W9-BVE-926

with contributions by

Sheldon Berger
Edward W. Bermes, Jr.
Robert V. Blanke
Sati C. Chattoraj
Ralph D. Ellefson
Emanuel Epstein
Vivian Erviti
Virgil F. Fairbanks
Willard R. Faulkner
Ermalinda A. Fiereck
Donald T. Forman
Gregor H. Grant
G. Phillip Hicks
Nai-Siang Jiang
John W. King
Robert F. Labbé
Maurice L'Heureux
Harold Markowitz
Donald W. Moss
Hipolito V. Nino
James L. Quinn III
Joseph I. Routh
William Shaw
Ole Siggaard-Andersen
Robert H. Spitzer
Donald S. Young

fundamentals of Clinical Chemistry

Edited by

NORBERT W. TIETZ, Ph.D.

Director of Clinical Chemistry and Professor, Department of Pathology,
University of Kentucky Medical Center, Lexington, Kentucky.
Formerly: Director of Clinical Chemistry, Mount Sinai Hospital Medical Center;
Professor of Clinical Chemistry, Department of Pathology,
The University of Health Sciences/The Chicago Medical School;
Professor of Biochemistry and Pathology, Rush Medical College, Chicago, Illinois.

Editorial Committee:

WENDELL T. CARAWAY, Ph.D.

McLaren General Hospital, Flint, Michigan

ESTHER F. FREIER, M.S., M.T. (ASCP)

University of Minnesota, Minneapolis, Minnesota

JOHN F. KACHMAR, Ph.D.

Rush-Presbyterian-St. Lukes Medical Center, Chicago, Illinois

HOWARD M. RAWNSLEY, M.D.

Dartmouth Medical School, Hanover, New Hampshire

W. B. SAUNDERS COMPANY • Philadelphia • London • Toronto

W. B. Saunders Company: West Washington Square
Philadelphia, PA 19105

1 St. Anne's Road
Eastbourne, East Sussex BN21 3UN, England

1 Goldthorne Avenue
Toronto, Ontario M8Z 5T9, Canada

Fundamentals of Clinical Chemistry ISBN 0-7216-8866-7

Library of Congress catalog card number 73-91279.

Last digit is the print number: 9 8 7 6

In Thankfulness Dedicated to

MY WIFE GERTRUD

and to my children

MARGARET ANN, KURT RICHARD,

ANNETTE MARIE *and* MICHAEL GERHARD

Contributors

SHELDON BERGER, M.D.

Associate Professor of Medicine, Northwestern University Medical School. Attending Physician, Department of Medicine, Northwestern Memorial Hospital, Chicago, Illinois.
Thyroid Function

EDWARD W. BERMES, JR., Ph.D.

Professor, Departments of Pathology and Biochemistry and Biophysics, Stritch School of Medicine, Loyola University. Director of Clinical Chemistry, Loyola Medical Center, Maywood, Illinois.
Basic Laboratory Principles and Procedures; Statistics, Quality Control, and Normal Values

ROBERT V. BLANKE, Ph.D.

Professor of Clinical Pathology, Medical College of Virginia, Virginia Commonwealth University. Director, Toxicology Laboratory, Medical College of Virginia Hospitals. Consultant, McGuire Veterans Administration Hospital, Richmond, Virginia.
Analysis of Drugs and Toxic Substances

WENDELL T. CARAWAY, Ph.D.

Biochemist, McLaren General Hospital, St. Joseph Hospital, and the Flint Medical Laboratory, Flint, Michigan.
Photometry; Carbohydrates; Lipids

SATI C. CHATTORAJ, Ph.D.

Associate Professor, Departments of Biochemistry, Obstetrics and Gynecology, Boston University School of Medicine. Director, Research Laboratories, Department of Obstetrics and Gynecology, Boston University School of Medicine, Boston, Massachusetts.
Gas Chromatography; Endocrine Function

RALPH D. ELLEFSON, Ph.D.

Assistant Professor of Biochemistry and Laboratory Medicine, Mayo Medical School. Director, Lipids and Lipoproteins Laboratory, Section of Clinical Chemistry, Mayo Clinic, Rochester, Minnesota.
Lipids and Lipoproteins

EMANUEL EPSTEIN, Ph.D.

Clinical Chemist, William Beaumont Hospital, Royal Oak, Michigan. Clinical Assistant Professor, School of Medicine, Wayne State University.
Electrophoresis

VIVIAN ERVITI, Ph.D.

Associate Director, Department of Research and Development, National Board of Medical Examiners, Philadelphia, Pennsylvania. Formerly: Associate Professor and Chairperson, Department of Educational Psychology and Statistics, University of Health Sciences, The Chicago Medical School, Chicago, Illinois.
Statistics, Quality Control, and Normal Values

VIRGIL F. FAIRBANKS, M.D.

Associate Professor, Mayo Medical School. Consultant, Hematology and Laboratory Medicine, Mayo Foundation, Rochester, Minnesota.
Hemoglobin, Hemoglobin Derivatives, and Myoglobin

WILLARD R. FAULKNER, Ph.D.

Associate Professor of Biochemistry, Vanderbilt University School of Medicine. Director, Research and Development, Clinical Chemistry Laboratories, Vanderbilt University Hospital, Nashville, Tennessee.
Renal Function

ERMALINDA A. FIERECK, M.S.

Associate Clinical Chemist, Mount Sinai Hospital Medical Center, Chicago, Illinois.
Analysis of Calculi; Amniotic Fluid

DONALD T. FORMAN, Ph.D.

Associate Professor of Biochemistry and Pathology, Northwestern University Medical School. Director, Clinical Biochemistry, Evanston Hospital, Evanston, Illinois.
Basic Laboratory Principles and Procedures; Statistics, Quality Control, and Normal Values

ESTHER F. FREIER, M.S., M.T.(ASCP)

Professor, Department of Laboratory Medicine and Pathology, University of Minnesota Medical School. Hospital Chemist and Associate Director of Clinical Chemistry Laboratory, University of Minnesota Hospitals, Minneapolis, Minnesota.
Osmometry

GREGOR H. GRANT, M.A., B.Sc., B.M., B.Ch., F.R.C.Path.

Consultant Chemical Pathologist, Royal Salop Infirmary, Shrewsbury, U.K.
Chemistry of Amino Acids and Proteins; Methods for the Analysis of Proteins and Amino Acids; Amino Acids and Related Metabolites

G. PHILLIP HICKS, Ph.D.

Clinical Associate Professor of Medicine, University of Wisconsin Medical School. President, Laboratory Computing, Inc., Madison, Wisconsin.
Laboratory Computers

NAI-SIANG JIANG, Ph.D.

Associate Professor of Biochemistry and Laboratory Medicine, Mayo Medical School. Consultant, Department of Laboratory Medicine, Mayo Clinic and Mayo Foundation, Rochester, Minnesota.
Immunochemical Principles

JOHN F. KACHMAR, Ph.D.

Associate Professor of Biochemistry, Rush University. Senior Biochemist, Rush-Presbyterian-St. Luke's Medical Center, Chicago, Illinois.
Chemistry of Amino Acids and Proteins; Methods for the Analysis of Proteins and Amino Acids; Amino Acids and Related Metabolites; Enzymes

JOHN W. KING, M.D., Ph.D.

Clinical Pathologist, Cleveland Clinic Foundation, Cleveland, Ohio.
Renal Function

ROBERT F. LABBÉ, Ph.D.

Professor of Laboratory Medicine, University of Washington, Seattle, Washington.
Porphyrins and Related Compounds

MAURICE V. L'HEUREUX, Ph.D.

Professor of Biochemistry, Stritch School of Medicine, Loyola University of Chicago, Maywood, Illinois.
Basic Principles of Radioactivity and Its Measurement

HAROLD MARKOWITZ, Ph.D., M.D.

Associate Professor of Microbiology and Laboratory Medicine, Mayo Medical School. Consultant, Department of Laboratory Medicine, Mayo Clinic and Mayo Foundation, Rochester, Minnesota.
Immunochemical Principles

DONALD W. MOSS, M.A., M.Sc., Ph.D.

Reader in Enzymology, Royal Postgraduate Medical School, University of London. Honorary Biochemist, Hammersmith Hospital, London, U.K.
Enzymes

HIPOLITO V. NINO, Ph.D.

Director, Clinical Chemistry, Department of Laboratory Medicine, The Christ Hospital, Cincinnati, Ohio. Formerly: Chief, Nutritional Biochemistry Section, Center for Disease Control, Atlanta, Georgia.
Vitamins

JAMES L. QUINN, III, M.D.

Professor of Radiology, Northwestern University Medical School. Director, Nuclear Medicine, Northwestern Memorial Hospital, Chicago, Illinois.
Thyroid Function

HOWARD M. RAWNSLEY, M.D.

Professor of Pathology, Dartmouth Medical School, Hanover, New Hampshire. Director of Diagnostic Laboratories, Mary Hitchcock Memorial Hospital, Hanover, New Hampshire.

JOSEPH I. ROUTH, Ph.D.

Professor of Biochemistry and Pathology, University of Iowa College of Medicine. Director, Special Clinical Chemistry Laboratory, University Hospitals, Iowa City, Iowa.
Liver Function

WILLIAM SHAW, Ph.D.

Supervisory Research Chemist, Immunochemistry Section, Center for Disease Control, Atlanta, Georgia.
Vitamins

OLE SIGGAARD-ANDERSEN, M.D., Ph.D.

Professor of Clinical Chemistry, University of Copenhagen. Head, Department of Clinical Chemistry, Copenhagen County Hospital, Herlev, Denmark.
Electrochemistry; Blood Gases; Acid-Base and Electrolyte Balance

ROBERT H. SPITZER, Ph.D.

Professor, Department of Biochemistry, The Chicago Medical School, Chicago, Illinois.
Chromatography

NORBERT W. TIETZ, Ph.D.

Director of Clinical Chemistry and Professor, Department of Pathology, University of Kentucky Medical Center, Lexington, Kentucky. Formerly: Director of Clinical Chemistry, Mount Sinai Hospital Medical Center, Chicago, Illinois. Professor of Clinical Chemistry, Department of Pathology; and Associate Professor of Biochemistry, The Chicago Medical School, University of Health Sciences, and Professor of Biochemistry and Pathology, Rush Medical College, Chicago, Illinois.
Electrolytes; Acid-Base and Electrolyte Balance; Gastric Function, Pancreatic Function, and Intestinal Absorption

DONALD S. YOUNG, M.B., Ph.D.

Chief, Clinical Chemistry Service, Clinical Pathology Department Clinical Center, National Institutes of Health, Bethesda, Maryland.
Automation

Foreword

By Michael Somogyi, Ph.D.

The clinical chemistry laboratory a few decades ago was a modest handmaiden supplying a few chemical tests that appeared serviceable in diagnostic procedures. These tests were developed in Otto Folin's laboratory at Harvard University, in D. D. Van Slyke's laboratory in the hospital of the Rockefeller Institute, and in a number of biochemistry departments, which were themselves new in some medical schools. These methods, which filled small booklets, were in most instances like cookbook recipes, and were frequently in the hands of semiskilled persons with no great competence in any branch of chemistry. A great change was inaugurated when a few progressive hospitals appointed to their laboratories highly trained professional chemists, men who were qualified to carry on investigative work in addition to supplying reliable data for diagnostic use. These chemists faced great challenges in refining existing methods and in the development of a rapidly increasing number of new tests and techniques. Today, in place of small "cookbooks," the chemist is confronted with a burgeoning amount of literature on clinical chemistry, the "fundamentals" of which are presented in this volume by the cooperative and coordinated effort of a group of competent and erudite clinical chemists.

With the well organized, rich content of this volume before me, and with the picture of instrumentation and equipment of an up-to-date clinical laboratory in mind, I cannot help but hark back to the humble beginnings in this field. In 1926, when I moved from academic work to the newly constructed Jewish Hospital of St. Louis, this was the first hospital in this city (priding itself on two medical schools!) to place a chemist in charge of its clinical laboratory. The space originally allotted to chemistry consisted of a single room of moderate size, with $1200 available for equipment. As my technical assistant I trained a young, small-town school teacher with no previous experience in chemistry. True, the routine work comprised at that time but a dozen or so analytical procedures, involving exclusively simple manual operations. Thus we found time for some investigative and educational activities, which year after year attracted the interest of the interns, residents, and younger members of the medical staff. They learned to use and appreciate the services of the laboratory so that the erstwhile modest handmaiden has developed into "laboratory medicine." My participation in this process, my collaboration with progressive, spirited, mostly youthful, clinicians was the most rewarding experience in my professional life. I am confident that these progressive young clinicians will find the study of the *Fundamentals of Clinical Chemistry* as useful as will workers in the laboratory. By so doing they will narrow the gap between medical science and medical practice.

Michael Somogyi

(1883–1971)

Preface

Clinical chemistry, by its nature, is a discipline that draws upon all fields of chemistry (inorganic, analytical, organic, physical, and biological). In addition, it requires a certain familiarity with many aspects of related fields, such as biology, physiology, hematology, immunology, physics, instrumentation, and electronics. The editor and authors, who are engaged in teaching clinical chemistry to students of biochemistry, medicine, and medical technology, believe that there is a need for a book that conveys the fundamentals of clinical chemistry and guides the novice in initial studies of this complex and rapidly growing field. It is assumed that the reader has had two or three years of basic college science courses. The auxiliary subjects are included only to the extent that they are necessary to explain and understand the clinical chemical aspects.

The text presents not only analytical procedures, but also the chemical principles upon which these methods are based. It is our hope that this will help in training individuals who are prepared to stand up to the challenge of the rapidly growing number of manual and automated analytical techniques for the detection and quantitation of, at times, complex metabolites or other compounds of interest to contemporary medicine. The concerted efforts of clinical chemists, pathologists, and public health officials to upgrade the quality of analytical work produced in clinical laboratories cannot succeed without such a group of well trained laboratory technicians.

The sections related to the clinical significance of laboratory tests, although limited, are of value not only to the clinician but also to the technician because they provide the latter with a realization of the usefulness of his data as well as the importance of accurate laboratory work. For a comprehensive coverage of these aspects, the student is referred to appropriate texts, professional journals, and references.

Occasionally the material presented goes beyond the scope of basic teaching programs and thus provides information for advanced students; an effort was made to present such material in a form understandable to the undergraduate student. In some sections, such material has been presented in smaller type.

A list of references and reading assignments is included at the end of each chapter. The references pertaining to methods given in detail are listed directly after the procedure.

The work on this book started several years ago. Throughout this time various national and international scientific organizations and many individuals have expended a great amount of effort to standardize various units applicable to clinical chemistry so that they may be more meaningful and also consistent throughout the world. The S.I. system has been adopted by the authors, in principle, in conformity with the practices in this country. In many places the older and the S.I. units are used side by side to aid in the transition.

Each of the chapters presented in this textbook has been reviewed by members of the editorial committee. Thus, it was hoped to incorporate the viewpoint's of several individuals and thereby to improve the quality of the material and promote continuity and intellectual uniformity in this multiauthor book. If the authors or editors have failed in certain objectives, they would greatly appreciate receiving comments from instructors and students so that these suggestions may be incorporated in any future edition of the text.

NORBERT W. TIETZ, PH.D.

Acknowledgments

Preparation of a multiauthor book requires close cooperation between the contributors and the editor. I am most grateful to all contributors for submitting so willingly to this principle in preparing their manuscripts and for cooperating so well in the process of reviewing the manuscripts. My special thanks go to the Editorial Committee, for their significant effort in editing the various manuscripts. I am most appreciative to Dr. H. M. Rawnsley for reviewing all chapters in regard to clinical application and for his many helpful suggestions. Thanks are expressed to Drs. W. Dasler, F. Fiorese, and A. Weinstock, as well as to Mrs. M. Rettaliata, who proofread some manuscripts and made many useful comments.

The cooperation of the staff of W. B. Saunders Company and especially of Jack Hanley, Medical Editor, is also gratefully acknowledged.

NORBERT W. TIETZ, PH.D.

CONTENTS

CHAPTER 1

BASIC LABORATORY PRINCIPLES AND PROCEDURES

by Edward W. Bermes, Jr., Ph.D., and Donald T. Forman, Ph.D.

The function of the clinical chemistry laboratory is to perform qualitative and quantitative analyses of body fluids, such as blood, urine, and spinal fluid, as well as feces, calculi, and other materials. If the results of these tests are to be useful to the physician in the diagnosis and treatment of disease, they must be done as accurately as possible. This involves using sound analytical methods and good instrumentation. Thus, the persons performing these analyses should not only be thoroughly familiar with the technique involved in the test being done, but they should also be well grounded in general methods of analytical chemistry. A good background in instrumentation is also desirable and, indeed, is becoming essential. In addition to a complete knowledge of the technical aspects of the tests involved, the technologist should know the principles of the method. This involves a knowledge of the chemical reactions and the effect of physical variables on them and the purpose of each reagent used. An understanding of normal values and the extent of deviation from these normal values, which are found in health and disease, is also necessary.

Before a new procedure is introduced it must be evaluated in the laboratory. The variables of the method must be thoroughly explored. For a colorimetric technique, adherence to Beer's law should be investigated. The time for maximum color development and stability of color, as well as reagent and sample stability, must be known. The effect of various anticoagulants should be known if plasma or whole blood is to be used in the procedure. Finally, the accuracy and precision of the procedure have to be evaluated using statistical techniques (see Chap. 2).

GENERAL LABORATORY SUPPLIES

Clinical chemistry laboratories of today are still served in the main by the classic material glass in spite of the modern introduction of plastics and stainless steel. The advantages as well as the limitations imposed by the properties of these materials should be appreciated in light of their possible effects upon analysis.

GLASSWARE

About the middle of the eighteenth century, Lavoisier discovered that glassware might be attacked by its contents. Stas[26] also found that his glassware was attacked by reagents to such an extent that there was interference with his atomic weight determinations. Many of these difficulties were associated with the use of glasses that had a high alkali content. This problem was decidedly improved after World War I when borosilicate glasses became available.

Clinical laboratory glassware can be divided into five general types of glass:[18] (1) high thermal resistant glass, (2) high silica glass, (3) glass with high resistance to alkali, (4) low actinic glassware, and (5) standard flint glass.

High Thermal Resistant Glass. This type is usually a borosilicate glass that has a low alkali content. It is free from magnesia-lime-zinc group elements, heavy metals, arsenic, and antimony. The borosilicate glass resists heat, corrosion, and thermal shock.[1] Since its dimensions change very little with temperature (relatively low coefficient of expansion), this type of glassware should be used whenever heating or sterilization by heat is employed. Some borosilicate glassware, if properly supported and not under internal pressures, can be heated to about 600°C for a relatively short period of time. If the glass is cooled too quickly, however, it will acquire strains that may affect its future serviceability. The highest safe operating temperature for this glass is its strain point. Table 1–1 shows the thermal durability of borosilicate glass at varying temperatures.

Pyrex and *Kimax* brand glass are the most common thermal resistant borosilicate glassware found in the laboratory. Laboratory apparatus such as beakers, flasks, and pipets are usually made of this type of borosilicate glass. Because it contains so few elements, contamination of liquids by the vessel is minimal, even when liquids are hot. *Exax* brand glassware is a lower grade borosilicate glass and may be used when it is not necessary or desirable that high quality borosilicate glass be used.

Several years ago a special alumina-silicate glass was developed that is at least six times stronger than borosilicate glass.[21] This *Corex* brand laboratory glassware has been strengthened chemically rather than thermally. Corex pipets have a typical impact strength of 30,000 psi compared to a rating of 2000 to 5000 psi for borosilicate pipets. Corex laboratory glassware is harder than conventional borosilicates and better able to resist clouding due to alkali and scratching. This glass is also used in higher temperature thermometers, graduated cylinders, and centrifuge tubes.

Vycor brand laboratory glassware is recommended for use in applications involving high temperatures, drastic heat shock, and extreme chemical treatment with acids and dilute alkalies. This transparent glassware is resistant to attack by all acids except hydrofluoric, and even in the upper temperature range it is more resistant to alkalies than is borosilicate glass. Vycor ware is used primarily in ashing and ignition techniques. It can be heated to 900°C and can withstand downshocks from 900°C to ice water.

High Silica Glass. The high silica content of this glass (over 96 per cent) makes it comparable to fused quartz in its thermal endurance, chemical stability, and electrical characteristics. The glass is made by removing almost all elements except silica from borosilicate glass. It is radiation resistant and has good optical qualities and temperature capabilities; it is used for high precision analytical work and can also be used for optical reflectors and mirrors.

Glass with High Resistance to Alkali. This boron-free glassware was developed particularly for use with strong alkaline solutions. Its thermal resistance is much less than that of borosilicate glass and it must therefore be heated and cooled very carefully. Its primary use should be whenever solutions or digestions with strong alkali are made. This glass is often referred to as soft glass.

TABLE 1–1 THERMAL DURABILITY OF BOROSILICATE GLASS

Strain point*	510°C
Annealing point†	555°C
Softening point	820°C

* Temperature at which deformation of glass may result due to heat stress.

† Temperature to which glass must be heated to relieve strains.

Low Actinic Glassware. This glassware contains materials that usually impart an amber or red color to the glass and reduce the amount of light passing through to the substance within the glassware. It was developed to provide a highly protective laboratory glassware for handling materials sensitive to light in the 300 to 500 nm range (i.e., bilirubin, carotene, and vitamin A).

Standard Flint Glass. This is a soda-lime glass composed of a mixture of the oxides of silicon, calcium, and sodium. This type of glass is lowest in cost of all glasses and is readily fabricated in a wide variety of shapes. Such glass has poor resistance to high temperatures and sudden changes of temperature, and its resistance to attack by chemicals is only fair. Since this glass is relatively easy to melt and shape, it has been used for bottles and some disposable laboratory glassware. Certain manufacturers supply soft glass (in contrast to low alkali borosilicate glass) disposable pipets to the clinical laboratory. Users of such pipets should rinse the pipets before use if they have not been water rinsed by the manufacturer. Soda-lime pipets may release alkali into the pipeted liquid and cause considerable errors in certain critical assay procedures. Organic contaminants found in soft glass and borosilicate glass disposable pipets can also interfere with analytical procedures using UV-absorption or fluorescence techniques.

Special glasses

Colored and Opal Glasses. These are made by adding small amounts of coloring materials to glass batches. These glasses are used in light filters, lamp bulbs, and lighting lenses.

Coated Glass. This glass has a thin, metallic oxide permanently fire bonded to the surface of glass. The film, unlike glass, can conduct electricity. It has electronic applications as heat shields to protect against infrared light and as electrostatic shields to carry off charges.

Optical Glass. It is made of soda-lime, lead, and borosilicate and is of high optical purity. It is used in making prisms, lenses, and optical mirrors.

Glass-ceramics (Pyroceram). This glassware has high thermal resistance, chemical stability, and corrosion resistance (like borosilicate glass). This material is useful in hot plates, table tops, and heat exchangers.

Radiation-absorbing Glass. This is made of soda-lime and lead and is useful in preventing transmission of high energy radiation (i.e., gamma ray, x-ray).

PLASTICWARE[13,25]

The introduction of plasticware to clinical chemistry laboratories has greatly enhanced laboratory analysis. Beakers, bottles, flasks, graduated cylinders, funnels, centrifuge tubes, tubing, and pipets now have unique qualities that make them ideal for use when high corrosion resistance and unusual impact and tensile strength are required. Table 1–2 describes the physical and chemical properties of various resins that are used in the preparation of laboratory ware.

Polyolefins (Polyethylene, Polypropylene). These are a unique group of resins with relatively inert chemical properties. Generally, the polyolefins are unaffected by acids (however, concentrated sulfuric acid slowly attacks polyethylene at room temperature), alkalies, salt solutions, and most aqueous solutions; aromatic, aliphatic, and chlorinated hydrocarbons cause moderate swelling at room temperature; organic acids, essential oils, and halogens slowly penetrate these plastics. Strong oxidizing agents attack this group of resins at elevated temperatures only. Polypropylene plasticware is slightly more vulnerable to attack by oxidizing agents, but it has the advantage of withstanding higher temperatures. Polyethylene and polypropylene are used primarily to fabricate bottles, beakers, jars, carboys, jugs, funnels, pipet jars, pipet baskets, tanks, buret covers, check valves, dis-

TABLE I–2 PHYSICAL AND CHEMICAL PROPERTIES OF LABORATORY PLASTICS

Material	Special Properties or Uses	Clarity	Heat Resistance °C*	Effect of: Weak Acids / Strong Acids	Weak Alkalies / Strong Alkalies	Organic Solvents / Sunlight
Polystyrene (styrene)	Excellent optical properties, hard, biologically inert	Transparent	65–80	None	None	Soluble in aromatic and chlorinated hydrocarbons
				Attacked by oxidizing acids	None	Yellow
High impact polystyrene	For molding, where impact resistance important	Translucent to opaque	60–80	None	None	Soluble in aromatic and chlorinated hydrocarbons
				Attacked by oxidizing acids	None	Some strength loss
Polyethylene, regular (PE)	Tough, chemically and biologically inert, pliable	Translucent to opaque	80–99	Resistant	Resistant	Resistant below 60°C
				Attacked by oxidizing acids	Resistant	Surface cracking
Polyethylene, high density	Tough, chemically and biologically inert, high heat resistance, harder than conventional polyethylene	Translucent to opaque	105–115	Very resistant	Very resistant	Resistant below 80°C
				Attacked by oxidizing acids	Very resistant	Surface cracking
Polypropylene (PP)	Very tough, heat resistant, autoclavable, inert	Transparent to opaque	135–160	Very resistant	None	Resistant below 80°C
				Attacked by oxidizing acids	Very resistant	Surface cracking

Methyl methacrylate (Plexiglas)	Good for casting, fabricating, machining, polishing, fairly hard, tough, good optical properties	Transparent	60–90	Practically none	Practically nil	Soluble in ketones, esters, aromatic hydrocarbons
				Attacked by oxidizing acids	Practically nil	Very slight
Cellulose acetate (acetate)	Clear, easily fabricated (sheet), quite tough	Translucent to opaque	60–105	Slight	Slight	Soluble in ketones and esters, softens in alcohol
				Decomposes	Decomposes	Slight
Polyamide (nylon)	Tough, heat resistant machinable, autoclavable	Transparent to opaque	130–150	Resistant	None	Resistant to common solvents
				Attacked	None	Slight discoloration
Rigid vinyl (PVC)	Tough, clear, ideal for vacuum forming, plasticized for tubing	Transparent to opaque	50–73	None	None	Soluble in ketones, esters, swells in aromatic hydrocarbons
				None to slight	None	Slight
Polytetrafluoro-ethylene (Teflon)	Extremely high heat resistance, waxy slippery surface, machinable, chemically inert	Opaque	255	None	None	None
				None	None	None
Polycarbonate	Very tough, heat resistant, biologically inert	Transparent	135–160	Resistant	Slight	Soluble in aromatic and chlorinated hydrocarbons
				Attacked by oxidizing agents	Attacked	Surface cracking

* Heat resistance (continuous) = maximum temperature at which no distortion occurs during continuous exposure (not under tension).

connect valves, twistcock connectors, needle valves, hollow stoppers, dropping pipets, hydrometer jars, stirring rods, tubing and reagent dispensers. Polyethylene is less expensive than polypropylene and is used in most disposable plasticware. Polypropylene has a distinct advantage in that it can be sterilized; however, it absorbs pigments and tends to become discolored.

Polycarbonate Resin. It is twice as strong as polypropylene and may be used at temperatures ranging from −100 to +160°C; however, its chemical resistance is not as wide as that of the polyolefins. This resin is unsuitable for use with bases such as amines, ammonia, and alkalies, as well as oxidizing agents. It is dissolved by chlorinated aliphatic and aromatic hydrocarbons. Polycarbonate resin is insoluble in aliphatic hydrocarbons, some alcohols, and dilute aqueous acids and salts. Since labware molded from this resin is glass-clear and shatterproof, it is used extensively in centrifuge tubes and graduated cylinders.

Tygon. A nontoxic, clear plastic of modified plasticized polyvinyl chloride, it is used extensively for the manufacture of tubing (i.e., tubing used in AutoAnalyzers). It is very flexible and can be used to handle most chemicals, although it should not be subjected to prolonged immersion in aliphatic or aromatic hydrocarbons, ketones, and esters. It is flexible at −30°C (brittle at −45°C), resists dry heat to 95°C, and can be steam autoclaved or chemically sterilized. This tubing is soft and flexible and quickly slips over tubulatures, gripping tightly on glass or metal.

Teflon Fluorocarbon Resins. They have unique qualities that make them almost chemically inert and ideal when high corrosion resistance at extreme temperatures is essential. Bottles and beakers made of Teflon resist severe temperatures ranging from −270 to +255°C, which allows use in cryogenic experiments or work at high temperatures over extended periods.

This labware is pure translucent white and inert to such corrosive reagents as boiling aqua regia, nitric and sulfuric acids, boiling hydrocarbons, ketones, esters, and alcohols. Because of its unique antiadhesive properties and its nonwettable surface, Teflon is used for self-lubricating stopcocks, stirring bars, bottle cap liners, and tubing. It is also quite easy to clean and is fast drying, but it can be scratched and misshaped.

Whenever possible, plastics should be used in place of glass. Plasticware has the advantage of being unbreakable, and it does not release ions into a solution as does glass. A disadvantage of plasticware is its tendency to bind various solutes and its potential leaching of the surface-bound constituents into subsequent solutions.[6] Polyethylene is also permeable to water vapor, and evaporation may occur through breathing of the plastic with concomitant increase in concentration of reagents and standards even in tightly stoppered bottles. As the volume of remaining liquid approaches a small fraction of the total capacity, the ratio of the surface area to volume increases. With the increased surface area of polyethylene available for diffusion, evaporation becomes a significant factor. Therefore, small volumes of reagent should never be stored in oversized plastic bottles for long periods of time.

Polyethylene is also not completely inert and may bind or adsorb proteins, dyes, stains, iodine, and picric acid. Colorless reagents may bind to plasticware and be undetected only to manifest themselves by erratic results in subsequent chemical analyses. Slow reduction of ceric and cuprous ions has been observed[6] in solutions stored in polyethylene bottles. Some solutions have developed significant fluorescence after being stored in plastic containers for extended periods.

VOLUMETRIC EQUIPMENT AND ITS CALIBRATION[19]

Most clinical chemistry procedures require accurate measurement of volumes. In very precise work it is never safe to assume that the volume contained or delivered by any piece of equipment is exactly that amount indicated. Ideally all volumetric apparatus for precise work should be designated Class A, which means that it conforms to the specifications in NBS (National Bureau of Standards) circular C-602 for Class A volumetric ware. Manufacturers no longer sell glassware with a NBS certificate of calibration.

PIPETS

In general, two main types of pipets are used in clinical chemistry (Fig. 1–1). The *volumetric* or *transfer pipet* (Fig. 1–1*A*) is designed to deliver a fixed volume of liquid and consists of a cylindrical bulb joined at both ends to narrower glass tubing. A calibration mark is etched around the upper suction tube and the lower delivery tube is drawn out to a fine tip. The more important requirements of the volumetric pipet are that the calibration mark should not be too close to the top of the suction tube, that the bulb should merge gradually into the narrower tubes, and that the tip should have a gradual taper. The orifice should be such that the outflow of liquid is not too rapid in order to reduce drainage errors to negligible proportions.[9]

Volumetric or transfer pipets are calibrated to deliver the volume specified. The most commonly used sizes are 1, 2, 3, 4, 5, 10, 25, 50, and 100 ml. Less frequently used sizes, delivering 6, 8, and so on, ml can also be obtained. They are used for accurate measurements of aliquots of nonviscous samples, filtrates, and standard solutions. These pipets should be of Kimax or Pyrex grade. The reliability of the calibration of the volumetric pipet decreases with a decrease in size and, therefore, special micropipets have been developed for chemical microanalysis.

Ostwald-Folin pipets (Fig. 1–1*B*) are similar to volumetric pipets, but have their bulb closer to the delivery tip. These pipets are found in 0.5, 1.0, 2.0, and 3.0 ml sizes. They are used for measuring viscous fluids such as blood or serum. The Ostwald-Folin pipet has an etched ring near the mouthpiece, indicating that it is a blow out pipet. The pipet is not blown out, however, until the blood or serum has drained to the last drop in the delivery tip. With opaque fluids such as blood, the top of the meniscus must be read when filling, and controlled slow drainage is required so that no residual film of blood is left on the walls of the pipet. All the previously mentioned pipets have been calibrated to deliver (TD) a specific volume.

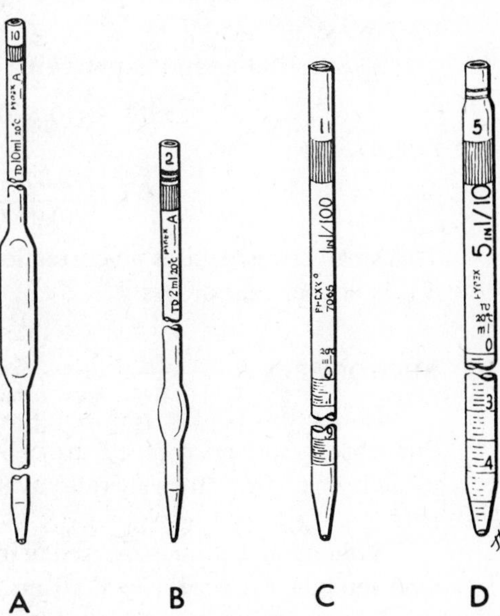

Figure 1–1 Pipets. *A*, Volumetric (transfer). *B*, Ostwald-Folin (transfer). *C*, measuring (Mohr). *D*, Serological (graduated to the tip).

The second type of pipet is the *graduated* or *measuring pipet* (Fig. 1–1*C*). This is a plain, narrow tube drawn out to a tip and graduated uniformly along its length. Two types calibrated for delivery are available; one is calibrated between two marks on the stem (*Mohr*) and the other has graduation marks down to the tip (*serological*). These pipets are intended for the delivery of predetermined volumes. The serological type of pipet (Fig. 1–1*D*) must be blown out to deliver the entire volume of the pipet. It has an etched ring (or pair of rings) near the mouth end of the pipet signifying that it is a blow out pipet. Measuring pipets are commonly used in 0.1, 0.2, 0.5, 1.0, 5.0, and 10.0 ml sizes. They are principally used for the measurement of reagents and are not generally considered accurate enough for measuring samples and standards. The serological pipet has a larger orifice than the measuring (Mohr) pipet and thus drains faster.

Calibration of to-deliver pipets is usually performed by measuring the amount of water delivered by the pipet. This measurement may be made by weighing the water delivered and calculating the volume from its density. It is important to note the flow out time and to maintain the pipet and receiving vessel in the same position for a 10 second drainage period. The receiving vessel is stoppered and weighed and the equilibrated water and air temperature recorded. The weight of water delivered from the to-deliver pipet should be corrected for temperature to obtain the true delivery volume. Water is commonly used as a calibration liquid because it is readily available and because it is similar in viscosity and speed of drainage to the dilute solutions ordinarily employed in clinical chemical analysis.

The following formula is used for the calibration of a to-deliver pipet:

$$\begin{array}{c}\text{Apparent weight of}\\ \text{water in g at}\\ t^{\circ}\text{C (equilibrated}\\ \text{temperature)}\end{array} + \left[\begin{array}{c}\text{stated capacity of}\\ \text{pipet in ml}\end{array} \times \begin{array}{c}\text{correction at}\\ t^{\circ}\text{C}\end{array}\right] = \begin{array}{c}\text{actual capacity}\\ \text{in ml at 20°C}\end{array}$$

One determines the actual capacity of a volumetric pipet made of borosilicate glass and marked to deliver 10 ml at 20°C in the following manner:

1. First determine the weight of the empty receiving vessel, e.g., 22.0391 g.

2. Determine the weight of receiving vessel and water delivered with temperature of water at 24.1°C, e.g., 31.9961 g. In this example the weight of water, therefore, is 31.9961 − 22.0391 = 9.9570 g. Correcting per ml of stated capacity, using a table to correct for determining true capacities of borosilicate glass vessels from weight of water in air at 24.1°C, we get 0.00369 (24.1°C).

3. Substituting in the preceding formula:

$$9.9570 + (10 \times 0.00369) = 9.9939 \text{ ml actual capacity at } 20^{\circ}\text{C}$$

Finally

$$\frac{0.0061}{10.0000} \times 100 = 0.061\,\%\text{ error}$$

This small error requires no correction. A correction need not be applied in general analysis if it is 0.1 per cent or less.[15]

MICROPIPETS

In microwork, the remaining volume left in a pipet can cause significant error. For this reason most micropipets are calibrated to contain (TC) the stated volume rather than to deliver it. They are generally available in small sizes, such as 20, 50, 100, and 200 μl (Fig. 1–2).

Volumes are often expressed in lambdas, although this term is no longer recommended. One lambda (λ) = 1 μl = 0.001 ml. Clinical laboratory personnel are familiar with the

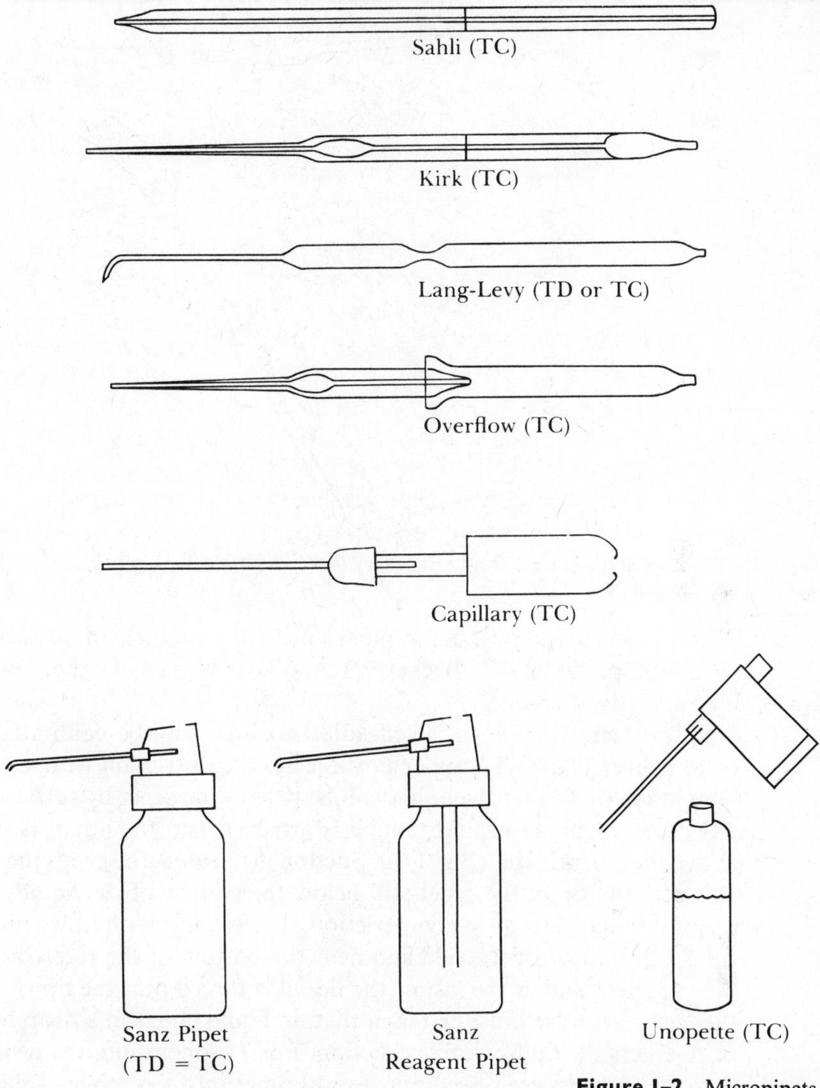

Figure 1–2 Micropipets.

Sahli hemoglobin pipet calibrated to contain 0.02 ml (20 μl). When using these or other micropipets, the sample is drawn just above the mark, the outer surface of the pipet is wiped clean with tissue paper, and the sample is adjusted to the mark by touching the tip repeatedly to the paper. The contents are then delivered by rinsing several times into a diluent. The diluent should not be drawn above the calibration mark because this will add a film of sample in excess of the nominal volume. Similar pipets calibrated to contain (TC) are available in various sizes. Larger sizes have a bulb to decrease the surface area to volume ratio and a small bore to facilitate accurate adjustment to the calibration mark.

To sample serum from a capillary blood collecting tube, the tube is etched with a file above the cells, held horizontally, and broken in two. The tip of the pipet is held against the cut end of the tube of serum and the pipet is filled by gentle suction (Fig. 1–3). The Sahli type of pipet is convenient because no new techniques need be introduced; however, the rather large blunt tip requires extra care in wiping all excess serum from the outer surface. In addition, the tip may be too large to sample serum from blood collected in 6 × 50 mm test tubes. The *Kirk pipet* has the tip drawn out to a finer point and thus over-

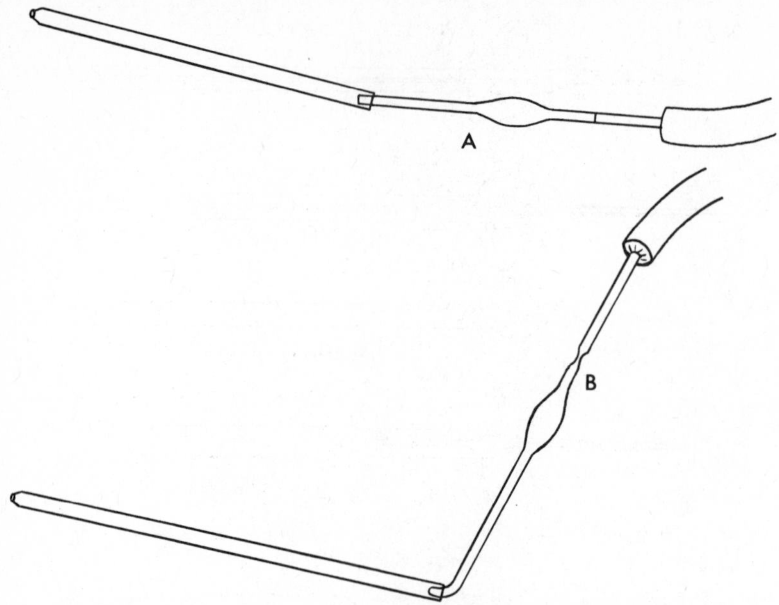

Figure 1–3 Sampling of serum from a capillary tube, *A*, with a Kirk pipet; *B*, with a Lang-Levy pipet.

comes these objections. Kirk pipets may be obtained in sizes as small as 1 μl (Microchemical Specialties Co., Berkeley, CA.); however, 10 μl is a practical lower limit for microliter analysis of serum.

The *Lang-Levy pipet* is self-adjusting and may be calibrated either to contain (TC) or to deliver (TD). The upper portion has a constriction with a bore slightly less than the bore at the tip so that the liquid will be held in the pipet by surface tension. When used as a TD pipet, a rubber aspirator tube is attached and the liquid is drawn into the pipet just above the constriction (Fig. 1–3). Suction is released and, with the pipet held nearly vertical and with the tip of the pipet still below the surface of the liquid, the solution in the pipet stops automatically at the constriction. The pipet is withdrawn and wiped off, and the bent tip is placed against the side and near the bottom of the receiving tube. Gentle pressure is now applied, and as the last of the liquid is forced out, the tip is removed from the wall of the test tube. Care must be taken that no liquid clings in a drop to the outside of the pipet or re-enters the tip by capillary action. For TC operation, the procedure is the same except that the contents are rinsed out several times into a receiving solution. Polyethylene Lang-Levy pipets are also available. They have a nonwetting surface and deliver the entire contents without rinsing.

The advantage of using the Lang-Levy pipet TD (i.e., without rinsing into the diluent) is that multiple samples of the same serum may be pipetted without washing and drying the pipet between samples. Similarly, in micro-operations a reagent may be pipetted into a number of tubes by reusing the same pipet. In practice, a single set of pipets, sizes 10, 20, 25, 50, 100, and 200 μl, is adequate for most operations. The pipet may be rinsed once with serum before sampling for analysis. Between different reagents the pipet is first rinsed with water, and then with the succeeding reagent. Pipets are rinsed thoroughly with water immediately after use and stored in a horizontal position in a closed container. Lang-Levy pipets, calibrated TC, may also be used TD provided the standard and unknown are delivered from the same pipet, since any volume error is then canceled out.

Overflow type pipets are calibrated TC (Fig. 1–2). Smaller sizes fill automatically by capillary action; gentle suction is applied to fill larger sizes. Contents are rinsed out in the manner just described. This type of pipet is more difficult to clean when sample has overflowed into the holder around the upper tip. Precision-bore disposable capillary tubes

(Fig. 1–2) calibrated TC are also available (Drummond Scientific Co., Broomall, PA.). The capillary is inserted into a holder attached to a rubber bulb that has a small opening in the top. The bulb is squeezed gently, the opening closed with the fingertip, the tube filled with sample, and suction released by removing the fingertip. To dispense, the opening is again closed and the contents expelled by gentle pressure. Rinsing is performed in the same manner.

An interesting variation of the overflow pipet has been described by Sanz.[23] A calibrated polyethylene pipet is fitted into a stopper inserted into the side of the cap of a small polyethylene bottle (Fig. 1–2). The cap of the bottle is a transparent dome with an opening in the top. The operation is comparable to that described for the Drummond capillary tubes. By inserting a vertical tube into the bottle, the device is converted into a reagent pipet. To use, the bottle is squeezed until the liquid rises into the dome and fills the pipet. On release of pressure the remaining liquid flows back into the bottle. The opening at the top is then closed with a fingertip and gentle pressure is applied to expel the reagent. The polyethylene is nonwetting, when thoroughly cleaned, and contents are delivered completely. Thus, the TD and TC volumes are the same. The polyethylene pipetting devices are an integral part of the Spinco ultramicroanalytical system (Beckman Instruments, Inc.).

A disposable self-filling pipet attached to a polyethylene reagent reservoir has also been described (Fig. 1–2).[29] This unit is sold under the trade name Unopette (Becton, Dickinson & Co., Rutherford, N.J.). A glass capillary pipet fitted in a plastic holder fills automatically with serum or whole blood by capillary action. The reagent bottle is squeezed slightly while the pipet is inserted. On release of pressure the sample is aspirated into the diluent or precipitating reagent. Intermittent squeezing fills and empties the pipet to wash out the contents. The unit has been adapted for a number of chemical and hematological determinations.

Calibration of micropipets

Inaccurate calibration of micropipets constitutes an important source of error in clinical chemistry. A gravimetric procedure employing mercury as the substance delivered or contained by the micropipet can be used to confirm proper calibration.

A pipet may be calibrated TC as follows: Grease the plunger of a 2 ml syringe with stopcock grease to provide an airtight seal. Insert the hub of the barrel into a one-hole No. 0 rubber stopper (Fig. 1–4). Insert the upper end of a clean, dry micropipet into the other side of the stopper. Pour some clean, dry mercury into a small beaker. Draw up the plunger slightly; then raise the beaker until the tip of the pipet is slightly below the surface of the mercury. Raise the plunger slowly until the mercury meniscus is adjusted exactly to the calibration mark; then withdraw the beaker downward in a clean sharp motion. Bring a small tared beaker under the tip and depress the plunger to expel all the mercury. Obtain the weight of mercury and divide by the density of mercury at its observed temperature to find the volume of the pipet. The density of mercury at 25 ± 5°C is 13.5340 ± 0.0123 or ± 0.091 per cent; therefore, for temperatures between 20 and 30°C it is sufficiently accurate to use the density at 25°C.

Micropipets may be calibrated TD (or TC) by various titrimetric or spectrophotometric procedures. Thus, a concentrated solution of sodium chloride may be sampled, diluted, and analyzed for either sodium or chloride and compared with an accurate macrodilution. Similarly, 1 volume of a 4 g/100 ml solution of potassium ferricyanide may be delivered (for TD) or washed out (for TC) into 200 volumes of water to provide a 1/201 microdilution. Use 20 μl of solution plus 4.00 ml of water, 50 μl of solution plus 10.00 ml of water, and so on. A 1/201 macrodilution is prepared by adding 1.00 ml of the ferricyanide solution to a 200 ml volumetric flask previously filled to the mark with water.

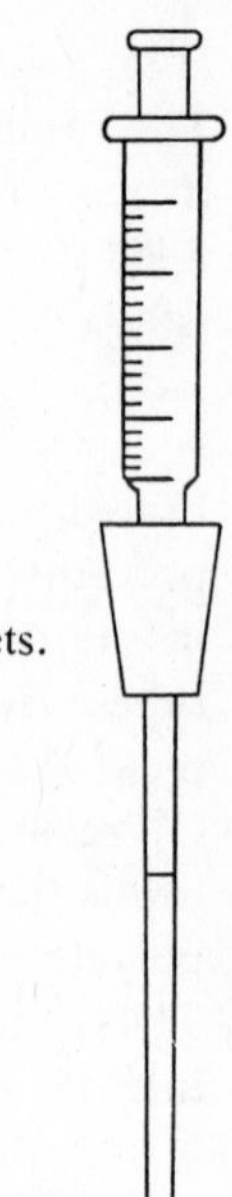

Figure 1–4 Apparatus for calibration of micropipets.

The absorbance of each dilution is then measured in a spectrophotometer at 420 nm against a water blank. An average of triplicate determinations should be used.

$$\frac{\text{Average } A\text{, micro}}{\text{Average } A\text{, macro}} \times \text{nominal capacity } (\mu\text{l}) = \text{actual capacity } (\mu\text{l})$$

Similar micro- and macrodilutions may be made from radioiodinated serum albumin. Equal aliquots are pipetted and counted in a scintillation detector to the desired degree of precision. Calculations are the same as with the spectrophotometric method.

AUTOMATIC PIPETTING AND DILUTING SYSTEMS

Semiautomatic versions of pipets are obtainable in sizes ranging from 0.005 to 20 ml and are very useful in the routine clinical laboratory. Figure 1–5 shows a Seligson pipet.[24] This system consists of a buret tip calibrated to contain a stated volume and a 2-way stopcock, one side attached to a suction device and the other side attached to a buret. In operation the tip is filled by suction and the stopcock closed. The tip is then wiped, and by reversing the direction of the stopcock a measured volume of diluent or reagent washes the serum into a receiving vessel. The stopcock is closed when the desired volume of diluent plus the capacity of the tip have been measured on the buret (i.e., 5 ml of diluent must be measured on the buret to deliver 4.8 ml for a 25-fold dilution, when a 0.2 ml sample is being aspirated). This device is self-cleaning and has the advantage that all of the specimens and standards are measured with the same pipet, giving good precision. Pipets are available in 0.02, 0.1, 0.2, and 0.5 ml sizes.

Figure 1–6 illustrates a single or multirange micropipetting device. This device is designed to operate as a to-deliver pipet. It will dispense its prerated sample (from 5 to 1000 microliters) when the plunger is moved through its complete stroke. The fluid containment tip utilizes a disposable plastic material, the tendency of which to retain an inner surface film is less than that of glass. This enables the technician to eliminate pipet cleaning and cross-contamination, and reduces error due to variation in pipetting. There is a macro version of this device which can be used to "deliver" or transfer larger volumes (e.g.,

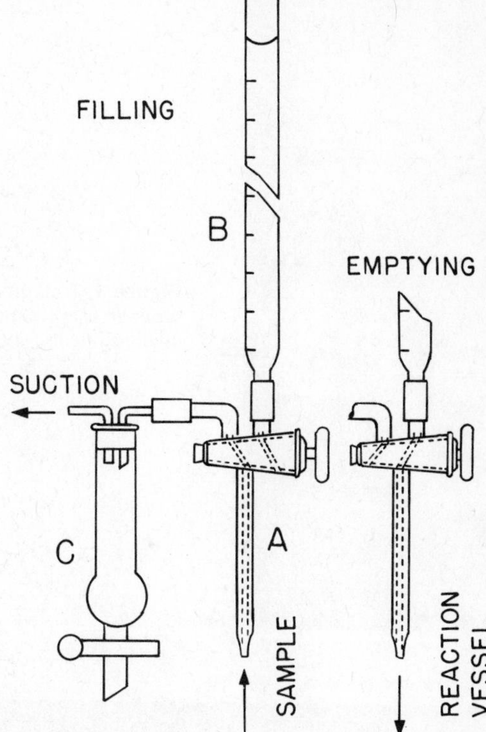

Figure 1-5 Diagram illustrating the significant features of a Seligson automatic pipet. *A*, Pipet and 2-way stopcock. *B*, Calibrated tube. *C*, Waste receptacle.

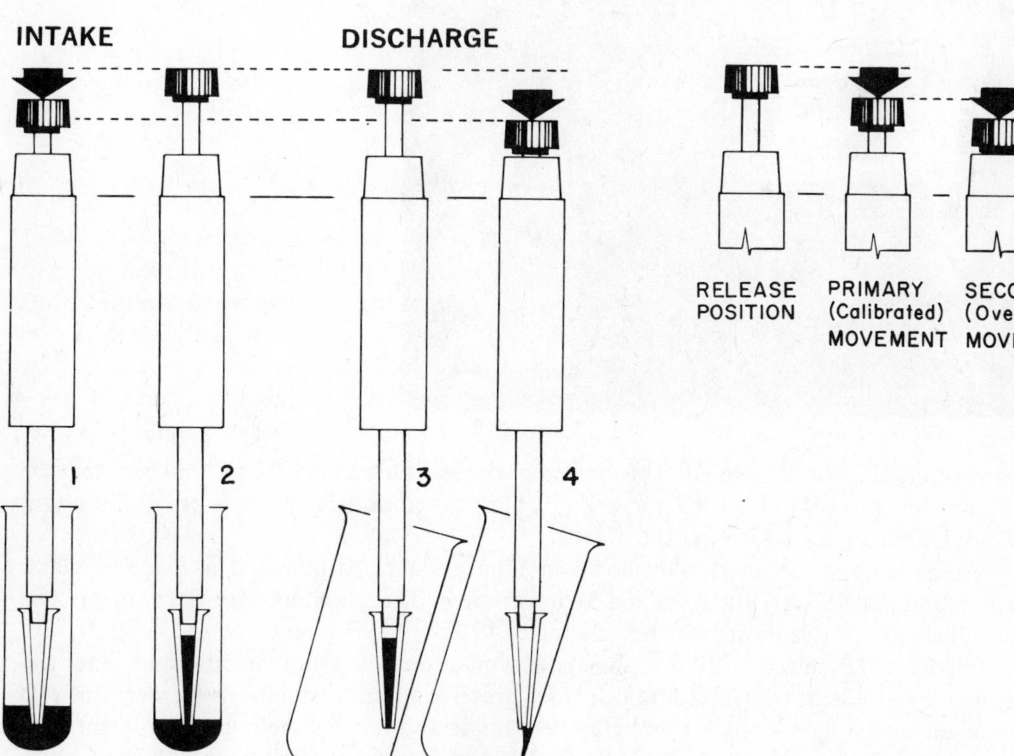

Figure 1-6 A semiautomatic micropipetting device which is designed to operate as a TD pipet.

Figure 1–7 Automatic pipettors. *A*, Manual Teflon glass syringe type. *B*, A semiautomatic sampling diluting and dispensing apparatus. (*B* courtesy Micromedic Systems, Inc., Independence Mall West, Philadelphia, PA.)

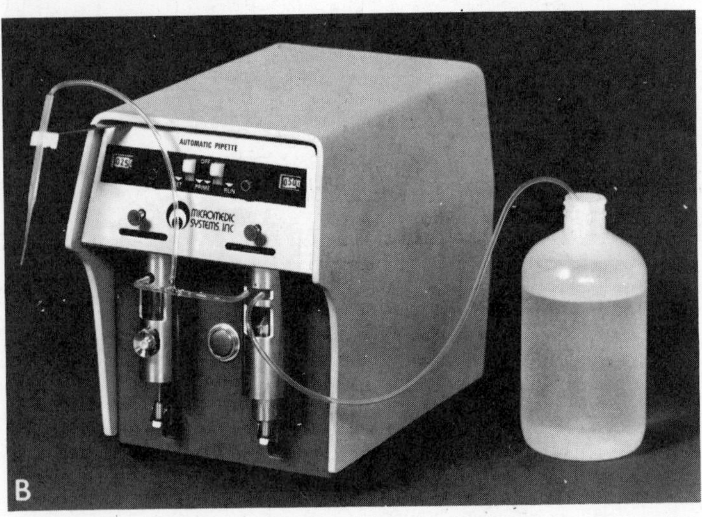

protein-free filtrates, dilutions). This device has a useable range of 1 to 5 ml and may be adjusted for any volume within those limits. The latter transfer pipet is especially useful in radioimmunoassay techniques.

A semiautomatic pipettor is shown in Figure 1–7*A*. Although it is made of glass, Teflon, and plastic, only the glass and Teflon come in direct contact with the reagent. The reproducibility of this device has been reported to be ± 1.0 per cent.

Figure 1–7*B* shows a dilutor that is a semiautomatic sampling, diluting, and dispensing apparatus. It measures and dispenses preset volumes of solutions by means of two motor-driven syringes—one for metering the sample and one for metering the diluent. The dilutor can be adjusted to accept as little as 10 μl of sample and deliver it with as much as 10 ml of diluent. The reproducibility of the metered pumps has been claimed to be ± 0.5 per cent.

These automatic and semiautomatic devices must be correctly calibrated by the analyst before they can be accepted as a useful tool. The volume dispensed by the diluent syringe can be determined by gravimetric calibration, and the volume of sample aspirated can be determined by using a dye and measuring the resultant dispensed solution by spectrophotometry.

Some of these devices come precalibrated. If the absolute calibration is slightly off, this would be of no concern if standards were diluted with the same dilutor as unknowns, e.g., in flame photometry. However, in performing a hemoglobin analysis an accurate dilution is required, since the standard certified by the College of American Pathologists is a prediluted cyanmethemoglobin solution and is therefore not treated in the same way as the unknown.

Problems encountered using automatic pipets and dilutors depend to a large degree on the nature of the solution being delivered. Strong bases and acids, and undiluted organic solvents cause some problems. Reagents (acids or bases) may cause bubbling, and reproducibility can be quite poor. Some organic solvents tend to "creep" in the system and adversely affect reproducibility. The different degrees of viscosity of samples and the degrees of washout of samples also affect the accuracy of these devices. In either case, autodilutors must receive daily maintenance and periodic recalibration in order to insure their reproducibility and accuracy.

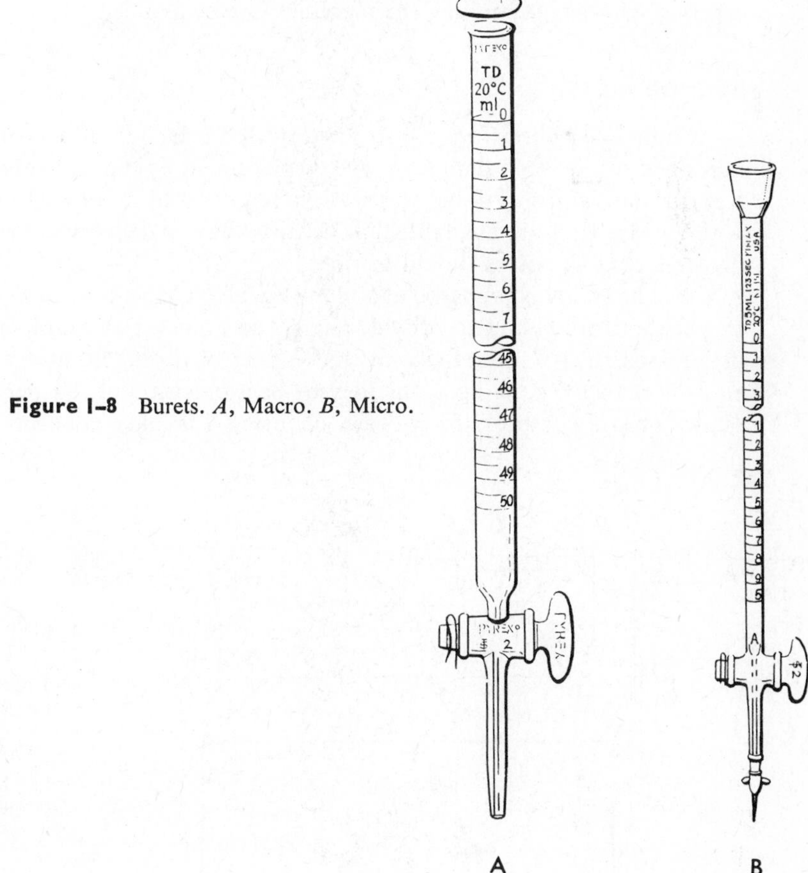

Figure I–8 Burets. *A*, Macro. *B*, Micro.

BURETS

The standard buret (Fig. 1–8) is found in sizes varying from 1 to 100 ml. A buret of 50 ml capacity is subdivided at 0.1 ml intervals. Burets having a capacity of 10 ml or less are classified as microburets.

The outflow of liquid from the buret is usually controlled by an all-glass or all-Teflon stopcock. The latter type does not require any lubricant and is especially useful when the titrant is an alkali. The all-glass stopcock should be lightly greased with petrolatum-like lubricants. Silicone-containing lubricants are not recommended since they "creep" along the length of the buret with subsequent contamination of the walls. Some burets are equipped with a reservoir and a 2-way stopcock for self-filling.

Burets (Fig. 1–8) are calibrated by first filling them to a point just above the zero line with freshly boiled and cooled (to room temperature) distilled water, and then carefully adjusting the meniscus to the zero line. The drop of water adhering to the buret tip is removed by touching the tip to the inside of a glass vessel. A tared vessel is placed beneath the tip and the buret is fully opened. Delivery should proceed freely with the buret tip not in contact with the vessel wall until the meniscus is about 1 cm above the graduation line to be tested. The buret tip should then be touched to the wall of the container and the content allowed to drain into the solution, until the meniscus reaches the graduation mark. The tared vessel is then stoppered and reweighed, and the delivered volume corrected for temperature. As in the case with graduated pipets, the precision of calibrations using water decreases with a decrease in buret size.

Burets used in macroanalysis have major graduation marks completely around the long cylindrical tube and minor graduation marks at least halfway around. By this means, errors of parallax in reading the meniscus can be avoided.

MICROBURETS

Some microburets are constructed on the principle of a syringe. The plunger can be machined to precision diameter. The displacement by the plunger is indicated on a micrometer dial and is proportional to the volume delivered. A typical model is shown in Figure 1–9. Various sizes are available that permit accurate delivery of as little as 0.01 μl per scale division. Several points should be noted:

1. The titrant is prepared at a relatively high concentration to reduce the volume necessary for a titration and to provide sharp end points. For example, if an end point can be determined to ± 1 scale division, the volume of titrant in an average titration should be equivalent to 100 scale divisions to provide a precision of ± 1 per cent. The concentration of titrant is a compromise between choosing a weaker concentration, which provides a

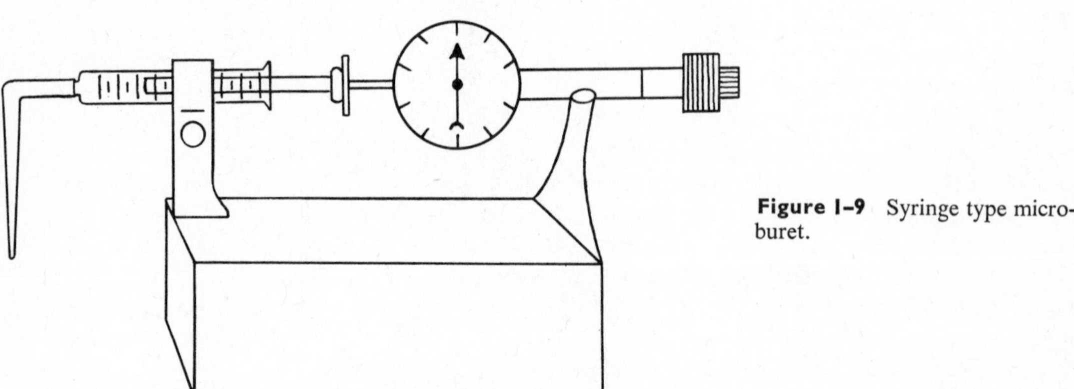

Figure 1–9 Syringe type microburet.

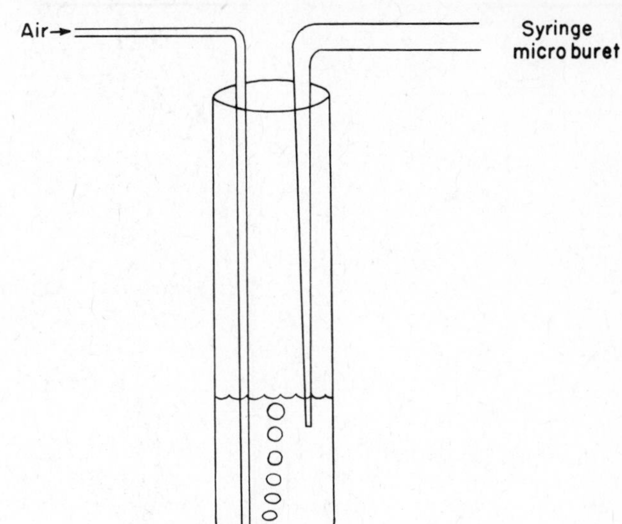

Figure 1–10 Microtitration in a test tube. Mixing is accomplished with a stream of air bubbles.

larger volume and therefore a smaller reading error, and a stronger concentration, which minimizes end point error and avoids overdilution of the sample.

2. The tip of the pipet is placed beneath the surface of the solution to be titrated to permit even dispensing of reagent and to avoid discrete drop formation. The tip should be drawn to a fine point to prevent diffusion of liquid back into the pipet.

3. Titrations may be performed in white porcelain spot plates to permit easier detection of color changes. The solution is stirred with a small glass rod as the titration proceeds. For titration in test tubes the solution may be mixed with a fine stream of air bubbles or an inert gas (Fig. 1–10). Other units employ vibrators or small cups that are rotated electrically to obtain mixing during the titration. Magnetic stirrers are also convenient. Very small stirring bars (e.g., 1 × 5 mm) are available commercially.

4. Blanks and standards should be included routinely and titrated to the same end point as the unknown. Indicator error is more significant in microtitrations; it is therefore advisable to pipet a fixed volume of indicator rather than to add it dropwise. Duplicate determinations are recommended.

VOLUMETRIC FLASKS

Volumetric flasks (Fig. 1–11) are found in the following sizes: 1, 2, 5, 10, 25, 50, 100, 200, 250, 500, 1000, 2000, and 4000 ml. They are primarily used in preparing solutions of known concentration.

An important factor in the use of volumetric apparatus is the need for accurate setting or reading of the meniscus level. A small piece of card that is half-black and half-white is most useful. The card is placed 1 cm behind the apparatus with the white half uppermost and the top of the black area about 1 mm below the meniscus. The meniscus then appears as a clearly defined thin black line. This device is also useful in reading the meniscus of a buret.

Volumetric equipment should be used with solutions equilibrated at room temperature. Solutions diluted in volumetric flasks should be repeatedly mixed during dilution so that the contents are quite homogeneous before making up to volume. In this way, errors due to expansion or contraction of liquids in mixing are negligible.

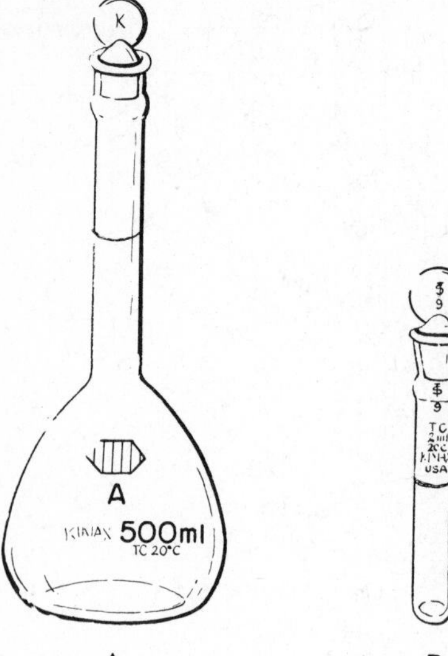

Figure 1–11 Volumetric flasks. *A*, Macro. *B*, Micro.

Volumetric flasks should be thoroughly cleaned and dried with purified ethanol or acetone before calibration. The flask is weighed and then filled with carbon dioxide-free distilled water until just above the graduation mark. The neck of the flask just above the water level should be kept free of water. The meniscus mark is set at the graduation line by removing excess water and the flask is reweighed. The final weight is corrected for the equilibrated water and air temperature to obtain the volume of the flask.

SEPARATORY FUNNELS

Separatory funnels (Fig. 1–12) are employed in the clinical chemistry laboratory for simple extraction procedures. This involves bringing a given volume of solution into con-

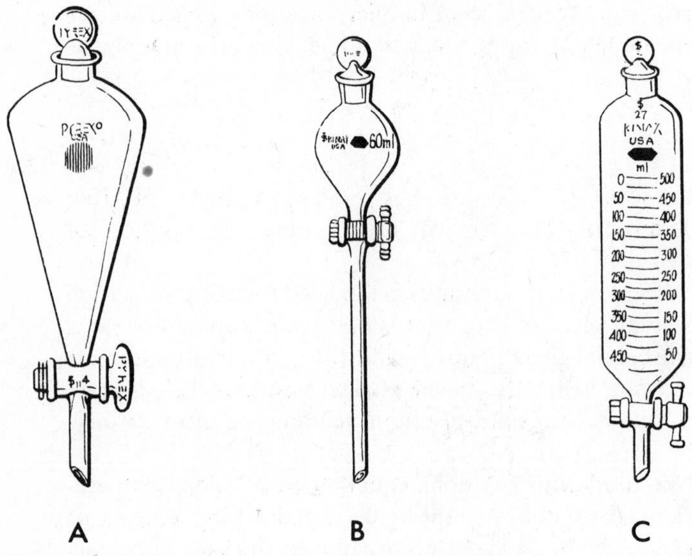

Figure 1–12 Separatory funnels. *A*, Pear shaped. *B*, Globe shaped. *C*, Cylindrical, graduated.

tact with a given volume of solvent (immiscible in the solution) and shaking until equilibrium has been attained, followed by separation of the liquid layers. Most separatory funnels taper off to a narrow bottom, which contains a sealed stopcock. Thus, it is relatively easy to separate two phases for further analysis. The capacity of the funnel should be such that the liquids occupy no more than approximately one-half the volume of the funnel. If necessary, the extraction procedure may be repeated after the addition of fresh solvent. This type of extraction gives rapid, simple, and clean separations. When extracting from a liquid to a lighter solvent, it is necessary to remove the lower phase from the funnel after each extraction before removing the extraction solvent, as in ethyl ether extractions of aqueous solutions.

MAINTENANCE AND CARE OF GLASSWARE

It is essential that volumetric glassware and glass apparatus be absolutely clean; otherwise, volumes measured will be inaccurate and chemical reactions affected adversely. One gross method generally used to test for cleanliness is to fill the vessel with distilled water and then empty it and examine the walls to see whether they are covered by a continuous thin film of water. Imperfect wetting or the presence of discrete droplets of water indicates that the vessel is not sufficiently clean. Residual detergent may be detected by measuring the pH of water added to the glassware or more simply by using a dilute solution of BSP dye.

A wide variety of methods has been suggested for the cleaning of volumetric glassware. Of the various cleaning agents in common usage, the National Bureau of Standards prefers fuming sulfuric acid and a chromic-sulfuric acid mixture. The latter is used by most laboratories. It is imperative that glassware cleaning be as mild as possible and appropriate to the type of contamination present. Fats and grease are the most frequent causes of severe contamination and it is advisable to dissolve these contaminants by a lipid solvent (water-miscible organic solvent) followed by water washing before removing the last traces with an oxidizing agent (chromic acid). The most widely used oxidant is a solution of sodium dichromate in concentrated sulfuric acid. Because of its oxidizing power, the solution, particularly when hot, removes grease quickly and completely. Cleaning solution, as the mixture is called, is not a general solution for cleaning all apparatus, but only for cleaning borosilicate glassware, including volumetric ware. Glassware is generally left in contact with the mixture for 1 to 24 hours, depending upon the amount of grease or lipid present. After removal of the acid and draining, the glassware should be washed out at least four times with tap water and then three times with distilled or deionized water. Glassware contaminated with chromic and sulfuric acid cleaning solution is unsuitable for enzyme analyses because even minute traces of these acids will inactivate enzymes. Soap, especially when used in calcium-containing water, leaves a film on the glassware which cannot be washed away even with many rinsings. Alkaline detergents, such as are used in commercial dishwashing apparatus, will etch glass surfaces and also strongly inhibit enzymes. The use of neutral detergents with adequate rinses (at least three) of deionized water are therefore recommended for glassware washing. If possible, commercially available laboratory dishwashers should be employed. It should be noted, however, that many detergents contain large amounts of sodium and phosphorus which must be rinsed from all glassware.

Ultrasonic cleaning devices with specially adapted cleaning solutions have also been useful in cleaning glassware and equipment by a combination of chemical and physical forces. The technique is especially advantageous when standard cleaning techniques are ineffective (for small orifices, glass coils).

Cleaning microglassware

Because of the small sample size used in microwork, the effect of possible contamination is much greater; thus, all glassware must be scrupulously cleaned. An entire sample of

20 μl of serum contains only about 64 μg of sodium. Detergents frequently contain trisodium phosphate or other sodium salts and preferably should be avoided. A satisfactory procedure is to immerse glassware in, or fill it with, a 100-fold dilution of ammonium hydroxide and let it stand to dissolve protein films or precipitates. The glassware is then rinsed several times, in turn, with water, dilute hydrochloric acid (100-fold), water, and distilled water before drying at 100°C. Pipets, after cleaning, may be rinsed with acetone, then placed on an aspirator for a few minutes and dried by a stream of air. Polyethylene pipets and other ware may be cleaned in a similar manner, but they are never oven dried since softening and distortion occur above 80°C.

For more thorough cleaning, glassware is soaked in chromic acid solution followed by rinsing with water, dilute ammonium hydroxide, and dilute hydrochloric acid. The glassware is then soaked for two days in distilled water to remove all traces of chromic acid. Dilute solutions of pepsin (1 g/100 ml in 0.1 molar HCl) have also been recommended for soaking glassware to digest protein films and precipitates.

Preparation of Chromic Acid Cleaning Solution. Stir about 20 g of powdered technical sodium dichromate with just enough water to make a thick paste. Slowly and carefully add 300 ml of technical grade concentrated sulfuric acid, stirring well. This preparation is best carried out in a sink. Store in a glass-stoppered bottle or covered glass jar. Clear supernatant solution should be decanted from the bottle each time it is used. The solution may be used repeatedly until the reddish color of dichromate has been replaced by the green color of the chromic ion. Do not allow this cleaning solution to come into contact with you or your clothing because it will burn the skin severely and destroy clothing. If any acid is spilled on the floor or bench top, neutralize it immediately with commercial grade sodium bicarbonate; then wash completely with water. If it is spilled on the skin, wash under running water as soon as possible.

ALTERNATIVE METHODS OF CLEANING. These include the use of fuming nitric acid, 10 per cent alcoholic potassium hydroxide, hydrochloric-nitric acid (1/1), 50 per cent potash solution, and acid, neutral or alkaline permanganate solutions. In case of the latter, manganese dioxide separates at the grease sites and its subsequent removal by concentrated hydrochloric acid causes the evolution of chlorine, which destroys the grease. Precipitated material may also be removed by aqua regia, fuming sulfuric acid, or an ultrasonic device. When chemical cleaning is inadequate, mechanical cleaning using warm soapy water or synthetic alkaline detergents can be used to good effect. Care is necessary to insure complete removal of these cleaning agents, since even low concentrations can result not only in a chemical contamination but also in a marked lowering of the surface tension of aqueous solutions with consequent change in meniscus shape. In extreme cases, steaming out glassware has been suggested, but the thermal retardation effect on the glass can cause measurable changes in its volume.

Borosilicate laboratory glassware is extremely resistant to acidic cleaning solutions, with the exception of hydrofluoric acid; however, strongly alkaline solutions will attack any glass over a period of time. Therefore, borosilicate glassware should generally be washed in a cleaning solution that is neutral or acid in reaction. It is important that scratching of glass be avoided since scratching diminishes the thermal shock resistance of glassware and can lead to breakage.

Care of Absorption Cells (Cuvets). Absorption cells must be scrupulously clean. Optical surfaces should not be touched since grease smudges are difficult to remove. As soon as possible after use, absorption cells should be rinsed and soaked in distilled water. When cleaning cells, a mild detergent should be used. Ultrasonic cleaning devices can also be an effective way to clean cells. Stubborn contaminants can be removed by soaking the cells in dilute sulfuric acid. Absorption cells should never be allowed to soak in hot concentrated acids, alkalies, or other agents that may etch the optical surfaces. When drying cuvets, high temperatures and unclean air should be avoided. A low to medium temperature oven (not to exceed 100°C) or vacuum, or a combination of the two, can be used to dry cuvets rapidly. Drying cuvets with acetone is to be discouraged since a film may be left on the optical surface.

Cleaning Pipets. Pipets should be placed in a vertical position with the tips up in a jar of cleaning solution. A pad of glass wool is placed at the bottom of the jar to prevent breakage. After soaking for several hours, the pipets are drained and rinsed with tap water until all traces of cleaning solution are removed. The pipets are then soaked in distilled water for at least an hour. A gross test for cleanliness is made by filling with water, allowing the pipets to empty, and observing whether drops form on the side within the graduated portion. Formation of drops indicates greasy surfaces. After the final distilled water rinse, the pipets are dried in an oven at not more than 110°C. Most laboratories that use large numbers of pipets daily use a convenient automatic pipet washer. These devices are made of metal or polyethylene and can be connected directly to hot and cold water supplies. Polyethylene baskets and jars may be used for soaking and rinsing pipets in chromic acid cleaning solution.

Cleaning and Care of Burets. Inspect the stopcock plug before cleaning. If it is well greased, the plug will turn easily and the surface between the plug and barrel will appear transparent. If the plug needs greasing, remove it and wipe it clean; also wipe out the inside of the barrel. Both parts must be dry before greasing. Apply a thin layer of good grade stopcock grease. Heat about 100 ml of chromic acid cleaning solution to a temperature of 60 to 70°C. After clamping the buret in an inverted position with the opening reaching nearly to the bottom of the beaker, draw the cleaning solution into the buret by suction until the level is slightly past the final graduation mark. Do not allow the cleaning solution to reach the stopcock area where it will remove the grease. After closing the stopcock, allow the filled buret to stand 3 to 5 minutes. The stopcock is then opened and the buret raised above the liquid level and allowed to drain thoroughly. The buret should then be flushed out well with tap water and finally rinsed at least three times with distilled water. When all traces of cleaning solution have been removed and the buret has been properly rinsed, drops of water will not adhere to the inner wall surface. The buret should be filled with clean distilled water and left in this state until needed. If the buret is left empty, it will quickly become contaminated with a film of grease.

Cleaning Flasks, Beakers, Cylinders, Volumetric Equipment, and Other Glassware. Pour warm cleaning solution into each vessel and stopper or cover carefully. Each vessel should be manipulated so that all portions of the wall are repeatedly brought into contact with the solution. This procedure should be followed for at least 5 minutes. The cleaning solution can be poured from one vessel to another and then returned to its original container. The vessels should then be rinsed repeatedly with tap water (four rinses) and finally rinsed three times with distilled water. It is important that the necks of volumetric flasks above the graduation mark be clean because, when solutions are diluted in the flask, drops of water may adhere to an unclean wall and may invalidate the measurement of volume.

FILTER PAPERS

A properly stocked clinical chemistry laboratory must have on hand filter papers in a variety of sizes and characteristics. These differ in thickness, porosity, and wet strength.

Filter papers can be classified into various categories. Within each category there are papers with slow, medium, and rapid filtering speeds, which correspond to fine, medium, and coarse retentions. A grade of filter paper is selected on the basis of precipitate size and method of filtration (gravity or suction). The success or failure of a separation depends a great deal on the method of handling the filtration. The majority of analytical filtrations are made with a 60° funnel using a gravity filtration. Many analysts have their own individual methods of folding filter paper and, in general, paper for use in a funnel should be just large enough to reach the edge of the funnel or slightly below and should never extend over the top. Filter papers with diameters of 9.0 and 12.5 cm are in common use in the clinical laboratory because they fit funnels generally used for preparation of blood and urine filtrates. Precipitates are poured into the paper with the aid of a stirring rod to avoid splashing and to keep the precipitate near the apex of the filter paper cone. Suction and other filter aids (accelerators, ashless floc, ashless clipping) may be used to enhance the filtration process.

Table 1–3 describes the grade, flow rate, and retention characteristics of commonly used filter papers. Retention refers to the type of precipitate the grade will retain; speed refers to relative mean flow rates.

Protein precipitates are best removed on a smooth surfaced, medium speed filter paper. Gelatinous precipitates such as the hydroxides of many multivalent metals are quantitatively retained on a rapid filtering paper. The granular crystalline precipitates of metallic halides and phosphates are best collected on a slower paper of a medium texture. Finely divided precipitates such as calcium oxalate or barium sulfate should be filtered using a very slow, fine pore paper.

Filter paper circles are made in several diameters, of which the most common are 5.5, 7, 9, 11, 12.5, and 15 cm. The size taken for a particular filtration depends on the bulk of the precipitate to be caught, and the volume of the solution to be filtered. The total insoluble matter after filtration should occupy less than one-third of the volume of the paper cone.

An additional filtering aid is Whatman *Phase Separating Paper* No. 1 PS. This paper is a water repellent cellulose acetate that acts as a phase separator in place of conventional separatory funnels. It separates aqueous solutions from water-immiscible solutions quickly and conveniently. Because this paper is hydrophobic, it retains the aqueous layer but allows the organic layer to go through. It also filters out any solid that may be present in the system at the time of filtration. This phase separating paper can be used with organic solvents that are lighter or heavier than water. If the organic solvent is lighter than water, it will gradually migrate to the edge of the water meniscus and pass through the paper, leaving the water phase behind. If heavier than water, it will pass directly through the paper, and the water will be retained.

Glass fiber filters are also available for clinical techniques. These filters are produced entirely from borosilicate fibers and when used in a Gooch, Büchner, or similar filtering apparatus they give a combination of fine retention with extremely rapid filtering speed not usually found in any cellulose grade. Glass fiber filters are well suited for filtration of heavy

TABLE 1–3 SELECTION OF USEFUL FILTER PAPERS—PROPERTIES OF THE VARIOUS GRADES

S and S Grade	*Whatman No.*	*Retention*	*Speed**	*Thickness (inches)*	*Surface*
595	1	Medium	75	0.005	Smooth
597	2	Fine	110	0.006	Smooth
604	4	Coarse	23	0.007	Smooth
602	5	Very fine	275	0.007	Rough
588	12	Whatman No. 2 ready folded			
402	30	Medium	95	0.007	Rough
589-white	40	Medium	95	0.006	Rough
589-black	41	Coarse	22	0.007	Smooth
589-1H	41H	Coarse	17	0.005	Smooth
589-blue	42	Very fine	300	0.007	Rough
589-red	44	Very fine	250	0.006	Rough

* Relative flow values; the larger the number, the slower the filtration rate.

viscous solutions or gels since they do not clog as quickly as open cellulose papers and yet can retain very fine particles.

FILTER MEDIA FOR SPECIAL APPLICATIONS

Cellulose products for chromatography and electrophoresis can be separated into four general classes: (1) papers, (2) powders, (3) ion-exchange celluloses, and (4) specialties.

Papers. The papers for chromatography and electrophoresis vary with respect to flow rate, loading capacity, and adsorption. The flow rates are relative and are classified as fast, medium, and slow. In general, the loading capacity is related to the thickness of the paper, and the thicker the grade, the more heavily it can be loaded. Adsorption on the cellulose fibers of one or more of the constituents being separated occurs occasionally because of the nature of the cellulose itself. The thin grades like Whatman No. 1 and No. 44 and the hardened grades show the least adsorption.

Powders. Cellulose powders for column chromatography meet the demand for a cellulose material capable of larger scale separations than is possible on paper. The powders are fibrous in nature and, in this respect, differ from many of the chromatographic adsorbents, which are powdered inorganic compounds. The grades can be of a coarse material with no consistently measurable particle size and of a powder ground to pass a 200 mesh sieve and give a closely packed, slow running column.

Ion-exchange Celluloses. These materials include phosphorylated cellulose, CM (carboxymethyl) cellulose, DEAE (diethylaminoethyl) cellulose, ECTEOLA (epichlorohydrin triethanolamine), and other celluloses.

Specialties. These include the many individual precut papers and membranes of special design useful for chromatographic and electrophoretic techniques. In this group are barium-impregnated strips for the removal of bilirubin in the estimation of urobilinogen, and cation- and anion-exchange paper useful for chromatography of amino acids.

Cellulose acetate membranes made of homogeneous cellulose acetate with uniform pores of less than 2 μm diameter are used in electrophoresis systems to separate proteins, hemoglobins, immunoglobulins, and various isoenzymes. The porosity of these membranes when carefully controlled permits a precisely predictable flow of buffer solution between the cell reservoirs.

The Millipore filters (Millipore Corp., Bedford, MA.) are cellulose acetate porous membranes ranging in pore size from 8 μm downward to 10 nm. These filters enable the analyst to concentrate efficiently any particles by pore size. The Millipore filter is free of biological inhibitors and provides an optimum collection environment for microorganisms.

Centriflo membrane filter cones (Amicon Corp., Lexington, MA.) have recently been introduced in ultrafiltration techniques for concentration and purification of multiple small samples of biological fluids. The Centriflo cones retain molecules with a molecular mass greater than 50,000 and are intended essentially for protein concentration.

These anisotropic membranes, highly retentive for the serum protein components, permit passage of low molecular-mass moieties and ionic species. Since the isolation of the diffusible (nonprotein-bound) calcium fraction of serum by more conventional ultrafiltration is laborious and must be done under rigorous control, the Centriflo membrane has been helpful in separating diffusible from protein-bound serum calcium by centrifugation through high flux ultrafiltration membrane cones.[12] These membranes have also been useful for rapidly concentrating cerebrospinal fluid or urine before protein electrophoresis.[31] Recently, Centriflo membrane cones, which can retain molecules with a molecular mass greater than 20,000, have been used to concentrate Bence-Jones proteins in urine prior to electrophoresis. The advent of fine hollow fibers[2] with walls of well-defined porosity has also provided the means for incorporating massive amounts of available surface for ultrafiltrate production or concentration of protein-containing solutions.

MEASURES AND WEIGHTS

Ultimately, most of the determinations in clinical chemistry are concerned with measurements of volume and mass. Every measurement includes both a number and a unit. The unit identifies the kind of dimension(s) of the property being measured, and the number indicates how many of the reference units are contained in the quantity being measured. A coherent system of units is formed by first defining ten units, which by convention are regarded as dimensionally unrelated and independent. These are listed in Table 1–4 and include seven "basic" units, two supplementary units, and a provisional unit for the quantity of enzyme activity. All other units (coherent derived units) are defined in terms of products and/or quotients of these basic units. Several noncoherent units, such as the "litre" (liter) not derivable from the basic units, are permitted and included among the SI units. The term "metric system," still in common use, should now be interpreted to mean the SI system.

TABLE 1–4 BASIC UNITS IN THE INTERNATIONAL SYSTEM OF UNITS (SI SYSTEM)

Type	*Quantity*	*Name*	*Symbol*
Basic	Length	metre*	m
Basic	Mass	kilogram	kg
Basic	Time	second	s
Basic	Electric current	ampere	A
Basic	Temperature	kelvin	K
Basic	Luminous intensity	candela	cd
Basic	Amount of substance	mole	mol
Provisional	Amount of enzyme activity	unit†	U†
Supplementary	Plane angle	radian	rad
Supplementary	Solid angle	steradian	sr

* The spelling "meter" is more commonly used in the United States.
† The use of U is now discouraged. Enzyme *activity* is to be expressed in mol/s, while enzyme *concentration* is to be expressed in katal (kat)/l.

Nomenclature for prefixes

Prefixes approved by the CGPM (Conférence Générale des Poids et Mésures), 1964, and the International Congress of Clinical Chemistry, 1966, for use in clinical chemistry that denote the approved decimal factors are shown in Table 1–5. It is recommended that the clinical laboratory have the book by Dybkaer and Jorgensen,[10] the article by Dybkaer,[11] or NBS Special Publication 330[28] as available references.

TABLE 1–5 PREFIXES DENOTING DECIMAL FACTORS

Factor	*Name*	*Symbol*	*Factor*	*Name*	*Symbol*
10^{12}	tera	T	10^{-1}	deci	d
10^{9}	giga	G	10^{-2}	centi	c
10^{6}	mega	M	10^{-3}	milli	m
10^{3}	kilo	k	10^{-6}	micro	μ
10^{2}	hecto	h	10^{-9}	nano	n
10^{1}	deka	da	10^{-12}	pico	p
			10^{-15}	femto	f
			10^{-18}	atto	a

LENGTH

The SI unit for length is the meter, although the centimeter (cm = 10^{-2} m) and millimeter (mm = 10^{-3} m) are the subunits most often used in clinical work. Other multiples and fractional units are as follows:

1 kilometer (km) (10^3 m) = 1000 meters (m)

1 meter (m) = 1000 millimeters (mm)

1 millimeter (10^{-3} m) = 1000 micrometers (μm)

1 micrometer (10^{-6} m) = 1000 nanometers (nm)

1 nanometer (10^{-9} m) = 1000 picometers (pm)

The use of Angstrom unit (Å), a noncoherent unit (= 10^{-10} m = 10^{-1} nm) is permitted, but not encouraged. The micron (10^{-6} m = μ) is replaced by the coherent unit, the micrometer (μm) and is no longer to be used.

Wavelength (λ)

Radiant energy can be defined in terms of (1) the wavelength of the electromagnetic waves carrying the energy, (2) the frequency of the radiation (in cycles per second), or (3) the wave number (cycles/cm or cycles/m). The visible portion of the electromagnetic spectrum extends over the range of approximately 400 to 700 nm (4000 to 7000 Å). The infrared region (short infrared and long infrared) covers the range of about 700 to 25,000 nm (0.70 to 25 μm, 13.3 to 0.4 $\times$ 10^5 cycles/m).

MASS

Mass refers to the quantity of matter; the unit is the kilogram, equal to the mass of the platinum-iridium international prototype of the kilogram. The ambiguous term *weight* reflects the force of gravity, and therefore cannot be used in the sense of mass. The subunits of mass are gram (g), milligram (mg), microgram (μg), and so forth. The term gamma (γ) for microgram is not to be used. The use of the term parts per million (ppm) is also discouraged and should be replaced by a coherent unit such as μg/g.

VOLUME

The coherent derived unit for volume is the cubic meter (m^3). This unit is quite large, and the derived unit, the cubic decimeter (dm^3 = 10^{-3} m^3) is a unit of more practical magnitude for the clinical laboratory. In 1964 the CGPM accepted the use of "litre" (liter) as a special name for the cubic decimeter (dm^3). Commonly used subunits are milli-, micro-, and nanometer (ml, μl, nl). It follows that 1 cm^3 = 1 ml, and 1 mm^3 = 1 μl.

AMOUNT OF SUBSTANCE

The unit for the amount of (chemical) substance is the mole (mol). This is defined as the quantity of chemical matter equal to that present in 0.0120 kg of pure carbon-12; or, more precisely, as that amount of substance which contains as many elementary entities as there are atoms in 0.0120 kg (one Avogadro's number = 6.0225 $\times$ 10^{23}). It is important that the nature of the elementary particles be clearly specified; they may be atoms, molecules, ions, electrons, chemical radicals, chemical residues, other chemically reactive particles, or specified groups of such particles. The dimension of the mole is the *relative atomic* or *molecular mass unit* (dalton). These terms replace the commonly used terms atomic and molecular weight.

TIME

The standard unit of time is the second (s) which is the duration of 9,192,631,770 periods of the radiation corresponding to the transition between two hyperfine levels of the ground state of the cesium-133 atom.[28] One hour (h) is equal to 60 minutes (min) or 3600 seconds. The symbol for 24 hours is d (diem).

INTERCONVERSION BETWEEN NONMETRIC AND SI UNITS

Until the use of SI units becomes established and legally obligatory, many quantities of clinical chemical interest will be measured in nonmetric units, and conversion from the common units to SI units may on occasion be required. Some of the more important conversion factors are listed in the following table (see also table in the Appendix).

Number Values In	*Multiply By* $\longrightarrow$ $(\longleftarrow)$	*Number Values In*
inch	2.54 (0.3937)	cm
quart	0.9463 (1.057)	l
cubic inch	16.39 (0.0610)	ml
fluid ounce	29.57 (0.0338)	ml
pound	453.6 (0.00220)	g
ounce	28.35 (0.0353)	g
grain	65 (0.0154)	mg

GAS CONCENTRATION AND PRESSURE

The common units for gas concentrations in blood are volume per cent (v/v) and, preferably, mmol/l. For example, the CO_2 content of blood is expressed in mmol/l. If all the carbon dioxide in the blood were in the form of bicarbonate, mEq/l of bicarbonate and mmol/l of carbon dioxide would be identical. Since this is not the case (carbon dioxide in the blood is in the form of free CO_2, HCO_3^-, and H_2CO_3), it is best to report CO_2 levels in mmol/l (mEq/l is no longer preferred, CGPM, 1964).

Until recently the common unit of expression for partial pressure of a gas, e.g., carbon dioxide, was mm Hg; however, it is now recommended that "pascal" (Pa) be substituted for the conventional millimeter of mercury (1 mm Hg = 133.3224 Pa). At the present time this recommendation has not been followed in the United States, and mm Hg is still widely used. (See also under Blood Gases, Chapter 15, Part I.)

IONIC STRENGTH

The concentration of buffer solutions is often given in terms of ionic strength since the effect of a buffer solution is a function of the concentration and the charge of all ions present. The ionic strength is one-half the sum of the numbers obtained by multiplying the concentration (mol/l) of each ion by the square of its valence.

Example: 0.2 mol/l NaCl

$$\text{ionic strength} = \frac{0.2 \times (1)^2 + 0.2 \times (1)^2}{2}$$

$$\text{ionic strength} = 0.2$$

0.5 mol/l Na_2SO_4

$$\text{ionic strength} = \frac{0.5 \times 2 \times (1)^2 + 0.5 \times (2)^2}{2}$$

$$\text{ionic strength} = 1.5$$

CHEMICALS

Laboratory chemicals are supplied in about six grades. The differentiation of these is quite unofficial and there is no agreement among manufacturers concerning the designation of the various degrees of purity. The most common designations follow.

REAGENT GRADE OR ANALYTICAL REAGENT GRADE (AR)

This degree of purity belongs to several hundred chemicals called reagents; these chemicals meet specifications designed to permit use in quantitative and qualitative analyses. These are found in two forms: (1) *lot analyzed reagents*, in which each individual lot is analyzed and the actual amount of impurity reported (e.g., arsenic—0.0005 per cent) and (2) *maximum impurities reagents*, for which maximum impurities are listed (e.g., arsenic—maximum 0.001 per cent). In the latter instance, the arsenic may only be 0.0004 per cent, but, of course, the analyst has no tangible figure to put his finger on other than the guaranteed maximum limit of 0.001 per cent.

In this reagent group the specifications have been prepared by the Committee on Analytical Reagents of the American Chemical Society for many of the most commonly used items.[22] Establishment of the ACS specifications marked a turning point in the development of chemical purity. Manufacturers of ACS chemicals check each lot in a control laboratory and only place ACS on the labels of those lot chemicals that meet the Society's published specifications. These reagent grade chemicals are of very high purity and are recommended for trace metal analyses and for standardization of reference methods. In addition, spectroanalyzed solvents and chromatographically pure reagents (e.g., amino acids) are now commercially available.

CHEMICALLY PURE GRADE (CP)

The degree of purity of materials bearing this label is shared by the terms "highest purity" and "chemically pure." The designation chemically pure fails to reveal what limits of impurities are tolerated, and the practice followed by different manufacturers in the use of this designation is not uniform. Chemicals in this category are probably not dependable for research and various clinical chemical techniques unless the chemist has analyzed the materials to assure himself of the absence of impurities that may cause trouble. The term highest purity is used by manufacturers for organic chemicals that they have purified to as great a degree as they find practical. The purity is usually determined by measurement of melting points or boiling points. Again, for research purposes, this grade of purity may be unsatisfactory without further purification. This group of chemicals may be used in clinical chemical analysis when higher purity biochemicals are not available.

USP AND NF GRADE

These chemicals are produced to meet specifications set down in the *United States Pharmacopeia* (USP) or the *National Formulary* (NF). These designations are of primary concern to the pharmaceutical chemist, and the tolerances specified for impurities are those that will not be injurious to health. In many cases these compounds may be very pure and can be used in chemical analysis and in the preparation of various reagents, but this cannot be assumed to be the case in all instances. The important thing to remember is that in these categories chemical purity is only incidental.

PURIFIED, PRACTICAL, OR PURE GRADE

These chemicals can be used as starting materials for laboratory synthesis of other chemicals of greater purity, but probably require purification and analysis before they can be used as analytical reagents. For certain analyses when reagents are not available, practical chemicals can be used if a blank is also run. In general these chemicals should not be used in clinical chemical analysis.

TECHNICAL OR COMMERCIAL GRADE

These chemicals are generally used only in manufacturing. The degree of purity varies widely and depends on the ease with which contaminants can be removed.

PURITY OF ORGANIC REAGENTS

The purity of commercially obtained organic reagents for clinical chemistry purposes is generally inferior to that of inorganic reagents. The majority of impurities in these compounds will have been introduced in their synthesis either with the starting materials or as by-products, and these are presumably more difficult or more expensive to remove than impurities in inorganic substances. In addition, some organic compounds oxidize or decompose on standing and the amount of impurities from this cause will depend on how long the bottle of reagent has been opened or stored. A well-known example of this sort of deterioration is provided by solutions of the chloride indicator s-diphenylcarbazone. This orange-yellow solution must be protected from light, since it turns orange-red when exposed to daylight and cannot be used thereafter. Its stability is much improved when it is stored in amber bottles and refrigerated; however, phenols and amines oxidize on standing and tend to darken even when refrigerated. Sugars have been shown to be quite hygroscopic and absorb moisture rapidly unless they are properly stored.

The presence of impurities in an organic reagent may be a source of difficulty in its use. If the contaminant does not react with the substance being determined, interference will not occur as long as there are no interfering side reactions and there is enough of the original reactant remaining. If a reagent is impure, the net final color developed by a reaction may be considerably less than ideal because of a high blank due to the impurity. The existence of isomers and their presence in a particular lot of an organic reagent may be a source of difficulty, since in the rather specific geometrical requirements of a chelate ring only certain isomers may produce the desired colored complex,[17] or in enzyme reactions only one of the isomers may be suitable as substrate.

PRIMARY STANDARDS

These highly purified chemicals may be weighed out directly for the preparation of solutions of selected concentration or for the standardization of solutions of unknown strength. They are supplied with an analysis for contaminating elements and an assay for each lot (see Table 1–6). The assay should not be less than 99.95 per cent. The specifications for primary standards have been prepared by the Committee on Analytical Reagents of the American Chemical Society.[22] These chemicals must be stable substances of definite composition, which can be dried, preferably at 104°C to 110°C, without a change in composition. They must not be hygroscopic, so that water is not adsorbed during weighing (see discussion on preparation of standard solutions, pp. 32, 39).

Secondary standards are solutions whose concentration cannot be determined directly from the weight of solute and volume of solution. The concentration of secondary standards is usually determined by analysis of an aliquot of the solution by an acceptable reference method, using a primary standard.

REFERENCE STANDARDS

The purity of some chemicals used in the clinical chemistry laboratory has been assured by the availability of standard reference materials (SRM) from the National Bureau of Standards (Table 1–6). In the near future, additional compounds will be available.

TABLE 1-6 CLINICAL CHEMISTRY STANDARDS FROM NBS

SRM No.	*Constituent*	*Purity* (%)
911	Cholesterol	99.4
912	Urea	99.7
913	Uric acid	99.7
914	Creatinine	99.8
915	Calcium carbonate	99.9
916	Bilirubin	99.0
917	D-Glucose	99.9
918	Potassium chloride	99.9
919	Sodium chloride	99.9
921	Cortisol	98.9
	(Total steroids)	(99.9)
924	Lithium carbonate	100.5

CERTIFIED STANDARDS

There are a few standards that are certified by the College of American Pathologists. Examples of these are bilirubin and cyanmethemoglobin.

The National Committee for Clinical Laboratory Standards (NCCLS) has made available a standardized protein solution. Others are in preparation.

DESICCANTS AND USE OF DESICCATORS

Most published information concerning the comparative efficiency of drying agents is based upon experiments that measure the amount of moisture absorbed from air flowing through a bed of desiccant.[4] The limited information that is available concerning the behavior of drying reagents in a desiccator suggests that their comparative usefulness for this purpose may not be the same as in flow measurements. Every effort should be made to avoid desiccants that produce dust when used in desiccators (granular calcium chloride frequently carries a large amount of dusty "fines" when fresh). Drying agents that incorporate cobalt chloride or other indicators to show when they are exhausted are much preferred to those that do not. Silica gel and anhydrous calcium sulfate (Drierite) are sold in indicating forms.

Desiccators should be opened carefully. Ordinary desiccators may contain air at less than atmospheric pressure; this is a result of heating of the air when hot specimens are inserted and then cooling of the air with the lid on. If the desiccator is not opened slowly, the inrush of air may create draughts sufficient to dislodge materials from open vessels or to stir up dust particles from the drying agent that may subsequently settle in the vessels that are being stored. Vacuum desiccators should be provided with a curved inlet tube to deflect incoming air against the lid, and the stopcocks on these should be opened very carefully when restoring the internal pressure to that of the atmosphere.

Table 1–7 describes various desiccants. From this table it is apparent that several are distinctly alkaline and one is strongly acidic. The choice of the drying reagent required for the quantitative absorption of moisture depends on the composition of the gases or materials to be dried, convenience, efficiency, and sometimes cost. Magnesium perchlorate quantitatively absorbs ammonia gas, and anhydrous calcium chloride (technical grade) absorbs carbon dioxide and ammonia. These facts should be kept in mind when choosing a desiccant intended for the quantitative removal of water from gases that may also contain ammonia, carbon dioxide, or other reactive substances. Certain drying reagents are deliquescent and, when liquefaction of the drying agent occurs, a decline in drying efficiency results. Calcium chloride and magnesium perchlorate are examples; however, both have a considerable capacity before deliquescence sets in. Phosphorus pentoxide is one of the most powerful drying agents in use, but its effective capacity is rapidly reduced by formation of metaphosphoric acid. Drying agents prepared with moisture-sensitive salts, which indicate exhaustion of drying capacity by a change in color (such as silica gel, activated alumina, and anhydrous calcium sulfate), are advantageous but considerably more costly.

Some drying agents can easily be regenerated, and this is an important consideration when comparing costs. Silica gel can be regenerated by heating in a drying oven at 120°C, but anhydrous calcium sulfate and activated alumina require temperatures of 275°C and 175°C, respectively. Magnesium perchlorate can be regenerated by heating to 240°C in a partial vacuum.

TABLE 1–7 CHEMISTRY AND ACTIVITY OF DESICCANTS

Drying Agent	*Activity**	*Capacity*	*Deliquescence*	*Easy Regeneration*	*Chemical Reaction*
Phosphorus pentoxide	0.02	v. low	yes	no	acidic
Barium oxide	0.6–0.8	moderate	no	no	alkaline
Alumina	0.8–1.2	low	no	yes	neutral
Magnesium perchlorate (anhydrous)	1.6–2.4	high	yes	no	neutral
Calcium sulfate (Drierite)	4–6	moderate	no	yes	neutral
Silica gel	2–10	low	no	yes	neutral
Potassium hydroxide (stick)	10–17	moderate	yes	no	alkaline
Calcium chloride (anhydrous)	330–380	high	yes	no	neutral

* Micrograms residual water per liter of air at 30°C.

DISTILLED AND DEIONIZED WATER

Distilled or deionized water is necessary for the preparation of all reagents and solutions in the laboratory. Even such water is not entirely pure, however, for it is contaminated by dissolved gases and by material dissolved from the container in which it has been stored. The dissolved gases may be removed by boiling the water for a short time. Occasionally distilled water is found to be contaminated by nonvolatile impurities that have been carried over by steam, in the form of a spray. The substances most likely to be carried over by spurting or spray are sodium, potassium, manganese, carbonates, and sulfates. The kinds and types of impurities introduced into the distilled water from the distilling apparatus also depend on the material used in the construction of the equipment. The most common impurities of this type are copper and glass products. Also, a still may froth while in operation and badly contaminate the distillate.

When water of the highest purity is required, distilled water is redistilled from an alkaline permanganate solution, in a silica or block tin apparatus. The permanganate solution oxidizes nitrogenous matter present. Redistilled water prepared in this manner is known as conductivity water. Many distillation units are now equipped with deionization resin beds, which feed deionized water into the still and thereby make the redistillation from an alkaline permanganate solution unnecessary.

The storage vessel employed may have a marked effect on the purity of water and reagents. Solutions stored in soda glass vessels are much more easily contaminated by trace metals then those stored in borosilicate glass vessels. Thus, when reagents or water are to be stored for long periods of time, it is desirable to use Pyrex bottles.

The simplest overall check on the quality of distilled and deionized water is the measurement of specific conductance. This is most conveniently monitored by a conductivity warning light installed in the effluent water line. Laboratories should endeavor to keep the resistance of their water over 1 million ohms/cm. However, the measurement of conductance does not take into account the nonionized substances (organic contaminants) that may be present in the water, and the conductivity of a given sample may in part be due to dissolved gases ($CO_2 \rightleftharpoons H_2CO_3$). CO_2-free water may be prepared by boiling the water immediately prior to use or by collecting water from an all-glass still and avoiding contact with air.

In a recent publication of the College of American Pathologists, the specifications were given for a higher grade of reagent water that must be used for atomic absorption spectrophotometry and may be needed (depending on methodology and sensitivity) for such procedures as enzyme measurements, pH determinations and flame photometry. This water, in addition to meeting the general requirements listed below, should have a resistance over 10 million ohms/cm.[22]

The following properties meet the general analytical requirements of distilled or deionized water.[22,27]

Property	*Requirement*
Residue after evaporation	Not more than 1 mg/l after drying for 1 h at 105°C
Chloride content	Not more than 0.1 mg/l
Ammonia content	Not more than 0.1 mg/l
Heavy metals (e.g., lead)	Not more than 0.01 mg/l
Consumption of permanganate	When 0.03 ml of 0.1 molar potassium permanganate is added to 500 ml of water containing 1 ml of concentrated surfuric acid, the color should not completely disappear on standing for 1 h at room temperature.
Silicate	Not more than 0.01 mg/l
Sodium	Not more than 0.1 mg/l
Carbon dioxide	Not more than 3 mg/l

Deionized water is water from which the mineral salts have been removed by a process of ion exchange. It is most conveniently prepared by using commercial deionizing equipment. The quality of the deionized water is dependent upon the quality of the water supply and the type of ion-exchange resins utilized. An arrangement which has been found to give good quality water for most laboratory tests consists of a primary filtration to remove particulate matter, followed by a charcoal bed to remove organics and then two mixed bed deionizers. A conductivity light is installed between the two ion-exchange tanks. The light is set to go out when the resistance of the effluent from the first tank goes below 200,000 ohms/cm resistance. The second tank then "polishes" the water to give water with a resistance of 15 to 18 megohms/cm. When the deionizers are exchanged, the first tank is removed for regeneration, the second tank is put in its place, and the new tank is put in the second position. This system can be expected to give water that has a total solid content of less than 1 part per million and a pH near 7.

For specialized applications or in situations where the source water is bad, a combination of distillation and deionization may be employed. These have been used in either order depending on the circumstances.[32] If sterile water is desired, an in-line ultraviolet light, available from suppliers of deionizing equipment, may be installed.

PREPARATION AND CONCENTRATIONS OF SOLUTIONS

Solute and solvent

In a solution of one substance in another substance, the dissolved substance is called the solute. The substance in which the solute is dissolved is called the solvent. When the relative amount of one substance in a solution is much greater than that of the other, the substance present in a greater amount is generally regarded as the solvent. When the relative amounts of the two substances are of the same order of magnitude, it becomes difficult, in fact arbitrary, to specify which substance is the solvent.

DILUTE, SATURATED, CONCENTRATED, AND STANDARD SOLUTIONS

A *dilute* solution contains a relatively small proportion of solute, and a *concentrated* solution contains a relatively large proportion of solute. Concentrated solutions are possible only when the solute is very soluble.

A *saturated* solution exists when the molecules of the solute in solution are in equilibrium with the excess undissolved molecules. Since temperature affects solubility, the exact temperature of the solution should be specified.

A *supersaturated* solution exists when there is more solute in solution than is present in a saturated solution of the same substance at the same temperature and pressure. Supersaturated solutions are unstable and cannot exist in equilibrium with the solid phase.

Standard solutions are an integral part of every quantitative analysis in the clinical chemistry laboratory and are used during the assay of an unknown sample. This is a solution whose precise concentration is known and which is used as a reference to assign concentrations to the assayed unknown samples. The process of determining or adjusting the concentration of the standard solution is known as *standardization*, which may be carried out in one of three ways:

1. Direct preparation of the standard solution by dissolving a weighed amount of a pure, dry chemical and diluting the solution to an exactly known volume. For example, a 1.0 molar sodium chloride solution may be prepared by dissolving 58.50 g of dry NaCl and diluting the solution in a volumetric flask to exactly 1 liter.

2. Titration of a solution of a weighed portion of a pure, dry chemical by the solution to be standardized. The weighed material used for this purpose is known as a primary

standard. If a solution of sodium hydroxide is standardized by determining the volume of the solution required to react with a known weight of pure, dry sulfamic acid ($NH_2{\cdot}SO_3H$), the latter serves as a primary standard. The equation of the reaction is:

$$NH_2{\cdot}SO_3H + NaOH = NH_2SO_3Na + H_2O$$

3. Titration against a primary standard such as hydrochloric acid that has been made up from constant boiling hydrochloric acid. The solution thus standardized is known as a secondary standard. The concentration of the base to be standardized in this titration cannot be known to any greater degree of accuracy than that of the standard hydrochloric acid solution.

A well-equipped clinical chemistry laboratory will maintain a supply of primary reference standard solutions for use, when necessary, in the standardization of other secondary standard solutions (working standards). The solutions used as reference standards must be stable on storage; i.e., their concentrations must not change appreciably with time. The solution should not react with glass or with constituents of the atmosphere and it should not be affected by light. Many standard solutions that can be used safely within a few days of their preparation are not sufficiently stable for use as permanent reference standards.

As a general rule one should not risk contamination of a standard by introducing a pipet into the container. It is better technique to pour off the necessary amount into a beaker or tube just before use. If there is a dried residue present on the mouth of the container, it should be rinsed off with a gauze moistened with distilled water and wiped with a dry gauze before pouring any standard. Condensed moisture that forms on the upper inside of the standard container on storage should also be swirled back into the solution before removing any standard solutions.

Expressing concentrations of solutions in physical units

The concentrations of solutions are generally expressed in the following ways:

1. Weight of Solute Per Unit Volume of Solution (Mass Concentration). This is the mass of the component divided by the volume of the mixture. The use of kilogram per liter (kg/l) and fractions of kilograms is preferred. Thus, kg/l, g/l, mg/l, μg/l, and ng/l are the preferred units. Consequently, 20 g of KCl per liter of solution should be expressed as 20 g/l, and not as 2.0 g/100 ml, 2 g per cent, or 2 per cent. However, due to common usage, the term mg/100 ml will probably be used for some time to express concentrations of substances in biological samples.

2. Amount (Mass) of Solute Per Mass of Solution. The use of kilograms (and fractions of kilograms) per kilogram is preferred (kg/kg). Thus, kg/kg, g/kg, mg/kg, μg/kg, and ng/kg, are the preferred units. If 10 g of NaCl are dissolved in 90 g of water to form 100 g of solution, this would be expressed as 100 g/kg, and not as 10 per cent (w/w) or as 10 g/100 g.

3. Volume of Solute Per Volume of Solution. When the solution and solvent are both liquids, as in alcohol solutions, the concentration of such a solution is frequently expressed in terms of volume per volume (v/v). By adding 70 ml of alcohol to a flask and filling it up to 100 ml with water, one would achieve a solution whose concentration could be expressed as 700 ml/l, thus replacing the terms volume per cent or 70 per cent (v/v).

Expressing concentration in chemical units

1. A molar solution contains 1 mole of the solute in 1 liter (1 mol/l) of solution; i.e., a 1 molar solution of H_2SO_4 contains 98.08 g of H_2SO_4 per liter of solution, since the molecular weight of H_2SO_4 is 98.08. A 0.50 molar solution contains 0.50×98.08 g = 49.04 g H_2SO_4 per liter of solution. The symbol *M* for molar is no longer acceptable.

2. A molal solution contains 1 mole of solute in 1 kg of solvent. In comparison, a molar solution has a final volume of 1000 ml whereas the molal solution has a volume that exceeds 1000 ml.

3. A normal solution contains 1 gram equivalent weight of the solute in 1 liter of solution; i.e., 1 mole of HCl, 0.5 mole of H_2SO_4, and 0.333 mole of H_3PO_4, each in 1 liter of solution, give normal solutions of these substances. A normal solution of HCl is also a molar solution. A normal solution of H_2SO_4 is also a 0.5 molar solution. Although normality is no longer a suggested way to express concentration, it is included since it will probably be in common usage for some time and is related to the equivalent concept, which will also probably be used to report serum electrolyte concentrations for some time to come. The following equations define the expressions of concentration:

$$\text{Molarity of a solution} = \frac{\text{number of moles of solute}}{\text{number of liters of solution}}$$

$$\text{Moles} = \frac{\text{mass (grams)}}{\text{gram molecular mass}}$$

$$\text{Molality of a solution} = \frac{\text{number of moles of solute}}{\text{number of kilograms of solvent}}$$

$$\text{Normality of a solution} = \frac{\text{number of gram equivalents of solute}}{\text{number of liters of solution}}$$

$$= \frac{\text{number of milligram equivalents of solute}}{\text{number of milliliters of solution}}$$

$$\text{Number of milligram equivalents (milliequivalents)} = \text{number of milliliters} \times \text{normality}$$

$$\text{Normality (in oxidation-reduction reaction)} = \text{molarity} \times \text{oxidation state change}$$

$$\text{Gram equivalent weight (as oxidant or reductant)} = \frac{\text{formula mass (g)}}{\text{oxidation state change}}$$

A limitation of the normality scale is that a given solution may have more than one normality, depending on the reaction for which it is used. *The molarity of a solution, however, is a fixed number, since there is only one molecular mass for any substance.*

A milligram equivalent of a substance is its equivalent weight expressed in milligrams. The equivalent *mass* of H_2SO_4 is 49.04 g. Then 1 gram equivalent of H_2SO_4 = 49.04 g H_2SO_4, and 1 milligram equivalent of H_2SO_4 = 49.04 mg H_2SO_4. Since substances may react on the basis of their valence, 1 mole of calcium (atomic mass = 40), which is bivalent, has twice the combining power of 1 mole of sodium. Therefore, 1 mole of calcium has the combining power of two equivalents or 2 moles of sodium, or 40 g of Ca is equivalent to two times 23 g of Na.

The unit of measure commonly used to express the concentration of electrolytes in plasma is the milliequivalent (mEq), which is one-thousandth of an equivalent.

$$\text{Milliequivalents (mEq)} = \frac{\text{mass (g)}}{\text{milliequivalent mass (g)}}$$

Milligrams per 100 ml can be converted to mEq per liter using the following formula:

$$\text{mEq/l} = \frac{\text{mg/100 ml} \times 10 \times \text{valence}}{\text{atomic mass}}$$

Example. If serum sodium is 322 mg/100 ml, then there are 3220 mg/l. The equivalent mass of sodium is 23, and the valence is 1, therefore

$$\text{mEq/l} = \frac{322 \times 10 \times 1}{23} = 140$$

The suggested new terminology for sodium concentration in plasma is millimole per liter (mmol/l). In the example above, the concentration of sodium is

$$\text{mmol/l} = \frac{\text{mg/l}}{\text{molecular mass}}$$

$$\frac{322 \times 10}{23} = 140$$

Dilution problems

All of the solutions considered thus far are volumetric solutions that contain a definite amount of solute in a fixed volume of solution. In percentage, molar, normal, or molal solutions, the amount (mass) of solute contained in a given volume of solution is equal to the product of the volume times the concentration. Whenever a solution is diluted, its volume is increased and its concentration decreased, but the total amount of solute remains unchanged. Hence, two solutions of different concentrations but containing the same amounts of solute will be related to each other as follows:

$$\text{amount of solute}_1 = \text{volume}_1 \times \text{concentration}_1$$

$$\text{amount of solute}_2 = \text{volume}_2 \times \text{concentration}_2$$

The volume and concentration on both sides of the equation must be expressed in the same units.

Example. It is desired to make 500 ml of a 0.12 molar solution of HCl from a 1.00 molar stock solution. Substituting into the equation

$$\text{ml}_1 \times \text{molarity}_1 = \text{ml}_2 \times \text{molarity}_2$$

$$500 \times 0.12 = X \times 1.00$$

$$X = \frac{500 \times 0.12}{1.00} = 60$$

Diluting 60 ml of 1.00 molar HCl to 500 ml will give a 0.12 molar HCl solution.

BUFFER SOLUTIONS AND THEIR ACTION

Buffers are defined as substances that resist changes in the pH of a system. All weak acids or bases, in the presence of their salts, form buffer systems. The action of buffers and their role in maintaining the pH of a solution can best be explained with the aid of the Henderson-Hasselbalch equation. The Henderson-Hasselbalch equation may be derived as follows:

If we have a weak acid, HA, and a salt of that acid, BA, the ionization of each can be represented as

$$HA \rightleftarrows H^+ + A^-$$

$$BA \rightleftarrows B^+ + A^-$$

The dissociation constant for a weak acid may be calculated from the following equation:

$$\frac{[H^+][A^-]}{[HA]} = K_a$$

Thus

$$[H^+] = K_a \times \frac{[HA]}{[A^-]}$$

$$\text{or } \log [H^+] = \log K_a + \log \frac{[HA]}{[A^-]}$$

Now multiplying thru by "−1":

$$-\log [H^+] = -\log K_a - \log \frac{[HA]}{[A^-]}$$

Since $pH = -\log [H^+]$ and $pK_a = -\log K_a$, we may write

$$pH = pK_a + \log \frac{[A^-]}{[HA]}$$

Since A^- is derived principally from the salt, the equation may, for practical purposes, be written

$$pH = pK_a + \log \frac{[\text{salt}]}{[\text{undissociated acid}]}$$

or simply

$$pH = pK_a + \log \frac{[\text{salt}]}{[\text{acid}]}$$

where [salt] = $[A^-]$ = dissociated salt

and [acid] = [HA] = undissociated acid

Consequently, the pH of the system is determined by the pK of the acid and the ratio of $[A^-]$ to [HA]. The buffer has its greatest buffer capacity at the point where the $[A^-] = [HA]$ and the $pH = pK_a$.

This entered into the preceding equation gives

$$pH = pK_a + \log 1$$
$$pH = pK_a + 0$$

The capacity of the buffer decreases as the ratio deviates from 1. If the ratio is beyond 50/1 or 1/50, the system is considered to have lost its buffer capacity. This point is approximately 1.7 pH units to either side of the pK of the acid since

$$pH = pK_a + \log 50/1$$
$$pH = pK_a + 1.7$$

Table 1–8 demonstrates the relationship between pH and the ratio of CH_3COONa to CH_3COOH (A^-/HA).

The chemical mechanisms by which buffers exert their effect may be seen by considering the reactions involved upon the addition of base to a buffer solution containing acetate ions, CH_3COO^-, and acetic acid molecules, CH_3COOH.

On addition of NaOH: CH_3COOH

$$+ Na^+OH^- \rightarrow 2CH_3COONa + H_2O$$

CH_3COONa

OH^- is removed by combining with the hydrogen ion dissociated from acetic acid.

TABLE 1–8 SALT/ACID RATIO AND pH

Concentration CH_3COONa (mol/l)	CH_3COOH (mol/l)	*Ratio* *Salt/Acid*	pH
0.00	0.20	0.00	2.7
0.01	0.20	0.05	3.4
0.05	0.20	0.25	4.1
0.10	0.20	0.50	4.4
0.20	0.20	1.00	4.7
0.40	0.20	2.00	5.0
1.00	0.20	5.00	5.4
2.00	0.20	10.00	5.7

The addition of alkali decreases the CH_3COOH in the buffer and increases the CH_3COONa. The pH of the solution increases in proportion to the change in ratio of salt to acid in the buffer solution.

On addition of HCl: CH_3COOH

$$\begin{matrix} CH_3COOH \\ CH_3COONa \end{matrix} + H^+Cl^- \rightarrow 2CH_3COOH + NaCl$$

H^+ is removed. (The hydrogen ion has combined with acetate to form poorly dissociated acetic acid.)

In this case the HCl reacts to decrease CH_3COONa and increase CH_3COOH in the buffer. The pH of the solution falls in proportion to the change in ratio of salt to acid in the solution; however, since the pH is related to the logarithm of the A^-/HA ratio, only a small change in pH occurs.

Other buffers commonly used and important in clinical chemistry are phosphate (Sørensen), citrate (Sørensen), carbonate-bicarbonate (Delory-King), acetate, phthalate, boric acid–borate (Palitzsch), veronal-sodium veronal, glycine-glycinate, diethanolamine, and tris(hydroxymethyl)aminomethane.

Tables 1–9 to 1–12 describe the preparation of commonly used buffers in clinical chemistry laboratories.

TABLE 1–9 PHOSPHATE BUFFER (0.1 mol/l)

pH range: 5.29 to 8.04; 20°C (Sørensen); 10 ml mixtures of *X* ml of 0.1 molar Na_2HPO_4 (14.2 g/l) and *Y* ml of 0.1 molar KH_2PO_4 (13.6 g/l). The pH values are about 0.03 pH unit less at 37°C.

X ml Na_2HPO_4	Y ml KH_2PO_4	pH *at* 20°C
0.25	9.75	5.29
0.5	9.5	5.59
1.0	9.0	5.91
2.0	8.0	6.24
3.0	7.0	6.47
4.0	6.0	6.64
5.0	5.0	6.81
6.0	4.0	6.98
7.0	3.0	7.17
8.0	2.0	7.38
9.0	1.0	7.73
9.5	0.5	8.04

TABLE I–10 TRIS(HYDROXYMETHYL)-AMINOMETHANE BUFFER

pH range: 7.20 to 9.10 at 23°C; 0.5057 g of tris-(hydroxymethyl)aminomethane dissolved in 50 ml of distilled water and mixed with the indicated amounts of *X* ml of 0.1 molar HCl and diluted to 100 ml give the pH values shown in the table. The pH values are approximately 0.15 pH unit lower at 37°C.

X ml 0.1 molar HCl *Added*	pH *at* 23°C
5.0	9.10
7.5	8.92
10.0	8.74
12.5	8.62
15.0	8.50
17.5	8.40
20.0	8.32
22.5	8.23
25.0	8.14
27.5	8.05
30.0	7.96
32.5	7.87
35.0	7.77
37.5	7.66
40.0	7.54
42.5	7.36
45.0	7.20

TABLE I–11 CARBONATE-BICARBONATE BUFFER

pH range: 9.1 to 10.6 at 25°C (Delory and King); 10 ml mixtures of *X* ml of 0.1 molar Na_2CO_3 (10.6 g/l) and *Y* ml of 0.1 molar $NaHCO_3$ (8.4 g/l). The pH values are about 0.1 pH unit less at 37°C.

X ml Na_2CO_3	Y ml $NaHCO_3$	pH *at* 25°C
1.1	8.9	9.1
1.4	8.6	9.2
2.2	7.8	9.4
2.7	7.3	9.5
3.9	6.2	9.7
5.1	4.9	9.9
6.4	3.6	10.1
7.4	2.5	10.3
7.9	2.1	10.4
8.3	1.6	10.5
8.8	1.2	10.6

TABLE 1-12 ACETIC ACID-SODIUM ACETATE BUFFER

pH range: 3.6 to 5.8 at 25°C; Mixtures of *X* ml of 0.2 molar CH_3COOH (dilute 11.5 ml AR grade glacial acetic acid to 1 liter) and *Y* ml of 0.2 molar CH_3COONa (16.4 g/l). The pH values are approximately 0.05 pH unit lower at 37°C.

X ml CH_3COOH	Y ml CH_3COONa	pH *at* 25°C
92.5	7.5	3.6
88.0	12.0	3.8
82.0	18.0	4.0
73.5	26.5	4.2
63.0	37.0	4.4
52.0	48.0	4.6
41.0	59.0	4.8
30.0	70.0	5.0
21.0	79.0	5.2
14.0	86.0	5.4
9.0	91.0	5.6
6.0	94.0	5.8

PREPARATION OF VARIOUS SOLUTIONS

Standard acids and bases

Accurately prepared standard solutions of hydrochloric and sulfuric acids, potassium hydrogen phthalate, and sodium hydroxide are necessary in a clinical chemistry laboratory. These standard solutions are used to establish the molarity of all acids and bases in the laboratory. The standard solutions are usually not prepared directly, but are made approximately and standardized by titration. The final concentration most frequently required is 0.1 mol/l or less. Table 1–13 shows the strengths of various concentrated solutions of acids and bases.

Potassium Hydrogen Phthalate. A primary standard for acidimetry, this chemical may be obtained from the National Bureau of Standards in very pure form. An accurate normality can be obtained by carefully weighing out the desired amount on an analytical balance. The molecular mass of potassium hydrogen phthalate ($HKC_8H_4O_4$) is 204.22. Since there is one dissociable hydrogen, the molecular mass is also the equivalent mass.

TABLE 1-13 STRENGTHS OF CONCENTRATED SOLUTIONS OF ACIDS AND BASES

Acid or Base	*Specific Gravity*	*% by Weight*	*g/l*	*Approximate Molarity*	*Milliliters Required to Make 1 l of 1 Molar Solution*
Hydrochloric acid (HCl)	1.19	37	440	12.1	83
Sulfuric acid (H_2SO_4)	1.84	96	1730	18	56
Nitric acid (HNO_3)	1.42	70	990	15.7	64
Acetic acid (CH_3COOH)	1.06	99.5	1060	17.4	57
Ammonium hydroxide (NH_4OH)	0.880	29	250	15–17	57–67
Sodium hydroxide (saturated solution) (NaOH)	1.50–1.53	about 50	600–700	15–18	57–67
Potassium hydroxide* (saturated solution) (KOH)	1.55	about 50	800	14	70

* Saturated solutions made from the usual CP potassium hydroxide will vary in strength, chiefly because of the variable amount of carbonate that such solutions contain.

For a 0.1 molar solution, weigh 20.422 g of the dry substance and dissolve in distilled water, diluting volumetrically to 1 liter. *Bases* can be standardized against this or other acidimetry standards, and other acids standardized by titration against the *base*. It must be remembered that the normality of alkaline solutions may change as a result of absorption of CO_2 or reaction with the glass container; hence, they must be restandardized often. Acids are relatively stable.

Hydrochloric Acid. Standard solutions of HCl may be prepared from constant boiling HCl. When an aqueous solution of hydrochloric acid stronger than 20 per cent by weight is distilled, it soon approaches a constant boiling point at which the vapor phase is constant in composition and a distillate of constant composition is obtained. Table 1–14 gives the concentration of the hydrochloric acid in aqueous solutions with constant boiling points at different barometric pressures as determined by Bonner and Branting.[3] This constant boiling HCl, prepared by distillation, has a constant concentration of 20.22 per cent of HCl (760 mm Hg) by weight and is used to prepare the dilute primary standard acid (0.1000 molar $\pm$ 0.0004) by diluting the proper weight (not volume) of the acid to the required volume. Further standardization is unnecessary. Acid prepared in this manner is used for the standardization of alkali and other reagents.

Sodium Hydroxide (Concentrated). Add 100 g of NaOH to 100 ml of water warmed to 55°C. Mix until the NaOH is dissolved and allow to stand overnight in a covered container. The NaOH solution will contain carbonate that will settle out and the supernatant can be removed by decanting or centrifugation. This concentrated solution is 16 to 18 molar and can be diluted to any strength desired. Its molarity can be established by titration against standard potassium acid phthalate, standard HCl, or standard sulfamic acid. The NaOH standard should be stored in a plastic bottle or a paraffin-lined container. The standard should be well protected from CO_2 of the air. Rubber or plastic stoppers should be used with alkaline solutions since alkali will cause a glass stopper to freeze. It is advantageous to use a soda-lime trap in the container in order to prevent atmospheric CO_2 from entering and weakening the solution.

pH reference solutions

For the standardization of pH meters, the following standards are useful.

1. *Potassium hydrogen phthalate*, 0.05 mol/l. It is made by dissolving 10.2 g per liter in water, and has pH values of 4.001 at 20°C and 4.025 at 37°C.

2. *Equimolar phosphate buffer*, contains 0.025 mol dihydrogen phosphate and 0.025 mol monohydrogen phosphate/l. It is made by dissolving 3.402 g of anhydrous potassium

TABLE 1-14 CONCENTRATION OF HYDROCHLORIC ACID IN AQUEOUS SOLUTIONS WITH CONSTANT BOILING POINTS AT DIFFERENT BAROMETRIC PRESSURES*

Barometric Pressure at Time of Distillation (mm Hg)	*HCl Concentration by Weight in Distillate (%)*	*Distillate Required to Make 1 Liter of 0.1 Molar HCl (g)*
720	20.317	17.935
730	20.293	17.956
740	20.269	17.977
750	20.245	17.998
760	20.221	18.019
770	20.197	18.041

* Reprinted from Bonner, W. D., and Branting, B. F.: J. Am. Chem. Soc., *48*:3093, 1926. Copyright by the American Chemical Society.

TABLE I-15 NBS pH STANDARDS

SRM #	*Material*	*pH at 25°C*
185d	Acid potassium phthalate	4.004
186Ic	Potassium dihydrogen phosphate	6.865
186IIc	Disodium hydrogen phosphate	7.413
187b	Borax	9.183
188	Potassium hydrogen tartrate	3.557
189	Potassium tetroxalate	1.679
191	Sodium bicarbonate	10.01
192	Sodium carbonate	
922	Tris(hydroxymethyl)aminomethane	7.699
923	Tris(hydroxymethyl)aminomethane hydrochloride	

dihydrogen phosphate (KH_2PO_4) and 3.549 g of anhydrous disodium hydrogen phosphate (Na_2HPO_4) per liter in water. The pH values are 6.88 at 20°C and 6.81 at 37°C.

3. *Standard phosphate buffer, ionic strength*, 0.1. It is made by adding 1.264 g of potassium dihydrogen phosphate and 4.30 g disodium hydrogen phosphate to 1 liter of freshly boiled distilled water at 25°C. This buffer has a pH of 7.38 at 25°C and 7.36 at 37°C.

4. *Borax* (*sodium tetraborate*) 0.05 *molar*. It is made by dissolving 19.07 g of $Na_2B_4O_7 \cdot 10H_2O$ per liter in water. It has pH values of 9.22 at 20°C and 9.08 at 37°C.

The National Bureau of Standards has a number of primary pH standards available. These Standard Reference Materials are to be used in admixtures as directed on the certificate of analysis accompanying the material to give a desired pH. A partial list of these materials is given in Table 1–15.

LABORATORY MATHEMATICS

EXPONENTS

Any number may be expressed as an integral power of 10, or as a product of two numbers, one of which is an integral power of 10.

Examples:

$$100 = 10^2$$

$$600 = 6 \times 10^2$$

$$460 = 4.6 \times 10^2$$

$$0.46 = 4.6 \times 10^{-1}$$

In multiplication, exponents of like bases are added and in division exponents of like bases are subtracted.

Examples:

$$20 \times 600 = 2 \times 10^1 \times 6 \times 10^2 = (6 \times 2) \times 10^{1+2} = 12 \times 10^3 = 1.2 \times 10^4 = 12{,}000 \quad (1)$$

$$\frac{20}{400} = \frac{2 \times 10^1}{4 \times 10^2} = \frac{2}{4} \times 10^{1-2} = 0.5 \times 10^{-1} = 5 \times 10^{-2} = 0.05 \quad (2)$$

LOGARITHMS

Logarithms may be used in the laboratory to carry out multiplications and divisions, to extract square roots, square numbers, and so on, but their value in biochemistry is more than merely a convenience. A number of basic phenomena are described by equations or formulas that utilize logarithms. Two obvious examples are pH and absorption of light.

Because their use is necessary, a brief review will be given of the system of logarithms to the base 10 (called the common or Briggsian system). The "common" logarithm of a number is the exponent or power to which 10 must be raised to give that number.

Example:

$$\text{Log } 10 = 1.0, \text{ since } 10^1 = 10$$
$$\text{Log } 100 = 2.0, \text{ since } 10^2 = 100$$
$$\text{Log } 2 = 0.301, \text{ since } 10^{0.301} = 2$$

A common logarithm always consists of two parts, an integer, called the characteristic, and a decimal number, called the mantissa.

The mantissa of the log of the number is found in logarithm tables or from a slide rule and is independent of the position of the decimal point in the original number. Thus 352, 0.352, and 0.00352 all have the same mantissa.

The characteristic indicates the position of the decimal point in the set of figures represented by the mantissa and is determined by inspection of the original number according to the following rules:

1. For a number greater than 1, the characteristic is positive and is one less that the number of digits to the left of the decimal point.

Example:

The characteristic for 167.0 = 2
16.7 = 1
1.67 = 0

2. For a number less than 1, the characteristic is negative and is numerically one more than the number of zeros *immediately* following the decimal point. The negative sign of the characteristic may be expressed in two ways:

a. by placing a bar above the characteristic as $\bar{1}$, $\bar{2}$, and so on
b. as 9. −10, 8. −10, and so on.

Examples:

The characteristic for 0.0741 is $\bar{2}$ or 8. −10
0.00067 is $\bar{4}$ or 6. −10

The antilogarithm is the number corresponding to a given logarithm. Thus, "the antilog of 2" really means "the number whose logarithm is 2," which in this case would be 100. Suppose it is required to find the antilog of 1.5020. The characteristic is 1 and the mantissa is 0.5020, which would be looked up in a table and found to be 3177. Since the characteristic is 1, there are two digits to the left of the decimal point and the number becomes 31.77.

Logarithm tables of mantissas can be found in various handbooks and antilogs can be determined either by working backward in a log table or by utilizing a separate antilog table.

Multiplication

When multiplying two numbers, the logarithm of each number is determined, the logarithms are added, and the antilogarithm of the sum is found.

Examples:

2.2 × 4.8 =
Log of 2.2 = 0.3424
Log of 4.8 = 0.6812
Sum 1.0236
Antilog of 1.0236 = 10.56 (answer)

Division

When carrying out divisions, the logarithm of each number is determined and then the logarithm of the divisor is subtracted from the logarithm of the dividend.

Examples:

$$\begin{aligned} \text{Log of } 4.8 &= 0.6812 \\ \text{Log of } 2.2 &= \underline{0.3424} \\ \text{Difference} \quad & \ \ 0.3388 \\ \text{Antilog of } 0.3388 &= 2.18 \text{ (answer)} \end{aligned}$$

Logarithms are very useful when carrying out multiple multiplications or when solving problems involving both multiplication and division.

Example:

$$\frac{4.8}{2.2} \times 2.5 =$$

$$\begin{aligned} \text{Log } 4.8 &= 0.6812 \\ \text{Log } 2.2 &= \underline{0.3424} \\ \text{Difference} \ & \ \ 0.3388 \\ \text{Log } 2.5 &= \underline{0.3979} \\ \text{Sum} \quad & \ \ 0.7367 \end{aligned}$$

$$\text{Antilog of } 0.7367 = 5.45$$

The logarithmic relationship used daily in all clinical chemistry laboratories is that of pH, which may be defined as the negative logarithm of the hydrogen ion concentration:

$$\text{pH} = \log \frac{1}{[\text{H}^+]} = -\log\,[\text{H}^+]$$

In the living organism, the concentration of acids and bases is very dilute. For example, the hydrogen ion concentration of the living organism is about 0.00000005 molar. Sørensen believed that laboratory workers would have difficulty in expressing $[H^+]$ of acid and basic solutions and devised the term "pH." The relationship is very simple if the $[H^+]$ is exactly $1 \times 10^{-\text{power}}$, because the pH is then equal to the value of the minus power of 10.

For example, in water

$$K_w = 10^{-14}$$

and

$$[\text{H}^+] \times [\text{OH}^-] = 10^{-14}$$

Since

$$[\text{H}^+] = [\text{OH}^-]$$

therefore

$$[\text{H}^+] = 10^{-7}$$

or

$$\text{pH} = 7$$

To calculate the pH of solutions with a $[H^+]$ more complex than $1 \times 10^{-\text{power}}$, logarithms must be used. Thus the pH of a solution with a $[H^+]$ of 0.004 may be calculated as follows:

$$0.004 \text{ may be expressed as } 4 \times 10^{-3}$$

Then

$$\begin{aligned} \text{pH} &= -\log\,(4 \times 10^{-3}) = -(\log 4 - 3) \\ &= -\log 4 + 3.00 \\ &= 3.00 - 0.602 = 2.398 \end{aligned}$$

A valuable working relationship between the pH of a buffer solution and the dissociation constant of the weak acid in the presence of its salt is given by the Henderson-Hasselbalch equation, which may be stated as follows:

$$\text{pH} = \text{pK} + \log \frac{[\text{salt}]}{[\text{acid}]}$$

The pK is defined as $-\log K$ and may be found in tables such as the table in the appendix or in suitable handbooks. The concentration of salt and acid is commonly expressed as moles or equivalents per liter. From this equation it can be seen that if the pK of a buffer pair is known, the amount of salt and acid needed to prepare a buffer of a given pH and concentration may be calculated.

Example. It is desired to prepare 1 liter of a 0.1 molar acetate buffer with a pH of 4.9. The pK is 4.76. Substituting in the equation

$$\text{pH} = \text{pK} + \log \frac{[\text{A}^-]}{[\text{HA}]}$$

$$4.9 = 4.76 + \log \frac{[\text{A}^-]}{[\text{HA}]}$$

$$\log \frac{[\text{A}^-]}{[\text{HA}]} = 0.14$$

$$\frac{[\text{A}^-]}{[\text{HA}]} = \text{antilog}\ 0.14 = 1.38$$

If mol/l of salt + mol/l of HA = 0.1, then mol/l of A^- = (0.1 − mol/l of HA) or

$$[\text{A}^-] = 0.10 - [\text{HA}]$$

Since

$$\frac{[\text{salt}]}{[\text{acid}]} = \frac{[\text{A}^-]}{[\text{HA}]} = 1.38$$

we may now substitute into this equation:

$$\frac{0.1 - [\text{HA}]}{[\text{HA}]} = 1.38$$

$$0.1 - [\text{HA}] = 1.38 \times [\text{HA}]$$

$$0.1 = [\text{HA}] + 1.38 \times [\text{HA}]$$

$$0.1 = 2.38[\text{HA}]$$

$$[\text{HA}] = \frac{0.1}{2.38} = 0.042$$

$$[\text{A}] = 0.1 - [\text{HA}]$$

$$[\text{A}] = 0.1 - 0.042 = 0.058$$

The buffer then should consist of

0.042 moles acetic acid per liter
and 0.058 moles sodium acetate per liter

To prepare 1 liter of this buffer, the following amounts are used:

$$\begin{array}{rl} 60.0\text{ g} & \text{(molecular mass of acetic acid in grams)} \\ \times\ 0.042 & \\ \hline 2.52 & \text{grams of acetic acid} \end{array}$$

and

$$\begin{array}{rl} 82.0\text{ g} & \text{(molecular mass of sodium acetate in grams)} \\ \times\ 0.058 & \\ \hline 4.75 & \text{grams of sodium acetate} \end{array}$$

These amounts dissolved and diluted to 1 liter will yield a 0.1 molar acetate buffer with a pH of 4.9.

EXPRESSING AND CALCULATING DILUTIONS

Dilution of solutions, which involves making a weaker solution from a stronger one, is a frequent laboratory procedure. The following are some of the most common reasons for dilutions: (1) if the concentration of the material in solution (usually the specimen to be analyzed) is too great to be accurately determined; (2) if in the removal of undesirable substances (e.g., proteins), solutions are added to precipitate the proteins and a dilution has taken place; and (3) in the preparation of working solutions from stock solutions.

Dilutions are usually expressed as a ratio, such as 1/10. This refers to 1 unit of the original solution diluted to a final volume of 10 units, i.e., 1 volume of solute plus 9 volumes of solvent. This would yield a solution which is 1/10 the concentration of the original solution. To calculate the strength of the dilute solution, multiply the concentration of the original solution by the dilution, expressed as a fraction.

Example. A 500 mg/100 ml solution is diluted 1/25. The concentration of the final solution is

$$500 \times 1/25 = 20 \text{ mg/100 ml}$$

If more than one dilution is made with a given solution, the concentration of the final solution is obtained by multiplying the original concentration by the product of the dilutions.

Example. A 1000 mg/100 ml solution is diluted 1/10 and then this diluted solution is further diluted 1/100. The concentration of the final solution would be

$$1000 \times 1/10 \times 1/100 = 1 \text{ mg/100 ml}$$

Large dilutions can conveniently be made in two steps. For example, if a 1/1000 dilution is required but only 100 ml is needed or the diluent is in short supply, this dilution could be accomplished, as in the previous example, without having to resort to micropipets.

The systematic dilution and redilution of a solution is called "serial dilution." This technique is commonly used in serology when the test may actually be reported in terms of the number of times a solution had to be serially diluted to still give a positive or negative reaction. In serial dilutions, to find the concentration in any one tube, the dilution in that tube is multiplied by each of the preceding dilutions, including the original tube.

Example. Into each of eight tubes is placed 0.2 ml of diluent. Then 0.2 ml of serum is added to the first tube and mixed. Next 0.2 ml of the mixture is removed and placed in the second tube and mixed. Then 0.2 ml is removed and placed in the third tube and so on until tube eight, when the

0.2 ml removed is discarded. The dilutions are calculated as follows:

Tube	*Dilution*
1	$\frac{0.2}{0.2 + 0.2} = \frac{0.2}{0.4} = 1/2$
2	$1/2 \times 1/2 = 1/4$
3	$1/4 \times 1/2 = 1/8$
4	$1/8 \times 1/2 = 1/16$
5	$1/16 \times 1/2 = 1/32$
6	$1/32 \times 1/2 = 1/64$
7	$1/64 \times 1/2 = 1/128$
8	$1/128 \times 1/2 = 1/256$

In the preparation of a so-called protein-free filtrate, protein-precipitating reagents are added and thus a dilution has taken place. If an aqueous standard is used in the procedure, this standard is sometimes used directly since deproteinization is unnecessary. In this case the dilution of the serum must be calculated so that the standard may be appropriately diluted.

If the standard is always used this way, it is usually made up to contain an amount that is less, by a factor of the dilution, than the amount it is equivalent to in the procedure. Its true concentration is thus less (by a known factor) than the serum concentration it represents.

Example. Serum is diluted 1/10 in the preparation of a protein-free filtrate. Then 1 ml of this filtrate is added to a cuvet, followed by the appropriate reagents for colorimetric analysis. Also, 1 ml of standard is added to another cuvet. If the standard is to be equivalent to a 100 mg/100 ml concentration of the substance, it is obvious that the standard cannot contain 100 mg/100 ml since the serum was diluted and the standard was not diluted. Thus, this standard must be diluted 1/10 before use or more practically it would be made up to contain $100 \times 1/10 = 10$ mg/100 ml. The standard therefore would contain 10 mg/100 ml, but would be equivalent to a serum concentration of 100 mg/100 ml in this particular test.

SIGNIFICANT FIGURES

Some degree of error is involved in all chemical determinations and, indeed, in all measurements taken whether reading a spectrophotometer, making a weighing, or measuring a distance. None of these measurements would be absolutely correct and the accuracy would be limited by the reliability of the measuring instrument. In a chemical determination, the accuracy would be limited by all the steps involved, such as pipetting. In reporting results, some indication of the reliability of the measurement is given by the number of significant figures that are given in the result. A significant figure is one that is known to be reasonably reliable. As an example, consider that the length of an object is recorded as 12.8 cm. By convention, this means that the length was measured to the nearest tenth of a centimeter, and that its exact value lies between 12.75 and 12.85 cm. If this measurement were exact to the nearest hundredth of a centimeter, it would have been reported as 12.80 cm. Thus in the clinical laboratory a result reported as 14.5 means that the result is accurate to the nearest tenth and that the exact value lies between 14.45 and 14.55. If this result were reliable to the nearest hundredth it would be reported as 14.50. The figure 14.5 contains three significant figures while 14.50 contains four significant figures. The table below gives the implied maximal limits of a result reported to various significant figures.

Reported Result	*Number of Significant Figures*	*Implied Limits*
14	2	13.5–14.5
14.5	3	14.45–14.55
14.50	4	14.495–14.505

Zeros to the right of the decimal point are always significant if they follow the digits and not significant if they precede a digit and the total number is less than one. Thus, 0.072 contains two significant figures, 0.720 contains three significant figures and 1,072 contains four significant figures. Zeros to the left of the decimal point may or may not be significant. Thus, a report of 220 does not indicate whether the measurement was to the nearest ten or to the nearest one. If there are digits on both sides of the zero, as 202, it is a significant figure. If the zero is followed by a decimal point and digits or zeros, as 220.1, the zero is a significant figure.

As was mentioned earlier, a result reported as 14.5 implies that the result is accurate to the nearest tenth and that the exact value lies between 14.45 and 14.55. Although the result may not be that accurate, three significant figures may still be used if the result is significantly more accurate than would be indicated by only two significant figures. In other words, if the result is not as good as between 14.45 and 14.55 but much better than 13.5 to 14.5, there is good reason to report the result as 14.5.

The dropping of one or more digits to the right to give the desired number of figures is called "rounding off." When the first digit dropped is less than 5, the last digit retained remains unchanged; i.e., 4.571 = 4.57. When it is more than 5 or is followed by digits greater than 0, the last digit is increased by 1; i.e., 3.788 = 3.79. When it is exactly 5, the last digit is increased by 1 if that digit is odd, otherwise the digit remains unchanged; in other words, it is rounded to the nearest even number. Thus, 6.85 becomes 6.8 and 6.75 also becomes 6.8.

COLLECTION AND HANDLING OF SPECIMENS

COLLECTION OF BLOOD

Blood may be obtained from veins or arteries, or by capillary puncture. Most tests in the clinical chemistry laboratory are performed on venous blood, while arterial blood is primarily used for blood gas determinations. The use of a tourniquet will aid in the identification of veins and in the rapid removal of blood. However, the tourniquet should occlude only the superficial veins and not the deep veins or arteries. Excessive fist clenching by the patient is to be avoided. Usually, disposable needles are used whether the blood is drawn into a syringe or into a vacuum tube. (For collection of blood for electrolyte analysis, see Chap. 15.)

The *preservation* of the chemical integrity of the specimen from the time of collection to the time of testing is of utmost importance if the results are to be meaningful. All tubes and syringes must be chemically clean. In general, tubes for chemical analysis do not have to be sterile, but one should not assume that a sterile tube is chemically clean. When blood is drawn from a patient by using a syringe, it is immediately transferred to a clean dry tube after the needle has been removed. The blood is then allowed to clot for at least 10 to 15 minutes at room temperature, and longer if allowed to stand in a refrigerator. The clot may adhere to the wall of the tube, so that "ringing" (making a gentle sweep around the inside walls of the tube with a wooden applicator stick) should be performed before centrifugation. Excessive ringing is unnecessary and can produce hemolysis. By allowing the clot to retract for a longer period of time, hemolysis is minimized and the yield of serum is greater; however, during this time glycolysis takes place and there can be a shift of substances from cells to serum. Thus, the time allowed for clot retraction is dependent upon the procedure to be done.

An alternative procedure is to draw blood into a vacuum tube (Vacutainer). Vacuum tubes may be siliconized to minimize hemolysis and prevent the clot from adhering to the wall of the tube.

After the blood has clotted, the tube is centrifuged and the supernatant serum removed. The individual handling of the serum will now depend on the analysis that is to be done and the time that will elapse before the analysis is started. The serum may either remain at room temperature, be refrigerated and protected from the light, or be frozen, depending on circumstances. The stability of individual components of serum will be discussed later in the respective chapters. The serum or plasma should be removed from the clot or cells soon after collection. Glucose changes most rapidly of all the commonly measured chemical constituents when left in contact with the cells. Since glycolysis is an enzyme reaction, it is very sensitive to temperature. Many laboratories collect the blood for glucose analysis in a tube containing fluoride, which inhibits the enzyme enolase involved in glycolysis. Glucose in serum or plasma (which is essentially free of cells that contain the glycolytic enzymes) is quite stable, especially in the refrigerator. The stability of glucose in serum at room temperature has been reported to be variable, but it is much greater than it is in whole blood.

Most enzymes are stable in serum for at least 24 hours under refrigeration and longer if frozen. The inorganic ions are stable for at least 8 hours at room temperature and for days under refrigeration. Bilirubin (particularly unconjugated) is very sensitive to light so that the serum must be assayed immediately or protected from direct light. Urea nitrogen, creatinine, and uric acid are stable for at least 24 hours without refrigeration, and longer if refrigerated.

Collecting tubes for microspecimens

Capillary collecting tubes approximately 150 mm in length with a 1.5 mm bore and tapered ends are available commercially or can be made as follows: Clean, dry, borosilicate tubing (Pyrex or Kimax), 3 mm OD (1.5 to 1.8 mm ID), is cut into approximately 27 cm lengths. The middle portion of the tubing is heated in a gas flame, drawn out to form a bore of about 0.8 mm, and broken in two, and the ends fire polished. The tubes are cleaned thoroughly and dried in an oven at 110°C. Each tube has a capacity of approximately 0.25 ml. After filling with blood, the larger end may be sealed either with sealing wax, by heating in a small flame to close the larger end, provided care is taken to avoid overheating and hemolyzing the blood; or by covering with a special plastic cap (Critocap). The latter technique is preferred. As a general rule anticoagulants are not used because fibrinogen interferes with electrophoretic separations and other protein fractionation techniques, and ammonium ions from ammonium heparinate interfere with the determination of urea by techniques that employ urease. Heparinized tubing, however, will provide a somewhat greater volume of plasma, and such specimens are easier to centifruge and are preferred by some workers for certain determinations.

Similar collecting tubes, 75 to 100 mm long, can be made from 4 mm OD tubing with a bore of 2.5 to 2.7 mm. Blood may also be collected directly in 6 × 50 mm glass test tubes or in small polyethylene centrifuge tubes (Beckman Instruments Inc., Spinco Div., Palo Alto, CA., and Brinkman Instruments, Westbury, N.Y.).

Collection of blood microspecimens

The site selected for puncture is cleaned with 70 per cent ethyl or isopropyl alcohol and allowed to dry. In adults, the fingertip is the usual site although some prefer to sample from the ear lobe; in infants, either the heel, the big toe, or the sole of the foot is satisfactory. Better flow of blood is obtained if the extremity is first placed in warm water or wrapped with a warm towel. The puncture is made with a Bard-Parker No. 11 scalpel blade or other type of lancet and should be sufficiently deep to permit a free flow of blood without squeezing. The first drop of blood is wiped off with dry gauze. When a large drop of blood has appeared, the tip of the collecting tube is touched to the drop while the tube is held slightly downward from a horizontal position. The blood enters the tube by capillarity. The tube is filled to within about 1 cm of the larger end with care to avoid the introduction of air bubbles. Several tubes may be filled, if required, from a single puncture. Excessive squeezing of tissues must be avoided because this can cause hemolysis and contamination with tissue juices.

The tubes are capped and placed in a test tube for transporting and centrifuging. A free-swinging, horizontal type of centrifuge is recommended since the capillary tubes tend to break or leak in an angle head centrifuge. When analyses must be delayed, the tube is scratched with a file and broken above the junction of clot and serum. The serum tube is then recapped and refrigerated.

To collect blood in small test tubes or centrifuge tubes, the surface of the puncture site is held in a nearly vertical position and the drops of blood are permitted to flow freely into the tube. The lip of the tube should not be scraped over the surface as this may cause hemolysis. Up to 1 ml of blood may be collected from a satisfactory puncture. The tubes are placed in larger tubes for centrifuging and sampling or may be centrifuged directly in special microcentrifuges. As with macromethods, the serum should be removed from the clot and refrigerated when analysis must be delayed.

If desired, certain determinations, e.g., glucose and urea, may be performed on whole blood. In this event, blood may be sampled directly into a micropipet, then washed out into a measured volume of one component of a protein-precipitating reagent that changes the pH sufficiently to prevent glycolysis.

Anticoagulants

If whole blood or plasma is desired, an anticoagulant must be added to the specimen immediately after it is drawn or placed into the tube into which the blood is collected. In the case of heparin it is more advantageous to coat the walls of the syringe with the anticoagulant.

The technologist must be certain that the anticoagulant used does not affect the chemical analysis. This may happen in a number of ways. If the anticoagulant is present as the sodium or potassium salt and electrolytes are being analyzed, a significant error can occur. In this case, use of the lithium or ammonium salt would obviate the problem. The anticoagulant may also remove the constituent to be measured, as in the case of oxalate, which removes calcium from the serum by forming an insoluble salt. The action of anticoagulants on enzymes is a function of anticoagulants as well as enzymes. Oxalate has been reported to inhibit lactate dehydrogenase, acid phosphatase, and amylase. Fluoride, if present in very high concentrations, inhibits urease but may activate amylase.[5]

Heparin is the anticoagulant that least interferes with clinical chemical tests. It is present in most of the tissues of the body, but in concentrations less than that required to prevent the clotting of blood. Heparin exists in the highest concentrations in the liver and the lungs. It is a mucoitin polysulfuric acid and is available as the sodium, potassium, and ammonium salt. This anticoagulant is believed to act as an antithrombin, preventing the transformation of prothrombin into thrombin, and thus preventing the formation of fibrin from fibrinogen. Several workers have shown that this action of heparin needs the presence of a cofactor that seems to be associated with the albumin fraction of plasma. Heparin has also been shown to possess antithromboplastin activity and to inhibit the lysis of platelets.

Usually about 20 units of heparin are used per milliliter of blood. Since heparin is not readily soluble, it is frequently used as a solution. If it is to be dried, it is better to dry it on the walls of the tube so that when the blood is added, solution may be as rapid as possible. Some disadvantages of heparin are high cost, temporary action and the fact that it gives a blue background on a blood smear stained with Wright's stain.

Oxalates, such as sodium, potassium, ammonium, or lithium, inhibit blood coagulation by forming rather insoluble complexes with calcium ions, which are necessary for coagulation. Of these, potassium oxalate at a concentration of about 1 to 2 mg/ml of blood is the most widely used. If the laboratory prepares its own tubes, the potassium oxalate may be added as a 30 g/100 ml solution and dried in an oven. The clotting of 15 ml of blood may be prevented by 30 mg (0.1 ml) of potassium oxalate. Temperatures in excess of 150°C should be avoided during the drying since at elevated temperatures the oxalate will be converted to carbonate, which has no anticoagulant activity.

Sodium fluoride, although usually considered as a preservative for blood glucose determinations, also acts as a weak anticoagulant. When used as a preservative along with an anticoagulant such as potassium oxalate, it is effective in a concentration of about 2 mg/ml of blood. It exerts its action by inhibiting the enzyme system involved in glycolysis. When sodium fluoride is used as an anticoagulant, the concentration must be much greater, i.e., 6 to 10 mg/ml of blood. As a general rule fluoride should not be used when collecting specimens for enzyme determinations or when using an enzyme as a reagent in a test, e.g., the urease method for the determination of urea.

Ethylenediaminetetraacetic acid (*EDTA*) is a chelating agent that is particularly useful for hematologic examination since it preserves the cellular constituents of the blood. It is used as the disodium or dipotassium salt, the latter being more soluble. It is effective in a final concentration of 1 to 2 mg/ml of blood. This chelating agent derives its anticoagulant activity from the fact that it binds calcium, which is essential for the clotting mechanism.

Specimen variables

The major chemical difference between *arterial* and *venous blood* is the degree of oxygen saturation of the hemoglobin. Blood is oxygenated while passing through the lungs and is then pumped from the left ventricle of the heart into the arterial circulation. A slight hydrostatic pressure still exists in the arterial end of the capillaries; hence, blood collected by capillary puncture is essentially arterial blood. Oxygen is partially removed from the blood by tissues surrounding the capillary bed. The blood then returns through the venous circulation to the heart.

For most practical purposes there is little difference in the chemical composition of plasma obtained from arterial or from venous blood. Glucose is removed from arterial blood by the tissues and metabolized in part to lactate, which then appears in venous blood in higher concentration than in arterial blood. In the fasting state, the concentration of glucose in capillary (arterial) blood is approximately 5 mg/100 ml higher than in venous blood. During a glucose tolerance test, however, this difference may approach 30 to 50 mg/100 ml. For this reason the source of the specimen should be consistent throughout the test.

In general, blood for chemical analysis should be drawn while the patient is in the *postabsorptive state*. An overnight fast is the usual procedure although a 6 hour fast is ample. During this time there is no need to restrict water intake. The common procedures that are affected most significantly by eating are those involving glucose (elevated), inorganic phosphorus (decreased), thymol turbidity (increased), and triglycerides (increased). In addition, lipemia, which is caused by a transient rise in chylomicrons following a meal containing fat, causes interference with a large number of chemical analyses because of turbidity.

There is *diurnal variation* (not related to eating) on certain constituents such as iron and corticosteroids. Dietary habits influence constituents such as uric acid and lipids. A proper glucose tolerance test is performed when the patient has had an adequate carbohydrate intake (250 g per day) for three days before the test.

Hemolysis may interfere with a number of chemical procedures and should be avoided. Several constituents, such as glutamic oxalacetic transaminase, lactic dehydrogenase, acid phosphatase, and potassium, are present in large amounts in erythrocytes so that hemolysis will significantly elevate the values for these substances in serum. Hemoglobin may directly interfere in a chemical determination by inhibiting an enzyme such as lipase, by interfering with a reaction such as the diazotization of bilirubin, or by yielding a significant amount of color and thus interfering with a colorimetric analysis. This is particularly true when the reading is taken in the blue portion of the spectrum. In grossly hemolyzed serum a dilution of the serum components occurs if the concentration of the metabolite in the red cell is less than in plasma; thus, the sodium and chloride concentration of serum would be falsely low in a grossly hemolyzed serum. Table 1–16 shows the concentration of some common constituents in erythrocytes and plasma.

Some common causes of hemolysis are moisture in the syringe and the mechanical destruction of cells by forcing the blood from the syringe into a tube without removing the needle. The blood should be allowed to slowly run down the side of the tube after the needle has been removed from the syringe. The use of vacuum tubes has largely taken care of these problems. Too vigorous mixing of the tube after drawing the blood may also lead to hemolysis.

Elevated levels of bilirubin may interfere with certain methods for measuring cholesterol and protein. In severe uremia, the elevated levels of creatinine and uric acid can cause false high levels of glucose as measured by the alkaline ferricyanide method or the neocuproine procedure.

Drug Interference. Although our knowledge on the effect of medications on chemical tests is still somewhat limited, information on this important aspect is rapidly becoming available. An extensive compilation of drug effects may be found in recent publications.[7,8,14,33] Caraway has stated that medication may alter the results of chemical tests through either pharmacological, physical, or chemical mechanisms. The pharmacological effect may be either primary (allopurinol, a xanthine oxidase inhibitor, causes a lowering of uric acid levels) or secondary (estrogens cause an increase in binding proteins such as ceruloplasmin and thyroxine-binding globulin and thereby increase copper and thyroxine levels). A drug or its metabolite may also interfere chemically with the determination of a particular constituent. Thus, organic iodines interfere with the determination of protein-bound iodine, and antiovulatory drugs interfere with the fluorometric determination of plasma cortisol.

Some drug manufacturers now supply information on the effect of their drug on laboratory tests. When a medication is suspected of giving a false elevation or depression of a

TABLE 1–16 CONCENTRATION OF SUBSTANCES IN ERYTHROCYTES AND PLASMA*

Substance	*Erythrocytes*	*Plasma*	*Erythrocytes/Plasma*
Glucose, mg/100 ml	74.0	90.0	0.82
Nonsugar-reducing substances, mg/100 ml	40.0	8.0	5.00
Nonprotein N, mg/100 ml	44.0	25.0	1.76
Urea N, mg/100 ml	14.0	16.0	0.88
Creatinine (Jaffe), mg/100 ml	1.8	1.1	1.63
Uric acid, mg/100 ml	2.5	4.6	0.55
Total cholesterol, mg/100 ml	139.0	194.0	0.72
Cholesterol esters, mg/100 ml	0.0	129.0	0.00
Sodium, mEq/l	16.0	140.0	0.11
Potassium, mEq/l	100.0	4.4	22.70
Chloride, mEq/l	52.0	104.0	0.50
Bicarbonate, mmol/l	19.0	26.0	0.73
Calcium, mEq/l	0.5	5.0	0.10
Inorganic P, mg/100 ml	2.5	3.2	0.78
Acid β-glycerophosphatase, units	3.0	0.25	12.00
Acid phenylphosphatase, units	200.0	3.0	67.00
Lactate dehydrogenase, units	58,000.0	360.0	160.00
Transaminase, AST, units	500.0	25.0	20.00
Transaminase, ALT, units	150.0	30.0	5.00

* After Caraway, W. T.: Am. J. Clin. Path., *37*:445, 1962.

chemical test result, it should be withdrawn and the test repeated in a few days. As a general rule, it is preferred to have the patient off medication when endocrine studies are performed.

CEREBROSPINAL FLUID

Cerebrospinal fluid is usually collected in three or four sterile tubes. The first and second tubes of spinal fluid are frequently contaminated with blood and are sometimes not suitable for chemical analysis. Even a small amount of plasma may cause a significant error in the protein analysis. If an analysis for glucose is to be performed on the spinal fluid, the filtrate should be prepared immediately after collection in order to avoid glycolysis, or sodium fluoride (0.5 mg/ml) may be added to a portion of the fluid and the specimen refrigerated.

COLLECTION OF URINE

Many of the chemical analyses performed on urine specimens must be carried out on 24 h urine collections. One of the reasons for this is that many urine constituents, including most of the hormones, exhibit a diurnal variation. In order to eliminate the variability of these peak excretion times, it is much more definitive to perform the analysis on a 24 h specimen. Collection of this specimen requires the utmost cooperation of the patient, nursing staff, and laboratory. At the beginning of the collecting period (usually when the patient wakens) the bladder should be emptied, that specimen discarded, and the time noted. All urine specimens passed thereafter are collected in an appropriate container. At the end of the 24 h collecting period the bladder is emptied and this urine specimen added to those already collected.

Inadequate preservative, loss of voided specimens, or the inclusion of two morning specimens in a 24 h period are common errors encountered in the collection of a 24 h urine specimen. Determination of total creatinine excretion in the 24 h period may be used as a guide to the adequacy of the 24 h collection. This is particularly useful if several 24 h urine specimens are collected from the same person, since the 24 h creatinine excretion, largely a function of muscle mass, is relatively constant from day to day in the same individual.

Most laboratories measure the volume of the 24 h urine and store an appropriate aliquot in the refrigerator until the test can be performed. Some laboratories use containers of known weight and weigh the 24 h collection to determine the volume. Measurement of the specific gravity of each specimen is not necessary. Assuming a specific gravity of 1.020 and calculating the urine volume on this basis is satisfactory for routine purposes. Whether the specimen is weighed or measured prior to aliquoting, occasionally the volume of a 24 h urine will be so large that two (or more) containers are necessary. The contents of the individual containers must be mixed prior to aliquoting. A bottle having a capacity of several gallons is convenient for this purpose.

It is the responsibility of the clinical chemical laboratory to provide an appropriate chemically clean container, containing the proper preservative, for the collection of a 24 h urine specimen. The container should hold at least 3 to 4 liters. One-gallon bottles are convenient. These should be properly labeled. The label should have space for the name and room number of the patient and the test desired as well as the time the collection was started and the time the collection was finished. In addition the bottle should contain a warning "Do Not Discard" since this is a 24 h urine collection. If at all possible, all 24 h urine collection bottles should be refrigerated during collection; the label on the bottle should also state this information.

The *preservative* for a 24 h urine specimen will depend not only on the procedure to be carried out, but to a certain degree also on the methodology to be used. In general, no preservatives are used when a bioassay is to be performed on the specimen. Since catecholamines and VMA (3-methoxy-4-hydroxymandelic acid) are stable only in acidic solution, 10 to 15 ml of concentrated hydrochloric acid is the most common preservative. If a strong acid is added to the bottle prior to collection, a warning label should be placed on the container. For the chemical analysis of the urinary steroids, in general, refrigeration is adequate. If the specimen is to be mailed to a reference laboratory, the laboratory will supply information about the proper preservative to be added (such as hydrochloric or boric acid). This preservative must not only insure stability during collection but also during the transit of the specimen to the reference laboratory. The directions of the reference laboratory should be followed carefully. Specimens for certain analyses (e.g., creatine) must be transported frozen. Styrofoam containers are available, and when used with dry ice they will keep the specimen frozen for 48 hours, which is long enough in most cases. Specimens should be mailed early enough during the week so that they will not be in transit over a weekend.

Porphyrins are in a class by themselves because, in general, they are most stable in alkaline urine. This alkalinization is usually accomplished by adding 5 g of sodium carbonate to the collection bottle. Two additional precautions are the use of a brown bottle and the addition of 100 ml of petroleum ether to retard oxidation by formation of a protective layer.

SAFETY

Although laboratories are generally considered very safe places to work, it is only because necessary safety rules and procedures have been incorporated in the daily activities of the workers. In general, safety rules are similar from one laboratory to another, and should be learned early in the training of a chemist or technologist. Future practices in a laboratory are then governed almost automatically by these safety habits.

ELECTRICAL HAZARDS

With the ever increasing trend toward instrumentation, it must be remembered that wherever there are electrical wires or connections, there is a potential shock or fire hazard. Worn wires should be replaced immediately on all electrical equipment; all apparatus should be grounded using either three-prong plugs or pigtail adapters. If grounded receptacles are not available, check with an electrician for proper grounding techniques. A convenient device called a Receptacle Analyzer (Instrutek, Inc., Annapolis, MD.) is available for checking receptacles. With this device one may check for poor resistance in ground and neutral wiring (impedances above 1000 ohms are detected) or excess voltage in the neutral wiring.

Electrical equipment and connections should not be handled with wet hands. Electrical apparatus should not be used after liquid has been spilled on it. It must be turned off immediately and dried thoroughly; a fan or hair dryer will speed up the drying process. In the case of a wet or malfunctioning electrical instrument, pulling out the plug is not enough if the instrument is used by several people; a note cautioning coworkers not to use it should be left on the instrument.

Electrical apparatus, especially motors, that are to be operated in an area where there are flammable vapors should be explosion-proof, and air-driven stirrers should be considered. Induction-driven motors are well suited for these areas. The laboratory, particularly the bench top, is no place for numerous extension cords.

HAZARDS FROM CHEMICALS

The storage and use of chemicals may be attended by a variety of dangers such as burns, explosions, fires, and toxic fumes. A knowledge of the properties of these substances and proper handling procedures will virtually eliminate any danger. Bottles of chemicals and solutions should always be handled carefully, and a cart used to transport two or more containers or one heavy container from one area to another. An excellent manual on the handling of chemicals is available.*

Spattering from acids, caustic materials, and strong oxidizing agents probably represents the greatest hazard to clothing and eyes and is a potential source of chemical burns. Bottles should never be grasped by the neck, but firmly around the body with one or both hands, depending on the size of the bottle. When diluting acids, always add acid slowly to the water with mixing; water should never be added to concentrated acid. When working with acid or alkali solutions, particularly when making up reagents in large amounts, safety glasses should be worn. Thought should be given to the possibility of breakage, and if possible this mixing should be done in a sink, which would provide water for cooling as well as for confinement of the reagent in the event the flask or bottle were to break.

All bottles containing reagents should be properly labeled. Before adding the reagent, it is good practice to label the container; in this way the possibility of having an unlabeled reagent is avoided. The label should bear the initials of the person who made up the reagent and the date the reagent was prepared. If appropriate, the expiration date should also be included. The labeling may be color coded to designate specific storage instructions, such as the need for refrigeration or special storage. All reagents found in unlabeled bottles should be discarded. The reagent should be flushed down the drain with large amounts of water. As a general rule, do not spare the water when pouring reagents down the drain. In the case of an acid, it must be diluted sufficiently in order not to harm the plumbing. Other reagents that are not harmful in the drain in themselves may present a real hazard if followed by an acid. An example of such a chemical is cyanide, which results in the release of hydrogen cyanide if brought in contact with acid. Large amounts of water must be used after each reagent that presents a danger, either actual or potential, to health.

Strong *acids*, *caustic* materials, and strong *oxidizing* agents should, whenever possible, be dispensed with automatic dispensing devices. When it is necessary to pipet these materials, there are a number of excellent safety devices commercially available to eliminate the necessity of pipetting by mouth.

Most of the general precautions just mentioned also hold for *automatic chemistry equipment* such as AutoAnalyzers. With this equipment the disposing of reagents should be considered carefully in view of what was mentioned earlier. In certain methods rather strong acids are pumped into the drain and should be accompanied by a steady flow of water from the faucet. Many of these instruments pump hazardous reagents. The use of safety glasses is recommended when inspecting a plugged tube or a malfunctioning manifold when these reagents are involved.

Perchloric acid, because it often explodes when it comes in contact with organic materials, requires careful handling procedures. Do not use perchloric acid around wooden bench tops. Bottles of this acid should be stored on a glass tray. Disposal may be accomplished by adding the acid *dropwise* (use a splatter shield) to at least 100 volumes of cold water and pouring down the sink with large amounts of additional cold water. Special perchloric acid hoods with special wash-down facilities are available when one is using large amounts of this acid.

Special care is needed in dealing with *mercury*, a rather common element in a clinical laboratory. Many think of mercury as a material used to fill a gasometer and forget that it

---

* Safety in Handling Hazardous Chemicals. Matheson, Coleman, & Bell, Norwood, Ohio 45212 (available without charge).

is also a potential hazard. Small drops of mercury on bench tops and floors may poison the atmosphere in a poorly ventilated room. Mercury toxicity is cumulative. The element's ability to amalgamate with a number of metals is well known. After an accidental spillage of mercury, the area should be gone over carefully until there are no globules remaining. All containers of mercury should be kept well stoppered.

A chemical *fume hood* is a necessity for every clinical chemistry laboratory. A container of any material giving off a harmful vapor should be opened with caution only in the fume hood. Reagent preparation that results in fume production should also be done in a hood. The heating of all flammable solvents should be done in a hood with a sealed electric hot plate, water bath, or heating mantle. In the event of an explosion or fire, the hood may be closed and the fire confined.

SAFETY EQUIPMENT

There are a large number of safety items available for the laboratory. Most of these are shown in a manual on laboratory safety.* A few will be mentioned briefly and some of these should be considered in areas where they are appropriate. *Eyewashers* or facewashers should be available in every chemistry laboratory. Some of these may merely be connected to existing plumbing. *Asbestos gloves* should be available for handling hot glassware and for handling dry ice. Safety *goggles*, glasses, and visors are available in many sizes and shapes. Some fit conveniently over regular eye glasses. Shatterproof laminated *safety shields* should be used in front of systems that represent a potential danger because of vacuum collapse or pressure explosions. Desiccator guards should be used with vacuum desiccators. Hot beakers should be handled with *tongs*. Floor standing gas cylinders should be strapped to the bench or wall with special *cylinder supports*. Inexpensive polyethylene pumps are available to pump acids from large bottles.

FIRE SAFETY

Every laboratory should have the necessary equipment for general fire fighting to put out or confine the fire to a given area, as well as to put out the fire on the clothing of an individual. Easy access to *safety showers*, which may be near the door of the laboratory, is desirable. These showers should have either a pull chain attached to the wall at a convenient height, or a large ring attached to the chain so that the shower may easily be activated even with the eyes closed. *Fire blankets* in an easily accessible wall-mounted case should be available for smothering clothing blazes. By taking hold of the rope that is attached to the blanket and turning the body around, the blanket is unrolled from the case and rolled around the body, thus smothering the flames.

There are various types of *fire extinguishers* available. Their use depends on the type of fire. Almost any type of fire extinguisher may be used on a fire where only wood, paper, textiles, and similar materials are involved. For fires involving grease or oils, foam, dry chemical or vaporizing liquid extinguishers should be used. In fires involving electrical equipment or in areas where live electricity is present, water or soda type extinguishers are not to be used. These materials will conduct electricity and will lead to a serious electrical hazard in addition to the existing fire. For these fires, carbon dioxide, dry chemical, or vaporizing liquid extinguishers should be used. It is rather impractical to have several types of extinguishers present in every area. Dry chemical fire extinguishers are among the best all-purpose extinguishers for laboratory areas. They should be provided near every laboratory door and, in a large laboratory, also at the opposite end of the room. The assumption should not be made that all the laboratory personnel know how to use this equipment. Everyone in the laboratory should be instructed in the use of these extinguishers and any

* Manual of Laboratory Safety. Fisher Scientific Co., Pittsburgh, PA., 15219 (available without charge).

other fire fighting equipment. All dry chemical extinguishers should be tested twice a year by qualified personnel.

BIOLOGICAL HAZARDS[20]

Frequent causes of biological infections are (1) accidental oral aspiration of infectious material through a pipet, (2) accidental inoculation with syringes and needles, (3) animal bites, (4) sprays from syringes, and (5) centrifuge accidents. Together, these caused about 12 per cent of some 3700 laboratory infections. Some other commonly recognized causes of laboratory infections are (1) cuts or scratches from contaminated glassware, (2) cuts from instruments used during animal autopsy, and (3) spilling or spattering of pathogenic samples on floors, table tops, and other surfaces. Laboratory technicians, students, trained professional personnel, and animal handlers, those most closely associated with the infectious operations, are in the greatest danger of being infected.

SAFE TECHNIQUES AND PROCEDURES FOR PREVENTION OF LABORATORY INFECTIONS

Sound fundamental laboratory techniques, well supervised and conscientiously carried out, can do much to achieve environmental control and reduce the hazards of infection. A list of procedural rules that are widely applicable in clinical laboratories follows:[30]

1. Never do direct mouth pipetting of infectious or toxic biological fluids, sera, and reconstituted control sera; use a pipettor.
2. Plug pipets with cotton when an automatic pipettor is not applicable.
3. Do not blow infectious material out of pipets.
4. Do not prepare a mixture of infectious materials by bubbling expired air through the liquid with a pipet.
5. Use only needle-locking hypodermic syringes. Avoid using syringes whenever possible.
6. Expel excess fluid and bubbles from a syringe vertically into a cotton pledget moistened with disinfectant.
7. Before and after injecting an animal, swab the site of injection with a disinfectant.
8. Reusable pipets and syringes should be sterilized immediately after use by placing in a pan containing a phenolic compound, e.g., Staphene.
9. Before centrifuging, inspect tubes for cracks. Inspect the inside of the trunnion cup for rough walls caused by erosion or adhering matter. Carefully remove all bits of glass from the rubber cushion. A germicidal solution added between the tube and the trunnion cup not only disinfects the surfaces of both of these, but also provides an excellent cushion against shocks that otherwise might break the tube.
10. Avoid decanting centrifuge tubes; if you must do so, afterwards wipe off the outer rim with a disinfectant. Avoid filling the tube to the point that the rim becomes wet with contaminated specimen.
11. Never leave a discarded tube or infected material unattended.
12. Autoclave all contaminated material that is to be discarded (e.g., cerebrospinal fluid pipets and Vacutainers).
13. Periodically, clean out deepfreeze and dry-ice chests in which sera are stored to remove broken ampules or tubes. Use rubber gloves and respiratory protection during this cleaning.
14. Handle with rubber gloves diagnostic serum specimens carrying a risk of infectious hepatitis.
15. Develop the habit of keeping your hands away from your mouth, nose and eyes and any other part of the face. This may prevent self-inoculation.

16. Avoid smoking, eating, and drinking in the laboratory.

17. Do not store food or beverages in the laboratory refrigerator.

18. Evaluate the extent to which the hands may become contaminated. With some specimens and analytical techniques, forceps or rubber gloves are advisable.

19. Change into a different set of clean laboratory clothing when going to the dining room, library, and other nonlaboratory areas.

ISOTOPES

The use of radioactive materials in the clinical chemistry laboratory is becoming increasingly more popular with the introduction of methods which utilize the principle of radioimmunoassay. Since the actual amount of radiation present in the clinical chemistry laboratory is usually quite small, relatively few special precautions are needed.

Among the several radiation units of measurement, two should be of interest to those in the clinical laboratory. The curie (Ci) is the basic unit used to describe the amount of radioactivity. One curie is equal to 37 billion disintegrations per second ($3.7 \times 10^{10}\ s^{-1}$). Most applications in the clinical laboratory deal in microcuries, and most commercial radioimmunoassay kits contain 1 to 2 microcuries of radioactivity.

The term rem (roentgen equivalent man) is the amount of energy dissipated and the biological damage derived from such energy dissipation. This term is used whether the radiation effect is caused by electromagnetic radiation or radiation from particulate bodies, such as alpha, beta, positron, proton, or neutron emission. For the general population the radiation dose to the whole body should not exceed 0.5 rem per year, while for the radiation worker the maximum level is 5 rem per year or 100 millirem per week.

The basic factors to be kept in mind in dealing with isotopes are exposure time, distance, and shielding. Since the radiation effects to humans are cumulative, one should not be in the presence of unshielded sources any longer than necessary. The isotope storage facilities should be located as far from personnel as possible. Radiation can be reduced by proper shielding. Normally, gamma sources exceeding 10 μCi are packaged in small, lead containers and should be kept in these containers except when material is being withdrawn.

RADIATION SAFETY

Each laboratory should have a radiation safety officer who has the responsibility of knowing and enforcing the safety rules, which will vary with the type and amount of isotope used. This person would be responsible for monitoring radiation areas with a portable radiation detector (a portable Geiger-Mueller counter). Persons who are working with gamma emitters (e.g., ^{131}I) or with beta emitters having a maximum energy greater than 0.2 MeV (e.g., ^{125}I) should wear film badges. This badge, provided by various companies, is a sealed photographic film to measure exposure to ionizing radiation. The badges are then sent to the company and the exposure in rems is calculated and reported each month. Badges are not needed if one is working with low energy beta emitters (e.g., ^{3}H).

Some general rules for clinical laboratory radiation safety

1. Radioactive material should be appropriately labeled and stored in a specially designated area.
2. Radioactive materials should be used only at designated work areas.
3. Eating or storing food in these areas is prohibited.
4. Radioactive material should not be pipetted by mouth.
5. All spills of radioactive materials should be contained and the surfaces involved washed with an alkali detergent. Repeat this procedure until the count on a portable monitor is below 300 counts per minute.
6. Solid waste (test tubes) should be rinsed in water before disposal.

Disposal of radioactive waste

The level of radioactivity encountered in the usual radioimmunoassay procedures is low enough so that liquid wastes may be disposed of in the sink with running water. To calculate the amount of radioactive material that may be disposed of via the sewage system from one building, one must know the water usage. This may be obtained from the water bill or by calling the water company. The allowable quantity of ^{125}I in sewage is 4.0×10^{-5} μCi/ml of water usage. This is equal to 1.13 μCi per cubic foot of water usage.

Radioisotope licensing

This rule varies from state to state. Some materials are exempt from Atomic Energy Commission or state licensing requirements. Most institutions already have an institutional license which specifies the safety officer. It would be well for the clinical chemistry laboratory to check with this individual before beginning to use radioactive materials. If there is no license, many manufacturers of isotope materials will assist the laboratory to obtain the proper license.

REFERENCES

1. American Society for Testing Materials: Standards. Bull. #C-225-45T, pt. 3, Philadelphia, 1949, p. 354.
2. Blatt, W. F.: Am. Lab., *4*:78, 1972.
3. Bonner, W. D., and Branting, B. F.: J. Am. Chem. Soc., *48*:3093, 1926.
4. Bower, J. H.: J. Res. Nat. Bur. Standards, *33*:199, 1944.
5. Caraway, W. T.: Am. J. Clin. Path., *37*:445, 1962.
6. Caraway, W. T.: Sources of error in clinical chemistry. *In* Standard Methods of Clinical Chemistry. S. Meites, Ed. New York, Academic Press, 1965, vol. 5, pp. 19–30.
7. Caraway, W. T., and Kammeyer, C. W.: Clin. Chim. Acta, *41*:395, 1972.
8. Constantino, N., and Kabat, H.: Am. J. Hosp. Pharm., *30*:24, 1973.
9. Dean, G. A., and Herringshaw, J. F.: Analyst, *86*:434, 1961.
10. Dybkaer, R., and Jorgensen, K.: Quantities and Units in Clinical Chemistry. Baltimore, The Williams & Wilkins, Co., 1967.
11. Dybkaer, R.: *In* Standard Methods of Clinical Chemistry. R. MacDonald, Ed. New York, Academic Press, Inc., 1970. vol. 6, p. 223.
12. Farese, G., Mager, M., and Blott, W. F.: Clin. Chem., *16*:226, 1970.
13. Handbook of Chemistry and Physics, R. C. Weast, Ed. Cleveland, The Chemical Rubber Co., 1971, pp. C757–C771.
14. Hansten, P. D.: Drug Interaction. Philadelphia, Lea & Febiger, 1971.
15. Henry, R. J.: Clinical Chemistry: Principles and Technics. New York, Harper & Row, Publishers, 1964.
16. Kirsten, W.: Anal. Chem., *25*:1137, 1953.
17. Martell, A. E., and Calvin, M.: Chemistry of the Metal Chelate Compounds. Englewood Cliffs, New Jersey, Prentice-Hall, Inc., 1952.
18. Morey, G. W.: The Properties of Glass. American Chemical Society Monograph, New York, Reinhold Publishing Co., 1954.
19. National Bureau of Standards, Circular C-602, 1959.
20. Philips, C. B.: *In* Handbook of Laboratory Safety N. V. Steere, Ed. Cleveland, The Chemical Rubber Co., 1967.
21. Properties of Selected Commercial Glasses: Pyrex Corning Vycor, B-83. Corning, New York, Corning Glass Works, 1959.
22. Reagent Chemicals. The American Chemical Society, Fourth Edition Washington, D.C., 1968.
23. Sanz, M. C.: Ultramicro methods and standardization of equipment. Clin. Chem., *3*:406, 1957.
24. Seligson, D.: Clin. Chem., *5*:320, 1959.
25. Simonds, H. R., Weith, A. J., and Bigelow, M. H.: Handbook of Plastics. 2nd ed. New York, D. Van Nostrand Co., Inc., 1949.
26. Stas, J. S.: Chem. News, *17*:1, 1868.
27. Stier, A. R., Miller, L. K., and Smith, R. J.: Water. Commission on Laboratory Inspection and Accreditation, College of American Pathologists, 1972.
28. The International System of Units: NBS Special Publication 330, 1972.
29. Walter, A. R., and Gerarde, H. W.: A rapid semiautomatic system of chemical analysis using true microspecimens. Clin. Chem., *10*:509, 1964.
30. Wedum, A. G.: Public Health Rep., *76*:619, 1964.

31. Windisch, R. M., and Bracker, M. M.: Clin. Chem., *16*:416, 1970.
32. Winstead, M.: Reagent Grade Water, How, When and Why. Monograph. Published by American Society of Medical Technologists. 1967.
33. Young, D. S., Thomas, D. W., Friedman, R. B., and Pestaner, L. C.: Clin. Chem. *18*:1041, 1972.

ADDITIONAL READINGS

Davidsohn, I., and Henry, J. B., Eds.: Todd-Sanford Clinical Diagnosis by Laboratory Methods. 15th ed. Philadelphia, W. B. Saunders Company, 1974.

Henry, R. J.: Clinical Chemistry, Principles and Technics. 2nd ed. New York, Harper & Row, Publishers, 1974.

Manufacturing Chemists Association: Guide for Safety in the Chemical Laboratory. 2nd ed. New York, Van Nostrand Reinhold Company, 1972.

CHAPTER 2

STATISTICS, NORMAL VALUES, AND QUALITY CONTROL

by Edward W. Bermes, Jr., Ph.D., Vivian Erviti, Ph.D., and Donald T. Forman, Ph.D.

STATISTICS IN THE CLINICAL CHEMISTRY LABORATORY

Clinical chemistry laboratories produce data that are used to aid in making clinically important decisions. All data generated are associated with variations that interfere with data evaluation and interpretation. Although some variation is within and some outside of the control of the analyst, he must be able to evaluate the extent of variation and determine its significance for decisions to be based on the data.

Both the refinement of laboratory techniques and the application of better experimental design procedures are essential in reducing variation. Statistical methods are helpful in providing estimates of and in controlling variation which remains. The next section provides examples of some commonly encountered sources of variation. Later sections of the chapter describe some statistical tools useful in the estimation and control of variation.

SOURCES OF VARIANCE (ERROR)

Variance is a general term used to describe the many fluctuations in data that interfere with the measurement of the phenomenon of interest. One can classify most variance as either systematic variance or error (random) variance.

Systematic variance

Systematic variance is defined as variation that influences observations consistently in one direction or another. Systematic variance may arise from either natural or man-made events, and be known or unknown to the investigator. There are many sources of systematic variance.

Aging phenomena are sources of systematic variance. Chemicals or reagents may decompose during storage or be greatly altered in composition by distillation, crystallization, or microbial attack. Biological specimens may also change over time, or traces of glass containers can slowly dissolve in alkaline solutions. Impurities can be absorbed from the air. Deviations in results can occur due to mechanical changes of apparatus or periodic variations of such factors as temperature and humidity.

Another type of systematic variation is the *personal bias of the analyst*. The analyst may unconsciously introduce a constant variation into the results. This variation may differ from that introduced, equally unconsciously, by a second analyst using an identical procedure in the same laboratory. For example, one analyst may read the meniscus mark on a pipet consistently lower than another. Such technician variation may be even greater if a new or less skilled operator performs the test.

Laboratory bias may also exist. Such bias may be due to differences arising from variations in standards, reagents, environment, methods employed, or apparatus used. These differences can result in either an upward or downward bias relative to results that would be obtained in a different laboratory by the same analyst using similar techniques and equipment.

Inter- and intraindividual bias refers to variation between and within patients. Interindividual variation occurs because of differences between races, ages, sexes, creeds, and health status. Variation within an individual patient from hour to hour, day to day, and seasonally is known as intraindividual variation.

Experimental error results from systematic differences between groups of measurements. It is associated with manipulation or change in methods, instruments, or analysts. For example, a given method may yield results consistently lower or higher than another analytical procedure. This may be due to variations in methods based on the same principle (e.g., glucose by copper reduction or ferricyanide reduction) or methods based on different principles but measuring the same constituent (e.g., glucose by *o*-toluidine or glucose oxidase).

Error variance (random error)

Error variance is variation whose source(s) cannot be completely identified and controlled. It is considered to be due to chance. The fluctuations may be in either direction rather than in one direction only. The chance for random error may be affected by the quality of reagents, the number and complexity of the steps in a procedure, and the skill of the analyst performing the test.

An example of error variance occurs when replicate analyses of the same specimen are made in exactly the same way by the same analyst and produce a variety of results. An analyst, for example, may make duplicate analyses of serum calcium using the same technique and the same specimen and obtain results of 9.7 and 10.0 mg/100 ml.

A system is said to be in a state of *statistical control* when systematic variance, error variance, and the interaction between the sources of variance have been evaluated and are judged to be within acceptable limits. This does not eliminate all variability in the data, but remaining variance is considered unimportant and reasons for its occurrence are not sought.

The concept of variation is so fundamental to the theory, application, and interpretation of statistics that it will be discussed in greater detail in the following pages.

POPULATION AND SAMPLE

The branch of mathematics called statistics is concerned with the properties of large collections of persons, things, or observations. Such large collections are called *populations*. The exact meaning of the word "population" varies according to the context in which it is used. A statistician would call the urea nitrogen values of all adult males living in the United States during a stated time period a population of values. He might also call the glucose values determined on all infants born in a given hospital during a specified period of time a population.

Usually, it is not possible or feasible to study entire populations of values. Sometimes information is available or can be obtained for only a subgroup of the population. The subgroup of the population of values is called a *sample*.* Sample values can be used to study the characteristics of the population from which they were selected if they are selected properly. For example, we might not be able to analyze specimens for glucose from all newborns in a given hospital during a one week period of time. However, we might be able to make 25 or 50 determinations and use the results to make inferences about the

* It should be noted that this use of the word sample differs from its use in many clinical chemistry laboratories. In many laboratories the word sample is used to refer to a specimen. In statistical terminology, sample is used to refer to a group of specimens, i.e., a subgroup of the population.

entire population of newborns of that hospital during the specified time period. We could not generalize from results based on the sample to all infants in the hospital during the entire year, or to all infants born in all hospitals during the specified year. Thus, while the definition of a population may change, its exact specification is necessary if results based on a sample are to be interpretable.

FREQUENCY DISTRIBUTIONS

A frequency distribution is a representation of the relationship between categories of data and frequencies. It may be a list of categories and the frequency associated with each, a graph, or a rule for pairing observations with frequency in mathematical form.

If determinations of glucose values were made on 105 apparently healthy infants, the values might be similar to those shown in Table 2–1. Plotting the glucose values against the frequency of occurrence results in the histogram shown in Figure 2–1. In order to better comprehend the meaning of data, it is desirable to describe a distribution of values in a concise way. A measure of the center of the distribution and a measure of the scattering of values about the center are used for this purpose. Such measures can be used in reference to populations or samples. Measures of the characteristics of populations are called parameters and are represented by Greek letters. Measures of the characteristics of a sample are called statistics and are represented by Roman letters.

Measures of location

Central Tendency. One measure of the center of a distribution is the *arithmetic mean.* To compute the arithmetic mean, the sum of all obtained values is divided by the number of values. The mean of a population is represented by the Greek letter μ. The formula for

TABLE 2–1 CUMULATIVE DISTRIBUTION OF 105 BLOOD GLUCOSE VALUES FROM INFANTS (mg/100 ml)

Glucose Value	*Frequency*	*Cumulative Frequency*	*Per Cent* Cumulative Frequency*	*Glucose Value*	*Frequency*	*Cumulative Frequency*	*Per Cent* Cumulative Frequency*
30	1	1	0.95	65	2	59	56.19
33	1	2	1.90	66	2	61	58.10
36	2	4	3.81	68	2	63	60.00
39	1	5	4.76	69	5	68	64.76
42	1	6	5.71	70	1	69	65.71
43	1	7	6.67	71	2	71	67.62
44	2	9	8.57	72	1	72	68.57
45	2	11	10.48	73	2	74	70.48
47	1	12	11.43	74	3	77	73.33
48	3	15	14.29	75	1	78	74.29
49	5	20	19.05	76	4	82	78.10
50	2	22	20.95	77	1	83	79.05
51	1	23	21.90	78	3	86	81.90
52	2	25	23.81	79	2	88	83.81
53	1	26	24.76	80	1	89	84.76
54	4	30	28.57	82	5	94	89.52
55	2	32	30.48	84	2	96	91.43
56	1	33	31.43	86	1	97	92.38
57	1	34	32.38	87	1	98	93.33
58	6	40	38.10	91	3	101	96.19
59	4	44	41.90	99	1	102	97.14
61	1	45	42.86	108	1	103	98.10
62	3	48	45.71	113	1	104	99.05
63	4	52	49.52	121	1	105	100.00
64	5	57	54.29				

* The per cent cumulative frequency is found by dividing the cumulative frequency by the total number of observations and multiplying by 100.

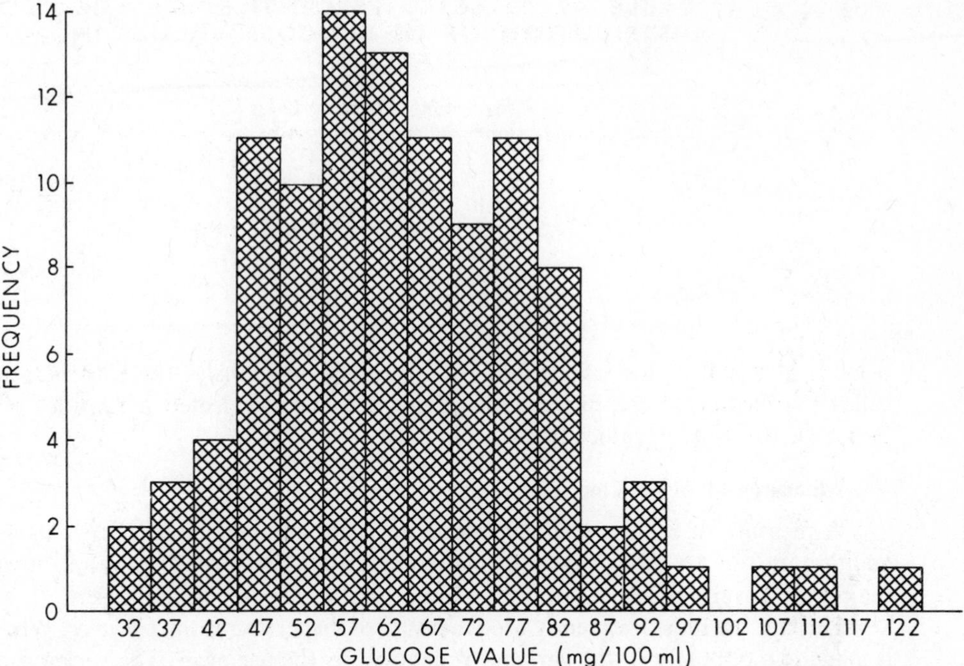

Figure 2–1 Histogram of 105 blood glucose values from newborns. The values along the horizontal axis represent the midpoints of the intervals. The limits of the intervals extend from 2.5 mg/100 ml below the midpoint to 2.5 mg/100 ml above it.

calculating the arithmetic mean of a sample of values is

$$\bar{X} = \frac{\sum X}{n} \tag{1}$$

where

$\bar{X}$ (pronounced X bar) = the mean
$\sum X$ = the sum of the observed values, from the first through the last observation
n = total number of observations

Another frequently used measure of central tendency is the *median*. If the observed values are arranged in increasing order, the median is the middle observation. In other words, the median is the value of the measured variable below which half the observations fall. If the number of obtained values, n, is odd, there will be a unique median. The value can be found by taking the $(n + 1)/2$ value from either end in the ordered sequence. When the number of values is even and there is no middle observation, the median is defined as the arithmetic mean of the two middle observations.

The *mode* is the value of the variable which occurs most frequently.

The *geometric mean* may also be used as a measure of central tendency. The geometric mean is calculated by finding the logarithm of each value of a variable, summing the logarithmic values, dividing the sum of the logarithms by the number of observations (n), and then taking the antilog of the result. The geometric mean is usually less than the arithmetic mean of the same observations. The only circumstance in which the geometric mean equals the arithmetic mean arises when all values observed are identical.

Percentiles. A percentile is defined as the value of a variable below which a certain proportion of the observations fall; i.e., percentiles are general measures of location rather than measures of central tendency. For example, the 10th percentile, P_{10}, is defined as the value below which 10 per cent of the values fall. The 50th percentile, P_{50}, is the point

TABLE 2–2 SELECTED PERCENTILES OF THE DISTRIBUTION OF 105 GLUCOSE VALUES IN TABLE 2–1

Percentile	*Glucose Value*
2.5	34
5	40
10	44.5
50	63.5
90	83
95	88
97.5	102.5

below which half of the values fall. As mentioned previously, the 50th percentile is also called the median of the distribution. Table 2–2 shows selected percentiles of the distribution of 105 glucose values from Table 2–1.

Measures of dispersion

A measure of central tendency alone does not completely describe a distribution. In addition, it is desirable to have information regarding the spread or scatter of values about the center. Several measures of dispersion or scatter are in common use.

Perhaps the most frequently used measure of dispersion is the *variance*. When reference is made to a population, the variance is denoted by the use of σ^2. The variance of a sample is denoted by s^2 and is defined as the sum of squares of the deviations of the observations from the mean divided by the total number of observations less 1. The greater the scatter of values from the center of the distribution, the greater the magnitude of the deviations and the larger the variance. The formula for the variance of a sample is

$$s^2 = \frac{\sum (X - \bar{X})^2}{n - 1} \tag{2}$$

where

s^2 = variance of a sample of values

$\sum (X - \bar{X})^2$ = the squares of the deviations of the observations from the mean of the observations, summed over the n observations

n = total number of observations

The numerator of Formula (2) is called the sum of squared deviations. Any sum of squared quantities has a number called the *degrees of freedom* associated with it. For n observations there are n squares of deviations, but one restriction has been imposed. The restriction results from the fact that the sum of the deviations of observations from the mean equals zero (see Example 1, Block A). If the values of $n - 1$ deviations are known, the value of the one remaining deviation is not free to vary but is determined by subtracting the sum of the $n - 1$ known deviations from zero. The degrees of freedom is defined as the number of squares minus the number of restrictions. Thus, the divisor of Formula (2) is the number of degrees of freedom associated with the sum of squares of the numerator. The reason for dividing the sums of squares by its associated degrees of freedom rather than by n when calculating the variance of a sample of values will be explained later.

The numerator of Formula (2) is a squared quantity. Since it is often inconvenient to talk about the scatter of a distribution in terms of squared units, a measure of dispersion expressed in the same units as the mean is desired. The *standard deviation* is such a measure of scatter. The standard deviation is defined as the positive square root of the variance. Thus, the standard deviation of a sample is

$$s = \sqrt{\frac{\sum (X - \bar{X})^2}{n - 1}} \tag{3}$$

EXAMPLE 1: SAMPLES FROM 105 GLUCOSE VALUES PRESENTED IN TABLE 2-1

BLOCK A

Computation of Some Measures of Central Tendency and Dispersion Using a Sample of Ten Observations from the Glucose Data

Glucose Value (mg/100 ml) (X)	*Deviation from Mean* ($X - \bar{X}$)	*Deviation Squared* ($(X - \bar{X})^2$)	*Absolute Difference of Value from Median* $\lvert X - \text{median}\rvert$	*Values Squared* (X^2)
33	−26.5	702.25	28	1089
43	−16.5	272.25	18	1849
44	−15.5	240.25	17	1936
59	−0.5	0.25	2	3481
59	−0.5	0.25	2	3481
63	3.5	12.25	2	3969
69	9.5	90.25	8	4761
71	11.5	132.25	10	5041
74	14.5	210.25	13	5476
80	20.5	420.25	19	6400
$\sum X = 595$	$\sum (X - \bar{X}) = 0$	$\sum (X - \bar{X})^2 = 2080.50$	$\sum \lvert d\rvert = 119$	$\sum X^2 = 37483$

Measures of central tendency

$$\text{Mean} = \bar{X} = \frac{\sum X}{n} = \frac{595}{10} = 59.5 \text{ mg/100 ml}$$

$$\text{Median} = \left(\frac{n+1}{2}\right) \text{ value} = 61 \text{ mg/100 ml}$$

Mode = 59 mg/100 ml
Geometric mean = 57.53 mg/100 ml

Measures of dispersion

$$\text{Variance} = s^2 = \frac{\sum (X - \bar{X})^2}{n-1} = \frac{2080.50}{9} = 231.17$$

Standard deviation $= s = \sqrt{s^2} = \sqrt{231.17} = 15.20$ mg/100 ml
Average deviation $= 119 \div 10 = 11.9$ mg/100 ml
Range $= w = 47$ mg/100 ml

BLOCK B

Application of the Computational Formula (4) for the Calculation of the Variance and Standard Deviation of the 10 Glucose Values of Block A

$$\sum X = 595$$
$$(\sum X)^2 = 354025$$
$$\sum X^2 = 37483$$

$$\text{Variance} = s^2 = \frac{n\sum X^2 - (\sum X)^2}{n(n-1)} = \frac{10(37483) - 354025}{10(9)} = \frac{20805}{90} = 231.17$$

Standard Deviation $= \sqrt{s^2} = s = 15.20$ mg/100 ml

BLOCK C

Computation of the Standard Error of the Mean, the 95 Per Cent Confidence Interval of the Mean, the 90 Per Cent Parametric Tolerance Interval To Include 95 Per Cent of the Population, and the Coefficient of Variation for the 10 Observations of Glucose Presented in Block A.

1. Standard error of the mean: $\sigma_{\bar{X}} = \frac{s}{\sqrt{n}} = \frac{15.20}{3.16} = 4.81$ mg/100 ml
2. 95 per cent confidence interval of the mean $= \bar{X} \pm 2\sigma_{\bar{X}} = 59.5 \pm 2(4.81)$

$$= 59.5 \pm 9.62$$

The 95 per cent confidence interval extends from 49.88 to 69.12 mg/100 ml

3. Tolerance interval to include 95 per cent of the population at a 90 per cent

confidence level $= \bar{X} \pm Ks = 59.5 \pm 3.018(15.20)$

$$= 59.5 \pm 45.87$$

Tolerance interval extends from 13.63 to 105.37 mg/100 ml

4. Coefficient of variation $= CV = \frac{s}{\bar{X}}(100) = \frac{1520}{59.5} = 25.55$ per cent

BLOCK D

Comparison of Population Parameters with Sample Statistics

STATISTICS COMPUTED ON 12 SAMPLES EACH OF SIZE 10 FROM THE 105 GLUCOSE VALUES IN TABLE 2–1

Sample No.	*Mean*	*Standard Deviation**	*Percentiles†* 2.5th	97.5th	*Tolerance Limits‡* LOWER	UPPER
1	67.3	19.88	27.54	107.06	7.30	127.30
2	61.1	18.41	24.28	97.92	5.54	116.66
3	56.4	8.83	38.74	74.06	29.75	83.05
4	55.7	12.99	29.72	81.68	16.50	94.90
5	63.6	22.32	18.96	108.24	−3.76	130.96
6	63.9	15.83	32.24	95.56	16.13	111.67
7	73.1	22.57	27.96	118.24	4.98	141.22
8	67.2	21.85	23.50	110.90	1.26	133.14
9	60.2	10.39	39.42	80.98	28.84	91.56
10	69.2	17.89	33.42	104.98	15.21	123.19
11	55.6	10.43	34.74	76.46	24.12	87.08
12	66.0	20.13	25.74	106.26	5.25	126.75
Population Parameter§	64.87	16.41	34	102.5		

* Calculated using $n - 1$ in denominator.

† Calculated by using $\bar{X} \pm 2s$.

‡ 90 Per cent confidence level to include 95 per cent of the population of 105 glucose values.

§ Variance calculated using n in denominator, percentiles calculated directly from Table 2–1.

Consider the 105 glucose values presented in Table 2–1 as the population of interest. A random sample of 10 values may be selected from the population and the mean, variance, and standard deviation of the sample computed.* The sample may result in values such as those in Example 1, Block A.

In Example 1, Block A, the calculation of the variance and standard deviation are fairly simple because the sample size is small. As the sample size is increased, the computation becomes more tedious and the chance for an arithmetic error increases. Whenever the mean is not a whole number, there is also the possibility of some rounding error in each deviation computed. The formula presented as Formula (2) and used in the above example is frequently called the *definitional formula for the variance.* An alternative formula that is algebraically equivalent to Formula (2) but more convenient for routine and machine calculations is presented below.

$$s^2 = \frac{n \sum X^2 - (\sum X)^2}{n(n-1)} \tag{4}$$

where

n = number of observations
$\sum X^2$ = sum of squares of the original observations
$\sum X$ = sum of the original observations

The *computational formula* for the variance, Formula (4), will be used throughout the remainder of the chapter.

Although the variance and standard deviation are the most frequently used measures of dispersion, it is sometimes desirable to use other statistics. Two other measures are the *average deviation* and the *range.*

The average deviation (AD) is calculated by taking the average of the absolute values of the deviations of the individual measurements from either the mean or the median. When the median is used as the measure of central tendency, the average deviation from the median is often used as the measure of dispersion instead of the variance and standard deviation which are more appropriate to the use of the mean.

In addition, there are some instances when the mean is used as the measure of central tendency, and it is still desirable to calculate the AD. For certain data it may be easier to compute the AD about the median than to compute s. The AD can then be used to estimate the standard deviation of the population. However, since the AD is smaller than the population standard deviation for a given set of data, it must be multiplied by a factor. For large samples (i.e., $n \geqq 25$), the AD must be multiplied by 1.25. The magnitude of the multiplying factor increases as sample size decreases and reaches 1.77 when $n = 2$. Dixon and Massey[8] provide a table of factors for sample sizes from 2 through 10 as well as short-cut formulas for computation of deviations from the median. For the data presented in Example 1, Block A (i.e., $n = 10$), the multiplying factor is 1.353. The AD computed by finding the mean of the absolute deviations from the median was found to be 11.9. The resulting estimate of the standard deviation is 16.1. This estimate compares favorably with the value of 15.2 obtained on a sample of 10 and the value of 16.4 calculated for the standard deviation of all 105 glucose values.

The range (w) is simply the difference between the highest and lowest values observed. Although there is no formula relating the range and the standard deviation, the range can be used as a rough estimate of the population standard deviation if a correction factor is applied. The correction factor varies with the number of observations upon which the range is based. For $n = 5$ the range is about twice σ; for $n = 10$ the range is approximately 3σ; while for $n = 25$ the range will be between 4σ and 6σ. Dixon and Massey give a table of

* In actual practice, a sample of this size is too small to provide much useful information. An $n = 10$ was chosen to simplify calculations only, and is *not* an endorsement of the use of small sample sizes in the laboratory.

correction factors. For the data in Example 1, Block A ($w = 47$, $n = 10$), the correction factor given by Dixon and Massey is 0.325 and the estimate of σ is 15.3, an estimate of σ that is not quite as good as that provided by the AD.

CONTINUOUS FREQUENCY DISTRIBUTIONS

If a large number of values of a variable are determined and a histogram constructed, the shape of the histogram will approach a smooth curve. Many types of smooth curves can be described by a mathematical expression known as a frequency function, which relates observations with their associated frequency. The function is defined for all values of the variable in such a way that the total area under the curve and above the axis equals one.

The Gaussian (normal) distribution

For some variables the smooth curve approximated by the histogram is a symmetrical, bell-shaped curve that can be described by a frequency function called the Gaussian* or normal distribution. A Gaussian distribution is a theoretical distribution in which the arithmetic mean, median, and mode coincide. Thus, the number of values that exceed the mean by a given amount equals the number that fall short of it by the same amount. Results close to the mean occur more frequently than results further removed from it, and results equal to the mean value occur with the greatest frequency.

Since the standard deviation is in the same units as the original measurements, it can be used to determine the proportion of values falling in a given area under the normal curve. If the total area under the curve represents all the values in the population, the area from the point $1s$ below the mean ($-1s$) to $1s$ above the mean ($+1s$) includes about 68 per cent of the values. The area from minus $2s$ to plus $2s$ includes approximately 95 per cent of the values, and the area from minus $3s$ to plus $3s$ includes approximately 99 per cent of the values. The scatter of results is shown graphically in Figure 2–2.

In Gaussian distributions the mean and the standard deviation can be used to identify specified percentiles of the distribution. The mean of a Gaussian distribution is the 50th percentile of the distribution. The 2.5th and 97.5th percentiles can be estimated by calculating the value of the mean minus and plus two standard deviations, respectively.

The use of *normal probability paper* is another method useful in obtaining the percentiles of a distribution that is Gaussian or approximately Gaussian. Normal probability paper is scaled so that a plot of the cumulative frequencies of values of a variable against the values results in a straight line if the distribution is Gaussian. Percentiles of the distribution can be read directly from the graph (see Fig. 2–3) by determining the per cent of values above the point of interest and locating that point on the left-hand scale. Then a horizontal line is drawn from the specified point on the left-hand scale to meet the

* This was first described by the French mathematician Abraham Demoivre in his treatise of 1733, and further developed by the astronomer-mathematician, Karl Friedrick Gauss in the 19th century.

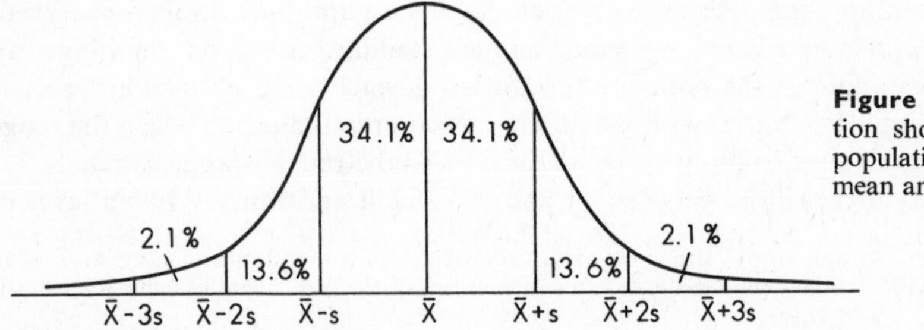

Figure 2–2 Gaussian distribution showing proportion of the population falling between the mean and $\pm s$, $\pm 2s$, and $\pm 3s$.

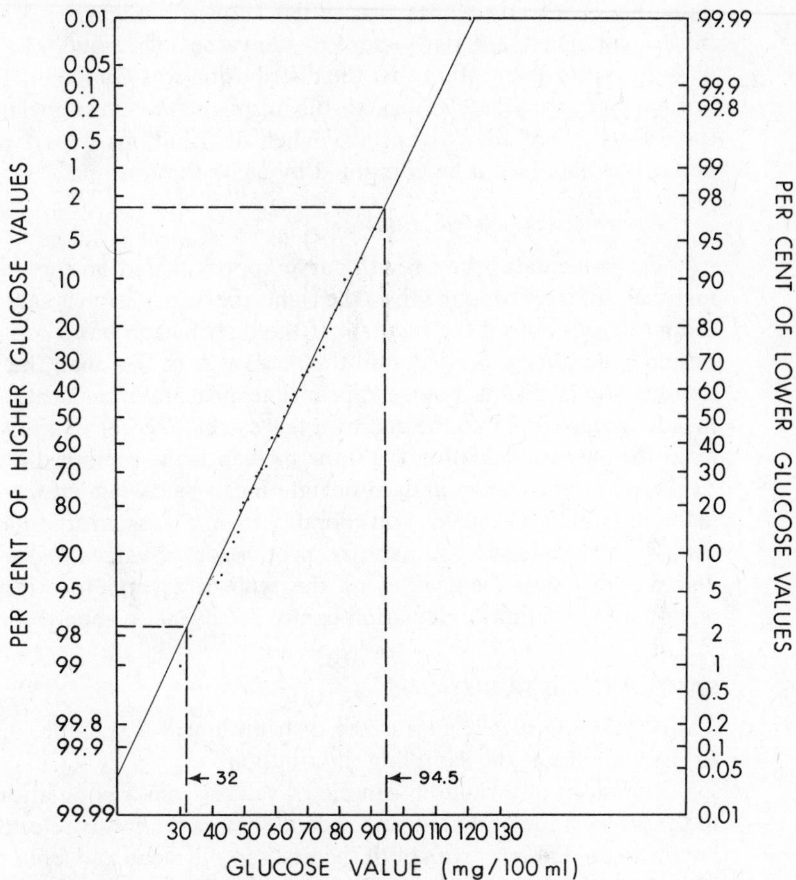

Figure 2–3 Cumulative frequency distribution of 105 glucose values plotted on normal probability paper.

cumulative distribution line. The value of the variable corresponding to the specified percentile is read from the horizontal axis directly below the intersection.

One could also use the right-hand scale to determine percentiles. If the point corresponding to the per cent of values below the percentile of interest is located on the right-hand scale, a horizontal line can be drawn from the point to the cumulative distribution line and the value of the variable read from the abscissa directly below the point of intersection.

The distribution of the 105 glucose values referred to previously is approximately normal (see Figure 2–1). If the mean and median of the values presented in Table 2–1 are calculated, they will be found to be close together: 64.87 and 63.5, respectively. In this distribution, the values for $\bar{X} \pm 2s$ are 32 and 98 to the nearest integer.

In Figure 2–3 the 49 different glucose values from Table 2–1 ($n = 105$) are plotted on the horizontal axis of normal probability paper, and the per cent of the distribution exceeding a given glucose value is plotted on the left-hand vertical axis. A straight line is drawn by inspection so that it passes through or falls close to as many of the points as possible. The points are equally distributed on both sides of the line. Horizontal lines are drawn from the 97.5 and 2.5 per cent points on the left-hand scale to the straight line. The horizontal line from the 97.5 point on the left-hand scale represents the point above which 97.5 per cent of the values fall, i.e., the 2.5th percentile. Vertical lines are dropped from the intersections to the horizontal axis and glucose values are read from the horizontal axis. Since the accuracy of such a subjective approach depends on the skill and judgment of the analyst drawing the line, results may differ considerably among analysts for the same set of data. The 2.5th and 97.5th percentiles shown in Figure 2–3 are seen to be 32 and 94.5 mg/100 ml. These are fairly close to the values of 32 and 98 that were obtained by using $\bar{X} \pm 2s$ on the assumption that the distribution was Gaussian. Thus, when distributions are normal or approximately normal, the mean and standard deviation can be used to facilitate the estimation of the percentiles. When distributions depart markedly from the normal, percentiles should not be computed by using the formula $\bar{X} \pm 2s$.

Asymmetrical distributions

For some data, the smooth curve approximated on the basis of a histogram is asymmetrical. If there is a long tail to the right, the distribution is said to be *positively skewed*, and the mean will exceed the median. If the distribution has a long tail on the left, the distribution is *negatively skewed*, and the mean will be less than the median. For skewed distributions, the median is a more appropriate measure of the center of the distribution than the mean, because it is less affected by a few extreme scores in a tail of the distribution. In such cases the average deviation from the median is the preferred measure of dispersion.

The variance and standard deviation can be calculated for any distribution. However, if the distribution deviates considerably from Gaussian, the mean and standard deviation should not be used to estimate the proportion of values under an area of the curve or to derive percentiles. Depending on the type of asymmetry present, calculations using the mean and the standard deviation can either over- or underestimate the percentile.[9]

SAMPLING DISTRIBUTIONS

A *sampling distribution* is the distribution of a statistic, and *sampling variance* refers to the variance of the sampling distribution.

The mean of a random sample of values from a population may not be exactly equal to the mean of the population. A second sample randomly selected from the population may have a mean different from both the first sample mean and from the mean of the population. If we continued to select samples of the same size from the population and compute the mean of each sample, we would obtain a distribution of sample means called the *sampling distribution of the mean.* Statisticians have shown that if *all* possible random samples of a given size are drawn from the population and the mean computed for each sample, the sampling distribution of the mean will approach the Gaussian form as sample size is increased. The mean of all the individual means will equal the mean of the population, and the standard deviation of the distribution of the means will be smaller than the standard deviation of the population. These concepts are extremely important in making inferences about a population from a sample.

The *population variance* is defined as the sum of the squares of the deviations of the observations from the mean, divided by the total number of observations. In most applications, the population variance is unknown. The variance computed for a sample of observations must be used to estimate the variance of the population as well as to describe the variance of the sample. If samples are drawn from a population and the variance of each sample computed, we would not expect every sample variance to be exactly equal to the population variance. Some samples would have a variance greater than σ^2 while others would have a smaller variance. In fact, these sample variances form a distribution of variances. Statisticians have shown that if the sample variances are computed by dividing the sum of the squares of deviations by the number of observations, the mean of the *sampling distribution of the variance* will not equal, but will be less than, the variance of the population. However, if the denominator is decreased by 1 and the sample variance computed, the mean of the distribution formed by all the sample variances will equal the population variance. The formulas for the variance presented earlier (2 and 4) used the number of degrees of freedom ($n - 1$) associated with the sum of squares as the denominator rather than the total number of observations. These formulas are appropriate for use when estimating the population variance from a sample, especially when the sample size is 30 or less.

There are a great many different samples, each of size 10, that might be drawn from a population of 105 observations. Block D of Example 1 presents the means and standard deviations of 12 such samples randomly selected from the population of 105 glucose values. Although the mean of the 12 sample means (63.28) is not equal to the population mean (64.87), it is a better estimate than 9 of the 12 individual sample means (except samples 5, 6, and 12).

The sample standard deviations presented in Example 1, Block D, were computed using $n - 1$ in the denominator. The mean value of the 12 sample standard deviations is 16.79, a difference of only 0.38 from the population value. If n had been used in the denominator for each of the 12 samples, each standard deviation would have been smaller and their mean would be equal to 15.92, a difference of 0.49 from the population value. The means of the sampling distributions in this example do not equal the corresponding population parameters because the sampling distributions are composed of only a few (12) of all possible samples of size 10.

Standard error of the mean ($\sigma_{\bar{X}}$)

Each mean in the sampling distribution of the mean is an estimate of the population mean. The mean of the sampling distribution of the mean is equal to the mean of the population. However, the standard deviation of the sampling distribution of the mean is smaller than the standard deviation of the population. The standard deviation of the sampling distribution of the mean is called the *standard error of the mean:*

$$\sigma_{\bar{X}} = \frac{\sigma}{\sqrt{n}} \tag{5}$$

where

$\sigma_{\bar{X}}$ = standard error of the mean
σ = standard deviation of the population
n = sample size

Thus, the standard error of the mean depends on the standard deviation of the population from which the samples were drawn and also on the size of the sample. The larger the sample size and the smaller the standard deviation of the population, the smaller the standard error of the mean. The smaller the standard error, the more closely the sample means will cluster about the mean of the population.

In practice, usually only a single sample is selected and its mean computed. The standard deviation of the population is usually unknown. However, the standard deviation calculated for a single sample and the size of the sample can be used to estimate the standard error of the mean. The formula is:

$$\hat{\sigma}_{\bar{X}} = \frac{s}{\sqrt{n}} \tag{6}$$

where

$\hat{\sigma}_{\bar{X}}$ = the sample estimate of the standard error of the mean
s = standard deviation of the sample

Thus, $\hat{\sigma}_{\bar{X}}$ is a measure of expected variation in sample means. The larger the number of values of the variable selected in a sample, the smaller the standard error, and the more reliable the value of the sample mean, i.e., it is likely to approximate more closely the mean of the population.

Confidence interval of the mean

Sometimes it is desirable to obtain an estimate of the interval within which a parameter lies, as well as an estimate of the parameter itself. The parameter may be the mean, the median, a standard deviation, a proportion, or any other estimate of a point. The estimate of the interval is called the *confidence interval.* In this chapter, only the confidence interval of the mean will be considered.

Confidence intervals of the mean ($CI_{\bar{X}}$) can be calculated to provide estimates of the interval in which the mean of the population lies (the unknown "true" mean). The mean obtained for a single sample and the standard error of the mean, calculated by using the standard deviation of a single sample, can be used to estimate the $CI_{\bar{X}}$. Since the sampling distribution of the mean is Gaussian, knowledge of the percentiles of the normal distribution is used to obtain the $CI_{\bar{X}}$. It is to be expected, if the sampling were repeated indefinitely and a $CI_{\bar{X}}$ calculated for each sample, that in 95 per cent of the samples selected from the population, the interval between the sample mean and $\pm 2 \frac{s}{\sqrt{n}}$ will include the value of the population mean. The formula for a 95 per cent confidence interval is

$$CI_{\bar{X}} = \bar{X} \pm 2\hat{\sigma}_{\bar{X}} \tag{7}$$

For the 99 per cent confidence limits, the interval would be $\bar{X} \pm 3\hat{\sigma}_{\bar{X}}$.

For example, in a random sample of 100 white males between the ages of 20 and 29, the mean urinary nitrogen excretion was found to be 8.0 g/d with a standard deviation of 1.6 g. On the basis of figures from this single sample, the following statements can be made about the true mean of the population. First, the value 8.0 g is considered to be the best estimate of the true mean. Second, since the standard deviation of the sample is the best estimate of the standard deviation in the population, the standard error of the mean can be calculated by use of the formula

$$\hat{\sigma}_{\bar{X}} = \frac{s}{\sqrt{n}} = \frac{1.6}{\sqrt{100}} = 0.16 \text{ g}$$

For the 95 per cent confidence interval of the mean

$$\begin{aligned} CI_{\bar{X}} &= \bar{X} \pm 2\hat{\sigma}_{\bar{X}} \\ &= 8.0 \pm 2(0.16) \\ &= 7.68 \text{ to } 8.32 \text{ g/d} \end{aligned}$$

The numbers 7.68 and 8.32 are called confidence limits for the mean. The percentage level, 95 per cent, is called the confidence level. If *all* possible samples of size 100 were drawn from the population and the confidence interval of the mean for each sample computed, 95 per cent of the calculated intervals would be expected to include the mean of the population. For any single sample there is no guarantee that the interval computed does in fact bracket the true mean. Neither is there any guarantee that of every 100 intervals calculated, 95 will cover the mean of the population.

If for a given confidence level, the confidence interval is too wide to be practically useful, an increase in sample size will narrow the interval. Thus, in the above example, if n were 225 instead of 100, $\hat{\sigma}_{\bar{X}} = 1.6/15 = 0.11$. The 95 per cent confidence interval would be from 7.78 to 8.22 g/d, a shorter interval than that calculated for $n = 100$.

Tolerance limits

In many practical situations it is desirable to obtain an interval that will include a fixed proportion of the population with a specified confidence. This may be more appropriate than an interval about a single point such as the mean. For example, clinical chemists are often interested in estimating the range within which a constituent is likely to vary in a group of apparently healthy individuals. In this case, the chemist is not interested in merely describing the sample of individuals with which he is working, nor in making statements about the mean value in the population from which the sample was drawn. He wishes instead to estimate the percentiles of a population. Intervals which cover a proportion of the population are called tolerance intervals. The end points of the intervals are called tolerance limits and can be calculated as

$$\bar{X} \pm Ks \qquad (8)$$

where

$\bar{X}$ = mean of the sample
K = factor determined so the interval computed will cover a given proportion, P, of the population with specified confidence, γ
s = standard deviation of the sample

Tolerance limits are wider than confidence limits, because tolerance limits cover a large portion of values in a population while confidence limits cover a single value.

Table 2–3 gives values of K for different sample sizes that will include 95 per cent of the values in the population 90 per cent of the time. The values of K can be seen to decrease as sample size increases and, therefore, the tolerance interval becomes narrower with increasing sample size.

Bowker[6] gives complete tables of K for populations up to 1000, and for additional confidence levels (other than $\gamma = 0.90$) and additional proportions of the population (other than $P = 95$ per cent). Condensed tables giving K values also appear elsewhere.[14,15] The reader should keep in mind that appropriate use of K factors requires the assumption that the population of values is normally distributed.

TABLE 2–3 TOLERANCE FACTORS FOR 0.90 CONFIDENCE LEVEL, γ, TO INCLUDE 95 PER CENT OF THE POPULATION FOR VARIOUS SAMPLE SIZES*

n	*K Factor*
5	4.152
6	3.723
7	3.452
8	3.264
9	3.125
10	3.018
15	2.713
20	2.564
25	2.474
30	2.413
40	2.334
50	2.284
60	2.248
70	2.222
80	2.202
90	2.185
100	2.172

* From Bowker, A. H.: *In* Selected Techniques of Statistical Analysis. C. Eisenhart, M. W. Hastay, and W. A. Wallis, Eds. Copyright 1947 by McGraw-Hill Book Company. Used with permission of McGraw-Hill Book Company.

The 90 per cent tolerance interval to include 95 per cent of the population is given in Example 1, Block C, for a sample of 10 glucose values. It is 13.63 to 105.37. This is considerably wider than the 95 per cent confidence interval of the mean, and wider than the 2.5th to 97.5th percentiles actually computed for the 105 glucose values (34 to 102.5). In Block C the tolerance limits calculated for a sample include almost the entire population. Block D presents tolerance limits for each of the twelve samples of size 10. Sample 3 provides limits that include only 94 per cent of the population. Such discrepancies are to be expected because values of K have been determined on the basis of statistical theory which predicts that 90 per cent of the tolerance intervals computed for all possible samples of size 10 will include 95 per cent of the population. Just as with the confidence interval of the mean, there is no guarantee that for a single sample the calculated interval will include the specified population proportion.

STATISTICAL TESTS

The investigator may wish to describe the characteristics of a sample of values or make inferences about populations from which the samples were derived. Many statistical tests are available for these purposes. A few tests will be described and illustrated in the following pages. Additional information concerning the applicability of these tests, as well as descriptions of other statistical techniques, can be found in standard textbooks of statistics, such as those in the list of Additional Readings at the end of this chapter.

Coefficient of variation (relative standard deviation)

Frequently it is necessary to compare the relative variability between different sets of data. The data may result from the use of different series of specimens, different methods, or different analysts. Comparison of different data sets cannot be made by directly comparing their means and standard deviations because the magnitude of the statistics and units in which the results are expressed may vary from one set of data to another. Such comparisons can be made, however, if the standard deviation of each set is expressed as a percentage of the mean. The statistic appropriate for such comparisons is the coefficient of variation (CV). The CV provides a measure of relative variability and is calculated as

$$CV = \frac{s}{\bar{X}}(100) \tag{9}$$

As an example, suppose that for a group of individuals the mean glucose value ($\bar{X}$) determined by one method was 98.5 mg/100 ml with a standard deviation (s) of 2.5 mg/100 ml, and the mean glucose determined by a second procedure was 78 mg/100 ml with a standard deviation of 2.0 mg/100 ml. The relative variability for the first method would be

$$CV = \frac{2.5(100)}{98.5} = 2.5 \text{ per cent}$$

and that for the second method would be

$$CV = \frac{2.0(100)}{78} = 2.6 \text{ per cent}$$

Thus, the relative variability of the two methods is about the same although the absolute variabilities differ by 0.5 mg/100 ml.

Since the coefficient of variation is a dimensionless quantity, it can also be used to compare the variation in sets of data measured in different units. For example, we may wish to perform a comparative study between a kinetic method and the Somogyi method for measuring amylase activity. The kinetic method results in data expressed in mU/ml, while the Somogyi method provides results in terms of SU/100 ml. Suppose our tests by the kinetic method resulted in values of 124 mU/ml for the mean and 20 for the standard deviation. The Somogyi method provided a mean of 138 SU and a standard deviation of 31. We cannot compare 31 SU with 20 mU/ml directly because of the differences in units. However, calculating the coefficients of variation we find 16.1 per cent for the kinetic method and 22.5 per cent for the Somogyi method. Thus, the Somogyi method yielded data of greater variability, i.e., less precision.

Linear regression and correlation

The clinical chemist often obtains more than one value of a variable or has information concerning more than one variable for each of the individuals in a group. For example, different methods are applied to analyze aliquots of a single specimen, or laboratory values for two or more constituents per patient are obtained.

When two values are available for each object or individual in the sample to be studied, it is possible to investigate the degree of association between the two sets of data by using linear regression and correlation.

If data are plotted, the resulting scatter plot may show a random scatter of points across the graph, a trend of points from the lower left to the upper right corner of the graph, or a trend from the upper left to the lower right. A random scatter indicates the lack of a relationship between the variables. A scatter from lower left to upper right with high values of one variable associated with high values of the second is called a direct relationship. A scatter of results from upper left to lower right indicates an indirect or inverse relationship, in which high values of one variable are associated with low values of the second.

It is possible to calculate a line of best fit going through the plotted points. The line of best fit is defined as the least squares regression line, i.e., the line for which the sum of squares of the deviations of the observed points from the line is smaller than the sum of squares of the deviations from any other line that might be drawn through the scatter of points. The equation which describes this line is called the regression equation. It is also possible to obtain a numerical measure of the strength of linear association between two variables called the *product-moment coefficient of correlation.*

In applying these statistical methods, the first step is to plot the points. If there appears to be a linear trend, an attempt is made to describe the least squares regression line through the points algebraically. The equation for a regression line of Y on X is in the general form

$$\hat{Y} = a + bX \tag{10}$$

where

$\hat{Y}$ = the best estimate of Y for a given value of X, based on a sample of X and Y values
a = Y intercept; i.e., the point at which the line crosses the Y axis, the value of Y when $X = 0$
b = slope of the line, the rate of change in Y for a unit increase in X
X = value of the preselected variable, the variable used to make the prediction

Values for the regression coefficients b and a can be calculated by substituting in the following formulas:

$$b = \frac{\sum (X - \bar{X})(Y - \bar{Y})}{\sum (X - \bar{X})^2} \tag{11}$$

where

$\sum (X - \bar{X})(Y - \bar{Y})$ = sum of the products of the deviations of X and Y from their respective means

$\sum (X - \bar{X})^2$ = sum of the squares of the deviations of the observations of the preselected variable from the mean of the observations

Formula (11) requires that deviations about the means be found. As stated previously in reference to the variance, deviations from the mean are laborious to calculate and may introduce error. A computational formula for b is available:

$$b = \frac{\sum XY - \frac{(\sum X)(\sum Y)}{n}}{\sum X^2 - \frac{(\sum X)^2}{n}} = \frac{n \sum XY - (\sum X)(\sum Y)}{n \sum X^2 - (\sum X)^2} \qquad (12)$$

The Y intercept (a) can then be found by

$$a = \bar{Y} - b\bar{X} \qquad (13)$$

Once values of a and b have been obtained, the investigator selects three values of X arbitrarily and substitutes them, along with the calculated values of a and b, into Formula (10) to obtain three values of Y. The resulting pairs of X, Y values are plotted on the graph. Although two pairs of points determine a straight line, the third is recommended as a check on the calculation.

In situations where neither X nor Y may be considered as preselected, it may be desirable to compute the regression of X on Y as well as the regression of Y on X, and to plot a second regression line on the same graph. Generally, these lines will not coincide. (See Figure 2–4.)

The regression line has been defined as the line for which the sum of squares of the deviations of the observed points from the points calculated using the regression equation is a minimum. The standard deviation of the differences between the observed and the calculated values is called the *standard error of estimate*. The formula is:

$$s_{X \cdot Y} = \sqrt{\frac{\sum(Y - \hat{Y})^2}{n - 2}} \qquad (14)$$

where

Y = the observed value of Y for a given value of X

$\hat{Y}$ = the calculated value of Y (using the regression equation) for a given value of X

$n - 2$ = the degrees of freedom associated with the deviations from regression

The calculated Y values are based on estimates of two constants, a and b, and thus the squares of deviations are associated with $n - 2$ degrees of freedom.

For computational purposes the following formula is easier to use because some of the quantities have already been calculated in computing b:

$$s_{Y \cdot X} = \sqrt{\frac{\sum(Y - \bar{Y})^2 - b \sum (X - \bar{X})(Y - \bar{Y})}{n - 2}} \qquad (15)$$

The product-moment correlation coefficient provides a measure of the extent to which every point on a scatter plot falls exactly on a straight line. It is thus a measure of the extent to which the rank ordering of observations remains the same for the two sets of data. The correlation coefficient can have any value from -1 to 1. The correlation coefficient will equal ± 1 only when every point on the graph falls exactly on the regression line. In most practical applications the correlation coefficient lies somewhere between these extremes. It will be zero when there is just a random scatter of points across the graph. The formula for the correlation coefficient is

$$r = \frac{\sum XY - \frac{(\sum X)(\sum Y)}{n}}{\sqrt{\left[\sum X^2 - \frac{(\sum X)^2}{n}\right]\left[\sum Y^2 - \frac{(\sum Y)^2}{n}\right]}} \tag{16}$$

or

$$r = \frac{n\sum XY - (\sum X)(\sum Y)}{\sqrt{[n\sum X^2 - (\sum X)^2][n\sum Y^2 - (\sum Y)^2]}} \tag{17}$$

It is often useful to obtain the proportion of total variance in one variable (e.g., Y) that is accounted for by its association with the second variable. This measure is called the *coefficient of determination*. It is calculated as the ratio of the sum of squares associated with regression to the total sum of squares in Y, and is equal to the square of the correlation coefficient:

$$r^2 = \frac{b\sum(X - \bar{X})(Y - \bar{Y})}{\sum(Y - \bar{Y})^2} \tag{18}$$

The variance in Y that is unexplained by its association with X would then be equal to $1 - r^2$. This quantity is referred to as the *coefficient of alienation.*

Calculations of the regression of Y on X, the regression of X on Y, the correlation coefficient, and the standard error are illustrated in Example 2 and in Figure 2–4 for a series of 20 specimens analyzed for serum amylase by both the Somogyi and a kinetic method.

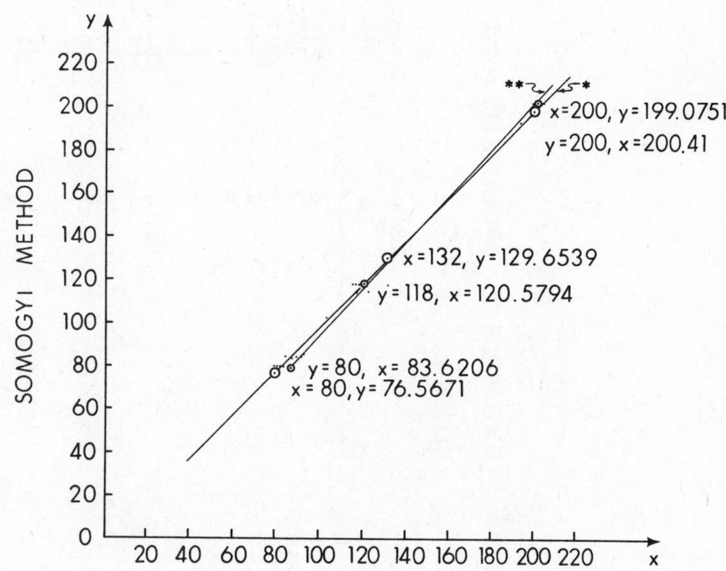

Figure 2–4 Regression lines for serum amylase data. The regression of Y on X has the equation $\hat{Y} = 1.021X - 5.10$ and is indicated on the graph by $*$. The regression of X on Y has the equation $\hat{X} = 0.973Y + 5.81$ and is indicated by $**$.

EXAMPLE 2: SPECIMENS ANALYZED BY TWO METHODS FOR SERUM AMYLASE

BLOCK A

Specimen No.	*Somogyi* (Y)	*Kinetic Method* (X)	D (Y − X)	D^2	Y^2	X^2	XY
1	118	132	−14	196	13924	17424	15576
2	80	84	−4	16	6400	7056	6720
3	202	203	−1	1	40804	41209	41006
4	84	90	−6	36	7056	8100	7560
5	118	116	2	4	13924	13456	13688
6	80	81	−1	1	6400	6561	6480
7	85	87	−2	4	7225	7569	7395
8	202	200	2	4	40804	40000	40400
9	84	92	−8	64	7056	8464	7728
10	118	119	−1	1	13924	14161	14042
11	80	82	−2	4	6400	6724	6560
12	85	85	0	0	7225	7225	7225
13	115	118	−3	9	13225	13924	13570
14	202	202	0	0	40804	40804	40804
15	84	88	−4	16	7056	7744	7392
16	80	80	0	0	6400	6400	6400
17	115	123	−8	64	13225	15129	14145
18	118	118	0	0	13924	13924	13924
19	102	104	−2	4	10404	10816	10608
20	193	193	0	0	37294	37249	37249
$n = 20$	$\sum Y = 2345$	$\sum X = 2397$	$\sum D = -52$	$\sum D^2 = 424$	$\sum Y^2 = 313429$	$\sum X^2 = 323939$	$\sum XY = 318472$

BLOCK B

***Regression of* Y *on* X**

Means

$$\bar{Y} = \frac{\sum Y}{n} = \frac{2345}{20} = 117.25$$

$$\bar{X} = \frac{\sum X}{n} = \frac{2397}{20} = 119.85$$

Slope

$$b = \frac{\sum XY - \frac{(\sum X)(\sum Y)}{n}}{\sum X^2 - \frac{(\sum X)^2}{n}} = \frac{318472 - \frac{(2397)(2345)}{20}}{323939 - \frac{(2397)^2}{20}} = 1.0209$$

Intercept

$$a = \bar{Y} - b\bar{X} = 117.25 - 1.0209(119.85) = -5.1049$$

Regression equation of Y on X

$$\hat{Y} = -5.1049 + 1.0209X$$

BLOCK C

***Regression of* X *on* Y**

Slope

$$b = \frac{\sum XY - \frac{(\sum X)(\sum Y)}{n}}{\sum Y^2 - \frac{(\sum Y)^2}{n}} = \frac{37423.75}{313429 - \frac{(2345)^2}{20}} = 0.9726$$

Intercept

$$a = \bar{X} - b\bar{Y} = 119.85 - 0.9726(117.25) = 5.8126$$

Regression equation of X on Y

$$\hat{X} = 5.8126 + 0.9726Y$$

BLOCK D

***Standard Error of Estimate of* Y *on* X**

Specimen No.	$Y - \bar{Y}$	$(Y - \bar{Y})^2$	$X - \bar{X}$	$(Y - \bar{Y})(X - \bar{X})$
1	0.75	0.56	12.15	9.11
2	−37.25	1387.56	−35.85	1335.41
3	84.75	7182.56	83.15	7046.96
4	−33.25	1105.56	−29.85	992.51
5	0.75	0.56	−3.85	−2.89
6	−37.25	1387.56	−38.85	1447.16
7	−32.25	1040.06	−32.85	1059.41
8	84.75	7182.56	80.15	6792.71
9	−33.25	1105.56	−27.85	926.01
10	0.75	0.56	−0.85	−0.64
11	−37.25	1387.56	−37.85	1409.91
12	−32.25	1040.06	−34.85	1123.91
13	−2.25	5.06	−1.85	4.16
14	84.75	7182.56	82.15	6962.21
15	−33.25	1105.56	−31.85	1059.01
16	−37.25	1387.56	−39.85	1484.41
17	−2.25	5.06	3.15	−7.09
18	0.75	0.56	−1.85	−1.39
19	−15.25	232.56	−15.85	241.71
20	75.75	5738.06	73.15	5541.11
$n = 20$		$\sum = 38477.70$		$\sum = 37423.70$

$$s_{Y \cdot X} = \sqrt{\frac{\sum(Y - \bar{Y})^2 - b\sum(Y - \bar{Y})(X - \bar{X})}{n - 2}}$$

$$= \sqrt{\frac{38477.70 - 1.0209(37423.70)}{18}}$$

$$= 3.89$$

BLOCK E

Correlation Coefficient Between Two Methods

$$r = \frac{\sum XY - \frac{(\sum X)(\sum Y)}{n}}{\sqrt{\left[\sum X^2 - \frac{(\sum X)^2}{n}\right]\left[\sum Y^2 - \frac{(\sum Y)^2}{n}\right]}}$$

$$= \frac{318472 - \frac{(2397)(2345)}{20}}{\sqrt{\left[323939 - \frac{(2397)^2}{20}\right]\left[313429 - \frac{(2345)^2}{20}\right]}} = \frac{37423.75}{37557.14} = 0.9964$$

Coefficient of determination

$$r^2 = 0.9928$$

Coefficient of alienation

$$1 - r^2 = 0.0072$$

BLOCK F

t Test for Difference Between Means of Paired Samples

Standard error of the difference

$$\sum(D - \bar{D})^2 = \sum D^2 - \frac{(\sum D)^2}{n} = 424 - \frac{(-52)^2}{20} = 288.8$$

$$\sigma_{\bar{Y}-\bar{X}} = \sqrt{\frac{\sum(D - \bar{D})^2}{n(n - 1)}} = \sqrt{\frac{288.8}{20(19)}} = 0.8718$$

***t* Test**

$$t = \frac{\bar{Y} - \bar{X}}{\sigma_{\bar{Y}-\bar{X}}} = \frac{117.25 - 119.85}{0.8718} = -2.9823$$

***F* test for homogeneity of variances**

Questions concerning the difference in precision between two different methods or between two analysts using a single method arise quite frequently. We can test the hypothesis that the precisions are equal against the hypothesis that the precision of one method or analyst is greater than that of the second by computing and comparing the variances of the two sets of values obtained. The appropriate statistical test for such a comparison is called the F test for homogeneity (equality) of variance. The test consists of forming a ratio of the variances such that

$$F = \frac{\text{larger variance}}{\text{smaller variance}} \quad (19)$$

Since each variance is associated with some number of degrees of freedom, a ratio of two variances would be associated with two such quantities, one for the numerator and another for the denominator.

If the ratio is equal to 1, there is no difference in the precision of the results. However, if the ratio exceeds 1, the investigator is faced with the problem of deciding just how much in excess of 1 the ratio must be before he can state with some confidence that there is a difference in precision. To answer this question, reference must be made to a table presenting the critical values of F for various levels of significance. Such tables are available in many statistics textbooks.* The table can be entered by locating the number of degrees of freedom associated with the numerator in a column, and the number of degrees of freedom associated with the denominator in a row. If the calculated value of the F ratio does not equal or exceed the critical value found in the F table, the two sample variances are said to be homogeneous; i.e., there is not enough evidence to conclude that the variances differ significantly. If, however, the calculated value for F equals or exceeds the tabular value, the precision of the data corresponding to the numerator is said to be poorer than the precision of the sample corresponding to the denominator.

An example may help to make the interpretation of the F ratio clearer. Suppose two analysts perform cholesterol determinations. Analyst A submits values on 25 specimens while analyst B submits values on 20 specimens. The variances of the results are 9.0 and 7.8 for analysts A and B, respectively.

The investigator should first decide on the magnitude of the chance he is willing to take of deciding that the numerator is less precise than the denominator when, in fact, there is no difference in precision. It is common to set this level of error (called the significance level) at 5, 2.5, 1.0, or 0.5 per cent. If 5 per cent is chosen, it means that the investigator is willing to be in error 5 per cent of the time when he decides that the data corresponding to the numerator have poorer precision.

The degrees of freedom for both numerator and denominator of the ratio are determined next. The degrees of freedom are calculated by taking one less than the number of specimens in each sample. The appropriate critical value of F can then be obtained from an F table by entering it with the value of the selected significance level, the degrees of freedom for the numerator, and the degrees of freedom for the denominator.

Let's say that a 5 per cent level was chosen by the investigator of this problem. He would then enter the table with a 5 per cent significance level, 24 degrees of freedom for the numerator, and 19 degrees of freedom for the denominator. The critical value listed in the table is 2.1. The calculated value ($9.0 \div 7.8$) is 1.2. Since the calculated value is less than the critical value found from the table, the investigator would decide that he has insufficient evidence to support a difference in precision between the analysts.

The F test for equality of variances is based on the assumption that the populations are Gaussian in form. Unless the sample size is large, failure of the distributions to meet the normality assumption can lead to serious errors in inferences.[8]

t test for equality of means

It is frequently necessary to test the hypothesis that the means of two samples are equal. One situation in which such a test may be desirable arises when a new method is compared with an older, well-established method. A single analyst would be supplied with a number of specimens and be asked to determine the value of a constituent on two aliquots of each specimen, one determination by each method. The appropriate statistical tool to determine whether the mean value obtained with one method differs from the mean ob-

* The tables to be found in many texts are those of Snedecor for the 0.05 and 0.01 levels of significance, and those of Merrington and Thompson for the 0.25, 0.10, 0.025, and 0.005 levels. These are one-tailed tables which provide the probabilities that F exceeds 1.00 when the hypothesis of no difference is true. One- and two-tailed tables and one- and two-tailed tests are discussed in greater detail in the sections which follow.

tained by the second method is the t test for paired data. The formula for the t test is

$$t = \frac{\bar{Y} - \bar{X}}{\hat{\sigma}_{\bar{Y}-\bar{X}}} \tag{20}$$

where

$\bar{Y}$ = mean of constituent determined by one method
$\bar{X}$ = mean of constituent determined by a second method
$\hat{\sigma}_{\bar{Y}-\bar{X}}$ = standard error of the difference between the means, estimated from sample values

The standard error of the difference between means can further be defined as

$$\hat{\sigma}_{\bar{Y}-\bar{X}} = \sqrt{\frac{\sum (D - \bar{D})^2}{n(n-1)}} \tag{21}$$

where

D = difference between values of the constituent determined by two different methods
$\bar{D}$ = mean of the differences
n = number of specimens in the sample, i.e., the number of differences computed

The t value obtained must be evaluated in terms of a critical value of t. The critical value of t for various levels of significance and degrees of freedom can be found in a table of values for the t distribution. Such tables are available in many statistics books. If the calculated value of t does not equal or exceed the critical value found in the t table, the two means cannot be said to differ. If the calculated value equals or exceeds the value found in the tables, the means are said to differ to a statistically significant degree.

In order to use the table, the investigator must decide on the significance level he wishes to maintain, and determine the number of degrees of freedom for his data. While the significance level helps the investigator to make a decision, its use means that the null hypothesis (the hypothesis that the means are equal) will be wrongly rejected a certain proportion of the time. In the t test for paired data, the standard error of the difference between means is the standard deviation of the differences divided by the square root of n [see Formula (21)]. The degrees of freedom associated with the sum of squares of differences is one less than the number of differences.

There is yet another decision to be made before the t test can be applied. The t test may be used as either a one-tailed or a two-tailed test. If the hypothesis that the means are different is tested against the null hypothesis of no difference, the test is two-tailed. Sometimes, however, one may wish to use the t test to test a directional hypothesis (e.g., that the mean obtained by the Somogyi method is less than the mean obtained by the kinetic method) against the null. The latter test would be called a one-tailed test because there is no interest in testing whether the mean by the Somogyi method is of greater magnitude than the mean by the kinetic method. When a two-tailed test is used and the significance level is set at 5 per cent, there is a 2.5 per cent probability of deciding, incorrectly, that one mean is less than the other; and a 2.5 per cent probability of deciding that the first mean is greater than the second. When a one-tailed test is performed at the 5 per cent level, there is a 5 per cent probability of deciding, incorrectly, in favor of the directional hypothesis.

Example 2, Block F illustrates the application of the t test to the amylase data. There are 20 differences and 19 degrees of freedom. A two-tailed test is performed. The critical value of t for 19 degrees of freedom at the 5 per cent level of significance is 2.093.* Since the absolute value calculated in this example (2.9823) exceeds the critical value, the mean of

* The exact type of t table varies from text to text. Some provide one-tailed tables while others provide two-tailed tables. If a 5 per cent level of significance is chosen for a two-tailed test, one enters the column in a one-tailed table marked 2.5 per cent, or the column in a two-tailed table marked 5 per cent.

the specimens analyzed by the Somogyi method can be said to be significantly different from the mean of the specimens analyzed by the kinetic method.

If a one-tailed test had been performed on the same data, the critical value at the 5 per cent level would have been 1.729, a value considerably smaller than that used for the two-tailed test. The choice of a one- or two-tailed test depends on the information sought and the amount of knowledge the investigator has beforehand.

The same data were used in calculating both the t value and the correlation coefficient of 0.9964. Thus, the rank order of aliquots analyzed by the Somogyi method is almost the same as the rank order by the kinetic method, but the means obtained by the two methods differ by an amount statistically significant at the 5 per cent level.

The comparison of correlation and t test results illustrates the point that a single set of data can be analyzed with different statistical tools, depending on the purpose of the investigator and the nature of the question he wishes to answer. It should *not* be construed as advocacy of performing myriad statistical tests on a single set of data in order to arrive at a conclusion predetermined by the investigator. It should be interpreted as a caution against indiscriminate data collection and analysis without a well-defined purpose.

An introduction to some basic statistical tools has been presented in this section. The following sections will further describe and illustrate the application of these tools to problems arising in the clinical chemistry laboratory.

NORMAL VALUES

The normal range for a given constituent of clinical interest is considered to be the concentrations of the constituent which are found in the body fluid or excretions of a group of clinically normal (apparently healthy) persons.* Values thus obtained are considered "normal data"; however, it is important to differentiate between these "normal data" and those that are physiologically "desirable" or "ideal" for any given constituent. If we were to determine the serum cholesterol values and the body weight of 20 clinically healthy American businessmen between the ages of 40 and 50 years, there would be general agreement that the "normal range" obtained after statistical evaluation of these data would be higher than that which we would like to consider as "ideal." The Metropolitan Life Insurance Company uses terminology that rather clearly points out this distinction. In 1912 its standards of weights for a given body height were called "average" weights. The term "average" was altered in the 1942–1943 period to "ideal," and again in 1959 to "desirable." The weights listed in the respective tables are considerably below the present average weight for Americans.[4]

When establishing the range of normal values for chemical constituents of body fluid and excretions, one should take into consideration physiological variations, which may include diurnal variation, true day to day variation, and the effect of the environment. One should also consider that there may be differences due to race, age, sex, weight, nutritional and absorptive states, degree of physical activity, position of the body during blood sampling, stage of the menstrual cycle, ovarian status, emotional state, climate, geographic location, cultural habits, and the time of day at which the sample was taken.[4]

ESTABLISHMENT OF NORMAL VALUES

The clinical chemistry laboratory is frequently confronted with the need to establish normal values for a body constituent of clinical interest. Such situations may arise when a

* The term "clinically normal" or "apparently healthy" is not easily defined; however, it is generally considered to be a group of individuals who are judged to be in a state free from any obvious or overt abnormalities. It should be emphasized that a possible biochemical abnormality may not have progressed far enough to be clinically detectable, so that the subject is still an "apparently" healthy person.

new method has been developed, (e.g., for a new enzyme) or if the laboratory merely wishes to compare its range of normal values with that found in the literature (an approach that is certainly recommended).

In practice, when establishing normal values, it is customary to analyze the body fluid of a group of apparently healthy individuals, as previously defined, for the level of the constituent of interest. Values thus obtained have to be treated in a way that will result in a "meaningful" normal range. The procedure to be selected for analyzing data will depend on the number of data available, on the type of data obtained, on the way in which they are distributed, and on the "type" of normal range that is desired.

Ideally, the range of normal values should be determined by analysis of specimens from every individual in the population of interest. However, in most cases it is not possible to investigate the entire population, and a sample must be selected from the population for study. The larger the sample size, the more likely the statistical values calculated on the sample will approximate those of the population. In practice, it is usually considered desirable to obtain at least 100 data. Frequently, however, even 100 are not available, and the statistical technique employed may have to be selected accordingly.

The number of specimens to be analyzed also depends on the specific constituent under study. For some constituents, such as the cholesterol mentioned previously, variables such as age and sex are known to affect the distribution. For such constituents the variable affecting the distribution should be categorized and a sample of appropriate size used to determine the normal range of constituent values for each category of the associated variable.

Another issue in the selection of subjects on which the normal range is to be based is whether the subject pool is to be selected from the population of "walking normals" (e.g., volunteers, employees, or voluntary blood donors) or from a hospitalized population. Normal values are defined in terms of apparently healthy individuals. However, it is very expensive (in terms of both time and money) and difficult to identify large groups of "healthy" people at each age-sex combination and then solicit their cooperation in the collection of urine and blood specimens. Therefore, a number of methods have been suggested whereby hospitalized individuals undergoing routine analyses can be used to establish normal values. Hospital patients are apt to form a heterogeneous sample; i.e., a mixture of two distributions, one of healthy individuals and one of sick individuals.

Neumann[19] suggested a graphical method applicable to a truncated* distribution that is Gaussian or lognormal in form. More recently, Becktel[3] has suggested a way in which hospitalized subjects can be used to establish normal ranges. Best *et al.*[5] validated the Becktel procedure and concluded that it is a workable method, at least for the SMA-12 (Technicon Instruments, Corp., Tarrytown, N.Y. 10591). These methods require several times as many patients to achieve the same precision as can be obtained for a given number of normal subjects.[21]

Aside from the selection of an appropriate number and type of subjects, the most important point in determining a normal range is ascertaining the shape of the distribution of values obtained. The construction of a histogram will frequently aid in the determination. For some data, the histogram will approach a smooth curve that is approximately Gaussian. For other data the curve approached may exhibit positive or negative skewness or have greater peakedness than is characteristic of the Gaussian distribution. Such data should be treated differently from those approximating a Gaussian distribution.

The interval most frequently used to establish normal values is the interval into which the central 95 per cent of the apparently healthy persons fall. It is important to remember that 5 per cent, or one of 20, of these apparently healthy individuals will have values outside the "normal" range. Similarly, a value just inside the normal range should not be viewed as absolutely normal.

* A truncated distribution is one in which one or both tails are cut off.

For most constituents tested in clinical chemistry laboratories there is an overlap of values obtained from the normal and abnormal populations. If the normal range is established in such a way that it would include the entire normal population, some of the abnormal values would be in the "normal" range and go undetected (type II error). If, on the other hand, a narrower normal range is used, a number of normal individuals would have an "abnormal" laboratory result (type I error). In the first case a disease state may be overlooked, while in the latter case undue concern over a normal individual could develop. It has therefore been proposed that two ranges be used.[24] The first range would include about 98 per cent of the apparently normal population, and any value outside this range would be considered as definitely abnormal. The second range would include 80 per cent of the apparently normal population. All values that fall into this range would be considered as definitely normal. All values falling within the range of 80 to 98 per cent constitute a "gray zone," and these values should be viewed with suspicion. The concept of two normal values is appealing, but it increases the complexity of computation and data interpretation and, hence, has not found widespread use.

DETERMINATION OF NORMAL VALUES BASED ON A GAUSSIAN DISTRIBUTION

The values obtained on apparently healthy individuals are first plotted as a histogram. If the histogram approaches a Gaussian distribution, it may be possible to establish the limits of the normal range by visual inspection of the distribution or by a cumulative frequency graph or table (see previous section on statistics and Fig. 2–3). A statistical approach can also be used to establish normal limits. The mean ($\bar{X}$) and the standard deviation (s) of the sample values are calculated and all values falling between the mean and $\pm 2s$ would constitute the "normal" range and would include approximately 95 per cent of the group sampled.*

If the distribution of glucose values (shown in Table 2–1) is considered to be the population, samples of size 10 selected from the population might be expected to yield values of sample statistics that differ from the population values and from each other. Values of the parameters for the population of 105 values and of corresponding statistics for 12 random samples, each of size 10, are presented in Example 1, Block D. Note that the percentiles of the population calculated by using $\bar{X} \pm 2s$ (32 and 98) differ from the percentiles obtained by graphical means (Fig. 2–3) and from those calculated directly from Table 2–1. The difference is due to the slight departure of the distribution from Gaussian. Percentiles based on samples of size 10 (computed with the formula $\bar{X} \pm 2s$) provide estimates of population percentiles that may be either narrower or wider than the population values.

The accuracy of an interval as an estimate of the normal range of the population is affected by the number of data in the sample analyzed, as previously mentioned. Since, in practice, the sample size is usually relatively small, Henry[14] has suggested that a tolerance interval is a more appropriate estimate of the population range than the percentiles of the sample distribution. As previously stated, a tolerance interval provides an estimate of the interval into which a specified proportion of the population may be expected to fall and is wider than the sample percentiles.

If a great many samples of size 10 were drawn and the $P = 95$ per cent, $\gamma = 90$ per cent tolerance limits for each sample calculated, 90 per cent of the intervals may be expected to include at least 95 per cent of the values in the parent population. It should be remembered, however, that there is no guarantee that any single interval does, in fact, cover at least 95 per cent of the population values. Although the tolerance intervals presented in

* As previously mentioned, estimation of the 2.5th and 97.5th percentiles based on the use of $\bar{X} \pm 2s$ is acceptable only if the distribution is Gaussian or approximates a Gaussian distribution.

Example 1, Block D differ from the actual 2.5th and 97.5th percentiles of the 105 glucose values, they provide a better fit to population values than that provided by $\bar{X} \pm 2s$ in some cases. In other cases the tolerance interval may be considerably wider than the population percentiles and may even exceed the range of the population (see sample 5).

It is important to note that the calculation of tolerance intervals is appropriate only when the population of values can be assumed to be Gaussian or approximately Gaussian. Because of this restriction, Elveback and Taylor,[9] and Reed *et al.*[20] suggest that percentile estimates accompanied by a confidence interval of each percentile estimate be used. While these latter methods may provide the best estimates of the normal range and its precision, they require a sample of 120 or more values for a 90 per cent confidence interval. Since samples of 120 may not be readily available, many workers may still have to rely on the calculation of sample percentiles or tolerance limits to estimate the normal range.

DETERMINATION OF NORMAL VALUES FROM NON-GAUSSIAN DATA

The previous discussion dealt with data that were distributed in a Gaussian or approximately normal manner. In reality, much biological data do not follow such a distribution, and calculations performed using either the tolerance interval or the $\bar{X} \pm 2s$ approach to estimate the percentiles of the population may in these cases be erroneous. Thus, use of these formulas could result in normal ranges extending below zero or normal ranges extending too far on the high side, depending on the distribution obtained.

Inspection of a histogram may reveal whether the distribution is symmetrical or skewed, although the shape of the distribution is not always easy to recognize, especially if the sample size is small. Distributions of biological data with positive skewness can sometimes be converted to Gaussian form by the application of suitable transformations. In cases where a logarithmic transformation tends to normalize the distribution, the positively skewed distribution of the original values is called a *lognormal distribution.*

Figure 2–5 shows a histogram of data for urea nitrogen values from 103 apparently healthy infants. This histogram shows the scatter of the data, but from direct observation it is difficult to decide if this data could be better fit by a Gaussian or a lognormal distribution. A graphical approach may be used to answer the question.

If the urea nitrogen values are arranged in order of increasing magnitude and the

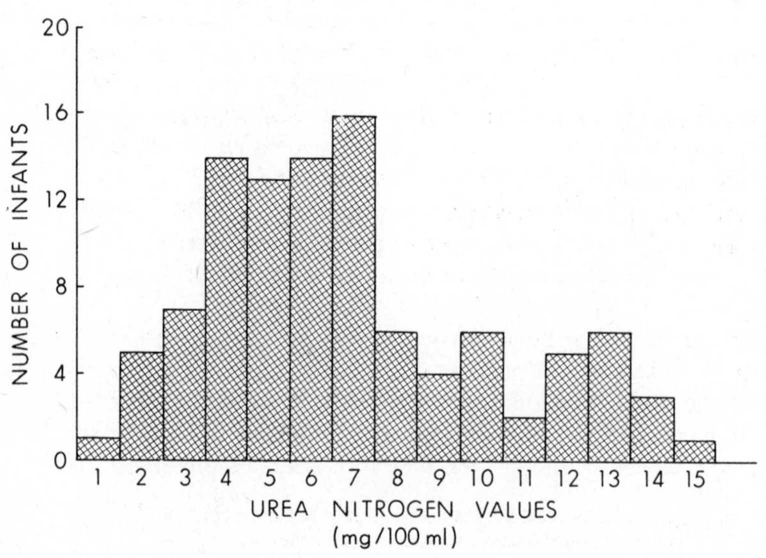

Figure 2–5 Histogram showing the distribution of urea nitrogen values from 103 apparently healthy infants.

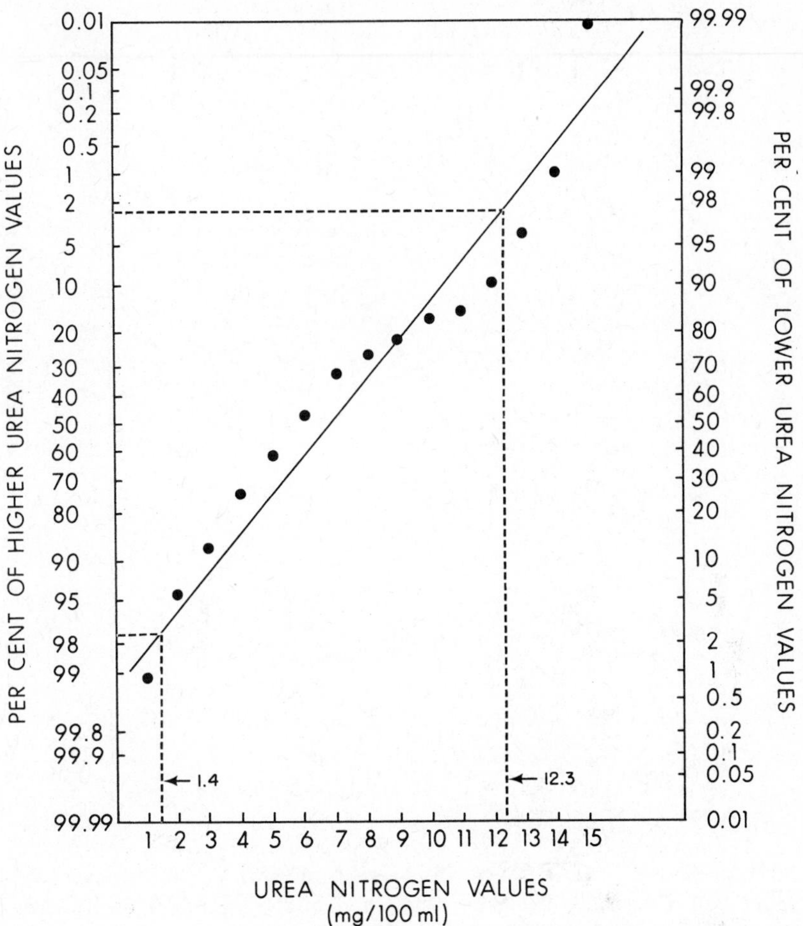

Figure 2–6 Cumulative frequency distribution of 103 urea nitrogen values plotted on normal probability paper.

cumulative frequency calculated (see Example 3), the data can be plotted on normal probability paper, with urea nitrogen values on the horizontal axis and the cumulative percentages on the vertical axis (Fig. 2–6). If the points appear to be on a straight line, a line is drawn so that it passes through or falls near as many of the points as possible. The 95 per cent interval is read directly from the graph. The 95 per cent interval obtained from Figure 2–6 extends from 1.4 to 12.3 mg/100 ml of urea nitrogen, but it should be obvious that the straight line does not provide a very good fit to the scatter of points.

If the distribution of plotted points appears to be a curve, the log of each urea nitrogen value is obtained and plotted on the horizontal axis against the cumulative frequency on the vertical axis of probability paper. If the resulting plot can be fitted by a straight line, the distribution is considered to be lognormal. A quick method of determining whether a log transformation is likely to be of value is to calculate the *CV* for the data. If the *CV* is greater than 15 or 20 per cent, a logarithmic transformation may be useful.[3,10] In our example, the *CV* for the urea nitrogen data is 48 per cent, indicating the desirability of the transformation.

Figure 2–7 shows the log of the urea nitrogen values plotted on normal probability paper. The straight line drawn in Figure 2–7 provides a better fit to the data than the line in Figure 2–6. This means that the data appear to have a lognormal distribution. The 95 per

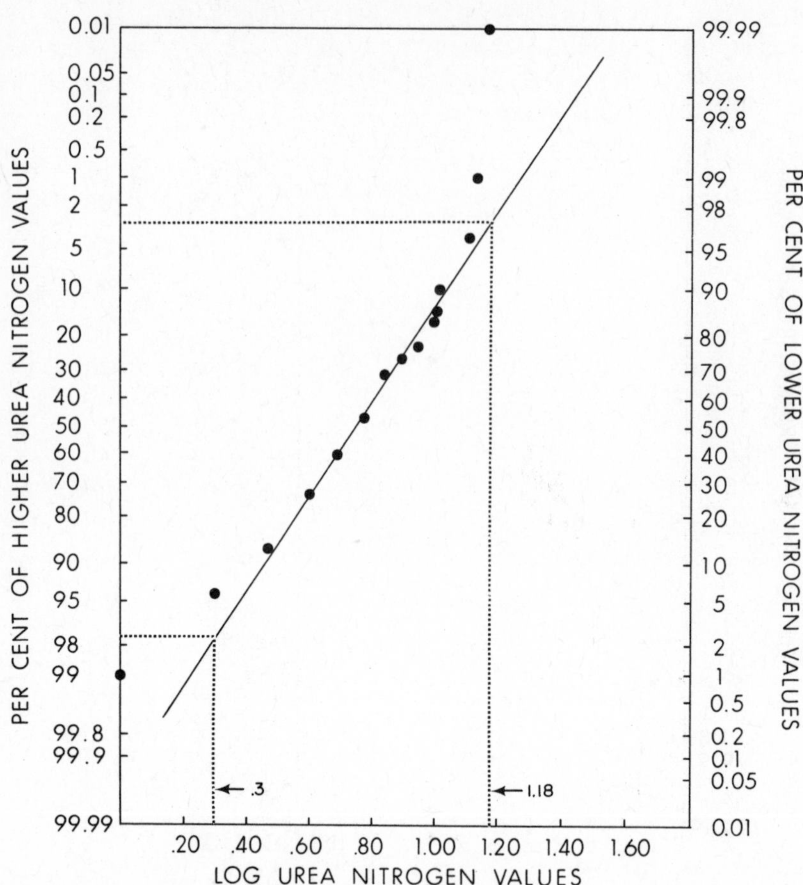

Figure 2–7 Cumulative frequency distribution of the logarithms of 103 urea nitrogen values plotted on normal probability paper.

cent interval can be read directly from Figure 2–7 as 0.3 to 1.18 in logs. Taking the antilogs of these values, the normal range extends from 2 to 15.14 mg urea nitrogen/100 ml.

Once it has been ascertained that a distribution is lognormal, the mean and standard deviation of the transformed distribution may be calculated and used to estimate the normal limits. The log of each urea nitrogen value is taken, and the logarithms are now handled in exactly the same manner as any value would be handled in computing the mean, variance, and standard deviation. The mean of the log values of the urea nitrogen data (see Example 3) was found to be 0.78 and the standard deviation was 0.23, expressed as logs. These values are now used to establish an interval within which approximately 95 per cent of the sample falls. The interval ($\bar{X} \pm 2s$) extends from 0.32 to 1.24. Taking the antilogs, the mean is 6.03 and the normal range is from 2.09 to 17.38 mg/100 ml. These calculated values are similar to the values of 2 and 15.14 obtained from Figure 2–7.

If the data for urea nitrogen had been handled as if they followed a Gaussian distribution, the mean would have been calculated as 6.89 with a standard deviation of 3.32. Thus, based on $\bar{X} \pm 2s$, the range for urea nitrogen would have been $6.89 \pm 2(3.32)$, or 0.25 to 13.53 mg/100 ml. These values are close to those read from the graph shown in Figure 2–6, namely, 1.4 to 12.3 mg/100 ml. Both approaches yield values that differ considerably from those obtained using a log transformation. This illustrates what was emphasized earlier: the formula $\bar{X} \pm 2s$ without transformation should only be used when the distribution approaches Gaussian.

EXAMPLE 3: CUMULATIVE DISTRIBUTION OF 103 VALUES OF UREA NITROGEN FROM INFANTS (mg/100 ml)

Urea Nitrogen Values (X)	*Frequency* (f)	*Per Cent Cumulative Frequency*	*Log Urea Nitrogen Values* (X')	fX'	fX'^2
1	1	0.97	0.00	0.00	0.0000
2	5	5.82	0.30	1.50	0.4500
3	7	12.62	0.47	3.29	1.5463
4	14	26.21	0.60	8.40	5.0400
5	13	38.83	0.69	8.97	6.1893
6	14	52.43	0.77	10.78	8.3006
7	16	67.96	0.84	13.44	11.2896
8	6	73.79	0.90	5.40	4.8600
9	4	77.67	0.95	3.80	3.6100
10	6	83.50	1.00	6.00	6.0000
11	2	85.44	1.04	2.08	2.1632
12	5	90.29	1.07	5.35	5.7245
13	6	96.12	1.11	6.66	7.3926
14	3	99.03	1.14	3.42	3.8988
15	1	99.99	1.17	1.17	1.3689
	$n = 103$			$\sum fX' = 80.26$ $(\sum fX')^2 = 6441.6676$	$\sum fX'^2 = 67.8338$

Calculation of the mean, variance, and standard deviation of log values using grouped data

$$\text{Mean} = \bar{X}' = \frac{\Sigma fX'}{n}$$

where

$$f = \text{frequency of occurrence of each value}$$

Substituting the value in the table for the sum of the frequency times the logarithms of the original urea nitrogen values, we find

$$\bar{X}' = \frac{80.26}{103} = 0.7792$$

The appropriate formula for the variance of grouped data is

$$\text{Variance} = s^2 = \frac{n\,\Sigma fX'^2 - (\Sigma fX')^2}{n(n-1)}$$

Substituting in this formula, we find

$$s^2 = \frac{103(67.8338) - (80.26)^2}{103(102)} = \frac{545.2138}{10506} = 0.0519$$

$$s = \sqrt{s^2} = 0.2278$$

$$2s = 0.4556$$

Using $\bar{X} \pm 2s$ to estimate the 2.5th and 97.5th percentiles of the distribution of urea nitrogen values in logs (0.78 ± 0.46) we find the 2.5th percentile to be 0.32 and the 97.5th percentile to be 1.24. Taking the antilogs of these values, the percentiles are 2.09 and 17.38 mg/100 ml.

The methods described in this section are applicable to Gaussian or positively skewed distributions. However, positive skew does not account for all departures from normality that are likely to be observed. Some distributions may have more or less peakedness (kurtosis) than a Gaussian distribution, and some may be negatively skewed.

Transformations other than the log transformation exist and are described elsewhere.[8,11] It should be remembered that a single type of transformation is appropriate to only one type of deviation from normality. Normalization of a distribution may require the application of more than one transformation. A more general approach that would reduce concern over the exact specification of the underlying distributional form and simplify computations would be highly desirable. As mentioned previously, Elvebach and Taylor have suggested such an approach.[9] While a full explanation of the statistics involved in calculating the confidence intervals about the percentile estimates is beyond the scope of this text, the method is applicable to distributions of any form. If a sample of 120 or more values is available, these nonparametric tools are very helpful.

The purpose of this section was to describe some of the problems involved in obtaining and interpreting normal values. So many problems exist, and so few solutions are possible, that Mainland[18] has suggested abandoning the term normal range in favor of something like cutoff points. Regardless of the appellation, the important thing to remember is that all numerical values, no matter how precise, must be interpreted. Intelligent interpretation can occur only when the interpreter is knowledgeable concerning the source of the data and the methods by which the numbers were produced.

QUALITY CONTROL

As the field of clinical chemistry has developed, there has been a corresponding increase in the number of tests requested per patient. Many hospitals are now performing chemical tests on almost all admissions and there is greater use of the "profile" tests for diagnostic purposes. The net result is that chemistry laboratories are faced with an ever-increasing workload. As more chemical data are generated there is a greater need for effective quality control programs in the laboratory.

During the past few years clinical biochemists and pathologists have become increasingly aware of the lack of agreement seen in assays performed by different laboratories.[7,23] In the surveys, gross differences were observed in the values obtained from various laboratories. Thus, the data produced in many laboratories are not as accurate and precise as may be desired.

ACCURACY, PRECISION, AND RELIABILITY

The terms accuracy, precision, and reliability are often used loosely, almost as synonyms, although they have distinct and different meanings. *Accuracy* refers to the extent to which measurements agree with the true value of the quantity being measured. Determination of accuracy consists of comparing observed results with actual "true" values. In the laboratory the true value is unknown (except for reference standards) and one usually compares measurements made by the method in question with those obtained by an acceptable reference method. When there is a constant systematic error or bias in a clinical method, such as that due to air oxidation in a colorimetric reaction, the mean of a series of determinations will vary from the actual true value and the method is considered inaccurate. Various methods for determining glucose may give different results because of differences in specificity. In the Folin-Wu glucose method, for example, nonglucose reducing substances normally present in blood (ascorbic acid, glutathione, ergothioneine, and so forth) are determined along with glucose. Such a test has poor accuracy, but this does not necessarily negate the value of the test if one takes into consideration the normal values associated with the procedure.

Precision refers to the magnitude of the random errors and the reproducibility of the measurements. The precision of a clinical method is measured by its variance or standard deviation. The smaller the variance, the greater the precision, and if two methods are being compared, the method with the smaller variance is more precise.

The *reliability* of a method is measured by its capacity to maintain both accuracy and precision. If a procedure has maintained a steady state of accuracy and precision over a considerable period, the method can be considered reliable. The reliability of a procedure can only be established by checking the method, using appropriate primary standards and control specimens.

CONTROLLING LABORATORY ERROR

The primary concern of the analyst is to ascertain that his results are as accurate and precise as necessary. The accuracy of an analytical result depends on the selection of a method that provides results which most nearly match those obtained by a reference method.* The accuracy of results may be evaluated by comparing the values obtained by the particular method under study, with the values derived from a reference method having acceptable accuracy.

The precision of an analytical result depends to a large extent on the following:

1. Use of a well-tested, suitable analytical procedure.
2. The availability of conscientious and well-trained technologists who maintain proper analytical skills from start to finish.
3. The reduction of the number of manipulations to a minimum and the simplification of such manipulations.
4. The ability to maintain the required quality of reagents.
5. The ability to maintain equipment in optimal functioning condition.

The routine use of accurate standards, quality control sera, and statistical methods to evaluate the results obtained are necessary to monitor precision on a day to day basis.

Each set of unknowns analyzed in a laboratory should contain at least one standard or set of standards. Use of standards alone, however, falls short of controlling the complete procedure, since they frequently control only some steps of a test. Thus, although the value of a standard and the instrument calibration may be correct, it is possible for the final value of the unknown sample to be considerably in error.

In order to control all the steps of a procedure, it is necessary to use preanalyzed control material which is similar in composition to the unknown specimen. The control specimen is then carried through the entire test procedure in parallel with the unknown, affected by any and all variables that affect the unknown. Use of control material is one way to check on the procedure, techniques, reagents, and instrument calibration.

The control material does not meet the requirements of a primary standard and should not be used as such. No control serum can be assigned a definite value for all constituents. Even a "weighed in" constituent of a control specimen is, as a rule, only a part of the total of that component. Thus, the values assigned to any constituent must be, eventually, based on chemical analyses.

THE USE OF QUALITY CONTROL SPECIMENS (POOLS)

A constructive and economical way to study and control the variations of methods is the inclusion of pooled, preanalyzed sera with each run of unknowns followed by the tabulation of data obtained on control charts and statistical evaluation of the data.

The procedures necessary for this phase of the quality control program can be summarized as follows:

1. Use of serum pools or commercial control sera with each run of unknowns.
2. Determination of mean values for each component of the control specimen, by a current laboratory method.
3. Determination of standard deviations for each component of the control specimen (day to day and batch to batch precision).

* A reference or referee method is the method which will give a result that is regarded as being as close as possible to the true value. Such methods are frequently too time consuming or cumbersome to be utilized in routine laboratory work.

4. Choosing acceptable limits of variation.

5. Setting up quality control charts.

6. Setting up specific trouble-shooting procedures for "out of control" results for each method.

7. Postgraduate education of technical staff in regard to quality control and advances in clinical chemistry, in order to increase skills and technical ability.

Preparation of controls from pooled serum or plasma

1. Pool volume. A pool should be sufficient for at least a three month need. Less than this needlessly increases the necessary work to establish means; much more than this may introduce the possibility of deterioration of some constituents during storage in the freezer.

2. Sources of serum or plasma.

a. The most practical source is salvaged excess serum collected at the end of each day.

b. Blood bank plasma.

3. Collection of serum for pool. Each day collect only clear serum samples free of hemolysis, lipemia, jaundice, or dyes. Place in a plastic bottle in the deep freeze (−20°C). Subsequent samples are added directly to the frozen mixture until an appropriate amount is collected for processing.

4. Preparation of pool.

a. Serum. Allow frozen serum to thaw completely at room temperature; mix thoroughly (e.g., with a magnetic stirrer) for 1 hour without producing any foam. Centrifuge at 3000 rpm for 30 minutes to eliminate fibrin and other debris. Thrombin may be added to enhance fibrin formation. Although filtration through filter paper has been used, it is relatively slow and bacterial growth and consequent alteration of ingredients may occur. Filtration through a Seitz or Millipore filter removes bacteria and is recommended. Mix well again.

b. Plasma salvaged from out-dated blood. Control plasma may be prepared from outdated blood bank blood. Plasma to be used is carefully chosen; only the plasma showing no hemolysis and containing a minimum amount of chyle is selected. The salvaged plasma is centrifuged to remove cells and clots. Since the blood bank plasma has been diluted approximately 20 per cent by the addition of anticoagulants and preservatives, it is restored to normal concentration by dialysis. Smaller molecules and ions lost during the dialyzing process may be restored by adding a calculated amount of these materials to the plasma after dialysis.

Dialysis is carried out in a cellulose casing suspended in a 25 per cent solution of polyvinylpyrrolidone (PVP) in barbital buffer at pH 8.6. The barbital buffer is prepared by dissolving 2.76 g of diethylbarbituric acid and 15.4 g of sodium barbital in 1 liter of distilled water.

5. Adjustment of serum values to levels meeting the needs of the laboratory. Pool values are best set at clinically critical levels, e.g., bilirubin values at 1 mg and 20 mg, and glucose at 100 mg/100 ml. Examples of adding supplemental amounts of pure materials follow:

a. Bicarbonate. Add 84 mg of $NaHCO_3$/l for each desired millimolar increase in the concentration of CO_2/l.

b. Glucose. If the raw pool contains 60 mg/100 ml and the desired value is 100 mg/100 ml, add 40 mg of glucose/100 ml of pool.

c. Bilirubin. If the raw pool contains 0.4 mg/100 ml and the desired value is 1.0 mg/100 ml, add 0.6 mg of bilirubin/each 100 ml of pool. Dissolve crystalline bilirubin in 0.1 molar NaOH (approximately 1 ml 0.1 molar NaOH for each 10 mg of bilirubin). Minimize exposure to light! Dilute to 3 to 5 ml with distilled water and add slowly with gentle stirring to the serum pool. The pH of the pool should be readjusted to pH 7.40 using 0.1 molar HCl.

6. Aliquoting.

a. Size of aliquots. The aliquots should be large enough to provide for the needs of a complete day's work (e.g., 10 ml) or aliquots may contain smaller amounts to provide material sufficient for individual tests or groups of tests.

b. Storage. Place tubes in deep freeze and keep them stoppered or covered with Parafilm until use.

Preparation of pool for thyroid function tests

This pool must be prepared separately in order to avoid contamination from sera containing iodides or mercury compounds. After the unknown sera have been analyzed for PBI or T_4, the remainder of all sera with values in the normal range are pooled until about 250 ml have been accumulated. Pooled sera are stored in a flask in the refrigerator and processed as for the regular pool, aliquoting 2.5 ml per tube.

Preparation of enzyme pools

A pool for amylase, transaminase, lactate dehydrogenase, or other enzymes should be prepared separately, since it is desirable to have an increased level of these enzymes to check properly the quality of the reagents.

Transaminase. Pool the remainder of all sera with transaminase values at least 3-fold above normal. Do not freeze. When 200 ml have been collected, the pool is assayed. If it is lower than the desired range, save and add only very high transaminases. If it is higher, calculate the amount of normal serum needed to reduce its level to the desired range. Filter through a Seitz or Millipore filter, and place small aliquots into 5 ml capacity screw-top vials. Place vials in freezer. Defrost one vial each week, and keep in the refrigerator. It has been found that freezing lowers the transaminase value of the pool considerably, but once it has dropped it remains stable.

Commercial control preparations

An important source of stable control material is commercial preparations, which usually are available in the lyophilized form. Some of this material may be purchased "unassayed," in which case the inter-vial variability is of greatest concern to the user. Others are purchased as "assayed" controls, i.e., with stated values for each of the constituents. In the latter case, information on the analytical methods and statistical procedures used in arriving at the assigned value for a given constituent should be made available to the user.

In 1967 a CDC (Center for Disease Control) committee was appointed to recommend general standards for control materials. In 1972, the National Committee for Clinical Laboratory Standards (NCCLS) drew up a tentative standard to aid manufacturers in the development of control materials that would best satisfy the present and future needs of the users. This committee stated that control materials should be sufficiently well characterized in regard to accuracy and precision, so that the use of these materials will aid in the quality control of clinical laboratories. It was realized that the accuracy and precision attainable in the large-scale production of control materials is limited by the performance of bulk processing equipment. As an example, a small amount of residual moisture in lyophilized serum samples will result in instability of glucose and some of the enzymes, especially lactate dehydrogenase. This would affect both the accuracy and precision of the lot.

These products should be from human sources. In some instances, i.e., enzymes, it may be necessary to "spike" the pool with enzymes other than human in origin. This source of enzyme should be stated. In the process of lyophilization some changes in the proteins occur which may result in a slight cloudiness in the reconstituted material. The manufacturer should provide instructions for the user to insure homogeneity of the final reconstituted material. For certain tests, appropriate blank correction procedures might be necessary.

Inter-vial precision is necessary for any control material. The NCCLS Committee recommended that intervial differences of the final reconstituted product not exceed ±1 per cent of the mean value in more than 5 per cent of the vials. Assuming a normal distribution, this represents a coefficient of variation of 0.5 per cent. As an example, if the mean value for a lot of vials of control sera for glucose is 100 mg/100 ml, then 95 per cent of the vials should contain a concentration of glucose between 99 mg/100 ml and 101 mg/100 ml when reconstituted.

The use of commercial control materials represents the only practical way to carry out a large survey of laboratories because (1) a large pool is needed and (2) these materials are disseminated via the mail and it would be impractical to mail out frozen sera. The daily use of commercial controls in the laboratory may present a serious cost factor that must be evaluated by each laboratory. In addition, these controls require reconstitution with diluent, thus adding another variable not present in frozen sera. However, one may purchase enough of this material of the same lot to be used for one year, which represents a convenience. Some of the lots of commercial sera are large enough so that many laboratories would be using the same lot over an extended period of time. This provides the basis for comparison of the data with those of a peer group.

Caution in the preparation and use of serum pools

When a control serum is prepared from a large pool of sera, there is always the danger that this pool may be positive for hepatitis B antigen. While it is impractical to test each serum before adding it to the pool, every effort should be made to exclude sera from patients with hepatitis. This is of particular importance in the preparation of pools with high enzyme activities, since these patients will usually have high levels of certain enzymes, i.e., transaminases. The final pool should then be tested for hepatitis B antigen by a radioimmunoassay technique. We have recently found that many lots of various commercial control sera give a positive test for hepatitis B antigen by this technique. Unless control sera have given a negative test for hepatitis B antigen, they should be handled with proper precautions.

ESTABLISHING ACCEPTABLE LIMITS OF VARIATION

Any analytical procedure, regardless of how carefully performed, is subject to error, since it is impossible to duplicate all conditions from day to day or run to run. Some causes of such variation may be discovered and corrected and are therefore called assignable (determinate) errors. Other variations occur to which no cause can be attributed, over which the analyst has no control, and are called random errors (see previous section on statistics). A process without assignable variations (great enough to be of importance) is said to be in statistical control.

The expression used to indicate the permissible range by which the obtained results may vary is "acceptable limits of error." Repeated analysis of the control or an unknown specimen should give results within this limit. If the value of a control is outside these limits, the run is considered to be out of control and reasons for the aberrant result are sought.

Quality control charts are used as the basis for inferences about laboratory procedures from samples of data. To establish limits of acceptability, a minimum of 20 (preferably more) independent analyses should be performed and the mean and standard deviation calculated. These statistics are used to establish "acceptable limits of error." For example, a laboratory may obtain 20 values for cholesterol and calculate the mean and standard deviation to be 175.5 and 2.8, respectively. Thus, the interval that will include about two-thirds of the values is approximately 173 to 178 mg/100 ml ($\bar{X} \pm s$); the interval that includes approximately 95 per cent of the values is 170 to 181 mg/100 ml ($\bar{X} \pm 2s$); and the interval that will include about 99 per cent of the values is 167 to 184 mg/100 ml ($\bar{X} \pm 3s$). Any one of these intervals could be accepted as the control limits against which future values could be evaluated, but all are subject to limitations. Thus, there is a need for guidelines to indicate the permissible range by which the obtained results may vary.

For any set of control limits established on a sample of values, there is a probability that a future value of the constituent will fall outside the established limits when the process is in control (type I error). There is also a concomitant probability that an obtained value will fall inside the established limits when the process is actually behaving erratically (type II error). The two error types are related. A decrease in type I error will result in an increase in type II error. Either type of error, but not both, can be decreased for a single constituent and a single procedure; i.e. as the probability of seeking trouble where none exists (type I error) decreases, the probability that erroneous results will remain unidentified (type II error) increases.

The use of $1s$ to establish the "acceptable limit of error" can readily be seen to have little value. About one-third of the values could be expected to fall outside the established control limits when the procedure is functioning normally. The use of $2s$ is helpful and is widely used, but it should be used with discretion. It should be remembered that occasionally (in about 5 of 100 assays) results may be expected to fall outside the limits when the procedures and apparatus are functioning normally. Since assay values that fall outside the limits should normally be examined, the investigator may seek trouble when none, in fact, exists. Such searches involve needless time and expense. Although the use of $3s$ would eliminate virtually any possibility of a value being beyond the acceptable range without actual error, the use of $3s$ is discouraged in most instances because of the increased probability that actual errors will go undetected.

There are other considerations in the establishment of control limits besides the two possibilities just discussed. For certain constituents, the range of values assigned to the control may be as great as the range of normal values. The clinical usefulness of such a procedure is questionable.

The precision of a laboratory is another factor to be considered in establishing control limits and evaluating results from different laboratories. In a laboratory with rather poor precision (i.e., obtaining results with larger variances) the use of $2s$, for example, to establish control limits would allow the acceptance of a large number of results that would be considered unacceptable by a laboratory with greater precision.

Since at present there is no definite system that can be used to establish the desirable control limits, respective laboratories will have to set their own limits based on their own experience and needs.[22] The values published in the literature can be used as a guide for the selection of desirable levels of precision to be attained.

Barnett[2] has synthesized the opinions of clinicians and laboratory specialists and has proposed medically critical ranges for 16 of the most commonly used blood tests. He lists some of the data at more than one decision level, which is the dividing point at which medical decisions concerning diagnosis and treatment are frequently made. The data are shown in Table 2–4. For some constituents, a "low level" is also listed. This is a level below which the same degree of accuracy and precision is no longer required.

These figures could be considered minimal standards for precision. With the instrumentation and methodology available at this point in time, most laboratories should have little difficulty in attaining this degree of precision. As an example, two university affiliated hospital laboratories report their coefficient of variation for sodium to be 0.6 per cent and 0.9 per cent, and calcium to be 1.1 per cent and 1.4 per cent. Each laboratory should strive for the best precision possible under their circumstances.

For the above reasons, the term "control" should be interpreted cautiously. Its meaning depends on the variance of values for a process that was determined during a prior series of runs. It also depends on the random errors that are operative.

TABLE 2–4 MEDICALLY CRITICAL RANGES FOR VARIOUS CONSTITUENTS OF BLOOD*

Component	*Decision Level*	*s at Same Level†*	*Calculated CV (%)*	*Low Level*
Hemoglobin	10.5 g/100 ml	0.5	4.76	5 g/100 ml
Hematocrit	32%	1.0	3.12	10%
Glucose	50 mg/100 ml	5.0	10.00	20 mg/100 ml
Glucose	100 mg/100 ml	5.0	5.00	
Glucose	120 mg/100 ml	5.0	4.17	
Blood urea nitrogen	27 mg/100 ml	2.0	7.41	4 mg/100 ml
Uric acid	6.0 mg/100 ml	0.5	8.33	4 mg/100 ml
Total protein	7.0 g/100 ml	0.3	4.28	2 g/100 ml
Albumin	3.5 g/100 ml	0.25	7.14	1.5 g/100 ml
Globulin	3.5 g/100 ml	0.25	7.14	1.5 g/100 ml
Cholesterol	250 mg/100 ml	20.0	8.00	80 mg/100 ml
Bilirubin	1.0 mg/100 ml	0.2	20.00	0.4 mg/100 ml
Bilirubin	20.0 mg/100 ml	1.5	7.50	
Calcium	11.0 mg/100 ml	0.25	2.27	5.0 mg/100 ml
Sodium	150 mEq/l	2.0	1.33	
Sodium	130 mEq/l	2.0	1.54	100 mEq/l
Phosphorus	4.5 mg/100 ml	0.25	5.56	1.5 mg/100 ml
Potassium	3 mEq/l	0.25	8.33	1.5 mEq/l
Potassium	6 mEq/l	0.25	4.17	
Chloride	90 mEq/l	2.0	2.22	50 mEq/l
Chloride	110 mEq/l	2.0	1.82	
CO_2	20 mEq/l	1.0	5.00	8 mEq/l
CO_2	30 mEq/l	1.0	3.33	

* From Barnett, R. N.: Am. J. Clin. Path., *50*:671, 1968. © by the Williams & Wilkins Company, Baltimore, Maryland.

† Same units as corresponding decision level.

Control charts

The control chart is the simplest way to present daily control data. The control chart is a sheet of rectangular coordinate graph paper, marked by days on the horizontal axis; the units of the vertical axis are in terms of the assay concerned. The mean value, as well as the "limits of acceptable error" indicated on the chart, are based on past laboratory experience. The daily value is plotted on the control chart, and any point falling beyond the "limits of acceptable error" can easily be seen.

For example, suppose that during a specified period of time, controls were run and the mean and standard deviation of all control values were found to be 40.00 and 3.36, respectively. Using these values and the criterion of $\pm 2s$, the limits of acceptable error are calculated to be approximately 33 to 47. These values are shown in Figure 2–8. Control values obtained subsequently are also plotted on the chart. Any value falling outside the limits is considered as an indication that the process is out of control and a reason is sought. As can be seen, the data on the 19th day should have been the subject of close scrutiny since the value is beyond the limit of acceptability.

The control chart for cholesterol shown in Figure 2–9 illustrates the situation in which several batches of unknowns were analyzed on some days. A control was included in each of these runs and the result for each of these controls is included in the plot shown in Figure 2–9. The mean for all controls for the total month is 176 mg/100 ml, and the limits have been set at 168 and 184 mg/100 ml.

Interpretation of Quality Control Charts. In a quality control program, one measures the magnitude of the variation and then establishes the variation that might reasonably be expected in the future. If something goes wrong, points will begin to fall outside the established limits. When points begin to fall outside, one should examine not only the process but the limits themselves.

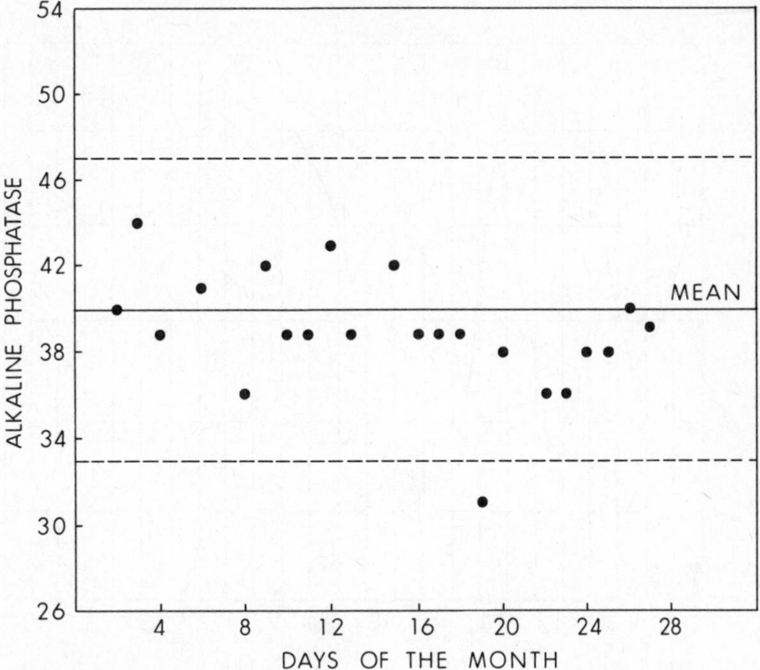

Figure 2–8 Quality control chart for alkaline phosphatase. Each point represents an alkaline phosphatase value for the control serum.

A *trend* in a quality control chart refers to a series of values for the control that continue either to increase or to decrease over a period of several consecutive days (Fig. 2–10). Trends may be the result of gradual deterioration of one or more of the reagents, changes in standards, incomplete protein precipitation, failure of instrumental components, and so on.

When values on several consecutive days distribute themselves on one side of the mean value line, but are remaining at a constant level, the trend is referred to as a *shift* (Fig.

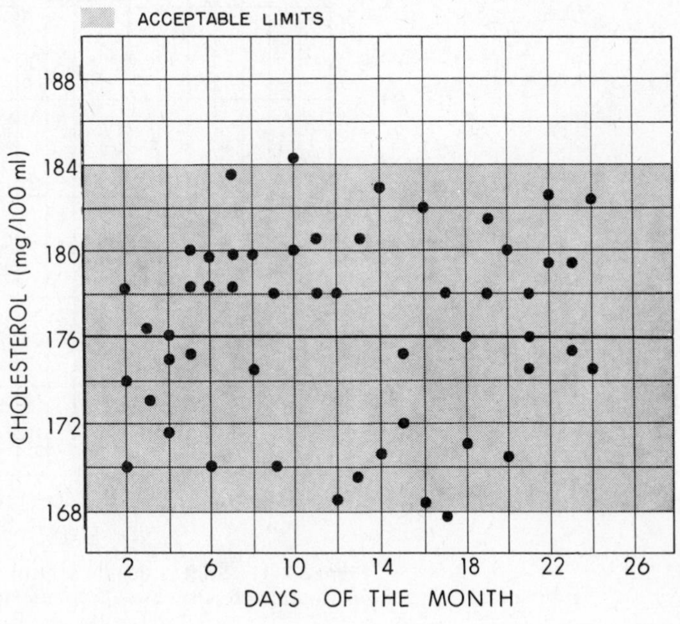

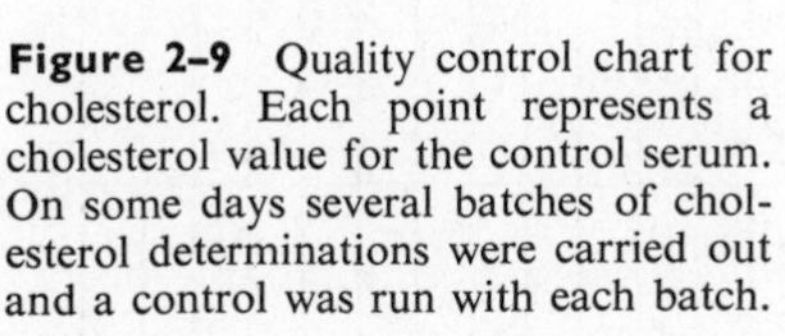
Figure 2–9 Quality control chart for cholesterol. Each point represents a cholesterol value for the control serum. On some days several batches of cholesterol determinations were carried out and a control was run with each batch.

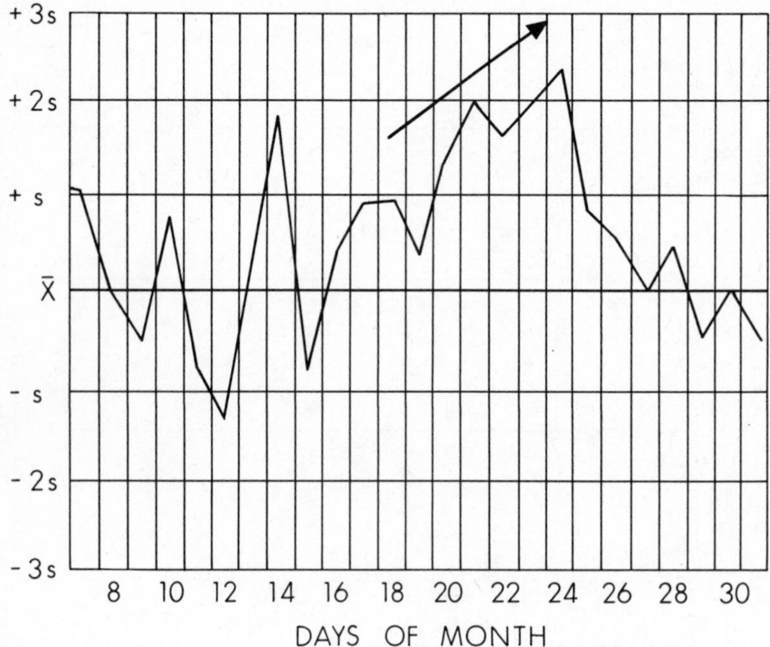

Figure 2–10 Trend in quality control values. The arrow denotes the upward trend.

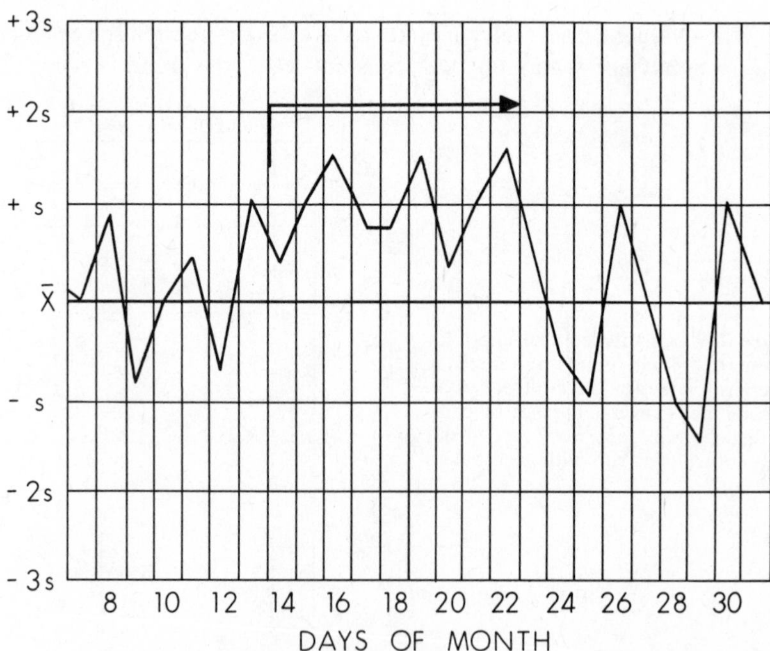

Figure 2–11 Shift in quality control values. The arrow denotes the shift in the distribution of values away from the mean.

2–11). An upward shift might indicate, for example, that a standard has deteriorated but is maintaining a constant level, a new standard has been prepared at a lower concentration than that required for a test, or a reagent has shifted to a new level of sensitivity. In test procedures in which solutions are boiled, an upward shift could indicate a prolonged boiling time due to an inaccurate timer, or an indicator that has lost sensitivity, thereby prolonging the end point reading. As a rule, downward shifts are caused by conditions that are opposite to those causing upward shifts.

NOTES REGARDING THE QUALITY CONTROL PROGRAM

A series of controls are run with each day's group of determinations. The frequency and total number of controls used are variable, depending upon the total number of assays being carried out and the complexity of the procedures. If the values are outside the acceptable range, the data are carefully evaluated before any results are reported. If the source of the apparent error cannot be found, a new control specimen, along with several of the unknowns chosen at random, is reassayed. If the new control gives an appropriate value and the unknowns check the previous run, the values are reported; if not, the data are held back and the reagents and procedural steps checked one by one until the source of trouble is determined.

In recent years there has been considerable shift from manual methods to automated procedures. This requires some change in quality control procedures. It is recommended that standards and controls be run not only at the beginning of the test run but also between the unknowns and at the end of the run. In this way, any changes in instrument performance, such as change in light source or weakening of photocells, and any changes in environment (temperature) which might occur during one run can be detected.

In addition to the routine controls which are included in each group of analyses, the use of "blind" controls provides another method of quality control. In this method, which many laboratories occasionally use, the person performing the test does not know that this sample is a control, i.e., it is numbered or identified as a patient's sample would be coded. This can be achieved in a number of ways. Commercial control sera may be used, in which case this blind control could be used once a week for a long period of time. Another way is to split a sample into two or three aliquots and process it as two or three different patient samples. A third method, providing the constituent under study is stable, is to process the specimen a day later and compare results. These supplemental methods of quality control are used in conjunction with the primary quality control program to eliminate any bias.

One of the purposes of a quality control program is to insure that every group of tests is in statistical control. The assumption is made that this can be interpolated to each specimen. However, there is no substitute for looking at and reflecting on the patient data being sent out. Certain constituents have a relationship to each other in health and disease. For example, there is a mathematical relationship between the anions and cations in an electrolyte report. Also, in certain diseases there is a reciprocal relationship between serum calcium and phosphorus levels. To inspect the data for such relationships is an important part of the quality control program. Also, the accuracy of the result on an individual specimen might be affected by the specimen itself, i.e., the serum may be hemolyzed or lipemic. Looking at the specimen is also a part of the overall quality control program. In addition, the technologist should look at the final color in the cuvet to detect turbidity or off color. This latter form of quality control may be difficult or impossible to achieve with automated equipment.

INTERLABORATORY QUALITY CONTROL PROGRAMS (PROFICIENCY SURVEYS)

Recently, a number of national and state agencies have instituted programs to help laboratories assess and maintain quality control. The College of American Pathologists

(CAP), the Center for Disease Control (CDC), the American Association of Bioanalysts (AAB), some state health departments, and so forth, send specimens to participating laboratories periodically. After the data that are submitted by the laboratories are compiled and analyzed, reports are sent to the participating laboratories. Data sent by the CAP, for example, include the laboratory's result for each constituent reported and a Standard Deviation Index (SDI). In addition, the number of other laboratories reporting results using the same method, and the mean and standard deviation of the distribution of results from all laboratories are reported.

The SDI indicates the position of the value reported by a specific laboratory in the distribution of all such values reported. For example, data reported by CAP to a laboratory may show the following:

Constituent	*Method*	*Your Result*	*Mean*	*s*	*No. of Labs*	*Your SDI*
glucose	ferricyanide	132	129.0	4.6	393	+0.7

The SDI of +0.7 was obtained by a computer program using the formula

$$\text{SDI} = \frac{\text{lab result} - \text{mean of all labs using same method}}{s \text{ of all labs}}$$

$$= \frac{132 - 129}{4.6} = +0.7$$

An SDI of +0.7 means that the laboratory in question reported a glucose value 0.7 s above the mean of all laboratories reporting results obtained by the ferricyanide method. Approximately 68 per cent of the laboratories would report results within $\pm 1\ s$ of the mean (see Fig. 2–2). Such results are considered "good." Approximately 18 per cent of the laboratories would be expected to report values between 1 and 1.5 standard deviations from the mean. Their results are considered "acceptable." Values that are more deviant than $\pm 1.5\ s$ are judged "nonacceptable." Thus, the central 86 per cent of the values are used as the basis of acceptability.

Data on accuracy are also provided by the proficiency service of CAP. Mean, standard deviation, and number of laboratories reporting data resulting from use of a well-accepted method are also reported. Data may appear as follows:

Constituent	*Well-accepted Method*	*Mean*	*s*	*No. of Labs*	*Your SDI*
glucose	*o*-toluidine	124.0	6.8	259	+1.2

The SDI as a measure of accuracy is computed as

$$\text{SDI} = \frac{\text{your result} - \text{mean of accepted method}}{s \text{ of accepted method}}$$

$$= \frac{132 - 124}{6.8} = +1.2$$

The SDI of 1.2 means that the laboratory reported a glucose value determined by the ferricyanide method that was slightly more than 1 s above the mean of the distribution of 259 laboratories reporting data obtained by the *o*-toluidine method. We know (from Fig. 2–2) that about 84 per cent of the values may be expected to fall below a value of 130.8 ($\bar{X} + s$) and about 16 per cent above this value. Although a value of 132 is among the upper 16 per cent of the values, it is not deviant enough from the mean to be considered unsatisfactory.

Even though a laboratory has an internal quality control program involving the use of control sera on a daily basis, participation in at least one external quality control program is wise. In some areas, laws now require such participation.

EVALUATION OF A NEW PROCEDURE

If the results of an analytical procedure performed in the laboratory are to be of use to the physician, they must be precise, reasonably accurate, and available soon enough to be medically useful. For many tests, the time to perform the test must be considered in terms of the accuracy needed by the physicians. Thus, a physician would rather receive a report in 1 hour that the blood glucose is between 40 mg/100 ml and 45 mg/100 ml than receive a report 8 hours later that the blood glucose is 40.3 mg/100 ml.

In the evaluation of a new procedure, some time should be spent analyzing sets of standard solutions until one thoroughly understands the parameters of the test, such as linear absorbance range. It is also recommended that known quantities of a constituent be added to the biological specimens and that the recovery be calculated. This shows, for example, that there is no loss of quantities during protein precipitation or at some other stage of the method. Analysis of serum pools containing the constituent of interest preferably in low, normal, and high amounts is also advisable. This should be done over a period of about one month, during which time a reasonable amount of data may be accumulated. The pools should also be analyzed by a reference method or at least a "generally accepted method." From these data one can calculate the mean, standard deviation, and coefficient of variation and compare these with data obtained by the reference method. By analyzing many replicate specimens in the same run, one can obtain useful data on the within-day or within-run precision. These within-run or within-day precision data indicate how well the method will perform under optimal conditions and will give a "target value" for the day to day precision of the method. If the method under evaluation is to replace a method already in use in the laboratory, the new method should have at least as good a precision as the method it is going to replace. If the precision data indicate that this new method is reproducible enough to be used by the laboratory, paired analyses by both methods of a large number of patient specimens should be performed to obtain information concerning the agreement of the two methods. If the pair of results on one individual specimen differs much more than the results on other specimens, the tests should be repeated. If the difference still exists, interference by medication should be considered, since it is possible that a medication may interfere with one method and not with the other.

If the new procedure has superior precision but gives values slightly higher or lower than the existing method, the normal range will have to be determined by analyzing specimens from normal individuals. Also, the clinical usefulness of the range should be confirmed. It will then be necessary to inform the clinical staff of the change in method and the difference in normal values.

In recent years there has been an ever-increasing workload in the clinical laboratory. Automation has helped to cope with part of this; however, requests for more specialized tests and more "off hour" tests cannot be handled by automation. Simplification of many procedures has made it possible to handle this type of increased workload. Manufacturers have responded to this need by supplying the necessary reagents, standards, and directions for carrying out a procedure in a single package. This is now referred to as a kit. Evaluation of such kit procedures poses an especially important problem to the laboratory director. An ad hoc committee on kits met at the Center for Disease Control in Atlanta, Georgia, in 1969 and formulated the following definition of a kit:

"A kit reagent set or diagnostic aid is a collection or assembly of reagents, devices, or equipment, or a combination thereof, offered for sale or distribution and containing all of the major components and written instructions to perform one or more designated tests or procedures."[17]

In 1966 the American Association of Clinical Chemists published a policy statement regarding reagent sets and kits.[1] In this policy it was recommended that the manufacturers supply to the user or perspective buyer the following information:

1. Adequate literature documentation of the method and any modifications by the manufacturer.
2. A statement of the principle involved in the test.
3. Typical values obtained in health and disease.
4. Data on precision, accuracy, and specificity using biological samples.
5. Statistical data showing the degree of correlation with other methods, and recovery data.
6. Stability of reagents, instructions for storage, and expiration dates.
7. Reagent composition and concentration.
8. Calibration procedures, use of blanks, and source of standards should be carefully described.
9. The instrumental requirements and limitations should be clearly stated.

It is recommended that one evaluate a kit in the same manner as a new method would be evaluated. In addition, one would have to make certain that the kit procedure had clear, concise instructions applicable to the instrumentation in that laboratory.

REFERENCES

1. American Association of Clinical Chemists, Committee on Standards and Controls, AACC Policy Regarding Reagent Sets and Kits: Clin. Chem., *12*:43, 1966.
2. Barnett, R. N.: Am. J. Clin. Path., *50*:671, 1968.
3. Becktel, J. M.: Clin. Chim. Acta, *28*:119, 1970.
4. Beeler, M. F.: Postgrad. Med., *43*:67, 1968.
5. Best, W. R., Mason, C. C., Bar, S. S., and Shepherd, H. G.: Clin. Chim. Acta, *28*:127, 1970.
6. Bowker, A. H.: *In* Selected Techniques of Statistical Analysis. C. Eisenhart, M. W. Hastay, and W. A. Wallis, Eds. New York, McGraw-Hill Book Company, 1947, p. 97.
7. Copeland, B. E., Skendzel, L. P., and Barnett, R. N.: Am. J. Clin. Path., *58*:281, 1972.
8. Dixon, W. J., and Massey, F. J.: Introduction to Statistical Analysis. New York, McGraw-Hill Book Company, 1969.
9. Elveback, L. R., and Taylor, W. F.: Ann. N.Y. Acad. Sci., *161*:538, 1969.
10. Grasbeck, R., and Fellman, J.: Scand. J. Clin. Lab. Invest., *21*:193, 1968.
11. Harris, E. K., and DeMets, D. L.: Clin. Chem., *18*:605, 1972.
12. Henry, R. J.: Clin. Chem., *5*:309, 1959.
13. Henry, R. J.: Am. J. Clin. Path., *34*:326, 1960.
14. Henry, R. J.: Clinical Chemistry: Principles and Techniques. New York, Harper & Row, Publishers, 1964.
15. Henry, R. J., and Dryer, R. L.: *In* Standard Methods of Clinical Chemistry. New York, Academic Press, Inc., *4*:205, 1963.
16. Hoffman, R. G.: J.A.M.A., *185*:865, 1963.
17. Logan, J. E.: CRC Critical Reviews in Clinical Laboratory Sciences, p. 271, September, 1972.
18. Mainland, D.: Clin. Chem., *17*:267, 1971.
19. Neumann, G. J.: Clin. Chem., *14*:978, 1968.
20. Reed, A. H., Henry, R. J., and Mason, W. B.: Clin. Chem., *17*:275, 1970.
21. Shepherd, H. G.: Lab. Med., *1*:24, 1970.
22. Tietz, N. W.: Hosp. Prog., *45*:140, 1964.
23. Tonks, D.: Clin. Chem., *9*:217, 1963.
24. Wootton, I. D. P., and King, E. J.: Lancet, *1*:470, 1953.

ADDITIONAL READINGS

Armitage, P.: Statistical Methods in Medical Research. Dorking, Great Britain, Blackwell Scientific Publications, 1971.

Castle, D. M.: Statistics in Small Doses. London, Churchill Livingstone, 1972.

Hoffman, R. G.: New Clinical Laboratory Standardization Methods. New York, Exposition Press, 1974.

Hollander, M., and Wolfe, D. A.: Nonparametric Statistical Methods. New York, John Wiley and Sons, 1973.

Huntsberger, D. V., and Leaverton, P. E.: Statistical Inference in the Biomedical Sciences. Boston, Allyn and Bacon, 1970.

CHAPTER 3

ANALYTICAL PROCEDURES AND INSTRUMENTATION

SECTION ONE

PHOTOMETRY

by Wendell T. Caraway, Ph.D.

Many determinations made in the clinical laboratory are based upon measurements of radiant energy emitted, transmitted, absorbed, or reflected under controlled conditions. The principles involved in such measurements will be considered in this section. Details of operation and maintenance for a particular instrument are typically supplied by the manufacturer and should be consulted for further information.

The term photometric measurement was defined originally as making a measurement of light intensity independently of wavelength. Most instruments used at present, however, have some means of isolating a narrow wavelength range of the spectrum for measurements. Those that use filters for this purpose are referred to as filter photometers, while those that use prisms or gratings are called spectrophotometers. Both types will be considered. Older colorimetric procedures, involving visual comparison of the color of an unknown to that of standards, have now been replaced for the most part by more specific and accurate photometric procedures.

Nature of light

Electromagnetic radiation includes radiant energy from short wavelength gamma rays to long wavelength radio waves. The term "light" will be used to describe radiant energy with wavelengths visible to the human eye and with wavelengths bordering on those visible to the human eye. The wavelength of light is defined as the distance between two peaks as the light travels in a wavelike manner. This distance is preferably expressed in nanometers (nm) for wavelengths commonly used in photometry. Other obsolete units that may be encountered are Ångstroms (Å) and millimicrons (mμ).

$$1 \text{ nm} = 1 \text{ m}\mu = 10 \text{ Å} = 10^{-9} \text{ meter}$$

The human eye responds to radiant energy with wavelengths between about 400 and 750 nm, but modern instrumentation permits measurements at both shorter wavelength (ultraviolet, UV) and longer wavelength (infrared, IR) portions of the spectrum.

Sunlight, or light emitted from a tungsten filament, is a mixture, or spectrum, of radiant energy of different wavelengths that the eye recognizes as "white." Table 3–1 shows approximate relationships between wavelengths and color characteristics for the UV, visible, and short IR portions of the spectrum. Thus, a solution will appear green when viewed against white light if it transmits light maximally between 500 and 580 nm but absorbs

TABLE 3–1 UV, VISIBLE, AND SHORT IR SPECTRUM CHARACTERISTICS

Wavelength (nm)	*Region Name*	*Color Observed**
180–220	ultra short UV	not visible
220–320	short UV	not visible
320–400	long UV	not visible
400–440	visible	violet
440–500	visible	blue
500–580	visible	green
580–600	visible	yellow
600–620	visible	orange
620–750	visible	red
750–2000	short IR	not visible

* Owing to the subjective nature of color, the wavelength ranges shown are only approximations.

light at other wavelengths. Similarly, a solid object appears green if it reflects light in this region (500 to 580 nm) but absorbs light at other portions of the spectrum. In general, if we compare the intensity of light transmitted by a colored solution to that of a blank or reference solution over the entire spectrum, we obtain a typical spectral-transmittance curve characteristic for that specimen. Such curves are shown in Figure 3–1 for solutions of nickel sulfate (A) and potassium permanganate (B). Inspection of the curves should lead us to predict that the color of solution A is green inasmuch as light is transmitted maximally near the green portion of the spectrum. Curve B, on the other hand, illustrates the spectrum of a solution that transmits light maximally in the blue, violet, and red portions of the spectrum. The eye recognizes this mixture of colors as purple.

The chief advantage of filter photometry or spectrophotometry is that we can isolate and use discrete portions of the spectrum for purposes of measurement. Thus, in the case of the permanganate solution, we may use a relatively pure green light source which will be absorbed selectively, rather than use a white (composite) light source. This results in improved specificity as well as better sensitivity of the measurement.

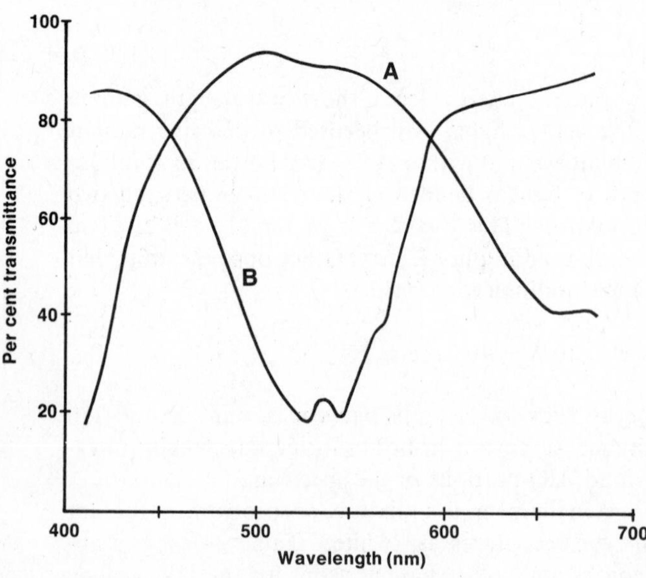

Figure 3–1 Spectral-transmittance curves of nickel sulfate (*A*) and potassium permanganate (*B*). Arbitrary concentrations, read versus water (Beckman DB-G spectrophotometer).

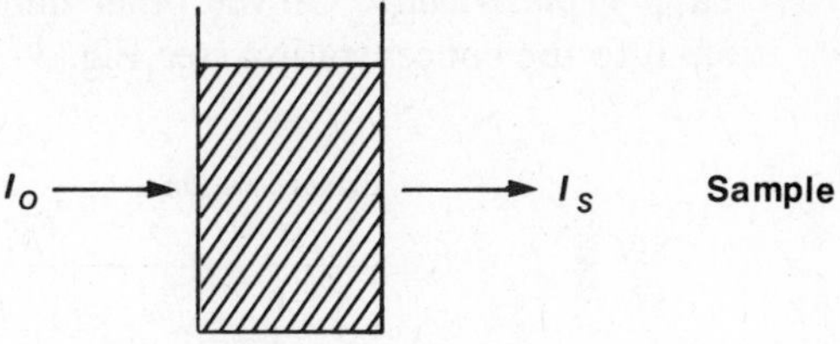

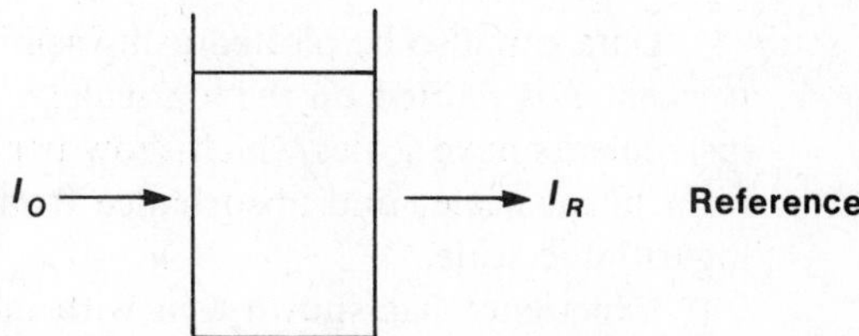

Figure 3–2 Transmittance of light through sample and reference cells. Transmittance of sample versus reference $= I_S/I_R$.

Beer's law

Consider an incident light beam with intensity I_O passing through a square cell containing a solution of a compound that absorbs light of a certain wavelength (lambda = λ) (Fig. 3–2). The intensity of the transmitted light beam I_S will be less than I_O and we may define the transmittance (T) of light as I_S/I_O. Some of the incident light, however, may be reflected by the surface of the cell or absorbed by the cell wall or by the solvent. In order to focus attention on the compound of interest, it becomes necessary to eliminate these factors. This is done by using a reference cell identical to the sample cell, except that the compound of interest is omitted from the solvent. The transmittance through this reference cell is I_R/I_O; the transmittance for the compound in solution is then defined as I_S/I_R. In practice, the reference cell is inserted and the instrument adjusted to an arbitrary scale reading of 100 (corresponding to 100 per cent transmittance), following which the per cent transmittance reading is made on the sample. As we increase the concentration of the compound in solution, we find that transmittance varies inversely and logarithmically with concentration. Consequently, it is more convenient to define a new term, absorbance (A), that will be directly proportional to concentration.* Hence

$$A = -\log I_S/I_R = -\log T = \log \frac{1}{T} = \log \frac{100 \text{ per cent}}{\text{per cent } T}$$
$$= \log 100 - \log \text{per cent } T = 2 - \log \text{per cent } T$$

To clarify these relationships more fully, assume that we have a solution of a compound at a concentration of 1 g/l that transmits only half of the incident light ($T = 0.5$) compared to a reference setting of 100, i.e.,

$$T = 50 \text{ per cent, and } A = 2 - \log 50 = 0.301$$

If we increase the concentration to 2 g/l, the light transmitted becomes $0.5 \times 0.5 = (0.5)^2 = 0.25 = 25$ per cent T. At 3 g/l, transmittance $= (0.5)^3 = 0.125 = 12.5$ per cent T, and so on. It can be seen that as the concentration increases *linearly*, the per cent T de-

* Another, now obsolete term sometimes used for absorbance is optical density (OD). Note that absorbance is *not* 100 − per cent T.

creases *geometrically*. On the other hand, absorbance is *linearly* related (directly proportional) to the concentration (see Fig. 3–3, *A* and *C*):

Concentration (g/l)	*Per Cent* T	A	A/C
0	100	0.000	
1	50	0.301	0.301
2	25	0.602	0.301
3	12.5	0.903	0.301

Data can also be plotted using semilogarithmic paper as shown in Figure 3–3*B*. When per cent *T* is plotted on the log scale, a straight line with negative slope is obtained. Most instruments have scales which show per cent *T* from zero to 100, reading from left to right on a linear scale; and absorbance from zero to infinity, reading from right to left on a logarithmic scale.

Experience has shown that with most photometers the response of the detector to a signal of transmitted light is such that any uncertainty in *T* is constant over the entire *T* scale. The uncertainty derives from electrical and mechanical imperfections in the instrument and from individual variations in use of the instrument.

A fixed distance on the linear scale (e.g., 1 per cent *T*) represents a greater change in absorbance for low values of per cent *T* than for high values of per cent *T*. For this reason, the *absolute* concentration error or uncertainty is greater when taking readings at high absorbance. However, the *relative* concentration error is greater for readings at either low *or* high absorbances. An uncertainty of 1 per cent in readings on the linear *T* scale results in changes in absorbance (and in apparent concentration) as follows:

Per Cent T	*Absorbance*	*Absolute Error*	*Per Cent Relative Error*
10	1.000		
11	0.959	0.041	4.1
45	0.347		
46	0.337	0.010	2.9
90	0.046		
91	0.041	0.005	10.9
95	0.022		
96	0.018	0.004	18.2

The relative error is actually *minimal* at 36.8 per cent *T*, corresponding to an absorbance of 0.434. Consequently, methods should be designed such that readings fall near the center

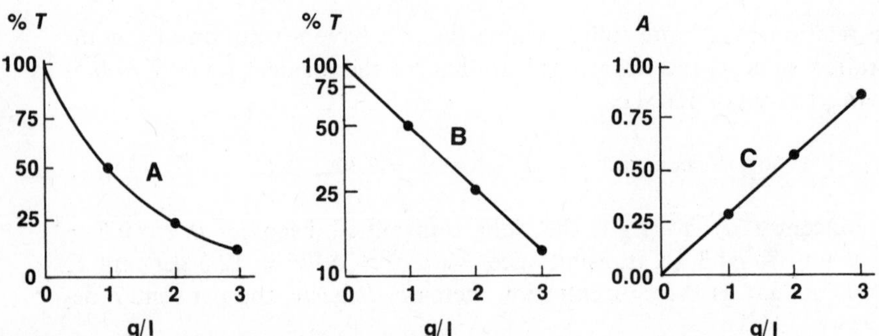

Figure 3–3 Transmittance and absorbance as a function of concentration. *A*, Per cent *T*, linear scale. *B*, Per cent *T*, logarithmic scale. *C*, Absorbance, linear scale.

of the scale, preferably between 20 and 80 per cent T (A between about 0.7 and 0.1). Judgment must be exercised as to the importance of absolute and relative errors in a particular determination.

When we hold the concentration constant, say at 1 g/l, and double the inside diameter of the cell, the effect on absorbance is the same as doubling the concentration, since we have introduced twice as many absorbing molecules in the light path. From this it follows that absorbance is also directly proportional to the light path through the cell. This relationship is often referred to as Bouguer's law or Lambert's law.

The overall equation relating these variables may be expressed as

$$A = abc \tag{1}$$

where A is the absorbance, a is a proportionality constant defined as absorptivity, b is the light path in cm, and c is the concentration of the absorbing compound. This equation is called *Beer's law* and forms the basis of quantitative analysis by absorption photometry. Absorbance values have no units; hence, the units for a are the reciprocal of those for b and c. Note also that since $A = 2 - \log$ per cent T, that $2 - \log$ per cent $T = abc$ and therefore log per cent $T = 2 - abc$.

When $b = 1$ cm and c is expressed in moles per liter, the symbol ε (epsilon) is substituted for the constant a. The value for ε is a constant for a given compound at a given wavelength under prescribed conditions of solvent, temperature, pH, and so forth, and is called the *molar absorptivity*. Values for ε are useful to characterize compounds, to establish their purity, and to compare sensitivities of measurements obtained on derivatives. Bilirubin, for example, when dissolved in chloroform at 25°C, should have a molar absorptivity of 60,700 ± 1600 at $A_{453\text{nm}}$. The molecular weight of bilirubin is 584. Hence, a solution containing 5 mg/l (0.005 g/l) should have an absorbance of

$$A = (60{,}700)(1)(0.005/584) = 0.520$$

Conversely, a solution of this concentration showing an absorbance of 0.490 could be assumed to have a purity of 0.490/0.520 or 94 per cent.

The molar absorptivity of the complex between ferrous iron and *s*-tripyridyltriazine is 22,600, while that with 1,10-phenanthroline is 11,000. This illustrates that, for a given concentration of iron, the former reagent produces a complex which absorbs about twice as much light energy as the complex with the latter reagent. Hence, it imparts greater sensitivity to the measurements of iron.

In toxicological work, it is customary to list constants based on concentrations in g/100 ml rather than in mol/l. This may also be necessary when the molecular weight of a substance is unknown. For $b = 1$ cm and $c = 1$ g/100 ml (1 per cent), $A = a$. This constant is often written as $A^{1\%}_{1\text{cm}}$, called the *absorption coefficient*, or as $E^{1\%}_{1\text{cm}}$, usually called the *extinction coefficient*. The latter symbol and term are now obsolete.

The direct proportionality between absorbance and concentration must be established experimentally for a given instrument under specified conditions. Frequently there is a linear relationship up to a certain concentration or absorbance. We then say that the solution obeys Beer's law up to this point. Within this limitation, a calibration constant or "factor" may be derived and used to calculate the concentration of an unknown solution by comparison to a standard. From Formula (1) ($A = abc$)

$$a = \frac{A}{bc} \tag{2}$$

Therefore

$$\frac{A_1}{b_1c_1} = \frac{A_2}{b_2c_2} \tag{3}$$

The light path (b) remains constant in a given method of analysis with a fixed cuvet size, and $b_1 = b_2$. Formula (3) then becomes

$$\frac{A_1}{c_1} = \frac{A_2}{c_2} \quad \text{or} \quad \frac{A_s}{c_s} = \frac{A_u}{c_u} \tag{4}$$

where s and u represent standard and unknown, respectively. Solving for the concentration of unknown, we obtain

$$c_u = \frac{A_u}{A_s} \times c_s \tag{5}$$

or the equivalent expression

$$c_u = A_u \times \frac{c_s}{A_s} = A_u \times F \tag{6}$$

where $F = c_s/A_s$. The value of the factor F is obtained by measuring the absorbance (A_s) of a standard of known concentration (c_s).

Certain precautions must be observed in using such factors or calibration constants. Under no circumstances should the factor be used when either the standard or unknown readings exceed the linear portion of the calibration curve, i.e., if the readings no longer obey Beer's law. Whenever possible, two or more standards should be included in each series of determinations to permit direct comparison of unknown to standard, or to calculate the calibration constant, since variations in reagents, working conditions, cell diameters, deterioration or change in instruments, and so on, may result in day to day changes of the absorbance value for the standard. A nonlinear calibration curve may be used, of course, if a sufficient number of standards of varying concentration are included to cover the entire range encountered for readings on unknowns.

In some cases, pure standard may not be readily available and constants may be provided that were obtained on pure standards by others and reported in the literature (see section on porphyrins, p. 463). In general, the use of published constants should be discouraged unless the method is followed in detail and readings are made on a spectrophotometer capable of providing light of high spectral purity at a verified wavelength. Use of broader band light sources usually leads to some decrease in absorbance. The absorbance of NADH at 340 nm, for example, is frequently used as a reference for the determination of enzyme activity, based on an assigned molar absorptivity of 6.22×10^3. Again, this value is acceptable only under the carefully controlled conditions just described and should not be used unless these conditions are met. In summary, published values for molar absorptivities and absorption coefficients should be used only as guidelines until they can be verified by readings on pure standards for a given piece of equipment.

ABSORPTION SPECTROPHOTOMETRY

The major components of a simple photometer are shown schematically in Figure 3–4. Light from the lamp is passed through a filter, or monochromator, to provide selection of the desired region of the spectrum to be used for measurements. A slit is used to isolate a narrow beam of the light source and improve its chromatic purity. The light next passes through an absorption cell (cuvet), where a portion of the radiant energy is absorbed, depending on the nature and concentration of the solution. Any light not absorbed is trans-

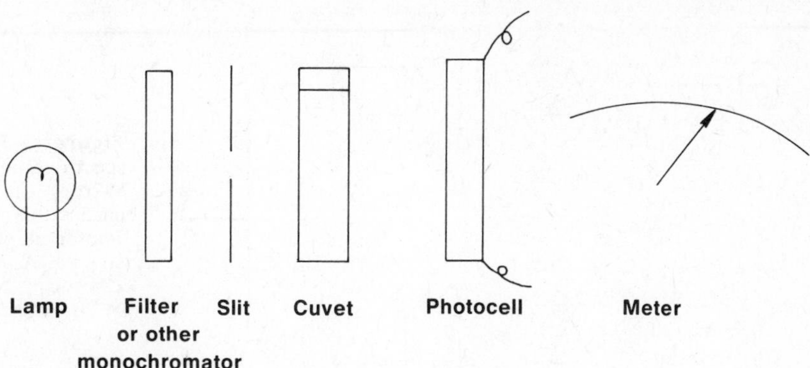

Figure 3–4 Major components of a simple photometer.

mitted to a photocell or phototube which converts light energy to electrical energy that can be registered on the meter or recorder.

In operation, an opaque block is substituted for the cuvet, so that no light reaches the photocell, and the meter is adjusted to read 0 per cent *T*. Next, a cuvet containing a reagent blank is inserted and the meter is adjusted to read 100 per cent *T*. The composition of the reagent blank should be identical to that of standard or unknown solutions except for the substance to be measured. Standard solutions containing various known concentrations of the substance are inserted and scale readings are recorded. Finally, a reading is made on the unknown solution and its concentration is determined by interpolation between or by comparison to the readings obtained on the standards.

Since numerous modifications of this simple approach have been made, component parts will next be considered in more detail.

The light source

The light source for measurements in the visible portion of the spectrum is usually a tungsten light bulb. The tungsten lamp is acceptable for making measurements of moderately dilute solutions in which the change in color intensity varies significantly with small changes in concentration. A common disadvantage with some photometers is that a considerable amount of heat is generated by the light source; this may cause problems in measurement either by changing the geometry of the optical system or by changing the sensitivity of the photocell.

At the temperature of the tungsten filament, some of the metal vaporizes and condenses on the cooler glass surface of the bulb, thus reducing the intensity of radiant energy. Such coatings may also change the spectrum sufficiently to alter instrument response. Similar changes may be brought about by deposits of chemical fumes or oily particles on the outside of the bulb. These coatings are often uneven and are particularly likely to cause trouble in instruments (such as the AutoAnalyzer colorimeter) that utilize two light beams and two photocells at right angles to the lamp. Periodic lamp inspection and replacement is recommended to avoid these difficulties.

One such type of double beam system is shown schematically in Figure 3–5, in which all components are duplicated except the light source. Another approach is to use a light beam chopper (a rotating wheel with alternate silvered sections and cut out sections) inserted after the exit slit. A system of mirrors passes the reflected portions of the light off the chopper alternately through the sample and a reference cuvet onto a common detector. Just as a single beam instrument is adjusted to 100 per cent *T* with the blank before and between sample readings, the double beam system makes these adjustments automatically.

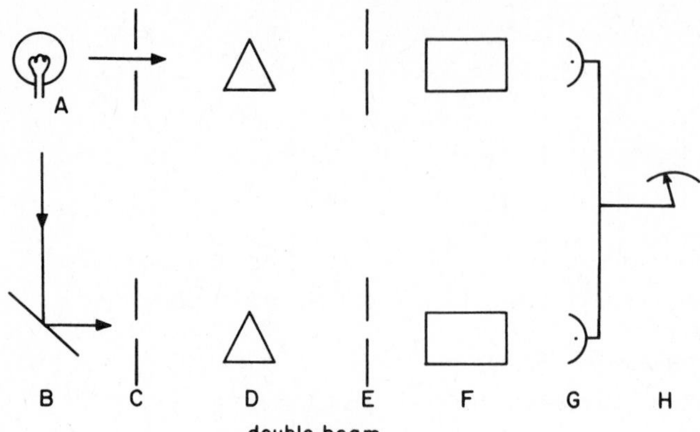

Figure 3–5 Double beam in space spectrophotometer. *A*, Light source. *B*, Mirror. *C*, Entrance slits. *D*, Monochromators. *E*, Exit slits. *F*, Cuvets. *G*, Photomultipliers. *H*, Meter. From Evenson, M. A. *In:* Fundamentals of Clinical Chemistry. N. W. Tietz, ed. Philadelphia, W. B. Saunders Company, 1970.

The chopped beam approach, using one detector, compensates for light source variation as well as for sensitivity changes of the detector.

A tungsten light source does not supply sufficient radiant energy for measurements below 320 nm. In the ultraviolet region, a low pressure mercury vapor lamp, which emits a discontinuous or line spectrum, is useful for calibration purposes but is not very practical for absorbance measurements. Hydrogen and deuterium lamps provide sources of continuous spectra in the ultraviolet region with some sharp emission lines, as do high pressure mercury and xenon arc lamps. These sources are more commonly used in ultraviolet absorption measurements. A deuterium lamp is more stable and has a longer life than a hydrogen lamp.

Electric current is the energy source for the lamp. Central power stations may vary in their ability to maintain a constant voltage in the lines; in addition, emergency generators may be put into use in case of power failures. Variations in light intensity from the lamp will result in unstable readings on the meter; consequently, suitable voltage regulation devices are essential for optimum stability. Transformers may also be used to reduce the line voltage for some instruments. A storage battery provides a stable source of current but presents maintenance problems and is susceptible to a slow drift with time.

Spectral isolation

A system for isolating radiant energy of a desired wavelength and excluding that of other wavelengths is called a monochromator. There are various ways of accomplishing this, including the use of filters, prisms, and diffraction gratings. Combinations of lenses and slits may be inserted before or after the monochromatic device to render light rays parallel or to isolate narrow portions of the light beam. Variable slits may be used to permit adjustments in total radiant energy reaching the photocell.

The simplest type of filter is a thin layer of colored glass. Certain metal complexes or salts, dissolved or suspended in glass, produce colors corresponding to the predominant wavelengths transmitted. Strictly speaking, a glass filter is not a true monochromator since it transmits light over a relatively wide range of wavelengths. The spectral purity of a filter or other monochromator may be described in terms of half bandwidth. This is defined as the width, in nm, of the spectral transmittance curve at a point equal to one-half the peak transmittance (Fig. 3–6). Commonly used glass filters have half bandwidths of approximately 50 nm and are entirely adequate for many purposes. These are referred to as wide bandpass filters.

Other glass filters include the narrow bandpass and sharp-cutoff types (Fig. 3–6). As implied, the latter filter typically shows a sharp rise in transmittance over a narrow portion

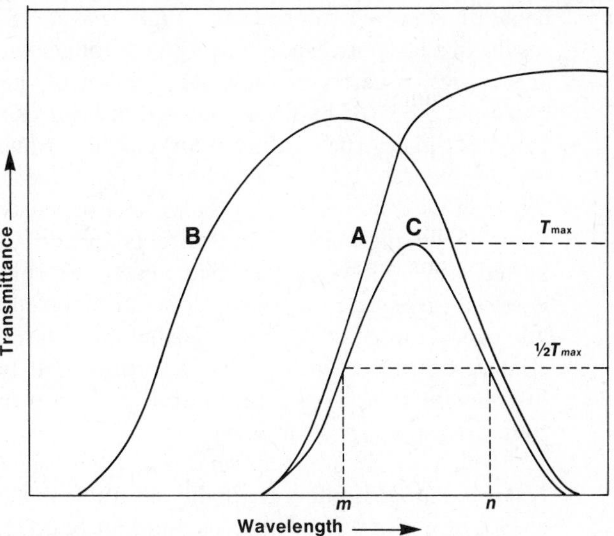

Figure 3–6 Spectral characteristics of a sharp-cutoff filter (*A*) and a wide bandpass filter (B). The narrow bandpass filter (*C*) is obtained by combining filters *A* and *B*. The half bandwidth of filter *C* (distance $m - n$) is defined as the width of the spectral-transmittance curve at a point equal to one-half of maximum transmittance.

of the spectrum and is used to eliminate light below a given wavelength. The narrow bandpass filters are usually constructed by combining two or more sharp-cutoff filters or regular filters.

Another approach for construction of narrow bandpass filters is to use a dielectric material of controlled thickness sandwiched between two thin-silvered pieces of glass. The thickness of the layer determines the wavelength of energy transmitted. Energy of wavelengths that are multiples of this thickness stay in phase as they reflect back and forth through the dielectric and finally emerge, whereas other wavelengths will cancel due to phase differences. Light striking the semitransparent film at a given angle θ will be partly reflected and partly passed (Fig. 3–7, point a). The same process occurs at a′, b, b′, and so on. For reinforcement to occur at point b, the distance traveled by the beam reflected at a′ must be some multiple of its wavelength in the medium. For purposes of clarity, the incident beam is shown as arriving at an angle θ from the perpendicular. In ordinary use, however, θ approaches zero and $n\lambda' = 2t$, where λ' is the wavelength of radiation in the dielectric material, t is its thickness, and n is an integer called the order of interference. These filters have narrow half bandwidths, usually from 5 to 10 nm, and are referred to as *interference filters*. Since they also transmit harmonics, or multiples, of the desired wavelength, accessory glass filters are required to eliminate these undesired wavelengths. Thus, an interference filter designed for 620 nm will also transmit some radiation at 310 and 1240 nm. The AutoAnalyzer is an example of an instrument equipped with interference filters.

Prisms and diffraction gratings are also widely used as monochromators. A *prism*

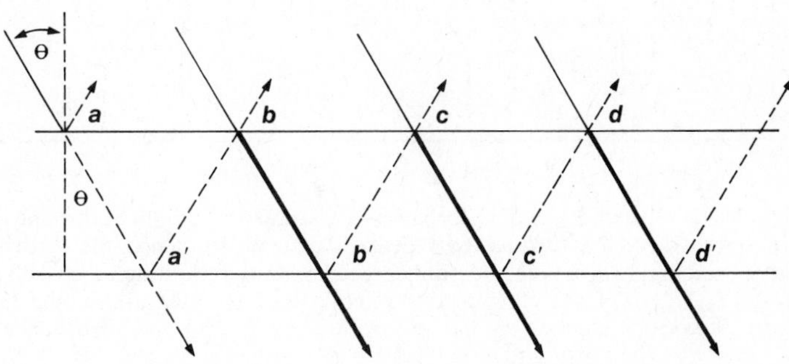

Figure 3–7 Passage of light through an interference filter. The heavier lines indicate reinforcement of entering light rays.

separates white light into a continuous spectrum by refraction, i.e., shorter wavelengths are bent, or refracted, more than longer wavelengths as they pass through the prism. This results in a nonlinear spectrum with the longer wavelengths closer together, but with suitable accessories, a narrow bandwidth portion of the spectrum may be isolated. A *diffraction grating* is prepared by depositing a thin layer of aluminum-copper alloy on the surface of a flat glass plate, then ruling many small parallel grooves into the metal coating. Better gratings contain 1000 to 2000 lines per millimeter and must be made with great care. These are then used as molds to prepare less expensive replicas for general use in instruments. Rays of radiant energy bend (refract) around a sharp corner, and the extent of refraction varies with the wavelength. Thus, each line ruled on the grating, when illuminated, gives rise to a tiny spectrum. Wave fronts are formed which reinforce those wavelenths in phase and cancel those not in phase. The net result is a uniform linear spectrum. Some instruments contain diffraction gratings that produce half bandwidths of 20 to 35 nm; higher priced instruments may have a resolution of 0.5 nm or less. Good gratings are generally better than prisms for spectral isolation.

The type of monochromator chosen will depend on the objectives of the analyst. Narrow bandpass filters are highly desirable in flame photometry in order to isolate emission energy at a given wavelength. Small half bandwidths are required in spectrophotometers if one is interested in resolving and identifying sharp absorption peaks that are closely adjacent. Lack of agreement with Beer's law will occur when a part of the spectral energy transmitted by the monochromator is not absorbed at all by the substance being measured. This is more commonly observed with the wide bandpass instruments.

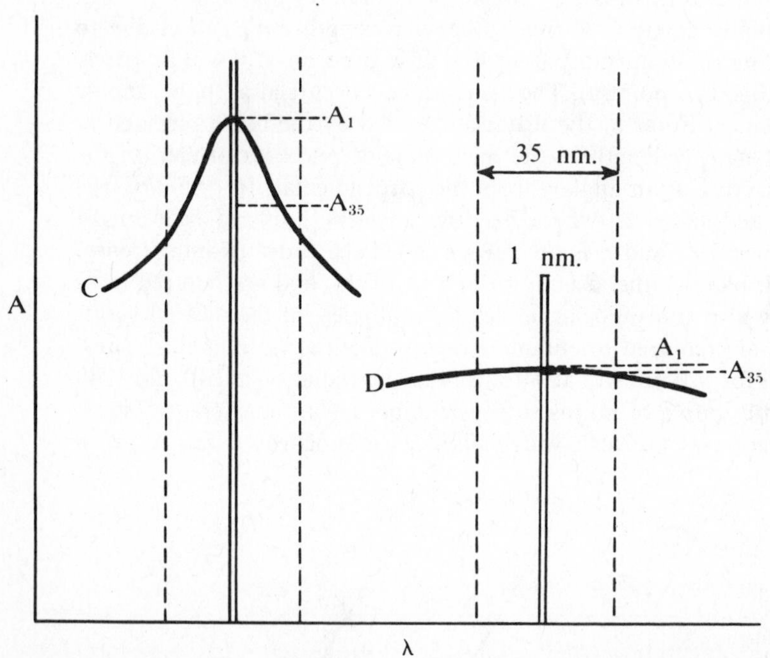

Figure 3–8 Dependence of absorbance on bandwidth of light and shape of absorption curve. Curve *C* exhibits a sharp peak of absorption: hence, the absorbance (A_1) with a 1 nm bandwidth is considerably greater than that obtained with a 35 nm bandwidth (A_{35}). (The A_{35} reading corresponds to the integrated area under the curve.) Curve *D* exhibits a relatively flat absorption curve, and the absorbance readings are essentially the same with either a 1 nm or 35 nm bandwith.

Some increase in absorbance as well as improved linearity with concentration are usually observed with instruments that operate at narrower bandwidths of light. This is especially true for substances that exhibit a sharp peak of absorption. In Figure 3–8, curves are shown for two solutions, C and D, with markedly different spectral absorbance curves. It is apparent that solution C will show a much higher absorbance in a spectrophotometer that supplies monochromatic light with a 1 nm bandwidth, as compared to an instrument with a 35 nm bandwidth. On the other hand, solution D would have essentially the same absorbance with either instrument.

The wavelength selected is usually at the peak of maximum absorbance in order to achieve maximum sensitivity; however, it may be desirable to choose another wavelength to minimize interfering substances. For example, turbidity readings on a spectrophotometer are greater in the blue region than in the red region of the spectrum, but the latter region is chosen for turbidity measurements to avoid absorption of light by bilirubin (460 nm) or hemoglobin (417 and 575 nm). The color developed in the alkaline picrate procedure for creatinine produces a relatively flat peak in the visible region at approximately 480 nm, but the reagent blank itself absorbs light strongly below 500 nm. A compromise is made by selecting a wavelength at 520 nm to minimize the contribution of the blank. Blank readings should, of course, be kept to a minimum. A small difference between two large numbers is subject to greater uncertainty; hence, minimizing absorbance of the blank improves precision and accuracy. The linear working range of a method can be expanded also by not measuring at the peak absorbance. This approach is used in some methods for glucose determinations. Measurements should preferably not be taken on the steep slope of an absorption curve, since a slight error in wavelength adjustment would introduce a significant error in absorbance readings.

Multiple wavelength readings

In addition to including blanks, background interference can often be eliminated or minimized by reading absorbance at two or three wavelengths. In one approach, currently used on the DuPont ACA system, absorbance is measured at two wavelengths, one corresponding to peak absorbance and another at a point near the base of the peak to serve as a baseline. The difference in absorbance at the two wavelengths is related to concentration. In effect, this provides a blank reference point for each individual specimen. Another method to correct for background interference is to measure absorbance at the peak wavelength and at two other wavelengths equidistant from the peak. Values for the latter are averaged to obtain a baseline under the peak, which is then subtracted from the peak reading. The value thus obtained is known as a "corrected" absorbance and can be related to the concentration, provided that the background absorbance is linear with wavelength over the region in which readings are made. This technique of making corrections for interfering substances is called the *Allen correction* and is illustrated in Figure 3–9. The corrected absorbance at 300 nm is obtained from the Allen equation:

$$A_{\text{corr}} = A_{300} - \frac{(A_{280} + A_{320})}{2}$$

Similar corrections are applied in procedures for spectrophotometric determinations of salicylates, porphyrins, steroids, and other compounds. The correction must be applied similarly to standards, since the Allen corrected absorbance is typically less than the total absorbance, even in the absence of interfering background.

Before using the Allen correction, knowledge of the shape of the absorption curve for the substance of interest and of the interference must be obtained. The linearity of the baseline shift should be verified by measuring the absorption spectrum of commonly encountered interferences. Care should be exercised in the use of the Allen correction because if it is not

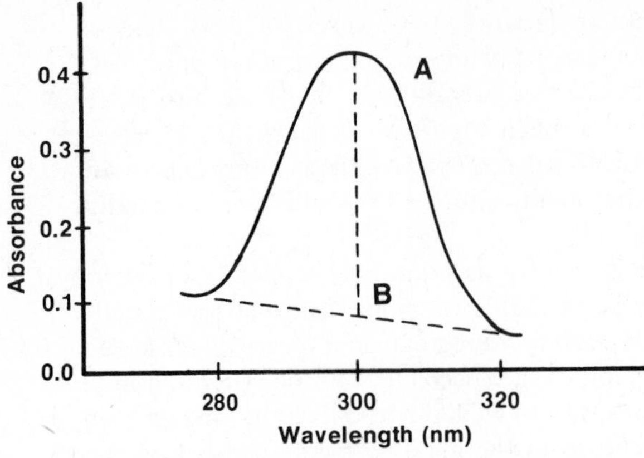

Figure 3–9 Example of an Allen correction. *A* represents the absorption curve of the test mixture. Absorbance readings are taken at the absorption peak (300 nm) and at equidistant points from the peak (280 and 320 nm). *B* represents the background absorbance, which is linear over the measured range. From Evenson, M. A. *In:* Fundamentals of Clinical Chemistry. N. W. Tietz, ed. Philadelphia, W. B. Saunders Company, 1970.

properly used, it may introduce larger errors than would be observed without correction. For example, such a situation may occur if the background reading is not linear over the region measured.

Wavelength calibration

For many analytical purposes, the wavelength chosen may be approximate, provided it is reproducible. Most filters fall into this category and are quite satisfactory since unknowns are compared to standards at a fixed wavelength and half bandwidth. With prisms and diffraction gratings, however, a continuous choice of wavelengths is available, and it becomes necessary to verify their accuracy and reproducibility. Knowledge of exact wavelength becomes critical when using published molar absorptivities, for identification of substances in toxicological studies, and in the use of differential absorption techniques. Enzyme assays employing the NAD-NADH reaction, for example, are based on a molar absorptivity constant for NADH of 6.22×10^3 at 340 nm. This implies that the wavelength setting must be accurate and reproducible and the instrument must show spectrophotometric accuracy if this constant is to be used in calculating results.

For the narrow half bandwidth instruments, a holmium oxide glass may be scanned over the range of 280 to 650 nm. This material shows very sharp absorbance peaks at well-defined wavelengths, and the operator may compare the wavelength scale readings which produce maximum absorbance with established values. Should these not coincide, a calibration curve can be constructed to relate scale readings to true wavelengths.[1] A typical spectral-transmittance curve for holmium oxide glass is shown in Figure 3–10. Selected absorption peaks for this filter, suitable for calibration purposes, occur at the following wavelengths (nm):

279.3	418.5
287.6	536.4
333.8	637.5
360.8	

With broader bandpass instruments, such as the Coleman Jr. II or 6A, a didymium filter is used to verify wavelength settings. This filter is specified to read a given per cent transmittance (e.g., 45 per cent *T*) against an air blank at 610 nm (Fig. 3–11). At this wavelength, per cent *T* changes sharply with change in wavelength. A simple adjustment is made to correct the wavelength setting and provide uniformity of operation. Such checks should be made periodically and are mandatory after replacement of the exciter lamp. Note that it is possible to obtain the correct per cent *T* reading also at a wavelength setting near 570 nm.

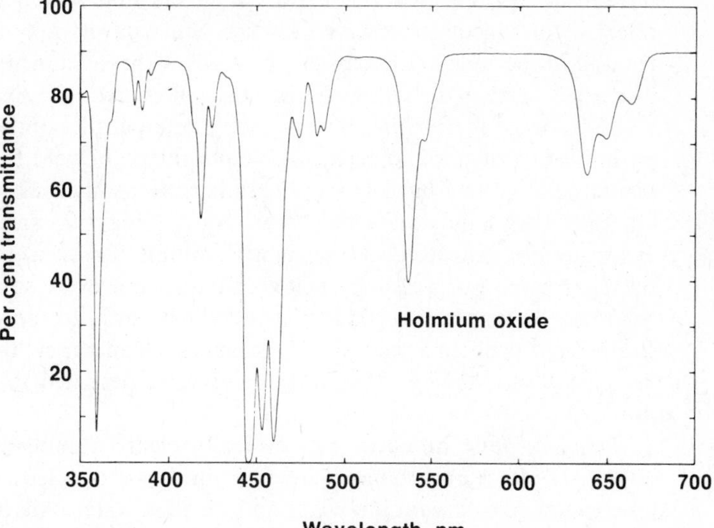

Figure 3-10 Spectral-transmittance curve of holmium oxide filter. (Courtesy of Beckman Instruments, Inc.)

Visual inspection of the color of the light beam can prevent such errors. The light at 610 nm appears orange; that near 570 nm appears greenish-yellow. A second reading of the didymium filter near 590 nm (point of minimal transmittance) is also of value. Increases in per cent T with time are observed with aging or darkening of bulbs, an increase in stray light, or sagging of bulb filaments. Significant changes in this reading should alert the user to replace the exciter lamp.[3]

Cuvets

Cuvets, also known as absorption cells, may be round, square, or rectangular, and may be constructed from glass, silica (quartz), or plastic. (Depending on the material used, the amount of light transmitted by the cell may differ significantly.) For routine colorimetric work, the round, glass, test tube type is most common. These are inexpensive and satisfactory, provided they are matched, i.e., to be used interchangeably, the cuvets must be of uniform inside diameters so that the absorbance of a solution will be within specified tolerances when measured in different cuvets. This criterion is readily checked by preparing a stable solution, such as cyanmethemoglobin, to read approximately 50 per cent T

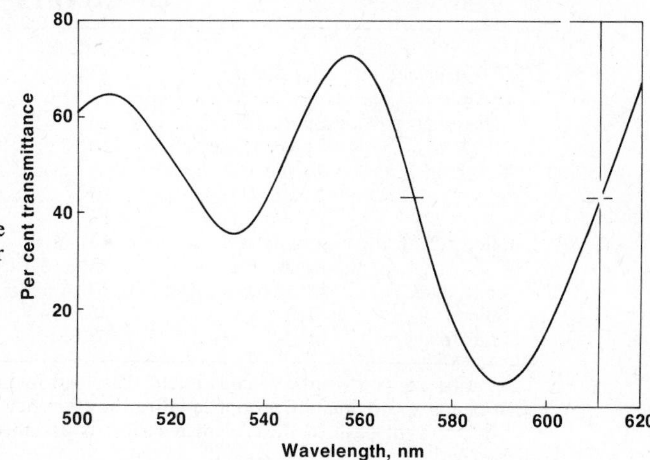

Figure 3-11 Spectral-transmittance curve of didymium filter read on a Coleman Junior II spectrophotometer, Model 6/20.

at 540 nm against a reference blank. This solution is then added to each dry cuvet to be tested and readings are taken. Those cuvets that match within 0.5 per cent *T* (or other selected tolerance) are reserved for use. Since cuvets may not be perfectly round, they are rotated in the well to observe any changes with position; those showing changes are either discarded or etched to indicate the position of use.

Square or rectangular cuvets have plane-parallel optical surfaces and a constant light path. The most popular have a 1.0 cm light path, held to close tolerances. Compared to round cuvets, they have few optical aberrations such as lens effect and refractive errors. On the other hand, these cuvets are more expensive and are less convenient to handle. Ordinary borosilicate glass cuvets are suitable for measurements in the visible portion of the spectrum. For readings below 340 nm, however, silica absorption cells are usually required. Some plastic cells have good clarity in both the visible and ultraviolet range but can present problems relating to tolerances, cleaning, etching by solvents, and temperature deformations. Many of the plastic cuvets are designed for disposable, single use applications.

Cuvets must be clean and optically clear. Etching or deposits on the surface will obviously affect absorbance values. Round cuvets, used in the visible range, are cleaned by copious rinsing with tap water and distilled water. Alkaline solutions should not be left standing in cuvets for prolonged periods. Both round and square cuvets may be cleaned in mild detergent or soaked in a mixture of concentrated HCl-water-ethanol (1:3:4). Cuvets should never be soaked in dichromate cleaning solution since the solution tends to adsorb onto and discolor the glass. Cuvets used for measurements in the ultraviolet region should be handled with special care. Invisible scratches, fingerprints, or residual traces of previously measured substances may be present and absorb significantly. A good practice is to fill all such cuvets with distilled water and measure the absorbance for each against a reference blank over the wavelengths to be used. This should be essentially zero. Startling and disturbing inconsistencies will be avoided if this procedure is practiced routinely.

Correctness of photometric measurements for narrow bandwidth instruments, using square or rectangular cuvets, may be verified as described by Rand.[2] A solution is prepared which contains exactly 0.0500 g of high purity potassium dichromate dissolved in, and diluted to 1000 ml with, 0.005 molar sulfuric acid. The absorptivity of this solution, determined at 350 nm, should be 10.70 liter $g^{-1}cm^{-1}$, with an uncertainty of less than 0.3 per

TABLE 3–2 EFFECTIVE LIGHT PATH AND MINIMUM VOLUME OF CUVETS

Instrument	*Cuvet*	*Light Path (mm)*	*Minimum Volume (ml)*	*Efficiency* (mm/ml)*
Beckman	Standard	10	3.0	3.3
	Lowry-Bessey	10	0.1	100.0
Coleman Jr.†	10 mm *OD*	8	1.0	8.0
	12 mm *OD*	10	1.5	6.7
	19 mm *OD*	17	5.5	3.1
Klett, Clinical	Standard	12	5.0	2.4
	Micro	12	2.0	6.0
Leitz, B & L	Standard	10	3.0	3.3
Spinco	Micro	6.4	0.1	64.0
Gilford	Micro	10	0.5	20.0

* For a given weight of constituent dissolved in the minimum volume shown, the absorbance is approximately proportional to the efficiency.

† The light path in the Coleman cuvets is assumed to be 2 mm less than the outside diameter.

cent. Since $A = abc$, the absorbance of the solution in a 1.0 cm cell should be

$$A = (10.70)(1.00)(0.0500) = 0.535$$

The solution should also show an absorbance peak at 350 nm.

Glass filters, made from a moderately neutral glass, may also be used to check photometric accuracy. These are available from the National Bureau of Standards (Schott NG-4 SRM 930) and are calibrated on the NBS high accuracy spectrophotometer at approximately 10, 20, and 30 per cent transmittance levels. A certificate is supplied which describes the filters and gives the calibration values at four wavelengths ranging from 440 to 635 nm.

In microchemical procedures, one is often concerned with minimal volumes of sample and suboptimal absorbance values. According to Beer's law [Formula (1)], absorbance can be increased by increasing the light path (*b*) or by increasing the concentration (*c*). For a limited sample, the latter may be achieved by decreasing the final volume of solution required for absorbance measurements in the spectrophotometer. Thus, a measure of efficiency may be defined for any cuvet by relating maximum light path with minimal volume required, i.e., mm/ml. Table 3–2 shows typical values for some instruments currently in use. Other factors, such as ease of manipulation, enter into the choice of a particular system.

Detectors

The two most commonly used devices for measuring light intensity in the UV and visible regions of the spectrum are barrier layer cells and photomultiplier tubes. The barrier layer cells are rugged and are used in inexpensive instruments; photomultipliers are almost always used in the higher quality, more expensive spectrophotometers.

Barrier Layer Cells. These cells operate on the principle that when light falls on certain metals or semiconductors, electrons will flow in proportion to the intensity of the light. The barrier layer cell consists of a thin layer of silver on a layer of the semiconductor selenium. The silver and selenium metals are then mounted on an iron backing or support. The iron backing is deficient in electrons and is therefore the positive electrode. The silver mounted on top of the selenium where the light will be incident is the negatively charged electrode. When light falls on the thin semitranslucent silver metal, electron flow from the semiconductor selenium into the iron backing occurs, but not in the reverse direction. The electron flow is measured and is directly proportional to the light intensity incident on the photocell.

The sensitivity of the barrier layer cell to light of different wavelengths is similar to the human eye. The maximum sensitivity of both occurs at 550 nm. Barrier layer cells are usually used at high levels of illumination, and the output from these photocells is generally not amplified. Barrier layer cells are very stable, but are slow in responding to changes of light intensity. Because of their slow response time, they are not suitable as detectors in instruments that employ interrupted (chopped) light beams falling on the detectors. Another disadvantage of this photocell is that it tends to show fatigue. Fatigue occurs in a barrier layer cell when, at a constant, continuous high level of intensity, the electrical output of the photocell decreases with time. Therefore, barrier layer cells should not be used at extremely high illumination.

The Coleman II spectrophotometer and the Technicon AutoAnalyzer interference filter colorimeter contain photocells of this type. The cells are very rugged, last for years, and perform well as inexpensive detectors. A potential problem with barrier layer cells is that their electrical output is very temperature-dependent. If a heat producing, high intensity lamp or flame is used for the light source, instrument design must be such that thermal stability and rapid temperature equilibrium of the photocell are achieved. This is accomplished in the AutoAnalyzer colorimeter by having the light source far removed from the photocell. Use of heat shields or plastic materials that do not conduct heat readily are other ways of improving thermal stability of instruments.

Photomultiplier Tubes. A photomultiplier is an electron tube that is capable of significantly amplifying a current. It is constructed by using a light-sensitive metal as the cathode that is capable of absorbing light, and emitting electrons in proportion to the radiant energy that strikes the surface of the light-sensitive material. The electrons produced by this first stage go to a secondary stage (surface) where each electron produces between four and six additional electrons. Each of these electrons from the second stage go on to another stage, again producing four to six electrons. Each electron produced cascades through the photomultiplier stages; thus, the final current produced by such a tube may be one million times as much as the initial current. As many as 10 to 15 stages or dynodes are present in common photomultipliers.

When operating such a tube, voltage is applied between the photocathode and each successive stage. The normal increment of voltage increase of each photomultiplier stage is from 50 to 100 volts larger than that of the previous stage. A common photomultiplier tube will have in the neighborhood of 1500 volts applied to it.

Photomultiplier tubes have extremely rapid response times, are very sensitive, and do not show as much fatigue as other detectors. Because of their excellent sensitivity and rapid response, all stray light and daylight must be carefully shielded from the photomultiplier. A photomultiplier with the voltage applied should never be exposed to room light because it will burn out. Because of the fast response time of the photomultiplier, this detector is applicable to interrupted light beams, such as those produced by choppers, and this provides significant advantages when used as a UV-visible detector in spectrophotometers. The rapid response times are also needed when a spectrophotometer is being used to determine an absorption spectrum of a compound. The photomultiplier also has adequate sensitivity over a wider wavelength range than photocell detectors.

When voltage is applied to photomultipliers and all light has been blocked from them, some current will usually be produced. This current is called dark current. It is desirable to have the dark current of photomultipliers at their lowest level, since it would also be amplified and would appear as background noise.

Readout devices

Electrical energy from a detector is displayed on some type of meter or readout system. These may be direct reading or null point systems. In direct reading systems, such as the Coleman Jr. or Leitz photometer, the output of the photocell is used to drive a sensitive meter directly with no further amplification. Other instruments, such as the Spectronic 20, utilize an amplifier to increase the output of the detector. In the null point system, the output of the detector is balanced against the output of a reference circuit. This may be adjusted manually, as in the Klett photoelectric colorimeter or the Beckman DU spectrophotometer. The meter may be replaced by a servometer activated by an imbalance of current which stops when the two circuits are balanced. Direct digital readouts are thus obtained. An example of the latter is the Instrumentation Laboratory Model 143 flame photometer.

Also popular are digital readout devices that provide a visual numerical display of absorbance or converted values of concentration. These operate on the principle of selective illumination of portions of a bank of light-emitting tubes or diodes, controlled by the voltage signal generated. Typical examples include the NIXIE tube (Burroughs Corp.), the Digivac tube (Motorola Semiconductor Products) and visible-light-emitting diodes (LEDs). The latter incorporate gallium as the major component. At present, $GaAs_xP_x$ diodes that emit red light are most widely used, but newer materials are being developed that emit green, yellow, or blue light. Compared to meters, the digital readout devices have faster response, are easier to read, and decrease operator fatigue.

Recorders

Certain spectrophotometers are equipped with recorders in addition to or instead of the digital display of values. These are synchronized to provide line traces of transmittance or absorbance, either as a function of time or wavelength. Change of absorbance with time is widely used to measure enzyme activity. Thus, in the kinetic determination of lactate dehydrogenase, the rate of disappearance of NADH can be monitored at 340 nm as pyruvate is converted to lactate. In the AutoAnalyzer, line tracings of transmittance are produced sequentially as standards and samples pass through the flowcell. When a continuous tracing of absorbance versus wavelength is recorded, the resultant figure is called an absorption spectrum. This type of procedure is especially useful for identification of drugs that absorb in the ultraviolet. Several criteria are used, including determination of those wavelengths showing maximum and minimum absorbance in both dilute acid and alkaline solutions; absorptivity at the wavelength of maximum absorbance; and ratios of absorbance at two wavelengths. Finally, the entire spectrum is compared to that of a known sample of the suspected drug. This sort of information is available in textbooks of toxicology.

FLAME PHOTOMETRY

Flame emission photometry is most commonly used for the quantitative measurement of sodium and potassium in body fluids. Lithium, while not normally present in serum, may also be measured in connection with the therapeutic use of lithium salts in the treatment of some psychiatric disorders.

Atoms of many metallic elements, when given sufficient energy such as that supplied by a hot flame, will emit this energy at wavelengths characteristic for the element. A specific amount or quantum of thermal energy is absorbed by an orbital electron. The electrons, being unstable in this high energy (excited) state, release their excess energy as photons of a particular wavelength as they change from the excited to their previous or ground state. If the energy is dissipated as light, the light may consist of one or more than one energy level and therefore of different wavelengths. These line spectra are characteristic for each element. Sodium, for example, emits energy primarily at 589 nm, along with other, much less intense emissions (Fig. 3–12). The wavelength to be used for the measurement of an element depends upon the selection of a line of sufficient intensity to provide adequate sensitivity as well as freedom from other interfering lines at or near the selected wavelength.

Alkali metals are comparatively easy to excite in the flame of an ordinary laboratory

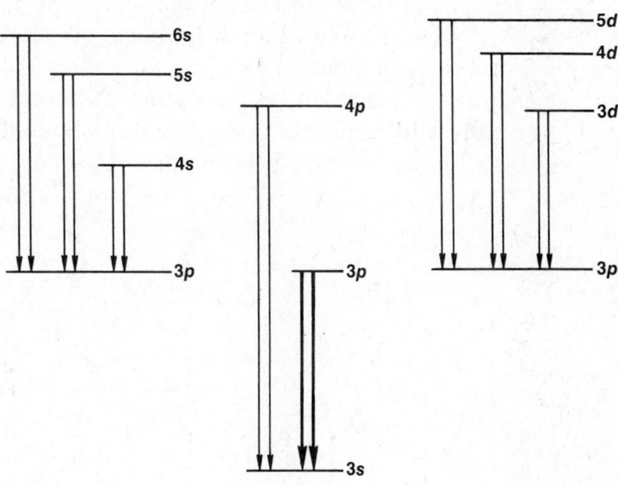

Figure 3–12 Schematic diagram showing energy levels for certain lines of the sodium spectrum. The major doublet at 589 nm (shown in heavy lines) results when the excited valence electron returns from the 3*p* orbital to the ground state 3*s* orbital.

burner. Lithium produces a red, sodium a yellow, potassium a violet, rubidium a red, and magnesium a blue color in a flame. These colors are characteristic of the metal atoms that are present as cations in solution. Under constant and controlled conditions, the light intensity of the characteristic wavelength produced by each of the atoms is directly proportional to the number of atoms that are emitting energy, which in turn is directly proportional to the concentration of the substance of interest in the sample. Thus, flame photometry lends itself well to direct concentration measurements of some metals.

Other cations, such as calcium, are less easily excited in the ordinary flame. In these cases, the amount of light given off may not always provide adequate sensitivity for analysis by flame emission methods. The sensitivity can be improved slightly by using higher temperature flames.

Of the more easily excited alkali metals like sodium, only 1 to 5 per cent of those atoms present in solution become excited in a flame. Even with this small percentage of excited atoms, the method has adequate sensitivity for measurement of alkali metals for most bioanalytical applications. Most other metal ions are not as easily excited in a flame, and flame emission methods are not as applicable to their measurement.

Components of a flame photometer

Figure 3–13 shows a schematic diagram of the basic parts of a flame photometer. A tank of gas is required and a two-stage pressure regulator. High pressure tubing must be used to lead the gases to the flame. An atomizer is needed to spray the sample as fine droplets into the flame. The monochromator, entrance and exit slits, and detectors are similar to those discussed previously for spectrophotometers. In effect, the light source for the spectrophotometer has been replaced with an atomizer-flame combination, and one is measuring emission of light rather than absorption.

Various combinations of gases and oxidants have been proposed and are being used in flame photometry. These include acetylene and oxygen for the hottest flame, and natural gas, acetylene, and propane in combination with either oxygen or compressed air. The choice of flame depends largely on the temperature desired; for sodium and potassium determinations, a propane-compressed air flame appears entirely adequate. Typical flame temperatures are shown in Table 3–3.

The atomizer and the flame are critical components in a flame photometer. The atomizer provides a means of drawing the sample through the aspirator and converting it into a fine mist which then enters the flame. This can be done by passing a gas of high velocity over the upper outlet of a capillary tube, the lower end of which is inserted into the sample. Liquid is then drawn up into a chamber and dispersed into small droplets. The larger droplets settle to the bottom and go to waste. The most important variable in the flame itself is the temperature. Frequent standardization of flame photometers is essential because thermal changes do occur and affect the response of the instrument. In addition, temperature changes affect the output of photocell detectors; for this reason, a period of warm-up, with aspiration of water and standards, is required to establish thermal equilibrium for the flame and the atomizer chamber before measurements are taken.

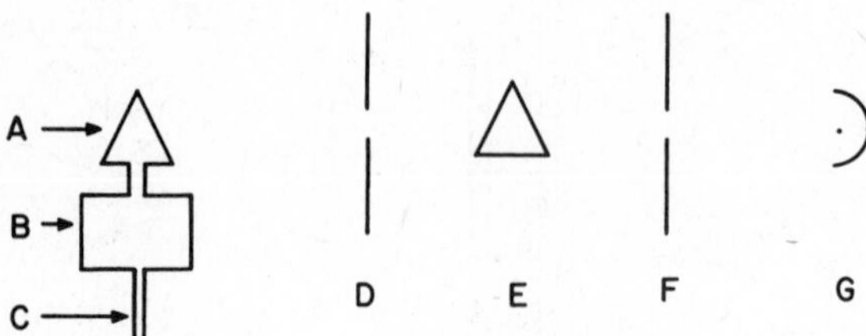

Figure 3–13 Essentials of a flame photometer. *A*, Flame. *B*, Atomizer. *C*, Aspirator. *D*, Entrance slit. *E*, Monochromator. *F*, Exit slit. *G*, Detector. From Evenson, M. A. *In:* Fundamentals of Clinical Chemistry. N. W. Tietz, ed. Philadelphia, W. B. Saunders Company, 1970.

TABLE 3–3 FLAME TEMPERATURES FOR VARIOUS GAS MIXTURES

Gas Mixture	*Flame Temperature, °C*
Natural gas-air	1840
Propane-air	1925
Hydrogen-air	2115
Acetylene-air	2250
Hydrogen-oxygen	2700
Natural gas-oxygen	2800
Propane-oxygen	2850
Acetylene-oxygen	3110

Ideally, monochromators in flame photometers should be of higher quality than those found in absorption spectrophotometers. When nonionic materials are burned, light of varying wavelength is given off. This is known as *continuous emission* and will be added to the *line emission* of the element being measured. For this reason, the narrowest bandpath that is achievable should be used to eliminate as much of the extraneous, continuous emission as possible, but still permit a maximum of the line emission to pass through to the detector. The detectors used in flame photometers operate by the same principle and in the same way as those described for spectrophotometers.

Direct and internal standard flame photometry

In some of the instruments of earlier designs, standard solutions of sodium or potassium were atomized or aspirated directly into the flame to provide a series of meter readings against which an unknown solution could be compared. This approach, referred to as the direct reading method, presents certain problems: (1) Minor fluctuations in air or gas pressure cause unstable response in the instrument and lead to errors and operator fatigue. (2) Separate analyses and sometimes separate dilutions must be made for sodium and potassium. (3) The potassium signal is enhanced by the sodium concentration in the specimen. The latter effect, known as *mutual excitation*, results from the transfer of energy from an excited sodium atom to a potassium atom. Consequently, more potassium atoms are excited and light emission is increased. Ideally, then, the concentration of sodium and potassium in the standards should closely approximate those in the unknown, a situation that is difficult to achieve when analyzing urine in which these electrolytes show wide variation in concentration.

In the "internal standard" method, another element, lithium, is added to all standards, blanks, and unknowns in equal concentrations. Lithium has a high emission intensity, is normally absent from biological fluids, and emits at a wavelength sufficiently removed from sodium and potassium to permit spectral isolation. The flame photometer makes a comparison of the emission of the desired element (sodium or potassium) with the emission of the reference lithium element. By measuring the ratios of emissions in this way, small variations in atomization rates, flame stability, and solution viscosity are compensated for.

In a true sense, lithium does not function as a "standard" under these conditions but as a reference element. Variable concentrations of sodium and potassium, in the lithium diluent, must be used to establish calibration curves or to verify linearity of response.

Lithium also acts as a *radiation buffer* to minimize the effects of mutual excitation. The final working concentration of lithium is so high, compared to either sodium or potassium, that the same percentage of potassium becomes excited regardless of the sodium concentration in the sample. Serum lithium concentrations in patients receiving lithium salts are maintained at approximately 1 mEq/l. This will produce no significant change in final lithium concentrations in samples containing lithium in the diluent.

A wetting agent is frequently recommended for inclusion in standards and sample dilutions. This minimizes changes in atomizer flow rates due to differences in viscosity of the samples and provides more uniform flow rates. Viscosity effects are further reduced by dilutions of samples of the order of 1:100 to 1:200.

ATOMIC ABSORPTION SPECTROPHOTOMETRY

Atomic absorption spectrophotometry in some respects is the inverse of flame emission photometry. In all emission methods, the sample is excited in order to measure the radiant energy given off as the element returns to its lower energy level. Extraneous radiation must be filtered out from the energy of interest if interference by these signals is to be avoided.

In atomic absorption spectrophotometry the element is not appreciably excited in the flame, but is merely dissociated from its chemical bonds and placed in an unexcited or ground state. This means that the atom is at a low energy level in which it is capable of absorbing radiation at a very narrow bandwidth corresponding to its own line spectrum. A hollow cathode lamp, made of the material to be analyzed, is used to produce a wavelength of light specific for the kind of metal in the cathode. Thus, if the cathode were made of sodium, sodium light at predominantly 589 nm would be emitted by the lamp. When the light from the hollow cathode lamp enters the flame, some of it is absorbed by the ground state sodium atoms in the flame, resulting in a net decrease in the intensity of the beam from the lamp. This process is referred to as atomic absorption.

The process is analogous to absorption spectrophotometry. A specific hollow cathode lamp serves as the light source, and the sample, heated in the flame, replaces the sample in the cuvet. The pathlength of the flame is analogous to the light path through the cuvet. As noted previously, only a small fraction of the sample in the flame contributes emission energy, and only a fraction of this will be transmitted to the detector. Hence, most of the atoms are in the ground state and able to absorb light emitted by the cathode lamp. In general, atomic absorption methods are approximately 100 times more sensitive than flame emission methods. In addition, owing to the unique specificity of the wavelength from the hollow cathode lamp, these methods are highly specific for the element being measured.

Components of an atomic absorption spectrophotometer

Figure 3–14 shows the basic components of an atomic absorption spectrophotometer. The hollow cathode lamp is the light source. A nebulizer sprays the sample into the flame; the monochromator, slits, and detectors have the functions described for flame photometry. The cathode is made of the metal of the substance to be analyzed and is different for each metal analysis. In some cases, an alloy is used to make the cathode, producing a multi-element lamp.

On most instruments, an electrical beam chopper and a tuned amplifier are incorporated. The power to the hollow cathode lamp is pulsed so that the light is emitted by the lamp at a certain number of pulses per second. On the other hand, all of the light coming from the flame is unpulsed. When light leaves the flame it is composed of pulsed unabsorbed light from the lamp and a small amount of unpulsed flame spectrum and sample emission. The

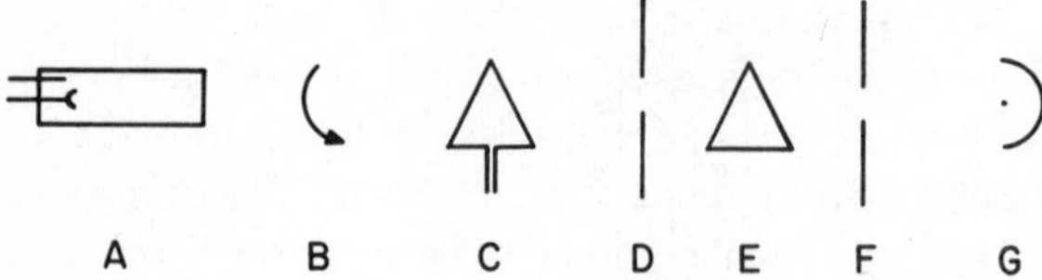

Figure 3–14 Essentials of an atomic absorption instrument. *A*, Hollow cathode. *B*, Chopper. *C*, Flame. *D*, Entrance slit. *E*, Monochromator. *F*, Exit slit. *G*, Detector. From Evenson, M. A. *In:* Fundamentals of Clinical Chemistry. N. W. Tietz, ed. Philadelphia, W. B. Saunders Company, 1970.

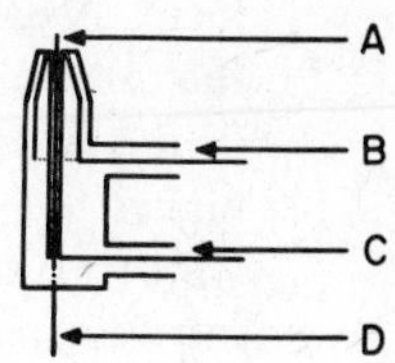

Figure 3–15 Total consumption burner. *A*, Capillary tip. *B*, Fuel inlet. *C*, Oxidant inlet. *D*, Sample capillary. From Evenson, M. A. *In:* Fundamentals of Clincial Chemistry, N. W. Tietz, ed. Philadelphia, W. B. Saunders Company, 1970.

detector detects all light, but the amplifier is electrically tuned to accept only pulsed signals. In this way, the electronics in conjunction with the monochromator remove all of the flame spectrum and sample emission.

Neon or argon gas at a few millimeters of mercury pressure is usually used as a filler gas for the hollow cathode lamp. A neon-filled lamp during operation produces a reddish-orange glow, and the argon produces a blue to purple glow inside the hollow cathode lamp. Quartz, or special glass that allows transmission of the proper wavelength, is used as a window. A current is applied between the two electrodes inside the hollow cathode lamp, and metal is sputtered from the cathode into the gases inside the glass envelope. When the metal atoms collide with the gases, neon or argon, they lose energy and emit their characteristic radiation. Calcium has a sharp, intense, analytical emission line at 422.7 nm, which is most frequently used for calcium analysis. In an interference-free system, only calcium atoms will absorb the calcium light from the hollow cathode as it passes through the flame.

Two kinds of burners have been used in most clinical applications. One is a *total consumption burner*, as illustrated in Figure 3–15. With this burner, the gases, hydrogen and air, and the sample mix within the flame. One disadvantage is that relatively large droplets are produced in the flame, which cause signal noise by light scattering. Also, the amount of acoustical noise produced is very high, and may become uncomfortable after a few hours of operation. An advantage of this type of burner is that the flame is more concentrated and it can be made hotter, causing molecular dissociations that may be desirable for some chemical systems.

Figure 3–16 shows a *premix burner* (laminar flow burner) and illustrates how the sample is aspirated, volatilized, and burned. Note that the gases are mixed and the sample atomized before being burned. An advantage of this system is that the larger droplets go to waste and not into the flame, thus producing a less noisy signal. Another advantage is that the pathlength through the flame of the burner is longer than that of the total consumption burner. This produces a greater absorption and increases the sensitivity of the measure-

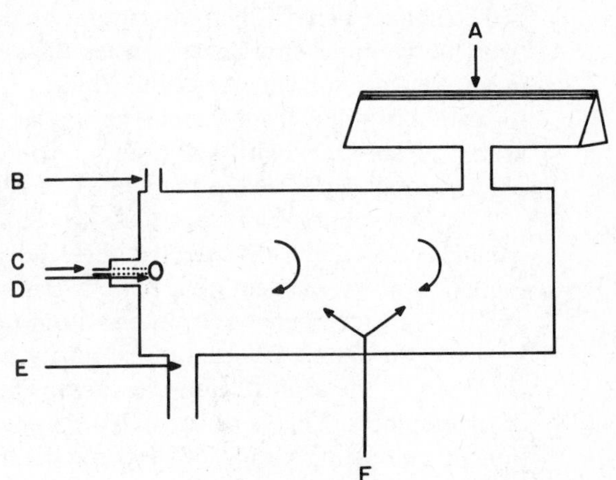

Figure 3–16 Laminar flow burner. *A*, Bolling head. *B*, Fuel inlet. *C*, Sample capillary and nebulizer (atomizer). *D*, Oxidant inlet. *E*, Drain. *F*, Spoilers. From Evenson, M. A. *In:* Fundamentals of Clinical Chemistry, N. W. Tietz, ed. Philadelphia, W. B. Saunders Company, 1970.

ment. A disadvantage of the premix burner is that the flame is usually not as hot as that of the total consumption burner, and it cannot sufficiently dissociate certain metal complexes in the flame (e.g., calcium phosphate complexes). *Nitrous oxide premix burners* produce higher temperatures and will dissociate some calcium complexes; however, at these higher temperatures, calcium becomes excited to a significant extent and emits in the flame, thus introducing another problem.

In *flameless atomic absorption* techniques the sample is placed in a depression on a carbon rod in an enclosed chamber. Strips of tantalum metal may also be used. In successive steps, the temperature of the rod is raised to dry, char, and finally atomize the sample into the chamber. The atomized element then absorbs energy from the corresponding hollow cathode lamp. This approach is more sensitive than the conventional flame methods and permits determination of trace metals in small samples of blood or tissue. As discussed later, phosphate interferes with calcium determinations by atomic absorption methods. This presents more problems in flameless analysis than with conventional flame atomization; hence, the latter method appears more suitable for the determination of calcium in serum and urine.

Interference in atomic absorption spectrophotometry

There are three general types of interferences in atomic absorption spectrophotometry. These are chemical, ionization, and matrix effects.

Chemical interference refers to the situation when the flame cannot dissociate the sample into free atoms so that absorption can occur. An example of this is the phosphate interference in the determination of calcium, caused by the formation of calcium phosphate complexes. These complexes do not dissociate in the flame unless a special high temperature burner is used. The phosphate interference is overcome by adding a cation that will compete with calcium for the phosphate. Usually, in the determination of calcium, lanthanum or strontium is added to the dilute sample to replace and release calcium from its phosphate complex. The freeing of calcium occurs because lanthanum and strontium form more stable complexes with phosphate than does calcium. The free calcium is then capable of absorbing the calcium light from the hollow cathode.

Ionization interference results when atoms in the flame become excited, instead of only being dissociated, and then emit energy of the same wavelength that is being measured. This effect can be overcome by adding an excess of a more easily ionized substance that will absorb most of the flame energy so that the substance of interest will not become excited. Ionization interference can also be decreased by reducing the flame temperature.

A third type of interference is the *matrix interference*. One example of a matrix effect is the enhancement of light absorption by organic solvents. An atom may absorb between two and five times more energy when dissolved in an organic solvent than when dissolved in an aqueous solvent. A second kind of matrix effect is the light absorption caused by formation of solids from sample droplets as the solvent is evaporated in the flame. This will usually occur only with solutions of concentrations greater than 0.1 mol/l. Refractory oxides of metals formed in the flame can also lead to matrix interferences.

Atomic absorption spectrophotometry is sensitive, accurate, precise, and highly specific. One of the reasons for these advantages is that the method does not require excitation of the element and, thus, is less affected by temperature variations in the flame and the transfer of energy from one atom to another. The high specificity results from the fact that the light used has an extremely narrow bandwidth (0.01 nm) and is selectively absorbed by the atoms being measured. The most significant disadvantage is the problem of interferences. This is being solved, however, by extraction techniques and the introduction of competing cations to release the element to be measured from complexing or chelating anions.

FLUOROMETRY

A substance is said to fluoresce when it absorbs light at one wavelength and emits light of a longer wavelength and lower energy. The most important advantage of fluorometry is its extreme sensitivity. Sometimes the sensitivity may be 1000 times that of colorimetric methods. Some molecules fluoresce directly, but a larger percentage must be complexed or reacted chemically to transform them into fluorescent compounds. The fluorescence of the new compound is then measured. By selection of different complexing agents, fluorescence at different wavelengths may be produced, thus making it possible to work at a wavelength significantly removed from interferences.

In the fluorescent process, a delay time of between 10^{-8} and 10^{-4} seconds occurs between the absorption of energy and the releasing of part of the energy in the form of light. If the length of time is longer than 10^{-4} seconds from the time the chemical species absorbs the energy until the light is emitted, this process is called phosphorescence.

Fluorometric instrumentation

In a photofluorometer, ultraviolet light (exciting source) is passed through a primary filter to screen out light above the desired wavelength. Part of the light energy is absorbed by the substance in solution. The emitted light is passed through a secondary filter at right angles to the ultraviolet beam and is measured photometrically. (By measuring the emitted light at the right angle, interference by transmitted light is avoided.) A combination of secondary filters may be used to screen out undesired fluorescence from other compounds. More elaborate spectrofluorometers are equipped with prisms or gratings and have facilities for dialing specific wavelengths for the exciting and secondary beams.

Figure 3–17 shows a schematic diagram of the components of a fluorometer. The energy source of a fluorometer is generally a mercury arc lamp or other suitable UV-visible light source that will produce sufficient energy so that, when absorption occurs, electron transitions to a higher energy level within the molecule will occur. In a fluorometer, the entrance and exit slits are similar to those used in spectrophotometers. Fluorometers that use a continuous source, such as a xenon source, have a monochromator so that isolation of the desired wavelengths is achieved before excitation of the substance occurs (primary monochromator). All fluorometers have another (secondary) monochromator system that will selectively remove unwanted wavelengths of light before they fall upon the detector. The secondary monochromator and detector are placed at right angles to the incident beam to prevent light from the high energy source of the lamp from reaching the detector.

Fluorometric methods

Fluorometric methods have been developed for hundreds of compounds of potential clinical interest, including drugs and drug metabolites. The more common determinations include porphyrins, catecholamines, calcium, magnesium, salicylate, quinidine, tyrosine,

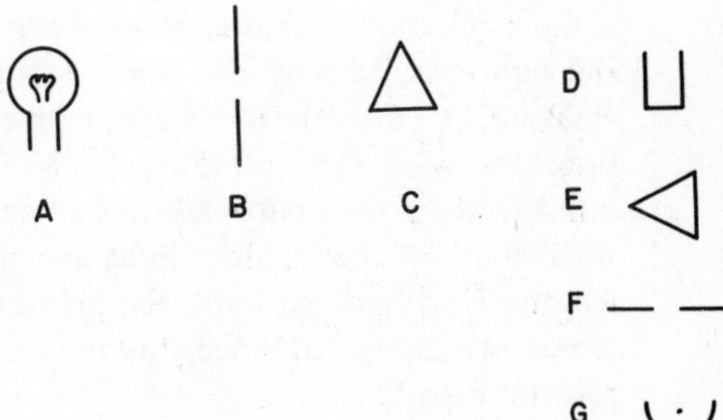

Figure 3–17 Essentials of a fluorometer. *A*, Light source. *B*, Slit. *C*, Primary monochromator or filter. *D*, Cuvet. *E*, Secondary monochromator or filter, *F*, Slit. *G*, Detector. From Evenson, M. A. *In:* Fundamentals of Clinical Chemistry. N. W. Tietz, ed. Philadelphia, W. B. Saunders Company, 1970.

and phenylalanine. Dehydrogenase enzymes may be assayed in small amounts of serum by measuring the increase or decrease of the fluorescent coenzyme NADH.

Methods must be evaluated thoroughly to insure adequate specificity inasmuch as many compounds exhibit fluorescence under appropriate conditions. These interfering substances can be of endogenous or exogenous origin. In the latter group are a number of drugs and their metabolites. When fluorometric methods are applied to measure products in an enzymatic procedure, it is advisable to include serum blanks to correct for interfering substances. Fluorescence may also be inhibited when the substance under study complexes with proteins or phosphates, or when the solution has a color that can act as a filter to reduce the intensity of the emitted light. Such "quenching" effects may be overcome or evaluated by use of an internal standard added to a second aliquot of sample and carried through the entire procedure.

TURBIDIMETRY AND NEPHELOMETRY

Turbidimetric measurements determine the amount of light blocked by particulate matter in the sample as light passes through the cuvet. Measurements can be made with either a filter photometer or a spectrophotometer.

Several problems are inherent in making turbidimetric measurements, and these problems are associated mostly with sample and reagent preparation rather than with the operation of the instrument. The amount of light that is blocked by a suspension of particles in a cuvet depends not only upon the number of particles present, but also upon the cross-sectional area of each particle. If the particle size of the standards is not the same as the particle size of the samples being measured, errors in turbidimetric measurements result. Another problem with the turbidimetric measurement is the need to keep the length of time between sample preparation and measurements as constant as possible. Particles may aggregate or settle out of solution while the measurements are being made, thus producing an error. Control of the rate of settling is usually accomplished by using gum arabic or gelatin, which provides a viscous medium that retards particle settling. Preparation of very fine suspensions also helps to minimize these problems.

Turbidimetric measurements are acceptable provided the number of particles and their size are in a reasonably narrow range. A high intensity light or a very low intensity light should not fall on the photodetector, because errors in instrumentation would then augment other errors in the measurement. Light of shorter wavelengths will produce apparent absorbance readings that are greater than those obtained with light of longer wavelengths; however, longer wavelengths (e.g., 620 or 660 nm) may be more desirable to minimize absorption effects encountered in the presence of such pigments as bilirubin and hemoglobin. With a colorless, turbid solution, a plot of log A versus logarithm of the wavelength normally produces a straight line. This relationship has proved helpful in making corrections for absorbance of colored solutions in which slight turbidity is present by taking readings at two wavelengths, one of which is well removed from the peak of absorbance.

Nephelometric measurements are similar to turbidimetric measurements except that the light scattered by the small particles or colloids in the sample cuvet is measured at right angles to the beam incident on the cuvet. The arrangement is similar to that used in fluorometry shown in Figure 3–17. The amount of scatter that occurs is related to the number and size of particles in the light beam. The size of the particles, their shape, and the wavelength of the incident light are important variables to control. The shorter the wavelength of the incident light, the greater the degree of dispersion. However, ultraviolet light is not recommended since this may produce fluorescence in the solution and introduce a positive error.

Nephelometry has an advantage over turbidimetry in that nephelometric measurements are usually capable of somewhat greater precision. Turbidimetric and nephelometric methods are not frequently used because they are inherently incapable of high precision. Specific chemical reactions are being sought to achieve higher accuracy and precision for the compounds for which these measurements are currently being used.

REFERENCES

Photometry

1. MacDonald, R. P.: Clin. Chem., *10*: 1117, 1964.
2. Rand, R. N.: Clin. Chem., *15*: 839, 1969.
3. Winstead, M.: Am. J. Med. Technol., *28*: 67, 1962.

ADDITIONAL READINGS

Bender, G. T.: Chemical Instrumentation: A Laboratory Manual Based on Clinical Chemistry. Philadelphia, W. B. Saunders Company, 1972.
Meloan, C. E.: Instrumental Analysis Using Spectroscopy. Philadelphia, Lea & Febiger, 1968, vol. 1.
Meloan, C. E.: Instrumental Analysis Using Physical Properties. Philadelphia, Lea & Febiger, 1968, vol. 2.
Robinson, J. W.: Atomic Absorption Spectroscopy. New York, Marcel Dekker, Inc., 1966.
Udenfriend, S.: Fluorescence Assay in Biology and Medicine. New York, Academic Press, Inc., 1962.
Willard, H. H., Merritt, L. L., Jr., and Dean, J. A.: Instrumental Methods of Analysis. 4th ed. Princeton, New Jersey; D. Van Nostrand Co., Inc., 1965.

SECTION TWO

ELECTROPHORESIS

by Emanuel Epstein, Ph.D.

Electrophoresis refers to the migration of charged solutes or particles in an electric field. While the expression electrophoresis applies to the migration of all kinds of particles, *iontophoresis* specifically refers to the migration of small ions.

The first electrophoretic method used in the study of proteins was the free solution or moving boundary method described by Tiselius in 1937. This technique is still used by many scientists for the measurement of electrophoretic mobility and the study of protein-protein interaction, but rarely by clinical chemists in routine work. A complex apparatus is needed, the technique is difficult, and samples of the order of 0.5 ml of serum are required. Results similar to those obtained with moving boundary electrophoresis can be obtained by gel electrophoresis. This technique, which requires only a small sample size and simple apparatus, will be described later in this section.

The term *zone electrophoresis* is used to refer to the migration of charged macromolecules within the confines of a nonconvective (or solid) supporting medium (e.g., cellulose paper, cellulose acetate, agar gel, and so forth). The solutes of interest in clinical chemistry are mainly macromolecular in size and colloidal in nature and include proteins in serum, erythrocytes, urine, CSF, and other physiological fluids, as well as the proteins in tissues.

THEORY

Molecules carrying an electric charge by virtue of proton ionization will move either to the cathode or to the anode of an electrophoretic system depending on the nature of the charge on the molecules. In a solution more acid than the isoelectric point (pI) of the solute, an ampholyte (a molecule which can be either positively or negatively charged) takes on a positive charge and migrates toward the cathode. In the reverse situation, the ampholyte is in the anionic form and migrates toward the anode. The rate of migration is dependent on such factors as (1) net electric charge of the molecule, (2) size and shape of the molecule, (3) electric field strength, (4) nature of the supporting medium, and (5) the temperature of operation. The equation expressing the driving force in any electrophoretic system is given by

$$F = (X)(Q) = \frac{(E)(Q)}{(d)}$$

where

$F =$ the force exerted on an ion
$X =$ the current field strength (V/cm)
$Q =$ the net charge on the ion
$E =$ the voltage applied (V)
$d =$ the width of, or distance across, the electrophoretic medium (cm)

The steady acceleration of the ion is counteracted by a resisting force characteristic of the solution in which migration occurs. This force, expressed by Stokes law, is

$$F' = 6\pi r\eta\, v$$

where

$F' =$ the counter force
$r =$ the ionic radius of the solute
$\eta =$ the viscosity of the buffer solution in which migration is occurring
$\pi =$ 3.1416, a constant
$v =$ the rate of migration of the solute = velocity (cm/s)

The force F' counteracts the acceleration produced by F (as the ion approaches the electrode of opposite charge it tends to speed up or accelerate) and the resultant of the two forces is a terminal and constant velocity. Therefore, when

$$F = F'$$

$$6\pi r\eta\, v = (X)(Q)$$

or

$$\frac{v}{X} = \frac{Q}{6\pi r\eta} = \mu$$

where

$\frac{v}{X} =$ the rate of migration (cm/s) per unit field strength (E/cm), and is defined as the electrophoretic mobility. It is expressed by the symbol μ. The units of μ are $cm^2/(V)(s)$

It can be seen that electrophoretic mobility is directly proportional to the net charge and inversely proportional to the size of the molecule and the viscosity of the electrophoresis medium. As mentioned above, $\mu = cm^2/(V)(s)$, and this equation can be used to calculate the electrophoretic mobility in actual practice. To illustrate, suppose that albumin has traveled 3 cm on a cellulose acetate strip 10 cm in length. If it takes 75 min (or 75 × 60 s) to travel this length at a voltage of 250 V, then

$$\mu = \frac{(3)(10)}{(75)(60)(250)} = 2.6 \times 10^{-5}\ \text{cm}^2/(\text{V})(\text{s})$$

Since one mobility unit is defined as 10^{-5} cm^2/(V)(s), the above mobility then is 2.6 mobility units. Some literature sources show mobilities as having a negative sign. Assuming that protein migration occurs in a direction opposite to that of the electrophoretic field (assumed to be from the + to the − electrode), by convention the mobility will have a negative value. In the example just given, electrophoresis was performed at pH 8.6 where proteins have a negative charge; therefore, migration occurred from the − to the + electrode, and the mobility value is amended to read −2.6 mobility units.

Mobility values obtained from zone electrophoresis measurements may not exactly compare with values obtained by free electrophoresis because of possible differences in the temperature, ionic strength, rate of endosmotic flow, and point of sample application.

Another contributing factor to the electrophoretic counterflow is the so-called *wick flow*. When electrophoresis is in process, heat is evolved, resulting in evaporation of solvent from the electrophoretic support. The drying effect causes buffer to rise into the electrophoretic support from both buffer compartments. This flow of buffer from both directions influences the protein mobility.

BUFFER

The ionic strength and the pH of the buffer in which the solute is migrating significantly affect the direction and the speed of movement of the molecule. In addition, the ionic strength determines the thickness of the ionic cloud surrounding the charged molecules and, thus, also the rate of migration and the sharpness of the zones.

The effect of ionic strength on mobility of ions in the electrophoretic process is due to the concentration of ions clustered about the charged molecule. With increasing concentration of buffer ions, the ionic cloud increases, and the molecule becomes more hindered in its movement. High ionic strength buffers yield sharper band separation, but the benefits of sharper resolution are diminished by the Joule (heat) effect that leads to denaturation of heat-labile proteins.

Many buffer systems have been used in electrophoresis, but those of greatest utility are the barbital buffers and the tris-boric acid–EDTA buffer. Because the base in these buffers is monovalent, any association effect between buffer and protein is minimal.

ELECTROENDOSMOSIS OR ENDOSMOSIS

An electrophoretic support medium in contact with water takes on a negative charge because of adsorption of hydroxyl ions. Since the ions are fixed to the surface of the electrophoretic support, they are rendered immobile relative to the other ions in solution. Positive ions in solution cluster about the fixed negative charge sites, forming an ionic cloud of mostly positive ions. The number of negative ions increases with increase in distance from the fixed negative charge sites until, eventually, positive and negative ions are present in equal concentration (Fig. 3–18). The potential which exists between the fixed ions and the associated cloud of ions is termed the *electrokinetic potential* or the *zeta potential* (ζ). When a current is applied to such a system, the charges attached to the immobile support remain fixed, but the cloud of ions in solution is free to move to the electrode of opposite polarity.

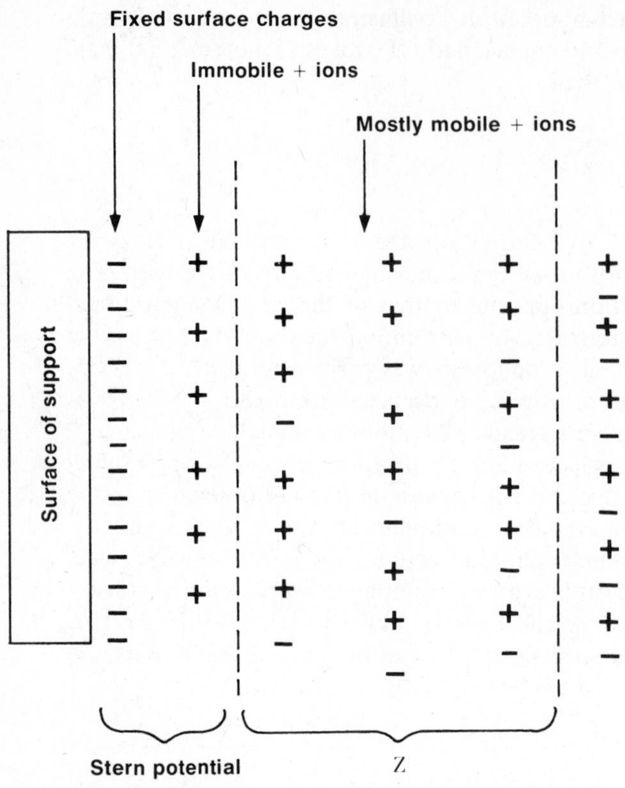

Figure 3–18 Distribution of + and − ions around the surface of an electrophoretic support. Adsorbed to the surface of the solid is a layer of − ions. (These may be + ions under suitable conditions.) A second layer of + ions is attracted to the surface. These two layers compose the Stern potential. The large, diffuse layer is the electrokinetic or zeta (ζ) potential. Extending further from the surface of the solid is homogeneous solution. The Stern potential plus the zeta potential equals the electrochemical potential or epsilon (ε) potential.

Since the ions in solution are highly hydrated, this movement of the ionic cloud results in movement of the solvent as well. This movement of solvent and its solutes relative to the fixed support is referred to as *endosmosis*. Macromolecules in solution which move in the opposite direction must "buck" this flow of hydrated positive ions. If the molecules are insufficiently charged, they may remain immobile or they may even be swept back toward the opposite pole. In media where endosmosis is strong, such as in paper, cellulose acetate, and agar gel, gamma globulins are swept behind the line of application. In electrophoretic media where surface charges are minimal, such as in starch gel or polyacrylamide gel, endosmosis is also minimal.

METHOD

Principle

The sample is applied to the electrophoretic support medium of choice and electrophoresis is conducted either under constant voltage or constant current. The support is then removed from the electrophoresis cell and rapidly dried or placed in a fixative to prevent diffusion of the sample. The support is then treated with a dye-fixative reagent to locate and stain the individual zones. After removal of excess dye, the support is dried (in the case of paper) or placed in a clearing agent (in the case of cellulose acetate membranes). (For details, see Chaps. 7, 10, and 12.) A representative electrophoresis system is shown in Figure 3–19. Actual apparatuses may vary in form, but Figure 3–19 illustrates the essentials.

Stains

The stains used to locate the separated fractions of the sample differ in accordance with type of application and personal choice. The stains utilized for a specific application

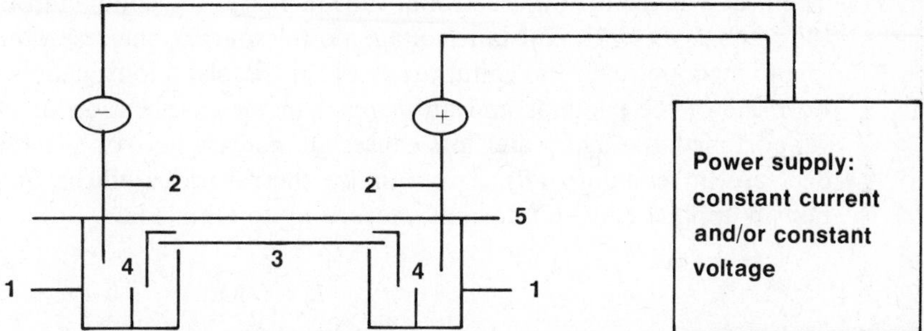

Figure 3–19 A schematic of a typical electrophoresis apparatus. Two buffer boxes (1) with baffle plates contain the buffer used in the process. In each buffer box is an electrode (2), either platinum or carbon, the sign of which is fixed by the mode of connection to the power supply. The electrophoresis support (3) on which separation takes place is in contact with buffer by means of wicks (4). The whole apparatus may be covered (5). Direct current power supply may be either constant current (adjustable) or constant voltage (adjustable), or both.

are given in connection with the respective procedures. The amount of dye uptake by the sample is affected by many factors, such as the type of protein and the degree of denaturation of proteins by fixing agents.

Quantitation

It is customary to give electrophoresis reports in terms of the percentage of each fraction present or in absolute concentration if the combined total of all fractions is known. Quantitation can conveniently be accomplished either by direct *densitometry* or by *elution* of the separated fractions followed by spectrophotometry of the dye associated with the protein zones. In densitometry, the electrophoretic strip (or other medium) is moved past the measuring optics, and each fraction, in accordance with its location and absorbance value, is presented on a strip chart. In some cases, the area under each peak is automatically integrated. The elution method involves cutting the support into zones containing the individual fractions and elution of the adsorbed dye by means of suitable solvents such as basic buffers or 0.1 molar NaOH.

In paper electrophoresis, the accuracy of densitometry is affected by the scattering of light by the surface of the paper. Also, the linearity of the photoelectric measurement is affected by the thickness of the dye-protein layer. In such cases, the elution method is likely to be more accurate.

In modern cellulose acetate techniques, the clearing of the membrane and the fact that a very small sample size is being used have essentially eliminated the shortcomings associated with densitometry. Thus, in such cases, both techniques are equally accurate.

Power supply

Commercially available power supplies allow operation at either constant current or constant voltage. The flow of current through a medium which offers electrical resistance involves production of heat:

$$\text{heat} = (E)(I)(t)$$

where

$$E = \text{voltage (V)}$$

$$I = \text{current (A)}$$

$$t = \text{time (s)}$$

The heat evolved during electrophoresis increases the conductance of the system (decreases the resistance). With constant voltage power sources, the resultant rise in current, which is due to increased thermal agitation of all the dissolved ions, causes an increase in the migration rate of the proteins and an increase in the evaporation of water from the stationary support medium. The water loss causes an increase in ion concentration and thus a further decrease in resistance (R). To minimize these various effects on the migration rate, it is best to apply a constant current. According to Ohm's law

$$E = (I)(R)$$

Therefore, if R decreases, the voltage (E) also decreases. This in turn decreases the heat effect and thus keeps the migration rate relatively constant. If a constant voltage source were used, the current, and therefore the migration rate, would progressively increase.

PAPER ELECTROPHORESIS (PE)

Paper electrophoresis, once used widely in clinical laboratories for the separation of serum proteins, has largely been replaced by cellulose acetate electrophoresis. Most PE procedures use either Whatman 1 or 3 MM paper, or the equivalent from other manufacturers, supplied in the form of precut paper strips.

A paper is chosen on the basis of distance of migration, separation quality, and qualities of handling, dyeing, destaining, and so forth. In general, a thick, soft paper (such as Whatman 3 MM) is best for resolution of proteins, but electroendosmosis is more prominent than with a hard paper. A disadvantage of the PE-technique is the long separation time needed (14 to 16 hours). Also, in case of protein electrophoresis, albumin is adsorbed by the paper, resulting in excessive background (trailing). The advantages of paper are its high tensile strength, its low cost, and its ease of handling.

AGAROSE GEL ELECTROPHORESIS (AGE)

Agar gel electrophoresis has been successfully applied to the analysis of serum proteins, hemoglobin, lactate dehydrogenase isoenzymes, lipoproteins, and other substances. In fact, this gel medium parallels cellulose acetate in versatility and convenience and generally competes with other media for applicability to routine clinical laboratory demands.

Impure or even purified agar is composed of at least two fractions; agaropectin and agarose. The former fraction contains sulfate and carboxylic acid groups and accounts for the considerable endosmosis and background color which is observed with unfractionated agar. A purified fraction of agar, called agarose, is essentially neutral and exhibits little endosmosis. It is becoming the agar medium of choice. The original literature by Hegenauer and Nace[3] may be consulted for the complete procedure for purification of agarose.

Advantages of AGE over paper electrophoresis are lower affinity for proteins and a native clarity after drying that allows for excellent densitometric examination. Usually 0.5 to 1.0 g/100 ml of agar is used to give a gel of desired strength with good migration properties. In AGE, the sample is applied either into a precut well or as a solution in warmed agar. The latter technique is less convenient but it has the advantage that the agar-sample solution solidifies to become part of the agar plate. The sample size employed in the technique is relatively small (1 to 3 μl), and the electrophoresis time is relatively short (30 to 90 min, depending on experimental conditions). Specific applications of this technique are mentioned later in this text.

CELLULOSE ACETATE ELECTROPHORESIS (CAE)

If the hydroxyl groups in cellulose are treated with acetic anhydride, cellulose acetate, the raw material for cellulose acetate membranes, is formed. The membranes commercially available are composed of about 80 per cent air space, forming pockets within interlacing cellulose acetate fibers. When purchased, the membrane appears as a dry, opaque, brittle film which may easily crack when stressed. However, when wetted with buffer, the air spaces fill with liquid and the film becomes quite pliable. The characteristics of the membrane will vary with the extent of acetylation, the prewashing procedure employed by the manufacturer, the additives used, and the pore size as well as the thickness of the membrane.

Cellulose acetate membranes may be made transparent for densitometry by treatment with a mixture of solvents. The mixture contains one solvent for cellulose acetate and another solvent which acts as vehicle (e.g., 95 parts methanol and 5 parts glacial acetic acid). The cellulose fibers are partially dissolved by the action of the solvent and coalesce so that the original air spaces are eliminated. The sample (0.2 to 2.0 μl) is generally applied with a twin-wire applicator. The edge of glass slides or micropipets have also been used for this purpose. An advantage of the technique is the speed of the separation (20 min to 1 h) and the ability to store the transparent membrane indefinitely. Application of this technique will be discussed in Chapters 7, 8, 10, and 12.

ACRYLAMIDE GEL ELECTROPHORESIS (AGE)

The resolution by ordinary electrophoretic techniques is incomplete in that the zones are composed of several proteins with the same electrophoretic mobility. Zones, in general, are also broad due to diffusion of the proteins during the procedure. A new type of electrophoresis, termed disc electrophoresis, was introduced in 1964[1,4] to overcome these deficiencies. The term derives from the *disc*ontinuities in the electrophoretic matrix and from the *disc*oid shape of the separated zones of ions (due to the tubular shape of the electrophoresis cell). While protein electrophoresis using paper, cellulose acetate, or agarose gel yields five zones, namely, albumin and alpha 1-, alpha 2-, beta-, and gamma globulins, polyacrylamide or starch gel electrophoresis yields 20 or more fractions. These media are therefore widely used in the study of individual proteins in serum, especially the study of genetic variants and enzymes.

The technique employs layers of gel which differ in composition and pore size. The individual gels are first prepared *in situ* by polymerizing a gel monomer and a cross-linking agent with the appropriate catalyst. The first gel to be poured into the tubular-shaped electrophoresis cell is the "separation gel." After 30 minutes, during which gelation takes place, a large-pore gel, the spacer gel, is cast on top of the separation gel. Finally, the sample in a gel with a pore size similar to the spacer gel (e.g., 3 per cent acrylamide) is cast above the spacer gel so that the finished product is composed of three different layers of gel. When electrophoresis begins, the protein ions migrate through the large-pore gels (which do not impair movement of protein in solution) and stack up on the separation gel in a very thin line. Separation of the ions then takes place in the bottom separation gel, not only on the basis of their charge but also on the basis of molecular size.

Since the sample concentrates at the border of the separation gel, preconcentration of solutions with low protein content (e.g., CSF) is not necessary.

Acrylamide gel is thermostable, transparent, strong, and relatively chemically inert, and it can be made in a wide range of pore sizes. Furthermore, these gels are uncharged, thus eliminating electroendosmosis.

In recent years, simplifications of this technique have been introduced. In one technique, the use of spacer and sample gel was discontinued and a continuous buffer system

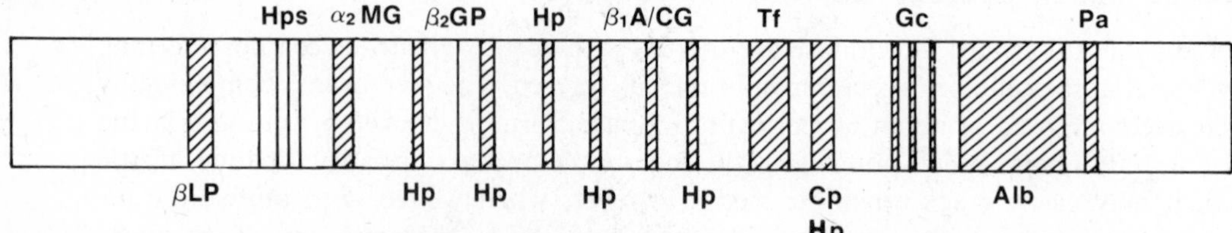

Figure 3–20 A simplified schematic of a protein pattern from the serum of a subject with haptoglobin type 2-1 (separation by acrylamide gel electrophoresis). Some zones contain more than the one protein shown, as demonstrated by immunological techniques. (Abbreviations: *Pa*, prealbumin; *Alb*, albumin; *Gc*, Gc globulins; *Cp*, ceruloplasmin; *Tf*, transferrin; $\beta_1 A/CG$, β_1-A/C globulin; *Hp* or *Hps*, haptoglobin(s); $\beta_2 GP$, β_2-glycoprotein (best shown by PAS staining); $\alpha_2 MG$, α_2 macroglobulin; *βLP*, β lipoprotein.

used with excellent results. Epstein *et al.*,[2] in their studies of alkaline phosphatase isoenzymes, used a similar technique. These authors introduced the serum sample not as a dilute solution as originally proposed, but as undiluted serum delivered by micropipet through the covering buffer layer to the top of the separation gel. A schematic representation of serum protein electrophoresis is shown in Figure 3–20.

STARCH GEL ELECTROPHORESIS

Starch gel electrophoresis also possesses the property of separating macromolecular ions on the basis of both charge and molecular size.

The starch used in starch gel electrophoresis is partially hydrolyzed, since native starch does not gel. As with agar ge, starch gel may be utilized in a horizontal process, but the sample must be applied in such a way that a very fine starting zone results. This can be accomplished by introducing the sample into a cut in the gel and loading the gel either with a liquid sample or a paper-loaded sample. The technique can also be applied with the migration taking place in a vertical direction. In this case, the liquid sample is applied to a cut or precast slot in the gel and then covered with a warmed wax-petrolatum mixture. The wax hardens and holds the liquid sample in place. Application of current results in compaction of proteins on the surface of the gel so that a thin uniform starting zone is produced. The gel preparation is a relatively difficult operation, requiring a skillful operator.

Starch used in the preparation of gels can be purchased from Merck & Co., Inc. (West Point, PA. 19486) or Connaught Medical Research Laboratories (Toronto, Ontario, Canada). The starch gels are used in a concentration of 10 to 16 g/100 ml. The pH of the buffer varies according to application between pH 3 and pH 11 but is generally between pH 8.6 and pH 9.0.

REFERENCES

1. Davis, B. J.: Ann. N.Y. Acad. Sci., *121*: 404, 1964.
2. Epstein, E., Wolf, P. L., Horwitz, J. P., and Zak, B.: Am. J. Clin. Pathol., *48*: 530, 1967.
3. Hegenauer, J. C., and Nace, G. W.: Biochim. Biophys. Acta., *111*: 334, 1965.
4. Ornstein, L.: Ann. N.Y. Acad. Sci., *121*: 321, 1964.

SECTION THREE

ELECTROCHEMISTRY

by Ole Siggaard-Andersen, M.D., Ph. D.

Several types of analytical methods are based on electrochemical phenomena. Those which have been applied to clinical chemical methods are potentiometry, polarography, amperometry, coulometry, and conductometry.

POTENTIOMETRY

Potentiometry is the measurement of the electric potential difference between two electrodes in an electrochemical cell. An *electrochemical* (galvanic) *cell* always consists of two electrodes (electron or metallic conductors) connected by an electrolyte solution (ion conductor). An *electrode* or *half-cell* consists of a single metallic conductor in contact with an electrolyte solution. The negative electrode (excess electrons) is called the *cathode*, and the positive electrode (deficient in electrons) is called the *anode*. The ion conductors can be composed of one or more phases in direct contact with each other, or they can be separated by membranes which are permeable only to specific cations or anions (Fig. 3–21). One of the electrolyte solutions is the unknown or test solution, which may be replaced by an appropriate standard solution for calibration or quality control purposes. A salt solution or bridge may be interposed in the cell to reduce any liquid-liquid junction potential present (see later). By convention, the cell is shown so that the left electrode (M_L) is the *reference electrode*, while the right electrode (M_R) is the *indicator* (*measuring*) *electrode*.[1] The flow of electrons between the two electrodes is governed by the electromotive force (EMF, E), which is defined as the maximal difference in potential between the two electrodes obtained when the current drawn from the cell is zero. By convention, the EMF of the cell is determined by measuring the potential at the right electrode [$V(M_R)$] and subtracting from this value the potential at the left electrode [$V(M_L)$]:

$$E = V(M_R) - V(M_L) \tag{7}$$

Measurements of this type are carried out with a *potentiometer*, of which the common pH meter is a special type. Two main classes of potentiometers are available. The *null-point potentiometer* (compensation or balancing potentiometer) applies a potential of increasing magnitude between the cell electrodes until that point at which no current flows through the cell (Fig. 3–21). The *direct-reading potentiometer* is a voltmeter which measures the potential across the cell (between the two electrodes), but in order that such potential measurement be accurate, it is necessary that no current flows through the cell. This is accomplished by incorporating a high resistance (impedance) within the voltmeter. Modern direct-reading potentiometers are equal in accuracy to compensation potentiometers, and have largely replaced the latter. For convenient reading and recording of data, these instruments may be equipped with accessories for direct digital display or printout.

Between points within any one given phase, the potential is constant as long as the current flow is zero. Between two different phases, however, a potential difference arises. The overall potential of an electrical cell is the sum of all the potential jumps existing between different phases of the cell (Fig. 3–21). The potential of a single electrode with respect to the surrounding electrolyte, as well as the absolute magnitude of the individual potential jumps between the phases, is actually unknown and cannot be measured. We can measure only the *potential differences* between two electrodes (half-cells). The potential

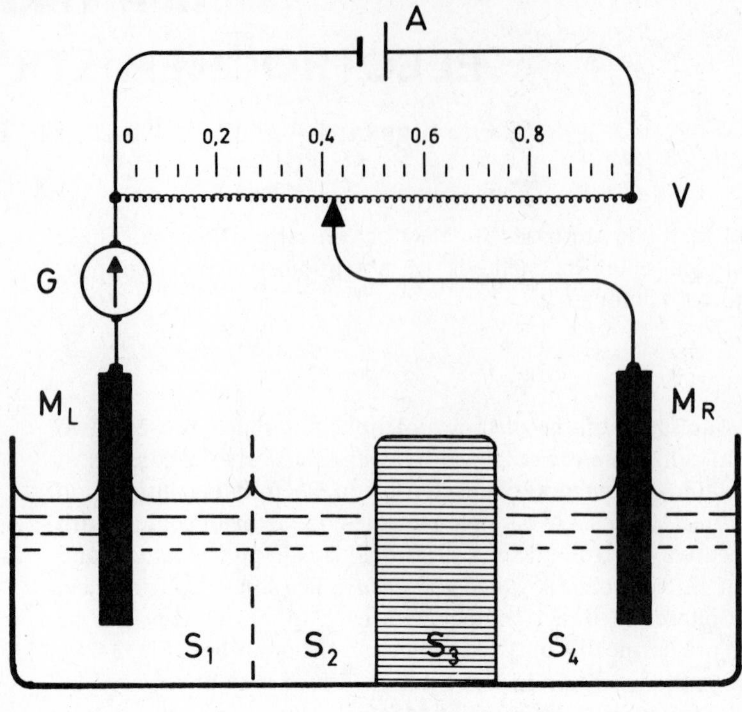

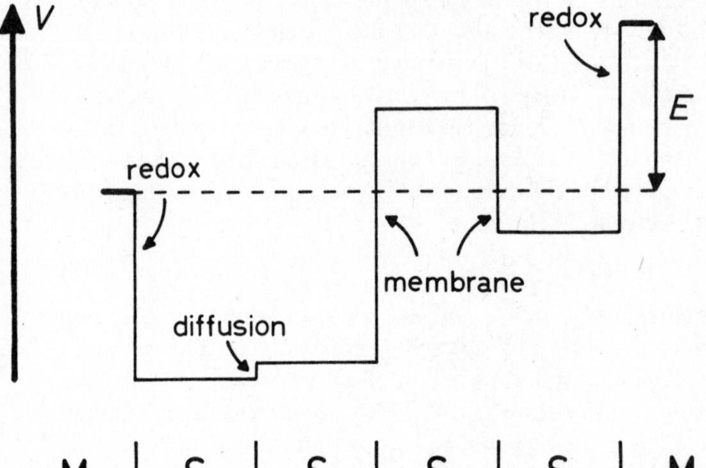

$M_L \mid S_1 \mid S_2 \mid S_3 \mid S_4 \mid M_R$

Figure 3–21 Schematic diagram of an electrochemical cell. M_L and M_R are two metallic conductors. S_1, S_2, S_3, and S_4 symbolize a variable number of ion conductive phases. S_1 may be a saturated KCl solution, S_2 a given test solution, S_3 an ion-selective membrane (which need not necessarily be a thin membrane), and S_4 a given reference solution. The liquid-liquid junction between S_1 and S_2 may be an open contact between the two solutions, or a porous membrane, or a fiber junction. The two electrodes (M_L and M_R) are connected externally via a potentiometer, e.g., a null-point potentiometer, consisting of a voltage source (A), a voltage divider (V), and a sensitive galvanometer (G). The voltage divider is adjusted so that the galvanometer shows zero current. The EMF of the cell is then read directly on the calibrated scale.

The lower half of the figure illustrates the potential jumps at the phase boundaries caused by (1) redox potentials, (2) membrane potentials, and (3) diffusion potentials (liquid-liquid junction potentials). At the bottom, the cell is written in symbols, vertical lines indicating phase boundaries.

jumps can be classified as (1) redox potentials, (2) membrane potentials, and (3) diffusion potentials. Generally, it is possible to devise a cell in such a manner that all the potential jumps except one are constant. This potential can then be related to the activity of some specific ion of interest (H^+, Na^+, and so forth).

DIFFUSION POTENTIALS (LIQUID-LIQUID JUNCTION POTENTIALS)

Diffusion potentials arise at a liquid-liquid junction ($S_L \mid S_R$) where two solutions of different ionic composition are in direct contact. If cations diffuse faster from left to right than anions, a positive liquid-liquid junction potential develops at the interface. Diffusion potentials are due to irreversible processes and are therefore more difficult to calculate than the true equilibrium potentials, but they can often be calculated with good approximation by means of the so-called Henderson equation (see Bates[1]). Generally, the aim is to reduce the liquid-liquid junction potential to a value as low and as reproducible as possible. This is achieved by using a concentrated KCl solution as the bridge solution, (3.5 mol/l, or saturated = 4.3 mol/l at 37°C). The potential is then dominated by the large excess of K^+ and Cl^- diffusing at almost the same rate into the test solution, and the potential becomes largely independent of the composition of the test solution.

It should be added that *thermal* diffusion potentials may arise at a liquid-liquid junction where two identical solutions of different temperature are in contact. Such thermal diffusion potentials should be avoided by maintaining the whole cell at the same, constant temperature.

MEASUREMENT OF ION ACTIVITY VERSUS ION CONCENTRATION

Most ions in biological fluids are present in free form and in complex-bound form, and a distinction must therefore be made between the concentration and activity of free ions and the concentration of total ions (free + bound).

Potentiometric measurements by means of ion-selective electrodes provide a means for the determination of the *activity* of the ion involved. This is the relevant quantity for most purposes because chemical equilibria and biological phenomena are dependent on the activity of the ions rather than the concentration. For some purposes, however, the concentration of the ions is the most relevant quantity, e.g., for balance studies where input and output of the total amount of substance are of interest.

The relationship between activity and concentration of free ion (I′) in an aqueous solution (S) is given by

$$a\mathrm{I}'(\mathrm{S}) = \gamma\mathrm{I}'(\mathrm{S}) \times m\mathrm{I}'(\mathrm{S})/(\mathrm{mol/kg}) \tag{8}$$

where

$m\mathrm{I}'(\mathrm{S}) = c\mathrm{I}'(\mathrm{S})/\rho\mathrm{H_2O}(\mathrm{S})$
$a\mathrm{I}'(\mathrm{S})$ = activity of I′ in S, (unit: 1).
$\gamma\mathrm{I}'(\mathrm{S})$ = activity coefficient of I′ in S (unit: 1).
$m\mathrm{I}'(\mathrm{S})$ = molality of I′ in S (unit: $\mathrm{mol} \times \mathrm{kg}^{-1}$).
$\rho\mathrm{H_2O}(\mathrm{S})$ = mass concentration of water in S (unit: $\mathrm{kg} \times \mathrm{liter}^{-1}$).
$c\mathrm{I}'(\mathrm{S})$ = concentration of I′ in S (unit: $\mathrm{mol} \times \mathrm{liter}^{-1}$).

In biological fluids it is often impossible to calculate the activity coefficients precisely, partly because the ionic strength is higher than 0.10 mol/kg, and partly because the contribution of protein ions to the ionic strength is highly uncertain. The activity coefficients for Na^+ and K^+ in human blood plasma have been found to be 0.73 and 0.75, respectively. These coefficients have been determined by measuring the activity by means of ion-selective electrodes, and by measuring the concentrations of Na^+ and K^+ by means of flame photometry. A thorough discussion of the problem inherent in the definition and measurement of single-ion activities and activity coefficients is given by Bates.[1]

In general, *activity determinations* are based on the comparison of the potential of the unknown solution with the potential of several standard solutions with known activity. The best example is the pH determination, as illustrated in the following equation:

$$\mathrm{pH(X)} = \mathrm{pH(S_1)} + \frac{\mathrm{pH(S_2)} - \mathrm{pH(S_1)}}{E(\mathrm{S_2}) - E(\mathrm{S_1})} \times [E(\mathrm{X}) - E(\mathrm{S_1})] \tag{9}$$

where $E(S_1)$ and $E(S_2)$ are the readings for two different standard solutions (NBS buffers). pH(X) and E(X) represent the pH and the reading, respectively, for the unknown.

The precision of potentiometric measurements is limited by the standard deviation for the measured potentials (about 50 μV). This uncertainty is equivalent to a standard deviation of about 0.001 for log aI′ for a monovalent ion. This corresponds to a coefficient of variation for aI′ of about 0.2 per cent. For divalent ions, the coefficient of variation will be twice as high.

For the determination of the concentration of total ion (free + bound), the sample must be diluted by a suitable diluent which liberates the complex-bound ion from its binding agent (e.g., by a pH adjustment). The dilution should at the same time serve to establish a constant ionic strength so that a constant activity coefficient is obtained independently of variations in ionic strength of the original sample. A different approach consists of *titration* utilizing a potentiometric end point detection. In this case, the electrode is used only to sense the sudden change in activity as the end point is reached. This technique is generally considered among the most accurate and precise analytical methods available.

REDUCTION (REDOX) POTENTIALS

Redox potentials are due to chemical equilibria involving electron transfer reactions:

$$\text{Oxidized form (Ox)} + ze^- \rightleftharpoons \text{Reduced form (Red)} \tag{10}$$

e.g.,

$$Fe^{3+} + e^- \rightleftharpoons Fe^{2+}$$

or

$$2H^+ + 2e^- \rightleftharpoons H_2$$

where z symbolizes the number of electrons involved in the reaction (the numerical stoichiometric number). Any substance which binds (accepts) electrons is an oxidant (Ox) and a substance which gives off electrons is a reductant (Red). The two forms, Ox and Red, represent a redox couple (conjugate redox pair). Usually, redox processes take place only between two redox couples, the electrons being transferred from a reductant (Red_1) to an oxidant (Ox_2). In this process, Red_1 is oxidized to its conjugate Ox_1, while Ox_2 is reduced to Red_2:

$$Red_1 + Ox_2 \rightleftharpoons Ox_1 + Red_2 \tag{11}$$

e.g.,

$$2S_2O_3^{2-} + I_2 \rightleftharpoons S_4O_6^{2-} + 2I^-$$

In an electrochemical cell, electrons may be accepted from, or donated to, an inert metallic conductor (e.g., platinum). A reduction process tends to charge the electrode positively (remove electrons), and an oxidation process tends to charge the electrode negatively. By convention, redox equilibrium (11) is represented by the cell

$$M_L\,|Red_1 - Ox_1\|Ox_2 - Red_2|\,M_R \tag{12}$$

A positive potential ($E > 0$) for cell (12) signifies that the cell reaction (11) proceeds spontaneously from left to right; $E < 0$ signifies that the reaction proceeds from right to left; and $E = 0$ indicates that the two redox couples are at mutual equilibrium.

The *electrode potential* (reduction potential) for a redox couple is defined as its potential measured with respect to the standard hydrogen electrode which is set equal to zero (see later). This potential, by convention, is the EMF of a cell, where the standard hydrogen electrode is the reference electrode (left electrode) and the given half-cell is the indicator electrode (right electrode). The reduction potential for a given redox couple is given by the

Nernst equation (for derivation, see textbooks of physical chemistry):*

$$E = E^\circ - \frac{N}{z} \times \log \frac{a\text{Red}}{a\text{Ox}} = E^\circ - \frac{0.0592 \times \text{V}}{z} \times \log \frac{a\text{Red}}{a\text{Ox}} \quad (13)$$

where

E = electrode potential of the half-cell.
E° = standard electrode potential when $a\text{Red}/a\text{Ox} = 1$.
z = number of electrons involved in the reduction reaction.
R = gas constant (= $8.3143 \text{ J} \times \text{K}^{-1} \times \text{mol}^{-1}$).
T = absolute temperature (unit: K, kelvin).
F = Faraday constant (= $96487 \text{ C} \times \text{mol}^{-1}$) (C = coulomb).
$N = R \times T \times \ln 10/F$ (= the Nernst factor if $z = 1$).
$N = 0.0592$ V if $T = 298.15$ K (= 25°C).
$N = 0.0615$ V if $T = 310.15$ K (= 37°C).
$\ln 10$ = natural logarithm of 10 = 2.303.
a = activity.
$a\text{Red}/a\text{Ox}$ = product of mass action for the reduction reaction.

The electrodes presently in use can be divided into three major groups or classes, namely, (1) inert metal electrodes immersed in solutions containing redox couples, (2) metal electrodes which function as a member of the redox couple, and (3) selective membrane electrodes.[13,17]

Inert metal electrodes

Platinum or *gold* are examples of inert metals used to record the redox potential of a redox couple dissolved in an electrolyte solution.†

Not all dissolved redox couples, however, are able to equilibrate with an inert metal like platinum or gold. In many cases the presence of a catalyst is required to establish a reproducible potential. The catalyst can be a small amount of another redox couple, a so-called mediator, which readily equilibrates with both the metal electrode and the more sluggish redox couple of interest. Examples of mediators are methylene blue and quinhydrone. Measurement of the redox potential of dissolved redox couples may be used for end point detection in redox titrations, but this procedure rarely finds application in clinical chemistry.

The **hydrogen electrode** is a special redox electrode for pH measurement. It consists of a

* In accordance with the recommendations of the International Organization for Standardization (International Standard, ISO 31 series; see McGlashan[9] and Dybkaer[4]), certain abbreviations are used in this section of the book which are now widely accepted but which are, at this time, not used by clinical chemists in the United States. Although this represents an inconsistency in style, as compared to other sections of the book, it was felt that it is important to introduce these terms since they may find wider use in the future:

Any quantity requires specification of (1) kind of quantity, (2) component, and (3) system. Examples: *concentration* of sodium ion in the plasma, *activity* of hydrogen ions in the erythrocytes, and *partial pressure* of carbon dioxide in the alveolar air. In symbols these examples may be written: $c\text{Na}^+(\text{P})$, $a\text{H}^+(\text{E})$, $p\text{CO}_2(\text{A})$.[16] The symbols of the kinds of quantities should be single letters (Latin or Greek) and written in italic type. A list of recommended symbols has been published by IUPAC (see McGlashan[9]). The symbols for the components are the familiar chemical symbols or suitable abbreviations (e.g., Hb for hemoglobin [Fe]). The symbols suggested for the systems relevant for clinical chemistry are single letter, roman-type symbols, e.g., B = blood, P = plasma, E = erythrocytes, A = alveolar air, and so forth.[4]

† Attempts have been made to measure the redox potential of blood or plasma. However, this potential is an undefined quantity as long as the redox couple is not specified. The different redox couples of blood or plasma are not in thermodynamic equilibrium, in contrast to the equilibrium between the different acid-base pairs. If it were not for this disequilibrium, life would be impossible because all organic substances would rapidly burn to CO_2 and H_2O. The redox potential in blood measured with a gold electrode (using a calomel reference electrode) appears to be the redox potential of the ascorbic acid/dehydroascorbic acid couple, which is of much less clinical interest than that of redox couples such as $NADH/NAD^+$.

platinum or gold electrode which is coated with highly porous platinum (platinized) to catalyze the electrode reaction

$$H^+ + e^- \rightleftharpoons \tfrac{1}{2}H_2$$

The electrode potential is given by

$$E = E^\circ - N \times \log \frac{(aH_2)^{1/2}}{aH^+}$$

or

$$E = E^\circ - N \times (\log aH_2^{1/2} - \log aH^+)$$

where

E° = 0 at all temperatures (by convention).
aH^+ = activity of hydrogen ions.
$-\log aH^+$ = negative log of the H^+ activity (paH^+ or pH).

When the partial pressure of H_2 in the solution (and hence aH_2) is maintained constant, by bubbling H_2 through the solution, the potential is a linear function of log aH^+ (= −pH). In the *standard hydrogen electrode* the electrolyte consists of an aqueous solution of HCl with $aH^+ = 1.000$ (or cHCl = 0.876 mol/l) in equilibrium with a gas phase with $aH_2 =$ 1.000 (or pH_2 = 101.3 kPa = 1 atm).

The **quinhydrone electrode** is another special redox electrode used for pH measurement. It consists of a platinum or gold electrode immersed in a saturated solution of quinhydrone, which is an equimolar mixture of quinone (Q) and hydroquinone (H_2Q). The electrode reaction is

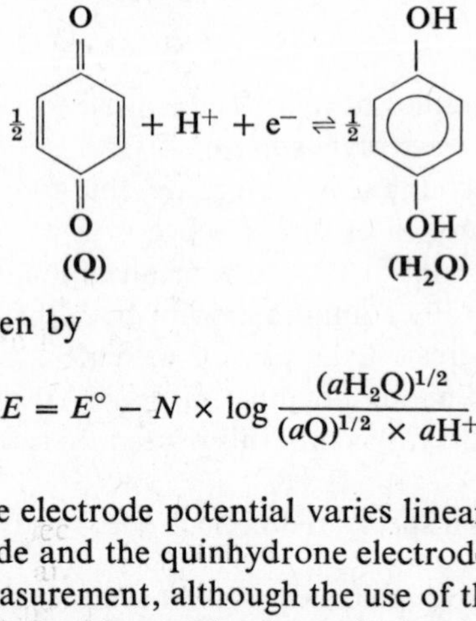

$$\tfrac{1}{2}\,\text{Q} + H^+ + e^- \rightleftharpoons \tfrac{1}{2}\,H_2Q$$

(Q) (H_2Q)

The electrode potential is given by

$$E = E^\circ - N \times \log \frac{(aH_2Q)^{1/2}}{(aQ)^{1/2} \times aH^+}$$

and because $aH_2Q = aQ$, the electrode potential varies linearly with the negative log aH^+ or pH. The hydrogen electrode and the quinhydrone electrode have both been replaced by the glass electrode for pH measurement, although the use of the quinhydrone electrode was recently proposed in connection with a pCO_2 electrode (see later).

Metal electrodes participating in redox reactions

The **silver electrode** consists of a silver wire immersed in a solution containing silver ions. The electrode process consists of the reduction of silver ions to metallic silver:

$$Ag^+ + e^- \rightleftharpoons Ag$$

The expression for the electrode potential reduces to:

$$E = E^\circ + N \times \log aAg^+$$

because the activity of pure silver is unity. According to this equation, the electrode measures the silver ion activity in the solution. An application for this electrode is the determination of chloride by titration with $AgNO_3$. In this method the silver electrode is coupled with a mercurous sulfate reference electrode ($\tfrac{1}{2}Hg_2^{2+} + e^- \rightleftharpoons Hg$) for the end point determination.

When all Cl^- ions are precipitated as AgCl, the sudden excess of Ag^+ causes a sudden change in the electrode potential, which indicates the end point of the reaction.

In an analogous fashion, other metal electrodes (e.g., Zn, Cu) measure the activity of their respective ions. When the metal is chemically unstable in pure form, an electrode consisting of an amalgam of the metal (e.g., calcium amalgam) can sometimes be used. Unfortunately, such electrodes are of little value for direct measurements in biological fluids because proteins and lipids tend to coat the amalgam surface, and various redox couples present in biological fluids tend to affect the electrode potential.

The **silver/silver chloride electrode** consists of a silver wire, electrolytically coated with AgCl, which dips into a solution containing chloride ions. The electrode process is

$$\text{AgCl (solid)} + e^- \rightleftharpoons \text{Ag (metal)} + Cl^-$$

Since aAgCl and aAg are both unity (because both components are present as pure substance), the expression for the electrode potential reduces to

$$E = E^\circ - N \times \log aCl^-$$

which shows that the electrode measures chloride activity. This electrode has been utilized for direct measurement of the chloride activity in serum. By placing it directly on the skin surface, it can also be used for the measurement of the chloride activity of sweat in connection with the diagnosis of mucoviscidosis (cystic fibrosis).

When the chloride activity is kept constant, e.g., with cKCl fixed at 0.1 mol/l, the electrode potential is constant, and therefore the electrode is frequently employed as a reference electrode or "inner electrode" in membrane electrodes (see Fig. 3–23).

In complete analogy, a silver/silver bromide electrode, a silver/silver iodide electrode, and a silver/silver sulfide electrode measure bromide activity, iodide activity, and sulfide activity, respectively. Unfortunately, these electrodes are quite sensitive to other redox couples in the test solution and therefore often fail in biological solutions.

The **calomel electrode** consists of mercury covered by a layer of calomel (Hg_2Cl_2), which is in contact with an electrolyte solution containing chloride. The electrode process is

$$\tfrac{1}{2}Hg_2Cl_2 \text{ (solid)} + e^- \rightleftharpoons \text{Hg (metal)} + Cl^-$$

Since aHg and aHg_2Cl_2 are both unity (present as pure substance), the electrode potential reduces to

$$E = E^\circ - N \times \log aCl^-$$

This shows that the electrode also functions as a chloride electrode, and its potential varies with the chloride activity. The chloride activity is generally maintained constant, either as saturated KCl or at a concentration of 3.5 (or 4.0) mol/l. Calomel electrodes are frequently employed as a reference electrode together with a glass electrode for pH measurement.

ION-SELECTIVE MEMBRANE ELECTRODES

For the discussion of these types of electrodes, it is necessary first to give some background information.

Membrane potentials are due to permeability of certain types of membranes to selected anions or cations. Biological membranes (e.g., the glomerular membrane in the kidney) are often impermeable to the high molecular protein ions; this gives rise to the so-called Donnan potential, which in turn leads to an uneven distribution of the diffusible ions on both sides of the membrane (see p. 949). For analytical applications, membranes are required which possess a selective permeability for a single ion species.

The ion-selective membrane, separating the solution on the left side (S_L) from the solution on the right side (S_R), can be illustrated as

$$S_L \,|\text{Membrane}|\, S_R \tag{14}$$

The membrane potential is conventionally defined as the potential on the right side minus the potential on the left, i.e.,

$$V(S_L \,|\, S_R) = V(S_R) - V(S_L) \tag{15}$$

The potentials of solutions S_L and S_R are recorded with any suitable reference electrode. In the case of a glass membrane, the potential of the reference solution (S_R) is generally recorded against an Ag/AgCl electrode, while the potential of the test solution (S_L) is recorded against a calomel electrode via a liquid-liquid junction between the saturated KCl and the test solution.

If the activity of a diffusible cation (I′) is higher in S_L than in S_R, a positive membrane potential develops. In most cases the mechanism is as follows: On the left side, where the cation activity is high, cations are bound to the membrane surface by specific binding groups and the membrane is thereby charged positively. This charging of the membrane causes a dissociation of cations from the other side of the membrane into the solution on the right side, which is thereby charged positively with respect to the solution on the left side. Thus, it appears as if the membrane were permeable to I′ only. This process proceeds until an equilibrium is established. This occurs when the electric potential difference across the membrane matches the difference in activity of the diffusible ions on the two sides of the membrane. Attempts to force I′ through the membrane by an electric current often damage the membrane, or at least make it behave differently from when it is at zero current. It might be mentioned that a thin silver membrane behaves as if it were a silver ion permeable membrane. However, the membrane is not permeable to ions and the real mechanism is a redox equilibrium ($Ag^+ + e^- \rightleftharpoons Ag$) at both surfaces.

The membrane potential is given by the Nernst equation (for derivation see textbooks of physical chemistry):

$$V(S_L \,|\, S_R) = V(S_R) - V(S_L) = -\frac{R \times T}{z \times F} \times \ln \frac{a\text{I}'(S_R)}{a\text{I}'(S_L)} \tag{16}$$

where z is the ion charge number (positive for cations, negative for anions), and I′ is the diffusible ion (cation or anion). If S_R is a reference solution with constant $a\text{I}'(S_R)$, the equation reduces to

$$V(S_L \,|\, S_R) = V' + \frac{N}{z} \times \log a\text{I}'(S_L) \tag{17}$$

where V' is constant. In other words, the *membrane potential is directly proportional to the logarithm of the activity of the diffusible ion in the test solution* (Fig. 3–22).

The sensitivity of the electrode, $dV/d \log a\text{I}'(S_L)$, is given by the so-called (theoretical) Nernst slope, N/z. However, the actual sensitivity often deviates slightly, being $s \times N/z$, where s is the relative sensitivity. For most pH-glass electrodes s is 0.98 to 1.00. For other ion-selective electrodes s may be considerably less than 1. It should be noted that the theoretical sensitivity for divalent ions is only half the value for monovalent ions.

The selectivity of the electrode for a single ion species is seldom absolute. If the membrane is "permeable" to several different ions, a diffusion potential may arise, and the theory for the membrane potential becomes more complicated. The following empirical equation (Nicolsky) describes the measurement potential as a function of the activity of the

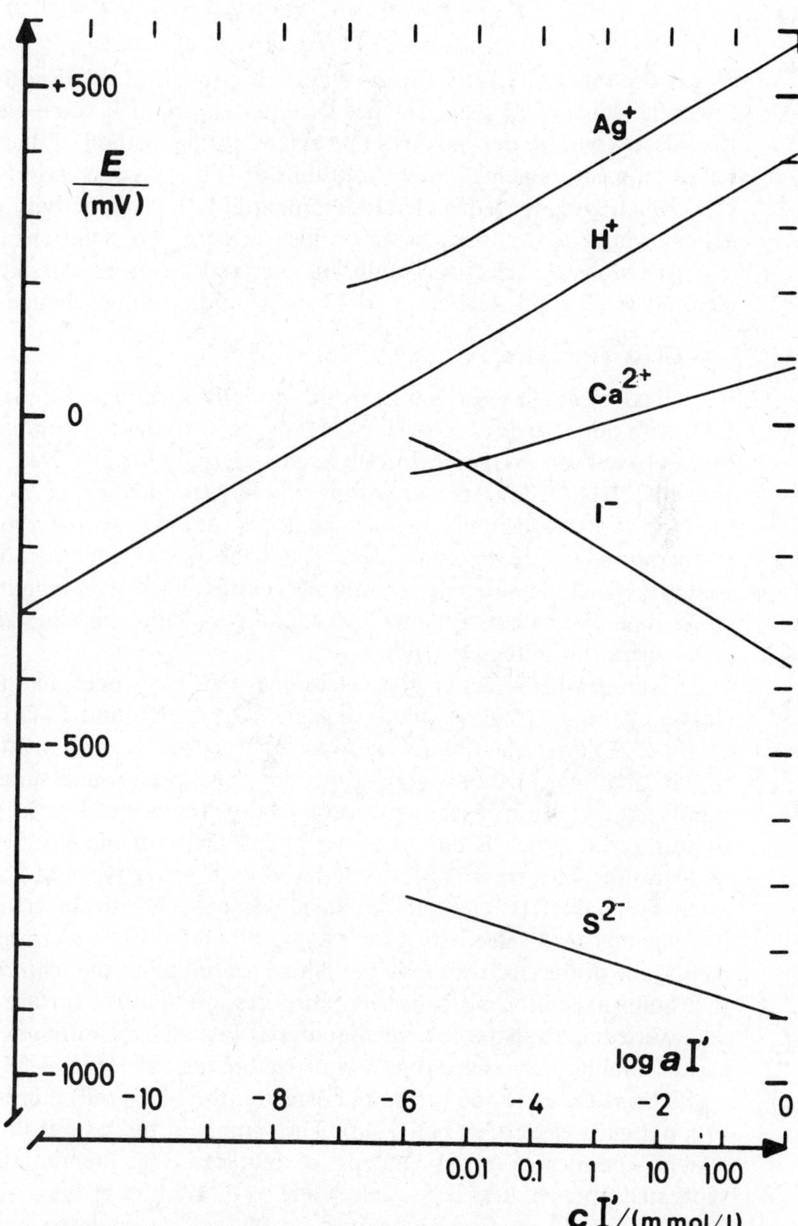

Figure 3-22 Illustration of the Nernst equation, i.e., the relationship between electromotive force of an ion-selective electrode chain (E) and the logarithm of the activity of the ion ($\log a\mathrm{I}'$):

$$E = E' + \frac{R \times T \times \ln 10}{z \times F} \times \log a\mathrm{I}'$$

where E' is constant at constant temperature, and z is the charge number for the ion (positive for cations, negative for anions). The slope for the monovalent ions is ± 59.16 mV; for divalent ions it is ± 29.58 mV ($T = 298.15$ K). The abscissa also indicates the concentration of I′ provided the activity coefficient is taken to be 1.

primary ion I_1' and an interfering ion I_2':

$$V = V' + \frac{N}{z} \times \log (aI_1' + K_{1,2} \times aI_2'^{z_1/z_2}) \quad (18)$$

$K_{1,2}$ is the selectivity ratio for the interfering ion I_2' in relation to the primary ion I_1'. The lower the value of $K_{1,2}$, the smaller the interference of I_2' when measuring I_1'. The value for the selectivity ratio depends to some extent on the method of measurement. One method is based on measurements in pure solutions of I_1' and I_2', respectively, and finding the activity of I_2' which gives the same electrode potential as a given activity of I_1'. In this case, $K_{1,2} = aI_1'/aI_2'$. Better methods are based on measuring $K_{1,2}$ in solutions containing both I_1' and I_2'.

Available ion-selective membrane electrodes can be classified arbitrarily as (1) glass electrodes, (2) solid-state electrodes, and (3) liquid ion-exchange electrodes.[3,10]

Glass electrodes

Glass electrodes are made from specially formulated glass consisting of a melt of SiO_2 with added oxides of various metals. By varying the composition of the glass, membranes have been prepared with directed selectivity for H^+, Na^+, K^+, Li^+, Rb^+, Cs^+, Ag^+, Tl^+, and NH_4^+.[5] The membranes generally have a thickness of 50 to 100 μm. Depending on the type of glass, their electrical resistance is very high, being about 10 to 800 MΩ at room temperature, increasing considerably with decrease in temperature. Recently, glass electrodes in which the reference solution is omitted have been manufactured, the glass surface being directly coated to a metallic conductor. This simplifies the electrode and permits miniaturization of the electrodes.

Hydrogen-ion selective glass electrodes (pH electrodes) can be manufactured from the classic Corning 015 glass, consisting of SiO_2, Na_2O, and CaO in the mass ratio of 72.2/21.4/6.4. A newer glass more selective to H^+ consists of SiO_2, Li_2O, and CaO in the ratio 68/25/7, but many other compositions have also been found suitable. With the older glass, significant error in measurement arose in the presence of Na^+ at pH values above 8.0. The sodium error, which is due to a loss in specificity of the electrode, increases greatly with temperature. This error is greatly reduced with newer types of glass which allow pH measurements in NaOH solutions up to about pH 13 with an error of less than 0.02. This corresponds to a selectivity ratio $K_{H,Na}$ of about 10^{-15} at room temperature. The glass membrane of the electrode may be shaped according to the requirements of the application. It is bulb-shaped for most titration purposes, and flat for surface measurements, but it has an inverted bulb-shape for microanalysis. For pH measurements in blood, the thermostatted capillary glass electrode has proved invaluable (Fig. 3–23).

The ***p*CO₂ electrode** (or more correctly, the pCO_2 cell) represents a special application of a pH glass electrode (Fig. 3–24). The sample in this case is in contact with a membrane which is permeable to gas but not to solutions. This membrane (e.g., silicone rubber) is separated from the actual glass electrode by a thin film of bicarbonate solution (5 mmol/l). The CO_2 gas diffuses from the sample (or test gas) through the membrane and rapidly enters CO_2 equilibrium with the bicarbonate solution altering its pH. The pH of the bicarbonate solution is a simple function of the pCO_2, obtained by rearrangement of the Henderson-Hasselbalch equation:

$$\text{pH} \doteq -s \times \log pCO_2 - \log \alpha + \text{p}K' + \log cHCO_3$$

where s is the relative sensitivity of the electrode, normally 0.95 to 1.00. α is the solubility coefficient of CO_2 in the bicarbonate solution, K' is the apparent, overall, first dissociation constant of carbonic acid.

Na^+-selective glass electrodes can be prepared from glass consisting of SiO_2, Na_2O, and Al_2O_3 in the ratio of 71/11/18. Lithium aluminum silicates have also been found suitable.

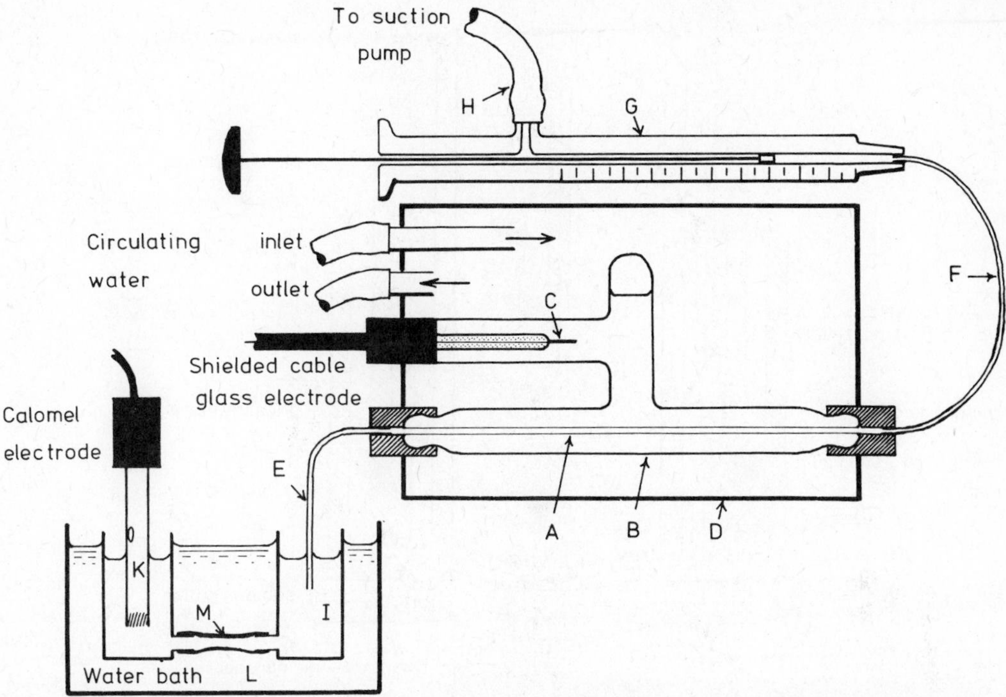

Figure 3-23 Capillary glass electrode for pH measurements in blood. A is a capillary tube manufactured from H^+-sensitive glass. Tube B contains the "inner" reference solution. C is the "inner" Ag/AgCl electrode. D is a water jacket for circulating water from a thermostatted bath. F is a polyethylene tube connecting the glass capillary with a Hamilton syringe, G, which is used for drawing the blood into the glass capillary through the polyethylene tubing E. K is a calomel electrode immersed in a thermostatted water bath. The liquid-liquid junction is formed between the saturated KCl solution in I and the blood in the tip of E.

The selectivity ratio $K_{Na,K}$ may be as low as 10^{-3}, and the electrode is insensitive to H^+ in the pH range 6 to 10. Electrodes with a flat surface have been used for the direct measurement of sodium ion activity ($a\text{Na}^+$) on the skin surface for the diagnosis of mucoviscidosis. Capillary electrodes have been constructed for measurements of $a\text{Na}^+$ in serum or serum dilutions and are employed in a recent multichannel analyzer (SMAC, Technicon Instruments Corporation, Tarrytown, N.Y. 10591).

K^+-selective glass electrodes are as yet less satisfactory than are the Na^+ electrodes, $K_{K,Na}$ being about 0.05. Attempts have been made to measure H^+, Na^+, and K^+ in serum simultaneously with three different glass electrodes, utilizing a computer for calculation of the correction for the different electrodes on the basis of the different selectivity constants.

The **NH_4^+-selective glass electrode** has been utilized to measure NH_4^+ concentrations. An interesting application of this electrode is in the determination of urea.[7] In this technique the urea in the sample diffuses to a membrane consisting of a polyacrylamide matrix containing immobilized urease fixed on a Dacron net. The urea is then hydrolyzed to NH_4^+ which is measured by the glass electrode. This is but one example of a potentiometric enzyme electrode. The use of enzyme membranes offers great possibilities for the development of electrodes "specific" to a variety of biochemical materials.

Solid-state electrodes

Solid-state membranes can be either homogeneous membranes consisting of a "single" crystal, or heterogeneous membranes consisting of an active substance imbedded in an inert matrix.[12,15]

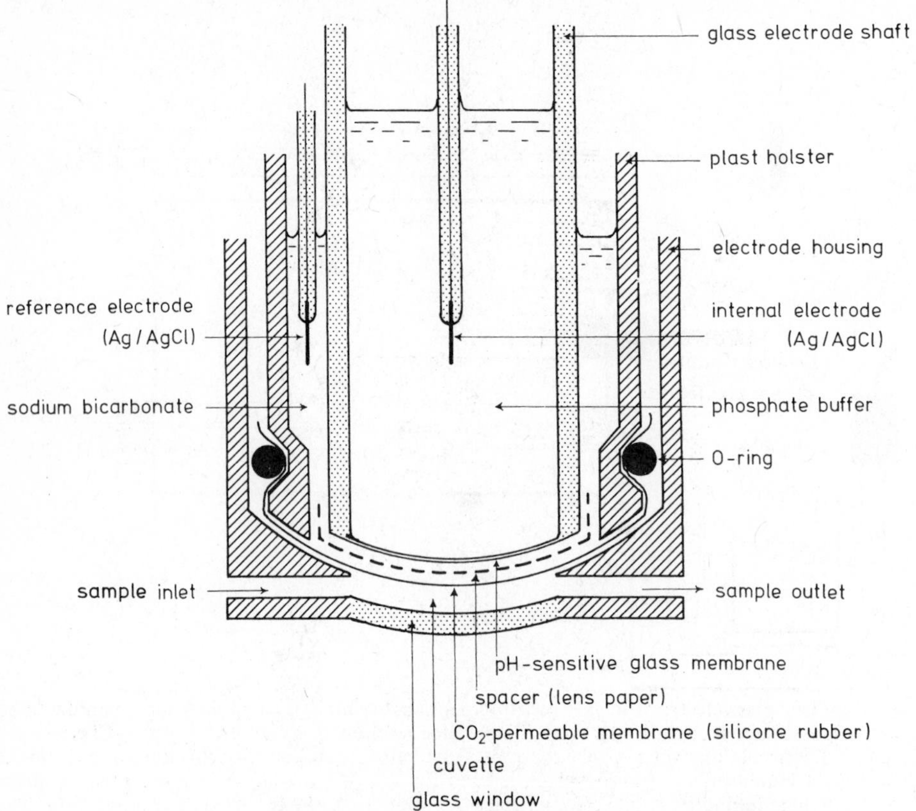

Figure 3–24 Schematic illustration of a pCO_2 electrode.

The homogeneous membrane electrodes include those for F^- (lanthanum fluoride crystal), Cl^- (AgCl crystal), Br^- (AgBr crystal), I^- (AgI crystal), S^{2-} (Ag_2S), Cu^{2+} (cupric selenide crystal), and others. The silver salt solid-state membrane electrodes are less susceptible to interference from redox systems than are the equivalent silver/silver salt redox electrodes. The AgCl membrane electrode is used for measurement of the activity of chloride in sweat by direct measurement on the skin surface (Orion electrode; Orion Co., Boston, MA.). The fluoride electrode has been utilized for measurements of the fluoride concentration in saliva.

Ion-exchange electrodes

Liquid ion-exchange membranes consist of an *inert solvent* in which an *ion-selective carrier* substance is dissolved. Both solvent and carrier should be insoluble in water. The membrane solution can be separated from the test solution by means of a collodion membrane, or a porous matrix can be soaked by the membrane solution. Many different membranes have been prepared with selectivity for specific cations or anions.

A calcium-selective membrane can be made by dissolving dioctyl phosphate (selective carrier) in dioctylphenylphosphonate (inert solvent).[14] The calcium electrode has been used for measurement of the calcium ion activity of serum. As the technique becomes perfected, this method will undoubtedly become a routine method in the clinical laboratory because of the clinical importance of the calcium ion activity (ionized calcium) in many physiological processes (see Chap. 15).

Potassium-selective membranes can be made by dissolving the antibiotic valinomycin in a suitable solvent.[6,11] Valinomycin is a neutral carrier which binds K^+ in the center of a

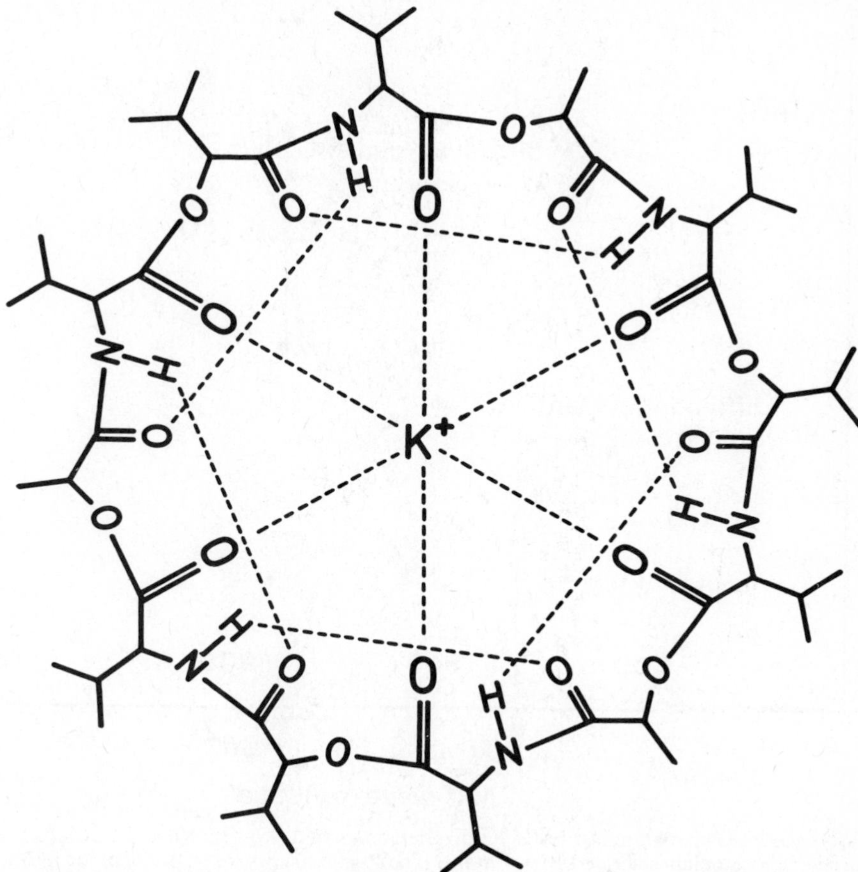

Figure 3–25 Valinomycin has a heterocyclic structure consisting of alternating peptide and ester linkages. The molecule is cylindrical, stabilized by six hydrogen bridges (from —NH to C═O). The outer surface of the cylinder is strongly hydrophobic, and the interior is strongly hydrophilic. The size of the central cavity nearly equals the diameter of an unhydrated potassium ion.

ring of oxygen atoms (Fig. 3–25). This membrane is highly selective for potassium, the selectivity ratio $K_{K,Na}$ being 2.5×10^{-4}. The electrode has already found application for measurement of potassium ions in serum.

POLAROGRAPHY

Polarography is based on the dual measurement of the current flowing through an electrochemical cell, and the electrical potential between the two electrodes when the potential is gradually increased at a constant rate by means of an external voltage source.[8] The function between current and potential is called a *polarogram* (Fig. 3–26).

A polarographic cell generally consists of a polarizable indicator electrode and a non-polarizable reference electrode. The indicator electrode can be a dropping mercury electrode, i.e., a glass capillary tube (50 μm ID) filled with mercury which slowly drips in small droplets out of the tip of the tubing into the test solution. In this way the electrode surface is constantly renewed. The reference electrode can be a large pool of mercury in the bottom of the vial.

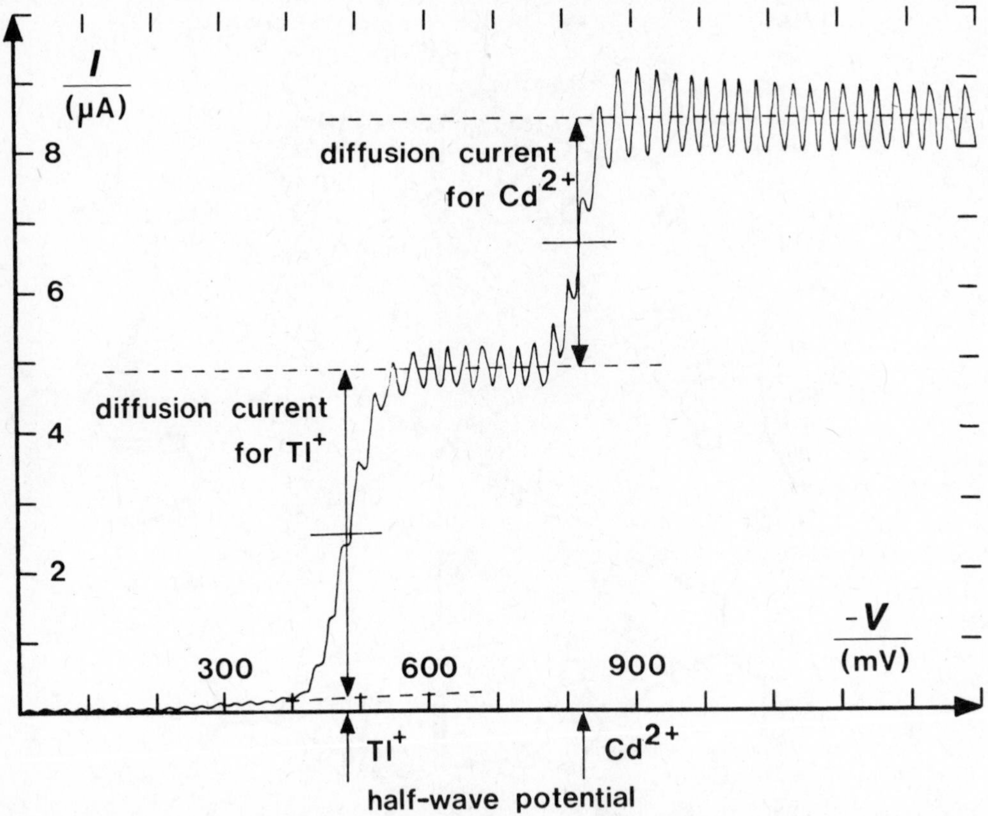

Figure 3–26 Polarogram of a test solution containing TlCl (1.0 mmol/l) and $CdCl_2$ (0.5 mmol/l) in an ammonia/ammonium chloride buffer (1 mol/l). The abscissa shows the potential of the indicator electrode. The ordinate shows the current through the cell. The current pulsates due to the dropping mercury cathode.

The cell may be symbolized as follows, where the indicator electrode is the cathode:

Hg	|Test solution|	Hg
(Anode)		**(Cathode)**

The cell is symmetrical and therefore shows an electromotive force of zero. When an increasingly negative potential is applied to the cathode at a constant rate, the current at first is almost zero, because the cathode is electrolytically polarized (i.e., the cell becomes asymmetrical and therefore generates a counterelectromotive force which balances the applied potential). If the test solution is oxygen-free water, the electrode processes are

Cathode $$H^+ + e^- \rightleftharpoons \tfrac{1}{2}H_2$$

Anode $$2Hg \rightleftharpoons Hg_2^{2+} + 2e^-$$

Hydrogen is formed at the cathode, which becomes a "hydrogen electrode," and mercury is oxidized to Hg_2^{2+} at the anode. However, due to the large surface of the anode, the concentration changes at the anode are small and insignificant in relation to the concentration changes at the cathode. The anode is therefore considered to be nonpolarized. A calomel electrode or Ag/AgCl electrode also behaves as a nonpolarizable electrode because the changes in chloride concentration when a current is flowing are relatively small, and the electrode potential is therefore relatively constant.

The maximal counterelectromotive force which can be obtained is reached when pH_2

at the cathode has reached atmospheric pressure. This is at a potential of about −1.7 V, although the electromotive force of a cell with a hydrogen electrode and a mercury reference electrode is only about −0.8 V. The difference is called the overpotential, which is ascribed to activation energy required for the formation of H_2. If the cathode potential is increased above this so-called *decomposition potential*, a current will flow and bubbles of hydrogen develop at the cathode. When H^+ is continuously reduced, H^+ is said to depolarize the cathode.

If the test solution contains substances which are reduced more easily than H^+, the cathode will be depolarized by these substances and a current will flow through the cell. For example, if the test solution contains oxygen, this will be reduced at the cathode according to the reaction

$$O_2 + 2H_2O + 4e^- \rightarrow 4OH^-$$

When the decomposition potential for this reaction (about −0.3 to 0.4 V) is reached, the current increases until it reaches a plateau, the so-called diffusion current. The height of the plateau is dependent on the rate at which O_2 can diffuse from the surrounding solution to the surface of the cathode where pO_2 is zero. The diffusion current is therefore directly proportional to the pO_2 of the test solution. Any substance which is reduced at the cathode is characterized by two parameters: (1) the *half-wave potential* (i.e., the potential where the current is equal to one-half the plateau current) characterizes the kind of substance being reduced, and (2) the *diffusion current* is proportional to the concentration of the substance in the test solution. Polarography may therefore be used for identification as well as for quantitation.

A variant of polarography is the so-called *anodic stripping voltammetry*. One of the electrodes is a mercury coated graphite rod. A negative potential is applied to the electrode. The trace metal ions of the sample are then reduced and plate the electrode. The plating time is usually from 1 to 30 min, depending on the concentration. A polarogram is then recorded with the plated electrode as the anode and a nonpolarizable cathode. The metals are stripped off the anode by oxidation to the respective cations. The order in which they are stripped off is a function of the metal's unique redox potential. The current flow during the stripping of a given metal is a function of the amount of this metal. The method therefore provides identification as well as quantitative measurement of trace metals. The preconcentration (plating) step permits the analysis of extremely dilute samples.

AMPEROMETRY

This technique is based on measurement of the current flowing through an electrochemical cell when a constant electric potential is applied to the electrodes. Some examples of this analytical principle are described in the following text:

The ***p*O₂ electrode** (Clark electrode, Fig. 3–27) is actually a complete electrochemical cell consisting of a small platinum cathode (area about 300 μm²) and an Ag/AgCl anode in phosphate buffer with added KCl. The platinum cathode, covered by a thin film of electrolyte, is separated from the test solution by a gas-permeable membrane (e.g., polypropylene). The cathode potential is adjusted to −0.65 V. In the absence of oxygen in the test solution, the current is almost zero because the cathode is polarized. In the presence of oxygen, a current is observed which is due to a diffusion of O_2 from the test solution through the membrane to the cathode, where it is reduced. The current is directly proportional to the pO_2 in the test solution. The sensitivity of commercial pO_2 electrodes is on the order of $dI/dpO_2 = 10^{-12}$ A/mm Hg, depending on the size of the cathode area and the thickness of the gas-permeable membrane.

The purpose of the membrane is two-fold: first, to prevent proteins and other (dissolved) oxidants from gaining access to the cathode surface and, second, to limit the diffusion zone

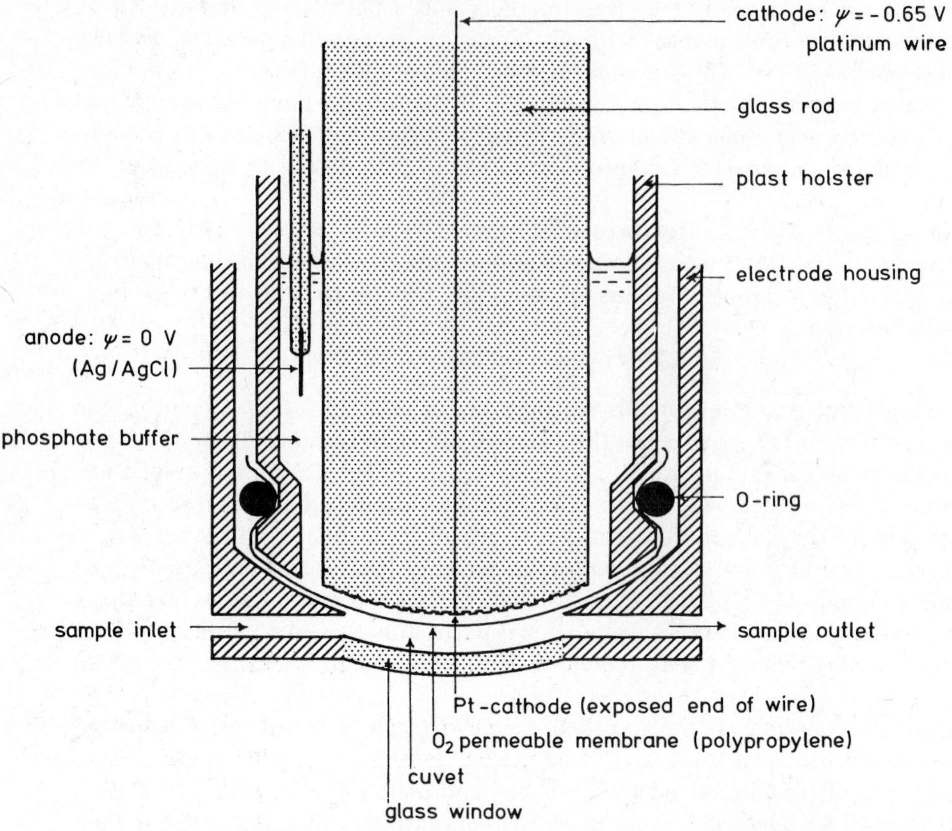

Figure 3-27 Schematic illustration of a pO_2 electrode.

to the membrane and hence prevent variations in the diffusion coefficient of O_2 in the test solution (or gas) from influencing the result.

The pO_2 electrode has found widespread application for measurements of pO_2 in arterial or capillary blood. It has also been applied for the measurement of the concentration of *total oxygen* in the blood after liberation of hemoglobin-bound oxygen with ferricyanide or carbon monoxide (by forming methemoglobin and carboxyhemoglobin, respectively).

Glucose determination by means of the pO_2 electrode is based on the glucose oxidase catalyzed reaction:

$$\text{Glucose} + O_2 \rightarrow \text{Gluconic acid} + H_2O_2$$

The rate of fall in pO_2 under standardized conditions is a measure of the glucose concentration. The glucose oxidase may also be trapped in a gel around the cathode so that a diffusion equilibrium involving both glucose and oxygen is established. Under standardized conditions, the electrode response is then directly related to the glucose concentration.

The **peroxidase electrode** consists of *a polarized platinum anode* and a nonpolarized silver/silver chloride cathode. The anode potential is +0.6 V. In the presence of peroxide, a current flows owing to oxidation of peroxide at the anode:

$$H_2O_2 \rightarrow 2H^+ + 2e^- + O_2$$

The peroxide electrode covered by an enzyme membrane or enzyme layer has been proposed by Clark[2] for the determination of any substance for which a suitable oxygen oxidoreductase is available. Important examples are glucose, uric acid, and ethanol, in which case the membrane contains glucose oxidase, uricase, and alcohol dehydrogenase, respectively. The substrate diffuses to the enzyme layer where it is dehydrogenated, and H_2O_2 is formed. The H_2O_2 diffuses to the anode and causes a current proportional to the rate of formation of H_2O_2, i.e., proportional to the diffusion of substrate to the membrane. Further applications of this principle will probably lead to other useful methods in the future.

Amperometric end point determination may be applied in the titration of Cl^- with Ag^+. The sample is acidified by means of nitric acid (or HNO_3 plus acetic acid) and the chloride ions are titrated with Ag^+. The silver ions may be added as a solution of $AgNO_3$ (volumetric titration), or Ag^+ may be generated from a silver electrode by means of an electric current (coulometric titration; see later). During titration, the Ag^+ concentration remains low due to the reaction $Ag^+ + Cl^- \rightarrow AgCl$, which causes the precipitation of AgCl. At the end point, Ag^+ appears in excess, and the increase in Ag^+ activity may be detected either potentiometrically (with an Ag electrode and a mercury/mercurous sulfate reference electrode), or the Ag^+ may be detected amperometrically. In the latter case, two silver electrodes are employed and a negative potential of 0.15 to 0.25 V is applied to the cathode. During titration, the cathode is polarized and the current is very low. At the end of the titration, the excess of Ag^+ depolarizes the cathode by the reaction $Ag^+ + e^- \rightarrow Ag$, and a current flows which is proportional to the excess of Ag^+. When the current has reached a preset value, the titration can be automatically stopped. This indicator principle is applied in the Cotlove chloride titrator, which in addition uses coulometric generation of Ag^+.

COULOMETRY

Coulometry is the technique used to measure the *amount* of electricity passing between two electrodes in an electrochemical cell. The amount of electricity is directly proportional to the amount of substance produced or consumed by the redox process at the electrodes. This is called Faraday's first law and may be expressed as

$$z \times n \times F = Q \tag{19}$$

where z is the numerical stoichiometric number of electrons involved in the reduction (or oxidation) reaction (unit: 1), n is the amount of substance reduced or oxidized (unit: mol), F is the Faraday constant ($= 96487\ C \times mol^{-1}$), and Q is the amount of electricity (unit: C = coulomb = ampere × second) passing through the cell.

An example of an application is the coulometric titration of chloride (Cotlove titrator), where silver ions are generated by electrolysis from a silver wire used as anode. At the cathode, H^+ is reduced to H_2. The amount of silver ions generated is measured coulometrically. When the current is kept constant, the measurement is reduced to a measurement of time according to

$$Q = I \times t \tag{20}$$

where Q is the amount of electricity (unit: coulomb), I is the electric current (unit: ampere), and t is time (unit: s, second). The current may also be gradually decreased as the titration approaches the end point. In this case the amount of electricity is calculated as the integral: $Q = \int_0^t I \times dt$, a calculation performed electronically by the coulometer. The end point of the titration may be detected either amperometrically or potentiometrically.

Acid-base titrations can also be performed coulometrically using a platinum generator electrode in the test solution separated from the other electrode by a sintered glass filter. If the generator electrode is the cathode, H^+ is removed ($H^+ + e^- \rightarrow \frac{1}{2}H_2$), which is equivalent to the addition of base. If the generator electrode is the anode, H^+ is added ($OH^- \rightarrow \frac{1}{2}O_2 + H^+ + 2e^-$). In either case, two sets of electrodes are necessary: the generator electrodes and the indicator electrodes. The latter may be used for direct potentiometric determination of the end point or in connection with amperometric end point determination. The position of the indicator electrodes should be specially adjusted in relation to the generator electrodes so that the current through the latter does not disturb the performance of the former. Coulometric titrations are among the most accurate analytical determinations available.

CONDUCTOMETRY

Conductometry is the measurement of the current flow between two nonpolarized electrodes between which a known electrical potential is established. In order to avoid polarization of the electrodes, an alternating potential is applied with a frequency between 100 to 3000 Hz. With increase in conductivity* of the solution, there is less impedance (resistance) and, therefore, increased current flow. (Since the applied potential is alternating, the resulting current is also alternating.) The alternating current flowing is directly proportional to the conductivity of the solution.

The conductivity of aqueous solutions is dependent on the concentration of electrolytes and is closely related to the ionic strength. In the purest available water, the conductivity is $\kappa = 4.9 \times 10^{-6}$ S × m^{-1} at 18°C. In ordinary distilled or deionized water the value is $\kappa < 2 \times 10^{-4}$ S × m^{-1}. A higher conductivity indicates the presence of electrolytes, and conductivity measurements therefore serve to contol the performance of deionizers and to provide a warning that the ion-exchange resin should be regenerated. Conductivity measurement can also be used for end point detection in many kinds of titrations (acid-base, precipitation, compleximetric).

The conductivity of whole blood is greatly dependent on the volume fraction of erythrocytes, and attempts have been made to utilize conductivity measurements for determination of the erythrocyte volume fraction (hematocrit).

REFERENCES

1. Bates, R. G.: Determination of pH. Theory and Practice. 3rd ed. New York, John Wiley & Sons, Inc., 1973.
2. Clark, L. C., Jr.: A family of polarographic enzyme electrodes and the measurement of alcohol. Biotechnol. Bioeng., *3:* 377, 1972.
3. Durst, R. A. (Ed.): Ion-Selective Electrodes. Washington D.C., National Bureau of Standards Publication, 314, 1969.
4. Dybkaer, R.: Nomenclature for quantities and units. Stand. Methods Clin. Chem., *6:*223, 1970.
5. Eisenman, G. (Ed.): Glass Electrodes for Hydrogen and Other Cations. New York, Marcel Dekker Inc., 1967.
6. Frant, M. S., and Ross, J. W., Jr.: Potassium ion specific electrode with high selectivity for potassium over sodium. Science, *167:*987, 1970.
7. Guilbault, G. G.: Analytical uses of immobilized enzymes. Biotechnol. Bioeng., *3:*361, 1972.
8. Heyrovsky, J., and Kuta, J.: Principles of Polarography. Prague, Publ. House of the Czechoslovak Acad. Scí., 1965.
9. McGlashan, M. L.: Physicochemical Quantities and Units. The Grammar and Spelling of Physical Chemistry. 2nd ed. London, Royal Institute of Chemistry, 1971.
10. Moody, G. J., and Thomas, J. D. R.: Selective Ion Sensitive Electrodes. Watford, England, Merrow Publishing Co., 1971.
11. Pioda, L. A. R., Stankova, V., and Simon, W.: Highly selective potassium responsive liquid membrane electrode. Anal. Letters, *2:*665, 1969.
12. Pungor, E., Havas, J., and Tóth, K.: Membranes of heterogeneous structure for the determination of the activity of anions, I. Acad. Sci. Hung. Acta Chim., *41:*239, 1964.

---

* *Conductance* is defined as the current (unit: A) divided by the potential difference (unit: V), when no EMF is present, and the unit therefore is ohm^{-1} = siemens, (Ω^{-1} = S).

Conductivity is defined as the current density (unit: A × m^{-2}) divided by the electric field strength (unit: V × m^{-1}), when no EMF is present, and the unit therefore is: $\Omega^{-1} \times m^{-1} = S \times m^{-1}$.

13. Purdy, W. C.: Electroanalytical Methods in Biochemistry. New York, McGraw-Hill Book Company, 1965.
14. Ross, J. W.: Calcium-selective electrode with liquid ion exchanger. Science, *156:*1378, 1967.
15. Růžička, J., Lamm, C. G., and Tjell, C.: Selectrode™, the universal ion selective electrode. III. Concept, constructions and materials. Anal. Chim. Acta, *62:*15, 1972.
16. Siggaard-Andersen, O.: Names and symbols for quantities (letter). Clin. Chem., *18:*1443, 1972.
17. Weyer, E. M. (Ed.): Bioelectrodes. Ann. N.Y. Acad. Sci., *148:*1, 1968.

SECTION FOUR

OSMOMETRY

by Esther F. Freier, M.S.

Osmometry is a technique for measuring the concentration of solute particles which, in turn, is related to the osmotic pressure of a solution. The osmotic pressure governs the movements of water (or solvent) across membranes separating two different solutions. Those membranes which are permeable only to water are referred to as being *strictly semipermeable*. The term *partially semipermeable* is applied to those membranes which permit passage of some small molecules and ions in addition to water. Membranes such as those enclosing the glomerular and capillary vessels are permeable to water and to essentially all small molecules and ions but not to macromolecular colloids such as proteins.

Consider an aqueous solution of sucrose placed within a sac made up of a strictly semipermeable membrane, with an open vertical glass tube manometer attached to this sac. If this device is placed into a beaker of distilled water, water will move from the beaker across the membrane into the sucrose solution. The sucrose solution will rise some distance up the manometer tube, the height of the rise in the tube being a measure of the *osmotic pressure* of the sucrose solution. Osmotic pressure can be defined as the pressure that is required to maintain equilibrium between the sucrose solution on one side of the membrane and the water on the other side of the membrane; i.e., the pressure that must be applied to the sucrose side of the membrane to prevent the influx of water into the monometer tube.

Osmosis

Osmosis[3,4] is that process which constitutes the movement of water across the membrane in response to differences in osmotic pressure on the two sides of a membrane. Water generally migrates from that side of the membrane with the more dilute solute to that containing the more concentrated solute. The water is more concentrated on the pure (100 per cent) water side of the membrane and, therefore, water moves down a concentration gradient from its higher concentration to a lower concentration in the solution side, analogous to the diffusion of molecules and ions.

If the sucrose solution within the membrane sac were replaced with a sodium chloride solution of the same molarity, the solution in the manometer would reach equilibrium at a point almost twice as high as that previously observed. Sodium chloride dissociates into two ions per molecule, and if the ion activity were unrestricted, it would have twice as many osmotically active particles for the same molecular concentration as does the sucrose solution. In reality, the number of active particles is less, as explained later. The total number of individual (solute) particles present in the solution per given mass of solvent, irrespective of their molecular nature (i.e., nonelectrolyte, ion, or colloid), determines the total osmotic pressure of the solution. In blood plasma, for example, nonelectrolytes such as glucose and urea, as well as the "electrolytes" present as free ions, and even the proteins contribute to the osmotic pressure of this body fluid.

If a solute is dissolved in a solvent (water, in the case of biological systems), the following four related phenomena occur:

1. The *osmotic pressure* is increased, as just explained.
2. At any given temperature, the *vapor pressure* of the solution is *lowered* below that of the pure solvent.
3. The boiling point of the solution, i.e., the temperature at which the vapor pressure equals atmospheric pressure, is *raised* above that of the pure solvent.
4. The *freezing point* of the solution, i.e., the temperature at which the vapor pressure of the solid equals that of the solvent, is *lowered* below that of the pure solvent.

These four properties of solutions, that is, an increase in osmotic pressure and boiling point and a lowering of vapor pressure and freezing point, are called *colligative properties* because they are tied together and mathematically interconvertible. They all are directly related to the total number of solute particles per mass of solvent. The term *osmolality* expresses concentrations of solutes in terms of *mass* of solvent (1 osmola*l* solution is defined to contain 1 osmol/*kg H_2O*). In contrast, the term osmola*r*ity expresses concentrations per *volume of solution* (1 osmolar solution is defined to contain 1 osmol/*liter solution*). In biological solution, the milliosmol is used as a more convenient unit.

In the ideal case, 1 mole of a nondissociated solute, dissolved in 1 kg of water is a 1 molal solution and contains 6.023×10^{23} particles (Avogadro's number) per kg of water. This solution boils at 0.52°C higher and freezes at 1.858°C lower than pure water. The vapor pressure of this solution is 0.3 mm Hg lower than the vapor pressure of pure water, which is 23.8 mm Hg at 25°C. The osmotic pressure of the same solution is increased to 17,000 mm Hg or to 22.4 atmospheres.

A solution of an electrolyte dissociates into two (in the case of NaCl) or three (in the case of $CaCl_2$) particles, and therefore the colligative effects of such solutions are multiplied by the number (n) of dissociated ions formed per molecule. Because of incomplete electrolyte dissociation, as well as association between solute molecules and solute molecules and solvent, many solutions do not behave as expected in the ideal case, and a 1 osmolal solution may give an osmotic pressure greater or less than that theoretically expected. The osmotic activity coefficient is a factor for the correction of the deviation from the "ideal" behavior of the system:

$$\text{osmolality} = \text{osmol/kg } H_2O = \phi(n)(C)$$

where

ϕ = osmotic coefficient
n = number of particles into which each molecule in the solution potentially dissociates
C = concentration in mol/kg H_2O

The total osmotic pressure of a solution is equal to the sum of the osmotic pressures or osmolalities of all solutes present.

A table of osmotic coefficients of most solutes of biological interest has been compiled by Wolf.[5] The osmotic coefficients were derived experimentally by measuring osmolality of solutions of known concentration. While glucose and ethanol have osmotic coefficients of 1.00, the ϕ for urea is 0.94 and the ϕ for NaCl is 0.93 at the concentration found in plasma and 0.91 at the higher concentrations that can occur in urine. Potassium dihydrogen phosphate (KH_2PO_4) in a 0.5 per cent solution has an osmotic coefficient of only 0.46.[5]

Because of their high molecular weights, serum proteins contribute less than 0.3 per cent of the total osmolality of serum. The electrolytes, Na^+, Cl^-, and HCO_3^-, which are present in a relatively high concentration, make the greatest contribution to serum osmolality, whereas the nonelectrolytes, such as glucose and urea, which are present at low concentrations, contribute only to a small extent. For example, the molality of osmolutes (osmotically active particles) in blood plasma is about 307 mmol/kg H_2O because of the presence of approximately 148 Na^+, 4 K^+, 2 Ca^{2+}, 1 Mg^{2+}, 108 Cl^-, 27 HCO^{3-}, 2 HPO_4^{2-}, 1 SO_4^{2-}, 5 organic$^-$, 5 glucose, and 4 urea (all values being molalities; unit: mmol/kg). The plasma proteins contribute only about 1 mmol/kg H_2O.

THE OSMOMETER

Theoretically, any of the four colligative properties could be used as a basis for the measurement of the osmotic pressure or the osmolality. Of these, the freezing point depression is the one most commonly used to determine the concentration of osmotically active particles in biological fluids. Unlike vapor pressure and osmotic pressure measurements, the determination of freezing point depression is independent of changes in ambient temperature. The instrument used is a freezing point depression osmometer, but it is usually referred to as an osmometer or, less commonly, as a cryoscope. The components of a freezing point depression osmometer (see Fig. 3–28) are as follows:[1]

1. A thermostatically controlled cooling bath maintained at −7°C.
2. A rapid stir mechanism to initiate ("seed") freezing of the specimen.
3. A thermistor probe connected to a Wheatstone bridge circuit to measure the temperature of the sample (the thermistor is a glass bead attached to a metal stem whose resistance increases with decrease in temperature).
4. A galvanometer used to null the current in the Wheatstone circuit.
5. A measuring potentiometer used to balance the circuit.

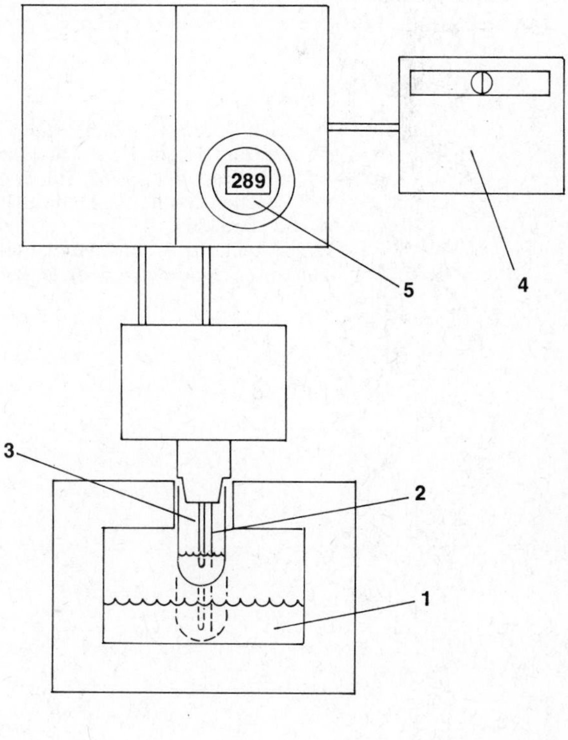

Figure 3–28 Block diagram of freezing point depression osmometer. 1, Cooling fluid. 2, Stirring rod. 3, Thermistor. 4, Galvanometer. 5, Potentiometer with direct readout. Test tube is shown above the liquid in the cooling bath (solid line) and inside the cooling liquid (dashed line).

PROCEDURE[2]

The sample, into which a thermistor probe and a stirring wire have been inserted, is lowered into the bath and with gentle stirring is supercooled to a temperature several degrees below its freezing point (−7°C). When the galvanometer movement indicates that sufficient supercooling has occurred, the sample is raised to a point above the liquid in the cooling bath and the wire stirrer is changed from a gentle rate of stir to a momentary vigorous amplitude which initiates freezing of the supercooled solution. The galvanometer reverses direction, as the released heat of fusion warms the solution, and then remains stationary, indicating the equilibrium temperature at which both freezing and thawing of the solution is occurring. During this equilibration period of 2 to 3 minutes, the balancing potentiometer is adjusted, thus changing the variable resistance of the Wheatstone bridge to bring the galvanometer to the null position. At the end of the equilibrium temperature plateau, the galvanometer again indicates decreasing temperature as the sample freezes to a complete solid.

The balancing potentiometer readings can be related to the degree of freezing point depression in °C, but more commonly they are calibrated with suitable standards to read directly in milliosmols. If an instrument is used which measures directly the freezing point depression, the osmolal concentration may be calculated as follows:

If the observed freezing point is −0.53°C,

$$\text{the mosmol/kg } H_2O = \frac{-0.53}{-1.86} \times 1000 = 285,$$

where −1.86 is the molal freezing point depression of pure water.

If the sample volume available for analysis is restricted, an 0.2 ml sample tube may be used instead of the 2 ml sample tube routinely used. After analysis, the thawed sample may be used for other tests. A precision of at least ±2 mosmol/kg H_2O can be obtained in a routine laboratory (see also Chap. 17).

REFERENCES

1. Abel, J.: Am. J. Med. Electronics, *Jan.–Mar.*: 32, 1963.
2. Johnson, R. B., Jr., and Hoch, H.: *In* Standard Methods of Clinical Chemistry. S. Meites, Ed. New York, Academic Press, Inc., 1965, vol. 5, p. 159.
3. Lifson, N., and Visscher, M. B.: *In* Medical Physics. O. Glasser, Ed. Chicago, Year Book Publishers, Inc., 1961, vol. 1, p. 869.
4. Warhol, R. M., Eichenholz, A., and Mulhausen, R. O.: Arch. Int. Med., *116*: 743, 1965.
5. Wolf, A. V.: *Aqueous Solutions and Body Fluids.* New York, Hoeber Medical Division, Harper and Row, 1966.

SECTION FIVE

CHROMATOGRAPHY

by Robert H. Spitzer, Ph.D.

Chromatography, because of its great versatility, efficiency, and convenience, finds an increasing number of applications in the clinical laboratory. The wide variety of chromatographic techniques makes it possible to separate molecular species of all sizes and forms, including (1) simple chemical entities such as gases (CO_2, N_2, D_2) and ions (Ca^{2+}, NH_4^+, Pb^{2+}); (2) more complex inorganic and organic molecules such as amino acids, sugars, lipids, vitamins, drugs, and steroids; (3) macromolecular species, including proteins, polysaccharides, nucleic acids, and soluble viruses; and (4) particulate materials such as subcellular components and bacteria. In addition to its use as a separatory tool, chromatography can also be used to establish the identity of one or more substances in a mixture, to estimate the molecular mass* of a compound, to concentrate solute molecules, to separate a substance of particular interest from other interfering compounds in a sample, and to aid in structure elucidation.

Separation of a mixture of solute molecules or particles by chromatography is achieved by virtue of differences in migration rates on or through a porous supporting medium called an *adsorbent* or, simply, *sorbent*. The migration is effected by the flow of a liquid or a gas (moving phase) which percolates through the sorbent medium (stationary phase) or a second liquid phase. Solute molecules are separated by one or a combination of physicochemical processes such as surface adsorption, solvent partition, molecular sieving (gel filtration), and ion exchange. Although the various types of chromatography are sometimes classified on this basis, a degree of ambiguity may arise because in a given chromatographic separation more than one process may be operative and there may be uncertainty as to which one is predominant. Other more practical classifications are based on the nature of the adsorbent employed (e.g., paper, Sephadex, or DEAE cellulose) or on the physical characteristics of the moving and stationary phases, respectively (e.g., gas-liquid chromatography).[9] A third classification is based on the type of technique employed (e.g., column chromatography or thin-layer chromatography). One view is that all chromatographic procedures can be categorized as either column chromatography or thin-layer chromatography (TLC). Paper chromatography can be considered a technique closely related to TLC.

In this section, the three main types of chromatography (column, paper, and thin layer) will be described, followed by a discussion of the principles and clinical applications of the four main physicochemical processes operative in chromatography (adsorption, ion exchange, gel filtration, and partition). The technique of gas chromatography is considered in Section 6 of this chapter. Applications of chromatography to specific analytical procedures are described in the individual chapters. For further details on chromatography, the reader is referred to the monographs listed in References 1 through 8, and Reference 10.

COLUMN CHROMATOGRAPHY

Column chromatography is the oldest, and was for many years the most common, form of chromatography. The pioneer work of the Russian botanist Mikhail Tswett and the

* In accordance with recent recommendations, the term molecular weight should be replaced by the term molecular mass (unit: dalton).

American chemist-geologist David Day at the turn of the century established the basis of chromatography as an analytical tool. Tswett, in 1906, introduced the terms *chromatographic* method and *chromatogram* during studies of plant pigments (chlorophyll, xanthophyll, and carotenes), since the compounds separated by column chromatography appeared as colored bands on the column.

In a simple form of column chromatography (Fig. 3–29), the sorbent is packed in a glass or plastic column supported at the bottom by a cotton or glass-wool plug or by a sintered glass filter. Depending on the nature of the sorbent, the separation of the sample mixture may be achieved by adsorption, ion exchange, gel sieving, or partition. The sample, dissolved in a suitable solvent, is added at the top of the column and caused to flow through the sorbent to allow the separation of the solute molecules. This process is called *column development* (Fig. 3–30), and the pattern of solute separation obtained on the column is called a chromatogram. In adsorption chromatography, the column may be extruded from the container (e.g., glass tube), and after localization of the solute zones the appropriate segments may be separated (e.g., by cutting the sorbent into sections) and the solute molecules eluted from the separated segments with suitable solvents. More commonly, the column adsorbent is not removed, but the solute molecules are successively eluted by passing more of the same solvent through the column.

Separation of sample components can also be achieved by adding different solvents at appropriate intervals (*stepwise elution*) or by adding solvents with progressively changing composition (*gradient elution*). In either case, appropriate and often serial volumes of eluate are collected and analyzed to determine the amount and kind of solute present.

The entire process of column chromatography can be performed in a completely automated manner. Important components in such an apparatus include a gradient device to change progressively the composition of the solvent, a pump to convey the gradient to the column, an injection unit to apply the sample, a means to keep the column temperature

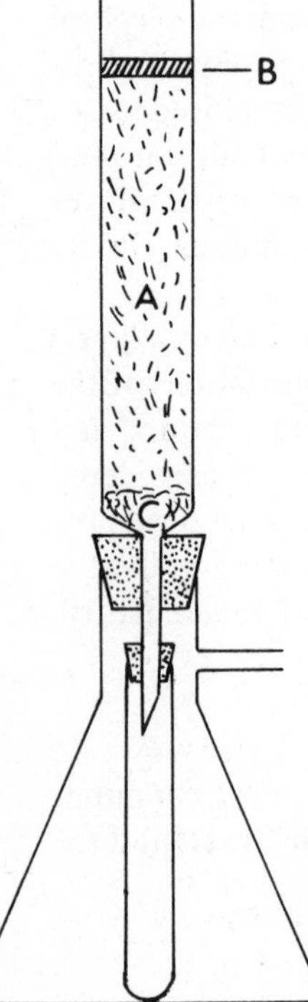

Figure 3–29 A simple column chromatographic apparatus. Column *A* is filled with adsorbent which is selected to separate solute *B* on the basis of adsorption, partition, gel filtration, or ion-exchange processes. *C* is a cotton or glass-wool plug. An Erlenmeyer flask containing a test tube is used to collect effluent fractions.

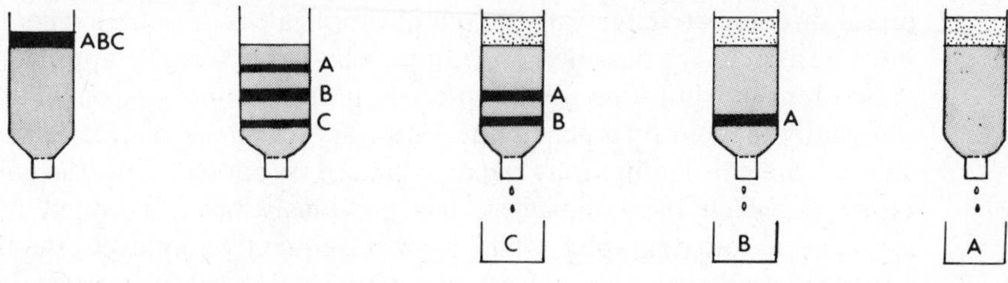

Figure 3–30 Representation of a basic analytical-preparatory procedure of column chromatography. The solutes *A*, *B*, and *C* are separated on the adsorbent (development) to obtain the chromatogram, followed by stepwise removal of the solutes (elution). (Modified from Olsen, R. E.: Methods in Medical Research, Chicago, Year Book Medical Publishers, Inc., 1970, vol. 12, p. 13.)

constant, an eluate fraction collector, and a detector (e.g., ultraviolet or visible absorption spectrophotometer with recorder) to measure solute concentration in the effluent.

Column chromatography can also be used for preparative purposes. Since the desired compounds may be present in one or more fractions of the column eluate, all fractions must be analyzed by an appropriate method (spectrophotometry, wet chemistry, or immunochemistry), and those fractions containing the desired material are combined and concentrated, if necessary.

Many forms of column chromatography using differing physicochemical processes have been developed to meet specific needs. Some of these are discussed in subsequent sections of this chapter and in other chapters.

PAPER CHROMATOGRAPHY

The sorbents used in this technique are special grades of filter paper selected for homogeneity, wet strength, thickness, level of impurities, and development rate.

The sample is applied to the paper in the form of a streak or spot using calibrated micro-

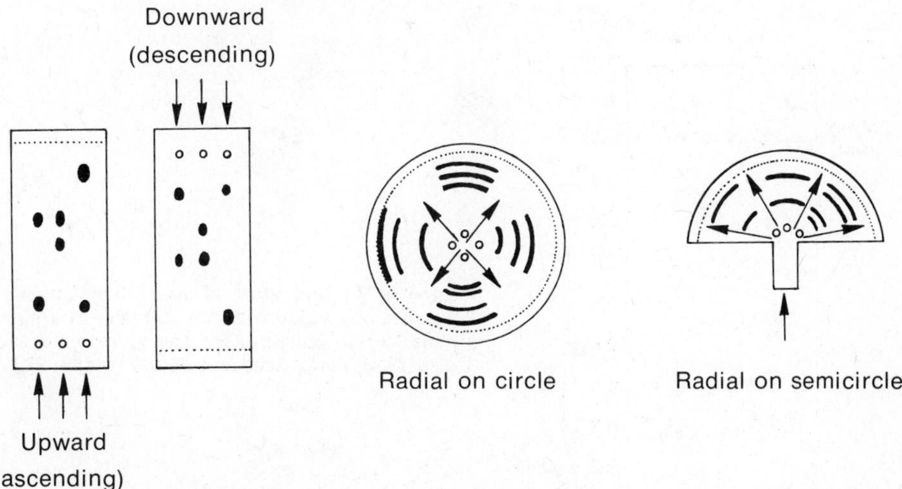

Figure 3–31 Examples of paper chromatograms. Sites of sample are indicated by open circles, direction of solvent development by arrows, and solvent front by dotted lines. (Modified from Long, C.: Biochemist's Handbook. Princeton, N.J., D. Van Nostrand Company, Inc., 1961, p. 132.)

pipets or commercially available sample applicators. The area where the sample is applied must be as small as possible. The sample may, therefore, be applied repeatedly to the same area after allowing time for the preceding application to dry. The area of application is frequently marked by a pencil line. After drying of the sample at the application site, one edge of the chromatography paper is placed in contact with the solvent in a closed container in which the atmosphere has previously been saturated with solvent vapor. In *ascending* chromatography, which is most frequently employed, the solvent migrates up the paper by capillary action. Other directions of solvent movement (development) are also effective (Fig. 3–31). Depending on the nature of the solvent and paper, any of the four physicochemical processes (adsorption, ion exchange, gel filtration, partition) can be used to achieve separation of the sample.

The chromatogram shown in Figure 3–32 illustrates the use of paper chromatography as a screening procedure for the detection and identification of amino acids. For the separation of more complex mixtures, two-dimensional chromatography may be required. After development, the amino acids may be visualized with a spray reagent containing ninhydrin or some other appropriate chemical, and the distance of each spot from the point of application measured. The ratio of this distance to the distance the solvent moved during development time is called the R_f value. The R_f value of each substance is then calculated and compared with the R_f value of known amino acids, preferably determined concurrently.

The R_f value of a substance is valid only for the experimental conditions used (e.g., type of solvent, sorbent, temperature, development time) and is not a true physical constant. Thus, identification is not established by the determination of one R_f value alone. Usually, supporting R_f values must be obtained using several other solvent systems. In addition, structural corroboration may be established using group-specific spray reagents.

Often it is necessary to utilize *two-dimensional chromatography* (development in two directions) to separate complex mixtures containing compounds with similar R_f values. This is done by removing the paper (or TLC plate) after chromatography, drying it, turning it 90°, and rechromatographing the substance in a different solvent mixture (Fig. 3–33). An alternative to two-dimensional chromatography is a technique which combines chromatography with electrophoresis; the sample is subjected to electrophoresis in one direction and then to chromatography after rotation of the paper by 90°.

Quantitative analysis of substances after separation on paper is often accomplished by

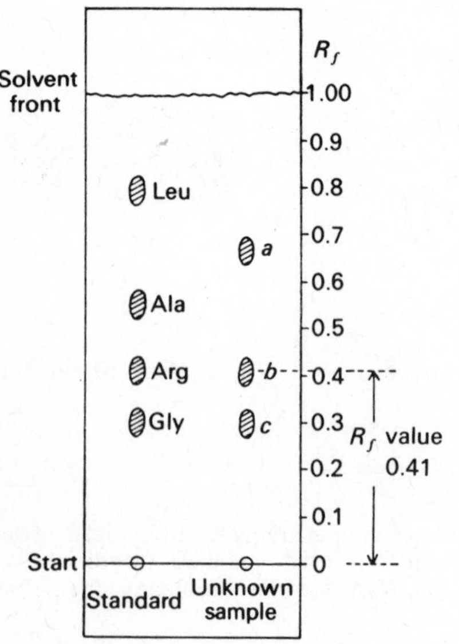

Figure 3–32 Separation of amino acids by paper chromatography. Substance *a* is unknown; *b* is identical to arginine; *c* is identical to glycine. (From Karlson, P.: Introduction to Modern Biochemistry. 3rd ed. New York, Academic Press, 1968, p. 32.)

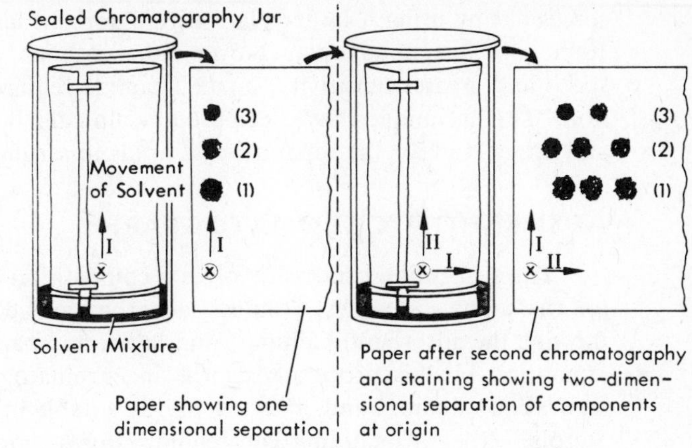

Figure 3–33 Two-dimensional paper chromatography. Two different solvents are used for development. (From Bennett, T. P.; Graphic Biochemistry. Vol. 1: Chemistry of Biological Molecules. New York, The Macmillan Company, 1968.)

reacting the unknown with appropriate color reagents and scanning the paper with a densitometer. An alternate but time-consuming procedure is to elute the substance or its colored derivative from the paper and then subject the eluate to analysis.

THIN-LAYER CHROMATOGRAPHY

TLC and paper chromatography are closely related techniques using similar procedures for application, development, detection, and analysis. In TLC, a thin layer of sorbent, such as silica gel (usually only 0.2 mm in thickness), is applied uniformly to a glass plate or a plastic sheet, using applicators designed for this purpose. Prepared plates coated with a variety of sorbents (silica gel, microcellulose, alumina, Sephadex, and so forth) are commercially available. The sample to be analyzed is added as a concise spot or streak near an edge of the plate which is then placed in a closed glass container with the lower edge of the plate in contact with an appropriate solvent mixture (Fig. 3–34). As in the case of column chromatography, solvents and sorbents can be chosen to achieve separation on the basis of any of the four types of separatory processes. After the solvent front has reached a desired height, the plate is removed from the tank and dried. The separated sample components are located and identified using appropriate procedures such as ultraviolet illumination, specific spray reagents, or autoradiography. If it is necessary to recover the separated substances for quantitation or for further experiments, the appropriate sorbent region is scraped from the plate and extracted to remove the solute.

TLC is both a convenient analytical tool and micropreparative tool. It can be scaled up

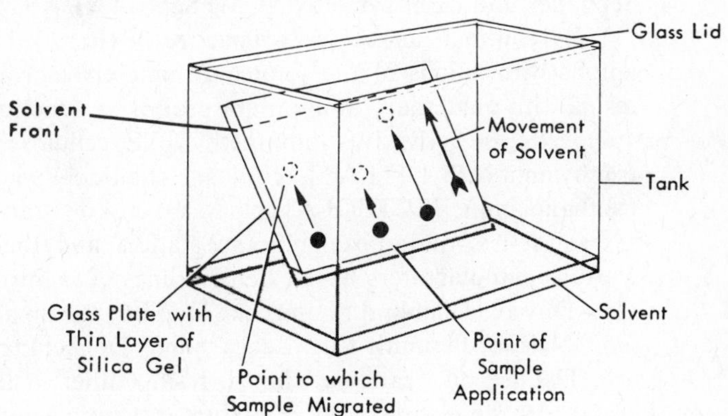

Figure 3–34 Illustration of thin-layer chromatography. The solvent is drawn up the thin layer of adsorbent by capillary action. Ascending development is commonly used in TLC. (Modified from Bennett, T. B.: Graphic Biochemistry. Vol. 1: Chemistry of Biological Molecules. New York, The Macmillan Company, 1968.)

somewhat by using a larger plate with a sorbent layer thickness of 2 to 5 mm, or even 10 mm. Its major advantages over either column or paper chromatography are greater speed and greater sensitivity, a small sample requirement, and frequently sharper separations. TLC techniques have been used in clinical laboratories for the identification of sugars and drugs, and for the separation of lipids in serum, amniotic fluid, and feces.

ADSORPTION CHROMATOGRAPHY

The adsorption process involves a competition between the solvent and the adsorbent for the solute molecules. Solute is adsorbed and desorbed repeatedly during percolation through the adsorbent. Column, thin-layer, and paper techniques can be used to facilitate separation by this process. Adsorption chromatography is particularly suitable for the separation of compounds that are poorly soluble in water but soluble in organic solvents. Sterols, terpenes including carotenoids, lipids, and fat-soluble vitamins are among the substances that have been separated by this method.

The most widely used adsorbents are silica (also called silica gel and silicic acid), alumina, and magnesium silicate (e.g., Florisil). Diatomaceous earth (e.g., Celite, Kieselguhr), sucrose, and charcoal are employed less frequently but are effective for some purposes. A solvent is chosen which will not only dissolve the solute molecules but also facilitate separation of solutes (development) on the chosen adsorbent. In certain applications, a nonpolar solvent is used for development and a more polar solvent for elution. Selection of the appropriate organic solvent for use with a particular combination of solute and sorbent can be made with the aid of tables summarizing the elution characteristics of solvents (*eluotropic series*).[3,5,8]

ION-EXCHANGE CHROMATOGRAPHY

In ion-exchange chromatography, solutes in a sample mixture are separated by virtue of their differences in sign and magnitude of ionic charge. This technique is especially applicable for compounds soluble in aqueous systems, such as inorganic ions, amino acids, proteins, and nucleotides. Ion-exchange materials are insoluble polymeric substances containing ionic groups as part of their structure. They are available in a number of particle sizes suitable for use in either column or thin-layer techniques.

A *cation-exchange resin* has many negatively charged functional groups (Fig. 3–35*A*) that are covalently bound to it. Associated with these groups are loosely bound cations which are available for exchange with cations of the solute. Some commercially available cation-exchange materials include resin copolymers (Dowex 50, Amberlite), silicates, cellulose derivatives (carboxymethyl [CM] cellulose, phosphoryl [P] cellulose, sulfomethyl [SM] cellulose, and sulfoethyl [SE] cellulose), and dextran derivatives (sulfopropyl [SP] Sephadex and carboxymethyl [CM] Sephadex).

By contrast, an *anion-exchange resin* (Fig. 3–35*B*) can exchange its loosely bound anions with anions of the solute. Commercially available anion-exchange resins include resins with quaternary ammonium groups as functional groups (Dowex 1 and Amberlite 400), cellulose derivatives (aminoethyl [AE] cellulose, diethylaminoethyl [DEAE] cellulose, triethylaminoethyl [TEAE] cellulose, guanidoethyl [GE] cellulose, and epichlorohydrin triethanolamine [ECTEOLA] cellulose), and dextran derivatives [DEAE, Sephadex]. More extensive lists of ion-exchange substances and their application are found elsewhere.[3] Various manufacturers have extensive lines of ion-exchange materials (e.g., Dow Chemical Co.: Dowex; Diamond Shamrock: Duolite; Rohm and Haas: Amberlite; Nalco Chemical Co.: Nalcite; Permutit Co.: Zeocarb and De Acidite).

The use of a cation-exchange resin (either in the form of columns or on thin-layer plates) for the separation of a mixture of amino acids at an acid pH is illustrated in Figure

A

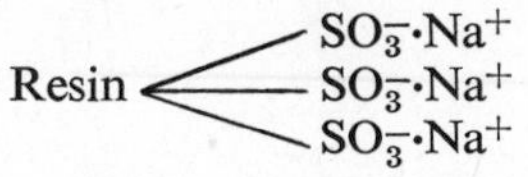

Resin — $SO_3^-{\cdot}Na^+$
Resin — $SO_3^-{\cdot}Na^+$
Resin — $SO_3^-{\cdot}Na^+$

A cation (sulfonate) exchange resin in its Na^+ form

B

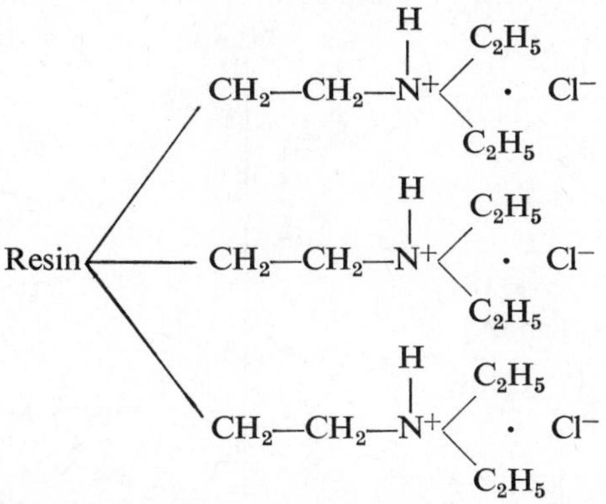

Resin — CH_2—CH_2—$N^+H(C_2H_5)_2 \cdot Cl^-$
Resin — CH_2—CH_2—$N^+H(C_2H_5)_2 \cdot Cl^-$
Resin — CH_2—CH_2—$N^+H(C_2H_5)_2 \cdot Cl^-$

An anion (diethylaminoethyl-) exchange resin in its Cl^- form.

Figure 3–35 Examples of anion- and cation-exchange materials. The dot represents a loose electrostatic bond between the resin and ions. The ions are capable of being exchanged with charged solute particles.

3–36. The positively charged amino acids in a strongly acidic solution will exchange with the loosely bound sodium ions and become bound to the negatively charged resin. By use of a buffer with increased pH, the amino acids become more negatively charged and are eluted from the resin by interchanging once again with sodium ions. The acidic amino acids, such as aspartic and glutamic acids, are eluted from the column more quickly than are the basic amino acids, such as arginine, histidine, and lysine. The separation of amino acids in this manner using a column of cation-exchange resin can be accomplished automatically by means of commercially available instruments called amino acid analyzers. This type of instrument is used to quantitate amino acids present in biologic extracts, protein hydrolysates, or urine.

Ion exchangers of the cellulose type (e.g., DEAE, ECTEOLA, CM) and dextran type (e.g., DEAE, SP) are especially useful in the separation of proteins. Column fractionation of plasma proteins, hemoglobins, hormones, enzymes, and isoenzymes are examples of such application.

In the clinical laboratory, ion-exchange resins are also used to remove substances which may interfere in a given analytical procedure. For example, an anion-exchange resin (Dowex 2) is used to separate porphobilinogen (PBG) from δ-aminolevulinic acid (ALA) in assays for these two hemoglobin precursors, and the DuPont Automatic Clinical Analyzer (ACA) uses a column in the reaction packet to remove enzyme inhibitors.

GEL-FILTRATION CHROMATOGRAPHY

Gel-filtration (gel permeation, molecular exclusion, molecular sieve chromatography) facilitates separation of solute particles mainly on the basis of differences in their molecular size, although molecular shape and hydration are contributing factors. The stationary phase most commonly consists of gel-forming hydrophilic beads, such as cross-linked dextran (Sephadex), polyacrylamide (Bio-Gel), and agarose (Sepharose). The beads are porous and contain openings and microchannels. By varying the degree of cross-linking in the course of

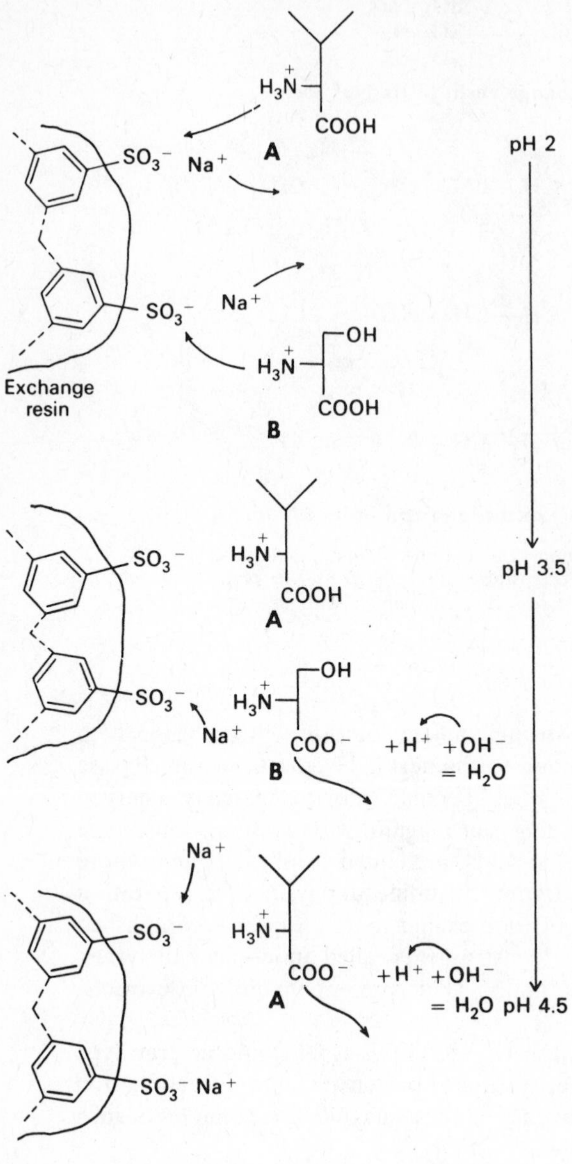

Figure 3–36 Separation of amino acids with a cation-exchange resin under acidic conditions. The amino acid cations exchange positions with cations (Na^+) of the resin at pH 2 (both amino acids *A* and *B* are adsorbed). By increasing the pH of the buffer, the amino acids are differentially eluted, reflecting differences in the dissociation constant K_a' of the ionizable groups from the resin (at pH 3.5 amino acid *A* remains adsorbed and amino acid *B* is eluted; at pH 4.5, amino acid *A* is also eluted). (From Karlson, P.: Introduction to Modern Biochemistry. New York, Academic Press, 1968, p. 33.)

manufacture, the size of these channels and pores can be designed to pass or retard molecules of different sizes. The solvent and all solute molecules with diameters less than that of the pore will enter the gel beads and be entrapped within the channels. Large solute molecules which cannot enter the pores must by-pass the beads and flow along the spaces between the beads, thus appearing in the effluent first (Fig. 3–37). The smaller, entrapped solute molecules will be delayed in traversing through the beads and therefore appear in the column effluent much later in time.

Gel-filtration chromatography can be used for the separation of either hydrophilic or hydrophobic materials, depending on the type of gel used. *Hydrophilic gels* are particularly suitable for the separation of solutions in aqueous systems (e.g., enzymes, antibodies, proteins, hemoglobins, polysaccharides, nucleic acids). *Hydrophobic gels*, e.g., methylated Sephadex or hydroxypropylated Sephadex (LH-201), polystyrene beads (Styragel), and cross-linked polymethacrylate have been developed for the separation of solutes which are

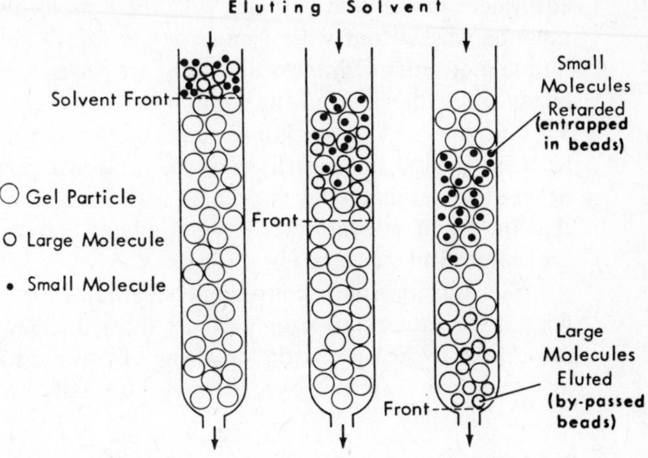

Figure 3–37 Representation of gel-filtration chromatography using a column technique. (From Bennett, T. P.: Graphic Biochemistry. Vol. 1: Chemistry of Biological Molecules. New York, The Macmillan Company, 1968.)

soluble only in organic solvents. The latter gels have been useful in the separation of triglycerides and fatty acids.

A packed column contains beads and a suspending fluid (buffer, saline, water, and so forth) which fills the channels inside and the spaces between and about the beads. The volume surrounding the beads is referred to as outer volume or *void volume* (Vo), whereas Vi is the volume inside the gel beads. In a separation process, the larger molecules which cannot enter the gel will emerge from the column in the first portion of the eluate, which is

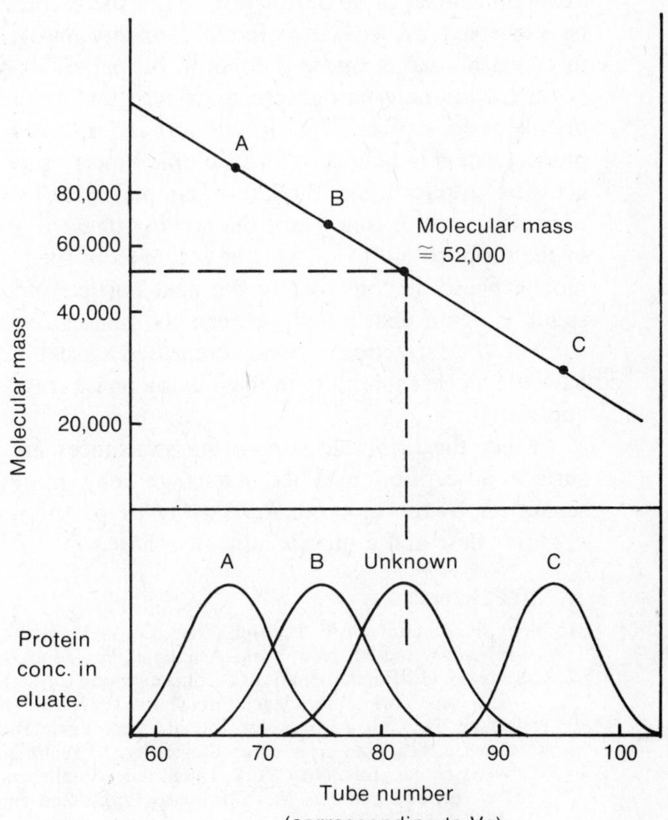

Figure 3–38 Molecular mass determination by gel filtration. To calibrate the column, two or more proteins (*A*, *B*, and *C*) of known molecular mass are allowed to pass through the column, and their peak elution volumes (Ve) are plotted as a function of molecular mass. On the basis of the elution volume of an unknown protein, its molecular mass can be interpolated from the calibration graph. This relationship holds true only for spherical proteins. In the case of nonspherical particles, the elution volume is directly related to the Stokes radius, i.e., the radius of a spherical particle of equivalent hydrodynamic properties. (From Lehninger, A. L.: Biochemistry. New York, Worth Publishers, Inc., 1970, p. 143.)

equivalent to the void volume. Very small molecules which freely enter the pores of the gel must be washed out with a much larger liquid volume before they appear in the effluent. Solute molecules of intermediate size are restricted but not prevented from entering the gel beads and will exit from the column in an *elution volume* (Ve) that is intermediate between Vo and (Vo + Vi). Within limits, and for a given column, a linear relationship exists between Ve and the logarithm of the molecular mass of compounds. This relationship can be used to estimate the molecular mass of a compound from its measured Ve provided that the Ve of several other compounds of known molecular mass is also determined (see the calibration curve in Fig. 3–38).

Gel filtration is a convenient technique for separating molecules which differ significantly in molecular size, such as inorganic ions and biopolymers (e.g., proteins). An important application is the desalting of protein solutions which, by this technique, is fast and therefore preferred over dialysis, especially when working with labile substances.

PARTITION (LIQUID-LIQUID) CHROMATOGRAPHY

Separation by this method is based on differences in the solubility of individual solute molecules in two immiscible liquids that are in contact with one another. The degree to which the individual solutes distribute between the two solvents is expressed by the partition (distribution) coefficient K_d. This coefficient is defined as the ratio of the solubility of the solute in the (receiving) phase to the solubility in the original solvent. In general, polar solutes extract better into polar solvents, and nonpolar solutes move preferably into nonpolar solvents (see also p. 712). The greater the value of the partition coefficient, the more rapid and complete is the movement of solute from the original into the extracting solvent. If the coefficient has a value only slightly above 1.0, many repeated consecutive extractions need to be performed before the extraction will be complete. Separations may be performed in a separatory funnel (nonchromatographic), or the two solvents can be made to come in contact inside a column, on paper, or on a thin-layer chromatography plate.

Partition chromatography as conventionally used in the clinical laboratory employs a mobile organic phase (e.g., *n*-butanol) and a stationary phase (e.g., water). The stationary phase (water) is bound to an insoluble "inert" support such as cellulose, starch, or silica gel. After application of the solute (sample) to the hydrated supporting substance in the form of a small spot or thin layer, the moving phase is allowed to percolate through this region of the support and to extract the solute from the stationary phase. The solute, now in the mobile phase, is conveyed to the next particle of the sorbent (stationary phase) and the solute is again distributed between the phases. As the mobile phase moves through the sorbent, the extraction process is repeated a great number of times. Solute molecules which have the highest solubility in the moving phase travel the greatest distance from the point of application.

Since the insoluble supporting substances are rarely inert, other processes such as surface adsorption and ion exchange may frequently contribute to chromatographic separation by this method. *Reversed-phase* partition chromatography employs a stationary organic phase and a mobile aqueous phase.

REFERENCES

1. Block, R. J., Durrum, E. L., and Zweig, G.: A Manual of Paper Chromatography and Paper Electrophoresis. 2nd ed. New York, Academic Press, Inc., 1958.
2. Cassidy, H. G.: Fundamentals of chromatography. *In* Technique of Organic Chemistry. A. Weissberger, Ed. New York, Interscience Publishers, 1957, vol. 10.
3. Heftmann, E.: Chromatography. 2nd ed., New York, Reinhold Publishing Co., 1967.
4. Kirchner, J. G.: Thin layer chromatography. *In* Technique of Organic Chemistry. E. S. Perry and A. Weissberger, Eds. New York, Interscience Publishers, 1967, vol. 12.
5. Lederer, E., and Lederer, M.: Chromatography. 2nd ed. Amsterdam, Elsevier Publishing Company, 1957.

6. Shellard, E. J.: Quantitative Paper and Thin-Layer Chromatography, New York, Academic Press, Inc., 1968.
7. Stahl, E.: Thin-Layer Chromatography. 2nd ed. New York, Springer-Verlag, 1969.
8. Strain, H. H.: Chromatographic Adsorption Analysis. New York, Interscience Publishers, 1942.
9. Szepesy, L.: Gas Chromatography. Translated by E. D. Morgan. Budapest/London, Iliffe Books, Ltd., 1970.
10. Zweig, G., and Whitaker, J. R.: Paper Chromatography and Electrophoresis. New York, Academic Press, Inc., 1971.

SECTION SIX

GAS CHROMATOGRAPHY

by Sati C. Chattoraj, Ph.D.

Since the first report by James and Martin[4] on the separation of volatile fatty acids by gas chromatography (GC), the application of this technique has been extended to various industries as well as to the biomedical field. In clinical chemistry, the extent of utilization of gas chromatography has been thwarted by the fact that many substances to be measured are ionic in nature and are, therefore, nonvolatile at the operating temperatures of a gas chromatograph. In addition, the volatility of high molecular mass compounds (e.g., palmitic ester of cholesterol, molecular mass 624) is sufficiently low to prevent elution from the gas chromatographic column in a reasonable time. The improvement of instrumental design, the availability of thermally more stable stationary phases, and the improvement of techniques for derivative formation (to increase volatility) have made gas chromatography a powerful analytical tool for the separation of many compounds of clinical interest. (see Chaps. 13, 14, and 21).

GC has great advantages over conventional chromatography because of its high resolution, sensitivity, rapid separation, and ability to provide simultaneous quantitation. The following is a brief discussion on the principles, apparatus, and requisites for gas chromatography. For detailed information, the reader is referred to pertinent reviews and monographs.[1–3,5–10]

BASIC PRINCIPLES OF TECHNIQUE

Gas chromatography is a process by which a mixture of compounds in volatilized form is separated into its constituent parts by moving a mobile (gas) phase over a stationary phase (sorbent or liquid phase). Under controlled conditions, the individual components of the sample, in accordance with their vapor pressure, will be present partially in the stationary phase and partially in the mobile phase. The ratio of the weight of solute per milliliter stationary phase to the weight of solute per milliliter mobile (gas) phase is termed the partition coefficient K. A compound with a high vapor pressure will have a low partition coefficient, i.e., it will be present to a greater extent in the gas phase. Thus, this compound will be eluted more rapidly than compounds with lower vapor pressure. If there is selective interaction between a compound in the sample and the stationary phase, the order of elution from the column may be different.

Gas chromatography is divided into two major categories: (1) gas-solid chromatography (GSC), in which the sorbent is a solid of large surface area, and (2) gas-liquid chromatography (GLC), in which a nonvolatile liquid (stationary phase) is coated on an inert solid support. The mobile phase in both cases is an inert gas (nitrogen, helium, or argon) which carries the solute molecules through the gas chromatographic column—hence the name *carrier gas*. The effluent leaving the column carries the separated sample constituents to the detector according to their time of elution. The response signal of the detector, after amplification in an electrometer, is fed into a recorder for display. The graphic representation usually has a straight baseline on which is superimposed a series of peaks corresponding to the number of components in the sample and the quantity of the components in the sample mixture. The *retention time* (the time of emergence of the peak maximum from the injection point), by comparison with known materials, provides a means for a qualitative identification. The measure of the *peak size* (area or height) allows for quantitation.

THE GAS CHROMATOGRAPH AND ITS COMPONENTS

A gas chromatograph, in general, consists of six basic components: (1) a carrier gas supply with flow control; (2) a sample introduction system; (3) the column, column oven, and its temperature control; (4) the detector; (5) the electrometer (amplifier); and (6) the recorder (Fig. 3–39). "Ancillary" components are devices for collecting the eluate, or may be a mass spectrometer which aids in the identification of the eluted components of the sample mixture.

Carrier gas supply with flow control

The type of carrier gas is determined by the type of detector utilized in the system. The most frequently used carrier gas is nitrogen, which can be used with flame ionization, electron capture, or thermal conductivity detectors. Helium may be used with flame ionization and thermal conductivity detectors, while argon is used with the argon ionization detector.

Precise control of the carrier gas flow rate through the column is important, since the retention time of the individual sample components is in part determined by the flow rate. Thus, any variations in the flow rate will lead to a change in retention time and uncertainty as to the identity of the chromatographic peaks. When operating under isothermal con-

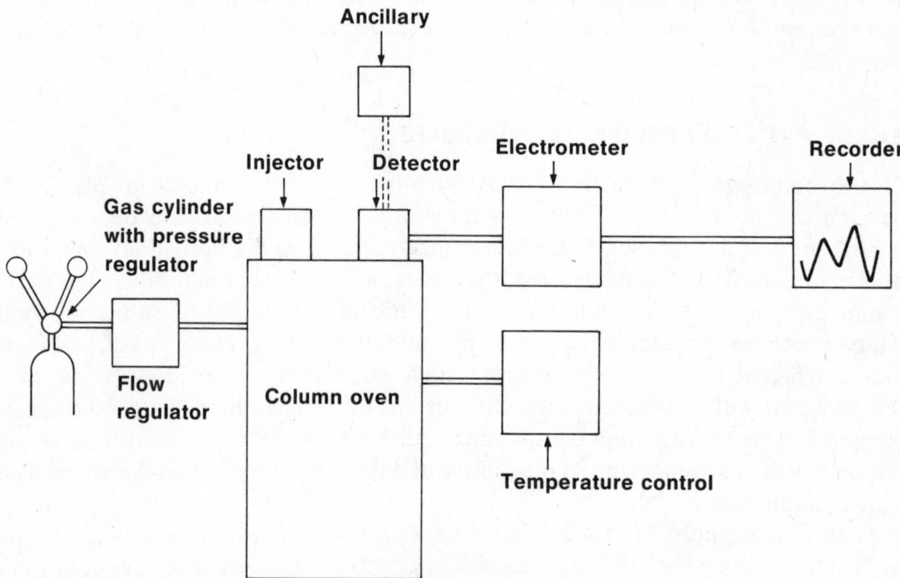

Figure 3–39 Gas chromatographic system.

ditions, the use of a good flow-control device and freedom from leaks in the system will insure adequate constancy of flow rate. When, however, constant flow is required at different column temperatures, as in a temperature-programmed operation, some form of differential flow controller is required to compensate for changes in column pressure with temperature.

Sample introduction system

The sample to be introduced may be a solid, liquid, or gas. Liquids are injected into the carrier gas stream with a microliter syringe through a septum in the apparatus. Solids may be dissolved in a suitable solvent and injected in this form, or they may be placed into capsules or onto a metal grid. (see Chap. 13). The sample introduction system is kept at a temperature which will result in quick vaporization of the sample components so that the sample mixture may move onto the column in the form of a narrow band. Excessive dead volume in the injection port results in diffusion of the sample and therefore increased peak width and tailing. The efficiency of separation is also affected by the sample size, which should be kept as small as possible (2 to 10 μl).

The high boiling point of some compounds requires that they be injected in the form of a more volatile derivative so that the desired instantaneous vaporization can be accomplished. Derivative formation has the additional advantages of reducing adsorption to the column, improving the gas chromatographic properties of the sample, and, in some cases, protecting heat-labile chemical groups of the sample compounds. (Procedures for derivative formation are discussed in Chap. 13.)

Column oven and its temperature control

The column oven of the gas chromatograph is designed to maintain constancy and uniformity of the column temperature. This is necessary for reproducibility of retention times and for maintaining a constant bleed rate of the column (evaporation of components of the liquid phase itself). Any "hot spots" along the column may cause localized deterioration of the column packing which may result in undesirable adsorption of compounds. Any temperature drop in either the column itself or the connection between the column and the detector (detector line) may lead to partial condensation and, consequently, to reduced efficiency of separation and decreased sensitivity.

Gas chromatographic column

Separation of the sample mixture takes place in the gas chromatographic column. Thus, extensive consideration must be given to every detail of column construction to insure efficiency and sufficient sensitivity for quantitative analysis. The column itself may vary from 1/8 to 1/4 inch in diameter and from 4 to 12 feet in length. The material used for the construction of the column is mainly glass or stainless steel tubing. The choice of material depends upon the substance being analyzed. Cholesterol and, particularly, samples containing halogens may react with hot metal surfaces; thus, the use of a glass column in such cases is essential. *Stainless steel* columns are inexpensive and durable, and they have a nonadsorptive surface with excellent heat-transfer properties. *Glass columns*, on the other hand, allow visual inspection of the support during the packing process and detection of discontinuity of packing material or the deposition of nonvolatile carbonaceous residues in the top of the column. However, the fragility of the column and the need to deactivate the active sites inside the tubing (silanization) are disadvantages of this type of column.

Solid support

Ideally, the solid support should be inert, and its sole purpose should be to hold the thin layer of the stationary phase. If the support is not inert, interactions of the vapor solute will take place and asymmetrical peaks or partial loss of the injected material due to irreversible adsorption or decomposition will result. The supports generally used for GLC are silicates (diatomaceous earths).

The interaction of polar compounds with these active groups may be prevented by treatment of the support with dichlorodimethylsilane or hexamethyldisilazane (silanization).

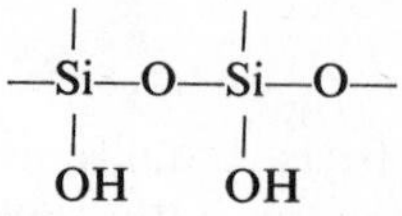

```
   |     |                      |     |
 —Si—O—Si—O—                  —Si—O—Si—
   |     |                      |     |
   OH    OH                     O     O
                                 \   /
                                  Si
                                 /  \
                              CH3    CH3
```

Examples of active sites 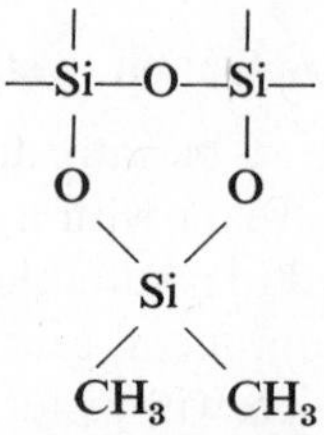 **After silanization**

During the silanizing process, some of the hydroxyl groups are replaced by chlorine instead of silane radicals. If the support comes in contact with water, the chlorine atoms would again be replaced with hydroxyl groups. The support, therefore, should be treated after silanization with absolute ethanol or methanol to replace the chlorine with ethyl or methyl radicals (for detailed procedure see p. 751).

The choice of supporting material and its preliminary preparation depend largely upon the nature and type of substance to be analyzed. Among the commercially available supports, diatomaceous earths (e.g., Anakrom U, Celite 545, Gas-Chrom P, Chromosorb W) are most widely used. Fluorocarbon supports, porous polymer beads, and glass beads are other examples.

The efficiency of the column is in part determined by the size and distribution of the supporting particles. By decreasing the particle size of a support of uniform distribution, the multiple-path effect on the transport of the solute is minimized, and thus the separatory power of the column is enhanced. Reduction of particle size, however, adversely affects the permeability of the column for the carrier gas, and therefore the practical limit of particle diameter is about 150-mesh. Solid supports in the range of 80- to 100-mesh and 100- to 120-mesh generally give adequate columns for routine analysis.

Stationary phase (liquid phase)

The proper choice of a liquid phase is of great importance for the separation, quantitation, and stability of the sample compounds. The nature of the stationary phase to be used is primarily determined by the physical and chemical properties of the compounds to be analyzed. The amount of stationary liquid phase used in relation to the amount of inert support may vary from less than 1 to 50 per cent by weight. In general, diffusion phenomena are reduced in columns with relatively low amounts of a liquid phase. However, at extremely low concentrations, the support may possess sufficient residual adsorptivity to cause sample adsorption or tailing of the peaks.

Although a wide variety of compounds have been used as liquid phases, a few carefully selected compounds probably suffice to fill the needs of the analytical procedures carried out in a clinical laboratory. The ideal liquid phase, under experimental conditions, should be nonvolatile, thermally stable, and chemically inert toward the solutes (i.e., no chemical transformation) under study. Methyl silicone polymers (e.g., SE-30, OV-1, DC-200, UCL-45), substituted silicone polymers (e.g., SE-52, OV-17, F-60, XE-60, QF-1, UCW-98) and silicone polyesters (e.g., EGSS-X, EGSP-A, ECNSS-S) fulfill these criteria fairly well. All methyl silicones have the following basic structure:

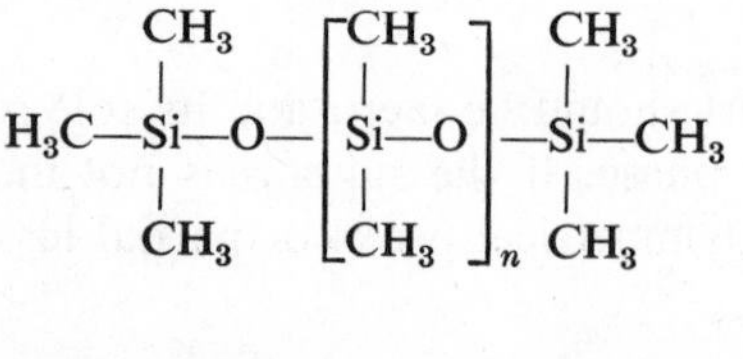

$$H_3C-\underset{CH_3}{\overset{CH_3}{\underset{|}{\overset{|}{Si}}}}-O-\left[\underset{CH_3}{\overset{CH_3}{\underset{|}{\overset{|}{Si}}}}-O\right]_n-\underset{CH_3}{\overset{CH_3}{\underset{|}{\overset{|}{Si}}}}-CH_3$$

The greater the value of n, the higher the molecular weight and the greater the viscosity. As n increases, the compounds change from oils to gums. When the molecular weight is of the order of 10,000, the vapor pressure of the gum is negligible, even at temperatures of 350°C. The methyl silicone polymer SE-30 is an excellent example of a thermally stable stationary phase and is therefore used widely. The partition property of the silicone polymers can be modified by replacing the methyl group or silicone part to varying degrees by other groups. Substitution by phenyl (e.g., SE-52), cyanoethyl (e.g., XE-60), and trifluoropropyl groups (e.g., QF-1) or the introduction of a polyester ethylene glycol succinate (e.g., EGSS-X) has yielded extremely useful stationary phases.

Stationary phases can be separated into two types: the *nonselective* phase (e.g., SE-30) and the *selective phase* (e.g., NGS, XE-60). The former type is suggested for the separation of compounds with differing molecular weight (and differing boiling points), and for the separation of samples of a relatively nonpolar nature. The selective type phases are more frequently employed for the separation of isomers and compounds of varying polarity. In general, the selective phases are less thermostable than the nonselective phases.

The procedure of coating the solid support with the stationary phase is one of the important facets of column preparation (p. 751). The stationary phase should consist of a thin uniform film around the solid support. Fragmentation of the very brittle support particles must be avoided to prevent exposure of adsorptive sites.

Column conditioning

Freshly prepared gas chromatographic columns, especially the stationary liquid phase, may contain impurities such as low molecular weight compounds of shorter polymers of the liquid phase. Such impurities are removed by heating the column, generally to 30 to 40°C above the intended operating temperature for the column. This procedure is called column conditioning. In order to avoid contamination of the detector with these impurities, detector lines should be disconnected during this process. The conditioning period varies with different liquid phases, but it is generally 12 to 24 hours. During this time the carrier gas is permitted to sweep through the column to carry off the impurities. Satisfactory conditioning of the column has occurred if a gas chromatogram shows a steady baseline without any peaks or spikes.

Detectors

The column effluent, containing the individual, separated compounds, enters the detector through the detector lines. As the compounds pass through the detector, an electrical signal is generated proportional to the amount of substance present in the carrier gas stream. Many types of detectors have been invented, but the thermal conductivity detector, the flame ionization detector, and the electron capture detector are most frequently employed.

Thermal Conductivity (TC) Detector. As the name implies, the operational principle of this detector is based on the changes in thermal conductivity of the carrier gas caused by the admixture of sample components. Changes in conductivity are proportional to the amount of substance eluted from the column. The detector consists of a metal block with two separate channels through which the column effluent and the pure gas, respectively, flow (Fig. 3–40*A*). Each channel carries a wire of tungsten or tungsten-rhenium alloy; these wires are connected to opposite arms of a Wheatstone bridge circuit which is fed from a constant current supply. Both filaments are heated electrically to raise their temperature above that of the detector block.

Since most substances analyzed have low thermal conductivities, sensitivity of the detector is increased by using a gas with high thermal conductivity. Helium is most frequently used. As the column effluent containing the mixture of carrier gas and separated sample

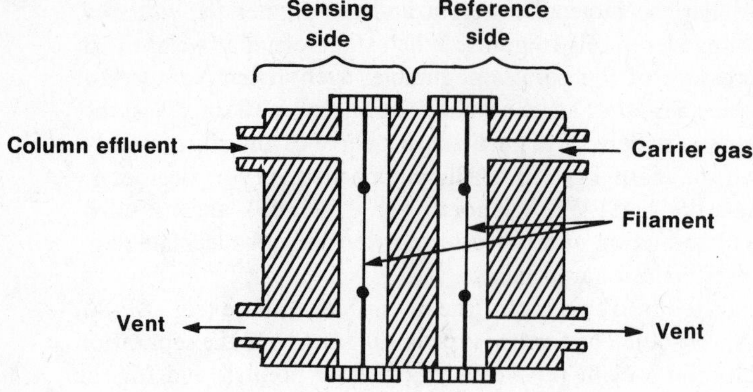

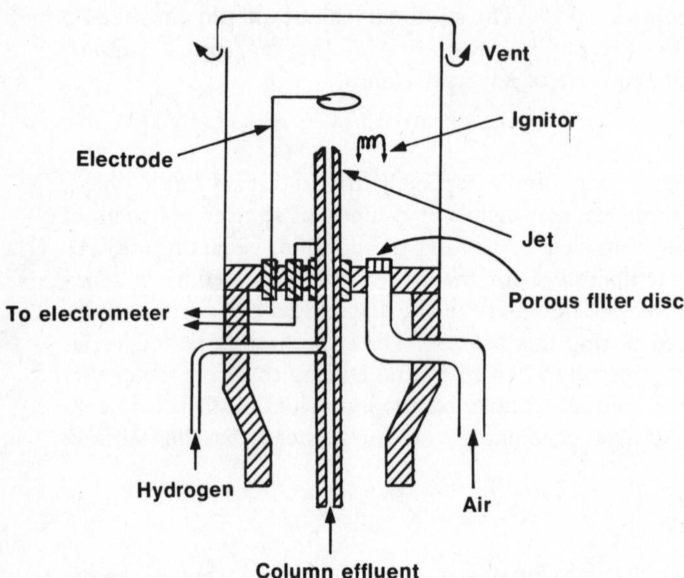

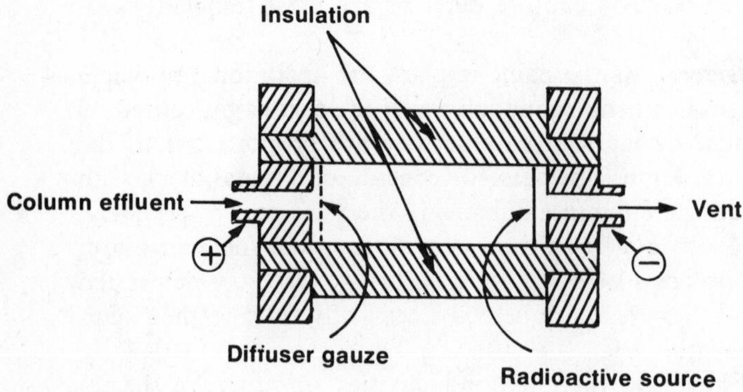

Figure 3–40 *A*, Schematic diagram of a thermal conductivity detector. *B*, Schematic diagram of a flame ionization detector. *C*, Schematic diagram of an electron capture detector.

passes through the sample side of the detector, the temperature and therefore the electrical resistance of the sample filament rise, while the reference filament remains unchanged. This causes a current flow which is amplified by the electrometer and transmitted to the recorder for graphic display.

The TC detectors are nondestructive, and thus sample components may be collected for further studies. Another advantage of TC detectors is their ruggedness and their ability to respond to a wide variety of samples, including respiratory gases such as O_2, N_2, CO_2, CO, and others which can not be detected with the flame ionization detector. Limitations of the TC detector are its relatively low sensitivity (up to 10^{-6} g) and its tendency to give baseline drifts.

Flame Ionization (FI) Detector. This is the most widely used detector because of its simple construction, reliable performance, wide range of linear response, high sensitivity, and ease of operation. FI detectors are more sensitive than TC detectors (up to 10^{-9} g), they are not subject to corrosion, and they can be operated at high temperatures (the stability of the liquid phase is generally the limiting factor). FI detectors can essentially sense all organic substances and are therefore highly suitable for analytical work in a clinical laboratory. Inorganic gases and water, however, cannot be detected. The sample is destroyed during detection in the FI detector, and therefore sample collection is possible only if the column effluent is split prior to entering the detector.

In most FI detectors (Fig. 3–40*B*), the effluent from the column is mixed with hydrogen gas prior to its admission to the detector. The mixture then enters a jet, where it is burned in either air or oxygen atmosphere to produce a small flame. As the sample burns in the flame, thermal ionization occurs and the transmitted electrons are collected either by a collector loop placed around the flame or by collector plates located on both sides of the flame. A potential (e.g., 300 V) is applied to the jet and the collector loop, each of which serves as electrodes. The degree of thermal ionization, and therefore the current flow, is directly proportional to the amount of sample burned in the flame. The current thus produced is linearly amplified by the electrometer and transmitted to a strip chart recorder for display.

In addition to the applied potential and the type of construction of the detector, optimal performance is dependent on the ratio of hydrogen to carrier gas flow rate. This ratio determines the flame temperature and, consequently, the efficiency of ionization. It is generally recommended, for maximum response, that the ratio of hydrogen to carrier gas flow should be about 1:1, and that of air to hydrogen flow 10:1. It is obvious, then, that carrier gas, hydrogen, and air flow rates should be optimized for each experimental condition if maximal detector response is expected.

Electron Capture (EC) Detector. These detectors are the most sensitive (up to 10^{-12} g), but they are also the most selective, since they detect only compounds with affinity for electrons (i.e., compounds which capture electrons). Thus, its application is essentially limited to halogenated compounds. If the compound itself does not contain halogens, appropriate derivatives of the sample compounds can be formed.

The EC detectors contain two electrodes separated by an insulator. A radioactive source such as tritium or nickel-63 is attached to the cathode. The column effluent containing the carrier gas and the separated sample components enters the detector through a tube leading to the anode, while the gas is exhausted through a hole in the cathode. As the carrier gas (e.g., nitrogen admixed with argon-methane) goes through the detector, it is ionized by the radioactive source. A fixed potential, or a short pulse of potential, just sufficient to collect all ions and electrons, is applied to the electrode, yielding a very small but constant current across the detector (standing current). If compounds having an electron capturing ability enter the detector, some electrons are captured, and as a result the standing current is diminished. This change in current can be amplified and displayed as negative peaks on the recorder. At normal operation, however, the signal from the detector is inverted in the electrometer so that a positive peak is depicted on the recorder chart.

EC detectors have the great advantage of sensitivity and selectivity, but unfortunately they are also extremely sensitive to impurities in the sample and to changes in the operating conditions. Thus, their use should be considered only when the FI detector proves inadequate.

REFERENCES

1. Cram, S. P., and Juvet, R. S.: Gas chromatography. Anal. Chem., *44*: 213R, 1972.
2. Eik-Nes, K. B., and Horning, E. C.: Gas Phase Chromatography of Steroids. New York, Springer-Verlag New York Inc., 1968.
3. Gas chromatography in clinical chemistry. Clin. Chim. Acta, *34*: 129, 1971. 9th West European Symposium on Clinical Chemistry.
4. James, A. T., and Martin, A. J. P.: Gas-liquid partition chromatography: the separation and micro-estimation of volatile fatty acid from formic acid to dodecanoic acid. Biochem. J., *50*: 679, 1952.
5. Littlewood, A. B.: Gas Chromatography: Principles, Techniques and Applications. 2nd ed. New York, Academic Press, 1970.
6. Porter, R. (Ed.): Gas Chromatography in Biology and Medicine, London, J. & A. Churchill Ltd., 1969.
7. Street, H. V.: The use of gas-liquid chromatography in clinical chemistry. Adv. Clin. Chem., *12*: 217, 1969.
8. Szymanski, H. (Ed.): Biomedical Applications of Gas Chromatography. New York, Plenum Press, 1964.
9. Weinstein, B.: Separation and determination of amino acids and peptides by gas-liquid chromatography. *In* Methods of Biochemical Analysis, D. Glick, Ed. New York, Interscience, 1966, vol. 14, p. 203.
10. Wotiz, H. H., and Chattoraj, S. C.: The role of gas-liquid chromatography in steroid hormone analysis. J. Chromatogr. Sci., *11*: 167, 1973.

SECTION SEVEN

BASIC PRINCIPLES OF RADIOACTIVITY AND ITS MEASUREMENT

by Maurice V. L'Heureux, Ph.D.

The wide use of artificially produced radioisotopes as diagnostic, therapeutic, and research tools in the basic medical sciences necessitates this discussion of some of the basic principles of radioactivity and its measurement.

Atomic nomenclature

The *atom* is the simplest unit into which an element can be divided and still retain the properties of the original element. Atoms of all elements are made up of three primary building blocks: protons, neutrons, and electrons. A useful model of the structure of the atom is that of an entity composed of a central core, the *nucleus*, consisting of *protons* and *neutrons,* and a cloud of much lighter electrons in motion about the nucleus in well-defined orbits. All elements are composed of various combinations of neutrons and protons within the nucleus, and these subatomic particles make up practically all of the mass of the atom. The atom as a whole is electrically neutral. The number of the positively charged particles, the protons, in the nucleus is equal in number to the negatively charged particles, the *orbital electrons*. The neutron is a nuclear particle with about the same mass as the proton, but it has no charge.

The number of protons in the nucleus is called the *atomic number* and is often designated by the letter Z. Each element has an atomic number which is characteristic of the element and is synonymous with the chemical symbol. The *mass number*, designated by the letter A, is the sum of the number of protons and neutrons in the nucleus and is the nearest whole number equal to the sum of the atomic weights of the protons and neutrons (1.0076 and 1.0090 daltons, respectively).

The term *nucleons* is a collective term for protons and neutrons. A *nuclide* is an atom with a specific combination of neutrons and protons.

A generally accepted way of representing a nuclide is to place the mass number (A) as a left superscript to the chemical symbol, and the atomic number (Z) as a left subscript, e.g., ${}^{32}_{15}P$, ${}^{14}_{6}C$, ${}^{131}_{53}I$. Since both the atomic number and the chemical symbol identify the chemical species, in most circumstances the left subscript is omitted from the representation, i.e., ${}^{32}P$, ${}^{14}C$, ${}^{131}I$.

Isotopes are nuclides with the same atomic number but different mass numbers. These represent various nuclear species of the same element. Most elements occurring in nature are mixtures of isotopes, although one particular isotope usually predominates. Natural carbon is a mixture of ${}^{12}C$ (carbon of mass 12), small amounts of ${}^{13}C$, and a trace of ${}^{14}C$. The other known isotopes of carbon are man-made. Hydrogen has three isotopes: ${}^{1}H$, ${}^{2}H$ (deuterium), and ${}^{3}H$ (tritium).

Chemical properties of all isotopes of an element are identical, since chemical properties depend upon the number and arrangement of the orbital electrons surrounding the nucleus. This fact is the basis for the specialty known as isotopic tracer methodology, the fundamental principle of which is that the living system does not differentiate between isotopes of the same element. Both stable and radioactive isotopes are available as tracer atoms. Stable isotopes, e.g., ${}^{15}N$ and ${}^{18}O$, are used when no radioactive isotopes are available which are suitable for a particular purpose. Since the use of stable isotopes has little application in clinical chemical operations, they will not be discussed further.

RADIOACTIVITY

Radioactivity is a property of the nucleus and is evidence of nuclear instability. It is manifested by a spontaneous change within unstable atomic nuclei, resulting in the emission of energetic radiations. Many structural criteria are involved in determining nuclear instability—e.g., the types of attractive and repulsive forces acting on the particles in the nucleus. Nuclear stability is related to the *neutron-proton* (n/p) *ratio* in the nucleus. Only certain combinations of neutrons and protons form stable nuclei. An excess of either one or the other leads to a redistribution of the particles. The unstable atomic nucleus tends to adjust its n/p ratio by emission of the excess energy and particles, producing a stable configuration. The series of transformations by which an unstable atomic nucleus converts to a stable nucleus is known as *radioactive decay*. Radioactive decay is a spontaneous and random event. The rate of the process is not affected by known chemical or physical agents, e.g., temperature, pressure, or concentration.

Types of radioactive emissions

There are three types of radiations emitted by radioactive nuclides: alpha particles, beta particles, and gamma rays. Many radioisotopes may emit more than one type.

Alpha particles, designated by the Greek letter α, are nuclei of helium atoms. They are composed of two protons and two neutrons. Thus, they bear two units of positive charge ($Z = 2$) and a mass number (A) of 4.

Beta particles, represented by the Greek letter β, can be negatively charged electrons called negatrons, or positively charged electrons called positrons. Usually the term beta particle without qualification is understood to mean a negative electron. Both negatrons and positrons are of nuclear origin and are emitted from radioactive atoms at high speeds almost equal to the speed of light.

A striking characteristic of beta radiation is that beta particles from a given emitter are not all of the same energy. Rather, emitted beta particles encompass a spectrum of energies varying from practically zero up to a maximum energy (E_{max}), and the energy is characteristic of the beta-emitting radionuclide.

Gamma rays, designated by the Greek letter γ, are electromagnetic radiations of very short wavelength. They are similar in their essential nature to x-rays, light waves, and radio waves, but they differ from these in that they originate from unstable atomic nuclei.

Modes of radioactive decay

Each radioactive nuclide has its own characteristic pattern of decay. In any process of decay, the radioelement which decays is called the parent, and the product nucleus of this transformation is called the daughter. The daughter element may be stable or radioactive. Essential to the meaningful use of radioisotopes is knowledge of two aspects of this pattern: the type and energies of the emissions, and the rate of decay.

With few exceptions, *alpha particle emission* is characteristic of heavy radioactive nuclides. The release of one alpha particle removes four nucleons from the nucleus, two of them protons. Hence, the daughter element has an atomic number two less and a mass number four less than the parent. For example, radium-226 decays by alpha emission to produce radon as follows:

$$^{226}_{88}\text{Ra} \rightarrow {}^{222}_{86}\text{Rn} + {}^{4}_{2}\text{He} \ (\alpha \text{ particle})$$

Most alpha emitters are naturally occurring radioisotopes and, for the most part, have little clinical chemical application.

Beta particle emission gives rise to two types of disintegration. In one type, called *negatron decay*, the parent nucleus emits a negatively charged electron. The atomic number of the product element is increased by one but the mass number remains unchanged. For example, radioactive carbon-14 disintegrates by the scheme

$$^{14}_{6}\text{C} \rightarrow {}^{14}_{7}\text{N} + \beta^-({}^{0}_{-1}\text{e})$$

Negative beta particle emission is characteristic of nuclei which exhibit a high neutron-proton ratio, too high for stability. In this transformation, a neutron is converted into a proton and a practically weightless, negatively charged electron, thus decreasing the n/p ratio. The negative beta particle is emitted with a kinetic energy characteristic for this particular transformation.

Occasionally, protons are in excess in a nucleus and the n/p ratio is too low for stability. Here, a different mode of stabilization results—*positron decay*. In this case, a proton is converted into a neutron with the emission of a positron, thus increasing the n/p ratio. The atomic number decreases by one and the mass number remains unchanged, as in the following example:

$$^{13}_{7}\text{N} \rightarrow {}^{13}_{6}\text{C} + \beta^+({}^{0}_{+1}\text{e})$$

Positron emission occurs only with artificially produced radionuclides.

Under certain circumstances, an unstable nucleus with an excess of protons does not decay by positron emission but by an alternate pathway called *orbital electron capture*. In this process, the nucleus captures an electron from an orbital shell of the atom, usually the K shell, and the electron combines with a proton to form a neutron, thus increasing the n/p ratio. Iodine-125, which is often used in radioimmunoassay procedures, exhibits this mode of decay:

$$^{125}_{53}\text{I} + {}^{0}_{-1}\text{e} \rightarrow {}^{125}_{52}\text{Te} + \gamma\text{-ray and x-ray}$$

The vacancy in the inner electron orbit resulting from the K capture is filled by an electron from an outer orbit, and this rearrangement of orbital electrons is reflected by the emission of an x-ray characteristic of the process.

Following alpha decay, beta decay, or electron capture reactions, the product nucleus may be left in a more or less excited state. This *excitation energy*, if present, is given off in the form of gamma rays. Therefore, the emission of gamma rays reflects the transition between two energy levels of the same nucleus. Normally, there is no significant delay in the release of the gamma rays. However, there are some excited states of product nuclei which do not emit their gamma rays immediately but show a delay in the disposition of excess energy. Nuclides of this type are known as metastable, and their mode of transformation to a stable state is called isomeric transition.

In any radioactive decay process, a fixed total amount of energy is released with each disintegration. Most or all of this energy release appears as the kinetic energy of the emitted particle(s) or gamma ray photons. The energies associated with these radiations are described in terms of a unit called an electron volt (eV), which is defined as the amount of kinetic energy acquired by an electron when it is accelerated in an electric field which is produced by a potential difference of 1 volt. Since the electron volt is a very small amount of energy, larger multiple units are commonly used: KeV for thousand or kilo electron volts, and MeV for million electron volts. Typically, the energy of the alpha particles varies from 2 to 10 MeV, that of beta particles from 0.018 to 4.8 MeV, and that of gamma rays from 10 KeV to 10 MeV.

Rate of radioactive decay

One cannot predict when a given radioactive atom will decay. But in a group of millions or billions of radioactive atoms present in any measurable amount of a radioactive element, the number that will disintegrate during a given time interval can be predicted rather closely. Observation verifies the principle that the number of disintegrations which occur in a unit of time is proportional to the number of atoms present at the beginning of the time interval. The proportionality constant is referred to as the decay constant, λ. The mathematical formula that expresses this relationship is

$$\frac{-dN}{dt} = \lambda N$$

The *d*'s are the symbols of differential calculus and the minus sign indicates that the number of radioactive atoms present decreases with time. Upon integration, the following expression is obtained:

$$N = N_0 e^{-\lambda t}$$

The intensity or strength of a radioactive source is often called the activity (A).

$$\mathrm{A} = -dN/dt = \lambda N$$

Accordingly, one may rewrite the above equation as follows:

$$\mathrm{A} = \mathrm{A}_0 e^{-\lambda t}$$

Stated in words, the activity of a sample remaining after a time t is equal to the initial activity A_0 times $e^{-\lambda t}$, where e is the base of the natural logarithm series, and λ is the decay constant. The decay constant λ is a measure of the intrinsic probability of decay. The greater the value of λ, the faster the decay rate.

Instead of describing the rate of decay in terms of the decay constant, it is more convenient to describe it in terms of the *half-life* ($t_{1/2}$). The half-life is the time required for a given amount of radioactivity to decrease to one-half its original value. Radioisotopes used in medicine have half-lives ranging from a few hours to several years. The half-life can be shown to be related to the radioactive decay constant in the following way:

$$t_{1/2} = \frac{0.693}{\lambda}$$

Thus, the amount of activity remaining at time (t) is given by the following expression:

$$A_t = A_0 e^{-\frac{0.693 \times t}{t_{1/2}}}$$

This equation is commonly modified to a more convenient form by using logarithms to the base 10:

$$\log A_0/A_t = \frac{0.301 \times t}{t_{1/2}}$$

This relationship is a useful working equation for calculating the activity of a sample after it has undergone decay for some time interval. This is conveniently done by plotting the log of per cent activity remaining, against time, and reading the per cent activity at any desired time interval (Fig. 3–41).

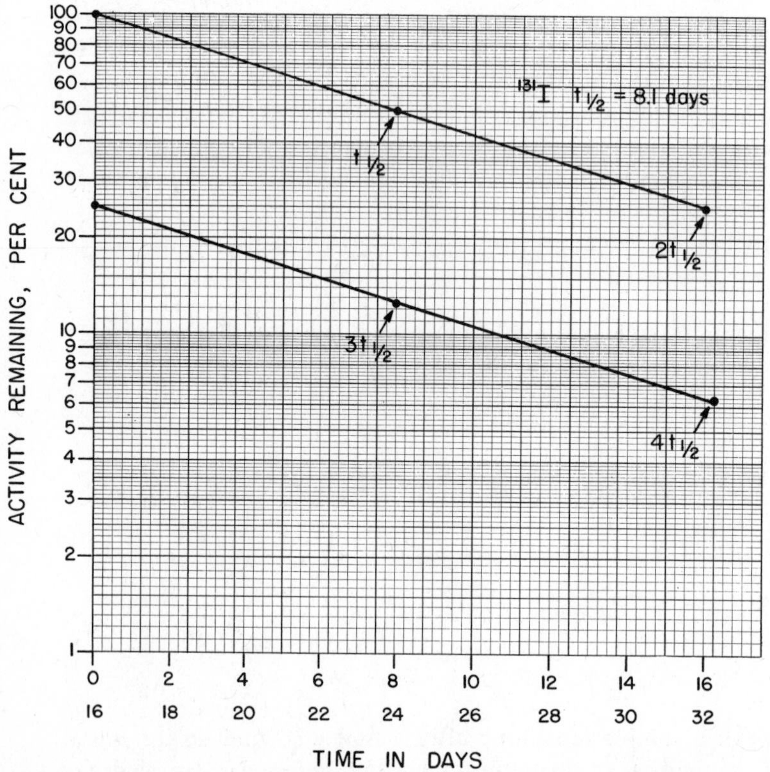

Figure 3–41 Radioactive decay plot for ^{131}iodine (half-life = 8.1 days). The per cent of the original radioactivity is plotted against time in days on semilog graph paper. The 100 per cent activity value is that present when the preparation was first assayed. Thus, 12 days after the original assay, 35.7 per cent of the original activity still remains in a preparation such as ^{131}I-labeled serum albumin.

TABLE 3–4 RADIATION PROPERTIES OF COMMONLY USED RADIONUCLIDES*

Nuclide	Half-life	Type of Decay	*Maximum Energy of Radiation (MeV)*† BETA	GAMMA
^{3}H	12.3 years	β^-	0.018	
^{14}C	5730 years	β^-	0.156	
^{24}Na	15 hours	β^-, γ	1.39	1.37, 2.75
^{32}P	14.3 days	β^-	1.71	
^{35}S	87.1 days	β^-	0.167	
^{42}K	12.4 hours	β^-, γ	2.00, 3.52	1.52
^{51}Cr	27.8 days	EC‡		0.32
^{59}Fe	45 days	β^-, γ	0.273, 0.475	1.095, 1.292
^{57}Co	270 days	EC		0.122, 0.136
^{60}Co	5.26 years	β^-, γ	0.313	1.17, 1.33
^{99m}Tc	6 hours	IT		0.141
^{125}I	60 days	EC‡		0.036
^{131}I	8.1 days	β^-, γ	0.61	0.364
^{198}Au	2.7 days	β^-, γ	0.962	0.411
^{197}Hg	65 hours	EC		0.077
^{203}Hg	46.9 days	β^-, γ	0.213	0.279

* Data taken in part from Dillman, L. T.: Radionuclide decay schemes and nuclear parameters for use in radiation-dose estimation, J. Nucl. Med., Supplement Number 2, March, 1969.

† Only the principal energies are given here.

‡ EC refers to electron capture.

IT refers to isomeric transition.

Each half-life reduces the activity by one-half and the effect is cumulative. In the general case,

$$A_n = A_0(\tfrac{1}{2})^n$$

where n is the number of half-lives which have elapsed. This expression is also useful in the planning of experimentation with radioisotopes and in the disposal of radioactive wastes. In the latter case, a useful rule of thumb is that seven half-lives will reduce any activity to less than 1 per cent of its original value ($(1/2)^7 = \frac{1}{128} < 0.008 \approx 0.8\%$).

The unit of radioactivity is the *curie* (Ci), which is the quantity of radioactive material in which 3.7×10^{10} disintegrations (dis) occur per second. Smaller units in common use are the millicurie, mCi (3.7×10^7 dis/sec), and the microcurie μCi (3.7×10^4 dis/sec). This definition of the curie is independent of the quantity of the radionuclide. For example, a curie of pure cobalt-60 would weigh less than a milligram, whereas a curie of uranium-238 would weigh over 2 metric tons.

A list of commonly used radionuclides and some of their radiation properties are presented in Table 3–4.

Interactions of radiations with matter

Radioactive emissions possess energy either in the form of kinetic energy of motion, as is the case with alpha particles and beta particles; or in the form of electromagnetic radiation, as is the case with gamma rays. The interactions of these radiations with matter reflect the transfer of energy to the atoms of the material through which they pass. This transfer takes place chiefly through processes of excitation and ionization.

In excitation, part of the energy of the incident radiation is absorbed by orbital electrons, and these are raised to higher energy levels. The excited atom then subsequently returns to its normal state with the emission of electromagnetic radiation. In ionization, sufficient energy may be transferred to one of the orbital electrons to remove it from the atom. This results in the production of a positive ion and an electron. This combination is called an ion pair.

The transfer of energy from alpha particles, and from beta particles of either sign, is accomplished by interacting electrical fields. Gamma rays, being uncharged, do not cause ionization directly, but they do so indirectly as will be noted presently.

The ability of radiations from radioactive decay processes to cause excitation and ionization is a most important property, for it is the basis for the detection of radioactivity, it is responsible for the biological effects of ionizing radiation, and it dictates the requirements for radiation protection.

The magnitude of the ionization produced is expressed in terms of specific ionization, which is defined as the number of ion pairs formed per unit length of path in a medium. Roughly, the specific ionization varies directly as the square of the charge and inversely as the square of the velocity of the ionizing radiation.

Alpha Particles. An alpha particle, by virtue of its properties—its large mass, its charge, and its velocity—produces dense, localized ionization in an absorbing medium. Accordingly, it dissipates its kinetic energy over a short distance, its range in matter is short, and its track is straight. Alpha particles are thus weakly penetrating. Alpha particles of a few MeV or less from a radioactive source near or on the body penetrate only the most superficial layers of the skin. In contrast, alpha emitters are a very serious internal hazard because of their high specific ionization.

Beta Particles. When beta particles interact with matter, they lose energy by excitation and ionization, as do the alpha particles. Since the beta particles have very small mass, they are easily deflected in passing through matter, and scattering effects are prominent in beta particle interactions. Due to the smaller size and charge and the high velocity of the beta particle, there is a smaller probability of interaction with atoms, and thus beta particles will travel much further through matter before expending their energy than will alpha particles of comparable energy. Another type of interaction of beta particles with matter is that of Bremsstrahlung (braking radiation). This is a highly penetrating electromagnetic radiation which results from the rapid deceleration of fast-moving beta particles in the field of the nucleus. Bremsstrahlung effects are low in tissues because of the low atomic number of tissue constituents. Like alpha particles, beta particles are considered to be internal radiation hazards, but beta radiation of high energy may also be an external hazard, since it can penetrate the outer layers of the skin.

Gamma Rays. Three processes are mainly responsible for the absorption of gamma rays: photoelectric absorption, Compton scattering, and pair production (Fig. 3–42).

In the *photoelectric effect*, the gamma ray photon imparts all of its energy to an electron, usually one in an inner orbit, in an atom of the absorbing material. The electron is ejected, and it in turn interacts with other atoms and causes secondary ionization. This type of interaction is most probable for gamma rays with energies below 0.5 MeV interacting with absorbers of high atomic number.

In the *Compton effect*, the photon imparts part of its energy to an orbital electron, which is then ejected. The gamma ray is deflected in the process and is emitted at a longer wavelength and a lower energy. The electron may produce secondary ionization in the same manner as the photoelectron. The scattered gamma ray is further attenuated by subsequent interactions. The relative probability of this type of interaction is quite high for gamma rays of intermediate energies, e.g., between 0.5 and 1.0 MeV with absorbers of medium to low atomic number.

In *pair production*, the entire energy of the photon is absorbed in the production of a positron and an electron in the vicinity of the nucleus. The negative electron produces secondary ionization. The positron also produces secondary ion pairs until it has lost most of its kinetic energy. It then combines with a free negative electron. Their mass is anni-

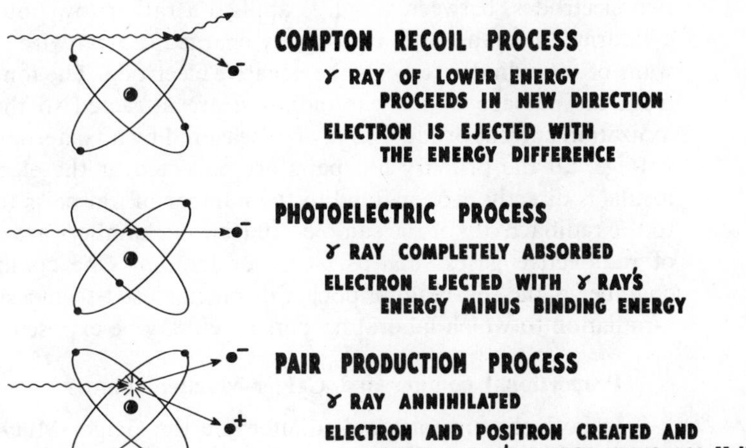

Figure 3-42 Interactions of gamma rays with matter.

hilated with the production of two gamma ray photons of 0.51 MeV, each emitted at 180° with respect to each other. The minimum gamma ray energy necessary for pair production is 1.02 MeV. Any gamma ray energy in excess of 1.02 MeV appears as the kinetic energy of the particles involved. The probability of the process increases with gamma rays of high energies and absorbers of high atomic number.

The emission of gamma rays reflects transition between two energy levels of the same nucleus. Gamma rays have no detectable mass, are uncharged, and travel with the velocity of light. The low specific ionization of gamma rays implies low energy loss per centimeter of path and, thus, high penetrating power. Externally, high energy gamma rays can irradiate the whole body. Internally, gamma rays present less of a hazard than alpha and beta radiation.

DETECTION AND MEASUREMENT OF RADIOACTIVITY

The determination of radioactivity requires a 2-fold approach in instrumentation: first, detection, which is concerned with the interaction of the radiation with some form of matter; and second, measurement, which is the quantitative assessment of the interaction.

Various types of media are used for the interaction. These include gases, liquids and solids. In general, the interaction involves one of the following effects:

1. The ionization of gases.

2. The production of scintillations or flashes of light in certain media called fluors, either solids or solutions.

3. The darkening of photographic emulsions.

Instrumentation assemblies for measuring ionizing radiation differ in the medium in which the interaction takes place, and on the method by which the interaction is measured.

When a rapidly moving charged particle enters a gas, it has the ability to remove electrons from the atoms of the gas molecules and to form ion pairs, a process called ionization. The various types of detectors based upon ionization effects in gases are differentiated on the basis of the behavior of these ion pairs in an electrical field. Three major groups of such detectors include ionization chambers, proportional counters, and Geiger-Mueller counters.

Ionization chambers

An *ionization chamber* consists of an enclosed chamber containing air or other gas and two electrodes, between which is applied a rather low potential. In its simplest form, it is cylindrical in shape with a positively charged central wire running through its length. The walls of the cylinder serve as the negative electrode. The ion pairs produced in the chamber from the passage of nuclear radiation are attracted to the collecting electrodes, and the ionization current produced may be measured by an external meter. At the proper operating voltage, all the primary ion pairs are collected at the electrodes, and the current which results is directly proportional to the number of ion pairs formed in the chamber and thus to the radioactivity of the sample. Ionization chambers are often used for the measurement of radioactive gases such as $^{14}CO_2$ or tritium. One common example of an ionization chamber is the self-reading pocket dosimeter used to measure the amount of gamma and x-radiation to which laboratory personnel may be exposed.

Proportional counter and Geiger-Mueller counter

In both the proportional counter and the Geiger-Mueller counter, detection of radiation is accomplished by taking advantage of a process called *ion multiplication.* The design of the counter and the voltage applied are such that a very high voltage gradient exists in the vicinity of the positive electrode. Electrons formed within the chamber through ionization are accelerated to high speeds and are capable of causing secondary ion-pair formation in the gas contained in the chamber. Thus, from a few primary ion pairs, a veritable torrent of ions moves toward the collecting electrodes. Provided the voltage is maintained at a constant value, the amplification remains constant and the size of the pulse produced is directly proportional to the number of primary ion pairs produced by the entering ionizing particle. Thus, the proportional counter derives its name from the proportionality between the size of the output pulse and the primary ionization. The advantage afforded by this type of detector is that it makes possible discrimination between types of radiation which differ in the intensity of the primary ionization, e.g., between alpha and beta radiation.

The *Geiger-Mueller counter* is characterized by the fact that at very high potential gradients, the electrical field between the electrodes becomes so strong that any single ionizing event produces ionization that spreads along the entire anode wire. Incident radiation of strength sufficient to release one single electron can result in a cumulative action which produces maximal possible ionization for a given voltage. The size of the charge collected on the anode is now independent of the number of primary ions collected. Hence, the Geiger-Mueller counter does not differentiate between different types of ionizing radiation which enter the detector.

Scintillation counting

In recent years, scintillation counting has become a very important tool for the measurement of radioactivity. It is based on the principle that exposure of certain materials to ionizing radiation results in the conversion of the kinetic energy of the particles or photons into flashes of light or scintillations. Materials which exhibit this property are called fluors or scintillators. Excitation of the fluor leads to light emission. The scintillation is directed to another unit called a photomultiplier tube, which converts the light photons into an electrical pulse whose magnitude is proportional to the energy lost by the incident radiation in the excitation process. This voltage pulse is then amplified and counted.

Two main types of scintillation counting are in use: solid scintillation counting and liquid scintillation counting. *Solid scintillation counting* (*in vitro* and *in vivo*) is concerned with the interaction of the radiation from a source physically external to the scintillator. *In vitro* applications in clinical chemistry involve, for example, measurement of radioactivity of samples of material outside of the living organism or patient, e.g., blood, urine, or feces.

In vivo counting refers to the direct measurement of radioactivity within the living organism or patient, or in a particular organ. *Liquid scintillation counting* is concerned with the measurement of radiation from a source intimately in contact with the fluor, either dissolved or suspended in it. Liquid scintillation media constitute a very important radiation detector in biochemical research applications and in nuclear medicine laboratories, especially in the laboratory using radioimmunoassay techniques.

Solid Scintillation Counting. Solid scintillation counters have the great advantage that different types of fluors are available to serve as detectors for different types of radiation. The fluors may be inorganic or organic in nature. Inorganic fluors are normally used in the form of single large crystals of inorganic salts containing small amounts of added material called activators. The activator is added during the process of crystal production, and it is essential for maximum scintillation production. The fluor most commonly used as a gamma ray detector is sodium iodide containing small amounts of thallium as the activator, NaI(Tl). Silver activated zinc sulfide, ZnS(Ag), is commonly used for alpha particle detection, and organic crystals of anthracene or naphthalene are useful for beta radiation detection. However, solid scintillation counting of alpha and beta radiation has little application in the clinical laboratory.

A solid scintillation detector applicable for gamma ray detection is shown in Figure 3–43. It consists of a solid crystal of sodium iodide, thallium activated, which is optically coupled to the face of a photomultiplier tube. As the gamma ray passes into the fluor, it dissipates its energy by the three main processes of interaction mentioned previously. Electrons are ejected, which in turn produce excitation or ionization of other portions of the crystal. The secondary excitation is the important event. Once excited, the more loosely bound electrons quickly return to a more stable energy state by emitting light photons. The photons pass through the transparent fluor and impinge on the photocathode of the photomultiplier tube, causing emission of electrons from the cathode. Amplification in the number of electrons occurs by means of a series of electrodes called dynodes spaced along the length of the photomultiplier tube. Each dynode is maintained at a potential higher than the preceding one by some 100 volts. The dynodes are so arranged that the electrons ejected from one are focused onto the next. The electron multiplying process continues until, at the last dynode, 10^5 to 10^6 electrons are collected for each electron ejected from the photocathode. The magnitude of the output pulse from the photomultiplier tube is proportional to the amount of energy dissipated in the fluor by the incident gamma ray.

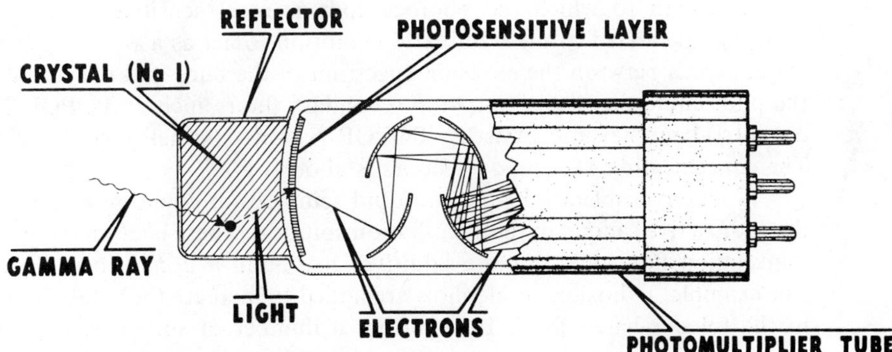

Figure 3–43 Schematic diagram of a solid scintillation detector.

The pulses are amplified and counted by means of associated electronic components. Usually, the complete system includes a preamplifier, a linear amplifier, a discriminator, and an electronic scaler. The *preamplifier* shapes the pulse and the *amplifier* strengthens the signal from the photomultiplier. The *discriminator* is a device that rejects pulses smaller than a predetermined size. It is used primarily to block out electric noise pulses from being counted. The *scaler* is a totalizing unit which acts as an electronic adding machine. It is commonly used together with an electronic register and timer, and each pulse passed by the discriminator is recorded as a count on the scaler. Most scalers employ a timer which will automatically stop the scaler from counting at the end of a preset time of operation or when a preset number of counts have been recorded.

Liquid Scintillation Counting. Liquid scintillation counting has become a very useful method for measuring radioisotopes in biomedical applications, because it is regarded as the most efficient system for counting low energy beta emitters such as carbon-14 and tritium. However, the method is also applicable to other types of radiation, including low energy x-rays emitted by electron capture decay processes, e.g., from iodine-125; and beta particles of high energy, e.g., from phosphorus-32.

In liquid scintillation counting, the radioactive sample is dissolved in or intimately mixed with a suitable solvent in a glass vial together with an organic liquid scintillator called a primary fluor. The solvents in general use are the alkyl benzenes such as toluene and xylene. A common characteristic of primary fluors is that they contain conjugated aromatic rings. Primary fluors used most often include PPO (2,5-diphenyloxazole), BBOT [2,5-bis-2(5-*t*-butyl-benzoxazolyl)-thiophene], p-terphenyl, and PBD [2-phenyl-5-(4-biphenyl)-1,3,4 oxadiazole]. The concentration of these fluors in the scintillation medium is of the order of 4 g/l.

The energy transfer process which takes place in the scintillation mixture is complex. Radiation emitted by the sample excites the solvent molecules and some of this excitation energy is transferred to molecules of the primary fluor. The primary fluor in turn re-emits this excitation energy in the form of light photons. The light so produced is directed toward two photomultiplier tubes (see Fig. 3–43). The signal from each photomultiplier tube is transmitted via a preamplifier to a coincidence circuit. The coincidence circuit will pass an output pulse to the scaler for counting, provided pulses reach it simultaneously from both photomultiplier tubes. Main contributors to the background count rate are the spurious pulses which arise within the photomultiplier tube itself and not from impinging light photons derived from a radioactive source. Photomultiplier tube noise is emitted randomly and may occur in one tube and not in the other. In contrast, the probability is high that radiation emitted by the sample will produce a signal in both photomultiplier tubes simultaneously. This arrangement of two photomultiplier tubes together with a coincidence circuit is an effective means of reducing the background count rate.

The wavelength of light emitted by the scintillation mixture may not match the range of wavelength to which the photocathode is most sensitive. Accordingly, a secondary fluor may be added to the scintillation solution to act as a wavelength shifter to achieve a better match between the emission spectrum of the fluor and the sensitivity of response of the photomultiplier tube. Common secondary fluors include POPOP [1,4-bis-2(5-phenyloxazolyl)-benzene] and dimethyl POPOP [1,4-bis-2(4-methyl-5-phenyloxazolyl)-benzene]. The concentration of secondary fluors is about 1 g/l.

A major problem inherent in liquid scintillation counting is the matter of dissolving the radioactive sample in the scintillation solution. There is often the need to include in the scintillation fluid one or more additives to obtain a homogeneous scintillation mixture. For example, *p*-dioxane or alcohols are added to increase the solubility of aqueous samples in the toluene-based fluor. In addition, a number of solubilizers are now commercially available for the purpose of getting the radioactive sample into true solution before counting.

Radioactive samples which cannot be dissolved in suitable solvents may be counted as suspensions if they can be reduced to a fine, uniform particle size. For example, gelling agents such as aluminum stearate or Cab=O=Sil have been used to stabilize insoluble material in toluene-PPO scintillation solutions. Counting of samples impregnated on filter paper discs, paper chromatograms or Millipore filters can be done by liquid scintillation. Such approaches provide meaningful results only if the activity remains either completely on the paper or comes completely off when the paper is immersed in the counting solution. It is important to be sure that the sample is not partially eluted into the scintillation mixture. Any fraction in solution will be counted with an efficiency different from that of the fraction in the solid phase.

Another precaution that must be taken before assay of any sample is undertaken is to allow the sample to become "dark-adapted" before counting. Such a procedure is necessary to exclude or minimize the contribution of phosphorescence to the count rate.

Quench Correction. With all sample preparation techniques, some degree of an adverse process called *quenching* invariably occurs. Quenching is defined as any process taking place within the sample container which results in decreased number or intensity of the light flashes produced, and thus in a decrease in counting efficiency. It may occur by several mechanisms. The sample introduced into the scintillation vial may interfere with the energy transfers among the components of the fluor before the production of light, or it may interfere with the transmission of the light after it has been produced. In either case, the problem of quenching is solved not so much by avoiding it as by devising means to correct for it. Several methods of quench correction have been developed, and three of these will be described briefly.

METHOD OF INTERNAL STANDARDIZATION. In this method, the samples are counted before and after a known quantity of the same isotope in a nonquenching form is added as an internal standard. It is assumed that the internal standard is subject to the same quenching conditions as the original sample. The decrease in count rate of the added standard will be a measure of the degree of quenching. The count rate of the sample in the absence of the internal standard is subtracted from the count rate with the internal standard. The difference in the count rate divided by the known disintegration rate of the standard provides the counting efficiency for that sample. The true disintegration rate for the sample in disintegrations per minute (dpm) may then be obtained for each sample by dividing the net count rate of the sample by the corresponding counting efficiency.

THE CHANNELS-RATIO METHOD. This method attempts to correct for quenching by following the change in the ratio of the sample counts directed by pairs of discriminators to two separate counting scaler channels. When quenching occurs, the output pulses are decreased in intensity, giving rise to a shift in the pulse height spectrum to lower energies. The ratio of the count rates in the two channels varies with the amount of quench and reflects this pulse height shift. In practice, one adds increasing amounts of quenching agent to a series of standard samples containing a known amount of radioactivity, and the channels ratio and detecting efficiency is determined for each sample. A plot is constructed on linear coordinate paper relating counting efficiency to the ratio of the count rates in the two channels. This relationship may then be used to correct a particular series of experimental samples for quenching.

METHOD OF EXTERNAL STANDARDIZATION. Liquid scintillation assemblies are generally equipped with an external standard facility whereby a standard gamma-emitting source such as cesium-137 or radium may be placed automatically at a fixed position adjacent to the sample vial and used instead of an internal standard to measure counting efficiency for each sample and, by extension, for quench correction.

A series of counting vials are prepared, containing the same amount of radioactivity but different amounts of quenching agent in the liquid fluor. As with the channels-ratio method, care is taken to make this set of standards as representative as possible of the composition of experimental samples to be counted. The samples are counted in the presence of the gamma source in a selected counting window and, subsequently, alone. The counting efficiency for each of the standard samples is plotted against the net count rate for the external standard to obtain a calibration curve for this system. Experimental samples are counted under the same operating conditions. The detection efficiency is obtained from the plot, and the activity of each sample corrected for quench can then be calculated.

ADDITIONAL READINGS

Faires, R. A., and Parks, B. H.: Radioisotope Laboratory Techniques. Halsted Press Division, John Wiley and Sons, Inc., 3rd ed. New York, 1973.

Hendee, W. R.: Medical Radiation Physics. Chicago, Yearbook Medical Publishers, Inc., 1970.

Simmons, Guy H.: A Training Manual for Nuclear Medicine Technologists (publication BRH/DMRE 70-3). U.S. Department of Health, Education and Welfare, Public Health Service Bureau of Radiological Health, Rockville, Maryland, 1970.

Wang, C. H., and Willis, D. L.: Radiotracer Methodology in Biological Science. Englewood Cliffs, New Jersey, Prentice Hall, Inc., 1965.

CHAPTER 4

AUTOMATION

by Donald S. Young, M.B., Ph.D.

When used in an industrial context the term *automation* implies mechanical or electronic control of a process. In clinical chemistry the same term is applied, although not correctly, to the performance of analytical tests by an instrument or combination of components with only minor involvement of an analyst. Mechanization of a procedure by the use of mechanical pipets has been described as *semiautomation. Partial automation* refers to procedures in which the initial preparation of a specimen is done manually, but in which the analysis proceeds without human intervention (e.g., chromatography).

With physicians becoming increasingly dependent on clinical laboratory data for the diagnosis of disease and the monitoring of therapy, automation has become essential to process the increasing workload in hospital laboratories. The alternatives to automation are either a much larger laboratory staff or a much more judicious selection of appropriate laboratory tests by physicians. Automation provides a means by which an increased workload can be processed rapidly and reproducibly, although it does not necessarily improve the accuracy of results. Indeed, limitations in the design of some automated instruments make it difficult to achieve results of acceptable quality. Where the accuracy of results is influenced by the analytical methodology (e.g., lack of specificity), automation can not compensate for inherent deficiencies in the procedure.

Whether an analysis is performed manually or automatically, it can be broken down into several individual steps. Although not all of these need to be mechanized to achieve increased efficiency, it is generally desirable to perform as many steps as possible without manual intervention. Full automation also reduces the possibility of human errors that arise from technicians making repetitive and boring manipulations, such as pipetting and analyzing many serum specimens for sodium and potassium by flame photometry.[3] An automated instrument may be designed for a specific task or may be built with sufficient flexibility to perform a variety of analyses. On other instruments (multichannel analyzers), it is possible to perform many different tests simultaneously.

Several steps are common to most analyses. These can be summarized as follows: (1) sample pickup; (2) sample delivery with or without subsequent washout of the sample probe with reagent or diluent; (3) protein separation by precipitation, filtration, centrifugation, chromatographic technique, or dialysis; (4) the addition of one or more reagents, mixing, and incubation; (5) reaction detection by visible, UV, or flame photometry, fluorometry, or nephelometry; and (6) data presentation on a digital readout, strip-chart recorder, printed tape, or computer terminal.

Automation was initially applied to those tests in the clinical laboratory that were requested most often. Although many of the same analytical principles are used for the chemical analysis of serum and urine, automation of urine analysis has not played an important role. This is partially because of the greater spread of observed values in urine specimens, requiring sensitivity to detect analytes at a low concentration while also requiring extended calibration linearity to permit analysis of samples of high concentration without dilution. The requirements for precision and accuracy in urine analyses are not quite as high as those for serum. Therefore, many laboratories determine values for calcium, phosphate, sodium, potassium, uric acid, creatinine, and urea nitrogen in urine on mechanized instruments which were originally designed for the analysis of serum constituents.

DEVELOPMENT OF AUTOMATED SYSTEMS

Automation of clinical chemistry analyses has evolved in three major directions. The first employs the *continuous-flow principle* conceived by Skeggs and first introduced commercially in 1957 as the AutoAnalyzer by the Technicon Instruments Corporation. This system has undergone considerable refinement and modification, so that the latest units have considerably greater versatility than the initial instruments.

The second group of instruments employs the principle of *discrete-sample processing*. Each specimen is processed separately, generally by steps mimicking the conventional manual procedure. This approach is an outgrowth of the development of the semiautomated pipets, in which the manual transfer stages between pipetting steps and the transfer into the photometer cuvet have been mechanized. The flexibility inherent in this approach has resulted in the development of many different instruments of this type.

The third and most recently developed principle makes use of *centrifugal force* to transfer and mix samples and reagents. A single light source and detector is used to record data from many cuvets.

It is, of course, possible to combine two of the above concepts in one analytical procedure. The initial transfer of serum into the centrifugal fast analyzer, for example, requires a discrete pipetting unit. On the other hand, an AutoAnalyzer may be used to analyze aliquots which were first prepared by manual dilutions.

AUTOANALYZER

Although the designation AutoAnalyzer has often been applied to any instrument capable of making chemical analyses automatically, the name is a registered trademark of Technicon Instruments Corporation, Tarrytown, New York, and should be reserved for the products of that company. The name AutoAnalyzer is used now to describe the simplest variant of the chemical processing instruments manufactured by the company. Those instruments, designed initially to process many tests at the same time, are designated *S*equential *M*ultiple *A*nalyzers (SMA) with a numerical qualification to indicate the number of tests that may be made simultaneously at a specified rate per hour (e.g., SMA 12/60).

The AutoAnalyzer was conceived around four important principles:

1. The use of continuous tubing of different diameters and a peristaltic pump to meter samples and reagents through the tubes. This process replaces the pipetting steps in manual procedures.

2. The introduction of air bubbles to separate sample and reagent streams into segments so that cross-contamination ("carryover") of the liquid stream by residual liquid from a preceding segment of the stream is minimized.

3. Dialysis through a semipermeable membrane to separate proteins from analytes so that interference of protein with the chemical reactions is avoided. This procedure eliminates the need for manual preparation of protein-free filtrates.

4. Modular construction of the analyzer to enable the convenient replacement of faulty components and to allow for substitution of one component for another if this is required for a different procedure (e.g., the replacement of a photometer with a fluorometer).

If an analysis is to be performed on the AutoAnalyzer, the required instrument modules are connected by plastic "transmission tubing." Specimens are then placed into cups in the sampler unit, from which an aliquot is removed through the sample pickup tube of the manifold. Samples and reagents are brought together in fixed-volume proportions and the liquid stream is segmented by air bubbles. The product of the chemical reaction formed after mixing of dialyzed sample and reagent, and after heating where applicable, is followed in a colorimeter or other appropriate device. Results are displayed as a series of peaks on a strip-chart recorder. It is not necessary for the reaction to be complete before measurement, but it is essential that all specimens, including standard solutions, be treated identically if accurate results are to be obtained.

COMPONENTS OF THE AUTOANALYZER

Sampler

The sampler is the device that holds the cups containing the standards and specimens for analysis. The specimen cups are placed in a circular tray mounted on a spindle that rotates so that each specimen is presented in turn for analysis at a predetermined time. The tray holds 40 disposable polystyrene cups of different sizes. In the first model of the sampler, the crook through which the sample was aspirated into the analyzer moved relatively slowly in and out of the specimens so that differences in depth of specimen in the cups affected the volume of sample analyzed, and this contributed to the variability of results.[6]

In the model II sampler, the action of the sample aspiration probe is such that it moves rapidly in and out of each specimen. Thus, it is not necessary to place identical volumes of specimens into all cups as long as the sample volume exceeds the minimum required for analysis. Aspiration of the sample is initiated by a programing cam. The appropriate cam is selected to determine the *rate of analysis*, the dwell time in the specimen, and the sample-to-wash ratio (assuming constant motor speed). As with the sampler I, the *volume of sample* aspirated and analyzed is dependent both on the diameter of the sample tubing of the manifold and on the dwell time in the specimen.

In the sampler II, the sample probe aspirates water between samples from a wash receptacle. The most commonly used cams provide a sample-to-wash ratio of 2:1, although cams with ratios from 9:1 to 1:6 are commercially available. Analysts may alter rates of analysis and sample-to-wash ratios for their procedures by changing the cam in the cam well, or by the use of their own design of cam. Imperfections in the cams, especially those of earlier design, have resulted in variability in the volume of samples aspirated.[8] The present plastic-molded cams exhibit much less variability.

The tray of the sampler II is designed so that it can accept cups that contain as little as 0.5 ml, as well as those holding 8.5 ml. The sample aspiration unit may be modified to accept twin probes or a paddle to agitate the specimen before analysis. For all the sampler units, evaporation of specimens can be minimized and exposure to direct light reduced by placing an amber plastic cover over the sample cups.

The identity of the specimens in the different cup locations of the basic AutoAnalyzer may be recorded in washable ink directly on the cups, on top of the sampler tray, or in a separate directory in which the identity of the sample is recorded against the number of the position in the tray. For operation of an AutoAnalyzer on-line to a computer or data-acquisition device, sequential identification may be used, i.e., the results obtained are matched with the proper specimen according to the sequence in which specimens were loaded on the sample tray. Greater certainty of identity is insured when the identity of each specimen is recorded as the sample is aspirated, and this information is stored in a memory device for correlation with the results as they are generated.

The sampler IV is the sampler unit recommended for the AutoAnalyzer II systems. It is interchangeable with the unit used in the SMA 12/60, and is employed in the Technicon Idee system for positive sample identification. The sampler can accept either Vacutainers (Becton-Dickinson & Company, Rutherford, N.J.), from which serum may be aspirated directly from above the cells after centrifugation, or conventional AutoAnalyzer sample cups. A preprinted label attached to the specimen container is read optically as the sample is aspirated.

The sampler developed for the *S*equential *M*ultiple *A*nalyzer *C*omputer (SMAC) multitest system differs from the other models in that it accepts racks of up to eight specimens. Thus, there is not the same limit to the number of specimens that can be processed before intervention by an analyst is required. The identity of each specimen is contained in human- and machine-readable form on the same coded label used with the sampler IV. To insure minimal contamination of one specimen by another in the SMAC Analyzer, the probe is moved up and down rapidly to introduce four air segments in the leading part of the sample as it enters the analytical system. This reduces contamination from the specimen analyzed previously and the water used in the rinse cycle, and enables a much higher rate of analysis to be achieved. Most analyses were made at a rate of 40 per hour when the sampler I was used, and there was significant contamination of adjacent samples, but with the SMAC sampler it is possible to achieve a rate of 150 samples per hour with less contamination.

Pumps and manifolds

The pumps of the AutoAnalyzer function on the principle of proportional addition of reagents and this depends on uniform delivery of solutions. The delivery of samples, reagents, and air into the analytical system occurs through pump tubing. The amount of sample, air, or reagent advanced into the system depends on the diameter of the tubing, which can be obtained with nominal delivery rates from 0.015 ml per minute to 3.90 ml per minute. If an analytical method is critically dependent on a precise ratio of sample to reagents, it is necessary to measure the actual delivery rate of the tubing in the system to be used because of differences between tubing and proper pumping action. The precision pump tubing that is now available insures less variability.

Tubes with the desired diameter are stretched between plastic holders (end blocks), which are placed on the pump. Since the walls of all tubes are of uniform thickness, tubing of different diameter may be used on the same pump. Some analysts, however, prefer not to place tubes of small diameter directly adjacent to those with large diameter. The total assembly of tubing and fittings is generally referred to as the *manifold.*

Reproducibility of delivery necessitates absence of change in the diameter of the pump tubing, and constant speed of the pump motor during the period of analysis of a batch of specimens. While only negligible changes occur in the tubing during one set of analyses, the polyvinyl (Tygon) tubing stretches, and the rate of delivery increases gradually after some time. Variability in delivery is reduced if all tubes are changed at the same time and before they lose their elasticity. As tubes begin to stretch, proper tension can be restored by moving the end blocks of the manifold further apart.

For most analytical procedures on the AutoAnalyzer, polyvinyl tubing is satisfactory. Siliconized rubber tubing (Acidflex) must be used when pumping strong acids, and solvent-resistant Tygon tubing (Solvaflex) is required when pumping most organic solvents. Judicious selection of tubing enables the analyst to perform on the AutoAnalyzer most procedures carried out in the clinical laboratory employing good analytical principles.

For the transfer of solutions between components of the AutoAnalyzer, polyvinyl tubing is usually satisfactory, but glass tubing is widely used in the SMA 12/60, AutoAnalyzer II, and SMAC systems to minimize contamination. For the mixing of solutions, all AutoAnalyzer systems utilize glass coils, which are mounted in such a way as to produce a tumbling action of a heavier liquid falling through a lighter until the liquid bubble is uniform. The differential velocity of liquid in the center of a stream from that at the edges, as well as the scrubbing action of the air bubbles, also contributes to the very effective mixing that occurs in a short time.

Four models of pumps are in use with AutoAnalyzer systems, and all employ the same principle of action. In the earliest model, five steel rollers are mounted transversely on a parallel bicycle-chain type of assembly. The tubing in the pump is compressed against a solid platen from above. Although the model I pump can accommodate 15 tubes, it is not advisable to exceed 12. The model II pump is built to accept 23 tubes without stacking and is arranged so that the same type of roller assembly compresses the tubing from below. The model II pump insures greater uniformity of delivery of fluid by the use of more rollers of larger diameter and by the elimination of the rocking motion that characterizes the model I pump when a roller lifts from the platen.

The pump III holds up to 28 pump tubes and incorporates an air-bar device that acts as a valve to occlude completely some of the air tubes. When the air bar lifts from the tubing, a more precisely controlled volume of air is delivered into the manifold than is possible without the device. This pump is used in the SMA 12/60 and AutoAnalyzer II system. For the SMAC analyzer, the pump consists of small replaceable modules dedicated to one analytical cartridge. Air segmentation of the sample diluent is performed to minimize sample interaction by the introduction of 90 bubbles per minute at a uniform rate through a simple valve mechanism.

Dialyzers

Dialysis is a process by which sample constituents of low molecular mass are separated from compounds with high molecular mass (proteins) with the aid of a semipermeable membrane. In the basic AutoAnalyzer, a Cuprophan sheet is stretched between two plates with matched spiral grooves, creating a channel on either side of the membrane. The diluted sample stream ("donor stream") moves on one side of the membrane, while the "recipient stream" (generally one of the reagents or saline) circulates through the other channel. For optimum transfer of solute, with minimal contamination of specimens, the two streams should flow concurrently at equal rates.

The quantity of solute that passes through the membrane is influenced by the duration of the contact of the two solutions. The area of contact, the temperature at which the dialysis occurs, and the thickness and porosity of the membrane are additional factors that determine the quantity of solute transferred. A further factor is the concentration gradient across the membrane. As solutes move across the membrane, the rate of transfer decreases so that even if the membrane area is doubled, the quantity of solute is not increased proportionately. The size and shape of molecules, their electrical charge, and the composition of fluids on either side of the membrane are further factors that influence the rate of dialysis across a membrane. Babson[2] has demonstrated, for example, that the rate of dialysis of solute from a solution containing a relatively high concentration of protein is different from that of a solution of lower protein concentration. This becomes important when a large volume of serum is slightly diluted only, and compared against a standard solution containing no protein.

In the AutoAnalyzer II, increased sensitivity of the analytical system is obtained by decreasing the speed of the liquid streams and by decreasing the dilution of the samples. As a result, the length of the dialyzer grooves has been reduced from 88 inches in the basic system to 3, 6, or 12 inches, depending on the requirements of the procedure. In the SMAC Analyzer, even greater sensitivity is obtained by the use of a new thin cellulose membrane that has twice the efficiency of the Cuprophan membrane. The grooves of the new dialyzer plates are not as deep, leading to decreased interaction of specimens.

In the AutoAnalyzer I, dialyzer plates are maintained in a water bath at 37°C to increase the rate of dialysis and maintain constant temperature during the dialysis process. In the AutoAnalyzer II, temperature control of the dialyzer is not provided, since it is assumed that changes of temperature will be slight during the short transit time of solutions through the dialyzer. Significant temperature fluctuations, however, affect the accuracy.

For CO_2 analyses with the AutoAnalyzer II system, a silicone-rubber membrane is used for the separation of the gas, thus eliminating the need for the debubbler that is required in the basic system and which contributes to sample cross-contamination.

Dialysis in the continuous-flow AutoAnalyzer system never reaches equilibrium, and reproducibility of results depends on the dialysis process occurring to the same extent in all standard solutions and specimens. The quantity of solute crossing the membrane rarely exceeds 20 per cent of the initial quantity, and for many constituents it is less than 10 per cent. This proportion is even less in the AutoAnalyzer II. With aging of a membrane, protein deposition reduces the quantity of material that crosses the membrane, and in routine laboratories it is usually necessary to replace membranes at intervals of one week or less.

Heating baths

Heating baths are used to provide the elevated temperature and time delay required for the development of a colored reaction product or for utilization of substrate by an enzyme. For the basic AutoAnalyzer, a 40 foot glass coil immersed in mineral oil is used most often as a heating bath. The temperature is maintained around 37° or 95°C, or in certain models the set point can be altered by means of an adjustable thermoregulator. The temperature in all models is controlled to within ± 0.1°C of the set point. The dwell time of a solution in the average glass coil is approximately 5 minutes with a typical flow rate. The dwell time may be increased by connecting coils in series, by using coils of larger diameter (and hence volume), or by decreasing the flow rate of the solution.

In the AutoAnalyzer II, SMA 12/60, and SMAC systems, shorter coils are used and each is tailored to the analytical procedure. This demands great precision in temperature control to insure that reproducible conditions are achieved for all samples. In these systems a solid metal heating block is used for transfer of heat to the solutions. A tightly wound glass coil inside the heating block allows rapid equilibration of temperatures. In the SMAC system the heating bath is coupled to a heat exchanger to reduce rapidly the temperature of the solution prior to further manipulations.

MEASUREMENT OR DETECTION DEVICES

Colorimeters

Several different models of colorimeters are used with the Technicon AutoAnalyzer. They have many features in common. A flow-through cuvet is used in all, and the color of the reagent stream is monitored continuously. The colorimeters employ a dual-beam principle to reduce the effect of variation in the output of the light source due to voltage fluctuations. The light from a single tungsten filament lamp is split and collimated into two beams, each of which strikes a barrier-layer photocell. On one side, the cuvet is interposed between the collimating unit and the photocell. The two photocells are arranged in a null-balancing circuit so that the ratio of the voltage from the test cell to that from the reference cell is displayed on a strip-chart recorder. Initial baseline conditions are achieved by means of a series of apertures placed in the reference beam and a "100 per cent *T*" control potentiometer. With normal use of the AutoAnalyzer system, the strip-chart recorder is calibrated so that its limits are 0 and 100 per cent *T*, and the output of the colorimeter is displayed on the recorder as per cent transmittance (%*T*). Wavelength selection is made by the insertion of appropriate narrow bandpass interference filters on both the reference and the test sides of the colorimeter.

In all AutoAnalyzer I and II colorimeters, air bubbles must be removed from the reagent stream before they enter the cuvet; otherwise, the electrical noise produced interferes with the interpretation of the strip-chart recording. The debubbling unit consists of an inverted glass "T" (or similar device) placed close to the inlet of the cuvet. This allows all bubbles and part of the reagent stream to be discarded, and only a small part is pulled through the cuvet in a continuous stream by the AutoAnalyzer pump. The most common cuvet in use with the AutoAnalyzer is tubular-shaped and has a pathlength of 15 mm. The same cuvets are incorporated into the SMA systems. In the SMA 12/30, 12 are mounted in a ring around a common light source. A single reference photocell is used for all cuvets. In the SMA 12/60, four are mounted in an arc around the light source. Other models of the AutoAnalyzer have been built in which essentially two colorimeters are combined in the same housing and arranged so that the per cent transmission displayed on the recorder indicates the difference in output between the two units (differential colorimeter). This is used when the analytical procedure requires a blank correction. With correct phasing of solutions through the two cuvets, it is possible to eliminate the calculation of the blank values.

The colorimeter of the AutoAnalyzer II is a modification of that of the SMA 12/60. While the early models were capable of processing up to three tests simultaneously, the current model is used for only one test at a time. A single phototube is used in place of the selenium-barrier photocells employed in the AutoAnalyzer I colorimeters. The colorimeter has a differential linear output, so that a calibration curve appears linear on a strip-chart tracing and calibration of the curve is feasible with a single standard.

The improved electronics incorporated into each generation of the Technicon colorimeters have enabled several unique features to be introduced into the SMAC system. Unlike all the other systems, the air bubbles in the reagent stream are not removed before it enters the cuvet. Elimination of noise is accomplished electronically. The sudden large changes in voltage caused by an air bubble is differentiated from the more gradual increase, associated with the development of a peak, and is filtered out by computer monitoring of the absorbance. The passage of frequent air bubbles through the cuvet enhances its cleaning and reduces sample interaction. A lesser volume of solution is required, since none is lost as waste in the process of bubble removal; in addition, more attention can be given to optical requirements and less to flow requirements of the cuvets. The cuvet of the SMAC system has a pathlength of 10 mm but a volume of only 2 μl. The squared-off "U" design reduces stray light to a minimum, with the horizontal part of the "U" forming the cuvet. Sapphire rods are fused flush against the optically flat ends of the cuvet, and the entire cuvet is encased in a glass housing. The cuvets in the SMAC system are mounted in such a way that a single photomultiplier tube is used. It has the capability of interrogating each cuvet four times per second. Interrogation is determined by a scanning disc, so that a signal is recorded from each cuvet in turn. Fiber optics are used to transmit light to and from the cuvets. Interference filters and a reference channel are used as in other AutoAnalyzer colorimeters. The SMAC colorimeter has the ultimate capability of being able to monitor up to 23 cells in the visible wavelength range, and up to 12 cells in the ultraviolet range.

Other detectors

Flame Photometers. The Technicon flame photometer has undergone many modifications and changes in design to obtain improved efficiency and safety. The principle of flame emission is employed with an internal reference so that sodium and potassium are measured independently against lithium present in the sample diluent. This minimizes errors due to variations in the pumping rate or changes in the character of the flame. The early flame photometer models used propane and oxygen as fuel and were equipped with a total consumption burner. With the latest model IV flame photometer, the fuel is methane or propane, which is mixed with compressed air. An atomizer type of burner is now used. The design of the burner is such that it is up to 10 per cent efficient, and sample cross-contamination has been reduced. This enables steady state conditions to be achieved rapidly, and 80 samples per hour may be analyzed without loss of reproducibility of results.

Fluorometer. A fluorometer may be substituted for the colorimeter in the AutoAnalyzer system to make use of the increased sensitivity of fluorometric technique. Although Technicon markets a filter fluorometer for fluorometric and nephelometric measurements with the AutoAnalyzer, most of the instruments that are commercially available may be used if the conventional cuvet is replaced by one adapted for continuous flow of solution.

Spectrophotometer. Spectrophotometers may be interfaced with an AutoAnalyzer manifold to measure substances that absorb in the ultraviolet region. In many instruments, such as the Gilford 300-N, no modifications are required since the cuvet has been designed initially for flow-through applications. Double-beam instruments, e.g., the Beckman-DBG, may also be used, but they require special manifolds so that an appropriate blank passes through the reference cuvet at the time that the test sample is in the other cuvet.

Other measurement devices

The Technicon AutoAnalyzer can be used as an input device in many different types of measuring instruments. Essentially any instrument that can accommodate a flow-through measuring unit is compatible with the AutoAnalyzer concept. Thus, atomic-absorption flame photometers may be used for calcium, magnesium, or trace-element analysis. Flow-through ion-specific electrodes are used in the SMAC system for measurement of sodium and potassium activity.

DATA PRESENTATION

Although physically discrete from the colorimeter, the AutoAnalyzer recorder is in effect an integral part of the measuring system. In the early models of the AutoAnalyzer, the recorder provided the only means of display of data. The strip-chart tracing indicates continuously the difference in voltage between test and reference photocells in the colorimeter. After the initial set-up of the colorimeter and recorder for a series of analyses, the reference voltage is essentially constant, and movement of the recorder pen represents an amplification of the difference in the analog voltages between the two photocells as the absorbance of the solution passing through the cuvet changes. The speed of the strip chart is constant, so that with the constant input of specimens into the analytical system, a series of peaks and valleys are produced on the paper at uniform intervals.

In a correctly set-up and calibrated system, the height of a peak is readily correlated with the concentration of the constituent measured in a specimen. Normally, transmission is displayed, so that on visual inspection the linearity of a calibration curve appears to fall off with increasing concentration although it is linear with respect to absorbance. By use of an appropriate slide wire, the calibration curve can be displayed in an absorbance mode. The straight-line calibration curve obtained in this way enables a calibration curve to be constructed from the analysis of only one standard solution and a reagent baseline when Beer's law is followed.

Some of the Technicon recorders have been modified for use with two colorimeters, so that two tracings are produced on a single strip chart. This enables two different tests, or a test and its blank value, to be displayed at the same time. When the voltage signal is low, as with methods of low sensitivity, a range expander may be used with the recorder to amplify the signal.

A digital printer can be incorporated into the AutoAnalyzer II system so that raw data from the linear output photometer may be converted directly into concentration units. This involves an analog to digital voltage converter. With this system the identity of a sample may be stored so that the identification, together with the test result, may be printed on paper tape. Digital printers were used with some AutoAnalyzer I recorders to print the result in concentration units.

AutoAnalyzers have been interfaced with computers to obtain the convenience of automatic data acquisition and the elimination of errors in chart reading. Usually a retransmitting slide wire is incorporated into the recorder so that the same signal is available to the recorder-pen drive mechanism and the computer. The analog voltage, however, must first be amplified and digitized before it can be used by a digital computer. Although the strip-chart tracing is redundant when an AutoAnalyzer is interfaced with a computer, it provides a useful visual indication of the quality of data being acquired by the computer, and a back-up in case of computer failure.

In the SMA 12/60, a Teletypewriter may be interfaced with the system so that all data from one sample may be presented together. The mechanical phasing of the system enables this to be done without the requirement of a memory-storage device. Punched paper tape may also be generated for input to a computer at a later time in an off-line mode. For the SMA 12/60 (and other models of the multitest systems), a preprinted strip chart is used to display graphically the results of all 12 tests in concentration units.

A process control computer is an integral part of the SMAC system. It governs the phasing of the analytical process and also regulates the display of results. The SMAC system provides a printed output of a numerical result as well as a graphical display relating the measured value to the normal range. Abnormal results are identified by a symbol to differentiate them from those falling within the normal range.

CALIBRATION OF THE AUTOANALYZER

With the models I and II AutoAnalyzers, the baseline is determined while water is aspirated continuously through the sample line. The colorimeter is adjusted so that the baseline lies between 90 and 95 per cent *T* on the recorder scale. Then a series of four to six standards, ranging in concentration from below the normal range to above that observed in most pathological samples, is run immediately preceding the specimens from patients. The standards contain the major analyte in water or appropriate diluent, although some chemists prefer to use preanalyzed serum. For most tests the standards are analyzed again after 20 to 30 patient specimens, and after all the unknowns either in the same sequence or in reverse order. This permits the analyst to make corrections for drift in the baseline or alteration of volume of sample aspirated during the run. Instead of having to use a calibration curve constructed from the peaks of the single set of standards, the analyst can connect the corresponding pairs of standards throughout the run by straight lines when the chart is removed from the recorder. This improves the precision of the calculation of the concentration of the unknowns if the drift that has occurred is linear.

Although the calibration curve constructed from the standards is smooth in the transmission mode, it is conventional practice when calculating results to connect adjacent points by straight lines. This produces a slight, but unnecessary, error when the concentration of the unknown lies midway between two standards. In the case of glucose when measured by the alkaline ferricyanide procedure, the error may reach 5 mg/100 ml. This error is obviated when a French curve is used to prepare the calibration curve or when the procedure is used in an absorbance mode so that the calibration curve is a straight line (assuming that it obeys Beer's law).

With the multichannel analyzers, serum specimens of known concentration are used to calibrate the analytical procedures. All these instruments display data in the absorbance mode so that a single calibration solution and a reagent baseline suffice to construct the calibration curve. The results are drawn on the strip chart, which is preprinted in appropriate concentration units, thus providing an immediate indication of the concentration of constituents in serum specimens. Accuracy of the results is dependent on the quality of the analyses that the manufacturer uses to assign the label values of the calibration materials, and on the skill of the analyst in adjusting the recorder pen to the correct settings on the strip chart. It is usual for the analyst using an SMA 12/60 to recalibrate the system manually, usually after every tenth unknown specimen. This reduces drift that might otherwise cause results to be quite erroneous. The analyst knows that he should suspect a fault in the analytical process if much readjustment is required. The normal adjustment required should not change a glucose value by more than 3 mg/100 ml or a urea nitrogen value by 1 mg/100 ml if performed after every tenth unknown. A large drift during an AutoAnalyzer run indicates a lack of stability in either the mechanical or electrical components of the system or in the chemistry. Unsatisfactory reagents or formation of a clot in the system can also be the cause of gradual or sudden drift.

DYNAMICS OF PEAK FORMATION

Thiers[5] has demonstrated that the rise curves of the same solution aspirated for different periods of time may be superimposed (Fig. 4–1). For the same solution, the peak height is determined by the duration of aspiration of the sample, and in all cases is part of the same curve before it reaches steady state. If the aspiration time is constant, the initial slope of the curve is related to the concentration of the sample. The fall curve also approximates first order kinetics with respect to concentration. Thus, the rise and fall curves of all normal AutoAnalyzer peaks comprise portions of the same curves that would have been present had steady state been achieved. During usual operation of an AutoAnalyzer the rise curve of one peak has commenced before the fall curve from the previous peak has reached the baseline. This carry-over from the previous peak causes the sample interaction, and it is apparent that if the interval between samples is increased, the sample interaction can be reduced. In this way, the duration of sample aspiration determines the percentage of steady state that is reached, and the rate of analysis (interval between samples) determines the interaction between specimens. This practical experience had been incorporated into the SMAC system to reduce sample interaction by mechanical means. The built-in computer ensures precise control of the duration of sample aspiration so that the same, albeit low,

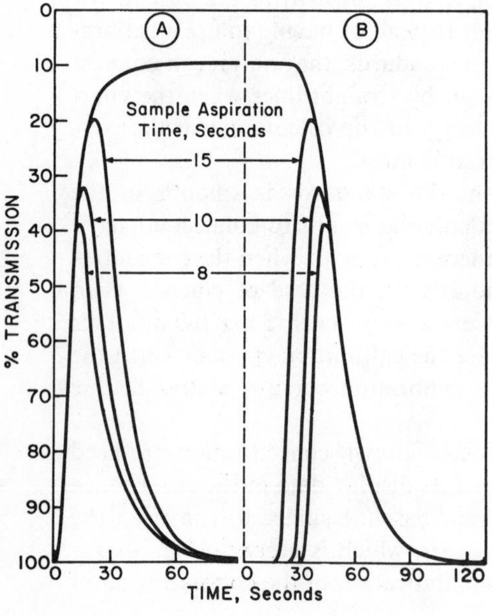

Figure 4–1 Shape of peaks with relation to aspiration time. (From Thiers, R. E., *et al.*: Kinetic parameters of continuous flow analysis. Clin. Chem., *13*:451, 1967.)

percentage of steady state is achieved for all samples. By this combination of mechanical and computer corrections, the SMAC system is able to analyze 150 samples per hour.

By considering the factors involved in the formation of a peak in the basic AutoAnalyzer system, it is possible to understand some of the abnormal peaks that may be observed in daily practice. When there is insufficient specimen for analysis, or when a clot occludes the sample line during an analysis, a situation analogous to a decrease in aspiration time occurs. The rise curve of an affected peak is normal, but it is curtailed early so that the peak is skinny. Differences in aspiration time, as with irregular cams, will alter peak heights but will not affect the slopes of the curves. While an air bubble produces a sharp spike on a recorder tracing due to the sudden change in voltage output from the colorimeter, the slope of the curve is unaffected. Dirt or precipitate in the sample produces noise that is most apparent at the tip of the peak but does not alter its rise and fall curves.

INTERACTION BETWEEN SAMPLES

Interaction is the term used in continuous flow analysis to refer to the contamination of one specimen by an adjacent specimen. This is the factor that has been most responsible for limiting the rate at which analyses can be performed on the AutoAnalyzer. If a specimen of high concentration is analyzed immediately preceding one of low concentration, then a certain proportion of the first specimen is left behind in the tubing and analyzed with the second. To obtain an accurate value for the second specimen, this contribution should be subtracted from the measured result. While it can be shown experimentally that it may require as long as a minute for the baseline to return to its former position following a sample of high concentration, it is assumed in practice that one sample only affects the specimen immediately after it. Since contamination is not related to the concentration of samples but to the flow of solutions through the system (which remains constant throughout a run), the extent of contamination can be calculated and a correction factor applied to all results to obtain greater accuracy. Unless the AutoAnalyzer recorder is used in the absorbance mode, all corrections for contamination can only be made after the data have been converted to concentration units. Normally, the correction factor is expressed as a percentage. To derive the percentage interaction ($\%I$), three specimens (often standard solutions) are analyzed after a set of standards. The first (S_1) and third (S_3) are of the same low concentration, while the middle (S_2) contains the same constituent at a high concentration. The concentrations are determined by reference to the standard solutions. If there is significant contamination between specimens, S_3 will be greater than S_1 and the correction factor is derived from the formula

$$\frac{S_3 - S_1}{S_2} \times 100 = \%I$$

This simple formula does not allow for the artifactual lowering of the peak of S_2 because it followed a sample of much lower concentration; nor does it allow for the same effect occurring in the standard solutions with which the unknowns are compared.

At high rates of analysis (e.g., 120/h), it is essential to derive and apply the correction factor to all results if they are to be meaningful. Although it is difficult to apply corrections to data derived by an SMA 12/60, the $\%I$ factor is usually small, as it is with other systems run at 60 samples per hour. The built-in computer of the SMAC system enables the factor to be derived rapidly and applied to all results before they are printed.

DUPONT ACA (AUTOMATIC CLINICAL ANALYZER)

The DuPont ACA is different from other automated systems in that it permits many different tests to be performed on the same specimen while retaining the capability of performing the same test on different specimens. The overall throughput is approximately 100 tests per hour. The instrument is well suited for processing a large portion of the workload in a small hospital or clinic, or for performing automatically those tests that would be performed manually even in most large hospitals. With the exception of sodium and potassium analyses, most of the tests commonly performed in the clinical laboratory have been adapted to the ACA. The simplicity of operation of the system and the short time for analysis enables emergency requests to be processed swiftly and accurately even by relatively unskilled staff.

For each test, the ACA employs a heat-sealed transparent plastic envelope containing analytical reagents in separate compartments (Fig. 4–2). The envelope (or pack) becomes the reaction chamber and, subsequently, also the cuvet for the photometric measurement.

When a test is performed, the serum specimen is placed in a cup to which an identifying label is attached. The cup is then inserted into the input tray of the ACA. A pack specific for each of the tests requested on the same specimen is placed immediately behind the specimen cup. All packs are marked with a human-readable identification and also a machine-readable binary code that programs the analyzer. As the machine code is read, the appropriate volume of sample is aspirated and delivered with diluent into the respective reagent pack. The pack is then transported on a continuous chain through two processing stations at which the diluted sample is mixed with the reagents first by breaking the reagent compartments, and then by vibrating the pack. Before the pack reaches the photometer, the machine code is read again and the correct filter is moved into place. Once in the photometer, the pack is compressed so that a cuvet with a 1.0 cm pathlength is formed, containing a portion of the reaction mixture. After completion of the photometric measurement, the packs are automatically discarded. Results, in appropriate concentration units, are printed on paper tape, together with the reproduction of the identifying label which was used to identify the specimen cup. Various codes may also be printed on the report form to indicate possible malfunctions of the system.

The volume of serum used for each test is between 20 and 500 μl. All reactions occur in an air bath in which the temperature is maintained at 37°C. Specificity of analyses is increased by the use of adsorption columns in the analytical packs to remove interfering

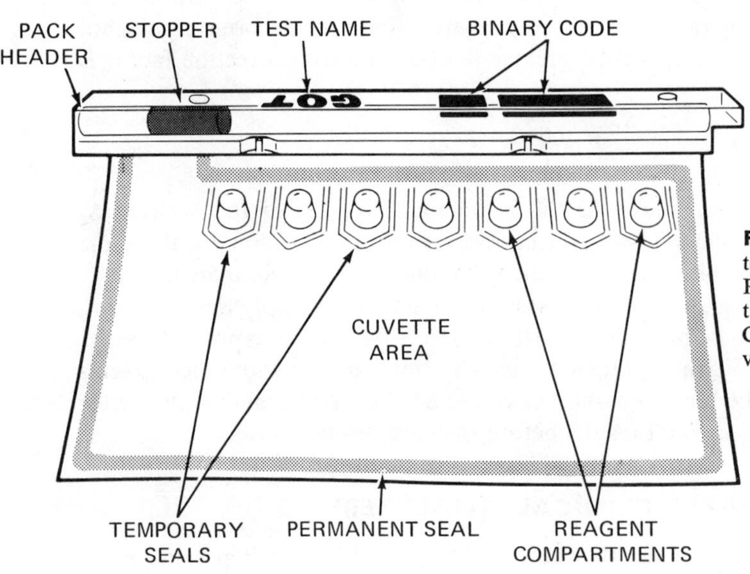

Figure 4–2 Reagent pack used for tests with the DuPont ACA. (From Perry, B. W., *et al.*: A Field Evaluation of the DuPont Automatic Clinical Analyzer. Birmingham, University of Alabama, 1970, p. 4.)

substances which otherwise might invalidate results. Polyacrylamide resin beads are used, for example, to eliminate inorganic salts from samples used in enzyme assays, whereas in some procedures a protein-precipitating agent in a column of glass beads is used to avoid interference by protein. The instrument employs two photometric readings for every test. For fixed end point reactions, the instrument employs for every test two photometric readings at two different wavelengths to minimize the influence of turbidity on the results. For enzyme activity measurements, the change in absorbance during 17.1 seconds is used to calculate results in International Units. Although the change in absorbance is small in many tests, the exceptional stability and extended linearity of the split-beam photometer of the ACA insures precise measurements.

Although 20 different tests may now be performed on the ACA, it has an ultimate capacity of more than 60 different tests. For the calibration of the system, standard solutions or specimens of known concentration are utilized. The slope and offset of the calibration curves can be adjusted independently for each test so that results from the ACA may be made comparable to those obtained with other instruments. Accuracy and reproducibility of results is critically dependent on the uniformity of the manufactured packs. Because separate packs are used for each test, there is no assurance that all the results from one batch of specimens will be valid even if a single quality control sample yields the anticipated results. Nevertheless, the typical day to day variation obtained for methods performed on samples of normal concentration is usually less than 3 per cent, even for enzyme assays.

VICKERS MULTICHANNEL 300 (M-300)

The Vickers M-300 is a modular analytical system in which tests are performed in parallel in up to 10 reaction consoles at a rate of 300 specimens per hour. Each reaction console can be considered as a discrete analytical instrument capable of performing either two tests simultaneously or one test with a blank determination. Indeed, individual reaction consoles combined with a specimen input unit are available for processing two tests at the same time. This instrument is the Vickers Dual 300 (D-300).

The complete M-300 analyzer consists of a control console and 10 reaction consoles. The entire analytical process is regulated by a small digital computer built into the control console. The specimens are placed into specially designed coded specimen containers which are placed into the control-console, from which they are transferred into cups which are transported on a conveyor belt past the individual reaction consoles. After use, the cups, while still attached to the conveyor, are rinsed and prepared for reuse.

The reaction console consists of a rotor with two concentric rings containing small reaction cups. An appropriate volume of specimen, previously diluted when transferred from the control console to the distributor, is dispensed by a double-transfer diluter into a cup in each of the two rings of the rotor. Reagents are added at predetermined intervals. If heat is required for the reaction to proceed, the temperature of the rotor may be varied from ambient to 50°C. For the photometric measurement, the contents of the cup are aspirated into the photometer mounted above the reaction unit. A separate photometer is used for each test. If sample blanking is used, both the blank and test are measured in the same photometer and the blank value is subtracted automatically before presentation of results. Signals from the individual photometers, together with the identity of the specimen, are received and collated by the computer and printed out by a Teletype or a low cost line printer.

A fluorometer or flame photometer may be substituted for the photometer of a reaction console. Two-point calibration of the instrument is effected by the initial analysis of aqueous standards, or serum specimens of known concentration, and proper adjustment of the instrument. The system may be recalibrated automatically during a run by insertion of the same calibration materials in certain coded containers. All specimens, including calibration materials and serum, are treated alike. Thus, it is not essential that reactions are complete before photometric measurements are made. The analytical system does not include provision for deproteinization and cannot measure enzyme activities by multipoint or continuous sampling techniques.

Positive identification of specimens in the Vickers M-300 is assured by the use of plastic, rectangular vials enclosed by foil paper on which a numerical matrix is printed, as on a Hollerith card. Machine-readable holes are produced when the prescored rectangles are pushed out, as with a Port-A-Punch card.

The Vickers analytical systems are still undergoing considerable development, but preliminary figures from field trials indicate that between-batch reproducibility is between 1 and 2 per cent for most tests in which samples of normal concentration are analyzed. For enzyme determinations, a between-batch reproducibility of less than 5 per cent is attainable for samples with activities in the normal range. The Vickers system incorporates sufficient flexibility in design so that a user may select the tests he wishes to run. A recently announced modification allows up to 20 tests to be made on 400 μl of serum.

HYCEL MARK X DISCRETIONARY MULTIPHASIC ANALYZER

The Hycel Mark X is a self-contained unit in which up to 10 different tests can be performed simultaneously. The Hycel Mark 17 is a further development of the same apparatus in which selectively up to 17 different analyses can be made on each specimen at the same time. On this instrument, in contrast to the SMA 12/60, the operator can determine which tests will be performed and reported. The selection of tests is effected through a push-button control panel with 10 rows of buttons (one per test) in 60 columns (one for each specimen). The reactions take place in test tubes mounted 10 to a row in 60 racks. Each test tube is in effect linked to one button in the control panel. The circular sample tray also has 60 numbered positions and thus the identity of a specimen is maintained and coordinated with the tests performed throughout the analytical system. Sequential numbering of specimens is employed to match sample identity and results.

The Hycel Mark X can process 40 specimens per hour to yield a maximum output of 400 test results. The dwell time of a specimen in the analyzer is 30 minutes, thereby limiting the usefulness of the apparatus to perform determinations in an emergency.

During the test procedure, a 0.1 ml sample is aspirated from the tube in the sample tray and dispensed into a series of reaction tubes in accordance with the number and type of tests requested.

A maximum of 1.0 ml of serum is required for analysis of all 10 tests, although 1.2 ml must be placed into the sample cups. The individual test tubes, contained in a rack, are transported by a conveyor-belt system at a fixed rate, and a rack with empty tubes is ready to accept samples every 90 seconds. Carry-over from one specimen to another is reduced by automatic flushing of the sample probe with distilled water between specimens. Reagents are added to the test tubes at preset intervals for those tests that have been programed into the system. No mixing stations are built into the instrument, but mixing occurs primarily by the force of reagent addition. Further mixing occurs with the movement of the test tubes through the heating baths of the apparatus. The configuration of reagent addition stations and heating baths is illustrated in Figure 4–3.

Measurement of the absorbance of the test solutions is made at a single readout station, but separate cuvets, lamps, filters, and photocells are used for each test. A peak-locking mechanism is employed to eliminate noise. The voltage from each photocell is recorded in turn and the output is displayed in concentration units on a precalibrated strip chart, but the same analog voltage signal can be transmitted via a slave potentiometer to a computer. Once the test tubes have traversed the readout station, the residual contents are

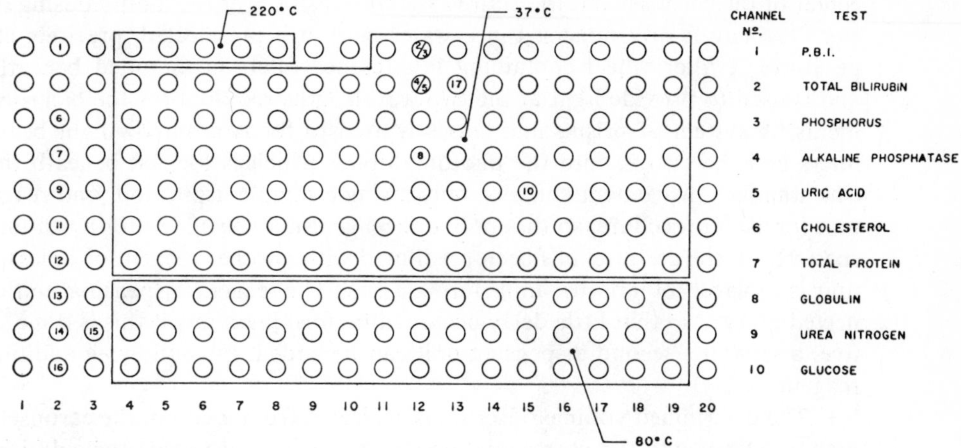

Figure 4–3 Layout of pipetting stations and reaction baths for Hycel Mark X. (From Young, D. S., and Gallup, J. D.: *In* Automation and Data Processing in the Clinical Laboratory. G. M. Brittin and M. Werner, Eds. Springfield, Illinois, Charles C Thomas, Publisher, 1970.)

tipped into a waste receptacle and the tubes are washed automatically first with a detergent solution and then with water, before being dried by air jets to prepare them for reuse.

Calibration of the test procedures is made by analysis of a control serum preparation containing the constituents to be analyzed at a high concentration. Reagent blanks are used to set the baseline. The calibration of the strip chart allows for the nonlinearity of those tests that do not conform to Beer's law. The apparatus does not include provision for blank assays (normally required for turbid or icteric specimens) or for the separation of protein, so that only procedures not affected by the presence of protein may be used for analysis. End point readings are used for enzyme assays since the system can not record the multiple data points required for kinetic assays. While it is theoretically possible for a user to set up his own analytical methods, the design of the instrument effectively precludes this because of the arrangement of heating baths and the format of the strip chart.

Accuracy of the data from the Mark X is critically dependent on the correctness of the labeled values of the calibration material and the precision with which the operator adjusts the calibration of the recorder. It is not possible to use solutions of pure chemicals in water to calibrate the system. Results obtained by the Mark X do not necessarily agree with those obtained by more traditional analytical methods. This is especially true for protein-bound iodine, total protein, and phosphate measurements. The variation of the system is between 3 and 5 per cent for most tests if samples are in the normal range.[7]

ABBOTT BICHROMATIC ANALYZER 100 (ABA-100)

The Abbot bichromatic analyzer is a simple mechanized instrument in which a sample is picked up and dispensed with a single reagent, and the absorbance of the solution is measured in a photometer. The instrument is capable of performing either end point or multiple reading type (kinetic) measurements on sample volumes of 5 to 10 μl. Only one test procedure can be performed at any time, although change-over from one procedure to another is simple and rapid. Measurement of absorbance at two wavelengths (bichromatic) is employed to compensate for the effects of turbidity in the solution.

Specimens are placed into the 32 available positions on a carousel. A disposable multiple-cuvet unit containing the same number of individual cuvets is inserted into the center of the carousel and the entire assembly is slid over the lamp housing of the photometer. The cuvets are immersed in a water bath at a preselected, precisely controlled temperature. A filter wheel containing interference filters is mounted beneath the carousel and is used to provide light at the two wavelengths required for the bichromatic measurements. A system of prisms and lenses is used to focus a narrow light beam through the filters and the cuvets onto the photomultiplier which is located beneath the filter wheel. The standard instrument contains only a single-sample aspiration, and reagent-dispensing unit for each procedure, so that only one reagent may be added for a test unless the serum specimen in the carousel has been prediluted with another reagent or unless the dispensing unit is replaced by another unit during a run. The reagent-dispensing unit can be disconnected quickly so that little delay occurs in performing a test if this is done. As an alternative, a separate, second dispensing unit can be added, thus allowing addition of a second reagent.

The determined volume of serum is aspirated from a cup in the carousel and delivered with a reagent into the corresponding cuvet. Air is used to mix the solution in the cuvet. The carousel advances at a fixed rate so that each specimen is presented for transfer at a uniform interval and the carousel makes one complete revolution in 5 minutes. Absorbance measurements are made during the second revolution of the carousel, and for end point measurements the absorbance of a reagent blank is subtracted automatically from the test values. Results are printed together with the identifying position number on human-readable paper tape. At the same time a Nixie-tube display on the instrument indicates the number of carousel revolutions, the cuvet number, and the absorbance. The instrument may be programed to print out data in concentration units if a standard of known concentration is analyzed initially and a conversion factor is entered into the instrument.

For continuous and multipoint type (kinetic) measurements, normal and fast modes of operation are available. In the normal mode, the change in absorbance for each cuvet is printed at intervals of 5 minutes (after one complete revolution of the carousel). In the fast mode, one cuvet is held in the light beam for 1 or 2 minutes and three measurements of the absorbance are printed at intervals of 15 or 30 seconds. In this mode, the throughput is reduced to 60 or 30 samples per hour, respectively.

Measurements can be made readily at several wavelengths, including 340 nm. Red colored printing on the paper tape is employed to indicate when zero order kinetics are not followed. Unlike many mechanized instruments, the ABA-100 can be calibrated with solutions of pure chemicals in water. Contamination between samples is claimed to be less than 0.1 per cent. No mechanism for the separation of protein has been developed, but the effect of turbidity is minimized by the bichromatic measurement. The most common application of the instrument is for enzyme analyses, although the relatively long lag time between reagent dispensing and absorbance measurement does not enable initial reaction characteristics to be examined in the normal mode of operation.

COULTER CHEMISTRY SYSTEM

The Coulter chemistry analyzer permits from one to 22 different tests to be performed at the same time, on up to 60 specimens per hour. Sample volumes range from 15 to 40 μl. Reagent volume is between 1 and 6 ml. Each test procedure is performed in its own set of modules so that there are separate sampling, reagent-addition, and mixing stations. Each test procedure also has its own dedicated colorimeter with an appropriate interference filter. A flame photometer with internal standardization is used for sodium and potassium analyses. Specimen identification is maintained throughout the instrument by mechanical interlocking of specimen and reaction containers. Analyses are performed without protein separation, using accepted methods. The reagents are available for use in manual back-up procedures if required. Specimens for emergency analyses may be inserted during a run.

Calibration of each test procedure is performed automatically and, in most cases, with aqueous solutions of pure chemicals. Serum blank corrections are performed automatically. Results are matched with the specimen identification previously entered into the system and are printed in concentration units on paper tape.

No data on performance of the instrument under routine laboratory conditions are yet available.

ACUCHEM MICROANALYZER

This instrument, developed by the Damon Corporation and distributed by Ortho Diagnostics Instruments, is a discrete analyzer that is still undergoing evaluation. It is expected to be introduced commercially in the near future. The instrument is capable of performing 16 tests simultaneously on 200 μl of serum at a rate of 60 samples per hour. Single or multiple tests may be performed on each specimen. Most of the tests commonly performed in the clinical chemistry laboratory may be performed on this instrument using traditional analytical procedures. Reagent delivery is effected by air displacement, eliminating the need for a complex syringe system. The photometric system uses a double-beam principle for each test and is capable of precise measurements in the visible and ultraviolet region. Enzyme analyses may be followed kinetically. Calibration of the instrument is made by use of a precalibrated control serum, since the system can not be calibrated with aqueous solutions. To reduce interaction between specimens, water and air rinses are employed between samples, and all reaction tubes are washed prior to reuse. Results are printed on paper tape. The instrument may be interfaced with most computer systems.

INSTRUMENTATION LABORATORY CLINICARD 368 ANALYZER

The Clinicard is an easy to operate, compact instrument intended for the performance of serum tests and enzyme tests requested frequently as "*stats*." Reagents for the different tests are prepackaged in disposable cuvets. The cuvets are supplied with a program card which determines the various measurement functions required for an analysis—such as the operating mode, the volume of sample (from 25 to 200 μl) to be used, and the appropriate interference filter. After the program card has been placed into the instrument, the correct volume of sample is aspirated and dispensed into the cuvet, which is then sealed and manually placed into one of the two built-in incubators preset at 37° or 90°C. After the appropriate incubation time, the instrument indicates that the operator should remove the cuvet and place it into the shuttle that automatically transports it to the photometer. Results are presented digitally by Nixie tubes in appropriate concentration units. A different test may be performed immediately afterward by inserting the appropriate program card and preloaded cuvet. Although each test must be processed sequentially, up to nine samples may be in various stages of analysis at one time in the analyzer.

The cuvets and photometer are so designed that one of three modes of operation can be selected by the program card. In the "standard" mode, readings of unknown, standard, and blank are made, and the computed concentration of the unknown is displayed within 4 seconds. In the "absolute-absorbance" mode, the blank is subtracted from the unknown value and the difference is displayed. In the "rate" mode, the absorbance of the cuvet is monitored continuously for 60 seconds and the change during the final 48 seconds is converted to International Units and presented. Built into the instrument are a series of warning indicators that alert the operator to malfunctions of the system or to abnormally high absorbance values.

There are approximately 30 different tests, some presently available and others under development. No published evaluations of the equipment are yet available, but the system incorporates many features that may be useful in the clinical laboratory. The simplicity of operation with built-in monitoring of error functions, accompanied by precise control of pipetting and automatic recalibration for each end point analysis should guarantee a reduction of the error rate in a clinical laboratory. The instrument appears to be well designed for processing small numbers of many different tests and for use as a *stat* analyzer.

ENZYME ANALYZERS

Several analytical systems have been built primarily to meet the demand for automation of enzyme analyses. A unique requirement of these systems is the ability to record the change in absorbance with time continuously or at frequent intervals so that kinetic measurements may be made. This mode of operation is difficult to combine into one instrument, which may also be used for single-point measurements. This is especially true of those analytical systems in which many different tests are performed on the same specimen at the same time. The delay required to make kinetic measurements would greatly reduce the throughput of the instruments. Nevertheless, the centrifugal analyzers, the DuPont ACA, and the Abbott ABA instruments are capable of making more than one measurement per sample in a cuvet and could be considered as kinetic enzyme analyzers. Since they are considered in detail elsewhere, they will not be discussed in this section.

GILFORD 3400 AEA (AUTOMATIC ENZYME ANALYZER)

This is a modular analytical system built around the 300-N spectrophotometer. The system consists of a sampler-transport device, reagent dispenser, spectrophotometer, data lister, and activity computer. Serum is pipetted manually into reaction vials contained in this unit. A reagent is dispensed automatically. Mixing is also performed automatically by a jet of air directed against fluting on the bottom part of the reaction vials, causing the tubes to rotate. After a variable delay of up to 3 minutes, during which preincubation occurs at ambient temperature, the solution is aspirated into the cuvet of the Gilford 300-N, which is maintained at one of the four selectable temperatures (25, 30, 32, or 37°C). Data are presented on paper tape, and for each assay the listing includes an initial absorbance reading followed by a specimen identification number (rack and vial), and then six values in International Units for each assay. Measurements are made at 10 second intervals during 2 minutes, and the six changes in absorbance for 1 minute are used to derive the reported values. Each procedure has a predetermined calculation factor (related to sample volume, temperature, and so forth) that is dialed into the system before the first measurement is made. The product of the change in absorbance and this factor converts the results to International Units. Nonlinearity is readily detected, as there will be large differences between the six printed values.

A sample size of 0.1 to 0.2 ml of serum is required for each analysis. The throughput of specimens is approximately 30 per hour, since the absorbance of each specimen is followed for 2 minutes. This rate of analysis is too slow if a single instrument is to be used for all analyses in a large hospital laboratory, although it may suffice for the small laboratory. In the latter situation, the role of the analyzer may be expanded by modifications that permit end point analysis, the addition of a second reagent, and longer preincubation at room temperature.

LKB-8600 REACTION RATE ANALYZER

This system consists of a unit which combines specimen transport, reagent dispensing, preincubation and photometric measurement functions in one unit with a separate strip-chart recorder. The reaction rate analyzer is limited to making measurements at 340 nm.

Samples are pipetted outside the instrument into polystyrene vials of 10 mm diameter, which subsequently serve as cuvets. Aluminum racks, holding the vials, are transported automatically through the system and pass through a hot-air tunnel to bring the solutions to the temperature at which absorbance measurements are made (usually 37°C). A preheated substrate is added to initiate the reaction at the measuring position and the cuvet is agitated automatically to mix the solution. The change in absorbance of the solution is followed continuously for 1 minute (although it is possible to increase this to 9 minutes) and displayed on the strip-chart recorder. The background absorbance is offset electronically, so that full-scale deflection is 0.05 absorbance units independent of background. The instrument may be interfaced on-line to a laboratory computer or to a hard-wired digital printer that is available as an accessory from the manufacturer.

This latter unit permits display of error messages, matching of sample identity with results, and conversion of raw data into International Units of activity. Activity is derived from the average change in absorbance over fixed time intervals. Nonlinearity of the reaction is displayed. The instrument is capable of processing 50 samples per hour when each specimen is held in the photometer for 1 minute. This analyzer differs from most others in that a separate cuvet is used for each specimen, thereby eliminating contamination of one specimen by another.

BECKMAN ENZYME ACTIVITY ANALYZER, SYSTEM TR

This recently announced system enables enzyme activities to be measured kinetically, rapidly, and with considerable assurance of instrumental precision. The throughput of specimens is up to 50 per hour, yet each reaction is followed continuously. As the reaction proceeds, after automatic sampling and rapid mixing of reagents and sample, both the rate of change of absorbance with time (reaction rate or enzyme activity; dA/dT) and the change of this reaction rate with time (second derivative) are monitored electronically. When the reaction rate is constant, the second derivative becomes zero. If this condition prevails for at least 17 seconds, the display of sample identification and reaction rate (in International Units) are triggered and the result is printed on paper tape. No display of results occurs until the temperature is constant and any lag phase has been passed. Analyses are rejected for which the second derivative is not zero within programed limits.

A differential double-beam spectrophotometer with a narrow bandpass (5 nm) grating monochromator is used in the system. Measurements are normally made with air in the reference cuvet, but a reagent or serum blank may also be placed in the reference cuvet. For tests requiring a serum blank, a dual sample probe is used to aspirate two precise aliquots of sample at the same time from the specimen cup. One aliquot is transferred to the sample cuvet and the other to the reference cuvet.

Reagent (525 μl) is added to the sample cuvet through a port in the bottom of the cell, and the sample (35 μl) is added through the top. A magnetic stirrer located beneath the light beam is used to mix the solutions. The cuvet has a pathlength of 1.0 cm and a volume of 0.5 ml. The solution inside the cuvet is heated by heating elements surrounding the cuvet. Temperatures of 25, 30, and 37°C may be selected.

An assay is initiated by inserting the respective reagent pack and depressing the appropriate test button. Reagent packs for six enzyme tests are available at present.

Reproducibility of instrumental performance is facilitated by the availability of an electronic calibration. Various parameters, such as actual reaction temperature or absorbance, as well as out-of-range conditions, can be displayed automatically.

The system demonstrates linearity over a broad range of activity—in many cases exceeding 1000 International Units. Little information is as yet available on the precision of the rate analyzer, although some of the manufacturer's own data indicate a within-run precision of less than ±2 per cent for most tests. The instrument to instrument variation is approximately ±4 per cent.

CENTRIFUGAL ANALYZERS

The use of centrifugal force to transfer liquids and the simultaneous monitoring of several chemical reactions in a centrifuge[1] marked the first radically new analytical concept for clinical chemistry since the introduction of the continuous flow technique of the AutoAnalyzer. At present, three commercial instruments utilizing this concept are manufactured in the United States: CentrifiChem (Union Carbide Corporation), Gemsaec (Electronucleonics, Inc.), and ROTOCHEM II (American Instrument Corporation).

The principle of operation of the system is shown in Figure 4–4. Chemical reactions take place and are measured in a rotating centrifuge head. The rotor consists of an outer ring containing a number of cuvets formed by compression of an inert spacer between two plates of optically transparent material (glass, quartz, or plastic) and a central removable transfer disc of Teflon containing three concentric rows of cavities. Each radially aligned set of cavities and the corresponding cuvet create a discrete reaction unit. A light source, monochromator or interference filter, and photodetector are arranged so that light passes through the cuvets perpendicularly to the rotor, thereby forming the photometric analysis system.

For operation, serum and reagents are placed in the innermost and center cavities of the transfer disc in proportions and dilutions appropriate for the assay to be performed. Acceleration of the rotor causes these liquids to move outward to the third cavity of the transfer disc, and from there to the cuvet. Mixing occurs with the transfer. More vigorous mixing can be accomplished by creating a vacuum at the center of the rotor and aspiration of air bubbles through the solution via the siphon drain opening in the distal end of the cuvet. In all the instruments, the reaction in the rotor may be controlled at a selected temperature. With rotation of the centrifuge, the cuvets pass sequentially through the light beam of the photometer. The transmittance of each reaction mixture is measured and referenced against a reagent blank in the first cuvet. Thus, the instrument functions as a multichannel double-beam spectrophotometer with a complete set of absorbance values produced with each revolution of the rotor (50 to 100 milliseconds). In practice, readings from eight consecutive revolutions are averaged for each cuvet. For monitoring a kinetic reaction, sets of eight readings are averaged every second so that values can be displayed or printed at intervals of 1 second.

Data acquisition, averaging, manipulation, and display are performed by an internal computation system. The degree of sophistication of the data processing system constitutes the major difference between the three commercial models, ranging from a hard-wired calculator using two-point kinetics, to a programable minicomputer system permitting nonlinear curve fitting and automatic control of the analytical system. After data measurements are made, the rotor is purged, washed, and dried automatically by the programed introduction of a mild detergent solution or water followed by air through the center of the rotor.

Most of the tests requested frequently in the clinical laboratory may be performed on the centrifugal analyzer, and many commercial test kits are adaptable to the system with little or no modification. The instrument may be used in several different modes. Either kinetic or end point reactions may be followed. The same type of test may be performed on many different specimens, or different tests may be performed on the same specimen. However, in the latter case all measurements must be made at the same wavelength and temperature.

Transfer discs are loaded outside the centrifugal analyzer. They may be prepared

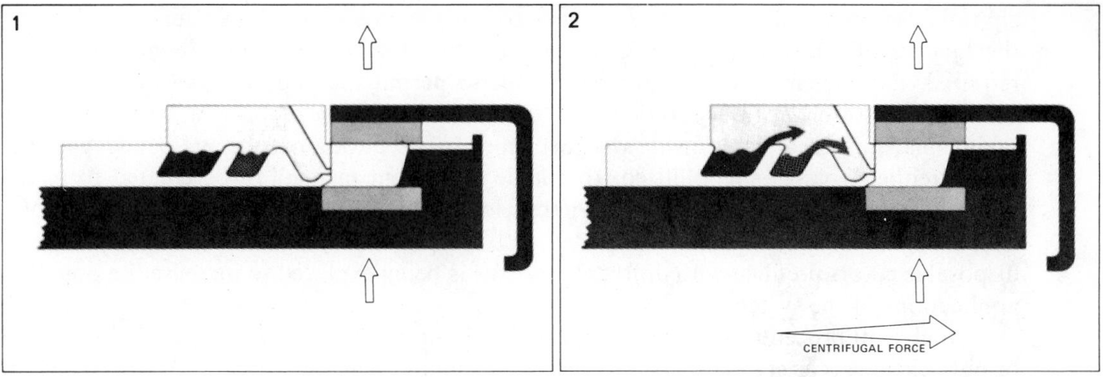

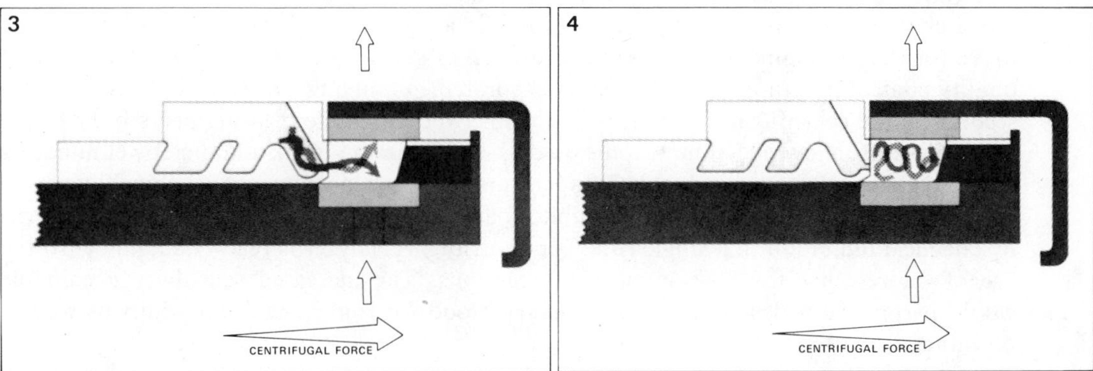

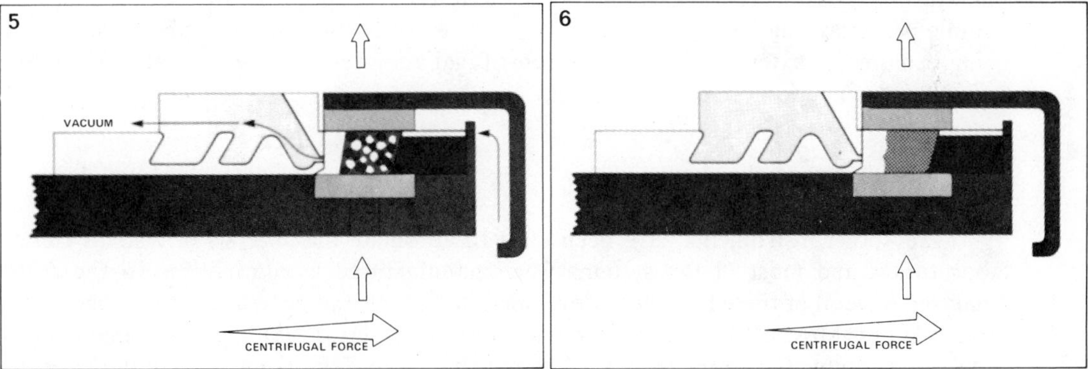

Figure 4–4 Sequence of transfer of solutions in rotor of centrifugal analyzers. Reagents and sample are placed in the appropriate wells in the transfer disc (1). Upon rotation, the centrifugal force causes the liquids to move into the cuvet in the rotor (4). Mixing is completed by drawing air bubbles through the solution (5) and absorbance of the solution is monitored as it passes through the vertical optical path (6). (Courtesy of the American Instrument Company.)

manually, but each of the commercial instruments is equipped with a separate dispensing module to insure uniform introduction of samples and reagents. The volume of serum employed is from 5 to 100 μl for each assay. The reaction volume required for measurement varies from 450 to 700 μl. The combination of reagents may be varied for different tests to permit preincubation of serum and reagents with the final addition of the activator or substrate upon centrifugation.

With the present design of rotors, it is impossible to perform assays requiring the precipitation of protein. However, it is possible to obviate interference by protein by following the kinetics of the reaction or by using a procedure for which protein separation is not required. Experimental rotors have been built to permit the addition of whole blood to the spinning rotor, with separation and transfer of aliquots of plasma into the cuvets taking place automatically inside the instrument. The cuvets can be preloaded with reagents (either frozen or lyophilized) to enable a "patient profile" to be created from the different tests. The same principle of dynamic loading could be applied to the addition of a single reagent to a rotor previously loaded with samples from different patients. The use of disposable rotors prefilled with different reagents is being explored to increase the potential applications of the system.

Analytical procedures in the centrifugal analyzer may be calibrated readily by the use of pure chemicals as standards when these are available. The precision of measurements in general equals or exceeds that obtainable in other instruments, with the major limitation being the reproducibility of the pipetting of serum and reagents.[4]

For those instruments with sophisticated data-processing modules, provision may be made for reagent blank correction and detection of substrate depletion. Maintenance and quality control programs may also be incorporated. As mixing can be completed within 3 seconds in the centrifugal analyzer, it is possible to extrapolate measurements back to the time of mixing, allowing a sample to be used as its own serum blank and thereby eliminating the need for a separate serum blank. Temperature equilibration may occur within 30 seconds. Incubation of serum with different inhibitors allows differentiation of isoenzyme patterns by chemical inhibition in a single run. The centrifugal analyzer is readily adaptable to the use of fluorescence for measurement of reactions. The increased sensitivity obtainable would permit the widespread use of capillary blood for routine tests on adults as well as on infants.

Development of the centrifugal analyzer concept and its applications is continuing at Oak Ridge National Laboratories with the production of an advanced prototype of a miniaturized system occupying less than 1 cubic foot of space. In this system the total volume per assay has been reduced to 100 μl. The use of fluorescent antibody reagents is being explored to extend the use of the centrifugal analyzer into the area of immunology and the determination of specific proteins.

OTHER ANALYTICAL SYSTEMS

Many other instruments have been built to automate the analysis of tests in clinical laboratories, and most of the systems were manufactured as alternatives to the AutoAnalyzer. Several of these have been developed in Europe and have had little impact in the United States. Most of the systems are "discrete" analyzers, in that each specimen is processed in an individual tube or cup, although the Carlo Erba Company of Italy has developed a system employing the continuous flow principle used in the Technicon AutoAnalyzers.

In the United Kingdom, the Joyce-Loebl Mecholab, the Baird and Tatlock Analmatic, and the Honeywell Digiclan have had some application in clinical laboratories, although none of these has been as widely used as the AutoAnalyzer. In Germany, the Braun Systematik has been used in several clinical laboratories. The Perkin-Elmer C4, designed in Europe, has also been used more successfully there than in the United States. All of these instruments were intended for operation in the typical hospital laboratory.

The Swedish AGA Autochemist was built to process 24 tests per specimen with a throughput of 135 specimens per hour. This workload would only arise in a large central hospital reference laboratory, a commercial bioanalytical laboratory, or in a large population screening center. It has been used in the latter two situations in this country, but the overall capacity of the system, and its cost and complexity, have precluded serious consideration of the system for use in a hospital laboratory.

Another simultaneous multitest analytical system has been introduced recently by Greiner Scientific in Switzerland. While only preliminary information has been released, the instrument appears capable of performing many of the routine tests done in a clinical laboratory. The system includes provision for automatic identification of the specimens and analysis of only those tests that have been marked on the request card.

Other analytical systems are under development. One of the most promising is being developed in the Batelle Institute in Switzerland and sold by Micromedic Systems, Inc. This is essentially a modular system in which very precise pipets (which have also found application in other analytical systems, e.g., the E.N.I. Gemsaec Centrifugal Analyzer) are used to pipet and dispense sample and reagents. A tube transport system and stable spectrophotometer enable accurate and precise results to be obtained with minimal operator intervention. Additional components are under development, so that most of the routine analyses performed in a clinical chemistry laboratory can be made on small volumes of sample.

SELECTION OF APPROPRIATE INSTRUMENTS

The number of different analyzers available attests to the fact that there is no single solution to the problems of the clinical laboratory that initially stimulated the development of automated instruments. Selection of equipment for a laboratory must be influenced by the funds available as well as by the requirements of the laboratory. Other factors affecting the choice include the skill of the staff, the number and variety of tests performed, the priorities of the laboratory director with regard to efficiency versus accuracy or precision, and the interest of the hospital staff in test batteries. Yet another determinant is the availability of back-up procedures or equipment. The laboratory director should consider automation only as one approach to accomplishing his workload. Judicious use of automatic pipets and prepackaged reagent kits may, in many cases, eliminate his need for an expensive instrument.

If any form of admission screening is undertaken, an automated analyzer must be considered essential. Although centrifugal analyzers, with further development, may be capable of undertaking this task, the most efficient means for doing this at present is a multichannel analytical system. Indeed, what is now accepted as an admission screening profile has largely been determined by the tests that are available on one of the multichannel analyzers. The optimal use of manpower, which a multichannel analyzer makes possible, must be balanced against the need to accept the analytical procedures decreed by the manufacturer and the inability to use primary standards to calibrate the system. Unless the values for the calibration materials provided by the manufacturer are confirmed, the laboratory becomes dependent on the manufacturer for the accuracy of the data generated. In neither of the two currently most popular multitest analytical systems is it possible to use primary standards or specific enzymatic procedures for uric acid or glucose (the procedures depending on reducing properties are known to be prone to interference from therapeutic drugs). It is also impossible to make kinetic enzyme analyses on these instruments. With the Hycel Mark X and XVII instruments, it is not possible to make corrections for samples that are, for example, lipemic, since no provision for blanking is incorporated into the systems. While these considerations are not necessarily very important for screening of a well population, they are important in the clinical laboratory in which many abnormal specimens must be processed. Greater freedom in choice of methodology for the SMAC analytical system will reduce the number of errors that may arise from the use of inappropriate methods.

A laboratory director should insure that he has back-up procedures for all the clinically important tests that are performed on his multichannel instrument, or a breakdown will have serious consequences. The modular construction of the multichannel instruments often enables only one test to be rendered nonusable when a breakdown occurs, so that the ability to process the workload is only mildly impaired. Nevertheless, the laboratory is especially dependent on the reliability of this equipment and the responsiveness of a manufacturer to service problems.

It is necessary to decide whether one analytical instrument will be dedicated to a single test or whether the same instrument will be used for many different tests. Frequently instruments are dedicated to a variety of tests for which the same analytical principle is used, e.g., enzyme analyses for which it is believed that kinetic measurements are more reliable. The laboratory director should remember, however, that reproducibility of results from the kinetic analyzers is often less than that from end point determinations as may be derived by the AutoAnalyzer. In the latter case, reproducibility may be achieved at the expense of accuracy, since the calibrating material may not be treated in the same manner as the unknown samples, and the calibration enzyme material may not be of human origin. It is probable that much of the variability of enzyme assays will be eliminated with the centrifugal analyzers which combine the qualities of precise temperature control, a multicuvet double-beam spectrophotometer, kinetic measurement capability, and precise metering of sample and reagents, with continuous monitoring of absorbances.

Experience in a laboratory reveals that best performance from instruments is obtained when the least number of changes are made with any instrument. This applies to the number of operators, as well as to the settings of the controls. Thus, instruments usually perform best when dedicated to one test, or to tests for which little adjustment of controls is required, if different assays are performed.

Great reliability of analytical equipment in a clinical chemistry laboratory is essential if results are to be returned to clinicians rapidly enough to be meaningful. Instruments for the clinical laboratory should require little service attention and should be sufficiently rugged to withstand, without breakdown, the misuse that often occurs in the laboratory. They should be so constructed that the laboratory staff can readily detect the location of a malfunction and make on-site repairs. Total dependence on service engineers is unwise in the clinical laboratory, since repairs often cannot be effected rapidly. Modular systems for which replacement parts are available within the laboratory are obviously more satisfactory for routine laboratory use than are uncommon instruments, especially of overseas manufacture, for which there is a relatively poor service organization. The ideal analytical system should employ the rapidity and simplicity of operation necessary for emergency tests, the small volume of specimens required for pediatric patients, and the high throughput required for routine analyses. When one instrument capable of these different challenges is used, the overall quality of work is improved. It is possible that the centrifugal analyzers and the Ortho/Damon Acuchem may meet these objectives, although experience is limited at present. The laboratory director should always weigh the use of mechanization and work-simplication procedures against automation. A well-motivated, conscientious staff can perform better with partial automation than can a poor staff with excellent equipment. Certainly instrumentation is no substitute for analytical competence in technical staff.

REFERENCES

1. Anderson, N. G.: Analytical techniques for Cell Fractions XII. A multiple-cuvette rotor for a new microanalytical system. Anal. Biochem., *28*: 545, 1969.
2. Babson, A. L., and Kleinman, N. M.: A source of error in an autoanalyzer determination of serum iron. Clin. Chem., *13*: 163, 1967.
3. Robinson, R.: The four-o'clock phenomenon. Lancet, *II*: 744, 1966.
4. Statland, B. E., Nishi, H. H., and Young, D. S.: Serum alkaline phosphatase: total activity and isoenzyme determinations made by use of the centrifugal fast analyzer. Clin. Chem., *18*: 1468, 1972.

5. Thiers, R. E., Cole, R. R., and Kirsch, W. J.: Kinetic parameters of continuous flow analysis. Clin. Chem., *13*: 451, 1967.
6. Thiers, R. E., and Oglesby, K. M.: The precision, accuracy, and inherent errors of automatic continuous flow methods. Clin. Chem., *10*: 246, 1964.
7. Winter, S. D., Skodon, S. B., Goldberg, M. H., and Barnett, R. N.: An evaluation of the performance of the Hycel Mark X Discretionary Multiphasic Analyzer. Am. J. Clin. Path., *56*: 526, 1971.
8. Young, D. S., Montague, R. M., and Snider, R. R.: Studies of sampling times in a continuous flow analytic system. Clin. Chem., *14*: 993, 1968.

ADDITIONAL READINGS

Bick, M., Lindridge, J. M., Mitchell, F. L., Pickup, J. F., Rideout, J. M., and Snook, M.: Assessment of the Vickers M-300 multichannel analysis system. Clin. Chim. Acta, *44*: 33, 1973.

Brittin, G. M., and Werner, M.: Automation and Data Processing in the Clinical Laboratory. Springfield, Illinois, Charles C Thomas, Publisher, 1970.

Henry, R. J., Cannon, D. C., and Winkelman, J. W., eds. Clinical Chemistry: Principles and Technics. 2nd ed. Hagerstown, MD., Haper & Row, Publishers, 1974.

Laessig, R. H.: The analytical chemist and multielement chemical testing in preventive medicine. Anal. Chem., *44*: 18A, 1971.

Perry, B. W., Hosty, T. A., Coker, J. G., Doumas, B., and Straumfjord, J. V.: A field evaluation of the DuPont Automatic Clinical Analyzer. Birmingham, University of Alabama, 1970.

Scott, C. D., and Burtis, C. A.: A miniature fast analyzer system. Anal. Chem., *45*: 327A, 1972.

White, W. L., Erickson, M. M., and Stevens; S. C.: Practical Automation for the Clinical Laboratory. 2nd ed. St. Louis, The C. V. Mosby Co., 1972.

CHAPTER 5

LABORATORY COMPUTERS

by G. Phillip Hicks, Ph.D.

A computer is an electronic system which can acquire, store, process, and retrieve information very rapidly. When applied in the clinical laboratory it becomes a powerful tool for improved management and services, allowing expansion without a loss of quality and proportionately increased cost.

Computerization of the laboratory, to a large extent, can be considered as an extension of automation with some similarities but also with significant differences in application. Inasmuch as the laboratory computer processes requests from various parts of the hospital and generates reports sent to locations outside the laboratory, the computer has an impact on almost every individual in the hospital directly involved in patient care. Thus, in discussing computers in the laboratory, it is necessary to consider the entire patient care process and the impact the computer has on all associated areas.

The laboratory has only begun to make a significant impact on the patient's care in the last two or three decades. In former times, the history (interview) of the patient and the information gained from direct physical examination were practically the total source of data for patient care. The role of laboratory data has been changing from simply confirming the diagnosis of the patient to providing information which is used directly in the diagnosis and treatment of patients. This increasing importance of laboratory data has placed a burden upon the laboratory to produce considerably more results of greater accuracy with higher speed.

Before the impact and benefits of the computer can be fully appreciated, one must thoroughly understand the laboratory's problem of managing specimens and data. It is not only the increase in laboratory tests *per se*, but also the increased tendency to centralize laboratory services which has led to a significant management problem of handling and identifying specimens submitted for analysis to the laboratory. Also, because of the critical nature of the data involved, efficient flow of laboratory information to other areas of the hospital, including nurses' stations, physicians, business offices, and patient record rooms, is of great importance. Since laboratory data are mostly numeric with only brief English statements, the clinical laboratory was probably one of the first major areas selected to computerize medical information.

THE COMPUTER SYSTEM

There are two sets of components to a computer system, the *hardware* and the *software*. These two components are analogous to the division of the analytical process into the instrument and the chemical method. Just as the best instrumentation in the laboratory cannot produce good results without a good method designed especially for clinical analysis, neither can a computer be used effectively without programs (software) specifically designed for clinical laboratory application.

Although it was recognized in the 1950s that computers could have a significant impact on health care, early investigators were hindered significantly by the types of hardware available. Computers were quite expensive and were unable to communicate directly with laboratory personnel. These limitations brought with them the need for punched cards for input of information and different types of personnel such as keypunch operators. In the early 1960's, a revolution in electronics, mainly the development of integrated solid state electronic components, made possible the production of very sophisticated computer equipment at low cost. This new generation of hardware, coupled with the experience gained with earlier equipment, made it possible to develop highly efficient "interactive systems" which allow for communication directly between the computer and laboratory personnel without the intervention of programers or other types of skilled personnel. The following portion of this chapter will deal with discussions of the major components of contemporary hardware and with the application of these new types of systems and concepts in the clinical laboratory. More detailed information dealing with basic terminology and the state of computers in the early 1970s can be found elsewhere.[1–5]

COMPUTER HARDWARE

Peripheral devices

The hardware can be divided into two basic categories—the Central Processing Unit (often called the CPU), which is the computer proper, and the *peripherals* which are external to the CPU. The various peripheral devices provide the means of transferring information into and out of the computer and storing information for the computer. Peripherals, combined with the CPU, make up what is known as a "configuration."

Peripherals can be considered to be of basically two types: those which allow the transfer of information from the computer operator or laboratory instrumentation to the computer, and those which store information.

Input and Output Devices. Devices which communicate with the user or instrumentation are frequently referred to as I/O (Input/Output) devices. One of the most common peripheral I/O devices is a *terminal*, which has a keyboard, similar to a typewriter, allowing keyboard entry of information or queries. The computer in turn responds by typing information back either on the same or another terminal. Such terminals have the distinct advantage of providing a permanent *hardcopy* (called audit trail) of every transaction, thus allowing retrospective examination of entries. For example, if the user were to type into the computer a serum potassium value which appeared on the report as 44, one could

Figure 5-1 CRT terminal. (Courtesy Digital Equipment Corp., Maynard, MA.)

examine the hardcopy to determine whether this was due to a human typing error or to a computer failure. There are a variety of hardcopy terminals which range in speeds from 10 characters/s up to 120 characters/s.

The second major type of terminal is the Cathode Ray Tube (CRT), which is very much like a television set with a keyboard. Figure 5–1 shows a CRT, known as the VTØ5 (Digital Equipment Corporation, Maynard, MA.).

The CRT can be used to transfer any information between the user and the computer, similar to the keyboard terminal. Its major asset is the higher speed of output from the computer on the screen (10 to 240 characters/s). The CRT is also very quiet since there are no moving parts associated with printing. Its major disadvantage is the lack of a hardcopy audit trail. The CRT instead allows observation of the keyboard entries and displays an instantaneous answer on a screen. As the screen is filled up, the information is either erased or disappears off the top of the screen. Consequently, there is no permanent record of the transaction between the user and the computer. It is for this reason that the laboratory often prefers to use the CRT for retrieving information rapidly, and to use the typewriter terminal to perform transactions where the accuracy of data input is critical. The costs of the CRT and typewriter terminals are essentially the same.

Another type of device that is suitable only for input of information is the *card reader* (Fig. 5–2). It can "read" both pencil marks and holes on a card (thus the name *mark-sense* reader).* Cards can be read at a rate of up to 200 or more cards/min, although the effective

* A mark-sense card reader "reads" both pencil marks and holes in a card by means of a series of photocells which correspond to the rows on a card. Light reflected from an unpunched or unmarked row generates a reference voltage. When there are marks or punches on a card which cause a change in reflected light, a change in voltage occurs. The voltage changes are converted to codes which the computer interprets as data.

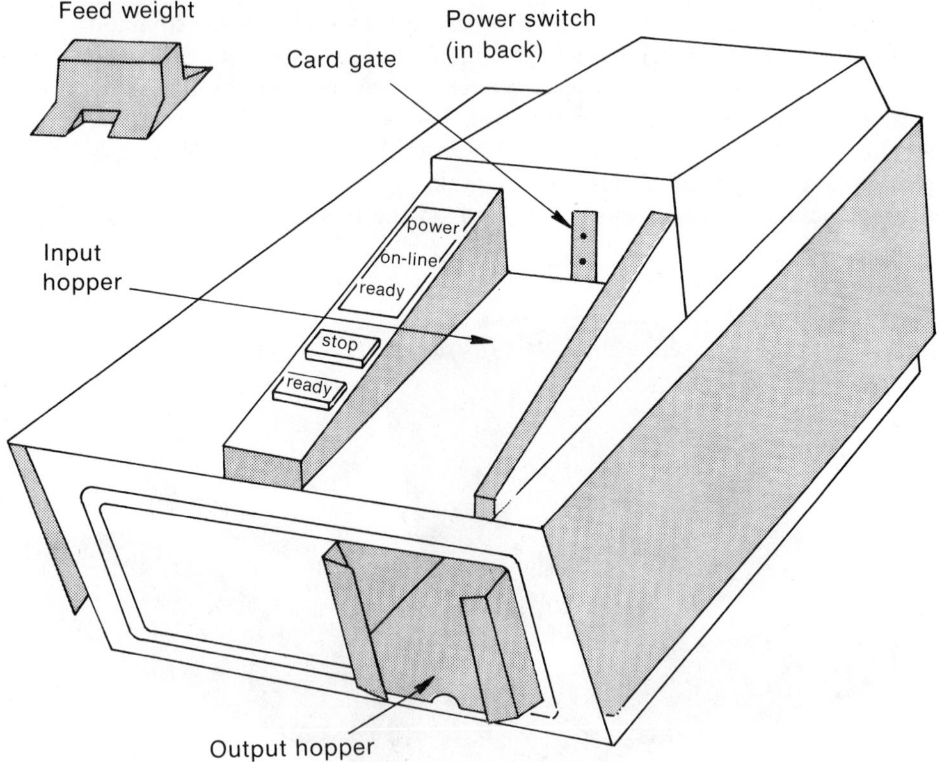

Figure 5–2 Mark-sense card reader.

Figure 5-3 High speed line printer. (Courtesy Digital Equipment Corp., Maynard, MA.)

rate of reading depends on how rapidly the computer can actually process the information received from the reader. The design of mark-sense cards for the entry of requests and laboratory data is discussed later in this chapter.

The *high speed printer* (Fig. 5-3) is an *output device* only, usually capable of printing information at speeds from 300 to 1100 lines/min depending on the amount of information that is to be printed on a given line. This device is often referred to as a *lineprinter* simply because it prints one whole line at a time. There is effectively a complete character set for each space on the lineprinter, and when a complete line has been received by the printer from the CPU, the entire line is printed in one process. Because of their high printing speed, lineprinters are suitable for the output of bulky reports such as cumulative or interim reports for all patients in the hospital, and long administrative reports such as patient directories and master work logs for the entire laboratory. Since the high speed lineprinter is expensive, compared to other printing terminal devices, usually only one such printer is used to handle the laboratory workload.

Finally, another element which can be included among the external peripherals is the *instrument interface*, which allows the direct electronic connection of a laboratory instrument to the CPU. There are two major types of interfaces, *analog* and *digital*. One must understand the different characteristics of an analog signal and a digital signal in order to understand these interfaces and how they are used with laboratory devices.

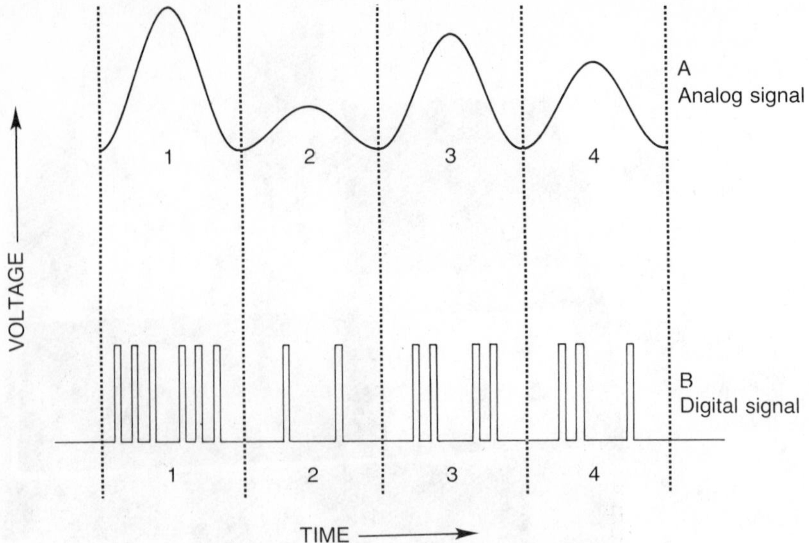

Figure 5–4 Representation of an analog and digital signal.

Figure 5–4 illustrates the two types of signals. An analog signal varies continuously with time. A recorder tracing of an AutoAnalyzer gives a good example of a graphic demonstration of an analog signal. In Fig. 5–4*A* the analog tracing might represent four peaks from an AutoAnalyzer. Since a *digital computer* can only process *discrete numbers*, it is necessary to convert the continuously changing voltage into a series of discrete numbers in order to interface such an instrument to the computer. This conversion is made by an *Analog to Digital Converter* (ADC) which "reads" the analog signal and converts the value of the signal height into a discrete number proportional to the voltage of the original signal. A series of numbers taken at sufficiently frequent intervals (e.g., each second for an AutoAnalyzer) permits the computer to construct a corresponding curve and to find the highest point as the sought-for value. Other examples of analog instruments are the continuous-recording densitometers for electrophoresis, recording spectrophotometers, and recording gas chromatographs.

A digital signal typically consists of a series of equal height voltage "pulses." The magnitude or "value" of the signal is indicated by the presence or absence of pulses in the pulse pattern, or sometimes by the time interval between pulses, but not by the height of the voltage pulses. Figure 5–4*B* illustrates four sets of digital pulses which correspond to the four analog peaks. Since the digital pulses are already discrete, a digital interface need only "decode" the pulse pattern to obtain the discrete number representing the signal value. Examples of laboratory instruments which produce digital signals and then use a digital interface are the Coulter Model S blood cell counter, the digital pH meter, and some densitometers for reading electrophoresis strips. Any instrument which has a printer output device requires a digital interface.

Because digital pulses from laboratory instruments are of high frequency (e.g., Coulter Model S) compared to the most common analog signals (e.g., AutoAnalyzer), digital instruments must be closer to the computer (25 to 200 feet) than must analog instruments (25 to 1000 feet) unless the pulses are amplified.

Storage Devices. The computer stores all information in a variety of storage devices. These peripherals are frequently classified by two characteristics—speed and capacity.

Perhaps the storage device external to the CPU with the highest speed is the **fixed**

head disc. A disc is literally a platter on which magnetized spots (bits)* can be deposited (or erased). A disc can be thought of as a phonograph record with grooves which correspond to magnetic tracks. The heads which read and write information on the discs may be compared to a phonograph pickup and stylus.

Figure 5–5 illustrates magnetic bits of data on the tracks of a fixed head disc. Each bit produces a pulse when it passes the read heads. The bit of data is detected as the presence or absence of an electronic pulse. A number of bits together produce a pattern of pulses called a "word." For example, a 12 bit word consists of a pattern of 12 pulses which give the numeric value of the word in a manner similar to that described for digital signals in Figure 5–4.

A 64 track fixed head disc has 64 corresponding heads to read and write information on the disc platter. Because the heads are fixed and do not have to move from one track to another, the time to access the information on the disc is dependent only upon the time the platter takes to make one revolution. A typical fixed head disc requires an average of from 8 to 16 milliseconds to find the desired information. Additional time required to transfer the information into the CPU depends upon the amount of information to be transferred at one time. High speed fixed head discs are used primarily for applications where the speed of accessing information is extremely important. For example, if 10 individuals were using a time-sharing system and each were to ask for information from the disc at the same time, the disc access speed must be very high in order to avoid significant delays. It must be realized that with the exception of large computers, the CPU can only process one program or one user's request at a time, even though there may be many such programs or requests in memory at one time. A fixed head disc usually has a smaller capacity than other discs, e.g. 250,000 to 500,000 words per platter. Several discs may be added to a system to increase the total storage capacity.

In practice a disc is usually divided into areas called "files." Such areas may be designated to store specific information such as programs (program file), patient laboratory results (patient files), and so forth. The size of a file is often expressed as the number of words it contains.

* All information handled by the computer is contained in numerical code. This is because the actual pieces of hardware contain information in a form called bits. Bits in the computer memory are stored on ferrite doughnuts (i.e., cores) which can be magnetized in a clockwise or counterclockwise direction. If a core is magnetized in one direction the computer gives it the value of zero; and if it is magnetized in the opposite direction the computer gives it the value of one. Since the values of a bit are limited to zero and one, all calculations and data are contained as coded numbers in base 2 or "binary." Our normal system of numeric calculation is base 10 or decimal.

In the computer the bits are grouped together into bunches which are called words. Word sizes for typical laboratory computers range from 8 to 16 bits. In core memory each "word" has a specific location assigned to it; this location is called the "address" of the word.

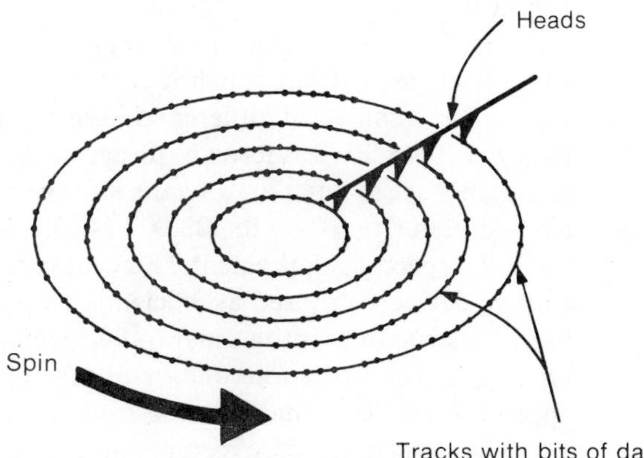

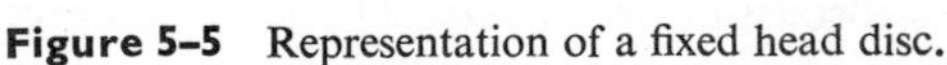
Figure 5–5 Representation of a fixed head disc.

The next highest speed device is the **moving head disc** which has a "platter" similar to that of the fixed head disc but which is equipped with only one read head. Since this head has to move to the appropriate track to find the information, the average access time is typically 50 to 80 milliseconds. The moving head disc, however, usually has a much higher capacity, namely in the range of 1.5 to 3 million words per platter. Since the cost for a moving head disc is essentially the same per platter as that for the fixed head disc, much more storage capacity per dollar is realized with this type of disc.

Another characteristic of most moving head discs is the necessity to read at one time an entire sector or block of information. Thus, if the name of only one patient is to be retrieved, the entire sector containing this information would have to be transferred to the CPU and the computer program would have to select the desired name. In contrast, the fixed head disc can usually read single words, referred to as "word addressable" capability.

In some systems, discs with both types of heads may be used for different applications For example, the fixed head disc may be used for the basic time-sharing system and for accessing information most frequently needed, such as data for hematology and chemistry. The moving head disc might be used for storage of outpatient data where access and sorting of information does not need to be accomplished at very high speeds.

The magnetic tape is the next slowest and least expensive storage device which contains information also in the form of magnetic bits. The retrieval time depends on the length of the tape. Small tapes, used on typical laboratory computers, contain only 128,000 words and the computer can scan such a tape within 20 to 30 s. While this is extremely slow compared to a disc, it may be quite acceptable for storing information which is to be retrieved infrequently, e.g., billing information generated once a day.

Large amounts of information may be stored on "industry-compatible tape" sometimes referred to as IBM-compatible tape. These tapes may be 2400 feet in length and may have 20 times or more the capacity of small tapes but the access time may be as long as 20 min. Use of industry-compatible tape has the great advantage that information generated by the laboratory computer can be transferred to any other computer equipped with an industry-compatible tape drive. This may be helpful for transmitting billing information, or information related to admissions, transfers, and discharges from one computer to another. Such transfer of information by tape is less expensive than direct computer interfaces by wire or telephone.

The **core memory** is another storage device. By definition, it is not a peripheral device since it is an actual part of the CPU. Thus, this form of memory is discussed in the following section.

The central processing unit (CPU)

A computer configuration, exclusive of the lineprinter and terminals, is shown in Fig. 5–6.

The CPU is the "electronic logic" of the computer configuration. Each CPU has its own set of instructions which it executes in the appropriate order as provided by the software. The number of different instructions (instruction set) range from six to several hundred. The CPU has its own storage device, the *core memory*. This device has an extremely fast access speed compared to the storage devices discussed previously. The core memory may contain from 4000 to 128,000 (4 to 128 K) or more words or instructions, depending upon the specific requirements. Since it is relatively more expensive than any other type of storage device, it is used as efficiently as possible (for example, to store on a "temporary basis" instructions [programs] to be executed by the CPU). The CPU electronic logic will typically fetch instructions one-by-one from the core memory to recognize and execute them appropriately. The time required for these functions is generally one microsecond or less.

The *speed* of the core memory and the CPU in general is very seldom the limiting factor

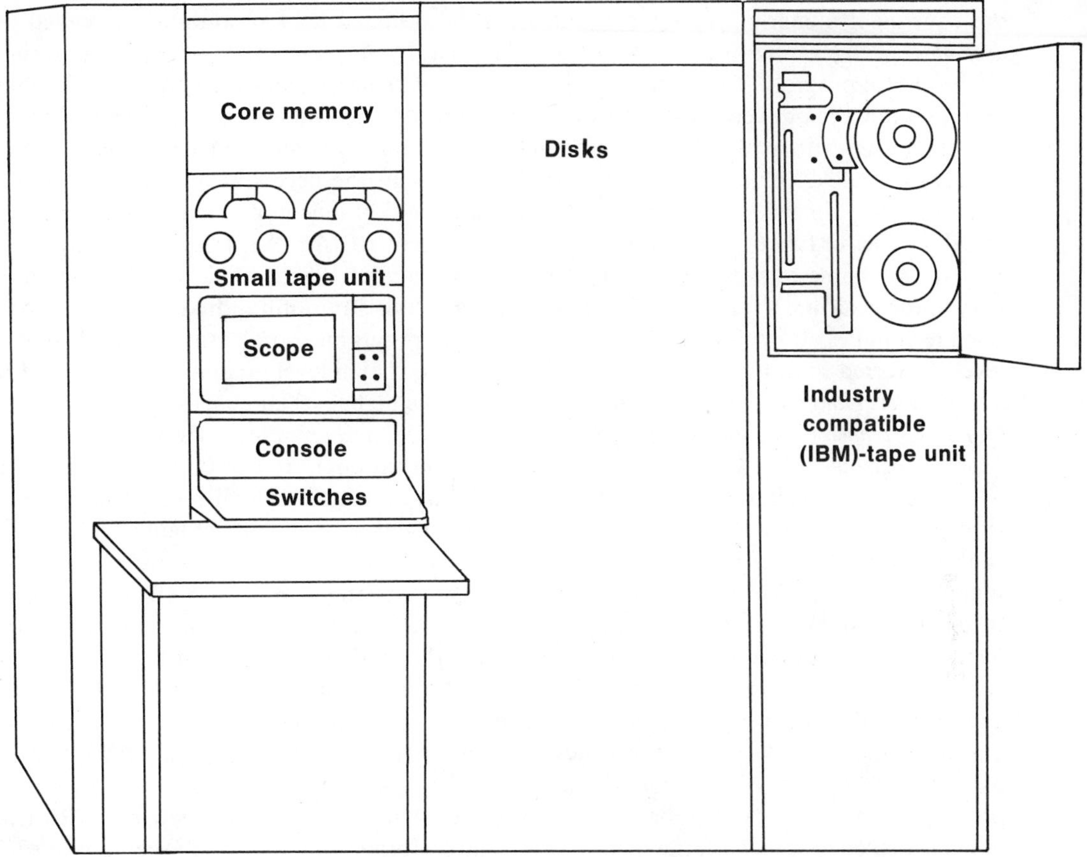

Figure 5–6 A computer system configuration.

in the speed of the overall system. The peripheral devices (discs, terminals, tape units, etc.) are many orders of magnitude slower than the CPU and thus determine the overall speed of the system. This can be even better appreciated if one realizes that the CPU does not limit the speed, even if it is interfaced with as many as 32 laboratory instruments, four tape drives, 24 terminals, eight discs, a card reader, and a lineprinter. In fact, the central processor is capable of executing 10,000 instructions or more in the same time in which it takes a terminal to print only one character. The purchase price for a CPU is typically 10 per cent or less of the total cost of the hardware.

A brief introduction to the concept of a computer *instruction* might be useful since it is often confused with a computer *language*. The machine instruction is the most fundamental form of telling the computer what function to perform. An instruction is expressed as one or more discrete numbers which can be decoded by the electronic logic in the computer. For example, the number 2000 may instruct the computer to ADD some number to another number. The address (location) of the number to be added might be indicated as an addition to the basic instruction. Thus, 2100 may instruct the computer to ADD the number in location 100 to the other number. A series of such discrete instructions which are executed in sequence constitute a *program*.

Constructing a program with such a series of discrete numbers is very tedious work, and therefore techniques have been developed to allow the programer to enter English-type mnemonic codes for instructions, such as ADD, LOAD, MULT, or STORE, instead of the numeric equivalents. The computer program which converts the mnemonic codes to the equivalent numeric instructions which can be decoded by the computer is called an *assembler*. Programing in this manner is called *machine language* or *assembly* programing.

A useful program, for example, to print a patient's name from the computer file, may require hundreds or even thousands of discrete machine instructions and can be tedious to construct even with assembly language. Once a function is written in machine instructions, it may be "called" or used by many programs without the necessity of writing the same function many times in one system. An even higher level of mnemonics can be implemented to call a series of functions. For example, the mnemonic *PRINT NAME, 178756* may mean to print the name of the patient with the hospital number 178756. This simple mnemonic may in turn call a corresponding machine program consisting of hundreds of instructions. Such techniques allow the programmer to construct programs more easily and at a "higher level," referred to as high level programing. A series of high level mnemonics is referred to as a *high level language*. Examples of some well-known high level languages are FORTRAN (*for*mula *tran*slation, which is especially useful for mathematics), COBOL (*co*mmon *b*usiness *o*riented *l*anguage, useful for business applications), and MUMPS (designed at Massachusetts General Hospital for medical applications). Both high level and machine level programing have their advantages and disadvantages. A high level language is easier to learn and to use, requires less knowledge of the computer, and is therefore preferred by many programers. While more difficult to use, machine language is more efficient in terms of speed and size of programs. A single high level instruction may use hundreds of machine instructions generalized to carry out a function with a wide range of conditions and applications, while a machine program can be tailored for a specific application. A well-designed system usually will mix the higher level functions with specific machine instructions to optimize the advantages of both types of programing. A more detailed description of programing languages is beyond the scope of this book, but one cannot simply rate a program or system as "good" or "bad" based on the language in which it is written. Rather, one must be concerned with the overall functions performed, the speed of execution of each function, the response times, and the ease with which laboratory functions are changed or developed with the system.

THE IMPACT OF LABORATORY COMPUTERS

The laboratory produces services for all patients in the hospital and therefore interacts with almost all personnel associated with patient care. The processes involved can be separated into four basic steps—request, laboratory work flow, analytical testing, and, finally, reporting. The laboratory personnel, aside from performing the actual analysis, are confronted with a complex administrative task which requires a considerable portion of working time and frequently represents a major bottleneck in the laboratory work flow. In addition, many of these manual tasks, such as transcription of doctors' orders onto laboratory requisition forms, or transcription of test data from the laboratory log to the report form leave considerable room for error. It is therefore not surprising that early attempts at computerization of the laboratory focused primarily on the clerical and administrative aspects. Figure 5–7 illustrates the manner in which a computer can have an impact on the flow of information and work produced by the laboratory, if the latest features of computer capability are available. It is apparent that, in general, the greatest degree of laboratory automation coupled with appropriate computer capability is most efficient and will have the greatest impact on the efficiency of health care delivery by laboratories. With few exceptions, the computer can essentially eliminate manual efforts in the work flow and reporting of laboratory data. In addition, the computer has a significant impact on the analytical process, if results are produced by automated equipment or if a significant amount of computations are required to produce results (see Fig. 5–7). A more detailed

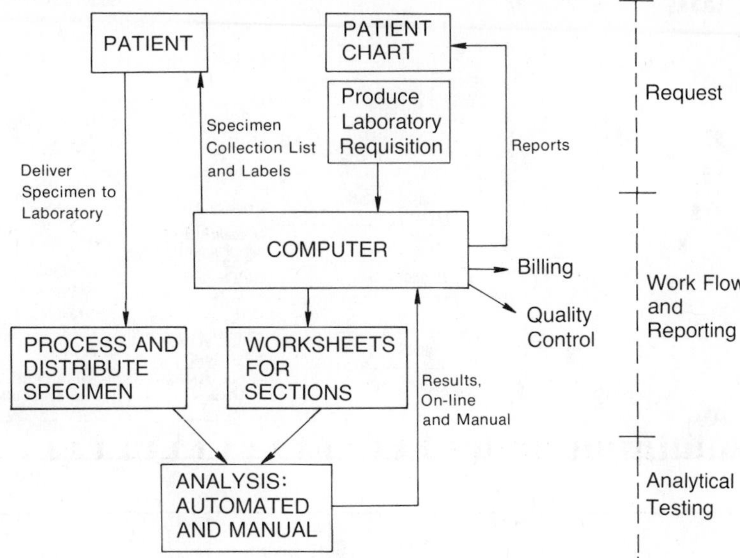

Figure 5-7 Laboratory information flow with a computer.

description of the effect of computers on specific laboratory functions is given in the following portions of this chapter.

BASIC ELEMENTS OF THE COMPUTER SOFTWARE SYSTEM

For the sake of continuity and understanding of the overall operation of a typical computer system, all basic elements from the requisitioning process to the final patient report will be described in this section in chronological order.

Admissions

Patients are "admitted to the computer" in one of a variety of ways. At Mount Sinai Hospital Medical Center in Chicago, Illinois, the data-processing center generates punched cards containing all vital patient demographic data, which are then read into the laboratory computer. Similar cards are issued if the patient is transferred during his hospital stay and at the time of discharge. This data can also be entered by means of magnetic tape or by manually using any of the terminals in the conversational mode.

Requisitions

The capability of the computer to read mark-sense cards makes it possible to consolidate many of the previously used manual requisition forms. Since these forms are used solely for entering requisitions into the computer, they no longer have to be designed for the convenience of certain laboratory sections but instead can be designed to simplify the ordering process at the nurses' station. In accordance with the needs of the individual institution, many types of mark-sense cards have been designed for use in hospitals with computer systems. Figure 5–8*A* is an example of a form generated at the nurses' station of Mount Sinai Hospital Medical Center in Chicago. In this institution, each nurses' station is provided with a Source Record Punch, which is capable of printing onto the requisition all pertinent administrative data in human-readable form (see upper right and upper left corner). The punched holes on the left side of the card give the patient's identification number and the date of the request. Subsequent to this procedure, the nurse or clerk marks the test desired by placing an "X" in the respective boxes on the card. This requisition, after delivery to the laboratory, is fed into the card reader.

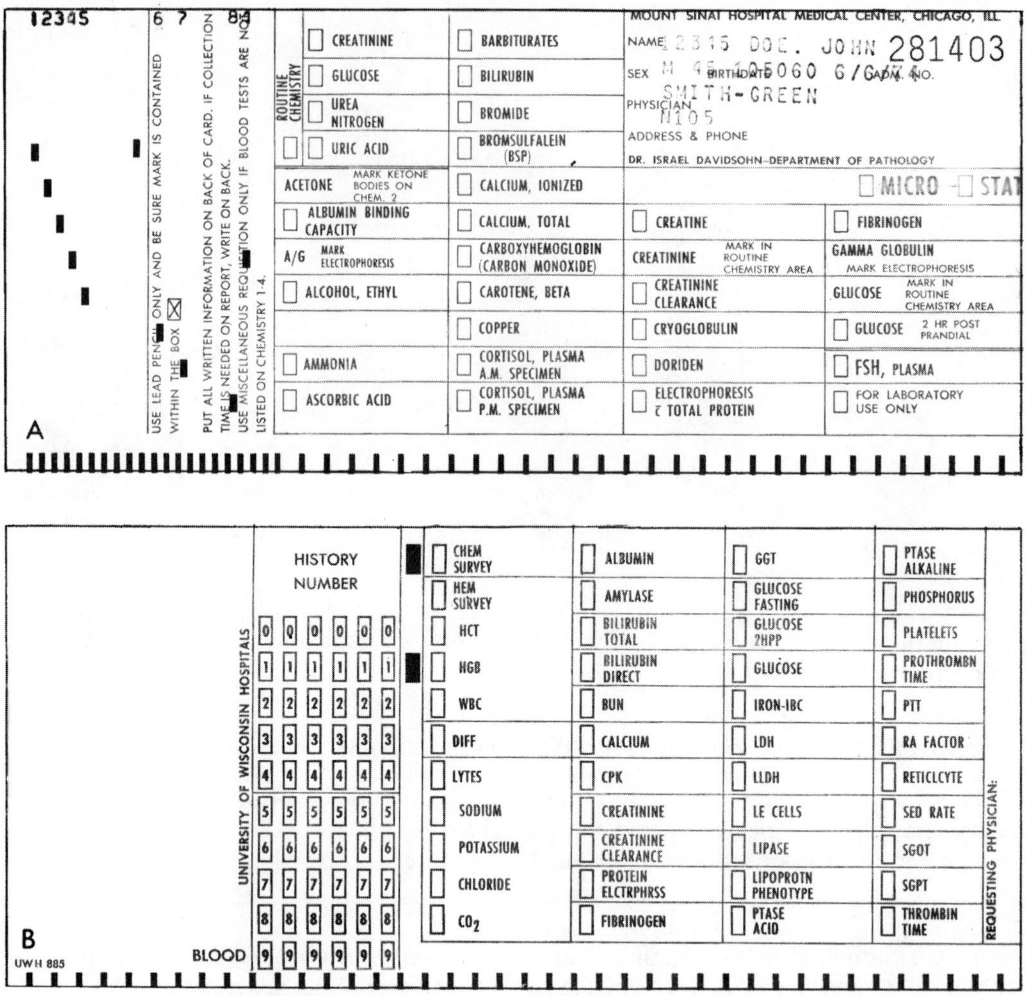

12345 6 7 8 9

USE LEAD PENCIL ONLY AND BE SURE MARK IS CONTAINED WITHIN THE BOX ☒

PUT ALL WRITTEN INFORMATION ON BACK OF CARD. IF COLLECTION TIME IS NEEDED ON REPORT, WRITE ON BACK.

USE MISCELLANEOUS REQUISITION ONLY IF BLOOD TESTS ARE NOT LISTED ON CHEMISTRY 1-4.

MOUNT SINAI HOSPITAL MEDICAL CENTER, CHICAGO, ILL.

NAME 2315 DOE. JOHN 281403

SEX M BIRTHDATE 05060 6/6/74 ADM. NO.

PHYSICIAN SMITH-GREEN N105

ADDRESS & PHONE

DR. ISRAEL DAVIDSOHN–DEPARTMENT OF PATHOLOGY

MICRO — STAT

ROUTINE CHEMISTRY: CREATININE, GLUCOSE, UREA NITROGEN, URIC ACID

BARBITURATES, BILIRUBIN, BROMIDE, BROMSULFALEIN (BSP)

ACETONE (MARK KETONE BODIES ON CHEM. 2) | CALCIUM, IONIZED

ALBUMIN BINDING CAPACITY | CALCIUM, TOTAL | CREATINE | FIBRINOGEN

A/G (MARK ELECTROPHORESIS) | CARBOXYHEMOGLOBIN (CARBON MONOXIDE) | CREATININE (MARK IN ROUTINE CHEMISTRY AREA) | GAMMA GLOBULIN (MARK ELECTROPHORESIS)

ALCOHOL, ETHYL | CAROTENE, BETA | CREATININE CLEARANCE | GLUCOSE (MARK IN ROUTINE CHEMISTRY AREA)

COPPER | CRYOGLOBULIN | GLUCOSE 2 HR POST PRANDIAL

AMMONIA | CORTISOL, PLASMA A.M. SPECIMEN | DORIDEN | FSH, PLASMA

ASCORBIC ACID | CORTISOL, PLASMA P.M. SPECIMEN | ELECTROPHORESIS ̄c TOTAL PROTEIN | FOR LABORATORY USE ONLY

A

HISTORY NUMBER

UNIVERSITY OF WISCONSIN HOSPITALS

0 1 2 3 4 5 6 7 8 9

CHEM SURVEY | ALBUMIN | GGT | PTASE ALKALINE

HEM SURVEY | AMYLASE | GLUCOSE FASTING | PHOSPHORUS

HCT | BILIRUBIN TOTAL | GLUCOSE 2HPP | PLATELETS

HGB | BILIRUBIN DIRECT | GLUCOSE | PROTHROMBN TIME

WBC | BUN | IRON-IBC | PTT

DIFF | CALCIUM | LDH | RA FACTOR

LYTES | CPK | LLDH | RETICLCYTE

SODIUM | CREATININE | LE CELLS | SED RATE

POTASSIUM | CREATININE CLEARANCE | LIPASE | SGOT

CHLORIDE | PROTEIN ELCTRPHRSS | LIPOPROTN PHENOTYPE | SGPT

CO_2 | FIBRINOGEN | PTASE ACID | THROMBIN TIME

REQUESTING PHYSICIAN:

B

UWH 885

BLOOD

Figure 5–8 Mark-sense requisition cards.

Figure 5–8*B* shows a card designed for use at the University of Wisconsin Hospitals in Madison. Notice that on this requisition a variety of laboratory tests are included to combine the most commonly ordered laboratory tests on one single form regardless of the section performing the test. The patient identification in this case is entered by marking the appropriate boxes in the "mark-sensefield." Although this approach of marking the patient's identification is slower and less convenient, it offers an acceptable alternative to the use of a Source Record Punch or similar machinery. The great divergency in the design of the requisition cards points out the usefulness of user control programs (see p. 231) which allow for the design of the requisition forms in the laboratory by laboratory personnel without the aid of an outside programer.

Requisition cards of the type shown not only represent a convenience to the laboratory but they also safeguard against errors, including duplicate orders. A single enzyme test, for example, may have been ordered both alone and as part of a liver profile. Such redundancy can be detected by the computer and called to the attention of the operator.

Requisitions can also be entered manually via any of the terminals in the conversational mode. When specimens are received intermittently (e.g., outpatient specimens), manual entry is actually more efficient than processing only a few cards with the card reader.

```
4M/S 405-2        #5010 | H#425510        #5010 |       |       | H#425510        #5010
OLIVIER LYNN            | OLIVIER LYNN          | #5010 | #5010 | OLIVIER LYNN
                   LAVH | LAVH: WBC  DIFF       | - - - | - - - | LAVH: WBC  DIFF
H#425510           6/14 | 6/14                  | #5010 | #5010 | 6/14
                   9AM  | 9AM                   |       |       | 9AM

                  #5062 | H#425510        #5062 |       |       | H#425510        #5062
                        | OLIVIER LYNN          | #5062 | #5062 | OLIVIER LYNN
                   GRN  | GRN : LYTE            | - - - | - - - | GRN : LYTE
                   6/14 | 6/14                  | #5062 | #5062 | 6/14
                   11AM | 11AM                  |       |       | 11AM

5ICU 525-A        #5000 | H#654123        #5000 |       |       | H#654123        #5000
JONES ROBERT A          | JONES ROBERT A        | #5000 | #5000 | JONES ROBERT A
                   GRN  | GRN : NA   K    CL    | - - - | - - - | GRN : NA   K    CL
H#654123           6/14 | 6/14                  | #5000 | #5000 | 6/14
                  1:20P | 1:20P.                |       |       | 1:20P
```

Figure 5–9 Blood-drawing labels.

Blood-drawing lists and labels

On the basis of the information which has been entered into the computer, either in the form of machine-readable requisitions or through the manual entry procedure, the computer can print a blood-drawing list together with self-adhesive labels for the identification of specimen containers. In the example shown in Fig. 5–9, the extreme left column of labels contains the name of the patient, sorted by nursing station and ascending room number within each nursing station. Such labels are used as the actual blood-drawing list. The set of labels in the extreme right column contains the patient's identification number, the patient's name, the collection container needed, and the codes for the tests ordered. In addition, the label provides the date and the laboratory accession number and time. Attached to this label are four smaller labels carrying the same laboratory accession number. The label in the right column, together with these four additional labels, can be peeled off and placed onto the specimen container. The labels in the center column serve as additional labels in case a series of samples are drawn.

The use of labels with all pertinent information eliminates the need to match up requisitions with specimens and also eliminates the need to assign manually an accession number to each specimen. The accession number serves to identify the specimen and is carried through all section worksheets and logs produced by the computer. This same accession number may also be used instead of the patient name to query the computer for information concerning the status of test results obtained on a particular specimen.

Logs

In a typical laboratory, pertinent information related to patients, to specimens received, and to results obtained is manually recorded. This significant amount of information can quickly be produced by the computer in the form of different types of *logs*. The log shown in Fig. 5–10 lists the laboratory accession number, the time, the date, the name and hospital number of each patient on whom laboratory work was ordered, and the codes for those tests and test combinations which have been requested. Test codes with no symbol affixed indicate that the test or the entire test package has not been completed, while a square bracket preceding the test code indicates that the test is requested and that the assay has already been completed. A per cent sign before the code for a test battery (e.g., %ELYT) indicates that such a battery has been ordered and that a portion of these tests, e.g., sodium and potassium, have been completed, while (e.g.) CO_2 and chloride determinations are still pending.

```
MASTER LOG                      11:33AM    3/23/1974      PAGE 1


ACC#  REQUISITIONS                   TIME   DATE NAME                       HOSP# N.S.

 365   CHOL   SGOT                   7AM    3/21 HAGGARG DOROTHY L 6783456 4WST
 401 [NA     K      CL               7AM    3/23 ABBEY CARL J              6543217 2WST
 455 %ELYT   BGAS                    7AM    3/23 BYRON JOHNATHAN L 3451234 5EST
 567   CA   [CHOL  [SGOT             9:20A  3/23 OLIVER WENDEL K           4567123 2WST
 600 [NA     CO2                     7AM    3/23 OLIVER WENDEL K           4567123 2WST
 605 %ELYT   PH                     10AM    3/23 JONES JOHN D              1234567 4W
 700 %ENZ    DBIL  [CHOL            12:10A  3/23 SHUTE DALE F              5437645 5EST
END OF LOG
```

Figure 5–10 Computer-generated laboratory log.

Incomplete tests are listed on such a log regardless of the day they were ordered and the request will re-appear on subsequent logs until the test is completed. This greatly reduces the chance that pending work from a previous day may be overlooked. Logs of this type may be printed for the entire laboratory (master logs) or for individual sections (section logs).

These and other computer printouts, such as a "patient directory" (containing all administrative data on all patients in the hospital), indicate the versatility of the computer, and its effectiveness as a management tool in a laboratory.

Worksheets

Laboratory computers, in general, are programed to produce two types of worksheets: *manual worksheets*, to be used for all tests which are performed manually, and *tray-loading worksheets*, to be used to load specimens on instrumentation which is connected on-line to the computer.

An example of a manual worksheet for urine electrolytes is shown in Fig. 5–11. Note that in each case the patient's name, accession number, date of request, nursing station, patient identification number, and time of specimen acquisition are included on the worksheet. In addition, the worksheet indicates how a test was requested. If, for example, a urine sodium were ordered individually, the code UNA is printed, while the code "UELY" indicates that a complete urine electrolyte set is ordered.

Tray-loading sheets for fully automated equipment on-line to the computer differ in several aspects from manual worksheets. A previously designed "control/schedule block"

```
MANUAL WORKSHEET, ALL INC. 11:48AM      3/23/1974          PAGE 1

ACC#    DATE   TIME    NAME                  HOSP# N.S.
                                                   UNA    UK     UCL    VOL   HRS


5001    3/23   8AM     ABBEY CARL J         6543217 2WST
                                                   UELY   UELY   UELY   4280  24.0


5002    3/23   8AM     BYRON JOHNATHAN L    3451234 5EST
                                                          UK            1225  24.0


5003    3/23   8AM     OLIVER WENDEL K      4567123 2WST
                                                   UNA                   345  12.5


5010    3/23   9AM     WREN CAROL R         2345678 3E
                                                   UELY   UELY   UELY   1350  24.0
```

Figure 5–11 Computer-printed manual worksheet.

LOADING WORKSHEET FOR GLUC 11:41AM 3/23/1974 PAGE 1

CUP	ACC#	DATE	TIME	NAME	HOSP#	N.S.	GLUC	BUN
1-MP								
2-STD								
3-STD								
4-STD								
5-CTL								
6	7Ø5	3/23	8AM	JONES JOHN D	1234567	4W	GLUC	
7	71Ø	3/23	9AM	OLIVER WENDEL K	4567123	2WST	GLUC	BUN
8	78Ø	3/23	8AM	HAGGARG DOROTHY L	6783456	4WST		BUN
9	791	3/23	8AM	ABBEY CARL J	6543217	2WST	GLUC	BUN
1Ø-CTL								
11	8ØØ	3/23	8AM	BYRON JOHNATHAN L	3451234	5EST	GLUC	BUN
12	9ØØ	3/23	8AM	SHUTE DALE F	5437645	5EST	GLUC	BUN

Figure 5–12 Tray-loading worksheet.

contains all information related to the position on a loading tray of the individual standards, quality control specimens, and so forth. Since the computer also has all information regarding the tests ordered on the individual patient, the computer can automatically assign positions for standards (STD), controls (CTL), and specific patient's specimens (see Fig. 5–12). Alternatively, a tray can be loaded with standards, controls, and patient's specimens in any desired order and the computer can then be informed as to the position of the individual specimens. Some computers also allow a mixture of these approaches where all but a few positions are assigned by the computer, leaving the empty positions for emergency requests.

Either manual or tray-loading worksheets can be generated at any time, thus providing information related to requests which arrive throughout a working day and helping to organize work flow.

Data entry

The most sophisticated form of data entry is *on-line data acquisition.* In this case, the electrical signals generated by the analytical instrument are transmitted directly to the computer. For example, analog signals (e.g., from an AutoAnalyzer) are converted into digital form by the analog-digital converter. Such data are generated for all standards, controls, and test specimens, allowing the computer to calculate the final results. In a well-designed system, such test data are retained in a temporary file and are transferred into the permanent patients' computer file only with the specific approval of an authorized technologist. A significant advantage of on-line data acquisition is the ability of the computer to perform a number of quality control functions during instrument operation (in real time) such as linearity checks of the standard curve (lack of linearity is indicated by the code "L" which is attached to the voltage reading of the standard). An example of such an on-line report is shown in Fig. 5–13. The computer can also automatically check whether a quality control specimen is within predetermined limits. Quality control results which are

```
GLUC,BUN,CRET,URIC                                10:05AM 4/15/1972 PAGE 1

CUP ACC#
           GLUC      BUN     CRET     CREA      CRE      VOL     URIC

PLATE 1
  1 MP

  2 STD
           129V      97V       MP       NC       NC       NC       MP
  3 STD
           238V     170VL      MP       NC       NC       NC       MP
  4 STD
           397V#    397V#      MP       NC       NC       NC       MP
  5 STD
             MP       MP     109V       NC       NC       NC     102V
  6 STD
             MP       MP     189V       NC       NC       NC     192V
  7 STD
             MP       MP     376V#      NC       NC       NC     325V
  8 MP

  9 QCI
            115       25      2.3       NC       NC       NC      4.3C
GLUC
PLATE 1
  1 Ø111111
  4 1111ØØØ
BUN
PLATE 1
  4 111111Ø
CRET
PLATE 1
  1 Ø111111
  7 111111Ø
END OF REPORT
```

Figure 5–13 On-line report with quality control information.

outside of such acceptable limits (as specified by the laboratory) are flagged by a "C," as shown in Fig. 5–13 for uric acid. Since such an on-line report can be generated automatically as soon as the analysis of the quality control specimen has been completed, the technologist has the option to decide on the appropriate action before completing a run which may be "out of control," thereby saving valuable technician time.

Immediately following the results, the on-line report provides "diagnostic" comments, consisting of "1's" and "Ø's," for each of the results given in the report. In Fig. 5–13, a series of seven "1's" (not shown) indicates to the technologist that the recorder peak (e.g., from an AutoAnalyzer) met all specified criteria (i.e., that all seven points taken by the computer are taken from the highest point of the curve, and that these points do not differ from each other by more than a predetermined amount). A "Ø" indicates that a particular reading differs by more than a preset amount from the other readings. This is the case if one of the readings is taken before the peak has reached its maximum, or after the peak maximum has passed. "Ø's" dispersed between "1's" are obtained in case of noisy peaks with the noise level exceeding a predetermined level. If any of the test results has been based on peaks falling into this category, a diagnostic sign "#" is attached to the test result.

Some instruments in the clinical laboratory which are interfaced directly with the computer generate digital signals rather than analog signals. The computer can receive such signals readily through the digital interface as previously described, but there is no way of evaluating the significance of a particular number. There are, however, calculations which the computer can readily perform to help detect incorrect laboratory results. If, for example, a pH and a pCO_2 are performed on a patient's blood sample, the computer will not only print out the actual values transmitted, but in addition will calculate the theoretical CO_2 content value, using the Henderson-Hasselbalch equation. If this calculated CO_2 value compares with the experimentally obtained CO_2 value, it may be assumed that the experimental pH, pCO_2, and CO_2 values are correct. Similar calculations may be performed for other applications in chemistry, hematology, and other parts of the laboratory.

The **mark-sense card** is another form of data input, as shown in Fig. 5–17 and explained in more detail later in the text. Such cards may be used to record the accession number and the test result by simply marking an "X" in the appropriate boxes with a pencil. Such entry cards are directly computer-readable when fed through a card reader. As the computer accepts data, it can perform a number of quality control tasks, e.g., by checking that not more than one value is entered for a given test, or by checking that all counts in a differential cell count add up to 100.

Manual entry cards can be designed to meet the requirements for any number of applications. The format of such cards, in a well-designed computer system, can be designed by the laboratory without the modification or addition of programs.

Keyboard entry of laboratory data is convenient if only a small number of data have to be entered, or if on-line entry or card reader entry is not available. Such manual entry of data can be simplified if the computer, on the basis of information already available to it, automatically prints the accession number, patient name, test codes, and so forth, and requires the technologist to enter only the final result. Where applicable, the computer can also perform a series of calculations which are frequently done by a desk calculator. Examples of such calculations include computation of electrophoresis results, creatinine clearances, and blood gases.

Finally, laboratory data can be entered through special single-purpose terminals. In such terminals, which have been designed for hematology, urinalysis, and other areas, each key on the keyboard has a specific designation. Although convenient, such terminals can be used only in that area for which they are designed. This subject has been treated in more detail by others.[1,2]

Patient reports

One of the great advantages of a laboratory computer system is its ability to generate reports of laboratory data for each patient at any desired time and in an organized form which is most meaningful to the physician. There are basically two types of reports: the interim report and the cumulative report.

The function of the *interim report* (also referred to as the ward report) is to provide laboratory data on an interim basis, throughout the day, to keep nurses and physicians informed of the status of various laboratory requests. Such reports may be issued at any given time of the day, or in response to a telephone query. If a laboratory test has been ordered but has not been completed, the word "pending" will appear on the report. Normal ranges are generally included, together with the laboratory results, and all abnormal results are flagged (Fig. 5–14). Various other versions of interim reports can be generated, e.g., those which list in concise form selected results ordered for patients on one given floor.

The second major type of computer report is the *cumulative report* (also called the summary report). This report is perhaps one of the most discussed benefits of the laboratory computer and, at the same time, one of the most controversial. The main purpose of the cumulative report is to summarize all of the laboratory data on a given patient (not limited to seven days, etc.), with the data organized in a fashion most convenient for the physician to review. There are several ways in which laboratory data can be organized. The test names can be printed horizontally across the top of the page, and the dates and laboratory results printed vertically. Alternatively, the dates and laboratory tests may be lined up horizontally side-by-side, with the test names listed vertically. Each of these formats has

```
LCI LABCOM EXAMPLE              INTERIM REPORT             PRINTED  8:29AM 10/24
DOE JOHN HOWARD         H#   123456789  6E   618
                        R#   987654321   JONES             TESTS FOR 10/22   PG 1

                                              NORMAL
ACC#                                          LO  HI     TECH

5025              ENZYME BATTERY
SGOT                      38 U/L                           56
SGPT                      56 U/L                           56
LDH                       97 U/L                           56
CPK                       60 U/L                           56

5026
UREA NITROGEN             26 MG/DL        *    10 22       45

4067    8AM
GLUCOSE             PATIENT NOT FASTING                    67

5027  10:50AM
CREATININE               2.0 MG/DL        *    .8 1.6      23
                      MOD HEMOLYSIS

5029    8:30PM    ELECTROLYTE
SODIUM                   140 MEQ/L            136 142      45
POTASSIUM                4.2 MEQ/L            3.5 4.8      45
CO2                       30 MM/L         *    24 28       45
```

Figure 5–14 Interim report.

its merits and its disadvantages. The horizontal format is at first glance (Fig. 5–15) much easier to read than the vertical format, since tests performed appear immediately under each other, and the results for a single test are easy to follow. In this approach, the test names, however, must be listed across the top of the page where there is only limited space available. Thus, the name of the laboratory test must be fitted into only a few characters, usually four to seven in number. This makes it difficult at times to assign meaningful abbreviations to some laboratory tests, especially for those tests which are more esoteric. Furthermore, if different tests of a different category are to be examined together, it is more difficult to relate them to each other.

In contrast, in the vertical format (Fig. 5–16), the test names are listed under each other, and test results are placed side-by-side across the page. This restricts the number of results which can be listed on one page, thus possibly requiring generation of additional pages. This limitation has caused some manufacturers to actually limit the cumulative report to seven days, or to the number of columns which conveniently fit across the page.

Regardless of the format of the cumulative report, there is still the significant advantage that the printed report is easier to read than hand-written reports, and that the data are well organized in regard to the sequence in which tests were performed and in regard to test groupings which are generally ordered together in a given disease state. The cumulative report has the additional advantage in that abnormal results can be flagged, and normal ranges for each test are supplied, stratified by age and sex. Furthermore, symbols can be used to indicate the areas where new test data have been added, reducing the time required for a physician or nurse to locate new laboratory data.

ADDITIONAL COMPUTER APPLICATIONS

In addition to the direct application of the computer to the flow of information from the requisition process to the reporting process, the computer provides a tool for other functions related to the laboratory. The computer may provide, for example, administrative data, perform a variety of quality control functions, and generate billing reports. Such applications are discussed in the following section.

LCI LABCOM EXAMPLE CUMULATIVE REPORT PG 2
NAME: DOE JOHN HOWARD NS: 6E ROOM: 618
H# : 123456789 DR: JONES AGE: 36YR 4MO 5:29PM 10/24/1973
R# : 987654321

H E M A T O L O G Y

	WBC 4.8-10.8 K/CUMM	RBC 4.6-6.2 M/CUMM	HGB 14-18 GM/DL	HCT 42-52 %	MCV 82-92 CU U	MCH 27-31 UUGM	MCHC 32-36 %	PLT EST
OCT 20 7A	7.6	3.53♦	11.0♦	33.1♦	92	31.2♦	33.1	ADEQ
OCT 21 7A			9.4♦	28.3♦ SEE				
OCT 23 7P			10.0♦	30.3♦				
◊ OCT 24 ◊ 2:54P			10.8♦	33.1♦				
OCT 25 7A			PENDING	PENDING				

D I F F E R E N T I A L

	BAND UP TO 3 %	NEUT 35-65 %	EOS UP TO 6 %	BASO UP TO 2 %	LYMPH 25-45 %	MONO UP TO 10 %
OCT 20 7A	0	72♦	6	0	12♦ CORR	10
OCT 23 7A	0	91♦	0	0	5♦	4
◊ OCT 24 ◊ 7A	0	90♦	0	0	6♦	4

B A S I C U R I N A L Y S I S

	APPEAR	UR SG 1.003- 1.038	UR PH 4.5-8.0	UR PROT UP TO .02 GM/DL	UR GLUC UP TO .02 GM/DL	KETONES	BILE	BLOOD
OCT 20	YEL CLDY	1.021	7	.06♦	NEG	NEG	NEG	MOD

M I C R O S C O P I C U R I N A L Y S I S

	UR WBC UP TO 3 /HPF	UR RBC UP TO 2 /HPF	CAST UP TO 2 /LPF	CAST #2 /LPF	CRYSTAL	MUCUS	CELLS	BACTERA
OCT 20 6:14P	2	1	HYAL <10	GRAN <10	NONE	MOD	SQU FEW	NONE
OCT 22 9:30P	1	NONE	HYAL <10		NONE	NONE	SQU FEW	FEW
OCT 23 12:14A	4♦	1	HYAL <10		NONE	NONE	SQU FEW	NONE

<10 5-10
ADEQ ADEQUATE
CLDY CLOUDY
CORR CORRECTED REPORT
GRAN GRANULAR
HYAL HYALINE
MOD MODERATE
NONE NONE DETECTED
SEE SEE FREE TEXT COMMENTARY
SQU SQUAMOUS EPITHELIAL
YEL YELLOW

DOE JOHN HOWARD PG 2

Figure 5–15 A horizontal cumulative report format.

LCI LABCOM EXAMPLE CUMULATIVE REPORT PG 4A
NAME: DOE JOHN HOWARD NS: 6E ROOM: 618
H# : 123456789 DR: JONES AGE: 36YR 4MO 5:29PM 10/24/1973
R# : 987654321

	NORMALS	OCT 20	----OCT 21----		OCT 22	OCT 23	OCT 24 ◊◊◊◊◊◊	OCT 25
CHEMISTRY - ELECTROLYTES								
GLUCOSE	70-	7AM						
MG/DL	120	118						
UREA NITROGEN	10-	7AM					7AM	
MG/DL	22	24♦	26♦		26♦		38♦	
SODIUM	136-	5PM	7AM		8:30P		7AM	7AM
MEQ/L	142	153♦	145♦		140	141	137	PENDING
POTASSIUM	3.5-	5PM	7AM	7:02A	8:30P		7AM	7AM
MEQ/L	4.8	3.5	2.9♦	4.5	4.2	4.1	4.3	PENDING
			REPT	CORR				
CO2	24-	5PM	7AM		8:30P		7AM	
MM/L	28	26	30♦		30♦	30♦	32♦	
CHLORIDE	95-							7AM
MEQ/L	105						90♦	PENDING

Figure 5–16 A vertical cumulative report format.

Administrative data

In a busy laboratory, where many specimens are being processed each day, it is very time-consuming to determine the exact workload and to distribute that workload appropriately. A computer, through its various administrative data programs, can readily provide statistics in regard to requests received, results produced, and the source of such request. The ability to document the arrival time of specimens in the laboratory, the time the reports were issued, and a list of stat requests received in a day, is another indication of the usefulness of the computer in the management of the laboratory and in the documentation of workload and personnel requirements.

Quality control

The significant increase in laboratory requests and the need for greater accuracy and precision of laboratory data have created the demand for more effective and efficient quality control programs (see Chap. 2). The availability of computers in the laboratory has significantly altered some of the older approaches. Some of these quality control aspects have already been discussed in preceding sections. These procedures include proper checking of request cards (p. 222), positive verification of patient identification (p. 223), and inspection of proper peak formation and linearity checks on standard curves in automated procedures (p. 225). Inspection and evaluation of the result of quality control specimens (p. 226), and verification of correctness of test results entered by the card reader (p. 227) must also be considered as part of the quality control program.

A series of statistical procedures, helpful in a quality control program, can readily be performed by the computer. Such procedures include tabulation of results on various quality control specimens, and calculation of standard deviation, the median, or the cumulative sum for quality control specimens.

The "limit check" and the "delta check" are further tools by which the computer can help identify incorrect laboratory data. The limit check program searches the patient file for a specified number of days and looks for values which are outside of reasonable predetermined limits. For example, the computer can be asked to search for all laboratory data which are known to be incompatible with life.

The delta check is in many ways more sensitive than the limit check. The delta check program compares consecutive values for a given test in a patient's file for abrupt changes. The deltas for each test procedure can be specified by the laboratory. The computer, for example, may be asked to identify any patient with an electrolyte value which has changed by more than 10 per cent in two consecutive values.

Billing and crediting procedures

Accurate billing information can be provided by generation of a billing list which identifies the patient by an appropriate hospital number and lists the tests which were performed, together with billing codes and charges. In more sophisticated systems, the computer can communicate billing information on magnetic tape directly from the laboratory to a central data-processing computer. The computer, through its billing procedure, can essentially eliminate two administrative problems in the laboratory: namely, double billing and missed billing.

If a laboratory test, once requested, cannot be performed, proper *credit* can be given to the patient or the charge can be deleted if the billing report has not yet been issued. Certain codes, such as "QNS" (quantity not sufficient) or "ACC" (laboratory accident) are "crediting statements" which cause the computer to generate automatically a credit to the patient. Such control over the billing (and crediting) process, without human intervention, greatly improves the laboratory operation, while guaranteeing that the patient is charged only for services rendered.

SYSTEM DESIGN AND USER CONTROL

Clinical laboratories differ in the way they are operated, depending upon the idiosyncrasies or needs of a particular laboratory, its staff, and users of the laboratory services. As a result, certain computer applications found useful in one laboratory may not be suitable in the same form for another. All laboratories, however, will want the general capability of the computer to generate worksheets, logs, cumulative reports, and so forth. It is in respect to the exact composition and format of these various reports that the differences may occur. It is therefore important to make a clear distinction between the *computer program* and the *parameters* in the system. The computer program, as previously stated, is a series of instructions which tell the computer step-by-step how to perform various functions. The exact manner in which the functions are carried out, however, is determined by parameters. Information regarding parameters, e.g., the exact tests that are done in a given laboratory section, can be kept in tables in the computer and can be easily altered without an actual change in the computer program. Similarly, the format in which the cumulative report is printed by the cumulative report program can be changed simply by changing parameters, without changing the actual program. Changes of this type are made by the *user control program*, which operates in the "conversational mode," i.e., the computer communicates interactively with laboratory personnel. No programing knowledge is required for such actions.

Another example which can illustrate the concept of user control is the design of a new mark-sense card for requisition or data entry. The capability to design such cards without the help of a programer is important, since most laboratories must design these cards to meet their specific needs, and these needs may change with time. Each card is divided into columns and each of the columns is numbered across the top of the card between the black marks, which are called "clock marks." Each of the rows down the card is identified as "A" through "L" (Fig. 5–17). This makes it possible to designate clearly any particular area on the card, and to define the meaning if one of these areas is marked. For example, in Fig. 5–17, the box indicating "uric acid" occurs in column 28, row E, and the word "few" occurs in column 30, row F. If the technologist makes marks in these two boxes, the report will indicate that few uric acid crystals were seen. The computer, of course, will accept only marks which are made exactly in the area of the box. Thus, any print or marks in areas outside the boxes are not "read" by the computer.

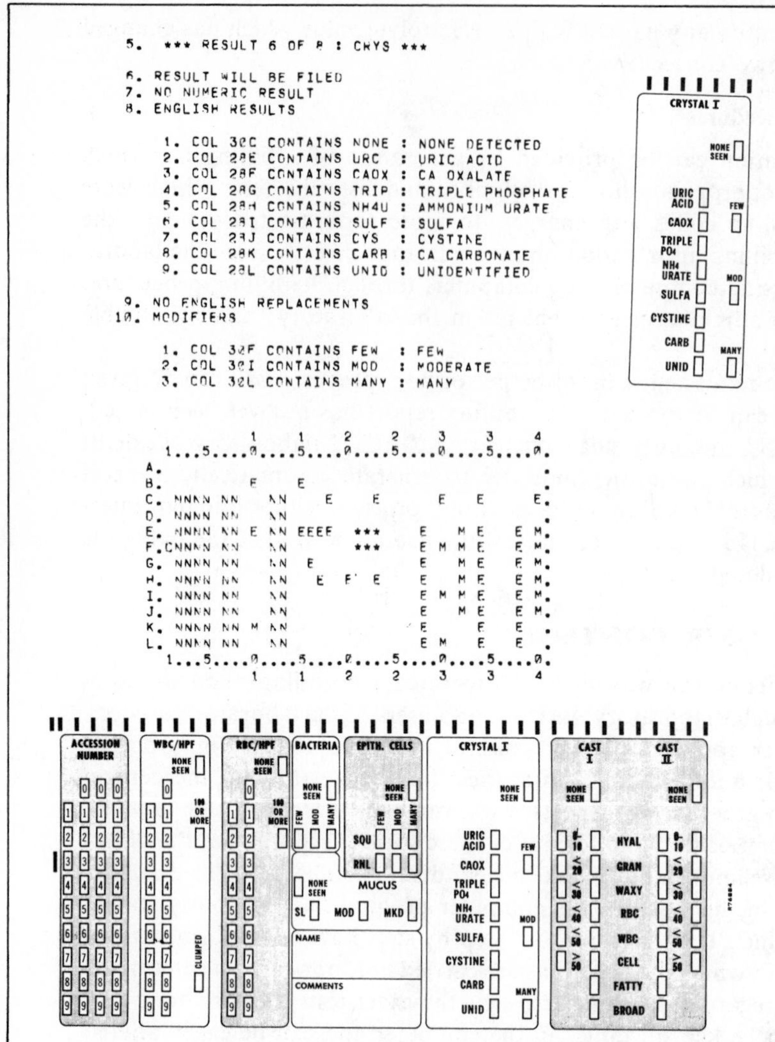

Figure 5–17 Procedure for user-designed mark-sense card

Figure 5–17 illustrates the user design process for a mark-sense card. The top of Fig. 5–17 is a computer printout showing the parameters of the corresponding section of the mark-sense card at the side. The middle of Fig. 5–17 shows a computer "printed picture" of the overall mark-sense card as designed by the laboratory personnel who entered the appropriate parameters. The card at the bottom is the final printed form of the mark-sense document as designed and used by the laboratory.

The ability of the user to change the parameters in the system is especially important during the early stages of a computer installation in the initial phase of operation. Parameters which need to be defined and frequently changed during these periods as well as after the installation are the following:

1. Test names with codes, units, prices, tube type for blood drawing lists, and normal ranges.
2. English statements used on reports.
3. Composition of worksheets and laboratory sections.
4. Format for cumulative reports and order of printing for laboratory tests on such reports.

5. Instrument protocol as to the number of standards, positions of controls, limits of acceptability for controls, frequency of automated on-line reports, and so forth.
6. Design of (mark-sense) requisitions or data entry cards.

FUTURE USE OF COMPUTERS

All applications described in this chapter have dealt essentially with passive use of the computer. The computer is one giant clerk capable of accepting requests, collecting the data, and generating many types of reports. The computer has not actively participated in analytical procedures or in the activities on the nursing stations. This is a natural application for initial computer installations because the computer must be justified by the benefits it provides. However, the computer can play a far more active role in the laboratory in the future.

Small computers, for example, are now being used in laboratory instruments to actually control the analytical process. By placing sample aspiration, reagent pipetting, and measuring devices under direct control of the computer, research is now possible to develop methods for the computer to make decisions instantaneously and carry out procedures which are only impractically performed on a manual basis.

It is also being speculated that the interpretation of laboratory data by a computer could ultimately lead to the diagnosis of disease by a computer. It should be pointed out, however, that clinical laboratory data are only a portion of the total data base required for making a rational diagnosis. This suggests that the computer must have access to many and various sources of data.

Many hospitals and companies are in the process of developing total information systems where computers will control the information flow throughout the entire hospital. Such information systems can be interfaced with the laboratory computer, so that laboratory data can be automatically transmitted to the central communications system. Such development is certainly technically feasible, and routine implementation can be expected in the very near future. It is unlikely, however, that large central computers will replace smaller laboratory computers. The performance of functions unique to the laboratory and the integration of laboratory instrumentation with small computers make the presence of the latter in the laboratory highly desirable, if not essential.

As a result of new and significant developments in electronics, the price of the computer hardware has decreased steadily. Thus, it becomes apparent that the most limiting and most costly aspect of a computer system is the software. In order to reduce the high cost of individualized programs, and to avoid staffing each laboratory with its own programing experts, new approaches in the development of more flexible software are also necessary and can be anticipated.

REFERENCES

1. Ball, M. J.: Selecting a Computer System for the Clinical Laboratory. Springfield, Illinois, Charles C Thomas, Publisher, 1971.
2. Krieg, A. F., Johnson, T. J., Jr., McDonald, C., and Cotlove, E: Clinical Laboratory Computerization. Baltimore, University Park Press, 1971.
3. Krieg, A. F.: Clinical Laboratory Computerization. *In* Todd-Sanford Clinical Diagnosis by Laboratory Methods. I. Davidsohn and J. B. Henry, Eds. 15th ed. Philadelphia, W. B. Saunders Company, 1974, pp. 1340–1358.
4. Johnson, J. Lloyd, and Assoc.: Clinical Laboratory Computer Systems, A Comprehensive Evaluation. Northbrook, Illinois, J. Lloyd Johnson Assoc., 1975.
5. Thiers, R. E.: Automation and On-Line Computers. *In* Clinical Chemistry, Principles and Technics. J. R. Henry, D. C. Cannon and J. M. Winkleman, Eds. 2nd ed. New York, Harper & Row, Publishers, 1974, pp. 209–282.

CHAPTER 6

CARBOHYDRATES

by Wendell T. Caraway, Ph.D.

Carbohydrates are the major food supply and energy source for the people of the world. Depending on dietary habits, 50 to 90 per cent of the carbohydrates consumed come from grain, starchy vegetables, and legumes. Typical items in this group include rice, wheat, corn, and potatoes. Other important sources are sucrose (cane and beet sugar, molasses), lactose (milk and milk products), glucose (fruits, honey, corn syrup), and fructose (fruits, honey). Meat products and sea food contain less than 1 per cent glycogen and do not contribute appreciably to the total carbohydrate intake. Fresh liver contains approximately 6 per cent glycogen, but this content decreases rapidly to less than 1 per cent on standing at room temperature. Cellulose occurs in stalks and leaves of vegetables but this carbohydrate is not digested by humans. Although a small portion may be split to glucose by bacterial action in the large bowel, this contribution is relatively insignificant.

Despite the major utilization of carbohydrate for energy, only a small amount is stored in the body. The average adult reserve is about 370 g, stored chiefly as liver and muscle glycogen. Since 1 g of carbohydrate supplies 4 Calories, the total body store of carbohydrate would provide only 1480 Calories, or approximately half the average daily caloric needs. When total caloric intake exceeds the daily expenditure, the excess carbohydrate is readily converted to fat and stored as adipose tissue. This results in increased efficiency of food storage, since 1 g of fat yields 9 Calories, i.e., more than twice that obtained from 1 g of carbohydrate.

The most common disease related to carbohydrate metabolism is *diabetes mellitus*, which is characterized by insufficient blood levels of active insulin. Deficiency of insulin results in inability of glucose to enter muscle and liver cells, with a corollary hyperglucosemia, impaired metabolism of proteins and fats, and secondary changes in fat metabolism, leading eventually to ketosis and possible diabetic coma (see Chap. 10). Other long-term complications include hypercholesterolemia, atherosclerosis, and kidney disease. There are approximately two million recognized diabetics in the United States, and probably two million more as yet undiagnosed cases. The overall incidence, therefore, is nearly 2 per cent of the entire population. The condition shows a strong familial tendency; the probability that an individual will develop diabetes is several times greater when there is a family history of the disease.

Early recognition of diabetes will permit earlier management and perhaps delay or minimize the complications of the disease. Overproduction or excess administration of insulin causes a decrease in blood glucose to levels below normal. In severe cases the resulting extreme *hypoglycemia* is followed by muscular spasm and loss of consciousness, known as *insulin shock*. Measurements of blood glucose, therefore, assume considerable significance, and the clinical laboratory must be prepared to furnish results rapidly and with a high degree of reliability.

The determination of blood glucose is the procedure most frequently performed in the hospital chemistry laboratory. In addition, many glucose determinations are performed in clinics, independent laboratories, and physicians' offices as an aid in the diagnosis and treatment of diabetes. The determination may be performed on patients in the fasting state, in the postprandial state, or in conjunction with glucose tolerance tests, in accordance with the physician's request. As we shall see later, various factors other than insulin affect the blood glucose level, and it is a mistake to assume that an elevated level is diagnostic for diabetes in the absence of other supporting information or confirmatory tests.

CHEMISTRY OF CARBOHYDRATES

The term carbohydrate refers to hydrates of carbon and is derived from the observation that the empirical formulas for these compounds contain approximately one molecule of water per carbon atom. Thus glucose, $C_6H_{12}O_6$, and lactose, $C_{12}H_{22}O_{11}$, can be written as $C_6(H_2O)_6$ and $C_{12}(H_2O)_{11}$, respectively. These compounds are not hydrates in the usual chemical sense, however, and noncarbohydrate compounds such as lactic acid, $CH_3CH(OH)COOH$ or $C_3(H_2O)_3$, can have similar empirical formulas. In more descriptive terminology, the carbohydrates are defined as the aldehyde and ketone derivatives of polyhydric alcohols. The simplest carbohydrate is glycol aldehyde, the aldehyde derivative of ethylene glycol. The aldehyde and ketone derivatives of glycerol are, respectively glyceraldehyde (glycerose) and dihydroxyacetone (Fig. 6–1).

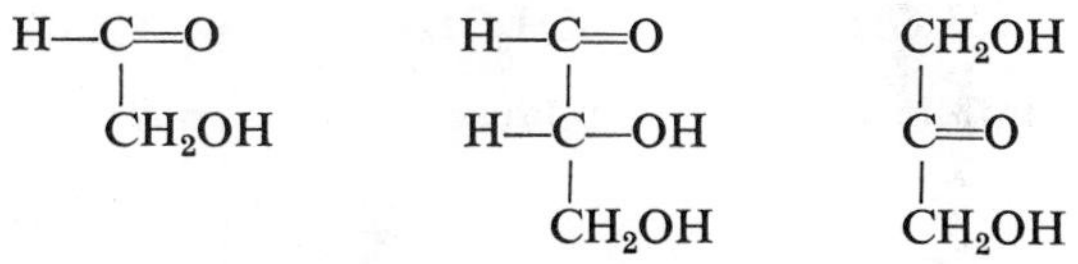

Glycol aldehyde	Glyceraldehyde (glycerose)	Dihydroxyacetone
H—C=O	H—C=O	CH_2OH
CH_2OH	H—C—OH	C=O
	CH_2OH	CH_2OH

Figure 6–1 2- and 3-carbon carbohydrates.

Monosaccharides

Sugars containing three, four, five, and six or more carbon atoms are known, respectively, as trioses, tetroses, pentoses, hexoses, and so on, and are classified as monosaccharides. Aldehyde derivatives are called aldoses and ketone derivatives are called ketoses. Typical examples are the six-carbon sugars glucose (an aldose) and fructose (a ketose), as shown in Figure 6–2.

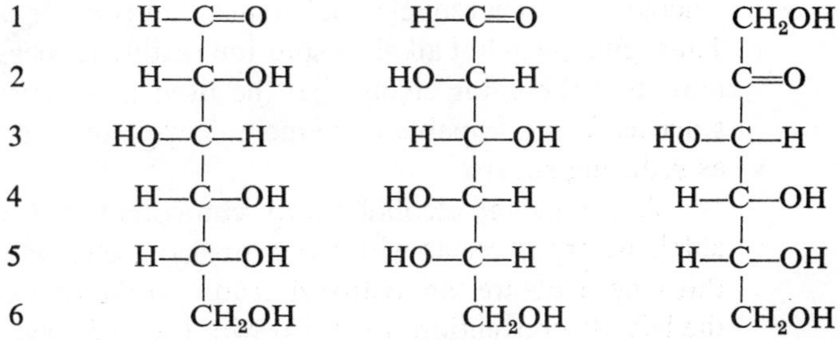

	D-Glucose	L-Glucose	D-Fructose
1	H—C=O	H—C=O	CH_2OH
2	H—C—OH	HO—C—H	C=O
3	HO—C—H	H—C—OH	HO—C—H
4	H—C—OH	HO—C—H	H—C—OH
5	H—C—OH	HO—C—H	H—C—OH
6	CH_2OH	CH_2OH	CH_2OH

Figure 6–2 Typical 6-carbon sugars.

The carbon atoms in the chain are numbered 1 to 6 as shown by the numbers at the left of the formula for D-glucose. The designation D- or L- refers to the position of the hydroxyl group on the carbon atom next to the last (bottom)—CH_2OH group. By convention, the D-sugars are written with the hydroxyl group on the right and the L- sugars are written with the hydroxyl group on the left. Compounds that are identical in composition and differ only in spatial configuration are called stereoisomers. The majority of the sugars occurring in the body are of the D-configuration. A number of different structures exist,

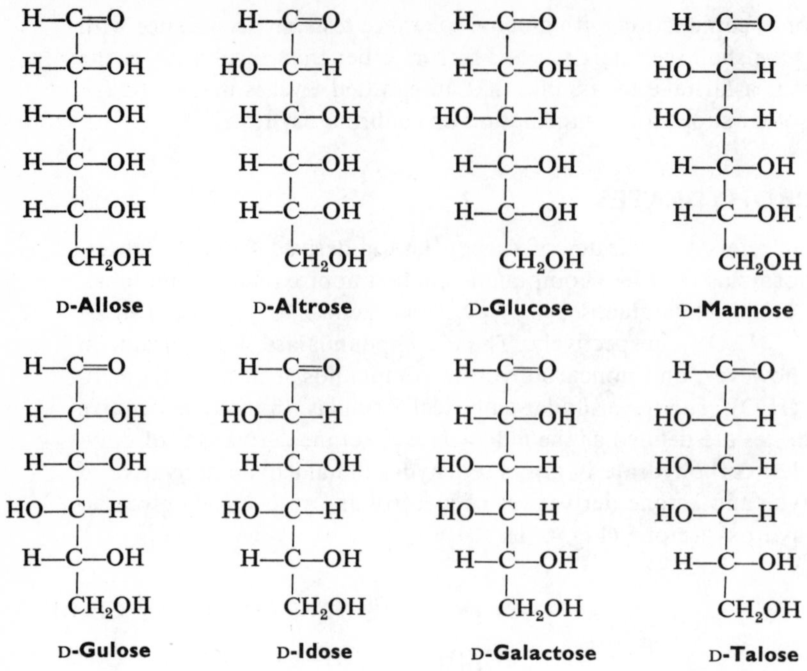

D-Allose	D-Altrose	D-Glucose	D-Mannose
H—C=O	H—C=O	H—C=O	H—C=O
H—C—OH	HO—C—H	H—C—OH	HO—C—H
H—C—OH	H—C—OH	HO—C—H	HO—C—H
H—C—OH	H—C—OH	H—C—OH	H—C—OH
H—C—OH	H—C—OH	H—C—OH	H—C—OH
CH_2OH	CH_2OH	CH_2OH	CH_2OH

D-Gulose	D-Idose	D-Galactose	D-Talose
H—C=O	H—C=O	H—C=O	H—C=O
H—C—OH	HO—C—H	H—C—OH	HO—C—H
H—C—OH	H—C—OH	HO—C—H	HO—C—H
HO—C—H	HO—C—H	HO—C—H	HO—C—H
H—C—OH	H—C—OH	H—C—OH	H—C—OH
CH_2OH	CH_2OH	CH_2OH	CH_2OH

Figure 6–3 The D-hexose sugar series.

depending on the relative positions of the hydroxyl groups on the carbon atoms. The D-hexose series is shown in Figure 6–3.

The formula for glucose can be written in either aldehyde or enol form. Shift to the latter structure is favored in alkaline solution:

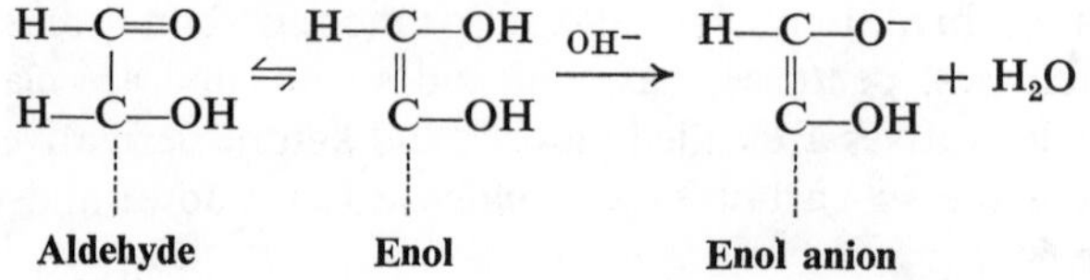

$$\begin{array}{c}\text{H—C=O}\\ |\\ \text{H—C—OH}\\ \vdots\\ \textbf{Aldehyde}\end{array} \rightleftharpoons \begin{array}{c}\text{H—C—OH}\\ \|\\ \text{C—OH}\\ \vdots\\ \textbf{Enol}\end{array} \xrightarrow{OH^-} \begin{array}{c}\text{H—C—O}^-\\ \|\\ \text{C—OH}\\ \vdots\\ \textbf{Enol anion}\end{array} + H_2O$$

The presence of a double bond and a negative charge in the enol anion form make glucose an active reducing substance and provide a basis for its analytical determination. Thus, glucose in hot alkaline solution readily reduces metallic ions such as cupric to cuprous ions, and the color change can be used as a presumptive indication for the presence of glucose. Sugars capable of reducing cupric ions in alkaline solution are commonly known as reducing sugars.

Aldehyde and alcohol groups can react to form hemiacetals. In the case of glucose the aldehyde group reacts with the hydroxyl group on carbon 5 as shown in Figure 6–4. With this ring structure the hydroxyl group on the first carbon can be written to the right or to the left. By convention, the form with the hydroxyl group on the right is called α-D-glucose and the form with the hydroxyl group on the left is called β-D-glucose. The common anhydrous crystalline glucose is in the α-D-form. The β-D-form is obtained by crystallization from acetic acid. The two forms differ with respect to optical rotation of polarized light. The specific rotation, $[\alpha]_D^{25}$, for the α-D-form is +113° and for the β-D-form it is +19.7°. Either form in aqueous solution gives rise to an equilibrium mixture that has a specific rotation of +52.5° (mutarotation). The equilibrium established at room temperature is such that about 36 per cent of the glucose exists in the α-form and 64 per cent in the β-form; only a trace remains in the free aldehyde form. The enzyme glucose oxidase reacts only with β-D-glucose. For this reason, standard solutions to be used in glucose oxidase methods for

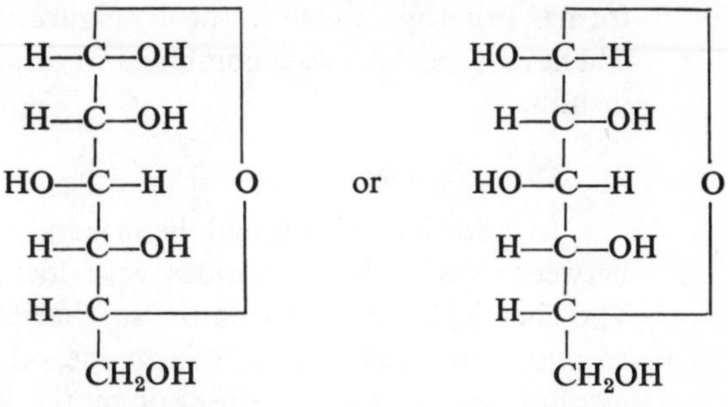

α-D-**Glucose** or β-D-**Glucose**

Figure 6–4 Structure of D-glucose (hemiacetal form). Note that carbon 1 is asymmetric.

glucose determinations should be permitted to stand at least 2 hours in order to obtain equilibrium comparable to that in the test samples to be analyzed.

From the ring structures shown in Figure 6–4, it is not apparent why the aldehyde group should react with the distant hydroxyl group on carbon 5. The spatial arrangement of the atoms is better represented by a symmetrical ring structure, depicted by the Haworth formula, in which glucose is considered as having the same basic structure as pyran (Fig. 6–5). In this formula the plane of the ring is considered as perpendicular to the plane of the paper, with the heavy lines pointing toward the reader. Hydroxyl groups in position 1 are then below the plane (α configuration) or above the plane (β configuration). A six-membered ring sugar, containing five carbons and one oxygen, is a derivative of pyran and is called a pyranose. When linkage occurs with formation of a five-membered ring, containing four carbons and one oxygen, the sugar has the same basic structure as furan and is called a furanose. Representative formulas are shown in Figure 6–5. Fructose is shown in two cyclic

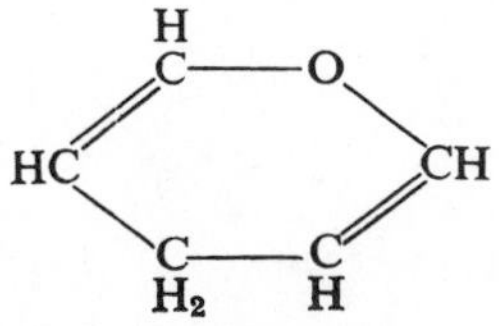

Pyran

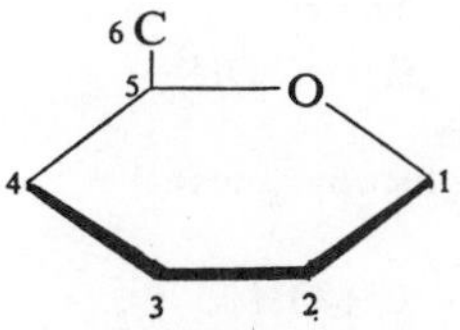

Pyranose structure

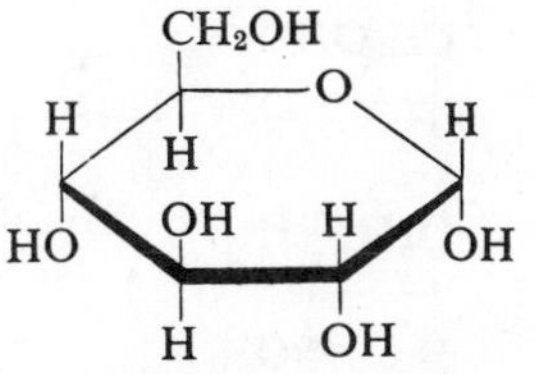

α-D-**Glucose**
α-D-**Glucopyranose**

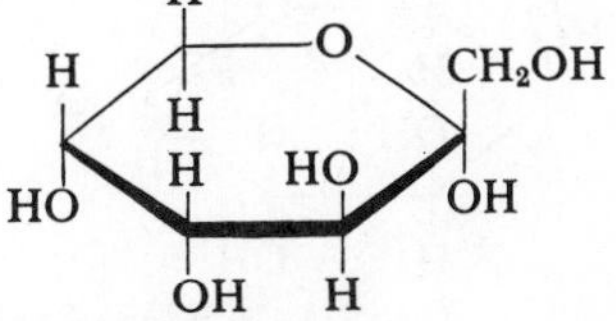

α-D-**Fructose**
α-D-**Fructopyranose**

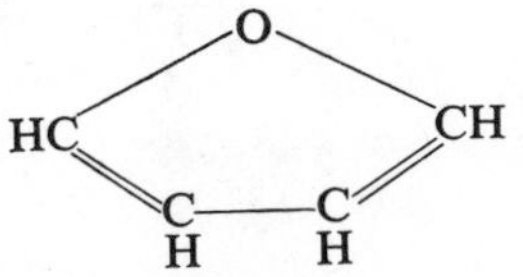

Furan

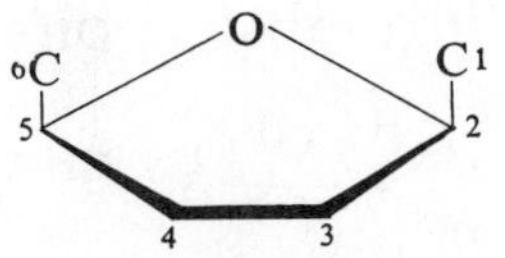

Furanose structure

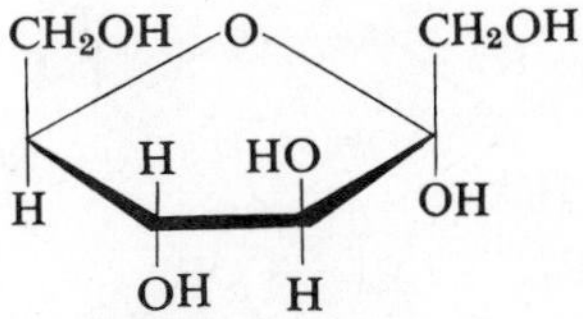

α-D-**Fructose**
α-D-**Fructofuranose**

Figure 6–5 The Haworth formula for sugars.

forms. Fructopyranose is the configuration of the free sugar and fructofuranose occurs whenever fructose exists combined in disaccharides and polysaccharides, as in sucrose and insulin.

Disaccharides

In addition to internal hemiacetal formation, we may have interaction of groups between two monosaccharides with loss of a molecule of water to form disaccharides. The chemical bond between the saccharides always involves the aldehyde or ketone group of one monosaccharide joined to the other monosaccharide either by one of the latter's alcohol groups (e.g., maltose) or by the latter's aldehyde or ketone group (e.g., sucrose). The linkage of an —OH group on C-1 of sugars (C-2 in the case of fructose) with an alcohol is called a glycosidic linkage. The most common disaccharides are

Maltose (glucose + glucose)
Lactose (glucose + galactose)
Sucrose (glucose + fructose)

Structural formulas and chemical names are shown in Figure 6–6.

If the linkage between the two monosaccharides is between the aldehyde or ketone group of one molecule and the hydroxyl group of another molecule (as in maltose and lactose), there will remain one potentially free ketone or aldehyde group on the second monosaccharide. Consequently, the sugar will be a reducing sugar. The reducing power, however, is only approximately 40 per cent of the reducing power of the two single monosaccharides together, since one of the reducing groups is not available. If the linkage between the two monosaccharides involves the aldehyde or ketone groups of both molecules (as in sucrose), a nonreducing sugar results since there is no remaining free aldehyde or ketone group.

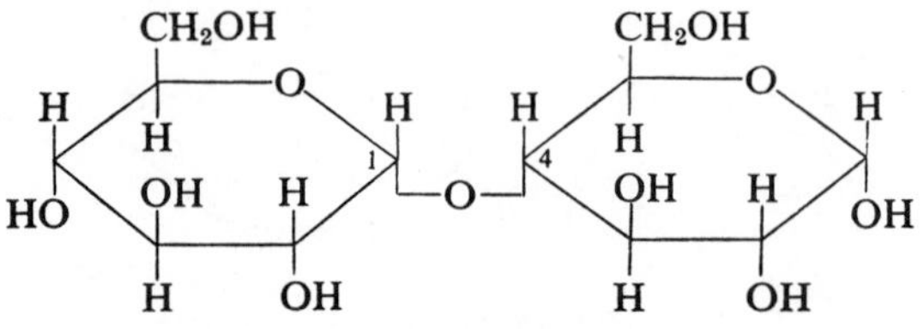

α-D-Maltose
α-D-glucopyranosyl-4-α-D-glucopyranose

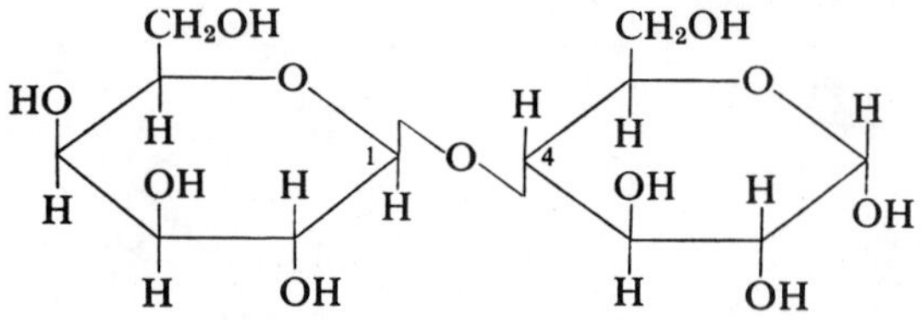

α-D-Lactose
β-D-galactopyranosyl-4-α-D-glucopyranose

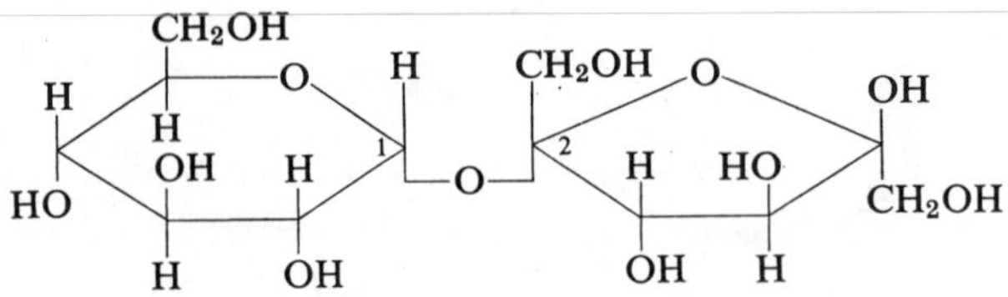

Sucrose
α-D-glucopyranosyl-β-D-fructofuranose

Figure 6–6 Structural formulas of disaccharides.

Polysaccharides

The linkage of many monosaccharide units together results in the formation of polysaccharides. In starch and glycogen, the chief reserve carbohydrates of plants and animals, respectively, one molecule may contain from 25 to 2500 glucose units. The suffix *-an* attached to a name of a monosaccharide indicates the main type of sugar present in the polysaccharide. Starch and glycogen, for example, are glucosans, since they are composed of a series of individual glucose molecules. Inulin, found in the tubers of the dahlia and the Jerusalem artichoke, is a polysaccharide consisting largely of fructose units and is known as a fructosan.

Nearly all starches are composed of a mixture of two kinds of glucosans called amyloses and amylopectins. The relative proportions of these two glucosans in a starch vary from approximately 20 per cent amylose and 80 per cent amylopectin in wheat, potato, and ordinary corn starch, to nearly 100 per cent amylopectin in the starch of waxy corn. On the other hand, a few corn starches are known that contain up to 75 per cent amylose. Iodine gives a typical deep blue color with amylose and a red to violet color with amylopectin. These characteristic colors disappear when the respective glucosans are hydrolyzed to smaller units such as dextrins and maltose. This disappearance of the starch-iodine color is utilized in some quantitative methods for amylase determination.

Although both amylose and amylopectin are made up of glucose molecules, there is one significant difference in their structure. Amylose, which has a molecular mass from 4000 to about 50,000, consists of one long unbranched chain of 25 to 300 units of glucose. These units are linked together by a 1,4-linkage with only the terminal aldehyde group free. In amylopectin the majority of the units are also connected with 1-4 α links, but, in

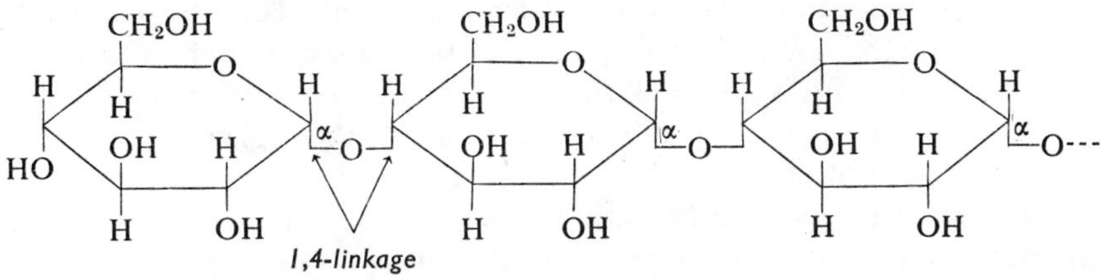

1,4-linkage

Chain of glucose molecules linked by 1,4-linkages as found in amylose

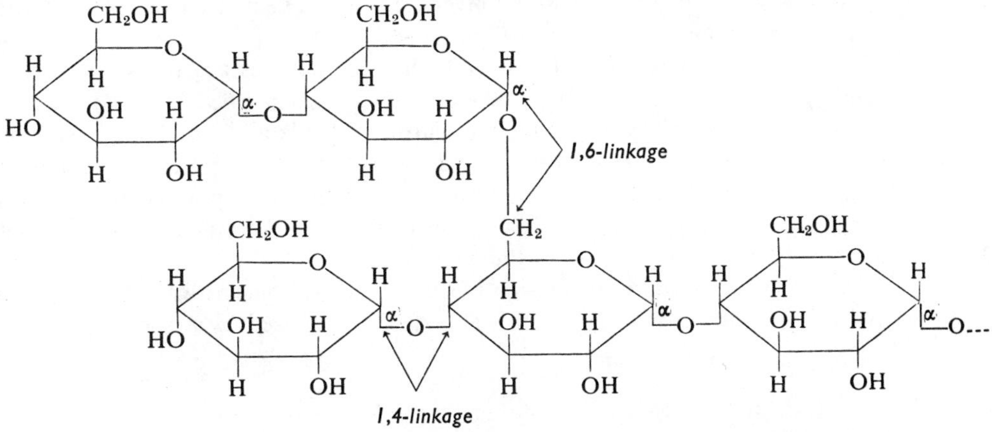

1,6-linkage

1,4-linkage

Chain of glucose molecules linked by 1,4-linkages and side chains linked by 1,6-linkage as found in amylopectin

Figure 6–7 Structures for amylose and amylopectin.

addition, there are 1,6-α-glycosidic bonds (amounting to about 4 per cent of the total) that form side chains. The structure is a branch-on-branch arrangement of 1000 or more D-glucopyranose units with a molecular mass for amylopectin ranging from 50,000 to about 1,000,000. Glycogens have structures similar to amylopectins except that branching is more extensive. The average length of a branch in a glycogen molecule is usually 12 or 18 D-glucopyranose units, compared to about 25 units in amylopectin. Examples of polysaccharide linkages are shown in Figure 6–7.

The difference in structure between amylose and amylopectin becomes important in the proper selection of the starch substrate for amylase determination (see p. 626). Any differences in the structure of starch will affect the rate of hydrolysis. The so-called α-amylase of pancreatic origin hydrolyzes the 1,4-glucoside linkage with special preference for the more central internal linkages. This results initially in the production of some maltose and a mixture of dextrins, which are subsequently also hydrolyzed to maltose. The 1,6-glucoside linkages are not attacked by α-amylase, and relatively large molecules of so-called residual dextrins are left after the action of the enzyme on amylopectin.

Dextrins are the products of partial hydrolysis of starch. They are a complex mixture of molecules of different sizes. Those formed from amylose are unbranched chains, and those formed from amylopectins are branched chains of glucose molecules. Erythrodextrins are larger, branched dextrins that produce a reddish color with iodine.

METABOLISM OF CARBOHYDRATES

Starch and glycogen ingested as food are partially digested by the action of salivary amylase to form intermediate dextrins and maltose. Amylase activity is inhibited at the acid pH of the stomach. In the small intestine the pH is increased by alkaline pancreatic juice, and the amlyase of the pancreas effects digestion of starch and glycogen to maltose. The latter, along with any ingested lactose and sucrose is split by the disaccharidases (maltase, lactase, and sucrase) in the intestinal mucosa to form the monosaccharides glucose, galactose, and fructose.

Absorption of these monosaccharides is fairly complete and appears to occur by an active enzymatic transfer process. This is inferred because the rate of absorption for glucose and galactose is several times greater than that for xylose, which is thought to be absorbed by passive diffusion. Some conversion of fructose to glucose apparently occurs during the process of absorption, and the interconversion can be visualized in terms of the enediol form common to both, as shown in Figure 6–8.

Following absorption into the portal vein, the hexoses are transported to the liver. Depending on the needs of the body, the carbohydrates may be converted to and stored as liver glycogen; metabolized completely to carbon dioxide and water to provide immediate energy; converted to keto acids, amino acids, and protein; or converted to fat and stored as adipose tissue. The complete picture of intermediary metabolism of carbohydrates is rather complex and interwoven with the metabolism of lipids and amino acids. For details, the reader should consult textbooks of biochemistry.

Factors involved in the regulation of blood glucose concentration are illustrated in the partial outline shown in Figure 6–9. Each step is catalyzed by a specific enzyme and, in

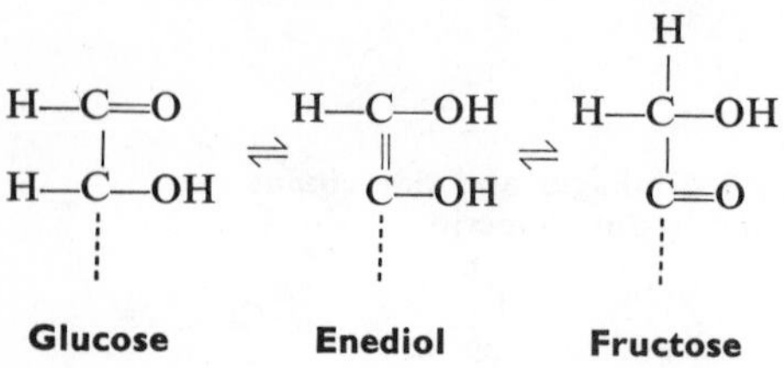

```
                                      H
                                      |
H—C=O          H—C—OH           H—C—OH
  |       ⇌       ||        ⇌       |
H—C—OH           C—OH              C=O
  :               :                 :
  :               :                 :
Glucose         Enediol          Fructose
```

Figure 6–8 Interconversion of glucose and fructose.

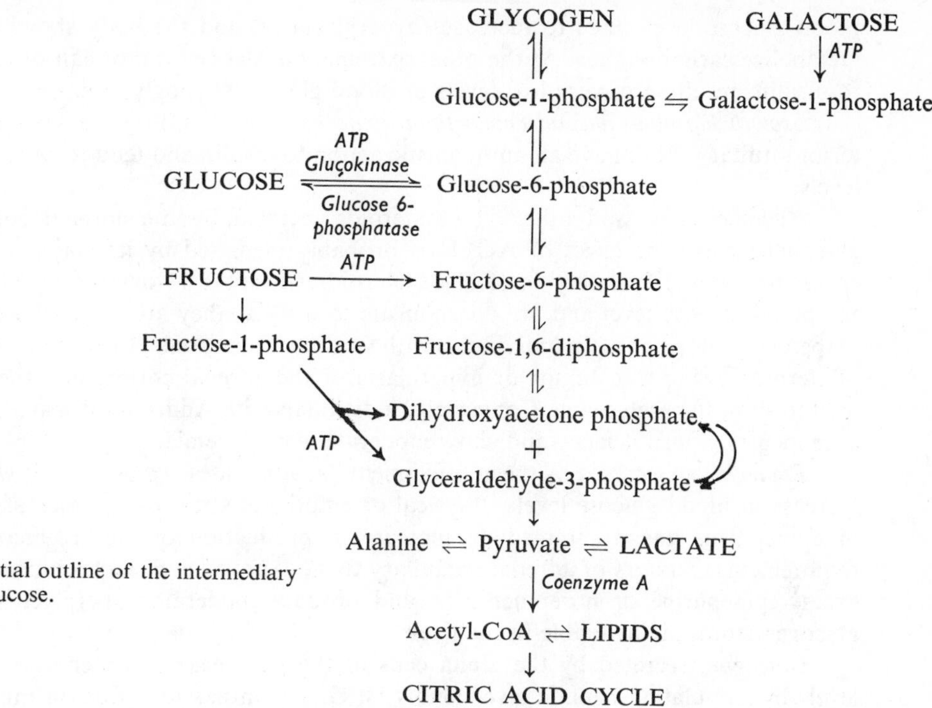

Figure 6–9 Partial outline of the intermediary metabolism of glucose.

some cases, different enzymes may be responsible for a given step, depending on which way the reaction proceeds. For example, the initial phosphorylation of glucose is catalyzed by glucokinase, but the reverse reaction depends upon glucose-6-phosphatase. Fructose and galactose are phosphorylated and eventually enter the same metabolic pathway as glucose.

Various terms are applied to describe general processes in carbohydrate metabolism. *Glycogenesis* refers to the conversion of glucose to glycogen, and *glycogenolysis* refers to the breakdown of glycogen to form glucose and other intermediate products. The formation of glucose from noncarbohydrate sources, such as amino acids, glycerol, or fatty acids, is *gluconeogenesis*. The conversion of glucose or other hexoses into lactate or pyruvate is called *glycolysis*. The net result of factors affecting these various processes determines the level of glucose in the blood.

Regulation of blood glucose concentration

In the fasting state the level of blood glucose is maintained by drawing upon the glycogen stores of the liver, and a slight amount may also be derived from the kidney. Both of these organs contain the specific enzyme, glucose-6-phosphatase, necessary for the conversion of glucose-6-phosphate to glucose. Skeletal muscle, although it stores glycogen, is lacking in this enzyme and cannot directly contribute glucose to the blood. As blood glucose levels increase, usually by absorption of carbohydrates from the intestine, glycogenolysis is replaced by glycogenesis, whereby excess blood glucose is converted into liver and muşcle glycogen.

A number of hormones are important in the regulation of blood glucose concentration. *Insulin*, produced by the beta cells of the islets of Langerhans in the pancreas, promotes glycogenesis and lipogenesis (formation of fat from carbohydrate), with a resultant decrease in blood glucose levels. Another important action of insulin is to increase the permeability of cells to glucose. With deficiency in effective insulin (diabetes), the fasting blood glucose level tends to increase (hyperglycemia) and the body shows less ability to metabolize carbohydrates. At the other extreme, an islet cell tumor can produce an excess of insulin, resulting in very low levels of blood glucose (hypoglycemia).

Growth hormone and adrenocorticotrophic hormone (ACTH) are secreted by the anterior pituitary. Both have an antagonistic action to insulin and tend to raise blood glucose levels.

Hydrocortisone and other 11-oxysteroids secreted by the adrenal cortex stimulate gluconeogensis. The effect of ACTH is probably mediated by its action on the adrenal cortex to stimulate production of 11-oxysteroids. Since the 11-oxysteroids tend to increase the blood glucose level and are antagonistic to insulin, they are sometimes referred to as diabetogenic hormones. In individuals with Cushing's syndrome, there is an overproduction of steroids owing to a tumor or hyperplasia of the adrenal cortex, and these individuals tend to show hyperglycemia. Conversely, individuals with Addison's disease have a primary adrenocortical insufficiency and show moderate hypoglycemia.

Epinephrine, secreted by the adrenal medulla, stimulates glycogenolysis with a resultant increase in blood glucose levels. Physical or emotional stress causes increased production of epinephrine and an immediate increase in production of blood glucose for energy requirements. Tumors of adrenal medullary tissue, known as pheochromocytomas, secrete excess epinephrine or norepinephrine and produce moderate hyperglycemia as long as glycogen stores are available in the liver.

Glucagon, secreted by the alpha cells of the pancreas, also increases blood glucose levels by stimulating hepatic glycogenolysis. Glucagon has no effect on muscle glycogen, as shown by the lack of elevation of blood lactate and pyruvate levels following its administration.

Thyroxine, secreted by the thyroid, also appears to stimulate glycogenolysis. Thyrotoxic individuals may show symptoms of mild diabetes and almost complete absence of liver glycogen. Thyroxine also increases the rate of absorption of glucose from the intestine.

METHODS FOR THE DETERMINATION OF GLUCOSE IN BODY FLUIDS

The current status of methodology for the determination of glucose has been reviewed by Cooper[3] and by Martinek.[9]

Normal values

The normal glucose concentration in human plasma or serum, determined by highly specific methods, ranges from 70 to 105 mg/100 ml. The normal true glucose concentration in whole blood is somewhat less, ranging from 65 to 95 mg/100 ml.

In the fasting state, the arterial and capillary blood glucose concentration is approximately 5 mg/100 ml higher than the venous blood glucose level.

The normal cerebrospinal fluid glucose concentration is about 60 to 70 per cent of the plasma level and ranges from 40 to 70 mg/100 ml.

Specimen

In the past, glucose determinations were usually performed on whole blood, often with relatively nonspecific methods which provided higher ranges of normal values. As we shall see, there are several good reasons for measuring glucose in plasma or serum rather than in whole blood.

Glucose is uniformly distributed in the water phase of the body. If we know the average water content of erythrocytes and plasma, we should be able to calculate the water content of a specimen of whole blood with a given hematocrit. Erythrocytes contain about 73 ml of water per 100 ml of cells; plasma contains about 93 ml of water per 100 ml of plasma. Assuming a hematocrit of 45 per cent, the water content of whole blood would be (0.45)(73) + (0.55)(93) = 84 ml of water per 100 ml of blood. If glucose were uniformly distributed in the water phase, then the ratio of plasma glucose to whole blood glucose would be 93/84 or 1.11. This agrees with the work of Tustison *et al.*,[12] who developed regression equations relating glucose concentrations in plasma versus that in whole blood. Other workers have reported widely varying ratios. The figures suggest that the method used to estimate glucose can have considerable influence on the apparent difference between the plasma glucose and the whole blood glucose concentration, and that much of the variation can be attributed to interfering substances in the red cells. Time of separation of plasma, type of preservative, method of precipitation of proteins, and variations in hematocrit are also factors. As an approximation, plasma glucose can be considered to be 10 to 15 per cent higher than corresponding whole blood glucose concentrations.

The following reasons are given to support the use of plasma (or serum), rather than whole blood, for the determination of glucose:

1. Whole blood is a 2-component system; interpretations are less hazardous when applied to a single-component system, such as plasma.
2. Whole blood must be mixed thoroughly before sampling. This is especially inconvenient with automated analyses.
3. Specificity for glucose is improved with most methods when plasma is used for analysis. Results of determinations of glucose in whole blood show greater variation with methodology and methods of protein precipitation than do results of determinations on plasma. In some methods, no deproteinization is required for plasma samples.
4. Values on whole blood tend to vary with the hematocrit. Again, this varies with the method of analysis.
5. Glucose is more stable in separated plasma than in whole blood.
6. Plasma is generally easier to handle and store than whole blood.

Probably the one major advantage in using whole blood is the convenience of measuring glucose directly on capillary blood from infants. In such cases, appropriate methodology will provide accurate and meaningful results. Capillary blood is essentially arterial blood. In the fasting state, capillary blood glucose concentration is approximately 5 mg/100 ml higher than that in venous blood. When capillary blood is used for a glucose tolerance test, however, levels may be 20 to 70 mg/100 ml higher than those in venous blood. For proper interpretation, either capillary or venous blood must be used consistently throughout the test, and a notation should be made if capillary blood is used so that results may be interpreted properly.

Erythrocytes contain sulfhydryl compounds, glutathione and ergothioneine, which act as reducing substances. These and other materials, such as uric acid and creatinine, interfere in procedures for glucose based upon its reducing action in alkaline solution unless they are removed before analysis. These nonglucose reducing substances, called saccharoids, increase the apparent glucose values by about 10 to 30 mg/100 ml when whole blood is analyzed by the Folin–Wu copper reduction method. The resulting normal range, which includes saccharoids, becomes 80 to 120 mg/100 ml and is reported as blood sugar, although total reducing substances is the more appropriate term. Some copper-reduction methods (e.g., Somogyi–Nelson) eliminate most saccharoids by suitable precipitation techniques and provide values in good agreement with highly specific enzymatic procedures.

Stability of glucose in body fluids

When *blood* is drawn, permitted to clot, and to stand uncentrifuged at room temperature, the average rate of decrease in serum glucose is approximately 7 per cent in 1 hour.[13] In separated unhemolyzed *serum* the glucose concentration is generally stable up to 8 hours at 25°C and up to 72 hours at 4°C. Variable stability, related to bacterial contamination, is observed for longer storage periods. Plasma, removed from the cells after moderate centrifugation, contains leukocytes that also metabolize glucose. Cell-free plasma shows no glycolytic activity. From these data it follows that whole blood glucose determinations should be performed promptly after collection of the specimen. Plasma or serum from blood without preservative must be separated from the cells or clot within a half-hour after the blood is drawn if glucose values within 10 mg/100 ml of the original value are to be obtained consistently.

Glycolysis can be prevented and glucose stabilized up to 24 hours at room temperature by adding sodium fluoride to the specimen. Fluoride ion also inhibits coagulation by precipitating calcium; however, clotting may occur after several hours, and it is advisable to use a combined fluoride-oxalate mixture, such as 2 mg of potassium oxalate and 2.5 mg of sodium fluoride per ml of blood. Either the whole blood or plasma may be analyzed, but plasma is preferred. Fluoride ion in high concentration inhibits the activity of urease and certain other enzymes; consequently, the specimens are unsuitable not only for determination of urea in some procedures that require urease, but also for direct assay of some serum enzymes.

Cerebrospinal fluids are frequently contaminated with bacteria or other cellular constituents and should be analyzed for glucose without delay. Glucose may be preserved in 24 hour collections of urine by adding 5 ml of glacial acetic acid to the container before starting the collection. The final pH is usually between 4 and 5 and bacterial activity is inhibited at this level of acidity. The use of 5 g of sodium benzoate per 24 hour specimen is also effective in preserving the urine.

ALKALINE FERRICYANIDE METHODS

PRINCIPLE

In hot alkaline solutions ferricyanide ion (yellow) is reduced by glucose to ferrocyanide ion (colorless):

$$\underset{\substack{\textbf{Ferricyanide}\\ \textbf{(Ferric iron)}}}{Fe^{III}(CN)_6^{3-}} \xrightarrow[\text{(glucose)}]{+e^-} \underset{\substack{\textbf{Ferrocyanide}\\ \textbf{(Ferrous iron)}}}{Fe^{II}(CN)_6^{4-}}$$

The decrease in color of ferricyanide ion, measured at 420 nm, is proportional to the glucose concentration (inverse colorimetry). The reagent blank, without glucose, has the greatest absorbance, and measurements in the low or normal range are inherently less accurate since readings are based on a small difference between large absorbance values. More precision is obtained by automation, and the alkaline ferricyanide method is currently in wide use where glucose determinations are performed on the AutoAnalyzer (see Chap. 4). Other reducing substances interfere; e.g., 1 mg of creatinine reacts the same as 1 mg of glucose, and 1 mg of uric acid reacts equivalently to 0.5 mg of glucose. Thus, serum from a patient with uremia, in which both creatinine and uric acid are markedly elevated, would show a falsely elevated value for glucose. High concentrations of creatinine and uric acid render the alkaline ferricyanide method unsuitable for determination of glucose in urine.

COPPER REDUCTION METHODS

PRINCIPLE

In hot alkaline solution, glucose readily reduces cupric ion to cuprous ion with formation of (mainly) cuprous oxide (Cu_2O). The reaction is not stoichiometric, but depends on the alkalinity, the time and temperature of heating, and the concentration of reagents.

Under carefully controlled conditions the reaction is reproducible and provides quantitative results when standards are analyzed in the same manner as protein-free filtrates. Reoxidation of cuprous ion by oxygen from the air is prevented by using a constricted tube to minimize surface area or by incorporating sodium sulfate (18 g/100 ml) in the reagent to decrease the solubility of oxygen. Added phosphomolybdic (or arsenomolybdic) acid (Mo^{6+}) is reduced by the cuprous ion to form compounds with lower oxidation states of molybdenum, which have a blue color and are suitable for photometric measurements.

In the *Folin-Wu procedure*, proteins are precipitated with tungstic acid and the water-clear, protein-free filtrate is used in the reaction. The method lacks specificity, owing to the presence of glutathione and other nonglucose reducing substances that appear in the filtrate, and should now be considered obsolete.

In the *Somogyi-Nelson procedure*, proteins are precipitated by the addition of barium hydroxide and zinc sulfate. Protein is removed as zinc proteinate, sulfhydryl compounds as zinc salts, and the remaining zinc and barium ions as zinc hydroxide and barium sulfate:

$$ZnSO_4 + Ba(OH)_2 \rightarrow Zn(OH)_2\downarrow + BaSO_4\downarrow$$

Uric acid and some creatinine are also precipitated and adsorbed on barium sulfate so that the resultant filtrate is virtually free of nonsugar reducing substances. When the reagents are properly balanced, the filtrate has a pH of approximately 7.4. In spite of its relative specificity, this method is also considered obsolete.

In the *neocuproine method*, used at present in the Technicon SMA 12/60 system, cupric ion is reduced to cuprous ion by glucose in hot alkaline solution. Cuprous ion forms an orange-colored complex with 2,9-dimethyl-1,10-phenanthroline (neocuproine) suitable for photometric measurement. The procedure is highly sensitive but is relatively nonspecific. Major problems encountered are high blanks, increasing baseline values in automated systems, and interference by uric acid and ascorbic acid. Careful control of temperature and heating time are also important, and this has limited the use of the procedure in manual methods. Serum glucose values obtained with the SMA neocuproine system are slightly higher than those with the AutoAnalyzer ferricyanide method.

DETERMINATION OF GLUCOSE USING GLUCOSE OXIDASE

PRINCIPLE

Glucose oxidase is an enzyme that catalyzes the oxidation of glucose to gluconic acid and hydrogen peroxide;

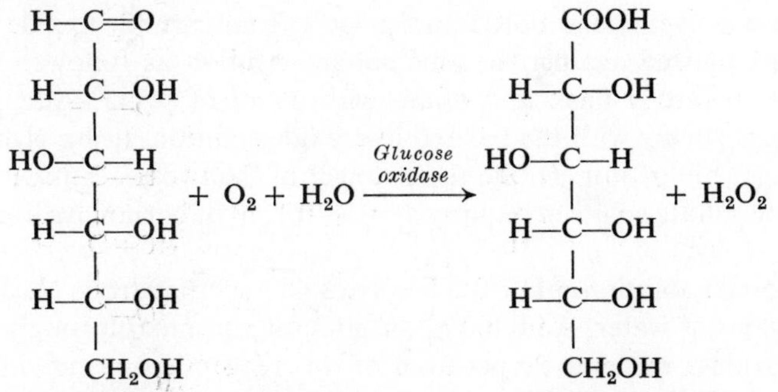

$$\begin{array}{c} \text{H—C=O} \\ | \\ \text{H—C—OH} \\ | \\ \text{HO—C—H} \\ | \\ \text{H—C—OH} \\ | \\ \text{H—C—OH} \\ | \\ CH_2OH \\ \textbf{Glucose} \end{array} + O_2 + H_2O \xrightarrow[\textit{oxidase}]{\textit{Glucose}} \begin{array}{c} \text{COOH} \\ | \\ \text{H—C—OH} \\ | \\ \text{HO—C—H} \\ | \\ \text{H—C—OH} \\ | \\ \text{H—C—OH} \\ | \\ CH_2OH \\ \textbf{Gluconic Acid} \end{array} + H_2O_2$$

Introduction of the enzyme peroxidase and a chromogenic oxygen acceptor, such as *o*-dianisidine, results in formation of color which can be measured:

$$o\text{-Dianisidine} + H_2O_2 \xrightarrow{\textit{Peroxidase}} \underset{\textbf{(Color)}}{\text{Oxidized } o\text{-dianisidine}} + H_2O$$

Glucose oxidase is highly specific for β-D-glucose. As noted earlier, glucose in solution exists as 36 per cent α and 64 per cent β form. Complete oxidation of glucose, therefore, requires mutarotation of α to β form. Some commercial preparations of glucose oxidase contain an enzyme, mutarotase, that accelerates this reaction.

The second step, involving peroxidase, is less specific. Various substances, such as uric acid, ascorbic acid, bilirubin, and glutathione, inhibit the reaction, presumably by competing with the chromogen for hydrogen peroxide. For this reason, results obtained directly on serum tend to be lower than true glucose values. Morin and Prox,[10] however, by decreasing the pH of the reaction mixture to 5.5 and by greatly increasing the concentration of peroxidase, have developed a 2 min procedure for glucose that appears to be relatively free from interferences when applied directly to serum. Most interfering substances are eliminated by use of a Somogyi zinc filtrate. Peroxides may be released in acid filtrates and cause positive errors. Some glucose oxidase preparations may contain catalase as a contaminant that decomposes peroxides and decreases the final color obtained. Standards and unknowns should be analyzed simultaneously under conditions such that the rate of oxidation is proportional to glucose concentration. The reaction is stopped and color developed after a standard incubation period.

In some methods the final mixture is acidified slightly to stop the reaction, and the yellow color is measured at 395 nm. In stronger acid solution the color becomes pink, with maximum absorbance at 540 nm, and both sensitivity and stability are improved. The method to be described can be used to determine glucose in the presence of other sugars.

One instrument, the Beckman Glucose Analyzer, employs a polarographic oxygen electrode which measures the rate of oxygen consumption after addition of the sample to a buffered solution containing glucose oxidase. Since this measurement involves only the first reaction shown above, interferences encountered in the peroxidase step are eliminated. Owing to the binding of oxygen by hemoglobin, the instrument cannot be used for the determination of glucose in whole blood.

The glucose oxidase methods are not directly applicable to urine specimens, owing to the high concentration of enzyme inhibitors present. A method has been described in which the urine is first treated with an ion-exchange resin to remove interfering substances.[8]

REAGENTS

1. Zinc sulfate solution, 5 g/100 ml. Dissolve 50 g of $ZnSO_4 \cdot 7H_2O$ in water and dilute to 1 liter.

2. Barium hydroxide solution, approximately 0.15 mol/l. Dissolve 50 g of $Ba(OH)_2 \cdot 8H_2O$ in water and dilute to 1 liter. Let stand for 2 days in a covered container; then decant or filter. Store in a polyethylene bottle and protect from carbon dioxide of the air. This solution must be balanced against the zinc sulfate solution as follows: Pipet 10.0 ml of zinc sulfate solution into a flask and dilute with 25 ml of water. Add phenolphthalein indicator and titrate slowly with the barium hydroxide solution, using vigorous mixing, to a definite permanent pink color. Dilute the stronger of the two reagents, if necessary, such that 10.0 ml of zinc sulfate solution requires 10.0 $\pm$ 0.1 ml of barium hydroxide solution for neutralization.

3. Glycerol-buffer solution, pH 7.0. Dissolve 3.48 g of anhydrous Na_2HPO_4 and 2.12 g of KH_2PO_4 in 600 ml of water. Add 400 ml of glycerol and mix thoroughly.

4. Glucose oxidase reagent. Preparation of this reagent from individual constituents has been described by Fales. The enzymes and chromogens may be obtained separately from a number of supply houses. A suitable packaged mixture known as Glucostat is available (Worthington Biochemical Corp., Freehold, N.J. 07728). The smaller vial contains the chromogen (10 mg of *o*-dianisidine), and the larger vial contains approximately 125 mg of glucose oxidase and 5 mg of peroxidase. Completely dissolve the soluble chromogen in 1.0 ml of distilled water and drain into an amber bottle. Dissolve the contents of the

larger vial in, and dilute to 100 ml with, glycerol-buffer solution. Add this solution to the amber bottle and mix well. This solution is stable up to at least 1 month when stored in the refrigerator.

5. Sulfuric acid, 3 mol/l. Add slowly, with mixing, 200 ml of concentrated sulfuric acid to 1000 ml of distilled water.

6. Benzoic acid solution, 2 g/l. Dissolve 2 g of benzoic acid in water, with warming, and dilute to 1 liter.

7. Stock standard glucose, 1 g/100 ml. Transfer 1.000 g of dry reagent grade glucose to a 100 ml volumetric flask. Add benzoic acid solution, mix to dissolve, then dilute to the mark with benzoic acid solution. Mix thoroughly and store in a tightly stoppered bottle in the refrigerator. This solution appears to be stable indefinitely.

8. Working standard. Warm a portion of the stock standard to room temperature. Pipet 10.0 ml of stock standard to a 100 ml volumetric flask and dilute to the mark with benzoic acid solution to provide a concentration of 100 mg/100 ml. Additional working standards may be prepared by suitable dilution of the stock standard. Standard solutions prepared from dry glucose should stand at least 2 h to insure that mutarotation has reached a state of equilibrium. Solutions to be administered for glucose tolerance tests should also be prepared at least 2 h before use.

PROCEDURE

1. Pipet 0.5 ml of blood, serum, plasma, or cerebrospinal fluid into a small Erlenmeyer flask. An Ostwald-Folin pipet is convenient for this measurement. Heparin, oxalate, or EDTA are satisfactory anticoagulants. Sodium fluoride (2.5 mg/ml of blood) may be used as preservative, but not thymol, which inhibits the reaction.

2. Add 7.5 ml of water and 1.0 ml of barium hydroxide solution. Mix and let stand at least 30 s.

3. Add 1.0 ml of zinc sulfate solution, mix well, let stand 2 min or more, and filter or centrifuge. This provides a 1:20 protein-free filtrate.

4. Prepare blank and standard in the identical manner as described for the sample, using water and 100 mg/100 ml of glucose standard, respectively, as starting materials.

5. Pipet 0.20 ml of each filtrate into respective test tubes.

6. Add 1.0 ml of glucose oxidase reagent (previously warmed to room temperature) to the tubes, one at a time in timed sequence. Mix each tube immediately and place in a 37°C water bath.

7. After exactly 30 min, remove the tubes one at a time, add 5.0 ml of sulfuric acid, 3 mol/l, and mix. In this manner the incubation time for all tubes is kept constant.

8. After 5 min, but within an hour, measure the absorbance at 540 nm against the blank.

$$\frac{A_{\text{sample}}}{A_{\text{standard}}} \times 100 = \text{mg glucose/100 ml of sample.*}$$

COMMENTS ON THE METHOD

The absorbance is usually linear with glucose concentrations up to 400 mg/100 ml; however, this should be checked with each new lot of glucose oxidase reagent, using standards of 100, 200, 300, and 400 mg of glucose/100 ml. When the sample glucose concentra-

* Because of long-standing usage, results in this text will be expressed in mg/100 ml. In the SI system, however, results are expressed in mmol/liter.

$$\text{mmol/l} = \frac{\text{mg/100 ml} \times 10}{\text{molecular mass}}$$

Glucose has a molecular mass of 180; hence, a serum glucose of 100 mg/100 ml is equivalent to 5.56 mmol/l.

tion is too high for accurate measurement, the filtrate is diluted 2-fold or 4-fold and the analysis repeated.

Rubber tubing, used for dispensing deproteinizing solutions or distilled water, has been found to interfere with adequate color development; therefore, rubber tubing connections on dispensers or automatic sampling and diluting devices should be kept to a minimum and should be suspect if standards yield unexpectedly low color values.

Many of the common drugs likely to be encountered in blood have been tested and found to have no appreciable effect on this procedure. Potential inhibitors or color-producing compounds are apparently either removed with the protein precipitate or are diluted to the point that interference becomes negligible.

Commercially, there is a similar product, Galactostat (Worthington Biochemical Corp., Freehold, N.J. 07728), suitable for the specific determination of galactose in the presence of other sugars.

REFERENCE

Fales, F. W.: *In* Standard Methods of Clinical Chemistry. D. Seligson, Ed. New York, Academic Press, Inc., 1963, vol. 4, p. 101.

HEXOKINASE METHODS

PRINCIPLE

Hexokinase is an enzyme that catalyzes the phosphorylation of glucose by adenosine triphosphate (ATP) to form glucose-6-phosphate and adenosine diphosphate (ADP). To follow the reaction, a second enzyme, glucose-6-phosphate dehydrogenase (G-6-PD) is used to catalyze the oxidation of glucose-6-phosphate by nicotinamide adenine dinucleotide phosphate (NADP) to form NADPH in direct proportion to the amount of glucose originally present. The absorbance of NADPH at 340 nm provides a measure of glucose content.

$$\text{Glucose} + \text{ATP} \xrightarrow{\textit{hexokinase}} \text{Glucose-6-phosphate} + \text{ADP}$$

$$\text{Glucose-6-phosphate} + \text{NADP}^{+} \xrightarrow{\textit{G-6-PD}} \text{6-phosphogluconate} + \text{NADPH} + \text{H}^{+}$$

Since most methods described in the literature rely on commercially prepared reagents supplied in lyophilized form, only a general discussion of the procedure will be presented. Lyophilized enzyme preparations stored in a desiccator at 5°C are stable. After reconstitution, the reagent is stable at 2°C for at least 24 hours but is not usable after 5 hours when kept at 26°C.

A Somogyi-Nelson protein-free filtrate should be prepared for determinations on whole blood. Direct determinations (without deproteinization) are feasible if serum or plasma is used as a specimen. However, specimens containing more than 0.5 g of hemoglobin per 100 ml are unsatisfactory owing to the release from red cells of phosphate esters and enzymes that may react to produce changes in NADP concentration and thus interfere with the measurement. No interference is observed with heparin, fluoride, oxalate, and EDTA anticoagulants at usual concentrations. The reaction is not affected by ascorbic acid or uric acid.

Absorbance is measured after completion of the reaction because reaction rates during the initial stages are fast and variable. Completeness of the reaction is checked by repeatedly reading absorbance near the selected time of measurement. Although glucose concentration may be calculated directly, based on the molar absorptivity of NADPH, a set of standards should always be included to detect possible deterioration of enzymes, ATP, or NADP.

Reagents may also contain substances that react with NADP. This can be evaluated by substituting water for sample in the test procedure.

Direct analyses on serum, performed either manually or by automation, should include serum blanks to correct for background absorption. In one study, the background absorbance was shown to have a mean equivalent to a glucose concentration of 12 mg/100 ml with a range of 8 to 29 mg/100 ml. The DuPont ACA instrument uses a resin column to remove interfering substances prior to analysis. The hexokinase method has also been shown to be an excellent procedure for the determination of glucose in urine. Fructose and mannose do not interfere if the corresponding isomerases which convert their phosphates to glucose-6-phosphate are absent.

Colorimetric hexokinase procedures have also been developed, so that absorbance may be measured in the visible range. An oxidation-reduction system containing phenazine methosulfate (PMS) and a substituted tetrazolium compound, 2-*p*-iodophenyl-3-*p*-nitrophenyl-5-phenyl tetrazolium chloride (INT) is reacted with NADPH formed in the reaction. Reduced INT produces a color with maximum absorbance at 520 nm. The PMS-INT color developer must be refrigerated when not in use and must be protected from excessive exposure to light.[3]

Glucose determinations by the hexokinase method, performed with or without deproteinization, show good agreement. Results are similar to those obtained with the glucose oxidase method when using Somogyi-Nelson protein-free filtrates of plasma or whole blood.

o-TOLUIDINE METHODS

PRINCIPLE

Various aromatic amines react with glucose in hot acetic acid solution to produce colored derivatives. Among those used are aniline, benzidine, 2-aminobiphenyl, and *o*-toluidine. The latter condenses initially with the aldehyde group of glucose to form an equilibrium mixture of a glycosylamine and the corresponding Schiff base, as illustrated in Figure 6–10. Further rearrangements and reactions take place after the original condensation to produce a green chromogen mixture with an absorption maximum at 630 nm.[14]

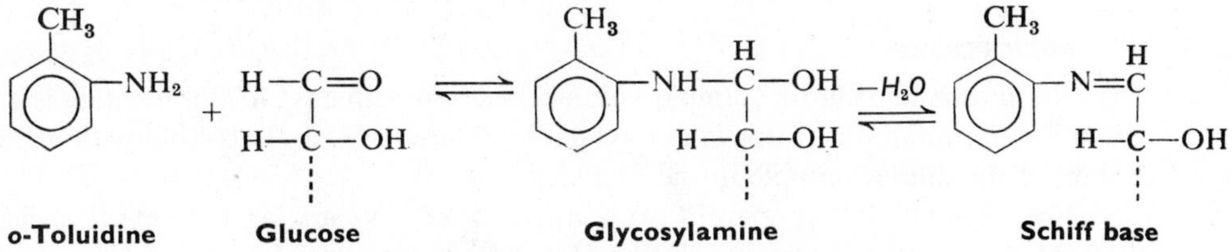

CH_3, NH_2 + H—C=O / H—C—OH $\rightleftharpoons$ CH_3, NH—C(H)—OH / H—C—OH $\underset{}{\overset{-H_2O}{\rightleftharpoons}}$ CH_3, N=C(H) / H—C—OH

o-Toluidine **Glucose** **Glycosylamine** **Schiff base**

Figure 6–10 Reaction of glucose with *o*-toluidine.

Dubowski[4] applied this reaction to trichloroacetic acid filtrates of serum and demonstrated good specificity for glucose. Negligible values were obtained on serum, cerebrospinal fluid, and urine, following yeast fermentation to destroy glucose. Other workers have noted that the reaction could be applied directly to serum without removal of protein.

Sugars other than glucose produce variable amounts of color in the reaction. The relative absorbance observed, compared to the same quantity of glucose, appears to depend somewhat on reaction conditions and should be determined by the individual laboratory interested in applying such data. The following absorbance ratios relative to glucose were found in our laboratory, using the method to be presented, for standards containing 200 mg of the respective sugar per 100 ml of solution. Ratios are also included for the same solutions analyzed by the AutoAnalyzer automated ferricyanide procedure.

Sugar	*o-Toluidine*	*Ferricyanide*
Glucose	1.00	1.00
Fructose	0.06	1.03
Mannose	0.96	0.91
Galactose	1.42	0.82
Sucrose	0.16	0.00
Maltose	0.09	0.48
Lactose	0.39	0.42
Xylose	0.12	1.20

Pentoses react with *o*-toluidine to produce an orange color with maximum absorption near 480 nm. By reading at two wavelengths, xylose may be measured in the presence of glucose. A wavelength is chosen (near 680 nm) where the absorbance produced by glucose is the same as that observed at 480 nm, and the color produced by xylose is negligible. The absorbance at 680 nm is then subtracted from the total absorbance at 480 nm to provide the net absorbance contributed by xylose.

REAGENTS

1. *o*-Toluidine, 5 ml/100 ml. Transfer 3.0 g of thiourea to a 3 liter Erlenmeyer flask. Add 1900 ml of glacial acetic acid and 100 ml of *o*-toluidine. Mix until the thiourea is dissolved and store in a brown bottle at room temperature. Contact with the skin should be avoided. The reagent should be dispensed from an all-glass automatic pipet such as a Repipet (Labindustries, Berkeley, CA. 94710).

This reagent may be used for months, although some variability occurs between batches and older reagents tend to produce increased absorbance values with glucose. Thus, a standard should be included with each series of determinations. Thiourea decreases the color of the reagent blank to an absorbance of about 0.01. Blanks can usually be omitted and the standards and unknowns measured against a distilled water blank; however, each new lot of *o*-toluidine should be checked by substituting water for serum in the method. If the absorbance of the blank is appreciable, it will be necessary to include a blank with each series of determinations.

2. Standard glucose solution, 200 mg/100 ml in benzoic acid solution. Prepare by dilution from stock 1.000 g/100 ml standard described for the glucose oxidase procedure.

PROCEDURE

An automatic dilutor (sample volume: 0.10 ml; diluent: 0.90 ml water) is convenient for this determination. If this is not available, dilute the sample 10-fold with water and use 1.0 ml of the diluted sample in the first step.

For whole blood, prepare a protein-free filtrate by mixing 0.20 ml of specimen with 1.80 ml of trichloroacetic acid solution (3 g/100 ml). Allow to stand 5 to 10 min, centrifuge, and pipet 1.0 ml for analysis. This procedure should also be followed for grossly hemolyzed or lipemic sera. If protein-free filtrates are used, the 200 mg/100 ml standard must also be diluted 10-fold with the trichloroacetic acid solution, since the latter reduces the absorbance obtained in the reaction.

1. Sample 0.10 ml of serum, plasma, cerebrospinal fluid, urine, or standard solution and dilute with 0.90 ml of water in a test tube (or pipet 1.0 ml of a 10-fold dilution or protein-free filtrate). Large test tubes (19 × 150 mm) are used for ease of mixing.
2. Add 7.0 ml of *o*-toluidine reagent, mix, and place in a 100°C bath for 10 min.
3. Cool in a cold water bath for 2 or 3 min; remix.
4. Measure the absorbance within the next 30 min against a water blank at 630 nm.

$$\frac{A_{\text{sample}}}{A_{\text{standard}}} \times 200 = \text{mg glucose/100 ml of sample}$$

COMMENTS ON THE PROCEDURE

Occasionally the final solution may show some turbidity. This can occur if the cooling bath is too cold, but usually is encountered with serum or plasma specimens having a high lipid content. In this event, add 2.0 ml of isopropanol to the 8 ml of final reaction mixture, mix, measure the absorbance, and multiply the answer by 10/8. If the mixture is still cloudy, repeat the test on a protein-free filtrate. The color follows Beer's law with most spectrophotometers but this should always be checked for a given instrument by analyzing standards ranging from 100 to 500 mg/100 ml. Sufficient reagent is present to permit simple dilution of the final reaction mixture with acetic acid for values up to 2000 mg/100 ml.

Moderate hemolysis does not interfere significantly. Each 100 mg/100 ml of hemoglobin increases the apparent glucose concentration by 2 mg/100 ml. Bilirubin does not react under the above conditions, and interference is negligible. In some modifications of this method, undiluted serum is added directly to the reagent. More intense color is obtained with glucose under these relatively anhydrous conditions, but the interference from bilirubin becomes significant because of its conversion to the green pigment biliverdin.

EDTA in concentrations greater than 1 mg/ml and sodium fluoride at levels greater than 5 mg/ml in the specimen will cause some increase in color. Thymol preservative should be avoided since this inhibits color formation. Dextran, used as a plasma expander, produces turbidity in the reaction and leads to falsely elevated values.

Normal ranges for the *o*-toluidine method are essentially the same as those for the glucose oxidase and hexokinase procedures, since all three methods give similar results. In patients with uremia, however, higher values are obtained by the *o*-toluidine method, though they are not as high as those obtained with the ferricyanide method.

The *o*-toluidine method has been applied to the AutoAnalyzer continuous-flow system. The major disadvantage is the deleterious effect of the reagent on tubing, thus necessitating frequent changes of the tubing. To overcome this, formulas have been devised to reduce the concentration of acetic acid in the reagent or to eliminate it. Boric acid, added directly to the *o*-toluidine reagent prepared in glacial acetic acid, increases the sensitivity over 2-fold. Citric acid, malic acid, and glycolic acid also enhance color. Some formulations substitute ethylene glycol or propylene glycol for acetic acid, but such reagents tend to have undesirably high viscosities. For continuous flow applications with sample dialysis, a solution containing boric acid, citric acid, and thiourea in approximately 50 per cent aqueous acetic acid solution has been proposed,[3] but others have not obtained uniformly consistent results with such reagents. The latter reagent cannot be applied directly to serum, since protein precipitates under these conditions and produces turbidity.

REFERENCE

Feteris, W. A.: Am. J. Med. Tech., *31*:17, 1965.

GLUCOSE TOLERANCE TESTS

Patients with mild or diet-controlled diabetes may have fasting blood glucose levels within the normal range, but be unable to produce sufficient insulin for prompt metabolism of ingested carbohydrate. As a result, blood glucose rises to abnormally high levels and the return to normal is delayed. In other words, the patient has decreased tolerance for glucose. Therefore, glucose tolerance tests are most helpful in establishing a diagnosis of a mild case of diabetes.

When a standard dose of 100 g of glucose is given orally, absorption occurs rapidly and the blood glucose concentration increases. This stimulates the pancreas to produce more insulin, with the result that after 30 to 60 minutes the blood glucose level begins to decrease. Since there now exists more insulin than necessary, the blood glucose tends to drop below the fasting level after 1.5 to 2 hours, and then returns to normal levels in

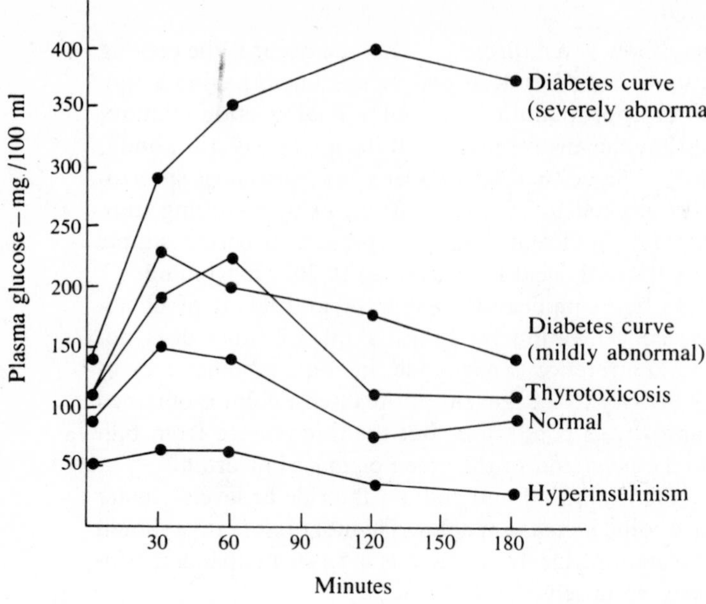

Figure 6–11 Plasma glucose values in response to an oral glucose load (glucose tolerance test).

approximately 3 hours. Response to glucose in various conditions is shown in Figure 6–11. Values refer to serum or plasma glucose concentrations. As noted earlier, these values are approximately 10 to 15 per cent greater than true glucose levels in whole blood.

The significance and interpretation of glucose tolerance tests have been reviewed by Duffy *et al.*[5] The major problem is to define criteria that provide both sensitivity and specificity. Sensitivity is the extent to which the test identifies diseased individuals. Specificity on the other hand, is the extent to which nondiseased individuals are classified as normal. When limits are set too low we achieve good sensitivity but poor specificity. Conversely, with high limits we lose sensitivity but attain good specificity.

Criteria for the diagnosis of diabetes have been evaluated by the Committee on Statistics of the American Diabetes Association[2] and are shown in Table 6–1. The Wilkerson Point

TABLE 6–1 VARIOUS CRITERIA FOR THE STANDARD ORAL GLUCOSE TOLERANCE TEST

	Hour	*Whole Blood**	*Plasma**	*Points*
Wilkerson Point System	0	110	130	1
	1	170	195	½
	2	120	140	½
	3	110	130	1
Values equal to or more than those listed are given points as shown. Two points or more are judged diagnostic of diabetes.				
Fajans-Conn Criteria	1	160	185	
	1½	140	165	
	2	120	140	
All levels must be equal to or greater than values shown at the times specified to make a diagnosis of diabetes. Criteria apply to ambulatory individuals under the age of 50.				
University Group Diabetes Program				
The subject is judged diabetic if the sum of values obtained at 0, 1, 2, and 3 hours equals 500 or more for whole blood, or 600 or more for plasma.				

* Values for whole blood or plasma glucose are given in mg/100 ml.

System provides the same interpretation as the criteria adopted by the United States Public Health Service. Other investigators are of the opinion that plasma values of more than 175 mg/100 ml after 1 hour and more than 120 mg/100 ml after 2 hours should be considered abnormal.

Attempts have been made to correlate results of glucose tolerance tests on diabetics with a fundamental defect in insulin secretion. Ketosis-prone diabetics show low to zero levels of plasma insulin, both fasting and after glucose administration. The most consistent abnormality in diabetics appears to be a blunting of the early insulin peak in the first few minutes after intravenous glucose injection. Results of one study are shown in Figure 6–12. However, 15 to 20 per cent of "normal" patients also have absence of the early insulin peak. Whether this group is really normal or potentially diabetic is not clear at the present time.

Growth hormone inhibits glycolysis and glucose uptake by muscle cells, and causes a rise in blood glucose. Growth hormone secretion is stimulated by hypoglycemia. Hence, growth hormone and insulin levels tend to vary inversely. As glucose is absorbed from the gastrointestinal tract, blood glucose levels rise. Feedback control normally results in a 10- to 15-fold rise in insulin levels and almost complete disappearance of growth hormone from the plasma. This insures storage of glycogen. After 2 to 4 hours, growth hormone rises to near basal levels, and those of insulin fall, although remaining at several times those of the initial concentrations. If there is a relative excess of insulin at this time, hypoglycemia may occur. If fasting continues, insulin almost disappears from the plasma and growth hormone rises to very high levels. This stimulates oxidation of fat and release of free fatty acids while minimizing metabolism of glucose and protecting glycogen stores for potential stress situations. Both insulin and growth hormone assays in plasma may be requested of the laboratory as aids in the interpretation of glucose tolerance tests.

The severe diabetic is strongly disposed to develop ketosis and pass into diabetic coma. In diabetic ketosis, plasma glucose levels, derived mostly from gluconeogenesis, are usually

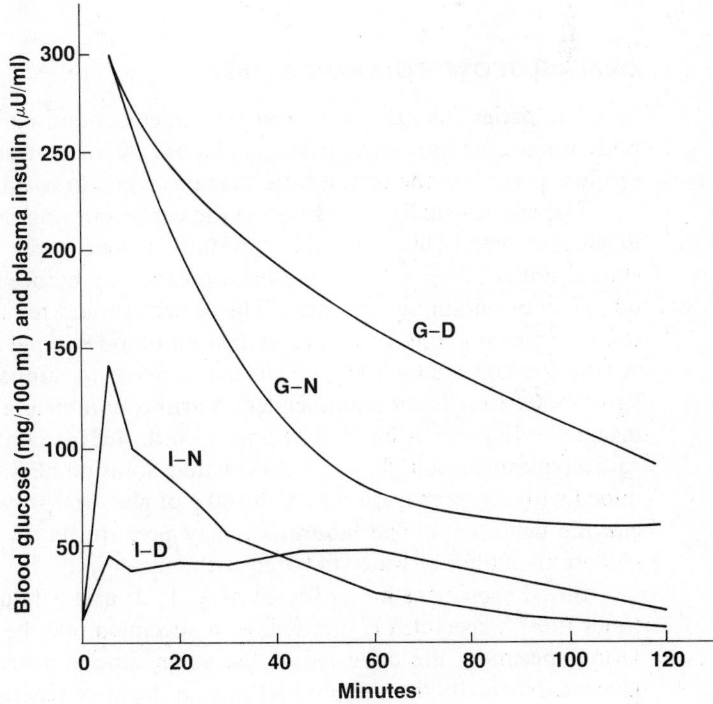

Figure 6–12 Mean values for blood glucose and plasma insulin in nonobese subjects following intravenous injection of glucose (0.5 g/kg of body weight) during the first 5 minutes. G–N: glucose, normals. G–D: glucose, mild diabetics. I–N: insulin, normals. I–D: insulin, mild diabetics. Curves are based on data of Seltzer *et al.*[11]

significantly elevated. Resultant severe glycosuria produces osmotic diuresis with fluid loss and depletion of electrolytes. Vomiting is frequently present and adds to the fluid and electrolyte depletion. Ketosis develops as a result of reduced glucose metabolism. Increased utilization of depot fat results in increased release of free fatty acids which in turn form acidic ketone bodies. These react with part of the plasma bicarbonate, resulting in a lower blood pH, a lower plasma bicarbonate, and a decrease in urinary pH. The developing metabolic acidosis stimulates the respiratory center and breathing becomes deeper, followed by a lowering of pCO_2. In summary, the major findings in diabetic ketosis are hyperglycemia and glycosuria, hyperventilation, dehydration (hemoconcentration and mild uremia), ketone bodies in blood and urine, low plasma bicarbonate, low blood pH, and hyperkalemia.

Severe symptoms may develop also in hypoglycemia, i.e., with plasma glucose levels below 40 mg/100 ml. The clinical symptoms are related to the rate of decrease of plasma glucose levels; if levels have dropped rapidly, a person may appear clinically hypoglycemic with higher glucose levels. If the levels have fallen gradually, the individual may show no symptoms, even with a plasma glucose as low as 30 mg/100 ml. Cerebral metabolism is dependent on an adequate supply of glucose from the blood, and symptoms of hypoglycemia resemble those of cerebral anoxia. These include faintness, dizziness, or lethargy which may progress rapidly into coma. If untreated, death or permanent cerebral damage may result. Rapid restoration of blood glucose concentration is essential.

A number of conditions may cause or precipitate hypoglycemia. Among these are overdosage with insulin, drug administration (sulfonylureas, phenformin, antihistamines), functional hypoglycemia (sensitivity to glucose), depleted glycogen stores in the liver, ingestion of large amounts of alcohol, islet cell tumors (insulinoma) or islet cell hyperplasia, galactosemia, and glycogen storage disease.

Plasma glucose levels in newborn infants are typically less than those for adults. In the low-birth-weight neonate, hypoglycemia may be defined as levels of whole blood glucose below 20 mg/100 ml. In the full-sized infant, blood glucose levels less than 30 mg/100 ml in the first 48 h of life and less than 40 to 50 mg/100 ml thereafter may be considered hypoglycemic.

ORAL GLUCOSE TOLERANCE TEST

The patient should be placed on a diet containing 1.75 g of carbohydrate per kg of body weight for three days before a glucose tolerance test. If carbohydrate intake has been too low preceding the test, a false diabetic-type curve may be obtained.

The test is usually performed in the early morning because there is a diurnal variation in glucose metabolism and the "normal" values apply to this time of day. The patient should not eat after the evening meal on the day before the test. Water may be taken but no caffeine-containing beverages. The patient should remain at rest during the test and also refrain from smoking or eating. A fasting blood sample and urine specimen are obtained. A solution containing 100 g of glucose is given to adults; for children, 2 g of glucose per kg of body weight is recommended. Various commercial preparations are available. One product (Glucola; Ames Co., Elkhart, Ind. 46514) contains 75 g of partially hydrolyzed carbohydrate in cola-flavored carbonated solution. Essentially the same response is obtained with this preparation as with 100 g of glucose (provided there is no intestinal mucosal enzyme deficiency). The laboratory may prepare its own solution by dissolving 100 g of glucose in 200 ml of water flavored with lemon juice.

Blood specimens are collected at $\frac{1}{2}$, 1, 2, and 3 hours after glucose ingestion. Sometimes the $\frac{1}{2}$ h specimen is omitted, or a specimen may be obtained at $1\frac{1}{2}$ h (see Table 6–1). Urine specimens are collected at the same time and are analyzed semiquantitatively for glucose. Normally these should all show a negative reaction. The level of plasma glucose at which glucose appears in the urine is called the renal threshold and is approximately 180 mg/

100 ml. Some individuals exhibit lower renal thresholds and excrete glucose in the urine even when the glucose tolerance curve is normal. This is usually considered a benign condition.

A 3 h test is usually adequate for routine evaluation of impaired glucose tolerance. If hypoglycemia is suspected, additional specimens are taken at 4 and 5 hours. Patients with adrenal insufficiency or with islet cell tumors of the pancreas tend to have low fasting levels of blood glucose. Response to a glucose tolerance test may appear normal over the first 3 hours, but values continue to fall during the fourth and fifth hour (hypoglycemic tail). Some patients with latent diabetes tend to show hypoglycemia during this period also, probably associated with a delayed secretion of insulin.

In certain potential diabetics (i.e., patients with near relatives who are diabetic), the oral glucose tolerance test may be normal but cortisone priming will yield an abnormal response. The procedure is to give 50 mg of cortisone $8\frac{1}{2}$ and 2 h before the start of the oral glucose tolerance test. The criterion for a positive test is a level of 160 mg/100 ml or more, 2 h after glucose ingestion. This test may have value in large scale studies, but it lacks sufficient sensitivity and specificity to be of much use in single cases.

INTRAVENOUS GLUCOSE TOLERANCE TEST

Poor absorption of orally administered glucose may result in a "flat" tolerance curve. On the other hand, patients with a history of gastrointestinal surgery may have accelerated intestinal absorption of glucose, as in the "dumping syndrome." The latter may result in a diabetic-type oral glucose tolerance curve. In either case, an intravenous glucose tolerance test may be performed to eliminate factors related to rate of absorption. As noted previously, plasma insulin assays may also be performed in conjunction with intravenous glucose tolerance tests to aid in the diagnosis of diabetes (see Fig. 6–12).

The dose of glucose is 0.5 g per kg of body weight, given as a 25 g/100 ml solution. The dose is administered intravenously within 2 to 4 min and blood is collected every 10 min for 1 h. If insulin assays are performed, a specimen is also obtained 5 min after the start of the injection. Blood glucose levels decrease in an exponential manner and the rate of glucose disappearance can be calculated from the formula $K = 70/t_{1/2}$, where $t_{1/2}$ is the number of minutes required for the blood glucose to fall to one-half of the 10 min level, and K is the rate of disappearance of blood glucose, expressed as per cent per minute of the 10 min level. In normal individuals, K usually exceeds 1.5 per cent; values below 1.0 per cent are considered diagnostic of diabetes.

TWO HOUR POSTPRANDIAL GLUCOSE

Since the 2 hour specimen in a glucose tolerance test has considerable significance in evaluating diabetes, the test may be shortened for screening purposes by determining glucose on the 2 hour specimen only. The patient consumes a breakfast, lunch, or glucose solution, containing 100 g of carbohydrate. Two hours after the meal, blood is drawn for a glucose determination. The patient should be instructed to consume the required amount of carbohydrate and to remain at rest during the 2 hour period following the meal. Many physicians now request 2 hour postprandial glucose determinations routinely in lieu of fasting glucose levels as guides to insulin requirements. Under usual hospital conditions it is often difficult, however, to control the 2 hour time interval very accurately, since timing may start at the beginning, midway, or end of the test meal. To insure uniformity of carbohydrate intake and accurate timing, it is recommended that 100 g of glucose in solution be used routinely as a test meal for postprandial blood glucose determinations.

Interpretation, as with other glucose determinations, varies with the method and depends on whether whole blood or plasma is used for the analysis. In accordance with the criteria of Table 6–1, plasma or serum glucose at 2 h postprandially should be less than 140 mg/100 ml. However, values between 120 and 140 mg/100 ml may warrant further investigation with a complete glucose tolerance test.

INSULIN TOLERANCE TEST

This test is useful in evaluating patients with insulin resistance or certain endocrine disorders. The patient is placed on a diet containing at least 300 g of carbohydrate daily for two or three days before the test. With the patient in the fasting state, blood is taken for a baseline glucose level, after which regular insulin is injected intravenously by a physician in an amount corresponding to 0.1 unit per kg of body weight. Blood specimens are then taken for glucose determinations at 20, 30, 45, 60, 90, and 120 min after the insulin is given. A syringe containing 50 ml of 50 g/100 ml glucose should be available for intravenous injection. The patient should be observed closely and a physician should be available to make the injection and terminate the test should a hypoglycemic reaction occur.

Normally the blood glucose decreases to about 50 per cent of the fasting level by 30 min, and then returns to normal fasting limits by 90 to 120 min. There are two types of abnormal response. The insulin-resistant type shows only slight or delayed decrease in blood glucose and occurs with adrenal cortical hyperfunction (Cushing's syndrome), in acromegaly, and in some cases of diabetes. In the second type of response the blood glucose falls normally, but the subsequent rise is delayed or does not occur at all. This situation occurs with hypofunction of the anterior pituitary (Simmond's disease) or of the adrenal cortex (Addison's disease), and in hyperinsulinism. In cases of suspected Simmond's disease it is recommended that half the usual dose of insulin be given and that the patient be watched carefully for signs of hypoglycemia. Glucose solutions or fruit juice should normally be given to patients at the end of insulin tolerance tests.

OTHER TOLERANCE TESTS

Various other tests have been proposed that require serial determinations of blood glucose. Some of these will now be described briefly.

Tolbutamide (1-butyl-3-*p*-tolylsulfonylurea, Orinase) is a compound that stimulates the pancreas to produce insulin. Following intravenous injection, the normal response is similar to that observed with the insulin tolerance test: the blood glucose decreases to about 50 per cent of the fasting level by 30 minutes, then returns to normal. If the blood glucose level at 20 minutes is between 80 and 84 per cent of the fasting value, the patient is said to have a 50 per cent probability of having diabetes.[7] In more severe cases the response will be even less, inasmuch as the pancreas is unable to secrete adequate insulin. The test has also proved to be valuable in evaluating hypoglycemic states caused by insulinomas.[6] In this condition, injection of tolbutamide results in marked decrease in blood glucose to values in the range of 20 to 30 mg/100 ml and persistent hypoglycemia up to 3 h. As with the insulin tolerance test, patients must be watched carefully for hypoglycemic reactions and the test terminated, if necessary, by intravenous administration of glucose.

The *epinephrine tolerance* test is used to evaluate one form of glycogen storage disease (type I, von Gierke's), a condition in which there is a deficiency or absence of the enzyme glucose-6-phosphatase in the liver. This enzyme is the catalyst for the final step in the formation of blood glucose from hepatic glycogen. Individuals with von Gierke's disease have low glucose levels in the blood, increased liver glycogen, but decreased *availability* of liver glycogen as shown by less than normal or no increase in blood glucose following administration of epinephrine. In a normal person, after intramuscular injection of 1 ml of a 1/1000 solution of epinephrine hydrochloride, the blood glucose increases 35 to 45 mg/

100 ml in 40 to 60 min and returns to the fasting level by 2 h. Blood specimens are taken at 30, 45, 60, 90, and 120 min after injection.

Deficiency of small bowel mucosal lactase has been found to be a rather common condition in healthy adults. Such deficiency may be associated with intolerance to lactose manifested by diarrhea and other symptoms following ingestion of milk. The diarrhea will usually disappear if lactose is eliminated from the patient's diet. A *lactose tolerance test* can be done to evaluate this condition.[1] A standard oral glucose tolerance test is performed first to provide a basis for comparison. On the following day the test is repeated, except that 100 g of lactose is substituted for glucose. If lactase activity is present, the lactose will be split to glucose and galactose and the resultant tolerance curve will be similar to that observed with glucose. With lactase deficiency the lactose tolerance curve will be flat, with a rise not exceeding 20 mg/100 ml over the fasting level. As discussed previously, galactose may react to a greater or lesser extent than glucose, depending on the method of analysis, but this difference is unlikely to affect interpretation of results when either the *o*-toluidine or ferricyanide method is used. Glucose oxidase and hexokinase methods do not detect galactose.

URINARY SUGARS

Occurrence of sugars in urine

Urine is examined routinely to detect or determine the presence or amount of glucose; this is done either as a screening procedure or as a guide to insulin therapy. Other sugars may also appear in the urine in certain conditions and interfere with the detection and determination of glucose. The sugars of clinical interest are all reducing sugars; that is, they readily reduce cupric ion in hot alkaline solution. Except for the very rare cases of galactosuria, glucose is the only sugar found in urine that is of pathological significance.

Galactose appears in the urine of infants with galactosemia, a condition characterized by inability to metabolize galactose. Such infants fail to thrive on milk, since half of the milk sugar, lactose, is converted to galactose. Lactose is sometimes found in urine of women during lactation and occasionally toward the end of pregnancy. The laboratory may be required to differentiate this sugar from glucose. Fructose may appear in the urine after eating fruits, honey, and syrups, but has no significance. Fructosuria is a rare and harmless congenital defect that should not be confused with diabetes. Pentoses may occur in urine after eating such fruits as cherries, plums, or prunes, or as a harmless congenital anomaly and, as with fructose, must be distinguished from glucose. Maltose has been reported to occur along with glucose in the urine of some patients with diabetes.

Many reducing substances other than sugars may also occur in urine. A partial list

TABLE 6-2 REDUCING SUBSTANCES IN URINE

Fructose	Ketone bodies
Lactose	Sulfanilamide
Galactose	Oxalic acid
Maltose	Hippuric acid
Arabinose	Homogentisic acid
Xylose	Glucuronic acid
Ribose	Formaldehyde
Uric acid	Isoniazid
Ascorbic acid	Salicylates
Creatinine	Cinchophen
Cysteine	Salicyluric acid

of the more important reducing substances is shown in Table 6–2. Ascorbic acid, especially, may be ingested in large amounts or be present in antibiotic preparations administered intravenously. In either case, excess concentrations usually appear in the urine and contribute significantly to the total reducing substances present.

QUALITATIVE METHODS FOR TOTAL REDUCING SUBSTANCES

PRINCIPLE

Benedict's qualitative reagent contains cupric ion complexed with citrate in alkaline solution. Glucose, or other reducing substances, reduces cupric ion to cuprous ion with resultant formation of yellow cuprous hydroxide or red cuprous oxide.

REAGENT

Dissolve 17.3 g of $CuSO_4 \cdot 5H_2O$ in 100 ml of hot water. Dissolve separately, with heating, 173 g of sodium citrate ($Na_3C_6H_5O_7 \cdot 2H_2O$) and 100 g of Na_2CO_3 in 800 ml of water. Allow to cool, then add the citrate-carbonate solution, with mixing, to the copper sulfate solution. Dilute to 1 liter with water. This reagent is stable.

PROCEDURE

Add 8 drops (0.4 ml) of urine to 5 ml of reagent in a test tube. Mix and place in a boiling water bath for 3 min. Remove and examine immediately. Report as 0 to 4+ according to the following criteria:

Appearance	*Report*	*Approximate Glucose Concentration* (g/100 ml)
Blue to green, no precipitate	0	0–0.1
Green with yellow precipitate	1+	0.3
Olive green	2+	1.0
Brownish-orange	3+	1.5
Brick red	4+	2.0 or more

A convenient adaptation of the preceding procedure is marketed in tablet form (Clinitest; Ames Co., Elkhart, Ind. 46514). The tablets contain anhydrous cupric sulfate, sodium hydroxide, citric acid, and sodium bicarbonate. Five drops (0.25 ml) of urine are mixed with 10 drops of water in a test tube. One tablet is added and the mixture is allowed to stand undisturbed for 15 seconds, remixed, and observed for color. A chart provided by the manufacturer is used to interpret the result. Heat is generated by contact of sodium hydroxide and water. The initial reaction between citric acid and sodium bicarbonate causes the release of carbon dioxide, which blankets the mixture and reduces contact with oxygen from the air.

QUANTITATIVE METHODS FOR TOTAL REDUCING SUBSTANCES

Although quantitative measurement of total reducing substances in urine provides information of limited diagnostic value, the test is still performed in a number of hospital laboratories. The Folin-Wu or Somogyi-Nelson methods may be used for this purpose. The urine usually needs diluting to below about 300 mg/100 ml to bring the concentration of glucose within the range of the method. The dilution necessary can be estimated from the qualitative test. The preparation of protein-free filtrate is omitted; instead, the urine is further diluted with water and analyzed in the same manner as a protein-free filtrate. Results

TABLE 6-3 RESULTS OBTAINED BY FOUR GLUCOSE PROCEDURES ON 20 URINE SPECIMENS ALL NEGATIVE FOR REDUCING SUBSTANCES

Method	*Apparent g Glucose/g Creatinine* RANGE	MEAN
Ferricyanide	1.07–2.77	1.80
Folin-Wu	0.83–2.16	1.24
Somogyi-Nelson	0.26–1.26	0.53
o-Toluidine	0.10–0.37	0.20
Fermentation	0–0.18	0.06

are corrected for the initial dilution. A small amount of reducing substances are found in urine specimens that are negative with the qualitative tests. Expressed as glucose, the concentration is less than 150 mg/100 ml of urine. The qualitative tests described above actually provide semiquantitative results and should, in most cases, obviate the need for quantitative measurements of total reducing substances.

QUANTITATIVE METHODS FOR GLUCOSE IN URINE

The amount of glucose excreted normally, as determined by highly specific enzymatic methods, is less than 500 mg/d. Random specimens show an upper limit of normal of approximately 30 mg/100 ml. The hexokinase procedure is recommended for greatest accuracy and specificity for determination of glucose in urine.

Other procedures vary widely in specificity for glucose. In one study 20 morning urine specimens were selected that were all negative for reducing sugar by the copper reduction test. These were analyzed by an automated alkaline ferricyanide method, the Folin-Wu method, the Somogyi-Nelson method, and the *o*-toluidine method. Creatinine was also determined. After yeast fermentation (see below), the specimens were again analyzed by the *o*-toluidine method and the decrease in value recorded as fermentable sugar. Yeast blanks were included. Results, shown in Table 6–3, are expressed as apparent g of glucose per g of creatinine. Of the four procedures, the *o*-toluidine method was found to be the most specific and the ferricyanide method the least specific for the estimation of glucose in urine. Obviously, neither the ferricyanide nor copper reduction methods should be used for quantitative measurements of glucose in urine. The *o*-toluidine procedure is recommended as a satisfactory alternative to the hexokinase methods for the quantitative determination of glucose in urine.

SEPARATION AND IDENTIFICATION OF SUGARS

Techniques for separating and identifying sugars include fermentation, osazone formation with phenylhydrazine, specific chemical tests, and paper or thin-layer chromatography. The availability of glucose oxidase test strips has greatly simplified the differentiation of glucose from the many other reducing substances.

Glucose, fructose, maltose, and mannose are fermentable with yeast, but lactose, galactose, and pentoses do not ferment. Of the fermentable sugars, only glucose and, rarely, fructose are likely to occur in urine. The fermentation test can be used, therefore, to differentiate glucose from lactose or other nonfermenting sugars.

FERMENTATION TEST

Bring a portion of the urine to boiling to destroy *E. coli*, which can ferment lactose. Cool to room temperature and reserve a portion for a qualitative test for reducing sugar.

Add about 0.3 g of dry active baker's yeast to 10 ml of the boiled specimen and mix with a stirring rod until the yeast is dispersed into a homogeneous mixture. Transfer to a large test tube and incubate unstoppered with occasional mixing for 1 h in a 37°C water bath. Centrifuge the incubated specimen and perform a qualitative test for reducing sugar on the supernatant. Compare the results with a similar test performed on the unfermented specimen.

If the fermented specimen is negative, all the sugar in the urine is probably glucose. If the fermented sample is the same as the unfermented sample, some sugar or reducing substance other than glucose is present. When the fermented sample is positive, but lower than the untreated sample, the difference is considered to be glucose. It is good practice to include a control to check the activity of the yeast. Dissolve 0.1 g of glucose in 10 ml of urine previously shown to be negative for reducing substances. This specimen, when carried through the fermentation test, should become negative for reducing sugar. A blank may also be included by mixing 0.3 g of yeast with 10 ml of water to rule out the possible presence of nonfermenting reducing substances in the yeast. If necessary, the yeast may be suspended in saline, centrifuged, and rewashed to remove reducing substances.

QUALITATIVE TESTS FOR INDIVIDUAL SUGARS

Glucose

A convenient paper test strip is commercially available (Clinistix; Ames Co., Elkhart, Ind. 46514). The filter paper is impregnated with glucose oxidase, peroxidase, and *o*-tolidine and provides a simple color test according to principles discussed in an earlier section. The test end is moistened with urine and examined after 10 seconds. A blue color develops if glucose is present. The sensitivity of the strip has been adjusted to take into account the presence of enzyme inhibitors normally occurring in urine. Thus, a positive test will be obtained with lower concentrations of glucose in water as compared to urine. For the same reason, a false positive test may be obtained with very dilute specimens.

In one study of 2000 urine specimens, 11 false negative enzyme paper tests were encountered. Among the inhibitors identified were ascorbic acid, dipyrone and meralluride sodium (mercuhydrin). Several antibiotics contain ascorbic acid as a preservative. The acid is largely excreted unchanged and can cause false negative results. Contamination of urine with hydrogen peroxide or a strong oxidizing agent, such as hypochlorite, produces false positive results. For routine examinations, a negative result by the strip test is usually interpreted to mean that the urine specimen is negative for glucose. To rule out false negatives, however, a better approach would be to test all specimens routinely by one of the copper reduction methods, such as a Clinitest tablet, then confirm the presence of glucose with a paper strip test. Discrepancies in the two results would require further investigation as described below.

Another strip test (Diastix; Ames Co., Elkhart, Ind. 46514) allows for the semiquantitative estimation of both glucose and ketone bodies. The glucose portion of the strip utilizes the glucose oxidase-peroxidase method. The peroxide produced oxidizes iodide to iodine, yielding varying intensities of brown color which correspond to the concentration of glucose in the urine. Compared to Clinistix, the Diastix glucose test is much less sensitive to inhibition by ascorbic acid.

Galactose, pentoses, and other reducing sugars do not react with the glucose oxidase strips. It is recommended that urine from infants and children be tested routinely by both the glucose oxidase and copper reduction tests, to identify individuals with congenital anomalies who might otherwise be diagnosed as diabetics. Nonglucose reducing substances should be further identified by chromatographic procedures.

REFERENCE

Free, A. H., and Free, H. M.: Crit. Rev. Clin. Lab. Sci., *3*:481, 1972.

TABLE 6–4 DIFFERENTIATION OF REDUCING SUGARS

Sugar	*Fermentation*	*Clinistix*	*Special Tests*
Glucose	+	+	
Fructose	+	−	Seliwanoff
Galactose	−	−	Galactose oxidase
Lactose	−	−	
Maltose	+	−	
Pentoses	−	−	Bial

Seliwanoff's test for fructose

Hot hydrochloric acid converts fructose to hydroxymethyl furfural, which links with resorcinol to produce a red-colored compound. To make the reagent, dissolve 50 mg of resorcinol in 33 ml of concentrated hydrochloric acid and dilute to 100 ml with water. Add 0.5 ml of urine to 5 ml of reagent in a test tube and bring to a boil. Fructose produces a red color within ½ minute. The test is sensitive to 100 mg fructose/100 ml provided excess glucose is absent. A 2 g/100 ml solution of glucose will produce about the same color as 100 mg/100 ml of fructose after ½ minute of boiling. A solution of fructose (0.5 g/100 ml) should be used as a control. With high concentrations of fructose, a red precipitate forms, which may be filtered and dissolved in ethanol to produce a bright red-colored solution.

Bial's test for pentose

By heating with hydrochloric acid, pentoses are converted to furfural, which reacts with orcinol to form green-colored compounds.

Dissolve 300 mg of orcinol in 100 ml of concentrated hydrochloric acid and add 0.25 ml of ferric chloride solution (10 g/100 ml). Glucose, if present in the urine, should be removed by fermentation. Add 0.5 ml of urine to 5 ml of reagent in a test tube and bring to a boil. Pentoses produce a green color. The test is sensitive to 100 mg pentose/100 ml. A solution of xylose (0.5 g/100 ml) should be used as a control. Glucuronates will produce a similar color if the boiling is prolonged. Fructose, as with Seliwanoff's reagent, produces a red color.

A combination of the preceding tests will usually suggest the nature of the reducing sugar, as summarized in Table 6–4. Lactose and galactose can be differentiated from each other by galactose oxidase procedures. A more definitive identification can be achieved by use of paper or thin-layer chromatography.

IDENTIFICATION OF URINARY SUGARS BY PAPER CHROMATOGRAPHY

PRINCIPLE

Sugars can be separated by ascending or descending chromatography on paper and located after color development with dinitrosalicylic acid. The variable rates of migration depend upon the solubility of the sugars in the particular solvent system. Presumptive identification is made by comparison of the R_f value of the unknown with those of authentic samples. The following procedure may be performed conveniently in a 6 × 18 inch Pyrex jar with a tightly fitting glass cover.

REAGENTS

1. Solvent. Perform the following procedure under a hood. Mix 60 ml of *n*-butanol, 40 ml of pyridine, and 30 ml of water. The mixture is completely miscible. Pour into the bottom of the jar, cover, and allow to equilibrate at least 30 min before use.

2. Spray reagent. Dissolve 0.5 g of 3,5-dinitrosalicylic acid in 100 ml of sodium hydroxide solution (4 g/100 ml).

3. Reference sugar solutions. Prepare solutions of glucose, fructose, galactose, maltose, lactose, and xylose to contain 1.75 g each in 100 ml of benzoic acid solution (2 g/l). These solutions are stable for months when refrigerated.

PROCEDURE

1. Determine the concentration of sugar in the urine by means of one of the qualitative copper reduction tests. Dilute the specimen, if necessary, to a sugar concentration of approximately 1 g/100 ml. If the concentration is only 0.5 g/100 ml, use twice as much sample in the test.

2. Draw a pencil line 1 inch from and parallel to the 10 inch side of a 10 × 14 inch section of Whatman No. 1 filter paper. Place pencil marks at least 1 inch apart on the line to indicate starting positions for each reference sample and for the urine specimen.

3. Apply approximately 0.01 ml of each solution to its respective point from a microhematocrit tube or a 10 μl pipet. Half of this amount should be applied and permitted to dry before adding the remainder in order to keep the diameter of the spots as small as possible. Allow all samples to dry completely.

4. Staple the sheet into a 14 inch high cylinder so that the line of application is at the bottom. Insert the paper into the chromatography jar, tape on the cover, and allow to stand undisturbed for about 16 h (overnight) at room temperature.

5. Remove the sheet and mark with a pencil along the solvent front. Allow to air dry under the hood for 4 h.

6. Spray the sheet with dinitrosalicylic acid reagent from an atomizer and allow to air dry for 4 h.

7. Heat the paper at 100°C for 10 min in a drying oven. The reducing sugars appear as brown spots against a yellow background.

8. Measure the distance from the starting line to the center of each spot; similarly, measure the distance from the starting line to the edge of the solvent front. Calculate the ratio of fronts, R_f:

$$R_f = \frac{\text{distance traveled by solute spot}}{\text{distance traveled by solvent front}}$$

The R_f values vary slightly from run to run, and for this reason known reference samples should be included each time. Average approximate values are as follows:

Sugar	R_f
Lactose	0.22
Maltose	0.28
Galactose	0.36
Glucose	0.41
Fructose	0.46
Xylose	0.52

A typical run is shown in Figure 6–13.

The unknown sugar in the urine is presumed to be the same as a known reference standard when both migrate the same distance under the same test conditions. For confirmation, the urine specimen may be mixed with an equal volume of the known standard and rechromatographed. Only one sugar spot will appear on the paper if the two sugars are identical. Dinitrosalicylic acid is a highly specific reagent for reducing sugars.

Identification of urine sugars can also be made by using thin-layer chromatographic techniques as described by Young and Jackson.[15] When frequent chromatographic separa-

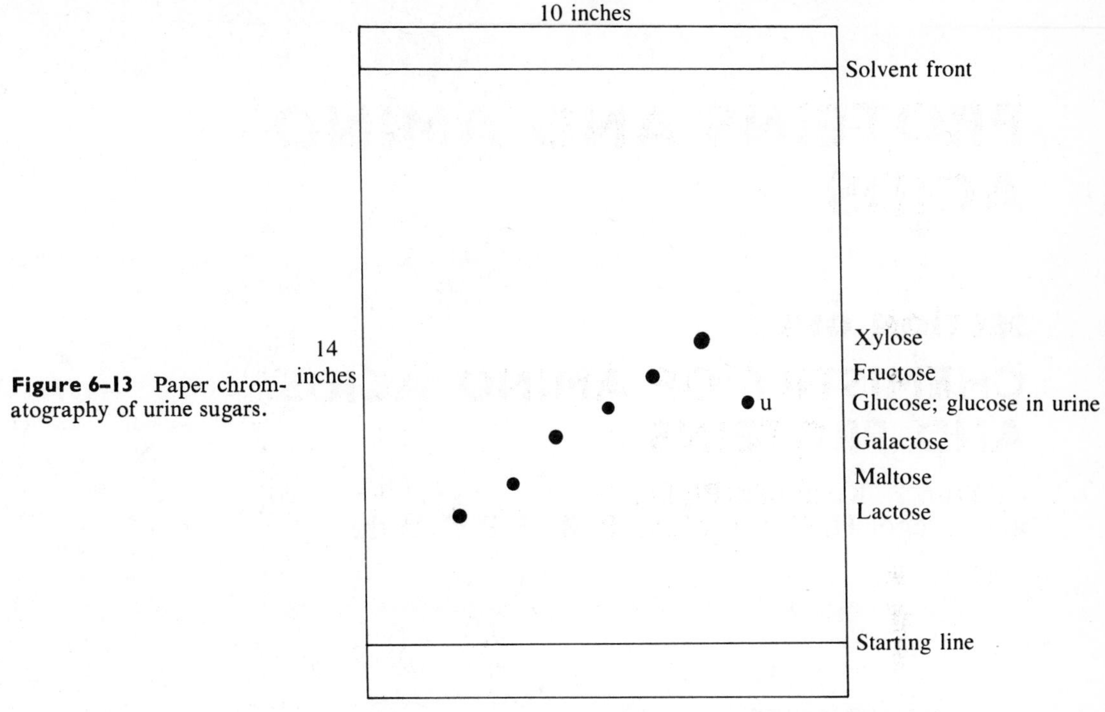

Figure 6–13 Paper chromatography of urine sugars.

tions are necessary, this method is preferred over paper chromatography, because of the shorter time period required. If such studies are performed infrequently, paper chromatography is simple, is adequate for most separations, and requires little actual working time.

REFERENCE

Sophian, L. H., and Connolly, V. J.: Am. J. Clin. Path., *22:* 41, 1952.

REFERENCES

1. Basford, R. L., and Henry, J. B.: Postgraduate Med., *41*:A-70, 1967.
2. Committee on Statistics of the American Diabetes Association: Diabetes, *18*:299, 1969.
3. Cooper, G. R.: Crit. Rev. Clin. Lab. Sci., *4*:101, 1973.
4. Dubowski, K. M.: Clin. Chem., *8*:215, 1962.
5. Duffy, T., Phillips, N., and Pellegrin, F.: Am. J. Med. Sci., *265*:117, 1973.
6. Fajans, S. S., Schneider, J. M., Schteingart, D. E., and Conn, J. W.: J. Clin. Endocrinol. & Metab., *21*: 371, 1961.
7. Kaplan, N. M.: Arch. Int. Med., *107*:212, 1961.
8. Logan, J. E., and Haight, D. E.: Clin. Chem., *11*:367, 1965.
9. Martinek, R. G.: J. Am. Med. Technol., *31*:530, 1969.
10. Morin, L. G., and Prox, J.: Clin. Chem., *19*:959, 1973.
11. Seltzer, H. S., Allen, E. W., Herron, A. L., Jr., and Brennan, M. T.: J. Clin. Invest., *46*:323, 1967.
12. Tustison, W. A., Bowen, A. J., and Crampton, J. H.: Diabetes, *15*:775, 1966.
13. Weissman, M., and Klein, B.: Clin. Chem., *4*:420, 1958.
14. Yee, H. Y., and Goodwin, J. F.: Anal. Chem., *45*:2162, 1973.
15. Young, D. S., and Jackson, A. J.: Clin. Chem., *16*:954, 1970.

ADDITIONAL READING

Zilva, J. F., and Pannall, P. R.: Clinical Chemistry in Diagnosis and Treatment. Chicago, Year Book Publishers, Inc., 1971, Chap. 6.

CHAPTER 7

PROTEINS AND AMINO ACIDS

SECTION ONE

CHEMISTRY OF AMINO ACIDS AND PROTEINS

by John F. Kachmar, Ph.D.,
and Gregor H. Grant, M.A., B.M., F.R.C. Path.

AMINO ACIDS

Proteins constitute that class of biochemical compounds most characteristic of protoplasm and life. Many individual carbohydrates, lipids, and nucleotides are encountered in all animal and plant forms. Proteins, however, are specific; each species of organism is associated with a large number of proteins typical of itself and itself alone. Indeed, there occur proteins unique for a given organ or tissue, and even for an individual organism. All proteins contain carbon, hydrogen, oxygen, nitrogen, and sulfur; in addition, individual proteins may contain phosphorus, iodine, iron, copper, zinc, or other elements. The presence of *nitrogen* in all proteins sets them apart from carbohydrates and lipids. The average nitrogen content is approximately 16 per cent.

When proteins are broken down into individual elementary units by acid, alkali, or enzymatic hydrolysis, it is found that these basic units consist of *alpha-amino acids*. Some 40 different amino acids have been isolated from various proteins, but only about 20 are present in all proteins in varying amounts. These amino acids are linked together by *peptide bonds* (see Formula [4]) into long chains, which contain from 50 to many thousands of amino acids. The number of amino acids, the order in which they are joined together in the long chain, and the manner in which the chain is coiled, folded, and cross-linked make possible the many millions of unique proteins present in the multitude of living organisms.

Alpha-amino acids constitute that class of organic acids which contain an amino group located on the carbon atom adjacent to the carboxyl group, as illustrated in the general formula

```
          O     H          H     H
           \\  /            \   /
    H       C—O               N       O
     \      |                 |     //
      N—C—H               R—C—C                (1)
     /      |                 |     \
    H       R                 H      O—H
         (A)                    (B)
```

The R-group represents the remainder of the molecule, and varies from H in *glycine* to the complex indole ring system in *tryptophane*. With the exception of the simplest amino acid, *glycine*, the central (α) carbon atom is asymmetric because all four groups linked to it are different. Thus, mirror image D and L stereoisomeric forms are possible; however, in natural proteins all amino acids have the L configuration shown in Formula (1), in which the structure (*A*), with the NH_2 group to the left of the α-carbon atom, is the conventional method of designating the L configuration, and the structure (*B*) is the form more convenient for writing and printing.

Amino acids contain both the acidic *carboxyl* group (proton donating) and the basic *amino* group (proton accepting) within the molecule. In the pH range encountered physiologically (bordering neutrality), the carboxyl group is dissociated, and the NH_2 group binds a proton to give the following structure:

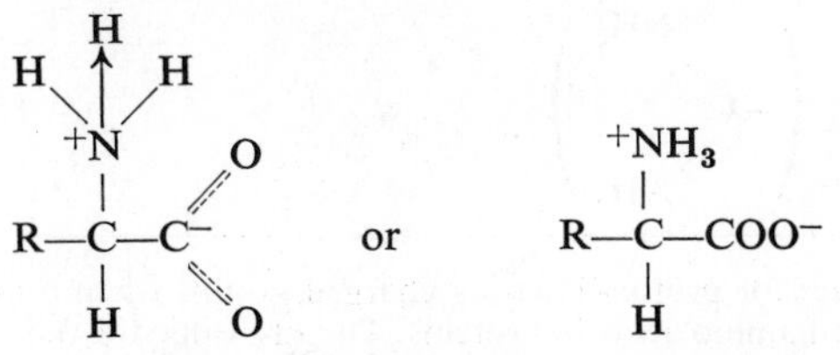

$$R{-}CH(N^{+}H_3){-}C(=O)O^{-} \quad \text{or} \quad R{-}CH(^{+}NH_3){-}COO^{-} \tag{2}$$

This type of ionized molecule with negative and positive charges is referred to as a dipolar ion or zwitterion. Proteins themselves are complex zwitterion type structures containing many negatively and positively charged groups in each molecule. Although the structure shown in Formula (2) is the more precise way to present the structure of an amino acid, for purposes of convenience the simpler formulation, R-CH(NH_2)COOH, will be used most often in this chapter.

Because amino acids can react with either acids or bases, their solutions can be used as pH buffers. This is illustrated in the following equations:

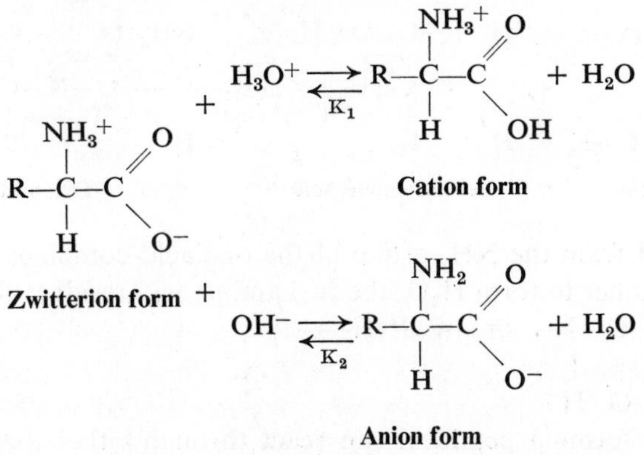

$$\underset{\text{Zwitterion form}}{R{-}CH(NH_3^{+}){-}C(=O)O^{-}} + H_3O^{+} \underset{K_1}{\rightleftharpoons} \underset{\text{Cation form}}{R{-}CH(NH_3^{+}){-}C(=O)OH} + H_2O \tag{3a}$$

$$\underset{\text{Zwitterion form}}{R{-}CH(NH_3^{+}){-}C(=O)O^{-}} + OH^{-} \underset{K_2}{\rightleftharpoons} \underset{\text{Anion form}}{R{-}CH(NH_2){-}C(=O)O^{-}} + H_2O \tag{3b}$$

For glycine, the value of $pK_1 = 2.34$ and that of $pK_2 = 9.60$. Inasmuch as buffering is effective only over a pH range of 1.5 units on either side of the pK value, glycine may be used as a buffer in the pH intervals of about 1.0 to 3.8 and 8.1 to 11.0 only. The ionization constants, expressed as pK values, of the various ionizable groups present in amino acids and proteins are presented in Table 7–1.

There are a number of amino acids that are repeatedly encountered in clinical chemistry. Some of these are among the 20 occurring in all proteins. Others are found only in certain special proteins, in free form, or as constituents of nonprotein compounds. The laboratory worker should be familiar with the formulas of these compounds and the nature of the R-groups that characterize them. These are presented in Table 7–2. Attention is directed to the conventional abbreviations for amino acids. These are used to describe the amino acid sequence in proteins.

TABLE 7–1 IONIZATION CONSTANTS OF IONIZABLE GROUPS IN FREE AMINO ACIDS AND IN PROTEINS*

Ionizing Group	*Range of pK Values* FREE AMINO ACIDS	PROTEINS
Principal carboxyl (−COOH) = pK_1	1.7– 2.6	3.0– 3.2
α-Amino (—NH_3^+) = pK_2	9.1–10.8	7.6– 8.4
Second carboxyls of Glu and Asp (—COOH)	3.8– 4.3	3.0– 4.5
Imidazole nitrogen of His (imidazole N^+—H)	6.0	5.6– 7.0
Sulfhydryl of cysteine (−SH)	8.3	9.1–10.8
Phenolic hydroxyl of Tyr (−OH)	10.1	9.2– 9.8
ε-Amino of Lys (—NH_3^+)	10.5	9.4–10.6
Guanido group of Arg (−C(NH_2)=NH_2^+)	12.5	11.5–12.6

* Given are the pK values for proton-donating charged groups when present in free amino acids and when present in peptide-linked amino acids in proteins. The pK value for the primary carboxyl varies from 1.71 for Cys to 2.63 for Thr. Similarly, the pK for the α-ammonium group varies from 8.95 for Lys to 10.78 for Cys. In protein chains the proximity of other amino acid residues and charged groups may modify the pK for any given ionizable group. The amino acid symbols are those listed in Table 7–2.

The peptide bond

Amino acids can react with one another to form peptides by the linking of the amino group of one acid with the carboxyl group of another acid:

$$\underset{\text{First amino acid}}{R_1\text{—}\overset{NH_2}{\underset{H}{C}}\text{—}C(=O)\text{—OH}} + \underset{\text{Second amino acid}}{H\text{—}\underset{H}{N}\text{—}\overset{C(=O)\text{—O—H}}{\underset{R_2}{C}}\text{—H}} \rightarrow \underset{\text{Dipeptide}}{R_1\text{—}\overset{NH_2}{\underset{H}{C}}\text{—}\overset{O}{\overset{\|}{C}}\text{—}\underset{H}{N}\text{—}\overset{C(=O)\text{—OH}}{\underset{R_2}{C}}\text{—H}} + H_2O \quad (4)$$

In this reaction the H from the NH_2 group of the one acid combines with the OH of the COOH group of the other to form H_2O, the two amino acid residues then being linked via the *peptide linkage*, $\text{—}\underset{O}{\underset{\|}{C}}\text{—}\underset{H}{\underset{|}{N}}\text{—}$ or —CONH—.

Amino acids, in forming peptides, can react through either the amino or carboxyl ends. For example, glycine and alanine can react to form two different dipeptides, *glycyl-alanine* and *alanyl-glycine:*

$$\underset{\text{Glycine}}{NH_2CH_2COOH} + \underset{\text{Alanine}}{CH_3CHNH_2COOH} \rightarrow \underset{\text{Glycyl-alanine}}{NH_2CH_2\text{—}\overset{O}{\overset{\|}{C}}\text{—}\underset{H}{\underset{|}{N}}\text{—}\overset{H}{\overset{|}{\underset{CH_3}{\underset{|}{C}}}}\text{—COOH}} \quad (5)$$

$$\underset{\text{Alanine}}{CH_3CHNH_2COOH} + \underset{\text{Glycine}}{NH_2CH_2COOH} \rightarrow \underset{\text{Alanyl-glycine}}{CH_3CHNH_2\text{—}\overset{O}{\overset{\|}{C}}\text{—}\underset{H}{\underset{|}{N}}\text{—}CH_2COOH} \quad (6)$$

TABLE 7-2 IMPORTANT AMINO ACIDS

Name	*Symbol*	*Formula*	*Comments*
		ALIPHATIC, UNCHARGED R-GROUPS	
Glycine	Gly	NH_2CH_2COOH	Simplest amino acid; optically inactive; used as buffer.
Alanine	Ala	CH_3CHNH_2COOH	Substrate for alanine transaminase (GPT,ALT).
Serine	Ser	$HOCH_2CHNH_2COOH$	Involved in "active center" of many enzymes. hydroxyl may be phosphorylated.
Leucine	Leu	$CH_3CHCH_2CHNH_2COOH$ | CH_3	Branched chain R-group, which is hydrophobic; faulty metabolism in "maple syrup" disease; ketogenic; essential.
Isoleucine	Ile	$CH_3CH_2CHCHNH_2COOH$ | CH_3	See leucine; partly ketogenic; essential.
Valine	Val	$CH_3CHCHNH_2COOH$ | CH_3	See leucine; partly ketogenic; essential.
Threonine	Thr	$CH_3CHOHCHNH_2COOH$	Essential.
Cysteine	Cys	$HSCH_2CHNH_2COOH$	Functional sulfhydryl (—SH) group; free —SH necessary for activity of many enzymes.
Cystine	Cys | Cys	$S—CH_2CHNH_2COOH$ | $S—CH_2CHNH_2COOH$	Oxidized form of cysteine; insoluble at neutral pH; forms one type of kidney stone; may link two peptide chains, as in insulin.
Methionine	Met	$CH_3—S—CH_2CH_2CHNH_2COOH$	Contains sulfur; methyl group transfer agent; essential.
		AROMATIC AND HETEROCYCLIC, UNCHARGED R-GROUPS	
Phenylalanine	Phe	$C_6H_5—CH_2CHNH_2COOH$ (benzene ring)	Metabolism deficient in phenylketonuria; essential.
Tyrosine	Tyr	$HO—C_6H_4—CH_2CHNH_2COOH$ (benzene ring)	Usually nonessential; intermediate in synthesis of catecholamines, thyroxine; functional phenolic hydroxyl; reacts with Folin reagent in quantitative protein assay.
Tryptophane	Trp	CH_2CHNH_2COOH on indole ring (C—H, N—H)	Essential; contains indole ring system; metabolites involved in carcinoid disease.

Table 7-2 *continued on following page.*

TABLE 7–2 *(Continued.)*

Name	*Symbol*	*Formula*	*Comments*
		DICARBOXYLIC, ACIDIC R-GROUPS	
Aspartic acid	Asp	$HOOC—CH_2CHNH_2COOH$	Substrate for aspartate transaminase (GOT).
Glutamic acid	Glu	$HOOC—CH_2CH_2CHNH_2COOH$	Substrate for both GOT and GPT.
		BASIC AMINO ACIDS, BASIC R-GROUPS	
Lysine	Lys	$H_2NCH_2CH_2CH_2CH_2CHNH_2COOH$	Terminal NH_2 referred to as ε (epsilon) NH_2; essential.
Arginine	Arg	$H_2N—CNH—CH_2CH_2CH_2CHNH_2COOH$ ‖ NH	Involved in urea synthesis; the basic group is guanidine.
Histidine	His	H—N—CH H—C⁺=N—C—CH_2CHNH_2COOH H	R-group represents imidazole; ring $\overset{+}{>}$NH ionizes at physiological pH; in hemoglobin it ionizes on oxygenation.
		IMINO ACIDS (RING >NH REPLACES NH_2)	
Proline	Pro	CH_2 CH_2 CHCOOH CH_2 N H	Important constituent of collagen and gelatin.
Hydroxyproline	Hyp	H CH_2 HO—C CHCOOH H_2C N H	Present in collagen; urine output used as an index of bone matrix metabolism.
		AMINO ACID AMIDES	
Glutamine	Gln	O ‖ $H_2NC—CH_2CH_2CHNH_2COOH$	Storage form of ammonia in tissues.
Asparagine	Asn	$H_2NC—CH_2CHNH_2COOH$ ‖ O	

MISCELLANEOUS AMINO ACIDS

Thyroxine	T_4	$HO-C_6H_2I_2-O-C_6H_2I_2-CH_2CHNH_2COOH$ (two I on each ring)	Thyroid hormone; contains 4 iodine atoms.
Triiodothyronine	T_3	$HO-C_6H_3I-O-C_6H_2I_2-CH_2CHNH_2COOH$ (one I on the first ring, two on the second)	Thyroid hormone; more active than T_4; contains only 3 iodine atoms.
β-Alanine		$H_2N-CH_2CH_2COOH$	Constituent of pantothenic acid (a vitamin).
Dihydroxyphenylalanine	DOPA	$(HO)_2C_6H_3-CH_2CHNH_2COOH$	Intermediate in synthesis of catecholamines.
γ-Aminobutyric acid	GABA	$H_2NCH_2CH_2CH_2COOH$	Metabolite of Glu; regulates nerve impulses in brain.
Ornithine		$H_2N(CH_2)_3CHNH_2COOH$	Intermediate in urea synthesis.
Phosphoserine		$H_2O_3P-O-CH_2CHNH_2COOH$	In casein and other phosphoproteins
Pyrrolidone carboxylic acid	PCA	Five-membered ring: $H_2C-CH_2-C(H)(COOH)-N(H)-C(=O)-$ (closing to H_2C)	Cyclicized form of Glu; rare, used to terminate peptide chains, as at N-terminal end of L-chains in γ globulins

An important tripeptide is *glutathione*, γ-glutamyl-cysteinyl-glycine, in which the terminal COOH of glutamic acid is linked with the NH_2 of cysteine, and the COOH of the latter is linked with the NH_2 of glycine:

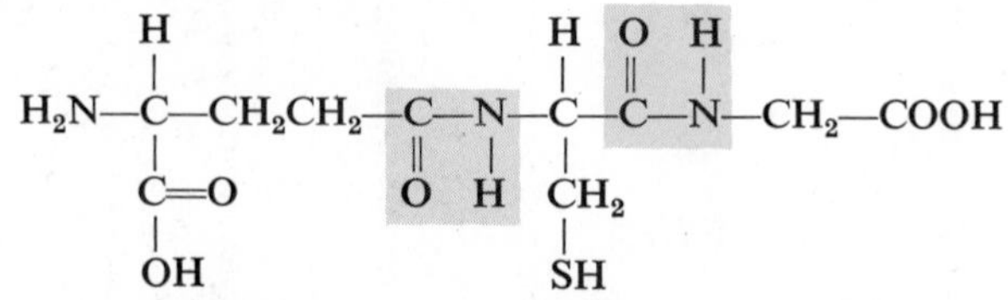

$$
\begin{array}{ccccccccccccc}
 & H & & & & & & H & & O & & H & \\
 & | & & & & & & | & & \| & & | & \\
H_2N- & C & -CH_2CH_2- & C & - & N & - & C & - & C & - & N & -CH_2-COOH \\
 & | & & \| & & | & & | & & & & & \\
 & C{=}O & & O & & H & & CH_2 & & & & & \\
 & | & & & & & & | & & & & & \\
 & OH & & & & & & SH & & & & & \\
\end{array}
\tag{7}
$$

γ-Glutamyl **Cysteinyl** **Glycine**

Glutathione, reduced form

Short chains of amino acids (6 to 30 residues) linked together by peptide bonds are referred to as *polypeptides*. The hormone *oxytocin* is an octapeptide. The terms *proteose* and *peptone* refer to protein breakdown products containing large polypeptides, which differ from true proteins in that they are not coagulated by heat. When the number of amino acids linked together reaches 40 or more (molecular mass of 5000), the chain takes on the physical properties associated with proteins.

CLASSIFICATION AND PROPERTIES OF PROTEINS

Proteins differ from one another in the number of the various amino acids linked together in the chain and in their arrangement or order in the chain. The molecular weight can vary from 5738 for the 51 amino acids in insulin to as high as several million for certain structural proteins.

In solution the polypeptide chains (primary structure) twist into coils or helices (secondary structure), and these are bent into folds (tertiary structure) to form complex globular, ellipsoidal, or needle-shaped forms. This folded chain constitutes the final *native protein*. The coils (helices) and folds are held together by hydrogen bonding, by weak covalent or electrostatic bonds between different R-groups, and by the disulfide (—S—S—) bonds of cystine. Many physical and chemical properties of proteins derive from this tertiary (folded) structure of the protein. When these tertiary bonds are broken, the chain may be partly or completely unfolded, and the protein is said to be *denatured*, with a resulting change in properties. The most striking change is a decrease in solubility and, in the case of enzymes, partial or total loss of catalytic properties. Egg white is a native protein; it is easily soluble in water and gives a clear water-white solution. If the solution is vigorously stirred, however, such as with an egg beater, or heated to above 60 to 70°C, the mechanical or heat energy, respectively, will alter the tertiary structure, causing the egg white to coagulate or set into a white insoluble form. *Protein solutions must be treated with extreme care to avoid the occurrence of denaturation.*

The larger protein moieties consist of two or more similar or related chains linked together by weak chemical bonds (quaternary structure). The five isoenzymes of lactate dehydrogenase (4 chains) and the copper-containing protein, ceruloplasmin (8 chains), are examples of such associations.

CLASSIFICATION BY STRUCTURE AND COMPOSITION

Proteins can be classified as *simple* or as *conjugated* proteins. In the conjugated proteins, some type of nonprotein moiety is linked with the folded amino acid chain. This added *prosthetic group* conveys new and characteristic properties to the complex formed. The colored *chromoproteins*, such as hemoglobin, contain an organic prosthetic group that is linked to some metal ion. In the *metalloproteins* some metal ion is attached directly to the protein. Examples of metalloproteins are *ferritin*, containing iron, and *ceruloplasmin*, containing copper. *Lipoproteins* contain bound cholesterol, phospholipids, and/or triglycerides, whereas *glycoproteins*, such as *mucins* and *orosomucoid*, contain complex carbohydrates in their structure. In the *nucleoproteins* (cellular chromatin material), the protein is associated with chains of deoxyribonucleic acids.

A satisfactory chemical classification of the very large number of individual proteins will not be possible until much more is known about their detailed structures. In the meantime, proteins can be conveniently differentiated into a few classes by laboratory techniques based on their special chemical and physical properties. For example, the oldest method of classification is based on solubility differences.

CLASSIFICATION OF PROTEINS BY SOLUBILITY PROPERTIES

Proteins which are soluble in water and also in dilute and moderately concentrated salt solutions are referred to as *albumins*. Human serum albumin is an example; egg albumin is another. Albumins are differentiated from protein derivatives such as peptones by being insoluble in saturated ammonium sulfate solution and other highly concentrated salt solutions. *Globulins*, as a class, are defined as proteins which are insoluble in water, soluble in weak salt solutions (2 to 10 g/100 ml $\sim$ 0.6 mol/l), but insoluble in concentrated salt solutions (12 to 60 g/100 ml $\sim$ 0.9 to 4.5 mol/l). Ammonium sulfate is the salt most frequently used, but Na_2SO_4, Na_2SO_3, $MgSO_4$, and others are also used.

The *albuminoids* are a special group of proteins, characterized by being essentially insoluble in most common reagents. They comprise the various fibrous proteins, which have a supporting or protecting function in the organism. Among these are the *collagens* (skin, cartilage), *elastins* (tendons), and *keratins* (hair, feathers).

LABORATORY TECHNIQUES USED FOR THE SEPARATION AND CLASSIFICATION OF PROTEINS

The methods currently used and the protein properties that are utilized in these methods are as follows:

1. Salt or solvent fractionation: changes in solubility in the presence of charged ions or in the presence of dehydrating solvents, or both.
2. Electrophoresis: differences in surface electrical charge density.
3. Ultracentrifugation: variations in molecular mass and molecular shape.
4. Chromatography: differences in size, shape, and charge as affecting rate of flow through various types of chromatographic media.
5. Gel filtration: differences in molecular volume and shape.
6. Immunochemical analysis: differences in localized molecular surface structure (determinants) giving rise to reactions with specific antisera.

Each of these techniques separates protein mixtures into individual fractions by use of a specific protein property. A fraction that may appear as a single homogeneous electrophoretic band may, on ultracentrifugation, give rise to several sedimentation fractions. Similarly, an albumin fraction obtained by salt precipitation will, invariably, when subjected to electrophoresis, form two or more individual bands in the albumin area of the electrophoretogram. Both large and small protein molecules may have the same net surface electrical charge density. Immunochemical techniqies, on the other hand, do not usually separate proteins, but instead make possible identification and quantitation of individual proteins in a protein mixture. Thus, these various protein separation techniques complement each other.

SALT AND SOLVENT FRACTIONATION OF PROTEINS

Protein solubility is governed by several factors and phenomena:

Proteins are colloidal solutions. The individual molecules are charged because of the presence of ionized chemical groups on their surfaces. Included among such charged groups are carboxyl (—COO^-), NH_3^+, and also, if the pH is favorable, the *guanidine*, *imidazole*, or ε-NH_3^+ groups of the amino acids *arginine*, *histidine*, and *lysine*, respectively.

In solution, a protein molecule is associated with a layer of water molecules (hydration shell, bound water), reflecting interaction between the water molecules and charged groups. The thickness of the hydration shell depends on the distribution and relative number of polar and nonpolar groups on the protein surface. The water shell and the Helmholtz electrical double layer associated with it enhance solution forces by inhibiting aggregation of protein molecules.

Solubility is promoted by a high concentration of free water, as well as by the high dielectric constant of this solvent. The activity of water is reduced by the addition of inorganic salt ions because they also bind water molecules (ion hydration). The addition of organic solvents such as ethanol or acetone similarly reduces free water concentration, and, in addition, these solvents also effect a sizeable drop in the dielectric properties of the solvent mixture.

The charges on the protein molecule tend to promote intermolecular electrostatic association or aggregation. This tendency is increased with decrease in the dielectric constant of the solvent, decreased quantity of free water, and contraction of the bound water shell. Proteins insoluble in pure water, such as euglobulins, dissolve on addition of low concentrations of inorganic ions ("salting in") because of attraction between the protein molecules and the salt ions. However, at high salt levels, all the various factors operate to cause diminishing or loss of solubility ("salting out"). The individual globulins in plasma and tissue extracts differ considerably in structure and size, and as a result require different concentrations of ions before they will precipitate out. In general, the more the charge on an ion, the more effective it is as a protein precipitant. By varying the type of salt and, especially, salt concentration, an entire spectrum of globulins can be obtained. Salt fractionation is of little value for isolating individual pure proteins, but it is a useful tool for separating a mixture into broad groups.

In the use of solvents for protein precipitation and separation, denaturation of the proteins is avoided by working at low temperatures (-5 to -30°C). Often changes in pH and the addition of certain metal ions are used to improve specific separations. This technique is used commercially to prepare plasma protein fractions (γ globulins, albumin) for clinical use.

DIFFERENCES IN SURFACE ELECTRICAL CHARGE PROPERTIES OF PROTEINS

Proteins are amphoteric compounds similar to the amino acids, having both positive and negative charges on the same molecule, the number depending upon the pH of the solution. The pH at which the numbers of positive and negative charges are equal is called the *isoelectric* point (pI). For most proteins this pH is on the acid side, varying from 4.6 for albumins to 5.1 to 6.2 for globulins. At a pH below the pI, the positive charges outnumber the negative charges and the protein behaves like a positive ion (cation); solutions contain $(Pr)^{+n}$ ions, analogous to the Na^+ ion. At physiological pH values, however, the negative charges will be in the majority and the protein will behave like an anion $(Pr)^{-n}$, analogous to Cl^- and HCO_3^-. Different proteins will vary in the net positive or negative charge per unit area of surface. Thus, if a solution containing proteins is placed between a pair of electrodes connected to a voltage source, the proteins will migrate to either cathode or anode, depending on their net charge, at speeds depending on the amount of charge and the size and shape of the individual protein molecules. This technique is referred to as *electrophoresis*. At pH 8.6, all serum proteins are negatively charged, but the albumins migrate most rapidly and are separated from the globulins. The latter can be split into three or more fractions, called alpha (α), beta (β), and gamma (γ) globulins, the γ being the slowest moving. Under proper conditions, each of these may be separated into subfractions.

An exquisite technique for protein separation based on surface charge properties is referred to as *isoelectric focusing*. The protein mixture is caused to migrate through a medium in which a continuous pH gradient has been established. The individual proteins migrate under the impetus of the electric field until they reach a pH equal to their isoelectric point, at which they precipitate out or cease to move (p. 434).

DIFFERENCES IN MOLECULAR MASS AND DENSITY OF PROTEINS

Because proteins are made up of different numbers and kinds of amino acids, individual proteins differ in their molecular mass. If a protein solution is placed in a centrifuge operating at a very high speed, the high centrifugal force will force the heavier molecules to sediment or settle out faster than the lighter molecules. The necessary high speeds can be attained only by use of an *ultracentrifuge*, usually not available in a routine clinical laboratory. This technique is particularly useful in separating various light lipoproteins from other (heavier) serum proteins.

SEPARATION OF PROTEINS BY ION-EXCHANGE CHROMATOGRAPHY

Because of differences in size, shape, and chemical composition, different proteins will be adsorbed on, and can be eluted from, various adsorbents at different rates. By control of pH, buffer type, and concentration, and by the proper choice of adsorbents and eluting solutions, individual proteins can be separated from each other and obtained in a highly purified form. This technique is not yet important in routine work (except to separate thyroxine-binding protein from contaminating iodine-containing material), but it will probably become a very important tool in the clinical laboratory in the near future. Modified celluloses such as diethylaminoethyl (DEAE) cellulose and carboxymethyl (CM) cellulose are commonly used in protein work.

SEPARATION OF PROTEINS BY GEL FILTRATION (MOLECULAR SIEVES)

These techniques separate proteins and other solutes by virtue of differences in molecular size and shape. The solutes are passed through columns of beads containing pores of fixed size. Small molecules become entrapped in the pores and are held back, whereas molecules too big to enter the pores pass through the column rapidly. By varying the sieve openings, different sizes of molecules can be entrapped, permitting separation of proteins and solutes into molecular mass classes. The shape of a molecule will also determine the ease with which it will enter or leave a bead channel. Dextran (Sephadex) and Agarose beads are available commercially (pp. 163 and 332).

SEPARATIONS OF PROTEINS BY IMMUNOCHEMICAL TECHNIQUES: DIFFERENCES IN ANTIGENICITY

The many differing amino acid chains in proteins allow an enormous variety of molecular shapes and surface structures; even a change of one amino acid, while often not affecting the basic structure, may significantly alter charge density and chemcial properties in a localized region of the protein. Thus, to distinguish proteins from one another, especially if they are closely related, reagents are required which are sensitive to such changes.

Fortunately, animals are capable of making just such reagents: these reagents are the antisera described in Section Two of this chapter. The proteins to be studied are injected into appropriate animals, who react by forming antibodies which can specifically react with and neutralize the antigen protein.

Immunochemical methods are very specific, and only in the case of proteins in very low concentrations, such as immunoglobulin E or protein hormones, is their sensitivity inadequate. In such cases, radioimmunoassay, which combines the sensitivity of radioactive labeling with the specificity of the immune reaction, is used, permitting detection of a few pg/ml.

ADDITIONAL READING

1. Bezkorovainy, A.: Basic Protein Chemistry. Springfield, Illinois, Charles C Thomas, Publisher, 1970.
2. Harper, H. A.: Review of Physiological Chemistry. 14th ed. Los Altos, California, Lange Medical Publishers, 1973.
3. Henry, R. J., Cannon, D. C., and Winkelman, J. W.: Clinical Chemistry, Principles and Technics. 2nd ed. Hagerstown, Maryland, Harper and Row, Publishers, 1974.
4. Hoffman, W. S.: The Biochemistry of Clinical Medicine. 4th ed. Chicago, Illinois, 1970, pp. 13–28, 41–56.
5. Lehninger, A. L.: Biochemistry. New York, Worth Publishers, 1970, pp. 55–146.
6. Mazur, A. and Harrow, B.: Textbook of Biochemistry. Philadelphia, W. B. Saunders Company, 1971, pp. 17–95.
7. Peters, T., Jr.: Serum albumin. *In* Advances in Clinical Chemistry. O. Bodansky and C. P. Stewart, Eds. New York, Academic Press, Inc., 1970, vol. 13, pp. 37–111.
8. Watson, D.: Albumin and "Total Globulin" Fractions of Blood. *In* Advances in Clinical Chemistry. H. Sobotka and C. P. Stewart, Eds. New York, Academic Press, Inc., 1965, vol. 8, p. 237.
9. White, A., Handler, P., and Smith, E. L.: Principles of Biochemistry. 5th ed. New York, McGraw-Hill Book Company, 1973, pp. 89–179, 629–636.

SECTION TWO

IMMUNOCHEMICAL PRINCIPLES

by Harold Markowitz, Ph.D., M.D., and Nai-Siang Jiang, Ph.D.

At the beginning of the 20th century, immunology could be defined simply as the study of immunity in infectious disease. Since then, growth has been explosive and many new subdisciplines have arisen, including immunochemistry, immunobiology, immunopathology, and immunogenetics. Immunochemistry originated with the detailed investigations by Landsteiner on the interactions of haptens and antigens with antibodies and the detailed studies by Heidelberger and others in which quantitative chemical techniques were applied to immunologic systems. Hand in hand with the development of more powerful tools for protein separation and analysis has come better understanding of the structures of antigens and immunoglobulins, of the interaction between antigen and antibody, of the structure of antigenic determinants, of the size of the antibody-combining site, and of the role of complement and its components in reactions *in vitro* and *in vivo*.

This section of Chapter 7 will outline some of the fundamental concepts that are the bases of immunologic techniques and will present some of the mechanisms involved in specific tests.

DEFINITIONS AND PROPERTIES OF IMMUNOGENS, ANTIGENS, AND ANTIBODIES

It is extremely difficult to provide a rigid definition of these terms, and almost any definition becomes circular. Thus, an *immunogen** (or an immunogenic material) is a protein or a substance coupled to a protein that, when introduced into a foreign host, is capable of inducing the formation of another protein, an *antibody*, in the host. This definition is neither precise nor complete. The route of introduction is usually, but not always, parenteral, and the antibody produced may either be circulating (*humoral*) or tissue-bound (*cellular*), as in delayed hypersensitivity reactions or graft reactions.

The simplest demonstration of an antigen-antibody reaction is the precipitin reaction detected by the appearance of turbidity or flocculation when an immune serum (serum containing antibody) is mixed with antigen. Such a reaction usually does not occur with normal serum (serum not containing antibody). Methods have been developed by which the amount of precipitate formed can be measured, and application of radioimmunoassay techniques (see p. 286) makes possible the detection of even nanogram quantities of antigen or antibody. Also, methods for the detection of antibodies are available that do not involve formation of precipitates (for example, reactions involving haptens and univalent antibodies).

The types and sources of immunogens vary greatly. If we include those small molecules (*haptens*) that are capable of inducing specific antibody formation when coupled to a carrier protein, even molecules such as L-tyrosine, penicillin, D-glucose, and arsanilic acid are immunogenic. In addition to proteins and polypeptides, lipids (such as the Wassermann antigen, cardiolipin), nucleic acids, and many other materials can function as antigens. Some general properties requisite for immunogenicity are as follows: (1) stable structure or areas of stability within the molecule, (2) randomness of structure, (3) a minimal molecular mass of 4000 to 5000, (4) ability to be metabolized (a necessary but not sufficient criterion for some classes of antigens), (5) accessibility of a particular immunogenic configuration to the antibody-forming mechanism, and (6) foreignness of structure to the host. The overall shape and charge of the molecule in most cases have very little influence on the ability to induce antibody formation.

Injection of an immunogen into a host animal induces a series of biologic changes leading to markedly increased production of some serum globulins with specific antibody activity—the *immunoglobulins*— and to changes in the capability of certain cellular elements to give an immune response. Continuing stimulation by immunogen can result in increasing production of immunoglobulins of different types and of different specificities and avidities† for antigen. In some cases, excessive stimulation can lead to immunologic paralysis, in which no antibody response is mounted until a major fraction of the immunogen is cleared from the organism. After the first exposure to an immunogen, a latent period (*induction*) occurs during which no antibody is present in serum; this period may last from five to ten days. Because antibody formation by sensitized (immunogen-coated) cells can be demonstrated *in vitro* in a period of minutes, the long delay in antibody build-up *in vivo* is due to the time required for transport of immunogen to the site of antibody production, transmission of information, and actual synthesis of antibody.

Serum sickness can be induced by a single injection of a large amount of immunogen.

* The term "immunogen" (instead of the term "antigen") is now used when referring to materials capable of eliciting antibody formation when injected into a host. The term "antigen" is used for any material capable of reacting with an antibody without *necessarily* being capable of inducing antibody formation. For example, egg albumin is an immunogen, since it is capable of inducing formation of anti-egg albumin antibody. Morphine is an antigen, since it will react with antimorphine antibodies, but it is not an immunogen since it does not induce antibody formation (unless first conjugated to a protein). In connection with immunochemical laboratory tests it is current practice to use the broader term "antigen."

† *Avidity* describes the capacity of an antibody to react rapidly and firmly with a specific antigen.

In about five to eight days, sufficient antibody forms so that some appears in the blood. Circulating antibody combines with antigen already present, causing the production of histamine as well as other amines and peptides. The antigen-antibody complexes formed also bind *complement* (*C*) and so deplete the circulating *C* level (see p. 291). The aggregates are removed by the kidney, and this causes an *immune-complex nephritis* and subsequent albuminuria. This protean disease, more frequent in the past as a result of antitoxin therapy, is only one example of the involvement of immunologic reactions in disease.

Antibodies (immunoglobulins) are proteins or carbohydrate-containing proteins of various molecular masses and structures. The feature common to all immunoglobulins is their immunologic specificity—namely, their ability to combine with the specific immunogen that led to their production. In many cases, an antibody is also capable of cross-reacting with compounds that are structurally similar but not identical to the eliciting immunogen. The circulating immunoglobulins are collectively referred to as humoral antibodies. Another large group of immunoglobulins, about which much less is known, are referred to as cellular antibodies; these are involved in delayed hypersensitivity reactions, rejection of transplants, allergic reactions, tumor immunity, and other phenomena.

Five distinct immunoglobulin (Ig) classes have been described: IgG, IgA, IgM, IgD, and IgE (Table 7–3). The simplest immunoglobulin molecule is IgG; it consists of two identical heavy (*H*) polypeptide chains (molecular mass 50,000 daltons) and two identical light (*L*) chains (molecular mass 25,000 daltons). The *H* chains vary with the Ig type, and five chains have been described—γ, α, μ, δ, and ε—associated with IgG, IgA, IgM, IgD, and IgE, respectively. Only two types of *L* chains are found, kappa (κ) and lambda (λ), and one or the other is found in all Ig classes. Thus, a given Ig may be designated on the basis of its chains—for example, $\gamma_2\kappa_2$ or $\alpha_2\lambda_2$. *H* and *L* chains are linked by a disulfide bridge. Intrachain disulfide bridges are also present on both *H* and *L* chains (Fig. 7–1).

The molecule can be cleaved by papain into a pair of *Fab* fragments (containing the *L* chain and one-half of an *H* chain) and an *Fc* fragment (consisting of the remaining halves of the *H* chains). The *Fab* portion is variable in amino acid composition. It accounts for the uniqueness of the antibody in regard to combining with its specific antigen. The *Fc* fragment is of constant composition and functions in such reactions as complement fixation and skin fixation. IgA, IgD, and IgE are similar in structure to IgG, all having two *L* chains and two *H* chains of appropriate type. The macroglobulin IgM (molecular mass 900,000) is a pentamer containing five $\mu_2\kappa_2$ units linked by short peptide fragments. Subclasses of the immunoglobulins also exist—four for IgG and two for IgA (IgA_1 and IgA_2) (Table 7–3).

TABLE 7–3 PROPERTIES OF HUMAN IMMUNOGLOBULINS

Class	*Mol. Mass*	*Carbo-hydrate (%)*	*Heavy Chain Type*	*Serum Level*	*Complement Fixing*	*Placental Transfer*	*Skin Sensitizing*	*In CSF*
IgG	150,000	2.5	γ	11.36*	...	...	...	+
IgG_1	...	...	...	66 ± 8†	2+	3+	3+‡	...
IgG_2	...	...	...	23 ± 8†	+	3+	−‡	...
IgG_3	...	...	...	7.3 ± 3.8†	3+	3+	3+‡	...
IgG_4	...	...	...	4.2 ± 2.6†	−	3+	3+‡	...
IgA	180,000–500,000	7.5	α	1.76*	−	−	−	+
IgM	900,000	12	μ	1.06*	+	−	−	−
IgD	180,000	11	δ	...	−	−	−	−
IgE	200,000	11	ε	...	−	−	+	−

* Mean value (mg/ml) in normal adults by automated nephelometric procedure.
† Per cent distribution.
‡ Passive cutaneous anaphylaxis (PCA).

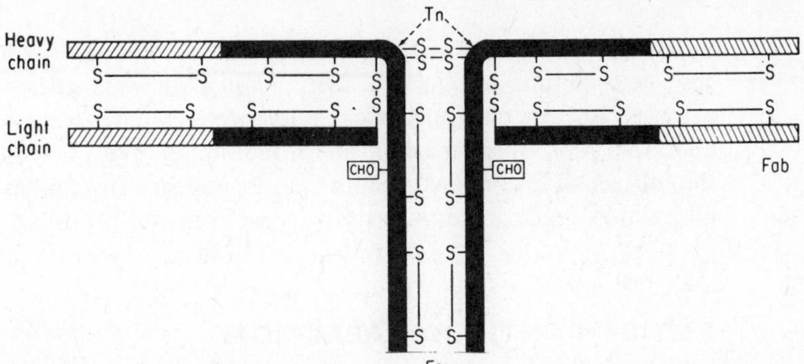

Figure 7-1 Model of the structure of human IgG_1, based on various data; details are based on the structure of the myeloma protein, Eu. Site of cleavage by trypsin (Tn) to produce Fab and Fc fragments is shown by broken lines. Interchain and intrachain disulfide bonds are indicated, and CHO marks the approximate position of the carbohydrate moiety. Variable regions of L and H chains are indicated by hatching. (Reproduced, with permission, from Edleman, G. M., and Gall, W. E.: The antibody problem. Annual Review of Biochemistry, vol. 38, p. 418. Copyright © 1969 by Annual Reviews Inc. All rights reserved.)

The distribution of Ig's in the body varies. IgM and IgD are primarily present intravascularly; the other Ig's are distributed about evenly between blood and tissues. The half-life of IgG is 23 days while that of other Ig's varies from two to six days.

IgA

This is different from the other Ig's in regard to its sites of synthesis and its relationship to secretory IgA (S-IgA). It seems to be particularly adapted to function on secretory surfaces and possesses enhanced ability to adhere to these surfaces and resist proteolytic digestion. Saliva, tears, and gastrointestinal secretions have the bulk of their Ig's in the form of IgA ($7S$* + $11S$); 30 to 40 per cent may be $7S$ IgA. S-IgA is the principal exocrine Ig found in tears, colostrum, saliva, gastrointestinal secretions, and other secretions. It appears to be synthesized locally. This is a dimeric form of IgA, consisting of two molecules bound together by a small peptide (*J* chain) and linked to a "secretory piece" (*T* component) with a molecular mass of about 50,000 daltons. The molecular mass of the *J* chain is about 23,000 daltons so that the total molecular mass is about 400,000 daltons (equivalent to $(\alpha_2\kappa_2)_2TJ$).

IgM

In part because of its large size, IgM enhances agglutination and is more effective in fixing complement. IgM appears very early after an immunogenic stimulus and this, along with its major intravascular distribution, may enable it to serve as a first line of defense. In rheumatoid arthritis and some associated diseases, a $19S$ IgM, known as *rheumatoid factor* (*RF*), is formed and has specificity for $7S$ IgG. *RF* is thought to be an autoantibody and may function in a protective role (an autoantibody is one that is produced by the host against some of its own tissue antigens).

IgE

The most recent of the Ig's to be identified, IgE represents a distinct class of antibodies involved in allergic reactions. IgE is present in the serum of healthy persons in concentrations usually well below 1 μg/ml. Increased amounts may be found in persons with hay fever, asthma, or parasitic diseases and in patients with IgE multiple myeloma. The levels are inconsistently higher in allergic persons and, as a group, their mean serum IgE levels are higher during the season in which they have maximal symptoms. Desensitization, which

* The symbol S refers to the Svedberg number (see Chapter 10, pp. 529 and 535).

leads to higher levels of IgG-blocking antibody, does not lower IgE levels. IgE has been shown to contain antibody activity against ragweed allergen *E* and penicillin, as well as against parasitic antigens such as *Ascaris lumbricoides*, *Necator americanus*, and mite allergens. A proposed mechanism for IgE action involves stimulation of histamine release by the antigen-IgE complex, leading to an increase in capillary permeability that facilitates migration of leukocytes into the affected area and produces an inflammatory response.

ANTIGEN-ANTIBODY REACTIONS

Various qualitative and quantitative immunologic techniques have been developed that are capable of detecting, identifying, and measuring antibody and antigen in the milligram and even picogram range. Because of the exquisite sensitivity and specificity of immunologic procedures, these are being used in diagnostic laboratories for the identification and quantitation of certain biologic compounds (see Chapter 7, Section Three, and Chapters 11, 13, 14, 20, and 21).

The **precipitin reaction** appears to be a two-stage process in which the primary combination of antigen and antibody occurs in seconds or minutes. The secondary stage is characterized by aggregation of the antigen-antibody complex to visible size; this may take minutes to days. The speed of this reaction depends on factors such as electrolyte concentration, pH, temperature, and lipid concentration in the serum, as well as on antigen and antibody types and the binding affinity of the antibody. The solubility of the complex and the effect of the initial volumes vary with the type of antigen (carbohydrate, protein, and so forth), the source of antibody, and the ratio of reacting antibody to antigen in the final mixture. The concentration of NaCl is also important; in most cases, 0.15 molar NaCl is optimal. Greater concentrations of NaCl can lead to smaller amounts of precipitate, which is due not to increased solubility of the antigen-antibody complex but to an equilibrium shift causing a given amount of antigen to combine with smaller amounts of antibody. Decreasing the NaCl concentration can lead to increased precipitation of non-specific serum euglobulins. Generally, the presence of divalent cations has no effect on precipitation.

An appropriate amount of antiserum (usually 0.5 to 1.0 ml) is placed in each of a number of test tubes in an ice bath, and antigen is added to each tube in increasing amounts. The volume is held constant by addition of 0.15 molar NaCl. The tube contents are then mixed. After 48 h, with occasional mixing, the precipitates are separated by centrifugation, washed several times with cold 0.15 molar NaCl, and recentrifuged. The supernatant is saved for the study of antibody or antigen excess (see later). The nitrogen in the precipitate is then estimated by micro-Kjeldahl or other suitable techniques. As shown in Figure 7–2*A*, increasing amounts of precipitate are obtained when increasing amounts of antigen are added. Curve IV represents total N precipitated, while curve III shows only antibody N. The difference between curves III and IV represents the amount of antigen added. Quantitative data with regard to antibody content of the serum and antigen-antibody binding can be obtained from the general equation

$$A = 2RS - \left(\frac{R^2}{A}\right)S^2$$

in which A = mg antibody N precipitated, S = mg antigen N added, and R = A at equivalence point per mg antigen N in precipitate at equivalence point. Plots based on this equation are shown by curves I and II of Figure 7–2*A*.

The supernatant solution is divided into two portions and tested by adding more antigen to one part and more antibody to the other. In the initial phases of the reaction, antibody can still be found in the supernatant (as shown by additional precipitation after

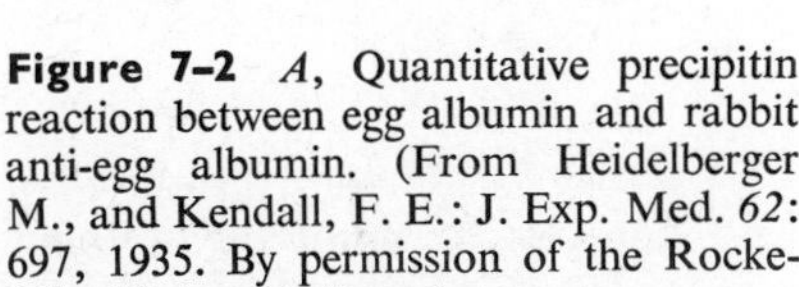

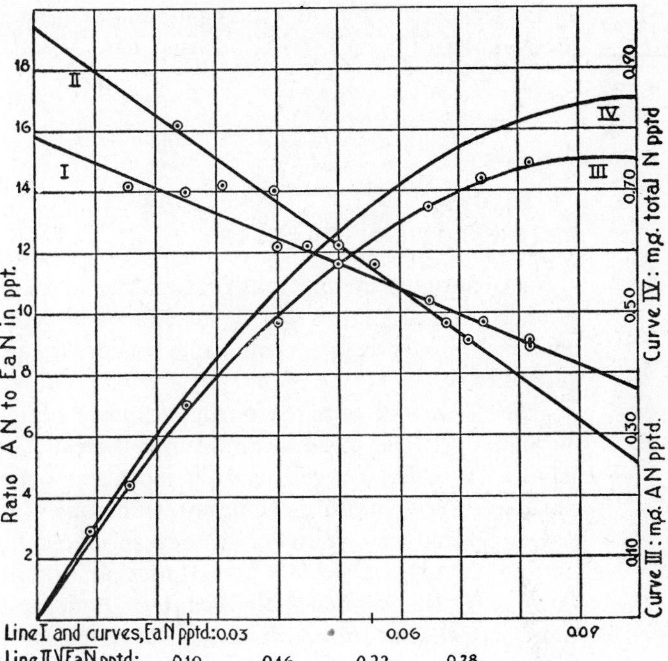

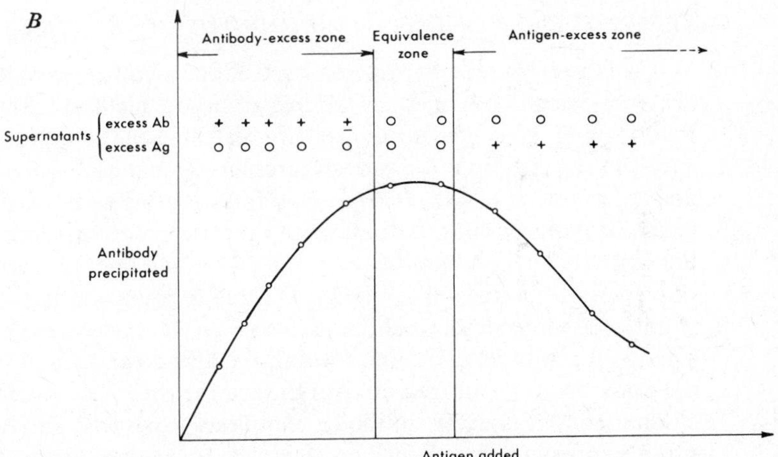

Figure 7-2 *A*, Quantitative precipitin reaction between egg albumin and rabbit anti-egg albumin. (From Heidelberger M., and Kendall, F. E.: J. Exp. Med. *62*: 697, 1935. By permission of the Rockefeller University Press.)

B, Precipitin curve for a monospecific system: one antigen and the corresponding antibody. (From Davis, B. D., *et al.*: Microbiology. 2nd ed. Hagerstown, Maryland, Harper & Row, Publishers, 1973, p. 371.)

antigen addition) but no antigen can be detected (Fig. 7-2*B*). This region is called the *antibody excess zone*. When more antigen is added, an area of the curve is reached in which neither excess antigen nor excess antibody can be detected in the supernatant (*equivalence zone*). The equivalence point is that point at which maximal precipitation occurs. Further addition of antigen leads to the zone of *antigen excess*, in which free antigen can be detected in the supernatant. Finally, very large excesses of antigen may lead to either decreased or no precipitation. This represents the *inhibition zone*, in which soluble antigen-antibody complexes are formed.

A number of theories have been proposed to explain the precipitin reaction. The simplest assumes sets of simultaneous bimolecular reactions between antigen (Ag) and antibody (Ab), each substance having more than one reactive group. This is a dynamic reaction and the composition of the precipitate varies continuously with the proportion of the reactants. Thus, in the egg albumin (Ea)-anti-egg albumin (Ab) system, used as an illustration in Figure 7–2,

Ab_5Ea	→	Ab_3Ea	→	Ab_2Ea	→	$AbEa_2$
Antibody excess		**Equivalence zone**		**Antigen excess**		**Inhibition zone (soluble complexes)**

PRECIPITATION IN GELS

Extension of the precipitin reaction to interaction in a semisolid medium has greatly extended the usefulness of the reaction and has allowed analysis of complex biologic mixtures, comparison of antigenic materials, and rapid quantitation of microgram quantities.

The technique is based on the diffusion of antigen and antibody toward each other through a gel. The speed of diffusion is basically governed by the concentration of the reactants. However, antigens vary in molecular mass, structure, and surface charge, all of which affect the diffusion coefficient. Consequently, individual antigens will separate from each other and will interact with their specific antibodies at different distances from their starting point; this produces arcs of precipitation in the gel. Two general types of tests are used: *single immunodiffusion*, which usually depends on diffusion of an antigen into agar impregnated with antibody; and *double diffusion*, in which both antigen and antibody migrate through the semisolid medium.

TECHNIQUE OF SINGLE IMMUNODIFFUSION IN TUBES

This procedure, first devised by Oudin, is still used to some extent, although it has generally been supplanted by double diffusion methods. Short pieces of small-diameter Pyrex tubing, precoated with agar, are sealed at one end. Antiserum and agar (0.5 to 1.0 g agar/100 ml) are mixed (at temperatures low enough to avoid denaturation of the antibody), and the tubes are filled about one-half full with this mixture. After the agar has set, appropriate dilutions of antigen are layered over the agar to a depth of 2 to 3 cm, the tops of the tubes are sealed with Parafilm or sealing wax, and the tubes are kept at a constant temperature (usually between 20° and 37°C). The appearance of bands of precipitation in the agar is observed after an interval that depends on the system used and the concentration of reactants. Rings of precipitation form in the agar as antigen diffuses into and reacts with the antibody. As more antigen diffuses toward the ring, the concentration of antigen increases, causing soluble antigen-antibody complexes to form, and the ring dissolves. As these soluble complexes move further down the tube, they meet additional antibody, the ratio of antigen to antibody changes, and reprecipitation occurs. If the concentration of antigen is too low in relation to the antibody, the antibody can diffuse out of the agar and form a precipitate in the overlaid antigen solution. Precipitation patterns of various systems are shown in Figure 7–3.

When mixtures of antigens and antibodies are used, multiple bands are usually obtained. The procedure is designed so that the number of bands seen indicates the minimal number of reactive systems present. Occasionally, bands are formed by normal serum proteins because of temperature artifacts or nonspecific interaction with antigen. These are minimized by maintaining constant temperatures, and they can be identified readily because the positions of these bands do not change with time.

Bands can be identified in various ways. Adsorption of the antiserum with pure antigen will remove the precipitin ring caused by that antigen. Alternatively, augmentation of an antigen solution with one of the purified antigens will lead to a higher mobility of one of the bands.

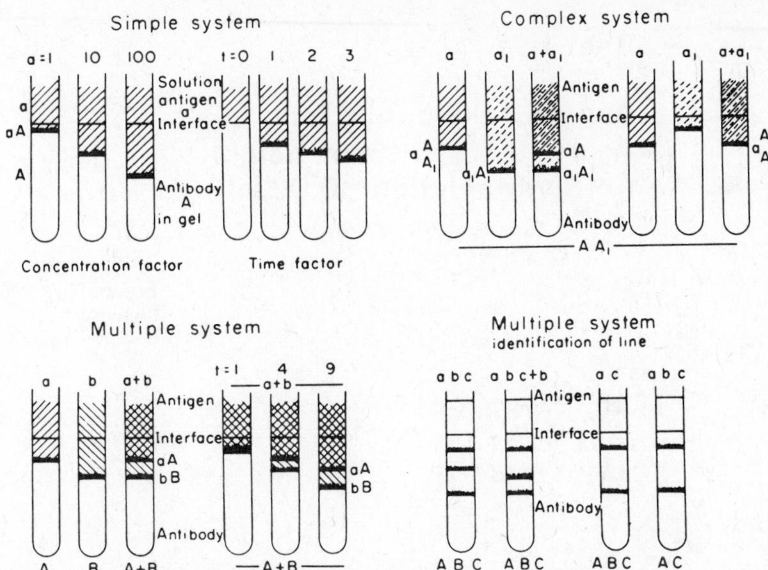

Figure 7–3 Precipitation patterns of simple, multiple, and complex serologic systems tested with one-dimensional simple diffusion (Oudin tube technique).

Increasing concentration of antigen or longer time leads to increase in mobility of precipitin ring (simple system). Complex systems, in which cross-reacting antigens are present, will give different patterns depending on antigen concentration. Native antigen *a*, when present in higher concentrations, moves more rapidly through agar, exhausting both native and cross-reacting antibodies. In this case, cross-reacting antigen *a* will not be detected. (From Weir, D. M.: Handbook of Experimental Immunology. 2nd ed. Oxford, Blackwell Scientific Publications, 1973, p. 668.)

This system is amenable to quantitation, although errors of up to 25 per cent may occur. With tubes held at constant temperature, the distance each band moves from the interface at specific time intervals can be measured. For a single band, the distance traveled from the meniscus is proportional to $t^{1/2}$. The slope of the plot, distance versus $t^{1/2}$, represents the diffusion rate (*d*), which is proportional to the log of the antigen concentration at low antibody-antigen ratios.

DOUBLE IMMUNODIFFUSION TECHNIQUE

The double diffusion technique has been used with diffusion in one or in two dimensions. In the *one-dimension technique* described by Oakley-Fulthorpe and Preer, antibody and antigen are separated from each other by a semisolid column of agar (Fig. 7–4*A*). Antibody and antigen diffuse toward each other through the agar, leading to band formation within hours or days, depending on the concentration and avidity of the antiserum. The distance of band migration from the antigen-agar interface, expressed as per cent (*p*) of the total length of the agar column, is used to identify the band position. The sharpest band is seen when the proportions of antigen and antibody are close to the equivalence point. At this ratio of concentrations, the band remains stationary with time. The sensitivity of the method can be illustrated with the diphtheria toxin–horse antitoxin system in which as many as 17 bands can be seen under optimal conditions. Semilog plots of *p* against antigen dilution will give straight lines.

Another form of the one-dimension double diffusion technique is the *two-basin plate procedure*, in which antigen and antibody are placed in separate wells in an agar plate and allowed to migrate toward each other (Fig. 7–4*B*). The initial concentrations of antibody and antigen at the source of diffusion are critical. Because heavier molecules diffuse more slowly, the position of the band is in part a function of the molecular masses of both antigen

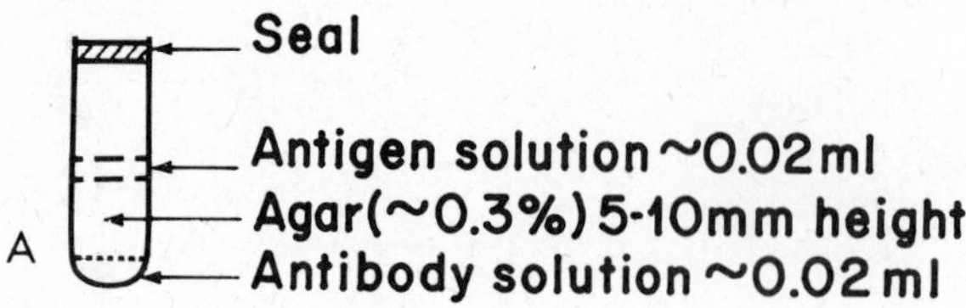

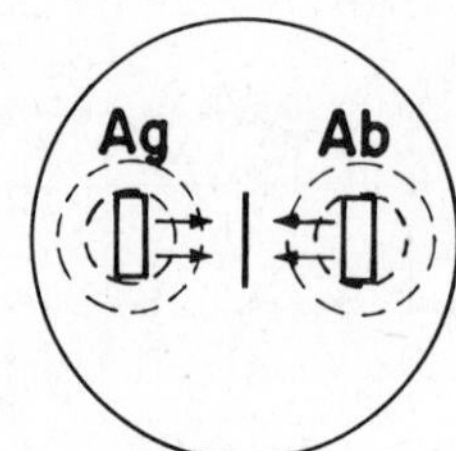

B

Figure 7-4 *A*, Double immunodiffusion in one dimension. *B*, Two-basin plate technique for double immunodiffusion.

and antibody. The precipitation band acts as a specific barrier—neither specific antigen nor antibody can penetrate without being precipitated by the other, but unrelated molecules can cross the band of precipitation freely. The rules which govern the reactivity with cross-reacting materials are the same as in the Oudin procedure. Specific bands of precipitation can grow as more antigen and antibody diffuse into the band zone. Excess antigen or antibody diffusing into the band may dissolve portions of the band which are then reprecipitated at a different position on the plate.

Double immunodiffusion in two dimensions (Ouchterlony technique) is now perhaps one of the most widely used techniques in immunology. It allows direct comparison of two or more test materials and provides a simple and direct method for determining if the materials are identical, cross-reactive, or nonidentical (see also p. 327).

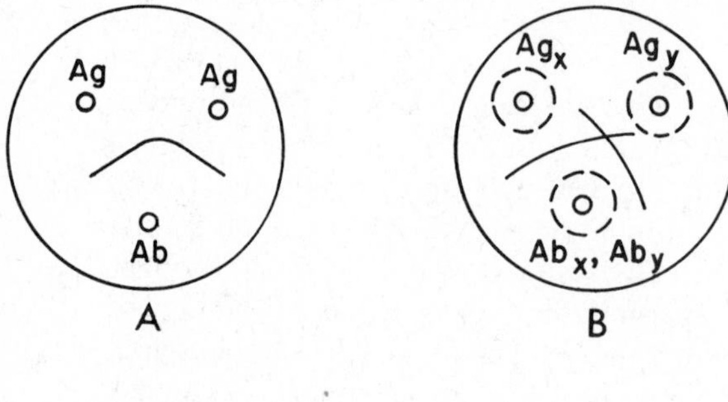

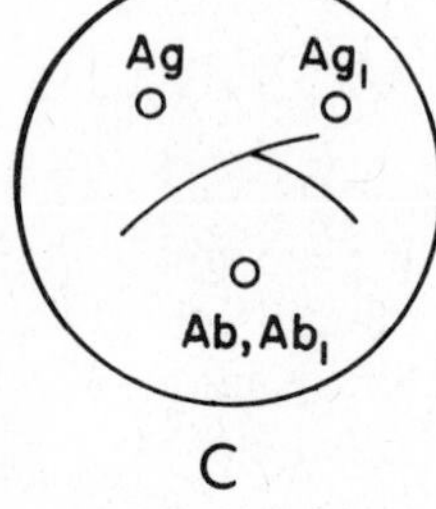

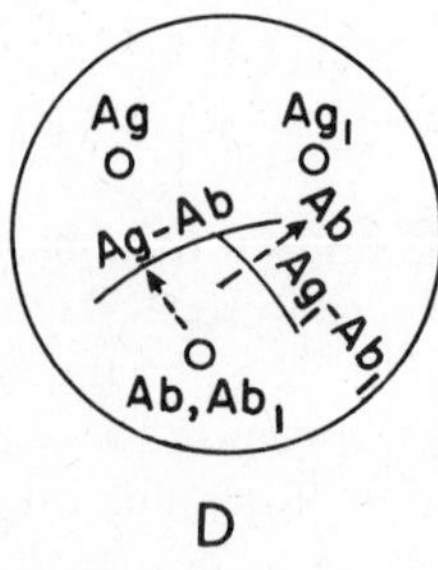

Figure 7-5 Double immunodiffusion in two dimensions by Ouchterlony technique. Ag = antigen; Ab = antibody. *A*, Reaction of identity. *B*, Reaction of nonidentity. *C*, Reaction of partial identity. *D*, Scheme for spur formation.

The simplest method uses a standard Petri dish filled with agar in saline (2 to 4 g/100 ml) to a depth of about 3 mm. Holes are cut in the agar plate with a punch or a sharp cork borer (Fig. 7–5). The plugs of agar are removed by suction, and the bottom of each well is sealed with a drop of agar in saline (1 g/100 ml) to prevent diffusion under the agar. When the same antigen is in both wells, the lines of precipitation fuse and are continuous—the *reaction of identity*. When the precipitin bands cross each other, it is a *reaction of non-identity;* in this case the two antigens are not identical and the antibody well contains antibody to both antigens. If the antigens are related but not identical, a *reaction of partial identity* is observed. Here, the cardinal point is that the precipitate serves as a barrier that does not block unrelated diffusing reactants. As shown in Figure 7–5*D*, when two related antigens, Ag and Ag_1, are in separate wells and the respective antibodies, Ab and Ab_1, are in the third well, an Ag-Ab precipitate forms on one side and blocks further diffusion of Ab from the antibody well, but on the other side the Ag_1-Ab precipitate does not stop Ab from migrating further and forming an Ag-Ab spur.

Although the concentrations of antigen and antibody can vary over a wide range, the production of dense, heavy precipitin bands can block passage of materials nonspecifically. In an unbalanced, nonequivalence system, formation of a second band is possible while the first is still dissolving in the region of antigen excess. These two problems may be resolved by observing the reaction at various time intervals and by setting up a series of plates with different concentrations of both reactants. Still another artifact may be due to precipitation of antibody by organ extracts or, as reported in some cases, by phosphate buffer.

A special modification of this technique is the slide microtechnique described by Wadsworth and Ouchterlony; it has a number of advantages. Only a small sample (1 μl) is required, and the reaction time is markedly decreased from as long as a week to 24 h or less in some cases. Sharp lines are produced because the agar layer is thin and the area from which the reactants diffuse is minute. Finally, the slides can be stained fairly rapidly after the template has been removed or, alternatively, photographed unstained with an oblique light source illuminating the precipitin bands. In some cases, the intensity and number of bands have been measured by using a photoelectric microphotometer and recorder.

The sensitivity of the method depends on the antigen-antibody system and the way in which the system is set up. As little as 1 μg of antibody N may be detected. Precipitin bands that are not visible to the naked eye may be detected by shining monochromatic light through the slide and observing the pattern of the interference fringes (Ambrose and Easty). It should be emphasized that a negative reaction does not necessarily imply absence of antibody or antigen. A negative reaction can result from using amounts of material too small for the sensitivity of the method or, alternatively, the antibody may be nonprecipitating.

RADIAL IMMUNODIFFUSION TECHNIQUE

This is probably the immunologic technique now most widely used for quantitation of immunoglobulins and other serum proteins. Various kits containing all the supplies needed for conducting the tests (including standards) are available commercially. The antigen is allowed to diffuse from a well into agar containing antibody; the antigen-antibody interaction is manifested by a well-defined ring of precipitation around the antigen well (Fig. 7–6*A*). Occasionally, more than one ring can be seen, indicating the presence of more than one reacting antigen-antibody system (in the case of IgM, the presence of two rings usually indicates the presence of both 7S and 19S IgM in the serum being tested). Standards are run at the same time as the sample, and a standard curve of ring area or diameter versus concentration is plotted (Fig. 7–6*B*). A characteristic of this procedure is that ring diameter continues to increase until equilibrium is reached, and a straight-line plot is not obtained until this time. If speed is important, the nonlinear plot may be used if a greater number of standards are run. Plates are read by oblique back-lighting; if necessary, the plates may be

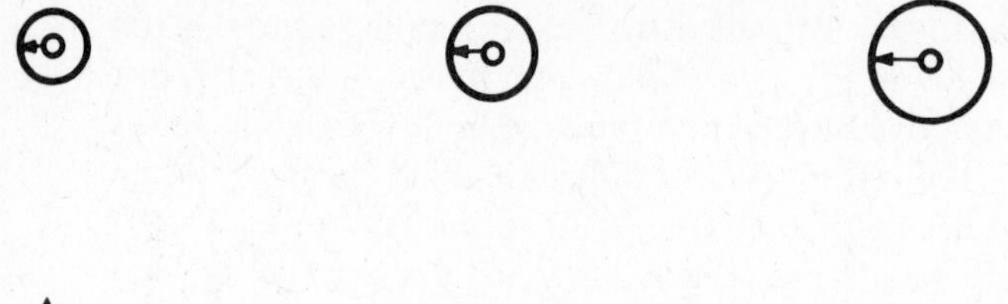

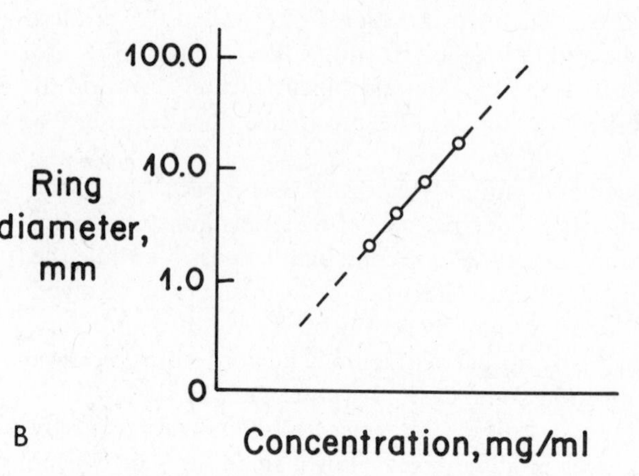

Figure 7-6 Radial immunodiffusion (quantitative technique). *A*, Increased diameter of ring of precipitation as a result of increased antigen concentration. *B*, Typical semilog plot of maximal ring diameter versus concentration.

washed briefly and treated with trichloroacetic acid or Ponceau S dye to detect small quantities of precipitate.

Plates may be prepared or purchased with a variety of wells punched and with various concentrations of antibody present so that different quantities of antigen can be detected. Although the method is simple in principle, one prerequisite is purity of antibody. Contamination with other antibodies can cause confusing and invalid results. The simplicity of the technique is somewhat deceptive because the choice of buffer, incubation times, extent of well filling, and other variables can greatly influence the results. Nevertheless, immunoglobulins in serum can be quantitated with a fair degree of sensitivity and reproducibility. This method serves as a screening procedure for the detection of immunologic deficiency states, macroglobulinemia, multiple myeloma, the dysgammaglobulinemias, and hypergammaglobulinemia secondary to various diseases (see p. 321 and Fig. 7-14).

IMMUNOELECTROPHORESIS

The technique of immunoelectrophoresis (IEP) developed by Grabar and Williams combines zone electrophoresis of proteins with diffusion of antigen toward antibody. The method utilizes an initial electrophoretic separation of the protein components in agar gel followed by diffusion of these components into an oppositely diffusing zone of antibody to produce a series of precipitin arcs. The number of arcs produced and their positions can be used to determine the individual components (antigens) in a complex mixture. By comparison with a known control on the same plate, tentative identification can be made. In some cases the characteristics of the precipitin arc, such as bowing, brushing at the end of the arc, or thickening, may suggest to the experienced observer the presence of special types of proteins (see pp. 315-316).

In addition to agar gel, various other support media have been used, including cellulose acetate, hydrolyzed starch, and dextran (Sephadex). In one of the modifications of the procedure, electrophoresis is first conducted on cellulose acetate or starch, and then sections of the support medium with the separated material are placed on agar containing antibody so that the antigen and antibody can diffuse toward each other. By using the microscope slide technique of Scheidegger, antigen requirements can be decreased to the microliter and microgram level and the time interval for electrophoresis and diffusion can be decreased to 1 h and 12 to 24 h, respectively.

This combination of electrophoretic separation in one dimension followed by separation based on diffusion in a second dimension, along with the great discriminating power resulting from the use of specific antibodies, has been applied to the analysis of antigen mixtures, the study of genetic variants, the comparison of antigens from different species, the study of growth and development by comparison of antigens at different embryonic stages, the comparison of serums from different species, and in purification procedures.

In performing immunoelectrophoresis, a number of factors must be carefully controlled to take full advantage of the sensitivity of the method. For example, the buffer pH may be selected to coincide with the isoelectric point of an antigen to be separated. The choice of the buffer anion may be more important than selection of the actual pH; e.g., borate buffer promotes electrophoretic migration of polysaccharides with *cis*-hydroxyl groupings and also alters the supporting medium (agar) to increase electroendosmotic flow. Improved resolution is obtained by decreasing buffer ionic strength at constant current, by varying the duration of electrophoresis, and by adjusting the distance between the lateral antibody trough and the antigen locations after electrophoresis. Prolonged electrophoresis results in heating of the agar and limits the resolution.

Perhaps the greatest problem is the availability of the desired antibody at the optimal concentration. This situation is improving with the increasing commercial availability of well-defined reagents against an increasing variety of antigens.

There is another similar technique, known as *counter-electrophoresis*, in which antigen and antibody are placed in two wells close together and caused to migrate toward one another under the impetus of an electric field. The interaction is again characterized by the formation of a band of precipitation. This procedure is somewhat more sensitive than simple diffusion of reaction components and is much faster, so that reactions may be read in 1 h, compared with the 16 or more hours required for diffusion techniques. This method has been used in studies of Australia (hepatitis) antigen (HbAg) but is now being supplanted by the more sensitive radioimmunoassay technique.

The technique of *electroimmunodiffusion* (also called "rocket" technique) has the proper sensitivity to detect antigens in biologic fluids with relatively low protein content such as cerebrospinal fluid, tears, saliva, and hypogammaglobulinemic serums. In this technique, electrophoretic migration of antigens takes place in antibody-containing agar on a glass slide. Migration of antigen toward the anode is visualized by its reaction with the antibody in the agar. The resulting pattern resembles a rocket; hence the name of the technique. The electrophoretic migration requires only a few hours, and the "rockets" may be seen without staining if concentrations are high enough. The sensitivity of the method stems from the antigen being concentrated in a very narrow migration path with some enhancement of the precipitin arc as a result of electrophoretic migration of the antigen. Quantitation is effected by using standards on the same plate along with the unknowns and then estimating the concentrations of unknowns from the heights of the "rockets" obtained. The calibration curve is linear only over a narrow concentration range, so samples may have to be diluted or concentrated as needed.

RADIOIMMUNOASSAY

Because of its specificity, sensitivity, and technical simplicity, radioimmunoassay has found increased use in clinical laboratories in recent years. The rapid multiplication of the literature in this field is evidenced by the fact that a recent review on radioimmunoassay listed 591 references. No attempt is made here to give a detailed review of all the aspects; instead, the basic principles, the reagents used, the conditions affecting sensitivity, and the separation of free and bound antigen will be discussed in a general way.

PRINCIPLE

The basic principle of radioimmunoassay is shown in the following equations:

$$\begin{array}{rcl} Ag^* + Ab & \rightleftharpoons & Ab\text{-}Ag^* \\ & & + \\ & & Ag \\ & & \downarrow\uparrow \\ & & Ab\text{-}Ag \end{array}$$

In the absence of unlabeled antigen (Ag), a certain amount of labeled antigen (Ag*) will bind to the antibody (Ab) according to the law of mass action. In the presence of unlabeled antigen, however, the amount of antibody-bound labeled antigen (Ab-Ag*) will decrease because the unlabeled antigen competes with it for the binding sites on the antibody. The amount of labeled antigen displaced from the antibody is proportional to the amount of unlabeled antigen present in the system.

The competitive principle of radioimmunoassay can be extended to nonimmunologic systems. Any substance that can be bound specifically to a macromolecule can be measured by methods based on the principle of competitive protein binding. The specific binding macromolecule can be a serum protein (such as thyroxine-binding globulin), a specific tissue receptor (such as estradiol receptor from the uterus), or an antibody. The assays that use serum protein or tissue receptor as the binding agent generally are referred to as competitive protein-binding assays, radioligand binding assays, receptor binding assays, or saturation analysis.

The advantages of the method are its specificity and sensitivity. For example, of the hundreds of different species of protein molecules in the serum, only human growth hormone will react with human growth hormone antibody. The method can easily detect nanogram quantities of protein and picogram quantities of low-molecular-mass substances. The degree of specificity and sensitivity achieved by radioimmunoassay or competitive protein-binding assay cannot be obtained by any other method now available. Table 7–4 lists a variety of compounds that have been measured by radioimmunoassay or competitive protein-binding assay.

REAGENTS

Antigen Used to Generate Antiserum. The antigen injected into animals to produce antibody need not be pure. It can be a crude extract or a semipurified preparation. Proteinaceous material may be used without any prior special treatment because protein itself is antigenic. Low-molecular-mass substances such as steroids, which are not immunogenic by themselves (haptens), need to be conjugated to an inert protein to render them immunogenic. Most commonly used for this purpose is bovine or human serum albumin. One of the following two reactions is generally used for conjugation.

In the *carbodiimide method*, free carboxyl groups in the protein react with carbodiimide to give a hypothetical intermediate which then reacts with free amino groups in the hapten to form the conjugate via an amide linkage:

$$\underset{\textbf{Protein}}{\text{Prot—C}\begin{matrix}\text{O}\\ /\!\!/\\ \\ \backslash\\ \text{OH}\end{matrix}} + \underset{\textbf{Carbodiimide}}{\text{R—N=C=N—R}'} \rightarrow \left[\text{Prot—}\overset{\text{O}}{\overset{/\!\!/}{\text{C}}}\text{—O—}\overset{\text{HNR}}{\overset{|}{\text{C}}}\underset{\text{NR}'}{\backslash\!\!\backslash}\right] \xrightarrow{Hapten—NH_2}$$

$$\underset{\textbf{Conjugate}}{\text{Prot—}\overset{\text{O}}{\overset{/\!\!/}{\text{C}}}\text{—NH—hapten}} + \text{R}'\text{NH—}\overset{\text{O}}{\overset{\|}{\text{C}}}\text{—NHR}$$

The intermediate can also react with hydroxyl groups to form an ester linkage.

The *mixed anhydride method* can be used to make conjugates fiom haptens with free carboxyl groups. If the compound to be conjugated does not have a carboxyl group, one can be introduced in the molecule. For example, an alcohol treated with succinic anhydride forms a semisuccinate. The free carboxyl group of the semisuccinate reacts with isobutyl chloroformate to form the mixed acid anhydride, which in turn reacts with the free amino group in the protein molecule to form the conjugate:

$$\underset{\textbf{Hapten}}{\text{ROH}} + \underset{\textbf{Succinic anhydride}}{\begin{matrix}\text{CH}_2\text{—C(=O)}\\ | \qquad\qquad \text{O}\\ \text{CH}_2\text{—C(=O)}\end{matrix}} \rightarrow \underset{\textbf{Semisuccinate}}{\text{RO}\overset{\text{O}}{\overset{\|}{\text{C}}}\text{CH}_2\text{CH}_2\text{COOH}} \xrightarrow[\textbf{Isobutyl chloroformate}]{(CH_3)_2CHCH_2\overset{O}{C}Cl}$$

$$\underset{\textbf{Mixed anhydride}}{\text{RO}\overset{\text{O}}{\overset{\|}{\text{C}}}\text{CH}_2\text{CH}_2\text{—}\overset{\text{O}}{\overset{/\!\!/}{\text{C}}}\text{—O—}\overset{\text{O}}{\text{C}}\text{CH}_2\text{CH(CH}_3)_2} \xrightarrow{Prot—NH_2} \underset{\textbf{Conjugate}}{\text{RO}\overset{\text{O}}{\overset{\|}{\text{C}}}\text{—CH}_2\text{CH}_2\overset{\text{O}}{\overset{/\!\!/}{\text{C}}}\text{—NH—Prot}}$$

Antiserum. The antiserum used in the assay need not be pure—that is, it need not contain only one population of antibody. However, the antibody that actually participates in the assay should not cross-react with compounds other than the compound to be measured. If it does, interfering substances must be removed (e.g. by adsorption) before the measurement can be made. Furthermore, the antibodies that do not participate in the assay but are present in the antiserum should not cross-react with the substance to be measured. Because the specificity and the sensitivity vary with different antisera, every antiserum should be carefully characterized as to its specificity and sensitivity before it is used in the assay.

Labeled Antigen. The labeled antigen must be chemically pure and should have a high specific activity. The higher the specific activity of the labeled antigen, the more sensitive the assay. Carbon-14 is ruled out for labeling because of its low specific activity. Iodine-131, iodine-125, and tritium are generally the isotopes of choice. For labeling proteins and peptides, only iodine isotopes are used, and specific activities of 50 to 300 μCi/μg are obtained. The iodination of protein is achieved by oxidizing the free iodide ion (I^-) to iodine (I_2) with chloramine-T. The I_2 generated will react with the tyrosyl residues. The I_2 also can be generated from I^- enzymatically or electrolytically. The reaction is stopped by reducing the excess I_2 to I^- with sodium metabisulfite. The final product is purified on a gel

TABLE 7–4 COMPOUNDS MEASURED BY RADIOIMMUNOASSAY AND COMPETITIVE PROTEIN-BINDING ASSAY

Peptide Hormones		
Growth hormone (GH)		Human chorionic gonadotropin (HCG)
Luteinizing hormone (LH)		Insulin
Follicle-stimulating hormone (FSH)		Secretin
Thyroid-stimulating hormone (TSH)		Gastrin
Adrenocorticotropic hormone (ACTH)		α-Melanocyte-stimulating hormone
Prolactin		β-Melanocyte-stimulating hormone
Oxytocin		Parathyroid hormone
Vasopressin		Calcitonin
TSH-releasing factor		Angiotensin
LH-releasing factor		Proinsulin
		Bradykinin
Steroids		
Estrogens		Cortisol
Testosterone		Aldosterone
Androstenedione		11-Deoxycortisol
Progesterone		Deoxycorticosterone
Drugs		
Digoxin		Morphine
Digitoxin		Barbiturates
Lysergic acid diethylamide		Ouabain
Miscellaneous		
Prostaglandins	Folic acid	Vitamin B_{12}
Cyclic AMP	Cyclic GMP	Triiodothyronine
Thyroxine	IgE	Plasminogen
Hepatitis-associated antigen	Carcinoembryonic antigen	Messenger RNA

filtration column to separate the protein-bound iodine from the free I^-. For most low-molecular-mass substances, tritium is used for labeling purposes. A large variety of tritium-labeled compounds can be purchased from commercial sources. Their specific activities range from 10 to 100 μCi/mmol.

Standard. The standard used in the assay must be immunologically identical to the substance to be measured. This is checked by plotting the dose-response curves for both the standard and the unknown on semilog paper; if the two curves are parallel, the standard and the compound measured are immunologically identical. For example, in Figure 7–7, which shows the dose-response curves of pituitary LH standard, urinary LH standard, and urinary LH extract, it is apparent that urinary LH and pituitary LH are not immunologically identical. Therefore, when measuring urinary LH, urinary LH standard should be used; if pituitary LH standard is used for measuring urinary LH, the result obtained is not valid.

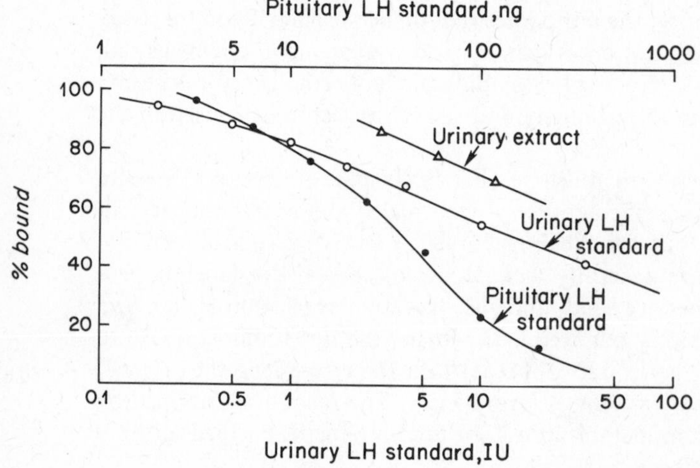

Figure 7–7 Radioimmunoassay dose-response curves of pituitary and urinary LH standards and urinary extract, with rabbit antiserum against human pituitary LH as antibody.

CONDITIONS AFFECTING THE SENSITIVITY OF RADIOIMMUNOASSAYS

Figure 7–8 shows three standard curves obtained by using the same antiserum at three different final concentrations. Decreasing the concentration of the antiserum increases the sensitivity. However, if the antiserum concentration is too low, the amount of labeled antigen that is bound is so small that a long time is required to accumulate a statistically significant number of counts, and this makes the assay impractical. In practice, the optimal concentration of antiserum is the amount that can bind 30 to 50 per cent of the labeled antigen added.

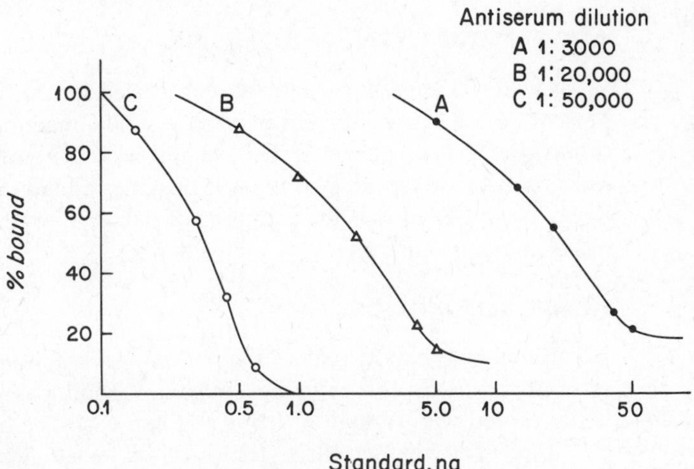

Figure 7–8 Standard curves for a radioimmunoassay. Antiserum dilutions: *A*, 1:3000; *B*, 1:20,000; *C*, 1:50,000.

The sensitivity also can be improved by using a minimal mass of labeled antigen; i.e., by using labeled antigen with the highest specific activity possible (radioactive iodine and tritium are available in carrier-free form and are the isotopes of choice for labeling purposes). The amount of labeled antigen in the assay should be as small as possible and yet sufficient to permit a reasonably short counting time; a practical counting rate is 2000 to 20,000 cpm.

The sensitivity of a radioimmunoassay is ultimately determined by the affinity of the antibody for the antigen. This is an intrinsic property of the antibody and cannot be manipulated.

SEPARATION OF FREE AND BOUND ANTIGEN

At the end of the incubation period, the reaction mixture contains both free and bound antigen, and these must be separated before radioisotope counting. Many methods have been used for this separation. Whatever method of separation is chosen, it must be reproducible, simple to perform, fast, and economically feasible. The methods now most commonly used can be classified into three types.

In the first type the removal of the free antigen is achieved by adsorption. Many adsorbents are used for this purpose, including activated charcoal, dextran-coated charcoal, ion-exchange resin, magnesium silicate (Florisil), Fuller's earth, and talc. The disadvantage of this method is that the time of contact between the adsorbent and the incubation mixture is critical, especially for the more active adsorbents such as activated charcoal. Sometimes timing has to be controlled to within seconds in order to get reproducible results.

The second type of method involves precipitating the bound antigen from the solution by using a protein precipitant such as $(NH_4)_2SO_4$, ethanol, dioxane, or polypropylene glycol. The bound antigen can also be precipitated immunologically by using a second antibody. For example, if the primary antibody is derived from rabbits, the second antibody can be an antiserum against rabbit γ-globulin derived from goats or sheep. This method has the advantage that it can be used for practically any radioimmunoassay. It has the disadvantage, however, that it usually requires 24 to 48 h of additional incubation time.

The third type of method utilizes solid phase antibodies. In this method, the binding and competition of antigens occur on a solid surface to which the antibody is attached. The solid surface may be the inside surface of a plastic tube, a polymerized antibody, or antibody conjugated by a covalent bond to an insoluble inert material, such as cellulose. At the end of incubation, the solid phase antibody, along with bound antigen, can be sedimented at the bottom of the tube and the supernatant containing the free antigen can easily be siphoned off or decanted.

AGGLUTINATION REACTIONS

The direct agglutination technique is used for the direct or indirect detection of antigens present on the surface of cells or particles. The reaction end point is evidenced by aggregation of the cells or particles. This technique is generally more sensitive than the precipitin reaction because the bulk of the cell provides added mass for visualization of the reaction. Specific agglutination is the clumping of bacteria, erythrocytes, and so forth in the presence of homologous antibody.

HEMAGGLUTINATION

Hemagglutination refers to agglutination reactions in which the antigen is located on an erythrocyte. Erythrocytes not only are good passive carriers of antigen, but are also easily coated with foreign proteins and can easily be obtained and stored.

Direct testing of erythrocytes for blood group, Rh, and other antigenic types is widely used in blood banks; specific antisera, such as anti-A, anti-C, and anti-Kell, are used to detect such antigens on the erythrocyte surface.

In *indirect* or *passive* hemagglutination, the erythrocytes are used as a particulate carrier of foreign antigen (and, in some tests, of antibody); this technique has wider applications. Other materials available in the form of fine particles, such as bentonite and latex, also have been used as antigen carriers, but they are more difficult to coat, standardize, and store. In a related variation of this technique, known as *hemagglutination inhibition*, the ability of antigens, haptens, or other substances to specifically inhibit hemagglutination of sensitized (coated) cells by antibody is determined. An application of this technique is the pregnancy test described on page 799.

The classic agglutination test can be conducted either on a slide or in a test tube. Micro procedures using very small quantities of the costly reagents are being used with increased frequency. These are easier to set up, are more sensitive, and produce settling patterns that can be read easily. The choice of animal as the source of erythrocytes is important; one should use the species that produces the least degree of cross-reaction. In practice, sheep erythrocytes are used most frequently. Erythrocytes are relatively unstable after depletion of substrate and accumulation of metabolic products. However, if the cells are collected directly into Alsever's solution (acid-citrate-dextrose solution, pH 6.1, sometimes containing a broad-spectrum antibiotic) and stored at 4°C, the erythrocytes are usable for up to 30 days. Cells treated with formalin (formalinized cells) can be used for months and may be stored for even longer periods in the frozen or lyophilized state. However, settling patterns may not be as satisfactory as with wet cells stored in the cold.

Erythrocytes to be used in indirect agglutination tests can be coated (sensitized) in various ways. The simplest method is by direct adsorption of antigen onto the cell, usually by direct incubation with the desired antigen. The degree of sensitization of the cells thus treated is a function of concentration of the antigen used and time of exposure but seems to be unrelated to the temperature during adsorption. The erythrocyte membrane possesses many reactive sites, and it is possible to coat the cell surface sequentially with a number of different antigens, without interference from those initially on the surface. Substances such as polysaccharides and lipopolysaccharides are easily adsorbed onto the cells, whereas more complicated methods are usually necessary for the adsorption of many, but not all, proteins. However, some nonhemagglutinating viruses and antibiotics such as penicillin may be coupled directly to the cells. Some proteins will bind only if the cells are pretreated; the most widely used method for this is the *tanned red cell technique* in which erythrocytes are treated with tannic acid. The exact mechanism by which tannic acid sensitization works is unknown.

The procedure for coating antigens onto tannic acid–treated cells is much like that described for simple adsorption. The optimal concentration of antigen to be used may have to be determined in some cases, although only a very small proportion of the antigen (around 1 per cent) is taken up by the cell. Tannic acid also increases cell agglutinability. In some cases, the tanned erythrocytes are so sensitive that they will clump without addition of antibody. If normal serum or serum albumin is added as a stabilizer (if the antigens being used do not cross-react), the suspension not only will give specific agglutination but also will be extremely sensitive to even small amounts of antibody.

The most specific method of sensitization involves the direct, covalent attachment of the antigen to the erythrocyte via a chemically reactive grouping. The great hazard in these procedures is the possibility of drastically altering the erythrocyte surface or denaturing or immunologically altering the antigen. There are three generally accepted techniques for the direct attachment of the reactive groupings, based on use of *bis*-diazotized benzidine (BDB), chromic chloride, or 1,3-difluoro-4,6-dinitrobenzene (DFDNB). The BDB technique couples erythrocyte and antigen through stable, covalent azo bonds. Cells prepared in this fashion are fragile, but if the erythrocytes are first formalinized and then coupled, they can be coated more effectively with larger quantities of antigen and can be stored for longer periods.

Coupling with heavy metal cations—for example, chromic chloride—is simple. Chromic chloride, erythrocytes, and antigen are mixed and incubated at room temperature for 1 h; then the reactants are removed by centrifugation. Chromium-treated cells settle poorly. Thus, when using such cells, the tube agglutination technique is preferred.

In another technique, DFDNB reacts with amino groups of the protein and the antigen. When sera are tested with this reagent, they should first be pretreated with DFDNB-treated (non-antigen-conjugated) cells to effect a nonspecific reaction between serum and DFDNB itself. After removal of any precipitate and addition of the coated cells, agglutination is due to action between the antibody and the coated erythrocyte. Erythrocytes sensitized by this or the previous techniques can be used on a micro scale.

COMPLEMENT

CHEMISTRY

Complement (*C*) is a complex of proteins present in serum which, when combined with antigen-bound antibody, forms several important products involved in mediating immune and allergic reactions. It is classically detected by its ability to cause hemolysis of sheep erythrocytes sensitized by coating with small quantities of anti-sheep cell antibody. Complement is composed of nine interacting proteins and enzymes, referred to as components *C*1 through *C*9. The first component of complement, *C*1, has three subunits (*q*, *r*, *s*) and the process of fixation in the classic pathway is initiated by the formation of a complex between *C*1 and the antigen-antibody complex (EA) in the presence of calcium. This complex possesses esterase activity (via activation of *C*1, which is a proesterase) and reacts with and modifies *C*4 and *C*2 to form a new enzyme *C*4*b*2*a* (*C*3 convertase). This enzyme complex transforms a large number of *C*3 molecules to a shower of *C*3*b* fragments. Some of these combine with *C*4*b*2*a* to form another enzyme, which cleaves *C*5. The product, *C*5*b*, then combines with *C*6 and *C*7, and this complex is subsequently joined by *C*8 and *C*9. Addition of *C*9 leads to a functionally and physically damaged cell that undergoes rapid lysis.

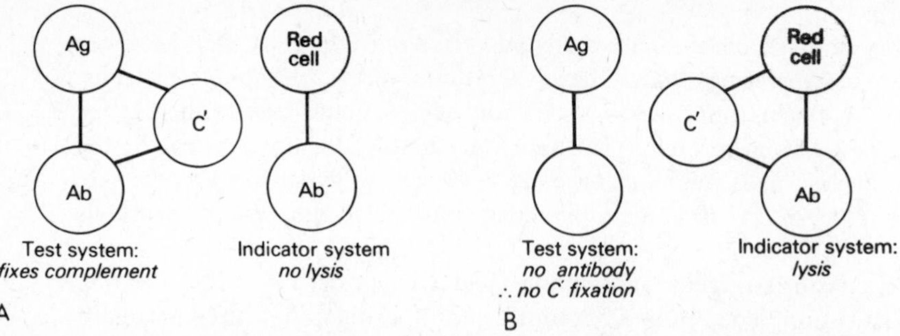

Figure 7–9 Complement-fixation test. Antigen (Ag) and antibody (Ab) are incubated in the presence of guinea pig serum which acts as a source of complement (C′). Then the indicator system, consisting of sheep red cells coated with antibody, is added.

A, When antigen and antibody are present in the test system, they fix complement and none remains to lyse the added indicator system. *B*, When antigen or antibody is lacking in the test system, complement is available to lyse the sensitized indicator cells. (From Roitt, I. M.: Essential Immunology. Oxford, Blackwell Scientific Publications, 1971, p. 97.)

Products of this reaction sequence interact with the clotting, fibrinolytic, and kinin-generating systems present in blood. Some are biologically active, such as the anaphylatoxins (derived from *C*3 and *C*5), which are low-molecular-mass materials that cause contractions in smooth muscle and increases in capillary permeability. Complement-dependent mechanisms for histamine release also have been described. The inherited and acquired abnormalities of the complement system that have been described in humans include hereditary angioedema (deficiency or inactivity of serum inhibitor of activated *C*1), *C*2 deficiency, *C*3 and *C*4 polymorphism, and familial *C*5 dysfunction.

COMPLEMENT FIXATION

The simple complement-fixation test (CFT) illustrated in Figure 7–9 can be used to test for either the presence of antigen-antibody interaction or the presence or absence of complement. If antibody is present and reacts with antigen, the complex will bind any complement added to the test system; and if sensitized erythrocytes are then added (indicator system), no lysis will be observed because no free complement is available to effect lysis. If lysis is observed, this indicates that the unknown specimen contained active complement, and thus no antigen-antibody interaction occurred.

Manual and automated procedures available for quantitating complement are based on the colorimetric measurement of the hemoglobin released, on turbidimetric measurement before and after lysis of cells, or on the determination of the number of erythrocytes remaining by particle counting. Antisera to all nine components of *C* are now available, and individual *C* components can now be assayed by using radial immunodiffusion.

IMMUNOFLUORESCENCE

Small quantities of antigen or antibody can be labeled or "tagged" by fluorescent dyes. This method is particularly useful for the morphologic localization of antigens in or on various types of bacterial or tissue cells. A number of fluorescent dye derivatives are available, but fluorescein isothiocyanate (FITC) is the one used most often. Another useful dye is lissamine rhodamine B (RB 200); this conjugates to the protein via a sulfonamide bond. Because tissues exhibit a blue-gray autofluorescence, the dye used must emit a color that contrasts sharply with this background. Fluorescein gives off a yellow-green color, and the rhodamine dye is orange-red. A good fluorochrome should conjugate easily and firmly to antibody protein but should not interfere with antibody reactivity toward the antigen.

The conjugated fluorochrome should be stable, have a good quantum yield of fluorescence energy, and show a maximal spread in wavelength between absorbed (primary) and emitted (secondary) light. The quantum yield is simply defined as quanta emitted per quanta absorbed and is 0.70 for fluorescein. With FITC, the exciting wavelength, 490 nm, is very close to the emission (fluorescence) wavelength, 520 nm.

In preparing the protein-dye conjugate, only the purest possible crystalline FITC should be used. The protein to be conjugated is usually a mixture of immunoglobulins and should be free of other serum protein fractions because albumin and other globulins label more rapidly than γ-globulin. Thus, if whole serum is used, certain protein fractions will successfully compete with the antibody protein for the conjugating dye, rendering the preparation nonspecific and therefore useless. After labeling, the unreacted FITC is removed by passage through a Sephadex G-25 column, and the conjugated protein is purified by fractionation on a DEAE cellulose column, using different buffers with increasing pH gradient or ionic strength. Ideally, the conjugated protein should contain 2 moles of fluorophor per mole of γ-globulin and its solubility should be similar to that of the original protein.

Direct staining is used to demonstrate the presence of a given antigen in or on a given cell or tissue preparation. Areas or structures that contain antigen specific to the fluorophor-labeled antibody exhibit a brilliant fluorescence (yellow-green in the case of FITC). In the more widely used indirect staining technique for an autoimmune antibody, the antigen (tissue, cell, or nuclear material) on a slide is treated with the patient's serum. This coats the antigen with the specific autoantibody present in the patient's serum. The preparation is then treated with FITC-labeled antihuman globulin, and the presence of bound fluorescence demonstrates the presence of autoimmune antibodies.

Fluorescent antibody techniques are used to aid in the diagnosis of autoimmune diseases. In the course of liver disease, a host of antibodies are formed, including antibodies against mitochondria, smooth muscle, and bile canaliculi. The reason for the appearance of these antibodies and their role in the disease process are unknown, although the correlation between them and the disease is quite high—for example, there is a high correlation between antimitochondria antibody and primary biliary cirrhosis. Systemic lupus erythematosus (SLE) is an autoimmune disease in which antibodies to a variety of nuclear antigens, including deoxyribonucleic acid, are produced. Antinuclear antibody (ANA) is found in SLE and in a wide variety of what are now termed immune complex diseases (formerly called collagen diseases). The occurrence of autoantibodies in thyroid disease (Hashimoto's thyroiditis) is also well known.

DETECTION OF ANTINUCLEAR ANTIBODIES

The procedure most commonly requested is a test for presence of antinuclear antibodies (ANA). A partial list of the disease states associated with the presence of these antibodies is shown in Table 7–5.

Sections (2 to 4 μm thick) of livers from young rats are prepared, placed on microscope slides, and air dried. A 1/32 or 1/64 dilution of the test serum with saline is then placed gently on top of the tissue. The slide with tissue and serum is incubated at room temperature in a humidifying dish for 30 min, washed twice for 10 min periods in 0.15 molar NaCl with gentle agitation, and then wiped dry (except for the tissue section). The appropriate fluorescein-conjugated antiglobulin is then added and allowed to remain in contact with the tissue for 30 min, after which the washing procedure is repeated. When dry, the slides are examined with a fluorescence microscope for bound fluorescent dye, and the preparations are reported as either negative or positive at the dilution tested. If positive at a 64-fold dilution, the serum may be titered so that the highest dilution still giving a positive result can be reported. In addition, the morphologic appearance of the fluorescent nuclear pattern is often ex-

TABLE 7-5 DISORDERS ASSOCIATED WITH ANTINUCLEAR ANTIBODIES

Collagen Vascular Diseases
- Systemic lupus erythematosus
- Rheumatoid arthritis
- Sjögren's syndrome
- Polymyositis

Hepatic Disease

Lymphoproliferative and Myeloproliferative Diseases
- Waldenström's macroglobulinemia
- Lymphoma
- Chronic leukemia

Pulmonary Disease

Miscellaneous Disorders
- Discoid lupus
- Hashimoto's thyroiditis
- Chronic ulcerative colitis
- Infectious mononucleosis
- Autoimmune hemolytic anemia
- Drug reactions

amined and is reported as homogeneous, speckled, shaggy, nucleolar, or mixed (most common).

QUANTITATION OF IMMUNOGLOBULINS

For the measurement of immunoglobulin levels in large numbers of specimens, some form of automated procedure is desirable. One such procedure has been developed for the AutoAnalyzer and is known as the automated immunoprecipitin procedure (AIP). This method, a continuous-flow procedure, is based on the classic precipitin reaction between antibody and antigen, but with the components at concentrations much lower than those used in the manual procedure. The antibody-antigen reaction produces a turbidity that is proportional to the antigen concentration under the test conditions and that is measured by means of a fluoronephelometer. Comparative studies have shown excellent correlation between the reference immunodiffusion technique and the AIP procedure.

One of the many problems encountered in the quantitation of immunoglobulins is in the selection of appropriate standards. The average human serum contains a heterogeneous population of immunoglobulins. Therefore, serum containing the monoclonal immunoglobulins is not suitable for use as a standard, even though the antibody may be present in a high concentration. A normal human serum pool has been prepared by the World Health Organization (in collaboration with national agencies such as the National Institutes of Health) to serve as an international standard. The total immunoglobulin content of this pool has been determined by a number of reference laboratories and the concentration is expressed in International Units to avoid confusion inherent in the use of weight units. Some standards for IgG, IgA, and IgM are commercially available and have been calibrated against these international reference preparations (see also p. 321).

PREPARATION OF ANTISERA

Many antisera are available from commercial sources. Occasionally, however, antiserum of the desired specificity or avidity may not be available commercially and will have to be made. Antisera for use in immunologic work must be selected carefully with respect to their suitability and purity. Even though a given antiserum has been shown to be homogeneous by the usual chemical and physical tests, it may still be unsuitable for immunochemical application because of the presence of trace amounts of nonspecific antibodies. Removal of these contaminants by adsorption may be necessary.

INDUCTION OF ANTIBODY

Animals most commonly used for antiserum preparation include rats, mice, guinea pigs, rabbits, chickens, goats, and horses. The last two are used commercially for preparation of large batches of some antisera (for example, anti-immunoglobulins). Animals to be used should be tested for the possible presence of "natural" antibodies that may interfere with the planned use of the antiserum.

The immunogen to be used for inducing antibody formation may be processed and used in a variety of ways. Soluble immunogens may be used as solutions in saline or in buffered saline; bacterial or other particulate material may be used as a suspension in saline. The immunogen can be introduced into the host animal by intravenous, intramuscular, or subcutaneous injection.

Immunogen to be injected intravenously may be incorporated into an alumina suspension. It is believed that the alumina particles containing bound immunogen are trapped by the lymph nodes, and the subsequent slow release of the immunogen provides sustained immunogenic stimulation. When combined with immunogen, certain chemical preparations (*adjuvants*) increase the amount of antibody production. Such preparations require fewer injections and smaller amounts of immunogen to provide a given amount of antibody. The adjuvant used most commonly is based on that devised by Freund 33 years ago and modified by Uhr. It is a stable emulsion containing mineral oil, a monooleate, and lyophilized *Mycobacterium butyricum*. Enhanced formation of antibodies is attributed to focal inflammation as a result of mobilization of macrophages and immunogen uptake and to retardation of absorption of immunogen from the depot emulsion at the injection site. The gain in antibody production should be balanced against the possible introduction of contaminating antibodies derived from the mycobacteria.

Immunization by the intravenous route is commonly used, even though this technique requires greater amounts of immunogen. Many different immunization schedules have been published. Most commonly, host animals are given gradually increasing doses of immunogen preparations, 2 to 5 mg per injection, on 3 to 4 days per week for 2 to 3 weeks. In many cases, it may be necessary to repeat this schedule once or twice to obtain adequate antibody levels.

SEPARATION OF ANTISERUM

After it is withdrawn from the inoculated animal, the blood is allowed to clot and to retract with a minimum of handling. The serum is removed by centrifugation at 4 to 6°C, at 1000 × *g* for 30 min. If any erythrocytes or leukocytes remain they are removed by a second centrifugation. The resultant serum should be a straw-colored, clear liquid. Chylous serum (with high lipid concentrations) can prove troublesome, but lipids usually can be removed either by filtration through coarse filter paper or by centrifugation and skimming off of the floating lipid. The serum then can be stored with preservatives or divided into small portions and stored frozen.

In some cases, antibody may be desired at a concentration higher than that present in the whole serum, or it may be desired to remove or isolate some special proteins. In these cases, the antibody-containing serum will have to be fractionated to separate either the entire globulin fraction or the designated specific immunoglobulin class from the other serum proteins present. In the majority of instances, simple precipitation of all γ-globulin by a salting-out procedure is all that is needed. The isolated globulins are then dissolved in a volume of saline calculated to yield the desired antibody concentration, and the solution is dialyzed against cold 0.15 molar NaCl until the dialysate is free of SO_4^{2-} or other ions from the precipitant. The antibody solution can then be assayed and either diluted or further concentrated as required. Concentration can be achieved by negative-pressure dialysis or by use of pressure filtration in a micropore apparatus using a membrane (PM-10) of pore size such that all γ-globulin is retained.

Any of the individual immunoglobulin classes (IgG, IgA, or IgM) may be separated by use of a combination of chromatographic methods or by use of specific immunosorbents. These methods are primarily used in the preparation of pure antisera monospecific against the heavy chains of IgG, IgA, or IgM. IgG may be isolated on diethylaminoethyl cellulose (DEAE cellulose) columns. Elution with dilute phosphate buffers (0.001 to 0.01 mol/l) results in a "fall-through" fraction that is almost entirely IgG. Pure IgM and IgA are more difficult to prepare, and this requires a combination of euglobulin precipitation, ultracentrifugation, and chromatography on Sephadex G-200.

The use of immunosorbents is the most specific technique available for preparation of antibodies. These agents consist of a specific antibody (or antigen) bound by cyanogen bromide to an insoluble support such as cellulose or agarose. In practice, whole antiserum or a crude fraction is added to the column, and then the column is washed with isotonic saline until no more protein is eluted. The only serum component remaining is the one specifically bound to the antibody (or antigen), which in turn is covalently linked to the agarose. It can be eluted in many cases by use of acidic buffers, about pH 2.5 to 4.

PURIFICATION OF ANTISERUM BY ADSORPTION

Because even minute amounts of immunogenic material can give rise to antibody production in a host organism, the use of a "highly purified" immunogen does not guarantee production of antibody directed only against this particular immunogen. In a classic example, eight-times recrystallized hen egg albumin gave rise to six antibodies representing a group of trace antigenic components present in the "highly purified" hen egg albumin preparation. Thus removal of unwanted antibodies from the preparation should be considered, even when highly purified immunogens had been used. Adsorption techniques are varied and usually must be tailored to the individual problem. Use of excess adsorbent may lead to contamination of the antibody preparation with adsorbent and should be avoided by the use of solid adsorbents.

Antibodies to human serum components other than immunoglobulins may be adsorbed by using serum from a patient with agammaglobulinemia. The antialbumin and anti-α- and anti-β-globulins will react with the respective serum proteins and can thus be removed, leaving only the anti-γ-globulins. Similarly, antiserum to ceruloplasmin may be purified with serum from a patient with Wilson's disease which contains only small amounts of ceruloplasmin. Thus, the antibodies to all other serum proteins except ceruloplasmin will be removed.

ASSAY OF ANTIBODY PREPARATIONS

Details of assay procedures will vary considerably, depending on the kind of antibody and the properties of the antigen. The two criteria of most concern are potency and specificity. In practice, potency is expressed in terms ranging from qualitative statements regarding reactivity to a quantitative expression in terms of mass units of antibody per unit volume of antiserum (mg/ml or mg N/ml). Semiquantitative expressions (for example, reacts out to 512-fold dilution) are based on titrations using serial dilutions of antiserum.

Precise details as to the specificity of an antiserum are often lacking for many commercial preparations, although a significant change for the better is now in progress.

Assays for purity should include not only physical and chemical tests but also tests involving immunoelectrophoresis and reverse radial immunodiffusion.

PRESERVATION OF ANTISERUM PREPARATIONS

Antiserum or fractionated products are susceptible to bacterial contamination or enzymatic degradation even when stored in the cold. This can be avoided by addition of suitable preservatives, by sterilization using micropore filtration (Millipore or Amicon), by lyophilization, or by storage in the frozen state. The addition of preservatives is the simplest method; the two preservatives used most commonly are thimerosal (Merthiolate) at a concentration of 0.01 g/100 ml and sodium azide at a concentration of 0.10 g/100 ml. Other preservatives that have been used are borates, phenol (0.25 g/100 ml), benzethonium chloride, and methyl or ethyl esters of *p*-hydroxybenzoic acid.

ADDITIONAL READING

Alpert, M. E., Uriel, J., and de Nechaud, B.: Alpha$_1$ fetoglobulin in the diagnosis of human hepatoma. N. Engl. J. Med., *278:* 984, 1968.

Austen, K. F.: Inborn errors of the complement system of man. N. Engl. J. Med., *276:* 1363, 1967.

Beck, J. S.: Variations in the morphological patterns of "autoimmune" nuclear fluorescence. Lancet, *1:* 1203, 1961.

Berson, S. A., and Yalow, R. S.: General principles of radioimmunoassay. Clin. Chim. Acta, *22:* 51, 1968.

Beutner, E. H.: Defined immunofluorescent staining: Past progress, present status, and future prospects for defined conjugates. Ann. N.Y. Acad. Sci., *177:* 506, 1971.

Campbell, D. H., *et al.*: Methods in Immunology: A Laboratory Text for Instruction and Research. 2nd ed. New York, W. A. Benjamin, 1970.

Freedman, S. O.: Clinical Immunology. New York, Harper & Row, Publishers, 1971.

Goldman, M.: Fluorescent Antibody Methods. New York, Academic Press, Inc., 1968.

Good, R. A., and Fisher, D. W.: Immunobiology: Current Knowledge of Basic Concepts in Immunology and Their Clinical Applications. Stamford, Connecticut, Sinauer Associates, 1971.

Hong, R.: The immunoglobulins. Clin. Immunobiol., *1:* 29, 1972.

Humphrey, J. H., and White, R. G.: Immunology for Students of Medicine. 2nd ed. Philadelphia, F. A. Davis Company, 1964.

Kabat, E. A., and Mayer, M. M.: Experimental Immunochemistry. 2nd ed. Springfield, Illinois, Charles C Thomas, Publisher, 1961.

Kwapinski, J. B. G.: Methodology of Immunochemical and Immunological Research. New York, Wiley-Interscience, 1972.

Mancini, G., Carbonara, A. O., and Heremans, J. F.: Immunochemical quantitation of antigens by single radial immunodiffusion. Immunochemistry, *2:* 235, 1965.

Markowitz, H., and Tschida, A. R.: Automated quantitative immunochemical analysis of human immunoglobulins. Clin. Chem., *18:* 1364, 1972.

Mayer, M. M.: The complement system. Sci. Am., *223:* 54, 1973.

Metzger, H.: Structure and function of γM macroglobulins. Adv. Immunol., *12:* 57, 1970.

Nairn, R. C.: Fluorescent Protein Tracing. 3rd ed. Baltimore, The Williams & Wilkins Company, 1969.

Natvig, J. B., and Kunkel, H. G.: Human immunoglobulins: Classes, subclasses, genetic variants, and idiotypes. Adv. Immunol., *16:* 1, 1973.

Odell, W. D., and Daughaday, W. H.: Principles of Competitive Protein-Binding Assays. Philadelphia, J. B. Lippincott Company, 1971.

Olitzky, I.: Standardized, automated procedure for measurement of serum hemolytic complement activity. Appl. Microbiol., *16:* 1635, 1968.

Ouchterlony, O.: Handbook of Immunodiffusion and Immunoelectrophoresis. Ann Arbor, Michigan, Ann Arbor Science Publishers, 1968.

Péron, F. G., and Caldwell, B. V.: Immunologic Methods in Steroid Determination. New York, Appleton-Century-Crofts, 1970.

Reimer, C. B., *et al.*: Comparative evaluation of commercial precipitating antisera against human IgM and IgG. J. Lab. Clin. Med., *76:* 949, 1970.

Roitt, I. M.: Essential Immunology. Oxford, Blackwell Scientific Publications, 1971.

Skelley, D. S., Brown, L. P., and Besch, P. K.: Radioimmunoassay. Clin. Chem., *19:* 146, 1973.

Smith, M., and Williams, R.: Immunology of the Liver. Philadelphia, F. A. Davis Company, 1971.

Stiehm, E. R., and Fudenberg, H. H.: Serum levels of immune globulins in health and disease: A survey. Pediatrics, *37:* 715, 1966.

Whittingham, S.: Serological methods in autoimmune disease in man. Res. Immunochem. Immunobiol. *1:* 121, 1972.

Williams, C. A., and Chase, M. W.: Methods in Immunology and Immunochemistry. New York, Academic Press, 1967–1971, vols. 1–3.

SECTION THREE

THE PROTEINS OF BODY FLUIDS

by Gregor H. Grant, M.A., B.M., F.R.C. Path.,
and John F. Kachmar, Ph.D.

PLASMA AND SERUM PROTEINS

Blood is made up of particulate cell forms suspended in a fluid medium called *plasma*, a very complicated mixture of inorganic and simple and complex organic materials dissolved in water. If the blood is collected without anticoagulant, in vessels with siliconized or nonpolar surfaces, the fluid separating is referred to as *native plasma.* This approximates very closely the plasma actually present in circulating blood. It is the practice, however, to use such anticoagulants as oxalate, citrate, EDTA, and heparin to prepare specimens of plasma for study or analysis. This differs from native plasma because it is modified by the presence of the added chemicals and by the partial loss of calcium bound by oxalate, if this is used. If blood is permitted to clot, the fluid separating is referred to as *serum.* It lacks the protein fibrinogen present in plasma, the fibrinogen having been transformed into insoluble fibrin in the clotting process. The fibrinogen constitutes only some 3 to 6 per cent of the total plasma proteins. It is satisfactory and much more convenient to use serum rather than plasma in most clinical chemical studies.

About 93 per cent of plasma or serum is solvent water. The remaining 7 per cent, the solutes, consists almost entirely of proteins, but because of their large molecular masses their molecular concentration is only about 1 mmol/l, compared with a mass concentration of 70 g/l. In clinical work the concentration of proteins is given in grams per 100 ml or per liter of serum volume. It would be more meaningful to express it in terms of serum water, a value about 1.07 times greater.

FUNCTIONS AND ORIGINS OF PLASMA PROTEINS[77]

As described in this and later chapters, many individual plasma proteins have specific functions, for example as enzymes, as antibodies, as factors in blood coagulation, or as transport proteins: many vital metabolites, metal ions, hormones, and lipids are transported around the body bound to carrier proteins.

Plasma proteins collectively also serve a number of different functions. They play a nutritive role inasmuch as they constitute a portion of the amino acid pool of the body; thus, the proteins are a form of storage amino acids. If needed, these proteins can be broken down in the liver to produce amino acids for use in building other proteins. Alternatively, they can be deaminated to give keto acids which can be mobilized to provide caloric energy, or they can be transformed into carbohydrates and lipids.

Another important function of proteins is a physicochemical one. The plasma proteins, being large, colloidal molecules, are nondiffusible; i.e., they cannot move through the thin capillary wall membranes as can most other plasma solutes. They are thus entrapped in the vascular system and exert a *colloidal osmotic pressure*, which serves to maintain a normal blood volume, and a normal water content in the interstitial fluid and the tissues. The albumin fraction is most important in maintaining this normal colloidal osmotic or *oncotic pressure* in blood. If the albumin falls to low levels, water will leave the blood vessels and enter the interstitial fluid and the tissues, thus producing edema.

The maintenance of the acid-base balance in blood also involves the plasma proteins. As amphoteric compounds, they function as buffers to minimize sudden, gross changes in the pH of the blood.

Despite their presence together, the various proteins of plasma do not all originate from the same source. Albumin and most of the α- and β-globulins, including the blood clotting and transport proteins, are synthesized by the liver, but antibodies (immunoglobulins, γ-globulins) are synthesized and secreted by plasma cells in the lymph nodes, bone marrow, spleen, and elsewhere.

Total Serum Protein

NORMAL VALUES

The level of total serum proteins found in healthy young and middle-aged adults is 6.0 to 8.0 g/100 ml* when recumbent. It is higher when ambulatory (6.5 to 8.5). This effect of posture varies with the individual, and the difference may be as much as 1.2 g/100 ml.

In plasma, fibrinogen increases the total protein by an additional 0.2 to 0.4 g/100 ml.

If the patient is on drugs, this must be taken into account, as some drugs (e.g., hormonal) may affect the level significantly.

CLINICAL SIGNIFICANCE[36,40,59,63,84,90]

In disease states, both the total protein and the ratio of the individual protein fractions may change independently of one another. In states of dehydration, total protein may increase some 10 to 15 per cent, the rise being reflected in all protein fractions. Dehydration may result either from a decrease in water intake, as occurs in frank water deprivation, or from excessive water loss, as occurs in severe vomiting, diarrhea, Addison's disease or diabetic acidosis. The absolute quantity of serum proteins is unaltered, but the concentration is increased because of the decreased volume of solvent water. In multiple myeloma, the total protein may increase to over 10 g/100 ml, the increase being due to the presence of markedly elevated levels of myeloma proteins (monoclonal immunoglobulins or paraproteins, p. 349). The quantities of other proteins are essentially unaltered.

Hypoproteinemia, characterized by total protein levels below 6.0 g/100 ml, is encountered in many unrelated disease states. In the nephrotic syndrome large masses of albumin may be lost in the urine as a result of leakage of the albumin molecules through the damaged kidney. In salt retention syndromes, water is held back to dilute out the retained salt, resulting in the dilution of all protein fractions. Large quantities of proteins are lost in patients with severe burns, extensive bleeding, or open wounds. Water is replaced by the body more rapidly than is protein, effecting a decreased total protein concentration. A long period of low intake or deficient absorption of protein may affect the level and composition of serum proteins, as in sprue and in other forms of intestinal malabsorption, as well as in acute protein starvation (kwashiorkor). In these conditions the liver has inadequate raw material to synthesize serum proteins to replace those lost in the normal turnover (wear and tear) of proteins and amino acids.

In general, changes in total proteins may occur in one, several, or all fractions. It is also possible for significant changes to occur in different directions in different fractions, without changes in the total protein concentration. Consequently, in a study of a patient's serum proteins, a determination of total protein is first made, followed by an estimation

* Values are given here in g/100 ml in accordance with present usage. The SI unit system, however, requires reporting proteins in g/l.

of total albumin and/or total globulins. These can now be carried out conveniently with automated techniques. A cellulose acetate electrophoretogram can be obtained next (or in place of albumin/globulin fractionation). If necessary these initial studies[48] may be followed by (1) estimation of individual serum (plasma) proteins by chemical or immunochemical procedures, (2) investigation of abnormal electrophoretic bands by immunoelectrophoresis and other techniques, and/or (3) investigation of the molecular size by use of gel filtration or ultracentrifugation.

Anabolic steroids, androgens, corticosteroids, corticotrophin, growth hormone, progestin, insulin, and thyroxine all increase protein synthesis and the serum protein level, whereas estrogens, including contraceptive pills, lower the serum protein, as in pregnancy. Transport proteins, however, such as transferrin and thyroxine-binding globulin, increase in concentration in pregnancy and during estrogen therapy.

Other drugs said to decrease serum protein levels are rifampin (rifamycin), which impairs protein synthesis; trimethadione, which decreases protein levels by urinary loss; and the long-term use of laxatives, which results in fecal malabsorption and thus in decreased serum protein levels. Clofibrate, digitalis, and epinephrine are reported to raise the protein level.

METHODS FOR THE DETERMINATION OF TOTAL PROTEIN IN SERUM[31,90]

Because the proteins of serum are a complex, variable mixture, certain assumptions are normally made when total serum protein is estimated. It is assumed that:

1. The lipid and carbohydrate constituents of protein may be regarded as if they were not part of the protein molecules.
2. The average percentage of nitrogen in serum proteins is 16 per cent.
3. The differences in the composition of different sera do not influence the analytical result.

These assumptions are arbitrary, but they allow estimates of total serum protein which have proved to be of great value clinically.

In practice, a standard reference method based on nitrogen determination is required (the Kjeldahl technique), as well as a routine method based on peptide bond estimation (the biuret method); occasionally, when speed is all-important, a rough but rapid method for determination of total serum protein is required (the refractive index method).

The biuret method is not very sensitive, but it is sensitive enough to estimate protein concentrations in serum; and it has the great advantage that it estimates peptide bonds and so is least affected by variations in protein composition. The more sensitive methods, such as those depending on ultraviolet light absorption, or on the use of Folin and Ciocalteu's phenol reagent, either are unnecessarily complex or depend on the estimation of a side chain such as tyrosine, and thus may give rise to serious errors if the composition of a serum is abnormal. Except when used for the estimation of single proteins (e.g., fibrinogen after its isolation as fibrin clot), these latter methods should be confined to the estimation of protein solutions so dilute that their concentrations are beyond the sensitivity of the biuret reaction (see p. 354).

DETERMINATION OF PROTEIN NITROGEN (KJELDAHL TECHNIQUE)[2]

Inasmuch as all proteins contain nitrogen derived from their constituent amino acids, proteins can be determined by measuring the nitrogen present in the isolated protein. It is conventionally assumed that 16 per cent of the mass of serum proteins is nitrogen (N), and from this it follows that

$$\text{protein} = \frac{1.00}{0.16} \times N = 6.25 \times N$$

More recent careful studies have indicated that this factor, 6.25, for converting N values to protein values, is probably too low and that the true value may be closer to 6.45. The old factor of 6.25, however, continues to be used pending acceptance of a more accurate value by a consensus of clinical scientists. In any case, whichever factor is used, the figure will remain to some extent arbitrary. Even when lipid and carbohydrate conjugates are excluded, the nitrogen content of the major individual protein fractions in serum varies considerably, ranging from 15.1 to 16.8 per cent, with corresponding protein/N factors ranging from 6.00 to 6.65[82,90], so that the exact factor in any given case will depend on the composition of the serum and will differ from one patient to another. The error or uncertainty is negligible for sera of normal composition, but this may not be true for sera from patients with protein abnormalities.

Protein nitrogen is usually measured by any one of many semimicro modifications of the classic *Kjeldahl nitrogen* technique. For precise work, trichloracetic acid or tungstic acid is employed to precipitate the proteins, the NPN (nonprotein nitrogen) being removed with the filtrate-supernatant. The washed precipitate is transferred to Folin-Wu or Kjeldahl digestion tubes and the organic matter oxidized by hot refluxing H_2SO_4, with or without added catalysts such as $CuSO_4$, $HgSO_4$, selenium, or SeO_2 to speed up the oxidation reactions. For efficient oxidation, the temperature should be at about 340 to 360°C; K_2SO_4 is added to raise the boiling point of the sulfuric acid. In this procedure, the carbon, hydrogen, and sulfur in the protein are oxidized to CO_2, CO, H_2O, and SO_2, and the nitrogen is converted to NH_4HSO_4:

$$\underset{[C,H,N,O,S,P]}{\text{Protein}} \xrightarrow[H_2SO_4 + \textit{Catalyst}]{\textit{Heat}} CO_2 + H_2O + SO_2 + P_2O_7^{4-} + CO + NH_4^+ + HSO_4^- \quad (8)$$

The NH_4HSO_4 formed is then measured by adding excess alkali and distilling the liberated NH_3 into excess standard acid and back-titrating with standard NaOH (Equations 9 and 10), or by distilling the NH_3 into boric acid, and measuring the NH_3 entrapped as $NH_4H_2BO_3$, by titrating with standard HCl (Equations 9, 11, and 12).

$$NH_4HSO_4 + 2\,OH^- \xrightarrow{\textit{Distill}} NH_3\uparrow + 2\,H_2O + SO_4^{2-} \quad (9)$$

$$NH_3 + H^+ \xrightarrow[\textit{HCl}]{\textit{Excess}} NH_4^+ + H^+ \text{ (excess), measured with Std. NaOH} \quad (10)$$

$$NH_3 + H_3BO_3 \xrightarrow[H_3BO_3]{\textit{Excess}} NH_4H_2BO_3 \longrightarrow NH_4^+ + H_2BO_3^- \quad (11)$$

$$H_2BO_3^- + H^+ \xrightarrow[\textit{HCl}]{\textit{Std.}} H_3BO_3 \quad (12)$$

Alternatively, the NH_3 formed in the Kjeldahl digestion may be measured by direct nesslerization of the digestate or of the distilled NH_3. Neither of these alternatives is to be recommended; their precision is much inferior to the previous titration methods, and the Cu^{2+} or Hg^{2+} catalysts interfere with the Nessler reaction. In practice, the Kjeldahl reaction is performed directly on a serum aliquot and the total nitrogen value obtained is corrected for nonprotein nitrogen by a separate assay of the NPN.

The Kjeldahl procedure is used primarily as a standard method against which all other methods are calibrated, because it is capable of high precision and accuracy. It is too slow for routine analysis of a large number of specimens. Owing to the uncertainty about the correct average factor for converting nitrogen into protein, the Kjeldahl nitrogen value cannot be used as an absolute standard for quantitating serum proteins, if indeed any such standard is possible. It is conventional practice, however, to calibrate other protein methods against the Kjeldahl value as a standard.

BIURET METHOD FOR THE DETERMINATION OF TOTAL PROTEIN IN SERUM AND EXUDATES[42,71]

All proteins contain a large number of peptide bonds. When a solution of protein is treated with Cu^{2+} ions in a moderately alkaline medium, a colored chelate complex of unknown composition is formed between the Cu^{2+} ion and the carbonyl (—C═O) and imine (═N—H) groups of the peptide bonds. An analogous reaction takes place between the cupric ion and the organic compound biuret, $NH_2{-}\underset{\underset{O}{\|}}{C}{-}NH{-}\underset{\underset{O}{\|}}{C}{-}NH_2$ and, therefore, the reaction is referred to as the *biuret reaction*. The reaction takes place between the cupric ion and any compound containing at least two NH_2CO—, NH_2CH_2—, NH_2CS—, and similar groups joined together directly, or through a carbon or nitrogen atom. Amino acids and dipeptides cannot give the reaction, but tri- and polypeptides and proteins react to give pink to reddish violet products. In the biuret reaction one copper ion is linked to between four and six nearby peptide linkages by coordinate bonds; the more protein present, the more peptide bonds available for reaction. The intensity of the color produced is proportional to the number of peptide bonds undergoing reaction. Thus, the biuret reaction can be used as a basis for a simple and rapid colorimetric method for determining protein.

Kingsley introduced the first procedure simple enough to be practical in the clinical laboratory. Since then, the biuret method has become the method of choice for measuring proteins because of its simplicity and adequate precision and accuracy. The reaction is performed directly, using either undiluted or diluted serum. Peptides are present in serum, but their concentration is so low that they contribute no increment to the biuret color. Ammonium (NH_4^+) ions interfere by buffering the pH at too low a level, but the NH_4^+ level in serum is too low to interfere. Most biuret methods can determine about 1.0 to 15 mg of protein in the aliquot being measured. A large number of modifications of the procedure are available;[42] all are alike in principle, but vary in details and in the composition of the biuret reagent, which must be stabilized so that the $Cu(OH)_2$ does not precipitate and is not too susceptible to reduction. The method to be presented is based on the procedure of Reinhold,[71] using the Weichselbaum biuret formulation, in which sodium potassium tartrate is used as a complexing agent to keep the copper in solution, and potassium iodide to prevent autoreduction. It will serve as an example of the biuret technique.

REAGENTS

1. Biuret diluent. 5 g KI in 0.25 ($\pm$0.02) mol NaOH/l (= 10 g NaOH/l). The NaOH solution may be prepared in 8 liter quantities by dilution of 2.50 mol/l (100 g/l) stock NaOH with CO_2-free water. Water fresh from a still or a good ion-exchange column is essentially CO_2 free, as is freshly boiled water. Store the reagent in stoppered polyethylene bottles.

2. Stock biuret reagent, Weichselbaum formula. Dissolve 15.0 g of finely pulverized $CuSO_4{\cdot}5H_2O$ in 70 to 80 ml of water. Prepare a solution of 45.0 g Rochelle salt (potassium sodium tartrate, tetrahydrate) in 600 to 700 ml of biuret diluent and slowly add the $CuSO_4$ solution with stirring. Both solutions must be at room temperature when mixed to prevent reduction of Cu^{2+} by the tartrate. Add biuret diluent to a volume of 1000 ml. Before use, filter the solution through a qualitative paper to remove any deposited cuprous oxide. Store in a polyethylene bottle, away from strong, direct light.

3. Working biuret. Dilute the stock biuret 5-fold with biuret diluent. Store the same way as the stock biuret.

4. Alkaline tartrate, 0.9 g Rochelle salt in 100 ml biuret diluent. The solution is

identical to working biuret solution except that Cu^{2+} is omitted. This is used to correct for serum pigment error as discussed under Procedure and Comments.

5. Saline, NaCl, 0.15 mol/l (0.85 g/100 ml).

6. Standard protein. It is most convenient to use Armour Bovine Serum Albumin Standard (BSA) solution, available in 3.0 ml ampules containing pure bovine albumin. The analysis in terms of protein N is imprinted on the ampule and is in the range of 9.7 to 10.3 mg of protein N/ml. This is equivalent to 60.6 to 64.4 mg of protein/ml (6.06 to 6.44 g/100 ml), if the customary protein/N ratio of 6.25 is used (see the discussion under the Kjeldahl procedure). Dilute the ampule solution 10-fold with saline by pipetting 2.50 ml into a 25.0 ml volumetric flask containing 20 ml of saline; the albumin is delivered down the sides of the flask to avoid foaming. After diluting to the mark with saline, gently but thoroughly mix the flask contents.

The National Committee for Clinical Laboratory Standards (Los Angeles, CA.) provides an Approved Protein Standard containing 2.2 ml of protein solution having a mass concentration of approximately 70 g protein/l (known within ± 0.2 g/l) contained in glass ampules under an inert gas. A pamphlet describes the specifications and the procedure for preparing a protein solution,[62] as well as a recommended assay procedure.[24]

7. Control serum. Prepare this by pooling leftover normal sera, which are free of hemolysis, significant icterus, turbidity, or chyle. Mix the pool, pass it through several layers of Pyrex wool to remove fine clots, transfer to 15×120 mm test tubes (polycarbonate is best) in 1 ml aliquots, and store frozen. Before assaying, remove a tube, thaw in a bath at 37°C, and carefully mix. The protein value may be measured by the Kjeldahl procedure or by assaying it against another serum of known protein value. Alternatively, a commercial control material may be used. If lyophilized, the reconstituted solution must be clear and free of turbidity. Caution should be used in pipetting such pools because of frequent contamination with Australian hepatitis antigen. The same precaution also applies to most commerical control sera.

PROCEDURE

The reaction is carried out directly in appropriate cuvets.

1. Set up two cuvets, each containing 2.0 ml of saline for each unknown, and pipet 0.10 ml of specimen into each, using 100 μl micropipets. If TC micropipets are used, rinse the pipets three or four times with saline. Carefully wipe the tips and ends of the pipets free of adherent serum before delivering the serum aliquot to the saline. An automatic dilutor is a convenience if a large number of specimens are to be analyzed. Also set up a saline blank cuvet containing 2.10 ml of saline only.

2. To one of each pair of cuvets add 8.0 ml of working biuret; to the other (the serum blank) add 8.0 ml of alkaline tartrate. It is convenient to use automatic pipets. Thoroughly mix the tube contents and allow the color to develop for 30 min.

3. Measure the absorbances of the solutions at 550 nm, using a spectrophotometer with the saline blank set at 100 per cent T. After deducting the serum blank reading, refer to the calibration curve and convert the absorbance value to g of protein/100 ml of serum. Alternatively, if the calibration is linear, $C = F \times A$, where F is the calibration factor (see next section), A is the absorbance, and C is the protein concentration in g/100 ml.

CALIBRATION

1. Transfer the following volumes, 0.50, 1.00, 1.50, and 2.00 ml, of the 10-fold dilution of Armour BSA standard to cuvets. Add saline to make the volume in each equal to 2.10 ml. Then add 8.0 ml of working biuret to each, as in the serum procedure, and measure the absorbance of the standard solutions after 30 min of color development.

2. In the 10.1 ml cuvet volume, 1.00 ml of the diluted standard will contain

$$S = 6.25 \times N \times 1/10, \quad \text{or}$$

$$S = (0.625) \times N \text{ mg of protein} \tag{13}$$

where N represents the mg N/ml value marked on the ampule.

3. If the unknown serum contains T g of total protein/100 ml, the 0.10 ml used for the determination will contain T mg of protein, also dissolved in a cuvet volume of 10.1 ml. The absorbance of this serum must equal that of some standard, containing S mg of protein. Thus, T mg $= S$ mg, or $S = T$, and the numerical value of the protein in the cuvet in milligrams is identical with the serum protein concentration expressed in g/100 ml. The standards represent 0.5, 1.0, 1.5, and 2.0 × 0.625 × N g protein/100 ml. Depending on the actual concentration in the Armour ampule, the standards represent approximately 3.1, 6.3, 9.5, and 12.6 g of protein/100 ml. The calibration factor F is equal to the slope of the calibration line and is given by C_s/A_s. The individual absorbances of the respective standards are used to calculate the slope by the following approximation formula:

$$F = \frac{\Sigma C}{\Sigma A} = \frac{\text{sum of concentrations represented by standards}}{\text{sum of absorbances of the standards}} \tag{14}$$

COMMENTS

1. At the time the blood sample is taken, it should be recorded whether the patient is ambulatory or recumbent; he need not be fasting. The use of a tourniquet may lead to falsely high results; hemolysis must be avoided or the added protein will give falsely high results.

2. The biuret reagent should be filtered at least weekly.

3. The precision (repeatability) expected for this procedure is about ±0.30 g/100 ml, or about ±3.5 per cent (95 per cent confidence interval). The main source of error is in pipetting the initial sample aliquot of 0.10 ml. The calibration curve is reproducible for any given spectrophotometer, small differences in calibration factor F arising from small differences in individual biuret reagents.

4. Serum blanks are particularly important if the serum is highly pigmented (hemoglobin, bilirubin), milky (lipemic), or turbid in appearance. Hemoglobin and, to a much lesser extent, bilirubin absorb light at 550 nm, and this absorbance will be added to the true protein biuret color. Less satisfactory alternatives to using blanks are to extract the turbidity with ether or to proceed as follows: After the biuret color of the unknown has been read, add 50 to 70 mg of dry KCN to the tube and to the biuret blank. The CN ion will remove the Cu^{2+} from the red-violet copper protein complex by forming the nonionized $Cu(CN)_2$, thus clearing the tube of any biuret color. Read the absorbance of the cyanide-treated solution and subtract this absorbance, representing pigment and turbidity, from the biuret color reading. Utmost care must be the rule when working with the extremely poisonous KCN. Correction may also be made by reading the solution at a second wavelength (bichromatic analysis). Nevertheless, the presence of pigments in the specimen, even with the corrections applied, is a source of error.

5. Presently, in most hospital laboratories an automated version of the method is used; satisfactory modifications for both discrete and continuous-flow analysis are available.

6. Human serum albumin may be used as a standard in place of bovine serum albumin. The same standard can then be used for both albumin and total protein determinations.[42]

REFERENCE

Reinhold, J. G.: Total protein albumin and globulin. *In* Standard Methods of Clinical Chemistry. D. Seligson, Ed. New York, Academic Press, Inc., 1953, vol. 1, p. 88.

DETERMINATION OF TOTAL PROTEIN BY MEASUREMENT OF REFRACTIVE INDEX

The refractive index η of water at 20°C is 1.3330. If a solute is added to the water, η will be increased by an amount directly proportional to the concentration of solute in dilute solutions and nearly proportional to concentration over a 2- or 3-fold range in more concentrated solutions (5 to 20 g/100 ml). This proportionality still holds for mixtures of solutes if their refractive indices are similar in magnitude. The relation between refractive index and concentration is usually obtained by empirical calibration.

The increment in refractive index can be used to measure total solids present in urine. The solutes consist mainly of urea, Na^+, Cl^-, K^+, and $H_2PO_4^-$ ions, the combined concentration ranging from 2.5 to 7.0 g/100 ml. Similarly, refractive index measurements can be used to obtain clinically useful, though not always accurate, measurements of serum protein concentrations. In this application, a reasonable assumption is made that the concentrations of individual inorganic electrolytes and simple organic metabolites do not vary appreciably from serum to serum, and that differences in refractive index are primarily a reflection of differences in protein concentration. Studies have shown that this is generally true, and refractive index measurements of clear, nonpigmented, nonturbid sera provide a very rapid and direct method for measuring serum protein when extreme accuracy is not essential. Samples from patients with azotemia, hyperlipemia, or hyperbilirubinemia or hemolyzed samples are subject to gross errors.

The TS Meter, Model 10401 (American Optical Co., Scientific Instrument Div., Buffalo, NY.), is a simple and convenient instrument of the refractometer type, designed for measurement of serum protein. Only slightly more than one drop of protein solution is required, the sample being spread by capillary action as a thin film between the measuring prism and a cover plate. The refractive index of a solution has an appreciable negative temperature coefficient; therefore, there must be either precise measurement of temperature or some form of compensation for change of η with temperature. The TS Meter provides for such compensation so that the readings are valid for the entire temperature range of 15 to 37°C. After placing the sample drop on the prism, the cover plate is pressed down, and the instrument is pointed at a source of reasonably intense light. The light beam enters parallel to the prism, is refracted by the protein solution, and is then projected against the eyepiece, which contains scales calibrated in both refractive index and grams of protein per 100 ml. The refracted rays light up a segment of the field viewed in the eyepiece, the field being separated by a sharp demarcation line into a light area and a dark area. The reading of the scale at the boundary line separating the two areas gives the protein concentration directly.

Methods for the Identification and Determination of Individual Protein Fractions

The most useful techniques[48] for the investigation of protein fractions are electrophoresis, immunodiffusion, and molecular filtration. In addition, use can be made of special reactions of lipoproteins (Chap. 10) and mucoproteins, involving their attached lipid or carbohydrate moieties.

Salt fractionation, although simple to carry out and essential in the past for the separation of albumin from globulins, is being used less frequently. Now that macroglobulins can be investigated by gel filtration, the expensive technique of ultracentrifugation can be restricted to the investigation of lipoproteins (Chap. 10). Ion-exchange chromatography has only a few applications at present: DEAE cellulose is used, for example, for the isolation of γ-globulin or for the purification of fluorescein-labeled proteins (Section Two, p. 293).

TABLE 7-6 ELECTROPHORETIC FRACTIONS

Electrophoretic Bands	*Normal Range (depth of staining of electrophoretic fractions), g/100 ml*	*Main Constituents and Their Molecular Masses in Daltons*	
Albumin	3.5–5.0	*Albumin* (Alb)	65,000
α_1	0.17–0.33	*α_1-anti-trypsin* (α_1ATr) Cause of band on protein staining.	45,000
		α_1-lipoprotein (α_1Lp) (apoprotein A) Cause of band on fat staining.	200,000
		Orosomucoid (α_1AGp) (α_1-acid-glycoprotein, seromucoid) Stains only faintly with protein stain.	44,000
α_2	0.42–0.90	*α_2-macroglobulin* (α_2M)	820,000
		Haptoglobin (Hpt)	From 85,000 up to more than 1,000,000 in higher polymers.
		Pre-β-lipoprotein (VLDL) (apoprotein A, B, and C) Seen only on fat staining.	Very large
β_1	0.52–1.05	*Transferrin* (Tr) (siderophilin)	80,000–90,000
		Hemopexin (Hpx) Stains poorly.	80,000
		β-lipoprotein (βLp) (apoprotein B) Requires fat stain.	About 3,000,000
β_2	0.08–0.14 (only present in fresh serum)	β_{1C}	
γ	0.71–1.65	Immunoglobulin G (IgG, γG)	150,000
		Immunoglobulin A (IgA, γA)	180,000 (higher when polymerized)
		Immunoglobulin M (IgM, γM)	900,000 (higher when polymerized)

* Some levels vary with season, age, or sex.[51] Values may also differ in accordance with the method employed (e.g., electrophoresis vs. immunochemical techniques).

AND THEIR MAIN CONSTITUENTS[47,77]

Average Values and Approximate Normal Ranges of Individual Proteins g/100 ml*	*Biological Function*	*Main Pathological Abnormalities*
4.4 (3.5–5.0)	Transport of fatty acids, etc. Regulation of plasma volume by colloid osmotic pressure. Provides main protein reserve.	Reduced in cirrhosis and nephrotic syndrome, malnutrition, protein losing enteropathy and hemodilution.
0.25 (0.2–0.4)	Proteinase inhibitor.	Increased in acute phase reaction. Its congenital deficiency leads to emphysema and neonatal liver necrosis.
0.36 (0.29–0.77)	Transport of lipids, fat-soluble vitamins, and hormones.	Reduced in liver disease.
0.09 (0.05–0.14)	Not established.	Increased in acute phase reaction, chronic inflammation, rheumatoid arthritis and malignancy.
Males: 0.24 (0.15–0.35) Females: 0.29 (0.17–0.42)	Plasmin inhibitor.	Increased in nephrosis, liver disease, and diabetes.
0.1–0.3 (depending on type)	Binds hemoglobin, preventing loss of iron.	Reduced in hemolysis and liver failure. Increased in acute phase reaction.
	Transports lipids, especially triglycerides.	Increased in primary and secondary hyperlipidemias Fredrickson Type IV.
0.29 (0.2–0.4)	Iron transport.	Increased in iron deficiency. Reduced in nephrotic syndrome.
0.10 (0.07–0.13)	Heme binding.	Reduced in hemolytic anemia.
0.53 (0.29–0.95) (depending on age and sex)	Lipid transport, especially cholesterol, fat-soluble vitamins, and hormones.	Increased in primary and secondary hyperlipoproteinemia Fredrickson Type II.
0.11 (0.08–0.14)	Complement Factor C3. On standing converted to β_{1A}.	Reduced in autoimmune diseases, e.g., glomerulonephritis, and SLE.
1.16 (125 IU/ml) (0.8–1.4)	Antibodies.	Polyclonal increase in liver disease and chronic infections. Monoclonal increase in myeloma. Reduced in immune paresis.
0.21 (116 IU/ml) (0.09–0.45)	Antibodies, especially in secretions.	Increased in portal cirrhosis. Otherwise similar to IgG.
0.1 (140 IU/ml) (0.06–0.25)	Antibodies formed first; confined to blood.	Monoclonal increase in Waldenström's macroglobulinemia. Otherwise similar to IgG.

SEPARATION OF PROTEINS BY ELECTROPHORESIS[77,78]

A mixture of protein molecules of differing net charges can be separated into fractions by an electric field. Two categories of supporting media are used. The first group includes paper (Durrum, 1950), agar gel (Gordon, 1949), and cellulose acetate (Kohn, 1957); these all have pores wide enough for the protein molecules to move freely. Methods employing such supports give four to seven fractions similar to those of the original moving boundary method (Tiselus, 1937). The second group of support media includes starch gel (Smithies, 1955) and acrylamide (Ornstein and Davis, 1964); they have pore diameters so small that they act as molecular sieves. Techniques employing these materials fractionate proteins according to their molecular volumes as well as their charges and are able to separate serum proteins into some 20 or more fractions. Their resolving power has proved of great value in the study of genetic protein variants and some isoenzymes, as these cannot be distinguished immunochemically. In routine work, however, the many bands tend to cause confusion rather than provide useful additional information.

In the clinical laboratory, paper has now been superseded by cellulose acetate[44] or agar gel,[95] which have a more uniform structure and provide faster and sharper fractionation. These media can be made transparent, permitting better scanning. Agar gel[47] is probably best for lipoprotein electrophoresis, but for routine investigation of serum proteins, cellulose acetate is simpler and more popular. An example of its use is described in detail below.

Barbital (veronal) buffer at pH 8.6 is the standard buffer for routine serum protein electrophoresis. If calcium lactate is added, a total of seven bands (prealbumin, albumin, α_1, α_2, β_1, β_2, and γ) is obtained with fresh serum. Other buffers are only used for special work, such as the separation of glycoproteins or of hemoglobin variants.

There are many alternative dyes available for staining the protein bands, such as Amido Black 10B, Ponceau S, Nigrosin, or Lissamine green. Glycoproteins can be detected by staining a duplicate strip for carbohydrate by the periodic acid Schiff's reagent method (p. 334), but this is seldom required in routine work. Lipoproteins can be demonstrated with fat stains: four bands are seen which do not correspond exactly to the protein-stained bands (see below and Chap. 10).

Only 15 proteins in normal serum are present in sufficient concentration to influence the electrophoretic patterns seen after staining for protein or for fat.[47] With the exception of prealbumin, they are listed in Table 7–6 in order of mobility, with molecular masses, serum concentrations, functions (if known), and principal alterations in disease.

Because several of them have similar mobilities, they give rise to only six clear bands when stained for protein. These bands are listed in the table; one of them (the γ band) is broad and ill-defined because of the heterogeneity of the immunoglobulins. The β_2 band is peculiar in that it is clearly separated from the β_1 band only when the buffer contains Ca

---

Figure 7–10 A set of cellulose acetate electrophoretic patterns obtained with the Beckman microzone system. A, albumin; B, α_1-globulin; C, α_2-globulin; D, β-globulin; and E, γ-globulin region.

Pattern 1, *Normal* serum (Reconstituted Metrix Control Serum). Pattern 2, *Chronic infection* (hepatitis): a relative decrease in albumin with a notable elevation in γ globulin. Pattern 3, *Destructive lesion* (acute reaction): pattern associated with a myocardial infarction with relative increased levels of α_2 and α_1-globulins. Pattern 4, *Hypogammaglobulinemia*: the considerable drop in γ-globulin is readily observed. A slight increase in α_2-globulin is also seen on this pattern, but is not a common finding in hypogammaglobulinemia. Pattern 5, *Nephrotic syndrome*: the very low level of albumin is striking, as is the extremely elevated peak for α_2-globulin, the moderate rise in β-globulin and the fall in γ-globulin. Pattern 6, *Cirrhosis*: the elevation in both γ- and β-globulin, with a partial fusion of these bands, is present, along with a decrease in albumin. Pattern 7, *Multiple myeloma*: the clone of myeloma proteins is associated entirely with the β-globulins. Pattern 8, *Multiple myeloma*: the myeloma protein is entirely in the γ-globulin area. The fall in albumin is only apparent. It is interesting to compare this γ-globulin region with those in patterns 2, 4, 6, and 7. Note the homogeneous peak in patterns 7 and 8 and the heterogenous peak in pattern 2.

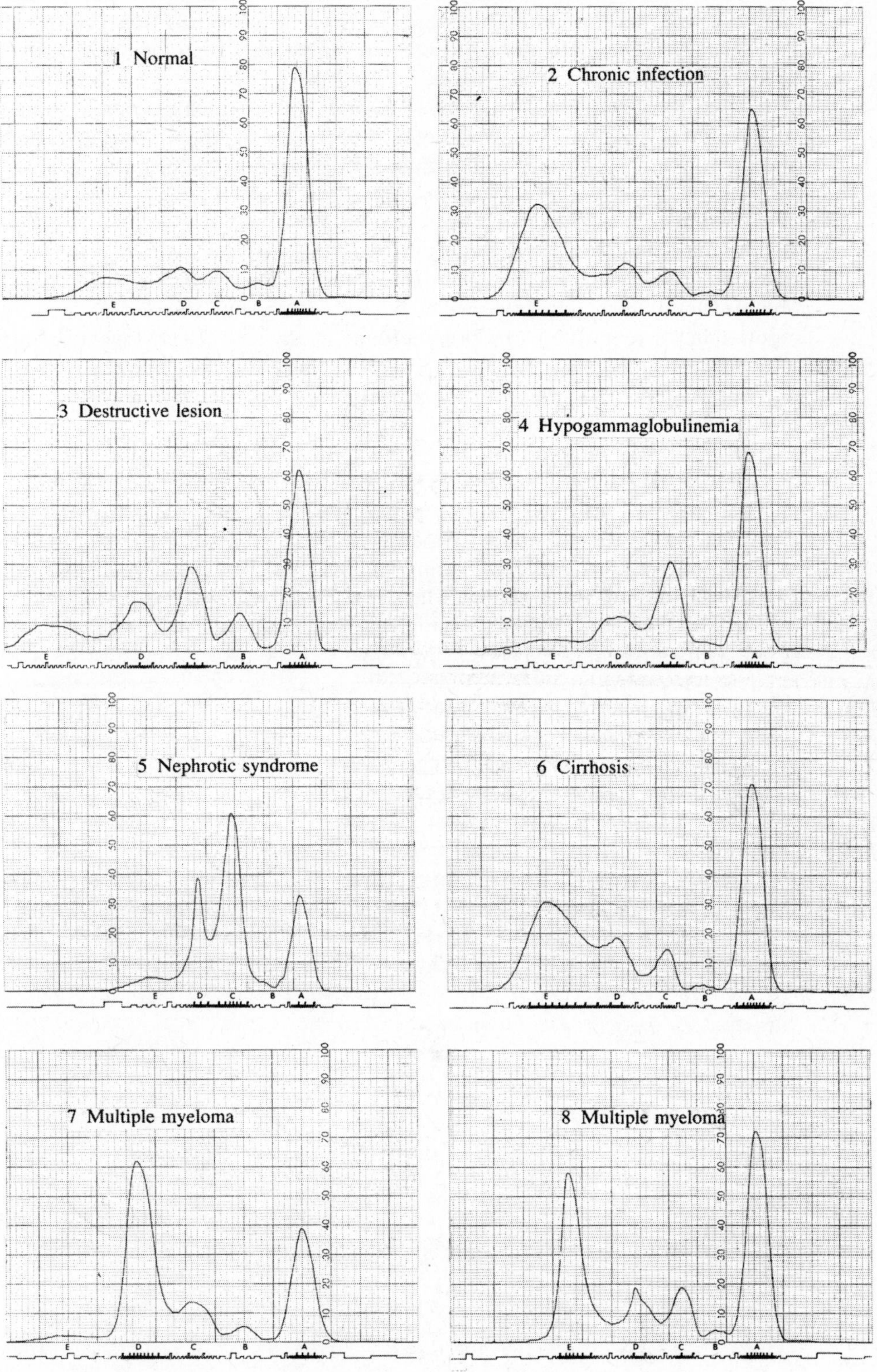

Figure 7–10 *See opposite page for legend.*

ions, and because it is seen only in fresh serum: it is due to the complement factor C3 (also called β_{1C}). On standing, this breaks down to form the faster β_{1A}, which is found in the β_1 band. The prealbumin band is often visible only by transmitted light. In plasma, a fibrinogen band is seen in the β-γ region.

When stained for fat, three bands—poorly demonstrated by protein staining and somewhat variable in mobility—can be seen, in the fast α_1, the α_2 (usually known as pre-β), and the β_1 regions, together with a chylomicron band at the origin.

NORMAL VALUES

A representative sample of results for the normal range of values for the various fractions is presented in Figure 7–10, pattern 1, and in Table 7–7. These results were obtained by different investigators or laboratories, using different methods. The data are reported in per cent of the total protein found in any given electrophoretic fraction. Despite some evident differences, there is a general agreement among the various sets of data. As one would expect, the values for the globulin fractions, being mixtures, vary more than do those for the albumin fraction.

CLINICAL INTERPRETATION OF ELECTROPHORETIC PATTERNS[47]

A number of healthy persons produce genetic variants of proteins[23] which give rise to altered electrophoretic patterns. Among the more common of such altered patterns are those due to variants of haptoglobin. The 2-1 and 2-2 types (see p. 343) join α_2-macroglobulin in forming the α_2 band, whereas the 1-1 type is faster and extends the α_2 band forward toward the α_1. Other common genetic variations are the replacement of the single α_1 band (due to α_1-anti-trypsin) or the single β_1 band (due to transferrin) by fainter double bands (as described later).

In diseases with tissue destruction, the concentration of C-reactive protein may be sufficient to produce an additional band in the γ region. In some leukemias the lysozyme (muramidase) concentration is sufficient to give a band in the post-γ region. More commonly observed is an intense extra band somewhere in the γ, β, or α_2 zones, due to an isolated

TABLE 7–7 VALUES FOR PROTEIN FRACTIONS OF NORMAL SERUM OBTAINED BY VARIOUS TYPES OF ELECTROPHORETIC TECHNIQUES (RELATIVE CONCENTRATION IN PER CENT OF TOTAL PROTEIN)

Protein Fractions						
Albumin	*Globulins*					
	α_1	α_2	β	γ	*Method*	*Reference*
52.2–67.0	2.8–4.6	6.6–13.6	9.1–14.7	9.0–20.6	C.A.-scan	Kaplan and Savory[38]
55–64	3.6–7.2	7.6–10.1	11.6–15.2	10.1–17.2	M.B.	Ehrmantraut[14]
47–71	2.5–5.8	5.1–12.0	4.5–15.7	11.3–24	Paper-scan	Ehrmantraut[14]
55–70	2.0–4.0	7.1–12.9	7.2–10.8	9.8–17.2	C.A.-scan	Grunbaum *et al.*[28]
55.7	7.8	9.9	11.1	15.3	M.B.	von Frijtag Drabbe and Reinhold[88]
58.9	4.3	9.0	10.3	17.3	Paper-scan	von Frijtag Drabbe and Reinhold[88]
58.6	3.9	8.0	12.5	17.6	Paper-scan	Klatskin *et al.*[43]
55–65	1.0–5.7	4.9–11.2	7–13	9.8–18.2	C.A.-elution	Kachmar
54–70	2–5	7–11	8–14	10–20	C.A.	Sunderman[81]
53–65	2.5–5.0	7–13	8.0–14.0	12.0–22.0	C.A.-Microzone	Tietz
Absolute concentration in g protein/100 ml serum. total globulin = 2.5–3.5.					Abbreviations:	
3.5–5.0	0.17–0.33	0.42–0.90	0.52–1.05	0.71–1.65	C.A. = cellulose acetate M.B. = moving boundary	

increase in a single member of the immunoglobulin family of proteins. Such proteins are synthesized by a "clone" of many identical plasma cells, which arise from the abnormal rapid multiplication of a single cell. These cells all produce identical immunoglobulin molecules with the same mobility, giving a sharp "monoclonal" or "M" band (a "paraprotein") on electrophoresis.

There remain many proteins of clinical importance whose concentrations cannot be assessed from the appearance of the electrophoretic pattern. Individual bands may be overshadowed by larger amounts of other proteins with the same mobility (e.g., ceruloplasmin in the α_2 band, or immunoglobulins other than IgG in the γ region), or the proportion of protein in the molecule may be so low that the protein stains poorly (e.g., lipoproteins contain considerable amounts of lipid, orosomucoid carbohydrate), or their concentrations in blood may simply be too low. Those proteins which do not show up as visible electrophoretic bands may be just as important clinically as those which do. The concentration at which a serum protein occurs in health bears no relation to its clinical importance (gram changes in albumin may be less important than nanogram changes in the glycoprotein hormones).

In interpreting the results of electrophoretic fractionations, very little clinical significance can be associated with small variations from the normal range of values. Serial patterns obtained at intervals of several days may increase discrimination. These may show regular, progressive changes or relative constancy and thus be more meaningful than a single examination. Results must, however, be read in conjunction with a normal control serum run at the same time.

Sometimes changes in the electrophoretic pattern may be very useful in suggesting further investigations. For example, reduction of the prealbumin band occurs in the acute phase reaction and in liver disease and is an indication for further investigation. In a few diseases the changes are very characteristic, as illustrated in Figure 7–10.

Nephrotic Syndrome. In this condition, albumin and other lower molecular mass proteins[77] are excreted in the urine. Some immunoglobulin-G is also lost, and this, coupled with a rise in concentration of larger molecules such as α_2-macroglobulin and haptoglobins type 2-1 or 2-2, leads to faint albumin and γ bands associated with an intense α_2 band (see Fig. 7–10, pattern 5).

Monoclonal Immunoglobulin Disorders (see p. 349). These disorders are usually characterized by an intense extra band anywhere in the γ, β, or even α_2 region, designated "M" band (see Fig. 7–10, patterns 7 and 8). This presence in the patient's serum of one specific kind of immunoglobulin in high concentration suggests the existence of either a benign or a malignant proliferation of plasma cells (e.g., multiple myeloma). These cells are genetically identical, and secrete an identical kind of immunoglobulin. The process is fundamentally similar to the proliferation of other kinds of tumor cells, but almost unique in that these tumor plasma cells make their presence known by secreting into the blood easily detectable specific proteins.

Primary or Secondary Immune Deficiency. This condition is characterized by a deficiency of immunoglobulin G (see p. 347), and is suggested by the presence of a faint or absent γ band (see Fig. 7–10, pattern 4).

Acute Phase Reaction. The change in serum protein composition which follows tissue damage, as occurs in injury, acute glomerulonephritis, myocardial infarction, or acute infection, is referred to as the *acute phase reaction*. The most marked change is in the increase in α_2-globulin which is mainly due to a 2- or 3-fold increase in the level of haptoglobin (see Fig. 7–10, pattern 3). Other less frequent changes are a fall in prealbumin and albumin and a rise in α_1 due to increases in α_1-antitrypsin and orosomucoid. γ-Globulins may increase at a later stage of the disease (see Fig. 7–10, pattern 2). The most sensitive indication of the acute phase reaction, however, is the enormous increase in the concentration of C-reactive protein, a protein almost absent from normal serum.

Increased Immunoglobulin (Antibody) Synthesis. Increased concentration of serum immunoglobulins causes a general increase in the intensity of the γ band, as is observed in chronic infections, cirrhosis of the liver, sarcoidosis, and autoimmune diseases, including lupus erythematosus.

In *chronic liver disease*, e g., portal cirrhosis, the concentration of immunoglobulin A is most frequently increased. Since IgA is predominantly of β mobility, this causes a characteristic β-γ bridging or fusion, a finding very suggestive of chronic liver disease (see Fig. 7–10, pattern 6, and p. 348).

Combinations of these responses may occur; for example, a monoclonal immunoglobulin disorder may be combined with an immune deficiency, or a reaction to tissue destruction may be combined with increased synthesis of polyclonal immunoglobulins, as in rheumatoid arthritis.

Table 7–8 summarizes the changes usually observed in various disease states.

SERUM ELECTROPHORESIS USING CELLULOSE ACETATE

REAGENTS

1. Barbital buffer, pH 8.6, ionic strength 0.05. Dissolve 103 g of sodium diethylbarbiturate, 18.4 g diethylbarbituric acid, 6.2 g calcium lactate,[47] and 10 g sodium azide (preservative) in water and make up volume to 10 liters. Check the pH, using a pH meter; if it is not 8.6 ± 0.1, carefully add concentrated HCl or NaOH to bring the pH to the desired value.

2. Staining solution, 0.2 g Ponceau S dye/100 ml of dilute aqueous trichloroacetic acid (3 g/100 ml).

PROCEDURE

1. Fill the electrophoresis tank with buffer. Set the bridge gap to a width of about 8 cm.

2. Soak the cellulose acetate strips (e.g., Oxoid 12.5 × 2.0 cm) with buffer for at least 15 min in a separate plastic pan; to avoid entrapping air, first float the strips over the buffer and allow them to fall into the buffer when saturated.

TABLE 7–8 ELECTROPHORETIC PATTERNS IN SELECTED DISEASE STATES

Disease	*Albumin*	α_1	α_2	β		γ	*Mechanism*
Nephrotic syndrome	D, DD	—	E, EE	—		D	Loss of low molecular weight proteins in urine.
Myeloma	—	—	—	E, EE	or	E, EE	Monoclonal Ig production.
Waldenström's macro-globulinemia	—	—	—	E	or	E	
Cryoglobulinemia	—	—	—	—		E	
Hypogammaglobulinemia	—	—	—	—		D, DD	May be primary or secondary deficiency.
Acute glomerulonephritis	D±	E±	E	—		—	Acute phase reactions.
Acute infection	D±	E±	E±	—		—	
Chronic infection	D	—	—	—		E	Polyclonal IgG increase and reduced albumin synthesis.
Sarcoidosis	—	—	—	E		E	
Rheumatoid arthritis	D±	—	E	—		E	Autoimmune diseases: as above + acute phase reaction.
Lupus erythematosus	D	E	E±	—		E	
Hepatic cirrhosis	D	—	E±	E±		E, EE	Reduced albumin synthesis + polyclonal Ig increase.
Obstructive jaundice	D±	—	E	E		—	Reduced albumin synthesis and increased lipoproteins.
α_1-Anti-trypsin deficiency	—	D	—	—		—	Primary deficiency.

Symbols: D or E = decrease or elevation up to 25 per cent from normal.
DD or EE = decrease or elevation over 25 per cent from normal.
± = marginal decrease or increase from normal.

3. Blot the strips between sheets of filter paper to remove excess buffer.

4. Place single wicks along the bridge of the tank. Then place the cellulose acetate strips onto the wicks, across the bridges, making sure they are parallel to each other and to the edges of the tank. Take care that the strips and wicks are in intimate contact with each other to insure good current flow. Several strips can be run at one time. If less than a full complement of strips is used, they are distributed uniformly over the width of the tank.

5. Place the curved holders against the outside edges of the bridges to insure contact between the strips and wicks to hold the strips level and taut.

6. Apply about 3 μl of serum to each strip approximately 3 cm from the cathode end of the strip. Streak the serum contained in a micropipet across the strip with a steady but rapid motion, starting about 0.5 cm from the side of the strip to within 0.5 cm of the other side. With some practice it is possible to draw a smooth, uniform streak with no difficulty. Alternatively, use a special applicator. Each serum is run in duplicate, and a standard normal serum is always included as a control.

7. Place the lid over the tank immediately after completing application of the serum specimens onto the strips. A slanting lid assures that any moisture that may have evaporated during the course of the run and condensed on the lid will flow to the edge of the tank, and not drop onto the strips.

8. Connect the electrical leads from the tank to the power supply, taking care to insure the correct polarity of the connections. It is wise to alter the direction of current flow on alternate runs to minimize chemical changes in the two buffer compartments. The point of application of the serum must be governed by the projected direction of current flow.

9. Adjust the current on the power supply to give 1.5 mA/2 cm width of the cellulose acetate strips carrying the current. For a complement of eight strips, the total current would be 12 mA. This will require an initial voltage of about 200 volts, which will slowly drop to about 160 volts during the 90 min run.

10. After the electrophoresis has proceeded for 90 min, turn off the power supply, rapidly remove the strips from the tank, and place them in the Ponceau S dye solution for about 5 to 10 min. During all processing, handle the strips with plastic-tipped forceps.

11. Place the strips in 5 per cent (v/v) acetic acid to remove excess dye. The trichloroacetic acid will have fixed the stained protein onto the strips.

12. Place the strips on glass plates and dry in the oven at 90 to 100°C until they appear brittle (about 40 min); or, if time allows, dry at room temperature between filter papers. Examine strips by transmitted as well as reflected light, preferably on an x-ray viewing screen; this shows up the prealbumin band well.

13. The duplicate-stained strips should look identical; if they do not, the procedure should be repeated. Since changes, if significant, are detectable to the naked eye, one strip should be sent to the physician. The other is kept by the laboratory. If quantitative results are required, the relative intensities of the bands are first read in a strip scanner and densitometer (the Photovolt Corporation densitometer or any other suitable instrument may be used), either by measuring reflectance or, after clearing, by measuring absorbance. Their concentrations can then be calculated from the total protein concentration previously estimated by the biuret method. The results are sent with the tracing to the physician.

In the scanning procedure, the strip is made to move past a slit at a constant speed, and the absorbance of the dye along its entire length is determined. As the readings are made, the amplified signal from the detector governs the motion of a pen, which records a curve showing changes in absorbance with distance along the paper. This procedure can best be performed with a densitometer coupled with a built-in integrator, which will measure the area under each peak of the scan. In the mechanism of the densitometer there may be included mechanical means (cams) to compensate for lack of linearity and for the unequal dye binding of the various protein fractions. In the actual scan, the pattern formed is very similar to that produced with the moving boundary apparatus, a series of peaks and valleys, the areas under the peaks being proportional to the quantity of protein in that fraction. (See Fig. 7–10.)

If quantitative results are required and no scanner is available, the following elution method may be used:

Apply a sample of 10 or 20 μl for electrophoresis. After staining, examine the strips, mark off the protein bands, and cut them out with a pair of scissors. Also cut out a piece of the strip about the same width to serve as a background blank. Insofar as possible, cut the fraction strips to about the same width, although the rather diffuse γ-globulin zone may be considerably wider than the other bands.

Place the cut, stained fraction zones into cuvets, and add exactly 3.0 ml of 0.10 molar NaOH. Tap the tubes to insure complete wetting of the cellulose acetate strip by the NaOH, and then allow them to stand for 20 min to permit elution of the dye from the strips into the alkali.

Finally, add 0.3 ml of 6 molar acetic acid (40 ml/100 ml) to neutralize the NaOH partially. Mix the cuvet contents and read absorbances in a spectrophotometer against the blank tube at 520 nm. The eluted strips can be removed with a stirring rod or allowed to rest at the bottom of the cuvet, causing no interference in the absorbance readings. Calculate the results as follows:

Add the absorbance readings of all the protein fractions or zones to give the total absorbance, A_T, of all the protein on the strip.

Divide the absorbance, A_U, of each zone by the value for total absorbance and multiply by 100; this gives the per cent of the total protein in that zone.

Divide the per cent value of each fraction by 100 and multiply by the value of the total protein concentration; this gives the protein concentration in that fraction in g/100 ml.

COMMENTS

1. Since changes that are not visually assessable are unlikely to be clinically significant, it is felt by some that quantitation may be dispensed with. Furthermore, uncertainties arise from the unequal dye binding of different proteins, the "trailing" of albumin being left behind, and from the nonlinearity between absorbance and dye concentration.

2. A micro modification may be used when large numbers of sera are to be analyzed. With a special multi-applicator, small identical volumes of the sera are applied simultaneously at fixed intervals across a single large sheet of cellulose acetate. For example, the Shandon 10 sample applicator has teeth 7 mm long, each holding 0.4 μl. It is dipped simultaneously into small pools (drops) of the 10 samples and then applied across a 78 × 150 mm Oxoid cellulose acetate sheet. Samples 1, 4, 7 and 10 are normal serum controls. Other examples of microtechniques are the Microzone method (Beckman Instruments, Inc.) and the Gelman technique.

3. Results should always be read in conjunction with a normal control serum run at the same time and, if quantitative results are reported, each laboratory should establish its normal range and the 95 per cent confidence limits. If more than one technologist is entrusted with performing the separations, it should be established that all technologists can obtain essentially identical results; that is, approximately the same average value with the same standard deviation.

4. There are few good studies of the precision attainable in doing replicates of the same serum by the same procedure. In one such study, Kaplan and Savory,[38] using cellulose acetate, found an average value of 61 per cent for the albumin fraction, based on 40 replicates, with a standard deviation, s, of 1.46 per cent, giving a coefficient of variation (CV) of 2.4 per cent. These data suggest a range of 58.0 to 64.0 per cent for the 95 per cent confidence limits for the albumin fraction. As can be seen by referring to the Table 7–7, this roughly corresponds to the range of values obtained for the serum albumin fraction in normal sera by the seven laboratories for which ranges are given. Two conclusions are suggested by these data: the degree of precision of the various methods is about the same, and the ob-

served range in albumin values may reflect the inherent degree of imprecision in the methods as much as it does the true physiological variation in the population.

For the α_1-globulin fraction, Kaplan and Savory[38] obtained a coefficient of variation (CV) of 14.2 per cent for an average value of 3.2 per cent of total protein. This corresponds to a 95 per cent confidence interval of 2.2 to 4.2 per cent protein. The CV values for the other three globulin fractions (α_2, β, and γ) were 6.0, 6.1, and 5.1 per cent, respectively. These correspond to 95 per cent confidence limits of 8.8 to 11.2 per cent for the α_2-globulins, and 9.7 to 12.4 and 13.1 to 16.2 per cent for the β and γ fractions. With abnormal sera the variation found was somewhat greater.

Reproducible values for the *lipoprotein* and *glycoprotein* fractions are even more difficult to obtain. Methods are far from standardized and the staining procedures are more difficult to perform.

REFERENCE

Kohn, J.: Cellulose acetate electrophoresis. *In* Chromatographic and Electrophoretic Techniques. I. Smith, Ed. London, Heinemann, 1968, vol. 2, p. 84.

IMMUNOCHEMICAL ANALYSIS OF PROTEINS[8,27,47,58]

The application of gel immunodiffusion techniques to the study of the proteins in blood serum and other body fluids has revolutionized our knowledge of their protein compositions. These simple, specific, and sensitive techniques require very small samples and seldom cause protein denaturation. Antisera can be designed to react with all, some, or only one of the proteins in a complex mixture such as human plasma. A single individual protein can be identified without previous separation and its concentration determined if a respective standard solution is available.

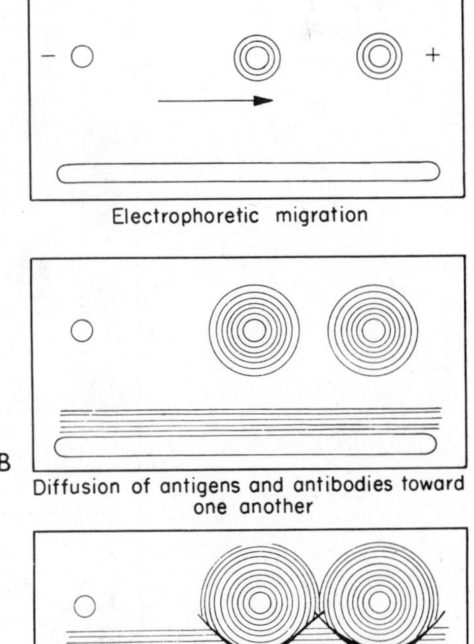

Figure 7–11 The principle of immunoelectrophoresis.

A rectangular glass plate is covered by a thin sheet of agar. The mixture to be analyzed is placed in a small hole (shown top left) cut out of the agar. A potential difference is applied across the plate for several hours, so that the different proteins separate as in cellulose acetate electrophoresis. (In A, two proteins are shown separating and the arrow indicates the direction of migration.)

Then, instead of staining the agar sheet, antiserum is placed in a trough cut parallel to the direction of the current (shown along bottom of plate), and the plate is incubated so that the separated proteins and the antiserum diffuse towards one another (B). Curved lines of precipitate form wherever a protein meets its homologous antibody (C).

The principle of gel immunofiltration is similar, except that the initial electrophoresis is replaced by thin layer gel filtration.

(From Grant, G. H. and Butt, W. R.: *In* Advances in Clinical Chemistry. O. Bodansky and C. P. Stewart, Eds. New York, Academic Press Inc., *13*: 401, 1970.)

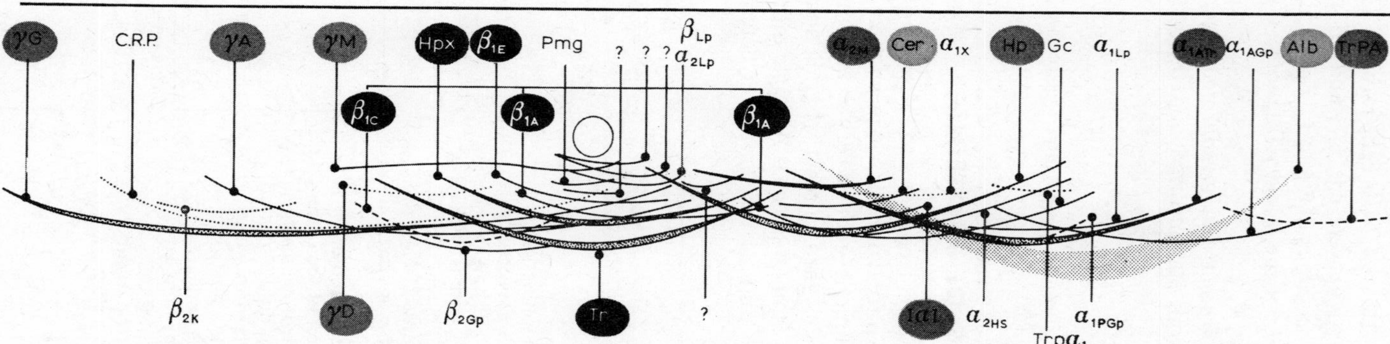

Figure 7–12 Immunoelectrophoresis pattern of normal serum. Abbreviations are those given in Table 7–6, with the addition of CRP for C-reactive protein, Gp for glycoprotein, Pmg for plasminogen, Cer for ceruloplasmin, IαI for inter-α-trypsin inhibitor, α_2HS for α_2 HS-glycoprotein, Trpα_1 for tryptophan-poor α_1-glycoprotein, α_1PGp for α_1 easily precipitable glycoprotein, and TrPA for tryptophan-rich prealbumin. (From Schultze, H. E., and Heremans, J. F.: Molecular Biology of Human Proteins. Amsterdam, Elsevier Publishing Co., 1966, p. 175.)

Whereas specific antisera are required for the identification or determination of *single* individual proteins, a "polyvalent" antiserum (made by injecting whole human plasma and designed to react with as many plasma proteins as possible) is required for detecting *all* the individual proteins simultaneously. When human plasma is allowed to react with such a polyvalent antiserum on a double-diffusion plate (p. 281), the proteins are separated by their differing diffusion rates, but a confusingly large number of lines of precipitate result. The differing diffusion rates alone, like the electrophoretic mobilities alone, do not allow discrimination between all the serum proteins. However, when preliminary electrophoresis was combined with double diffusion at right angles (immunoelectrophoresis, Fig. 7–11)[25] there was effective discrimination between the serum proteins. Immediately it became apparent that the bands seen on electrophoresis were mixtures, that the main γ-globulin band was formed by a family of antigenically similar proteins with mobilities extending from α_2 to slow γ, and that plasma contained many more proteins than was previously supposed. With the best antisera, about 30 different precipitin lines could be distinguished (Fig. 7–12). Those corresponding to known proteins were identified by the use of antisera made specifically against them, or by noting which line disappeared when the polyvalent antiserum was absorbed with the known protein, or by the reaction of identity (p. 283). Protein lines which could not be identified were then labeled by a suffix added to the Greek letter of the original electrophoretic band.

The most important of the newly discovered proteins (originally called β_{2A}) was immunoglobulin A; immunogobulin M, being a macroglobulin, was already known from ultracentrifugal studies. Later the realization that myeloma proteins are pure *normal* immunoglobulins in very high concentration and the finding that some of them were not classifiable as G, A, or M led to the discovery of immunoglobulins D and E in normal serum.

Antisera made against concentrated urine or other body fluids and "adsorbed" with plasma were used to detect and measure those proteins specific to the fluid but not present among the plasma proteins.[26]

In routine clinical chemistry, quantitative techniques, which use single diffusion (p. 280) and specific antisera, are of even more importance than these qualitative techniques. Radial immunodiffusion (p. 283) is presently the most widely used method; electro-immunodiffusion (p. 285) is faster but more complex. Both methods are capable of estimating protein concentrations down to a few μg/ml. If greater sensitivity is required, radioimmunoassay permits detection down to a few pg/ml (p. 286). This is seldom required for any plasma protein estimation except for the assay of hormone glycoproteins, IgE, and fetal proteins such as α-fetoprotein and carcinoembryonic antigen (CEA). The clinical significance of the last is still being evaluated and debated. Raised levels of α-fetoprotein (p. 354) and IgE (p. 277) can be determined with sufficient sensitivity for clinical work by radial immunodiffusion, modified, if necessary, to increase its sensitivity (p. 325).

Attempts to reduce the work in multiple analyses have led to techniques capable either of determining several proteins at once or of determining a single protein rapidly and automatically. The first approach is exemplified by the quantitative immunoelectrophoretic technique of Laurell as modified by Clarke and Freeman[7] (Fig. 7–13). The serum proteins are first separated by electrophoresis in agar, and the separated proteins are then driven into a sheet of agar containing polyvalent antiserum by an electric field applied at right angles to the first. The different proteins are, as in electro-immunodiffusion (p. 285), precipitated as "rockets" whose lengths are proportional to their concentrations, on comparison with a standard human serum. In practice the method is less complex and more useful if a smaller number of proteins are determined by using an antiserum against such individual proteins only, but these antisera are more difficult to make.

For the determination of plasma proteins, such as immunoglobulins, on a large number of specimens, some form of automated procedure is desirable, and one such has been developed for the AutoAnalyzer and is known as the Automated Immunoprecipitin

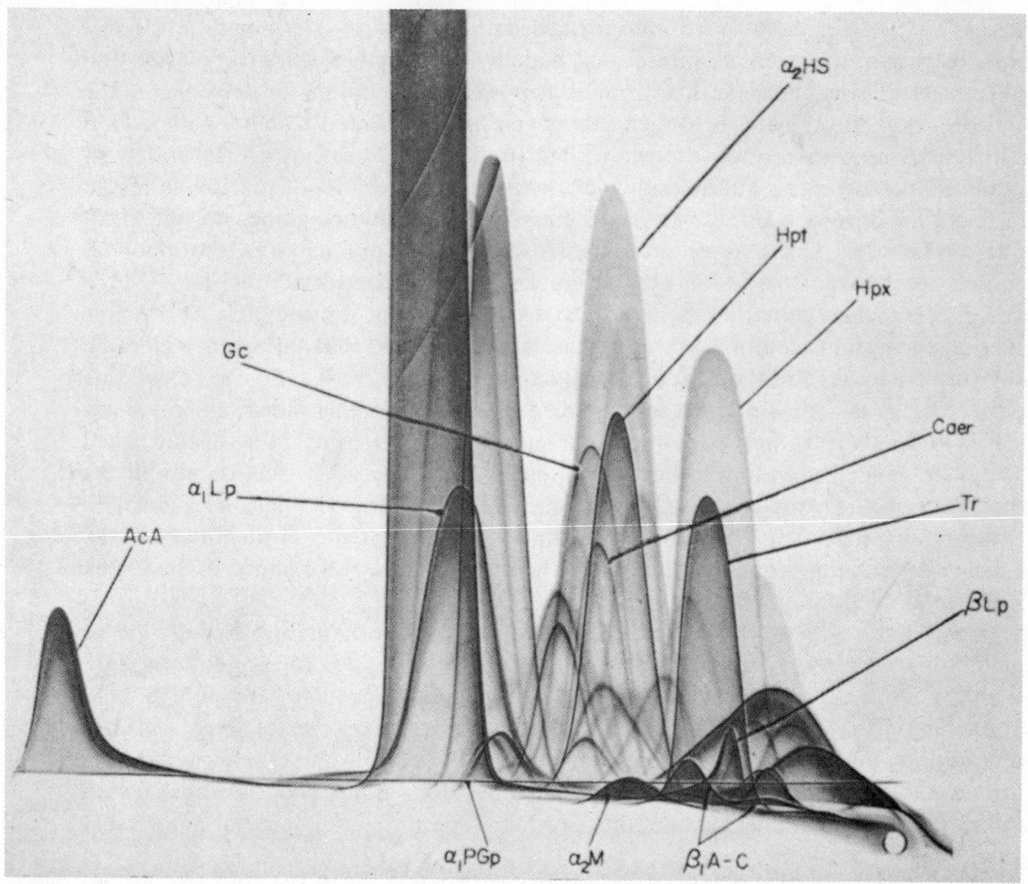

Figure 7–13 Quantitative immunoelectrophoresis. Initially the serum proteins were separated on an agar plate from right to left in an electric field. Then they were driven upward into antiserum-containing agar by an electric field at right angles to the first. For abbreviations see Figure 7–12; AcA, acetylated albumin control. (From Clarke, H. G. M., and Freeman, T.: Clin. Sci., *35*: 403, 1968.)

Procedure (AIP) (p. 294). The antibody-antigen reaction produces a turbidity which is proportional to the antigen concentration under the test conditions. The method, however, is very liable to hidden errors from other causes of turbidity, such as lipids. The antiserum must be very specific so that other proteins will not contribute to the turbidity. The protein antigen concentration must not exceed the equivalence ratio if re-solution of the precipitate is to be avoided.

In all quantitative techniques, errors may arise from using unsatisfactory standards or unsatisfactory antisera. Like colorimetric methods, immunochemical methods depend on comparing unknowns with standards. It is often impossible, however, to obtain protein pure enough to serve as a standard because it cannot be isolated without denaturation which changes its antigenic properties. In such cases a pooled normal serum has to be used as a standard. Immunoglobulins pose a particularly difficult problem because they occur in serum as heterogeneous mixtures, so that even normal serum standards used in different laboratories may vary. Consequently, the World Health Organization[74] has prepared a freeze-dried normal human serum pool to serve as an international standard, which is defined to contain 100 IU each of IgG, IgA, and IgM.

When using antisera as analytical reagents,[56] it is important to remember that no batch of antiserum can be assumed to be identical to any other. Even if made by identical procedures, they may differ in antibody titer, in specificity, and in the firmness or "avidity" with which they combine with the protein antigen used for their production. This is because an antiserum contains a complex mixture of antibodies which combine with different parts (determinants) and with differing degrees of firmness to the protein, and because individual animals vary greatly in the population of such antibodies they make. Antisera may also contain human proteins which have been added to "adsorb" unwanted antibodies. Antisera, especially against immunoglobins (p. 296), should therefore be obtained only from manufacturers who provide full details on how they have been raised: their high cost usually prevents their being tested adequately by the user for quality.

In the clinical laboratory single-diffusion techniques are used for the quantitative analysis of individual plasma proteins, especially immunoglobulins, while double-diffusion techniques are used for the detection of proteins not normally demonstrable in plasma, such as C-reactive protein, α-fetoprotein, immunoglobulin D, or Bence-Jones protein. Electrophoresis followed by double diffusion is employed for the identification of abnormal bands seen on cellulose acetate electrophoresis.

Dilute buffered agar or agarose is the gel of choice for use in diffusion techniques. Cellulose acetate membranes require less antiserum, but they are difficult to use and have the disadvantage that the precipitin lines cannot be seen until the membrane is stained at the end of the test.

Because of the high cost of antisera, microtechniques on microscope slides or small glass plates are preferred. Whether single or double diffusion is used, the reactants either may be allowed to diffuse toward one another or, to make the test more rapid, may be made to move in an electric field, the protein antigen moving toward the anode and the antibody immunoglobulins moving, if at all, toward the cathode as a result of electro-endosmosis.

QUANTITATIVE ANALYSIS OF A SINGLE PROTEIN (MANCINI AND LAURELL TECHNIQUES)

PRINCIPLES

The single-diffusion principle is used, and the amount of antiserum required to precipitate the relevant protein in the patient's serum is compared with the amount required

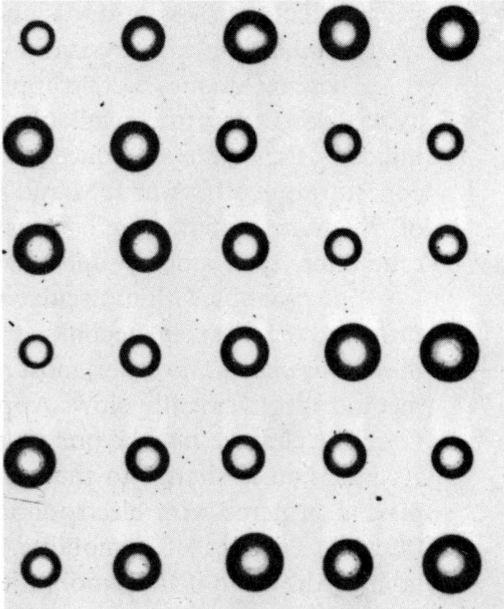

Figure 7–14 Specific protein determination by radial immunodiffusion. The first four holes in the top and fourth rows contain the standards followed by the normal serum control. The remaining holes contain the unknowns with randomly arranged duplicates. (The protein being determined in this example is transferrin.) For further details, see text.

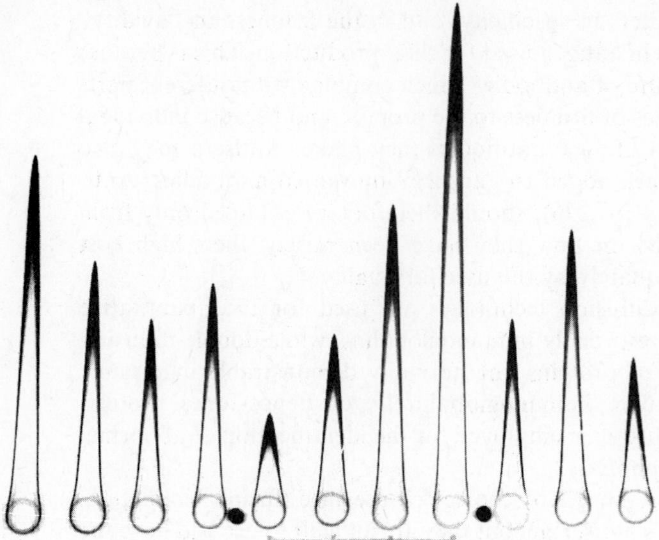

Figure 7–15 Determination of specific proteins by electro-immunodiffusion. The four central holes contained the four standards; the other holes contained unknowns. (Albumin was being estimated.) (From Laurell, C. B.: Scand. J. Clin. Lab. Invest., *29* (suppl. 124): 22, 1972.)

to precipitate a standard solution of it. Standards and unknowns are placed in holes cut in a thin layer of agar, into which a suitable quantity of the specific antiserum has already been incorporated. The antigen protein is then allowed to diffuse into the agar (single radial immunodiffusion: Mancini technique;[52] Fig. 7–14), or it is moved into the agar more rapidly by an electric field (electro-immunodiffusion; single crossed immunoelectrophoresis; Laurell "rocket" technique;[47] Fig. 7–15). As the antigen moves into the antiserum-containing gel, it begins to precipitate when it has combined with sufficient antibody. This visible line of precipitate then moves gradually outwards, because as more antigen arrives and redissolves the earlier precipitate, more precipitate is forming ahead. This process stops and the circle (Mancini technique) or rocket (Laurell technique) becomes stationary only when the antigen has reacted with an equivalent proportion of antibody. The area bounded by the precipitate gives a measure of the antiserum used and, thus, of the amount of protein antigen added. The reaction must be run to completion before reading; until then the positions of the lines will depend not only on the protein concentration but also on other factors which affect the rates of diffusion or the electrophoretic mobilities of the plasma proteins, such as molecular polymerization and heterogeneity as well as temperature.

These techniques can be applied for the determination of any serum protein for which a suitable antiserum is available. The intensity of the γ band on the electrophoretic strip indicates the combined concentrations of all immunoglobulins; it throws no light on the concentration of IgA or IgM and gives only an approximate indication of the concentration of the more abundant IgG. Consequently, the Mancini technique is widely used for their estimation; their mobility and heterogeneity make them unsuitable for the Laurell technique.

For proteins with molecular masses small enough to diffuse rapidly, e.g., Bence-Jones proteins, the Mancini technique is always the method of choice. Because of their large molecular masses, however, most plasma proteins diffuse slowly, and this makes the Mancini method inconveniently slow. Approximate results may be read earlier; or the much faster Laurell technique may be preferable although it is more complex and uses more antiserum. Its precision is similar to that of the Mancini method, and it is particularly suitable for plasma proteins with electrophoretic mobilities equal to or greater than that of the α_2 fraction. Proteins of β mobility, such as transferrin, may also be estimated by electro-immunodiffusion if their mobilities are first increased by carbamylation.

ESTIMATION OF IMMUNOGLOBULINS G, A, AND M BY SINGLE RADIAL IMMUNODIFFUSION[3,52] (MODIFIED FROM MANCINI, CARBONARA, AND HEREMANS)

REAGENTS

1. Antisera. Specific antisera against the heavy chains of IgG, IgA, and IgM may be obtained commercially.* The exact volume to be used per plate must be determined by trial. (When the antisera are in optimum concentration in the agar, the stock human serum standard should produce precipitin rings of 6 to 8 mm on the anti-IgG plate when tested at 21-fold dilution, and, on the anti-IgA and anti-IgM plates, when tested undiluted.) The most sensitive (and economical) antiserum concentration to use is the least amount capable of producing clear rings.

2. Borate buffer, pH 8.6, ionic strength 0.05. Dissolve 6.7 g of boric acid and 13.4 g of sodium borate (decahydrate) in distilled water and dilute to 1 liter.

3. Buffered agar gel. Drop one tablet (Oxoid 1 D Agar, Code BR27) into 50 ml of borate buffer contained in a flask. On bringing the solution to a boil, the agar will dissolve. Distribute in 7.5 ml volumes in small, capped glass containers. Cool in an incubator to 48 to 52°C.

4. Diluent. Dissolve 5 g sucrose and 4 g crystalline bovine albumin in distilled water and dilute to 100 ml. Patients' sera are stable for two weeks at 4°C in this medium but will deteriorate overnight in saline[35].

5. Standards. *Stock:*. A human serum standardized in IU against the International Reference Preparation of the World Health Organization should be used, e.g. Behringwerke's Accuracy Control Serum OTVB 17: for IgG estimations it is used diluted 21-fold; for IgM it is used undiluted; and for IgA it is diluted 2-fold with diluent (reagent 4).

Results are reported in International Units as well as in mg/100 ml, for there is no general agreement yet as to the exact concentrations of the immunoglobulins which correspond to the international standard (p. 294).

IgG Working Standard (100 per cent): Dilute the stock standard with albumin-sucrose diluent to give a 21-fold dilution (use a 2 ml volumetric pipet and 100 μl pipets; e.g., Eppendorf or Oxford pipets). Further dilute this working standard to yield 75, 50, and 25 per cent working standards using a 200 μl pipet. Make fresh every week.

75 per cent standard: 3 vol 100 per cent working standard + 1 vol diluent
50 per cent standard: 1 vol 100 per cent working standard + 1 vol diluent
25 per cent standard: 1 vol 50 per cent working standard + 1 vol diluent

IgM Working Standard: Prepare in a similar way 100, 75, 50, and 25 per cent dilutions of undiluted stock standard in albumin-sucrose diluent.

IgA Working Standard:

100 per cent standard: Use the 50 per cent working standard made for the IgM determinations.

75 per cent standard: Dilute the 75 per cent working standard from IgM standards 1/1 with diluent, using a 200 μl pipet.

50 per cent standard: Use the 25 per cent working standard made for the IgM determinations.

25 per cent standard: Dilute 50 per cent standard 1/1 with diluent, using a 200 μl pipet.

Control Serum: Pool aliquots of fresh serum from 50 normal adults, preferably from 25 males and 25 females. Split the pool into 0.5 ml aliquots, place into polypropylene

---

* Commercial sources include: Behring Diagnostics, Department of Hoechst Pharmaceuticals, Inc., Route 202–206 North, Somerville, N.J. 08876; Meloy Laboratories, Biological Products Division, 6715 Electronic Drive, Springfield, VA. 22151; Wellcome Reagents Ltd., Wellcome Research Laboratories, Beckenham, Kent, England, BR3 3BS; and Hyland Div., Travenol Laboratories, Inc., 3300 Hyland Avenue, Costa Mesa, CA. 92626.

microtubes, and store at −20°C. Such a pool is satisfactory for at least two years. When control serum is required, one tube is removed, thawed and used; any excess is discarded.

To use as a control for IgA or IgM determinations, dilute the pool 2-fold, and when used for IgG determinations dilute the pool 32-fold, with sucrose/albumin diluent.

METHOD

Preparation of plates

Make a mold from two 10 × 10 cm glass plates separated by a U-shaped Perspex or Lucite frame 1.0 mm thick and 10 mm wide, as shown in Figure 7–16. Press them firmly together with suitable clamps (e.g., bulldog paper clips). Prewarm the plates, pipets, test tubes, and molten agar in an incubator at 48 to 52°C. Pipet the requisite amount of antiserum into one of the tubes containing 7.5 ml buffered agar gel. Mix rapidly but thoroughly; pipet down side of mold, which is standing vertically on two of the clips. When agar has set, store in the refrigerator at 4°C after removing clips from mold.

When a plate is required for use, remove the top glass, supporting the lower glass against a transverse stop which is thin enough to allow the top glass to be slid off over it. Alternatively, use one of the small plates of this type which are commercially available (see comment 3).

Dilution of the patients' sera

Make 2- to 64-fold serial dilutions of the patients' sera with albumin-sucrose diluent; if a high IgG is suspected, make a 128-fold dilution. Generally, use the following:

For an IgG plate: 32-fold and 64-fold dilutions.
For an IgA plate: 4-fold and 8-fold dilutions.
For an IgM plate: 2-fold and 4-fold dilutions.

When low values are expected, correspondingly lower dilutions should be used.

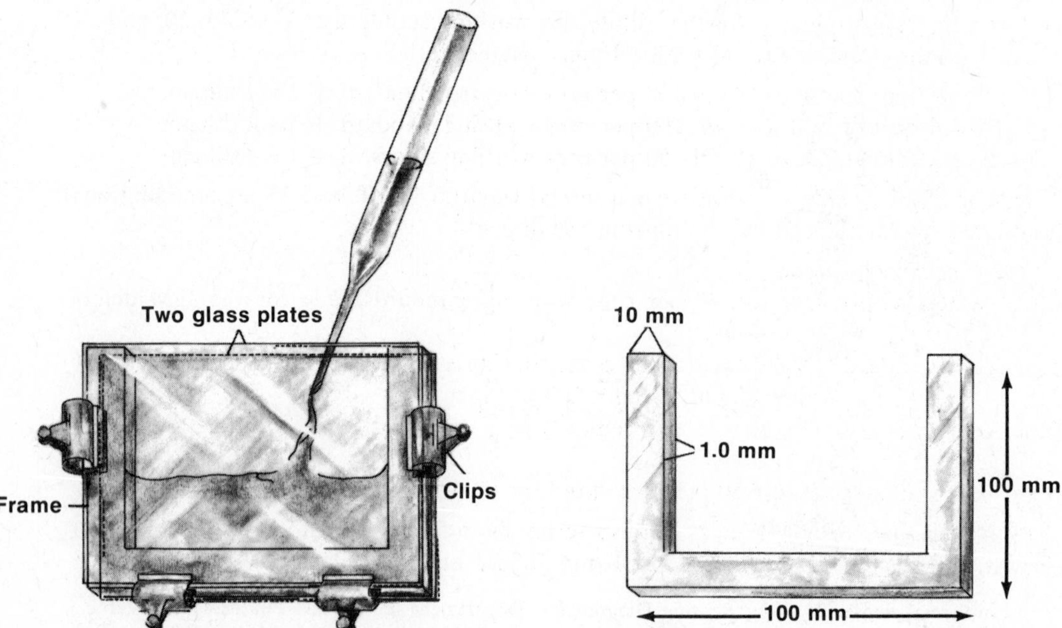

Figure 7–16 Preparation of immunodiffusion plate. See text for explanation. (Modified from Mancini, G., *et al.*: Immunochemistry, *2*:235, 1965.)

Cutting the wells

This should be done at least 24 h after pouring, and after standards, controls, and unknowns have been appropriately diluted. A 2 mm diameter needle is cut square, and then sharpened to give it an internal bevel. Place the agar plate over a piece of graph paper, with points marked 1 cm apart to serve as a template, and, using the needle, cut a rectangular pattern of 30 holes 1 cm apart (Fig. 7–14). Alternatively, use a specially made cutting jig.

Remove the agar plugs by suction with a Pasteur pipet attached to a vacuum line. Add the samples with a 2 μl micropipet or fill the cups exactly to the brim with a drawn-out capillary. The latter is preferable if Petri dishes are used, since it corrects for any variation in depth of the agar (see comment 2). Oblique light is best for this operation.

Duplicate sets of the four standards are placed at either end of the plate. The control serum and unknowns should all be tested in duplicate.

Incubation of the plates

Incubate for three days in a level, moist chamber at room temperature or, if urgent, in a closely airtight chamber at 37°C.

Reading the plates

Readings of the ring diameters are normally made after 60 h at room temperature. At 37°C, IgG may be read after one day, and IgM or IgA after two days. If urgent, preliminary approximate results may be reported much earlier. A measuring eyepiece is held against the bottom of the inverted plate and focused on the rings. Alternatively, a Berthon (vernier) caliper rule may be used.

Prepare a standard curve by plotting the square of the ring diameters of the standards against their concentrations. (This should be a straight line.) Then square the ring diameters of the unknowns and read their concentrations from the curve. Some sera may have to be rediluted and remeasured, since it is not permissible to extrapolate beyond the calibration curve.

NORMAL VALUES

See Tables 7–9A and B.

COMMENTS

1. The method can be applied in a similar way to the estimation of other plasma proteins; e.g., albumin, transferrin, α_1-anti-trypsin, ceruloplasmin, or β_1C-globulin. Specific antisera for each of these proteins are available commercially.

2. Plastic Petri dishes may be used instead of glass plates, provided their bases are perfectly flat and scratch-free. Unfortunately, most batches of Petri dishes on the market cannot be relied on in these respects. Use about 8 ml agar for a 9 cm dish or 2½ ml agar for a 5 cm dish. After mixing with the antiserum, pour at once into the prewarmed Petri dishes adjusted to the horizontal on a leveling table. When set, store inverted in a refrigerator.

3. Plates are also available commercially (see firms listed on p. 321) and are especially useful when only a few determinations are required. However, for the large number of tests required in routine clinical work, the high cost and small size of these plates render them less suitable. Only those commercial plates designed to be read at equilibrium should be used.

4. Results read before equilibrium has been reached are only approximate, both because they depend on the rate of diffusion, which is very susceptible to variable conditions, and because there is no straight-line relationship between concentration and ring size. When the precipitate first becomes visible, the antigen concentration is usually still so high that the ring diameter is directly proportional to the logarithm of the concentration, as in the Oudin technique (p. 281), while at the end at equilibrium it is proportional to the square root of the concentration. No constant relationship holds between these times.

TABLE 7–9A NORMAL IMMUNOGLOBULIN CONCENTRATIONS IN SERUM AT DIFFERENT AGES

(Cejka, J., Mood, D. W., and Kim, C. S.: Clin. Chem., *20*: 656, 1974.)

		IgG		*IgA*		*IgM*		*IgE*[b]
Age	*No. Subjects*	IU/ML[a] (RANGE)	% OF ADULT LEVEL	IU/ML[a] (RANGE)	% OF ADULT LEVEL	IU/ML[a] (RANGE)	% OF ADULT LEVEL	IU/ML (RANGE)
Cord blood	29	135.9 (92.7–199.2)	99	0.4 (0.03–5.3)	0.4	12.5 (5.0–31.1)	8	30 (Cord blood to 6–12 mo)
½–3 mo	7	60.5 (36.5–100.2)	44	9.2 (2.1–40.2)	8	60.5 (18.9–193.9)	39	
3–6 mo	9	44.8 (17.3–116.2)	33	12.9 (3.0–54.8)	11	58.0 (23.7–141.7)	37	
6–12 mo	13	82.8 (51.0–134.4)	61	22.8 (9.1–57.6)	20	123.0 (56.1–268.9)	78	
1–2 yr	22	78.4 (43.4–141.7)	57	25.0 (8.8–71.7)	22	121.4 (47.5–270.2)	77	50 (20–120)
2–3 yr	16	94.6 (60.0–149.3)	69	35.7 (15.4–82.8)	32	125.2 (63.7–246.2)	80	50 (20–150)
3–6 yr	74	105.7 (68.8–162.5)	77	54.2 (23.2–126.7)	48	131.0 (66.5–257.9)	83	70 (20–225)
6–9 yr	32	120.4 (80.3–180.5)	88	70.8 (19.6–156.1)	63	133.6 (65.1–274.3)	85	80 (23–275)
9–12 yr	20	119.7 (76.2–188.0)	88	85.1 (40.3–180.0)	75	166.7 (82.9–335.1)	106	90 (25–290)
12–16 yr	14	123.0 (83.0–182.1)	90	90.4 (53.9–154.4)	80	133.6 (57.9–308.4)	85	100 (27–350)
Adults	22	136.7 (81.8–228.5)		113.0 (50.3–253.8)		157.0 (47.5–310.2)		105 (30–400)

[a] Geometric mean $\pm 1.96\ SD_{log}$, converted to antilogs.
[b] Provisional figures.

TABLE 7–9B NORMAL IMMUNOGLOBULIN CONCENTRATIONS IN SERUM OF INDIVIDUALS IN DIFFERENT COUNTRIES

(Rowe, D. S.: Lancet, *ii*: 1232, 1972.)

Town and country	*Number of individuals examined*	*Geometric mean of estimates of concentrations of immunoglobulins in IU/ml with 95% confidence limits*		
		IgG	IgA	IgM
Algiers, Algeria	100	143(97–213)	164(84–317)	190(84–429)
Perth, Australia	94*	143(94–219)	127(56–286)	191(86–425)
Santiago, Chile	100	156(83–292)	163(73–365)	158(109–228)
Birmingham, England	51	123(73–207)	115(46–289)	133(47–372)
Offenbach, Germany	45	124(86–178)	108(48–244)	133(59–298)
Osaka, Japan	98	146(102–210)	129(70–237)	144(68–308)
Mexico City, Mexico	100	127(82–196)	97(29–327)	63(12–333)
Utrecht, Netherlands	100	116(65–206)	94(40–223)	127(48–334)
Ibadan, Nigeria	100	287(146–567)	80(21–207)	211(34–1413)
Uppsala, Sweden	94	126(90–177)	126(57–282)	135(52–345)
Lausanne, Switzerland	100	135(87–208)	136(56–334)	176(81–380)

* 95 for IgA.

5. The CV of the method is about ±8.5 per cent.

6. The above technique will determine protein concentrations down to about 10 mg/100 ml. Its sensitivity can be increased 20-fold by staining with Coomassie Brilliant Blue R (see below). If this is combined with larger samples applied by refilling the holes in the agar (p. 373) or by cutting larger holes, concentrations down to about 100 μg/100 ml can be determined. This is approximately the upper level of the normal serum IgE concentration (mean: 25 μg/100 ml). Normal levels of IgE can be determined with antiserum labeled with ^{125}I.[75] Alternatively, the more complex, but very sensitive, radioimmuno-adsorbent technique is available for protein concentrations down to 10 ng/100 ml.

7. In order to use antisera for immunoglobulin estimation intelligently, their method of preparation should be known, for their use involves two special problems.

a. The different classes of immunoglobulins cross-react because of the light chains they have in common. Since their characteristic differences are in the "constant" Fc segments of their chains, antisera should be made against these Fc segments. This is possible for IgG and IgM, but not for IgA. In order that all subclasses are represented, pooled normal serum should be the source of the antigen, rather than a pure paraprotein mixture. Unfortunately, this is not possible for IgA since its Fc fragment is too susceptible to further digestion by the enzymes used to split it from the intact Ig molecule. IgA whole molecules, isolated from normal serum, are used as antigen, and the antiserum obtained is then absorbed with light chains and $F(ab)_2$. For IgD and IgE antisera, the use of paraproteins cannot be avoided initially. The resulting antisera can then be used for the isolation of a greater range of Ig molecules by affinity chromatography, and this preparation is then used to provide a more satisfactory antiserum.

b. Both the serum immunoglobulin being measured (for example, IgG) and the reagent being used to measure it are complex mixtures.[74] The more different the mixture in the patient is from the mixture used to make the antiserum, the more inaccurate the result will be. In particular, estimations of a monoclonal immunoglobulin using antisera made against normal serum will inevitably be unreliable.

REFERENCE

Mancini, G., Carbonara, A. O., and Heremans, J. F.: Immunochemistry, *2*: 235, 1965.

QUANTITATIVE ESTIMATION OF INDIVIDUAL PLASMA PROTEINS BY SINGLE ELECTROIMMUNODIFFUSION (LAURELL ROCKET TECHNIQUE)

REAGENTS

1. Antiserum specific against the protein to be estimated. This is usually available commercially (p. 321).

2. Barbiturate buffer, pH 8.6, ionic strength 0.05. Dissolve 10.3 g sodium barbital, 1.84 g diethyl barbituric acid, and 1 g sodium azide (preservative) in distilled water and dilute to 1 liter.

3. Buffered agarose gel (1 g/100 ml). Add 1 g agarose to 100 ml barbiturate buffer contained in a flask. On bringing to a boil, the agarose will dissolve. Cool in an incubator to 48 to 52°C.

4. Stain. Dissolve 0.2 g Coomassie Brilliant Blue R (Searle Diagnostic, Inc.) in a mixture of 45 ml methanol, 10 ml glacial acetic acid, and 45 ml distilled water.

5. Washing fluid. Five parts water, 1 part glacial acetic acid, and 4 parts ethanol.

STANDARDS

If the pure individual undenatured protein is available, prepare a standard solution with a concentration of the same order as that expected in the unknown. Alternatively, use a commercial standard, such as Behringwerke's Standard Human Serum, stabilized, containing known concentrations of many individual protein components.

WORKING STANDARDS

Include and analyze at least four standards in duplicate to cover the likely range of values.

ELECTROPHORESIS APPARATUS

The apparatus must provide water cooling to the lower surface of the plate as in the Laurell tank.[47]*

TECHNIQUE

A glass plate (200 × 110 mm) is coated with a thin layer of antiserum-containing agarose at a uniform thickness of about 1.5 mm (see under Mancini technique). About 30 ml are required to fill the mold (see Fig. 7–16). The antiserum concentration used is the lowest shown by preliminary trial to give a clear precipitin rocket 3 to 5 cm high, when used to assay approximately 0.5 to 2 μg samples of antigen (Fig. 7–15).

Holes 2 mm in diameter are cut in the agarose along a line parallel with and 2 cm from one of the long edges. The distances between the centers of the holes must be at least 8 mm, which will allow about 20 holes to be cut. They are filled with exactly 2 μl of the antigen solutions (up to six unknowns and four standard solutions of varying dilution in duplicate), using an Oxford disposable tip micropipet. Electrophoresis is started as soon as possible to minimize diffusion. The current is adjusted to provide conveniently fast movement without overheating and is maintained until the rockets reach equilibrium. The experimental parameters should be so selected that the size of the rockets is between 2 and 5 cm.

Remove the plate, cover it with a piece of filter paper, dry in a current of air, and stain with Coomassie Brilliant Blue for 10 to 15 min. Wash with the aqueous ethanol-acetic acid mixture for about the same time interval.

Measure the heights of the rockets. Plot height, which is assumed to be proportional to the area enclosed by the precipitate, against concentration of standards and read the concentration of the unknowns from the standard curve.

REFERENCE

Laurell, C. B.: Electrophoretic and electro-immunochemical analysis of proteins. Scand. J. Clin. Lab. Invest., *29* (suppl. 124): 21, 1972.

QUALITATIVE TESTS FOR SPECIFIC PROTEINS[27,58]

This method employs the principle of double diffusion in plain agar, since it would be wasteful to incorporate the antiserum in the agar. The antiserum and the test samples are placed in holes in the agar. They may be allowed to diffuse toward one another across the gel to form lines of precipitate (Ouchterlony double diffusion; see p. 281), or if a rapid result is required, movement can be accelerated by electrophoresis (double-crossed or counter-immunoelectrophoresis).

If the protein to be tested for is one not normally detectable in serum, a specific antiserum is used, as in the three examples described below. If a protein normally present is suspected to be deficient, a "polyvalent" antihuman antiserum (made by injecting whole human serum into an animal) is placed in a central hole surrounded by holes containing the unknowns and the positive and negative controls. Opposite the patient's serum, lines corresponding to the different plasma proteins will form. If one of these which normally

* Obtainable from Tekniker Otto Olofsson, Kronetorpsgatan 58C, 212 26 Malmö, Sweden.

fuses with the line opposite the pure protein standard is missing, it indicates that the patient's serum is deficient in this protein (see Fig. 7–5), a conclusion which may be checked by immunoelectrophoresis (p. 317).

OUCHTERLONY DOUBLE-DIFFUSION METHOD

REAGENTS

1. Buffered agar gel. See Radial Immunodiffusion, p. 321.

2. Antisera. "Polyvalent" antihuman antiserum. Most of the specific antisera required are available commercially. Any other antiserum can be made by injection of the appropriate protein into animals such as the rabbit, provided a pure sample of the protein is available; however, this is a procedure requiring several months and at least two animals.[8]

TECHNIQUE

The molten buffered agar is poured into Petri dishes chosen for their plane bases and freedom from scratches (about 8 ml agar for a 9 cm plate). When the agar has set, a suitable pattern of holes, such as those illustrated below, is cut with cork borers, using a drawing underneath as a template. (Diameter of central holes = 8 mm; diameter of peripheral holes = 6 mm; distance between them = 4 mm.) The cores of agar are removed with a small scalpel or platinum wire loop, or sucked out with a Pasteur pipet. The plate is then dried at 37°C to remove excess moisture.

The antisera, unknowns, and controls are added to the holes with a pipet delivering 0.02 ml drops. The plates are then incubated at 37°C and examined for precipitin lines at 24 h and again at 48 h.

To ensure that the protein antigen in the patient's serum and the corresponding antibody in the antiserum reach equivalence and form a line somewhere between their holes, serial dilutions of the patient's serum are arranged around the antiserum in the central cup or, if the antiserum is potent, serial dilutions of it are arranged around the patient's serum in the central cup.

Typical arrangements for detecting C-reactive protein, α_1-fetoprotein, and IgD are as follows (amounts of antiserum will vary with antiserum titers; the diagrams are not to scale):

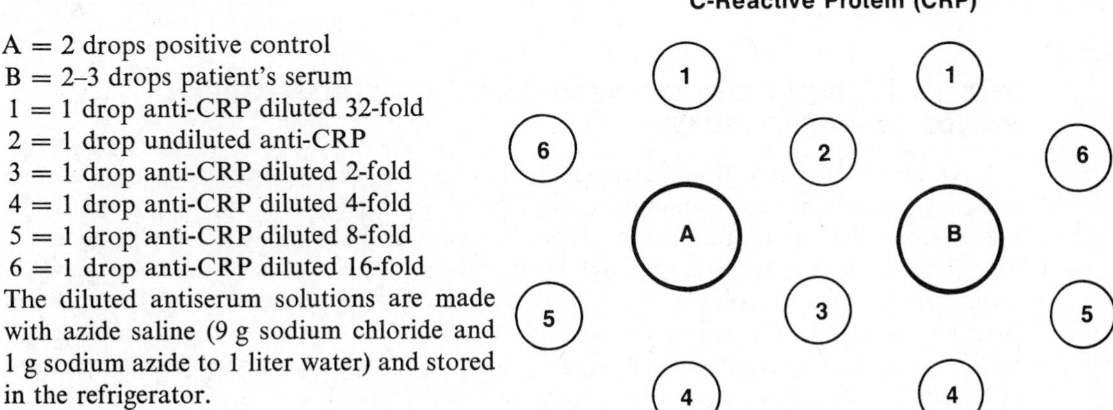

A = 2 drops positive control
B = 2–3 drops patient's serum
1 = 1 drop anti-CRP diluted 32-fold
2 = 1 drop undiluted anti-CRP
3 = 1 drop anti-CRP diluted 2-fold
4 = 1 drop anti-CRP diluted 4-fold
5 = 1 drop anti-CRP diluted 8-fold
6 = 1 drop anti-CRP diluted 16-fold

The diluted antiserum solutions are made with azide saline (9 g sodium chloride and 1 g sodium azide to 1 liter water) and stored in the refrigerator.

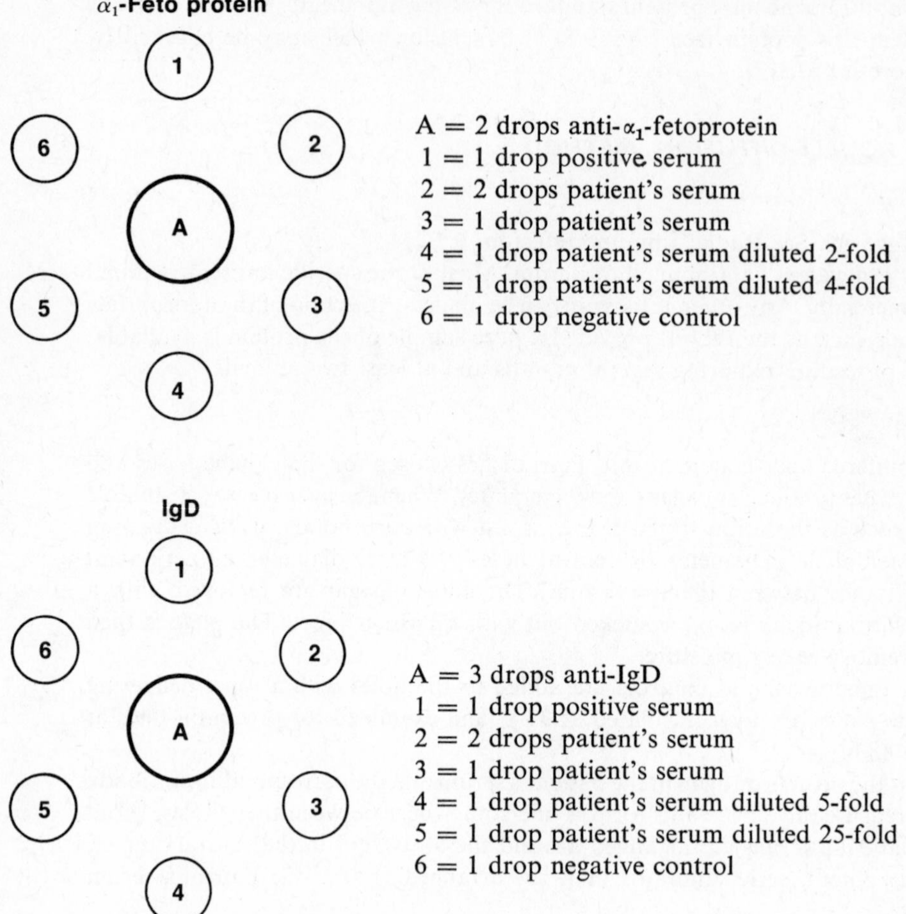

A similar pattern of holes is used for detecting and typing Bence Jones proteins in urine (p. 367).

THE IDENTIFICATION OF ABNORMAL ELECTROPHORETIC BANDS (PARAPROTEINS)

An abnormal band in an electrophoretic pattern is nearly always due to a normal protein present in excessive amount rather than to a protein not normally present. Thus, for its identification, an antihuman plasma antiserum capable of reacting with as many plasma proteins as possible is required. The positions and densities of the lines found on immunoelectrophoresis (see Fig. 7–12) depend more on the characteristics of the antiserum than on the protein antigen concentrations; but if normal serum is run simultaneously against the same antiserum, the technique is semiquantitative, and is much simpler than the corresponding quantitative technique (see p. 317 and Fig. 7–13).

Abnormal bands found by electrophoresis are usually due to monoclonal immunoglobulins. Their detection and typing with polyvalent and specific antisera is a good example of the application of immunoelectrophoresis in clinical chemistry.

THE TYPING OF MONOCLONAL IMMUNOGLOBULINS BY IMMUNOELECTROPHORESIS[45]

REAGENTS

1. Antisera. Sera against the heavy chains of IgG, IgA, and IgM and against kappa and lambda light chains when part of the complete (intact) Ig molecule (p. 367).
2. Barbiturate agarose. See Laurell technique, p. 325.

APPARATUS

Electrophoresis tank and power supply.
Microscope slides.
Microscope slide holder to fit across electrophoresis tank and adequate to support at least five slides.
Jig to cut pattern of holes and troughs.
Leveling table.
(All this apparatus is available commercially, e.g., from Shandon Scientific Co. Ltd.)

PROCEDURE

Preparation of agarose-coated slides

Clean a slide holder with distilled water and then alcohol. Block any holes in the holder with polyethylene tape from the back. Clean slides and number with a diamond marker. Place slide holder on a leveling table, and adjust to the horizontal. Pour sufficient barbiturate agarose at 60°C into the base of the slide holder so that the slides when placed in it are sealed in. Avoid inclusion of air bubbles if possible. Allow to set and remove excess gel from top with a clean slide. Pour further barbiturate agarose on top of the slides to cover them to a depth of 2 mm.

If possible, prepare the slides the day before they are required.

Cutting of plate

Use a well and trough cutter (e.g., Shandon) to cut two troughs and three holes on each microscope slide.

Line up cutter with slides on plate, and cut right through the gel to the glass slide. Widen wells by overpunching with 2 mm needle (p. 323). Remove the round plugs of gel, using a narrow bore pipet and vacuum. Remove the troughs with a small scalpel by cutting the ends of the troughs and easing up the piece of agar on the blade. At all times avoid touching the rest of the gel with the fingers.

Application of samples

Five slides will be required to investigate one M-protein. Set up the first three (for plasma proteins in general, for IgG, and for IgA) as follows:

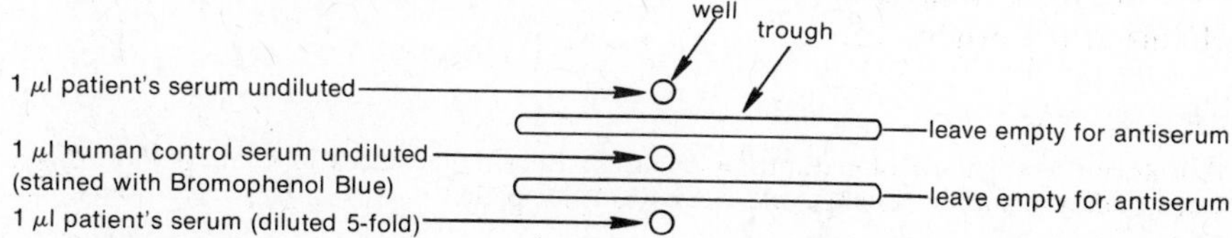

Set up the fourth slide (for IgM) similarly but with 2 μl of each sample. For the fifth (for light chains) use 1 μl samples and undiluted patient's serum in both outer holes. The samples are applied using a Partigen Dispenser or a Hamilton type of syringe. The control serum is stained with Bromophenol Blue so that the progress of the albumin can be observed. Two dilutions of the patient's serum should always be used against the heavy chain antisera, because if the concentration of paraprotein in the serum is very high, precipitation may occur in the trough and be missed.

Electrophoresis

Place the holder with its slides into the electrophoresis tank and apply two fresh Whatman 3 MM filter paper wicks saturated in buffer, so as to bridge the gap between tank and gel.

Adjust the power pack to give a steady current of 20 mA for each of the five slides. Allow separation to continue for about 2 h.

Development of the patterns

Remove the plate from the tank and pipet antiserum into the troughs as follows:
Slide 1: 50 μl "polyvalent" antihuman antiserum to both troughs
Slide 2: 50 μl anti-IgG to both troughs
Slide 3: 50 μl anti-IgA to both troughs
Slide 4: 30 μl anti-IgM to both troughs
Slide 5: 50 μl anti-κ to one trough and 50 μl anti-λ to the other.

Leave the slides overnight in the moist chamber and then examine for precipitation pattern. Continue incubation for another 24 to 48 h and re-examine. Wash and stain slides for permanent preparations (see below).

Interpretation

Look for a localized displacement toward an antiserum trough, a distortion or a thickening of the IgG, IgA, or IgM lines in slide 1 and in the specific antiserum slides. A localized duplication of a line, especially the IgG line, is another common finding. Changes may be seen at only one of the two dilutions of the patient's serum. The main value of slide 5 is to check for the presence of an M-protein. An abnormal bulge of similar mobility to that seen in previous slides should be seen against anti-κ or anti-λ, but not against both. This confirms the presence of an M-protein and shows to which light chain class it belongs. Such a finding in the absence of abnormalities in slides 2, 3, and 4 suggests the presence of either a rare class of M-protein, probably IgD, or free light chains. Such findings should be confirmed by the Ouchterlony double-diffusion technique as described previously.

The presence of a Bence Jones (light chain) protein will be suspected from its rapid diffusion rate and is easily distinguished, if necessary, from the larger Ig molecules by thin-layer gel filtration (p. 332). Bence Jones proteins also react specifically with antisera designed to react only with those determinants of a light chain which are hidden when it is part of an intact four chain Ig molecule (p. 367).

If one of the heavy chain antisera gives positive results but the light chain antisera are negative, *heavy chain disease* should be suspected. It may be confirmed by repeating the immunoelectrophoretic run against the heavy chain antiserum using agar into which has been incorporated the light chain antiserum. This will precipitate all intact immunoglobulins and free chains at the origin.

REFERENCE

Kohn, J.: The laboratory investigation of paraproteinemia. *In* Recent Advances in Clinical Pathology. S. C. Dyke, Ed. London, Churchill Livingstone, Series 6, 1973, p. 363.

PREPARATION OF PERMANENT IMMUNOCHEMICAL RECORDS

Permanent, stained records of the results of any of the above gel diffusion tests may be made as follows:

REAGENTS

1. Azide saline. Dissolve 9 g sodium chloride and 1 g sodium azide in 1 liter distilled water.
2. Stain. Dissolve 0.5 g Amido Black 10B in a mixture of 400 ml ethanol, 100 ml glacial acetic acid, and 500 ml distilled water.
3. Wash solution. Prepare as above, omitting the stain.

METHOD

1. Wash the plate for 3 days in azide saline to remove proteins in solution. Change wash solution each day.
2. Remove the plate from the wash solution and cover with a piece of wet filter paper.
3. Dry in 37°C incubator.
4. Place in Amido Black 10B solution, remove filter paper, and leave to stain for 15 to 30 min.
5. Rinse with wash solution until the background is clear.
6. Dry in air.

COMMENT

If precipitin lines are faint, use Coomassie Brilliant Blue R (p. 325) in place of Amido Black 10B solution.

MOLECULAR FILTRATION

Ordinary filtration is used to separate small but visible particles suspended in some solvent or solution. The original suspension is poured into a funnel, which holds a filter paper or a membrane containing pores or openings small enough to prevent passage of the suspended particles, but large enough to allow rapid flow-through of solvent or solution. The smaller the particles to be separated, the smaller must be the average size of the pores in the paper or membrane. In molecular sieving, the pore size is made so small that macromolecular solutes can be separated from solvent, ions, and other solutes of small molecular mass. In this way protein solutions may be concentrated and/or desalted. Furthermore, by varying the pore size, proteins can be fractionated according to their molecular volumes. Two approaches are used: the macromolecules may be held back by a membrane (ultrafiltration); or the small molecules may be held up within the pores of gel "beads" impermeable to the large protein molecules (gel filtration).

ULTRAFILTRATION

Ultrafilter membranes are usually prepared from collodion or viscose. Such membranes are permeable to all small solutes but impermeable to proteins of a molecular mass of 20,000 to 40,000 or higher. These are now available from a number of manufacturers; a large variety of pore sizes are offered so that proteins of different molecular size can be separated from one another. Because of the small pore size, the flow of water or solution through the ultrafilter is very slow; thus, to accelerate the process, filtration is usually carried out under positive or negative pressure, or a concentrated macromolecular solution (e.g., Carbowax, sucrose, PVP) is used to exert an osmotic force.

The membranes may be in the shape of disks as used for the removal of small particles, bacteria, and viruses from solutions or sera (sterilization) or they may be in the form of Visking tubing as used for renal dialysis, or they may be thimble-shaped. Recently, bundles of very fine cellulose or cellulose acetate tubules have been introduced; their enormous surface area leads to greatly increased speed of ultrafiltration ("hollow fiber devices," Dow Chemical Company).

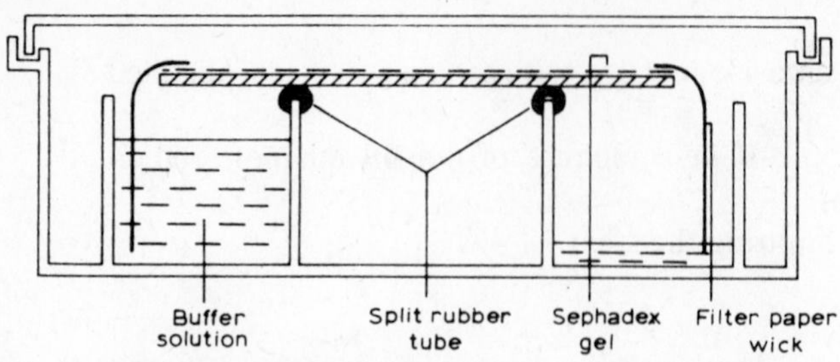

Figure 7–17 Separation of proteins by thin-layer gel filtration. See text for details. (From Grant and Everall: *In* Protides of the Biological Fluids, 1965; Section B. Amsterdam, Elsevier Publishing Co., 1966, p. 322.)

GEL FILTRATION[11,17]

Unlike the immunochemical techniques which use continuous gels, gel filtration makes use of the properties of "beads" of gel suspended in fluid (p. 163). It provides a simple way of removing small solute molecules from a protein solution, and allows the separation of proteins of different molecular volumes from one another by a much simpler and cheaper technique than ultracentrifugation. For the separation of the plasma proteins, the most loosely linked of the dextran gels (Sephadex G-200) is the most useful. This technique fractionates the serum proteins into three groups, an excluded large molecular group comprising the macroglobulins and lipoproteins, an intermediate group that is mainly IgG, and a small molecular group dominated by albumin.

Two examples of the use of gel filtration columns are the separation of proteins labeled with radioisotopes or fluorescent dyes from the labeling reagent (p. 293) and the isolation of CSF proteins prior to estimation (p. 373).

For analytical work it is more convenient to use a simple thin-layer modification of

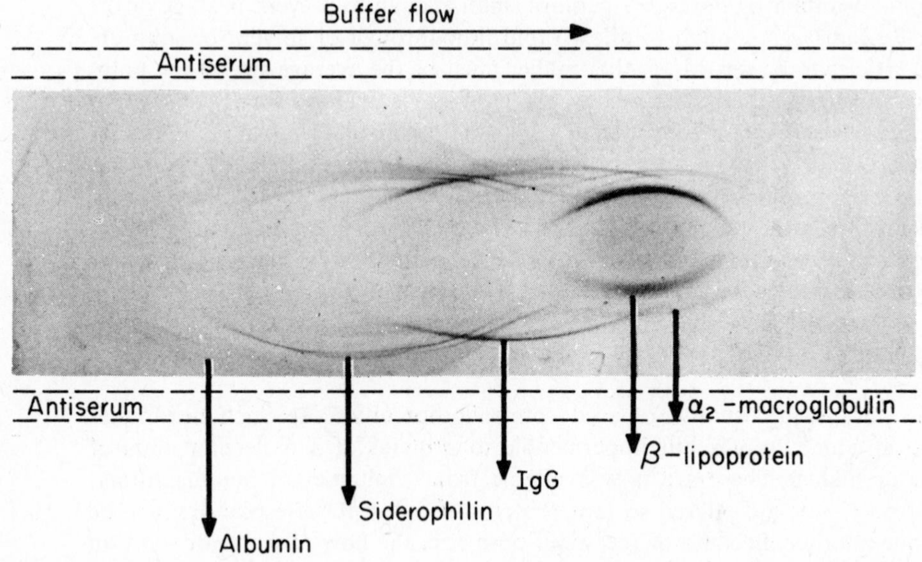

Figure 7–18 Gel immunofiltration of human serum. The serum proteins are separated on a thin layer of Sephadex G 200. After this preliminary separation, a strip of cellulose acetate is placed on the gel to soak up the separated antigens. When this strip is saturated, it is transferred to an agar plate, and thin paper strips soaked in antiserum are placed parallel to the cellulose acetate strip. The precipitin lines form after diffusion in the usual manner. (From Grant, G. H. and Everall, P. H., Lancet, *ii*: 368, 1967.)

this technique.[1] A glass plate is covered by a thin layer of the gel and suspended between two reservoirs of buffer solution. The fluid level of one reservoir is maintained at a level higher than in the other, so that when wicks dipping into the buffer reservoirs are connected to the ends of the plate, buffer flows gently through it (Fig. 7–17) (a disconnected, slightly tilted, electrophoretic apparatus with more buffer in one compartment than in the other is very suitable). The system is covered and allowed to reach equilibrium, and the sample is applied through a hole in the lid. After allowing time for separation of the fractions, the plate is removed, dried, and stained with any convenient protein stain. For serum, Sephadex G-200 superfine is used and the serum is first stained with Bromophenol Blue. The position of the stained albumin fraction, although the slowest, gives some indication of the progress of the fast-moving macroglobulin fraction. Examples of the use of this technique are the detection of monomer fragments in the investigation of macroglobulinemia sera and the separation of Bence-Jones proteins from immunoglobulins.

The method may also be combined with immunodiffusion to give a very delicate technique for the qualitative analysis of protein mixtures according to molecular volume. A pattern of precipitin lines is produced, similar to that in immunoelectrophoresis, but these are separated by their differing molecular volumes instead of their differing electrophoretic mobilities (Fig. 7–18).[1,27]

REFERENCE

Fischer, L.: An introduction to gel chromatography. *In* Laboratory Techniques in Biochemistry and Molecular Biology. T. S. Work and E. Work, Eds., Amsterdam, North-Holland Publishing Co., 1969, vol. 1, part II, p. 151.

SPECIFIC TESTS FOR MUCOPROTEINS

Conjugated proteins containing a complex carbohydrate moiety linked to the peptide chain are referred to as *glyco- and mucoproteins*. The carbohydrate chain linked to the protein is made up of the *hexoses*, galactose or mannose; the *hexosamines* (amino sugars), glucosamine or galactosamine; the *methylpentose*, fucose; and a *sialic acid*, such as N-acetylneuraminic acid. The sialic acid is located at one end of the carbohydrate chain; the other end of the chain is linked by a peptide bond to one of the peptide chains of the protein molecule.

(15)

D-2-amino, 2-desoxy-galactose D-galactosamine

D-6-desoxy-galactose D-fucose

N-acetylneuraminic acid (one of several related forms of sialic acid)

In a useful classification, those carbohydrate-proteins containing less than 4 per cent of hexosamines are termed *glycoproteins*.[41] They contain from a trace up to about 15 per cent of total carbohydrate. Those with over 4 per cent of hexosamines and containing from

10 to 75 per cent of carbohydrate are termed *mucoproteins*. The expression *mucoids* is also used to characterize those mucoproteins in which the bond linking the carbohydrate chain and the protein is split only with difficulty, in contrast to another group containing acidic mucopolysaccharides, in which the linkage is polar and easily split.

The glyco- and mucoproteins differ from the other serum proteins in giving a positive staining reaction with Schiff's reagent and in not being precipitated by perchloric and sulfosalicylic acids. These properties allow them to be estimated separately, either electrophoretically or chemically. In the latter case,[96] the other serum proteins are precipitated by perchloric acid; then the mucoproteins are precipitated by phosphotungstic acid and assayed by measuring either their total carbohydrate content with orcinol or diphenylamine, or their protein content by the Folin, biuret, or Kjeldahl methods, or by a turbidimetric technique. Methods are also available for assaying serum protein–bound hexose and serum hexosamines.

Unfortunately, unlike the lipoproteins, which are few in number and have a common function as transporters of lipids, the glyco- and mucoproteins are numerous and have no function in common. Most serum proteins are either glycoproteins (e.g., transferrin, ceruloplasmin, prothrombin, or immunoglobulins) or mucoproteins (e.g., orosomucoid, haptoglobin, hemopexin). As a result, estimations of total serum mucoprotein have been disappointing clinically. Orosomucoid and haptoglobin, however, contain a large proportion of the serum protein–bound carbohydrate, and both increase in the acute phase reaction. Consequently, total serum mucoproteins increase in inflammatory and proliferative diseases such as acute infections, rheumatoid arthritis, tuberculosis, carcinoma, and lymphosarcoma. This increase, however, presupposes normal liver and adrenal function, inasmuch as serum mucoprotein levels are depressed in acute liver disease, cirrhosis, and adrenal insufficiency. As a result, normal levels may be encountered even in patients with demonstrated acute reactive disease. Values can range from 100 to 400 mg/100 ml in rheumatoid arthritis, and from 120 to 450 mg/100 ml in acute infections, compared with a normal range of 75 to 135 mg/100 ml (average: 120 mg/100 ml).

Because of the overlap between normal and abnormal values, the nonspecific nature of the acute phase reaction, and its ease of demonstration by routine electrophoresis, total serum mucoproteins as such are rarely estimated. Instead, it is clinically more useful to determine specific serum proteins, such as orosomucoid, haptoglobin, and α_1-anti-trypsin.

SERUM ELECTROPHORESIS FOR GLYCO- AND MUCOPROTEINS[44]

REAGENTS

1. Periodic acid solution. Dilute 2.5 ml of 50 per cent periodic acid and 25.0 ml of 0.2 molar sodium acetate to 250 ml with distilled water.

2. Potassium iodide solution. Dissolve 40.0 g potassium iodide in 400 ml distilled water. Before use add 10.0 ml of 0.1 molar HCl.

3. HCl, 1 mmol/l.

4. Saturated ammonium thiosulfate, $(NH_4)_2S_2O_3$, solution.

5. Schiff's stain. Dissolve 2 g of basic fuchsin in 400 ml of boiling water, cool to 50°C, filter, and add 10 ml of 2 molar HCl and 4 g K-metabisulfite. Stopper and let stand in a cool, dark place overnight. Add 1 g activated charcoal, shake, and filter. Add 1 molar HCl until the reagent fails to turn filter paper pink. Stopper and store in a cold, dark place. Discard when solution turns pink.

6. Nitric acid, 0.1 mol/l (3.2 ml concentrated HNO_3/500 ml H_2O).

PROCEDURE

1. Refer to the procedure for electrophoresis on page 312 and proceed to step 9, but apply 5 to 10 μl of serum.

2. After the electrophoresis, dry the strips for about 10 min in a hot-air oven at 80 to 100°C.

3. Wash the strips in denatured ethanol for about 10 min.

4. Transfer to the periodic acid solution for about 10 min.

5. Rinse in 1 mmolar HCl and transfer to potassium iodide solution. The solution and strips will turn brown.

6. Add a few drops of saturated ammonium thiosulfate solution from a capillary pipet until brown color disappears.

7. As soon as the stain has just cleared from the iodide-stained bands, rinse in 1 mmolar HCl and transfer into Schiff's stain.

8. When the bands have an intense purple color (after 15 to 30 min), place in three changes of 0.1 molar nitric acid, each for 10 to 15 min.

9. Rinse again in 1 mmolar HCl. Blot and dry at room temperature. Then place between blotting paper sheets under pressure.

REFERENCE

Kohn, J.: Cellulose acetate electrophoresis. *In* Chromatographic and Electrophoretic Techniques. I. Smith, Ed. London, Heinemann, 1968, vol. 2, p. 101.

DETERMINATION OF SERUM MUCOPROTEINS

The main bulk of serum proteins are precipitated with perchloric acid at a final concentration of 0.60 mol/l. The mucoproteins remaining in the filtrate are then precipitated with phosphotungstic acid, washed free of nonprotein nitrogen, and after solution in sodium carbonate, treated with Folin-Ciocalteu phenol reagent (see p. 354). The assay is standardized against solutions of tyrosine, assuming a value of 0.042 g as the tyrosine fraction in 1.0 g of mixed serum mucoproteins.

NORMAL VALUES AND COMMENTS

Different investigators have reported somewhat differing normal values. Most normal sera will fall within the range of 75 to 135 mg/100 ml. The precision expected is about ± 15 per cent. The results are probably low because a fraction of the mucoproteins is lost by coprecipitation with other proteins in the perchloric acid treatment. The separation of mucoproteins from the other proteins is at best a rough, empirical procedure, and reproducible results are obtained only by strict adherence to experimental protocol.[87]

REFERENCE

Winzler, R. J.: Determination of serum glycoprotein. *In* Methods of Biochemical Analysis. D. Glick, Ed. New York, Interscience Publishers, Inc., 1955, vol. 2, p. 279.

THE ESTIMATION AND CLINICAL SIGNIFICANCE OF THE INDIVIDUAL PLASMA PROTEINS

ALBUMIN[63]

PROPERTIES AND CLINICAL SIGNIFICANCE

Albumin is a carbohydrate-free molecule with a mass of 65,000 daltons (Table 7–6). Unlike most plasma proteins, it has several different functions. They include the transport of large organic anions normally insoluble in aqueous fluids (in particular, long-chain fatty acids and bilirubin), the binding of toxic heavy metal ions, the transport of poorly soluble hormones such as cortisol, aldosterone, and thyroxine when the capacities of their more specific binding proteins are exceeded, the maintenance of plasma colloidal osmotic pressure, and the provision of a reserve store of protein.

Normal serum from a healthy *ambulatory* adult contains about 3.5 to 5.0 g of albumin per 100 ml, averaging about 4.5. The other serum proteins (the total globulins) average about 2.9 g/100 ml, corresponding to an albumin/globulin ratio of 1.3 to 1.8 (average, 1.6). The hemodilution on recumbency lowers the serum albumin and other proteins by about 0.5 g/100 ml without affecting the albumin/globulin ratio significantly. In newborn infants, albumin and globulin average about 3.5 and 2.0 g/100 ml, respectively. These levels increase slowly until about the third year, when adult values for each are reached.

Although the total serum protein concentration may be above normal in dehydration and below normal in water-loading, the ratio of albumin to globulins in these states may be essentially unchanged.

Hypoalbuminemia is very common in disease. It results from four main factors, often acting together: (1) an impairment in synthesis, as in malnutrition or liver disease; (2) an increase in catabolism, as in the acute phase reaction (p. 311); (3) a loss of protein through the urine or feces; or (4) a change in distribution between the intravascular and extravascular compartments.

In most infections, serum albumin levels are moderately depressed (3.2 to 4.2 g/100 ml). In conditions such as liver disease, carcinomatosis, or congestive heart failure, levels of 2.3 to 3.2 g/100 ml are common. In liver disease they are often accompanied by a rise in immunoglobulins and a decrease in the A/G ratio to 0.7 to 1.0. The lowest albumin levels are due to proteinuria of 10 to 30 g/d as seen in the nephrotic syndrome. Serum levels in this condition may decrease to 1.5 to 2.0 g/100 ml or even lower. Significant decreases in serum albumin are also caused by protein loss through the intestinal tract (enteropathies). Low levels of albumin are important contributory causes of edema.

METHODS FOR THE DETERMINATION OF ALBUMIN

The procedures currently being used for the estimation of serum albumin fall into four types: (1) assay of albumin by the biuret technique after removal of globulins by salt precipitation; (2) measurement of albumin directly, by virtue of the tendency of albumin to bind certain dyes, such as bromcresol green, methyl orange, and 2-(4′-hydroxyazobenzene)-benzoic acid (HABA); (3) assay of albumin after separation by electrophoretic procedures; and (4) assay by immunochemical techniques.

Removal of globulins by salt precipitation, followed by the determination of protein by the biuret or Kjeldahl techniques, was for many years the standard procedure. Sodium sulfate (22 g/100 ml) was introduced by Howe to replace ammonium sulfate (p. 271), which interferes with the biuret reaction and cannot, of course, be used with the Kjeldahl nitrogen technique. When electrophoresis was introduced, it was found that concentrations of sodium sulfate as high as 27 g/100 ml were required to make the results of the two methods consistent. As a result, a temperature of 35 to 37°C was needed to keep this quantity of salt in solution; alternatively, the sulfate may be replaced by the more soluble sulfite. In the Reinhold-Kingsley procedure[71] a mixture of the sulfate and sulfite (total concentration, 27 g/100 ml) is used and ether is added so that the globulins separate between the ether and aqueous layers, thus avoiding tedious filtration.

Currently, albumin is frequently estimated directly by its dye-binding capacity, because this technique is simpler and easy to automate.[13] The results compare fairly well with the electrophoretic technique. At low levels of serum albumin, dye binding methods may give results that are too high and immunochemical assays may be preferred (p. 317).

REFERENCES

Peters T. J.: Serum albumin. *In* Advances in Clinical Chemistry. O. Bodansky and C. P. Stewart, Eds., New York, Academic Press, Inc., 1970, vol. 13, p. 37.

Watson, D.: Albumin and "total globulin" fractions of blood. *In* Advances in Clinical Chemistry. H. Sobotka and C. P. Stewart, Eds. New York, Academic Press, Inc., 1965, vol. 8, p. 238.

Owen, J. A.: Effect of injury on plasma proteins. *In* Advances in Clinical Chemistry. H. Sobotka and C. P. Stewart, Eds. New York, Academic Press, Inc., 1967, vol. 9, p. 2.
Hoffman, W. S.: The Biochemistry of Clinical Medicine. 4th ed. Chicago, Year Book Medical Publishers, Inc., 1970.

DETERMINATION OF ALBUMIN BY DYE BINDING

All proteins, and especially albumins, tend to react with many chemical species by means of electrostatic and tertiary van der Waals' forces and by virtue of hydrogen bonding. Bilirubin, fatty acids, most hormones, and many drugs are transported about the body bound to albumin. Many colored dyes in the anionic form also possess this protein-binding property. This property has been used in attempts to devise methods by which albumin could be measured directly, without previous separation from globulins.

In dye-binding methods, only those dyes or indicators can be used that bind very tightly to the albumin molecule, so that practically 100 per cent of the albumin present is bound to dye. The binding must be unaffected by small changes in ionic strength and pH. Furthermore, the color of the protein-bound dye should be different from that of the free dye; i.e., there should be a substantial shift in the wavelength of light at which maximum absorption occurs in the two forms. The albumin-dye concentration can then be measured in the presence of excess dye. Finally, dye binding to other protein fractions (globulins) must be negligible if the dye binding is to be the basis for a valid assay of albumin. For practical reasons, the color characteristics of the dye should be such that it can be measured at wavelengths of light where bilirubin and hemoglobin will give negligible or minimal interference. The use of *methyl orange* was proposed by Bracken and Klotz and Watson and Nankiville; HABA [2-(4′-hydroxyazobenzene)-benzoic acid] by Rutstein, Ingenito, and Reynolds and Martinek, and *bromcresol green* by Rodkey. The use of bromcresol green is most free of pigment interference. Bilirubin at levels over 5 mg/100 ml interferes with HABA procedures and to a lesser degree with methyl orange methods. Rodkey's original method, in which the bromcresol green was buffered at pH 7.05, gave inconveniently high blank absorbance values; thus, lower pH buffers are now used, as in the method presented below.

ASSAY OF SERUM ALBUMIN BY BROMCRESOL GREEN BINDING

REAGENTS

1. Succinate buffer, 0.10 mol/l, pH 4.15. Dissolve 11.8 g of succinic acid and 100 mg of sodium azide in 800 ml water. Adjust pH with 0.25 molar NaOH (10 g/l), transfer to a 1 liter volumetric flask and dilute to volume with water.

2. Bromcresol green solution, Stock (0.60 mmol/l). Transfer 419 mg of bromcresol green, indicator grade, and 100 mg sodium azide to a 1 liter volumetric flask. Dissolve in 10 ml of 0.1 molar NaOH and dilute to volume with water.

3. Working buffered dye solution. Mix one volume stock with three volumes of 0.1 molar succinate buffer. Adjust pH to 4.20 ± 0.05; if too high, adjust with succinic acid (5 g/100 ml).

4. Succinate buffer, 0.075 mol/l, pH 4.20. Dilute 750 ml of 0.1 molar succinate buffer to 1 liter with water and adjust pH to 4.20 ± 0.05.

5. Stock albumin standard, 10.0 g/100 ml. Dissolve 10.00 g of human serum albumin Fraction V, corrected for moisture content, and 50 mg of sodium azide in about 70 ml of water. Transfer into a 100 ml volumetric flask and dilute to volume with water. Store in refrigerator. Confirm protein concentration by Kjeldahl estimation.

6. Albumin working standards. Dilute stock appropriately with 50 mg/100 ml sodium azide to give solutions containing 2.0, 3.0, 4.0, 5.0, and 6.0 g albumin/100 ml and store in refrigerator.

PROCEDURE

1. Pipet 5.0 ml of the working dye solution into a series of numbered cuvets, one for each of the five albumin working standards and one for each unknown.
2. Pipet 5.0 ml of 0.075 molar succinate buffer pH 4.20 into a similar series of numbered cuvets to serve as blanks.
3. Add 25 μl of each standard and each unknown to the appropriately numbered tube in each series.
4. Mix thoroughly and allow to stand 10 min at room temperature.
5. Using a spectrophotometer at 628 nm or a colorimeter with a 630 nm interference filter, measure the absorbance of the first set of tubes after adjusting the instrument to zero absorbance with working dye solution. Read the second (blank) set of tubes after adjusting the spectrophotometer to zero with water. Deduct blank values from first readings.
6. Plot absorbance (A) against concentration of standards, and read unknowns from curve or, if linear, calculate as follows:

$$\text{Albumin g/100 ml} = \frac{A \text{ unknown}}{A \text{ standard}} \times \text{concentration of standard}$$

COMMENTS

1. The method is easy to modify for automation either for discrete or for continuous flow analysis. To prevent turbidity in continuous-flow methods, 4 ml of 30 per cent Brij-35 (Fisher Scientific Company, Pittsburgh, PA. 15219) should be added to each liter of working dye solution and of 0.075 molar pH 4.20 succinate buffer, respectively, before adjusting the pH to 4.20 ± 0.05. Brij-35 causes a slight change in the absorbance maximum of the dye-albumin complex, but does not appear to affect the results.
2. Citrate buffer (0.075 mol/l, pH 4.20 ± 0.05) may be used instead of succinate.
3. Addition of electrolytes decreases the absorbance. Neither bilirubin nor salicylate interferes.
4. At low serum albumin levels, results tend to be too high owing to attachment of dye to other proteins. In such cases, radial immunodiffusion or electro-immunodiffusion gives more accurate results.[91]

REFERENCE

Doumas, B. T., and Biggs, H. G.: Determination of serum albumin. *In* Standard Methods of Clinical Chemistry. G. A. Cooper, Ed. New York, Academic Press, Inc., 1972, vol. 7, p. 175.

FIBRINOGEN

PROPERTIES AND CLINICAL SIGNIFICANCE

Fibrinogen is the precursor of fibrin, the insoluble protein of blood clots. In the process of blood coagulation, the trypsin-like enzyme thrombin is formed from the serum glycoprotein prothrombin. It acts on fibrinogen to split off a peptide fragment, converting the soluble fibrinogen into insoluble fibrin. The fibrinopeptide split off from fibrinogen by the thrombin contains some 3 per cent of the original protein nitrogen. The residual fibrinogen fibrils (fibrin monomers) then aggregate and enmesh to form a three-dimensional, cross-linked fibrin gel. A translucent colorless gel is produced from plasma, but in blood clotting the gel is red because of entrapped blood cells. On standing, the gel contracts (syneresis) to form a tougher mass, extruding serum in the process.

The plasma concentration of fibrinogen is about 200 to 400 mg/100 ml; none is present in serum, since fibrinogen is removed as fibrin in the clotting process. Elevations of fibrinogen levels up to 700 mg/100 ml are encountered in many inflammatory diseases, such as *rheumatic fever*, *pneumonia*, *septicemia*, and *tuberculosis*. An increase in the erythrocyte sedimentation rate (ESR) is intimately associated with increased plasma fibrinogen levels.

Decreased levels of plasma fibrinogen are rather uncommon. In *congenital hypofibrinogenemia*, very little fibrinogen is present because of a rare genetic defect that is reflected by the inability of the liver to synthesize fibrinogen. The condition of *acquired hypofibrinogenemia* is of more interest in clinical work. Low levels of fibrinogen may be observed in severe, usually terminal, liver disease, which is also associated with low levels of several other clotting factors, such as prothrombin and factor VII.

More important clinically are cases of low fibrinogen levels in pregnancy. In premature separation of the placenta (antepartal hemorrhage), the high levels of thromboplastic agents present in the placenta are released into maternal blood, resulting in a rapid conversion of fibrinogen to fibrin in the blood, placenta, and other organs. This eventually depletes the fibrinogen stores and results in severe hemorrhage and often shock and death. A similar situation occurs as a result of fetal death *in utero*, although the process is slower. These conditions are emergencies, requiring prompt and appropriate treatment, which may include intravenous administration of fibrinogen and other clotting factors in large doses. Plasma fibrinogen may fall to concentrations as low as 50 to 70 mg/100 ml. Severe fibrinogen depletion can also occur as a complication of lung and prostatic surgery.

METHODS FOR THE DETERMINATION OF FIBRINOGEN

The most accurate quantitative methods[70,83] resort to isolating the fibrinogen (as fibrin) from the other plasma proteins. Clotting is carried out by diluting the plasma with saline and adding either exogenous thrombin or excess Ca^{2+} ion in a quantity sufficient to overcome the anticoagulant present. The isolated, washed fibrin clot is then assayed by one of the available protein methods: measurement of Kjeldahl nitrogen and conversion to protein by the factor 5.95 (16.9 per cent N); or re-solution of the clot in NaOH and measurement of protein with either the biuret, Folin-Ciocalteu, or Lowry copper-phenol methods; or measurement of ultraviolet absorbance at 280 nm (p. 355). There is considerable disagreement over the tyrosine equivalent of fibrinogen as measured by the Folin reagent; therefore, the biuret and copper-phenol methods are preferred.

An example of a biuret method is presented below. A similar procedure with a somewhat simpler technique for collecting the clot has been described by Swaim and Feders.[83]

Such methods are too slow in obstetric emergencies, the situation in which fibrinogen estimations are most often required. Rapid, if less accurate, methods have to be used. One group of methods[17a,54] is based on adding sufficient salt solution to cause precipitation of the fibrinogen without precipitating the other plasma proteins, and then measuring the turbidity of the suspension. For example, the plasma may be diluted 10- or 20-fold with one of the following solutions: 12.5 g Na_2SO_3/100 ml, 10.5 g Na_2SO_4/100 ml, 13.4 g $(NH_4)_2SO_4$/100 ml, or 1.2 molar phosphate, as described below.

In the other group of tests thrombin is added, and either the greatest plasma dilution at which clotting occurs or the time required to cause the plasma to clot is measured. The latter technique is fast and of great value clinically, but it is liable to error in the presence of any factor which interferes with the action of thrombin or with fibrin polymerization. Such factors may be found in obstetric emergencies, in which the extremely complex processes of fibrinolysis (solution of clot) and of clotting occur at the same time.[4] Clinically, however, the rate of fibrin formation is probably more important than the concentration of fibrinogen.

DETERMINATION OF FIBRINOGEN BY FIBRIN ISOLATION AND ASSAY WITH BIURET REAGENT

REAGENTS

1. NaCl, 0.15 mol/l (saline).
2. Thrombin solution, 500 NIH units/ml. Add sterile saline with a sterile syringe and

needle to a vial of 5000 units Thrombin-Topical. Mix gently to dissolve. The solution is stable for three to four weeks if refrigerated. Withdraw aliquots, using sterile technique.

3. NaOH, 0.75 mol/l (30 g/l).

4. Phosphate buffer, 0.20 mol/l, pH 6.3 to 6.4. Dissolve 1.85 g of KH_2PO_4 and 0.91 g of Na_2HPO_4 in water and dilute to 100 ml.

5. Ground Pyrex glass. Obtain about 30 to 50 cm of 3 to 4 mm diameter Pyrex tubing and cut into 1 to 2 cm pieces, place in a mortar, and grind with a pestle until pieces about 0.5 to 1.0 mm in size are obtained. Suspend the ground glass in about 400 ml of water, stir vigorously to suspend the particles, let settle for several seconds, and decant the very fine particles. Repeat the removal of the fine particles several times. Wash the glass particles with 0.05 molar NaOH and 0.10 molar HCl and then rinse with water by decantation until free of acid. Collect on a filter paper, air dry, and transfer to a clean, dry test tube.

6. Standard. Determine the protein content of a pooled serum specimen with an established method after diluting it 5-fold with saline. A commercial control serum of verified protein content may also be used.

PROCEDURE

1. The specimen consists of plasma, obtained by centrifuging blood containing oxalate or citrate as anticoagulant. Run unknown plasmas in duplicate.

2. Place about 0.50 g of ground glass into two 15 × 125 mm test tubes, using the end of a spatula. Add 8.0 ml of saline and 1.0 ml of phosphate buffer to each of the tubes, followed by 0.50 ml of the plasma to be analyzed. If a low level of fibrinogen is expected, use 1.0 ml of plasma. Mix well by vigorous tapping.

3. Rapidly add 2 drops (0.10 ml) of 500 unit thrombin solution to each tube, using a 1 ml sterile syringe and needle. Mix immediately by tapping vigorously enough to suspend most of the ground glass several centimeters up into the solution in the tube. If more than one unknown is being analyzed, treat only two tubes at a time with thrombin.

4. Let the tubes set for 5 min; clotting is completed when a solid gel forms. Tap to loosen the gel from the sides of the tube and let set for another 5 to 10 min.

5. Centrifuge for 5 min at 2500 rpm. The glass particles enmeshed in the clot will force the clot to sediment to the bottom of the tube in the form of a clot-glass pellet. Decant the supernatant. With care, practically all the fluid can be poured off without disturbing the clot-glass particles. If some gel is still present, insert a thin glass rod and squeeze and press out any fluid in the gel and press the gel against and into the glass particles.

6. Wash the clot-glass particles by adding saline and resuspending the particles by vigorous shaking. Saline is added down the sides of the tubes while they are hand-rotated to wash the sides. Centrifuge, decant supernatant, and repeat the wash procedure a second time. Decant the saline as completely as possible.

7. To each tube add 2.0 ml of 0.75 molar NaOH. Place the tubes into a boiling water bath and heat for 15 min to dissolve the fibrin. Shake the tubes every few minutes to aid solution. After heating, cool the tubes to room temperature and mix.

8. Working standards and blank. Pipet 0.20 ml and 0.40 ml of the diluted standard to tubes and add 1.80 ml and 1.60 ml, respectively, of the 0.75 molar NaOH. The blank tube contains the powdered glass plus 2.0 ml of the NaOH.

9. To all tubes add 4.5 ml of working biuret reagent (p. 302), mix, and allow the color to develop for 15 to 20 min. Read the absorbances of the samples, standards, and blank against an instrument blank containing 2.0 ml of NaOH and 4.5 ml of working biuret reagent at a wavelength of 550 nm. If any opalescence is present, centrifuge the tubes before reading their absorbances.

CALCULATIONS AND COMMENTS

If the diluted serum standard contains Z g/100 ml of protein, the two working standards will represent $Z/500$ and $Z/250$ g of protein in the cuvet volume of 6.5 ml. The unknown cuvet will contain $P/100$ g of fibrinogen if the unknown has P g fibrinogen/100 ml and if 1.0 ml of plasma is used for the assay. If the unknown is compared with the 0.20 ml standard, then

$$\frac{P}{100} = \frac{A_u}{A_s} \times \frac{Z}{500} \qquad \text{or} \qquad P = \frac{A_u}{A_s} \times Z \times \tfrac{1}{5}$$

where

P = fibrinogen concentration in g/100 ml

A_u and A_s = absorbances of unknown and standard, respectively

Similarly, if 0.50 ml of unknown is used and the absorbance is compared with the 0.40 ml standard, then

$$P = \frac{A_u}{A_s} \times Z \times \tfrac{4}{5}$$

The reproducibility of the procedure is fairly good; the coefficient of variation is about 8 per cent. Some other proteins are partly occluded in the clot despite the washing procedures. The error here has been estimated at anywhere from zero to 20 per cent (in extreme cases). On the other hand, fibrinolysins may destroy (solubilize) some fibrin, and there is some loss on heating the fibrin clot with the dilute alkali. Occasionally, losses occur because of incomplete clotting, especially when low levels of fibrin are present. In such situations, when no real emergency exists, allowing a longer time in the clotting step will improve the fibrin recovery. Fibrinogen is stable at room temperature for several days, and for several weeks in the refrigerator. If a patient is on anticoagulant therapy, additional thrombin may have to be added to ensure complete clotting.

REFERENCE

Reiner, M., and Cheung, H. C.: Fibrinogen. *In* Standard Methods of Clinical Chemistry. D. Seligson, Ed. New York, Academic Press, Inc., 1961, vol. 3, p. 114.

DETERMINATION OF FIBRINOGEN BY PHOSPHATE SALT PRECIPITATION

REAGENTS

1. Phosphate reagent, 1.20 mol/l (0.60 mol KH_2PO_4 and 0.60 mol Na_2HPO_4/l). In a 250 ml volumetric flask, dissolve 20.414 g of reagent-grade KH_2PO_4, 21.297 g of reagent-grade Na_2HPO_4, and 0.25 g of reagent-grade potassium sorbate in distilled water. Add the distilled water with constant shaking to prevent as much of the Na_2HPO_4 as possible from caking on hydration and mix thoroughly until all solute is dissolved. Dilute to the 250 ml mark with distilled water. The pH checked with a glass electrode will be 6.5 ± 0.02, but the pH is not critical within the range 6.4 to 7.5. The reagent is stable at least two years at room temperature. The potassium sorbate serves as a preservative.

2. Sodium chloride, 0.15 mol/l. Dissolve 9.0 g NaCl, AR, in 1 liter of distilled water.

3. Standard fibrinogen. Commercially available fibrinogen may be used after first assaying for nitrogen by a Kjeldahl method. The factor 5.95 may be used to convert nitrogen to fibrinogen. The solution is prepared according to the manufacturer's directions. It is stable approximately 12 h in the refrigerator (approximately 5°C). Alternatively, an aliquot of the fibrinogen solution may be clotted with thrombin as described in the previous procedure, and the washed clot assayed by the biuret method to get a measure of true "clottable" fibrinogen in the commerical preparation.

Two further working standards are made by diluting 1 vol of standard fibrinogen with 1 vol (2-fold dilution) and with 3 vol (4-fold dilution) of 0.15 molar sodium chloride solution, respectively.

PROCEDURE

1. Into pairs of cuvets at room temperature pipet 0.25 ml of each of the three standards (undiluted, 1/2, 1/4) and of each plasma (any of the usual anticoagulants are satisfactory).

2. Into one cuvet of each pair, marked "sample," add 3.0 ml phosphate reagent; into the other, marked "blank," add 3.0 ml of 0.15 molar sodium chloride solution.

3. Mix at once by gentle shaking and read sample within 9 min in a spectrophotometer at 450 nm against a corresponding blank.

4. Plot absorbance against fibrinogen concentrations for the three standards and read concentrations of unknowns from the curve.

COMMENT

In order to obtain reproducible results, the order of adding the reagents and their mixing must be rigidly standardized.

REFERENCE

Martinek, R. G., and Berry, R. E.: Clin. Chem., *11*, 10, 1965.

TRANSFERRIN

PROPERTIES AND CLINICAL SIGNIFICANCE[9]

Transferrin (or siderophilin) has a molecular mass of about 80,000 daltons. It is the principal component of the β_1-globulin band and has the vital function of transporting iron (see p. 922).

There are about 20 genetic variants of transferrin[23], all apparently compatible with good health. The fastest and slowest are clearly separated from the usual transferrin (TrC) on cellulose-acetate electrophoresis, so that heterozygotes give two bands of half strength. For most variants, however, including the three most common, the molecular sieving provided by starch gel or acrylamide gel electrophoresis is required to separate them from the TrC band.

At a normal blood pH, one or two atoms of iron (Fe^{3+}) are attached by ionic bonding to the colorless protein to form a pink complex, which transports iron from old red cells, broken down by the reticuloendothelial cells, to the bone marrow for resynthesis of hemoglobin. Dietary iron is similarly transported from the gut to the bone marrow.

The average concentration of transferrin in plasma is about 290 mg/100 ml (about 30 μmol/l), a concentration capable of transporting about 300 μg Fe/100 ml (about 60 μmol/l). Normally it is only about one-third saturated with iron.

Increased serum transferrin levels are observed in iron deficiency anemia and in the last months of pregnancy. Low levels, usually accompanied by low albumin levels, are found in many diseases, due either to impaired synthesis (e.g., cirrhosis, starvation, chronic infection) or to increased excretion as in the nephrotic syndrome (p. 311).

DETERMINATION OF TRANSFERRIN

Transferrin may be estimated directly by an immunochemical method or indirectly by measuring the maximum amount of iron the serum can bind (total iron-binding capacity). This indirect technique (see Chap. 15) overestimates transferrin by 10 to 20 per cent, because the metal attaches to proteins other than transferrin when the latter is more than half saturated.[86]

Transferrin may be estimated immunochemically by the radial diffusion or the electrodiffusion methods. Standards are made from a pure preparation of the protein, or a standard serum containing a proved concentration of transferrin may be used, e.g., Behringwerke's Standard Human Serum, stabilized. Satisfactory antisera may be made in animals with transferrin isolated with Rivanol (2-ethoxy-6,9-diaminoacridine lactate) at pH 9. This precipitates all serum proteins except transferrin and the immunoglobulins; the latter are removed by 55 per cent saturation with ammonium sulfate.

Electrodiffusion can be made to proceed more rapidly if the electrophoretic mobility of the transferrin is first increased by carbamylation as follows:[92] One vol of serum is mixed with 1 vol of 2 molar potassium cyanate (KOCN), incubated at 40°C for 30 min and then cooled in an ice bath. The standards must be treated in *exactly* the same way.

HAPTOGLOBIN[65]

PROPERTIES AND CLINICAL SIGNIFICANCE

If breakdown of red cells should occur in the blood (intravascular hemolysis), the body has several mechanisms for preventing loss of iron from liberated free hemoglobin. The first line of conservation is the serum haptoglobin, which combines with hemoglobin to form stable complexes. In addition, heme from broken-down hemoglobin combines with the β_1-globulin hemopexin or with albumin to form methemalbumin. Iron is conserved because all these complexes are too large to pass through the glomeruli. Even when hemolysis is severe enough to cause hemoglobinuria, some of the hemoglobin filtered is reabsorbed by the renal tubules.

Haptoglobin (Hp) is an α_2-glycoprotein containing about 20 per cent carbohydrate. It has been suggested that Hp and the immunoglobulins may have had a common evolutionary origin, since both contain two light chains (α) and two heavy chains (β) (to which the carbohydrate section is attached), and both can combine bivalently with another protein. The α-chains are of two main kinds, giving rise to three genetically determined normal types of serum haptoglobins $\alpha_2^1\beta_2$ (known as type 1-1), $\alpha^1\alpha^2\beta_2$ (type 2-1), and $\alpha_2^2\beta_2$ (type 2-2). Type 1-1 is monomeric with a molecular mass of 85,000; type 2-2 is represented by a series of at least 10 polymers, the molecular mass of the largest component being more than one million.[23]

The concentration of serum haptoglobin is usually expressed in terms of hemoglobin-binding capacity. It has a normal range of 40 to 180 mg Hb bound per 100 ml serum. The Hb-binding capacity is related to the haptoglobin type and remains fairly constant for a given individual. In about 4 per cent of Blacks, no haptoglobin is present (congenital anhaptoglobinemia).

The concentration rises in the acute phase reaction and in the nephrotic syndrome. In Caucasians after the first year of life, absence or a low level indicates hemolysis or acute hepatocellular damage.

An isolated estimation of haptoglobin is of limited value because of the wide normal range. However, if the patient's normal haptoglobin is known (baseline), a further estimation is useful in confirming an acute hemolytic episode, since in such cases haptoglobin is saturated and metabolized. It takes several days for it to be resynthesized and for its concentration to return to a normal level.

DETERMINATION OF HAPTOGLOBIN

In most methods an excess of hemoglobin is added to the serum to saturate the Hp. The amount of HpHb may then be estimated from the peroxidase activity of the bound Hb. Alternatively, the complex may be separated from the free hemoglobin electrophoretically, taking advantage of its greater mobility (see Chap. 8), or by gel filtration,[68] taking advantage of its greater molecular size. The proportion of Hb bound to Hp is then calculated as described below.

Direct immunochemical estimation is not often used because the molecular compositions of standard and unknown are often different owing to the heterogeneity of haptoglobin.

If radial diffusion is used, the final reading of the plate must be postponed for several days to assure that highly polymerized haptoglobin type 2-2 has had time to complete its diffusion and to react with the specific antihaptoglobin antiserum.

ESTIMATION OF HAPTOGLOBIN BY HEMOGLOBIN BINDING AND GEL FILTRATION

REAGENTS

1. Sephadex G-100, 140-400 mesh (Pharmacia Fine Chemicals, Inc.).
2. Sodium chloride, solid.
3. Sodium chloride solution, 2 g/100 ml, containing chloroform as a preservative.
4. Standard hemoglobin solution. Collect about 10 ml of normal blood, using any common anticoagulant. Centrifuge. Wash cells three times with 0.15 molar sodium chloride. Shake packed cells vigorously for 2 min with an equal volume of distilled water and half this volume of chloroform. Centrifuge at 3000 rpm for 20 min. Pipet off the upper layer of hemoglobin solution, filter if necessary, and estimate the hemoglobin concentration by one of the standard methods (Chap. 8). Adjust to 13.5 g hemoglobin/100 ml. Store at −20°C.

PROCEDURE

1. Suspend about 10 g Sephadex G-100 in an excess of sodium chloride solution (2 g/100 ml) and let stand for 48 h to swell. Pour into a glass column, 40 × 1.5 cm, and allow to pack under its own weight. Wash at least 12 h with the salt solution.
2. Add 20 μl hemoglobin solution (2.7 mg Hb) to 0.5 ml of the patient's serum and mix. Add about 5 mg solid sodium chloride to increase the specific gravity and, without disturbing the column, deliver this solution onto the top surface of the gel but below the layer of the salt solution. Elute the column with the 2 g/100 ml NaCl solution at about 1 ml/min and collect 1 ml eluates in a fraction collector until both colored bands are eluted. (The column can be used repeatedly if, after use, the 2 g/100 ml salt solution is used to wash out the free hemoglobin and then left *in situ* with about 3 ml above the gel surface.)
3. Add 2 ml of 2 g NaCl/100 ml to each tube, transfer to a cuvet, and read in a spectrophotometer at 415 nm, using the NaCl solution as blank.
4. Plot absorbance against the tube number. There should be two peaks, the first corresponding to the complex and the second corresponding to the excess free hemoglobin. Pool all fractions from the first peak and measure the volume and absorbance.
5. Prepare from the hemolysate (standard) a range of solutions containing 0.01 to 0.1 mg hemoglobin/ml. Read their absorbances and construct a standard curve or determine a calibration factor. Calculate the total amount of hemoglobin in the first peak:

$$\text{haptoglobin concentration (expressed as hemoglobin binding capacity in mg/100 ml serum)} = \text{hemoglobin (in first peak)} \times \frac{100}{0.5}$$

COMMENTS

1. Use fasting blood samples to avoid hyperlipidemia, which may cause the first peak to be turbid and give too high results. Hemolysis in the sample causes no error.
2. Sephadex G-200 should not be used since it will separate the complexes of the different haptoglobin polymers, whereas G-100 does not resolve proteins in this molecular range and all proteins of the molecular size of the monomer complex and above are eluted in the first peak.

REFERENCE

Ratcliff, A. P., and Hardwicke, J.: J. Clin. Path., *17*: 676, 1964.

α_1-ANTI-TRYPSIN

PROPERTIES AND CLINICAL SIGNIFICANCE[15,49]

α_1-Anti-trypsin has a molecular mass of 45,000 and is the main plasma protein among the α_1 globulins, giving rise to the α_1 band in stained electrophoretograms. It is synthesized by the liver and is responsible for nearly all the protease inhibiting capacity of serum; hence its name. Its normal serum concentration is about 250 mg/100 ml, and it is one of the proteins whose concentration increases in the acute phase reaction and in pregnancy, or when oral contraceptives are taken. It falls in infants with idiopathic hyaline membrane disease.

Its congenital deficiency is associated clinically with the development of emphysema at an unusually early age, and an increased incidence of neonatal hepatitis, usually progressing to cirrhosis.

Many genetically determined variants[23] of α_1-anti-trypsin (identified by letters) have been demonstrated as bands of varying mobility on starch gel electrophoresis. Normal individuals have M bands and thus have the phenotype MM. α_1-Anti-trypsin deficiency is usually associated with the autosomal recessive gene Z, demonstrable as a slow moving Z band, which stains only faintly. Homozygotes ZZ are likely to be affected clinically, but whether heterozygotes MZ, which have about half the normal concentration of protein, are at risk is not yet clarified. The only other genotypes affected are SS and SZ.

The frequency of deficient enzyme inhibition as a cause of emphysema is still disputed, probably owing to varying criteria used to diagnose chronic bronchitis and emphysema and to extraneous factors such as smoking. Some smokers with α_1-anti-trypsin deficiency escape emphysema. As the condition becomes more widely recognized, the reported incidence among emphysematous patients is growing. It is probably about 5 per cent. The disease is primary in type, mainly affecting the lung bases, and may possibly be due to the action of proteolytic enzymes, normally inhibited by α_1-anti-trypsin, which are released by leucocytes and bacteria when lung tissue is infected.

ESTIMATION

Because α_1-lipoprotein and orosomucoid (α_1-acid-glycoprotein) stain so poorly, the α_1 band seen in protein electrophoresis is composed almost entirely of α_1-anti-trypsin. About 95 per cent of Caucasians are of the MM phenotype and show the normal α_1 band. Other alleles result in the synthesis of α_1-anti-trypsin with faster or slower mobility and roughly the same concentration. Heterozygotes possessing one of these genes and the normal M give an α_1 band of normal mobility and an extra adjacent faster or slower band, each about half as intense as the normal α_1 band. Since the Z gene results in the synthesis of so little protein, homozygotes appear to have no α_1 band, while heterozygotes have a band of normal mobility but of half normal intensity.

If no α_1 band is seen on electrophoresis, the α_1-anti-trypsin should be estimated immunochemically (pp. 321 and 325) or by enzymatic assay. Either the radial immunodiffusion or the electrodiffusion technique described above may be used. The latter is faster and very suitable for determining protein with α mobility. Suitable standards are available commercially (e.g., Behring Diagnostics).

The enzymatic assay method[12] is rapid and precise, but it is more complex and estimates the total anti-trypsin activity of serum, only about 80 per cent of which is due to α_1-anti-trypsin. The rates of hydrolysis of a mixture of trypsin and substrate are measured with and without the previous addition of the patient's serum. A convenient substrate is α-N-benzoyl-DL-arginine-*p*-nitroanilide (BAPNA). After a suitable time the reaction is stopped by adding acetic acid, and the absorbance of the liberated *p*-nitroaniline (PNA) is read at 400 nm. For linearity, protein must be present in the control and conditions must be rigidly standardized.

$$\text{Serum Trypsin Inhibitory Capacity } (\mu\text{mol/min/ml}) = \frac{\text{Assay volume}}{\text{Serum sample volume}} \times \frac{\Delta A}{M \times T}$$

where ΔA is the difference between the absorbance of the sample assay and that of an albumin-trypsin control after time T; M is the millimolar absorptivity of PNA.

Values for normal subjects are 2.1 to 3.8; for MZ heterozygotes, 1.05 to 2.1; and for ZZ homozygotes, 0.5 to 0.7 μmol/min/ml.

REFERENCE

Dietz, A. A., Rubinstein, H. M., and Hodges, L.: Clin. Chem., *20*: 396, 1974.

CERULOPLASMIN

PROPERTIES

Ceruloplasmin has a molecular mass of about 160,000, and has an α_2-globulin mobility. Since the main components of the α_2 band are α_2-macroglobulin and haptoglobin, the electrophoretic pattern gives no indication of the concentration of ceruloplasmin. Ceruloplasmin is blue in color, contains more than 90 per cent of the plasma copper (see Chap. 15, Section Three) and behaves as an oxidase (see Chap. 12).

CLINICAL SIGNIFICANCE

Normal values for the concentration of ceruloplasmin in adult sera are about 25 to 50 mg/100 ml. They are lower in babies, about 8 to 23 mg/100 ml for the first four months. The level is *reduced* in hepatolenticular degeneration (Wilson's disease) as well as in the nephrotic syndrome, sprue, and scleroderma of the small intestine. It is *increased* in the acute phase reaction, in pregnancy, during treatment with estrogens, and in certain chronic infections (see Chap. 12).

ESTIMATION

Ceruloplasmin may be estimated either on the basis of its oxidase activity, using, for example, *p*-phenylenediamine as substrate and measuring the color produced[69,76b] (see Chap. 12), or as an antigen, using either the electrodiffusion rocket technique or the radial diffusion technique as described previously. Either is suitable, the former technique being faster, the latter simpler.

IMMUNOGLOBULINS AND THEIR RELATION TO DISEASE[35]

Immunoglobulins are produced by plasma cells and their precursors, the B lymphocytes. There are five classes which have been discussed in Section Two (see Table 7–3). The B lymphocytes carry the different antibodies (immunoglobulins) on their surfaces; when one of them is stimulated by the corresponding antigen, it proliferates to form plasma cells able to secrete the appropriate specific immunoglobulin.[8a]

Probably, an individual plasma cell can produce only one species of antibody[5], that is, a single kind of immunoglobulin. The injection of a bacterium with its numerous protein antigens, or the injection of even a single pure immunogen with its numerous determinants, leads normally to the production of a great variety of immunoglobulins. Numerous different plasma cells are stimulated to produce antibodies of different classes against different determinants and with different degrees of fit. Their varying electrophoretic mobilities give rise to the normal, wide, and diffuse γ band. The intensity of the band increases on repeated immunogenic stimulation, as in chronic infections. On the other hand, if a single plasma cell precursor continually divides to form a clone of cells, all the daughter cells will produce the same single immunoglobulin and their molecules will be identical. On electrophoresis they will have the same mobility and, therefore, produce a single narrow band.

TABLE 7–10 CHANGES IN ELECTROPHORETIC PATTERNS OF SERUM RESULTING FROM ABNORMALITIES IN IMMUNOGLOBULIN CONCENTRATIONS

Abnormality	*Electrophoresis*	*Clinical Syndromes*
1. Deficiency of IgG	No γ band	Multiple infections (immunoparesis)
2. Increase in normal heterogeneous immunoglobulins (polyclonal)	Diffuse increase in γ band	Secondary, e.g., to cirrhosis or chronic infections
3. Increase in a single homogeneous immunoglobulin (monoclonal) in serum or urine.	Narrow M-band (± Bence Jones proteinuria)	Myelomatosis; Waldenström's macroglobulinemia

When tested immunochemically, they will react with antisera made against only one class of heavy chain and one class of light chain. These characteristics allow such monoclonal proteins (M-proteins, paraproteins) to be identified with certainty in the laboratory.

In disease there may be deficiencies or increases in the normal heterogeneous immunoglobulins (which produce the diffuse γ band) or there may be increases in monoclonal immunoglobulins (sharp "paraprotein" bands) (see Table 7–10). Cryoglobulinemia and amyloid disease appear also to be disorders of immunoglobulins.

IMMUNOGLOBULIN DEFICIENCIES[35,80]

Decreased production of plasma immunoglobulins is clinically associated with poor resistance to infection owing to failure of antibody production. This may lead, for example, to recurrent respiratory infections, malabsorption, or unexplained splenomegaly.

Deficiency of IgG (hypogammaglobulinemia) is indicated by marked reduction or absence of the γ band on electrophoresis. It is usually secondary to protein loss or production failure (see Table 7–11) but may be a primary congenital abnormality. It may or may not be accompanied by IgM and IgA deficiencies.

At birth a normal infant has practically no serum immunoglobulin apart from the IgG transferred across the placenta from the mother, although the fetus has a full complement of B lymphocytes with surface immunoglobulins.[8a] On contact with antigens after birth, the B lymphocytes give rise to plasma cells and the production of the three main immunoglobulins gradually increases, as shown in Figure 7–19. As the maternal IgG disappears, the serum IgG level reaches a minimum at about 3 months, which is the danger period. Two groups of babies are at risk—premature babies and those who have a transient delay in the synthesis of IgG. In such cases IgG estimates are invaluable, since the level may fall dangerously low unless they are given injections of γ globulin.

TABLE 7–11 CAUSES OF IgG DEFICIENCY SYNDROME (IMMUNOPARESIS)

1. In the newborn only
   (i) Physiological
   (ii) Transient delay in production
2. Abnormal loss of protein
   Nephrotic syndrome; protein-losing enteropathy
3. Failure of production
   (i) Primary
      (a) Failure of antibody production only
      (b) Failure of cellular immunity as well[8a,80]
   (ii) Secondary
      Myelomatosis, leukemia, reticuloses, drug induced

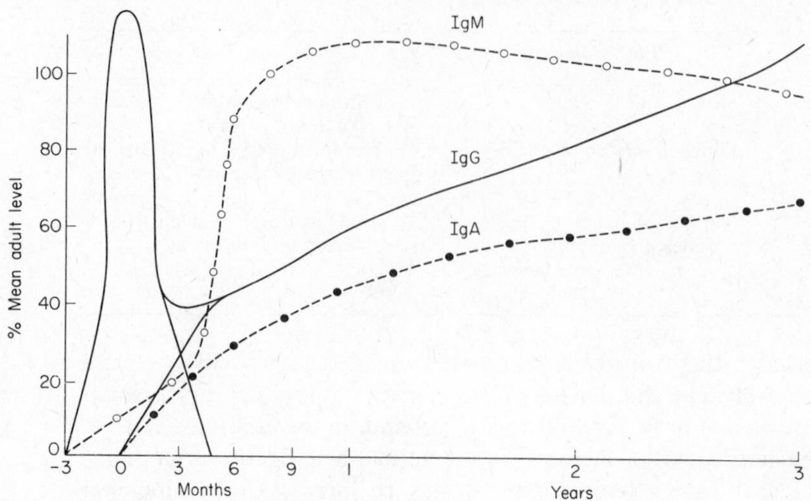

Figure 7–19 Serum immunoglobulins in the newborn and young child. ———, IgG (first curve, maternal; second curve, child); o - - - - o, IgM; ● - - - - ●, IgA. (From Grant, G. H., and Butt, W. R.: *In* Advances in Clinical Chemistry. O. Bodansky and C. P. Stewart, Eds. New York, Academic Press, Inc., *13*:419, 1970.)

Antibody deficiency may also be present in spite of a γ band that appears to be normal (dysgammaglobulinemia). This is usually due to severe deficiency of IgM, IgA, or both, and can be diagnosed only by immunochemical analysis. In the "toxic" type of antibody deficiency secondary to renal failure, celiac disease, diabetes mellitus, or severe infection, the level of IgM usually falls first, then IgA, and finally IgG.

Congenital deficiencies of every possible combination of the three main immunoglobulin classes have been reported, but they are very rare.[35,80]

POLYCLONAL OR SECONDARY HYPERGAMMAGLOBULINEMIA

The "normal" levels of serum immunoglobulins (Table 7–9) depend on the extent of environmental antigenic stimulation. Germ-free animals, for example, have very low levels. Natives of undeveloped countries usually have higher "normal" levels of IgG, IgM, and IgE than those from countries with better public health services.

In sarcoidosis and most generalized infections, including subacute endocarditis, multiple antigenic stimulation by several routes results in a non-specific increase in concentration of all the immunoglobulins. If the bloodstream is especially affected, as in tropical parasitemia, IgM will predominate (secondary macroglobulinemia). If mucous surfaces are mainly affected, IgA is likely to be the predominant form. Such a rise is characteristic of malabsorption states; it is of little value in differential diagnosis but may be useful as an index of response to treatment. An isolated rise in IgG is especially characteristic of autoimmune diseases. Increases in IgE occur in parasitic infections and in allergic diseases such as extrinsic asthma, hay fever, and atopic eczema.

Since the fetus is normally protected from foreign antigens, it is born with very low levels of IgA and IgM (see Fig. 7–19). Their estimation in cord blood and during the first six weeks of life is, therefore, useful for the detection of intrauterine and neonatal infections.

In the differential diagnosis of liver disease,[35] immunoglobulin estimations are of considerable value when disease elsewhere in the body can be excluded. In chronic active hepatitis (juvenile macronodular cirrhosis), the increase is mainly in IgG; in portal cirrhosis it is mainly in IgA; while primary biliary cirrhosis is characterized by an isolated rise in serum IgM. In acute hepatitis, IgM is also raised followed later by IgG. In drug-induced jaundice the Ig levels are normal.

PARAPROTEINEMIA[35,45]

The finding of a paraprotein (monoclonal immunoglobulin, M-component) in a patient's serum indicates that he has an immunocytoma, a proliferating clone of immunoglobulin-producing cells. Its associated clinical symptoms and prognosis will depend on two factors: first, the number, site, and rate of growth (degree of malignancy) of the proliferating cells, and second, any secondary conditions such as the viscosity syndrome, amyloidosis, cryoglobulinemia, or renal failure caused by the paraproteinemia itself.

Unlike most neoplasms, an immunocytoma has a specific product, its paraprotein, whose serum concentration and characteristics provide a means of assessing the extent of the disease[76a] and its degree of malignancy, respectively (p. 352).

The incidence of immunocytoma increases with age. Sometimes it starts as a frankly malignant myelomatosis; sometimes it is slow growing, and a secondary effect of the paraprotein itself causes death, e.g., the viscosity syndrome in Waldenström's macroglobulinemia. Sometimes it is benign, and occasionally it is transient.

Myelomatosis. This disease of the middle-aged and elderly is a cancer of antibody-forming cells (plasma cells) proliferating in the bone marrow and elsewhere. Occasionally (in about 4 per cent of cases), they first multiply locally to form an isolated tumor. The fact that each paraprotein appears to be different is due to the malignant change occurring randomly in one of the very large number of different kinds of normal plasma cells, each producing a different, although normal, antibody. The relative frequencies of the various types of myeloma proteins (IgG, IgA, etc.) seem to correspond to the relative numbers of plasma cells normally producing these proteins (Table 7–12).

TABLE 7–12 MYELOMATOSIS[a]

Type of Immunoglobulin	*Approximate Incidence (Percent)*	*Approximate Mean Age Incidence*	*Average Doubling Times of M Protein Conc.*	*Common Clinical Findings*[b]
IgG	50	65	Slowest (10 months)	More liable to immune paresis; higher level of M protein; 60% with Bence Jones proteinuria
IgA	25	65	Intermediate (6.3 months)	Hypercalcemia; 70% Bence Jones proteinuria; amyloidosis
Bence Jones protein only	20	56	Fastest (3.4 months)	Associated with more osteolytic lesions, hypercalcemia, renal failure, amyloidosis; worse prognosis
IgD	1.5	57	—	Frequently have osteolytic lesions, Bence Jones proteinuria, hypercalcemia and renal failure; nearly all λ type; also tendency to have extraosseous tumors
IgM	0.5	—	—	Represents about 20% of cases of monoclonal IgM increase. May or may not be associated with the viscosity syndrome
IgE	3 cases	—	—	
Biclonal	2	—	—	
Nil	1	—	—	Usually severe reduction of normal Igs. Appears to have poor prognosis

[a] After Hobbs.[34] Data from about 500 patients.
[b] No significant difference found for incidence of anemia or infection.

Myelomatosis is often difficult to diagnose clinically, since it may be associated with many other symptoms besides pain in the back, a fracture, or an obvious tumor. More often one finds an ill-defined illness characterized by vague generalized pain, weakness, loss of weight, anorexia, uremia, hemorrhages, repeated infections, or diarrhea and vomiting, symptoms unlikely to suggest the diagnosis without laboratory help. Anyone who could conceivably have the disease, including all those found by hematological studies to have an obscure anemia or very high erythrocyte sedimentation rate, should be screened for abnormal serum protein bands by cellulose acetate electrophoresis. In addition, concentrated urine must be tested routinely for Bence Jones protein (p. 366); otherwise, those patients with Bence Jones proteinuria but without serum paraproteins will be missed. Such routine investigation has led to a remarkable apparent increase in the incidence of the disease.

In myelomatosis there are almost always M-components in the serum and/or urine, and a malignant plasmacytosis in the bone marrow. The majority of cases also exhibit typical bone changes, a normocytic normochromic anemia, and a reduction in normal serum immunoglobulins. The clinical findings depend to some extent on the immunoglobulin class. The paraprotein is usually IgG, often IgA or free light chains (Bence Jones protein), only occasionally IgM or IgD, and very rarely IgE (see Table 7–12). A small proportion of malignant immunocytomas arise in lymph nodes, but their enlargement, or the blood changes, usually make the diagnosis clear without the investigation of the serum proteins (lymphosarcoma, reticulosarcoma, chronic lymphatic leukemia).

An uncommon related disorder is *heavy chain disease*[2a] in which free *Fc* fragments of Ig heavy chains are present and produce wide diffuse bands on the electrophoretogram of serum or urine. IgG heavy chain disease has been observed in middle-aged men with a soft tissue plasmacytoma or enlarged lymph nodes, weakness, and loss of weight. In the IgA type (α chain disease) there is intestinal malabsorption secondary to plasma cell infiltration of the gut and mesenteric lymph nodes (Arabian lymphoma of the small gut). The IgM form is least common and is usually associated with chronic lymphatic leukemia. In all types the *Fc* fragments vary, and they produce broader bands than other paraproteins because of the heterogeneity of their carbohydrate moieties.

SECONDARY EFFECTS OF HYPERGAMMAGLOBULINEMIA[35]

The Viscosity Syndrome. The viscosity syndrome is caused directly by paraproteinemia. It occurs in those cases (about 6 per cent) in which the abnormal protein has a very large molecular mass ("primary macroglobulinemia"),[32] causing an increase in serum viscosity. The most common cause is an IgM paraprotein (molecular mass about one million) with a serum concentration of 3 g/100 ml or more ("Waldenström's macroglobulinemia"). Less common are those unusual forms of IgG, IgA, or Bence Jones protein which polymerize or form complexes, for example, as cryoglobulins (q.v.). Waldenström's macroglobulinemia is usually accompanied by Bence Jones proteinuria, but is much less malignant than myelomatosis and is treatable by exchange plasma transfusions.

The other disorders which may result directly from hypergammaglobulinemia occur as often in polyclonal immunoglobulin abnormalities as in paraproteinemias.

Amyloid Disease.[19] In amyloid disease, an abnormal proteinaceous material is deposited in the blood vessels and in the matrix of a number of organs such as the liver, spleen, adrenals, and kidney. Because of its starch-like reactions, this material is called amyloid. Histologic specimens of such tissue stain brown with aqueous iodine, and then turn blue on acidification. Two forms of amyloid disease are recognized: primary and secondary amyloidosis. The primary form is not associated with any previous disease. Secondary amyloidosis, however, is the more common form in which amyloid degeneration occurs as an aftermath of long term suppurative disease, rheumatoid arthritis, syphilis, tuberculosis, or Hodgkin's disease, as well as myelomatosis. Amyloid material has the property of absorbing dyes such as Congo Red or Evans Blue. This property was formerly utilized for its chemical detection, to confirm diagnoses made by histologic examination of biopsy specimens; and it is the basis of the Congo Red test described later.

Recently it has been shown by electron microscopy that amyloid is almost entirely composed of characteristic fibrils, which can be isolated in pure form. Their insolubility in saline makes it possible to separate them from salt soluble protein contaminants. The fibrils can then be dissolved in pure water to form a colloidal solution. They are found to be built of protein subunits varying in molecular mass from 5000 to 30,000, and are often closely related to immunoglobulins, in particular to the variable half of the light chain (see Fig. 7–1, Section II). It seems possible that they are produced from plasma cells. Specific antisera against them can be prepared, and may well prove to be diagnostically useful.

Many urinary Bence Jones proteins, on treatment with pepsin or trypsin, turn into fibrils which are apparently identical to amyloid and are built from their variable parts. Although there is some evidence that amyloid is not always composed of this type of protein, we are beginning to have a much clearer understanding of this disease. It seems to occur either when the immune system has escaped from normal regulation as in myelomatosis, or when the size of the immunogenic stimulus is excessive compared with the capacity of the immune system. This would explain its occurrence both when the system is defective, as in immune deficiency diseases, and when it is normal but exposed to prolonged immunogenic stimulation, as in chronic infections.

Cryoglobulinemia.[35] The term cryoglobulin refers to certain abnormal globulins, occasionally encountered in serum, which can precipitate or gel out when serum is cooled to a low temperature but redissolve when the serum is rewarmed to 37°C. Whether this causes clinical symptoms seems to depend on whether precipitation occurs above 21°C (p. 353). The amount of cryoglobulin varies from a few mg to several grams per 100 ml, and it may be monoclonal (essential cryoglobulinemia) or polyclonal (secondary cryoglobulinemia). The monoclonal cryoglobulins are due to polymerization and may be of the IgG or IgM class or can even be rheumatoid factors. Polyclonal cryoglobulins are usually immune complexes containing more than one class of immunoglobulins, such as IgM rheumatoid-factor-antibody bound to IgG. These complexes lodge in vessel walls and fix fibrinogen and complement, causing a vasculitis. They may occur in systemic lupus erythematosus, rheumatoid arthritis, and other autoimmune diseases.

Clinically, cryoglobulinemia may present with the features of Raynaud's syndrome—intolerance to cold, purpura, gangrene of the extremities, and skin sores. The monoclonal cryoglobulins can cause skin lesions long before the usual symptoms of immunocytoma occur. Death may result from blockage of key blood vessels in such vital organs as the kidneys, brain, and lungs. The specific symptoms depend on the amount present and the degree of anti-inflammatory response of the body. Similar symptoms may be associated with cryofibrinogens (which precipitate from plasma but not serum) and with cold agglutinins.

Secondary Renal Disease. Paraproteinemia may cause renal damage secondary to amyloid disease or cryoglobulinemia, or as a result of formation of urinary casts which block the distal tubules of the kidney. In myelomatosis, renal failure is usually due to hypercalcemia and pyelonephritis.

LABORATORY INVESTIGATION OF IMMUNOGLOBULINS[33,45]

If possible, separate the serum at 37°C to avoid missing cryoproteins that precipitate out between room temperature and 37°C. Cellulose acetate electrophoresis is an essential screening test, even though it gives little information on the concentrations of IgM or IgA. Marked reduction of the γ band or a polyclonal increase of unknown origin should be followed up by quantitative estimations of IgG, IgM, and IgA by radial immunodiffusion (p. 321).

The IgG level can also be deduced approximately from the total γ-globulin concentration. This may be estimated by electrophoresis or by the turbidity produced on adding serum to suitable zinc sulfate[46] or ammonium sulfate solutions,[21] or it may be determined chemically by precipitation with Wolfson's reagent (19.3 g ammonium sulfate and 4 g NaCl/100 ml) followed by re-solution of the precipitate and its estimation by the biuret reaction.[21]

Sera containing high IgM concentrations usually cause a precipitate when one drop is placed into water or a dilute salt solution. This simple "Sia" test[18] is positive, for example, in secondary macroglobulinemia which accompanies malaria or kala-azar. Changes in IgM concentrations are also reflected in the level of the total macroglobulins, which include α_2-macroglobulin. They may be estimated by measuring the protein in the fastest moving fraction obtained by Sephadex G 200 gel filtration (p. 332) or from the 19*S* fraction obtained by ultracentrifugation. If, on cellulose acetate electrophoresis, there is an abnormal band thought to be monoclonal, this should be confirmed by immunoelectrophoresis to make certain that it is not caused, for example, by fibrinogen or by old or uremic serum. Tests should be carried out with antisera against serum proteins in general, against the three main immunoglobulin heavy chains, and against the two light chains as described on page 329, in order to type the paraprotein and to determine whether the other immunoglobulins are so deficient that the patient is likely to have lost his resistance to infection. Estimate the concentration of the paraprotein by cellulose acetate electrophoresis and densitometry. It cannot be estimated by radial immunodiffusion, since there will be no standard solution of the paraprotein for comparison (see p. 325). A concentrated sample of urine should be subjected to electrophoresis parallel with the serum specimen to check for excreted monoclonal proteins, usually free light chains (Bence Jones proteins, p. 366). The concentrations of the normal immunoglobulins are determined by radial immunodiffusion (p. 321).

Four pieces of evidence help to diagnose malignancy:

(a) a progressive rise in serum paraprotein level,

(b) serum paraprotein level greater than 1 g/100 ml,

(c) suppression of normal immunoglobulins, and

(d) the findings of immunoglobulin fragments, such as free light chains (Bence Jones proteins), free heavy chains or monomeric (7*S*) IgM, especially in the absence of whole immunoglobulin molecules.

If the paraproteinemia appears to be benign, the patient's serum should be retested at intervals, in case a malignancy has developed. A sample of the patient's serum, obtained when the paraproteinemia is first diagnosed, is used as a reference for the assessment of subsequent changes in paraprotein level by radial immunodiffusion.

Tests for Cryoglobulins[35]

It is important to keep blood specimens at 37°C during the procedure. After the specimen has been drawn from the patient, allow it to clot in a water bath at 37°C for at least one hour to ensure removal of all fibrinogen. Centrifuge the tube containing the clot in cups containing water at 37 to 40°C, and store the separated serum at 37°C until tested. To test for the presence of cryoglobulins, cool the serum specimen overnight in a refrigerator at 4 to 10°C. If no precipitation or gelling is observed overnight at 4 to 10°C, keep the serum chilled for at least another 48 h. Check it on the second and third days to be certain that no latent precipitation has occurred. Turbidity and lipemia present in the specimen make it difficult to detect low degrees of gelling. If such a serum must be tested, divide it into two portions; chill one and leave the other at room temperature or preferably 37°C as a control.

The temperature at which the precipitate appears should be measured, since if this is not above 21°C, the lowest natural skin temperature, then exposure to cold is unlikely to cause symptoms unless the cryoglobulins are immune complexes.

Monoclonal cryoglobulins can be distinguished from polyclonal (mixed) cryoglobulins by immunoelectrophoresis of both whole serum and the cold supernatant. Their concentrations can be assessed from the difference between the protein concentrations of whole serum and cold supernatant, both estimated by the biuret method.

The Congo Red Test for Amyloid Disease

This test is essentially a measure of the rate of removal of Congo Red dye from blood by amyloid material deposited in the blood vessels and in the matrix of various tissues. The patient receives an intravenous injection of pure, parenteral grade Congo Red dye at a dosage of 4.5 mg/kg of body weight up to a total of 100 mg. At 3 to 4 min and again at exactly 60 min after the injection, specimens of blood are obtained. Serum proteins are removed from the specimens with acetone, and the dye concentration in the 60 min specimen is compared with that in the 4 min specimen. The latter is assumed to represent the initial dye level (= 100 per cent). A level of 60 per cent (usually 65 to 70 per cent) of the dye in the 60 min specimen is considered normal. Serum levels of 40 per cent are suggestive of amyloid disease, and those of 25 per cent or less are highly so, provided no dye has been lost into the urine as a result of proteinuria, which may invalidate the test or make interpretation difficult.[87]

β_{1C}-GLOBULIN, THE C3 COMPONENT OF COMPLEMENT

PROPERTIES AND CLINICAL SIGNIFICANCE

β_{1C}-Globulin forms the β_2 band seen on electrophoresis of *fresh* serum, and the intensity of this band is a good indication of its concentration. On standing, β_{1C} changes to the faster moving β_{1A} and the β_2 band disappears.

β_{1C} is one of the components of complement (see Section Two, p. 291). Some of these components have been identified with precipitin bands seen on immunoelectrophoresis of normal serum: e.g., C3 has been identified as β_{1C}, C4 as β_{1E}, and C5 as β_{1F}. Of these forms, β_{1C} (normal serum level 70 to 160 mg/100 ml) is present in greatest concentration, and so its estimation is often used as an indication of the level of complement as a whole. Its concentration increases in the acute phase reaction, but more important diagnostically, its level falls in certain renal diseases. In acute poststreptococcal glomerulonephritis, if the initial fall is not followed by a return to normal as the patient recovers, a change from the acute to the chronic form is suggested. In diseases in which "soluble immune complexes" are deposited in the glomeruli, such as systemic lupus erythematosus (SLE), serum sickness, or certain cases of drug reactions, the initial low β_{1C} level rises as the patient recovers, and a second fall indicates an imminent relapse. Patients with most other renal diseases have normal β_{1C} levels; the exceptions are those with membrano-proliferative glomerulonephritis, about half of whom have low β_{1E} levels but normal or raised β_{1C} levels. Low serum β_{1C} levels are also found in severe hemolytic anemia, septicemia, and some liver diseases.

Determination of β_{1C}

A rough estimate of the concentration of β_{1C} can be made on the basis of the intensity of the β_2 band in *fresh serum*. For quantitative estimation the serum is left at room temperature for 24 h so that all the β_{1C} is converted to β_{1A}; the latter is then estimated against anti-β_{1A} by the radial immunoassay method.

α_1-FETOPROTEIN

If blood from a 12 week old fetus is examined by cellulose acetate electrophoresis, a prominent extra band is seen between the albumin and α_1-anti-trypsin bands. This α_1-fetoprotein (AFP) is produced by the fetal liver and yolk sac; its serum concentration reaches a maximum at about 12 to 15 weeks and then falls until it is no longer detectable, about five weeks after birth.

In most patients with primary carcinoma of the liver (hepatoma) and in about one-third of those with teratomas, the tumor cells revert to the fetal function of synthesizing and excreting α_1-fetoprotein. Its concentration rises in the patient's serum and can be used as a diagnostic aid, in following the course of treatment, and for indicating the occurrence of secondary growth after operation. The α_1-fetoprotein may be detected by double diffusion (see p. 328) and can be estimated by radial immunodiffusion or electro-immunodiffusion using a standard solution available commercially. A suitable specific antiserum can be bought commercially (Behringwerke) or can be prepared by injecting rabbits with heart blood serum from a fresh 20 week old human fetus and then "absorbing" the resulting rabbit antiserum with normal adult serum. The test is highly specific but may occasionally be positive in patients with upper intestinal tract tumors and in patients with carcinoma of the stomach with liver metastases, and temporarily in hepatitis.

Recently it has been shown that fetuses with open spina bifida or with anencephaly leak so much α_1-fetoprotein into the amniotic fluid that its level reaches many times the normal and can be estimated similarly. If spina bifida is suspected, α_1-fetoprotein estimations in amniotic fluid are the only satisfactory tests available to detect the disease early enough in pregnancy (18 weeks) to allow termination. The only other condition causing high levels of AFP is intrauterine death. The low levels present in maternal serum, measurable only by radioimmunoassay, may also be elevated.

ESTIMATION OF PROTEINS IN OTHER BODY FLUIDS

With the exception of some exudates and of urine from patients with gross proteinuria, the concentrations of proteins in other body fluids are much less than in serum and require a different analytical approach. The volume of specimen available for analysis may also be inconveniently small. Samples of cerebrospinal fluid, for example, are necessarily limited in volume and normally have a concentration only about one or two hundredths of that in serum. An analytical technique more sensitive than the biuret method and/or preliminary concentration by ultrafiltration, gel filtration, or precipitation and re-solution is required. Methods available include the copper-phenol method of Lowry, direct UV spectrophotometry and, for rapid approximate determination, turbidity measurement following protein precipitation.

TOTAL PROTEIN DETERMINATION USING THE FOLIN-CIOCALTEU REAGENT[87]

Solutions of various types of phenols behave as weak reducing agents and can be oxidized to quinone-type compounds by many oxidizing agents. A complex phosphotungsto-molybdic acid reagent, *phenol reagent*, devised by Folin and Ciocalteu, oxidizes phenolic compounds under alkaline conditions; it is reduced from its initial golden yellow color to a deep blue. This reagent is used to measure phenolic compounds and to assay tyrosine, the amino acid that contains a phenol side chain. Inasmuch as all proteins contain some tyrosine, this reaction has been used to quantitate plasma, serum, and spinal fluid proteins. The indole and imidazole groups of tryptophan and histidine also react with the reagent, but the reaction is weaker than that given by the phenol structure.

The method is some five to ten times as sensitive as the biuret reaction but is unsatisfactory for serum protein mixtures because the individual proteins vary in their tyrosine, tryptophan, and histidine contents and therefore give different degrees of color intensity with the reagent. Lowry and his associates,[50] however, have developed a technique both more specific and more sensitive, by carrying out the biuret reaction and then adding the phenol reagent. The biuret complex as well as the aromatic side chains of the proteins reduces the reagent. The reaction is about one hundred times as sensitive as the biuret method; also, it determines mainly peptide bonds and it is far less affected by variations in the tyrosine content. However, a number of buffers containing amino groups or substituted amino groups (e.g., TRIS, glycine amide) interfere by producing a color.

MEASUREMENT OF PROTEIN BY ULTRAVIOLET LIGHT ABSORPTION

Protein solutions show strong ultraviolet light absorption bands at 279 nm and at 210 to 215 nm. The absorption peak at 280 nm is due to the aromatic rings of tyrosine and tryptophan, and that at 215 nm is due to the peptide bond.

Unfortunately, precise measurement of absorbance in the short ultraviolet range (215 to 220 nm) presents many problems of technique that are sufficient to discourage its use in a routine laboratory except when there is no other option. Measurement of the aromatic rings at 280 nm has the same limitations as the use of phenol reagent, namely, the varying tyrosine content of the different serum proteins.

Urine Proteins[53,77]

PHYSIOLOGY AND NORMAL VALUES

The basement membranes of the glomeruli (see Fig. 17–2) behave as ultrafilters and probably have structures somewhat similar to sheets of conjugated dextran gel; they exclude proteins of molecular masses greater than 100,000 daltons but allow the passage of increasing amounts of proteins of lesser molecular mass down to about 30,000; they are freely permeable to smaller molecules of proteins. However, only a very small proportion of the proteins filtered are excreted in the urine. Most are reabsorbed and broken down by the cells lining the tubules, so that normal urine contains very little plasma protein, even of low molecular mass. It contains, however, small quantities of proteins of renal origin, of which the mucoprotein of Tamm and Horsfall (uromucoid) is much the most abundant (see p. 367).[26] The total quantity of protein excreted by healthy individuals in a 24 h period and the relative proportions of the main groups of protein excreted are approximately as follows:[76]

	Approximate Mean Concentration mg/100 ml	*Approximate Diurnal Output* mg/d
Total Protein	12	100*
Mucoprotein[30]	5	70*
Albumin	2.5	10
Globulins	4.5	20

* Includes carbohydrate moiety of mucoprotein.

There remain considerable differences between the findings of different authors;[53] the above figures give only a rough indication of the normal levels.

The quantities of some of the individual plasma proteins excreted daily are listed in Table 7–13. After considerable muscular exertion, the total protein excreted may be as

TABLE 7-13 QUANTITIES OF SELECTED SERUM PROTEIN COMPONENTS IN NORMAL HUMAN URINE DETERMINED IMMUNOCHEMICALLY[53,93]

Protein	Concentration in Serum (mg/100 ml)	Approximate Molecular Mass (daltons)	Some Estimates of Quantity in Urine (mg/d): Number of Individuals Tested	Average	Range	Notes
Albumin	3500—5000	65,000	12	7.8	3.6—14.1	
			20	10.0	3.9—24.4	
			29	5.0	2.5—28.8	
Immunoglobulins						
IgG	800—1400	160,000	12	3.2	1.2—6.5	
			29	3.1	0.9—6.3	
Fc fragments	—	58,000	12	0.2	0.1—0.4	IgG breakdown product
IgA (7*S*)	90—450	160,000	12	1.4	0.7—2.7	Includes secretory
		or more	29	0.25	0.1—0.4	IgA
Ig light chains	Trace	45,000	10	3.3	1.1—6.8	Partly monomeric
		(dimer)	12	3.4	1.2—7.2	
Transferrin	200—400	85,000	5	0.38	0.32—0.47	
			29	0.17	0.04—0.45	
α_1-Anti-trypsin	200—400	45,000	29	0.15	0.04—0.63	
Orosomucoid	50—140	44,000	5	0.37	0.18—0.67	
			29	0.45	0.21—2.32	
GC globulin	30—55	51,000	5	0.05	0.03—0.07	
α_2 HS glycoprotein	40—85	49,000	6	1.1	0.60—1.35	
			29	0.23	0.07—1.12	
β_2-Glycoprotein I	20—25		5	0.33	0.21—0.44	
3S γ_1-globulin		25,000	6	0.23	0.14—0.35	
β_2-Microglobulin	0.12—0.18[64]	11,800	20	0.12	0.06—0.21	

high as 500 mg/100 ml. Most random normal specimens contain under 10 mg/100 ml, although in the more concentrated overnight specimens the level may be as high as 20 mg/100 ml. Proteinuria is said to be present whenever the urinary protein output is greater than that reflected in these normal values. Not all proteinuria is clinically significant, but persistent abnormal levels of protein in the urine are an indicator of the presence of kidney or urinary tract disease. Thus, the detection of urinary protein and the quantitative assessment of the degree of proteinuria are very important laboratory procedures.

CLINICAL SIGNIFICANCE OF PROTEINURIA

Mechanisms of proteinuria

The type and quantity of proteins excreted in disease depend on four factors, of which the first is by far the most important:

Glomerular Leakage of Plasma Proteins. The amount of protein excreted, which may be very large in the nephrotic syndrome, is of less clinical significance than the degree of selectivity,[29] that is, the extent to which the glomerulus can still distinguish between larger and smaller molecules. Because albumin, present in such high concentration in serum, is the principal protein excreted, the condition is sometimes called albuminuria.

SELECTIVE LEAKAGE. In this condition, the molecular cross linkages in the basement membrane are intact but wider than normal. This allows larger molecules than usual, especially albumin, to pass through, but there is still selection according to size, so that macroglobulins are excluded.

NON-SELECTIVE LEAKAGE. The damaged glomerulus allows proteins of any size to pass through, indicating greater damage to the glomerulus and usually a poorer prognosis. The estimation of selectivity is of value in deciding which patients with the nephrotic syndrome, especially children, should be treated with steroids. Only those with selective proteinuria are likely to respond.

Overflow of Filtered (Low Molecular Mass) Proteins. This occurs when low molecular mass proteins are present in the serum and the glomerular filtrate at such high concentrations that they cannot be completely reabsorbed by the tubules (e.g., Bence Jones protein in myelomatosis, lysozyme in monocytic leukemia).

Failure of Tubular Reabsorption of Filtered Proteins. If the glomeruli are undamaged but the tubules are damaged, the small molecular weight proteins normally filtered will not be reabsorbed by the tubules but will appear in the urine and give a characteristic electrophoretic pattern (tubular proteinuria) completely unlike that seen in glomerular proteinuria. The albumin band is relatively faint, the α_2- and β_2-globulin bands are prominent, and there may be a post-γ band. Such isolated tubular damage is uncommon; it occurs in a variety of clinical conditions affecting the kidney tubules, such as *cadmium poisoning*, *chronic hypokalemia*, and *congenital tubular diseases* like Fanconi's syndrome.

Far more commonly, tubular damage accompanies glomerular damage, as in *chronic pyelonephritis*. It cannot then be diagnosed from the electrophoretic pattern, since the changes caused by glomerular leakage obscure those due to tubular damage. It may be assessed by estimating immunochemically the clearance of a convenient low molecular mass protein such as β_2-microglobulin and comparing its clearance with that of albumin.[64]

In *severe renal failure*, as more glomeruli are destroyed, the glomerular filtration rate falls and the degree of albuminuria falls with it. Tubular proteinuria, however, increases, both because the serum levels of the small molecular mass proteins rise above their usual renal thresholds and because there are fewer tubules left to reabsorb them.

Renal Proteins Derived from Kidney Tissue. Kidney breakdown products found in urine include such materials as glomerular basement membrane fragments in nephritis, and fibrin degradation products. Renal secretions include Tamm-Horsfall mucoprotein, which can be estimated immunochemically after depolymerization,[30] and probably secretory IgA, the form in which most urinary IgA occurs.

Degree of proteinuria

Proteinuria, mainly glomerular, is often a manifestation of primary renal disease, although transient proteinuria may occur with fevers, thyroid disorders, and in heart disease, in the absence of renal disease. Proteinuria may be evident very early in the course of various renal disease states. With such conditions as pyelonephritis, reflecting bacterial infection in the kidney, and acute glomerulonephritis, often associated with recent streptococcal infections, the degree of proteinuria is slight, usually amounting to less than 2 g per day. In chronic glomerulonephritis and in the nephrotic syndrome, including lipoid nephrosis, and in some forms of hypertensive vascular disease (*nephrosclerosis*), protein loss may vary from a few grams to as much as 30 g/d. Proteinuria is encountered in certain other disease entities when and if they give rise to kidney lesions. Among these are lupus erythematosus, amyloidosis, toxemia of pregnancy, septicemia, and certain forms of drug and chemical poisoning.

In healthy individuals, transitory elevations in urine protein output are encountered after intense exercise or work, and after exposure to cold. *Orthostatic proteinuria* is a benign condition in which protein excretion is normal when the patient is lying down (prone), but is elevated when the patient walks or stands erect for any period of time. Persons subject to orthostatic proteinuria are often embarrassed during medical examinations for insurance because their urines are found to be positive when tested for protein after they have been on their feet and active for a good part of the day. On rechecking the urine in the morning after a night's rest, the urine is usually found to be negative for protein by the usual routine tests.

Tests for Proteins in Urine

QUALITATIVE TESTS

In devising qualitative tests for urine proteins, it is desirable that the test be negative whenever protein is present in normal concentrations, but that the test be positive whenever the protein concentration is greater than 20 to 25 mg/100 ml. The test should not be sensitive to mucins and other proteins of the urinary tract and it should be capable of rough quantitation. A large number of such tests are available. The older tests are based on precipitating the proteins either by heat or with anionic protein precipitants. A more recent test depends on a change in the color of an indicator in the presence of protein.

Qualitative tests are best done on fresh morning specimens of urine. Such specimens are usually fairly concentrated, thus making possible the detection of proteinuria early in disease. They will also be free of the orthostatic effect discussed previously. If urines must be held for examination at a more convenient time, and if timed specimens are to be examined, they should be refrigerated and examined within 48 h. Layering with toluene may be used to minimize microbial growth. Bacterially contaminated specimens (pyelonephritis, lower genitourinary tract infections) should be examined only when fresh.

Heat Coagulation Test

The pH of the urine is checked with bromcresol green pH indicator paper (pH range 3.8 to 5.5), and if it is above pH 5.0, 5 molar acetic acid is added dropwise until the pH is between 4.0 and 4.6. If any turbidity or insoluble matter is present or is formed (mucoproteins may precipitate on adding the acid), it is removed by centrifugation. About 7 to 10 ml of clear urine are placed in a 15 × 120 mm test tube and the upper one-third is gently heated over a Bunsen flame until boiling just begins. The heated part is compared in ordinary light with the lower unheated portion of the specimen. Any turbidity, cloudiness, or precipitate demonstrates the presence of protein. The degree of the reaction is customarily graded 0 to 4+. False positive results may occur and are discussed in the section on interpretation of results. The grading of the turbidity observed and its relation to the protein concentration is given in Table 7–14.

REFERENCE

Kark, R. M. *et al.*: A Primer of Urinalysis. 2nd ed. New York, Hoeber Medical Div., Harper and Row, Publishers, 1963.

TABLE 7–14 INTERPRETATION OF THE HEAT COAGULATION TEST FOR URINE PROTEIN*

Appearance	*Reading*	*Approximate Protein Concentration* (mg/100 ml)
No turbidity	negative	0–4
Slight, distinct turbidity	± (trace)	4–10
Definite turbidity; light print readable	1+	15–30
Light cloud; heavy print readable	2+	40–100
Moderate cloud with slight precipitate	3+	200–500
Heavy cloud with flocculent precipitation and coagulation	4+	over 800–1000

* After Kark, R. M. *et al.*: A Primer of Urinalysis. 2nd ed. New York, Hoeber Medical Division, Harper and Row, Publishers, 1963.

Sulfosalicylic Acid Test

Place 5 ml of clarified urine (pH 4.5 to 6.5) into a 15 × 120 mm test tube. Then carefully add 0.50 to 0.80 ml of 0.8 molar sulfosalicylic acid (20 g/100 ml) down the sides of the tube to layer underneath the urine. After 1 min examine the interface for the presence or absence of turbidity or cloudiness. A barely evident turbidity (5 to 10 mg/100 ml) is graded as ±, and increasing degrees of turbidity, cloudiness, or precipitation are graded 1+ to 4+. False positives, due to nonspecific precipitation, may be encountered and are discussed in the section on interpretation of results.

The reagent is prepared by dissolving 100 g of reagent grade sulfosalicylic acid (2-hydroxybenzoic, 5-sulfonic acid dihydrate) in about 350 ml of water with the aid of heat. The warm solution is filtered through two layers of filter paper and the total filtrate diluted to 500 ml when cool. The solution should be crystal clear and colorless. It is discarded when turbidity or darkening develops.

Albustix Test

Albustix[72] is the name of a commercial product from the Ames Company (Div. Miles Laboratories, Elkhart, IN.). It consists of strips of special adsorbent paper, the ends of which have been treated with citrate buffer of pH 3.0, containing bromphenol blue indicator (pH 3.0, yellow; pH 4.2, blue). The test is based on the so-called "protein error" of indicators. At the pH of the indicator strip, most of the indicator is in the yellow, nonionized acid form, HI, with only a small fraction in the blue ionized form, I^-. If protein is present, it will bind with the anion form, I^-, since the anion form has a greater affinity for the protein than for the hydrogen ion. The removal of I^- will cause more of the indicator acid to ionize and the concentration of HI will decrease, resulting in a change in the ratio of I^-/HI. The color presented by the paper strip will depend on the relative proportions (ratio) of the two forms. The quantity of indicator is fixed and, if sufficient protein is present, the larger fraction of the indicator will be in colored form.

If the strip is dipped momentarily into a urine containing no or only traces of protein, it will show a yellow color. With increasing levels of protein, the strip will present colors ranging from yellow-green, to green, to increasingly deeper shades of blue. The strips come in a vial, to which is attached a chart relating color to protein concentration. A yellow-green color is graded as a trace and reflects 20 to 30 mg of protein/100 ml of urine. Various shades of blue are graded 1+ to 4+ and reflect levels of 30 to 500 mg/100 ml. This test is a rapid and convenient laboratory aid, although it is not as sensitive as the heat and sulfosalicylic acid tests.

Interpretation of Results of Qualitative Tests

All three techniques are positive in the presence of albumin, globulins, and Bence Jones proteins.[39] False positive results are obtained when certain organic x-ray contrast media are present in solution in the urine specimens. They interfere with the SSA test by producing a turbidity. Similar false positives may be obtained if patients are receiving tolbutamide, certain sulfa drugs (Gantricin), and medications such as benzoin. These materials may show up as an initial turbidity on acidifying the urine in the heat test, but such turbidity is usually dissolved on subsequent boiling. The Albustix test is not affected by the presence of such materials; however, the Albustix strips may give false positive reactions if urines are alkaline and well buffered. Dialysis of such urine specimens for several hours will often decrease interfering materials to negligible concentrations. A carefully performed heat test is the most sensitive of the three tests discussed, while the other two are more convenient for the rapid testing of large numbers of urine specimens.

QUANTITATIVE TESTS FOR URINARY PROTEINS

Quantitative tests for urine proteins are generally carried out on 24 h urine specimens, although 12 and 8 h specimens are occasionally more informative. Protein clearances are sometimes measured on the basis of specimens collected over a short time (e.g., 15 or 30 min).

Turbidimetric methods[57,89] using trichloracetic acid, sulfosalicylic acid, or naphthalene sulfonic acid as precipitants are most frequently used. Such a method is described in detail. If the protein concentration is too low for such methods, a dye elution method may be used similar to that used after electrophoresis (p. 314).[66,85]

For most accurate results the protein is precipitated (for example, with trichloracetic acid), redissolved in weak alkali, and estimated by the Lowry[50,76] or biuret methods. The latter is described on p. 363.[20]

URINE PROTEIN DETERMINATION BY TURBIDITY WITH PROTEIN PRECIPITANTS

REAGENTS

1. Alternative precipitating reagents:

(a) Trichloracetic acid (TCA), 0.19 mol CCl_3COOH/liter of solution (3.0 g/100 ml of water). Keep at room temperature but prepare fresh once a month.

(b) Sulfosalicylic acid–sodium sulfate reagent. Dissolve 3.0 g of sulfosalicylic acid (SSA) in a solution containing 7.0 g Na_2SO_4/100 ml water (see p. 371).

(c) BNSA reagent, 0.04 mol β-naphthalenesulfonic acid in 0.43 molar acetic acid solution. Dissolve 4.5 g of the sulfonic acid (available from Eastman Kodak Co.) in a solution containing 12 ml of glacial acetic acid diluted to 500 ml. The solution should be filtered.

2. Saline, NaCl, 0.15 mol/l (0.85 g/100 ml).

3. Standard solutions. Dilute a composite of normal sera, clear, nonicteric, and free of hemolysis, 3:100 with saline in a volumetric flask. Then prepare working standards by diluting 1.0, 2.0, 4.0, 6.0, 8.0, and 10.0 ml of the diluted serum to 10.0 ml with saline. These are then treated exactly as are the unknowns. The protein content of the original serum is determined by the biuret method in quadruplicate and the average value is calculated. If this value is P g/100 ml, the working standards will have concentrations of $3P$, $6P$, and so on, up to $30P$ mg/100 ml or $0.03P$, $0.06P$ and so on, up to $0.30P$ g/l. If a serum containing 6.85 g/100 ml is used, the standards will represent concentrations of 0.21, 0.41, . . . , 2.06 g protein/l.

4. Daily control. Urine protein controls are available from several commercial sources. Alternatively, a laboratory can prepare its own urine protein control as follows: Dilute 3.0 ml of a normal serum to 100 ml with benzoic acid (0.05 g/l water) Store in the refrigerator. Daily, warm an aliquot to room temperature and dilute 2.0 ml to 5.0 ml with saline. Assay this dilution along with the unknowns. A value within ±5 per cent of the concentration established for that preparation should be obtained.

PROCEDURE

1. Preparation of urine: Gently, but thoroughly, mix the entire urine specimen; record volume. Filter an aliquot of 70 to 100 ml through a fluted filter paper (e.g., Whatman No. 1) into a dry flask, discarding the first 10 to 15 ml. If still turbid, refilter through Whatman No. 44. If available, bacteriological membrane filters may be used to clarify the urine. If specimens have been refrigerated, they should be warmed to room temperature (23 to 28°C) before analysis.

2. Check the urine by qualitative tests (Albustix is convenient) to estimate roughly the amount of protein present. If over 150 to 200 mg/100 ml, dilute the urine with saline to bring the concentration into the range of 50 to 100 mg/100 ml.

3. Carry out the turbidity reaction in appropriate (e.g., Coleman 12 × 75 mm) cuvets. One pair is needed for each unknown urine, and an additional pair is used to correct for pigment (if any) in the reagent. Arrange the cuvets in parallel rows in a rack.

4. Pipet 0.50 ml of urine or diluted urine into each pair of cuvets. Add 2.50 ml of saline to the control cuvets (C) in the front row. (They serve to correct for pigment in the urine specimen.) Then, at 1/2 minute intervals, add 2.50 ml of one of the precipitating reagents to the back row cuvets (T), one at a time. Cover with Parafilm, and mix immediately. After 7 min, at 1/2 min intervals, mix each tube again gently to avoid air entrapment, and read its apparent absorbance immediately in a spectrophotometer at 620 nm against its (C) tube mate as blank.

5. Correct for any color present in the precipitating reagent by reading the absorbance of 2.50 ml of reagent plus 0.50 ml saline against a blank of saline. The BNSA reagent may develop a color which may require a correction. The reagent is discarded when the absorbance exceeds 0.040 units.

CALCULATIONS

1. Subtract the reagent absorbance reading (if any) from the absorbance of the unknowns.

2. Convert the corrected absorbance reading to concentration by the use of a calibration graph, or less accurately, by the use of a calibration factor for the procedure. If the urine was diluted, multiply this value by the appropriate dilution factor. This gives the protein concentration in grams per liter. The 12 or 24 h outputs of protein are obtained by multiplying the concentration value by the volume expressed in liters.

$$Q = (A_U - A_R) \times F \times D \times V \qquad (16)$$

where Q = gram protein output during period of collection
A_U, A_R = absorbances of unknown and of reagent, respectively
F = calibration factor
D = dilution factor
V = volume of urine in liters collected during period.

3. The results are reported in grams, if over 1 gram per volume, and in milligrams if under that value.

CALIBRATION

1. Calibration graph: Plot the corrected absorbances of the standards against their protein value in g/l. The graph is slightly S-shaped, but the curve is essentially linear in the range of 0.20 to 2.0 g/l.

2. The calibration factor F can be obtained by calculating the slope of the line from the measured absorbances of the standards.

$$F = \frac{\text{sum of protein concentrations of all standards}}{\text{sum of absorbances of all standards}} \qquad (17)$$

The factor should not be used with high absorbance values, where deviation from linearity is apparent.

COMMENTS

1. The reproducibility of the method is about ±5 per cent except with very low or very high absorbance readings, in which case it falls to ±10 per cent. Overall accuracy is of the order of 10 to 15 per cent. To exclude collection errors, the protein output should be determined in duplicate over two similar collection periods.

2. Although different precipitating agents give slightly different degrees of turbidity

with individual classes of proteins, results obtained are very similar. Simplicity of preparation favors the use of TCA, and stability favors the use of SSA-sulfate reagent. All reagents react with Bence Jones proteins. Large polypeptides are probably less sensitive to the TCA than to the sulfonic acid reagents. No two reagents can possibly precipitate all the protein types present in urine to the same degree.

3. *In turbidity measurements*, true light absorbance is not measured. The particles in suspension, by diffracting and refracting light incident on them, decrease the light energy that reaches the detector. Although some light energy is also absorbed, it is primarily the loss of light energy that has been dispersed that is being measured. Any wavelength may be used, but the dispersion will increase as the wavelength is decreased. For uniformity of dispersion, the size and shape of the particles must be uniform. The latter is affected by temperature and by the rate and intensity of the mixing of the reagent and the specimen. Narrow cuvets are preferred because over a short light path a lesser degree of variability in particle geometry is encountered. Temperature control is important; standardizations and determinations should be done at the same temperature, 25 ± 2°C. Particle sizes change with time and, therefore, time control is also important. It takes 5 min after reagent addition to obtain a reasonably uniform particle size at 23 to 28°C, and most investigators report that size dispersion is stable for up to about 10 min. With longer periods of time of reaction, agglomerates of particles begin to develop. At low levels of protein, turbidity develops more slowly; at high levels particles tend to sediment. Turbidity readings are most reproducible in the range of 25 to 90 mg/100 ml.

4. *Significant figures* in reporting results: Absorbances can be read to three significant figures, but the last figure is subject to considerable doubt. If $A = 0.165$, we are certain of the 1 and 6, but in replicate readings the 5 may be replaced by a 3, 6, 2 or even 9 ($A = 0.165 \pm 0.004$). Similarly, measurements of urine volumes—for example, in 1 or 2 liter cylinders graduated at 10 or 20 ml intervals respectively—are accurate only to two or three significant figures. A reading of 1124 ml could just as well be 1122 or 1130 ml or worse, if the meniscus is read at the wrong angle. This degree of accuracy assumes that the smallest cylinder able to hold the volume of urine is used, that no urine is lost during collection and in transferring of the urine contents from bottle to cylinder, and that, during the measurement of volume, sufficient time is allowed to permit complete drainage of bottle contents to the cylinder. The uncertainty in measuring 1124 ml is at least 10 ml.

The number of significant figures in the product of two numbers can never be more than the least number in either of the two factors. The following is a sample calculation:

V = urine volume = 1125 ± 10 ml; 3 significant figures; the per cent uncertainty is $\frac{10}{1125} \times 100 = \pm 0.9$ per cent.

A = absorbance = 0.124 ± 0.004; 2 significant figures; the uncertainty in the value = ±3.2 per cent.

D = dilution factor = 5.0; 2 significant figures; uncertainty = 1/50 or ±2 per cent.

F = calibration factor = 6.5; 2 significant figures; uncertainty = ±5 per cent, the reproducibility of the procedure.

Q = quantity of protein in grams

$$Q = (1.12\mathit{5}) \times (0.12\mathit{4}) \times (5.\mathit{0}) \times (6.\mathit{5}) = 4.\mathit{53375} \text{ g/volume.}$$

In the preceding formula, the uncertain digits are printed in italics. At least two factors have uncertainties in the second figure; hence, the product of all factors can be given only to two significant digits.

$$Q = 4.\mathit{5} \text{ g/volume*}$$

* The per cent uncertainty of a product of several factors is given as the square root of the sum of the squares of the per cent uncertainties of each factor. In the sample calculation:

$$\text{per cent uncertainty in } Q = \sqrt{(0.9)^2 + (3.2)^2 + (2)^2 + (5)^2} = 6.3 \text{ per cent;}$$

6.3 per cent of 4.53 = 0.29. Thus $Q = 4.53 \pm 0.29$, or $Q = 4.5 \pm 0.3$ g/volume.

The procedure has a reproducibility of ±5 per cent which is the minimum uncertainty expected in the final answer, and equals ±0.23 g. The total output then can be expressed as 4.5 ± 0.2 g/volume.

Urine protein outputs should never be reported to more than three significant digits; in most cases only two will be warranted. Use of more will imply pretensions to a level of accuracy not possible with the methods now available. Repeated collection of the 24 h specimen or determination of creatinine in urine may detect gross collection errors. Obviously, the preceding comments apply to the reporting of any laboratory results.

REFERENCES

Waldman, R. K. *et al.*: J. Lab. Clin. Med., *42*: 489, 1952.
Meulemans, O.: Clin. Chim. Acta, *5*: 757, 1960.

DETERMINATION OF URINE PROTEIN WITH THE BIURET REACTION

PRINCIPLE

The protein in urine is precipitated with trichloracetic acid, redissolved in alkali, and measured colorimetrically, using the biuret reaction.

REAGENTS

1. Trichloracetic acid, 20 g/100 ml.
2. Sodium hydroxide, 0.20 mol/l (8.0 g NaOH/l).
3. Biuret reagent (alkaline copper reagent). Dissolve 9.0 g Rochelle salt, $NaKC_4H_4O_6 \cdot 4\,H_2O$, in about 400 ml of 0.2 molar NaOH. Add 3.0 g of $CuSO_4 \cdot 5H_2O$; when that is dissolved, add 5.0 g KI. Dilute to 1.0 liter with 0.2 molar NaOH solution. The solution is stored in a polyethylene bottle.
4. Stock standard protein (bovine albumin) solution. It is convenient to use the approximately 30 per cent (30 g/100 ml) BSA solution available from the Armour Laboratories, Div., Armour and Co., Chicago, IL.
5. Dilute and working standards. The stock BSA is diluted 5-fold with 0.15 molar NaCl (0.85 g/100 ml) containing 50 mg/100 ml sodium azide as preservative, to make a dilute standard of approximately 6.0 g BSA/100 ml. The actual concentration is determined by the biuret procedure for serum proteins (see p. 302). The dilute standard is further diluted 25-fold to make the working standard. (If the dilute standard has been found to contain 6.08 g protein/100 ml, the working standard will contain 2.43 mg/ml.)

PROCEDURE

1. Measure and record the volume of the timed (24 h) urine specimen. Centrifuge to remove particulate matter.
2. Perform a careful qualitative test for urine on the specimen. If the result is ±, 1+, 2+, 3+, or 4+, use 8.0, 8.0, 2.0, 1.0, and 0.50 ml, respectively, of the centrifuged specimen for the assay (step 3). If the test is negative for protein, a report of "negative" is issued.
3. Pipet the indicated volume of urine into a 12 ml centrifuge tube. Perform the test in duplicate.
4. Pipet 1.0, 2.0, 3.0, and 4.0 ml of working standard into additional tubes.
5. Add water to all tubes (except those with 8 ml) to bring volume to 4.0 ml.
6. Add 1.0 ml of TCA (20 g/100 ml) to each tube containing 4.0 ml and mix well. Add 2.0 ml of the TCA to the tubes with 8.0 ml of specimen.
7. Allow the protein to precipitate for 30 min, and then centrifuge for 10 min (see note 1). Pour off the supernatant, invert the tubes over a piece of filter paper, and permit them to drain completely (5 min).
8. Add 5.0 ml of the 0.20 molar NaOH to each tube. Promote solution by immersing

the tubes in a boiling water bath for 1 min, and by agitation with a fine-pointed stirring rod. Cool to *room temperature*.

9. Add 5.0 ml of biuret reagent to all tubes, mix well with the stirring rods, and transfer to cuvets.

10. Prepare a reagent blank containing 5.0 ml of NaOH and 5.0 ml biuret reagent.

11. Place all cuvets into a bath at 37°C, and let color develop for 30 min.

12. Read absorbances of solutions in all cuvets at 550 nm. Construct a calibration curve by plotting absorbance vs mg protein/ml solution, and calculate the protein content of each urine sample (in mg). Multiply these values by the appropriate dilution factor and report protein concentration in g/liter, or preferably g/d. Report results to two significant figures only.

Quantity of urine protein excreted =

$$\begin{array}{l}\text{mg protein in}\\ \text{cuvet as read}\\ \text{from graph}\end{array} \times \frac{\begin{array}{c}\text{total volume}\\ \text{of urine spec.}\\ \text{in ml/d}\end{array}}{\begin{array}{c}\text{ml urine in}\\ \text{aliquot for}\\ \text{assay}\end{array}} \times \frac{1\ \text{g}}{1000\ \text{mg}} = \text{g protein/d}$$

Concentration of urine protein =

$$\begin{array}{l}\text{mg protein}\\ \text{in cuvet}\end{array} \times \frac{1000\ \text{ml}}{\begin{array}{c}\text{ml urine in}\\ \text{assay aliquot}\end{array}} \times \frac{1\ \text{g}}{1000\ \text{mg}} = \text{g protein/liter}$$

NOTES AND COMMENTS

1. Use a centrifuge with a radial head rather than an angle head so that the tube will be spun in a horizontal position, permitting packing of the protein precipitate in the bottom of the tube.

2. If the NaOH solution of the protein (step 8) is significantly colored, carry a second aliquot of urine specimen through to this point, and then add 5.0 ml (additional) NaOH in place of the 5.0 ml of biuret reagent. This "color blank" is read against 0.2 molar NaOH, and the absorbance is subtracted from the value obtained with the biuret reagent. The corrected absorbance is then used to calculate the protein content in the cuvet.

REFERENCE

E. Freier, unpublished.

ELECTROPHORETIC SEPARATION OF PROTEINS IN URINE USING CELLULOSE ACETATE

PROCEDURE

Centrifuge urine (preferably a complete early morning specimen), and decant supernatant. Estimate total urine protein by turbidimetry. Concentrate the urine, if necessary, to 2 or 3 g protein/100 ml, using negative pressure ultrafiltration and Visking dialysis tubing as in Figure 7–20.[16] Sufficient urine should be used so that enough specimen of the desired concentration will be available. If necessary the urine is further concentrated by an Amicon concentrator (p. 374) which, like the narrow 8/32 inch Visking tubing, concentrates proteins with molecular masses down to 40,000 daltons;[49a] this includes the usual dimeric form of immunoglobulin light chains (Bence Jones proteins). The 23/32 inch tubing is required for proteins with molecular masses down to 15,000 to 10,000 daltons.

Other techniques, commercially available, using negative pressure ultrafiltration include the hollow fiber technique (Bio-Rad 50, Bio-Rad Laboratories) and collodion bags

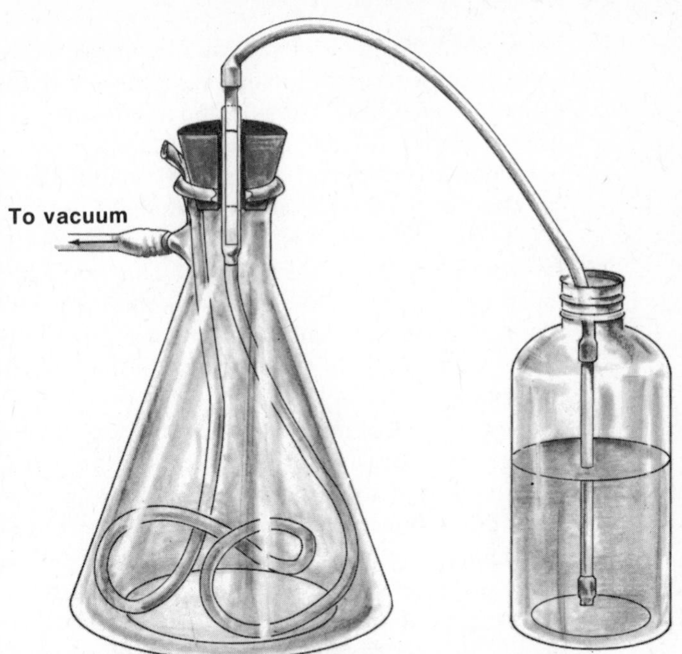

Figure 7-20 Apparatus for concentration of urine protein by negative pressure ultrafiltration. The bottle on the right contains the urine sample to be concentrated (with some preservative such as azide added). One end of the dialysis tubing is closed by being compressed between the neck of the left flask and its stopper. A glass tube is inserted into the other end, and then both of them are pushed together through the rubber stopper to make a tight joint. When the flask on the left is evacuated, urine is sucked into the dialysis tubing. Filtrate collects at the bottom of the flask and concentrate collects within the tubing.

The volume of urine to be processed depends on the degree of concentration required. The length of Visking dialysis tubing is adjusted accordingly and the vacuum flask may, if necessary, be replaced by a container of any other size from a test tube to a large aspirator bottle. (Modified from Everall, P. H. and Wright, G. H.: J. Med. Lab. Tech., *15*:209, 1958.)

(Schleicher and Schuell). Electrophoresis is then carried out exactly as outlined in the procedure for serum. The patient's serum should be analyzed simultaneously.

CLINICAL SIGNIFICANCE

The pattern in disease depends on three factors: the extent of glomerular leakage, the degree of impairment of tubular reabsorption, and the serum level of low molecular mass proteins (see p. 356).

Albuminuria is characteristic of glomerular leakage. Higher molecular mass proteins are also filtered in accordance with the selectivity properties of the glomeruli (see below). These are best judged by comparing the urine and serum patterns; the more similar they are, the less the selectivity.

In tubular failure (tubular proteinuria) the albumin band is faint, but excretion of the low molecular mass proteins—α_2-microglobulin, β_2-microglobulin, and γ-trace (post-γ protein)—produces α_2 and β_2 bands and sometimes a post-γ-globulin band.

Proteinuria may also be the result of high serum levels of low molecular mass proteins (Bence Jones protein) (see p. 366).

In terminal uremia the urine contains small serum protein molecules as well as large ones, because there are not enough functioning tubules left to absorb all the small protein molecules and the damaged glomeruli allow large ones to leak through (see p. 357).

ESTIMATION OF GLOMERULAR SELECTIVITY IN THE NEPHROTIC SYNDROME[29]

In this syndrome the "selectivity" of the kidney in distinguishing between molecules of different sizes is a better indication of the patient's prognosis and of the likely response to steroid therapy than the total quantity of protein excreted. It may avoid the necessity for a renal biopsy. Patients with high selectivity, especially children, usually respond to steroids, whereas those with poor selectivity have irreversible chronic renal disease.

Selectivity is assessed by comparing the clearances of proteins of different molecular sizes. The measurement of UV/P involves the ratio of urine to plasma concentration (see Chap. 17). Standard protein solutions are therefore not required: the concentrations of a protein in urine and in serum can be compared directly on the same radial immunodiffusion plate. Ideally, several proteins providing a wide range of molecular sizes, such as orosomucoid (molecular mass 40,000), albumin (67,000), transferrin (85,000), IgG (150,000), and α_2-macroglobulin (840,000), are measured; their clearances (usually relative to another protein, e.g., transferrin) are plotted against their molecular masses on log-log paper, and the slope of the line through the results is used as a measure of selectivity.

In routine work it is sufficient to measure the ratio of the UV/P value of albumin or transferrin relative to that of IgG.[6] This is much simpler and does not require the use of a timed specimen of urine, since V cancels out, an important practical advantage with children. The ratio of the concentrations of each protein in plasma and urine is estimated from the ratio of the squares of the diameters of their precipitin rings in the Mancini radial diffusion method.

An IgG/albumin clearance ratio less than 0.16 indicates high selectivity, minimal change nephritis, and probably satisfactory response to steroids. Ratios of 0.16 to 0.30 (medium selectivity) occur in membranous glomerulonephritis and pre-eclamptic toxemia, while ratios above 0.30 (low selectivity) occur in proliferative glomerulonephritis. Errors in determining the IgG clearance may occasionally arise as a result of excess of light chains in the urine or because of polymerization of IgG in the serum.

An indication of the degree of selectivity can also be obtained by comparing the patterns obtained by electrophoresis or thin layer gel filtration when serum and concentrated urine are run in parallel (pp. 312 and 332).

BENCE JONES PROTEINURIA[35,45]

This is the only common form of overflow proteinuria. Normally only very small quantities of κ and λ immunoglobulin light chains are excreted in the urine. They pass straight through the glomerular filters but are reabsorbed by the kidney tubules and broken down in their lining cells. In Bence Jones proteinuria, the concentration of light chains (Bence Jones proteins) exceeds the kidney tubule threshold, and proteinuria results just as glucosuria occurs in diabetes. This happens when the single clone of plasma cells of an immunocytoma is producing large quantities of identical free light chains. The situation is exactly similar to the excess production of a single complete immunoglobulin molecule, which is demonstrable as an M-band on electrophoresis of serum. However, in this case the smaller protein molecules pass through the glomeruli and cause an overflow proteinuria; on electrophoresis, the urine then shows a sharp extra band which immunochemically is either entirely κ or entirely λ in type.

In myelomatosis, Bence Jones proteinuria may be associated with the monoclonal immunoglobulin band in the serum, but it may also occur alone. For this reason, electrophoresis of both concentrated urine and serum must always be carried out in parallel in suspected cases.

Most urine proteins, when heated, do not coagulate and precipitate out of solution until the temperature reaches 56° to 70°C. Bence Jones proteins coagulate at a much lower temperature, 40 to 60°C; furthermore, unlike other proteins, the precipitated Bence Jones proteins redissolve as the temperature of the solution is increased to between 85 and 100°C and often reprecipitate as the temperature is again decreased to 45 to 50°C. Since they are precipitated by sulfosalicylic and trichloracetic acids, they are included among proteins measured by the common methods used to quantitate urinary proteins.

The molecular mass of immunoglobulin light chains is about 25,000 daltons (see p. 276). Bence Jones proteins occur as monomers with this molecular mass or as dimers with a molecular mass of about 50,000.

Tests for Bence Jones proteinuria are essential for the investigation of all suspected cases of myelomatosis. Until recently the peculiar behavior of Bence Jones proteins on heating was used as a test, but the advent of electrophoresis and immunoelectrophoresis has shown that this fails to detect about a third of the cases.[35] The concentrated hydrochloric acid ring test (Bradshaw's test) is the best of the simple screening tests, but even this gives about 5 per cent false negatives. More reliable screening requires the electrophoresis of concentrated urine and the demonstration of a narrow (monoclonal) band. This should be done whenever there is other evidence of myelomatosis as well as when Bradshaw's test is positive. Finally, the narrow electrophoretic band has to be shown immunochemically to be due to a single light chain class.

Bradshaw's test[87]

The urine specimen is treated with 1 or 2 drops of dilute acetic acid (23 ml/100 ml) to bring the pH down to 4.8 to 5.0, centrifuged clear, and diluted 2-fold with water. The diluted urine is layered carefully over concentrated HCl in a test tube. If Bence Jones protein is present, it will be precipitated by the HCl and will form a fine or heavy ring at the interface of urine and HCl. A positive test should be confirmed by electrophoresis because occasionally albumin may precipitate, if present in large quantity.

If Bradshaw's test is positive and the protein concentration is greater than 100 mg/100 ml, proceed as follows:

(a) Perform electrophoresis on unconcentrated urine. Look for a Bence Jones protein band, usually in the β or γ regions.

(b) Test with antisera specific for κ or λ chains respectively by double diffusion on two Ouchterlony plates (see p. 327 and the note below on antisera) as follows: Prepare two agar plates. Place 2 drops of anti-κ in the center well of one and 2 drops of anti-λ antiserum in the center well of the other. Place into the peripheral cups of both plates: (i) 1 drop patient's urine undiluted, (ii) 1 drop of patient's urine diluted 15-fold, (iii) 1 drop of patient's urine diluted 25-fold, (iv) positive κ control, (v) positive λ control, (vi) negative control. Read plate after two hours.

If Bradshaw's test is negative or total protein concentration is less than 100 mg/100 ml, proceed as follows:

(a) Perform electrophoresis of urine concentrated to 3 g/100 ml as previously described.

(b) If an M-band is present, test for the presence of κ and λ chains by immunoelectrophoresis (see p. 329). Note that normal polyclonal light chains may also be present at this concentration.

Note: Anti-κ and anti-λ sera are made from pools of κ and λ Bence Jones proteins, respectively. These pools should be as large as possible so that the antisera react with the widest possible range of Bence Jones proteins of the same class.

These antisera will react with light chains whether they are free (Bence Jones proteins) or parts of a complete immunoglobulin molecule and are, therefore, suitable for the investigation of serum myeloma proteins (p. 329).

Antisera which react only with *free* Bence Jones proteins can be prepared by adsorbing the above antisera with intact IgG molecules, so that they react only with those determinants of the light chains which are hidden within the intact immunoglobulin molecule. They are useful for identifying Bence Jones proteins in the presence of intact immunoglobulin molecules (p. 330).

URINARY MUCOPROTEIN

Uromucoid (mucoprotein of Tamm and Horsfall) is a normal excretion product of the cells lining the kidney tubules and is excreted at a rate of about 70 mg/d. It is a filamentous mucoprotein which contains 28 per cent carbohydrate and occurs in urine in two forms, a polymeric form with a molecular mass of about seven million daltons and a monomeric one with a molecular mass of about 80,000 daltons. The former precipitates with salt, while the latter remains in solution. Uromucoid may be estimated by radial immunodiffusion or electro-immunodiffusion if first changed completely into the monomeric form with sodium dodecyl sulfate.[30] The antiserum for the test is easily prepared by injecting into rabbits uromucoid prepared from normal human serum by salt precipitation.

Urinary casts are formed of uromucoid.[55] The presence of "hyaline casts" as such is not pathological; but when, as a result of tubular disease, cells or cellular debris are admixed with them, such "cellular" or "granular" casts in the urine sediment provide striking evidence that the cellular exudation is from the tubules.

The urinary excretion of uromucoid appears to be unchanged in patients with renal calculi unless there is tubular damage.

Qualitative test for increased urinary mucins

Filter the urine specimen to be tested to remove particulate matter, and dilute with an equal volume of water. Into each of three tubes put 2 or 3 ml of diluted urine. Put 5 to 6 drops of saline into the first tube, which serves as a control. Dropwise add acetic acid (33 ml/100 ml) into the second tube, until a precipitate or opalescence is observed, and then add more acid dropwise as long as additional precipitation occurs. Put this amount of acid plus 5 to 7 additional drops into the third tube. Compare the quantity of precipitate in the third tube with that in the second. If the quantity of precipitate formed in the third tube appears not to have decreased with the addition of excess acid, excess mucin (mucus) is present. Urine containing normal amounts of mucin will give no precipitate with the acid.

Proteins in Cerebrospinal Fluid[77]

The cerebrospinal fluid (spinal fluid, CSF, Csf*) is a clear, colorless fluid that fills the nontissue spaces in and around the brain and spinal cord. The fluid serves to maintain constancy of intracranial pressure and to provide a mechanical, water-jacket-like protective coating for the delicate nerve tissue. It is formed as a secretion by cells in the cerebral ventricles. The total volume is about 150 ml. The fluid is a slightly modified ultrafiltrate of plasma. Its composition varies slightly, depending on where it is sampled. There is a gradual change in concentration of its components as the fluid flows down from the brain ventricles to the cisternal space and then to the lumbar area of the spinal cord, where the usual laboratory spinal fluid specimen is obtained by means of a spinal needle puncture.

Chemical changes of diagnostic importance occur in CSF in many diseases of the central nervous system.

Usually the physician will be interested only in the total protein, γ-globulin (immunoglobulins), glucose, and perhaps chloride,[79,87] but there will be occasional requests for protein electrophoresis, sodium, calcium, bilirubin, and the enzymes lactate dehydrogenase (LDH) and creatine kinase (CK). In addition to these chemical data, the clinician will be interested in the CSF pressure, the color, turbidity, and cell count of the fluid, and the presence of old or new blood. Serological tests for syphilis and bacteriological studies are also often requested. Therefore, only part of the specimen collected may be available for chemical analysis.

---

* Newly recommended abbreviation. See Appendix.

SPECIMEN

Normally a spinal puncture will supply a 5 to 10 ml specimen, sometimes less. Because of the trauma to the patient and because of the effort and skill demanded in obtaining good specimens, repeat punctures are to be avoided. Spinal fluids are, therefore, very precious specimens; they are handled carefully to avoid loss and every effort is made to get as much information as possible from the volume of sample available. Any leftover portion is stored refrigerated or frozen for 7 to 10 days for possible repeat determinations or for other tests that might be dictated by the clinical status of the patient.

It is important that every CSF specimen be centrifuged before it is subjected to chemical study. Any erythrocytes or leukocytes that are present must be removed in order not to include cellular protein in the determination of the true CSF protein level. If, because of a traumatic puncture, a considerable amount of fresh blood is present in the specimen, the results for protein (and LDH) may be invalid because they will reflect the proteins of blood more than those originally present in the spinal fluid. (Sugar and chloride assays, however, will still be meaningful.) Serum contains 200 to 400 times more soluble protein than does the CSF; therefore, as little as 0.20 ml of blood can add to the CSF specimen four times the quantity of soluble protein present in 5 ml of normal CSF. Gross blood, when present, should always be mentioned in the laboratory report. Specimens from patients with a recent brain hemorrhage may possess a light yellow color. Such yellow pigmentation is referred to as *xanthochromia*, and is due to bilirubin, present as a result of the breakdown of old hematoporphyrin. When xanthochromia is evident, this also should be noted in the laboratory report. Occasionally, surgical talc may contaminate the specimen; this must be removed, although often it is difficult to centrifuge down the light talc grains. Talc turbidity will invalidate any turbidimetric assay for protein. Other occasional contaminants are starch granules and x-ray contrast media.

NORMAL VALUES

A normal spinal fluid specimen is clear, colorless, free of turbidity, contains a few lymphocytes, and has a total protein content in the range of 15 to 45 mg/100 ml. Ventricular fluids have the least protein (10 to 15 mg/100 ml), and lumbar fluids have the highest level of protein (15 to 45 mg/100 ml).

As with the plasma and urine proteins, our knowledge of individual CSF proteins has been revolutionized by the introduction of electrophoretic and immunochemical techniques. Since CSF is fundamentally an ultrafiltrate, one would expect macromolecules such as β-lipoprotein, α_2-macroglobulin, and IgM to be absent, α-lipoprotein, haptoglobins 2-1 and 2-2 and IgA to be present only in traces, and the IgG/albumin ratio to be less than in serum. This has been confirmed by immunoelectrophoresis, and it causes the electrophoretic pattern to differ from that of serum in having less intense α_2 and γ bands. The approximate normal ranges for total protein, albumin, and the immunoglobulins in mg/100 ml are as follows (serum levels are given for comparison):

	Total Protein	*Albumin*	*IgG*	*IgA*	*IgM*
CSF	15–45	10–30	0.8–6.4	Trace	Nil
Serum	6000–8200	3500–5000	800–1400	90–450	60–250

CSF also differs from serum in having two extra bands, a conspicuous pre-albumin band and a slow β_2 (or T) band due to a second carbohydrate-deficient form of transferrin. Possibly these two proteins are secreted locally; certainly, additional immunoglobulins can be secreted locally by plasma cells within the central nervous system.

CLINICAL SIGNIFICANCE

The level of total CSF proteins rises in many diseases of the central nervous system; alterations in the concentrations of its individual components, such as the immunoglobulins, are useful for differential diagnosis (see Table 7–15). Allowance must be made for the fact that the CSF levels of all proteins with molecular masses up to 150,000 daltons (e.g., IgG) will be affected by their concentrations in serum. For this reason, serum should always be examined in parallel with the CSF. In neurological diseases, three kinds of changes are found.[77] Firstly, there may be an admixture of proteins from blood owing to increased permeability of the cerebral capillaries or to cerebral hemorrhage. This causes an increase in total protein, albumin, and IgG. Electrophoresis shows a relative decrease of the usually prominent pre-albumin and transferrin components and an increase in the usually faint α_2 and γ bands by proteins of larger molecular weight. Such changes occur in acute inflammation, especially meningitis, in which case the acute phase proteins (including C-reactive protein) will be prominent. They also sometimes occur in chronic inflammation, in nearly half of the patients with intracranial tumors, in cerebrovascular disease (even without hemorrhage), and in the CSF below the level of any obstruction of the spinal canal.

Secondly, there may be an isolated rise in the level of IgG in the CSF due to an increase in local synthesis. On electrophoresis, the γ-band is more intense than usual and may be banded (oligoclonal).[47] On immunoelectrophoresis, IgM may be seen and the IgG line may be duplicated. Such changes are associated with a local infiltration of lymphocytes and plasma cells, and are found in chronic and subacute infections of the central nervous system, such as syphilis, brain abscess, encephalitis and chronic forms of meningitis. They also occur in multiple sclerosis, subacute panencephalitis, and some degenerative diseases.

Thirdly, there may be a decrease in the mobility of several proteins: in particular, the β_1-globulins, transferrin and hemopexin, acquire β_2 mobility, so that the β_1 band appears less intense and the β_2 or T band more intense than normal. This is found in some cases of degenerative disease and seems to indicate progressive damage to the nervous system, possibly by the release of enzymes.

TABLE 7–15 CEREBROSPINAL FLUID PROTEIN IN VARIOUS DISEASES[77,79,87]

Clinical Condition	*Appearances*	*Total Protein (mg/100 ml)*	*Pandy Globulin Test*
Normal	Clear, colorless	15—45	Negative
Increased admixture of plasma proteins			
Increased capillary permeability:			
bacterial meningitis	Purulent, cells, turbid, opalescent	100—500	1+ to 3+
viral meningitis	Clear, colorless, cells	30—100	Usually negative
encephalitis	Clear, colorless, cells	15—100	Usually negative
poliomyelitis	Clear, colorless	10—300	Usually negative
brain tumor	Usually clear	15—200 (usually normal)	Usually negative
Mechanical obstruction:			
spinal cord tumor	Clear, colorless or yellow	100—2000	1+ to 3+
Hemorrhage:			
cerebral hemorrhage	Colorless, yellow or bloody	30—150	Negative or 1+
Local immunoglobulin production:			
neurosyphilis	Clear, colorless	50—150	3+ to 4+
multiple sclerosis	Clear, colorless	25—50	Negative or 1+
Both increased capillary permeability and local immunoglobulin production:			
tuberculous meningitis	Colorless, fibrin clot, cells	50—300 (occ. up to 1000)	1+ to 3+
brain abscess	Clear or slightly turbid	20—120	Usually negative

LABORATORY INVESTIGATIONS

The three kinds of pathological change just noted could be detected most easily if an initial estimation of total protein were followed by electrophoretic analysis, but unfortunately electrophoresis of CSF raises difficulties not found with serum or even with urine: the protein concentration is so low that a 100-fold concentration is frequently necessary. At least 5 ml of fluid may be required, an inconveniently large proportion of the amount available.

For the detection of the first two kinds of change, however, it is sufficient to estimate the CSF albumin and immunoglobulin levels. This requires only 0.1 ml of unconcentrated fluid. Then, taking into account any changes in the serum proteins, the proportion of IgG to albumin in CSF can be used to detect and distinguish those diseases in which Ig's are made locally within the brain from those diseases in which blood proteins leak into the CSF[10,22] (see p. 370). A qualitative test, the Lange colloidal gold curve flocculation test, was until recently used for this purpose; it depends on the tendency of immunoglobulins to cause flocculation of metastable colloidal gold sols and on the tendency of albumin to inhibit such flocculation. Nowadays, after estimation of total protein, and perhaps a rough qualitative test for globulins such as the Pandy test, albumin and IgG are estimated by radial immunodiffusion. Where sufficient, the remaining fluid may be concentrated and analyzed electrophoretically.

METHODS FOR THE DETERMINATION OF TOTAL CEREBROSPINAL FLUID PROTEIN

The most common procedures used to estimate CSF proteins employ measurement of turbidity[57,61] when proteins are precipitated with reagents such as trichloracetic and sulfosalicylic acids. This has been discussed briefly in the section on protein determination in urine (p. 360). These procedures are simple and, in contrast to the Lowry and UV methods, do not suffer interference from amino acids, peptides, and other non-protein substances; but they do require 0.5 or 1 ml of CSF.

The main difficulty with the turbidimetric procedures is the selection of the ideal precipitating reagent. Such a reagent should react similarly with both albumin and globulins. The turbidity should be stable long enough to permit its convenient estimation, and it should be reproducible from day to day and from sample to sample, normal or abnormal (see p. 362). The details of the technique must, therefore, be rigidly adhered to if reproducible results are to be obtained. The precipitant least affected by changes in albumin/globulin ratio appears to be 3 per cent sulfosalicylic acid in 7 per cent sodium sulfate.[61]

Of the more precise methods, ultraviolet light absorption has the merit of using only 0.1 to 0.2 ml CSF, but requires preliminary removal of interfering low molecular mass substances.[37,60] A method using gel filtration to do this is described in outline on p. 373.

DETERMINATION OF TOTAL PROTEIN BY SULFOSALICYLIC ACID–SODIUM SULFATE TURBIDITY

REAGENTS

1. Sulfosalicylic acid (SSA). Dissolve 3.0 g sulfosalicylic acid, AR, in a solution of 7.0 g sodium sulfate/100 ml. After preparation, filter the reagent until clear and store away from direct light. The reagent is colorless and should be discarded when any color or turbidity develops.

2. Standard protein solutions. Prepare a composite of normal, hemolysis-free, nonicteric and nonturbid sera. Determine the total serum protein in quadruplicate by the biuret procedure. Carefully dilute 3.0 ml of the composite serum to 100 ml with 0.15 molar NaCl. Then dilute aliquots of this dilute serum (of approximately 180 to 200 mg/100 ml concen-

tration) with 0.15 molar NaCl to form a series of standards of the order of 20, 40, 80, 120, 160, and 200 mg protein/100 ml. Do all mixing gently to avoid protein denaturation. These standards are then treated in the same way as CSF specimens.

3. NaCl, 0.15 mol/l (0.85 g/100 ml).

PROCEDURE

1. Check the turbidity of the reagent by diluting 2.0 ml of reagent with 0.50 ml of saline and measuring its absorbance against saline at 620 nm. If the absorbance is over 0.010, prepare new reagent.

2. Centrifuge the CSF specimen to sediment any cells or other particulate matter. Use small centrifuge tubes or test tubes; 10 × 75 mm tubes are convenient for specimens of 1 to 5 ml.

3. Use a pair of cuvets for each CSF specimen. Add the reagents and specimen directly. Place the cuvets in parallel rows in a test tube rack in a water bath at 25 ± 2°C. The front row (control) cuvets will serve as pigment controls for the back row (turbidity) cuvets.

4. Place 0.50 ml of a CSF specimen into each of a pair of cuvets. If a protein value over 150 mg/100 ml is expected (turbid or purulent specimens), prepare a 3-, 5-, or 10-fold dilution with 0.15 molar saline and use 0.50 ml of this diluted specimen.

5. Place 2.0 ml of saline into all front row (control) cuvets. At one-half minute intervals, add 2.0 ml of sulfosalicylic acid reagent to each of the back row cuvets and mix.

6. After a specific time interval (e.g., 10 min[61]), mix each turbidity (back row) cuvet again by gentle inversion and read its absorbance at 620 nm against its front row mate, (which is set at 100 per cent transmittance).

CALCULATIONS

The calibration curve is linear in the range from 0 to about 150 to 200 mg protein/100 ml of CSF. For the interval of linearity, a Beer's law calibration constant or factor may be calculated and used, or the data may be graphed and unknowns read directly from the calibration graph. The calibration constant F is obtained by calculating the average slope of the line relating absorbance to concentration, either by least squares statistical treatment or by the approximation formula presented in equations (14) and (17); then

$$C_u = F \times A_u \times D$$

where C_u and A_u represent concentration of protein in the unknown and absorbance of the unknown, and D is the dilution factor, if the CSF was diluted before being assayed.

COMMENTS

1. Reproducibility is about ±5 per cent except at levels under 15 mg/100 ml (±10 per cent). Overall accuracy is about 7 to 10 per cent, being least accurate at the very low and very high extremes.

2. The control tube corrects for slight xanthochromia. The use of 620 nm light further minimizes error due to the presence of yellow pigment.

3. Interference by contaminating hemoglobin (blood) pigment is not corrected for; hemoglobin, being a protein, reacts with the SSA reagent to form a turbidity.

REFERENCES

Meulemans, O.: Clin. Chim. Acta, *5*: 757, 1960.
Pennock, C. A., Passant, L. P., and Bolton, F. G.: J. Clin. Path., *21*: 518, 1968.

DETERMINATION OF CSF PROTEIN BY ULTRAVIOLET SPECTROPHOTOMETRY

PRINCIPLE

Protein solutions exhibit strong ultraviolet light absorption bands at 277 to 280 nm and at 210 to 220 nm. The band at 280 nm is due to aromatic rings (tyrosine, etc.), whereas that at 215 nm reflects the presence of the peptide bond. The molar absorptivity at 215 nm is quite high, permitting assay of protein in specimens as small as 5 to 10 μl, provided that interfering substances of low molecular mass, such as peptides, have first been removed.

This can be done by gel filtration through a small chromatographic column designed to retain a constant volume of fluid and containing Sephadex G-50 suspended in 0.15 molar NaCl solution. A sample of 0.20 ml of CSF is applied to the top of the column, followed by one wash of 3.0 ml of saline. The column dimensions are such that the 3.0 ml of eluate draining from the bottom of the column contains all the CSF protein. Interfering materials are retained by the column. The absorbance of the eluate (after appropriate dilution, if needed) is then measured at 220 nm. A bovine serum albumin solution (Armour), treated similarly, serves as a calibrator. The procedure yields satisfactory results but requires special care in technique. It uses much less CSF but is less practical for routine use than the turbidity method.

REFERENCES

Patrick, R. L., and Thiers, R. E.: Clin. Chem., *9*: 283, 1963.
Igon, P.: Am. J. Med. Tech., *33*: 501, 1967.

DETERMINATION OF CSF IMMUNOGLOBULINS

QUALITATIVE TESTS

The presence of increased levels of globulins in a CSF specimen can be detected by several simple turbidity tests. In the *Pandy test* the reagent is 1.0 ml of a saturated aqueous solution of phenol, to which a drop (0.04 ml) of CSF is added. The degree of turbidity obtained after a prescribed time is read as 1+ to 3+. The results of the Pandy test obtained in a number of clinical conditions are given in Table 7–15. This test is simple and requires only a drop of fluid, but the availability of precise immunochemical methods is making such once useful tests obsolete.

DETERMINATION OF IMMUNOGLOBULINS (AND OTHER SPECIFIC PROTEINS) IN CSF BY RADIAL IMMUNODIFFUSION

The same technique and standards described for the serum procedure (see p. 321) can be used, except that the specimen is tested undiluted in single and double quantities. In applying double quantities, the sample is first added with a 2 μl micropipet, or the cup is filled to the brim as usual. After waiting until it has drained, the application is repeated. Alternatively, larger holes can be used.

In meningitis the presence of IgM at a concentration of more than 3.0 mg/100 ml suggests a bacterial rather than a viral infection.[77a] Normal levels are found in meningism.

TEST FOR LOCALLY SECRETED IgG

PRINCIPLE

The level of CSF albumin is used as a reference to indicate whether an increase of CSF IgG is due to local secretion or to increased permeability of the blood capillaries (or both). Account is taken of the influence of their serum concentrations.

METHOD

The concentrations of IgG in CSF and in serum are determined by radial immunodiffusion on the same plate. The concentrations of albumin in CSF and in serum are determined similarly. The ratio

$$\frac{\text{CSF IgG}}{\text{Serum IgG}} \bigg/ \frac{\text{CSF albumin}}{\text{Serum albumin}}$$

is calculated. The normal value is approximately 0.5. Increased values indicate local IgG secretion.

REFERENCES

Delpech, B., and Lichtblau, E.: Clin. Chim. Acta, *37*: 15, 1972.
Ganrot, K., and Laurell, C. B.: Clin. Chem., *20*: 571, 1974.

ELECTROPHORESIS OF CSF PROTEINS[47,77]

Because of the relatively low concentration of proteins usually present in CSF, the fluid must generally be concentrated up to 100-fold. This process is most conveniently done by arranging a semipermeable membrane between the fluid and an absorbent material as in the Minicon-B Macro-concentrator (Amicon Corp., Lexington, MA.). These concentrators are small, disposable, and rapid, and automatically shut off when only a small volume of fluid remains. Alternatives are ultrafiltration through collodion membranes, by the use of either positive pressure (N_2 gas) or negative pressure (vacuum). The latter can be conveniently carried out using special collodion bag holders available commercially (Schleicher and Schuell Co.). Dialysis against concentrated solutions of synthetic polymers (or even against thick sucrose syrups) may be used, but may entail some contamination of the specimen by these agents.

Electrophoresis of the concentrated CSF is then carried out exactly as for serum (see p. 312). Increases in locally secreted IgG frequently give rise to several oligoclonal γ-globulin bands placed closely together, instead of the usual diffuse polyclonal band. A sample of normal CSF will show the following average distribution of proteins: prealbumin, 4.5 per cent; albumin, 64.5 per cent; α_1-globulin, 3.6 per cent; α_2-globulin, 6.5 per cent; β-globulin, 12.6 per cent; γ-globulin, 7.9 per cent.

REFERENCES

1. Andrews, P., and Male, C. A.: Sephadex thin layer gel filtration. *In* Chromatographic and Electrophoretic Techniques. 3rd ed. I. Smith, Ed. London, W. Heinemann Medical Books Ltd., 1969, vol. 1, p. 823.
2. Archibald, R. M.: Nitrogen by the Kjeldahl method. *In* Standard Methods of Clinical Chemistry. D. Seligson, Ed. New York, Academic Press, Inc., 1958, vol. 2, p. 91.
2a. Association of Clinical Pathologists: Symposium on Disorders of Protein Metabolism. J. Clin. Path. Supplement, 1975.
3. Berne, B. H.: Clin. Chem., *20*: 61, 1974.
4. Bonnar, J.: Blood coagulation and fibrinolysis in obstetrics. *In* Clinics in Hematology. A. S. Douglas, Ed. Philadelphia, W. B. Saunders Company, 1973, vol. 2, No. 1, p. 213.
5. Burnet, M.: Cellular Immunology. London, Cambridge University Press, 1970.
6. Cameron, J. S., and Blandford, G.: Lancet, *ii*: 242, 1966.
7. Clarke, H. G. M., and Freeman, T.: Clin. Sci., *35*: 403, 1968.
8. Clausen, J.: Immunochemical techniques for the identification and estimation of macromolecules. *In* Laboratory Techniques in Biochemistry and Molecular Biology. T. S. Work and E. Work, Eds. Amsterdam, North-Holland Publishing Co., 1969, vol. 1, p. 397.
8a. Cooper, M. D., and Lawton, A. R.: Am. J. Path., *69*: 513, 1972.
9. Dagg, J. H., Cumming, R. L. C., and Goldberg, A.: Disorders of iron metabolism. *In* Recent Advances in Haematology. A. Goldberg and M. C. Brain, Eds. London, Churchill Livingstone, 1971, p. 77
10. Delpech, B., and Lichtblau, E.: Clin. Chim. Acta, *37*: 15, 1972.
11. Determan, H.: Gel Chromatography. New York, Springer-Verlag, Inc., 1968.

12. Dietz, A. A., Rubinstein, H. M., and Hodges, L.: Clin. Chem., *20*: 396, 1974.
13. Doumas, B. T., and Biggs, M. G.: Determination of serum albumin. *In* Standard Methods in Clinical Chemistry. G. A. Cooper, Ed. New York, Academic Press, Inc., 1972, vol. 7, p. 175.
14. Ehrmantraut, H. C.: Clinical Significance of Paper Electrophoresis. Palo Alto, Calif., Beckman Instruments Inc., Spinco Div., 1958.
15. Eriksson, S.: Studies in α_1-anti-tryspin deficiency. Acta Med. Scand., *117* (Suppl. 432), 1965.
16. Everall, P. H., and Wright, G. H.: J. Med. Lab. Tech., *15*: 209, 1958.
17. Fischer, L.: An introduction to gel chromatography. *In* Laboratory Techniques in Biochemistry and Molecular Biology. T. S. Work and E. Work, Eds. Amsterdam, North-Holland Publishing Co., 1969, vol. 1, p. 151.
17a. Fowell, A. H.: Am. J. Clin. Path., *25*: 340, 1955.
18. Franglen, G.: Clin. Chim. Acta, *14*: 559, 1966.
19. Franklin, E. C., and Zucker-Franklin, D.: *In* Advances in Immunology. F. J. Dixon and H. G. Kunkel, Eds. New York, Academic Press, Inc., 1972, vol. 15, p. 249.
20. Freier, E.: Unpublished procedure.
21. Friedman, H. S.: Gamma globulin in serum. *In* Standard Methods of Clinical Chemistry. D. Seligson, Ed. New York, Academic Press, Inc., 1958, vol. 2, p. 40.
22. Ganrot, K., and Laurell, C-B.: Clin. Chem., *20*: 571, 1974.
23. Giblett, E. R.: Genetic Markers in Human Blood. Oxford, Blackwell Scientific Publications, 1969.
24. Gornall, A. G., Bardawill, C. J., and David, M. M.: J. Biol. Chem., *177*: 751, 1949.
25. Grabar, P., and Burtin, P. (Eds.): Immuno-electrophoretic Analysis. Amsterdam, Elsevier, 1964.
26. Grant, G. H.: J. Clin. Path., *12*: 510, 1959.
27. Grant, G. H., and Butt, W. R.: Immunochemical methods in clinical chemistry. *In* Advances in Clinical Chemistry. O. Bodansky and C. P. Stewart, Eds. New York, Academic Press, Inc., 1970, vol. 13, p. 383.
28. Grunbaum, B. W., Lyons, M. F., Carroll, N., and Zec, J.: Microchem. J., *7*: 54, 1963.
29. Hardwicke, J., Cameron, J. S., Harrison, J. F., Hulme, B., and Soothill, J. F.: Proteinuria studied by clearances of individual macromolecules. *In* Proteins in Normal and Pathological Urine. Y. Manuel, J. P. Revillard, and H. Betuel, Eds. Basel, S. Karger, 1970, p. 111.
30. Haupt, H., Bichler, K-H., and Becker, W.: Characterization and immunochemical quantitation of human uromucoid. *In* Protides of the Biological Fluids, 21st Colloquium. H. Peeters, Ed. New York, Pergamon Press, 1974, vol. 21, p. 529.
31. Henry, R. J., Cannon, D. C., and Winkelman, J. W. (Eds.): Clinical Chemistry, Principles and Technics. 2nd ed. Hagerstown, Maryland, Harper and Row, Publishers, 1974.
32. Heremans, J. F., Laurell, A. H., and Waldenstrøm, J.: Acta Med. Scand., *170* (Suppl. 367): 1, 1961.
33. Heremans, J. F., and Masson, P. L.: Clin. Chem., *19*: 294, 1973.
34. Hobbs, J. R.: Brit. J. Haematol., *16*: 599, 1969.
35. Hobbs, J. R.: Immunoglobulins in clinical chemistry. *In* Advances in Clinical Chemistry. O. Bodansky and A. L. Latner, Eds. New York, Academic Press, Inc., 1971, vol. 14, p. 219.
36. Hoffman, W. S.: The Biochemistry of Clinical Medicine. 4th ed. Chicago, Year Book Medical Publishers, Inc., 1970.
37. Igou, P. C.: Am. J. Med. Tech., *33*: 354, 1967.
38. Kaplan, A., and Savory, J.: Clin. Chem., *11*: 937, 1965.
39. Kark, R. M., Lawrence, J. R., Pollak, V. E., Pirani, C. L., Muerke, R. C., and Silva, H.: A Primer of Urinalysis. 2nd ed. New York, Hoeber Medical Division, Harper and Row, Publishers, 1963.
40. Kawai, T.: Clinical Aspects of the Plasma Proteins. Berlin, Heidelberg, New York, Springer, 1973.
41. Kent, P. W.: Structure and function of glycoproteins. *In* Essays in Biochemistry. P. N. Campbell and G. D. Greville, Eds. New York, Academic Press, Inc., 1967, vol. 3, p. 105.
42. Kingsley, G. R.: Procedure for serum protein determinations. *In* Standard Methods of Clinical Chemistry. G. R. Cooper, Ed. New York, Academic Press, Inc., 1972, vol. 7, p. 199.
43. Klatskin, G., Reinmuth, O. M., and Barnes, W.: J. Lab. Clin. Med., *48*: 476, 1964.
44. Kohn, J.: Cellulose acetate electrophoresis. *In* Chromatographic and Electrophoretic Techniques. I. Smith, Ed. London, Heinemann, 1968, vol. 2, p. 84.
45. Kohn, J.: The laboratory investigation of paraproteinemia. *In* Recent Advances in Clinical Pathology. S. C. Dyke, Ed. London, Churchill Livingstone, 1973, Series 6, p. 363.
46. Kunkel, H. G.: Proc. Soc. Exp. Biol. Med., *66*: 217, 1947.
47. Laurell, C. B.: Electrophoretic and electro-immunochemical analysis of proteins. Scand. J. Clin. Lab. Invest., *29* (Suppl. 124): 21, 71, 1972.
48. Laurell, C. B.: Clin. Chem., *19*: 99, 1973.
49. Leading Articles: Brit. Med. J., *i*: 1, 758, 1973.
49a. Lindstedt, G., and Lundberg, P. A.: Clin. Chim. Acta., *56*: 125, 1974.
50. Lowry, O. H., Rosebrough, N. J., Farr, A. L., and Randall, R. J.: J. Biol. Chem., *193*: 265, 1951.
51. Lyngbye, J., and Krøll, J.: Clin. Chem., *17*: 495, 1971.
52. Mancini, G., Carbonara, A. O., and Heremans, J. F.: Immunochemistry, *2*: 235, 1965.
53. Manuel, Y., Revillard, J. P., and Betuel, H. (Eds.): Proteins in Normal and Pathological Urine. Basel, S. Karger, 1970.
54. Martinek, R. G., and Berry, R. E.: Clin. Chem., *11*: 10, 1965.
55. McQueen, E. G.: J. Clin. Path., *15*: 367, 1962.
56. Medical Research Council Working Party on the characterization of antisera as reagents. Immunology, *20*: 3, 1971.
57. Meulemans, O.: Clin. Chim. Acta, *5*: 757, 1960.

58. Ouchterlony, O.: Immunodiffusion and immunoelectrophoresis. *In* Handbook of Experimental Immunology. D. M. Weir, Ed. Oxford, Blackwell Scientific Publications, 1967, p. 655.
59. Owen, J. A.: Effect of injury on plasma proteins. *In* Advances in Clinical Chemistry. H. Sobotka and C. P. Stewart, Eds. New York, Academic Press, Inc., 1967, vol. 9, p. 2.
60. Patrick, R. L., and Thiers, R. E.: Clin. Chem., *9*: 283, 1963.
61. Pennock, C. A., Passant, L. P., and Bolton, F. G.: J. Clin. Path., *21*: 518, 1968.
62. Peters, T.: Clin. Chem., *14*: 1147, 1968.
63. Peters, T., Jr.: Serum albumin. *In* Advances in Clinical Chemistry. O. Bodansky and C. P. Stewart, Eds. New York, Academic Press, Inc., 1970, vol. 13, p. 37.
64. Peterson, P. A., Evrin, P., and Berggärd, I.: J. Clin. Invest., *48*: 1189, 1969.
65. Pintera, J.: The biochemical, genetic and clinicopathological aspects of haptoglobin. *In* Series Haematologica. K. G. Jansen and S-A. Killmann, Eds. Baltimore, The Williams and Wilkins Co., 1971, vol. IV, p. 2.
66. Poortmans, J., and van Kerchove, E.: Clin. Chim. Acta, *8*: 485, 1963.
67. Fowell, A. H.: Am. J. Clin. Path., *25*: 340, 1955.
68. Ratcliff, A. P., and Hardwicke, J.: J. Clin. Path., *17*: 676, 1964.
69. Ravin, H. A.: J. Lab. Clin. Med., *58*: 161, 1961.
70. Reiner, M., and Cheung, H. C.: Fibrinogen. *In* Standard Methods of Clinical Chemistry. D. Seligson, Ed. New York, Academic Press, Inc., 1961, vol. 3, p. 114.
71. Reinhold, J. G.: Total protein albumin and globulin. *In* Standard Methods of Clinical Chemistry. D. Seligson, Ed. New York, Academic Press, Inc., 1953, vol. 1, p. 88.
72. Rice, E. W., and Cicone, M. A.: Am. J. Clin. Path., *29*: 90, 1958.
73. Rodkey, F. L.: Clin. Chem., *11*: 478, 1965.
74. Rose, D. S., Anderson, S. G., and Grab, B.: Bull. World Health Org., *42*: 535, 1970.
75. Rowe, D. S.: Bull. World Health Org., *40*: 613, 1969.
76. Saifer, A., and Gerstenfeld, S.: Clin. Chem., *10*: 321, 1964.
76a. Salmon, S. E., and Smith, B. A.: J. Clin. Invest., *49*:1114, 1970.
76b. Schosinsky, K. H., Lehmann, H. P., and Beeler, M. F.: Clin. Chem., *20*:1556, 1974.
77. Schultze, H. E., and Heremans, J. F.: Molecular Biology of Human Proteins. Amsterdam, Elsevier Publishing Co., 1966.
77a. Smith, H., Bannister, B., and O'Shea, M. J.: Lancet, *ii*:591, 1973.
78. Smith, I. (Ed.): Chromatographic and Electrophoretic Techniques. 3rd ed. London, William Heinemann Medical Books Ltd., 1969.
79. Stewart, C. P., and Dunlop, D.: Clinical Chemistry in Practical Medicine. 6th ed. Edinburgh, Livingstone, 1962.
80. Stiehm, E. R., and Fulginite, V. A.: Immunologic Disorders in Infants and Children. Philadelphia, W. B. Saunders Company, 1973.
81. Sunderman, F. W., Jr., and Sunderman, F. W.: Am. J. Clin. Path., *27*: 125, 1957.
82. Sunderman, F. W., Jr., Sunderman, F. W., Falvo, E. A., and Kallick, C. J.: Am. J. Clin. Path., *30*: 112, 1958.
83. Swaim, W. R., and Feders, M. B.: Clin. Chem., *13*: 1026, 1967.
84. Thompson, R. H. S., and Wootton, I. D. P.: Biochemical Disorders in Human Disease. 3rd ed. London, Churchill, and New York, Academic Press, Inc., 1970.
85. Tidstrøm, B.: Scand. J. Clin. Lab. Invest., *15*: 167, 1963.
86. van der Heul, C., van Eijk, H. G., Wiltink, W. F., and Leijnse, B.: Clin. Chim. Acta, *38*: 347, 1972.
87. Varley, H.: Practical Clinical Biochemistry. 4th ed. London, Heinemann Medical Books Ltd., and New York, Interscience Books Inc., 1967.
88. von Frijtag Drabbe, C. A. J., and Reinhold, J. G.: Anal. Chem., *27*: 1092, 1955.
89. Waldman, R. K., Krause, L. A., and Borman, E. K.: J. Lab. Clin. Med., *42*: 489, 1952.
90. Watson, D.: Albumin and "total globulin" fractions of blood. *In* Advances in Clinical Chemistry. H. Sobotka and C. P. Stewart, Eds. New York, Academic Press, Inc., 1965, vol. 8, p. 237.
91. Webster, D., Bignall, A. H. C., and Attwood, E. C.: Clin. Chim. Acta, *53*: 101, 1974.
92. Weeke, B.: Scand. J. Clin. Lab. Invest., *21*: 351, 1968.
93. Weeke, E. O. B.: Urinary serum proteins. *In* Protides of the Biological Fluids, 21st Colloquium. H. Peeters, Ed. New York, Pergamon Press, 1974, vol. 21, p. 363.
94. West, C. D., Hong, R., and Holland, N. H.: J. Clin. Invest., *41*: 2054, 1962.
95. Wieme, R. J.: Agar Gel Electrophoresis. Amsterdam, London and New York, Elsevier Publishing Co., 1965.
96. Winzler, R. J.: Determination of serum glycoprotein. *In* Methods of Biochemical Analysis. D. Glick, Ed. New York, Interscience Publishers Inc., 1955, vol. 2, p. 279.

SECTION FOUR

AMINO ACIDS AND RELATED METABOLITES

by Gregor Grant, M.A., B.M., F.R.C. Path., and John F. Kachmar, Ph.D.

Next to urea, amino acids constitute the second largest source of the non-protein nitrogen (NPN) in serum. In general, urea-N and NPN parallel each other in serum; as the former rises, the latter increases in about the same proportion. The difference between the two includes nitrogen contained in a number of different metabolites; however, the largest portion (75 per cent) of nitrogen other than that included in urea-N is accounted for as amino acid nitrogen (AAN).

There is no convenient method for measuring total AAN, although procedures for measuring α-amino-N are available. Despite the importance of amino acids in many aspects of human biochemistry, determinations of AAN have not been as useful in clinical diagnosis as one would have expected. It is now appreciated that it is not the total of all amino acids that is important clinically, but rather the changes in concentrations of certain individual amino acids or groups of related amino acids.

A large number of inborn errors of metabolism associated with abnormal metabolism of individual amino acids are now known, some of which result in mental retardation (see Table 7–17, p. 387). Furthermore, it has been established that the deleterious effects of such conditions may sometimes be prevented by a diet deficient in the non-metabolized amino acids. Such conditions must be diagnosed and treated before they cause irreversible brain damage. Programs for mass screening of newborns to detect such genetic defects have been developed, and microbiological and chemical screening tests are available for the detection of individual amino acids present in high concentrations. When investigation of several amino acids is required, chromatography[23] is the preferred technique. Many different amino acids can then be detected at the same time by their positions on the chromatogram and by their color reactions with various reagents.

One-dimensional ascending paper chromatography, as illustrated in Figure 3–32, is simple, but two-dimensional chromatography gives much better resolution. The two-dimensional method using paper takes 24 to 36 h, but a small scale thin layer chromatographic technique, such as that described below,[21] can be carried out in half a day. Alternatively, high voltage electrophoresis may be combined with chromatography, carried out at right angles to it. The initial electrophoretic separation takes only half an hour and eliminates the need for desalting of the specimen before performing the chromatographic separation. The equipment used, however, is expensive and potentially dangerous for routine work.

Commercially available automatic amino acid analyzers separate the amino acids on a column of cation exchange resin by gradient elution and then quantitate the individual amino acids spectrophotometrically after employing the ninhydrin reaction.

CLINICAL SIGNIFICANCE[2,19,22a]

The normal total *plasma* or *serum* concentration of amino acids (as AAN) ranges from 2.5 to 3.5 mg/100 ml (lower limit) to 4.5 to 7.0 mg/100 ml (upper limit), depending on the method of assay. Low values are rarely encountered; the deficiency in amino acid intake in starvation is countered by the breakdown of plasma and tissue proteins into their constituent amino acids. In the liver, amino acids are converted into protein or deaminated to keto acids. The amino group is excreted as urea, which is synthesized in the liver by means of the Krebs-Henseleit cycle.[16] These processes tend to keep the amino acid nitrogen at a fairly constant level for any one individual. A transient elevation of 1 to 2 mg/100 ml is observed after a meal that is relatively heavy in proteins. A significantly large increase in total AAN is encountered only in very severe liver disease, particularly in fulminating hepatic necrosis ("acute yellow atrophy" of the liver) caused by chemical poisoning by such agents as phosphorus, chloroform, and carbon tetrachloride. A slight increase is also found accompanying diabetes mellitus, and in patients with impaired renal function, along with elevations in all components of the NPN fraction.

The measurement of the *urinary output* of amino acids has been of more clinical interest than plasma AAN levels. The daily output of amino acid nitrogen is found to be of the order of 50 to 200 mg, some 1 to 2 per cent of the total daily urinary nitrogen excretion. The individual amino acids are freely filtered by the glomeruli, but the urinary concentration is quite low because the acids, like peptides and low molecular mass proteins, are actively reabsorbed by the renal tubules. Only a few amino acids are present in the urine in quantities of over 50 mg/d. The remainder are present in small amounts, varying from traces to 5 to 20 mg/d. Roughly one-third are present in the free form; the rest occur as peptides or in other bound or conjugated forms. In a chromatogram of normal urine,[21,23] glycine is usually the dominant fraction, with alanine, serine, glutamine, and (in heavy meat eaters) histidine and 1-methyl histidine present in smaller quantities; in some normal urines taurine is prominent, in others β-aminoisobutyric acid. The renal threshold for many substances, including amino acids, is lowered in pregnancy, so that in this condition histidine, phenylalanine, lysine, and tyrosine spots may also be seen. Premature babies, especially during the first week, have a physiological generalized renal aminoaciduria, and even in full term infants, aminoaciduria is more marked than in normal adults.

Increases in urinary amino acids may arise by two mechanisms: the plasma amino acid level may rise above the renal threshold as a result of some metabolic disorder (overflow aminoaciduria), or the plasma level may be normal but the renal threshold low (renal aminoaciduria). A comparison of the blood and urine chromatograms will distinguish between these two mechanisms. Either may be primary or secondary: a primary defect in amino acid metabolism, for example, may raise the blood level of one (or more) amino acid, as in phenylketonuria; or a primary defect of an active transport system in the kidney tubules may allow excretion of a group of amino acids in spite of the fact that their blood levels are normal, as in cystinuria. More commonly, aminoaciduria is generalized and secondary to some underlying disease. Again, it may be of the overflow type, as in liver failure; or the renal tubules may be affected, as in heavy metal poisoning, in acute tubular necrosis, in severe wasting (from starvation or disease), and in some congenital diseases such as galactosemia (see p. 257), Wilson's disease (see p. 651), or the Fanconi syndrome. In the last, reabsorption of amino acids, glucose, phosphate, and other metabolites may be impaired.

ANALYTICAL PROCEDURES FOR AMINO ACIDS

DETERMINATION OF TOTAL α-AMINO ACID NITROGEN

The most precise method for assaying amino acids employs the gasometric ninhydrin reaction.[4] Ninhydrin (triketohydrindene hydrate) reacts with amino acids as follows:

$$\text{Ninhydrin} + \text{R—C(NH}_2\text{)(H)—COOH (Amino acid)} \longrightarrow \text{Hydrindantin} + \text{R—C(=O)H (Aldehyde)} + CO_2 + NH_3 \quad (19)$$

Ninhydrin Amino acid

Hydrindantin Aldehyde

In this reaction the amino acid is oxidized to an aldehyde containing one less carbon atom, resulting in the release of ammonia and carbon dioxide. Exactly 1 mole of carbon dioxide is liberated per mole of each amino acid, with the exception of aspartic acid, which gives off 2 moles. All true amino acids undergo the ninhydrin reaction, as do peptides and proteins and any compounds containing free amino groups. Proline and hydroxyproline do not react. The carbon dioxide formed can be determined manometrically, and is a measure of the amino acids present in a given sample. Protein interference is eliminated by using picric acid filtrates.

If the pH of the reaction environment is about 3 to 4, the ammonia, ninhydrin, and the reaction product, hydrindantin, react to form a bluish colored reaction product. This color

$$\text{Ninhydrin} + NH_3 + \text{Hydrindantin} \longrightarrow \text{Colored Product (C=N—C)} + 3H_2O \quad (20)$$

Ninhydrin Hydrindantin Colored Product

reaction is also used for both detection and quantitation of amino acids in chromatographic procedures.

Sodium β-naphthoquinone-4-sulfonate will react with amino acids under weakly alkaline pH conditions to form a brownish-orange colored product. Frame[5] developed this Folin reaction into a useful method for measuring total AAN in plasma and, with care, in urine. Ammonia interferes and must be removed (urine). Uric acid and sulfonamides give a weak reaction, which can be corrected for. A modification of this procedure is outlined below.

Alternatively, amino acids in serum can be quantitated using *1-fluoro-2,4-dinitrobenzene (FDNB)*.[6,20] Free amino groups in proteins, peptides, and amino acids react with the reagent to form yellow derivatives with the following basic structure:

$$\text{R—C(H)(COOH)—NH—C}_6\text{H}_3\text{(NO}_2\text{)}_2$$

Protein interference is eliminated by use of tungstic acid filtrates. An aliquot of the filtrate is heated with the FDNB reagent at pH 10.0 for 15 min at 56°C, and the color of the amino acid derivative is measured at 420 nm. Creatine, urea, and uric acid do not interfere.

DETERMINATION OF PLASMA AND URINARY AMINO ACID NITROGEN BY THE NAPHTHOQUINONE SULFONATE PROCEDURE

SPECIMENS

Plasma specimens are preferable to serum. Values for AAN in serum are some 10 per cent higher than those found in plasma, partly because of the amino acids released by the clotting process. Urine should be collected in bottles containing 10 ml of concentrated HCl to prevent microbial destruction.

PROCEDURE FOR PLASMA[5]

The reaction is carried out with tungstic acid filtrates of plasma. The filtrate is neutralized, buffered with borate, treated with a fresh aqueous solution of sodium β-naphthoquinone-4-sulfonate, and then heated for 10 min at 100°C. Standards, consisting of a mixture of glycine and glutamic acid, are determined concurrently with the unknowns. All tubes are cooled, treated with acid formaldehyde, and diluted with water. The brownish-orange color formed is measured spectrophotometrically at 470 nm.

PROCEDURE FOR URINE[24]

Interfering materials such as uric acid are removed by passing the urine through a Dowex-50 × 8 cation-exchange resin, and washing the resin column with dilute HCl. The isolated amino acids are eluted with NaOH, NH_3 is removed by aeration, and the eluate is then treated the same way as the filtrates in the plasma procedure.

COMMENTS

The normal range for plasma AAN in adults by the naphthoquinone procedure has been reported to be 4.0 to 6.0 mg/100 ml, as against 3.2 to 5.5 mg/100 ml for the ninhydrin methods, and 3.4 to 7.0 mg/100 ml for the FDNB method. The upper limit for children may be somewhat lower. The difference between plasma and serum (higher) values is probably less than the uncertainty of present methods.

The value for urine output in normal individuals is not well defined and varies with the reagent used and the effort used to remove interfering materials. Apparently the lowest values (50 to 200 mg/d or about 0.85 to 2.3 mg/kg/d) are obtained with the naphthoquinone procedure when urine specimens are passed through Dowex resin. Ninhydrin methods give values of 80 to 250 mg/d (about 1.2 to 2.9 mg/kg/d) and FDNB methods give values of about 2.0 to 4.4 mg/kg/d for children and probably also for adults.

About 85 per cent of the amino acid nitrogen in plasma represents amino acids present in free form. In urine, only one-third of the amino acids are free; the remainder are present as peptides, or as conjugates with various other metabolites.

CHROMATOGRAPHY

There are three stages in a chromatographic analysis of amino acids in blood or urine specimens: preparation of the sample, chromatographic separation, and quantitation of the separated amino acids. Chromatography is a versatile technique: details can be varied to suit any specific problem. In particular, a great variety of solvent mixtures are available for separating different amino acids, and many different reagents may be used for locating and identifying them. Conditions suitable for routine preliminary testing are described below, but any abnormality found must always be confirmed by further work. Changes caused by bacterial contamination or drug therapy are first excluded by retesting a fresh specimen; if the presence of abnormalities is confirmed, reference is made to a standard work on chromatography[23] and the most suitable solvent mixtures and reagents are chosen for further investigations.

Blood plasma requires preliminary treatment to remove proteins, lipids, and salts which interfere with the chromatographic separation. In the case of urine samples, salts and proteins should be removed and the amino acid concentration suitably adjusted. Fortunately, amino acids are very stable and soluble in water and aqueous alcohol, and proteins and lipids can be easily removed by precipitation or extraction. Salts may be removed by electrolysis or ion exchange resins; the former requires special apparatus, and the latter requires reconcentration. For the small samples used in thin layer chromatography, simple extraction procedures are usually adequate (see procedure).

The amount of urine sample to be applied may be standardized in relation to a definite amount of amino acid nitrogen, urea-N, or creatinine, or to the volume excreted in a timed interval. These methods are only approximate and do not allow for concentration changes of individual amino acids. In the clinical laboratory it is usually best to determine the urine creatinine.

The most popular solvent mixture for *one-dimensional chromatography* is *n*-butanol–acetic acid–water (12/3/5). The acetic acid maintains miscibility of water in the *n*-butanol up to a concentration which gives the best balance of hydrophobic and hydrophilic properties to effect maximum separation. Although one-dimensional chromatography gives limited resolutions of amino acids, it provides a useful screening test owing to the consistency of the normal plasma amino acid pattern [22] (see p. 160).

In *two-dimensional chromatography*, one solvent mixture usually contains small amounts of ammonia or other amine to increase the mobility of basic amino acids, and the other solvent mixture contains acid. The selection of effective solvent mixtures is achieved primarily empirically (by trial and error), rather than on the basis of theoretical considerations.

Standard mixtures of pure amino acids are run at the same time as the unknowns, and the positions of the spots for standard and unknown are compared (see Fig. 7–23). The amino acids present are detected by spraying with either ninhydrin or isatin, which are general staining reagents (see Table 7–16). Individual amino acid fractions may then be identified by specific stains. Some of these may be applied on top of the general stains.[23] Alternatively, several samples of the unknown are run and stained in different ways. The identity of any individual amino acid may be confirmed by repeating the run after adding some of the pure amino acid to the sample. Other special techniques for identification of individual amino acids involve pretreatment, such as oxidation for the identification of sulfur-containing amino acids, conversion to copper salts to distinguish α-amino acids from other ninhydrin-staining compounds, or preparation and chromatographic separation of dinitrophenyl or other derivatives.

Paper chromatography has been used widely in the past for screening patients for possible amino acid abnormalities. Whatman 3MM filter papers, 25 cm square, mounted on a frame are commonly used (Fig. 7–21).[23] Eight samples and the necessary standard mixtures can be run overnight at room temperature on each sheet, using the butanol–acetic acid–water solvent. When studying the newborn, blood plasma is used because it is easier to obtain and because urine chromatograms may be difficult to interpret. The separated amino acids are detected by dipping the dried chromatograms into a solution containing 0.2 g ninhydrin/100 ml acetone.

Thin layer chromatography has now largely replaced paper because of its greater

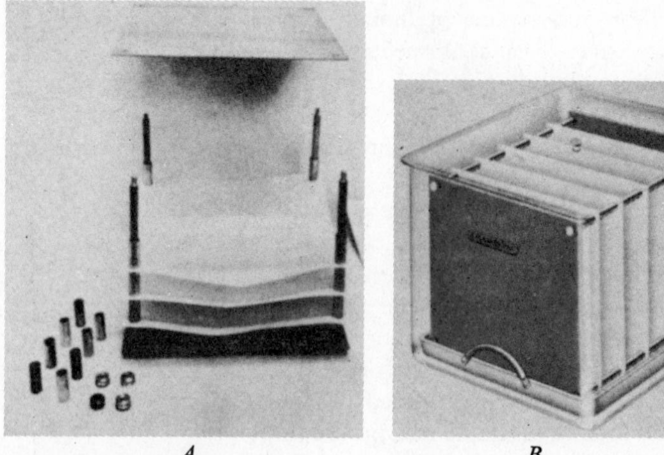

A B

Figure 7–21 Apparatus for paper chromatography. (*A*) The frame is being assembled. The papers are placed horizontally on the frame and are held apart by collars; finally, the end plates are placed in position and held by the nuts. The solutions to be chromatographed are applied to the origins with the frame still in the horizontal position shown. (*B*) Assembled for ascending chromatography. The frame, holding from one to five papers, sits in the solvent in the bottom of the tray. (From Smith, I.: Chromatographic and Electrophoretic Techniques, 3rd Ed. London, W. Heinemann, 1969, vol. 1, p. 9.)

speed, sensitivity, and resolution. A technique using thin layers of cellulose powder is described below.

QUALITATIVE ANALYSIS OF BLOOD AND URINE FOR AMINO ACIDS BY THIN LAYER CHROMATOGRAPHY (SAIFER'S MICROMODIFICATION OF THE METHOD OF WHITE)[21,30,33]

SPECIMENS

Blood. With the patient in a fasting state, collect blood into heparinized tubes or capillaries and separate the plasma from the cells.

Urine. Collect an early morning specimen. If analysis cannot be carried out soon, preserve the specimen by adding one part of a solution of 10 g thymol in 100 ml isopropanol to 200 parts urine.

PRINCIPLE

Interfering substances, such as salts and proteins, are removed from plasma by alcoholic precipitation of protein, followed by chloroform extraction of the supernatant, or from urine by electrolytic desalting. The amino acids are then separated by two-dimensional thin layer ascending chromatography using cellulose powder plates. The positions of the amino acids on staining are compared with those of standard mixtures run at the same time. The amount of urine used is adjusted according to the creatinine concentration.

REAGENTS

Solvent mixtures

First dimension: Mix 225 ml pyridine, 150 ml acetone, 25 ml NH_4OH (28 per cent, w/w) and 100 ml water.

Second dimension: Mix 375 ml isopropanol, 62.5 ml formic acid (80 per cent aqueous), and 62.5 ml water.

Standard solutions

Solution A. 5 mmol/l each of leucine, phenylalanine, tryptophan, valine, proline, hydroxyproline, threonine, glycine, aspartic acid, and lysine in 0.1 molar HCl. Either the L or DL forms may be used.

Solution B. 5 mmol/l each of isoleucine, methionine, tyrosine, alanine, glutamic acid, serine, arginine, histidine, and cystine in 0.1 molar HCl.

Store the standard solutions in the refrigerator when not in use. Neutralize before use by adding an equal volume of 0.1 molar NaOH.

Staining reagents

Ninhydrin-collidine reagent. Ninhydrin, 0.2 g/100 ml ethanol. Store in a refrigerator. Just prior to use, add 0.5 ml of 2,4,6-collidine to each 50 ml of the ninhydrin solution.

Isatin-cadmium reagent. Dissolve 0.2 g of cadmium acetate in 20 ml of water to which 5.0 ml glacial acetic acid has been added (Solution 1). Dissolve 0.2 g isatin in 100 ml of acetone (Solution 2). Just prior to use, mix 1 volume of Solution 1 with 8 volumes of Solution 2.

PROCEDURE

Preparation of amino acid extract of plasma

To 1 volume of plasma (minimum 100 μl) in a centrifuge tube, add 3 volumes of absolute ethanol. Mix well and allow to stand for 10 min. Centrifuge at 2000 rpm for 10 min to separate precipitated protein. Decant off the ethanolic extract and add 3.0 volumes of chloroform. Mix well and allow to stand for 5 min. The aqueous layer above the chloroform-alcohol mixture contains the free amino acids and is pipeted into a separate tube. Lipids and many other organic compounds are removed by the chloroform. Most salts have previously been removed with the precipitate.

Note. In preparing the extract, about 20 per cent of the total amino acids are lost, including approximately half the phenylalanine, leucine, and isoleucine.

Preliminary treatment of urine

An aliquot of urine is set aside for determination of creatinine concentration (p. 994). Another aliquot, about 3 ml, is desalted electrolytically as illustrated in Figure 7–22. (A similar apparatus is available commercially.) Although not always essential, this gives the best separations.

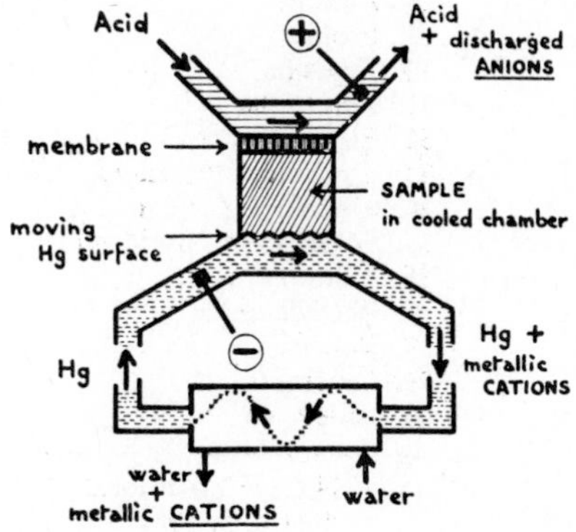

Figure 7–22 The principles of electrolytic desalting. Metallic cations discharge at and are carried away by the flowing mercury cathode, which is then washed free of the cations and returned to the cell. Anions, inorganic and some organic, pass through the ion-permeable membrane, discharge, and are washed away by the flowing acid anode, which is run to waste. Most organic cations and neutral molecules remain in the cooled sample chamber. (From Smith, I.: Chromatographic and Electrophoretic Techniques. 3rd ed. London, W. Heinemann, 1969, vol. 1, p. 41.)

Thin-layer chromatography

Cut 6.0 × 6.0 cm TLC plates from standard 20 × 20 cm aluminum TLC sheets coated with 100 μm cellulose (E. Merck, Brinkmann Instr., Westbury, New York, No. 5537/9H). Scrape off about 1 mm of the cellulose layer from each edge of the small plates; arrange them so that the grain (seen by examining the aluminum backing) runs vertically: spot 2.5 μl of the aqueous extract with a 5 μl Hamilton syringe, or with a Drummond micropipete, at a point about 1 cm interior to both the bottom and left edges. Keep the spot as small as possible by applying the sample in small amounts and drying with a heat

TABLE 7–16 COLOR REACTIONS OF AMINO ACIDS* ON CELLULOSE TLC PLATES

Amino acid	Color		
	NINHYDRIN-COLLIDINE	ISATIN-Cd, IMMEDIATE	ISATIN-Cd, > 12 h AT ROOM TEMP
1. Cystine	Gray	Pale pink	Light blue
2. Cysteine	Gray	Pale pink	Violet
3. Cystathionine	Gray	Pale pink	Violet
4. Cysteic acid	Violet	Yellow	Pink
5. Phosphoethanolamine	Violet	N.C.	Very faint pink
6. Argininosuccinic acid (major anhydride)	Violet	Pink	Pink
6′. Argininosuccinic acid (minor anhydride)	Violet	Pink	Pink
7. Ornithine HCl	Violet	Deep pink	Deep pink
8. Asparagine	Brown	Pink	Deep purple
9. Homocystine	Violet	Pale violet	Gray/brown
10. Arginine	Violet	Pink	Deep pink
11. Lysine	Violet	Deep pink	Deep pink
12. 1-Methylhistidine	Gray/brown	Violet	Blue
13. Histidine	Gray/brown	Violet	Purple
14. Carnosine	Light brown	N.C.	Faint brown
15. Taurine	Violet	N.C.	Very faint pink
16. Glutamine	Violet	Deep pink	Deep pink
17. Aspartic acid	Green	Pink	Violet
18. Citrulline	Violet	Purple	Pink
19. Methionine sulfoxide	Violet	Light pink	Violet
20. Methionine sulfone	Violet	Gray	Pink
21. Serine	Violet	Pink	Pinkish purple
22. Glycine	Golden brown	Pink	Pink
23. Hydroxyproline	Yellow	Pink	Blue
24. Glutamic acid	Violet	Purple	Pink ring
25. Sarcosine	Dark gray	Pale violet	Pale pink
26. Threonine	Violet	Deep pink	Deep pink
27. β-Alanine	Green	Bluish gray	Blue
28. Alanine	Violet	Violet	Purple
29. Proline	Yellow	Blue	Deep blue
30. Tyrosine	Gray blue	Bluish gray	Green
31. Tryptophan	Light violet	Blue	Gray
32. Ethanolamine	Violet	Violet	Pale pink
33. γ-Aminobutyric acid	Violet	Pink	Pink
34. α-Aminobutyric acid	Violet	Bluish gray	Violet
35. β-Aminoisobutyric acid	Violet	N.C.	N.C.
36. Methionine	Violet	Pink	Pink
37. Phenylalanine	Brown	Violet	Deep blue
38. Valine	Violet	Bluish violet	Deep pink
39. Leucine	Violet	Pink	Pinkish violet
40. Isoleucine	Violet	Pink	Pink

N.C. = no color.
* After Saifer.[21]

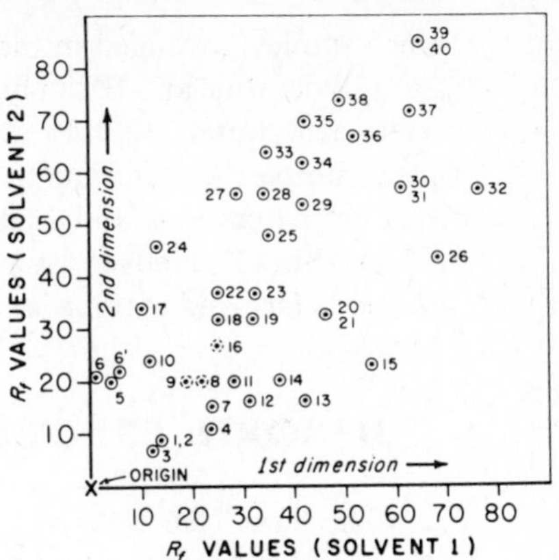

Figure 7–23 Chromatogram of amino acids commonly found in blood and urine, after separation by two-dimensional chromatography on thin layer cellulose microplates as described in text. First dimension pyridine-acetone-ammonia, second dimension isopropanol-formic acid. Numbering as in Table 7-16.

R_f values of unknowns will vary slightly with each individual run. They are best determined by relating their positions to more easily recognized spots, such as proline or hydroxyproline, and to standard mixtures run at the same time. (From Saifer, A.: Rapid screening methods for the detection of inherited and acquired aminoacidopathies. *In* Advances in Clinical Chemistry. O. Bodansky and A. L. Latner, Eds. New York, Academic Press, Inc., 1971, vol. 14, p. 146.)

gun between applications. In the case of urine, apply a volume equivalent to 1 μg creatinine but do not apply more than 3 μl.

Also spot 1 to 5 μl of each standard solution on separate plates. Run standards and unknowns in duplicate so that both stains may be used.

Line two glass covered dishes (e.g., staining dishes, Fisher No. 8-812A) with Whatman 3 MM paper and fit with low notched racks capable of supporting the plates and holding the filter paper in position. Add the first solvent mixture to one dish and the second solvent mixture to the other. About 25 ml is required to saturate the paper and cover the rack. Cover the dishes for at least 30 min prior to chromatography to allow equilibration to take place. Quickly insert the thin layer plates in the notches of the rack in the first dish, cover the dish and place a weight on its glass cover. Allow the solvent to ascend to the top of the plates, which takes about 10 to 15 min. Remove the plates and allow them to air dry at room temperature until solvent-free. Rotate the plates 90° and place into the rack in the second dish, which is then closed tightly. This solvent also takes about 15 min to reach the top of the plate. Remove the plates and dry them with a hot air gun. Repeat second run.[33] Dry again to remove all traces of the solvent. (It is advisable to leave the solvents in the dishes between runs and to refill with fresh solvent before each run.)

Hold the plates at a corner with forceps. Dip one into the ninhydrin-collidine reagent and heat with a warm air gun to develop the color. Dip the duplicate plate into the isatin-cadmium reagent and heat at 70°C for 10 min to develop the color. Compare unknowns and standards, using Table 7–16 and Figure 7–23 as guides. If the concentration of a particular amino acid appears to be increased, this may be confirmed by repeating the run after first spotting the pure acid on top of the unknown at the origin.

COMMENTS

1. This rapid microtechnique is essentially a screening test. If an abnormal pattern is found, repeat the procedure in larger tanks with full size (20 × 20 cm) sheets, before carrying out specific tests for the suspected amino acids. About 10 μl of serum extract and a quantity of urine equivalent to 5 μg creatinine will be required.
2. Patients being studied should be off all medications.
3. If the patient has proteinuria, remove proteins from the sample by ultrafiltration, using a small apparatus of the type illustrated in Figure 7–20. Then desalt some of the ultrafiltrate collected. An electrolytic desalter will remove small amounts of protein but may become clogged if larger amounts are present.
4. The isatin—cadmium stain fades after 24 h, but the ninhydrin stain lasts longer and can be made permanent by spraying with nickel sulfate solution (7 g/100 ml water).
5. For rapid screening of large numbers, use one-dimensional chromatography.[33] The plasma samples require extraction, but the urine samples need not be desalted. Run

nine samples (arranged in the order Standard A, seven unknowns, Standard B) on a TLC plate 20 cm wide and 10 cm high (half a standard plate cut across the grain) using *n*-butanol–acetic acid–water (12/3/5). Apply the standards (5 μl) and the plasma extracts (10 μl) or urine samples (volumes containing 4 μg creatinine) 1.5 cm from the bottom of the plate as streaks 1 cm long and 1 cm apart. Run until solvent front is about 0.5 cm from the top (35 to 40 min), remove, dry with hot air gun, and run again until solvent reaches the same position. Dry and stain as above.

INBORN ERRORS OF AMINO ACID METABOLISM[9,11,26]

The term "inborn errors of metabolism" was first used by A. E. Garrod 65 years ago following his study of alkaptonuria, a rare inherited condition in which the urine darkens on standing owing to the presence of a substance, homogentisic acid, not normally found in urine. He suggested that this and other related congenital conditions might be due to the absence of one of the enzymes in a series of reactions forming part of the normal course of metabolism; and he was able to explain their various biochemical, pathological, and clinical manifestations as secondary to the missing enzyme. Alkaptonuria was also the first example of "recessive inheritance" in man. Fifty years later the enzyme deficiency was directly demonstrated.

This idea that an enzyme deficiency could result from a double dose of an abnormal gene implied that the normal gene was necessary for the formation of the enzyme. This was the first clue to the now well established generalization that genes exert their effects by directing the synthesis of enzymes and other proteins. With the discovery of the DNA structure of genes and their base triplet coding for protein synthesis, the differences between human beings became explicable as inherited differences in the enzymes and other proteins which their DNA is coded to synthesize.

Although an enormous number of congenital differences in the enzyme composition of individuals are possible, most such differences are harmless and often not recognized. Other combinations are lethal, their presence being incompatible with life processes. An increasing number of diseases are being described in which the deficiency of an enzyme or the presence of an abnormal enzyme causes a block in the biochemical processing of an amino acid or of some other metabolite (e.g., a carbohydrate in galactosemia), resulting in elevated blood levels and increased renal excretion. The symptoms and prognosis associated with such amino acid disorders may vary from being almost benign and harmless, as in *alkaptonuria*, to being usually acute and lethal, as in *maple syrup disease*. The more important of these abnormalities are listed in Table 7–17. Phenylketonuria, alkaptonuria, and tyrosinosis are described below in more detail.

Prenatal tests are now being developed to test enzyme deficiencies in amniotic fluid so that pregnancies highly likely to produce affected children can be terminated by therapeutic abortion where necessary (see Chap. 22).

A second related but less well understood group of disorders (e.g., cystinuria) results from congenital defects in amino acid transport systems, systems which control the passage of small molecular mass substances in and out of cells across their membranes. The precise molecular basis of these defects is not yet understood, but they seem to involve defects or absence of particular enzymes or "carrier" proteins situated locally in the membranes, causing metabolic blocks, essentially analogous to those in the first group of disorders, and hence a lowering of the renal threshold for one or more amino acids (see Table 7–18).

CYSTINURIA

On occasion a laboratory is asked to test urine specimens or kidney stones for the presence of the sulfur-containing amino acid, cystine. Both cystine and cysteine are normally

TABLE 7–17 PRIMARY OVERFLOW AMINOACIDURIAS
(Based on references 16, 19, 21, 23, and 26.)

Disease	*Deficient Enzyme*	*Amino Acids with Increased Blood Levels*	*Clinical Features*
Phenylketonuria	L-Phenylalanine hydroxylase	Phenylalanine	Mental retardation.*
Tyrosinosis	*p*-Hydroxyphenyl-pyruvic acid oxidase	Tyrosine	Hepatic cirrhosis;* renal damage.
Histidinemia	Histidine α-deaminase	Histidine	Speech defects; mental retardation in half the cases.
Maple syrup disease	Branched chain α-ketoacid decarboxylase	Valine, leucine, isoleucine, alloisoleucine	Mental retardation,* toxic encephalopathy, odor of maple syrup in urine. Usually fatal in weeks or months.
Valinemia	Valine transaminase	Valine	Retarded development*—physical and mental. Hypotonia, vomiting.*
Homocystinuria	Cystathionine synthetase	Homocystine, methionine	Mental retardation,* skeletal abnormalities, paraplegia, thromboembolism.
Cystathioninuria	Cystathionase	Cystathionine	Mental retardation, psychoses.
Prolinemia	Proline oxidase	Proline	Renal abnormalities and mental retardation in some cases.
Hydroxyprolinemia	Hydroxyproline oxidase	Hydroxyproline	Severe mental retardation, thrombopenia.*
Hyperglycinemia		Glycine	Two forms—ketotic and non-ketotic. Mental retardation and other symptoms.
Citrullinemia	Argininosuccinate synthetase	Citrulline	Mental retardation, hepatomegaly. (These urea cycle enzyme deficiencies raise the blood ammonia, but the blood urea is unchanged.)
Argininosuccinicaciduria	Argininosuccinase	Argininosuccinic acid	
Lysinemia	Lysine dehydrogenase	Lysine	Mental retardation and neurological symptoms.
Sarcosinemia	Sarcosine oxidase	Sarcosine	Mental retardation, retarded growth.
β-Alaninemia	β-Alanine transaminase	β-Alanine, γ-amino-butyric acid	Mental retardation, somnolence.
Carnosinemia	Carnosinase	Carnosine	Mental retardation.

* Responds to diet deficient in the amino acids whose concentrations are raised in the blood.

TABLE 7–18 PRIMARY RENAL AMINOACIDURIAS (CONGENITAL RENAL TUBULAR TRANSPORT DEFECTS)
(Based on references 19, 21, 23, and 26.)

Disease	*Amino Acids with Increased Concentration in Urine*	*Clinical Features*
Cystinuria	Cystine, arginine, lysine, ornithine	Cystine renal calculi. (Note: There are three subtypes of cystinuria, which vary in the degree of intestinal transport defect.)
Prolinuria or iminoglycinuria	Proline, hydroxyproline, glycine	Mild form: probably benign. Severe form (Joseph's syndrome): Mental retardation, epilepsy.
Hartnup disease	Thirteen neutral amino acids	Mental retardation and neurological symptoms.
Oasthouse urine disease	Methionine	Mental retardation, burnt sugar odor.
Hyperglycinuria	Glycine	Probably benign (compare with hyperglycinemia, Table 7–17).

present in small amounts in urine. The combined total 24 h output of both is about 10 to 100 mg. Cysteine constitutes about 10 per cent of the total. The sulfhydryl or thiol, —SH, group is easily oxidized to the disulfide (—S—S—), 2 moles of cysteine forming 1 mole of cystine. The reduction of cystine to cysteine is more difficult to achieve. The reactions by

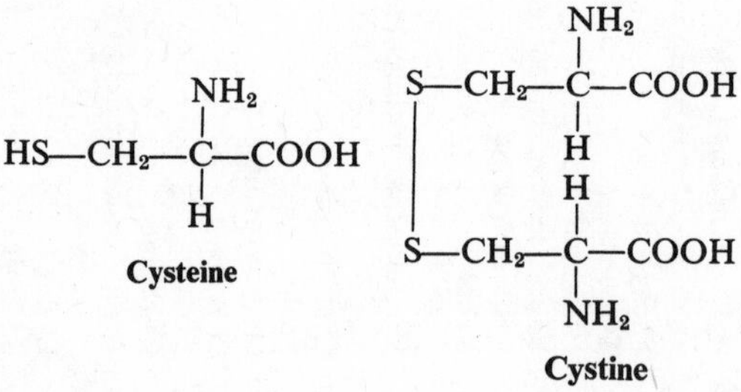

$$\underset{\textbf{Cysteine}}{HS{-}CH_2{-}\overset{\displaystyle NH_2}{\underset{\displaystyle H}{C}}{-}COOH} \qquad \underset{\textbf{Cystine}}{\begin{array}{l} S{-}CH_2{-}C(NH_2)(H){-}COOH \\ | \\ S{-}CH_2{-}C(H)(NH_2){-}COOH \end{array}} \tag{21}$$

which this occurs physiologically are not well understood, although one known pathway includes an enzymatic reduction involving glutathione. The usual chemical procedure for cystine analysis is to add NaCN to the cystine preparation, which results in the following reaction:

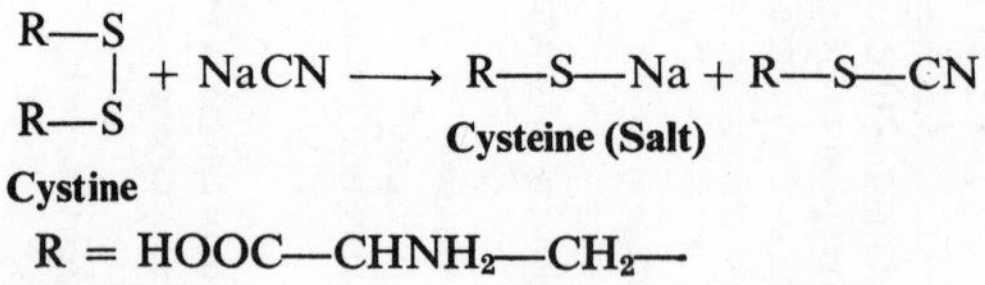

$$\underset{\textbf{Cystine}}{\begin{array}{l} R{-}S \\ \quad | \\ R{-}S \end{array}} + NaCN \longrightarrow \underset{\textbf{Cysteine (Salt)}}{R{-}S{-}Na} + R{-}S{-}CN \tag{22}$$

$$R = HOOC{-}CHNH_2{-}CH_2{-}$$

Since there is no convenient chemical way to detect cystine as such, all routine procedures use this reaction to convert cystine to cysteine, the —SH group of which is reactive chemically. Cystine is relatively insoluble at near neutral pH, but is soluble in mineral acids and in alkalis, including NH_4OH. The limit of solubility in urine is about 30 mg/100 ml. In patients with cystinuria, this limit is surpassed and cystine is precipitated out in the urine, and even in the renal pelvis, forming cystine stones (calculi). The cystine in urinary sediment can be identified by the characteristic appearance of the crystals—clear, flat hexagonal plates—soluble in HCl and NH_4OH, but insoluble in acetic acid.

Cystinuria is a disease characterized by the excretion of large quantities of the basic amino acids, lysine, arginine, ornithine, and cystine, and by a tendency to form and deposit cystine calculi (stones). The disease[1,19,22a] involves a genetic defect in the renal tubular reabsorption of lysine, arginine, and ornithine, which indirectly affects cysteine excretion and results in the excretion of increased levels of cystine. Several related forms of the disease exist, and some also include a defect in intestinal absorption of cysteine. The disease *cystinosis* is unrelated to cystinuria, and its physiologic basis is even less well understood. A genetic defect in the metabolism or transport of cystine is present, resulting in the deposition of cystine in the body tissues, although the urinary output may be normal, or only slightly elevated.

Cystinuria is best diagnosed by chromatography. The blood pattern is normal, but analysis of urine shows the presence of increased quantities of the basic amino acids lysine, arginine, ornithine, and cystine. The increase in cystine may be confirmed by adding to the dry sample spot 0.3 μl of 30 per cent H_2O_2 containing 0.2 mg of ammonium molybdate per ml, and then redrying. By this reaction cystine and cysteine are oxidized to cysteic acid ($HO_3SCH_2CHNH_2COOH$), so that on repeating the chromatographic separation there will be a characteristic change in the pattern (see Fig. 7–23 and Table 7–16). Any methionine present is oxidized to methionine sulfone.

Chemical tests are also used to identify cystine in stones and urinary sediments. In the *nitroprusside test*, the unknown is treated with NaCN (see equation 22), followed by a few drops of a fresh solution of nitroprusside. If cystine or cysteine is present, a deep reddish violet (magenta) color is formed.

In the *lead acetate test*, the specimen is heated with 10 molar NaOH to split the sulfide group from the amino acid. The free sulfide ion is then detected by the addition of lead acetate, the Pb^{2+} ion reacting with the sulfide ion to form a black precipitate of PbS.

NITROPRUSSIDE TEST[27]

Pipet about 3 ml of fresh urine into a test tube or a small crucible. In the case of a renal stone, grind the stone to a fine powder, transfer a small amount to a crucible, and heat with 2 to 3 ml of H_2O; allow the crucible to cool. Add 1.0 ml of fresh NaCN solution, 5 g/100 ml (handle with care), and allow the tube or crucible to stand for 10 min. It is wise to use a control consisting of 3 ml of normal urine to which a few milligrams of cystine have been added. Then add dropwise freshly prepared sodium nitroferrocyanide solution (5 g $Na_2FeNO(CN)_5 \cdot 2H_2O$/100 ml). If cystine or cysteine is present, a magenta color develops. Normal urines give a faint brown or reddish brown color. The test is also positive (purple) with homocystine. *Homocystinuria*, unlike cystinuria, gives a positive reaction with silver nitroprusside as well.[25]

LEAD ACETATE TEST

Boil about 5 to 7 ml of urine with 1.0 to 1.5 ml of 10 molar NaOH. Cool the tube or crucible and add 3 to 5 drops of saturated lead acetate ($Pb(CH_3COO)_2 \cdot 2H_2O$) solution. The formation of a brownish black to black precipitate of PbS indicates organically bound sulfhydryl. If the urine contains more than 2+ protein, the protein will interfere because the heating with alkali will split off sulfhydryl from the protein molecules. Protein is removed by heating the urine to 70 to 80°C, followed by filtering.

DEFECTS IN AROMATIC AMINO ACID METABOLISM

Metabolic blocks may occur in the conversion of phenylalanine to tyrosine (phenylketonuria), or in the further metabolism of tyrosine to homogentisic acid (alkaptonuria), or to dihydroxyphenylalanine (DOPA) and melanin (albinism). In contrast, melanin-producing tumors (melanomas) may produce an excess of melanin.

PHENYLKETONURIA

Phenylketonuria[9,19,26,32] is an inheritable disease with an incidence of about 1 in 10,000 live births, characterized by a defect in the ability to metabolize the amino acid phenylalanine. The disease is caused by the absence or deficiency of the enzyme L-*phenylalanine hydroxylase*, which converts phenylalanine to the amino acid tyrosine. As a result, phenylalanine accumulates in the tissues of the body. A portion of this phenylalanine is excreted in the urine as such, but a larger fraction is deaminated to phenylpyruvic acid and excreted in that form.

$$C_6H_5{-}CH_2{-}C({=}O){-}COOH \underset{\text{Transaminase}}{\rightleftharpoons} C_6H_5{-}CH_2{-}CH(NH_2){-}COOH \xrightarrow[\text{hydroxylase}]{\text{Phenylalanine}} HO{-}C_6H_4{-}CH_2{-}CH(NH_2){-}COOH \qquad (23)$$

↑ Location of genetic block (at phenylalanine hydroxylase)

Phenylpyruvic acid (PPA) — **Phenylalanine** — **Tyrosine**

The phenylalanine accumulating in the tissues is apparently toxic, especially to developing brain tissue, with the result that brain damage and progressive mental retardation occur. The injury to brain tissue begins within the second and third week of life and progresses with time, becoming maximal at about 8 to 9 months. Brain damage can be minimized if the newborn child is placed on a low-phenylalanine diet soon after birth. Even if diagnosis is made late, if phenylalanine is then removed from the diet, the rate of further mental deterioration can be decreased, provided that this is done before 4 to 6 months have passed. If no therapy is instituted, the child may develop as an idiot with an I.Q. of 20 to 30. Because the effects of the disease may be minimized by very early diagnosis and the use of phenylalanine-free diets, a number of states in the U.S. now require, by statute, that all newborns be tested for the presence of this genetic defect.

It is useful to have both a simple screening test to detect increases in serum phenylalanine in newborns and a more accurate method to monitor the dietary treatment of known cases. A simple urine test for phenylpyruvic acid (a transient blue-green color given with ferric chloride) is available, but unfortunately it is unsatisfactory as a screening test, as it is insensitive and thus often fails to detect cases early enough after birth.[18]

Normal full-term infants have serum phenylalanine levels ranging from about 1.2 to 3.4 mg/100 ml, averaging 2.1 mg/100 ml. In premature infants the level will often be higher but will fall into the range for normal full-term infants within 7 to 20 days. Phenylketonuric infants if fed a normal diet have levels of over 4.5 mg/100 ml by the third or fourth day. This value steadily increases as long as the infant is receiving a normal (phenylalanine-containing) diet, and levels of 20 to 30 mg/100 ml or more are found after 7 to 10 days. If a phenylalanine-free diet formula is instituted, the level falls to near normal values within a short time, and remains there as long as phenylalanine intake is restricted.

Chromatography of blood amino acids as described above is the best screening method for the detection of amino acid defects in general, but if only screening for phenylketonuria is desired, the Guthrie microbiological blood test[8] or the chemical determination of serum phenylalanine is simpler and adequately reliable. The latter is the method of choice for dietary monitoring. The most commonly used chemical method is that of McCaman and Robins.[17] This procedure is a fluorometric method in which phenylalanine is reacted with ninhydrin and the fluorescence formed is measured after enhancement by the addition of the dipeptide, L-leucylalanine. In the procedure of Udenfriend and Cooper[28] the phenylalanine is decarboxylated to phenylethylamine, which is measured by reaction with methyl orange. LaDu and Michael[14] used L-amino oxidase to convert phenylalanine to phenylpyruvic acid, which is then converted to a borate complex that absorbs in the ultraviolet range (308 nm).

THE GUTHRIE SCREENING TEST FOR INCREASED SERUM PHENYLALANINE[8]

This microbiological procedure is based on the ability of phenylalanine to counteract the effects of a metabolic antagonist on the growth of a special strain of *B. subtilis* which requires phenylalanine as growth factor. An agar plate is used which contains a mixture of a suspension of the *subtilis* spores and a solution containing a minimum amount of growth nutrients plus a fixed amount of the metabolic antagonist β-2-thienylalanine. The similarity of its structure to phenylalanine is evident from their formulae:

H—C——C—H
H—C   C—CH_2CHNH_2COOH
S

β-2-Thienylalanine (metabolic antagonist to phenylalanine)

—CH_2CHNH_2COOH

Phenylalanine

(24)

When the infant reaches the age of six days (or as soon as possible afterward if the baby is being treated with antibiotics which interfere with the test), blood is taken by heel-prick onto filter paper, dried, and sent to the laboratory. Uniform discs are cut from such blood spots for testing and are compared with similar discs containing a series of known quantities of phenylalanine.

If the serum has an elevated level of phenylalanine, enough will diffuse from the sample disc into the bacterial medium to counteract the inhibitory effect of the thienylalanine, and a ring of bacterial growth will surround the disc. If the phenylalanine level is not more than 2 mg/100 ml, the inhibition to growth will not be overcome, and no bacterial growth will occur. False positives can occur, and any positive test should be verified by a chemical determination of phenylalanine, or by chromatography. Tests should not be done on specimens obtained from premature infants or from full-term babies immediately after birth; such specimens often give false positive results, owing to immaturity of the liver. Also, the infant should be on a phenylalanine-containing diet for at least 24 h before the sample is taken.

FLUOROMETRIC DETERMINATION OF SERUM PHENYLALANINE

REAGENTS

1. Succinate buffer, 0.6 mol/l, pH 5.88 $\pm$ 0.03. Dissolve 70.9 g of succinic acid in about 600 ml of water; using a pH meter, carefully add about 200 ml of 5 molar NaOH with thorough stirring, until a pH of 5.88 is obtained. Dilute to 1 liter. Store the reagent in a refrigerator; recheck its pH frequently.

2. Ninhydrin solution, 0.03 mol/l. Dissolve 2.67 g of the best grade ninhydrin in water to make 500 ml of solution. Keep this solution refrigerated.

3. L-Leucyl-L-alanine, 5 mmol/l. Dissolve 101 mg in 100 ml of water. This reagent has limited stability, even when refrigerated.

4. Buffered ninhydrin-peptide mixture. This is prepared fresh daily by mixing 5 vol of succinate buffer, 2 vol of ninhydrin solution, and 1 vol of the L-leucyl-L-alanine peptide solution.

5. Copper sulfate solution, 0.8 mmol/l. Dissolve 50 mg of $CuSO_4 \cdot 5H_2O$ in 250 ml of water. This reagent is stable.

6. Alkaline tartrate solution. Dissolve 1.33 g of Na_2CO_3 and 57 mg Rochelle salt, $KNaC_4H_4O_6 \cdot 4H_2O$, in 500 ml of water. This reagent is stable; store in a polyethylene bottle.

7. Alkaline copper reagent. Mix 3 volumes of alkaline tartrate with 2 volumes of copper solution. Prepare this reagent fresh daily.

8. Trichloracetic acid, 0.6 mol/l. Use fluorometric grade CCl_3COOH. The solution contains 98.0 g of TCA/1 and should be stored refrigerated.

9. Phenylalanine-in-serum standard. Prepare a stock standard containing 200 mg phenylalanine/100 ml water, using only L-phenylalanine of the purest grade available. For the preparation of the working standards, use as diluent a composite of nonhemolyzed, nonicteric, and nonchylous sera. Five working standards are formulated as follows:

Standard No.	*Stock Standard* (ml)	*Composite Serum* (ml)	*Concentration of Phenylalanine* (mg/100 ml)
1	none	5.00	A
2	0.10	4.90	0.98 A + 4.0
3	0.20	4.80	0.96 A + 8.0
4	0.30	4.70	0.94 A + 12.0
5	0.40	4.60	0.92 A + 16.0

The A stands for the concentration of phenylalanine in the composite serum. This is obtained by analyzing the composite serum against a previous set of standards or against a set of phenylalanine-in-water standards. If A = 2.1 mg/100 ml, the five standards will contain 2.1, 6.1, 10.0, 14.0, and 17.9 mg/100 ml. These working standards are analyzed exactly as are the unknown sera. The standards can be kept frozen for 4 to 6 weeks. The calibration curve is linear through 14 to 16 mg phenylalanine/100 ml.

10. Standards-in-water. Dilute 25 and 50 μl of the 200 mg/100 ml stock standard to 4.0 ml with 0.3 molar TCA. Treat the standards in the same way as the sample filtrates; they represent 2.5 and 5.0 mg/100 ml. These standards are used only to assay the normal composite serum used to prepare the in-serum standards.

SPECIMEN

Draw specimens of infant blood into microcapillary tubes. If the assay is done immediately, whole blood may be used; however, if the assay is delayed, centrifuge the blood, separate the plasma or serum, and freeze until analysis is undertaken. (50 μl of sample, as required in the procedure, may be stored in plastic centrifuge tubes.) Whole blood values appear to increase on standing; this increase may be as much as 30 to 50 per cent in the course of 5 to 7 days, when phenylalaline levels are in the normal range. Such an increase does not occur with frozen stored serum or plasma.

PROCEDURE

This is a microprocedure, and a set of microliter pipets, a microcentrifuge, as well as plastic microcentrifuge tubes (0.3 to 0.5 ml capacity), should be available.

1. Into microcentrifuge tubes, pipet 50 μl of sample or standards, respectively, and add 50 μl of 0.6 molar trichloracetic acid. Mix the tubes by vigorous hand-tapping or with a mechanical mixer. Allow them to set for 10 min and then centrifuge in a microcentrifuge.

2. Pipet 50 μl of each filtrate and transfer to a set of 13 × 100 mm test tubes. Also set up a blank containing 50 μl of 0.30 molar trichloracetic acid.

3. Add 0.80 ml of the buffered ninhydrin-dipeptide mixture to each tube and thoroughly

mix the tubes. Cover the tubes with small marbles, or otherwise cap, transfer to a water bath at 60°C ± 1°C, and heat for 2 h.

4. While the tubes are incubating, turn on the fluorometer. The activating (primary) wavelength used is 365 nm; the emission (secondary) wavelength is 515 nm.

5. After the 2 h incubation period, remove the tubes from the bath and place into a cooling bath at about 20°C.

6. Add 5.0 ml of fresh alkaline copper reagent to each tube. Mix the contents of the tubes and transfer to fluorometer cuvets. Zero the fluorometer and choose the aperture or slit settings so that the 20 mg/100 ml or highest standard will give a scale reading of 60 to 80. Take the fluorescence readings of the blank, standards, and samples. Correct all readings for the reading of the TCA blank.

7. Draw a calibration curve relating the corrected fluorometer scale reading to the concentration of phenylalanine, and read the concentrations of the unknowns from the curve.

8. Run all standards in duplicate. Repeat any test with a value over 4 mg/100 ml to verify the result. The pediatrician should be immediately informed and asked to provide a fresh, new specimen to confirm the result.

COMMENTS

Any specimens received between runs should be spun down, serum or plasma separated, and the microtube portion containing the plasma or serum kept frozen until the assay is begun. The procedure can be scaled down, in which case filtrates are made with 20 μl of serum and 20 μl of trichloracetic acid, and 20 μl of filtrate used for the reaction; 300 μl of buffered ninhydrin-peptide and 2.0 ml of alkaline copper reagent are used in place of the quantities previously mentioned.

The pH of the buffer used must be 5.88 ± 0.03. If the pH is lower, a loss of sensitivity occurs, and at higher pH values, other serum amino acids will give significant fluorescence. At pH 5.88, only tyrosine, leucine, and perhaps arginine give measurable readings, but these are less than 5 per cent of an equimolar level of phenylalanine. Obviously, if these amino acids are present at elevated levels, they will give an apparent increase in the phenylalanine value. The use of in-serum standards corrects for the presence of normal levels of other amino acids and assures the same pH in unknowns and standards.

NORMAL VALUES FOR SERUM PHENYLALANINE

Adults:	0.8 to 1.8 mg/100 ml
Full-term, normal-weight newborns:	1.2 to 3.4 mg/100 ml
Prematures, low-weight newborns:	2.0 to 7.5 mg/100 ml
Phenylketonuric newborns:	over 4.5 mg/100 ml after 2 to 3 days, and if not treated, rising to 15 to 30 mg/100 ml in a period of 10 days.

REFERENCES

Wong, P. W. K., O'Flynn, M. E., and Inouye, T.: Clin. Chem., *10*: 1098, 1964.
Faulkner, W. R.: Phenylalanine. *In* Standard Methods of Clinical Chemistry. S. Meites, Ed. New York, Academic Press, Inc., 1965, vol. 5, p. 199.

QUALITATIVE TESTS FOR PHENYLPYRUVIC ACID (PPA)

REAGENTS

1. $FeCl_3$ solution. Dissolve 10 g $FeCl_3 \cdot 6H_2O$ in water to give 100 ml solution.

2. Phosphate precipitating agent. This reagent contains 2.2 g $MgCl_2 \cdot 6H_2O$, 1.4 g NH_4Cl, and 2.0 ml concentrated NH_4OH in 100 ml of solution.

PROCEDURE

Add with mixing 1.0 ml of phosphate precipitant to 4.0 ml of fresh urine. Filter the specimen and acidify the filtrate with 2 to 3 drops of concentrated HCl. Add 2 to 3 drops of $FeCl_3$; observe the filtrate after each drop for any color formation. A green to blue-green color that persists for 2 to 4 min indicates a positive test. Very rapidly fading greens suggest homogentisic acid (HGA) or *p*-hydroxyphenylpyruvic acid (PHPPA), and should be read as negative. Imidazolepyruvic acid gives a positive test, identical with that for PPA, but it is encountered only rarely. Bilirubin will give a false positive reaction, but it normally would be absent. Other color hues are negative. The sensitivity is about 10 mg PPA/100 ml. By using serial dilutions of the urine filtrate, a rough quantitative measure of the PPA concentration may be made.

PPA usually cannot be detected in infant urines until the serum phenylalanine concentration has reached levels between 12 and 15 mg/100 ml. Infants affected with phenylketonuria will excrete up to 2 g PPA/d, and urinary concentrations will be from 50 to 100 mg/100 ml.

A commercial impregnated test paper called *Phenistix* (Ames Co., Div. Miles Laboratories, Elkhart, IN.) may also be used to test for PPA. In this procedure, also based on the ferric chloride reaction, the presence of PPA will change the color of the paper from a light gray to a gray-green or green-blue. Light, medium, and high readings correspond to PPA levels of 15, 40, and 100 mg/100 ml.

TEST FOR THE HETEROZYGOSITY OF THE PHENYLKETONURIA GENE

Serum phenylalanine levels in persons who are heterozygotes (carriers) for the phenylketonuric gene are frequently within the range encountered in normal persons free of the genetic defect. At times, it is desirable to be able to establish which parent and which siblings of an affected child are carriers of the genetic defect. Since the serum level is noninformative in this respect, advantage can be taken of the fact that such carriers have a limited capacity to metabolize phenylalanine. The carrier to be tested is given an oral dose of phenylalanine, 100 mg/kg of body weight. In a noncarrier, the phenylalanine concentration will rise from a fasting value of about 1.4 mg/100 ml to a value of 9 mg/100 ml at 1 and 2 h, and then drop to 5 mg/100 ml at 4 h. In a person who is a heterozygote, the phenylalanine concentration will rise from normal to a value of about 19 mg/100 ml at the first hour and fall much more slowly than in the normal person. A *phenylalanine tolerance index* is calculated by adding up the 1, 3, and 4 h values. The values of the index for normal individuals and carriers are sufficiently disparate that there is no difficulty in distinguishing between the two. In phenylketonurics the rise in serum level is higher and more prolonged, reaching 30 mg/100 ml at 1 h and 40 to 59 mg/100 ml at 2 to 5 h.

REFERENCES

Hsia, D. Y.-Y., Driscoll, K. W., Troll, W., and Knox, W. E.: Nature, *178*: 1239, 1956.
Hsia, D. Y.-Y., and Inouye, T.: Inborn Errors of Metabolism. Chicago, Year Book Medical Publishers, Inc., 1966, pp. 69, 71, 222.

TYROSINOSIS AND ALKAPTONURIA

Tyrosine is derived partly from dietary proteins and partly from hydroxylation of phenylalanine to tyrosine *in vivo* (equation 23). The further metabolism of tyrosine may follow a number of different pathways. In a minor pathway, tyrosine is converted to dihydroxyphenylalanine (DOPA) and then to the pressor hormones, norepinephrine and epinephrine. The reaction pathway is as follows:

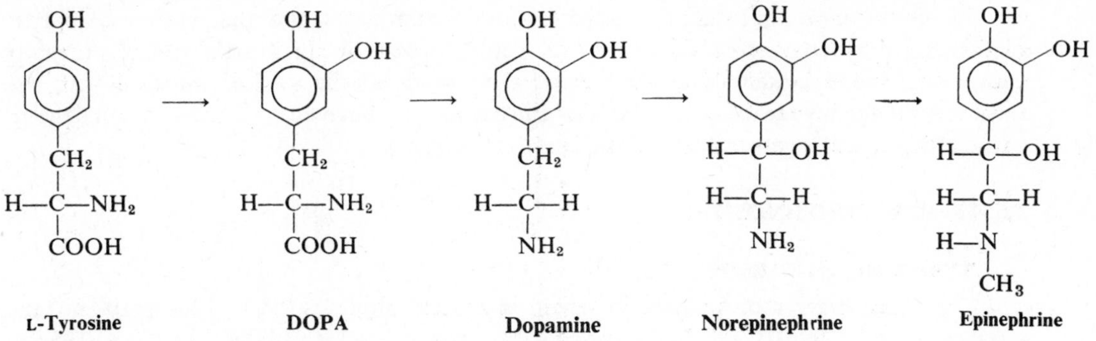

L-Tyrosine → DOPA → Dopamine → Norepinephrine → Epinephrine (25)

The major pathway leads, by deamination (see equation 26), to *p*-hydroxyphenylpyruvic acid (PHPPA) and then to homogentisic acid (HGA).[9,19,26] The aromatic ring in the latter is opened by homogentisic acid oxidase to form maleylacetoacetic acid, which is then hydrolyzed into fumarate and acetoacetate. These four-carbon compounds are then metabolized via the tricarboxylic acid cycle. Deficiencies of PHPPA-oxidase and HGA-oxidase, of genetic origin, give rise to two rare but well recognized biochemical diseases, *tyrosinosis* and *alkaptonuria*.

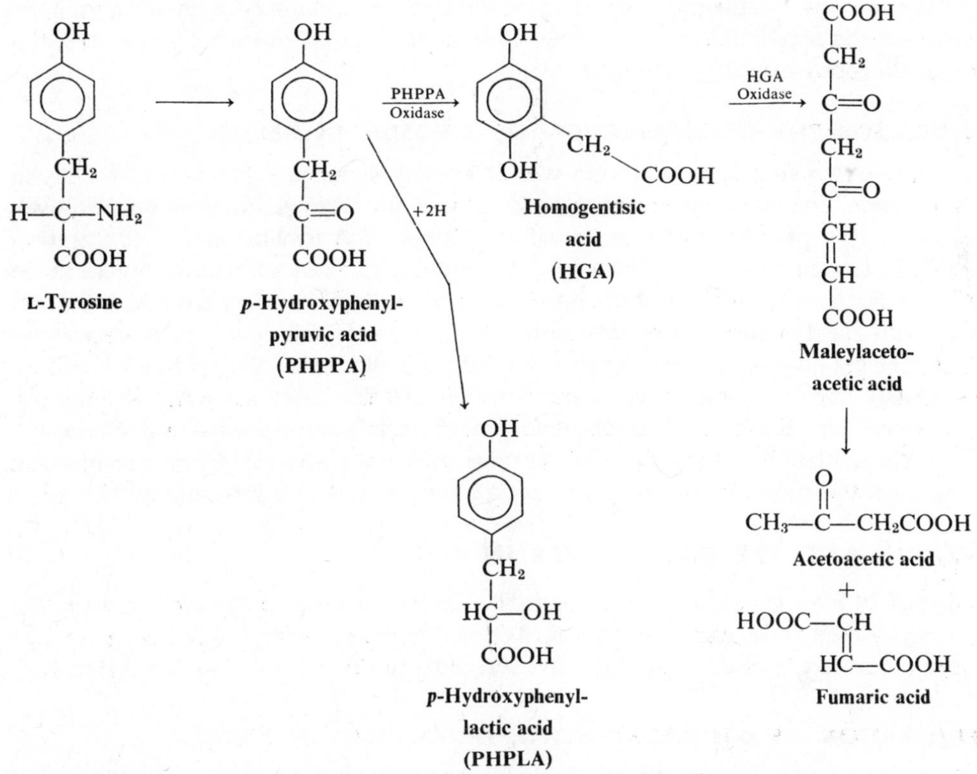

L-Tyrosine → *p*-Hydroxyphenylpyruvic acid (PHPPA) —PHPPA Oxidase→ Homogentisic acid (HGA) —HGA Oxidase→ Maleylacetoacetic acid → Acetoacetic acid (CH_3—CO—CH_2COOH) + Fumaric acid (HOOC—CH=CH—COOH) (26)

p-Hydroxyphenylpyruvic acid (PHPPA) —+2H→ *p*-Hydroxyphenyllactic acid (PHPLA)

In *tyrosinosis*, little or no HGA is formed and the serum tyrosine level rises. Tyrosine and *p*-hydroxyphenylpyruvic acid as well as DOPA and *p*-hydroxyphenyllactic acid (PHPLA) accumulate and are found in significantly elevated amounts in the urine.

Severe liver damage accompanies tyrosinosis; the condition may be fatal in infancy or lead to cirrhosis later in life. Other associated phenomena include renal tubular damage and resistant rickets. Most of the cases have been found in an isolated French Canadian population in Quebec.

In *alkaptonuria*, homogentisic acid (HGA) accumulates and the further oxidation of the para-dihydroxy aromatic ring by oxygen to brown or black melanin-like pigments causes the urine to darken on standing, especially if made alkaline. Alkaptonuria is harmless until middle age is reached, when sufficient pigment may have accumulated in cartilage to cause arthritis and pigmentation of the ears (ochronosis).

TESTS FOR TYROSINOSIS

TYROSINE IN SERUM

The concentration of tyrosine in serum in normal adults is 0.8 to 1.3 mg/100 ml, the average being 1.1 mg/100 ml. In tyrosinosis the tyrosine level is usually 4 to 10 mg/100 ml, whereas in phenylketonuric adults it is below normal (0.55 ± 0.2 mg/100 ml).

In full-term newborns of normal weight, the normal serum levels for tyrosine range from 1.6 to 3.7 mg/100 ml; however, considerably higher concentrations, 7 to 24 mg/100 ml, are found in the sera of premature infants and in full-term infants of low birth weight, owing to the immaturity of the liver and its decreased ability to synthesize the appropriate enzymes. As the liver matures, the accumulated tyrosine is metabolized and serum levels will decrease to adult levels within 4 to 8 weeks. An immaturity of the liver may also cause a phenylalanine hydroxylase deficiency and, therefore, a (temporary) increase in serum phenylalanine levels. Furthermore, the enzyme phenylalanine hydroxylase may be inhibited as a result of the accumulation of tyrosine. Care must be taken in such cases to avoid an erroneous diagnosis of phenylketonuria.

METHODS FOR THE DETERMINATION OF TYROSINE IN SERUM

Routine chromatography of serum will detect any significant increase in the serum level of tyrosine. The most convenient method for assaying tyrosine in serum quantitatively is the fluorometric procedure of Udenfriend and Cooper.[28] A trichloracetic acid filtrate of the serum is treated with a mixture of α-nitroso-β-naphthol (ANBN) and nitrite in the presence of nitric acid. Initially the ANBN and nitrite react to form a pinkish colored complex with the tyrosine. Treatment with HNO_3 converts this to a yellow fluorescing material. The yellow pigment is extracted into ethylene dichloride to separate it from the excess ANBN and its fluorescence is measured at 570 nm, after activation by light of 460 nm. Hsia[11] has adapted the Udenfriend procedure to simple micro- and semimicro-methods. The method is not specific, and tyramine and other *p*-hydroxyphenyl compounds also react, but these are present in insufficient quantity to interfere in serum assays.

TYROSINE AND ITS DERIVATIVES IN URINE

Normal 24 h urine output of tyrosine is 8 to 20 mg for both adults and children. The assay method used for serum is not so satisfactory for urine, owing to interfering tyrosine derivatives, but may be used as a qualitative screening test for para-substituted phenols.[27]

DETERMINATION OF p-HYDROXYPHENYLPYRUVIC ACID IN URINE

Tyrosinosis can be diagnosed most reliably by the isolation and identification of *p*-hydroxyphenylpyruvic acid (PHPPA) from urine by chromatographic means. Urine of affected patients may contain up to 1.6 g of PHPPA, some 25 times the quantity present in normal urine. More conveniently, the *Millon reaction* for tyrosine can be used as a means of detecting tyrosinosis urine. A positive reaction is given by tyrosine and other tyrosine derivatives, as well as by PHPPA, and is therefore not specific. Nevertheless, the test is useful because elevated levels of tyrosine and hydroxyphenyllactic acid accompany the increased urinary output of PHPPA.

Method. Mix 2.5 ml of urine and 2.0 ml of a solution of 15 g $HgSO_4$ in 2.5 molar H_2SO_4, allow to set for one hour, and then centrifuge. Transfer the supernatant to a 50 ml beaker, add 10 ml of 1 molar H_2SO_4, and gently boil the mixture for 10 to 15 min. Add another 2 ml of 1 molar sulfuric acid and permit the beaker to cool for 30 min. Then add 1 ml of fresh $NaNO_2$ (2 g/100 ml). The formation of an orange color indicates a positive test.

TESTS FOR THE DETECTION OF ALKAPTONURIA

The characteristic darkening of the urine on exposure to air may take many hours, but it is usually seen on the baby's diaper. It must be distinguished from melanuria (see below) and from darkening due to phenols, gentisic acid (a salicylate metabolite), methyldopa, and indoxyl sulphate (indican), which is an intestinal decomposition product of tryptophan accumulating under conditions of intestinal stagnation.

Reduction tests

HGA will reduce an ammoniacal solution of silver nitrate very rapidly. Add 5.0 ml of $AgNO_3$ solution (3 g/100 ml) to 0.50 ml of urine, followed by a few drops of dilute ammonia solution (10 ml NH_4OH/100 ml). If HGA is present, a brown-black to black precipitate of reduced, elemental silver will be formed immediately, often even before the addition of the NH_4OH. With melanogens the reaction is very slow and excess NH_3 must be present. On heating the urine with Benedict's qualitative sugar reagent, the supernatant becomes dark and the usual precipitate of yellow cuprous oxide is observed. Melanogens do not reduce Benedict's reagent except if present in very large quantities. The supernatant is not darkened.

Ferric chloride test[27,29]

If a ferric chloride solution (10 g $FeCl_3 \cdot 6H_2O$ in 100 ml of 0.1 mol/l HCl) is added dropwise to 5 ml of urine, a transient blue color will form, as with phenylpyruvic acid, but PPA-containing urines will not react positively with any of the reduction tests just described. Melanogens give a brownish black color with $FeCl_3$. Methods for the quantitative assay of HGA are available, but quantitation is seldom requested.

ALBINISM AND MELANURIA

Melanins, abnormal pigments occasionally observed in urine specimens, are end products of the metabolism of tyrosine. These complex pigments and their colorless precursors, melanogens, are formed as the result of the oxidation of DOPA (dihydroxyphenylalanine) via pathways that are still obscure. The pigments occur normally in the skin and are found in most tissues to some degree. More is present in dark-skinned individuals than in light-skinned persons. Absence of the pigment gives rise to the condition of *albinism*.[26] *Melanomas* are melanin-producing tumors of the skin, sometimes developing in the retina of the eye. These uncommon tumors may produce sufficient pigment to be detected in the urine. If metastasis to the liver has occurred, easily demonstrable quantities of the pigments or their colorless precursors are often present in the urine. Melanogens in urine darken when exposed to air and sunlight just like homogentisic acid.

TESTS FOR MELANOGENS IN URINE

Ferric chloride test

Add about 1 ml of a solution of 10 g $FeCl_3 \cdot 6H_2O$ in 100 ml of 1.2 molar HCl to about 5 ml of urine. If melanogens are present, the urine color changes rapidly to a dark brown.

Thormählen test

Add 5 to 6 drops of fresh sodium nitroferricyanide solution (5.0 g/100 ml) to 5.0 ml of urine in a test tube, then add 0.5 ml of 10 molar NaOH, and mix vigorously. Rapidly cool the tube in cold tap water and acidify with acetic acid solution (33 ml/100 ml). With a normal urine, an olive or brownish green color is obtained; melanogens give a color varying from a greenish blue to bluish black, depending on the quantity present. This is probably the most specific and sensitive test available. High levels of creatinine give a brown color.

Ammoniacal silver nitrate test

Add to 0.5 ml of urine 5.0 ml of $AgNO_3$ solution (3 g/100 ml) followed by dilute ammonia solution (2 ml concentrated NH_4OH/100 ml) until the AgCl precipitate is almost dissolved. The solution will darken as a result of the formation of both melanins and colloidal silver. The reaction develops slowly. In contrast, homogentisic acid darkens the silver solution rapidly, even before the addition of the ammonia.

Hydroxyproline, Collagen Metabolism, and Bone Disease

Collagen is the only human protein that contains appreciable amounts of hydroxyproline. Estimations of urinary hydroxyproline can therefore be used as a measure of the daily turnover of collagen, provided dietary sources of hydroxyproline are first reduced to a minimum. Since bone contains about 40 per cent of the total body collagen, determinations of the 24 h urinary hydroxyproline output have proved to be of value in the clinical study of bone disease.

In blood and urine, hydroxyproline occurs in three forms: as free hydroxyproline, or combined in peptides or in proteins. In urine it is mainly (90 per cent) in the peptide form; in blood it is mainly (80 per cent) in the protein form (hypro-protein).[15]

CLINICAL SIGNIFICANCE

The level of hydroxyproline in plasma or urine is extremely sensitive to dietary intake: a single large intake of gelatin based ice cream is said to be enough to shift the 24 h output into the abnormal range. If, however, the patient is on a diet sufficiently low in collagen, the 24 h output provides a useful index of the turnover of bone matrix, complementing the evidence on mineral turnover provided by estimations of serum calcium, phosphate, and alkaline phosphatase. Urinary hydroxyproline probably indicates the degree of bone destruction by osteoclasts, and serum alkaline phosphatase indicates the degree of bone formation by osteoblasts.

High levels of urinary hydroxyproline are found in bone diseases associated with a high rate of matrix turnover, such as Paget's disease of bone or metastatic bone disease. Determinations are more often carried out for the control of treatment than for diagnosis, as for example in the control of calcitonin therapy for Paget's disease.[3] In this disease the rates of both bone formation and bone destruction are greatly increased; on treatment with calcitonin, osteoclastic bone resorption is reduced, leading usually to clinical improvement and a slow fall in urinary hydroxyproline and serum alkaline phosphatase.

In endocrine disease, increased levels of thyroxine, parathormone, or growth hormone in serum are accompanied by increased urinary hydroxyproline outputs. High levels are also found inconsistently in some connective tissue diseases. In children, the more rapid and variable growth rate is paralleled by much higher and more variable hydroxyproline excretion. A "hydroxyproline index" (hydroxyproline/creatinine × weight in kg) has been used as a test of growth when assessing the treatment of malnourished children.

Estimations of plasma free hydroxyproline are required for the diagnosis of congenital hydroxyprolinemia (see Table 7–17). As one would expect, plasma peptide hydroxyproline levels rise in uremia, especially when accompanied by bone disease. The hypro-protein level has been reported to increase in diseases associated with the acute phase reaction.

DETERMINATION OF URINARY HYDROXYPROLINE

The patient must be on a low collagen diet during the test and for the previous 24 h. He may eat eggs and milk but no meat, fish, or poultry and no jelly or other foods or sweets containing gelatin.

An aliquot of a 24 h urine specimen is hydrolyzed with acid to free hydroxyproline from peptides. The hydroxyproline is then oxidized to pyrrole, usually with chloramine T, and the concentration of the red compound formed by the reaction between pyrrole and Ehrlich's reagent (*p*-dimethylaminobenzaldehyde) is determined colorimetrically.

Unfortunately, urine contains substances such as tyrosine and tryptophan which also form colored compounds. Various methods for eliminating this interference have been recommended. Preliminary to oxidation, a cation exchange resin may be used to isolate the hydroxyproline either on a column[2a,13] or by a batch method.[7] Alternatively, the pyrrole produced on oxidation may be removed by distillation or by extraction with toluene[12] before the colorimetric reaction. An automatic amino acid analyzer may also be used.

To verify that the patient has been on a low collagen diet, the free hydroxyproline, which is very sensitive to dietary collagen, may also be determined. For this procedure an aliquot of unhydrolyzed urine is used.

NORMAL LEVEL

Adult on low collagen diet: 10 to 50 mg/d (total)
0.2 to 1.0 mg/d (free)

REFERENCE

LeRoy, E. C.: The techniques and significance of hydroxyproline measurement in man. *In* Advances in Clinical Chemistry. O. Bodansky and C. P. Stewart, Eds. New York, Academic Press, Inc., 1967, vol. 10, p. 213.

REFERENCES

1. Berlow, S.: Abnormalities in the metabolism of sulfur containing amino acids. *In* Advances in Clinical Chemistry. H. Sobotka and C. P. Stewart, Eds. New York, Academic Press, Inc., 1967, vol. 9 p. 165.
2. Bigwood, E. J., Crokaert, R., Schram, E., Soupart, P., and Vis, H.: Aminoaciduria. *In* Advances in Clinical Chemistry. H. Sobotka and C. P. Stewart, Eds. New York, Academic Press, Inc., 1959, vol. 2, p. 201.
2a. Cleary, J., and Saunders, R. A.: Clin. Chim. Acta., *57*: 217, 1974.
3. Foster, G. V.: Calcitonin. *In* Recent Advances in Clinical Pathology, Series 6. S. C. Dyke, Ed. London, Churchill Livingstone, 1973, p. 43.
4. Frame, E. G.: Free amino acids in plasma and urine by the gasometric ninhydrin-carbon dioxide method. *In* Standard Methods of Clinical Chemistry. D. Seligson, Ed. New York, Academic Press, Inc., 1963, vol. 4, p. 1.
5. Frame, E. G., Russel, J. A., and Wilhelmi, A. E.: J. Biol. Chem., *149*: 255, 1943.
6. Goodwin, J. F.: Spectrophotometric quantitation of plasma and urinary amino nitrogen with fluorodinitrobenzene. *In* Standard Methods of Clinical Chemistry. R. P. MacDonald, Ed. New York, Academic Press, Inc., 1970, vol. 6, p. 89.
7. Goverde, B. C., and Veenkamp, F. J. N.: Clin. Chim. Acta, *41*: 29, 1972.
8. Guthrie, R., and Susi, A.: Pediatrics, *32*: 338, 1963.
9. Harris, H.: Frontiers of Biology: The Principles of Human Biochemical Genetics. A. Neuberger and E. L. Tatum, Eds. London, North-Holland Publishing Co., and New York, American Elsevier Publishing Co., 1970, vol. 19.
10. Hsia, D. Y.-Y., Driscoll, K. W., Troll, W., and Knox, W. E.: Nature, *178*: 1239, 1956.

11. Hsia, D. Y.-Y., and Inouye, T.: Inborn Errors of Metabolism. Part 2. Chicago, Yearbook Medical Publishers, Inc., 1966, pp. 69, 71, 222.
12. Kivirikko, K. I., Laitinen, O., and Prockop, D. J.: Anal. Biochem., *19*: 249, 1967.
13. Klein, L.: Hydroxyproline in urine and tissues. *In* Standard Methods of Clinical Chemistry. R. P. MacDonald, Ed. New York, Academic Press, Inc., 1970, vol. 6, p. 41.
14. LaDu, B. N., and Michael, P. J.: J. Lab. Clin. Med., *55*: 491, 1960.
15. LeRoy, E. C.: The technique and significance of hydroxyproline measurement in man. *In* Advances in Clinical Chemistry. O. Bodansky and C. P. Stewart, Eds. New York, Academic Press, Inc., 1967, vol. 10, p. 213.
16. Levin, B.: Hereditary metabolic disorders of the urea cycle. *In* Advances in Clinical Chemistry. O. Bodansky and A. L. Latner, Eds. New York, Academic Press, Inc., 1971, vol. 14, p. 66.
17. McCaman, M. W., and Robins, E.: J. Lab. Clin. Med., *59*: 885, 1962.
18. Medical Research Council Working Party on Phenylketonuria: Brit. Med. J. 1968 (4): 7.
19. Milne, M. D.: Some abnormalities of amino acid metabolism. *In* Biochemical Disorders in Human Disease. 3rd ed. R. H. S. Thompson and I. D. P. Wootton, Eds. London, Churchill, 1970, p. 553.
20. Rapp, R. D.: Clin. Chem., *9*: 27, 1963.
21. Saifer, A.: Rapid screening methods for the detection of inherited and acquired aminoacidopathies. *In* Advances in Clinical Chemistry. O. Bodansky and A. L. Latner, Eds. New York, Academic Press, Inc., 1971, vol. 14, p. 146.
22. Scriver, C. R., Davies, E., and Cullen, A. M.: Lancet, 1964 (ii), 230.
22a. Scriver, C. R., and Rosenberg, L. E.: Major Problems in Clinical Pediatrics: Amino Acid Metabolism and its Disorders. A. L. Schaffer, Ed. Philadelphia, W. B. Saunders Company, 1973.
23. Smith, I.: Chromatographic and Electrophoretic Techniques. 3rd ed. London, Heinemann, 1969, vol. 1.
24. Sobel, C., Henry, R. J., Chiamori, N., and Segalove, M.: Proc. Soc. Exp. Biol. Med., *95*: 808, 1957.
25. Spaeth, G. L., and Barber, G. W.: Pediatrics, *40*: 586, 1967.
26. Stanbury, J. B., Wyngaarden, J. B., and Fredrickson, D. S.: The Metabolic Basis of Inherited Disease. 3rd ed. New York, McGraw-Hill Book Company, 1972.
27. Thomas, G. H., and Howell, R. R.: Selected Screening Tests for Genetic Metabolic Diseases. Chicago, Year Book Medical Publishers, Inc., 1973, p. 81.
28. Udenfriend, S., and Cooper, J. R.: J. Biol. Chem., *203*: 953, 1953.
29. Varley, H.: Practical Clinical Biochemistry. 4th ed. London, Heinemann and New York, Interscience Books, Inc., 1967.
30. White, H. H.: Clin. Chim. Acta, *21*: 297, 1968.
31. Wong, P. W. K., O'Flynn, M. E., and Inouye, T.: Clin. Chem., *10*: 1098, 1964.
32. Woolf, L. I.: Inherited metabolic disorders: errors of phenylalanine and tyrosine metabolism. *In* Advances in Clinical Chemistry. H. Sobotka and C. P. Stewart, Eds. New York, Academic Press, Inc., 1963, vol. 6, p. 98.
33. Wootton, I. D. P.: Microanalysis in Medical Biochemistry. 5th ed. London, Churchill Livingstone, 1974.

CHAPTER 8

HEMOGLOBIN, HEMOGLOBIN DERIVATIVES, AND MYOGLOBIN

by Virgil F. Fairbanks, M.D.

Structures

Hemoglobin is a spheroidal molecule with a central cavity and a molecular mass of 64,456 daltons. It is divisible radially into four tetrahedrally arranged segments, each segment consisting of loops of globin and each presenting to the exterior surface of the sphere a heme (iron-porphyrin) group. The heme group is the site of O_2 bonding.

The spheroidal hemoglobin tetramer is composed of two pairs of identical globin chains. In hemoglobin of adults (hemoglobin A), these chains are designated α and β. The two α and two β chains are arranged about the central cavity as shown in Figures 8–1 and 8–2. There are two narrow areas of contact between adjacent α and β chains (designated $\alpha_1\beta_2$ contact points) and two broad areas of contact (designated $\alpha_1\beta_1$ contact points).[55] As hemoglobin passes from the oxygenated to the deoxygenated state, there is rotational movement about the $\alpha_1\beta_2$ contact points such that the central cavity enlarges and accommodates a molecule of 2,3-diphosphoglycerate (DPG), a product of glucose metabolism. Conversely, with oxygenation, rotation of the two halves of the hemoglobin molecule results in a smaller central cavity and loss of the DPG.

Normal adult human blood contains two other hemoglobins in addition to hemoglobin

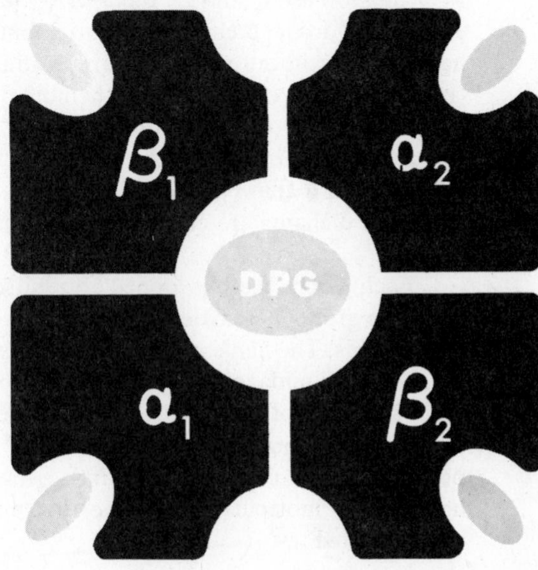

Figure 8–1 Hemoglobin is a tetramer composed of two pairs of identical globin chains. For hemoglobin A, there are two α and two β chains. Each globin chain has an attached heme group (light gray oval). 2,3-Diphosphoglycerate (DPG), a product of glucose metabolism, occupies the central cavity when hemoglobin is in the deoxygenated state (deoxyhemoglobin).

Figure 8–2 Three-dimensional representation of the hemoglobin molecule based on x-ray diffraction studies. The α chains (light gray) lie over the β chains (dark gray). The central cavity (site of DPG binding) is apparent. The disc-shaped objects represent heme groups, one in each globin chain. The amino and carboxyl ends of the α globin chains are indicated by letters *N* and *C*, respectively. The identical globin chains (for example, both α chains) are linked at their carboxyl and amino ends. (From: The hemoglobin molecule, by M. F. Perutz. Copyright © 1964 by Scientific American, Inc. All rights reserved.)

A ($\alpha_2\beta_2$). These are hemoglobin $A_2(\alpha_2\delta_2)$, composed of two α and two δ chains, and hemoglobin F ($\alpha_2\gamma_2$), composed of two α and two γ chains. Hemoglobin F is the major hemoglobin of fetal life and small amounts of this hemoglobin normally persist in adult life. In hemoglobins A_2 and F, the δ or γ chains have the same structural relationship to the α chains as do the β chains in hemoglobin A. Hemoglobin Gower II ($\alpha_2\varepsilon_2$) is an embryonic hemoglobin without a counterpart in adult life. In certain disorders, tetramers of β chains (hemoglobin H) or of γ chains (hemoglobin Barts) may occur.

Each of the globin chains is a long strand of amino acids, 141 in each α chain and 146 in each β, δ, or γ chain (Fig. 8–3).[16] Each chain contains helical (coiled) segments (seven in an α chain and eight in a β chain) separated by short noncoiled segments. The β, γ, and δ chains are closely homologous, the β and δ chains differing in only 10 amino acids.

Each globin chain is looped about itself so as to form a pocket or cleft in which nestles the heme group. Normally, this heme pocket is formed entirely by nonpolar (hydrophobic) amino acids. The heme moiety is suspended within this pocket by an attachment of the iron atom to the imidazole group of the proximal histidine (position 92 of the β chain [β^{92}] or position 87 of the α chain [α^{87}]). The imidazole group of the distal histidine (β^{63} or α^{58}) is also in contiguity with the iron of heme, but it appears to swing in and out of this position to permit the ingress and egress of O_2. The four iron atoms of the tetrameric hemoglobin molecule are in the divalent state, whether hemoglobin is oxygenated or deoxygenated.

Function

Hemoglobin is uniquely adapted to take up and discharge O_2 within a relatively narrow range of O_2 pressure (pO_2). This is a function of a chain-to-chain interaction whereby an alteration in spatial relationships of one portion of the molecule to another facilitates the uptake or release of O_2. The O_2 dissociation curve of normal blood hemoglobin is sigmoidal (Fig. 8–4). The shape of this curve implies that, once the hemoglobin molecule has taken up a molecule of O_2 at one heme site (and given up DPG), the other three sites are oxygenated with only a modest increment in oxygen pressure. Conversely, a molecule with all four heme sites oxygenated readily gives up O_2 at three of these sites. This highly coordinated response to slight changes in oxygen tension, which is absolutely vital for effective O_2

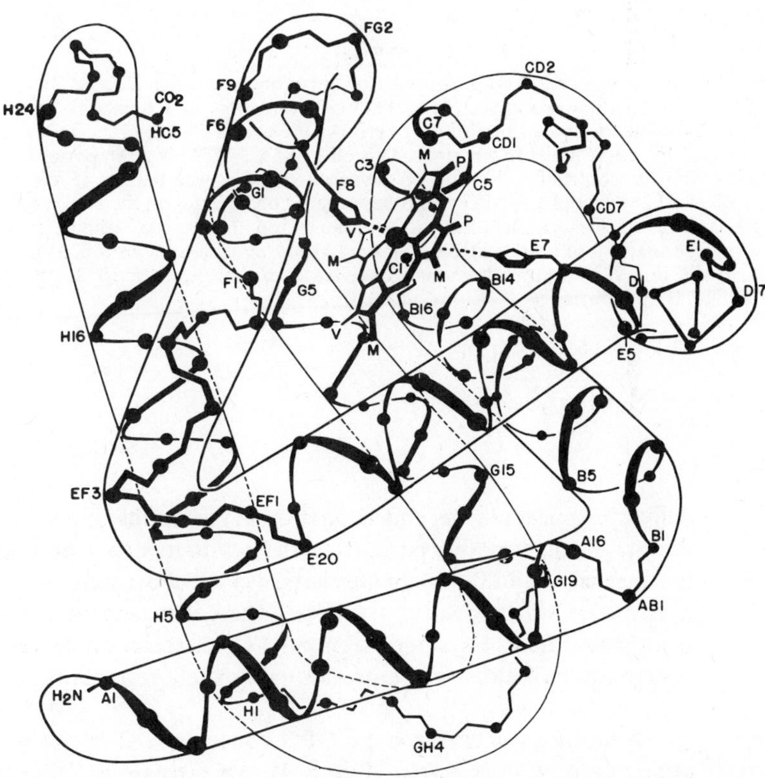

Figure 8–3 The basic unit of both hemoglobin and myoglobin is a long chain of amino acids so looped to provide a cleft that is occupied by the heme group. The globin chain is composed of several segments of amino acids in a spiral or helical configuration linked by short nonhelical chains of amino acids. The helical segments are designated A through H. In this illustration, amino acids are designated in accordance with the helical or nonhelical segment in which they occur.

Some of the important sites in this molecule to which reference is made in this chapter are as follows: for the β chain, A2 is the site of substitution in hemoglobins S and C; F8 is the position of "proximal histidine," which links the globin chain with iron in the heme group. Nonhelical segments (indicated by irregular lines) are designated according to the adjacent helical segments—for example, CD2 is the second amino acid in the nonhelical region between segments C and D. (From Dickerson, R.E.: X-ray analysis and protein structure. *In* The Proteins: Composition, Structure, and Function. 2nd ed. H. Neurath, Ed. New York, Academic Press, Inc., 1964, vol. 2, pp. 603–778. By permission.)

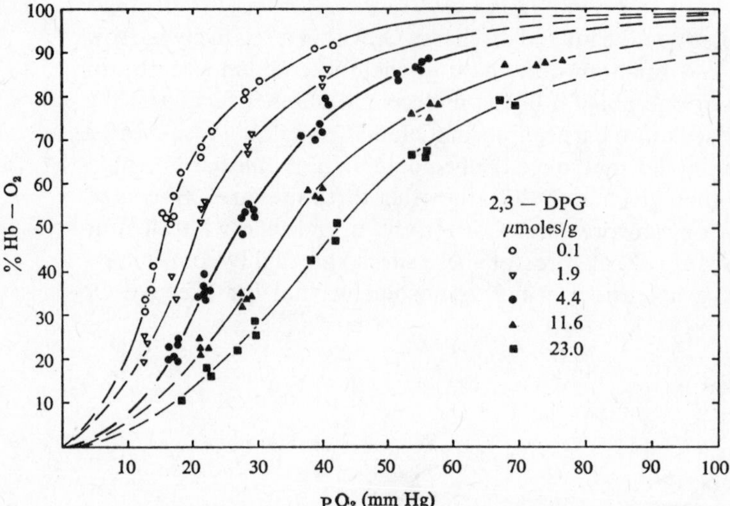

Figure 8–4 The normal oxygen dissociation curve of hemoglobin is sigmoidal. Changes in 2,3-diphosphoglycerate (DPG) concentration in the erythrocyte markedly influence the position of the curve. As the concentration of 2,3-DPG increases, the curve shifts to the right. This effect of 2,3-DPG is an important mechanism for controlling the rate of oxygen delivery to the tissues. (From Duhm, J.: The effect of 2,3-DPG and other organic phosphates on the Donnan equilibrium and the oxygen affinity of human blood. *In* Oxygen Affinity of Hemoglobin and Red Cell Acid Base Status [Alfred Benzon Symposium, IV]. M. Rørth and P. Astrup, Eds. New York, Academic Press, Inc., 1972, pp. 583–594. By permission.)

delivery, appears to depend in part on (1) reversible binding of certain amino acids across the $\alpha_1\beta_2$ contact point, permitting the $\alpha\beta$ dimers to snap back and forth as the molecule is oxygenated and deoxygenated, and (2) the movement of DPG in and out of the central cavity (this small molecule stabilizes the deoxygenated configuration). Pure solutions of hemoglobin do not manifest sigmoidal O_2 dissociation curves, but the addition of DPG in nearly equimolecular concentration completely restores the normal sigmoidal dissociation curve.

Alterations in erythrocytic DPG concentration *in vivo* markedly affect the O_2 dissociation curve of hemoglobin (Fig. 8–4). An increase in DPG concentration causes the dissociation curve to shift to the right, indicating decreased O_2 affinity of hemoglobin and increased O_2 delivery to tissues.[18] This is an important compensatory mechanism in anemia, and it has been shown that the erythrocytic DPG concentration varies reciprocally with the blood hemoglobin concentration. Conversely, low DPG concentration is associated with shift of the O_2 dissociation curve to the left, indicating increased O_2 affinity of hemoglobin and decreased O_2 delivery to tissues. This is a problem in blood banking, because blood preserved with ACD (acid-citrate-dextrose) solution shows a progressive decrease in DPG concentration.

Changes in pH and pO_2 of blood also influence the affinity of hemoglobin for oxygen. A decrease in blood pH or an increase in pCO_2 causes the dissociation curve to shift to the right (acid Bohr effect), and an increase in blood pH or a decrease in pCO_2 causes the curve to shift to the left (alkaline Bohr effect).

Synthesis and degradation

Hemoglobin synthesis involves the independent formation of *heme* and *globin* chains. Heme synthesis is considered elsewhere (see Chap. 9). Globin chains are synthesized by cytoplasmic ribosomes in the cells as directed by the nucleotide sequence of the chromosomal DNA. Abnormal hemoglobins arise by errors in the chromosomal DNA sequence. For example, a change, at a single site in the DNA base sequence, from cytosine-thymine-cytosine to cytosine-adenine-cytosine would be translated into a messenger RNA that would direct the synthesis of hemoglobin S (the abnormal hemoglobin responsible for sickle cell disorders). This "mis-sense" type of mutation is the cause of the vast majority of hemoglobinopathies. The thalassemias, on the other hand, appear to be due to a decrease in the rate of synthesis of normal globin chains rather than to the formation of abnormal globin chains. This appears to be the result of an actual decrease in the rate of formation of messenger RNA.

Within the intact erythrocyte, hemoglobin iron is normally oxidized very slowly to the trivalent state to form methemoglobin. As a result, the molecule can no longer bind and transport oxygen. Methemoglobin formation is counterbalanced by a continual conversion of methemoglobin back to hemoglobin, a process requiring the action of the NADH-dependent enzyme, methemoglobin reductase. If there is a deficiency of methemoglobin reductase, methemoglobin accumulates within the erythrocyte to a concentration of 10 to 30 per cent of the total pigment (hemoglobin + methemoglobin).

Hemoglobin degradation normally occurs within phagocytic cells of the reticuloendothelial system, when the aged and metabolically run-down erythrocyte can no longer provide the energy to sustain the K^+-Na^+ cation pump and therefore leaks cations at an increasing rate. The sequence of this extravascular erythrocyte destruction is (1) phagocytosis of aged cells, (2) opening of the heme ring to form verdohemoglobin, and (3) removal of iron and protein from the tetrapyrrole chain. The tetrapyrrole chain is further metabolized to bilirubin. (The details of hemoglobin catabolism are presented in more complete detail in Chapter 19.) The iron is reutilized in the formation of new hemoglobin. The protein is digested, and its component amino acids enter the general amino acid pool. It is not certain whether methemoglobin is an obligatory intermediate in this sequence. When there is accelerated extravascular erythrocyte destruction, as in most autoimmune hemolytic disorders and hypersplenic states, this process is accentuated.

Under certain circumstances (for example, mechanical hemolysis due to faulty intracardiac valvular prostheses or paroxysmal nocturnal hemoglobinuria), hemolysis may be predominantly intravascular. In this case, the free plasma hemoglobin concentration may be transiently increased. Free plasma hemoglobin is normally bound by the plasma protein, *haptoglobin*. In severe hemolytic episodes, the unbound haptoglobin concentration may be moderately decreased or it may be completely depleted. The hemoglobin-haptoglobin complex formed is rapidly removed by parenchymal cells of the liver. Free hemoglobin, in excess of the binding capacity of haptoglobin, is excreted in the urine, to which it imparts a chocolate-brown color. Hemoglobinuria may be identified by the occult blood tests described later in this chapter. The plasma haptoglobin concentration returns to normal after two to three days unless the hemolysis is a continuing process. Some of the heme that dissociates from hemoglobin in plasma may be taken up by *hemopexin*, a heme-binding protein present in normal plasma.[48] Some is taken up by serum albumin to form an albumin-heme complex called *methemalbumin*.

Hemoglobin variants

Approximately 400 structural variants of the hemoglobin molecule have been described.[41,56] Many of these have been discovered incidentally and are not associated with any abnormal symptoms or physical abnormalities. Some of the abnormal hemoglobins and their clinical manifestations are listed in Table 8–1.

TABLE 8–1 CLINICALLY IMPORTANT HEMOGLOBINOPATHIES*

Clinical Manifestation	*Hemoglobin Designation*	*Substitution*	*Comments*
Hemolytic anemia	S	$\beta^{6\text{glu}\rightarrow\text{val}}$	Sickling
	C	$\beta^{6\text{glu}\rightarrow\text{lys}}$	Target cells
	E	$\beta^{26\text{glu}\rightarrow\text{lys}}$	Target cells
	O_{Arab}	$\beta^{121\text{glu}\rightarrow\text{lys}}$	Target cells
	Köln	$\beta^{98\text{val}\rightarrow\text{met}}$	Unstable hemoglobin
	Zürich	$\beta^{63\text{his}\rightarrow\text{arg}}$	Unstable hemoglobin
	H	β_4	Unstable hemoglobin
Cyanosis	$M_{\text{Iwate (Kankakee)}}$	$\alpha^{87\text{his}\rightarrow\text{tyr}}$	This group of rare abnormal hemoglobins is characterized by abnormal absorption spectra and by normal methemoglobin values when tested by the usual spectrophotometric measurement of methemoglobin (see methemoglobin assay)
	$M_{\text{Boston (Osaka)}}$	$\alpha^{58\text{his}\rightarrow\text{tyr}}$	
	$M_{\text{Saskatoon (Chicago)}}$	$\beta^{63\text{his}\rightarrow\text{tyr}}$	
	$M_{\text{Milwaukee 1}}$	$\beta^{67\text{his}\rightarrow\text{glu}}$	
	$M_{\text{Hyde Park}}$	$\beta^{92\text{his}\rightarrow\text{tyr}}$	
	Kansas	$\beta^{102\text{asn}\rightarrow\text{thr}}$	Rightward displacement of O_2 dissociation curve
Erythrocytosis	Chesapeake	$\alpha^{92\text{arg}\rightarrow\text{leu}}$	This group of abnormal hemoglobins is characterized by marked leftward displacement of the O_2 dissociation curve
	J_{Capetown}	$\alpha^{92\text{arg}\rightarrow\text{gln}}$	
	Malmö	$\beta^{97\text{his}\rightarrow\text{gln}}$	
	Ypsilanti	$\beta^{99\text{asp}\rightarrow\text{tyr}}$	
	Kempsey	$\beta^{99\text{asp}\rightarrow\text{asn}}$	
	Yakima	$\beta^{99\text{asp}\rightarrow\text{his}}$	
	Rainier	$\beta^{145\text{tyr}\rightarrow\text{cys}}$	
	Bethesda	$\beta^{145\text{tyr}\rightarrow\text{his}}$	

* This is a selected list of abnormal hemoglobins with significant clinical effects. Complete lists of known hemoglobin variants are published in several recent textbooks of hematology and in a review by Lehmann and Carrell.[41] Abbreviations for amino acids are: glu, glutamic acid; val, valine; lys, lysine; met, methionine; his, histidine; arg, arginine; tyr, tyrosine; gln, glutamine; asn, asparagine; leu, leucine; thr, threonine; asp, aspartic acid; cys, cysteine.

Genetics

Inheritance of hemoglobinopathies that are associated with structural abnormalities follows simple mendelian laws. Most abnormal hemoglobins cause little or no clinical manifestations in the heterozygote, and thus are autosomal recessive traits. These include hemoglobins S, C, and E. On the other hand, the person who is heterozygous for an unstable hemoglobin (such as hemoglobin Köln), for a hemoglobin M, or for a high-O_2–affinity hemoglobin (such as hemoglobin Malmö) has the clinical manifestations indicated in Table 8–1. Therefore, these are transmitted as autosomal dominant traits.

The structure of α chains is probably controlled by four gene loci, two each on homologous chromosomes.[40] Therefore, abnormal hemoglobins that have substitutions in the α chain generally make up about 25 per cent of the total hemoglobin (one mutant α chain gene, three normal α chain genes). Hemoglobinopathies due to α chain abnormalities are uncommon in comparison with those caused by substitutions in the β chain.

The amino acid sequence of β chains is determined by two gene loci located on a single pair of homologous chromosomes. If one of the two β chain genes contains an altered sequence of DNA (a mutation), approximately 50 per cent of the β chains formed will be abnormal. In the formation of the complete tetrameric hemoglobin molecule, there is a preferential matching of like chains. For example, if β^A chains (normal) and β^S chains are both being formed at the same rate, hemoglobin A ($\alpha_2\beta_2{}^A$) and hemoglobin S ($\alpha_2\beta_2{}^S$) both will be present in equal amounts.* The presence of both hemoglobin A and hemoglobin S in nearly equal quantities is called *sickle cell trait*, a condition that almost never gives rise to any symptoms or clinical abnormalities.

In fact, β^S chains are not as rapidly produced as β^A chains, and the ratio of hemoglobin A to hemoglobin S in a person with sickle cell trait is usually about 60/40. If there is a mutation in only one of a pair of genes controlling β chain structure, such that an unstable hemoglobin is formed (for example, hemoglobin Köln), the proportion of hemoglobin A to unstable hemoglobin in the blood is often 5/1 or greater. This may be due in part to a slow rate of formation of the unstable hemoglobin and in part to accelerated denaturation and removal of the unstable hemoglobin from circulating erythrocytes.

The inheritance of an identical mutation in both of the genes that determine β chain structure results in a homozygous hemoglobinopathy. Thus, if both genes direct the synthesis of β^S chains, only hemoglobin S ($\alpha_2{}^A\beta_2{}^S$) can be formed. No hemoglobin A will be found. This gives rise to *sickle cell disease*, a very serious disorder that usually leads to death by the time of puberty. The simultaneous inheritance of different mutations in both of the β chain genes may also lead to serious disease, as exemplified in *hemoglobin S-C disease*. In this disorder, which exhibits all of the features of *sickle cell disease*, both β^C and β^S chains are formed, leading to production of hemoglobin S ($\alpha_2\beta_2{}^S$) and hemoglobin C ($\alpha_2\beta_2{}^C$). Again, no hemoglobin A is formed.

Like the β chains, δ chains are controlled by a single pair of gene loci. Mutations affecting δ chains are of no clinical significance. There are believed to be four γ chain loci, two on each of the homologous chromosomes. Mutations in the γ chain loci are of no clinical significance.

Studies of families in which there are both α chain and β chain mutant hemoglobins have indicated that these abnormalities are transmitted independently. This implies that the genes for the α and β chains are not closely associated on the chromosome and might even be on different chromosomes. In contrast, β, δ, and also γ genes occur very close to each other on the chromosome. In addition, there is a switching gene which, shortly before birth, decreases the expression of the γ genes (thereby decreasing the formation of hemoglobin F) and reciprocally increases the expression of the β genes (thereby increasing the formation of hemoglobin A). The role of the switching gene is to bring about a transition from nearly 100 per cent hemoglobin F in fetal blood to nearly 100 per cent hemoglobin A in postnatal blood. Failure of normal function of the switching gene gives rise to a condition known as hereditary persistence of fetal hemoglobin (HPFH),[29] a benign disorder in which 10 to 20 per cent of the total blood hemoglobin is hemoglobin F, the remainder being hemoglobin A (and a small amount of normal hemoglobin A_2). Recent studies indicate that HPFH may be due to deletion of one of the gene loci.

* Comments on notation. In the description of the structure of normal hemoglobins, component chains were designated α, β, γ, and δ. Hemoglobin A is, therefore, $\alpha_2\beta_2$. However, in discussing mutations that affect the β chain, confusion is avoided by designating the normal β chain as β^A, and abnormal β chains as β^S (for the β chain abnormality of hemoglobin S), β^C (for the β chain abnormality of hemoglobin C), and so forth. A similar convention is used for the α, γ, and δ chains. Thus, the normal α chain may be designated α^A, and α^I would be the abnormal α chain of hemoglobin I ($\alpha_2{}^I\beta_2{}^A$); the normal γ chain may be designated γ^F, and γ^{Hull} would be the abnormal γ chain of hemoglobin F^{Hull} ($\alpha_2{}^A\gamma_2{}^{Hull}$); the normal δ chain may be designated δ^A, and $\delta^{Flatbush}$ would be the abnormal δ chain of hemoglobin Flatbush ($\alpha_2{}^A\delta_2{}^{Flatbush}$).

An additional notational convention followed in this chapter is the designation of known amino acid substitutions by superscripts following the chain symbol. Thus, the following symbols for hemoglobin S are equivalent: $\alpha_2{}^A\beta_2{}^S$, $\alpha_2\beta_2{}^{6glu\rightarrow val}$, and $\alpha_2\beta_2{}^{6val}$. The last of these expressions indicates that valine is the sixth amino acid in the β chain, a site normally occupied by glutamic acid. This is the structural abnormality responsible for hemoglobin S.

It would seem logical that, because there is a hemoglobin A_2 and a hemoglobin A_3, the major normal adult hemoglobin should be A_1; however, convention decrees that this be called simply hemoglobin A.

Thalassemias and related disorders

Thalassemia is a disorder in which there are anemia and usually characteristic morphologic changes in the erythrocytes, such as hypochromia, microcytosis, target cells, and poikilocytosis. Once thought to be a rare disorder of those whose ancestors lived on the Mediterranean littoral, thalassemia has become recognized as a common disorder throughout the world.

Thalassemia major is a serious affliction. The onset is in early infancy and the clinical features include severe anemia, jaundice, liver and spleen enlargement, skeletal malformations, and cutaneous ulcers. Thalassemia major fortunately is a rare disorder except in populations with a very high frequency of the thalassemia gene (for example, Italians and Greeks).

Thalassemia minor usually causes a mild hypochromic anemia but may occur without any manifestations. It is most common in people derived from the Mediterranean area, Africa, or Asia. However, it is not rare in people derived from northern Europe and is a relatively common cause of mild anemia in the United States.

In the thalassemias there is no structural abnormality of hemoglobin, but there is a decreased rate of hemoglobin synthesis. This is due to failure to translate the DNA code into messenger RNA at a normal rate. It is generally held that this is the result of a repressor gene that inhibits (or represses) the formation of messenger RNA. If α chain synthesis is inhibited, an α thalassemia is produced; if β chain synthesis is inhibited, a β thalassemia is produced.*

As noted above, the structure and synthesis of the α chains of hemoglobin are probably determined by two pairs of α chain genes (two genes on each of two chromosomes). Failure of translation (or repression) of one of these genes would lead to a mild anemia, an α thalassemia minor, with no other identifying features. Failure of translation of two of these α chain loci would give rise to a mild hypochromic anemia because not enough hemoglobin would be formed to fill the erythrocyte. Failure of translation of three of these loci would give rise to hemoglobin H disease, a form of α thalassemia. Failure of translation of all four α loci would result in intrauterine death of the fetus due to extreme anemia.[40]

An alternative hypothesis[69] for the α thalassemias is that there are two α thalassemia genes, designated α-thal$_1$ and α-thal$_2$; α-thal$_1$ is assumed to cause a greater inhibition of α chain synthesis than does α-thal$_2$. According to this hypothesis, an adult who carries one α-thal$_1$ gene (heterozygote) is asymptomatic, is not anemic, and has no abnormal hemoglobins. However, the newborn who is heterozygous for α-thal$_1$ thalassemia has a minor hemoglobin anomaly in that hemoglobin Barts (γ_4) is present in a concentration of 3 to 10 per cent of the total hemoglobin. An adult who is heterozygous for the α-thal$_2$ gene is also asymptomatic and nonanemic. However, the newborn heterozygote for α-thal$_2$ will have 1 to 2 per cent hemoglobin Barts. Homozygosity for α-thal$_2$ would, by this hypothesis, cause a hypochromic anemia without other hemoglobin abnormalities. Double heterozygosity (α-thal$_1$ and α-thal$_2$) would cause moderate to severe hypochromic anemia with hemoglobin H as 10 to 15 per cent of the total hemoglobin. Homozygosity for α-thal$_1$ would lead to fetal death, with 100 per cent of the hemoglobin of the fetus being hemoglobin Barts.

These hypotheses for the α thalassemias may be reconciled by the assumption that the gene α-thal$_2$ represents the repression of one α chain gene, and that the gene α-thal$_1$ represents the repression of two α chain genes. Then, double heterozygosity (α-thal$_1$ and α-thal$_2$) would cause repression (or deletion) of three α chain loci, giving rise to hemoglobin H disease.

There is a repressor gene, believed to be in close proximity to the β and δ gene regions of the chromosome, that may repress the expression of the β gene or of both the β and δ genes. Abnormal function of this repressor gene gives rise to the β thalassemias. If an abnormal repressor gene is inherited from each parent, the result most commonly is thalassemia major. However, if the abnormal repressor gene is inherited from only one parent the result

*More recent evidence indicates that α thalassemia is due to deletion of the gene loci determining α chain structure. Gene deletion may also account for some cases of β thalassemia. However, β thalassemia may be the result of any of several genetic mechanisms: the "repressor" mechanism stated above is clearly a simplification which cannot explain all β thalassemias.

is thalassemia minor. The two main types of β thalassemia minor are the high A_2 type, which accounts for about three-fourths of all β thalassemia minor, and the high F type, which accounts for most of the rest. The common (high A_2) type of β thalassemia is due to inheritance, on one chromosome, of a gene that causes repression of one β chain gene, with normal expression of the other β chain gene (because it is on the other chromosome) and normal expression of both δ chain genes. As a result, the total quantity of hemoglobin produced is diminished, erythrocytes are not well filled with hemoglobin, and a mild anemia is usually present. The proportion of hemoglobin A_2 ($\alpha_2\delta_2$) is increased, but that of hemoglobin F ($\alpha_2\gamma_2$) is normal. Thalassemia minor of the high F type (F-thalassemia, β-δ thalassemia) is due to inheritance, on one chromosome, of a gene that causes repression of both β and δ genes on the same chromosome. The γ chains are not repressed and indeed may be reactivated to compensate for decreased hemoglobin synthesis. Thus, formation of β and δ chains is decreased, a mild hypochromic anemia is observed, a moderate increase in hemoglobin F is demonstrated, and a lower than average proportion of hemoglobin A_2 may be found. Inheritance of the genes for both high A_2 β thalassemia and for high F β-δ thalassemia may cause an anemia of severity intermediate between that of thalassemia major and that of thalassemia minor, or it may cause a clinical disorder virtually indistinguishable from severe thalassemia major.[63] Homozygosity for the high F β-δ thalassemia gene causes only a mild anemia resembling thalassemia minor.

Of the children of one parent with β thalassemia minor and one parent with sickle cell trait, some will have received the β thalassemia minor gene from one parent and the hemoglobin S gene from the other parent. In this case, since the normal β chain gene will be repressed by an abnormal repressor gene on the same chromosome, the predominant hemoglobin formed will be hemoglobin S ($\alpha_2\beta_2{}^S$). Because repression of the normal β chain is usually not complete, a small amount of hemoglobin A will be formed and, in addition, there will be a slight increase in either hemoglobin A_2 or hemoglobin F.

This interaction between a gene for hemoglobin S and, on the opposite chromosome, a gene for β thalassemia minor gives rise to the serious disease known as *sickle-thalassemia* or S-thalassemia. S-thalassemia may have all of the clinical manifestations and the serious prognosis of sickle cell disease, or it may be a milder form. Of the clinically significant sickling disorders in American Blacks, approximately one-fifth are due to S-thalassemia. S-thalassemia may be difficult to distinguish from classic sickle cell disease or from the combined effects of the sickle gene and the gene for HPFH (S-HPFH). The demonstration of small quantities of hemoglobin A is evidence of S-thalassemia, if transfusion can be excluded as a cause. Pedigree studies or the acid elution test of Kleihauer and Betke, described later in this chapter, differentiates S-thalassemia from S-HPFH.

The β and δ gene loci are so close to each other that several abnormal hemoglobins have arisen from fusion of portions of the β and δ genes during meiotic divisions of germinal cells. These are the Lepore hemoglobins, $\alpha_2(\delta\text{-}\beta)_2$. Persons heterozygous for Lepore hemoglobinopathy have both the features of thalassemia minor and the presence of a slow-moving band (indistinguishable from hemoglobin S) on electrophoresis that constitutes 10 to 15 per cent of the total hemoglobin.

The features of the various thalassemia syndromes are given in Table 8–2.

Hemoglobin derivatives

The term hem*o*globin implies that the iron atom is in the reduced or divalent (ferr*o*us) state. This is true whether hemoglobin is in the oxygenated form (oxyhemoglobin) or deoxygenated form (deoxyhemoglobin). The terms reduced hemoglobin and ferrohemoglobin, which have been used as synonyms for deoxyhemoglobin, are redundant (since the iron in hemoglobin is, by definition, in the reduced or "ferro" state) and should be avoided. When the iron is oxidized to the trivalent state, the resultant brown pigment is methemoglobin, also (rarely) called hem*i*globin.

TABLE 8–2 THE THALASSEMIAS AND ASSOCIATED HEMOGLOBIN ABNORMALITIES

Abnormality	*Target Cells*	*Deoxyhemoglobin Solubility*	*Hypo-chromia*	*Anemia*	*Hemoglobin F*	*Hemoglobin A_2*	*Hemoglobin H*	*Hemoglobin F Cellular Distribution*	*Other Features*
α Thalassemia									
1 α chain repressed or deleted	0	Normal	0	0	Normal	Normal	0	Not applicable	Small quantities (~2%) hemoglobin Barts (γ_4) detectable at birth
2 α chains repressed or deleted	0	Normal	0	0	Normal	Normal	0	Not applicable	Hemoglobin Barts (~5%) detectable at birth
3 α chains repressed or deleted	+	Normal	+	Mild to severe	May be increased	Normal	10–15%	Nonuniform	Unstable hemoglobin hemolytic anemia (hemoglobin H)
4 α chains repressed or deleted	0	Abnormal	0	Severe	0*	0*	0*	—	Fetal death with severe pallor and edema (hydrops fetalis)
β Thalassemia (High A_2) and βδ Thalassemia (High F)									
Heterozygous									
High A_2	±	Normal	+	+	Normal	4–10%	0	Not applicable	—
High F	±	Normal	+	+	5–15%	Decreased	0	Nonuniform	—
Homozygous									
High A_2	±	Normal	Marked	Marked	80–100%	Decreased or 0	0	Uniform	Skeletal abnormalities, cutaneous ulcers, growth retardation
High F (βδ)	+	Normal	Moderate	±	100%	0	0	Uniform	—
Doubly heterozygous									
Hemoglobin S and thalassemia	Present	Abnormal	Present	Moderate to marked	Increased	Increased	0	Nonuniform	May exhibit all typical features of sickle cell disease†
Hemoglobin C and thalassemia	Present	Normal	Present	Moderate	Increased	Increased	0	Nonuniform	Often marked splenic enlargement†
Hemoglobin E and thalassemia	Present	Normal	Present	Moderate to severe	Increased	Increased	0	Nonuniform	†
Hereditary Persistence of Fetal Hemoglobin									
Heterozygous	0	Normal	0	0	10–20%	Decreased	0	Uniform	Asymptomatic
Homozygous	0	Normal	0	0	100%	0	0	Uniform	Asymptomatic
Hemoglobin S-HPFH	0	Abnormal	0	0	10–20%	Decreased	0	Uniform	Usually asymptomatic

* Hemoglobin Barts, 100 per cent.
† Either A_2 or F may be increased; hemoglobin A may also be present.

TABLE 8–3 ABSORBANCE MAXIMA OF HEMOGLOBIN AND HEMOGLOBIN DERIVATIVES

Compound	*Absorbance Maxima (nm)*				
Deoxyhemoglobin	428–430		555		
Oxyhemoglobin	412–415		541	576–578	
Carboxyhemoglobin	417–418		537	568–572	
Methemoglobin (neutral pH)	404–407	500	540	578	630
Cyanmethemoglobin	413–418	480	541	580–590	

Carboxyhemoglobin is a hemoglobin-carbon monoxide complex. Carbon monoxide is normally generated in small quantities from the catabolism of heme, although the amount of carboxyhemoglobin formed is too small to be readily measured (measurement of carboxyhemoglobin is reviewed in Chap. 21). In strongly basic or strongly acidic solution, hemoglobin is denatured to *alkaline hematin* or *acid hematin*, respectively. Obsolete methods for the determination of blood hemoglobin were based on the measurement of these compounds. *Sulfhemoglobin*, *verdohemoglobin*, and *choleglobin* are degradation products of hemoglobin.

Hemoglobin functions physiologically not only in the transport of oxygen to the tissues but also in the transport of carbon dioxide from the tissues to the lungs (see also Chap. 16). The hemoglobin-CO_2 complex is designated carbonyl hemoglobin (or carbaminoyl hemoglobin, $RNHCOO^-$).

Spectral absorbance maxima for the more important hemoglobin derivatives are given in Table 8–3.

MEASUREMENT OF HEMOGLOBIN CONCENTRATION IN WHOLE BLOOD

The measurement of hemoglobin concentration in venous or capillary blood is one of the most frequently performed clinical laboratory tests. The principle of the method given here has been almost universally adopted for either manual procedures or automated systems.

CLINICAL SIGNIFICANCE

When the hemoglobin value is less than the normal value, the patient is said to be anemic; when it is higher than the normal value, the patient may have polycythemia or erythrocytosis (a higher than normal concentration of erythrocytes in the blood). If there is anemia, concomitant measurements of hemoglobin concentration, hematocrit, and erythrocyte count (and, from these data, calculation of the mean corpuscular volume and mean corpuscular hemoglobin concentration) usually provide an important clue to the likely causes of the anemia.

PRINCIPLE

Hemoglobin is oxidized to methemoglobin by ferricyanide, and the methemoglobin is converted into the stable cyanmethemoglobin by addition of KCN. The absorbance of cyanmethemoglobin is measured at 540 nm, where cyanmethemoglobin exhibits a broad absorbance peak (Fig. 8–5). The method is standardized against certified cyanmethemoglobin standards.

SPECIMEN

The procedure requires a minimum of 0.02 ml of whole blood. Blood may be anticoagulated with disodium ethylenediaminetetraacetate (EDTA) or may be taken directly from a finger (or heel) puncture without use of anticoagulant.

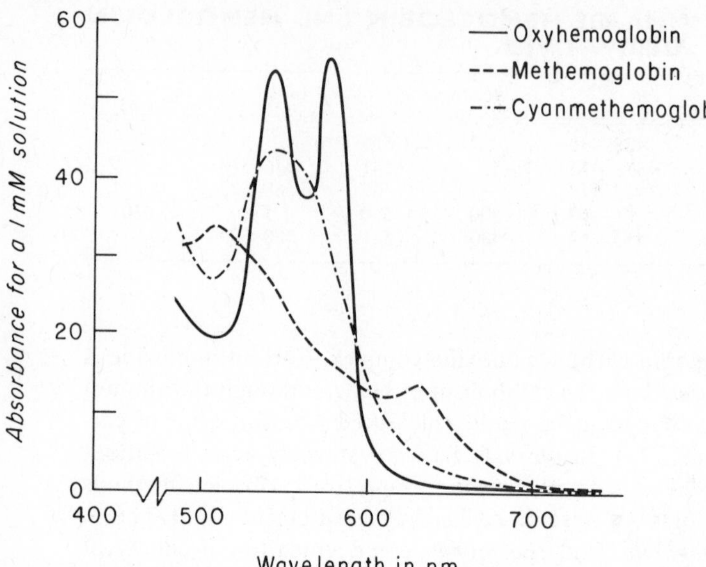

Figure 8–5 Spectrophotometric absorption curves for oxyhemoglobin, methemoglobin, and cyanmethemoglobin (author's data). Oxyhemoglobin and cyanmethemoglobin are used in measuring the hemoglobin concentration. The peak at 630 nm, which is distinctive for methemoglobin, is abolished by addition of cyanide, and the resultant decrease in absorbance is directly proportional to the methemoglobin concentration. All heme proteins exhibit their maximal absorbance in the region of 400 to 440 nm (Sorét band). This band is exhibited by all porphyrin compounds. Because the absorbance of hemoglobin in the Sorét region is approximately 10 times the absorbance at 540 nm, the Sorét peaks have been omitted from this diagram. The high absorbance of hemoglobin and of its derivatives in the Sorét region has found limited application. However, in the method for quantitation of hemoglobin A_2 by elution from a DEAE column (pp. 428–431), absorption measurements are made at the Sorét maximum for oxyhemoglobin, 418 nm. The absorption curve for methemoglobin is markedly influenced by small changes in pH. The curve given here is at pH 6.6.

REAGENTS

1. Drabkin's solution. Dissolve, in succession, 0.20 g of $K_3Fe(CN)_6$, 0.05 g of KCN, and 1.0 g of $NaHCO_3$ in distilled water and dilute to 1000 ml. The reagent must be stored in a dark bottle in the refrigerator (do not freeze).

2. Cyanmethemoglobin standard (usually 80 mg/100 ml). The standard preparation must be certified by an appropriate certifying agency.[9,19]

PROCEDURE

1. Set spectrophotometer to wavelength of 540 nm.

2. Set up a series of 16 mm (ID) cuvets so that there are three more cuvets than the number of samples to be tested. Label cuvets as follows: blank, standard, control, sample 1, sample 2, sample 3, and so forth.

3. Pipet 6.0 ml of standard into the standard cuvet. Pipet the same volume of Drabkin's solution into all other cuvets.

4. Thoroughly mix all blood specimens to be tested by repeated inversion immediately before testing.

5. Use a Sahli or calibrated capillary pipet to transfer 0.020 ml of whole blood from each sample to its respective cuvet. Blow out the Sahli or capillary pipet, and rinse to the mark with Drabkin's solution from the cuvet.

6. Mix each sample cuvet by rapid rotary swirling and then let stand for approximately 5 min to ensure complete cell lysis and conversion of hemoglobin to cyanmethemoglobin.

7. During the 5 min specified in step 6, place the blank in the light path of the spectrophotometer and adjust the galvanometer to indicate 100 per cent transmittance (0 absorbance). Then replace the blank cuvet with the standard cuvet and read the absorbance. This reading may be designated A_s, and the concentration of the standard (usually 80 mg/100 ml) may be designated C_s.

8. Measure the absorbance of each of the sample cuvets. Each sample cuvet reading should immediately be preceded by a recheck of the blank and a resetting of the galvanometer to 100 per cent transmittance with the blank if necessary.

CALCULATIONS

The dilution of blood in Drabkin's solution is then $\frac{0.02}{(V + 0.02)}$, and the hemoglobin concentration of each sample is calculated as

$$C_b \text{ (g/100 ml)} = \frac{\left(\frac{V + 0.02}{0.02}\right) \cdot \left(\frac{A_b \times C_s}{A_s}\right)}{1000}$$

in which C_b is the concentration of hemoglobin in a given blood specimen, A_b is the absorbance of the corresponding sample cuvet, C_s is the cyanmethemoglobin concentration of the standard, V is the volume of Drabkin's solution used, and A_s is the absorbance of the standard cyanmethemoglobin solution.

NORMAL VALUES

For persons residing at or near sea level, the normal values of hemoglobin concentration are 15.5 ± 1.1 g/100 ml (mean ±1*SD*) in adult males, and 13.7 ± 1.0 g/100 ml in adult females.[70] (As a practical rule, anemia is generally defined as a hemoglobin concentration of less than 14 g/100 ml in a man or less than 12 g/100 ml in a woman.)

COMMENTS AND PRECAUTIONS

The cyanmethemoglobin method has been adopted internationally as the approved standard method for hemoglobin measurement.[9,13,19,65] Although cyanide is a well-known lethal chemical, its concentration in Drabkin's solution is so low that it does not constitute any significant hazard to personnel, but reasonable care must be exercised in handling the solution. The preparation of Drabkin's solution from KCN powder must be performed with much caution, and the disposal of any solutions containing KCN into a sink should be preceded and followed by copious flushing of the sink with water to prevent the generation of lethal HCN gas in case the sink has been used for disposal of acids.

Convenient packets of dry Drabkin's reagent mix are commercially available (the contents of the packet are added to 1000 ml of distilled water).

The method outlined here permits fairly rapid manual measurement of hemoglobin concentration. However, automated methods, based on the same principle, may be more convenient when large numbers of specimens are being processed. In the Coulter Model S automated hematological apparatus, a precalibrated whole blood standard is provided by the manufacturer and is used at least daily in the calibration of the instrument for hemoglobin concentration, mean corpuscular volume, erythrocyte count, and leukocyte count.

REFERENCES

van Kampen, E. J., and Zijlstra, W. G.: Adv. Clin. Chem., **8**:141–187, 1965.
Eilers, R. J.: Am. J. Clin. Path., **47**:212–214, 1967.

DETERMINATION OF METHEMOGLOBIN AND SULFHEMOGLOBIN

When the iron in hemoglobin is oxidized to the trivalent state, the resulting brownish pigment is methemoglobin.

CLINICAL SIGNIFICANCE

The *methemoglobin* concentration in blood is increased to 10 to 30 per cent of the total heme pigment in congenital methemoglobinemia due to methemoglobin-reductase deficiency.[10,36] It is increased to a variable degree in persons who have ingested nitrates[11] or certain drugs or chemicals such as sulfones or aniline dyes.* *Sulfhemoglobin* is not a normal constituent of blood. It is sometimes found in concentrations of 1 to 10 per cent of the total hemoglobin after excessive use of certain drugs such as acetophenetidin.

PRINCIPLE

The absorption spectrum of methemoglobin exhibits a small, characteristic peak at 620 to 640 nm[17,20,46] (Fig. 8–5). This peak decreases as methemoglobin is converted to cyanmethemoglobin by addition of cyanide and the decrease in absorbance is proportional to the methemoglobin concentration.

The normal absorption spectrum of oxyhemoglobin shows very little absorbance above 600 nm. However, if certain poorly defined hemoglobin denaturation products are present in a hemolysate, there is a broad increase in the absorption curve in the range of 600 to 620 nm. This sulfhemoglobin plateau is not affected by treatment with cyanide. The abnormal pigments that give rise to the 620 nm absorption band were first observed when H_2S gas was bubbled through hemoglobin solutions. It was then thought that a hemoglobin-sulfur complex had formed, hence the term sulfhemoglobin. However, sulfhemoglobin has never been isolated or prepared in pure form. Thus, it is not possible to measure the sulfhemoglobin concentration of a specimen in terms of a standard of known concentration. In calculating sulfhemoglobin concentration, a mathematically derived, hypothetical, molar absorptivity for sulfhemoglobin is used, based on studies by Drabkin and Austin.[17]

SPECIMEN

Blood should be freshly obtained and may be anticoagulated with heparin, EDTA, or ACD (acid-citrate-dextrose) solution. No fluid or food restriction is needed.

REAGENTS

1. Potassium ferricyanide. Dissolve 2.0 g of $K_3Fe(CN)_6$ in distilled water and dilute to 10.0 ml. If stored in a brown bottle at 4°C, this solution is stable for at least one year.

2. Potassium cyanide solution (**caution**: lethal poison). Dissolve 500 mg of KCN in distilled water and dilute to 10 ml. Label "**Poison.**"

3. Potassium phosphate buffer, 0.15 mol/l, pH 6.6 (20°C). Dissolve 17.1 g of K_2HPO_4 $3H_2O$ (or 13.2 g anhydrous) and 10.2 g of KH_2PO_4 separately, each in 500 ml of distilled water; then mix in a ratio of approximately 9 vol to 16 vol, respectively, to adjust pH to 6.6. Store at 4°C. All phosphate buffers are likely to develop mold growth. Discard whenever solution appears turbid. New buffer should be prepared at least once every three months.

---

* A few cases of congenital methemoglobinemia have been reported in which the methemoglobin concentration of peripheral blood has been as high as 40 to 55 per cent of the total heme pigment.[36] Methemoglobin concentrations greater than 60 per cent may be lethal regardless of cause.

PROCEDURE FOR THE DETERMINATION OF METHEMOGLOBIN

1. Deliver 0.1 ml of whole blood into a 10 ml test tube containing 3.9 ml of distilled water; swirl to mix.
2. Add 4.0 ml of potassium phosphate buffer; mix thoroughly.
3. Prepare a blank cuvet containing 1.5 ml of phosphate buffer and 1.5 ml of water. This cuvet is designated C_1.
4. Transfer 3 ml of hemolysate (step 2) to each of two cuvets, designated C_2 and C_3.
5. To cuvet C_3, add 0.1 ml of $K_3Fe(CN)_6$ solution. Cover with Parafilm and mix by inverting three times. Allow this cuvet to stand at least 2 min before measurement of absorbance.
6. Measure the absorbance at 630 nm for cuvets C_2 and C_3; use C_1 as blank. Record values as A_{2a} and A_{3a}.
7. Add 0.1 ml of KCN to all cuvets (use a safety pipet or add 2 drops from a transfer pipet fitted with a rubber bulb). Cover with Parafilm, mix by inverting three times, and allow to stand at least 5 min.
8. Measure absorbance at 630 nm for cuvets C_2 and C_3 with C_1 as blank. Record these values as A_{2b} and A_{3b}.
9. Calculate methemoglobin concentration as follows:

$$\text{Methemoglobin (per cent of total pigment)} = 100\left(\frac{A_{2a} - A_{2b}}{A_{3a} - A_{3b}}\right)$$

The hemoglobin solution is slightly turbid; however, because the turbidity does not change with addition of the $K_3Fe(CN)_6$ or KCN, the absorbance caused by it is the same for both readings with each cuvet and therefore is compensated for in these calculations.

PROCEDURE FOR THE DETERMINATION OF METHEMOGLOBIN AND SULFHEMOGLOBIN

1. Hemolyze 0.1 ml of whole blood with 3.9 ml of distilled water and add 4.0 ml of potassium phosphate buffer as in the preceding section.
2. Centrifuge at 1600 × *g* for 30 min at 5°C to remove stroma.
3. Set up blank cuvet as in preceding section.
4. Transfer 3.0 ml of clear supernatant to a second cuvet, designated C_2, and 1.0 ml of clear supernatant to a third cuvet, designated C_3.
5. Add 2.0 ml of a mixture of equal volumes of distilled water and phosphate buffer to cuvet C_3.
6. Measure the absorbance of cuvet C_2 at 630 nm. Designate this value A_1.
7. Add 0.1 ml of KCN (use safety pipet or dropper as noted above) to cuvets C_2 and C_3. Cover with Parafilm and invert three times to mix thoroughly. Allow to stand 5 min.
8. Measure absorbance of cuvet C_2 at 630 nm and at 620 nm against blank cuvet C_1. Designate these values A_2 and A_3, respectively. Further readings will not be performed on cuvet C_2, and this may be set aside.
9. Add 0.1 ml of $K_3Fe(CN)_6$ solution to cuvets C_1 and C_3. Cover with Parafilm and invert three times to mix. Allow to stand for at least 2 min.
10. Reset spectrophotometer to 540 nm and adjust to give zero absorbance with cuvet C_1. Measure absorbance of C_3. Designate this value A_4.
11. The following calculations are made:

$$\text{Total hemoglobin (g/100 ml)} = F_1A_4$$

$$\text{Methemoglobin (g/100 ml)} = F_2(A_1 - A_2)$$

$$\text{Sulfhemoglobin (g/100 ml)} = 80F_3[A_3 - F_4(A_1 - A_2) - F_5A_4]$$

$$\text{Methemoglobin as per cent of total} = \frac{\text{methemoglobin (g/100 ml)} \times 100}{\text{total hemoglobin (g/100 ml)}}$$

$$\text{Sulfhemoglobin as per cent of total} = \frac{\text{sulfhemoglobin (g/100 ml)} \times 100}{\text{total hemoglobin (g/100 ml)}}$$

In these calculations, absorptivity constants (F_1, F_2, and so forth) are often taken directly from the work of Drabkin and Austin[17] or of Michel and Harris.[46] They may be obtained independently in the following manner.

Transfer 3 ml of commercial cyanmethemoglobin standard (80 mg/100 ml) to a cuvet with a 1.0 cm light path and measure the absorbance at 540 nm. Designate this reading A_s. With a properly calibrated spectrophotometer of narrow band pass (approximately 1 nm or narrower), an absorbance of 0.546 should be obtained for A_s (a reading between 0.540 and 0.550 is acceptable). Thus, a cyanmethemoglobin concentration of 1.0 g/100 ml would have an absorbance of 6.83 $\left(\text{from the ratio } \frac{0.546}{0.080}\right)$. Because the whole blood specimen was diluted 80-fold in the hemolyzing solution and 3-fold again in the cuvet,

$$F_1 = \frac{240}{A_s} \times \text{cyanmethemoglobin concentration of standard}$$

$$F_1 = \frac{240}{6.83} \left(\text{from } \frac{240 \times 0.080}{0.546}\right)$$

$$F_1 = 35.14$$

Prepare a 100 per cent methemoglobin standard as follows. Hemolyze normal blood, centrifuge to remove stroma as previously described, and prepare a blank as in procedure for methemoglobin (step 3). Transfer 3.0 ml of clear hemolysate to a second and third cuvet, and 1.0 ml to a fourth cuvet. To the third cuvet add 0.1 ml, and to the fourth cuvet 1.0 ml of a mixture of equal volumes of distilled water and phosphate buffer. Add 0.1 ml of $K_3Fe(CN)_6$ solution to second and fourth cuvets. Cover with Parafilm and invert three times to mix. Set the spectrophotometer to 630 nm, and use the blank to set zero absorbance. Measure the absorbance of the second cuvet and designate this value as $A_{1(s)}$. Add 0.1 ml (2 drops) of KCN solution to each cuvet. Cover with Parafilm and mix by inverting three times. Allow to stand 2 min. Then read absorbance of the second cuvet at 630 nm; designate this value $A_{2(s)}$. Reset the spectrophotometer to 540 nm, and read the absorbance of the second cuvet again. Designate this value $A_{3(s)}$. Read at 620 nm the absorbance of the third cuvet. Designate this value $A_{4(s)}$. Reset the spectrophotometer to 540 nm, and read the absorbance of the fourth cuvet Designate this value $A_{5(s)}$. The constant F_2 is the ratio of absorbance at 540 nm of cyanmethemoglobin solution at 1 g/100 ml concentration to the decrease in absorbance at 630 nm. Thus,

$$F_2 = \frac{A_{5(s)} F_1}{A_{1(s)} - A_{2(s)}}$$

F_3 may be taken from the data of Drabkin and Austin,[17] which indicate an absorbance of 0.157 for a hypothetical concentration of sulfhemoglobin of 1 g/100 ml.

$$F_4 = \frac{A_{3(s)}}{A_{1(s)} - A_{2(s)}}$$

$$F_5 = \frac{A_{4(s)}}{3A_{5(s)}}$$

We have found the following values for these coefficients:

$$F_1 = 35.2$$
$$F_2 = 33.6$$
$$F_4 = 0.0294$$
$$F_5 = 0.0125$$

A value for F_3 of 0.157, when applied to data from hemolysates of normal blood specimens, results in calculated sulfhemoglobin concentrations that are predominantly less than zero. We have empirically obtained coefficients for the sulfhemoglobin calculation by using the above values of F_4 and F_5 as coefficients for $(A_1\text{-}A_2)$ and A_4 respectively, assuming a normal mean sulfhemoglobin value of zero, and solving for the coefficient F_3. The coefficient which we determined in this manner on 30 hemolysates from normal blood specimens was 0.169.

Inserting the empirically obtained absorptivity constants in the formulas previously given, we have derived the following formulas:

$$\text{Total hemoglobin (g/100 ml)} = 35.2A_4$$
$$\text{Methemoglobin (g/100 ml)} = 33.6(A_1 - A_2)$$
$$\text{Sulfhemoglobin (g/100 ml)} = 80[0.169A_3 - 0.0294(A_1 - A_2) - 0.0125A_4]$$

NORMAL VALUES

The data (expressed as percentage of the total hemoglobin concentration) from consecutive specimens from 30 healthy persons were as follows:

	Methemoglobin	*Sulfhemoglobin*
Mean	0.78	0.084
Standard deviation	0.37	0.26
Median	0.85	0.00
95th percentile	1.28	0.74

Methemoglobin concentrations greater than 1.5 per cent and sulfhemoglobin values greater than 1.0 per cent should be regarded as abnormal. Because sulfhemoglobin concentrations in normal blood specimens are distributed about a median value of zero per cent, negative values are often calculated. All such negative values should be reported as zero per cent sulfhemoglobin.

COMMENTS AND PRECAUTIONS

Because the equilibrium between hemoglobin and methemoglobin is influenced by many factors, it is essential that the methemoglobin concentration be determined shortly after the blood has been drawn. Hemolysates must not be permitted to stand too long before the absorbance is measured. However, the 30 minutes of centrifugation of the hemolysate, as specified in the methemoglobin-sulfhemoglobin procedure, is without significant effect on methemoglobin concentration.

Sulfhemoglobin at present is purely a spectrophotometric concept—no specific chemical compound has been identified. The hemoglobin degradation products, verdohemoglobin and choleglobin, also absorb near 620 nm, and it is likely that measurement of sulfhemoglobin only provides an index, albeit a useful index, of the presence of these compounds. A more direct spectrophotometric measurement of choleglobin has been described by Mills and Randall.[47] It is based on absorbance differences between choleglobin and carboxyhemoglobin at 627 and 570 nm.

The theoretical formula for sulfhemoglobin is not entirely accurate in that corrections are not made for the contributions of sulfhemoglobin to absorbance at 630 and 540 nm. Because the sulfhemoglobin concentration normally is near zero and rarely is as high as 10 per cent of the total heme proteins, the error would be slight at most. Furthermore it can be shown mathematically that these corrections become automatic when the coefficient F_3 is derived empirically as described above. Despite the undefined nature of sulfhemoglobin, its measurement is likely to remain the most practical clinical laboratory index of the presence of hemoglobin denaturation products. Attempts to devise more precise measurements of sulfhemoglobin seem pointless as long as the nature of this substance remains in doubt.

It must be stressed that, in the rare cases of congenital methemoglobinemia that are due to the presence of an abnormal hemoglobin (for example, hemoglobin M_{Boston}), the test for methemoglobin by the methods outlined above will not be positive. This is because such abnormal hemoglobins have absorption spectra which, at acidic pH, differ markedly from the spectrum of normal methemoglobin and may not give absorbance changes on addition of KCN. The spectral anomaly that characterizes most of these abnormal hemoglobins is an absorption peak at or near 600 nm at acidic pH (pH 6.5).[4,62] Betke[4] has shown that by measuring the absorbance of hemolysates at 500, 600, and 630 nm, methemoglobinemia due to abnormal hemoglobins may be identified. For example, for methemoglobin derived from hemoglobin A and for methemoglobin due to hemoglobin $M_{Milwaukee\ I}$, at pH 6.5 the absorbance ratio A_{630}/A_{600} is not less than 1.25, and the absorbance ratio A_{500}/A_{600} is not less than 2.8. Methemoglobin due to hemoglobins M_{Boston} and $M_{Saskatoon}$ ($M_{Chicago}$) have relatively higher absorbance at 600 nm, and if one of these pigments is present in a hemolysate, these absorbance ratios will be substantially lower. Methemoglobin derived from hemoglobin $M_{Milwaukee\ I}$ resembles that derived from hemoglobin A in its spectral absorption curve, in its reaction with KCN, and in the absorbance ratios. However, careful analysis of the absorption spectrum of methemoglobin $M_{Milwaukee\ I}$ at acidic pH reveals a peak at 623 nm, rather than the peak at 630 to 635 nm characteristic of methemoglobin derived from hemoglobin *A*.[62]

REFERENCES

Evelyn, K. A., and Malloy, H. T.: J. Biol. Chem., **126**:655–662, 1938.
van Kampen, E. J., and Zijlstra, W. G.: Adv. Clin. Chem., **8**:141–187, 1965.

SOLUBILITY TEST FOR HEMOGLOBIN S

CLINICAL SIGNIFICANCE

The procedure described below is a simple and rapid method for detection of the presence of hemoglobin S in blood. It does not differentiate sickle cell trait from sickle cell disease.

PRINCIPLE

Hemoglobin S, when deoxygenated, is insoluble in concentrated phosphate buffer and produces a visible turbidity. Almost all other hemoglobins are soluble in such solutions, including hemoglobins A, F, C, and D. Thus, this test quickly differentiates blood containing hemoglobin S from that containing most other hemoglobins. A reducing substance, sodium hydrosulfite ($Na_2S_2O_4$, sodium dithionite), is used to deoxygenate the hemoglobin and saponin is used to lyse the erythrocytes.

SPECIMEN

The test requires a minimum of 0.05 ml of whole blood; the blood may be anticoagulated with heparin, EDTA, or ACD (acid-citrate-dextrose) solution. Specimens that have been kept for as long as one to two weeks at 4 to 5°C are satisfactory. No preliminary restriction of food or fluid is required.

REAGENTS

1. Potassium phosphate-saponin buffer, 2.3 mol/l, pH 7.0 (20°C). Dissolve 303.74 g $K_2HPO_4 \cdot 3H_2O$ (or anhydrous K_2HPO_4, 231.9 g) in 200 ml distilled water. Separately, dissolve 131.86 g KH_2PO_4 (crystalline) in 700 ml distilled water, warming if necessary to hasten solution. When solution is complete, mix both solutions, make up to 1000 ml with distilled water, and add 0.143 g of saponin. The buffer-saponin solution is stable at room temperature for three months.

2. Sodium hydrosulfite, dry powder.

PROCEDURE

1. On the day of the test, mix the saponin-buffer solution thoroughly and warm, if necessary, to ensure that all crystals are dissolved. Add sodium hydrosulfite to the phosphate buffer-saponin solution in the proportions of 100 mg to 10 ml. Shake thoroughly until the sodium hydrosulfite has dissolved.

2. Transfer 3 ml of this solution to each of several appropriately labeled tubes (8 mm ID); there should be at least one tube for a normal control, one tube for an abnormal control containing hemoglobin S, and one tube for each of the unknown specimens.

3. Add 0.05 ml of whole blood (control and unknown specimens) to the respective tubes. Cover each tube with Parafilm and invert twice.

4. Wait for at least 5 min, and then examine the tubes, held against a page of print, for turbidity. Good illumination is necessary.

INTERPRETATION

The test is positive if there is sufficient turbidity to prevent reading of newsprint (Fig. 8–6). Normal specimens will show only a faint haziness; many will be completely transparent. A positive test is presumptive evidence for hemoglobin S. However, the rare sickling

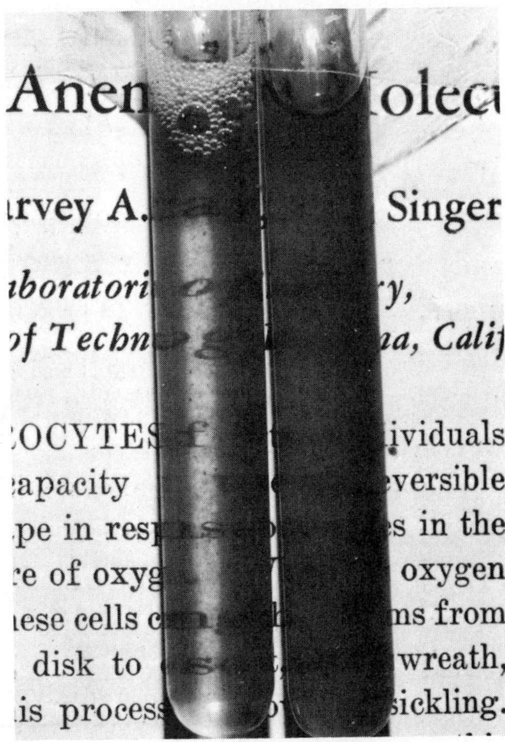

Figure 8–6 Solubility test for hemoglobin S. Deoxyhemoglobin S (right tube) is insoluble in 2.3 mol/l phosphate buffer. By contrast, normal hemolysate (left tube) is sufficiently transparent that print is easily read through it.

hemoglobin C_{Harlem} ($C_{Georgetown}$) will give positive solubility tests. Hemoglobin Barts, if present in the blood of a fetus with hydrops fetalis due to α thalassemia, will also give a positive solubility test. In patients with unstable hemoglobins who have undergone splenectomy, the test may also be positive because of numerous insoluble erythrocyte inclusions. Positive tests have also been reported in patients with marked hyperglobulinemia (for example, in multiple myeloma). Thus, every positive solubility test should be confirmed by electrophoresis.

COMMENTS AND PRECAUTIONS

The procedure outlined above closely follows the original method of Itano,[34] who first observed the poor solubility of deoxyhemoglobin S in concentrated phosphate buffer. Several minor modifications of this procedure have been published[26,42,51] and presumably give comparable results. A quantitative measurement of deoxyhemoglobin solubility has little, if any, practical value.

Reagent kits for hemoglobin S screening are commercially available from several sources. These are believed to be fundamentally similar to the procedure described above. In a carefully controlled comparison of seven commercially available kits, Schmidt and Wilson found both false negative and false positive results to be very frequent. They commented, "Accuracy of commercial solubility test kits varies greatly, and frequent incorrect diagnoses can be expected."

One cause of false negative results by this method is severe anemia. When the hemoglobin concentration is 5 g/100 ml or lower, false negative tests are the rule. If a patient has anemia (7 g/100 ml or less) doubling the volume of blood added in the test should result in a positive test if hemoglobin S is present. It may be noted that, although the reagents used in this test are quite inexpensive and can be readily prepared in most clinical laboratories, the cost of commercial kits is relatively high (as high as $50 per 100 tests). The solubility test has also been adapted for automated large-scale screening.[52] False negative results have also been observed with this method.

Use of this procedure for screening for hemoglobin S has been properly criticized on the basis that it does not identify hemoglobin C, which is prevalent in the same ethnic group. Screening for abnormal hemoglobins can also be done quickly and inexpensively by electrophoresis on cellulose acetate, which permits identification of both of these common hemoglobinopathies and also allows distinction among sickle cell trait, sickle cell disease, and S-C disease. Cases of β thalassemia minor and sickle-thalassemia also may be identified by electrophoresis. Thus, the best use of the solubility test may be not in screening programs but in determining whether a hemoglobin variant that appears to be hemoglobin S or D or G by electrophoresis is, in fact, hemoglobin S.

REFERENCES

Itano, H. A.: Arch. Biochem. Biophys., **47**:148–159, 1953.

Nalbandian, R. M., Nichols, B. M., Camp, F. R., Jr., Lusher, J. M., Conte, N. F., Henry, R. L., and Wolf, P. L.: Clin. Chem., **17**:1028–1032, 1971.

Schmidt, R. M., and Wilson, S. M.: J.A.M.A., **225**:1225–1230, 1973.

TESTS FOR UNSTABLE HEMOGLOBINS

More than 50 unstable hemoglobin variants have been described, and most of these are associated with hemolytic anemia. Of these, the most frequently reported unstable hemoglobins are hemoglobin H and hemoglobin Köln (Table 8–1).

CLINICAL SIGNIFICANCE

The major clinical manifestation caused by an unstable hemoglobin is hemolytic disease. Some unstable hemoglobins cause severe hemolytic anemia; others, such as hemoglobin Köln, are associated with characteristic manifestations of hemolysis, such as hyperbilirubinemia, reticulocytosis, and polychromatophilia in the peripheral blood film, despite normal or nearly normal values for hemoglobin concentration in the peripheral blood. This group of disorders can be differentiated from other causes of congenital hemolytic disease by demonstration of the characteristically increased thermolability of the unstable hemoglobin.

PRINCIPLE

If an unstable hemoglobin is present in a hemolysate, heating produces marked turbidity and, occasionally, a flocculent precipitate. Hemolysates that contain normal, stable hemoglobins remain transparent for a longer time at higher temperatures than do hemolysates containing unstable hemoglobins. Many, slightly different heat stability tests have been described. In general, when higher temperatures are used, shorter time intervals are required to effect precipitation or turbidity. For example, normal hemoglobin does not precipitate when heated at 50°C for 1 or 2 hours,[14] or at 60°C for 30 minutes, and only slight precipitation occurs when normal hemoglobin, in the form of cyanmethemoglobin, is heated at 68°C for 1 minute.[38] In contrast, unstable hemoglobins show distinct turbidity under each of these sets of conditions. It is useful to perform these tests routinely at both 50°C for 60 minutes and 68°C for 1 minute.

SPECIMEN

This procedure requires 2 ml of whole blood; it may be obtained with any common anticoagulant. Specimens that have been refrigerated at 4 to 5°C for up to one week are also satisfactory. No prior restriction of food or fluid is required.

REAGENTS

1. Tris buffer, 0.19 mol/l, pH 8.0 (25°C). Dissolve 23 g of tris(hydroxymethyl)aminomethane in 800 ml of distilled water. Adjust to pH 8.0 by addition of 1 mol/l HCl with constant stirring. Bring volume up to 1000 ml with distilled water.

2. Potassium cyanide. Dissolve 5 g of KCN in distilled water to a final volume of 100 ml. Label "**Poison.**" Keep in a safe place at room temperature.

3. Potassium ferricyanide. Dissolve 20 g of $K_3Fe(CN)_6$ in distilled water to a final volume of 100 ml. Store in a dark bottle at 5°C. This reagent is stable for at least one year.

4. Potassium phosphate buffer, 2 mol/l, pH 6.8 (25°C). Dissolve 45.6 g of $K_2HPO_4 \cdot 3H_2O$ (or 34.8 g anhydrous) and 26.2 g of KH_2PO_4 (crystalline) separately, each in 100 ml of distilled water; warm flasks in a hot water bath to accelerate dissolution. For use, mix in a ratio of 9 vol of K_2HPO_4 solution to 10 vol of KH_2PO_4 solution.

PROCEDURE

1. *Processing of blood.*

a. Centrifuge for 15 min at 700 × *g* (2000 rpm when radius of rotation is 16 cm) at 5°C. Discard plasma.

b. Resuspend erythrocytes in 10 vol of cold 0.15 mol/l NaCl solution (0.9 g/100 ml) and centrifuge at 700 × *g* for 15 min at 5°C. Discard supernatant saline.

c. Repeat step b.

2. *1 hour test, 50°C, pH 8.0.*

a. Transfer 0.5 ml of washed, packed erythrocytes to a 15 ml conical centrifuge tube.

b. Add 4.5 ml of distilled water; mix to hemolyze.

c. Add 5 ml of tris buffer; mix.

d. Centrifuge for 30 min at 1600 × *g* (3000 rpm when radius of rotation is 16 cm) at 5°C to remove erythrocyte stroma.

e. Transfer 2 ml of clear supernatant hemolysate to a test tube (10 mm ID).

f. Place test tube in 50°C water bath for exactly 60 min.

g. Examine for turbidity by holding against light. Hemolysates with normal hemoglobin remain completely clear. Unstable hemoglobins cause variable turbidity.

3. *1 min test, 68°C, pH 6.8.*

a. Transfer 0.1 ml of washed packed erythrocytes to a 15 ml conical centrifuge tube.

b. Add 9.9 ml of distilled water; mix.

c. Add 5 drops of KCN solution and 5 drops of $K_3Fe(CN)_6$ solution; mix and permit to stand for 2 min at room temperature. (The cyanmethemoglobin concentration of the hemolysate should be approximately 200 mg/100 ml; this may be checked spectrophotometrically and adjusted if necessary.)

d. Centrifuge for 30 min at 1600 × *g* at 5°C to remove erythrocyte stroma.

e. Transfer 2 ml of the supernatant to each of two test tubes (10 mm ID); to each, add 0.2 ml of potassium phosphate buffer and mix thoroughly.

f. Place tubes simultaneously in a water bath preheated to 68 ± 1°C and start a timer or stopwatch; agitate the tubes rapidly in the hot water bath.

g. At exactly 1 min, remove tubes from hot water bath and inspect for turbidity by holding in front of a strong light. Hemolysates with normal hemoglobin show a slight haziness. Turbidity is marked when an unstable hemoglobin is present.

COMMENTS AND PRECAUTIONS

1. Care must be exercised in preparing and handling concentrated cyanide solutions.
2. Thermostability tests are crude and not easily quantitated. Yet, they are extremely useful in establishing the presence of an unstable hemoglobin. We have found instances of hemoglobin Köln disease that did not show precipitation with the customary test—60 minutes at 50°C—but precipitated readily in 1 minute at 68°C. For this reason, it is our practice always to do both tests.

REFERENCE

Dacie, J. V., Grimes, A. J., Meisler, A., Steingold, L., Hemsted, E. H., Beaven, G. H., and White, J. C.: Brit. J. Haemat., **10**:388–402, 1964.

ELECTROPHORETIC SEPARATION OF HEMOGLOBINS ON CELLULOSE ACETATE

CLINICAL SIGNIFICANCE

The identification of hemoglobin variants such as hemoglobin S requires demonstration of their characteristic electrophoretic mobility. This is easily and conveniently accomplished by the following technique. On the other hand, many clinically important hemoglobin variants have identical electrophoretic mobilities.

PRINCIPLE

Hemoglobins are negatively charged ions at pH 8.6 and move toward the anode (positive electrode) in an electrical field. However, hemoglobins differ in the amount of charge per molecule. Hemoglobin S has two more and hemoglobin C has four more positive charges per molecule than does hemoglobin A. Thus, hemoglobin S moves toward the anode more slowly than does hemoglobin A, and hemoglobin C moves even more slowly. Conversely, hemoglobins H, I, and J, which have additional negative charges, all move toward the anode more rapidly than does hemoglobin A.

Various supporting media, such as potato starch gel, paper, and polyacrylamide gel, have been used in electrophoresis. Cellulose acetate has come into general use because it is easily available, provides sharp resolution of hemoglobin bands in a relatively short time, and permits clearing, densitometric quantitation, and permanent storage of the transparent film.[2,8,24]

SPECIMEN

This procedure requires a minimum of 0.5 ml of whole blood. Any anticoagulant may be used. Fasting and fluid restrictions are unnecessary. A specimen will keep at 4 to 5°C for one week if the anticoagulant is heparin, and for three to four weeks if collected with ACD (acid-citrate-dextrose) solution. However, if an unstable hemoglobin (for example, H or Köln) is suspected, electrophoresis should be carried out within 24 h.

REAGENTS

1. Tris-EDTA-borate (TEB) buffer, pH 8.6 (25°C). Dissolve 12.0 g of tris(hydroxymethyl)aminomethane (Tris), 1.56 g of disodium ethylenediaminetetraacetate (EDTA), and 0.92 g of boric acid in deionized water and dilute to 1000 ml. Store at 5°C. This solution is stable indefinitely.
2. Stain. Dissolve 20 g of Ponceau S, 30 g of trichloroacetic acid, and 30 g of sulfosalicylic acid in distilled water and dilute to 1000 ml. Make a new solution at least once every two months.

PROCEDURE

1. If the blood has sedimented overnight, remove the supernatant plasma before proceeding.
2. Place 0.5 to 1 ml of whole blood or, if the plasma has been removed, 0.2 to 0.5 ml of sedimented cells into a centrifuge tube (the volume of packed cells is not important). Suspend the cells in 10 to 15 ml of cold isotonic saline.
3. Centrifuge at 5°C for 10 to 15 min at 900 × *g*.
4. Aspirate the supernatant saline and discard.
5. Add 9 vol of cold distilled water per 1 vol of packed cells.
6. Add 0.2 ml of xylene or toluene. Resuspend the packed cells with a wooden applicator stick or a glass stirring rod, and then thoroughly mix with a vortex mixer.*
7. Centrifuge at 1600 × *g* for at least 15 min at 5°C.
8. Aspirate and discard the supernatant xylene or toluene.
9. The clear hemolysate lying between the supernatant and the stromal precipitate may be removed with a transfer pipet for immediate use in electrophoresis, or may be transferred to another tube and stored for short intervals at 5°C (for use the same day) or at less than 0°C for use in subsequent studies.
10. Prepare a cellulose acetate membrane as described on page 312. Apply samples at a template position near the cathode. A battery of known hemoglobin types is alternated in the template positions with the unknown hemolysate. Place TEB buffer in both buffer chambers, making sure that the ends of the membrane are in contact with buffer and that the buffer level is the same in each chamber.
11. Electrophoresis is carried out at room temperature for 30 min at 450 V.
12. Stain with Ponceau S as described on page 313.
13. Pick the membrane up on a glass plate of appropriate size. Drain off excess solution. Place the glass plate and membrane in an oven, preheated to 100°C, for 15 min, remove, and allow to cool to room temperature; alternatively, allow the membrane to air dry for approximately ½ hour. With a razor blade, carefully peel the membrane from the glass plate. Place the membrane in a plastic envelope and trim off the ends.

INTERPRETATION

The relative order of increasing electrophoretic mobility of the more commonly encountered hemoglobin variants, from cathode (negative side) to anode (positive side), is as

---

* Hemoglobin H is denatured and removed by this step. If an unstable hemoglobin is suspected, omit step 6.

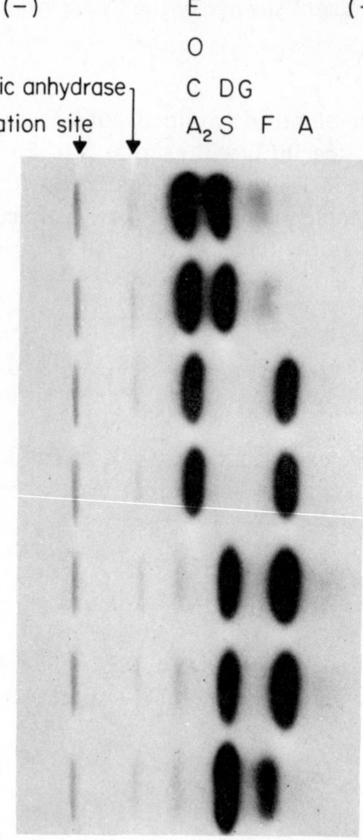

Figure 8–7 Hemoglobin electrophoresis on cellulose acetate at pH 8.6, showing the relative positions of various hemoglobins. Hemoglobin F often is poorly resolved from A, which it trails slightly. Hemoglobins D, G, and Lepore have the same, or nearly the same, mobility as hemoglobin S; hemoglobins O, E, and C have the same, or nearly the same, very slow mobility as hemoglobin A_2. Carbonic anhydrase has a slower (more cathodal) mobility than any of the common hemoglobins. It is seen when the cellulose acetate membrane is stained with nonspecific protein stains.

follows: hemoglobins C, A_2, E, and O_{Arab}; hemoglobins S, D, and G; hemoglobin F; hemoglobin A. These positions are indicated in Figure 8–7. Usually, hemoglobin A is preceded by a fainter band that merges with it. This has been called A_3 and is believed to be a hemoglobin-glutathione complex. Hemoglobins H and I have identical mobilities, substantially more anodal than hemoglobin A. Hemoglobins J, K, and N have an anodal mobility intermediate between that of H or I and A.

COMMENTS AND PRECAUTIONS

Hemoglobins A_2, C, O_{Arab}, and E cannot be differentiated on the basis of electrophoretic mobility in this medium, nor can hemoglobins D, G, and Lepore be distinguished from hemoglobin S. Hemoglobin F is poorly resolved from hemoglobin A. Hemoglobins H and I are indistinguishable. However, hemoglobin A_2 is rarely more than 10 per cent of the total hemoglobin. Thus, if a very slowly moving band constitutes more than 20 per cent of the total hemoglobin, it may be presumed to be hemoglobin C, O_{Arab}, or E. Hemoglobins C and O_{Arab} are virtually limited to persons of Central African ancestry, whereas hemoglobin E is virtually limited to those of Southeast Asian ancestry. Hemoglobin Lepore constitutes only 10 to 15 per cent of the total hemoglobin, whereas hemoglobin S, D, or G constitutes 25 to 45 per cent. Similarly, hemoglobin H constitutes approximately 10 to 15 per cent of the total hemoglobin, whereas hemoglobin I (an α chain variant) is usually about 25 per cent. Hemoglobin H is most commonly seen in persons from Southeast Asia.

Thus, by measuring the proportions of these hemoglobin bands densitometrically, and by knowing the ethnic origin of a patient, one may determine the probable nature of the abnormal hemoglobin. However, additional confirmatory tests are often necessary before the hemoglobin type can be established. A hemoglobin solubility test will establish the presence of hemoglobin S; a thermostability test will establish the presence of an unstable hemoglobin (for example, Köln or H). Agar gel electrophoresis at pH 6.2 is valuable in differentiating hemoglobin C from E and O_{Arab} or hemoglobin S from D and G.

It is mandatory that a circuit breaker be an integral part of the electrophoresis system and that no attempt be made to bypass this safety feature. Electrocution is a potential hazard of electrophoresis, and death has resulted from failure to observe safety precautions.

REFERENCES

Bartlett, R. C.: Clin. Chem., **9**:325–329, 1963.
Briere, R. O., Golias, T., and Batsakis, J. G.: Am. J. Clin. Path., **44**:695–701, 1965.
Graham, J. L., and Grunbaum, B. W.: Am. J. Clin. Path., **39**:567–578, 1963.

ELECTROPHORETIC SEPARATION OF HEMOGLOBINS IN AGAR GEL AT pH 6.2

CLINICAL SIGNIFICANCE

Electrophoresis at pH 6.2 aids in the identification of the clinically important hemoglobin variants, E, D, G, and O_{Arab}.

PRINCIPLE

Electrophoresis demonstrates differences in mobility as the result of differences in charge and molecular shape. The conditions provide far superior separation of hemoglobins F and A than is obtainable on cellulose acetate at pH 8.6.

SPECIMEN

This procedure requires the same specimen and conditions described in the preceding section.

REAGENTS

1. Citrate buffer, stock solution, 0.5 mol/l, pH 5.8 (25°C). Dissolve 147 g of trisodium citrate ($Na_3C_6H_5O_7{\cdot}2H_2O$) in deionized water and dilute to 800 ml. Dissolve 10.5 g of citric acid ($H_3C_6H_5O_7{\cdot}H_2O$) in deionized water and dilute to 100 ml.

Add approximately 68 ml of citric acid solution to the trisodium citrate solution and carefully adjust the mixture to pH 5.8. Add deionized water to a final volume of 1000 ml. Store at 4°C. The solution is stable for approximately one year.

2. Citrate buffer, working solution, 0.05 mol/l, pH 6.2 (25°C). Prepare by diluting 1 vol of stock solution with 9 vol of deionized water. The pH will increase to 6.2.

3. Bacto-agar (Difco Laboratories, Detroit, MI 48232).

4. Stain (any of the following may be used; for general purposes, we prefer the Naphthol Blue-Black stain).

Light Green Stain (A General Protein Stain). Dissolve 0.5 g of Light Green dye in 5.0 ml of glacial acetic acid, 25.0 ml of absolute methanol, and 75.0 ml of distilled H_2O. Prepare fresh solution once a month.

o-Dianisidine Stain (A Peroxidase Stain, More Specific for Hemoglobin). Dissolve 40 mg of *o*-dianisidine in 60.0 ml of acetate buffer, pH 4.7 (1.46 g of sodium acetate and 0.52 ml of glacial acetic acid, diluted to 500 ml with distilled water). Add 0.4 ml of 30 per cent hydrogen peroxide (Superoxol) and 100.0 ml of absolute ethanol. Prepare fresh daily. (**Caution**: *o*-Dianisidine is a benzidine derivative and excessive exposure may lead to carcinoma of the urinary bladder. Hydrogen peroxide is extremely caustic and may cause severe burns of the hands or permanent corneal injury. Wear glasses or protective goggles and rubber gloves.)

Naphthol Blue-Black Stain. Mix 0.6 g of Naphthol Blue-Black dye (Amido Black 10B) (Sigma Chemical Co., St. Louis, MO), 6.0 g of trichloroacetic acid, 50.0 ml of 95 per cent ethanol, and 0.75 ml of glacial acetic acid; dilute with distilled H_2O to 200.0 ml. Prepare fresh solution once a month.

5. Acetic Acid Rinse, 2 Per Cent. Dilute 10 ml of glacial acetic acid with deionized water to 500 ml.

EQUIPMENT

There is no uniform practice with respect to the dimensions of the glass gel supports. Some use microscope slides (2.5 by 7.6 cm) for this purpose. The thin glass plates (8.2 by 10.0 cm) commonly used in making lantern slides are preferred by this author. These have the advantages that they are readily available in standard size, they make it possible to analyze six to eight specimens simultaneously in a single gel, and the gels, when dried, are conveniently filed. Much larger gel supports (for example, 13 by 22 cm) are also sometimes used in this procedure when a large number of specimens must be examined. The use of supports of various sizes (and thereby gels of various sizes) results in different specifications for voltage, current, and duration of electrophoresis. The procedure described here assumes the use of 8.2 by 10 cm glass plates.

PROCEDURE

1. Preparation of gel. To 50 ml of working buffer in a 250 ml Erlenmeyer flask, add 0.5 g of agar and suspend it by vigorous swirling. Loosely cap the Erlenmeyer flask (for example, with a 100 ml beaker) to prevent inadvertent contamination and to decrease evaporation. Place the flask in a water bath at 90 to 100°C, containing enough water to cover the lower two-thirds of it. At approximately 30 min intervals, remove the flask from the water bath and swirl it to resuspend the agar. After 2 to 3 h of heating, the agar suspension will begin to clarify. Inspection against a strong light will reveal an extremely fine granularity. Return it to the water bath and continue to heat. In another 30 to 60 min, the flask will be completely transparent and free of granularity when inspected against a strong light (it may be necessary to bring the agar momentarily to the boiling point). Remove the flask from the hot water bath and allow it to cool to 50 to 60°C. The cooling may be accelerated by holding the flask at an angle under running cold water while constantly rotating it.

2. Pouring of agar. Place the glass supports on a perfectly level surface (use of a spirit level will help determine what surfaces are suitable). Fill a prewarmed 10 ml pipet to the 9 ml mark with molten agar. Place the tip on the glass plate near one corner and, with a back-and-forth movement of the pipet, allow the agar to run out on the plate. An even coating, 1 mm deep, will result.

If any bubbles are seen on the surface of the agar, they may be eliminated by aspirating them into a Pasteur pipet. Allow the agar to cool undisturbed until solid; this usually requires 15 to 20 min. Cover with a sheet of polyethylene film or Saran Wrap (or place in a tightly sealed container to decrease evaporation) and store in a refrigerator for at least 24 h prior to use.

3. Preparation of hemolysates.

a. Centrifuge 1 ml of whole blood at 700 × *g* for 10 min.
b. Remove and discard plasma.
c. Resuspend erythrocytes in 10 ml of 0.9 per cent NaCl solution.
d. Centrifuge at 700 × *g* for 10 min. Remove and discard supernatant.
e. Lyse erythrocytes by alternately freezing and thawing twice.
f. Measure the hemoglobin concentration of the hemolysate and adjust it to 10 g/100 ml by adding working buffer.

Example: The hemoglobin concentration of the hemolysate is found to be 18.5 g/100 ml. Since the volume of the hemolysate remaining is quite small, it is desired to expend only 0.1 ml. How much working buffer must be used for dilution? Since $10/(10 + 8.5) \times 18.5$ g/100 ml = 10 g/100 ml, the correct dilution would be achieved by diluting 10 vol of hemolysate with 8.5 vol of working buffer, or 0.1 ml hemolysate with 0.085 ml of buffer. This volume is easily approximated with an 0.1 ml pipet graduated in 0.01 ml units. Precise measurement is not required.

4. Electrophoresis.

a. Cut a 2 mm wide strip of Whatman No. 1 filter paper into 10 segments, each approximately 8 mm long.

b. Remove cover from agar gel. Electrophoresis is to be performed along the long axis of the gel. Select which end is to be anodal (positive). Cut a tiny notch in one corner of the anodal end of the gel as a marker; then, lay a straightedge on top of the gel to define a line 4 cm from the notched end and perpendicular to the long axis of the gel. Using a razor blade or other very sharp edge, make a series of incisions approximately 1 cm long and separated from each other by approximately 0.5 cm of unbroken gel.

c. With tweezers, pick up a filter paper strip and immerse it in diluted hemolysate. Remove, blot lightly, and place it on a spatula, about 0.5 cm from the end. Very carefully insert the spatula into the slot cut in the gel. Retract the gel slightly and, using the tips of tweezers, slide the filter paper strip into the slot and remove the spatula. Repeat this for each hemolysate. Allow approximately 10 min for the hemoglobin to be absorbed into the gel. Cautiously remove the filter paper from each slot, being careful not to tear the gel. (As an alternative to the use of filter paper wicks, the tip of the spatula may be moistened with hemolysate prior to insertion into the slot; this will leave a very fine line of hemoglobin.) If the slot remains separated, a drop of working buffer may be placed over it and, with gentle finger pressure on either side, the slot may be closed. Blot any excess buffer or hemolysate from the surface. Note the order of the specimens in reference to the notched corner of gel.

d. Position the gel plate above the buffer chambers, which contain working buffer. Establish the circuit by a bridge composed of several layers of filter paper moistened with working buffer. This filter paper should overlap approximately 1 cm of gel at each end. Cover the gel with polyethylene film or Saran Wrap, or spray with Krylon. Attach electrodes so that the positive electrode (anode) is in electrical contact with the notched end of the gel. Carry out electrophoresis in a cold room or refrigerator at 5°C, applying 50 mA and approximately 100 V (10 V/cm of gel) for 90 min.

e. The gel should be stained immediately after termination of electrophoresis if a permanent record is desired. Immerse the gel for 20 min in one of the stain solutions listed under Reagents. Wash briefly in distilled water and then destain with 2 per cent aqueous acetic acid. Repeated changes of destaining solution are required for the Napthol Blue-Black or Light Green stain, and destaining usually is not complete within 24 h. Destaining should be continued until the background is virtually colorless.

f. After completion of staining and destaining, the gel may be dried to a transparent, paper-thin film on the glass plate. It may be dried at room temperature, in a 90°C oven, or under an infrared heat lamp.

INTERPRETATION

Hemoglobins A, D, E, O_{Arab}, and G migrate slightly toward the cathode. Hemoglobin F shows the most marked cathodal movement of the common hemoglobins. Conversely, hemoglobin C shows anodal migration. Hemoglobin S remains close to the site of application.[25,44,58,68]

The relative order of hemoglobin migration in this medium is shown in Figure 8–8. Hemoglobins D and G migrate with A and thus separate well from hemoglobin S. Hemoglobin E also travels with A and thus is easily differentiated from C. Hemoglobin O_{Arab} has a slightly slower cathodal migration than A, and this serves to distinguish it from C as well as from A, D, G, and E. Hemoglobin A_2 has mobility like that of hemoglobin A. Hemoglobin Lepore also exhibits an agar gel mobility identical to that of hemoglobin A, thus differentiating it from hemoglobin S.

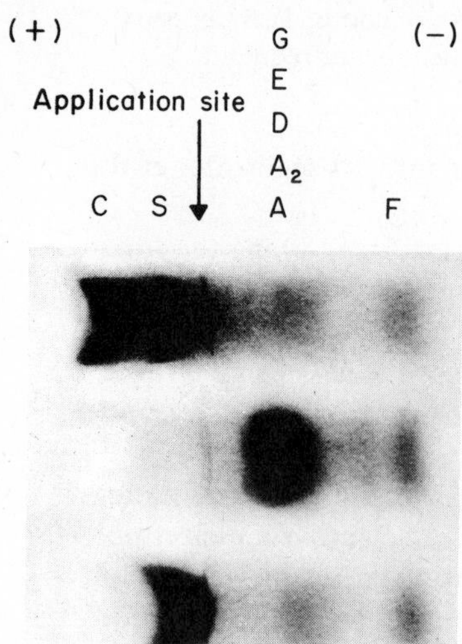

Figure 8–8 Hemoglobin electrophoresis on agar gel at pH 6.2, showing relative positions of various hemoglobins. The common hemoglobins are found in the order: C, S, A, F, proceeding from most anodal (C) to most cathodal (F). Hemoglobins A_2, E, G, and D have the same mobility as hemoglobin A.

COMMENTS AND PRECAUTIONS

The general precautions described for cellulose acetate electrophoresis should be observed. Note the special precautions concerning the use of *o*-dianisidine stain. Known standards must be included with every group of unknowns because apparently minor variations in technique may cause quite variable results. The specified voltage and current will give a maximal rate of electrophoretic mobility without overheating, provided that electrophoresis is carried out at 5°C and with gels of the specified dimensions. Electrophoresis may be performed at room temperature for 16 hours with approximately 25 mA and a voltage drop not in excess of 2 V/cm of gel. It is the author's experience that more dependable separations are attained by using shorter runs at higher voltage with adequate cooling.

The procedure just outlined gives satisfactory results with the agar specified. Agar preparations vary markedly and in poorly understood ways. If a different type of agar or a more highly purified agar is used, altogether different results may be obtained, even though the buffer and pH are identical. This is presumed to be due to differences in reactive groups of the agar.

The concentration of hemoglobin applied to the gel affects mobility. Thus, if there is a larger quantity of hemoglobin A, the apparent electrophoretic mobility will be increased.[44] For this reason, care is required in adjusting the hemoglobin concentration of each sample.

REFERENCES

Robinson, A. R., Robson, M., Harrison, A. P., and Zuelzer, W. W.: J. Lab. Clin. Med., **50**:745–752, 1957.

Marder, V. J., and Conley, C. L.: Bull. Johns Hopkins Hosp., **105**:77–88, 1959.

MEASUREMENT OF HEMOGLOBIN A_2

CLINICAL SIGNIFICANCE

Beta thalassemia minor is a common cause of mild anemia and is usually accompanied by a slight increase in the hemoglobin A_2 concentration in the blood.

PRINCIPLE

Diethylaminoethyl cellulose (DEAE cellulose) is an ether of cellulose and a substituted, strongly basic amine:

$$\begin{matrix} C_2H_5 \\ C_2H_5 \end{matrix} \!\!>\! \overset{+}{N}\text{—}CH_2\text{—}CH_2\text{—}O\text{—}R$$

in which R is the fourth carbon of a glucose in the cellulose chain. Proteins may be bound to the DEAE cellulose and then eluted by changes in pH or salt concentration. In the procedure which follows, hemoglobin is eluted from DEAE cellulose by tris-phosphate buffer, pH 9.5. Hemoglobins C, E, and A_2 are eluted at a lower molarity than are hemoglobins A, F, and S, which in this method remain on the DEAE cellulose column.

SPECIMEN

The procedure requires 1 ml of whole blood. However, because the specimen is prepared as for electrophoresis on cellulose acetate, both procedures can be performed with the same 1 ml specimen. Fasting or fluid restriction is not necessary. Any anticoagulant is satisfactory. Maintain the specimen at 5°C until the procedure is begun.

REAGENTS

1. Diethylaminoethyl cellulose, Selectacel type 40 (Brown Company, Berlin, N.Y.).
2. Tris-phosphate buffer No. 1. Dissolve 121.14 g of tris(hydroxymethyl)aminomethane (Tris) in distilled water and dilute to 2000 ml. Dissolve 103.5 g of $NaH_2PO_4 \cdot H_2O$ (or 90.0 g anhydrous) in distilled water and dilute to 1500 ml. To 2000 ml of Tris solution, add approximately 1000 ml of NaH_2PO_4 solution and adjust the pH to 8.5 (20°C). Refrigerate at 5°C. Make fresh buffer every three to four months.
3. Tris-phosphate buffer No. 2. To 100 ml of buffer No. 1, add 500 mg of KCN and enough distilled water to bring the volume to 5000 ml. Check the pH and adjust with dilute H_3PO_4 or NaOH to pH 8.5. (This volume is sufficient for preparation of the gel and, in addition, for several chromatographic analyses of hemoglobin A_2. For subsequent analyses, it may be convenient to use only one-fifth of the volume specified.)
4. Tris-phosphate buffer No. 3. Dilute 7 vol of buffer No. 2 with 3 vol of distilled water. (This buffer is prepared only when needed for measuring hemoglobin A_2 in a sample that contains hemoglobin S, G, D, Lepore, or any other hemoglobin with similar electrophoretic mobility.)
5. Xylene or toluene, AR.

SPECIAL EQUIPMENT

1. Disposable columns, 0.8 cm ID by 20 cm long. These may be made by cutting off the narrow tops of plastic disposable 10 ml pipets. Pack the tips with glass wool. Attach 21 gauge disposable needles to the tips, and slide polyethylene tubing over the needle.
2. Rack or stand to hold columns.
3. Pump (Polystaltic pump, Buchler Corp., Fort Lee, N.J.). Alternatively, gravity may be used to ensure an adequate flow rate, with the buffer reservoir placed approximately 1 m above the upper end of the column.
4. Spectrophotometer, with good sensitivity at 418 nm.

PROCEDURE

1. Preparation of DEAE cellulose suspension. Suspend 50 g of DEAE cellulose in 2 liters of distilled water. Stir, and allow to sediment for 30 min. Decant the supernatant to remove the light particles (fines). Repeat this step two or more times until all fines are removed. Resuspend in 1 liter of buffer No. 1 for 24 h with constant stirring. Filter on a Büchner filter and discard the filtrate. Resuspend the DEAE cellulose in 1 liter of buffer No. 2 and stir for 1 h. Repeat the filtration and washing in buffer No. 2 three more times.

Finally, resuspend in 500 ml of buffer No. 2. Store at 4°C. This material is stable for one month. Determine the amount of DEAE cellulose in an aliquot of thoroughly resuspended material by centrifuging for 20 min at 1400 × *g* in a graduated conical centrifuge tube. From this, calculate the volume of suspension that will contain sufficient material to give 3.5 ml of packed DEAE cellulose.

2. Preparation of hemolysate. Prepare the hemolysate as for electrophoresis on cellulose acetate (see p. 423) and further dilute it by adding 2 vol of hemolysate to 1 vol of distilled water. This should provide a hemolysate containing 10 to 13 mg of hemoglobin/ml.

3. Preparation of the column. The column to be used is clamped to a rigid support, and a spirit level is used to ensure that the column is in a vertical position. Pour sufficient DEAE cellulose suspension into the column to provide 3.5 ml of packed adsorbent. Overlay this with enough buffer No. 2 to fill the column. Clamp the polyethylene outlet tubing, and allow the column to stand for 20 to 30 min while the DEAE cellulose sediments. It is crucial that the sedimentation be even and that no air pockets form. Select a rubber stopper to fit the inlet (upper) end of the column, insert a 20 gauge needle through it, and attach 1 m of polyethylene tubing to the hub of the needle. Lead the tubing through the pump and then attach it to the buffer reservoir. Make sure that the stopper fits the column snugly and that the tubing is properly secured in the pump. Fill the buffer reservoir with buffer No. 2.

4. Chromatographic elution of hemoglobin A_2. When the DEAE cellulose is fairly well sedimented, open the clamp on the outlet tubing and allow the buffer in the column to run out until only a fine meniscus of clear buffer overlies the adsorbent. Then, carefully overlay the DEAE cellulose with 1 ml of hemolysate (containing 10 to 13 mg of hemoglobin) while the outlet is unclamped. Allow the hemolysate layer to descend into the DEAE cellulose until only a fine meniscus remains. Wash down the sides of the column with 1 ml of buffer No. 2 and wait until only a fine meniscus remains. Then, carefully rinse down sides of the column with buffer No. 2, gradually filling the column completely. Turn on the pump and allow buffer to fill the tubing completely. Seat the stopper firmly into the upper end of the column. Regulate the pump speed to deliver buffer solution at a rate of 2 ml/min. (**Caution**: If the pump generates excessive fluid pressure, the cellulose will become so tightly packed that it impedes the flow through the column.) Collect only the first 20 ml of effluent from the column. This contains 95 to 100 per cent of the hemoglobin A_2 applied to the column.

If the blood sample is known to contain a hemoglobin with S mobility (S, G, D, or Lepore), use buffer No. 3 in the above chromatographic procedure in place of buffer No. 2. Collect and discard the first 10 ml of effluent. Collect and retain the next 20 ml, which contains hemoglobin A_2. Do not reuse the column or the DEAE cellulose.

5. With buffer No. 2 as blank, set the spectrophotometer to 418 nm. Measure the absorbance of the eluate and designate this value A_{A2}. Dilute 0.1 ml of hemolysate to 20 ml with buffer No. 2 and measure the absorbance at 418 nm. Designate this value A_{total}.

CALCULATION

$$\text{Hemoglobin A}_2 \text{ (as per cent of total hemoglobin)} = \frac{100\, A_{A2}}{10\, A_{total}} = \frac{10\, A_{A2}}{A_{total}}$$

NORMAL VALUES

Normal values range from 1.6 to 3.2 per cent (mean, 2.4; *SD*, ±0.3). In β thalassemia minor, hemoglobin A_2 values range from 3.4 to 6.4 per cent (mean, 5.1; *SD*, ±0.7). (These values were reported by Bernini[3] in a study of 473 individuals. They are consistent with other reported studies and with the author's experience.)

COMMENTS AND PRECAUTIONS

In contrast to other hemoglobins (hemoglobin F excepted), hemoglobin A_2 must be measured with considerable precision because the diagnosis of β thalassemia minor may hinge on a hemoglobin A_2 value only 1 to 2 per cent above the normal range.

Several alternative methods have been used. These include densitometric scanning of cellulose acetate strips and elution from segments cut from starch gel or cellulose acetate, followed by spectrophotometric measurement. These alternative procedures suffer from considerable imprecision. Efremov *et al.* have reported a still simpler chromatographic method using microgranular DEAE cellulose and Pasteur pipets as columns. It should be stressed that hemoglobins C, E, and O cannot be distinguished from hemoglobin A_2 by these methods.

REFERENCES

Bernini, L. F.: Biochem. Genet., **2**:305–310, 1969.
Efremov, C. D., *et al.*: J. Lab. Clin. Med., **83**:657–664, 1974.

DETERMINATION OF FETAL HEMOGLOBIN (ALKALI-RESISTANT HEMOGLOBIN)

CLINICAL SIGNIFICANCE

In the high F or the β-δ type of thalassemia minor, the fetal hemoglobin concentration is generally in the range of 2 to 10 per cent of the total hemoglobin. In β thalassemia major, however, the fetal hemoglobin may be 90 per cent or more of the total hemoglobin. Slight increases of hemoglobin F concentration are found in a variety of unrelated hematological disorders such as aplastic anemia and acute leukemia. In homozygous sickle cell disease, the hemoglobin F concentration is often slightly increased. Higher concentrations of hemoglobin F occur in sickle-thalassemia and in patients who are doubly heterozygous for the hemoglobin S gene and for a gene for hereditary persistence of fetal hemoglobin (HPFH). These disorders may be differentiated by family studies or by the acid elution test for fetal hemoglobin (see below), which distinguishes between the uniform intraerythrocytic distribution of hemoglobin F in HPFH and the irregular distribution occurring in sickle-thalassemia. The presence of small quantities of hemoglobin A, demonstrated by electrophoresis in a patient who has mostly hemoglobin S and a moderate increase in hemoglobin F, is strong evidence for sickle-thalassemia.

PRINCIPLE

Most human hemoglobins denature readily at alkaline pH and can then be precipitated by addition of ammonium sulfate to 40 per cent saturation.[61] However, fetal hemoglobin is resistant to denaturation by alkali and remains soluble. These differences permit a rapid and relatively simple quantitative measurement of the amount of fetal hemoglobin present in human blood. In the method which follows, all hemoglobins are first converted to cyanmethemoglobins.[5]

SPECIMEN

Fresh blood anticoagulated with any standard anticoagulant may be used; 1 ml is required. No prior fasting or fluid restriction is needed.

REAGENTS

1. Cyanide-ferricyanide solution. Dissolve 0.2 g of KCN and 0.2 g of $K_3Fe(CN)_6$ in distilled water, and dilute to 1000 ml.

2. Saturated solution of ammonium sulfate. Add 750 g of $(NH_4)_2SO_4$ to 1000 ml of distilled water. Warm and agitate to bring into solution (the final volume of this solution is expected to be well in excess of 1000 ml). A few crystals may appear and settle to the bottom of the flask as the solution cools to room temperature.

3. Sodium hydroxide solution, 1.2 mol/l. Dissolve 48.0 g of NaOH, AR (carbonate-free), in distilled water, and dilute to 1000 ml.

PROCEDURE

1. Place approximately 1 ml of whole blood in a centrifuge tube and add cold isotonic saline to a volume of approximately 10 ml.
2. Centrifuge at 5°C at 700 × g for 15 min.
3. Aspirate and discard the supernatant.
4. Mix the packed cells with a wooden applicator stick or glass rod. Then pipet 0.2 ml of packed cells into a tube containing 4.0 ml of cyanide-ferricyanide solution.
5. Mix the cells in this solution with a glass stirring rod.
6. Using a safety pipet, transfer 2.8 ml of the hemoglobin-cyanide-ferricyanide mixture to a test tube. Blow in 0.2 ml of sodium hydroxide solution, 1.2 mol/l, and rapidly mix with a glass stirring rod. Begin timing, with a stopwatch or other accurate timer, at the moment the sodium hydroxide solution is blown in.
7. Exactly 2 min after the addition of sodium hydroxide, blow in 2 ml of saturated ammonium sulfate solution and mix well with a stirring rod. Allow to stand for 5 min to coagulate the protein.
8. Filter through a small filter paper (Whatman No. 42 is recommended). If the filtrate is not absolutely clear, refilter through the same paper. Read absorbance of the filtrate at 540 nm, using a cuvet with a 1.0 cm light path. If the absorbance exceeds 1.0, dilute 2-fold or 3-fold with distilled water and reread. The absorbance of the filtrate is designated A_f.
9. Using a safety pipet, transfer 0.4 ml of the original blood-cyanide-ferricyanide solution mixture (not treated with sodium hydroxide) to another tube and add 6.75 ml of distilled water.* Agitate to mix, and read the absorbance of this in the same manner as above. Designate the absorbance A_b.

CALCULATION

$$\text{Per cent alkali-resistant hemoglobin} = \frac{100 \times A_f}{10 \times A_b}$$

in which A_f is the absorbance of the filtrate, and A_b is the absorbance of the diluted blood-cyanide-ferricyanide mixture. The numerator contains a factor of 100 to convert from a decimal value to percentage of the total hemoglobin.

NORMAL VALUES

In normal adults or children more than 1 year old, hemoglobin F by this method is usually less than 1 per cent and never greater than 2 per cent of the total hemoglobin.

COMMENTS AND PRECAUTIONS

There are several published variations of the alkali denaturation test. Two important advantages of the procedure presented above are that it is the most sensitive method for measurement of concentrations of fetal hemoglobin less than 10 per cent of total hemoglobin and that it does not give erroneously high values when CO-hemoglobin is present, as do other alkali-denaturation tests.

The cyanide-ferricyanide solution has approximately four times the cyanide concentration of Drabkin's solution. It should be handled with appropriate care, including the

* This provides a 375-fold final dilution of the original washed, packed erythrocytes. In contrast, the final dilution of the NaOH-treated, washed, packed erythrocytes is 37.5-fold. The ratio of dilutions is 10/1. Therefore, in the calculations, the absorbance of the diluted, untreated hemolysate (A_b) is multiplied by 10.

use of safety pipets. **Do not** pipet by mouth. Other precautions noted under Measurement of Hemoglobin Concentration in Whole Blood (p. 413) apply to the handling of this reagent. Note that the same packed cells can be used for this procedure as for hemoglobin electrophoresis by removing the necessary volume of cells immediately prior to hemolyzing the sample in the hemoglobin electrophoresis procedure.

It must be recognized that this procedure, and other rapid tests for fetal hemoglobin, give erroneously low results when hemoglobin F makes up 30 per cent or more of the total hemoglobin in a specimen. For example, this technique underestimates, by about 50 per cent, the actual hemoglobin F concentration of umbilical cord blood. When it is known, or seems likely, that hemoglobin F is present in high concentration, the more cumbersome but more accurate hemoglobin F assay of Jonxis and Visser[37] should be used.

REFERENCE

Betke, K., Marti, H. R., and Schlicht, I.: Nature (Lond.), **184**:1877–1878, 1959.

TEST FOR FETAL HEMOGLOBIN DISTRIBUTION BY ACID ELUTION

In this test, blood smears prepared on glass slides are fixed in 80 per cent methanol and then immersed for 5 minutes at 37°C in a pH 3.3 buffer (citric acid [0.075 mol/l] and disodium phosphate [0.050 mol/l]). During this period, the slide is gently agitated. Under these conditions, hemoglobin A is eluted relatively rapidly while hemoglobin F elutes relatively slowly. The slide is then rinsed with water and stained with Ehrlich's acid hematoxylin (Harleco Laboratories, Philadelphia, PA 19143) and counterstained with erythrosin (0.2 g/100 ml). The stained slide is then examined microscopically at ×300 to ×200 magnification.

INTERPRETATION AND COMMENTS[39,59]

Normal adult erythrocytes show complete elution of hemoglobin (Fig. 8–9). Erythrocytes containing hemoglobin F stain darkly. When fetal erythrocytes are admixed with maternal erythrocytes (as commonly occurs during labor), the mother's blood shows occasional darkly stained erythrocytes containing hemoglobin F. In the genetic trait, hereditary persistence of fetal hemoglobin (HPFH), all erythrocytes show nearly uniform partial retention of hemoglobin. In thalassemias with increased hemoglobin F, there is nonuniform distribution of retained hemoglobin from cell to cell. For satisfactory results, the elution time and the pH must be very carefully controlled.

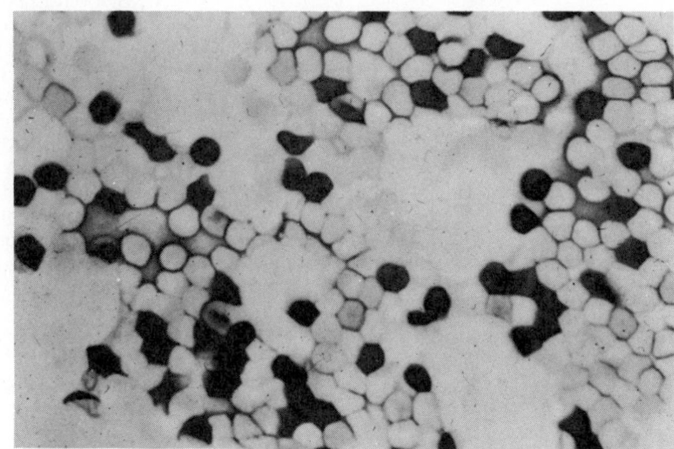

Figure 8–9 Acid elution test for cellular distribution of fetal hemoglobin. Hemoglobin F is more slowly eluted from erythrocytes than is hemoglobin A when exposed to a solution of citric acid. The dark cells in this illustration contain hemoglobin F which has not been eluted. The clear cells contain only hemoglobin A. (Counterstained with hematoxylin and erythrosin, ×300.)

REFERENCES

Kleihauer, E., Braun, H., and Betke, K.: Klin. Wochenschr., **35**:637–638, 1957.
Shepard, M. K., Weatherall, D. J., and Conley, C. L.: Bull. Johns Hopkins Hosp., **110**:293–310, 1962.

ISOELECTRIC FOCUSING IN POLYACRYLAMIDE GEL

When migrating under the influence of an electric field along a pH gradient, a charged protein, such as hemoglobin, will concentrate, or focus, at the pH that corresponds to its isoelectric point. The pH gradient is established by incorporating into a supporting medium (such as polyacrylamide gel) a mixture of molecules carrying positive and negative charges (ampholytes). Commercially available ampholyte preparations are mixtures of arginine and aspartic and glutamic acids with synthetic polyamino-polycarboxylic acids. One end of the supporting medium is immersed in acid; the other end is immersed in a basic solution. Electrodes are connected so that the anode is in contact with the acidic end of the gel and the cathode is in contact with the basic end. Passage of a weak current (for example, 1 mA/5 ml of gel) causes the ampholytes to migrate into such positions that they establish a gradual pH gradient from one end of the gel to the other. Hemoglobin, or any other protein, also moves within the gel until it becomes concentrated at its isoelectric point.[22,57]

The relative order of hemoglobins (and myoglobin) observed after isoelectric focusing is, beginning at the acidic anodal end: myoglobin, hemoglobin A, hemoglobin S, and hemoglobin C (Fig. 8–10). Each hemoglobin may appear as a double band. Additional bands are easily produced artifactually.

Isoelectric focusing has only begun to be used systematically as a clinical laboratory tool in the identification of abnormal hemoglobins. It is a relatively simple technique, requiring basically the same equipment used for electrophoresis. Because it provides more powerful resolution of hemoglobin bands than does electrophoresis, it seems inevitable that it will have general application in the future.

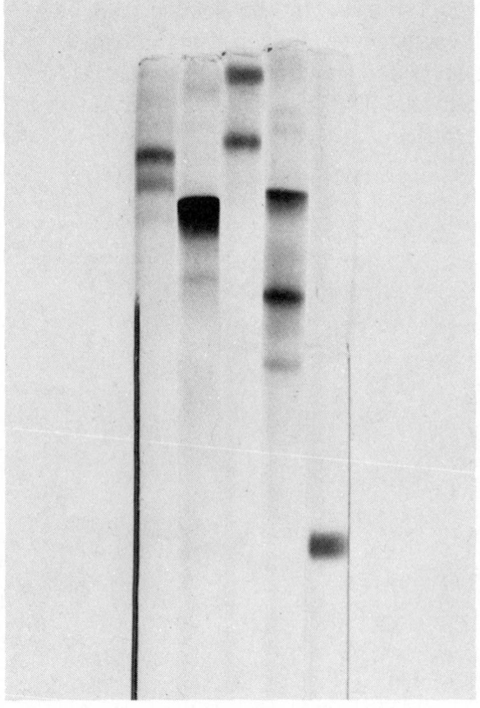

Figure 8–10 Isoelectric focusing of hemoglobins and myoglobin in polyacrylamide gel columns. Upper end of gel column is basic cathode and lower end is acidic anode. From left to right, the gel columns contain: hemoglobins S (upper) and F (lower); hemoglobin A; hemoglobins C (upper) and S (lower); hemoglobins A (upper) and Malmö (lower prominent band); and human myoglobin. Myoglobin is widely separated from hemoglobins S and C. Hemoglobins A and Malmö are easily separated, a separation which is not possible by electrophoresis. Additional light bands are seen with the hemoglobin specimens. The nature of these is not known; possibly they are degradation products. (Based on a collaborative study with Dr. David N. Fass.)

REFERENCE

Righetti, P., and Drysdale, J. W.: Biochim. Biophys. Acta, **236**:17–28, 1971.

DETECTION OF HEMOGLOBIN IN URINE AND FECES

The methods to be described in this section are useful in detecting occult blood—that is, blood present in such low concentration that it is not detectable visually. Another important use is in establishing whether the dark color of a specimen of urine or stool is due to the presence of heme. A black or red discoloration of excreta can be due to the presence of any of several abnormal pigments other than heme. Heme proteins, such as hemoglobin and myoglobin, or their derivatives (methemoglobin, metmyoglobin, and sulfhemoglobin) give positive reactions in these tests. Ingested plant peroxidases or peroxidases in pus cells in urine also may give positive reactions.

CLINICAL SIGNIFICANCE

Hemoglobin is not normally present in sputum, urine, or feces except in minute amounts; the demonstration of the presence of hemoglobin in these materials is an important diagnostic clue that usually implies bleeding within the respiratory, urinary, or gastrointestinal tract, respectively. The demonstration of the presence of hemoglobin on weapons or clothing has obvious importance in forensic pathology.

In considering tests for hemoglobin in the urine, a distinction must be made between hematuria (the presence of intact erythrocytes in the urine) and hemoglobinuria (the presence of free hemoglobin in the urine). Hematuria implies a bleeding lesion in the urinary tract. Hemoglobinuria, in the absence of hematuria, usually implies severe intravascular hemolysis. Microscopic examination of the urine is the appropriate and most sensitive test for hematuria. Hemoglobinuria is detected by the occult blood tests outlined in this section.

PRINCIPLE

The tests for the detection of hemoglobin depend on the fact that heme proteins act as peroxidases, catalyzing the reduction of hydrogen peroxide to water. This reaction requires a hydrogen donor. If guaiac, benzidine, or a benzidine derivative is used as a hydrogen donor, the oxidation of the donor chromogen results in a blue color, and the intensity of the final color is primarily a function of the hemoglobin concentration:

$$\underset{\textbf{(Reduced, Colorless)}}{H_2O_2 + \text{benzidine}} \xrightarrow{\textit{Heme protein}} \underset{\textbf{(Oxidized, Blue)}}{2\,H_2O + \text{benzidine}}$$

Actually, the benzidine reaction is more complex than indicated here and has not been completely elucidated.

In addition to benzidine, other compounds are used in testing for the presence of heme proteins. Most widely used is *o*-tolidine, a benzidine derivative (Fig. 8–11), and what has

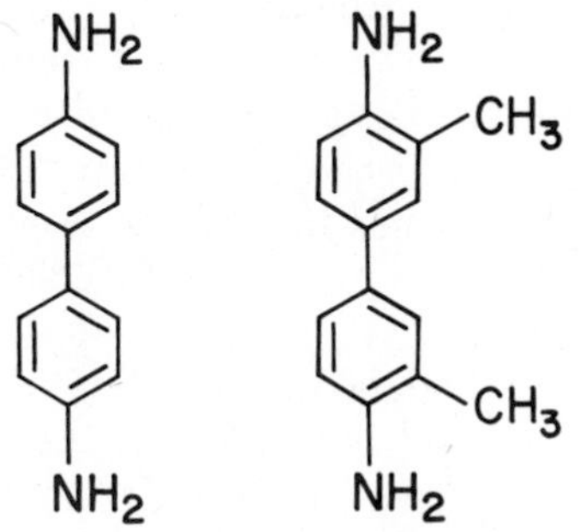

Figure 8–11 Structures of compounds used in testing for occult blood. *Left*, Benzidine base. *Right*, *o*-Tolidine.

NH2 NH2 CH3 CH3 NH2 NH2

been stated for benzidine applies also to *o*-tolidine. Gum guaiac extracts are also used for establishing the presence of heme proteins. As with benzidine and *o*-tolidine, the guaiac reaction product is an intense blue. Guaiac extracts are crude, and the chemical composition of the hemoglobin-reactive component(s) is not definitely known.

SPECIMEN

With urine, the procedure requires 10 ml. The specimen should be kept at 5°C until the test is made and should be studied as soon as possible on the day it is received (otherwise, if there is hematuria, lysis of the erythrocytes in the specimen may lead to an erroneous finding of hemoglobinuria).

With feces, a 24 hour specimen is preferred to a random specimen. However, 1 ml of stool is a sufficient volume for the test.

REAGENTS

1. Saturated solution of chromogen (benzidine or *o*-tolidine). Dissolve 1.0 g of benzidine dihydrochloride (do not use benzidine base) in distilled water and dilute to 100 ml. Mix until dissolved. Do not heat. Filter and store in a dark bottle at 5°C. *Or*, dissolve 1.08 g of *o*-tolidine in absolute methanol and dilute to 100 ml.
2. Hydrogen peroxide, 3 per cent, USP.

PROCEDURE FOR TEST TUBE METHOD FOR HEMOGLOBINURIA OR MYOGLOBINURIA[15]

1. Centrifuge 10 ml of urine at 1700 × *g* for 10 min in a conical centrifuge tube.
2. Transfer 0.2 ml of supernatant to a test tube; add 0.2 ml of distilled water to a second test tube labeled "negative control."
3. To approximately 5 ml of supernatant, add one drop of whole blood and mix; transfer 0.2 ml of this to another test tube labeled "positive control."
4. Decant and discard the remainder of supernatant. This will leave approximately 0.1 to 0.2 ml of sediment. Resuspend this by gentle tapping. Remove a sample for microscopic examination and retain the residual sediment for the occult blood test.
5. To the three test tubes—negative control, unknown, and positive control—add 4 drops of *o*-tolidine, followed by 2 drops of hydrogen peroxide and 1 drop of glacial acetic acid. Do the same with the urinary sediment. Inspect for color development after 1 min. The negative control should appear yellow. The positive control should show a distinct and intense blue color. The unknown will turn blue if there is hemoglobinuria or myoglobinuria; otherwise it will be yellow or amber. The sediment will turn blue if there is hematuria. This will be confirmed by microscopic examination. A green or greenish-blue color may appear when there is marked pyuria (excessive number of leukocytes in the urine).

PROCEDURE FOR TEST TUBE METHOD FOR BLOOD IN FECES

1. With a spatula, thoroughly mix the stool specimen in the container in which it is received. Then, with an applicator stick, transfer a small amount to the bottom of each of two graduated 15 ml centrifuge tubes (to approximately the 0.3 ml mark). To one of these, designated "positive control," add 0.05 ml of whole blood. Add 5 ml of distilled water to each tube and mix thoroughly to emulsify. Add distilled water to the 10 ml mark and again mix thoroughly, using an applicator stick.
2. Centrifuge at 700 × *g* for 10 min.
3. Decant approximately 1 ml of supernatant from each tube into appropriately labeled test tubes.
4. To a third test tube add 1 ml of distilled water and label this "negative control."
5. Place all three tubes in a boiling water bath for 5 min (to destroy bacterial peroxidase and catalase).

6. Cool by transferring the tubes to a beaker containing water at room temperature.

7. When the tubes have cooled to room temperature, add to each tube 1 ml of chromogen solution followed by 1 ml of hydrogen peroxide solution.

8. At exactly 1 min, inspect for color development. The negative control should show, at most, a yellowish-green color. The positive control should show a distinct blue color. If there is blood in the feces, there will be a distinctly blue color in the specimen tube; otherwise, it will be a yellow, tan, or yellowish-green.

COMMENTS AND PRECAUTIONS

Benzidine base is carcinogenic, and excessive absorption through the skin, orally, or by inhalation of powder may lead to cancer of the urinary bladder. This risk probably exists also for benzidine dihydrochloride and *o*-tolidine, although this is disputed. Care in handling these compounds is mandatory.

The tests given here for occult blood in the stool are sensitive to about 100 mg of hemoglobin per 100 g of feces, the equivalent of approximately 0.5 to 1.0 ml of blood per 100 ml of feces. Therefore, these are extremely sensitive tests and are well known to give false positive results.[30,31,33,54] The tests do not distinguish myoglobin from hemoglobin and also are sensitive to peroxidases which occur in all plants (particularly in large amounts in celery and horseradish). It is, therefore, essential to have the patient refrain from ingestion of any meat, fowl, fish, celery, radishes, or horseradish for at least three days before collection of fecal specimens for the test for occult blood. Negative tests do not rule out bleeding lesions, because these may bleed only intermittently. Inorganic iron in the feces, originating from medicinal iron, may cause positive tests for occult blood with *o*-tolidine or benzidine but rarely with guaiac unless the guaiac is contaminated with tannic acid.[32]

Commercially available reagent kits for testing for fecal blood include the Hematest (Ames Company, Elkhart, IN 46514), which is based on the *o*-tolidine reaction, and the Hemoccult test (Smith, Kline & French Laboratories, Philadelphia, PA 19101), which is based on the guaiac reaction. Each provides a special filter paper onto which the stool specimen is smeared. Reagents are then applied to the fecal streak to develop the color. The Hemoccult filter paper is guaiac-impregnated. The sensitivity of these tests is claimed to be about 1 ml of blood per 100 ml of feces for the Hematest, and about 2 ml of blood per 100 ml of feces for the Hemoccult test. By simultaneously measuring blood loss with a radioisotope tag (^{51}Cr), the author has confirmed that 10 ml or more of blood per 100 ml of feces almost always gives a positive Hemoccult test.

Similarly, proprietary reagent tablets and impregnated strips, such as Hematest, Occultest, and Hemastix, based on the *o*-tolidine reaction, are available for testing for hemoglobinuria-hematuria-myoglobinuria.

REFERENCE

Derman, H., and Pauker, S.: *In* Hemoglobin: Its Precursors and Metabolites. F. W. Sunderman and F. W. Sunderman, Jr., Eds. Philadelphia, J. B. Lippincott Co., 1964, pp. 70–80.

MEASUREMENT OF PLASMA HEMOGLOBIN

Virtually all of the hemoglobin in blood is contained within the erythrocytes. A minute quantity of hemoglobin normally is released into plasma in the destruction of erythrocytes, and this is promptly bound by haptoglobin. The haptoglobin-hemoglobin complex is rapidly removed by parenchymal cells of the liver. Thus, the normal plasma hemoglobin concentration is close to zero.

CLINICAL SIGNIFICANCE

An increase in plasma hemoglobin concentration is indicative of acute destruction of erythrocytes (hemolysis) within the vascular system. Therefore, the practical value of the measurement is virtually limited to circumstances in which acute intravascular hemolysis is believed to have occurred as a result of transfusion reactions, or to the evaluation of degree of hemolysis occurring in extracorporeal treatment of blood. The measurement of free hemoglobin in plasma is of no practical value in the diagnosis of chronic hemolytic disorders.

PRINCIPLE

The benzidine-heme color reaction,* described in the preceding section (Detection of Hemoglobin in Urine and Feces) in which the blue color of oxidized benzidine indicates the presence of heme protein, is used.[12] There is a sequence of color changes from green to an intense blue to a final brown which reaches maximal intensity in 20 minutes. These changes may be followed spectrophotometrically: the gradual increase in brown color is reflected in a progressive increase in absorbance at 515 nm to a plateau at about 20 minutes. The brown pigment is measured at 515 nm with the standard method for plasma hemoglobin.

Turbidity of the plasma contributes to the absorbance, and therefore a reagent blank or water blank is not suitable. This problem is circumvented by using plasma treated with hydrogen peroxide as the matrix for both blank and standard. The hydrogen peroxide denatures plasma hemoglobin, thus markedly decreasing its peroxidase activity; the chromogen is not oxidized when added subsequently. In this manner, turbidity of the peroxide-treated plasma blank can be subtracted from the turbidity of the "test."

Hanks *et al.*[27] pointed out that the test has far greater sensitivity if the initial green color is measured at 700 nm. Because this color is transitory, exact timing is crucial and the absorbance measurement must be made $3\frac{1}{2}$ minutes after addition of the chromogen. This technique is necessary when measurement of hemoglobin concentrations less than 2 mg/100 ml is desired. For routine clinical use, measurement of the absorbance of the stable brown color seems preferable.

SPECIMEN

This procedure requires a minimum volume of 8 to 10 ml of whole blood. No prior dietary or fluid restriction is needed. The blood must be freshly drawn by the following technique.

1. Attach a 30 cm length of polyvinyl tubing to the hub of a sterile 18 gauge needle (or use an 18 gauge needle with attached tubing from an infusion set).

2. Prepare two plastic (or silicone-coated glass) centrifuge tubes, labeled 1 and 2, for collection of blood from the free end of the tubing.

3. Lightly coat the interior of tube 2 by spraying with a fine mist of sodium heparin solution, 1000 units/ml. Only a small amount (0.05 ml) of heparin solution should be applied. Any large droplets or puddles should be wiped away.

4. Place a tourniquet lightly around the upper arm. With minimal delay, perform a

---

* Since this chapter was written, severe Federal restrictions on the marketing and use of benzidine preclude the use of methods dependent upon benzidine and its derivatives for the foreseeable future. Fortunately, the clinical indications for measuring plasma hemoglobin are very limited. For those rare clinical situations in which a plasma hemoglobin measurement is demanded, the method of Harboe (Harboe, M., Scand. J. Lab. Clin. Invest., **11**: 66–70, 1959) may suffice. This method is based upon measurement of the absorbance of oxyhemoglobin at 415 nm. The author has verified Harboe's finding that this method is essentially linear and of sufficient sensitivity to detect plasma hemoglobin in concentrations as low as 3 mg/100 ml. The method is, thus, approximately one-tenth as sensitive as benzidine methods. Precision is poor below a hemoglobin concentration of 5 mg/100 ml. On the other hand, this degree of precision may well be sufficient for most clinical purposes. The major error in plasma hemoglobin measurement will still depend on the care with which the sample is obtained.

clean puncture of the antecubital vein. Release the tourniquet, and permit blood to flow freely into tube 1 until approximately 4 ml has accumulated; then, transfer the free end of the tubing into tube 2 and collect 8 ml of blood. Be careful that droplets of blood do not adhere to the test tube above the meniscus. Discard the blood in tube 1.

5. Centrifuge tube 2 at 1000 × *g* (2400 rpm when the radius of rotation is 16 cm) for 10 min; carefully draw off the supernatant plasma with a transfer (Pasteur) pipet and deliver it into a second centrifuge tube.

6. Recentrifuge at 1600 × *g* for 20 min; carefully draw off the supernatant plasma and transfer it into a third tube.

7. Proceed promptly to hemoglobin measurement or freeze at −20°C until the determination is to be completed.

REAGENTS

1. Benzidine. Dissolve 0.1 g of benzidine base* in 9.0 ml of glacial acetic acid and 1.0 ml of sulfate-free distilled water. Store in a brown bottle. Prepare weekly.

2. Hydrogen peroxide. Dilute 0.3 ml of 30 per cent H_2O_2 (Superoxol) with 9.7 ml of sulfate-free distilled water. Prepare fresh at least every three days. Handle Superoxol with care. It is best to wear rubber gloves and goggles or glasses.

3. Diluent. Dilute 1.0 ml of glacial acetic acid with 9.0 ml of sulfate-free distilled water.

4. Standard hemoglobin solution. Centrifuge 1 ml of normal whole blood at 1000 × *g* for 10 min. Remove and discard the plasma. Resuspend the erythrocytes in 10 ml of isotonic (9 g/l) sodium chloride and centrifuge at 1000 × *g* for 10 min. Remove and discard the supernatant. Resuspend the erythrocytes again in 10 ml of isotonic sodium chloride and centrifuge at 1000 × *g* for 10 min. Remove and discard the supernatant. Lyse the packed, washed erythrocytes by alternately freezing and thawing twice. Measure the hemoglobin concentration by the cyanmethemoglobin method. Dilute this hemolysate with isotonic sodium chloride to a concentration of 0.2 mg of hemoglobin/ml. Alternatively, the method may be standardized against a certified commercial hemoglobin preparation.

Example: Hemoglobin concentration of hemolysate is found to be 18 g/100 ml, or 18,000 mg/100 ml. How much sodium chloride solution must be added to 0.1 ml of hemolysate to adjust the concentration to 20 mg/100 ml? With the basic formula $C_1V_1 = C_2V_2$, in which C_1 = initial concentration, V_1 = initial volume, C_2 = final concentration desired, and V_2 = final volume,

$$\frac{18{,}000 \text{ mg}}{100 \text{ ml}} \times 0.1 \text{ ml} = \frac{0.2 \text{ mg}}{1.0 \text{ ml}} \times V_2$$

$$V_2 = \frac{18}{0.2} \text{ ml}$$

$$V_2 = 90 \text{ ml}$$

Since 90 ml is to be the final volume, then 90 ml − 0.1 ml, or 89.9 ml, of sodium chloride solution must be added.

EQUIPMENT

A Coleman Junior or similar spectrophotometer is suitable.

All glassware (cuvets, test tubes, pipets, centrifuge tubes) must be acid-washed with

---

* Because of the carcinogenicity of benzidine, its manufacture and distribution have been markedly curtailed for use in clinical laboratories. See footnote on preceding page. At the time of publication it remained unclear whether other compounds with comparable sensitivity could be used.

hydrochloric or nitric acid to remove sulfate (sulfate inhibits the color reaction). Chromic acid, commonly used for acid washing of glassware, contains excessive amounts of sulfate.

PROCEDURE

1. Dilute 1 ml of plasma with 2 ml of sulfate-free distilled water; mix.
2. Set up three cuvets for each plasma specimen to be tested. Label these "standard," "blank," and "unknown." Into each, pipet 0.25 ml of diluted plasma.
3. Into the standard and blank cuvets, pipet 0.25 ml of hydrogen peroxide solution (to inactivate the hemoglobin). Shake and allow to stand for 10 min before adding any other reagents.
4. Add 0.25 ml of standard hemoglobin solution to the standard cuvet. Add 0.25 ml of sulfate-free distilled water to the blank and unknown cuvets.
5. Add 0.5 ml of benzidine solution to all cuvets.
6. Add 0.25 ml of hydrogen peroxide solution to the unknown and standard cuvets.
7. Shake all tubes and allow to stand until development of a mahogany-brown color is maximal (about 20 min).
8. Add 10 ml of diluent and mix by inversion. Allow to stand for another 10 min.
9. Set spectrophotometer to 515 nm, and set the blank cuvet at zero absorbance. Measure absorbance of the standard (A_s) and unknown (A_u) cuvets.

CALCULATIONS

$$\text{Plasma hemoglobin concentration (mg/100 ml)} = \frac{A_u}{A_s} \times 20$$

in which 20 represents the concentration of the standard in mg/100 ml.

NORMAL VALUES

Normal values for plasma hemoglobin in carefully obtained plasma specimens are less than 2.5 mg/100 ml.[53] Strenuous physical exertion may result in increased plasma hemoglobin values.

COMMENTS AND PRECAUTIONS

It is likely that hemoglobin does not normally exist free in plasma in any measurable concentration. The amounts that are measured in plasma from normal subjects appear to be a result of the trauma to which blood is subjected during or after venipuncture. Thus, very careful attention to technique in drawing and handling the blood specimen is essential in order to avoid, as far as possible, artifactual increases of the plasma hemoglobin concentration. Furthermore, since measured concentrations of hemoglobin are almost invariably higher in serum than in plasma, it appears that coagulation results in hemolysis with liberation of hemoglobin. Therefore, the test should only be performed on plasma specimens obtained with meticulous "atraumatic" techniques.

When plasma specimens are pink-tinged, it must be assumed that the hemoglobin concentration is at least 30 mg/100 ml. Such specimens must be diluted; otherwise, absorbance will be out of the range of the spectrophotometer scale.

Sulfates form insoluble complexes with benzidine, and therefore the stipulated precautions must be observed to avoid sulfate contamination. Glassware that is acid-washed with chromic acid (a mixture of sodium chromate and sulfuric acid) is unsuitable. All reagents must be prepared with sulfate-free water. The presence of ferric ion in reagents or on glassware may increase the apparent plasma hemoglobin concentration, since ionic iron has a weak peroxidase-like activity. Such contamination may be suspected when the blank exhibits a brownish color (the blank should be nearly colorless).

Vanzetti and Valente[67] have described a variation of this method in which the twice-centrifuged plasma is treated with acetic acid (to convert hemoglobin to acid hematin) and is then extracted with ethyl ether. The ethyl ether extracts the acid hematin, and the aqueous phase is discarded. The ethyl ether phase is then extracted with ethanol, and the acid hematin is taken up in the ethanol phase. Benzidine and hydrogen peroxide are added to the ethanol extract, and after 15 minutes the absorbance is measured at 500 nm. A principal advantage of this method is that it eliminates the turbidity which contributes to the absorbance measured by other techniques. By this method, normal plasma hemoglobin concentrations should not exceed 0.5 mg/100 ml.[35] Thus, this method may be preferred should it be necessary to measure very low levels of plasma hemoglobin precisely.

All methods for plasma hemoglobin measurement demand the most meticulous technique and rigid standardization to ensure reliable results. It also bears repeating that the hazardous reagents, hydrogen peroxide and benzidine base, must be handled with the greatest care.

REFERENCE

Naumann, H. N.: *In* Hemoglobin: Its Precursors and Metabolites. F. W. Sunderman and F. W. Sunderman, Jr., Eds. Philadelphia, J. B. Lippincott Co., 1964, pp. 40–48.

MEASUREMENT OF OXYGEN SATURATION OF HEMOGLOBIN

The prime function of hemoglobin is to carry oxygen from the alveoli of the lungs to all the body tissues. Blood that has been oxygenated in the lungs and is being distributed through the arteries contains hemoglobin which under normal circumstances is at least 95 per cent oxyhemoglobin. During passage of the blood through the capillary network, oxygen is extracted by the tissues; in the venous blood that returns to the heart, the hemoglobin normally is approximately 75 per cent oxyhemoglobin.

CLINICAL SIGNIFICANCE

Measurement of oxygen saturation of blood may be indicated when a patient has either cyanosis (a bluish discoloration of skin, lips, eyelids, and nail beds) or erythrocytosis (increase in the number of erythrocytes in the blood). Among the possible causes of cyanosis or erythrocytosis are severe pulmonary disease (resulting in diminished oxygenation of blood) and admixture of venous and arterial blood (for example, from an arteriovenous shunt, usually a developmental abnormality of the heart or great vessels). In either of these conditions, both cyanosis and erythrocytosis may coexist, and a low value for oxygen saturation of arterial blood will be obtained. Measurement of oxygen saturation does not differentiate between these disorders. Cyanosis without erythrocytosis may be due to the presence of an abnormal hemoglobin (for example, methemoglobinemia due to the presence of hemoglobin M); methemoglobinemia due to the ingestion of nitrates or various drugs; or sulfhemoglobinemia, usually due to drug ingestion. When there is cyanosis due to methemoglobinemia or sulfhemoglobinemia, measurement of arterial oxygen saturation may give either a normal value or a lower than normal value. Erythrocytosis occurs without cyanosis in polycythemia vera, in the presence of certain neoplastic diseases, as a result of abnormalities of the kidney, or as a result of the presence of a hemoglobin with high oxygen affinity. In these disorders, measurement of arterial oxygen saturation gives a normal value.

PRINCIPLE

The oxygen saturation of hemoglobin may be measured by using either of two basic principles.

1. The oxygen contained in a sample of blood may be released and measured volumetrically in a Van Slyke apparatus. The total oxygen binding capacity of a blood specimen may be measured in the same manner. Then, the oxygen saturation is the ratio of the actual oxygen content to the total oxygen binding capacity of the specimen.[28,66]

2. The ratio of oxyhemoglobin to deoxyhemoglobin may be measured on the basis of difference in spectral (color) characteristics between the two pigments. Spectrophotometers or reflectance meters may be used for this purpose, spectrophotometers being more generally used. Among applications of the spectrophotometric method are measurement of oxygen saturation *ex vivo*, as described below,[49,50] and *in vivo* measurement by cuvet oximeters that may be applied to the pinna of the ear for measurement of the absorbance of light by blood in the capillary bed, or by cuvet oximeters through which arterial blood may be passed to permit continuous monitoring of oxygen saturation.[71] The spectrophotometric method presented below has the advantages that it is at least as accurate as other methods, is relatively simple to perform, requires no unusual or costly equipment, and is the basis of a widely used semiautomated (CO-Oximeter) method.[43]

With the absorption spectrum of oxyhemoglobin as an example, for any solution of the pure pigment the absorbances at any two wavelengths will have a constant ratio irrespective of the concentration of the pigment (for example, the ratio of absorbance at 577 nm to that at 548 nm is always 1.2/1 for a pure solution of oxyhemoglobin). However, if another pigment with a different spectral absorption curve is also present in the solution, the absorbance ratio will change, and the extent to which the ratio changes is proportional to the concentration of the other pigment. It was recognized long ago that by measuring the ratio of absorbances at two wavelengths for a solution containing two hemoglobin derivatives (oxyhemoglobin and deoxyhemoglobin or oxyhemoglobin and carboxyhemoglobin), the relative proportion of each could be precisely ascertained. The absorbance ratios became known as Hüfner's quotients.

Because the relationship of absorbance ratios to concentration ratios (concentration of oxyhemoglobin to concentration of deoxyhemoglobin) is not a straight-line arithmetic function, it was necessary to construct tables or graphs displaying the absorbance ratio for every concentration ratio of pigments. Then, for any specimen of blood known to contain only two hemoglobin pigments, the relative proportions could be determined by making absorbance measurements at two wavelengths and using the ratio of absorbances to find the concentration ratio on a published graph. One drawback of this method is that much reliance must be placed on data published by others and obtained with different instruments. It is very tedious to reconstruct graphs relating absorbance ratios to concentration ratios.

The method described below is derived from the absorbance ratio method. It differs only in that one of the absorbance measurements is made at an isobestic point of the two pigments. An isobestic point is a wavelength at which the spectral absorption curves for the two pigments intersect (identical concentrations of the two pigments exhibit identical absorbance values). The absorbance at the isobestic point, then, is proportional to the total concentration of the two pigments. The other absorbance measurement is made at a wavelength at which the two pigments have markedly different absorbances. For a mixture of oxyhemoglobin and deoxyhemoglobin, a convenient isobestic point is 548 nm (Fig. 8–12). The spectral absorption curves of these pigments differ markedly at 577 nm, and this wave-

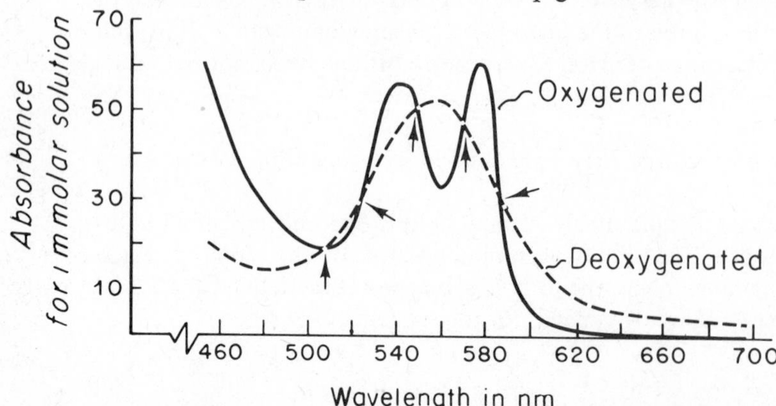

Figure 8–12 Spectral absorption curves for oxyhemoglobin and deoxyhemoglobin. The data were obtained by the author with whole blood hemolyzed with Triton X-100 in a Nahas cuvet with 0.1 mm light path, as described in the text. Arrows indicate isobestic points.

length is suitable for the second absorbance measurement. A major advantage of this method is that the ratio of concentrations of the two pigments is a straight-line arithmetic function of the absorbance ratio. Therefore, a graph or table need not be constructed; the concentration ratios can be calculated by a relatively simple equation. Furthermore, it is a simple matter to obtain and verify independently the constants that are required in the calculations.

The derivation of the formula for oxygen saturation is as follows: C_t represents the total concentration of hemoglobin (oxyhemoglobin plus deoxyhemoglobin), C_o is the concentration of oxyhemoglobin, and C_r is the concentration of deoxygenated (reduced) hemoglobin. A_{548} and A_{577} are the absorbances at 548 and 577 nm, respectively. K_1, K_2, and K_3 are the absorptivity constants for oxyhemoglobin and deoxyhemoglobin at 548 nm, oxyhemoglobin at 577 nm, and deoxyhemoglobin at 577 nm, respectively. Then, when only oxyhemoglobin or deoxyhemoglobin is present

$$A_{548} = K_1C_t = K_1C_o = K_1C_r$$
$$A_{577} = K_2C_o$$
$$A_{577} = K_3C_r$$

Let S = fractional oxygen saturation of hemoglobin. Then

$$S = \frac{C_o}{C_t}$$

or

$$C_o = SC_t$$

For any mixture of oxyhemoglobin and deoxyhemoglobin,

$$A_{577} = K_2C_o + K_3C_r$$

or

$$A_{577} = K_2SC_t + K_3C_r$$

However, $C_r = C_t - C_o$ (assuming presence of only two hemoglobin pigments). Therefore,

$$\begin{aligned} A_{577} &= K_2SC_t + K_3(C_t - C_o) \\ &= K_2SC_t + K_3(C_t - SC_t) \\ &= C_tK_2S + C_tK_3 - C_tK_3S \\ &= C_t(K_2S + K_3 - K_3S) \\ &= C_t[K_3 + S(K_2 - K_3)] \end{aligned}$$

$$\begin{aligned} \frac{A_{577}}{A_{548}} &= \frac{C_t[K_3 + S(K_2 - K_3)]}{C_tK_1} \\ &= \frac{K_3}{K_1} + \frac{S(K_2 - K_3)}{K_1} \end{aligned}$$

$$\frac{A_{577}}{A_{548}} - \frac{K_3}{K_1} = \frac{S(K_2 - K_3)}{K_1}$$

$$S(K_2 - K_3) = K_1\left[\frac{A_{577}}{A_{548}} - \frac{K_3}{K_1}\right]$$

$$S = \left[\frac{K_1}{K_2 - K_3}\right]\left[\frac{A_{577}}{A_{548}} - \frac{K_3}{K_1}\right]$$

$$S = \left[\frac{K_1}{K_2 - K_3}\right]\left[\frac{A_{577}}{A_{548}}\right] - \left[\frac{K_3}{K_2 - K_3}\right]$$

$$\text{Per cent saturation} = 100\left(\left[\frac{K_1}{K_2 - K_3}\right]\left[\frac{A_{577}}{A_{548}}\right] - \left[\frac{K_3}{K_2 - K_3}\right]\right)$$

Since the fractions $\frac{100K_1}{K_2 - K_3}$ and $\frac{100K_3}{K_2 - K_3}$ are also constants, designate the former K_4 and the latter K_5. Then,

$$\text{Per cent saturation} = K_4 \frac{A_{577}}{A_{548}} - K_5$$

This is an equation for a straight line, in which per cent saturation represents the ordinate, $\frac{A_{577}}{A_{548}}$ the abscissa, K_4 the slope, and K_5 the ordinate-intercept. When only deoxyhemoglobin is present, $S = 0$, and therefore

$$\frac{A_{577}}{A_{548}} = \frac{K_3}{K_1}$$

Similarly, when only oxyhemoglobin is present, $S = 1.0$, and

$$\frac{A_{577}}{A_{548}} = \frac{K_2}{K_1}$$

Thus, the absorbance ratios $\frac{A_{577}}{A_{548}}$ for pure solutions of either oxyhemoglobin or deoxyhemoglobin are constants, irrespective of concentration. If the ratio $\frac{A_{577}}{A_{548}}$ for oxyhemoglobin is designated R_o and the ratio $\frac{A_{577}}{A_{548}}$ for deoxyhemoglobin is designated R_r, then

$$K_4 = \frac{100}{R_o - R_r}$$

and

$$K_5 = \frac{100R_r}{R_o - R_r} = R_r K_4$$

Thus, K_4 and K_5 can be calculated from measured values of the ratios R_o and R_r. The constants thus obtained permit subsequent calculation of oxygen saturation in any specimen containing mixtures of oxyhemoglobin and deoxyhemoglobin, by measurement of absorbance at 548 and 577 nm.

The same basic equations are valid with other wavelengths. Nahas[50] used the isobestic point at 805 nm and the oxyhemoglobin absorption measurement at 660 nm (where absorbance is very low). Similarly, van Kampen and Zijlstra[65] used absorbance values at 560 and 506 nm, the value at 506 nm being isobestic. The values of the constants K_4 and K_5 will vary depending on the wavelengths used.

By measuring absorbance at a third wavelength, oxygen saturation and carbon monoxide content also can be determined. The commercially available CO-Oximeter (Instrumentation Laboratory, Inc., Lexington, MA 02173) performs simultaneous measurements at 548, 578, and 568 nm. From these data are derived hemoglobin concentration in g/100 ml, oxygen saturation of hemoglobin, and carboxyhemoglobin concentration. Oxygen saturation and carboxyhemoglobin concentration are expressed as a percentage of the total hemoglobin. The computations are performed automatically, based on the principles and formulas outlined above.

REAGENTS

1. Hemolyzing reagent. Dilute 1.0 ml of Triton X-100 (Rohm and Haas Company, Philadelphia, PA 19105) with distilled water to 100 ml.
2. Blank reagent. Dilute 1 ml of hemolyzing reagent with 1 ml of distilled water.
3. Sodium hydrosulfite ($Na_2S_2O_4$, dithionite).
4. Heparin sodium. Draw 0.5 ml of solution containing 1000 units/ml into a sterile syringe and eject into a sterile vial containing 10 ml of isotonic saline. The final concentration of heparin is 48 units/ml.

SPECIAL EQUIPMENT

1. Spectrophotometer with narrow (1 nm or less) band-pass.
2. Nahas cuvets with 0.1 mm light path, 0.01 ml volume (available as NAC-10 cuvet, Waters Corp., Rochester, MN). Because these do not fit many standard spectrophotometers, the sides may need to be ground to produce a 1 by 1 cm cross-section.*

PROCEDURE FOR DETERMINATION OF CONSTANTS IN FORMULA FOR OXYGEN SATURATION

1. Obtain approximately 5 ml of fresh heparinized blood from a fasting normal subject who is a nonsmoker.
2. Rotate in tonometer for 15 min to equilibrate with room air. Air in tonometer may be either continually or intermittently replaced. Aspirate 1 ml of blood into a syringe containing a glass bead (steel ball or mixing ring) and 1 ml of hemolyzing reagent. Invert several times until clear hemolysate is obtained.
3. Fill one Nahas cuvet (blank) with distilled water and one with hemolysate from step 2. Enough hemolysate should be injected so that 1 or 2 drops of the hemolysate pass through the tiny orifice at the bottom of the cuvet. This flushes out hemolysate that has absorbed oxygen in passing through the cuvet. Wipe away these droplets.
4. Set spectrophotometer to 548 nm and determine absorbance of hemolysate. Designate this $A_{548(o)}$.
5. Reset spectrophotometer to 577 nm and determine absorbance of hemolysate. Designate this $A_{577(o)}$.
6. Add 6 mg of $Na_2S_2O_4$ to remaining hemolysate; mix.
7. Measure absorbance of treated hemolysate at 577 and 548 nm. Designate these readings $A_{577(r)}$ and $A_{548(r)}$, respectively. $A_{548(r)}$ should be identical, or nearly identical, with $A_{548(o)}$.
8. Calculate ratios R_o and R_r:

$$R_o = \frac{A_{577(o)}}{A_{548(o)}}\ ;\ R_r = \frac{A_{577(r)}}{A_{548(r)}}$$

9. Calculate constants K_4 and K_5:

$$K_4 = \frac{100}{R_o - R_r}\ ;\quad K_5 = \frac{100R_r}{R_o - R_r}$$

MEASUREMENT OF OXYGEN SATURATION IN UNKNOWN SPECIMEN

1. Prepare a 5 ml glass syringe by placing a 1 mm glass bead (or a 1 mm steel ball or mixing ring) inside the syringe (to facilitate mixing) and inserting the plunger as far as it will go. Clearly label and autoclave.
2. Prior to drawing a blood sample, aspirate heparin solution (48 units/ml) into syringe. Allow entire area of contact between barrel and plunger to become moistened as a seal. Point syringe upward and eject excess heparin solution, making certain that all bubbles have been ejected and that heparin solution completely fills the small amount of "dead space" and the needle.
3. Prepare blank cuvet by filling a Nahas cuvet with blank reagent.
4. Draw 1 ml of arterial blood specimen into the heparinized syringe (arterial blood specimens should be drawn only by a physician).
5. Immediately insert needle of syringe into a vial containing hemolyzing solution.

---

* The short light path of the Nahas cuvet makes it possible to obtain a satisfactory absorbance reading from minimally diluted hemolysates at the specified wavelengths. Other cuvets with special adapters may be used.

Eject approximately 0.1 ml of blood into hemolyzing solution to clear needle of any blood exposed to ambient air. Then aspirate 1.0 ml of hemolyzing solution into the syringe.* Seal needle point by forcing it into a rubber stopper or Plasticine or replace needle with a steel cap. Mix by repeated inversion. Contents of syringe will become darker and clearer as blood hemolyzes.

6. Carefully and rapidly replace the needle (or cap) with a 25 gauge needle. Eject 2 drops of blood to eliminate hemoglobin exposed to air in the syringe nipple and in the needle. Then, inject the solution into a Nahas cuvet until the cuvet is full and a drop of blood has emerged from the outlet at the bottom. Wipe off the effluent blood.

7. Place the blank cuvet and the cuvet with the hemolysate into the spectrophotometer and read absorbance at 548 and 577 nm.

CALCULATIONS

$$\text{Per cent saturation} = K_4 \times \frac{A_{577}}{A_{548}} - K_5$$

NORMAL VALUES

Normal values are 95 to 99 per cent.

COMMENTS AND PRECAUTIONS

In determining the constants K_4 and K_5, inadvertent exposure of the blood to ambient air will not influence the results. On the other hand, the greatest care must be exerted to ensure that the arterial blood specimen (unknown) is not inadvertently exposed to room air or air bubbles. Furthermore, studies of the arterial blood specimen must be performed speedily; otherwise, the oxygen saturation may decrease. Use of a spectrophotometer with narrow band-pass and a narrow slit width (0.01 mm) is essential.

It must be borne in mind that this technique is valid only when the two normal forms of hemoglobin (oxyhemoglobin and deoxyhemoglobin) are present in blood in appreciable quantities. Methemoglobin, sulfhemoglobin, carboxyhemoglobin, high concentration of bilirubin, or lipemia introduce errors. However, methemoglobin at a concentration of less than 5 per cent does not appreciably influence the results. The studies should be undertaken only in fasting subjects. (A correction can be made for lipemia by measuring absorbance ratios in plasma, as described by Nahas.[50])

Blood concentrations of carboxyhemoglobin in excess of 7 per cent of the total hemoglobin significantly decrease the accuracy of the measurement of oxygen saturation by the method described here. Persons who consume more than one package of cigarettes daily may have carboxyhemoglobin concentrations as high as 15 per cent of total hemoglobin. Higher carboxyhemoglobin concentrations may occur in those who have occupational or accidental exposure to carbon monoxide. Whenever there is reason to suspect carbon monoxide exposure, the carboxyhemoglobin concentration should be measured (see Chap. 21); if it is increased, a correction may be made in the oxygen saturation measurement in the following manner.

For any mixture containing only oxyhemoglobin, deoxyhemoglobin, and carboxyhemoglobin

$$A_x = [a_{x(o)}C_o + a_{x(c)}C_c + a_{x(r)}C_r]d$$

in which A_x is the total absorbance of the solution at any given wavelength (x); C_o is the concentration of oxyhemoglobin; C_c is the concentration of carboxyhemoglobin; C_r is the concentration of

* When the blood specimen has a hemoglobin concentration greater than 16 g/100 ml, the absorbance at 577 nm may be too high to be read accurately. In this case, dilute 1 ml of blood with 2 ml of hemolyzing solution.

deoxyhemoglobin; d is the length of the light path through the cuvet; and $a_{x(o)}$, $a_{x(c)}$, and $a_{x(r)}$ are the absorbances, at wavelength x, of pure solutions of oxyhemoglobin, carboxyhemoglobin, and deoxyhemoglobin, respectively, in equimolar concentrations. Then, at 577 and 548 nm

$$A_{577} = [a_{577(o)}C_o + a_{577(c)}C_c + a_{577(r)}C_r]d$$

$$A_{548} = [a_{548(o)}C_o + a_{548(c)}C_c + a_{548(r)}C_r]d$$

Furthermore, at 568 nm (an isobestic point of carboxyhemoglobin and deoxyhemoglobin),

$$A_{568} = [a_{568(o)}C_o + a_{568(c)}C_c + a_{568(r)}C_r]d$$

Therefore, if absorbance is measured at the three wavelengths, 577, 548, and 568 nm, solution of these three simultaneous equations will give the concentrations of oxyhemoglobin, deoxyhemoglobin, and carboxyhemoglobin. These calculations are performed automatically by a component computer in the instrument known as the CO-Oximeter. Oxygen saturation, then, is

$$\frac{C_o}{C_o + C_c + C_r} \times 100$$

The same results may be achieved manually by measurement of absorbance at two wavelengths (for example, 577 and 548 nm), if the concentration of carboxyhemoglobin has been determined separately (see p. 1105). Thus, in such cases, only the first two of these equations are required. The equations are rewritten and solved for oxygen saturation (per cent oxyhemoglobin) as follows,[43] wherein S_o = per cent oxyhemoglobin, S_r = per cent deoxyhemoglobin, and S_c = per cent carboxyhemoglobin:

$$\frac{A_{577}}{A_{548}} = \frac{S_o a_{577(o)} + S_r a_{577(r)} + S_c a_{577(c)}}{S_o a_{548(o)} + S_r a_{548(r)} + S_c a_{548(c)}}$$

However, $S_r = 100 - S_c - S_o$. Therefore

$$\frac{A_{577}}{A_{548}} = \frac{S_o a_{577(o)} + (100 - S_c - S_o)a_{577(r)} + S_c a_{577(c)}}{S_o a_{548(o)} + (100 - S_c - S_o)a_{548(r)} + S_c a_{548(c)}}$$

$$S_o = \frac{\frac{A_{577}}{A_{548}}[100a_{548(r)} - S_c a_{548(r)} + S_c a_{548(c)}] - S_c a_{577(c)} - 100a_{577(r)} + S_c a_{577(r)}}{\frac{A_{577}}{A_{548}}[a_{548(r)} - a_{548(o)}] + a_{577(o)} - a_{577(r)}}$$

Since 548 nm is an isobestic point for oxyhemoglobin and deoxyhemoglobin, $a_{548(o)} = a_{548(r)}$. Therefore,

$$S_o = \frac{\frac{A_{577}}{A_{548}}[100a_{548(r)} + S_c(a_{548(c)} - a_{548(r)})] - S_c(a_{577(c)} - a_{577(r)}) - 100a_{577(r)}}{a_{577(o)} - a_{577(r)}}$$

The absorptivity constants required for solution of this equation are

$$a_{548(r)} = 49.84;\ a_{548(c)} = 48.40;\ a_{577(c)} = 40.4;\ a_{577(r)} = 38.0;\ a_{577(o)} = 61.48^*$$

By substituting these absorptivity constants in the equation, one obtains the following:

$$S_o = \frac{\frac{A_{577}}{A_{548}}[4984 - 1.44S_c] - 2.4S_c - 3800}{23.48}$$

$$S_o = \frac{A_{577}}{A_{548}}[212.26 - 0.061S_c] - 0.102S_c - 161.8$$

* The constants given are for 1.0 mmol/l concentrations of these hemoglobin pigments. Absorbance values obtained at any other concentrations are equally valid, provided only that the concentrations be identical for all three pigments.

REFERENCES

Nahas, G. G.: J. Appl. Physiol., **13**:147–152, 1958.
van Kampen, E. J., and Zijlstra, W. G.: Adv. Clin. Chem., **8**:141–187, 1965.

MYOGLOBIN

Myoglobin is the oxygen-binding protein of striated (cardiac and skeletal) muscle. It resembles the hemoglobin subunits (α, β, γ, and δ chains) so closely that it has served as the prototype in studies of the molecular structure of hemoglobin. Unlike hemoglobin, myoglobin exists only as a monomer, and thus its molecular weight is approximately one-fourth that of hemoglobin. Also, unlike hemoglobin, it is unable to release oxygen except at extremely low oxygen tensions. Its physiological role in man is uncertain. It may serve as an oxygen reservoir of last resort, accessible only under circumstances of extreme hypoxia. Such conditions conceivably may exist when there is strenuous physical exertion, as in athletic events. In contrast to the small amounts of myoglobin contained in the muscle of terrestrial mammals, aquatic mammals have relatively large amounts of myoglobin in their muscles, possibly as an adaptation to the need for oxygen during protracted periods of submersion.

CLINICAL SIGNIFICANCE

When there has been injury to skeletal or cardiac muscle, myoglobin may be released and is then excreted in the urine (myoglobinuria). There are numerous causes of this condition.[45] In many, the degree of myoglobinuria is so slight that it is very rarely perceived by either the patient or the physician or by routine examination of the urine. More pronounced myoglobinuria, sufficient to cause obvious discoloration of urine, may occur after violent physical exercise or as the result of a severe accident in which there is crushing of muscle (with crush injuries, copious amounts of myoglobin may appear in the urine and may contribute to impairment of renal function in badly injured patients; postmortem studies show myoglobin apparently occluding renal tubules in some instances).

When present in large quantity, myoglobin imparts to the urine a color similar to that of a cola drink or black coffee. Myoglobinuria cannot be distinguished from other causes of dark urine simply on the basis of visual inspection. Essentially the same color may be observed in hemoglobinuria. Patients with hemolytic anemia due to unstable hemoglobin also may have coffee-colored urine, but this color is due to a nonprotein, heme-free degradation product of hemoglobin that has not yet been chemically defined.

The possibility of myoglobinuria may be considered when a urine specimen gives both a positive test for protein by the sulfosalicylic acid test (the heat and acetic acid tests may be negative when there is hemoglobin or myoglobin in the urine) and a positive occult blood test. It then becomes necessary to determine whether the heme protein in the urine is hemoglobin or myoglobin.

IDENTIFICATION OF MYOGLOBIN

The physicochemical properties that permit differentiation of hemoglobin and myoglobin are (1) hemoglobin is less soluble than myoglobin in ammonium sulfate solutions, (2) the absorption spectra are different for analogous forms of derivatives of hemoglobin and myoglobin,[23,64] (3) myoglobin is more electropositive then hemoglobin A and can therefore be differentiated electrophoretically,[21,45,60] and (4) the molecular sizes of hemoglobin and myoglobin are different, allowing for differentiation by ultrafiltration or gel filtration.[21] Hemoglobin and myoglobin also can be differentiated by immunodiffusion or immunoprecipitation.[1] These are extremely sensitive techniques capable of measuring minute quantities of myoglobin. However, antihuman-myoglobin serum is not yet commercially available. Even if it were, it would not be practical for most laboratories to keep it on hand for the infrequent and unpredictable situation in which identification of urinary myoglobin is needed.

ELECTROPHORETIC METHOD FOR THE IDENTIFICATION OF MYOGLOBIN

PRINCIPLE

At pH 8.6, myoglobin has an electrophoretic mobility resembling that of hemoglobin C.[7] This is demonstrable with paper, starch, or cellulose acetate as the supporting medium. When myoglobinuria is pronounced, two bands may be observed: a minor, more anodal band, which is myoglobin; and a more prominent, slightly slower-moving band, which is metmyoglobin (Fig. 8–13). When the urine is strongly colored by heme protein, myoglobinuria can be rapidly and conveniently confirmed by direct application of the urine sample to the cellulose acetate membrane. Otherwise, it may be necessary to dialyze the sample overnight against distilled water or tris-EDTA-borate buffer and then concentrate it (by evaporation or by water-absorbing resins) before applying it to the membrane.

REAGENTS

See electrophoretic separation of hemoglobins on cellulose acetate (p. 422).

PROCEDURE

See electrophoretic separation of hemoglobins on cellulose acetate (p. 422). In addition to the urine sample, hemolysate from the patient's peripheral blood and known hemoglobin S, hemoglobin C, and metmyoglobin standards should be applied to the cellulose acetate membrane.

INTERPRETATION

If the urinary heme protein is hemoglobin, the electrophoretic pattern will be the same or nearly the same as that of the patient's hemolysate (usually, only a hemoglobin A band will be observed in both). If the urinary heme protein is myoglobin, it will appear to have the same or nearly the same mobility as hemoglobin C. On unstained cellulose acetate strips, one may see a faint pink band (myoglobin) slightly preceding a more pronounced brown band (metmyoglobin).

COMMENTS AND PRECAUTIONS

If a human myoglobin standard is not available, commercially available horse myoglobin (Sigma Chemical Co., St. Louis, MO) may be used. The commercial preparation

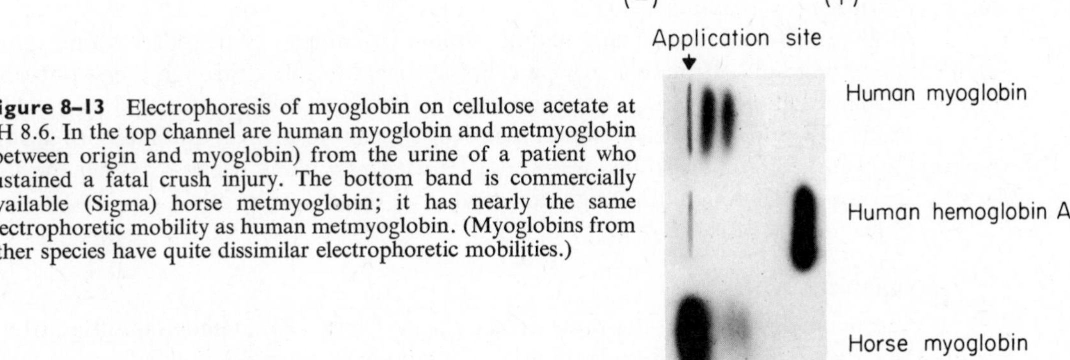

Figure 8–13 Electrophoresis of myoglobin on cellulose acetate at pH 8.6. In the top channel are human myoglobin and metmyoglobin (between origin and myoglobin) from the urine of a patient who sustained a fatal crush injury. The bottom band is commercially available (Sigma) horse metmyoglobin; it has nearly the same electrophoretic mobility as human metmyoglobin. (Myoglobins from other species have quite dissimilar electrophoretic mobilities.)

contains only horse metmyoglobin, which shows slightly less anodal migration than human metmyoglobin. If a person with sickle cell trait (or hemoglobin C trait) has hemoglobinuria, hemoglobins A and S (or C) will be demonstrable in the urine specimen tested by the above method. The presence of a pigment with the mobility of hemoglobin A is a clue that the patient has hemoglobinuria, not myoglobinuria. This potential error may be avoided by concomitant electrophoretic study of the hemoglobin of the patient's blood.

Electrophoresis or isoelectric focusing in polyacrylamide gel provides a separation of myoglobin from hemoglobins that is superior to that attainable with cellulose acetate. Therefore, this medium should be used if the patient has a variant hemoglobin in his blood.

REFERENCE

Boulton, F. E., and Huntsman, R. G.: J. Clin. Path., **24**:816–821, 1971.

SPECTROPHOTOMETRIC METHOD

PRINCIPLE

In alkaline solution, methemoglobin and metmyoglobin form hydroxides with sufficiently different optical absorption in the wavelength range of 580 to 600 nm so that these pigments can be readily distinguished. The ratio of absorbances at 600 and 580 nm is substantially higher for metmyoglobin than for methemoglobin.

SPECIMEN

This procedure requires 10 ml of urine.

REAGENTS

Potassium ferricyanide, 0.6 mol/l, is prepared by dissolving 20 g of $K_3Fe(CN)_6$ in distilled water to a volume of 100 ml. A sodium hydroxide solution, 1 mol/l (4.0 g/100 ml), and a sodium chloride solution, 0.15 mol/l (0.9 g/100 ml), are needed.

PROCEDURE

1. Prepare a hemoglobin control by diluting 1 vol of whole blood with 9 vol of distilled water. Centrifuge at 1500 × *g* for 30 min to remove the stroma. Then, dilute 1 ml of clear supernatant with 9 ml of NaCl solution.
2. Centrifuge the urine specimen at 1700 × *g* for 20 min to remove erythrocytes and other particulate material.
3. Transfer 10 ml of diluted hemoglobin control to a 50 ml beaker. To another 50 ml beaker add 10 ml of supernatant urine sample. To each add 0.1 ml (2 drops) of potassium ferricyanide. With constant stirring, add the NaOH, 1 drop at a time, until the pH is 10.0.
4. Transfer the hemoglobin control and the urine sample to 15 ml conical centrifuge tubes; centrifuge for 30 min at 1500 × *g*.
5. After centrifugation, the supernatants should be completely transparent on examination in a strong light. If there is any perceptible turbidity, filter through a fine-porosity filter (a Millipore filter, type VM, has been used for this purpose).
6. Transfer 3 ml of each clear supernatant (or filtrate) to a cuvet and measure absorbance at 600 and 580 nm. (Distilled water may be used as a blank because neither the salt nor the ferricyanide has significant light absorption at these wavelengths.)
7. Calculate the ratio, A_{600}/A_{580}.

INTERPRETATION

If there is myoglobinuria, the ratio will be greater than 0.85. Hemoglobinuria gives a ratio of less than 0.8. Observed ratios in urine containing metmyoglobin have ranged from

0.90 to 0.96, and ratios for urines with methemoglobin have ranged from 0.67 to 0.75. Urine free of either pigment has a ratio of approximately 0.5 to 0.8. Therefore, a ratio below 0.8 can be taken as evidence of the presence of hemoglobin only if the benzidine test indicates the presence of a heme protein.

COMMENTS AND PRECAUTIONS

This technique is reputed to be sensitive to concentrations of myoglobin as low as 30 mg/100 ml; lower concentrations will not perceptibly discolor the urine. Thus, it appears to be the most sensitive test that is practical for a clinical laboratory. As yet, experience with this procedure is limited, and it is not known whether the presence of nonheme pigments may give rise to false positive results. The test cannot be interpreted, and should not be undertaken, unless the urine gives a positive benzidine or *o*-tolidine test indicating the presence of heme protein.

REFERENCE

Glauser, S. C., Wagner, H., and Glauser, E. M.: Am. J. Med. Sci., **264**:135–139, 1972.

OTHER METHODS FOR THE IDENTIFICATION OF MYOGLOBIN

Ultrafiltration of clear, centrifuged urine provides a means of differentiating hemoglobin from myoglobin on the basis of molecular size. Hemoglobin does not pass through a Millipore filter 8 to 12 nm in pore diameter, whereas myoglobin does. Thus, the presence of heme protein in the ultrafiltrate is indicative of myoglobin. This procedure appears to be satisfactory in reported studies.[21]

Myoglobin is readily distinguished from hemoglobins by isoelectric focusing in polyacrylamide gel. In this medium, myoglobin focuses farthest to the acidic anode end of the gel, whereas hemoglobins S and C focus at the basic cathode end of the gel (Fig. 8–10). This may be of value in patients whose blood contains hemoglobin S or C and in whom hemoglobinuria would be difficult to differentiate from myoglobinuria on the basis of electrophoresis.

A solubility test for myoglobin[6] is based on the fact that hemoglobin is completely precipitated in 80 per cent saturated ammonium sulfate solution, whereas myoglobin often remains in solution. To perform this test, add 2.8 g of crystalline ammonium sulfate, $(NH_4)_2SO_4$, to 5 ml of urine in a test tube. Mix gently until no crystals are seen at the bottom of the tube. Filter. Test the filtrate for heme protein by the benzidine or *o*-tolidine method (see preceding section; Hemastix is convenient for this purpose). If the test for heme protein is positive, this is presumptive evidence for the presence of myoglobin. Unfortunately this simple and widely used test frequently gives erroneous results.[1,21,45]

REFERENCES

1. Adams, E. C.: Differentiation of myoglobin and hemoglobin in biological fluids. Ann. Clin. Lab. Sci., **1**:208–221, 1971.
2. Bartlett, R. C.: Rapid cellulose acetate electrophoresis. II. Qualitative and quantitative hemoglobin fractionation. Clin. Chem. **9**:325–329, 1963.
3. Bernini, L. F.: Rapid estimation of hemoglobin A_2 by DEAE chromatography. Biochem. Genet., **2**:305–310, 1969.
4. Betke K.: Hämoglobin M: Typen und ihre Differenzierung (Übersicht). *In* Haemoglobin-Colloquium (Wien, 1961). H. Lehmann and K. Betke, Eds. Stuttgart, Georg Thieme Verlag, 1962, pp. 39–47.
5. Betke K., Marti, H. R., and Schlicht, I.: Estimation of small percentages of foetal haemoglobin. Nature (Lond.), **184**:1877–1878, 1959.
6. Blondheim, S. H., Margoliash, E., and Shafrir, E. A.: A simple test for myohemoglobinuria (myoglobinuria). J.A.M.A., **167**:453–454, 1958.
7. Boulton, F. E., and Huntsman, R. G.: The detection of myoglobin in urine and its distinction from normal and variant haemoglobins. J. Clin. Path., **24**:816–821, 1971.
8. Briere, R. O., Golias, T., and Batsakis, J. G.: Rapid qualitative and quantitative hemoglobin fractionation: cellulose acetate electrophoresis. Am. J. Clin. Path., **44**:695–701, 1965.

9. Cannan, R. K.: Proposal for a certified standard for use in hemoglobinometry—second and final report J. Lab. Clin. Med., **52**:471–476, 1958.
10. Cawein, M., and Lappat, E. J.: Hereditary methemoglobinemia. *In* Hemoglobin: Its Precursors and Metabolites. F. W. Sunderman and F. W. Sunderman, Jr., Eds. Philadelphia, J. B. Lippincott Co., 1964, pp. 337–349.
11. Comly, H. H.: Cyanosis in infants caused by nitrates in well water. J.A.M.A., **129**:112–116, 1945.
12. Crosby, W. H., and Furth, F. W.: A modification of the benzidine method for measurement of hemoglobin in plasma and urine. Blood, **11**:380–383, 1956.
13. Crosby, W. H., Munn, J. I., and Furth, F. W.: Standardizing a method for clinical hemoglobinometry. U.S. Armed Forces Med. J., **5**:693–703, 1954.
14. Dacie, J. V., Grimes, A. J., Meisler, A., Steingold, L., Hemsted, E. H., Beaven, G. H., and White, J. C.: Hereditary Heinz-body anaemia: a report of studies on five patients with mild anaemia. Brit. J. Haemat., **10**:388–402, 1964.
15. Derman, H., and Pauker, S.: Detection of occult blood in feces and urine. *In* Hemoglobin: Its Precursors and Metabolites. F. W. Sunderman and F. W. Sunderman, Jr., Eds. Philadelphia, J. B. Lippincott Co., 1964, pp. 70–80.
16. Dickerson, R. E.: X-ray analysis and protein structure. *In* The Proteins: Composition, Structure, and Function. 2nd ed. H. Neurath, Ed. New York, Academic Press, Inc., 1964, vol. 2, pp. 603–778.
17. Drabkin, D. L., and Austin, J. H.: Spectrophotometric studies. II. Preparations from washed blood cells: nitric oxide hemoglobin and sulfhemoglobin. J. Biol. Chem., **112**:51–65, 1935.
18. Duhm, J.: The effect of 2,3-DPG and other organic phosphates on the Donnan equilibrium and the oxygen affinity of human blood. *In* Oxygen Affinity of Hemoglobin and Red Cell Acid Base Status (Alfred Benzon Symposium, IV). M. Rørth and P. Astrup, Eds. New York, Academic Press, Inc., 1972, pp. 583–594.
19. Eilers, R. J.: Notification of final adoption of an international method and standard solution for hemoglobinometry specifications for preparation of standard solution. Am. J. Clin. Path., **47**:212–214, 1967.
20. Evelyn, K. A., and Malloy, H. T.: Microdetermination of oxyhemoglobin, methemoglobin and sulfhemoglobin in a single sample of blood. J. Biol. Chem., **126**:655–662, 1938.
21. Farmer, T. A., Jr., Hammack, W. J., and Frommeyer, W. B., Jr.: Idiopathic recurrent rhabdomyolysis associated with myoglobinuria: report of a case. New Eng. J. Med., **264**:60–66, 1961.
22. Fawcett, J. S.: Isoelectric fractionation of proteins on polyacrylamide gels. F.E.B.S. Letter, **1**:81–82, 1968.
23. Glauser, S. C., Wagner, H., and Glauser, E. M.: A rapid simple accurate test for differentiating hemoglobinuria from myoglobinuria. Am. J. Med. Sci., **264**:135–139, 1972.
24. Graham, J. L., and Grunbaum, B. W.: A rapid method for microelectrophoresis and quantitation of hemoglobins on cellulose acetate. Am. J. Clin. Path., **39**:567–578, 1963.
25. Gratzer, W. B., and Beaven, G. H.: Electrophoretic behaviour of haemoglobins in agar gel. *J. Chromat.*, **5**:315–329, 1961.
26. Greenberg, M. S., Harvey, H. A., and Morgan, C.: A simple and inexpensive screening test for sickle hemoglobin. New Eng. J. Med., **286**:1143–1144, 1972.
27. Hanks, G. E., Cassell, M., Ray, R. N., and Chaplin H., Jr.: Further modification of the benzidine method for measurement of hemoglobin in plasma: definition of a new range of normal values. J. Lab. Clin. Med., **56**:486–498, 1960.
28. Harington, C. R., and Van Slyke, D. D.: On the determination of gases in blood and other solutions by vacuum extraction and manometric measurement. II. J. Biol. Chem., **61**:575–584, 1924.
29. Huisman, T. H. J., Schroeder, W. A., Charache, S., Bethlenfalvay, N. C., Bouver, N., Shelton, J. R., Shelton, J. B., and Apell, G.: Hereditary persistence of fetal hemoglobin: heterogeneity of fetal hemoglobin in homozygotes and in conjunction with β-thalassemia. New Eng. J. Med., **285**:711–716, 1971.
30. Humphery, T. J., and Goulston, K.: Chemical testing of occult blood in faeces: "Haematest," "Occultest," and guaiac testing correlated with ^{51}chromium estimation of faecal blood loss. Med. J. Aust., **1**:1291–1293, 1969.
31. Huntsman, R. G., and Liddell, J.: Paper tests for occult blood in faeces and some observations on the fate of swallowed red cells. J. Clin. Path., **14**:436–440, 1961.
32. Illingworth, D. G.: Influence of iron preparations on occult blood tests. J. Clin. Path., **18**:103–104, 1965.
33. Irons, G. V., Jr., and Kirsner, J. B.: Routine chemical tests of the stool for occult blood: an evaluation. Am. J. Med. Sci., **249**:247–260, 1965.
34. Itano, H. A.: Solubilities of naturally occurring mixtures of human hemoglobin. Arch. Biochem. Biophys., **47**:148–159, 1953.
35. Jacobs, S. L., and Fernandez, A. A.: Hemoglobin in plasma. *In* Standard Methods ofClinical Chemistry. R. P. MacDonald, Ed. New York, Academic Press, Inc., 1970, vol. 6, pp. 107–114.
36. Jaffé, E. R.: Hereditary methemoglobinemias associated with abnormalities in the metabolism of erythrocytes. Am. J. Med., **41**:786–798, 1966.
37. Jonxis, J. H. P., and Visser, H. K. A.: Determination of low percentages of fetal hemoglobin in blood of normal children. Am. J. Dis. Child., **92**:588–591, 1956.
38. Kleihauer, E.: Unpublished data.

39. Kleihauer, E., Braun, H., and Betke, K.: Demonstration von fetalem Hämoglobin in den Erythrocyten eines Blutausstrichs. Klin. Wochenschr., **35**:637–638, 1957.
40. Lehmann, H.: Different types of alpha-thalassaemia and significance of haemoglobin Bart's in neonates. Lancet, **2**:78–80, 1970.
41. Lehmann, H., and Carrell, R. W.: Variations in the structure of human haemoglobin: with particular reference to the unstable haemoglobins. Brit. Med. Bull., **25**:14–23, 1969.
42. Loh, W.-P.: Evaluation of a rapid test tube turbidity test for the detection of sickle cell hemoglobin. Am. J. Clin. Path., **55**:55–57, 1971.
43. Maas, A. H. J., Hamelink, M. L., and de Leeuw, R. J. M.: An evaluation of the spectrophotometric determination of HbO_2, HbCO and Hb in blood with the CO-Oximeter IL 182. Clin. Chim. Acta, **29**:303–309, 1970.
44. Marder, V. J., and Conley, C. L.: Electrophoresis of hemoglobin on agar gels: frequency of hemoglobin D in a Negro population. Bull. Johns Hopkins Hosp., **105**:77–88, 1959.
45. Medical Grand Rounds: Myoglobinuria. Am. J. Med. Sci., **261**:351–358, 1971.
46. Michel, H. O., and Harris, J. S.: The blood pigments: the properties and quantitative determination with special reference to the spectrophotometric methods. J. Lab. Clin. Med., **25**:445–463, 1940.
47. Mills, G. C., and Randall, H. P.: Hemoglobin catabolism. II. The protection of hemoglobin from oxidative breakdown in the intact erythrocyte. J. Biol. Chem., **232**:589–598, 1958.
48. Muller-Eberhard, U.: Hemopexin. New Eng. J. Med. **283**:1090–1094, 1970.
49. Nahas, G. G.: Spectrophotometric determination of hemoglobin and oxyhemoglobin in whole hemolyzed blood. Science, **113**:723–725, 1951.
50. Nahas, G. G.: A simplified Lucite cuvet for the spectrophotometric measurement of hemoglobin and oxyhemoglobin. J. Appl. Physiol., **13**:147–152, 1958.
51. Nalbandian, R. M., Nichols, B. M., Camp, F. R., Jr., Lusher, J. M., Conte, N. F., Henry, R. L., and Wolf, P. L.: Dithionite tube test: a rapid, inexpensive technique for the detection of hemoglobin S and non-S sickling hemoglobin. Clin. Chem., **17**:1028–1032, 1971.
52. Nalbandian, R. M., Nichols, B. M., Heustis, A. E., Prothro, W. B., and Ludwig, F. E.: An automated mass screening program for sickle cell disease. J.A.M.A., **218**:1680–1682, 1971.
53. Naumann, H. N.: The measurement of hemoglobin in plasma. *In* Hemoglobin: Its Precursors and Metabolites. F. W. Sunderman and F. W. Sunderman, Jr., Eds. Philadelphia, J. B. Lippincott Co., 1964, pp. 40–48.
54. Paver, W. K. A., and Goldman, P.: The detection of occult blood in faeces. Med. J. Aust., **1**:669–670, 1966.
55. Perutz, M. F.: The hemoglobin molecule. Sci. Am., **211**:64–79, 1964.
56. Perutz, M. F., and Lehmann, H.: Molecular pathology of human haemoglobin. Nature (Lond.), **219**: 902–909, 1968.
57. Righetti, P., and Drysdale, J. W.: Isoelectric focusing in polyacrylamide gels. Biochim. Biophys. Acta, **236**:17–28, 1971.
58. Robinson, A. R., Robson, M., Harrison, A. P., and Zuelzer, W. W.: A new technique for differentiation of hemoglobin. J. Lab. Clin. Med., **50**:745–752, 1957.
59. Shepard, M. K., Weatherall, D. J., and Conley, C. L.: Semi-quantitative estimation of the distribution of fetal hemoglobin in red cell populations. Bull. Johns Hopkins. Hosp., **110**:293–310, 1962.
60. Singer, K., Angelopoulos, B., and Ramot, B.: Studies on human myoglobin. I. Myoglobin in sickle cell disease. Blood, **10**:979–986, 1955.
61. Singer, K., Chernoff, A. I., and Singer, L.: Studies on abnormal hemoglobins. I. Their demonstration in sickle cell anemia and other hematologic disorders by means of alkali denaturation. Blood, **6**:413–428, 1951.
62. Smith, M. H.: Spectral properties of the M haemoglobins. *In* Haemoglobin-Colloquium (Wein, 1961). H. Lehmann and K. Betke, Eds. Stuttgart, Georg Thieme Verlag, 1962, pp. 49–52.
63. Stamatoyannopoulos, G., Fessas, P., and Papayannopoulou, T.: F-thalassemia: a study of thirty-one families with simple heterozygotes and combinations of F-thalassemia with A_2-thalassemia, Am. J. Med., **47**:194–208, 1969.
64. Theil, G. B.: Separation and identification of myoglobin and hemoglobin. Am. J. Clin. Path., **49**: 190–195, 1968.
65. van Kampen, E. J., and Zijlstra, W. G.: Determination of hemoglobin and its derivatives. Adv. Clin. Chem., **8**:141–187, 1965.
66. van Slyke, D. D., and Neill, J. M.: The determination of gases in blood and other solutions by vacuum extraction and manometric measurement. I. J. Biol. Chem., **61**:523–573, 1924.
67. Vanzetti, G., and Valente, D.: A sensitive method for the determination of hemoglobin in plasma. Clin. Chim. Acta, **11**:442–446, 1965.
68. Vella, F.: Acid-agar gel electrophoresis of human hemoglobins. Am. J. Clin. Path., **49**:440–442, 1968.
69. Wasi, P.: Alpha thalassaemia (abstract). XIV International Congress of Hematology, São Paulo, July 16 to 21, 1972, pp. 36–37.
70. Williams, W. J., and Schneider, A. S.: Examination of the peripheral blood. *In* Hematology. W. J. Williams, E. Beutler, A. J. Erslev, and R. W. Rundles, Eds. New York, McGraw-Hill Book Company, Inc., 1972, pp. 10–22.
71. Wood, E. H., Sutterer, W. F., and Cronin, L.: Oximetry. *In* Medical Physics. O. Glasser, Ed., Chicago, Year Book Publishers, Inc., 1960, vol. 3, pp. 416–445.

ADDITIONAL READINGS

Fairbanks, V. F.: Introduction to discussion of Heinz body anemias: unstable hemoglobinopathies. Exp. Eye Res., **11**:365–372, 1971.

Finch, C. A.: Methemoglobinemia and sulfhemoglobinemia. New Eng. J. Med., **239**:470–478, 1948.

Graubarth, J., Bloom, C. J., Coleman, F. C., and Solomon, H. N.: Dye poisoning in the nursery: a review of seventeen cases. J.A.M.A., **128**:1155–1157, 1945.

Haglund, H.: Isoelectric focusing in pH gradients: a technique for fractionation and characterization of ampholytes. *In* Methods of Biochemical Analysis. D. Glick, Ed. New York, John Wiley & Sons, Inc., 1971, vol. 19, pp. 1–104.

Harris, J. W., and Kellermeyer, R. W.: The Red Cell: Production, Metabolism, Destruction; Normal and Abnormal. Cambridge, Harvard University Press, 1970.

Huehns, E. R.: Diseases due to abnormalities of hemoglobin structure. Ann. Rev. Med., **21**:157–178, 1970.

Huehns, E. R., and Bellingham, A. J.: Annotation: diseases of function and stability of haemoglobin. Brit. J. Haemat., **17**:1–10, 1969.

Jonxis, J. H. P., and Huisman, T. H. J.: A Laboratory Manual on Abnormal Haemoglobins. 2nd ed. Oxford, Blackwell Scientific Publications, 1968.

Lehmann, H., and Huntsman, R. G.: Man's Haemoglobins: Including the Haemoglobinopathies and Their Investigation. 2nd ed. Philadelphia, J. B. Lippincott Co., 1974.

Motulsky, A. G.: Frequency of sickling disorders in U.S. Blacks. New Eng. J. Med., **288**:31–33, 1973.

Sonnet, J., and de Noyette, J. P.: Gel isoelectric focusing of fetal and adult hemoglobin M Iwate. Science Tools, The LKB Instrument Journal, **18**:12–14, 1971.

Stamatoyannopoulos, G., Bellingham, A. J., Lenfant, C., and Finch, C. A.: Abnormal hemoglobins with high and low oxygen affinity. Ann. Rev. Med., **22**:221–234, 1971.

Sunderman, F. W., and Sunderman, F. W., Jr.: Hemoglobin: Its Precursors and Metabolites. Philadelphia, J. B. Lippincott Co., 1964.

Weatherall, D. J.: The Thalassaemia Syndromes. 2nd ed. Philadelphia, F. A. Davis Company, 1973.

Williams, W. J., Beutler, E., Erslev, A. J., and Rundles, R. W.: Hematology. New York, McGraw-Hill Book Company, Inc., 1972.

CHAPTER 9

PORPHYRINS AND RELATED COMPOUNDS

by Robert F. Labbé, Ph.D.

Chemistry of porphyrins

Porphyrins are derivatives of porphin, a macrocyclic, highly unsaturated structure composed of four pyrrole rings linked by four methene bridges (=CH—) (Fig. 9–1). While porphin itself has no side chains, the porphyrins are differentiated on the basis of the kind and order of substituents in the eight peripheral positions on the pyrrole rings of porphin. The type III isomer series occurs in normal metabolism. Type I porphyrins form in significant amounts only in the presence of a rare biochemical defect. Other isomers do not occur naturally.

The *porphyrinogens* are forms of porphyrins in which the tetrapyrrole ring is reduced by the addition of six hydrogen atoms, one at each of the four methene bridge carbons, and one at each of the two nonhydrogenated pyrrole nitrogens (Fig. 9–1). The porphyrinogens, but not the porphyrins, undergo biochemical alterations in their substituents.

Many kinds of porphyrins are known; however, very few are found in nature, and only three of these, *uroporphyrin* (URO), *coproporphyrin* (COPRO), and *protoporphyrin* (PROTO), are of clinical significance. Free porphyrins play no biological role in man; they are active only in the form of metal chelates. The iron chelate of a porphyrin is termed a *heme*, *protoheme* being by far the most common and, in a quantitative sense, the most significant (Fig. 9–2). Heme always functions as a prosthetic group of a protein. Hemoproteins participate in a variety of biochemical processes, all of which are associated with some aspect of biological oxidation such as oxygen transport (hemoglobin), cellular respiration (cytochromes), or hydrogen peroxide utilization (catalase). A cobalt chelate (vitamin B_{12}) and a magnesium chelate (chlorophyll) are other forms of naturally occurring tetrapyrroles, though greatly modified in structure from the porphyrin from which they are derived. In the nonchelated state some porphyrins are found as pigments, e.g., in egg shells, bird feathers, and worm integument.

Crystalline porphyrins are dark red or purple in color. In acid solution porphyrins are well known for their intense orange-red fluorescence (620 to 630 nm) on exposure to long wavelength ultraviolet light (397 to 408 nm), a property that is used in porphyrin analysis. The intense color and fluorescence are due to the high degree of conjugated unsaturation in the tetrapyrrole ring. The porphyrinogens, by contrast, are colorless and nonfluorescent; they also are quite unstable and readily oxidized to porphyrins, especially in an acid medium. Among the three porphyrins of clinical significance, stability of the methyl esters on dried chromatograms increases with the number of carboxyl groups as follows, URO > COPRO > PROTO.[6] In general, free porphyrins tend to be more stable in acid solution

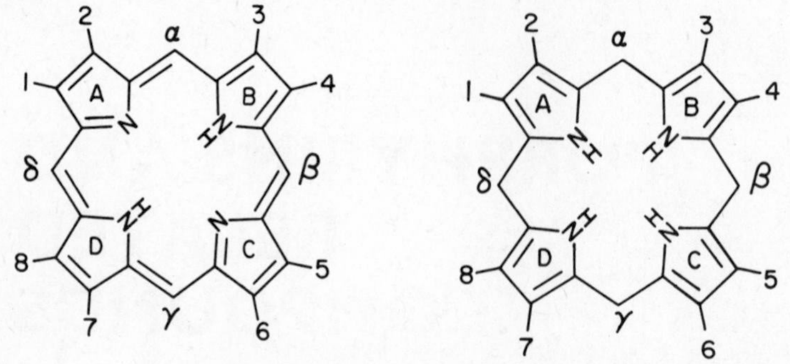

PORPHIN SUBSTITUENTS IDENTIFYING DIFFERENT PORPHYRINS

Pyrrole Ring	Porphin Position	URO	COPRO	PROTO
			Porphyrin III or Porphyrinogen III	
A	1	$-CH_2COOH$	$-CH_3$	$-CH_3$
	2	$-CH_2CH_2COOH$	$-CH_2CH_2COOH$	$-CH{=}CH_2$
B	3	$-CH_2COOH$	$-CH_3$	$-CH_3$
	4	$-CH_2CH_2COOH$	$-CH_2CH_2COOH$	$-CH{=}CH_2$
C	5	$-CH_2COOH$	$-CH_3$	$-CH_3$
	6	$-CH_2CH_2COOH$	$-CH_2CH_2COOH$	$-CH_2CH_2COOH$
D	7	$-CH_2CH_2COOH$	$-CH_2CH_2COOH$	$-CH_2CH_2COOH$
	8	$-CH_2COOH$	$-CH_3$	$-CH_3$
			Porphyrin I or Porphyrinogen I	
D	7	$-CH_2COOH$	$-CH_3$	Type I isomer of PROTO does not occur in nature
	8	$-CH_2CH_2COOH$	$-CH_2CH_2COOH$	

Figure 9–1 Structures of porphyrin and porphyrinogen, and a listing of the porphin substituents that identify the different porphyrins.

and in the dark. Coproporphyrin in dilute hydrochloric acid is sufficiently stable to be used as a convenient fluorescence standard.

The solubility of the porphyrins in water is influenced by the number of carboxyl groups in the pyrrole substituents. URO, having eight carboxyl groups, is the most soluble in aqueous media. PROTO, having only two carboxyl groups, is quite insoluble in aqueous media at physiological pH, but it is very soluble in lipid solvents. COPRO, with its four

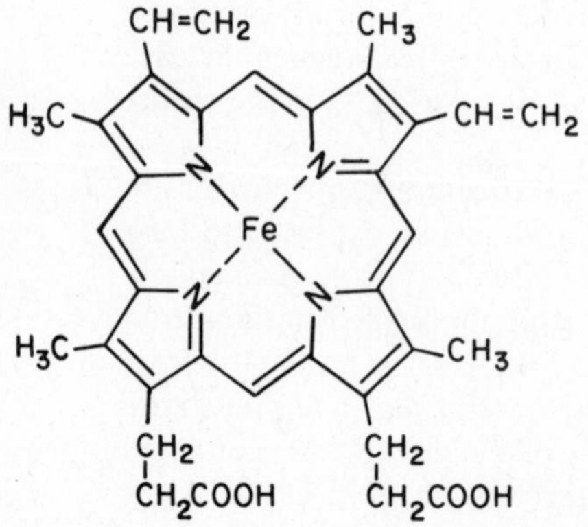

Figure 9–2 The structural formula of heme or, more precisely, protoheme, a chelate of ferrous iron and protoporphyrin IX. The vinyl groups occur in positions 2 and 4 of rings A and B, and the propionic acid (carboxyethyl) groups occur in positions 6 and 7 of rings C and D (compare with Fig. 9–1).

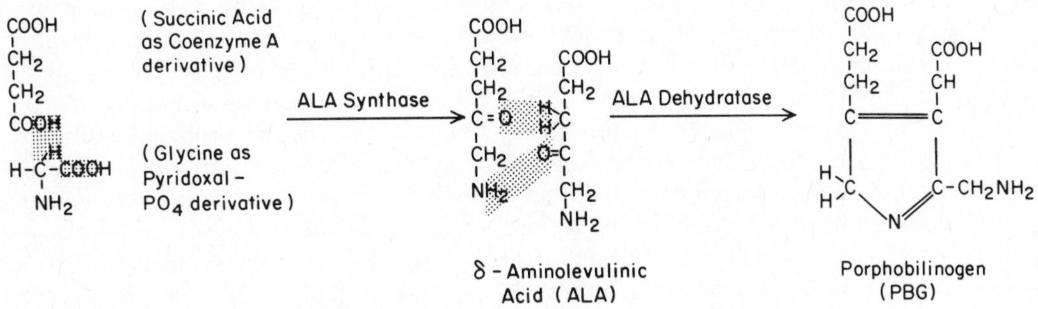

(Succinic Acid as Coenzyme A derivative) — ALA Synthase — (Glycine as Pyridoxal-PO_4 derivative) → δ-Aminolevulinic Acid (ALA) — ALA Dehydratase → Porphobilinogen (PBG)

Figure 9–3 Biochemical reactions leading to the formation of porphobilinogen.

carboxyl groups, has intermediate solubilities. These differing solubility properties are the basis for the separation and assay of the individual porphyrins. As can be inferred, URO is excreted for all practical purposes exclusively in the urine, PROTO exclusively in the feces, and COPRO by either route depending upon the amount formed and the pH of the urine, alkalinity favoring COPRO excretion in the urine.[2] The clinical symptoms in those diseases that are characterized by porphyrin deposition in the skin and by photosensitivity are related to the photochemical properties, stability, and solubility of porphyrins.

Biochemical synthesis of porphyrins

Porphyrin and heme biosynthetic activity is most prominent in the marrow of the long bones and in the liver. However, porphyrins and heme are synthesized in all mammalian cells; there is no evidence of heme formation at any other site than in the cell where it is utilized. The series of reactions leading to heme begins with the condensation of succinyl-coenzyme A and glycine-pyridoxal phosphate. The product of this condensation is decarboxylated to produce *δ-aminolevulinic acid* (ALA), a reaction catalyzed by ALA synthase. Two molecules of ALA then condense and cyclize through the action of ALA dehydratase to yield *porphobilinogen* (PBG), the monopyrrole precursor of porphyrins (Fig. 9–3). Four

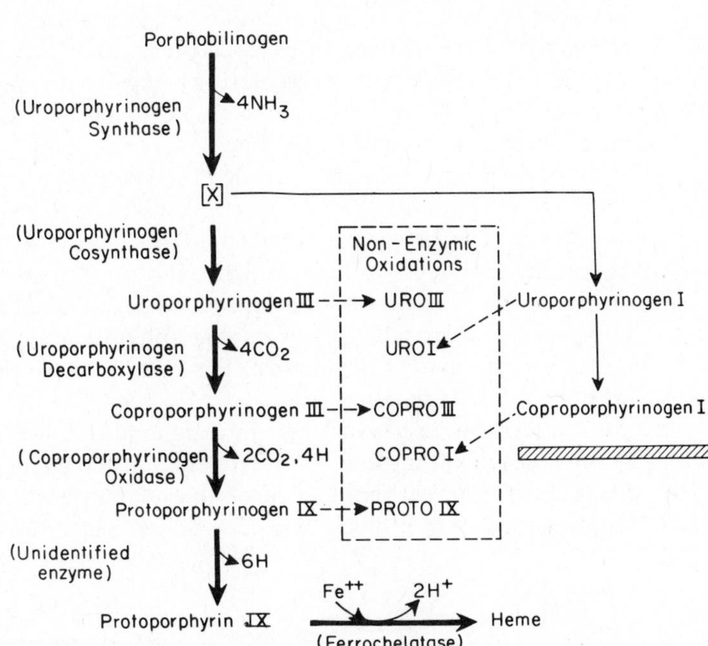

Figure 9–4 The biosynthetic pathway from porphobilinogen to heme. The reaction(s) and intermediate(s) between porphobilinogen and uroporphyrinogen III are not yet clearly defined. Since coproporphyrinogen oxidase cannot utilize coproporphyrinogen I as a substrate, the latter is not further metabolized.

molecules of PBG condense and cyclize to *uroporphyrinogen III* through a complex reaction sequence involving two enzymes: uroporphyrinogen synthase and uroporphyrinogen cosynthase. The four acetate (carboxymethyl) substituents of uroporphyrinogen III are decarboxylated to methyl groups with the formation of *coproporphyrinogen III*. By action of the enzyme coproporphyrinogen III oxidase, two specific propionate (carboxyethyl) groups are decarboxylated and oxidized (dehydrogenated) to yield *protoporphyrinogen IX*. This is oxidized to *protoporphyrin IX*,* which finally chelates iron to become heme. The insertion of iron is catalyzed by the enzyme ferrochelatase (Fig. 9–4).

The free porphyrins that are normally found in body fluids and tissues arise as by-products of this heme biosynthetic pathway. Only trace amounts of porphyrins escape this synthesis because various control mechanisms maintain the pathway in a delicate state of balance in regard to the requirements of each cell. In normal erythrocytes, for example, about one molecule of excess PROTO accumulates for each 30,000 molecules of heme that are formed in the process of hemoglobin synthesis. However, a number of pathological conditions are associated with stimulated or inhibited heme biosynthesis leading to abnormal excretion rates or tissue levels of the porphyrins and their precursors. From a diagnostic standpoint, abnormalities are considered only in terms of increased levels of these metabolites. Impaired heme synthesis at the point of ALA synthase may result in levels of porphyrins or their precursors being below normal, but also below the capability of accurate analysis using the routine methodology available.

Clinical significance of porphyrins

Increased tissue concentrations or excretion rates of porphyrins are found in three general kinds of pathological conditions that differ in basic etiology. The *porphyrias* are inherited diseases that involve the heme biosynthetic pathway. They lead to an overproduction or accumulation of porphyrins, and often of their precursors, resulting in markedly increased levels in the tissues and/or excreta. The *porphyrinurias* and *porphyrinemias* designate conditions characterized by moderately increased porphyrin excretion rates or blood porphyrin levels that are due to secondary effects on porphyrin metabolism.

Congenital erythropoietic porphyria (Günther's disease) is of historical interest in that it was one of the first inborn errors of metabolism to be described, even though it is quite rare. This disease appears in infancy and is characterized by a photo reaction with bullous eruptions, severe scarring of exposed areas, and hypertrichosis. A unique feature of the disease is erythrodontia, the discoloring being due to porphyrin deposition which imparts an intense orange-red fluorescence to the teeth on exposure to long wavelength ultraviolet light. There is a pattern of very high levels of porphyrins in erythrocytes, urine, and feces. Urinary porphyrins frequently occur in such a high concentration as to impart a burgundy color and exhibit a red fluorescence directly upon exposure to long wavelength ultraviolet irradiation. The biochemical defect in this disease is, partially at least, a deficiency in uroporphyrinogen cosynthase, which catalyzes formation of the type III isomer. A consequence of this deficiency is an overproduction of URO I and COPRO I. Although the I and III isomers are separable by chromatography, the procedure is too complex for other than research purposes. The diagnosis of this type of porphyria, however, is not dependent upon an isomer differentiation.

Protoporphyria has its onset in childhood and is manifested by an immediate photosensitive reaction that leads to scleroderma of the hands, nose, and cheeks. Screening tests for porphyrins reveal very high levels in erythrocytes and feces, with normal urine findings. On examination of a blood smear by fluorescence microscopy, one can frequently detect

* There are 15 possible isomers of protoporphyrin. The naturally occurring isomer has been designated IX, although it is now known to be derived biochemically from precursors of the isomer type III.

a portion of the erythrocytes giving off a transient fluorescence due to PROTO. The nature of the biochemical defect has not been unequivocally established.

Acute intermittent porphyria usually appears about the third decade of life and is rarely seen in children. It is characterized by neurological symptoms that include severe abdominal pain, peripheral neuropathy, psychosis, and a variety of other less frequent symptoms. The hallmark of this disease is an increased urinary excretion of the porphyrin precursors ALA and PBG. A definitive diagnosis of the disease requires a demonstration of increased PBG levels in urine, a point that demands emphasis because the symptomatic picture of acute intermittent porphyria is seen so commonly in other conditions. It is noteworthy also that there is only poor correlation at best between the degree of symptomatology and the rate of PBG excretion. For these reasons, the diagnosis can be a difficult problem, especially in the absence of any clear evidence of increased urinary PBG; that is, the disease may occur in a biochemically latent state. Although an exacerbation of the disease may be readily precipitated, there is no sure way to cause a biochemical expression of the latent form without high risk to the patient. The many metabolic abnormalities observed in acute intermittent porphyria leave the nature of the primary, inherited biochemical defect(s) in doubt. Increased activity of ALA synthase and decreased activity of uroporphyrinogen synthase in the liver have both been reported; either contributes to the overproduction or accumulation of ALA and PBG.

Porphyria variegata (South African genetic porphyria) usually appears later in life than the other porphyrias. The disease exhibits no immediate photo reaction, but there may be hyperpigmentation, increased mechanical fragility of the skin, bullous lesions, and hypertrichosis (hirsutism) in sun-exposed areas. There may or may not be neurological symptoms similar to those found in acute intermittent porphyria. The urinary findings, like the symptoms, can be quite variable, although there is usually an increase in porphyrins. The distinguishing feature of this type of porphyria is the presence of markedly elevated fecal porphyrins with essentially normal erythrocyte porphyrins. The nature of the inherited biochemical defect is unknown.

Porphyria cutanea tarda, acquired type (symptomatic cutanea tarda, Bantu porphyria, Turkish porphyria), has some features in common with the inherited type, but the precipitating factor can often be traced to chronic alcoholism or a liver toxin. Other clinical features include skin fragility with bullous lesions in areas exposed to sunlight, hyperpigmentation, and hypertrichosis. Iron loading is frequently associated with this disease. Typical laboratory findings will include elevated urinary URO as the predominant feature, with all other porphyrin and porphyrin precursors being within the normal range.

Coproporphyria is often asymptomatic, but may on occasion appear as a variant of acute intermittent porphyria. Most often this disease is characterized by marked increases in COPRO in the urine and feces. Following an exacerbation that may have been precipitated by porphyrinogenic drugs (many hypnotics, anticonvulsants, tranquilizers) a patient may excrete PBG and URO in the urine and appear biochemically to have acute intermittent porphyria.

Porphyrinurias refer to conditions with moderate increases (usually 2- to 3-fold) in urinary COPRO excretion. These abnormalities in porphyrin metabolism may be associated with liver damage, accelerated erythropoiesis, infection, lead intoxication, and many other conditions. Obviously, porphyrinuria has little diagnostic value *per se*, but it must be recognized in order to differentiate the condition from one of the porphyrias. Frequently, coproporphyrinuria is accompanied by increases in urinary urobilinogen, which can be confused with porphobilinogen. This situation calls for the careful use of laboratory methods capable of differentiating between urobilinogen and porphobilinogen to avoid a false diagnosis of porphyria. A biochemical explanation of porphyrinuria is not known.

Porphyrinemias refer to conditions accompanied by moderate increases (usually 2- to 3-fold) in erythrocyte protoporphyrin concentration. As in the porphyrinurias, the etiology is nonspecific. A variety of factors, including iron deficiency, chronic infection, impaired iron absorption, and lead intoxication (see discussion of ALA and ALA dehydratase assays below), will lead to porphyrinemia. However, in most cases an explanation for the development of the disease lies in impaired iron availability as a substrate for the ferrochelatase reaction; that is, an imbalance in the iron and protoporphyrin available at the last step in heme biosynthesis. Presently, the porphyrinemias have limited diagnostic application, but they are mentioned because recent investigations suggest that blood PROTO analysis may soon come into common use as a test for iron deficiency and lead intoxication.

A SYSTEMATIC APPROACH TO THE LABORATORY DIAGNOSIS OF PORPHYRIN DISEASES

Laboratory procedures for the diagnosis of inborn errors in porphyrin metabolism (porphyrias) are frequently limited to qualitative or quantitative examination of the urine for PBG and porphyrins. A much more elementary procedure is the exposure of a urine specimen to sunlight, which causes specimens containing considerable amounts of PBG and porphyrinogens to darken. Only rarely are examinations of blood or feces considered. A review of the porphyrias, porphyrinurias, and porphyrinemias shows that there are many overlapping clinical and biochemical features requiring a more thorough laboratory work up. The diagnosis of these pathological conditions requires the analysis and differentiation of porphyrins and porphyrin precursors in blood, urine, and feces. With only a few exceptions, screening tests are adequate and will suffice, making the assay of porphyrins and porphyrin precursors seldom necessary. The *pattern of changes* in porphyrin metabolism are most important.[7,12] These, combined with a medical history, will nearly always permit a definitive diagnosis of the nature of the defect in porphyrin metabolism, the principal exceptions being those cases in which the metabolic defect is in a latent state.

The approach outlined here requires the collection of 1 to 2 ml of blood with anticoagulant, a random urine specimen, and a small fecal specimen. These specimens are analyzed according to the following protocol:

1. Whole blood is examined for elevated porphyrin levels by a screening test only.
2. Urine is examined for PBG using the Watson-Schwartz screening test. If the test yields a positive or even questionable result, this should be confirmed by quantitative analysis.
3. Urine is examined for elevated porphyrin levels by screening tests. If abnormal, the porphyrin extract is subjected to thin-layer chromatography to determine whether URO or

TABLE 9–1 TYPICAL PATTERN OF SCREENING TEST RESULTS FOUND IN VARIOUS DISEASES OF PORPHYRIN METABOLISM*

Diagnosis	*Porphyrins*			*Porphobilinogen*
	BLOOD	URINE	FECES	URINE
Congenital erythropoietic porphyria	increase	increase	increase	normal
Erythropoietic protoporphyria	increase	normal	increase	normal
Acute intermittent porphyria	normal	increase	normal	increase
Porphyria variegata	normal	variable	increase	variable
Porphyria cutanea tarda (acquired type)	normal	increase	normal	normal
Hereditary coproporphyria	normal	variable	variable	variable
Lead poisoning	variable	variable	variable	normal
Porphyrinurias	normal	increase	normal	normal
Porphyrinemias	increase	normal	normal	normal

* The patterns of changes shown will be found in most cases. However, several are particularly prone to deviate. This is especially true of porphyrias that can occur in latent forms and of those diseases that involve porphyrin precursor changes.

COPRO, or both, are present in excess. Only with the most difficult diagnostic problems may it become necessary to quantitate URO and COPRO in a 24 hour urine collection.

4. Feces are examined for elevated porphyrin levels by a screening test only.

5. In the case of suspected lead intoxication showing porphyrinuria or porphyrinemia, an assay of the erythrocytes for ALA dehydratase will exclude other causes of these changes in porphyrin metabolism.

6. Reference to Table 9–1 will aid in interpreting the pattern of changes in laboratory findings that typically accompany the different types of diseases of porphyrin metabolism. This information, combined with the symptomatology and family history, will in most cases reveal an abnormality in porphyrin metabolism and permit a differential diagnosis to be made. Latent forms of the porphyrias pose a diagnostic problem for which there is presently no satisfactory answer.

METHODS FOR PORPHOBILINOGEN ANALYSIS

PRINCIPLE

PBG condenses with *p*-dimethylaminobenzaldehyde in acid solution (Ehrlich's reagent) to form a magenta-colored product (Fig. 9–5). Since reaction can occur with other urinary constituents, the screening test uses pH adjustment and solvent extractions to remove interfering substances and make the test reasonably specific. For quantitative analysis, the PBG is purified by adsorption to an ion exchange resin from which interfering substances can be removed. This part of the procedure removes not only color-producing substances but also indole and related compounds which react with the chromophore to produce colorless derivatives.[13]

SPECIMENS

For diagnostic purposes, only urine is analyzed for PBG. The screening test is preferably carried out on a morning specimen. Quantitative analyses should be made on 24 hour urine collections. If the pH of the urine is adjusted and maintained near neutrality (pH 6 to 7), it can be stored for periods up to two weeks in the refrigerator, although assays should preferably be performed as soon as possible.[1]

REAGENTS

1. Ehrlich's reagent, qualitative (Watson-Schwartz test).* Dissolve 0.7 g of *p*-dimethylaminobenzaldehyde in 150 ml of concentrated HCl and add 100 ml water. Store in a brown

* Several modifications of Ehrlich's reagent have been used, each varying in terms of the kind and concentration of acid solvent and the concentration of *p*-dimethylaminobenzaldehyde. This is the result of attempts to obtain a stabilized chromophore and to enhance its absorbance. The particular Ehrlich's reagent indicated is that which is described in the original method.

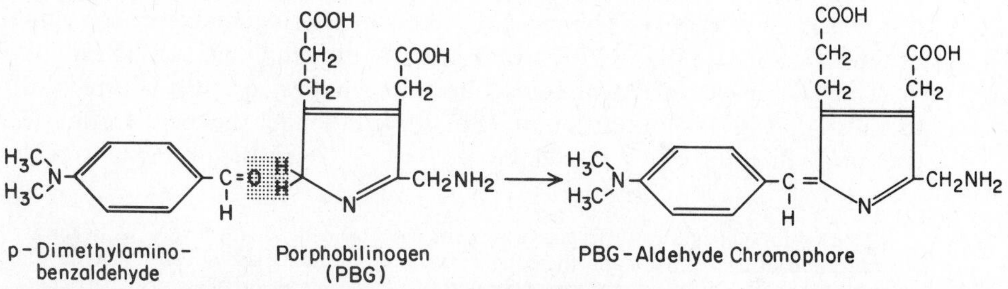

Figure 9–5 The reaction of porphobilinogen with Ehrlich's reagent.

bottle. Although the reagent becomes increasingly yellow with storage, it can be used over a period of months.

2. Ehrlich's reagent, quantitative (PBG assay).* Dissolve 2.0 g of *p*-dimethylaminobenzaldehyde in 25 ml of concentrated HCl plus 75 ml of glacial acetic acid. The reagent becomes increasingly yellow on prolonged storage. This can be retarded by refrigeration in the dark.

3. Sodium acetate, saturated. An excess of sodium acetate·$3H_2O$ is added to water (approximately 1 g/ml) and the reagent is stored at room temperature.

4. Chloroform, AR.

5. *n*-Butanol, AR.

6. Dowex 2-X8 resin (200 to 400 mesh), acetate form. Suspend 100 to 200 g of Dowex 2-X8 resin in about 4 vol of sodium acetate solution (10 g/100 ml). Mix the suspension well and allow it to settle by gravity until the supernatant solution can be easily decanted. Wash the resin 6 to 10 times in this manner, using about 4 vol of distilled water and decanting each time to remove both the fines and the excess sodium acetate. To hasten the procedure, the suspension may be centrifuged briefly between washings. A stock suspension of resin is prepared by adding an equal volume of distilled water to the washed, packed resin. Immediately before use, the suspension should be shaken thoroughly, since the resin settles rapidly. This resin preparation is stable for several months.

SCREENING TEST FOR PORPHOBILINOGEN IN URINE

PROCEDURE

1. To 2.0 ml of urine in a test tube, add 2.0 ml of Ehrlich's reagent (qualitative) and mix.

2. Add 4.0 ml of saturated sodium acetate and mix again. Check with pH indicator paper to confirm that the solution is in the range of pH 4 to 5.

3. Add 5 ml of chloroform, stopper, and shake vigorously for 1 min. Permit the phases to separate, centrifuging if necessary.

4. Transfer the upper (aqueous) phase with a pipet to another tube, to which is added 2 ml of *n*-butanol. Stopper, and shake vigorously for 1 min. Allow the phases to separate, centrifuging if necessary.

5. Examine the lower (aqueous) layer for a magenta color. If color is present, the test is positive for PBG, indicating a concentration several times normal.

COMMENTS

This procedure is commonly known as the Watson-Schwartz test. The extractions with chloroform and butanol are essential to remove frequently occurring substances that interfere with the test. However, if no magenta color is observed in the upper (aqueous phase) following the chloroform extraction, the *n*-butanol extraction may be omitted and the test considered negative. The most common interfering substance is urobilinogen, which produces a color similar to PBG with Ehrlich's reagent. Other substances sometimes present in urine can give a variety of colors, including yellow, orange, and red. All of these tend to make a positive identification of PBG difficult. Under these circumstances, the quantitative procedure for PBG should be performed because this method removes the most

---

* Several modifications of Ehrlich's reagent have been used, each varying in terms of the kind and concentration of acid solvent and the concentration of *p*-dimethylaminobenzaldehyde. This is the result of attempts to obtain a stabilized chromophore and to enhance its absorbance. The particular Ehrlich's reagent indicated is that which is described in the original method.

common interfering Ehrlich chromogens, as well as the decolorizing indoles, to permit positive identification of PBG. A variation of the Watson-Schwartz test incorporating a purification step using a disposable ion-exchange column has been proposed as a means of achieving increased reliability.[3]

REFERENCE

Schwartz, S., Berg, M. H., Bossenmaier, I., and Dinsmore, H.: Determination of porphyrins in biological materials. *In* Methods of Biochemical Analysis. D. Glick, Ed. New York, Interscience Publishers, Inc., 1960, vol. VIII, p. 249.

QUANTITATIVE DETERMINATION OF PORPHOBILINOGEN IN URINE

PROCEDURE

1. Pipet 4 ml of Dowex 2-X8 resin suspension into a 12 ml centrifuge tube and centrifuge about 1 min; discard the supernatant solution. Pipet 1 ml of urine onto the packed resin, followed by 0.1 ml of concentrated ammonium hydroxide (28 g NH_3/100 g). Stir the mixture thoroughly and centrifuge about 1 min; discard the supernatant solution.

2. Wash the resin to which the PBG has been adsorbed 4 times with 5 ml aliquots of distilled or deionized water to remove interfering substances. Centrifuge about 1 min after each wash.

3. Elute the PBG from the washed resin with four 2 ml aliquots of 1 molar acetic acid. After each acetic acid addition, centrifuge the resin for 1 min and decant the supernatant solution into a 12 ml graduated centrifuge tube, adjusting the final volume to exactly 10.0 ml with 1 molar acetic acid. Mix the solution and, finally, centrifuge briefly to remove any traces of resin which may have been carried over during decanting.

4. Treat 2 ml of acetic acid eluate with 2 ml of Ehrlich's reagent (quantitative). Likewise prepare a reagent blank, substituting 2 ml of 1 molar acetic acid for the eluate. Thoroughly mix the reactants. After allowing the unknowns to stand 6 to 8 min, read them against the reagent blank at 555 nm and 525 nm in a spectrophotometer having a resolution of 1 to 2 nm or better.

CALCULATION

Porphobilinogen concentration is calculated as follows:

$$\mu\text{g of PBG/ml of urine} = \frac{A_{555} \times 10}{0.114}$$

where

0.114 = the absorbance, in a 1 cm light path, which the reaction product would give when the concentration of PBG is 1 μg/ml.
10 = the dilution of the original urine aliquot with acetic acid.

COMMENTS

The factor 0.114 has been determined empirically using the conditions outlined and δ-aminolevulinic acid as a standard. ALA can be condensed with acetylacetone to give a pyrrole which reacts like PBG in this test. (Alternatively, PBG can now be obtained commercially.) To calculate the 24 hour excretion, multiply the concentration of PBG by the total urine volume.

If PBG is present in significant amounts, the color developed by Ehrlich's reagent should be rose to crimson, and the ratio of A_{525}/A_{555} should be 0.83. When the ratio $A_{525}/A_{555} > 1.00$, interfering substances are present and the result should not be interpreted as an abnormal concentration of PBG. With a PBG concentration two to three times normal, which is clinically significant, no difficulty is likely to be encountered in its quantitation.

Alternatively, an Allen correction can be applied by measuring the absorbance at 535, 555, and 575 nm, and then applying the formula $A_{corrected} = 2A_{555} - (A_{535} + A_{575})$. This eliminates the effects of interfering chromogens and gives generally lower values for PBG.[9]

A seemingly large amount of resin is used for the amount of PBG to be adsorbed; however, the optimum ratio for the batch process has been found to be 1 vol of urine to 2 vol of packed resin. The concentration of acetic acid required to elute PBG from the resin is not critical, but to obtain quantitative recovery of the PBG, four elutions are essential.

The reaction of PBG with Ehrlich's reagent yields a product whose absorbance follows Beer's law from the lower limits of detection through an absorbance of at least 0.750. If the absorbance is too high, an appropriate dilution of the acetic acid eluate should be made, or preferably less original sample should be used and a correction made in the calculation.

Upon addition of the Ehrlich's reagent, maximum color develops within 6 min. It remains stable for 2 to 3 min and then begins to fade slowly. Within 20 min, absorbance decreases by about 10 per cent.

NORMAL VALUE

The normal urinary excretion rate of PBG is <1 mg/24 h, or about 1 to 2 μg/ml.

REFERENCE

Moore, D. J., and Labbé, R. F.: Clin. Chem., **10**:1105, 1964.

METHODS FOR PORPHYRIN ANALYSES

PRINCIPLE

All porphyrin analyses are based upon the isolation of the porphyrin(s) from the specimen, quantitative separation of the individual porphyrins, and observation or measurement of the porphyrin(s) present by fluorometry or spectrophotometry. Porphyrins are isolated from body excreta or tissues by acidification and extraction into an organic solvent. For screening purposes, extracts can usually be examined at this step. For quantitation, the individual porphyrins are separated by selective solvent extraction or by chromatography. Their characteristic orange-red fluorescence (620 to 630 nm) on irradiation with long ultraviolet light (398 to 408 nm) allows acid solutions of porphyrins to be detected fluorometrically at concentrations below 10^{-8} mol/l. Alternatively, when the concentrations of porphyrins are sufficiently elevated, they can be measured spectrophotometrically.

SPECIMENS

Whole blood specimens are collected using any common anticoagulant.

Urines for screening purposes should preferably be morning specimens, but random specimens can be used. Since the assay of single urine specimens provides less meaningful data, quantitative analyses should be performed on 24 hour collections. In this case, the urine should be collected in a container to which 4 to 5 g of sodium carbonate had been added; if PBG also is to be assayed, the urine must be kept at pH 6 to 7 (see above). Insuring that the urine is neutral or alkaline will retard the spontaneous condensation and cyclization of PBG to URO, a reaction that can in some circumstances give falsely elevated porphyrin results.

Feces are used only for qualitative tests, and a small specimen (1 g) is adequate.

When porphyrin analyses cannot be performed soon after collection of the specimens, the specimens should be kept in a darkened container and refrigerated.

REAGENTS

1. HCl, 3 mol/l and 8 mol/l. Dilute 25 ml or 67 ml, respectively, of concentrated HCl, AR, to 100 ml with water.
2. Acetic acid, glacial, AR.
3. Diethyl ether, AR.
4. *n*-Butanol, AR.
5. Glycerol solution. Dilute 2 parts of glycerol, AR, with 1 part of water.
6. Chromatographic solvent. Mix chloroform/methanol/ammonium hydroxide/water in a ratio of 12/12/3/2.
7. Glacial acetic acid/ethyl acetate (HAc/EtAc). Mix 1 part of HAc with 4 parts of EtAc.
8. Thin-layer chromatography plates. These are prepared by making a slurry of 2 g of silica gel G (Brinkmann Instruments, Inc., Westbury, N.Y. 11590) in 5 ml of water. Pipet 1 ml of the slurry onto a clear, nonfrosted microscope slide; carefully and quickly agitate to give a uniform distribution. Dry the slides either at room temperature or in a warm oven. They can be prepared at any convenient time and stored in a dust-free container until needed. The slides should not be stored with a desiccant because a silica coating that has become too dry will cause some streaking of the COPRO. The gel layer can be carefully scored lengthwise with a pencil or sharp instrument to permit application of two specimens or standards on each slide.
9. Coproporphyrin standard. Either isomer I or III may be used; the biological source from which the porphyrin is isolated is unimportant. To prepare a standard solution, dissolve 10.0 mg of coproporphyrin tetramethyl ester (Calbiochem, La Jolla, CA 92037) in 18.4 ml of 6 molar HCl; the ester hydrolyzes in about one day. Stored refrigerated and in the dark, this stock standard solution of 500 μg coproporphyrin/ml HCl will be stable for many months. It can be diluted for use usually to a concentration of 5.0 μg coproporphyrin/100 ml in 1.5 molar HCl; this diluted standard should be made from the stock solution monthly. As an alternative, dry standardized vials, each containing 5.0 μg of coproporphyrin, are available (Sigma Chemical Co., St. Louis, MO 63118). These are simply reconstituted with 1.5 molar HCl.

SCREENING TESTS FOR PORPHYRINS

All screening tests are performed in glass apparatus. *Plastic tubes must be avoided* because optical properties of some synthetic materials will interfere with observance of the porphyrin fluorescence.

Blood porphyrins

Place 1 ml of whole blood into 3 ml of HAc/EtAc in a centrifuge tube. Stir the mixture thoroughly, centrifuge, and decant the supernatant solution into a second centrifuge tube containing 0.5 ml of 3 molar HCl. Shake the two phases thoroughly and allow to separate. Illuminate the lower HCl layer and observe with an ultraviolet light* for an orange-red fluorescence indicative of porphyrins. With these volumes of solutions, normal blood will usually show only a trace of porphyrin fluorescence. This will be equivalent to 50 to 60 μg of porphyrin/100 ml whole blood, or near the upper end of the normal range. A distinct

* For all of these screening tests (blood porphyrins, urinary porphyrins, fecal porphyrins), this can be any ordinary Wood's lamp of adequate intensity; a 100 W mercury spot bulb with a filter is particularly useful.

porphyrin fluorescence is seen at concentrations of two to three times normal, becoming deep red as the concentration increases further.

For practical purposes, all of the porphyrin found in blood is located within the erythrocytes, the plasma containing only traces of porphyrins. Therefore, an analysis of whole blood actually reflects the porphyrin content of the erythrocytes. Increased cellular porphyrin concentrations can be quite easily observed also by fluorescence microscopy of an unstained blood smear. One problem with this latter procedure is the fact that the intense ultraviolet irradiation focused on the slide quickly destroys the PROTO, causing its fluorescence to fade rapidly in the field of view. By contrast, URO and COPRO are quite stable under these conditions.

Urinary porphyrins

Place 5 ml of urine plus 3 ml of HAc/EtAc into a centrifuge tube and shake thoroughly. Allow the phases to separate by standing, or hasten the process by a brief centrifugation. Irradiate the upper (organic) layer and observe with an ultraviolet lamp for porphyrin fluorescence. This layer fluoresces lavender with moderately elevated concentrations of porphyrins, then pink, and finally red with markedly increased porphyrin concentration.

Occasionally drugs, abnormal metabolites, or unknown materials will impart a fluorescence or color to the organic layer that will make interpretation difficult. When such interfering substances are encountered or when increased sensitivity is desired, this upper layer may be removed carefully with a pipet and transferred to a second tube to which is added 0.5 ml of 3 molar HCl. With shaking, the porphyrins are extracted into the lower acid phase while most interfering substances remain in the organic phase. In an acid medium, porphyrin fluorescence is greatly intensified; moreover, the bluish tint of common urine components is separated out by the extraction to give the more typical orange-red fluorescence identified with porphyrins.

If urinary porphyrins are abnormally elevated, it is usually advisable to determine whether the increased porphyrins are URO or COPRO, or both. This distinction is made very simply and quickly by thin-layer chromatography. When a screening test is positive for porphyrins, or when a result is difficult to interpret, 10 μl of the upper HAc/EtAc layer containing the porphyrins is spotted onto a silica gel slide, either at room temperature or more rapidly with a stream of warm air from a blower. In a covered staining jar containing 3 ml of chromatographic solvent, the chromatogram is developed until the front has moved about halfway up the slide, or until the COPRO band has clearly separated from the origin (5 to 10 min). Viewed under ultraviolet light, a normal urine specimen will reveal a sharp COPRO band or spot to have moved almost two-thirds of the distance from the origin to the solvent front, with very little or no fluorescence remaining at the origin. Specimens containing elevated URO will have a clearly visible fluorescent spot at the origin. In the more common case of coproporphyrinuria, the COPRO band will be significantly more intense than that seen in normal urine, which can be run simultaneously as a reference.

An alternate procedure that is convenient utilizes commercially available disposable chromatographic columns (Bio-Rad Laboratories, Richmond, CA 94804).[4] The principal advantage is the adsorption of the porphyrins to an anion-exchange resin followed by the removal of many interfering substances. Final elution with HCl then gives a relatively pure solution of porphyrins that permits easier interpretation.

Fecal porphyrins

Place a small portion of feces (about 50 mg or a volume about the size of a large drop of water) on the end of a glass rod in a 12 ml centrifuge tube. Add 3 ml of HAc/EtAc and stir vigorously. Centrifuge briefly and decant the supernatant solution from the residue into a second centrifuge tube containing 0.5 ml of 3 molar HCl. Shake the solutions well for 1 min and allow to separate, or centrifuge briefly. Observe the lower (aqueous) layer in ultraviolet light for porphyrin fluorescence. Normally, this acid layer shows little or no fluorescence; porphyrins, when present in increased concentration, will impart their clearly distinguishable orange-red fluorescence.

The occasional presence of chlorophyll of dietary origin in fecal matter may cause confusion because it also has a red fluorescence that is not easily distinguishable from that of porphyrins. Even though chlorophyll is insoluble in aqueous solution and remains in the upper organic layer, it can exhibit a deep red or yellowish-brown fluorescence that might interfere with interpretation of the test. When high levels of chlorophyll are present, the upper organic layer can be drawn off and discarded and a fresh 3 ml aliquot of HAc/EtAc used to extract the remaining HCl layer. This procedure will remove the chlorophyll interference, and any remaining orange-red fluorescence can be considered due to porphyrins.

REFERENCE

Haining, R. G., Hulse, T., and Labbé, R. F.: Clin. Chem., **15**:460, 1969.

Urinary uroporphyrin and coproporphyrin

Screening tests for urinary URO and COPRO can be performed also by solvent extraction. Place 10.0 ml of urine into a 60 ml Squibb type of separatory funnel. Acidify with 4.0 ml of glacial acetic acid, and then add 20 ml of diethyl ether. Shake vigorously for 1 min and allow the phases to separate. If an emulsion forms, it can usually be broken by the addition of about 1 ml of ethanol. This step extracts the COPRO into the upper (ether) phase. Transfer the urine (lower phase) to another funnel and extract a second time with 20 ml of ether. Combine the two ether extracts in one funnel and add 2.0 ml of 3 molar HCl. Shake vigorously for 1 min to extract the COPRO into the HCl. Observe this lower phase in an ultraviolet light for the orange-red fluorescence of porphyrins.

The test for URO is performed on the same urine specimen following removal of the COPRO. Adjust the ether-extracted urine to pH 3 using 3 molar HCl. Transfer the specimen to a 60 ml Squibb type of separatory funnel and extract with 20 ml of ethyl acetate. Shake vigorously for 1 min and allow the phases to separate. Transfer the urine (lower phase) to another funnel and extract a second time with 20 ml of ethyl acetate. Combine the two ethyl acetate extracts in one funnel and add 2.0 ml of 3 molar HCl. Shake vigorously for 1 min to extract the URO into the HCl. Observe this lower phase in an ultraviolet light for the orange-red fluorescence of porphyrins.

A normal urine specimen can be carried through the screening procedure as an aid in interpretation of the results. Normal urine should show no porphyrin fluorescence in either the COPRO or URO extract. A moderately elevated porphyrin concentration will appear as a distinctly orange-red fluorescence. As the porphyrin concentration in the urine increases, the fluorescence observed in the test becomes increasingly intense, reaching a brilliant red in cases where the porphyrin concentration is greater than about 10 times normal.

QUANTITATIVE METHOD FOR URINARY URO AND COPRO

PROCEDURE

1. Place 8.0 ml of urine in a 12 ml centrifuge tube. Add 0.2 ml of 8 molar HCl to effect solution of any precipitate. Adjust the urine to pH 3, as determined with pH indicator paper, by adding saturated sodium acetate (0.2 to 0.3 ml).
2. Add 2.0 ml of *n*-butanol and shake the mixture vigorously, then centrifuge for 10 to 15 min to break the emulsion and to separate the phases cleanly. This procedure removes and concentrates the porphyrins and separates them from water-soluble fluorescing substances. The upper butanol layer will be 1.5 ml in volume.

3. Spot 60 μl samples from the butanol layer at 10 mm from one end of a silica gel chromatography slide. Dry the spot with a stream of warm air.

4. Chromatography is carried out in a histological staining jar containing 3 ml of chromatography solvent. After the solvent front has nearly reached the top of the slide ($<$10 min), air dry the slide for a few minutes. By observing the plate with a filtered Wood's lamp or with transillumination using a viewing box, the red spots of porphyrin fluorescence can be located and marked with a pencil or sharp instrument.

5. Scrape the silica gel containing the individual porphyrin spots (about a 5 mm section) from the plate and place into individual tubes for fluorometric measurement. Remove a similarly sized area of silica gel containing no fluorescence from below the origin for use as a blank.

6. Add 2 ml of glycerol-water solution, shake the tubes vigorously to disrupt all silica particles, and then allow them to stand for 15 to 20 min. This waiting period before reading the fluorescence is important because it allows the background contributed by the silica gel to stabilize at a minimum value.

7. Just before placing in the fluorometer, invert the tubes to mix without entrapping air bubbles. Read the fluorescence against a porphyrin standard. COPRO at a concentration of 0.05 μg/ml in 1.5 molar HCl is commonly used for this purpose.

CALCULATIONS

$$\mu\text{g COPRO/d} = \frac{\text{sample reading}}{\text{standard reading}} \times \frac{5}{100} \times 2 \times \frac{1.5}{8.0} \times \frac{24\text{ h vol (in ml)}}{0.06}$$

$$\mu\text{g URO/d} = \frac{\text{sample reading}}{\text{standard reading}} \times \frac{5}{100} \times 2 \times \frac{1.5}{8.0} \times \frac{24\text{ h vol (in ml)}}{0.06} \times 0.81$$

where

5/100 = concentration of COPRO standard (5 μg/100 ml).
2 = vol (in ml) of the solution measured.
1.5/8.0 = concentration factor for the porphyrin after its extraction into butanol.
0.06 = vol (in ml) of the sample chromatographed.
0.81 = an empirically determined conversion factor for URO using a COPRO standard.

COMMENTS

Urine specimens known to contain a high concentration of porphyrins should be used in smaller amount and the volume made up to 8.0 ml with water. Potential difficulties in subsequent steps will be eliminated, and complete recovery of the porphyrins will be insured by making this initial dilution.

This procedure is described because quantitative information on urinary porphyrins is often requested. However, a preferred approach to the laboratory diagnosis of porphyrin diseases has been outlined above. An alternate procedure for urinary porphyrin analysis requires COPRO separation by solvent extraction and URO separation by alumina adsorption. The porphyrins are subsequently taken into HCl and measured fluorometrically.[14]

Quantitative analyses of feces and blood are not normally done except in research studies because of the difficulty of these procedures and the fact that the data have been usually of limited diagnostic value. While fecal porphyrin assays can be a diagnostic aid in differentiating the types of porphyria,[7,12,16] screening test results combined with other porphyrin laboratory data will usually suffice. Blood or erythrocyte porphyrin concentrations are not usually needed for the diagnosis of porphyrias.

Recently, protoporphyrinemia has received increasing attention as an index of lead intoxication,[5] which has resulted in the development of simplified methods for blood porphyrin analysis.[8,11] Increased erythrocyte PROTO concentration results also from numerous abnormalities in iron metabolism, including nutritional deficiency, but these can be differentiated by secondary tests.

NORMAL VALUES

The rate of urinary URO excretion is $<40\ \mu g/d$. The rate of COPRO excretion is $<200\ \mu g/d$ for adults and $2\ \mu g/kg$ body weight for infants and young children.

REFERENCE

Scott, C. R., Labbé, R. F., and Nutter, J.: Clin. Chem., **13**:493, 1967.

DETERMINATION OF δ-AMINOLEVULINIC ACID CONCENTRATION AND δ-AMINOLEVULINATE DEHYDRATASE ACTIVITY

Lead intoxication poses a special problem to the laboratory. First, neither the rate of lead excretion nor the blood lead concentration necessarily reflects the total body load of the metal, which includes also that in bone. Nevertheless, these two parameters are generally considered as indicative of the effective load of the metal, or that which is capable of producing tissue damage. The most frequently accepted upper limit for blood lead is $40\ \mu g/100$ ml whole blood; yet concentrations below this level can cause detectable biochemical changes. Whether these changes are accompanied by subclinical tissue damage remains to be answered. Second, the heme biosynthetic pathway, which appears to be uniquely sensitive to the presence of lead, is altered through several mechanisms[15] (Fig. 9–4): ALA dehydratase activity in erythrocytes, and probably in other cells, is decreased, leading to an increase in ALA concentration in serum and urine; a block in the conversion of coproporphyrinogen III to protoporphyrinogen IX probably accounts for the coproporphyrinuria of lead intoxication; and the incorporation of iron is inhibited to cause an accumulation of PROTO, which is most easily seen in the erythrocytes. In addition to the direct measurement of lead concentration in the blood or lead excretion rate in the urine, these changes in porphyrin metabolism offer several dependable biochemical indicators of lead intoxication.

An increase in erythrocyte PROTO concentration appears to be a very sensitive indicator of the presence of lead in the body,[5] and it can now be determined on a micro scale.[11] The major drawback to the use of erythrocyte PROTO concentration is the fact that it is also increased by iron deficiency and a variety of other derangements in iron metabolism.

The measurement of COPRO in urine has been, and still is, used as a screening test in the detection of lead intoxication. However, this method has only limited diagnostic value because, first, coproporphyrinuria does not closely correlate with the effective body load of lead, especially in early stages of intoxication; and, second, coproporphyrinuria, having so many etiologies (as previously discussed) is the least specific test used for lead intoxication.

The excretion of ALA in urine can be a suitable indicator for lead intoxication providing that an accurate 24 h urine collection is made with precaution to preserve the ALA, which rapidly decomposes in even slightly alkaline media. These difficulties may account for the poorer correlation reported in several studies between ALA excretion and blood lead concentration. An alternative is assay of serum ALA concentration,although this apparently has not found wide acceptance as a test for lead intoxication.

The most sensitive biochemical response to lead intoxication is a decrease in erythrocyte ALA dehydratase activity, an effect that can be detected at blood lead levels in the subclinical range.[10] This enzyme appears to offer several advantages: studies have indicated a close correlation with blood lead concentration, the enzyme appears to be specific in its response to lead, and a simple micro assay for enzyme activity is available. ALA dehydratase activity provides a convenient means also of differentiating lead intoxication and iron deficiency when PROTO analysis is used as a screening test for either condition.

THE DETERMINATION OF δ-AMINOLEVULINIC ACID CONCENTRATION IN SERUM

PRINCIPLE

Interfering thiol compounds present in serum are eliminated by adding iodoacetamide. The serum is freed from protein by precipitation with trichloroacetic acid and subsequent centrifugation. The supernatant solution containing ALA is heated with acetylacetone to bring about quantitative formation of a pyrrole derivative (Fig. 9–6). This compound is reacted with Ehrlich's reagent and measured spectrophotometrically.

REAGENTS

1. Iodoacetamide, 0.06 mol/l. Prepare fresh by dissolving 60 mg in 5 ml of distilled or deionized water.
2. Trichloroacetic acid. Dissolve 20 g/100 ml of distilled or deionized water.
3. Acetate buffer, pH 4.6. Dissolve glacial acetic acid (57 ml) and sodium acetate trihydrate (136 g) in water and dilute to 1000 ml.
4. Acetylacetone (2,4-pentanedione; Eastman Organic Chemicals, Rochester, N. Y. 14650).
5. Ehrlich's reagent (ALA assay) (see footnote, p. 462). Dissolve 1 g of *p*-dimethylaminobenzaldehyde in about 30 ml of glacial acetic acid and 8 ml of 70 per cent perchloric acid, then dilute to 50 ml with acetic acid. This reagent is stable for about 6 h.

PROCEDURE

1. To 3.0 ml of serum add 0.2 ml of 0.06 molar iodoacetamide. After 2 min add 1.0 ml of trichloroacetic acid (20 g/100 ml). Shake the mixture vigorously and centrifuge to remove the precipitate.
2. To 1.0 ml of the supernatant solution placed in a glass-stoppered test tube, add 1.0 ml of acetate buffer, 0.1 ml of 2.5 molar sodium hydroxide, and 0.05 ml of acetylacetone. Mix and place the solution in a boiling water bath for 10 min. Cool to room temperature.
3. To 1.5 ml of this solution, add 1.5 ml of Ehrlich's reagent (ALA assay). Read the absorbance at 553 nm after exactly 15 min using a spectrophotometer having at least 1 to

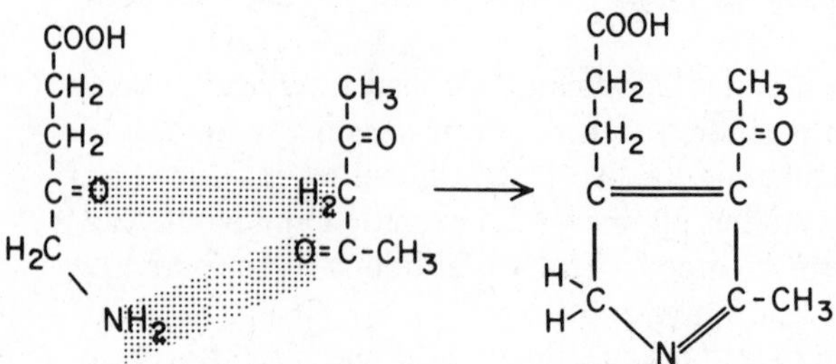

Figure 9–6 The conversion of δ-aminolevulinic acid to an Ehrlich-reacting pyrrole.

2 nm resolution. The same serum is used as a blank; it is treated identically but with the exclusion of acetylacetone in the heating process.

CALCULATIONS

$$\mu g \text{ ALA/100 ml of serum} = 1060 \times A_{553}$$

where

1060 = a calculation constant determined empirically from a standard curve, and indicates the slope of the curve obtained by plotting A_{553} versus μg ALA/100 ml serum.

COMMENTS

Beer's law applies up to an absorbance of at least 0.750. Since pure δ-aminolevulinic acid hydrochloride is commercially available from several suppliers of biochemicals, any laboratory can and should prepare its own standard curve. This is done by adding ALA to serum in concentrations of 100, 200, 400, 600 μg/100 ml. The assay is carried out as described and a correction is made for the original ALA contained in the serum.

NORMAL VALUES

For adult males, the mean ALA concentration in serum is reported as 19 μg/100 ml ($SD \pm 4$); children may have slightly lower values.

REFERENCE

Haeger-Aronsen, B.: Scand. J. Clin. Lab. Invest., *12*, Suppl. 4:33, 1960.

DETERMINATION OF δ-AMINOLEVULINATE DEHYDRATASE (EC 4.2.1.24) ACTIVITY IN ERYTHROCYTES

PRINCIPLE

The enzyme ALA dehydratase converts ALA to PBG. In peripheral blood, the reaction stops at this monopyrrole stage rather than proceeding to porphyrin. Therefore, ALA-dehydratase activity in erythrocytes can be determined simply by incubating a sample of peripheral blood with ALA and measuring the PBG formed.

REAGENTS

1. Triton X-100 (Sigma Chemical Company, St. Louis, MO 63178). Dissolve 0.2 g in 100 ml of distilled or deionized water.
2. δ-Aminolevulinic acid hydrochloride (substrate). Prepare a 0.01 molar solution of ALA in a citrate-phosphate buffer, pH 6.65. To prepare the buffer, adjust a 0.25 molar solution of $Na_2HPO_4 \cdot 7H_2O$ (6.7 g/100 ml) to the proper pH by adding a 0.25 molar solution of citric acid monohydrate (5.2 g/100 ml). Enzyme substrate is made by dissolving 16.8 mg of ALA·HCl/10 ml of this buffer.
3. Protein precipitant. To 0.250 g of *N*-ethylmaleimide dissolved in 30 to 40 ml of warm water, add 10 g of trichloroacetic acid and dilute to 100 ml.
4. Ehrlich's reagent (ALA dehydratase assay). Dissolve 10 g of *p*-dimethylaminobenzaldehyde, reagent grade, in 420 ml of glacial acetic acid. On the day of the test, mix 8.0 ml of 70 per cent perchloric acid with a sufficient volume of the above solution to make a final volume of 50 ml.

PROCEDURE

1. Draw a heparinized blood specimen and keep refrigerated until the time of assay. Determine the hematocrit on an aliquot of the blood.

2. To prepare the incubation mixture, add 0.10 ml of blood to 1.40 ml of Triton X-100 reagent, which assures immediate hemolysis. Add 1.0 ml of buffered ALA substrate and mix. Remove 1.0 ml immediately for the blank and promptly add it to a tube containing 1.0 ml of protein precipitant.

3. Cover the tube containing the remainder of the reaction mixture (1.5 ml) and incubate at 38°C for 1 h. At the end of the incubation period, immediately add 1.5 ml of protein precipitant.

4. After mixing and centrifuging, remove 1.5 ml of each of the supernatant solutions (blank and test) and add 1.5 ml of Ehrlich's reagent.

5. Mix the solutions and allow the color to develop for 13 min.

6. Within the next 10 min read the absorbance against water at 555 nm. Use a spectrophotometer having at least 1 to 2 nm resolution.

CALCULATIONS

$$\text{Units ALA dehydratase/ml erythrocytes} = (A_{test} - A_{blank}) \times \frac{100 \times 25}{\text{hematocrit}} \times \frac{1}{0.1}$$

where

100/hematocrit = correction for the erythrocyte fraction of whole blood.
25 = the dilution factor of the blood.
0.1 = correction for the definition of enzyme activity, one unit being defined as an increase in A_{555} of 0.100 (1.0 cm light path) per h at 38°C.

COMMENTS

Chelating anticoagulants such as EDTA should not be used because they may interfere with enzyme activity. Several assays for ALA dehydratase in erythrocytes have been described. These are generally based upon the same principle and differ only in methodological details. Since enzyme activity is very sensitive to pH, this must be rigidly controlled. The enzyme assays should be performed on freshly drawn blood specimens if possible. Storage at 4°C for 24 h results in an average decrease in activity of 4 per cent, but may be as high as 10 per cent. Blood stored overnight at 23°C has a loss in ALA dehydratase activity of about 20 per cent.

NORMAL VALUES

Normal ALA dehydratase activity is 175 ($SD \pm 36$) units for children and 176 ($SD \pm 43$) units for adults, the units being those specifically defined by the authors. To convert these activities into U/ml erythrocytes, the values should be multiplied by the factor $4 \times \frac{2}{61} \times \frac{1}{60}$, which reduces to 2.19×10^{-3}, where $4 \times \frac{2}{61}$ converts A of PBG in Ehrlich's reaction to micromoles ALA utilized, and $\frac{1}{60}$ corrects for the incubation time.

REFERENCE

Burch, H. B., and Siegel, A. L.: Clin. Chem., **17**:1038, 1971.

REFERENCES

1. Bossenmaier, I., and Cardinal, R.: Clin. Chem., *14:*610, 1968.
2. Bourke, E., Copeman, P. W., Milne, M. D., and Stokes, G. S.: Lancet, *1:*1394, 1966.
3. Castrow, F. F., 2nd: South. Med. J., *63:*541, 1970.
4. Castrow, F. F., 2nd, Mullins, J. F., and Mills, G. C.: J. Clin. Invest., *50:*340, 1968.
5. Chisholm, J. J.: Pediatrics, *51:*254, 1973.
6. Doss, M., and Ulshöfer, B.: Biochim. Biophys. Acta, *237:*356, 1971.
7. Eales, L., Dowdle, E. B., Saunders, S. J., and Sweeney, G. D.: S. Afr. J. Clin. Lab. Med., *9:*126, 1963.
8. Kammholz, L. P., Thatcher, L. G., Blodgett, F. M., and Good, T. A.: Pediatrics, *50:*625, 1972.
9. Loriaux, L., Delena, S., and Brown, H.: Clin. Chem. *15:*292, 1969.

10. Morgan, J. M., and Burch, H. B.: Arch. Intern. Med., *130:*335, 1972.
11. Piomelli, S., Davidow, B., Guinee, V. F., Young, P., and Gay, G.: Pediatrics, *51:*254, 1973.
12. Rimington, C.: S. Afr. J. Clin. Lab. Med., *9:*255, 1963.
13. Taddeini, L., Kay, I. T., and Watson, C. J.: Clin. Chim. Acta, *7:*890, 1962.
14. Talman, E. L.: Porphyrins in urine. *In* Standard Methods of Clinical Chemistry. D. Seligson, Ed. New York, Academic Press, Inc., 1958, vol. 2, p. 137.
15. Vallee, B. L., and Ulmer, D. D.: Biochemical effects of mercury, cadmium, and lead. *In* Annual Review of Biochemistry. E. E. Snell, Ed. Palo Alto, California, 1972, vol. 41, p. 91.
16. Wetterberg, L., Haeger-Aronsen, B., and Stathers, G.: Scand. J. Clin. Lab. Invest., *22:*131, 1968.

ADDITIONAL READINGS

Burnham, B. F.: Metabolism of porphyrins and corrinoids. *In* Metabolic Pathways. 3rd ed. M. Greenberg, Ed. New York, Academic Press, Inc., 1969, vol. III, pp. 403–537.

Falk, J. E.: Porphyrins and Metalloporphyrins. New York, Elsevier Publishing Co., 1964.

Goldberg, A.: Porphyrins and porphyrias. *In* Recent Advances in Haematology. A. Goldberg and M. C. Brain, Eds. Churchill Livingstone, London, 1971, pp. 302–336.

Heilmeyer, L.: Disturbances in Heme Synthesis. Springfield, Illinois, Charles C Thomas, Publisher, 1966.

Levere, R. D., and Kappas, A.: Biochemical and clinical aspects of the porphyrias. *In* Advances in Clinical Chemistry. O. Bodansky and C. P. Stewart, Eds. New York, Academic Press, Inc., 1968, vol. 11, pp. 133–174.

Marks, G. S.: Heme and Chlorophyll—Chemical, Biochemical and Medical Aspects. Princeton, N. J., D. Van Nostrand Co., 1969.

Marver, H. S., and Schmid, R.: The porphyrias. *In* The Metabolic Basis of Inherited Disease. 3rd ed. J. B. Stanbury, J. B. Wyngaarden, and D. S. Fredrickson, Eds., New York, McGraw-Hill Book Co., 1972, pp. 1087–1140.

Schwartz, S., Berg, M. H., Bossenmaier, I., and Dinsmore, H.: Determination of porphyrins in biological materials. *In* Methods of Biochemical Analysis. D. Glick, Ed. New York, Interscience Publishers, Inc., 1960, vol. VIII, pp. 221–293.

CHAPTER 10

LIPIDS AND LIPOPROTEINS*

by Ralph D. Ellefson, Ph.D., and Wendell T. Caraway, Ph.D.

The term "lipid" is applied to those *fatty*, *oily*, and *waxy substances* of animal or vegetable origin that are practically insoluble in water but that dissolve freely in nonpolar solvents, such as chloroform, ether, hexane, and benzene. These properties stem from the characteristics of the relatively large *hydrocarbon* portions in lipid molecules. In addition to the hydrocarbon portions most lipid molecules contain hydroxyl or ester groups. Some of the more complex lipids contain sugars, amino, phosphoryl, or other structural forms that tend to confer solubility in more polar solvents such as alcohol and water. In physiologic fluids and in tissues, most lipid molecules are present in combination with proteins; such lipid-protein complexes are referred to as *lipoproteins*. This combination promotes solubility of the lipids in an aqueous medium.

As defined above, the lipids include sterols, vitamins A, D, E, and K, bile pigments, waxes, carotene, and related dietary pigments, as well as the fatty acids, triglycerides, and phosphatides. For our purposes, we shall consider only those lipids that can be put into the following subgroups: non-esterified fatty acids, triglycerides, glycerophosphatides, and sphingolipids, as well as cholesterol and cholesteryl esters. Triglycerides, phosphatides, and sphingolipids can be degraded by digestive processes or metabolic reactions to several intermediate split products.

In the body, lipids can serve as structural and functional elements of biomembranes, as precursors to other essential substances, as sources of biochemical fuel, as energy storage depots, and as insulators. Major clinical interest has focused on changes in blood lipid and lipoprotein components in various diseases, on the association of blood lipids and lipoproteins with *atherosclerosis*, and on studies of *lipid storage diseases*.

CHEMISTRY OF LIPIDS

FATTY ACIDS

The fatty acids comprise one of the simpler molecular forms of the lipids. Those of importance in human metabolism and nutrition are virtually all monocarboxylic acids containing even numbers of carbon atoms in straight chains. The principal chain lengths range from 16 (C_{16}) to 24 (C_{24}) carbon atoms and may be characterized further as saturated, monounsaturated, or polyunsaturated. Unsaturated acids contain double bonds connecting two adjacent carbon atoms. When more than one double bond exists, they are usually separated by three carbon atoms (=CH—CH_2—CH=). Typical examples are shown schematically in Figure 10–1.

Although two carbon atoms may freely rotate about a single bond between them, a

* Some portions of this chapter have been retained from Chapter 7 of the First Edition. This contribution from Dr. Robert L. Dryer is thankfully acknowledged.

```
   H  H  H  H  H  H  H  H  H  H  H  H  H  H  H     O
   |  |  |  |  |  |  |  |  |  |  |  |  |  |  |   //
H—C—C—C—C—C—C—C—C—C—C—C—C—C—C—C—C
   |  |  |  |  |  |  |  |  |  |  |  |  |  |  |   \
   H  H  H  H  H  H  H  H  H  H  H  H  H  H  H     OH
```

Palmitic Acid

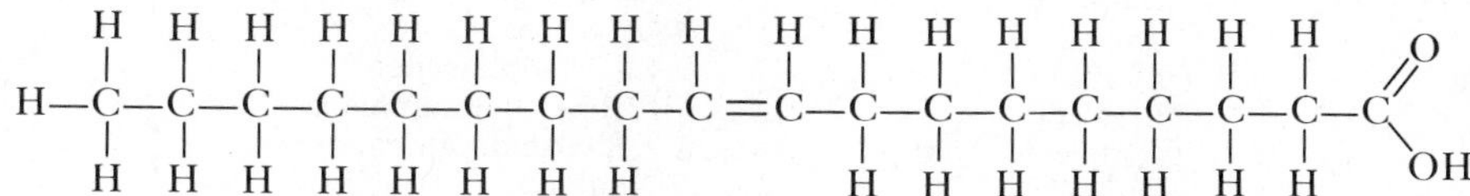

Oleic Acid

```
   H  H  H  H  H  H  H  H  H  H  H  H  H  H  H  H  H     O
   |  |  |  |  |  |  |  |  |  |  |  |  |  |  |  |  |   //
H—C—C—C—C—C—C=C—C—C=C—C—C—C—C—C—C—C—C
   |  |  |  |  |        |        |  |  |  |  |  |  |   \
   H  H  H  H  H        H        H  H  H  H  H  H  H     OH
```

Linoleic Acid

Figure 10–1 Examples of common fatty acids. Although shown schematically in linear form, the carbon atoms actually exist in wavy or "kinked" configurations as shown in Figure 10–2.

double bond more or less rigidly fixes the relative positions of the two carbon atoms in space so that the remaining bonds of each carbon have specific spatial orientations. When a double bond occurs along the length of a carbon chain, the pair of hydrogen atoms on the participating carbon atoms may lie on the same side of the double bond, in which case they are described as *cis*, or they may lie on opposite sides of the double bond, in which case they are described as *trans*. The double bonds of the naturally occurring fatty acids are of the *cis* configuration as shown in Figure 10–2.

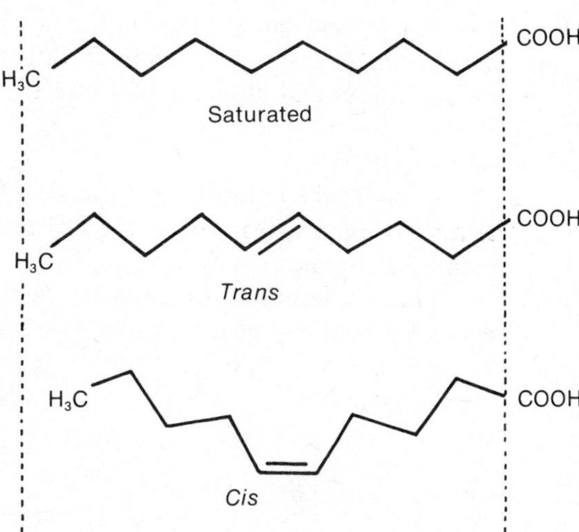

Figure 10–2 Representation of fatty acid structures by conventional line segments. The dotted vertical lines show that the saturated chain is the longest, that the *trans*-unsaturated chain is a little shorter, and that the *cis*-unsaturated chain is the shortest and the most bent.

Table 10–1 FATTY ACIDS COMMONLY FOUND IN HUMAN LIPIDS

Common Name	*Systematic Name*	*Numerical Designation*
Lauric	Dodecanoic	12:0
Myristic	Tetradecanoic	14:0
Palmitic	Hexadecanoic	16:0
Palmitoleic	9-Hexadecenoic	$16{:}1^{9}$
Stearic	Octadecanoic	18:0
Oleic	9-Octadecenoic	$18{:}1^{9}$
*Linoleic	9,12-Octadecadienoic	$18{:}2^{9,12}$
*Linolenic	9,12,15-Octadecatrienoic	$18{:}3^{9,12,15}$
γ-Linolenic	6,9,12-Octadecatrienoic	$18{:}3^{6,9,12}$
Arachidic	Eicosanoic	20:0
Arachidonic	5,8,11,14-Eicosatetraenoic	$20{:}4^{5,8,11,14}$
Behenic	Docosanoic	22:0
Lignoceric	Tetracosanoic	24:0

* Obtained from foods; not synthesized in human cells.

The common saturated and unsaturated fatty acids are shown in Table 10–1. A shorthand designation is used. The first number refers to the number of carbon atoms in the straight chain molecule, the number following the colon represents the number of double bonds in the molecule, and the superscript shows the double bond position, the carboxyl carbon always being designated carbon atom number one. According to this convention, linoleic acid can be described as $18{:}2^{9,12}$ which means that one double bond lies between carbons 9 and 10, counting from the carboxyl end of the chain, and the second lies between carbons 12 and 13. The systematic name, referring to the formal Geneva system of nomenclature, is 9,12-octadecadienoic acid.

Most of the fatty acids of human materials have even numbers of carbon atoms because the fatty acids are biosynthesized from two-carbon acetyl units. The fatty acids have K_a' values similar to those of acetic acid and propionic acid, namely 1.4×10^{-5} to 1.7×10^{-5} ($pK_a' = 4.85$ to 4.77). In human blood, lymph, and tissues, the fatty acids exist almost entirely in the form of anions (salts). This circumstance is based on two factors. First, in an aqueous medium at neutral pH, more than 99.4 per cent of the molecules of an acid with a K_a' of 1.7×10^{-5} would be ionized. The second factor is that nearly all the fatty acid molecules (or anions) in human fluids and tissues are bound to protein molecules, and this protein binding favors the anions rather than the undissociated acids. The concentrations of fatty acid anions in human materials are quite low. Most of the lipids in human tissues and blood consist of esters or amides of fatty acids.

GLYCEROL

Many of the natural lipids are *glyceryl esters* of fatty acids. Therefore, a brief consideration of certain features of the glycerol molecule and of glyceride molecules in general seems appropriate.

The two terminal (or primary) carbon atoms in the glycerol molecule are chemically equivalent and can be designated by α and α'; the middle carbon atom is designated by β.

$$\begin{array}{l} H_2C_{\alpha}\text{—}OH \\ \quad | \\ HC_{\beta}\text{—}OH \\ \quad | \\ H_2C_{\alpha'}\text{—}OH \end{array}$$

If one of the primary (α) carbon atoms is part of an ester group or some other functional group and hydroxyl groups remain intact in the α′ and β positions, the molecule becomes asymmetric; the β carbon atom is the center of asymmetry, and D and L optical antipodes are possible.

In order to designate systematically the position of a substituent in an asymmetric molecule that is a derivative of glycerol, the IUPAC-IUB Commission adopted the following rule. All of the glyceryl lipids that possess molecular asymmetry are considered to exist in the L form. Using the Fisher projection formula and showing the L configuration with the glyceryl β hydroxyl group (or substituent) on the left, the carbon atom above is designated number 1 and the carbon atom below is designated number 3. Systematic names that are derived from this rule should include the designation "sn-" to indicate "stereospecific numbering." By this rule, the following formulas **A** and **B**, which represent enantiomers, are designated sn-1-acylglycerol and sn-3-acylglycerol, respectively.

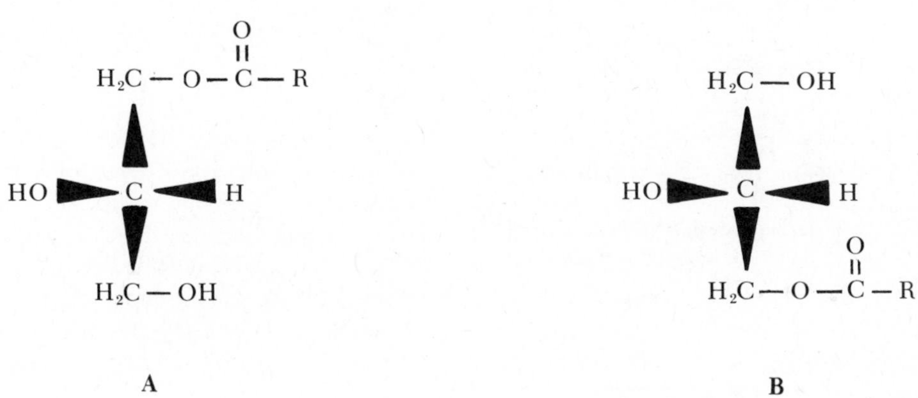

TRIGLYCERIDES (TRIACYLGLYCEROLS)

The *neutral* glycerides are the simple esters of the fatty acids and glycerol, and the triglycerides are the most abundant of the neutral glycerides.

Triglycerides can be hydrolyzed readily by strong alkalis or acids or by enzymes known as *lipases*, which are found in all tissues, blood plasma, pancreatic juice, other body fluids, and feces. Under drastic conditions, the products include the ensemble of component fatty acids and free glycerol, but under milder conditions incomplete hydrolysis produces a mixture of free fatty acids and mono- and diglycerides. Equations describing these reactions may be written as shown on page 478.

The mono- and diglycerides exist in blood and tissues only in small concentrations. They are metabolic intermediates in the formation and degradation of the triglycerides and glycerophosphatides.

The triglycerides constitute most of the mass of adipose tissue. In adipose cells they exist mainly in molecular clusters in the form of micro-droplets and ultramicro-droplets. The concentrations of triglycerides in tissues other than adipose tissue are relatively small, except in intestinal cells soon after the ingestion of fat. Fatty acids from food fats are absorbed by the intestinal mucosal cells and incorporated into triglycerides. However, soon after being formed, these triglycerides are released from the mucosal cells into the intestinal lymph.

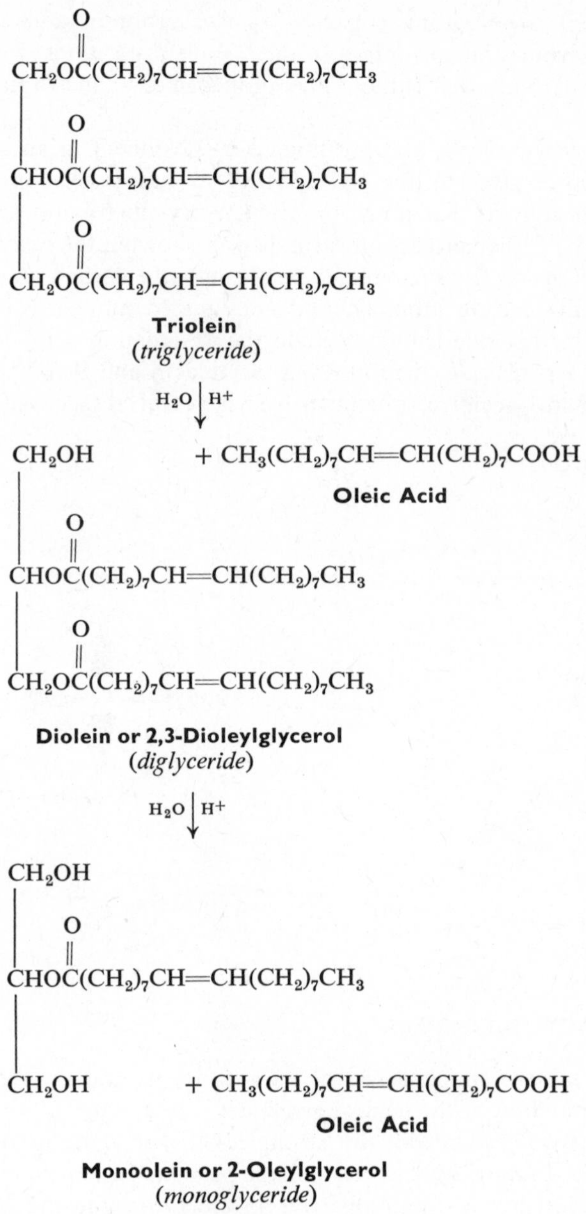

$$
\begin{array}{l}
\quad\;\; O \\
\quad\;\; \| \\
CH_2OC(CH_2)_7CH{=}CH(CH_2)_7CH_3 \\
| \\
|\quad\;\; O \\
|\quad\;\; \| \\
CHOC(CH_2)_7CH{=}CH(CH_2)_7CH_3 \\
| \\
|\quad\;\; O \\
|\quad\;\; \| \\
CH_2OC(CH_2)_7CH{=}CH(CH_2)_7CH_3
\end{array}
$$

Triolein
(triglyceride)

$$\xrightarrow[H_2O]{H^+}$$

$$
\begin{array}{ll}
CH_2OH & + \; CH_3(CH_2)_7CH{=}CH(CH_2)_7COOH \\
| & \\
|\quad\;\; O & \\
|\quad\;\; \| & \\
CHOC(CH_2)_7CH{=}CH(CH_2)_7CH_3 & \\
| & \\
|\quad\;\; O & \\
|\quad\;\; \| & \\
CH_2OC(CH_2)_7CH{=}CH(CH_2)_7CH_3 &
\end{array}
$$

Oleic Acid

Diolein or 2,3-Dioleylglycerol
(diglyceride)

$$\xrightarrow[H_2O]{H^+}$$

$$
\begin{array}{ll}
CH_2OH & \\
| & \\
|\quad\;\; O & \\
|\quad\;\; \| & \\
CHOC(CH_2)_7CH{=}CH(CH_2)_7CH_3 & \\
| & \\
CH_2OH & + \; CH_3(CH_2)_7CH{=}CH(CH_2)_7COOH
\end{array}
$$

Oleic Acid

Monoolein or 2-Oleylglycerol
(monoglyceride)

GLYCEROPHOSPHATIDES

These are the simplest members of the complex lipids, so called because they contain atoms of P and frequently N in addition to C, H, and O. The glycerophosphatides and the sphingomyelins which we shall consider soon, were commonly described in the past as "phospholipids," but aside from the indication of phosphorus content, the term phospholipid conveys very little structural information. For this reason we shall use the more precise terms glycerophosphatides and sphingomyelins to describe the respective classes. The glycerophosphatides may be regarded as derivatives of phosphatidic acid, the structure of which is shown at the end of the paragraph. Phosphatidic acid is converted to a glycerophosphatide by the addition of one of a variety of alcohols; three examples are choline, ethanolamine, and serine. The last of these is not only an amino alcohol but an amino acid as well.

```
            O                                   O
            ||                                  ||
   O  CH2OCR                           O  CH2OCR
   ||   |             amino            ||   |
 RCOCH     O        ---------->      RCOCH     O
           ||         alcohol               |     ||
      CH2O—P—OH                           CH2O—P—O—CH2—CH2—NH3+
           |                                   |
           OH                                  O−
```

L-α-Phosphatidic acid → **L-α-Glycerophosphatide**

$HOCH_2CH(NH_3^+)—COO^-$ — **Serine**

$HOCH_2CH_2N^+H_3 \cdot Cl^-$ — **Ethanolamine chloride**

$HOCH_2CH_2N^+(CH_3)_3 \cdot Cl^-$ — **Choline chloride**

Typical amino alcohols

Phosphatidylcholine is frequently termed *lecithin*, and phosphatidylserine and phosphatidylethanolamine are frequently termed, collectively, *cephalins*. Since the fatty acids associated with these substances may be the same or may be different, designation as a lecithin is still only a class distinction.

A few phosphatides with more complicated molecular structures are known; one example is cardiolipin or 1,3-*bis*-phosphatidylglycerol.

```
              O                  OH
              ||                 |
          CH2OCR        CH2O—P—O—CH2
           |             |    ||     |
   O       |             |    O  O   |
   ||      |             |       ||  |
  RCOCH                 HCOH    RCOCH
   |                     |         |
   |        O            |         |      O
   |        ||           |         |      ||
  CH2O—P—O—CH2                   CH2OCR
            |
            OH
```

Cardiolipin

SPHINGOSINE AND CERAMIDES

Instead of glycerol, the *sphingolipids* contain the amino alcohol, sphingosine.

```
                       H
                      /
 CH3(CH2)12C=C—CH—CH—CH2OH
           /       |    |
          H        OH   NH2
```

Sphingosine

Note that in sphingosine the configuration around the double bond is *trans*. In sphingolipids only one molecule of fatty acid is found per molecule of lipid; it is attached in amide form to the nitrogen of sphingosine. The term ceramide is used to designate the N-acyl derivatives of sphingosine. Ceramides are intermediates in the biosynthesis and in the physiologic degradation of the sphingomyelins and other sphingolipids.

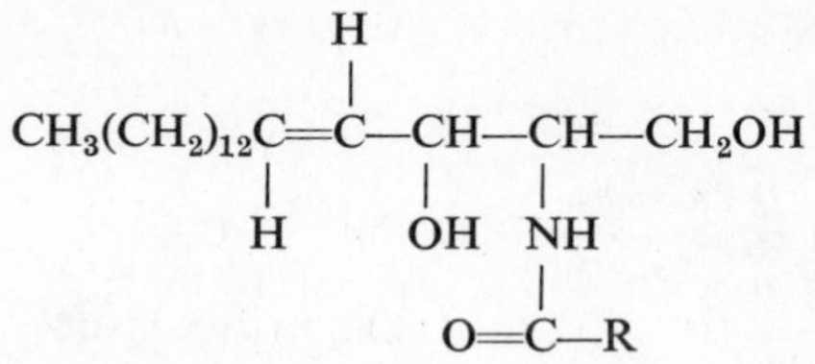

$$CH_3(CH_2)_{12}CH{=}CH{-}CH(OH){-}CH(NH{-}C({=}O){-}R){-}CH_2OH$$

Example of a Ceramide

SPHINGOMYELINS

In the sphingomyelins the primary alcohol function serves as a point of attachment for a molecule of phosphoryl choline. The structure of sphingomyelin is, therefore, as follows:

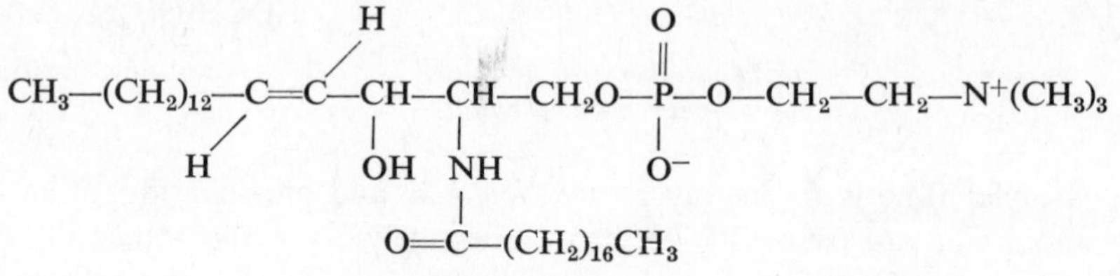

$$CH_3{-}(CH_2)_{12}{-}CH{=}CH{-}CH(OH){-}CH(NH{-}C({=}O){-}(CH_2)_{16}CH_3){-}CH_2O{-}P(=O)(O^-){-}O{-}CH_2{-}CH_2{-}N^+(CH_3)_3$$

Sphingomyelin

CEREBROSIDES

Several *lipid storage diseases* (see p. 488) are characterized by accumulations of derivatives of ceramides. For example, a ceramide glucoside accumulates in tissues of patients with *Gaucher's disease.* Such substances, in which the primary alcohol of sphingosine forms a glycosidic linkage with carbon atom 1 of a hexose, are usually referred to as cerebrosides or ceramide monohexides; note that cerebrosides contain no phosphorus. Some cerebrosides form sulfuric acid esters at C_6 of the hexose ring; these cerebroside sulfates are known commonly as *sulfatides.*

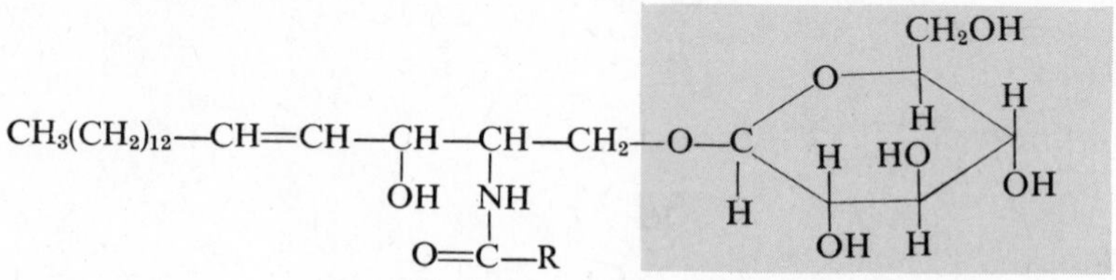

$$CH_3(CH_2)_{12}{-}CH{=}CH{-}CH(OH){-}CH(NH{-}C({=}O){-}R){-}CH_2{-}O{-}\text{[glucose, } C_1\text{; ring substituents: } CH_2OH,\ H,\ HO,\ OH,\ OH,\ H\text{]}$$

Example of a Glucocerebroside

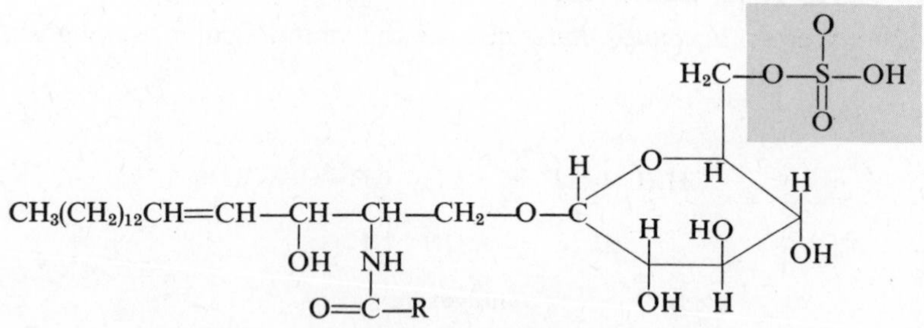

$$CH_3(CH_2)_{12}CH{=}CH{-}CH(OH){-}CH(NH{-}C({=}O){-}R){-}CH_2{-}O{-}\text{[hexose, } C_1\text{; } C_6\text{: } H_2C{-}O{-}S(=O)_2{-}OH\text{]}$$

Example of a Sulfatide

Most of the sphingolipids (sphingophospholipids and ceramide hexosides) are found in the central nervous system and in cell membranes. The fatty acids involved usually have long carbon chains (C_{20} to C_{24}), and they contain most of the hydroxy fatty acids found in human lipids.

CHOLESTEROL AND CHOLESTERYL ESTERS

The term sterol describes the alcoholic lipids that include in their molecular structures the tetracyclic *perhydrocyclopentanophenanthrene* skeleton. Cholesterol is the most abundant sterol in human tissues and fluids. The most descriptive name for cholesterol is Δ^5-cholestene-3β-ol. The symbol Δ^5 indicates the presence and position of a double bond between carbons 5 and 6, and the term 3β-ol indicates the presence and direction of orientation (as explained below) of the hydroxyl group. Identification of the positions of substitution in the sterol molecule is by means of the numbering system shown here for the structure of cholesterol.

Sterane

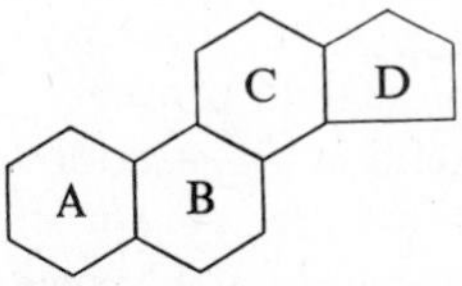

Perhydrocyclopentanophenanthrene (sterane) skeleton

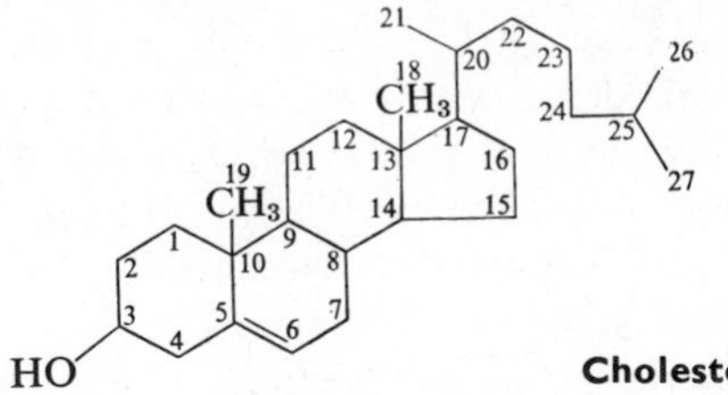

Cholesterol

If the double bond of cholesterol is reduced, two different products, cholestanol and coprostanol, can be formed. In the *trans* product, cholestanol, the atoms lie very nearly in a slightly pleated sheet; in the *cis* product, coprostanol, the A ring is folded at nearly a right angle to the remainder of the molecule. The difference in molecular form between cholestanol and coprostanol is illustrated below. The structural formula for cholesterol that is shown above represents a view from above the average plane of the tetracyclic structure and from directly above the molecule; the kinked structures shown below for cholesterol, cholestanol, and coprostanol also represent a view from above the plane of the tetracyclic structures, but at an angle of approximately 45 degrees toward the reader

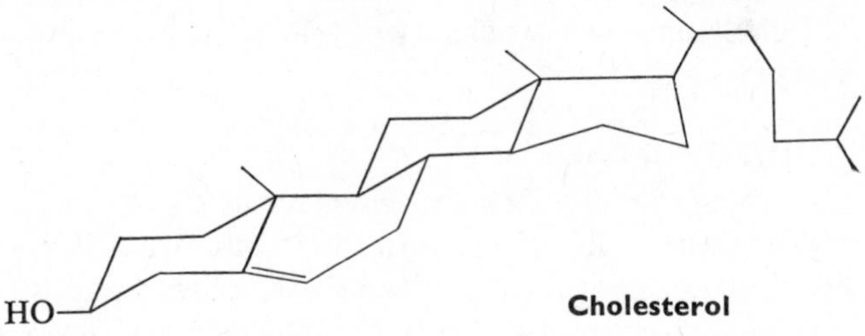

Cholesterol

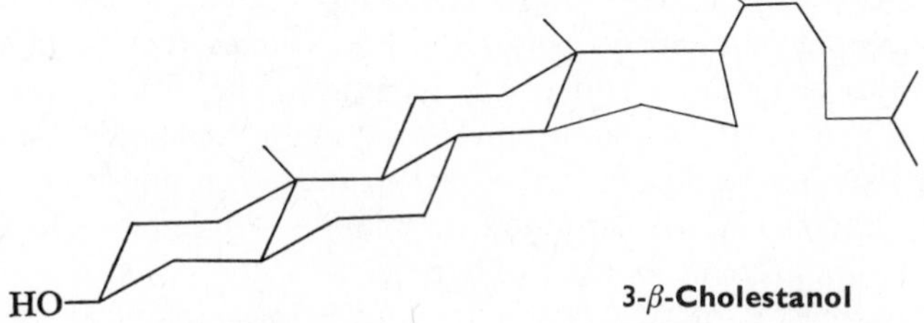

3-β-Cholestanol

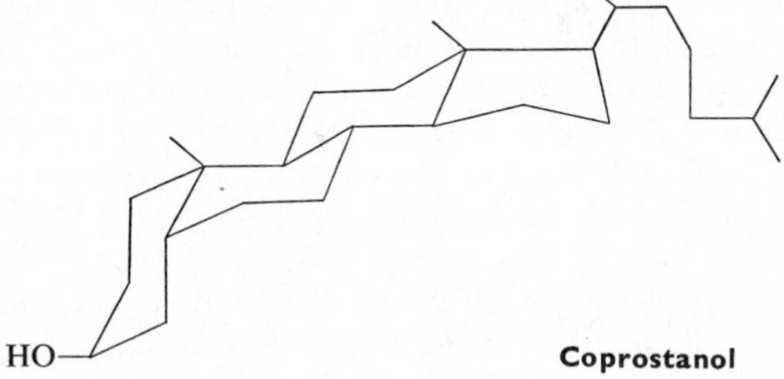

Coprostanol

instead of from directly above the molecule. The difference in molecular structure is responsible for large differences in physical and biochemical properties between the two cholesterol derivatives. Coprostanol is formed by the reductive action of microorganisms that live in the lumen of the gut.

In a sterol molecule, the substituents around a ring may lie above or below the plane of the ring. In the drawings, solid lines linking substituents to a ring refer to bonds above the ring plane, and dotted lines indicate substituents below the plane. In cholesterol, for example, the —OH group is above the plane; in this condition we speak of a β-oriented group. If the group lies below the ring plane, it is termed α-oriented. This difference affects the chemical reactivity of the molecule. For example, only β-oriented OH groups will form insoluble derivatives with *digitonin*, a most useful reagent for the analytical isolation of serum cholesterol and other β-hydroxylated steroids from α-hydroxylated steroids.

The double bond in cholesterol has many of the general properties of double bonds. It can be brominated readily, and the dibromide is useful in preparation of pure cholesterol for analytical standards.[28] Pure cholesterol slowly oxidizes, even at refrigerator temperature, because the double bond is vulnerable to molecular oxygen. Some workers therefore prefer to prepare and store quantities of the pure dibromide, which can then be easily debrominated in small lots as needed.

As an alcohol, cholesterol can form esters with fatty acids; normally, about two-thirds of the cholesterol in plasma exists in such form. These cholesteryl esters are formed from the free alcohol and phosphatidylcholine. The enzyme involved is best described as a *transferase*, since it transfers fatty acids from phosphatidylcholine to the alcohol function of the sterol. This is of analytical significance, since this enzymatic reaction continues after blood has been drawn; i.e., the ratio of free and esterified cholesterol in collected blood samples will change with time, even at refrigerator temperature. For analyses in which correct values for free and esterified cholesterol are important, specimens should be processed promptly after collection. In most individuals, the ratio of the quantity of esterified cholesterol to the quantity of total cholesterol is rather constant, except in severe liver disease.

LIPOPROTEINS

All the lipids in plasma circulate in combination with proteins. Most of the fatty acid molecules, in anionic form, are bound to albumin, while the other lipids are combined with other proteins in complexes called lipoproteins. The latter range in size from small lipoproteins of about 10 nm diameter, with molecular masses of less than 400,000, to particles having diameters of approximately 1 μm and having molecular masses that are proportionately greater. The lipid component causes the lipoproteins to have densities that are less than the densities of albumin and of the globulins. The differences in the relative proportions of lipids and proteins among the several lipoproteins give them different densities and permit their separation in the ultracentrifuge. The lipoproteins also differ in density of electrical charge. This property, in combination with differences in size, also permits separation of the lipoproteins by electrophoresis.

Complex lipid-protein aggregates of two types—the *chylomicrons* and the *very low density lipoproteins* (VLDL)—are primarily triglyceride transport vehicles. The chylomicrons (diameter 0.1 to 0.5 μm, density <0.95 g/ml) are the largest of the physiologic fat particles in the blood, and they consist mostly of triglycerides formed from food lipids. The chylomicrons originate in the intestinal epithelial cells, and they carry triglycerides to adipose cells for storage and to working cells for catabolism. The liver also clears chylomicrons from the blood, incorporates the triglycerides into lipoproteins, and releases them back into the bloodstream as VLDL (density <1.006 g/ml). These VLDL particles are primarily carriers for transporting triglycerides from the liver to other organs.

In addition to the VLDL, the plasma lipoproteins include *low density lipoproteins*

(LDL, density = 1.006 to 1.063 g/ml) and *high density lipoproteins* (HDL, density = 1.063 to 1.21 g/ml). The VLDL, LDL, and HDL differ in terms of both lipid composition and protein composition (see Table 10–3, p. 529). While the major lipid components of the VLDL are triglycerides, the lipids of the LDL and HDL consist mostly of cholesterol, cholesteryl esters, and phospholipids.

The major forces holding together the lipids and proteins of the lipoproteins are mostly noncovalent. Chylomicrons, VLDL, and LDL are composed of less than 30 per cent protein; they are micellar and consist of hydrophobic centers of triglycerides and cholesteryl esters surrounded by hydrophilic coats of protein and phospholipid.[45]

LIPID METABOLISM[23,57,63]

A discussion of lipid metabolism must encompass both the disposition of preformed food lipid and the conversion and storage of excess calories as fat. The average American diet provides nearly 40 per cent of total calories as fat; in addition, the total daily caloric supply is often above the total expenditure, and the surplus is stored largely as fat. Careful studies have shown that the adult human can absorb as much as 150 g of food fat per day, and that liquid fats are digested and absorbed more quickly than solid fats.

Digestive enzymes degrade ingested lipids to fatty acids, monoglycerides, water-soluble phosphates, free cholesterol, and a variety of intermediate products. Triglycerides are converted to fatty acids and monoglycerides, which are absorbed as such. This process of hydrolytic digestion, which takes place in the lumen of the small intestine, is accomplished mainly by enzymes from the pancreas, and it is aided considerably by the emulsifying properties of the *bile salts*, themselves derivatives of cholesterol. Therefore, in the absence of bile, for whatever reason, the digestion of fats is hindered and the stools passed contain significant amounts of unabsorbed fats. These are partly split by colonic microorganisms and produce foul-smelling and highly irritating stools. Fat that is not taken up by the small intestine contains dissolved fat-soluble vitamins; malabsorption of lipids, therefore, can result in deficiencies of vitamins A and E, and possibly D and K.

In the small intestine, the liberated fatty acids exist as soaps and, in the presence of bile salts as micellar complexes. In this form they are absorbed by the intestinal mucosa, along with the other digestion products (including monoglycerides). Within the mucosal cells the fatty acids are reassembled to form triglycerides, glycerophosphatides, and cholesteryl esters. It is important to note, however, that the glycerol employed in resynthesis of lipids is not that freshly absorbed from the intestine. The triglyceride-forming enzymes can transfer activated fatty acids only to *glycerophosphate* and to monoglycerides; the glycerophosphate is formed mainly from *glucose.*

Within the mucosal cells, the reformed lipids are incorporated into the chylomicrons. These particles are released from the mucosal cells, are transferred to the lymphatic vessels, and ultimately reach the blood via the thoracic lymph duct. It is well established that the chylomicron content of blood and lymph rises sharply after fat ingestion while the chylomicron content of blood in normal fasting persons is very small. The presence of chylomicrons produces a turbidity of the serum (postprandial lipemia), and this phenomenon has been made the basis of a test for measuring fat absorption efficiency.

Fatty acids with twelve carbons or less are incorporated into glycerides and chylomicrons only to a very small extent. These substances are transferred to the portal blood and are transported directly to the liver. In the liver cells they are degraded and/or used for biosynthetic processes. A small amount of the fat reaching the liver is probably degraded by that organ for the purpose of releasing energy; normally the liver does not store lipids, and nearly all lipids in the liver are structural lipids.

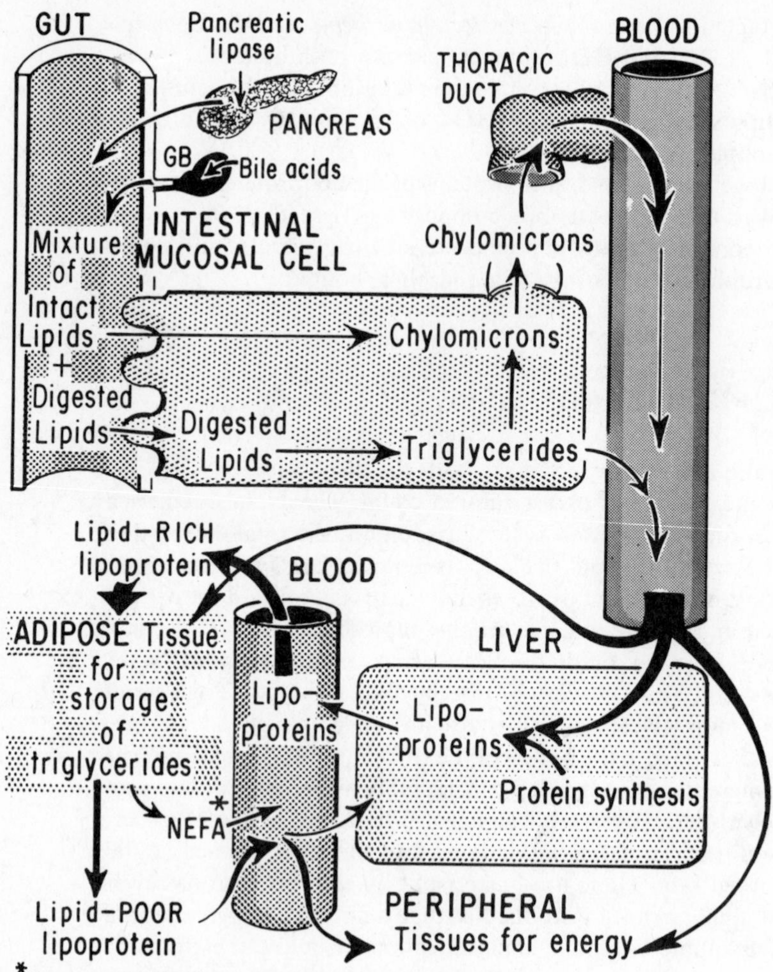

Figure 10–3 Factors involved in lipid transport in the body. The liver is central in the formation of lipid-rich lipoproteins, which carry triglycerides to the adipose tissue for storage. NEFA are released as needed to furnish available energy elsewhere. The lymphatic system cooperates in the absorption process. (After Dryer, modified.)

Lipoprotein transport is available not only for preformed lipids, but also for lipids made from excess carbohydrate or protein in the diet. The liver converts these energy forms to triglycerides and other lipids, and then handles them in the same way as the preformed lipids. Some details of lipid transport are summarized in Figure 10–3.

FATTY ACIDS

Fatty acids (FA) sometimes are referred to as nonesterified fatty acids (NEFA), free fatty acids (FFA), or unesterified fatty acids (UFA). In the blood, this important lipid form is carried mostly by *plasma albumin*; however, small amounts of fatty acids are carried by the larger lipoproteins, also. The best evidence indicates that one molecule of albumin can carry as many as 20 molecules of fatty acid. The normal level of FA in human blood is only 0.30 to 1.10 mmol/l, amounting to about 8 to 31 mg/100 ml of plasma, but the flux is very large and quite sensitive to exercise and physical work, to the level of blood glucose, and to excitement or other psychologic stress that liberates epinephrine. The FA are readily taken up by most tissues for satisfying energy requirements.

The oxidation of a mole of palmitic acid produces carbon dioxide and water plus a rather large quantity of energy, as shown in the following equation.*

$$C_{15}H_{31}COOH + 23\ O_2 \rightarrow 16\ CO_2 + 16\ H_2O + 2{,}330{,}500\ \text{cal}$$

Of this energy, nearly 40 per cent can be trapped in the body as useful chemical energy. The chemical energy is contained in specialized molecules, which, by their structure, have the ability to hold unusually large quantities of energy. By means of suitable enzyme reactions the chemical energy may be released for metabolic processes or stored in the form of high energy compounds. One of these specialized molecules is known as adenosine triphosphate (ATP). The structure of ATP and other information about this and related compounds may be found in any biochemistry text. Keeping in mind that a triglyceride molecule is an ester containing three fatty acid molecules, it is clear that fats make an excellent storage form for reserve energy. Among their advantages are their high intrinsic energy content, their low density (less than 1.0 g/ml), and their hydrophobic property. The amount of energy produced by metabolizing one mole of palmitic acid (16 carbon atoms) is approximately twice that produced by metabolizing an equivalent amount (2.5 moles) of glucose (6 carbon atoms per molecule). Sugar storage requires water of hydration; triglyceride storage does not.

Long chain fatty acids are oxidized for the production of energy in the *mitochondria* of cells by a series of reactions that operate in a repetitive manner to shorten the chains by two carbon atoms at a time. This process has become known as *β-oxidation.* As a result, each C_{16} molecule, for example, is converted to eight molecules of an intermediate known as acetyl coenzyme A (acetyl CoA). Acetyl CoA does not normally accumulate in the cell, but becomes enzymatically condensed with *oxaloacetate*, a substance derived largely from carbohydrate metabolism. The product of the condensation reaction is *citrate*, which is a major component of another cyclic series of reactions known variously as the Krebs cycle, the citric acid cycle, or the tricarboxylic acid cycle. The Krebs cycle serves as a common pathway for the final oxidation of nearly all food material, whether derived from carbohydrate, fat, or protein. It is important to bear in mind that the smooth operation of the metabolic machinery depends on the availability of sufficient oxaloacetate to serve as acceptor for acetyl CoA.

KETOSIS

In acute starvation or in impaired carbohydrate metabolism, represented by uncontrolled diabetes mellitus, the supply of acetyl CoA is greater than the supply of oxaloacetate. The cause of this problem is an excessive degradation of fatty acids by β-oxidation in liver cells, resulting from an excessive mobilization of fatty acids from adipose cells. As a result, acetyl CoA must be handled by an alternative pathway, which forms increased amounts of acetoacetic acid, β-hydroxybutyric acid, and acetone (a condition known as ketosis). These three metabolites are known collectively as *ketone bodies*, and since the first two are acidic substances, they may cause a metabolic acidosis as they accumulate.

Ketosis can, therefore, be regarded as developing from an excessive production of acetyl CoA as the body attempts to obtain necessary energy from stored fat in the absence of an adequate supply of carbohydrate metabolites. The body liberates large amounts of free fatty acids to override the glucose deficit. As fat flows from the depot stores, and as the long chain fatty acids are received by working cells (including liver cells), they become converted into CoA derivatives in preparation for degradation; the cellular levels of palmityl CoA and other long chain CoA derivatives may increase. Long chain fatty acid CoA

* The unit employed in discussing the energy value of food is the Calorie, equal to 1000 calories or 1 kilocalorie. In the SI system, the unit of energy is the joule (J), and 1 calorie = 4.1868 J.

derivatives inhibit the enzymes that produce oxaloacetic acid from glucose or glycogen; therefore, in conditions that promote excessive release of fatty acids from adipose cells, ketogenesis becomes augmented.

Much of an excess of fatty acids released by adipose cells is cleared from the blood by liver cells. Liver cells, then, are largely responsible for converting fatty acids to acetoacetate. Unlike muscle cells, liver cells cannot metabolize acetoacetate. Ketogenesis occurs by a coupling of acetyl CoA with another molecule of acetyl CoA (instead of with oxaloacetate) in accordance with the following reactions:

$$2\ CH_3{-}\overset{O}{\overset{\|}{C}}{-}CoA \rightarrow CH_3{-}\overset{O}{\overset{\|}{C}}{-}CH_2{-}\overset{O}{\overset{\|}{C}}{-}CoA + \text{coenzyme A}$$

Acetyl CoA **Acetoacetyl CoA**

$$CH_3\overset{O}{\overset{\|}{C}}CH_2\overset{O}{\overset{\|}{C}}{-}CoA + CH_3\overset{O}{\overset{\|}{C}}{-}CoA \rightarrow {}^{-}O{-}\overset{O}{\overset{\|}{C}}CH_2\underset{CH_3}{\underset{|}{\overset{OH}{\overset{|}{C}}}}CH_2\overset{O}{\overset{\|}{C}}{-}CoA + \text{coenzyme A}$$

β-Hydroxy-β-methylglutaryl CoA

$$\swarrow$$

$$CH_3\overset{O}{\overset{\|}{C}}CH_2C\begin{smallmatrix}\nearrow O \\ \searrow O^- \end{smallmatrix} + CH_3\overset{O}{\overset{\|}{C}}{-}CoA$$

Acetoacetate **Acetyl CoA**

The first product, acetoacetyl CoA, condenses with a third molecule of acetyl CoA to yield β-hydroxy-β-methylglutaryl CoA. This product becomes cleaved enzymatically to yield acetoacetate and acetyl CoA. Some of the acetoacetate formed in liver cells usually is reduced to β-hydroxybutyrate.

$$CH_3{-}\overset{O}{\overset{\|}{C}}{-}CH_2{-}COO^- + NADH + H^+ \rightleftharpoons CH_3{-}\overset{OH}{\overset{|}{C}H}{-}CH_2{-}COO^- + NAD^+$$

Acetoacetate **β-Hydroxybutyrate**

Free acetoacetate is quite unstable, and some of it decomposes to form carbon dioxide and acetone. This decarboxylation reaction accounts for the formation of the third ketone body, acetone, frequently observed in severe untreated diabetes mellitus. *Muscle*, *heart*, and *brain* can use ketone bodies carried to them via the blood, because they can resynthesize the CoA derivatives of the acids and subsequently oxidize them for the production of energy.

The entire process of ketosis can be reversed by restoring an adequate level of carbohydrate metabolism. In starvation this consists of adequate carbohydrate ingestion; in diabetes mellitus this consists of insulin administration, which permits the circulating blood glucose to be taken up by the cells, a process that apparently cannot proceed in the absence of the hormone. When oxaloacetate, the acceptor of acetyl CoA, is again available the normal metabolism is restored, and the rush of fatty acids from adipose tissues slows and is finally reversed.

A more comprehensive view of the metabolic reactions we have discussed may be gained from the scheme outlined in Figure 10–4, which relates the main features of carbohy-

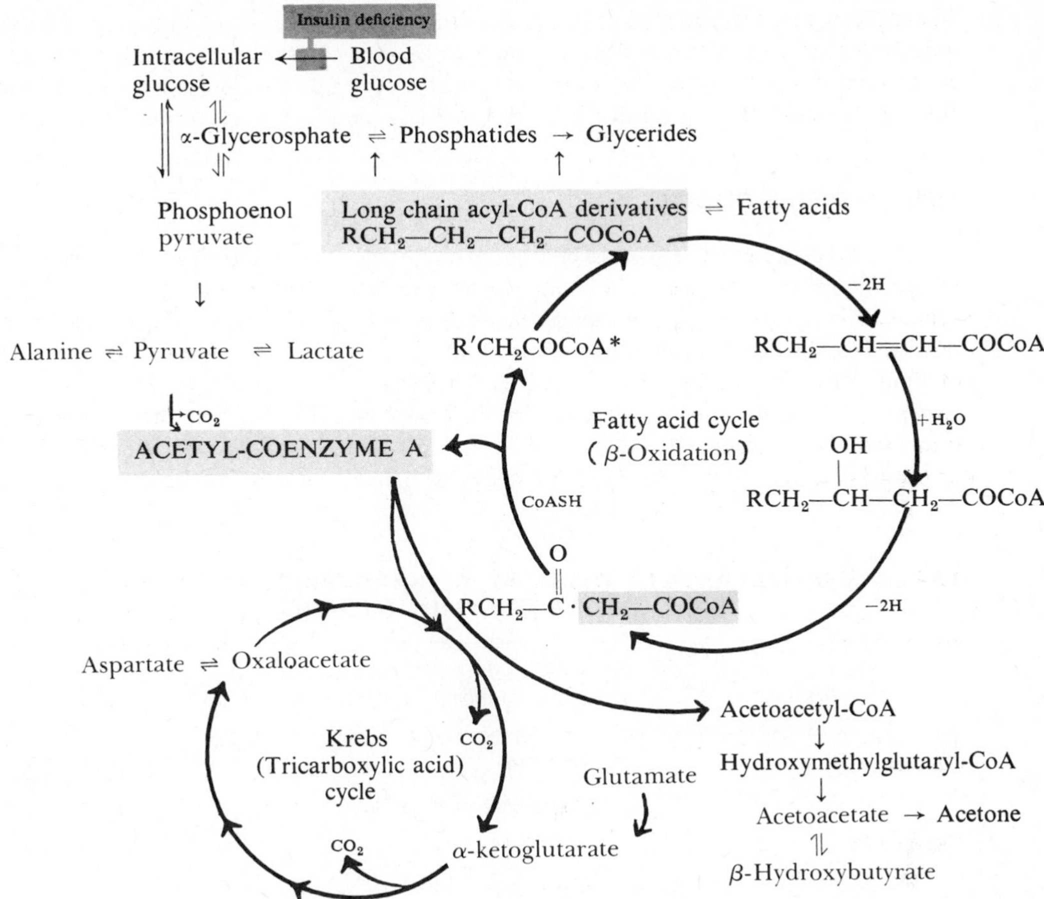

Figure 10–4 Metabolic relations of fats, proteins, and carbohydrates. Note that both fat and carbohydrate metabolism produce acetyl-coenzyme A. Amino acids from proteins may feed by transamination into the Krebs cycle. Other amino acids can be converted to acetyl-coenzyme A also.

drate and lipid metabolism, and which further shows how even amino acids derived from proteins may be incorporated. The student may also note that oxaloacetate can be obtained from aspartate by transamination; similarly, *glutamate* can yield *α-ketoglutarate*, another intermediate in the Krebs cycle.

FATTY LIVER

A second type of abnormality of lipid metabolism involves the accumulation of excessive fat in the liver. Fatty liver is often associated with *alcoholism*, *malnutrition*, *obesity*, and *diabetes mellitus*; however, it may also be caused by extremely high levels of fat in the diet or by such toxic agents as carbon tetrachloride and chloroform. Since these agents are frequently used carelessly, *intoxication by solvents* may occur as an occupational hazard in some industries. One underlying mechanism in the production of fatty liver is an inadequate synthesis of phospholipids that limits hepatic synthesis of lipoproteins and the secretion of lipids from the liver into the blood.

Appreciable amounts of fat absorbed from food are brought to the liver either directly through the portal blood or by the lymphatics as chylomicrons. The liver must then synthesize lipoproteins, which must contain phosphatides, to transport the fat from the liver to the adipose tissue for proper storage. If the phosphatides are lacking in lecithin, or if the total production of phosphatides is insufficient, then the formation of lipoproteins is incomplete, and transport of triglycerides from the liver becomes impaired. *Choline* and *methionine* are known as *lipotropic agents* because they can prevent the fatty liver caused by nutritional deficiency, perhaps by enhancing the formation of the phosphatides essential for the transport of triglycerides from the liver via the plasma lipoproteins.

LIPID STORAGE DISEASES

The lipid storage diseases are characterized by the accumulation of excessive quantities of specific fatty substances in various tissues with attendant malfunction of the involved organs. The nature of the metabolic defects in nine known sphingolipid storage diseases is now well established.[9] In each case, an enzyme deficiency is responsible for the accumulation of lipid. Diagnosis can be established in most cases by relatively simple enzyme assays. Except for Fabry's disease, which is an X-linked abnormality, all these conditions are transmitted as autosomal recessive mutations and are associated with mental retardation or progressive central nervous system impairment. The characteristic lipid that accumulates

TABLE 10–2 METABOLIC DISEASES CHARACTERIZED BY INABILITY TO DEGRADE SPHINGOLIPIDS

Disease	*Major Sphingolipid Accumulated and Position of Enzyme Block*	*Enzyme Defect*
Gaucher	Cer+β-Gluc *Ceramide glucoside* (*glucocerebroside*)	β-Glucosidase
Niemann-Pick	Cer+PChol *Sphingomyelin*	Sphingomyelinase
Krabbe	Cer+β-Gal *Ceramide galactoside* (*galactocerebroside*)	β-Galactosidase
Metachromatic leukodystrophy	Cer-β-Gal+OSO_3^- *Ceramide galactose*-3-*sulfate* (*sulfatide*)	Sulfatidase
Ceramide lactoside lipidosis	Cer-β-Gluc+β-Gal *Ceramide lactoside*	β-Galactosidase
Fabry	Cer-β-Gluc-β-Gal+α-Gal *Ceramide trihexoside*	α-Galactosidase
Tay-Sachs	Cer-β-Gluc-β-Gal+β-NAcGal | NAcNA *Ganglioside* GM_2	Hexosaminidase A
Tay-Sachs variant	Cer-β-Gluc-β-Gal-α-Gal+β-NAcGal *Globoside* (*plus Ganglioside* GM_2)	Total hexosaminidase
Generalized gangliosidosis	Cer-β-Gluc-β-Gal-β-NAcGal+β-Gal | NAcNA *Ganglioside* GM_1	β-Galactosidase

Cer = ceramide; Gluc = glucose; PChol = phosphorylcholine; Gal = galactose; NAcNA = N-acetylneuraminic acid; NAcGal = N-acetylgalactosamine.

in each of the lipid storage diseases is indicated diagrammatically in Table 10–2, adapted from Brady *et al.*[9]

METABOLISM OF CHOLESTEROL

Although the structure of cholesterol is complex, the biosynthesis of the total molecule is accomplished from relatively simple acetyl units. Consequently, many amino acids, carbohydrates, and fatty acids, when supplied in excess of other metabolic needs, can contribute to the cholesterol pool. The liver, skin, adrenals, gonads, intestine, and even the aorta carry out this biosynthesis. It is estimated that the liver can produce 1.5 g per day, and all other tissues together can produce 0.5 g per day. The total amount biosynthesized from acetate is, therefore, about two or three times the amount of preformed cholesterol consumed in the typical American diet.

Elevated levels of cholesterol in blood plasma are recognized as a risk factor toward *atherosclerotic disease*. Attempts to lower plasma cholesterol by decreasing the dietary cholesterol have led to only small changes—decreases of approximately 15 per cent. It is possible to lower the serum cholesterol by diets low in cholesterol if the amount of fat and simple sugars (hexoses and disaccharides) is also curtailed. Orientals, whose diets meet these standards, typically show cholesterol levels some 100 to 150 mg/100 ml lower than their American counterparts. If the Orientals switch to a typical Western diet, the levels of plasma cholesterol rise to typical American levels.

Circulating cholesterol can be eliminated through conversion by the liver to salts of bile acids. Figure 10–5 shows the structures of the major bile acids and their relationships to cholesterol. The conversion of cholesterol to the individual bile salts involves introduction of one or two additional hydroxyl groups, all in the α-configuration, and the partial oxidation of the aliphatic side chain to introduce a carboxyl group. The simple bile salts undergo further modification by condensation with glycine or with taurine to form the so-called conjugated bile salts. For simplicity, the figure shows only the structures of glycocholic and taurodeoxycholic acids. Chenodeoxycholate can undergo similar condensation reactions with glycine and taurine.

The bile salts not only require cholesterol in their own formation, but they also can form molecular complexes with unchanged cholesterol; in so doing, they facilitate the excretion of still more cholesterol through the bile. If the molecular complex between the bile salts and cholesterol is broken down within the gallbladder, as sometimes happens during infectious processes, the cholesterol may deposit about some microscopic nidus to form gallstones, which may grow to the size of large marbles, and which frequently contain 60 to 99 per cent, by weight, of cholesterol.

In the intestinal lumen, the bile salts serve to emulsify ingested fats and thereby promote digestion. During the absorptive phase of digestion, approximately 90 per cent of the bile salts are reabsorbed. Thus the bile salts, produced by the liver, pass to the intestine and are then largely returned to the liver. This process is termed the *enterohepatic circulation*, and, as we have mentioned, it is important in facilitating the excretion of sterols and the digestion of dietary lipids. Failure of the enterohepatic circulation of the bile salts results in malabsorption of lipids and of the fat-soluble vitamins.

PRINCIPLES OF LIPID ANALYSIS

The handling and quantitative analysis of lipids pose some problems not encountered in analyses of other types of materials. Since the lipids are usually water-insoluble, the use of organic solvents is necessary. Some of these present the hazard of flammability. Most lipid solvents evaporate so readily that the concentrations of solutions may change appreciably unless special care is exercised.

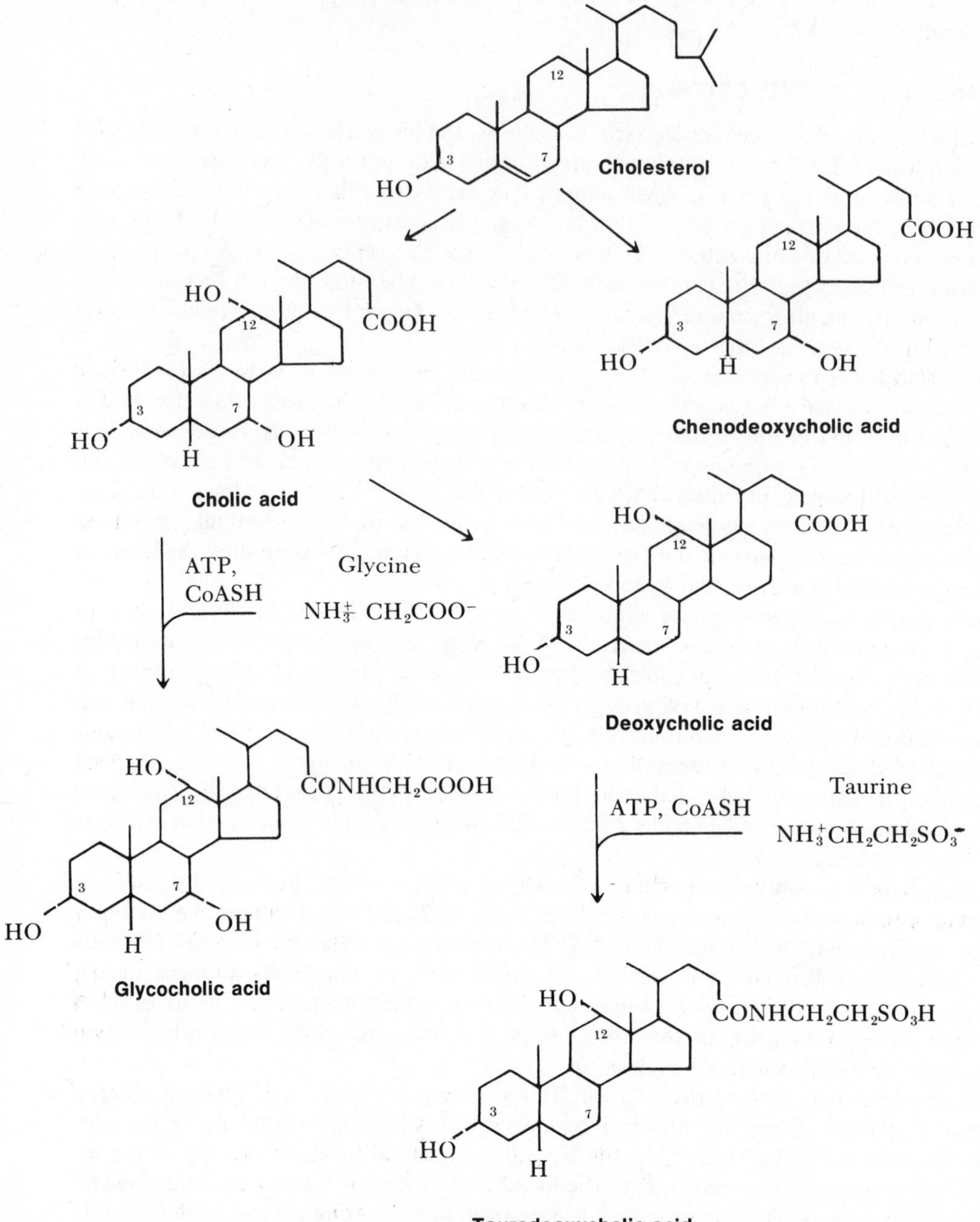
Cholesterol
COOH
Chenodeoxycholic acid
COOH
Cholic acid
ATP,
CoASH
Glycine
NH₃⁺ CH₂COO⁻
COOH
Deoxycholic acid
CONHCH₂COOH
ATP, CoASH
Taurine
NH₃⁺CH₂CH₂SO₃⁻
Glycocholic acid
CONHCH₂CH₂SO₃H
Taurodeoxycholic acid

Figure 10-5 Structures of the major bile acids of man. Note that the OH-group is on one or more of C-3, C-7, and C-12.

Standard solutions of purified lipids and extracts of lipids from biologic specimens can be stored satisfactorily for several days and even for weeks if certain precautions are observed. Losses of solvents by *evaporation* can be insignificant if the storage flasks are tightly capped, and if the solutions are refrigerated during storage. Two types of chemical reactions can affect lipids during storage. *Oxidation* of unsaturated lipids can occur to a significant degree during storage for a few days unless the solution contains an antioxidant; in the experience of one of the authors,[24] ethanol and isopropyl alcohol (but not methanol) have served well as antioxidants. The second type of problem arises from *transesterification.* Ester lipids in solutions that contain methanol can transesterify to form appreciable amounts of methyl esters in a period of a few days; the use of ethanol presents the same problem but to a lesser degree. Use of isopropyl alcohol as a lipid solvent causes less of a problem than does use of methanol and ethanol, because its involvement in transesterification of lipid esters seems to be hindered sterically.[24] The oxidation and transesterification reactions can change certain lipids to the extent that a definitive analysis of the original lipid profile of a specimen becomes impossible.

Lipid solutions that are stored at refrigerator temperature must be warmed to room temperature before aliquots are measured. At refrigerator temperature, the concentrations of solutes are increased as a result of contraction of solvent volumes.

All solvents, including ether, chloroform, methanol, and hexane, should be of at least analytical grade (AR) quality and preferably spectrochemically pure. Ether of suitable grade (analytical reagent or for anesthesia) is supplied in metal cans, which prevent peroxidation of the solvent. Generally other solvents are supplied in brown glass bottles; it is wise to procure these in containers of approximately 500 to 1000 ml so that the contents can be consumed before they deteriorate. Since the purest chloroform is highly unstable, commercial material usually contains a trace of alcohol to inhibit decomposition, and this small amount of additive will not interfere. Hydrocarbon solvents usually are mixtures of pentanes and hexanes (low boiling petroleum ether, b.p. 30 to 60°C) or of hexanes and heptanes (ligroin, b.p. 60 to 80°C). Because these mixtures may be of variable composition and purity, they must be washed with concentrated sulfuric acid and then rinsed with distilled water. The washed product should be dried with anhydrous sodium sulfate and then distilled before use in lipid extraction.

When it is necessary to concentrate solutions of the lipids, solvents should be removed under vacuum at moderate temperature with a rotary evaporator. For certain purposes an alternative method of solvent evaporation may be employed; if the volume is only a few milliliters, the solvent may be swept away with a fine, gentle stream of nitrogen or carbon dioxide. Gentle heating may facilitate this process. This maneuver does not always constitute a satisfactory method, however; it can cause losses of lipids from specimens by sublimation, especially in cases of small specimens.

Removal of precipitates or separation of phases should be performed by centrifugation rather than by filtration, because the evaporation of solvents from the large exposed surfaces of filter papers may lead to considerable change in the concentrations of solutions. Although this can be offset by repeated washing of the papers with fresh solvent, it is quicker and less expensive in the end to employ centrifugal separation. For precise work, collected precipitates can be washed quickly by dispersion with fresh solvent; the wash solution may be added to the original supernatant lipid solution. When separatory funnels are needed, they should be equipped with Teflon plugs, since these are not likely to leak and do not require lubricating grease which might otherwise contaminate the specimens.

A primary requisite for any analysis is a suitable standard substance of high purity and stability; unfortunately, most lipid materials are difficult to purify, and many of them are not as stable as might be desirable. It is currently possible to purchase some lipids of adequate purity from a few sources.* High purity cholesterol is now available from the U.S. National Bureau of Standards as Standard Reference Material No. 911 and is recommended for the preparation of primary standard solutions. Criteria for the purity of cholesterol have been reviewed by Williams *et al.*[66] Cholesterol may be purified by the method of Fieser,[28] which consists of preparing cholesterol dibromide, regenerating the cholesterol by treating the dibromide with zinc dust, and recrystallizing the product from methanol.

---

* Several major sources include: Applied Science Laboratories, Inc., P.O. Box 440, State College, PA. 16801; Sigma Chemical Co., P.O. Box 14508, St. Louis, MO. 63178; Nu Chek Prep, Inc., P.O. Box 172, Elysian, MN. 56028; Supelco, Inc., Supelco Park, Bellefonte, PA. 16823; and Analabs, 80 Republic Drive, North Haven, CT. 06473.

EXTRACTION OF LIPIDS FROM BIOLOGIC SPECIMENS

In preparation for analysis, lipids generally are extracted from specimens into organic solvents. By this process, the lipids become separated from associated proteins and from many other substances that would interfere with the analyses. Selection of the extraction solvent must be based on the types of analyses that are to be performed with the lipid extract. The most complete extraction of lipids can be accomplished with a mixture of an alcohol and either chloroform or an ether.[54] Solvents that have been used extensively for lipid extraction include chloroform-methanol in a variety of proportions (Folch and Lees[29]), ethanol-diethyl ether (Bloor[8]), isopropyl alcohol-diethyl ether (Ellefson[24]), and isopropyl alcohol-chloroform (Ellefson[24]).

Suitable extraction solvents and procedures will be described with the presentation of each of the analytical methods.

TOTAL LIPIDS

The term "total lipids" refers to all the lipid material that can be extracted readily from a specimen. This test can be used as a screening test for hyperlipidemia in place of more specific lipids tests.[27] However, the total lipids test can now be considered obsolete. Information gained from this test has limited clinical usefulness, since it does not indicate in which of the lipid fractions abnormalities may exist. Also, a modest elevation of one of the lipid fractions may not readily be reflected in the total lipid value.

Nevertheless, a method for the determination of total lipids is presented in this chapter for persons who wish to use this test for the screening of patients.

METHODS FOR THE DETERMINATION OF TOTAL LIPIDS IN SERUM

Several types of methods have been employed for the measurement of total lipids. The simplest, but least accurate and least precise method is based on the measurement of turbidity after treatment of the specimen with an appropriate aqueous reagent. Other methods require a preliminary extraction of the lipids and quantitation by micro-oxidation; saponification with titration of the resulting fatty acids; or weighing of the lipid residue from an aliquot of the extract. In addition, a colorimetric method has been recommended, and this procedure is presented here in detail.[18,33,70]

DETERMINATION OF TOTAL LIPIDS USING THE PHOSPHORIC ACID–VANILLIN REACTION

SPECIMEN

Either serum or plasma suffices. This test should be performed within a few hours of the time of drawing of the specimen. If the analysis cannot be performed on the same day, the serum (or plasma) should be separated from the cells (and clot) and frozen. For this test, frozen specimens may be stored for several days.

PRINCIPLE

In this method, the lipid specimen is heated with concentrated sulfuric acid; then vanillin and phosphoric acid are added to yield a pink colored product. The chemical reactions that form the basis of this method remain unknown. One assumption that has been stated in several published papers is that the unsaturated components of a lipid specimen first become oxidized to ketones, and then the ketones condense with vanillin or a derivative of vanillin under the influence of acid catalysis. Following the assumed condensation reaction, dehydration of an aldol-type intermediate is further assumed to yield a more highly unsaturated product that absorbs visible light. Saturated lipids are known to produce little or no color, and the color yields among unsaturated lipids vary. For example, the color yield from oleic acid is greater than the yield from linoleic acid, and both of those acids produce greater color yields than does cholesterol. The variation of color yield, of course, limits the accuracy and usefulness of this test method; however, by employing as a standard a natural, mixed lipid such as olive oil or lipid extracted from a pool of human blood serum, reasonable accuracy can be achieved.

REAGENTS

1. Sulfuric acid, concentrated, AR.
2. Phosphoric acid–vanillin reagent. Dissolve 1.0 g of vanillin in 160 ml of distilled water in a 500 ml volumetric flask. Add concentrated phosphoric acid to the mark, and mix thoroughly. Store this solution in an amber glass bottle in a dark cabinet. The reagent is stable for approximately 6 months.
3. Stock standard, 2400 mg/100 ml. Place 337 mg of tripalmitin (0.418 mmol) and 863 mg of triolein (0.974 mmol) in a 50 ml volumetric flask, dissolve in chloroform, dilute to the mark, and mix thoroughly. This solution must be stoppered tightly and stored in a freezer (−20°C).
4. Working standard, 240 mg/100 ml. Allow stock standard to warm to room temperature, and then transfer 1.0 ml to a 10.0 ml volumetric flask. Dilute to the mark with chloroform, and mix the solution thoroughly. Stoppered tightly and stored in a refrigerator, this solution should be usable for as long as 2 weeks.

PROCEDURE

1. Use a glass-stoppered 16 × 150 mm tube for each specimen, for each standard, and for a water blank.
2. Transfer 200 μl of serum to each specimen tube and 200 μl of water to the blank tube.
3. Transfer 0.5 ml of working standard to another tube. Vaporize the chloroform with a gentle stream of nitrogen; heat the tube in warm water to hasten the evaporation. Add 200 μl of water.
4. Add 5.0 ml of concentrated sulfuric acid to each tube; mix immediately and thoroughly with a vortex mixer.
5. Stopper the tubes loosely and heat at 100 °C for 10 min.
6. Allow the tubes to cool to room temperature. Then transfer 200 μl of the contents of each tube to a clean cuvette (12 × 75 mm).
7. Add 3.0 ml of phosphoric acid–vanillin reagent to each cuvette. Mix thoroughly with a vortex mixer.
8. Allow the cuvettes to stand in the dark for 60 min.
9. Set the absorbance reading at zero with the water blank, and determine the absorbances of the specimens and standard(s) at 520 nm.

CALCULATIONS

$$\text{mg total lipids/100 ml serum} = \frac{A_u}{A_s} \times C_s \times \frac{0.5}{0.2}$$

$$= \frac{A_u}{A_s} \times 600$$

COMMENTS

1. Specimens that have total lipid concentrations over 1000 mg/100 ml should be diluted 4-fold with 0.85 g/100 ml sodium chloride and re-analyzed.

2. Specimens that contain more than 600 mg total lipids/100 ml should be analyzed for total cholesterol and triglycerides concentrations.

NORMAL VALUES

Normal values range from 400 to 700 mg/100 ml.

REFERENCES

Zollner, N., and Kirsch, K.: Ztschr. Ges. Exper. Med., *135*:545–561, 1962.
Drevon, B., and Schmit, J. M.: Bull. Trav. Soc. Pharm. (Lyon), *8*:173–178, 1964.
Frings, C. S., and Dunn, R. T.: Am. J. Clin. Path., *53*:89–91, 1970.

FATTY ACIDS IN PLASMA

Knowledge of the level of nonesterified fatty acids (FA) in blood plasma can be helpful in the diagnosis and management of certain diseases and disorders of metabolism and in the evaluation of prospective causes of hyperlipoproteinemia. A variety of "lipoactive" hormones can stimulate the release of fatty acids into the blood from adipose cells. These hormones include *epinephrine*, *norepinephrine*, *ACTH*, *thyrotrophin*, *growth hormone*, and *glucagon*. Suppression of the release of fatty acids by adipose cells can be effected by *insulin* and *glucose* in combination. A disorder or disease that causes an excessive release of a lipoactive hormone can induce an elevation of the blood level of fatty acids. In persons afflicted with *diabetes mellitus*, deficiency in normal insulin may result in uncontrolled release of fatty acids from adipose cells and thus in an increased level of FA in the plasma. In acute fasting, or starvation, the FA levels may rise to as much as three times the normal value.

A sustained release of fatty acids from adipose cells in excess of energy needs can contribute to the development of secondary hyperlipoproteinemia. In such cases, excessive fatty acids are cleared from the blood by the liver and in part are secreted back into the blood as lipoprotein-bound glyceride lipids. An excessive release of fatty acids from adipose cells, therefore, can cause an increase in formation and secretion of lipoproteins by the liver. High levels of FA in plasma have also been reported in patients with serious cardiac arrhythmias; however, a direct role for fatty acids in the production of cardiac arrhythmias has not been established.[35]

METHODS FOR THE DETERMINATION OF FATTY ACIDS IN PLASMA

The levels of fatty acids in the serum or plasma can be determined simply by a microtitration with a dilute solution of a strong base. The method of Dole[19] is very satisfactory. Fatty acids are extracted from acidified plasma by a mixture of heptane and isopropyl alcohol. The purified extract is titrated with dilute alkali, using an ethanolic solution of thymol blue or thymolphthalein as an indicator. During the titration, the system must be protected from exposure to atmospheric carbon dioxide, which would also consume the dilute alkali. This is accomplished by bubbling a stream of alkali-washed nitrogen through the titration tube, which also stirs the system to ensure thorough mixing of the added alkali. The alcohol promotes the contact of fatty acid with the aqueous base, since it is a solvent for both.

Colorimetric methods, based on the formation of *cupric soaps* of FA, also have been described.[21] The soaps are extracted into a mixture of chloroform-heptane-methanol, and the amount of bound cupric ion is determined colorimetrically after reaction of the copper with a suitable reagent, such as sodium N,N-diethyldithiocarbamate.

SPECIMEN

Either serum or heparinized plasma constitutes a satisfactory specimen. The serum or plasma should be separated and analyzed without delay, since the presence of small amounts of lipase permits the occurrence of slow *in vitro* lipolysis with the evolution of newly formed free fatty acids. This process occurs more rapidly if heparin has been given to the subject from whom blood is drawn. Generally, specimens from heparinized persons are not satisfactory for this test. Heparin added as an anticoagulant *in vitro* has no effect. The values in paired plasma and serum samples indicate that the serum levels are substantially the same as plasma levels if sera are separated promptly.

Situational stress, of the kind that may arise during drawing of the blood sample, can cause a rapid increase in plasma FA. This may be an important factor in children, or in patients who are not yet accustomed to hospital procedures. The sample should be drawn in the fasting state; ingestion of food induces a decrease in plasma FA.

DETERMINATION OF FATTY ACIDS BY TITRATION

REAGENTS

1. Extraction mixture. Mix 40 vol of isopropyl alcohol, 10 vol of heptane, and 1 vol of 0.5 molar sulfuric acid.
2. Thymol blue indicator. Dissolve 0.1 g of thymol blue in 100 ml of redistilled ethanol.
3. Sodium hydroxide, 18 mmol/l. Dilute 0.9 vol of 0.10 molar sodium hydroxide with 4.1 vol of freshly boiled, distilled water. Check concentration by titrating 0.10 molar hydrochloric acid to the phenolphthalein end point.
4. Titration mixture. Mix 10 ml of 0.1 per cent thymol blue solution (ethanolic) in 90 ml of ethanol; add 18 mmolar sodium hydroxide until the blank (see Procedure, step 2) titrates between 5 and 10 microliters of titration mixture. This mixture will absorb CO_2 from the air during storage and, as a result, will increase in acidity.
5. Palmitic acid, stock standard, 3.33 mmol/l. Dissolve 85.47 mg of palmitic acid in heptane and dilute the solution to 100 ml.
6. Palmitic acid, working standard, 0.333 mmol/l. Dilute 10 ml of stock standard to 100 ml with heptane.

PROCEDURE

1. *Extraction.* Transfer 1.00 ml or 2.00 ml of serum to a 20 ml glass-stoppered tube and dilute with water to 3.00 ml. Add 5.0 ml of extraction mixture, shake thoroughly, and then let stand for 10 min. Add 3.0 ml of heptane and shake thoroughly.
2. *Titration.* Transfer 3.0 ml of the upper phase fluid (heptane phase) from the extraction tube into a 15 ml conical tube that contains 1.0 ml of titration mixture. Purge the mixture with a gentle stream of nitrogen from a capillary delivery tube. The tip of the capillary tube should extend to the bottom of the conical tube. With a microburette, titrate the mixture with 18 mmolar sodium hydroxide to a yellow-green end point. Titration blanks must be measured for each set of titrations. For this purpose, distilled water is substituted for the serum specimen, and the described extraction and titration procedures are followed.
3. *Standardization.* In the extraction step, substitute 3.0 ml of working standard for the 3.0 ml of heptane, substitute distilled water for the specimen and proceed as described above. The 3.0 ml of working standard contains 1.0 μmol of fatty acid.

CALCULATION

$$\text{mmol FA/l serum} = (V_u - V_b) \times F$$

$$F = \frac{1.00\ \mu\text{mol}}{(V_s - V_b)} \times \frac{1000\ (\text{ml/l})}{\text{ml specimen}} \times \frac{1\ \text{mmol}}{1000\ \mu\text{mol}}$$

$$F = \frac{1.00}{(V_s - V_b) \times \text{ml specimen}}$$

where V_u, V_s, and V_b represent titration values of unknown, standard, and blank, respectively.

COMMENTS

1. The ultramicroburet used for titrating FA must be graduated in 0.1 μl increments if satisfactory accuracy is expected. It is helpful to fit the outlet of the syringe with a fine hypodermic needle (26 gauge) over which thin-wall polyethylene tubing may be slipped. The tubing is cut to a length sufficient to reach approximately to the bottoms of the centrifuge tubes in which the FA samples are contained.

2. The nitrogen that stirs the solution and serves as an inert blanket must be washed with alkali to remove any traces of carbon dioxide. This is accomplished by a wash bottle of NaOH (10 g/100 ml) through which the gas is passed. The outlet of the wash bottle should be protected with a cotton or glass wool plug to entrap any chance droplets of alkali. The washed gas is then delivered into the FA solution via a fine polyethylene tube.

NORMAL VALUES

The normal adult range is 0.30 to 0.90 mmol/l. Values for children and obese adults are slightly higher, with an upper limit of normal of 1.10 mmol/l.

REFERENCE

Dole, V. P.: J. Clin. Invest., *35*:150, 1956.

TRIGLYCERIDES

Elevated levels of both cholesterol and triglycerides in plasma have been identified as risk factors related to atherosclerotic disease. The hyperlipidemias can be inherited traits, or they can be secondary to a variety of disorders or diseases including diabetes mellitus, nephrosis, biliary obstruction, and metabolic disorders associated with endocrine disturbances. The plasma levels of cholesterol and triglycerides can vary independently; therefore, evaluation of hyperlipidemias should include determinations for both of these lipids.

SPECIMEN

The level of triglycerides in plasma is slightly higher than that in serum, possibly because of entrapment of chylomicrons and lipoproteins during the clotting process, and perhaps also because of lipolysis during clot formation. For most purposes, however, either plasma or serum is satisfactory. Triglycerides reach a maximal level in the plasma approximately 4 to 6 h postprandially. Values will increase even if only pure carbohydrate is ingested. For reliable interpretation, blood specimens should be drawn after the patient has fasted for at least 12 h.

METHODS FOR THE DETERMINATION OF TRIGLYCERIDES

In one colorimetric method, lipids are extracted and the phospholipids are removed by adsorption onto zeolite, silicic acid, or magnesium silicate. The remaining esters are converted to hydroxamic acid derivatives by treatment with an alkaline solution of hydroxylamine.[53] The hydroxamates form a red to violet color when treated with ferric salts in a weak acid solution. A correction must be applied for the color contributed by cholesteryl esters. This method is difficult to use, and accuracy and precision are not readily achieved.

Another approach involves hydrolysis of the triglycerides, either enzymatically or with alkali, to liberate glycerol. The glycerol can then be measured by enzymatic, colorimetric, or fluorometric procedures.

Two *enzymatic* methods exist for the analysis of serum triglycerides. One method (E-1)[22] employs alkaline hydrolysis of the triglycerides and the following sequence of coupled enzymatic reactions:

$$\text{glycerol} + \text{ATP} \xrightleftharpoons{\text{glycerol kinase}} \text{sn-glycerol-3-phosphate} + \text{ADP}$$

$$\text{ADP} + \text{phosphoenolpyruvate} \xrightleftharpoons{\text{pyruvate kinase}} \text{ATP} + \text{pyruvate}$$

$$\text{pyruvate} + \text{NADH} + \text{H}^+ \xrightleftharpoons{\text{lactate dehydrogenase}} \text{lactate} + \text{NAD}^+$$

The amount of NADH consumed is equivalent to the amount of glycerol (free or derived from triglyceride) present in the specimen. The decrease in NADH in the test mixture is determined by measuring the absorbance at 340 nm or at 366 nm.

The second enzymatic method (E-2) employs lipase hydrolysis of the triglycerides and the following sequence of reactions:

$$\text{glycerol} + \text{ATP} \xrightleftharpoons{\text{glycerol kinase}} \text{sn-glycerol-3-phosphate} + \text{ADP}$$

$$\text{sn-glycerol-3-phosphate} + \text{NAD}^+ \xrightleftharpoons{\text{glycerol phosphate dehydrogenase}} \text{dihydroxyacetone phosphate} + \text{NADH} + \text{H}^+$$

In this method the amount of NADH formed is equivalent to the amount of glycerol (free or derived from triglyceride) present in the specimen.[65] Both enzymatic methods are available in kit form, and, in fact, the kit is the only form in which the lipase purified for triglyceride analysis is available at this time. Experience with the system using alkaline hydrolysis (E-1) has been more satisfactory than experience with the other system (E-2). The E-1 system produces values that are quite consistent with values obtained with colorimetric and fluorometric methods, but are uniformly higher by 10 to 20 mg/100 ml of serum. In principle, method E-1 is superior, because the equilibrium of each reaction is favorable. In method E-2, the equilibrium of the glycerol phosphate dehydrogenase reaction is unfavorable, and in order to drive the reaction to completion, a trapping agent (hydrazine) for the dihydroxyacetone phosphate must be used. Method E-1 will be described.

Two *colorimetric* methods have been used extensively for the determination of glycerol liberated from triglycerides by alkaline hydrolysis. Both methods require a preliminary separation of the triglycerides from the phospholipids, sugars, amino acids, and proteins. This separation can be accomplished by column chromatography or by extraction of lipids into an organic solvent and removal of the soluble interfering substances by selective adsorption. Also, direct partition of serum lipids with a mixture of nonane, isopropyl alcohol, and dilute sulfuric acid may be used; the triglycerides are transferred to the nonane phase, while the phosphoglycerides and other interfering substances remain in the aqueous-isopropyl alcohol phase.[43] Both methods include alkaline hydrolysis of the isolated triglycerides and oxidation of the glyceride glycerol to formaldehyde and formic acid.

In the method of Van Handel and Zilversmit,[59] serum lipids are extracted with chloroform-methanol, and the phospholipids are removed by adsorption onto a zeolite (a preparation of aluminosilicates). Triglycerides are hydrolyzed, and the glycerol is oxidized with periodic acid to produce formaldehyde. When formaldehyde is treated with a sulfuric acid solution of chromotropic acid (4,5-dihydroxy-2,7-naphthalenedisulfonic acid), a pink colored derivative of unknown composition is formed. The absorbance of the colored solution is proportional to the amount of formaldehyde. The procedure, with modifications, has been proposed as a *standard colorimetric method.* In general, the method requires meticulous attention to details and is quite time-consuming for routine manual determinations.

A second, and more widely used, colorimetric procedure involves the *Hantzsch condensation reaction* in which formaldehyde combines with ammonium ion and acetylacetone to produce a yellow colored diacetyl lutidine with maximal absorbance at 412 nm. The lutidine derivative is also fluorescent and forms the basis of a *fluorometric* procedure especially suitable for automated analyses.[40] Fluorescence can be measured by using a 400 nm primary filter and a 485 nm secondary filter.

$$H{-}\overset{O}{\overset{\|}{C}}{-}H + NH_4^+ + 2\,CH_3{-}\overset{O}{\overset{\|}{C}}{-}CH_2{-}\overset{O}{\overset{\|}{C}}{-}CH_3$$

Formaldehyde **Acetylacetone (2,4-pentanedione)**

$$\downarrow$$

$$\text{3,5-diacetyl-1,4-dihydrolutidine } (H_3C{-}CO{-}C{=}C(CH_3){-}NH{-}C(CH_3){=}C{-}CO{-}CH_3 \text{ ring with } CH_2) + 3\,H_2O + H^+$$

3,5-Diacetyl-1,4-dihydrolutidine

Both the colorimetric and fluorometric methods have been incorporated into commercially available kits designed for manual determinations.

DETERMINATION OF TRIGLYCERIDES BY AN ENZYMATIC METHOD (E–I)[22]

REAGENTS

1. Ethanolic KOH, 0.5 mol/l. Dissolve 3.3 g of KOH pellets (AR, assay 85 per cent) in 10 ml of water. Cool and dilute to 100 ml with absolute ethanol.

2. Magnesium sulfate, 0.15 mol/l. Dissolve 3.7 g of $MgSO_4{\cdot}7H_2O$ in water and dilute to 100 ml.

3. Triethanolamine buffer, 0.1 mol/l, containing 4 mmolar magnesium sulfate. Dissolve 14.9 g of triethanolamine and 0.98 g of $MgSO_4{\cdot}7H_2O$ in water and dilute to 1 liter.

4. NADH-ATP-PEP reagent: NADH, 0.006 mol/l; ATP, 0.033 mol/l; PEP, 0.011 mol/l. Dissolve 42.6 mg of NADH disodium salt, 180 mg of ATP disodium salt, and 21 mg of phosphoenolpyruvate monosodium salt in 10.0 ml of water.

5. LDH-PK reagent, lactate dehydrogenase, 2 mg/ml; pyruvate kinase, 1 mg/ml. Combine appropriate aliquots of respective stock suspensions and dilute with water to the concentrations indicated.

6. GK reagent, glycerol kinase, 2 mg/ml. Dilute an appropriate aliquot of stock suspension with water to the concentration indicated.

7. Triglyceride standard, 100 mg/100 ml. Dissolve 100 mg of triolein (99+% pure) in absolute ethanol and dilute to 100 ml.

PROCEDURE

1. *Alkaline hydrolysis.* Transfer 200 μl of serum into a glass-stoppered test tube and add 0.5 ml of ethanolic KOH. Stopper the tube, mix the contents, and incubate at 60°C for 30 min. Allow tubes to cool to room temperature.

2. *Hydrolysis of standards.* Transfer 200, 400, 600, and 800 μl aliquots of the standard solution to glass-stoppered tubes, and vaporize the ethanol under a gentle stream of dry nitrogen. Add 200 μl of water and 0.5 ml of ethanolic KOH. Stopper the tubes, mix gently, and incubate at 60°C for 30 min. Allow the tubes to cool. These standards correspond to 100, 200, 300, and 400 mg/100 ml of triglycerides in serum.

3. *Reagent blank.* Pipet 200 μl of water into a glass-stoppered tube and add 0.5 ml of ethanolic KOH. Treat the blank tube in the same way as the specimen and standard tubes.

4. *Specimen blank.* Pipet 200 μl of serum into a glass-stoppered tube and add 0.5 ml of 90 per cent ethanol. Do not add alkali; treat the specimen blank in the same way as the specimen and standard tubes from step 5 on.

5. Add 1.0 ml of magnesium sulfate solution to each tube, mix thoroughly, and centrifuge.

6. Transfer 0.50 ml aliquots of supernatant fluids to respective cuvets. To each cuvet add 2.5 ml of buffer, 0.10 ml of NADH-ATP-PEP solution, and 20 μl of the LDH-PK suspension. Mix gently but thoroughly, and let stand at room temperature for 10 min.

7. Measure absorbance (A_1) at 340* nm or at 366 nm after setting the spectrophotometer to zero with a water blank.

8. Pipet 0.02 ml of GK reagent into each cuvet, mix gently, and let the mixture stand 10 min or until there is no further decrease in absorbance. Measure the absorbance (A_2). When absorbance exceeds 0.400 at 366 nm or 0.800 at 340 nm, mix 0.10 ml of the supernatant fluid from step 5 with 0.90 ml of dilute NaCl solution (9 g/l) and repeat steps 6 to 8 with 0.5 ml of the diluted fluid. Final results are multiplied by 10 if diluted fluid is used.

CALCULATIONS

Calculate ΔA for the reagent blank (b'), specimen blank (b), standards (s), and unknown (u).

$$\Delta A = A_1 - A_2$$

$$\text{Triglycerides, mg/100 ml} = \frac{(\Delta A_u - \Delta A_b)}{(\Delta A_s - \Delta A_{b'})} \times C_s$$

where C_s is the concentration of the standard. Plot the blank-corrected ΔA against concentration for each of the standards. If the curve is a straight line, the calculations for the concentrations of unknowns can be based on the concentration and value of ($\Delta A_u - \Delta A_b$) for any one of the standards. If the curve is not straight, the concentrations of the unknowns should be determined directly from the curve.

* The wavelength of 340 nm is optimal, as it is the wavelength of maximal absorbance. Measurement may also be made at 366 nm, which in less expensive instruments is necessary if the light source is a mercury vapor lamp that has little radiant output at 340 nm. Reading at the longer wavelength involves some loss of sensitivity.

COMMENTS

1. The buffer solution is stable for approximately 1 year. The solution of NADH, ATP, and PEP can be used for approximately 2 weeks if stored at 4°C. The solution of LDH and PK and the solution of GK can be stored at 4°C for approximately 1 year.

2. Alcoholic KOH converts triglycerides to fatty acid salts and glycerol, but it converts phosphoglycerides to fatty acid salts and glycerophosphates. The glycerophosphates have been shown to be stable in alkaline solution. The enzyme glycerol kinase (GK) is specific for glycerol and does not act on glycerophosphates; therefore, the phosphoglycerides do not interfere with this test.

3. Magnesium sulfate is added to precipitate liberated fatty acids and prevent turbidity in subsequent steps of the procedure.

4. Reagent blanks must be used with the standards. For determinations of greatest accuracy, a serum blank for each specimen must be carried through the procedure. Without the use of serum blanks, the free glycerol of the specimen will contribute to the triglyceride-glycerol value and will be a cause of variable inaccuracy.

The content of free glycerol in most specimens is equivalent to 5 to 20 mg triglycerides/100 ml. However, certain diseases or drugs may cause higher serum free glycerol levels. Thus, corrections for free glycerol by subtracting a fixed amount, such as 5 per cent of the total triglyceride, may lead to significant inaccuracies. High levels of free glycerol have been observed frequently in control sera.

NORMAL VALUES

There is considerable disagreement over the normal values for serum triglycerides. For the above method, Eggstein[22] reports values of 40 to 164 mg/100 ml for males, and 35 to 235 mg/100 ml for females. Others, however, have reported that normal values are higher in males than in females. We suggest the use of the following values that are based on age and sex (lower 95 percentile):

	Total Triglycerides, mg/100 ml		
Age, Years	*Males and Females* (Fredrickson & Levy[31])	*Males* (Ellefson[24])	*Females* (Ellefson[24])
0–19	140		
20–29	140	157	101
30–39	150	182	111
40–49	160	193	122
50–59	190	197	133

Conversion of mg to mmoles

Results may vary slightly depending on the method used and the kind of standard employed. Expressing results in mmol/l, instead of mg/100 ml, makes the values independent of the molecular mass of the standard and of the considerable variation in the fatty acid composition of serum triglycerides. Assuming an average molecular mass of 882 for the triglycerides in serum, a triglycerides concentration of 1.00 mmol/l is equivalent to 88.2 mg/100 ml.

REFERENCES

Eggstein, M., and Kreutz, F. H.: Klin. Wochenschr., *44*:262, 1966.

Fredrickson, D. S., and Levy, R. I.: Familial hyperlipoproteinemia. *In* The Metabolic Basis of Inherited Disease, 3rd ed. Stanbury, J. B., Wyngaarden, J. B., and Fredrickson, D. S., Eds. McGraw-Hill Book Co., 1972, p. 546.

Ellefson, R. D.: Unpublished data.

DETERMINATION OF TRIGLYCERIDES BY A COLORIMETRIC METHOD

The following procedure is acceptable as a reference method; however, it is not recommended as a routine test procedure, because it lacks convenience and requires much time.[50]

REAGENTS

1. Chloroform, AR.
2. Methanol, absolute.
3. Chloroform-methanol mixture. Mix 64 ml of chloroform and 34 ml of methanol.
4. Sodium chloride, saturated aqueous solution. Add 400 g of NaCl, AR, to 1000 ml of distilled water; use the clear supernatant solution.
5. Silicic acid, Mallinckrodt, 100 mesh, suitable for chromatographic analysis. Activate by heating at 120°C for 3 h. Store in a tight container.
6. Ethanolic potassium hydroxide solution. Dissolve 0.4 g of KOH pellets, AR, in 95 per cent ethanol and dilute to 100 ml.
7. Sulfuric acid, 0.1 mol/l. Add 3.0 ml of concentrated H_2SO_4, AR, to 450 ml of water and dilute to 500 ml.
8. Sodium metaperiodate solution. Dissolve 0.5 g of $NaIO_4$, AR, in water and dilute to 100 ml.
9. Sodium bisulfite solution. Dissolve 5.0 g $NaHSO_3$, AR, in water and dilute to 100 ml.
10. Chromotropic acid reagent. Dissolve 1.14 g of chromotropic acid, disodium salt in 100 ml of water. In another vessel add carefully 300 ml of concentrated H_2SO_4, AR, to 150 ml of water; cool to room temperature. Cool the chromotropic acid solution in ice, and slowly add the solution of sulfuric acid; mix thoroughly during the addition of H_2SO_4. Store at 4°C in a dark bottle.
11. Thiourea solution. Dissolve 7 g of thiourea in water and dilute to 100 ml.
12. Triglyceride standard solution. Dissolve 100 mg of tripalmitin (123.9 μmol) or triolein (113.0 μmol) in chloroform and dilute to 100 ml. (Triolein has a molecular mass closer to that of serum triglycerides, while tripalmitin is not subject to hydroperoxide formation.)

PROCEDURE

1. Transfer 200 μl aliquots of serum to tubes of 12 to 15 ml capacity with Teflon-lined screw caps. Add 9.8 ml of the chloroform-methanol mixture, stopper the tubes, and mix thoroughly with a vortex mixer or by shaking. Re-mix the contents of the tubes after 10 min and again after 20 min. After 30 min centrifuge the tubes at approximately 2000 rpm for 10 min.

A reagent blank tube should be prepared by substituting 200 μl of 0.9 per cent sodium chloride solution (9 g/l) for the serum.

Prepare standard tubes with 0.20, 0.40, 0.60, 0.80, and 1.00 ml aliquots of standard solution, representing 100, 200 300, 400, and 500 mg/100 ml. Vaporize the chloroform with a gentle stream of nitrogen. Add 200 μl of 0.9 per cent sodium chloride solution and 9.8 ml of chloroform-methanol mixture to each tube; cap and treat in the same way as the specimen tubes.

2. Transfer 4.0 ml aliquots of the lipid extracts to 15 ml tubes with Teflon-lined screw caps, each containing 8.0 ml of the saturated sodium chloride solution. Cap the tubes, shake vigorously, let stand for 1 h, and centrifuge at 2000 rpm for 10 min. Draw off the upper (aqueous) layer of fluid.

3. Transfer the washed chloroform solutions to dry screw-cap tubes. Use transfer pipets and be careful to avoid transferring droplets of aqueous phase. Add 0.2 g of acti-

vated silicic acid to each tube, cap, a . snake gently. Mix again at 10 min, 20 min, and 30 min. Centrifuge at 2000 rpm for 10 min.

4. Transfer 0.50 ml aliquots of the supernatant fluids to 18 × 150 mm glass-stoppered test tubes, and vaporize the chloroform in a gentle stream of nitrogen or carbon dioxide.

5. Add 0.5 ml of ethanolic KOH to each tube. Stopper the tubes, mix, and heat at 65°C for 30 min.

6. Add 0.5 ml of 0.1 molar H_2SO_4, and heat in a boiling water bath for about 10 min.

7. Cool tubes to room temperature, and add 100 μl of sodium metaperiodate solution. Mix, and allow the mixtures to stand 10 min. Time this interval carefully.

8. Add 100 μl of sodium bisulfite solution. Mix immediately. Let the mixtures stand 10 min.

9. Add 5.0 ml of the chromotropic acid reagent to each tube, mix thoroughly with a vortex mixer, and heat the tubes at 100°C for 30 min (use a thermostated aluminum block heater). Cool to room temperature.

10. Add 0.50 ml of the thiourea solution. Mix thoroughly with a vortex mixer.

11. Determine the absorbance (A) at 570 nm. (The color is stable for at least 2 h).

CALCULATIONS

$$\text{mg triglycerides/100 ml serum} = \frac{A_u}{A_s} \times C_s \text{ (in mg/100 ml)}$$

or

$$\text{mmoles triglycerides/l serum} = \frac{A_u}{A_s} \times C_s \text{ (in mmol/l)}$$

COMMENTS

When the absorbance exceeds 0.700, the serum must be diluted 2, 5, or 10-fold with 0.15 molar NaCl solution before measuring the 200 μl aliquot for the test.

REFERENCE

Rice, E. W.: Triglycerides ("neutral fat") in serum. *In* Standard Methods of Clinical Chemistry. R. P. MacDonald, Ed. Academic Press, New York, 1970, Vol. 6, p. 215.

DETERMINATION OF TRIGLYCERIDES BY THE HANTZSCH CONDENSATION REACTION

The following procedure represents an abbreviated, manual, colorimetric method that is practical for routine testing.[43] It is based on a selective extraction of triglycerides and on the Hantzsch reaction for formaldehyde. Formaldehyde derived from triglyceride-glycerol has been reported to be the only significant chromogen in the colorimetric reaction; therefore, *specimen blanks* appear to be unnecessary.

REAGENTS

1. Nonane,* 99+ per cent pure.
2. Isopropyl alcohol, AR.†
3. Sulfuric acid, 0.04 mol/l. Add 4.5 ml of concentrated sulfuric acid to 500 ml of distilled water; dilute this solution to 2.0 liters.
4. Sodium methylate, 9.3 mmol/l. Dissolve 50 mg in 100 ml of isopropyl alcohol.

---

* Heptane may be substituted for the much more expensive nonane.

† See purification of isopropyl alcohol, p. 505.

Prepare this on the day of its use, and keep stoppered to protect it from carbon dioxide and atmospheric moisture.

5. Sodium metaperiodate, 0.06 mol/l in 0.875 molar acetic acid. Dissolve 1.283 g of $NaIO_4$ in 0.88 molar acetic acid (5.0 ml of glacial acetic acid diluted to 100 ml with distilled water) and dilute to 100 ml with additional 0.88 molar acetic acid. Store this reagent in an amber glass bottle.

6. Ammonium acetate, 2 mol/l. Dissolve 154 g of ammonium acetate in distilled water and dilute the solution to 1 liter.

7. Acetylacetone reagent. Mix 1.50 ml of acetylacetone (2,4-pentanedione) with sufficient 2 molar ammonium acetate to yield a volume of 200 ml.

8. Stock standard. Dissolve 2.500 g of triolein (99+ per cent pure) in isopropyl alcohol, and dilute to 250 ml.

9. Working standards. Prepare solutions containing 50, 100, 200, 300, and 400 mg of triolein/100 ml by diluting 2.5, 5.0, 10.0, 15.0, and 20.0 ml aliquots of the stock standard to 50.0 ml with isopropyl alcohol.

PROCEDURE

1. Place 0.5 ml of each specimen into a 16 × 150 mm tube with a Teflon-lined screw cap. Place 0.5 ml of water into a reagent blank tube and in each standard tube.

2. Place 0.5 ml of working standard into each standard tube.

3. Add 3.0 ml of isopropyl alcohol to each specimen tube, to the reagent blank tube, and to each standard tube.

4. Add an additional 0.5 ml of isopropyl alcohol to each specimen tube and to the blank tube, but *not to the standard tube.*

5. Add 1.0 ml of 0.04 molar sulfuric acid and then 2.0 ml of nonane to each tube.

6. Mix the contents of each tube for 20 s in a vortex mixer.

7. Transfer 200 μl of the nonane layer from each tube to 3.0 ml of sodium methylate solution. Mix with the vortex mixer, and heat the tubes at 60°C for 15 min with an aluminum block tube heater.

8. Add 100 μl of sodium metaperiodate reagent and 1.0 ml of acetylacetone reagent. Mix immediately with a vortex mixer, and continue to heat at 60°C for another 10 min.

9. Centrifuge the tubes, and then transfer the upper phase fluid to 10 × 75 mm cuvettes (or similar, small cuvettes).

10. Measure the absorbances at 412 nm; set the absorbance at zero with the reagent blank.

11. Plot a standard curve with absorbance (A) along the ordinate and mg triglyceride/100 ml along the abscissa. The triglycerides concentrations of the unknowns can be read directly from the standard curve.

CALCULATIONS

If the standard curve proves to be a straight line that passes through the origin, then the triglycerides concentrations of the unknowns can be calculated with the following expression:

$$\text{mg triglycerides/100 ml serum} = \frac{A_u}{A_s} \times C_s$$

A_s and C_s are the absorbance and concentration of *one* of the standard solutions.

COMMENT

1. Specimens in which the triglycerides concentrations exceed 300 mg/100 ml should

be diluted with aqueous sodium chloride (9 g/l) so that the final concentrations are 300 mg/100 ml or less.

DETERMINATION OF TRIGLYCERIDES BY A SEMI-AUTOMATED FLUOROMETRIC METHOD[24,40]

The following procedure is outlined to illustrate a semi-automated method for use with the *Technicon AutoAnalyzer* (Modified N-78 method). The instrument has a redesigned

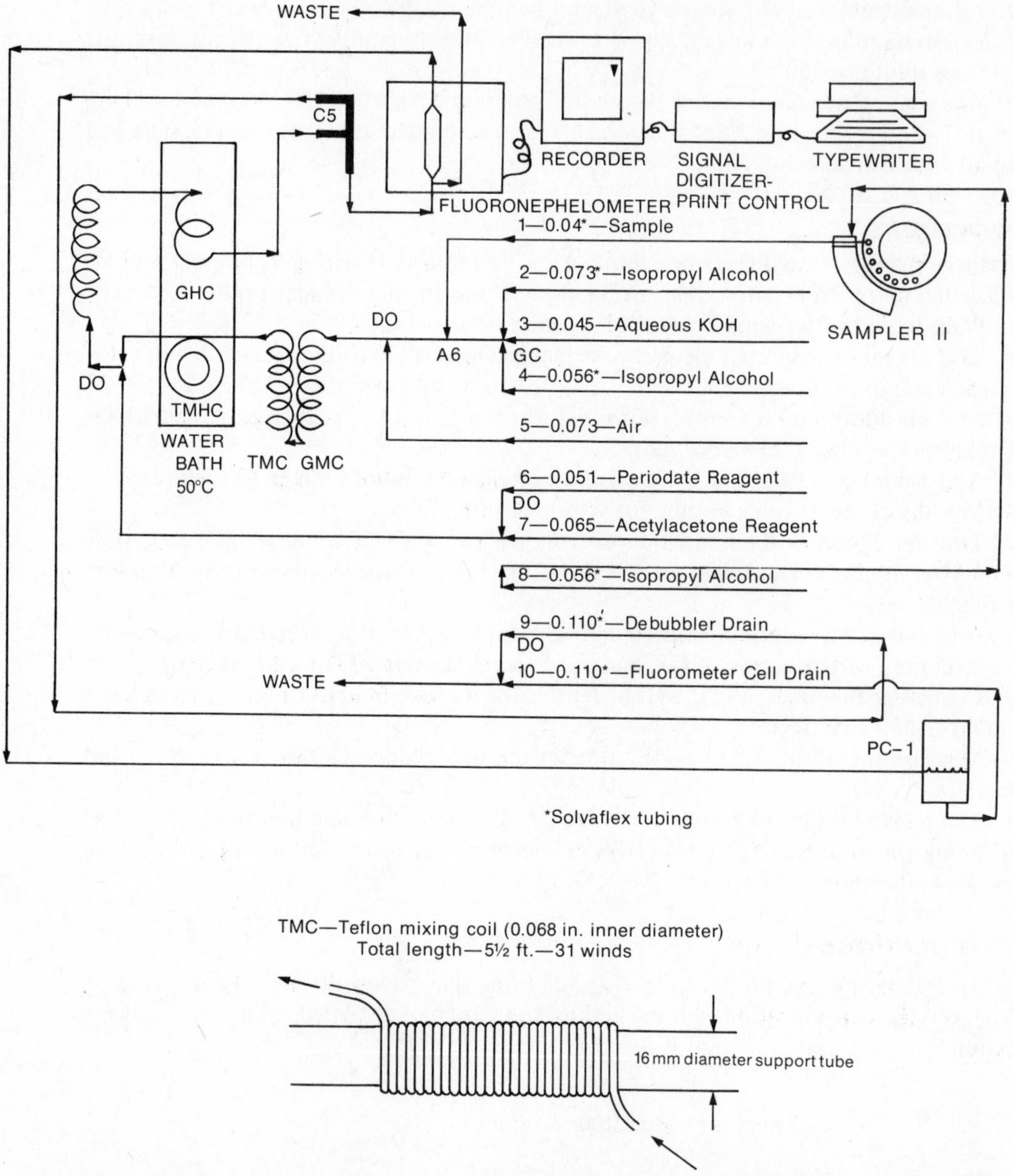

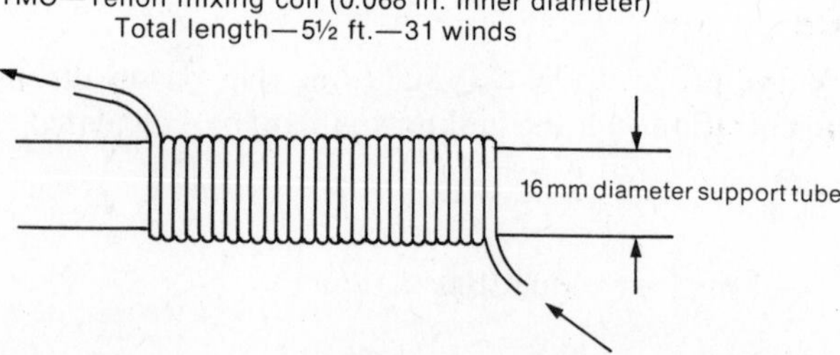

Figure 10-6 Flow diagram for triglycerides determination by the AutoAnalyzer procedure.[24,40] The symbols shown in the figure represent the following components: GMC—Glass mixing coil, long size (Technicon); GHC—Glass heating coil, 40 ft. long; TMHC—Teflon mixing and heating coil (0.068 in. ID), 26 ft. long, continuous with 18 in. transmission tube and TMC for a total length of 33 ft. of Teflon tubing. TMHC is wound in a coil with a diameter of 3 in.

manifold, Teflon mixing coils, Teflon heating coils and Teflon transmission tubing; Sampler II with 2–3 ml cups; Pump II; heating bath (50–60°C) with a 40-foot Teflon coil (0.068 inch ID), and a 20-foot glass coil; and a fluoronephelometer with a 400 nm interference filter as the primary filter and a 485 nm sharp cut-off secondary filter (see Fig. 10–6).

REAGENTS

1. Triglyceride standard solutions. Dissolve 100, 200, 300, and 400 mg of triolein respectively in chloroform and dilute to 100 ml.

2. Adsorbent. Mix gently 1000 g of zeolite (80 mesh), 100 g of Lloyd's reagent, 50 g of anhydrous $CuSO_4$, and 100 g of $Ca(OH)_2$ in a container with a capacity of 1.5 to 2 liters. Heat for 12 to 24 h in an oven at 100 to 110°C. Store in a moisture-proof container.

3. Isopropyl alcohol. Purify by distillation in the presence of KOH.[24] Place approximately 2.4 liters of isopropyl alcohol in a 3 liter round-bottom flask that is equipped with a good distillation column, a still head, a condenser, and a receiver. Add 10 to 15 g of KOH pellets and 3 or 4 boiling chips ("Hengar Granules," Hengar Co., 6825 Greenway Ave., Philadelphia, PA. 19142). Heat the distillation flask with an electric mantle so that the isopropyl alcohol boils gently and yields distillate at a rate of 200 to 300 ml/h. Discard the first 200 ml of distillate, and then collect the product until approximately 300 to 400 ml of undistilled liquid remains in the boiling flask. Discontinue distillation at that time. The boiling flask may be refilled with more isopropyl alcohol without adding more KOH. Fresh boiling chips must be added, however. Again, the first 200 ml of distillate must be discarded. A third batch of isopropyl alcohol may be distilled before cleaning of the boiling flask becomes necessary. After distillation of the third batch, however, the residual liquid in the boiling flask should be discarded, and the flask should be rinsed thoroughly with distilled water and then dried. The distillate should be very pure isopropyl alcohol, virtually free of aldehydes and other problem-causing substances.

4. KOH, aqueous, 25 g/l. In a 2 liter volumetric flask dissolve 50 g of KOH pellets, AR, in approximately 500 ml of distilled water. Cool to room temperature and dilute to 2 liters. Keep stoppered or covered with Parafilm to prevent absorption of carbon dioxide from the air.

5. Hydrochloric acid, 2 mol/l. Slowly add 331 ml of concentrated HCl to approximately 1 liter of water in a 2 liter volumetric flask. Cool to room temperature and dilute to 2 liters.

6. Ammonium acetate, 2 mol/l. Weigh 308 g of ammonium acetate, AR, dissolve in 1.8 liters of distilled water, and add 100 ml of 2 molar HCl. Adjust the pH of the solution to 6.0 with additional 2 molar HCl. Dilute the solution to 2 liters and mix.

7. Acetylacetone reagent. Measure 1.5 ml of acetylacetone (2,4-pentanedione) and 5.0 ml of purified isopropyl alcohol into a 200 ml volumetric flask. Dilute to the mark with 2 molar ammonium acetate solution and mix thoroughly. Transfer to a dark bottle. This reagent should be prepared on the day it is used.

8. Periodate reagent. Dissolve 5.4 g of sodium periodate ($NaIO_4$) in a mixture of 125 ml of water and 115 ml of glacial acetic acid. Dilute to 1.0 liter with water, mix thoroughly, and transfer the solution to a dark bottle.

PROCEDURE

1. *Extraction of Triglycerides from Serum.* A pipettor-diluter or syringe pipets should be used. Transfer a 200 μl aliquot of serum to a 15 to 20 ml tube (with Teflon-lined screw cap) and dilute with 4.8 ml of isopropyl alcohol. Mix the contents of the tube thoroughly with a vortex mixer.

2. *Removal of Phospholipids and Other Interfering Substances.* Add approximately 1 g of the adsorbent mixture to the lipid extract, and mix thoroughly. Allow the mixture

to stand for 10 to 20 min, and then mix thoroughly again. Allow the mixture to stand for another 10 min and then centrifuge for 10 min at 2400 rpm.

At this stage the extract should be perfectly clear. If the extract is turbid, it might be cleared by a longer period of centrifugation. If the extract does not clear during additional centrifugation, the turbidity is probably caused by very fine particles from the adsorbent mixture; in such case, a new adsorbent must be prepared using less vigorous mixing.

3. *Standards.* A standard curve is established by using solutions of triolein in chloroform with concentrations of 100, 200, 300, and 400 mg of triolein/100 ml. Aliquots of the standard solutions (0.2 ml) are measured with the same device used for measuring the serum specimens. The aliquots are diluted with isopropyl alcohol and are treated exactly according to the procedure for the processing of the serum specimens.

4. *Analysis.* Transfer the lipid extracts, following treatment with the zeolite adsorbent, to 2 or 3 ml cups, and process in the AutoAnalyzer at a rate of 40 specimens/h.

Blanks should be determined for all specimens. This determination of nonspecific fluorescence can be accomplished by substituting water for the KOH solution and reanalyzing the specimen extracts.

COMMENTS

In the original Technicon fluorometric triglyceride procedure, KOH in aqueous isopropyl alcohol was pumped with acid-flex tubing; that arrangement was unsatisfactory. Also, inadequate mixing occurred, and inadequate time was allowed for the saponification reaction. The baseline of the original N-78 AutoAnalyzer method was very erratic and the reproducibility of peak values was poor. (The same problems have been carried into the AutoAnalyzer II systems.) These problems were overcome through the development of the design described above[24] and shown in Figure 10–6.

REFERENCE

Kessler, G., and Lederer, H.: Fluorometric measurement of triglycerides. *In* Automation in Analytical Chemistry. L. T. Skeggs, Jr., Ed. New York, Mediad, 1965, p. 341.

CHOLESTEROL

CLINICAL SIGNIFICANCE

The serum cholesterol level is affected by a variety of factors.[31] Primarily, the level of cholesterol in the blood plasma, just as the level of any other lipid, reflects the concentrations of the lipoproteins, and, therefore, the factors that control the secretion of lipoproteins into the plasma and the clearance of lipids and/or lipoproteins from the plasma. The HDLs and LDLs are the cholesterol-rich lipoprotein fractions; therefore, high or low levels of cholesterol in the plasma may reflect high or low levels of those lipoproteins. Unfortunately, the factors that regulate the levels of LDL and HDL in the plasma have not been identified. Therefore, the specific causes of high levels of cholesterol in the plasma remain unknown. Generally, high plasma levels of cholesterol that reflect high levels of LDLs may be caused by an *inherited defect* in lipoprotein metabolism, by *disease of the endocrine system*, by *liver disease*, or by *renal disease*. Low levels of cholesterol in the plasma may reflect an inherited deficiency of either LDL or HDL, or they may reflect *impairment of liver function*.

Several hormones markedly affect the plasma level of cholesterol. An inverse relationship exists between the plasma levels of *thyroxine* and cholesterol. *Hypothyroidism* is associated with hypercholesterolemia to such a degree that measurement of serum cholesterol was once used to monitor thyroid status routinely, until the development of the more specific measurements of serum thyroxine. The increased plasma levels of cholesterol in hypothyroid persons is associated with an increase in levels of LDL and a decrease in levels of HDL. In *hyperthyroidism* the cholesterol levels may be in the low normal range, and a corresponding decrease in the lipoproteins can be seen. *Estrogens* also tend to lower cholesterol levels.

Postmenopausal women show higher cholesterol levels than do premenopausal women, and (parallel to some men of the same age groups) they exhibit an increased risk toward atherosclerotic disease. The effect of *estrogens* on plasma levels of cholesterol is sufficiently impressive that their administration to men with known predispositions to coronary artery disease has been suggested. As might be expected, estrogen treatment of men does lead to a lowering of the serum cholesterol that reflects a lowering of the LDL levels in their plasma, but side effects of estrogen therapy in men make this less than an ideal form of treatment.

Pregnancy may be accompanied by a moderate increase in cholesterol level, probably as a result of altered endocrine function. The change is usually regarded as entirely physiological, and the cholesterol values return to normal after parturition.

Early *hepatitis* produces an increase in serum cholesterol, but as the disease becomes increasingly severe, the level falls, probably because of decreased synthesis by the damaged or necrosing liver cells. In persons who have *obstructed bile canaliculi* and *ducts*, cholesterol excretion into the intestine via the bile is reduced. Abnormal loading of plasma lipoproteins with cholesterol and other lipids occurs. One abnormal lipoprotein complex that may be formed is called *LP-X*. This abnormal lipid-loading of lipoproteins may result in a marked hypercholesterolemia. Mechanical interference with the flow of cholesterol-containing bile is only part of the explanation for the rise of serum cholesterol, however.

In the *nephrotic* syndrome there is a rise in serum lipids, often of sufficient magnitude to give the serum a chylous or milky appearance. Much of the lipid increase is caused by elevated levels of VLDL or of VLDL and LDL together. Therefore, elevated levels of cholesterol, phospholipids, and triglycerides may be seen in this group of disorders.

Diabetes mellitus (effective insulin lack) is associated with hypercholesterolemia and hypertriglyceridemia; the more uncontrolled the diabetes, the greater the elevation of lipids. Adequate treatment of the disease will return the cholesterol level to essentially normal values. In addition, *emotional stress* has been shown to cause marked fluctuations in serum cholesterol in some individuals but not in others.

The implication of cholesterol in the development of *atherosclerosis* and *heart disease* has stimulated an enormous and growing literature, which we cannot review here. In the simplest terms, however, it appears that there is a statistically significant correlation between high serum cholesterol levels and the incidence of coronary artery disease. This suggests that it would be desirable to maintain normal levels of cholesterol in the plasma. Substitution of *polyunsaturated fat for saturated fats* in the diet tends to lower the serum cholesterol level in some persons. Certain medications are used widely to reduce plasma cholesterol levels. As with triglycerides, measurements of serum cholesterol are important in the diagnosis of *hyperlipoproteinemias*.

SPECIMEN

As in the case of the serum triglycerides test, the patient should fast for 12 h before specimen collection. In most persons, the plasma cholesterol level is affected very little by a recent meal; but in some persons, elevations to the extent of 100 mg/100 ml can be induced by a recent meal. The effect of a recent meal appears to depend on the types of food ingested and on undetermined characteristics of individuals.

METHODS FOR THE DETERMINATION OF CHOLESTEROL

A number of extensive reviews have appeared that deal with this subject.[56,60,68,69] Despite the number of procedures available for the determination of cholesterol and its

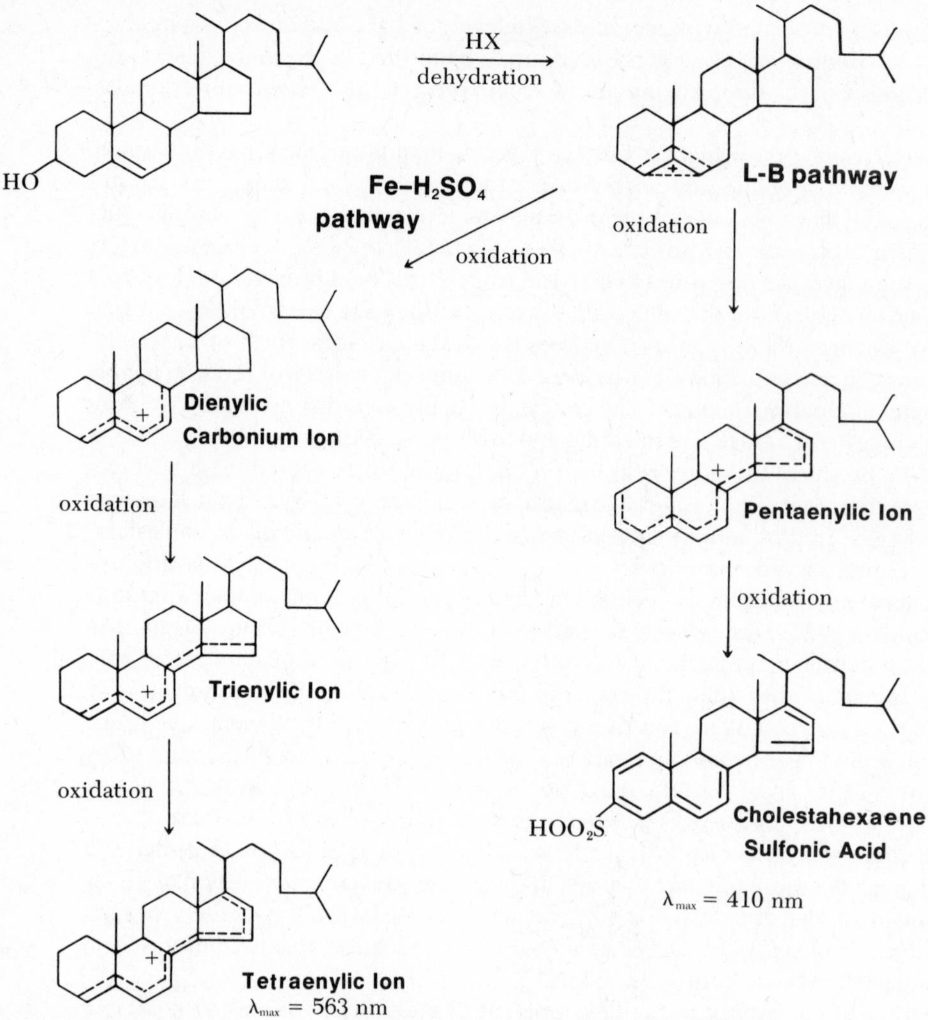

Figure 10–7 Proposed major intermediates and products from cholesterol in the iron-sulfuric acid reaction and in the Liebermann–Burchard reaction.

esters, there are just a few fundamental principles on which the majority of methods are based. The two reactive centers in the molecule are represented by the double bond and the hydroxyl group; both of these are important in the chemical test methods.

Cholesterol reacts with strong, concentrated acids as a typical alcohol, and the products are colored substances, chiefly cholestapolyenes and cholestapolyene carbonium ions.[11,61] In virtually all procedures, acetic acid and acetic anhydride are used as solvents and dehydrating reagents, and sulfuric acid is used as a dehydrating and oxidizing reagent. In some procedures, the reaction with these agents is enhanced by the addition of various metal ions, including iron. The general reaction sequence of the color tests for cholesterol is presented in Figure 10–7.

According to this scheme, cholesterol is first attacked by strongly acid reagents, generalized as HX, where X might stand for the sulfate ion. Such reagents first remove a molecule of water, then oxidize the intermediate to produce 3,5-cholestadiene (two double bonds) or its cation. The oxidizing agent is usually sulfuric acid, which is converted to sulfur dioxide. The cholestadiene reacts further to form cholestapolyene carbonium ions; the stabilities of these cations are dependent on the sulfuric acid concentration. In the

Liebermann-Burchard reaction, the polyenes are the main chromophores. In the presence of more concentrated sulfuric acid and ferric ion, the polyene cations are the main chromophores. Therefore, depending on the relative concentration of the sulfuric acid, and the presence or absence of ferric ion, one obtains either a green color (Liebermann-Burchard) due to a cholestapolyene sulfonic acid, or a red color due to the formation of a cholestapolyene carbonium ion. In the absence of iron or other metal ions, a red colored product will form if the reaction mixture is heated to 80 to 95°C (Salkowski reaction); however, the absorbance in the presence of iron is at least five times greater than the absorbance of the product that forms in the absence of iron.[24]

The reactions are not entirely specific, nor can they be controlled with sufficient precision to yield always the exact products shown; but with proper care, the reactions can be the basis of an accurate and reproducible analytical system. There is little reason to prefer a "red" color reaction over a "green" color reaction, except that the "red" reaction that includes iron generally yields a product with a higher absorbance than that of the "green." The choice of a specific procedure is, therefore, based on other considerations. These colorimetric reactions are subject to interferences by such substances as bilirubin; therefore, the test procedure should include steps to eliminate bilirubin from the colorimetric reaction mixture.

The following classification of procedures is based primarily on the nature of the preliminary treatment of the specimen.

Direct Procedures. In these methods, the colorimetric reaction is performed directly on serum or plasma. Serious errors may arise from the presence of protein, which may produce a variety of colored products. Direct procedures may also suffer from interferences from nonspecific chromogens[15] and chromophores including bilirubin, and instability of the final colored product is sometimes a problem. The Liebermann-Burchard reaction, applied directly to serum, will produce with 1 mg of bilirubin a color equivalent to 5 or 6 mg of cholesterol; the reaction is, therefore, unsuited for determination of cholesterol in serum from jaundiced patients unless suitable corrections or blanks are incorporated into the procedure. The direct procedures are rapid, require the least degree of manipulation, and are suited to automation. Acceptable results have been reported with the direct methods of Pearson, Stern, and McGavack[47] (which employs a serum blank), and of Wybenga *et al.*[67] However, direct colorimetric procedures are never entirely reliable.

Extraction Procedures. These methods introduce an extraction step, primarily to remove proteins prior to color development. Certain solvents or solvent mixtures that can dissolve small amounts of water and cause proteins to precipitate are necessary for a complete extraction of the lipids.

Following extraction of the lipids, the extracting solvent is evaporated, and the color reaction is applied to the lipid residue. With isopropyl alcohol extracts the iron-sulfuric acid reagent may be added directly without preliminary evaporation of the solvent. Unequal chromogenicity of the free and esterified forms of cholesterol has been observed in the Liebermann-Burchard reaction and in the reaction involving ferric ion.

Extraction and Hydrolysis Procedures. In this approach, cholesterol and cholesteryl esters are extracted and the latter are hydrolyzed or "saponified" before color development. Such procedures eliminate interference from protein and the problem of unequal rates of color development by the free and ester forms of cholesterol. The saponification step also tends to destroy some nonspecific chromogens. Extraction into a second solvent further increases specificity. One example of this approach is the method of Abell *et al.*,[1] which is widely accepted as a reference method.

Procedures Involving Extraction, Hydrolysis, and Precipitation. These methods are complicated and tedious but reliable. The cholesterol is extracted, the esters are saponified, and the total cholesterol is then further purified by isolation as the *digitonide*. The digitonide is decomposed by saponification, which again frees the cholesterol, and the product of this step is subjected to color development.

By introduction of the digitonin step, the effect of nonspecific chromogens is considerably reduced or eliminated. The method of Sperry and Webb[55] is a widely accepted reference method for determinations involving precipitation of cholesterol as the digitonide.

Many other methods in each category could be cited, most of which represent slight modifications of fundamental procedures that will be presented in detail. Before accepting modifications of the more standardized procedures, the modifications should be checked carefully against a method of accepted performance.

Fluorometric Method. A fluorometric method, based on the Liebermann-Burchard reaction, has been published.[4] It is a true micromethod, capable of measuring as little as 0.1 μg of cholesterol with high accuracy and precision. Free and esterified cholesterol produce nearly the same molar fluorescence, and the fluorescence readings are relatively stable.

Enzymatic Method. More recently an enzymatic method employing *cholesterol oxidase* has become available[5] and promises to improve the direct measurement of cholesterol in serum without a preliminary extraction.

The enzymatic reactions that form the basis of this method are:

$$\text{Cholesterol} + O_2 \xrightarrow[\text{oxidase}]{\text{Cholesterol}} \text{Cholest-4-ene-3-one} + H_2O_2$$

$$2\,H_2O_2 + \text{4-Aminoantipyrene} + \text{Phenol} \xrightarrow{\text{Peroxidase}} \text{Quinoneimine dye} + 4\,H_2O$$

The principle is analogous to the glucose oxidase method for glucose and is based on the determination of hydrogen peroxide formed in the reaction. Esterified cholesterol, however, must be hydrolyzed before it can react with cholesterol oxidase. This is accomplished by incorporating the enzyme *cholesteryl ester hydrolase* into the reaction mixture, or by alkaline hydrolysis.

The cholesterol oxidase method has been found to be specific for sterols that have 3β hydroxyl functions and a double bond in the 4–5 or 5–6 position. While this method is not absolutely specific for cholesterol, it has been found to be less subject to error than any of the colorimetric methods, including the method of Abell *et al.*[1]

DETERMINATION OF CHOLESTEROL BY THE METHOD OF ABELL et al.[1]

REAGENTS

1. Absolute ethanol, redistilled.
2. Petroleum ether, b.p. 68°C, redistilled.
3. Acetic acid, AR.
4. Acetic anhydride, AR, free from HCl.
5. Sulfuric acid, AR. Keep tightly stoppered.
6. Potassium hydroxide solution. Dissolve 10 g of KOH pellets, AR, in 20 ml of water.
7. Alcoholic KOH. Prepare immediately before use by adding 6 ml of the KOH solution (reagent 6) to 94 ml of absolute ethanol.
8. Liebermann-Burchard reagent. Chill 20 vol of acetic anhydride to a temperature below 10°C in a glass-stoppered container. Add 1 vol of concentrated sulfuric acid, mix thoroughly, and keep the mixture cold for 9 min. Add 10 vol of acetic acid and warm to room temperature. The reagent should be used within 1 h.*
9. Cholesterol standard solutions, 200, 400, and 600 mg/100 ml. Dissolve 200, 400,

* A modified Libermann-Burchard reagent, described by Huang *et al.*,[38] incorporates 20 g of anhydrous sodium sulfate per liter of solution. This reagent was found to be stable for at least 2 weeks at room temperature.

and 600 mg of cholesterol (NBS certified standard or material of comparable quality) in absolute ethanol, and dilute each solution to 100 ml.

PROCEDURE

1. Measure 200 μl serum or standard solution into 12 or 15 ml glass-stoppered centrifuge tubes, and add 2.0 ml of alcoholic KOH to each tube. Stopper the tubes, mix the contents with a vortex mixer, and incubate the mixtures at 37 to 40°C for 55 min.
2. Cool the tubes to room temperature, add 5 ml of petroleum ether to each tube, and mix thoroughly by shaking. Add 2.0 ml of water and mix thoroughly again by shaking. Centrifuge the stoppered tubes at slow speed, if necessary, to break the emulsion.
3. Transfer 4.0 ml aliquots of the petroleum ether layers to clean, dry tubes. Vaporize the solvent by heating the tubes to 60°C and directing gentle streams of nitrogen into the tubes.
4. Place the tubes in a water bath, sand bath, or aluminum block thermostated to 25°C. Allow 5 min for the temperature to stabilize.
5. Add 6.0 ml of the Liebermann-Burchard reagent to an empty reagent blank tube. Use this solution to set the spectrophotometer to zero absorbance.
6. At timed intervals of 15 s, add 6.0 ml of Liebermann-Burchard reagent to each of the specimen tubes and standard tubes. Mix the contents of each tube thoroughly with a vortex mixer immediately after the addition of the reagent, and return the tube to the 25°C environment.
7. Allow each tube to remain in the 25°C environment for exactly 30 min; then transfer the contents to a cuvette, and determine the absorbance (A) at 620 nm, in timed intervals of 15 s.

CALCULATIONS

$$\text{mg cholesterol/100 ml} = \frac{A_u}{A_s} \times C_s \text{ (mg/100 ml)}$$

(C_s = concentration of standard solution)

COMMENTS

1. Aliquots of standard solutions and aliquots of serum should be measured with the same pipetting device or with high quality pipets.
2. The color yield usually follows Beer's law; however, if it does not, plot a standard curve, or calculate the concentration using the standard with an absorbance reading closest to the absorbance of the specimen.
3. The accuracy and precision of this method are very good. Recoveries of added cholesterol have been quantitative. Usually, at least 99 per cent of the measured absorbance represents cholesterol. With a little experience, precision corresponding to a coefficient of variation of 2 per cent or less (variance within one day and, also, from day to day) can be achieved readily.
4. The full color develops only in an anhydrous medium. Therefore, contamination with any water from step 3 onward must be avoided.

NORMAL VALUES FOR SERUM TOTAL CHOLESTEROL

The normal range for serum total cholesterol at birth is 45 to 100 mg/100 ml. This level approximately doubles during the first few days of life. Normal values reached by 1 month hold fairly constant until the age of 20 years and range from 100 to 230 mg/100 ml.

Normal ranges for total serum cholesterol in adults vary with age, sex, and diet. A seasonal variation has also been observed, with levels higher in fall and winter than in spring and summer. Keys *et al.*[41] reviewed the literature and presented their own studies of over 2000 carefully selected healthy individuals. Reed *et al.*[49] examined cholesterol values in more than 1400 sera from subjects judged to be normal and found results similar to those of Keys. Typical ranges are shown below.

Age, Years	*Total Cholesterol*, mg/100 ml		
	90 *Per Cent Limits* (Fredrickson & Levy[31])	*Upper Limits* (Ellefson[24])	
		Males	*Females*
0–19	120–230		
20–29	120–240	235	220
30–39	140–270	265	240
40–49	150–310	280	265
50–59	160–330	300	320

Although the ranges shown have been established for apparently normal, healthy adults, these may not necessarily be "desirable" values. Some authorities, therefore, feel that it is desirable to maintain a cholesterol level below 250 mg/100 ml regardless of age or sex. Normally, 65 to 75 per cent of the total cholesterol in serum is present in the form of esters.

REFERENCE

Abell, L. L , Levy, B. B., Brodie, B. B., and Kendall, F. E.: Standard Methods Clin. Chem., *2*:26, 1958.

DETERMINATION OF FREE AND TOTAL CHOLESTEROL BY THE FERRIC CHLORIDE–SULFURIC ACID REACTION (LEFFLER,[42] MODIFIED)

REAGENTS

1. Iron reagent. Dissolve 2.5 g of ferric chloride, $FeCl_3 \cdot 6H_2O$, in 100 ml of phosphoric acid (assay 85 per cent). Store in a glass-stoppered bottle. This solution is stable indefinitely.
2. Digitonin solution. Dissolve 1.0 g of digitonin in 50 ml of absolute ethanol. Warm gently to dissolve; then add 50 ml of water and mix.
3. Absolute ethanol-acetone mixture. Mix equal volumes of each.
4. Glacial acetic acid, acetone, isopropyl alcohol, and concentrated sulfuric acid, all AR grade. Keep the sulfuric acid bottle tightly stoppered to prevent absorption of moisture from the air.
5. Cholesterol standards. Dissolve 100, 200, 300, and 400 mg of cholesterol (NBS certified standard or preparation of similar quality) in isopropyl alcohol and dilute each solution to 100 ml. The 200 mg/100 ml standard is used routinely. All four standards should be used to prepare a standard curve or to check linearity.

Procedure for Total Cholesterol

1. Pipet exactly 200 μl of serum into the bottom of a test tube.
2. Pipet exactly 200 μl of standard (200 mg/100 ml) into a second tube.
3. Add 5.0 ml of isopropyl alcohol rapidly from a pipet or automatic dispensing device.
4. Mix thoroughly and centrifuge. The standard does not require centrifugation.
5. Pipet 1.0 ml of each supernatant fluid into the bottom of a dry glass-stoppered 12 or 15 ml centrifuge tube.
6. Pipet 1.0 ml of isopropyl alcohol into a clean dry tube, which will hold the reagent blank.

7. Add 3.0 ml of glacial acetic acid to all tubes and mix.

8. Add 0.3 ml of iron reagent and mix well.

9. At 15 s intervals, add 3.0 ml of concentrated sulfuric acid to one tube at a time; let the acid flow freely down the side of the tube. Stopper, and mix immediately with a vortex mixer.

10. Let the tubes stand 10 min to cool. Do *not* hasten cooling by immersing in cold water.

11. Transfer solutions to respective cuvettes and let stand 2 or 3 min for any bubbles to rise.

12. Without delay, measure absorbances of standard (A_s) and unknown (A_u) against the reagent blank at 560 nm.

CALCULATION

$$\text{mg cholesterol/100 ml} = \frac{A_u}{A_s} \times 200$$

COMMENTS

1. It is important to mix each tube immediately after adding concentrated sulfuric acid in order to prevent possible charring of organic substances extracted from serum and to promote uniform heating and color development.

2. Calibration curves are linear on most instruments to an absorbance of 1.0; however, linearity should be checked by using standards over the range from 100 to 400 mg/100 ml.

3. Substances known to interfere in this reaction are bromides, nitrates, and nitrites. A serum bromide level of 10 mmol/l (103 mg NaBr/100 ml) increases the apparent cholesterol concentration by about 30 mg/100 ml. Nitrite, encountered in some lots of isopropyl alcohol, depresses the color reaction. This error is not significant, provided a standard is included with each series of determinations. To test for the presence of nitrites, dissolve about 1 g of KI in 10 ml of water, add 50 ml of isopropyl alcohol, and let stand 5 min. The solution should remain colorless. If nitrites are present, a yellow color develops.

4. Isopropyl alcohol may be redistilled over KOH as described on p. 505.

5. Corrections for errors caused by high levels of bilirubin may be necessary. Corrections may be accomplished by preparing a standard curve for the color produced by a bilirubin standard and the cholesterol reagent and subtracting the appropriate absorbance value from the total absorbance value. Alternatively, and preferably, the lipid extract can be treated with zeolite (p. 506) to remove bilirubin before the cholesterol tests are performed.

Procedure for Free Cholesterol

1. Analyze the serum for total cholesterol as described above, including the standard.

2. Pipet exactly 200 μl of serum into a test tube.

3. Add 3.0 ml of alcohol-acetone mixture rapidly from a pipet or automatic dispenser. Mix thoroughly and centrifuge.

4. Pipet 2.0 ml of clear supernatant fluid into a 12 or 15 ml glass-stoppered centrifuge tube.

5. Add 2 ml of digitonin solution, mix thoroughly, and place in a water bath at approximately 50°C for at least 15 min until flocculation occurs.* Cool to room temperature.

6. Centrifuge for 10 min. A slight turbidity in the supernatant fluid at this point should be ignored.

7. Carefully decant the supernatant fluid and dry the lip of the tube on absorbent paper.

* Others have used 37° for 1 h.

8. Add 3 ml of acetone to the tube from a pipet. First wash down the walls, and then force the remainder directly onto the precipitate. Break up the precipitate by tapping.

9. Centrifuge as before, decant, drain, and dry the lip of the tube.

10. Add 1.0 ml of isopropyl alcohol and 3.0 ml of glacial acetic acid directly onto the precipitate and mix until dissolved.

11. Proceed as described under total cholesterol, starting with step 8.

CALCULATIONS

$$\text{mg free cholesterol/100 ml} = \frac{A_u}{A_s} \times 200 \times 0.3077$$

$$= \frac{A_u}{A_s} \times 61.5$$

The factor 0.3077 is derived from the ratio of the volumes of serum present in the final solution for the total and free procedures respectively, i.e., 0.20 ml × 1.0/5.2 = 0.03846 ml and 0.20 ml × 2.0/3.2 = 0.125 ml.

$$\text{cholesterol in esterified form, mg/100 ml} = \text{total cholesterol} - \text{free cholesterol}$$

$$\text{per cent ester cholesterol} = \frac{\text{ester cholesterol}}{\text{total cholesterol}} \times 100$$

REFERENCE

Leffler, H. H.: Am. J. Clin. Path., *31*:310, 1959.

DETERMINATION OF TOTAL CHOLESTEROL BY A SEMI-AUTOMATED METHOD

The following procedure is outlined to illustrate a semi-automated method for use with a *Technicon AutoAnalyzer*. The instrument was redesigned[24] with Tygon manifold tubing, 22 feet of Teflon transmission tubing, including a 4-foot preliminary mixing coil and a 14-foot mixing coil within a 95°C heating bath, and a colorimeter with 550 nm filters (see Fig. 10–8).

REAGENTS

1. Isopropyl alcohol and adsorbent are the same as described for the semi-automated method for triglycerides, p. 505.

2. Acid-iron reagent. Dissolve 500 mg of ferric chloride hexahydrate, AR, in 2000 ml of glacial acetic acid, AR, and add 1000 ml of concentrated sulfuric acid.

3. Cholesterol standard solutions. Dissolve 100, 200, 300, and 400 mg of cholesterol (NBS certified standard or equivalent) respectively in isopropyl alcohol and dilute to 100 ml.

PROCEDURE

1. *Extraction and adsorption*. These steps are performed exactly as described in the semi-automated method for triglycerides. (See pp. 505 and 506.)

2. *Standards*. Establish a standard curve using the four standards described above. Aliquots are measured with the same device used for measuring the serum specimens. The aliquots are treated exactly the same as serum.

3. *Analysis*. Following treatment with the zeolite adsorbent (see p. 506), transfer the lipid extracts to 2 or 3 ml cups and process in the AutoAnalyzer at a rate of 40 specimens/h.

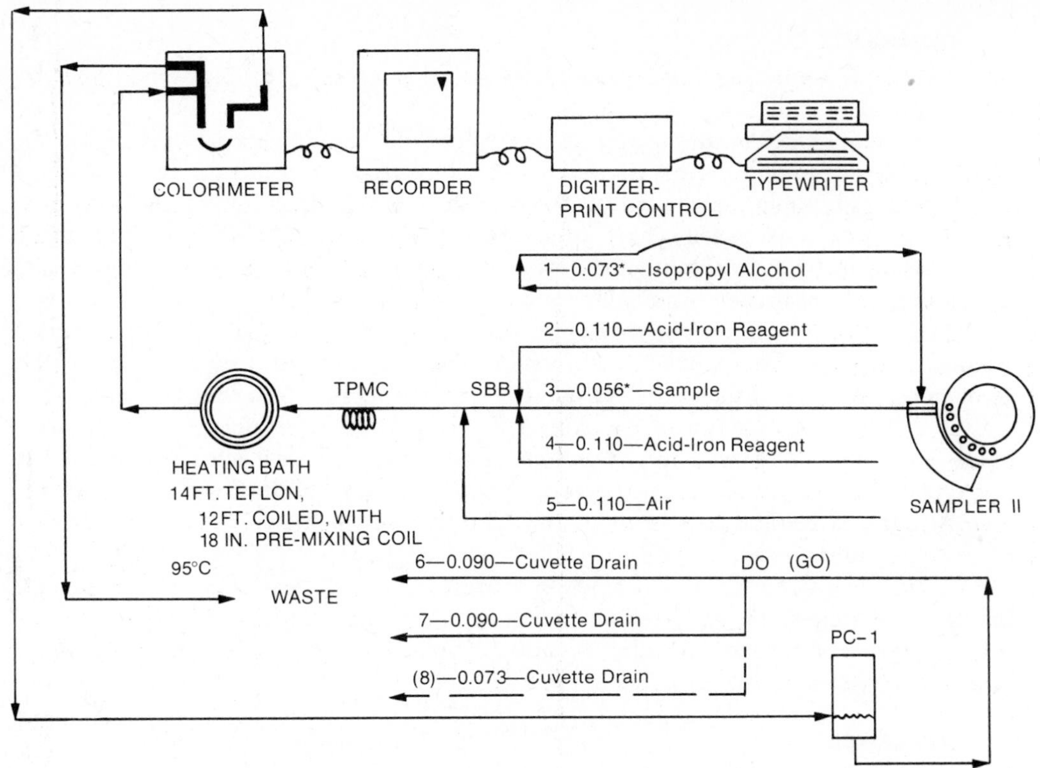

Special Components:

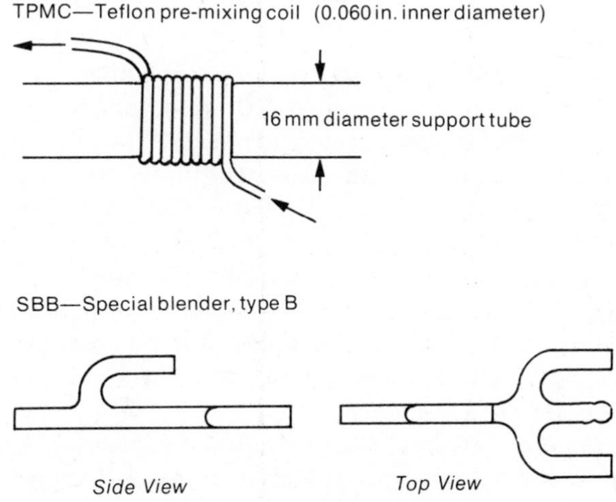

A satisfactory substitute for this blender can be prepared from one DO connector and one GO connector.

Figure 10–8 Manifold used in the AutoAnalyzer method for cholesterol.[24] For abbreviations, see Figure 10–6. The figures in parentheses and the broken line represent modifications of the manifold that should be used when the system includes the Technicon "Pump II." The Teflon mixing coil for the heating bath is wound from the Teflon transmission tubing in the same way as is the TPMC: it is wound in two sections, and the two sections are wound on separate tubes; each section is 6 ft. long and contains 30 turns of tubing. The Teflon tubing from the SBB through the TPMC and through the mixing coils in the heating bath to the debubbler ahead of the colorimeter is one continuous length, and the inner diameter is 0.060 in. One length of Teflon transmission tubing is used to deliver acid-iron reagent from the bottle to lines 2 and 4 of the pump manifold; this transmission tubing must have an I.D. of at least 0.125 in.

COMMENTS

1. The procedure described produces results that agree with those obtained by the Abell method.[1]

2. The use of Teflon tubing permits pumping of the viscous reaction mixture through the full 22 feet of tubing without the severe carry-over problems that would be encountered if the usual glass tubing were used. In the Technicon N-24a glass system, the path of the reaction mixture is necessarily short. In addition, the retarded flow of the mixture caused by the glass and by the viscosity of the mixture develops a hazardous fluid pressure in the manifold tubes and in the transmission tubing near the pump, and prevents uniform pumping of the reaction mixture. The redesigned AutoAnalyzer system for this method permits adequate mixing of the specimen extract with the reagent, adequate exposure of the cholesterol and esters to the reagent, and adequate heating for more complete color development.

3. The improved design of the AutoAnalyzer permits equivalent color yields from cholesterol and cholesteryl esters. Therefore, standardization with pure cholesterol solutions is possible. The original AutoAnalyzer N-24 (Technicon) method and the more recent AutoAnalyzer II method require standardization with *serum* standards in order to obtain agreement with the Abell method.

4. The corrosive reagent and reaction mixtures are quite destructive of the manifold tubing; nevertheless, Tygon manifold tubing functions more satisfactorily than any substitute known at this time, including Technicon "Acid-Flex" tubing. The manifold tubing should be replaced after approximately 24 h of use.

REFERENCE

Ellefson, R. D.: Unpublished information.

PHOSPHOLIPIDS

For most clinical purposes the glycerophosphatides and sphingomyelins are grouped together under the term "phospholipid." Until recently, a worthwhile approach to analysis of the serum lipid profile involved measurement of the concentration of total phospholipids in the blood serum together with measurements of the concentrations of the cholesteryl esters and "total fatty acids." Now, with the availability of accurate methods for triglycerides quantitation, the need for the determination of total phospholipids has diminished. The total phospholipids concentration in the blood plasma varies in much the same directon and to the same degree as does total cholesterol concentration. Therefore, knowing the serum concentration of total phospholipids adds very little, if any, additional useful information about a patient's serum lipid profile when the total cholesterol and total triglycerides concentrations are already known.

A graph that correlates serum concentrations of total phospholipids and total cholesterol, based on a study of approximately 1200 normal subjects between the ages of 2 and 77 years, is shown in Figure 10–9.[3] With concentrations expressed in mg/100 ml, the equation of the regression line is:

$$\text{phospholipids} = 68 + (0.89 \times \text{cholesterol})$$

Occasionally a need for quantitation of specific phospholipids may arise. For this purpose, in most cases fractionation of phospholipids into specific components can be accomplished by thin-layer chromatography (Figure 10–10).[44,48,62,71]

A clinical condition in which the quantitation of specific phospholipids can be helpful

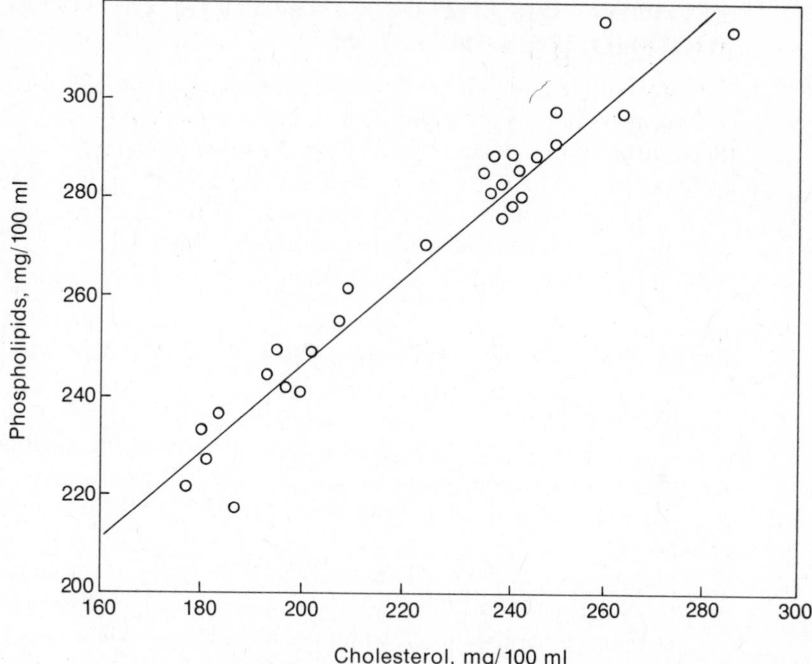

Figure 10–9 Correlation of phospholipid and cholesterol concentrations in serum. Each point represents a mean value for an age group. Data from Adlersberg, *et al.*[3]

is in the diagnosis of *Niemann-Pick disease*, a phospholipid storage disease caused by a deficiency of the enzyme *sphingomyelinase*. As a result, sphingomyelin accumulates in tissues—particularly in the nervous system, spleen, liver, and kidneys—and elevated concentrations of sphingomyelin may also be observed in blood serum.

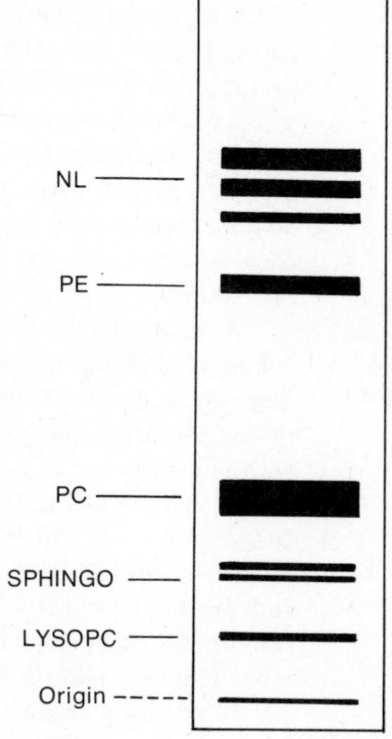

Figure 10–10 Illustration of a typical thin-layer chromatogram of phospholipids in human serum. LYSOPC: lysophosphatidyl choline; SPHINGO: sphingomyelin; PC: phosphatidyl choline (lecithin); PE: phosphatidyl ethanolamine; NL: neutral lipids (includes cholesteryl esters, triglycerides, and cholesterol) and fatty acids. Dimensions of the TLC plate are 5 cm × 20 cm. Chromatography of specimens in bands rather than in spots permits maximal separation of components with minimal streaking.[2]

METHODS FOR THE DETERMINATION OF TOTAL PHOSPHOLIPIDS IN SERUM

Most of the methods for determination of phospholipids involve the oxidative digestion of a purified lipid extract followed by a measurement of the inorganic phosphorus liberated. (Separation of the phospholipids from the other lipids present is not necessary.) Oxidizing agents used include hot concentrated sulfuric acid in conjunction with hydrogen peroxide, nitric acid, or perchloric acid. Virtually all of the analytical methods depend on the formation of the phosphomolybdate ion, which is then reduced by a suitable reagent to form the complex "heteropoly blue," a colloidal dispersion of unknown composition.

DETERMINATION OF PHOSPHOLIPIDS BY THE GOMORI PROCEDURE

REAGENTS

1. Standard. Potassium dihydrogen phosphate in demineralized water; 0.4393 g KH_2PO_4 diluted to 1000 ml (100 μg P per ml).

2. Oxidizing acid. Mix 10 ml of 70 per cent perchloric acid and 90 ml of 2.5 molar sulfuric acid.

3. Sodium molybdate. Dissolve 25.0 g of sodium molybdate, AR, in 142 ml of 2.5 molar sulfuric acid, and dilute to 1000 ml with demineralized water.

4. Elon (*p*-methylaminophenol sulfate). Dissolve 1.0 g of Elon and 3.0 g of sodium bisulfite, and dilute to 100 ml with demineralized water.

5. Extracting solvent. Combine 1 vol of absolute ethanol with 1 vol of diethyl ether. For convenience, prepare 1000 ml or more of the mixture at one time. Store the mixture in an amber, glass-stoppered bottle, and store away from heat sources and direct light.

PROCEDURE

1. Transfer 400 μl of serum to a Teflon-lined, screw-capped centrifuge tube calibrated for 10 ml; add sufficient extraction solvent to fill the tube to the calibration mark. Cap the tube, mix the contents of the tube vigorously, and then allow to stand for 10 min. Mix the contents of the tube again, and then centrifuge for 10 min to pack the protein precipitate.

2. Transfer 5.0 ml of the lipid extract to a 16 mm × 150 mm Pyrex tube, add one or two tiny carborundum boiling chips, and vaporize the solvent by heating the tube in a tube heater set at 100 to 110°C.

3. To the lipid residue, add 400 μl of oxidizing acid, and insert the tube halfway into a well of an aluminum block heater set at 240 to 250°C. At the end of the first 30 min of the digestion period, lower the tube to the full depth of the heater well; heat for an additional 30 min. Remove the tube from the heater and allow to cool; the acidic residue should be clear and colorless or slightly yellow. If the residue contains dark colored material, the procedure for oxidation must be repeated with a smaller aliquot of the lipid extract.

4. Add 7.0 ml of demineralized water to the acidic residue from the digestion. Add 2.0 ml of sodium molybdate solution, and mix. Add 0.5 ml of Elon solution and mix thoroughly. Transfer the reaction mixture to a cuvette (19 × 105 mm) and allow the color to develop for 1 h from the time of the last mixing. Then read the absorbance at 700 nm. Set the instrument to zero absorbance using the reagent blank.

5. A standard curve is prepared by the use of standard samples containing 10.0 μg, 20.0 μg, 30.0 μg, and 40.0 μg of phosphorus/test tube. The standard samples are prepared by measuring 100, 200, 300, and 400 μl aliquots of the standard solution into Pyrex tubes and processing exactly according to the procedure described for serum from step 3 (be sure to add carborundum chips). The standard curve should be a straight line through the range from 0 to 40 μg of phosphorus per sample (0 to 20 mg of lipid-bound phosphorus/100 ml of serum; each sample is equivalent to 200 μl of serum).

Once the linear standard curve has been established, the use of a single standard sample containing 10.0 μg of phosphorus should provide verification of continuing validity of the standard curve. At least two samples of this standard solution should be included with each set of 18 to 20 specimens. The absorbance value for each standard sample should agree within 3 per cent with the value from the standard curve.

CALCULATIONS

As an alternative to reading specimen values from a standard curve, lipid-bound phosphorus and phospholipid concentrations may be calculated with the following formulas.

$$\mu\text{g of lipid-bound phosphorus}/200\ \mu\text{l of serum} = \frac{A_u}{A_s} \times C_s$$

$$\text{mg of phospholipids}/100\ \text{ml of serum} = \frac{25 \times \frac{A_u}{A_s} \times C_s/(1000\ \mu\text{g/mg})}{(200\ \mu\text{l})/(1000\ \mu\text{l/ml})} \times 100\ \text{ml}$$

$$= 12.50 \times \frac{A_u}{A_s} \times C_s$$

where C_s = μg of phosphorus in standard
25 = average factor to convert phosphorus to phospholipids.

COMMENTS

1. The rate of color development is sensitive to pH. The optimal pH is near 0.90.
2. Solvent used for extraction of the lipids from the specimen must be completely vaporized before the oxidizing acid is added.
3. During the first 30 min of oxidative digestion, flecks of black carbonacious ash may form and ride on the upper edge of the condensing acid in the upper portion of the Pyrex tube. This ash becomes trapped and thoroughly digested when the tubes are lowered to the full depth of the heater wells at the end of the first 30 min.
4. The lipids from most specimens become completely oxidized within 45 min; therefore, the hour-long digestion period allows a time cushion of a few minutes. Prolonging the digestion period beyond 1 h permits excessive leaching of silicate from the tubes. Silicate, like phosphate, reacts with molybdate in an acidic medium to form a heteropoly acid, and heteropoly silicomolybdic acid, like heteropoly phosphomolybdic acid, can be reduced to form a "molybdenum blue."
5. While the described method employs "Elon" as the reducing agent, another agent, *p*-semidine, has been found to be very satisfactory.[20]

NORMAL VALUES

The total phospholipid level of serum increases with age from birth, as does the level of cholesterol. A distinct elevation occurs during pregnancy. Average normal values are summarized below.

Age	*Phospholipids*, mg/100 ml
Birth	29–93
Males, up to 65 years	175–275
Females, nonpregnant	158–232
Females, early pregnancy	205–291
Males or females, 65–88 years	196–366

REFERENCES

Gomori, G.: J. Lab. Clin. Med., *27*:955, 1942.
Silverstein, M. N., and Ellefson, R. D.: Sem. Hematol., *9*:299–307, 1972.

DETERMINATION OF THE LECITHIN–SPHINGOMYELIN RATIO IN AMNIOTIC FLUID (GLUCK, MODIFIED[34])

A determination of the lecithin in amniotic fluid, or the ratio of the lecithin to sphingomyelin concentration (L/S ratio), has been found to be of value in assessing fetal lung maturity.[34] Evidence suggests that *hyaline membrane disease* results from insufficient production of surface-active phospholipids in the fetal lungs. These are probably produced and stored within the cells lining the alveoli and are then secreted into the lumen where they coat the alveolar cells, thus decreasing surface tension at end-expiration, and thereby preventing alveolar collapse. The amount of pulmonary surfactant is insufficient in the lungs of prematurely born fetuses and of those newborn infants who develop hyaline membrane disease.

Prior to the 34th week of gestation, the lungs produce nearly equal amounts of sphingomyelin and lecithin. However, after this time, synthesis of lecithin increases sharply, and the L/S ratio increases accordingly. This change is reflected in the amniotic fluid. In a case of uncomplicated pregnancy, an L/S ratio in the amniotic fluid of 3/1 or greater indicates that the fetus has sufficiently mature lungs to be able to breathe normally at birth. A ratio of less than 2/1 suggests that the fetal lungs may not be sufficiently mature to avoid *respiratory distress syndrome*. The L/S ratio can also be a helpful test in the determination of optimal times for cesarean delivery. Maternal diseases can cause the L/S ratio of the amniotic fluid to be unreliable as an indicator of fetal lung maturity (see also Chapter 22).

SPECIMEN

Centrifuge at least 3 ml of amniotic fluid at 250 $\times$ *g* for 20 min to remove cellular components; decant and refrigerate the supernatant. Bloody specimens cannot be used, because small volumes of blood plasma can alter the L/S ratio enough to make interpretation difficult.[10] Specimens which stand at room temperature longer than 1 h may also have an altered L/S ratio. The cell-free specimen should be frozen if the analysis cannot be done the same day.

MATERIALS AND REAGENTS

1. Precoated TLC plates, 5 $\times$ 20 cm, coated with plain silica gel.
2. Screw-cap jar for developing solvent. A jar about 22 cm high with a 7 cm diameter cap is satisfactory.
3. Developing solvent. Mix 25 ml of methanol, 4 ml of water, and 65 ml of chloroform. Transfer 40 ml of the developing solvent to the screw-cap jar, cap it, and mix by swirling.
4. Sulfuric acid, 50 per cent (w/w). Carefully add 56 ml of concentrated H_2SO_4 to 100 ml of water. Swirl the mixture gently during the addition of the acid.
5. Standards. Dissolve 300 mg each of lecithin and sphingomyelin in 100 ml of chloroform. The standards should be used as markers to help locate the correct bands on the unknown, and to calculate the L/S ratio. Measured volumes of 10, 25, and 50 μl of the standard provide suitable bands for determinations of ratios.

PROCEDURE

1. Pipet 2.0 ml of amniotic fluid into a test tube.
2. Add 2 ml of methanol and mix.
3. Add 4 ml of chloroform, insert stopper, and shake vigorously for 1 min.
4. Centrifuge. The lower layer should be clear.
5. Tilt the tube, insert a 2 ml volumetric pipet past the protein button at the interface, withdraw 2 ml of the lower chloroform layer, and transfer it to a centrifuge tube.
6. Hold the centrifuge tube in a hot water bath, with continuous shaking, until the solvent has evaporated, or vaporize the solvent with a gentle, small stream of nitrogen

while warming the tube in a bath of warm water. Add about 50 μl of chloroform and rotate the tube to dissolve the residue.

7. With a 50 μl microsyringe add the chloroform extract to the silica plate in a thin line about 2 cm long and about 2 cm from the bottom of the plate. A standard mixture of lecithin and sphingomyelin should be chromatographed alongside the specimen on the same chromatoplate.[24]

8. Let the line of lipid dry for a few minutes; then insert the plate into the jar with developing solvent and screw on the cap.

9. Let the chromatogram develop until the solvent nearly reaches the top.

10. Remove the plate and place it on a paper towel. Let the plate dry in the air.

11. Spray the chromatogram uniformly with the sulfuric acid solution. Heat the plate at 250°C for 30 min.

12. For quantitation, scan the plate with a recording densitometer, and measure the areas under the peaks for lecithin and sphingomyelin.

13. The areas from the densitometer recordings must be converted to micrograms of lecithin and sphingomyelin with the aid of standard curves.

CALCULATIONS

$$\text{L/S ratio} = \frac{\text{lecithin}}{\text{sphingomyelin}} = \frac{A_{\text{PC}}}{A_{\text{SPHINGO}}} \times \frac{A_{\text{SSPHINGO}}}{A_{\text{SPC}}} \times \frac{Q_{\text{SPC}}}{Q_{\text{SSPHINGO}}}$$

A_{PC} and A_{SPHINGO} are areas under the lecithin (phosphoryl choline) and sphingomyelin peaks on the tracing from the densitometer.

A_{SPC} and A_{SSPHINGO} are areas under the peaks that correspond to lecithin and sphingomyelin standards.

Q_{SPC} and Q_{SSPHINGO} are the quantity of lecithin standard and of sphingomyelin standard applied to the chromatogram, in μg.

Values for the L/S ratio should be reported to the nearest 0.1.

COMMENTS

1. Alternatively, the lipid bands on the chromatogram may be stained nonspecifically by exposure to iodine vapor for a few minutes. This can be accomplished by allowing the chromatogram to stand in a sealed jar that contains a few crystals of iodine. Also, the phospholipid bands may be stained blue with the phospholipid-specific stain of Voskevsky and Kostetsky.[62]

2. For quantitation without recovery of the lipids from the chromatogram, optical scanning after charring the lipids as described is the only satisfactory method. Alternatively, the lipids may be eluted from the silica gel, and quantitative analysis may be accomplished with the colorimetric method described for serum phospholipids (p. 518). Quantitative elution of the phospholipids is somewhat difficult and time-consuming, and is not recommended for routine determinations of the L/S ratio of amniotic fluid.

For reviews of this topic, see references under "Additional Reading."

NORMAL VALUES

The L/S ratio in amniotic fluid at term normally ranges from 2.0 to 5.0.

REFERENCES

Gluck, L., Kulovich, M. V., Borer, R. C., Jr., Brenner, P. H., Anderson, G. G., and Spellacy, W. N.: Am. J. Obst. Gyn., *109*:440, 1971.

Gluck, L., and Kulovich, M.: Am. J. Obst. Gyn., *115*:539, 1973.

FATTY ACID PROFILES OF SERUM LIPIDS

The assortment of fatty acids present in the serum lipids can have specific diagnostic significance. In normal individuals, the fatty acid profiles of the serum lipids following a 12 h fast vary only within narrow limits.

Persons affected with *Refsum's disease* lack the enzyme *phytanate α-oxidase*;[39] the lack of this enzyme prevents the normal degradation of *phytanic acid*, a metabolite of phytol and chlorophyll. Phytanic acid then accumulates and becomes incorporated into a variety of lipids. The accumulation of phytanic acid appears to be pathogenic. Persons with Refsum's disease develop a polyneuropathy, but respond favorably to diets that contain little phytol, phytanic acid, and chlorophyll. One known metabolic effect of phytanic acid is its blockage of β-oxidation, the normal energy producing degradation of fatty acids. Phytanic acid is readily detected in blood specimens from persons who have Refsum's disease and may account for as much as 20 per cent of the fatty acids in the lipids of the serum.

Persons with *acrodermatitis enteropathica* comprise another group who may have an unusual fatty acid profile in their blood lipids.[16,64] Such persons have a true steatorrhea accompanied by impairment of absorption and metabolism of the essential fatty acids. The serum lipids, therefore, contain much less than the usual amounts of linoleic and arachidonic acids.

Fatty acid profiles of serum lipids can be determined by a method which includes alkaline hydrolysis and subsequent conversion of the resulting fatty acids to methyl esters with diazomethane. The fatty acid methyl esters are then quantitated by gas-liquid chromatography.

ANALYSIS OF FATTY ACID PROFILES BY GAS-LIQUID CHROMATOGRAPHY[25]

REAGENTS

1. Potassium hydroxide solution. Dissolve 33 g of KOH pellets, AR, in 66 g of distilled water.
2. Petroleum ether, b.p. 30 to 60°C, redistilled.
3. Sulfuric acid, 1 mol/l. Dilute 98 g of concentrated H_2SO_4 to 1.0 liter with distilled water. (*Caution*: Add the sulfuric acid to 500 ml of water in a 1 liter volumetric flask, and dilute to 1 liter when the solution has cooled to room temperature.)
4. *N*-Methyl-*N*-nitroso-*p*-toluenesulfonamide, AR (crystalline).
5. Diethyleneglycol monoethyl ether, AR.
6. Diethyl ether, AR.
7. Methanol, AR.
8. Diethyl ether–methanol (9/1, v/v). Mix 9 vol of diethyl ether with 1 vol of methanol.

PROCEDURE

1. Transfer 1.0 ml or 2.0 ml of serum to a 50 ml glass-stoppered Erlenmeyer flask. Add 3.0 ml of the KOH solution and 10 ml of methanol. Add a small carborundum boiling chip, and heat the mixture to gentle boiling for 30 min. Add 10 ml of water, and heat for another 30 min. Cool the mixture to room temperature, add 15 ml of water, and extract the nonsaponifiable lipids with three 10 ml portions of petroleum ether. Acidify the aqueous portion to pH 1 with 1 molar H_2SO_4, and add 1.0 ml of the acid in excess. Extract the fatty acids with two 10 ml portions of petroleum ether. Vaporize the petroleum ether in a rotary evaporator. Dissolve the fatty acid residue in 10 ml of the ether-methanol mixture, and transfer the mixture to a 15 ml test tube.

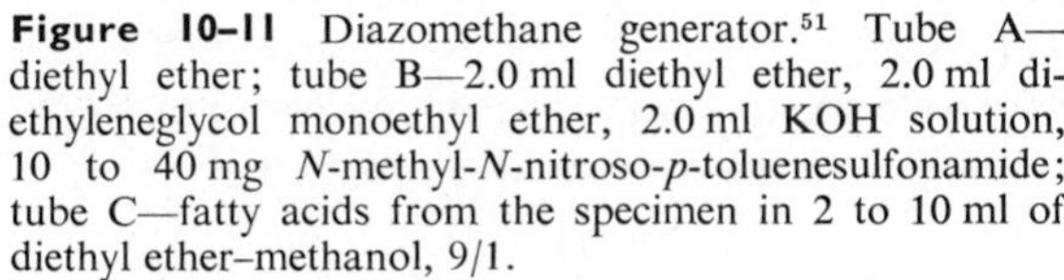

Figure 10–11 Diazomethane generator.[51] Tube A—diethyl ether; tube B—2.0 ml diethyl ether, 2.0 ml diethyleneglycol monoethyl ether, 2.0 ml KOH solution, 10 to 40 mg *N*-methyl-*N*-nitroso-*p*-toluenesulfonamide; tube C—fatty acids from the specimen in 2 to 10 ml of diethyl ether–methanol, 9/1.

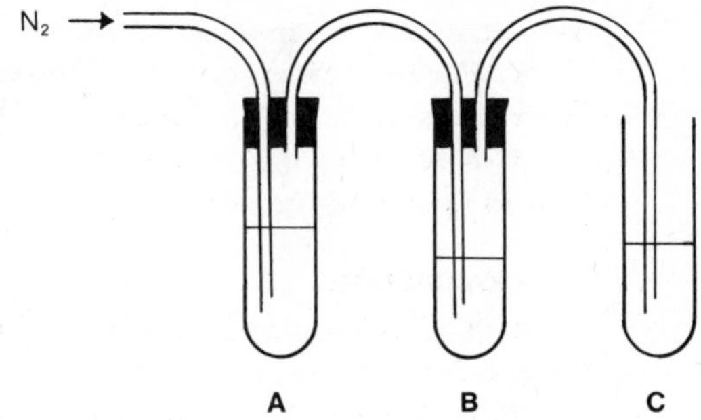

2. Into tube A of the diazomethane generator (Fig. 10–11), place 10 ml of diethyl ether. Into tube B, place 2.0 ml of diethyl ether, 2.0 ml of diethyleneglycol monoethyl ether, and 2.0 ml of 33 per cent KOH. Place the gas delivery tube of the diazomethane generator into the solution of fatty acids. Add 50 to 100 mg of *N*-methyl-*N*-nitroso-*p*-toluenesulfonamide to tube B. Close tubes A and B with rubber stoppers. Pass a gentle stream of nitrogen through the generator, carrying the diazomethane from tube B into the solution of fatty acids until the solution of fatty acids turns distinctly yellow. At that time, the fatty acids should have become methylated; the yellow color represents an excess of diazomethane.[51]

3. Transfer the solution of methyl esters to a 50 ml round-bottom flask, and vaporize the solvent with a rotary evaporator. Dissolve the residue of methyl esters in 1 ml of petroleum ether, and transfer the solution to a small vial. Vaporize the solvent in a gentle stream of nitrogen, and redissolve the residue in 5 to 15 μl of petroleum ether.

4. Inject 0.5 to 1.0 μl of the solution into the gas chromatograph. For analyses of fatty acid methyl esters, the gas chromatograph should be equipped with a 12 to 15 foot column and a hydrogen flame ionization detector. The liquid phase of the column should be diethyleneglycol succinate or ethyleneglycol succinate, and the solid support should be diatomacious earth—preferably acid washed and siliconized. The load of liquid phase in the column packing material may range from 4 per cent to 10 per cent. The column temperature should be 185 to 190°C.

STANDARDIZATION

For identification of the fatty acids present in the specimen, a standard mixture of methyl esters should be chromatographed immediately before or after the specimen. Preferably, the standard should contain known amounts of the following methyl esters: laurate, myristate, palmitate, palmitoleate, phytanate, stearate, oleate, linoleate, linolenate, arachidate, and arachidonate. The percentage of each fatty acid in the specimen can be approximated closely by determining the percentage of the total area comprised by the area under the appropriate peak on the recording. The instrument response per mass unit of each methyl ester should be constant over a wide range of specimen mass. However, the instrument response per mass unit of specimen will vary from one methyl ester to another. Therefore, a more exact percentage for each methyl ester can be determined by adjusting the peak area according to the instrument response per unit of mass.

Methyl esters from mixtures of fatty acids can be determined by gas-liquid chromatography more accurately through the incorporation of an *internal standard* into the biologic specimen (e.g., Ellefson and Mason[25]).

CALCULATIONS

1. For both specimen and standard, count from the recordings the number of area units (chart squares, integrator units, etc.) under each peak.

2. For each component of the standard, divide the number of area units by the weight per cent of that component in the standard mixture.

3. For each component of the specimen methyl esters, divide the number of area units under the peak by the respective quotient from step 2.

4. Add the quotients from step 3, and divide the quotient for each component by the total to determine the percentage of each methyl ester in the specimen mixture.

COMMENTS

1. A phytanic acid level of 1 per cent or more of the serum lipid fatty acids is diagnostic for Refsum's disease.

2. Following a 12 h fast, the ratio of per cent linoleate/per cent palmitate should be at least 0.70. A ratio much less than 0.70 indicates chronic malabsorption of linoleate.

3. A variety of methods for the preparation of fatty acid methyl esters from biologic lipids have been published. The method described here is quantitative, and the results are easily reproducible.

REFERENCES

Kahlke, W.: Klin. Wochenschr., *42*:1011, 1964.
Ellefson, R. D., and Mason, H. L.: Endocrinology, *75*:179, 1964.
Schlenk, H., and Gellerman, J. L.: Anal. Chem. *32*:1412, 1960.

FECAL LIPIDS

Fecal lipids include triglycerides, mono- and diglycerides, phospholipids, glycolipids, fatty acids and fatty acid salts (soaps), sterols, and cholesteryl esters. Most of the lipids arise from intestinal bacterial cells and epithelial cells that have sloughed normally from the intestinal mucosa. Bile carries appreciable amounts of a variety of lipids into the duodenum; however, these lipids may become absorbed by the intestinal mucosa, depending on the availability of the enzymes necessary for their digestion, which is a prerequisite for absorption.

The lipid content of feces is determined for the diagnosis of *steatorrhea*, which occurs in *pancreatic insufficiency*, *Whipple's disease*, and a small number of other *malabsorption* problems. In pancreatic disease, steatorrhea is caused primarily by deficiency of *pancreatic lipase*. This may occur with resection of the pancreas or with intrinsic pancreatic disease, such as *cystic fibrosis*, *chronic pancreatitis*, or obstruction due to *neoplasm* or *calculi*. *Enterogenous steatorrhea* may be caused by anatomic changes in the intestinal wall, such as occur in Whipple's disease, regional enteritis, tuberculous enteritis, or the atrophy of malnutrition. It may also occur in the absence of anatomic abnormalities of the intestinal wall, as in *gluten-induced enteropathy* (celiac disease) and *tropical sprue*. The former is thought to be based on allergy, while the latter may be due to folic acid deficiency or to changes in the intestinal flora, since it can be cured by antibiotics.

It was formerly believed that fractionation of the total lipids into neutral fat and free (nonesterified) fatty acids might aid in the differential diagnosis of pancreatic insufficiency from malabsorption due to other causes. Today most investigators agree that such an approach gives very little additional information. The extensive overlap in results, enzymatic and spontaneous hydrolysis of fats during collection and storage, and incorrect collection of specimens lessen the reliability of the test.

DETERMINATION OF TOTAL FECAL LIPIDS

Reliable quantitative analyses of total lipids in feces can be accomplished gravimetrically or by titrimetric measurement of the total fatty acids. In one approach, widely accepted as a standard reference method, a weighed sample is subjected to alkaline hydrolysis to convert the lipids of interest to fatty acid salts and sterols. The mixture is acidified, and the resulting fatty acids and the sterols are extracted into petroleum ether. The solvent is evaporated and the residue is weighed to obtain a measure of total lipids present. Alternatively, the residue is dissolved in ethanol and titrated with a standard solution of sodium hydroxide to obtain a measure of total fatty acids present.[58] On the average, the latter represent about 78 per cent of the total fecal lipids. Both the gravimetric and titrimetric methods provide reliable results and are recommended for metabolic or fat balance studies, and for clinical diagnostic purposes.

SPECIMEN

To obtain useful values, the patient must ingest a total of 50 to 150 g of fat/d. Also, for reliable results, all specimens should be collected over a period of 72 h and the fat excretion reported in g/d. Occasionally, analyses of random specimens or 24 h collections may be required. In this event, lipid content can be expressed as per cent of dry weight; however, owing to wide variation in the bulk content of diets, these results will be subject to more variation than those obtained from 72 h collections.

Specimens may be collected in individual glass or plastic containers or in one large tared container. Wax-coated containers should not be used. During the collection period, the fecal specimens should be refrigerated. Contamination of feces with urine should be avoided. Any obvious foreign matter should be removed before proceeding with the analysis.

REAGENTS

1. Potassium hydroxide, 6.2 mol/l. Dissolve 450 g of KOH pellets, AR, in 900 ml of distilled water.

2. Ethanol–isoamyl alcohol mixture. Add 4 ml of isoamyl alcohol to 1 liter of 95 per cent ethanol.

3. Petroleum ether, boiling point range 30 to 60°C. Mix approximately 2.5 liters of petroleum ether with 250 ml of concentrated sulfuric acid in a 3 liter flask. Agitate the mixture with a magnetic stirrer for 2 to 3 h. Transfer the petroleum ether to a second flask and wash three times by stirring with 300 ml portions of distilled water. Transfer the petroleum ether to a third flask and dry by storing with anhydrous sodium sulfate for a period of 4 h or longer. Finally, purify the dried petroleum ether by distillation; the fraction that boils between 30° and 60°C should be collected. This purification may not be necessary if the gravimetric method is used to determine fecal fat content.

4. Hydrochloric acid, 8.2 mol/l. Dilute 684 ml of concentrated hydrochloric acid to 1 liter with distilled water.

5. Sodium hydroxide, 0.1 mol/l. Dilute 6.3 ml of saturated NaOH solution to 1.0 liter with CO_2-free water. (Prepare CO_2-free water by boiling distilled water for 10 to 15 min and then protecting the water from atmospheric CO_2 as it cools. Protection from atmospheric CO_2 can be accomplished by allowing air to enter the cooling flask only through a tube filled with ascarite or soda lime.) Standardize this solution by titration against a standard solution of HCl or against a primary standard such as potassium acid phthalate.

6. Ethanol, 95 per cent.

7. Thymol blue, 0.2 g/100 ml in 50 per cent ethanol. Mix 55 ml of 95 per cent ethanol and 45 ml of distilled, CO_2-free water, and dissolve 0.2 g of thymol blue.

8. Nitrogen, CO_2-free.

PROCEDURE

1. Weigh the entire specimen. Transfer the specimen to a Waring Blendor and add a measured amount of water sufficient to allow blending of the specimen into a homogenate

of uniform consistency. Blend the specimen and water. Retain approximately 100 g of the homogenate, and discard the rest.

2. Transfer approximately 10 g of the homogenate to a tared foil pan. Weigh the sample. Heat gently on a steam plate, under a heat lamp, or on a hot plate that has a fine control; the specimen should be heated until it appears to be dry. Heating should be continued in an oven at 100°C for 12 h or until the sample is entirely dry. The dried residue is weighed.

3. Transfer a second sample of the homogenate (approximately 10 g) to a tared flask (125 to 150 ml), and weigh the sample.

4. Add 10 ml of 6.2 molar KOH; mix well. Add 40 ml of the ethanol–isoamyl alcohol mixture, and mix again. Add 2 or 3 boiling chips (Hengar Granules or carborundum chips). Attach the flask to a reflux system and boil gently for at least 30 min. Remove the flask from the reflux system, and cool the contents to room temperature.

5. Add 17 ml of 8.2 molar hydrochloric acid; mix thoroughly. Cool the contents of the flask again.

6. Add 50.0 ml of petroleum ether to the flask; stopper the flask tightly, and shake the flask vigorously for 1 min. Remove the stopper cautiously, and allow the liquid phases in the flask to separate.

7. *Gravimetric measurement.* Transfer a 25.0 ml aliquot of the petroleum ether solution into a tared beaker. Allow the petroleum ether to evaporate with gentle heating. Weigh the residue of fatty acids and sterols. *Normal values* by this method should be less than 7 g/d.

8. *Titrimetric measurement* (alternative to step 7). Transfer a 25.0 ml aliquot of the petroleum ether solution into a 50 ml Erlenmeyer flask. Add a small piece of filter paper, and gently boil off the petroleum ether. Add 10 ml of redistilled ethanol, and boil gently for a few seconds to expel CO_2. Add 250 μl of thymol blue solution. Titrate the fatty acids with 0.1 molar sodium hydroxide, using a slow stream of nitrogen for mixing during the titration. Determine a titration blank, using 10 ml of redistilled ethanol in another flask and following the same procedure as for the specimen. *Normal values* by this method should be less than 6.3 g of fatty acids/d (based on an average molecular mass of 265).

CALCULATIONS

Gravimetric measurement

$$\text{Weight of lipid in entire specimen } (W_l) = \frac{\text{Weight of lipid residue (step 7)} \times 2 \times 1.04}{\text{Weight of wet sample of specimen (step 3)}} \times \text{weight of entire homogenate.}$$

$$\text{Weight of lipid excreted in 24 h} = \frac{W_l}{\text{Number of h of collection}} \times 24\text{ h.}$$

$$\text{Weight of total solids in entire specimen } (W_s) = \frac{\text{Weight of solid residue (step 2)}}{\text{Weight of wet sample of specimen (step 2)}} \times \text{weight of entire homogenate.}$$

$$\text{Lipids as percentage of total solids} = \frac{W_l}{W_s} \times 100$$

Titrimetric measurement

$$\text{g of fatty acids per entire specimen } (W_{fa}) = \frac{(V_u - V_b) \times C \times 2 \times 1.04 \times 265}{\text{Weight of wet sample of homogenate} \times 1000} \times \text{weight of entire homogenate.}$$

$$\text{g of fatty acids excreted in 24 h} = \frac{W_{fa}}{\text{Number of h of collection}} \times 24 \text{ h.}$$

where V_u = titration volume for unknown,
V_b = titration volume for blank,
C = concentration of base.

COMMENTS

1. An entirely definitive diagnosis of lipid malabsorption may require a carefully controlled balance study. In such a case a "coefficient of fat retention" can be determined. This is simply the number of grams of lipid excreted in the feces expressed as a percentage of the number of grams of lipid ingested. A study period of 72 h or more is necessary for the determination of a reproducible "coefficient."

2. The fat content of foods can be determined with this method, and in balance studies this method should be used for that purpose, also.

3. Generally it has been found that the determination of total lipids is the most useful test in detecting steatorrhea. The fractionation of the total lipids into fatty acids and neutral fats is of little additional value. If desired, the neutral fat alone can be determined quite simply. The stool is acidified with H_2SO_4 or HCl; the lipids are extracted with petroleum ether after the addition of ethanol. The lipid extract is made alkaline with sodium hydroxide in aqueous alcohol. The fatty acids form soaps and pass entirely into the alcohol phase, leaving the neutral lipids in the petroleum ether phase. For greatest accuracy, the extracted neutral lipids should be carried through the saponification and re-extraction into petroleum ether before quantitation. (The fatty acid salts, or soaps, are not soluble in petroleum ether.)

NORMAL VALUES

The normal adult daily excretion of all fats in feces is no more than 8 g/d. This includes lipids which are not detected by the methods described in detail, e.g., sterols, sphingolipids, and glyceride-glycerol. The normal range, expressed as percentage of dry weight, is 15 to 25 per cent. The upper limit for normal values determined by the described gravimetric method is 7 g per day, and the upper normal limit by the titrimetric method is 6.3 g per day.

REFERENCE

Van de Kamer, J. H.: Total Fatty Acids in Stool. *In* Standard Methods of Clinical Chemistry. D. Seligson, ed. Academic Press, New York, 1958, Vol. II, pp. 34–39.

FAT IN URINE (CHYLURIA)

Normally, urine contains small amounts of complex lipids; however, no more than traces of triglycerides can be found. Ordinarily, even persons who have hyperlipidemia excrete no more than traces of triglycerides in urine.

Extensive trauma, especially the type associated with crushing lesions or multiple fractures of the long bones, frequently results in the mobilization of lipid particles into the circulating blood. Some such particles, known as *lipid emboli*, are large enough to block the circulation through the capillary arterioles, and the destruction of nearby cells occurs. Lipid emboli can cause brain damage, pneumonia, and necrosis of bone. The emboli are composed predominantly of triglyceride. Following the appearance of lipid emboli in the blood, an excretion of triglycerides in urine may occur, so the appearance of measurable amounts of triglycerides in urine is an indication of possible fat embolism.

Filariasis can be a second cause of appearance of triglycerides in urine.[14] In this condition, a block may occur between the intestinal lymph vessels and the thoracic duct, causing renal lymph varices which rupture into the renal tubules. Lymph that is rich in chylomicrons from intestinal fat absorption then may flow through the renal tubules and cause chyluria.

Tissue damage resulting from *kidney stones*, also, can cause the appearance of triglycerides in the urine.[24]

Triglycerides in urine that exist as fat droplets will tend to float to the top of an undisturbed specimen; the process can be hastened by centrifugation. The particles may be observed as spherical objects or grape-like clusters under the microscope. Visualization and confirmation can be improved by staining the fat globules with Sudan III.

PROCEDURE FOR EXAMINING FAT GLOBULES IN URINE[20]

1. Centrifuge about 15 ml of well-mixed urine in a clean tube for 15 min.
2. Use a medicine dropper to take off a small amount of urine from the surface of the centrifuged specimen, and transfer two separate drops to a slide. Add a drop of Sudan III stain (saturated solution in 70 percent ethanol) to one drop and cover both drops with coverslips.
3. Examine the unstained specimen under a microscope for fat globules. Examine the stained specimen for round droplets of fat which are colored red-orange with the stain. The staining reaction may require 4 or 5 min. Mineral oil, sometimes used as a lubricant, may be found in urine of some patients; it will also stain with Sudan III.

DETERMINATION OF TRIGLYCERIDES IN URINE

Quantitative determination of triglycerides in urine usually is not necessary; however, this test can be helpful in establishing the presence of emboli, and it may be accomplished by extracting the urine with petroleum ether, evaporating the solvent, and analyzing the residue.[24] The triglycerides excreted in the urine after a bone fracture are not bound by protein and are readily soluble in the solvent. Petroleum ether is not a satisfactory extractant for lipids bound in the form of lipoproteins.

Procedure

1. Transfer about 200 ml of well-mixed urine (single specimen or 24 h collection) to a separatory funnel.
2. Add 100 ml of petroleum ether (boiling point range 30 to 60°C). Shake vigorously for 1 min, let the phases separate, and discard the lower aqueous phase.
3. Add 50 ml of water to the extract, and shake vigorously again; let the phases separate, and discard the lower aqueous phase.
4. Transfer the extract to a clean round-bottom flask and evaporate the solvent with a rotary evaporator.
5. The residue may be dissolved in the appropriate solvent and processed in a manner similar to that used for the standards in one of the methods described for the quantitative determination of triglycerides in serum.
6. The triglycerides concentration of the urine should be converted to μg of triglycerides excreted per diem.

NORMAL VALUES

Normal values for triglycerides in urine are less than 20 μg/d.

LIPOPROTEINS IN SERUM

The value of the lipoproteins tests lies in the recognition and identification of *hyperlipoproteinemias*,[31] identification of *abnormal lipoproteins*,[24,31] and recognition of *subnormal amounts* and *deficiencies of specific normal lipoproteins*.[30] Unusual and abnormal serum lipoprotein profiles may be the result of inherited traits, they may reflect disorders and diseases of metabolism, or they may be the result of infectious diseases.

Numerous studies have demonstrated an association between serum lipids and atherosclerosis.[72] The major lipid fractions have been investigated in this respect, but the actual mechanism by which atheromatous plaques are formed remains obscure. The prospect that the physicochemical state of the plasma lipids might play an important role in this process has been one factor that has stimulated studies of the individual lipoproteins.

Plasma lipoproteins may be separated by *ultracentrifugation* into four major fractions, listed in Table 10–3. These fractions show differences in electrophoretic mobility and chemical composition. Studies by Fredrickson and co-workers demonstrated the value of separating lipoproteins by electrophoresis and led to a system of classification of hyperlipoproteinemias.[31] In 1970 a panel, under the auspices of the World Health Organization, concluded that the most rational basis for describing hyperlipidemias at present is in terms of the concentrations of the individual lipoproteins in serum.[6]

The *hyperlipidemias* include six types of inheritable hyperlipoproteinemias; these types frequently are referred to as lipoprotein phenotypes.[31] The prominent characteristics of these hyperlipoproteinemias are given in Table 10–4. Characteristics of representative specimens are illustrated in Fig. 10–12, and characteristic electrophoregrams of the blood lipoproteins are illustrated in Figs. 10–13, 10–15, and 10–16. The illustrations in Fig. 10–13 represent idealized lipoprotein electrophoregrams according to the original definitions of the hyperlipoproteinemias. Figures 10–14, 10–15, and 10–16 show the actual electrophoretic mobilities of the VLDLs, LDLs, and HDLs, and the several biochemical variants of the hyperlipoproteinemias which exist within the accepted types.[24] Recognition of the existence of these variants of the hyperlipoproteinemias is important for accurate analysis of lipoprotein profiles; without knowledge of the variants indicated, the analysis of lipoprotein electrophoregrams can readily become confusing. Classification of inherited

TABLE 10–3 PROPERTIES OF LIPOPROTEINS

	Chylomicrons	VLDL	LDL	HDL
Density, g/ml	<0.95	0.95–1.006	1.006–1.063	1.063–1.21
Diameter, nm	120–1100	30–70	17–26(β)	7.4–9.0 (α_1)
Molecular mass (in millions)	10^3–10^4	5–100	2–3	0.25
Flotation rate (S_f value)	>400	20–400	0–20	—
Electrophoretic mobility (paper or agarose)	origin	β, pre-β, α_2, α_1	β, pre-β, α_2, α_1	α_1, α_2
Approximate composition* (per cent of dry weight)				
triglycerides	80–95	40–80	10	1–5
total cholesterol	2–5	10–40	45	20
phospholipids	3–6	15–20	20	30
protein	1–2	5–10	25	45–50

Abbreviations: VLDL, very low density lipoproteins; LDL, low density lipoproteins; HDL, high density lipoproteins.

S_f values represent the negative sedimentation coefficient in Svedbergs in NaCl solution (1.063 g/ml) at 26°C.

* Approximate compositions refer only to chylomicrons, pre-β VLDL, β LDL, and α_1 HDL; compositions of other VLDLs, LDLs, and HDLs are not indicated.

TABLE 10-4 FAMILIAL HYPERLIPOPROTEINEMIA PHENOTYPES[31]

Type	I	IIa	IIb	III	IV	V
Age of detection	Early childhood	Any age	Any age	Adult	Adult	Early adult
Clinical features	Eruptive xanthomas; lipemia retinalis; hepatosplenomegaly; abdominal pain	Tendinous, tuberous xanthomata; arcus cornea; xanthelasma; premature ASVD	Obesity; premature ASVD; no xanthomas; xanthelasmas may occur	Planar, eruptive, tuberous xanthomata; accelerated ASVD; arcus cornea	Obesity; premature ASVD; hyperuricemia; hepatosplenomegaly	Abdominal pain; lipemia retinalis eruptive xanthomas; hepatosplenomegaly; premature ASVD
Criteria	Chylomicrons present VLDL normal or sl. ↑ LDL and HDL normal or ↓	LDL ↑, VLDL normal	LDL ↑, VLDL ↑	VLDL with increased chol. content and abnormal mobility. LDL normal or ↓	VLDL ↑, LDL normal, no chylomicrons	Chylomicrons present VLDL ↑
Chylomicrons	↑	No	No	Slight ↑	No	
Cholesterol	↑	↑	↑	↑	Normal or ↑	↑
Triglycerides	↑	Normal	↑	↑	↑	↑
Chol/TG	<0.2	>2.0	variable	0.6–1.9	<0.6	<0.6
Treatment	Restrict dietary fat.	Low cholesterol diet. Reduce weight to normal. Substitute unsaturated fat for saturated. Chemotherapy.	As for IIa. May not require chemotherapy	Restrict fat and carbohydrate. Chemotherapy.	Restrict carbohydrate, limit alcohol. Chemotherapy.	Restrict fat and carbohydrate, increase protein intake. Chemotherapy

Abbreviations: ASVD = arteriosclerotic vascular disease; ↑ = increased; ↓ = decreased; < = less than; > = greater than. Chol/TG = cholesterol: triglyceride ratio.

Chemotherapy: clofibrate, thyroxine, nicotinic acid, cholestyramine.

hyperlipoproteinemias according to phenotype is important, since dietary management and drug therapy are largely dependent on this information.

CLASSIFICATION AND CLINICAL SIGNIFICANCE

Type I hyperlipoproteinemia, characterized by *chylomicronemia*, results from a low level or a lack of *lipoprotein lipase activity* in the vascular system.[31] The biochemical defects responsible for the *type II* and *type III* profiles have not been identified. The type II and type III profiles appear to be primarily reflective of disorders of lipid metabolism. While

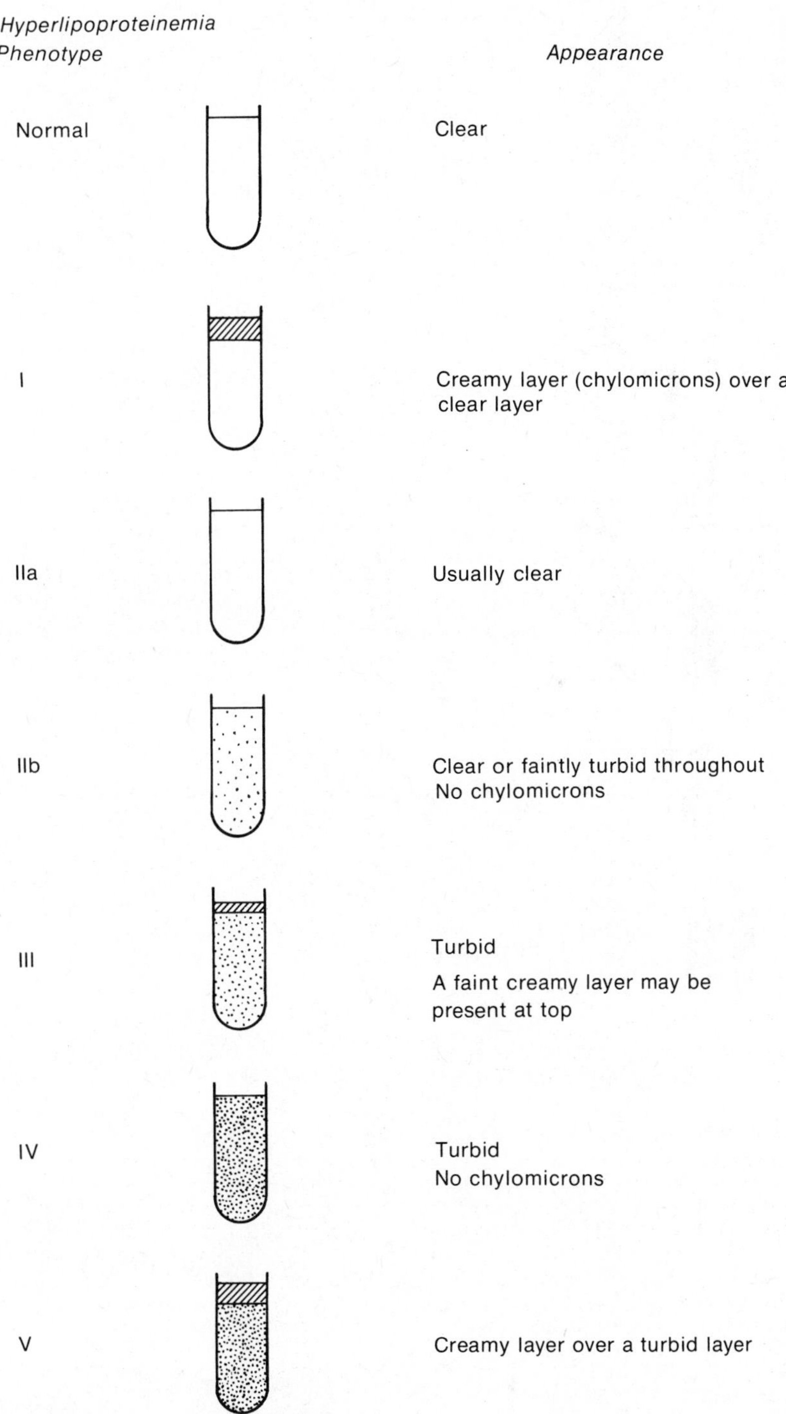

Figure 10–12 Typical appearance of serum after standing 18 h at 4°C.[31]
Creamy layer of chylomicrons.
Turbid layer associated with increased pre-β lipoprotein.

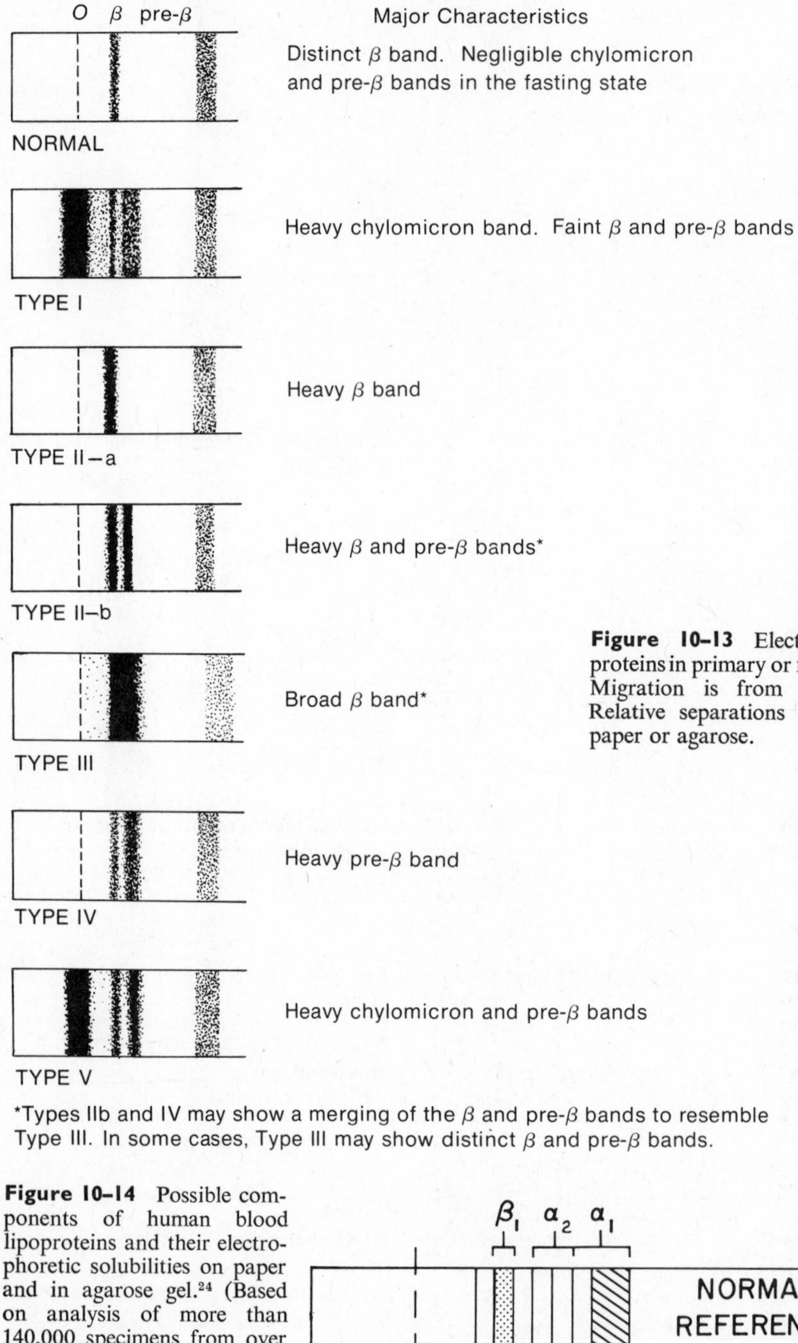

O β pre-β

Major Characteristics

NORMAL — Distinct β band. Negligible chylomicron and pre-β bands in the fasting state

TYPE I — Heavy chylomicron band. Faint β and pre-β bands

TYPE II–a — Heavy β band

TYPE II–b — Heavy β and pre-β bands*

TYPE III — Broad β band*

TYPE IV — Heavy pre-β band

TYPE V — Heavy chylomicron and pre-β bands

*Types IIb and IV may show a merging of the β and pre-β bands to resemble Type III. In some cases, Type III may show distinct β and pre-β bands.

Figure 10–13 Electrophoretic patterns of lipoproteins in primary or familial hyperlipoproteinemia.[3] Migration is from left to right. "O" = origin. Relative separations are those typically found on paper or agarose.

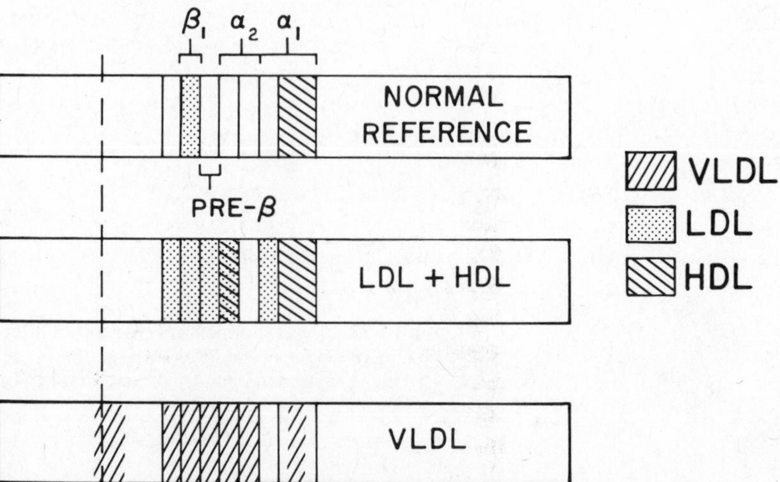

Figure 10–14 Possible components of human blood lipoproteins and their electrophoretic solubilities on paper and in agarose gel.[24] (Based on analysis of more than 140,000 specimens from over 80,000 patients.)

Top: Major components in normal serum after a 12 h fast—β_1 LDL and α_1 HDL; *middle:* forms of LDLs and HDLs that can exist in patients' specimens—β_2 LDL, β_1 LDL, pre-β LDL, α_2 LDL, α_1 LDL, α_2 HDL, α_1 HDL; *bottom:* forms of VLDLs that can exist in patients' specimens—chylomicrons (origin), β_2 VLDL, β_1 VLDL, pre-β VLDL, two α_2 VLDLs, α_1 VLDL.

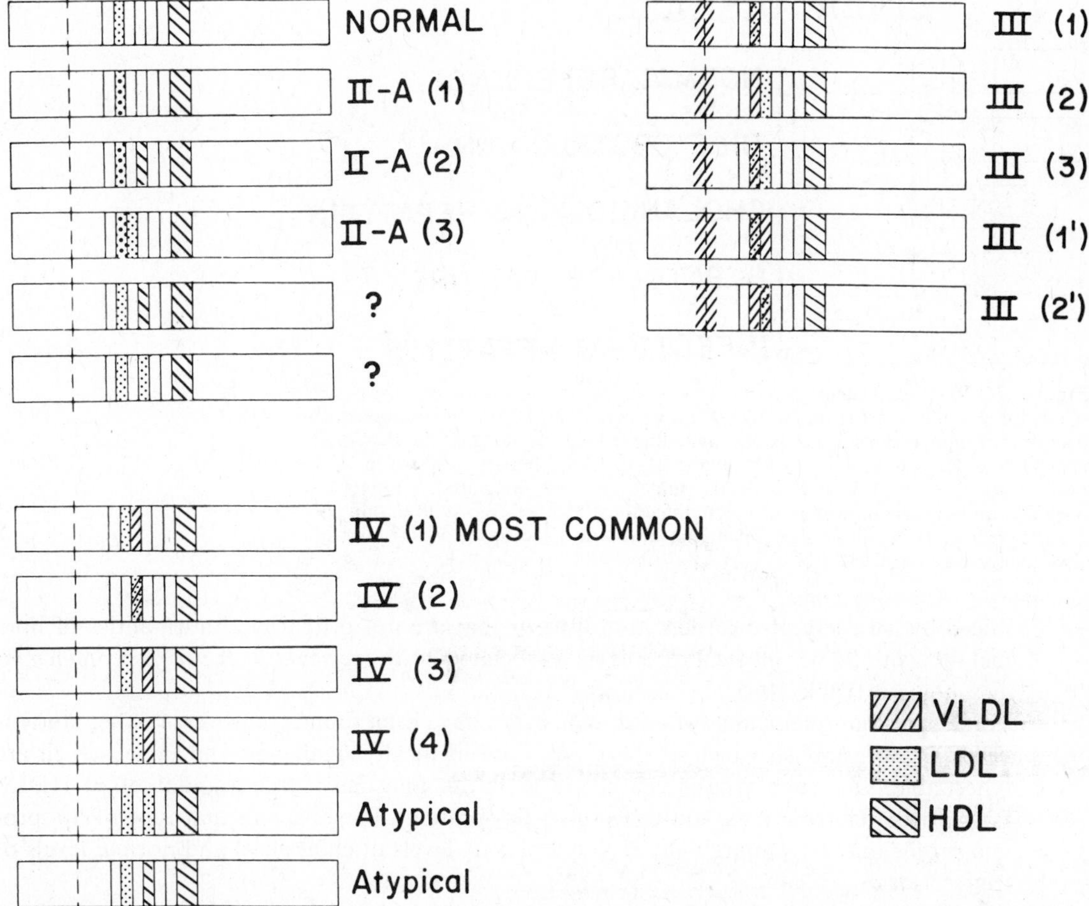

Figure 10–15 Major variants of whole serum lipoprotein electrophorograms that are characteristic of the hyperlipoproteinemias.

Type II-A: (1) Excess of β LDL, normal β_1 HDL; (2) excess of β LDL, excess of α_2 HDL, normal α_1 HDL; (3) excess of β LDL, presence of pre-β LDL, normal α_1 HDL; (?) atypical profile, hypercholesteremia resulting from excess of α_2 HDL; β LDL and α_1 HDL are normal; (?) atypical profile, hypercholesteremia resulting from excess of α_2 LDL; β LDL and α_1 HDL are normal.

Type III: (1) Typical β VLDL with normal β LDL and α_1 HDL; (2) typical β VLDL with pre-β LDL and normal α_1 HDL; (3) typical β VLDL with pre-β VLDL and normal β LDL and α_1 HDL; (1′) same components as (1) plus pre-β VLDL; (2′) same components as (2) plus pre-β VLDL. Combinations of β and pre-β bands in variants 2, 3, 1′, and 2′ are in appearance similar to the typical "broad β band."

Type IV: (1) Excess of pre-β VLDL with normal β LDL and α_1 HDL; (2) excess of pre-β VLDL with pre-β LDL and normal α_1 HDL; (3) excess of α_2 VLDL with normal β LDL and α_1 HDL; (4) excess of α_2 VLDL with pre-β LDL and normal α_1 HDL.

Atypical profiles that can be mistaken for type IV profiles: normal β LDL and α_1 HDL with excess of α_2 HDL (upper) and excess of α_2 LDL (lower).

hyperlipoproteinemias of *types IIa and IIb* can occur either as *inheritable* traits or *secondary* to a variety of diseases, *type III hyperlipoproteinemia* appears to be entirely *an inheritable trait*. The β *VLDL* that is the prominent characteristic of the type III profile has not been detected in normal plasma or serum; it contains a greater proportion of cholesteryl esters than does the normal VLDL, and it is thought to be related to β LDL by an abnormal loading with triglycerides or by an impairment of the unloading of triglycerides from a VLDL that would be a normal precursor to β LDL. The *type IV and V profiles* contain excesses of VLDLs that appear as prominent *pre-*β[31] *and* α_2[24] *bands* on lipoprotein electrophoregrams. The type IV and V profiles often reflect a sensitivity to dietary carbohydrates.

In addition to being reflective of inherited disorders that seem to be primarily disorders of lipid metabolism, the type IV and V profiles, as well as the type II profiles, can be in-

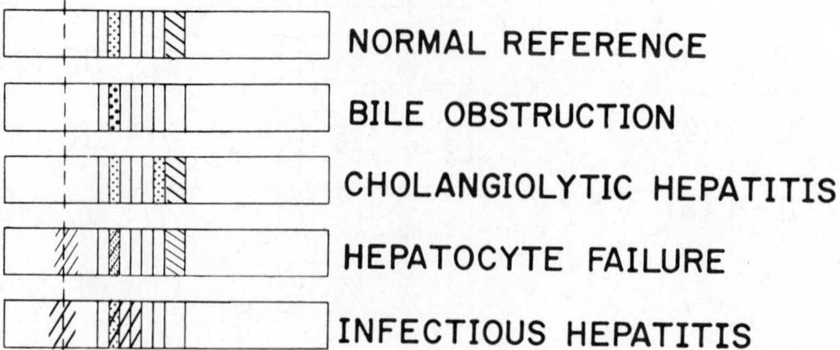

Figure 10–16 Lipoprotein profiles representative of some liver diseases.[24] (Code for shading is in Figure 10–15.) Bile obstruction: α_1 HDL absent, abnormal complex of lipoproteins in β band, including LP-X; cholangiolytic hepatitis: normal β LDL and α_1 HDL with abnormal α_1 LDL; hepatocyte failure: low concentration of β LDL and α_1 HDL, sometimes with increased amounts of β VLDL and chylomicron-like particles; infectious hepatitis: normal amount of β LDL with α_1 HDL diminished or absent, and with excesses of VLDL and of chylomicron-like particles.

duced by a variety of disorders and diseases that are not primarily abnormalities of lipid metabolism. Some medical problems associated with *secondary hyperlipoproteinemias* are shown in Table 10–5.

Hyperlipoproteinemias of additional types have been found; however, further study is necessary to establish which of these types are medically significant and, also, which are inheritable. One such type is characterized by an unusually large amount of α_2 HDL[24] (sometimes referred to as "sinking pre-β" lipoprotein). Persons with that type of lipoprotein profile may have marginally elevated plasma levels of cholesterol and normal levels of triglycerides.

Two inheritable forms of **lipoprotein deficiency** (hypolipoproteinemia) have been recognized. *Abetalipoproteinemia* occurs rarely, and is accompanied by *acanthocytosis* and in some cases by a *sensory neuropathy*; this condition is known also as *Bassen-Kornzweig disease*. A deficiency of α_1 HDL, known as *Tangier disease*,[30] is very rare. A striking feature of this condition is a massive enlargement of the tonsils with an abnormal storage of cholesteryl esters in tonsil tissue. *Sensory neuropathy* is a feature of Tangier disease, also.

An evaluation of the blood lipoprotein profile, including determinations of serum total cholesterol and total triglycerides values, is a more satisfactory test than is the measurement of serum cholesterol and triglycerides values only. In many cases, *quantitative analyses* of lipoprotein components are necessary to ascertain whether excesses or abnormally low

TABLE 10–5 DISEASES OR CONDITIONS ASSOCIATED WITH "SECONDARY" HYPERLIPOPROTEINEMIAS[24,31]

Disease or Disorder	*Types of Hyperlipoproteinemia*
Acute intermittent porphyria	IIa, IIb
Alcoholism	I, IV, V
Dysgammaglobulinemia	IIb, IV, V
Estrogen therapy, pregnancy	IV
Hypothyroidism	IIa, IIb, IV
Insulin-dependent diabetes (uncontrolled)	IIb, IV, V
Nephrotic syndrome	IIb, IV, V
Pancreatitis	IV, V
Steroid therapy	IV, V

levels of lipoproteins of specific classes may exist.[24] Usually, however, an adequate analysis of serum lipoproteins can be accomplished through electrophoresis and simple inspection of the electrophoregrams. In addition, the appearance of the plasma or serum after refrigerated storage for 12 to 24 hours can be helpful in determination of the type of hyperlipoproteinemia that is present. (Variations in specimen appearance are illustrated in Fig. 10–11.)

Quantitative analyses of the lipoproteins may be carried out with the analytical ultracentrifuge or by separation of the lipoprotein classes followed by spectrophotometric or colorimetric quantitations of lipid and/or protein components.

For analytic ultracentrifugation, a salt usually is added to plasma to raise its density to 1.063 or to 1.21 g/ml; the specimen then is centrifuged at high speeds at a constant temperature of 26°C. The moving boundaries of the floating lipoproteins are photographed at intervals to determine the rates of flotation (S_f values). Additional optical measurements allow for the quantitation of the fractions. Chylomicrons are not generally measured as such in the analytical ultracentrifuge, but are separated first and analyzed independently. Quantitative measurements during analytical ultracentrifugation usually are made of the LDL ($S_f = 0$ to 20), the VLDL ($S_f = 20$ to 400), and several subclasses of those components.

Plasma has a salt density of approximately 1.006 g/ml. Ultracentrifugation of plasma alone for about 1 min at 100,000 $\times$ g brings the chylomicrons to the top of the tube. Longer ultracentrifugation at this density allows the VLDL to be collected at the surface. LDLs are commonly isolated between densities of 1.006 and 1.063, and HDLs between densities of 1.063 and 1.21 (Table 10–4).

Lipoprotein fractions may be isolated by preparative ultracentrifugation, by selective precipitation, or by chromatography and electrophoresis; quantitation may be accomplished by spectrophotometry or chemical analysis of one or more of their components.

LIPOPROTEIN ELECTROPHORESIS

Electrophoresis of the serum lipoproteins can be accomplished very satisfactorily using either *paper*[26] or *agarose gel*[37,46] as the support medium. The relative positioning of the lipoprotein bands in agarose electrophoregrams is very similar to the positioning in paper electrophoregrams.[24]

Electrophoresis in paper is described, because it is simple and convenient, and the electrophoregrams are satisfactorily reproducible. Electrophoregrams of the lipoproteins are evaluated simply by visual analysis. The important features of lipoprotein electrophoregrams are illustrated in Figs. 10–14, 10–15, and 10–16.

A variety of papers have been evaluated, and the most satisfactory papers are Whatman No. 1 and Whatman 3MM. Because it is transparent, agarose gel allows for greater sensitivity in the detection of faint lipoprotein bands than does paper. Also, streaking in the lipoproteins occurs to a lesser extent in agarose electrophoregrams than in paper electrophoregrams. The quality of paper electrophoregrams depends on the concentration of the buffer, the period of time allowed for electrophoresis, and the choice of vehicle for the staining bath.[24]

Starch gel and *agar gel* are satisfactory support media, also.[24] *Cellulose acetate* has been promoted as an electrophoresis support medium and it is used in many laboratories for lipoprotein electrophoresis. Cellulose acetate presents certain problems, however. Lipids and lipoproteins can be lost easily from the cellulose acetate electrophoregrams into the staining medium. Also, with cellulose acetate, specimens that contain excesses of chylomicrons yield poor electrophoregrams; the chylomicrons move through the electrophoregrams and distort the electrophoretic mobilities of all of the lipoprotein bands.[24] Certain new preparations of cellulose acetate are supposed to eliminate the problems of staining and of chylomicron interference; nevertheless, the problems persist at this time.

References to electrophoresis of lipoproteins in agarose gel[37,46] and in cellulose acetate[7,17] are given at the end of this chapter.

Electrophoresis in *polyacrylamide gel* is a useful tool for special lipoprotein analyses;[24] however, it is *not* an appropriate method for the routine electrophoresis of lipoproteins. In polyacrylamide gel (3.75 g/100 ml), *atypical* VLDLs and VLDLs that are characteristic of the type IV profile do not migrate; therefore, those fractions cannot be distinguished from each other or from the chylomicrons, which also remain stationary. In polyacrylamide gels of lower concentrations (between 2 and 3.75 g/100 ml) the VLDLs move more slowly, but they fail to form bands that would be readily recognizable. Electrophoresis in polyacrylamide is useful in identifying the abnormal β VLDL that is characteristic of type III hyperlipoproteinemia, since these VLDLs move slowly as a band in 3.75 g/100 ml gel;[6] also, it is useful for special studies of LDL and HDL components.[24]

PROCEDURE FOR ELECTROPHORESIS OF LIPOPROTEINS[26]

SPECIMEN

For reliable classification of the lipoprotein profile, the patient should fast for at least 12 h before blood is drawn for analysis. Either plasma or serum may be used, although most workers prefer plasma. EDTA (1 mg/ml of blood) is the anticoagulant most commonly employed, because it binds divalent cations that promote changes of lipoproteins. Ideally, the plasma or serum should be separated promptly from the cells, stored at 0 to 4°C, and analyzed the same day it is drawn. Specimens stored longer than 7 days show variable results. Generally, frozen specimens should not be used because both VLDLs and chylomicrons aggregate and deteriorate during freezing and thawing.

REAGENTS

1. Barbital buffer, 0.058 mol/l, pH 8.6. Dissolve 82.4 g of sodium barbital, 11.2 g of barbital, and 2.97 g of EDTA disodium salt in 8.0 liters of distilled water. Mix well and store at room temperature.
2. Albumin-barbital buffer. Dissolve 10.0 g of albumin (Bovine Plasma Albumin, fraction V from bovine plasma) in 1 liter of 0.058 molar barbital buffer. Mix well, and store in a refrigerator.
3. Oil Red O stain. In a cylinder 6 inches in diameter and 18 inches high, dissolve 4 g of Oil Red O (Harleco) in 3800 ml of acetone. Stir gently with a magnetic stirrer and slowly add 3000 ml of distilled water. Stir gently with heating at 45°C until 1 h before use. This bath of stain will accommodate 100 electrophoregrams. Discard after using once. (For small batches of electrophoregrams, the quantities of dye and vehicle should be scaled down.)

PROCEDURE[26,36]

1. Fill the buffer reservoirs of the electrophoresis cells with 0.058 molar barbital buffer.
2. Load Whatman 3MM paper strips (1 inch wide) onto support racks and thoroughly soak them in albumin-barbital buffer. Then load the strips into the electrophoresis cells and allow them to drain for 2 h or longer, but not longer than 24 h.
3. Transfer 10 μl aliquots of serum or of serum fractions from ultracentrifugation to the paper strips with a double wire applicator.
4. Apply a DC potential of 200 volts to the strips for 5 to 6 h.
5. Remove the electrophoregrams from the cells and dry quickly in an oven at 110 to 115°C, or under heat lamps.
6. Stain the electrophoregrams by soaking in the saturated bath of Oil Red O for 2 to 3 h. Stop the heating and stirring of the staining bath 1 h before the bath is used. Gentle agitation of the bath is necessary during the staining period, and this can be accomplished

satisfactorily by suspending the electrophoregrams in the bath from a wire hanger and applying a gentle, dunking motion. The dunking motion can be provided by a crankshaft attached to a small, inexpensive gear-motor.

7. Rinse away the solvent of the staining bath and much of the excess stain by soaking the electrophoregrams in a cylinder of water for 30 to 60 min.

8. Allow the electrophoregrams to dry. Additional excess stain can be removed from the electrophoregrams, if desired, by soaking them in petroleum ether for 30 min; however, this step is unnecessary for routine evaluation of the lipoprotein profile.

9. The electrophoregrams can be evaluated by comparing them with the interpretive diagrams in Figures 10–14, 10–15, and 10–16.

COMMENTS

1. Chylomicrons form a distinct band on the electrophoregrams at the line of application of the specimen, and some of the chylomicrons streak out toward the β lipoproteins.

2. The VLDLs of the type IIb and type IV profiles usually are pre-β VLDLs; in some cases however, the VLDLs are α_2 VLDLs. Pre-β VLDLs and α_2 VLDLs are distinctly different complexes—not artifacts—and occasionally both forms of VLDL are found in a single specimen.[24] Apparently, excessses of the pre-β VLDLs and the α_2 VLDLs represent two biochemical variants of type IIb and of type IV profiles.

3. In electrophoregrams of whole serum, pre-β VLDL is indistinguishable from pre-β LDL; and the α_2 VLDL, α_2 LDL, and α_2 HDL are indistinguishable from each other.[24] Preliminary fractionation of the specimen by ultracentrifugation eliminates this problem of lipoprotein identification.

4. Cholesterol-rich β VLDLs are characteristic of type III hyperlipoproteinemia; however, atypical β VLDLs that are not rich in cholesterol are found in some specimens. For the diagnosis of type III hyperlipoproteinemia, the ratio of cholesterol/triglycerides in the VLDL should be determined; when the ratio is greater than 0.42, the type III profile is very likely, and when the ratio is less than 0.35, the β VLDL probably is atypical.[6] Ratios between 0.35 and 0.42 are equivocal. The cholesterol/triglycerides ratio of the type III β VLDL usually is 0.50 or greater. Frequently, however, the type III β VLDL is accompanied by pre-β VLDL;[24] the cholesterol/triglycerides ratio of the pre-β VLDL usually is 0.20 to 0.32. The amount of pre-β VLDL that accompanies the type III β VLDL, therefore, will determine whether the ratio for the total VLDL will be clearly indicative of the type III disease.

5. In cases of *bile obstruction*, the α_1 HDL band may stain lightly or may not be visible on the electrophoregram.

6. In cases of *cholangiolytic hepatitis*, a band of LDL is found on the electrophoregrams behind and contiguous with the α_1 HDL band;[24] in the diagrams of Figure 10–15 this band is designated as α_1 LDL. Confirmation of the presence of α_1 LDL requires a centrifugal separation of the LDL from the HDL at a density of 1.063 and electrophoresis of the LDL and of the HDL separately.

7. Some investigators have recommended that the serum or plasma specimens should be preserved with EDTA. In the author's experience, the use of EDTA has been unnecessary. However, prompt ultracentrifugation and electrophoresis of the specimens and refrigeration during any necessary periods of storage are essential for reliable analyses.

8. The barbital buffer in the electrophoresis cells is useable for a period of 3 to 4 weeks, but the polarity of the electrophoresis cells should be alternated in successive periods of use.

FRACTIONATION OF LIPOPROTEINS BY PREPARATIVE ULTRACENTRIFUGATION

An analysis of a lipoprotein profile through electrophoresis of the whole serum may not be entirely definitive, because β, pre-β, α_2, and α_1 bands in electrophoregrams of whole serum may contain either VLDLs or LDLs or both, and both α_2 and α_1 may contain HDLs (Fig. 10–13).[24] In such a case, ultracentrifugation of the sample at its natural density (1.006 g/ml) may be carried out; then, separate electrophoregrams for the VLDLs (top fraction), for the LDLs + HDLs (bottom fraction), and for the whole serum are prepared. For most specimens that require ultracentrifugation, fractionation in a medium of density 1.006 g/ml is adequate. In some cases, recentrifugation of the LDL + HDL fraction at a density of 1.063 is necessary.

REAGENTS

1. Sodium chloride solution, density of 1.006 g/ml (1.10 per cent w/w). Dissolve 11.0 g of sodium chloride, AR, in 989.0 g of distilled water.
2. Sodium chloride solution, density of 1.074 g/ml (10.4 per cent w/w). Dissolve 104.0 g of sodium chloride, AR, in 896.0 g of distilled water.

PROCEDURE

1. Dilute 2.0 ml aliquots of serum with 1.0 ml of sodium chloride solution, density 1.006, in polycarbonate centrifuge tubes, 11 mm × 77 mm. Dilution of the serum is not essential, but 2.0 ml of serum is a sufficient sample, and diluting the serum to the capacity of the centrifuge tube increases the distance of separation between the VLDLs and the LDLs at the end of the spin.
2. Centrifuge the specimens at forces of 100,000 to 120,000 × g for 12 to 15 h.* The tubes need not be capped.
3. Withdraw the floating layers of VLDLs from the centrifuge tubes with disposable transfer pipettes; approximately 0.5 ml of fluid containing the VLDLs must be removed from each tube.
4. Remove the LDL + HDL fraction from the bottom one-third of each tube. Analyze the VLDL and LDL + HDL fractions by electrophoresis, as described for whole serum.
5. In order to identify certain components of some lipoprotein profiles, a separation of the LDLs from the HDLs is necessary; this is best accomplished by re-spinning the LDL + HDL fraction from the first spin in a medium of density 1.063. Ordinarily the LDLs and HDLs are respun at density 1.063 only when the electrophoretic analysis of the specimen indicates the necessity to do so. In preparation for fractionation at density 1.063, dilute 0.5 ml of the LDL + HDL solution with 2.5 ml of saline of density 1.074. (A counterbalance tube must be prepared, using 0.5 ml of saline of density 1.006 and 2.5 ml of saline of density 1.074.)

REFERENCES

Fredrickson, D. S., and Levy R. I.: Familial hyperlipoproteinemia. *In* The Metabolic Basis of Inherited Disease. 3rd ed. Stanbury, J. B., Wyngaarden, J. B., and Fredrickson, D. S., Eds. McGraw-Hill, New York, 1972, pp. 545–614.

Fejfar, Z., Fredrickson, D. S., and Strasser, T.: Circulation, *45*:501–508, 1972.

Ellefson, R. D., Jimenez, B. J., and Smith, R. C.: Mayo Clin. Proc. *46*:328–332, 1971.

QUANTITATION OF LIPOPROTEINS

Quantitative determinations of individual lipoproteins are helpful when the serum total cholesterol and total triglycerides values are marginal and when the intensities of stain in the bands of the lipoprotein electrophoregrams are equivocal.

The technique of lipoprotein quantitation by an *optical scanning of electrophoregrams* has been popular. This approach to quantitation, however, has not proven to be satisfactory. The staining of lipoprotein electrophoregrams by the usual methods—with lipid stain or by ozonolysis followed by exposure to Schiff's aldehyde reagent—is too variable to allow accurate and reproducible quantitation, even with very careful control and standardization. The prospect of lipoprotein quantitation by optical scanning of electrophoregrams has been

* The Beckman Type 35 rotor with special adapters (Ellefson) will accommodate 42 specimens, and the IEC Type A-192 rotor with standard adapters will accommodate 24 specimens.

evaluated carefully, using a variety of electrophoresis support media, several approaches to staining, and a variety of optical instruments.[24]

Although the optical scanning of lipoprotein electrophoregrams has not proven to be a satisfactory basis for accurate quantitation, it can be a useful method for the communication and storage of information. The tracing from an optical scanner might be more convenient than the stained electrophoregram for addition to a patient's history file or for the storage of data in the laboratory.

While quantitative analysis using analytical ultracentrifugation can be accomplished with relative ease, the procedure is impractical for routine use in most, if not all, clinical laboratories. A more practical approach to the quantitation of lipoproteins in routine laboratories consists of isolation of the lipoproteins by the major classes—VLDL, LDL, and HDL—followed by quantitation of the cholesterol, triglycerides, phospholipids, and/or protein content of each fraction. Currently, the cholesterol content of each lipoprotein fraction is being used quite extensively as a basis for lipoprotein quantitation. The isolation of fractions of the lipoproteins for quantitation can be accomplished satisfactorily by a combination of preparative ultracentrifugation and selective precipitation. The VLDLs are separated from the LDLs and HDLs by ultracentrifugation; the LDLs are precipitated from the infranatant fluid by heparin and manganous ion or by dextran sulfate and calcium ion; and the HDLs are precipitated by dextran sulfate and manganous ion.[12,13,24] The LDL and HDL precipitates are dissolved in sodium chloride solution (5 g/100 ml), and these solutions and the VLDL fluid are diluted to standard volumes and analyzed for cholesterol and other lipoproteins by the methods used for whole serum.

METHODS FOR THE ESTIMATION OF LIPOPROTEIN CONCENTRATIONS WITHOUT ULTRACENTRIFUGATION[32]

Concentrations of HDL cholesterol can be measured, and concentrations of VLDL cholesterol and LDL cholesterol can be estimated, by a rather simple procedure.[13] This procedure can be quite helpful in interpretations of lipoprotein profiles, and it is based on the following relationships.

$$C_{\text{VLDL}} = \text{TG}/5$$
$$C_{\text{LDL}} = C_{\text{SERUM}} - C_{\text{HDL}} - \text{TG}/5$$

where C = concentration of cholesterol in the LDL, VLDL, HDL fractions or in serum, expressed in mg/100 ml.

TG = serum triglycerides concentration, expressed in mg/100 ml.

These equations are fairly accurate for specimens in which the triglycerides concentrations are no more than 400 mg/100 ml and which are *not* from persons who have type III hyperlipoproteinemia.

The equations are based on two observations: (1) The ratio of cholesterol/triglycerides in VLDL is approximately 1/4 and, therefore, the weight ratio of VLDL cholesterol/serum triglycerides is near 1/5. (2) When chylomicrons are not detected, most of the serum or plasma triglycerides are in the VLDL. Plasma cholesterol and triglyceride are readily determined by methods described previously. C_{HDL} can be determined in the supernatant after all other lipoproteins are precipitated by a mixture of heparin and manganous ions.

FRACTIONATION OF LIPOPROTEINS BY SELECTIVE PRECIPITATION

REAGENTS

1. Heparin, 5 g/100 ml aqueous solution.
2. Manganous chloride, 1 mol/l. Dissolve 19.8 g of $MnCl_2 \cdot 4\,H_2O$ in water and dilute to 100 ml.

PROCEDURE

To 1.0 ml of plasma or serum, add 40 μl of heparin solution; mix, and then add 50 μl of manganous chloride solution. Mix, let stand for 15 min, and then centrifuge for 10 min. Analyze the clear supernatant fluid for total cholesterol, and multiply the result by 1.09 to correct for dilution of the sample. This value represents C_{HDL}. Calculate the value for C_{LDL} as described above.

NORMAL RANGE

LDL cholesterol = 66 to 178 mg/100 ml plasma.

REFERENCE

Burstein, M., and Samaille, J.: Clin. Chim. Acta, *5*:609, 1960.
Friedewald, W. T., Levy, R. I., and Fredrickson, D. S.: Clin. Chem., *18*:499, 1972.

REFERENCES

1. Abell, L. L., Levy, B. B., Brodie, B. B., and Kendall, F. E.: Cholesterol in serum. *In* Standard Methods of Clinical Chemistry. D. Seligson, Ed. New York, Academic Press, Inc., 1958, Vol. 2, pp. 26–33.
2. Achaval, A., and Ellefson, R. D.: J. Lipid Res., *7*:329, 1966.
3. Adlersberg, D., Schaefer, L. E., Steinberg A. G., and Wang, C. I.: J. A.M.A., *162*:619–622, 1956.
4. Albers, W., and Lowry, O. H.: Anal. Chem. *27*:1829–1831, 1955.
5. Allain, C. C., Poon, L. S., Chan, C. S. G., Richmond, W., and Fu, P. C.: Clin. Chem., *20*:470–475, 1974.
6. Beaumont, J. L., Carlson, L. A., Cooper, G. R., Fejfar, Z., Fredrickson, D. S., and Strasser, T.: Circulation, *45*:501–503, 1972.
7. Beckering, R. E., and Ellefson, R. D.: Am. J. Clin. Path., *53*:84–88, 1970.
8. Bloor, W. R.: J. Biol. Chem., *77*:53–73, 1928.
9. Brady, R. O., Johnson, W. G., and Uhlendorf, B. W.: Am. J. Med., *51*:423–431, 1971.
10. Bryan, R.: Clin. Chem., *18*:1551–1552, 1972.
11. Burke, I. W., Diamondstone, B. I., Valapoldi, R. A., and Menis, O.: Clin. Chem., *20*:794–801, 1974.
12. Burstein, M., Scholnick, H. R., and Morfin, R.: J. Lipid Res., *11*:583–595, 1970.
13. Burstein, M., and Samaille, J.: Clin. Chim. Acta, *5*:609, 1960.
14. Cahill, K. M.: J. Trop. Med. Hyg., *68*:27–31, 1965.
15. Caraway, W. T., and Kammeyer, C. W.: Clin. Chim. Acta, *41*:395–434, 1972.
16. Cash, R., and Berger, C. K.: J. Ped., *74*:717–729, 1969.
17. Chin, H. P., and Blankenhorn, D. H.: Clin. Chim. Acta, *20*:305–314, 1968.
18. Drevon, B., and Schmit, J. M.: Bull. Trav. Soc. Pharm. (Lyon), *8*:173–178, 1964.
19. Dole, V. P.: J. Clin. Invest., *35*:150–154, 1956.
20. Dryer, R. L.: The lipids. *In* Fundamentals of Clinical Chemistry, 1st ed. N. W. Tietz, Ed. Philadelphia, W. B. Saunders Company, 1970, pp. 328, 343.
21. Duncombe, W. G.: Clin. Chim. Acta, *9*:122–125, 1964.
22. Eggstein, M., and Kreutz, F. H.: Klin. Wochenschr., *44*:262–273, 1966.
23. Ellefson, R. D.: Pathways of lipid metabolism: Mammals. *In* Biology Data Book, 2nd ed. P. L. Altman and D. S. Dittmer, Eds. Bethesda, Maryland, Federation of American Societies for Experimental Biology, 1974, Vol. 3, pp. 1538–1542.
24. Ellefson, R. D.: Unpublished data, observations, and methods.
25. Ellefson, R. D., and Mason, H. L.: Endocrinology, *75*:179–186, 1964.
26. Ellefson, R. D., Jimenez, B. J., and Smith, R. C.: Mayo Clin. Proc., *46*:328–332, 1971.
27. Epstein, E., Baginski, E· S., and Zak, B.: Ann. Clin. Lab. Sci., *2*:244–254, 1972.
28. Fieser, L. F.: J. Am. Chem. Soc., *75*:5421, 1953.
29. Folch, J., Lees, M., and Sloane-Stanley, G. H.: J. Biol. Chem., *226*:497–509, 1957.
30. Fredrickson, D. S., Gotto, A. M., and Levy, R. I.: Familial lipoprotein deficiency. *In* The Metabolic Basis of Inherited Disease, 3rd ed. J. B. Stanbury, J. B. Wyngaarden, and D. S. Fredrickson, Eds. New York, McGraw-Hill Book Co., 1972, pp. 493–530.
31. Fredrickson, D. S., and Levy, R. I.: Familial hyperlipoproteinemia. *In* The Metabolic Basis of Inherited Disease, 3rd ed. J. B. Stanbury, J. B. Wyngaarden, and D. S. Fredrickson, Eds. New York, McGraw-Hill Book Co., 1972, pp. 545–614.
32. Friedewald, W. T., Levy, R. I., and Fredrickson, D. S.: Clin. Chem., *18*:499–502, 1972.
33. Frings, C S., and Dunn, R. T.: Am. J. Clin. Path., *53*:89–91, 1970.
34. Gluck, L., Kulovich, M. V., Boerer, R. C., Jr., Brenner, P. H., Anderson, G. G., and Spellacy, W. N.: Am. J. Obstet, Gynecol., *109*:440–445, 1971.
35. Gupta, D. K., Young, R., Jewitt, D. E., Hartog, M., and Opie, L. H.: Lancet, ii:35, 1209–13, 1969.

36. Hatch, F. T., and Lees, R. S.: Practical methods for plasma lipoprotein analysis. *In* Advances in Lipid Research. R. Paoletti and D. Kritchevsky, Eds. New York, Academic Press, Inc., 1968, Vol. 6, pp. 1–68.
37. Houtsmuller, A. J.: Agarose-Gel-Electrophoresis of Lipoproteins. Assen, Netherlands, Royal Van Gorcum, Ltd., 1969.
38. Huang, T. C., Chen, C. P., Wefler, V., and Raftery, A.: Anal. Chem., *33*:1405–1407, 1961.
39. Kahlke, W.: Klin. Wochenschr., *42*:1011–1016, 1964.
40. Kessler, G., and Lederer, H.: Fluorometric measurement of triglycerides. *In* Automation in Analytical Chemistry. L. T. Skeggs, Ed. New York, Mediad, 1965, pp. 341–344.
41. Keys, A., Mickelsen, O., Miller, E. V. O., and Chapman, C. B.: Science, *112*:79–81, 1950.
42. Leffler, H. H.: Am. J. Clin. Path., *31*:310–313, 1959.
43. Levy, A. L.: Ann. Clin. Lab. Sci., *2*:474–479, 1972.
44. Mangold, H. K.: Aliphatic lipids. *In* Thin-Layer Chromatography. E. Stahl, Ed. New York, Springer-Verlag, 1969, pp. 363–421.
45. Margolis, S.: Structure of very low and low density lipoproteins. *In* Structural and Functional Aspects of Lipoproteins in Living Systems. E. Tria and A. Scanu, Eds. New York, Academic Press, Inc., 1969, pp. 269–424.
46. Papadoupoulis, N. M., and Kintzois, J. A.: Anal. Biochem., *30*:421–426, 1969.
47. Pearson, J., Stern, S., and McGavack, T. H.: Anal. Chem., *25*:813–814, 1953.
48. Randerath, K.: Thin-Layer Chromatography. New York, Academic Press, Inc., 1964, pp. 126–151.
49. Reed, A. H., Cannon, D. C., Winkelman, J. W., Bhasin, Y. P., Henry, R. J., and Pileggi, V. J.: Clin. Chem., *18*:57–66, 1972.
50. Rice, E. W.: Triglycerides ("neutral fat") in serum. *In* Standard Methods of Clinical Chemistry. R. P. MacDonald, Ed. New York, Academic Press, Inc., 1970, Vol. 6, pp. 215–222.
51. Schlenk, H., and Gellerman, J. L.: Anal. Chem., *32*:1412–1414, 1960.
52. Silverstein, M. N., Ellefson, R. D., and Ahern, E. H.: N. Eng. J. Med., *282*:1–4, 1970.
53. Skidmore, W. D., and Entenman, C.: J. Lipid Res., *3*:356–363, 1962.
54. Sperry, W. M.: Lipid analysis. *In* Methods of Biochemical Analysis. D. Glick, Ed. New York, Interscience Publishers, Inc., 1955, Vol. 2, pp. 83–111.
55. Sperry, W. M., and Webb, M.: J. Biol. Chem., *187*:97–106, 1950.
56. Tonks, D. B.: Clin. Biochem., *1*:12–29, 1967.
57. Tria, E., and Scanu, A. (Eds.): Structural and Functional Aspects of Lipoproteins in Living Systems. New York, Academic Press, Inc., 1969.
58. Van de Kamer, J. H.: Total fatty acids in stool. *In* Standard Methods of Clinical Chemistry. D. Seligson, Ed. New York, Academic Press, Inc., 1958, Vol. 2, pp. 34–39.
59. Van Handel, E., and Zilversmit, D. B.: J. Lab. Clin. Med., *50*:152–157, 1957.
60. Vanzetti, G.: Clin. Chim. Acta, *10*:389–405, 1964.
61. Velapoldi, R. A., Diamondstone, B. I., and Burke, R. W.: Clin. Chem., *20*:802–811, 1974.
62. Voskevsky, V. E., and Kostetsky, E. Y.: J. Lipid Res., *9*:396, 1968.
63. Wakil, S. J. (Ed.): Lipid Metabolism. New York, Academic Press, Inc., 1970.
64. White, N. B., Jr., and Montalvo, J. M.: J. Pediatr., *83*:999–1006, 1973.
65. Wieland, O.: Glycerol. *In* Methods of Enzymatic Analysis. H. U. Bergmeyer, Ed. New York, Academic Press, Inc., 1963, pp. 211–214.
66. Williams, J. H., Kuchmak, M., and Witter, R. F.: Clin. Chem., *16*:423–426, 1970.
67. Wybenga, D. R., Pileggi, V. J., Dirstine, P. H., and Digiorgio, J.: Clin. Chem., *16*:980–984, 1970.
68. Zak, B., and Ressler, N.: Am. J. Clin. Path., *25*:433–446, 1955.
69. Zak, B., Epstein, E., and Baginski, E. S.: Ann. Clin. Lab. Sci., *2*:101–125, 1972.
70. Zollner, N., and Kirsch, K.: Ztschr. Ges. Exp., *135*:545–561, 1962.
71. Zollner, N., and Wolfram, G.: TLC in clinical diagnosis. *In* Thin-layer chromatography. 2nd ed. E. Stahl, Ed. New York, Academic Press, Inc., 1969, pp. 592–599.
72. Inter-Society Commission of Heart Disease Resources: Circulation, *42*:A53, 1970.

ADDITIONAL READING

Mallikarjuneswara, V. R.: Lecithin-sphingomyelin ratio in amniotic fluid, as assessed by a modified thin-layer chromatographic method in which a commercial pre-coated plate is used. Clin. Chem., *21*:260–263, 1975.

Olson E. B., Jr., and Graven, S. N.: Comparison of visualization methods used to measure the lecithin/sphingomyelin ratio in amniotic fluid. Clin. Chem., *20*:1408-1415, 1974.

Tria, E., and Scanu, A. (Eds.): Structural and Functional Aspects of Lipoproteins in Living Systems. New York, Academic Press, Inc., 1969.

Velapoldi, R. A., Diamondstone, B. I., and Burke, R. W.: Spectral interpretation and kinetic studies of the Fe^{3+}-H_2SO_4 (Zak) procedure for determination of cholesterol. Clin. Chem., *20*:802–811, 1974.

Wakil, S. J. (Ed.): Lipid Metabolism. New York, Academic Press, Inc., 1970.

White, A., Handler, P., and Smith, E. L. (Eds.): Principles of Biochemistry. 6th ed. New York, McGraw-Hill Book Company, 1973.

CHAPTER 11

VITAMINS*

by Hipolito V. Nino, Ph.D., and William Shaw, Ph.D.

New social patterns as well as advances in the understanding of nutritional needs have led to an increased number of requests for laboratory tests for the evaluation of the nutritional status of individuals. In former times, laboratory tests available for this purpose were very limited in number and scope and were generally used only to confirm suspected deficiencies in undernourished individuals. Today, it is recognized that laboratory tests are needed in many other clinical situations. For example, it is important to know the clinical effects of extended intravenous nutrition, excessive use of vitamins, self-imposed or other dietary restrictions, oral contraceptives, massive losses of blood, and prolonged chronic illnesses. It has also been recognized that a great many people are deficient in many nutrients, even if these individuals do not exhibit any overt clinical symptoms of such deficiencies. A recent 10 state nutrition survey organized by the Nutrition Program of the Department of Health, Education and Welfare indicated the prevalence of less than desirable vitamin A levels (less than 20 μg/100 ml plasma) in a large number of young people of all ethnic groups.

Vitamins are a class of biochemical compounds essential for normal growth and development. The human organism is not capable of synthesizing these compounds, and thus each of the vitamins must be supplied in the diet either in the form of the vitamin itself or in the form of a precursor. Vitamin deficiency may lead to a series of pathological conditions, as shown in Table 11–1. There is no chemical similarity between any of the different vitamins, and the metabolic functions of each are different.

Vitamins are divided into two main categories on the basis of their solubility. Among the *fat-soluble vitamins* are included vitamins A, D, E, and K, while the *water-soluble vitamins* include thiamine (vitamin B_1), riboflavin (vitamin B_2), niacinamide (niacin, nicotinic acid), pyridoxine (vitamin B_6), pantothenic acid, lipoic acid, folic acid, inositol, cyanocobalamin (vitamin B_{12}), and ascorbic acid (vitamin C). Some of the vitamins serve directly as coenzymes (e.g., biotin), whereas others serve as precursors of coenzymes. Niacin, for example, can readily be converted to nicotinamide adenine dinucleotide (NAD). Other vitamins, such as vitamin A, E, and D, have no function as coenzymes. In fact, the exact metabolic role of the oil-soluble vitamins still has not been completely elucidated.

Methods presented in this chapter are limited to those which assess the levels of vitamins either by direct analysis or by measuring the level of other components in body fluids, the activity or concentration of which is directly affected by the level of certain vitamins in tissues. Examples of this latter approach are the measurement of red cell transketolase, which can be used to assess vitamin B_1 (thiamine) status, and the assay of glutathione reductase, which can serve as an indirect test for vitamin B_2 (riboflavin). A number of excellent textbooks are available, and the reader may consult these for further details.

* Use of trade names is for identification only and does not constitute endorsement by the Public Health Service or by the United States Department of Health, Education, and Welfare.

Table 11-1 THE VITAMINS

Chemical Name	*Common Name*	*Symptoms of Deficiency in Man*	*Methodology Available for Assay in Biological Fluids*	*References*
Retinol	vitamin A	xerophthalmia, keratomalacia	colorimetric assay, fluorometric assay	3, 4, 6, 7, 14
Calciferol Cholecalciferol	vitamin D_2 vitamin D_3	rickets	competitive protein-binding radioassay	2
Thiamine	vitamin B_1	beriberi, Wernicke-Korsakoff syndrome	microbiological assay, fluorometric assay	8, 12
Riboflavin	vitamin B_2	angular stomatitis, photophobia	microbiological assay, fluorometric assay	8, 13, 17
Pyridoxine, pyridoxal, pyridoxamine	vitamin B_6 group	convulsions, hypochromic anemia	microbiological assay	8
3-Pyridinecarboxylic acid	niacin, niacinamide	pellagra	microbiological assay	8
Pteroylglutamic acid	folic acid	megaloblastic anemia	competitive protein-binding radioassay, microbiological assay	5, 22
Cyanocobalamin	vitamin B_{12}	pernicious anemia, megaloblastic anemia, neuropsychiatric disorders	competitive protein-binding assay, microbiological assay	8, 11, 23
Biotin	—	dermatitis	microbiological assay	8
Ascorbic acid	vitamin C	scurvy	colorimetric assay	9, 15, 24
Pantothenic acid	—	burning feet syndrome	microbiological assay	8
Tocopherols	vitamin E	controversial at present	colorimetric assay	1
Naphthoquinones	vitamin K	increased clotting time, cerebral hemorrhage in infancy	colorimetric assay	16
Lipoic acid	—	none known	microbiological assay	10

CAROTENES AND VITAMIN A

The vitamin A required by the human organism is derived from plants, fish, and other animals, either as such or in the form of carotenes (provitamin A). The major precursors are a group of pigments called the carotenes, which are synthesized by all plants and by some animals, but not by man. The major precursor of vitamin A in humans is *β-carotene*, a symmetrical hydrocarbon molecule containing two β-ionone rings connected by an 18-carbon hydrocarbon chain with 11 conjugated double bonds (Fig. 11–1). Scission of β-carotene at the central double bond results in the formation of *retinal*, which is reduced to *retinol* (*vitamin A*) by an NADH-dependent enzymatic reaction. This conversion is believed to take place mainly in the liver.

Vitamin A occurs in two forms: vitamin A_1 and A_2. Vitamin A_2 (3-dehydro retinal) has an additional double bond in the β-ionone ring and is biologically less active than vitamin A_1; α- and γ-carotene, as well as cryptoxanthin, contain only one β-ionone ring and thus form only one molecule of retinal upon oxidative scission.

Fruits and vegetables containing carotenes provide over half of the vitamin A requirement in the American diet. Certain animal tissues or products, such as liver extracts or fish oils, are a direct source of retinal or retinol esters. Vitamin A and carotenes are subject to oxidation, and their stability depends on the presence of antioxidants.

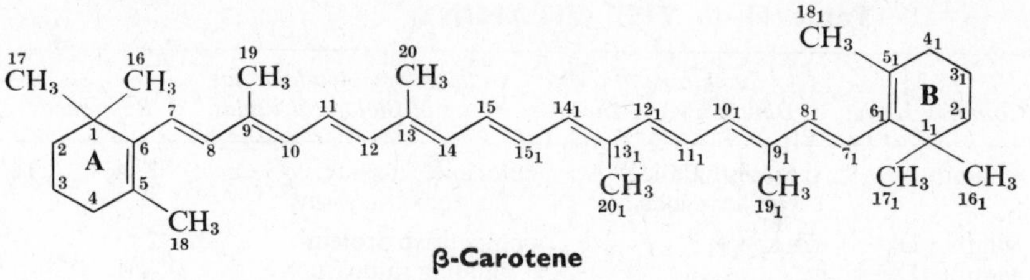

β-Carotene

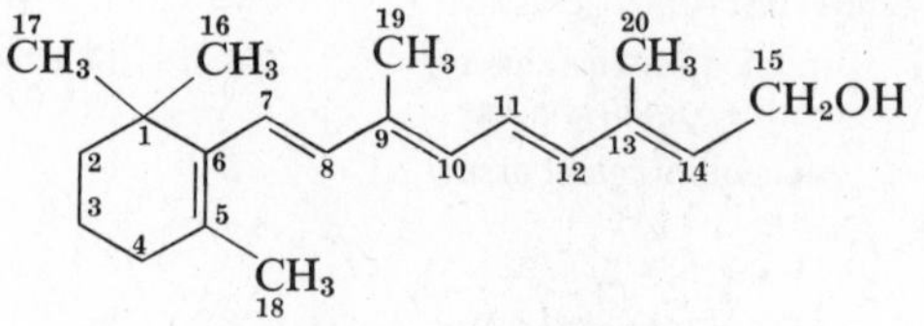

Vitamin A (retinol)

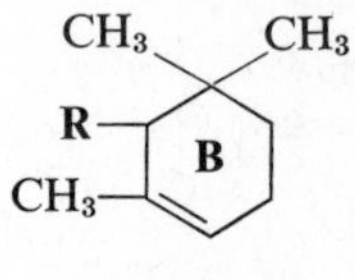

α-Carotene **Cryptoxanthin** **γ-Carotene**

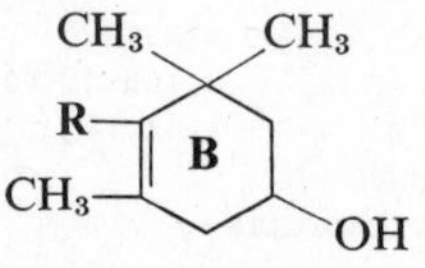

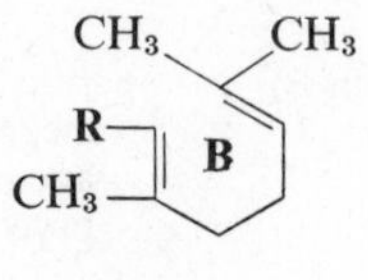

Figure 11-1 Structure of β-carotene, vitamin A, and other carotenoids. R represents β-carotene without the "B" ring system.

When the retinol esters are ingested, they are hydrolyzed in the intestine by pancreatic lipase. Free retinol is absorbed into the intestinal mucosal cells and re-esterified with fatty acids, predominantly palmitic acid. It then passes via the lymphatic system to the liver in association with chylomicra, and is transported in the general circulation by low density lipoproteins. Blood contains an esterase which is capable of hydrolyzing the vitamin A ester. This may explain why, in persons in a fasting state, approximately 80 to 95 per cent of the total vitamin A is in the form of the free alcohol, and only the remainder is in the form of the ester.

Vitamin A plays an important role in the chemical events associated with visual excitation. In addition, vitamin A is also required for normal growth, development, and reproduction. Vitamin A has also been implicated in protein synthesis, development of skin tissue and mucous membranes, electron transfer reactions, and maintenance of the integrity of cell membranes or cell organelles.

Hypovitaminosis A (vitamin A deficiency) leads to night blindness and, if protracted, to keratomalacia and xerophthalmia (dryness, and greasiness of the cornea). Hypovitaminosis A is a leading cause of blindness in children in tropical countries.

While decreased intake of vitamin A may lead to deficiency diseases, *hypervitaminosis A* (excess ingestion of vitamin A) produces toxic effects, such as loss of hair, joint pain, drowsiness, headache, vomiting, skin defects, increased intracranial pressure, abdominal pain, excessive sweating, and brittle nails. High serum vitamin A levels are found in women using oral contraceptives, although the significance of this finding at this time is not clear.

DETERMINATION OF CAROTENE AND VITAMIN A[4,14]

PRINCIPLE

Trifluoroacetic acid reacts with the π-electrons in the conjugated double bonds of vitamin A to form a chemical compound with a blue color (Neeld-Pearson procedure).

Trifluoroacetic acid (TFA) replaces the antimony trichloride-chloroform reagent used in the classic Carr-Price reaction, which suffered from the disadvantage of forming insoluble SbOCl in the presence of moisture. The Neeld-Pearson reaction also shows increased sensitivity and specificity. Since carotenes also react with trifluoroacetic acid, a correction for the absorbance contributed by carotenes is applied.

SPECIMEN

Blood for the determination of serum β-carotene and vitamin A should be collected when the patient is in the fasting state, and the specimen should be free from hemolysis and protected from light. If freshly separated serum is promptly frozen at −10°C, vitamin A is stable for at least two weeks.

REAGENTS

1. Ethanol, 95 per cent (v/v), AR.
2. Petroleum ether, reagent (pesticide) grade.
3. Chloroform, anhydrous, AR.
4. Trifluoroacetic acid reagent. Mix 1 vol of trifluoroacetic acid (CF_3COOH), AR, with 2 vol of anhydrous chloroform just prior to use. This solution is stable for 4 h at 25°C.
5. Vitamin A stock standard, 160 μg/ml. Transfer 16.0 mg of all-*trans*-retinyl acetate (Eastman Organic Chemicals, Rochester, N.Y.) to a 100 ml volumetric flask and dilute to volume with anhydrous chloroform.
6. Vitamin A working standard. Pipet 2.5, 5.0, 7.5, and 10.0 ml of the stock standard (160 μg/ml) into 100 ml volumetric flasks and dilute to volume with anhydrous chloroform to obtain working standards with vitamin A concentrations of 4.0, 8.0, 12.0, and 16.0 μg/ml. These standards are stable for one week at 4 to 8°C if protected from light. Preparation of standards and all analytical operations must be performed in nonactinic or very low intensity light.
7. β-Carotene stock standard, 200 μg/ml. Transfer 20.0 mg of synthetic crystalline β-carotene (Sigma Chemical Company, St. Louis, Mo.) to a 100 ml volumetric flask. Dissolve in approximately 4 ml of chloroform, and then dilute to volume with petroleum ether.
8. β-Carotene working standard. Pipet 10.0 ml of the 200 μg/ml standard into a 100 ml volumetric flask and dilute to volume with petroleum ether to obtain a concentration of 20.0 μg/ml. Then transfer 2.5, 5.0, 10.0, 15.0, and 20.0 ml of the 20.0 μg/ml standard to a 100 ml volumetric flask and dilute to volume with petroleum ether to obtain working standards with concentrations of 0.5, 1.0, 2.0, 3.0, and 4.0 μg/ml. These standards are stable for only a few hours at 25°C and should be made up fresh each time an analysis is performed. All standard solutions *must* be protected from direct light.

PROCEDURE

1. Pipet 1.0 ml of serum into respective 15 ml glass-stoppered centrifuge tubes.
2. Pipet 2.0 ml of 95 per cent (v/v) ethanol, stopper, and mix well with a vortex mixer.
3. Add 3.0 ml of petroleum ether and place tubes in a Kraft "Shaker in the Round" (Kraft Apparatus, Inc., Queens, N.Y.) or some other suitable shaker to extract vitamin A and the carotenes into the petroleum ether phase.
4. Centrifuge the stoppered tubes for 10 min at 2500 rpm.
5. Carefully pipet off 2.0 ml of the petroleum ether phase (upper layer) and transfer into a dry cuvet (e.g., 10 × 75 mm Coleman cuvet). Read the absorbance at 450 nm in a suitable spectrophotometer against a petroleum ether blank without delay to prevent evaporation of solvent and destruction of carotenoids by light. Mark this reading A_1.
6. Evaporate the contents of the cuvets to dryness in a 50°C water bath with the aid of a

fine stream of nitrogen. After removing the cuvets from the water bath, dry each one carefully with nonabrasive paper (to prevent scratching or marking them).

7. Add 0.1 ml of chloroform to each cuvet and mix briefly with a vortex mixer.

8. Add 1.0 ml of trifluoroacetic acid reagent to the cuvet serving as blank, mix, and set spectrophotometer to 0 absorbance at 620 nm.

9. Add forcefully (to facilitate immediate mixing) 1.0 ml of trifluoroacetic acid reagent to all other cuvets, and record the reading at 620 nm at exactly 2 s after addition of the reagent. (Add the trifluoroacetic acid reagent only to one cuvet at a time so that readings may be taken as required.) **Caution**: Trifluoroacetic acid is a strong acid and care should be exercised to prevent spilling or splattering it.

Best results are obtained if the trifluoroacetic acid reagent is added forcefully with an automated pipet. If a recorder is used, the absorbance value can be read at the peak or inflection point after the initial surging peak caused by the introduction of the trifluoroacetic acid reagent. This second peak or inflection point occurs about 2 s after the introduction of the color reagent. Mark this reading A_2.

CALCULATIONS

β-Carotene. Determine the amount of β-carotene per ml (see A_1 reading in step 5 of Procedure) from the carotene standard curve and carry out the following calculations:

$$\mu g\ \beta\text{-carotene}/100\ ml\ serum = \mu g\ \beta\text{-carotene}/ml \times 3.0 \times 100$$

where

3.0 = vol petroleum ether containing the β-carotene from 1.0 ml of serum after extraction
100 = conversion factor from μg/ml to μg/100 ml.

Vitamin A. For accurate calculation of the vitamin A content, it is necessary to correct for the absorbance contributed by carotenes at 620 nm:

$$A_3 = A_2 - (F \times A_1)$$

where

A_1 = absorbance of carotene at 450 nm (step 5).

A_2 = absorbance at 620 nm due to both carotene *and* vitamin A (step 9).

A_3 = absorbance at 620 nm of vitamin A (corrected for absorbance contributed by β-carotene).

F = factor which converts the carotene absorbance at 450 nm (step 5) into the equivalent absorbance at 620 nm in the color reaction (step 9).

$$F = \frac{A_{620}\ \text{of carotene using vitamin A procedure}}{A_{450}\ \text{of petroleum ether solution of carotene}}$$

There have been reports indicating considerable variation in the value of this factor. The authors have also found a factor different from that given in the original Neeld-Pearson report. Therefore, each laboratory *must* determine its own factor.

After A_3 has been found, the actual concentration of vitamin A per 100 ml of serum is calculated as follows:

μg vitamin A (free alcohol)/100 ml

$$= \frac{A_3 \times \mu g\ \text{retinyl acetate standard/cuvet}}{A_{620}\ \text{retinyl acetate standard}} \times \tfrac{3}{2} \times 100 \times 0.872$$

or

$$\mu g\ \text{vitamin A}/100\ ml = \frac{A_3 \times \mu g\ \text{retinyl acetate standard/cuvet}}{A_{620}\ \text{retinyl acetate standard}} \times 130.8$$

where

3 = volume of the petroleum ether extract of 1.0 ml serum.

2 = aliquot of the petroleum ether extract used for the assay.

100 = conversion of μg retinyl acetate/ml to μg retinyl acetate/100 ml.

0.872 = ratio of molecular mass of retinol to molecular mass of retinyl acetate. Thus, this factor corrects for the use of retinyl acetate instead of retinol as the standard.

CALIBRATION

β-Carotene. Place β-carotene working standards in concentrations of 0.5, 1.0, 2.0, 3.0, and 4.0 μg/ml into appropriate cuvets (e.g., 10 × 75 mm Coleman cuvets), and read the absorbance at 450 nm against a petroleum ether blank. Plot β-carotene concentration (μg/ml) against absorbance.

For the purpose of calculating the carotene correction factor F for the vitamin A procedure, 2.0 ml of each of the β-carotene standards is treated as a sample, beginning with step 6 of the vitamin A procedure. The average ratio of absorbance at 620 nm/concentration of β-carotene (in μg/ml) is then calculated and used in the computation of the β-carotene correction factor F.

Vitamin A. Pipet 0.1 ml of each retinyl acetate working standard into cuvets and treat as in Vitamin A procedure. Plot μg retinyl acetate/cuvet against A_{620}.

DISCUSSIONS OF METHODS

A variety of methods for the determination of vitamin A in serum have been proposed, but the reactions between vitamin A and antimony trichloride in chloroform (Carr-Price) and the reactions between vitamin A and trifluoroacetic acid (Neeld-Pearson) are most widely used in clinical laboratories.

Bessey and coworkers[3] irradiated the specimen for vitamin A and carotene assay with ultraviolet light and measured the ΔA_{328} before and after the irradiation of the specimen with ultraviolet light of a wavelength between 310 and 400 nm. Since vitamin A (and carotene) is destroyed under these conditions, ΔA values may be taken as a measure of the vitamin A concentration. The method may also be applied to the determination of β-carotene, in which case ΔA is measured at 460 nm. This method requires a preliminary saponification and appears to be too long and involved for routine clinical purposes but may be of use if only small sample volumes, such as 50 μl or less are available.

Several fluorometric micro and macro procedures have been proposed. Because of interference by fluorescent compounds in serum, Garry et al.[7] suggested removal of these interfering compounds by passing the sample through a silicic acid column. A similar but improved method has been described by Thompson et al.[19]

NORMAL VALUES

The normal range for vitamin A in serum is between 20 and 65 μg/100 ml. Values over 140 μg/100 ml may be found in patients with vitamin A toxicity. The normal range for carotenes (mainly β-carotene and xanthophyll) is between 60 and 200 μg/100 ml serum. Elevated serum carotene levels are frequently found in infants and in hypothyroid individuals and result from inability of the liver to convert carotenes to vitamin A. Other causes are excessive dietary intake and hyperlipemia associated with diabetes mellitus.

ASCORBIC ACID

The antiscorbutic vitamin (L-ascorbic acid, vitamin C) is a carbohydrate with strong acidic properties from the presence of two easily dissociable enolic hydroxyl hydrogens in the molecule. Ascorbic acid is a moderately strong reducing agent which can be reversibly oxidized to dehydroascorbic acid. All plants and animals, with the exception of man, other primates, and the guinea pig, possess the capacity to synthesize ascorbic acid from D-glucose via the metabolic pathway that involves D-glucuronic acid and L-gulonolactone as intermediates. Man and the animals listed above lack the ability to convert L-gulonolactone to the vitamin, and therefore ascorbic acid must be provided in the diet. Both L-ascorbic acid and its immediate oxidation product, dehydroascorbic acid, are biologically active. Products of further metabolism such as L-diketogulonic acid and oxalate have no vitamin activity.

Clinical significance

The exact function of L-ascorbic acid in cell metabolism is still far from understood. Ascorbic acid functions as a hydrogen donor in certain *in vitro* enzymatic reactions in which the hydroxyl group is introduced into a variety of substrates. It is not certain, however, that ascorbic acid performs this role *in vivo*. Ascorbic acid is also known to be required for the hydroxylation of such metabolites as proline, tyrosine, dihydroxyphenylalanine (DOPA), and some adrenal steroid hormone intermediates. Furthermore, it is involved in the ring-opening oxidation of homogentisic acid (HGA) to maleylacetoacetic acid. The urine of children deprived of ascorbic acid darkens when exposed to air (HGA accumulation). This condition resembles alkaptonuria, a condition also characterized by the accumulation of HGA. In the latter condition, however, HGA accumulation is caused by the deficiency of the enzyme HGA-oxidase. Administration of vitamin C alleviates the symptoms of the deficiency in an ascorbic acid-deficient, but not in the enzyme-deficient individual. Deficiency in the vitamin has also been implicated in certain anemias, often in association with either iron or folate deficiency.

Acute deficiency of ascorbic acid intake leads to the classic disease of *scurvy*. Symptoms of this disease are bleeding gums, loose teeth, and poor healing of wounds. These symptoms may be at least partially explained by the role of ascorbic acid in proline and collagen metabolism. Patients with scurvy often have no detectable plasma levels of vitamin C. However, a person must be on a diet severely deficient in ascorbic acid for three or four months before clinical symptoms can be detected.

Most cases of scurvy encountered today in the United States occur in individuals in the age range of 7 months to 2 years. The 10 state nutrition survey, previously mentioned, shows that the incidence of plasma levels of vitamin C below 0.20 mg/100 ml is low.

The Recommended Daily Allowance (RDA) of vitamin C for adults, as established by the National Research Council, is 70 mg/day, although many nutritionists recommend 100 to 200 mg/day. Recently, several scientists have recommended the use of massive doses (up to 1000 to 3000 mg or more/day) to increase the resistance against colds. The efficacy of this therapy is being challenged and is still under investigation.

SPECIMENS

The most frequently used specimen is plasma collected with oxalate as anticoagulant, although serum may be used. EDTA or heparin can also be used as an anticoagulant for the collection of plasma samples. The stability of ascorbic acid in whole blood is presumably increased by the presence of reducing agents such as glutathione and cysteine. However, specimens so treated are unsuitable for the use in assays employing the titration with 2,6-dichlorophenol-indophenol. Also, the iron in red cells may interfere with some modifications of the dinitrophenylhydrazine method. However, regardless of the type of specimen used, the test should be performed without delay.

If the dinitrophenylhydrazine method is used, the ascorbic acid content in whole blood remains stable for about 3 h if the specimen is refrigerated. Vitamin C is stable for two weeks if the plasma specimen is promptly diluted with 6 g/100 ml of metaphosphoric acid and stored at −20°C. Trichloroacetic acid (10.0 g/100 ml) can also be used in place of metaphosphoric acid to prepare a protein-free filtrate of plasma or serum.

METHODS[9,15,24]

Ascorbic acid may be determined by titration with the oxidation-reduction indicator (dye) 2,6-dichlorophenol-indophenol in acid solution. Ascorbic acid reduces the indicator to a colorless form. A protein-free filtrate is prepared from fresh blood or plasma by deproteinization with metaphosphoric acid or trichloroacetic acid. An aliquot of the filtrate is titrated directly, with the indicator reagent added from a microburet. The indicator solution is blue, but becomes pink in acid solution. The titration is continued until a persisting faint pink color is formed. Similar titrations are carried out on fresh standard solutions and on a water blank, both treated in the same manner as the blood specimen, including the addition of metaphosphoric acid to provide an acid medium.

Although the titration methods are rapid and convenient, some difficulty is encountered in assessing the end point because the indicator color is rather pale and tends to fade. Other methods have been developed in which excess of a colored indicator is added to a protein-free filtrate and to standard solutions. The decrease in absorbance, compared to standards, is related to the concentration of ascorbic acid. Some turbidity is usually encountered with protein-free filtrates, and this contributes to the absorbance. After readings are made, excess ascorbic acid is added to the cuvets to completely decolorize the indicator and to permit measurement of the background absorbance due to turbidity. A satisfactory method is the Henry[9] modification of the method by Owen and Iggo. Both indicator-dye methods measure only the reduced ascorbic acid and not dehydroascorbic acid; therefore, these methods are not suitable for urine specimens since urine contains a sizable fraction of the vitamin in the dehydroascorbic acid form. Metal ions such as cuprous, stannous, and ferrous ions interfere with the manual and automated modifications of the indicator-dye methods.

In another method commonly used for the determination of ascorbic acid (Fig. 11–2) the acid reacts with 2,4-dinitrophenylhydrazine after oxidation of vitamin C to dehydroascorbic acid. This procedure, however, does not distinguish between the biologically active ascorbic acid and dehydroascorbic acid, and the biologically inactive diketogulonic acid. In this approach, ascorbic acid is oxidized to dehydroascorbic acid by either cupric sulfate or 2,6-dichlorophenol-indophenol. The dehydroascorbic acid in a strongly acid solution reacts with 2,4-dinitrophenylhydrazine to form a dinitrophenylhydrazone. The hydrazone, in the presence of strong sulfuric acid solution, develops a red color which can be measured spectrophotometrically. Thiourea is added to the dinitrophenylhydrazine reagent to prevent the oxidation of the dinitrophenylhydrazine reagent by interfering substances.

Since normal blood specimens contain very little dehydroascorbic acid or diketogulonic acid, the indophenol and the hydrazone methods give similar results.

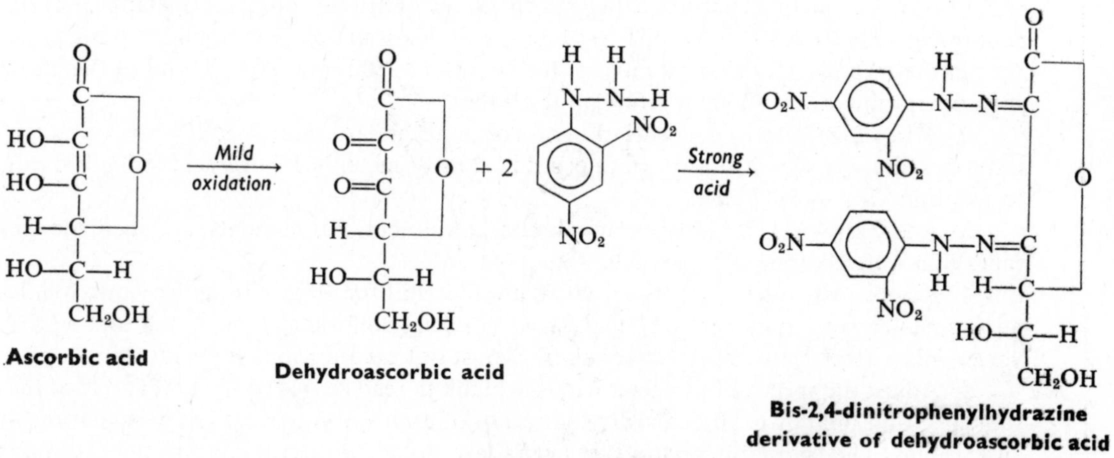

Ascorbic acid —*Mild oxidation*→ **Dehydroascorbic acid** + 2 (2,4-dinitrophenylhydrazine) —*Strong acid*→ **Bis-2,4-dinitrophenylhydrazine derivative of dehydroascorbic acid**

Figure 11–2 The dinitrophenylhydrazine reaction for vitamin C.

REAGENTS

1. Metaphosphoric acid solution, 6.0 g/100 ml. Dissolve 30.0 g of metaphosphoric acid (HPO_3) in distilled water and bring to a final volume of 500 ml. Prepare immediately before use.

2. Sulfuric acid, 4.5 molar. Add slowly 250 ml of concentrated sulfuric acid, AR, to 500 ml of cold water in a 1 liter flask and bring to a final volume of 1 liter with distilled water. **Caution:** Since significant heat is generated when concentrated sulfuric acid is diluted, the flask should be placed in an ice bath. The concentrated acid should be added slowly and the resulting solution mixed constantly.

3. Sulfuric acid, 12 molar. Add 650 ml of concentrated sulfuric acid to 300 ml of cold water in a 1 liter flask, cool, and bring to a final volume of 1 liter with distilled water. Refrigerate.

4. 2,4-Dinitrophenylhydrazine reagent, 2.0 g/100 ml in 4.5 molar sulfuric acid. Dissolve 10 g of 2,4-dinitrophenylhydrazine in 4.5 molar sulfuric acid and dilute to a final volume of 500 ml. Let stand in the refrigerator overnight, and then filter.

5. Thiourea solution, 5.0 g/100 ml. Dissolve 5 g of thiourea in glass-distilled water and dilute to a final volume of 100 ml. This reagent is stable for one month at 4°C.

6. Copper sulfate solution, 0.6 g/100 ml. Dissolve 0.6 g of anhydrous copper sulfate in glass-distilled water and dilute to a final volume of 100 ml.

7. Dinitrophenylhydrazine-thiourea-copper sulfate (DTCS) reagent. Combine 5 ml of the thiourea solution, 5 ml of the copper sulfate solution, and 100 ml of the 2,4-dinitrophenylhydrazine reagent. Store in a bottle at 4°C for a maximum of one week.

8. Standards. All ascorbic acid standards should be prepared fresh daily.

a. Ascorbic acid stock standard, 50.0 mg/100 ml. Dissolve 50 mg of ascorbic acid in metaphosphoric acid (6.0 g/100 ml) and bring to a final volume of 100 ml with metaphosphoric acid.

b. Intermediate ascorbic acid standard, 5.0 mg/100 ml. Pipet 10.0 ml of stock standard into a 100 ml volumetric flask and dilute to a volume of 100 ml with metaphosphoric acid (6.0 g/100 ml).

c. Working standards. In a series of 25 ml volumetric flasks, pipet the following amounts of intermediate standard: 0.5, 2.0, 4.0, 6.0, 10.0, 15.0, and 20.0 ml. Bring to a final volume of 25 ml with metaphosphoric acid (6.0 g/100 ml) to yield working standards with concentrations of 0.10, 0.40, 0.80, 1.20, 2.00, 3.00, and 4.00 mg/100 ml.

PROCEDURE

1. Add 0.5 ml of heparinized plasma to 2.0 ml of freshly prepared metaphosphoric acid (6.0 g/100 ml) in 13 × 100 mm test tubes, and mix well on a vortex mixer. Centrifuge the plasma-metaphosphoric acid mixture for 10 min at 2500 rpm. Pipet 1.2 ml of the clear supernatant into 13 × 100 mm screwcap test tubes.

2. Add 1.2 ml of each concentration of working standard into 13 × 100 mm screwcap test tubes. Prepare standards in duplicate. Add 1.2 ml of metaphosphoric acid (6.0 g/100 ml) to two tubes for use as blanks.

3. Add 0.4 ml of DTCS reagent to all tubes. Cap tubes, mix contents, and incubate the tubes in a water bath at 37°C for 3 h.

4. Remove the tubes from the water bath and chill for 10 min in an ice bath. While mixing, slowly add to all tubes 2.0 ml of *cold* 12 molar sulfuric acid, cap, and mix with a vortex mixer (the temperature of the mixture must not exceed room temperature).

5. Adjust the spectrophotometer with the blank to read zero A at 520 nm, and read the standards and unknowns. Plot the concentration of each working standard versus absorbance values. The standard curve obeys Beer's law up to an ascorbic acid concentration of 2.0 mg/100 ml.

CALCULATION

The concentration of the samples are obtained from the standard curve and are multiplied by 5 (to correct for dilution of the plasma by metaphosphoric acid) to give the concentration of ascorbic acid per 100 ml of plasma.

NORMAL VALUES

In persons with an adequate intake of vitamin C, plasma concentrations of total vitamin (ascorbic acid plus dehydroascorbic acid) are between 0.6 to 2.0 mg/100 ml. The lower limit value may be seen in some cases with subclinical vitamin C deficiency and in older individuals.

Values for plasma concentrations and body pool levels of vitamin C correlate very well down to a plasma level of 0.3 mg/100 ml. Low plasma vitamin C levels are not necessarily an indication of low tissue levels of vitamin C. It has therefore been suggested that vitamin C levels in leukocytes are a closer measure of tissue saturation with vitamin C.

FOLIC ACID

Folic acid (FA)* is one of the water-soluble vitamins. Chemically, it is pteroylmonoglutamic acid, and it is made up of a substituted pteridine joined to a para-aminobenzoic acid (PABA) residue, which in turn is linked to glutamic acid by a peptide bond (Fig. 11–3).

PABA is an essential growth factor for many microorganisms. If PABA is provided, these organisms can synthesize FA; however, microorganisms cannot synthesize PABA itself. Sulfa drugs are structurally similar to PABA, and their antibiotic function is due to competitive inhibition of PABA uptake by the microorganisms (Fig. 11–4).

In man and animals, ingested folic acid is reduced to dihydrofolic acid (FH_2) and then to the coenzyme, 5,6,7,8-tetrahydrofolic acid (FH_4). This conversion of FA and FH_2 to FH_4 takes place in the liver and several other tissues. Folic acid, along with vitamin B_{12} and several related derivatives of FH_4 such as N^5-formyl-FH_4 (folinic acid) and $N^{5,10}$-methylene-FH_4 (Fig. 11–3), is involved in the biochemical transfer of 1-carbon fragments such as CH_3— and —CHO. For example, FH_4 is involved in the conversion of serine to glycine; other folate coenzymes participate in the synthesis of purines, thymine, and methionine, and the metabolism of histidine.

There is some evidence that N^5-methyl-FH_4 may be the predominant folic acid form in human serum. Excess FA is excreted in the urine primarily in the form of folinic acid, but the evidence is conflicting. Approximately 20 per cent of the daily intake of FA (300 to 500 μg) is derived from dietary sources; the remainder is synthesized by intestinal microorganisms. Serious deficiency of FA may arise if such microorganisms are absent (gut sterilization), or if intestinal absorption is poor (e.g., after surgical resection or in celiac disease or sprue). Parenteral administration of folate will overcome these deficiencies. Other causes for folate deficiency are insufficient dietary intake, excessive utilization (as in pregnancy, liver disease, and malignancies) and antifolate drugs (e.g., amethopterin).

In man, folate deficiency is primarily associated with the development of a severe macrocytic, megaloblastic anemia, suggesting arrested red cell maturation. (Vitamin B_{12} deficiency will cause a similar type of anemia.) The required presence of folic acid in hemoglobin synthesis and erythrocyte maturation derives mainly from its role in the synthesis of thymine, which is required for DNA synthesis. The neurological symptoms of pernicious anemia cannot be cured by folate but require treatment with vitamin B_{12}.

* Although the term folic acid is used, it is understood that at physiological pH the compound will be present in part as folate ion.

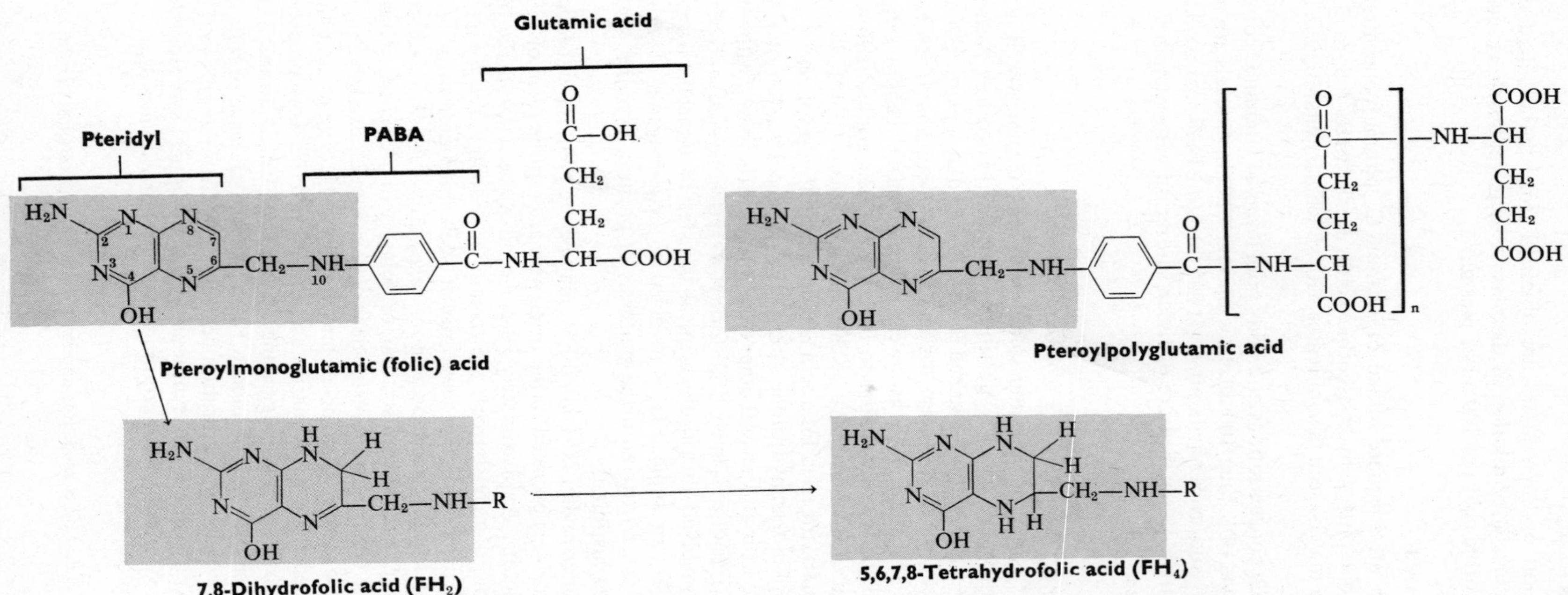

Figure 11–3 Folic acid and its derivatives. The light-shaded portion indicates the pteridine portion (2-amino, 4-hydroxy, 6-methylpterin). The position of the four hydrogens is shown in the formula for tetrahydrofolic acid (FH_4). The lower part of the figure (opposite page) shows the relation of the various FH_4 derivatives to each other. The dark-shaded areas indicate the one-carbon fragments.

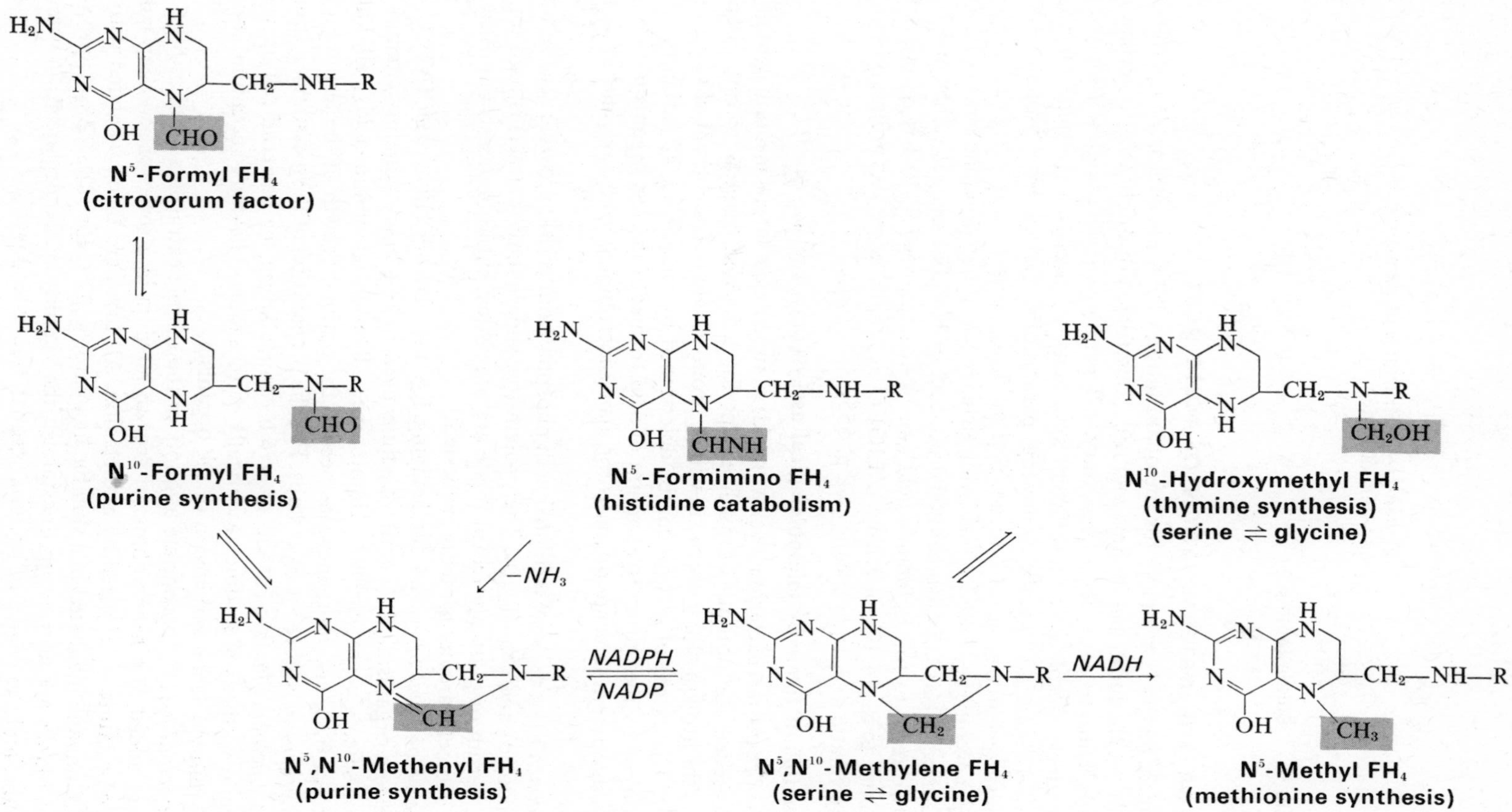

N^5-Formyl FH_4
(citrovorum factor)

N^{10}-Formyl FH_4
(purine synthesis)

N^5-Formimino FH_4
(histidine catabolism)

N^{10}-Hydroxymethyl FH_4
(thymine synthesis)
(serine $\rightleftharpoons$ glycine)

$-NH_3$

N^5,N^{10}-Methenyl FH_4
(purine synthesis)

NADPH / NADP

N^5,N^{10}-Methylene FH_4
(serine $\rightleftharpoons$ glycine)

NADH

N^5-Methyl FH_4
(methionine synthesis)

Figure 11-3 continued

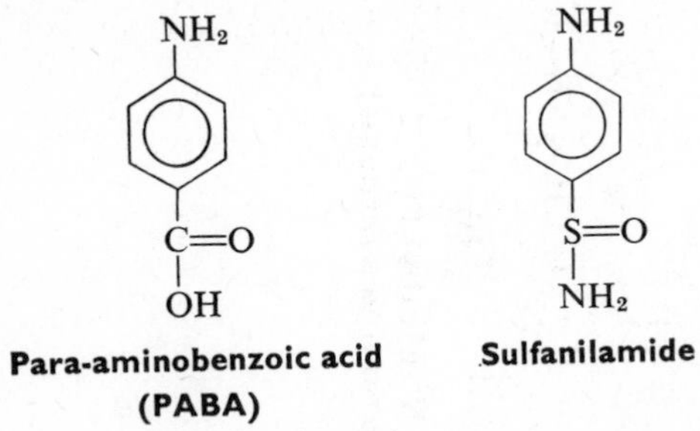

NH$_2$ / C=O / OH

Para-aminobenzoic acid (PABA)

NH$_2$ / S=O / NH$_2$

Sulfanilamide

Figure 11-4 Structures of para-aminobenzoic acid (PABA) and sulfanilamide.

METHODS FOR THE DETERMINATION OF FOLIC ACID

The chemical methods available for the quantitation of FA are subject to interference by all kinds of biochemical metabolites and are, therefore, not applicable to serum and urine specimens. Until recently, it was necessary to depend on microbiological assays which were both relatively simple and of sufficient specificity to give useful results. One such procedure will be outlined in the following paragraphs. Competitive protein-binding radioassays have recently been introduced, and are now in use.

An indirect technique, based on assaying urinary *form*i*mino*glu*tamic acid (FIGLU), has also been used. FIGLU is a breakdown product of histidine and, if sufficient FH$_4$ is present, the two react to form glutamic acid and N^5-formimino-FH$_4$. If FH$_4$ is present in limited amounts, (e.g., in FA deficiency), FIGLU accumulates in the urine and can be assayed by enzymatic and other means (see p. 555).

Determination of folate by microbiological assay (*Lactobacillus casei*)

If a microorganism (bacterium, yeast, mold) is inoculated into a medium containing all nutrients needed for optimum growth except for a limited amount of one essential growth factor, the growth of the organism will be proportional to the amount of this growth factor present. The extent of growth can be determined by measuring the turbidity of the cell suspension, or by measuring the amount of acid formed if the test organism is an acid producer. These observations are the basis for many microbiological assays, including that for folate.

The *specimen* to be assayed is diluted with phosphate buffer and ascorbic acid and then autoclaved, cooled, and centrifuged. The protein-free supernatant (extract) is used for the actual assay. Serum or plasma levels of folate are of greatest clinical interest, but assays of FA levels in red cells are occasionally requested.

One *microbiological procedure* for assaying FA uses *Lactobacillus casei* (ATCC No. 7469) as the test organism. The sterile medium used in this assay contains vitamin-free casein hydrolysate, glucose, acetate, glutathione, all essential amino acids, all known vitamins (except FA), and inorganic ions needed for optimal growth of the test organism.

The *test organism* is stored in agar stabs. For the preparation of the inoculum, a transfer is made into *Lactobacillus* broth, and after 6 h of incubation the bacterial cell harvest is centrifuged and washed three times aseptically with saline. The final inoculum consists of a suitably diluted saline suspension of the organism.

For the *actual assay*, appropriate aliquots of the sample extracts, as well as folic acid standards, are added to a fixed volume of growth medium in a series of screwcap culture tubes, and the mixture is sterilized by autoclaving. All tubes are then inoculated with one drop of inoculum and incubated in a suitable incubator at 37°C for 18 to 22 h. Appropriate controls and blanks are also set up to establish the viability of the organism, the sterility of the medium, and the degree of bacterial growth due to endogenous FA.

At the end of the incubation period, the turbidity of the test medium is measured at 620 nm. The absorbance of the "test" is compared to the absorbance of the "standards," and the concentration of FA (including most of the FH_4 intermediates) is calculated. Results are reported in terms of "total folate activity."

NORMAL VALUES

There is considerable disagreement as to the normal range. However, serum values between 3.0 and 25.0 ng/ml are considered by many laboratories as being within the normal range. Values below 3.0 ng/ml are usually found in persons with megaloblastic anemia that is responsive to folic acid therapy.

Determination of formiminoglutamic acid (FIGLU)

Although formiminoglutamic acid (FIGLU) is elevated in some patients with folate deficiency, such elevation is not consistently present. The reason for this may be due to a deficiency of histidine, the precursor of FIGLU in the diet of some of these patients. Consequently, to remove this source of uncertainty, patients are "loaded" with a 15 g dose of histidine, given in three doses of 5 g each, 4 h apart. After the initial dose of histidine, a 24 h urine collection is started. Under these loading conditions, a urine FIGLU output greater than 35 mg/d indicates folate deficiency. Both patients with B_{12} deficiency and normal subjects with low histidine intake will show a decreased output of FIGLU.

The most convenient and accurate assay for FIGLU is based on the enzymatic conversion of FIGLU to glutamate with the simultaneous conversion of tetrahydrofolate (FH_4) to N^5-formimino-tetrahydrofolate (N^5-CHNH-FH_4).[18] The N^5-CHNH-FH_4 is then quantitatively converted to N^5, N^{10}-methenyl-tetrahydrofolate (N^5, N^{10}-CH_2-FH_4) by the action of a second enzyme. Since N^5, N^{10}-CH_2-FH_4 is very unstable and is converted non-enzymatically to N^{10}-formyltetrahydrofolate (N^{10}-CHO-FH_4), it is necessary to reconvert the latter compound back to N^5, N^{10}-CH_2-FH_4 by acidification. Thus, the addition of perchloric acid at the end of the reaction incubation serves two purposes: (1) it stops the enzymatic reactions, and (2) it regenerates any decomposed N^5, N^{10}-CH_2-FH_4. The absorbance of N^5, N^{10}-methenyl-FH_4 is then read at 365 nm. Appropriate blanks are used to correct for the background absorbance contributed by the enzyme preparation and urine. The absorbance at 365 nm is linearly related to the concentration of FIGLU in urine. The necessary enzymes, FH_4, and the FIGLU standard are all commercially available. A summary of the enzymatic reactions involved is given below:

$$\text{Formiminoglutamate} + FH_4 \xrightarrow[]{\textit{FIGLU transferase}} \text{glutamate} + N^5\text{-formimino-}FH_4 \quad (1)$$

$$N^5\text{-formimino-}FH_4 \xrightarrow{\textit{Cyclodeaminase}} N^5,N^{10}\text{-methenyl-}FH_4 + NH_3 \quad (2)$$

$$N^5,N^{10}\text{-methenyl-}FH_4 + H_2O \underset{H^+}{\overset{OH^-}{\rightleftharpoons}} N^{10}\text{-formyl-}FH_4 \quad (3)$$

(*nonenzymatic*)

VITAMIN B_{12}

Vitamin B_{12} or cyanocobalamin was isolated and crystallized from liver in 1948. It consists of two main parts: a porphyrin-like corrin ring system, and a nucleotide with the unusual heterocyclic base 5,6-dimethylbenzimidazole (Fig. 11–5). Cobalt is chelated to the four nitrogens of the corrin ring, to a nitrogen of the benzimidazole, and to a cyanide ion. Cyanocobalamin itself is not coenzymatically active and must be enzymatically converted to the true B_{12} coenzymes, deoxyadenosyl B_{12} and methyl B_{12}. When serum or other natural products are extracted in the presence of cyanide ions, the B_{12} coenzymes are converted to cyanocobalamin.

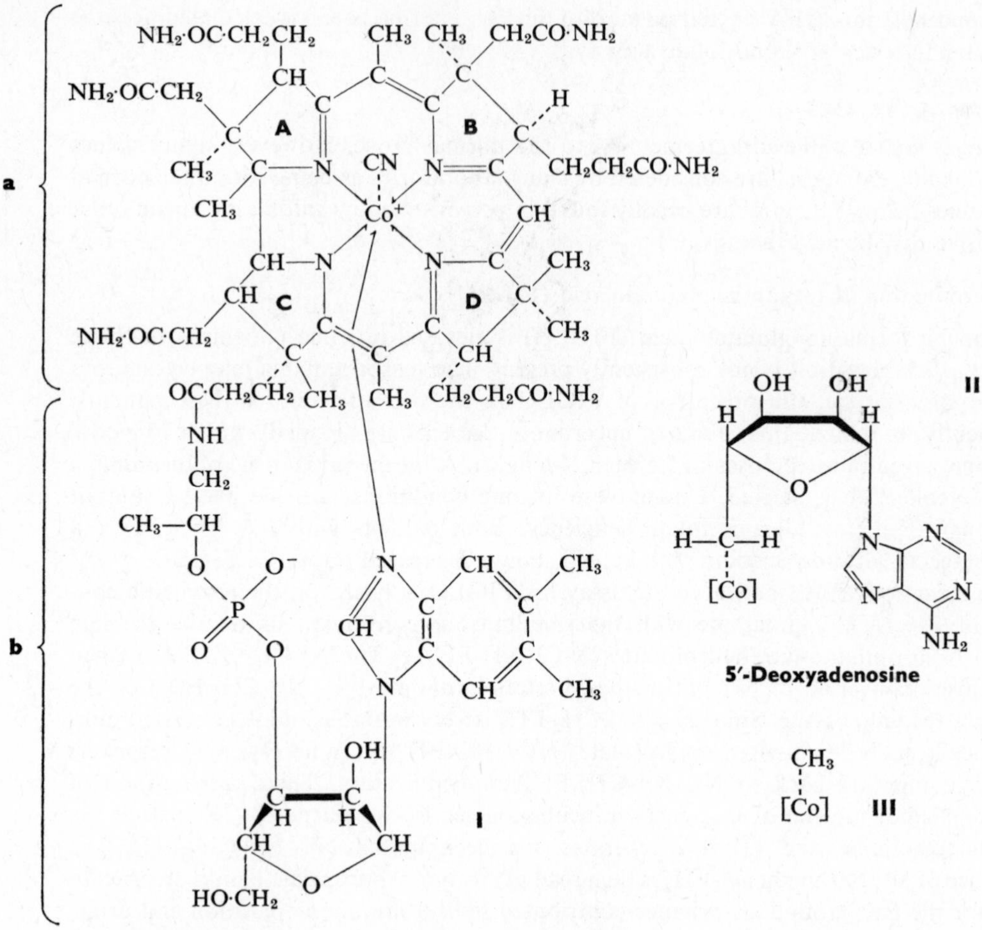

a = Corrin ring
b = 5,6-Dimethylbenzimidazole ribonucleotide

Figure 11–5 Vitamin B_{12} and derivatives. (I) Structure of vitamin B_{12} (cyanocobalamin). (II) In deoxyadenosyl-B_{12}, the CN of cyanocobalamin is replaced by 5′-deoxyadenosine. (III) In methyl-B_{12}, a methyl group replaces the CN of cyanocobalamin.

The intestinal absorption of vitamin B_{12} is highly unique in that a glycoprotein called intrinsic factor (IF) is necessary for its absorption in the ileum. Intrinsic factor has a molecular mass of about 50,000 and is secreted by the mucosal cells at the cardiac end of the stomach. Many other substances, such as saliva, milk, and serum, can bind vitamin B_{12}, but no other known substance has the property of transporting it across the intestinal wall. After intrinsic factor combines with B_{12}, the B_{12}-IF complex combines with calcium ions, is adsorbed onto the mucosal surface of the ileum, and is then transported into the ileal muscosal cells where it is converted predominantly to deoxyadenosyl B_{12}. The most important and common cause of B_{12} deficiency is due to a defect in the secretion of intrinsic factor. Antibodies to intrinsic factor can be demonstrated in a large number of such patients, suggesting that this condition might be an autoimmune disease. The term pernicious anemia is applied to a disease characterized by B_{12} deficiency resulting from inadequate secretion or absence of intrinsic factor. Pernicious anemia is by far the most common B_{12}-deficiency disease. B_{12} deficiency can also occur in vegetarians who eschew the intake of meat or dairy products, since these foods are the only major sources of B_{12}. B_{12} deficiency may also occur in individuals suffering from infestation with fish tapeworm, which has a very high B_{12} requirement.

The predominant form of vitamin B_{12} in serum is methyl-B_{12}. Most or all vitamin B_{12} present in the serum of normal individuals is protein bound, predominantly to the α globulin transcobalamin I, which is believed to be a storage protein. Transcobalamin II, a glycoprotein and β globulin, delivers B_{12} to the reticulocytes or bone marrow. A deficiency of transcobalamin II has been observed in persons with B_{12} deficiency and chronic myelogenic leukemia. B_{12} is secreted into the bile, and an enterohepatic circulation of the vitamin has been demonstrated.

Two definite enzymatic reactions in mammals have been established in which B_{12} participates as coenzyme: (1) the synthesis of methionine from homocysteine and methyltetrahydrofolate, and (2) the conversion of the methyl malonyl coenzyme A thioester to succinyl Co-A.

Reaction (1) is thought to be the reaction that links folic acid and B_{12} metabolism. If B_{12} is deficient, the transfer of a methyl group from methyltetrahydrofolate is retarded and methyl-FH_4, a folate derivative that is inactive in hemopoiesis, accumulates. Consequently, elevated serum folate levels may often be found in individuals who are deficient in B_{12}. Administration of high levels of folic acid can cause a temporary remission of anemias caused by B_{12} deficiency, but this administered folic acid is itself soon "trapped" as methyl-FH_4 unless B_{12} is also administered. In addition, the hematological remission induced by folate in cases of B_{12} deficiency is short and anemia may recur. As mentioned previously, administered folate has no effect on the neurological symptoms associated with B_{12} deficiency.

Reaction (2) plays an important role in the metabolism of fatty acids and the aliphatic amino acids, and it is possible that the neurological effects of pernicious anemia may be due to the accumulation of abnormal lipids in the nervous system. The measurement of methyl malonic acid in urine has been shown to be a good indicator of vitamin B_{12} status since the B_{12} coenzyme deoxyadenosyl B_{12} is required for the further metabolism of the Co-A derivative of this acid. Levels of methyl malonic acid are greatly elevated in persons with B_{12} deficiencies because of a lack of the B_{12} coenzyme for this reaction.

Methods available for B_{12} determination are either microbiological assays or competitive protein-binding radioassay. Since the B_{12} microbiological assay is very similar to that discussed for folate (p. 554), only the radioassay for B_{12} is discussed in the following section.

DETERMINATION OF VITAMIN B_{12} BY CPB RADIOASSAY

Competitive protein-binding radioassays require (1) a radioactive (labeled) form of the substance to be measured, which is of sufficiently high specific activity (amount of radioactive atoms per amount of total compound), (2) a binding protein possessing a relatively high affinity and specificity for the substance to be measured, and (3) an efficient means for separating the free and protein-bound radioactive substance.

Many modifications of the original method for B_{12} radioassay have been proposed; however, all of these methods have a number of steps in common: (1) The vitamin B_{12} in serum is released from serum-binding proteins by boiling or autoclaving serum in HCl, in HCl containing cyanide ions, or in acetate or glutamate buffers containing cyanide ions. Some of these procedures require a centrifugation step while others do not. (2) The extracts of serum and cyanocobalamin standards are then mixed with equal aliquots of ^{57}Co cyanocobalamin. (3) A suitable binding protein such as intrinsic factor or transcobalamin-rich serum from patients with chronic myeloid leukemia is added to the mixture of ^{57}Co cyanocobalamin and serum extract or standard. (4) After a suitable incubation period, free and bound vitamin B_{12} are separated. The procedures or agents used in separating the two fractions include: dialysis, ultrafiltration, albumin-coated charcoal, DEAE cellulose, zinc sulfate/barium hydroxide, Amberlite, uncoated charcoal, and intrinsic factor coupled to Sephadex. The best separating agent appears to be intrinsic factor coupled to Sephadex, since such a preparation serves both as binding and as separating agent. Other separating agents, especially charcoal, may not completely separate bound and free species of vitamin B_{12}. (5) After the separation of the bound and free vitamin B_{12}, the bound or free vitamin B_{12} form, or sometimes both, are counted in a suitable gamma scintillation counter or a beta counter using a special liquid scintillation fluid containing heavy atoms, such as lead, to increase the counting efficiency. Many different methods for plotting standard curves have been advocated, and a few investigators have relied only on a simple radioisotope dilution formula and have neglected to prepare an entire standard curve. However, most investigators insist that an entire standard curve should be performed with each assay. The normal range for serum vitamin B_{12} is from 200 to 1600 pg/ml.

RIBOFLAVIN

Riboflavin was first isolated in pure form from milk whey. It is an orange-colored material and its structure is a substituted isoalloxazine linked to D-ribitol. It is a constituent

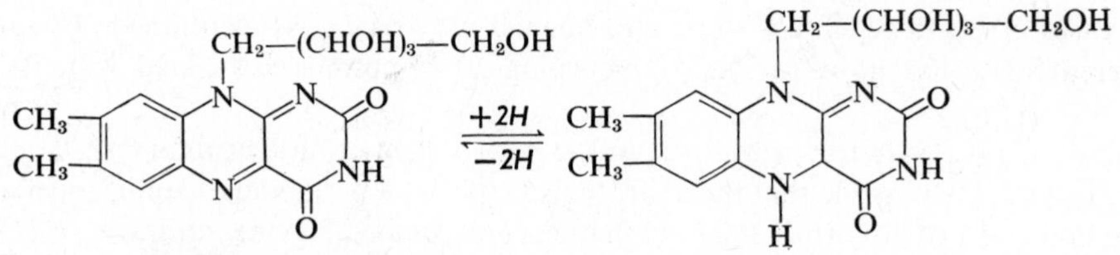

Riboflavin

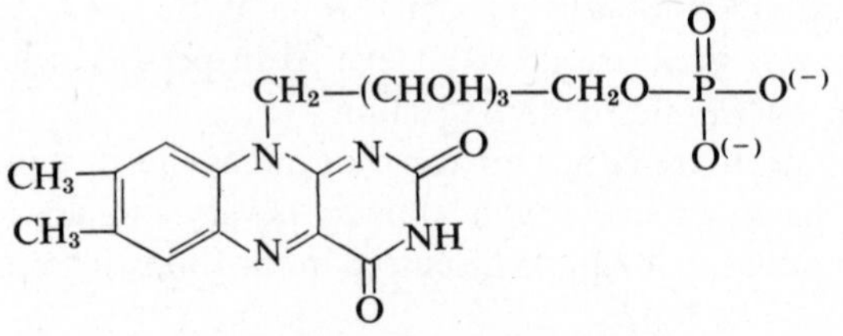

Flavin mononucleotide (FMN)

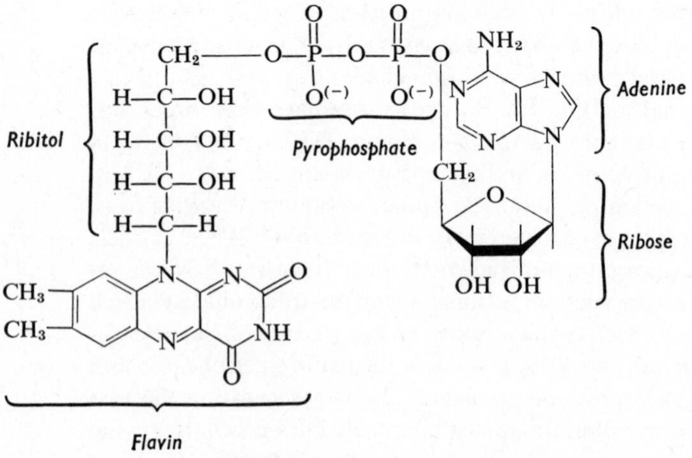

Flavin adenine dinucleotide (FAD)

Figure 11–6 Riboflavin and flavin coenzymes.

of two redox coenzymes, namely, flavin mononucleotide (FMN, riboflavin-5′-phosphate) and flavin adenine dinucleotide (FAD). In the latter, FMN is bonded to adenine ribonucleotide through an acid anhydride bond between the two phosphate groups (Fig. 11–6).

The flavin coenzymes take part in a large number of oxidation-reduction reactions. Enzymes whose function is associated with tightly bound flavin coenzymes are called flavoproteins. Among these are pyruvic acid, α-ketoglutaric acid, D-amino acid, and xanthine dehydrogenases. One of the important biochemical properties of these coenzymes is their ability to exist in a half-reduced (free radical), semiquinone form (FADH) containing an unpaired electron, or in a fully reduced form ($FADH_2$). Flavin-containing enzymes exhibit characteristic absorption spectra which are similar to those of the free flavins. Reduction of the quinone form is accompanied by decreased absorbances at 280 and 380 nm.

In humans, *riboflavin deficiency* symptoms include (1) angular stomatitis (painful fissures at the corners of the mouth), (2) disturbances in vision accompanied by itching and burning of the eyes and sensitivity to light, and (3) dermatitis of the nose, scalp, scrotum, or vulva. Riboflavin is found in meats, milk, cereals, and leafy vegetables, but no food is particularly rich in this vitamin.

DETERMINATION OF RIBOFLAVIN

As was the case with many of the other B vitamins, the microbiological assays were the first methods used to quantitate riboflavin in biological samples, such as urine. The subsequent introduction of fluorometric methods for riboflavin determination resulted in much greater accuracy and precision. The most widely used methods today are modifications of the procedure by Najjar,[13] and Slater and Morrell.[17] In these modifications, interfering substances in urine are destroyed by permanganate oxidation, the riboflavin is then extracted into a butanol-pyridine solution, and its fluorescence is determined. A separate blank must be run for each sample to correct for the residual fluoresence of those interfering substances which are not destroyed by the permanganate oxidation. Blanks are prepared by irradiating with two 100 watt light bulbs for 8 h an aliquot of the original urine sample in tightly capped test tubes rotating on a turntable-like device. This treatment destroys about 95 per cent of the riboflavin. The urine blanks are then treated like samples. To calculate the fluorescence due to riboflavin, the amount of fluorescence obtained with the blank is subtracted from the amount of fluorescence produced by the sample.

MEASUREMENT OF GLUTATHIONE REDUCTASE ACTIVITY

Since riboflavin itself is not metabolically active but is converted to the metabolically active coenzymes FMN or FAD, many investigators believe that the measurement of the biochemical activities of these coenzymes may be more important in ascertaining nutritional status than is the measurement of the level of riboflavin in urine itself. Two possible causes for biochemically inadequate riboflavin nutriture exist: (1) the individual ingests and absorbs enough riboflavin but fails to convert it adequately to the coenzyme derivative FAD, and (2) the individual does not ingest and/or absorb enough riboflavin.

The biochemical activities of the two active coenzymes can be determined by measuring the degree of activation of an FAD- (or FMN-) dependent enzyme on addition of the appropriate coenzyme. If the enzyme is saturated with FAD, little or no increase in activity will be found, indicating that the individual from which the enzyme sample was taken had adequate riboflavin nutriture. Marked activation upon addition of FAD, indicates inadequate riboflavin intake.

One enzyme that has been found to be a good indicator of riboflavin status is erythrocyte glutathione reductase. Glutathione (GSH) is a tripeptide which plays a role in maintaining the integrity of erythrocytes. The enzyme glutathione reductase replenishes the erythrocytes' stores of the reduced form of glutathione (GSH). The reaction which glutathione reductase catalyzes is written as follows:

$$\text{NADPH} + \text{H}^+ + \text{GSSG} \xrightarrow[\textit{(GSH-reductase)}]{\textit{FAD}} \text{NADP}^+ + 2\text{GSH}$$

The degree of stimulation by added FAD is measured in terms of the activity coefficient, which is defined as the decrease in absorbance of NADPH at 334 nm in the presence of FAD, divided by the decrease in absorbance of NADPH at 334 nm without added FAD. An activity coefficient greater than 1.2 is said to be indicative of riboflavin deficiency.

The specimen used in this assay is a hemolysate of red blood cells. A blank (without enzyme) is also measured, although Tillotson and Sauberlich[20] reported that a blank was not necessary under the conditions of their assay.

THIAMINE

Thiamine is a water-soluble vitamin consisting of a pyrimidine ring combined with a substituted thiazole by a methylene bridge. The active coenzyme thiamine pyrophosphate (TPP) is formed by the phosphorylation of a hydroxyethyl group on the thiazole ring of the vitamin (Fig. 11–7). The ultimate isolation of thiamine and its identification as the anti-beriberi factor resulted from the observation of Eijkman that hens with a disease similar to human beriberi were cured when fed rice polishings.

Approximately 20 to 22 different labeled compounds have been chromatographically separated from the urine of rats and man given thiamine labeled with ^{14}C in the pyrimidine or thiazole ring. Most of these compounds are present in minute amounts, but the metabolites present in larger amounts are pyramin (the pyrimidine moiety of thiamine), thiamine disulfide, thiochrome, thiazole, and 2-methyl-4-amino-pyrimidinecarboxylic acid.

Thiamine functions biologically as a coenzyme in enzymatic reactions involving the decarboxylation of α-keto acids, and the transfer of a 2-carbon fragment from one carbohydrate to another. Because of the involvement of this coenzyme in decarboxylase reactions, TPP has been termed cocarboxylase. In mammalian cells, the TPP requiring enzymes involved in decarboxylation of pyruvic and α-ketoglutaric acids are part of high molecular weight complexes associated with the mitochondrial membrane. In persons with thiamine deficiency, the blood levels of pyruvic and α-ketoglutaric acids increase, presumably because of a lack of TPP which is necessary to maintain the activity of the respective decarboxylases. Since pyruvic acid is derived from glucose, one test for thiamine deficiency

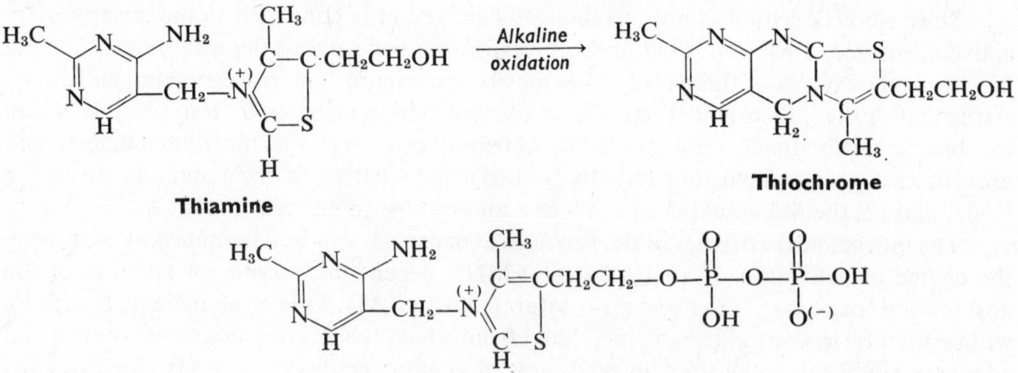

Alkaline oxidation

Thiamine → **Thiochrome**

Thiamine pyrophosphate (TPP)

Figure 11–7 Thiamine and derivatives.

involves the measurement of blood lactic and pyruvic acid levels after administration of glucose. If TPP is lacking, pyruvic acid derived from glucose is preferentially "shunted" to the formation of lactic acid. Thus, an elevated ratio of lactic to pyruvic acid is indicative of thiamine deficiency. Another index of thiamine deficiency is related to the coenzyme role of TPP in the reaction catalyzed by the enzyme transketolase. This enzyme is required in the metabolism of glucose by the pentose phosphate pathway.

Any *decrease in transketolase* activity due to thiamine deficiency results in a reduced catabolism of glucose 2-^{14}C in erythrocytes, as measured by the decreased production of $^{14}CO_2$. This radioassay serves as an indirect measurement of transketolase activity. However, the direct measurement of transketolase activity as an index of thiamine status is becoming increasingly popular, and the principle of this method is given on page 563.

Since all cells possess a requirement for TPP, thiamine deficiency affects all organ systems, but the effects on the neuromuscular system are especially severe. Clinical beriberi has been endemic in the Far East because of the prevalent diet of polished rice. In the Western countries, beriberi is seen mainly in alcoholics. Clinical symptoms of this deficiency disease are peripheral neuritis and extreme muscular weakness as well as anxiety and mental confusion. The heart may become enlarged, and the pericardium may become edematous. Congestive heart failure may develop because of weakening of the heart muscle itself. Less severe thiamine deficiency results in symptoms of a subjective nature: fatigue, vertigo, insomnia, irritability, and headache.

DETERMINATION OF THIAMINE

Determinations of thiamine were first performed by microbiological assay, but the most widely accepted method for the determination of thiamine in urine is the fluorometric thiochrome method. In this procedure, thiamine is oxidized to thiochrome by alkaline potassium ferricyanide (Fig. 11–7). The thiochrome formed is extracted into isobutanol, and its fluorescence is determined with a suitable fluorometer. Blanks are necessary and are determined for each sample by omitting the potassium ferricyanide oxidizing agent. However, the variety of interfering substances present in urine necessitate purification of the sample by passing it through special chromatographic (Decalso*) columns. Recently, Decalso has been replaced by the ion-exchange resin Amberlite.

Possibly the most promising approach is the method of Leveille.[12] In this method, three tubes containing an equal amount of the urine sample are prepared. Standard thiamine solution is added to tube A, water to tube B, and benzenesulfonyl chloride to tube C (benzenesulfonyl chloride is added to destroy thiamine). The ferricyanide oxidizing agent is then added to all tubes and the fluorescence is determined. The difference in fluorescence between tubes B and C represents the fluorescence due to the thiochrome produced from urinary thiamine corrected for interfering substances. The difference in fluorescence between tubes A and B represents the fluorescence due to the thiochrome from the added thiamine standard corrected for any quenching of fluorescence by components of individual urine samples. Tube A thus serves as an internal standard for each sample.

DETERMINATION OF TRANSKETOLASE ACTIVITY

Transketolase is an enzyme found in nearly all tissues, including the blood and liver of mammals. It catalyzes the reversible transfer of a 2-carbon moiety from a donor carbohydrate possessing a keto group (a ketose), to an acceptor carbohydrate possessing an aldehyde group (an aldose), as shown in Figure 11–8. The enzyme has a requirement for magnesium ions as well as for thiamine pyrophosphate.

* The Permutit Co. (Vendor: Fisher Scientific Co., Pittsburg, PA.)

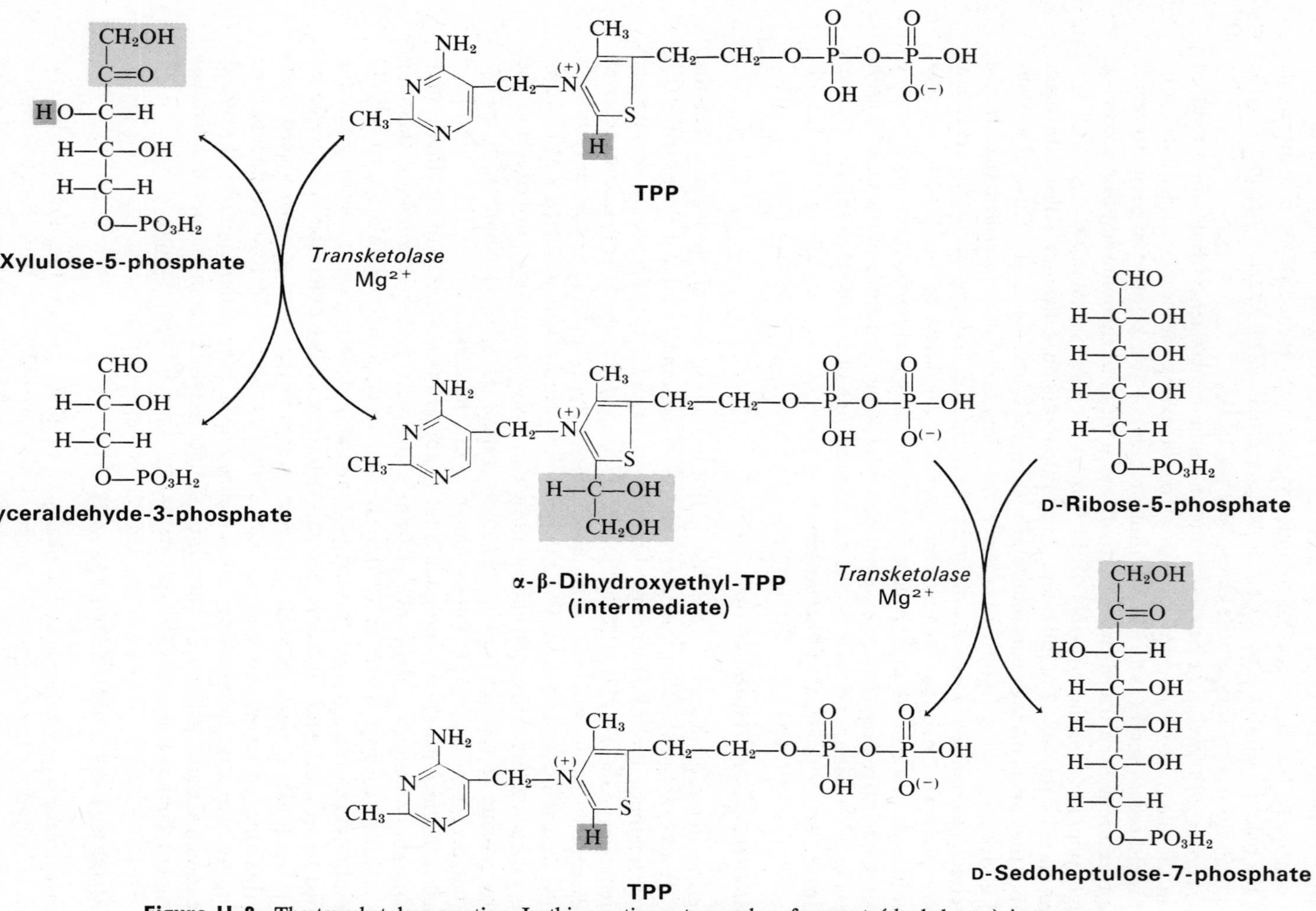

Figure 11–8 The transketolase reaction. In this reaction a two-carbon fragment (shaded area) is transferred from a donor ketose (D-xylulose-5-phosphate) to thiamine pyrophosphate (TPP), and then finally to an acceptor aldose (D-ribose-5-phosphate). The products of this reaction are D-glyceraldehyde-3-phosphate and D-sedoheptulose-7-phosphate.

Transketolase catalyzes two reactions in the pentose phosphate pathway:

Xylulose-5-phosphate + ribose-5-phosphate → sedoheptulose-7-phosphate
+ glyceraldehyde-3-phosphate (4)

Xylulose-5-phosphate + erythrose-4-phosphate → fructose-6-phosphate
+ glyceraldehyde-3-phosphate (5)

Transketolase activity in blood is usually measured by determining the rate of disappearance of D-ribose-5-phosphate with the orcinol reagent (orcinol and ferric chloride in concentrated hydrochloric acid). Transketolase activity in blood can also be measured by determining the amount of fructose-6-phosphate formed in reaction (5) utilizing the anthrone reagent.

The enzyme assay is performed by incubating a sample of heparinized blood (previously hemolyzed in a phosphate buffer containing sodium chloride and magnesium sulfate) with ribose-5-phosphate and removing aliquots of the incubation mixture at three successive time intervals. Trichloroacetic acid is immediately added to the aliquots to stop the reaction. After centrifugation, the supernatant is analyzed for the remaining level of ribose-5-phosphate by the orcinol reaction. The product of the transketolase reaction, sedoheptulose-7-phosphate, also reacts with the orcinol reagent to yield a colored product with an absorption spectrum ($\lambda_{max} = 670$ nm) different from but overlapping with the spectrum of the colored product formed when ribose-5-phosphate reacts with the orcinol reagent ($\lambda_{max} = 580$ nm). By measuring the absorbances of samples at both 670 nm and 580 nm, and solving simultaneous equations, one can measure transketolase activity in terms of both rate of substrate (ribose-5-phosphate) disappearance and rate of product (sedoheptulose-7-phosphate) formation.[21]

Although ribose-5-phosphate is used as the substrate for the transketolase assay, xylulose-5-phosphate is also needed for this reaction. Xylulose-5-phosphate is not included in the assay mixture because the erythrocytes contain nonlimiting amounts of two extremely active enzymes that can convert ribose-5-phosphate to xylulose-5-phosphate. These enzymes are *phosphoribose isomerase*, which forms ribulose-5-phosphate from ribose-5-phosphate, and *pentose phosphate epimerase*, which forms xylulose-5-phosphate from ribulose-5-phosphate. Ribose-5-phosphate added to an incubation mixture containing these enzymes is rapidly converted to an equilibrium mixture of ribose-5-phosphate, ribulose-5-phosphate, and xylulose-5-phosphate in a 1/9/1 ratio. There is a rapid decrease in ribose-5-phosphate during the first 5 min of incubation due to the formation of this equilibration mixture of pentoses. The decrease in ribose-5-phosphate concentration after the 5 min equilibration period is linear with time and is presumed to be due only to transketolase activity. Some previously proposed assays of transketolase activity measure ribose-5-phosphate disappearance before the equilibration of the pentoses takes place. Such methods give falsely high values for transketolase activity.

Normal values are expressed in terms of decrease in total pentose concentration. Values found are between 9 and 12 μmol/h/ml whole blood. If the hematocrit is abnormal, normal values are 2.1–2.4 μmol/h/10^9 red cells.

The sensitivity of the above procedure as a measure of thiamine deficiency can be improved by performing the assay before and after addition of thiamine pyrophosphate (TPP). Presence of TPP results in optimal enzyme activity. This increase in activity is known as the TPP effect. Results are interpreted as follows:

TPP Effect	
0–14 per cent	Normal, no apparent thiamine deficiency
15–24 per cent	Marginal thiamine deficiency
over 25 per cent	Thiamine deficiency

REFERENCES

1. Baker, H., and Frank, O.: Clinical Vitaminology. New Interscience Publishers, 1968, p. 169.
2. Belsey, R., Deluca, H. F., and Potts, J. T., Jr.: J. Clin. Endocrinol. Metab., *33:*554, 1971.
3. Bessey, O. A., Lowry, O. H., Brock, M. J., and Lopez, J. A.: J. Biol. Chem., *166:*177, 1946.
4. Bradley, D. W., and Hornbeck, C. L.: Biochem. Med., *7:*78, 1973.
5. Cooperman, J. M.: Am. J. Clin. Nutr., *20:*1015, 1967.
6. Dugan, R. E., Frigerio, N. A., and Siebert, J. M.: Anal. Chem., *36:*114, 1964.
7. Garry, P. J., Pollack, J. D., and Owen, G. M.: Clin. Chem., *16:*766, 1970.
8. Gyorgy, P., and Pearson, W. N. (Eds.): The Vitamins. 2nd ed. New York, Academic Press, Inc., 1967, vol. 7.
9. Henry, R. J.: Clinical Chemistry Principles and Technics. New York, Harper and Row, Publishers, 1964, p. 715.
10. Herbert, A. A., and Guest, J. R.: *In* McCormick, D. B., and Wright, L. D. (Eds.): Methods in Enzymology. New York, Academic Press, Inc., 1970, vol. 18, part A, p. 269.
11. Herbert, V., Gottlieb, C., and Lau, K. S.: Blood, *28:*130, 1966.
12. Leveille, G.: Amer. J. Clin. Nutr., *25:*273–274, 1972.
13. Najjar, V. A.: J. Biol. Chem., *141:*355, 1941.
14. Neeld, J. B., and Pearson, W. N.: J. Nutr., *79:*454, 1963.
15. Roe, J. H., and Kuether, C. A.: J. Biol. Chem., *147:*339, 1943.
16. Scudi, J. V., and Buhs, R. P.: J. Biol. Chem., *144:*599, 1942.
17. Slater, E. C., and Morrell, D. B.: Biochem. J., *40:*644, 1946.
18. Tabor, H., and Wyngarden, L.: J. Clin. Invest., *37:*824, 1958.
19. Thompson, J. N., Erdody, P., Crien, R., and Murry, T. K.: Biochem. Med., *5:*67, 1971.
20. Tillotson, J., and Sauberlich, H. E.: J. Nutr., *101:*1459, 1971.
21. Warnock, L.: J. Nutr., *100:*1057, 1970.
22. Waxman, S., Schreiber, C., and Herbert, V.: Blood, *38:*219, 1971.
23. Wide, L., and Killander, A.: Scand. J. Clin. Lab. Invest., *27:*151, 1971.
24. Wilson, S. S., and Guillan, R. A.: Clin. Chem., *15:*282, 1969.

ADDITIONAL READINGS

Albanese, A. A., and Hunter, B. J. (Eds.): Newer Methods in Nutritional Biochemistry. New York, Academic Press, Inc., 1963–1972, vols. 1–5.

Association of Vitamin Chemists, Inc. (Ed.): Methods of Vitamin Assay. 3rd ed. New York, Interscience Publishers, 1966.

Baker, H., and Frank, O.: Clinical Vitaminology. New York, Interscience Publishers, 1968.

Bogert, J., Briggs, G. M., and Calloway, D. H.: Nutrition and Physical Fitness. Philadelphia, W. B. Saunders Company, 1973.

Gyorgy, P., and Pearson, W. N. (Eds.): The Vitamins. New York, Academic Press, Inc., 1968, vols. 6 and 7.

Sauberlich, H. E., Dowdy, R. P., and Skala, J. H.: *In* J. W. King and W. R. Faulkner (Eds.): Critical Reviews in Clinical Laboratory Sciences. Cleveland, Ohio, CRC Press Inc., 1973, vol. 4.

Sebrell, W. H., Jr., and Harris, R. S. (Eds.): The Vitamins. 2nd ed. New York, Academic Press, Inc., 1968, vols. 1–5 (vol. 4 in preparation).

Shaefer, A., Exec. Dir. Interdepartmental Committee on Nutrition for National Defense (NIH): Manual for Nutrition Surveys. 2nd ed. U.S. Govt. Printing Office, Washington, D.C., 1963.

Tannahill, R.: Food in History. New York, Stein Day Publishers, 1973.

Ten-State Nutrition Survey, 1968–1970. 5 volumes, DHEW Publication Numbers (HSM) 72-8130, 72-8131, 72-8132, 72-8133, 72-8134 (1972).

CHAPTER 12

ENZYMES

by John F. Kachmar, Ph.D., and Donald W. Moss, Ph.D.

Clinical enzymology is the application of the science of enzymes to the diagnosis and treatment of disease processes. It is one of the most rapidly developing fields in contemporary clinical chemistry. Knowledge of enzymes and their actions goes back at least 140 years, and the word "enzyme" has been used for over a century. Although the role of enzymes in metabolism gradually became clear during the first 40 years of this century, systematic clinical enzymology itself is a comparatively recent development.

Measurements of digestive enzyme activity in body fluids as an aid to diagnosis date back to the early 1900's, and some of these earliest observations (e.g., those on amylase in urine, first studied by Wohlgemuth in 1908) are still useful. Measurements of enzyme activity in serum began in the 1920's and 1930's with the studies of Kay, King, Bodansky, and Roberts on alkaline phosphatase in bone and liver disease. Shortly afterward Kutscher and Wolbergs, as well as Gutman and Gutman, recognized the value of changes of acid phosphatase activity in serum in the diagnosis of prostatic cancer.

An important phase of serum enzymology began in 1943, with the observation of Warburg and Christian that increased activities of glycolytic enzymes were present in the sera of tumor-bearing rats. The measurement of cellular enzymes released into plasma as a consequence of tissue damage received an even greater stimulus in 1955 when La Due, Wroblewski, and Karmen reported the transitory rise of glutamic-oxaloacetic transaminase activity in serum following an acute myocardial infarction. This observation marks the beginning of the modern phase of diagnostic enzymology, which has been characterized by a search for methods of greater sensitivity and for enzymes whose changes are more specific for particular disease states and organs.

Today, in the larger hospital laboratories, enzyme assays may account for as much as 20 to 25 per cent of the total work load, with as many as 12 to 18 different enzymes being estimated routinely. This chapter outlines the main features of enzyme chemistry and surveys the principles which govern enzyme assay methods. The factors which influence the choice of the enzyme to be assayed and the procedure to be used are also discussed.

Enzymes as Catalysts

A catalyst may be defined as a substance which increases the rate of a particular chemical reaction without itself being consumed or permanently altered. In other words, at the end of a catalyzed reaction the catalyst appears unchanged in form and quantity, whereas the main reaction materials have undergone transformation into new products. Enzymes are protein catalysts of biological origin.

A catalyst changes only the *rate* at which equilibrium is established between reactants and products; it does not alter the equilibrium constant of the reaction. In a reaction in which only one set of products is chemically possible, the catalyst cannot effect any change in the nature of the products; but in a reaction in which several different possible pathways exist (as in many organic reactions), the catalyst may favor one pathway over other possible pathways, resulting in different yields of the various reaction products compared with the uncatalyzed process.

Only a small quantity of catalyst is needed to exert an effect on a reaction, and this small quantity can repeat its catalytic role over and over again. Thus, the mass of reactants consumed and products formed is typically many times greater than the mass of the catalyst present.

Catalysts are used widely in a variety of industrial chemical processes, for example, in the manufacture of sulfuric acid (catalysis of $SO_2 + O_2 \rightarrow SO_3$ by platinum), and in the synthesis of gasoline from high boiling oils. In the laboratory determination of calcium, the titration of oxalate with permanganate is catalyzed by a trace of manganous ion (Mn^{2+}), without which the reaction proceeds very slowly. The catalytic action of iodide ion in promoting the oxidation of arsenite by ceric ion is the basis for the determination of protein-bound iodine (PBI). However, compared with enzymes, these inorganic catalysts are usually nonspecific in their action, and will accelerate a wide range of reactions.

Enzymes are synthesized by all living organisms, and they accelerate the multitude of metabolic reactions upon which life depends. Without exception, all enzymes are proteins and, like all proteins, show varying degrees of susceptibility to denaturation by such factors and agents as heat, extremes of pH, concentrated solutions of urea, and organic solvents such as acetone. Since the catalytic activity of enzymes depends on the presence of a precise conformational structure in the folded polypeptide chains, even minor alterations in this structure may result in loss of activity. Denaturation may therefore be detected by the loss of enzyme activity long before other physical or chemical evidence of denaturation is demonstrated. In living organisms, enzymes are rapidly degraded, and their supply replenished by new synthesis. Enzyme preparations are generally unstable and must be handled with special care. Solutions are stored at refrigerator temperatures, often with added stabilizing agents such as ammonium sulfate or glycerol. For longer periods of storage, preparations may be frozen and stored either in this form or as lyophilized powders.

Over 700 enzymes have been isolated, many in highly pure or crystalline form. Each of these is capable of catalyzing a specific type of organic or inorganic reaction. Some enzymes are relatively small molecules, with molecular masses of the order of 10,000 daltons, whereas others are very large molecules, with molecular masses ranging from 150,000 to over a million daltons. Similarly, enzymes show a broad distribution of molecular charge, so that different enzymes may vary considerably in electrophoretic mobilities. Because of their prime importance in biochemical phenomena, enzymes have been the subjects of intensive study. Not only have the amino acid sequences of a number of enzymes been elucidated, but even the conformational structures of a few, such as ribonuclease A and carboxypeptidase B, have been established by X-ray crystallography.

The sequence of amino acids in the chain of ribonuclease A, the enzyme which catalyzes the depolymerization of ribonucleic acid (RNA), as worked out by Smyth, Stein, and Moore,[187] is shown in Figure 12-1. This relatively small enzyme consists of 124 amino acids and has a molecular mass of 12,700 daltons. Figure 12-2 depicts the three-dimensional structure (conformation) of the amino acid chain, as elucidated by Harker from X-ray studies.

An enzyme-catalyzed reaction can be depicted in a simplified or schematic way as follows:

$$\underset{S}{\text{Substrate}} \xrightleftharpoons{\text{Enzyme } (E)} \underset{P}{\text{Product}} \quad (1)$$

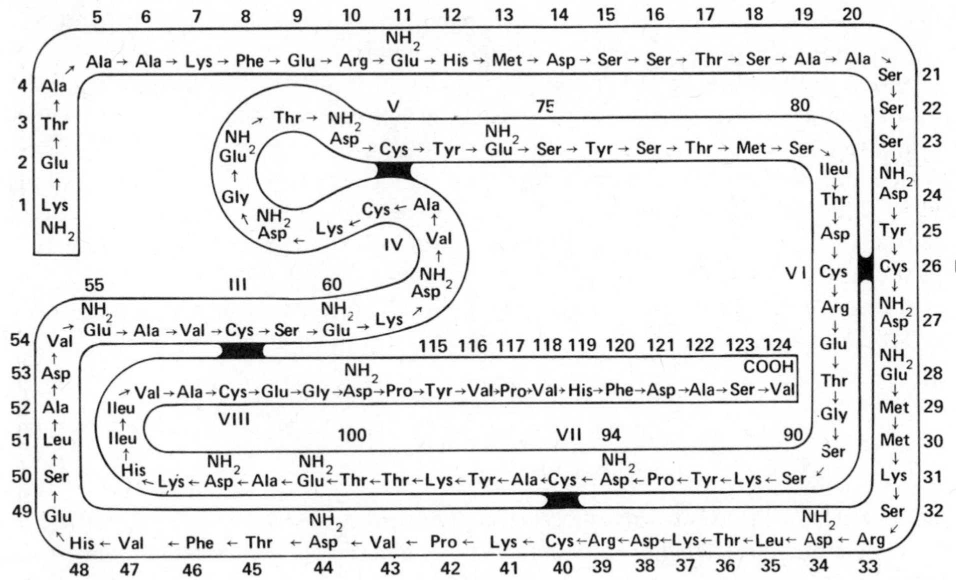

Figure 12-1 Sequence of the 124 amino acids in the peptide chain of ribonuclease A (EC 2.7.7.16), as formulated by Moore and his associates. Note the cystine bridges cross-linking different segments of the amino acid chain. The numbers along the sides of the chain refer to the numerical order of the amino acids in the chain. (From Smyth *et al.*: J. Biol. Chem., *238*: 227, 1963.)

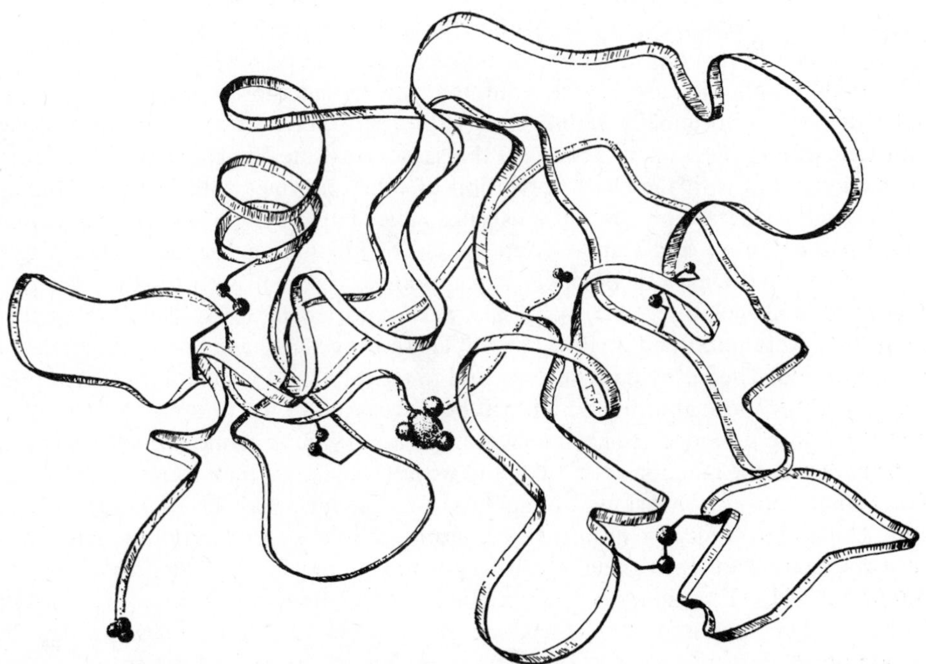

Figure 12-2 Plane projection of the three dimensional structure of ribonuclease A, as proposed by Harker from X-ray studies. Note the cystine bridges also shown in Figure 12-1. The phosphate group indicated in the figure lies in the "active center" of the enzyme, situated in a cleft in the three dimensional structure. (From Kartha: Accounts of Chemical Research, *1*: 374, 1968.)

Here it is assumed that only one substance participates in the reaction. This starting material, referred to as the *substrate*, S, undergoes chemical change to form P, the *product* of the reaction. In the absence of the enzyme, very little or no P will be formed; but in the presence of the catalyzing enzyme, E, the reaction will proceed and the product will be formed at a rate which will depend on the concentration of the enzyme and other factors such as substrate concentration, temperature, and pH. The reaction is reversible, and may proceed in either the forward or reverse directions, depending on the value of the equilibrium constant, $K_{eq} = P/S$, and the conditions prevailing. If the reverse reaction is being studied, the labels "substrate" and "products" are interchanged.

Most enzyme-catalyzed reactions involve two substrates, and usually two products are formed. The more general reaction, therefore, may be written as:

$$\underset{S_1}{\text{Substrate 1}} + \underset{S_2}{\text{Substrate 2}} \xrightleftharpoons{\text{Enzyme } (E)} \underset{P_1}{\text{Product 1}} + \underset{P_2}{\text{Product 2}} \tag{2}$$

Very often, water, the solvent, serves as one of the substrates. Many enzymes require the presence of one or more *activators* and *cofactors*. These, as well as *inhibitors*, will be discussed later in this chapter.

Enzyme Nomenclature

By the turn of the century it had become customary to identify individual enzymes by using the name of the substrate or group on which the enzyme acts, and then adding the suffix, *-ase*. Thus the enzyme hydrolyzing urea was called ure*ase*, that acting on starch, amyl*ase* (from amylum, starch), and that acting on phosphate esters, phosphat*ase*. In a few instances, for purposes of clarity, the type of reaction involved was also identified, as in carbonic *anhydrase*, D-amino acid *oxidase*, and succinic *dehydrogenase*. Enzymes known prior to this attempt at systemization had already been given empirical names. Some trivial names of this type are trypsin, diastase, ptyalin, pepsin, and emulsin.

This combination of a few trivial, common names and the larger number of semi-systematic names was originally found serviceable. However, with the rapid discovery of a multitude of enzymes, many acting on the same substrate but catalyzing different or related reactions, and with the characterization of other enzymes with a specific requirement for a single substrate or pair of substrates, this simple system became inadequate. The need for a definitive and standardized system of identifying enzymes was acute. In 1955 the International Union of Biochemistry appointed a commission to study the problem of enzyme nomenclature. The proposals were reported in 1961, and, after further discussion, they were published in 1964.[52,63,155] These proposals have been accepted by all workers in the field. They provide a rational and practical basis for identifying all enzymes now known and for those enzymes which will be discovered in the future.

Two names are provided for each enzyme: (1) a *systematic* name, which clearly describes the nature of the reaction catalyzed, and with which is associated a unique *numerical code designation*; and (2) a working or practical name, which may be identical with the systematic name or is a modification thereof, more suitable for everyday use. The unique numerical designation for each enzyme consists of four numbers, separated by periods, as for example 2.2.8.11. The number is often prefixed by the letters "EC," denoting "Enzyme Commission." The first number defines the class to which the enzyme belongs. All enzymes are assigned to one of six classes, characterized by the type of reaction they catalyze: (1) oxidoreductases, (2) transferases, (3) hydrolases, (4) lyases, (5) isomerases, and (6) ligases. The next two numbers indicate the subclass and sub-subclass to which the enzyme is assigned. These may differentiate the amino-transferring subclass from the phosphate-transferring group, and the ethanol acceptor sub-subclass from that accepting acyl groups. The last number is the specific serial number given each enzyme in its sub-subclass.

The name of each enzyme consists of two parts: the first gives the name of the substrate or substrates acted upon, and the second, a word ending in *-ase*, indicates the type of reaction catalyzed by all enzymes in the group. If two substrates are involved, both names are used and are separated by a colon; e.g., L-lactate:NAD oxidoreductase. Occasionally an expression in parentheses, such as (decarboxylating), may be inserted to further identify the reaction. Because of the precise rules concerning terminology, any enzyme can be identified by both its code number and its systematic name. In Table 12-1 are listed some selected enzymes of clinical interest, identified by trivial, practical, and systematic names and also by code numbers.

It has been a common and convenient practice to use capital letter abbreviations for the names of certain enzymes, such as GPT for glutamate pyruvate transaminase (EC 2.6.1.2); other examples are GOT, LDH, and CPK, as illustrated in Table 12-1. This practice is not recommended by Commission rules.[155] However, it is so well established, and the convenience in oral and printed communication is so real, that the practice persists, despite the formal disapproval.

Recently, a group of British workers[8,9] proposed the use of a set of "standard" letter abbreviations for the names of those enzymes most commonly referred to in the literature. This has received some support in the United States,[56] although it was vigorously challenged officially[43] (Commission on Biochemical Nomenclature). Examples of such abbreviations are LD, for lactate dehydrogenase (formerly: LDH); ALT, for alanine aminotransferase (formerly: GPT); and LPS, for lipase. Such tentative "standard" abbreviations will be used in this chapter, when appropriate, and after being clearly defined. A selected few of these abbreviations are listed in Table 12-2.

TABLE 12-1 CODE DESIGNATIONS, SYSTEMATIC NAMES, AND TRIVIAL OR PRACTICAL NAMES OF SOME ENZYMES OF CLINICAL INTEREST

Trivial or Common Name, Common Abbreviation	*Practical Name and "Standard" Abbreviation*	*EC Code Designation*	*Systematic Name*
Aldolase	Aldolase ALS	4.1.2.13	D-Fructose-1,6-diphosphate: D-glyceraldehyde-3-phosphate lyase
Alkaline phosphatase	Alkaline phosphatase ALP	3.1.3.1	Orthophosphoric monoester phosphohydrolase
Amylase, Diastase, Ptyalin	Amylase AMS	3.2.1.1	α-1,4-Glucan 4-glucanohydrolase
Creatine phosphokinase (CPK), (CK)	Creatine kinase CK	2.7.3.2	Adenosine triphosphate: creatine phosphotransferase
Glutamate oxalacetate transaminase (GOT)	Aspartate transaminase AST	2.6.1.1	L-Aspartate:2-oxoglutarate aminotransferase
Isocitrate dehydrogenase (ICD)	Isocitrate dehydrogenase ICD	1.1.1.42	L-Isocitrate:NADP oxidoreductase (decarboxylating)
Lactate dehydrogenase (LDH)	Lactate dehydrogenase LD	1.1.1.27	L-Lactate:NAD oxidoreductase
Lipase, Steapsin	Lipase, Triacylglycerol lipase LPS	3.1.1.3	Triacylglycerol acyl-hydrolase
Pseudocholinesterase, non-specific, Type II cholinesterase	Cholinesterase CHS	3.1.1.8	Acylcholine acyl-hydrolase
Trypsin	Trypsin TPS	3.4.21.4	None given. Subclass: peptide hydrolases; sub-subclass: peptide peptidohydrolases.

TABLE 12–2 LIST OF PROPOSED ABBREVIATIONS OF CLINICALLY IMPORTANT ENZYMES†

EC Number	*Recommended Name*	*Abbreviation*
1.1.1.1	Alcohol dehydrogenase	AD
1.1.1.14	L-Iditol dehydrogenase	ID
	(Sorbitol dehydrogenase)*	SDH**
1.1.1.27	Lactate dehydrogenase	LD
	Hydroxybutyrate dehydrogenase	HBD
1.1.1.49	Glucose-6-phosphate dehydrogenase	GPD
2.1.3.3	Ornithine carbamoyl-transferase	OCT
2.3.2.2	γ-Glutamyltransferase	GGT***
2.6.1.1	Aspartate transaminase	AST
2.6.1.2	Alanine transaminase	ALT
2.7.3.2	Creatine kinase	CK
3.1.1.3	Triacylglycerol lipase	LPS
3.1.1.8	Cholinesterase	CHS
3.1.3.1	Alkaline phosphatase	ALP
3.2.1.1	α-Amylase	AMS
3.2.1.31	β-Glucuronidase	GRS
3.4.21.4	Trypsin	TPS
3.4.23.1	Pepsin A = Uropepsin	PPS
4.1.2.13	Fructose bisphosphate aldolase	ALS
4.2.1.1	Carbonate dehydratase (Carbonic anhydrase)*	CA

† List as proposed by Baron, Moss, Walker, and Wilkinson.[9]
* Older, more commonly used name.
** Abbreviation used in this chapter, in place of the "semi-official" abbreviation.
*** Initially, GMT was proposed, but was replaced by GGT in 1975.

Elementary Aspects of Enzyme Catalysis

A chemical reaction involving the transformation of a chemical material, S, into a product, P, can proceed spontaneously only if there is a decrease in *free energy* or *chemical potential* in the course of the reaction. Stated simply, chemical reactions proceed downhill, from the energy viewpoint. The symbol for Gibbs' free energy is G, and ΔG represents the difference between the chemical potential of the products, P, and that of the initial reactants, S. Thus, a chemical reaction will tend to proceed spontaneously to completion or to equilibrium only if ΔG has a negative value. This concept is illustrated in Figure 12-3. If ΔG is zero, no reaction will occur—the system is in equilibrium, and P and S have the same free energy content. If ΔG is positive, P will tend to react to form S; the uphill reaction will take place only if energy is provided from the outside to push the reaction uphill. In the living organism this outside energy is often provided by adenosine triphosphate, ATP.

Even though a chemical reaction is thermodynamically possible (i.e., ΔG is negative), the reaction may not proceed spontaneously, since only those molecules that are "excited" or "activated" will undergo reaction. The "active" molecules are those which have absorbed extra energy in some way, with the result that the bonds linking some or all of the atoms in the molecule are weakened. Most chemical materials are stable at ordinary conditions because only a very small fraction of the molecules are in such an "activated" state, that is, have enough extra energy to cross the "activation barrier." In the spontaneous reaction

$$S \rightleftharpoons S^* \rightleftharpoons P^* \rightleftharpoons P \tag{3}$$

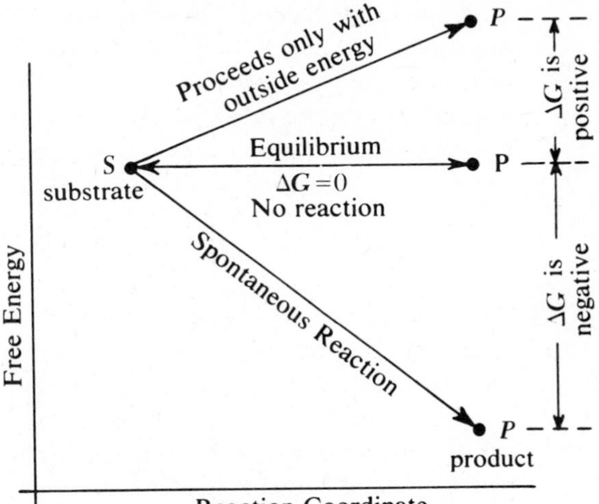

Figure 12-3 Free energy change and the course of chemical reactions.

the molecules of S must first absorb ΔA units of activation energy to form the "activated" molecules, S^*, as shown schematically in Figure 12-4 and in equation 3. The activated S^* molecules then undergo reaction to P^* and P. Many reactions require only a relatively low activation energy, and the energy of the thermal motion of the solute and solvent molecules is sufficient to activate enough molecules to initiate the reaction. The energy released in the reaction then makes possible the activation of all the reactants and allows the reaction to proceed to completion within a comparatively short time period. Many reactions with high activation barriers can be initiated by heating the reactants, as is often done in the chemical laboratory.

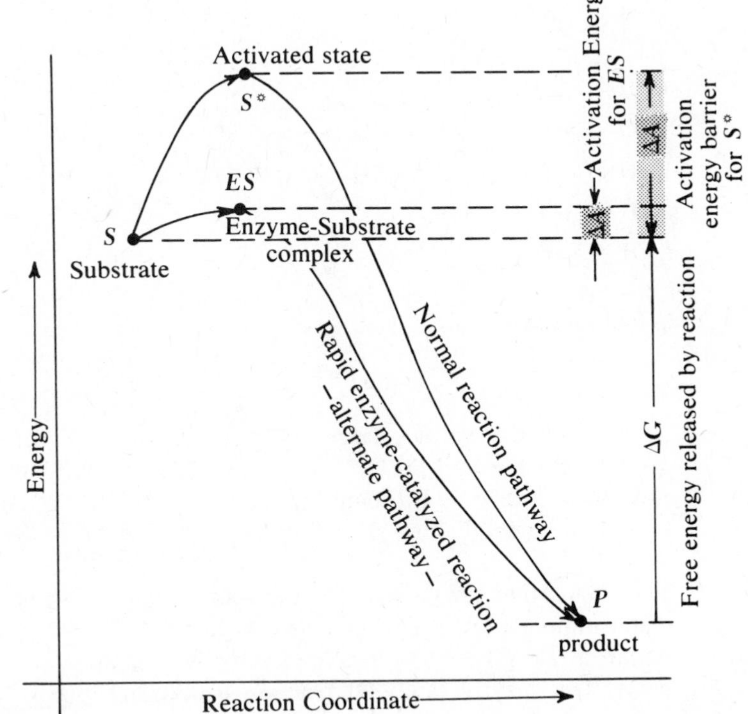

Figure 12-4 Activation energy barrier and reaction course, with and without enzyme catalysis.

If the activation energy needed for a given reaction pathway is too high to permit initiation of the reaction, the reactant may still be able to undergo chemical change if some other reaction pathway having a lower activation energy requirement is available or is provided. Catalysts provide such alternate paths. It is now known that enzymes act by forming an enzyme-substrate (*ES*) complex which has a higher reactivity and breaks down to give the reaction products and the free enzyme.

$$E + S \rightleftharpoons ES \rightleftharpoons P + E \qquad (4)$$

The activation energy required for both E and S to form ES is lower than that required to activate S alone, and the reaction will proceed via the alternate enzyme-substrate complex pathway. Such an alternate pathway is indicated in Figure 12-4.

When an enzyme acts on its substrate, only a single bond or chemical group in the substrate is involved in the reaction, even though the substrate may be a large molecule, perhaps also a protein. It is now well documented that only a small area of the enzyme peptide chain is actually involved in the catalytic activity. This area is referred to as the *active center* of the enzyme. Characteristically, this catalytic center is formed by three or four amino acids in the twisted, coiled polypeptide chain, separated considerably from each other in the amino acid sequence, but very near to each other spatially. Certain side chains of these amino acids hold or anchor the substrate at the site, while others modify the bonding forces at the reacting group of the substrate. Presumably, the remainder of the enzyme peptide chain is needed to maintain the unique conformation necessary to form the active center.

Glucose-6-phosphate is an important intermediate in the metabolism of glucose. In aqueous solution, it can be hydrolyzed to form glucose and phosphate ion (equation 5). The free energy decrease for the hydrolysis of this ester at 37°C and pH 8.5 is approximately

$$\text{α-D-Glucose-6-phosphate} + H\text{—}O\text{—}H \xrightleftharpoons[\text{pH} = 8.5]{\textit{Glucose-6-phosphatase}} \text{α-D-Glucose} + H\text{—}O\text{—}P(=O)(O^-)_2 \qquad (5)$$

α-D-Glucose-6-phosphate α-D-Glucose Phosphate ion

−3000 calories, which is large enough to permit rapid and spontaneous hydrolysis of the ester to glucose and HPO_4^{2-}. Yet a sterile solution of glucose-6-phosphate can be kept at room temperature for many weeks, with very little or no hydrolysis taking place. However, if a few milligrams of a phosphatase are added to the solution, the ester is split very rapidly.

This rather simple hydrolytic reaction requires the severing of an (H—O) bond in a water molecule and a (P—O) bond in the ester. The rupture of these chemical bonds demands a rather extensive degree of activation of the water and glucose-6-phosphate molecules. The enzyme-catalyzed reaction proceeds along a different pathway. The enzyme reacts with the glucose-6-phosphate, liberating the glucosyl residue and forming an enzyme-phosphate intermediate, which rapidly transfers the phosphate residue to a hydroxyl ion, freeing the enzyme to repeat the process over and over again. The enzyme-mediated pathway has a significantly lower activation energy requirement and, therefore, proceeds easily.

Enzyme Kinetics

Some of the techniques used in the analysis of inorganic and organic materials are not applicable to measuring enzymes, because the enzyme protein usually represents only a small part of the total mass of other proteins and compounds. Partial or total isolation is impractical, because suitable procedures for such separations are time-consuming and entail considerable loss of material. Fortunately, each enzyme provides us with a tool or "tag" for its measurement, namely its unique biochemical property of catalyzing a specific chemical reaction. We can thus quantitate the activity of a specific enzyme, even in complex mixtures containing other enzymes, by measuring what the enzyme can do, rather than by measuring the enzyme in terms of mass. Even a small quantity of enzyme can, over a period of time, catalyze the transformation of a large quantity of substrate. If the substrate changes are measured by sensitive procedures, it is possible to relate these changes quantitatively to the activity of minute amounts of the associated enzyme.

The degree of catalytic activity can be utilized as a precise measure of enzyme concentration only if such activity is measured under well-defined experimental conditions. Thus, the main need in working with enzymes is to define the conditions applicable to each specific enzyme reaction.

The acid phosphatase in serum can hydrolyze phenylphosphate (substrate) to phenol and phosphate ion (products). As the reaction proceeds, the substrate, phenylphosphate, is consumed, and its concentration in the reaction tube will decrease. At the same time the concentrations of the products, phenol and inorganic phosphate, increase over their initial values. The more enzyme present, the greater the rate of catalysis and, therefore, the greater the rate of change in the concentrations of the reaction components. Inasmuch as one mole of each product is formed from one mole of substrate, under ideal conditions the rate of increase in product concentration should be identical to the rate of decrease in substrate concentration. Therefore, the reaction may be followed by measuring either the drop in substrate concentration or the increase in concentration of either product. (In our example, shown in equation 6, there is no convenient method for determining phenylphosphate, whereas simple and precise methods are available for measuring either inorganic phosphate or phenol.) In general, measurement of the appearance of products is preferred over the disappearance of substrate, since it is easier analytically to measure small increases from zero than to detect small decreases from an initially high level.

$$C_6H_5\text{—O—}\overset{\overset{\displaystyle O}{\|}}{\underset{\underset{\displaystyle OH}{|}}{P}}\text{—}O^- + H\text{—O—}H \xrightarrow[\text{pH 4.9}]{\textit{Enzyme}} C_6H_5\text{—O—}H + HO\text{—}\overset{\overset{\displaystyle O}{\|}}{\underset{\underset{\displaystyle OH}{|}}{P}}\text{—}O^- \qquad (6)$$

Phenylphosphate ion **Phenol** **Dihydrogenphosphate ion**

Both the concentration of enzyme present and the time interval during which the reaction proceeds will determine the quantity of substrate consumed or product formed. This is expressed in the following equation:

$$Q = k_1 \times E \times t \qquad (7)$$

where Q is the quantity of product formed (or increase in concentration of product),

E is the concentration of active enzyme present, t is the interval of time during which the reaction has been permitted to proceed, and k_1 is a rate constant. If the time interval (t) is fixed, then $Q = (k_1t) \times E$; if, on the other hand, the enzyme concentration is kept constant, then $Q = (k_1E) \times t$. Thus, the quantity of product formed (or substrate consumed) will be a straight-line function either of enzyme concentration or of time, if the other is kept fixed. Figures 12-5 and 12-6 are graphs illustrating such straight-line functions. In the Ellman procedure for measuring cholinesterase, the quantity of enzyme present in the reaction tube is expressed in terms of the number of sulfhydryl groups (—SH of thiocholine) liberated in the hydrolysis of acetyl thiocholine; in a commonly used method for lactate dehydrogenase, the enzyme is assayed by measuring the decrease in absorbance as the reduced coenzyme I (NADH) is oxidized during the course of the reaction.

If the reaction time is fixed, the quantity of product formed is proportional to the concentration of enzyme present in the reaction tube. From equation 7 it follows that the *rate* of product formation $= Q/t = k_1 \times E$. Thus, the enzyme activity can be determined by measuring the *rate* of product formation or substrate consumption, over any convenient time period.

At the moment when the enzyme and substrate are mixed, the rate of the reaction is initially zero (or virtually so). Typically, it then rises rapidly to a maximum value, which remains constant for a period of time. During the period of constant reaction rate, which may be short or long in different circumstances, the rate depends only on enzyme concentration and is completely independent of substrate concentration. The reaction is said to follow *zero-order* kinetics, i.e., its rate is a function of the zeroth power ($S^0 = 1$) of the substrate concentration. Ultimately, however, as substrate is consumed, the reaction rate declines because of the effect of this and several other factors, such as accumulation of products which may be inhibitory, the growing importance of the reverse reaction, and even enzyme denaturation. As the reaction rate begins to decline, the reaction enters a phase of first-order dependence on substrate concentration.

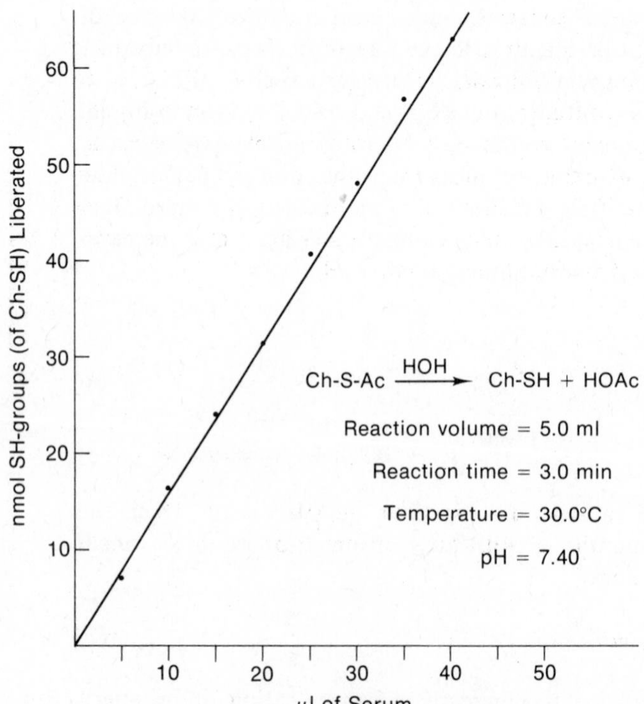

Figure 12-5 Relation between quantity of thiol groups liberated from acetyl thiocholine by cholinesterase and volume of serum containing the enzyme. Ch-SH = thiocholine, Ch-S-Ac = acetyl thiocholine; HOAc = acetic acid.

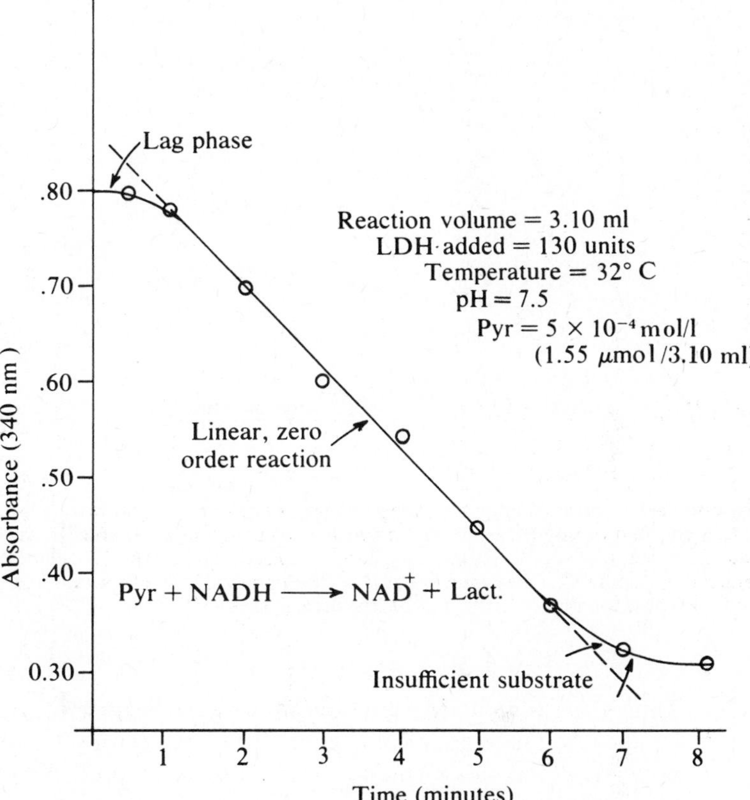

Figure 12-6 Rate of oxidation of NADH as a function of time in an LD assay. The concentration of NADH is expressed as its absorbance at 340 nm.

Although the rates of reaction produced by different amounts of an enzyme can be compared under first-order conditions, it is obviously easier to standardize such comparisons when the enzyme concentration is the only variable that influences the reaction rate. Therefore, enzyme assays are almost always made under conditions which are initially saturating with respect to substrate concentration (see next section). The rate of reaction during the zero-order phase (equivalent to the "initial rate") can be determined by measuring the product formed during a fixed period of incubation, but this assumes that the rate has, in fact, remained constant during this period. If it has not, the apparent rate of reaction will no longer be proportional to enzyme concentration. This is illustrated in Figure 12-7. Measurement of reaction rates at any portion of curve A will give results that will be identical to the true "initial rate." However, curve B deviates from linearity over its entire course, and rates fall off with time. At no time will it give a measure of the "initial rate." From curve C, correct results can be obtained only if the rate is measured along segment II. *In*correct results are obtained if the rate is measured during the lag phase (I), or during phase III. Methods in which the progress of the reaction is continuously monitored, e.g., in a recording spectrophotometer (often called "kinetic" methods or continuous sampling type techniques), are therefore preferable because any deviation from zero-order kinetics becomes apparent.

In the Wohlgemuth procedure for amylase, the enzyme activity is measured by the time required to *consume all of a fixed quantity of substrate.* In this case, equation 7 becomes

$$E = \frac{Q}{k_1} \times \frac{1}{t} \quad \text{or} \quad E = k \times \frac{1}{t} \quad \text{or} \quad E \propto \frac{1}{t} \tag{8}$$

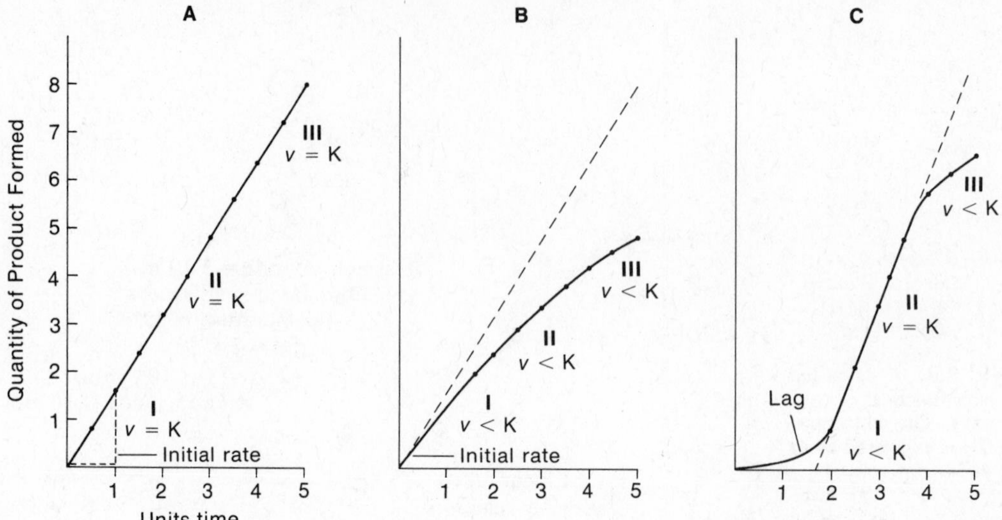

Figure 12-7 Forms of graphs showing change in enzyme reaction rate as a function of time. In **A**, the rate is constant during the entire run, and rates calculated at I, II, and III will be identical to the initial rate. In **B**, the rate falls off continuously; rates calculated at I, II, and III will be different, and less than the true initial rate. In **C**, a measurement at II will be representative of the maximal rate, but at I (lag period) and III (substrate depletion), it will be less than at II.

Thus, there is an inverse relationship between enzyme concentration and the time required for the reaction to go to completion, the curve relating E and t having the shape of a hyperbola. However, this method cannot be used for precise measurement of enzyme activity, because it depends on total consumption of substrate. Thus, for a considerable period of reaction time, the substrate concentration will be less than optimal and zero-order kinetics will not apply.

Rates of all chemical reactions are governed by the concentrations rather than the absolute quantities of the reactants participating in the reaction. Therefore, in the previous discussion it would have been more precise to have referred only to concentrations of substrates and enzymes, and not to the quantities of these materials. In practice, in any given assay, the volume of solution in which the enzyme, substrates, and products are contained is kept fixed at some convenient value, so that the quantities of the components present are proportional to their concentrations. Under these conditions, concentration and quantity can be used interchangeably. In precise studies of enzyme reactions, it is important to employ and measure only true concentrations of each of the components involved in the reaction.

Factors Governing the Rate of Enzyme Reactions

HYDROGEN ION CONCENTRATION (pH)

If an enzyme-catalyzed reaction is carried out at a series of different pH values with the concentrations of enzyme and substrate kept constant, no reaction occurs at certain pH ranges, whereas various rates of reaction are observed at other pH values. When enzyme activity (rate) is plotted against pH, curves of the types presented in Figures 12-8 and 12-9 are obtained. In the case of amylase (Fig. 12-8), no activity is measurable at values of pH below 4.0 or above 9.5, and maximum activity occurs at pH 7.0. Similar curves relating

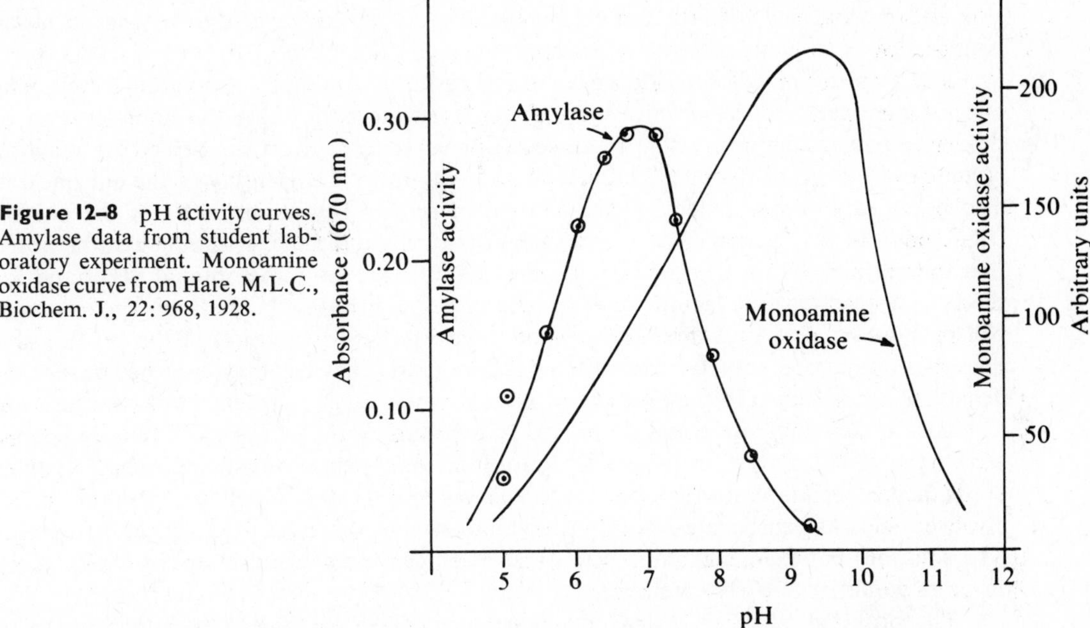

Figure 12–8 pH activity curves. Amylase data from student laboratory experiment. Monoamine oxidase curve from Hare, M.L.C., Biochem. J., *22*: 968, 1928.

reaction rate and pH can be obtained for all enzymes, although the pH at which maximum activity occurs and the shape of the curve vary from one enzyme to another and with the conditions of measurement. The curve for monoamine oxidase extends over a wide range of pH, with a sharp drop-off in activity on the alkaline side, reflecting enzyme denaturation. The curves for urease, investigated by Howell and Sumner (the latter is one of the early

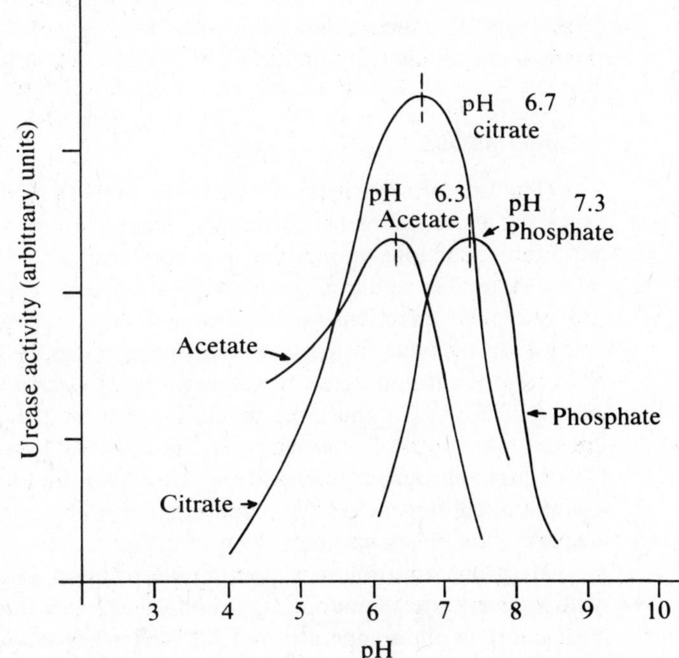

Figure 12–9 pH activity curves for urease, showing effect of buffer species on pH optimum. (Adapted from Howell and Sumner: J. Biol. Chem., *104*: 619, 1934.)

pioneers in enzyme science), show sharp optima, and also demonstrate that the type of ion (buffer) environment in which the enzyme is functioning has an effect on the reaction rate and the optimal pH for the enzyme (see Fig. 12-9). Many of the enzymes in blood plasma show maximum activity somewhere in the pH range from 7 to 8.

The form of the pH-dependence curve is a result of a number of separate effects. The enzyme may catalyze the reaction only if the substrate is in either the undissociated or dissociated form, and the extent of dissociation is dependent on the pH of the reaction system and the p*K* of the dissociating acid or base group. The activity of the enzyme will also be affected by the extent of dissociation of certain key amino acids in the protein chain, both at the "active center" and elsewhere in the molecule. It should also be noted that the optimal pH for a given reaction may be different from the optimal pH found for the reverse reaction (see lactate dehydrogenase, p. 652). Both pH and ionic environment will have an effect on the three-dimensional conformation (structure) of the protein and therefore on enzyme activity. At extreme values of pH, enzymes may even be irreversibly denatured, as is shown in the example of monoamine oxidase. The extent of denaturation depends on the temperature and the period of exposure to the extreme pH. Some enzymes are known to be associations of two or more individual peptide chains; at extreme values of pH these associations are disrupted, with ensuing loss of catalytic activity. Other enzymes, however, such as pepsin, are not only stable at extreme values of pH (e.g., in 0.12 molar HCl solution) but exert maximum activity at hydrogen ion concentrations at which many enzymes would be entirely inactivated.

The pronounced effects of pH on enzyme reactions emphasize the need always to control this variable by means of adequate buffer solutions or by use of a pH Stat. Enzyme assays should be carried out at the pH of optimal activity because (1) the sensitivity of the measurement is maximal at this pH, and (2) the pH-activity curve usually has a minimum slope near this pH so that a small variation in pH will cause a minimal change in enzyme activity.

In some reactions, acids or bases are formed during the reaction (e.g., formation of fatty acids by the action of lipase, or formation of ammonia from urea by urease). In these cases it is important to use buffers of sufficiently great buffer capacity to effectively counteract the changes in pH resulting from such acid or base formation. The p*K* of the buffer system should be within one pH unit or less of the optimal pH of the enzyme system. Unfortunately, these rules have often been ignored in devising procedures for assaying the activity of clinically important enzymes.

TEMPERATURE

The rate of any chemical reaction increases as the temperature at which the reaction is taking place increases. Enzymatic reactions are no exception. For most chemical and enzymatic reactions an increase in temperature of 10°C will approximately double the rate of reaction. The actual Q_{10} value (the relative increase in rate per 10°C temperature rise) for enzymatic reactions varies from 1.7 to 2.5. However, an additional factor must be taken into account in the case of enzymatic catalysis. As the temperature increases, the enzyme protein undergoes increasingly rapid *heat denaturation*, and this becomes marked above 40 to 45°C. Thus, the increasing rate of the reaction is counteracted by the even greater rate of loss of active enzyme. The observed rate of enzyme activity shows a maximum value at a particular temperature, as shown in Figure 12-10. The actual temperature optimum will depend on the reaction conditions, particularly the time interval over which enzyme activity is measured.

Heat denaturation is a continuous process, and the rate of denaturation increases with temperature. Some denaturation of enzymes occurs even at room temperature, and even more at the temperature of 37°C. To prevent or minimize loss of activity, enzyme

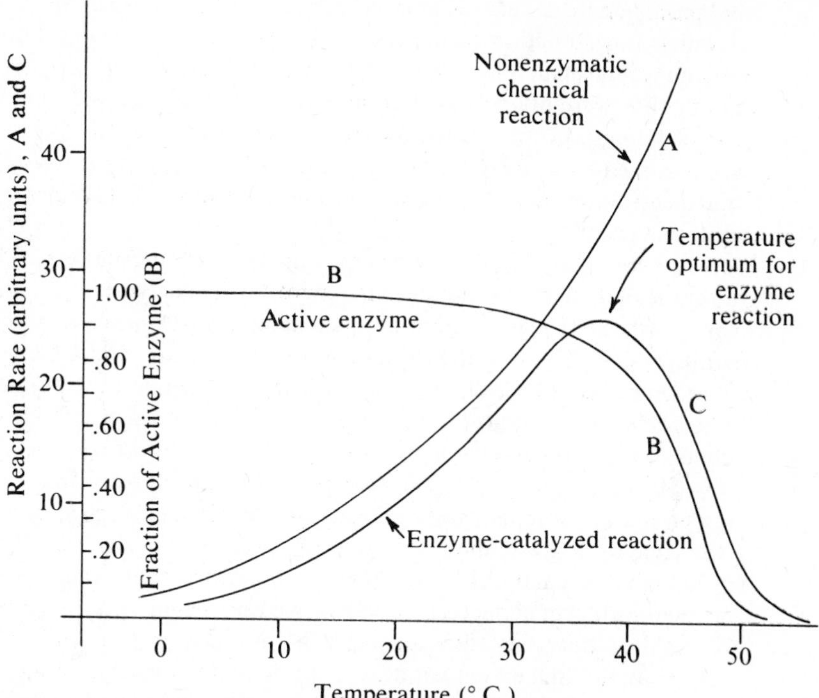

Figure 12–10 Schematic diagram showing effect of temperature on rate of nonenzyme-catalyzed and enzyme-catalyzed reactions.

preparations or specimens are stored at refrigerator temperatures (2 to 8°C) for short periods of time, and below freezing temperatures (−20 to −70°C) for long periods of time. *Repeated* freezing and thawing is itself denaturing and should be avoided. The presence of other proteins and other impurities tends to slow down enzyme denaturation, and therefore more highly purified preparations are often less stable. The protective mechanism is not well understood, but it is believed that proteins and other colloids provide a more natural ionic environment and protection against localized pH changes.

There is considerable variation in the stability of individual enzymes. As a general rule, enzyme specimens should be assayed immediately or, if relatively stable, before significant loss of activity occurs. Optimal storage conditions and times will vary with each enzyme or enzyme specimen. Amylase, for example, is stable at room temperature (22 to 25°C) for 24 hours, whereas acid phosphatase is exceedingly unstable, even when refrigerated, unless kept at a pH below 6.0. Alkaline phosphatase also exhibits an unusual property: the tendency for the activity of frozen, partially purified preparations of the enzyme to increase after thawing over a period of 24 hours or longer (this effect is shared by reconstituted, lyophilized preparations of the enzyme and affects their use for quality-control purposes). A few enzymes are inactivated at refrigerator temperatures; clinically important examples are the liver-type isoenzymes of lactate dehydrogenase (LD-4 and LD-5), which appear to be less stable at lower temperatures. As a result, sera for LD determinations *should not be refrigerated.*

As already mentioned, the optimal temperature for any given enzyme is also dependent on the duration of the exposure of the enzyme to the given temperature; the shorter the period of exposure, the higher the optimal temperature found. An enzyme may demonstrate an optimal temperature of 50 to 52°C (e.g., amylase) for a period of 4 to 5 minutes, but a temperature optimum of only 35°C over 3 to 4 hours. The large majority of cellular and plasma enzymes are reasonably stable at 37°C, and since this is close to the temperature at which they function in the body, the practice of carrying out enzyme assays at 37°C became established. When spectrophotometric assays (e.g., for transaminases) were introduced, they were often carried out at ambient (room) temperature because spectrophotometers with thermostatted cell-compartments were rare at that time. "Ambient temperature" was assumed to be 25°C, although 30 to 32°C is the more usual temperature prevailing in a spectrophotometer cuvet compartment because of heating from the nearby lamp and electrical circuits.

In 1961, the Commission on Enzymes of the International Union of Biochemistry[155] recommended that the standard temperature in enzyme work should be 25°C. Since in many parts of the world high ambient temperatures make it difficult to maintain a temperature of 25°C without cooling as well as heating, the Commission in 1964 amended its recommendation and proposed a standard temperature of 30°C.[63] The temperature to be preferred in clinical enzymology remains at present a subject of intensive study and debate. Most clinical enzymologists in Europe favor either 25 or 37°C, but many in the United States prefer 30°C. The strongest point in favor of the 37°C temperature is that a greater change in some rate-measuring parameter (e.g., absorbance) will be registered in a given time for a given amount of enzyme, and that the accuracy and precision of the assay will thereby be enhanced. The higher temperature will permit shorter recording intervals and therefore provide necessarily for a greater throughput of samples.

Against this, the advocates of 25°C argue that the stability of the enzyme will be reduced at the higher temperature. Although this effect may be slight or even negligible in the case of some enzymes, the lowest temperature compatible with an accurate assay should be chosen.

The debate at present is also concerned with the temperature to be specified for use in reference (referee) methods. For these types of assays, highly sensitive thermostats are needed which control the reaction temperature to within ±0.05°C of the set-point.

Whatever the temperature chosen for enzyme analysis, the important consideration is that this variable must be rigidly controlled. Tubes containing the reaction mixture must be equilibrated at the incubation-bath temperature before the enzymatic reaction is initiated, and the reaction must be stopped before the tubes are removed from the bath. so that the reaction mixture is at the controlled temperature for the *entire* timed period. When a spectrophotometer is used for rate assays, it must be equipped with means for measuring and controlling the temperature of the cuvet compartment. The temperature control device in routine work must maintain the set temperature to within ±0.1°C.

SUBSTRATE CONCENTRATION

If one gradually increases the concentration of substrate in an enzyme reaction system, keeping all other factors constant, the rate of reaction will increase with the increase in substrate concentration until a maximum value is reached. Any further increase in substrate concentration, no matter how great, will elicit no further increase in reaction rate. This phenomenon is illustrated in Figure 12-11. At low substrate concentrations, the rate of the reaction increases linearly with an increase in concentration of substrate; then, as moderate levels of substrate concentration are approached, the rate of increase falls off until the maximum rate of reaction is obtained.

The significance of substrate-rate curves was first emphasized by Michaelis and Menten, and such curves are referred to as Michaelis-Menten plots. The shape of the curve is explained by the assumption that the substrate and enzyme first associate to form an enzyme-substrate complex, already referred to on p. 572. The rate of reaction depends on the concentration of the complex, which is the moiety undergoing reaction, rather than on the concentration of the enzyme itself. At low concentrations of substrate, only a fraction of

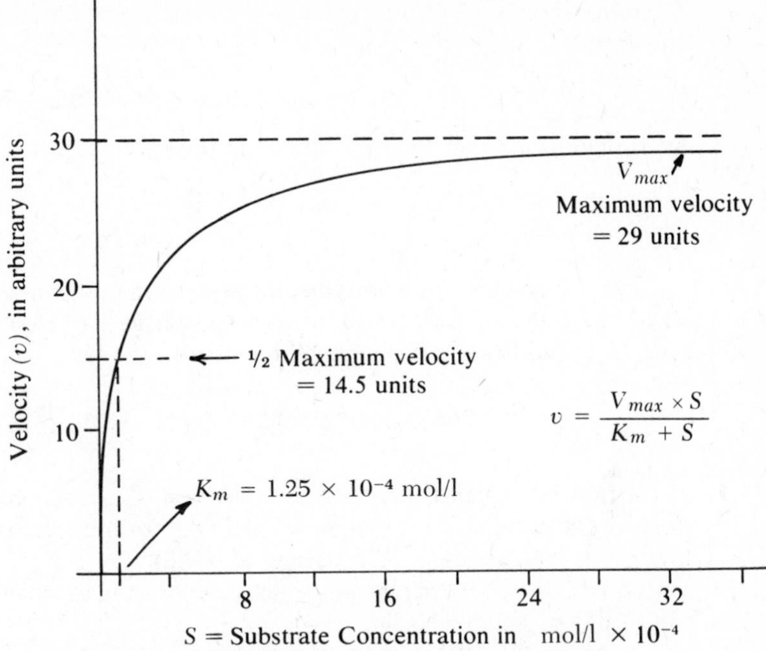

Figure 12–11 Michaelis-Menten curve relating enzyme reaction velocity (rate) to substrate concentration. The value of K_m is given by the substrate concentration at which one-half of the maximum velocity is obtained.

the enzyme is associated with substrate and the rate observed reflects the low concentration of enzyme-substrate complex. At very high substrate concentrations, all the enzyme is bound to substrate, and a much higher rate of reaction is obtained. Moreover, inasmuch as all the enzyme is now present in the form of the complex, no further increase in complex concentration and no further increment in reaction rate are possible. The maximum possible velocity for the reaction has been reached.

The enzyme reaction is formulated as follows:

$$E_f + S \underset{k_{-1}}{\overset{k_1}{\rightleftharpoons}} ES \xrightarrow{k_2} E_f + P \tag{9}$$

In these equations E_f refers to free enzyme, and k_1 and k_{-1} to the rate constants for the association and dissociation of the complex. The complex breaks down into P and free enzyme (E_f) at a rate governed by k_2. Michaelis and Menten assumed that equilibrium is attained rapidly between E, S, and ES; i.e., the effect of product formation ($ES \rightarrow P$) on the concentration of ES is negligible. If we denote the total enzyme concentration by $[E_t]$,* then at any time $[E_f] = [E_t] - [ES]$. The concentration of free substrate can be equated with total substrate concentration, since the amount combined as ES will be a negligible fraction of the total. (The concentration of E is very small compared to that of S.) Then the dissociation (Michaelis-Menten) constant of the equilibrium reaction, K_m, is given by:

$$K_m = \frac{[E_f] \times [S]}{[ES]} = \frac{([E_t] - [ES]) \times [S]}{[ES]} = \frac{k_{-1}}{k_1} \tag{10}$$

* In this section the square brackets [ ] are used in the conventional sense to denote concentration of the quantity included in the brackets; e.g., [S] = concentration of substrate, in mmol/l.

The measured velocity of the reaction (v) is determined by the rate of decomposition of ES into P and E_f, i.e.,

$$v = k_2 \times [ES] \tag{11}$$

Rearranging equation 10 and eliminating $[ES]$ from equation 11,

$$v = k_2 \frac{[E_t] \times [S]}{K_m + [S]} \tag{12}$$

For a given amount of enzyme, the maximum reaction velocity (V_{max}) is reached when all of the enzyme is saturated with substrate, i.e., $[ES] = [E_t]$, and therefore, $V_{max} = k_2 \times [E_t]$. Substituting this in equation 12 gives:

$$v = \frac{V_{max}[S]}{K_m + [S]} \quad \text{or} \quad \frac{V_{max} - v}{v} = \frac{K_m}{[S]} \tag{13}$$

A plot of v against S should give a section of a rectangular hyperbola, and this is the shape of the curve that is found experimentally for most enzymes. It can be seen that when $[S] = K_m$, manipulation of equation 13 yields $v = \frac{1}{2}(V_{max})$; i.e., the Michaelis-Menten constant K_m is that particular substrate concentration at which the reaction proceeds at one half of its maximum velocity.

In Michaelis and Menten's original scheme, K_m was a true equilibrium constant; a smaller value indicated a greater affinity of the enzyme for its substrate and vice versa. In introductory derivations such as that presented here, the symbol K_m for the Michaelis constant is usually used in this sense. However, now that it is possible to measure rates of formation and breakdown of ES in several cases (e.g., by fast-reaction absorption spectrophotometry), it has been found that the rate of breakdown of ES into products may not be negligible compared to that of the dissociation of ES into E and S. The Michaelis constant thus ceases to be a true equilibrium constant, and now contains a velocity term, becoming a steady-state constant,

$$K_m = \frac{k_{-1} + k_2}{k_1}$$

Nevertheless, the shape of the curve relating v and S remains the same. In precise demonstrations, it is now customary to restrict the term Michaelis-Menten constant (K_m) to the experimentally determined substrate concentration at which $v = \frac{1}{2}V_{max}$, and to use the symbol K_s to represent the true enzyme-substrate association constant, where this is known. While it is quite simple to set up an experiment to determine the variation of v with S, the exact value of V_{max} cannot be evaluated easily from the hyperbolic curve. Furthermore, many enzymes deviate from ideal behavior at high substrate concentrations and indeed may be inhibited by excess substrate, so that V_{max} is not achieved in practice. The Michaelis-Menten equation (13) is therefore usually transformed into one of several reciprocal forms (equations 14 and 15), and either $\frac{1}{v}$ is plotted against $\frac{1}{[S]}$, or $\frac{[S]}{v}$ is plotted against $[S]$.

$$\frac{1}{v} = \left(\frac{K_m}{V_{max}} \times \frac{1}{[S]}\right) + \frac{1}{V_{max}} \tag{14}$$

$$\frac{[S]}{v} = \left(\frac{1}{V_{max}} \times [S]\right) + \frac{K_m}{V_{max}} \tag{15}$$

Equation 14, for example, when plotted, gives a straight line with intercepts at $\frac{1}{V_{max}}$ on the ordinate and $-\frac{1}{K_m}$ on the abscissa (Lineweaver-Burk plot). For illustrative purposes, the data for Figure 12-11 are recast in Lineweaver-Burk form in Figure 12-12. The Michaelis-Menten curve and its linearized plots are altered in the presence of inhibitors in a way that is described in the next section.

The value of the Michaelis constant has been used to compare the binding of homologous or related substrates to the same enzyme. Also, if measured against the same substrate under defined conditions, the K_m value can be used to compare the properties of similar enzymes from different sources. Studies of this type have shown that the different isoenzymes of lactate dehydrogenase from the same organism but from different sites have significantly different K_m values. A knowledge of the K_m of an enzyme is also important in determining the substrate concentration to be used to measure enzyme activity at optimal substrate concentrations.

Zero-order kinetics will be maintained if all substrates (and all obligate activators) are present in large excess, i.e., at concentrations at least 20 and preferably 100 times that of the value of K_m. The K_m values for the majority of enzymes are of the order of 10^{-5} to 10^{-3} molar; therefore, substrate concentrations are chosen to be in the range of 0.001 to 0.10 mol/l. The actual concentration of substrate to be used for each enzyme system must be determined carefully by experiment. Henry and his associates have shown, for example, that the concentration of one substrate used in the original method for the assay of serum alanine transaminase was appreciably below the optimal level. If the determination of glutamate pyruvate transaminase (GPT, ALT) activity is run at the optimal substrate level, a considerably higher value for the enzyme level is obtained. There are occasions when optimal concentrations of substrate cannot be used (for example, when the substrate has limited solubility, or when the concentration of a given substrate will inhibit the activity of another enzyme needed in a coupled reaction system); one must then compromise and

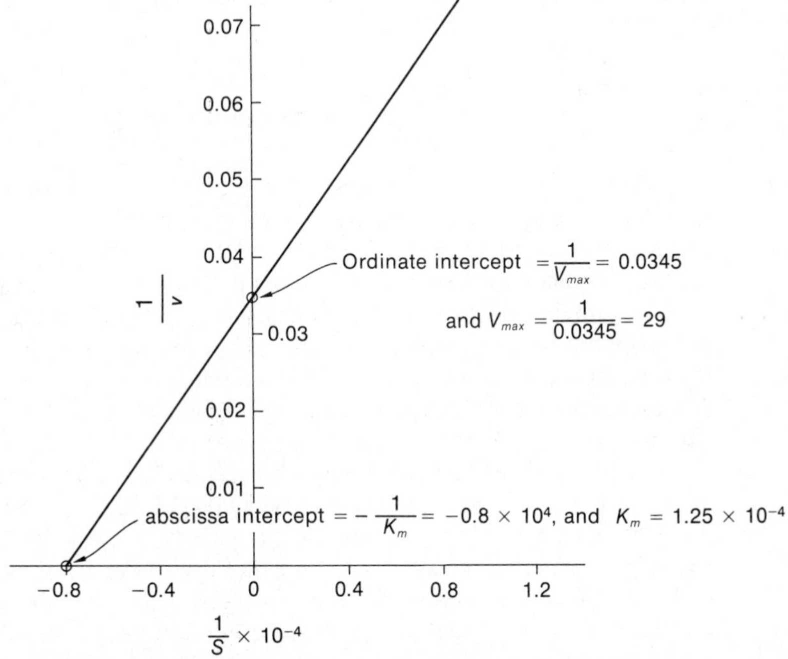

Figure 12–12 Lineweaver-Burk transformation of the curve in Figure 12-11, with $1/v$ plotted on the ordinate (Y-axis), and $1/S$ on the abscissa (X-axis). The indicated intercepts permit calculation of V_{max} and K_m. The units of v and S are those given (for the original graph) in Figure 12-11.

use the nonideal system. To obtain reproducible results, it is necessary to adhere *precisely* to *all* details of the procedure.

The role of substrate depletion as a factor which contributes to the decline in the rate of an enzymatic reaction with time has already been mentioned. For this reason, not more than 20 per cent of the substrate should be transformed in the course of the enzyme assay. If more than this fraction has been consumed, or if for any other reason nonlinear reaction curves prevent accurate measurement of the initial rate, it is necessary to repeat the assay, using a smaller quantity of enzyme specimen. Dilution of the specimen with water or saline may give disproportionate changes in activity and should be undertaken with caution. Normal or heat-inactivated serum and pure albumin solutions are often better diluents, though this also should not be assumed uncritically in every case. When fixed-time methods are in use, a shortened incubation period is preferable to dilution.

Enzyme Cofactors and Inhibitors

Although many enzymes (e.g., urease) are simple proteins with catalytic activity linked directly to the unique structure and conformation of their amino acid chains, other enzymes require in addition the presence of some *nonprotein* entity in the enzyme-substrate system before enzyme activity can be manifested. Such nonprotein materials are referred to as cofactors. Cofactors are often discovered because the activity of some enzyme is diminished or even entirely lost when the enzyme preparation is dialyzed, but its activity is restored by the addition of some chemical material present in the dialysate. For example, prolonged dialysis of an active transaminase results in loss of activity, owing to the removal of the dialyzable cofactor, pyridoxamine-5′-phosphate. If the cofactor is an organic compound, it is called a coenzyme; if it is some inorganic ion, it is referred to as an activator, except when it is part of the molecule (metalloenzyme).

Enzyme Activation

Many enzyme activators are metal ions and are essential or obligate; i.e., in their complete absence, no enzyme activity takes place. In some cases enzyme activators are nonessential; the enzyme can function without the activator being present, but the rate of reaction is enhanced when the activator is present. Most activators are weakly bound to the enzyme protein and can readily be removed by dialysis, but in a few instances (e.g., four of the copper atoms in ceruloplasmin), the ion is firmly bound to the protein. All phosphate transfer enzymes (kinases) require the presence of Mg^{2+} ions. Without magnesium, the enzymes are inactive; thus, the metal ion is an essential activator. Other common activating cations are Mn^{2+}, Fe^{2+}, Ca^{2+}, Zn^{2+}, and K^{+}. Anions may also act as activators. Amylase will function at its maximal rate only if Cl^{-} (at 0.01 mol/l) or other monovalent anions such as Br^{-} or NO_3^{-} are present. Some enzymes require the obligate presence of two activating ions: K^{+} and Mg^{2+} are essential for the activity of pyruvate kinase. The velocity of the reaction depends on the activator ion concentration in a fashion similar to its dependence on substrate concentration. A Michaelis constant can be determined from data relating enzyme activity to increasing metal ion concentration in the presence of excess substrate, and the value of K_a, the measure of the degree of affinity between enzyme and metal ion, can be calculated. It is important that activator-dependent enzyme reactions be performed in the presence of both excess activator and excess substrate. In many cases, however, the addition of activator beyond a certain optimal concentration may result in a decrease in reaction rate (inhibition by excess activator). In such cases the optimal concentration of activator must be used.

The mechanisms by which cations and anions activate enzymes vary from case to case. The ion may, for example, alter the spatial configuration of the protein to allow for proper binding of the substrate to the enzyme. This is the most likely explanation of the activity of monovalent ions like K^+ and Cl^-. Divalent ions such as Mg^{2+} may link substrate or coenzyme to the enzyme protein. Other metal ions are involved in electron transfer reactions and undergo oxidation and reduction, as in the case of iron ($Fe^{2+} \rightleftharpoons Fe^{3+}$) in cytochrome c oxidase.

In a broader sense the term "activation" can be applied to any process whereby an inactive enzyme is made catalytically active. The proteolytic enzymes are synthesized in the body in the form of inactive precursors, termed *proenzymes* or *zymogens*, which are then transformed by chemical agents into the active enzymes. Trypsin is formed from its precursor, trypsinogen, when a small hexapeptide fragment is split from the amino terminal end of the protein by the action of enterokinase or by active trypsin. Similarly, H^+ ions in the stomach convert pepsinogen to the active gastric enzyme, pepsin.

Apparent activation of an enzyme may be observed whenever some material is added which can counteract the presence of some inhibiting agent.

Coenzymes and Prosthetic Groups

The combination of enzyme plus coenzyme is referred to as a *holoenzyme*; *apoenzyme* refers to the enzyme protein without the cofactor. If the coenzyme is bound so tightly to the enzyme that it cannot be removed by dialysis (e.g., the heme protein in peroxidase), it is referred to as being a *prosthetic group*. Many coenzymes are only loosely bound to the enzyme, and indeed act as cosubstrates, forming complexes similar to those between enzyme and substrate.

Since coenzymes are cosubstrates in enzyme reactions, they must, like other substrates, be present in excess concentration for the reaction to proceed at the maximum rate. Furthermore, as is true for substrates, they undergo reaction, and their structure is altered as the reaction proceeds.

The majority of the water-soluble vitamins are components of important coenzymes and prosthetic groups. Vitamin B_1 in the form of the pyrophosphate ester, thiamine pyrophosphate, is the coenzyme associated with decarboxylases, enzymes which remove carbon dioxide from oxocarboxylic acids. Nicotinamide is part of the molecule of the coenzyme associated with hydrogen transfer reactions; pyridoxine, as pyridoxal and pyridoxamine phosphates, participates in amino transfer reactions. Not all coenzymes are associated with vitamin activity; e.g., adenosine di- and triphosphate (ADP, ATP) are involved in phosphate transfer reactions, and glucose-1,6-diphosphate is the coenzyme of phosphoglucomutase.

Dehydrogenases are a class of enzymes which catalyze such hydrogen transfer reactions as the oxidation (dehydrogenation) of an α-hydroxy acid to the corresponding oxo-acid. A clinically important enzyme of this class is lactate dehydrogenase. Two related coenzymes are involved in these hydrogen transfer reactions: one is *n*icotinamide *a*denine *d*inucleotide (NAD), in the past referred to as coenzyme-I or as DPN (*d*iphospho*p*yridine *n*ucleotide), and the other is NADP (coenzyme-II, TPN), a phosphorylated derivative of NAD.

A nucleotide is an organic compound composed of a heterocyclic nitrogen base linked to a ribose unit, which is in turn joined to a phosphate residue at the C-5 position. In NAD one nucleotide contains adenine and the other contains nicotinamide; both bases are linked to the sugar at the C-1 position of the ribose unit. The two nucleotides are joined through

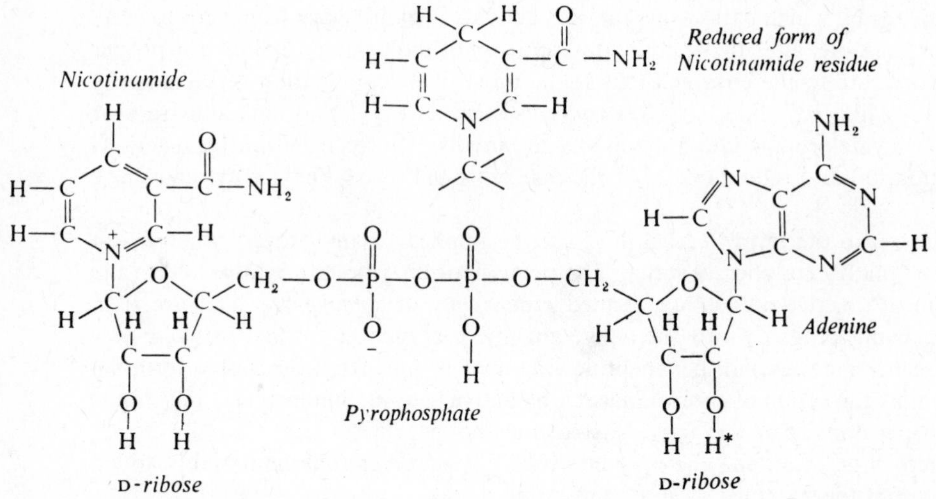

Nicotinamide

Reduced form of Nicotinamide residue

Pyrophosphate

Adenine

D-ribose

D-ribose

Figure 12-13 Structural formula for coenzyme I, nicotinamide adenine dinucleotide (NAD^+). In coenzyme II, $NADP^+$, a phosphoryl residue, $-P(=O)(OH)_2$, replaces the (H) at the position indicated by the star(*). The structure of the nicotinamide portion in the reduced form of the coenzyme is also shown.

the phosphate residues to form a diphosphate (pyrophosphate) group. The formula for NAD is presented in Figure 12-13. The star (*) indicates the position of the third phosphate

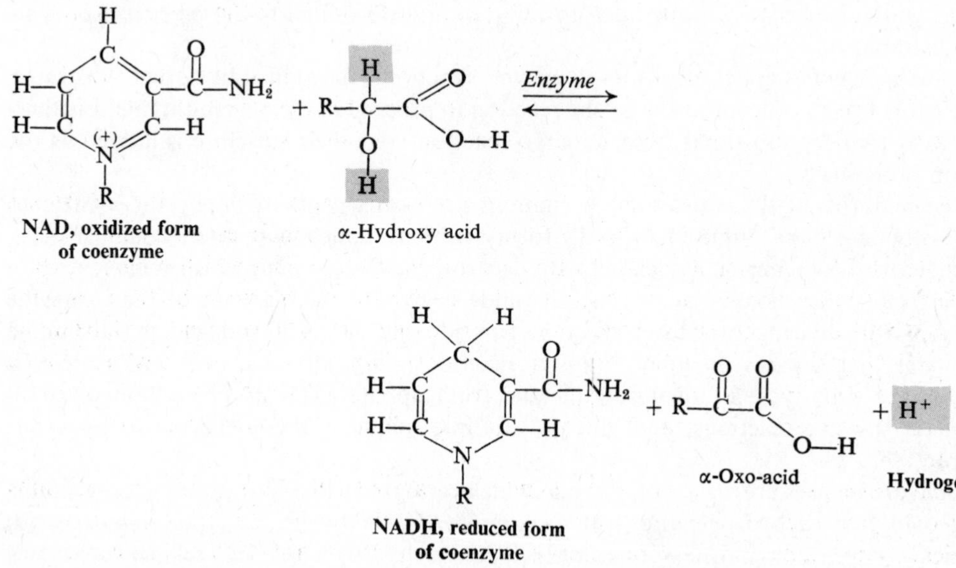

NAD, oxidized form of coenzyme + α-Hydroxy acid $\xrightarrow{\text{Enzyme}}$ **NADH, reduced form of coenzyme** + α-Oxo-acid + H^+ **Hydrogen ion** (16)

group in *n*icotinamide *a*denine *d*inucleotide *p*hosphate (NADP). Only the pyridine moiety in nicotinamide is involved in the hydrogen transfer reaction. When the ring is reduced, two electrons are transferred to the ring by the two hydrogens removed from the alcohol (α-hydroxy acid). One electron and one hydrogen go to the C-4 position of the pyridine ring; the other electron neutralizes the charge on the nitrogen, and its hydrogen remains as a proton. Reduced coenzyme is formed, the customary abbreviation for which is NADH. The reaction is reversible; NADH will reduce an oxo-acid to an alcohol, and NAD will oxidize the alcohol to the oxo-acid. The reaction proceeds to equilibrium, which is determined by the relative oxidation-reduction potentials of the NADH/NAD system and the oxo-acid-alcohol pair. Glucose-6-phosphate dehydrogenase and isocitrate dehydrogenase are examples of enzymes requiring the NADPH/NADP system as hydrogen transfer coenzyme.

The coenzymes for the phosphate-, amino-, and glucosyl-transfer enzymes function in an analogous way. In these transfer reactions, coenzymes serve as cosubstrates and are consumed, undergoing reaction to their complementary forms. If a coenzyme can be regenerated continuously to its original form by another reaction (enzyme system), then a very small amount can serve to promote the reaction of a large quantity of specific substrate. In the transaminase reaction, pyridoxal phosphate is converted to its complementary form, pyridoxamine phosphate, on receiving the amino group from the donor substrate (e.g., aspartate), and is reconverted to its original form on transferring the amino group to the acceptor substrate (oxoglutarate).

The first step in the metabolism of glucose is the phosphorylation of glucose to glucose-6-phosphate by hexokinase. Adenosine triphosphate (ATP) is the source of transferred phosphate, with adenosine diphosphate (ADP) being formed at the rate of one mole per mole of glucose. The reaction would cease the moment that the small quantity of ATP present was consumed, if it were not for the fact that, further along in the series of reactions by which glucose is converted to pyruvate, 1,3-diphospho-D-glycerate and phosphoenol-pyruvate are formed. The phosphate transfer potentials of these phosphorylated compounds are higher than that for ATP, and these compounds will transfer their phosphate groups to ADP. Consequently, ATP will be regenerated for re-use in the glucose-hexokinase reaction, thus permitting a very large quantity of glucose to be phosphorylated. This is illustrated in equation 17, which is also an example of the method devised by Baldwin to portray cyclic reactions.

$$\begin{array}{ccccc} \text{D-Gluc} & \longrightarrow & \text{ADP-P} & \longleftarrow & \text{Pyr} \\ & Mg^{2+} \; \text{HK} & & \text{PK} \; K^{+}, Mg^{2+} & \\ \text{D-Gluc-6-P} & \longleftarrow & \text{ADP} & \longrightarrow & \text{Pyr-P} \end{array} \qquad (17)$$

D-Gluc	= D-glucose	ADP	= Adenosine diphosphate
D-Gluc-6-P	= Glucose-6-phosphate	ADP-P	= ATP = Adenosine triphosphate
HK	= Hexokinase	Pyr	= Pyruvate
PK	= Pyruvate kinase	Pyr-P	= Phospho(enol)pyruvate

Enzyme Denaturation and Inhibition

DENATURATION

Catalytic activity of enzymes is dependent on the maintenance of a precise three-dimensional structure of the enzyme molecule. Disruption of this structure by such processes as exposure to elevated temperature, to extremes of acidity or alkalinity, to irradiation, or to phenomena occurring at air-liquid interfaces (e.g., frothing) will result in a partial or complete loss of enzyme activity. This alteration of structure with loss of activity is referred to as denaturation, the term usually being applied to those changes which occur as a result of physical forces such as temperature or mechanical energy. Denaturation may be wholly or partly reversible, depending on the nature of the process and the duration of its action, or it may be entirely irreversible. The presence of the substrate is often found to protect the enzyme against denaturation to some extent. Apparently, the substrate molecules bridge the active center and thus prevent any alteration of the spatial configuration at this key site on the enzyme. In general, denaturing processes affect all enzymes in a similar fashion and are nonspecific in nature, although the rate of inactivation by any given agent may vary with the individual enzyme concerned.

Many chemical materials are able to inactivate enzymes, and this action is termed *chemical inactivation* or denaturation. Protein precipitants, such as trichloracetic and tungstic acids, are examples of such chemical *inactivating* agents, as are many heavy metal ions, such as lead, zinc, copper, and mercury. Such chemical inactivation is differentiated from *inhibition*, although no sharp line can be drawn between these two processes. The action of inhibitors is usually more specific, and their effects may be restricted to a few enzymes, or even to one particular enzyme only.

ENZYME INHIBITION

Whenever an enzyme reaction is proceeding at a rate less than that expected for the existing conditions of pH, temperature, substrate concentration, and activator concentration, the enzyme is said to be inhibited. Inhibition may be partial or total, reversible or irreversible. The action of an irreversible inhibitor is progressive with time, and inhibition may be complete if enough inhibiting agent is present. Such inhibition is not readily reversed (e.g., by dialysis) but, since it usually involves a covalent chemical combination between enzyme and inhibitor, chemical reversal is sometimes possible. Organophosphorus compounds inhibit cholinesterases in this way.

Combination between an enzyme and a *reversible inhibitor* takes the form of an equilibrium reaction which is reversed on removal or dilution of the inhibitor (e.g., by dialysis). Two forms of reversible enzyme inhibition may be encountered, competitive and noncompetitive. In *competitive inhibition* the inhibiting agent binds to the enzyme at the same site as does the substrate, and the substrate and inhibitor compete for the same position on the enzyme surface:

$$\begin{gathered} E + S \underset{}{\overset{K_S}{\rightleftharpoons}} ES \xrightarrow{k_2} E + P \\ E + I \underset{}{\overset{K_I}{\rightleftharpoons}} EI \text{ (Inactive)} \end{gathered} \tag{18}$$

The competitive inhibitor resembles the substrate sufficiently to combine with the active center, but the *EI* complex cannot break down to form products. The more of the enzyme present in the *EI* form, the less will be available to bind with substrate. If enough inhibitor is present, it may saturate the enzyme so that no active enzyme-substrate complex can be formed, and no activity will be observed. If more substrate is added, it will displace the inhibitor from the enzyme and activity will begin to be restored. The degree of activity observed will depend on the ratio of inhibitor concentration to substrate concentration and on the relative degree of binding of each to the enzyme. In the presence of a fully competitive inhibitor, the same maximum velocity will be obtained as that observed in the absence of the inhibitor, but a higher substrate concentration will be needed to achieve it (*apparent* increase in K_m). A competitive inhibitor has a chemical structure very similar to that of the substrate. Hexokinase, which acts on D-glucose, is inhibited by D-xylose and 6-deoxy-D-glucose, which have the same D-pyranose ring skeleton. Enzymes requiring coenzymes can be inhibited by chemical analogs of the coenzymes. Thus, transaminases, requiring pyridoxal phosphate, will be competitively inhibited by deoxypyridoxal. If an enzyme acts on two similar substrates, each will act as a competitive inhibitor of the other. The catalytic reduction of pyruvate to lactate by lactate dehydrogenase is inhibited by α-ketobutyrate, a substrate containing an additional —CH_2 group.

Competitive inhibition provides further evidence for the presence of a highly specialized substrate-binding site on the enzyme molecule. Competitive inhibition by metal ions will be discussed in a separate paragraph.

In *noncompetitive inhibition*, the inhibitor binds to the enzyme at a locus other than that at which the substrate is bound. In some cases the enzyme remains active, but the inhibitor has so modified its structure that its rate of activity is diminished. Increasing the substrate concentration does not overcome this type of inhibition.

An interesting example of noncompetitive enzyme inhibition is the inactivation of enolase by fluoride (F^-) ion. In one step in the metabolism of glucose, enolase catalyzes the conversion of 2-phosphoglycerate to phosphopyruvate. By inhibiting this reaction, F^- stops the glycolytic breakdown of glucose, preventing formation of pyruvate. This is the biochemical rationale for the use of F^- as a preservative in specimens to be used for glucose determinations.

Competitive and noncompetitive inhibitors can be distinguished experimentally by their effects on the shape of the Michaelis curve or on one of its linearized forms such as the Lineweaver-Burk plot (Fig. 12-14). The linearized plots aid calculation of V_{max} and K_m in the presence or in the absence of inhibition. In noncompetitive inhibition there is a decrease in V_{max}, with no change in apparent K_m, whereas in competitive inhibition the V_{max} is unaffected, with an apparent increase in the K_m. These changes are reflected in changes in the slopes of the lines, and in the values of the intercepts on the ordinate and/or abscissa. Change in both V_{max} and apparent K_m indicates mixed or partially competitive inhibition.

A unique form of enzyme inhibition is manifested by *anti-enzymes*, as represented by a variety of trypsin inhibitors. These are proteins which bind to trypsin irreversibly, nullifying its proteolytic activity. One such inhibitor is present in the α_1-globulin fraction of serum proteins; others are found in soy beans and lima beans. Similar proteolytic inhibitors present in plasma prevent the accumulation of excess thrombin and other coagulation enzymes, thus keeping the coagulation process under control.

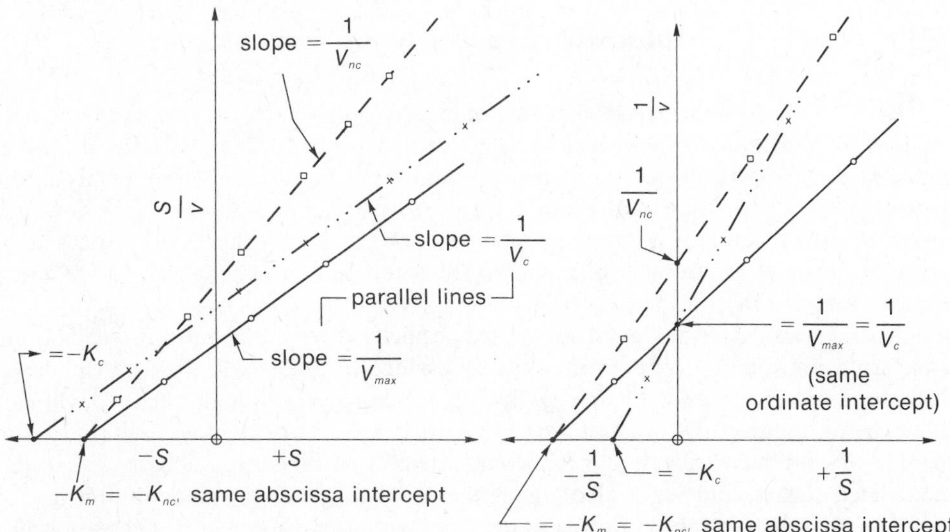

Figure 12-14 Use of linearized forms of the Michaelis-Menten equation to distinguish between competitive and non-competitive inhibition. Left: Hanes plot $\left(S \text{ vs. } \frac{S}{v}\right)$. Right: Lineweaver-Burk plot, $\left(\frac{1}{S} \text{vs.} \frac{1}{v}\right)$. K_{max}, K_c, and K_{nc} are values for the Michaelis constant with no inhibition, with competitive inhibition, and with non-competitive inhibition, respectively. The corresponding symbols for maximum velocity are V_{max}, V_c, and V_{nc}, respectively. 0——0, no inhibition; ×—···×, competitive inhibition; ◇— —◇, non-competitive inhibition.

ENZYME INHIBITION BY METAL IONS

In many enzymes the presence of free cysteine sulfhydryl groups (—SH) is essential for activity. Oxidation of these groups to disulfide groups (—S—S—) may alter the spatial structure of the enzyme, resulting in loss of activity. Papain and bromelin, for example, are activated by treatment with cysteine, which reduces disulfide links in the protein to the free —SH form. Very often the —SH groups on the enzyme will react with heavy metal ions present as contaminants in the enzyme system, resulting in loss of enzyme activity. Such heavy metal–poisoned enzyme preparations can frequently (but not always) be reactivated by treatment with or dialysis against a solution of EDTA or BAL (dimercaptopropanol), or some inert chelating agent having a greater affinity for the metal than the enzyme has. Competitive inhibition by metal ions can arise when two metal ions compete for the same binding site on the enzyme. Thus, Ca^{2+} is an inhibitor for some enzymes which depend on Mg^{2+} activation. Sodium and lithium are potent inhibitors of pyruvate kinase, for which potassium is an obligatory activator. Metal ions can also act as noncompetitive inhibitors; heavy metal poisoning of enzymes is usually of this type.

In order to avoid either inhibition or enzyme inactivation, glassware used in the laboratory for the assay of enzymes must be washed and rinsed very thoroughly. Most detergents are chemical protein denaturants; repeated rinsing is necessary to insure their complete removal. Chromic acid should not be used in cleaning glassware, since the last traces of chromic ions are difficult to remove from glass. Warm nitric acid (diluted tenfold) can be used to remove chromium and other metal ions such as copper, mercury, and iron. Because reagents containing these metal ions are used daily in the clinical laboratory, it is prudent to segregate equipment used in enzyme work from that used elsewhere in the laboratory. This applies even to spectrophotometer cuvets. Obviously, only redistilled or deionized water should be used both for rinsing glassware and for preparing all reagents. For enzyme *referee* methods, the quality of the water used becomes a matter for specification, as for all other reagents.

Substrate Specificity of Enzymes

Early in the study of enzymes it was appreciated that each enzyme catalyzes only one reaction or at most a limited range of chemical reactions. In fact, enzymes show a rather high degree of specificity for the substrates and types of reactions they catalyze, but the degree of specificity varies from one enzyme to another. Some enzymes show *absolute specificity*—they catalyze a single unique reaction, and no others. Pyruvate kinase, for example, mediates the transfer of a phosphate group between phosphopyruvate and ADP, and can function in no other reaction.

A somewhat lesser degree of substrate specificity is found in hexokinase. This enzyme transfers a phosphate group from ATP to D-glucose, but it will also phosphorylate D-fructose, D-mannose, and 2-deoxy-D-glucose at almost equivalent rates. It will not act, however, on D-galactose, or on a variety of other hexoses or pentoses, although some of these are bound to the enzyme and can (competitively) inhibit enzyme activity. Disaccharides, methylated sugars, and sugar alcohols neither bind to the enzyme nor inhibit its activity.

The phosphatases are examples of enzymes with *group* specificity. These enzymes split phosphate from any of a large variety of organic phosphate esters, although at somewhat different rates. Substances as varied as glucose-6-phosphate, phenylphosphate, and β-glycerophosphate can serve as substrates.

The esterases and proteinases are groups of enzymes with even less specificity. The former hydrolyze esters to alcohols and carboxylic acids. A considerable range of chain length in both the alkyl (alcohol) and acyl (acid) portions of the esters is permitted (see lipase, p. 633). The proteinases and peptidases attack peptide bonds, and will hydrolyze a variety of large or small proteins and polypeptides (and even simple compounds containing only one peptide bond) to form free amino and carboxyl groups. Some (such as trypsin) possess esterase and amidase action as well.

Stereoisomeric specificity is characteristic of many enzymes. The enzymes involved in glycolysis act only on the D- stereoisomers of glucose and its derivatives and never on the L- forms. The transaminases convert oxo-acids only to the L- isomers of the amino acids, and fumarase hydrates fumarate to the L- form of malate, rather than to the D-glucose-related mirror image (D-) form. Human α-amylase hydrolyzes only the linear segments of starches in which the D-glucose residues are linked by α-1,4-linkages. It is inactive to cellulose, in which the sugar residues are connected by β-1,4-linkages, and to the branch points (α-1,6-linkages) in glycogen and amylopectin.

Units for Measuring Enzyme Activity

Enzyme concentrations are measured in terms of *activity* units present in a convenient volume or mass of specimen. The unit of activity is the measure of the rate at which the reaction proceeds, e.g., the quantity of substrate consumed or product formed in an arbitrary or convenient unit of time. The quantity of substrate may be given in any convenient unit—milligrams, micromoles, change in absorbance, change in viscosity, or microliters of gas formed; time may be expressed in seconds, minutes, or hours. Since the rate of the reaction will depend on experimental parameters such as pH, type of buffer, temperature, nature of substrate, ionic strength, concentration of activators, and other variables, these parameters must be specified in the definition of the unit. Any methodological modification that involves a change in the values chosen for these variables, even if slight, may alter the significance of the units associated with that method.

In the course of many decades a multiplicity of units for expressing enzyme activity have been introduced. Even for the same or similar enzymes, each investigator defined his unit in terms of quantities analytically or otherwise convenient for him at the time. A classic example is encountered in the types of units used in measuring phosphatase activity in serum. Bodansky defined his unit as the amount of enzyme that will split one milligram of phosphate-phosphorus from β-glycerophosphate at pH 8.6 in 60 min at 37°C. Kind and King, in their alkaline phosphatase method, employed phenylphosphate as a substrate and defined their unit as the quantity of enzyme that would liberate one milligram of phenol at pH 10.0 in 15 min at 37°C. The Gutman and Gutman unit for acid phosphatase (pH 4.9) was defined in terms of the amount of phenol liberated during 1 h reaction time. The Bessey-Lowry-Brock (BLB) unit of alkaline or acid phosphatase activity is expressed in terms of one millimole of substrate (*p*-nitrophenylphosphate) hydrolyzed in 60 min. Bowers and McComb also used *p*-nitrophenylphosphate, but with a different buffer; they expressed their unit in terms of micromoles of *p*-nitrophenol liberated per minute at 30°C. Thus, if a given phosphatase preparation hydrolyzes one millimole of each of the above-mentioned substrates per minute, the following numerical values in terms of the individual units defined above would be obtained:

Bodansky	pH 8.6	31 mg phosphorus/mmol × 60 min	1860 units
Kind-King	pH 10.0	94 mg phenol/mmol × 15 min	1410 units
Bessey-Lowry-Brock	pH 10.2	1 mmol × 60 min	60 units
Bowers-McComb	pH 10.15	1000 μmol × 1 min	1000 units

The concentration of an enzyme is often expressed in terms of activity units per volume or mass of specimen (e.g., per mg of specimen [wet liver tissue], per mg of protein, or per mg of fat-free protein-nitrogen in the specimen). In clinical work, the concentration is generally reported in terms of some convenient unit of volume, such as activity per 100 ml or per liter of serum, or per 1.0 ml of packed erythrocytes.

Just as a multiplicity of units for reporting activity developed, a similar multiplicity of units of volume evolved. The Commission on Enzymes proposed that the *unit of enzyme activity* be defined as that quantity of enzyme that will catalyze the reaction of one micromole (μmol) of substrate per minute, and that this unit be termed the International Unit (U). Concentration is to be expressed in terms of U/ml or mU/ml (= U/l), whichever gives the more convenient numerical value. In this chapter the symbol U is used to denote the International Unit. In those instances in which there is some uncertainty as to the precise nature of the substrate or where there is difficulty in calculating the number of micromoles reacting (as with macromolecules such as starch, protein, and complex lipids), the unit is to be expressed in terms of the chemical group or residue measured in following the reaction (e.g., reducing sugar units, or amino acid units formed).

Although the proposals of the Enzyme Commission have been accepted by many scientists working with enzymes, it is not likely that laboratories using well-established enzyme procedures will cease to report enzyme values in terms of the older, empirical units to which they have become accustomed. But as new methods are devised, it is anticipated that the activity units will be established in accordance with the recommendations of the Commission.

In the presentation of specific enzyme procedures later in this chapter, the enzyme units used will frequently be those introduced by the originators of the methods, but for each procedure the units and ranges of normal values will be converted into International Units. However, it is important to remember that the International Unit itself is a function of the method used and is not an absolute quantity. Therefore, different methods will give different numerical values for activity even if they are expressed in International Units. Examples showing the conversion from empirical units to International Units follow.

1. *Shinowara-Jones-Reinhart* (SJR) unit for alkaline phosphatase activity: The quantity of enzyme that will split 1.0 mg of phosphate-phosphorus from β-glycerophosphate in 60 min at 37°C and pH 9.6, under specified conditions of barbital buffer and substrate concentration. The enzyme level is customarily reported in terms of units per 100 ml of serum. The 1.0 mg of phosphorus is equal to 1/30.98 mmol or 32.3 μmol of P. One mole of substrate will give rise to one mole of phosphorus. Thus, 32.3 μmol of substrate react in 60 min, which is equivalent to 0.538 μmol/min. Thus, 1.0 SJR unit is equivalent to 0.54 U, and 1.0 SJR unit per 100 ml is equal to 10 SJR units per liter or 5.4 U/l (= 5.4 mU/ml).

2. *Sibley-Lehninger* (SL) unit for aldolase activity: This is defined as the quantity of enzyme in 1 ml of specimen that will split 1.0 μl of fructose-1,6-diphosphate into two trioses, dihydroxyacetone phosphate and 3-phosphoglyceraldehyde, in 60 min under the conditions of the test. Recalling that 1 molar volume is 22.4 liters, then 1 μl of substrate is equal to 1/22.4 or 0.0446 μmol, and 1.0 SL unit is equal to 0.0446/60 or 0.743×10^{-3} U. Finally, 1.0 SL unit/ml is identical to 0.74 U/l = 0.74 mU/ml.

3. *Karmen-Wroblewski-LaDue* (spectrophotometric) unit for transaminase activity: This unit is defined as a change (decrease) of 0.001 in the absorbance of NADH/min, when the NADH is present in a total volume of 3.0 ml and is measured at 340 nm across a 1.0 cm light path.

The molar absorptivity (ε) of NADH is 6.22×10^3. A solution containing 1.0 μmol/ml will have an absorbance of 6.22, and a drop of 0.001 in the absorbance reflects a concentration change of $(1/6.22) \times 10^{-3}$ μmol/ml = 1.61×10^{-4} μmol/ml. In the total volume of 3.0 ml this will represent a decrease in the quantity of NADH (substrate consumed) of 4.82×10^{-4} μmol. The Karmen unit is defined in terms of absorbance change per minute, so that 1.0 Karmen unit per ml is equal to 0.48 mU/ml. This method of converting absorbance change to micromoles of substrate transformed, by the use of the known molar absorbance of NADH (or NADPH), is extremely useful since these coenzymes take part, directly or by means of coupled reactions, in many enzyme assays.

Discussions in progress at present indicate that the International Unit itself may soon be replaced by a new unit. It is currently recommended that enzyme activity be expressed in mol/s, given the name Katal, and that the enzyme concentration be expressed in terms of Katals per liter (Kat/l). This is consistent with the SI (Systeme Internationale) scheme of units in that it uses the *mole* as the measure of substrate transformed and the *second* as the unit of time. Thus, 1.0 U $= 10^{-6}$ mol/60 s $= 16.7 \times 10^{-9}$ mol/s, or 1.0 nKat/l = 0.06 U/l.

Use of Enzymes as Reagents

The relatively recent availability of pure enzyme preparations at reasonable cost has made it practical to use enzymes as laboratory reagents for analytical work. Their substrate specificity gives enzymes unique value in the determination of many biological compounds. An important application is the use of glucose oxidase for the assay of "true glucose," which has been discussed in the chapter on carbohydrates (see p. 245). Glucose can also be determined by using the hexokinase–glucose-6-phosphate dehydrogenase coupled enzyme system, also discussed in the same section. Similarly, galactose can be determined in the presence of other sugars by the use of the substrate-specific galactose oxidase (see p. 248).

The chemical assay of *serum* uric acid, based on the reduction of phosphotungstic acid, is subject to interferences. The use of uricase, which is almost substrate-specific for uric acid (oxidation to allantoin), makes it possible to measure "true" uric acid in serum. (The enzyme also slowly attacks 6-thiouric acid, a metabolite of 6-mercaptopurine, which may be present in sera from patients receiving this anticancer drug.) The enzyme is not as useful for *urine* uric acid assays, because many urines contain materials which interfere with or inhibit the enzyme. Similar inhibiting effects are encountered in the use of other enzymes, and the possibility of such interference must be kept in mind when using enzymes with materials as complex as urine.

Other important serum constituents that can be rapidly and accurately analyzed by use of enzymes are lactic and pyruvic acids (lactate dehydrogenase), ethyl alcohol (alcohol dehydrogenase), glycerol (glycerol dehydrogenase), creatine (creatine kinase), and ammonia (glutamate dehydrogenase).

In using enzymes for the assay of metabolites, the enzyme reaction system is so devised that all components are present in excess with the exception of the material to be determined. Sufficient enzyme is added to insure that the reaction will proceed to completion in a reasonably short period of time. If the equilibrium of the reaction is not favorable, other reagents (or another enzyme system) may be added to force the reaction in the desired direction. In measuring lactate, hydrazine or semicarbazide is added to "trap" the pyruvate formed, thus forcing the reaction to complete oxidation of the lactate (see Chapter 15, p. 937).

Because of the cost of reagent enzymes, such methods for determining compounds of clinical interest should be developed and used only when convenient or accurate chemical methods are not available.

Within the last several years enzymes attached to particulate material, so-called *immobilized enzymes*, have been prepared and are available commercially. These preparations consist of enzymes which have been chemically bonded to adsorbents such as microcrystalline cellulose, diethylaminoethyl cellulose, carboxymethyl cellulose, and agarose. Diazo, triazine, and azide groups are used to join the enzyme protein to the insoluble matrix, forming either (1) particles in contact with the substrate solution, (2) a membrane surface in contact with substrate solution, or (3) the inner surface of a vessel holding the substrate solution. Among enzymes available in such immobilized form are urease, α-amylase, glucose oxidase, trypsin, and leucine aminopeptidase. These are used as recoverable reagent enzymes. Substrate is passed over or through the insoluble enzyme preparation, and the product is recovered and further processed, while the enzyme is used to react with additional substrate. The immobilized enzymes can be used for batch processes or in continuous-flow systems. Stability to heat, pH, and oxidation is considerably increased compared to enzymes in solution. Immobilized proteolytic enzymes are not subject to autodigestion.

Enzymes in Plasma and Their Origins

Enzymes are the essential factors which enable the many biochemical reactions that constitute life to proceed in the cells of the body. Changes in enzyme concentrations in tissue cells should, therefore, reflect states of health and disease. Unfortunately, it is generally not practical to assay enzymes in tissues on a routine basis. As a convenience, changes in enzyme activities in the blood are followed, with the assumption that they may reflect changes that have taken place in specific organs or tissues. Serum and all the cellular elements in blood have been studied for possible value in diagnostic medicine. Erythrocytes contain a large number of enzymes, but alterations in the levels of only a few, such as glucose-6-phosphate dehydrogenase and glutathione reductase, have as yet been correlated with the presence of specific disease states. The concentration of alkaline phosphatase in leukocytes has been of some value in the differentiation of certain types of leukemias. However, most of the effort has been expended in studying enzymes in plasma and serum.

Hess[98] differentiates plasma enzymes into two classes: the *plasma-specific* enzymes and the *non-plasma-specific enzymes*. The first group includes those enzymes which have a very definite and specific function in plasma. Plasma is their normal site of action, and they are present in it at higher levels than in most tissue cells. Among these are such enzymes as plasmin and thrombin, which are involved in blood coagulation and clot lysis, and such enzymes as ceruloplasmin, cholinesterase, and lipoprotein lipase. These enzymes are synthesized in the liver and are constantly released into the plasma to maintain an optimal and effective steady-state concentration. They are of clinical interest when present in plasma at levels that are below normal as a result of impaired liver synthesis, or when entirely absent as a result of an inborn genetic defect. Hepatolenticular degeneration (Wilson's disease), for example, is associated with a deficiency of ceruloplasmin, a copper-containing oxidase, and sensitivity to the anesthetic, succinylcholine bromide, results from deficiency or absence of cholinesterase.

The second group, the *non-plasma-specific enzymes*, have no known physiological function in plasma. They are present in plasma at concentrations much lower than their concentrations in most tissues. In many cases plasma is deficient in activators or coenzymes necessary for the activity of these enzymes. This group can also be divided into two subclasses—the *enzymes of secretion* and the *enzymes associated with cellular metabolism.* Among the enzymes of secretion are such enzymes as amylase, lipase, and pepsin(ogen). Though secreted at high rates, they are rapidly disposed of through the usual excretory channels such as the urine and the intestinal tract, and normal plasma levels are relatively low and constant. If, however, any of the usual pathways of excretion are blocked, or the rate of liberation into the extracellular fluid is suddenly accelerated, or the rate of production is increased, a significant increase in plasma levels of these enzymes will occur.

The enzymes of cellular metabolism are located within the tissue cells, and are present there at very high concentrations. Some exist free in the cellular fluid, and others are contained in cellular structures such as the mitochondria and lysosomes. Some of the organelles

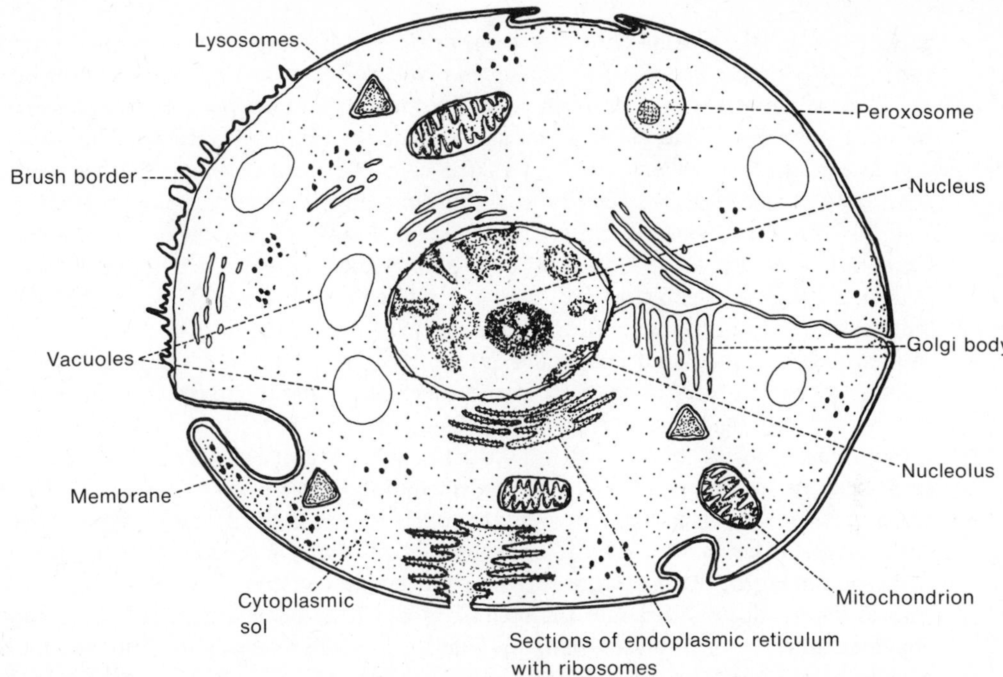

Figure 12–15 Schematic diagram of a living cell showing compartments or organelles in which enzyme activities are located. Some cells may lack one or more types of organelles. A list of some of the individual enzymes predominantly located in each of the various compartments follows (based on data from Dixon and Webb[52], and Lehninger[126]):

Cytosol
(Cytoplasmic supernatant)
Lactate dehydrogenase
Aspartate transaminase
Aldolase
Pyruvate kinase
Glucokinase
Leucineaminopeptidase
Isocitrate dehydrogenase (NADP dependent)

Lysosomes
Acid phosphatase
β-Glucuronidase
Cathepsin
Arylsulphatase

Peroxisomes
D-Amino oxidase
Urate oxidase
Catalase

Membrane Border
Alkaline phosphatase
Invertase
Maltase

Nucleus and Nucleolus
Adenyl transferase
Arginase
NAD pyrophosphorylase
NAD(P) nucleosidase

Endoplasmic reticulum
(Ribosomes, microsomes)
Cholesterol esterase
Cholinesterase
Vitamin A esterase
Glucose-6-phosphatase
Glutathione reductase

Golgi Bodies
Trypsinogen
Amylase

Mitochondria
Isocitrate dehydrogenase (NAD dependent)
Cytochrome oxidase
ATP-ase
Succinic dehydrogenase
D-3-Hydroxybutyrate dehydrogenase
Aspartate transaminase
Adenylate kinase
Sorbitol dehydrogenase

and structures contained in cells are illustrated in Figure 12-15. In the legend to the figure are listed some of the individual enzymes which have been found to be present in the several individual cellular compartments or structures. A given enzyme activity may be present in more than one organelle, but when this is so, the individual enzyme proteins differ to some degree in their structure. As long as the cell membrane is intact, such enzymes are contained within the cell walls, and the level of these enzymes in the extracellular fluid and the plasma is extremely low or zero. The cell membranes are impermeable to enzymes as long as the cells are metabolizing normally. If cell activity is impaired or destroyed as a result of deficiencies in the supply of oxygen or oxidizable metabolites such as glucose, or if the cell is damaged in some way (bacterial or viral infection), the membrane becomes permeable or ruptures. The cell contents, including their enzyme complement, are then released into the extracellular fluid, and eventually in part reach the plasma. If a large number of cells are so affected, the plasma level of the enzymes may increase very suddenly, to levels many times greater than normal. This happens in myocardial infarctions. A blood clot in a branch of the coronary artery may result in a localized insufficiency in perfusion of well-oxygenated blood to nearby cardiac tissue with ensuing damage to and death of cardiac muscle cells.

When diffusion of enzymes out of the cells has begun, several factors contribute to the rate of rise of enzyme activity in the plasma and the level of activity that is reached. The "driving force" for the diffusion is the concentration gradient that exists between the intra- and extra-cellular enzyme levels; the greater the gradient, the more readily the enzyme leaks out. Rates of diffusion are inversely related to molecular size, so that enzymes of smaller molecular mass will diffuse more rapidly through the damaged cell membrane into the extracellular fluid and eventually into the plasma. Enzymes that are located in subcellular organelles (such as the mitochondrial enzyme, glutamate dehydrogenase) leak from cells less readily than do the enzymes of the cytoplasm. These factors contribute in part to the different rates of increase of individual enzyme activities observed in plasma after an episode of tissue damage, such as a myocardial infarction.

Numerous experiments with animals, together with clinical observations, have demonstrated that the levels of enzyme activity reached in plasma are directly related to the amount of tissue damaged, but not necessarily to the severity of the clinical condition. Although little is known at present about the means by which enzymes are cleared from the plasma, enzymes differ in their plasma *half lives* and these differences contribute to the observed variations in rates of rise and fall of individual enzymes.

Most authors agree that leakage of enzymes from dying cells is the main cause of the observed rise in serum enzyme activity in such conditions as myocardial infarction and infectious hepatitis. However, some are less convinced that failure to maintain the semi-permeable cell membrane in its normal state is also a cause of enzyme leakage in conditions that stop short of actual cell death. They believe that increased synthesis of enzyme by the affected cells is also an important factor. In some conditions increased enzyme synthesis with overflow into the plasma is almost certainly the cause of part of the increased enzyme activity found in plasma. For example, the rise in plasma alkaline phosphatase in obstructive liver disease is now regarded as being the result of such a process.

Sometimes increased synthesis of a particular enzyme is due to a multiplication of the cells which produce that enzyme, rather than to an increased enzyme production per cell. This process may also be accompanied by an alteration in the relationship between the enzyme-producing cells and the blood. The cells of the normal prostate produce large quantities of acid phosphatase, but this does not enter the blood in significant quantities. However, when metastases of prostatic cancer appear in other tissues of the body, the characteristic acid phosphatase is more readily able to pass from these tissues into the plasma. This augments the normally low plasma activity of this enzyme in spite of the fact that each of the cancerous cells produces less acid phosphatase than does the normal prostatic cell.

If some given enzyme, such as alcohol dehydrogenase or sorbitol dehydrogenase, is present to any appreciable concentration in only one organ (the liver in the case of these two enzymes), an increase in the plasma level of that enzyme would pinpoint its source and thus identify the diseased or affected organ. Such enzymes are referred to as *tissue- or organ-specific enzymes*. Other enzymes of this type are the acetylcholinesterase of erythrocytes and neurons, the acid phosphatase present in the prostate gland, and the alkaline phosphatase of the osteoblasts, the bone-building cells. The enzymes associated with glycolysis, the pentose cycle, and the citric acid cycle are probably present in most body cells. They cannot be thought of as being organ-specific, although concentrations in different tissues do vary considerably. Aspartate transaminase, aldolase, lactate dehydrogenase, and transketolase are among the many such enzymes. An increase in plasma concentration of these enzymes can result from release of the enzyme from a variety of possible tissue sources.

Although the level of any one enzyme may not be informative about its tissue of origin, a comparison of the levels of several enzymes may be useful, since the individual enzymes are present in different tissues in different ratios. If they are derived from a single cell type, their ratio in the plasma will approximate that in the tissue of origin. An examination of the serum isoenzyme pattern (see the next section) for the specific enzyme may also aid in pinpointing the source of the released enzyme.

It is understandable that it is the dream of the diagnostic clinician to be able to associate one specific enzyme with each tissue or disease state. A large number of enzymes have been studied in plasma and other physiological fluids to ascertain whether variations in the activities of specific enzymes can be related to particular states of disease. Some have been found to be poor qualitative or quantitative indicators and have not passed into general use. Those found by two decades of experience to be most useful in clinical diagnosis are listed in Table 12–3.

Isoenzymes

Many enzymes that catalyze specific chemical reactions are widely distributed throughout the cells of different species, and even throughout different kingdoms of organisms, animals, plants, and bacteria. Although they possess identical or very similar activities, these enzymes from different sources are not identical proteins since they exhibit demonstrable differences in physical, biochemical, and immunological properties. It is now generally believed that the structure of the "active center" is identical (or at least very similar) for all

TABLE 12–3 LIST OF ENZYMES OF DEMONSTRATED USEFULNESS IN CLINICAL DIAGNOSIS

Enzyme	*Organ or Disease of Interest*	*Specimen**
Acid phosphatase	Prostate (carcinoma)	S
Alkaline phosphatase	Liver, bone	SUW
Amylase	Pancreas	SUF
Lipase	Pancreas	SF
Aspartate transaminase	Liver, heart	SFC
Alanine transaminase	Liver, heart	SF
Lactate dehydrogenase	Liver, heart, red blood cells	SFC
Creatine kinase	Heart, muscle, brain	S
Isocitrate dehydrogenase	Liver	S
Ceruloplasmin	Copper transport protein (Wilson's disease)	S
Aldolase	Muscle, heart	S
Glucose-6-phosphate dehydrogenase	Anemia (genetic defect)	R
Trypsin	Pancreas, intestine	FS
Sorbitol dehydrogenase	Liver	S
Ornithine carbamoyltransferase	Liver	S
Cholinesterase	Liver (insecticide poisoning, anesthetic apnea)	S
5′-Nucleotidase	Liver	S
Uropepsin	Stomach	U
Hexosephosphate isomerase	Liver	S
Pyruvate kinase	Liver	S
Hexose-1-phosphate uridyl transferase	Galactosemia	R
Malate dehydrogenase	Liver	S
Glutathione reductase	Anemia, cyanosis	R
Acetylcholinesterase	Insecticide poisoning	R
Plasmin	Blood coagulation	P

* Symbols: S-Serum P-Plasma U-Urine F-Body fluids C-Cerebrospinal fluid R-Erythrocytes W-Leukocytes

enzymes of a given specificity, regardless of origin, but that the amino acid sequences in many sections of the peptide chains may differ considerably.

The work of many investigators has established that even a given enzyme obtained from the same individual organism can exist in multiple forms.[119,122,123,221] If specimens of sera or extracts of different tissues are subjected to electrophoresis on a variety of media, specific enzyme activity can be demonstrated at several areas or bands along the electrophoretogram. Examples of such electrophoretic patterns are shown in Figure 12-16. Markert and Møller proposed that different proteins present in the same individual with similar or like enzymatic activity be referred to as isozymes, although most authors prefer the term *isoenzymes*, recommended by the International Union of Biochemistry. Although the isoenzymes of lactate dehydrogenase have been most thoroughly investigated, other enzymes of clinical interest have also been shown to exist in multiple forms. Among these are alkaline phosphatase, amylase, creatine kinase, ceruloplasmin, glucose-6-phosphate dehydrogenase, and aspartate transaminase. It is conceivable that all enzymes occur in multiple isoenzyme forms.[123]

Isoenzymes differ not only in their rates of electrophoretic migration but also in such properties as stability to heat denaturation, resistance to various chemical inhibiting agents, and affinity for substrates and coenzymes. Of considerable interest is the fact that the proportions of individual isoenzymes present vary in many instances from one tissue to another. Figure 12-16 shows that the LD pattern obtained from extracts of human heart tissue is characteristically different from that obtained from human liver.[122,221]

The lactate dehydrogenase isoenzyme distribution in extracts of erythrocytes and kidney is similar to that in heart, with LD-1 and LD-2 predominating, whereas the pattern of skeletal muscle lactate dehydrogenase resembles that of liver, with the electrophoretically slower fractions more abundant. A third pattern is seen in many tissues, including lung, adrenal, thyroid, lymph node, and spleen, and in various forms of malignancies, in which the intermediate zones are most marked. When an organ or tissue is injured, the intracellular enzyme released from the tissue cells diffuses into the plasma and imposes its pattern over that normally present. The distribution of serum isoenzyme bands obtained may then identify the tissue from which the increased level of the particular enzyme originated within the limits imposed by the three types of patterns. If two or more organs are exuding their enzymes into the plasma, the pattern will be composite.

Isoenzyme patterns can be of significant value in differential diagnosis. Unfortunately, only a few enzymes can be resolved into their individual isoenzymes as easily as can lactate

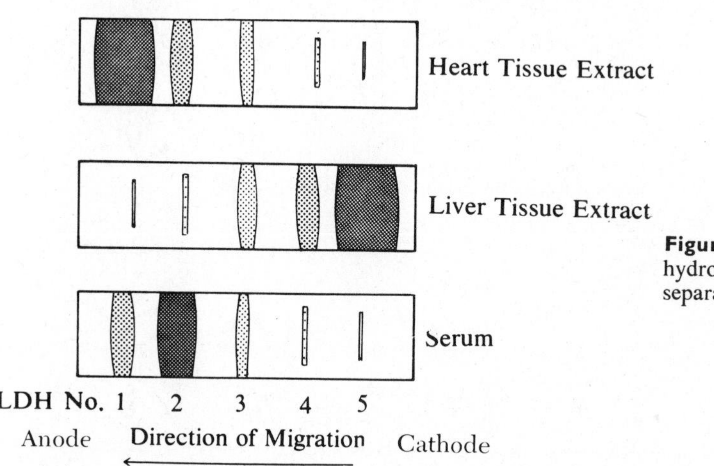

Figure 12–16 Examples of lactate dehydrogenase (LD) isoenzyme patterns as separated on cellulose acetate.

dehydrogenase, and only some isoenzymes show a clear tissue-specific distribution. Because of its importance, the development of convenient isoenzyme fractionation techniques is a field of intense investigative activity. The separation and identification of the isoenzymes of creatine kinase and of alkaline phosphatase, especially, are of considerable current interest.

PROCEDURES FOR SEPARATING ISOENZYMES

Resolution of isoenzyme mixtures (e.g., in serum) by zone electrophoresis, followed by visualization of the separated enzyme zones by specific staining procedures that depend on enzymatic activity (as shown for lactate dehydrogenase in Figure 12-16), is the procedure that is most generally useful in clinical chemistry laboratories. Often, visual inspection of the intensities of the several zones gives all the information that is needed for diagnosis, but densitometric scanning can be resorted to when a more quantitative estimate of the relative proportions of the different isoenzymes is needed. Gel support media (starch or polyacrylamide) are often chosen since these tend to give sharp enzyme zones, even when the isoenzymes do not greatly differ in molecular size.

Since electrophoresis is time-consuming and requires special equipment, alternative approaches to isoenzyme differentiation have been sought and developed. These procedures exploit distinctions in other physical properties (differences in elution from ion-exchange adsorbents) or in altered catalytic characteristics of the isoenzymes. The greater heat stability of LD-1, compared with LD-5, forms the basis of a very useful test in which the proportion of the total activity of a serum specimen which survives a fixed period of heating at 60 or 65°C is measured. Differences in the relative substrate specificities of isoenzymes can also be used to aid differentiation: e.g., LD-1 is relatively more active with α-hydroxybutyrate as substrate than is LD-5. Measurements of differences in response of isoenzymes to inhibitors provide a further useful quantitative procedure. This approach has been used to detect and estimate the fraction of the intestinal form of the enzyme in the total alkaline phosphatase activity in serum, since this isoenzyme is significantly more inhibited by L-phenylalanine than are the forms derived from bone and liver.

Sometimes the structural differences between isoenzymes are so marked that the molecules are antigenically distinct. Provided that enough of the isoenzymes can be prepared free of each other and of other antigens, antisera against the different isoenzymes can be induced by injection of the purified material into animals such as rabbits. These antisera can then be used to analyze isoenzyme mixtures since each antiserum will react with, and under appropriate conditions will inactivate or precipitate, the particular isoenzyme against which it was prepared. LD-1 and LD-5 proteins differ sufficiently for this technique to be effective. Placental alkaline phosphatase is also sufficiently different from other phosphatases to permit preparation of antisera specific to this enzyme form.

STRUCTURE OF LD ISOENZYMES

It has now been established that lactate dehydrogenase is a protein composed of four subunits (i.e., it is tetrameric). If the band from heart tissue (which moves fastest during electrophoresis) is treated with urea to dissociate it into its monomeric units, it is found that the units are all identical. Similarly, those units obtained from the slowest moving band (from liver tissue) are also identical, but they differ from those obtained from heart tissue. The other isoenzyme bands dissociate into mixtures of heart (H) and liver (M) units, the proportion of H units increasing with the increase in the electrophoretic mobility of the isoenzyme. The five isoenzymes of LD thus differ in the number of H and M units in their active tetrameric forms. The fastest moving form, designated LD-1, is composed of four H units, and can also be designated HHHH or H_4. The next fastest band is LD-2 (HHHM or H_3M). The other forms are similarly designated: LD-3 (H_2M_2), LD-4 (HM_3), and LD-5 (M_4), the slowest band. The designation M for the liver unit derives from the fact that it is also the predominant unit found in skeletal muscle. Syntheses of the H and M forms are under separate genetic control. The relative proportions of the H and M forms in different tissues are governed by the activities of the two controlling genes, so that cells in which the H-gene is more active have a greater content of that isoenzyme. Differences in rates of degradation of the isoenzymes also contribute to the observed tissue isoenzyme patterns: for example, the LD-5 is synthesized four times faster in rat liver than in heart, but it is degraded only one-tenth as fast; thus, the half life of this isoenzyme in the liver is 10 times longer than in heart.

Variants of both LD-1 and LD-5 have been described, arising presumably from mutations at the H- and M-gene loci, respectively. The variant subunits hybridize with the normal subunits so that very complex electrophoretic patterns are seen in tissues in which both normal and mutant genes are active. The relative activities of the normal H- and M-genes change during development so that the isoenzyme patterns of fetal or immature tissues differ in many cases from those of the adult. Malignant changes in tissues may be accompanied by a return to the fetal isoenzyme distribution, while the appearance of atypical isoenzymes has also been reported in some patients with cancers. A sixth LD isoenzyme (LD-X) occurs in the mature human testis and in spermatozoa. This is thought to be a tetramer of a subunit different from the H- and M-subunits. Hybrid isoenzymes consisting of mixtures of H- and M- and X-subunits can be formed *in vitro* but do not occur in living cells.

The existence of two different subunits which can assemble in groups of four to form the active enzyme molecule leads to the prediction of the formation of five isoenzymes. However, if the active enzyme is a dimer, three isoenzymes will result from the association of the two different subunits. Creatine kinase is such a dimeric enzyme, and the form that predominates in muscle is formed from subunits (M) which are different from those (B units) which make up the enzyme predominant in brain tissue. Mixed (MB) forms exist in smaller amounts in some tissues (heart), so that the three isoenzymes of creatine kinase may be represented as MM, BB, and MB. Aldolase shows a type of heterogeneity similar to LD, and three different subunits, A, B, and C, may combine to give rise to active, tetrameric aldolase molecules.

Not all examples of enzyme multiplicity can be attributed to combination of different subunits. Among other possible causes are variations in the amount of charged residues (e.g., sialic acid) associated with the enzyme molecules, nongenetic alterations in molecular structure, and even variations in the way that a single polypeptide chain may fold up, giving rise to conformational isomers, or "conformers."

Specimens for Enzyme Assay

Enzyme assays, unless otherwise specified, should be carried out with serum rather than with plasma as the test specimen. The common anticoagulants inhibit the activity of many enzymes, although the degree and type of inhibition encountered may vary with the specific enzyme under investigation and with the assay method used. Anticoagulants such as EDTA, which act by chelation of divalent metal ions, also inhibit those enzymes which depend on these ions for their activity (e.g., many phosphatases). Heparin is an inhibitor of some enzymes, including acid phosphatase. Often, a given enzyme activity is found both in erythrocytes and in serum, but the concentration of the activity in the two phases is quite disparate. The red cell contains the much higher level of enzymes of cellular metabolism, while the serum has the greater concentration of plasma-specific enzymes and enzymes of secretion. Most serum enzyme determinations thus require specimens free of any hemolysis. Serum contains the enzymes of lysed platelets, and it is fortunate that with the large majority of enzymes routinely measured in human serum, no significant fraction is derived from platelets. Acid phosphatase may be an exception. The need to store and maintain specimens to be used for enzyme assay under conditions which minimize the possibility of any enzyme denaturation or inactivation has already been mentioned.

The Phosphatases

The phosphatases belong to the class of enzymes called *hydrolases*. Hydrolytic enzymes catalyze the scission of compounds containing acyl or phosphoryl (phosphate) ester bonds, as well as compounds containing peptide, amide, hemiacetal and similar bonds. While these bonds are being split, a concurrent splitting of an O–H bond in a water molecule also takes place (see equation 19).

Hydrolytic reactions are exergonic, "down-hill," and proceed easily; *in vivo*, they are involved in the breakdown of metabolites, and not in their synthesis. Most of the early work in clinical enzymology was concerned with these enzymes. That they were usually enzymes of secretion, and that they were found in normal plasma (serum) at measurable concentrations, also made their study convenient. The several digestive enzymes are also included among the hydrolases.

The phophatases constitute a group of hydrolases of low specificity which are characterized by their ability to hydrolyze a large variety of organic phosphate esters with the formation of an alcohol and a phosphate ion. In this group are included only those enzymes which typically attack only monoesters of orthophosphoric acid.

$$\underset{\textbf{Organic ester}}{R{-}O{-}\overset{\overset{\displaystyle O}{\|}}{\underset{\underset{\displaystyle O^-}{|}}{P}}{-}O^-} + H{-}O{-}H \xrightleftharpoons[]{\substack{\text{Monoester}\\\text{Phosphohydrolase}}} \underset{\textbf{Alcohol}}{R{-}O{-}H} + \underset{\textbf{Phosphate ion}}{H{-}O{-}\overset{\overset{\displaystyle O}{\|}}{\underset{\underset{\displaystyle O^-}{|}}{P}}{-}O^-} \qquad (19)$$

However, certain enzymes in this class (e.g., alkaline phosphatase, ALP) will also hydrolyze monoesters of pyrophosphoric and metaphosphoric acids (e.g., ATP). Some monophosphate hydrolases with a more restricted specificity, such as 5′-nucleotidase, are also useful diagnostic aids. The diesterases (EC 3.1.4.X, orthophosphoric diester hydrolases) constitute a separate group which splits compounds in which H_3PO_4 is esterified at two positions, as shown in equation 20.

$$\begin{array}{l} CH_2{-}O{-}\overset{\overset{\displaystyle O}{\|}}{C}{-}R_1 \\ | \\ CH{-}O{-}\overset{\overset{\displaystyle O}{\|}}{C}{-}R_2 \\ | \\ CH_2{-}O{-}\overset{\overset{\displaystyle O^-}{|}}{\underset{\underset{\displaystyle O}{\|}}{P}}{-}O{-}CH_2CH_2{-}N^+(CH_3)_3 \end{array} + HOH \xrightleftharpoons{\textit{Phospholipase C}}$$

Phosphatidyl choline
(Lecithin)
Diester

$$\begin{array}{l} CH_2{-}O{-}\overset{\overset{\displaystyle O}{\|}}{C}{-}R_1 \\ | \\ CH{-}O{-}\overset{\overset{\displaystyle O}{\|}}{C}{-}R_2 \\ | \\ CH_2{-}O{-}H \end{array} + (CH_3)_3\overset{+}{N}CH_2CH_2{-}O{-}\overset{\overset{\displaystyle O^-}{|}}{\underset{\underset{\displaystyle O}{\|}}{P}}{-}O^- + 2H^+ \qquad (20)$$

1,2-Diglyceride
(1,2-Diacylglycerol)
Alcohol

Choline phosphate
(Phosphorylcholine)
Monoester

The alcohol esterified to the orthophosphoric acid, $(HO)_3P{=}O$ (equation 19), may be a simple aliphatic alcohol, a polyhydric alcohol such as glycerol or glucose, or any one of a variety of aromatic hydroxy compounds. The phosphatases are not one enzyme but a group of functionally related enzymes. Clinically, it has been useful to recognize three types: *alkaline phosphatase* (in serum, liver, bone, placenta, and intestines), with optimal activity at about pH 10; *acid phosphatase* (in serum, prostate, and liver), with optimal activity at pH 4.9 to 5.0; and *red cell phosphatase*, with optimal activity at pH 5.5 to 6.0. Phosphatases catalyze the transfer of the phosphate group from a donor substrate to an acceptor compound containing an (—OH) group.[20] If the acceptor is water, HOH, the net effect observed is hydrolysis. Schematically,

$$R_1{-}O{-}PO_3H_2 + R_2{-}O{-}H \underset{}{\overset{\textit{Phosphate transfer}}{\rightleftharpoons}} R_1{-}O{-}H + R_2{-}O{-}PO_3H_2 \qquad (21a)$$

$$\underset{\textbf{Donor ester phosphate}}{R_1{-}O{-}PO_3H_2} + \underset{\textbf{Acceptor alcohol}}{H{-}O{-}H} \overset{\textit{Hydrolysis}}{\rightleftharpoons} \underset{\textbf{Donor alcohol}}{R_1{-}O{-}H} + \underset{\textbf{Acceptor ester phosphate}}{H{-}O{-}PO_3H_2} \qquad (21b)$$

If simple hydrolysis occurs, one mole of substrate, $R_1{-}O{-}P$, should produce one mole of each of the two products, alcohol and phosphate ion. If an acceptor alcohol is present in the system, less than the stoichiometric quantity of inorganic phosphate is formed, the difference being accounted for by the phosphate transferred to the organic acceptor. Experimentally, it has been found that a good acceptor must contain a hydroxyl group plus a second hydroxyl or amino group.[1]

ALKALINE PHOSPHATASE IN SERUM

(EC 3.1.3.1; orthophosphoric acid monoester phosphohydrolase, ALP)

This group of enzymes has optimal activity at a pH of about 10, but the optimum observed varies with the nature and concentration of the substrate acted upon, the type of buffer or PO_4-acceptor present, the composition of the isoenzymes present in serum, and other experimental variables. Eight or more different isoenzyme forms may be detected, but there are rarely more than three or four in any one serum sample.[71] The serum isoenzymes have the properties of the specific forms present in intestinal, liver, bone, placental, and, rarely, renal tissue, but there is still uncertainty as to the source of the dominant forms present in normal serum. Alkaline phosphatases act on a large variety of physiologic and nonphysiologic substrates, but the natural substrates on which they act in the body are not known. The fact that individuals with an apparent inborn absence of the enzyme excrete large quantities of ethanolamine phosphate suggests that this (or perhaps phosphatidyl ethanolamine) may be one of the true physiologic substrates of the enzyme.

$$H_3N^+{-}CH_2CH_2{-}O{-}P(=O)(OH){-}O^-$$

Phosphorylethanolamine or ethanolamine phosphate

Some divalent ions such as Mg^{2+}, Co^{2+}, and Mn^{2+} are activators of the enzyme, optimal activity being obtained with Mg^{2+} at about 1.0 mmol/l. Phosphate ions are inhibitors of all forms of the enzyme, as are also borate, oxalate, and cyanide ions. The individual isoenzyme forms are inhibited to different extents by L-phenylalanine,[83] urea,[105] Zn^{2+}, or AsO_4^{3-}. Variations in Mg^{2+} and in substrate concentration cause variations in the pH optimum. The type of buffer present (except at low concentrations) affects the rate of enzyme activity.[1] Buffers can be classified as inert [carbonate, barbital], inhibiting [glycine, propylamine], or activating [2-methyl-2-aminopropanol-1 (MAP), tris(hydroxymethyl)-aminomethane (Tris), and diethanolamine (DEA)]. Glycine inhibits apparently by complexing the activating Mg^{2+} ion.

The enzyme is present in practically all tissues of the body, especially at or in the cell membranes, and it occurs at particularly high levels in intestinal epithelium, kidney tubules, bone (osteoblasts), liver, and placenta.[110] Although the nature of the precise metabolic function of the enzymes is not yet understood, it appears that the enzyme facilitates transfer of metabolites across cell membranes, and that it is associated with lipid transport and with the calcification process in bone synthesis. There is no known function for the enzyme present in milk, intestinal fluid, and bile. It may be present as a contaminant as a result of shedding from cell membranes in contact with these fluids.[110]

The form present in normal adult serum probably originates mainly in the liver or the biliary tract,[112] although some researchers maintain that a large component originates in the intestinal fluid.[71] A small amount of bone component may also be present. The enzyme found in urine is probably derived from renal tissue and does not represent serum enzyme cleared by the kidney. Alkaline phosphatase in serum is rapidly denatured at 56°C but is relatively stable at lower temperatures (the placental forms are most stable). Sera kept at room temperatures usually show a slight but real increase in activity, which varies from 1 per cent over a 6 h period, to 3 to 6 per cent over a 1 to 4 day period.[31,186] Even in sera stored at refrigerator temperature activity increases slowly (2 per cent/d). In some frozen sera, activity will also slowly increase with time, but in most others there is decrease in activity, which is recovered after thawing the serum.[132]

A similar enhancement of activity, but of greater magnitude, is encountered with reconstituted lyophilized preparations, such as those available as commercial "control sera" or "reference materials."[132,186] If the reconstituted material is then stored at 37°C, the increase in activity with some materials can be as high as 50 to 100 per cent over a 24 h period; and the increases with storage at 4 and 20°C are about 10 and 30 per cent, respectively. The enhancement of activity continues for several days, but at a decreasing rate. The cause of this phenomenon is not understood, but may be due to the formation, on freeze-drying, of a phosphate-lipoprotein complex or a multimeric form of the enzyme, which dissociates on warming to free the "bound" enzyme.

CLINICAL SIGNIFICANCE

Serum alkaline phosphatase estimations are of interest in the diagnosis of two groups of conditions: hepatobiliary disease, and bone disease associated with increased osteoblastic activity. For many years, it was believed that alkaline phosphatase reaching the liver from other tissues (bone) was excreted into the bile, and that the elevated serum enzyme activity found in hepatobiliary disease was a result of a failure to excrete the enzyme through the bile. However, more recent studies[112] suggest that the response of the liver to any form of biliary-tree obstruction is to synthesize more ALP. Some of this additional enzyme enters the circulation to raise the enzyme level in serum. The elevation tends to be more marked in extrahepatic than in intrahepatic obstruction (e.g., by stone or by cancer of the head of the pancreas) and is greater the more complete the obstruction. Serum enzyme activities may reach 10 to 12 times the upper limit of normal, returning to normal on surgical removal of the obstruction.

Intrahepatic obstruction of the bile flow (e.g., by invading cancer tissue, or by drugs such as chlorpromazine which affect the biliary tree) also raises serum alkaline phosphatase, but usually to a lesser extent (up to 2.5 times the upper normal limit). Liver diseases which principally affect parenchymal cells, such as infectious hepatitis, typically show only moderately elevated serum alkaline phosphatase levels, the degree of elevation depending on the degree of biliary stasis. These distinctions between extrahepatic and intrahepatic obstructive jaundice and parenchymal jaundice are clinically very useful, but it must be remembered that they are based on averages of many cases, and that in individual instances exceptions to the rules may be encountered.

Among the *bone diseases*, the highest levels of serum ALP activity are encountered in *Paget's disease* (*osteitis deformans*), as a result of the action of the osteoblastic cells as they try to rebuild bone that is being resorbed by the uncontrolled activity of osteoclasts. Values from 10 to 25 times the upper limit of normal are not unusual. Only moderate rises are observed in *osteomalacia*, the levels dropping slowly in response to vitamin D therapy. Levels are normal in *osteoporosis*. In *rickets*, levels two to four times the normal are obtained, and these fall slowly to normal on treatment with vitamin D. Slight to moderate elevations are seen in the *Fanconi syndrome*. *Primary hyperparathyroidism*, and often (but less regularly) *secondary hyperparathyroidism*, are associated with slight to moderate elevations of the enzyme, the degree of elevation reflecting the extent of skeletal involvement. Very high enzyme levels are present in patients with *bone cancer*. Transient elevations are often but not always found during healing of bone fractures. Physiological bone growth elevates alkaline phosphatase in serum, and this accounts for the fact that in the sera of growing children one finds enzyme activity some 1.5 to 2.5 times that present in normal adult serum. An increase of up to 2 to 3 times normal is observed in women in the third trimester of *pregnancy;* this additional enzyme is of placental origin.

Moderate elevations of ALP may be seen in several disorders that do not involve the liver or bone. Among these are Hodgkin's disease, congestive heart failure, ulcerative colitis, regional enteritis, and intra-abdominal bacterial infections.[111]

ASSAY OF ALKALINE PHOSPHATASE

Whenever the natural substrate of an enzyme is known, the assay of that enzyme is usually carried out with this substrate, even though it may be one to which the enzyme shows a low level of specificity. When the true substrate is not known, as is true for alkaline phosphatase, the analyst is justified in using any substrate that is analytically convenient and gives a reasonably rapid rate of reaction.

In 1930, Kay introduced the use of β-glycerophosphate as a substrate for alkaline phosphatase, and Bodansky used this as the basis for his classic procedure in 1932. The reaction can be followed by measuring the rate of liberation of inorganic phosphate ion as the enzyme hydrolyzes the glycerophosphate to phosphate and glycerol. In his assay, Bodansky used a weak barbital buffer which gave a resultant pH for the reaction mixture of approximately 8.8 at 37°C. He defined his unit of phosphatase activity as that quantity of enzyme which liberated 1.0 mg of phosphate-phosphorus in 1.0 h at 37°C under the conditions of his procedure. The activity was reported in terms of units per 100 ml of serum. Since serum contains inorganic phosphate, this assay entails two measurements of phosphate-phosphorus, namely, before and after the one hour incubation period. It was soon appreciated that the pH of the assay was below the optimal value, and Shinowara, Jones, and Reinhart modified the procedure by using a more alkaline barbital buffer, giving a pH of 9.8 for the enzyme reaction.

The concentration of buffer used was, however, too low to ensure the identical desired pH with all sera. The pH of serum increases rapidly on standing, from about 7.4 to as much as 8.5, as a result of loss of carbon dioxide to the air and from ammonia which is generated from the hydrolysis of metabolites. The buffer capacity of serum is appreciable, and any buffer used in enzyme assays must be of a high enough concentration to give a fixed and predetermined pH, regardless of the initial pH of the serum specimen to be assayed. On the other hand, it is known that a buffer of too high concentration may inhibit the reaction.

In 1934 King and Armstrong proposed the use of *phenylphosphate* as substrate. The rate of reaction was followed by measuring the phenol formed. This substrate has the advantage of being hydrolyzed more rapidly than is β-glycerophosphate; moreover, phenol is a reactive compound which can be determined by several sensitive colorimetric methods, thus permitting shorter incubation times. Furthermore, because there are only traces of phenol-like compounds in serum, only one analysis is needed, rather than the two phosphate

$$C_6H_5\text{—O—}\overset{\overset{\displaystyle O}{\|}}{\underset{\underset{\displaystyle O^-}{|}}{P}}\text{—}O^- + H\text{—}O\text{—}H \underset{pH=9.6}{\overset{P\text{-}ase}{\rightleftarrows}} C_6H_5\text{—O—H} + H\text{—}O\text{—}\overset{\overset{\displaystyle O}{\|}}{\underset{\underset{\displaystyle O^-}{|}}{P}}\text{—}O^- \qquad (22)$$

Phenylphosphate **Phenol** **Phosphate ion**

analyses required in the Bodansky method. The phenol which is formed can be assayed with the Folin-Ciocalteu reagent (King-Armstrong) or with 4-aminoantipyrine (Kind and King), or by reaction with a diazo reagent. The original method employing the Folin-Ciocalteu reagent required a 30 min incubation period followed by deproteinization of the incubation mixture. Using antipyrine as a chromogenic reagent, deproteinization is not needed, and a 15 min reaction period is sufficient.

In 1946 Bessey, Lowry, and Brock[20] proposed the use of *p*-nitrophenylphosphate (PNPP), a chromogenic, substituted phenylphosphate, as the substrate (BLB procedure). This ester is colorless, but the reaction product is colored at the pH of the reaction, and the enzyme reaction can be followed by observing the rate of formation of the yellow color of the *p*-nitrophenoxide ion (equation 23, p. 607). No time-consuming deproteinization or further chemical reaction is required, and the reaction is linear with time. The selection of glycine buffer is not the best of choices; the glycine tends to give a low sub-optimal reaction rate because it chelates some of the necessary activator ions.

Huggins and Talalay introduced the use of *phenolphthalein diphosphate* as a substrate; the indicator phenolphthalein is released in the reaction, being measured by its red color (at pH 10). Unfortunately, the hydrolysis of the substrate (a diphosphate ester) forms a product (the monophosphate ester) which itself is a substrate of the enzyme. The phosphate bonds in the two materials are split concurrently but at different rates, giving rise to complex kinetics and a nonlinear enzyme reaction. Babson and his associates[4] prepared *phenolphthalein monophosphate*, and it has been demonstrated that this substrate gives linear reaction kinetics and is a useful and sensitive substrate for assaying phosphatase. The Shinowara, King-Armstrong, Kind-King, and Bessey procedures have been widely used, and all have been adapted for automated analysis, but they may be considered obsolete at this writing.

These procedures, despite their improvements, were still two-point assays, with long incubation periods, and there has been an ever increasing interest in devising continuous monitoring procedures, based on using the more advantageous chromogenic substrates (e.g., *p*- nitrophenyl phosphate and thymolphthalein phosphate) and new buffers. Moss[141] has shown that α-naphthol monophosphate may also be used, the α-naphthol that is formed being measured at 340 nm. This is convenient, since it is the same wavelength used to measure reactions involving the NAD/NADH coupled dehydrogenase systems.

In all the alkaline phosphatase methods discussed so far, the liberated phosphate group is transferred to water, i.e., the reaction is primarily hydrolytic. The rate of phosphatase action is much enhanced, however, if certain amino alcohols are used as buffers. Among these apparent activators are compounds such as 2-methyl-2-aminopropanol-1 (MAP), diethanolamine (DEA), tris(hydroxymethyl)aminomethane (Tris), and ethylaminoethanol (EAE). These materials are derivatives of aliphatic amines, and function as buffers by binding protons at the nitrogen atom. Being hydroxyl compounds, however, they can act as phosphate group acceptors, and the enzyme rate enhancement observed in their presence derives from their participation in the PO_4 transfer reaction. Not all compounds containing

NH_2 and OH groups can serve as activating buffers[1]: the greatest degree of phosphorylation is observed when the OH group is separated from the N atom by two carbon atoms (e.g., in DEA and EAE) and a one- or two-carbon alkyl group is attached to one of the other two nitrogen bonds:

$$\text{—(C)—C—}\underset{\displaystyle \text{H}}{\overset{}{\text{N}}}\text{—C—C—OH}$$

Enzyme activity in the presence of optimal concentrations of these buffers can be three- to six-fold greater than in the presence of a non-activating buffer (e.g., carbonate). The ALP methods described here employ such phosphate-accepting buffers.

The enhanced phosphatase activity observed with a buffer such as MAP, first proposed by Lowry[130] to replace the inhibiting glycine in the BLB procedure, was soon appreciated to derive in part from its PO_4-acceptor properties. The improved sensitivity and the considerably shortened reaction period associated with the use of such a buffer were demonstrated by Bowers and McComb,[28] in their procedure using PNPP as substrate. Many other amino-alcohols, such as DEA, EAE, and MAE (2-methylaminoethanol), are phosphorylatable buffers, which are effective in the pH range from 8.5 to 10.5. However, use of such buffers complicates the kinetics of the enzyme reaction, since the reaction rate will depend not only on the substrate concentration but also on the concentration of the buffer-acceptor which behaves as cosubstrate. Rather high concentrations of buffers (approximately 1.0 to 2.2 mol/l) are needed for maximum enzyme activity.

A large number of manual, automated, and commercial "kit" procedures which use these buffers and chromogenic substrates have been introduced. The "standard" procedure for the assay of alkaline phosphatase proposed by clinical chemists in Germany[82] and the "recommended" method put forward by their Scandinavian counterparts[175] both use DEA buffer, pH 9.8, at 1.0 mol/l and PNPP as substrate at 10 mmol/l. The methods to be presented in detail employ MAP as buffer, PNPP as substrate, and temperatures of 30°C and 37°C, respectively. Optimal reaction conditions, including temperature, to be used in phosphatase activity assays are still under discussion. The increased rate of reaction at 37°C must be weighed against the evidence suggesting a slow but real denaturation of the enzyme at this temperature. For this reason a temperature of 30°C and a very short incubation time are probably most advantageous.

UNITS FOR REPORTING PHOSPHATASE ACTIVITY

In the past, alkaline phosphatase activity was expressed in a multiplicity of arbitrary units. The Bodansky unit has already been defined. The King-Armstrong unit is equal to 1.0 mg of phenol liberated from phenylphosphate (in carbonate buffer) in 30 min at 37°C, whereas the Bessey-Lowry-Brock (BLB) unit is equal to the enzyme activity which is able to release 1.0 mmol of *p*-nitrophenol (PNP) from PNPP in 60 min at 37°C.

As these older procedures are being replaced by the new group of methods referred to, activity is being reported uniformly in International Units (U/l). Recall that the International Unit is defined as that activity which acts on 1.0 μmol of substrate per minute (and thus releases 1.0 μmol of chromogen from the substrate per minute) under the conditions of the assay.

DETERMINATION OF ALKALINE PHOSPHATASE USING A CONTINUOUS MONITORING PROCEDURE (BOWERS AND McCOMB)[28,29,135]

PRINCIPLE

Para-nitrophenylphosphate (PNPP) is colorless. The enzyme splits off the phosphate group to form free *p*-nitrophenol (PNP), which in the acid form in dilute solution is also colorless. Under alkaline conditions this is converted to the *p*-nitrophenoxide ion, which assumes a quinoid structure with a very intense yellow color. At the pH of the reaction, most of the free nitrophenol is in the yellow-colored quinoid form. Thus, the course of the reaction can be followed by observing the increase in yellow color. At the pH of the reaction,

$$p\text{-Nitrophenylphosphate} + HOH \xrightarrow[\substack{Mg^{2+} \\ pH = 10.3}]{ALP} p\text{-Nitrophenoxide} \xleftrightarrow[\text{at alkaline pH}]{\text{Rearranges}} p\text{-Nitrophenoxide (quinoid)} + PO_4^{3-} \quad (23)$$

p-Nitrophenyl-phosphate (colorless) **p-Nitrophenoxide (colorless benzenoid form)** **p-Nitrophenoxide (yellow, quinoid form)**

not all of the phenoxide is in the quinoid form, but the fraction is constant for any one assay, and the rate of color formation can be used as a measure of the reaction.

The rate of formation of *p*-nitrophenol (PNP) by the action of the enzyme on *p*-nitrophenylphosphate at 30°C can be monitored colorimetrically with a recording spectrophotometer. Substrate, buffer/PO_4-acceptor, and Mg^{2+} are at the optimal concentrations of 1.4×10^{-3}, 0.75, and 1.0×10^{-4} mol/l, respectively.

In a manual two-point procedure described later, the reaction is permitted to proceed for a predetermined time, and is then stopped by adding NaOH to raise the pH to 11.5 to 12.0, which inactivates the enzyme, converts all the phenoxide to the colored quinoid form, and dilutes the yellow color to a measurable intensity.

SPECIMENS

Only serum or heparinized plasma, preferably free of hemolysis, should be used.

REAGENTS

1. Buffer, 2-methyl-2-aminopropanol-1 (MAP), 0.84 mol/l, pH 10.30 at 30°C. In a 2000 ml volumetric flask, mix 300 ml of 1.0 molar HCl solution and 150 g of MAP dissolved in 1000 ml of H_2O, and dilute to volume with additional water. Use CO_2-free water; in preparing and handling the buffer, take care that contact with air (CO_2 absorption) is minimal. The buffer should show a pH of 10.30 ± 0.02, measured at 30°C. It is stable at room temperature for about 60 days.

2. Stock $MgCl_2$ solution, 150 mmol/l. Dissolve 3.0 g of $MgCl_2{\cdot}6H_2O$ in water to make 100 ml of solution. This reagent is stable.

3. Substrate solution, 215 mmol/l of *p*-nitrophenylphosphate (PNPP) in 1.5 mmol/l $MgCl_2$ solution. Prepare a fresh working solution of $MgCl_2$ by diluting 1.00 ml of the stock $MgCl_2$ solution to 100 ml. Weigh out 800 mg of high purity disodium hexahydrate salt of PNPP (correct for hydration, if other than $6H_2O$) and dissolve it in 10.0 ml of this working (dilute) $MgCl_2$ solution. This solution should preferably be made and used fresh, but solutions will keep with negligible decomposition for about 7 days if kept at 4°C and in the dark.

4. PNP stock standard, 1.00 mmol *p*-nitrophenol/l. Dissolve 139.1 mg of high purity or recrystallized PNP in water to make 1000 ml of solution. This solution is stable if kept in the dark. High purity PNP is available commercially, but a batch of PNP may be purified by recrystallization from hot water, and drying overnight in a vacuum desiccator over silica gel.

5. PNP working standard solution, 4.00×10^{-2} mmol/l in 0.84 mol/l MAP buffer solution. Pipet 10.0 ml of the stock standard into a 250 ml volumetric flask and dilute to

volume with the MAP buffer. Mix thoroughly. Check the absorbance of this solution by measuring its value at 404 nm in a 1.00 cm cell at 30°C in a narrow beam spectrophotometer. The reading should be 0.751 ± 0.002 (ε = 18,700 to 18,800). A similar dilution of the stock PNP in 0.05 mol NaOH/l, measured at 401 nm at 25°C, should give a value of 0.735 (ε/mol × l = 18,300 to 18,400).[28]

If the instrument on which the absorbance readings are to be made is other than a standardized narrow beam spectrophotometer, a value other than 0.751 may be obtained. In such a case, use this measured value to calculate the "apparent molar absorptivity" for that instrument:

$$= \text{Reading} \times \frac{1.0\ \text{mol/l}}{4.0 \times 10^{-2} \times 10^{-3}\ \text{mol/l}} = \text{R} \times 25{,}000 \leq 18{,}750$$

This value is used to convert absorbance readings into U/l. See comment 2.

PROCEDURE

1. A spectrophotometer having a thermostatted cell compartment and connected to a recorder is required. Set the wavelength to 404 nm. Adjust the temperature of the cell compartment to 30°C ± 0.1°C. Place cuvets in the compartment or an incubator at 30°C to prewarm to temperature.
2. Pipet 2.70 ml of buffer into a 12 × 100 mm test tube.
3. Add 100 μl of serum to the tube containing the buffer. Mix gently but well. Deliver the serum with an Eppendorf or similar micropipet, with disposable tips. If a glass micropipet (TC) is used, rinse the pipet twice with buffer before proceeding to the next specimen.
4. Place the tubes containing buffer and serum into a 30°C water bath for at least 5 min to equilibrate to temperature. At the same time, place a tube containing an aliquot of substrate in the bath to bring it to temperature also.
5. Initiate the reaction by adding 200 μl of warmed substrate to the tube containing the buffered serum. Immediately transfer the tube contents to a prewarmed cuvet, and place the cuvet into the cell compartment of the spectrophotometer. In many instruments it is possible to read several (4 to 7) cuvets serially in a single assay run. The changes in the absorbances of the solutions in the cuvet(s) with time, measured at 404 nm, are recorded for a period of from 2 to 5 min. The trace of absorbance versus time should be a straight line, or should include a linear segment over a 2 to 3 min period.
6. The rate of change of absorbance with time $= \dfrac{\Delta A_{404}}{\text{time}} = \Delta A/\text{min}$ is calculated, using data obtained from a linear portion of the recorder trace.

CALCULATIONS

1. If ε = 18,750, then a solution containing 1.0 mmol/l will have an absorbance of 18.75, and the measured ΔA/min will correspond to $\dfrac{\Delta A/\text{min}}{18.75}$ μmol PNP/ml. The quantity of PNP is present in a 3.0 ml volume and the sample size used is 0.1 ml. Thus,

$$\text{U/l} = \frac{\Delta A_{404}/\text{min}}{18.75} \times 3.0 \times \frac{1000}{0.10} = (\Delta A/\text{min}) \times 1600$$

COMMENTS

1. If a recorder is not available, the absorbances of the cuvets can be read manually every 15 or 30 seconds and ΔA_{404} can be plotted against time on graph paper.

2. If the spectrophotometer does not give an absorbance of 0.751 for the 4.0×10^{-2} mmol/l working standard, use the "apparent molar absorptivity" (AMA) as calculated from the actually observed value in place of 18.75 in the formula for calculating activity in U/l.

$$\left(\text{U/l} = \frac{\Delta A}{\text{min}} \times 1600 \times \frac{18{,}750}{\text{AMA}}\right)$$

3. Temperature coefficient: The relative enzyme activities at several selected temperatures, taking that at 30°C as 1.00, have been reported to be as follows: 25°C = 0.76 and 37°C = 1.47. The Q_{10} value for the system is 1.7.

NORMAL VALUES FOR ALP ACTIVITY IN SERUM

1. Adults: 25 to 90 U/l.
2. Children: Upper limit is about 150 U/l; the lower limit is not yet established, but 40 U/l may be taken as a reasonable estimate.
3. Using 1.0 molar DEA (pH 10.10) as buffer in place of MAP, a PNPP concentration of 14 mmol/l, and a sample size of 0.02 ml, Tietz, Weinstock, and Wills[204] obtained a range of 70 to 220 U/l for ALP activity in normal sera of adults (30°C).

REFERENCES

Bowers, G. N., Jr., and McComb, R. B.: Clin. Chem., *12*:70, 1966.

Bowers, G. N., Jr., and McComb, R. B.: *In* Proceedings, International Seminar and Workshop on Enzymology. N. W. Tietz, Program Coordinator. Mt. Sinai Hosp. Med. Center, Chicago, 1972, p. 2–2.

McComb, R. B., and Bowers, G. N., Jr.: Clin. Chem., *18*:97, 1972.

DETERMINATION OF ALKALINE PHOSPHATASE BY A MANUAL TWO-POINT KIT PROCEDURE USING p-NITROPHENYLPHOSPHATE

The procedure to be outlined is an example of the many enzyme kit procedures developed by commercial concerns. The substrate is PNPP and the buffer/PO_4-acceptor is MAP, at final concentrations of 3.8 mmol and 0.81 mol/l, respectively.

REAGENTS

1. Buffered substrate, 4.0 mmol PNPP/l, 0.5 mmol Mg^{2+}/l, in 0.84 molar MAP buffer, pH 10.2. Prepare by adding 21 ml of buffer solution (available as a kit component) to a vial containing 84 μmol *p*-nitrophenylphosphate, disodium salt, and 10.5 μmol $MgCl_2$. Dissolve the salts by gentle mixing. The reagent is stated to be stable for two months, if kept refrigerated.
2. NaOH solution, 0.05 mol/l. Prepare by diluting the contents of a 25 ml vial of a 2 mol/l NaOH solution to 1000 ml, and transfer to a polyethylene bottle. The reagent is stable, but take care to minimize contact with air (absorption of CO_2).
3. Working standard solution, *p*-nitrophenol (PNP), 4.5×10^{-2} mmol/l (= 4.5×10^{-2} μmol/ml) in 0.05 molar NaOH solution. Prepare fresh daily by diluting 2.5 ml of a 3.6×10^{-4} mol/l stock standard (provided in the kit) to 20.0 ml with the 0.05 mol/l NaOH solution. (See comment 1.)

PROCEDURE

1. Pipet 1.0 ml of buffered substrate into each of two 15 × 125 mm test tubes, labeled T (test) and B (blank). Place these in a 37°C water bath for 5 to 7 min to equilibrate to temperature. Use one pair of tubes for each specimen.
2. With the timer set, add 50 μl of serum to the T tube and incubate at 37°C for exactly 30 min.

3. At the end of 30 min, add 10 ml of 0.05 molar NaOH to both tubes to stop the reaction and to dilute the PNP formed. Mix well.

4. Add 50 μl of serum to the B tube, and mix contents thoroughly.

5. Pour the contents of the B and T tubes into appropriate cuvets and read the absorbances of the solutions at 405 nm against water as an instrument blank. Calculate $\Delta A = (A_T - A_B)$, where A_T and A_B are the absorbances of the contents of the T and B tubes, respectively.

6. Referring to the calibration curve, convert the ΔA value to International Units (U/l) of enzyme activity.

CALIBRATION

1. Prepare the calibration solutions by diluting 1.0, 2.0, 3.0, 4.0, and 5.0 ml of working standard to 5.5 ml with 0.05 molar NaOH.

2. Transfer to cuvets and read the absorbances at 405 nm against 0.05 molar NaOH solution as instrument blank.

3. Plot the absorbance readings against the enzyme activity values of the standards. The latter are: 60, 120, 180, 240, and 300 U/l. (See calculation.)

CALCULATION

The 1.0 ml calibration standard contains 4.5×10^{-2} μmol of PNP in 5.5 ml. If this quantity of PNP is formed in 30 min by the enzyme that was present in 50 μl of specimen and diluted to 11.0 ml, then the rate of formation is

$$\frac{11.0 \text{ ml}}{5.5 \text{ ml}} \times (4.5 \times 10^{-2}\ \mu\text{mol}) \times \frac{1}{30 \text{ min}} \times \frac{1000 \text{ ml}}{0.05 \text{ ml}} = 60\ \mu\text{mol/min per liter}$$

or 60 U/l. This is the activity value which corresponds to the 1.0 ml standard.

COMMENTS

1. This procedure is a modification of that described in the actual kit directions. The latter uses 20 μl of specimen added to 0.50 ml of buffered substrate, or a ratio of 1/25. In the procedure described here, the ratio is 1/20. The concentration of the working standard has also been modified to make allowance for this change in concentration (from 3.6×10^{-5} mol/l to 4.5×10^{-5} mol/l).

2. The enzyme activity values of the standards in the kit procedure were arbitrarily designated to give values identical with those obtained by the Bessey-Lowry-Brock procedure (2, 4, . . . , 10 BLB units). These convert to 33, 66, . . . , 167 μmol/min per liter. However, in terms of the actual quantity of product formed under the actual assay conditions, the values in U/l for the standards are 60, 120, 180, 240, and 300 U/l as given under Calibration. The range of values found in normal sera as stated in the kit is from 13 to 38 U/l; these become $\frac{60}{33} \times (13 \text{ to } 38)$, or 24 to 69 U/l. The discussion in the instructions furnished with the kit does point out that with MAP the assay becomes 1.8 times more sensitive than in the original BLB procedure, and that the adjusted standard concentration compensates for this effect.

3. The use of the Blank (B) tube compensates for absorbance at 405 nm due to bilirubin and minimal hemolysis (hemoglobin).

NORMAL VALUES

Adults: 24 to 69 U/l.
Children: 58 to 140 U/l.

REFERENCE

Tekbulletin No. 15, AL-P, and Bulletin No. 15–1. Determination of Alkaline Phosphatase with Tekit AL-P, Searle Diagnostic, Inc., Columbus, OH.

ISOENZYMES OF ALKALINE PHOSPHATASE

Alkaline phosphatases prepared from different human tissues are not identical in all properties, although they are similar in many respects. The nature of and explanation for these tissue-specific differences are not understood, but in the case of the placental isoenzyme, there is evidence that a distinct structural gene (or genes) determines the synthesis of this particular form. Criteria which have been used to differentiate the isoenzymes of alkaline phosphatase include the following: differences in relative rates of reaction with various substrates; differences in response to the presence of selected inhibitors; variations in stability to denaturation by heat or urea; differences in electrophoretic mobility; and differences in immunochemical characteristics.

The tissue-specific characteristics of the alkaline phosphatase isoenzymes[71,123,214] are retained when the enzymes are released into the circulation, and this can be used to identify the tissue contributing the enzyme responsible for the elevation of the serum level of the enzyme. However, the differences in properties between alkaline phosphatase isoenzymes are in some cases quite small (this is particularly true for bone and liver phosphatases), so that a combination of more than one technique is needed to ensure reliable discrimination.

When serum samples containing different alkaline phosphatase isoenzymes are separated by electrophoresis at alkaline pH, the liver phosphatase moves rapidly toward the anode (Fig. 12-17). Bone phosphatase, which typically gives a more diffuse zone than does the liver isoenzyme, has a slightly lower anodal mobility, although the two zones usually overlap to some extent. Intestinal phosphatase also migrates diffusely but more slowly than the bone enzyme, and kidney phosphatase (which occurs very rarely in serum) migrates even more slowly. The placental isoenzymes have mobilities of the same order as those of liver and bone, depending on the phenotype. Minor, slow-moving phosphatase zones can also arise from these various tissues, but their diagnostic significance is uncertain. Variants of each form may be encountered, of which some are genetic and some are only artifactual, reflecting changes occurring through experimental procedures. The intestinal enzyme is especially unstable; a preparation which moves as a single zone may become altered during storage, giving three zones during electrophoretic migration.

Although the difference in mobility between bone and liver phosphatases is small, it can be exploited in favorable situations to distinguish between these two tissues as the

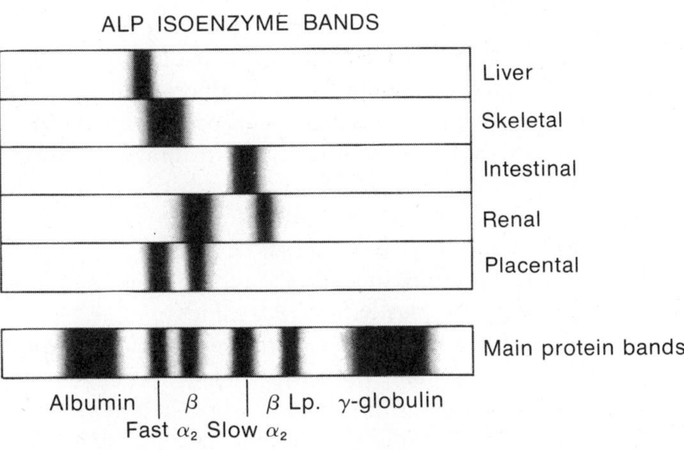

Figure 12–17 Alkaline phosphatase isoenzyme patterns visualized after starch gel electrophoresis of saline extracts of respective human tissues. (After Latner and Skillen.[123])

possible source of increased phosphatase levels (hyperphosphatasemia). However, it is advisable to run reference sera (markers) of known isoenzyme composition (e.g., samples from cases of proven bone or liver disease) along with the unknown specimens. Comparison is easier when the electrophoresis of sample and markers is done on the same support medium at the same time (e.g., on gel slabs rather than in individual tubes). It is also useful to compare the electrophoretic patterns obtained on specimens before and after they were subjected to heat denaturation (see below).

Starch,[196] agar,[152] and polyacrylamide[38] gels, as well as cellulose acetate,[76] have been used for the separation of alkaline phosphatase isoenzymes. The references quoted give typical examples of each method. The choice of a particular system depends on the interest and resources of individual laboratories, but it is important that the chosen method be used regularly and consistently to ensure repeatable and reliable results.

When electrophoresis suggests the presence of *intestinal* phosphatase, confirmatory evidence can be obtained by repeating the separation after incubation of the sample with neuraminidase. This treatment, by removing negatively charged sialic acid residues, reduces the anodal mobility of all phosphatase isoenzymes except those of intestinal origin, which are neuraminidase-resistant. The intestinal enzyme also contains sialic acid groups, but they are internal and not accessible to neuraminidase action.[71,140]

The most remarkable property of *placental* alkaline phosphatase is its pronounced stability to heat.[83,151] Incubation of the enzyme at a temperature as high as 65°C does not diminish its activity, whereas that of other phosphatase isoenzymes is completely destroyed at this temperature. The determination of the amount of alkaline phosphatase activity in serum surviving heating at 65°C for 30 min therefore provides the most convenient and specific test for the presence of this isoenzyme,[70,189] e.g., in late pregnancy. Pronounced heat stability is also shown by the "Regan isoenzyme," a placental-like fetal form of alkaline phosphatase which occurs in 5 to 15 per cent of specimens from patients with cancers of various types. Thus, heating at 65°C can also be used to detect this abnormal isoenzyme. The similarity between these two enzyme forms is evidenced by the fact that the "Regan isoenzyme" will react with antibodies made against the placental isoenzyme.[207]

Other isoenzymes of alkaline phosphatase can be differentiated on the basis of their stability to temperatures lower than 65°C. At 56°C, for example, *intestinal* ALP is more stable than liver ALP, which in turn is more stable than the bone isoenzyme. Thus, if after 10 min at 56°C, less than 20 per cent of the serum alkaline phosphatase activity remains, the enzyme present is probably largely of bone type; residual activities of between 25 and 55 per cent support electrophoretic evidence that liver phosphatase is the predominant isoenzyme. At present, however, there is no agreement on a standard temperature or time of heating. Many workers prefer 56°C, but time intervals vary from 10 to 30 min. For this test to be meaningful, however, careful control of the experimental conditions is essential. The temperature coefficients for inactivation of enzymes by heat are high, so that a slight variation in the temperature produces a large change in the rate of inactivation. The use of a water bath which is capable of precise and accurate control of temperature within narrow limits is therefore essential, and the duration of incubation must be timed exactly. At the end of the heating period, specimens must be immediately immersed in cold water.

The following procedure for heat inactivation of alkaline phosphatase in serum has been found to give reproducible results:

1. Put approximately 0.5 ml of serum in a small, thin-walled glass tube (Dreyer tube) and seal with plastic film (Parafilm).

2. Place the tube into a thermostatically controlled water bath that is already stabilized at 56.0°C, so that the surface of the serum in the tube is below the level of the water, and simultaneously start the stopwatch. The water bath should have good temperature stability and be of large volume and well stirred, to minimize temperature fluctuations on introducing the specimens.

3. After *exactly* 10 min, remove the serum tube rapidly and place it in ice.
4. Determine the alkaline phosphatase activity of the heated specimen; express it as a percentage of that in an unheated specimen of the same serum that has been kept in ice but not frozen.

Residual activities of 20 per cent or less suggest that the predominant isoenzyme is of the bone type. Values between 25 and 55 per cent are associated with sera in which the predominant isoenzyme(s) is of liver or intestinal origin, or both.

If the presence of placental or Regan isoenzymes is suspected, the incubation should be done at 65°C for 30 min.

Specific *chemical inhibitors* have also been applied to the characterization of alkaline phosphatase isoenzymes in serum. L-Phenylalanine[72,83] markedly inhibits intestinal, placental, and Regan isoenzymes when present at a concentration of 5 mmol/l, but has little effect on the isoenzymes of bone or liver.

Urea[7] inhibits the enzyme at high concentrations, the inhibition being irreversible and varying with the tissue origin of the ALP. Again, no agreement exists as to the concentration of urea to be used, or the duration of treatment. Bone isoenzyme is most susceptible to urea denaturation (16 per cent residual activity after treatment at 37°C for 18 min in 3 molar urea).[110] The liver enzyme has intermediate resistance (44 per cent residual activity), whereas the intestinal and placental enzymes are most resistant under these conditions (69 per cent activity left).

Another approach to aid discrimination between the alkaline phosphatase isoenzymes in serum includes the use of antisera to specific tissue phosphatases (e.g., placental phosphatase), but considerable cross-reaction is encountered.[122,208] The highest degree of certainty in the identification of ALP isoenzymes can be obtained if a combination of two or more of the following techniques is employed: electrophoresis, heat-stability measurements, and inhibition by L-phenylalanine or urea.

REFERENCES

Fishman, W. H., and Ghosh, N. K.: Isoenzymes of human alkaline phosphatase. *In* Advances in Clinical Chemistry, Vol. 10, O. Bodansky and C. P. Stewart, Eds. Academic Press, New York, 1967, pp. 256–370.

Latner, A. L., and Skillen, A. W.: *In* Isoenzymes in Biology and Medicine. Academic Press, New York, 1968, pp. 56–63, 113–115.

Warnock, M. L.: Clin. Chim. Acta, *14*:156, 1966.

ACID PHOSPHATASE IN SERUM

(EC 3.1.3.2; orthophosphoric monoester phosphohydrolase)

Under the name of acid phosphatase (ACP) are included all phosphatases with optimal activity below a pH of 7.0. Thus, the name refers to a group of similar or related enzymes rather than to one particular enzyme species. The greatest concentrations of acid phosphatase activity are present in liver, spleen, milk, erythrocytes, platelets, bone marrow, and the prostate gland. The last is the richest source, and it contributes about one-third to one-half of the enzyme present in serum of healthy males. The source of the remainder of the acid phosphatase in serum in healthy males and females is not known, but there is some evidence that it derives from disintegrated platelets, red cells, and the liver. The optimal pH for the individual acid phosphatases varies, depending on the tissue from which they are obtained. The prostatic enzyme, which is of greatest clinical interest, has a well-defined optimum pH at 4.8 to 5.1. The observed pH optimum also varies with the substrate on which the enzyme acts; the more acidic the substrate, the lower the pH at which maximum activity is obtained. The enzymes can hydrolyze a variety of phosphate esters, and indeed, every substrate utilized in measuring serum alkaline phosphatase has also been used in evaluating acid phosphatase activity.

The acid phosphatases are unstable, especially at temperatures above 37°C and at pH levels above 7.0.[54] Some of the enzyme forms in serum (especially the prostatic enzyme) are particularly labile, and over 50 per cent of the ACP activity may be lost in one hour's time at room temperature. Acidification of the serum specimen to a pH below 6.5 aids in stabilizing the enzyme.

Because of the clinical importance of elevated serum acid phosphatase levels in the diagnosis of prostatic cancer, it is desirable to be able to differentiate between an increase in the concentration of the specific prostatic enzyme and an increase in the activity of the nonspecific forms. Certain chemical agents can inhibit the activity of one or the other type of acid phosphatase. The prostatic enzyme is strongly inhibited by L-tartrate ions, whereas the red cell enzyme is inhibited by formaldehyde and by cupric ions. The most common procedure for differentiating between prostatic and nonspecific activities is to perform the assay in the presence and absence of L-tartrate ion. Tartrate-inhibited forms of acid phosphatase do occur in tissues other than prostate but they rarely, if ever, enter the serum. Attempts have been made to discover differential substrates (i.e., substrates which are hydrolyzed rapidly by the prostatic enzyme, but at a significantly slower rate by the other forms of the enzyme). Evidence has been presented that β-glycerophosphate, α-naphthyl phosphate, and thymolphthalein monophosphate are relatively more sensitive to the action of the prostatic enzyme than are such substrates as phenylphosphate and *p*-nitrophenylphosphate.

CLINICAL SIGNIFICANCE

Elevations of the prostatic acid phosphatase (and thus, generally, total acid phosphatase) are found in the sera of males with *prostatic cancer* with metastases.[88,226] Total activities may reach 40 to 50 times the upper limit of normal, nearly all of the activity being inhibited by tartrate. However, when the carcinoma remains localized in the prostate gland, normal or only slightly raised levels are found. After surgery or estrogen therapy, the levels slowly approach normal, with a subsequent rise if the treatment is unsuccessful. In patients with benign hypertrophy of the prostate, the enzyme level in serum is within normal limits, even after prostatic manipulation, except in rare instances.[219]

Slight or moderate elevations in total ACP activity often occur in Paget's disease, in hyperparathyroidism with skeletal involvement, and in the presence of malignant invasion of the bones by cancers such as female breast cancer. The enzyme in these cases is not inhibited by tartrate and is thought to come from osteoclasts. (In some cases of breast cancer, however, elevations of the tartrate inhibited ACP have been observed.) Elevated levels of nonprostatic acid phosphatase have also been observed in patients with *Gaucher's* and *Niemann-Pick* diseases, in a number of prepubertal conditions, in *myelocytic leukemia*, and in some other hematological disorders.

Acid phosphatase is present in very high concentrations in semen, a fact utilized in forensic medicine in investigations of rape and similar offenses. Acid phosphatase activity is also present in urine.[69]

DETERMINATION OF ACID PHOSPHATASE ACTIVITY IN SERUM

Methods for acid phosphatase activity measurements are adaptations of those developed for alkaline phosphatase. Thus, in 1938 Gutman and Gutman[88] used phenylphosphate as the substrate for serum acid phosphatase, and first demonstrated the great value of this determination in the diagnosis of prostatic cancer and in following treatment. Their procedure was a modification of the King-Armstrong method for alkaline phosphatase determination, in which the acid phosphatase activity was measured at a pH of 4.9. Shinowara also adapted his alkaline phosphatase procedure for acid phosphatase, and methods measuring the hydrolysis of *p*-nitrophenylphosphate at pH 4.9 were devised by others. Fishman and Lerner recommended the use of phenylphosphate but converted the liberated phenol into a diazo dye for measurement. Babson, Read, and Phillips[5,178] urged the use of

α-naphthylphosphate as substrate specific for the prostatic enzyme; they also measured the hydrolyzed α-naphthol as a diazo pigment. This approach was further developed by Hillman[101] into a continuous monitoring procedure. Thymolphthalein monophosphate (TMP) was proposed as a substrate by Roy and associates,[171] who claimed that TMP is the most specific of all substrates for assaying prostatic acid phosphatase. Several of these approaches have been adapted to automated analysis, e.g., α-naphthylphosphate (Centrifichem) and phenylphosphate (AutoAnalyzer).

The Bessey-Lowry-Brock (BLB)[20] procedure using *p*-nitrophenylphosphate (also studied by Andersch and Szcypinski) and the Roy[171] and Hillmann[101] procedures will be presented as examples of methods available for measuring acid phosphatase activity.

DETERMINATION OF TOTAL AND PROSTATIC ACID PHOSPHATASE ACTIVITY USING *p*-NITROPHENYLPHOSPHATE (PNPP)*

SPECIMENS

Hemolyzed serum specimens are contaminated with considerable amounts of red cell acid phosphatase and should be rejected. Chylous sera give analytical difficulties and should be avoided. Serum from all specimens should be immediately separated from red cells and stabilized by the addition of disodium citrate monohydrate at a level of 10 mg/ml of serum. Small 10 mg pellets are available from several commercial sources. Alternatively, two or three drops of 6 molar acetic acid per 5 ml of serum will also lower the pH to the level at which the enzyme is stable. Under these conditions activity will be maintained at room temperature for several hours, and for up to a week if the serum is refrigerated. Both fluoride and oxalate ions, as well as heparin, inhibit enzyme activity and should not be used as anticoagulants.

PRINCIPLE

The enzyme is permitted to act on PNPP in citrate buffer at pH 4.9, both in the presence and in the absence of 0.04 mol tartrate ion/l, using a 30 min reaction time. The liberated PNP is measured spectrophotometrically after conversion to the quinoid form by addition of NaOH. The reaction cannot be followed continuously as can that of alkaline phosphatase, since the spectral differences between PNP and PNPP are developed only in alkaline solution (pH = 11). The assay carried out without tartrate gives a measure of all forms of the enzyme (platelet, red cell, prostatic, etc.). The tartrate inhibits the activity of the prostatic form, and therefore the test containing tartrate measures only the activity of the nonprostatic forms. The level of the prostatic enzyme is obtained by difference. A calibration curve is prepared using standard solutions of *p*-nitrophenol. Results are reported in "total" and "prostatic" (= tartrate-inhibited) Bessey-Lowry-Brock (BLB) units, or in (International) U/l. The BLB unit is defined as that quantity of enzyme activity consuming 1 mmol of substrate/h.

REAGENTS

1. Citrate buffer, 0.09 mol/l, pH 4.85 ± 0.05. To a solution of 18.91 g of citric acid monohydrate in 500 ml of water, add 180 ml of 1.0 molar NaOH and 100 ml of 0.01 molar HCl. Check the pH (at 37°C), adjust to 4.85, and dilute the solution to 1 liter. Add a few drops of chloroform as a preservative. The reagent is stable for about six months if refrigerated.

2. Tartrate-citrate buffer, pH 4.85, 0.09 mol citrate, and 0.04 mol tartrate/l. Dissolve 1.50 g of L(+)-tartaric acid in 250 ml of citrate buffer. The reagent is stable for six months, if refrigerated.

---

* The procedure is essentially the same as that described in Bulletin No. 104, Sigma Chemical Co., St. Louis, MO.

3. Stock substrate: *p*-nitrophenylphosphate, disodium salt (PNPP) (4.0 mg/ml or 15.2 μmol/ml). The concentration will vary slightly, depending on the water content of the salt. The solution is unstable; prepare only as much as needed.

4. Buffered substrates. Mix one part of reagent 1 with one part of stock substrate; in another container, mix one part of reagent 2 with one part of stock substrate. Adjust the pH of each mixture to 4.9. Pipet 1.0 ml aliquots of each into 15 × 125 mm test tubes, stopper the test tubes, and store frozen. Identify the citrate-substrate tubes and the tartrate-citrate tubes appropriately. These substrates are less stable than those having an alkaline pH; thus, the volumes of the buffered substrates prepared should not exceed one week's needs.

5. Sodium hydroxide, 0.10 mol/l. Prepared by dilution from any available stock standard NaOH.

6. Standard solutions of *p*-nitrophenol (PNP). Stock standard, 0.010 mol/l. Dissolve 0.3479 g of pure crystals in 250 ml of water. Specially purified crystals are available from several commercial sources. The standard solution is stable for six months if kept refrigerated and stored in the dark (see also p. 607). Dilute standard, 0.050 mmol/l. Make fresh as needed. Dilute the 0.010 mol/l stock two hundred-fold with water. Use within 4 to 6 h.

PROCEDURE

1. Use two citrate-substrate tubes and one tartrate-substrate tube for each unknown serum to be assayed. Mark one of the citrate tubes "CB" (for blank), and the other "CT" (total). Mark the tartrate tube "NP" (for nonprostatic). In addition, obtain one pair of each type of substrate tube (citrate and tartrate) to serve as substrate quality controls and instrument blanks. Bring *all* tubes and the sera to be tested to 37°C.

2. Add 200 μl of the first unknown serum to one citrate "CT" tube and to one tartrate "NP" tube. At timed intervals, repeat this step for the other unknown sera. Mix contents by swirling the tubes rapidly, and permit them to incubate for exactly 30 min. Incubate the "CB" tubes and the two pairs of substrate control tubes without adding serum.

3. At the end of the 30 min reaction period, add 4.0 ml of NaOH (0.10 mol/l) to all tubes to stop any enzyme reaction.

4. Remove all tubes from the water bath, and then add 200 μl of each serum to the respective citrate "CB" tubes, and 200 μl of water to the four substrate control tubes. Mix the contents of all tubes, and transfer the contents to cuvets.

5. Compare the absorbances of the two citrate substrate controls against each other at 410 nm, and select the cuvet with the lowest absorbance reading to serve as an instrument blank for the citrate tubes, "CT" and "CB." Do the same for the two tartrate controls. (If the paired control tubes differ in absorbance by more than 0.040, the quality of the substrate is suspect!)

6. Read the absorbances of the "CT" and "CB" tubes against the selected citrate instrument blank, and convert the readings to enzyme units, using the calibration curve. Record these as T_u and B_u. Similarly, select a tartrate instrument blank, and read the absorbance of the "NP" tubes. Using the curve, convert the absorbance readings to enzyme units and record these as N_u. Then $(T_u - B_u)$ = total acid phosphatase activity in BLB units, and $(T_u - N_u)$ = prostatic acid phosphatase activity in the same units. The "CB" tubes correct for any serum pigments present and absorbing at 410 nm.

CALIBRATION

Prepare working standards by diluting 1.0, 2.0, 4.0, 6.0, 8.0, and 10.0 ml of dilute standard solution with water to a volume of 10.0 ml. The instrument blank consists of 10.0 ml of water. Add exactly 1.1 ml of 0.20 molar NaOH to all tubes, mix, pour into cuvets, and read absorbance at 410 nm.

The 1.0 ml standard contains 0.05/(1000 × 11.1) mmol of PNP/ml, and the 5.2 ml final test volume contains $(5.2 \times 0.05)/(1000 \times 11.1) = 2.34 \times 10^{-5}$ mmol of PNP. If this is the quantity of product produced by 0.20 ml of enzyme in a 30 min reaction, then the quantity produced by 1 liter in 1 hour will be $2.34 \times 10^{-5} \times (60/30) \times (1000/0.2) =$ 0.234 mmol/h/l = 0.234 BLB units. The working standards, thus, represent 0.23, 0.47, 0.94, 1.40, 1.8, and 2.3 BLB units. Inasmuch as 1 BLB unit = 16.7 (International) U/l, the standards represent similar multiples of 3.9 U/l.

If the absorbance obtained is over 0.600, the assay should be repeated using either a shorter incubation time (15 or 10 min) or a smaller aliquot of serum (200 μl of a five-fold or ten-fold dilution of the serum with saline). The identical volume of the same diluted serum must be used in all three tubes (CT, NP, and CB). The correction factor is $(30/t) \times D$, where t is the (shorter) incubation time (in minutes) and D is the dilution factor.

NORMAL VALUES

Total acid phosphatase	Males:	0.15–0.65 BLB units (2.5–11.7 U/l)
	Females:	0.02–0.55 BLB units (0.3–9.2 U/l)
Prostatic (tartrate labile) acid phosphatase	Males:	0.01–0.20 BLB units (0.2–3.5 U/l)
	Females:	0.00–0.05 BLB units (0–0.8 U/l)

REFERENCE

Technical Bulletin No. 104, Revised, 1971, Sigma Chemical Co., St. Louis, MO.

DETERMINATION OF ACID PHOSPHATASE USING THYMOLPHTHALEIN MONOPHOSPHATE (ROY, BROWER, AND HAYDEN[171])

The substrate, thymolphthalein monophosphate, disodium salt (TMP) (Worthington Biochemical Corp., Freehold, NJ.), is dissolved in 0.10 molar citrate buffer, pH 6.0, to give a final concentration of 2.2 to 2.4 mmol/l. The selected pH appears to be the pH optimum for the prostatic enzyme using this substrate. A reaction temperature of 37°C is used by the authors to enable sufficient product formation to permit convenient absorbance measurement. The enzyme appears to be stable at that temperature and pH during the 30 min reaction period. (Obviously, a reaction temperature of 30°C may be used, if preferred.)

For the assay, 200 μl of serum are added to 1.0 ml of buffered substrate at 37°C, and the reaction is stopped with 5 ml of alkaline carbonate solution, which also converts the liberated thymolphthalein into its blue form. The color is measured at 590 nm. TMP is hydrolyzed by red cell acid phosphatase at a very slow rate, so that the reaction measures essentially only the prostatic enzyme (plus some platelet enzyme). Thus, indirect measurement of the prostatic enzyme by its inhibition with tartrate is not required and is not used.

The range of values found by this method for enzyme activity in sera from healthy men of all ages is 0.11 to 0.60 U/l. This range is considerably lower than that found with procedures for total acid phosphatase using PNPP as substrate, which give 2 to 11 U/l as the upper limit of normal. There is poor correlation between values obtained by the Roy procedure and those found for total acid phosphatase by the BLB procedure. However, correlation is reasonably good, if the values by the Roy procedure are compared with values obtained by the BLB procedure for the prostatic (tartrate-inhibited) enzyme.

CONTINUOUS MONITORING TYPE TECHNIQUES FOR ACID PHOSPHATASE ACTIVITY MEASUREMENTS

Although the number of acid phosphatase determinations performed daily by clinical chemistry laboratories is never so great as to make manual methods impractical, there has always been interest in the development of a continuous sampling procedure for ACP. One such method was devised by Hillman,[101] and is available in kit form (Eskalab, Acid Phosphatase) from Smith Kline Instrument Co., Inc., Philadelphia, PA. 19101.[36,67] The substrate consists of a solution containing α-naphthylphosphate, Fast Red TR salt (diazotized 2-amino-5-chlorotoluene), citric acid, and sodium citrate at concentrations of approximately 3.0 mmol, 30 mmol, and 50 mmol/l, respectively, at a pH of 5.0. On addition of 200 μl of serum to 3.0 ml of the mixture, the enzyme liberates α-naphthol, which is converted *in situ* to a diazo pigment by the diazonium salt present. After a 5 min period, the reaction kinetics become linear; the absorbance is then read at 405 nm for 5 min, and the ΔA/min is calculated. In a parallel run, tartrate (final concentration = 17 mmol/l) is added to the substrate; thus, values for the activities of both the total and prostatic enzymes can be obtained.

The normal range for total acid phosphatase activity at 30°C is 0 to 9 U/l; the value for the prostatic enzyme is 0 to 3 U/l.

5′-NUCLEOTIDASE

(EC 3.1.3.5; 5′-ribonucleotide phosphohydrolase)

5′-Nucleotidase (NTP; 5′NT)* is a phosphatase which acts only on nucleoside 5′-phosphates,[73] such as adenosine 5′-phosphate (*a*denosine 5′-*mono*phosphate, AMP; adenylic acid), releasing inorganic phosphate. The enzyme is widely distributed throughout

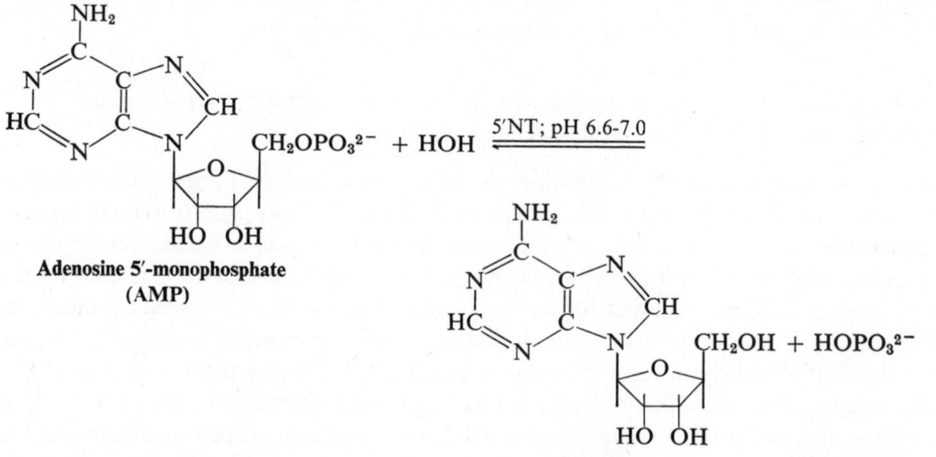

$$\text{Adenosine–}CH_2OPO_3^{2-} + HOH \xrightarrow{\text{5'NT; pH 6.6-7.0}} \text{Adenosine–}CH_2OH + HOPO_3^{2-} \quad (24)$$

Adenosine 5′-monophosphate (AMP) → **Adenosine** + **Phosphate**

the tissues of the body and appears to be localized in the cytoplasmic membrane of the cells in which it occurs. Its pH optimum is between 6.6 and 7.0.

CLINICAL SIGNIFICANCE

The activity of 5′-nucleotidase in serum is increased two- to six-fold in those *hepatobiliary diseases* in which there is interference with the secretion of the bile.[10,65,100] This may be due to extrahepatic causes (a stone or tumor occluding the bile duct), or it may arise from intrahepatic conditions such as *cholestasis* caused by chlorpromazine, *malignant infiltration* of the liver, or *biliary cirrhosis*. When parenchymal cell damage is predominant, as in early infectious hepatitis, serum 5′-nucleotidase is normal or only moderately elevated. Thus, both 5′-nucleotidase and alkaline phosphatase[10,13] behave similarly in hepatobiliary disease, except that elevations of 5′-nucleotidase are somewhat more pronounced and persist longer than those of alkaline phosphatase, particularly in cases of chronic liver disease.

In cases of *skeletal disease*, increases in the concentration of 5′-nucleotidase are rarely observed, and when they do occur, they are marginal. Thus, assays of 5′-nucleotidase are

* NTP is the unofficial "standard" abbreviation.[9] 5′NT is a more literal abbreviation form occasionally seen in the literature; we will use the former.

particularly valuable in differentiating between elevations in alkaline phosphatase caused by hepatobiliary diseases or obstruction, and those caused by disease involving the skeletal system. In fact, any rise in NTP activity that is more than marginal is virtually specific for hepatobiliary disease.[10]

Normal levels of 5′-nucleotidase are seen in all forms of *pancreatic disease*, provided the common duct is not involved and hepatic metastases are not present. No increase or, at best, only minimal increases are seen in Hodgkin's disease, lupus erythematosus, a large variety of infectious diseases, and diseases involving organs other than the liver. 5′-Nucleotidase is also normal in late pregnancy and during childhood, conditions in which alkaline phosphatase is generally elevated.

METHODS FOR THE DETERMINATION OF NTP ACTIVITY

The substrate most generally used in measuring the activity of 5′-nucleotidase is adenylate (adenosine 5′-monophosphate, AMP). However, this substrate is an organic phosphate ester and thus can also be hydrolyzed to an appreciable degree by other nonspecific (alkaline) phosphatases, especially at pH 7.5, the pH previously assumed to be optimal for NTP activity. Methods for the estimation of 5′-nucleotidase in serum, such as that of Dixon and Purdom, must therefore incorporate some means for correcting for the hydrolysis of the substrate by the nonspecific phosphatases.

5′-Nucleotidase differs from alkaline phosphatase in being inhibited by nickelous ions,[25] and this property is used to distinguish the two enzymes in the method of Campbell[37,176] described here. However, some isoenzymes of alkaline phosphatase may also be inhibited by Ni^{2+}. In another approach used by Belfield and Goldberg,[14] and by Persijn *et al.*,[146,209] a large excess of some phosphate ester, such as β-glycerophosphate or phenylphosphate, is added to the reaction mixture. These materials are substrates for alkaline phosphatase but not for 5′-nucleotidase, and by forming substrate complexes with the former enzyme, they reduce the proportion of the total alkaline phosphatase activity which is directed to the hydrolysis of the NTP substrate, adenosine 5′-monophosphate. This approach cannot be used when the measurement of the reaction rate is based on the determination of inorganic phosphate, since this is a product of the reaction catalyzed by both enzymes. However, methods have been described in which adenosine (the second product of the 5′-nucleotidase reaction) is determined spectrophotometrically by measuring the amount of ammonia formed by the action of added adenosine deaminase, as described by Persijn, van der Slik, and Bon[146] and by Belfield and Goldberg.[12]

Automated methods for the estimation of 5′-nucleotidase are based on nickel inhibition, or on the addition of an excess of a nonspecific substrate, as outlined above.

DETERMINATION OF 5′-NUCLEOTIDASE ACTIVITY IN SERUM (CAMPBELL METHOD)

PRINCIPLE

Serum is incubated with adenosine 5′-monophosphate at pH 7.5 and 37°C, with and without added nickel ions. After 30 min, the amount of inorganic phosphate liberated is determined. Phosphate produced in the absence of nickel represents the combined activities of alkaline phosphatase and 5′-nucleotidase, whereas that produced in the presence of nickel is due to the activity of ALP alone. Thus, the difference between these two values for liberated phosphate corresponds to the activity of 5′-nucleotidase in the serum sample. Manganese ions serve as an activator of 5′-nucleotidase. The presence of copper in the acetate buffer accelerates color development.

REAGENTS

1. Barbiturate buffer, 0.04 mol/l, pH 7.5.* Dissolve 8.25 g of sodium diethylbarbiturate (sodium barbital) in 140 ml of 0.2 molar HCl and dilute to 1 liter with water. Adjust to pH 7.5 at 37°C if necessary.

2. Substrate, 10 mmol/l. Dissolve 0.365 g of adenosine 5′-phosphoric acid monohydrate in 18 ml of 0.1 molar NaOH and dilute to 100 ml with water. Alternatively, use 0.391 g of disodium adenosine 5′-phosphate and dilute to 100 ml with water.

3. Manganese ion solution, 20 mmol/l. Dissolve 0.338 g of $MnSO_4 \cdot H_2O$ or 0.396 g of $MnCl_2 \cdot 4H_2O$ in 100 ml water.

4. Nickel chloride solution, 0.1 mol/l. Dissolve 2.4 g of $NiCl_2 \cdot 6H_2O$ in 100 ml water.

5. Trichloracetic acid, 100 g/l.

6. Acetate buffer, 2.4 mol/l, pH 4.0. Dissolve 2.5 g of $CuSO_4 \cdot 5H_2O$ and 46 g of sodium acetate trihydrate in 1 liter of 2 molar acetic acid (115 ml glacial acetic acid diluted to 1 liter). Adjust to pH 4.0 if necessary.

7. Ammonium molybdate solution, 5 g/100 ml. Dissolve 5 g of $(NH_4)_6Mo_7O_{24} \cdot 4H_2O$ in 100 ml water.

8. Reducing agent. Dissolve 2 g of *p*-methylaminophenol monohydrogen sulfate ("Rhodol," "Elon," or "Metol") in 80 ml of water. Add 5 g of Na_2SO_3, anhydrous, dilute to 100 ml with water and filter. Store in a dark bottle at 4°C.

9. Stock phosphate standard, 1.00 mg P/ml. Dissolve 2.193 g of dry KH_2PO_4, anhydrous, AR, in 500 ml water. Add 0.20 ml of concentrated sulfuric acid as preservative before diluting to the mark. Store at 4°C.

10. Working phosphate standard, 10 μg P/ml. Measure 1.00 ml of stock standard into a 100 ml graduated flask, and dilute to 100 ml with trichloracetic acid (100 g/l). Stable for 4 weeks if stored at 4°C.

PROCEDURE

1. The reaction is carried out in 12 ml conical centrifuge tubes (borosilicate glass). Two tubes are needed per specimen, labeled T (test) and C (control). To the T tube add 1.5 ml of buffer and 100 μl of $MnSO_4$ solution; to the C tube add 1.3 ml buffer, 100 μl $MnSO_4$, and 200 μl of $NiCl_2$ solution. Mix tube contents well.

2. Add 200 μl of serum to both reaction tubes, mix, and place them into a water bath at 37 ± 0.2°C. Allow 5 min for the tubes and contents to attain 37°C, and then add 200 μl of substrate solution to each tube and mix without removing the tubes from the water bath.

3. After exactly 30 min, stop the enzymatic reaction by the addition of 2.0 ml of trichloracetic acid (100 g/l) to each tube. Mix the tube contents thoroughly, remove tubes from the water bath, and centrifuge them at about 3000 rpm for 15 min.

4. Pipet 2.0 ml of the clear supernatant from the test (T) and control (C) mixtures into clean borosilicate tubes. All tubes should be washed with dilute HCl and thoroughly rinsed with distilled water, to insure freedom from detergent contamination; similarly, pipets should be rinsed with reagent before delivering same.

5. Prepare a standard (S) by pipetting 1.0 ml of working standard and 1.0 ml of water into an acid-washed test tube. Similarly, prepare a reagent blank (B) by mixing 1.0 ml water and 1.0 ml of the trichloracetic acid.

6. To all tubes (T, C, S, B) add 3.0 ml of acetate buffer, followed by 0.50 ml of ammonium molybdate solution and 0.50 ml of Elon solution. Mix well after each addition.

7. After 5 min color development, transfer the tube contents to cuvets and measure the

---

* Newer work[99] establishes the optimal pH to be in the range from 6.6 to 7.0. As alternatives to barbital ($pK_a' = 7.43$), PIPES [piperazine-*N*,*N*′-bis(2-ethanesulfonic acid)], with $pK_a' = 6.80$, and HEPES (*N*-2-hydroxyethylpiperazine-*N*-2-ethanesulfonic acid), with a $pK_a' = 7.55$, have been proposed as useful buffer materials. The last two are examples of Good[86] buffers.

absorbance of all solutions against the blank in a spectrophotometer at some wavelength between 680 and 880 nm. (The higher the wavelength, if available on the instrument, the higher the absorbance for any given quantity of phosphorus produced.) The color is stable for at least 30 min.

CALCULATION

The standard tube contains 10 μg of phosphate (as P). Therefore, the amount of phosphate-phosphorus produced by the action of the 5′-nucleotidase is given by

$$\mu\text{g P} = \frac{A_T - A_C}{A_S - A_B} \times 10; \qquad \text{or} \qquad \mu\text{mol P} = \frac{A_T - A_C}{A_S - A_B} \times \frac{10}{31}$$

This amount of PO_4-P is released in 30 min by the enzyme present in 0.10 ml of serum (since half of the total reaction mixture is taken for estimation of phosphate). Therefore, serum 5′-nucleotidase activity in μmoles of substrate hydrolyzed per min per liter of serum (U/l) is given by

$$\text{U/l} = \frac{A_T - A_C}{A_S - A_B} \times \frac{10}{31} \times \frac{1}{30} \times \frac{1000}{0.1}, \qquad \text{or} \qquad \text{U/l} = \frac{A_T - A_C}{A_S - A_B} \times 108$$

If the activity is greater than 150 U/l, repeat the assay using a shorter incubation time and make corresponding adjustments in the calculations.

NORMAL VALUES

The normal range for 5′-nucleotidase activity in sera from normal men and normal nonpregnant and pregnant women is from 2 to 17 U/l. Lower values have been reported in children.

REFERENCE

Campbell, D. M.: Biochem. J., *82*:34P, 1962.

GAMMA GLUTAMYLTRANSFERASE

(EC 2.3.2.2; γ-glutamyl-peptide:amino acid γ-glutamyltransferase)

Peptidases are enzymes which catalyze the hydrolytic cleavage of peptides to form amino acids or smaller-sized peptides or both. They constitute a broad group of enzymes of varied specificity, and some individual enzymes catalyze the transfer of amino acids from one peptide to another amino acid or peptide, i.e., they act as amino acid transferases. Included among the latter is γ-glutamyltransferase (GGT),[167] which transfers the γ-glutamyl group from peptides and compounds that contain it to some acceptor. This enzyme was originally termed a "transpeptidase," but authorities[63] suggest that this name be replaced by the more appropriate term, "transferase." The γ-glutamyl acceptor can be the substrate itself, some amino acid or peptide, or even water, in which case a simple hydrolysis takes place. The enzyme acts only on peptides or peptide-like compounds containing a terminal glutamate residue joined to the remainder of the compound through the terminal (-γ-) carboxyl (see equation 25). Beyond the fact that glycylglycine is five times more effective as an acceptor than is either glycine or the tripeptide (gly-gly-gly), little is known about the optimal properties of the acceptor cosubstrate. The rate of the peptidase transfer reaction is considerably faster than that of the simple hydrolysis reaction. An example of a reaction catalyzed by the enzyme is presented in equation 25. This is the reaction used in measuring enzyme activity.

$$\underset{\substack{\gamma\text{-Glutamyl-}\\ p\text{-nitroanilide}\\ \textit{Substrate (donor)}}}{\mathrm{HOOC{-}{}_{\alpha}CH(NH_2){-}{}_{\beta}CH_2{-}{}_{\gamma}CH_2{-}C({=}O){-}NH{-}C_6H_4{-}NO_2}} + \underset{\substack{\text{Glycylglycine}\\ \textit{Acceptor}}}{\mathrm{H_2N{-}CH_2{-}C({=}O){-}NH{-}CH_2{-}COOH}} \underset{\text{GGT}}{\overset{\text{pH 8.2}}{\rightleftharpoons}} \underset{\substack{\gamma\text{-Glutamyl-}\\ \text{glycylglycine}\\ \textit{Transfer product}}}{\mathrm{HOOC{-}CH(NH_2){-}CH_2{-}CH_2{-}C({=}O){-}NH{-}CH_2{-}C({=}O){-}NH{-}CH_2{-}COOH}} + \underset{\substack{p\text{-Nitro-}\\ \text{aniline}\\ \textit{Donor residue}}}{\mathrm{H_2N{-}C_6H_4{-}NO_2}} \qquad (25)$$

Our basic knowledge of the biochemistry of the transpeptidases is still fragmentary. The enzyme was first identified in kidney tissue, where its concentration is quite high, but was later shown to be present in serum and in all cells, except those in muscle. Some enzyme is present in the cytosol, but the larger fraction is located in the cell membrane, and may act to transport amino acids and peptides into the cell across the cell membrane in the form of γ-glutamyl peptides.[167] It may also be involved in some aspects of glutathione metabolism.

CLINICAL SIGNIFICANCE

Even though renal tissue has the highest level of the enzyme, the enzyme present in serum appears to originate primarily from the hepatobiliary system, and as would be expected, GGT activity is elevated in any and all forms of *liver disease*.[114,131] It is highest in cases of intra- or posthepatic *biliary obstruction*, reaching levels some 5 to 30 times normal. Along with 5′-nucleotidase and leucine aminopeptidase, it is one of the group of cholestasis-indicating enzymes. It is much more sensitive than alkaline phosphatase, the transaminases, and leucine aminopeptidase in detecting obstructive jaundice, cholangitis, and cholecystitis; the rise occurs earlier than with these other enzymes, and persists longer.[10] Only moderate elevations (2 to 5 times normal) are seen in infectious hepatitis, and in this condition GGT determinations are less useful diagnostically than are measurements of the transaminases. High elevations of GGT are also seen in patients with either primary or secondary (metastatic) *neoplasms*; again changes occur earlier and are more pronounced than those with the other liver enzymes. Small increases (2 to 5 times normal) of GGT activity are observed in patients with fatty livers, and similar, but transient, increases are seen in cases of drug intoxication. In acute and chronic *pancreatitis*, and in some pancreatic malignancies (especially if associated with hepatobiliary obstruction), enzyme activity may be 5 to 15 times the upper limit of normal.

Normal levels of the enzyme are seen in cases of *skeletal disease* (Paget's disease, bone neoplasms), in children above 1 year of age, and in healthy, pregnant women—conditions in which alkaline phosphatase (ALP) is elevated.[131] Thus, measurement of GGT levels in serum can be used to ascertain whether observed elevations of ALP are due to skeletal disease or reflect the presence of hepatobiliary disease. 5′-Nucleotidase (NTP) determinations can provide the same information, but GGT levels have the advantage that they become elevated earlier in liver disease and rise to greater levels than do the former. However, elevations of 5′-nucleotidase have been more useful clinically, because they are seen only in hepatobiliary disease.

Normal levels of GGT are encountered in various *muscle diseases* and in *renal failure*, but mild elevations may be seen in untreated lipoid nephrosis. In *myocardial infarctions*, GGT is usually normal; but if there is a rise, it occurs at about the fourth day and reaches a maximum value in another four days, and probably implies liver damage secondary to cardiac insufficiency.

Not only are elevated levels of GGT seen in the sera of patients with *alcoholic cirrhosis*, but they are also seen in the majority of sera from persons who are heavy drinkers. Rosalki[169] stresses the value of serum GGT levels in detecting alcohol-induced liver disease. In patients receiving *drugs* such as Dilantin and phenobarbital, raised levels of the enzyme are found in serum specimens, but not in specimens of spinal fluid. Such an increase of activity may reflect induction of new enzyme activity by the action of the anticonvulsant drugs. The release of GGT into serum reflects the toxic effects of alcohol and other drugs on microsomal structures in liver cells. The enzyme level found correlates well with the duration of the drug action. Hepatic complications occurring in mucoviscidosis (cystic fibrosis) also lead to elevations of GGT.

High levels of GGT are present in the prostate,[131] and this may account for the fact that the activity of GGT in sera of males is about 50 per cent higher than that seen in sera from females. *Prostatic malignancy* may, at times, be the source of the elevated GGT activity. It has been suggested that the enzyme may be of as much interest in forensic medicine as is prostatic acid phosphatase (p. 614). The irradiation of tumors in cancer patients may be accompanied by a rise in GGT activity, although lactate dehydrogenase activity in the course of such treatment remains unchanged.

The enzyme which is found in *urine* probably originates in the kidney. Elevated enzyme activity is found in the urine of patients with acute urorenal infections and with diseases involving renal tissue destruction. However, in chronic renal disease and in older individuals, urine enzyme levels may be depressed.

METHODS

Current procedures for measuring GGT activity use γ-glutamyl-p-nitroanilide (GGPNA, GLUPA) (equation 25) as the substrate, with glycylglycine serving as the γ-glutamyl residue acceptor. Both two-point and continuous monitoring methods have been described. These are based on the method developed by Szasz,[193] which itself was based on earlier work by Orlowski and by Dimov and Kulhanek. Other substrates that have been investigated include the γ-glutamyl derivatives of aminopropionitrile, α-naphthylamine, and aniline, but GGPNA has been found to be most convenient (it is most sensitive, the p-nitroaniline formed can be directly measured, it has reasonable stability, and so forth). Szewczuk and Orlowski noted the activating effect of glycylglycine, as a corollary to its being a γ-glutamyl acceptor, but very little work has been done in searching for and evaluating other possible peptide or nonpeptide acceptors or activators.

The substrate, GGPNA, unfortunately does present several negative features as well. Its solubility in water at pH 8.0 is limited; to approach concentrations in the reaction mixture of some ten or twenty times the value of the K_m requires that the concentration in the substrate reagent preparation be in the supersaturation range (Szasz), even for reactions carried out at 25°C. To obtain a substrate reagent concentration high enough so that the addition of 50 to 100 μl of the preparation could be used to initiate the transferase reaction, Rosalki and Tarlow[170] dissolved the GGPNA in HCl and used a stronger buffer to overcome the added acid. The substrate is somewhat unstable, and does undergo nonenzymatic hydrolysis, even at pH 8. Moreover, it is even more subject to rapid decomposition at 30 and 37°C and in the presence of HCl. An appropriate blank must be used to correct for this hydrolysis.

The pH optimum appears to vary with the nature of the buffer used. In the Szasz procedure a pH of 8.2 is employed; Rosalki and Tarlow used a pH of 8.0.

DETERMINATION OF γ-GLUTAMYL TRANSFERASE (GGT) (METHOD OF ROSALKI AND TARLOW[170])

The substrate employed in this method is γ-glutamyl-p-nitroanilide (GGPNA), with glycylglycine serving as acceptor. Buffering is provided by tris(hydroxymethyl)aminomethane

(Tris) and by glycylglycine ($pK_2 = 8.25$). Serum is added to the buffer-acceptor solution, and the reaction is initiated by addition of substrate in HCl solution. The increase in absorbance at 405 nm due to the *p*-nitroaniline formed in the reaction is measured spectrophotometrically. The temperature used is 30°C, although Rosalki and Tarlow used 37°C.

REAGENTS

1. Buffer-glycylglycine reagent, pH 8.5, 115 mmol Tris/l and 138 mmol glycylglycine/l. Dissolve 6.96 g of Tris (base) and 9.11 g of glycylglycine in about 400 ml of water. Adjust the pH to 8.5 at 30°C by the addition of 0.5 molar NaOH, and then dilute to 500 ml. The reagent is stable for at least three months if refrigerated.

2. GGPNA substrate, 104 mmol/l in 0.5 molar HCl. Prepare a volume sufficient for a day's work, by dissolving 29.5 mg of L-γ-glutamyl-*p*-nitroanilide, monohydrate (or 27.6 mg of nonhydrated GGPNA) per ml of 0.50 molar HCl. When prepared, use as soon as possible (within 2 h). Do not store; discard excess reagent.

3. Adjustment of buffer-glycylglycine concentration. When 0.10 ml of the substrate solution is added to exactly 1.00 ml of buffer-reagent, the pH should be exactly 8.0 at 30°C. If not, adjust the pH of the buffer to pH 8.0 with 0.5 molar HCl or 0.5 molar NaOH.

PROCEDURE

1. Pipet 2.0 ml of buffer-glycylglycine and 100 μl of specimen into a 1.0 cm cuvet. Prewarm the cuvet to 30°C. Each run should include a substrate blank, in which 100 μl of water replaces the serum. This is used to correct for nonenzymatic hydrolysis of the substrate.

2. After temperature equilibration has been achieved, initiate the enzyme reaction by adding 0.20 ml of substrate. Mix cuvet contents rapidly and then monitor the change in absorbance at 405 nm continuously, or at 1.0 min or 0.5 min intervals, for a period of several minutes. From these absorbance values calculate the ΔA/min, and the activity in the serum specimen in U/l.

The composition of the final reaction mixture is as follows: Tris buffer, 100 mmol/l; glycylglycine, 120 mmol/l; GGPNA, 9.0 mmol/l; pH 8.0; serum dilution = 1/23.

CALCULATIONS

The millimolar absorptivity for *p*-nitroaniline is $9.87 \times \text{liter} \times \text{mmol}^{-1} \times \text{cm}^{-1}$. Therefore, enzyme activity is given by

$$\text{U/l} = \frac{\Delta A}{\text{min}} \times \frac{10^3}{9.9} \times \frac{2.30}{0.10} = \frac{\Delta A}{\text{min}} \times 2330$$

where $\dfrac{\Delta A}{\text{min}}$ = observed rate of reaction

2.3 = total reaction volume in ml
0.10 = specimen volume in ml
9.9 = millimolar absorptivity (in cm^2/mmol)
10^3 = μmol in 1.0 mmol

COMMENTS

1. The relative enzyme activities at selected temperatures are as follows: 25°C: 0.77; 30°C: 1.00; 32°C: 1.10; 35°C: 1.25; and 37°C: 1.35.

2. 2-Methyl-2-aminopropane-1,3-diol, diethanolamine, and triethanolamine can also be used as buffers in place of Tris.

3. Bondar and Moss[26] have reported that glutamate acts as an activator, doubling activity when the enzyme concentration is in the normal range, and increasing it some 20 per cent when enzyme levels are elevated. Others have found much less or no activation as a result of added free glutamate ion.[162]

4. Commercial "kit" procedures are available through Bio-Dynamics/bmc, Indianapolis, IN. (Catalog #15794, #16312) and Worthington Biochemical Corp., Freehold, NJ. (as "Statzyme, γ-GTP"). Both use a 30°C incubation temperature, and the composition of the reaction mixture is the same in both procedures: Tris buffer, 185 mmol/l; glycylglycine, 40 mmol/l; and GGPNA, 4.0 mmol/l. The Worthington mixture also includes glutamate at 1 mmol/l. The ratios of serum to buffered substrate are 1/16 and 1/30 in the Bio-Dynamics/bmc and Worthington procedures, respectively; Szasz used a ratio of 1/11.

5. Glutathione and bromsulfophthalein are inhibitors of the enzyme, but the quantity derived from the specimen is too small to exert any measurable effect.

6. To minimize consumption of the expensive substrate (GGPNA), semi-micro cuvets (10 mm light path, 4.0 mm cuvet width) requiring only 1.0 ml of buffered substrate may be used.

NORMAL VALUES

Insofar as methodology is still in the developmental stage, no well-established values for the normal range of serum GGT are as yet available. Some values are available for all three temperatures suggested (25, 30, and 37°C). GGT values in the sera of adult males are some 50 per cent higher than those in sera of women or of children of both sexes.

Authors	*Reaction temperature*	*Activity in U/l*	
		Males	*Females*
Szasz 1969	25°	5–21	3–14
1975	25°	5–30	3–20
Bio-Dynamics/bmc	30°	8–37	5–24
Worthington	30°	12–38	9–31
Lum-Gambino	30°	3–61	3–56
Ewen-Griffiths	30°	5–40	5–40
Rosalki-Tarlow	37°	50 upper limit	30 upper limit

REFERENCES

Rosalki, S. B., and Tarlow, D.: Clin. Chem., *20*:1121, 1974.
Szasz, G.: Clin. Chem., *15*:124, 1969.

AMYLASE

(EC 3.2.1.1; α-1,4-glucan 4-glucanohydrolase)

Amylases are a group of hydrolases which split complex carbohydrates such as starch and glycogen. These polysaccharides are constituted of α-D-glucose units linked through carbon atoms 1 and 4 located on adjacent glucose residues. Both straight chain (linear) polyglucans (such as amylose) and branched polyglucans (such as amylopectin and glycogen) are hydrolyzed, but at different rates. The enzyme splits the chains at alternate α-1,4 hemiacetal (—C—O—C—) links, forming maltose and some residual glucose in the case of amylose, and both these sugars plus a residue of limit dextrins in the case of the branched chain polyglucans. The α-1,6 linkages at the branch points in the latter (see Fig. 6–7) are not attacked by the enzyme.

Two types of amylases are recognized; the *beta amylases* (e.g., plant and bacterial exoamylases) can act only at the terminal reducing end of a polyglucan chain, splitting off a segment of two glucose units (maltose) at a time. Animal amylases, including those present in human tissues, are *alpha amylases*. They are also referred to as endoamylases, because they can attack α-1,4 linkages in a random manner anywhere along the polyglucan chain. Large polysaccharide molecules are thus rapidly broken down into smaller units—dextrins, maltose, and some glucose units. Since both maltose and glucose are reducing sugars, the course of the hydrolytic reaction is paralleled by an increase in soluble reducing materials.

Linear starch chains (in helical form) react with molecular iodine to form the well-known, deep blue starch-iodine complex. The endo-amylolytic hydrolysis of starch to dextrins and oligosaccharides is thus also paralleled by a gradual loss of ability to bind iodine, and the hue of the glucan-iodine complex changes to a light blue, then to violet, and finally to red. No iodine color is formed when the chain size is six glucose units or less. As dextrinization proceeds, the turbidity and viscosity characteristics of starch sols fall off very rapidly, much faster than the increase in reducing groups. The turbidimetric and nephelometric procedures for measuring amylase are based on this decrease in turbidity as a result of the dextrinization (amyloclastic) reaction.

Amylase (AMS) in human serum has a moderately sharp pH optimum at 6.9 to 7.0. The enzyme is customarily assayed at 37 or 40°C, although it is active at 50°C, and some automated procedures employ this higher temperature. The temperature coefficient is +6 per cent/degree increase in temperature ($Q_{10} = 1.6$). α-Amylases are calcium metalloenzymes,[127] with the calcium absolutely required for functional integrity; saturation with Ca is obtained at about 1.0 mmol/1. However, full activity is displayed only in the presence of a variety of inorganic anions, such as chloride, bromide, nitrate, chlorate, or HPO_4^{2-}. The first two are the most effective activators; optimal activity is obtained at a Cl^- level of 10 mmol/l. The amylases normally occurring in human plasma are small molecules, with molecular masses varying from 40,000 to 50,000 daltons. The enzyme is thus small enough to pass through the glomeruli, and amylase is the only plasma enzyme normally found in the urine. Serum and urine amylases migrate electrophoretically with the β- and γ-globulins.

The enzyme is quite stable; activity loss is negligible at room temperature in the course of a week, or at refrigerator temperatures over a two month period. With the exception of heparin, all common anticoagulants inhibit amylase activity; citrate, EDTA, and oxalate inhibit it by as much as 15 per cent. As a consequence, amylase assays should be performed only on serum or heparinized plasma.

In the body, amylase is present in a number of organs and tissues. The greatest concentration is present in the pancreas, where the enzyme is synthesized by the acinar cells and then secreted into the intestinal tract for digestion of starches. The salivary glands secrete a potent amylase to initiate hydrolysis of starches while the food is still in the mouth and esophagus. The action of the salivary enzyme, once referred to as *ptyalin*, is terminated by the acid in the stomach. In the intestinal tract, effective action of pancreatic and intestinal amylase is favored by the mildly alkaline conditions in the duodenum. Intestinal maltase then further hydrolyzes maltose into glucose. Most of the pancreatic amylase is destroyed by trypsin activity in the lower portions of the intestinal tract, although some amylase activity is present in stool. Weak amylase activity is found in extracts from liver, fallopian tubules, striated muscle, and adipose tissue. The enzyme is also found in colostrum and in milk. The enzyme present in normal serum and urine is predominantly of pancreatic and salivary gland origin. Amylase activity is not markedly diminished after pancreatectomy. Thus, the remaining activity is derived from the salivary glands, and perhaps from some as yet unidentified source. Little or none is of liver origin, as once thought.[17] The enzyme found in urine is derived from the plasma.

The occurrence of at least seven amylase isoenzymes in human tissue can be demonstrated by gel filtration and electrophoretic techniques.[102] Multiple forms are found in salivary and pancreatic extracts, but usually only two forms, already referred to, are found in serum and three forms in urine. *Macroamylases*[17,72,75] are present in the sera of a few persons. These rare forms are probably complexes between ordinary amylase and IgA or other normal or abnormal high molecular mass plasma proteins. Because of their large size (relative molecular masses over 200,000 daltons), these macroamylases cannot be filtered through the glomeruli, and are thus retained in the plasma, where their presence may increase amylase activity to some six- to eight-fold over that observed in normal persons. In contrast, urine amylase activity is lower than normal, less being cleared by the kidneys. No clinical symptoms are associated with macroamylasemia.

CLINICAL SIGNIFICANCE

Assays of amylase activity in serum are of interest clinically in relation to diseases of the pancreas and the evaluation of pancreatic function.[177] In acute pancreatitis one sees a transient rise of amylase in serum from a normal range between 60 and 180 Somogyi units to values of about 550 units and sometimes to as high as 2000 to 3000 units. This takes place over a period of 8 to 72 h, with a peak at 24 to 30 h. The accompanying hyperamylasuria may persist for a number of days, even though the serum amylase will have reached normal values. AMS in clinical diagnosis is discussed further in Chapter 20.

ASSAY METHODS

Development of precise assay methods[180] for measuring amylolytic activity is confounded by the difficulties encountered in preparing a substrate of known and reproducible composition and concentration. Starches vary considerably in their proportion of amylose and amylopectin, and the average chain length of the starch molecule depends on the methods by which it is manufactured and by which the substrate solution is prepared. Starch does not disperse in water to form a true molecular solution, but forms instead a colloidal sol containing hydrated starch micelles of various sizes. The degree of dispersion varies with temperature; at lower temperatures amylose chains aggregate (retrograde) into large micelles. It is essential that starch substrates be prepared and treated in exactly the same manner if data from different assays are to be comparable.

Potato, corn, and Lintner's "soluble" starch are most commonly used, although pure amylose, amylopectin, and glycogen are preferred by some analysts. Starch sols deteriorate rather rapidly as the result of mold contamination; benzoic acid, sodium azide, or *p*-hydroxypropylbenzoate may be added as preservatives. Sterile (autoclaved) sols keep well for several months.

In "*saccharogenic*" assays[95,188] the course of the enzyme reaction is followed by measuring the quantity of reducing materials (sugars, dextrins) formed. Any of the common methods for measuring reducing substances (Folin-Wu or Somogyi-Nelson) may be used. Methods based on the reduction of picrate, ferricyanide, and 3,5-dinitrosalicylic acid[179] and on the use of the anthrone-sugar reaction have also been advocated. The results obtained with these methods are reported in terms of milligrams of "apparent glucose" formed, although the chief reducing sugar present is maltose, which, mass for mass, has about 40 per cent of the reducing capacity of glucose. Quantities of reducing sugars produced are determined in protein-free filtrates of the reaction mixtures. With some starch preparations, it is difficult to obtain clear, nonopalescent filtrates. This does not affect the assay, provided that the turbidity is due to starch and not due to incompletely precipitated proteins. (Turbidity due to starch clears up at a later stage of the procedure.)

Recently, attempts have been made to design "multi-point" or "continuous monitoring type" assays. In one such approach maltase (α-1,4-glucosidase, EC 3.2.1.20) and glucose oxidase are added to the starch-amylase system to form a coupled enzyme system, in which soluble starch is first hydrolyzed by the amylase in the sample to maltose (plus some glucose) (reaction 26), the maltose being then further hydrolyzed by α-glucosidase to glucose (reaction 27). The glucose is then oxidized in the presence of glucose oxidase to δ-gluconolactone and H_2O_2 (reaction 28).

$$\text{Starch} \xrightarrow{\textit{amylase}} \text{Maltose} + \text{Glucose} \tag{26}$$

$$\text{Maltose} \xrightarrow{\alpha\textit{-glucosidase}} 2\ \text{Glucose} \tag{27}$$

$$\text{Glucose} + O_2 \xrightarrow{\textit{glucose oxidase}} \delta\text{-Gluconolactone} + H_2O_2 \tag{28}$$

In the overall reaction, one mole of oxygen is consumed per mole of glucose oxidized, resulting in a decrease in pO_2 which is monitored during the reaction with a pO_2 electrode and recorded on a strip chart recorder.[202] Sodium azide is added to prevent decomposition of H_2O_2 by catalase and regeneration of O_2. Amylase activity is calculated as a function of the decrease in pO_2.

In an alternate approach, peroxidase can be added to the system described by equations 26, 27, and 28, to use the H_2O_2 formed to oxidize some colorless redox compound (oxygen acceptor), such as ABTS,* to its colored form, which can be measured colorimetrically.[205]

$$H_2O_2 + \underset{\textit{colorless}}{A{-}H_2} \xrightarrow{\textit{peroxidase}} 2\,H_2O + \underset{\textit{colored}}{A} \tag{29}$$

The hexokinase/glucose-6-phosphate dehydrogenase couple may be used with maltase in the same way, although the endogenous glucose in the serum gives rise to a high blank.

Several automated forms of saccharogenic procedures have been described. In one method[143] glucose oxidase and catalase are used to destroy endogenous glucose; after starch is added, the reducing sugars are determined using the cupric-neocuproine reaction (see Chapter 6).

A number of *dye-labeled* amylase substrate materials have also been investigated, and these have been introduced into use by several commercial concerns. Most of these are synthesized by linking amylose or amylopectin via ether or ester (covalent) bonds to a variety of reactive dyes. The dyes are derivatives of triazine, in which R_1 and R_2 are aryl chromophores, linked to the triazine ring through their amino groups. The starches are coupled to the dye at C-2 of the glucose residues.

[Structure: triazine ring with R_1—C, C—O—Starch, and C—R_2, ring nitrogens N]

Roche Diagnostics (Nutley, NJ.) makes available Amylochrome,[120] amylose bonded to Cibachron Blue F3GA. Amylose-Azure, available from Calbiochem (Los Angeles, CA.), was described by Rinderknecht, Wilding, and Haverback[159] and consists of amylose coupled

* ABTS = diammonium salt of 2,2′-azino-di-(3-ethylbenzthiazoline)-6-sulfonic acid (Bio-Dynamics/bmc).

to Remazol Brilliant Blue R. Phadebas is marketed by Pharmacia Laboratories, Inc. (Piscataway, NJ.); this substrate is made up of an unidentified dye bonded to a soluble starch insolubilized by cross-linking with diepoxide. All three dye substrates are water insoluble. Their suspensions, in buffer solutions, are attacked by the enzyme at the α-1,4 bonds to produce small sized, blue dye-containing fragments which are water soluble. These can be measured colorimetrically, after being separated by centrifugation or filtration from the insoluble, unreacted substrate.

Diamyl (General Diagnostics, Morris Plains, NJ.), on the other hand, is composed of a red dye, Reactone Red 2B, coupled to amylopectin, and is water soluble. The soluble red hydrolytic products are measured after precipitating the unreacted substrate with tannic acid and alcohol.

With the single exception of Phadebas,[40] the absorbance of the color of the soluble split products formed from these dye-labeled starches is proportional to enzyme activity, over a practical range of enzyme concentration. In the case of Phadebas, it appears that the logarithm of absorbance is proportional to the logarithm of enzyme activity. Even though insoluble "substrates" are used, in most of these procedures results obtained generally appear to be comparable with those obtained with the more classical procedures using natural glucans. Procedures using these dye-coupled starches either have been calibrated empirically or have been calibrated using enzyme preparations evaluated by the classical methods.

Amylase activity can also be evaluated by following the decrease in substrate (starch) concentration, rather than by measuring the product formed. Methods based on this approach are referred to as *amyloclastic* methods. In *chronometric* procedures the time required for amylase to hydrolyze completely all the starch present in a reaction mixture is measured. The endpoint is reached when there is absence of any substrate capable of forming the blue starch-iodine color. In the Wohlgemuth method serial dilutions of the enzyme preparation are added to a fixed quantity of starch, and that dilution is found which is just able to hydrolyze all the starch present in a fixed time period. The *amylometric* procedures of Van Loon and others measure the amount of starch hydrolyzed in a fixed period of time, using the intensity of the blue starch-iodine color as the means for quantitating unhydrolyzed starch.

Some sera and urines contain material that apparently inhibits formation of the starch-iodine color, thus erroneously indicating high amylase activity. The nature of this material is not known; some reports state that it is dialyzable, and others that it is some type of paraprotein. Because of this, caution suggests that the use of amylase procedures based on measuring starch by the starch-iodine color should be discouraged. However, procedures based on this approach are simple and rapid and are still in current use.[39]

In most assay methods the pH is maintained at about 6.9 to 7.0 by the use of 0.02 molar phosphate buffer; in some methods, Cl^- ion is added, but in others, the Cl^- added with the serum or urine is assumed to be sufficient (0.01 mol/l) to provide maximum activation. The calcium requirement has been ignored and it is possible that, at times, the Ca^{2+} in the reaction mixture dilution may not be at optimal levels.

About 30 per cent of the requests for serum amylase determinations are ordered on a "stat" basis to confirm or rule out acute pancreatitis as a possible cause of acute upper abdominal pain. Thus, the method of Peralta and Reinhold (see Ware *et al.*[213]), which measures the change in turbidity of a starch sol over a short reaction period, offers certain advantages.

Two methods will be presented as examples of procedures presently in use. The Henry-Chiamori[95] modification of Somogyi's procedure[188] is an example of a saccharogenic procedure, and the Amylochrome procedure[120] is included as an example of the dye-starch methods. The details of the various techniques for amylase assay are the same irrespective of whether the specimen being used is serum, plasma, urine, or duodenal aspirate. Urine specimens should be timed; 1, 2, or 24 h outputs are most often called for. The pH of urine and duodenal fluids should be checked and adjusted to between 6.5 and 7.5 if necessary.

The saccharogenic and amylometric procedures can measure only up to some 450 to 500 Somogyi units of activity. Specimens with greater activity must be diluted with saline or, preferably, with dilute albumin (5 mg/100 ml) in saline solution. Some of the dye-labeled starch methods can measure up to 700 to 1000 units before dilution of specimens becomes necessary.

UNITS FOR EXPRESSING AMYLASE ACTIVITY

Somogyi defined a unit of amylase activity in the saccharogenic assay as that quantity of enzyme that is able to liberate reducing substrates with a reducing value equivalent to 1 mg of glucose in the course of a 30 min reaction at 40°C and at a pH of 6.9 to 7.0. The concentration is expressed as the number of units per 100 ml of specimen. Most methods based on measuring the formation of reducing sugars report results in terms of Somogyi units, although not all measurements are performed at 40°C.

There is no such uniformity, however, with the amyloclastic methods. The Wohlgemuth *diastatic index* is defined as the number of milligrams of starch digested by the enzyme present in 1.0 ml of specimen at 37°C in 30 min. The Huggins and Russell unit is the quantity of enzyme that will hydrolyze 1.0 mg of starch to the dextrin stage in 60 min at 37°C, whereas one Street and Close unit is that amount which converts to dextrin 20 mg of starch in 15 min. There is no direct relationship between any of these units and the Somogyi unit. The latter has been so generally accepted in this country that values obtained by other types of procedures are usually multiplied by an appropriate factor to convert them into equivalent Somogyi units. In the chronometric methods, amylase units are calculated from the formula, units $= C/t$, where t is the time required to decolorize the starch preparation, and C is a factor so chosen that the values obtained are again comparable with those obtained with saccharogenic methods. Inasmuch as neither the substrate nor the products of the amylolytic reaction are single, well-defined chemical entities, and a variety of temperatures are used, there is no real advantage to be gained in converting common amylase units into International Units. To convert from Somogyi units measured at 40°C to International Units at 37°C, the appropriate combined factor is 1.6 (1.0 Somogyi unit/100 ml = 1.85 U/l; temperature factor = 0.85).

DETERMINATION OF SERUM AMYLASE BY A SACCHAROGENIC METHOD (SOMOGYI, MODIFIED BY HENRY AND CHIAMORI)

1. Phosphate buffer, 0.10 mol/l, pH 7.0. Dissolve 4.55 g of KH_2PO_4 and 9.35 g of Na_2HPO_4 in distilled water and dilute to one liter.

2. Buffered substrate. With stirring, suspend 2.15 g of Lintner's soluble starch (or corn starch) in 30 to 40 ml of water to form a smooth paste, free of clumps. To this add slowly, with stirring, 100 to 120 ml of boiling hot buffer solution. The starch granules will dissolve to form an opalescent, homogeneous starch sol. Heat the solution, with stirring, to boiling, and hold at that temperature for three minutes. Permit the sol to cool to 30 to 35°C, transfer to a volumetric flask, and dilute to 200 ml (starch concentration = 10.7 mg/ml). This solution is stable for three to four weeks when refrigerated. Discard the solution when any sign of microbial growth is evident. Before use, reheat a well-mixed aliquot of the solution to 90°C.

3. Glucose standards. Prepare two standards containing 200 and 400 mg glucose/100 ml in a saturated benzoic acid solution.

4. Folin-Wu protein-free filtrate reagents.

   a. H_2SO_4, 0.33 mol/l.

   b. Tungstate solution, 10.0 g $Na_2WO_4 \cdot 2H_2O$/100 ml.

5. Folin-Wu sugar reagents.

a. Alkaline copper reagent. Dissolve 40.0 g of Na_2CO_3, anhydrous, in about 400 to 500 ml of water. To this add 7.5 g of tartaric acid, with stirring, followed by a solution of 4.5 g of $CuSO_4 \cdot 5H_2O$ in approximately 100 ml of water. Then dilute the solution to 1000 ml. The reagent is stable.

b. Phosphomolybdic acid reagent. Add 400 ml of NaOH (2.5 mol/l), with stirring, to 70 g of molybdic acid (MoO_3) and 10 g of $Na_2WO_4 \cdot 2H_2O$ in a 2000 ml beaker, followed by 400 ml of water. Boil the solution for 30 to 40 min, until it is free of ammonia; then cool, and dilute with water to about 700 ml. Add 250 ml of concentrated H_3PO_4 and dilute the solution to 1000 ml. The reagent is stable.

PROCEDURE

1. Use two 15 × 120 mm tubes, marked T and C, for each serum specimen. To each, add 3.50 ml of buffered substrate and equilibrate to 37°C in a water bath. To the T tube add 0.50 ml of serum. (Avoid contamination with saliva.) After mixing, incubate both tubes for exactly 30 min.
2. During this period, prepare the blank and standards to contain 3.50 ml of buffered starch; 0.50 ml of H_2O or standards, respectively; 0.75 ml of 0.33 molar H_2SO_4; and 0.25 ml of tungstate solution.
3. At the end of the 30 min reaction time, add 0.75 ml of 0.33 molar H_2SO_4 and 0.25 ml of tungstate solution to the T and C tubes to stop the reaction. Then add 0.50 ml of serum to the C tube. After mixing, centrifuge the tubes and filter the supernatants.
4. Transfer 1.00 ml each of the blank, standards, and filtrates to properly labeled Folin-Wu sugar tubes marked at the 12.5 ml graduation. After adding 1.0 ml of alkaline copper reagent, heat the tubes in a heating block or water bath at 100°C for 8 min, and cool to room temperature.
5. Add 1.0 ml of phosphomolybdic acid to each tube, followed by vigorous mixing, to allow the escape of *all* effervescent gas. Dilute the tubes' contents to 12.5 ml with distilled water, mix thoroughly, and transfer the contents to cuvets.
6. Using a wavelength of 680 nm, read the absorbances of the solutions against the blank. In the calculations, use that standard which most closely matches the absorbance of the color in the T tube. Then

$$\frac{(A_T - A_C)}{A_S} \times C_S = \text{Somogyi units/100 ml}$$

where C_s represents the glucose concentration of the standard in mg/100 ml. If the amylase value is over 550 units, repeat the assay using 0.50 ml of a dilution of the serum in saline. If the photometric reading is outside the linear range of the sugar method, use less filtrate in the determination of reducing substances.

COMMENTS

1. The precision of the procedure is of the order of ±10 per cent in the range of 80 to 200 units/100 ml, somewhat better in the 200 to 450 unit/100 ml range, and a little worse at low or very high levels.
2. A control should be run with each batch of specimens. Commercial control sera may be used, or a control may be prepared in the laboratory by adding clear saliva to a composite of normal sera to raise the amylase activity to about 250 to 300 Somogyi units. Pipet 1.0 ml aliquots into tubes and store these in a freezer. Each day thaw out one tube by immersing it in water at 40°C. Mix well, and keep in a refrigerator when not needed.
3. Pipetting is done carefully to avoid getting any saliva into the pipets. The use of cotton-plugged pipets is a useful precaution.

NORMAL VALUES

It is generally accepted that the amylase level in the *serum* of healthy persons is 60 to 180 Somogyi units per 100 ml or from 95 to 290 (International) U/l. The actual lower and upper limits vary with the methods used, and reflect the poor precision and accuracy of current assay techniques. The normal range in terms of Street and Close units is 6 to 33 units.

In assays for *urinary amylase* on 1, 2, or 24 h timed specimens, the lower range of normal is between 1500 and 1800 Somogyi units/d and the upper range is between 6000 and 7500 units/d. The average daytime hourly output may be taken as between 70 and 275 units. Amylase levels in *duodenal fluid* specimens may vary in various conditions; concentrations of the order of 50,000 to 80,000 Somogyi units/100 ml are not uncommon for persons with normal pancreatic function.

REFERENCE

Henry, R. J., and Chiamori, N.: Clin. Chem., *6*: 434, 1960.

DETERMINATION OF AMYLASE BY A DYE-COUPLED STARCH METHOD (AMYLOCHROME PROCEDURE)

In the substrate Cibachron Blue-amylose, the dye is coupled to the amylose by ether linkage through the C-6 of the glucose residues. Even though the substrate is insoluble, amylase is able to degrade it, liberating small water-soluble fragments of various sizes, some of which may contain coupled dye residues. The intensity of color of the soluble starch fragments obtained is a measure of the enzyme activity. The procedure is calibrated against solutions of the free dye.

REAGENTS

A kit containing all necessary reagents is available from Roche Diagnostics, Nutley, NJ.

1. Substrate. This is furnished in the form of blue tablets or pellets containing 200 mg Cibachron Blue-amylose mixed with phosphate buffer and NaCl in such quantity that, when dissolved in 2.0 ml of H_2O, the concentrations of buffer and Cl^- are 0.04 mol/l and 0.02 mol/l, respectively.
2. $NaH_2PO_4 \cdot H_2O$, 3.45 g packets. The packet contents are dissolved in 250 ml of water to prepare the (acid) diluent, used to stop the enzyme reaction and to dilute the reaction products.
3. Cibachron Blue F3GA dye, standard solution, 0.200 mg/ml.

PROCEDURE

1. With forceps (not fingers), place one blue substrate pellet into a 15 × 125 mm test tube containing 1.9 ml of water. Disperse the substrate in the water by vigorous shaking (vortex mixer). One tube is needed per specimen, unless the specimen is turbid or highly pigmented, in which case a second tube is run to serve as a specimen control.
2. Place the tube(s) in a 37°C water bath to equilibrate to temperature (about 5 min).
3. Pipet 100 μl of the serum (or urine) specimen into the (assay) tube, mix again, start the timer, and incubate the tube in the water bath at 37°C for exactly 15 min. If a number of specimens are being run, allow 1 min intervals between individual tests.
4. Run a reagent blank each day. The pellet, dispersed in 2.0 ml H_2O, is incubated at 37°C for 15 min, followed by addition of 8.0 ml of diluent and mixing. This is a measure of any free dye present in the substrate. The absorbance reading should be low (less than 40 dye units), and reasonably constant, for any lot of pellets.
5. Stop the reaction by blowing 8.0 ml of the acid diluent into the reaction tube, followed by vigorous mixing.

6. If a specimen control is being run, pipet 100 μl of the specimen into the reaction tube after addition of the diluent, and then mix.

7. Centrifuge all tubes to sediment out the unreacted, insoluble substrate, and then filter through a high-flowrate filter.

8. Transfer filtrates into appropriate cuvets, and measure absorbance at 625 nm against water. Subtract reading of the reagent blank from all readings.

9. Refer to the calibration curve (or use factor), and convert absorbance readings into dye units of amylase activity.

CALIBRATION PROCEDURE

1. Dilute 1.0, 2.0, and 3.0 ml of the calibration dye solution to 10.0 ml with H_2O.

2. Measure the absorbance of these solutions at 625 nm against water as a blank.

3. There is some lot-to-lot variation in the reactivity of the Amylochrome substrate, perhaps caused by variations in dye-starch chain length and in the number of dye units bound per glucose residue. Hence, each lot of pellets is provided with a reactivity factor, expressed as a percentage of nominal standard activity. This value divided into 100 gives a standardization factor to convert results to standard activity.

4. The three standards represent uncorrected values of 200, 400, and 600 dye units of activity. These are multiplied by the standardization factor to give the corrected activity. For example, if the reactivity factor is 105, the 200 unit standard has a corrected activity of (100/105) × 200, or 190 units.

5. Plot absorbance readings for the three standards against their activity values. The points should fall on a straight line.

6. The dye unit of enzyme activity is defined as that quantity that will, under the conditions of the assay, split off, in 15 min, dye color equivalent to 1.0 mg of pure dye. Concentration is expressed in terms of dye units/100 ml of specimen. The first standard contains 0.200 mg of dye in 10.0 ml. If 0.100 ml of specimen gives the same intensity of color in the 10 ml (reaction volume plus diluent volume), then 100 ml of specimen will split off (100/0.100) or 1000 times 0.200 mg = 200 mg. This is equivalent to 200 units/100 ml.

7. An approximate value of activity in Somogyi units can be obtained by multiplying the dye unit value by the factor 0.63. (That is, 200 dye units/100 ml = 126 Somogyi units/100 ml.)

NORMAL VALUES (AMYLOCHROME PROCEDURE)

The following values for the range of activity in normal serum are provided by the Roche laboratories:

Serum: 45 to 200 dye units/100 ml
Urine: 40 to 330 dye units/h

REFERENCES

Klein, B., Foreman, J. A., and Searcy, R. L.: Clin. Chem., *16*:32, 1970.
Instructions, Amylochrome Kit, Roche Diagnostics, Div., Hoffman-La Roche Inc., Nutley, NJ.

LIPASE

(EC 3.1.1.3; triacylglycerol acyl-hydrolase)

Lipases are defined as that group of enzymes which hydrolyze the glycerol esters of long-chain fatty acids. The ester bonds at carbons 1 and 3 (α positions) are preferentially attacked, and the products of the reaction are two moles of fatty acids and one mole of 2-acylglycerol (β-monoglyceride) per mole of substrate. The latter is relatively resistant to hydrolysis, but it can be isomerized to the α-form (3-acylglycerol), permitting the third fatty acid to be split off. A scheme for the steps in the complete hydrolysis of a fat to glycerol and three fatty acids is given in equation 30.

Desnuelle[48] has demonstrated that the enzyme acts only at the interface between water and a water-insoluble substrate, and thus can act only on substrates which are present in an emulsified form. The same author has also demonstrated that the rate of enzyme action is related not to the absolute concentration of substrate present, but rather to the surface area of the dispersed substrate. The preparation of a reproducible, stable emulsion of uniform particle size requires considerable care; it is only within the last several years that satisfactory procedures for preparing such emulsions have been developed. The considerable disagreement regarding the properties of the enzyme and its utility in clinical diagnosis can in part be attributed to work with unsatisfactory substrates and inadequate methods. The observed optimal pH depends on the nature of the substrate, on the presence or absence of other reaction components, and on the length of the reaction period. Experiments with newer methods using short reaction periods indicate that the pH optimum is 8.8 at 30°C.[203]

$$\begin{array}{ccccccccc}
CH_2OFA' & & CH_2OH & & CH_2OH & & CH_2OH & & CH_2OH \\
| & & | & & | & & | & & | \\
CHOFA'' & \xrightarrow[HOH,\ OH^-]{Lipase} & CHOFA'' & \xrightarrow[HOH,\ OH^-]{Lipase} & CHOFA'' & \xrightarrow{Isomerization} & CHOH & \xrightarrow[HOH,\ OH]{Lipase} & CHOH \\
| & & | & & | & & | & & | \\
CH_2OFA''' & & CH_2OFA''' & & CH_2OH & & CH_2OFA'' & & CH_2OH \\
\textbf{Triglyceride} & & \boldsymbol{\alpha,\beta}\textbf{-Diglyceride} & & \boldsymbol{\beta}\textbf{-Monoglyceride} & & \boldsymbol{\alpha}\textbf{-Monoglyceride} & & \textbf{Glycerol} \\
 & & + & & + & & & & + \\
 & & \textbf{FA'OH} & & \textbf{FA'''OH} & & & & \textbf{FA''OH} \\
 & & \textbf{(Fatty acid I)} & & \textbf{(Fatty acid III)} & & & & \textbf{(Fatty acid II)}
\end{array} \qquad (30)$$

In attempting to measure lipase activity, some investigators have not properly differentiated between "true" lipases and related enzymes. Among the latter are *carboxylic-ester hydrolase* (EC 3.1.1.1), commonly referred to as aliesterase; *aryl-ester hydrolase* (EC 3.1.1.2); and *lipoprotein lipase*[161] (EC 3.1.1.34). The aliesterases hydrolyze glycerol esters of short chain fatty acids (e.g., tributyrin), as well as esters of monohydric alcohols (e.g., ethyl acetate) and esters of dibasic acids (diethyl adipate). The esterases are inhibited by atoxyl (sodium arsanilate), fluoride, and triorthocresylphosphate, and show no real or apparent activation by bile salts. The aryl-esterases hydrolyze such esters as phenyl acetate and β-naphthyl butyrate. Lipoprotein lipase[161] (clearing factor) is activated by heparin and hydrolyzes protein-bound triglycerides to form free fatty acids and monoglycerides, which are then transferred to an acceptor protein such as serum albumin. All three may be present in normal plasma, along with true lipase. By the use of various activators and inhibitors, the activity of one or several of these related enzymes may be partially potentiated or suppressed.

Past studies have indicated that serum lipase is activated by bile salts (e.g., sodium glycocholate), albumin, and calcium ions. The bile salts are excellent emulsifying agents, and probably promote the formation of a stable, finely dispersed fat-in-water emulsion.

Recent studies[74,203] suggest, however, that the "apparent" activating effect of bile salts and of albumin derives from their prevention of the denaturation of the enzyme at the surface of the emulsion micelle, rather than from a true activation process. In short-run rate assays, Ca^{2+} ions have no effect on the reaction; but in assays with long reaction periods, the Ca^{+} may potentiate activity by binding fatty acids which otherwise might inhibit the reaction (product inhibition). The enzyme is inhibited by heavy metals, quinine, and some esterase inhibitors such as eserine and diisopropylfluorophosphate, but not by fluoride or arsanilate.

Most of the lipase is produced in the pancreas, but some is also secreted by the gastric and intestinal mucosa. Lipase activity can be demonstrated in leukocytes, in adipose tissue cells, and in milk.

The lipase present in normal serum and that released into serum during a pancreatitis attack show some differences in properties and activities. This suggests that the lipase activity in normal serum is not entirely of pancreatic origin.

Despite reports in the past purporting to have measured lipase activity in urine, recent work using more specific and sensitive assay techniques appears to contradict this: either no enzyme is present, or its activity is repressed by inhibitors, or the enzyme is excreted in some bound, inactive form.

CLINICAL SIGNIFICANCE

The use of measurements of lipase activity in serum and in other fluids is limited, almost without exception, to the evaluation of conditions associated with the pancreas and will be discussed in Chapter 20.

METHODS FOR MEASURING LIPASE ACTIVITY

Cherry and Crandall first appreciated the clinical value of plasma lipase determinations. Their method employed as substrate a 50 per cent emulsion of olive oil in 5 per cent (w/v) gum acacia buffered with phosphate at pH 7.0. The amount of free fatty acid formed during a 24 h reaction period at 37°C was titrated with 0.05 molar NaOH to a phenolphthalein endpoint. A unit of lipase activity was defined as the quantity of enzyme which liberated acid equivalent to 1.0 ml of 0.05 molar NaOH (50 μmol). The enzyme concentration was reported in units per 1.0 ml of serum. It has been established that over 50 per cent of the hydrolytic activity measured by the original method occurs in the course of the first several hours, and various investigators have reduced the incubation time to six, four, three, or even one hour.[201] Tris [tris(hydroxymethyl)aminomethane] buffer and veronal buffer (pH 7.4 to 8.0) have been used in place of phosphate buffer, and thymolphthalein has been recommended as a superior indicator, since the endpoint at pH 10.5 is a better stoichiometric measure of the acid produced. Tietz and Fiereck[201] demonstrated the increased precision possible if the titration is performed potentiometrically.

In most methods described, either olive oil or corn oil is used as substrate. The use of tributyrin, as proposed by Goldstein, Einstein, and Roe, has been shown to be inappropriate, since that substrate is much more responsive to aliesterase activity than to lipase activity.[94] Since oils are mixtures of many triglycerides, and may also contain mono- and diglycerides, theoretically they should be less satisfactory as substrates than a material containing only one molecular species. With the availability of triolein of high purity, this triacylglycerol should become the substrate of choice for reference methods. Specially purified olive oil will, however, continue to remain in use for routine applications.

A considerable amount of effort, especially in the last several years, has been spent in trying to devise new methods for lipase activity measurements which would be more sensitive than the Cherry-Crandall type of procedure, but also specific and easy to perform. A number of different colorimetric approaches have been advanced. Seligman and Nachlas used β-naphthyl laurate as substrate and measured the liberated β-naphthol after diazotization. More recently, Whitaker[218] suggested using α-naphthyl palmitate, in the presence of eserine, to effectively inhibit esterase activity. Two other proposed chromogenic substrate materials are phenyl laurate (Saifer and Perle) and *p*-nitrophenyl stearate. These compounds, however, are probably not true lipase substrates; results obtained using phenol esters of fatty acids containing 16 to 18 carbon atoms suggest that even these substrates are hydrolyzed by an arylesterase, rather than by a true lipase.

A considerably different colorimetric technique was devised by Massion and Seligson,[133] who measured the fatty acid split off from an olive oil emulsion by the red color of its complex with methyl red. The method requires only 50 μl of serum and a 30 min reaction period. Similarly, Gindler[84] suggested the use of the dye Spectru Cationic Blue, which reacts with fatty acids (liberated by lipase action) to form blue-colored complexes (acid-base ion pairs). In a quite different approach, Yang and Biggs[227] added Cu^{2+} to form cupric salts of the fatty acids, isolated the salts by extraction into chloroform, and then measured the Cu^{2+} photometrically as the diethyldithiocarbamate chelate. An olive oil emulsion and a 100 μl specimen were used in a 10 min reaction with linear kinetics. They found the range for lipase in normal serum measured at 37°C to be from 21 to 171 ($\bar{x}$ = 63) U/l (μmol fatty acid formed/min per liter).

Suitable sensitivity with short reaction runs should be obtainable by the use of fluorogenic substrate materials. However, those compounds available and studied to date, such as monodecanoylfluorescein (Fleischer and Schwartz[73]), fluorescein dibutyrate, *N*-methylindoxyl myristate, and 4-methylumbelliferone laurate (Guibault[87]), are probably not true lipase substrates.

Emulsions of fats in water are milky in appearance. As lipase hydrolyzes the triglycerides in the fat-oil micelles, the latter are altered in structure, or even broken up, so that a decrease in the turbidity (milkiness) results. Vogel and Zieve[211] were the first to propose a procedure based on this principle, using a spectrophotometer to measure the change in turbidity (ΔA_{400nm}) of an olive oil emulsion over a 20 min reaction period. Shihabi and Bishop[182] refined the procedure: they used a specimen volume of 100 μl and two 1 min reaction periods, made measurements at 340 nm, and calibrated their procedure against a series of emulsified olive oil standards. Using a pH of 8.8 and a temperature of 37°C, they obtained a normal range for serum lipase of 7 to 120 U/l. A further refinement was introduced by Shipe and Savory[183] and Seligson,[229] who used a fluorometer to measure true light scattering by the micelles. It should be noted that it is probable that the rate of disintegration of the micellar structure, as measured by changes in turbidity (light scattering), may differ from the rate of triglyceride hydrolysis, as measured by free fatty acid formation. Several difficulties are encountered with using turbidimetric methods: Stable, reproducible substrates with suitable initial turbidity values are difficult to prepare and store, and the high initial absorbance readings have large uncertainties. Occasionally, turbidities have been observed to increase, rather than decrease, possibly owing to different rates of aggregation of micelles occurring in blank and test cuvets or to precipitation of proteins present in the reaction mixture. Attempts have also been made to overcome the shortcomings of the titrimetric (Cherry-Crandall) procedures, one of these being the poor precision obtained in the titration of the fatty acid hydrogen ions in the presence of a buffer. Bandi and Kenny were able to improve titration precision considerably by extracting the free fatty acid into hexane, evaporating to dryness, and titrating after re-solution into petroleum ether-ethanol. More basically, Tietz and Repique,[203] refining the method of Rick and Hausaman, used a pH- Stat apparatus to titrate the fatty acid continuously, as it was formed *in situ*. The addition of base (keeping the pH constant at 8.8) was recorded as a function of time. Improved precision is obtained, even though only 0.50 ml of specimen or less is used, and only a 3 to 8 min reaction time is necessary. Triolein was shown to be more sensitive and preferable to olive oil, although results using both substrates with the same sera correlate well. The authors reported a normal range for lipase in serum of up to 200 U/l using triolein, and up to 160 U/l using olive oil as substrate.

The kinetic method employing the pH-Stat is potentially capable of giving the most reliable results, but it requires skill and relatively expensive equipment. The Tietz-Fiereck modification of the Cherry-Crandall titrimetric procedure will be described in detail, since it is more practical for use in the average routine laboratory.

DETERMINATION OF LIPASE ACTIVITY IN SERUM

SPECIMENS

Lipase activity in serum is stable at room temperature for a week; sera may be stored for three weeks in the refrigerator, and for several months if frozen. Bacterial contamination may result in an increase in lipase activity.

PRINCIPLE

An aliquot of serum is incubated with a stabilized olive oil emulsion at a reaction pH of 7.8 for 3 h at 37°C. The liberated fatty acids are titrated to pH 10.5 with NaOH, 0.050 mol/l, either potentiometrically or to a light blue color with thymolphthalein as indicator.

REAGENTS

1. Purified olive oil. Add 300 ml of the best quality olive oil to 60 g of chromatographic grade alumina (Al_2O_3, Merck No. 71207) with stirring. Stir the suspension at frequent intervals over the course of an hour. Permit the alumina to settle out, and filter the supernatant through a qualitative filter paper (Whatman No. 1 or equivalent) using a Büchner funnel and suction. Use of pure, fresh olive oil is important; olive oil, U.S.P., is satisfactory as a starting material.

2. Oil emulsion. Add 7.0 g of gum acacia (emulsifier) and 0.2 g of sodium benzoate (preservative) to 100 ml of water in a high speed blender, and dissolve with gentle blender action. Then add 100 ml of purified olive oil, and emulsify the mixture by operating the blender at top speed for 10 min. During mixing, the emulsion may get warm, but do not permit it to get hot; if necessary, stop mixing, and let it cool. A water-cooled blender is available and is a real convenience. Preparation of larger volume batches is not recommended.

Store the emulsion at 4 to 10°C. Do not permit it to freeze, as this destroys the emulsion. On standing over a period of time, some creaming may occur, but the emulsion can be fully reconstituted by shaking it vigorously 10 times. If complete separation of the oil and water phases occurs, discard the emulsion. Emulsions are stable for a month at 4 to 10°C.

3. Buffer base, stock solution, 0.80 mol/l tris(hydroxymethyl)aminomethane (Tris). Dissolve 48.55 g of Tris in 500 ml of water. Keep the reagent refrigerated.

4. Tris-hydrochloride buffer, 0.20 mol/l, pH 8.0 at 27°C (pH 7.75 at 37°C). To 50 ml of buffer base in a beaker, add 21 ml of HCl (0.2 mol/l) and dilute with H_2O to 150 to 160 ml. Permit the solution to cool to 25°C, check the pH with a pH meter, and then adjust to pH 8.0 by careful addition of more acid. Adjust the volume to 200 ml. Note that Tris buffers have a large temperature coefficient,

$$\Delta pH/\Delta T = -0.25/10°C = -0.025 \text{ pH units/°C} \qquad (= -0.025 \text{ pH} \times K^{-1})$$

5. Sodium hydroxide standard, 0.050 mol/l. Prepare by diluting any laboratory stock NaOH to exactly 0.050 mol/l. Confirm titer.

6. Thymolphthalein indicator, 1 per cent (v/v) in 95 per cent (v/v) ethyl alcohol.

7. Ethanol, 95 per cent (v/v).

PROCEDURE

1. Use a pair of 40 to 50 ml centrifuge tubes, labeled T (test) and B (blank), for each unknown. Into each tube place 2.50 ml of water, 10.0 ml of olive oil emulsion, and 1.00 ml of Tris-HCl buffer. The water and emulsion may be added by automatic pipets or with serological pipets, but the buffer should be measured out precisely. The test should be done in duplicate, if sufficient serum is available.

2. Place the tubes in a water bath at 37°C and permit to warm up for 5 min.

3. Add 1.0 ml of unknown serum (or other specimen) to the T tube, cover the tube with Parafilm, mix vigorously, and return to the water bath. Permit the reaction to proceed at 37°C for 3 h.

4. At the end of the incubation period, remove the tubes from the water bath, and

add 3 ml of 95 per cent (v/v) ethanol to all tubes to stop the enzyme reaction (protein denaturation). Then add 1.00 ml of the unknown sera to the respective B tubes, and determine the fatty acids liberated in the lipase reaction by titration with 0.05 molar NaOH to an endpoint of pH 10.5, using a pH meter (potentiometric titration). Any pH meter capable of reading to within 0.01 pH unit may be used, but shielded electrodes are essential. It is important that the solution be continuously mixed during the titration. This can be conveniently done by using an electromagnetic stirrer and a stir-bar in the solution.

5. During the titration feed the NaOH rapidly, dropwise, until the pH approaches 10, and then more slowly until the endpoint is obtained. The B titration (V_B) measures the NaOH needed to bring the pH of the buffer and any acids in the serum to pH 10.5. The T titration (V_T) measures this value plus the acid liberated from the triglycerides in the enzyme reaction.

6. If a suitable pH meter is not available, and a visual endpoint titration must be performed, proceed as follows: At the end of the incubation period remove the tubes from the bath and pour their contents into 50 ml Erlenmeyer flasks, wash the tubes with 3 ml of 95 per cent (v/v) ethanol, and add the washings to their respective flasks. Then add 1.0 ml of sera to the respective B flasks and mix. Add 5 drops of thymolphthalein, and titrate the contents of the tubes with 0.05 molar NaOH. Add the NaOH rapidly, dropwise, until the first hint of blue is seen, and then more cautiously, until a definite blue color (greenish-blue with icteric sera) is obtained.

7. Calculate the difference $(V_T - V_B) = \Delta V$. This difference gives the value of the lipase activity in the customary Cherry-Crandall (Tietz-Fiereck) units. The unit of activity is defined as the quantity of enzyme in 1.0 ml of serum which will produce fatty acids equivalent to 1.00 ml of 0.05 mol/l NaOH (50 μmol), under the conditions of the test.

COMMENTS

1. The reported 95 per cent confidence limits for values in the normal range are ± 12 per cent; this approaches ± 4 per cent with sera containing increased levels of lipase activity.

2. The 10 ml of olive oil emulsion is sufficient to maintain maximum activity within the conditions of the procedure. The actual pH of the enzyme reaction mixture after mixing emulsion, buffer, and serum is about 7.65 at 37°C. This pH may be the optimum for the 3 h reaction period. As mentioned on p. 634, the optimum pH for short period reactions appears to be from 8.6 to 8.8. Over a 3 h period, considerable enzyme inactivation may be expected at 37°C, and therefore 30°C may be the preferable reaction temperature.

3. The enzyme reaction is not linear with time over the 3 h period, although the deviation is not pronounced, and is significantly less than that obtained in methods using 6 or 24 h incubation periods.

4. The results may be converted to International Units, if desired, as follows: One conventional unit represents 50 μmol of fatty acid split off in the 180 min reaction run, or 50/180 = 0.278 μmol/min. Thus, 1.0 Tietz-Fiereck unit/ml = 0.278 U/ml or 280 U/l.

NORMAL RANGE

Tietz and Fiereck report a normal range of 0.10 to 1.00 conventional unit, with an average value of 0.42 unit/ml. The equivalent values in International Units are 28 to 280 ($\bar{x}$ = 116) U/l (see Comments).

REFERENCE

Tietz, N. W., and Fiereck, E. A.: Measurement of Lipase in Serum. *In* Standard Methods of Clinical Chemistry, G. R. Cooper, Ed., New York, Academic Press, 1972, Vol. 7, pp. 19-31.

LIPASE IN URINE AND DUODENAL FLUID

Lipase activity in urine, using the Cherry-Crandall procedure, has been reported in the literature. However, the validity of these results has been questioned. As already pointed out, using some of the new, sensitive procedures, no lipase activity in urine could be demonstrated.[203]

The unsatisfactory nature of lipase methods has discouraged measurement of lipase activity in duodenal fluids in evaluating pancreatic function, although amylase values are routinely so measured. Duodenal specimens may need to be diluted before being measured. Physiological saline may be used as diluent, although a solution containing 50 mg albumin/100 ml is preferable. For further comments refer to Chapter 20.

TRYPSIN

(EC 3.4.21.4; no systematic name)

Trypsin is a proteinase, and more specifically an endopeptidase, with a specificity directed toward peptide bonds involving the carboxyl groups of the basic amino acids, arginine and lysine. However, the chemical bond attacked need not be a peptide bond: indeed, esters and amides are split more rapidly than are peptide bonds. This specificity is illustrated in Figure 12-18, which presents the structures of three synthetic substrates, *p*-toluenesulfonyl-L-arginine methyl ester (TAME), benzoyl-L-arginine-*p*-nitroanilide (BAPNA), and benzoyl-L-lysineamide (BLA), which are all hydrolyzed rapidly by the enzyme.

Trypsin is synthesized in the acinar cells of the pancreas in the form of the inactive proenzyme, trypsinogen. The latter is stored in the zymogen granules, and is secreted into the duodenum under the stimulus of either the vagus nerve or the intestinal hormone cholecystokinin-pancreozymin. In the intestinal tract the trypsinogen is converted to the active enzyme by the intestinal enzyme enterokinase, or by preformed trypsin. For more information the reader is referred to Chapter 20.

The molecular mass of trypsin is 23,800, and the enzyme contains a predominance of basic amino acids (isoelectric point = pH 10). The pH for optimal tryptic activity toward protein substrates is in the range of 8.0 to 9.0; but with the synthetic substrates such as TAME and BAPNA referred to above, the optimal pH is 7.8. The enzyme is stable in weak acid but is rapidly destroyed at an alkaline pH. The enzyme digests itself, and when it is functioning at its optimal pH of 9, its activity falls off rapidly.

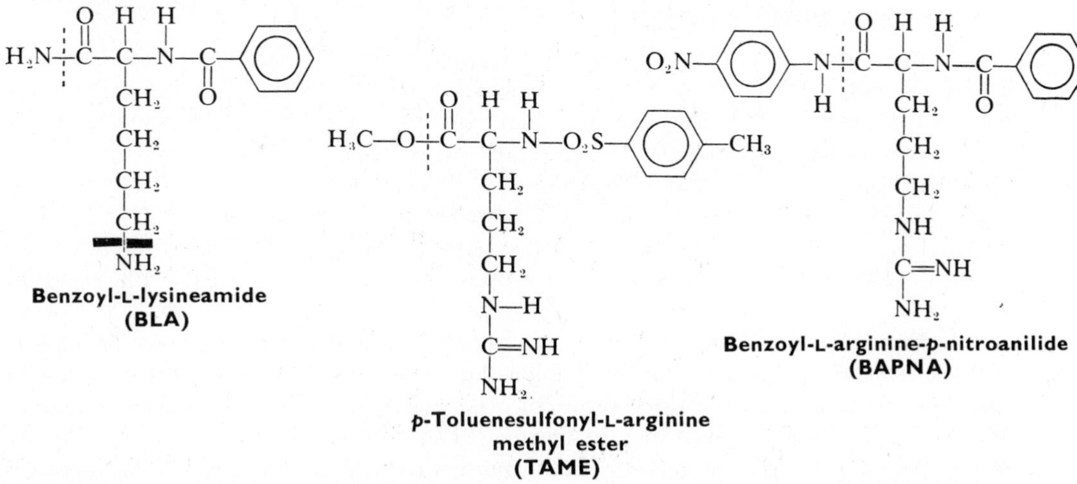

Benzoyl-L-lysineamide (BLA)

***p*-Toluenesulfonyl-L-arginine methyl ester (TAME)**

Benzoyl-L-arginine-*p*-nitroanilide (BAPNA)

Figure 12–18 Formulas for three synthetic substrates for trypsin, illustrating the bond specificity of the enzyme, which attacks bonds involving the —COOH of lysine and arginine. The dashed line indicates the bond which is hydrolyzed. Hydrolysis of TAME is an example of esterase activity; hydrolyses of BLA and of BAPNA are examples of amidase activity.

Activity is stimulated by calcium and magnesium ions, and to a lesser extent by cobalt and manganese ions and by aliphatic alcohols. Enzyme activity is inhibited by cyanide, sulfide, citrate, fluoride, and heavy metals, and by those organic phosphorus compounds which inhibit esterases.

Materials such as soy beans, lima beans, and egg-white contain trypsin inhibitors—small-sized proteins which combine irreversibly with trypsin and inactivate it by covering up the "active center." Similar nondialyzable trypsin inhibitors are present in pancreatic juice, in serum, and in urine. The plasma inhibitor α_1-antitrypsin and its immunochemical assay have been discussed in Chapter 7. This inhibitor presumably protects the plasma proteins against hydrolysis by trypsin, if for some reason any appreciable quantity of the enzyme enters the vascular system.

The enzyme attacks denatured proteins at a more rapid rate than it attacks native proteins, and therefore the former have been used as substrates for the enzyme.[157] However, the use of proteins (hemoglobin, casein, denatured serum) as trypsin substrates has the disadvantage that they are also hydrolyzed by chymotrypsin and other proteases and peptidases. Gelatin, however, still serves as a substrate in the convenient X-ray film test. These protein substrates of indefinite composition are being replaced by synthetic peptide substrates such as BAEE (benzoyl-L-arginine ethyl ester) and BLA, BAPNA,[64] and TAME, already alluded to—substrates of known composition which are specific for trypsin. These materials have the additional advantage of not being attacked by chymotrypsin and other peptidases, which are associated with trypsin in duodenal fluids and stool specimens.

Procedures using these substrate materials are simple and convenient, although on occasion specimen color may be a source of interference. Where possible, the change in absorbance at some fixed wavelength is followed during the course of the reaction. For example, since benzoyl-arginine, a reaction product, has a greater absorbance at 253 nm than does the substrate BAEE, the reaction may be followed by measuring the rate of increase in absorbance at 253 nm. If amides are used, the NH_3 formed can be measured by the Berthelot method. With the use of any synthetic substrate, the reaction can be monitored by measuring the carboxyl H^+ produced (pH-Stat apparatus). With BAPNA as a substrate, the yellow color of the *p*-nitroaniline split off in the reaction permits measuring trypsin activity by a continuous monitoring procedure. Only one chemical bond can be split in the synthetic substrates, permitting the expression of activity in International Units.

CLINICAL SIGNIFICANCE[106]

Although determinations of trypsin activity are of limited clinical value, they have been used as an aid in the evaluation of pancreatic function and as an aid in the diagnosis of chronic pancreatitis and fibrocystic disease, as discussed in Chapter 20 (Pancreatic Function). Trypsin assays can be performed either on specimens of duodenal juice aspirated from the intestinal tract[92] or on fresh stool specimens. Most investigators agree that determinations of trypsin in stool specimens from adults are usually of limited value. Much pancreatic trypsin may be destroyed as the stool passes through the intestinal tract, and it cannot easily be differentiated from trypsin and other proteinases associated with intestinal bacteria.

In hemorrhagic and other forms of acute pancreatitis, some trypsin may diffuse out into the interstitial fluid and eventually enter the plasma and be detected there. Attempts have been made to assay trypsin in serum or plasma, but the results have not been promising. All or part of the trypsin may be inactivated by the trypsin inhibitors present in the serum; furthermore, thrombin and plasmin (present to some extent in all plasmas) possess trypsin-like activity, and there is no simple method of differentiating this activity from true trypsin activity.

Fibrocystic disease (cystic fibrosis, mucoviscidosis) in children is accompanied by deficient secretion of trypsin by the pancreas. Although the disease is diagnosed clinically and by measurement of sweat electrolytes, the demonstration of little or no tryptic activity in stool specimens (or in duodenal fluid specimens) is useful supporting information.This can be done conveniently by use of the X-ray film test, or one of the new methods employing synthetic substrates, as discussed earlier and in Chapter 20.

SEMIQUANTITATIVE DETERMINATION OF TRYPSIN BY THE X-RAY FILM TEST[119]

The test specimen may be either duodenal fluid or the more readily available fresh stool specimen. Serial, two-fold dilutions of the specimen in pH 8.0 buffer are prepared. For a duodenal fluid specimen the dilution may range from 1/2 to 1/256, and for a stool specimen from 1/5 to 1/320. A drop of each dilution is spotted onto a small piece of unexposed X-ray film, along with drops of suitable controls. The X-ray film with its series of specimen drops is then incubated for 30 min at 37°C, cooled in a refrigerator to harden the softened gelatin, and washed with a light stream of water to remove any loose or broken-down gelatin. The piece of film is then examined for the presence of clear (digested) areas, indicating hydrolysis of the gelatin layer by the action of the enzyme. The greatest dilution of the specimen giving a cleared area is taken as the measure of the trypsin activity present.

NORMAL VALUES

The stool of a normal infant under one year of age will show tryptic activity through a dilution of 1/80 or higher. With older children, activity may be evident only through a dilution of 1/20 or 1/40. Infants with fibrocystic disease will rarely give positive tests beyond the 1/10 dilution. In the case of duodenal fluid specimens, normal infants give positive tests through a dilution of 1/32 or 1/64, fibrocystic children usually 1/4 or less.

DETERMINATION OF TRYPSIN IN STOOL OR DUODENAL FLUID USING TAME AS SUBSTRATE

A procedure using TAME in Tris buffer, and based on measuring tryptic activity by titrating the carboxyl hydrogen ion released on hydrolysis of the methyl ester, has been described by Haverback and associates.[92]

NORMAL VALUES

Normal children and adults give values of 40 to 760 μg trypsin/g stool; adults with (alcoholic) pancreatic insufficiency give values of 0 to 33 μg/g. Fibrocystic children have levels of less than 20 μg/g stool.

Both the Bio-Dynamics/bmc and Mann Research Laboratories[190] make available kits for trypsin assay, using BAPNA as substrate in Tris buffer. No normal values have as yet been established.

REFERENCE

Biochemica Test Combination, Catalog No. 15950 TTAA, Bio-Dynamics/bmc, Indianapolis, IN.

UROPEPSIN

(EC 3.4.23.1; no systematic name; pepsin)

That human urine contains an enzyme capable of splitting proteins under acid conditions was known as early as 1861. Bendersky (1890) was the first to use the name *uropepsin* in referring to the enzyme. It had already become conventional to use the term *pepsin* to differentiate those proteinases which hydrolyzed proteins under the acid conditions present in the stomach from other proteolytic enzymes. Several such endopeptidases may be found in urine, and at least two of these are true pepsins. As a rule, pepsins show less substrate specificity than do trypsins; they preferentially split peptide bonds which join two aromatic amino acids (tyrosine, phenylalanine) or which join glutamic acid to an aromatic acid. (Examples of bonds subject to pepsin action are illustrated in Figure 12-19.) Large protein molecules are broken down only to the polypeptide stage, rather than to free amino acids, although simple peptides can be completely hydrolyzed. The optimal pH for pepsin activity is 1.5 to 2.0, although pepsins will function over a wide pH range (1.5 to 5.0), and the optimal pH observed will vary with the nature of the substrate.[158]

The chief cells in the stomach (refer to Chapter 20) are the source of pepsin, which is synthesized and secreted in the form of an inactive precursor (proenzyme or zymogen) called *pepsinogen* (or, in the new terminology, *propepsin*). At acid pH, pepsinogen (molecular mass 42,500) is broken down into fragments consisting of six small polypeptides and the active enzyme, *pepsin*, with a molecular mass of 34,500. Once formed, the pepsin itself will convert more pepsinogen to pepsin; i.e., the conversion of the proenzyme to pepsin is autocatalytic. Pepsinogen and pepsin differ markedly in their stabilities in alkaline

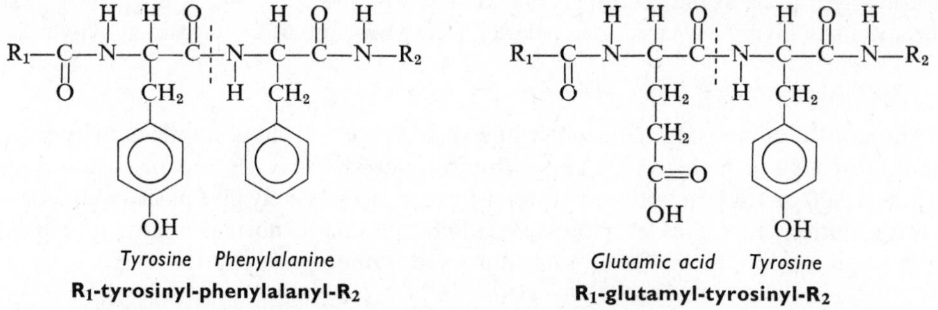

Tyrosine *Phenylalanine*

R_1-tyrosinyl-phenylalanyl-R_2

Glutamic acid *Tyrosine*

R_1-glutamyl-tyrosinyl-R_2

Figure 12–19 Two segments of peptide chains illustrating the peptide bonds split by pepsin. The dotted lines indicate the position of attack.

solution: free pepsin is rapidly and permanently destroyed at pH 12 even at room temperature, whereas pepsinogen at this pH is stable even at 50°C and can readily be converted to active pepsin on subsequent acidification.

Of the pepsinogen released from the gastric mucosa, about 99 per cent is secreted into the stomach to become part of the gastric fluid. The remaining 1 per cent diffuses into the interstitial fluid around the chief cells, and eventually reaches the blood. Any active pepsin that may get into the blood is rapidly inactivated by the pH conditions prevalent in blood. Pepsinogen, however, is stable and circulates in the blood as such. That portion which reaches the kidneys passes through the glomeruli and is excreted in the urine.[116] Under the mildly acid conditions present in urine most of the time, part or all of the proenzyme, now called uropepsinogen, may be converted (activated) to uropepsin. Daily excretion is fairly constant for any one individual, although there is a diurnal variation in output which parallels the diurnal variation in adrenal activity. The quantity excreted is independent of urine pH, volume, and specific gravity.

If the pH of the specimen is kept between 5.0 and 6.5, the urine enzyme is stable at room temperature for 2 or 3 days, and in the refrigerator for two weeks. No inhibitor of the enzyme is present in urine.

CLINICAL SIGNIFICANCE

Uropepsin determinations are used mainly as an aid in the diagnosis of diseases of the stomach. However, decreased levels are also found in patients with myxedema, Addison's disease, and hypopituitarism, and elevated levels are encountered in patients with hyperthyroidism or Cushing's syndrome, and in physiological stress situations.[197]

The rate of excretion of free HCl by the gastric mucosa and the rate of excretion of uropepsin in the urine roughly parallel each other, both in health and in disease. The greatest output is associated with duodenal ulcers,[116] and this increases with the duration of the disease. Considerably diminished outputs are found in patients with gastric ulcers and with gastric carcinoma. However, there is considerable overlap between the values seen in healthy persons and those observed in disease conditions. Patients with untreated pernicious anemia excrete very little or no uropepsinogen, and none of the activated enzyme or its proenzyme is present in urines from patients with total gastrectomies. Occasionally, however, uropepsin is excreted by patients with histamine-resistant *achlorhydria* (but not in true *achylia*). In patients with gastric resections, uropepsin output is related to the quantity of gastric tissue remaining.

Injection of cortisone or corticotrophin stimulates uropepsin output to levels two or three times the average normal value.

METHODS FOR UROPEPSIN ACTIVITY MEASUREMENTS

Three approaches have been used to assay uropepsin activity. In the first approach, the amount of some hydrolysis product is measured. Mirsky and Gray, for example, determined the amount of tyrosine split as a result of the action of the enzyme on hemoglobin. The tyrosine liberated was measured with the Folin-Ciocalteu reagent. In analogous procedures, serum proteins, edestin, or radio-iodinated serum albumin have been used as substrates, and activity was measured in terms of the amount of unhydrolyzed protein remaining. This could be done by precipitating the proteins with trichloroacetic acid and assaying them with biuret reagent, or by determining the liberated tyrosine by measuring the increase in absorbance at 280 nm.

In a second approach, the enzyme is quantitated by measuring the rate at which it clots milk casein under standard conditions (0.75 mol/l acetate buffer, pH 4.9; homogenized milk; 37°C). Pepsin possesses some rennin-like activity; i.e., it attacks soluble casein and modifies it to a form which is insoluble and which thus clots or precipitates out. This procedure was introduced by West and his associates,[216] and though probably not as precise as the Anson-Mirsky and similar methods, it is simple to perform, and the results are adequate for most clinical purposes.

The third approach utilizes synthetic substrates[158] of low molecular mass designed to contain bonds sensitive to peptic action.[52] One such substrate is *N*-acetyl-L-phenylalanyl-L-3,5-diiodotyrosine, which is rapidly hydrolyzed by (uro)pepsin to 3,5-diiodotyrosine; the latter can be measured colorimetrically after treating it with ninhydrin. Inasmuch as the substrate has only one peptide bond to be split, activity can be expressed in U/l.

REFERENCE

Keller, M. E., Agne, M., Mannebach, A., and Leppla, W.: *In* Enzymes in Urine and Kidney, U. C. Dubach, Ed., Baltimore, Williams and Wilkins Co., 1968, pp. 265–280.

SERUM CHOLINESTERASE

(EC 3.1.1.8; acylcholine acyl-hydrolase)

The term serum or plasma cholinesterase (CHS*) refers to a group of related enzymes found in serum, all of which have the ability to hydrolyze acetylcholine. However, many other substrates are also attacked, such as the butyryl and other acyl esters of choline and

---

* This unofficial standard abbreviation for the enzyme is used in the text, but one of the authors (JFK) prefers PCE, inasmuch as this directly alludes to plasma cholinesterase or to pseudocholinesterase. Bergmeyer[15] uses ChE.

thiocholine, as well as a number of aryl and alkyl esters of short-chain fatty acids. The hydrolysis of acetylcholine may be formulated as follows:

$$(CH_3)_3\overset{+}{N}{-}CH_2CH_2{-}O{-}\overset{O}{\overset{\|}{C}}{-}CH_3\ Br^- + H{-}O{-}H \xrightarrow{Cholinesterase} (CH_3)_3\overset{+}{N}{-}CH_2CH_2{-}O{-}H\ Br^- + CH_3\overset{O}{\overset{\|}{C}}{-}O^- + H^+ \quad (31)$$

Acetylcholine bromide → Choline bromide + Acetate ion + Hydrogen ion

Acetylcholine is very important physiologically. It is synthesized at nerve endings and acts to transmit impulses from nerve to muscle fiber. Cholinesterase destroys the acetylcholine after the impulse transmission has been mediated, so that additional impulses may be transmitted, if needed. Otherwise, the nerve would remain electrically charged and further conduction would not be possible.

Two enzymes with cholinesterase activity but with different substrate specificities and origin are present in human tissues. One enzyme or enzyme group (CHS), also referred to as *pseudocholinesterase*, is present in the liver, the pancreas, the heart, and the white matter of the brain, as well as in serum. The second enzyme occurs in erythrocytes, in the lungs and spleen, in nerve cell endings, and in the gray matter of the brain, and is referred to as "true" or *red cell cholinesterase*. This enzyme is listed by the Enzyme Commission as a separate enzyme (EC 3.1.1.7; acetylcholine acetyl-hydrolase). Pilz[147] has shown that small amounts of this enzyme are also present in serum and can be isolated from it. Although both enzymes have important physiological functions, only the serum enzyme will be discussed in detail because it has been of most interest clinically.

These two enzyme types differ in specificity toward some substrates but behave similarly toward others. The serum enzyme acts on benzoylcholine, but cannot hydrolyze acetyl-β-methylcholine; the red cell enzyme acts on the latter but not on the former. Only choline esters are split by the red cell enzyme; aryl or alkyl esters are not attacked. The red cell enzyme is inhibited by its substrate, acetylcholine, if present at concentrations above 10^{-2} mol/l; the serum enzyme is not inhibited by this substrate.

$$C_6H_5{-}\overset{O}{\overset{\|}{C}}{-}O{-}CH_2CH_2{-}\overset{+}{N}(CH_3)_3 \qquad (H_3C)_3\overset{+}{N}{-}CH_2{-}\underset{H}{\overset{CH_3}{C}}{-}O{-}\overset{O}{\overset{\|}{C}}{-}CH_3$$

Benzoylcholine **Acetyl-β-methylcholine**

Both enzymes are inhibited by the alkaloids prostigmine and physostigmine, both of which contain quaternary nitrogen (also present in choline) in their structures. These are typical competitive inhibitors, competing with the choline residue of acetylcholine for its binding site on the enzyme surface. Both enzymes are irreversibly inhibited by some organic phosphorus compounds, such as diisopropylfluorophosphate. The phosphoryl group binds

$$C_6H_4(\overset{+}{N}(CH_3)_3){-}O{-}\overset{O}{\overset{\|}{C}}{-}N(CH_3)_2 \qquad O{=}P(F)(O{-}CH(CH_3)_2)_2$$

Prostigmine **Diisopropylfluorophosphate**

very tightly to the enzyme at the site at which, normally, binding of the acetyl group occurs, thus preventing attachment of the acetylcholine. Both enzymes are also inhibited by a large variety of other compounds, among which are morphine, quinine, tertiary amines, phenothiazines, pyrophosphate, bile salts, citrate, fluoride, and borate.

The cholinesterase present in normal sera can be separated by electrophoresis into from seven to twelve bands, the number obtained depending on the experimental technique used.[108] The isoenzymes of CHS differ in molecular size, and appear to be aggregates of different numbers of the same basic unit.[121] The various forms hydrolyze the different substrates at different rates, but over 90 per cent of the activity against acetylcholine is contained in three fractions. Of more interest are the atypical (genetic) variants of the enzyme, characterized by diminished activity against acetylcholine,[125] which are found in the sera of a small fraction of apparently healthy persons. These forms possess increased resistance to inhibition by fluoride ion and by dibucaine[42,77] (see below).

The gene controlling the synthesis of CHS can exist in at least four allelic forms, designated as E_1^u, E_1^a, E_1^f, and E_1^s. Several other allelic forms may also exist, and another gene locus is recognized (E_2). The normal, most common phenotype is designated as E_1^u E_1^u, or UU. The gene E_1^a is referred to as the atypical gene; the sera of persons homozygous for this gene (E_1^a E_1^a = AA) are only weakly active toward most substrates for cholinesterase, and possess increased resistance to inhibition of enzyme activity by dibucaine. The E_1^f gene also gives rise to a weakly active enzyme, but with increased resistance to fluoride inhibition. The E_1^s gene (*s* for silent) is associated with absence of enzyme or the presence of a protein with minimal or no catalytic activity. The diminished activity associated with the variant alleles probably reflects very weak association between the modified enzyme protein and substrate. The homozygous forms, AA or FF, are found in only 0.3 to 0.5 per cent of the white population; their incidence among blacks is even lower.[121,125]

The cholinesterase activity in serum is quite stable, being unaltered for six hours at room temperature, and for several weeks at 0 to 5°C. Frozen preparations retain activity for many months, but repeated freezing and thawing of a preparation may result in 30 per cent loss in activity.

CLINICAL SIGNIFICANCE

The determination of cholinesterase levels in serum may be requested as a test of liver function, as an indicator of possible insecticide poisoning, or for the detection of patients with atypical forms of the enzyme. The spread of values encountered in apparently healthy people is rather wide, ranging between 4 and 12 U/ml (37°C). The level in any given person is fairly constant; a fall in enzyme level of greater than 0.67 U/ml is significant at the 95 per cent confidence level.[217] Levels at birth are low (one-fourth that of adults), but these levels increase rapidly, reaching adult levels by the second month of life. No enzyme is found in urine.

Measurements of serum cholinesterase activity can serve as sensitive measures of *liver* function if the patient's normal (baseline) level is known, which, unfortunately is rarely the case. In the absence of known inhibitors, any decrease in activity in serum reflects impaired synthesis of the enzyme by the liver. A 30 to 50 per cent decrease in level is observed in acute hepatitis, and in chronic hepatitis of long duration. Decreases of 50 to 70 per cent occur in advanced cirrhosis and carcinoma with metastases to the liver. Essentially normal levels are seen in chronic hepatitis, mild cirrhosis and obstructive jaundice.

Decreased levels of serum enzyme are also found in patients with *acute infections*, *pulmonary embolisms*, and *muscular dystrophy*, and after surgical procedures. After *myocardial infarctions*, the enzyme level decreases until the fifth day and then begins a slow rise to normal. Decreased levels are also seen in *chronic renal* disease and in *pregnancy*.

A marginal increase in enzyme levels may be observed in patients with *nephrotic syn-*

drome. Synthesis of albumin to replace that lost in the urine is accompanied by some synthesis of additional cholinesterase, although syntheses of the enzyme and albumin by the liver are independent of each other. Marginal increases in enzyme are also seen in thyrotoxicosis and hemochromatosis, in obese diabetics, and in patients with anxiety and other psychiatric states.

Among the organic phosphorus compounds which *inhibit* cholinesterase activity are many organic insecticides, such as Parathion, Sarin, and tetraethyl pyrophosphate. Workers engaged in agriculture, and those working in organic chemical industries, are subject to poisoning by inhalation of these materials or by contact with them.[139] Obviously, if enough material is absorbed to inactivate all the red cell acetylcholinesterase, death will result. Both types of enzyme are inhibited, but the activity of serum enzyme falls more rapidly than does that of the red cell enzyme. A 40 per cent drop in serum enzyme activity occurs before the first symptoms are felt, and a drop of 80 per cent is required before serious neuromuscular effects become apparent. Near zero or zero levels of enzyme activity require emergency treatment of the patient with such enzyme reactivators as pyridine-2-aldoxime.

Succinyldicholine (Suxamethonium) is a drug used in surgery as a muscle relaxant. Because it is very similar to acetylcholine, it is also hydrolyzed by cholinesterase, and its physiological effect persists only long enough (30 to 50 min) to meet the needs of the surgical procedure. In patients with low levels of enzyme activity, or in those with the atypical, weakly active variant, this destruction of the drug will not occur rapidly enough, and the patient may enter a period of prolonged apnea. It is therefore advisable to determine cholinesterase levels on patients for whom Suxamethonium anesthesia is planned, to be certain that use of the drug is safe. The presence of abnormal enzyme activity may be confirmed by determining either the dibucaine[109] (Nupercaine) or fluoride numbers.[51,89] These parameters indicate the percentage inhibition of enzyme activity toward specified substrates in the presence of a standard concentration of these reagents. The average values of the dibucaine numbers (DN) for normals, heterozygotes, and homozygotes ($E_1{}^a$ gene) are 78, 60, and 16 per cent, respectively, when *benzoylcholine* is used as substrate. Dietz and associates reported values of 81 to 86, 67 to 80, and 8 to 35 per cent, respectively, with *proprionyl thiocholine* (PTC) as substrate. Analogous fluoride numbers (FN) with PTC as substrate are 77 to 81, 67 to 69, and 54 to 65 per cent, respectively. The phenotypes most susceptible to apnea after succinylcholine administration are: AA, AS, FF, FS, SS, AF, and UA.[51]

Cholinesterase levels in *cerebrospinal fluid* are very low. The total activity present in fluids from healthy individuals is about 17 ± 4 mU/ml. Both the serum and red cell enzyme forms can be detected with the serum form present in the greater concentration. Appreciable elevations of CHS activity are seen in diseases involving destruction of brain parenchyma, in brain tumors and brain abscesses, in hydrocephalus (to 40 mU/ml), in Guillain-Barre disease (to 60 mU/ml), and often in meningitis and multiple sclerosis. In other neurological and non-neurological diseases, enzyme levels are normal or any changes seen are too non-specific and inconsistent to be of much value in diagnostic work. CHS activity is independent of protein concentration and of leukocyte count, the leukocytes containing no enzyme activity.

METHODS FOR THE DETERMINATION OF CHOLINESTERASE ACTIVITY

The hydrolysis of acetylcholine results in the formation of one mole of hydrogen ion for each mole of substrate reacted. In manometric methods this hydrogen ion reacts with HCO_3^- buffer to release CO_2, which is then measured. In the potentiometric procedure of Michel, the decrease in pH during the course of a fixed reaction period is measured. The acid formed can be titrated with a pH-Stat apparatus. If pH indicators which change color in the pH range of 6.0 to 8.5 (bromthymol blue, *m*-nitrophenol) are incorporated into the reaction mixture, the change in color, measured spectrophotometrically, can serve as a measure of enzyme activity. A number of "kits" and test papers devised for testing for enzyme activity in the field are based on this principle.

In the procedure of de la Huerga *et al.*, the quantity of unreacted acetylcholine remaining after a 60 min reaction period at 37°C is measured. The ester is reacted with hydroxylamine to form the hydroxamate derivative, which forms an orange-brown complex with ferric ion in acid solution which can be measured colorimetrically at 540 nm. Kalow and Genest[109] used benzoylcholine as substrate, and measured the decrease in absorbance of the substrate at 240 nm.

At the present time, however, all these methods have been replaced, to a large degree, by methods employing acetylthiocholine esters as substrates. These react at approximately the same rate as do choline esters, and the thiocholine formed can be measured by reaction with chromogenic disulfide agents such as 5,5′-dithiobis(2-nitrobenzoic acid) (DTNB = Ellman's reagent)[79] and 4,4′-dithiodipyridine.[35] The iodide salts of acetyl-, proprionyl-, and butyrylthiocholine have all been used as substrates.[78] The reactions for butyrylthiocholine, with DTNB as the chromogen, are given in equations 32 and 33.

$$(CH_3)_3N^+\text{–}(CH_2)_2\text{–}S\text{–}C(=O)\text{–}(CH_2)_2\text{–}CH_3 + HOH \xrightarrow{PCE} (CH_3)_3N^+\text{–}(CH_2)_2\text{–}S\text{–}H + {}^-O\text{–}C(=O)\text{–}(CH_2)_2\text{–}CH_3 + H^+ \quad (32)$$

Butyrylthiocholine ion; Thiocholine ion; Butyrate; Hydrogen ion

$$(CH_3)_3N^+\text{–}(CH_2)_2\text{–}S\text{–}H + (HOOC)(NO_2)C_6H_3\text{–}S\text{–}S\text{–}C_6H_3(HOOC)(NO_2) \rightleftharpoons (CH_3)_3N^+\text{–}(CH_2)_2\text{–}S\text{–}S\text{–}C_6H_3(HOOC)(NO_2) + HOOC(O_2N)C_6H_3\text{–}SH \quad (33)$$

Thiocholine ion; DTNB (colorless); Mixed disulfide; 5-Mercapto-2-nitrobenzoic acid (5-MNBA) (colored)

The DTNB reaction is an example of a color-producing reaction (absorbance at 410 nm). If 4,4′-dithiodipyridine is used as a chromogen, mercaptopyridine is formed; this immediately tautomerizes to thiopyridone, which absorbs at 344 nm.

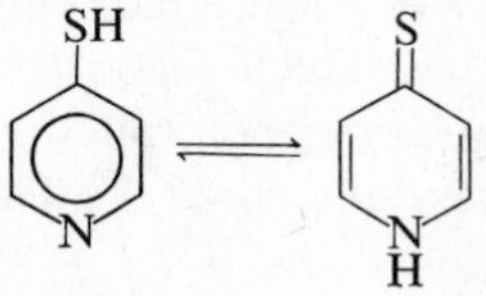

DETERMINATION OF SERUM CHOLINESTERASE BY THE PROPRIONYLTHIOCHOLINE-DITHIOBIS(NITROBENZOIC ACID) PROCEDURE (DIETZ, RUBINSTEIN, AND LUBRANO[51])

PROCEDURE (ABBREVIATED)

1. Pipet 3.0 ml of color reagent (0.423 mmol DTNB/l in phosphate buffer, pH 7.6, ionic strength 0.1) into each of three test tubes. Allow all tubes to equilibrate at 37°C for 5 min.

2. Add 1.0 ml of substrate (10 mmol proprionylthiocholine iodide/l in water) to each of the tubes.

3. Add 1.0 ml of a 100-fold dilution of the serum specimen in water to two of the tubes (duplicates), start timer, and permit the reaction to proceed for exactly 3.0 min.

4. Stop the reaction by adding 1.0 ml of quinidine sulfate solution (0.5 g/100 ml water) to all three tubes, and then add 1.0 ml of diluted specimen to the third tube (blank).

5. Read the absorbances of the solutions in the duplicate tubes against the blank at a wavelength of 410 nm, within the next 30 min. Calculate the increase in absorbance, ΔA, during the 3 min reaction period.

CALCULATIONS

1. If a narrow band-pass instrument is used, calculate the activity by using 13.6 as the millimolar absorptivity value for 5-mercapto-2-nitrobenzoic acid (5-MNBA). If other instruments are used, add a small crystal of cysteine to each of several tubes containing increasing volumes of color reagent. These solutions will then serve as a series of standards[58] with increasing levels of 5-MNBA. Measure the absorbance of the 5-MNBA color formed and plot the absorbance against concentration to prepare a calibration curve.

COMMENTS

1. To obtain the dibucaine and fluoride numbers, use the above procedure except that the 1.0 ml of substrate will contain, in addition, either 0.15 mmol of dibucaine (Nupercaine hydrochloride)/l in water or 20 mmol NaF/l in water. The DN and FN can be calculated from the fraction inhibition of enzyme activity.

2. There is no consensus among investigators as to the preferred assay temperature; most determinations have been carried out at 25°C or 37°C. Representative values are listed in Table 12-4.

NORMAL VALUES

Activity values obtained vary with the substrate used, with the type of buffer, and with buffer concentration. There is a sex difference, values for women being some 12 per cent lower than those for men. See Table 12-4.

TABLE 12–4 REPRESENTATIVE VALUES FOR NORMAL RANGES FOR CHOLINESTERASE ACTIVITY IN SERUM

Investigator	*Substrate*	*Temperature*	*Activity, U/ml*	*Sex*
Dietz[51]	PTC	37°C	4.9 to 11.9	M,F
Garry[77]	BTC	25°C	2.1 to 5.25	M,F
Szasz[194]	BTC	25°C	2.2 to 5.0	M
			1.8 to 4.6	F
Szasz[194]	ATC	25°C	1.3 to 3.5	M,F
E.M. Diagnostics[42]	BTC	30°C	3.6 to 9.5	M,F
Garry-Routh[79]	ATC	37°C	2.6 to 5.5	M
			1.9 to 4.6	F

REFERENCE

Dietz, A. A., Rubinstein, H. M, and Lubrano,T.: Clin. Chem., *19*:1309, 1973.

CERULOPLASMIN

(EC 1.16.3.1; Iron (II):oxygen oxidoreductase; Ferroxidase I)

Serum contains an enzyme which catalyzes the oxidation of *p*-phenylenediamine by oxygen to a purple-colored compound. This enzyme is a blue-colored metalloprotein (hence the name, *cerulo*plasmin) with an absorbance maximum at 605 nm and specific absorptivity of 0.068 × liter × g^{-1} × cm^{-1}. The protein migrates electrophoretically as an α_2-globulin (see p. 346) and contains 0.355 per cent copper. The green hue seen in some plasma specimens may be due to a high content of this blue-colored protein modifying the normal yellowish color of the plasma. The protein can be decolorized by reduction with dithionite ($S_2O_4^{2-}$), ascorbate, thioglycollate, and hydroxylamine. Ceruloplasmin is a glycoprotein, containing 8 to 10 sialic acid residues per mol, for which relative molecular masses varying from 132,000 to 162,000 have been reported. Before the enzyme properties of ceruloplasmin were discovered by Holmberg and Laurell, the protein was of interest as one of the "acute phase reaction" proteins present in serum.[145]

There are eight atoms of copper per molecule of ceruloplasmin, four of which are irreversibly bound to the protein moiety. The other four are linked to the protein by weak bonds, and are readily removable by copper chelating agents and by mild enzymatic hydrolysis of the protein. About 94 to 95 per cent of the total copper in plasma is present in the ceruloplasmin molecule, and the remainder is bound to albumin; only a trace is present as free Cu^{2+} (see p. 929).

The enzyme properties of ceruloplasmin are those of an oxidase. This term is applied to those enzymes which catalyze the oxidation of some substrate with molecular oxygen as obligate hydrogen acceptor. Although hydrogen peroxide may be the reduction product in many oxidase reactions, in the case of ceruloplasmin the reduction product is water.

$$2\,H{-}R{-}H + O_2 \xrightarrow{\textit{Oxidase}} 2R + 2H_2O$$

$$\tfrac{1}{2}O_2 + \text{Ascorbic acid} \xrightarrow{\textit{Ceruloplasmin}} H_2O + \text{Dehydroascorbic acid} \qquad (34)$$

Oxygen; Ascorbic acid (R = —CHOHCH₂OH); Water; Dehydroascorbic acid

Ascorbic acid can serve as hydrogen donor, and most or all of the ascorbate oxidase activity in serum can be attributed to the ceruloplasmin molecule. The following compounds can also serve as substrates for the enzyme: benzidine, *p*-phenylenediamine (PPD), *N,N*-dimethyl-*p*-phenylenediamine (DMPD), catechol, guaiacol, epinephrine, serotonin, and dihydroxyphenylalanine (DOPA). Although the last three and ascorbic acid are normal components of serum, most investigators agree that the *in vivo* physiological function of ceruloplasmin as an enzyme is not to oxidize these metabolites, but rather to catalyze the oxidation of ferrous ion to ferric ion[165] (ferroxidase activity) and to promote the saturation of apo-transferrin with iron (see also Iron Metabolism, p. 923).

Cyanide (CN^-) and azide (N_3^-) form complexes with the loosely bound copper and dissociate it from the protein molecule, thereby completely inhibiting enzyme activity. Many other anions, including Cl^-, Br^-, CNS^-, and a variety of organic ions (citrate, maleate), also inhibit enzyme activity if present at concentrations greater than 0.05 to 0.10 mol/l. Some anions (Cl^-) act as activators at low concentrations, as does Fe^{2+}; high levels of iron are inhibitory. As a result, the activity exhibited by any ceruloplasmin preparation will depend on the nature and ionic strength of the buffer used, and the kinds of ions present in the enzyme reaction solution.

CLINICAL SIGNIFICANCE

By virtue of its copper content, ceruloplasmin is an important vehicle for transport of copper in the body (see Copper, p. 929). The average level of ceruloplasmin in serum is about 31 mg/100 ml (0.31 g/l).[174] There is a significant sex difference, with the level in women some 20 per cent higher than that in men. The liver is the source of the protein-enzyme present in plasma. The level at birth is about 20 per cent of that seen in adults. The level then increases rapidly, reaching a maximum somewhat above the adult level at the second year of life. This is followed by a slow drop through adolescence. The data in Table 12-5 show how serum ceruloplasmin and copper levels change in parallel fashion and how they vary with age and disease state.

TABLE 12-5 CERULOPLASMIN AND COPPER CONCENTRATIONS IN SERUM IN A NUMBER OF CLINICAL CONDITIONS (AFTER SASS-KORTSAK)[174]

Conditions	*Ceruloplasmin (mg/100 ml)* Mean Value and Range	*Copper (μg/100 ml)* Mean Value and Range
Healthy newborns	7 (2–13)	16 (12–26)
Two-year-olds	43 (31–54)	140 (95–186)
Ten-year-olds	34 (22–45)	117 (72–162)
Young adults	31 (21–41)	109 (69–150)
Adults	33 (25–43)	114 (89–147)
Pregnancy at term	55 (39–89)	216 (118–302)
Infectious disease	68 (60–83)	
Nephrosis	30 (12–54)	
Wilson's disease	9 (2–19)	Usually very low in homozygotes; rarely low in heterozygotes

Inasmuch as ceruloplasmin is an "acute reactive" protein, its concentration in serum is moderately increased (50 to 100 per cent) in conditions of physiological stress. Increased levels of the protein are encountered in acute infections, in chronic conditions such as rheumatoid arthritis, lupus, and cirrhosis, after myocardial infarctions (tissue necrosis), and after the stress of surgery. Somewhat greater elevations (up to two-fold) are observed in hepatitis, Hodgkin's disease, metastatic carcinoma, and hyperthyroidism. Estrogens stimulate ceruloplasmin synthesis; as much as three-fold increases are encountered in pregnancy, and in women taking oral contraceptives.

Diminished levels of the enzyme are encountered in several conditions: nutritional protein deficiency (malabsorption as well as malnutrition), severe liver disease, nephrosis, and Wilson's disease.[11,174] In nephrosis some 50 to 75 mg of the enzyme may be lost daily in the urine, along with albumin; but blood levels, although usually decreased, may be normal.

Ceruloplasmin is of most interest clinically because of its relation to Wilson's disease, also known as *hepatolenticular degeneration*.[212] This disease was first described in 1912 and is characterized by low levels of serum copper and ceruloplasmin, the deposition of copper in many tissues of the body, a marked increased excretion of copper and amino acids, and muscular rigidity and lack of coordination as a result of injury to the cerebral basal ganglia. The copper found in the liver, brain, and kidney may be some 20 times that normally present. Copper is also deposited in the corneas of the eyes, giving rise to so-called Kayser-Fleischer rings. The neurological symptoms and the aminoaciduria are the result of the toxic effects of the copper deposited in the brain and kidneys. The nature of the biochemical defect which gives rise to the symptoms of the disease is not known. Some abnormality in copper metabolism is obviously involved, and it was first postulated that the defect was associated with the inability of the body to synthesize ceruloplasmin and regulate absorption of copper from the intestinal tract. It is now hypothesized that the metabolic defect may be in the mechanism by which copper is made available for synthesis into ceruloplasmin and for transport into and out of liver cells for excretion into the bile. The disease is treated by giving the patient penicillamine or other agents which complex copper and facilitate its removal from the tissues in which it has been deposited. Although the disease is congenital, the clinical features usually do not develop until adolescence or early adulthood.

METHODS FOR THE DETERMINATION OF CERULOPLASMIN

The spectrophotometric method for the determination of ceruloplasmin is based on the measurement of ΔA at 605 nm, before and after bleaching the blue color of the protein-copper complex with ascorbic acid or cyanide. This procedure measures the intact metalloprotein (but not necessarily the active enzyme) and is capable of reasonable precision; but it has the defects that a large volume of specimen is required, and that even small amounts of serum opalescence (turbidity) may give rise to falsely high values. Ceruloplasmin may also be measured immunochemically (see p. 346). Such methods also measure protein and they are suitably sensitive, but they have the defect that both the physiologically active and inactive proteins are measured.

Two approaches for measuring the enzymatic activity of ceruloplasmin have been developed. In one approach amine oxidase activity is measured, using either PPD or DMPD as substrate. Non-enzymatic oxidation also takes place, and this has to be inhibited or corrected for. The reaction is therefore carried out with and without the addition of azide (NaN_3). The assay carried out with azide (enzyme poison) measures the nonenzymatic oxidation; the difference between the two assays is a measure of the rate of the enzyme-catalyzed oxidation of the diamine. Carrying out the reaction in the dark minimizes the non-enzymatic reaction. The oxidase reaction is complex; a number of products are formed and the precise nature of these is still under discussion.[46,156] A semiquinone type of free radical is first formed, which is then further oxidized or undergoes dismutation to other products. Plasma specimens generally contain ascorbate (and often other O_2 acceptors of suitable redox potential) whose oxidation precedes that of PPD. Thus a "lag" phase of variable length is observed in the ceruloplasmin oxidase assay. Oxidase activity is essentially linear once the lag phase has been overcome.

In the newer approach, ceruloplasmin is measured by virtue of its ferroxidase activity. The reaction mixture contains Fe^{2+} and apo-transferrin, in addition to ceruloplasmin. The enzyme oxidizes the iron to Fe^{3+}, which then binds to the apo-protein to form transferrin, which is measured by its absorbance at 460 nm.[144] Only one reaction product is formed, a short reaction time is required, the reaction is linear, and only a 10 μl volume of serum is needed. However, the low absorptivity of transferrin limits the precision. Either the PPD or the Fe^{2+} oxidation may also be followed by measuring the O_2 consumed, using the Warburg apparatus, or by using an O_2 electrode.

DETERMINATION OF CERULOPLASMIN BY THE PPD OXIDASE METHOD

In the two-point procedure proposed by Sunderman and Nomoto,[191] the reaction is carried out at 37°C in acetate buffer, 0.10 mol/l, pH 5.45. The substrate is crystalline *p*-phenylenediamine dihydrochloride at a final concentration of 8.9 mmol/l. Two tubes are used, each containing buffer, serum (0.10 ml), and PPD in a total volume of 3.1 ml. After allowing a 5 min lag period, NaN_3 is added to the control tube to inhibit the enzyme. The time is noted and the reaction is permitted to proceed for 30 min, at which time azide is added to the assay tube. The ΔA_{530} is a measure of the ceruloplasmin activity. The procedure is calibrated against solutions of pure ceruloplasmin, which have been standardized in g/l by measuring their blue color absorbance at 605 nm. Results are reported in mg ceruloplasmin/100 ml (or in g/l).

An alternative procedure is that of Henry *et al.*[46] These investigators also use the hydrochloride salt of PPD, and acetate buffer, but at a pH of 6.0. The reaction is carried out at 37°C in 1.0 cm cuvets for a 30 min period, after allowing a 10 min lag period. Results are reported in arbitrary units: 1000 × ΔA_{530} between the assay and control tubes.

NORMAL VALUES

Sunderman and Nomoto[191] report a normal range of 0.315 ± 0.049 g/l (approximately equal to a 95 per cent range of 21.7 to 41.3 mg/100 ml) for a population of both sexes. Henry and associates, for a similar population, give values of 270 to 570 units, equivalent to about 16.8 to 34.2 mg/100 ml. Other values culled from the literature are: immunochemical methods, 29 ± 7 mg/100 ml; after isolation by column chromatography, 26.3 ± 3.7 mg/100 ml.

REFERENCES

Demetriou, J. A., Drewes, P. A., and Gin, J. B.: *In* Clinical Chemistry. 2nd ed. R. J. Henry, D. C. Cannon, and J. W. Winkelman, Eds. Hagerstown, MD., Harper and Row, 1974, pp. 857–864.

Sunderman, F. W., Jr., and Nomoto, S.: Clin. Chem., *16*:903, 1970.

LACTATE DEHYDROGENASE

(EC 1.1.1.27; L-lactate:NAD oxidoreductase)

Lactate dehydrogenase (LD) is a hydrogen transfer enzyme which catalyzes the oxidation of L-lactate to pyruvate with the mediation of NAD as hydrogen acceptor. The reaction is reversible and the reaction equilibrium strongly favors the reverse reaction, namely the reduction of pyruvate to lactate (P → L).

$$\underset{\text{L-Lactate}}{\mathrm{H{-}C(CH_3)(COO^-){-}O{-}H}} + \underset{\text{Oxidized coenzyme I}}{\mathrm{NAD^+}} \underset{\text{pH 7.4–7.8}}{\overset{\text{LD, pH 8.8–9.8}}{\rightleftharpoons}} \underset{\text{Pyruvate}}{\mathrm{CH_3{-}C(=O){-}COO^-}} + \underset{\text{Reduced coenzyme I}}{\mathrm{NADH}} + \mathrm{H^+} \quad (35)$$

The pH optimum for the lactate to pyruvate (L → P) reaction is 8.8 to 9.8; for the P → L reaction it is 7.4 to 7.8. The optimal pH varies with the source of enzyme and depends on the temperature as well as on substrate and buffer concentrations. The specificity of the enzyme extends from L-lactate to a variety of related α-hydroxy acids and α-hydroxy, γ-oxo-acids, although only α-hydroxybutyrate, the next higher homolog, reacts at a rate approximating that for lactate. This catalytic oxidation of α-hydroxybutyrate to α-oxobutyrate is referred to as α-hydroxybutyrate dehydrogenase (HBD) activity. The enzyme does not act on D-lactate, and only NAD will serve as coenzyme. The enzyme has a molecular mass of 134,000, and is composed of four peptide chains of two types, M and H, each under separate genetic control. (Refer to Isoenzymes, p. 597; and LD Isoenzymes, p. 599.)

Lactate dehydrogenase is inhibited by sulfhydryl reagents such as mercuric ions and *p*-chloromercuribenzoate, the inhibition being reversed by the addition of cysteine or glutathione. Borate and oxalate inhibit by competing with lactate for its binding site on the enzyme; similarly, oxamate competes with pyruvate for its binding site. Both pyruvate and lactate in excess inhibit enzyme activity, although the effect of pyruvate is greater. Inhibition by either substrate is greater for the H form than for the M form. Substrate inhibition decreases with increase in pH. EDTA inhibits, perhaps by binding Zn^{2+}, but the postulated activator role for zinc ions is not fully established.

DISTRIBUTION AND CLINICAL SIGNIFICANCE

LD Content in Various Tissues and Serum. LD activity is present in almost all the tissues of the body and is invariably found only in the cytoplasm of the cell. Enzyme levels in various tissues (in U/g) are very high compared to those in serum: liver, 260,000; heart, 160,000 to 240,000; kidney, 250,000 to 300,000; skeletal muscle, 133,000; and whole blood, 16,000 to 62,000. Thus, tissue levels are about 500-fold higher than those normally found in serum, and leakage of the enzyme from even a small mass of damaged tissue can increase the observed serum level of LD to a significant extent.

Serum Levels of Lactate Dehydrogenase in Different Disease States.[49] *Myocardial infarctions* may be associated with elevations of total LD to as much as ten times normal, but as a rule, the elevations seen are only three to six times the average normal value (normal range for the P → L reaction at 30°C is 85 to 190 U/l). The rise in serum level begins about 8 to 12 h after onset of pain, reaches a maximum 48 to 60 h later, and remains elevated for some 7 to 12 days. This contrasts with aspartate aminotransferase values (p. 674), which begin to rise at 6 to 8 h, peak at 24 to 36 h (10 to 20 times normal), but are back within the normal range by the fourth or fifth day. LD values higher than 7 to 8 times normal suggest a poor prognosis. Values may be moderately elevated in myocarditis and in cardiac failure with hepatic congestion, but are normal in angina and in pericarditis. Enzyme levels may be moderately elevated in severe shock and in anoxia.

Elevations of LD activity are observed in *liver disease*, but these, again, are not as great as the increases seen in aminotransferase activity. Elevations are especially high (ten times normal) in toxic jaundice; slightly lower values are observed in viral hepatitis and in infectious mononucleosis. LD activity is normal, or at most only twice the upper limit of normal, in cirrhosis and in obstructive jaundice.

Marked elevations of serum LD activity are seen in patients with untreated *pernicious anemia*. The levels seen may approach some 50 times normal, although they usually range from 1200 to 7500 U/l. With treatment of the anemia, the LD values approach normal. Large elevations (20 times normal) are also observed in *megaloblastic anemia* resulting from folate deficiency. It has been established that the megaloblastic bone marrow is the source of the increased serum LD.[62] A large fraction of the new erythrocytes in the marrow are destroyed while they are maturing or developing.

Increased levels of the enzyme in serum are also found in about a third of patients with *renal disease*, especially those with tubular necrosis or pyelonephritis. However, these elevations do not correlate well with proteinuria and other parameters of renal disease.

About 50 per cent of patients with some form of *malignant disease* show increased LD activity in serum. Especially high values are associated with Hodgkin's disease and with abdominal and lung cancers. The isoenzyme pattern of the elevated LD may occasionally reflect the organ affected by the malignancy, but most often it shows only a nonspecific increase in the slow-moving forms (LD-4 and LD-5), suggesting that the tissue has regressed to synthesizing the more embryonic, anaerobic LD types. As a rule, the elevations of LD seen in cancer patients are too erratic to be of use in clinical diagnosis. More helpful are measurements of LD activity in effusions obtained from areas near or adjacent to malignancies. In these fluids LD levels are often higher than in serum, whereas the opposite is true for fluids bathing healthy tissues. Leukemias are associated with only moderate elevations of serum LD.

Moderately increased LD activity is found in the sera of all patients affected by *progressive muscular dystrophy*, especially in the early and middle stages of the disease. The observed increase is confined to LD-5, the isoenzyme form characteristic of striated muscle. In the later stages of the disease, after a large mass of the LD-5 has been lost, the observed LD levels in serum may even drop to normal levels, with the predominant isoenzyme forms now being LD-1 and LD-2. Only occasionally is an elevated serum LD level found in persons with other forms of neuromuscular disorders.

Elevated values of the enzyme are also seen in cases of *pulmonary emboli*; on occasion, a raised LD level may be the only evidence to suggest the presence of a hidden embolus.

Lactate Dehydrogenase in Urine. Elevations of LD activity in urine to three to six times normal are associated with *chronic glomerulonephritis*, *lupus*, *diabetic nephrosclerosis*, and *bladder and kidney malignancies*. Determination of LD activity in urine is affected by uncertainties arising from the presence of inhibitors (urea, small peptides) and from the possible inactivation of LD under adverse pH conditions in the urine.

Lactate Dehydrogenase in Spinal Fluid. Elevations of LD activity in spinal fluid are associated with *subarachnoid hemorrhage* and with *cerebrovascular thrombosis and hemorrhage*. CSF-LD is usually normal in patients with brain or meningeal tumors, but may be elevated in cases of invasive cancers originating from primary sources elsewhere.

LD ISOENZYME PATTERNS IN DIAGNOSIS[49]

Where only a single tissue is known to be involved in the disease process, the change in LD level in serum will reflect the severity of the insult to that tissue or organ. However, frequently it is not possible to determine clinically which organ is contributing to the rise in LD, particularly in the early stage of a disease process, when the LD may be only marginally elevated. In such cases, a study of the change in LD isoenzyme pattern in serum may be very helpful in suggesting a diagnosis. Each tissue has its specific pattern of enzymes and, therefore, diffusion from a given tissue may impress its pattern onto the pattern found in the serum. LD isoenzyme patterns found by various investigators in some selected tissues and human serum are presented in Table 12-6. The isoenzyme pattern for any given tissue reflects the type of isoenzyme synthesized and the difference in rate between synthesis and degradation. The observed variations in isoenzyme pattern reflect both real variations in sera and differences in methods used to separate and quantitate the isoenzyme fractions. LD-4 and LD-5 form a small fraction of serum because they have a much shorter half-life than do the H forms.

Sera obtained from patients with *myocardial infarctions* will usually show only bands 1 and 2 (and occasionally 3), with the proportion of total LD activity present in the first two bands greater than that present in normal serum. The relatively high proportions of LD-1 and LD-2 in heart tissue will augment those normally present in these bands in serum. In the case of *hepatic disease*, the high proportion of LD-5 in liver will add to the normally low serum level of this isoenzyme and give a pattern in which the first two bands are rela-

TABLE 12-6 ISOENZYME PATTERN IN SOME SELECTED TISSUES IN PERCENTAGE OF TOTAL LD ACTIVITY

Organ or Tissue	*Lactate Dehydrogenase Isoenzyme*				
	LD–1	*LD–2*	*LD–3*	*LD–4*	*LD–5*
Heart	35–70	28–45	2–16	0–6	0–5
Kidney	28	34	21	11	6
Liver	0–8	2–10	3–33	6–27	30–85
Skeletal muscle	1–10	4–18	8–38	9–36	40–97
Brain	21–25	21–26	36–54	15–20	2–8
Erythrocytes	39–46	36–56	11–15	4–5	2
Lung	10	20	30	25	15
Spleen	6–10	11–15	35–40	20–28	5–20
Normal serum	25–31	38–45	17–22	5–8	3–6
Normal serum	31–54	37–54	3–15	0–5	0–3
Normal serum	26–40	40–54	10–18	3–9	0
Normal serum	31–32	40–42	21–22	5–6	0
Normal serum	21–42	40–53	10–23	1–5	0–4

tively lower than normal; bands 4 and 5, normally just barely visible, become distinct and clearly elevated. Quantification of the fractions will show the actual changes in the proportions of the various isoenzymes.

Disease processes in other tissues will also alter the serum isoenzyme pattern, but the changes observed are more difficult to correlate with the specific tissue involved. The pattern for untreated *pernicious anemia* resembles that for myocardial infarction, except for a slight decrease in LD-1 and a small increase in LD-2. The kidney pattern is very much like that for normal serum. In pulmonary infarcts the LD-3 band is elevated, and is almost equal to LD-2, whereas the level of LD-1 is clearly lower. But in general, except for *myocardial infarcts* and *hepatic disease*, isoenzyme patterns are of little value in pinpointing the tissue responsible for the increased serum LD level.

Rosalki and Wilkinson[220] advocated measurement of *α-hydroxybutyrate dehydrogenase* (HBD) activity, and calculation of the LD/HBD ratio. LD-1 and LD-2 (with greater proportions of monomer H) are relatively more active with α-ketobutyrate as substrate than with pyruvate. With normal serum, the LD/HBD ratio varies from 1.2 to 1.6; in *parenchymal liver disease*, the ratio is increased and ranges from 1.6 to 2.5. In *myocardial infarctions*, with elevated LD-1 and LD-2 activity (and greater HBD activity), the ratio is decreased to between 0.8 and 1.2. The exact ratio values will depend on the conditions chosen for the assay of the two activities.

METHODS FOR LACTATE DEHYDROGENASE ACTIVITY MEASUREMENTS

A multiplicity of individual procedures have been introduced over the last 20 years, using both the forward (L → P) and reverse (P → L) reactions. Wroblewski and LaDue adapted the classic assay of Kubowitz and Ott to the determination of LD in serum specimens. Their procedure is based on the P → L reaction and uses a pH of 7.4 and a temperature of 25°C. The reaction is followed by measuring the decrease in absorbance at 340 nm as NADH is oxidized to NAD. The unit of activity was defined as that quantity of enzyme which causes an absorbance change of 0.001 per min if present in a total volume of 3 ml and measured in a cuvet with a 1.0 cm light path. This method was subsequently modified and improved by Henry *et al.*[96]

The determination of LD activity by this method was probably the first *continuous monitoring* ("kinetic") enzyme assay to become widely used in clinical laboratories. The reaction rate for the P → L reaction is relatively rapid with serum specimens, and zero-order kinetics may not be maintained for long because of exhaustion of NADH supply and possible enzyme inhibition or inactivation. However, with adequate initial levels of NADH, reliable estimates of the initial rate can be made, provided that the rate measurement begins almost immediately after initiation of the reaction.

Amador and his associates[2] argued strongly for the use of the L → P reaction at a pH of 9.0 to 9.5. These workers maintained that both lactate and NAD were more stable than pyruvate and NADH and that, in particular, the use of NAD avoids the difficulties that may be encountered as a result of inhibitor formation in solutions of NADH. Furthermore, the inhibitory effect of pyruvate (substrate in the reverse reaction) tends to prevent attainment of true maximum rates. The measured rate by the L → P reaction is only one-fourth that obtained with the P → L reaction.

Both the L → P and P → L procedures are being used and are equally popular. Sufficient knowledge of the LD reaction is not yet available to make possible a decision as to which is the most desirable from the point of accuracy, precision, cost, and technical convenience.

Colorimetric methods available are of two types. In the first group, pyruvate is reacted with 2,4-dinitrophenylhydrazine (2,4-DNPH) to form the corresponding phenylhydrazone, which has a golden-brown color at alkaline pH. The colorimetric procedure of Cabaud and Wroblewski, based on this P → L reaction, is carried out at 37°C, and at a pH of 7.8 to 8.0.

$$\underset{\text{Pyruvic acid}}{CH_3-C(=O)-C(=O)-OH} + \underset{\text{2,4-Dinitrophenylhydrazine}}{H_2N-NH-C_6H_3(NO_2)_2} \xrightarrow{H^+} \underset{\text{Pyruvate 2,4-dinitrophenylhydrazone}}{CH_3-C(=N-NH-C_6H_3(NO_2)_2)-C(=O)-OH} + H_2O \quad (36)$$

$$\downarrow OH^-$$

Golden-brown colored form

The color of the pyruvate–2,4-dinitrophenylhydrazone is measured at 440 or 525 nm, but the units are expressed in terms of the W-L "kinetic" unit. The normal range for this method is 200 to 600 W-L units/ml. Many commercial kits are based on this method, as well as on the similar procedure measuring pyruvate formed in the L → P reaction.

The second group of colorimetric procedures is based on the reduction of such dyes as 2,6-dichlorophenol-indophenol or 2-*p*-iodophenyl-3-*p*-nitrophenyl tetrazolium chloride (INT) by the NADH formed in the forward reaction. Phenazine methosulfate (PMS) serves as an intermediate electron carrier between the NADH and the dyes. The colorimetric methods using tetrazolium salts are frequently used to visualize the various LD isoenzyme fractions after electrophoretic separation. A number of kits based on these reaction schemes are also available.

Inasmuch as NADH shows strong fluorescence, both the L → P and P → L reactions can be monitored fluorometrically, providing a considerably increased sensitivity. These methods are two-point assays, and they are standardized against solutions of NADH or against enzyme solutions of known activity. The NADH formed in an L → P reaction can be determined by using diaphorase to reduce resazurin to resorufin, which also can be measured fluorometrically.

The colorimetric procedures give LD activity values which, in terms of accuracy and precision, are much inferior to those obtained with continuous monitoring methods. Automated versions for all types of procedures have been developed.

Only a continuous monitoring procedure using the P → L reaction will be described in detail. This incorporates suggestions directed toward "optimizing"[82] the measurement of lactate dehydrogenase activity, i.e., establishing those experimental conditions which will permit the reaction to proceed most rapidly and yet maintain linearity for a reasonable period of time.

DETERMINATION OF LD ACTIVITY BY MEASUREMENT OF NADH CONSUMPTION

SPECIMENS

Hemolyzed serum should not be used, since red cells contain 150 times more LD than serum. For the same reason, it is best to separate serum from the clot as soon as possible after the blood specimen has been drawn. Heparinized plasma is satisfactory, but plasma containing other anticoagulants, especially oxalate, should not be used.

Purified LD is unstable when dissolved in water, but it is quite stable in concentrated ammonium sulfate, in glycerol-water (1/1), and in the presence of other proteins. The different isoenzymes differ in their sensitivity to cold, LD-4 and LD-5 being especially labile. In tissue extracts, all activity of these two forms is lost if the extracts are stored at −20°C overnight. Loss of activity may be prevented by addition of NAD or glutathione. Both types of monomers bind a mole of NAD, but the binding of NAD to the M form is weaker and some dissociation occurs,[66] with concomitant exposure of sulfhydryl groups to oxidation. In serum, the sulfhydryl in albumin and other proteins retards inactivation of the M-rich isoenzymes (LD-4 and LD-5). Serum specimens should be stored at room temperature, at which no loss of activity will occur for two to three days. If specimens of sera must be stored for longer periods, they should be kept at 4°C with NAD (10 mg/ml) or glutathione added to decrease the rate of inactivation of LD-4 and LD-5.

PRINCIPLE

Pyruvate is reduced to lactate by LD at pH 7.4 and at 30°C. The progress of the accompanying oxidation of NADH to NAD^+ is monitored continuously by measuring the rate of absorbance decrease at 340 nm in a spectrophotometer. See equation 35.

REAGENTS

1. Tris buffer, pH 7.4, 500 mmol/l. Weigh out 60.6 g of tris(hydroxymethyl)aminomethane, dissolve it in about 700 ml of water, warm to 30°C, and carefully add concentrated HCl to a pH of 7.4, using a pH meter at 30°C for the pH adjustment. Dilute to 1000 ml. The reagent is stable.
2. Working buffer, Tris, 57.5 mmol/l, pH 7.4. Dilute 57.5 ml of the 500 mmol/l buffer to 500 ml with water. Check pH after warming to 30°C, and adjust to 7.4 with 1 molar HCl or 1 molar NaOH.
3. NADH solution, 5.58 mmol/l in Tris buffer, pH 7.4. Dissolve 21.8 mg of β-NADH (disodium salt, trihydrate) in 5.0 ml of 57.5 mmol/l Tris buffer. Stable for 72 h at 4°C. It is recommended that only a volume sufficient for a day's work be prepared.
4. Pyruvate, 14.0 mmol/l in Tris buffer, pH 7.4. Dissolve 155 mg of sodium pyruvate in 100 ml of the working buffer. Keep refrigerated; stable for 3 to 4 weeks.
5. Dichromate blanks. Stock, 150 mg $K_2Cr_2O_7$/500 ml of water containing 3 to 4 drops of concentrated H_2SO_4. This is diluted 4-fold, 7-fold, etc., as needed (see step 2 of procedure).

PROCEDURE

1. Use a narrow band-pass spectrophotometer equipped with a circulating constant temperature water bath or equivalent device. Set the thermostat to maintain the temperature of the cuvet compartment at 30°C (±0.1°C).
2. Pipet 2.70 ml of Tris buffer (57.5 mmol/l), 100 μl of the NADH solution, and 100 μl

of serum into a 1.0 cm cuvet. After mixing the contents, place the cuvet in a 30°C incubator or water bath for 10 to 20 min. This incubation permits a reduction by the NADH of any pyruvate and other oxo-acids present in the serum. During this period, measure the absorbance (at 340 nm) of the reaction mixture and select a dichromate blank that will give an absorbance reading for the test cuvet of between 0.55 and 0.70. (See comment 1.)

3. At the end of the incubation period, add 200 μl of pyruvate solution, which has also been prewarmed to 30°C, mix rapidly, and then take absorbance readings at 0.5 or 1.0 min intervals for 3 to 6 min, either manually or by means of a recorder. The fall in absorbance per min should be constant over the reaction period. If the ΔA/min begins to fall off either gradually or sharply before more than four readings have been made, exhaustion of NADH may have occurred and the run should be repeated with 50 μl of serum or a dilution (5- or 10-fold) of the serum with buffer (or preferably albumin solution, 5 g/100 ml).

The actual concentration of reaction components in the cuvet is as follows: Tris buffer, pH 7.4, 50 mmol/l; NADH, 0.18 mmol/l; pyruvate, 0.90 mmol/l; ratio of specimen volume to total volume, 1/31.

CALCULATION

The LD activity in International Units at 30°C is obtained as follows:

$$\text{mU/ml} = \frac{\Delta A}{\text{min}} \times \frac{1000}{6.22} \times \frac{3.1}{0.1} = 4985 \times \frac{\Delta A}{\text{min}}$$

where 3.1 = total volume in cuvet, in ml

0.1 = volume of serum specimen, in ml

6.22 = millimolar absorptivity of NADH at 340 nm (unit = ml × mmol^{-1} × cm^{-1})

1000 = converts mmol to μmol

$\frac{\Delta A}{\text{min}}$ = average absorbance change (decrease) per min

The value of ΔA/min can be calculated by using total ΔA in t min/t min, or by plotting the data on graph paper (absorbance against time), drawing the best fitting line, and then calculating ΔA/min.

If the actual measurement of enzyme activity is made at some other temperature, this can be converted to U/l at 30°C by use of the following average temperature correction factors (after Demetriou *et al.*[46]): 25°C, 1.44; 30°C, 1.00; 32°C, 0.86; 35°C, 0.70; 37°C, 0.60. The use of these temperature factors involves some uncertainty, and is to be avoided, if possible.

COMMENTS

1. Reaction blank: The blank is used to compensate for the absorbance of serum pigments. Some analysts use water, or even omit the blank cuvet. Others have used a blank containing serum and buffer only. However, the dichromate blanks are convenient. With spectrophotometers with an absorbance offset mechanism, blanks are not necessary.

2. Some analysts omit the preincubation step, claiming that the endogenous reactions do not significantly alter the value of ΔA/min; others use very short preincubation periods (3 to 5 min).

3. In the Wroblewski-LaDue procedure, activity was expressed in the "Karmen" or "spectrophotometric" unit. Multiplying by 2.07 will convert U/l to W-L units/ml.

4. In their procedure, Henry *et al.*[96] used phosphate buffer at pH 7.4, 97 mmol/l; NADH, 0.22 mmol/l; and pyruvate, 0.60 mmol/l.

In the optimized procedure proposed by German clinical chemists,[82] the buffer is also phosphate, but at pH 7.5 and at 40 mmol/l; NADH and pyruvate are used at 0.18 and 0.60 mmol/l, respectively. The Scandinavian workers[175] propose Tris at 50 mmol/l, NADH and pyruvate at 0.15 and 1.2 mmol/l, respectively. Tris buffer is to be preferred because NADH is more stable in Tris than in phosphate buffer.

NORMAL VALUES

Values for the normal range of LD activity in serum vary considerably, depending on the direction of the enzyme reaction, the type of method used, and the experimental parameters. For the P → L reaction at 30°C, and at pH 7.4, a range of 95 to 200 U/l represents the experience of most workers in the field. This is equivalent to 205 to 425 W-L units/ml. For the L → P reaction at pH 8.8 to 9.0, the range of 35 to 88 U/l (at 30°C) represents the consensus of values proposed by various investigators.

The normal value for spinal fluid LD is 7 to 30 U/l for the P → L reaction at 30°C.

REFERENCE

Scandinavian Society for Clinical Chemistry and Clinical Physiology: Recommended Methods for the Determination of Four Enzymes in Blood. Scand. J. Clin. Lab. Invest., *33*:291, 1974.

METHODS FOR DEMONSTRATING OR MEASURING CHANGES IN LD ISOENZYME PATTERNS

Electrophoretic separation on cellulose acetate or agar or polyacrylamide gels[50,123] is still the procedure most commonly used, and many procedures and specialized apparatus are available commercially. A solution of thin agar gel containing the reaction mixture (NAD, lactate, buffer) is layered over the strip or gel that contains the separated enzymes. NADH is formed and can be visualized and quantitated, either by virtue of its fluorescence when excited by long wave ultraviolet light (365 nm) or by its reduction of a tetrazolium salt (MMT, NBT) with the formation of a colored formazan, which can be measured colorimetrically.

A less precise but often clinically useful evaluation of a serum LD pattern may be obtained by determining the proportion of LD activity destroyed by heating serum at 57°C for 30 min and the fraction which retains its activity after 30 min incubation at 65°C. If total activity (at normal assay temperature) is denoted by T, the activity remaining after heating at 57°C by L, and that remaining after heating at 65°C by H, then $(T - L)$ will represent the heat labile (LD-4 and LD-5) fractions, with a value of 10 to 25 per cent in normal sera, increasing to 33 to 80 per cent in patients with liver disease. The stable fraction is given by H, with a normal range from 20 to 40 per cent, rising to 45 to 65 per cent in patients with myocardial infarctions.

When α-oxobutyrate is used as enzyme substrate in place of pyruvate, the reduction of substrate proceeds at an appreciable rate only when LD-1 and LD-2 are present, the other LD isoenzymes being essentially inactive. α-Hydroxybutyrate dehydrogenase activity (HBD), as seen in serum, is considered by most experts not to represent a different specific enzyme, but to represent the LD activity of the LD-1 and LD-2 isoenzymes. Its measurement can then be a measure of the cardiac LD isoenzymes. British workers advocated the measurement of HBD activity and the calculation of the LD/HBD ratio (see p. 655). In general, the reaction systems are identical, except that five to ten times as much α-oxobutyrate is used in place of pyruvate. The reader is referred to the procedure of Ellis and Goldberg.[59]

Recently, efforts have been made to separate the LD isoenzymes by use of chromatography and by immunochemical techniques. These methods have been more successfully applied to the isoenzymes of creatine kinase, and are briefly discussed on page 689. The immunochemical approaches have been disappointing. Antisera can be prepared against both the M and H subunits, but they will precipitate not only the H_4(LD-1) and M_4(LD-5) isoenzymes, but also variable amounts of the intermediate forms (H_3M, H_2M_2, HM_3). Thus the LD-1 and LD-2 isoenzyme forms cannot be isolated cleanly from the other accompanying forms; the same is true for the liver forms, LD-4 and LD-5.

ISOCITRATE DEHYDROGENASE

(EC 1.1.1.42; threo-D_s-isocitrate:NADP oxidoreductase (decarboxylating))

Serum isocitrate dehydrogenase (ICD) catalyzes the oxidative decarboxylation of isocitrate to α-oxoglutarate (equations 37 and 38). The enzyme is substrate specific; only D_s-isocitrate is acted on by the enzyme (both *cis*-aconitate and L-isocitrate are inert), and only NADP can serve as the hydrogen transfer coenzyme in the reaction. It is not clear whether the oxidation and decarboxylation reactions occur simultaneously or in separate steps. The enzyme that is found in serum is primarily of liver origin, although it is found to some extent in all cells in soluble form in both the cytoplasm (cytosol) and the mitochondria. Another form of ICD (EC 1.1.1.41) is present in all cells, but it occurs only in the mitochondria (bound to particulate matter), and requires NAD as the hydrogen acceptor. The two enzymes further differ in the following properties: The serum (cytosol) enzyme is inert to ADP, which is a regulator (activator) of the mitochondrial enzyme. The accumulation of NADPH (product) has little effect on the activity of the serum (soluble) enzyme, whereas the mitochondrial enzyme is inhibited by its product NADH. The serum enzyme can decarboxylate added oxalosuccinate (equation 38), whereas the mitochondrial enzyme has no decarboxylating activity. It is the mitochondrial enzyme which participates in the tricarboxylic acid cycle; the soluble enzyme functions only in auxiliary biosynthetic reactions.

$$\underset{D_s\text{-Isocitrate}}{\begin{matrix} COO^- \\ | \\ H\text{—}C\text{—}OH \\ | \\ {}^-OOC\text{—}C\text{—}H \\ | \\ CH_2 \\ | \\ COO^- \end{matrix}} + \underset{\text{Coenzyme II}}{NADP^+} \underset{\longleftarrow}{\xrightarrow{\text{Isocitrate dehydrogenase}}} \underset{\text{Reduced coenzyme II}}{NADPH} + H^+ + \underset{\text{Oxalosuccinate}}{\begin{matrix} COO^- \\ | \\ C{=}O \\ | \\ {}^-OOC\text{—}C\text{—}H \\ | \\ CH_2 \\ | \\ COO^- \end{matrix}} \qquad (37)$$

Hydrogen Transfer Reaction

$$\underset{\text{Oxalosuccinate}}{\begin{matrix} COO^- \\ | \\ C{=}O \\ | \\ {}^-OOC\text{—}C\text{—}H \\ | \\ CH_2 \\ | \\ COO^- \end{matrix}} + H^+ \underset{Mn^{2+}}{\xrightarrow{\text{Isocitrate dehydrogenase}}} \underset{\alpha\text{-Oxoglutarate}}{\begin{matrix} COO^- \\ | \\ C{=}O \\ | \\ H\text{—}C\text{—}H \\ | \\ CH_2 \\ | \\ COO^- \end{matrix}} + \underset{\text{Carbon dioxide}}{CO_2} \qquad (38)$$

Decarboxylation Reaction

The serum type enzyme is found in high concentrations not only in the liver but also in heart, skeletal muscle, kidney, and adrenal tissue, as well as in platelets and red cells. It has a relative molecular mass of 64,000, and requires Mn^{2+} as an activator, the ion being essential for the decarboxylation reaction. The concentration giving optimal activation is in the range of 0.5 to 1.5 mmol/l. The Mn^{2+} can be replaced by Mg^{2+} or Co^{2+}; however, only 60 to 80 per cent of the activity possible with Mn^{2+} is obtained, even at the optimal concentration range of 1.7 to 3.0 mmol/l. The pH range for optimal activity is broad (pH 7.0 to 7.8).

Sulfhydryl binding reagents such as Cu^{2+}, Hg^{2+}, iodoacetate, and *p*-chloromercuribenzoate ($COOH\text{-}C_6H_4\text{-}HgCl$) are potent inhibitors, and the inhibition can be only partly reversed by addition of sulfhydryl compounds such as glutathione. Cyanide, azide, and *N*-ethylmaleimide are also potent inhibitors of enzyme activity. Recently, it has been shown that inhibition occurs in the presence of NaCl at concentrations as low as 50 mmol/l;[60] at 300 mmol/l the degree of inhibition is 50 per cent. The effect appears to be due to the Na^+, inasmuch as similar inhibition is obtained if other sodium salts are used in place of NaCl. Decrease in activity is also observed if the concentration of a variety of buffers is increased above 0.10 mol/l. This suggests that enzyme activity is depressed in the presence of increasing ionic strength. Anticoagulants such as oxalate, EDTA, and F^- also inhibit activity, perhaps by removing the activating Mn^{2+}.

Reports of different isoenzyme forms of ICD have been published[122] but definitive data are still lacking. Apparently four forms may be detected in serum, although only two can be demonstrated in liver and in cardiac tissue. The form with the fastest mobility is most abundant in liver tissue, whereas in heart extracts the predominant form has a slow mobility, and is also very heat-labile.

CLINICAL SIGNIFICANCE

The level of ICD activity found in the *sera* of healthy adults is quite low (1.2 to 7.0 U/l, measured at 30°C). Elevations of ICD activity have been found to be very sensitive indicators of parenchymal liver disease, and provide for early detection, inasmuch as they can be seen even in the early, incubation phase of the disease. As prognostic tools, they are of less value. The highest values are found in viral hepatitis (10 to 40 times the upper limit of normal). In chronic hepatitis, increased levels of enzyme activity can persist for several months, although serum ICD values usually return to within the normal range within 14 to 20 days. A sudden large drop in activity following an appreciable initial rise suggests that massive cell necrosis may have occurred, with a poor prognosis. Elevations as high as those seen in viral hepatitis may also be seen in cases of infectious mononucleosis, but in patients with serum or toxic hepatitis the elevations observed are only of the order of three to eight times normal.

Normal or slightly increased levels of ICD activity are encountered in uncomplicated obstructive jaundice, and the degree of increase is independent of the severity, duration, and etiology of the obstructive process.[215] However, very high levels of activity are observed in cases of acute inflammation of the biliary tract, the source of the enzyme probably being extrahepatic. High levels of the enzyme are also common in neonatal biliary duct atresia. In cirrhosis, serum enzyme levels will vary from normal to four times the upper normal value. If persistent and more elevated enzyme values are seen, the prognosis is poor. Hepatic neoplasms and carcinoma metastatic to the liver are also associated with slight to moderate elevations of serum ICD activity.

Normal enzyme values are encountered in a variety of diseases involving the heart, lungs, kidney, skeletal muscle, and other tissues. It is remarkable that despite the very high concentration of ICD activity in heart tissue, no elevations are observed in myocardial infarctions, unless the infarct is accompanied by congestive failure resulting in hepatic ischemia. Presumably the heart enzyme is either very labile, or is differentially cleared from the plasma more rapidly than are the other enzyme forms. However, increases in enzyme level are observed in the majority of cases of severe pulmonary infarctions.

Administration of alcohol[85] causes a rise in ICD activity in serum, and it has been suggested that ICD values be used as a measure of the reaction of the liver to oral ingestion of alcohol. Similarly, a number of drugs, such as *p*-aminosalicylic acid, may also cause a rise in the ICD level. This probably reflects the toxic effect of these drugs on the liver cells.

A sudden rise in the ICD level during pregnancy is suggestive of possible placental damage or degeneration. However, in uncomplicated pregnancies, ICD levels remain within normal limits.

Red cells contain some 100 times more enzyme than does serum. However, normal values for serum ICD activity are seen in all forms of anemias except those due to vitamin B_{12} or folate deficiency (the megaloblastic anemias), in which activity may range from normal to eight times the upper limit of normal.

Very low levels of ICD activity are encountered in *cerebrospinal fluid* (up to 0.3 U/l). These levels are too low to measure with the usual procedures used for serum. Moderate elevations of ICD in cerebrospinal fluid have been reported in cases involving acute bacterial meningitis, vascular cerebral lesions, and tumors primary or metastatic to the cerebrospinal system.

ASSAY METHODS

Wolfson and Williams-Ashman[225] first quantitated ICD activity in serum by using a manual spectrophotometric procedure and measuring the increase in NADPH concentration (ΔA_{340}) during the course of the reaction. This procedure was modified[27,46] to provide the continuous monitoring methods in current use. The various authors differ in the choice of buffer, temperature, and NADP and Mn^{2+} concentration; no "optimized" procedure has as yet been developed.

Some two-point, colorimetric methods have also been worked out. In one approach (Bell and Baron, and others), the oxoglutarate formed in the reaction is measured in the form of its colored 2,4-dinitrophenylhydrazone (ΔA_{390}). In an alternate approach (Nachlas), the NADPH formed was measured by its ability to reduce a tetrazolium dye (INT) to a colored formazan, with phenazine methosulfate (PMS) serving as electron carrier between the NADPH and the tetrazolium salt.

The continuous monitoring procedure of Ellis and Goldberg[60] will be presented in detail.

DETERMINATION OF ISOCITRATE DEHYDROGENASE ACTIVITY IN SERUM

SPECIMEN

Assays of enzyme activity are best done using a serum specimen which is free of lipemia and hemolysis. It is preferable to separate the serum from the clot as soon as possible, although contact of serum with cells for one to two hours does not significantly increase the serum ICD activity. Plasma prepared from either citrated or heparinized blood may also be used, although some reports suggest that heparin reacts with some amine buffers, causing the development of opalescence in the reaction cuvets. Other anticoagulants must be avoided.

Reports as to the stability of ICD in serum are contradictory. Most investigators agree that the activity is stable at 4°C for two to three weeks. Activity is unaffected by incubation at 55°C for one hour, or storage at room temperature for up to six hours. There is more disagreement as to stability when solutions are stored in the frozen state; but loss of activity, when it occurs, seldom exceeds 30 per cent.

REAGENTS

1. Buffer, 0.10 mol triethanolamine (TEA)/l, pH 7.3, 30°C. Dissolve 18.87 g of triethanolamine hydrochloride in approximately 750 ml of water. Add 15 ml of 1.0 molar NaOH, warm the solution to 30°C, and then add additional NaOH to a pH of 7.3, checking

the pH with a pH meter. Dilute with water to a volume of 1000 ml. Store in a dark cabinet at 25°C.

2. Activator, stock solution, 100 mmol Mn^{2+}/l. Dissolve 1.98 g of $MnCl_2 \cdot 4H_2O$ in water and dilute to a volume of 100 ml. Store at 4°C.

3. NADP solution, 13.3 mmol/l. Dissolve 20 mg of NADP, disodium salt, in 2.0 ml of water. Prepare just enough for one day's work. Any excess may be stored frozen for a few days only.

4. Substrate, D,L-isocitrate, 67 mmol/l. Dissolve 98.5 mg of trisodium D,L-isocitrate dihydrate in the TEA buffer and make up volume to 5.0 ml.

5. Composite reagent: When large numbers of assays are to be carried out, it is convenient to prepare a composite of reagents 1, 2, and 3, in the proportions 1.95, 0.05, and 0.10 ml, respectively, and to pipet 2.1 ml into the reaction cuvet.

PROCEDURE

1. Pipet 1.95 ml of TEA buffer, 0.05 ml of Mn^{2+} solution, and 0.10 ml of NADP (or 2.10 ml of the composite reagent) into each of two 1.0 cm cuvets ("Test" and "Blank").

2. Add 0.60 ml of the serum to be assayed to each cuvet and equilibrate the cuvets and contents at 30°C for 10 min.

3. Initiate the enzyme reaction by adding 0.30 ml of 67 mmolar isocitrate to the "Test" (reaction) cuvet. To the "Blank" cuvet add an additional 0.30 ml of buffer. Mix the cuvet contents rapidly, and then measure the change in absorbance of the test cuvet at 30°C continuously (or at 1 min intervals) for a period of 8 to 12 min. Readings are made against the blank at 340 nm. Alternatively, the change in absorbance in both cuvets may be read against water in the reference cuvet; the test readings are corrected for the blank readings.

4. The reaction shows a lag phase of 3 to 6 min, being longest in specimens with the least activity. After the reaction has become linear, take or record additional readings for a period of 4 to 5 min, and calculate ΔA/min for the linear phase.

CALCULATIONS

$$\text{ICD activity in serum in U/l} = \frac{\Delta A}{\text{min}} \times \frac{3.0}{0.6} \times \frac{1000}{6.22}$$

$$= \frac{\Delta A}{\text{min}} \times 804$$

where 3.0 = final volume in reaction cuvet, in ml

6.22 = mmolar absorptivity for NADPH at 340 nm

0.6 = volume of specimen used for assay, in ml

1000 = number of ml in 1.0 liter

$\frac{\Delta A}{\text{min}}$ = absorbance change per min, during linear phase of the reaction

COMMENTS

1. Tris and TEA are the most suitable buffers, although it has been reported that they react with heparin to form a fine opalescence in the reaction cuvet. Phosphate reacts with the Mn^{2+} and cannot be used; also, it promotes breakdown of the NADPH. Piperazine, *N*-ethylmorpholine, and cacodylic acid are satisfactory, but offer no advantages over Tris or TEA. No matter which chemical is used as buffer, the blanks become increasingly larger, as the pH is made more alkaline than 7.5.

2. Except for NADP, all reagents are present at optimal concentrations. About 10 per cent additional activity can be observed if the NADP level is doubled to 0.85 mmol/l. The concentrations of the various components in the reaction cuvet are as follows: TEA buffer, pH 7.3, 83 mmol/l; Mn^{2+}, 1.67 mmol/l; NADP, 0.43 mmol/l; D,L-isocitrate, 6.7 mmol/l; and serum dilution, 1/5. In procedures developed by other workers, NADP and isocitrate are used at much lower concentrations, whereas the level of Mn^{2+} is higher than necessary.

3. Because of the low level of ICD activity in serum, a relatively large volume of serum specimen must be used if a measurable reaction rate is to be obtained. With instruments capable of measuring accurately very small changes in absorbance, it is possible to decrease the volume of specimen; with less sensitive instruments, serum volume and reaction time may have to be increased.

4. With the majority of serum specimens, blank values rarely exceed 1.0 U/l, usually being much less. With some specimens, however, blanks can be much larger or erratic or both. As a result, some analysts prefer not to use blanks; this may be satisfactory with instruments permitting very short runs, but in the usual situation it is good policy to run the blank and to correct for it.

5. When Mn^{2+} is mixed with serum and buffer, a fine precipitate may occasionally be formed, giving rise to an opalescence in the specimen. In one procedure this precipitate is filtered out by passing the reaction mixture through glass wool, before initiating the reaction with isocitrate.

6. Some of the opalescence problems may be partly resolved by the addition of NaCl to the reaction mixture, since this prevents or retards turbidity formation. For example, Bernt and Bergmeyer[19] add NaCl to the mixture to a concentration of 42 mmol/l. However, Ellis and Goldberg[60] have demonstrated that the addition of sodium salts inhibits the ICD reaction, and the use of NaCl is to be avoided. As little as 50 mmol/l is inhibitory, and, depending on the volume of serum used, the specimen itself may introduce some 25 to 45 mmol Na^{+}/l.

7. Temperature-reaction rate factors: The following factors may be used to convert activity measured at the given temperature to that at 30°C: 25°C, 1.60; 28°, 1.18; 30°, 1.00; 32°, 0.85; 35°, 0.65; 37°, 0.56.

NORMAL VALUES

Ellis and Goldberg,[60] studying a mixed population of adults (male and female), obtained an average value for ICD activity in serum of 4.1 U/l, with a standard deviation of ±1.4 U/l. Measurements were made at 30°C. From these values we can deduce that the 95 per cent range of values is approximately 1.3 to 6.9 U/l. The normal range given by Bernt and Bergmeyer[19] is 1 to 4 U/l at 25°C. Multiplied by the appropriate temperature factor, this range becomes 1.6 to 6.4 U/l at 30°C. A similar treatment of the range given by Wolfson and Williams-Ashman (50 to 260 nmol/h/ml at 25°C) gives 1.3 to 6.9 U/l at 30°C. Thus, a range of 1.2 to 7.0 U/l (30°C) may be accepted as a reasonable composite value, and this is consistent with reports by other investigators.

Any differences in ICD values due to sex are not evident; racial and ethnic differences are still to be investigated. ICD activity in the sera of children is the same as that in adults. Cord blood values are roughly twice that for adults. In newborns, activity rises for a few days to values four times the adult normal; these then decrease to just above the upper normal by the end of the second week.

REFERENCE

Ellis, G., and Goldberg, D. M.: Clin. Biochem., *2*:175, 1971.

SORBITOL DEHYDROGENASE

(EC 1.1.1.14; L-iditol:NAD 5-oxidoreductase (polyol) dehydrogenase, SDH)

Sorbitol dehydrogenase (SDH)* catalyzes the reversible oxidation of D-sorbitol by NAD to D-fructose. The forward (sorbitol to fructose, S → F) reaction is favored by an alkaline pH (about pH 9), whereas the F → S reaction proceeds best at a near neutral pH.

$$\begin{array}{c}CH_2OH\\|\\H-C-OH\\|\\HO-C-H\\|\\H-C-OH\\|\\H-C-OH\\|\\CH_2OH\end{array} + NAD^+ \underset{\text{pH } 6.0\text{–}7.5}{\overset{\text{pH } 8.5\text{–}9.5}{\rightleftharpoons}} \begin{array}{c}CH_2OH\\|\\C=O\\|\\HO-C-H\\|\\H-C-OH\\|\\H-C-OH\\|\\CH_2OH\end{array} + NADH + H^+ \qquad (39)$$

D-Sorbitol (D-Glucitol) — D-Fructose

The actual pH permitting the fastest rate depends on the type of buffer and on the fructose (or sorbitol) concentration. The enzyme can oxidize a number of other sugar alcohols, such as L-iditol, D-xylitol, and D-ribitol; hence the practical name, polyol dehydrogenase, and the formal name, iditol oxidoreductase. However, only NAD can serve as H acceptor for the SDH found in serum, which originates in the cytoplasm of liver cells. An analogous enzyme present in mitochondria requires NADP as H acceptor.

CLINICAL SIGNIFICANCE

Sorbitol dehydrogenase is a good example of an enzyme with a high degree of organ specificity. Significant quantities of the enzyme are found only in the liver, with lesser quantities in the kidney and prostate. Cardiac and skeletal muscle and other tissues contain comparatively little SDH activity.

In *healthy persons* activity is barely detectable in serum, and the appearance of any measurable activity invariably indicates the presence of parenchymal liver cell damage. In *acute hepatitis* the level of serum SDH can increase to between 10 and 30 times the normal level. Similar elevations are seen in liver damage resulting from hypoxia following circulatory insufficiency or shock. Even greater elevations in activity are seen in cases involving poisoning by inhalation of organic solvents. However, in *chronic, stabilized hepatitis*, as well as in *cirrhosis* and *obstructive jaundice*, enzyme activity, after the initial rise, falls to normal or at most stays marginally elevated. Normal levels of activity are seen in myocardial infarctions, diabetes, and progressive muscular dystrophy. In cancer metastatic to the liver, serum activity is normal or slightly elevated.

METHODS FOR THE MEASUREMENT OF SORBITOL DEHYDROGENASE ACTIVITY[34,81,134]

SDH activity in serum may be quantitated colorimetrically by using the S → F reaction (equation 39) and measuring the amount of fructose formed (by reacting it with orcinol). However, the preferred procedure is to follow the increase in NADH spectrophotometrically. Most current procedures are based on the work of Gerlach and use the F → S reaction, which proceeds four to six times faster than the S → F reaction. Triethanolamine

* The unofficial abbreviation proposed is ID; one author (JFK) prefers the old form, SDH, which is used in the text.

(TEA) and tris(hydroxymethyl)aminomethane (Tris) are used as buffers at a concentration of about 125 mmol/l. Fructose and NADH are present in the reaction mixture in final concentrations of 400 and 0.25 to 0.40 mmol/l, respectively. A relatively large volume of serum is needed (serum/total reaction volume = 1/3 or 1/4) because of the low level of activity in normal serum. After allowing the endogenous reactions to proceed to completion (20 to 30 min), the reaction is initiated by the addition of fructose, and the ΔA/min is measured over a 5 to 8 min period.

COMMENTS

1. Hemolysis presents no problem, as there is no enzyme activity in erythrocytes.
2. The enzyme is very unstable; activity drops rapidly in sera stored at room temperature, and appreciable activity is lost in two days even at 0 to 4°C.
3. Serum is the preferred specimen; heparinized plasma is satisfactory, but other anticoagulants are inhibitory.
4. Because of the high initial absorbance, it is most practical to set the instrument against a serum blank.

NORMAL VALUE

Less than 1.0 U/l at 25°C; less than 1.3 U/l at 30°C.

REFERENCES

Bruns, F. H., and Werners, P. H.: Dehydrogenases. *In* Advances in Clinical Chemistry, Vol. 5. H. Sobotka and C. P. Stewart, Eds. New York, Academic Press, 1962, pp. 271–280.

Gerlach, U., and Hiby, W.: Sorbitol dehydrogenase. *In* Methods of Enzymatic Analysis, 2nd English Ed. H. U. Bergmeyer, Ed. New York, Academic Press, 1974, pp. 569–573.

GLUCOSE-6-PHOSPHATE DEHYDROGENASE

(EC 1.1.1.49; D-glucose-6-phosphate:NADP oxidoreductase)

Glucose-6-phosphate dehydrogenase (GPD) is a hydrogen transfer enzyme which mediates the reversible transfer of hydrogen from β-D-glucose-6-phosphate to coenzyme II (NADP).[34] The products of the reaction are 6-phosphogluconolactone and the reduced form of the coenzyme, NADPH. The enzyme has also been identified with the names *Zwischenferment* and *hexose phosphate dehydrogenase*. The following reaction is involved:

```
      H                                                     O
      |                                                     ‖
HO—C———————┐                                           C———————┐
      |          |                                          |          |
 H—C—OH         |            Glucose-6-phosphate     H—C—OH         |
      |          |               dehydrogenase              |          |
HO—C—H         O + NADP⁺  ——————————————————→   HO—C—H         O + NADPH + H⁺   (40)
      |          |                  (GPD)                   |          |
 H—C—OH         |    Coenzyme II                   H—C—OH         |      Reduced
      |          |                                          |          |    coenzyme II
 H—C———————┘                                        H—C———————┘
      |                                                     |
  H2C—OPO3 2−                                           H2C—OPO3 2−

β-D-Glucose-6-                                       D-Glucono-α-
  phosphate                                        lactone-6-phosphate
   (G6P)                                          (6-Phosphoglucono-
                                                    lactone) (PGL)
```

If a pure preparation of GPD is used, the reaction stops at the lactone stage. In tissues, however, a lactonase is present which catalyzes the rapid hydrolysis of the lactone to 6-phosphogluconate (PGA), which is the substrate of another enzyme, 6-phosphogluconate dehydrogenase (EC 1.1.1.44; PGD). This latter enzyme is also an NADP-mediated hydrogen transferase, which oxidizes 6PGA with decarboxylation to ribulose-5-phosphate and carbon dioxide. These two reactions are presented in equations 41 and 42.

$$\begin{array}{c} \text{C}{=}\text{O} \\ \text{H—C—OH} \\ \text{HO—C—H} \quad \text{O} \\ \text{H—C—OH} \\ \text{H—C} \\ \text{H}_2\text{C—OPO}_3^{2-} \end{array} + \text{HOH} \xrightarrow[\textit{lactonase}]{(\text{OH}^-)\ \textit{or}} \begin{array}{c} \text{O}{=}\text{C—O}^- \\ \text{H—C—OH} \\ \text{HO—C—H} \\ \text{H—C—OH} \\ \text{H—C—OH} \\ \text{H}_2\text{C—OPO}_3^{2-} \end{array} + \text{H}^+ \qquad (41)$$

6-Phosphogluconolactone (PGL) → **6-Phosphogluconate (PGA)**

$$\begin{array}{c} \text{O}{=}\text{C—O}^- \\ \text{H—C—OH} \\ \text{HO—C—H} \\ \text{H—C—OH} \\ \text{H—C—OH} \\ \text{H}_2\text{C—OPO}_3^{2-} \end{array} + \text{NADP}^+ + \text{H}^+ \xrightarrow[\substack{\textit{dehydrogenase} \\ \text{(PGD)}}]{\substack{\textit{6-Phospho-} \\ \textit{gluconate}}} \begin{array}{c} \text{H} \\ \text{H—C—OH} \\ \text{C}{=}\text{O} \\ \text{H—C—OH} \\ \text{H—C—OH} \\ \text{H}_2\text{C—OPO}_3^{2-} \end{array} + \text{NADPH} + \text{CO}_2 + \text{H}^+ \qquad (42)$$

6-Phosphogluconate (PGA) → **Ribulose-5-phosphate (Ru5P)**

Glucose-6-phosphate dehydrogenase is present in practically all mammalian cells. Normally, the highest levels are present in the adrenal glands and in adipose tissue, although even higher concentrations are found in the mammary glands during lactation. The enzyme present in erythrocytes is of most clinical interest, although GPD is also present in leukocytes and platelets. Only very small amounts are present in serum, skeletal muscle, heart, and kidney.

The enzyme present in normal red cells has a relative molecular mass of about 110,000 and is a dimer made up of two identical subunits; its greatest activity is in the pH range from 8 to 9. At appropriate values for pH, ionic strength, and protein concentration, there exist monomeric, tetrameric, or higher polymeric forms with either no or reduced activity. The enzyme acts preferentially on the β anomer of glucose-6-phosphate, although galactose-6-phosphate can be oxidized at a rate about one-tenth that for G6P, and 2-deoxy-D-glucose-6-phosphate is acted upon by one group of variant forms of the enzyme. The preferred coenzyme is NADP, although NAD can serve as a weak hydrogen acceptor. In the presence of NADP, all the enzyme is saturated with the NADP; this protects the enzyme from inactivation, especially during lysis of red cells. Mg^{2+} is an obligate activator; NADPH and ATP inhibit enzyme activity. Most heavy metals, except copper and zinc, are powerful inhibitors; this inhibition can be counteracted by EDTA, but not by sulfhydryl agents. The enzyme is also inhibited by many organics and drugs such as sulfonamides, primaquine (and other 8-aminoquinoline antimalarials), quinine, antipyretics, vitamin K analogs, and nitrofurantoin.

The oxidation of glucose-6-phosphate by GPD is the first step in the so-called pentose-phosphate shunt,[104] which accounts for about 10 per cent of glucose utilization. The main portion of glucose is metabolized by the Meyerhof pathway to pyruvate and lactate, and is eventually oxidized via the citric acid cycle. It appears that the function of the pentose-phosphate shunt is to produce NADPH and pentose phosphates; the former is needed for vital synthetic reactions, and the latter serve as starting materials for nucleotide and nucleic acid synthesis. By referring to equations 40 and 42, it can be seen that the successive GPD and PGD reactions produce 2 moles of NADPH for each glucose molecule oxidized to pentose phosphate. More specifically, high levels of NADPH are required by red cells to maintain sufficient glutathione (GSH) in the reduced state, without which cell integrity cannot be preserved.

Oxyhemoglobin in the cell oxidizes some metabolites and many drugs, with the formation of H_2O_2. In a reaction catalyzed by glutathione peroxidase, this peroxide is reduced to water, with the accumulation of oxidized glutathione (GSSG). The latter is then reduced to GSH by the action of glutathione reductase using NADPH as hydrogen donor. If GPD (and/or PGD) activity is deficient, too little NADPH is formed to maintain the necessary high cellular concentration of GSH, without which defects in the cell membrane structure occur and early senescence of the red cells results.[55]

CLINICAL SIGNIFICANCE[21,22,34,115]

Normal values for the level of GPD activity in serum and in erythrocytes are subject to some uncertainty. The serum level is affected by the degree of lysis of red cells (and to a lesser degree by the lysis of leukocytes and platelets). The level of activity in red cells is subject to hormonal factors, and depends on the genetic variants of the enzyme present in the cell; over 80 variants of the enzyme have been described. The normal range of GPD activity in serum is less than 3.0 U/1, measured at 30°C. The range of activity found in red cells may be given as 120 to 280 U/10^{12} cells, using the most appropriate unit for expressing activity. Young cells (less than 20 days old) have more enzyme activity than older or moribund cells. The level in red cells in newborns is 50 per cent greater than that in adults, and even greater in low-birth-weight and premature infants.

Although increased levels of GPD activity in *serum* may be observed in pulmonary and myocardial infarctions, the changes in enzyme activity in *red cells* are of most clinical interest. Elevations up to three times the average normal range are seen in untreated pernicious anemia, and somewhat smaller increases occur in *Werlhof's disease* (thrombocytopenic purpura); in the latter condition, levels decrease to normal after splenectomy. Similar increases are seen in hyperthyroidism, viral hepatitis, and myelogenous leukemia, and after myocardial infarctions.

A small percentage of individuals are born with low GPD activity in their red cells. The occurrence of such a deficiency of GPD or the presence of variant forms of the enzyme are the most common types of enzyme abnormality observed in the human population, affecting some one hundred million persons.[103,228] No clinical effects of the variant enzyme are manifested in individuals affected with more than half of the known variants. In these persons, GPD activity measured in red cells is usually found to be normal or, at most, marginally decreased; but in the case of certain enzyme variants, activity may be considerably depressed.

A second group of individuals, affected with some one-fourth of the many enzyme variants, may be subject to forms of drug-induced hemolytic disease. If these individuals are treated with antimalarials such as primaquine (or with any of the sensitizing drugs or chemicals already listed), they will suffer a severe hemolytic episode, males being much more severely affected than females. After the acute hemolytic attack is ended, the patient recovers fully; i.e., the disease is self-limiting. The patient remains well until another attack is precipitated by some new initiating agent, such as some drug or chemical, or some physiological crisis, such as severe acidosis in a diabetic. In the hemolytic episode, only the older cells are affected and destroyed, the younger cells being essentially untouched. Levels of red cell GPD activity, measured after a hemolytic attack, may be within normal limits. This hereditary defect affects some 10 to 15 per cent of black males, some Chinese, and Sephardic Jews as well as some other persons of Mediterranean origin. The defect is associated with the gene pair ($Gd\text{-}B^{+}$), which normally controls enzyme synthesis being replaced by the variant gene pair (Gd-A$^-$).[228] The (+) and (−) refer to genes governing high and low enzyme activity, respectively; the (B) refers to the gene controlling synthesis of an enzyme protein having normal electrophoretic mobility, and the (A) to one producing an enzyme protein having a mobility greater than that of (B). The (A$^+$) enzyme has 80 per cent of the activity of the normal (B$^+$) enzyme, but the (A$^-$) enzyme has less than 20 per cent of normal activity. Although in many blacks the B gene is replaced by the A gene, only those individuals belonging to the A$^-$ genotype are subject to hemolytic episodes; the low enzyme activity present causes the cells to be susceptible to lysis. The genetic defect is transmitted as a characteristic of intermediate dominance; it is sex-linked (and also linked to color blindness), with transmission through the female.

Another GPD variant (Gd-Mediterranean) is often associated with *favism*, a hemolytic anemia in which the hemolytic episode is triggered by ingestion of fava beans or by contact with the fava bean plant (*Vicia fava*). This enzyme variant has the mobility of the normal enzyme, but only a small percentage of the activity of the normal form. This phenotype is found among Italians and Sardinians, although not all persons harboring the favism gene are affected, suggesting that some other factor is also involved in setting off the hemolytic attack.

The type of defect with the most severe clinical effects is that encountered with those non-drug-dependent variants which are associated with a non-spherocytic hemolytic anemia. These may also cause a severe hemolytic jaundice of the newborn, not attributable to any ABO or Rh incompatibility.

METHODS FOR THE MEASUREMENT OF GLUCOSE-6-PHOSPHATE DEHYDROGENASE

Spectrophotometric methods are based on measuring the amount of NADPH formed in the GPD reaction (increase in absorbance at 340 nm), one mole of NADPH being formed by the oxidation of one mole of G6P. However, all clinical specimens (sera, red cell hemolysates, or tissue homogenates) will exhibit some 6-phosphogluconate dehydrogenase activity. This reaction (equations 41 and 42) uses as substrate the product of the GPD reaction, and also itself produces one mole of NADPH per mole of PGA oxidized. Thus, the NADPH formed will represent the combined activity of both enzyme reactions. Some 15 to 35 per cent of the NADPH may be derived from the PGD reaction, with the result that more than one mole of NADPH will be formed from each mole of the substrate G6P. The actual quantity will depend on the relative activities of the two enzymes under the conditions present in the reaction mixture.

A number of procedures have been proposed to correct for this contaminating PGD activity. Bishop[24] suggested that two parallel reactions be run; one with both G6P and PGA present as substrates, and the other with only PGA present. The former measures the combined activity of the two enzymes, whereas the latter measures only the PGD reaction; the difference between the two is a measure of the GPD activity. Extra PGA is added to a concentration high enough to maintain a constant, if not optimal, rate for the PGD reaction. (Otherwise, its rate would depend on the concentration of the substrate PGA, which would increase as the GPD reaction proceeds.) In another approach PGD is added in excess to the reaction mixture to promote complete oxidation of all the PGA formed in the GPD reaction to ribulose-5-phosphate.[142] Thus, each mole of G6P will give rise to two moles of NADPH and one-half of this value will be a measure of the NADPH arising from the GPD reaction.

If both a pH unfavorable for the PGD reaction and a short reaction time are used, the effect of any PGD activity can be minimized. Many screening or kit procedures utilize this approach. In another method NADP is added to the solution in which the red cells are to be lysed, and the conditions for lysis are chosen so that any PGD activity is fully inactivated, whereas the GPD is protected by the NADP from undergoing inactivation.

PGD activity is completely inactivated by maleimide,[128] when the latter is present in the system at a concentration of 4 mmol/l.

In specimens with low GPD activity, any PGD activity will also be very low, because very little PGA will be formed from G6P to serve as substrate for the enzyme. Thus, in those specimens containing the clinically important low-activity enzyme variants, the PGD activity error may be minimal, and GPD values will be valid even if uncorrected for PGD activity.

In fluorometric procedures,[23,129] the NADPH formed is measured by virtue of its fluorescence, instead of by its absorbance at 340 nm. The latter serves as the excitation wavelength, and the emitted light is measured at 460 nm. Fluorometric procedures are some hundred-fold more sensitive than UV absorbance methods, permitting the use of much smaller specimen volumes.

In the third group of procedures, devised mainly for convenient use in field or mass-screening studies, the NADPH formed in the GPD reaction reduces certain dyes to their colorless form and the color decrease is measured colorimetrically. In the procedure of Motulsky, the time required to decolorize a given quantity of brilliant cresyl blue is measured. In the Ellis and Kirkman[61] approach, an electron transfer intermediate, phenazine methosulfate (PMS), is added to mediate the reduction of the purple redox dye 2,6-dichlorophenolindophenol to its colorless form by the NADPH. The change in absorbance is measured at 620 nm.

DETERMINATION OF ERYTHROCYTE GLUCOSE-6-PHOSPHATE DEHYDROGENASE ACTIVITY (BISHOP, MODIFIED[24])

REAGENTS

1. Reagent T (total): Reaction mixture for measuring the combined GPD and PGD activity. Dissolve 1.82 g of tris(hydroxymethyl)aminomethane in 80 ml of water at 30°C, and adjust the pH to 7.5 at 30°C by the careful addition of 6 molar HCl. Weigh out 304 mg of $MgCl_2 \cdot 6H_2O$; 23 mg of NADP(Na); 30 mg of glucose-6-phosphate, monosodium salt; and 31 mg of sodium 6-phosphogluconate; and dissolve these chemicals in the Tris buffer. Dilute the solution to 100 ml. Pipet 4.2 or 10.5 ml aliquots into test tubes, freeze the tube contents, and store the aliquots in the freezer (−20°C). Stable for 6 to 8 weeks.
2. Reagent P (PGD): Reaction mixture for measuring only the PGD activity. Prepare this solution exactly as Reagent T, except omit the glucose-6-phosphate.
3. Saponin solution, 0.01 g/100 ml water.
4. Acid-citrate-dextrose blood collecting solution (ACD solution).

PROCEDURE

A. Preparation of hemolysate of blood specimen

1. Any of the common anticoagulants may be used to collect the blood specimen. However, if the assay is to be delayed for more than two days, the specimen should be collected in citrate or ACD solution. Keep specimen refrigerated until the assay can be performed.
2. Centrifuge specimen and remove the supernatant plasma and buffy coat.
3. Wash the packed cells twice with 10 volumes of saline (9 g NaCl/l) or ACD solution, and finally suspend the washed cells in an equal volume of saline or ACD.
4. Obtain a cell count of the red cell suspension in terms of $N \times 10^6/\mu l$.

B. Activity assay

1. Pipet 1.00 ml of the 0.01 per cent saponin solution into three 1.0 cm cuvets, and then transfer 10 μl of cell suspension to each cuvet. Mix, using a small plastic stirrer, and

let stand in a water bath or cuvet compartment at 30°C for 10 min for temperature equilibration and for cell lysis to occur.

2. Pipet 2.00 ml of Reagent P (pre-warmed to 30°C) into one cuvet, 2.00 ml of pre-warmed Reagent T into the second cuvet, and 2.0 ml of saline or ACD into the third (blank) cuvet. Mix contents of each cuvet rapidly.

3. Measure and record the change in absorbance in the T and P cuvets, measured against the blank at 340 nm, for a period of 10 to 15 min.

4. Calculate ΔA/min for the reactions in both cuvets P and T, and calculate the difference in ΔA/min between T and P (T minus P).

5. The concentrations of the components in the reaction mixture in mmol/l are as follows: Tris buffer, pH 7.5, 100; Mg^{2+}, 10; NADP, 0.20; G6P, 0.70; PGA, 0.60; specimen dilution, 1/301. All components are at optimal levels, except the NADP, which is present at a concentration which allows 90 per cent of the maximum activity.

CALCULATIONS

$$\text{Activity of GPD, in U}/10^{12}\text{ cells} = \frac{(R_T - R_P)}{N \times \dfrac{1}{0.001}} \times \frac{3.01}{6.22} \times \frac{1}{0.01} \times 10^6$$

$$= \frac{(R_T - R_P)}{N} \times 48{,}400$$

where R_T = ΔA/min in cuvet T (combined GPD + PGD reactions)

R_P = ΔA/min in cuvet P (PGD reaction only)

0.01 = ml of cell suspension used in assay

N = red cell count of the suspension, expressed in 10^6 cells/μl

0.001 ml = 1 μl = volume of suspension containing $N \times 10^6$ cells

10^6 = factor to give activity in 10^{12} cells

3.01 = total reaction volume in cuvet, in ml

6.22 = mmolar absorptivity of NADP at 340 nm

COMMENTS

1. Although the pH optimum for GPD is at pH 8.3, the pH activity curve is quite flat over the pH range from 7.3 to 8.6, and most workers have preferred to use a reaction pH between 7.4 and 7.6, which is less favorable to the PGA reaction. Triethanolamine or glycylglycine may replace Tris as buffer; phosphate should not be used, as it is inhibitory.

2. Saponin is the best lysing agent; it works most rapidly, and appears to release all the enzyme bound to the cell membrane. Digitonin may be used in place of saponin, but water lysis or lysis by freezing and thawing should not be used.

3. GPD in hemolysates is very unstable. If measurement of enzyme activity has to be delayed, blood should be stored intact, but even then it is best if the blood is drawn into citrate or ACD solution. Washed cells are best stored in ACD solution, where they are stable for 7 days at 4°C. Cells suspended in saline must be lysed and their PGD activity measured within 2 h. Long-term storage is to be avoided.

4. There is as yet no agreement as to the units for expressing red cell GPD activity. Inasmuch as it is the activity in the cell which is important, expressing activity in terms of a standard number of cells is preferred, although some investigators feel that cell counts are subject to too much uncertainty. The unit cell number employed has been either 10^9 or 10^{11} cells, but 10^{12} is to be preferred (10^{12} is a multiple of 10^3, and the absolute magnitude in units involves no decimal quantities).

Activity may be expressed in terms of U/g Hb, inasmuch as Hb can be measured accurately. However, the Hb content of cells may vary independently of the GPD activity, being subject to its own genetic and other factors. Assuming a mean corpuscular hemoglobin concentration of 0.35 g/1.0 ml packed cells, and a hematocrit of 0.45 (10^{10} cells/1.0 ml packed cells), then 10^{12} cells = (0.35) × ($10^{12}/10^{10}$) = 35 g Hb. Thus, 1.0 U/10^{12} cells = (1.0/35) U/g Hb = 0.0285 U/g Hb.

Activity has also been expressed in terms of U/100 ml packed red cells or U/100 ml blood. This has been especially convenient in field studies. If 1.0 U is present in 10^{12} cells (= 35 g Hb), and there is 15.7 g Hb in 100 ml of blood, then (15.7/35) = 0.45 U is present in 100 ml of whole blood. Since 100 ml of blood contains 45 ml of packed cells (average), 1.0 U/10^{12} cells = 1.0 U/100 ml packed cells.

For the sake of simplicity we shall report GPD activity in terms of U/10^{12} cells.

5. To assay *serum* activity, use the procedure described above for red cells, but use 1.0 ml of serum in place of the 1.0 ml of saponin plus 0.01 ml of packed cells.

6. The relative activity of GPD at several selected temperatures is: 25°, 0.72; 28°, 0.89; 30°, 1.00; 32°, 1.12; 35°, 1.33; 37°, 1.52.

NORMAL VALUES

An average, consensus value for the range of GPD activity found in the erythrocytes of "normal" individuals, i.e., in those who have not been subject to hemolytic episodes and other clinical manifestations owing to variant enzyme form, may be given as 120 to 280 U/10^{12} cells (or 3.4 to 8.0 U/g Hb), measured at 30°C. Additional values, culled from the literature, for both normal and variant cells are presented in Table 12-7.

TABLE 12–7 VALUES FOR ERYTHROCYTE GLUCOSE-6-PHOSPHATE DEHYDROGENASE REPORTED IN THE LITERATURE FOR CELLS CONTAINING BOTH NORMAL AND VARIANT ENZYMES*

Specimen, Source	*Assay Temperature, °C*	*GPD Activity, U/10^{12} cells*	*Reference*
Normal, mixed adult	25	189	134
Normal, mixed adult	30	125–280	129
Normal	30	140–490	142
Normal, adult	30	193–326	97
newborn		274–505	
Normal, black, male	25	161±14	90
Heterozygous, black, male	25	139±39	90
Homozygous, deficient, black, male	25	7.0±3.9	90
Normal, Egyptian, male	25	235±60	90
Normal, Italian, male	25	175±10	90
Homozygous, deficient, Italian, male	25	2.8±1.0	90

* Values are reported in U/10^{12} cells, although individual authors may have used other units. It is assumed that U/10^{12} cells = U/100 ml packed cells = (U/g Hb) × 35. This assumes an average hematocrit of 0.45, an average erythrocyte count of 4.5 × 10^6 cells/μl, and an average Hb concentration of 157.5 g/liter

REFERENCE

Bishop, C.: J. Lab. Clin. Med., *68*:149, 1966.

The Transaminases

ASPARTATE AMINOTRANSFERASE = ASPARTATE TRANSAMINASE (AST)

(Glutamate oxalacetate transaminase, GOT)
(EC 2.6.1.1; L-aspartate:2-oxoglutarate aminotransferase)

ALANINE AMINOTRANSFERASE = ALANINE TRANSAMINASE (ALT)

(Glutamate pyruvate transaminase, GPT)

(EC 2.6.1.2; L-alanine:2-oxoglutarate aminotransferase)

The transaminases constitute a group of enzymes which catalyze the interconversion of amino acids and α-oxoacids by transfer of amino groups. These enzymes are referred to as aminotransferases, and the "standard" letter abbreviations, AST and ALT, have been introduced for the two enzymes. However, the term "transaminase," and particularly the old abbreviations GOT and GPT, still remain in general use. Animal cells contain a variety of aminotransferases. AST is present in both the cytoplasm and the mitochondria of cells.[30] In conditions associated with a mild degree of tissue injury, the predominant form in serum is that from the cytoplasm, although some mitochondrial enzyme is also present. Severe tissue damage results in the release of much mitochondrial enzyme as well.

The α-oxoglutarate/L-glutamate couple serves as the amino group acceptor and donor pair in all amino transfer reactions; the specificity of the individual enzymes derives from the particular amino acid which serves as donor of the amino group. Thus, aspartate aminotransferase (AST, GOT) catalyzes the reaction shown in equation 43. Alanine aminotransferase (ALT, GPT) catalyzes the analogous reaction presented in equation 44.

$$\begin{array}{c} COO^- \\ | \\ H-C-NH_2 \\ | \\ CH_2 \\ | \\ COO^- \\ \\ \textbf{L-Aspartate} \end{array} + \begin{array}{c} COO^- \\ | \\ C{=}O \\ | \\ CH_2 \\ | \\ CH_2 \\ | \\ COO^- \\ \textbf{α-Oxoglutarate} \end{array} \underset{\text{P-5-P}}{\overset{\text{AST}}{\rightleftharpoons}} \begin{array}{c} COO^- \\ | \\ C{=}O \\ | \\ CH_2 \\ | \\ COO^- \\ \\ \textbf{Oxalacetate} \end{array} + \begin{array}{c} COO^- \\ | \\ H-C-NH_2 \\ | \\ CH_2 \\ | \\ CH_2 \\ | \\ COO^- \\ \textbf{L-Glutamate} \end{array} \qquad (43)$$

$$\begin{array}{c} COO^- \\ | \\ H-C-NH_2 \\ | \\ CH_3 \\ \\ \\ \\ \textbf{L-Alanine} \end{array} + \begin{array}{c} COO^- \\ | \\ C{=}O \\ | \\ CH_2 \\ | \\ CH_2 \\ | \\ COO^- \\ \textbf{α-Oxoglutarate} \end{array} \underset{\text{P-5-P}}{\overset{\text{ALT}}{\rightleftharpoons}} \begin{array}{c} COO^- \\ | \\ C{=}O \\ | \\ CH_3 \\ \\ \\ \\ \textbf{Pyruvate} \end{array} + \begin{array}{c} COO^- \\ | \\ H-C-NH_2 \\ | \\ CH_2 \\ | \\ CH_2 \\ | \\ COO^- \\ \textbf{L-Glutamate} \end{array} \qquad (44)$$

The reactions are reversible, but the equilibria of the AST and ALT reactions favor formation of aspartate and alanine, respectively.

Pyridoxal-5′-phosphate (phosphopyridoxal, P-5-P) and its amino analog, pyridoxamine-5′-phosphate, function as coenzymes in the amino transfer reactions.[68] The P-5-P is bound to the apoenzyme, and serves as a true *prosthetic* group. Only a small quantity of the coenzyme is required; that is, a concentration just sufficient to saturate the enzyme. The coenzyme is not consumed in the reaction, but is recycled over and over again. The pyridoxal-5′-phosphate bound to the apoenzyme accepts the amino group from L-aspartate or alanine to form enzyme-bound pyridoxamine-5′-phosphate and the reaction product, oxalacetate or pyruvate, respectively. The coenzyme in amino form then transfers its amino group to α-oxoglutarate to form L-glutamate and to regenerate pyridoxal-5′-phosphate. Thus, it is not necessary to add a large excess of P-5-P to the reaction mixture. This is in contrast to the dehydrogenase reactions, in which NAD/NADH or NADP/NADPH are consumed, and must be present in large excess.

Transaminases are widely distributed in animal tissues. Both AST and ALT are normally present in human plasma, bile, cerebrospinal fluid, and saliva, but none is found in urine unless a kidney lesion is present. Activities in various tissues, relative to that in serum, are shown in Table 12-8.

TABLE 12-8 TRANSAMINASE ACTIVITIES IN HUMAN TISSUES, RELATIVE TO SERUM AS UNITY[118]

	AST	*ALT*
Heart	7800	450
Liver	7100	2850
Skeletal Muscle	5000	300
Kidney	4500	1200
Pancreas	1400	130
Spleen	700	80
Lung	500	45
Erythrocytes	15	7
Serum	1	1

CLINICAL SIGNIFICANCE[220]

Following *myocardial infarctions*[117] an increased level of AST activity appears in serum, as might be expected from its relatively high concentration in heart muscle (Table 12-8). Serum levels do not become abnormal, however, until 4 to 6 h have elapsed after onset of chest pain. Peak values of AST activity are reached after 24 to 36 h, and the activity values fall to within the normal range by the fourth or fifth day, provided no new infarct has occurred. Levels some 10 to 15 times the upper limit of normal are usually associated with fatal infarcts. The peak values of AST activity are roughly proportional to the degree of cardiac damage. However, small elevations in serum levels do not necessarily indicate a favorable prognosis. Alanine transaminase (ALT) levels are within normal limits, or only marginally increased, inasmuch as the ALT activity in heart muscle is only a fraction of the quantity of AST activity present.

In *viral hepatitis* and other forms of liver disease associated with some degree of hepatic necrosis, serum levels of both enzymes will be elevated, even before the clinical symptoms of disease (such as jaundice) appear. AST and ALT levels may reach values as high as 100 times the upper limit of normal, although 30- to 50-fold elevations are most frequently encountered. Characteristically, ALT is higher than AST, and the ALT/AST (De Ritis) ratio,[47] which normally (and in myocardial infarctions) is less than 1.0, becomes greater than unity. This is the case especially in the later stages of the disease, and during the recovery phase. However, any changes introduced in the assay methods for the two enzymes may alter the activities of the two enzymes relative to each other, and hence the enzyme ratio in serum. In parenchymal liver disease, peak values of transaminase activity are seen between the seventh and twelfth days; activities then gradually decrease, reaching normal levels by the third to fifth week.

The picture in *toxic hepatitis* is similar to that seen in infectious hepatitis, with very high ALT and AST activities being observed in severe cases. Elevations up to 20 times the upper normal value may be encountered in *infectious mononucleosis*, and similar but somewhat lower values in *intrahepatic cholestasis.* The aminotransferase levels observed in *cirrhosis* vary with the status of the cirrhotic process, ranging from upper normal to some 4 to 5 times normal, with the level of AST activity higher than that of ALT activity. Slight or intermediate elevations of both enzymes are also seen in obstructive jaundice.

AST (and occasionally ALT) activity levels are increased in progressive *muscular dystrophy* and *dermatomyositis*, reaching levels up to eight times normal; they are usually normal in other types of muscle diseases, especially in those of neurogenic origin. Pulmonary emboli can raise AST levels to two to three times normal, and slight to moderate elevations (2 to 5 times normal) are seen in acute pancreatitis, crushed muscle injuries, gangrene, and hemolytic disease. Five-fold to ten-fold elevations of the two enzymes occur in patients with carcinoma of the liver, either primary or metastatic, with AST usually being higher than ALT. Slight elevations of both AST and ALT activities may be observed after intake of alcohol, in delirium tremens, and after administration of some medications (opiates, salicylates, ampicillin).

Although serum levels of both AST and ALT become elevated whenever disease processes affect liver cell integrity, ALT is the more liver-specific enzyme. Serum elevations of alanine transaminase activity are rarely observed except in parenchymal liver disease. Moreover, elevations of ALT activity persist longer than do those of AST activity.

Changes in the level of aspartate aminotransferase activity in *cerebrospinal fluid* have been found to be relatively nonspecific. Substantial elevations of AST activity are found in the spinal fluids of patients who suffer cerebrovascular accidents. If the enzyme level is elevated both in serum and in spinal fluid, massive parenchymal brain damage is suggested and the prognosis is poor; if only the spinal fluid level is raised, recovery is probable. Both primary and secondary malignancies involving the brain or spinal cord are accompanied by varied degrees of elevation of AST activity in the spinal fluid. Normal values are usually found in patients with bacterial meningitis; raised levels suggest complications.

METHODS FOR THE MEASUREMENT OF TRANSAMINASE ACTIVITY[96,166]

As is evident from equations 43 and 44, the assay system for measuring transaminase activity will contain two amino acids and two oxoacids. This presents certain problems in evaluating enzyme activity. There is no convenient method available for assaying either of the amino acids in the reaction system. One must therefore measure the oxoacids, either the acid consumed or the acid formed in the reaction. The oxoacids can be assayed colorimetrically by coupling with 2,4-dinitrophenylhydrazine, as discussed in connection with the colorimetric procedures for LD (equation 36). There are several disadvantages to this colorimetric approach, however. In each *aminotransferase* reaction, two oxoacids are present, one on each side of the equation. In the AST reaction, these are α-oxoglutarate and oxalacetate. Both can react to give dinitrophenylhydrazones. As the reaction proceeds one oxoacid is increasing in concentration, while the other is decreasing. The reaction begins with an oxoacid substrate, and hence reagent blanks have a high absorbance. If these are kept low by using low substrate concentrations, the latter may become a factor limiting the reaction rate. In the AST reaction, one of the products, oxalacetate, is relatively unstable. Furthermore, oxoacids such as pyruvate, which are normal components of serum, also produce phenylhydrazones, and contribute to the high blanks.

Despite these limitations, the colorimetric approach is still feasible, because the phenylhydrazones of the products of both the AST and ALT reactions (oxalacetate and pyruvate, respectively) are considerably more chromogenic than is the phenylhydrazone of α-oxoglutarate. Those colorimetric methods which are based on dinitrophenylhydrazone formation are relatively simple, and have limited but acceptable accuracy.[181,223] The assay methods based on this approach, particularly that described by Reitman and Frankel,[153] continue to be used and form the basis of many proprietary "kits."

In the case of AST, an alternative procedure was introduced by Babson and associates.[6] These workers measured the oxalacetate formed, by combining it in a specific reaction with Azoene Fast Violet B (6-benzamido-4-methoxy-*m*-toluidine diazonium chloride). Similar diazonium salts (e.g., Azoene Fast Red and True Ponceau L) have been introduced by Sax and Moore and other workers. This procedure is sensitive enough to permit use of a 20 min reaction time; reagent blank absorbances are minimal compared to those with the phenylhydrazone methods, and a wider range of concentrations can be assayed without specimen dilution. The procedure and reagents are marketed by General Diagnostics Laboratories, Morris Plains, NJ., as the TransAc Method. The method has the disadvantage that it must be standardized against pre-assayed serum specimens or the oxalacetate standard used must be assayed enzymatically.

Both of these types of colorimetric methods are two-point methods, and suffer from the fact that the enzyme is inhibited by its product, oxalacetate, which accumulates during the reaction, causing deviation from linearity. Although oxalacetate inhibits both enzymes, the mitochondrial enzyme is inhibited to a greater degree.

It is not possible to monitor transaminase reactions directly, but continuous-monitoring assays for *aminotransferases* can be devised by coupling the transaminase reactions to specific dehydrogenase reactions.[113] The oxoacids formed in the transaminase reaction are measured indirectly by reducing them with NADH to the corresponding hydroxy acids, and measuring the resultant change in NADH concentration spectrophotometrically. Thus, oxalacetate, formed in the AST reaction, is reduced to malate in the presence of malate dehydrogenase (MD) (equation 45); pyruvate formed in the ALT reaction is reduced to lactate by lactate dehydrogenase (LD). The substrate, NADH, and the auxiliary enzymes, MD and LD, should be present in large excess so that the reaction rate is limited only by the amounts of AST and ALT, respectively. As the reactions proceed, NADH is oxidized to NAD^+. The disappearance of NADH per unit of time is followed by measuring the decrease in absorbance for several minutes at 340 nm. The change in absorbance per minute (ΔA/min) may be related directly to micromoles of NADH oxidized and, in turn, to micromoles of substrate transformed per minute (International Units). The advantages

$$\begin{array}{ccccc} \alpha\text{-Oxoglutarate} \searrow & & \nearrow \text{L-Aspartate} \quad \text{L-Malate} \nwarrow & & \nearrow NAD^+ \\ & \text{AST} & & \text{MD} & \\ \text{L-Glutamate} \swarrow & & \searrow \text{Oxalacetate} \longrightarrow \nearrow & & \nwarrow \text{NADH} \end{array} \qquad (45)$$

Aminotransferase reaction
(Formation of oxalacetate)
Assay reaction

Dehydrogenase reaction
(Quantitation of oxalacetate)
Indicator reaction

this technique offers are that it is possible to use the initial, linear phases of the reaction to determine reaction rates, and that substrate concentrations can be set high enough so as not to be rate-limiting. Sufficient NADH and dehydrogenase can be added to drive the coupled reaction to completion and thus minimize reverse reactions.

Another type of coupled enzyme system has been proposed for use in assaying the transaminases. In this instance, the indicator enzyme is glutamate dehydrogenase, EC 1.4.1.3, which uses NAD to oxidatively deaminate the glutamate formed in the transaminase reaction to form NADH and NH_4^+ and to regenerate oxoglutarate. The rate of NADH formation serves as the measure of the transaminase reaction.

Continuous-monitoring procedures are definitely to be preferred for the assay of transaminases, as indeed for all enzymes where possible. Because of the large numbers of AST and ALT activity measurements performed daily in many laboratories throughout the world, the development of standard or reference methods for these two enzymes has engaged the attention of several national and international groups.[82,175,222] All of these have chosen the coupled reaction approach; however, the methods proposed differ in several details, such as substrate concentrations, nature of buffer (e.g., Tris, phosphate), and assay temperature. At the present time, several different transaminase units and normal ranges continue in use, and caution is required in interpreting results when methods are not specified in detail.

CONTINUOUS MONITORING METHOD FOR THE MEASUREMENT OF AMINOTRANSFERASE ACTIVITY

Measurements should be conducted in a spectrophotometer with established spectrophotometric and wavelength accuracy and with good resolution at 340 nm. Temperature of the reaction mixture in the cuvet must be controlled at a constant known level by means of a thermostatically controlled cuvet compartment. A preliminary incubation period should be included to destroy endogenous oxoacids in the serum (side reaction) prior to adding the oxoacid involved in the reaction. Following this, there is a brief lag phase as is seen with most coupled reactions. Subsequently, several readings (5 to 8, at 1 min intervals) are taken to establish the linear portion of the curve. A recorder which produces a curve directly related to absorbance is very convenient. Because the reaction mixture has a rather high initial absorbance, it is customary to use a solution such as potassium dichromate as a blank reference solution, so as to produce an initial absorbance for the test of about 0.7 to 0.9; this can also be accomplished with a zero offset (zero suppression) facility, if available with the spectrophotometer.

DETERMINATION OF ASPARTATE TRANSAMINASE (AST) ACTIVITY

In the procedure to be outlined, Tris is used as buffer, in place of the customary phosphate, because phosphate appears to increase the rate of NADH decomposition and to inhibit association of P-5-P with the transaminase apoenzyme.[154,200] The coenzyme is added to insure that all the transaminase is fully active. Lactate dehydrogenase is added to the coupled enzyme system to accelerate the endogenous side reactions,[163] and thus shorten the pre-incubation period. A pH of 7.8 is used[175] because the pH optimum for the coupled enzyme system appears to be between 7.7 and 7.9. Stock preparations of both MD and LD are diluted with glycerol rather than with $(NH_4)_2SO_4$ to avoid introducing ammonium ions, thereby eliminating a glutamate dehydrogenase side reaction and associated loss in NADH. The reaction temperature is 30°C.

REAGENTS

1. Tris base stock, 1.0 mol/l. Dissolve 121.1 g tris(hydroxymethyl)aminomethane in water to make 1000 ml. Store in a polyethylene bottle; the reagent is stable.

2. Tris buffer, pH 7.8, 0.10 mol/l. Dilute 50 ml of the 1.0 mol/l Tris base with about 300 ml of water. Add about 5.5 ml of HCl, 6 mol/l, adjust temperature to 30°C, and (with the aid of a pH meter) adjust pH to exactly 7.8 by the addition of 1 molar NaOH. Dilute with water to 500 ml. The reagent is stable, and may be stored in a polyethylene bottle at 4°C or at room temperature.

Note: Certain pH electrodes do not give correct readings in Tris solutions. Refer to the instructional brochure describing the properties of the glass and reference electrodes being used.

3. L-Aspartate, 228 mmol/l, in 0.10 mol/l Tris buffer, pH 7.8. Dissolve 7.59 g of L-aspartic acid in a solution containing 25.0 ml of Tris base and 175 ml of water, with the addition of 2.5 molar NaOH (25 ml) and the application of heat, if necessary. Cool to 30°C, and with the aid of a pH meter adjust the pH to 7.8 by the addition of 1 molar NaOH or HCl. Add water to make 250 ml of solution. Store in refrigerator; stable for 6 months.

4. 2-Oxoglutarate, 225 mmol/l, in 0.10 mol/l Tris buffer, pH 7.8. Dissolve 1.65 g of 2-oxoglutaric acid in about 25 ml of water, add 5.0 ml of Tris base, warm to 30°C, and, as before, with the aid of a pH meter, adjust the pH to exactly 7.8 by the addition of

2.5 molar NaOH. Add water to make 50 ml of solution. Keep refrigerated; stable for 2 weeks.

5. NADH solution, 6.5 mmol/l, in 0.10 mol/l Tris buffer, pH 7.8. Dissolve 5.0 mg of β-NADHNa$_2$·4H$_2$O, 98 per cent grade, in 1.0 ml of the 0.10 molar Tris buffer, pH 7.8. It is convenient to use the commercially available vials containing pre-weighed quantities of the reduced coenzyme. If stored at refrigerator temperature, the reagent is stable for one working week; it is best, however, to prepare enough reagent for one day's work. Keep cold during use.

6. Malate dehydrogenase suspension, 36,000 U/l in glycerol-water (1/1). Dilute the commercially available stock suspension of MD with glycerol-water (1/1) to a concentration of 36,000 U/l. The stock suspension should be free of aminotransferase, apo-aminotransferase, and L-glutamate dehydrogenase.

7. Lactate dehydrogenase suspension, 72,000 U/l in glycerol-water (1/1).[163] Prepare this reagent by a procedure similar to that described for MD. This enzyme suspension should also be essentially free of aminotransferase, apo-aminotransferase, and L-glutamate dehydrogenase activity. Both enzyme reagents are stable for about six months, if stored in a refrigerator and kept on ice during use.

8. Pyridoxal-5′-phosphate, 4.5 mmol/l, in 0.10 mol/l Tris buffer, pH 7.8. Dissolve 11.1 mg of pyridoxal-5′-phosphoric acid in 10 ml of 0.10 mol/l Tris buffer, pH 7.8. Keep refrigerated; stable for one week.

9. Dichromate solutions (blanks). Prepare a stock solution containing 30 mg of $K_2Cr_2O_7$/100 ml of water, containing 3 drops of concentrated H_2SO_4. Dilute this solution further with water as needed in the procedure. The reagent is stable.

PROCEDURE

1. Add the following reaction components to a cuvet that has a 1.0 cm light path:

L-Aspartate, 228 mmol/l	2.30 ml
NADH, 6.5 mmol/l	0.10
P-5-P, 4.5 mmol/l	0.10
MD, 36,000 U/l	0.05
LD, 72,000 U/l	0.05
Serum	0.20
Total Volume	2.80 ml

If many determinations are to be done, it is convenient to mix a batch of these five reagents, sufficient for one day's work, and to pipet 2.6 ml of the composite reagent into individual cuvets. The reagent mixture is stable for several days, if kept refrigerated. It should be kept in ice water on the bench, or in the refrigerator when not being used to prepare reaction tubes. The serum specimens are then pipetted into the cuvets, and the contents mixed.

2. Preincubate the cuvets and contents in a water bath or in the thermostatted cuvet compartment of the spectrophotometer at 30°C for 5 to 8 min to permit the endogenous side reactions to proceed to completion.

3. Initiate the AST enzyme reaction by adding 0.20 ml of the 225 mmol/l oxoglutarate solution that has been prewarmed to 30°C. Rapidly mix the cuvet contents; place the cuvet in the spectrophotometer, and measure the absorbance at 340 nm for a period of 5 to 8 min. The readings are made against a dichromate blank. Choose the blank or the zero suppression so that the initial absorbance readings are between 0.70 and 0.90.

The actual concentrations of the various components in the reaction mixture are as follows: Tris buffer, pH 7.8, 90 mmol/l; L-aspartate, 175 mmol/l; 2-oxoglutarate, 15 mmol/l; NADH, 0.21 mmol/l; MD, 600 U/l (25°C); LD, 1200 U/l (25°C); and P-5-P, 0.15 mmol/l. The ratio of serum volume to total reaction volume (serum dilution) is 1/15.

4. Calculate the average absorbance change per minute (ΔA/min), using either the recorder tracing or the measured absorbance values. Use only the linear portion of the rate curve, or those consecutive individual values of ΔA/min which are essentially constant.

CALCULATION

$$\text{AST activity in U/l} = \frac{\Delta A}{\text{min}} \times \frac{1}{6.22} \times \frac{3.0}{0.20} \times 1000$$

$$= \frac{\Delta A}{\text{min}} \times 2410$$

where 3.0 = total volume in cuvet, in ml
0.2 = volume of serum specimen, in ml
6.22 = millimolar absorptivity of NADH at 340 nm
1000 = number of ml/l

The average relative rates of the AST reaction system at various temperatures, relative to the rate at 30°C as unity, are as follows: 25°, 0.73; 32°, 1.12; 35°, 1.37; and 37°, 1.56. If the reaction is measured at some temperature other than 30°C, the activity obtained can be converted to that which would be obtained at 30°C by dividing the measured activity by the listed factor (or one obtained by extrapolation for temperatures not listed). Such conversions are only approximate, and their use is not encouraged.

COMMENTS

1. 2-Oxoglutarate protects AST from inactivation during the side-reaction incubation. Suggestions have been made to include oxoglutarate in the initial reaction mixture, in place of the aspartate, and to initiate the reaction by the addition of L-aspartate. For this purpose, the concentration of L-aspartate should be 1.05 mmol/l (in Tris buffer, pH 7.8), and that of the 2-oxoglutarate should be 22.5 mmol/l (in Tris buffer). The aspartate in the initial reaction mixture is replaced by 2.00 ml of the oxoglutarate, and after the 5 to 8 min pre-incubation, the reaction is initiated by the addition of 0.50 ml of the 1.05 molar aspartate.

2. Elevated levels of L-glutamate dehydrogenase (GMD, EC 1.4.1.3) may be encountered in some sera from patients with parenchymal liver disease. If ammonia is also present in the serum specimen or in one of the reagents, the GMD reaction will proceed and interfere with the AST determination by consuming oxoglutarate and NADH. Thus, use of ammonia-free reagents is recommended.

3. Occasionally, the side reactions in the pre-incubation stage may be quite extensive, thus consuming a large fraction of the available NADH. If the ΔA/min in the test decreases with time (non-linear rate), suboptimal NADH concentration is indicated. In this event, the test must be repeated with a higher initial concentration of NADH or a smaller volume of specimen (diluted specimen).

4. If the activity in a given serum specimen is high (ΔA/min over 0.10), the serum will need to be diluted. Saline may be used, but an albumin solution (5 g/100 ml) is preferred.

5. Hemolyzed serum specimens should not be used. As shown in Table 12-8, AST and ALT activities in erythrocytes are some 15 and 7 times, respectively, higher than those in sera.

6. Authorities disagree as to the stability of AST activity in serum. Specimens are best stored frozen if they are to be kept more than 3 to 4 days. Minimal loss of activity occurs at 0 to 4°C over 1 to 3 days.

NORMAL VALUES

Insofar as methods for the determination of AST activity are under intense investigation directed toward establishing conditions that will provide the most accurate results (optimal reference methods), any suggested range of normal values is, at best, tentative. The following values, in U/l at 30°C, reflect current and past experience; the values in parentheses are mean or modal values.

Specimen	*Values in U/l* — *Adults*	*Children*
Serum (no P-5-P added)	men 7–21 (12)	10–25
	women 6–18 (10)	
	both sexes 6–25 (18)[200]	
	both sexes 7–23 (13)	
Serum (P-5-P added)	both sexes 12–29 (19)[200]	
Cerebrospinal fluid (no P-5-P added)	both sexes 3–10	2–10 (?)

The upper limits are not firm, and they may be raised when truly optimal methods have been confirmed. The distribution of values is skewed to the right; the skewness may be real, but does reflect the inclusion in the normal listing of sera with non-visible hemolysis and sera from persons with possible occult tissue infarcts. Individuals in the fasting state may show slightly lower values. Values for men are slightly higher than those for women.[96,124] The inclusion of exogenous pyridoxal-5′-phosphate in the reaction mixture appears to increase AST values by about 15 to 40 per cent,[41,200] and at times even more.

MEASUREMENTS OF AST ACTIVITY BY DOUBLE-BEAM SPECTROPHOTOMETRY[163]

Investigators have shown that it is possible to effectively correct for the GMD (GDH), LD, and ALT side reactions by using a double-beam spectrophotometer in which the ΔA/min is measured in a specimen and in a reference cuvet at the same time. The specimen cuvet contains the complete reaction mixture, whereas the reference cuvet lacks the aspartate. The reaction is initiated with either serum or 2-oxoglutarate. The absorbance change in the specimen cuvet represents both the transaminase and the concurrent side reactions, whereas that in the reference cuvet represents the side reactions only. The difference in ΔA/min values in the two cuvets gives a more accurate measure of the AST activity. Alternatively, "dynamic blanking" may be used; the ΔA/min is measured in two separate cuvets in the usual spectrophotometer, with and without aspartate present. Oxoglutarate is preferably included in both the specimen and reference mixtures, in place of aspartate, because it is required for the GDH and ALT side reactions. If this technique is used, care must be taken that identical time intervals elapse between initiation of the reaction and the start of absorbance measurements. The rates of the side reactions, in general, are not linear in time.

In principle, this technique can also be applied to measuring the activity of other enzymes.

DETERMINATION OF ALANINE TRANSAMINASE (ALT) ACTIVITY

The procedure used is identical to that for measuring AST activity, except that L-alanine replaces aspartate as the amino group donor, and LD replaces MD as the indicator enzyme. The concentration of alanine needed to give an optimal rate of ALT activity is much greater than that of aspartate in measuring AST activity. Whether or not exogenous

P-5-P is required has as yet not been definitely established. The added LD speeds up the side reaction and serves as the indicator enzyme.

REAGENTS

1. Tris base, 1.0 mol/l, pH 7.8. This is identical to that used in the AST procedure.
2. NADH solution. Refer to the AST procedure.
3. L-Alanine, 525 mmol/l, in 0.10 mol/l Tris buffer, pH 7.8. Dissolve 11.8 g of L-alanine (99+ per cent purity) in about 200 ml of water and 25 ml of the 1.0 molar Tris base. Warm the solution to 30°C and, with the aid of a pH meter, adjust the pH to 7.8 with 1 molar HCl or 1 molar NaOH. Dilute with water to a volume of 250 ml. Stable for 4 to 6 weeks, if stored in a refrigerator.
4. 2-Oxoglutarate solution, 225 mmol/l, in 0.10 mol/l Tris buffer, pH 7.8. Refer to the AST procedure.
5. Pyridoxal-5′-phosphate, 4.5 mmol/l in 0.10 mol/l Tris buffer, pH 7.8. Refer to the AST procedure.
6. Lactate dehydrogenase solution, 72,000 U/l, in glycerol-water (1/1). This is the same preparation as used in the AST procedure.

PROCEDURE

1. Add the following components to a 1.0 cm cuvet:

L-Alanine, 525 mmol/l	2.30 ml
NADH, 6.5 mmol/l	0.10
LD, 72,000 U/l	0.10
P-5-P, 4.5 mmol/l	0.10
Serum	0.20
Total volume	2.80 ml

It is convenient to prepare a composite containing the four reagents, as suggested in the AST procedure, and to pipet 2.60 ml of this composite into a cuvet (or a 12 × 100 mm test tube), followed by 0.20 ml of the serum specimen(s). From this point on the procedure and calculations are identical to those outlined for the AST procedure.

With the large quantity of LD added, the endogenous (side) reactions are completed within a few minutes.

The ALT reaction is initiated by adding 0.20 ml of the 225 mmol/l oxoglutarate solution.

The composition of the enzyme reaction mixture is as follows: Tris buffer, pH 7.8, 90 mmol/l; L-alanine, 400 mmol/l; oxoglutarate, 15 mmol/l; NADH, 0.21 mmol/l; LD, 2400 U/l; and P-5-P, 0.15 mmol/l. The serum dilution is 1/15.

COMMENTS

1. As suggested in the procedure, the reaction components and serum may be added to a test tube, and the side reactions permitted to reach completion while the tube is coming to temperature (30°C). The oxoglutarate is rapidly added to the tube contents, which, after being mixed, are poured into cuvets pre-warmed to 30°C. The absorbance readings of the solutions are measured and recorded.
2. Very turbid or icteric sera may be diluted in order to avoid working with very high absorbances, although this problem is usually resolved by selecting a $K_2Cr_2O_7$ blank of appropriate concentration.
3. The comments discussed under the AST procedure are also pertinent to the determination of ALT.

NORMAL VALUES

Tentative values for the normal range for ALT activity in U/l in serum and spinal fluid measured at 30°C, without added P-5-P, may be stated to be as follows:

	Males	*Females*
Serum	6 to 21	4 to 17
Spinal fluid	none	

Values for assay systems utilizing added P-5-P have not as yet been established, but may be increased some 20 to 25 per cent. In most present procedures, the concentration of alanine is not optimal, and a range of 5 to 19 U/l of ALT activity at 25°C (= 7 to 29 U/l at 30°C) has been suggested for one proposed "optimized" method.

REFERENCES

Scandinavian Society for Clinical Chemistry and Clinical Physiology: Recommended methods for the determination of four enzymes in blood, report of the Committee on Enzymes, Scand. J. Clin. Lab. Inv., *33*:291, 1974.

Wilkinson, J. H., Baron, D. N., Moss, D. W., and Walker, P. G.: Standardization of clinical enzyme assays: A reference method for aspartate and alanine transaminases. J. Clin. Path., *25*:940, 1972.

CREATINE KINASE

(EC 2.7.3.2; adenosine triphosphate:creatine *N*-phosphotransferase)

Creatine kinase (CK), also referred to as creatine phosphokinase, catalyzes the reversible phosphorylation of creatine by adenosine triphosphate (ATP) (see equation 46). The optimal pH values for the forward (Cr + ATP → ADP + CrP) and reverse (CrP + ADP → ATP + Cr) reactions are 9.0 and 6.8, respectively. The equilibrium position for the reaction is dependent on pH. At neutral pH, CrP has a much higher phosphorylating potential than does ATP; this favors the reverse reaction, with ATP being formed from CrP. The reverse reaction proceeds six times faster than does the forward reaction, both measured at optimal conditions.

$$^{-}O{-}\overset{O}{\overset{\|}{C}}{-}CH_2{-}N(CH_3){-}\underset{\underset{NH}{\|}}{C}{-}NH_2 + AMP{-}O{-}\underset{O^-}{\overset{O}{\overset{\|}{P}}}{-}O{-}\underset{O^-}{\overset{O}{\overset{\|}{P}}}{-}O^- \underset{\text{pH, 6.7}}{\overset{\text{pH, 9.0}}{\underset{\text{CK, Mg}^{2+}}{\rightleftharpoons}}}$$

Creatine (Cr) — Adenosine triphosphate (ATP)

$$^{-}O{-}\overset{O}{\overset{\|}{C}}{-}CH_2{-}N(CH_3){-}\underset{\underset{NH}{\|}}{C}{-}NH{-}\underset{O^-}{\overset{O}{\overset{\|}{P}}}{-}O^- + AMP{-}O{-}\underset{O^-}{\overset{O}{\overset{\|}{P}}}{-}O^- + H^+ \qquad (46)$$

Phosphocreatine (Creatine phosphate, CrP) — Adenosine diphosphate (ADP)

As is true for all kinases, Mg^{2+} is an obligate activating ion, functioning in the form of its salts with ADP and ATP. The optimal concentration range for Mg^{2+} is quite narrow, and excess Mg^{2+} is inhibitory. Most heavy metal ions (Zn^{2+}, Cu^{2+}) inhibit enzyme activity, as do iodoacetate and other sulfhydryl binding reagents. Activity is inhibited by excess ADP and by citrate, fluoride, and L-thyroxine. Uric acid is a potent inhibitor of the enzyme in serum. Even Cl^- and SO_4^{2-} inhibit activity to some degree, and the concentrations of these ions should be kept low in any enzyme assay system based on the CrP + ADP (reverse) reaction. The enzyme in serum is relatively unstable, activity apparently being lost as a result of sulfhydryl oxidation or internal disulfide formation. Approximately full activity can be restored by incubating the enzyme preparation with sulfhydryl compounds such as cysteine, glutathione, mercaptoethanol, thioglycollic acid, and dithiothreitol (Cleland's reagent).

Creatine kinase is a dimer, composed of two units with a relative molecular mass of 50,000. Three isoenzymes are present in human tissues: CK_1 or BB, present predominantly in nerve tissue, but also found in the thyroid, kidney, and intestine; CK_2 or MB, present in heart muscle and in the diaphragm and esophagus (smooth muscle); and CK_3 or MM, the most common form, present in all tissues, and especially in skeletal muscle. On electrophoresis the CK_1 isoenzyme (BB) migrates most rapidly toward the anode and is found in the pre-albumin area. The slowest moving form is CK_3 (MM), which is found in the γ-globulin region, whereas the heart form, CK_2 (MB), with an intermediate mobility, is located in the α_2-globulin zone.

CLINICAL SIGNIFICANCE

CK activity is greatest in striated muscle, brain, and heart tissue, which contain some 2000, 700, and 350 U/g of protein, respectively. Other tissues, such as the kidney and the diaphragm, contain significantly less activity (2 and 12 U/g protein, respectively), and the liver and erythrocytes are essentially devoid of activity. The determination of CK activity has proved to be a tool more sensitive than many other laboratory tests in the investigation of skeletal muscle disease, and is also useful in the diagnosis of myocardial infarction and cerebrovascular accidents.

Serum CK activity is greatly elevated in all types of muscular dystrophy, and especially so in the Duchenne type, in which levels up to 50 times the upper limit of normal may be encountered. In *progressive muscular dystrophy* (PMD), enzyme activity in serum is highest in infancy and childhood (7 to 10 years of age), and may be elevated long before the disease is clinically apparent. CK activity characteristically falls as the patient gets older and the mass of functioning muscle is diminished with the progression of the disease. About 50 to 80 per cent of the asymptomatic female carriers of Duchenne dystrophy show three- to six-fold elevations of CK activity in serum, but values may be normal if specimens are obtained after patients have undergone a period of physical inactivity. Quite high values of serum CK are seen in *viral myositis*, *polymyositis*, and similar muscle diseases. However, in conditions such as myasthenia gravis, multiple sclerosis, poliomyelitis, and Parkinsonism—i.e., *neurogenic muscle diseases*—enzyme activity is normal. Very high serum activity is also seen in malignant hyperthermia, a familial disease characterized by high fever and brought on by administration of inhalation anesthesia to the affected individual. Apparently, MB isoenzyme replaces part of the MM form in the muscles, and this results in impaired storage of CrP in muscle, as well as other biochemical defects. In dystrophies and myosites, usually only the MM isoenzyme is seen in serum; but, if total CK activity is high, some MB can also be present, since muscle may contain some 20 to 30 per cent of the MB isoenzyme.

Following a myocardial infarction, CK activity in serum begins to rise within 4 to 6 h, reaches a peak value at between 18 and 30 h, and then rapidly returns to normal by the third day. The average maximum elevation is 7 to 12 times the upper limit of normal, and the test is the earliest and most sensitive enzyme indicator of myocardial infarction. Invariably, the rise in serum CK activity is accompanied by the appearance of an easily demonstrated MB isoenzyme band in electrophoretic patterns of the serum specimen.

Some experts feel that the demonstration of an MB isoenzyme band is more pathognomonic of an infarction than is the time-course of CK activity during the first 36 to 60 h after onset of pain. The MB protein is cleared from serum very rapidly, and usually is not demonstrable after the first 24 to 36 h, even after a severe infarction. Serum CK activity may be normal, or at best slightly elevated, in congestive failure. In cases of angina pectoris activity is normal, but in patients receiving high levels of digitoxin, a low-order elevation in serum CK is often found. However, in patients receiving injections of lidocaine, an anti-arrhythmia drug, serum CK activity is significantly increased, as it is in patients receiving direct current counter-shock therapy for the management of their arrhythmias.

Because liver contains a negligible quantity of CK, patients with *primary liver disease* and cirrhosis have normal CK activity in their sera. For the same reason hepatic congestion and hypoxia, which may accompany cardiac disease, do not effect an elevation in serum CK values, although they often contribute to the elevation seen in serum ALT and LD activity. Normal levels of CK activity are observed in the sera of patients in shock and in patients with azotemia, malignancies, and hemolytic disease.

Serum CK activity may be increased in patients with acute *cerebrovascular disease* and with cerebral ischemia. It is also considerably elevated just before and during the early phases of acute psychotic reactions; the cause of this increase is not understood, except that some stress phenomenon is involved.[136] Isoenzyme studies show that the increase is entirely in the MM isoenzyme; no BB isoenzyme increase is demonstrable.

Moderate increases in CK activity are found in *hypothyroidism*, in Doriden intoxication, in pulmonary infarctions, and in childhood disorders accompanied by convulsions or muscle spasms. In general, any trauma to muscle from bruises, fractures, or surgical procedures (and even severe exercise) may cause a marked elevation in serum CK levels, which may persist for a week or longer. Even the trauma to muscle caused by frequent intramuscular injections of medication may induce a sharp increase, ranging from two to four times the upper limit of normal.

In *newborn* infants, CK levels increase rapidly during the first 24 h after birth to levels considerably greater than those seen in cord blood, which themselves are two to four times greater than normal adult levels. This increase persists for about one year, and is in part attributable to birth trauma. The measurement of CK activity in amniotic fluid has been said to be of value in the diagnosis of possible intrauterine death. Whereas elevations in LD activity in amniotic fluid may reflect placental damage as well as fetal damage, elevations in CK reflect only fetal damage.

Assay of CK activity in *cerebrospinal fluid* is of only limited value in clinical diagnosis because any changes observed are irregular, and often non-specific in nature. Elevations can be seen in some specimens from epileptic patients, and from patients with brain tumors or cerebral infarcts, but the elevations are not consistent with the degree of pathologic change. Similarly, spinal fluid specimens from a majority (but not all) of patients with either bacterial or non-bacterial meningitis and those with autism have increased levels of CK.

METHODS FOR THE DETERMINATION OF CREATINE KINASE ACTIVITY

Numerous procedures have been developed for the assay of CK activity. Both the forward and reverse reactions have been used, and colorimetric, fluorometric, and coupled enzyme methods have been used to follow the rate of product formation during the course of the reaction. The use of the reverse reaction is preferred because it proceeds faster than does the forward reaction, although the cost of the starting chemicals, creatine phosphate and ADP, is greater than is the cost of creatine and ATP.

In one of the older two-point procedures, ATP and creatine are incubated with the specimen, and after a suitable time period, the reaction is stopped by addition of acid to a concentration of 1 mol/l. The CrP formed in the reaction is acid-labile under these conditions and is hydrolyzed to creatine and free phosphate,whereas the ATP and ADP present are not affected. The inorganic phosphate formed is determined colorimetrically, and serves as a measure of the CK reaction. The procedure provides a simple and direct measure of enzyme activity, but requires a large specimen volume to obtain enough PO_4 for accurate measurement. Also, as the ADP accumulates, it begins to inhibit the reaction.

Although the other reaction product, ADP, cannot be measured directly, Tanzer and Gilvarg[195] measured it by coupling the CK reaction to two other enzyme reactions, leading eventually to the oxidation of NADH, which could be measured spectrophotometrically at 340 nm (equation 47). The pH (9.0) is favorable for the CK reaction, but not

$$\begin{aligned} \text{Creatine} + \text{ATP} &\overset{\text{CK}}{\rightleftharpoons} \text{creatine phosphate} + \text{ADP} \\ \text{Phosphoenolpyruvate} + \text{ADP} &\overset{\text{PK}}{\rightleftharpoons} \text{pyruvate} + \text{ATP} \\ \text{H}^+ + \text{pyruvate} + \text{NADH} &\overset{\text{LD}}{\rightleftharpoons} \text{lactate} + \text{NAD}^+ \end{aligned} \tag{47}$$

for the auxiliary (pyruvate kinase) and indicator (LD) reactions. Because ADP is rephosphorylated to ATP, it does not accumulate to inhibit the CK reaction, and the regenerated ATP keeps the Mg^{2+}/ATP ratio constant at the desired optimal level. Reaction conditions are so chosen that only the CK reaction is rate limiting, and the rate of NADH oxidation ($-\Delta A$/min) is a measure of CK activity. Other investigators omitted the LD reaction and measured the pyruvate colorimetrically with 2,4-dinitrophenylhydrazine (see ALT and AST procedures).

If the reverse reaction is used for assay, one reaction product is creatine, which can be measured colorimetrically by combining it with diacetyl and α-naphthol to form a pink-colored product. Sax and Moore[173] developed a fluorometric method, based on measuring a fluorophor formed by reaction between creatine and ninhydrin in KOH solution. The other enzyme reaction product is ATP, which permits direct measurement by use of the highly specific and sensitive bioluminescence (firefly light) method. ATP oxidizes luciferin in the presence of luciferase (both obtained from fireflies) to an intermediate, which in the presence of O_2 decomposes and gives off a characteristic light, one quantum of energy being emitted for each molecule of ATP consumed. The reaction is carried out in a light-free system (modified photometer), and the rate at which the light impulses are given off is measured.[44,224]

In the CK procedure used most widely, the ATP produced in the reverse reaction is measured by coupling a hexokinase (HK) reaction and a glucose-6-phosphate dehydrogenase (GPD) reaction to the CK reaction (equation 48). The HK catalyzes the phosphorylation of glucose by the ATP to form glucose-6-phosphate (G6P) and regenerate ADP for the CK reaction. The G6P is then oxidized with NADP to form 6-phosphogluconic acid and produce NADPH. The rate of NADPH formation is a measure of the CK activity, provided that the concentrations of all other components in the three enzyme system

$$\begin{aligned} \text{Creatine phosphate} + \text{ADP} &\xrightarrow{\text{CK}} \text{creatine} + \text{ATP} \\ \text{ATP} + \text{glucose} &\xrightarrow{\text{HK}} \text{glucose-6-phosphate} + \text{ADP} \\ \text{Glucose-6-phosphate} + \text{NADP}^+ &\xrightarrow{\text{GPD}} \text{6-phosphogluconate} + \text{NADPH} + \text{H}^+ \end{aligned} \tag{48}$$

are present in suitable excess, so that the CK activity is the only limiting factor. The coupled enzyme system is completely "downhill," i.e., all reactions proceed in the favorable direction. The pH optimum for the entire system is 6.8 to 6.9.

When the source of CK is a specimen such as serum, certain "side reactions" may occur because of other enzymes introduced into the system by the specimen. If the introduced enzyme is adenylate kinase (AK; old name, myokinase), the reaction (2ADP → ATP + AMP) may proceed at a sufficient rate to consume enough ADP to cause the concentration of ADP to become rate-limiting. This effect is counteracted by adding AMP (adenylic acid) to the system. AMP is a competitive inhibitor of ADP in the AK reaction, and it forces the AK reaction to proceed in the direction of ADP formation. Any significant phosphatase or nucleotidase activity may also interfere by hydrolyzing some ATP, ADP, CrP, and the G6P needed for the GPD reaction. However, the level of enzymes in serum which are active at pH 6.8 is negligible.

This assay procedure, first used by Nielsen and Ludvigsen and by Oliver, was studied and modified by Rosalki,[168] Hess, and other workers. The "optimized" procedure developed by the German Society for Clinical Chemistry[82] is presented in detail.

DETERMINATION OF CREATINE KINASE

SPECIMEN

Serum or heparinized plasma are the preferred specimens, CK activity being inhibited by EDTA, citrate, and fluoride. CK activity in serum specimens is very unstable, and is rapidly lost during storage. Activity may persist at ambient temperature for 4 h, at 4°C for about 8 to 12 h, and, when frozen, for only 2 to 3 days; but the degree of stability varies with the individual specimen. A large fraction, if not all, of the original activity may be recovered by the addition of sulfhydryl agents, but the nature of the inactivation and re-activation processes is not understood. The addition of Cleland's reagent (or some other SH agent) to the serum specimen to a concentration of 5 mmol/l retards loss of activity significantly, but it is best if the blood specimen is collected directly into a tube already containing the SH agent.

A small degree of hemolysis can be tolerated because red cells have no CK activity. However, hemolyzed specimens should not be used, owing to the possible liberation from the red cells of enzymes and intermediates (adenylate kinase, ATP, glucose-6-phosphate), which may affect the lag phase and the side reactions occurring in the assay system.

A little-understood phenomenon is observed when specimens with very high activity have to be diluted to decrease the activity to the range compatible with the concentrations of reagents in the enzyme assay system. On dilution of the specimen one can observe an apparent increase in enzyme activity, which is greatest when the diluent is water or saline (NaCl, 8.5 g/l), but which also occurs if the diluent is heat-inactivated serum.[53] The augmentation of activity may vary from 1.5- to 3-fold but may be as high as 15-fold. It is greatest in specimens with the highest activity, and especially in those from patients with muscular dystrophy. The magnitude of this dilution effect appears to be as much a characteristic of the individual specimen as it is of the disease involved. The observed facts are consistent with the presence of some form of dissociable inhibitor which varies considerably from one serum to another.[199] Based on present knowledge, it appears best to avoid diluting a specimen, if at all practical, and to use inactivated serum as the diluent if dilution is unavoidable.

REAGENTS

1. Triethanolamine buffer (TEA), pH 6.8, 100 mmol/l. Dissolve 14.9 g of reagent grade triethanolamine in about 800 ml of water, warm the solution to 30°C, and carefully

adjust the pH to 6.8 by adding 6.0 molar acetic acid. Dilute the solution to 1000 ml. The solution is stable and may be stored at room temperature.

2. Magnesium acetate, 100 mmol/l, in 100 mmol/l TEA buffer. Dissolve 4.29 g of $Mg(C_2H_3O_2)_2 \cdot 4H_2O$ in 200 ml of the TEA buffer, pH 6.8. Stable, but keep refrigerated.

3. Glucose solution, 200 mmol/l, in TEA buffer. Dissolve 7.2 g of D-glucose in 200 ml of 100 mmol/l TEA buffer. Keep refrigerated.

4. Creatine phosphate, 350 mmol/l, in TEA buffer. Dissolve 1.27 g of creatine phosphate, disodium salt, hexahydrate, in TEA buffer, adjust pH to about 7.5, and dilute to 10 ml. Store frozen and keep chilled when being used.

5. NADP solution, 6.0 mmol/l, in TEA buffer. Dissolve 45 mg of NADP, disodium salt, in TEA buffer, adjust pH to 6.8, and dilute to 10 ml. Keep frozen.

6. ADP solution, 10 mmol/l, in TEA buffer. Dissolve 137 mg of ADP, trisodium salt, trihydrate, in TEA buffer, adjust pH to 6.8, and dilute to 25 ml. Keep frozen.

7. Glutathione, 90 mmol/l, in TEA buffer. Dissolve 278 g of (reduced) glutathione in TEA buffer, adjust pH to 6.8, and dilute to 10 ml. Prepare fresh as needed.

8. AMP solution, 100 mmol/l, in TEA buffer. Dissolve 500 mg adenosine-5′-phosphate, disodium salt, hexahydrate, in the buffer, adjust pH to 6.8, and dilute to 10 ml. Keep frozen.

9. Hexokinase, 3000 U/l, in TEA buffer. Prepare by diluting a stock crystalline suspension with TEA buffer to 3000 U/l. Keep frozen.

10. Glucose-6-phosphate dehydrogenase, 2000 U/l, in TEA buffer. Prepare from a stock suspension as described for HK.

11. Substrate mixture. Prepare just before use, and prepare enough for one day's work only. Warm the frozen reagents to 4°C, and then combine one part of each of the 10 reagents to form a composite reagent. Check pH and adjust to 6.8 (at 30°C), if needed. Keep in an ice bath, except when actually pipetting into tubes or cuvets. Use within 4 to 6 h after preparation.

PROCEDURE

1. Pipet 3.0 ml of substrate mixture into a 12 × 75 mm test tube.

2. Add 0.10 ml of serum specimen, mix gently, and place in a 30°C water bath for 6 min for temperature equilibration to occur, and for completion of the side reactions and lag phase. (See Comments.)

3. Pour the tube contents into a 1.0 cm cuvet, which has been pre-warmed in a water bath or cuvet compartment to 30°C. Insert the cuvet in the cell compartment of the spectrophotometer, and measure the increase in absorbance at 340 nm, against a water blank. Take readings at one minute (or half minute) intervals for a period of 5 to 7 min. The reaction should proceed at a linear rate. If necessary, take additional readings until at least five absorbance readings in the linear rate phase have been obtained.

CALCULATIONS

For each mole of CrP consumed, one mole of NADPH is formed (mmolar absorptivity = 6.22). The total reaction volume is 3.10 ml; specimen volume is 0.10 ml. Therefore,

$$\text{CK activity in U/l} = \Delta A/\text{min} \times \frac{3.1}{6.22} \times \frac{1000}{0.10} = \Delta A/\text{min} \times 5000$$

If the enzyme activity at 30°C is taken as 1.00, activities at some selected temperatures are as follows: 25°, 0.68; 28°, 0.86; 32°, 1.18; 35°, 1.47; 37°, 1.72.

COMMENTS

1. No serum blanks need be used. The pigment color in a highly icteric serum may be blanked out by using dichromate blanks (see AST procedure).

2. Tris, glycylglycine, and imidazole, at concentrations of 60 to 80 mmol/l, may be used as buffers in place of TEA. The use of MOPS (morpholinopropanesulfonic acid, $pK_a' = 7.2$) has also been suggested. Activity drops as the buffer concentration is increased above 0.10 mmol/l.

3. The actual concentrations of the components in the reaction mixture are obtained by dividing the concentrations of the reagents by 10 as noted under Reagents. These concentrations differ somewhat from those used by Rosalki[168]. The CrP must be at 35 mmol/l if maximum activity is to be observed; Rosalki uses only 10 mmol/l. On the other hand, he uses Mg^{2+} at the high concentration of 30 mmol/l; the German group[82] found that no more than 10 mmol/l was needed to effect maximum activity, and Nielsen and Ludvigsen found that levels over 10 mmol/l were increasingly inhibitory.

Smeaton[184] has shown that the lag phase can be considerably shortened by using very high levels of HK and GPD (5000 U/l), and by eliminating the AMP, which he found not only inhibits CK to some degree but also inhibits GPD activity. For this reason the HK and GPD concentrations in the procedure are greater than those proposed either by the German group (1200 U/l each)[82] or by Rosalki (600 and 300 U/l for HK and GPD, respectively).

The Statzyme-CPK kit (Worthington Biochemical Corp., Freehold, NJ.) incorporates the auxiliary enzymes at a level of 5000 U/l each. The concentration of AMP is reduced to 3.7 mmol/l; CrP is used at a concentration of 20 mmol/l.

4. In the procedure presented, and also in that of Rosalki, it is assumed that the 6 min allowed for the completion of the lag phase and side reactions is sufficient time to effect essentially complete activation of the CK by glutathione in the reagent mixture. Some workers feel that reactivation proceeds slowly over a period of time. They prefer to add the sulfhydryl agent to the serum, and to allow up to an hour's incubation time for complete activation of the enzyme in the serum.

NORMAL VALUES

Because of the large volume of work still being done to develop more accurate and definite procedures for measuring CK activity, and the many remaining unsolved problems relating to inactivation and activation of CK in serum, a range of normal values acceptable to all investigators in the field has yet to be established. Also, several population parameters which affect CK values are often not taken into account in establishing a normal range. A review of data in the literature suggests that the normal range for adult males may be taken as 12 to 65 U/l, and that for adult women as 10 to 50 U/l, with activity measured at 30°C.

There is considerable evidence that CK levels are lower by some 20 to 30 per cent when measured in patients who have been at complete bed rest for several days, compared to values obtained from patients who are ambulatory or otherwise exercising use of some body muscles. This fact is usually not taken into account when studies directed toward establishing normal values are being made, and the spread in normal values and the upper limit values reported may reflect populations containing both bed-rest and non-bed-rest patients.

That CK activity is lower in women than in men by some 25 to 40 per cent was evident even in the early work. Garcia[80] has shown that apparently healthy, muscular, heavy-framed men may have serum CK levels in the range of 100 to 120 U/l, and he suggested that an individual's serum CK level may be roughly proportional to his muscle mass.

The difference in CK levels in men and women may also reflect such a difference in muscle mass.

Recently, Meltzer[137] has directed attention to the observation that CK activity appears to be higher in Blacks than in Caucasians by some 30 per cent. He proposes 110 U/l as the upper limit of normal for Black males, as against 70 U/l for Caucasian males; the corresponding figures for females are 70 and 50 U/l, respectively. Such racial or genetic differences have been overlooked in previous work.

REFERENCES

German Society for Clinical Chemistry: Recommendations of the Enzyme Commission. Z. Klin. Chem. Klin. Biochem., *10*:281, 1972.

Swanson, J. R., and Wilkinson, J. H.: Measurement of Creatine Kinase Activity in Serum. *In* Standard Methods of Clinical Chemistry, Vol. 7, G. R. Cooper, Ed. New York, Academic Press, 1972, pp. 33–42.

METHODS FOR THE SEPARATION AND QUANTITATION OF CREATINE KINASE ISOENZYMES

Four different techniques have been developed for use in separating and measuring the three isoenzymes of CK. The first technique to be used was electrophoresis, and this is still the most commonly used procedure. The best results have been obtained with agarose[57] or polyacrylamide gels,[45,185] although many investigators have obtained satisfactory separations with cellulose acetate. The isoenzyme bands may be visualized colorimetrically, using the NADPH formed in the coupled CK-HK-GPD system to reduce a tetrazolium salt to a colored formazan. Such colorimetric methods can be used effectively only with sera with elevated CK levels (over 124 U/l); sera with normal activity must be first concentrated (see Protein chapter). Considerably more sensitivity is attained by using fluorometric detection or quantitation[32,164] of the NADPH (bluish-white fluorescence on excitation with long-wave ultraviolet light). By this technique, the CK bands can be detected even in sera with a total CK level of 20 U/l or less. Quantitation is most often carried out with a scanning accessory available with most fluorometers. The isoenzyme bands may be eluted from the cut strips of the film or gel, and the fluorescence of the eluates measured.[160]

The second approach to CK isoenzyme isolation involves the selective elution of the three isoenzyme fractions after adsorption on a chromatographic column.[138] The adsorbent used is DEAE-Sephadex-A-50, contained in minicolumns (0.5 × 6.0 cm) holding about 60 mg of the adsorbent. The CK and other proteins are adsorbed on the column, following which the column is washed and the MM isoenzyme eluted out by addition of several serial small-volume portions of Tris buffer, pH 8.0, 50 mmol/l, containing 100 mmolar NaCl. The MB fraction is next eluted with several portions of the same Tris buffer, but containing 200 mmolar NaCl. The BB isoenzyme is then eluted out with portions of Tris buffer, pH 7.0, containing 500 mmolar NaCl. The several published and kit procedures vary in minor details. The isoenzyme fractions in the eluates are then measured with the same procedure used to measure total CK; preliminary concentration of the eluates may be necessary before the quantitation is attempted. Experimental details are arranged so that little or no overlapping of isoenzyme fractions occurs in the eluates, particularly of MM and MB. This technique, though relatively new, is simple to carry out and takes little time, and may replace the electrophoretic procedures if experience shows that it is as reliable as the latter.

In the third method for measuring isoenzymes, advantage is taken of the fact that at very low substrate concentrations, the reaction rates for the three isoenzymes are quite different.[224] When creatine phosphate is present at a concentration of 2.5 mmol/l (optimal = 35 mmol/l) and ADP at 0.012 mmol/l (optimal = 1.0 mmol/l), the BB form reacts over four times faster than does the MM form, with the MB isoenzyme having an intermediate activity. The very small amount of ATP formed at these very low substrate concentrations is measured with the luciferin-luciferase (firefly light) system, using a very sensitive photometer (Chem-Glo-Photometer, American Instrument Co., Silver Spring, MD.).

The fourth technique for separating CK isoenzymes is based on precipitation of the BB and MM enzymes by antibodies to these isoenzymes.[107] Because the antibodies are directed against the B and M subunits of CK, the antibodies against MM and BB forms will also precipitate the MB isoenzyme. The quantity of BB present in a serum specimen is obtained by measuring the residual activity remaining after adding anti-MM antiserum and centrifuging down the resulting precipitate. The quantity of MB is obtained by subtracting the measured MM and BB activities from the total CK activity. Representative results are given in Table 12-9.

TABLE 12-9 CREATINE KINASE ISOENZYME PATTERNS IN SERA OF PATIENTS WITH A VARIETY OF CLINICAL CONDITIONS, AND OBTAINED BY SEVERAL DIFFERENT TECHNIQUES

Clinical Condition	*CK Activity, U/l*			*Per Cent MB Activity*	*Technique*† *& Reference*
	TOTAL	MM	MB		
Hospitalized, non-cardiacs	35 ± 13	33 ± 12	2 ± 1	3 to 9	E[160]
Myocardial infarctions	860 ± 120	797 ± 112	63 ± 43	2 to 12	E[160]
Patients receiving morphine IM	347 ± 89	345 ± 89	3 ± 2	0.3 to 0.9	E[160]
Dermatomyositis	1500	1410	90	6	E[32]
Polymyositis	1725	1052	673	39	E[32]
Normal, non-cardiacs	36–277	—	0–2.6	0 to 1	C[138]
Myocardial infarctions	84–236	—	4.6–28	6 to 11	C[138]
Patient A, myocardial infarct, 24 h	82	53	29	35	I[107]
Patient B, myocardial infarct, 48 h	416	449	67	16	I[107]
Myocardial infarctions (35 patients)					
Total CK, 35–100 U/l	60*	51	9	15	I[107]
Total CK, over 200 U/l	416*	339	67	16	I[107]

* Average value.
† Symbols: E = Enzymoelectrophoresis; C = Column chromatography; I = Immunoprecipitation.

ALDOLASE

(EC 4.1.2.13; D-fructose-1,6-diphosphate:D-glyceraldehyde-3-phosphate lyase)

The enzyme aldolase (ALS) belongs to the class of enzymes referred to as *lyases*, a miscellany of enzymes which reversibly cleave substrates without hydrolysis into two units, one or both of which contain a double-bonded carbon. Carbonic anhydrase and the many decarboxylases are included in the class. Aldolase catalyzes the splitting of D-fructose-1,6-diphosphate (FDP) to D-glyceraldehyde-3-phosphate (GLAP) and dihydroxyacetone-phosphate (DAP), one of the important reactions in the glycolytic breakdown of glucose to lactate. The reaction equilibrium favors the formation of fructose diphosphate. The

$$\begin{array}{c}CH_2OPO_3^{2-}\\ \text{(furanose ring: O, H—C, H, OH, C, } CH_2OPO_3^{2-}\text{, OH, HO, H)}\end{array} \underset{\longrightarrow}{\overset{Aldolase}{\longleftarrow}} \begin{array}{c}CH_2OPO_3^{2-}\\ |\\ C{=}O\\ |\\ CH_2OH\end{array} + \begin{array}{c}CH_2OPO_3^{2-}\\ |\\ H{-}C{-}OH\\ |\\ H{-}C{=}O\end{array} \qquad (49)$$

D-Fructose-1,6-diphosphate (FDP) **Dihydroxyacetonephosphate (DAP)** **D-Glyceraldehyde-3-phosphate (GLAP)**

enzyme shows absolute specificity only for DAP. The GLAP can be substituted with D- or L-glyceraldehyde or glycolaldehyde to form D-fructose-1-(mono)phosphate, L-sorbose-1-phosphate, and xylulose-1-phosphate, respectively. The optimal pH range is 6.8 to 7.2; but when the enzyme is coupled with triosephosphate isomerase (TPI) and glycerol-3-phosphate dehydrogenase (as in one assay method), the apparent range of pH for optimal activity is rather broad, 7.0 to 9.6 pH units.

Aldolase activity is present in all cells in the body, but the degree of specificity varies from organ to organ. Three forms are recognized, designated by Rutter[172] as aldolases A, B, and C. These may represent either different isoenzyme forms (A, C), or different enzymes (B). They are characterized by their different rates of activity toward the two substrates, FDP and fructose-1-phosphate (F1P). Isoenzyme A is the form predominant in skeletal muscle, and has very little activity toward F1P (FDP/F1P ratio of activities = 50/1). Isoenzyme B is the form present in liver, kidney, and leukocytes, and has more activity toward F1P (FDP/F1P = 1.5). The activity of form C, an embryonic form predominant in brain tissue,[206] lies between those of A and B. Both A and C have a tetrameric structure, and form hybrids AAAC, AACC, and so forth, analogous to the LD isoenzyme forms. The AAAA (A_4) form is present in muscle, and A_4 or hybrids of A and C are found in heart, lung, red cells, and serum.

CLINICAL SIGNIFICANCE

Serum aldolase determinations (normal: 1.0 to 7.5 U/l at 30°C) have been of greatest clinical interest in diseases involving muscle disintegration,[198] in which elevations of 10 to 50 times the upper normal level may be seen. The highest levels are found in progressive (Duchenne) muscular dystrophy. The greatest serum increases occur early in the disease, but as the capacity of the cells to synthesize enzyme decreases, serum levels also decrease. Slight to moderate increases in ALS activity can be demonstrated in the sera of female carriers of the gene causing the disease. Lesser degrees of ALS elevation are encountered in dermatomyositis, polymyositis, and limb-girdle dystrophy, but normal values are observed in polio, myasthenia gravis, multiple sclerosis, and diseases of muscle dysfunction of neurogenic origin.

Increases in serum aldolase activity are also observed in myocardial infarctions; the pattern of rise (5 to 8 times normal) and fall parallels that of AST. Increases of 7 to 20 times normal are associated with viral hepatitis, the pattern of rise and fall being similar to that for ALT; normal values are reached 15 to 20 days after the first rise occurred. Levels of ALS in chronic hepatitis, portal cirrhosis, and obstructive jaundice are normal or only marginally raised. In liver diseases it is the B enzyme form, the so-called fructose-1-phosphate aldolase, which is increased.

Other disease states in which increased levels of aldolase may be encountered are: trichinosis, moderate to high values; gangrene; prostatic tumors; some carcinomas metastatic to the liver; granulocytic leukemia, about 6 times normal; megaloblastic anemia, 10 to 13 times normal; delirium tremens; and 60 to 80 per cent of patients with acute psychoses and schizophrenia.[136] Activity is normal in hypothyroid patients, and normal or slightly depressed in persons affected with fructose intolerance.

Injections of cortisone and ACTH will raise serum ALS to levels between 10 and 18 U/l. This physiological response to hormone therapy must be kept in mind when interpreting elevated aldolase values.

In general, measurement of aldolase activity in serum does not provide information that is not available by measurement of some of the other, more routinely assayed enzymes, such as AST and LD. Even in muscular dystrophy, the determination of CK activity is more useful, because the elevations are greater and more easily differentiated from those due to other disease conditions. Because of this, no detailed method for measuring ALS will be presented.

METHODS FOR THE DETERMINATION OF ALDOLASE ACTIVITY

All assay methods are based on the forward reaction, as written in equation 49. Both photocolorimetric and continuous-monitoring procedures have been developed. In the colorimetric methods,[149] hydrazine is added to the reaction mixture as a trapping agent to force the reaction to completion, by forming the hydrazones of DAP and GLAP. These are then hydrolyzed with NaOH to form free dihydroxyacetone and glyceraldehyde, which are then treated with 2,4-dinitrophenylhydrazine. The colored dinitrophenylhydrazones are then measured at 540 nm.

Two continuous-monitoring procedures have been proposed. In the Bruns[16,33] procedure, on which all the commonly used procedures and kits are based, the aldolase reaction is coupled with two other enzyme reactions. Triosephosphate isomerase (TPI, EC 5.3.1.1) is added to insure rapid conversion of all GLAP to DAP. Glycerol-3-phosphate dehydrogenase (GPD, EC 1.1.1.8) is added to convert (reduce) the DAP to glycerol-3-phosphate, with NADH acting as hydrogen donor. The decrease in NADH concentration is then measured. A practical routine procedure based on this approach is that of Pinto, Kaplan, and Van Dreal.[148]

In the other coupled enzyme approach, the GLAP is oxidized by NAD^+ to 3-phosphoglyceric acid, and the rate of formation of NADH is measured. This reaction is catalyzed by the added enzyme, glyceraldehyde-3-phosphate dehydrogenase (GAD). Triosephosphate isomerase and arsenate are also added to force the reaction to completion. This coupled system is subject to many interferences, and has fallen into disfavor.

The ALS activity in serum is quite stable. Activity is unchanged at ambient temperatures for up to 48 h, and at 4°C for several weeks. Hemolyzed specimens should not be used, and plasma is preferred over serum because of the possible release of platelet enzyme during the clotting process.

NORMAL VALUES

The accepted range of values for the activity of ALS in serum in adults is 1.0 to 7.5 U/l, measured at 30°C. However, there is a definite sex difference. Pinto and coworkers[148] give 8.0 ± 4.0 U/l for men and 4.7 ± 3.2 U/l for women, for activity measured at 37°C (5.2 ± 2.6, and 3.0 ± 2.1 U/l, respectively, at 30°C). Experience has shown that ALS values for inactive persons and patients at bed rest are only some 50 to 70 per cent of the values obtained for active individuals.

The level at birth is twice as high as that of adults; this doubles in early childhood, and then slowly falls off to the adult range by 18 to 20 years of age.

REFERENCE

Pinto, P. V. C., Kaplan, A., and Van Dreal, P. A.: Clin. Chem., *15*:349, 1969.

Concluding Remarks

Enzyme assay techniques are changing rapidly. The slow, cumbersome, and inherently inaccurate two-point procedures are being replaced by rapid, automated or semi-automated continuous-monitoring methods. These are often associated or coupled with computer calculation of reaction rates, and direct print-out of results. The potential for greatly increased productivity and accuracy can be realized only if the machines are designed with the advice of clinical chemists, and used by informed technologists.

No analysis is better than the quality of the specimen to be analyzed and the quality of the reagents used. Hence, the emphasis in this chapter has been placed on the type of specimen to be used, its stability, the conditions governing storage and preservation, and on precise preparation of reagents and their stability. The use of certain high quality commercial "kits" and commercial reagents will increase, and is to be encouraged, but appropriate controls on reagent quality and the accuracy of the procedures must always be instituted. Increased governmental intervention in the setting up of standards for reagent quality and kit and instrumental performance is to be expected. There will be continued effort to develop "optimized" methods making possible more sensitive procedures. Accurate and precise "reference" methods will be developed in the effort to monitor the quality of kit and other routine procedures.

The enzymes discussed and the procedures presented were selected to acquaint the student with examples of all classes of enzymes, and with the large variety of enzyme assay methods available. The references include excellent books and monographs, as well as journal articles, many of which should be available in the laboratory library for repeated consultations.

The use of enzymology in diagnostic medicine is now well established, but much has still to be learned. Practical techniques for evaluating isoenzyme patterns for many enzymes are available, and methods for others are actively being sought. The natural history relating to the source, rise, and fall of enzyme activity in serum has only begun to be investigated. Accurate ranges for normal values under various conditions, differentiated by age, sex, race, and culture, are now being re-evaluated and established, using new methods and instruments.

It was not possible to discuss the large variety of more or less complex instruments devised to measure enzyme reaction rates. Some of these are discussed briefly elsewhere in this text. Even though many appear to be based on unique and exotic chemical and engineering principles, they all are still only applications of the few basic concepts discussed under enzyme kinetics.

REFERENCES

1. Amador, E.: Clin. Chem. *18*:94, 1972.
2. Amador, E., Dorfman, L. E., and Wacker, W. E. C.: Clin. Chem., *9*:391, 1963.
3. Amador, E., and Urban, J.: Am. J. Clin. Path., *57*:167, 1972.
4. Babson, A. L., Greeley, S. J., Coleman, C. M., and Phillips, G. E.: Clin. Chem., *12*:482, 1966.
5. Babson, A. L., Read, P. A., and Phillips, G. E.: Am. J. Clin. Path., *32*:83 and 88, 1959.
6. Babson, A. L., Shapiro, P. O., Williams, P. A. R., and Phillips, G. E.: Clin. Chim. Acta, *7*:199, 1962.
7. Bahr, M., and Wilkinson, J. H.: Clin. Chim. Acta, *17*:367, 1967.
8. Baron, D. N., Moss, D. W., Walker, P. G., and Wilkinson, J. H.: J. Clin. Path., *24*:656, 1971.
9. Baron, D. N., Moss, D. W., Walker, P. G., and Wilkinson, J. H.: J. Clin. Path. *28*:592, 1975.
10. Batsakis, J. G.: Ann. Clin. Lab. Science, *4*:255, 1974.
11. Bearn, A. G., and Cleve, H.: Wilson's Disease. *In* The Metabolic Basis of Inherited Disease, 3rd Ed. J. B. Stanbury, J. B. Wyngaarden, and D. S. Fredrickson, Eds. New York, McGraw-Hill, 1972.
12. Belfield, A., and Goldberg, D. M.: Clin. Chem., *15*:931, 1969; *16*:396, 1970.
13. Belfield, A., and Goldberg, D. M.: J. Clin. Path., *22*:144, 1969.
14. Belfield, A., and Goldberg, D. M.: Nature, *219*:73, 1968.

15. Bergmeyer, H. U., Ed.: Methods of Enzymatic Analysis, 2nd English Ed. New York, Academic Press, 1974.
16. Bergmeyer, H. U., Bernt, E., and Bechtler, G.: Fructose-1,6-diphosphate Aldolase. *In* Bergmeyer, H. U., op. cit., p. 1100.
17. Berk, J. E., Kizu, H., Geller, E., and Fridhandler, L.: Proc. Soc. Exp. Biol. Med., *131*:154, 1969.
18. Bernstein, R. E.: Clin. Chim. Acta, *8*:158, 1963.
19. Bernt, E., Bergmeyer, H. U., and King, J.: Isocitrate Dehydrogenase. *In* Bergmeyer, H. U., op. cit., p. 624.
20. Bessey, O. A., Lowry, O. H., and Brock, M. J.: J. Biol. Chem., *164*:321, 1947.
21. Beutler, E.: Am. J. Clin. Path., *47*:303, 1967.
22. Beutler, E.: Glucose-6-phosphate Dehydrogenase Deficiency. *In* The Metabolic Basis of Inherited Disease, 3rd Ed. J. B. Stanbury, J. B. Wyngaarden, and D. S. Fredrickson, Eds. New York, McGraw-Hill, 1972.
23. Beutler, E., and Mitchell, M.: Blood, *32*:816, 1968.
24. Bishop, C.: J. Lab. Clin. Med., *68*:149, 1966.
25. Bodansky, O., and Schwartz, M. K.: 5′-Nucleotidase. *In* Advances in Clinical Chemistry. O. Bodansky and C. P. Stewart, Eds. New York, Academic Press, 1968, Vol. 11, p. 277.
26. Bondar, R. J., and Moss, G. A.: Clin. Chem., *20*:317, 1974.
27. Bowers, G. N., Jr.: Clin. Chem., *5*:509, 1959.
28. Bowers, G. N., Jr., and McComb, R. B.: *In* Proceedings, International Seminar and Workshop on Enzymology. N. W. Tietz, Program Coordinator. Mt. Sinai Hosp. Med. Center, Chicago, 1972, pp. 2–2 and 3–8.
29. Bowers, G. N., Jr., and McComb, R. B.: Clin. Chem. *12*:70, 1966.
30. Boyd, J. W.: Clin. Chim. Acta, *7*:424, 1962.
31. Brojer, B., and Moss, D. W.: Clin. Chim. Acta, *35*:511, 1971.
32. Brownlow, K., and Elevitch, F. R.: J. A. M. A., *230*:1941, 1974.
33. Bruns, F.: Biochem. Z., *325*:156, 1954.
34. Bruns, F. H., and Werners, P. H.: Dehydrogenases. *In* Advances in Clinical Chemistry. H. Sobotka and C. P. Stewart, Eds. New York, Academic Press, 1962, Vol. 5. Glucose-6-phosphate Dehydrogenase, p. 243. Sorbitol Dehydrogenase, p. 271.
35. Bucolo, G., Chang, T. Y., and McCroskey, R.: Clin. Chem., *20*:881, 1974.
36. Bulletin ACP-BL-1, Eskalab Acid Phosphatase. Smith Kline Instruments, Inc., Palo Alto, CA.
37. Campbell, D. M.: Biochem. J., *82*:34P, 1962.
38. Canapa-Anson, R., and Rowe, D. J. F.: J. Clin. Path., *23*:499, 1970.
39. Caraway, W. T.: Am. J. Clin. Path., *32*:97, 1959.
40. Ceska, M., Birath, K., and Brown, B.: Clin. Chim. Acta, *26*:437, 1969.
41. Cheung, T., and Briggs, M. H.: Clin. Chim. Acta, *54*:127, 1974.
42. Cholinesterase (Kinetic Test), Catalog No. 3811, EM Laboratories, Inc., Elmsford, NY.
43. Cohn, W. E.: Letter, and King, J. S.: Reply, Clin. Chem., *18*:740, 1972.
44. CPK Kit, Catalog No. A1001, Antonik Laboratories, Elk Grove Village, IL.
45. Dawson, D. M., Eppenberger, H. M., and Kaplan, N. D.: J. Biol. Chem., *242*:210, 1967.
46. Demetriou, J. A., Drewes, P. A., and Gin, J. B.: *In* R. J. Henry, D. C. Cannon, and J. W. Winkelman, Eds., op. cit. Lactate Dehydrogenase, p. 819; Isocitrate Dehydrogenase, p. 837; Ceruloplasmin, p. 857.
47. De Ritis, R., Giusti, G., Piccinino, F., and Cacciatore, L.: Bull. World Health Org., *32*:59, 1965.
48. Desnuelle, P.: Pancreatic Lipase. *In* Advances in Enzymology. New York, Wiley-Interscience, 1961, Vol. 23, p. 129.
49. Diagnostic Application of Lactic Dehydrogenase to Disease. Seminar on Clinical Enzymology. B. M. Wagner, Ed. Department of Pathology, Columbia-Presbyterian Medical Center, New York, March, 1969.
50. Dietz, A. A., Lubrano, T., and Rubinstein, H. M.: LDH Isoenzymes. *In* Standard Methods of Clinica Chemistry. G. R. Cooper, Ed. New York, Academic Press, 1972, Vol. 7, p. 49.
51. Dietz, A. A., Rubinstein, H. M., and Lubrano, T.: Clin. Chem., *19*:1309, 1973.
52. Dixon, M., and Webb, E. C.: Enzymes, 2nd Ed. New York, Academic Press, 1964.
53. Dobosz, I.: Clin. Chim. Acta, *50*:301, 1974.
54. Doe, R. P., Mellinger, G. T., and Seal, U. S.: Clin. Chem., *11*:943, 1965.
55. Dousset, J., and Contu, L.: Ann. Rev. Medicine, *18*:55, 1967.
56. King, J. S., Jr.: Abbreviations of Names of Enzymes. Editorial. Clin. Chem., *18*:319, 1972.
57. Elevitch, F. R., and Brownlow, K.: Am. J. Clin. Path., *59*:133, 1973.
58. Ellin, R. I., Groff, W. A., and Kaminskis, A.: Clin. Chem., *18*:1009, 1972.
59. Ellis, G., and Goldberg, D. M.: Am. J. Clin. Path., *56*:627, 1971.
60. Ellis, G., and Goldberg, D. M.: Clin. Biochem., *2*:175, 1971.
61. Ellis, H. A., and Kirkman, H. N.: Proc. Soc. Exp. Biol. Med., *106*:607, 1961.
62. Emerson, P. M., Withycombe, W. A., and Wilkinson, J. H.: Brit. J. Haematology, *13*:656, 1967.
63. Enzyme Nomenclature (1972). Recommendations of the International Union of Pure and Applied Chemistry, and the International Union of Biochemistry. New York, American Elsevier, 1973.
64. Erlanger, B. F., Kokowsky, N., and Cohen, W.: Arch. Biochem. Biophys., *95*:271, 1961.
65. Eshchar, J., Rudzki, C., and Zimmerman, H. J.: Am. J. Clin. Path., *47*:598, 1967.

66. Everse, J., and Kaplan, N. D.: Lactate Dehydrogenase, Structure and Function. *In* Advances in Enzymology. A. Meister, Ed. New York, Wiley-Interscience, 1973, Vol. 37.
67. Fabiny-Byrd, D. L., and Ertingshausen, G.: Clin. Chem., *18*:841, 1972.
68. Fasella, P., and Turano, C.: *In* Vitamins and Hormones. R. S. Harris, P. L. Munson, and E. Duzfalusy, Eds. New York, Academic Press, 1970. Vol. 28, p. 157.
69. Fernandez, R., Seal, U. S., Mellinger, G. T., and Doe, R. P.: Invest. Urology, *2*:328, 1965.
70. Fishman, W. H.: Ann. N.Y. Acad. Science, *166*:745, 1969.
71. Fishman, W. H., and Ghosh, N. K.: Isoenzymes of Human Alkaline Phosphatase. *In* Advances in Clinical Chemistry. O. Bodansky and C. P. Stewart, Eds. New York, Academic Press, 1968, Vol. 10, p. 256.
72. Fishman, W. H., Inglis, N. I., and Krant, N. I.: Clin. Chim. Acta, *12*:298, 1965.
73. Fleisher, M., and Schwartz, M. K.: Clin. Chem., *17*:417, 1971.
74. Fraser, G. P., and Nicol, A. D.: Clin. Chim. Acta, *13*:552, 1966.
75. Fridhandler, L., Berk, J. E., and Ueda, M.: Clin. Chem., *17*:423, 1971; *18*:1493, 1972.
76. Fritsche, H. A., Jr., and Adams-Park, H. R.: Clin. Chem., *18*:417, 1972.
77. Garry, P. J.: Clin. Chem., *17*:183 and 192, 1971.
78. Garry, P. J., Owen, G. M., and Lubin, A. H.: Clin. Chem., *18*:105, 1972.
79. Garry, P. J., and Routh, J. I.: Clin. Chem., *11*:91, 1965.
80. Garcia, W.: J. A. M. A., *228*:1395, 1974.
81. Gerlach, U., and Hiby, W.: Sorbitol Dehydrogenase. *In* Bergmeyer, H. U., Ed., op. cit., p. 569.
82. German Society for Clinical Chemistry, Recommendations of the Enzyme Commission.: Z. Klin. Chem. Klin. Biochem., *8*:658, 1970; *10*:281, 1972.
83. Ghosh, N. K.: Ann. N.Y. Acad. Science, *166*:604, 1969.
84. Gindler, E. M.: Clin. Chem., *17*:633, 1971.
85. Goldberg, D. W., and Watts, C.: Gastroenterology, *49*:256, 1965.
86. Good, N. E., Winget, G. D., Winter, W., Conolly, T. N., Izawa, S., and Singh, R. M. M.: Biochemistry, *5*:467, 1966.
87. Guibault, G. G., Hulserman, J., and Sadar, M. H.: Anal. Letters, *2*:185, 1969.
88. Gutman, A. B., and Gutman, E. B.: J. Clin. Inv., *17*:473, 1938.
89. Harris, H., and Whittaker, M.: Ann. Human Genet., *26*:59, 1962.
90. Hartz, J., El Maghrabi, R., Namen, A., Gabr, M., Bowman, J., Carson, P., and Ajmar, F.: Clin. Chim. Acta, *48*:117, 1973.
91. Hausamen, T. U., Helger, R., Rick, W., and Gross, W.: Clin. Chim. Acta, *15*:241, 1967.
92. Haverback, B. J., Dyce, B. J., Gutentag, P. J., and Montgomery, D. W.: Gastroenterology, *44*:588, 1963.
93. Henry, R. J., Cannon, D. C., and Winkelman, J. W., Eds.: Clinical Chemistry, Principles and Technics, 2nd Ed. Hagerstown, MD., Harper and Row, 1974.
94. Henry, R. J.: Pancreatic Lipase. *In* Standard Methods of Clinical Chemistry. D. Seligson, Ed. New York, Academic Press, 1958, Vol. 2, p. 86.
95. Henry, R. J., and Chiamori, N.: Clin. Chem., *6*:434, 1960.
96. Henry, R. J., Chiamori, N., Golub, O., and Berkman, S.: Am. J. Clin. Path., *34*:381, 1960.
97. Herz, J., Kaplan, E., and Scheye, E. S.: Clin. Chim. Acta, *46*:147, 1974.
98. Hess, B.: Enzymes in Blood Plasma. K. S. Henley, Trans. New York, Academic Press, 1963.
99. Hill, P. G., and Sammons, H. G.: Enzyme, *12*:201, 1971.
100. Hill, P. G., and Sammons, H. G.: Quart. J. Med., *36*:457, 1967.
101. Hillman, G.: Z. Klin. Chem. Klin. Biochem., *9*:273, 1971.
102. Hobbs, J. R., and Aw, S. E.: Isoenzymes of Amylase. *In* Enzymes in Urine and Kidneys. U. C. Dubach, Ed. Baltimore, Williams and Wilkins, 1968, p. 291.
103. Hopkinson, D. A.: Genetically Determined Polymorphisms in Erythrocytes in Man. *In* Advances in Clinical Chemistry. O. Bodansky and C. B. Stewart, Eds. New York, Academic Press, 1968, Vol. 11, p. 37.
104. Horecker, B. L.: Am. J. Clin. Path., *47*:271, 1967.
105. Horne, M., Cornish, C. J., and Posen, S. J.: J. Lab. Clin. Med., *72*:905, 1968.
106. Howatt, H. T.: Gastroenterology, *42*:72, 1962.
107. Jockers-Wretou, E., Grabert, K., and Pfleiderer, G.: Z. Klin. Chem. Klin. Biochem., *13*:85, 1975.
108. Juul, P.: Clin. Chim. Acta, *19*:205, 1968.
109. Kalow, W., and Genest, K.: Canad. J. Biochem., *35*:339, 1957.
110. Kaplan, M. M.: Progress in Hepatology, Alkaline Phosphatase. Gastroenterology, *62*:452, 1972.
111. Kaplan, M. M.: Current Concepts, Alkaline Phosphatase. New Eng. J. Med., *286*:200, 1972.
112. Kaplan, M. M., and Righetti, A.: J. Clin. Inv., *49*:508, 1970.
113. Karmen, A., Wroblewski, F., and LaDue, J. E.: J. Clin. Inv., *34*:126 and 133, 1955.
114. Keane, P. M., Garcia, L., Gupta, R. N., and Walker, W. H C.: Clin. Biochem., *6*:41, 1973.
115. Keller, D. F.: CRC Crit. Rev. Clin. Lab. Science, *1*:247, 1970.
116. Keller, M. E., Agne, M., Mannebach, A., and Keppla, W.: Uropepsinogen. *In* Enzymes in Urine and Kidneys. U. C. Dubach, Ed. Baltimore, Williams and Wilkins, 1968, p. 265.
117. Kibe, O., and Nilsson, N. J.: Acta Med. Scand., *182*:597, 1967.
118. King, J.: Practical Clinical Enzymology. Princeton, D. Van Nostrand, 1965, p. 123.

119. King, J.: Practical Clinical Enzymology. Princeton, D. Van Nostrand, 1965, p. 275.
120. Klein, B., Foreman, J. A., and Searcy, R. L.: Clin. Chem., *16*:32, 1970.
121. La Motta, R. V., and Woronick, C. L : Clin. Chem., *17*:135, 1971.
122. Latner, A. L.: Isoenzymes. *In* Advances in Clinical Chemistry, H. Sobotka and C. P. Stewart, Eds. New York, Academic Press, 1967, Vol. 9, p. 57.
123. Latner, A. L., and Skillen, A. W.: Isoenzymes in Biology and Medicine. New York, Academic Press, 1968, pp. 56–63, 113–115, and 158–161.
124. Laudahn, G., Hartman, E., Rosenfeld, E. M., Weyer, H., and Muth, H. W.: Klin. Wochenschr., *48*:838, 1970.
125. Lehmann, H., and Liddell, J.: Prog. Med. Genetics, *3*:75, 1964.
126. Lehninger, A. L.: Biochemistry, 2nd Ed. New York, Worth Publishers, 1975, p. 28.
127. Levitzki, A., and Reuben, J.: Biochemistry, *12*:41, 1973.
128. Löhr, G. W., and Waller, H. D.: *In* Bergmeyer, H. U., Ed., op. cit., Vol. 2, p. 636, and Vol. 3, p. 1200.
129. Lowe, M. L., Stella, A. F., Mosher, B. J., Gin, J. B., and Demetriou, J. A.: Clin. Chem., *18*:440, 1972.
130. Lowry, O. H., Roberts, N. R., Wu, M., Hixon, W. S., and Crawford, E. J.: J. Biol. Chem., *207*:19, 1954.
131. Lum, G., and Gambino, S. R.: Clin. Chem., *18*:358, 1972.
132. Massion, C. G., and Frankenfeld, J. K.: Clin. Chem., *18*:366, 1972.
133. Massion, C. G., and Seligson, D.: Am. J. Clin. Path., *48*:307, 1967.
134. Mattenheimer, H.: Clinical Enzymology. Ann Arbor, MI., Ann Arbor Science Publ., 1971, pp. 54, 65, 99, 109.
135. McComb, R. B., and Bowers, G. N., Jr.: Clin. Chem., *18*:97, 1972.
136. Meltzer, H.: Science, *159*:1368, 1968.
137. Meltzer, H. Y.: J. A. M. A., *229*:1169, 1974.
138. Mercer, D. W., and Varat, M. A.: Clin. Chem., *21*:1088, 1975.
139. Milby, T. H.: J. A. M. A., *216*:2131, 1971.
140. Moss, D. W.: Ann. N. Y. Acad. Science, *166*:641, 1969.
141. Moss, D. W.: Enzymologia, *31*:193, 1966.
142. Nicholson, J. F., Bodourian, S. H., and Pesce, M. A.: Clin. Chem., *20*:1349, 1974.
143. O'Neal, W. R., and Gochman, N.: Clin. Chem., *16*:985, 1970.
144. Osaki, S., Johnson, D. A., and Frieden, E.: J. Biol. Chem., *241*:2796, 1966.
145. Owen, J. A.: Effect of Injury on Plasma Proteins. *In* Advances in Clinical Chemistry. H. Sobotka and C. P. Stewart, Eds. New York, Academic Press, 1967, Vol. 9, p. 1.
146. Persijn, J. P., Van der Slik, W., and Bon, A. W. M.: Z. Klin. Chem. Klin. Biochem., *7*:493, 1969.
147. Pilz, W.: Cholinesterases. *In* Bergmeyer, H. U., Ed., op. cit., p. 831.
148. Pinto, P. V. C., Kaplan, A., and Van Dreal, P. A.: Clin. Chem., *15*:349, 1969.
149. Pinto, P. V. C., Van Dreal, P. A., and Kaplan, A.: Clin. Chem., *15*:339, 1969.
150. Posen, S., Neale, F. C., and Clubb, J. S.: Ann. Int. Med., *62*:1234, 1965.
151. Quigley, G. J., Richards, R. T., and Shier, K. J.: Am. J. Obst. Gyn., *106*:340, 1970.
152. Rawstron, J. R., and Ng, S. H.: Clin. Chim. Acta, *32*:303, 1971.
153. Reitman, S., and Frankel, S.: Am. J. Clin. Path., *28*:56, 1957.
154. Rej, R., Fasce, C. F., Jr., and Vanderlinde, R. E.: Clin. Chem., *19*:92, 1973.
155. Report of the Commission on Enzymes of the International Union of Biochemistry. New York, Pergamon Press, 1961.
156. Rice, E. W.: Ceruloplasmin Assay in Serum. *In* Standard Methods of Clinical Chemistry. D. Seligson, Ed. New York, Academic Press, 1963, Vol. 4, p. 39.
157. Rick, W.: Trypsin. *In* Bergmeyer, H. U., Ed., op. cit., p. 1017.
158. Rick, W., and Fritsch, W. P.: Pepsin. *In* Bergmeyer, H. U., Ed., op. cit., p. 1046.
159. Rinderknecht, H., Wilding, P., and Haverback, B. J.: Experientia, *23*:807, 1967.
160. Roberts, R., Henry, P. D., Witteeveen, S. A. G. J., and Sobel, B. E.: Am. J. Cardiology, *23*:650, 1974.
161. Robinson, D. S., and Wing, D. R.: Studies on Tissue Clearing Factor Lipase Related to its Role in the Removal of Lipoprotein Triglyceride from the Plasma. *In* Plasma Lipoproteins. R. M. S. Smellie, Ed. New York, Academic Press, 1971, p. 123.
162. Rock, R. C., and Slickers, R. A.: Letter. Clin. Chem., *21*:166, 1975.
163. Rodgerson, D. O., and Osberg, I. M.: Clin. Chem., *20*:43, 1974.
164. Roe, C. R., Limbird, L. E., Wagner, G. S., and Nerenberg, S. T.: J. Lab. Clin. Med., *80*:577, 1972.
165. Roesser, H. P., Lee, G. R., Nacht, S., and Cartwright, G. E.: J. Clin. Inv., *49*:2408, 1970.
166. Rosalki, S. B.: A Review of Serum Glutamate-oxalacetate Transaminase (SGOT) Methods. Dade Div., Am. Hosp. Supply Co., Miami, FL., 1972.
167. Rosalki, S.: Gamma-Glutamyl Transpeptidase. *In* Advances in Clinical Chemistry. O. Bodansky and A. L. Latner, Eds. New York, Academic Press, 1975, Vol. 17, p. 53.
168. Rosalki, S. B.: J. Lab. Clin. Med., *69*:696, 1967.
169. Rosalki, S. B., and Rau, D.: Clin. Chim. Acta, *39*:41, 1972.
170. Rosalki, S. B., and Tarlow, D.: Clin. Chem., *20*:1121, 1974.
171. Roy, A. V., Brower, M. E., and Hayden, J. E.: Clin. Chem., *17*:1093, 1971.
172. Rutter, W. J., Rajkumar, T., Penhoest, E., and Kochman, M.: Ann. N.Y. Acad. Sci., *151*:102, 1968.

173. Sax, S. M., and Moore, J. J.: Clin. Chem., *11*:951, 1965.
174. Sass-Kortsak, A.: Copper Metabolism. *In* Advances in Clinical Chemistry. H. Sobotka and C. P. Stewart, Eds. New York, Academic Press, 1965, Vol. 8, p. 1.
175. Scandinavian Society for Clinical Chemistry and Clinical Physiology: Recommended Methods for the Determination of Four Enzymes in Blood. Scand. J. Clin. Lab. Inv., *33*:291, 1974.
176. Schwartz, M. K.: Measurement of 5′-Nucleotidase Activity in Serum. *In* Standard Methods of Clinical Chemistry. G. R. Cooper, Ed. New York, Academic Press, 1972, Vol. 7, p. 1.
177. Schwartz, M. K., and Fleischer, M.: Diagnostic Biochemical Methods in Pancreatic Disease. *In* Advances in Clinical Chemistry. O. Bodansky and C. P. Stewart, Eds. New York, Academic Press, 1970, Vol. 13, p. 114.
178. Seal, U. S., Mellinger, G. T., and Doe, R. P.: Clin. Chem., *12*:620, 1966.
179. Searcy, R. L., Hayashi, S., and Berk, J. E.: Am. J. Clin. Path., *46*:582, 1966.
180. Searcy, R. L., Wilding, P., and Berk, J. E.: Clin. Chim. Acta, *15*:189, 1967.
181. SGOT-Survey I, 1973. Proficiency Testing, Summary Analysis. U.S. Department of Health, Education, and Welfare, Center for Disease Control, Atlanta, GA., 1973.
182. Shihabi, Z. K., and Bishop, C.: Clin. Chem., *17*:1150, 1971.
183. Shipe, J. R., and Savory, J.: Clin. Chem., *19*:645, 1973.
184. Smeaton, J. R.: Clin. Chem. Newsletter, Perkin-Elmer Corp., Norwalk, CT., *4*:8, 1972.
185. Smith, A. F.: Separation of Tissue and Serum Creatine-kinase Isoenzymes in Polyacrylamide Gel Slabs. *In* Isoenzymes, Biochemical and Genetic Studies. J. A. Frelinger and R. T. Acton, Eds. New York, M.S.S. Information Group, 1973.
186. Smith, A. F., and Fogg, B. A.: Clin. Chem., *18*:1518, 1972.
187. Smyth, D. G., Stein, W., and Moore, G. J.: J. Biol. Chem., *238*:227, 1963.
188. Somogyi, M.: Clin. Chem., *6*:23, 1960.
189. Stolbach, L. L., Krant, M. J., and Fishman, W. H.: N. Eng. J. Med., *281*:757, 1969.
190. Stool Assay-Spectrophotometric, Kit No. 08001–7808, Mann Research Laboratories, Inc., New York, NY.
191. Sunderman, F. W., Jr., and Nomoto, S.: Clin. Chem., *16*:903, 1970.
192. Sussman, H. H., Small, P. A., and Cotlove, C.: J. Biol. Chem., *243*:160, 1968.
193. Szasz, G.: Clin. Chem., *15*:124, 1969.
194. Szasz, G.: Clin. Chim. Acta, *19*:191, 1968.
195. Tanzer, M. L., and Gilvarg, C.: J. Biol. Chem., *234*:3201, 1959.
196. Taswell, H. F., and Jeffers, D. M.: Am. J. Clin. Path., *40*:349, 1963.
197. Taylor, W. H.: Physiol. Rev., *42*:519, 1962.
198. Thompson, W. H. S.: The Clinical Chemistry of Muscular Dystrophies. *In* Advances in Clinical Chemistry. H. Sobotka and C. P. Stewart, Eds. New York, Academic Press, 1964, Vol. 7, p. 138.
199. Thompson, W. H. S.: Clin. Chim. Acta, *23*:105, 1969.
200. Tietz, N. W.: Personal communication.
201. Tietz, N. W., and Fiereck, E. A.: Measurement of Lipase Activity in Serum. *In* Standard Methods of Clinical Chemistry. G. R. Cooper, Ed. New York, Academic Press, 1972, Vol. 7, p. 19.
202. Tietz, N. W., Miranda, E., and Weinstock, A.: *In* Proceedings, International Seminar and Workshop on Enzymology. N. W. Tietz, Program Coordinator, Mt. Sinai Hospital Med. Center, Chicago, 1972, p. 2–9.
203. Tietz, N. W., and Repique, E. V.: Clin. Chem., *19*:1268, 1973.
204. Tietz, N. W., Weinstock, A., and Wills, D.: *In* Proceedings, International Seminar and Workshop on Enzymology, N. W. Tietz, Program Coordinator, Mt. Sinai Hospital Med. Center, Chicago, 1972, p. 2–4.
205. Tietz, N. W., Whang, S. J., and Miranda, E.: *In* Proceedings, International Seminar and Workshop on Enzymology, N. W. Tietz, Program Coordinator, Mt. Sinai Hospital Med. Center, Chicago, 1972, p. 2–11.
206. Tzvetanova, E.: Clin. Chem., *17*:926, 1971.
207. Usategui-Gomez, M., Yeager, F. W., and Fernandez de Castro, A.: Cancer Research, *33*:1574, 1973.
208. Usategui-Gomez, M., Yeager, F. W., and Fernandez de Castro, A.: Clin. Chim. Acta, *46*:355, 1973.
209. Van der Slik, W., Persijn, J. P., Engelsman, E., and Riethorst, A.: Clin. Biochem., *3*:59, 1970.
210. Varley, H.: Practical Clinical Chemistry, 4th Ed. New York, Wiley-Interscience, 1967.
211. Vogel, W. C., and Zieve, L.: Clin. Chem., *9*:168, 1963.
212. Walshe, J. M.: Brain, *90*:149, 1967.
213. Ware, A. G., Walberg, C. B., and Sterling, R. E.: Turbidimetric Measurement of Amylase. *In* Standard Methods of Clinical Chemistry. D. Seligson, Ed. New York, Academic Press, 1963, Vol. 4, p. 39.
214. Warnock, M. L.: Clin. Chim. Acta, *14*:156, 1966.
215. Watts, C.: Clin. Chim. Acta, *14*:177, 1966.
216. West, P. M., Ellis, F. W., and Scott, B. L.: J. Lab. Clin. Med., *39*:159, 1952.
217. Wetstone, H. J., and La Motta, R. V.: Clin. Chem., *11*:653, 1965.
218. Whitaker, J. F.: Clin. Chim. Acta, *44*:133, 1973.
219. Wiederhorn, A. R., and Pickens, R. L.: J. Urology, *109*:855, 1973.
220. Wilkinson, J. H.: Introduction to Diagnostic Enzymology. Baltimore, Williams and Wilkins, 1962.
221. Wilkinson, J. H.: Isoenzymes. Philadelphia, Lippincott, 1966, p. 104.

222. Wilkinson, J. H., Baron, D. N., Moss, D. W., and Walker, P. G.: Standardization of Clinical Enzyme Assays: A Reference Method for Aspartate and Alanine Transaminases. J. Clin. Path., *25*:940, 1972.
223. Winsten, S., Wilkinson, J. H., and Boutwell, J. H.: Clin. Chem., *15*:496, 1969.
224. Witteeveen, S. A. G. J., Sobel, B. E., and DeLuca, M.: Proc. Nat. Acad. Science USA, *71*:1384, 1974.
225. Wolfson, S. K., and Williams-Ashman, H. G.: Proc. Soc. Exp. Biol. Med., *96*:231, 1957.
226. Yam, L. T.: Am. J. Med., *56*:605, 1974.
227. Yang, J-S., and Biggs, H. G.: Clin. Chem., *17*:512, 1971.
228. Yoshida, A.: Science, *17*:532, 1973.
229. Zinterhofer, L., Wardlaw, S., Jatlow, P., and Seligson, D.: Clin. Chim. Acta, *44*:173, 1973.

CHAPTER 13

ENDOCRINE FUNCTION

by Sati C. Chattoraj, Ph.D.

Endocrinology is a science which deals with the products of a group of glands and their action in maintaining the chemical integrity of cell environment. These glands are specialized, structurally and functionally. Unlike other multicellular glands, they are devoid of ducts. Secretions are released directly into the blood stream; hence the designations *ductless* or *endocrine* (internally secreting). The glands secrete one or more specific types of active principles which are called *hormones* (Greek: *hormon*—exciting, setting in motion). For efficient transport of the secreted material, these glands are highly vascularized.

According to Bayliss,[6] a hormone is defined as any substance normally produced by specialized cells in some part of the body and carried by the blood stream to another part from which it affects the body as a whole. For example, adrenocorticotrophin (ACTH) is secreted by the pituitary, but it affects the functional activities of the adrenal cortex. Similarly, blood-borne hormones of the adrenal cortex regulate carbohydrate, fat, protein, and mineral metabolism of the body. Figure 13–1 shows the approximate location of the endocrine glands in the body.

NATURE AND ACTIONS OF HORMONES

The hormones vary widely in chemical composition, ranging from amines (thyroxine, epinephrine, etc.) through complex steroid ring structures (corticosteroids, androgens, estrogens, etc.) to proteins (adrenocorticotrophic hormone, chorionic gonadotrophin, thyrocalcitonin, etc.). Hormones possess a high degree of structural specificity. Any alteration in the molecular composition of a hormone brings dramatic changes in its physiological activity. For example, although the structural difference between the female sex hormones estradiol and estriol is only an additional α-hydroxyl group (- -OH) at the C-16 position, estradiol is the most potent estrogen, whereas estriol is almost inert as far as its effect on accessory sex organs is concerned. Similarly, when norepinephrine, a hormone of the adrenal medulla, is methylated on its amine-N to produce epinephrine, this minor structural change not only diminishes its potency but also alters the nature of its biological activity.

Actions of the hormones are complex and diverse. (The source, chemical nature, and possible site of the actions of individual hormones are summarized in Table 13–1.) They may, however, be broadly divided according to three general aspects.

*1. **Regulatory Function.*** One of the major functions of the endocrine system is to maintain constancy of chemical composition (homeostasis) of plasma and interstitial and intracellular fluids for proper and efficient function and growth of the organism. This homeostatic mechanism is maintained through the sensitively regulated metabolism of salt, water, carbohydrate, fat, and protein by secretion of appropriate hormones. When there is derangement in the salt and water balance, hormones such as vasopressin and aldosterone come into play, whereas if there is an increased concentration of blood glucose (hyperglycemia), as, for example, after a carbohydrate-rich meal, insulin is promptly secreted from the pancreas so that the glucose will be utilized at a faster rate until the concentration decreases to its normal level.

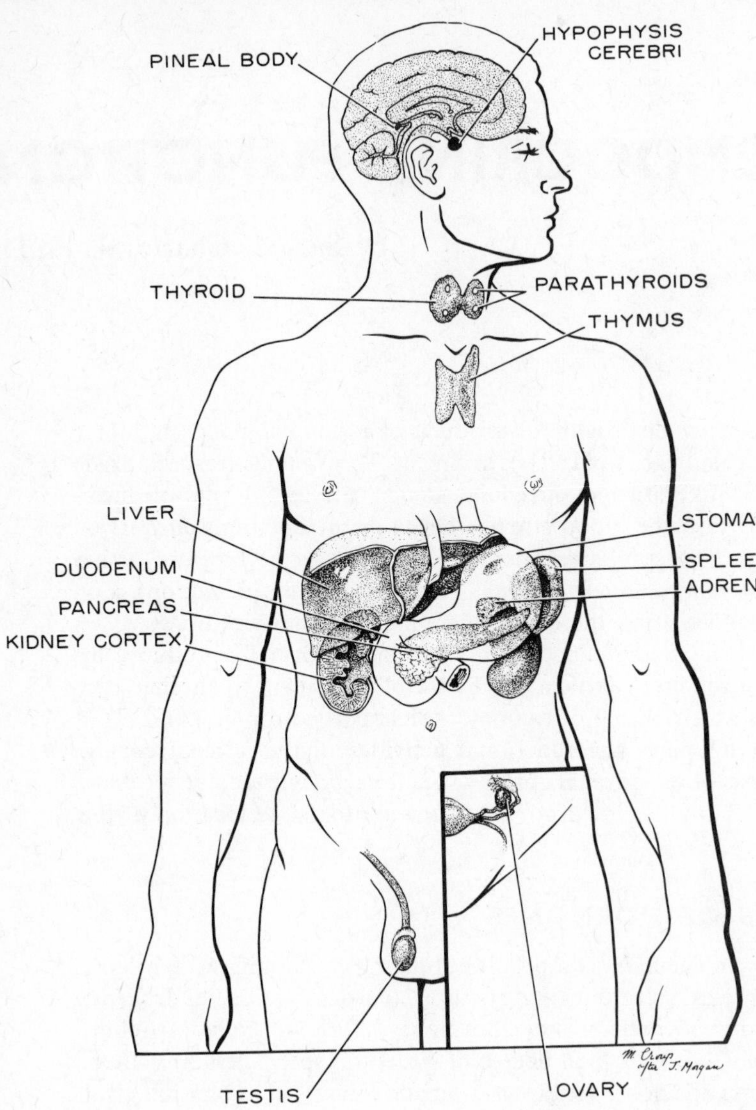

Figure 13–1 Approximate locations of the endocrine glands in man. Though the liver, kidneys, and spleen add important materials to the blood, they are not definitely known to be organs of internal secretion. (From Turner, C. D.: General Endocrinology, 4th ed., W. B. Saunders Company, Philadelphia, 1966.)

TABLE 13–1 THE HORMONES, THEIR SOURCE, AND A BRIEF DESCRIPTION OF THEIR ACTION

Endocrine Gland and Hormone	*Nature of Hormone*	*Site of Action*	*Principal Actions*
HYPOTHALAMUS			
Various releasing factors	Polypeptides	Anterior pituitary	Release of trophic hormones
ANTERIOR PITUITARY			
Somatotrophin, growth hormone (STH, GH)	Protein	Body as a whole	Growth of bone and muscle
Adrenocorticotrophin (ACTH)	Polypeptide	Adrenal cortex	Stimulates formation and secretions of adrenocortical steroids
Melanophore-stimulating hormone (MSH)	Polypeptide	Skin	Dispersion of pigment granules; darkening of skin
Thyrotrophin (TSH)	Glycoprotein	Thyroid	Stimulates formation and secretion of thyroid hormone
Follicle-stimulating hormone (FSH)	Glycoprotein	Ovary	Growth of follicles with LH, secretion of estrogens and ovulation
		Testis	Development of seminiferous tubules; spermatogenesis

TABLE 13–1 *(Continued)*

Endocrine Gland and Hormone	*Nature of Hormone*	*Site of Action*	*Principal Actions*
Luteinizing or interstitial cell-stimulating hormone (LH or ICSH)	Glycoprotein	Ovary	Formation of corpora lutea, secretion of progesterone
		Testis	Stimulation of interstitial tissue—secretion of androgens
Prolactin (lactogenic hormone, luteotrophin)	Protein	Mammary gland	Proliferation of mammary gland and initiation of milk secretion
POSTERIOR PITUITARY			
Vasopressin (ADH, antidiuretic hormone)	Nonapeptide	Arterioles	Elevates blood pressure
		Renal tubules	Water reabsorption
Oxytocin	Nonapeptide	Smooth muscle (uterus, mammary gland)	Contraction, action in parturition and in sperm transport; ejection of milk
THYROID			
Thyroxine and triiodothyronine	Iodoamino acids	General body tissue	Stimulates oxygen consumption and metabolic rate of tissues
Calcitonin (thyrocalcitonin)	Polypeptide	Skeleton	Inhibits calcium resorption; lowers plasma calcium and phosphate
PARATHYROID			
Parathyroid hormone (PTH, parathormone)	Polypeptide	Skeleton, kidney, gastrointestinal tract	Regulates calcium and phosphorus metabolism
ADRENAL CORTEX			
Adrenal cortical steroids—cortisol, aldosterone	Steroids	General body tissue	Carbohydrate, protein and fat metabolism; salt and water balance, inflammation, resistance to infection; hypersensitivity
ADRENAL MEDULLA			
Norepinephrine and epinephrine	Aromatic amines	Sympathetic receptor	Mimic sympathetic nervous system
		Liver and muscle	Glycogenolysis
		Adipose tissue	Release of lipid
OVARY			
Estrogens	Phenolic steroids	Female accessory sex organs	Development of secondary sex characteristics
Progesterone	Steroids	Female accessory reproductive structures	Preparation for ovum implantation; maintenance of pregnancy
Relaxin	Polypeptide	Symphysis pubis, uterus	Relaxation, aids in parturition
TESTIS			
Testosterone	Steroid	Male accessory sex organs	Development of secondary sex characteristics, maturation and normal function
PANCREAS			
Insulin	Polypeptide	Most cells	Regulation of carbohydrate metabolism; lipogenesis
Glucagon	Polypeptide	Liver	Glycogenolysis
PLACENTA			
Estrogens, progesterone, gonadotrophins (HCG) Placental lactogen, relaxin	Same as above	Same as above	Same as above
GASTROINTESTINAL TRACT			
Secretin	Protein	Pancreas	Secretion of alkaline fluid and digestive enzymes
Cholecystokinin-pancreozymin	Protein	Gallbladder	Contraction and emptying
Enterogastrone	Protein	Stomach	Inhibition of motility and secretion
Gastrin	Protein	Stomach	Secretion of acid

2. Morphogenesis. Some hormones play an important part in controlling the type and rate of growth of an organism. The development of the male and female sex characteristics under the influence of the respective sex hormones (testosterone and estradiol-17β) is perhaps the best example.

3. Integrative Action. This aspect of hormonal function is the most complex and the least understood. Broadly speaking, each hormone has a specific function. For example, the biological action of corticotrophin (ACTH), secreted by the pituitary gland, controls the functional status of the adrenal cortex; estrogens and progesterone, produced in the ovaries and called female sex hormones, regulate the development of secondary sex characteristics; the adrenal hormone, aldosterone, controls salt and water balance; the pancreatic hormone, insulin, regulates carbohydrate metabolism; and so on. However, even though a particular hormone dramatically influences a single biochemical event, or dramatically changes the morphology and rate of metabolism of a single organ, the presence of other hormones produced by different endocrine glands is also important for efficient functioning. Insulin alone would not be adequate to maintain the balance of carbohydrate metabolism. The concerted action of glucagon (from the pancreas) and of other hormones from glands such as the pituitary (corticotrophin, somatotrophin), adrenals (glucocorticoids, epinephrine), thyroid (thyroxine), and even gonads (estrogens) is also important. This interrelation is not limited to the endocrine glands but extends to the nervous system as well. While it is true that the mineralocorticoids (deoxycorticosterone, aldosterone) have a profound influence on the maintenance of salt and water balance, this control mechanism would fail without the simultaneous adjustment of the rate of blood flow, blood pressure, and vasoconstriction by the autonomic nervous system. Therefore, under normal circumstances, there exists an integrative functioning of the endocrine and nervous systems which is reflected by the maintenance of a constant body environment. Derangements of such interdependence give rise to disease states.

CONTROL OF HORMONE SECRETION

There are several mechanisms for maintaining the delicate balance between the products of hormones and the need of the organism for hormones. Detailed discussions of these mechanisms are beyond the scope of this book and students are referred to other texts.[89,99] The following is a brief outline, covering only the salient features of these mechanisms. The anterior pituitary occupies a central position in the control of hormone secretion. It secretes several trophic* hormones, which stimulate and maintain certain other endocrine glands. In the absence of these trophic hormones, the target glands† are unable to maintain a normal rate of secretion. The main target organs of the trophic hormones are the thyroid gland, adrenal cortex, testis, and ovary. The specific trophic hormones have been described elsewhere in the text (p. 787).

To understand the interplay of hormone secretion control, whether directed by the pituitary or not, a description of certain basic concepts is warranted. Most important is the *feedback* loop or *servo* mechanism. (The concept was originally derived from the operation of an electrical network; specifically, a relay device consisting of input and output signals to actuate the automatic control of a complex machine, instrument, or operation.) In simplified terms, the feedback system may be described as one in which the degrees of function of two variables, A and B, are interdependent. When A is directly proportional to B (i.e., when B

---

* There is some controversy about the spelling of the anterior pituitary hormones. Many workers prefer to use the ending "tropic" meaning "a turning," whereas others prefer the ending "trophic" meaning "to nourish." In the present writing the latter ending will be used.

† The term "target" is often used to refer to the site of action of any hormone, whether "trophic" in nature or not. For example, the thyroid gland and the uterus are the target organs for TSH and estrogens, respectively.

increases, A also increases), the relationship is described as *positive feedback*, whereas if A is inversely proportional to B (i.e., when B increases, A decreases), then a *negative feedback* exists. It is important to note in this connection that feedback regulation, especially negative feedback, seems to be a cardinal feature not only in the control of hormone secretion but also in the growth and metabolism of unicellular bacteria as well as highly complex multicellular organisms. The feedback relationship may be represented diagrammatically as follows:

1. Negative feedback:

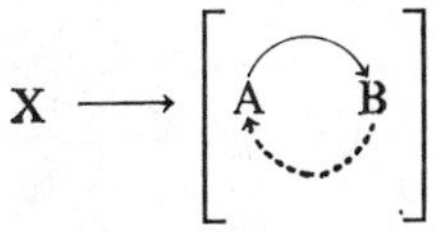

X ⟶ [A B]

X (external stimuli or any other demanding factor) increases A which in turn increases (solid line) the concentration of B to a level at which B decreases (dashed line) the concentration of A. The net effect of this control mechanism is the maintenance of an optimal constant level of both variables A and B. It should be noted that the effect of negative feedback control is characteristically opposite to the initial stimulus. A diminished level of calcium (initial stimulus) in the blood, for example, acts directly on the parathyroid glands to increase the secretion of the parathyroid hormone, which in turn raises the circulating calcium level (final response). Conversely, a rise in the amount of calcium in the circulation causes the parathyroid glands to secrete less of the hormone, resulting in a decline of the calcium level. A high calcium level also acts on the so-called clear cells of the thyroid gland which secrete the calcium-lowering hormone, calcitonin. Thus, the plasma calcium level is defended by two negative feedback control mechanisms involving two different hormones.

2. Positive feedback:

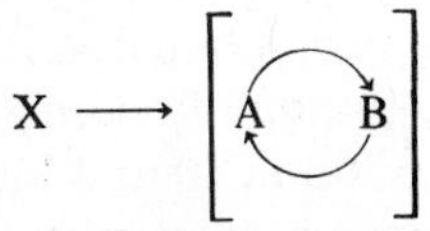

X ⟶ [A B]

X increases A, leading to the increase of B which in turn increases A. As opposed to negative feedback, positive feedback does not operate in isolation but constitutes an integral part of a control system and *confers oscillatory behavior upon the system*. The cyclicity in the hormonal secretion during the menstrual cycle (see p. 793) is considered to be due to an integrated control system consisting of the negative feedback–positive feedback relationship between the pituitary gonadotrophins (FSH and LH) and the ovarian hormones, estradiol-17β and progesterone. The secretion of FSH and estradiol-17β is controlled by the negative feedback mechanism, whereas secretion of LH during the preovulatory phase depends on the positive feedback relationship with circulating estradiol-17β. Thus, at the beginning of the menstrual cycle, a high level of FSH stimulates estradiol-17β secretion, which in turn decreases FSH (*negative feedback*). The increasing concentration of estradiol-17β, however, promotes LH secretion, which culminates in the so-called preovulatory *LH-surge* (*positive feedback*—as the estradiol level increases, the LH level also increases). From the physiological point of view, such a positive feedback relationship depends on an optimal level of estradiol-17β. Very high doses of estradiol-17β diminish rather than increase the secretion of LH. In the positive feedback mechanism, the initiating stimulus causes more of the same, and unless this is a part of an overall control system terminating in a negative feedback, it does not lead to control but to instability. In a normal menstrual cycle, the ovulated follicle under the influence of LH is transformed into the corpus luteum, which secretes large amounts of progesterone. Progesterone constitutes part of the negative feedback loop for the control of LH secretion.

STEROID HORMONES

Steroids are compounds containing the cyclopentanoperhydrophenanthrene ring system (Fig. 13–2). The three six-sided rings (A, B, C) constitute the phenanthrene nucleus

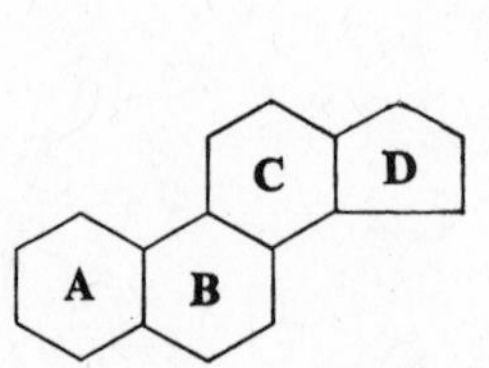

Cyclopentanoperhydrophenanthrene

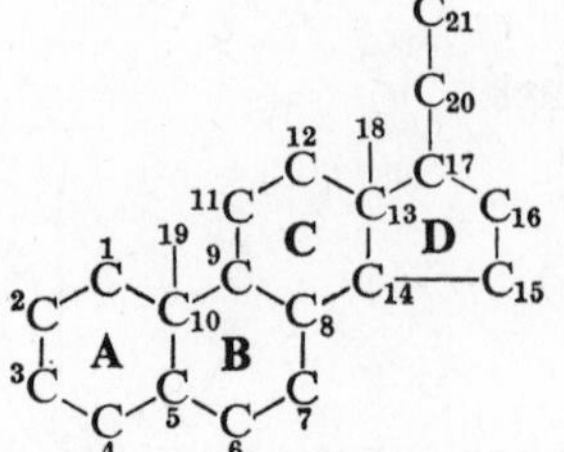

Figure 13–2 Common features and numbering system of steroids.

to which is attached a five-sided ring (D), cyclopentane. The prefix "perhydro" refers to the fact that all the necessary hydrogen atoms have been added to the compound to make it fully saturated. This class of compounds includes such natural products as sterols (e.g., cholesterol), bile acids (e.g., cholanic acid), sex hormones (e.g., estrogens, androgens), corticosteroids, cardiac glycosides (e.g., digitoxigenin), sapogenins (e.g., tigogenin), and some alkaloids (e.g., solasodine). The steroid hormones with which we are concerned contain up to 21 carbon atoms (C_{21} steroids), numbered as shown in Fig. 13–2. Each carbon atom of a ring bears two hydrogen atoms, except when it is common to two rings, in which case it bears only one hydrogen atom (i.e., at C-5, C-8, C-9, and C-14). C-17 bears one hydrogen atom, while C-10 and C-13 are bound only to other carbons (i.e., C-19 and C-18, respectively). The carbon atoms composing the rings and the hydrogen atoms attached to them are not usually written into the structure unless it is required to draw special attention to configuration. Furthermore, in all naturally occurring steroid hormones, the projected solid line from the carbon atom at position 10 or 13 usually designates the presence of an angular (—CH_3) group, unless otherwise indicated.

Steroids consist of tetracyclic rings and are three-dimensional. Thus, the constituent carbon atoms and the hydrogen atoms or their substituents lie in different planes, giving rise to *isomers*. The direction of the hydrogen atoms, the substituents, and the side chain play a much more important role in the distinction of various isomers of the steroid compounds than does the relative position of the carbon atoms in the rings. Thus, the isomers resulting from fusion of two rings are decided on the basis of the spatial relationship between the hydrogen atoms or the substituents at common carbon atoms. When rings A and B are fused, two isomers are possible, depending upon whether the hydrogen atom at C-5 and the methyl group at C-10 are on the same or opposite side of the plane of the rings. If the hydrogen atom points in the same direction as the angular methyl group at C-10, the compound is said to be the *cis* or *normal* form. If, however, they are on the opposite sides, the compound is said to be the *trans* or *allo* form.

While the rings A and B may be either *cis* or *trans*, the rings B/C and C/D have *trans* configuration in all naturally occurring steroid hormones.

The two methyl groups attached to C-10 and C-13 lie above the plane of the molecule and are customarily the points of reference for describing the spatial orientation of other substituents in the steroid nucleus. Substituents on the same side as these two methyl groups are said to possess a β configuration, which is indicated by a solid line (—) joining them to the appropriate carbon atoms in the nucleus. Substituents on the opposite side are attached by a broken line (- -) to denote an α configuration. Thus, in the structures shown in Fig. 13–3, when the hydrogen substituent at C-5 is *cis*, the isomer is the 5β isomer, and when it is

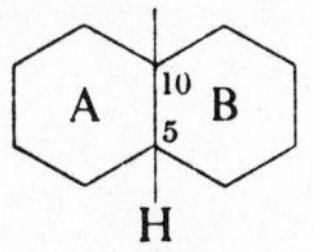

Figure 13–3 Fusion of rings A and B in naturally occurring steroids.

A 10 B 5
H
Cis-

Hydrogen atom and methyl group are on the same side (above the plane of the paper —5β-isomer).

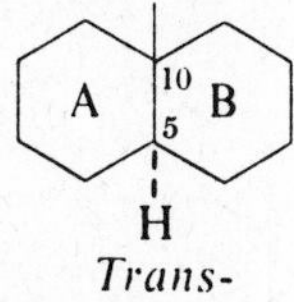

Trans-

Hydrogen atom is on the opposite side (below the plane of the paper—5α—isomer).

trans the isomer is accordingly the 5α isomer. Similarly, the substituents at C-3, C-11, C-17, or any other carbon atoms are indicated as either α or β configuration, depending on their spatial orientation relative to these methyl groups (C-10 and C-13).

The innumerable steroids containing the cyclopentanoperhydrophenanthrene nucleus differ from one another by the introduction of double bonds between certain pairs of carbon atoms, by the introduction of substituents for the hydrogen atoms, or by the addition of a specific type of side chain. On the basis of such structural characteristics, the steroidal compounds are classified as derivatives of certain parent hydrocarbons, namely *estrane* (for estrogens, Fig. 13–4), *androstane* (for androgens, Fig. 13–5), and *pregnane* (for corticosteroids and progestins, Fig. 13–6).

It should be noted that the parent substance *estrane* lacks one methyl group at C-10, and hence it is a C_{18} compound. Furthermore, estrogens are actually derivatives of the compound estratriene, since the benzenoid ring structure is a common feature of all the naturally occurring estrogens.

The parent substance *androstane* is a C_{19} compound and possesses 5α or 5β configuration. Naturally occurring *androsterone* and *etiocholanolone* are the examples of the respective derivatives of these isomers.

The special feature of the hydrocarbon *pregnane* is an ethyl side chain (—CH_2—CH_3) attached to C-17, making it a C_{21} compound. The side chain is in *cis* relationship to the methyl groups at C-10 and C-13 and is therefore β oriented. In all naturally occurring steroids or pregnane derivatives, the side chain is β oriented and the hydrogen or its substituent at this position is always α oriented. Consequently, in recent steroid nomenclature, the spatial orientation of the substituent at this position of the pregnane derivative is no

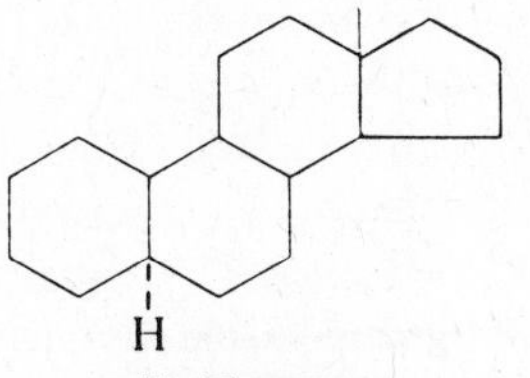

Figure 13–4 Parent hydrocarbons of estrogens.

H
5α-Estrane

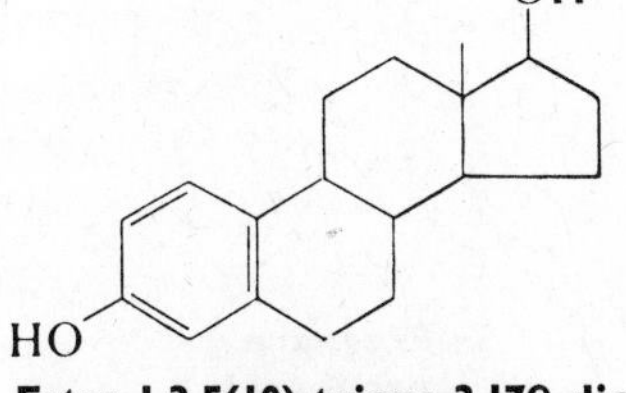

OH
HO
Estra-1,3,5(10)-triene-3,17β-diol (Estradiol-17β)

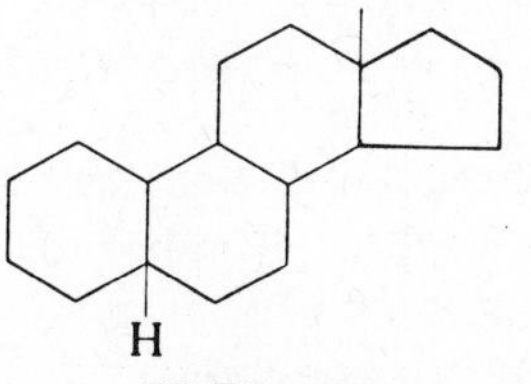

H
5β-Estrane

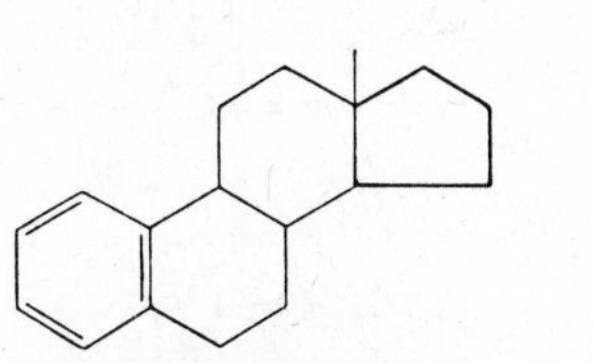

Estra-1,3,5(10)-triene (Estratriene)

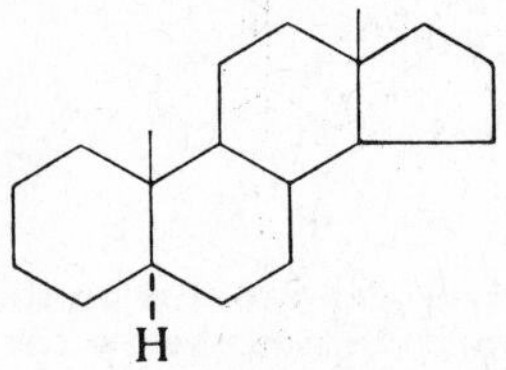

5α-Androstane

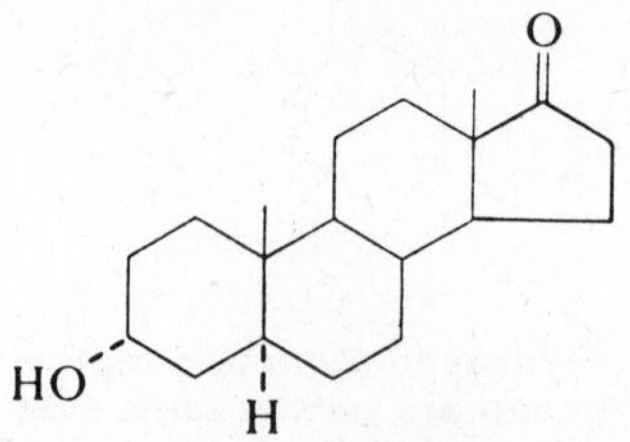

3α-Hydroxy-5α-androstan-17-one (Androsterone)

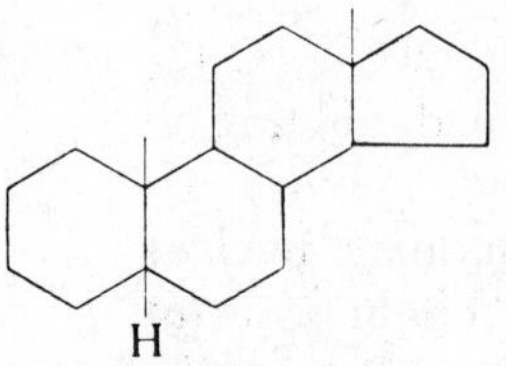

5β-Androstane (Etiocholane)

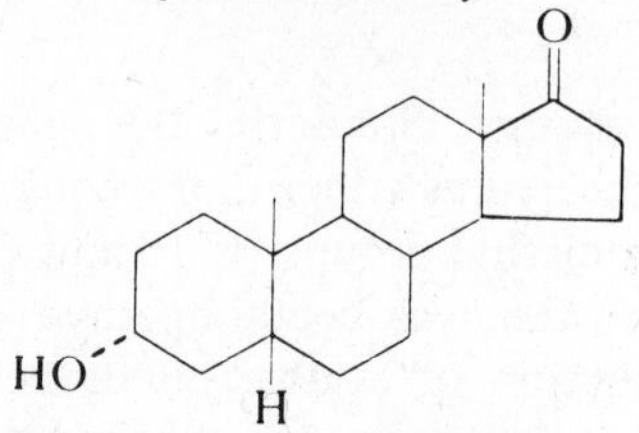

3α-Hydroxy-5β-androstan-17-one (Etiocholanolone)

Figure 13–5 Parent hydrocarbons of androgens.

longer specified as α (e.g., cortisol; 1β,17,21-trihydroxy-4-pregnene-3,20-dione). When the configuration at position 20 in the side chain of a pregnane derivative is as depicted in the projection formula (e.g., pregnanediol), substituents shown to the right of C-20 are termed α, and those to the left are termed β. Like androstane, this parent substance also has two isomers, 5α- and 5β-pregnane. *Allopregnanediol* and *pregnanediol* are the respective derivatives of these two isomers.

It should be noted that the prefix *allo* refers only to stereoisomerism of the hydrogen atom at C-5. When the configuration differs at any other carbon atom, the prefix *epi* is used; for example, androsterone possesses a 3α-hydroxyl group, and epiandrosterone a 3β-hydroxyl group; testosterone possesses a 17β-hydroxyl group, and epitestosterone a 17α-hydroxyl group.

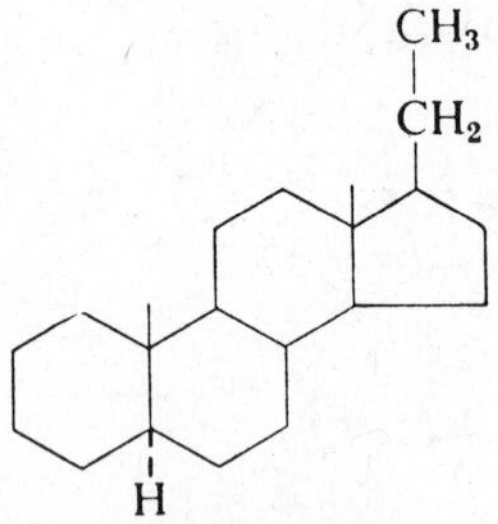

5α-Pregnane (Allopregnane)

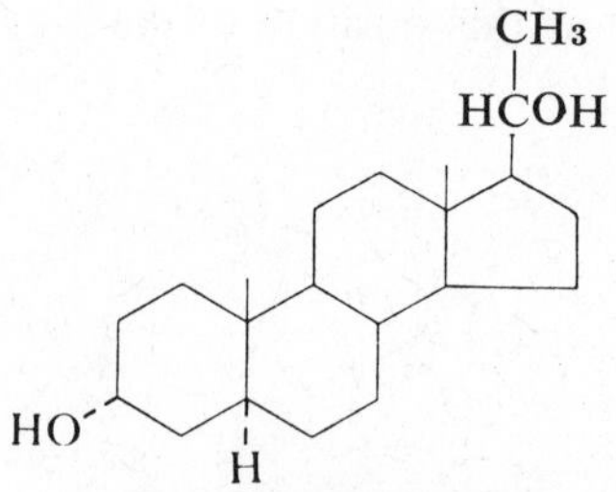

5α-Pregnane-3α,20α-diol (Allopregnanediol)

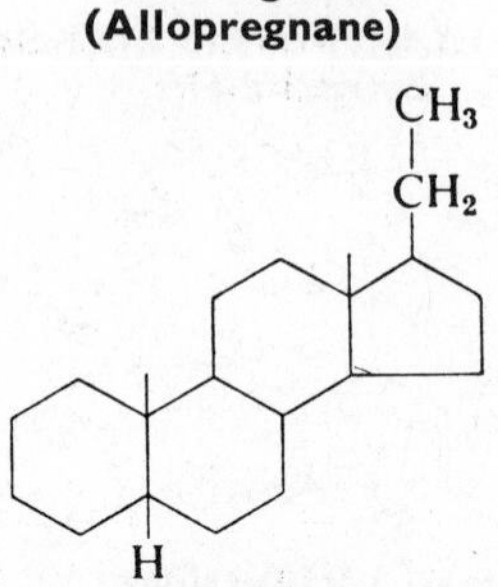

5β-Pregnane

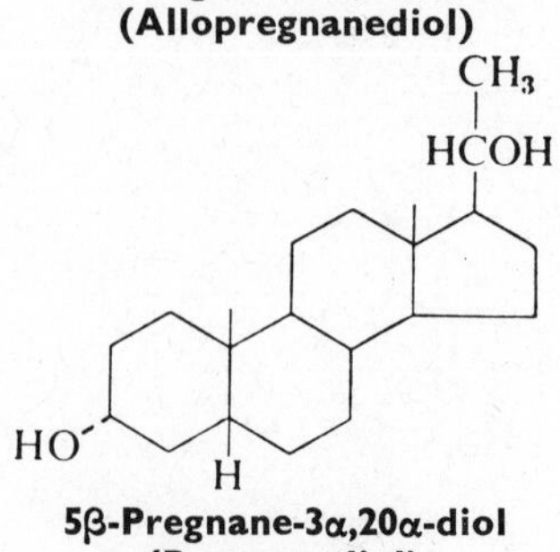

5β-Pregnane-3α,20α-diol (Pregnanediol)

Figure 13–6 Parent hydrocarbons of corticosteroids and progestins.

To describe a compound with chemical nomenclature, a variety of other *suffixes* and *prefixes* are used. The suffix *-ane* indicates a fully saturated compound (e.g., pregn*ane*); *-ene*, the presence of one double bond (e.g., pregn*ene*); *-diene*, two double bonds; *-triene*, three double bonds; the terminal "e" is omitted before a vowel, e.g., -4-en-3β-ol. The position of the double bond is indicated by the number of the carbon atom from which it originates, and it is understood to terminate at the next higher carbon atom (i.e., 4-ene means that a double bond lies between C-4 and C-5). However, when an alternative is possible (for example, a double bond originating at C-8 in the *estrane* nucleus can terminate at C-9 or C-14), the number of the carbon atom at which the bond ends is written in parentheses. Thus, a double bond at C-8 terminating at C-14 is designated as 8 (14) -ene. A formerly used prefix for a double bond is the symbol Δ with a superscript indicating the position of the double bond (e.g., Δ^5). An alcohol (—OH substituent of the nucleus) is indicated by the suffix *-ol*, (two alcohol groups as *-diol*, three as *-triol*, etc.), or by the prefix *hydroxy* or *oxy* (dihydroxy for two, trihydroxy for three, etc.). Ketones ($>C{=}O$) are identified by the suffixes *-one* for one keto group, *-dione* for two keto groups, etc., or by the prefix *oxo-* (see Table 13–2).

In naming a compound containing double bonds, hydroxyl groups, and ketones, priorities are given to the use of suffixes and prefixes. Thus, hydroxyl groups are indicated by the prefix followed by the suffixes for other substituents. Accordingly, the systematic name of dehydroepiandrosterone (Fig. 13–7) is written as 3β-hydroxy- androst-5-en-17-one. Note that to denote the bond of unsaturation, the first part of the parent hydrocarbon is followed by the position of the bond and the suffix (i.e., androst-5-en). If the prefix Δ for unsaturation is chosen, then suffixes for both the hydroxyl and the ketone groups are used, e.g., Δ^5-androsten-3β-ol-17-one. When there is only one kind of substituent, the use of a suffix is customary, e.g., pregnanediol (5β-pregnane-3α,20α-diol), androstenedione (androst-4-ene-3,17-dione).

The usefulness of the systematic name of a compound lies in the fact that it gives information about the parent substance, the position of unsaturation, and the nature, position, and orientation of substituents. The trivial name, as the term suggests, conveys little or no information about the chemical origin and characteristics of a compound, e.g., cortisol, progesterone, testosterone, etc. The trivial and systematic names of some of the important steroid hormones are given in Table 13–3.

TABLE 13–2 COMMON SUFFIXES AND PREFIXES FOR STEROIDS

Suffix or Prefix	*Definition*
SUFFIX	
-al	Aldehyde group
-ane	Saturated hydrocarbon
-ene	Unsaturated hydrocarbon
-ol	Hydroxyl group
-one	Ketone group
PREFIX	
hydroxy- (-oxy-)	Hydroxyl group
keto- (oxo-)	Ketone
deoxy- (desoxy-)	Replacement of hydroxyl group by hydrogen
dehydro-	Loss of two hydrogen atoms from adjacent carbon atoms
dihydro-	Addition of two hydrogen atoms
cis-	Spatial arrangement of two substituents on the same side of the molecule
trans-	Spatial arrangement of two substituents on opposite sides of the molecule
α-	Substituent which is *trans* to the methyl group at C-10
β-	Substituent which is *cis* to the methyl group at C-10
epi-	Isomeric in configuration at any carbon atom except at the junction of two rings
Δ^n-	Position of unsaturated bond

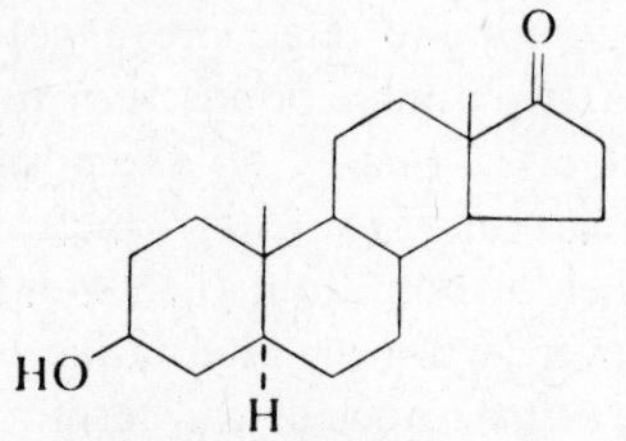

3β-Hydroxy-5α-androstan-17-one
(Epiandrosterone)

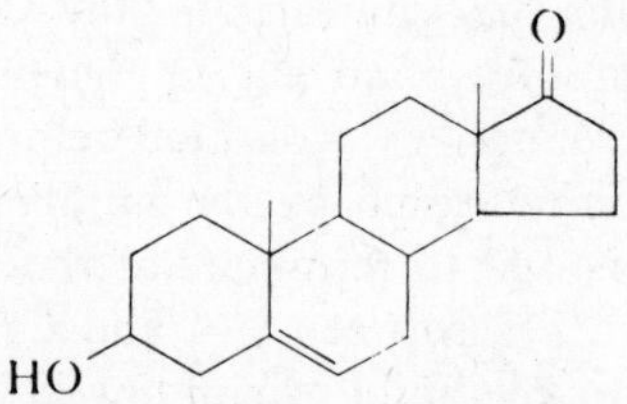

3β-Hydroxyandrost-5-en-17-one
(Dehydroepiandrosterone)

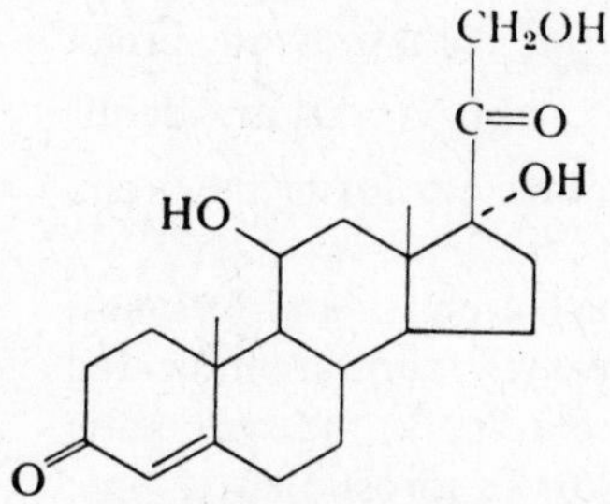

11β,17,21-Trihydroxypregn-4-ene-3,20-dione
(Cortisol)

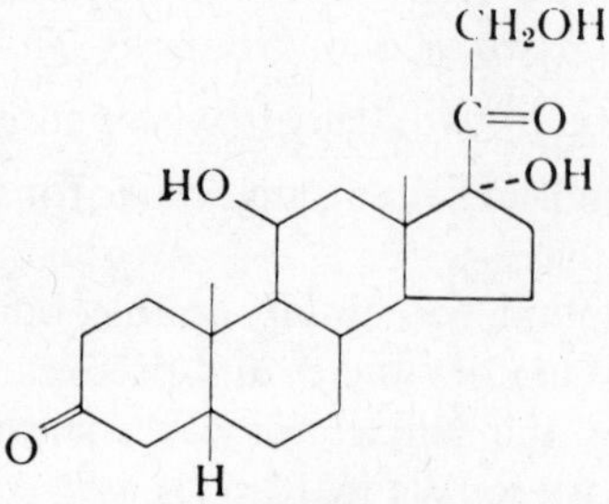

11β,17,21-Trihydroxy-5β-pregnane-3,20-dione
(Dihydrocortisol)

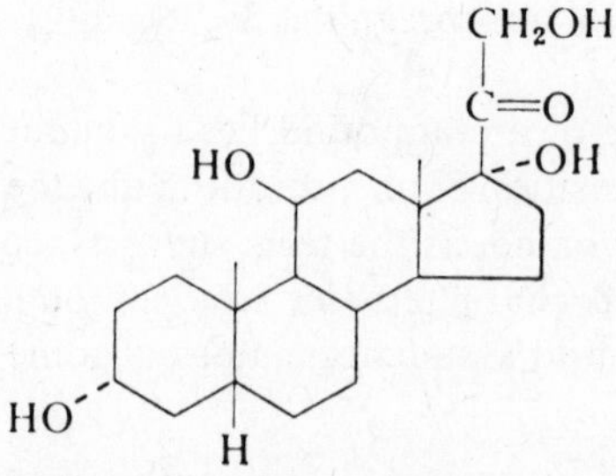

3α,11β,17,21-Tetrahydroxy-5β-pregnan-20-one
(Tetrahydrocortisol, Urocortisol)

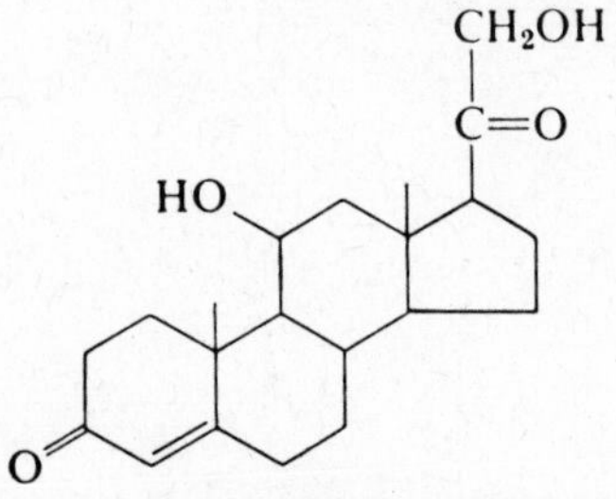

11β,21-Dihydroxypregn-4-ene-3,20-dione
(Corticosterone)

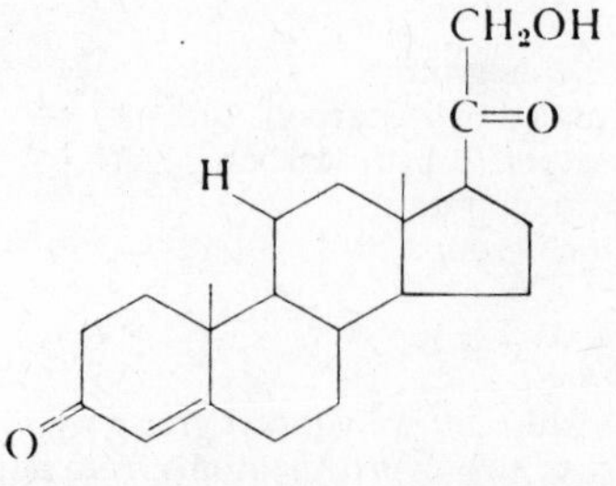

21-Hydroxypregn-4-ene-3,20-dione
(11-Deoxycorticosterone)

Figure 13–7 Illustration of semitrivial names.

TABLE 13–3 TRIVIAL AND SYSTEMATIC NAMES OF SOME IMPORTANT STEROID HORMONES

Trivial Name	*Systematic Name*
Estrone	3-Hydroxyestra-1,3,5(10)-trien-17-one
Estradiol-17β	Estra-1,3,5(10)-triene-3,17β-diol
Estriol	Estra-1,3,5(10)-triene-3,16α,17β-triol
Testosterone	17β-Hydroxyandrost-4-en-3-one
Androsterone	3α-Hydroxy-5α-androstan-17-one
Etiocholanolone	3α-Hydroxy-5β-androstan-17-one
Dehydroepiandrosterone	3β-Hydroxyandrost-5-en-17-one
Adrenosterone	Androst-4-ene-3,11,17-trione
Progesterone	Pregn-4-ene-3,20-dione
Pregnanediol	5β-Pregnane-3α,20α-diol
Cortisol	11β,17,21-Trihydroxypregn-4-ene-3,20-dione
Urocortisol (tetrahydro F)	3α,11β,17,21-Tetrahydroxy-5β-pregnan-20-one
Aldosterone	11β,21-Dihydroxy-3,20-dioxopregn-4-en-18-al

In addition to the usual suffixes and prefixes just discussed, there are some special prefixes which are generally used for the semitrivial names of the compounds. Thus, the prefix *dehydro* is used to indicate the loss of two hydrogen atoms from adjacent carbon atoms with the formation of a double bond, e.g., dehydroepiandrosterone (Fig. 13–7). The prefix *dihydro-* or *tetrahydro-* indicates the addition of two or four hydrogen atoms to the molecule, respectively, as in dihydrocortisol and tetrahydrocortisol (Fig. 13–7). The replacement of a hydroxyl group by hydrogen (C—OH→CH) is prefixed by *deoxy-* (or *desoxy-*), for example, 11-deoxycorticosterone.

GENERAL ASPECTS OF BIOSYNTHESIS AND METABOLISM OF STEROID HORMONES

The advent of radiolabeled compounds has played a very important role in the elucidation of biogenesis and metabolism of steroid hormones. The use of radioactive acetic acid and cholesterol for the study of steroidogenesis *in vivo* and *in vitro* has produced radioactive steroid hormones, lending support to the concept that both compounds are precursors of steroid hormones. Similarly, the administration of radioactive steroid hormones followed by the separation and identification of radioactive metabolites in the urine has helped to delineate the metabolic pathways. Even though acetate and cholesterol are both precursors of steroid hormones, they do not, in all probability, constitute a separate pathway of biosynthesis but follow the general sequence of acetate → cholesterol → steroid hormones. It has been amply documented that acetate is the sole precursor of cholesterol; among the 27 carbon atoms constituting cholesterol, 12 originate from the carboxyl carbon (C) and 15 from the methyl carbon (M) of acetic acid (CH_3COOH), as shown in Fig. 13–8. Stepwise degradations of radioactive steroid hormones (e.g., cortisol, corticosterone) synthesized from ^{14}C-labeled acetate indicate that the individual carbon atoms of steroid hormones correspond to those of cholesterol originating from carboxyl and methyl carbon atoms of acetic acid. There are, however, 30 separate biochemical reactions involved in the

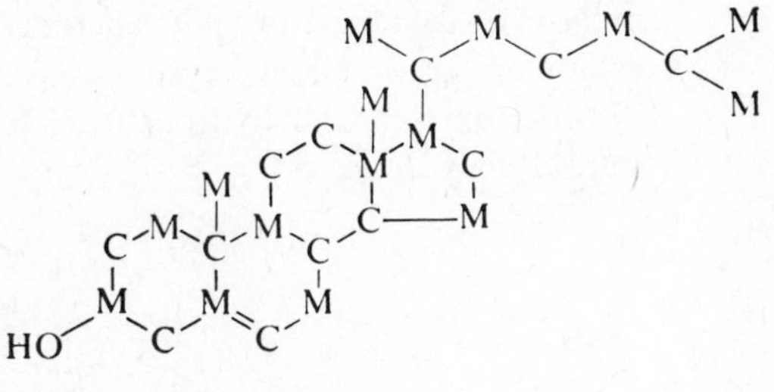

Figure 13–8 Carbon atoms of cholesterol derived from carboxyl carbon (C) and methyl carbon (M) of acetic acid.

biosynthesis of cholesterol from acetate, and many more enzymatic reactions come into play to convert cholesterol to a variety of steroid hormones. For a detailed discussion of these topics students are referred to an excellent authoritative monograph by Dorfman and Ungar.[20]

In normal men and women, steroid hormones are produced in the adrenals, ovaries, and testes. All these glands utilize the same precursors, such as acetate and cholesterol, but the quality and the quantity of steroid hormone produced by each gland are different. The difference is inherent in the degree of activity and the presence or absence of certain enzymatic systems. For example, the enzymes 11β-hydroxylase and 21-hydroxylase are uniquely present in the adrenals to synthesize the characteristic hormones of the glands, the corticosteroids. Similarly, the enzymatic distinction between the ovaries and the testes lies in the fact that the ovaries contain an active aromatizing enzyme system (necessary to convert male sex hormones, e.g., testosterone, to female sex hormones, e.g., estradiol), in addition to the enzymes found in the testes.

The different enzymes participating in the biogenesis of steroid hormones may be broadly classified into the following functional groups.

1. Hydroxylases. These enzymes catalyze the substitution of the hydroxyl group (—OH) for hydrogen (—H). For example, 21-hydroxylase introduces a hydroxyl group at C-21. Similarly, 11β-hydroxylase introduces an OH-group at the β-position of C-11. There are numerous examples of other important hydroxylases, such as 20α-hydroxylase, 19-hydroxylase, 17-hydroxylase, etc. The cofactors are NADPH and molecular oxygen. The reaction is irreversible.

$$\text{H—}^{21}\text{C(H}_2\text{)—}^{20}\text{C=O} \xrightarrow[\textit{21-Hydroxylase}]{\textit{NADPH, O}_2} {}^{21}\text{CH}_2\text{OH—}^{20}\text{C=O}$$

2. Desmolases. These enzymes are required for splitting off the side chain. There are two desmolases, 20,22-desmolase and 17,20-desmolase, which are very important in steroidogenesis. The former participates in the conversion of the C_{27} carbon compound, cholesterol, to a C_{21} compound, pregnenolone, whereas the latter transforms C_{21} steroid hormones to C_{19} steroid hormones. The required cofactors are thought to be NADPH and molecular oxygen.

$$\underset{\textbf{Cholesterol}}{\text{R—}^{20}\text{C(}^{21}\text{CH}_3\text{)—}^{22}\text{C—(CH}_2\text{)}_2\text{CH(CH}_3\text{)}_2} \xrightarrow[\textit{20, 22-Desmolase}]{} \underset{\textbf{Pregnenolone}}{\text{R—}^{20}\text{C(}^{21}\text{CH}_3\text{)=O}} + \underset{\textbf{Isocaproic aldehyde}}{\text{H—C(=O)—(CH}_2\text{)}_2\text{CH(CH}_3\text{)}_2}$$

3. Dehydrogenases. This group of enzymes catalyzes the transfer of hydrogen (oxidation and reduction). The reaction is generally reversible. The cofactor is either NAD or NADP (oxidized or reduced form, depending on the direction of the reaction). Some examples are 3β-hydroxysteroid dehydrogenase, 11β-hydroxysteroid dehydrogenase, 17β-hydroxysteroid dehydrogenase, Δ^5-3β-hydroxysteroid dehydrogenase, and 3α-hydroxysteroid dehydrogenase.

$$\text{(OH, H)} \underset{\textit{NADH or NADPH}}{\overset{\textit{NAD or NADP}}{\rightleftarrows}} \text{(=O)} \quad \textit{17β-Dehydrogenase}$$

4. Isomerases. These enzymes catalyze the migration of a double bond. The most important enzyme of this group that is involved in steroidogenesis is Δ^5-ketosteroid isomerase. The concerted action of Δ^5-3β-hydroxysteroid dehydrogenase and Δ^5-ketosteroid isomerase on pregnenolone produces progesterone through the oxidation of the 3β-hydroxyl group and the migration of a double bond ($\Delta^5 \rightarrow \Delta^4$).

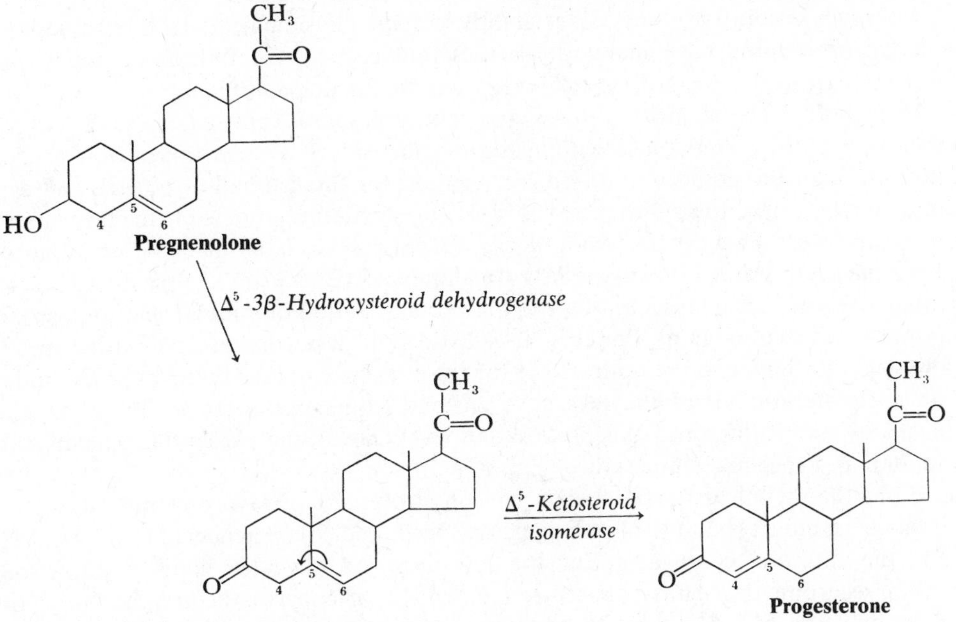

Pregnenolone

Δ^5-*3β-Hydroxysteroid dehydrogenase*

Δ^5-*Ketosteroid isomerase*

Progesterone

The liver is the major site of steroid metabolism. There is some evidence, however, that the kidney and the gastrointestinal tract may also carry out some of the metabolic transformations of steroids. Physiologically active steroid hormones have very high structural specificity. Any alterations in the number or nature of the substituents, or in the steroid nucleus, are liable to make the hormones inactive, or they may change their specific activity. Introduction of a new hydroxyl group (e.g., estradiol → estriol), dehydrogenation (e.g., testosterone → androstenedione), reduction of a double bond (e.g., cortisol → dihydrocortisol), or conjugation of an essential hydroxyl group(s) with a chemical moiety such as glucuronic acid (glucosiduronic acid) (e.g., testosterone → testosterone glucuronide) are important biochemical steps not only for neutralizing the effectiveness of hormones but also for their rapid elimination from the systemic circulation. The conjugation of these hormones and their metabolites with sulfuric or glucuronic acid is, by far, the most efficient single metabolic process for their excretion in the urine, by virtue of the high water solubility of these conjugates. Almost all the steroid hormones and their metabolites are excreted as either glucuronides or sulfates.

GENERAL COMMENTS ON THE METHODS OF STEROID DETERMINATION

The clinical significance of the determination of steroid hormones and their metabolites in urine or in plasma is the assessment of the secretory activity of the glands producing these hormones. Until recently, very few methods for plasma determination were suitable for use in a routine clinical laboratory. As a result, for clinical purposes, urinary measurements have been a widely accepted practice. In most instances the urinary excretion of a hormone, its metabolites, or both does not account for the total amount of hormone secreted by the gland, but it does represent an approximate proportion of the amount secreted during the period of urine collection. Thus, urinary assays indirectly reflect the secretory activity of the endocrine glands. However, factors such as incompleteness of collection, altered renal function, and contribution by more than one gland to the same hormone(s) or metabolites (e.g., both adrenals and gonads contribute to the urinary 17-ketosteroids) may warrant special precaution in the interpretation of urinary values.

Although steroid hormones differ greatly in their physiological activities in the body, the assay procedures have many similarities and require the following general steps: *hydrolysis*, *extraction*, *purification and separation*, and *final quantitation.*

Hydrolysis. The majority of the steroid hormones and their metabolites are excreted in the urine as the water-soluble conjugates of glucosiduronic acid (glucuronic acid) and sulfuric acid. In the absence of any direct method for the estimation of such conjugated steroids in the urine, the splitting (hydrolysis) of such ether and ester linkages is an obligatory first step. Two general types of procedures are available, namely, *acid hydrolysis* and *enzyme hydrolysis*. In acid hydrolysis, an aliquot of a 24 h urine sample is boiled, generally under reflux, in the presence of a specified concentration of mineral acid for a specified length of time (10 to 60 min). For enzymatic hydrolysis, a portion of a 24 h urine specimen is adjusted with buffer to the optimal pH for the enzyme employed, and after the addition of an adequate amount of the respective enzyme (β-glucuronidase to hydrolyze glucosiduronates, and sulfatase to hydrolyze sulfate conjugates), the test sample is incubated for 18 to 76 h at a specified temperature (e.g., 37 °C).

From the technical point of view, acid hydrolysis is always preferable (except for acid-labile hormones) because of its simplicity, speed, and completeness of reaction irrespective of the nature of conjugates. Enzyme hydrolysis, on the other hand, requires special attention regarding the optimal concentration and type of enzyme, the pH, the temperature, and the duration of incubation. In addition, the possible presence of enzyme inhibitors, varying in amount and nature with different urine samples, will always cast some doubt on the degree of completeness of hydrolysis. In spite of such drawbacks, enzymatic hydrolysis is employed in many procedures, particularly when the steroids are labile in strong acid solution (e.g., pregnanetriol, corticosteroids).

Extraction. Following hydrolysis, the free steroids become sparingly soluble in aqueous solution. Thus, when an immiscible organic solvent in which steroids are highly soluble is added to the hydrolyzed urine and is shaken, the vast majority of steroids are extracted into the organic layer. Repeating the process of extraction with a fresh volume of the organic solvent ensures the maximum recovery of steroids from the urine. The selection of the organic solvent is based on the polarity of the steroid hormones in question. One should bear in mind that even though the nonpolar component, the 4-ring system, is common to all steroids, the polarity increases as the number of oxygen groups (i.e., ketone and hydroxyl groups) and double bonds increases. Steroids with one or two oxygens (e.g., androgens, estrogens) are of low polarity, and the best choice of solvent for their extraction would be one of a nonpolar nature, such as ether or benzene. Similarly, steroids with three or more oxygens (e.g., corticosteroids and their metabolites) are quite polar, and for their extraction polar organic solvents such as chloroform, dichloromethane, or ethyl acetate would be most suitable.

The property of relative solubility of substances in two immiscible solvents not only has been exploited for the extraction process but has also been the basis for the separation and purification of substances (partition chromatography). The ratio of the concentration of a substance in a nonpolar phase to the concentration of the same compound in the polar phase is known as the partition coefficient (K). Substances with high K values will mostly be in the nonpolar phase, whereas those with low K values will preferably move into the polar phase. In the extraction process, the solvent system is composed of an organic solvent (a relatively nonpolar phase) and the hydrolyzed urine (polar phase). Steroids of high partition coefficients will consequently be extracted into the organic layer. A better recovery of polar compounds (e.g., urinary corticosteroids, estriol) from the hydrolyzed urine can be conveniently achieved by adding ammonium sulfate or sodium chloride prior to extraction. The addition increases the partition coefficient by decreasing the solubility of the steroid in the urine.

Purification and Separation. Although a proper choice of solvent improves the selectivity of extraction, nonetheless a large number of urinary pigments, chromogenic substances, and other nonspecific materials are invariably extracted along with the steroids. Removal of such contaminants, especially those which will interfere in the final estimation, is very important.

The *solvent partition* method is the most simple and suitable method for a clinical laboratory. Thus, it is widely used for preliminary purification and separation of compounds of interest. The basic principle is the same as that of the aforementioned extraction. The organic solvent containing steroids and other urinary impurities is treated with weakly *basic solution* (sodium bicarbonate, sodium carbonate). By virtue of their greater solubility, strongly acidic components migrate into the aqueous layer to be discarded. The separation of neutral and phenolic steroids can be achieved in a similar way. Because of the acidic nature of the phenolic steroids (estrogens) they are readily extractable with an aqueous solution of sodium hydroxide. After lowering the pH of the alkaline solution, estrogens are re-extracted with a suitable solvent (diethyl ether) and are processed further for final estimation. Most often the organic extract is washed to neutrality with water to insure complete removal of alkali which, if allowed to remain, might interfere in subsequent work-up of the materials.

The degree of purification and separation needed prior to quantitative measurement will depend, of course, on the method used for final quantitation. For example, if the final mode of estimation is a color reaction which is very specific for an individual steroid or for a group of steroids, further purification of the steroid extract may be omitted. In fact, many colorimetric assays of steroid hormones are performed on crude urinary extracts. Although such determinations yield fairly adequate information for most clinical purposes, occasionally, specific measurement of an individual steroid or group of steroids necessitates further purification and separation of the extracts. There are various techniques available for this purpose. For detailed description of theory and application of these methods to steroid analysis, the reader is referred elsewhere.[18,26] It suffices to mention here that these methods are based either on some selective chemical reaction (e.g., Girard derivative formation for the separation of ketonic and nonketonic steroids, digitonide formation to separate 3β-hydroxy and 3α-hydroxy steroids) or on physical techniques such as countercurrent distribution (CCD), paper and column partition chromatography, adsorption chromatography (column and thin layer), or gas chromatography (GC).

Quantitation. Methods for the quantitative estimation of steroids may be divided into four categories: *colorimetric*, *fluorometric*, *gas chromatographic*, and *radioisotope methods* (including double isotope derivative formation, competitive protein binding and radioimmunoassays).

COLORIMETRIC PROCEDURES. Colorimetric methods are by far the most commonly used clinical methods for the quantitation of steroid hormones. In these methods a certain functional group of a steroid is made to react with a particular chemical reagent to form a specific colored product. For example, steroids containing a keto $\left(\gt C{=}O\right)$ group in position 17 (17-ketosteroids) react with *meta*-dinitrobenzene in alcoholic alkali to produce a reddish-purple compound. The intensity of the color is proportional to the concentration of the steroid and is measured with a spectrophotometer or a colorimeter, using the wavelength of maximum absorption. The principal source of error in this procedure is the interference from nonspecific chromogens derived from other steroidal and nonsteroidal components of urinary extracts. Various measures have been suggested to correct for such interference.

In later sections some specific methods for steroidal hormone measurement are discussed. It is important to note that in some instances the nonspecific chromogens may account for as much as 90 per cent of the actual value.

FLUOROMETRIC ESTIMATION OF STEROIDS. Many steroids are known to produce a characteristic fluorescence if present in a suitable medium such as sulfuric acid or phosphoric acid. The activation and emission wavelengths, under specified experimental conditions, are relatively specific for a given substance. Although this technique offers certain advantages (sensitivity and, at times, increased specificity), from the practical point of view it is afflicted with many drawbacks, such as nonspecific fluorescence, and quenching effects caused by solvent residues, urinary contaminants, and improperly purified reagents. The need for delicate reaction conditions and the instability of fluorescence have further limited the usefulness of this technique in routine clinical laboratories. Nevertheless, the fluorometric assays of certain hormones appearing in microgram quantities in biological fluids (e.g., cortisol in plasma, estrogens in the urine of men and nonpregnant women) have proved quite useful. In some instances they have replaced previous colorimetric procedures.

GAS CHROMATOGRAPHIC ESTIMATION. This technique is one of the latest and most promising additions to steroid methodology. Speed, sensitivity, accuracy, precision, and specificity are demanding criteria for selection of procedures to measure compounds of clinical interest. Gas chromatographic techniques appear to fulfill these requirements. As a result, within a short period of time this technique has become an important tool for clinical laboratories and has replaced many of the time-consuming and occasionally nonspecific chemical methods. In recent years, gas chromatographic methods for the analysis of most of the steroid hormones have been reported.[22,44,101,102] Many of them have been found to be quite suitable for routine clinical use (e.g., pregnanediol, pregnanetriol, 17-ketosteroids, estriol), and these methods are described in greater detail in the appropriate sections of this chapter.

The theory and the instrumentation involved in gas chromatography have been described in Chapter 3. The following paragraphs describe briefly the features pertinent to steroid analysis.

A glass or stainless steel column of any size and shape may be used, limited only by the ease of packing and operation. The analytical columns for steroid work vary from $\frac{1}{8}$ to $\frac{1}{4}$ inch in diameter and 4 to 12 feet in length. The *solid supports* are deactivated (acid and base washed, silanized) diatomaceous earths of small and uniform particle size with great surface area. The mesh size of these supports is in the range of 80 to 100 and 100 to 120, and gives adequately efficient columns for steroid analysis. Commercially available supporting materials such as Diatoport S (F and M Scientific Company, Avondale, PA.), Gaschrom P, Gaschrom Z, and Gaschrom Q (Applied Science Labs, Inc., State College, PA.) are widely used. Both selective and nonselective liquid (stationary) phases are employed, depending on the nature of the steroids being analyzed. The separation with a *nonselective* phase depends primarily on the molecular size and shape of the compounds. Most of the useful nonselective phases are methyl-substituted silicone polymers which are marketed under various trade names, such as SE-30, OV-1, F-60 and JXR. Among them, SE-30 and OV-1 appear to be the most suitable because of their excellent thermal stability and chemical inertness toward the steroids. The separation of compounds which vary in the nature, number, or stereochemical arrangement of the functional group(s) is best achieved with *selective phases*. The selective phases that have been recommended for steroid separations include

XE-60 (cyanoethyl methyl silicone polymer), QF-1 (fluoroalkyl silicone polymer), OV-17 (methyl phenyl silicone polymer), and the polyester NGS (neopentyl glycol succinate). The nonselective phases are more stable thermally, the operation temperature limit being in the vicinity of 320°C in comparison to 235°C for most selective phases. The silicone polymer OV-17, however, may be used up to about 300°C. For steroid analysis the most suitable concentration of the stationary (liquid) phase is generally found to be between 1 and 4 per cent (w/w) of the solid support. Different techniques of coating the stationary phase onto the solid support, and packing and conditioning of the column will be described in a later section (17-ketosteroids).

The most suitable and most versatile *detection system* for routine steroid analysis is the flame-ionization detector (for principle of operation see Chapter 3). Since the rate of ion production is proportional to the concentration of combustible substances, the detector response depicted as a peak for an individual compound on the chart paper is related to the amount present in the sample. The linearity of response in the expected concentration range of analysis is pre-established with pure reference compounds. The quantitation is carried out by comparing the peak size of the unknown to that obtained for a known concentration of its reference standard.

Optimal performance of a flame-ionization detector is dependent on the ratio of hydrogen to carrier gas flow rate. The efficiency of a specific column also varies with the operating temperature and the flow rate of carrier gas. For this reason, whenever a new column is used, the optimal conditions of these parameters should be determined by trial using a set of reference standards.

There are two methods of *introduction of samples* into a gas chromatograph. The first is the so-called *solution method*. Depending on the anticipated concentration, the materials to be chromatographed are dissolved in a known volume of an organic solvent (benzene, iso-octane, acetone, ethyl acetate, carbon tetrachloride, carbon disulfide, etc.). Usually 1 or 2 μl of the final solution is drawn into a Hamilton microliter syringe and is injected into the gas chromatograph.

The other method for sample introduction is the *solid injection technique*. Generally, the sample is dissolved in a suitable solvent and a desired amount of this solution is applied onto an inert stainless steel or platinum gauze, which is located in a dimpled Teflon plate. The solvent evaporates, leaving a solid residue on the gauze which can then be introduced either manually or by automatic solid injection techniques.[78] The technique has the advantage of greater sensitivity, since the sample size can be increased without having an undesirable large solvent peak.

Irrespective of the methods of sample introduction, the use of an *internal standard* is desirable for several reasons. With the solution method it is difficult to inject exact aliquots of dissolved substances on each occasion. The variation may arise from either the mode of injection or the continual changes of sample concentration caused by evaporation of the solvent. With the solid injection technique, variation may also be caused by the loss of an unknown quantity of the materials left on the Teflon plate. The presence of an internal standard of known concentration compensates for such variations. It also corrects for the effect of variation in the detector response. Furthermore, the retention time of compounds relative to the internal standard may be used as a means of their identification in an unknown sample. The selection of a compound is generally based on the following considerations: it should approximate the physicochemical properties of the component or components to be measured; it must yield a completely resolved peak; and it must not be present in the original sample. The method of computation of results using internal standards is described elsewhere. (See methods for 17-ketosteroids, estriol.)

The *derivative formation* of steroids for the gas chromatographic methods has been found to play an important role in quantitative analysis. The purposes of such a procedure are numerous and may be summarized as follows: (1) decrease of polarity minimizing tailing and adsorption (e.g., trimethyl silyl ether or acetate derivative of estriol); (2) stabilization of thermally reactive structural arrangements preventing thermal breakdown (e.g., O-methyloxime-trimethyl silyl ether derivative of 17-hydroxycorticosteroids and their metabolites), dihydroxyacetone side chain of C_{21} steroids breaking down to C_{19} 17-ketosteroids; (3) increase in volatility so that the compounds will be eluted faster under allowable experimental conditions (e.g., trimethyl silyl ether derivatives of steroids); and (4) alteration of physicochemical property to improve gas chromatographic separation of closely related compounds (e.g., trimethyl silyl ether derivatives of 17-ketosteroids, estrogens, and progesterone metabolites as well as acetate derivatives of estrogens and progesterone metabolites).

DOUBLE-ISOTOPE DERIVATIVE ASSAY. In the double-isotope derivative assay, a steroid labeled with an isotope of high specific activity and negligible mass (e.g., ^{3}H or ^{14}C) is added to the specimen to be analyzed. Following the extraction and purification of the sample, a suitable reagent labeled with a second isotope (e.g., ^{35}S) is added to yield a chemically stable derivative. After further purification, the measurement of the isotope ratio of the derivative provides the amount of steroid originally present in the sample. Two isotopes (e.g., ^{3}H and ^{35}S) which emit particles of different energy spectra are chosen so that the radioactivity of both isotopes can be measured simultaneously in a liquid scintillation counter. The radioactive steroid, added prior to processing of the samples, serves as an indicator for calculating losses during the extraction and purification, whereas the formation of a derivative with a radiolabeled reagent aids in the specific detection and the quantitation. Double-isotope derivative assays are highly sensitive, specific, and accurate. Methods using the double-isotope derivative for the estimation of biologically active steroid hormones in human peripheral blood have been greatly curtailed owing to the complexity, tediousness, and expense involved.

RADIOIMMUNOASSAY. The development of radioimmunoassays has provided the field of endocrinology with highly sensitive and reliable methods.

The theoretical and practical aspects of the radioimmunoassay of steroid and other hormones have been discussed in recent monographs.[17,62,64,81] (See also Chap. 7.)

Steroid hormones are not immunogenic by themselves (i.e., they cannot elicit antibodies). Therefore, antibodies against steroids are produced by active immunization of sheep and rabbits with steroid protein conjugates. If the assay employs a binding protein which is not an antibody, the test becomes a competitive protein-binding assay.

Plasma proteins widely used as binding protein include corticosteroid-binding globulin (CBG), human sex steroid–binding globulin (SSBG, testosterone-estradiol–binding globulin, TeBG), and progesterone-binding protein (PBP) of guinea pig plasma. Target tissues of the respective hormones may also be used for the same purpose, e.g., uterus (for estrogens); prostate and seminal vesicles (for dihydrotestosterone); liver, hepatoma cells, and thymus (for glucocorticoids); and kidney and toad bladder (for mineralocorticoids). Although the potential for utilizing tissue receptors for competitive protein-binding assay is great, the unstable nature and complexity of the preparation of these tissue-binding proteins have limited their routine use.

Radioimmunoassays and competitive protein-binding methods are the most sensitive methods for steroid determination available at the present time. However, since plasma-binding proteins and antibodies are not absolutely specific for their interaction with steroids, extraction of steroids from the plasma and preliminary purification are required to increase the degree of specificity. The separation of steroids from their binding proteins in plasma samples is also necessary to prevent additional interaction in the assay system. Since protein hormones do not appear to circulate bound to carrier proteins, unextracted serum can be employed in radioimmunoassays for these hormones.

A major methodological problem with the competitive protein-binding, and radioimmunoassays of steroids is that the value of the blank tends to cause either under- or overestimation of a steroid. The blank value is defined as an "apparent" amount of the steroid under study in a sample free of the steroid. This "blank value" has been ascribed to several sources. Competing steroidal or nonsteroidal materials present in the plasma extract or unknown materials in the solvent or in the eluate from the chromatograms may either compete for the binding sites or denature part of the binding protein. This will decrease the binding of the radioactive steroid and cause overestimation. If the interfering substances adsorb the steroid or lower the adsorbing activity of the nonspecific adsorbent, then it will lead to an apparent increase in the bound fraction and result in an underestimation of the steroid.

RELIABILITY CRITERIA FOR A METHOD OF STEROID HORMONE ASSAY

Over the past three decades a very large number of methods for the quantitative determination of steroids in biological fluids have appeared in the literature. It is therefore important to establish, for a method, certain reliability criteria such as accuracy, precision, specificity, and sensitivity. Whenever a method is to be introduced for routine analysis, it should first be examined along these guidelines to determine whether it is suitable for its intended use. It is also necessary to repeat these evaluations at frequent intervals for the maintenance of good quality control.

Accuracy. This term refers to the degree to which measurements approach the true value of the quantity being measured. Evaluation of this criterion is generally performed by determining the percentage of recovery of added steroids. Ideally, the compound to be recovered should be added in the form in which it occurs in the sample. Because of lack of complete knowledge of the nature of the conjugates, and because of their unavailability, free steroids are usually added to the sample for such evaluations. Although experiments of this type are not entirely satisfactory, they still give some information about the losses incurred during processing of the sample following hydrolysis.

The accuracy for the methods based on the radioimmunoassay or competitive protein-binding assay is generally determined by adding a small amount of tritiated internal standard of high specific activity (consequently, negligible mass) to the specimen. After completion of the sample preparation, an aliquot is removed and recovery is calculated on the basis of the radioactivity found.

Precision. This term refers to the magnitude of the random errors and thus demonstrates the reproducibility of the measurements. The precision is frequently evaluated by performing replicate determinations on the same specimen or by carrying out multiple recovery experiments, each time adding the same concentration of the desired compound.

Usually for a particular method there is an intrinsic range of concentration which can be measured with maximum reproducibility (minimum variation between replicate analyses). According to Marrian,[48] a standard deviation of ± 10 per cent from the mean at the concentration optimal for the method is considered to be quite acceptable for a method of steroid assay. In hormone assays using antibodies, tissue proteins, or plasma proteins, the precision is also of the order of ± 10 per cent. However, higher precision can be achieved by automation.

Specificity. This refers to the exclusive measurement of a compound or compounds for which the method has been designed. In other words, when a method is designed to measure 17-ketosteroids, it should measure only those steroids and nothing else. The clinical usefulness of the determination of hormones, their metabolites, or both in blood or urine lies in the proper assessment of their production in the body. A method without specificity defeats this purpose.

Generally, the specificity of steroid assay hinges on selective color reactions (e.g., Zimmermann color reaction for 17-ketosteroids), the chromatographic behavior of the compounds (e.g., gas chromatographic measurement of 17-ketosteroids, pregnanediol, estriol), or the selective interaction of steroids with binding proteins (e.g., competitive protein-binding assay of cortisol using CBG; radioimmunoassay of estradiol-17β using antibodies). However, the chromatographic property of a compound, the selectivity of the color reaction, or the selective interaction of the binding protein does not necessarily impart specificity. The interference from many drugs, reagents, and other materials in the biological extract may yield spurious results. Antisera or binding proteins may also interact with steroids of similar structure. For this reason, a method needs to be given careful consideration as to its inherent specificity and sources of error.

Sensitivity. This term is defined as the minimum amount of a substance in biological medium that can be determined with accuracy, precision, and specificity by a particular method. This is largely limited by the degree of sensitivity of the final method of quantitation. Generally, the methods involving fluorescence are more sensitive than those based on color reactions. In some cases sensitivity can be further increased by the use of gas chromatographic methods. However, the highest sensitivity seems to be obtained by radioligand binding assays, which include radioimmunoassays and competitive protein-binding assays of steroid hormones. By judicious choice of the binding proteins and other experimental parameters, hormone levels as low as 30 pg can be quantitated.

ADRENOCORTICAL STEROIDS

The human adrenal cortex secretes a variety of steroid hormones that are intimately concerned with a wide range of metabolic processes. More than 40 different steroids have been isolated from the adrenals. These are the corticosteroids, which are formed exclusively by the adrenals, as well as the androgens, progestogens, and estrogens, which are also secreted by the gonads.

Adrenal activity is regulated by an anterior pituitary hormone, adrenocorticotrophic hormone (corticotrophin, ACTH). Of the various hormones released as a result of corticotrophin stimulation, only cortisol has a feedback inhibitory effect. The control of aldosterone secretion is not entirely dependent on corticotrophin; agents derived from the liver (angiotensinogen) and kidney (renin) also come into play here. A detailed discussion of the control mechanism of aldosterone secretion may be found elsewhere.[99]

Corticosteroids

The corticosteroids are, from the physiological as well as the quantitative point of view, the most important groups of adrenal steroids. The structural formulas of some of the biologically most active corticosteroids are shown in Fig. 13–9, along with their trivial and systematic names. These compounds all possess a Δ^4-3-keto group (unsaturation between carbon atoms 4 and 5 and a keto group at carbon atom 3); a side chain $(H_2\text{—}\overset{\text{HO}}{\overset{|}{\text{C}}}\text{—}\overset{\text{O}}{\overset{\|}{\text{C}}}\text{—})$ substituted at C-17 in the β position (above the plane of the paper); and, with the exception of compound S and deoxycorticosterone, an oxygen function (keto or β-hydroxyl) at C-11. Cortisone and hydrocortisone (cortisol) also have a 17α-hydroxyl group (below the plane of the paper).

The corticosteroids show maximum structural specificity. Structural alterations, especially the reduction of Δ^4-3-keto group, make them biologically inactive.

The major corticosteroids—namely, cortisol, corticosterone, and aldosterone—are secreted by the adrenals at the rate of 25 mg, 2 mg and 200 μg/d, respectively. There is

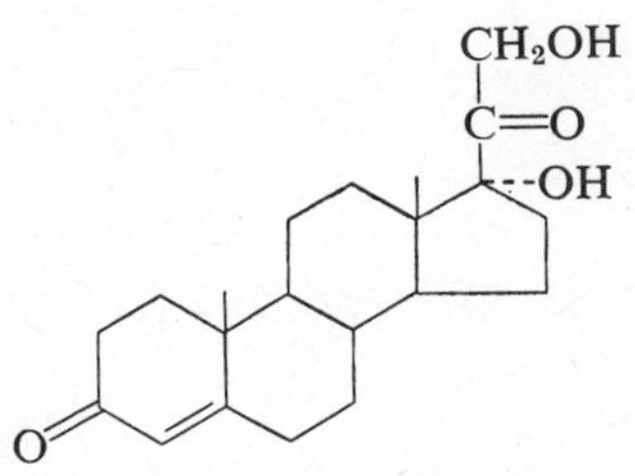

11-Deoxycortisol
(Cpd S)
(17,21-dihydroxy-pregn-4-ene-3,20-dione)

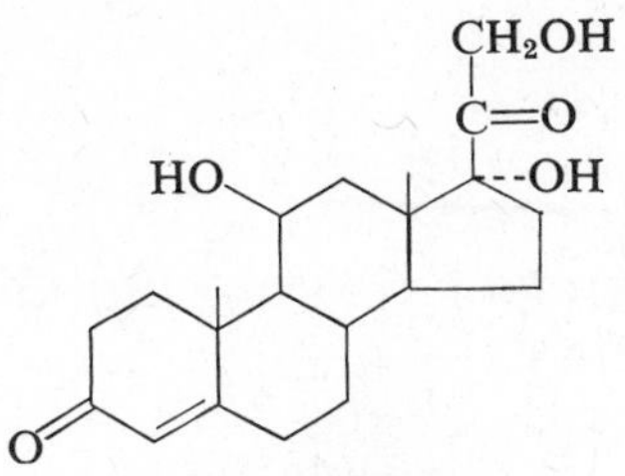

Cortisol
(Cpd F)
(11β,17,21-trihydroxy-pregn-4-ene-3,20-dione)

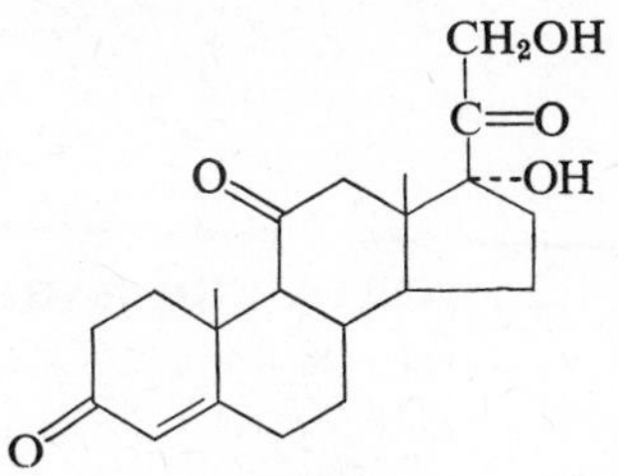

Cortisone
(Cpd E)
(17,21-dihydroxy-pregn-4-ene-3,11,20-trione)

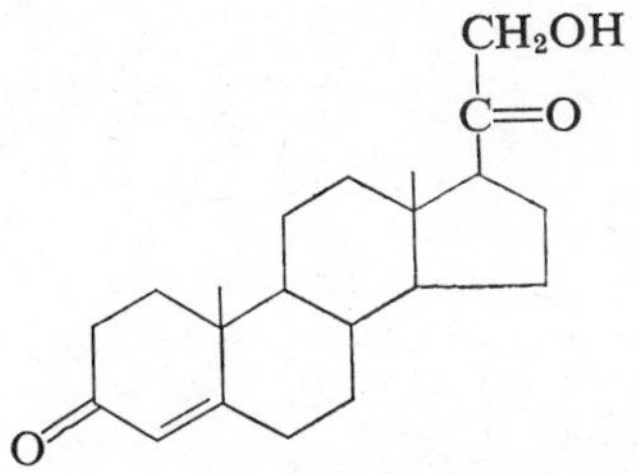

Deoxycorticosterone
(DOC)
(21-hydroxy-pregn-4-ene-3,20-dione)

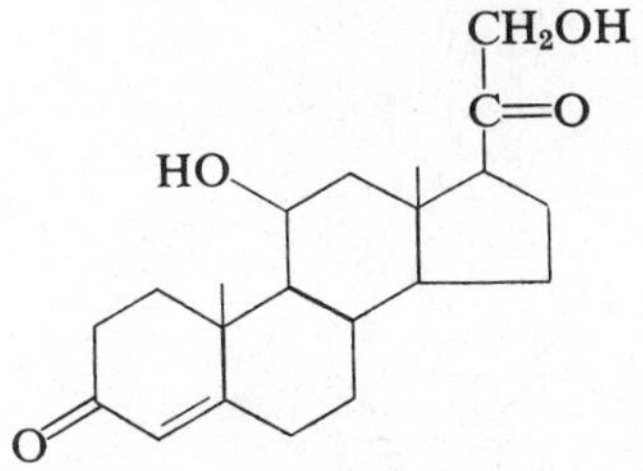

Corticosterone
(Cpd B)
(11β,21-dihydroxy-pregn-4-ene-3,20-dione)

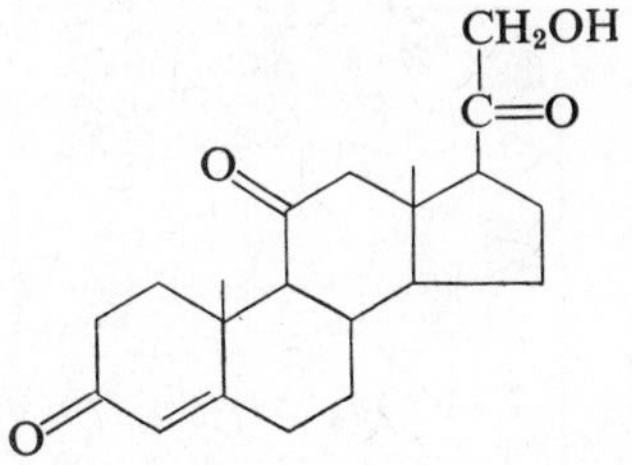

11-Dehydrocorticosterone
(Cpd A)
(21-hydroxy-pregn-4-ene-3,11,20-trione)

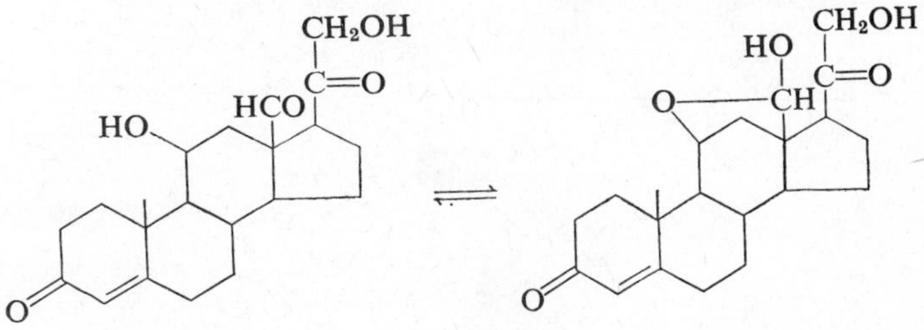

Aldehyde ⇌ **Hemiacetal**

Aldosterone
(11β,21-dihydroxy-3,20-dioxo-pregn-4-en-18-al)

Figure 13–9

diurnal variation in the secretion of cortisol and corticotrophin. Soon after midnight the blood level of cortisol starts to rise, reaching a maximum at about early morning, after which there is a gradual decline to the lowest level between early evening and midnight. In certain adrenal diseases, diurnal rhythmicity of blood levels of cortisol is absent, e.g., in Cushing's syndrome. Functionally, the adrenal cortical steroids may be subdivided into *glucocorticoids* and *mineralocorticoids*. Glucocorticoids participate in controlling carbohydrate metabolism and include the compounds cortisol, cortisone, corticosterone, and 11-dehydrocorticosterone. Mineralocorticoids regulate salt and water metabolism and include the compounds 11-deoxycorticosterone and aldosterone. It should be noted that aldosterone, in addition to having the general structural characteristic of corticosteroids, uniquely possesses at C-18 an aldehyde substituent which remains in equilibrium with the hemiacetal form (Fig. 13–9).

Other adrenal hormones

Besides corticosteroids, the adrenals also secrete androgens, progesterone, and estrogens, all of which are known to be produced by the gonads as well. The major androgens

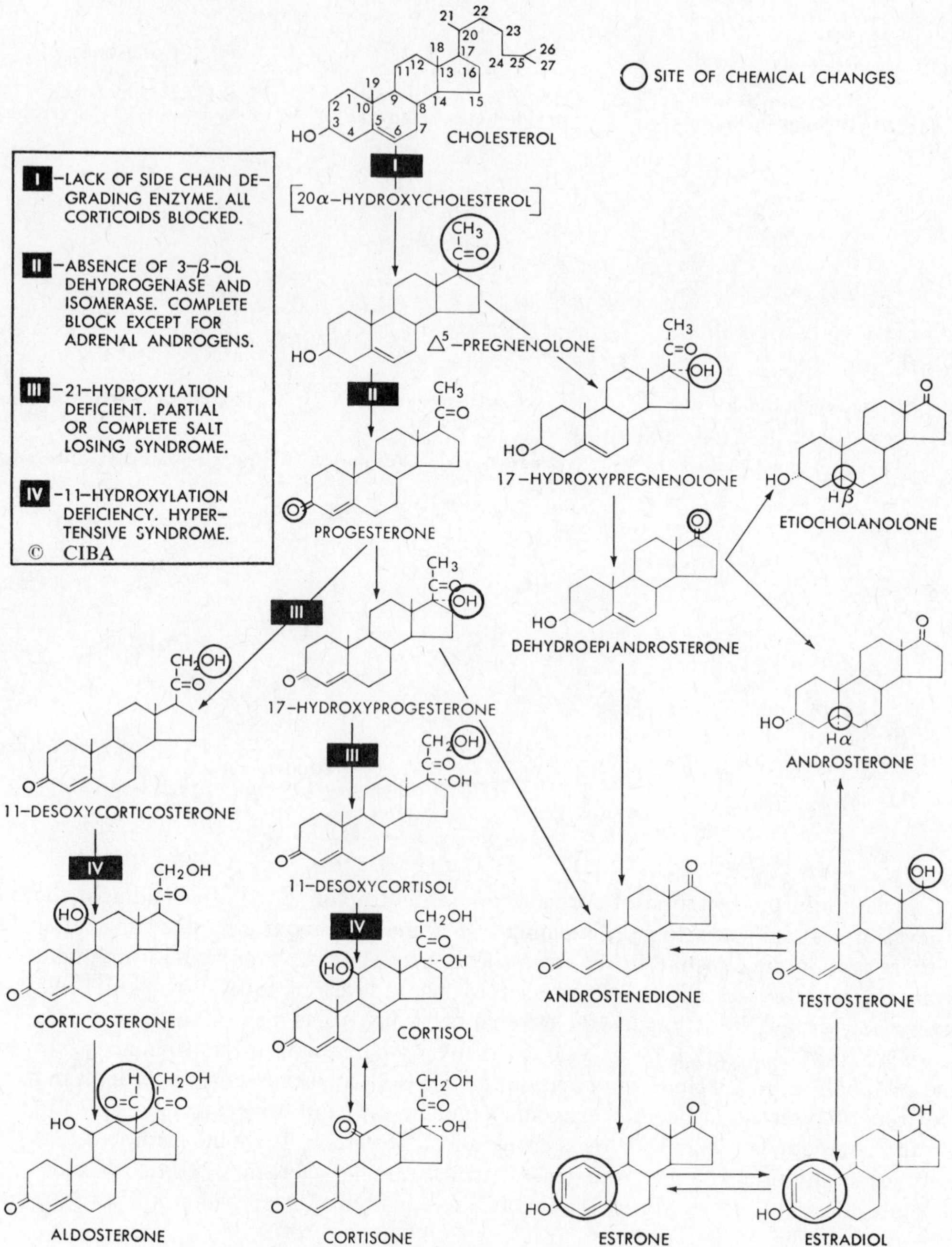

Figure 13–10 Biogenesis of corticosteroids.

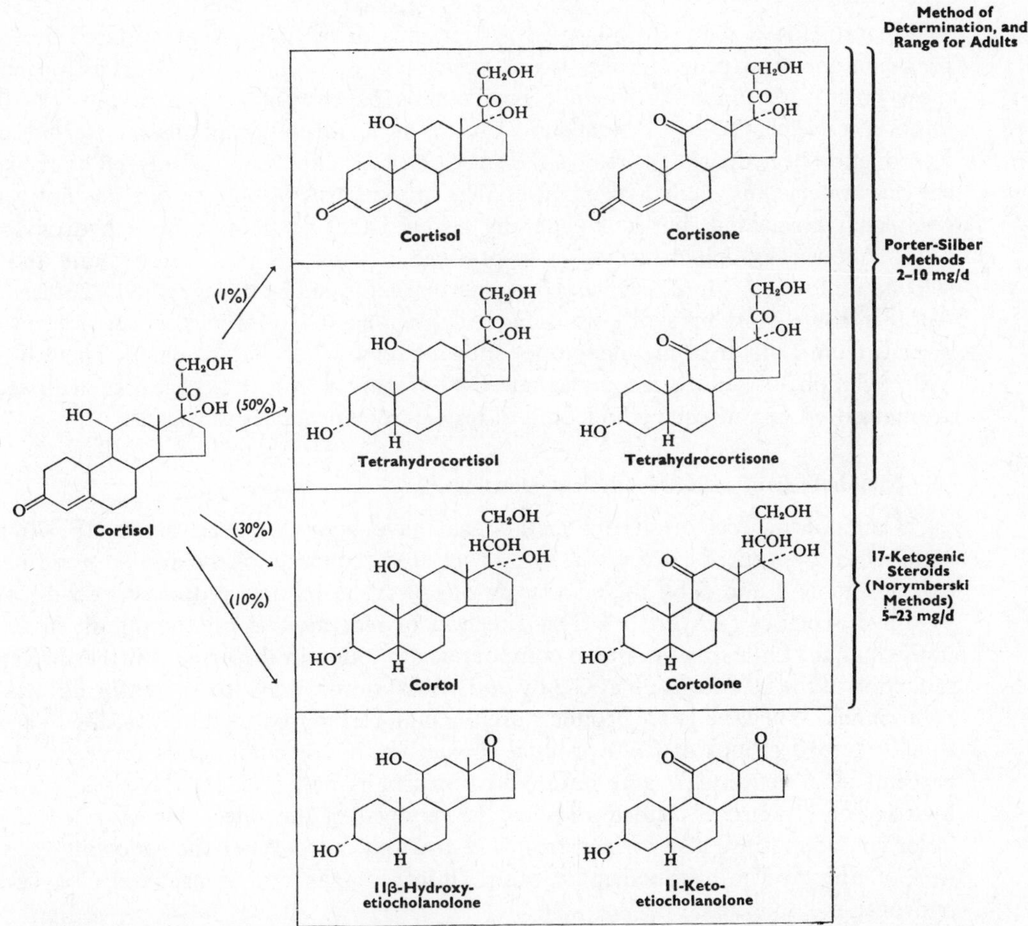

Figure 13–11 Major urinary metabolites of cortisol showing the approximate extent of conversion, and the metabolites as determined by different methods. (After James, V. H. T., and Landon, J.: *In* Recent Advances in Endocrinology. 8th ed. Little, Brown and Co., Boston, 1968.)

are androsterone, etiocholanolone, androstenedione, testosterone, dehydroepiandrosterone (DHA or DHEA), and 11β-hydroxyandrostenedione (Figs. 13–10 and 13–11). DHEA and androstenedione are, from the quantitative standpoint, the most important adrenal 17-ketosteroids and the former DHEA is believed to be produced exclusively by this gland. The rate of secretion of DHEA has been calculated to be as much as 25 mg/d. The estrogens (estrone, 18-hydroxyestrone) and progesterone of adrenal origin are quantitatively very insignificant.

Biogenesis of adrenal corticosteroids

Investigations with radio-labeled compounds have shown that acetate, cholesterol, pregnenolone, and progesterone are all precursors of corticosteroids. In all probability, these precursors constitute a single biosynthetic pathway as shown in Fig. 13–10. The important and characteristic biochemical events in the formation of adrenal steroids are the introduction of hydroxyl groups (—OH) at C-21, C-17, and C-11, catalyzed by the specific enzyme systems known as hydroxylases. The biosynthesis of aldosterone has not yet been completely elucidated, but it is generally believed that corticosterone is hydroxylated at C-18, followed by dehydrogenation to produce aldosterone. Androstenedione and testosterone are formed from 17-hydroxyprogesterone catalyzed by the enzyme 17,20-desmolase. Another route of their synthesis is from dehydroepiandrosterone, which is synthesized directly from 17-hydroxypregnenolone following removal of the side chain. The subsequent hydroxylation of androstenedione at C-11 forms 11β-hydroxyandrostenedione. The aromatization of androstenedione and of testosterone produces estrogens.

Metabolism of adrenal cortical steroids

Common aspects of steroid metabolism have been discussed before (p. 709). The metabolism of cortisol is shown in Fig. 13–11. The reduction of the double bond between carbon atoms 4 and 5 by the enzyme Δ^4-5β- or Δ^4-5α-reductase produces 5β-dihydro or 5α-dihydrocortisol (dihydro F). The direction of reduction is predominantly toward 5β in the human. Little or no dihydro compounds are found in the urine. Further preferential reduction of the ketone group at C-3 by 3α-hydroxysteroid dehydrogenase forms tetrahydro compounds. While the latter products are the major excretory metabolites, the hydrogenation of ketone groups at C-20 produces some hexahydro compounds (cortols). A small percentage of tetrahydro and hexahydro cortisol is converted to 11-oxygenated (11β-hydroxy or 11-keto) 17-ketosteroids by the removal of the side chain (Fig. 13–11). The latter compounds are also formed from 11β-hydroxyandrostenedione following the reduction of ring A. The metabolism of other androgens has been considered elsewhere (see androgen metabolism).

Corticosterone, aldosterone, and other minor C_{21} steroids such as 11-deoxycorticosterone and 11-deoxycortisol (compound S) follow the same sequence of reductive catabolism as cortisol. However, the compounds devoid of 17-hydroxyl groups (corticosterone, aldosterone, 11-deoxycorticosterone) are not metabolized to C_{19} 17-ketosteroids. The transformation of cortisol to cortisone, corticosterone to 11-dehydrocorticosterone, and 11β-hydroxyandrostenedione to 11-ketoandrostenedione is reversible. Consequently, the reduced metabolites with corresponding keto groups at C-11 are also excreted in the urine. The bulk of the C_{21} metabolites are excreted as conjugates of glucuronic acid, and under normal circumstances approximately 1 per cent of the total amount of cortisol secreted appears unchanged in urine.

Adrenogenital syndrome

The biosynthesis of cortisol from cholesterol requires the action of specific enzymes for the chemical modification and introduction of different functional groups on specific sites

(Fig. 13–10). The synthesis of these enzymes is genetically dictated, and any defect in a genome is manifested by an abnormal steroid synthesis. Such abnormalities are, therefore, familial and/or hereditary. *Adrenogenital syndrome* is characterized by the inherited absence or deficiency of biosynthetic enzymes which lead to the formation of cortisol (Fig. 13–10). Deficiencies of almost all of the enzymes of the corticosteroid biosynthetic pathway have been described. By far the most common deficiency in adrenogenital syndrome is in the hydroxylation at the C-21 position (type III block, Fig. 13–10). Functional activity of the adrenal cortex largely depends on the negative feedback control existing between the plasma levels of free cortisol and corticotrophin (ACTH). Thus, in the absence of cortisol, there is constant unrestrained secretion of corticotrophin, causing stimulation of the adrenal cortex, which leads to hyperplasia. The pituitary-adrenal relationships in the adrenogenital syndrome and in other disorders which cause hyperactive adrenals are illustrated in Figure 13–12. Since ACTH stimulates the conversion of cholesterol to pregnenolone, there is an increased output of those cortisol precursors synthesized prior to the enzyme block. These precursors lead to the formation of excess androgenic steroids which cause virilization. The effects are most marked in the female, and the affected female infants are usually born with various degrees of abnormalities of the external genitalia. Male infants appear normal at birth but subsequently develop signs of sexual and somatic precocity. In the absence of 21-hydroxylase action, there is a buildup of 17-hydroxyprogesterone and 21-deoxycortisol (Fig. 13–13). These two cortisol precursors are further metabolized to produce increased amounts of pregnanetriol, 17-hydroxypregnanolone, 11β-hydroxy- and 11-ketopregnanetriol. From the quantitative point of view, the increase of urinary pregnanetriol is most dramatic, and thus its excretion values in the urine have diagnostic importance. Indeed, the increased level of 17-ketogenic steroids in such cases is primarily due to the presence of this metabolite (see method for 17-ketogenic steroid). Most cases of congenital adrenal hyperplasia show also a high 17-ketosteroid excretion, but in certain cases values may be normal or only moderately elevated.

In the absence of Δ^5-3β-hydroxysteroid dehydrogenase-isomerase (type II block, Fig. 13–10), the oxidation of the hydroxyl group in the 3 position of the ring A and the shift of the double bond from C-5 to C-4 do not take place. As a result, progesterone cannot be formed from pregnenolone. In this deficiency, the excretion of pregnenolone metabolites such as Δ^5-pregnenetriol and dehydroepiandrosterone is increased. When a defect occurs in 11β-hydroxylation (type IV block, Fig. 13–10), 11-deoxycortisol (compound S) accumulates and the excretion of the urinary metabolite, tetrahydro S, becomes significant.

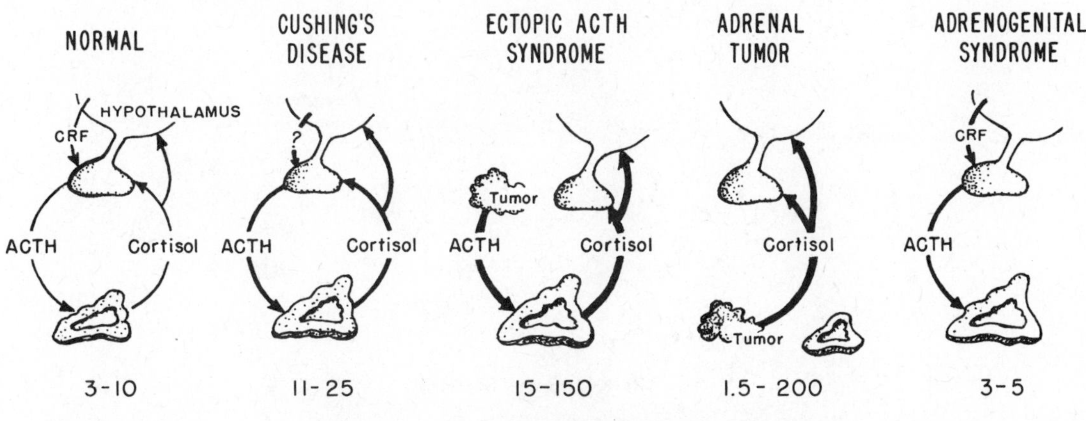

Figure 13–12 Pituitary-adrenal relations in various adrenal disorders. (After Lipsett, M. B. *et al.*: Ann. Int. Med., *61*:733, 1964.)

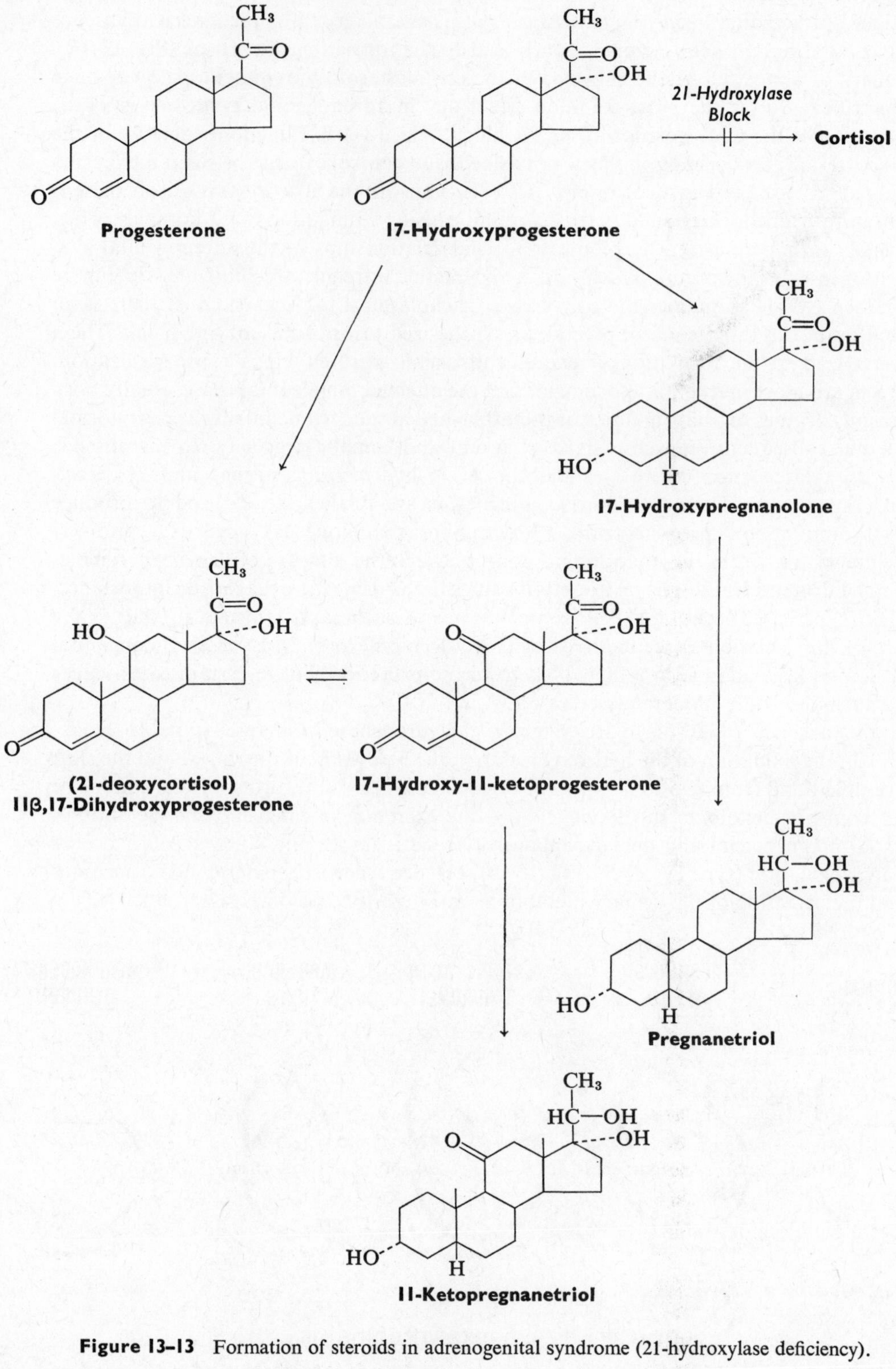

Figure 13–13 Formation of steroids in adrenogenital syndrome (21-hydroxylase deficiency).

THE ESTIMATION OF CORTICOSTEROIDS IN URINE

Since cortisol is the principal corticosteroid secreted by the adrenal cortex, the excretion values of its metabolites in the urine are used as an index of the functional status of this gland. The urinary metabolites derived from cortisol may be grouped as follows:

1. Tetrahydro metabolites: tetrahydrocortisol (THF), tetrahydrocortisone (THE), and allotetrahydrocortisol.
2. Hexahydro metabolites: α and β cortols, and cortolones.
3. 11-oxygenated 17-ketosteroids: 11β-hydroxyetiocholanolone, 11-ketoetiocholanolone, 11β-hydroxyandrosterone, and 11-ketoandrosterone.

In addition to these cortisol metabolites, the urinary products of clinical importance also include tetrahydro 11-deoxycortisol (THS), pregnanetriol, 11-keto- and 11β-hydroxypregnanetriol, and 17-hydroxypregnanolone.

The methods for estimation of 11-oxygenated 17-ketosteroids and pregnanetriol are considered in the respective sections. In the present section, the methods related to the determination of corticosteroids and their C_{21} metabolites are described.

The methods based on the Porter-Silber reaction given by the steroids containing the dihydroxyacetone side chain are the methods most widely used in a routine clinical laboratory.[18] It should be noted, however, that this color reaction does not include all the C_{21} metabolites of cortisol (e.g., α and β cortols, α and β cortolones). The assay procedures based on the chemical oxidation to 17-ketosteroids (see 17-ketogenic steroids) measure all the major C_{21} metabolites. Both these methods are described in detail. There are, however, many additional chemical methods for the quantitative determination of corticosteroids, but because of their limited application to the clinical field they are not discussed here. The analysis by gas chromatographic methods[22] appears to be very promising but cannot be considered presently as a routine assay procedure. The determination of aldosterone and its metabolites by chemical methods is too involved and complex for a common routine laboratory.

CLINICAL SIGNIFICANCE

The usefulness of the estimation of these steroids in blood or in urine will be discussed in the appropriate sections. It may suffice to say here that the measurements are used primarily for the evaluation of adrenal or pituitary dysfunction.

Decreased values are found in adrenal insufficiency, as exemplified by Addison's disease. Similar values can also be noted in panhypopituitary states, such as Sheehan's syndrome.

Increased values are found in numerous conditions which clinically can present as Cushing's syndrome. The latter entity was originally considered to be the result of a pituitary basophil adenoma but it is now known to be associated with ACTH-like secreting tumors (bronchial adenoma, carcinoma, islet cell adenoma of the pancreas, Zollinger-Ellison syndrome, multiple endocrine adenomas). In addition, markedly elevated values are found in cases of adrenal carcinoma.

In the adrenogenital syndrome, total urinary 17-ketogenic steroids are elevated, most frequently because of the presence of excess pregnanetriol which is measured by this method. The Porter-Silber chromogens are low to normal when 21-hydroxylation is blocked, and high when 11β-hydroxylation is blocked since the color reaction is characteristic of the dihydroxyacetone side chain in the corticosteroids. (See principle of the Porter-Silber method.)

It should be noted here that high or low values are not diagnostic as to primary or secondary dysfunction of the adrenal cortex. More sophisticated tests based on the stimulation and suppression of adrenal function using various agents such as ACTH, dexamethasone, and Metopirone are needed to differentiate between the primary and secondary sites of the adrenal dysfunction.

Stimulation and Suppression Tests. Stimulation tests using ACTH (corticotrophin) are most useful in differentiating between Addison's disease and hypopituitarism. The ACTH is administered either intramuscularly or intravenously. Most frequently, 25 units of aqueous ACTH in 500 ml normal saline is administered intravenously over an exact 8 or 6 h interval on two successive days. In normal patients, a 2- to 5-fold increase in 17-hydroxycorticosteroid and a 2-fold increase in 17-ketosteroid excretion levels are noted. In Addison's disease (primary adrenal insufficiency), ACTH stimulation causes little or no rise in 17-OH-corticosteroid, 17-ketosteroid, or 17-ketogenic steroid excretion. In patients with adrenal insufficiency secondary to pituitary hypofunction, a gradual rise (*staircase* response) in steroid excretion is seen on successive days of ACTH stimulation. Such a staircase response signifies that the adrenals are responsive to exogenous ACTH and that the cause of adrenal insufficiency is due to pituitary or hypothalamic dysfunction. In normal subjects, plasma cortisol values increase rapidly to more than double the basal level over the first 30 min and then rise more slowly. The criteria for a normal response include a basal level of more than 5 μg/100 ml and a value after 5 h of between 36 and 60 μg/100 ml. ACTH infusion tests involve the inconvenience of setting up an intravenous drip and of taking multiple blood samples. A *short ACTH stimulation test* which can be used for screening purposes on an outpatient basis is more convenient. It is commonly carried out by determining plasma cortisol before, and 30 min after a single intramuscular injection of 250 μg of a synthetic polypeptide, Cortrosyn (Organon), which has steroidogenic activity similar to natural corticotrophin. The criteria for a normal test on adults include an initial value of more than 5 μg/100 ml with a plasma cortisol increment at 30 min of at least 7 μg/100 ml.

The stimulation tests have also some diagnostic importance in hyperadrenalism. Thus, when Cushing's syndrome is present because of adrenocortical hyperplasia, the excretion of corticosteroids is increased 3- to 5-fold over control values following ACTH stimulation. In Cushing's syndrome secondary to adrenal carcinoma, on the other hand, no or little response to stimulation is noted because of the inherent functional autonomy of the tumor.

The use of *suppression tests* using either urinary steroids or plasma cortisol as an index of suppression is presently the most satisfactory procedure for differential diagnosis of adrenocortical hyperplasia and adrenal carcinoma. Hyperplasia results from increased output of ACTH from the pituitary. The underlying principle of the test is suppression of ACTH secretion by administration of cortisol or its synthetic analogs, resulting in decreased output of adrenal steroids. Because of the increased rate of ACTH production in adrenal hyperplasia, the required administered dose of cortisol or its synthetic analogues is higher than normal. The production of adrenal steroids in adrenal carcinoma is autonomous and excessive, and therefore the feedback control mechanism is ineffective. Consequently, even after the administration of large doses of cortisol or its synthetic analogs, the adrenal secretion remains unchanged in this condition.

In clinical practice, a potent cortisol analogue such as dexamethasone (9α-fluoro-16α-methylprednisolone) is utilized in order that the administered compound may be given in such small amounts that it will not contribute significantly to the analysis of urinary steroids. A standard method of testing is to administer 0.5 mg of dexamethasone every 6 h for 48 h. If no suppression is obtained, the test is repeated with consecutive doses of 2 mg and 8 mg. In normal patients the administration of the 0.5 mg dose reduces the basal excretion level of 17-hydroxycorticosteroids and total 17-ketogenic steroids. The increased steroid excretion which occurs in patients with adrenocortical hyperplasia is generally suppressed by 2 mg doses, whereas the elevated steroid excretion of patients with self-sustaining adrenal carcinoma persists even with higher doses (8 mg). Plasma cortisol analyses are of value in the diagnosis of Cushing's syndrome to establish the lack of diurnal variation and a fall of plasma levels by at least 70 per cent following the administration of a single dose of dexamethasone (2 mg); such determinations do not provide, however, an insight into the etiology of this syndrome.

The *metyrapone test* is employed to assess the pituitary reserve for ACTH. It is thus valuable for the differential diagnosis of adrenocortical insufficiency. Metyrapone (SU 4885, Metopirone, metapyrone) is a drug that selectively inhibits the enzyme action of 11β-hydroxylase. Consequently, when this drug is administered, the conversion of 11-deoxycortisol (compound S) to cortisol is inhibited (see biogenesis of cortisol). The secretion of ACTH is controlled solely by cortisol through the feedback mechanism. The inhibition of cortisol synthesis caused by the drug allows unrestricted release of ACTH, which in turn stimulates the adrenal cortex to produce increased amounts of 11-deoxycorticosteroids. Since the test demands an endogenous supply of ACTH, the degree of adrenocortical stimulation demonstrated by the increased urinary excretion of 11-deoxy C_{21} (metabolites of compound S and 17-hydroxyprogesterone) and of C_{19} steroids (17-ketosteroids) reflects the reserve capacity of the pituitary to release ACTH.

In common practice, 750 mg of metyrapone is administered orally every 4 h over a 24 h period. The urinary steroid excretion in normal subjects shows an increase of twice the amount of the basal level. In patients with hypoadrenocorticalism due to pituitary deficiency, no elevation in urinary excretion is noted. Since the net effect of this test is to increase the endogenous level of ACTH, the results in Cushing's syndrome are similar to the stimulation test as previously described. In adrenocortical hyperplasia, the urinary excretion of steroids increases as much as 2-fold, whereas with adrenal carcinoma no such change occurs. It is noteworthy that in Cushing's syndrome, caused by the ACTH-secreting nonendocrine tumors, the suppression test using dexamethasone and the metyrapone test do not elicit any response, because the site of excessive production of ACTH is beyond the pituitary-hypothalamic axis.

Urinary steroid excretion can be determined either by the Porter-Silber method (corticosteroids) or by measuring 17-ketogenic steroids. The latter method seems preferable since the chromogenicity of urinary tetrahydrodeoxycortisol (tetrahydro S) is less in the Porter-Silber reaction. Alternatively, more specific measurements of the urinary or plasma deoxycorticosteroids can be employed.

DETERMINATION OF URINARY CORTICOSTEROIDS BY A MODIFIED PORTER-SILBER METHOD[85]

PRINCIPLE

In 1950 Porter and Silber described a color reaction based upon the formation of a yellow pigment (absorption maximum at 410 nm) when certain corticosteroids react with phenylhydrazine in the presence of alcohol and sulfuric acid. They demonstrated that this color reaction is given primarily with corticosteroids that possess a dihydroxyacetone side chain as illustrated below:

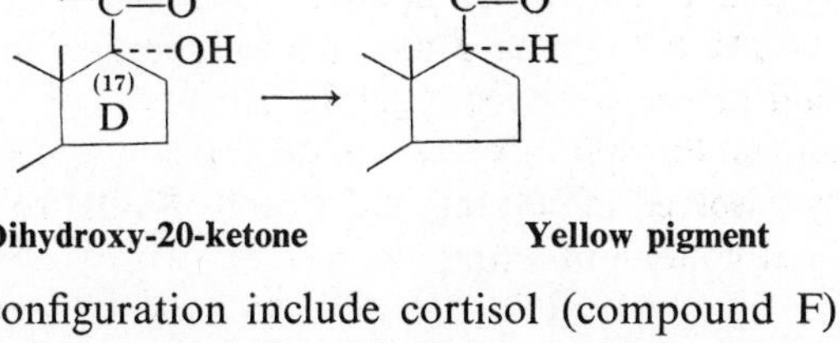

17,21-Dihydroxy-20-ketone **Yellow pigment**

Corticosteroids with this configuration include cortisol (compound F), cortisone (compound E), 11-deoxycortisol (compound S), and their tetrahydro derivatives (Figs. 13–10 and 13–11). In urine, major corticosteroids reacting with the Porter-Silber reagent are tetrahydro E and F; in certain adrenogenital syndromes (e.g., 11β-hydroxylase deficiency) and during the Metopirone test, which blocks 11β-hydroxylase activity, tetrahydro S comprises the bulk of the Porter-Silber chromogens.

The basic steps of the procedure consist of hydrolysis of conjugates by β-glucuronidase; extraction with chloroform; washing the chloroform extract with dilute alkali to remove estrogens, bile acids, and interfering chromogens; and color reaction with alcoholic phenylhydrazine–sulfuric acid reagent.

REAGENTS

1. Chloroform, AR. The solvent should be freshly distilled from anhydrous potassium carbonate (K_2CO_3). Store freshly distilled chloroform in an amber bottle. To prevent formation of phosgene, 1 per cent ethanol should be added.
2. Sodium hydroxide, 0.1 mol/l. Dissolve 4 g of sodium hydroxide pellets in 1 liter of distilled water.
3. Ethanol, purified. Absolute ethanol is purified as follows: To 1 liter of absolute ethanol add 2 g 2,4-dinitrophenylhydrazine hydrochloride and 0.5 ml concentrated hydrochloric acid (HCl). Let stand for approximately 48 h. Distill through a 10 inch Vigreaux column, discarding the first and last 100 ml. Redistill through the same column, again discarding the first and last 100 ml. The purity is determined by mixing this ethanol with the phenylhydrazine–sulfuric acid reagent. No color should develop on standing overnight at room temperature.
4. Sulfuric acid, 64 per cent (v/v). To 360 ml distilled water slowly add 640 ml of concentrated sulfuric acid, AR, with constant swirling. Prepare only in a Pyrex container (2 liter Erlenmeyer flask) immersed in an ice water bath; the solution becomes extremely hot.
5. Alcoholic–sulfuric acid reagent (blank reagent). Mix 100 ml 64 per cent sulfuric acid with 50 ml absolute ethanol. The reagent is stable indefinitely.
6. Phenylhydrazine hydrochloride, recrystallized. A commercially available, chemically pure grade of phenylhydrazine hydrochloride is purified further as follows: add 100 g of phenylhydrazine hydrochloride to 500 ml of warm water at 70°C. Add 1 g activated charcoal. Heat 1 liter ethanol to boiling and add to the dissolved phenylhydrazine in the water. Quickly filter while hot through Whatman No. 2 filter paper. Cool the filtrate in the refrigerator and collect the crystals in a sintered glass filter with medium porosity. Repeat the procedure of recrystallization, dissolving the crystals in proportionally less water. Wash the last collection of crystals with cold ethanol and dry thoroughly. Store in a tightly stoppered brown bottle in a desiccator over anhydrous calcium chloride. The purified material should have a melting point of 240°C to 243°C.
7. Alcoholic phenylhydrazine–sulfuric acid reagent. Dissolve 50 mg recrystallized phenylhydrazine hydrochloride in 50 ml alcoholic–sulfuric acid reagent. The reagent should be prepared fresh before use.
8. β-Glucuronidase. The optimal pH and buffer to be used will vary with the source of the enzyme. Beef liver β-glucuronidase (Ketodase, Warner-Chilcott Laboratories, Morris Plains, N.J.) is incubated in the presence of 0.1 mol/l acetate buffer at pH 5. Bacterial β-glucuronidase (Sigma Chemical Company, St. Louis, MO.) is incubated in 0.1 mol/l phosphate buffer at pH 6.8. Prepare the enzyme solution in the concentration of 1000 units/ml. This should be prepared fresh before use.
9. Buffer solutions. Phosphate buffer, pH 6.8, 0.5 mol/l. To 500 ml 1.0 mol/l solution of KH_2PO_4 (68.0 g dissolved in 500 ml) add 1 mol/l NaOH to bring the pH to 6.8. Adjust the solution to a final volume of 1 liter.

Acetate buffer, 1.0 mol/l, pH 5. Dissolve 95 g of sodium acetate·$3H_2O$ and 17.2 ml glacial acetic acid in water, and dilute to a volume of 1 liter.

10. Steroid standard, stock solution. Transfer 25 mg cortisol (compound F) or tetrahydrocortisone to a 250 ml volumetric flask, and dilute to the mark with absolute ethanol. This stock standard solution contains 100 μg/ml.

11. Working standard solution. Transfer 5 ml of stock standard solution to a 100 ml volumetric flask, and dilute to the mark with distilled water. This working standard solution contains 5 μg/ml.

COLLECTION OF SPECIMEN

Collection of a complete 24 h urine specimen is very important. Creatinine determination is believed to be a fair check for completeness of the specimen. The urine may be stored without any preservative in the refrigerator for a few days. Alternatively, the addition of 1 g boric acid per liter of urine will preserve the specimen at room temperature without any bacterial decomposition of the steroids.

PROCEDURE

Hydrolysis, Extraction, and Washing

1. Transfer 10 ml urine to a 250 ml glass-stoppered cylinder. Adjust the pH of the urine to 6.8 using indicator paper. Add 1 ml β-glucuronidase (bacterial) solution (1000 units), 2 ml 0.5 mol/l phosphate buffer, and 0.1 ml chloroform.
2. In a similar manner, prepare the water blank and standard using 10 ml of distilled water and 10 ml of working standard solution, respectively, instead of urine.
3. Mix the samples well and incubate at 37°C for 18 to 24 h.
4. To each tube add approximately 3 g ammonium sulfate, and mix. Add 100 ml chloroform to each glass-stoppered cylinder and mix the contents by repeated inversion for 30 s. Let the cylinders stand for 5 min in order to separate the aqueous and the organic phases.
5. Remove the aqueous supernatants by aspiration.
6. Add 10 ml of 0.1 mol/l NaOH to each cylinder and shake for 30 s. Allow to stand for 5 min. Aspirate off the alkali layer.
7. In a similar manner wash the chloroform extracts twice with 10 ml of distilled water.

Porter-Silber Reaction

1. Transfer 40 ml aliquots of each chloroform extract to properly labeled 50 ml glass-stoppered centrifuge tubes as follows:

Blank-Blank	*Phenyl-Blank*	*Standard-Blank*	*Standard-Phenyl*	*Test-Blank*	*Test-Phenyl*
40 ml blank extract + 5 ml blank reagent (alcoh. H_2SO_4)	40 ml blank extract + 5 ml phenylhydrazine reagent	40 ml standard extract + 5 ml blank reagent	40 ml standard extract + 5 ml phenylhydrazine reagent	40 ml urine extract + 5 ml blank reagent	40 ml urine extract + 5 ml phenylhydrazine reagent

2. Tightly stopper all tubes, shake vigorously for 30 s and allow to stand for 15 to 20 min. Alternatively, centrifuge the tubes at 2000 rpm for 10 min.
3. Transfer approximately 2.5 ml of the supernatant phase from each tube into correspondingly labeled 10 × 75 mm Coleman cuvets.
4. Incubate the tubes in a water bath at 60°C for 30 min, or overnight in the dark at room temperature.
5. Measure the absorbance (A) with a spectrophotometer at wavelength 410 nm as follows:

Adjust the photometer to zero absorbance using the blank-blank, and read the standard and test blanks. Similarly, set the phenyl blank at zero absorbance and read the standard and test phenyl tubes.

CALCULATION

The standard sample contains 0.05 mg of cortisol. Incorporating this value and the appropriate dilution factor (10) to calculate the concentration of corticosteroids/100 ml of urine, the following equation is derived:

$$\text{Corticosteroids (mg/100 ml)} = \frac{A\text{ test} - A\text{ test blank}}{A\text{ standard} - A\text{ standard blank}} \times 10 \times 0.05$$

$$= \frac{A\text{ test} - A\text{ test blank}}{A\text{ standard} - A\text{ standard blank}} \times 0.5$$

The excretion of corticosteroids per diem (d) is calculated as follows:

$$\begin{matrix}\text{Corticosteroids}\\ \text{(mg/d)}\end{matrix} = \frac{\text{conc (mg/100 ml)} \times \text{urine volume (ml/d)}}{100}$$

COMMENTS

Acid hydrolysis is unsuitable because the free corticosteroids are labile in a strongly acidic medium. The metabolites of cortisol contain numerous hydroxyl and keto groups, making them relatively hydrophilic. The use of a polar organic solvent such as chloroform insures quantitative extraction of these steroids from hydrolyzed urine. To remove acidic components and phenols including estrogens, the solvent extract is washed with *dilute* alkali. The use of a strong alkali (stronger than 0.1 mol/l) destroys the corticosteroids. The alkali-washed extract, termed the *neutral fraction*, contains metabolites of cortisol and of all other steroids excreted as glucuronides, as well as any other neutral lipid-soluble materials of urine. The selectivity of the color reaction toward the steroids with dihydroxyacetone side chains obviates the need for further purification. The impurities present in the extract form nonspecific brown chromogens in the presence of sulfuric acid. The use of a "urine blank" corrects for such background interference.

Various nonsteroidal substances, including acetone, fructose, and dehydroascorbic acid, also form a colored complex with the Porter-Silber reagent. In addition, the following drugs and their metabolites have been reported to cause interference with the colorimetric estimation: iodides, paraldehyde, chloral hydrate, Furadantin, bilirubin, colchicine, coffee, most sulfa drugs, chlorophenothiazines, spirolactones, quinine, and Darvon. Administration of these drugs should be withheld for several days prior to determination of corticosteroids.

NORMAL VALUES

Children (up to 1 year): 0.5–1.0 mg/d
Adult male: 3–10 mg/d
Adult female: 2–8 mg/d

REFERENCE

Sunderman, F. W., Jr.: *In* Lipids and the Steroid Hormones in Clinical Medicine. F. W. Sunderman and F. W. Sunderman, Jr., Eds. Philadelphia, J. B. Lippincott Co., 1960, p. 162.

DETERMINATION OF TOTAL 17-HYDROXYCORTICOSTEROIDS (TOTAL 17-KETOGENIC STEROIDS)[82]

PRINCIPLE

In 1952 Norymberski reported that sodium bismuthate oxidizes several groups of 17-hydroxycorticosteroids to 17-ketosteroids, which can then be measured by the Zimmermann reaction (see determination of total 17-ketosteroids). He termed these steroids "17-ketogenic steroids." The characteristic side chains which are oxidized by sodium bismuthate are shown below:

I. $^{21}CH_2OH$ / $^{20}C{=}O$ / ---OH (17) $\xrightarrow{NaBiO_3}$ 17-one

17,21-diol-20-one
(dihydroxyacetone)

II. $^{21}CH_2OH$ / $^{20}CHOH$ / ---OH (17) $\xrightarrow{NaBiO_3}$ 17-one

17,20,21-triol
(glycerol)

III. $^{21}CH_3$ / $^{20}CHOH$ / ---OH (17) $\xrightarrow{NaBiO_3}$ 17-one

(17,20-glycol)

IV. CH_3 / $C{=}O$ / ---OH (17) $\xrightarrow{NaBH_4}$ CH_3 / $CHOH$ / ---OH (17) $\xrightarrow{NaBiO_3}$ 17-one

17,20-ketol

V. O (17-one) $\xrightarrow{NaBH_4}$ OH $\xrightarrow{NaBiO_3}$ / No oxidation

17-ketosteroids

Group I includes cortisol, cortisone, their tetrahydro derivatives, 11-deoxycortisol (compound S), and tetrahydro S; group II includes cortols and cortolones; group III constitutes pregnanetriol and its 11-oxygenated derivatives; group IV includes 17-hydroxyprogesterone and 17-hydroxypregnanolone.

It should be noted that the first two groups consist of active corticosteroids (cortisol, cortisone) and their metabolites, whereas groups III and IV comprise mainly the metabolites of the precursors of active corticosteroids (e.g., 17-hydroxyprogesterone). The excretion of the latter is quantitatively very significant in certain forms of the adrenogenital syndrome. Sodium bismuthate does not oxidize the 17-hydroxy compounds containing a ketone at C-20 and a methyl group at C-21 as shown in IV. In later modifications, a reduction step, using sodium borohydride prior to bismuthate oxidation, was introduced. This made it possible to measure the metabolites containing a 21-deoxy keto side chain (e.g., 17-hydroxy-pregnanolone together with the compounds included in groups I, II, and III). Following borohydride reduction, the 17-hydroxy-20-keto-21-deoxy steroids are reduced to 17,20-dihydroxy-21-deoxysteroids, and naturally occurring urinary 17-keto steroids are reduced to C_{19} 17-hydroxysteroids. Subsequent treatment of the urine with sodium bismuthate produces 17-ketosteroids from all four groups of C_{21} 17-hydroxysteroids. Since sodium bismuthate does not reoxidize the C_{19} 17-hydroxysteroids, the 17-ketosteroids originally present in the urine become negative to the Zimmermann reaction. As a result, a determination of 17-ketosteroids after borohydride reduction and sodium bismuthate oxidation provides a direct measure of the total C_{21} 17-hydroxycorticosteroids.

REAGENTS

1. Ethylene dichloride. Distill commercially available AR solvent from sodium carbonate (2 g/l) in an all-glass distilling apparatus. Collect the fraction distilling between 83° and 84°C.
2. Sodium bismuthate, Merck, AR.
3. Sodium bisulfite solution, 5 g/100 ml. Prepare fresh before use.
4. Sodium borohydride (Metal Hydrides, Inc., Beverly, MA.).
5. Tes-Tape (Eli Lilly & Co., Indianapolis, IN.).

Other reagents are as those described for urinary 17-ketosteroids determination.

APPARATUS

Special glassware: glass-stoppered heavy-walled centrifuge tubes of 35 ml and 50 ml capacity.

Mechanical shaker: Burrel, wrist-action shaker.

PROCEDURE

1. Test urine with pH paper. If alkaline, acidify with glacial acetic acid (to dissolve phosphate precipitate if present).

2. Using Tes-Tape, determine the approximate concentration of glucose in the sample. If the specimen contains less than 0.5 g glucose/100 ml, proceed to step 3. If the specimen contains more than 0.5 g glucose/100 ml, separate the glucose from the steroids as follows: Transfer 20 ml of urine to a glass-stoppered centrifuge tube, add 10 g ammonium sulfate, and mix to dissolve the salt. Extract three times with 20 ml portions of solvent (ether-ethanol, 3:1). Evaporate the combined extracts to dryness under nitrogen in a water bath at 50°C. Add 10 ml ethanol to the residue and warm the solution in hot water to dissolve the steroids. (Ignore the insoluble material.) Cool, centrifuge, and transfer two 4 ml aliquots (equivalent to 8 ml urine) of the supernatant fluid (for duplicate analysis) to 50 ml centrifuge tubes. Evaporate the ethanol to dryness. Redissolve the residue in 0.5 ml methanol and dilute to 8 ml with water. Proceed to step 3, beginning with addition of sodium borohydride.

Reduction, Oxidation, Hydrolysis, and Extraction

3. Place 8 ml of urine in a 125 ml Erlenmeyer flask. Add 100 mg sodium borohydride. Check the pH. If the pH is not over 8, add an additional 25 mg borohydride. Let it stand for 2 h or overnight at room temperature. (Preferably, instead of adding solid borohydride, 0.8 ml of 10 g/100 ml of freshly prepared solution of sodium borohydride may be used.)

4. Add 8 ml glacial acetic acid and allow to stand for 15 min. (The acid decomposes the excess borohydride.)

5. Transfer to a 50 ml centrifuge tube. Add 2 g sodium bismuthate. Stopper and shake mechanically for 30 min away from direct sunlight. (The samples may be covered with a heavy black cloth during the treatment with bismuthate.) Add 2 g fresh sodium bismuthate and shake for an additional 15 min. Leave the samples overnight at room temperature. The following morning shake the tubes for 15 min.

6. Centrifuge for 10 min at 2000 rpm, and transfer 6.0 ml of the supernatant fluid to 35 ml glass-stoppered centrifuge tubes containing 1.5 ml of freshly prepared sodium bisulfite solution. Mix the solution and allow to stand for 5 min.

7. Add 5 ml distilled water and 3.6 ml concentrated hydrochloric acid. Let stand for 15 min.

8. Place in a boiling water bath for 10 min. Remove and cool the samples in a cold water bath.

9. Add 12 ml ethylene dichloride and shake mechanically for 15 min. Centrifuge for 2 min at 2000 rpm.

10. Aspirate off the upper phase as completely as possible without losing any organic solvent.

11. Add to the organic extract 25 to 30 pellets of sodium hydroxide. Place in shaking machine for 15 min, centrifuge, and filter through 7 cm Whatman No. 1 filter paper into a test tube.

12. Transfer 4 ml of filtrate (≡ 1 ml of urine) to a test tube and evaporate to dryness under nitrogen in a water bath at 50 to 55°C. (In the case of a 24 h collection of large volume, use 8 ml of filtrate.)

Color Reaction

13. Perform the Zimmermann color reaction and measure the absorbance as described in the method for total 17-ketosteroid determination.

CALCULATIONS

$$\text{Total 17-ketogenic steroids (mg/d)} = \frac{\text{corrected } A \text{ of sample}}{\text{corrected } A \text{ of standard}} \times 0.05 \times \text{total urine volume (ml)}$$

COMMENTS

Since the 17-ketosteroids formed from the 17-hydroxycorticosteroids are fairly stable in a hot acid medium, the hydrolysis of steroid conjugates can now be performed with acid as opposed to the enzymatic hydrolysis used in the direct method based on the Porter-Silber reaction. The presence of glucose in urine interferes with the bismuthate oxidation. All urine specimens should therefore be routinely tested with Tes-Tape, and the glucose removed before the determination is begun.

The most suitable means to rid the sample of the glucose appears to be the procedure outlined in step 2. Errors due to the presence of glucose may, however, also be avoided by increasing the amount of sodium bismuthate (1 g for each gram of glucose above 0.5 g/100 ml). The presence in urine of varying amounts of reducible substances other than glucose makes the use of a large excess of borohydride necessary. Addition of sufficient borohydride is indicated by effervescence on the addition of acetic acid (step 4). The absence of effervescence is suggestive of an insufficient amount of borohydride, which may yield misleading results because of the incomplete reduction of different ketone groups. Instead of sodium bismuthate, the oxidizing agent sodium meta periodate (10 vol per cent of 10 g/100 ml solution in 0.1 mol/l NaOH) may also be used. The advantage of using this reagent lies in the fact that in addition to oxidizing 17-hydroxycorticosteroids to 17-ketosteroids, it oxidizes glucuronides to the free steroids or to their formates, which are easily hydrolyzed in alkaline solution; thus, the need for acid hydrolysis is eliminated. Necessary precautions and drug interference in the color reaction will be discussed elsewhere (see determination of 17-ketosteroids).

Since a greater number of metabolites of cortisol (cortol and cortolone) are estimated by the ketogenic method, the normal urinary excretion values are generally higher than those obtained by the Porter-Silber method. This method also yields high values in the adreno-genital syndrome because of the presence of excessive amounts of urinary metabolites of the cortisol precursors, which go undetermined by the latter procedure.

NORMAL VALUES

Children (up to 1 year): < 1 mg/d
Children (1–10 years): 2.3–3.8 mg/d
Adult male: 5–23 mg/d
Adult female: 3–15 mg/d

REFERENCE

Sobel, C. S., Golub, O. J., Henry, R. J., Jacobs. S. L., and Basu, G. K.: J. Clin. Endocrinol., *18*:208, 1958.

THE ESTIMATION OF CORTICOSTEROIDS IN PLASMA

The main purpose of the estimation of corticosteroids in blood or urine is to evaluate the rate of secretion of cortisol by the adrenal cortex as well as the actual level of the hormone to which the tissues are exposed. While the estimation of the urinary excretion of metabolites renders indirect information regarding the overall activity of the gland (i.e., secretion rate), the blood estimation appears to be more useful to ascertain whether the tissues are exposed to proper amounts of cortisol. It should be noted in this connection that the urinary excretion may be elevated by the increased rate of production and metabolism of the hormones without the physiological level in the blood being enhanced. For example, in obesity and hyperthyroidism, the urinary excretion of 17-hydroxycorticosteroids is elevated even though the plasma level of cortisol is within the normal range. The measurement of plasma cortisol is of value in studying the existence of normal diurnal variation and in obtaining quick information regarding the response to functional tests employing stimulation and suppression of the adrenals.

Cortisol represents almost 80 per cent of the total 17-hydroxycorticosteroids in the blood, and the majority circulates in its original form along with small amounts of unconjugated reduced derivatives. The biologically active unconjugated cortisol in the plasma is bound to some extent by albumin and to an α-globulin derived mainly from the liver. This latter protein is called *transcortin*, or corticosterone-binding globulin (CBG). While the precise function of such protein binding is still obscure, it is generally suggested that this mechanism assures a ready source of available circulating hormone and protects it from inactivation and conjugation in the liver. The concentration of CBG in the plasma rises during pregnancy and during estrogen therapy, with a concomitant increase of total 17-hydroxycorticosteroids in plasma. However, since the amount of free hormone, unassociated with protein, remains at physiological levels, there are no untoward effects. It should be noted here that the unbound portion is biologically most significant, because this is the amount which is available for immediate physiological action. Customary procedures involving organic solvent extraction or protein precipitation estimate both free and protein-bound corticosteroids. Thus, with the procedure discussed, total unconjugated cortisol is measured.

Some authors[16] have suggested that determination of urinary cortisol levels may be a more reliable index of adrenocortical hyperfunction. It is reasoned that normally, only about 1 per cent of the total amount of cortisol secreted appears unchanged in urine, since most cortisol is protein-bound and therefore not filtered through the glomeruli. At concentrations in excess of 20 μg cortisol/100 ml plasma, transcortin becomes saturated. Consequently, the amount of unbound cortisol, and thus the amount available for filtration, increases rapidly. The result is a relatively large increase of urinary free cortisol. The free urinary cortisol, therefore, represents the excretion of the circulating unesterified, biologically active, nonprotein-bound cortisol responsible for the signs and symptoms of the disease.

There are three general methods for the estimation of blood corticosteroids in a routine clinical laboratory. These are the methods based on the Porter-Silber color reactions,[65] measurement of the sulfuric acid–induced fluorescence,[51] and the methods based on the principle of competitive protein-binding (CPB) radioassay.[58] In recent years the fluorometric technique and the CPB assay methods have found wider clinical application than the colorimetric procedures. The fluorometric method is more sensitive than the colorimetric technique and thus requires relatively small volumes of plasma. Several steroids, such as prednisone, prednisolone, and dexamethasone, if administered to the patient, interfere in the colorimetric but not in the fluorometric method. The same applies to ketones, hexoses, and some commonly used drugs (see Comments, p. 730). The major drawback of the fluorometric technique is an overestimation of cortisol by about 2.5 μg/100 ml plasma because of the presence of nonspecific fluorogens.[51] It has been observed that the nonspecific fluorogens correlate closely with the true plasma cortisol level and are dependent upon the functional status of the adrenal glands. Therefore, the nonspecific plasma fluorescence does not negate the clinical usefulness of the method. The fluorometric procedure has also the disadvantage that it does not measure 11-deoxycortisol (compound S), which is increased in patients undergoing the metyrapone test.

The CPB assay appears to be most promising. It is rapid, fairly simple to perform, sensitive, and specific. As opposed to the fluorometric or colorimetric procedures, the results obtained with this method are not affected by the presence of nonspecific substances, drugs, and synthetic analogs (e.g., Aldoctone, dexamethasone). In addition, the method can easily be modified to measure both cortisol and 11-deoxycortisol by the introduction of a selective solvent-extraction step.[58]

FLUOROMETRIC ESTIMATION OF 11-HYDROXYCORTICOSTEROIDS IN PLASMA[51]*

COLLECTION OF BLOOD

Blood should be withdrawn at a fixed time of the day (e.g., 8 A.M.) because of diurnal variation of circulating cortisol. Collect 10 ml of blood in a test tube containing heparin as an anticoagulant. Separate the plasma as soon as possible, because the uptake of corticosteroids by red cells increases on standing. If necessary, plasma can be stored for 72 h at 0 to 4°C. Do not freeze plasma, since this may produce a precipitate which may trap or adsorb steroids, causing low results.

PRINCIPLE

Free and protein-bound cortisol and corticosterone are extracted from the plasma with dichloromethane. The organic extract is shaken with a sulfuric acid–ethanol reagent. After removing the supernatant dichloromethane, a resulting fluorescence of the acid solution is read in a fluorometer at a specified time, and is compared with that of a known concentration of cortisol treated in the same manner. Maximum fluorescence of corticosteroids is produced by excitation at 470 nm (436 nm when a mercury lamp is used as the exciting

* It should be borne in mind that the use of this fluorometric technique is not suitable in patients undergoing the metyrapone test because 11-deoxycortisol (compound S), which is increased following the drug administration, does not produce fluorescence. Under these circumstances, the plasma method based on the Porter-Silber reaction for 17-hydroxycorticosteroids (Peterson[65]), or the urinary determination by the methods described here will have to be employed.

source). By exciting with light of 470 nm, results are more specific and correlate well with those of CPB methods. Maximum emission of fluorescence occurs at 530 nm.

REAGENTS

Use only reagents of AR quality.

1. Dichloromethane, purified. Purify commercially available dichloromethane as follows: Let the solvent stand for several days over concentrated sulfuric acid, shaking occasionally. Wash further by shaking with concentrated sulfuric acid (100 ml/l), followed by 1 mol/l NaOH (100 ml/l), and twice with distilled water (200 ml/l). Dry over anhydrous sodium sulfate for 24 h. Distill in an all-glass apparatus and collect the fraction boiling between 39° and 40°C. Store in an amber bottle.
2. Sulfuric acid (concentrated), AR.
3. Sodium hydroxide (pellets), AR.
4. Sodium sulfate (anhydrous), AR.
5. Ethyl alcohol, purified, as described for Porter-Silber reagent (p. 728).
6. Fluorescence reagent. Slowly add 7 vol. of concentrated sulfuric acid to 3 vol. of ethyl alcohol in a flask which is kept cold in iced water. The solution obtained should be colorless. If the ethyl alcohol is not purified enough, a brown color may develop. This reagent remains stable for a month at room temperature.
7. Cortisol standards. Dissolve 50 mg cortisol in 50 ml purified ethyl alcohol. Take 1 ml of this solution and dilute to 100 ml with distilled water (10 μg/ml). These solutions remain stable for months at 4°C.

For the working standard solution, dilute 1 ml of the 10 μg/ml standard to 10 ml, with distilled water. This solution contains 1 μg/ml.

APPARATUS

All glassware should be cleaned with chromic acid, followed by thorough washing with tap water and finally distilled water. If a rubber Propipette is used to pipet solvent and reagents, plug all pipets with cotton wool to prevent rubber particles from contaminating the solutions.

PROCEDURE

Extraction of Free Steroids from Plasma

1. Pipet 2 ml of plasma into a 25 ml glass-stoppered centrifuge tube. Into two separate tubes place 2 ml distilled water (water blank) and 1 ml working standard solution diluted with 1 ml distilled water (standard), respectively.
2. Add 15 ml dichloromethane to each tube. Stopper and shake very gently by hand, or place in a slow-moving mechanical shaker for 10 min (vigorous shaking causes emulsion formation).
3. Centrifuge for 2 min at 2000 rpm. Remove the supernatant plasma by suction.
4. (Fluorometry should be performed in batches of not more than six plasma extracts, a blank, and a cortisol standard. Careful timing is necessary to keep nonspecific fluorescence as low and as uniform as possible.) At 1 min intervals, beginning with the blank, add 10 ml of the extract to 5 ml of the fluorescence reagent in a suitable glass-stoppered tube. Shake vigorously for 20 s.
5. Carefully aspirate the supernatant dichloromethane from each tube in the same order as before, starting with the blank.
6. Thirteen minutes after mixing with the fluorescence reagent, read the fluorescence at 530 nm (emitted wavelength) following excitation at 470 nm. Set the water blank to read zero on sensitivity range 1 of the instrument. One minute later set the standard to read

100 on sensitivity range 2. Then read the samples on the same sensitivity range at 1 min intervals in the order in which the fluorescence reagent was added.

CALCULATION

$$\mu\text{g of cortisol/100 ml plasma} = \frac{\text{reading of sample}}{\text{reading of standard}} \times 1.0 \times \frac{100}{2}$$

COMMENTS

Scrupulous cleaning of glassware and rigorous purification of dichloromethane are very important, because extraneous materials from the solvent and glassware may quench the fluorescence. The nonspecific fluorogens in plasma increase linearly with time relative to the fluorescence of the cortisol standard. For this reason, careful timing is necessary to insure the constancy of this nonspecific fluorescence from estimation to estimation. The principal corticosteroids measurable by the method are cortisol, corticosterone, and 20-dihydrocortisol. However, the last two steroids normally are not quantitatively as significant as cortisol in blood. Estrogens, especially estradiol, produce a considerable amount of fluorescence, but again their presence—even in the plasma of pregnant women—is negligible in comparison to the amount of cortisol. Therapy with triparanol or spironolactone may lead to falsely high results, and therapy with sulfadimine to falsely low values. Aldactone also interferes with the fluorometric method. Spuriously high results may be obtained in patients taking contraceptive drugs, which increase CBG.

NORMAL VALUES

The values range from 6.5 to 26.3 μg (average 14.2 μg) per 100 ml plasma between 8 and 10 A.M, and 2 to 18 μg (average 8 μg) per 100 ml plasma at 4 P.M.

There is no significant difference between values according to age and sex in adults.

REFERENCE

Mattingly, D.: J. Clin. Pathol., *15*:374, 1962.

*MEASUREMENT OF PLASMA CORTISOL BY COMPETITIVE PROTEIN-BINDING**[58,60]

PRINCIPLE

When tritiated cortisol (or corticosterone) is added to a suitable dilution of binding protein (e.g., corticosteroid-binding globulin, CBG, in normal human plasma), some of the cortisol is taken up by CBG while the rest remains free in solution. If unlabeled cortisol (e.g., extract of serum sample) is now added, this will compete with the labeled steroid for the binding sites on CBG. Eventually a new equilibrium between labeled and unlabeled steroid will be reached which is dependent upon the amount of unlabeled steroid present (the higher the amount of unlabeled cortisol, the lower the amount of labeled steroid bound to CBG). The attainment of the equilibrium can be hastened by heating the solution to 45°C, at which temperature CBG releases most of the bound cortisol. This is followed by cooling to 5°C, when CBG recombines with both labeled and unlabeled cortisol. Protein-bound and free cortisol are then separated by adsorption of the free cortisol to Florisil (or Fullers' earth). The proportion of the tritiated steroid remaining bound to protein is then determined and used as an index of the amount of unlabeled cortisol in the sample.

The cortisol content of the sample is quantitated by comparing its displacement of tracer with that caused by known amounts of the steroid. A standard curve may be drawn, by plotting the cpm (counts per minute) bound, the percentage of bound tracer, or its

* This assay procedure is discussed in significant detail as an example of a hormone assay by competitive protein-binding. The same principles discussed here also apply to many other similar assays.

reciprocal, against the amount of steroid. If the reciprocal of the per cent bound steroid is used, essentially a straight-line relationship is obtained over the range suitable for the analysis. Since the time required to count the protein-bound fraction to a preset number of counts (e.g., 4000 cpm) is proportional to the reciprocal of the per cent bound steroid (i.e., the lower the per cent binding, the higher the time required), it is convenient to plot the time required to reach the preset count *versus* the amount of steroid added. In this case, or in the case of cpm bound against concentration plot, no additional calculation for the per cent bound is necessary.

REAGENTS

1. Corticosterone-1,2-^{3}H or cortisol-1,2-^{3}H (New England Nuclear Corp., Boston, MA.) of high specific activity (e.g., 23.5 Ci/mmol). Check the purity of the radioactive steroid by conventional thin-layer or paper chromatography. If the compound is 95 per cent pure, no further purification is necessary. Make solution of 10 μCi/ml in purified ethanol and store at -10°C. Avoid freezing radioactive solutions since it increases decomposition. (Cortisol-^{3}H should be used if Fullers' earth is employed as adsorbent. See explanation under Comments.)

2. Cortisol standards (Sigma Chemical Co., St. Louis, MO.). Prepare the stock solution (10 μg/ml in purified ethanol) as before. For the working standard solution (0.1 μg/ml) dilute 0.1 ml of the stock standard solution to 10 ml with purified ethanol. The working standard should be made up in small (10 ml) quantities since it deteriorates after several months if subjected to the frequent rapid changes of temperature which occur with daily use.

3. Binding protein (CBG). Collect and pool plasma from normal male subjects. The pooled plasma may be stored in small aliquots at -20°C. The plasma protein is stable indefinitely when stored frozen but cannot withstand repeated freezing and thawing.

4. CBG-isotope solution. To 5 ml pooled human plasma, add 0.4 ml cortisol-^{3}H solution (10 μCi/ml) and dilute to 100 ml with distilled water. To avoid precipitating the protein with ethanol, the plasma should be added after adding most of the water. The solution is made up fresh daily.

5. Ethanol, purified. See Porter-Silber reagent (p. 728).

6. Florisil, 60-100 mesh (Fisher Scientific Co., Pittsburgh, PA.) or Fullers' earth, acid washed, Cat. #134-F (BDH Ltd., Montreal, Canada). Florisil is washed with ethanol to remove the fines, air dried, reactivated at 100°C for 12 h, and kept in a vacuum desiccator at room temperature. Fullers' earth is used without further preparation.

7. Scintillation fluid (Bray's). Dissolve 120 g naphthalene, 8 g PPO, and 0.4 g dimethyl POPOP in 200 ml methanol, 40 ml ethylene glycol, and 1760 ml of *p*-dioxane.

8. Dispo-culture tubes, 12 × 75 mm (Arthur H. Thomas Co., Philadelphia, PA.).

COLLECTION OF BLOOD

Follow the method described on page 735.

PROCEDURE

Deproteinization of the Plasma Sample

1. Pipet 0.1 ml of heparinized plasma into 1 ml of ethanol in a centrifuge tube. Mix and centrifuge at 3000 rpm for 5 min. Transfer the supernatant to a small test tube (Dispo-culture tube).

2. Repeat the procedure by adding 0.5 ml ethanol to the centrifuge tube and combine supernatant with that in the small test tube.

3. Place the small tube into a water bath at 45°C and evaporate the ethanol to dryness under nitrogen. (Dry, filtered air can also be used, according to Murphy.[58]) It is convenient and economical to process 10 or more samples at a time.

Competitive Protein-Binding Assays

1. Prepare standard tubes in duplicate by pipetting 0, 0.1, 0.2, 0.3, and 0.4 ml of solution containing 0.1 μg cortisol/ml and evaporate to dryness.

2. Set the standard and sample tubes in a rack. To each tube add 1 ml of CBG-isotope solution, shake the tube rack well, and incubate in a water bath at 45°C for 5 min.

3. Transfer the rack to an ice bath (5°C) and let stand for 10 min. To each tube add 80 mg Florisil or 15 mg Fullers' earth (a precalibrated glass or plastic scoop may be conveniently used). Shake the rack for exactly 2 min (longer shaking increases the uptake bv the adsorbent) and return to the ice bath for a further 10 min. Florisil settles to the bottom. If Fullers' earth is used, centrifuge the tubes.

4. Pipet 0.5 ml supernatant from each tube into counting vials.

5. Add 10 ml Bray's scintillation fluid. Shake well. Count each vial to 4000 cpm (preset count) in a liquid scintillation counter. To obtain the standard curve, plot time required for each standard tube to reach the preset count against corresponding nanograms of cortisol (i.e., 0, 10, 20, 30, 40 ng). (Alternatively, cpm versus concentration may be plotted, provided there is enough radioactivity to achieve high precision in the counting.)

CALCULATION

Compare sample data (time or cpm) with the standard curve to obtain the steroid content. Since 0.1 ml plasma was used, the amount in nanograms in the sample is equivalent to that in μg/100 ml plasma.

COMMENTS

The plasma from pregnant women or women on contraceptive pills (increased CBG concentration), and from male dogs (CBG has higher affinity for most corticosteroids) has also been used as a source of binding protein.[17] The presence of endogenous cortisol in the binding plasma decreases the sensitivity of the method. Therefore, endogenous cortisol in the donor of CBG is decreased by suppression of the adrenal function by dexamethasone, by treatment of the pooled plasma with charcoal adsorbent, by gel filtration under conditions which favor dissociation of the protein-steroid complex, or by exposure of plasma to elevated temperatures (up to 45°C) or to low pH (5.0 to 5.5). The reason tritiated corticosterone is used as a radioligand to measure cortisol is that the affinity of both cortisol and corticosterone for human CBG is very similar. In addition, Florisil employed for separation of the unbound fraction adsorbs corticosterone and most other steroids much better than it adsorbs cortisol. An inconveniently large amount of Florisil would therefore be required if cortisol were to be used as the tracer.

The routine use of Fullers' earth is less convenient because it remains suspended in the solution until centrifuged. The time of exposure of the assay medium to the adsorbent then becomes very critical and therefore all samples must be centrifuged at the same time. Florisil, on the other hand, quickly settles to the bottom of the tube where further adsorption is negligible (maximum adsorption takes place during shaking) and obviates centrifugation. For this reason, Florisil is a commonly used adsorbent in the measurement of almost all steroids by competitive protein-binding assays. However, the use of Fullers' earth increases the specificity of the method for cortisol, provided tritiated cortisol is used as a radioligand. This is due to the fact that most steroids other than cortisol are readily and effectively removed by fullers' earth, rendering them unable to compete with the radioactive cortisol for binding sites on CBG. In a pediatric unit, where congenital adrenal hyperplasia (adrenogenital syndrome) is an important consideration, the use of Fullers' earth as an adsorbent and tritiated cortisol as a ligand may be advisable. The presence of relatively large amounts of progesterone, 17-hydroxyprogesterone, 11-deoxycortisol, or 21-deoxycortisol, which also bind well with human CBG, may lead to erroneous results and false diagnosis if Florisil is used.

The standardization of the amount of Florisil or Fullers' earth to be added for a particular batch of the adsorbents and assay system is very important. The optimal amount of adsorbents will vary with the quality and adsorptivity of the adsorbent, the concentration of the binding proteins, and the duration of shaking.

The method as outlined is suitable for the measurement of cortisol in the range of 0 to 40 μg/100 ml plasma. The useful range for a particular assay depends on the specific activity of the radioligand, and the concentration and quality of CBG employed.

The *sensitivity* of the method is increased if a lower concentration of CBG with high affinity for the steroid is employed. The efficiency of the agent used to separate bound and free steroid also has some effect. The more complete the separation, the greater the sensitivity. The problems related to the *blank* are minimal, since there are few manipulative procedures, and no organic solvent other than ethanol is used for the preparation of the sample (see radioisotope methods).

The *specificity* of the method largely resides in the selective interaction of the CBG with cortisol. However, steroids such as progesterone, 17-hydroxyprogesterone, 11-deoxycortisol (Cpd S), 21-deoxycortisol, and corticosterone have high affinity for human CBG. Under normal circumstances, the validity of measuring cortisol depends upon the premise that these competing steroids are present in negligible amounts in the assay system. This assumption, however, does not apply in late pregnancy (progesterone is high), in the adrenogenital syndrome (17-hydroxyprogesterone, 21-deoxycortisol, or 11-deoxycortisol may be high), or after the administration of metyrapone (11-deoxycortisol is high). Consequently, under these conditions, the measurement of cortisol cannot be valid without prior separation of the competing steroids, e.g., by solvent partition. A preliminary treatment of the plasma with petroleum ether to extract progesterone and 17-hydroxyprogesterone,[36] and with carbon tetrachloride to extract 11-deoxycortisol,[58] has been used for the removal of these steroids.

Most synthetic corticosteroids (except prednisolone) as well as numerous other compounds, including androgens, estrogens, thyroid hormones, tranquilizers, vitamins, and assorted pharmacological and natural products, have no significant effect on this assay system. The *normal* values for cortisol by this method are similar to those obtained by the fluorometric method.

The measurement of urinary free cortisol can be accomplished by the described method with a minor modification which includes the extraction of an aliquot of urine (2 ml) with methylene chloride.[59] Following evaporation of the organic extract, the competitive protein-binding assay is carried out as described for plasma cortisol. The relatively large amounts of cortisol metabolites (e.g., tetrahydro derivatives) do not interfere since they are excreted as conjugates which are soluble in water but not in methylene chloride. Both Fullers' earth[59] and Florisil[7] can be used for the separation of the bound and unbound fractions.

According to Murphy,[59] values in normal subjects range from 0 to 108 μg/d with a mean value of 48 ($SD \pm 32$) μg/d. As opposed to plasma levels, the excretion values in obese, chronically ill, or hypertensive patients do not differ significantly from those found in normal subjects. In the case of Cushing's syndrome the value exceeds 120 μg/d. In subjects whose adrenocortical function is suppressed by dexamethasone administration or who have adrenocortical hypofunction, very low values (less than 10.0 μg/d) are obtained.

REFERENCES

Murphy, B. E. P.: J. Clin. Endocrinol., *27*:973, 1967.
Murphy, B. E. P.: Recent Progr. Horm. Res., *25*:563, 1969.

MEASUREMENT OF ALDOSTERONE IN PLASMA AND URINE

The importance of measuring aldosterone lies in the diagnosis and treatment of patients with overproduction of this steroid hormone (*aldosteronism*). The continuous excessive secretion of aldosterone results in sodium retention, potassium loss, and eventually hypertension. This condition may result from adrenal disease (primary aldosteronism) due to single or multiple adenomas, adrenal carcinoma, or bilateral adrenal hyperplasia. Extra-adrenal diseases, such as congestive cardiac failure, the nephrotic syndrome, or cirrhosis with ascites, may also cause overproduction of aldosterone (secondary aldosteronism).

Until recently, the estimation of aldosterone in urine or in plasma was beyond the scope of a clinical laboratory because of the laborious and complex nature of the methods used to measure the minute quantities of this hormone normally present in body fluids. The techniques based on the double-isotope derivative formation[18] are extremely time-consuming, complex, and expensive. Consequently, their application has been limited to use in research and specialized laboratories. Radioimmunoassays appear to offer a simple and reliable method for measuring circulating aldosterone levels in human plasma.[64] The method of Bayard *et al.*[5] seems to be suitable for a routine laboratory, and a general outline of that technique follows.

Antibodies to aldosterone are produced in rabbits by injecting aldosterone-3-carboxymethoxime-18,21-diacetate coupled with rabbit albumin. The specificity of the antibody is determined by studying the per cent cross-reaction of other steroids compared with that of aldosterone. The cross-reactivity is negligible with most steroids except aldosterone and aldosterone-21-monoacetate. Despite excellent specificity, aldosterone must be separated from other plasma steroids, such as cortisol, because of the relatively high amounts in which they occur physiologically. Five to 20 ml plasma with ^{3}H-aldosterone added for recovery measurements is extracted and purified by paper chromatography. After 20 min incubation with antibody-steroid mixture at 37°C, bound and free fractions are separated using Florisil. The concentration of aldosterone in the plasma sample is determined by comparing the per cent bound with that of the standard curve. Healthy subjects on a normal sodium diet and lying down for 2 h have an aldosterone concentration in the range of 1 to 7 ng/100 ml of plasma; after standing for 4 h the values range from 3 to 28 ng/100 ml of plasma.

ANDROGENS

Androgens are a group of C_{19} steroids which exert profound influence on the male genital tract and are concerned with the development and maintenance of secondary male sex characteristics; hence they are called male sex hormones. They are secreted by the testes, the adrenals, and the ovaries. The principal and biologically most active naturally occurring androgen is testosterone, which is derived mostly from the Leydig cells of the adult testes. The structural characteristic of this steroid is an unsaturated bond between C-4 and C-5, a ketone group at C-3 (Δ^4-3-keto), and a hydroxyl group in the β position at C-17 (Fig. 13–14). The hydroxyl group in the β position (above the plane of the paper) is essential for biological activity, since epitestosterone (17α-hydroxyl) and androstenedione (17-keto) have very little biological activity.

In men, *testosterone* is undoubtedly the most important androgen secreted into the blood, and it or its metabolites (e.g., dihydrotestosterone) are responsible for genital development, beard growth, muscle development, and sexual drive. In normal women, the function of androgens, other than being estrogen precursors (see estrogen biogenesis), is less well defined. In pathological conditions, increased androgen production by ovaries and adrenals results in virilization. There are, however, in addition to testosterone, other potent androgenic steroids which may contribute significantly to the total blood androgen load in women.

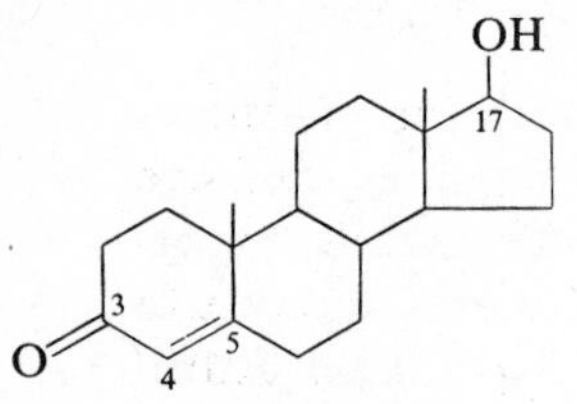

Testosterone
(17β-hydroxyandrost-4-en-3-one)

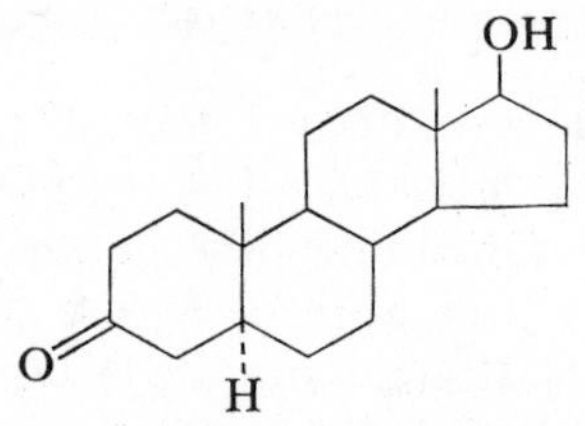

Dihydrotestosterone
(17β-hydroxy-5α-androstan-3-one)

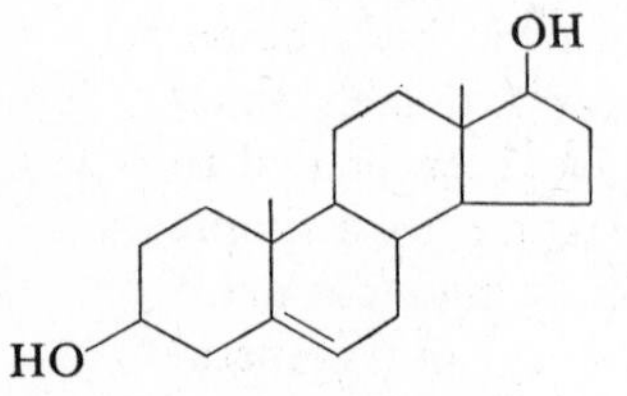

Androstenediol
(3β,17β-dihydroxyandrost-5-ene)

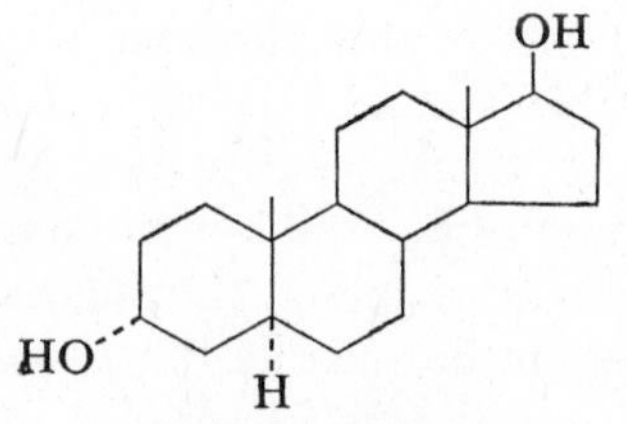

Androstanediol
(3α,17β-dihydroxy-5α-androstane)

Figure 13–14 Chemical structure of androgens.

From the bioassays in animals and the nitrogen balance studies in man, it would appear that the plasma androgens other than testosterone are *dihydrotestosterone*, *androstanediol*, and *androstenediol*. Their chemical structure is shown in Fig. 13–14. Interest in *dihydrotestosterone* was recently stimulated by studies[43] which suggest that this androgen may be an intracellular effector of testosterone action. Testosterone is reduced to dihydrotestosterone (addition of two hydrogens at C-4 and C-5) by a variety of tissues such as the prostate, skin and seminal vesicles. Dihydrotestosterone is thought to exert its biologic activity on these target tissues where it is formed, but in addition it may enter the blood and subsequently produce androgenic effects on other parts of the body. Indeed, the bioassay data suggest that dihydrotestosterone is a far more potent androgen than testosterone. *Androstanediol* is produced in peripheral tissue from testosterone and re-enters the blood. *Androstenediol* may be secreted by the adrenals or ovaries, and/or derived from the plasma dehydroepiandrosterone. In most textbooks, androstenedione and dehydroepiandrosterone are listed as androgens. Although in test animals they exhibit weak androgenicity, it has been shown in recent years that these two steroids are important *prehormones* for plasma testosterone. A prehormone is defined as a substance with little or no inherent biologic potency but which is converted to a more active product in peripheral tissue. From the quantitative point of view, androstenedione is a significant testosterone prehormone of the ovaries, and dehydroepiandrosterone is a testosterone prehormone of the adrenals. Consequently, in hirsute and/or virilized women, the functional tests for both ovaries and adrenals are of diagnostic value (see plasma testosterone—clinical significance). The 11-oxygenated C_{19} steroids derived from the adrenals are biologically inactive.

Biosynthesis

There are two pathways for the formation of androgens from pregnenolone (Fig. 13–15). The Δ^5 pathway includes the transformation of pregnenolone to 17-hydroxypregnenolone to dehydroepiandrosterone and thence to androstenedione and testosterone. The other route is from pregnenolone to progesterone and thence to androstenedione (Δ^4 pathway). Although both pathways are functional in the testis and adrenal, the Δ^5 route is probably the more active biosynthetic pathway. In the ovary, during the first half

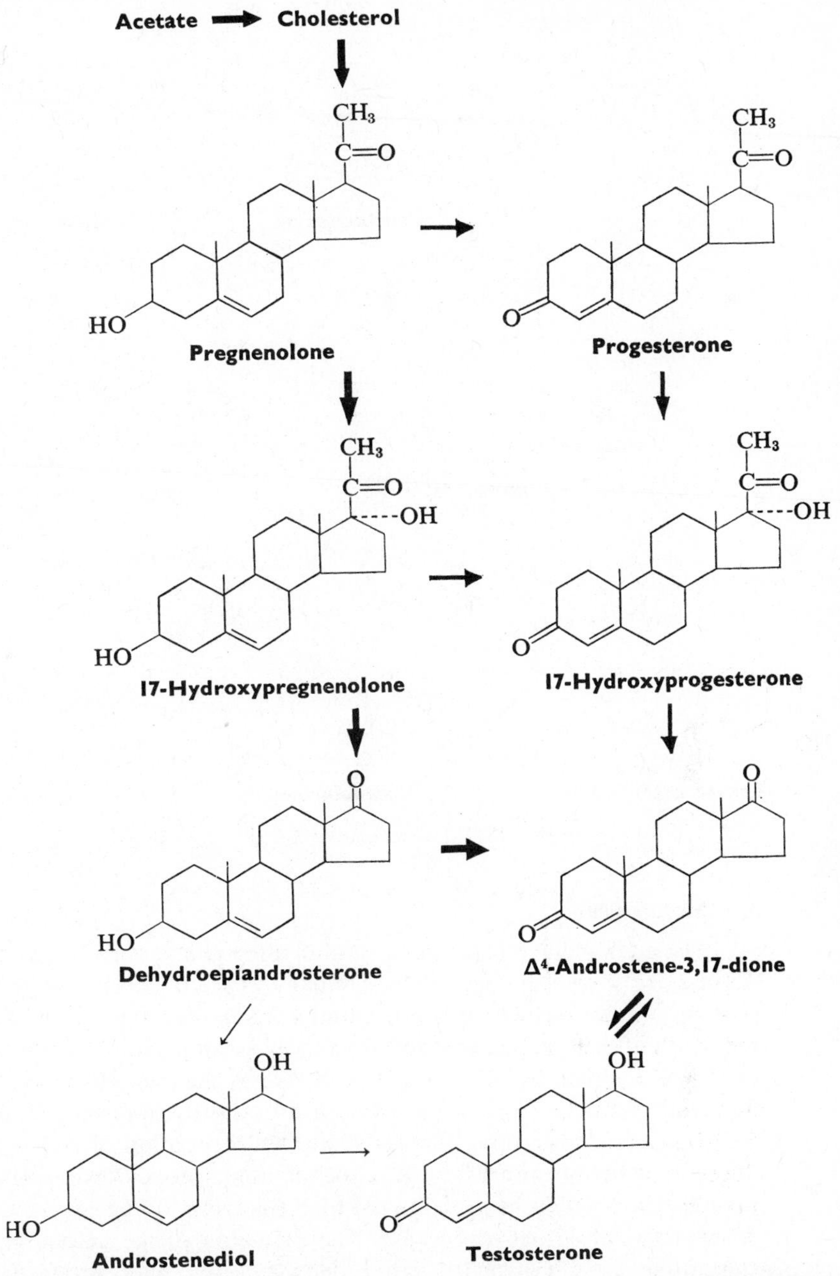

Acetate → Cholesterol

Pregnenolone → Progesterone

17-Hydroxypregnenolone → 17-Hydroxyprogesterone

Dehydroepiandrosterone → Δ^4-Androstene-3,17-dione

Androstenediol → Testosterone

Figure 13–15 Biogenesis of androgens (ovary, adrenal, testis). The heavy arrows indicate the preferred pathway. See text.

of the menstrual cycle (follicular phase), the estrogen precursors, androstenedione and testosterone, are mainly provided by the Δ^5 pathway, while in the second half of the cycle (luteal phase), these are preferentially formed from progesterone. Dehydroepiandrosterone may also be converted to testosterone via androstenediol, which does not involve androstenedione as an obligatory intermediate. The transformation of androstenedione into testosterone is a reversible reaction. In the ovary, the equilibrium lies far toward the formation of androstenedione, which is secreted. In the testes, on the other hand, the forward reaction is more active, thus favoring the production and secretion of testosterone. The two principal secretory products of the adrenals are dehydroepiandrosterone and androstenedione.

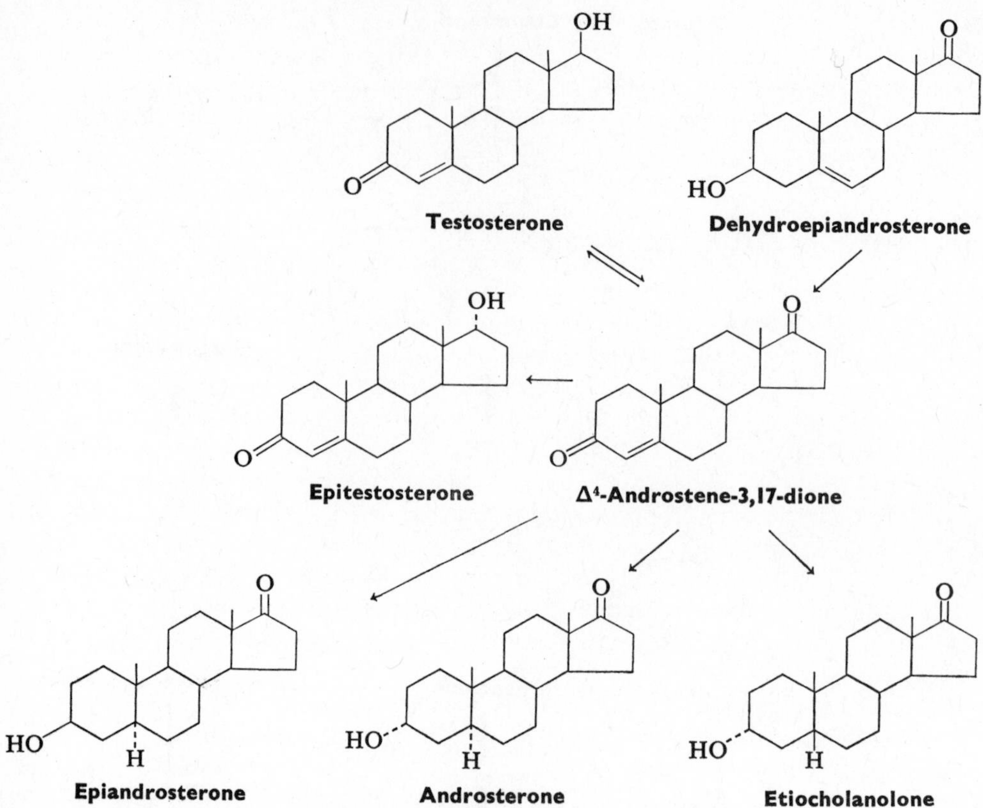

Figure 13–16 Catabolism of $C_{19}O_2$ androgens.

Metabolism

The main metabolic products of androstenedione, testosterone, and dehydroepiandrosterone are shown in Fig. 13–16. Similar to corticosteroid metabolism, the reduction of ring A, and subsequent oxidation and conjugation are the major metabolic steps. The reduction of the double bond gives rise to two isomers, differing in the spatial configuration of H at C-5. When the H is on the same side as the methyl group at C-10 (*cis*), then it is a 5β isomer (etiocholanolone), and when it is on the opposite side (*trans*), the isomer is a 5α isomer (androsterone). Similarly, the hydrogenation of the ketone group at C-3 produces 3α-hydroxy (androsterone, etiocholanolone) or 3β-hydroxy (epiandrosterone) compounds, the reaction being catalyzed by the corresponding enzymes, such as 3α-hydroxy or 3β-hydroxysteroid dehydrogenase. The reduction of the ketone group at C-17 of androstenedione by 17α-hydroxysteroid dehydrogenase can produce epitestosterone. Some workers[75] claim that the measurement of epitestosterone before and after stimulation of the ovary with gonadotrophin is useful in detecting primary ovarian failure and in confirming the ovarian stromal hyperfunction in polycystic ovary disease. However, the clinical importance of this test has not yet been firmly established.

Quantitatively, the 3α-hydroxysteroids are predominant in urine. Dehydroepiandrosterone, containing a β-hydroxyl group at C-3 and an unsaturated bond between C-5 and C-6 (3β-hydroxyandrost-5-ene), is first irreversibly converted to androstenedione, which in turn follows the same biochemical sequences as described before. Dehydroepiandrosterone is also excreted unchanged, as the sulfate conjugate. The metabolism of 11-oxygenated androgens has been described in a previous section (see cortisol metabolism). All these catabolites, except epitestosterone, constitute the group of steroids known as the 17-ketosteroids, by virtue of their ketone group at C-17. While the conjugation of all these steroids may occur with sulfuric acid and glucuronic acid, the glucuronide is predominant in androsterone, etiocholanolone, and 11-oxygenated 17-ketosteroids; dehydroepiandrosterone is present exclusively as the sulfate conjugate.

ESTIMATION OF URINARY 17-KETOSTEROIDS

The 17-ketosteroids are metabolites of precursors secreted by the adrenals, by the testes, and possibly to some extent by the ovaries. In men, approximately one-third of the total urinary 17-ketosteroids represent the metabolites of testosterone secreted by the testes, whereas most of the remaining two-thirds are derived from the steroids produced by the adrenals. In women, who usually excrete smaller quantities than men, the total 17-ketosteroids are derived almost exclusively from the adrenals.

The bulk of the urinary 17-ketosteroids consists of androsterone, epiandrosterone, etiocholanolone, dehydroepiandrosterone, 11-keto- and 11β-hydroxyandrosterone, and 11-keto- and 11β-hydroxyetiocholanolone. It should be recalled that dehydroepiandrosterone and 11-oxygenated 17-ketosteroids are the products of adrenals only, while the others also arise from the precursors (androstenedione, testosterone) elaborated by the gonads. Thus, the main purpose of the quantitation of these steroid metabolites is to assess gonadal and adrenal function.

Decreased values of 17-ketosteroids are generally obtained in primary hypogonadism (Klinefelter's syndrome, castration), secondary hypogonadism (panhypopituitarism), and primary hypoadrenalism (Addison's disease, especially in women). *Increased values* are obtained in testicular tumors (interstitial cell tumor, chorioepithelioma), adrenal hyperplasia, and adrenal carcinoma.

The level of total 17-ketosteroids in urine is not a good index of androgen production by the gonads. Urinary 17-ketosteroids are primarily derived from adrenal precursors (dehydroepiandrosterone and androstenedione), which have little or no biological activity. Moreover, the term 17-ketosteroid is strictly a chemical one; no specific biological activity, androgenic or otherwise, is implied. For example, etiocholanolone has no androgenic activity, and testosterone (which is a potent androgen) is not a 17-ketosteroid. Thus, in individual patients, the determination of urinary 17-ketosteroids does not indicate whether testosterone production is normal, excessive, or deficient. Direct measurements of circulating plasma testosterone in males and the production rate of testosterone in females (see plasma testosterone, clinical significance) represent the most accurate index of androgen production. Since suitable methods for plasma testosterone estimation have become recently available, the measurement of total urinary 17-ketosteroids for the evaluation of androgen production should be discontinued. However, the relevance of the 17-ketosteroids in adrenal disease and the popularity of this test in clinical laboratories require the description and discussion of a detailed procedure.

There are a number of chemical methods available for the estimation of total 17-ketosteroids.[18] The final quantitation in most of these is based on the color reaction originally described by Zimmermann. The method described by Drekter *et al.*,[21] with modifications by Sobel *et al.*,[82] has been shown to be most adequate for routine clinical use and is given below.

PRINCIPLE

The 17-ketosteroids are excreted as water-soluble conjugates of glucuronic acid and sulfuric acid. Cleavage of these conjugates with acid is followed by extraction, washing with alkali, and finally the color reaction. Estrone, which is a 17-ketosteroid, is removed by alkali treatment because of its phenolic nature (see p. 713) and thus is eliminated prior to the colorimetric reaction of the "neutral" 17-ketosteroids. The reaction is based on the treatment of 17-ketosteroids with *meta*-dinitrobenzene in alcoholic alkali to produce a reddish-purple color with maximum absorption at 520 nm. Marlow[47] has demonstrated that the development of color depends on the presence of an active methylene group adjacent to a carbonyl group, most likely giving the following product:

17-Ketosteroid + *m*-Dinitrobenzene $\xrightarrow{\text{Alcoholic Alkali}}$ Purple compounds ($=NO_2^-\ K^+$)

When the ketone group is situated at other positions (e.g., Δ^4-3-keto in testosterone, progesterone, cortisol) the color development is less intense and the absorption maxima differ. The quantitation is carried out by the comparison of the color density of the sample with that of a known amount of pure standard, such as dehydroepiandrosterone.

REAGENTS

1. Ethanol, purified. (See 17-hydroxycorticosteroid determination, p. 728.)
2. Ethanol, 70 per cent (v/v). Dilute 700 ml of purified ethanol to 1 liter with distilled water.
3. Ethylene dichloride, redistilled.
4. Potassium hydroxide, AR grade. Prepare a saturated aqueous solution.
5. Potassium hydroxide–ethanol solution. Add 1 vol of saturated solution of potassium hydroxide to 4 vol of purified ethanol. Centrifuge and use the supernatant. Prepare this reagent just before use.
6. *m*-Dinitrobenzene, 1.16 g/100 ml purified ethanol. Purify the commercially available material as follows: Dissolve 30 g of the substance in a minimal volume of ethanol by warming in a steam bath. Cool, add 30 ml of 20 g/100 ml sodium hydroxide in water and allow to stand for 30 min. Add 3 vol of distilled water with mixing and let it stand 15 min. Filter off the crystalline precipitate on a Büchner funnel. Wash the crystals on the funnel with distilled water and suck dry. Redissolve the crystals in a minimal volume of ethanol and add 3 vol distilled water as before. Wait for 15 min and filter on a Büchner funnel. Wash the crystals with distilled water and suck as dry as possible. Transfer crystals to a Petri dish and dehydrate in a desiccator over anhydrous calcium chloride. Store the final product in an amber bottle.
7. Dehydroepiandrosterone (DHEA) standard. Dissolve 10 mg of DHEA in 100 ml of purified ethanol. This solution contains 100 μg/ml.

COLLECTION OF SPECIMEN

See method for corticosteroids in urine, p. 729.

PROCEDURE

1. Test urine with pH paper. If alkaline, acidify with glacial acetic acid to dissolve phosphate precipitate if any is present.

Hydrolysis, Extraction, and Washing

2. Transfer 8 ml of urine to a 35 ml glass-stoppered centrifuge tube. Add 2 ml of glacial acetic acid and 3 ml concentrated hydrochloric acid.
3. Stopper the tube and place it in a 100°C bath for 10 min. Cool it under cold, running tap water.

4. Add 10 ml of ethylene dichloride and place in a shaking machine for 15 min.

5. Centrifuge for 2 min at 2000 rpm and aspirate off the urine as completely as possible.

6. Add to the solvent 25 to 30 pellets of sodium hydroxide and place in shaking machine for 15 min. Alternatively, the extract may be washed with 10 g/100 ml NaOH solution followed by two water washes. Centrifuge as before and filter the solvent through Whatman No. 1 filter paper.

7. Transfer 2.5 ml of the filtrate (equivalent to 2 ml urine) to a test tube and evaporate to dryness under nitrogen in a water bath at 50 to 55°C. (If very low concentrations are expected, use 5 ml of the filtrate.)

Color Reaction and Spectrophotometric Reading

8. Perform the Zimmermann reaction as follows: (a) To a blank tube, the sample tube, and a standard tube containing 50 μg DHEA, add 0.2 ml of *m*-dinitrobenzene solution. (b) Add 0.2 ml of freshly prepared alcoholic potassium hydroxide solution and mix. (c) Place the tubes in a water bath at 25°C in the dark for 30 min. (d) To each tube add 5 ml of 70 per cent ethanol and mix.

9. Measure the absorbance of the standard and the sample in a spectrophotometer or a colorimeter at 480, 520, and 560 nm, setting the instrument at 100 per cent transmission with the blank solution.

CALCULATION

Calculate the corrected optical density of the standard and sample using the following formula:

$$\text{Corrected absorbance} = (A_{520}) - \frac{(A_{480} + A_{560})}{2}$$

Calculate 24 h excretion of 17-ketosteroids as follows:

$$\text{mg 17-ketosteroids/d} = \frac{\text{corrected } A \text{ of sample}}{\text{corrected } A \text{ of standard}} \times 0.05 \times \frac{\text{24 h urine volume (ml)}}{2}$$

COMMENTS

Most of the urinary 17-ketosteroids are excreted as sulfate and glucuronide conjugates which are hydrolyzed by strong acid and heat. The duration of hydrolysis is very critical. Less than 10 minutes will cause incomplete hydrolysis and more than 10 minutes will lead to gradual destruction of the steroids and formation of an increased amount of non-steroidal chromogens. Addition of glacial acetic acid helps to minimize the formation of nonspecific chromogens during the hydrolytic procedure, particularly in the case of an alkaline urine specimen. Although solvents such as benzene, carbon tetrachloride, and ether are suitable for extraction, ethylene dichloride is aptly suitable because it can extract the steroid hormones from hydrolyzed urine more quantitatively at a relatively low ratio of the solvent to urine. This is technically advantageous for a routine laboratory method since it avoids the need for handling and evaporating large quantities of solvent.

According to Drekter *et al.*[21] the treatment of the extract with pellets of sodium hydroxide is superior to the customary treatment with aqueous sodium hydroxide because solid NaOH removes phenols and other urinary pigments more completely.

For the Zimmermann reaction two different alkaline reagents have been in common use, namely, aqueous and alcoholic KOH. The former yields colors of much less intensity than the alcoholic reagent and the latter has the disadvantage of being unstable. As suggested by Sobel *et al.*,[82] the saturated solution of KOH is stable and yields very low blank absorbance. According to the same authors, the time, temperature, and dilution with 70 per cent ethanol give maximum color development and stability. To avoid undue effects in the color reaction, the ethanol must be of highest quality and purified according to the procedure given in the text. However, in spite of meticulous care in the preparation of reagents and in the color development there is always the formation of nonspecific background chromophors arising from other ketonic steroids and nonsteroidal ketones. The reading of the absorbance at three wavelengths and the use of the correction formula serve to eliminate the effect of such background interference in the estimation. The correction is based on the assumption that the absorbances of nonspecific materials at the three chosen wavelengths lie on a straight line. In other methods the preparation of a urine blank and subtraction of its reading from the sample is supposed to serve the same purpose. The following drugs and their metabolites in urine are known to yield spurious results causing either under- or overestimation of 17-ketosteroids: ascorbic acid, Doriden, morphine, meprobamate, and penicillin G.

NORMAL VALUES

Up to 1 year	less than 1 mg/d
1–4 years	less than 2 mg/d
5–8 years	less than 3 mg/d
8–12 years	3 to 10 mg/d
13–16 years	5 to 12 mg/d
Young adult male	9 to 22 mg/d
Adult male	8 to 20 mg/d
Adult female	6 to 15 mg/d

The excretion values are the same for both sexes throughout childhood. After about age 60, the rate of excretion declines progressively in both sexes.

REFERENCES

Drekter, I. J., Heisler, A., Scism, G. R., Stern, S., Pearson, S., and McGavack, T. H.: J. Clin. Endocrinol., *12*:55, 1952.

Sobel, C. S., Golub, O. J., Henry, R. J., Jacobs, S. L., and Basu, G. K.: J. Clin. Endocrinol., *18*:208, 1958.

FRACTIONATION OF 17-KETOSTEROIDS

CLINICAL SIGNIFICANCE

The estimation of total urinary neutral 17-ketosteroids serves as a screening test for the diagnosis of an adrenal or gonadal disease. But to derive meaningful information, the determination of *individual* components of this group of steroids is very important.

When the 17-ketosteroids are measured as a group, no distinction is made between the metabolites derived mainly from the testes and those derived primarily from the adrenals. It is known that androsterone and etiocholanolone are primary metabolic products of testosterone. The increased excretion of these two compounds in a male, without proportionate changes of DHEA and 11-oxygenated 17-ketosteroids, will yield a positive indication of testicular dysfunction. When only total 17-ketosteroids are estimated, such specific changes will go unobserved. Similarly, in both adrenocortical hyperplasia and carcinoma, the excretion value of 17-ketosteroids is quite high. The fractionated estimation shows that whereas in hyperplasia all components of 17-ketosteroids are elevated, in carcinoma the increased excretion value is mainly the result of the presence of excessive DHEA. Thus, to differentiate adrenocortical carcinoma and hyperplasia, individual estimation of dehydroepiandrosterone, rather than the determination of the 17-ketosteroids as a group, is of considerable diagnostic value. Furthermore, the fractionation provides also the opportunity to determine the ratio of $C_{19}O_2$ (androsterone, etiocholanolone, DHEA) to $C_{19}O_3$ (11-oxygenated 17-ketosteroids). The importance of such study lies in differentiating the type of adrenogenital syndrome. For example, an increased ratio of urinary $C_{19}O_2$ to $C_{19}O_3$ will signify a block in 11β-hydroxylation (see also p. 720). The efficacy of the administration of metyrapone (metyrapone inhibits 11β-hydroxylase) may also be properly evaluated on the basis of changes in the ratio.

METHODS FOR FRACTIONATION

Before the modern techniques involving chromatography became available, the use of chemical reactions that were selective toward a characteristic group in the steroid molecule was the primary means of separation. Two such widely used procedures are the separation by Girard's Reagent T and separation by digitonin. The former reagent separates the ketonic neutral compounds containing the 17-ketosteroids from the nonketonic neutral compounds. The principle is based on the fact that Girard's Reagent T forms water-soluble derivatives with the 17-ketosteroids so that the ketones can be separated from the nonketonic material by distribution between water and ether. The treatment of the aqueous solution containing the ketone derivatives with mineral acid (e.g., HCl), followed by re-extraction with ether, will yield free ketonic compounds.

The separation of the 3α-hydroxy- and 3β-hydroxy-17-ketosteroids in urine is performed by the use of digitonin. This substance is capable of forming insoluble complexes with steroids containing a hydroxyl group with the β-configuration at C-3 (dehydroepiandrosterone), but it does not react with 3α-hydroxysteroids (androsterone, etiocholanolone). In patients with adrenocortical tumors, very high values for the 3β-hydroxy-17-ketosteroids, particularly dehydroepiandrostrone, are almost always obtained. Thus, in the investigation of patients with adrenal hyperfunction, the digitonin separation aids in the differential diagnosis. However, from the technical point of view, the quantitativeness of this approach is occasionally doubtful, and at present it is very seldom used. The separation of the ketonic and nonketonic fractions by Girard's Reagent is similarly of little practical value. During the past decade, the fractionation of individual 17-ketosteroids, employing adsorption and partition chromatography (see Dorfman[18]) followed by color reaction, has provided valuable clinical information. In spite of improved reproducibility and resolution, the time-consuming and sophisticated nature of such procedures make them unsuitable for the common, routine laboratory. In recent years, the advent of gas chromatography as an analytical tool appears to have overcome these methodological deficiencies. Speed, sensitivity, accuracy, and simplicity can be achieved with this technique.

The practical usefulness of gas chromatographic analysis has perhaps been realized to the fullest extent in the fractionation and quantitation of the 17-ketosteroids. Following the single injection of a urinary neutral extract, all the 17-ketosteroids can be analyzed individually. Other compounds of clinical interest, such as pregnanediol and pregnanetriol, can be analyzed as well.

Several procedures have been reported for the gas chromatographic analysis of 17-ketosteroids.[22,102]

The use of trimethylsilyl ether (TMSi) derivatives and a gas chromatographic column containing a selective phase (e.g., NGS, XE-60) has been found to be the most suitable means for adequate separation and reliable quantitation of these steroids. The following is a gas chromatographic procedure as described in the manual of a workshop on gas chromatography of steroids (Tietz). The method is claimed to be suitable for routine clinical use. A description of the preparation of a gas chromatographic column obtained from the same source is also given. The importance of the characteristics of the column in the gas chromatographic analyses of steroids has been discussed in the general comments on steroid determination.

GAS CHROMATOGRAPHIC SEPARATION AND DETERMINATION OF 17-KETOSTEROIDS, PREGNANEDIOL, AND PREGNANETRIOL

PRINCIPLE

The 17-ketosteroids are excreted in the urine as a mixture of glucuronides and sulfates. The hydrolysis is generally performed either by sulfuric acid in the presence of overlying benzene, to minimize the destruction particularly of dehydroepiandrosterone, or by the enzyme Glusulase, containing both β-glucuronidase and sulfatase. The organic extract (benzene) is washed with sodium hydroxide solution to remove strongly acidic components and phenolic steroids (estrogens). The benzene extract is further washed with water to remove excess alkali, is dehydrated, and is evaporated to dryness. The dried residue containing neutral steroids is dissolved in tetrahydrofuran (THF) and transferred to a centrifuge tube. Following evaporation of THF, the trimethylsilyl ether derivatives of 17-ketosteroids, pregnanediol, and pregnanetriol are formed by reacting the steroids in THF with hexamethyldisilazane and trimethylchlorosilane. The TMSi derivatives are then injected into a gas chromatographic column containing a selective stationary phase such as XE-60. The quantitation of individual steroids is carried out by comparing the ratio of peak area of individual steroids and the internal standard in the sample to that obtained from the known concentrations of the corresponding standard and internal standard.

I. Preparation of the Column

A. Silanizing the Column

Perform all manipulations with organic solvents and silanizing reagents under a hood.

REAGENTS

1. Acetone, AR.
2. Potassium hydroxide, 1 g/100 ml distilled water.
3. Methanol, AR.
4. Toluene, AR.
5. Silanizing reagent: Mix 5 ml of dimethyldichlorosilane (G.E. No. SC-3002) with 95 ml toluene.

Caution: Prepare silanizing reagent under a hood; use a rubber bulb to pipet; avoid introduction of moisture, as poisonous fumes may be liberated.

Reagent Check

All these reagents should be of highest purity, and all organic solvents should be redistilled unless there is evidence that their purity is sufficient. The purity may be checked by evaporating an appropriate amount of solvent (e.g., 50 ml) to approximately 0.5 ml and injecting this solvent onto the column. The presence of peaks would indicate impurities in the reagent.

PROCEDURE

1. With the aid of a syringe or a small funnel, fill and rinse the column with acetone. Remove traces of acetone with a vacuum.
2. Fill the column with 1 per cent KOH and let it stand approximately 5 min. Drain the KOH solution from the column.
3. Rinse the column 3 times with methanol and then once with toluene.
4. Place the silanizing reagent into the column and let it stand for 15 min.
5. Drain the column and then rinse it twice with toluene.
6. Rinse the column quickly with methanol using a vacuum.
7. Evaporate traces of methanol by placing the column into the heated column oven (at approximately 100°C) of the gas chromatograph.
8. Prior to filling the column with support, dry it thoroughly with a stream of nitrogen. This is necessary in case any moisture has condensed in the column during cooling.

B. Sieving the Support

Screen the support to decrease the variation in the sizes of individual particles by placing the solid support (e.g., Gas Chrom-P) into a set of standard sieves and shaking it on a mechanical shaker (U.S. Standard Sieve Series, W. S. Tyler Co., Cleveland, OH.).

To obtain a support of 100–120 mesh, use only the particles which passed through the 100 but not the 120 mesh size sieve.

C. Acid-Washing the Support*

(The support prepared by the following procedure is sufficient to fill two 2 m × 4 mm columns.)

1. Place 20 g of the support (100–120 mesh) into a 1 liter beaker and add approximately 400 ml of concentrated HCl, AR.
2. Let stand for 12 h; stir occasionally.
3. Remove the acid with a filter stick (Corning No. 39533).
4. Wash the support in this manner three more times. Allow 1 h for each of these washings.
5. After the fourth acid wash, pour 750 ml of water into the beaker with the support, stir, and let stand for several minutes. Then decant the water.
6. Wash the support again with water by repeating step 5.
7. Place the support into a Büchner funnel and wash it with water (pH of the wash should be the same as that of the water).
8. Dry the support by suspending it in methanol; decant the methanol along with the fragmented finer particles.
9. Preliminary drying is done at room temperature by spreading the support over the entire surface of a large watch glass, Petri dish, or porcelain dish at least 150 mm in diameter.
10. After the last traces of methanol vapors are gone, complete the drying in an oven at 80°C. Store the support in a desiccator.

D. Silanizing the Support

REAGENTS

1. Mix 5 ml dimethyldichlorosilane (G.E., No. SC-3002) with 95 ml toluene.
2. Toluene, AR.
3. Methanol, AR.

PROCEDURE

1. Place all the support from procedure C into a 500 ml filtering flask.
2. Add a 5 per cent solution of dimethyldichlorosilane to the support. The amount of solution added should be such that there is an adequate volume of liquid ($\frac{1}{2}$ to 1 inch) above the support.
3. Use a vacuum to reduce the pressure in the flask for several minutes, and swirl the flask to free any air bubbles trapped on the support.
4. After air bubbles cease to form, shut off the vacuum and remove the support from the flask by filtration through a Büchner funnel.
5. Wash the support well with toluene and dry it with methanol.
6. Air dry the support and then place it in an oven at 80°C as outlined under procedure C.

E. Coating the Support

1. Place the acid-washed and silanized support into a filtering flask.
2. Add the solution of the liquid phase in the appropriate solvent, in this case 4 to 5 g

* This step, as well as step D, can be omitted if the employed support has already been treated in this manner by the manufacturer.

XE-60 (Applied Science Lab.) in 100 ml acetone, AR. There should be an adequate level of the solution of the liquid phase above the support. *Note:* Many other liquid phases (NGS, Hi-Eff 8B, OV-17, etc.) have been used successfully for the separation of 17-ketosteroids. In our experience, XE-60 has given columns with the highest efficiency.

3. Reduce the pressure with a vacuum and swirl the flask to remove air bubbles from the surface of the support; disconnect the vacuum, and allow the suspension to stand for 5 min.

4. Swirl the flask again and pour the entire contents on a Büchner funnel. Reduce the vacuum and let the solution drain through the funnel until excess moisture is removed.

5. Carry out the preliminary drying at room temperature, and then dry in an oven at 80 to 100°C. Do not place the support into the oven until it is air dried.

F. PACKING THE COLUMN

1. Pour a small amount (4 to 6 inches) of the coated support into a 2 m × 3–4 mm diameter, U-shaped column, using a small funnel.

2. Pack the support firmly by tapping the walls of the column with a rubber-coated, heavy glass rod.

3. Pour 4 to 6 additional inches of support into the column, and pack the column by tapping until it is filled to within 1 inch of both the end and the side arm.

4. Place a glass-wool plug at each end of the column. The glass wool should first be silanized by a procedure identical to that outlined under procedure D.

Packing the Coil Columns. Place a silanized glass-wool plug into the lower opening of the column. Attach this end of the column to a vacuum. Then follow the procedure just described (F, 1–4).

G. CONDITIONING THE COLUMN

1. Place the column into the column oven and connect it to the carrier gas but not to the detector.

2. Heat the column overnight at a temperature of 250°C (for XE-60).

3. The carrier-gas flow should be set at 40 ml/min.

4. At the end of the conditioning period, connect the column to the detector and adjust the temperatures to the operating conditions (column 205–225°C, detector 260°C, injection port 260°C).

II. Acid Hydrolysis and Extraction

This procedure may be used for the determination of 17-ketosteroids and pregnanediol. Pregnanetriol cannot be determined by this procedure.

REAGENTS

1. Benzene, AR, redistilled
2. Tetrahydrofuran (THF), AR. Store over anhydrous Na_2SO_4.
3. Forty per cent (v/v) H_2SO_4, AR.
4. Concentrated H_2SO_4, AR.
5. NaOH, 10 g/100 ml water.
6. Na_2SO_4, AR, anhydrous.
7. Internal standard solution (20 mg/100 ml). Place 40 mg epicoprostanol into a 200 ml volumetric flask and make up the volume to the mark with tetrahydrofuran.

PROCEDURE

1. Mix, measure, and record the volume of the 24 h urine collection.
2. Pipet a 50 ml aliquot into a 100 ml beaker and adjust the pH to 0.8 with 40 per cent H_2SO_4. (It is recommended that the procedure be carried out in duplicate.)

3. Transfer the specimen into a 500 ml round-bottom flask. Add 50 ml of redistilled benzene and a few boiling chips.

4. Attach to a reflux condenser (with a glass fitting), and heat the mixture to boiling using a heating mantle. The heat should be just high enough to allow the urine to boil gently; too intense a temperature will char the organic layer.

5. After 15 min, remove the flask from the mantle and cool the mixture by immersion in an ice bath.

6. Pour the contents into a separatory funnel and drain the aqueous urine layer (bottom layer) into a round-bottom flask. Save the organic layer.

7. Using the urine layer, repeat this procedure (steps 4 to 6) twice; first refluxing at a pH of 0.8 for $\frac{1}{2}$ h, then refluxing at a pH of 0.2 for 1 h. Pool the benzene extracts from steps 6 and 7.

8. After the third refluxing, add 5 ml of concentrated H_2SO_4 to the urine and reflux without benzene for 20 min.

9. Cool, add 50 ml benzene, and shake in a separatory funnel. Discard the urine layer. Add the benzene extract to the previous three extracts (steps 6 and 7).

10. Wash the combined benzene extracts with three 15 ml portions of NaOH solution (10 g/100 ml).

11. Wash the benzene extract with water in a similar manner until the wash water is neutral.

12. Dry the benzene extract by pouring it quantitatively into a round-bottom flask through a glass funnel which has been fitted with a glass-wool plug and 2 g anhydrous Na_2SO_4. Wash the Na_2SO_4 with 10 ml benzene.

13. Add 1 ml of the internal standard solution. (In case of a very concentrated urine, or when high values are expected, use 2 ml of the internal standard.)

14. Evaporate the benzene under vacuum on a rotary evaporator at approximately 60°C.

15. Transfer the residue with two approximately 5 ml portions of THF to a conical, glass-stoppered centrifuge tube.

16. Evaporate the solvent to dryness under a stream of pure dry nitrogen by placing the tube into a water bath at 60°C.

17. Place the tube in a desiccator until the ether derivatives are prepared.

III. Enzymatic Hydrolysis and Extraction

May be used for 17-ketosteroid, pregnanediol, and pregnanetriol analysis.

REAGENTS

1. Glusulase (150,000 units/ml of β-glucuronidase and 75,000 units/ml of sulfatase). (Endo Labs., Inc., Garden City, N.Y.)

2. Benzene, AR, redistilled.

3. Tetrahydrofuran (THF), AR.

4. Glacial acetic acid, AR.

5. NaOH, 1 mol/l. Dissolve 40 g NaOH in water and dilute to 1 liter with distilled water.

6. Na_2SO_4, AR, anhydrous.

PROCEDURE

1. Mix, measure, and record the volume of the 24 h urine collection.

2. Pipet a 50 ml aliquot (duplicates) into a 100 ml beaker. Adjust the pH to 5 by adding 2 g sodium acetate and 0.5 ml glacial acetic acid. Check the pH with a pH meter, and add more glacial acetic acid if needed.

3. Transfer the sample to a 100 ml glass-stoppered flask, add 0.5 ml of Glusulase, and stopper.

4. Incubate at 37°C for at least 36 h.

5. Place the urine into a separatory funnel and extract it three times with 60 ml portions of benzene.

6. Wash the combined extracts twice with 15 ml 1 molar NaOH. Discard the aqueous layer.

7. Wash the extract with 15 ml portions of distilled water until the pH of the wash is the same as the pH of the distilled water. Discard the washes.

8. Proceed as in acid hydrolysis and extraction (II, 12–17).

IV. Preparation of Trimethylsilyl Ether Derivatives

Note: Precautions should be taken to exclude moisture from all equipment and reagents.

REAGENTS

1. Tetrahydrofuran (THF), AR.

2. Prepare a standard mixture by placing into a 200 ml volumetric flask 10 mg each of androsterone, etiocholanolone, dehydroepiandrosterone, pregnanediol, pregnanetriol, 11-ketoandrosterone, Δ^5-pregnenetriol, 11-ketoetiocholanolone, 11β-hydroxyandrosterone, 11β-hydroxyetiocholanolone, and 11-ketopregnanetriol. Dilute to volume with THF.

3. Hexamethyldisilazane.

4. Trimethylchlorosilane.

PROCEDURE

1. Remove the centrifuge tube with the urine extract (II, step 17) from the desiccator, and dissolve the residue in 2 ml of THF.

2. Set up a standard (in duplicate) by pipetting 1 ml of the standard mixture solution and 1 ml of internal standard solution (II, reagent 7) into a glass-stoppered centrifuge tube.

3. Add 0.3 ml of hexamethyldisilazane and 0.1 ml of trimethylchlorosilane to all tubes. Stopper, mix, and place in the desiccator overnight at room temperature. Alternatively, incubate the sample for ½ h at 60°C.

4. After incubation, mix and centrifuge.

5. Transfer the supernatant into another centrifuge tube. (This transfer does not have to be quantitative. Any loss of extract will be compensated by an equal loss of internal standard.) The supernatant may be cloudy at this point.

6. Place the tubes into a water bath at 60°C, and evaporate the solution under nitrogen.

7. Add 1 ml THF, mix, and centrifuge again.

8. Transfer the supernatant to another centrifuge tube. It should be clear (but may be colored) at this point. Repeat step 6.

9. Add 0.2 ml THF to the residue. If a high concentration of 17-ketosteroids, pregnanediol, or pregnanetriol is expected, use larger amounts of THF (e.g., 0.4 ml).

V. Operation of the Gas Chromatograph

Adjust the temperature of the injection port, column, and detector as follows: Injection port 260°C, column 205–225°C, detector 260°C. Adjust the flow of carrier gas (e.g., helium or nitrogen) to 45 to 65 ml/min. Adjust the hydrogen and air flow in accordance with the procedures provided by the manufacturer. For further details, see the manufacturer's instruction manual.

VI. Injection Technique

Many techniques for injection have been recommended, but for this application the following technique appears to provide the most reliable results.

1. Draw approximately 1 μl of THF into a 10 μl Hamilton syringe. Pull the needle out of the solution and withdraw the plunger slightly so as to introduce a very small air bubble.

2. Draw 1 to 1½ μl of sample into the syringe (depending on the expected concentration of the components).

3. Hold the plunger securely in position and place the needle through the septum of the injection port.

4. When the needle is in the proper position, inject the sample with one fast movement, and hold the plunger in position until the solvent front appears on the recorder.

5. Remove the syringe and wash it immediately with THF.

VII. Calculation

Determine the areas (peak height × peak width at ½ peak height) under the peaks corresponding in retention time to authentic steroids and the internal standard, and use the formula as described under the gas chromatographic analysis of estriol.

COMMENTS

The requisites for gas chromatographic analysis of steroids have been discussed before (see general comments on the methods of steroid determination). It should be emphasized here that the preparation of the column has much to do with the efficient separation and quantitation of steroids. The deactivation of active sites on the column tubing and the solid supports through acid-base washing and silanization, the use of a uniform mesh size, and the coating of the stationary phase onto the solid support are essential for the preparation of an efficient column. Solid supports are rather brittle. The support material must be handled as gently as possible at all times in order to minimize fragmentation. The coating of the support with the stationary phase as described in the text is sufficiently gentle. The only disadvantage of this particular method lies in the uncertainty of the final concentration of the stationary phase. Frequently this is of little significance, unless, as occurs with an occasional column, the phase concentration is too low. Another method of coating the solid support consists of evaporating the requisite amount of dissolved stationary phase onto the solid support while constantly stirring the mixture. This generally leads to significant fragmentation. It has been suggested that good results may be obtained by suspending the dry, coated support in absolute ethanol and gently floating the fragments off the top of the liquid.

The treatment of dehydroepiandrosterone sulfate with hot mineral acid leads to artifact formation. For colorimetric estimation involving the Zimmermann reaction, such structural rearrangements may not influence the results drastically as long as the reactive group in ring D is intact (see total 17-ketosteroid estimation). The gas chromatographic separation, on the other hand, requires the structural integrity of the compound, and hence proper precaution is needed to keep the structure intact while the sample is being processed. The mild acid hydrolysis with overlying benzene minimizes the losses caused by artifact formation. The enzyme hydrolysis is very specific and does not cause any undue structural rearrangements, but it is time-consuming. In addition, the presence of varying quantities of β-glucuronidase inhibitors in the urine may lead to incomplete hydrolysis. Several non-enzymic procedures for the apparently complete and safe hydrolysis of these steroid conjugates suitable for gas chromatographic methods may be found elsewhere.[22,102]

The fractionation and determination of individual 17-ketosteroids, pregnanediol, and pregnanetriol in crude urinary extracts most often yield clinically useful results. It should be noted, however, that the detection system in a gas chromatograph is nonspecific, and to achieve precise and specific analysis, adequate and efficient prepurification of the urinary extracts is very important. It is interesting to note in this connection that, even after considerable purification of the urinary extract, the gas chromatographic method yields lower values for etiocholanolone and dehydroepiandrosterone and a higher value for androsterone in comparison to the results obtained by the well-standardized gradient elution chromatographic methods followed by the Zimmermann reaction. The overestimation of androsterone by gas chromatography is attributed to the presence of epiandrosterone.

At present, no satisfactory explanations are known for the low yield of etiocholanolone and dehydroepiandrosterone. The precautions suggested in the text for the preparation of trimethylsilyl ether derivatives should be carefully followed. The use of tetrahydrofuran as the solvent may cause occasional problems if it is not freshly distilled. Other solvents, such as chloroform, hexane, or pyridine, may be used instead. Various commercially available silylation reagent mixtures (Tri-sil, BSA—Pierce Chemical Co., Rockford, IL.) seem to be conveniently suitable in a routine laboratory.

NORMAL VALUES DETERMINED BY GAS CHROMATOGRAPHY

	Males (mg/d)	*Females* (mg/d)
1. Pregnanediol	0.2–1.2	0.2–6.0
2. Androsterone	2.0–5.0	0.5–3.0
3. Etiocholanolone	1.4–5.0	0.8–4.0
4. Dehydroepiandrosterone	0.2–2.0	0.2–1.8
5. Epicoprostanol (internal standard)		
6. Pregnanetriol	0.5–2.0	0.5–2.0
7. Δ^5-Pregnenetriol		0.2–0.9
8. 11-Ketoandrosterone	0.2–1.0	0.2–0.8
9. 11-Ketoetiocholanolone	0.2–1.0	0.2–0.8
10. 11β-Hydroxyandrosterone	0.1–0.8	0–0.5
11. 11β-Hydroxyetiocholanolone	0.2–0.6	0.1–1.1
12. 11-Ketopregnanetriol		0–0.3
13. Total 17-ketosteroids	5.0–12.0	3.0–10.0

Note: Compounds are listed in the order in which they come off the XE-60 column. Blank spaces indicate that normal values have not yet been established.

REFERENCES

Vestergaard, P., and Clausen, B.: Acta Endocrinol, Suppl. 64, 1962.

Creech, B. G.: J. Gas Chromat. *2*:195, 1964.

Wotiz, H. H., and Clark, S. J.: Gas Chromatography in the Analysis of Steroid Hormones. New York, Plenum Press, 1966.

Eik-Nes, K. B., and Horning, E. C. (Eds.): Gas Phase Chromatography of Steroids. New York, Springer-Verlag, 1968.

Tietz, N. W.: Gas Chromatographic Separation and Determination of 17-Ketosteroids, Pregnanediol, and Pregnanetriol. Workshop Manual: "Gas Chromatography in Clinical Chemistry." The Chicago Medical School and Mount Sinai Hospital Medical Center, Chicago, Illinois. November 13–16, 1967.

MEASUREMENT OF PLASMA TESTOSTERONE

CLINICAL SIGNIFICANCE

Essentially all testosterone in the blood of males is secreted by the testes, and its level is regulated by the interstitial cell–stimulating hormone (ICSH, see gonadotrophins). Consequently, the plasma testosterone level in the male is usually a good index of Leydig cell function and correlates well with hypo- and hypergonadism. In females, testosterone and relatively nonandrogenic prehormones such as androstenedione and dehydroepiandrosterone (DHEA) are secreted into the blood by the adrenals and ovaries. As a result of metabolic processes in liver and extrahepatic tissues, a significant fraction of these precursors is converted to testosterone, which re-enters the blood. The origin of blood testosterone in women (65 per cent from androstenedione, 15 per cent from DHEA, 20 per cent from ovarian secretion) is thus not under strict control of the pituitary and, consequently, the concentration of testosterone in blood of females is not always a good index of the testosterone production rate. (*Production rate* is defined as the rate of entry into the blood of a hormone from *all* sources, while *secretion rate* denotes the rate of entry of the hormone from a specific endocrine gland.) The evaluation of androgen production in females warrants special consideration. In pathological conditions involving hirsutism and virilism, both the adrenals and ovaries may be the source of the excess androgens. Measurements of plasma testosterone before and after selective suppression or stimulation of these organs may aid in the elucidation of the sources of the androgens. Thus by administering exogenous glucocorticoids (e.g., dexamethasone), the pituitary-adrenal axis can be suppressed, and, by serial measurements of plasma testosterone, the adrenal contribution can be evaluated. Similarly, to delineate ovarian participation, estrogens alone or in combination with progestins can be administered to suppress the pituitary-ovarian axis, followed by analysis of androgenic components in blood. Stimulation of the ovaries by human chorionic gonadotrophin (HCG) while the adrenals are simultaneously suppressed can also be employed to establish which organ is responsible for the disorder. The clinical usefulness of urinary testosterone estimation in women is limited by the fact that testosterone glucuronide (biologically inactive), which is a major excretory product, may be formed directly from androstenedione and dehydroepiandrosterone in the liver. Therefore, by measuring urinary testosterone, one may seriously overestimate actual testosterone production.

In certain patients with virilism and idiopathic hirsutism, the plasma testosterone levels are within normal limits while the production rate of testosterone has always been shown to be elevated. It has therefore been suggested that the measurement of the production rate rather than the plasma level has more diagnostic usefulness.[4] As previously mentioned, in females testosterone is not maintained at a certain plasma level by an autoregulatory control as it is in men via the pituitary–Leydig cell axis. Therefore, the plasma concentration of testosterone can fluctuate not only with alterations of the production rate, but also with changes in the *metabolic clearance rate* (MCR). The *MCR* is defined as the volume of plasma cleared irreversibly of steroid per unit time, and is determined by the rate of disappearance of radioactivity from the blood following a single injection or a constant infusion of radioactive steroid. *The production rate* is equal to the MCR multiplied by the nonisotopic plasma steroid concentration. A variety of drugs and diseases can alter the testosterone MCR. (*Clearance is increased* in hypothyroidism and virilization; by androgen, medroxyprogesterone, and dexamethasone treatment; and by adrenal, ovarian, and testicular tumors. *Clearance is decreased* in hyperthyroidism and aging; by estrogen and barbiturate treatment; and by erect posture.) The MCR is inversely related to the plasma level, i.e., if the MCR increases, the plasma level decreases. The clearance of a given steroid is dependent upon its interaction with plasma protein and upon the ability of hepatic and extrahepatic tissues to extract and metabolize the steroid. Under physiologic conditions, 66 per cent of the unconjugated testosterone is bound with high affinity to a β-globulin which also binds estradiol (TeBG-testosterone estradiol–binding globulin, or HSBG-human sex steroid–binding globulin), one per cent is bound to CBG, 31 per cent is bound to albumin, and 2 per cent is free. The last fraction is generally considered to be available for physiological action. When TeBG increases (for example, during estrogen treatment), the proportion of bound testosterone is increased, resulting in a decrease of the metabolic clearance rate of testosterone (binding with protein protects testosterone from being metabolized). In this instance, the plasma level of testosterone will show an increase if the production rate does not concomitantly decrease. Similarly, the occurrence of a normal plasma testosterone level in virilized women in spite of increased production rates is explained by an increased metabolic clearance rate due to a decrease in TeBG concentration. The measurement of the production rate of testosterone is the best means of evaluating the androgenic state in females, and the assay of plasma testosterone is a much better diagnostic tool than the determination of urinary total 17-ketosteroids.

Methods

Earlier methods for measuring testosterone by fluorometric, double-isotope derivative and electron-capture gas-liquid chromatographic techniques[92] have been largely replaced by recently introduced competitive protein-binding assays using testosterone-binding globulin as binding protein.[17] The CPB methods differ mainly in the purification of the sample. The most simple and rapid method is the one described by Horton *et al.*,[32] which uses only extraction and washing of the extract for purification. It is an insensitive method and suitable only for the analysis of the plasma from men. Most procedures with adequate sensitivity for measuring testosterone in both male and female plasma require one or more chromatographic purifications to remove competing substances from the sample. The method reported by Mayes and Nugent[52] appears to possess great reliability and practicability. A brief outline of the procedure is given below.

Testosterone-binding globulin, TBG (testosterone-estradiol–binding globulin, TeBG), from human pregnancy plasma is used as a source of binding protein for the final assay. The procedure involves the extraction of the alkalinized plasma samples with methylene dichloride, purification by paper and thin-layer chromatography, and final estimation by competitive protein binding. A sufficient amount of radioactive testosterone is added to the plasma sample prior to extraction, so that an aliquot can be removed to check recovery. The remaining radioactivity is adequate to serve as a radio ligand for the binding assay. (Note that the radioactive steroid is premixed with the CBG solution used in the *cortisol* assay.)

The binding of TBG with testosterone is not absolutely specific. Other steroids (e.g., dihydrotestosterone, androstanediol, androstenediol, estradiol-17β) have fairly high affinity for TBG and are separated from testosterone by either paper or thin-layer chromatography. Small alumina columns are used during elution from paper and thin-layer chromatograms. Small particles of silica gel alter testosterone-TBG binding and are removed by the column. Ammonium sulfate is used to separate the bound (precipitate) from the unbound (supernatant) fraction. To compute the results, the per cent of unbound steroid in the sample is compared with a standard curve constructed by plotting the per cent of unbound steroid as a function of the unlabeled testosterone content of the standards. The standard curve is nearly linear for a range of 0 to 2 ng of testosterone.

NORMAL VALUES

Adult females: 40 ($SD \pm 14$) ng/100 ml plasma.
Adult males: 680 ($SD \pm 180$) ng/100 ml plasma.

PROGESTERONE AND ITS METABOLITES

Progesterone is a female sex hormone. This compound, in conjunction with estrogens, regulates the accessory organs during the menstrual cycle. The importance of this hormone also lies in preparing the uterus for the implantation of blastocysts and in maintaining pregnancy. In nonpregnant women, progesterone is secreted mainly by the corpus luteum (a yellow glandular mass in the ovary, formed by an ovarian follicle following the discharge of its ovum), whereas during pregnancy the placenta becomes the major source. It should be recalled that progesterone is also an obligatory precursor for the synthesis of corticosteroids and androgens. Consequently, the adrenals and the testes are also considered minor sources of this steroid hormone.

The structural formula of progesterone, a C_{21}-compound, is shown in Fig. 13–17. Similar to corticosteroids and testosterone, progesterone (pregn-4-ene-3,20-dione) contains a keto group (at C-3) and a double bond at C-4–5 (Δ^4) in ring A, and this structural characteristic is considered to be essential for progestational activity. The 2-carbon side chain

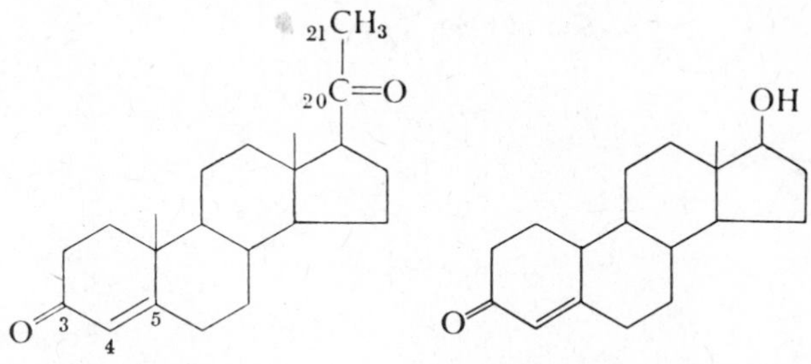

Figure 13–17 Structural formula of progesterone and 19-nortestosterone.

Progesterone (Pregn-4-ene-3,20-dione)

19-Nortestosterone 17β-Hydroxy-19-norandrost-4-en-3-one)

($CH_3—\overset{O}{\overset{\|}{C}}—$) on C-17 does not seem to be very important for its physiological action. Indeed, the synthetic compound 19-nortestosterone (absence of a methyl group at C-10) and its derivatives (Fig. 13–17) are (orally administered) more potent progestational agents than progesterone itself.

BIOGENESIS AND METABOLISM

The biosynthetic pathway of progesterone is believed to involve the same enzymatic sequences from acetate through cholesterol and pregnenolone to progesterone as described in earlier sections (see corticosteroids and general aspects of steroidogenesis). The important metabolic events leading to the inactivation of progesterone are reduction and conjugation. An examination of the chemical structure shows that there are three different sites in the molecule of progesterone which are liable to hydrogenation (reduction). These are the double bonds between the carbon atoms 4 and 5 and the keto groups at C-3 and at C-20. It should be further noted that the reduction of each site may give rise to two isomers differing in the spatial orientation of the hydrogen at C-5 or of the hydroxyl groups at C-3 and C-20. The formation of an α or β isomer is catalyzed by the specific enzymes (e.g., Δ^4-5α- or Δ^4-5β-reductase, 3α or 3β hydroxysteroid dehydrogenase, 20α- or 20β-hydroxysteroid dehydrogenase). The main metabolic pathway is outlined in Fig. 13–18.

The metabolites of progesterone may be classified into three groups based on the degree of reduction. (1) Pregnanediones. The double bond is reduced, producing two compounds: pregnanedione (H at C-5 is in β orientation—above the plane of the paper) and allopregnanedione (H at C-5 is in α orientation—below the plane of the paper). (2) Pregnanolones. The keto group at C-3 is reduced to a hydroxyl group. The hydroxy group may again be either α or β in orientation. However, the urinary metabolites are preponderantly in α configuration. (3) Pregnanediols. Further reduction of the keto group at C-20 gives rise to these metabolites. As in the case of the hydroxyl group at C-3, the metabolites containing the 20α-hydroxyl group are quantitatively most important. In fact, urinary measurement of pregnanediol (5β-pregnane-3α,20α-diol) is used as an index of endogenous production of progesterone because it has been shown that this metabolite is quantitatively most significant and correlates fairly well with a majority of clinical conditions. The reduced metabolites are finally conjugated with glucuronic acid and excreted as water-soluble glucuronides.

MEASUREMENT OF PLASMA PROGESTERONE

CLINICAL SIGNIFICANCE

Plasma progesterone levels remain consistently low (<0.02–0.18 μg/100 ml) during the follicular phase of the menstrual cycle. Progesterone in this phase is believed to be of adrenal

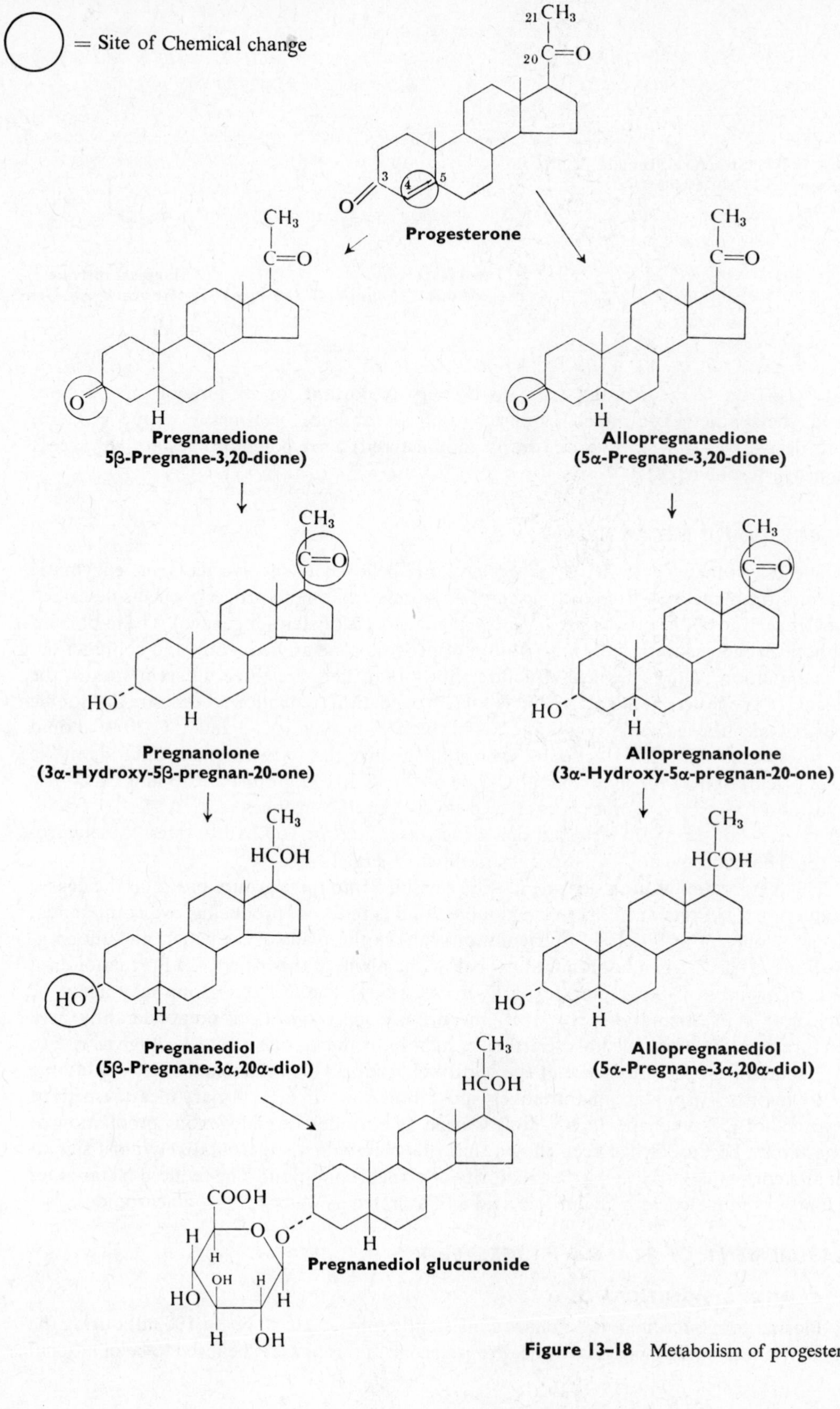

Figure 13–18 Metabolism of progesterone.

origin. Following ovulation, the levels increase rapidly, reaching maximal concentrations (1–2 μg/100 ml) in three to five days. Progesterone concentrations remain elevated for four to six days and then fall abruptly to the initial low levels approximately 24 h before the onset of menstruation. Since the rise and fall of progesterone parallels the activity of the corpus luteum, the clinical importance of measuring plasma progesterone is to confirm ovulation and proper functioning and normal duration (13 to 15 days) of the corpus luteum. If ovulation does not occur, there is no cyclic rise of plasma progesterone. Abnormal progesterone secretion has been implicated in premenstrual tension, irregular shedding of the endometrium, membranous dysmenorrhea, and so-called luteal insufficiency.

In pregnancy, progesterone is largely produced by the placenta. Plasma progesterone levels rise steadily with the progress of pregnancy, reaching a value as high as 20 μg/100 ml at term. Although abnormal excretion values of urinary pregnanediol (progesterone metabolite) in threatened abortions, toxemias of pregnancy, and intrauterine fetal deaths involving the placenta have been reported,[45] systematic studies of plasma progesterone in these conditions have not been made as yet.

There is considerable variation in progesterone levels between individuals, and from day to day in the same individual. Consequently, in gynecological disorders or in abnormal pregnancy, serial rather than single measurements of plasma progesterone are essential for proper interpretation of results.

Methods

The direct measurement of progesterone in plasma is the most reliable approach to assess its rate of production. Several techniques are now available for its estimation: the double-isotope derivative method, gas-liquid chromatography,[26] and most recently competitive protein-binding.[17] The latter technique is the best and most suitable for a routine clinical laboratory. It is extremely sensitive and capable of measuring subnanogram amounts of steroid, requiring less than 1 ml of blood. Relatively large numbers of samples can be measured in one assay, and analysis can be completed in one or two days. Among the numerous procedures described in the literature,[17] the method reported by Johansson[36,37] is the most simple, and it is specific enough for clinical purposes. The method is suitable for both pregnancy and nonpregnancy plasma and can be completed within one day.

MEASUREMENT OF PLASMA PROGESTERONE BY COMPETITIVE PROTEIN-BINDING[36,37]

PRINCIPLE

The competitive protein-binding assay of progesterone is based on the binding affinity of CBG toward this steroid. The principle of the procedure is similar to that described for cortisol measurement in plasma (p. 737). The method consists of a simple petroleum ether extraction of the plasma, followed by competitive protein-binding analysis. Petroleum ether extracts progesterone selectively, leaving important competing steroids such as cortisol, corticosterone, or 11-deoxycortisol in the plasma. Other steroids which are extracted to some extent include 17-hydroxyprogesterone and 20α-dihydroprogesterone. Their concentrations relative to progesterone in the plasma of both pregnant and nonpregnant women are negligible and do not cause any great error in the results. The plasma pool obtained from women on contraceptive pills as a source of CBG and corticosterone-1,2-^{3}H as radioactive ligand are utilized (see Comments on p. 739). Free and protein-bound steroids are separated by adsorbing the free steroids onto Florisil.

REAGENTS

1. Petroleum ether (Mallinckrodt) with a boiling point ranging from 30°–60°C.
2. Florisil (60–100 mesh). Wash four times with glass-distilled water and decant the finer particles. Dry the remaining particles at 120°C overnight.
3. Corticosterone-1,2-^{3}H, specific activity 57.2 Ci/mmol (New England Nuclear Corp., Boston, MA.).
4. Progesterone (Ikapharm, Israel) standard. For preparing standard and radioactive corticosterone solution, the procedures and care are similar to those described before (p. 738).
5. Binding protein (CBG) is the pooled plasma from healthy young women on contraceptive pills. The plasma is stored in 1 ml aliquots at −15°C.
6. CBG-isotope solution. Evaporate corticosterone-1,2-^{3}H, 15 ng (2.5 μCi), in ethanol (0.25 ml) to dryness under nitrogen and dissolve in 100 ml of glass-distilled water. Add 50 μl of the binding plasma protein and mix gently. Allow solution to stand for 1 h at room temperature or overnight at 6°C before use.
7. Liquid scintillation counting is done in 10 ml of scintillation solution containing 100 g of naphthalene, 7.0 g of 2,5-diphenyloxazole, and 0.3 g of 1,4-bis-2(5-phenyloxazolyl)-benzene per 1000 ml dioxane.

COLLECTION OF BLOOD

(See plasma cortisol.) Erythrocytes should be removed from the plasma immediately after the blood has been drawn because erythrocytes convert progesterone to 20α-dihydroprogesterone.

PROCEDURE

Extraction. Transfer 0.25 ml of nonpregnancy plasma (or 0.1 ml pregnancy plasma before 20th week of gestation and 0.05 ml thereafter) into 8 ml extraction tubes with well-fitted glass stoppers. Add 2.5 ml of petroleum ether and shake for 1 min. When the plasma has settled, withdraw the petroleum ether into culture tubes using Pasteur pipets and then evaporate to dryness. (Avoid getting any plasma over with the petroleum ether, as this will contain CBG and interfere with the binding system, resulting in a low progesterone value. The risk can be minimized by centrifuging the samples after shaking and by using Pasteur pipets in which plasma droplets can easily be detected.) Prepare similarly blank tubes in duplicate, using distilled water instead of plasma.

Competitive Protein-Binding Assays. Prepare standard tubes (0, 1, 5, and 10 ng) in duplicate by pipetting 0, 0.1, 0.5, and 1 ml of working standard solution containing 0.01 μg of progesterone per ml ethanol. Evaporate to dryness.

Follow the procedure as described for competitive protein-binding assays of cortisol using Florisil as adsorbent.

CALCULATION

Construct the standard curve by plotting the cpm of bound corticosterone-1,2-^{3}H as a function of the unlabeled progesterone content of the standard. Derive progesterone content in the plasma samples by comparing the cpm of bound steroid with the standard curve.

COMMENTS

The comments on the competitive protein-binding assay of cortisol should be reviewed. A greater specificity can be achieved by the use of progesterone-binding protein (PBP) of pregnant guinea pig plasma. The PBP, unlike CBG, does not bind most other steroids. The specificity of the method described here depends on the selective extraction of progesterone and the protein-binding system. Since the binding system is not absolutely specific for progesterone, the selectivity in the extraction of progesterone by the petroleum ether is very important. As petroleum ether is a mixture of hydrocarbons within a given boiling range, in this case 30–60°C, its ability to extract steroids will vary between distillers and also between batches from the same distiller. According to the author, the batch of petroleum ether to be selected for this progesterone method should extract less than 0.5 per cent corticosteroids (cortisol, corticosterone, or 11-deoxycortisol), no more than 20 per cent of 17-hydroxyprogesterone, and more than 80 per cent of progesterone.

The affinity of Florisil for steroids can be tested by determining the binding ability of a certain amount (e.g., 80 mg) of Florisil for corticosterone-1,2-^3H in a water solution. The test should be performed under the same conditions as the assay. If more than 80 per cent of the steroid is adsorbed, the batch is suitable for use in the assay. A decline of adsorptivity of Florisil may occur owing to hydration. Therefore, following standardization, Florisil should be stored in a screw-cap bottle and its exposure to air for long periods must be avoided.

It should be noted that in men the interference of testosterone makes reliable measurement of plasma progesterone impossible without further purification of the petroleum ether extract.

NORMAL VALUES

Menstrual cycle:

Follicular phase: 50–150 ng/100 ml
Luteal phase: 1000–2000 ng/100 ml

Pregnancy: From the 9th to the 32nd week of gestation, plasma progesterone increases steadily from 16.7 ($SD \pm 7.4$) μg/100 ml to 125 ($SD \pm 38$) μg/100 ml. During the last eight weeks of gestation, the plasma levels of progesterone do not show any significant rise.

REFERENCE

Johansson, E. D. B.: Acta Endocrinol., *61*:592, 607, 1969.

DETERMINATION OF URINARY PREGNANEDIOL

CLINICAL SIGNIFICANCE

Pregnanediol is a major metabolite of progesterone, the hormone of the corpus luteum and the placenta. Thus, the clinical significance of measuring urinary pregnanediol is to indicate ovulation and normal function of the corpus luteum in nonpregnant women and to evaluate the placental function in pregnancy. In children, little or no pregnanediol can be found in the urine and in men it is derived mainly from progesterone and 11-deoxycorticosterone secreted by the adrenals. The excretion values in men never exceed 1 mg/d. In normal menstrual cycles, the excretion pattern of urinary pregnanediol mirrors the secretion of progesterone by the corpus luteum. During the *follicular phase*, the level of urinary pregnanediol is similar to that found in men. During the *luteal phase*, which follows ovulation and corpus luteum formation, the excretion of pregnanediol increases gradually to its maximum (luteal peak) between the 21st and 24th days, depending on the interval of the menstrual cycle. In the absence of ovulation, there is no formation of corpus luteum. Consequently, in anovulatory menstrual cycles, no cyclical rise of urinary pregnanediol is observed.

During the first three months of pregnancy, the amount of urinary pregnanediol is little higher than that found during the luteal phase of the menstrual cycle. As the secretion of progesterone by the placenta increases, there is a steady rise of pregnanediol excretion until about the 32nd week of gestation, when the excretion curves level off. Within 24 h of delivery, excretion values start to drop, reaching nonpregnancy levels in four or five days.

In threatened abortions, toxemias of pregnancy, and intrauterine fetal deaths involving the placenta, values lower than those normally observed at corresponding weeks of gestation have been reported. For detailed information on the excretion of pregnanediol in abnormal pregnancies and other gynecological disorders, the reader is referred to the monograph by Loraine and Bell.[45]

Methods

Until recently, no suitable method for the analysis of progesterone was available. Consequently, the measurement of urinary pregnanediol, which is excreted in milligram amounts, has been used as an index of progesterone secretion. Studies on the recovery of urinary pregnanediol after the administration of progesterone have shown that this metabolite does indeed bear a proportionality to the total amount of circulating progesterone. It is expected, however, that since simple and sensitive methods for measuring plasma progesterone have become available, the determination of urinary pregnanediol will be discontinued wherever possible.

Since pregnanediol is an inactive metabolite, its measurement has to be performed by chemical methods rather than by bioassay techniques.[18,45] A detailed chemical method and the principle of the gas chromatographic method suitable for routine clinical use are described in the following section.

CHEMICAL DETERMINATION OF URINARY PREGNANEDIOL[41]

The main steps in the procedure consist of acid hydrolysis, toluene extraction, permanganate oxidation of the extract, first chromatography on alumina column, acetylation, second chromatography on alumina, color reaction with sulfuric acid, and measurement of the chromogen at 430 nm in a spectrophotometer.

APPARATUS

All glassware should be cleaned thoroughly. Contamination with cork or rubber should be avoided.

1. *Chromatographic column:* glass tubes 12 cm long, 10 mm in diameter with a 75 ml reservoir and a sealed-in sintered disc of porosity No. 3 to support the column of alumina. An alumina column is prepared as follows: Partly fill the chromatographic tube with specified solvent (benzene or petroleum ether) and pour 3 g of deactivated alumina (see below) in a thin stream into the tube. Allow the column to settle, tap to level its surface, and add a protective 5 mm layer of silver sand. Do not let the column dry at any time.
2. Spectrophotometer.

REAGENTS

All reagents should be of AR grade.

1. Toluene, redistilled.
2. Hydrochloric acid, concentrated.
3. Sodium chloride, 25 g/100 ml solution in 1 molar sodium hydroxide.
4. Potassium permanganate, 4 g/100 ml solution in 1 molar sodium hydroxide prepared freshly before use.
5. Ethanol, absolute, purified as described before (p. 728).
6. Deactivated alumina (100/150 mesh): deactivate by exposing alumina for 10 days in layers with occasional mixing to an atmosphere saturated with water vapor at room temperature. The convenient way of doing this is to place the alumina in a Petri dish in a desiccator containing water at the bottom. Store the deactivated alumina in an air-tight

container. Before use, each batch should be standardized with authentic standard solution using the solvent systems as described in the text.

7. Benzene redistilled, water-saturated.
8. Silver sand (40/100 mesh), purified by boiling with 30 per cent hydrochloric acid and then washing thoroughly with tap water and then distilled water.
9. Light petroleum ether (b.p. 40 to 60°C) redistilled, water-saturated.
10. Acetyl chloride.
11. Sodium bicarbonate, 8 g/100 ml distilled water.
12. Sodium sulfite, solid.
13. Sulfuric acid, concentrated.
14. Standard pregnanediol diacetate solution (20 mg/100 ml) in ethanol.

PROCEDURE

Collect a 24 h urine specimen, without using a preservative, and store at 4°C.

Acid Hydrolysis, Extraction, and Permanganate Oxidation. Measure one-twentieth of the total volume into a 500 ml round-bottom flask and dilute to 150 ml with distilled water. (This aliquot of urine is adequate for specimens from nonpregnant women. Smaller aliquots, such as 1 or 2 per cent of the total volume, should be used when urine from pregnant women is used.) Add 50 ml of toluene and a few glass beads to prevent bumping. Heat to boiling under a reflux condenser. Add 15 ml concentrated hydrochloric acid through the condenser and continue boiling for 10 min. Cool the flask rapidly under running tap water and transfer its contents to a 500 ml separating funnel. Remove the urine layer and re-extract with an additional 50 ml of toluene. Combine the toluene extracts and discard any urine which has settled out from the emulsion.

Shake the toluene extract with 25 ml of 25 g/100 ml sodium chloride in 1 molar sodium hydroxide. Discard the aqueous layer including the curdy precipitate at the liquid interface. Add 50 ml of freshly prepared 4 g/100 ml potassium permanganate in 1 molar sodium hydroxide, and shake for 10 min. (This step may be carried out conveniently in a wrist-action mechanical shaker.) The toluene extract is transferred to a 250 ml Erlenmeyer flask. To minimize losses, the separating funnel and the flask should be adequately rinsed with fresh toluene. Discard the permanganate layer and wash the toluene layer 4 to 5 times with successive 50 ml quantities of distilled water to remove all permanganate color. Filter the toluene through a Whatman No. 1 filter paper into a 250 ml round-bottom flask, pouring from the separating funnel so as to leave behind a few drops of emulsion. Evaporate the toluene under reduced pressure to a volume of approximatly 10 ml, and cool to room temperature.

First Alumina Chromatography. Prepare a column of deactivated standardized alumina using benzene as the solvent, as described previously. Apply the toluene extract to the column. Rinse the flask with 5 ml of benzene and add this to the column. When all the solvent has percolated through, add to the column 25 ml of 0.8 per cent ethyl alcohol in benzene. Discard the eluate, which contains some pigmented material and relatively nonpolar steroids. Elute with 12 ml of 3 per cent ethanol in benzene. Collect the eluate (which contains all the pregnanediol) into a 6 × 1 inch test tube. Evaporate the solvent to dryness under nitrogen.

Acetylation and Second Chromatography. Dissolve the residue in the tube in 2 ml of benzene, and add 2 ml of acetyl chloride. Loosely stopper the tube, swirl briefly, and let stand at room temperature for approximately 1 h. Add 25 ml of light petroleum ether and transfer to a separating funnel. Wash the solvent mixture once with 50 ml of distilled water, once with 25 ml of 8 g/100 ml sodium bicarbonate solution, and twice with 25 ml distilled water. Carefully run off the last drop of water and then pour the light petroleum ether onto an alumina column prepared as above, but using petroleum ether instead of benzene. When the solution has percolated through, elute the pregnanediol diacetate with 15 ml of benzene into a 6 × 1 inch test tube. Evaporate the solvent to dryness as before, and place the tube for at least an hour in a desiccator over anhydrous calcium chloride.

Colorimetry. Add approximately 10 mg of sodium sulfite to the residue in the test tube and then 10 ml of concentrated sulfuric acid. Stopper the tube, shake well, and place in a water bath at 25°C for 17 h. Prepare a standard tube containing 100 μg of pregnanediol diacetate (0.5 ml of a 20 mg/100 ml standard solution in ethanol). Evaporate the ethanol to dryness, and add to the residue 15 ml of benzene, and evaporate to dryness. Desiccate the tube and develop the color as above. Read the color density of the standard, as well as the unknown, against the sulfuric acid blank at 430 nm.

CALCULATION

If one-twentieth of the 24 h urine specimen was taken,

$$\text{mg of pregnanediol/24 h} = \frac{\text{reading of unknown}}{\text{reading of standard}} \times 20 \text{ (aliquot of urine)} \times 0.1 \text{ (concentration of standard in mg)} \times 0.8 \text{ (ratio of molecular mass of free and acetylated pregnanediol)}$$

The color density of the unknown may also be converted to mg of pregnanediol diacetate by means of a calibration curve constructed by employing standard concentrations which cover the ranges expected for the analyses of the samples. This will also verify the proportionality of color density to the concentration.

COMMENTS

Pregnanediol is excreted in urine almost exclusively as pregnanediol glucuronide. Acid hydrolysis liberates pregnanediol, which can be extracted easily with an organic solvent such as toluene. The exposure of free pregnanediol in hot acid solution increases its destruction. The overlaying toluene facilitates the removal of pregnanediol from acid solution as quickly as it is liberated. Exact timing of hydrolysis is very important if destruction of pregnanediol is to be minimized. A second extraction of the hydrolyzed urine will insure quantitative extraction.

Emulsions are likely to form during extraction with toluene. Washing the extract with sodium chloride in sodium hydroxide solution not only breaks the emulsion but also removes the acidic and phenolic components. The potassium permanganate oxidation step removes some of the steroidal artifacts formed during acid hydrolysis, which otherwise would be eluted from the column along with pregnanediol, producing falsely high sulfuric acid chromogens. In fact, the color reaction with sulfuric acid is very nonspecific. Almost all steroids, and for that matter many organic compounds, are liable to form a yellow color in the presence of concentrated sulfuric acid. Thus, to achieve specificity, the rigorous purification of pregnanediol by two alumina chromatography steps is very important.

The presence of oxidizing agents in commercially available sulfuric acid and impurities from the solvents interfere with the color development. Addition of a reducing agent such as sodium sulfite helps to overcome the untoward effects of oxidizing agents. The effects of solvent impurities are minimized by adding the same amounts of solvents to the standard tubes that are used in the final stage of the method (15 ml of benzene).

REFERENCE

Klopper, A., Michie, E. A., and Brown, J. B.: J. Endocrinol., *12*:209, 1955.

GAS CHROMATOGRAPHIC DETERMINATION OF URINARY PREGNANEDIOL[15,78]

A general discussion of the gas chromatographic analysis of steroids has already been presented (p. 714). The basic steps involved in this technique are as follows: acid hydrolysis, extraction with toluene, washing with alkali and water, addition of cholesterol propionate as internal standard, acetylation with acetic anhydride and pyridine, and injection into a gas chromatograph containing a 3 per cent SE-30 glass column, 2 m × 4 mm (ID). The quantitation is carried out by comparing the ratio of peak height or peak area obtained from the known concentrations of authentic standard and internal standard with that of pregnanediol and internal standard in the sample. A rigorous purification of the urinary extract prior to gas chromatography is not required, since a gas chromatographic column of relatively high efficiency separates pregnanediol diacetate adequately from other peaks arising from urinary contaminants, with the exception of allo-pregnanediol diacetate. However, the amount of the latter compound in the urine is minute, and thus it does not greatly affect the quantitative results. For detailed discussions of accuracy, precision, specificity, sensitivity, and the applicability of this technique to clinical conditions, the reader is referred to the original publications.[15,78,79]

NORMAL VALUES

Normal menstrual cycle:

Proliferative (follicular) phase : 0.10–1.3 mg/d
Luteal phase : 1.2–9.5 mg/d

During pregnancy: In pregnant women there are wide variations in the excretion of pregnanediol. The following is the normal range of excretion at different weeks of gestation:

Weeks	*Range (mg/d)*
10–12	5–15
12–18	5–25
18–24	13–33
24–28	20–42
28–32	27–47

ESTROGENS

Estrogens, like progesterone, are female sex hormones. They are responsible for the development and maintenance of the female sex organs and secondary sex characteristics. They also participate in the regulation of the menstrual cycle and in the maintenance of pregnancy. In women, estrogen is secreted mainly by the ovarian follicles and, during pregnancy, by the placenta. Adrenals and testes are also believed to secrete estrogens, but only minute quantities. Until recently, only three estrogens—estrone, estradiol-17β, and estriol—were known to be produced in the body. Estrone and estradiol-17β, which are interconvertible, were considered to be the secretory products of the ovary, and estriol was considered to be the ultimate end product of both (see Fig. 13–19). While estrone and estradiol-17β are still considered to be the only estrogens secreted by the ovary, their metabolism appears to follow a more complicated biochemical sequence, giving rise to more than a dozen metabolites. In Fig. 13–19 the names and structural formulas of some of the newly discovered estrogens are shown. It should be noted, however, that the three classic estrogens, i.e., estrone, estradiol-17β, and estriol, are quantitatively most important, and the discussion will mainly be confined to these estrogens.

Among the three major estrogens, estradiol-17β is by far the most potent and is considered to be the true ovarian hormone in women. Structurally, estrogens are derivatives

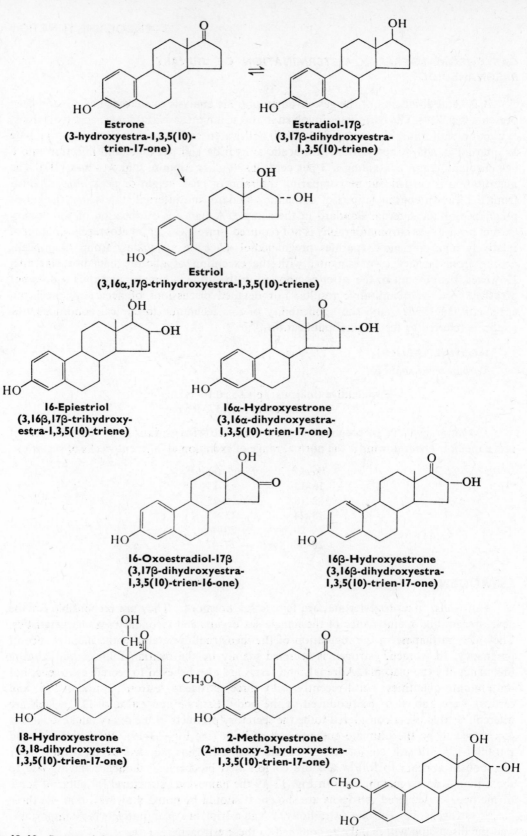

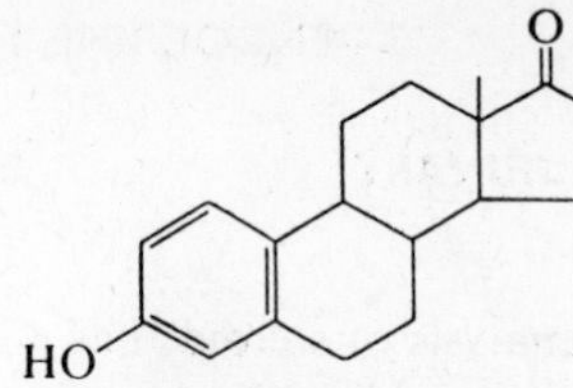

O

OH

HO

HO

Estrone
(3-hydroxyestra-1,3,5(10)-trien-17-one)

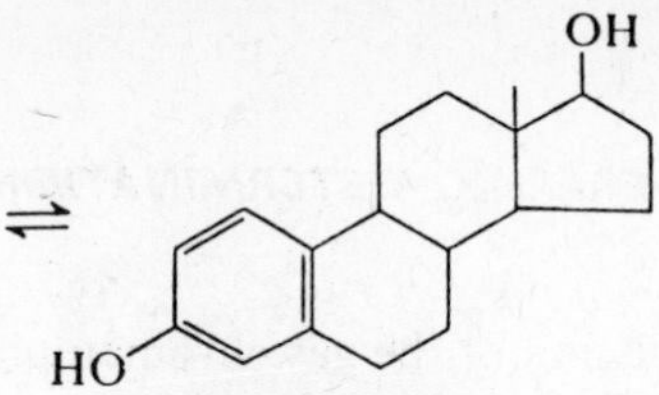

Estradiol-17β
(3,17β-dihydroxyestra-1,3,5(10)-triene)

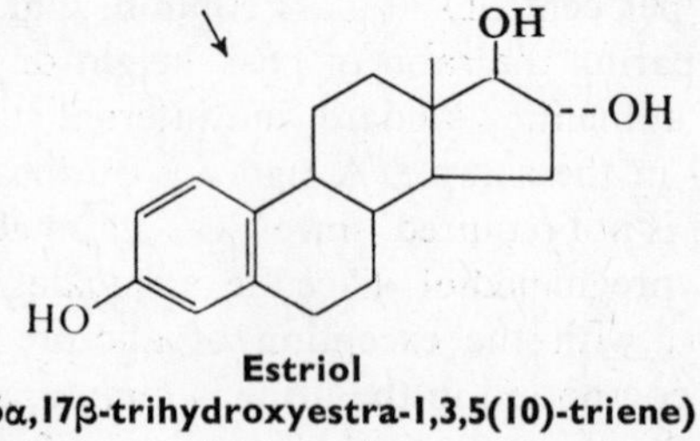

OH
OH
HO

Estriol
(3,16α,17β-trihydroxyestra-1,3,5(10)-triene)

OH
OH
HO

O
OH
HO

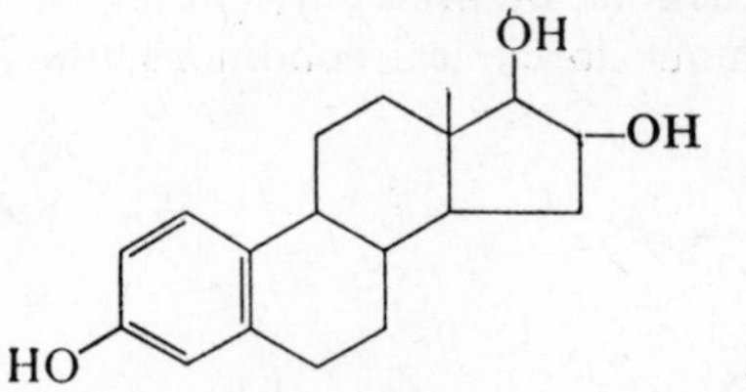

16-Epiestriol
(3,16β,17β-trihydroxy-estra-1,3,5(10)-triene)

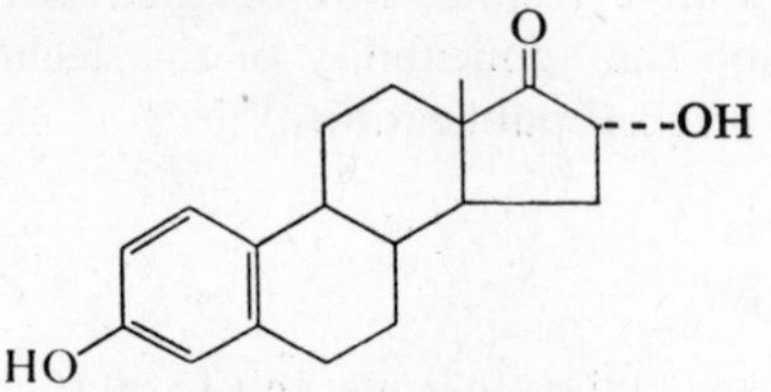

16α-Hydroxyestrone
(3,16α-dihydroxyestra-1,3,5(10)-trien-17-one)

OH
O
HO

O
OH
HO

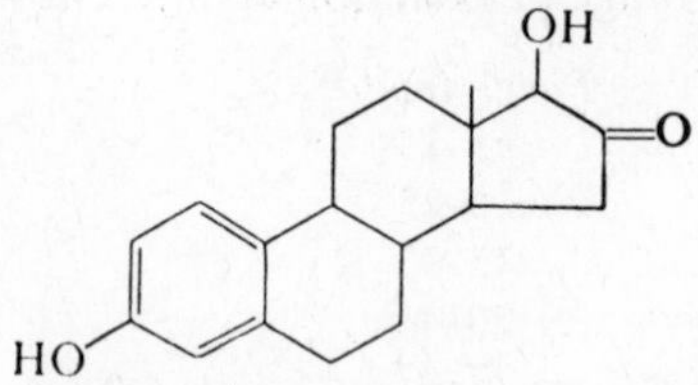

16-Oxoestradiol-17β
(3,17β-dihydroxyestra-1,3,5(10)-trien-16-one)

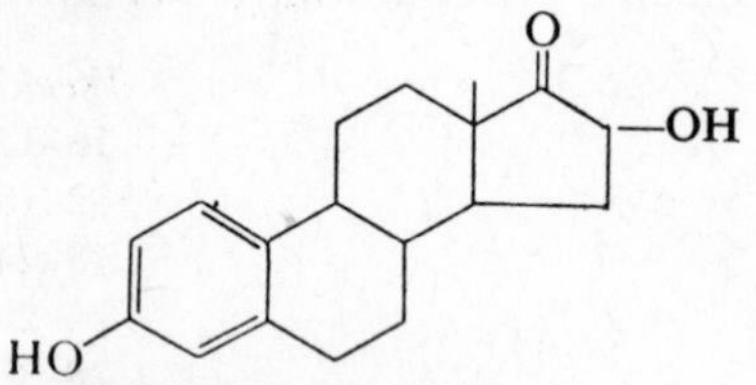

16β-Hydroxyestrone
(3,16β-dihydroxyestra-1,3,5(10)-trien-17-one)

OH
CH_2 O
HO

O
CH_3O
HO

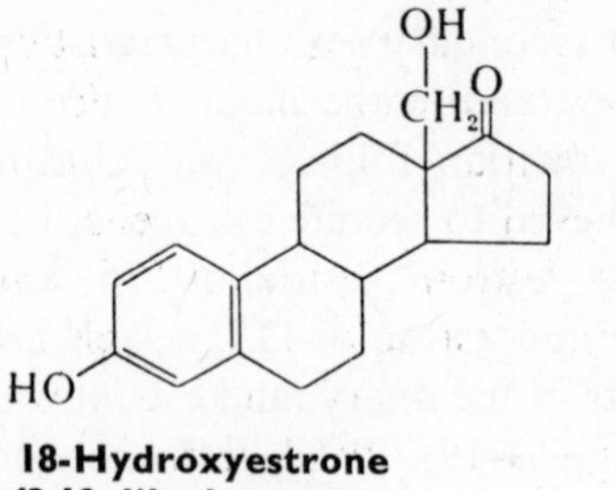

18-Hydroxyestrone
(3,18-dihydroxyestra-1,3,5(10)-trien-17-one)

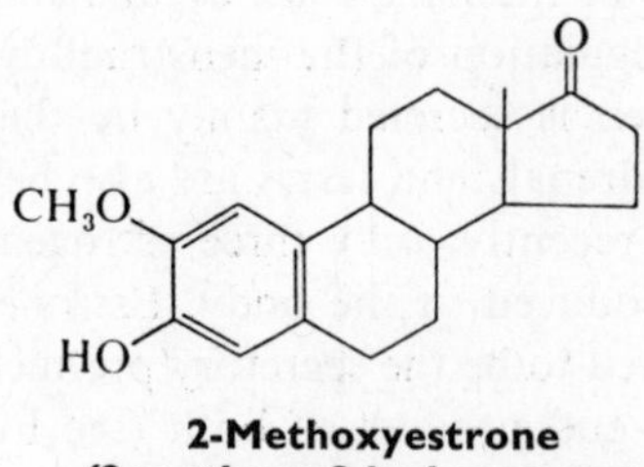

2-Methoxyestrone
(2-methoxy-3-hydroxyestra-1,3,5(10)-trien-17-one)

OH
OH
CH_3O
HO

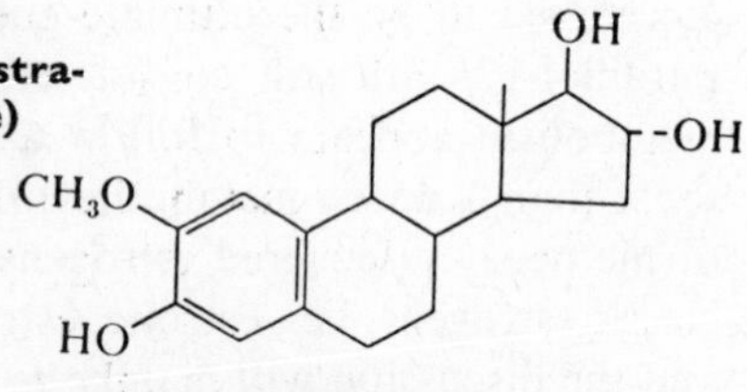

2-Methoxyestriol
(2-methoxy-3,16α,17β-trihydroxyestra-1,3,5(10)-triene)

Figure 13–19 Structural formula of the three classic estrogens (estrone, estradiol-17β, and estriol) and some of the newly discovered estrogens.

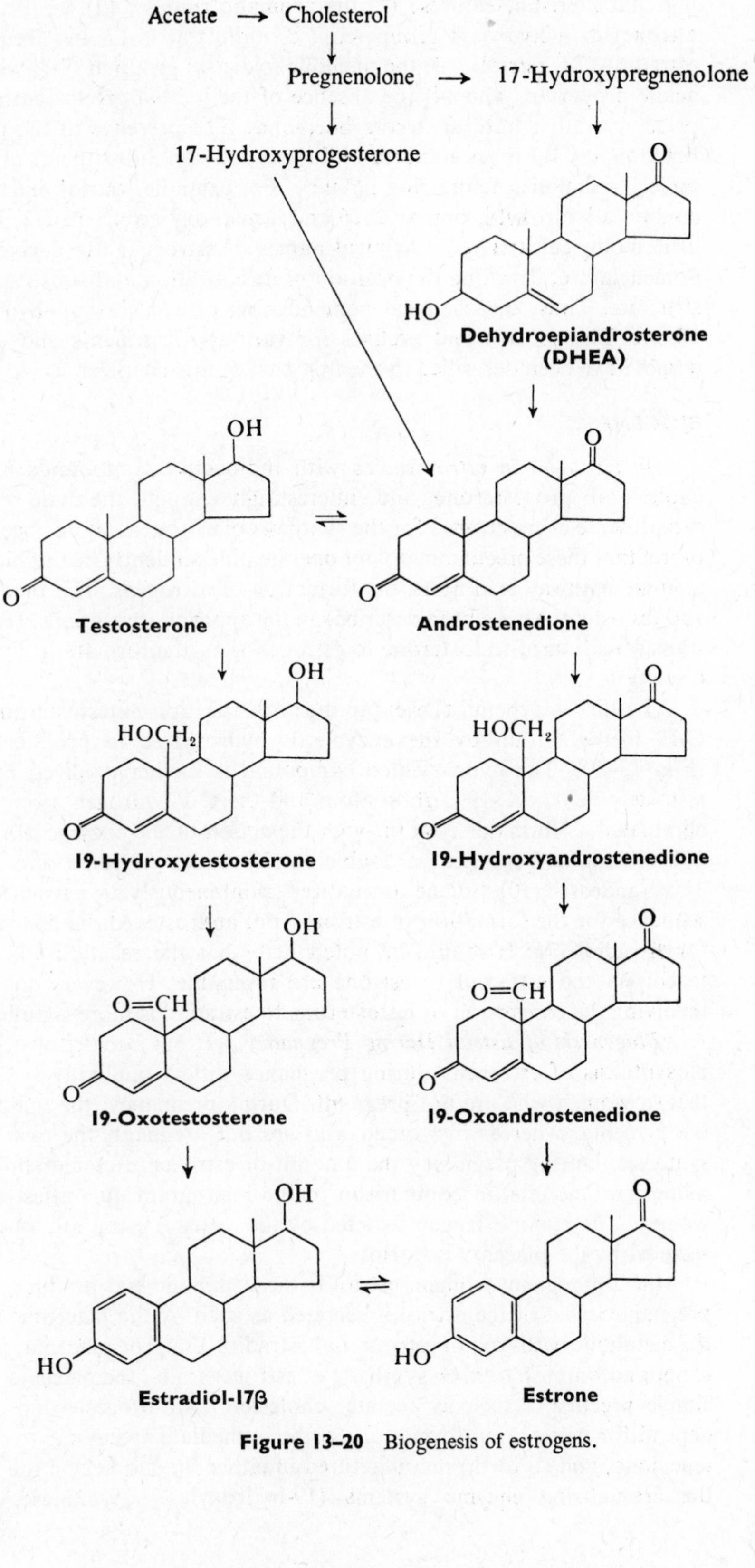

Acetate → Cholesterol

↓

Pregnenolone → 17-Hydroxypregnenolone

↓ ↓

17-Hydroxyprogesterone

O

HO

Dehydroepiandrosterone (DHEA)

↓

OH O

O ⇌ O

Testosterone **Androstenedione**

↓ OH ↓ O

$HOCH_2$ $HOCH_2$

O O

19-Hydroxytestosterone **19-Hydroxyandrostenedione**

↓ OH ↓ O

O=CH O=CH

O O

19-Oxotestosterone **19-Oxoandrostenedione**

↓ OH ↓ O

⇌

HO HO

Estradiol-17β **Estrone**

Figure 13–20 Biogenesis of estrogens.

of the parent hydrocarbon estrane. They consist of 18 carbon atoms and possess the following characteristic features: (1) the aromatic ring A; (2) the presence of a ketone (e.g., estrone) or a hydroxyl group (e.g., estradiol) at C-17 and frequently at C-16 (16-keto-estradiol-17β, estriol); (3) the phenolic hydroxyl group at C-3, which gives the compounds acidic properties; and (4) the absence of the methyl group (carbon 19) at C-10, which is present in other natural steroid hormones. The presence of the phenolic ring and oxygen function at C-17 is essential for biological activity. Substituents at any other position in the molecule diminish feminizing potency. For example, estriol and 2-methoxyestrone, which contain a hydroxyl group at C-16 and a methoxy group at C-2, respectively, possess very little biological activity. Chemical names of estrogens are derived by means of standard nomenclature, denoting the position of unsaturation and substituents based on the estrane structure. Thus, the chemical nomenclature of estrone is 3-hydroxyestra-1,3,5(10)-trien-17-one. The suffixes and prefixes for various substituents and the systems for chemical names have been described in the first part of this chapter.

BIOGENESIS

In vivo and *in vitro* studies with radioactive compounds have shown that acetate, cholesterol, progesterone, and, interestingly enough, the male sex hormone testosterone, can all serve as precursors for the synthesis of estrogens. It has been demonstrated, furthermore, that these precursors do not operate independently in the biosynthesis but constitute a single pathway leading to the formation of estrogens. The biosynthesis of progesterone and testosterone have been described in the appropriate sections. Here, only the biochemical transformation of testosterone to estrogen (i.e., the formation of aromatic ring A) will be discussed.

The first biochemical event in the aromatization of testosterone is hydroxylation of the C-19 methyl group by the enzyme 19-hydroxylase to produce 19-hydroxytestosterone (Fig. 13–20). The hydroxylated compound is further oxidized by 19-oxidase to 19-oxo-testosterone. The C-19 carbon atom and the C-1 hydrogen atom of this intermediate are eliminated as formaldehyde through the action of the enzyme 10,19-desmolase, leaving a C-1(10) double bond (i.e., a double bond between carbon atoms 1 and 10). The resulting 3-oxo-androst-1(10),4-diene aromatizes spontaneously to estradiol-17β. The biochemical sequence for the formation of estrone from androstenedione is the same as that described for estradiol-17β. It should be noted here that the reactions of testosterone to androstenedione and estradiol to estrone are reversible. However, quantitatively the pathway involving the conversion of testosterone to estradiol is more significant in the ovary.

Biogenesis of Estriol During Pregnancy. It has been shown in recent years that the biosynthesis of estrogens during pregnancy differs qualitatively and quantitatively from that in women who are not pregnant. During pregnancy, the major source of estrogens is the placenta, whereas in women who are not pregnant, the ovaries are the main site of synthesis. During pregnancy the amount of estrogen excreted shows a progressive rise to milligram amounts, in comparison to the microgram quantities excreted in nonpregnant women. The major estrogen secreted by the ovary is estradiol, whereas the major product secreted by the placenta is estriol.

In nonpregnant women, estriol is the metabolic end product of estradiol-17β. During pregnancy most of the estriol is secreted as such by the placenta, without involvement of the metabolic pathways of estrone and estradiol-17β. The placenta, as opposed to the ovary, cannot accomplish *de novo* synthesis of estrogens; i.e., the placenta is incapable of utilizing simple precursors such as acetate, cholesterol, or progesterone. Consequently, it must depend for its estrogen formation on the immediate precursors (e.g., testosterone, androstenedione, and so forth) manufactured in either the mother or the fetus. In the placenta, the aromatizing enzyme systems (19-hydroxylase, 19-oxidase, and 10,19-desmolase)

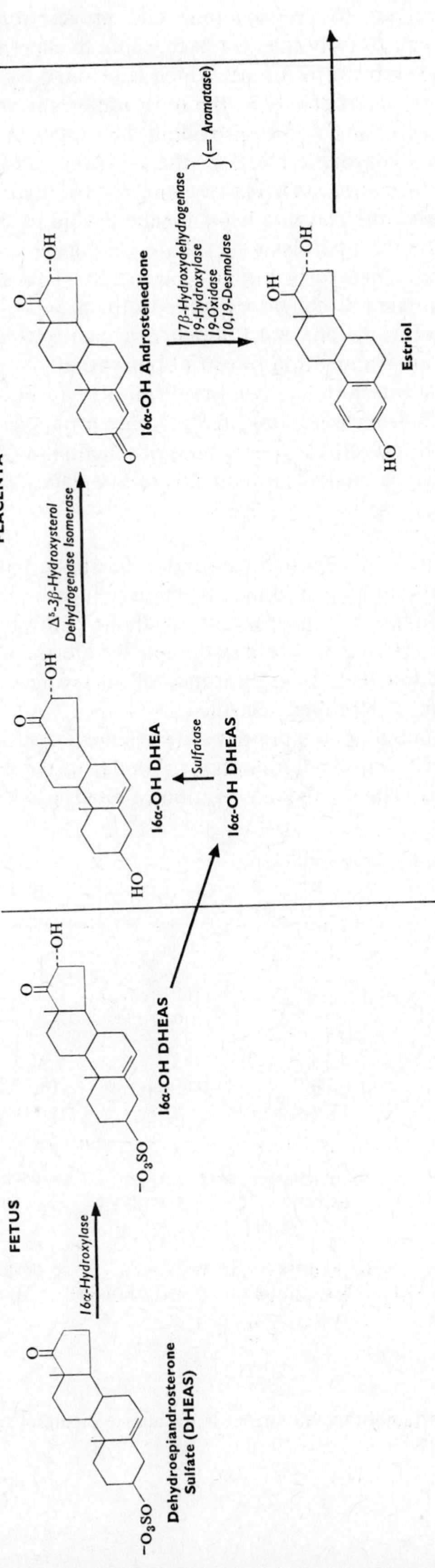

Figure 13–21 Biogenesis of estriol in the fetoplacental unit.

actively convert C_{19} steroids to estrogens. However, other enzyme systems necessary for the conversion of acetate to pregnenolone and progesterone to androgens (C_{19} compounds) are absent or relatively inactive. According to experimental and clinical evidence, the C_{19} steroid dehydroepiandrosterone, which is secreted by the fetal adrenals as the sulfate conjugate, serves quantitatively as the most important precursor for placental estriol synthesis (Fig. 13–21). In addition to providing the supply of the precursor, the fetus also performs an important enzymatic reaction, the α-hydroxylation at C-16, which is a necessary step for estriol formation. Whereas the enzyme 16α-hydroxylase is very active in fetal tissues (liver, adrenals), the placenta is practically devoid of this enzyme. But the enzymes such as sulfatase (for the hydrolysis of sulfate conjugate) and Δ^5-3β-hydroxysteroid dehydrogenase-isomerase (necessary for the conversion of Δ^5-steroid to Δ^4-steroid), whose actions on DHEA-sulfate are obligatory prior to the aromatization, are very active in the placenta. It also contains the enzyme 17β-dehydrogenase, which reduces the ketone group at C-17. Therefore, an examination of estriol biosynthesis in pregnancy reveals that both the fetus and the placenta join in a concerted action. The fetus supplies the precursors and carries out the 16α-hydroxylation, and the placenta aromatizes the C_{19} neutral steroids. A defect in either will be reflected in a decreased production of estriol. For this reason, estriol excretion in pregnancy is used as an indicator of fetoplacental status.

METABOLISM

With the discovery of new estrogens, the metabolic pathway has become far more complex than the hitherto known simple reaction sequence: estradiol ⇄ estrone → estriol. An extensive discussion of the most recent metabolic pathway is beyond the scope of this book. The reader is referred to the monograph by Dorfman and Ungar.[20] It should be noted in this connection that the significance of all the newly discovered metabolites in health and disease has not yet been clarified.

Figure 13–22 summarizes the present state of knowledge concerning the metabolism of estrone and estradiol-17β in the human body. Similar to the metabolism of other steroids, the liver is the primary site for the inactivation of estrogens. The main biological chemical

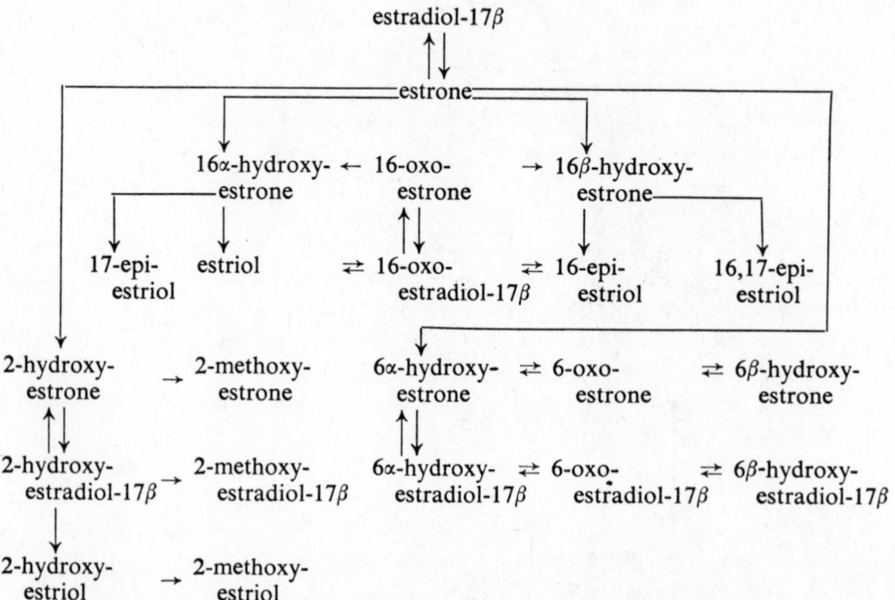

Figure 13–22 Metabolism of estradiol-17β and estrone in the human organism as evidenced by *in vivo* and *in vitro* experiments.

reactions are hydroxylation, dehydrogenation and hydrogenation, methylation, and conjugation.

The following is a brief outline of the salient features of estrogen metabolism. Estradiol-17β is rapidly oxidized to estrone which, in turn, is hydroxylated to form 16α-hydroxyestrone. The subsequent reduction of this compound to estriol (Fig. 13–22) appears to be the most important reaction quantitatively. It is to be noted further that the formation of estriol is achieved exclusively via this pathway; direct hydroxylation of estradiol does not appear to contribute to its formation significantly. It should be recalled, however, that in pregnancy estriol can be synthesized directly from 16α-hydroxylated dehydroepiandrosterone arising from the fetus followed by subsequent aromatization in the placenta (Fig. 13–21). The next important sequence of reactions is the oxidation of estriol to 16-oxo(-keto) estradiol-17β as well as the formation of 16β-hydroxyestrone with the subsequent reduction to 16-epiestriol. There is evidence of the formation of 17-epiestriol (17α-hydroxy group) and 16,17-epiestriol (16β-hydroxy and 17α-hydroxy groups) from their respective precursors (16α-hydroxyestrone and 16β-hydroxyestrone). From a quantitative point of view, these two epimeric estriols are not considered to be significant metabolites. *Conjugation* is the final step in the metabolic processes. Estrogens are conjugated with glucuronic and/or sulfuric acid. It should be recalled that the conjugation imparts more water solubility to these lipids, allowing them to be eliminated rapidly through the kidney.

DETERMINATION OF PLASMA ESTRONE AND ESTRADIOL-17β IN NONPREGNANT WOMEN

CLINICAL SIGNIFICANCE

Estradiol-17β, secreted largely by the ovaries, is biologically the most potent estrogen and is known to be interconvertible with estrone. A significant portion of estrone is also derived from circulating C_{19} neutral steroids (e.g., androstenedione). The major secretory product of the ovary is estradiol-17β, and its measurement is considered sufficient to evaluate the status of ovarian function. During the early part (six or seven days) of the menstrual cycle, the plasma level of estradiol-17β remains virtually constant and then begins to rise slowly, is followed by a rapid increase, and reaches a maximum the day before or the day of the LH surge. It is generally believed that the rise of estradiol-17β levels is the factor which triggers LH release (see gonadotrophins). Following ovulation, there is a drop in estradiol-17β secretion followed by a second rise which corresponds to the formation of the corpus luteum. The levels of estradiol-17β during the luteal phase of the cycle are higher than in the proliferative phase. Estrone, on the other hand, does not show a definite midcycle peak or a consistent pattern throughout the cycle. This has been ascribed in part to factors influencing the extragonadal sources of estrone. Except in pregnancy, measurements of plasma estriol have very little clinical significance, since this metabolite is almost exclusively derived from estradiol-17β. Until recently, methods suitable for routine analyses of plasma estrone and estradiol-17β were not available. As a result, the plasma concentrations and the clinical importance of these estrogens in various gynecological disorders are not yet established. Plasma levels of estrone, estradiol-17β, and the gonadotrophins (FSH and LH) during the normal menstrual cycle are shown in Fig. 13–23.

Methods

The circulating estrogens are present in plasma in the form of free, conjugated, and protein-bound steroid. Most of the plasma methods do not include hydrolysis and, therefore, measure free and protein-bound estrogens. Double-isotope derivative methods[3] or those using gas chromatography combined with electron-capture detection[22] appear to possess the necessary sensitivity and specificity, but the complexity of processing the sample is so

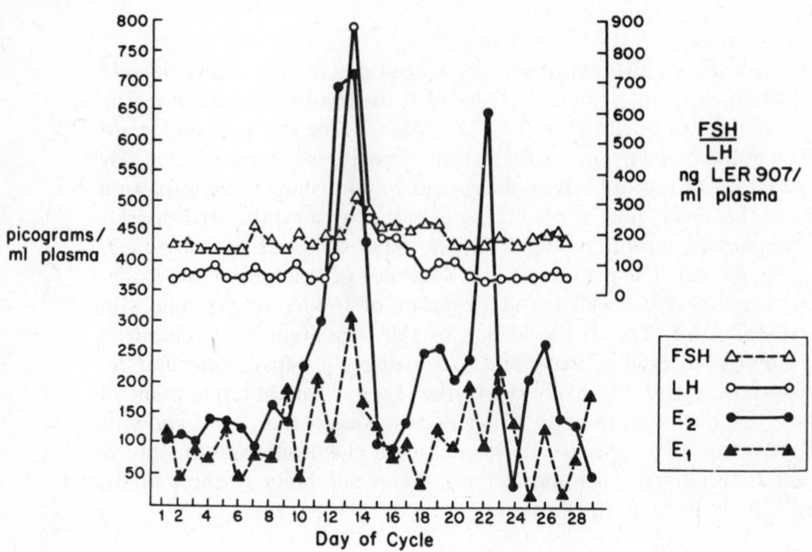

Figure 13–23 Gonadotrophins and estrogens in plasma during the normal menstrual cycle.

great that their usefulness in a clinical laboratory is limited. Recent introduction of the radioimmunoassay to steroid hormone analysis shows great potential, and a number of methods for the analysis of plasma estradiol-17β and estrone have already been described.[64] While these procedures rely on the specific interaction of estrogens with their antibodies for final quantitation, they differ mainly in the mode of purifying samples and of separating the antibody-bound and free steroids. Antibodies against a specified estrogen are as yet not absolutely specific. As a result, most of the procedures require a preliminary isolation of other cross-reacting estrogens. Methods for separating free from antibody-bound hormone include Sephadex column chromatography; precipitation of the bound form by ammonium sulfate or double antibody; adsorption of the free form by dextran-coated charcoal; or the solid-phase system which involves the use of polymerized antibodies, antibodies trapped in polyacrylamide gel, or antibody-coated polystyrene tubes. The radioimmunoassay of plasma estrone and estradiol-17β described by Mikhail *et al.*[56] is a method which has been applied to a comparatively large number of normal and abnormal gynecological cases,[91] and is discussed below.

Antibodies to estradiol-17β are produced by immunization of ewes with estradiol-17β hemisuccinate coupled to bovine serum albumin and are polymerized with ethyl chloroformate before use. Since estrone cross-reacts strongly with the antibodies, the latter are employed for the assay of both estrone and estradiol-17β. In the method, 1 to 5 ml of alkalinized plasma is extracted with diethyl ether. The organic extract is evaporated to dryness, redissolved, and further purified and separated into individual estrogens by a Sephadex LH-20 column. Predetermined amounts of polymerized antibodies and tritiated estrone or estradiol-17β are added to the tubes containing the standards and unknown. Incubation is then performed at room temperature for 2 h, followed by centrifugation in the cold. An aliquot of the supernatant, containing the unbound steroid, is pipetted into counting vials, mixed with scintillation fluid, and counted. The standard curves are plotted as cpm free steroid or per cent bound as a function of nanograms of estrone or estradiol-17β. The concentration of the steroid in samples is determined by comparing the free or per cent bound radioactivity with the standard curve. The daily concentrations of circulating gonadotrophins and estrone and estradiol-17β throughout the menstrual cycle as reported by

Mikhail *et al.*[56] are shown in Fig. 13–23. The values obtained by this method in normal men are

Estrone : 7.4 (*SD* ± 1.3) ng/100 ml
Estradiol-17β : 1.7 (*SD* ± 0.3) ng/100 ml

DETERMINATION OF ESTRIOL IN PREGNANCY PLASMA

CLINICAL SIGNIFICANCE

The use of urinary estriol assays (see p. 781) in the assessment of the fetoplacental function is fraught with many shortcomings. First, urinary collections are time-consuming and there is always uncertainty whether the urine specimen represents a complete 24 h collection. Furthermore, a decreased urinary estriol might be the expression of decreased renal clearance and not necessarily the reflection of decreased estriol production. A more rapid and direct approach to the estimation of placental estriol production is the determination of estriol in plasma. As opposed to 24 h urine collections, plasma can readily be obtained. Moreover, because of the short half-life of estrogens, a decrease in estriol may be in evidence one day earlier in blood than in urine.[61]

Methods

The fluorometric method of Nachtigall *et al.*[61] and the radioimmunoassay described by Gurpide *et al.*[29] appear to be simple, rapid, and adequately sensitive for the measurement of estriol in a minimum volume of plasma.

The main steps in the method of Nachtigall *et al.* are as follows: addition of a mixture of radioactive estriol conjugates (estriol-16-glucosiduronate, estriol-3-sulfate, and estriol-3-sulfate-16-glucosiduronate) of high specific activity to 1 ml of pregnancy plasma, followed by acid hydrolysis, extraction, separation of estriol by partition between benzene-hexane (1/1) and water, reextraction of estriol from the aqueous solution with diethyl ether, evaporation of the organic solvent, and fluorometric measurement of the residue by the Ittrich procedure.[35] Before fluorometry, an aliquot of the sample is removed for radioactivity assay to correct for procedural losses. The presence of glucose up to a concentration of 600 mg/100 ml plasma does not interfere with the test. The estriol levels obtained in 34 cases of normal pregnancy at term showed a range of 9.3 to 56 μg/100 ml plasma with an average of 27 and a median of 25.

The *radioimmunoassay* for estrogens in pregnancy urine, plasma, and amniotic fluid reported by Gurpide *et al.*[29] is based on the reaction of a sheep antiserum against an antigen prepared by linking estriol-16,17-dihemisuccinate to bovine serum albumin. It was observed that the antibody reacted not only with estriol but also with estriol-16-glucosiduronate, estradiol-17-glucosiduronate, estrone, estradiol, and other estrogens. Estrogens conjugated with sulfuric or glucosiduronic acid at the phenolic group did not cross-react. Nonphenolic steroids, such as cortisol, corticosterone, dehydroepiandrosterone, pregnanediol, progesterone, testosterone, and androstenedione, did not appreciably compete with labeled estriol for the antibodies even when 60 ng of these steroids was added to the reaction mixture. Since estriol and its ring D conjugates are quantitatively most significant in pregnancy urine, plasma, and amniotic fluid, the characteristics of this antiserum allowed its use in the measurement of estriol in pregnancy urine, plasma, and amniotic fluid. The general outline for the assay method is as follows: To each tube the following solutions are added successively: estriol standard or sample, diluted with buffer to a volume of 0.2 ml; 0.05 ml tritiated estriol corresponding to 30,000 cpm, and 0.5 ml diluted antiserum solution. Following gentle mixing, the tubes are incubated for a minimum of 10 min in an ice-water bath. An equal volume (0.75 ml) of a saturated solution of ammonium sulfate is then added to each tube to precipitate antibody-bound estriol. After mixing, the tubes are centrifuged and 0.5 ml of the supernatant is transferred to a vial to count radioactivity. The estriol content in the sample is calculated by comparing the per cent of unbound estriol in the test sample with that of a standard curve obtained for known concentrations of estriol. The range over which the method appears readily applicable is found to be 0.2 to 5 ng of estriol per tube. For the determination of plasma estriol, 0.2 ml of pregnancy plasma is treated with 1.8 ml of methanol. Following centrifugation, 0.1 ml and 0.2 ml aliquots of the supernatant are transferred to the assay tubes, dried, dissolved in 0.2 ml of buffer, and assayed for estriol as described above. A similar procedure, involving methanolic extraction, is followed for the measurement of estrogens in amniotic fluid. For urinary estriol, 0.1 and 0.2 ml of pregnancy urine is diluted 1000-fold with the buffer and used directly for the assay.

DETERMINATION OF URINARY ESTROGENS

DETERMINATION OF URINARY ESTROGENS IN NONPREGNANT WOMEN

CLINICAL SIGNIFICANCE

Measurement of estrogens in women during reproductive life is important to evaluate the functional status of the ovaries. Although there is wide individual variation, the excretion of estrogens in normally menstruating women has a definite pattern. During the first seven days of a 28 day cycle, the excretion is generally low. The estrogen level then rises steadily to a well-defined peak on or about the 14th day of the cycle. This peak is called the *ovulatory peak* and is assumed to coincide with ovulation. Following the peak excretion of estrogens, there is a rapid fall, succeeded by another rise around the 21st day of the cycle (*luteal peak*). The excretion value then declines gradually to the lowest level at the onset of the next menstrual cycle. The length of a normal menstrual cycle may vary between 23 and 35 days, and these variations are due to alterations in the duration of the *follicular phase*. Consequently, the ovulatory peak excretion of estrogens may occur as early as four days before, or as late as six days after, the so-called midcycle. Once the ovulation has taken place, the duration of the *luteal phase* is relatively constant from individual to individual, usually being 13 to 15 days. Such constancy is due to the limited life span of the corpus luteum. A luteal phase of less than 12 days is considered abnormal, and one which lasts longer than 16 days is almost diagnostic of pregnancy. The luteal peak excretion of estrogens generally occurs at or around the seventh day after ovulation.

Decreased excretion of estrogens is found in women with amenorrhea due to agenesis of the ovaries, primary ovarian malfunction, dysfunction of the pituitaries, and other metabolic disorders. *Increased* excretion of estrogens is generally associated with ovarian tumors (solid or cystic), hyperovarianism in females, and testicular tumors in males.

In assessing the etiology of amenorrhea, the administration of human gonadotrophins (FSH and LH; *ovarian stimulation test*) is useful. An increase in the excretion of estrogens following treatment indicates a deficiency in the pituitary or hypothalamic-pituitary function. Indeed, infertile women with amenorrhea or anovular menstrual cycles who demonstrate an adequate response to the stimulation test and who desire pregnancy are selected for the induction of ovulation by therapy with human chorionic gonadotrophin (HCG).[24] Stimulation of ovarian function and subsequent induction of ovulation can also be accomplished by nonsteroidal, nonhormonal drugs such as clomiphene citrate.[27] Recent evidence indicates that clomiphene acts on the pituitary-hypothalamic axis with resultant release of gonadotrophins which in turn stimulate ovarian function. The successful use of synthetic or naturally occurring hypothalamic LH- and FSH-releasing hormone (LH-RH/FSH-RH) for the induction of ovulation in women with amenorrhea due to suspected hypothalamic dysfunction has also recently been reported.[74] One prevailing problem of induction of ovulation by exogenous administration of gonadotrophins or clomiphene citrate is hyperstimulation of the ovaries and subsequent multiple pregnancy. Measurements of urinary total estrogens as a guide to such therapy have been suggested as a method of circumventing this complication.[87]

Methods

Studies based on the administration of tracer doses of radioactive estrogens have shown that the urine is the principal route of estrogen excretion; the urinary determination of three estrogens—namely, estrone, estradiol-17β, and estriol, either individually or together—renders adequate information about the endogenous production of estrogenic hormones. In recent years a variety of chemical methods of urinary estrogen determination have appeared in the literature.[45] In most methods the final estimation is carried out by either fluorometry or colorimetry. Estrogens form an orange-yellow color with intense greenish fluorescence when heated with sulfuric acid. Such acid-induced fluorescence is extremely sensitive and permits detection of as little as 0.005 μg of estrogens. However, the nonspecificity of the fluorescence and a number of other variables, such as length of exposure of acid and steroids to elevated temperature, the amount of water added, and the presence or absence of solvents, have always been causes of difficulty in utilizing this kind of reaction for routine analysis. Of all the colorimetric reactions, the Kober reaction is the best known and most accepted for the quantitative determination of estrogens. The specific groups in the steroid molecule required for the development of the pink color in the Kober reaction are the phenolic or phenolic ether groups at C-3 of ring A, and an intact ring D oxygenated at C-17. Thus, the reaction is specific for estrogens. If the specimen, however, contains urinary contaminants, a nonspecific yellow-brown color, produced during the reaction sequence, is superimposed on the pink color caused by estrogens. For this reason, extensive purification of the urinary extract prior to color reaction, in addition to the background color correction, becomes a necessity for reliable measurement of estrogens.

The modification of the Kober reaction by Brown[10] is presently the most widely applied method. This method has the distinction of being the first published procedure that is sensitive and specific enough to measure, individually, estrone, estradiol-17β, and estriol in the urine of nonpregnant women, in whom the excretion values lie in the microgram range. The main steps in this method are acid hydrolysis, extraction, separation into neutral and phenolic fractions, formation of 3-methyl ethers of the three estrogens, purification and separation by alumina chromatography, and final quantitation by the Kober color reaction using the Allen correction. Although Brown's method has been used successfully in many laboratories to obtain valuable information regarding the urinary excretion of estrogens in normal menstrual cycles and in menstrual disorders,[45] a method which is simple, rapid, and adequately sensitive to be suitable for routine clinical use is still to be achieved.

Ittrich[35] has introduced an ingenious method for avoiding the nonspecific portions of chromogens. The products of the second stage in the Kober reaction are diluted with water to an acid concentration of 20 to 30 per cent and are shaken in the cold with chloroform containing 2 per cent *p*-nitrophenol and 1 per cent ethanol. The color resulting from the estrogens is extracted into the chloroform layer, whereas the yellow-brown nonspecific color remains in the acid layer. The novelty of this modification lies in the fact that the color complex in chloroform can be measured either colorimetrically or fluorometrically, since it emits an intense yellowish-green fluorescence when excited with visible light. When it is measured colorimetrically, an amount equivalent to 0.2 μg of estrogens can be determined. Fluorometrically, as little as 0.005 μg of estrogens can be detected. This procedure has been incorporated into various methods for the estimation of estrogens in urine and blood.[45] Indeed, following the Ittrich extraction, the fluorometric measurement of total estrogens in crude urinary extracts without chromatographic purification has been shown to be highly suitable and fairly specific for routine analysis. In the following paragraphs, a method based on such a principle for urinary determination of total estrogens in nonpregnant women is described.

DETERMINATION OF URINARY TOTAL ESTROGENS[12,76]

PRINCIPLE

The following basic steps are involved: acid hydrolysis, extraction with diethyl ether, washing the ether extract with carbonate buffer, separation into phenolic and neutral steroids by partition between sodium hydroxide solution and the organic extract, re-extraction of phenolic steroids with diethyl ether from aqueous solution, development of the Kober color followed by extraction of the color complex with chloroform containing 2 per cent *p*-nitrophenol, and measurement of the fluorescence of the organic phase in a fluorometer using 530 nm as the wavelength for excitation and 550 nm as the wavelength for emission. The amount of estrogens is calculated by comparing the intensity of fluorescence of the sample with that obtained from standard mixtures (estrone, estradiol-17β, and estriol) of known concentration.

REAGENTS

All reagents should be of AR grade.

1. Hydrochloric acid, concentrated.
2. Diethyl ether, freshly distilled in all-glass apparatus.
3. Sodium carbonate buffer, pH 10.5, prepared by mixing 150 ml of 20 per cent (w/v) sodium hydroxide solution with 1 liter of 8 per cent (w/v) sodium bicarbonate.
4. Light petroleum ether (b.p. 40 to 60°C), redistilled in all-glass apparatus.
5. Sodium bicarbonate, 8 per cent (w/v) solution.
6. Sodium hydroxide, 1 mol/l. Dissolve 40 g sodium hydroxide pellets in 1 liter of distilled water.
7. Sodium sulfate, anhydrous.
8. Ethanol, absolute, purified (see corticosteroid estimation).
9. Hydroquinone, recrystallized from ethanol, 4 g/100 ml ethanol.
10. *p*-Nitrophenol, recrystallized from benzene; dissolve 2 g *p*-nitrophenol in 98 ml of chloroform containing 1 per cent (v/v) ethanol.
11. Sulfuric acid, concentrated.
12. Chloroform, freshly distilled in all-glass apparatus.
13. Standard solution: Weigh 8 mg each of estrone and estriol, and 4 mg of estradiol-17β and dissolve in 100 ml absolute ethanol. Dilute 0.1 ml of this stock solution with absolute ethanol to 100 ml in a volumetric flask. (This working standard contains 0.008 μg of estrone and estriol and 0.004 μg estradiol-17β/0.1 ml.) The stock solution should be stored at 4°C.

PROCEDURE

Collect the specimen of urine, and test the urine for glucose as described for the total 17-ketogenic steroids (p. 730). If the glucose concentration is more than 0.5 g/100 ml, follow the procedure as described there, using 1 per cent of the total volume of urine. Dilute the glucose-free residue to 25 ml with distilled water and proceed as follows.

Hydrolysis and Extraction. Transfer 1 per cent of the total volume of urine into a 250 ml round-bottom flask and dilute to 25 ml with distilled water. Add several glass beads (to prevent bumping) and heat to boiling under a reflux condenser. Add 5 ml of concentrated hydrochloric acid through the condenser and continue boiling for 30 min. Cool the flask rapidly under running tap water and transfer the contents to a separatory funnel. Extract the hydrolyzed urine once with 25 ml of ether and twice with half the volume (12.5 ml) of ether. Then shake the combined ether layers with 10 ml of sodium carbonate buffer (pH 10.5) solution. Discard the aqueous layer.

Separation of Phenolic Steroids. Add 50 ml of petroleum ether to the ether extract in the separatory funnel. Extract the organic solvent mixture two times with 25 ml of 1 molar NaOH, collecting the alkali layer in an Erlenmeyer flask. Partly neutralize the alkaline solution by adding solid $NaHCO_3$ in portions until the pH is 10. Transfer the aqueous solution into a separatory funnel. Extract the solution three times with ether—once with an equal volume (50 ml) and twice with half the volume (25 ml). Then shake the combined ether layers with 20 ml of sodium bicarbonate (8 g/100 ml). Discard the aqueous layer. Wash the ether extract with 10 ml of distilled water and drain off the water as completely as possible. Transfer the ether extract to an Erlenmeyer flask containing 5 g anhydrous sodium sulfate. Rinse the separatory funnel with a few ml of fresh ether and add to the flask. Filter the ether extract through a Whatman No. 1 filter paper into a 250 ml round-bottom flask. Evaporate the ether to dryness at 30°C under reduced pressure.

Fluorometry. Dissolve the residue in the flask in 5 ml of absolute ethanol. Transfer 2 ml aliquots to two glass-stoppered tubes. Prepare standard tubes (in duplicate) containing 0.1 ml ($\equiv$ 0.02 μg total steroids; estrone = 0.008 μg, estradiol = 0.004 μg, and estriol = 0.008 μg) of working standard solution and a blank tube containing pure ethanol. Add 0.5 ml of 4 per cent hydroquinone solution to each tube, and evaporate to dryness under nitrogen in a water bath at 50°C. Place the tubes in an ice-water bath. Add 0.4 ml of distilled water and 0.75 ml of concentrated sulfuric acid. Stopper the tubes and heat in a boiling water bath for 40 min with occasional shaking. Place the tubes in an ice bath. Add to the tubes 1.5 ml of distilled water and mix thoroughly. Allow the tubes to stand in ice not less than 5 min and not more than 25 min. Add 2.5 ml of 2 per cent *p*-nitrophenol in chloroform. Stopper the tubes and shake vigorously for 30 s. Centrifuge the tubes for 3 min at 2500 rpm. Aspirate off the upper layer. Transfer the organic phase into the cuvets. Read the fluorescence at 550 nm following excitation at 530 nm. Set the instrument to zero absorbance with the blank, and read the standards and the samples.

CALCULATION

If 1 per cent of the total volume of urine is used, then

$$\mu\text{g of total estrogens/d} = \frac{\text{reading of sample}}{\text{reading of standard}} \times 0.02 \times 5/2 \times 100$$

COMMENTS

Since the bulk of the estrogens are excreted as water-soluble conjugates of glucuronic and sulfuric acids, hydrolysis is necessary to allow extraction of the steroids with an organic solvent such as ether. As a matter of expediency, acid hydrolysis is generally used. However, during acid hydrolysis the presence of excess glucose, hydrochlorothiazide, and a wide variety of formaldehyde-generating drugs (e.g., methenamine mandelate) can seriously decrease the recovery of estrogens, particularly estriol. The interference is also observed in the presence of phenolphthalein, cascara, senna, and diethylstilbestrol. The acid hydrolysis of urine from patients with liver disease results in negative values for estrogens. In the presence of significant amounts of glucose, the destruction may be greater than 50 per cent, and is over 95 per cent in the presence of sucrose, fructose, and inulin. A change of specific gravity of urines from 1.010 to 1.020 also causes a linear decrease from 90 per cent to 60 per cent in the recovery of estriol. In recent years, urinary estriol determinations in pregnant women have been increasingly employed as an index of the fetal well-being. Thus, it is important that changes in estriol values truly reflect changes in excretion and not losses incurred during hydrolysis. Measures suggested to eliminate the effects of interfering substances include dilution of the urine, isolation of estrogen conjugates by ammonium sulfate precipitation, solvent extraction, gel filtration, and extraction of conjugates by neutral polystyrene resin.

Washing the specimen with sodium carbonate buffer (pH 10.5) removes strongly acidic components. As described before, estrogens are slightly acidic in nature because of the presence of a phenolic hydroxyl group at C-3. This acidic property has been utilized for the separation of phenolic estrogens and neutral steroids. Shaking the organic solvent with sodium hydroxide solution moves the estrogens into the alkali layer while the neutral steroids (17-ketosteroids, pregnanediol, corticosteroids, etc.) stay in the organic phase. Petroleum ether is added to achieve better recovery of some of the estrogens (estrone, estradiol-17β) which would otherwise be left in some quantity in the ether layer when partitioned with sodium hydroxide solution. Since estrogens are quite soluble at a pH above 11, the adjustment of the pH to below 10.5 is very important for quantitative re-extraction of estrogens with ether.

The addition of hydroquinone protects the estrogens from oxidation and facilitates the formation of the Kober-color complex. The directions for addition of exact aliquots of different reagents for color development, the duration of heating, and the extraction of the color complex with *p*-nitrophenol solution after the specified length of time should be followed as closely as possible. The tubes should be kept in ice at all times. The intensity of fluorescence decreases with time; thus, the fluorometric reading should be completed within half an hour after extraction of the color complex. Since the relative intensities of fluorescence of the three estrogens—estrone, estradiol-17β, and estriol—are different, determination of the total estrogen content without a separation of the individual estrogens may result in some error. The use of a standard solution of three estrogens mixed in a ratio generally excreted in the urine of a normal nonpregnant woman (estrone, estradiol-17β, estriol, 2/1/2) reduces the possibility of such error.

NORMAL VALUES

The excretion of estrogens in children is very low, being generally less than 1 μg/d. In men, a constant amount of estrogens excreted derives from the adrenals and probably also from the testes. The average value is approximately 11 μg, with a range of from 5 to 18 μg/d.

Normal menstrual cycle: Brown[10] has studied extensively the excretion of estrone, estradiol-17β, and estriol during normal menstrual cycles. The following are the excretion values of total estrogens, computed from his data:

Onset of menstruation:	4–25 μg (mean 13 μg)/d
Ovulation peak:	28–99 μg (mean 56 μg)/d
Luteal peak:	22–105 μg (mean 43 μg)/d
Menopausal women:	1.4–19.6 μg (mean 6.4 μg)/d

REFERENCES

Ittrich, G.: Acta Endocrinol., *35*:34, 1960.
Scholler, R., Leymarie, P., Heron, M., and Jayle, M. F.: Acta Endocrinol., *52* (Suppl. 107), 1966.
Brown, J. B., Macnaughton, C., Smith, M. A., and Smyth, B.: J. Endocrinol., *40*:175, 1968.

DETERMINATION OF URINARY ESTRIOL IN PREGNANCY

CLINICAL SIGNIFICANCE

In view of the intimate involvement of the fetus and placenta in the production of estriol (see biogenesis of estriol in pregnancy), the excretion rate of this hormone is a sensitive indicator of the fetoplacental condition. Chronic fetal metabolic distress or chronic alteration in the fetoplacental exchange can affect estriol synthesis at any of its critical biosynthetic steps, resulting in a decreased maternal urinary estriol excretion and decreased estriol levels in blood and amniotic fluid.[28,40,77] The estriol level during the first trimester is relatively low, and estriol assays during this period have not been used to any great extent for diagnostic purposes. From the clinical point of view, estriol determinations during the second and third trimester of pregnancy, to forecast fetal viability, have been found to be most useful. In pregnancies complicated by toxemia, hypertension, diabetes mellitus, and postmaturity where the fetus may be in jeopardy, estriol determination can be used by obstetricians as an aid in deciding whether to allow a pregnancy to continue or to terminate it and remove the fetus from a hostile intrauterine environment. An isolated single estimation of estriol in a high risk pregnancy and comparison of that value with one found in a normal population has limited clinical importance. Serial measurements to evaluate the trend of estriol excretion are more meaningful. A dramatic fall in estriol levels most frequently indicates fetal jeopardy. Therefore, speed and reliability of the estriol assay are of the utmost importance.

Methods

From the methodological point of view, the estimation of estriol in the second half of pregnancy is not a great problem since relatively large amounts of this steroid are excreted in the urine. A variety of reliable methods involving colorimetry, fluorometry,[18] and gas chromatography[22] for the analysis of estriol in blood, amniotic fluid, and urine have become available. Unfortunately, very few techniques are as fast and reliable as required for the clinical management of obstetrical patients. Recently, chemical procedures have been simplified for routine use,[11,35,88] but some degree of purification is still necessary and any shortcuts in methodology are accompanied by a decrease in reliability. The technique of choice will depend on the availability of the necessary instrumentation. The chemical methods described by Brown *et al.*[11] and by de la Torre and his associates,[88] and the gas chromatographic method reported by Scommegna *et al.*,[78] appear to be suitable for the analysis of estriol in a clinical laboratory. The chemical methods discussed here do not measure specifically estriol but total estrogens. Still, the analyses serve the clinical purpose, because estriol is the predominant estrogen in pregnancy urine.

The method of Brown *et al.*[11] is essentially the same as described earlier for the estimation of total estrogens in urine, except that the final measurement is carried out by colorimetry using the modified Kober reaction.[12] The reliability compares favorably with that of established methods[10] if urine contains at least 1 mg estrogen/24 h specimen.

The method of de la Torre *et al.*[88] involves enzymatic hydrolysis at 55°C for 60 min, ether extraction, solvent partition, and colorimetric estimation by means of a modified Kober reaction. The method is simple, rapid, and reliable when applied to pregnancy urine specimens containing 2.0 μg or more of estriol per ml (about 2 mg/d). Since enzyme hydrolysis is employed, the destruction of estrogens found to occur during acid hydrolysis in the presence of glucose and certain drugs is obviated.

GAS CHROMATOGRAPHIC ANALYSIS OF ESTRIOL[78,100]

PRINCIPLE

The main steps include acid hydrolysis, ether extraction, separation of the neutral and phenolic fractions by NaOH partition, addition of cholesterol as an internal standard to the phenolic fraction, derivative formation (acetate or TMSi) and finally gas-liquid chromatography on a 2 m × 4 mm, 3 per cent SE-30 (or XE-60) column. Quantitation is carried out by comparing the ratio of peak heights obtained for the known concentrations of authentic standard and internal standard with the peak heights of the unknown and internal standard in the sample.

REAGENTS

All reagents should be of AR grade. Reagents 1 to 7 are the same as described for the estimation of total estrogens (p. 778).

8. Pyridine, redistilled in all-glass apparatus under anhydrous conditions.
9. Acetic anhydride, redistilled as above.
10. Estriol standard, 20 mg estriol (gas chromatographic purity) dissolved in 100 ml absolute ethanol (purified).
11. Internal standard, 20 mg cholesterol (gas chromatographic purity) dissolved in 100 ml absolute ethanol.

PROCEDURE

Collection of urine specimen, test for glucose, and removal of glucose from the urine are followed as described in the procedure for determination of 17-ketogenic steroids (p. 730). Use 2.5 per cent of the total volume of urine obtained in the first and second trimesters of pregnancy, and 1 per cent of the total volume obtained in the last trimester.

Hydrolysis and Extraction. The procedure is the same as described for the determination of total estrogens.

If the urine aliquot to be used is more than 25 ml, dilute to 50 ml with distilled water and use double the amount of all reagents described before.

Separation of Phenolic Steroids. Follow the procedure as described for determination of total estrogens, including the step of sodium bicarbonate and water wash. Add 20 μg of cholesterol (0.1 ml of 20 mg/100 ml ethanol) to the ether extract. Transfer the ether extract to an Erlenmeyer flask containing 5 g of anhydrous sodium sulfate. Filter the ether extract through a Whatman No. 1 filter paper into a 250 ml round-bottom flask. Evaporate the ether to dryness at 30°C under reduced pressure. Dissolve the residue in 5 ml of acetone. Transfer the solution with a Pasteur pipet into a 15 ml centrifuge tube. Rinse the flask two more times with 2 ml of fresh acetone and add to the centrifuge tube. Evaporate the acetone to dryness at 50°C under a stream of nitrogen. At this point, prepare a standard tube by adding 20 μg each of estriol and cholesterol (0.1 ml of 20 mg/100 ml standard solution) to a centrifuge tube. Evaporate the ethanol to dryness.

Acetylation. To the dried residue add 0.1 ml of redistilled anhydrous pyridine and 0.5 ml of distilled acetic anhydride. Shake gently to dissolve the contents of the tube. Stopper. Incubate for 1 h at 68°C, or let stand overnight at room temperature. Treat the tube containing the standard and the internal standard in a similar manner. Evaporate the acetylating reagent mixture to dryness at 60°C under a stream of nitrogen. Alternatively, trimethylsilyl ether derivatives can be formed, as described on page 754.

Gas Chromatography. The operating conditions will vary with the column characteristics. The following is a typical description.

Column: 2 m × 4 mm (ID) glass tube packed with 3 per cent SE-30 or OV-1 on 80–100 mesh acid washed and silanized diatomaceous earth.

Temperature: 250°C.
Carrier gas (nitrogen): 40 ml/min.
Detector (flame ionization): 280°C.

Dissolve the contents of the standard tube in 100 μl of acetone. Draw 2 μl of solution into a 10 μl Hamilton syringe and inject into the gas chromatograph. Dissolve the acetylated sample residue in 100 μl of acetone and inject 2 μl as before. Repeat the injection of the standard at the end of the sample run.

CALCULATION

Measure the peak heights of the estriol triacetate and the internal standard in the standard and sample gas chromatograms. Calculate the amount of estriol in the unknown sample according to the following formula:

$$\text{mg of estriol/d} = \frac{R_{ST} \times R_U \times I \times \dfrac{\text{total volume of urine}}{\text{volume of urine used}}}{1000}$$

where

$$R_{ST} = \frac{\text{peak height of 20 } \mu\text{g acetylated cholesterol}}{\text{peak height of 20 } \mu\text{g acetylated estriol}}$$

$$R_U = \frac{\text{peak height of acetylated estriol in sample}}{\text{peak height of internal standard in sample}}$$

$$I = \mu\text{g of internal standard added to sample (20 } \mu\text{g)}$$

COMMENTS

Regarding the acid hydrolysis, extraction, and separation of phenolic steroids, see the method for determination of total estrogens. For relevant discussions on the gas-chromatographic procedure, refer to the sections on *gas chromatography of steroids* (p. 714).

Although cholesterol is present in the urine, the use of exogenous cholesterol as an internal standard is possible since the urinary cholesterol is removed during phenolic partition.

A constancy of peak height ratio of standard and internal standard over a concentration range of analysis should be established for individual columns. The amount of sample to be injected depends largely on the concentration of estriol expected. It has been our experience that injection of more than 5 μl of solution of crude urinary extract causes overlapping of peaks and often obscures minor peaks. It also prevents the detector response from approaching the baseline between peaks, thus preventing readings at low attenuation. (It is generally more desirable to inject small volumes and decrease the attenuation.)

Since the calculation is based on the comparison of known mass (20 μg) of free estriol, the relative change of molecular weight following acetylation is compensated, and thus the need for correction for molecular weight change is unnecessary.

NORMAL VALUES

In normal pregnancy, the values of urinary estriol excretion increase with each week of gestation. The range and the average excretion values (mg/d) from 19 to 42 weeks are given in Table 13–4. Because of individual variation, the range of normal values is very wide.

TABLE 13–4 NORMAL VALUES FOR ESTRIOL EXCRETION DURING PREGNANCY*

Weeks (of gestation)	*Average (mg/d)*	*Range (mg/d)*
19	3.45	1.90–5.00
20	5.00	2.80–7.25
21	5.40	3.00–8.00
22	6.00	3.30–8.60
23	6.40	3.60–9.50
24	7.00	3.80–10.50
25	7.50	4.20–11.25
26	8.20	4.55–12.25
27	9.00	5.00–13.50
28	9.85	5.40–14.50
29	10.50	5.80–16.00
30	11.50	6.30–17.50
31	12.75	6.90–19.00
32	13.50	7.50–20.80
33	14.80	8.15–22.50
34	16.50	8.80–24.60
35	17.50	9.70–27.00
36	19.00	10.50–29.00
37	21.00	11.50–32.00
38	22.80	12.40–35.00
39	24.60	13.50–38.00
40	27.00	14.50–41.20
41	29.00	16.00–44.00
42	32.00	17.25–49.00

*Data from Scommegna, A., and Chattoraj, S. C.: Am. J. Obstet. Gynecol., *99*:1087, 1967.

REFERENCES

Wotiz, H. H., and Chattoraj, S. C.: *In* Gas Chromatography of Steroids in Biological Fluids. M. B. Lipsett, Ed. New York, Plenum Press, 1965, p. 195.

Scommegna, A., and Chattoraj, S. C.: Obstet. Gynecol., *32*:277, 1968.

PROTEIN HORMONES

The hormones secreted by the pituitary, the pancreas, and the parathyroids are all protein in nature. The number and the type of hormones secreted by each gland are enumerated in Table 13–1. These hormones differ from each other not only in their physiological action but also in their chemical structure, ranging from a simple nonapeptide (e.g., oxytocin) to complex protein molecules (e.g., FSH, LH, TSH). Their assay involves associated problems. The hormones of low molecular weight, such as steroids and catecholamines, possess distinctive chemical groups toward which a specific color reaction may be directed as a means of assay (see chemical methods for the determination of 17-ketosteroids, catecholamines, etc.). In the case of protein hormones, no such singularity in chemical composition exists. Consequently, the direct chemical determination of these hormones in biological fluids such as blood and urine cannot be made. In the absence of any chemical assay procedures, determinations of protein hormones are limited to biological and immunological techniques. The basic principles and the prerequisites of these two techniques are discussed briefly in the following paragraphs.

BIOASSAY TECHNIQUES

In this type of procedure a concentrate of the hormone to be measured is injected into suitably prepared animals to elicit a specific physiological response, which is then compared with the response produced by a reference standard preparation. However, until very recently there was no recognized standard preparation available for comparative assay of the protein hormones. The results were therefore expressed in terms of arbitrary animal units. A unit is defined as the amount of a specific hormone required to produce a specific effect. Thus a "mouse uterine unit" is the dose of total gonadotrophins (FSH and LH) necessary to cause a 100 per cent increase in uterine weight of immature mice. Further examples of arbitrary animal units are "rat ovarian unit" for human pituitary gonadotrophin (HPG), "rat ovarian hyperemia unit" for human chorionic gonadotrophin (HCG), and "Junkmann-Schoeller unit" for thyroid-stimulating hormone (TSH).

Because of the nonreproducibility of quantitative bioassay results in different laboratories, it soon became apparent that the degree of response to a minimum dose of hormones differs not only with the animal colony but also between individual animals in the same colony maintained under rigidly controlled environmental conditions. It was, therefore, necessary to establish some kind of international standard or international reference preparation (IRP) with specified biological activity which could be utilized throughout the world for the purpose of comparing bioassay results. In recent years, international standards have become available for the following protein hormones: insulin, TSH, ACTH, prolactin, HCG, posterior pituitary hormones, HPG, human growth hormone (HGH), follicle-stimulating hormone (FSH), and luteinizing hormone (LH). The use of such international standards for the bioassay technique has greatly mitigated ambiguity and confusion of bioassay results.

Since the bioassay technique is based on the physiological response specific for the hormone being measured, this is the most direct means of analysis. To obtain reliable and reproducible results by this procedure, proper precautions and careful design of experiments are very important. The factors causing variation of biological response in experimental animals are numerous. Among them, the strain and the number of animals used for each experiment (four-point assay using two groups of animals for standard and two groups of animals for unknown is generally recommended), the number of injections (the same amount of hormone injected in two doses usually gives a larger response), the time of the day the assay is made (the diurnal variation in response, e.g., of the adrenals to ACTH), and the medium used to administer the hormone are perhaps the most noteworthy.

Other problems exist as well. It has been shown that in biological fluids there are substances other than the test material which may augment, decrease, or abolish the specific response. In addition, the probability of chemical change of the hormone before or after injection is always a cause of uncertainty. It is believed, however, that the establishment of a dose-response curve for the standard preparation and the test material is very important to validate the experiment. A significant deviation from parallelism between the slopes of the dose-response curves of the standard and the unknown preparations makes the results invalid. In fact, the computation of bioassay results in animal units without reference to a recognized standard has doubtful quantitative significance even though, for clinical purposes, such practice is still in vogue (see urinary gonadotrophins). The sources of errors and statistical treatment of the bioassay data have been discussed in a recent monograph.[19]

IMMUNOASSAY TECHNIQUES

The principle of immunoassay of protein hormones depends on the fact that these hormones are immunogenic, i.e., when introduced into an animal they cause the production of specific proteins known as antibodies (antihormones). Antigen and antibody combine in a specific manner to form an antigen-antibody complex, which can be detected and measured by various serological methods, such as agglutination, precipitation, complement fixation, lysis, etc. The hemagglutination inhibition, complement fixation, and radio-immunoassay are by far the most commonly used serological techniques for the qualitative and quantitative determination of the protein hormones in biological fluids. Technical details and authoritative discussion of these assay methods may be found in monographs and review articles.[8,62,81] General principles and characteristics of these immunochemical procedures have also been described in Chapter 7. Of all the immunological techniques, *the radioimmunoassay* (RIA) has found its widest application in the quantitative determination of protein hormones. Compared with bioassays, these methods are more specific and sensitive, and easier to perform. Indeed, the sensitivity of RIA is so high that it has made possible the analyses of almost all protein and polypeptide hormones with antigenic properties without prior extraction of the hormone from the sample. Antibodies or complete commercial kits are now available and sources for these have been listed in a recent review article.[81] Although the procedures involving RIA for different hormones vary in minor detail as to the mode of sample preparation and separation of free and antibody-bound hormone, the basic steps may be described as follows:

1. The optimal dilution (titer) of antibodies which gives a 50 per cent binding of ^{125}I- or ^{131}I-labeled highly purified protein hormone is determined. This dilution is generally considered to render adequate sensitivity and precision in a radioimmunoassay.

2. Varying concentrations of the unlabeled standard preparation or samples are added to the system containing predetermined optimal amounts (step 1) of radiolabeled hormones and antibodies, followed by incubation. During the incubation period, the radiolabeled and unlabeled hormones compete for binding sites on the antibody and reach equilibrium. The duration of incubation varies with the nature of the hormone being assayed. Protein hormones of large molecular size require days to attain equilibrium, while for low molecular mass substances such as steroids, the same can be accomplished within hours. Prolonged incubations may cause damage to the protein hormones because of their exposure to free radicals (caused by radiation), oxidants, and/or proteolytic enzymes present in the assay system. Glucagon, calcitonin, and ACTH are more susceptible to this type of effect than are HGH, TSH, LH, and insulin. The damage can be avoided or minimized by incubating at low temperatures, using low plasma concentrations, or adding proteinase inhibitors such as mercaptoethanol.

3. After equilibration, the antibody-bound and free hormones are separated. A wide variety of methods is available for this purpose. They include adsorption of free hormone to paper (chromatography or chromatoelectrophoresis), ion exchange, activated charcoal, silicate granules, talcum powder; precipitation of the antigen-antibody complex with anti-gamma-globulin (double-antibody), salts, or organic solvents; bonding of antigen or antibody to solid phase (Sephadex, beaded agarose, polystyrene tube); and gel filtration. The suitability of these methods depends on the nature of the hormone and the efficiency of the method for complete and rapid separation of the antibody-bound and free hormones so that the radioactivity of either or both of these fractions can be accurately measured. A separation procedure that permits further association and dissociation of the reacting components will seriously impair the quality of the assay.

4. Depending on the method of separation, the radioactivity of the antibody-bound and/or free hormone is determined. To obtain a standard calibration curve, the ratio of radioactivity of bound (B) and free (F) fractions or the per cent bound is plotted against known concentrations of the standard reference preparation. An illustrative standard curve is shown in Fig. 13–24. More than one dilution of a sample is generally utilized for an assay, so that the concentrations of the unknown fall within the range of the calibration curve.

The selection of the biological fluids used for the assay primarily depends on the lower limits of sensitivity for a specific assay technique. The circulating levels of these hormones are very minute—in the ng or pg/ml range. Accordingly, for bioassays, urine analyses following the concentration and extraction of a 24 h urine collection have been a common prac-

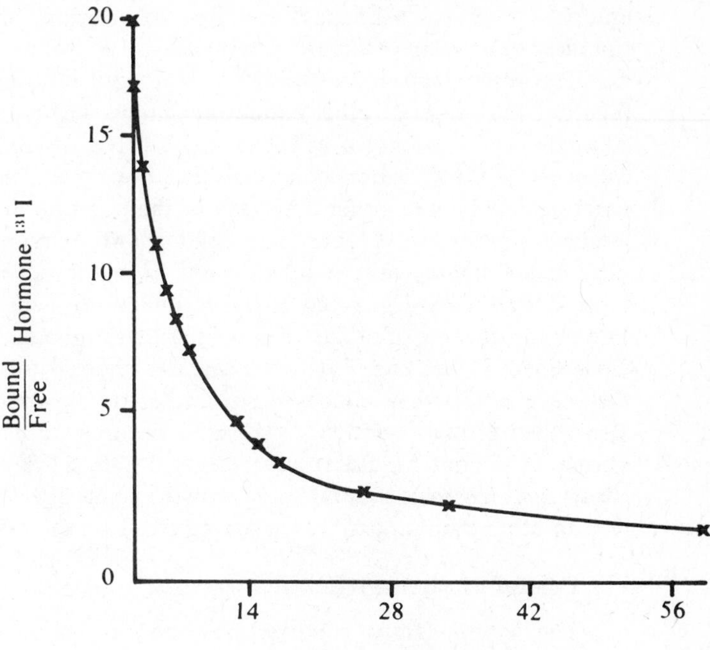

Figure 13–24 Illustrative standard curve for radioimmunoassay. Ratio of bound to free ^{131}I = labeled protein hormone plotted against the concentration of unlabeled standard preparation.

tice. The ease of blood collection, along with the fact that most of the physiologic events causing protein hormone fluctuations result in more pronounced changes in blood than in urine, favors the analysis of these hormones in plasma or serum.

Most of the protein hormones (e.g., insulin, growth hormone, gonadotrophins) can be analyzed in the plasma without preliminary extraction and concentration. These procedures may be necessary, however, if plasma levels of the hormones to be determined are below the sensitivity of the assay available (e.g., ACTH, thyrotrophin) and if the concentrations of degradative enzymes in plasma samples are high enough to cause significant inactivation of the hormone (e.g., angiotensin II, ACTH) *in vitro*. It was generally accepted that the polypeptide hormones do not circulate bound to specific plasma proteins as is the case with other hormones (e.g., steroid hormones, thyroxine), but recently the existence of specific binding proteins for oxytocin and vasopressin (e.g., neurophysin) and for ACTH has been reported. Consequently, for the analyses of these polypeptide hormones, the preliminary treatment of plasma is required. (See radioimmunoassay of steroids, p. 716.)

PITUITARY HORMONES

The pituitary gland (hypophysis) secretes at least nine hormones, all of which are proteins or peptides (see Table 13–1). Anatomically, the gland is divided into two lobes: the anterior pituitary (adenohypophysis) and the posterior pituitary (neurohypophysis). Of the nine hormones, seven are produced in the anterior pituitary (growth hormone, prolactin, TSH, FSH, LH, ACTH, α- and β-MSH). The other two hormones, vasopressin (also known as the antidiuretic hormone, ADH) and oxytocin, are produced in the hypothalamic nuclei (the supraoptic and paraventricular) and carried through the neurohypophyseal nerve axons to the posterior pituitary. The posterior lobe of the pituitary is therefore not in itself a discrete endocrine organ but acts as a reservoir for these two hormones. For a full account of each of the pituitary hormones the reader is referred to textbooks and other monographs.[19,25,99] This section will mainly be confined to the discussion of the anterior pituitary hormones. Detailed description, methods of analysis, and clinical significance of vasopressin and oxytocin may be found elsewhere.[19,25]

The anterior pituitary secretes several trophic hormones which regulate and maintain functional activities of other endocrine glands. The trophic hormones for the thyroid and adrenal cortex are known as *thyrotrophin* (TSH, thyroid-stimulating hormone) and *corticotrophin* (ACTH, adrenocorticotrophic hormone), respectively; those involved in the functional activity of either the ovary or the testes are follicle-stimulating hormone (FSH), luteinizing hormone (LH) or interstitial cell–stimulating hormone (ICSH), and prolactin (also called luteotrophin or luteotrophic hormone, LTH; lactogenic hormone). These last three hormones are grouped together under the generic term *gonadotrophic* hormones. In humans, the role of prolactin in regulating the gonadal function has not been established. Consequently, LH and FSH are generally referred to as gonadotrophins. In addition to these five pituitary hormones, there are somatotrophin (STH, somatotrophic hormone, also called growth hormone, GH) and melanocyte-stimulating hormone (MSH). These hormones do not have a specific target organ. STH exerts a wide variety of metabolic effects and promotes general body growth, while MSH does not have a clear-cut function in man other than a possible darkening effect on the skin.

Control of pituitary function

The hypothalamus manufactures specific polypeptides, known as releasing factors which regulate the secretion of anterior pituitary hormones, i.e., growth hormone, ACTH, TSH, FSH, LH, and prolactin. (Because these substances appear to qualify as hormones, i.e., are transported in the blood stream to produce a specific effect on the activities of a target organ, it has been suggested that they should be termed "releasing hormones." However, to avoid confusion with pituitary hormones, these substances will be referred to as "releasing factors" in this discussion.)

The anterior pituitary is not innervated. The functional connection between the hypothalamus and the anterior pituitary is due to the intimate contact between the hypothalamic nerve endings and the source of the pituitary blood supply. The neurosecretory materials (releasing factors) are released into the primary capillary loops of the hypophyseal portal system and are thus distributed to the anterior pituitary. In this manner, the hypothalamus controls the synthesis and release of the pituitary trophic hormones. Most of the hypothalamic releasing factors are stimulatory in action, including growth hormone–releasing factor (GHRF), luteinizing hormone–releasing factor *and* follicle-stimulating hormone–releasing factor (LH/FSH–RF), thyroid-stimulating hormone–releasing factor (TSH–RF, or thyrotrophin-releasing factor, TRF), and corticotrophin-releasing factor (CRF). The hypothalamic factors for prolactin and MSH are inhibitory in action and are termed prolactin-inhibiting factor (PIF) and melanocyte-inhibiting factor (MIF). In addition, the possible existence of prolactin- and MSH-releasing factors has also been reported. The isolation, elucidation of molecular structure, and chemical synthesis of LH/FSH–RF, TRF and MIF have already been accomplished. The TRF and MIF are tripeptides, whereas the LH/FSH–RF is a decapeptide. Information on the chemistry, assay, and mechanism of action of releasing factors may be obtained elsewhere.[50,73]

The releasing factors, as opposed to some trophic hormones (e.g., gonadotrophins), are not species-specific. The LH/FSH–RF and TRF isolated from the pig are equally effective in the human. The availability of highly purified and synthetic releasing factors has very important *diagnostic and therapeutic usefulness*. Until recently there have been no definitive tests available to distinguish hypothalamic and hypophyseal trophic hormone failure. This differentiation is now facilitated by administering individual releasing factors and measuring the response by the changes of the plasma concentration of the respective trophic hormones. In the case of pituitary dysfunction, the response is minimal. If, however, the defect is in the hypothalamus, the level of trophic hormone increases significantly. The therapeutic use of releasing factors may be indicated in those conditions where an insufficient secretion of anterior pituitary hormones is due to hypothalamic dysfunction. The clinical use of synthetic TRF and LH/FSH–RF has been discussed at length in recent review articles.[73,74]

The functional relationship between the pituitary and the target glands (adrenal cortex, gonads, thyroid) is based on feedback regulation, primarily negative feedback to the hypothalamus and pituitary (Fig. 13–25). As stated earlier, the effect of negative feedback control is characteristically opposite to the initial stimulus. Thus, a diminished level of cortisol in the blood (initial stimulus) triggers the release of hypothalamic CRF to cause increased secretion of ACTH, which in turn stimulates the adrenal cortex to increase secretion of cortisol (final response). Conversely, an elevated level of blood cortisol diminishes the synthesis and release of CRF, resulting in decreased secretion of ACTH and, consequently, cortisol. The importance of such feedback control lies in the delicate maintenance of an optimal concentration of hormones in the blood, which in effect maintains the constancy of other blood constituents and the functions of the target tissues. Although the hormones produced by the adrenals (cortisol) and gonads (estradiol-17β, progesterone, testosterone) primarily interact with the hypothalamic receptors, studies with highly purified and synthetic releasing factors indicate that these steroids may modulate directly the response of pituitary cells to respective releasing factors. For example, it has been shown that progesterone and testosterone suppress the response of gonadotrophin-producing cells to threshold doses of LH/FSH–RF, while small doses of estrogen potentiate the response (*positive feedback*).[74] The rhythmic release of gonadotrophins from the pituitary during the menstrual cycle has been related to the characteristics of the female hypothalamus. In

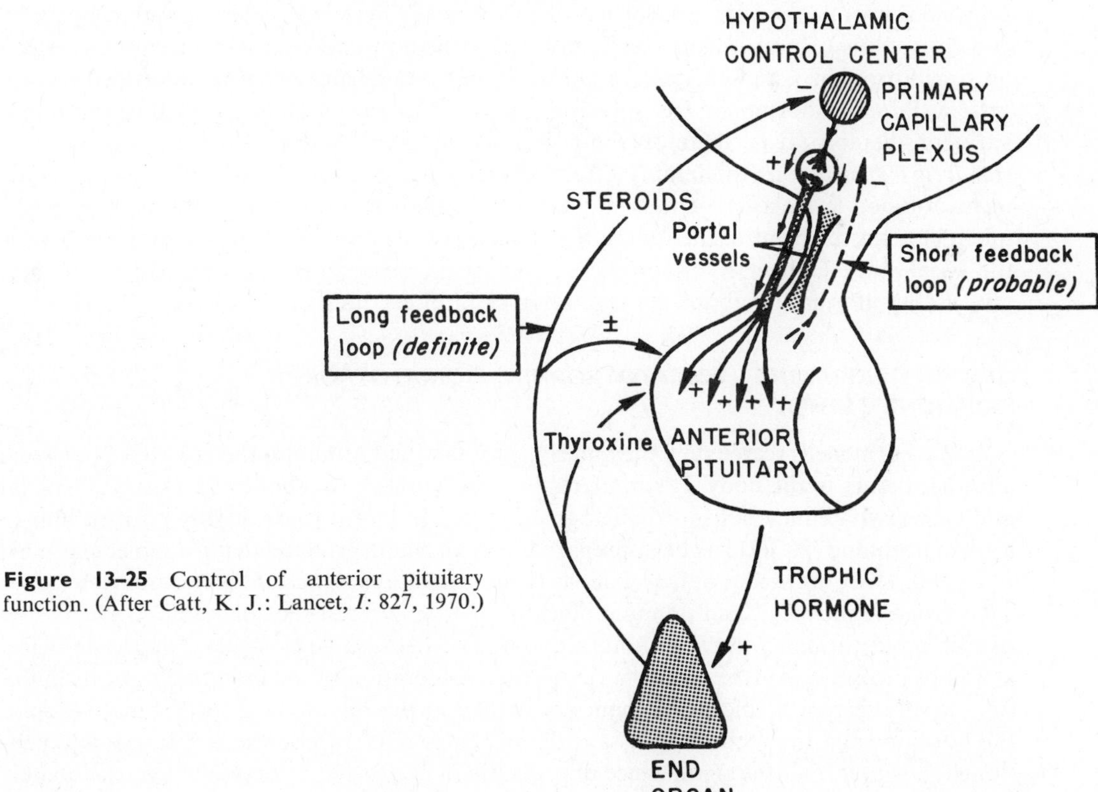

Figure 13–25 Control of anterior pituitary function. (After Catt, K. J.: Lancet, *1:* 827, 1970.)

view of the differential action of sex steroid hormones on the pituitary, such concept may require modification. The precise nature of the control of growth hormone (GH) secretion has not been elucidated, except that glucose inhibits the release of GH via the hypothalamus. In addition to the long feedback loops (see Fig. 13–25), there may exist a short feedback loop by which pituitary hormones can act directly upon the hypothalamus and modify their own secretions, but definite proof is still lacking.

CLINICAL SIGNIFICANCE

The pituitary gland is intimately involved in the maintenance of the metabolic as well as the reproductive function. Consequently, evaluation of pituitary function by determining the levels of trophic hormones is an important diagnostic approach. In clinical medicine, pituitary function has often been assessed indirectly by measuring the hormones secreted by the gonads, thyroid, and adrenal glands. However, this approach is not feasible for those trophic hormones (e.g., growth hormone) which do not have a specific target organ. Growth-hormone secretion cannot be assessed accurately from its metabolic effects. Direct measurement of pituitary trophic hormones is also important in the evaluation of the secretory capacity of the target gland. For example, an increased level of ACTH will indicate a deficiency of the adrenal cortex, whereas a low level will be indicative of the dysfunction of the pituitary gland itself. Indeed, impaired function of the target gland (e.g., thyroid) may be detected by raised levels of trophic hormones (e.g., TSH) even before the production of the target gland has significantly decreased. The measurement of pituitary hormones may also prove to be very useful in assessing the response of the pituitary gland to the administration of hypothalamic releasing factors.[74]

There are many factors besides endocrine disease which can influence the release of pituitary hormones; for example, emotional or physical stress (ACTH, GH), time of the day (diurnal variation, ACTH), time of month and time of life (menstrual cycle and menopause, gonadotrophins), food intake (GH), hormonal therapy (steroidal contraceptives, effects on gonadotrophins), and opiates, tranquilizers, and other drugs affecting the central nervous system. Pituitary hormones are rapidly removed from plasma, and circulating levels may vary considerably. It is, therefore, important to measure these hormones under *carefully standardized conditions*, preferably on more than one occasion. In addition, the measurement of blood levels after various standardized stimuli (e.g., metyrapone test) may give a much better assessment of the functional capacity of the pituitary than the measurement in the basal state. Conversely, suppression tests (e.g., dexamethasone administration) may give useful information about the regulation of pituitary function.

GROWTH HORMONE (SOMATOTROPHIN, SOMATOTROPHIC HORMONE [STH])

The hormone is secreted by the anterior pituitary and promotes the growth of bone and all other tissues in the body. It stimulates protein synthesis (anabolic effect) and affects fat and carbohydrate metabolism (diabetogenic effect). In recent years, highly purified human growth hormone (HGH) has been prepared. It is an acidic protein having a molecular mass of 21,500. There are some controversies as to the separate existence of the hormone *prolactin*. The physical, chemical, and immunological properties suggest that the biological activities of both prolactin and growth hormone are contained in the same molecule. The physiological action of prolactin is involved primarily in the proliferation of the mammary gland, initiation of milk secretion, and maintenance of the rat corpus luteum. For further discussion of the hormone and the method of assay, the reader is referred elsewhere.[45] It may be mentioned, however, that the significance of prolactin in health and disease is not yet adequately known.

The determination of growth hormone in blood and urine has been reviewed by Loraine.[45] It suffices to say that available biological assay methods are not adequately sensitive to detect the presence of this hormone in human blood or urine. The method of choice has become the radiologic procedures which are generally based on the one reported by Hunter and Greenwood.[34] The basic steps involve the labeling of the pure growth hormone (antigen) with ^{131}I, incubation of the optimal amount of labeled antigen with the test sample and with a predetermined amount of anti-HGH (antibody) for seven days, followed by the separation of the free antigen and the antigen-antibody complex by electrophoresis and the determination of radioactivity of the two fractions. The amount of unlabeled antigen (HGH) in the test sample is then calculated by comparison with a standard curve.

Recently, a method using a solid phase system (antibody-coated tube) for the separation of bound and free fractions has been reported.[14] From the practical point of view, this technique may become superior to the previously described methods. The plasma levels of GH(ng/ml) by this method in 21 fasting ambulatory adult males and 35 females were found to be 4.2 (SE $\pm$ 1.2) and 5.1 (SE $\pm$ 1.0) ng/ml, respectively. Plasma GH levels are generally higher in children than in adults. A comparison of levels by different methods in normal and abnormal conditions has been reported by Skelley *et al.*[81] It should be emphasized, however, that the half-life of GH is very short and the plasma GH levels show a wide fluctuation throughout the day and also vary with such factors as exercise, nutritional status, sleep, and stress. Because of these facts, reliable information in regard to the clinical significance of plasma GH assays will only be obtained when serial determinations are made at frequent intervals under controlled conditions. In addition, the level of GH in plasma is markedly influenced by the blood sugar level, the plasma concentration of nonesterified fatty acids (NEFA) and amino nitrogen, and blood levels of other hormones, notably insulin and glucagon. Consequently, for proper clinical interpretation of the GH results, simultaneous measurements of all these parameters may become important. A *decreased* level of GH is found in dwarfism, and an *increased* level is found in gigantism and acromegaly. The absolute amount of GH excreted in the urine and its clinical significance have not yet been established.

THYROID-STIMULATING HORMONE (TSH, THYROTROPHIN, THYROTROPHIC HORMONE)

This is an anterior pituitary hormone which controls the functional activity of the thyroid gland. In recent years it has become evident that TSH is not the only thyroid stimulator. Substances such as long-acting thyroid stimulator (LATS), detected in the serum of thyrotoxic patients, are produced outside the pituitary and mimic the biological action of TSH. The properties of this material have been discussed in detail by McKenzie.[53] Although highly purified preparations of TSH have been obtained, the hormone has not been isolated in pure form. According to the present evidence, it is a glycoprotein with a molecular mass of approximately 25,000, 8 per cent of which is represented by carbohydrates.

A great number of bioassay methods of varying sensitivity, reliability, and ease have been developed for the measurement of TSH. Unfortunately none of them is suitable for routine clinical application. The methods that appear to be most sensitive are radioimmunoassays. A comparison of the data obtained from the bioassay and the radioimmunoassay has recently been reported.[38] While the TSH values by both methods could be related to clinical status, there was little or no correlation between the absolute levels detected by the bioassay and the radioimmunoassay in individual subjects. Since the radioimmunoassay of TSH has become the method of choice, the values obtained by this technique are considered valid. For normal values and clinical significance, the section on thyroid function should be consulted.

ADRENOCORTICOTROPHIC HORMONE (ACTH, CORTICOTROPHIN, ADRENOCORTICOTROPHIN)

This is a trophic hormone which regulates the functional activity of the adrenal cortex. It has also some extra-adrenal metabolic effects in the body. The administration of adrenocorticotrophin to adrenalectomized animals has shown that it mobilizes fat depots, increases the rate of fatty acid oxidation, and enhances muscle-glycogen synthesis, resulting in hypoglycemia. The isolation of ACTH in pure form from the pituitary of several species, including man, has been accomplished. The primary structure of the hormone, including the amino acid sequence, has also been established. ACTH of all species examined so far consists of a single, unbranched polypeptide chain with 39 amino acid residues, having a molecular mass of approximately 4500. Among species, the difference in the sequence of amino acids lies mainly between residues 25 to 33; the biologically active portion of the molecule comprising the amino acid residues from 1 to 25 is identical in all species. The synthesis of biologically active ACTH has also been achieved. For further information the review article by Li[42] is suggested.

The so-called **melanophore-stimulating hormone** (MSH), has a close structural relationship to ACTH. Two general types of MSH (α and β) are generally found. Beta-MSH is comparatively less active than α-MSH. It has been shown that the first 13 amino acids in α-MSH and amino acids 11 to 17 in β-MSH are identical to those found in the same numbered sequence in ACTH. It should be noted that the considerable pigmentation associated with Addison's disease of the adrenals is most likely caused by one or both of these portions of the ACTH molecule. Apart from the possible darkening effect of the skin, the biologic effect of MSH in man has not been completely investigated.

Even though ACTH is available in pure form, chemical procedures for its measurement are not yet available. Accordingly, all assays are confined to biological or immunological techniques. Bioassays for this hormone suffer, however, from the lack of sensitivity necessary to detect its presence in blood and urine. As a result, until recently very little information has been available about the concentration of ACTH in body fluids in health and disease.

Among the *bioassays* employed so far, the *Sayers test*, depending on the depletion of adrenal ascorbic acid, is one of the most sensitive, specific, and widely used methods.[72] The test is conducted in hypophysectomized rats. The procedure is as follows: After approximately 24 hours following hypophysectomy of the rats (the purpose of hypophysectomy is to remove the endogenous source of ACTH), the left adrenal is removed, and immediately an intravenous injection of standard ACTH or the test material is given. One hour later the right adrenal is removed. On the following day the concentration of total ascorbic acid (dehydro and reduced) of both adrenals is determined. The response is evaluated on the basis of the difference between the ascorbic acid content of the right (test) and left (control) adrenal glands. With this technique, as little as 0.2 mIU (milli-International Unit) can be detected. One *international unit* is defined as the activity contained in one milligram of the first International Standard for ACTH.

The *radioimmunoassay* originally reported by Yalow *et al.*[103] has become the method of choice for the analysis of plasma ACTH. Various modifications as to the mode of sample preparation and separation of the antibody-bound and free hormone have been instituted.[23] Because of very low concentrations, and degradation of ACTH by proteolytic enzymes in the plasma, preliminary extraction has been employed for the majority of assay methods. The amount of ACTH excreted in the urine is still uncertain, while reports on the normal levels of ACTH in blood are quite variable. According to a recent report,[23] serum concentration of ACTH in normal subjects at 8 AM is 102 ($SD \pm 26.6$) pg/ml. (Because of diurnal variation, the values are generally lower in the afternoon.) No significant difference in serum ACTH levels was found between hospitalized and ambulatory persons, between the sexes, or at different age levels. In pathological conditions, such as Addison's disease, congenital adrenal hyperplasia, bilateral adrenalectomy, and "stress" conditions, the plasma levels are *high*, whereas in adrenal carcinoma and in panhypopituitarism, the levels are *low*. In induced hypoglycemia (e.g., insulin administration) and in metyrapone treatment, ACTH concentrations are elevated.

GONADOTROPHINS

The pituitary gonadotrophins consist of three hormones: follicle-stimulating hormone (FSH), luteinizing or interstitial cell–stimulating hormone (LH or ICSH), and prolactin (otherwise known as luteotrophic hormone or luteotrophin—LTH, lactogenic hormone, mammotrophin). The latter hormone is elaborated by acidophilic cells, which are also believed to produce growth hormone; the former two hormones are elaborated by the basophilic cells of the pituitary. The function of prolactin and its structural relationship with growth hormone have already been briefly outlined (see growth hormone) and will not be discussed further. Here, the discussion will mainly be confined to FSH and LH, henceforth to be denoted together as HPG (human pituitary gonadotrophin); human chorionic gonadotrophin (HCG) of placental origin; and human chorionic somatomammotrophin (HCS; human placental lactogen—HPL), also of placental origin.

HUMAN PITUITARY GONADOTROPHIN (HPG)

In the female, follicle-stimulating hormone (FSH) causes growth of the ovarian follicles, and in the presence of small amounts of LH it promotes the secretion of estrogens by the maturing follicles. In the male, the same (pituitary) gonadotrophin, FSH, stimulates spermatogenesis. Luteinizing hormone (LH) in the female causes ovulation of the follicle, which has previously been ripened by FSH. An optimum ratio of FSH and LH is believed to be important in producing ovulation, even though the effective concentration of each in the human has not yet been established. Following ovulation, the transformation of the ruptured follicle into the corpus luteum and its secretion of progesterone are also believed to be under the influence of LH. In the male, LH is called the interstitial cell–stimulating hormone (ICSH) and is responsible for the production of testosterone by the interstitial cells of the testes.

The control of the secretion of gonadotrophins (LH and FSH) has been discussed before (see control of pituitary hormone secretion). The availability of highly purified and synthetic LH/FSH-RF has provided an important tool for studying the relationship between the pituitary and the hypothalamus and the nature of the control mechanism during the reproductive cycle. For an in-depth discussion, the review article of Schally *et al.*[74] may be consulted. Highly purified bovine or porcine LH-RF and its synthetic duplicate cause the release of both LH and FSH in humans, and thus it has been suggested that the same factor (LH-RF) is responsibile for the release of both LH and FSH. The variable amounts of FSH and LH during the menstrual cycle have been ascribed to the differential responsiveness of gonadotrophin (LH, FSH) secreting cells to the stimulation by the releasing factor. Prolonged infusion of the synthetic releasing factor increases FSH secretion. The secretion of FSH is primarily controlled by the negative feedback existing between plasma estrogen level and the hypothalamus. In the absence of ovarian production of estrogen—for example, in postmenopausal women—the release of FSH is high and relatively large amounts are found in blood and urine. The control of FSH secretion in men has not been definitely explained. It has been suggested that a small quantity of estrogen secreted by the adrenals and/or the testes may suffice for the regulation of FSH secretion. Recent evidence indicates that the positive feedback of estrogen at certain concentrations to the hypothalamus and/or to the pituitary is responsible for the *LH surge* obligatory for ovulation (see positive feedback control, p. 703). Large doses of estrogen and progesterone inhibit the release of LH, resulting in the failure of ovulation. In women taking oral contraceptives (estrogen-progestogen), the midcycle peak of LH is absent and, hence, no ovulation occurs. In the male, testosterone regulates LH secretion by the negative feedback mechanism; there is no cyclicity in the production of gonadotrophins and testosterone. It is a continuous process, much like the control of adrenal secretion by a continuously operating feedback mechanism.

Although FSH and LH have not yet been isolated in absolutely pure form, the preparations so far obtained from animal and human pituitaries and from the urine of postmenopausal women indicate that these are glycoproteins with molecular masses of approximately 30,000. Aggregation and dissociation into subunits have been shown to occur and complicate the estimation of the molecular mass. The glycoprotein hormones (FSH, LH, HCG, TSH) are composed of two kinds of subunits, usually referred to as α subunits (which appear to be common to several of these hormones) and β subunits (which are hormone-effect-specific). Detailed discussions on this and other topics, including assay methods, have been presented in a recent monograph.[70]

MEASUREMENTS OF LH AND FSH

CLINICAL SIGNIFICANCE

The excretion of gonadotrophins in children is very low. Detectable amounts of HPG are present in the urine of normal men and normally menstruating women, and higher levels are found in postmenopausal women. Wide individual variation is generally observed, even though a midcycle peak in women with a normal menstrual cycle is a consistent pattern. Because of the nonfunctioning ovaries in postmenopausal women, relatively large amounts of gonadotrophins, consisting mainly of FSH, are excreted in the urine. The clinical usefulness of the measurement of gonadotrophins lies mainly in the differential diagnosis of primary gonadal failure and gonadal failure secondary to lesions of the anterior pituitary in both men and women.

Increased excretion is generally associated with ovarian agenesis, Klinefelter's syndrome (prepubertal or pubertal seminiferous tubule failure), and male climacteric. *Decreased* excretion may be found in patients with hypogonadotrophic eunuchoidism, panhypopituitarism, and anorexia nervosa.

The majority of bioassay methods currently available are not sensitive enough to detect FSH and LH in blood. Consequently, the application of this technique has been applied predominantly to the analysis of urinary FSH and LH. The advent of radioimmunoassays for FSH and LH has made it possible to estimate these hormones in small quantities of blood. The concentrations of FSH and LH tend to be higher in the preovulatory phase than in the luteal phase, and there is a marked surge of LH at midcycle which precedes the event of ovulation by about 24 to 36 hours. (See Fig. 13–23.) A smaller peak of FSH coincides with the LH peak. Following ovulation, both FSH and LH levels fall and remain low until the next cycle starts.

In addition to the evaluation of gonadal function by measurements of FSH and LH, the plasma analyses of LH has been found to be very useful in the diagnosis and treatment of infertility in women. A midcycle rise in LH levels is a good indication that ovulation will occur approximately 24 hours later. Subfertile couples and also women being treated with clomiphene for infertility can then be informed that ovulation is about to take place. Apart from the essential diagnostic significance of LH and FSH in gonadal dysfunction, LH measurement also has clinical importance because growth hormone and LH are frequently the first hormones to be affected by pituitary disease.

ASSAY OF HUMAN PITUITARY GONADOTROPHIN (HPG) IN URINE

Bioassay methods appear to be most suitable for the determination of *total* gonadotrophins. Immunological assays,[54,55] which are more sensitive, can be applied for the specific analysis of FSH and LH in the plasma (serum) and urine. The analysis of untreated urine by this technique, however, is not reliable, and extraction and concentration of urine prior to the assay are suggested.

The determination of urinary total gonadotrophic activity, rather than the measurement of FSH and LH separately, appears to render adequate clinical information, and a procedure which is currently in wide use will be described in detail. It should be noted, however, that there are several assay methods available which are claimed to be specific for the individual determination of FSH and LH. Among them, the *augmentation test* for the FSH by Steelman and Pohley[84] and the ovarian ascorbic acid depletion test (OAAD) for LH by Parlow[63] are those most widely used. Further information on these topics may be derived from the monograph by Albert.[1]

BIOASSAY FOR THE DETERMINATION OF TOTAL GONADOTROPHIC ACTIVITY IN URINE[9,39]

PRINCIPLE

The method depends on the increase in uterine weight of immature mice injected subcutaneously with a urinary concentrate of gonadotrophin twice a day for two and one-half days. The result is expressed in mouse uterine (MU) units. One MU unit is defined as the amount of gonadotrophins required to cause a 100 per cent increase in uterine weight. Since the amount of gonadotrophins excreted in the urine is very low, an extraction and concentration of the sample is required. During this procedure the unwanted toxic and interfering substances are eliminated. The main steps are as follows: Adsorption of gonadotrophins onto Kaolin at pH 4.5, followed by elution with 2 mol/l ammonium hydroxide (NH_4OH). The eluate is adjusted to pH 5.5 and the HPG is precipitated with 2 vol of cold acetone. The precipitated material is further washed with ether and ethanol. The dry material is dissolved in normal saline in varying dilutions and injected into immature mice. The mice which serve as controls receive concurrently a similar aliquot of saline.

REAGENTS

1. Glacial acetic acid, AR.
2. Kaolin, Tamms Industries, Chicago, IL. Oxford-English brand.
3. Ammonium hydroxide, 2 mol/l.
4. Acetone, AR.
5. Ether, AR.
6. Ethanol, purified.
7. NaCl, 0.85 g/100 ml.

PROCEDURE

A complete 24 h collection of urine is needed. During collection and prior to analysis, the urine should be kept refrigerated, but not longer than 24 to 48 h. Prolonged standing even at 5°C may destroy HPG. It is best to start processing the urine as soon as it is received, at least through the step of Kaolin treatment, after which the preparation may be kept overnight in the refrigerator.

Extraction. Adjust the pH of the 24 h urine specimen to 4.5 with glacial acetic acid. Add 20 g of Kaolin per liter of urine. Mix well with a stirring rod. Pour the mixture onto a Büchner funnel and filter under mild suction. (Alternatively, refrigerate overnight, aspirate off the urine layer on the following morning, add acidified water, mix, transfer onto the

Büchner funnel, and proceed as follows.) Discard the filtrate. Wash the Kaolin cake in the Büchner funnel with 2 liters of water acidified to pH 4.5 with glacial acetic acid.

Discard the wash. Elute the gonadotrophins from the Kaolin cake with 100 ml of 2 mol/l ammonium hydroxide. Again pour the eluate over the Kaolin cake to completely elute the HPG. Transfer the eluate to a 500 ml centrifuge bottle. Wash the Kaolin cake with 50 ml of distilled water and add the water wash to the ammonium hydroxide eluate.

Precipitation. Adjust the pH of the eluate to 5.5 with glacial acetic acid. Add 2 vol (300 ml) of cold acetone, and refrigerate for 1 h. Centrifuge for 15 min at 2000 rpm. Pour off the supernatant. Add 20 ml of ethanol, stir to suspend the precipitate, and centrifuge as above. Discard the supernatant. Repeat the process of washing with 20 ml of ether. Let the precipitate dry completely under a stream of air, taking care not to lose any flakes of the precipitate. (The precipitate may be stored overnight or longer under desiccation in a freezer.)

Preparation of Solution for Injection. Pulverize the precipitate with a glass rod and dissolve in 10 ml of normal saline. Centrifuge for 10 min at 2000 rpm. Aspirate the supernatant with a 10 ml syringe and transfer to a clean glass vial. Prepare two different dilutions, namely, 1 to 5 and 1 to 10, as follows: transfer 1 ml aliquots of the supernatant to two separate glass vials containing 4 ml and 9 ml of normal saline respectively. Refrigerate all three solutions.

Injection. The immature mice to be used should be 21 to 23 days old and weigh 8 to 10 g. Using three animals for each solution (undiluted, diluted 1/5, and diluted 1/10), inject each animal with 0.2 ml twice a day (AM and PM) for two days, and once the last day. Inject simultaneously three more mice with normal saline to serve as controls. Sacrifice the animals on the fourth day. Dissect out the uterus from each animal, avoiding the adjoining fatty tissues, and weigh on a torsion balance.

Result. The response is considered to be positive when the average uterine weight of the test animals is at least twice the average uterine weight of the control animals (i.e., one MU unit). For example, if 1 ml (5 × 0.2 ml) of the extract dissolved in 10 ml saline (undiluted supernatant) gives a positive response, the excretion of gonadotrophins/d is 10 MU units. Similarly, the positive response to the injection of 1 ml of the solution diluted to 1/5 or 1/10 indicates the excretion of 50 or 100 MU units of gonadotrophins/d respectively.

COMMENTS

Mice (or rats) are chosen as the experimental animals for gonadotrophin assays because they are more sensitive to stimulation by these hormones than are other animals, such as rabbits, guinea pigs, and amphibia. The increase in the weight of the uterus depends on the stimulation of the ovaries by gonadotrophins to secrete estrogens, which in turn act on the uteri. Previously, it was erroneously believed that the mouse uterus test was specific for FSH activity, but recent evidence suggests that this test measures both FSH and LH. However, in case of elevations of LH, the ovaries of the injected mice will be increased out of proportion. Thus, the result is designated "total gonadotrophic activity" in the urine.

It should be pointed out that the quantitation of gonadotrophic activity without reference to the international standard is not very accurate. A suitable standard for the mouse uterine test (e.g. NIH-HPG-UE) has recently become available from the Endocrinology Study Section, National Institutes of Health, Bethesda, Maryland. It is highly important to establish frequently, and preferably routinely, the parallelism of the dose-response curve for a suitable range of concentration between the reference standard and the test material prepared in a specific laboratory. Since gonadotrophins are glycoproteins, the experimental precautions concerning pH, temperature, ion concentration, heavy metal contamination, and so forth are very important. Otherwise, denaturation of these hormones may occur, resulting in spurious results. The presence of toxic substances in the extract affects biological response and occasionally causes death of the experimental animals. Among the various measures employed to reduce toxicity, additional purification of the crude Kaolin extract, using ammonium acetate and ethanol[1] or tricalcium phosphate,[45] appears to be quite helpful.

NORMAL VALUES

Adults: 10–50 Mouse Uterine units/d
Menopausal: 50–190 Mouse Uterine units/d
Children (before puberty): Less than 6 Mouse Uterine units/d

REFERENCES

Borth, R., Lunenfeld, B., and Manzi, A.: *In* Human Pituitary Gonadotrophins. H. Albert, Ed. Springfield, Illinois, Charles C Thomas, Publisher, 1961, p. 13.
Klinefelter, H. F., Jr., Albright, F., and Griswold, G. C.: J. Clin. Endocrinol., *3*:529, 1943.

RADIOIMMUNOASSAY FOR THE DETERMINATION OF HUMAN PITUITARY GONADOTROPHINS

A large number of methods for the determination of FSH and LH in blood and urine have been described[45,70] and the methods of Midgley for LH[54] and FSH[55] have been applied widely in clinical laboratories. It has been difficult to obtain highly specific antibodies for FSH and LH because these two hormones possess biologic and immunologic similarities not only to each other but also to HCG and TSH. FSH possesses antigenic groups common to the three glycoprotein hormones (LH, HCG, TSH). LH contamination in FSH preparations may be eliminated by using α-chymotrypsin or by adding HCG.[55] In view of the highly similar antigenic structures of LH and HCG, most investigators have used HCG antisera for the radioimmunoassay of LH.

The gonadotrophins (LH and FSH) show alterations in their immunochemical behavior as a result of both metabolic transformation and extraction from urine. In view of these differences in immunologic behavior, depending on the biologic fluid from which the gonadotrophins are extracted, it is highly desirable to employ a homogeneous assay system. For example, in assaying urinary LH, it is best to use a urinary reference preparation of LH as both tracer and standard, as well as an antiserum raised against urinary LH. Similarly, the radioimmunoassay of serum LH should ideally be performed with labeled and unlabeled LH of pituitary origin and antiserum to pituitary LH. When such homogeneous systems are used, the results obtained by radioimmunoassay and bioassay are concordant.

HUMAN CHORIONIC GONADOTROPHINS (HCG)

Chorionic gonadotrophin is secreted by the placenta and is found in urine, blood, amniotic fluid, colostrum, milk, and fetal tissues. This hormone appears within a few days after conception, and because of this, early confirmation of pregnancy is possible through its detection in urine and blood (pregnancy test). The level of HCG is highest during the first trimester and stabilizes at a lower level as the pregnancy progresses, becoming almost undetectable following parturition. The physiological activity of this hormone mimics that of the pituitary gonadotrophin LH, except that it prolongs the life of the corpus luteum (six to eight weeks), which secretes estrogen and progesterone until the placenta assumes the secretory activity. Chemically, it is a glycoprotein with an approximate molecular mass of 30,000.

DETERMINATION OF HCG IN PLASMA AND URINE

CLINICAL SIGNIFICANCE

The *qualitative* estimation of HCG is primarily used for the detection of pregnancy. The usefulness of *quantitative* HCG assays lies in the diagnosis of abnormal pregnancies (e.g., pre-eclamptic toxemia) and trophoblastic diseases. Patients with hydatidiform mole or choriocarcinoma excrete large amounts of HCG (100,000 units or more). A very high hormone level several weeks after parturition is strong evidence of choriocarcinoma. During pregnancy, especially during the early weeks, it is not always possible to differentiate between the normal HCG excretion of pregnancy and the abnormal excretion of moles or choriocarcinoma, because values overlap. Serial determination may, however, aid in the diagnosis, since the normal sequence in pregnancy is that of a decreasing level following the first two to three months, whereas levels will remain high or even increase with a mole or choriocarcinoma. Serial determinations of HCG excretion in these pathological conditions also provide an excellent indicator of response to treatment by surgery or chemotherapy. In an abnormal pregnancy, such as one complicated by pre-eclamptic toxemia, the urinary and plasma levels of HCG are significantly higher than those during a normal pregnancy. In threatened abortion and ectopic pregnancy, the excretion is generally below the expected value. Patients with ovarian and testicular teratomas have also been reported to excrete HCG in large quantities. For detailed information on the clinical importance of HCG assays, the reader is referred to the monograph by Loraine and Bell.[45]

Mishell *et al.*[57] reported data obtained by immunologic measurements of HCG in serum and urine obtained throughout pregnancies. Values for serum (IU/ml) or urine voided in the morning (IU/ml) were similar. There was generally a rapid rise in the HCG level to a peak between 60 and 80 days of pregnancy, with a mean value of about 100 IU of HCG/ml, followed by a somewhat more gradual fall to a mean level of about 20 IU/ml after 120 days. Curves showing HCG levels in serum and urine during normal pregnancy are shown in Fig. 13–26.

Methods

There are a number of assay procedures available involving both biological and immunological techniques. Of late, the routine application of bioassays has diminished significantly and their utility lies mainly in the preparation and standardization of HCG

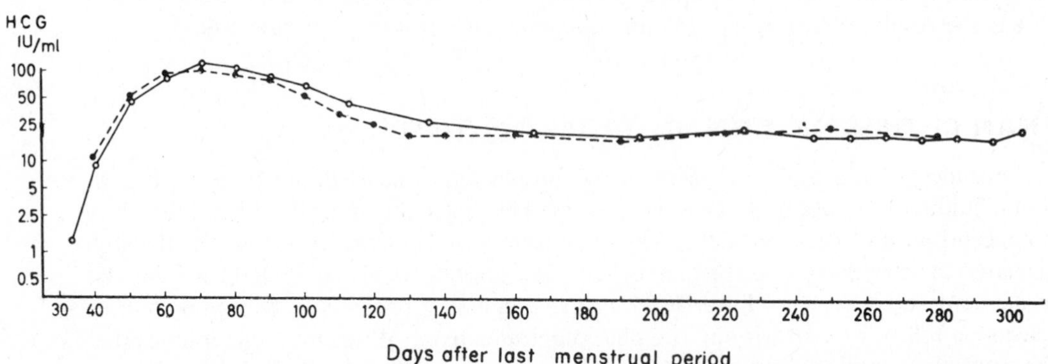

Figure 13–26 Concentration of HCG in serum compared with the concentration in morning urine in normal women during pregnancy. The closed circles represent the means of 311 determinations of serum and the open circles represent the means of 522 determinations of urine. The patients in the 2 groups were not identical. (From Mishell, D. R., *et al.*: J. Clin. Endocrinol., *23*:125, 1963.)

from biological sources. In spite of the sensitivity of the biologic procedures, there are several disadvantages to their routine use, such as cost and maintenance of animal colonies, time required to perform tests, variability in sensitivity of animals during various seasons, and the fact that toxic urine may cause death of the animal. Because of simplicity, speed, accuracy, and low cost, the immunologic methods for qualitative and quantitative analyses of HCG have largely replaced the bioassay techniques. For a comprehensive review on the biologic and immunologic methods, the reader is referred to the article by Hobson.[31]

QUALITATIVE ESTIMATION OF HCG

Biologic pregnancy tests

Since HCG closely resembles LH of the pituitary in its physiological activities, the underlying principle of HCG bioassay is the same as that of the LH assay. Thus, injection of HCG produces corpora lutea and hemorrhagica in the ovaries of mice and rabbits, hyperemia in the ovaries of rats, and secretions of sperm and ova in toads and frogs.[45] Until recently, methods depending on the production of ovarian hyperemia in immature rats[2] and expulsion of spermatozoa in amphibia[30] were most commonly used in clinical laboratories.

Immunologic pregnancy tests

Tests depending upon *hemagglutination inhibition* and *complement fixation* have been used for pregnancy tests and for the quantitative determination of HCG. The procedures based on hemagglutination inhibition are highly sensitive, rapid, simple, inexpensive, and well suited for a routine clinical laboratory. Limitations of the method include poor reproducibility, occasional nonspecificity, and the possibility of hemolysis of the antigen-coated red cells. The use of *latex particles* instead of tanned red cells as a carrier of hormone antigen overcomes the problem of hemolysis but is less reliable. In techniques based on the principle of complement fixation, the absence of hemolysis using a single undiluted sample is confirmatory of pregnancy (positive test). The method is considered to be highly sensitive and very specific. However, factors such as the concentrations of the reactants (antigen, antibody, complement, erythrocytes, hemolysin, and so forth), temperature, volume, ionic strength, pH, and duration of the reaction must be rigidly controlled. Because of these difficulties, the technique has as yet found little application in the routine laboratory.

By far the most popular and widely used pregnancy tests are based on the hemagglutination (or latex-agglutination) inhibition. In recent years, a number of commercial preparations for pregnancy tests have become available: *UCG Test*, Wampole Laboratories, Stamford, CT.; *Pregnosticon Tube and Slide Test*, Organon, Inc., West Orange, N.J.; and *Gravindex Test*, Ortho Pharmaceutical Company, Raritan, N.J. Because of wide use of commercially available test kits in clinical laboratories, the procedure is not described in detail here. The specific protocols for these test kits are supplied by the manufacturers. The following is a general description of these kit procedures.

Commercially available test kits, based on hemagglutination or latex-agglutination inhibition may be divided into two categories—*tube tests* (Pregnosticon and UCG Test) and *slide tests* (Gravindex and Pregnosticon).

1. Tube tests. A plasma extract or urine is first mixed with an optimum amount of anti-HCG (step 1) in a test tube. If HCG is present in the unknown sample, it combines with and neutralizes the antibodies even though no visible reaction occurs. When HCG-coated red blood cells are added to the system (step 2) and left undisturbed for 2 h, no hemagglutination will occur, since all antibody has been neutralized during step 1. As a result, the cells will settle to the bottom of the tube in a "doughnut" pattern, as do normal nonagglutinated cells (*positive test*). In the absence of HCG, the anti-HCG is not neutral-

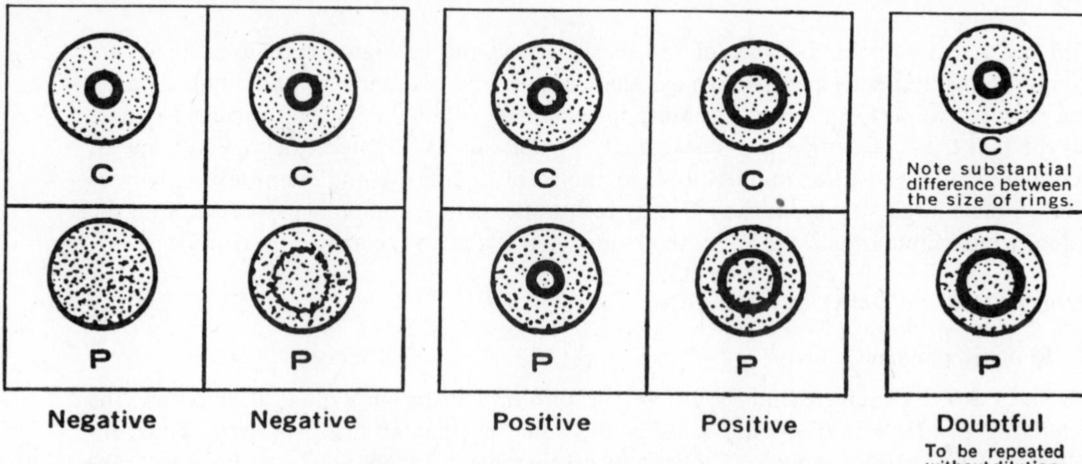

Figure 13–27 Interpretation of readings: After 2 h at room temperature, the cell patterns at the bottom of the tubes are interpreted and recorded as illustrated in the diagram. The best way to view the cell patterns is through the bottom of the test tube. Do this carefully, lifting up rack, then read from bottom, or by the mirror under the UCG test rack holder. Positives are when the pattern in both the control and patient tubes are comparable. Occasionally, the smaller "ring" tends to fill in and may appear as an opaque circle. C = control tube, P = patient tube. (Courtesy of Wampole Laboratories, Stamford, Conn.)

ized; it will react with HCG-coated red cells (i.e., agglutination) and settle out to form a mat of cells on the bottom of the test tube (*negative test*). The size and shape of the ring formation is generally compared with a control tube prepared simultaneously with the sample (Fig. 13–27).

2. Slide test. Similar to the tube test, the slide test is also an agglutination-inhibition reaction, but latex particles coated with HCG are used instead of red cells. The test is performed on a slide and the results are obtained within 2 min. Antiserum is added to the urine followed by the addition of HCG-coated latex particles. If there is no HCG in the urine, the antiserum will react with the HCG-coated latex, giving rise to a clumping effect (agglutination = *negative test*). In the presence of HCG in the urine, the agglutination will be inhibited and no clumping effect is observed (*positive test*).

In the *direct agglutination* slide test (DAP Test, Wampole Laboratories, Stamford, CT.) the latex particles are *coated with anti-HCG* and this becomes the only reagent to be added to the sample. The presence of HCG in the test specimen results in an agglutination of the anti-HCG–coated latex particles (clumping effect), which is read as a *positive test*.

COMMENTS

In immunologic methods, the drugs and toxic substances present in urine are less likely to interfere than in bioassays. For qualitative pregnancy tests, preferably the first morning void of urine should be collected in a clean container. The patient should be instructed to restrict intake of fluids from 8 PM until the morning collection. Urine with a specific gravity less than 1.010 may yield a false-negative reaction. The urine specimens should be clear; presence of turbidity or urine sediments requires filtration or centrifugation. HCG antisera do not possess absolute specificity. Consequently, interfering substances, especially in proteinuria, may neutralize the antisera and cause a false-positive pregnancy test. This may also occur if the sensitivity of the test exceeds 0.7 IU/ml because of the cross-reactions with pituitary gonadotrophins. Conditions which may cause denaturation of HCG (e.g., temperature, pH, contamination with heavy metals, and so forth) may yield false-negative results. Denaturation of the antisera enhances sensitivity with a concomitant increase of false-positive reactions (with loss of activity, relatively less HCG is required to neutralize the antibody). The use of known positive and negative controls and the occasional restandardization of the reagents are therefore extremely important if reliable results are to be obtained.

The hemagglutination-inhibition kits currently available are not suitable for undiluted or untreated serum because of the high incidence of false-positive reactions. These methods can be adapted to serum, provided an acetone extraction of the serum is performed for the purpose of removing inhibiting substances.[57] With the DAP Test, detection of HCG in serum is possible without any manipulation other than simple dilution. However, the greatest problem in the DAP Test is the occurrence of the prozone phenomenon—inhibition of agglutination in the presence of excess antigen. Consequently, in the case of hydatidiform mole, choriocarcinoma, and during certain stages of pregnancy where large amounts of HCG are present in blood and urine, false-negative results may be expected.

Recently, the results of a study involving five commercially available pregnancy test kits have been published.[33] The accuracies of various tests, based on the total numbers of correct test results from uncomplicated first trimester pregnancies and a control group of nonpregnant patients were as follows: Pregnosticon Tube Test, 98.8 per cent; UCG Tube Test, 98.6 per cent; Pregnosticon Slide Test, 96.7 per cent; Gravindex Test, 93.8 per cent; and DAP Test, 93.2 per cent. According to these investigators, the tube tests which have greater sensitivity and accuracy should be used for reliable results, particularly for patients having elective gynecologic surgery.

QUANTITATIVE DETERMINATION OF HCG

The procedure is the same as described for the pregnancy tests except that serially diluted test specimens are used. In the case of a bioassay (e.g., rat hyperemia test), serially diluted urine is injected into groups of animals and the first dilution which gives no response is determined. The results obtained in the test groups of animals are compared with the groups receiving International Standard preparation of HCG. The total amounts of HCG are expressed in terms of international units (IU).

For the quantitative assay by hemagglutination inhibition, the urine or processed serum specimen is serially diluted until further dilution will not cause inhibition of hemagglutination (i.e., the end point). A standard of known concentration is tested in the same manner and at the same time. Since the dilution of a sample that just prevents agglutination (end point) is expected to contain the same quantity of hormone as that of the standard preparation producing the same effect, a direct comparison gives the amount of hormone present in the unknown. A commercial kit (UCG Titration set, Wampole Laboratories, Stamford, CT.) is available for quantitation of human chorionic gonadotrophin.

The results obtained by the hemagglutination-inhibition technique have generally been reported to be considerably higher than those obtained by bioassays. A critical review of the current status of both types of methods in clinical practice has been made by Hobson.[31] Since antisera to HCG cross-react strongly with LH, the same *radioimmunoassay* can be used for the two hormones (see radioimmunoassay of gonadotrophins) following separation of HCG and LH by gel filtration on a Sephadex column.[55]

HUMAN PLACENTAL LACTOGEN (HPL)

Another protein hormone formed by the trophoblast is human placental lactogen (HPL) or human chorionic somatomammotrophin (HCS). In contrast to HCG, it does not appear to have a carbohydrate component. Although it has not yet been isolated in absolutely pure form, the estimated molecular mass of the hormone lies between 30,000 and 38,000. This substance possesses chemical properties similar to those of pituitary prolactin and immunologically cross-reacts with pituitary growth hormone. In addition, it possesses lactogenic and somatotrophic biological properties. A variety of metabolic changes in pregnancy, including insulinogenic and diabetogenic effects, have been ascribed to the action of HPL. Assays for HPL in clinical practice have so far only been conducted using radioimmunoassays, and the method of Saxena *et al.*[71] has been widely applied.

The HPL plasma level in normal pregnancy appears to be measurable between the seventh and ninth week and increases steadily throughout gestation, reaching its highest value near term. At present, an International Reference Standard for HPL is not available. As a result, there is considerable variation in normal ranges between laboratories. It has been postulated that both placental mass and the integrity of placental function are reflected in the serum HPL level.[71] Consequently, in high risk pregnancies (e.g., threatened abortion, toxemia, diabetes), the serial analysis of plasma or serum HPL has been proposed as a potentially useful diagnostic and prognostic index of the functional status of the placenta. A fall of the circulatory level of HPL is thought to indicate placental dysfunction and potential fetal insufficiency. It must be emphasized, however, that the clinical usefulness of HPL analysis has not yet been firmly established. The prevalent opinion is that the serial determination of urinary or plasma estriol levels is a more reliable index of fetoplacental well-being.

INSULIN

This hormone, elaborated by the beta cells of the islets of Langerhans in the pancreas, is responsible for the control of carbohydrate metabolism and maintenance of the blood glucose level (see Chap. 6). Insulin is the protein hormone whose structure was first elucidated. It is composed of two long polypeptide chains, A and B, with specific amino acid sequences bound together by two disulfide linkages. There is also an intrachain disulfide bridge linking the amino acids at positions 6 and 11 in chain A. Although no definite part of the insulin molecule can be designated as the active center, it has been shown that the intact disulfide bridge is essential for full biological activity. The excellent review article by Sanger[69] provides further information.

METHODS OF ASSAY

The determination of plasma levels of insulin is a necessary test in medical practice. A discussion of the methods for determining levels of insulin in blood may be found in the article by Taylor.[86]

As with other protein hormones, the assay of this hormone is carried out by biological and immunological methods.

The *biological methods* are of two types, *in vivo* and *in vitro*. The *in vivo* methods depend on the ability of the hormone to produce hypoglycemic convulsions in mice and to cause hypoglycemia in rabbits. Although previously widely used, the quantitative measurement of plasma insulin with both of these procedures is unsatisfactory, and their use in the clinical laboratory has been very limited.

Among the *in vitro* methods, the rat diaphragm test and the rat epididymal fat pad test are the most noteworthy. The *rat diaphragm technique* consists of incubating isolated diaphragms in a suitable glucose-containing medium and measuring the resultant glucose uptake during a given period of time. Although the technique is capable of detecting approximately 100 μIU (micro International Unit) of insulin per ml of plasma, the lack of specificity is its greatest drawback. It has been shown that substances other than insulin present in the plasma fraction interfere with the measurement. In addition, the degradation of insulin by proteolytic enzymes in the diaphragm and adsorption of the hormone onto glassware are other important sources of error.

The principle of the *epididymal fat pad test* depends on the ability of insulin to affect the glucose metabolism in the fatty tissue. The procedure consists of incubating the fat pad of a rat with ^{14}C-labeled glucose and determining either the rate of glucose uptake by the tissue or the rate of production of $^{14}CO_2$. The measurement of either of the end points correlates with the concentration of insulin in the medium. This method has greater sensitivity (10 μIU/ml) than the rat diaphragm test. As with other methods, it also suffers from lack of specificity, and thus measures *insulin-like* activity (ILA) of nonpancreatic origin.

Immunoassay

Because of severe limitations of biological methods in regard to sensitivity and specificity, in recent years the radioimmunoassay has become the technique of choice for measurement of insulin in blood. The subject has been reviewed by Berson and Yalow.[8] Briefly, the main steps of the insulin method are as follows: pork insulin is labeled with ^{131}I, and an antibody to pork insulin is produced in guinea pigs (a guinea pig antipork insulin serum reacts identically with pork and human insulin). A serial dilution of the standard human insulin and the plasma are prepared and incubated with a predetermined amount of radiolabeled insulin and anti-insulin for three to five days at 4°C. At the end of the incubation period, the aliquots are removed and subjected to electrophoresis in order to separate free insulin and antibody-bound insulin. The radioactivity of each of these two fractions is determined, and the ratio of the radioactive count (^{131}I content) of bound and free insulin is plotted against the concentrations of the standard, as shown in Fig. 13–24. The amount of insulin in the plasma is determined by comparing the results with the standard curve.

The *normal values* for plasma levels of insulin obtained by bioassays are higher than those obtained by immunoassays. Using the isolated rat diaphragm test, the average level of insulin in fasting plasma of normal subjects is 37 μIU per ml with a range from 11 to 240 μIU per ml, whereas that obtained by radioimmunoassay is 21 μIU per ml with a range from 0 to 60 μIU per ml.[8] In diabetes mellitus the plasma level is generally *low*, and the pattern of insulin secretion following administration of glucose is different from that found in normal subjects. In the former, the highest insulin levels are reached 2 hours after glucose intake, whereas in normal subjects the corresponding time is 30 to 60 min. In patients with *islet cell tumor of the pancreas*, the values are *high*, which in effect causes hypoglycemia. Acromegaly and gigantism are also associated with a high level of plasma insulin.

CATECHOLAMINES

Catecholamines are amines which are distinguished by the presence of two hydroxyl groups on a benzene ring (catechol). The most important endogenously produced compounds of this group are epinephrine (adrenaline), norepinephrine (noradrenaline), and dopamine [β(3,4-dihydroxyphenyl)ethyl amine]. The structural formulas and the numbering system of these compounds are shown in Fig. 13–28. Norepinephrine differs from dopamine in that it possesses a hydroxyl group on the β carbon atom; epinephrine is distinguished from norepinephrine by the presence of a methyl group on the nitrogen of the terminal amino group. Dopamine and norepinephrine are primary amines; epinephrine, on the other hand, is a secondary amine because of the substitution of a hydrogen atom in the amino group by a methyl group. Epinephrine and norepinephrine are dihydroxylated phenyl (catechol) β-ethanolamine, and therefore exhibit the chemical properties of phenols, alcohols, and amines. Chemically, epinephrine is rapidly oxidized in neutral and alkaline solutions. Norepinephrine is much more resistant to oxidation. These differences in properties of the two compounds have been utilized for their individual estimations in biological fluids. For further information on the chemical properties of catecholamines the review article by von Euler[95] is suggested.

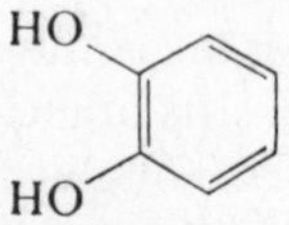

Catechol (Dihydroxybenzene)

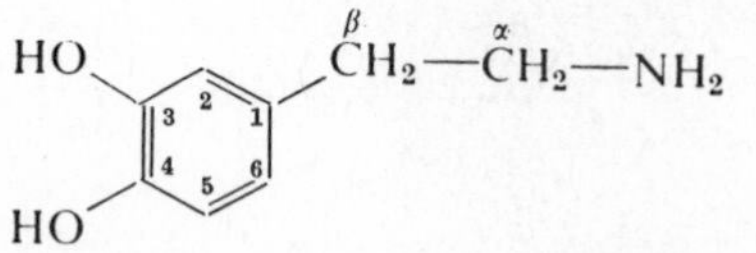

Dopamine[β(3,4-dihydroxyphenyl)ethyl amine]

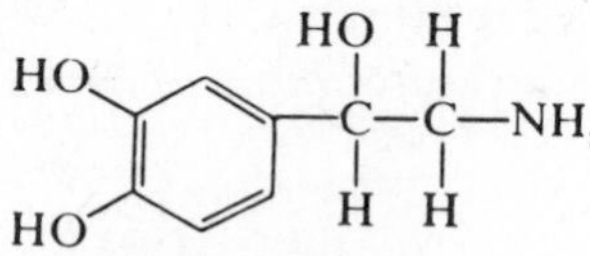

Norepinephrine (Noradrenaline)

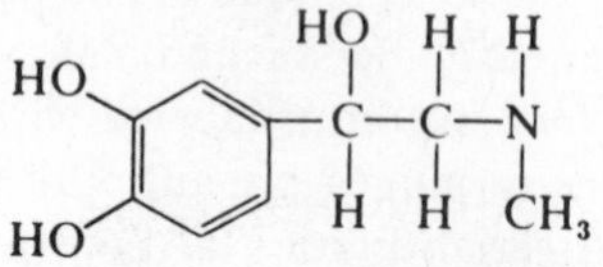

Epinephrine (Adrenaline)

Figure 13–28 Structural formulas and the numbering system of catecholamines.

The main sources of catecholamines in the body are the chromaffin cells in the adrenal medulla, the heart, lung, liver, intestines, and prostate glands, the chromaffin bodies of the fetus, the brain, and the adrenergic nerves. The chromaffin cells are so designated for their distinctive brown or black color on staining with chromic acid. Epinephrine is quantitatively the most important substance produced by the adrenal medulla. Norepinephrine is the major substance liberated by the postganglionic sympathetic nerves. In addition, norepinephrine is the predominant catecholamine produced in fetal life of mammals. Dopamine is considered to be produced significantly in the brain and other viscera such as the lungs, liver, and intestines. The hormones are produced in mitochondria and are stored in special granules of the chromaffin cells. Administration of reserpine or other pharmacological agents, insulin-induced hypoglycemia, and stressful situations have all been shown to cause depletion of the catecholamine content of storage granules. Each of the three catecholamines, dopamine, norepinephrine, and epinephrine, has characteristic physiological functions and pharmacologic actions. Detailed consideration of these topics may be found in the textbooks of endocrinology listed at the end of this chapter.[89,99] It may suffice to mention here that the first two hormones have, in general, a marked influence on the vascular system, whereas epinephrine, which is considered to be the true adrenal medullary hormone, exerts a profound influence on the metabolic processes, especially carbohydrate metabolism.

BIOSYNTHESIS AND METABOLISM

In vivo and *in vitro* studies have shown that the aromatic amino acid L-tyrosine plays the most important role as precursor for the biogenesis of catecholamines (Fig. 13–29). This amino acid is found in abundance in plasma and tissues. Normally, tyrosine is derived from the diet or synthesized in the liver following hydroxylation of the amino acid phenylalanine. The first step in catechol synthesis is the hydroxylation of tyrosine to produce dihydroxyphenylalanine (DOPA). The decarboxylation of DOPA gives rise to the first catecholamine, dopamine. Dopamine is then hydroxylated to produce norepinephrine. The conversion of norepinephrine to epinephrine is mediated by phenylethanolamine-N-methyl transferase, which is present almost exclusively in the adrenal medulla. Because of such

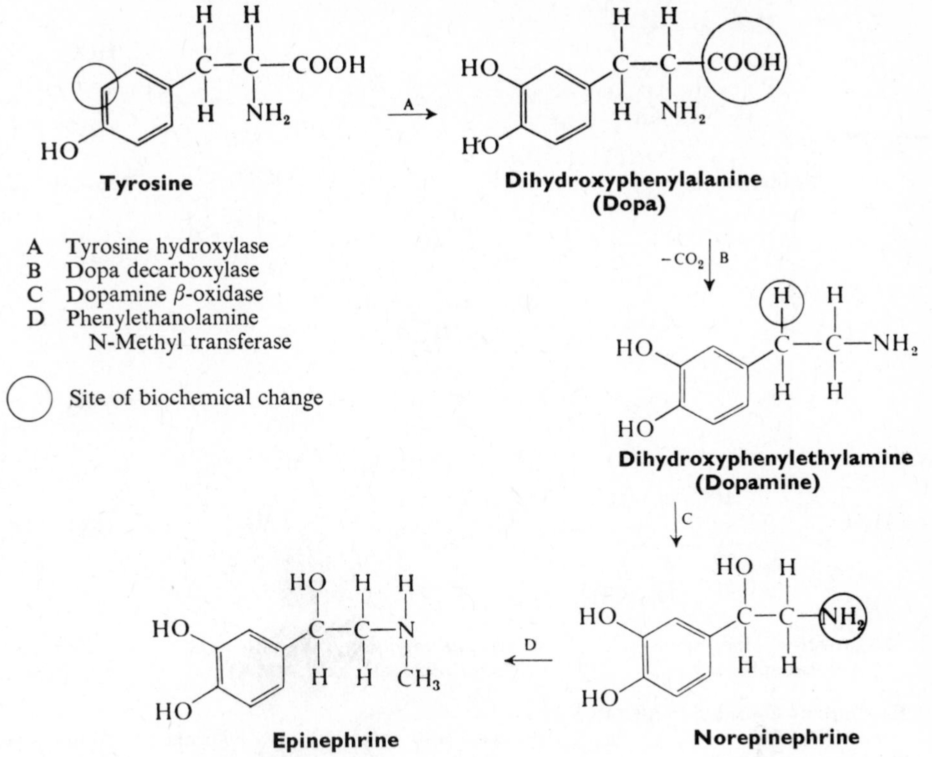

Tyrosine → (A) → **Dihydroxyphenylalanine (Dopa)** → ($-CO_2$, B) → **Dihydroxyphenylethylamine (Dopamine)** → (C) → **Norepinephrine** → (D) → **Epinephrine**

A Tyrosine hydroxylase
B Dopa decarboxylase
C Dopamine β-oxidase
D Phenylethanolamine N-Methyl transferase

○ Site of biochemical change

Figure 13–29 Biogenesis of catecholamines.

limited distribution of methyl transferase, a pheochromocytoma (tumor of chromaffin tissue) secreting a large amount of epinephrine is generally associated with the adrenal medulla.

There are two important metabolic events for the ultimate disposition of catecholamines in the body: catechol-O-methylation and oxidative deamination (Fig. 13–30). A certain percentage of the methylated amines are excreted either in the free form or conjugated with sulfuric or glucuronic acid. The majority of 3-methoxy derivatives, however, undergo deamination and are then oxidized to 3-methoxy-4-hydroxy-mandelic acid (vanilmandelic acid, VMA), which is excreted in the free form. When oxidative deamination occurs first, epinephrine and norepinephrine are converted to a common metabolite, 3,4-dihydroxymandelic acid (DHMA), which is subsequently O-methylated to VMA. There are some controversies about the sequence of the enzymatic inactivation of catecholamines. According to present evidence, O-methylation is considered to be quantitatively the most significant step and is followed by oxidative deamination. It should be noted, however, that irrespective of the nature of initial enzymatic attack, the ultimate end product of both epinephrine and norepinephrine is VMA. For this reason, the urinary measurement of the latter reflects the total rather than the differential production of these two catecholamines in the body. The final metabolite of dopamine following O-methylation and oxidative deamination is homovanillic acid, HVA (see Fig. 13–31).

CLINICAL SIGNIFICANCE

The normal human adrenal medulla produces about 80 per cent of epinephrine and 20 per cent of norepinephrine. Since only a small amount of epinephrine is released by other

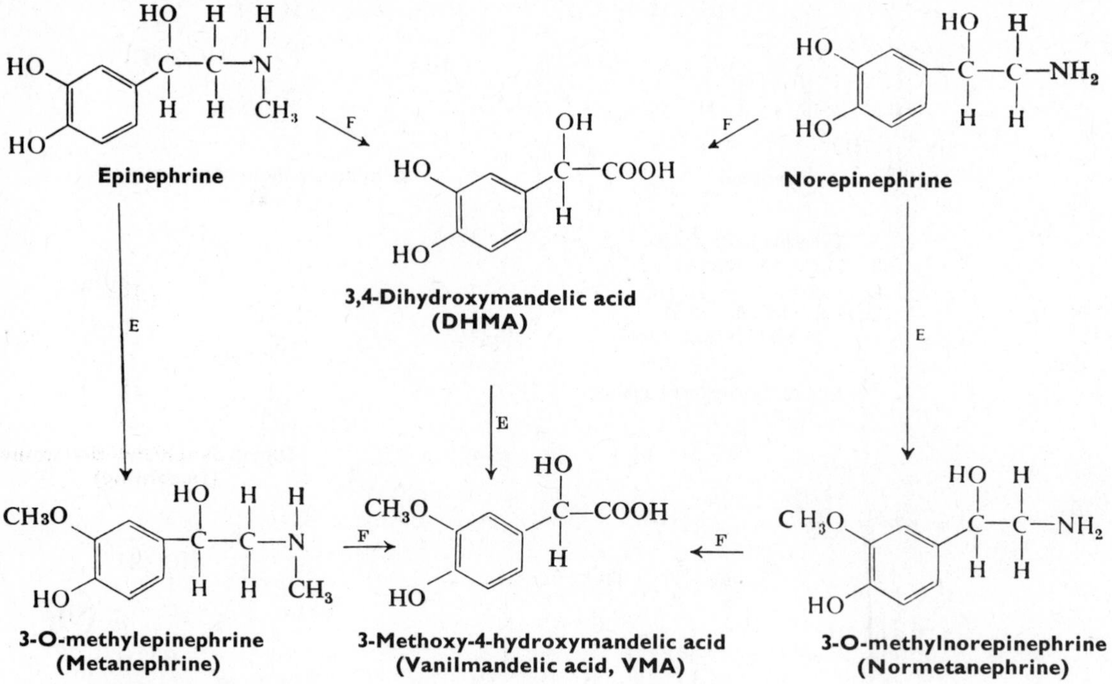

Epinephrine

Norepinephrine

3,4-Dihydroxymandelic acid (DHMA)

3-O-methylepinephrine (Metanephrine)

3-Methoxy-4-hydroxymandelic acid (Vanilmandelic acid, VMA)

3-O-methylnorepinephrine (Normetanephrine)

E—Catechol O-methyl transferase (COMT)
F—Monoamine oxidase (MAO)

Figure 13–30 Metabolism of epinephrine and norepinephrine.

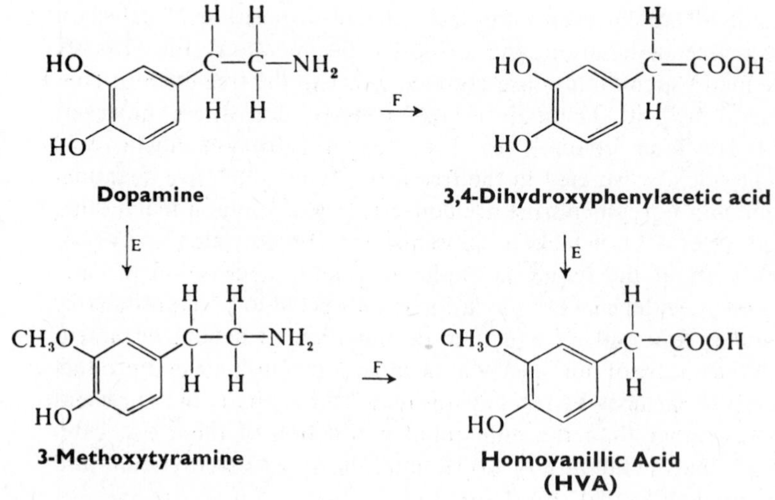

Dopamine

3,4-Dihydroxyphenylacetic acid

3-Methoxytyramine

Homovanillic Acid (HVA)

E—Catechol O-methyl transferase (COMT)
F—Monoamine oxidase (MAO)

Figure 13–31 Metabolism of dopamine.

tissues in the sympathetic nervous system, the plasma levels of epinephrine or the urinary excretion of this amine and its metabolites are useful in the assessment of the functional status of the adrenal medulla. It should be noted, however, that the adrenal medulla is not essential for the maintenance of life, as is the adrenal cortex. Therefore, the consideration of deficient activity (hypofunction) of the gland is of little clinical interest; medullary *hyperfunction*, on the other hand, has great diagnostic importance particularly in patients with hypertension. Pheochromocytoma (chromaffinoma), a tumor of the chromaffin tissue, is associated with the presence of greatly increased plasma and urinary levels of catecholamines and their urinary metabolites such as VMA. Some 90 per cent of pheochromocytomas (PCC) originate in the adrenal medulla and are characterized by an increased secretion of epinephrine. Extra-adrenal PCC's, which secrete largely norepinephrine, may develop anywhere along the course of the sympathetic chain from neck to pelvis, but often the paired para-aortic bodies of Zuckerkandl are affected. Children are more prone than adults to development of extra-adrenal, multiple PCC.

Norepinephrine, although secreted in small amounts by the adrenal gland, is produced mostly at the adrenergic nerve endings. Therefore, the norepinephrine level of plasma or urine is an indication of overall peripheral sympathetic activity. Patients with heart failure have higher than normal plasma levels of norepinephrine. Elevated values of catecholamines have also been found in patients with tumors of neural origin, such as neuroblastomas and gangliomas. However, the findings in these pathological conditions are not consistent, and for proper diagnosis the differential determinations of the individual catecholamines and their metabolites, including homovanillic acid (HVA), are most helpful.

It is now possible to measure epinephrine and norepinephrine in the plasma and to measure epinephrine, norepinephrine, VMA, HVA, and metanephrines in the urine. It is therefore important to consider the advantages and disadvantages of selecting the source—blood or urine—and the type of substances to be analyzed for any particular objective. For most physiologic and pathologic studies, measurements of catecholamines (total or individual) or a metabolite such as VMA appears to be adequate. However, in situations where the metabolism rather than the secretory rate is affected, causing asymmetrical alterations in concentrations of the individual amines and various metabolites, the analysis of just one individual product may not be satisfactory. The situation is best exemplified in patients receiving monoamine-oxidase and DOPA decarboxylase inhibitors for the treatment of hypertension. Under these circumstances, it would be advantageous to measure all products (catecholamines and metabolites) for the proper interpretation of changes in secretion and/or metabolism.

The *plasma levels* of catecholamines reflect the balance between biosynthesis, release, uptake, catabolism, and excretion *at the time of sampling*. The concentration of catecholamines and their metabolites in the *urine*, on the other hand, represents an overall estimate of sympathetic activity, during relatively *long periods of time*, and does not reflect the influence of transient changes in plasma levels. Therefore, in some cases of pheochromocytoma when paroxysmal attacks are very short, urine values may not be raised above the normal limits. In such patients, the measurement of plasma catecholamine levels during a spontaneous paroxysm or during an attack provoked by histamine or glucagon is an effective way of confirming the diagnosis. However, under conditions of relatively stable sympathetic activity, or when the induced changes are prolonged, the urinary levels of catecholamines will be directly proportional to those in the blood. The assay methods for catecholamines in urine are somewhat simplified because the amines (including dopamine) are present in relatively large amounts. Catecholamines in plasma samples are present in nanogram quantities and, consequently, methodological refinements and additional precautions are essential.

DETERMINATION OF PLASMA CATECHOLAMINES

The concentrations of epinephrine and norepinephrine in blood plasma are normally very low, necessitating the use of highly sensitive assay techniques. For many years the only methods available for the assay of catecholamines in body fluids were biological. In spite of their high sensitivity, the bioassay techniques pose many problems and have been replaced by simpler, less expensive, and more precise and specific fluorometric methods. For a comprehensive discussion of this topic, review articles[26,46,97] should be consulted. Fluorometric methods for the estimation of epinephrine and norepinephrine in plasma depend upon the conversion of the amines into derivatives which emit a characteristic fluorescence greater than that of the parent amine. The methods employed most extensively have been modifications of the ethylenediamine (EDA) method and the trihydroxyindole (THI) method. Discussions on the specificity of these two procedures may be found in the review article of Callingham.[26] A comparative study of these two fluorometric procedures for the analyses of plasma catecholamines has been recently made by Manger *et al.*[46] Although both methods are very accurate in quantitating epinephrine and norepinephrine in plasma, from the practical point of view, the EDA method has been considered to be more suitable for use in a routine clinical laboratory.[46] More recently, a variety of automated and semi-automated procedures based on these methods have been devised, with improvement in sensitivity and reproducibility.[49,94]

The principle and the reaction sequences for the THI method are described on page 812. In the EDA method, epinephrine is first oxidized to adrenochrome, and this compound then condenses with one mole of ethylenediamine with elimination of 2 H_2O and 2 H. The reaction of norepinephrine with ethylenediamine is more complex and not yet entirely resolved. Epinephrine gives but one fluorescent product, and norepinephrine gives two major products and a third minor product. According to the present concept, norepinephrine is oxidized to noradrenochrome; EDA then condenses with this compound, resulting in ring formation and the elimination of the side chain of norepinephrine and 2 H. The compound then condenses with another molecule of EDA to produce the compound shown below as one of the two principal fluorescent products formed; the second product has not been identified. Please note: norepinephrine reacts with 2 mols of EDA and loses its side chain, whereas epinephrine reacts with 1 mol of EDA and retains its side chain as part of the final product. The fluorescent products of epinephrine and norepinephrine have different emission spectra, allowing simultaneous measurement of both hormones without prior separation.

(1)

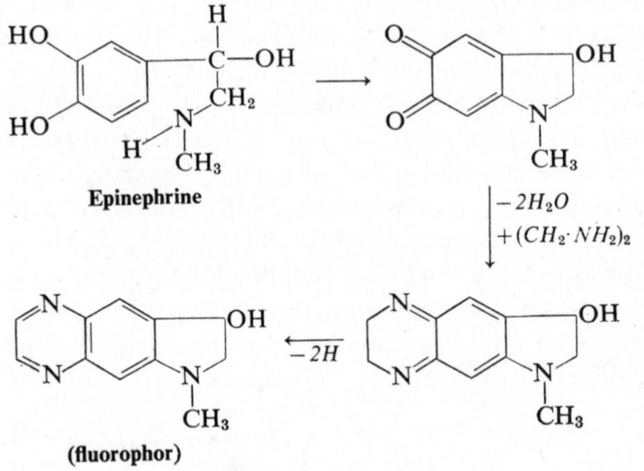

Epinephrine

$-2H_2O$
$+(CH_2 \cdot NH_2)_2$

$-2H$

(fluorophor)

(2)

Norepinephrine $\longrightarrow$ $CHOH—CH_2—NH_2$

$-2H$
$+(CH_2NH_2)_2$

$CHOH—CH_2—NH_2$
$NH—CH_2—CH_2—NH_2$

$-2H$
$-NH_2CH_2CHO$

$-2H$ $+(CH_2NH_2)_2$
$-2H_2O$

(fluorophor)

The most satisfactory EDA method is that described by Weil-Malherbe.[97] A general outline of the method is given below. Before setting up the method in a laboratory, the original reference and the article by Manger *et al.*[46] should be consulted.

Venous blood (20 ml) is collected through plastic tubing in a glass-stoppered 25 ml graduated cylinder containing 5 ml of anticoagulant solution (1 g EDTA and 2 g sodium thiosulfate/100 ml at pH 7.4). Essentially the same results are obtained with heparinized plasma. The plasma is separated by centrifugation, and a measured volume of plasma is adjusted to pH 8.4 with 0.5 mol sodium carbonate/l and passed through a column (5 mm, ID) prepared with 0.5 g of acid-washed, activated alumina. The column is washed with water (10 ml) and the catechols are eluted with 3.0 ml of 0.2 mol/l acetic acid followed by 3.0 ml water. The eluate is mixed with 0.6 ml EDTA (1g/100 ml) and adjusted to pH 6.0. The mixture is then transferred onto a column containing Amberlite CG-50 (0.4 g moist weight). The column is washed with 20 ml water and eluted with 3.5 ml of 1 mol/l acetic acid. The condensation reaction is performed in red glassware to exclude the effects of light upon the fluorophors (the product formed from norepinephrine is unstable in light). The eluate is mixed with 0.5 ml of redistilled ethylenediamine (EDA), shaken vigorously, and kept in a water bath at 55°C for 25 min. The mixture is cooled, saturated with sodium chloride (approximately 1.0 g), and extracted with 1.8 ml of isobutanol by shaking for 4 min. Standard solutions containing 0.05 μg epinephrine and norepinephrine are prepared together with two blanks; one blank is processed through the entire procedure (column blank), and the other through the condensation reaction only (reagent blank). The fluorescence is measured at wavelengths (emission spectra) of 510 nm (reading b) and 580 nm (reading y) using a single activation wavelength of 420 nm. The readings are corrected for column blank and the standards for reagent blank.

The fluorescence of the product formed from norepinephrine at 510 nm is twice as great as that from epinephrine; at 580 nm the ratio is reversed. The content of epinephrine and norepinephrine is calculated using the following equations:

$$E + \frac{N}{m} = y$$

$$E + \frac{N}{n} = b$$

where m = ratio of fluorescence of E/N standards at 580 nm
n = ratio of fluorescence of E/N standards at 510 nm
E = amount of epinephrine in sample
N = amount of norepinephrine in sample
y = amount of apparent epinephrine in sample indicated by the reading at 580 nm
b = amount of apparent epinephrine in sample indicated by the reading at 510 nm

Hence, $N = mn\,(b - y)/(m - n)$ and $E = y - N/m$.

Blood samples should be taken as quickly as possible and the plasma separated and frozen immediately, since small delays at this stage cause significant losses in catecholamine content. Prolonged storage of frozen plasma samples and repeated thawing and freezing also result in losses. Venipuncture can raise blood catecholamine content. For this reason, when multiple estimations are necessary, samples should be taken through an indwelling venous catheter. Certain common drugs (e.g., Ampicillin, methyldopa, promethazine, protamine, sulfonamides, vitamin B complex) and beverages (e.g., cocoa, coffee, tea) interfere with this fluorometric method for catecholamine estimation. Failure to standardize rigorously factors such as posture and avoidance of interfering substances may invalidate the most careful biochemical analysis and confound interpretation of the data.

With the procedure outlined above, the following mean values ($\pm SE$) in the plasma from 37 *normal* men and women were obtained: epinephrine, 0.22 ± 0.04 ng/ml; and norepinephrine, 0.58 ± 0.11 ng/ml. *Increased* levels are observed after electroshock and shocks caused by hemorrhage, anaphylaxis, endotoxin, or tumbling; after the administration of nicotine, KCl, and other drugs such as chlorpromazine; after surgery and other forms of trauma; after muscular work and tilting; in hypoglycemia; and in pheochromocytoma. *Decreased* levels are observed during Pentothal or Nembutal anesthesia and after the administration of reserpine and ganglionic blocking agents.

The normal levels of dopamine in plasma have not yet been established, although suitable fluorometric methods are now available.[26,97]

DETERMINATION OF CATECHOLAMINES AND THEIR METABOLITES IN URINE

Individual determinations of urinary catecholamines, vanilmandelic acid (VMA), 3-methoxy metabolites (metanephrine and normetanephrine), and homovanillic acid (HVA) are indicated for some clinical investigations. A large number of methods involving colorimetry, fluorometry and gas chromatography for measuring each of these substances have been reported.[13,18,96–98] Colorimetric methods for catecholamines are not sufficiently sensitive and the methods of choice are those based on fluorometry as discussed earlier.[96]

Fluorometric, colorimetric, isotope dilution (see review article[18]), and gas-chromatographic[13] methods for *metanephrine and normetanephrine* estimation have been described. Among them, the colorimetric method described by Pisano[66] has found wide application in clinical laboratories. After acid hydrolysis, the metanephrines are adsorbed on Amberlite CG-50 and eluted with 3 mol/liter NH_4OH. The eluted compounds are then converted to vanillin and reacted with periodate (see VMA estimation). The resulting compound is then assayed spectrophotometrically at 360 nm. This method does not distinguish between metanephrine and normetanephrine but serves as a rapid, easy, and reliable method for screening urine for the diagnosis of pheochromocytomas. The *normal values* of total metanephrines were reported to be 0.6 ($SD \pm 0.3$) mg/d. Occasional interfering substances have been noted with this method.

A recently reported gas-chromatographic method[13] which allows estimation of metanephrines and VMA in the same urine sample appears to be very simple, accurate, and suitable for clinical application. Highly acidified urine (pH 0.9) is extracted with ethyl acetate, which is saved for VMA estimation. Hydrolysis of the aqueous phase by hydrochloric acid is followed by oxidation of the metanephrines to vanillin using periodate at pH 10–11. The vanillin is extracted with methylene dichloride containing the internal standard, *p*-hydroxybenzoic ethyl ester and is then estimated by gas chromatography (3 per cent OV-1 column) as a trimethylsilyl (TMS) derivative. The ethyl acetate phase, which contains VMA, is extracted with acetate buffer, pH 6.2. VMA is oxidized to vanillin by sodium periodate at pH 11–12. After being extracted into methylene dichloride, vanillin is estimated by gas chromatography as above. Results obtained by gas chromatography were compared with those of Pisano's spectrophotometric procedure.[13] While the values for VMA by both methods were very similar, the quantities of metanephrines measured by GLC were substantially lower than those obtained by chemical methods. Such a disparity has prompted the authors to conclude that Pisano's procedure overestimates metanephrines because of the presence of interfering contaminants.

Homovanillic acid (HVA) has been identified as the principal urinary metabolite of dopa and dopamine. A lack of suitable methods for measuring dopamine in blood and urine and the importance of this amine in neuroblastoma have aroused interest in the estimation of urinary HVA, which is excreted in free form in comparatively large amounts. Various methods involving colorimetry, fluorometry, and gas chromatography have been described.[18,96] Methodological complications and deficiencies in most of these techniques have so far prevented the estimation of HVA from becoming a routine clinical test. However, the application of gas chromatography promises to circumvent many technical difficulties, and the gas chromatographic method described by Williams and Greer[98] merits consideration for clinical application. With this technique both HVA and VMA can be analyzed simultaneously. It should be recalled that HVA is the major terminal metabolite of the dopamine pathway, and VMA is the major terminal metabolite of the norepinephrine pathway (see Figs. 13–30 and 13–31). It is therefore important for the diagnosis of neuroblastoma to carry out simultaneous determinations of HVA and VMA, since dopamine and/or norepinephrine may be elevated in these patients.

The general outline of the method is as follows: A volume of urine containing 10 mg of creatinine is acidified to pH 1–2, saturated with NaCl, and extracted five times with equal volumes of diethyl ether. The ether extracts are pooled and evaporated to dryness under a stream of air. The residue is acetylated with 1 ml of acetic anhydride catalyzed by 0.5 ml of trifluoroacetic acid at 75°C in a water bath for 30 min. After evaporation of the acetylating mixtures to dryness, it is then methylated with ethereal diazomethane to form methylacetoxy esters of HVA and VMA. The sample (200 μg creatinine equivalent) is injected into a gas chromatograph containing a nonpolar 3 per cent OV-1 column. Either argon or flame-ionization detectors can be used. If a large peak is observed at the exact retention time of either HVA standard or the VMA standard, the identity of the peak is further checked by repeating the analytical run on a polar column (10 per cent EGA). The quantitation is carried out by comparing the peak area of the standard with that of the sample gas chromatogram. The specificity of measurements has also been confirmed by analyzing the sample in a combined gas chromatograph–mass spectrometer.

Normal values vary between 1 to 40 μg HVA/mg creatinine, and 1 to 12 μg VMA/mg creatinine. In the clinical field, the urinary aromatic acid excretion is expressed in terms of μg/mg creatinine because random specimens of urine may be used when complete collections are difficult to obtain. In addition, variations based on sex and age are minimized.

While it is evident from the above discussion that methods are available for the individual analyses of the catecholamines and their important urinary metabolites, very few of them have found wide application in the clinical field. At the present time, measurements of urinary total catecholamines and VMA are common routine clinical procedures. Two methods which are widely used for their estimation will be described.

ESTIMATION OF TOTAL CATECHOLAMINES[83,96]

PRINCIPLE

Catecholamines are excreted partly in free form and partly in conjugated form; in the latter form they are conjugated predominantly as ethereal sulfates which are easily hydrolyzed by heating with acid. Because of the high solubility of free catecholamines in aqueous media, customary extraction with organic solvents cannot be employed. They may, however, be adsorbed selectively on alumina (or appropriate resins), either by passing the hydrolyzed urine adjusted to pH 8.5 through a column of the adsorbent, or by mixing it with a batch of suspended alumina. The adsorbed catecholamines are then eluted with acid solution. The final method of quantitation by fluorometry is based on the oxidation of catecholamines (epinephrine and norepinephrine) to stable fluorescent derivatives. Epinephrine and norepinephrine are oxidized to corresponding adrenochromes, which rearrange in alkaline medium to the fluorescent compounds adrenolutin and noradrenolutin, respectively. Because the side chain closes to form an indole ring, and because there are three hydroxyl groups present, procedures based on this reaction are referred to as trihydroxy indole (THI) methods. These lutins are very unstable unless they are protected from oxidation by a suitable reducing agent such as ascorbic acid. The reaction sequence for epinephrine is as follows:

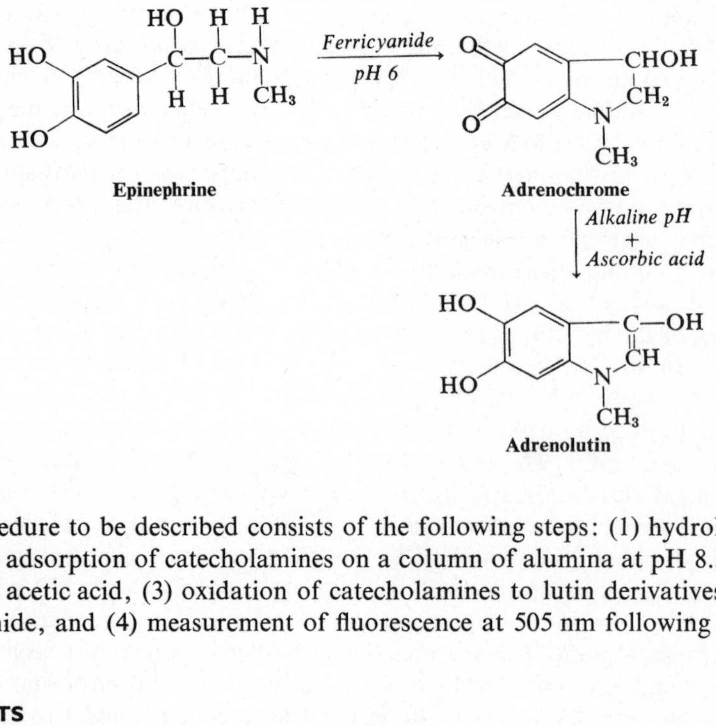

Epinephrine —(*Ferricyanide*, *pH 6*)→ Adrenochrome

Adrenochrome —(*Alkaline pH* + *Ascorbic acid*)→ Adrenolutin

The procedure to be described consists of the following steps: (1) hydrolysis of urine at pH 1.5, (2) adsorption of catecholamines on a column of alumina at pH 8.5 and elution with 0.2 mol/l acetic acid, (3) oxidation of catecholamines to lutin derivatives with potassium ferricyanide, and (4) measurement of fluorescence at 505 nm following excitation at 400 nm.

REAGENTS

1. H_2SO_4, 3 mol/l.

2. Aluminum oxide. (Alcoa activated alumina, grade F-20, 60–200 mesh.) Store tightly stoppered. If the activity has declined because of water adsorption, heat in a muffle furnace for 4 h at 400°C.

3. Stock standard (noradrenaline). Dissolve 20.1 mg levarterenol bitartrate, USP, in 100 ml of 0.1 mol/l hydrochloric acid. This solution contains 100 μg/ml. This remains stable at least six months when refrigerated. (The reaction condition described here is optimal for the formation of a fluorescent derivative from noradrenaline. For this reason,

noradrenaline rather than adrenaline is suitable for a reference standard. Also, noradrenaline is normally present in urine in larger quantity.)

4. Working standard. Dilute the stock standard 1/100 with distilled water. Prepare fresh before each use. (1 ml ≡ 1 μg norepinephrine.)

5. Sodium carbonate, 14 g/100 ml distilled water.

6. Sodium acetate, 0.2 mol/l. Dissolve 27.2 g $CH_3COONa{\cdot}3H_2O$ or 16.41 g CH_3-COONa (anhydrous) in water and dilute to 1 liter with water. Adjust the pH to 8.5 with 0.5 mol/l Na_2CO_3.

7. Acetic acid, 0.2 mol/l. Dilute 6 ml glacial acetic acid to 500 ml water.

8. Sodium hydroxide, 10 mol/l. Dissolve 40 g NaOH pellets and make volume up to 100 ml with water.

9. Sodium acetate, saturated solution. Add 500 g anhydrous salt to 900 ml water.

10. Acetate buffer solution. To 100 ml saturated sodium acetate add 6.0 ml of 10 mol/l NaOH. Do not keep this solution more than one month.

11. Potassium ferricyanide solution, 0.25 g/100 ml H_2O. Prepare only small quantities and store in a dark bottle. Do not store for more than one month.

12. Sodium hydroxide, 20 g/100 ml H_2O.

13. Ascorbic acid. Weigh out 20 mg quantities and place into test tubes. Cover with Parafilm until required for use.

14. Ascorbic acid–NaOH. Dissolve the ascorbic acid above in 1 ml water, and add 9.0 ml of 20 g/100 ml NaOH. Prepare fresh immediately prior to use.

COLLECTION OF SPECIMEN

The stability of catecholamines declines as the pH rises, with destruction becoming extremely rapid in an alkaline medium. If the pH of the urine is maintained below 2 during collection and kept at 4°C, there is virtually no loss of the amine content of the urine even over periods of several months.[18] The addition of 4 ml of 5 mol/l H_2SO_4 per 400 ml urine being collected has been shown to be adequate.[96]

PROCEDURE

Hydrolysis and Alumina Chromatography

1. Adjust two 10 ml aliquots of urine to pH 2 with 3 mol/l H_2SO_4.

2. Reflux the acidified urine for 20 min.

3. Cool and adjust the pH to 8.5 with 14 g/100 ml Na_2CO_3. Centrifuge at 2000 rpm for 5 min.

4. Prepare two alumina columns by adding an aqueous suspension of alumina to two chromatography tubes (8 × 130 mm) until a height of approximately 4 cm is reached. Wash the columns with 5 ml of 0.2 mol/l sodium acetate.

5. Allow all the urine to run through the columns and discard the eluate. Never allow the column to run dry.

6. Wash the column with 20 ml of 0.2 mol/l sodium acetate followed by 5.0 ml of water.

7. Elute the catecholamines with 50 ml of 0.2 mol/l acetic acid, and collect the initial 20 ml of eluate.

Oxidation with Ferricyanide and Fluorometry

8. Transfer 4 ml aliquots of urine eluate and the reagents to each of three test tubes as shown in the following table.

Reagent added	Standard + Sample	Sample	Blank
Eluate from column	4.0 ml	4.0 ml	4.0 ml
Standard (1 μg/ml)	0.5 ml	—	—
Distilled water	—	0.5 ml	0.5 ml
Acetate buffer	1.0 ml	1.05 ml	1.0 ml
Potassium ferricyanide	0.1 ml	0.1 ml	0.1 ml

Mix and allow to stand for 2 min.

9. To all tubes *except the blank* add 1.0 ml freshly prepared ascorbic acid–NaOH. Mix and centrifuge for 2 min. To the blank add 1.0 ml 20 g/100 ml NaOH, and allow to stand for 20 min; centrifuge.

10. Read the fluorescence at 505 nm following excitation at 400 nm.

CALCULATION

$$\mu\text{g of catecholamines/d} = \frac{A_t - A_b}{A_{st} - A_t} \times 0.5 \times \frac{20}{4} \times \frac{\text{T.V. urine}}{10}$$

where A_t = reading of the test
A_b = reading of the faded blank
A_{st} = reading of the internal standard (standard + test)

COMMENTS

Fluorometric methods are highly sensitive, but their susceptibility to extraneous factors demands scrupulous care in the cleaning of glassware and the preparation of reagents. Two common sources of contamination causing nonspecific fluorescence are introduced by the use of detergents and lubricants. Whenever these are used, measures should be taken to remove them completely. Other deleterious effects arise from the presence of heavy metals. Water to be used for the preparation of reagents must be deionized and distilled twice over alkaline potassium permanganate in an all-glass apparatus. Thorough washing of alumina with hydrochloric acid and water is also very important to remove alkaline, fluorescent, and heavy metal impurities. The use of an internal standard is suggested because pure noradrenaline standard produces greater fluorescence than the same standard added to urine eluates from alumina. This has allegedly been caused by the presence of urinary quenching substances.

Although the present method correlates fairly well with the clinical picture, certain inherent drawbacks of this procedure need to be mentioned. The single alumina-column treatment of hydrolyzed urine does not achieve the necessary purification. As a result, nonspecific fluorescence in the so-called *faded-blank* of these preparations is high. The quenching of the fluorescence of the alumina eluates is also liable to cause serious error. Furthermore, the experimental conditions employed for the formation of the fluorescent compounds do not give maximum fluorescence. In addition, the reaction does not include dopamine. Weil-Malherbe[96] advocates the use of a second purification step, consisting of adsorption of amines on a cation-exchange resin to eliminate interfering impurities. According to the same author, the addition of cupric ions catalyzes the formation of lutins and increases the intensity of fluorescence to a considerable extent. The method described appears to improve the intensity of fluorescence, the stability of the blank, and the discrimination between epinephrine and norepinephrine. Since the rates of oxidation of epinephrine and norepinephrine are different at different pH values, a differential oxidation at pH 3 and pH 6 will enable one to measure epinephrine and norepinephrine individually, even if they are present in a mixture. The determination of norepinephrine in the presence of epinephrine may also be carried out by naphthoquinone condensation.

Since the method described here is not absolutely specific, falsely elevated results may be obtained because of the presence of urinary metabolites of medications such as adrenaline, adrenaline-like drugs, tetracyclines, quinidine, and Aldomet (methyldopa). High values may also occur in patients with progressive muscular dystrophy, myasthenia gravis, and widespread burns. Individuals undergoing vigorous exercise may also show elevated urinary catecholamines. Malnutrition, transection of the cervical spinal cord, and familial dysautonomia may cause decreased excretion of catecholamines.

NORMAL VALUES

Normal values vary with the method of analysis. The excretion of total catecholamines in random urine samples ranges from 0 to 14 $\mu g/100$ ml, and up to 100 $\mu g/d$.

REFERENCES

Sobel, C., and Henry, R. J.: Am. J. Clin. Pathol., *27*:240, 1957.

Weil-Malherbe, H.: *In* Methods of Biochemical Analysis. D. Glick, Ed. New York, Interscience Publishers, 1968, vol. *16*, p. 293.

URINARY ESTIMATION OF VMA[67] *(Vanilmandelic Acid; 3-Methoxy-4-Hydroxymandelic Acid)*

Since urinary VMA is quantitatively the most important metabolite of catecholamines (epinephrine and norepinephrine), it has served as a useful index of endogenous production of these amines. Various methods depending on chromatography, isotope dilution, and colorimetry have been used for its quantitative measurement. Among them, colorimetric estimations based on the oxidation of VMA to vanillin have found widest application in routine clinical laboratories. Since the first introduction of such a procedure, a great number of methods with various modifications have appeared in the literature.[96] The method of Pisano *et al.*,[67] which is considered to be the most simple and best suitable for routine clinical use, is described.

PRINCIPLE

VMA, along with other phenolic acids, is extracted from acidified urine with ethyl acetate. It is then extracted from the organic solvent with aqueous potassium carbonate solution. The potassium carbonate extract is treated with sodium metaperiodate to oxidize VMA to vanillin. To separate it from contaminating urinary phenolic acids, vanillin is selectively extracted into toluene and is determined spectrophotometrically at a wavelength of 360 nm.

$$\text{VMA (3-}CH_3O\text{, 4-HO-phenyl)}-\underset{H}{\overset{OH}{C}}-COOH \xrightarrow{NaIO_4} \text{Vanillin (3-}CH_3O\text{, 4-HO-phenyl)}-CHO$$

REAGENTS

All reagents should be of AR quality.

1. Hydrochloric acid, 6 mol/l. Slowly add 500 ml concentrated HCl to a 1 liter volumetric flask containing approximately 300 ml water, and dilute to mark with water. Water to be used for the preparation of all reagents should be distilled twice in an all-glass apparatus.

2. Sodium chloride.
3. Ethyl acetate.
4. Potassium carbonate, 1 mol/l. Dissolve 138 g potassium carbonate in 1 liter of distilled water.
5. Sodium metaperiodate, 2 g/100 ml distilled water. Make fresh weekly and store in an amber bottle.
6. Sodium metabisulfite, 10 g/100 ml distilled water.
7. Acetic acid, 5 mol/l. Dilute 286 ml glacial acetic acid with distilled water to 1 liter.
8. Phosphate buffer, 1 mol/l, pH 7.5. *Solution A:* Dissolve 178 g disodium phosphate ($Na_2HPO_4 \cdot 2H_2O$) in distilled water and dilute to 1 liter. Store in a refrigerator. *Solution B:* Dissolve 27.22 g potassium dihydrogen phosphate (KH_2PO_4) in 200 ml distilled water. Mix 168.2 ml of solution A with 31.8 ml of solution B. Check pH on pH meter and make any necessary adjustment to obtain a pH of 7.5.
9. Hydrochloric acid, 0.01 mol/l. Dilute 0.83 ml of concentrated HCl to 1 liter with distilled water.
10. Standard solutions. Stock solution (1 mg/ml): Accurately weigh 100 mg of VMA and dissolve in 100 ml of 0.01 mol/l HCl in a volumetric flask. The solution is stable approximately three months under refrigeration. Working solution (10 μg/ml): Dilute 1 ml of the stock solution to 100 ml with 0.01 mol/l HCl. Prepare fresh before use.

COLLECTION OF SPECIMEN

To preclude false elevations of urinary VMA, the intake of chocolate, coffee, bananas, foods containing vanilla, citrus fruits, and drugs such as aspirin and antihypertensive agents (e.g., Aldomet) must be restricted three days prior to and during collection of the urine specimen. The pH of the urine should be kept at approximately 2 during the collection by placing 10 ml of 6 mol/l HCl into a suitable container (dark-brown bottle). After measurement of the total volume, 100 ml aliquots may be stored at 4°C for subsequent analysis. The specimen so preserved is stable for several weeks.

It has recently been reported that no dietary restrictions during urine collection are necessary if the VMA method based on the oxidation of VMA to vanillin is used.[68] However, for other methods which employ a reaction of the phenolic acids with diazotized *p*-nitroaniline, a rigid control of diet and drugs is still necessary.

PROCEDURE

Pipet 0.2 per cent of the 24 h volume into 50 ml glass-stoppered (or screw-cap) centrifuge tubes marked previously as "tests," "internal standards," and "unoxidized blanks" in duplicate. To the internal standard tubes add 1 ml of the working standard. Dilute the contents of all these tubes to 5.5 ml with distilled water, and further acidify with 0.5 ml of 6 mol/l hydrochloric acid. Add a saturating amount of sodium chloride (approximately 3 g), mix, and extract with 30 ml of ethyl acetate by shaking on a mechanical shaker for 30 min. Centrifuge for 5 min. Transfer 25 ml of the organic extract (upper layer) to a second glass-stoppered centrifuge tube containing 1.5 ml of 1 mol/l potassium carbonate. Shake mechanically for 3 min and centrifuge for 5 min. Pipet 1 ml of the carbonate phase (lower layer) to a third glass-stoppered centrifuge tube. To the test and standard tubes, add 0.1 ml of 2 g/100 ml sodium metaperiodate, mix, and stopper loosely; place all tubes including the tubes marked "unoxidized blank" (metaperiodate solution is omitted at this stage) into a water bath of 50°C for 30 min. At the end of the incubation period, remove the tubes and cool to room temperature. To the unoxidized blanks, add 0.1 ml of sodium metaperiodate and mix. Without delay add to all tubes 0.1 ml of metabisulfite solution to reduce residual periodate. Neutralize with 0.3 ml of 5 mol/l acetic acid, and add 0.6 ml of 1 mol/l phosphate buffer at pH 7.5. (The pH can be checked at this point by adding one drop of aqueous cresol red (0.04 g/100 ml). The solution should be yellow, indicating a pH of less than 8.8.) Shake mechanically for 3 min with 20 ml of toluene to extract vanillin, the oxidized product of VMA. Centrifuge for 5 min, and transfer 15 ml of the toluene extract into a fourth glass-stoppered centrifuge tube containing 4.0 ml of 1 mol/l potassium carbonate. Shake mechanically for 3 min and centrifuge for 5 min. Transfer the carbonate layer containing vanillin into a microcuvet, and determine the absorbance at 360 nm against a water blank.

CALCULATION

$$\text{mg VMA/d} = \frac{A_t - A_b}{A_{st} - A_t} \times \frac{10}{1000} \times \frac{100}{0.2} = \frac{A_t - A_b}{A_{st} - A_t} \times 5$$

where A_b = absorbance of "unoxidized" urine blank
A_t = absorbance of test
A_{st} = absorbance of internal standard (standard + test)

COMMENTS

Necessary care in the collection and preservation of the urine specimen as outlined in the text is very important. Diets and drugs contributing to the excretion of related phenoxy acids which may be oxidized to vanillin will yield falsely elevated results. However, as a precautionary measure, it is advisable to prepare an unoxidized blank for every sample to correct for the presence of vanillin in urine even when the dietary restrictions prior to and during collection of the specimen have been followed. The absorbance may be measured against the unoxidized blank instead of the water blank, and in that case the need for subtraction of the absorbance of the urine blank from the absorbance of the test samples is obviated. The internal standard (addition of a known amount of VMA to the test specimen) compensates for procedural losses, for decomposition of vanillin, and for the relative inhibition of its formation because of the presence of unknown urinary factors. Indeed, at room temperature the oxidation of VMA by periodate proceeds smoothly in pure solutions, whereas an elevated temperature (50°C) is required for the oxidation of VMA in urinary extracts. In occasional urinary samples, the oxidation may be strongly inhibited even at 50°C.[96]

The oxidation of VMA to vanillin is also sensitive to hydrogen ion concentration. In neutral and acidic solutions, oxidation results in the formation of a yellow pigment; strongly alkaline solutions, on the other hand, delay the formation of vanillin and cause its decomposition. Optimal conditions are obtained in (1 to 15 g/100 ml) sodium or potassium carbonate solution at an approximate pH of 11.[67] The maximum absorption of vanillin occurs at 348 nm. However, at this wavelength considerable absorbance of the oxidation product (*p*-hydroxybenzaldehyde) of *p*-hydroxymandelic acid, a normal constituent of urine, necessitates measurement at 360 nm, where the absorbance of vanillin is 80 per cent of its peak value and interference is minimal. It should be pointed out that the absorbance of vanillin drops sharply from 350 to 380 nm, and it is important that the wavelength setting remain exactly at 360 nm.[67]

NORMAL VALUES

Normal values range from 1.8 to 7.1 mg of VMA/d, or 1.5 to 7.0 μg/mg creatinine.

REFERENCE

Pisano, J. J., Crout, R. J., and Abraham, D.: Clin. Chim. Acta, *7*:285, 1962.

SEROTONIN AND ITS METABOLITE: 5-HYDROXYINDOLE ACETIC ACID (5-HIAA)

Serotonin (5-hydroxytryptamine, 5-HT), a powerful smooth muscle stimulant and vasoconstrictor, is a derivative of the amino acid tryptophan (Fig. 13–32). This compound is formed predominantly in the enterochromaffin cells (otherwise known as argentaffin cells, because of their affinity for silver salts) of the gastrointestinal tract. It is transported in the blood by the platelets and is present in the brain and other tissues. In recent years, interest in this substance and other related hydroxy indoles has grown considerably because of the discovery that they are excreted in large amounts by patients with metastatic carcinoid syndrome (argentaffinoma).

The formation and breakdown of serotonin is depicted in Fig. 13–32. The essential amino acid tryptophan is hydroxylated to form 5-hydroxytryptophan (5-HTP). Approximately 1 to 3 per cent of dietary tryptophan is normally metabolized by this pathway, but as much as 60 per cent of this amino acid is converted to 5-HTP in carcinoid tumors. The 5-hydroxytryptophan is decarboxylated to serotonin (5-hydroxytryptamine). While the enzymatic decarboxylation is most active in carcinoid tumors, it may also take place in the liver, kidney, lung, and brain.

Serotonin is pharmacologically the most active indole amine; however, its biological activity is apparently lost when it is bound to tissues or platelets. It may rapidly undergo oxidative deamination in a tumor or in the blood after release from a tumor. The oxidative deamination of serotonin by the enzyme monoamine oxidase (MAO) leads to the formation of 5-hydroxyindole acetic acid (5-HIAA), which is quantitatively the most significant metabolite of the 5-hydroxyindole pathway. The majority of the 5-HIAA is excreted in the free form, although a small amount may be conjugated as the O-sulfate ester before excretion.

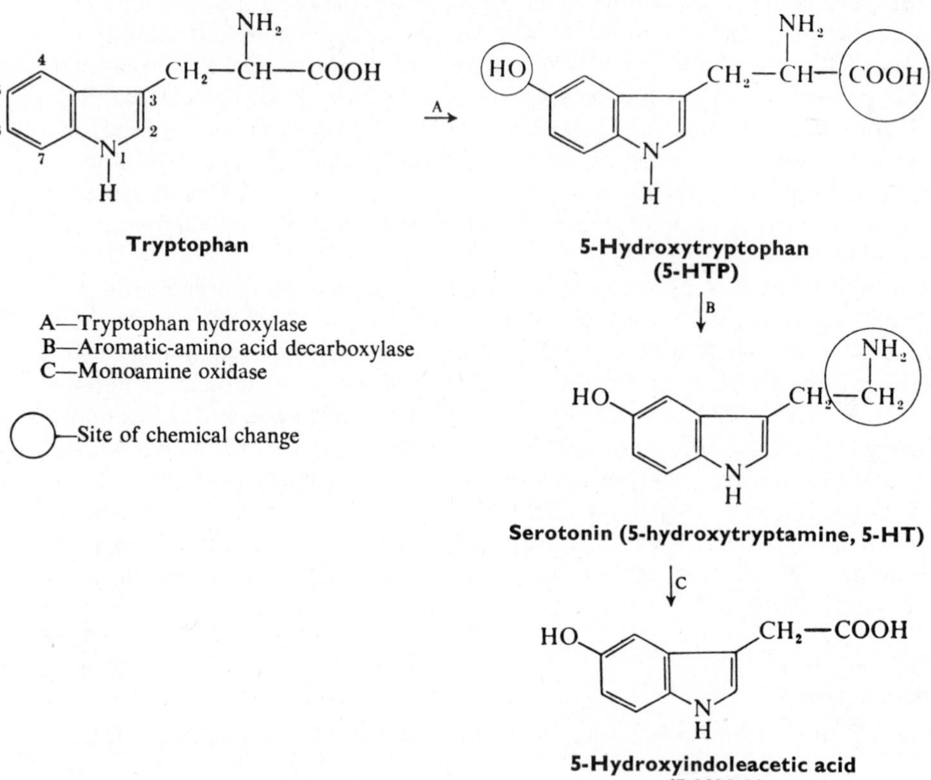

Figure 13–32 Biogenesis and metabolism of serotonin.

CLINICAL SIGNIFICANCE

Metastatic carcinoid tumors (argentaffinoma), arising from the argentaffin cells, produce excessive amounts of serotonin. Since the bulk of urinary 5-hydroxyindole acetic acid (5-HIAA) is derived from serotonin (5-hydroxytryptamine), the amount of this acid excreted reflects the secretion of the amine. Urinary levels of 5-HIAA ranging up to as much as 1 g/d have been reported in patients with carcinoid tumors. While marked *increases* are always associated with carcinoidosis, a slight elevation may be observed in some patients with nontropical sprue, or transiently following the administration of reserpine. In renal insufficiency and in some instances of phenylketonuria, *decreased* excretion of 5-HIAA has been noted.

URINARY DETERMINATION OF 5-HYDROXYINDOLE ACETIC ACID (5-HIAA)

Estimation of the parent hormone, serotonin, in blood and urine has been severely limited in the clinical laboratory because of its very low concentration and because of methodological complications. As a result, the urinary determinations of 5-HIAA continue to be the most useful means for the diagnosis of carcinoid tumors. In such cases, this metabolite of serotonin is excreted in very large amounts, often exceeding 350 mg/d, and a positive result is obtained on simple qualitative (screening) tests. However, for early diagnosis, when tumors are small and have not metastasized, and in some carcinoid tumors where the excretion values barely exceed 8 mg, the more sensitive and specific quantitative test is required. Here, both qualitative and quantitative procedures based on the methods reported by Udenfriend and his associates[90] are described.

Screening Test[80]

The test is based on the development of a purple color, specific for 5-hydroxyindoles, on the addition of 1-nitroso-2-naphthol and nitrous acid. Other interfering chromogens are extracted in ethylene dichloride.

COLLECTION OF SPECIMEN

A random specimen is usually suitable for the screening test. For quantitative analysis, a complete collection over 24 h should be used. The specimen should be collected in a bottle containing 25 ml of glacial acetic acid. Acidification is important to prevent decomposition of 5-hydroxyindole acetic acid. Falsely negative results may occur in patients taking phenothiazine drugs. The ingestion of bananas, pineapples, walnuts, other foods containing serotonin, or cough medication containing glycerol guaiacolate may produce falsely positive results. Therefore, these drugs and diets should be restricted three to four days prior to the collection.

REAGENTS

1. 1-Nitroso-2-naphthol, 0.1 g/100 ml 95 per cent ethanol.
2. Sulfuric acid, 1 mol/l.
3. Sodium nitrite, 2.5 g/100 ml water. Prepare freshly at frequent intervals. Refrigerate.
4. Nitrous acid reagent. Prepare fresh before use by adding 0.2 ml sodium nitrite (2.5 g/100 ml) to 5 ml of 1 mol/l sulfuric acid.
5. Ethylene dichloride, redistilled.

PROCEDURE

Pipet into a test tube 0.2 ml of urine, 0.8 ml of distilled water, and 0.5 ml of 1-nitroso-2-naphthol. Mix. Similarly prepare another tube with *normal* urine to serve as a negative control. Add 0.5 ml of freshly prepared nitrous acid reagent to both tubes and mix again. Let the tubes stand at room temperature for 10 min. Add 5 ml of ethylene dichloride and shake. If turbidity results, centrifuge. A *positive test* shows a purple color in the top aqueous layer. The negative control with normal urine produces a slight yellow color.

COMMENTS

Dietary and drug restrictions as outlined are important if falsely negative or positive results are to be avoided. The substance *p*-hydroxyacetanilide derived from acetanilide or related drugs reacts similarly and adds to the color. Color formation may be inhibited in clinical conditions accompanied by excretion of large amounts of keto acids.

A purple color (positive test) will be seen at levels of 5-HIAA excretion as low as 40 mg/d. At higher levels, the color is more intense and is almost black at levels above 300 mg/d. A positive result should be verified with a quantitative method.

Quantitative Test[90]

PRINCIPLE

This method is considered to be specific for 5-hydroxyindole acetic acid (5-HIAA). The procedure involves preliminary treatment of the urine with dinitrophenylhydrazine to react with any keto acids which may interfere. The urine is then extracted with chloroform to remove indole acetic acid. After saturation of the aqueous phase with sodium chloride, the 5-hydroxyindole acetic acid is extracted into ether and is then returned to a buffer of pH 7.0 for the color reaction, as described previously. The absorbance of the reaction product is measured at 540 nm in a spectrophotometer.

COLLECTION OF SPECIMEN

Follow the procedure described for the screening test.

REAGENTS

Prepare reagents 1 to 4 as described for the screening test.

5. 2,4-Dinitrophenylhydrazine, 0.5 g/100 ml in 2 mol/l HCl.
6. Chloroform, reagent grade, freshly redistilled.
7. Sodium chloride.
8. Diethyl ether, reagent grade. Wash with a saturated solution of acidified ferrous sulfate to destroy peroxides, then with water (twice). Distill in an all-glass apparatus.
9. Ethyl acetate, redistilled.
10. Phosphate buffer, 0.5 mol/l, pH 7.0. Mix 61.1 ml of a solution of $Na_2HPO_4 \cdot 2H_2O$ (89.07 g/l) and 38.9 ml of a solution of KH_2PO_4 (68.085 g/l). Check the pH on the pH meter and make any necessary adjustments to obtain pH 7.0.
11. Standard solutions. *Stock solution:* Weigh accurately 5 mg of pure 5-hydroxyindole acetic acid and dissolve in 20 ml of glacial acetic acid. *Working solution:* Transfer 1 ml of stock solution and dilute to 25 ml with distilled water. This solution contains 10 μg/ml.

PROCEDURE

Transfer 6 ml of urine to a 50 ml glass-stoppered centrifuge tube. Prepare blank and standard tubes using 6 ml of distilled water and 6 ml (60 μg) of working standard solution, respectively. Add 6 ml of 2,4-dinitrophenylhydrazine reagent. (If a large precipitate occurs following addition of the reagent, centrifuge and use the supernatant for the subsequent steps.) Allow to stand for 30 min. Add 25 ml of chloroform, shake for a few minutes, and centrifuge. Remove the organic layer, add another 25 ml of chloroform, and repeat the extraction. After centrifuging, transfer a 10 ml aliquot of the aqueous layer to a 40 ml glass-stoppered centrifuge tube. Add approximately 4 g of solid sodium chloride and 25 ml of ether. Shake for 5 min and centrifuge. Transfer a 20 ml aliquot of the ether to another 40 ml glass-stoppered centrifuge tube. Add 1.5 ml of phosphate buffer at pH 7.0. Shake for 5 min, centrifuge, and aspirate off the ether layer. Transfer 1 ml of the aqueous phase to a 15 ml glass-stoppered centrifuge tube containing 0.5 ml of nitroso-naphthol reagent. Add 0.5 ml of nitrous acid reagent, mix well, and warm at 37°C for 5 min. Add 5 ml of ethyl acetate, shake, allow the layers to separate, and remove the organic layer. Add another 5 ml of ethyl acetate and repeat the above process. Transfer the final aqueous layer to a microcuvet. Read the absorbance of the standard and test samples against the blank at 540 nm.

CALCULATION

$$\text{mg 5-hydroxyindole acetic acid/l urine} = \frac{A \text{ of unknown}}{A \text{ of standard}} \times \frac{60}{1000} \times \frac{1000}{6}$$

$$= \frac{A \text{ of unknown}}{A \text{ of standard}} \times 10$$

or

$$\text{mg/d} = \frac{A \text{ of unknown}}{A \text{ of standard}} \times \frac{60}{1000} \times \frac{\text{T.V. (ml)}}{6}$$

COMMENTS

The important points relevant to the urinary determination of 5-hydroxyindole acetic acid have already been discussed. The salient feature of the quantitative method is the removal of keto acids through the formation of phenylhydrazones. The presence of excessive amounts of keto acids interferes with color formation. Reaction of 5-hydroxyindole acetic acid with nitroso-naphthol to form a violet chromophore is claimed to be very specific. Serotonin and 7-hydroxyindoles do not respond to this reaction.[90] The use of organic solvents such as chloroform and ethyl acetate at different steps of the procedure insures the removal of other indole acetic acids and nonspecific substances. The distribution of 5-HIAA between ether and water is very low; the addition of a saturating amount of sodium chloride aids in the quantitative transfer of 5-HIAA into the ether phase. The pH of the buffer solution should be exactly 7, since at higher pH values the compound becomes progressively more unstable.

NORMAL VALUES

The normal values for urinary excretion of 5-HIAA range from 2 to 8 mg/d.

REFERENCES

Sjoerdsma, A., Weissbach, H., and Udenfriend, S.: J.A.M.A., *159*:397, 1955.
Udenfriend, S., Titus, E., and Weissbach, H.: J. Biol. Chem., *216*:499, 1955.

REFERENCES

1. Albert, A. (Ed.): Human Pituitary Gonadotrophins. Springfield, Illinois, Charles C Thomas, Publisher, 1961.
2. Albert, A., Mattox, V. R., and Mason, H. L.: *In* Textbook of Endocrinology. R. H. Williams, Ed. Philadelphia, W. B. Saunders Company, 1968, p. 1181.
3. Baird, D. T.: J. Clin. Endocrinol., *28*:244, 1968.
4. Bardin, C. W., and Mahoudeau, J. A.: Ann. Clin. Res., *2*:251, 1970.
5. Bayard, F., Beitins, I. Z , Kowarski, A., and Migeon, C. J.: J. Clin. Endocrinol., *31*:1, 1970.
6. Bayliss, W. M., and Starling, E. H.: J. Physiol., *28*:325, 1902.
7. Beardwell, C. G., Burke, C. W., and Cope, C. L.: J. Endocrinol., *42*:79, 1968.
8. Berson, S. A., and Yalow, R. J.: *In* The Hormones. G. Pincus, K. V. Thimann, and E. B. Astwood, Eds. New York, Academic Press, Inc., 1964, Vol. 4, p. 557.

9. Borth, R., Lunenfeld, B., and Menzi, A.: *In* Human Pituitary Gonadotrophins. A. Albert, Ed. Springfield, Charles C Thomas, Publisher, 1961, p. 13.
10. Brown, J. B.: Biochem. J., *60*:185, 1955.
11. Brown, J. B., MacLeod, S. C., Macnaughton, C., Smith, A. M., and Smyth, B.: J. Endocrinol., *42*:5, 1968.
12. Brown, J. B., Macnaughton, C., Smith, M. A., and Smyth, B.: J. Endocrinol., *40*:175, 1968.
13. Calseyde, J. F. van de, Scholtis, R. J., Schmidt, N. A., and Leyten, C. J.: Clin. Chim. Acta, *32*:361, 1971.
14. Catt, K. J., Tregear, G. W., Burger, H. G., and Skermer, C.: Clin. Chim. Acta, *27*:267, 1970.
15. Chattoraj, S. C., and Wotiz, H. H.: Fertil. Steril., *18*:342, 1967.
16. Cope, C. L., and Black, E. G.: Br. Med. J., *2*:117, 1959.
17. Diczfalusy, E. (Ed.): Steroid Assay by Protein Binding. Karolinska Symposia on Research Methods in Reproductive Endocrinology. Stockholm, 1970.
18. Dorfman, R. I. (Ed.): Methods in Hormone Research. 2nd. ed. New York, Academic Press, Inc., 1967, Vol. I.
19. Dorfman, R. I. (Ed.): Methods in Hormone Research. 2nd ed. New York, Academic Press, Inc., 1969, Vol. IIA.
20. Dorfman, R. I., and Ungar, F.: Metabolism of Steroid Hormones. New York, Academic Press, Inc., 1965.
21. Drekter, I. J., Heisler, A., Scism, G. R., Stern, S., Pearson, S., and McGavack, T. H.: J. Clin. Endocrinol., *12*:55, 1952.
22. Eik-Nes, K. B., and Horning, E. C. (Eds.): Gas Phase Chromatography of Steroids. New York, Springer-Verlag, 1968.
23. Galksov, A.: Acta Endocrinol., *69*:Suppl. 162, 1972.
24. Gemzell, C., and Johansson, E. D. B.: In Control of Human Fertility. E. Diczfalusy and U. Borell, Eds. New York, Wiley Interscience, 1971, p. 241.
25. Gray, C. H., and Bacharach, A. L. (Eds.): Hormones in Blood. 2nd ed., New York, Academic Press, Inc. 1967, Vol. I.
26. Gray, C. H., and Bacharach, A. L. (Eds.): Hormones in Blood. 2nd ed. New York, Academic Press, Inc., 1967, Vol. II.
27. Greenblatt, R. B., Zarate, A., and Mahesh, V. B.: *In* Clinical Endocrinology. E. B. Astwood and C. E. Cassidy, Eds. New York, Grune and Stratton, 1968, Vol. II.
28. Greene, J. W., Smith, K., Kyle, G. C., Touchstone, J. C., and Duhring, J. L.: Am. J. Obstet. Gynecol., *91*:684, 1965.
29. Gurpide, E., Giebenhain, M. E., Tseng, L., and Kelly, W. G.: Am. J. Obstet. Gynecol., *109*:897, 1971.
30. Henry, J. B., Krieg, A. F., and Davidsohn, I.: *In* Clinical Diagnosis by Laboratory Methods. 14th ed., I. Davidsohn and J. B. Henry, Eds. Philadelphia, W. B. Saunders Company, 1969, p. 1181.
31. Hobson, B. M.: J. Reprod. Fertil., *12*:33, 1966.
32. Horton, R., Kato, T., and Sherins, R.: Steroids, *10*:245, 1967.
33. Horwitz, C. A., Garmezy, L., Lyon, F., Hensley, M., and Burke, M. D.: Am. J. Clin. Pathol., *58*:305, 1972.
34. Hunter, W. M., and Greenwood, F. C.: Biochem. J., *91*:43, 1964.
35. Ittrich, G.: Z. Physiol. Chem., *312*:1, 1958.
36. Johansson, E. D. B.: Acta Endocrinol. *61*:607, 1969.
37. Johansson, E. D. B.: Acta Endocrinol., *61*:592, 1969.
38. Kirkham, K. E., Hunter, W. M., Jeffrey, F. H., and Bennie, J. G., *In* In Vitro Procedures with Radioisotopes in Medicine. Vienna, International Atomic Energy Agency, 1970, p. 597.
39. Klinefelter, H. F., Jr., Albright, F., and Griswold, G. C.: J. Clin. Endocrinol., *3*:529, 1943.
40. Klopper, A.: Am. J. Obstet. Gynecol., *107*:807, 1970.
41. Klopper, A., Michie, E. A., and Brown, J. B.: J. Endocrinol., *12*:209, 1955.
42. Li, C. H.: Recent Progr. Horm. Res., *18*:1, 1962.
43. Liao, S., and Fang, S.: Vitam. Horm., *27*: 18, 1969.
44. Lipsett, M. B. (Ed.): Gas Chromatography of Steroids in Biological Fluids, New York, Plenum Press, 1965.
45. Loraine, J. A., and Bell, E. T.: Hormone Assays and Their Clinical Application. 3rd ed. Baltimore, Williams & Wilkins Co., 1971.
46. Manger, W. M., Steinsland, O. S., Nahas, G. G., Wakin, K. G., and Dufton, S.: Clin. Chem., *15*:1101, 1969.
47. Marlow, H. W.: J. Biol. Chem., *183*:167, 1950.
48. Marrian, G. F.: *In* Proceedings of Third International Congress in Biochemistry, Brussels, 1955, p. 205.
49. Martin, L. E., and Harrison, C.: Anal. Biochem., *23*:529, 1968.
50. Martini, L., and Ganong, W. F. (Eds.): Frontiers in Neuroendocrinology, New York, Oxford University Press, 1971.
51. Mattingly, D.: J. Clin. Pathol., *15*:374, 1962.
52. Mayes, D., and Nugent, C. A.: J. Clin. Endocrinol., *28*:1169, 1968.
53. McKenzie, J. M.: Physiol. Rev., *48*:252, 1968.

54. Midgley, A. R., Jr.: Endocrinol., *79*:10, 1966.
55. Midgley, A. R. Jr.: J. Clin. Endocrinol. *27*:295, 1967.
56. Mikhail, G., Wu, C. H., Ferin, M., and Vande Wiele, R. L.: Steroids, *15*:333, 1970.
57. Mishell, D. R., Wide, L., and Gemzell, C. A.: J. Clin. Endocrinol., *23*:125, 1963.
58. Murphy, B. E. P.: J. Clin. Endocrinol., *27*:973, 1967.
59. Murphy, B. E. P.: J. Clin. Endocrinol., *28*:343, 1968.
60. Murphy, B. E. P.: Recent Progr. Horm. Res., *25*:563, 1969.
61. Nachtigall, L., Bassett, M., Hogsander, U., Slagle, S., and Levitz, M.: J. Clin. Endocrinol., *26*:941, 1966.
62. Odell, N. D., and Daughaday, W. H. (Eds.): Principles of Competitive Protein-Binding Assays. Philadelphia, J. B. Lippincott Co., 1971.
63. Parlow, A. F.: Fed. Proc., *17*:402, 1958.
64. Peron, F. G., and Caldwell, B. V.: Immunologic Methods in Steroid Determination. New York, Appleton-Century-Crofts, 1970.
65. Peterson, R. E.: *In* Lipids and Steroid Hormones in Clinical Medicine. F. W. Sunderman and F. W. Sunderman, Jr. Eds. Philadelphia, J. B. Lippincott Co., 1960, p. 164.
66. Pisano, J. J.: Clin. Chim. Acta, *5*:406, 1960.
67. Pisano, J. J , Crout, R. J., and Abraham, D.: Clin. Chim. Acta, *7*:285, 1962.
68. Rayfield, E. J., Cain, J. P., Casey, M. P., Williams, G. H., and Sullivan, J. M.: J.A.M.A., *221*:704, 1973.
69. Sanger, F.: Science, *129*: 1340, 1959.
70. Saxena, B. B., Beling, C. G., and Gandy, H. M. (Eds.): Gonadotropins. New York, Wiley Interscience, 1972.
71. Saxena, B. N., Retetoff, S., Emerson, K., and Selenkow, H. A.: Am. J. Obstet. Gynecol., *101*:874, 1968.
72. Sayers, M. A., Sayers, G., and Woodbury, L. A.: Endocrinol., *42*:379, 1948.
73. Schally, A. V., Arimura, A., and Kastin, A. J.: Science, *179*:341, 1973.
74. Schally, A. V., Kastin, A. J., and Arimura, A.: Am. J. Obstet. Gynecol., *114*:423, 1972.
75. Schneider, G. T., Weed, J. C., and Rice, B. F.: Am. J. Obstet. Gynecol., *113*:176, 1972.
76. Scholler, R., Leymarie, P., Heron, M., and Jayle, M. F.: Acta Endocrinol., *52*: Suppl. 107, 1966.
77. Scommegna, A., and Chattoraj, S. C.: Am. J. Obstet. Gynecol., *99*:1087, 1967.
78. Scommegna, A., and Chattoraj, S. C.: Obstet. Gynecol., *32*:277, 1968.
79. Scommegna, A., Chattoraj, S. C., Wotiz, H. H.: Fertil. Steril., *18*:342, 1967.
80. Sjoerdsma, A., Weissbach, H., and Udenfriend, S.: J.A.M.A., *159*:397, 1955.
81. Skelley, D. S., Brown, L. P., and Besch, P. K.: Clin. Chem., *19*:146, 1973.
82. Sobel, C. S., Golub, O. J., Henry, R. J., Jacobs, S. L., and Basu, G. K.: J. Clin. Endocrinol., *18*:208, 1958.
83. Sobel, C. S., and Henry, R. J.: Am. J. Clin. Path., *27*:240, 1957.
84. Steelman, S. L., and Pohley, F. M.: Endocrinol., *53*:604, 1953.
85. Sunderman, F. W., Jr.: *In* Lipids and Steroid Hormones in Clinical Medicine. F. W. Sunderman and F. W. Sunderman, Jr., Eds. Philadelphia. J. B. Lippincott Co., 1960, p. 162.
86. Taylor, K. W.: *In* Hormones in Blood. 2nd ed. C. H. Gray and A. L. Bacharach, Eds. New York, Academic Press, Inc. 1967, Vol. I, p. 47.
87. Taymor, M. L., Yussman, M. A., and Gminski, D.: Fertil. Steril., *21*:759, 1970.
88. de la Torre, B., Johanisson, E., and Diczfalusy, E.: Acta Obstet. Gynecol. Scand., *49*:165, 1970.
89. Turner, C. D.: General Endocrinology. 4th ed. Philadelphia, W. B. Saunders Company, 1966.
90. Udenfriend, S., Titus, E., and Weissbach, H.: J. Biol. Chem., *216*:499, 1955.
91. Vande Wiele, R. L., Bogumil, J., Dyrenfurth, I., Ferin, M., Jewelewicz, R., Warren, M., Rizkallah, T., and Mikhail, G.: Recent Progr. Horm. Res., *26*:63, 1970.
92. Vander Molen, H. J.: *In* The Androgens of the Testis. K. B. Eik-Nes, Ed. New York, Marcel Dekker, 1970, p. 206.
93. Vestergaard, P., and Clausen, B.: Acta Endocrinol., Suppl. 64, 1962.
94. Victora, J. K., Baukal, A., and Wolff, F. W.: Anal. Biochem., *23*:513, 1968.
95. von Euler, U. S.: *In* Hormones in Blood, C. H. Gray and A. L. Bacharach, Eds. New York, Academic Press, Inc., 1961, p. 515.
96. Weil-Malherbe, H.: *In* Methods of Biochemical Analysis. D. Glick, Ed. New York, Interscience Publishers, 1968, Vol. 16, p. 293.
97. Weil-Malherbe, H.: *In* Methods of Biochemical Analysis. (Suppl. volume on Biogenic Amines.) D. Glick, Ed. New York, Interscience Publishers, 1971, p. 119.
98. Williams, C. M., and Greer, M.: *In* Methods in Medical Research. R. E. Olson, Ed. Chicago, Year Book Medical Publishers, Inc., 1970, p. 106.
99. Williams, R. H. (Ed.): Textbook of Endocrinology. 5th ed. Philadelphia, W. B. Saunders Company, 1974.
100. Wotiz, H. H., and Chattoraj, S. C.: *In* Gas Chromatography of Steroids in Biological Fluids. M. B. Lipsett, Ed. New York, Plenum Press, 1965, p. 195.
101. Wotiz, H. H., and Chattoraj, S. C.: J. Chromatogr. Sci., *11*: 167, 1973.
102. Wotiz, H. H., and Clark, S. J.: *In* Gas Chromatography in the Analysis of Steroid Hormones. New York, Plenum Press, 1966.
103. Yalow, R. S., Glick, S. M., and Berson, S. A.: J. Clin. Endocrinol., *24*:1219, 1964.

CHAPTER 14

THYROID FUNCTION

by Sheldon Berger, M.D., and James L. Quinn, III, M.D.

The fully developed human thyroid gland, as viewed from the front, has been likened to a butterfly. The "wings of the butterfly" are the two lobes, and their connecting piece is the isthmus. The gland is wrapped tightly around the anterior and lateral surfaces of the trachea and larynx, the isthmus crossing the trachea just below the cricoid cartilage. Each of the lobes measures approximately 4.0 cm in height, 1.5 to 2.0 cm in width, and 2.0 to 4.0 cm in thickness; the isthmus is about 2.0 cm in both length and width and 0.2 to 0.6 cm in thickness. The average gland weighs 25 to 30 g.

The secretory unit of the gland is the follicle. Each follicle is a sphere having a diameter of approximately 300 μm, and its walls consist of an epithelial cell monolayer. These cells manufacture and secrete the two thyroid hormones, L-thyroxine (3,5,3′,5′-L-tetraiodothyronine) and L-triiodothyronine (3,5,3′-L-triiodothyronine). These hormones, commonly designated T_4 and T_3, respectively, are stored within the lumina of the follicles.

Hormone synthesis and *release* is controlled by a polypeptide which originates in the anterior lobe of the pituitary, the thyroid-stimulating hormone, or TSH. Pituitary secretion of TSH is, in turn, regulated by two factors: a hormone released by the hypothalamus into the hypophyseal-portal venous system, called thyrotrophin-releasing hormone (TRH); and the concentration of free thyroid hormones (FT_3 and FT_4) in the interstitial fluid which bathes both the pituitary and hypothalamus.

Since an increased concentration of hormone suppresses secretion of TSH, and a decreased concentration augments secretion of TSH, this regulatory arrangement has been designated a negative feedback system. TSH secretion, in turn, is regulated by a hypothalamic tripeptide called thyrotrophin-releasing hormone (TRH). Recent studies indicate that thyroxine blocks the pituitary response to TRH. As thyroid hormone levels fall below a critical level, the TSH-secreting cells of the anterior pituitary become responsive to TRH and secrete TSH which, in turn, raises T_3 and T_4 levels in the blood. When thyroid hormone levels rise excessively, these cells become refractory to TRH and cease production of TSH.

Thyroid hormone *biosynthesis* involves five distinct steps (Fig. 14–1):

1. Thyroidal trapping of serum iodide, catalyzed by a trapping enzyme. This step, thought to be oxygen-dependent, is inhibited by certain anions, including iodide in high concentration and perchlorate (ClO_4^-).

2. Enzymatic oxidation of iodide to some reactive intermediate ("active I"). The enzyme responsible for iodide oxidation is a peroxidase, and the reaction has been written as follows:

$$H_2O_2 + 2\,I^- + 2\,H^+ \xrightarrow[\textit{peroxidase}]{\textit{Iodine}} 2\text{ active I} + 2\,H_2O$$

Active I may be the iodinium ion (I^+).

3. Iodination of tyrosyl residues present in follicular thyroglobulin by active I to produce mono- and diiodotyrosine (MIT and DIT). Steps 2 and 3 are closely coupled. The product of the peroxidase reaction, active I, is quickly incorporated into the 3-, or the 3- and 5-positions of the tyrosyl residues by a tyrosine iodinase.

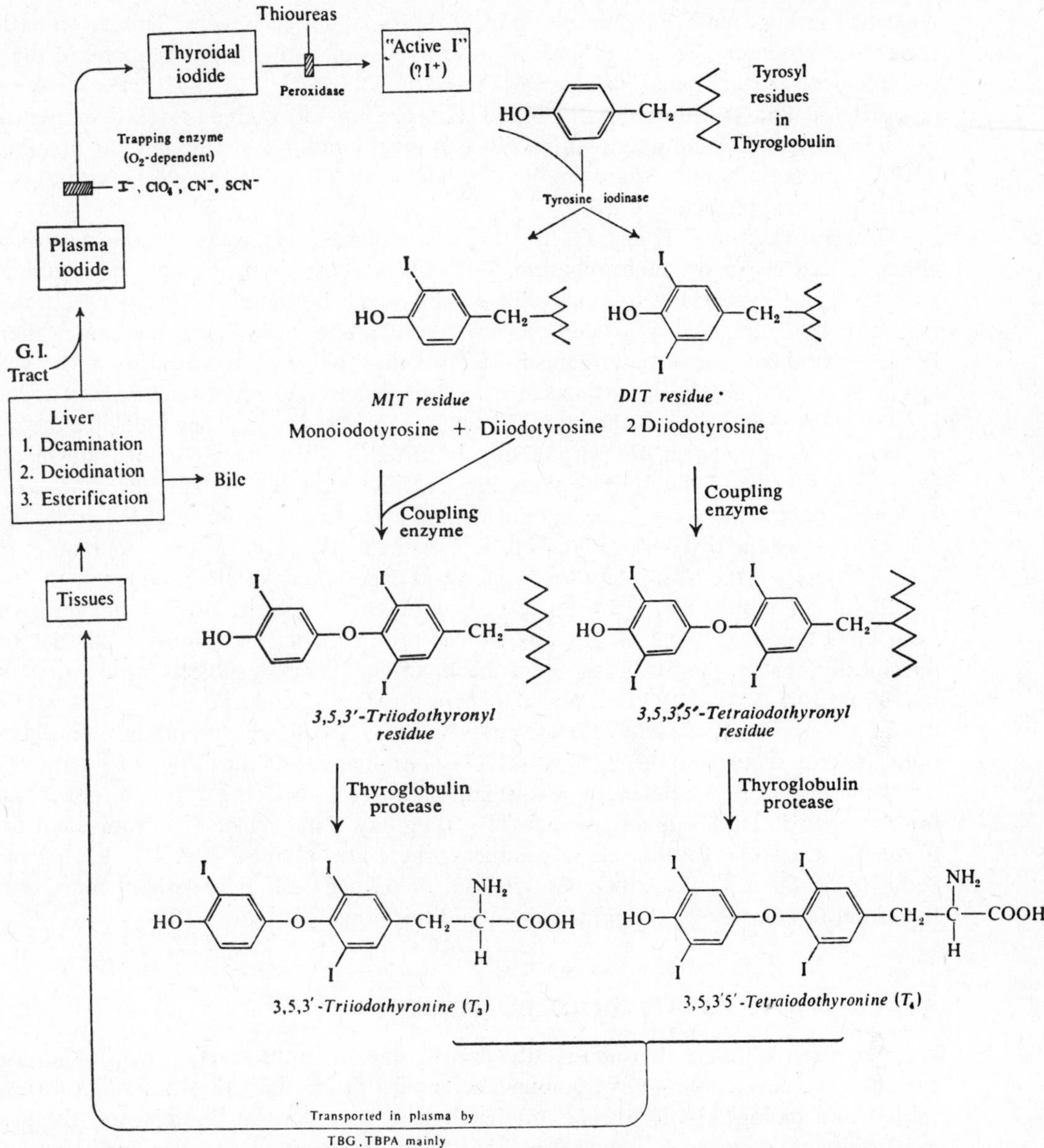

Figure 14–1 The metabolism of iodine, emphasizing formation and secretion of the thyroid hormones.

4. Oxidation and condensation of MIT and DIT to form the iodothyronines, T_3 and T_4. This step is presumed to represent an enzymatic coupling of two molecules of iodotyrosine with extrusion of an alanine side chain. It is important to appreciate that this coupling occurs while the iodotyrosines are bound in peptide linkages to other amino acids, and not when they are in the free form. Thus, the T_3 and T_4 formed as a result of iodotyrosyl coupling are also stored as peptide-linked iodothyronyl residues in thyroglobulin.

5. Proteolytic cleavage of follicular thyroglobulin to "free" the MIT, DIT, T_3, and T_4, which are incorporated, by peptide bonds, into the parent protein. The MIT and DIT are promptly deiodinated, and the iodide is recycled.

Proteolytic cleavage is achieved by thyroglobulin protease. MIT and DIT are deiodinated by a specific deiodinase (free iodotyrosine deiodinase), which is an NADP-dependent enzyme found in the microsomes of thyroidal epithelial cells. This deiodination is considered a conservation step, since it prevents loss of iodine from the gland in the form of biologically inactive material (i.e., MIT and DIT).

After T_3 and T_4 diffuse into the blood, they are bound to three proteins: an inter-alpha globulin (thyroxine-binding globulin, TBG), a prealbumin (thyroxine-binding prealbumin, TBPA), and an albumin (thyroxine-binding albumin, TBA). More will be said about the carrier proteins, particularly TBG.

The *metabolism* of T_3 and T_4, like that of all amino acids, involves their deamination, either oxidatively or by transamination. The resulting thyropyruvates are then decarboxylated to form thyroacetates, which still exhibit some hormonal activity. The hormones may also be conjugated as glucuronides in the liver, in which form they enter the bile. Proper iodine balance in the organism requires that iodine be removed from the various iodine-containing catabolites by specific deiodinases and returned to the thyroid as iodide.

The *biological effects* of the thyroid hormones are profound. They influence the rate of oxygen consumption and heat production in virtually all tissues. The mechanism of this effect may involve uncoupling of oxidation and phosphorylation. The hormones are also indispensable for the growth, development, and sexual maturation of growing organisms. The capacity of the thyroid hormones to trigger the metamorphosis of amphibian tadpoles is, in fact, one of the classic bioassay systems in experimental endocrinology.

T_4 is commonly but erroneously considered the principal thyroid hormone. T_3 is assigned a secondary role because its concentration in plasma is only 1/30 that of T_4. Specifically, mean plasma T_4 concentration among normal subjects approximates 7.0 μg/100 ml (4.5 μg T_4-I/100 ml), and the corresponding T_3 concentration is 0.24 μg/100 ml (0.12 μg T_3-I/100 ml). To equate relative physiologic importance with plasma concentration is always dangerous, and the T_4-T_3 situation is a dramatic example of this easy error.

In terms of relative potency (on a weight basis) $T_3/T_4 = 3/1$. Further, disposal kinetics for T_4 approximate 80 μg/day, and for T_3 50 μg/day. Fifty μg of T_3 is equivalent to 150 μg of T_4. A normal thyroid gland produces, therefore, about 230 (150 + 80) μg of "T_4 equivalent." Of this, T_3 contributes 150/230, or 65 per cent. It is perhaps more correct, therefore, to consider T_3 the principal thyroid hormone.[27]

EVALUATION OF THYROID FUNCTION

Characterization of thyroid function has become, in recent years, an elegant laboratory exercise. The development of ingenious biochemical procedures for estimating the concentration and biological effectiveness of the thyroid hormones in plasma, and the general availability of radioactive iodine for clinical use, have broadened the understanding of thyroid physiology to a degree which few would have imagined 20 years ago. Still, the newer tests which have been developed, though increasingly specific in their capacity to measure thyrometabolic status, are not without important limitations. None of the tests can be interpreted casually, since correct interpretation often depends on a subtle point in iodine metabolism. The modern tests now require an order of awareness of basic matters which was quite unnecessary in the (not too remote) days when measurement of the basal metabolic rate (BMR) was the "court of final appeal." But progress is not without its associated problems and responsibilities, and one simply must learn what must be known.[7]

Unlike the BMR, none of the tests which have supplanted it assesses directly the metabolic impact of thyroid hormones on peripheral tissues, and, after all, it is this impact which ultimately determines clinical thyroid status. A word about the BMR is perhaps in order, then, at the outset, to explain why it has been largely superseded.

Following the classic studies of Magnus-Levy, Benedict, the DuBois, and Boothby, the BMR became the test for assessing the calorigenic effect of thyroid hormones, and in

TABLE 14–1 EXTRATHYROIDAL FACTORS WHICH AFFECT THE BMR*

Factors Which Increase the BMR

1. Faulty preparation for the test
   a. anxiety
   b. inadequate rest
   c. recent food ingestion
   d. calorigenic drugs (xanthines, sympathomimetics, thyroid preparations)
2. Errors in test performance
   a. increased environmental noise
   b. extremes in room temperature
   c. uncomfortable table or bed
   d. insufficient elevation of patient (patients with cardiac failure, obesity)
   e. tight nose piece, girdle, or collar
   f. oxygen leaks (test system, perforated ear drum)
   g. improper breathing pattern
   h. beginning test with overexpansion of chest
3. Systemic disorders which increase oxygen consumption
   a. involuntary motor disorders
   b. cardiac failure
   c. pulmonary disease
   d. leukemia
   e. skin disease
   f. fever of any cause

Factors Which Decrease the BMR

1. Errors in test performance
   a. saturated soda lime
   b. beginning test with underexpansion of chest
2. Systemic disorders which decrease oxygen consumption
   a. malnutrition
   b. Addison's disease
   c. nephrotic syndrome
   d. shock

* After Ingbar, S. H., and Woeber, K. A.: *In* Textbook of Endocrinology, 4th ed. R. H. Williams, Ed. Philadelphia, W. B. Saunders Company, 1968.

general—when performed with due care—it served reasonably well. The great problems connected with the test involved the need for careful preparation of the patient, meticulous performance of the test, and the frequency of nonspecific results caused by interference by a variety of extrathyroidal systemic disorders. Extensive experience with the procedure permits one to classify the many factors which influence, and therefore obscure, interpretation of the BMR. These are shown in Table 14–1.

A search for laboratory alternatives was clearly in order. This chapter will consider those alternatives which most clinical laboratories either perform or are capable of performing in routine fashion. Each of the tests rests on important principles of thyroid physiology and iodine metabolism, as these are now understood.

SERUM PROTEIN-BOUND IODINE (PBI)

Iodine circulates in plasma in two forms: first, as a constituent of the two thyroid hormones, T_3 and T_4; and second, as free iodide. The average concentration of the former, i.e., hormonal iodine, in sera from normal subjects is 5.4 μg/100 ml; free iodide concentration is generally 1.0 μg/100 ml or less.

Unlike free iodide, both T_3 and T_4 are almost completely bound in dissociable linkage to various carrier proteins. Approximately 99.96 per cent of total serum T_4 and 99.6 per cent of total serum T_3 are, in fact, now known to be protein-bound. If, therefore, one measures the iodine content of a protein precipitate of serum, one will have measured the concentration of virtually all hormonal iodine in the sample.

Barker and his coworkers,[2] in 1951, were the first to report a practical technique for measuring serum PBI. Their method, and all subsequent modifications, involves four basic steps: (1) precipitation of serum proteins, (2) repeated washing of the precipitate to remove trapped iodide, (3) oxidation of the serum protein-thyroid hormone complexes to liberate free iodine, and (4) measurement of the iodine so liberated on the basis of its ability to catalyze the reduction of ceric sulfate by the arsenite ion.

Serum PBI values range between 3.5 or 4.0 and 8.0 μg/100 ml (mean = 5.4 μg/100 ml) in normal subjects. There is no significant difference in PBI values between adult men and women; both sexes, however, tend to show a slight fall beyond the age of 50. Duplicate analyses of samples should agree to within 0.6 μg/100 ml, and the PBI value of different sera from the same individual should agree to within 1.0 μg/100 ml. The manual method of Barker, Humphrey, and Soley is very satisfactory, and its modification by Henry[3] will be described in principle.

Specimens should be collected carefully in thoroughly acid-cleaned test tubes. Since red cells are essentially devoid of thyroxine, gross hemolysis lowers the PBI level slightly because of dilution of serum by hemolysate. Sera, once separated from red cells, are extremely stable, even at room temperature.

PRINCIPLE

Serum proteins are precipitated with zinc hydroxide. The washed precipitate is then alkalinized by the addition of sodium carbonate to minimize iodine loss during incineration, dried, and incinerated at 620°C. Iodide present in the alkaline ash is measured by the ceric-arsenite reaction.

The digestion of organic material by a mixture of sulfuric, nitric, and perchloric acids has been widely employed in the Technicon procedure for the automated determination of serum PBI. Complete digestion requires only 3 minutes at 280°C, and the conditions of the automated system are extremely well controlled. These factors confer acceptable accuracy to the procedure.

The iodine-catalyzed reaction between ceric ions (Ce^{4+}) and arsenious acid (As^{3+}) is a two-step reaction:

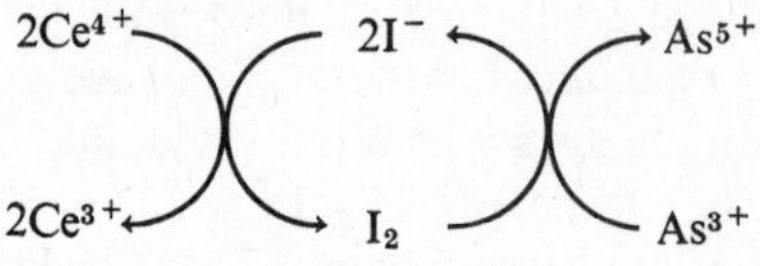

$$\begin{array}{ccccc} 2Ce^{4+} & & 2I^- & & As^{5+} \\ & \times & & \times & \\ 2Ce^{3+} & & I_2 & & As^{3+} \end{array}$$

Ce^{4+} reacts with I^- to form Ce^{3+} and elemental iodine (I_2). Elemental iodine then reacts with As^{3+} to form $As^{5+} + I^-$. The I^- thus generated can react again with Ce^{4+}, and so on. The reduction of the yellow-colored ceric ion (Ce^{4+}) to the colorless cerous ion (Ce^{3+}) is used to measure the rate of reaction, which is dependent, obviously, on the concentration of iodide present.

Chloride ions are known to enhance the catalytic action of iodide in the ceric-arsenite reaction. Thus, to eliminate interference by exogenous chlorides, an excess of chloride ions is added to the arsenious acid reagent.

The yellow color of the ceric ion exhibits maximum absorption in the ultraviolet range at about 317 nm but is measured, in most procedures, at 415 or 420 nm. Use of the longer wavelength has several advantages: it obviates the need for an ultraviolet light source and quartz cuvets, and it extends the working range for the procedure. Beer's law is followed over a wide range only if spectrophotometers with a narrow bandwidth are used.

Extrathyroidal factors affecting the serum PBI level

The excitement which followed the development of the PBI was both considerable and justified, since serum PBI was observed to be rather consistently elevated among patients with hyperthyroidism, and depressed among hypothyroid subjects. The eclipse of the BMR had begun. Increased experience with the PBI, however, led to a more restrained enthusiasm, and the reasons for this transition must be understood very clearly in order to appreciate the subsequent history of thyroid methodology.

Under a variety of conditions a discrepancy between the serum PBI and the patient's clinical thyrometabolic status was observed. Sometimes the PBI was disproportionately elevated, and sometimes it was disproportionately depressed. The causes of such discrepancies, as presently understood, are classified in Table 14–2.

The circumstances which produce elevation of the PBI without a proportional increase in metabolic rate will be detailed below. Rationalization of the mechanisms of disproportionate depression of the PBI will then require only inverse reasoning.

Both euthyroid and hypothyroid subjects commonly receive a variety of thyroid preparations for a variety of reasons. The extent of, and direction of, their effect on the PBI depends upon their relative content of T_4 and T_3. This is so for the following reasons: The normal thyroid secretes approximately 80 μg of T_4 and 50 μg of T_3 each day. T_3 is, however, far more potent metabolically on the basis of weight comparison; the potency ratio T_3/T_4 is approximately 3/1. Should a physician elect to restore a hypothyroid individual to a euthyroid state using T_4 alone, he must replace 50 μg of T_3 with its T_4 equivalent, i.e., 150 μg, in addition to replacing the 80 μg of T_4 normally secreted. He prescribes, therefore, at least 230 μg of T_4. (In fact, he prescribes 300 μg of T_4, since approximately one-third of an orally administered dose of T_4 is not absorbed.) By substituting T_4 for T_3 he is prescribing three molecules where one would do. Further, T_4 contains 33 per cent more iodine per mole-

TABLE 14–2 CAUSES OF DISCREPANCY BETWEEN SERUM PBI AND THYROMETABOLIC STATUS

I. *Disproportionate Elevation of the PBI*
   - A. Administration of iodine–containing drugs or radiographic media
   - B. Increased number of available thyroxine-binding sites* on serum TBG
     1. estrogen induced (birth control pills)
     2. idiopathic or familial
   - C. Thyroid replacement therapy (L-thyroxine)
   - D. Circulating iodoproteins
   - E. Circulating iodotyrosines

II. *Disproportionate Depression of the PBI*
   - A. T_3 the predominant, or only, circulating thyroid hormone
   - B. Thyroid replacement therapy (desiccated thyroid, purified thyroglobulin, L-triiodothyronine)
   - C. Decreased number of available thyroxine-binding sites on serum TBG
     1. idiopathic or familial
     2. nephrotic syndrome
     3. drug-induced
        - a. androgens
        - b. salicylates (large doses)
        - c. diphenylhydantoin
        - d. O-p′-DDD

* Present TBG assays measure the number of active binding sites rather than the protein concentration. It is not possible, therefore, to distinguish increased TBG binding due to an increased concentration of this carrier protein in serum from increased binding activity of the individual TBG molecules.

TABLE 14–3 EFFECT OF VARIOUS THYROID HORMONE PREPARATIONS ON THE PBI*

Hormone Preparation	*Daily Dosage (mg)*	*Expected Serum PBI* (μg/100 ml)
Desiccated thyroid (Armour)	120–180	normal to low (2.9–7.2)
Desiccated thyroid (Warner-Chilcott Spec. Prep.)	120–180	low (3.3–3.7)
Purified thyroglobulin (Proloid, Warner-Chilcott)	120–180	low to low-normal (1.6–4.8)
L-Thyroxine (Synthroid, Flint)	0.2–0.3	normal to high (5.6–11.2)
L-Triiodothyronine (Cytomel, SKF)	0.05–0.1	low (0.4–1.4)

* After Sisson, J. C.: J. Nucl. Med., *6*:853, 1965.

cule than does T_3 and, in addition, T_4 molecules are cleared from the serum far more slowly. For these reasons, a large amount of thyroxine is administered, and thus T_4 replacement therapy produces a disproportionate elevation of the PBI (Table 14–2, I-C.) Replacement by pure T_3, since much less is required, would obviously produce disproportionate lowering of the PBI. The effect of commonly used thyroid hormone preparations on the PBI in euthyroid or hypothyroid subjects receiving normal replacement dosage is shown in Table 14–3. If one modifies the normal range in relation to both preparation and dosage, the PBI will still provide useful information.

The remaining four causes of spurious elevation (Table 14–2, I-A, B, D, and E) are not so easily dismissed. Three involve matters of nonspecificity, and the fourth reflects the paradox of hyperthyroxinemia without hyperthyroidism.

The PBI: Problems of nonspecificity

T_4 in serum is for the most part loosely bound to various carrier proteins. The principal carrier protein is an α-globulin called the thyroxine-binding globulin, or TBG. A very small fraction of total serum T_4 (about 0.05 per cent) circulates in the free state. More will be said about both TBG and FT_4 later in this discussion, but for the moment it will suffice to indicate that the principal iodine-containing components in sera from normal subjects are T_4-serum protein complexes (Fig. 14–2). Under certain circumstances, however, other iodine-containing components may appear in serum and falsely elevate the PBI.

Intrafollicular T_4, i.e., T_4 stored within the thyroid follicles, is protein-bound as is T_4 in serum, but the nature of the intrafollicular T_4-protein bond is quite different from the serum T_4-protein complex (Fig. 14–1). T_4 is an amino acid—an iodoamino acid—and it is but one of many amino acids within the thyroid follicles which associate to constitute the several follicular proteins. (The principal follicular protein is called thyroglobulin.) T_4 is, in fact, but one of four iodoamino acids which are incorporated into the follicular proteins; the others are T_3, monoiodotyrosine (MIT), and diiodotyrosine (DIT). The amino acid residues which form these proteins are linked to one another by means of peptide bonds (—CONH—), and one property of such bonds is important to the thyroidologist. Peptide bonds are considerably stronger than adsorption bonds, and the various thyroxine-responsive tissues cannot cleave them. The tissues cannot respond to T_4 when it is supplied to them in this form, i.e., as iodoprotein. In several thyroidal disorders, Hashimoto's thyroiditis in particular, the follicles may "leak" and release iodoprotein into the circulation. Since iodoprotein will coprecipitate with the plasma proteins to which T_4 is normally

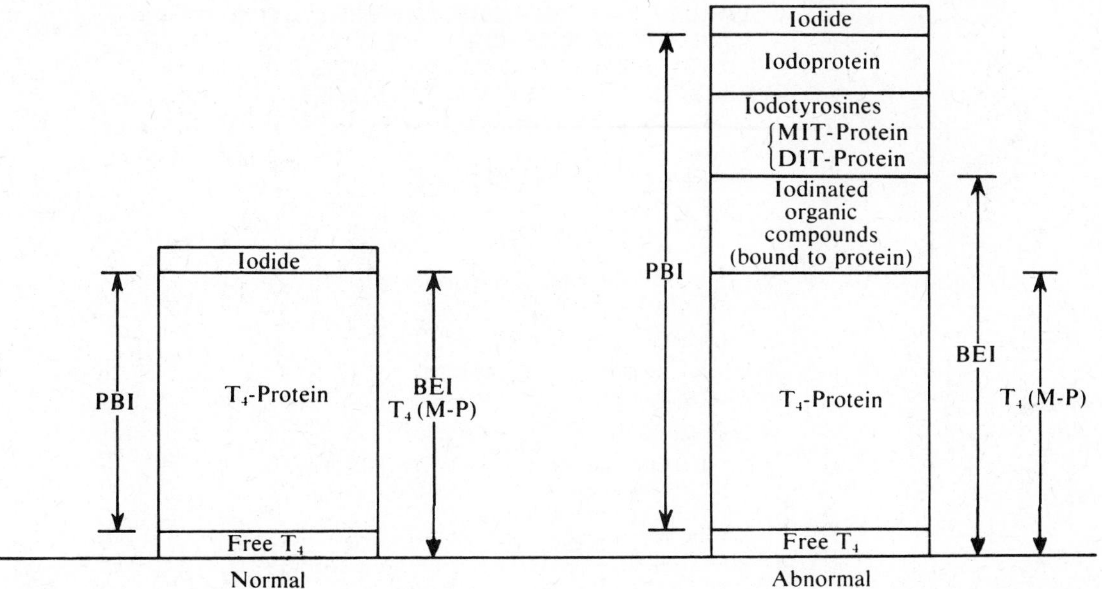

Figure 14–2 Principal forms of serum iodine. Arrows indicate those iodine-containing moieties which are measured by the respective laboratory procedure.

adsorbed, the four iodoamino acids represented in the follicular iodoproteins will be measured as PBI although they are hormonally inert. The PBI thus suffers what has been designated an *iodoprotein problem.*

Under other pathological circumstances, thyroid follicles may release the hormonally inactive T_3 and T_4 precursors, MIT and DIT. This situation is probably less common than iodoprotein escape. Since both MIT and DIT in serum are adsorbed to various proteins, they too contribute to the PBI, and this unwelcome contribution has been called the *iodotyrosine problem.*

The most common cause of spurious elevation of the PBI, however, is exposure to a variety of iodine-containing drugs and radiographic contrast media. Most contrast media in common use are iodine-containing organic compounds which exhibit two annoying characteristics as far as the thyroid diagnostic laboratory is concerned. First, they adsorb to the serum proteins and thus raise the PBI. Second, their rate of clearance from serum is slow and they therefore interfere for long periods of time. These compounds have been classified, in fact, according to the time required for their biological removal from the serum (Table 14–4).

Several drugs in common clinical use also contain iodine. These include diiodohydroxyquin (Diodoquin), isopropamide iodide (Darbid), the topical preparation povidone-iodine (Betadine), and inorganic iodine itself. Iodine-containing organic compounds raise the PBI because they adsorb to serum proteins and release iodide in the course of their metabolism.

Increased concentrations of inorganic iodide in serum are thought to raise the PBI by two mechanisms: iodide trapping and nonspecific iodination of serum proteins. Routine washes of precipitated serum proteins in the various PBI methods remove about 97 per cent of iodide trapped in the protein precipitate. If one assumes then that 3 per cent remains trapped, at serum iodide concentrations of approximately 1.0 μg/100 ml, a 3 per cent washing inefficiency will elevate the PBI by only 1.0×0.03, or 0.03 μg/100 ml. At iodide concentrations of 100 μg/100 ml, on the other hand, the situation is serious, since $100 \times 0.03 = 3.0$ μg/100 ml. Second, increased levels of serum iodide, by a mechanism which is

TABLE 14-4 CLASSIFICATION OF IODINATED RADIOGRAPHIC CONTRAST MEDIA BASED ON RATE OF BIOLOGICAL REMOVAL (DURATION OF PBI ELEVATION)

Short-Lived (Less than 6 Weeks)

1. Diatrizoate (Hypaque, Renografin)
2. Iothalamate (Conray)
3. Ipodate (Oragrafin)
4. Iodohippurate (Medopaque)
5. Iodopyracet (Diodrast)

Intermediate-Lived (6–12 Weeks)

1. Iopanoic acid (Telepaque)
2. Bunamiodyl (Orabilex)

Long-Lived (Greater than 12 Weeks)

1. Ethyl iodophenylundecylate (Pantopaque)
2. Meglumine iodipamide (Cholografin)
3. Propyliodone (Dionosil)
4. Iodized poppy seed oil (Lipiodol)

not well understood, produce some iodination of all the serum proteins. In other words, hyperiodidemia produces iodoproteinemia. This consequence of iodide ingestion may be more important, in fact, than iodide trapping. In any case, iodide trapping may be essentially obviated by pretreatment of the serum sample with resin.

In doses of 125 mg per day or less, for periods of up to seven weeks, iodides do not raise the PBI by either mechanism. Daily doses of 200 to 600 mg interfere from two to ten weeks, and massive doses (i.e., daily doses exceeding 3000 mg) raise the PBI for up to four months after they are discontinued.

Three developments in recent years have contributed greatly to the solution of the foregoing problems of nonspecificity.

Improved specificity of the PBI

1. Serum Butanol-Extractable Iodine (BEI).[5,19,20] The first of these was advanced by Man *et al.* in 1951.[19] These investigators appreciated that *n*-butyl alcohol, or *n*-butanol, has two properties on which an improved serum T_4 method might rest. First, it is an excellent solvent for the iodothyronines T_3 and T_4. Second, although it can cleave an adsorption bond, it cannot cleave a peptide bond. If, then, one were to measure the iodide content of a butanol extract of a protein precipitate of serum, one would have excluded iodoprotein from such an extract and thus solved the iodoprotein problem. Since, however, *n*-butanol extracts of serum also contain both iodides and iodotyrosines (MIT and DIT), there was work yet to be done.

Solution of both the iodide and iodotyrosine problems, based on the differential solubility of iodotyrosines and iodothyronines in alkali, was at hand. Whereas iodotyrosines are quite soluble in alkali, iodothyronines are not, and aqueous alkali washes also remove iodide.

Iodoprotein, iodotyrosine, and inorganic iodide interference could be circumvented by measuring the iodide content of alkali-washed butanol extracts of protein precipitates of serum. Butanol extracts of acidified serum were washed, therefore, with 3.8 molar sodium hydroxide containing 5 per cent sodium carbonate. The resulting extracts, "purified" of both iodides and iodotyrosines, were expected to contain only T_4 and T_3. (The extracts were evaporated to dryness and analyzed for iodide content by procedures identical to those described under PBI.) These expectations were confirmed, and the butanol-extractable iodine, or BEI, in normal subjects was found to range between 3.2 and 6.4 μg/100 ml.

When the same serum sample is examined, the difference between the PBI and BEI should be negligible, theoretically, in the absence of contaminating iodoprotein, iodotyrosine, and/or iodide. In fact, however, the BEI is, on the average, about 20 per cent lower. If the observed difference exceeds 20 per cent, one must suspect the presence of abnormal iodo-compounds.

The BEI has never achieved great popularity despite its theoretical appeal, because it is both cumbersome and imprecise.

2. Determination of Thyroxine by Column $[T_4(C)]$.[17,24] The usefulness of ion-exchange resins for the separation of serum iodoamino acids and inorganic iodide has been known for many years, but the method of Pileggi *et al.* was the first T_4 method suitable for the routine clinical hospital laboratory.[24] The method found wide acceptance and came to be considered by many the method of choice for the determination of serum T_4. Several modifications which have been reported confer a slight increase in accuracy and a considerable increase in both speed and convenience. The principles common to all methods will be presented, but no specific method detailed. Such information can be obtained either from the original reports or from any of the procedures now supplied commercially in kit form.

PRINCIPLE

1. Serum is diluted with an alkaline solution, generally 0.1 molar sodium hydroxide, to raise the pH above 12.6. *Rationale:* At a pH above 12.6, T_4 dissociates completely from its carrier proteins. Ionization of the carboxyl group (and perhaps the hydroxyl group, too) of the T_4 molecule is thought to facilitate its release from proteins and enhance its attraction to the resin.

2. The diluted specimen is then poured onto a strongly basic anion-exchange resin (generally of the Dowex AG-1, X-2 type). Since preparation of the resin, as well as selection of a suitable and well-controlled mesh size, appears to be extremely critical, the use of commercially prepared, disposable columns is recommended for those operators inexperienced in preparation of resin columns. Most methods currently employ resin columns in either the acetate or hydroxyl form. *Rationale:* As diluted serum passes through the column, T_4 and related compounds, as well as proteins and iodide, are adsorbed by the ion exchange resin. The affinity of the column for iodide, and the iodine moiety of the T_4 molecule, exceeds its affinity for the acetate or hydroxyl ions.

3. The column is washed with an acetate-alcohol solution. Either acetate-isopropanol (pH 8.0) or acetate-methanol (pH 5.5) is generally employed. *Rationale:* Washing at pH 8.0 completely removes all serum proteins from the column, and also removes both carbonate and bicarbonate, thus preventing CO_2 gas from disrupting the column when acetic acid is added in the next step. T_4, T_3, and inorganic iodide are retained by the column. Methods which incorporate the alkaline-wash step also retain thyronines and most iodinated organic compounds on the resin and thus require an additional wash with glacial acetic acid at a pH of approximately 5.5 in order to remove these compounds. The wash with acetate-methanol, recommended by some, removes proteins, iodotyrosines, and many iodinated organic compounds with a single elution, but does not remove bicarbonate and carbonate prior to acidification.

4. The columns are primed with a small and accurately measured amount of glacial acetic acid. *Rationale:* T_4 and T_3 are held in the upper portion of the resin bed. Addition of glacial acetic acid moves these down the column. It is important to add a precise amount of glacial acetic acid, since an excess would cause a loss of T_3 and T_4.

Note: In steps 1 through 4 all eluates are discarded.

5. The column is permitted to drain completely and a precisely measured amount of 50 per cent acetic acid is added to the column (in most cases 3.0 ml) to provide a pH of about

1.4. *Rationale:* Under these conditions, 80 to 95 per cent of the T_3 and T_4 is eluted from the column. The exact fraction is rather constant for a given method and depends on both the type of resin used and the nature of the wash solutions. The column effluent is saved in an appropriate container. Inorganic iodide, in concentrations up to 1000 μg/100 ml, is retained by the column and will not interfere with the procedure.

6. An identical amount of 50 per cent acetic acid is added to the column a second time, and once more the effluent is saved in a separate cup. *Rationale:* The second acetic acid wash removes the remainder of the T_3 and T_4 from the column. Separate collection and analysis of two eluates provides a pattern of elution which is useful in the detection of contaminated samples. If, for example, eluate No. 2 contains more than the standard percentage of T_3 and T_4, contamination by exogenous iodide is suspected and the result disregarded.

Note: The procedure outlined provides eluates which are sufficiently concentrated to permit direct analysis for iodide; the tedious evaporation of acetic acid eluates which was required in earlier procedures is now unnecessary.

7. The T_3 and T_4 content of the two eluates is determined directly by any of the standard procedures employing the ceric-arsenite reaction without preliminary wet or dry digestion. The eluates are treated with a solution of $KBrO_3$ and KBr, followed by a precisely timed addition of arsenious acid and then ceric reagent. *Rationale:* The catalytic activity of T_4 in the ceric-arsenite reaction is thought by some to depend on the release of iodine from the T_4 molecule by the action of the ceric ion. Although pure T_4 has the ability to catalyze the ceric-arsenious acid reaction (though to a lesser degree than iodide), T_4 eluted from a column has little or no such catalytic activity. This lack of activity may be due to the presence of inhibiting substances in the eluate which either bind to, or are inactivated by, Br_2. Thus, pretreatment of the eluate with Br_2 could result in the binding of those compounds which would otherwise have removed iodide from the ceric-arsenite catalytic cycle (see principle of the PBI method, p. 828). Some authors feel that treatment with Br_2 releases iodine from thyroxine and thus makes it available for catalytic action. However, no direct evidence has been adduced in support of this hypothesis.

Pretreatment of the eluate with Br_2 can be done using bromine water, but most procedures recommend a combination of $KBrO_3$ and KBr, which in an acid medium generates Br_2 according to the following equation:

$$5\ Br^- + BrO_3^- + 6\ H^+ \rightarrow 3\ Br_2 + 3\ H_2O$$

Formation of elemental bromine can be recognized by the appearance of a yellow color in the reaction mixture. After addition of arsenious acid, Br_2 is immediately reduced to Br^-, which results in decolorization of the solution. Br^- has little or no catalytic activity in the ceric-arsenite reaction and thus causes no interference.

Note: Procedures which do not require digestion were originally designed as manual methods, but recent modifications permit analysis of the eluate by the AutoAnalyzer.

STANDARDIZATION

Some authors recommend the use of aqueous KI standards. Others, considering these unsatisfactory, employ instead commercial control sera for preparation of the standard curve. Both approaches now appear to be undesirable, and the use of thyroxine in protein solution for standards is recommended.

CONCLUSIONS

Procedures which employ the principles just outlined are now sufficiently simple, rapid, and well-standardized to be suitable for the routine clinical chemistry laboratory.

TABLE 14–5 EFFECT OF ORGANIC IODINE COMPOUNDS ON T_4-COLUMN TEST

Compound	100 μg/100 ml* (*as iodine*)	1000 μg/100 ml* (*as iodine*)
Cholorografin	None	None
Dionosil	None	Yes
Floraquin	Yes	Yes
Hypaque	None	None
Lipiodol	None	None
Orabilex	None	Yes
Oragrafin	None	None
Renografin	None	None
Salpix	None	None
Skiodan	None	None
Telepaque	Yes	Yes

* *In vitro* additions of the compounds to serum.

The T_4 by column method offers greater specificity than the classic PBI method. The increased specificity is due to the elimination of many troublesome iodine-containing organic compounds in the first wash (step 3) and the retention of inorganic iodide by the column. Thus, the T_4 column method is not affected by inorganic iodine in concentrations below 1000 μg/100 ml. The influence of certain organic iodine compounds has been studied *in vitro* and the results are summarized in Table 14–5.

NORMAL VALUES

The normal values for the various T_4 by column methods range between 2.8 and 6.4 μg T_4-I/100 ml (4.3 to 9.8 μg T_4/100 ml or 55 to 126 nmol T_4/l). Nonincineration techniques are, on the average, 0.13 μg/100 ml higher.

3. Determination of Serum T_4 by Displacement [$T_4(D)$].[6,7,21,22] In 1964 Murphy and Pattee reported an ingenious method for measuring serum T_4 which for the first time did not depend on iodide analysis.[22] Their method, an extension of Ekins' studies of competitive protein-binding,[6] is based on a property of the intact T_4 molecule, specifically its capacity to displace radioactive T_4 from a standard radio-T_4-protein complex, i.e., T_4-TBG.

PRINCIPLE

T_4 in a serum specimen is dissociated from its various carrier proteins (TBG, TBPA, and albumin) by extraction with an appropriate T_4 solvent [e.g., ethanol, an ethanol-methanol-isopropanol mixture (90/5/5), dimethoxypropane, and so forth]. The efficiency of this extraction is 80 to 90 per cent. An aliquot of the extract is evaporated to dryness in a water bath at a temperature of 45 to 56°C, a standard radio-T_4-TBG mixture is added to the residue, and the mixture is incubated in an ice bath for 5 min. The radio-T_4-TBG mixture contains barbital buffer to inhibit any binding of T_4 to TBPA which may be present in the mixture in small quantities. The amount of radio-T_4 in this mixture is calculated to be adequate to saturate all T_4-binding sites on TBG.

After the initial incubation period, an anion-exchange resin is added, the mixture is incubated in an ice bath for 1 h, and the radioactivity in this mixture is measured in a well counter (= total count). During this incubation (equilibration) time, unlabeled T_4 derived from the sample replaces radio-T_4 from the radio-T_4-TBG mixture, i.e., the radio-T_4 displacement is directly proportional to the quantity of unlabeled T_4 in the sample.

The amount of radio-T_4 which has been displaced from TBG by unlabeled T_4 in the sample is adsorbed to the resin and can be counted in a well counter after centrifugation of the mixture and washing of the resin (= resin count). The per cent resin uptake of radio-

T_4 is determined by dividing the resin count by the total count in the sample. The total amount of T_4 in the sample can then be calculated by applying this value to a standard curve prepared by plotting per cent resin uptake of radio-T_4 against concentration of unlabeled T_4 standards. Serum T_4 levels, expressed as T_4-iodine, generally range from 2.9 to 7.0 μg/100 ml (4.4 to 10.7 μg T_4/100 ml or 57 to 137 nmol T_4/l).

Neither iodoprotein, iodotyrosines, inorganic iodide, iodinated drugs, nor iodinated contrast media apparently interfere with this procedure. The anticonvulsant Dilantin lowers the result by reducing the binding affinity of the subject's TBG for T_4. On the other hand, the D-isomer of T_4, D-T_4 (Choloxin), now commonly used clinically to lower serum cholesterol concentration, displaces radio-T_4 from TBG in the reaction mixture and thus produces a falsely elevated result. Despite these instances of drug interference, the serum T_4(D) method appears to be a very specific serum T_4 measurement. In any case, the results of this method, along with those of the PBI, BEI, and T_4(C) methods, need not correspond with clinical thyrometabolic status for the reason discussed below.

THE PARADOX OF HYPERTHYROXINEMIA WITHOUT HYPERTHYROIDISM (AND HYPOTHYROXINEMIA WITHOUT HYPOTHYROIDISM)

In 1959 Beierwaltes and Robbins described a clinically euthyroid 48 year old man who, during a routine periodic health examination, was found to have PBI levels which ranged between 11.8 and 16.0 μg/100 ml.[4] There was no history of exposure to either iodine-containing drugs or radiographic contrast media. Serum BEI determinations were 12.3 and 13.0 μg/100 ml. The BMR (−20 per cent), serum cholesterol (212 mg/100 ml), and thyroidal uptake of radioiodine (22 per cent/d) were normal. The man had three children; one, a 15 year old daughter, also exhibited elevated PBI and BEI levels. These were two apparently euthyroid people with increased concentrations of T_4 in their serum.

Ingbar, in 1961, observed the reverse situation. He reported a euthyroid man with serum PBI concentrations of 2.0 μg/100 ml.[14] This man had extrathyroidal disease, but none known to depress the PBI. Again the BMR (+3 per cent), serum cholesterol (210 mg/100 ml), and radioiodine uptake (28 per cent/d) were normal. This man was apparently an example of a person who was euthyroid despite subnormal concentrations of T_4 in his serum.

These reports suggested that clinical thyrometabolic status evidently depends on factors other than total serum T_4 concentration; perhaps, rather, it depends on that fraction of total serum T_4 which circulates in the free state and is, therefore, more accessible to the peripheral T_4-responsive tissues. Although it was not possible then to measure free T_4 in serum, it was possible to measure accessibility of serum T_4 to tissues, since this could be inferred from the T_4 degradation rate. Both authors proceeded to do just that.

Though measurement of T_4 degradation rate by the peripheral tissues was not difficult, it did require a clear understanding of the kinetics of T_4 metabolism and the terminology generally applied to these kinetics. These terms are as follows:

TDS = thyroxine distribution space: the volume of body fluids which would be required to contain exchangeable thyroxine were it present throughout at the same concentration at which it exists in the plasma.

ETT = Extrathyroidal thyroxine: the quantity of exchangeable thyroxine in terms of its content of iodine, believed to coincide closely with the quantity of extrathyroidal hormone.

k = Fractional rate of turnover of thyroxine: the fraction of hormone within the TDS which is degraded and replaced per unit time, calculated as $0.693/t_{1/2}$, where $t_{1/2}$ = the thyroxine half-time, or time required for half the exchangeable thyroxine to be degraded and replaced.

C = Thyroxine clearance rate: the volume of the TDS which contains a quantity of thyroxine equal to that being degraded per unit time.

HI = The concentration of hormonal iodine in the plasma.

D = Thyroxine degradation rate: the quantity of thyroxine undergoing degradation per unit time, in terms of its content of iodine.

The equations which express the relationships among these terms are:

$$
\begin{aligned}
\text{ETT} &= \text{HI} \times \text{TDS} && (1)\\
\text{C} &= \text{TDS} \times \text{k} && (2)\\
\text{D} &= \text{HI} \times \text{TDS} \times \text{k} && (3)\\
\text{D} &= \text{ETT} \times \text{k} && (4)\\
\text{D} &= \text{HI} \times \text{C} && (5)
\end{aligned}
$$

Thus, by determining the PBI, TDS, and $t_{1/2}$ one may calculate D [Equation (3)]. Consideration of the techniques normally employed to determine TDS and $t_{1/2}$ is beyond the scope of this chapter. Such measurements were made, however, both in normal subjects and in the patients reported above, and the results are shown in Table 14–6.

Note the similarity of the D values in the three columns. One patient (that of Beierwaltes and Robbins) had a pool size approximately twice normal, which was turning over half as fast; the other had a contracted pool, about half normal, with a turnover rate approximately twice normal. Both subjects had similar T_4 degradation rates and both were, therefore, euthyroid. The cause of the expanded ETT pool in the first instance was thought to be a genetically determined increase in T_4-binding sites on the thyroxine-binding α-globulin, TBG. The second case displayed a contracted ETT pool because of a decrease (idiopathic) in such binding sites.

The several messages to be gleaned from these patients are very important ones: (1) Serum T_4 concentration [whether measured as PBI, BEI, T_4(C), or T_4(D)] is determined mainly by the number of binding sites on TBG which are occupied by T_4 molecules. (2) Thyrometabolic status, on the other hand, is evidently determined by the concentration of free hormone in serum, and this very small fraction of the total also regulates the hypothalamic and pituitary "sensors" of serum thyroid hormone concentration.

The interaction between serum free T_4 (FT_4) and the unoccupied or available binding sites on its principal carrier protein TBG conforms to the law of mass action and can be described by the following equation (see Fig. 14–3):

$$[T_4] \times [TBG] = k \times [T_4 \cdot TBG]$$

where

$$
\begin{aligned}
[T_4] &= \text{concentration of free } T_4 \text{ in serum}\\
[TBG] &= \text{concentration of unoccupied binding sites on serum TBG}\\
k &= \text{an association constant}\\
[T_4 \cdot TBG] &= \text{concentration of } T_4\text{-occupied binding sites on serum TBG}
\end{aligned}
$$

The relationship among these moieties has important diagnostic implications. A primary increase in either $[T_4]$ or $[TBG]$ would drive this reaction to the right, increasing serum

TABLE 14–6 STUDIES OF THYROXINE KINETICS IN NORMALS AND TWO SUBJECTS WITH "PBI-THYROMETABOLIC DISSOCIATION"

Parameter	*Normals*	*Case of Beierwaltes and Robbins*[4]	*Case of Ingbar*[14]
BMR	−15 to +15	−20	+3
PBI (μg/100 ml)	3.5–8.0	11.8–16.0	2.0
Chol (mg/100 ml)	140–250	212	210
RaI uptake (% dose/d)	15–45	22	28
BEI (μg/100 ml)	3.4–6.8	12.3–13.0	—
TDS (liter)	9.4	7.5	13.2
k (%/d)	10.6	5.7	23.1
C (liter/d)	1.0	0.43	3.05
ETT (μg I)	508	1099	264
D (μg I/d)	54	63	61

$$[T_4]\ [TBG] = k\,[T_4 \cdot TBG]$$

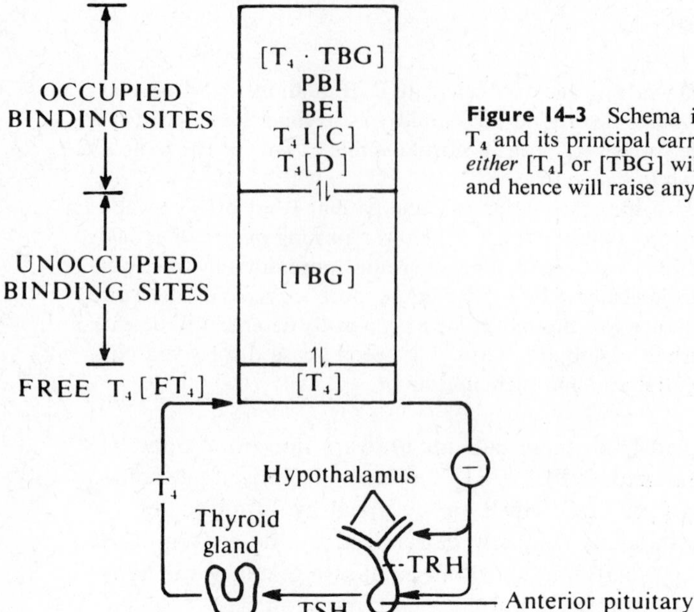

Figure 14–3 Schema illustrating the relationship between serum T_4 and its principal carrier protein, TBG. Note that an increase in *either* [T_4] or [TBG] will shift the equilibrium towards [$T_4 \cdot$ TBG] and hence will raise any of the serum T_4 tests.

TRH = Thyrotropin – releasing hormone
TSH = Thyroid – stimulating hormone

[T_4·TBG], and hence the PBI, BEI, T_4(C), and T_4(D). Hyperthyroidism produces a primary increase in [T_4], while estrogens and idiopathic or genetic influences produce a primary increase in [TBG]. In the former case the patient is ill and requires treatment; in the latter case the patient is euthyroid and requires nothing but an understanding of the equation. It is obviously necessary to be able to distinguish these two circumstances.

The differential diagnosis may be achieved in four ways:

1. Measurement of the concentration of unoccupied binding sites on TBG ([TBG]) in serum.
2. Measurement of FT_4 in serum.
3. Measurement of thyroid gland activity (in terms of avidity for radioiodine).
4. Measurement of metabolic impact of the thyroid hormones on the peripheral tissues (i.e., BMR).

The BMR has been discussed. The other diagnostic options will now be considered.

MEASUREMENT OF UNSATURATED [TBG]

If an elevated serum [T_4·TBG] were due to a primary increase in [T_4], the increased [T_4·TBG] would occur at the expense of [TBG] and, since serum [TBG] is not regulated by a feedback loop such as the hypothalamic-pituitary system which protects [T_4], [TBG] will decrease. On the other hand, a primary increase in [TBG] will generate an increase in [T_4·TBG] at the expense of [T_4]. The fall in [T_4] is, however, only transient since this moiety is "protected" (Fig. 14–3). Further, the increased [TBG] is not fully saturated by combination with T_4, and the serum will exhibit an increased [TBG]. So [TBG] is reduced in the first instance and elevated in the second.

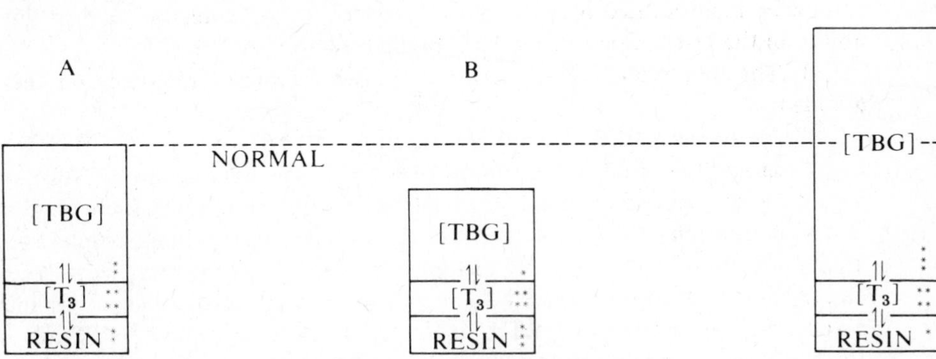

* = molecules of radiothyroxine added to a serum-resin mixture

Figure 14–4 Estimation of [TBG] by use of a "competing" inert T_4 (or T_3) receptor.

THE RESIN T_3 UPTAKE TEST (RT_3U)

[TBG] may be estimated indirectly by a group of tests which have been designated T_3 tests. All are based on *in vitro* competition for thyroid hormone between serum TBG and an added inert receptor (Fig. 14–4).

In 1957, Hamolsky and coworkers[11] developed the first of the T_3 tests, the *in vitro* red blood cell uptake of ^{131}I-T_3, on the basis of the following considerations. The red blood cell was known to be capable of binding thyroid hormones on the basis of experiments in which red blood cells were incubated in saline-radio-T_3 and saline-radio-T_4 mixtures. Red cells were therefore considered possible models for all the thyroid hormone-responsive tissues. A sample of whole blood was considered, then, to house two binding systems: the various carrier proteins in serum, and the red blood cells. These compete, in a sense, for secreted hormone, and each system was thought to possess a characteristic hormone-binding affinity. The binding affinity of the carrier proteins was thought to be variable, depending upon the number of unoccupied binding sites on TBG, i.e., [TBG]. The greater the number of unoccupied sites, or the higher the [TBG], the poorer the competitive position of the red cells for added hormone. The smaller the number of available sites, or the lower the [TBG], the better the competitive position of the red cells. The partition of both endogenous and added thyroid hormone in blood between carrier proteins and red cells was thought to depend upon these considerations, and since tracer amounts of labeled hormone added to blood were known to mix thoroughly with the endogenous hormone pool, the partition of added labeled hormone was expected to mirror the partition of endogenous hormone between carrier proteins and red cells. Thus, the fractional uptake of added labeled hormone by red cells was expected to increase in hyperthyroidism, where [TBG] is reduced, and to decrease in hypothyroidism, where [TBG] is increased. A large clinical experience with the procedure validated these predictions in general.

Radio-T_3 was considered preferable to radio-T_4 for purposes of this test for two reasons. First, because TBG has a lesser affinity for T_3, a greater fraction would attach to red cells and thus confer greater accuracy to the counting procedure. Second, radio-T_3 is more stable, i.e., less photosensitive. In the T_3 procedure, however, the red blood cells were soon replaced by resins, in either granular or sponge form. Sisson[26] has summarized the several advantages of resins over red blood cells:

1. Either serum or plasma, unlike whole blood, may be frozen for extended periods prior to analysis.

2. Pooled sera for use as a control for each run may also be frozen. Such standards cannot be incorporated into the red cell procedure as protection against day to day variability in the test because of red cell fragility.

3. The influence of intrinsic abnormalities in the erythrocytes on the procedure is avoided.

4. No hematocrit correction is necessary.

Combination of the T_4(D), or any of the tests which measure [T_4·TBG], with the RT_3U would appear to represent an ideal laboratory tandem, since such combination would permit elegant physiologic interpretations. A patient who exhibits an elevation in both the T_4(D) and RT_3U tests evidently has an increased [T_4·TBG] but a decreased [TBG]. This circumstance suggests increased T_4 secretion, i.e., hyperthyroidism. An increased T_4(D) associated with a decreased RT_3U indicates an increase in both [T_4·TBG] and [TBG] in serum. This suggests a primary increase in [TBG], and such patients are euthyroid. Such a circumstance is seen in pregnancy, in exposure to exogenous estrogen (commercial estrogen preparations, ovulation suppressants), or in genetic or idiopathic increase in [TBG]. A reduction in both the T_4(D) and RT_3U indicates a decrease in [T_4·TBG] in the face of an increased [TBG]. Such a situation suggests a primary decrease in T_4, i.e., hypothyroidism. Finally, a low T_4(D) but elevated RT_3U suggests a reduction in both [T_4·TBG] and [TBG]. This circumstance suggests a primary decrease in [TBG], due to either increased loss (e.g., nephrotic syndrome), decreased production (androgen excess, genetic or idiopathic), or successful competition for T_4-binding sites by certain drugs (large doses of salicylates, diphenylhydantoin, and the adrenolytic agent o-p′-DDD).

Thus, concordant variance of the T_4(D) and RT_3U tests suggests altered thyroid function; discordant variance suggests altered [TBG] and a euthyroid state.

These generalizations represent the basis of the *thyroxine-resin T_3 index* (*T_4-RT_3 index*). In this procedure a mathematical product is calculated using either PBI, T_4(C), T_4(D), or T_4(RIA) and RT_3U. The resultant value is assumed to be proportional to the concentration of free thyroxine (FT_4) in the blood. Thus, hyperthyroidism elevates and hypothyroidism depresses the T_4-RT_3 index, whereas pregnancy, estrogen therapy, nephrotic syndrome, and so forth generally confer normal results. This attempt to simplify thyroid diagnosis is commendable but dangerous, since many factors may influence only one member of the "index team" and thus eliminate the usefulness of the unaffected procedure.

The binding capacity of TBG for T_3 is influenced by a variety of extrathyroidal factors, and for this reason the RT_3U test often provides too many false positives and false negatives. One may take exception to use of the adjective "false" in this connection, but in any case the old problem of nonspecificity definitely haunts the RT_3U test. The principal extrathyroidal causes of abnormal RT_3U tests are classified in Table 14–7.

Additional comment is necessary concerning the effects of treatment with the various commonly used thyroid hormone preparations on the RT_3U test. When the normal or hypothyroid individual is given full replacement or suppressive doses of either desiccated thyroid (3 grains, or 180 mg) or L-thyroxine (0.3 mg), the RT_3U test is generally normal. Administration of comparable doses of T_3, on the other hand, produces low results in about 50 per cent of cases so treated. (The influence of these preparations on the PBI was shown in Table 14–3.)

Neither inorganic nor organic iodides interfere with the RT_3U test, with the exception of the oral cholecystographic contrast medium, sodium ipodate (Oragrafin).

Normal values for the RT_3U procedure have been expressed in many ways, but "per cent resin uptake" is perhaps the most common parameter and normal sera generally range from 25.0 to 35.0 per cent.

TABLE 14-7 EXTRATHYROIDAL FACTORS WHICH MAY AFFECT THE RT_3U TEST

Factors Which Elevate the Result

1. Competition by drugs for T_4-binding sites
   a. on TBG-diphenylhydantoin (Dilantin sodium)
   b. on TBPA (causing endogenous T_4 to "shift" to TBG and thus displace T_3)
      (1) salicylates (large doses)
      (2) phenylbutazone
2. Decrease in TBPA production due to acute illness, surgery
3. Reduced TBG synthesis
   a. hereditary-familial (idiopathic)
   b. hormone mediated (androgens, anabolic steroids after 7–21 days of therapy; effect persists for 7–21 days after cessation of treatment)
4. Loss of TBG in urine (nephrotic syndrome)
5. Mechanism uncertain
   a. anticoagulants (both heparin, coumarins)
   b. cardiac arrhythmias, supraventricular
   c. elevated arterial pCO_2
   d. metastatic carcinoma
   e. hepatic disease

Factors Which Lower the Result

1. Increased TBG synthesis
   a. hereditary-familial (idiopathic)
   b. estrogen mediated
      (1) endogenous: pregnancy (by 2–3 weeks gestation; returns to normal 1–2 weeks postpartum)
      (2) exogenous (after 7–21 days of treatment)
         (a) all estrogenic preparations (including stilbestrols)
         (b) ovulatory suppressants (oral contraceptives)
2. Mechanism uncertain
   a. drug induced: perphenazine (Trilafon), prolonged
   b. hepatic disease, acute and chronic

SERUM FREE THYROXINE (FT_4)

The free thyroid hormones in serum are the moieties which probably determine clinical thyrometabolic status, and it is now possible to measure their concentration directly. Estimation of serum free thyroxine (FT_4) was reported first. This development, truly an important one, is somewhat disappointing as well, since the principal methods, those of Ingbar *et al.*,[15] Sterling and Brenner,[28] and Lee *et al.*,[18] are hardly routine procedures.

The method of Ingbar and his coworkers is a two-stage "double dialysis" procedure. The first stage involves equilibrium dialysis of a test serum-radiothyroxine mixture against standard phosphate buffer, producing a dialysate which contains both free radiothyroxine and a radioiodide contaminant. A second dialysis, to remove this contaminant, is therefore necessary. Accordingly, a mixture of first stage dialysate and pooled plasma is dialyzed against phosphate buffer containing Amberlite IRA-400 anion-exchange resin. This resin adsorbs only the radioiodine contaminant and thereby maintains an effective diffusion gradient for iodide until all the contaminant is removed. Because of the strong binding affinity of the proteins in the pooled plasma for T_4, little if any radiothyroxine is available to the resin. All radioactivity which remains within the dialysis bag is assumed, therefore, to represent free radiothyroxine.

This technique measures per cent FT_4. Absolute $[FT_4]$ is determined from the following equation:

$$[FT_4]\ (\text{ng}/100\ \text{ml}) = \frac{T_4\ (\mu\text{g}/100\ \text{ml}) \times \text{per cent } FT_4}{0.65} \times 1000$$

TABLE 14–8 PER CENT FREE THYROXINE (PER CENT FT_4), PROTEIN-BOUND IODINE (PBI), AND ABSOLUTE CONCENTRATION OF FREE THYROXINE (FT_4) IN SERA OF NORMAL SUBJECTS AND SUBJECTS WITH VARIOUS ABNORMAL STATES*

Diagnosis	% FT_4	PBI (μg/100 ml) $\bar{x} \pm SD$	FT_4 (as T_4) ng/100 ml $\bar{x} \pm SD$
Normal	0.050 ± 0.009	5.4 ± 0.8	4.03 ± 1.08
Myxedema	0.037 ± 0.010	1.6 ± 0.5	0.88 ± 0.52
Pregnancy	0.026 ± 0.006	8.0 ± 0.9	3.21 ± 0.56
Thyrotoxicosis	0.110 ± 0.072	12.9 ± 2.9	20.56 ± 13.07
General extrathyroidal illness	0.078 ± 0.033	4.6 ± 1.4	5.12 ± 2.99

* After Ingbar, S. H. *et al.*: J. Clin. Invest., *44*:1679, 1965.

The factor 0.65 in the denominator represents the contribution of the four iodine atoms to the molecular weight of thyroxine, i.e., 508/778.

Typical FT_4 studies in normals and subjects having a variety of abnormalities are shown in Table 14–8.

Sterling and Brenner[28] simplified the FT_4 method significantly by replacing the second dialysis with a much simpler precipitation procedure. Their method involves (1) dialysis of a serum-radiothyroxine mixture against phosphate buffer (as above), and (2) precipitation of the radiothyroxine in the dialysate with $MgCl_2$ after addition of carrier T_4. This method is often called the magnesium precipitation method (after its second stage), and the method of Ingbar is called the resin dialysis technique.

The method of Lee *et al.*[18] is perhaps the simplest of the three. In this procedure radiothyroxine is added to serum, and the free and protein-bound fractions are separated on microcolumns of Sephadex G-25.

FT_4 correlates better than the PBI, T_4(C), or T_4(D) with clinical thyroid status, since FT_4 is less affected by states of altered TBG capacity. Like the PBI, T_4(C), and T_4(D), however, there is no invariable connection between FT_4 and clinical thyrometabolic status.

HYPERTHYROIDISM WITHOUT HYPERTHYROXINEMIA: T_3 HYPERTHYROIDISM, OR T_3 TOXICOSIS[13,29,30]

For a good many years after its discovery by Gross and Pitt-Rivers in 1952, the role of T_3 in clinical thyroid disease was not fully appreciated. It was commonly dismissed as "the other thyroid hormone." Since 1957, however, a succession of papers has documented a clinical syndrome characterized by thyrotoxicosis attributable to increased serum levels of T_3. The work of Hollander, in which gas chromatographic methods were used to measure serum T_3 and T_4, was especially convincing.[12] Sterling *et al.* soon developed an easier competitive protein-binding technique for T_3,[29] and Gharib and coauthors, a useful radioimmunoassay.[9] It is now possible to measure T_3 in serum with sufficient accuracy and specificity to permit the clinician to evaluate its contribution to clinical thyroid problems. Recent estimates of the incidence of "T_3 hyperthyroidism" approximate 5 to 10 per cent of all cases of hyperthyroidism, and the incidence may be considerably higher in areas of iodine deficiency.[13] Further, there is evidence now that certain hyperthyroid patients exhibit entirely normal thyroid tests with the exception of elevated T_3 levels during the early stages of their disease.

The diagnosis of T_3 hyperthyroidism rests on five criteria:

1. Conventional clinical signs of hyperthyroidism. These signs may be associated with either diffuse goiter, solitary nodules, or multinodular goiter. The gland may be small or

very large. Indeed, metastatic carcinoma of the thyroid may be the cause with no evident thyroid abnormality on palpation.

2. Normal (or low) serum PBI, BEI, T_4(C), or T_4(D) levels.

3. Normal serum TBG-binding capacity. Criterion 2 would obviously lose its significance were TBG capacity reduced. Ideally, this determination should be made by paper electrophoresis or by direct measurement rather than inferred from a RT_3U test.

4. Autonomous thyroid function as determined by any standard T_3 suppression test. This criterion is necessary because serum T_3 levels may be increased in circumstances other than hyperthyroidism, e.g., iodine deficiency, exogenous T_3.

5. Increased serum T_3 levels. Regardless of method, these range in general from two to five times the normal mean values.

MEASUREMENT OF T_3 IN SERUM

Hollander reported a sensitive gas-liquid chromatographic method for measuring T_3 in serum in 1968.[12] This method involves four basic steps: (1) extraction of T_3 from serum by passage through a cation-exchange resin column; (2) preparation of a stable, volatile derivative of T_3; (3) purification of this derivative using an anion-exchange resin column; and (4) gas chromatographic separation and quantitation using a nickel-63 electron capture detector. An additional step was added later to prevent deiodination of endogenous T_4 to T_3 (approximately 1.7 per cent). This method established the syndrome of T_3 hyperthyroidism unequivocally, but it was unsuitable for routine use by reason of its complexity.

Sterling, and later Wahner, reported simpler methods for estimating T_3 in serum using competitive protein-binding.[27,31] These, too, involved four basic steps: (1) removal of the thyroid hormones from serum by either chemical extraction or resin columns; (2) separation of T_3 from T_4 by either paper or thin-layer chromatography; (3) elution of the separated T_3; and (4) measurement of the eluted T_3 by competitive protein-binding. These methods provided spuriously elevated T_3 levels for two reasons: *in vitro* conversion of T_4 to T_3 during the extraction and purification steps, and contamination of chromatographically purified T_3 with residual T_4. These problems could introduce a 50 to 100 per cent error, and they have defied easy solution.

Gharib *et al.*,[9] and later others, have successfully introduced radioimmunoassays for T_3 which made routine analysis of T_3 in serum feasible. Radioimmunoassay involves five basic steps (Fig. 14–5): (1) injection of either thyroglobulin or a T_3 conjugate (e.g., T_3-poly-L-lysine or T_3-bovine serum albumin) into rabbits to elicit anti-T_3 antibodies; (2) harvesting of rabbit antisera and screening for anti-T_3 antibody using labeled T_3; (3) testing antisera for immunologic specificity, specifically for quantitative distinction of T_3 from T_4 (identical procedures may be used for T_4 radioimmunoassay—T_4[RIA]); (4) construction of standard curves using antisera with the highest anti-T_3 antibody titers and known concentrations of T_3 in buffer or a specially prepared low T_3 serum; and (5) assay of unextracted sera from patients for T_3 concentration using, in addition, some "blocking agent" to eliminate interference by native serum proteins in the antigen-antibody reaction. Tetrachlorothyronine, ANS (antineutrophilic serum), dilantin, and salicylates have been used for this purpose, i.e., to displace endogenous T_3 from TBG and inhibit binding of labeled T_3 to native TBG. When T_3 conjugates are used as antigens, this procedure can detect serum T_3 concentrations as low as 50 ng/100 ml or less, and the method is most sensitive in the serum T_3 zone between 50 and 500 ng/100 ml. Sera with higher concentrations must be diluted and reassayed. Among euthyroid subjects, serum T_3 concentration approximates 215 ng/100 ml ($SD \pm 55$). Hypothyroid patients average 100 ng/100 ml ($SD \pm 290$).

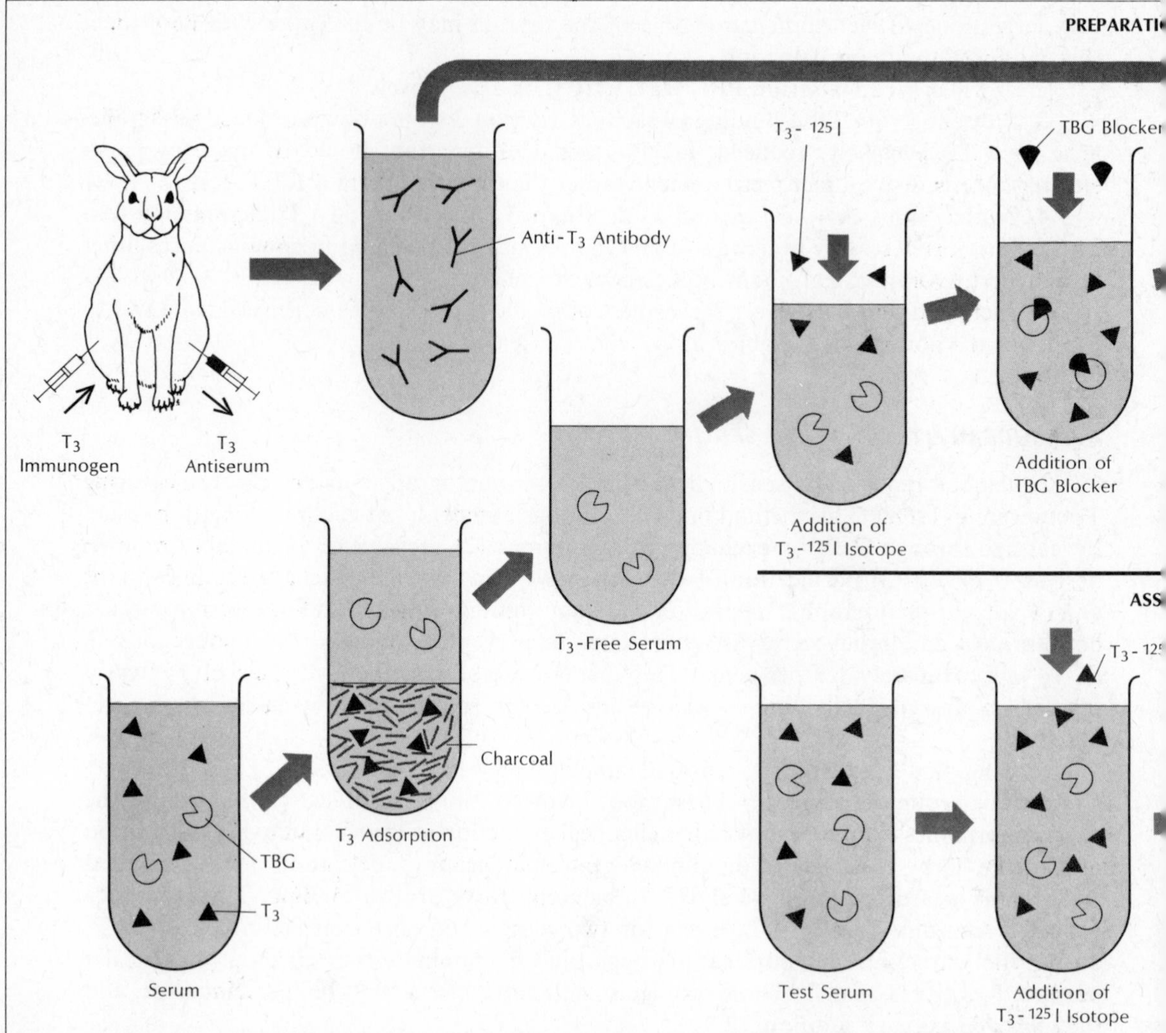

Figure 14–5 The techniques involved in the radioimmunoassay of T_3 are depicted above, starting with the preparation of a T_3-free serum for use in establishment of the assay standard (lower left), and with the raising of a T_3 antiserum in the rabbit. In preparing the standard, labeled T_3 is added to the T_3-free serum, then a blocker of thyroxine-binding globulin is introduced to prevent TBG competition with specific antibody. The antiserum and a known quantity of unlabeled T_3 are next added; the material is allowed to incubate, dextran-coated charcoal is

IMPROVED DEFINITION OF THE HYPOTHYROID STATE: MEASUREMENT OF TSH IN SERUM[23]

In general, measurement of serum levels of the thyroid hormones by either displacement analysis, column chromatography, or iodine content of serum protein precipitates is far more useful in cases of suspected hyperthyroidism than in hypothyroidism. Accordingly, the development by Odell, Wilber, and Utiger of a radioimmunoassay for thyroid-stimulating hormone (TSH) in serum has represented a major laboratory contribution.[23] This procedure is as useful as it is intellectually exciting.

If a diseased thyroid gland secretes insufficient thyroid hormone to support a eumetabolic state, the normal anterior pituitary quickly "senses" this insufficiency and responds by secreting increased amounts of TSH. Eventually the thyroid becomes incapable of

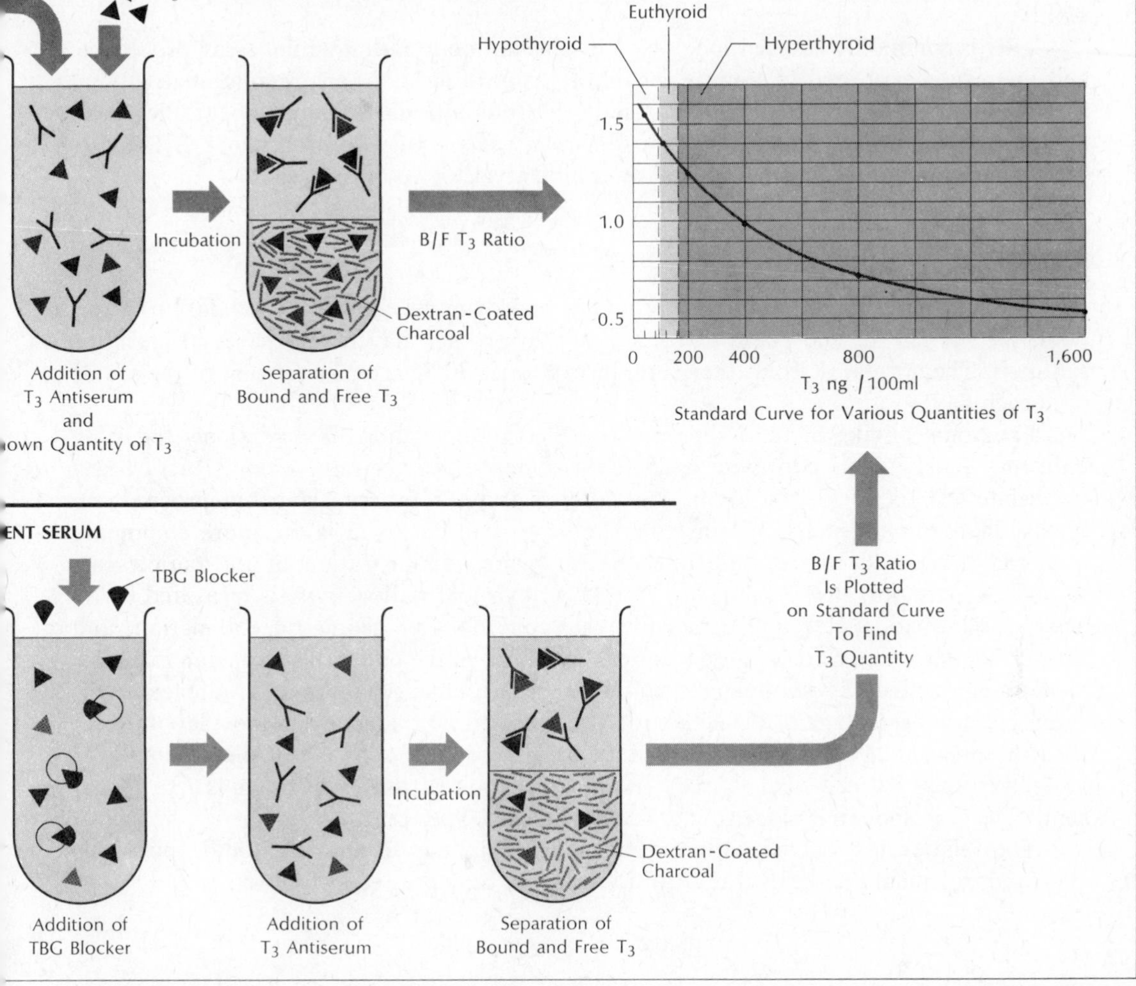

used to adsorb the free T_3, and measurement is made of the bound and free (B/F) T_3. The ratio is then plotted to establish a standard curve on which hypothyroid, euthyroid, and hyperthyroid ranges can be delineated. In the assay of patient serum, the steps are essentially the same as those taken with the T_3 serum in order to derive the standard curve. The B/F T_3 ratio is then located on the standard curve to find T_3 concentration. (From Hollander, C. S.: Newer aspects of hyperthyroidism. J. Hosp. Prac., May, 1972, p. 92.)

responding to this increased stimulation and hypothyroidism develops. This condition is termed *primary* hypothyroidism, and it is characterized by reduced levels of T_3 and T_4 in serum, but elevated levels of TSH.

If, on the other hand, hypothyroidism is due to disease of the anterior pituitary which interferes with TSH production or release, this circumstance is termed *secondary* hypothyroidism, and it is characterized by low serum levels of TSH as well as of T_3 and T_4. Finally, a syndrome called *tertiary* hypothyroidism has been recently described in which the abnormality resides in the patient's hypothalamus and leads to insufficient secretion of thyrotrophin-releasing hormone (TRH).[25] Consequently, insufficient TSH is released and thyroid function fails. The serum findings are identical to those in *secondary* hypothyroidism, i.e., low T_3, T_4, and TSH levels. *Secondary* and *tertiary* hypothyroidism may be distinguished by administering synthetic TRH parenterally and monitoring the patient's serum TSH response at appropriate intervals.[10] In *tertiary* hypothyroidism a significant rise in TSH is seen at 40 to 60 min, whereas in *secondary* hypothyroidism no such rise is observed.

TSH is commonly assayed today by a double antibody radioimmunoassay procedure, and the results expressed in μ units per ml of serum in terms of an International Human Thyrotrophin Standard obtained from the WHO International Laboratory for Biological Standards (Mill Hill, London, England). *Normal values* range from 1.5 to 8.0. Subjects with primary hypothyroidism consistently exhibit levels of 20 or greater.

THYROIDAL UPTAKE OF RADIOIODINE

The iodine pool in man normally contains approximately 350 μg of iodide. From this pool, the normal thyroid gland extracts about 70 μg each day for purposes of hormone synthesis. The normal thyroid, therefore, incorporates 70/350, or 20 per cent of the iodide pool each day.

Fractional uptake of the iodide pool by the thyroid is measured routinely today by uniformly labeling this pool with a small tracer dose of either radioiodine (RaI) or pertechnetate ($^{99m}TcO_4^-$) (Table 14–9). The latter is trapped by the gland but proceeds no further in hormonogenesis. ^{131}I in either liquid or capsule form is the more commonly used radiotracer. At appropriate intervals following administration of the radiotracers (6 and 24 h for RaI and 30 min for $^{99m}TcO_4^-$), thyroidal radioactivity is measured by an external scintillation detector. Under identical conditions of geometry, collimation, and time, a standard (i.e., a tracer dose identical to that given the patient) is also counted. After background activity is subtracted from both measurements, thyroidal radioactivity is expressed as a percentage of the standard. The accepted normal range in most laboratories for RaI uptake at 24 h has been falling in the past decade due to increased dietary intake of iodine (bread is the principal source). In the Chicago area, the usual range is 7 to 33 per cent.[8] The corresponding range for $^{99m}TcO_4^-$ is 0.4 to 3.0 per cent.

The relationship between thyroidal uptake of iodine (I uptake), iodide pool size (P_I), and fractional uptake of RaI (RaI uptake) may be expressed as follows:

$$\text{I uptake} = P_I \times \text{RaI uptake}$$

Under normal circumstances, a gland which accumulates 70 μg of iodine each day for hormonogenesis, in the presence of a P_I of 350 μg, will exhibit an RaI uptake of 20 per cent, as we have said. Both hyper- and hypothyroidism naturally alter the iodine requirements of the gland, and if the P_I is normal, the RaI uptake will reflect these changed requirements. Thus, a hyperthyroid gland which doubles its rate of hormonogenesis will have an RaI uptake of 140/350, or 40 per cent, and a hypothyroid gland with one-tenth the normal rate of hormone synthesis will have an uptake of 7/350, or 2.0 per cent.

These very simple calculations highlight the first of the two major limitations of the RaI uptake. Since P_I is not measured routinely, there is no necessary connection between RaI uptake and rate of hormone synthesis. For example, a patient with an iodide pool expanded to 1000 μg, but having normal thyroid function, will exhibit an RaI uptake of 70/1000, or 7 per cent, and a subject with a contracted pool (100 μg) but normal thyroid function will have an uptake of 70/100, or 70 per cent. The importance of these considerations varies directly, of course, with the frequency of abnormal iodide pools in the population, and, sad to say, abnormal pools are all too frequent. The many contemporary sources of iodine contamination have been considered in connection with previously described methods. These compounds all expand the P_I and thus decrease the values for RaI uptake.

TABLE 14-9 RADIONUCLIDES COMMONLY USED FOR IN VIVO THYROID EVALUATION STUDIES

Radionuclide	$t_{1/2}$	*Predominant γ-Emission (KeV)*	*Approximate Rads per μCi Administered*		
			INFANT	CHILD	ADULT
^{123}I	13 h	159	0.06	0.04	0.02
^{125}I*	60.6 d	27	0.50	0.80	1.10
^{131}I	8.05 d	364	5.00	2.50	1.50
^{132}I†	2.3 h	670, 780	0.05	0.03	0.01
^{99m}Tc	6 h	140	<0.001	<0.001	<0.001

* Used for scanning only.
† Used for thyroidal uptake only.

Contraction of the iodide pool, because of deficient iodine intake, is a far less common circumstance in the United States, since the average diet includes at least 100 to 300 μg of iodine in food and water. Although iodine intake may fluctuate moderately from day to day, these changes exert little effect on the RaI uptake. Iodized salt, to which most of us subscribe, is "iodized" in this country in a KI/NaCl weight ratio of 1/10,000; that is, each 10 g of iodized salt contains 1 mg of KI, or 760 μg of iodide, ten times the normal daily requirement. Thus, iodine prophylaxis very effectively eliminates "contracted poolers." Most of the high mountainous districts of the world (e.g., the Alps, Himalayas, Andes) and some nonmountainous areas as well (the Uele region of the Congo, Holland, and interior Brazil) still harbor numerous iodine-deficient inhabitants.

Just as it is potentially hazardous to assume a necessary connection between RaI uptake and thyroidal iodide accumulation, it is equally dangerous to link iodide accumulation with hormonogenesis. Iodide trapping is but the first step in hormone synthesis; there are five others. Since the post-trapping steps may be interfered with on either a congenital or acquired basis (thyroiditis, foods, drugs), trapped iodide may leave the gland in an incompletely metabolized state, much like $^{99m}TcO_4^-$. It is possible, therefore, to observe a normal or even elevated RaI or $^{99m}TcO_4^-$ uptake in hypothyroid subjects. Further, hyperthyroidism may result from excessive ingestion of thyroid hormone preparations (thyrotoxicosis factitia) or hormone production by a functioning tumor of the ovary (struma ovarii). In such instances pituitary secretion of TSH would be suppressed, thyroid function would compensatorily diminish, and the RaI or $^{99m}TcO_4^-$ uptake would be correspondingly depressed. Thus, one may observe hyperthyroidism in association with a reduced radiotracer uptake.

SUMMARY

Those tests of thyroid function which seem most relevant to the contemporary clinical laboratory have been reviewed. Their rationale and pitfalls have been emphasized. It is evident that despite the many remarkable advances which have led to the development of these procedures, there is as yet none sufficiently specific to permit facile interpretation. Care and caution must characterize both the performance and interpretation of these studies. In general, *in vivo* studies are being supplanted by increasingly sophisticated and specific *in vitro* procedures.

REFERENCES

1. Andros, G., Harper, P. V., Lathrop, K. A., and McCardle, R. J.: Pertechnetate-99m localization in man with application to thyroid scanning and the study of thyroid physiology. J. Clin. Endocrin., *25*:1067, 1965.

2. Barker, S. B., Humphrey, M. J., and Soley, M. H.: The clinical determination of protein-bound iodine. J. Clin. Invest., *30*:55, 1951.
3. Barker, S. B, as modified by Henry, R. J.: Clinical Chemistry. New York, Hoeber Medical Division, Harper and Row, Publishers, 1964, p. 937.
4. Beierwaltes, W. H., and Robbins, J.: Familial increase in the thyroxine binding sites in serum alpha globulin. J. Clin. Invest., *38*:1683, 1959.
5. Benotti, J., and Pino, S.: Simplified method for butanol-extractable iodine and butanol-insoluble iodine. Clin. Chem., *12*:491, 1966.
6. Ekins, R. P.: The estimation of thyroxine in human plasma by an electrophoretic technique. Clin. Chim. Acta, *5*:453, 1960.
7. Ekins, R. P.: Radioimmunoassay, protein-binding assay, and other saturation assay techniques. Year Book of Nuclear Medicine, 1973, pp. 5–37.
8. Ghahreman, G. G., Hoffer, P. B., Oppenheim, B. E., and Gottschalk, A.: New normal values for thyroid uptake of radioactive iodine. J.A.M.A., *217*:337, 1971.
9. Gharib, H., Ryan, R. J., and Mayberry, W. E.: Triiodothyronine (T_3) radioimmunoassay: A critical evaluation. Mayo Clin. Proc., *47*:934, 1972.
10. Hall, R. *et al.:* The thyrotropin-releasing hormone test in diseases of the pituitary and hypothalamus. Lancet, *1*:759, 1972.
11. Hamolsky, M. W., Stein, M., and Freedberg, A. S.: The thyroid hormone-plasma protein complex in man. II. A new in vitro method for study of "uptake" of labelled hormonal components by human erythrocytes. J. Clin. Endocrin., *17*:33, 1957.
12. Hollander, C. S.: On the nature of circulating thyroid hormone. Clinical studies of triiodothyronine and thyroxine in serum using gas chromatographic methods. Trans. Assoc. Am. Physicians, *81*:76, 1968.
13. Hollander, C. S.: Newer aspects of hyperthyroidism. J. Hospital Practice, *7*:87, 1972.
14. Ingbar, S. H.: Clinical and physiological observations in a patient with an idiopathic decrease in the thyroxine-binding globulin of plasma. J. Clin. Invest., *40*:2053, 1961.
15. Ingbar, S. H., and Woeber, K. A.: The thyroid gland. *In* Textbook of Endocrinology. 4th ed. R. H. Williams, Ed. Philadelphia, W. B. Saunders Company, 1968.
16. Ingbar, S. H., Braverman, L. E., Dawber, N. A., and Lee, G. Y.: A new method for measuring the free thyroid hormone in human serum and an analysis of the factors that influence its concentration. J. Clin. Invest., *44*:1679, 1965.
17. Lee, M., Tietz, N. W., and Martinez, C. J.: Clinical evaluation of a modified "Oxford T_4-by-Column" method for serum thyroxine. Clin. Chem., *18*:422, 1972.
18. Lee, N. D., Henry, R. J., and Golub, O. J.: Determination of the free thyroxine content of serum. J. Clin. Endocrin., *24*:486, 1964.
19. Man, E. B., Kydd, D. M., and Peters, J. P.: Butanol-extractable iodine of serum. J. Clin. Invest., *30*:531, 1951.
20. Masen, J. M.: A simplified procedure for serum butanol extractable iodine. Am. J. Clin. Path., *48*:561, 1967.
21. Murphy, B. E. P.: In vitro tests of thyroid function. Sem. Nucl. Med., *1*:301, 1971.
22. Murphy, B. E. P., and Pattee, C. J.: Determination of thyroxine utilizing the property of protein-binding. J. Clin. Endocrin., *24*:187, 1964.
23. Odell, W. D., Wilber, J. F., and Utiger, R. D.: Studies of thyrotropin physiology by means of radioimmunoassay. Recent Progr. Hormone Res., *23*:47, 1967.
24. Pileggi, V. J., Lee, N. D., Golub, O. J., and Henry, R. J.: Determination of iodine compounds in serum. I. Serum thyroxine in the presence of some iodine contaminants. J. Clin. Endocrin., *21*:1272, 1961.
25. Shenkman, L., Mitsuma, T., Suphavai, A., and Hollander, C. S.: Hypothalamic hypothyroidism. J.A.M.A., *222*:480, 1972.
26. Sisson, J. C.: Principles of, and pitfalls in, thyroid function tests J. Nucl. Med., *6*:853, 1965.
27. Sterling, K.: The importance of circulating tri-iodothyronine. N. Eng. J. Med., *284*:271, 1971.
28. Sterling, K., and Brenner, M. A.: Free thyroxine in human serum: simplified measurement with the aid of magnesium precipitation. J. Clin. Invest., *45*:153, 1966.
29. Sterling, K., Refetoff, S., and Selenkow, H. A.: T-3 thyrotoxicosis: thyrotoxicosis due to elevated serum triiodothyronine levels. J.A.M.A., *213*:571, 1970.
30. Wahner, H. W.: T_3 hyperthyroidism. Mayo Clin. Proc., *47*:938, 1972.
31. Wahner, H. W., and Gorman, C. A.: Interpretation of serum T-3 levels measured by the Sterling technique. N. Eng. J. Med., *284*:225, 1971.

CHAPTER 15

BLOOD GASES AND ELECTROLYTES

SECTION ONE

INTRODUCTION

by Norbert W. Tietz, Ph.D.

Electrolytes are classified as either anions or cations, depending upon whether they move in an electric field toward the anode or the cathode, that is, whether they have a negative or positive charge. They are essential components of all living matter and include the major electrolytes Na^+, K^+, Cl^-, HCO_3^-, $HPO_4^{-,2-}$ Ca^{2+}, and Mg^{2+}, as well as the trace elements $Fe^{2+,3+}$, $Cu^{+,2+}$, Mn^{2+}, Co^{2+}, $Cr^{3+,6+}$, Cd^{2+}, Zn^{2+}, Br^-, and I^-. Although amino acids and proteins in solution also carry an electrical charge, in clinical chemistry they are usually classified separately from electrolytes. The major electrolytes occur primarily as free ions, while the trace elements occur primarily in some special combination with proteins and thus are also frequently classified separately.

The dietary requirement for electrolytes varies widely; most need to be consumed only in small amounts or at rare intervals, and are retained when in short supply. Some, like calcium and potassium, are continuously excreted and must be consumed regularly in order to prevent deficiency. Excessive consumption leads to corresponding increased excretion, mainly in the urine. Abnormal loss of electrolytes, which occurs through excessive perspiration, vomiting, or diarrhea, is readily assessed by laboratory tests and can be corrected by administration of salts.

The role of electrolytes in the human body is manifold. There are almost no metabolic processes which are not dependent on or affected by electrolytes. Among other functions of the electrolytes are maintenance of osmotic pressure and hydration of the various body fluid compartments, maintenance of the proper body pH, regulation of the proper function of the heart and other muscles, involvement in oxidation-reduction (electron transfer) reactions, and participation as an essential part or cofactor of enzymes. Thus, it becomes quite apparent that abnormal levels of these electrolytes and trace elements may be either the cause or the consequence of a variety of disorders.

Determination of electrolytes is one of the most important functions of the clinical laboratory. Progress in this field, and especially in the field of trace elements, was hampered by the lack of suitable methods for their determination. In recent years, however, a number of instruments have been made available to facilitate analytical determination, and it is to be expected that much more work will be done and much knowledge will be gained in the next few years. Among the tools that we use to determine these elements are a variety of micro- and macromethods based on spectrophotometry, emission spectrography, flame spectrophotometry, neutron activation analysis, atomic absorption spectroscopy, and coulometric analysis and other electrochemical techniques (specific ion electrodes).

More specific information about the role of individual electrolytes, trace elements, and their determination is given in the respective paragraphs that follow.

DEFINITION OF TERMS AND ABBREVIATIONS USED FOR VARIOUS ACID-BASE PARAMETERS

Bicarbonate. The bicarbonate fraction is the second largest fraction of the anions in plasma. It is customary to include in this fraction the ionized bicarbonate (HCO_3^-) and the carbonate (CO_3^{2-}), as well as the carbamino compounds ($RCNHCOO^-$). At the pH of blood, the concentration of carbonate is only 1/1000 that of bicarbonate. The carbamino compounds are also present in such small amounts (in erythrocytes and plasma about 1.5 and 0.2 mmol/l respectively) that they are generally not mentioned specifically. The newly recommended abbreviation to express the concentration of bicarbonate is $cHCO_3^-$.

Carbonic acid. This fraction of blood, plasma, or serum includes the undissociated carbonic acid ($HHCO_3$)* and the physically dissolved (anhydrous) CO_2. Since the concentration of dissolved CO_2 is significantly higher than that of $HHCO_3$, the symbol $cdCO_2$ (concentration of dissolved CO_2), as opposed to $HHCO_3$, is frequently used; indeed, this is now the preferred symbol. This quantity is not experimentally measured but is calculated from pCO_2 by multiplication with α, the solubility coefficient of CO_2. In case of a blood plasma sample at 37°C:

$$\alpha dCO_2 = 0.0306 \text{ mmol} \times \text{liter}^{-1} \times \text{mm Hg}^{-1}$$
$$= 0.0306 \text{ mmol/l per mm Hg}$$

***pCO*$_2$.** The pressure of a mixed gas, such as air, is the sum of the partial pressures of the individual gases (see Section 2 on blood gases). That part of the pressure which is contributed by CO_2 is called the partial pressure of $CO_2(pCO_2)$. It is usually expressed in mm of Hg. The only place in the body where the blood is in contact with a gas phase is in the lung alveoli. The pCO_2 of the blood not in contact with a gas phase (e.g., arterial, capillary, and venous blood) refers to the pCO_2 in a hypothetical gas phase with which the blood *would be* in equilibrium.

Total CO*$_2$ *(formerly CO*$_2$ *content). The concentration of total $CO_2(ctCO_2)$ of blood, plasma, or serum consists of an ionized fraction that contains HCO_3^- (and CO_3^{2-}, as well as carbamino compounds) and a non-ionized fraction that contains $HHCO_3$ and physically dissolved (anhydrous) CO_2 (see Carbonic acid).

$$ctCO_2 = cHCO_3^- + cdCO_2\dagger$$

CO*$_2$ *combining power. The value of the CO_2 combining power is an index of the total amount of CO_2 that can be bound by serum, plasma, or whole blood at a pCO_2 of 40 mm Hg at 25°C.

Standard bicarbonate of blood. This indicates the concentration of bicarbonate in the plasma phase from blood which is equilibrated with a gas with $pCO_2 = 40$ mm Hg, $pO_2 > 100$ mm Hg $\Rightarrow sO_2 \approx 100$ per cent, at 37°C.

* It is customary to use the symbol $HHCO_3$ for carbonic acid to indicate that the first hydrogen atom is not ionized.

† The concentrations of all listed compounds are expressed in mmol/l.

Buffer base. The concentration of buffer base in the blood or plasma is defined as the sum of the concentrations of bicarbonate and net protein anion. It may therefore be defined "operationally" as the concentration of titratable base when titrating to the apparent isoelectric pH of the proteins at a pCO_2 of zero. Hence, the fundamental difference between the "buffer base" and the "base excess" is the choice of endpoint of titration.

Base excess. The base excess concentration is defined as the concentration of titratable base minus the concentration of titratable acid when titrating the blood or plasma with a strong acid or base to a plasma pH of 7.40 at a pCO_2 of 40 mm Hg at 37°C. Positive values indicate a relative deficit of non-carbonic acid, and negative values indicate a relative excess of non-carbonic acid, in the blood or plasma.

pH. The pH is the negative logarithm of the hydrogen ion activity (pH $= -\log aH^+$). Thus, the average pH of blood (7.40) corresponds to a hydrogen ion activity of 0.00000004. Assuming that the activity coefficient of H^+ is 1, the corresponding hydrogen ion concentration is 4×10^{-8} mol/l = 40 nmol/l.

COLLECTION OF BLOOD FOR THE DETERMINATION OF pH, CONCENTRATION OF TOTAL CO_2 ($ctCO_2$), AND pCO_2

The pCO_2 of air (about 0.2 mm Hg) is much less than that of blood (about 38 mm Hg). Thus, when blood is exposed to air, the CO_2 content and the pCO_2 decreases and the pH increases correspondingly. For the determination of pH, total CO_2($ctCO_2$) or pCO_2, it is, therefore, necessary to collect, transfer, and manipulate blood (serum or plasma) under conditions in which exposure to air is avoided or kept at a minimum. (See also the respective comments on Specimen in the sections on pH, pCO_2, and pO_2.) Before performing these techniques, one should be aware of certain other pertinent properties of blood.

CHOICE BETWEEN VENOUS, ARTERIAL, AND CAPILLARY BLOOD SAMPLES

The $ctCO_2$ of *venous* blood (usually collected from the radial vein of the arm) is approximately 2 mmol/l higher and the O_2 content about 2 mmol/l less than that of arterial blood simultaneously drawn. This arterial-venous difference varies with the metabolic activity of the organ or tissue from which the venous blood is obtained. For this reason, arterial blood is of more uniform composition than venous blood and is, therefore, preferred for many studies. The arterial-venous pH difference, however, is extremely small, 0.01 to 0.03 pH, and is minimized by the compensatory effect of *increased* acidity due to increased CO_2 (carbonic acid concentrations) and *decreased* acidity due to decreased O_2 content as arterial blood is changed to venous blood. Under the extreme conditions of an elevated respiratory quotient of 0.95 (in case of high carbohydrate and low fat diet) and an elevated blood pH of 7.6, the arterial-venous pH difference reaches a maximum of 0.03. In conditions of low respiratory quotient, e.g., 0.70 (low carbohydrate diet or decreased carbohydrate utilization), and a pH of 7.2 (e.g., in diabetic acidosis), venous blood may be more alkaline than arterial blood by 0.01 pH unit. The important fact is that the arterial-venous pH difference in the majority of patients does not exceed 0.01 pH unit; therefore, for clinical purposes, the pH of venous blood has the same clinical significance as that of arterial blood.

Arterialized blood in respect to pO_2 and $ctCO_2$, as well as pH, approximates arterial blood. It may be obtained by collecting the blood specimen from a limb that has been warmed to 45°C for several minutes immediately before the blood is drawn. The entire forearm, including the hand and fingers, is immersed in a water bath at 45°C, or a towel wetted with water of 45°C is wrapped around the forearm. The heat dilates the capillaries and, by accelerating the rate of blood flow, decreases the changes in blood composition caused by tissue respiration. Procurement of arterialized blood is rarely resorted to because of the inconvenience.

Capillary blood obtained by pricking the fingertip, earlobe, heel, or toe with a small sharp blade (Bard-Parker blade No. 11 is suitable) resembles arterial blood in its composition more closely than it resembles venous blood. Capillary blood may also be arterialized by warming the skin in the area of the puncture, but this arterialization procedure is rarely used. Speedy manipulation during specimen collection is required to avoid clotting and to prevent significant loss of carbon dioxide and gain of oxygen in capillary blood samples.

COLLECTION OF ARTERIAL BLOOD

Arterial blood should be collected by a skilled and experienced physician, since puncture of an artery is not only difficult but, in fact, may be dangerous. The blood should be collected in a Luer-Lok or other suitable syringe, which has previously been coated with a solution of heparin (1000 USP units/ml). The heparin not only will serve as an anticoagulant but will also fill the air spaces in the syringe and the needle. After collection of the blood, the tip of the syringe is sealed either with a metal or other suitable cap, or with a mercury button. A slight excess of mercury may be introduced into the syringe to facilitate the mixing of the specimen after collection and before analysis. This is accomplished by swirling the syringe rapidly and, thus, moving the mercury drop within the syringe. If whole blood is used for analysis, it is important that the specimen be mixed adequately before analysis.

COLLECTION OF VENOUS BLOOD

Most technologists apply a tourniquet to the patient's arm before performing the venous puncture. If this is the case, the patient should *not flex his fingers*; the blood for pH and gas studies should be drawn immediately after applying the tourniquet, and blood required for other tests should be drawn thereafter. This will prevent any changes in the composition of the blood caused by stasis and accompanying accumulation of metabolites and decrease in oxygen in the blood. If, for any reason, the tourniquet is left on for more than a few seconds, it should be removed after successful venous puncture and the blood permitted to flow freely for at least 1 min before withdrawing the sample. Aspiration of the sample immediately after removing the tourniquet introduces substantial errors, since compounds accumulated during stasis will flush into the blood and rapidly change its composition. These adverse changes may also be circumvented by performing the venous puncture without applying a tourniquet.

There are three containers in which the venous specimen can be collected.

1. The blood may be collected in a heparinized Luer-Lok or other suitable syringe as outlined under the collection of arterial blood.

2. The specimen may be drawn into a heparinized Vacutainer, a technique which has become very popular. In this case, however, the tube must be filled completely to avoid pH changes due to loss of CO_2 into the vacuum. If the pH is measured in plasma, the tube must be centrifuged with the stopper in place and determinations must be made immediately after removing the stopper. Once the stopper has been removed for more than a few seconds, the specimen is unsuitable for any repeat analysis of pH, $p\mathrm{CO_2}$ or $c\mathrm{tCO_2}$.

3. The blood may also be collected in a tube containing "light mineral oil," which is used to provide a barrier between blood and air to retard the exchange of O_2 and CO_2. Many investigators have expressed objections against the use of such oil because CO_2 is slightly soluble in this medium. The error introduced is negligible, however, if a thin oil

layer is used, if the blood is not agitated excessively, if it is centrifuged for only 3 min, and if it is analyzed within 30 min after collection. Contamination of electrodes by mineral oil can be prevented by withdrawing the sample from under the oil with the aid of a 1 ml tuberculin syringe, fitted with a 26 gauge needle. After removal of the needle, the sample may be aspirated or delivered into the electrode directly from the syringe.

COLLECTION OF CAPILLARY BLOOD

Thoroughly cleanse the skin area where the puncture is to be performed and remove any excess moisture to prevent dilution of the sample and hemolysis. The puncture should be made deep enough to provide a good blood flow. (Sterile Bard-Parker blades No. 11 are satisfactory for this purpose.) The first drop of blood should be wiped off because it frequently contains some tissue juice that tends to hemolyze the red cells and alter the pH and other blood constituents. The blood is then collected with the aid of a heparinized capillary tube, by placing the end of the capillary tube into the freshly formed drop of blood. Spreading of the blood over a wide area should be avoided, since this will expose a large surface of blood to air, causing a loss of CO_2 and a change in pH. It should always be remembered that it is better not to do a determination at all than to do a determination on an unsatisfactory specimen.

After collection of the specimen, a small metal bar is placed into the capillary and the latter is sealed with a suitable sealing substance. Mixing of blood is provided by moving a magnet along the capillary. Thus, the metal bar will move up and down the capillary, mixing the blood with heparin while keeping the red cells suspended. The mixing should be repeated shortly before the actual measurement.

COLLECTION OF SPECIMENS FOR THE DETERMINATION OF ANIONS, CATIONS, AND TRACE ELEMENTS

Specimens for the determination of various electrolytes in serum, such as Na^+, K^+, Cl^-, Fe^{2+}, Cu^{2+}, and Mg^{2+}, should be collected in tubes that are free of even trace contamination. If necessary, tubes should be washed with three-fold diluted nitric acid, rinsed in deionized water, and dried well. (For the determination of trace elements, the resistance of the deionized water should be about 10 megohms.) Hemolysis must be avoided, since this will interfere to some degree with all tests measuring electrolyte concentration. In some cases, for example for the K^+ analysis, a hemolyzed specimen is unsuitable.

Except in emergencies, all specimens, including those for pH and $ctCO_2$ analysis, should be collected when the patient is in a fasting state.

REFERENCES

Gambino, S. R., Astrup, P., Bates, R. G., Campbell, E. J. M., Chinard, F. P., Nahas, G. G., Siggaard-Andersen, O., and Winters, R.: Am. J. Clin. Path., *46*:376, 1966.

Fleischer, W. R., and Gambino, S. R.: Blood pH, Po_2 and Oxygen Saturation. Commission on Continuing Education, American Society of Clinical Pathologists, Chicago, 1972.

SECTION TWO

BLOOD GASES

by Ole Siggaard-Andersen, M.D., Ph.D.

THE PARTIAL PRESSURE OF OXYGEN IN ARTERIAL BLOOD, $pO_2(aB)$ *

This quantity refers to the oxygen partial pressure in an imaginary gas phase in equilibrium with oxygen dissolved in arterial blood. It is sometimes called the tension of oxygen in the blood and is traditionally expressed in mm Hg. The newly recommended unit for pressure is the pascal (Pa = N/m^2). The proportionality factor is 0.133 kPa/mm Hg.

The following quantities are directly related to the pO_2 of arterial blood:

1. The pO_2 is equal for the plasma phase and the erythrocyte phase of whole blood, and hence the pO_2 of the arterial blood is equal to *the* pO_2 *of the arterial plasma.*

2. *The* pO_2 *of the alveolar air* (A) is slightly higher than the pO_2 of arterial blood [$pO_2(aB)$]. Normally, O_2 diffuses rapidly through the alveolar membrane; however, small amounts of blood are always shunted through the lungs without passing functioning alveoli. $pO_2(A)$ is defined as the substance fraction of oxygen in the alveolar air times the total pressure of the alveolar air: $pO_2(A) = xO_2(A) \times p(A)$, and generally the substance fraction of oxygen in a gas mixture is taken to be equal to the volume fraction†: $xO_2(A) = \varphi O_2(A)$.

In gas mixtures such as atmospheric or alveolar air, the total barometric pressure [$p(Atm)$] equals the sum of the individual partial pressures of the different components (Dalton's law):

$$p(\text{Atm}) = pO_2 + pCO_2 + pN_2 + pH_2O$$

or

$$p(\text{Atm}) - pH_2O = pO_2 + pCO_2 + pN_2 \quad (1)$$

Dalton's law can also be expressed as

$$pG(M) = xG(M) \times p(M) \quad (2)$$

which defines the partial pressure of a component (G) in a gas mixture (M) as the substance fraction (mol fraction) of component G in M, times the total pressure of the gas mixture.

Dry atmospheric air has been found to have the following volume fractions (essentially equal to substance fractions because molar volumes of these gases differ by less than 1 per cent) of component gases: oxygen, 20.93 per cent; carbon dioxide, 0.03 per cent; and nitrogen, 79.04 per cent. The nitrogen value is generally measured by difference and includes small amounts of other inert gases. In the alveolar air, the sum of pO_2 and pCO_2 is normally constant at about 150 mm Hg, and an inverse relationship exists between these two quantities.

In connection with measurement of respiration and with clinical problems, the concentrations of gases are generally expressed in terms of their partial pressures (unit: mm Hg or, preferably, kPa) rather than in terms of substance fraction or volume fraction. The formula for calculating the partial pressure of a gas in a humidified gas mixture (hM) in terms

* Regarding the use of the abbreviation $pO_2(aB)$, see Chap. 3, p. 139 and the appendix.

† The volume fraction (φ) is defined as the volume of the pure substance divided by the volume of the mixture at constant pressure.

of the substance fraction of the gas in the dry gas mixture (dM) is shown in Equation 3 for oxygen, but the same formula is applicable to all other gases:

$$pO_2(\text{hM}) = (p(\text{hM}) - pH_2O(\text{hM})) \times xO_2(\text{dM}) \quad (3)$$

example

$$pO_2(\text{hM}) = (760 \text{ mm Hg} - 47 \text{ mm Hg}) \times 0.15 = 107 \text{ mm Hg}, \quad (4)$$

where 47 mm Hg is the partial pressure of water vapor at 37°C, 760 mm Hg represents the total pressure of the humidified air in the alveoli, which is equal to the existing atmospheric pressure, and 0.15 represents the substance fraction (or volume fraction) of oxygen in dry alveolar air. This formula is used in the standardization of instruments for blood gas analysis.

3. *The concentration of free dissolved oxygen* in the blood is proportional to the pO_2 in the blood (Henry's law): $cO_2(B) = \alpha O_2(B) \times pO_2(B)$, where $\alpha O_2(B)$ is the solubility coefficient of oxygen in the blood, which is 0.00140 mmol × liter^{-1} × mm Hg^{-1} (0.00140 mmol per liter per mm Hg) at 37°C.*[19] For plasma the value is $\alpha O_2(P) = 0.00126$ mmol × liter^{-1} × mm Hg^{-1} at 37°C.[4] With a normal arterial pO_2 value of 93 mm Hg, the concentration of free dissolved oxygen in the arterial blood is therefore 0.00140 × 93 mm Hg = 0.13 mmol/l.† With pure oxygen treatment, the arterial pO_2 may rise to about 640 mm Hg, depending upon the degree of shunting of blood in the lungs. The concentration of free dissolved oxygen in the arterial blood is then 0.9 mmol/l. With treatment in a hyperbaric chamber, the pO_2 may rise to several atmospheres and the concentration of free dissolved oxygen may then rise significantly, e.g., $pO_2(\text{aB}) = 2500$ mm Hg $\Rightarrow cO_2(\text{aB}) = 3.5$ mmol/l. Under these circumstances the physically dissolved oxygen contributes significantly to the amount of oxygen transported by the blood.

4. *The oxygen saturation of the hemoglobin* (sO_2) increases with increasing pO_2 as illustrated by the oxygen dissociation curve. (See Fig. 15–1, p. 860.)

5. *The concentration of total oxygen in the blood* (often called the oxygen content) is the sum of the concentrations of free dissolved oxygen and oxygen bound to hemoglobin. The latter value is dependent on several factors, including the pO_2, the hemoglobin concentration, and the concentration of 2,3-diphosphoglycerate in the erythrocytes. (See also pp. 404 and 861.)

NORMAL VALUES

For adults of 12 to 40 years of age, the 95 per cent tolerance limits for pO_2 in arterial blood are 83 and 108 mm Hg; for the 40 to 80 year group the limits are 72 and 104 mm Hg. The relationship between the normal mean $pO_2(\text{aB})$, measured in mm Hg, and age of the patients, measured in years, can be approximated as follows:[12,17]

$$\{pO_2\}‡ = -0.27 \times \{\text{age}\} + 104$$

---

* The so-called Bunsen solubility coefficient (α_B) of a gas (e.g., O_2) in a fluid (e.g., plasma, P) is defined as $\alpha_B O_2(P) = cO_2(P) \times V_m O_2(\text{STP})/pO_2(P)$, where $V_m O_2(\text{STP})$ is the molar volume of oxygen at standard temperature and pressure (22.4 liters). When the unit for pO_2 is atmospheres, the unit for the Bunsen solubility coefficient becomes atm^{-1} or (ml/ml)/atm.

The solubility coefficient decreases with increasing temperature and with increasing concentration of other solutes.

† The molar volume (V_m) of an ideal gas at STP (standard temperature and pressure, i.e., 0°C and 1 atm) is $V_m = 22.4$ liters/mol. For O_2 its value is 22.38 liters/mol; for CO_2 it is 22.26 liters/mol. The volume of 0.13 mmol of O_2 at STP is therefore 2.91 ml. It has been common practice to express the amount of dissolved oxygen in terms of "volumes per cent" (= ml/100 ml), but it is preferable to use the substance concentration (unit: mmol/l).

‡ The symbol of a quantity in braces, e.g., $\{pO_2\}$, indicates the pure number value, whereas the symbol in brackets, e.g., $[pO_2]$, indicates the unit of the quantity. In other words, $pO_2 = \{pO_2\} \times [pO_2]$.

These values refer to sea level. There is no significant sex difference for the pO_2 values (unlike that for pCO_2 values).

The pO_2 of venous blood from the right atrium is about 40 mm Hg. For peripheral venous blood the pO_2 varies greatly with skin temperature, duration of venous stasis, and muscular pumping.

CLINICAL SIGNIFICANCE

Decreased arterial pO_2 (hypoxemia) can be caused by the following:

1. Decreased pO_2 in the inspired air (high altitude).
2. Decreased diffusion capacity of the alveolar lining membranes for oxygen (e.g., respiratory distress syndrome in the newborn).
3. Decreased alveolar ventilation either of *peripheral* origin (suffocation, submersion, paralysis, emphysema, and so forth) or of *central* origin (depression of the respiratory center by barbiturates or morphine).
4. Increased venoarterial shunting of the blood (congenital heart failure with cyanosis).

The physiological effects of moderately *decreased* pO_2(aB) are (1) stimulation of the peripheral chemoreceptors leading to hyperventilation, and (2) stimulation of erythropoietin production leading to polycythemia. Both effects are observed during acclimatization to high altitude. Further reduction of the arterial pO_2 leads to tissue hypoxia with increased production of lactic acid and signs of cerebral hypoxia (confusion, coma). The dangers of a low arterial pO_2 are aggravated by a low hemoglobin concentration and by a high oxygen affinity (low $pO_2(0.5)$ value, see p. 862). The lowest pO_2 value that is tolerated for long periods is about 40 mm Hg.

Increased pO_2 *in the arterial blood* is caused by breathing O_2-enriched air. The toxic effects of oxygen are especially pronounced in the newborn, where oxygen treatment may cause fibrosis in the eye with blindness (retrolental fibroplasia). Therefore, it is essential to control the pO_2 in the newborn during oxygen therapy to avoid values higher than 110 mm Hg for long periods.

In the adult, exposure to pure oxygen at atmospheric pressure is tolerated for maximally four days and causes substernal pain, cough, and a progressive deterioration of pulmonary function. However, at reduced pressure (1/3 atm) pure oxygen is well tolerated. Oxygen at high partial pressures damages the pulmonary capillaries, causing a fibrinous exudation. Hyperbaric oxygen causes cerebral vasoconstriction and, eventually, convulsions. Inhibition of enzymes containing sulfhydryl groups has also been demonstrated.

When a large fraction of oxygen for the metabolism is transported in the form of free dissolved oxygen, a smaller fraction of the oxyhemoglobin is converted to deoxyhemoglobin in the tissues. Deoxyhemoglobin is the major vehicle for the transport of CO_2 in the form of carbamate (R—NH—COO⁻). The tissue pCO_2 therefore tends to rise slightly, but this effect is probably of minor importance.

DETERMINATION OF pO_2 *WITH THE* pO_2 *ELECTRODE*

pO_2 is generally measured directly, with the Clark pO_2 electrode. Alternatively, the pO_2 of the blood can be measured by gasometric analysis of a small gas bubble aspirated into a syringe with a large volume of blood, but this technique has proved too complicated for routine application.[1,15]

pO_2 can also be calculated from the oxygen saturation, pH, and temperature by means of the standard oxygen dissociation curve. Several nomograms have been published for this purpose, and a very practical slide rule has been constructed by Severinghaus.[20] However, the calculation is only an approximation, due to the significant effect of CO_2 as well as that of 2,3-diphosphoglycerate on the oxygen dissociation curve (see p. 404).

SPECIMEN

Arterial blood is collected anaerobically in a well-sealed heparinized syringe. Freely flowing capillary blood originates from the arterioles and therefore provides a substitute for arterial blood in many cases. However, a certain risk of admixture of blood from the venules always exists, causing erroneously low pO_2 values. This risk is especially pronounced in the following situations:[22]

1. In patients in shock, where the arterial blood pressure is low, the venous pressure high, and the arteriovenous blood gas differences are large.
2. In newborn infants in the first hours or even days after birth.
3. In newborn infants with respiratory distress syndrome.
4. In patients receiving oxygen therapy. Capillary blood is not suitable for estimation of high arterial pO_2 values. Even slight tissue deoxygenation of arterial blood, which causes a fall in oxygen saturation of only a few per cent, causes a very marked fall in pO_2 owing to the horizontal course of the oxygen dissociation curve at high pO_2 values.

If the measurement cannot be performed immediately, the blood should be stored in ice water, but not longer than 1 h. A syringe of blood should be completely immersed in ice water to cool it rapidly. Placing the syringe on the tips of ice cubes is not adequate. Capillary tubes should be placed horizontally to facilitate subsequent resuspension of the erythrocytes. When using cooling pads from the freezer, care must be taken to avoid hemolysis, since the temperature of such pads is often below zero.

Sources of error are the exposure of the blood specimen to air, and O_2 consumption by metabolic processes during storage. If an air bubble equal to 10 per cent of the blood volume is present in the syringe, the error at a pO_2 level of about 30 mm Hg may amount to +15 per cent. Oxygen consumption in blood causes a fall in pO_2 of about 3 to 12 mm Hg/h at 37°C if the pO_2 is in the normal range. At high pO_2 the rate is greater: at $pO_2 \approx 400$ mm Hg the pO_2 fall is 2 mm Hg/min. At 2 to 4°C the rates are reduced by a divisor of about 10. Leukocytes, thrombocytes, and reticulocytes account for virtually all of the oxidative metabolism. Erythrocytes lack mitochondria with the necessary enzymes of the tricarboxylic acid cycle, but they do contain glucose-6-phosphate dehydrogenase and are able to oxidize some glucose to CO_2 via the pentose phosphate pathway.

PRINCIPLE

The determination of pO_2 using a pO_2 electrode is based on amperometric (polarographic) measurement of oxygen as described in the section on Electrochemistry (p. 149).

EQUIPMENT

The pO_2 electrode is shown schematically in Fig. 3–27.

The *platinum cathode* is the exposed end of a platinum wire with a diameter of about 20 μm sealed in a glass rod. The Ag/AgCl anode can be shaped as a ring around this glass rod. The electrolyte consists of a phosphate buffer (0.5 mol/l, pH 7.0) saturated with AgCl, containing 0.13 mol/l KCl. The rough surface of the glass tip of the electrode insures that a thin small layer of electrolyte solution stays between the electrode and a membrane covering the tip of the electrode.

The membrane consists of a gas-permeable material, e.g., polypropylene, with a thickness of about 20 μm, which is tightly fitted over the platinum cathode. A shorter response time is obtained with Silastic, but the electrode then becomes more sensitive to the diffusion coefficient of O_2 in the test solution. If the membrane is too thin, the diffusion zone extends into the blood sample, whereas a thicker membrane or a less permeable membrane ensures that the diffusion zone is limited to the membrane itself.

The electrode chamber or cuvet should require as little blood as possible for filling, preferably only 50 to 75 μl for complete filling and rapid equilibration. The cuvet may be a closed chamber with an inlet and outlet tube. The tubings leading to the chamber should be made from a material with a low permeability to gases, e.g., nylon or vinyl. Open, upside-down electrodes have also been described.[11]

The pO_2 meter is a sensitive amperometer which provides a constant polarizing potential of -0.6 V irrespective of the current. Older instruments are provided with a mercury cell (1.34 V), which must be periodically replaced. The sensitivity of the pO_2 electrode with a microcathode and a polypropylene membrane is $\Delta I/\Delta pO_2 \approx 20$ pA/mm Hg. This value is almost constant up to about 500 mm Hg. This corresponds to a reduction of oxygen of 5.2 fmol/s at pO_2 values of 100 mm Hg. The zero current is about 1 to 50 pA for pO_2 values of 0. The pO_2 meter is provided with the following controls:

1. The zero adjustment allows adjustment of the meter reading to zero for a pO_2 of zero in the calibrating gas (see Calibration, below).
2. The sensitivity adjustment allows the adjustment of the meter reading to a pO_2 value corresponding to that of the calibrating gas or solution.
3. The membrane leakage test allows a check of the resistance over the membrane,which falls markedly if the membrane is ruptured.
4. The battery test allows a check of the potential of the mercury battery (if used).

CALIBRATION AND MEASUREMENT

Zero calibration is generally made with oxygen-free gases, e.g., pure N_2, He, or N_2O. The use of "zero solution" [an aqueous solution of sodium borate (10 mmol/l) with added sodium sulfite (50 to 100 mg/5 ml solution) freshly prepared] is not recommended because sulfite penetrates the membrane and may increase a background current. It also causes wear on the membrane. N_2 is safer since it reduces the chance of contamination. Zero calibration needs to be done only once a day.

Slope adjustment can be carried out with atmospheric air or with a calibration gas with a known volume fraction of oxygen. The volume fraction of oxygen in atmospheric air is 0.2093, and the pO_2 of the humidified air (h) is therefore $pO_2(h) = 0.2093 \times [p(h) - 47$ mm Hg], where 47 mm Hg is the saturated water vapor pressure at 37°C and $p(h)$ equals the ambient pressure.

The *blood/gas sensitivity ratio* (r) must be known. Owing to the greater diffusion coefficient for O_2 in gases than in fluids, the sensitivity of the pO_2 electrode tends to be greater with gases than with fluids, r being usually between 0.95 and 1.0.[7] In order to establish the blood/gas sensitivity ratio, a blood sample must be equilibrated with a known pO_2 in a tonometer (see p. 870). When measuring blood pO_2 values, the calibration value for humidified gas or air (h) should therefore be $pO_2(h)/r$. For practical purposes the water/gas sensitivity ratio may be assumed to be equal to the blood/gas sensitivity ratio. The thermostat water enters equilibrium with the atmospheric air after the thermostat has been in operation for 1 h, provided the water is well aerated and uncontaminated by microorganisms. Reading the pO_2 of the thermostat water and of the humidified air allows calculation of the water/gas sensitivity ratio.

PROCEDURE

The blood pO_2 is measured by injecting the blood slowly into the electrode chamber, after prior rinsing of the electrode chamber with physiological saline. Enough blood should be available to flush the chamber with 1 to 2 vol of blood. After measurement, the electrode chamber is rinsed gently with physiological saline, and the reading of atmospheric air is checked. After a series of measurements on blood, rinsing with a biodetergent is recommended, and at the end of the day rinsing should be done with an aqueous Zephiran chloride solution (0.05 g/100 ml). The electrode is left filled with saline.

SOURCES OF ERROR AND PROCEDURE NOTES

The analytical coefficient of variation when measuring pO_2 repeatedly on the same blood sample is about 2 per cent if the chamber is rinsed with saline and the calibration checked with air between samples. The analytical accuracy depends on a number of factors:

1. The electrode should ideally be calibrated by means of equilibrated blood. The use of air, and a correction by means of the blood/gas or water/gas sensitivity ratio, may introduce a systematic error of 1 to 2 per cent.

2. The electrode sensitivity is normally constant up to a $pO_2 \approx 500$ mm Hg. Beyond that, a deviation from linearity results in values being too low unless the electrode is calibrated by means of a gas with a pO_2 value in the high measuring range.

3. The temperature of the electrode should be accurate within ± 0.1°C. A considerable temperature difference may exist between the water bath and the electrode if the tubings leading to the electrode are too long or the water circulation is too slow. The pO_2 of the blood increases 8 per cent for a temperature rise of 1°C. The electrode error due to increasing current with increasing temperature adds an additional 3 per cent. The total bias of the measurement is therefore +11 per cent/°C.

4. pO_2 values measured at 37°C must be corrected to the actual temperature of the patient. The exact relationship between pO_2 and temperature is complicated: At $pO_2 <$ 100 mm Hg, the determining factor is the change in the position of the oxygen dissociation curve, which is displaced to the left by a fall in temperature as well as the concomitant rise in pH. The combined effect can be expressed as $\Delta \log pO_2/\Delta T = +0.031 \times K^{-1}$ (= +0.031 per degree temperature change in Kelvin). This equation is represented by the nomogram shown in Fig. 15–3. At very high pO_2 values, hemoglobin is fully saturated with oxygen irrespective of temperature. The determining factor is then the change in solubility of oxygen, i.e., $\Delta \log pO_2/\Delta T = -\Delta \log \alpha/\Delta T = +0.005 \times K^{-1}$. For pO_2 values between 100 and 400 mm Hg, no accurate formula can be given.

REFERENCES

Clark, L. C., Jr.: Monitor and control of blood and tissue oxygen tensions. Trans. Am. Soc. Artif. Intern. Organs, *2*:41, 1956.

Severinghaus, J. W.: Measurements of blood gases: pO_2 and pCO_2. Ann. N.Y. Acad. Sci., *148*:115, 1968.

THE HEMOGLOBIN OXYGEN DISSOCIATION CURVE AND THE pO_2 AT HALF SATURATION [pO_2 (0.5) OR P_{50}]

The relationship between the hemoglobin oxygen saturation (sO_2) and the pO_2 is illustrated by the hemoglobin oxygen dissociation curve (Fig. 15–1A) or the so-called Hill plot (Fig. 15–1B). The latter is obtained by plotting $\log [sO_2/(1 - sO_2)] \equiv \text{logit } sO_2$* on the ordinate and $\log pO_2$ on the abscissa. The oxygen dissociation curves in a Hill plot are virtually parallel, straight lines with a slope of $n_{Hill} = 2.7$. However, theoretical considerations indicate that the slope must decline toward unity at very high as well as at very low sO_2 values.

The pO_2 at half saturation of the hemoglobin, $pO_2(0.5)$, also symbolized P_{50}, is a measure of the affinity of hemoglobin for oxygen: A low $pO_2(0.5)$ signifies a high affinity, and vice versa. A better indicator of the affinity is $-\log pO_2(0.5)$, which is directly proportional to the affinity of hemoglobin for oxygen at half saturation.

The value of $pO_2(0.5)$ is influenced by the following main factors: (1) temperature, (2) pH, (3) pCO_2, (4) concentration of 2,3-diphosphoglycerate in the erythrocytes (often

---

* The logit $(x) = \log \dfrac{x}{(1 - x)}$, where $0 < x < 1$.

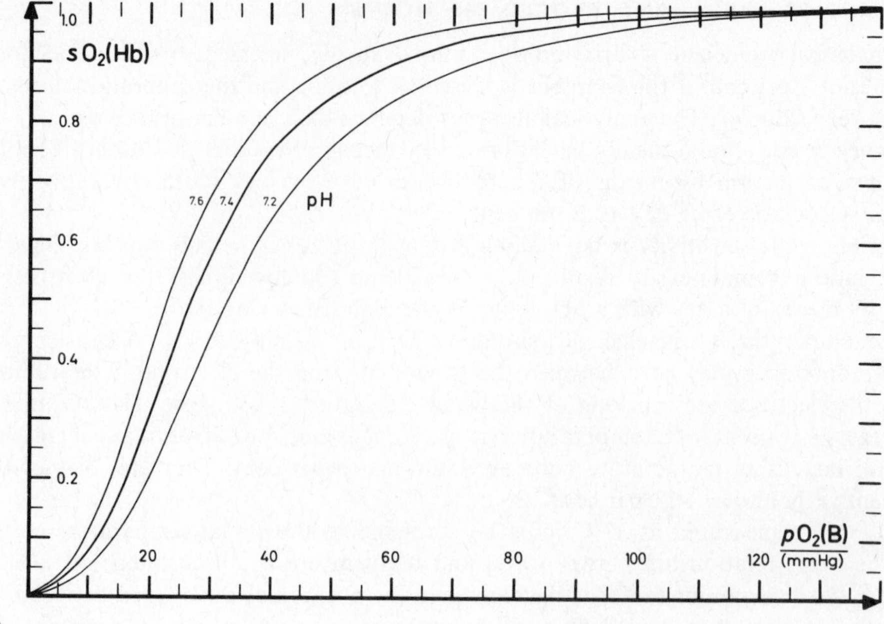

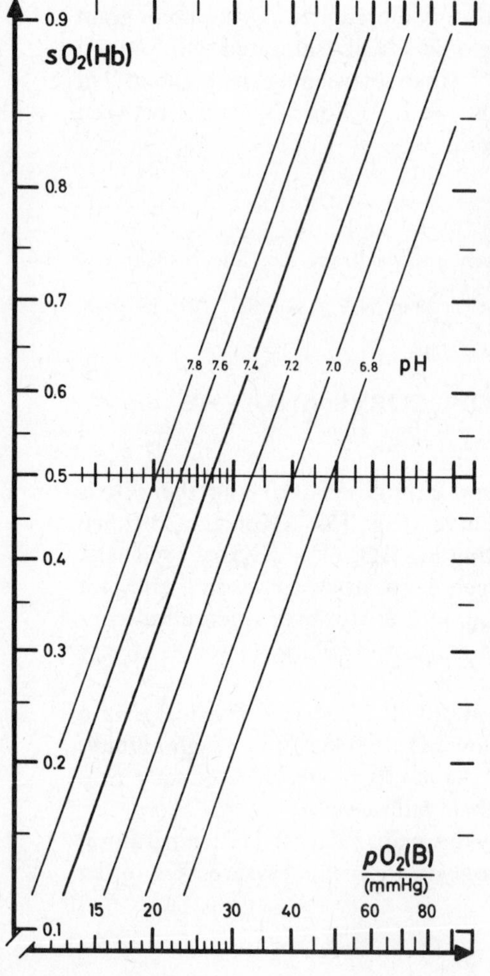

Figure 15–1 Oxygen dissociation curves for human blood with different pH(P) but with constant $pCO_2 = 40$ mm Hg, normal 2,3-DPG concentration in the erythrocytes (4.8 mmol/l), and temperature at 37°C. The displacement of the curve with changing pH(P) is called the Bohr effect. The oxygen dissociation curves are shown both in a coordinate system with pO_2 and sO_2 on linear scales (A), and in a coordinate system with pO_2 on a logarithmic scale and sO_2 on a logit scale [logit $sO_2 \equiv \log(sO_2/(1\text{-}sO_2))$] (B). In the latter plot (a so-called Hill plot) the oxygen dissociation curves are virtually parallel straight lines over a wide sO_2 range. The lines are not completely equidistant because the Bohr effect varies somewhat with the pH, falling off at high as well as low pH values, the maximum effect being at pH(P) $\approx$ 7.1.

expressed as the ratio 2,3-DPG concentration/hemoglobin concentration), and (5) type of hemoglobin (see also Chap. 8). The quantitative effect of each of these factors is as follows, if other variables are kept constant at their normal values:

1. $\frac{\Delta \log pO_2\,(0.5)}{\Delta T} = +0.024\ K^{-1}$

2. $\frac{\Delta \log pO_2(0.5)}{\Delta\ pH(P)} = -0.38$

3. $\frac{\Delta \log pO_2\,(0.5)}{\Delta \log pCO_2} = +0.10$

4. $\frac{\Delta \log pO_2\,(0.5)}{\Delta\left(\frac{c2,3\text{-DPG}}{cHb}\right)} = +0.44$

For a description of the molecular mechanisms by which H^+, CO_2, and 2,3-DPG influence the oxygen affinity of hemoglobin, the reader is referred to Chapter 8 and the specialized literature.[16,18]

The effect of pH on the $pO_2(0.5)$ is called the *Bohr effect* [coefficient (2)]. This coefficient applies to conditions when the pCO_2 is 40 mm Hg, and changes in the pH are caused by changes in the concentration of noncarbonic acids or bases. If, however, the concentration of noncarbonic acid or base is constant, and pH changes are being caused by changes in the pCO_2, the coefficient includes the pH and CO_2 effect and its absolute value is therefore higher: $\Delta \log pO_2(0.5)/\Delta\ pH(P) = -0.53$. It should be noted that both coefficients refer to whole blood, and the pH values refer to the plasma phase. It must be emphasized that all the coefficients vary with the values of the different variables, and any calculation based on these coefficients is therefore only approximate.[25]

NORMAL VALUES

For adults the 95 per cent tolerance limits for $pO_2(0.5)$, corrected to pH (P) = 7.4, are about 25 and 29 mm Hg. For newborn infants the limits are about 18 and 24 mm Hg.

CLINICAL SIGNIFICANCE

Elevated $pO_2(0.5)$ *values* indicate a displacement of the oxygen dissociation curve to the right, i.e., *a decreased oxygen affinity of hemoglobin.* The main causes are:

1. High temperature (hyperthermia).
2. Low plasma pH (acute acidemia).
3. High pCO_2 (hypercapnia).
4. High concentration of 2,3-DPG. If the $pO_2(0.5)$ value is corrected to 37°C, pH 7.4, and a pCO_2 of 40 mm Hg, the $pO_2(0.5)$ becomes essentially an indirect parameter of the 2,3-DPG concentration. The latter value tends to increase when the plasma pH increases (i.e., in states of chronic alkalemia) as well as in anemia.
5. Presence of abnormal hemoglobins which show a decreased oxygen affinity (e.g., hemoglobin Seattle).

The physiological effects of a decreased oxygen affinity are small. Generally the affinity is still high enough for the hemoglobin to bind adequate amounts of oxygen in the lungs. A low oxygen affinity therefore merely results in a facilitated dissociation of oxyhemoglobin in the tissues, which, in anemia, is a desirable compensatory mechanism. Patients with Hb Seattle have a $pO_2(0.5)$ value of about 41 mm Hg and a low hemoglobin concentration, but the subjects are otherwise unaffected.[13]

Decreased $pO_2(0.5)$ *values* signify a displacement of the oxygen dissociation curve to the left, i.e., *an increased oxygen affinity of the hemoglobin.* The main causes are:

1. Low temperature (hypothermia).
2. High plasma pH (acute alkalemia).
3. Low pCO_2 (hypocapnia).
4. Low concentration of 2,3-DPG. The 2,3-DPG concentration generally falls in states of acidemia, which have persisted for several hours. Thus, the increase in pO_2 (0.5), caused by the low pH, is gradually compensated by a fall in 2,3-DPG concentration, which causes a decrease in $pO_2(0.5)$ back toward normal. In bank blood collected with ACD (acid-citrate-dextrose), the 2,3-DPG concentration falls to very low values during storage, thus decreasing the efficiency of the hemoglobin to deliver oxygen. This fall can be prevented by adding inosine, pyruvate, and inorganic phosphate to the storage medium.
5. Carbon monoxide causes a slight increase in the oxygen affinity of the remaining heme groups (the Haldane-Smith effect). Methemoglobin has a similar effect. Both effects are small and without clinical significance.
6. Some abnormal hemoglobins show an increased oxygen affinity, e.g., hemoglobin Yakima. The higher oxygen affinity of fetal blood as compared to that of adult blood is due to the greater effect of 2,3-DPG on HbA than on HbF.[14]

The physiological effect of an abnormally high oxygen affinity is an inhibited dissociation of HbO_2 in the tissues and hence a lower tissue pO_2. Patients with Hb Yakima have a pO_2 (0.5) value of about 12 mm Hg and a compensatory increase in hemoglobin concentration.[13] Their ability to perform strenuous muscular exercise is diminished compared to normal individuals, but otherwise they are relatively unaffected.

Determination of $pO_2(0.5)$ or P_{50}

The simplest method consists of measuring both the pO_2 (with a Clark electrode) and sO_2 (with a spectrophotometric method) on a venous blood sample drawn anaerobically in a heparinized syringe.[18] Another method consists of establishing a known pO_2 by equilibrating the blood with a gas mixture of known pO_2 and measuring the resulting sO_2 spectrophotometrically. Alternatively, a known sO_2 can be established by mixing equal volumes of fully oxygenated and fully deoxygenated blood and measuring the resulting pO_2 with a Clark electrode. In all three methods the pH(P) and pCO_2 should be measured simultaneously.

The $pO_2(0.5)$ is then calculated from the measured pO_2 and the sO_2 by means of the following equation:

$$\log pO_2\,(0.5) = \log pO_2 - \frac{1}{2.7} \times \text{logit}\, sO_2 \qquad (5)$$

where 2.7 is the Hill slope, which is assumed to be independent of the other variables. The calculated (uncorrected) $pO_2(0.5)$ pertains to the actual values for temperature, pH(P), pCO_2, and 2,3-DPG concentration in the venous blood of the patient. The value for $pO_2(0.5)$ can be corrected to $T = 37°C$, pH(P) = 7.40, and pCO_2 = 40 mm Hg using the previously mentioned coefficients.

Example: The values for the venous blood are T = 37°C, pH(P) = 7.20, and pCO_2 = 80 mm Hg. The corrected value for pO_2 (0.5, 7.4) is then calculated as follows:

$$\log pO_2\,(0.5, 7.4) = \log pO_2\,(0.5) - 0.38 \times (7.20 - 7.40) + 0.10 \times (\log 80 - \log 40).$$

It is essential to specify whether the pO_2 (0.5) value pertains to the actual pH and pCO_2, or whether it has been corrected to pH 7.4 and pCO_2 40 mm Hg. As previously mentioned, the corrected value, which can be symbolized pO_2 (0.5, 7.4), is mainly an indicator of the 2,3-DPG concentration of the erythrocytes.

SOURCES OF ERROR

The sources of error are the same as those mentioned in the discussion of the pO_2 and sO_2 determinations. Assuming that the analytical coefficients of variation for these quantities are at best 1 per cent, the coefficient of variation for the derived pO_2 (0.5) value is at least 1.5 per cent. Extrapolation to $sO_2(0.5)$ by means of a standard Hill slope of 2.7 also involves a source of error due to a certain biological variation in the Hill slope. Therefore, the measured oxygen saturation should be as close as possible to 0.5 (or 50 per cent). Large corrections to pH 7.4 and pCO_2 of 40 mm Hg may lead to significant errors because of the variation of the correction factors with changes in pH, pCO_2, and 2,3-DPG concentration.

REFERENCES

See those for pO_2 and sO_2 determination, pages 441 and 859.

THE PARTIAL PRESSURE OF CARBON DIOXIDE IN ARTERIAL BLOOD, $pCO_2(aB)$

This quantity refers to the CO_2 partial pressure of an imaginary gas phase in equilibrium with the blood. It is sometimes called the *tension* of carbon dioxide in blood. The traditional unit is mm Hg, while the newly recommended unit is the pascal (Pa) (see pO_2, p. 854).

The following quantities are directly related to the pCO_2 of the arterial blood:[23]

1. The pCO_2 is equal for the plasma phase and the erythrocyte phase of whole blood, and therefore the pCO_2 of the arterial blood is equal to *the* pCO_2 *of the arterial plasma.*

2. The pCO_2 *of the alveolar air* [$pCO_2(A)$] is slightly less than the pCO_2 of the arterial blood [$pCO_2(aB)$]. Normally, CO_2 diffuses rapidly through the alveolar membrane; however, small amounts of blood are always shunted through the lungs without passing functioning alveoli.

3. *The alveolar ventilation*, $\dot{V}(A)$, i.e., the ventilation of the functioning alveoli of the lungs, is inversely related to the pCO_2 of the alveolar air and hence of the arterial blood (the symbol $\dot{V}$ denotes volume rate, i.e., volume divided by time):

$$pCO_2(A) = \frac{\dot{V}CO_2}{\dot{V}(A)} \times p(A) + pCO_2(I) \qquad (6)$$

where $\dot{V}CO_2$ is the volume rate of CO_2 produced in the metabolism and eliminated in the lungs, and $pCO_2(I)$ is the pCO_2 of the inspired air. According to this equation, an increase in $pCO_2(A)$ can be caused by an increase in $\dot{V}CO_2$, an increase in $pCO_2(I)$, or a decrease in $\dot{V}(A)$.

4. *The concentration of dissolved carbon dioxide* in the plasma (including H_2CO_3) is proportional to the pCO_2 (Henry's law):

$$cdCO_2(P) = \alpha CO_2(P) \times pCO_2(P)$$

α, the solubility coefficient for CO_2 (including H_2CO_3), varies with the temperature and composition of the solution. For pure water at 37°C, the solubility coefficient is $\alpha =$ 0.0329 mmol × liter^{-1} × mm Hg^{-1}. The presence of ions in the solution decreases the solubility coefficient. For plasma, the presence of proteins also decreases the solubility, whereas lipids increase the solubility of CO_2. The normal mean value for plasma at 37°C is 0.0306 mmol × liter^{-1} × mm Hg^{-1} with a biological standard deviation of about 0.0003 mmol × liter^{-1} × mm Hg^{-1}.[2,3] The temperature variation can be expressed approximately as

$$\Delta \log \alpha\, CO_2(P)/\Delta T = -0.0092 \times K^{-1}.$$

The equilibrium constant for the hydration of carbon dioxide, $CO_2 + H_2O \rightleftharpoons H_2CO_3$, is $K' = cH_2CO_3/cdCO_2 = 0.0023$ at 37°C, i.e., only a small fraction of dissolved CO_2 is hydrated.

5. *The concentration of total* CO_2 *in the plasma* increases with increasing pCO_2.

6. *The* pH *of plasma* falls with increasing pCO_2 of the blood when the concentration of noncarbonic acid or base is kept constant. The relationship between pH(P) and log pCO_2 is then approximately a straight line (Fig. 15–2).

NORMAL VALUES

The 95 per cent tolerance limits for the arterial pCO_2 are 35 to 45 mm Hg for men, and 32 to 43 mm Hg for women.[23] During pregnancy the value falls for unknown reasons to a mean value of 28 mm Hg, shortly before delivery.

The above-mentioned values refer to sea level. The mean value decreases by about 3 mm Hg per kilometer (or 5 mm Hg/mile) above sea level.

With the patient in the sitting or standing position, the values are slightly lower (2 to 4 mm Hg) than in the supine position.

The pCO_2 of venous blood from the right atrium is about 6 to 7 mm Hg higher than

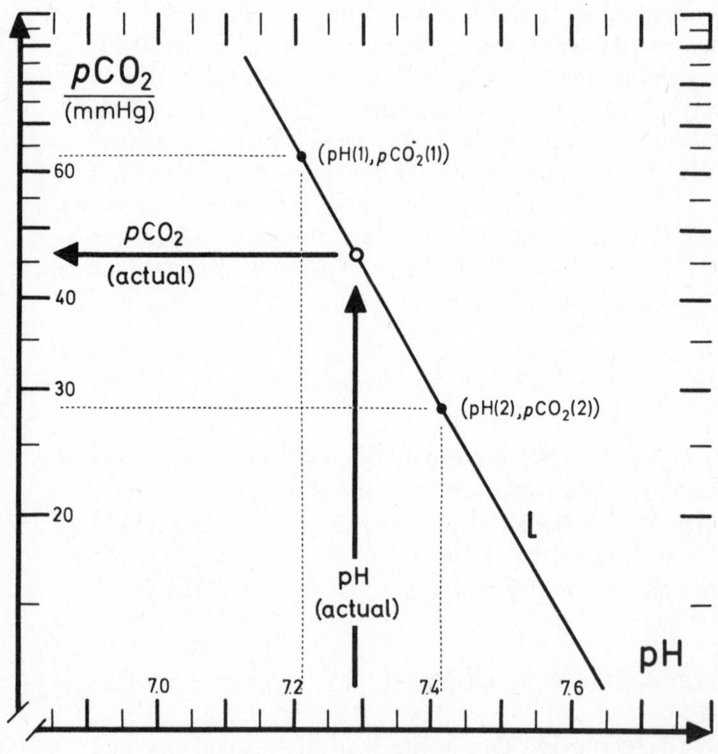

Figure 15–2 *Principle of the* CO_2 *equilibration method for* pCO_2 *determination.* The CO_2 equilibration curve (line *L*) is obtained by equilibrating the sample with two gases (1 and 2) with different but known pCO_2 values and measuring the two corresponding pH values. When the CO_2 equilibration curve is plotted as shown with pH on the abscissa and log pCO_2 on the ordinate, an approximately straight line is obtained on which the unknown pCO_2 can be read by interpolation using the measured actual pH value of the specimen as an entry.

the arterial pCO_2. The difference between arterial blood and peripheral venous blood varies considerably, depending upon the skin temperature, length of stasis, and muscular pumping.

CLINICAL SIGNIFICANCE

Elevated arterial blood pCO_2 (hypercapnia) is almost synonymous with *respiratory acidosis*, although the latter is restricted to clinical conditions with a *primary* increase in pCO_2. Hypercapnia may result from an increased pCO_2 in the inspired air (e.g., rebreathing air). An increased production of CO_2 in the organism also tends to elevate the blood pCO_2 (see Equation 1). In either case the rise in pCO_2 stimulates ventilation (see later), increasing the rate of CO_2 exhalation, and thus minimizing the rise in pCO_2. Significant hypercapnia is generally the result of *decreased alveolar ventilation.* This may arise either from *peripheral pathology* such as the presence of a foreign body in the airways, allergic laryngeal edema, asthmatic shock, chronic bronchitis, pulmonary tuberculosis, emphysema, poliomyelitis, curare intoxication, or *depression of the respiratory center* by drugs such as opiates or barbiturates. An elevated plasma pH, due to a *metabolic alkalosis*, causes a decrease in the normal respiration (compensatory hypoventilation), resulting also in hypercapnia.

The following physiological effects of a primary increase in pCO_2 (respiratory acidosis) are closely associated with the concomitant fall in plasma pH and intracellular pH:

1. Stimulation of the peripheral and central chemoreceptors.
2. Augmented urinary excretion of acid (renal compensation, see p. 962).
3. Vasodilatation in the brain (increased cranial pressure, and in extreme cases coma) and skin (warm, pink appearance).
4. Vasoconstriction in the viscera and extremities.
5. Increased levels of plasma catecholamines as well as plasma potassium.

The highest levels of pCO_2 observed in patients at physiological steady state breathing atmospheric air are about 90 mm Hg. Any levels higher than this will entail pO_2 values below 40 to 50 mm Hg which cannot be tolerated for long periods (see pO_2). Higher values (as high as 250 mm Hg) can be reached in patients breathing pure oxygen. During total apnea (cessation of breathing), the pCO_2 rises 4 to 6 mm Hg/min as a result of the continuous production of CO_2 in the organism.

Decreased arterial blood pCO_2 (hypocapnia) is almost synonymous with *respiratory alkalosis*, although the latter is restricted to clinical conditions with a *primary* decrease in pCO_2. The only relevant cause of hypocapnia is *increased alveolar ventilation* caused by mechanical artificial ventilation or by stimulation of the respiratory center due to one or more of the following factors:

1. A fall in plasma pH due to a *metabolic acidosis* causes a stimulation of the peripheral and central chemoreceptors leading to hyperventilation and a fall in pCO_2. This respiratory compensation of a metabolic acidosis may increase the pH approximately halfway towards normal.
2. *Hypoxia* stimulates the peripheral chemoreceptors. The low pO_2 at high altitude thereby causes a hyperventilation resulting in a respiratory alkalosis.
3. Several *pharmaceuticals* stimulate the respiratory center, e.g., salicylate which may lead to a severe respiratory alkalosis especially in children (this, however, may be followed by a metabolic acidosis).
4. *Anxiety*, for instance due to the blood sampling procedure, may cause hyperventilation and a fall in pCO_2.

The physiological effects of a primary fall in pCO_2 (respiratory alkalosis) are probably secondary to the concomitant rise in plasma pH or intracellular pH:

1. Augmented urinary excretion of base (renal compensation, see p. 962).
2. Vasoconstriction in the brain (fall in cerebral perfusion and pressure) and skin (pale, cyanotic appearance).

3. Fall in the activity of calcium ions in the plasma, possibly causing tetany.
4. Rise in the concentration of plasma lactate and fall in plasma potassium.

A rapid fall in pCO_2 may cause convulsions and even death, especially in older, arteriosclerotic patients. The lowest values obtained by vigorous spontaneous hyperventilation are about 18 mm Hg. The compensatory hyperventilation in metabolic acidosis may lead to values as low as 10 mm Hg.

DETERMINATION OF pCO_2 WITH THE pCO_2 ELECTRODE

pCO_2 is now generally measured directly with the pCO_2 electrode. Alternatively pCO_2 can be calculated as follows:

1. The Henderson-Hasselbalch equation allows the calculation of pCO_2 from the measured values for pH and total CO_2 concentration (see p. 895). However, the use of normal mean values for pK' and α may introduce significant errors in values for patients with abnormal composition of the plasma, especially abnormal ionic strength or increased lipid concentration.

2. The pCO_2 may be calculated from a measured pH value of the specimen by interpolation on the CO_2 equilibration curve (Fig. 15–2). However, the use of a straight line as an approximation for the CO_2 equilibration curve may introduce significant errors, especially with very high or very low pCO_2 values obtained by extrapolation rather than interpolation. Furthermore, a variable correction applies when the blood is partly desaturated. (For detailed procedure, see p. 868.)

SPECIMEN

Arterial blood is collected anaerobically in a heparinized syringe. Freely flowing capillary blood can generally be used as an adequate substitute for arterial blood. A slight admixture of blood from the venules does not affect the pCO_2 as much as the pO_2. Anaerobically collected venous blood (plasma) may also be used, although results are clinically less useful.

Sources of error are exposure of the blood to air, and acid formation by metabolic processes in the erythrocytes and leukocytes during storage. A small air bubble in the syringe is actually of little significance: a volume of air amounting to 10 per cent of the blood volume will cause a fall in the total CO_2 concentration of only 0.2 mmol/l, and a fall in pCO_2 of only 1 to 2 mm Hg. For this reason heparinized vacuum tubes (Vacutainers) may be used for anaerobic blood collection. Glycolysis and oxidative metabolism cause a rise in the pCO_2 of about 5 mm Hg/h at 37°C. At 2 to 4°C this rate is reduced by a divisor of about 10. Therefore, specimens which are not analyzed immediately should be stored in ice water, but not for longer than 1 h.

PRINCIPLE

The determination of pCO_2 using a pCO_2 electrode is based on a pH measurement of a stationary sodium bicarbonate solution which is in CO_2 equilibrium with the test solution or test gas via a CO_2-permeable membrane (see Chapter 3, Section 3 on Electrochemistry).

The pCO_2 electrode allows rapid and accurate determination of the pCO_2 of small volumes of blood including capillary blood.

EQUIPMENT

The pCO_2 electrode is shown schematically in Fig. 3–24.

The glass electrode consists of a slightly convex pH-sensitive glass membrane, mounted on the end of nonsensitive standard glass tubing. The inner solution bathing the inside of the glass membrane is a phosphate buffer with added NaCl. The inner electrode is an

Ag/AgCl electrode. The reference electrode can be an Ag/AgCl electrode, shaped as a ring around the stem of the glass electrode.

The sodium bicarbonate solution is an aqueous solution with cNaHCO$_3$ = 5 mmol/l, cNaCl = 20 mmol/l, and saturated with AgCl. A sodium bicarbonate concentration of 5 mmol/l has been found to be optimal. A lower concentration gives a faster but less reproducible response. The "spacer" is a piece of thin lens paper or a nylon net which covers the glass electrode membrane and insures a layer of bicarbonate solution between the glass membrane and the CO_2-permeable membrane.

The CO_2-*permeable membrane* generally consists of silicone rubber of a thickness of 25 μm, which provides a 98 per cent response time of 30 s.

The *electrode chamber* or cuvet should require as little blood as possible for filling (see pO_2 electrode, p. 149).

The pCO_2 *meter* is a sensitive galvanometer with a special pCO_2 scale. The sensitivity of the pCO_2 electrode ($-\Delta$pH/$\Delta \log pCO_2$) is generally between 0.96 and 1.0, and constant for a pCO_2 range from about 5 to 200 mm Hg. The sensitivity is independent of the nature of the sample, whether liquid or gaseous (unlike the pO_2 electrode).

Modern pCO_2 meters are provided with the following controls:

1. A *zero-adjustment control*, to adjust the electrical zero point to the pCO_2 value of the first calibrating gas.
2. A *standard adjustment control*, to adjust the reading of the first calibration gas to its known value.
3. A *sensitivity adjustment control*, to increase or decrease the sensitivity of the galvanometer and to adjust the reading to the pCO_2 value of the second calibrating gas.
4. A *membrane leakage control*, which allows measurement of the resistance across the CO_2-permeable membrane, which falls markedly if the membrane is ruptured.

CALIBRATION AND MEASUREMENT

The pCO_2 electrode is calibrated by means of two gases with known pCO_2 values (e.g., 40 and 100 mm Hg) at least once a day. The volume fraction of CO_2 in the dry calibrating gases [φCO_2(d)] must be accurately known (e.g., by gasometric analysis). The pCO_2 in the humidified gas is then calculated as

$$pCO_2(h) = \varphi CO_2(d) \times [p(h) - 47 \text{ mm Hg}],$$

where p(h) is the pressure of the humidified gas (which is equal to the atmospheric pressure) and 47 mm Hg is the saturated water vapor pressure at 37°C. If φCO_2(d) = 5.6 per cent and p(h) = 760 mm Hg, then pCO_2(h) = 40 mm Hg.

Recalibration is performed more frequently by means of one of the gases only, preferably before each measurement. The cuvet is rinsed with water; the calibration gas is flushed through the cuvet for 1 s every 30 s until no further change in the indicated pCO_2 occurs. The water droplets remaining in the cuvet are sufficient for saturating the gas with water vapor in the cuvet. Preferably, the gas should be flushed into the cuvet from the top, while solutions should be injected from the bottom. If the gas is adequately thermostatted and humidified in advance, it may also be flushed continuously through the cuvet until the appropriate reading is obtained.

The electrode chamber should be completely clean, and rinsed with physiological saline, before the blood is introduced. The blood is injected slowly into the chamber. Care should be taken to avoid pressure gradients and temperature gradients. Enough blood should be available to allow flushing of the cuvet with 1 to 2 vol of blood. After measurement, the electrode chamber is rinsed gently with physiological saline (preferably at 37°C), and the reading of the calibrating gas is checked. The procedure for rinsing and cleaning the electrode is identical to that described for the pO_2 electrode (p. 858).

SOURCES OF ERROR

The analytical standard deviation, when measuring repeatedly the same blood sample, rinsing with saline and checking the calibrating gas in between, is about ± 0.5 to 1.0 mm Hg, or the coefficient of variation is about 1 to 2.5 per cent. The analytical accuracy depends on a number of sources of error:

1. The pCO_2 values of the calibrating gases must be accurately known, preferably by gasometric analysis in the laboratory (Haldane apparatus[10] or Scholander apparatus),[19] although calibrated gases are also available commercially.
2. The temperature of the electrode chamber must be accurate within ± 0.1°C.
3. pCO_2 values measured at 37°C must be corrected to the actual temperature of the patient. The solubility of gases decreases with increasing temperature. In a closed system the pCO_2 therefore increases with increasing temperature. In the blood the change in dissociation of various buffer groups, primarily of the proteins, also affects the result. Empirically, the blood pCO_2 changes with temperature as follows: $\Delta \log pCO_2/\Delta t = +0.021\ K^{-1}$. In other words, the pCO_2 of the blood increases 5 per cent/°C. The formula is represented by the nomogram shown in Fig. 15–3.

REFERENCES

Severinghaus, J. W., and Bradley, A. F.: Electrodes for blood pO_2 and pCO_2 determination. J. Appl. Physiol., *13*:515, 1958.

Stow, R. W., Bear, F. R., and Randall, B. F.: Rapid measurement of the tension of carbon dioxide in the blood. Arch. Phys. Med. Rehabil., *38*:646, 1957.

DETERMINATION OF pCO_2 BY THE CO_2 EQUILIBRATION METHOD

PRINCIPLE

The pCO_2 may be calculated from a measured pH value of the specimen by interpolation on the CO_2 equilibration curve which relates pH and pCO_2. This is done by equilibrating blood with two gases with different (but known) pCO_2 values. When the CO_2 equilibration curve is plotted with pH on the abscissa and $\log pCO_2$ on the ordinate (see Fig. 15–2), an approximately straight line is obtained, and the formula for calculating pCO_2 can therefore be expressed:

$$\log pCO_2(\mathrm{P}) = \log pCO_2(1) + \frac{(\log pCO_2(2) - \log pCO_2(1))}{\mathrm{pH}(2) - \mathrm{pH}(1)} \times (\mathrm{pH}(\mathrm{P}) - \mathrm{pH}(1))$$

where 1 and 2 represent the blood samples equilibrated with the two different gas mixtures.

The pCO_2 values are essentially derived from three pH measurements: the measurement of the sample (without equilibration), and measurements of the sample after equilibration with each of two gases with different pCO_2 values. The values for $pCO_2(1)$ and $pCO_2(2)$ are the same for all determinations if the same set of gases is used.

TECHNIQUE

Two specimens of blood are equilibrated with two gases of known but different pCO_2 values, preferably containing volume fractions of CO_2 in oxygen of about 0.040 and 0.090 (= pCO_2 about 28 and 63 mm Hg, respectively) as outlined in the following section on tonometry. A CO_2 equilibration line is then drawn as outlined in Figure 15–2.

The volume fraction of CO_2 in the equilibrating gas (φCO_2) must be carefully determined in the laboratory by gasometric analysis (Lloyd-Haldane apparatus or Scholander

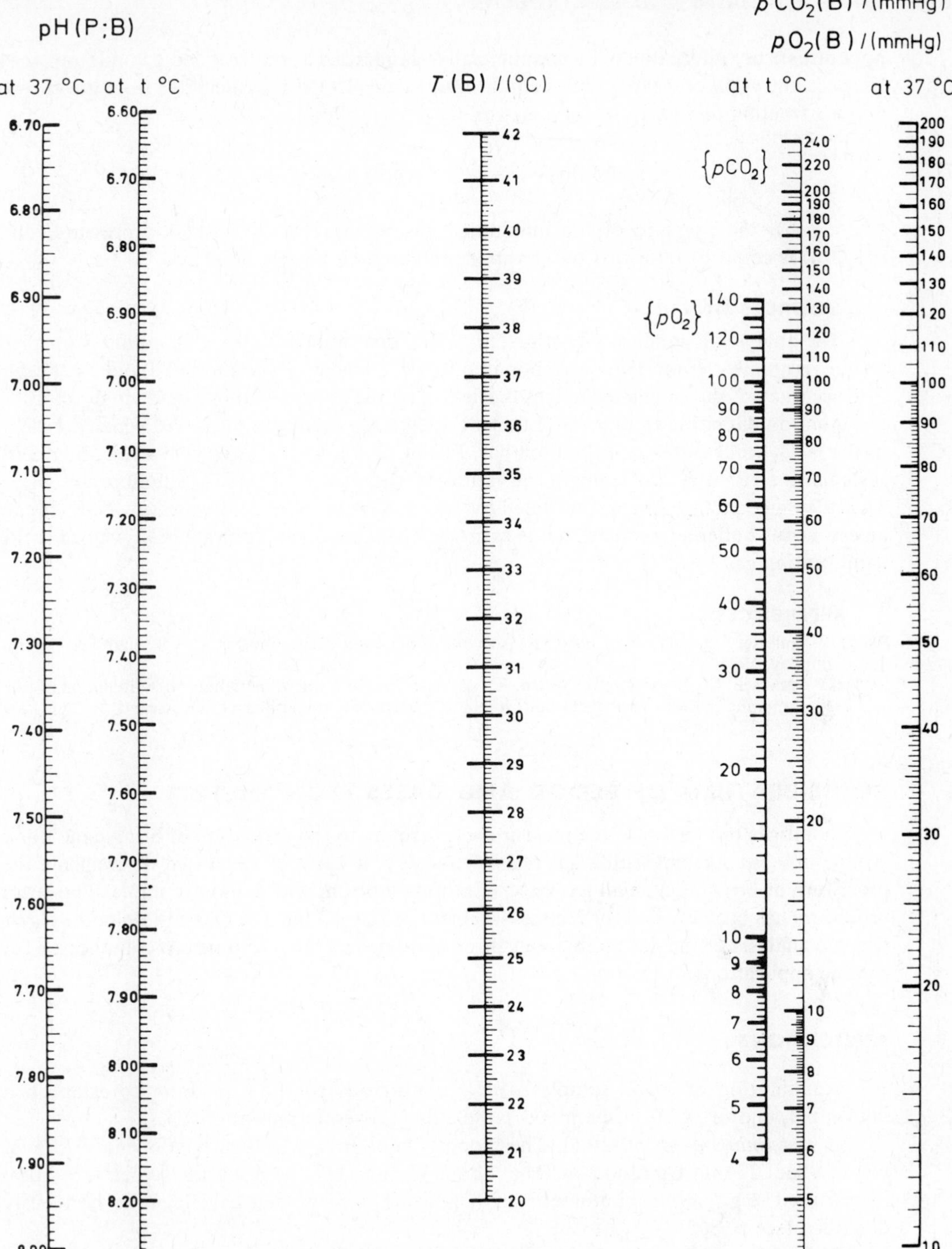

Figure 15-3 Nomogram for the conversion of pH, pCO_2 and pO_2 values in a blood sample measured at 37°C to values corrected for the body temperature of the patient. The nomogram is based on the following formulas: $\Delta pH(P)/\Delta T = -0.015\ K^{-1}$; $\Delta \log pCO_2/\Delta T = 0.021\ K^{-1}$; $\Delta \log pO_2/\Delta T = 0.031\ K^{-1}$.

Example: Measured: $pCO_2 = 60$ mm Hg at 37°C, temperature of patient = 31°C. Find 60 on the scale for pCO_2 at 37°C. Find 31 on the temperature scale. Place line between the two points and read pCO_2 on the scale for pCO_2 at t°C (here 45 mm Hg). The procedure is similar for pO_2 and pH.

apparatus), or, alternatively, a commercially standardized and certified gas mixture may be used. The pCO_2 of the humidified equilibrating gas ($pCO_2(h)$) is calculated from the known volume fraction of CO_2 in the dry gas ($\varphi CO_2(d)$) as follows:

$$pCO_2(h) = \varphi CO_2(d) \times (p(h) - pH_2O(h)),$$

where $p(h)$, the pressure of the humidified gas, is equal to atmospheric pressure, and $pH_2O(h)$ is equal to saturated water vapor pressure (47 mm Hg at 37°C).

SOURCES OF ERROR

The analytical standard deviation for pCO_2, determined by the CO_2 equilibrium procedure, is about ±1 mm Hg, corresponding to a coefficient of variation of about 2 per cent. The accuracy of the calculated values is affected by the approximations made in developing the equation. Greatest accuracy is obtained when pCO_2 values are measured which are close to the pCO_2 value of the equilibration gas. For pCO_2 values above or below this, there is a systematic error due to the slight curvature of the pH vs. $\log pCO_2$ equilibration lines. The calculated pCO_2 value is unaffected by any systematic error of the measured pH due to an erroneous buffer adjustment, but is entirely dependent on an accurate pCO_2 value for the equilibrating gas.

REFERENCES

Siggaard-Andersen, O.: The Acid-Base Status of the Blood. 4th ed. Baltimore, The Williams & Wilkins Company, 1974.

Siggaard-Andersen, O., Engel, K., Jørgensen, K., and Astrup, P.: A micro method for determination of pH, carbon dioxide tension, base excess and standard bicarbonate in capillary blood. Scand. J. Clin. Lab. Invest., *12*:172, 1960.

EQUILIBRATION OF BLOOD AND GASES (TONOMETRY)

Blood-gas tonometry in the present context refers to the procedure of bringing a blood sample into equilibrium with a gas phase. However, it has also been used to designate the procedure of bringing a small gas bubble into equilibrium with a blood sample. The latter procedure has been utilized for the measurement of the pO_2 and pCO_2 of the blood by gasometric analysis of the gas bubble, but this technique has proved to be too complicated for routine applications.[1]

APPLICATIONS

Equilibration of blood samples with a continuous gas flow in order to establish a known pO_2 and/or pCO_2 in the blood sample finds several applications:

1. *Determination of* $pO_2(0.5)$. The blood is equilibrated with a gas having a known pO_2 of about 27 mm Hg (and a pCO_2 of about 40 mm Hg), the resulting oxygen saturation is measured (e.g., spectrophotometrically), and $pO_2(0.5)$ is then calculated as previously described (see p. 862).

2. *Determination of the actual* pCO_2 *of the blood* by interpolation on the CO_2 equilibration curve using the actual pH as an entry (see Fig. 15–2).

3. *Determination of the so-called standard bicarbonate*, i.e., the concentration of bicarbonate in the plasma phase of blood which is equilibrated at 37°C with a pCO_2 of 40 mm Hg (and $pO_2 > 100$ mg Hg $\Rightarrow sO_2 > 0.9$), or the so-called *CO_2 combining power of the plasma*, i.e., the concentration of total CO_2 in the plasma after equilibrating the separated plasma with a pCO_2 of 40 mm Hg.

4. *Calibration of blood* pCO_2 *and* pO_2 *electrodes* by means of equilibrated blood has been advocated, especially for pO_2 electrodes owing to the variable blood/gas sensitivity ratio

of the latter. Alternatively, an equilibrated blood sample can be used for *analytical control purposes*.

EQUIPMENT

Many different types of tonometers (or saturators) have been described, the construction depending upon the volume of blood required. For applications (1) to (3), the aim is generally to reduce the sample volume; for preparation of control samples for pO_2 or pCO_2 electrodes, a volume of about 10 ml is generally suitable.

The general requirements are: (1) rapid equilibration, preferably in less than 3 min in order to minimize glycolysis in the blood, (2) temperature control to $\pm 0.1°C$, (3) humidity control, to prevent evaporation or condensation of water, (4) minimal hemolysis in the specimen, preferably with a concentration of hemoglobin in plasma of less than 50 μmol/l, and (5) ease of operation and cleaning of the tonometer.

In order to obtain rapid equilibrium, the gas-liquid interphase must be as large as possible. Two main principles are used, bubbling and rotation. Bubbling is the simplest method, requiring the least equipment (Fig. 15–4), but it is unsuitable for blood volumes of less than 1 ml. Rotation,[5,6] swirling,[9] or mechanical agitation,[24] whereby the blood is spread out to a thin film on the walls of the tonometer, may have some advantages (by avoiding the overpressure in gas bubbles; see later), but it requires special equipment.

A microtonometer especially designed for pCO_2 determinations in capillary blood is available from Radiometer (London Company, 811 Sharon Drive, Cleveland, OH 44145). It allows simultaneous equilibration of four aliquots of blood (30 to 50 μl) with two different gas mixtures. The thermostatted and humidified gas mixture flows over the blood, which is agitated in the bottom of a small disposable vial. The whole tonometer is thermostatted to 37°C. The time required for CO_2 equilibrium is about 2 to 3 min, depending on the degree of agitation and the viscosity of the sample. After equilibration, the blood is drawn into a glass capillary electrode for pH measurement.

SOURCES OF ERROR

1. *Erroneous gas analysis.* The volume fraction of O_2 and/or CO_2 in the dry gas should be measured gasometrically by means of the Scholander apparatus[19] or the Haldane

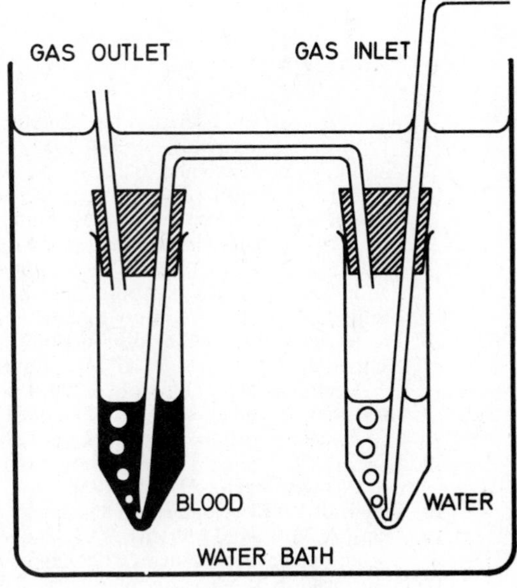

Figure 15–4 Schematic illustration of a simple blood-gas tonometer. The thermostatted and humidified gas is led to the bottom of a test tube containing about 10 ml blood and bubbles through the blood at a rate of about 20 to 40 ml/min. Foaming is avoided by greasing the walls of the upper portion of the tonometer chamber with silicone grease. The equilibration time with large bubbles ($d >$ 2 mm) is typically in the order of 10 to 15 min for 95 per cent equilibrium. A simple catheter can be inserted through the gas outlet tube.

apparatus (Lloyd modification[10]). Accurately analyzed gas mixtures are generally commercially available in tanks. Alternatively, pure gases can be mixed in the desired proportions by means of gas mixing pumps. The partial pressures in the humidified gas phase are calculated as exemplified for pCO_2 on page 867.

2. *Overpressure in the gas phase.* If the pressure of the gas phase is elevated over ambient pressure, the partial pressure of the equilibrating gas will also be erroneously high. This may happen if the gas exit tubing from the tonometer is too narrow. Also, the pressure inside bubbles tends to be slightly higher than ambient pressure, partly due to the hydrostatic pressure but especially due to the surface tension. The latter effect is more pronounced with small bubbles than with large bubbles and therefore the bubble size should not be less than 2 mm in diameter. Consequently, a fritted glass filter is not recommendable as the source of bubbles.

3. *Temperature difference.* The equilibrating gas should preferably be preheated and humidified in order to avoid cooling of the liquid phase (Fig. 15–4). The temperature of the tonometer should be accurate within $\pm 0.05°C$.

4. *Transfer of sample from tonometer to the measuring instrument.* The blood should be aspirated from the tonometer with the least possible vacuum (large diameter of sampling catheter), and the syringe should be previously flushed with the equilibrating gas. The transfer time should be as short as possible.

5. *Changes in the composition of the blood during equilibration.* Such changes are especially unwarranted when the purpose is to construct oxygen dissociation curves or CO_2 equilibration curves. To avoid metabolic changes (glycolysis, changes in 2,3-DPG concentration), as well as evaporation or dilution effects, the equilibration should be as rapid as possible, preferably less than 5 min. Sedimentation of erythrocytes in the tonometer or the aspirated sample must be carefully avoided. Hemolysis should be kept at a minimum, preferably less than 1 per cent of the erythrocytes [cHb(P) $<$ 100 mmol/l], and for this reason certain antifoams which tend to cause hemolysis (e.g., caprylic alcohol) must be avoided.

REFERENCE

Ravin, M. B., and Briscoe, W. A.: Blood-gas transfer, hemolysis, and diffusing capacity in a bubble tonometer. J. Appl. Physiol., *19*:784, 1964.

REFERENCES

1. Asmussen, E., and Nielsen, M.: A bubble method for determination of pCO_2 and pO_2 in blood. Scand. J. Clin. Lab. Invest., *10*:267, 1958.
2. Austin, W. H., Lacombe, E., Rand, P. W., and Chalterjee, M.: Solubility of carbon dioxide in serum from 15 to 30°C. J. Appl. Physiol., *18*:301, 1963.
3. Bartels, H., and Wrbitzky, R.: Bestimmung des CO_2-Absorptions-koeffizienten zwischen 15 und 38°C in Wasser und Plasma. Pfluegers Arch., *271*:162, 1960.
4. Christoforides, C., Laasberg, L. H., and Hedley-Whyte, J.: Effect of temperature on solubility of O_2 in human plasma. J. Appl. Physiol., *26*:56, 1969.
5. Farhi, L. E.: Continuous duty tonometer system. J. Appl. Physiol., *20*:1098, 1965.
6. Flenley, D. C.: A rotating disc tonometer. Resp. Physiol., *3*:256, 1967.
7. Heitmann, H., Buckles, R. G., and Laver, M. B.: Blood pO_2 measurements: performance of microelectrodes. Resp. Physiol., *3*:380, 1967.
8. Jørgensen, K., and Astrup, P.: Standard bicarbonate, its clinical significance, and a new method for its determination. Scand. J. Clin. Lab. Invest., *9*:122, 1957.
9. Laue, D.: Ein neues Tonometer zur raschen Äquilibrierung von Blut mit verschiedenen Gasdrucken. Pfluegers Arch., *254*:142, 1951.
10. Lloyd, B. B.: Development of Haldane's gas analysis apparatus. J. Physiol. (Lond.), *143*:5, 1958.
11. Maas, A. H. J., and Mertens, P. J.: The measurement of the pCO_2 and pO_2 of blood with electrodes in an open cuvette system. Clin. Chim. Acta, *28*:443, 1970.
12. Mellemgaard, K.: The alveolar-arterial oxygen difference: its size and components in normal man. Acta Physiol. Scand., *67*:10, 1966.

13. Metcalfe, J., and Dhindsa, D. S.: The physiological effects of displacements of the oxygen dissociation curve. Alfred Benzon Symp., *4*:613, 1972.
14. Novy, M. J.: Alterations in blood oxygen affinity during fetal and neonatal life. Alfred Benzon Symp., *4*:696, 1972.
15. Payne, J. P., and Hill, D. W. (Eds.): Oxygen Measurement in Blood and Tissues. London, Churchill, 1966.
16. Perutz, M. F.: Stereochemistry of cooperative effects in haemoglobin. Haem-haem interaction and the problem of allostery. The Bohr effect and combination with organic phosphates. Nature (Lond.), *228*:726, 1970.
17. Raine, J. M., Bishop, J. M.: A-a difference in O_2 tension and physiological dead space in normal man. J. Appl. Physiol., *18*:284, 1963.
18. Rørth, M.: Hemoglobin interactions and red cell metabolism. Ser. Haematol., *5*:1, 1972.
19. Scholander, P. F.: Analyzer for accurate estimation of respiratory gases in one-half cubic centimeter samples. J. Biol. Chem., *167*:235, 1947.
20. Severinghaus, J. W.: Blood gas calculator. J. Appl. Physiol., *21*:1108, 1966.
21. Severinghaus, J. W., Roughton, F. J. W., and Bradley, A. F.: Oxygen dissociation curve analysis at 98.7–99.6% saturation. Alfred Benzon Symp., *4*:65, 1972.
22. Siggaard-Andersen, O.: Acid-base and blood gas parameters. Arterial or capillary blood? (Editorial). Scand. J. Clin. Lab. Invest., *21*:289, 1968.
23. Siggaard-Andersen, O.: The Acid-Base Status of the Blood. 4th ed. Baltimore, The Williams & Wilkins Company, 1974.
24. Siggaard-Andersen, O., Engel, K., Jørgensen, K., and Astrup, P.: A micro method for determination of pH, carbon dioxide tension, base excess and standard bicarbonate in capillary blood. Scand. J. Clin. Lab. Invest., *12*:172, 1960.
25. Siggaard-Andersen, O., and Garby, L.: The Bohr effect and the Haldane effect. (Editorial). Scand. J. Clin. Lab. Invest., *31*:1, 1973.
26. Van Slyke, D. D., and Cullen, G. E.: Studies of acidosis. I. The bicarbonate concentration of the blood plasma, its significance, and its determination as a measure of acidosis. J. Biol. Chem., *30*:289, 1917.

SECTION THREE
ELECTROLYTES

by Norbert W. Tietz, Ph.D.

Sodium

Sodium is the major cation of extracellular fluid. It plays a central role in the maintenance of the normal distribution of water and the osmotic pressure in the various fluid compartments. The normal daily diet contains approximately 8 to 15 g ($\approx$130 to 260 mmol) of sodium chloride, which is nearly completely absorbed from the gastrointestinal tract. Except for a series of factors listed in the following paragraphs, the excess is excreted by the kidneys, which, therefore, are the ultimate regulators of the sodium content of the body. Sodium is a threshold substance with a normal renal threshold* of 110 to 130 mmol/l. Sodium is initially filtered by the glomeruli, but 80 to 85 per cent is reabsorbed in the proximal portion and an additional amount in the distal portion of the tubules. The total amount of sodium reabsorbed may reach as much as 99 per cent. The reabsorption is greatly controlled by some adrenal cortical hormones, mainly aldosterone (which enhances the tubular reabsorption of sodium and indirectly also of Cl^-, but decreases the tubular reabsorption of K^+). The exchange of Na^+ for H^+ in the renal tubules is an important mechanism of the acidification of urine and is discussed in detail in Chapter 16.

* Threshold is the level in serum below which any sodium present in the glomerular filtrate is completely reabsorbed by the tubules (see also Chapter 17, Renal Function).

Hyponatremia (low serum sodium level) is found in a variety of conditions, including the following: (1) severe polyuria, as seen in diabetes insipidus; (2) metabolic acidosis (e.g., diabetic acidosis), in which cations are excreted along with the anions (see Chapter 16); (3) Addison's disease, in which decreased secretion of corticosteroids, mainly aldosterone, causes decreased reabsorption of sodium by the tubules and, thus, loss of serum sodium; (4) diarrhea, in which an excessive amount of sodium is lost through the stool as a result of insufficient absorption of dietary sodium and sodium of pancreatic juice; and (5) renal tubular disease, in which there may be a defect in either the reabsorption of sodium or the Na^+–H^+ exchange.

Hypernatremia (increased serum sodium level) is found in the following conditions: (1) hyperadrenalism (Cushing's syndrome), in which there is an increased production of mineralocorticosteroids and, thus, an increased absorption of sodium by the renal tubules; (2) severe dehydration due to primary water loss, in which there is a relative increase of sodium in serum; (3) certain types of brain injury; (4) diabetic coma after therapy with insulin, in which it is believed that the removal of excess glucose from the serum causes a retransfer of cellular sodium into the extracellular fluid in order to maintain equal osmotic pressure in both compartments; (5) excess treatment with sodium salts.

SPECIMENS

Although serum is used as specimen most often, plasma has been claimed by some to be preferable. Lithium heparinate and ammonium heparinate are the preferred anticoagulants. Urine determinations are generally performed on a 24 h specimen. In cases where the output of urine from each kidney is studied, specimens are collected over a short time period.

METHODS FOR THE DETERMINATION OF SODIUM IN BODY FLUIDS

Although a number of gravimetric and titrimetric methods for the determination of sodium in body fluids are available,[16] these are outdated, slow, and require large amounts of specimen. Almost all laboratories today employ either emission flame photometry or atomic absorption spectrophotometry for this test because of the speed and simplicity of these methods. Since the principle of both techniques has been discussed in detail in Chapter 3, and since most manufacturers supply detailed procedures with their equipment, no attempt will be made here to outline a method. (See, however, p. 878.)

Sodium determinations, even in the routine laboratory, may be performed with an accuracy of at least ± 2 per cent if an appropriately designed instrument is used. As a general rule, a flame photometer, employing an internal standard, is slightly more accurate and precise than a direct reading instrument. Serum specimens for sodium analysis are generally diluted 100- or 200-fold; thus, the dilution error may be as great or greater than the instrumental error, especially if very small volumes of sample are used (e.g., 25 μl diluted to 5.0 ml with a mechanical dilutor).

A new and promising approach to the determination of sodium in body fluids makes use of ion-selective electrodes (see Chapter 3, Section 3). It should be remembered, however, that these electrodes measure ion-activity and not concentration. Changes in the concentration of sodium in biological fluids are related to, but do not parallel, changes in the activity of sodium ions. Thus, special calibration procedures or corrections must be employed if results obtained with ion-selective electrodes are to be reported in terms of concentration units.

The determination of Na in feces or tissues requires prior digestion with HNO_3 or with H_2SO_4 and H_2O_2 or, preferably, dry ashing.

NORMAL VALUES

The range of normal values for *serum* sodium is from 135 to 148 mmol/l. There is less agreement as to the value for normal *urinary* sodium excretion. In part this is due to the great effect of dietary intake of sodium on urine levels. Values given in the literature for normal individuals on an average diet vary from 40 to 90 mmol/d to 43 to 217 mmol/d and even 27 to 287 mmol/d. The rate of sodium excretion during the night is only one fifth of the peak rate during the day, indicating a large diurnal variation.

The sodium concentration in *cerebrospinal fluid* is almost identical to that found in serum, namely 138 to 155 mmol/l. Erythrocytes contain approximately 1/10 of the sodium found in serum. The concentration of sodium in sweat will be discussed in connection with the sweat test (see Chapter 20).

COMMENTS

An approximate check on the accuracy of sodium, chloride, and carbon dioxide determinations can be made by calculating the difference between the sodium value and the combined chloride and carbon dioxide values, in mmol/l. (For the purpose of this approximation, $ctCO_2$ is set equal to $cHCO_3^-$.) This difference is normally 7 to 14 mmol/l but may be as small as 5 mmol/l. Table 16-2 shows that in an average serum specimen there are, in addition to sodium ions, 12 mmol/l of other cations of which Ca^{2+}, Mg^{2+}, and "others" are not normally determined as part of an "electrolyte" study (= unmeasured cations). There are also 24 mmol/l of "undetermined anions" (HPO_4^{2-}, SO_4^{2-}, organic acids$^-$, proteins$^-$) present in serum. (See also Chapter 16, p. 970.)

Therefore:

Na^+	142	Cl^-	103
		HCO_3^-	27
unmeasured cations, incl. K^+	12	unmeasured anions	24
	154		154

or:

$$Na^+ - (Cl^- + HCO_3^-) = 12$$

or:

$$142 - (103 + 27) = 12 \text{ (range: 5 to 14)}$$

An alternative calculation is:

$$(Na^+ + K^+) - (Cl^- + CO_2) = 16$$

or:

$$(142 + 4) - (103 + 27) = 16 \text{ (range: 9 to 18)}$$

Note: All values are given in mmol of total ion charge/l.

These calculations are based on the assumption that all unmeasured ions are within the normal range. This is true in many cases, but may not be true in conditions such as diabetes and renal impairment, in which the organic acid fraction is increased as a result of increased production of β-hydroxybutyric acid and acetoacetic acid or retention of metabolites, respectively (see also under Organic Anions in this chapter and in Chapter 16). Inspection of all test results obtained on blood (serum), especially results for glucose, urea nitrogen, or acetone performed on the same specimen, may help to detect those situations in which the preceding equation does not hold true.

Potassium

Potassium is the major intracellular cation, having an average cellular concentration in tissue cells of 150 mmol/l and a concentration in red cells of 105 mmol/l. This is approximately 23 times higher than the concentration of potassium in extracellular fluid. The high intracellular concentration, in the presence of the low extracellular concentration, is believed to be maintained by an active transport mechanism that utilizes oxidative energy of the cells. In addition, the permeability of cell membranes for potassium is extremely low, and rapid shifts of potassium in or out of cells by diffusion have not been observed. There is, however, some movement of potassium ions out of the cells. This is important to the practicing clinical chemist, since shifts of potassium from the cells to the serum may invalidate measured serum potassium levels if the serum is not separated from the cells shortly after collection of blood (see under specimens).

The requirements of the body for potassium are satisfied by a normal dietary intake of 80 to 200 mmol/d. Potassium, once absorbed by the intestinal tract, is partially removed from the plasma by glomerular filtration and is then nearly completely reabsorbed in the proximal tubules. Unlike sodium and chloride, however, it is then effectively reexcreted by the distal tubules. There is no threshold level for potassium.

Any potassium absorbed by the intestinal tract causes only a slight and temporary increase in serum potassium levels. Only a fraction of this potassium moves into the red cells (and tissue cells), and the excess is rapidly excreted by the kidneys as just explained. It is believed that this mechanism protects the body against high serum potassium levels, which could cause serious changes in muscle irritability, respiration, and myocardial function as well as characteristic electrocardiographic changes. Such symptoms may appear with potassium levels above 7.5 mmol/l. Levels of 10 mmol/l may be fatal, although patients with slightly higher levels have recovered.

Low potassium levels cause excitatory changes in muscle irritability and myocardial function, which are also accompanied by characteristic electrocardiographic changes. For these reasons the serum potassium determination has become a most important diagnostic tool in situations in which extremely high or low serum potassium levels are suspected. Prompt performance of the test by the laboratory is mandatory.

The mechanism for potassium excretion described above and the absence of a threshold level for K^+ have the disadvantage that the body has no effective mechanism to protect itself from excessive loss of potassium. Even in potassium deficiency, the kidney continues to excrete potassium. For maintenance of normal potassium levels, a regular daily intake of potassium is therefore essential.

Hypokalemia or *hypopotassemia* (low serum potassium levels) may be seen in cases of body potassium deficiency, although the large intracellular potassium reserves may maintain normal serum potassium levels despite an actual potassium deficiency. Aside from inadequate intake of potassium, deficiencies in this ion may occur as a result of excessive loss of potassium through the feces (in prolonged diarrhea) or through vomitus (after prolonged periods of vomiting). Increased secretion of mineralocorticosteroids, especially aldosterone, results in a decreased reabsorption of potassium (the reverse of the behavior of sodium, which under such conditions is reabsorbed at an increased rate) and, thus, results in decreased serum potassium levels. Patients on a regular diet who exhibit serum potassium levels below 3 mmol/l for several days are likely to have hyperaldosteronism. Urine potassium output in these cases is also increased. In alkalosis there is a K^+ movement into the cell that is matched by a movement of H^+ from the cell into the extracellular fluid. Thus, all other things being equal, alkalosis in itself results in decreased serum potassium levels; the opposite is the case in acidosis. Intracellular potassium deficiency may lead to alkalosis, a process which is explained in more detail in Chapter 16.

Hyperkalemia or *hyperpotassemia* (increased serum potassium level) is generally observed in cases of oliguria, anuria, or urinary obstruction. Renal failure due to shock results in decreased removal of potassium from serum; renal tubular acidosis interferes with the Na^+ - H^+ exchange and, thus, also results in a retention of potassium in serum. In renal failure, one of the important purposes of renal dialysis is the removal of accumulated potassium from plasma.

SPECIMENS

Specimens for serum (plasma) potassium analysis must be free from hemolysis, since any release of potassium from red cells can significantly increase the serum levels and thus invalidate the test results. Hemolysis of 0.5 per cent of the erythrocytes can increase serum K^+ levels by 0.5 mmol/l.

It has been shown that potassium levels in plasma are about 0.1 to 0.7 mmol/l lower than those found in serum. It is believed that the higher level found in serum is due to release of potassium from ruptured platelets during the coagulation process.

Opening and closing of the fist prior to venous puncture should be avoided, since this muscle action may result in an increase in plasma potassium levels of 10 to 20 per cent.

After collection of the specimen, the blood should be allowed to clot, but any further delay in separating the plasma or serum from the erythrocytes should be avoided. Potassium is actively transported into the cell by a mechanism which involves phosphorylation of glucose. After consumption of intracellular glucose, or if the specimen is stored at refrigerator temperatures so that phosphorylation is inhibited, there is a shift of potassium from the cell into the plasma or serum.

DETERMINATION OF POTASSIUM IN BODY FLUIDS

A wide range of methods has been proposed for the determination of potassium in body fluids.[16] These include gravimetric, turbidimetric, emission flame spectrophotometric, and atomic absorption spectrophotometric procedures. Almost all modern laboratories use one of the last two procedures because they are fast, convenient, and reliable, as explained previously under the determination of sodium in serum. A modern, well designed flame photometer in a routine laboratory is capable of an accuracy of at least ± 0.2 mmol/l. For serum samples in the physiological range (≈ 4.0 mmol/l), this amounts to an error of approximately 5 per cent. The principles of flame photometry and atomic absorption spectrophotometry have been explained in Chapter 3, and detailed methodology is supplied with the respective instruments.

Potassium ion activity (aK^+) may be determined by an ion-selective electrode. The high selectivity of this electrode for potassium makes this method a highly useful tool. Measured values, however, have to be related to the potassium concentration as explained under Sodium and in Chapter 3, Section 3.

NORMAL VALUES

Normal serum potassium levels range from 3.5 to 5.3 mmol/l, with plasma levels being slightly lower. Levels in the newborn are somewhat higher than those of adults, and most authors give a range of 4.0 to 5.9 mmol/l. Higher normal values for this group have been reported by some authors, but errors in specimen collection or handling may have been involved in these cases.

Urinary excretion of potassium varies greatly with the potassium intake, but commonly observed levels of persons on an average diet are 25 to 120 mmol/d with a mean of 40 mmol/d.

PRACTICAL CONSIDERATIONS RELATED TO THE DETERMINATION OF SODIUM AND POTASSIUM BY EMISSION FLAME PHOTOMETRY

1. Sodium and potassium are the two metal ions that are analyzed by emission flame photometry most frequently in a clinical laboratory. Lithium is generally used as an internal standard, although analysis of lithium (e.g., in the blood of psychiatric patients under treatment with lithium compounds) may also be performed by flame photometry. It is more practical to determine metal ions other than these three by atomic absorption spectrophotometry or other techniques (see respective procedures). Each of these three ions (Na^+, K^+, Li^+) emits light of a variety of distinctive wavelengths, but the spectral lines most frequently used in clinical determinations are 589, 768, and 671 nm, respectively. The emitted light is isolated (separated from interfering light) by either filters, gratings, or prisms.

2. When analyzing biological material it becomes necessary to dilute the samples before analysis. The extent of dilution depends on the type of instrument, the type of specimen, and the concentration of the ions to be analyzed. Such dilution is necessary to adjust the concentration of the measured ion so that the intensity of emitted light will be linearly related to the concentration and within the range of optimum sensitivity and accuracy of the photometer. In the Technicon flame photometer, dilution and removal of protein are accomplished by the dialysis step in the procedure. By decreasing the protein concentration and viscosity, dilution promotes a uniform flow of sample and decreases the likelihood of plugging of the atomizer. Dilution also decreases or eliminates interference by other sample constituents, which otherwise may enhance or depress light emission.

3. Reagents and standards, as well as dilutions of the sample, must be made with highest purity water, preferably with deionized water with an electrolyte content of less than 0.1 ppm or a resistance of more than 1,000,000 ohms.

4. Sodium and potassium standards are generally prepared from their chloride salts since chloride ions do not affect the analysis. Lithium solutions are generally prepared from nitrates or sulfates. Most lots of lithium nitrate have an assay of less than 100 per cent; this must be considered when weighing out the salt for the preparation of the standards. Lithium standards for lithium analysis should be prepared from Li_2CO_3.

Most manufacturers recommend that sodium be added to the potassium standard in a concentration comparable to that in normal serum (e.g., 140 mmol/l). Addition of sodium compensates for the slight enhancement of potassium emission by sodium in the measurement of potassium. The presence of lithium in the sample dilution reduces this positive interference from Na^+ (see Chapter 3, Section 1).

Standard solutions are best stored in polyethylene squeeze bottles to eliminate the leaching of ions from glass containers into the standard solution. The bottles should be kept relatively full, since this type of bottle "breathes," which means that water vapor escapes, causing a slow increase in the concentration of the standard.

5. Accuracy of results depends on following the manufacturer's instructions explicitly in order to secure uniformity with respect to flame size, rate of introduction of the sample, and patency of the analyzer. Variations in these three conditions cause corresponding variations in temperature or in the size of atomized droplets, which in turn alter the sensitivity of the instrument and may lead to erroneous results.

6. Both burner and atomizer must be cleaned frequently with a thin wire or by other appropriate means, and both parts must be rinsed well with deionized water between samples and after use. Such a precaution is necessary to prevent or remove protein and salt deposits. Failure to do so will result in an uneven flow of the sample into the flame and thus in erroneous results. Addition of nonionic surface active agents such as Acationox, Sterox SE, or Brij to reagents and standards helps to provide an even flow into the burner and promotes the formation of small, uniform droplets of the sample solution during atomization.

7. The flow of fuel and oxidant into the flame must be strictly controlled, especially in direct reading instruments. Any change in the flow rate of either or both of these components results in a different flame temperature and flame size and therefore affects the sensitivity and reproducibility of the test procedure.

Internal standard instruments are less affected by these changes because of the compensatory effect of the internal standard, but even under these circumstances needle valves and high quality flow regulators should be used to prevent any gross fluctuations in the gas flow. Difficulties have been reported with the use of city gas, as a result of the varying concentrations of the different components of this gas or of gross fluctuations in the gas pressure. To avoid this, many laboratories use gas tanks, if this is permitted by local fire laws.

8. Most manufacturers recommend that the electronic circuit of the flame photometer be left on at all times to eliminate warm-up periods and, in some instances, to prolong the life of the equipment. The sensitivity of photo-sensing devices will change with temperature; thus, insufficient warm-up periods affect the accuracy of the results.

9. At the beginning of each test run, a standard or a lithium solution must be aspirated for a certain time period (e.g., 2 to 5 min) in order to bring the burner and the chamber which houses the atomizer to thermal equilibrium, which is necessary for obtaining reproducible results. Evaporation of water in the atomizer chamber and in the flame causes a drop in temperature.

10. If an internal standard such as lithium is used, the quantitation of the unknown ion is based on the ratio between the signal obtained from the internal standard and that obtained from the unknown. Thus, it is important that the lithium concentration in the standards and in the unknowns be identical. If a new batch of lithium reagent is prepared, old standards must be discarded and a new set prepared with the new lithium solution.

11. The use of combustible gases in the clinical laboratory requires special procedures and precautions. This applies not only to flame emission and flame absorption photometry, but also to other techniques such as gas chromatography. Acetylene can form an acetylide in the presence of copper; therefore, copper tubing should never be used in connection with this gas. Acetylene tanks should be changed before they are completely empty, since the last portion of gas contains appreciable amounts of solvent (e.g., acetone). Propane, which is a common fuel for use in flame photometers, is heavier than air and therefore collects easily in pockets (low spots) in the laboratory if there is a leak in the cylinder. Therefore, work areas should be well ventilated. Some newer instruments employ small propane tanks which greatly reduce the risk of an explosion. Propane is frequently supplied in liquefied form (LPG). Such gas contains appreciable amounts of butane and small amounts of other hydrocarbons. The relative proportions of these gases increase as the cylinder empties, causing a change in the flame characteristics. Thus, these tanks should be changed before they are completely empty.

Chloride

Chloride is the major extracellular anion, constituting about 103 mmol/l of the total anion concentration of approximately 154 mmol/l; thus, this ion is significantly involved in maintaining proper water distribution, osmotic pressure, and normal anion-cation balance in the extracellular fluid compartment. The concentration of chloride in red cell fluid is 49 to 54 mmol/l, and that of whole blood is 77 to 87 mmol/l. Tissue cells contain approximately 1 mmol/l.

Chloride ions ingested with food are almost completely absorbed by the intestinal tract. They are removed from the blood by glomerular filtration and are passively reabsorbed by the proximal tubules.

Chloride, to some extent, may be lost through excessive sweating during hot weather periods. However, this stimulates aldosterone secretion and thus causes the sweat glands to secrete sweat of lower sodium and chloride concentration than that excreted during normal temperatures. Thus, the loss of Cl^- under these conditions is minimized.

Low serum chloride values are observed in salt-losing nephritis, as associated with chronic pyelonephritis. This loss is probably due to deficient tubular reabsorption despite a body deficit of chloride. In Addison's disease, chloride values are generally maintained close to normal except during Addisonian crisis, in which chloride (and Na^+) levels may drop significantly. Low serum chloride values may also be observed in those types of metabolic acidosis which are caused by excessive production or diminished excretion of acids (e.g., diabetic acidosis and renal failure). In these cases chloride ions are partially replaced by the accumulated anions such as β-hydroxybutyrate, acetoacetate, lactate, and phosphate. Prolonged vomiting, from any cause, may result in a significant loss of Cl^- and ultimately a decrease in serum and body chloride.

High serum chloride values are observed in dehydration and in conditions causing decreased renal blood flow, such as congestive heart failure. Hyperchloremic acidosis may be a sign of severe renal tubular pathology. Excessive treatment with or intake of Cl^- obviously also results in high serum levels.

An increase in the chloride concentration to 106 ± 5 mmol/l has been observed in individuals with hypercalcemia due to primary hyperparathyroidism.[64] In the same study, serum chloride values in patients with hypercalcemia due to other causes were found to be 100 ± 3.0 mmol/l. It is believed that this difference in chloride concentrations is caused by the effect of parathyroid hormone on distal tubular function.

METHODS FOR THE DETERMINATION OF CHLORIDE IN BODY FLUIDS

MERCURIMETRIC TITRATION (SCHALES AND SCHALES, MODIFIED)

PRINCIPLE

A tungstic acid protein-free filtrate of the specimen is titrated with mercuric nitrate solution in the presence of diphenylcarbazone as the indicator. The free mercuric ions combine with chloride ions to form soluble but essentially nonionized mercuric chloride.

$$2Cl^- + Hg^{2+} \rightarrow HgCl_2$$

After all chloride ions have reacted with mercuric ions, any excess Hg^{2+} combines with the indicator diphenylcarbazone to form a blue-violet colored complex. The first appearance of this blue-violet color is considered the titration end point.

REAGENTS

1. Sulfuric acid, 0.36 mol/l. Add 20 ml of concentrated sulfuric acid, AR, to approximately 700 ml of distilled water in a 1000 ml volumetric flask. Cool and dilute to volume.
2. Sodium tungstate, 10 g/100 ml. Place 700 ml of deionized water into a 1000 ml volumetric flask and add 100 g of sodium tungstate ($Na_2WO_4 \cdot 2H_2O$). After all sodium tungstate is dissolved, dilute to volume with deionized water. Allow to stand for several days and decant or filter if any precipitate has formed.
3. Mercuric nitrate solution, 5 mmol/l. Place 1.0833 g of HgO (red) into a small beaker or flask; add 3 ml of concentrated HNO_3 and 20 ml of deionized water. Stir until dissolved and transfer to a 1000 ml volumetric flask. Dilute to volume with deionized water. Alternatively, 1.6681 g of $Hg(NO_3)_2 \cdot \frac{1}{2}H_2O$ may be used instead of HgO.
4. Diphenylcarbazone solution. Dissolve 250 mg of *s*-diphenylcarbazone in 100 ml

of methanol or ethanol. Store in a dark bottle in the refrigerator. The solution should have an orange-red color. If the color changes, e.g., to dark cherry red or to yellow, discard the solution. The reagent is generally stable for 2 months.

5. Chloride standard, 100 mmol/l. Place 5.845 g of NaCl (dried at 110°C) into a 1000 ml volumetric flask; add 3 ml of concentrated HNO_3 and approximately 100 ml of deionized water. After all the salt is dissolved, dilute to volume with deionized water.

METHOD

1. Place 0.5 ml of serum or standard, respectively, into an Erlenmeyer flask or test tube.* Add, in succession, 3.5 ml of water, 0.5 ml of 0.36 molar sulfuric acid, and 0.5 ml of sodium tungstate reagent. Mix well, allow to stand for 5 min, and centrifuge. (This procedure gives a modified Folin-Wu protein-free filtrate.)

2. Transfer 2 ml of clear supernatant fluid into a suitable titration vessel and add 0.1 ml (2 drops) of diphenylcarbazone solution. Titrate with mercuric nitrate solution from a microburet, calibrated in intervals of 0.05 ml and capable of delivering drops equal to not more than 0.02 ml.† Burets with a fine glass tip are satisfactory, but hypodermic needles should not be used as tips, since the metal will react with the mercuric nitrate solution.

3. When approaching the end point (first appearance of a faint blue-violet color), add the mercuric nitrate solution in amounts not greater than 0.02 ml at a time. Approximately 2.0 ml of reagent will be used to titrate a serum sample with normal chloride concentration.

CALCULATION

$$\text{Chloride concentration in mmol/l} = \frac{\text{titration of unknown}}{\text{titration of standard}} \times 0.02 \times \frac{1000}{0.2}$$

or

$$\frac{\text{titration of unknown (in ml)}}{\text{titration of standard (in ml)}} \times 100$$

where 0.02 = mmol chloride/ml of standard

0.2 = ml of sample used

1000 = ml/liter

COMMENTS ON THE METHOD

1. The titration end point is sensitive to pH. There is some disagreement as to the best pH to be used, but most authors use a pH between 3 and 4.5, which is obtained if the protein-free filtrate is prepared as just described. Therefore, it is important that all tests and standards be titrated at this pH range.

2. Many laboratories titrate specimens directly with $Hg(NO_3)_2$ without preparing a protein-free filtrate. Such an approach yields rapid results, but has an inherent positive

* If sweat is to be analyzed as discussed and outlined in Chapter 20, the following modification is applied: Titrate 2.5 ml of the extract from the gauze directly without making a protein-free filtrate. Also titrate 200 μl of standard directly, and enter both results into the formula provided with the procedure for the sweat test (p. 1096).

† If spinal fluid or urine with low protein content is to be analyzed, 200 μl of unknown and standard may be titrated directly without making a protein-free filtrate. One drop of approximately 0.07 molar HNO_3 should be added before titration in order to provide an acid pH. (At alkaline pH, the indicator has a color similar to that observed at the titration end point.)

error of approximately 2 per cent owing to the reaction of Hg^{2+} with —SH groups of proteins. Pigments in the sample may mask the end point, which is especially difficult to detect in highly icteric or hemolyzed sera. Some authors have reported positive errors of as much as 15 mmol/l.

3. The reagent reacts with any halogen, not only Cl^-; thus, ions such as Br^-, and I^-, as well as CN^-, CNS^-, and —SH groups, will also react and thus give rise to positive errors. In bromide poisoning, the bromide replaces some chloride and the titration may give an apparent normal level, since it measures the sum of the two halogens. In spite of these limitations, this is a simple, rapid, and useful method for a clinical laboratory and is especially useful to supplement an automated method.

REFERENCE

Schales, O., and Schales, S. S.: J. Biol. Chem., *140*:879, 1941.

COLORIMETRIC MEASUREMENT OF CHLORIDE WITH MERCURIC THIOCYANATE (AUTOANALYZER METHOD)

A procedure by Zall, Fisher, and Garner, which was adapted to the AutoAnalyzer by Skeggs, is based on the following principles: The specimen is mixed with a solution of $Hg(SCN)_2$. As a result of the high affinity of Cl^- for Hg^{2+}, undissociated $HgCl_2$ is formed, resulting in the release of free SCN^-. The SCN^- reacts with Fe^{3+} in the ferric nitrate reagent to form the highly colored, reddish colored complex of $Fe(SCN)_3$ with an absorption peak at 480 nm.

$$Hg^{2+} + 2SCN^- + 2Cl^- \rightarrow HgCl_2 + 2(SCN)^-$$
$$3(SCN)^- + Fe^{3+} \rightarrow Fe(SCN)_3$$

The useful analytical range of this procedure is limited to between 80 and 120 mmol/l, and even in this range it lacks linearity. In order to increase linearity and accuracy, the addition of a certain amount of $Hg(NO_3)_2$ to the reagent stream (e.g., an equivalent to 60 mmol chloride/liter) has been recommended. These Hg ions bind a fixed amount of chloride ions and thus make them unavailable for reaction with $Hg(SCN)_2$. As a result, only the chloride ions in excess of those bound by the Hg^{2+} from $Hg(NO_3)_2$ will react with $Hg(SCN)_2$ and produce a color. The associated base line shift allows greater sensitivity (wider spread of per cent T values for a given amount of Cl^-) and a more linear response over a wider portion of the physiological range. In the AA II methodology, a linearizer is available, which also improves linearity over a wider range.

The AutoAnalyzer method is not always suitable for the analysis of chloride in urine and sweat, depending on the concentration of chloride present in the sample. Thus, an alternative method for chloride must be available in a clinical laboratory. The AutoAnalyzer technique also suffers from base line drift, making use of drift controls and frequent standardization imperative.

COULOMETRIC-AMPEROMETRIC TITRATION OF CHLORIDE

In the coulometric-amperometric determination of serum chlorides, the sample is diluted in an acid solution (HNO_3–CH_3COOH mixture) containing a small amount (e.g., 25 mg/100 ml) of gelatin. The nitric acid provides good electrolytic conductivity; the acetic acid renders the solution less polar, thus reducing the solubility of silver chloride and providing a sharper end point; and the gelatin provides for a smoother and more reproducible titration curve by being adsorbed preferentially to high spots of the electrode and, thus, equalizing the reaction rate over the entire electrode surface. In addition to this, the acid solution aids in preventing reduction of precipitated silver chloride at the indicator cathode.

Excess of protein, however, may introduce some error owing to reaction of silver ions with the sulfhydryl groups of protein.

If the titration is carried out with the Chloridometer (American Instrument Co., Inc., Silver Spring, MD. 20910, and Buchler Instruments Inc., Fort Lee, N.J. 07024), the volume of the solution must be at least 2.5 ml so that the electrodes are fully immersed into the solution. When the titration is started, a silver generator electrode is fed a constant current, causing oxidation of the Ag of the silver wire anode to Ag^+ and a reduction of H^+ to H_2 at the cathode. The silver ions, which are generated by the current at a constant rate, combine with the chloride ions to form insoluble AgCl and are thus immediately removed from solution. After the equivalence point is reached (sufficient Ag^+ has been generated to react with all chloride present), any additional generation of Ag^+ will result in a sudden increase in electroactivity (conductivity) of the titrant, which is detected amperometrically by a set of silver indicator electrodes. The increase in potential between the electrodes activates a relay, which in turn stops both an automatic timer and the generation of additional Ag^+ at the anode. The Ag^+ generated is a function of both time and current; inasmuch as the current is constant, the quantity of Ag^+ generated becomes a function of time only. Thus, the time necessary to reach the titration end point can be taken as a measure of the chloride concentration. The titration time for a standard solution analyzed in the same manner can be used to calculate the chloride concentration of the unknown by the following formula:

$$\text{chloride concentration (mmol/l)} = \frac{T_u - T_b}{T_s - T_b} \times C_s$$

where

T_u = the titration time of the unknown

T_s = the titration time of the standard

T_b = the titration time of the blank

C_s = the concentration of standard in mmol/l

The principle of coulometry has been explained in more detail on page 151.

COMMENTS ON THE PROCEDURE

The procedure just described is probably the most accurate method available for the determination of chlorides in routine clinical chemistry laboratories. Results of the titration technique were 99.7 per cent of those found with the isotope dilution technique. The method is suitable for determining the chloride concentration over the entire normal and abnormal physiological range. Other halogens, as well as CN^-, CNS^-, and —SH, interfere with the determination.

Greatest accuracy is obtained if the titration time is held between 70 and 160 seconds. If solutions with extremely low chloride content are to be analyzed (e.g., in micro chloride determinations), the titration speed is reduced by reducing the current through the silver generator electrodes. Standard solutions used in the determination of specimens with low chloride content should be diluted to a range that corresponds to the approximate concentration of the unknown, or, as in microanalysis, the amount of standard used should be reduced (e.g., 20 μl, if the serum sample used is also 20 μl).

The electrodes must be rinsed well after each analysis. Before use of the instrument, polish the electrodes with silver polish and rinse well. The silver wire electrodes should be adjusted to the same length and should not be bent or pointed.

REFERENCES

Cotlove, E., Trantham, H. V., and Bowman, R. L.: J. Lab. Clin. Med., *51*:461, 1958.

Cotlove, E.: Determination of chloride in biological materials. *In*: Methods of Biochemical Analysis. D. Glick, Ed. New York, Interscience Publishers Inc., 1964, vol. 12.

Cotlove, E.: Chloride. *In*: Standard Methods of Clinical Chemistry. D. Seligson, Ed. New York, Academic Press, Inc., 1961, vol. 3, pp. 81–92.

COLORIMETRIC MEASUREMENT OF CHLORIDE WITH FERRIC PERCHLORATE

An automated procedure for serum chloride based on formation of a complex between ferric perchlorate and chloride ion has been reported by Fingerhut.[13] The colored complex formed is thought to be a chloro-complex of the ferric ion with an absorbance maximum of about 340 nm.

Although the method is said to be linear throughout the biological range, and although it eliminates the use of mercuric salts, it has not yet found wide acceptance in laboratories.

NORMAL VALUES

Normal values for chloride in *serum or plasma* range from 98 to 106 mmol/l, with relatively few values being above 104 mmol/l. The serum chloride values vary little throughout the day, although there is a slight decrease in chloride ions after meals because of the chloride required for the formation of gastric HCl. Values for *spinal fluid* are 118 to 132 mmol/l. *Urine* values vary greatly with Cl^- intake, but generally range between 110 and 250 mmol/d. *Sweat* chloride values will be discussed in connection with the sweat test (Chapter 20).

Carbon Dioxide

THE DETERMINATION OF TOTAL CO_2 ($ctCO_2$)

The concentration of total CO_2 in blood, plasma, or serum may be determined by several different techniques. The two most commonly employed in clinical chemistry laboratories today are the gasometric method[37,40] and the AutoAnalyzer procedure.[56]

The determination of the $ctCO_2$ is most frequently performed on serum specimens, but whole blood or plasma may also be used. As explained in the section Definition of Terms, the method for CO_2 determines not only the amount of physically dissolved CO_2 ($cdCO_2$) but also that released from HCO_3^-, CO_3^{2-}, and carbamino compounds. The amount of CO_2 gas released is measured either volumetrically (volume of gas at atmospheric pressure) or manometrically (pressure of gas at fixed volume). Initially, each of these techniques was carried out with the respective Van Slyke apparatus. Today, both Van Slyke gasometers have been replaced by the more convenient Natelson microgasometer, which allows for manometric measurement of the total CO_2.

CLINICAL SIGNIFICANCE

Knowledge of the $ctCO_2$ of serum (plasma, blood), together with other clinical and laboratory information (pH, pCO_2), is necessary for the evaluation of the acid-base status; however, the determination of the $ctCO_2$ alone, without additional information, is of limited value. A high $ctCO_2$ may be observed in compensated respiratory acidosis (retention of CO_2) as well as in metabolic alkalosis (increase in HCO_3^-). A low $ctCO_2$ may be observed in compensated respiratory alkalosis (loss of CO_2 due to hyperventilation) or in metabolic acidosis (decrease of HCO_3^-). Additional laboratory determinations, such as pH and pCO_2, will permit differentiation between metabolic and respiratory conditions. Examples of disorders associated with abnormal $ctCO_2$ are described in Chapter 16.

NORMAL VALUES

Specimen	*Range in mmol/l*
Venous plasma (serum)	23–29
Capillary plasma (serum)	21–28
Venous (whole) blood	22–26
Arterial (whole) blood	19–24

MANOMETRIC DETERMINATION OF ctCO$_2$ WITH THE NATELSON MICROGASOMETER

An anaerobically collected sample of serum, plasma, or whole blood is introduced anaerobically into the pipet attached to the microgasometer, followed by lactic acid, an antifoam reagent, and deionized water. Lactic acid releases CO_2 from HCO_3^-, CO_3^{2-}, and other sources of bound CO_2, the antifoam reagent (e.g., caprylic alcohol) prevents foaming, and the water washes the sample and reagents into the reaction chamber. The sample and all of the reagents are separated by mercury buttons; water is drawn into the reaction chamber with the aid of mercury, which at the same time prevents the introduction of air and seals the gasometer. The reaction chamber is then sealed by closing the stopcock, and a vacuum is applied to the reaction chamber. This will cause a diffusion of the liberated CO_2 gas and the physically dissolved CO_2 from the liquid phase into the vacuum above the liquid. After 1 min of agitation to assure complete release of the gas, the liquid level is advanced to a predetermined position and the pressure of the gas (which is now compressed into a fixed volume) is measured manometrically (= P_1). Alkali, such as NaOH, is now introduced into the reaction chamber to cause a total reabsorption of the CO_2 gas as Na_2CO_3. The pressure required to confine the residual gases (O_2, N_2) in a fixed volume is now measured (= P_2). The difference between P_1 and P_2 is a measure of the amount of CO_2 present. The manometer of the Natelson microgasometer is calibrated in mm Hg, but this figure must be corrected for the temperature and must then be converted to mmol/l by multiplication with the conversion factor supplied with the instrument. Correction for temperature is necessary since the pressure of a gas is directly related to the temperature if the volume is kept constant.

REFERENCES

Natelson, S.: Am. J. Clin. Path., *21*:1153, 1951.

Natelson, S.: Techniques of Clinical Chemistry. 3rd ed. Springfield, IL., Charles C Thomas, Publisher, 1971.

ALTERNATIVE METHOD FOR THE DETERMINATION OF ctCO$_2$

A new analyzer (Beckman Instruments, Inc., Fullerton, CA.) offers a different approach to the micro measurement of $ctCO_2$ which does not involve the use of mercury. The principle of operation involves release of CO_2 gas when the sample is added to sulfuric acid, and subsequent monitoring of the CO_2 released with a pair of pCO_2 electrodes (reference and sample electrode). The rate of change in pH of the buffer inside the pCO_2 electrode is said to be a measure of the concentration of total CO_2 ($ctCO_2$) in the sample. The technique requires 10 μl of sample and allows simultaneous measurement of the chloride concentration by a modified coulometric technique. Both tests can be completed within 30 seconds. For the discussion of the operation of pCO_2 electrodes and coulometric determinations of chloride, see p. 144 and p. 151, respectively.

THE DETERMINATION OF TOTAL CO_2 (ctCO$_2$) BY THE AUTOANALYZER (AA)

In the AutoAnalyzer I (AA I) methodology, the specimen is treated with sulfuric acid to release the CO_2 gas from HCO_3^- and $RCNHCOO^-$. A fixed aliquot of the total CO_2 collected is removed from a "gas trap" and is reabsorbed into a weak alkaline carbonate-bicarbonate buffer solution containing phenolphthalein. As CO_2 is reabsorbed, the pH of the buffer solution decreases, resulting in a *decrease* in the intensity of the red color. This change in the color of the phenolphthalein indicator is proportionate to the amount of CO_2 released from the sample.

This procedure has the disadvantage that it requires a freshly prepared or frequently restandardized color reagent. In addition, the technique results in a decrease in absorbance with an increase in concentration of CO_2 (inverse colorimetry).

In the AA II methodology, the color reagent has been changed to contain cresol red in TRIS buffer. Sulfuric acid is again used to liberate CO_2 gas. The CO_2 gas may be trapped as in the AA I methodology or, alternatively, the stream containing the CO_2 gas is passed by a silicone rubber membrane. Gases, but no other constituents of the sample stream, then diffuse through the membrane and mix with the recipient stream (color reagent). The color produced is a direct measure of the concentration of total CO_2 present in the sample.

Although the color reagent used in this method is more stable than is the older reagent, it still requires frequent adjustment to keep the pH of the color reagent within the

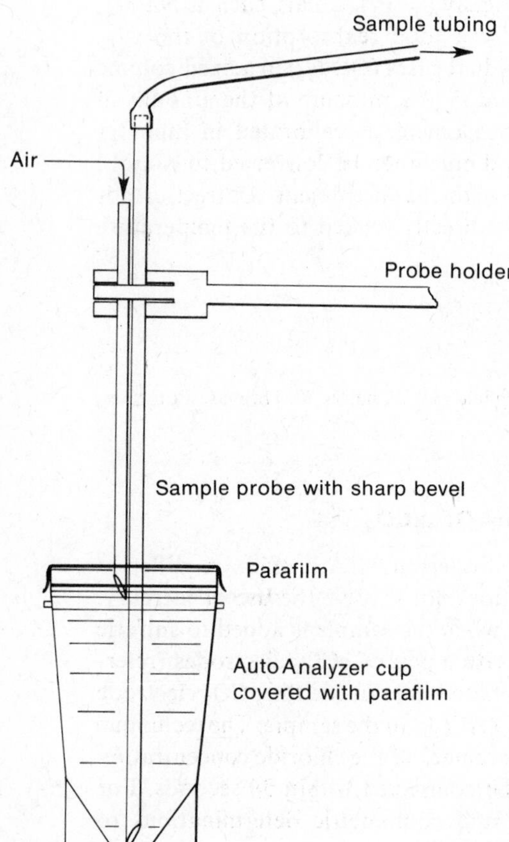

Figure. 15–5 Modified sample probe and Auto Analyzer cup, used in the determination of $ctCO_2$ by the Auto-Analyzer.

narrow range of 9.2 ± 0.2. In the new SMAC methodology, the color reagent is again phenolphthalein.

All AA methods for measuring the concentration of total CO_2 share the common problem of base line drift, making use of frequently redetermined standard curves and drift controls mandatory.

COMMENTS

Specimens used for determination of the concentration of total CO_2 must be handled anaerobically at all times (see Collection of Blood for the Determination of pH, $ctCO_2$, and pCO_2). If the specimen is collected in a heparinized Vacutainer, the top of the container should not be removed until the specimen is to be withdrawn and analyzed.

In the standard AutoAnalyzer procedure, the sample in the sample cup is exposed to atmospheric air, resulting in a loss of CO_2. As a result, values obtained with this technique may be low by as much as 6 mmol/l if samples are exposed to air for 1 h. This significant source of error can be eliminated by use of the following technique:

Transfer an adequate amount of an anaerobically collected specimen into a micro or macro AutoAnalyzer cup so that the cup is completely filled. Cover the cup immediately with a small piece of Parafilm, stretching the film only slightly (to prevent loss of CO_2 through the Parafilm). Load specimens and standards, treated identically, onto the AA tray and cover with the tray cover.

In order to facilitate piercing of the Parafilm by the sample probe, the latter should be modified as follows: File the tip of the probe to produce a sharp bevel. Prepare a second probe in the same way and mount it slightly higher, as shown in Figure 15–5. Connect the first probe to the sample tubing of the manifold and leave the second probe open. This will allow for replacement of the sample volume that has been withdrawn from under the Parafilm layer. Results obtained employing this technique compare favorably (within ±1 mmol/l) with results obtained with the Natelson microgasometer.

REFERENCE

Skeggs, L. T. Jr.: Am. J. Clin. Path., *33*:181, 1960.

THE DETERMINATION OF CO_2 COMBINING POWER

The CO_2 combining power is an index of the amount of CO_2 that can be bound by plasma as HCO_3^- at a pCO_2 of 40 mm Hg at 25°C. The test is performed by placing the specimen (serum, plasma) into a container such as a separatory funnel (suitable to increase the surface area of the specimen) and equilibrating it with a gas mixture containing CO_2 at a pCO_2 of 40 mm Hg, the average CO_2 tension of alveolar air. This can be realized by using a gas mixture containing 5.2 per cent (v/v) CO_2 in air at 760 mm Hg (or 5.4 per cent CO_2 at 740 mm Hg). After equilibration, a portion of the sample is removed and analyzed for total CO_2 in the same way as outlined under Determination of $ctCO_2$. This method determines not only bicarbonate but also the $cdCO_2$. Thus, the value obtained must be corrected by subtracting 1.2 mmol/l if the equilibration was performed at 38°C or 1.8 mmol/l if the equilibration was performed at 25°C. (The solubility of CO_2 increases with a decrease in temperature.) Although this test has some value in evaluating the acid-base status in metabolic disturbances, it has been replaced by the more accurate procedures measuring $ctCO_2$. The test for CO_2 combining power has less clinical value and results may even be misleading in respiratory disturbances, since the sample is equilibrated with a pCO_2 that corresponds to the average *normal* alveolar air rather than to the condition existing in the patient under investigation, who may have a low pCO_2 (in respiratory alkalosis) or a high pCO_2 (in respiratory acidosis).

The blood specimen for this determination can be collected aerobically; although, at times, this may be a convenience, the technique also requires equilibration, which is inconvenient and a potential source of error.

THE DETERMINATION OF PLASMA BICARBONATE

Bicarbonate is the second largest anion fraction in plasma (21 to 28 mmol/l). The small amounts of carbamino compounds present in plasma are customarily included in this fraction. Bicarbonate is important as a component of the bicarbonate buffer system, and it also serves as a transport form for CO_2 from the tissues to the lungs (see Chapter 16). Although bicarbonate can be determined by direct titration,[40,52] the determination of HCO_3^- as such is rarely requested in clinical medicine since determination of the total CO_2 serves as an approximate measure of bicarbonate concentration ($c\text{HCO}_3^- = ctCO_2 -$ 1.2 mmol/l).

As indicated in the section entitled The Interrelationship Between $ctCO_2$, $c\text{HCO}_3^-$, $cdCO_2$ and pH, the bicarbonate concentration can also be calculated from the Henderson-Hasselbalch equation, or it can be obtained with the aid of nomograms constructed on the basis of this equation.

DETERMINATION OF BLOOD, PLASMA, OR SERUM pH

The blood (plasma, serum) pH is the most valuable single factor in the evaluation of the acid-base status of a patient. As a result of the combined effort of the buffer system, the respiratory system, and the renal mechanism, the blood pH is also one of the most stringently controlled parameters in the human body. Disturbances in the acid-base status are classified into (a) metabolic acidosis* (primary bicarbonate deficit), (b) metabolic alkalosis (primary bicarbonate excess), (c) respiratory acidosis (primary dCO_2 excess), and (d) respiratory alkalosis (primary dCO_2 deficit). Conditions leading to an acid-base imbalance are discussed in detail in the following chapter on Acid-Base and Electrolyte Balance.

The determination of $ctCO_2$ alone is often insufficient to evaluate existing conditions, especially when dealing with respiratory disorders or respiratory disorders superimposed on metabolic disorders. Therefore, it is recommended that a pH determination be performed whenever the acid-base status of a patient is being investigated.

The blood pH is not necessarily an indicator of the acid-base status of the body as a whole. Changes in the pH of extracellular fluid are quickly reflected in the pH of the intracellular fluid only if the pH change is due to a change in the pCO_2. However, if the change in the pH of extracellular fluid is due to an increase or decrease in acids or bases other than HCO_3^-/CO_2, changes in intracellular pH occur more slowly (and frequently to a much lesser degree). The same situation applies also to spinal fluid. The reason for this behavior lies in the fact that gases, such as CO_2, can penetrate cell membranes much easier and faster than ionized particles.

Formerly it was very difficult to measure the pH of body fluids accurately; however, in recent years pH meters have been developed that make possible the measurement of the blood pH with the necessary degree of accuracy and with relatively little effort. Today, it is not the pH meter but the way in which the specimen is collected that limits the accuracy of the pH determinations.

---

* The terms acidosis and alkalosis refer to an abnormal *balance* of H^+ in the body or a fluid compartment; i.e., *acidosis* refers to an abnormal gain of acid or loss of base, while *alkalosis* signifies an abnormal gain of base or loss of acid. When referring to pH values of blood, the terms *acidemia* and *alkalemia* are preferred.[55]

SPECIMEN COLLECTION

Special care must be taken not to expose the sample to air during collection or when transferring it from the collection tube to the pH meter. If a pH measurement is to be confirmed, the sample should be withdrawn from the center of the plasma phase, as distant as possible from the surface and from the erythrocytes (see also Collection of Blood for Determination of pH, $ctCO_2$, and pCO_2).

If an anticoagulant is to be used, heparin is preferred. Oxalate, citrate, and EDTA should not be used, since they cause shifts of electrolytes and water between plasma and cells.

If blood is drawn and stored in a plastic syringe, it should be ascertained that the plastic is gas-tight and does not permit diffusion and exchange of gases.

EFFECT OF SPECIMEN STORAGE ON BLOOD pH

The pH of freshly drawn blood decreases on standing at a rate of 0.04 to 0.08 pH unit/h at 37°C, by about 0.03/h at 25°C, but by only 0.008/h at 4°C. The decrease in pH is accompanied by a corresponding decrease in glucose and an equivalent increase in lactate. The primary cause of these changes is thought to be glycolysis taking place in leukocytes, thrombocytes, and reticulocytes and only to a minimal degree in erythrocytes. In leukemia, the pH drop during the first 30 min after specimen collection may be as high as 0.6 pH unit.

Respiration in freshly drawn blood (protected from air) causes a decrease in total O_2 content of 0.1 mol $\times$ liter^{-1} $\times$ h^{-1} (= 0.1 mol/l per h) at room temperature and twice this value at 37°C. This decrease in O_2 is accompanied by a nearly equivalent increase in $ctCO_2$. Because of compensation between the alkalinizing (proton binding) effect of oxygen decrease (conversion of $HHbO_2$ to the weaker acid, HHb) and the simultaneous acidifying effect of the increase in CO_2 (carbonic acid), the effect of spontaneous respiration *in vitro* on the pH of blood is negligible (less than 0.01 pH unit) in the first hour, even at 37°C.

The adverse effects of glycolysis and respiration on pH, $ctCO_2$, and pCO_2 of blood (plasma, serum) can best be avoided by making all measurements within 30 min after the blood is drawn. If the analysis is to be delayed, the syringe or tube containing the freshly drawn blood (with appropriate anticoagulant) should be immediately placed in ice water at 0 to 4°C. Under these conditions any changes in blood pH are negligible.

If the pH measurement is to be performed on whole blood, the sample should be adequately mixed and prewarmed to 37°C prior to measurement. (Slow injection of the specimen through the narrow passage leading to the microelectrode of the blood pH meter allows the sample to come to 37°C prior to the pH measurement.) If the analysis is to be performed on plasma, the blood must be warmed to 37°C before separation of red cells from plasma by centrifugation, or else false high pH values are obtained. This precaution is necessary, since increases in temperature affect the plasma pH less than the pH of whole blood.

BUFFERS FOR THE STANDARDIZATION OF pH METERS

The pH meter must be standardized with buffers that are accurate to within ±0.005 pH units. Some manufacturers, such as Instrumentation Laboratories, Inc. (Watertown, MA. 02173) and Radiometer, Inc. (Cleveland, OH. 44145), make buffers available that are standardized to 0.001 pH unit, but such accuracy is not attainable nor can such solutions be maintained at their listed pH value under routine laboratory conditions. As buffers are exposed to air, their pH usually decreases gradually because of absorption of atmospheric CO_2. This is especially true if buffers are purchased in 500 ml stock bottles. Precision buffers, marketed in ampules containing approximately 3 ml of buffers, are less likely to change during storage; thus, these sealed ampules serve as good storage containers for primary standards.

The National Bureau of Standards has recently released instructions for the preparation of acceptable pH buffers. These buffers can be prepared as follows:

Phosphate buffer, pH 7.386 at 37°C, ionic strength 0.01 (KH_2PO_4, 0.008695 mol/l; Na_2HPO_4, 0.03043 mol/l). Dissolve 1.179 g of dried potassium dihydrogen phosphate, KH_2PO_4, and 4.302 g of dried disodium hydrogen phosphate, Na_2HPO_4, in distilled water, and dilute to 1 liter at 25°C.

The pH values at the respective temperatures are

20°C: 7.429	35°C: 7.389	40°C: 7.380
25°C: 7.413	37°C: 7.386	
30°C: 7.400	38°C: 7.384	

Phosphate buffer, pH 6.841 at 37°C, ionic strength 0.1 (KH_2PO_4, 0.025 mol/l; Na_2HPO_4, 0.025 mol/l). Dissolve 3.388 g KH_2PO_4 and 3.549 g Na_2HPO_4 in distilled water and dilute to 1 liter at 25°C.

The pH values at the respective temperatures are

20°C: 6.881	35°C: 6.844	40°C: 6.838
25°C: 6.865	37°C: 6.841	
30°C: 6.853	38°C: 6.840	

The phosphate buffer according to Sørensen is similar to the above NBS buffers, but it has a molar ratio of KH_2PO_4 to Na_2HPO_4 of 1/4 (instead of 1/3.5 and 1/1, respectively, for the two NBS buffers). This buffer is commercially available (e.g., Radiometer, Inc. and Instrumentation Laboratories, Inc.). It is prepared by dissolving 1.816 g KH_2PO_4 and 9.501 g $Na_2HPO_4 \cdot 2H_2O$ in 1000 g of water. The buffer has an ionic strength of about 0.17. The respective pH values of the buffer at different temperatures are

20°C: 7.426	35°C: 7.386	40°C: 7.377
25°C: 7.410	37°C: 7.383	
30°C: 7.397	38°C: 7.381	

These buffers have a temperature coefficient of $\Delta pH/\Delta t = -0.0015/°C$ ($= -0.0015 \times K^{-1}$). This is approximately one-tenth that of whole blood. Tris buffer, on the other hand, has a temperature coefficient of $\Delta pH/\Delta t = -0.026 \times K^{-1}$. Use of a buffer with such a high temperature coefficient could provide a means to detect and indeed quantitate variations in instrumental temperatures.[12] Tris is available as a Standard Reference Material from the National Bureau of Standards, as SRM 922 (Tris) and SRM 923 (Tris-HCl).

All these buffers are stable for several months if kept unopened and at refrigerator temperature in a bottle made of resistant glass or in a tightly stoppered plastic bottle. It has been suggested that addition of thymol prolongs the useful life of the buffer. The primary cause of deterioration is growth of microorganisms, primarily fungi. All buffers should be prepared with freshly double-distilled water that is free of CO_2 and ammonia.

OPERATION OF THE pH METER

Modern pH meters are accompanied by manuals that contain detailed instructions for blood pH measurements. Therefore, only those procedural steps that are generally applicable will be given.

1. Turn on the pH meter at least 15 min prior to use. Check whether the temperature of the electrode is 37°C ± 0.05°C.
2. Fill the electrode with a fresh aliquot of buffer, e.g., pH 7.384. The buffer should

never be aspirated directly from the stock bottle; instead, an aliquot should be transferred to a smaller bottle provided with a tight lid. Plastic bottles are preferred to glass bottles, since there is no release of ions from the container. At the end of the day, the buffer should be discarded. Adjust the pH meter with the adjustment knob to read the pH value of the buffer (e.g., 7.384).

3. Wash the electrode with distilled water or saline and air and introduce the second buffer, e.g., pH 6.841. Adjust the pH meter with the slope control (if the instrument is provided with such a device). If an adjustment of the slope control was necessary, re-check the calibration with the first buffer.

Standardization of the pH meter with the two reference buffers is generally not adequate. In addition it is recommended that a quality control specimen with a protein matrix be used. Such a sample is more likely to show errors due to protein build-up on the surface of the glass electrode or to temperature variations of the electrode. Controls of this type are presently commercially available in three ranges, Versatol Alkalosis, Acidosis, and Normal (General Diagnostics, Morris Plains, N.J. 07950).

4. Wash the electrode with saline, followed by air.

5. Introduce the specimen into the electrode. Make certain that no gas or air bubbles are present in the blood column or at the junction between the sample and the KCl solution.

6. Wait for 20 to 30 s before taking a reading. At this point the reading should be stable and no further drift should occur.

7. Rinse the electrode with saline and air and re-check the calibration after about every 5 specimens. The readings should agree to within ± 0.01 pH unit. Rinse the electrode with saline and air before introducing the next sample.

8. When all pH measurements have been completed, wash the electrode with a 100-fold diluted solution of a regular household detergent, rinse with water, and leave either saline or buffer in the electrode. From time to time, the electrode may be soaked in a solution containing 1 g pepsin/l of 0.1 molar HCl. This will aid in the removal of any protein film which may have formed on the electrode. A weak solution of sodium hypochlorite ($\approx$ 0.01 mol/l) is also an effective wash solution.

If the capillary electrode is blocked by a clot, use a horse hair, but *not* a wire, to remove the clot from the capillary.

NORMAL VALUES

For venous serum and plasma collected under routine conditions, the normal range of pH is generally considered to be 7.35 to 7.45, with an average of 7.40. This figure allows for the small experimental error experienced in routine determinations. The true range for serum or plasma pH is probably slightly narrower. The pH of whole blood is 0.01 to 0.03 pH unit lower than that for plasma and serum because of an effect of red cells on the liquid junction potential ("suspension effect")[54] and not because of an inherent pH difference (see also Table 16–1).

SOURCES OF ERROR

Maintenance of Electrodes. *a. Glass Electrode.* Internal and external surfaces of the electrode must be kept clean at all times. Rinsing of the electrodes as described in the procedure is extremely important to prevent or minimize protein build-up on the H^+-sensitive glass, which causes slow electrode response and drift. A check for proper response of the electrode may be performed by measuring the pH of two buffers which differ by more than 2 pH units. The readings should be correct within 0.02 pH unit. As a rule, pH meters do not give a perfect linearity, and a deviation of up to 0.02 pH unit may be expected.

It is therefore important that pH measurements be done with a pH meter that has been calibrated with a buffer of a pH as close as possible to the pH of the sample.

b. Reference Electrode. The mercury in the reference electrode should be shiny and the saturated KCl solution should cover the mercury. If refilling with KCl is indicated this should be done with prewarmed, saturated KCl, so that some crystals will form inside the electrode on standing. Some pH meters have a KCl liquid salt bridge between the reference and glass electrodes. This KCl must be exchanged at regular intervals, since red cells will accumulate at the junction and create a drift due to a changing potential. The reference electrode should be covered with a tight-fitting cap, to prevent evaporation and creeping of the KCl solution. Greasing the surface with silicone stopcock grease may also prevent creeping.

Temperature Control. The specimen pH and the electrode response slope are temperature dependent. Thus, it is important that the electrode be precisely thermostatted at 37 ± 0.05°C. The temperature should be verified at the place where the circulating water leaves the electrode jacket. Significant drops in temperature can occur while the water circulates from the water bath through the electrode, especially if the circulation rate is decreased. Good circulation is aided by preventing mold growth in the circulating water. This can be done by adding a bacteriostatic agent, which is available commercially, or by adding a few drops of caprylic alcohol.

Shielding and Grounding of the Electrode. The pH meter should be adequately grounded at the electrical outlet with a "true zero ground." However, the instrument should not be grounded through the outlet *and* a water pipe, and two grounded instruments should not be connected to each other through a ground wire. This will create ground loops, a phenomenon that can inject electrical "noise" into the system.

Synthetic clothing (Nylon, Dacron, etc.) worn by laboratory workers can accumulate electrostatic charges, causing meter movement. Such interference can be prevented by use of electrodes with cables which are adequately "shielded." Addition of sodium or potassium nitrate (0.1 g/l) to the circulating water will adequately "shield" the electrode itself.

Temperature Corrections. Blood pH measurements are customarily performed at 37°C (formerly 38°C) regardless of the actual body temperature of the patient. Since pH values decrease with increase in temperature, a correction factor must be applied to those pH measurements performed on patients with a blood temperature other than 37°C. The $\Delta pH/\Delta t$ has been reported by Rosenthal[48] to be -0.0146 (≈ 0.015)/°C ($= -0.0146 \times K^{-1}$) for whole blood; it is about -0.012/°C for plasma. However, this factor varies slightly with pH and has also been reported to vary slightly with the type of electrode employed. A convenient chart with correction factors for a wide range of temperatures has recently been published[6] and a convenient nomogram is given in Figure 15–3. Failure to correct for temperature may lead to gross errors if the body temperature deviates significantly from 37°C.

COMMENTS

The pH has been the customary unit with which to express the activity of hydrogen ions (aH^+). This has been done mainly for reasons of convenience and habit. With the introduction of SI units, it has been proposed that the aH^+ be expressed in terms of absolute concentration, nmol/l. While less convenient, this unit would reflect absolute changes in aH^+ more obviously than do the customary pH units. For example, a drop in pH of 0.30 unit from 7.40 to 7.10 is caused by an increase in aH^+ from 40 nmol/l to 80 nmol/l (= +100 per cent). An equally large increase in pH of 0.30 unit from 7.40 to 7.70, however, is caused by a decrease in aH^+ from 40 nmol/l to 20 nmol/l (= −50 per cent). A scale relating pH values to values for aH^+ is given in Figure 16–9.

THE INTERRELATIONSHIP BETWEEN $ctCO_2$, $cHCO_3^-$, $cdCO_2$, AND pH*

THE HENDERSON-HASSELBALCH EQUATION

Carbon dioxide and water react to form carbonic acid, which in turn dissociates to hydrogen ions and bicarbonate ions:

$$CO_2 + H_2O \underset{}{\overset{K_{\text{hydration}}}{\rightleftharpoons}} H_2CO_3 \underset{}{\overset{K_{\text{dissociation}}}{\rightleftharpoons}} H^+ + HCO_3^-$$

According to the law of mass action,

$$K_{\text{hydration}} = \frac{aH_2CO_3}{aCO_2 \times aH_2O} = 0.00229 \text{ (p}K = 2.64 \text{ at } 37°\text{C)}$$

and

$$K_{\text{dissociation}} = \frac{aH^+ \times aHCO_3^-}{aH_2CO_3} = 2.04 \times 10^{-4} \text{ (p}K = 3.69 \text{ at } 37°\text{C)}$$

These can be combined to give

$$K_{\text{combined}} = \frac{aH_2CO_3}{aCO_2 \times aH_2O} \times \frac{aH^+ \times aHCO_3^-}{aH_2CO_3} = \frac{aH^+ \times aHCO_3^-}{aCO_2 \times aH_2O} = 4.68 \times 10^{-7} \text{ (p}K = 6.33)$$

In the classical formulation, Henderson (1908) used concentrations (c) rather than activities (a) for bicarbonate, CO_2, and H^+; the concentration of water was assumed to be constant and was therefore incorporated into the constant K'.

$$K' = \frac{cH^+ \times cHCO_3^-}{cdCO_2}$$

The symbol $cdCO_2$ stands for the concentration of dissolved CO_2 *including* the small amount of undissociated (dissolved) H_2CO_3. It can be expressed as $cdCO_2 = \alpha \times pCO_2$, where α is the solubility coefficient for CO_2. The symbol $cHCO_3^-$ represents the concentration of total CO_2 ($ctCO_2$) minus the concentration of dissolved CO_2 ($cdCO_2$), which includes H_2CO_3.

$$cHCO_3^- = ctCO_2 - (\alpha \times pCO_2)$$

The "bicarbonate" concentration by definition includes carbonate (CO_3^{2-}) and carbamate (carbamino-CO_2; $RCNHCOO^-$), which are present only in small amounts.

If the Henderson equation is rearranged, and $cdCO_2$ is replaced by $\alpha \times pCO_2$, the following equation results:

$$cH^+ = K' \times \frac{\alpha \times pCO_2}{cHCO_3^-}$$

Hasselbalch (1916) showed that a logarithmic transformation of the equation was a more useful form and employed the symbols pH ($= -\log cH^+$) and pK' ($= -\log K'$). [Currently, the pH is defined as the negative log of the activity of H^+ (aH^+), which is the entity actually measured with pH meters.] The resulting Henderson-Hasselbalch equation becomes:

$$\text{pH} = \text{p}K' + \log\frac{cHCO_3^-}{\alpha \times pCO_2} = \text{p}K' + \log\frac{ctCO_2 - (\alpha \times pCO_2)}{\alpha \times pCO_2}$$

* This section (pp. 893 to 899) has been prepared in collaboration with O. Siggaard-Andersen.

The "constants," pK' and α, are defined as follows: (1) K' is the first, apparent, overall (combined) dissociation constant for carbonic acid; apparent, because concentrations are employed rather than activities and because it includes both the hydration and dissociation constants; overall, because both the concentration of dissolved CO_2 and the true concentration of H_2CO_3 are used. K' depends not only on the temperature but also on the ionic strength of the solution. For an aqueous sodium bicarbonate solution at 37°C the following approximate relationship exists between K' and ionic strength (I) measured in mol/kg H_2O:

$$pK' = 6.33 - 0.5\sqrt{\{I\}}$$

For blood plasma at 37°C the normal mean value is $pK'(P) = 6.103$ with a normal biological standard deviation of about ± 0.0015, mainly due to normal variations in ionic strength. (The normal average pK' at 38°C is 6.100). In pathological cases with markedly deviant ionic strength, the standard deviation for pK' may be significantly greater. Changes in ionic strength of ± 20 per cent cause changes in pK' between 6.08 and 6.12. The variation of pK' of plasma with temperature can be expressed approximately as $\Delta pK'(P)/\Delta T = -0.0026 \times K^{-1}$ or a decrease of 0.0026/°C increase in temperature. Owing to the inclusion of carbonate and carbamate in the bicarbonate concentration ($cHCO_3^-$), $pK'(P)$ appears to vary with pH, decreasing slightly with increasing pH. For most clinical purposes this variation of pK' with pH change can be ignored.

(2) α, the solubility coefficient for CO_2 gas (including that present in its hydrated form, H_2CO_3), varies with temperature and composition of the solution. For pure water at 37°C the solubility coefficient is $\alpha = 0.0329$ mmol/liter × mm Hg ($= 0.0329$ mmol × liter^{-1} × mm Hg^{-1}). The presence of salts or proteins in the solution decreases the solubility coefficient, while lipids increase it. The mean value for normal plasma at 37°C is 0.0306 mmol × liter^{-1} × mm Hg^{-1} with a biological standard deviation of about ± 0.0003 mmol × liter^{-1} × mm Hg^{-1}. In lipemic plasma the value of α may be 0.033 or even higher. The temperature variation, expressed as $\Delta \log \alpha CO_2(P)/\Delta T$, is approximately $-0.0092 \times K^{-1}$.

Inserting the "constants" for normal plasma at 37°C, the Henderson-Hasselbalch equation takes the following form:

$$pH = 6.103 + \log \frac{\{cHCO_3^-\}}{0.0306 \times \{pCO_2\}}$$

or

$$pH = 6.103 + \log \frac{\{ctCO_2\} - 0.0306 \times \{pCO_2\}}{0.0306 \times \{pCO_2\}}$$

where pCO_2 is measured in mm Hg, and $cHCO_3^-$ and $ctCO_2$ are measured in mmol/l.

Taking the antilogarithm, and combining constants, the equation becomes:

$$aH^+ = 24.1 \times \frac{\{pCO_2\}}{\{cHCO_3^-\}} \times 10^{-9}$$

or, taking the activity coefficient of hydrogen ions in plasma to be 1.00 (the true value is probably about 0.80), the equation can be written:

$$\{cH^+\} = 24.1 \frac{\{pCO_2\}}{\{cHCO_3^-\}}$$

where cH^+ is now given in nmol/l, pCO_2 in mm Hg, and $cHCO_3^-$ in mmol/l. If normal values are substituted in the equation,

$$cH^+ = 24.1 \times \frac{40}{25.4}\ \text{nmol/l} = 38.0\ \text{nmol/l}$$

APPLICATION OF THE HENDERSON-HASSELBALCH EQUATION

The Henderson-Hasselbalch equation was originally used for the calculation of the pH of the plasma from values for the alveolar pCO_2 and the concentration of total CO_2 in plasma, both determined gasometrically. Later, when pH measurements could be made more easily by means of the glass electrode, the equation was used to calculate the pCO_2 of the plasma from the measured pH value and the gasometrically determined total CO_2 concentration. The equation can also be used to calculate the bicarbonate or total CO_2 concentration of plasma from measured values for pH and pCO_2.

In other words, with the aid of the H-H equation one can calculate any of the following parameters if two of them are known: pH, pCO_2 (or $cdCO_2$), $ctCO_2$, and $cHCO_3^-$.

Example 1: A freshly drawn blood sample has an experimentally determined plasma pH of 7.44 and a $ctCO_2$ of 26.7 mmol/l at 37°C. Calculate pCO_2, $cHCO_3^-$, and $cdCO_2$.

Answer:

$$\text{pH} = \text{p}K' + \log \frac{ctCO_2 - (0.0306 \times pCO_2)}{(0.0306 \times pCO_2)}$$

Therefore:

$$7.44 = 6.10 + \log \frac{26.7 - (0.0306 \times pCO_2)}{(0.0306 \times pCO_2)}$$

or

$$7.44 - 6.10 = \log \frac{26.7 - (0.0306 \times pCO_2)}{(0.0306 \times pCO_2)}$$

$$\text{antilog of } 1.34 = \frac{26.7 - (0.0306 \times pCO_2)}{(0.0306 \times pCO_2)}$$

$$21.88 = \frac{26.7 - (0.0306 \times pCO_2)}{(0.0306 \times pCO_2)}$$

$$21.88(0.0306 \times pCO_2) = 26.7 - (0.0306 \times pCO_2)$$

$$0.6695 \times pCO_2 = 26.7 - (0.0306 \times pCO_2)$$

$$(0.6695 \times pCO_2) + (0.0306 \times pCO_2) = 26.7$$

$$0.7001 \times pCO_2 = 26.7$$

$$\boldsymbol{pCO_2} = \frac{26.7}{0.7001} = \mathbf{38.1\ mm\ Hg}$$

The $cdCO_2$ and the pCO_2 at 37°C are related as follows:

$$cdCO_2 = 0.0306 \times pCO_2$$

Therefore:

$$cdCO_2 = 0.0306 \times 38.1 \text{ mg Hg}$$

$$\mathbf{cdCO_2 = 1.17\ mmol/l}$$

The concentration of HCO_3^- is the difference between $ctCO_2$ and $cdCO_2$:

$$cHCO_3^- = ctCO_2 - cdCO_2$$

Therefore:

$$\boldsymbol{cHCO_3^-} = 26.7 - 1.17 = \mathbf{25.5\ mmol/l}$$

Example 2: Experimentally determined data for a blood plasma specimen at 37°C are $ctCO_2$ = 25.2 mmol/l and pCO_2 = 37.5 mm Hg. Calculate the values for the pH, $cdCO_2$, and $cHCO_3^-$ for this plasma.

Answer:

$$cdCO_2 = 0.0306 \times pCO_2$$

$$cdCO_2 = 0.0306 \times 37.5$$

$$\mathbf{cdCO_2 = 1.15\ mmol/l}$$

$$cHCO_3^- = ctCO_2 - cdCO_2$$

$$cHCO_3^- = 25.2 - 1.15$$

$$\mathbf{cHCO_3^- = 24.05\ mmol/l}$$

$$\text{pH} = \text{p}K' + \log \frac{cHCO_3^-}{cdCO_2}$$

$$\text{pH} = 6.10 + \log \frac{24.05}{1.15}$$

$$\text{pH} = 6.10 + \log 20.91$$

$$\text{pH} = 6.10 + 1.320$$

$$\mathbf{pH = 7.42}$$

Various nomograms and slide rules have been constructed to facilitate the calculations. An example of an alignment nomogram is shown in Fig. 15–6. Various desk-top calculators can also be programmed to solve the Henderson-Hasselbalch equation for any desired parameter.

SOURCES OF ERROR

The overall uncertainty in calculated pCO_2 values is about ±5 per cent, but the accuracy is dependent on the accuracy of the pH measurement as well as that of the measurement of $ctCO_2$. Furthermore, the use of normal mean values for pK' and α may introduce errors in patients with abnormal plasma composition, especially with increased ionic strength or increased lipid concentration. Assuming an ionic strength increase of 20 per cent causing a fall in pK' from 6.10 to 6.08, and assuming that the specimen is lipemic (giving a rise in $\{\alpha\}$ from 0.0306 to 0.033) and that the pH = 7.40 and $ctCO_2$ = 25.5 mmol/l, the pCO_2 would then become 35 mm Hg. But, using the normal mean values for pK' and α, the calculated pCO_2 is 40 mm Hg, i.e., a result with a bias of +14 per cent. Assuming the ionic strength falls by 20 per cent, causing a rise in pK' from 6.10 to 6.12, and assuming that $\{\alpha\}$ is slightly decreased to 0.0302 by a high plasma protein concentration, then pCO_2 would be 42 mm Hg. Using the standard values for pK' and α, the calculated value is 40 mm Hg, i.e., a result with a bias of −5 per cent.

In view of these sources of error, the method of choice for the most accurate determination of the blood pCO_2 is direct measurement by means of the pCO_2 electrode.

THE USE OF THE SIGGAARD-ANDERSEN ALIGNMENT NOMOGRAM FOR THE CALCULATION OF ACID-BASE PARAMETERS

The alignment nomogram (see Fig. 15–6) represents the Henderson-Hasselbalch equation in the manner originally proposed by Van Slyke and Sendroy (1928) and includes a representation of the CO_2 equilibration curves for plasma and whole blood at various

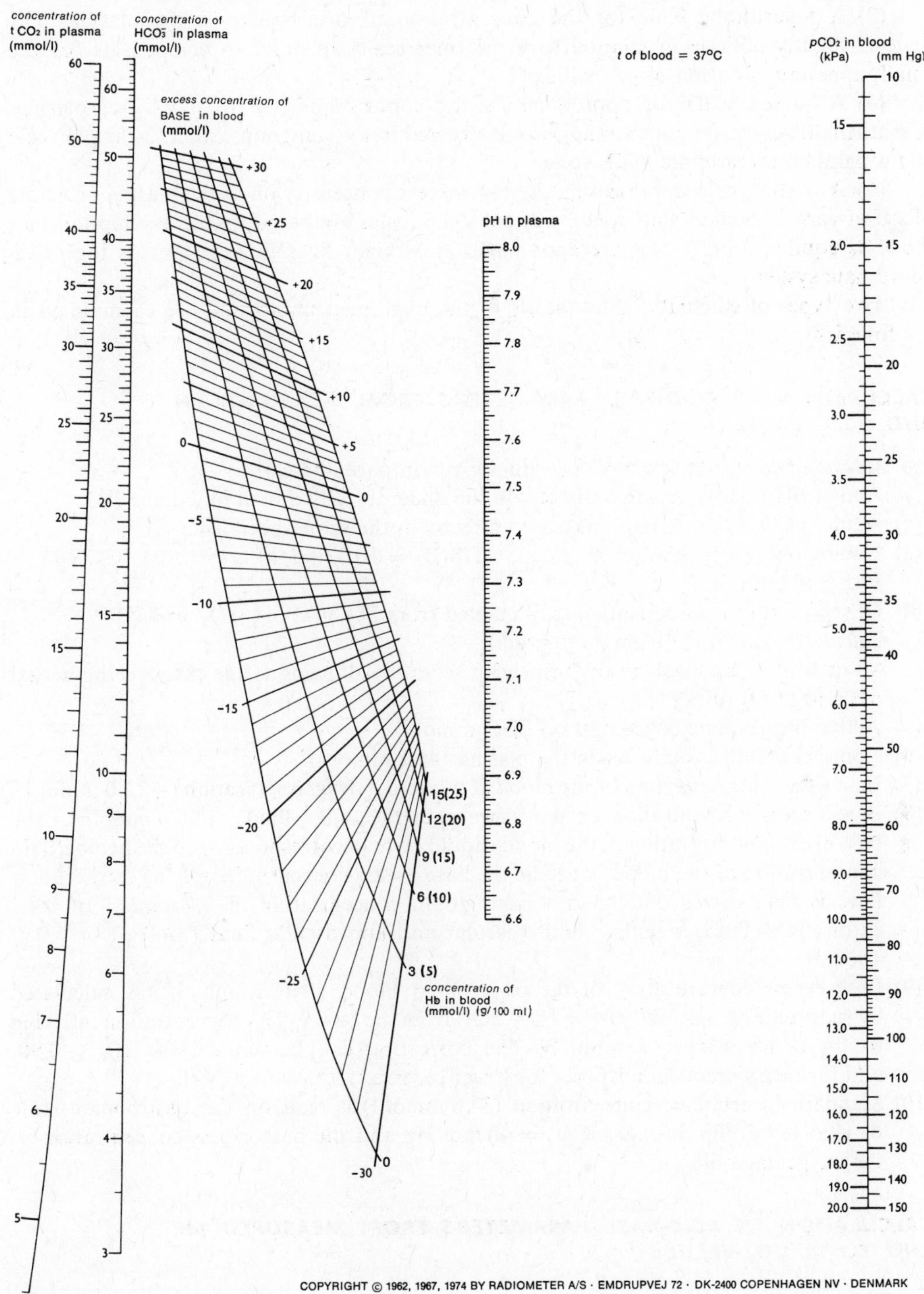

Figure 15–6. The Siggaard-Andersen alignment nomogram for the calculation of acid–base parameters.

hemoglobin concentrations in a manner originally proposed by Singer and Hastings (1948). The nomogram refers to human blood or plasma at 37°C and includes the following scales:

(1) A linear scale for pH of the plasma.

(2) A logarithmic scale for pCO_2 of the blood or plasma (unit: mm Hg), parallel to the pH scale.

(3) A logarithmic scale for the concentration of bicarbonate in the plasma. The scale is slightly oblique in relation to the former scales in order to compensate for the small, apparent variation of pK' with pH.

(4) A curved scale for approximating the concentration of total CO_2 of plasma. A mathematically exact scale cannot be constructed in a nomogram with a linear pH scale and a parallel, logarithmic pCO_2 scale.

(5) A grid of scales representing the base excess concentration of the plasma or whole blood at various hemoglobin concentrations. The scales are based on the assumption that the CO_2 equilibration curves are approximately straight lines in a pH versus log pCO_2 coordinate system.

Two types of calculations by means of the alignment nomogram are exemplified in the following.

CALCULATION OF ACID-BASE PARAMETERS FROM MEASURED pH AND pCO_2 VALUES

Measured quantities (example of values given in parentheses):

(1) Actual pH (7.34) measured directly in the anaerobically drawn blood at 37°C.

(2) Actual pCO_2 (82 mm Hg), measured directly in the same specimen.

(3) Hemoglobin concentration (e.g., cHb(B) = 10.9 mmol/l ≈ ρHb(B) = 175 g/l = 17.5 g/100 ml).

(4) Oxygen saturation, measured or calculated from pO_2 and pH ($sO_2 = 0.65$).

Calculated quantities (from nomogram):

All quantities are read on the respective scales by drawing a line through the actual pH and pCO_2 values.

(5) Actual bicarbonate concentration (43.5 mmol/l).

(6) Concentration of total CO_2 in the plasma (46.1 mmol/l).

(7) Base excess concentration of the blood at the actual oxygen saturation (+12.0 mmol/l).

(8) Base excess concentration of the average extracellular fluid (+16.0 mmol/l). The base excess concentration of the blood should be read corresponding to the hemoglobin concentration of the blood, whereas the base excess concentration of the extracellular fluid is read corresponding to a hemoglobin concentration of 3.7 mmol/l of total extracellular fluid, including both vascular and nonvascular fluid (≈60 g/l or 6.0 g/100 ml).

(9) Base excess concentration of the oxygenated blood (+10.86 mmol/l) is calculated by subtracting $0.3 \times c\text{Hb(B)} \times (1 - sO_2)$ from the base excess concentration referring to the actual oxygen saturation. The correction can be omitted for $sO_2 > 0.90$, and for many practical purposes for lower oxygen saturations as well.

(10) Standard bicarbonate concentration (34.6 mmol/l) is read on the bicarbonate scale by drawing a line through $pCO_2 = 40$ mm Hg and the base excess concentration of the oxygenated blood.

CALCULATION OF ACID-BASE PARAMETERS FROM MEASURED pH AND TOTAL CO_2 VALUES

This method is completely analogous to the preceding method except that the total CO_2 concentration of the plasma replaces the actual pCO_2 as a measured quantity [item

(2)] while the actual pCO_2 replaces the total CO_2 concentration as a calculated quantity [item (6)]. All other items listed remain unaltered.

NOTES

(1) The concentration of hemoglobin necessary for these calculations should be representative of the Hb concentration of the patient at the time the acid base status is evaluated. It may be determined directly, or it can be derived approximately from the volume fraction of the erythrocytes in the blood (hematocrit divided by 100) as follows:

$$c\text{Hb(B)} \approx \varphi\text{E(B)} \times 20 \text{ mmol/l}$$

or

$$\rho\text{Hb(B)} \approx \varphi\text{E(B)} \times 35 \text{ g/100 ml}$$

(2) The calculations with the nomogram should be performed with values referring to 37°C. The pH and pCO_2 values for blood at other temperatures can be calculated by means of the coefficients [$\Delta\text{pH(P)}/\Delta T = -0.015 \text{ K}^{-1}$ and $\Delta \log pCO_2/\Delta T = 0.021 \text{ K}^{-1}$] or from a nomogram (see Fig. 15–3).

(3) The base excess concentration of the average extracellular fluid is constant during acute respiratory changes (acidosis or alkalosis), whereas the base excess concentration of the blood (or the standard bicarbonate concentration) changes slightly owing to a redistribution of bicarbonate between blood and the weakly buffered, extravascular, extracellular fluid. An acute respiratory acidosis with an increase in pCO_2 from 40 to 80 mm Hg causes a fall in the base excess concentration of the blood of about 3 mmol/l, whereas the base excess concentration of the average extracellular fluid remains constant.

SOURCES OF ERROR

The *precision* (reproducibility) of the calculated values is entirely dependent on the precision of the measured values. If the precision of the two measured values is given by an analytical coefficient of variation of ±2 per cent, the coefficient of variation of the calculated quantity will be about ±2.8 per cent ($= \pm\sqrt{2^2 + 2^2}$), or even slightly greater if we add to this the imprecision in lining up the two points in the nomogram.

The *accuracy* (agreement with the true value) of the calculated values depends not only on the accuracy of the measured values but furthermore on the accuracy of the nomogram. The scale for $ctCO_2(P)$ is only approximate, and is most accurate for samples with pH ≈ 7.4. For specimens with pH ≈ 6.8, the greatest bias (about +6 per cent) in the calculated value for pCO_2 is obtained when the pCO_2 is high. The same limitations apply to the base excess grid, which is most accurate for pCO_2 values between 30 and 60 mm Hg. Variations in pK' and α affect the accuracy as described in the discussion of the Henderson-Hasselbalch equation on page 896.

REFERENCE

Siggaard-Andersen, O.: Blood acid-base alignment nomogram. Scales for pH, pCO_2, base excess of whole blood of different hemoglobin concentrations, plasma bicarbonate, and plasma total-CO_2. Scand. J. Clin. Lab. Invest. *15*: 211–217, 1963.

Lithium

Lithium ions in the form of Li_2CO_3 are widely used in the control of the manic stage of manic-depressive psychosis. Therapeutic levels during the initial treatment are 1.0 to 1.5 mmol/l serum until the threshold of effectiveness is reached after 6 to 10 days. The serum level during maintenance therapy is generally kept at 0.5 to 1.0 mmol/l. The *in vivo* half life of lithium is about 24 h, giving rise to urine levels of 2 to 12 mmol/l, which is about 5 to 25 times higher than those maintained in serum.

Li^+ penetrates cell membranes relatively easily and is therefore evenly distributed in both the extra- and intracellular fluid, where it replaces Na^+ and K^+. Increased concentrations of Li^+ are toxic. It is therefore mandatory to check the ability of the patient to clear Li^+ adequately by measuring urinary Li^+ excretion prior to initiation of treatment and to monitor serum Li^+ levels regularly. Toxic reactions are rarely seen with serum levels below 1.5 mmol/l but may be moderate to severe at levels of 2.0 to 2.5 mmol/l. Early signs of toxicity are drowsiness, slurred speech, and coarse tremors; this is followed by nausea, vomiting, diarrhea, sluggishness, thirst, polyuria, and eventually stupor and convulsions.[8] Li^+ is thought to interfere with cyclic AMP-mediated processes. It has also been reported to have teratogenic effects and should therefore not be given during pregnancy.

DETERMINATION OF LITHIUM BY ATOMIC ABSORPTION SPECTROPHOTOMETRY

SAMPLE

Determinations are most conveniently carried out on *serum* specimens. Hemolysis does not interfere since Li^+ levels in erythrocytes are similar to, although slightly lower than, those in plasma. The blood specimen is best drawn in the morning before the patient receives the first daily dose of the drug. Li^+ levels in serum are stable for at least 1 week at 4°C and longer when frozen. Determinations of Li^+ in *urine* are generally carried out on a 24 h specimen.

PRINCIPLE

The sample is diluted ten-fold (serum) or 100-fold (urine) with deionized water and the dilution is directly aspirated into the flame. The measurement is carried out using the Li-line at 670.8 nm. Na^+ and K^+ slightly enhance the lithium signal (about 2 per cent); to compensate for this effect, physiological concentrations of these ions are added to the lithium standards.

REAGENTS

1. Stock blank, 140 mmol sodium and 5 mmol potassium/l. Dissolve 32.7 g NaCl and 1.5 g KCl in water and dilute to 4 liters with deionized water. Store the reagent in a polyethylene bottle.
2. Stock lithium standard, 10.00 mmol Li^+/l. Dissolve 369.5 mg Li_2CO_3 (previously dried at 110°C and stored in a desiccator) in 20 ml water and 2 ml of 0.1 molar HCl. Dilute to 1000 ml with deionized water. Stable indefinitely. Store in polyethylene bottle.
3. Working standard, 1.00, 2.00, and 3.00 mmol Li^+/l. Add 10.0 ml of stock blank to separate 100 ml volumetric flasks and add 10.00, 20.00, and 30.00 ml of stock standard, respectively. Fill to mark with deionized water and mix. Store in polyethylene bottle.

INSTRUMENT SETTINGS

The instrument settings vary with the type of instrument used. The reader is referred to the operating manual supplied by the manufacturer.

PROCEDURE

1. Transfer 1.0 ml of the stock blank, of each standard solution, and of the patient's sample into separate test tubes (17 × 100 mm plastic-capped tubes are convenient).
2. Add 9.0 ml of deionized water and mix well.
3. Aspirate the solutions directly into the flame.
4. Prepare a standard curve and calculate the concentrations of the unknowns from this curve

COMMENTS

The procedure has a relative uncertainty of less than 3 per cent. Atomic absorption spectrophotometry is generally considered to be the method of choice. Reasonably comparable results, however, can also be obtained by flame emission photometry. In this case, K^+ is used as the internal standard. This method is also subject to interference by other ions such as Na^+; therefore, appropriate amounts of this ion must be added to the standard to compensate for this effect.

Urine specimens may be analyzed by this method in a similar way. In general it will be necessary to dilute urine 100-fold. Compensation for the urine matrix effect is necessary and is accomplished by adding a proportionate amount of stock blank to the greater urine dilution. Alternatively, the stock blank in the standard is replaced by urine that is free of lithium (e.g., 1 ml urine/100 ml standard solution).

REFERENCE

Pybus, J., and Bowers, G. N. Jr.: Clin. Chem., *16:*139, 1970.

Calcium and Phosphorus

The metabolism of calcium and that of phosphorus are so closely related that it is justifiable to discuss these elements in the same section. More than 99 per cent of the calcium in the body and 80 per cent of the phosphorus are present in the bones as calcium fluorophosphate apatite; the remainder of the calcium and phosphorus have varied and significant functions in the body. For example, calcium ions decrease neuromuscular excitability, participate in blood coagulation, and activate some enzymes, such as succinate dehydrogenase and adenosine triphosphatase. Furthermore, calcium ions and cyclic AMP (cAMP) may play a role in the transfer of inorganic ions across cell membranes and in the release of neurotransmitters. Phosphorus, on the other hand, is involved in the intermediary metabolism of carbohydrates (see Chapter 6) and is a component of other physiologically important substances, such as organic phosphate esters, phospholipids, nucleic acids, and nucleotides (e.g., ATP). It also plays a critical role in the mineralization of bone.

Both calcium and phosphorus are absorbed in the upper small intestine; calcium absorption is maximal in the duodenum, while that of phosphate is maximal in the jejunum. The absorption of both ions is favored at an acid pH and is greatly decreased at an alkaline pH, under which conditions both ions form insoluble compounds. Presence of vitamin D is essential for calcium absorption. Increased levels of the vitamin group promote calcium absorption, and decreased levels reduce it.

Essentially all of the calcium of blood is present in the serum; however, phosphorus is present mainly in the cells as organic phosphate, with only a small but significant amount occurring in serum as inorganic phosphate.

Calcium is present in serum in two distinct forms, the *nondiffusible protein-bound calcium*, which constitutes approximately 40 to 50 per cent of the total serum calcium, and the *diffusible free calcium fraction*. The latter can be further subdivided into *complexed calcium* (by citrate, phosphate, bicarbonate, and sulfate), and *ionized calcium*, which is the physiologically active form (see section on Ionized Calcium).

The degree to which calcium is bound to protein varies significantly with the type of protein. In the normal individual, about 81 per cent of the nondiffusible fraction is bound to albumin via the imidazole of histidine and the remainder to α-, β- and γ-globulins. However, the degree of binding may vary in certain disease states and with changes in pH. Any increase in pH decreases the ionized calcium by about 5 per cent for each 0.1 unit increase in pH, while a decrease in pH increases the ionized calcium.

FACTORS INFLUENCING SERUM CALCIUM LEVELS

Of the many factors that are known to influence serum calcium levels, the following are most important:

Parathyroid hormone (PTH) is a polypeptide with a molecular mass of 9000 and a half life of 20 to 30 min. It is synthesized and secreted by the chief cells of the parathyroid gland in response to decreased serum levels of ionized calcium. It is most conveniently determined by RIA methods. This hormone maintains normal serum calcium levels by (a) mobilization of calcium from bones; (b) increasing the synthesis of 1,25-dihydroxycholecalciferol (1,25-$(OH)_2D_3$), a vitamin D derivative, which also increases bone resorption and causes an increase in the intestinal absorption of calcium; (c) decreasing the clearance of calcium up to three-fold by increasing its reabsorption in the renal tubules. As normal serum calcium levels are restored, the negative feedback mechanism will shut off PTH secretion.

The effect of PTH on bone and kidney cells appears to be associated with stimulation of adenyl cyclase, leading to increased production of cAMP. These changes presumably mediate changes in calcium and phosphorus transport. The effect of 1,25-$(OH)_2D_3$ on the intestinal mucosa is due to an increase in the synthesis of both calcium binding protein and a Ca^{2+}-dependent ATPase, both of which are involved in the enzyme-dependent process of calcium absorption by the intestinal epithelium. The former is probably involved in ion-selection, while the latter is involved in supplying the energy requirements. PTH also inhibits reabsorption of phosphates by the renal tubules. Thus, high PTH levels are associated with increased urine phosphate levels.

Calcitonin is a 32-member peptide with a molecular mass of 3590. It is produced, and in large amounts stored, by the parafollicular cells of the thyroid gland (therefore, the former name thyrocalcitonin). It is also produced, but to a much lesser degree, by the parathyroid and thymus glands. The half life of the hormone is 2 to 15 min. The level in serum is best determined by RIA methods.

The main action of calcitonin is its inhibition of bone resorption by regulating the number and activity of osteoclasts. Thus, in this respect, calcitonin is an antagonist to PTH. The hormone is secreted in direct response to high serum calcium levels and thus may prevent large oscillations in serum calcium levels and excessive loss of body calcium. The hormone is also said to increase phosphate clearance by decreasing its absorption in the tubules.

Vitamin D is the term applied to a group of pro-vitamins and active vitamins that includes the biologically inactive calciferol (vitamin D_2) and cholecalciferol (vitamin D_3). Hydroxylation of C-25 converts both compounds into the active vitamins 25-hydroxycalciferol (25-OHD_2; 25-HC) and 25-hydroxycholecalciferol (25-OHD_3; 25-HCC), respectively. This process takes place in the liver and is controlled by a negative feedback mechanism involving 25-OHD_2 and 25-OHD_3. A second hydroxylation at C-1 converts 25-OHD_3 to 1,25-dihydroxycholecalciferol (1,25-$(OH)_2D_3$; 1,25-DHCC), which has 10 times the activity of 25-OHD_3 and which is thought to be the final active form of vitamin D_3. It is postulated that vitamin D_2 may undergo a similar hydroxylation.

1,25-$(OH)_2D_3$ has been shown to increase intestinal absorption of calcium and phosphates as well as resorption of bone (see PTH). The formation of 1,25-$(OH)_2D_3$ is catalyzed by a renal enzyme, and is inhibited by increase in Ca^{2+}. Because of its interaction with calcium and because of the mode of its action, 1,25-$(OH)_2D_3$ has been classified by some as a hormone.

Phenobarbital and phenytoin (diphenylhydantoin) can induce hypocalcemia by stimulating the microsomal oxidase system in the liver, which oxidizes vitamin D to inactive metabolites, thus leading to decreased intestinal absorption of calcium.

Plasma Proteins. Since approximately 50 per cent of plasma calcium is bound to protein, any decrease in serum proteins frequently results in a decrease in the total serum calcium level. However, this decrease affects mainly the nondiffusible fraction and, therefore, tetany is rarely associated with hypoproteinemia. Similarly, an increase in protein, as occurs in multiple myeloma, may increase the total serum calcium without any significant change in the diffusible calcium fraction. Simultaneous determinations of total and ionized calcium will readily allow differentiation of abnormal serum calcium levels due to primary causes from those due to secondary causes (see Ionized Calcium).

Serum (Plasma) Phosphates. There appears to be a reciprocal relationship between calcium and phosphorus. Any increase in serum inorganic phosphorus is associated with a decrease in serum calcium. The best example of such a situation is the increased serum phosphorus level in renal retention of phosphorus (e.g., in uremia), which is associated with decreased serum calcium levels. It has been established that, in chronic renal disease, the synthesis of 1,25-$(OH)_2D_3$ (and thus intestinal reabsorption of Ca^{2+}) is decreased.

MECHANISMS FOR MONITORING CALCIUM HOMEOSTASIS

The mechanisms involved in maintaining normal calcium levels in serum are quite complex, but the most important and typical events are as follows: If serum calcium levels decrease below normal levels, secretion of parathyroid hormone (a) stimulates bone resorption, (b) decreases calcium clearance by the kidney, and (c) stimulates synthesis of 1,25-$(OH)_2D_3$ by the kidney. Increased levels of 1,25-$(OH)_2D_3$ also cause an increase in bone resorption and, in addition, increased absorption of calcium from the intestines as explained in the section on Vitamin D. As calcium levels increase above normal, the release of PTH and the synthesis of 1,25-$(OH)_2D_3$ are halted by the negative feedback mechanism. In addition, calcitonin is released, which inhibits bone resorption.

CLINICAL SIGNIFICANCE OF CALCIUM AND PHOSPHORUS MEASUREMENTS

Hypercalcemia (increased serum calcium levels) is observed in primary hyperparathyroidism along with decreased serum phosphorus levels, increased urine calcium and phosphorus excretion, as well as increased serum levels of PTH and ionized calcium levels. However, if calcium stores are exhausted, serum calcium levels in this condition may be normal or even low. Hypervitaminosis D, multiple myeloma, and some neoplastic diseases of bone may also be accompanied by increased serum calcium levels, but, unlike the situation in hyperparathyroidism, serum phosphorus levels are normal or even elevated, PTH levels are normal or low, and ionized calcium levels are usually normal.

Hypocalcemia may be observed in hypoparathyroidism together with normal or increased serum phosphorus levels and decreased urinary calcium and phosphorus excretion. The low serum ionized calcium levels in this disease may lead to an increase in neuromuscular irritability and thus to tetany. Serum calcium levels may be low in steatorrhea (due to decreased absorption), in nephrosis (due to loss of protein), in nephritis (due to decreased absorption), and in pancreatitis (due to formation of calcium soaps), but in these conditions ionized calcium levels are normal and PTH levels may be elevated. Hypocalcemia due to hypoproteinemia is generally not associated with clinical signs of tetany, since the decrease in calcium is mainly in the protein-bound fraction. The clinical significance of ionized calcium is discussed in a separate section (p. 911).

Hyperphosphatemia (increased serum phosphorus levels) may be found in hypervitaminosis D, hypoparathyroidism, and renal failure. *Hypophosphatemia* (low serum phosphorus levels) may be seen in rickets (vitamin D deficiency), in hyperparathyroidism, and in the Fanconi syndrome, a disease associated with a defect in reabsorption of phosphorus and other metabolites from the glomerular filtrate.

METHODS FOR THE DETERMINATION OF CALCIUM IN BODY FLUIDS

SPECIMEN

Calcium determinations are generally performed on *serum* or heparinized plasma, collected in the fasting state. Blood collected with oxalate or from patients receiving EDTA treatment is unsuitable for analysis; the former will remove calcium from serum by precipitation as oxalate,[20] and the latter will chelate calcium and thus make it unavailable for analysis by many procedures.

Blood should be drawn without venous stasis, as this leads to hemoconcentration, and therefore to an increase in calcium levels of 0.1 to 0.15 mmol $Ca^{2+}_{1/2}$/l* above the normal value. This effect is mainly on the protein-bound calcium. The level of ionized calcium will change under these conditions only if the blood pH changes significantly.

Calcium values change with the posture of the patient, with values being about 0.1 to 0.15 mmol $Ca^{2+}_{1/2}$/l lower in patients in the recumbent position. For the analysis of calcium in *stool or urine*, a timed collection, such as a 24 h specimen, is recommended.

DETERMINATION OF SERUM CALCIUM BY OXALATE PRECIPITATION AND REDOX TITRATION

PRINCIPLE

A solution of ammonium oxalate is added to a diluted serum sample, and the calcium is precipitated as calcium oxalate (equation 2). The precipitate is washed with dilute ammonium hydroxide to remove the excess of ammonium oxalate and is then dissolved in sulfuric acid (equation 3). The oxalic acid thus formed from the calcium oxalate is titrated with a standardized solution of potassium permanganate (equation 4). The manganese atom is reduced by oxalic acid from the +VII to the +II valent state. The titration end point is indicated by the first appearance of a purple color (excess potassium permanganate), which occurs when all oxalic acid has reacted with the $KMnO_4$:

$$Ca^{2+} + C_2O_4^{2-} \longrightarrow CaC_2O_4\downarrow \quad (2)$$

$$CaC_2O_4 + H_2SO_4 \longrightarrow H_2C_2O_4 + CaSO_4 \quad (3)$$

$$2KMn^{(VII)}O_4 + 5H_2C_2O_4 + 3H_2SO_4 \xrightarrow{70°C} K_2SO_4 + 2Mn^{(II)}SO_4 + 10CO_2 + 8H_2O \quad (4)$$

This method gives highly reliable results and therefore has served for many years as a reference method for serum calcium determinations. It has now been largely replaced by the more convenient and potentially more accurate methods employing atomic absorption spectrophotometry.

REFERENCE

Kramer, B., and Tisdall, F. F.: J. Biol. Chem. *47*:475, 1921; modified by Clark, E. P., and Collip, J. B.: J. Biol. Chem., *63*:461, 1925.

THE DETERMINATION OF SERUM CALCIUM BY PRECIPITATION WITH CHLORANILIC ACID

PRINCIPLE

Calcium is precipitated from the specimen as calcium chloranilate by adding a saturated solution of sodium chloranilate. The precipitate is washed with isopropyl alcohol

* In the United States it is presently customary to report calcium values either in mg/100 ml or mEq/l. In the SI system, expression of concentrations in mmol/l is recommended. This unit does not take into account ion charge, which allows one readily to recognize or calculate the electrical balance of a solution. Thus, the concentration of *calcium ion charge* expressed in mmol $Ca^{2+}_{1/2}$/l has been introduced. The following expressions are, therefore, equivalent: 4.84 mg Ca/100 ml = 2.42 mEq Ca^{2+}/l = 2.42 mmol calcium ion charge ($Ca^{2+}_{1/2}$)/liter = 1.21 mmol Ca^{2+}/l. (See also p. 946.)

to remove the excess of chloranilic acid and is then treated with EDTA, which chelates with calcium and releases chloranilic acid. The latter is colored and can be measured photometrically.

$$Ca^{2+} + \text{chloranilate} \rightarrow \text{Ca-chloranilate}\downarrow$$

$$\text{Ca-chloranilate} + \text{EDTA} \rightarrow \text{Ca-EDTA} + \text{chloranilic acid (purple color)}$$

REAGENTS

1. Sodium chloranilate. Place 6.13 g of sodium chloranilate in 500 ml of deionized water. Shake well to saturate the solution and filter through filter paper.
2. Isopropyl alcohol, 50 per cent (v/v). Dilute 250 ml isopropyl alcohol, AR, to 500 ml with deionized water.
3. EDTA. Dissolve 25 g tetrasodium ethylenediaminetetraacetate in deionized water and dilute to 500 ml.
4. Calcium standard, 5.00 mmol $Ca^{2+}_{1/2}$/l. Place 0.2479 g of dried calcium carbonate, AR, into a 1000 ml volumetric flask. Add approximately 9 ml deionized water and 1 ml concentrated HCl. Shake until dissolved. Fill to the mark with deionized water. Store in a Pyrex bottle. The solution is stable.

PROCEDURE

1. Pipet 2.0 ml of serum, plasma (heparinized), or aqueous standard into a 15 ml conical centrifuge tube (preferably in duplicate).
2. Add forcefully to all tubes 1 ml of saturated sodium chloranilate. Mix tubes well; do not invert! (It is important to blow in the chloranilate to achieve immediate mixing; otherwise, proteins may be precipitated.)
3. Place tubes into a 37°C water bath for at least 3 h.
4. Centrifuge for 10 min at approximately 2000 rpm, decant the supernatant immediately, and drain tubes for approximately 2 min. Wipe the mouth of the tube with cotton gauze or filter paper to remove any excess of chloranilate.
5. Add 1 drop of isopropyl alcohol to each tube and break up the precipitate by tapping the tube against your fingers or, preferably, with the aid of a Vortex mixer. Wash precipitate with 6 to 7 ml of 50 per cent isopropyl alcohol. The alcohol is conveniently dispensed from a squirt bottle by directing the stream directly into the tip of the tube.
6. Centrifuge and drain tubes on cotton gauze or filter paper for approximately 2 min; wipe the mouth of the tube dry.
7. Add 1 drop of EDTA to the precipitate and break up the precipitate as outlined in step 5.
8. Add to each tube exactly 6 ml of EDTA.
9. Allow all tubes to stand for approximately 10 min, and then read at 520 nm against an EDTA blank. Although the color should appear immediately, in the opinion of this author it is advantageous to allow the tubes to stand for the specified time to insure complete solution of the precipitate. The color is extremely stable.

CALCULATION

$$\text{mmol } Ca^{2+}_{1/2}/\text{l} = \frac{A_U}{A_S} \times \frac{1000}{2} \times \frac{0.01}{1}$$

$$= \frac{A_U}{A_S} \times 5$$

where

A_U = absorbance of unknown

A_S = absorbance of standard

1000 = ml/l

2 = ml of specimen used

0.01 = mmol $Ca^{2+}_{1/2}$ in 2.0 ml standard

If values are expressed in mmol Ca^{2+}/l or mg/100 ml, multiply by 2.5 or 10, respectively, instead of multiplying by 5

COMMENTS ON THE METHOD

Some samples tend to give a cloudy solution in the final step of the procedure (after the addition of EDTA). In some cases this is due to faulty washing of the precipitate with isopropyl alcohol; in other cases, it is thought to be due to lipids. In the latter instance, extraction of the final solution with ether may be helpful.

This method has also been largely replaced by the atomic absorption procedure. It is retained in this text since it is a convenient and reasonably reliable method for those laboratories not equipped with an atomic absorption spectrophotometer.

REFERENCE

Ferro, P. V., and Ham, A. B.: Am. J. Clin. Path., *28*:208, 1957; *28*:689, 1957.

DETERMINATION OF SERUM CALCIUM BY EDTA TITRATION (MICROMETHOD)

PRINCIPLE

A diluted serum sample is titrated with EDTA in the presence of calcein indicator and at an alkaline pH (to avoid magnesium interference). The initial yellow-green fluorescence caused by the calcium-calcein complex changes to a nonfluorescent salmon-pink color (of free calcein) when all calcium present has been chelated by EDTA (Fig. 15-7).

REAGENTS

1. Calcium standard. See the preceding method, based on precipitation with chloranilic acid.

2. Potassium hydroxide, 1.25 mol/l. Dissolve 8.3 g KOH, AR (assay 85%), in deionized water, add 0.050 g of KCN (poison!), and dilute to 100 ml with deionized water. Check concentration against standard acid.

3. Calcein indicator. Dissolve 0.025 g of calcein in 100 ml of 0.25 molar NaOH.

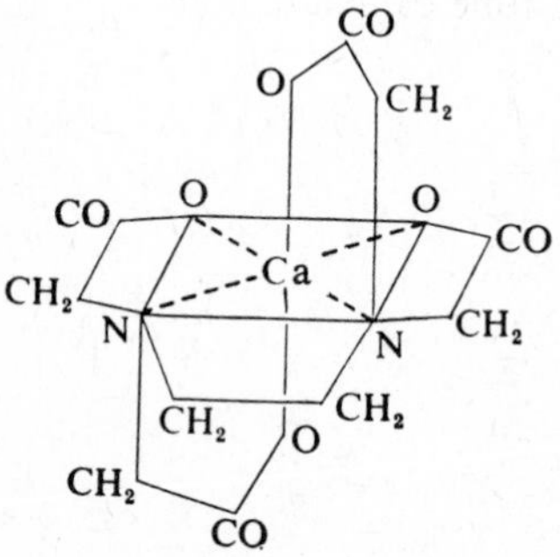

Figure 15-7 Chelate complex of EDTA with calcium.

Store in polyethylene bottles and keep refrigerated. Replace the indicator solution when it turns greenish.

4. EDTA, 0.010 mol/l. Dissolve 0.372 g of disodium dihydrogen ethylenediaminetetraacetic acid (dihydrate) in deionized water and dilute to 100 ml. Store in a small polyethylene bottle.

PROCEDURE

1. Add to microtitration cups (e.g., Beckman No. 314463) 40 μl of water, standard, and sample, respectively. Add to all cups 200 μl of KOH and 20 μl of indicator. The indicated amounts may be added conveniently with a Beckman micropipet. If such pipets are used, standards and unknowns should be pipetted with the same pipet.

2. Titrate blank, standards, and unknowns with 0.01 molar EDTA solution, using one of the commercially available microtitrators. Disappearance of the yellow-green fluorescence and the appearance of an orange-red color mark the end point. It has been found more convenient to carry out the titration in the dark (dark box with a viewing window or an Oxford Titrator) and with the aid of an ultraviolet light source. It is much easier to detect the disappearance of the fluorescence in the dark than to detect the appearance of the salmon-pink color.

CALCULATION

$$\text{mmol } Ca^{2+}_{1/2}/l = \frac{\text{reading of unknown} - \text{reading of blank}}{\text{reading of standard} - \text{reading of blank}} \times 5$$

For expressing values in mEq/l, mmol Ca^{2+}/l, or mg/100 ml, see pp. 904 and 906.

COMMENTS ON THE PROCEDURE

The sensitivity and speed of complexometric titration procedures are the main reasons for their use in the clinical laboratory. Unfortunately, all of these methods seem to have the great disadvantage that the titration end point is hard to detect, particularly in the presence of hemoglobin, jaundice, lipemia, and high phosphate concentration. In complexometric titrations of this type the slowly added complexing agent (EDTA) removes calcium little by little from the calcium-indicator-complex, resulting in a gradual color change from the yellow-green fluorescence to the nonfluorescent salmon-pink until all of the indicator is in the latter form.

A number of other indicators have been used, e.g., murexide (ammonium purpurate), Cal-Red, and Eriochrome Black T. In some of these determinations (Eriochrome Black T) magnesium ions interfere; in others (Cal-Red and calcein) magnesium interference can be eliminated by titration at a strongly alkaline pH, at which Mg is precipitated as $Mg(OH)_2$.

The accuracy of these methods has been improved by use of special instruments for photometric or fluorometric endpoint detection. Even with the use of these devices, the technique gives results which are inferior to those obtained with the atomic absorption methods or the chloranilate precipitation method.

REFERENCES

Appleton, H. D., West, M., Mandel, M., and Sala, A.: Clin. Chem., *5*:36, 1959.
Diehl, H., and Ellingboe, J. L.: Anal. Chem., *28*:882, 1956.
Technical Bulletin U M-TB-007 E, Beckman Instruments, Inc., Fullerton, CA.

DETERMINATION OF CALCIUM BY EMISSION FLAME PHOTOMETRY

Many attempts have been made to determine calcium in diluted serum samples by flame photometry. Although calcium has an arc line at 422.7 nm and two molecular bands at 554 and 662 nm, flame photometric procedures have not been very successful. Among the difficulties are: (1) the positive interference by sodium and potassium, (2) the inhibition of calcium emission by phosphates and sulfates, and (3) the fact that calcium cannot be excited easily even in a "hot" flame. Attempts have been made to eliminate interference by phosphorus, by adding an excess of PO_4 ions to standard and samples, which minimizes the effect of phosphorus in the sample. This has solved one problem but created another, namely, a further decrease in sensitivity due to the addition of phosphates.

Isolation of calcium by precipitation with oxalate before analysis has resulted in the elimination of interfering substances; however, here again, oxalate depresses the emission. This is possibly due to the formation of degradation products with low excitation potential.

Normal values for flame photometric methods are generally slightly higher than those obtained with chemical methods.

REFERENCE

Margoshes, M., and Vallee, B. L.: Flame photometry and spectrometry, principles and applications. *In* Methods of Biochemical Analysis. New York, Interscience Publishers, Inc., 1956, vol. 3.

DETERMINATION OF CALCIUM USING o-CRESOLPHTHALEIN COMPLEXONE

In a method adapted for use with the AutoAnalyzer, serum is mixed with 0.3 molar HCl to dissociate calcium from proteins. The calcium in the sample stream is then dialyzed into a reagent stream containing cresolphthalein complexone and 8-hydroxyquinoline in dilute HCl. A colored complex between calcium and cresolphthalein complexone is formed at a pH between 10 and 12, which is maintained by addition of diethylamine buffer. The absorbance is measured at 570 nm. 8-Hydroxyquinoline contained in the reagent binds magnesium and prevents interference from this ion.

Moorehead and Biggs[36] have recently improved this method by use of 2-amino-2-methyl-1-propanol at pH 10.0 to 10.5. The authors claim that this method has increased sensitivity, baseline stability, freedom from magnesium interference, and no blanking problems.

The method is said to give results which agree reasonably well with those obtained by atomic absorption spectrophotometric methods, although accuracy and precision are somewhat less. Strict adherence to experimental parameters is necessary if gross errors are to be avoided.

REFERENCES

Gitelman, H. J.: Anal. Biochem., *18*:521, 1967.
Kessler, G., and Wolfman, M.: Clin. Chem., *10*:686, 1964.

DETERMINATION OF CALCIUM BY ATOMIC ABSORPTION SPECTROPHOTOMETRY

PRINCIPLE

Calcium compounds, when introduced into a flame, dissociate to give free calcium atoms. Calcium in this form absorbs light of characteristic wavelengths (e.g., 422.7 nm) produced by a hollow cathode lamp with a calcium filament (see also Chapter 3). Under the described conditions only a small fraction of calcium atoms (about 1 out of 1000) will be raised to a higher energy level and will emit light on returning to the ground state (see flame photometry).

Some anions, such as phosphates, bind with calcium to form highly refractory compounds which do not dissociate in the flame, thus causing falsely low results. This interference is eliminated by the addition of La^{3+} or Sr^{2+}, which bind preferentially with phosphate and prevent the formation of calcium phosphates. Lanthanum is preferred, since it forms a tighter complex with phosphorus. Under the conditions of the following procedure, a 0.7 g/100 ml solution of La^{3+} is capable of preventing phosphate interference up to concentrations of 1.6 g phosphate-phosphorus/1000 ml. Proteins are precipitated by trichloroacetic acid to eliminate their interference (proteins reduce the apparent concentration of calcium) and to increase the reproducibility and accuracy.

A method based on atomic absorption spectrophotometry has recently been recommended as a reference method for determining total calcium in serum.[7] Pickup *et al.*[41] have subsequently pointed out some deficiencies in this method and have made additional recommendations.

Principles of atomic absorption spectrophotometry and major interferences have been discussed in Chapter 3.

REAGENTS

1. Stock standard, 100.0 mmol $Ca^{2+}_{1/2}$/l. Transfer quantitatively 2.497 g of dried $CaCO_3$ into a 500 ml volumetric flask. Dissolve in approximately 10 ml of deionized water and 5 ml of concentrated HCl. After all the $CaCO_3$ has dissolved, dilute with deionized water to the 500 ml mark.

2. Calcium working standards. Into six 100 ml volumetric flasks place, respectively, 2, 3, 4, 5, 6, and 7 ml of calcium stock standard and dilute with deionized water to the 100 ml mark. The respective concentrations are 2, 3, 4, 5, 6, and 7 mmol $Ca^{2+}_{1/2}$/l.

3. Diluent (0.7 g La^{3+} in 100 ml of 4 per cent TCA). Place 17.8 g $LaCl_3 \cdot 6H_2O$ and 40 g of trichloroacetic acid into a 1000 ml volumetric flask; dissolve in and dilute to the mark with deionized water. This reagent is used half-strength in regard to La^{3+}, if analysis is performed on an instrument which allows 20-fold or greater dilutions of the sample. (See instrument and instrumental parameters.) La_2O_3 may be substituted for $LaCl_3$ in the diluent as follows: Transfer 54.3 g of La_2O_3 into a 4 liter volumetric flask. Add approximately 50 ml of H_2O, followed slowly by 95 ml of concentrated HCl. Wash down the sides of the container with deionized water and mix well to dissolve the La_2O_3. After the La_2O_3 has dissolved, dilute the solution to approximately 2 liters with deionized water and then add 280 g of TCA. Mix well and dilute to volume with H_2O. Filter the reagent into a 7 liter plastic container and add an additional 3 liters of deionized water. Mix well. Note: Each new lot of La_2O_3 must be checked for calcium contamination.

INSTRUMENT

The following procedure has been developed for the Perkin-Elmer Atomic Absorption Spectrophotometer Model 303, which was equipped with an Intensitron hollow cathode lamp and a Perkin-Elmer burner with a Boling head. The instrument was connected to a Perkin-Elmer recorder readout and a Texas Instrument recorder (Servo/Riter II). The procedure may be adapted to other instruments which have sufficient sensitivity and stability, e.g., Perkin-Elmer Model 306 or 503.

Instrumental parameters. The Perkin-Elmer Atomic Absorption Spectrophotometer is operated at a wavelength of 422.7 nm, the slit is set to No. 4 (1 mm, 13 Å), and the source is set to 10 mA. The air and acetylene supply should be adjusted to a flow meter reading of "10." At this setting the flame should be blue with narrow yellow streaks. The sample tubing should be approximately 300 mm in length and 0.015 inch I.D. × 0.043 inch O.D. (Technicon polyethylene tubing No. 562–2002). The recorder is set to scale expansion No. 1, noise suppression No. 3, and a speed of 0.5 or 1.0 inch/min.

The Perkin-Elmer Atomic Absorption Spectrophotometer Model 306 has increased sensitivity for calcium and has a newly designed one-slot burner head (No. 303–0418 or 040–0266). In view of the different design, optimal air setting is "9" and acetylene setting is "7". The greater sensitivity allows for greater dilution of the sample (see Reagent 3, procedure step 1, and procedure note 1). The new design of the burner also decreases chemical interference.

PROCEDURE

1. Pipet 1.00 ml of each of the working standards, unknowns, and controls into appropriately labeled test tubes (duplicates if possible). (Pipet 0.5 ml only, if Model 306 is used).

2. Add forcefully 9.00 ml of diluent to each test tube. An automatic dispensing device may be used for this addition. Mix and allow to stand for 10 min. Centrifuge for approximately 10 min at 2000 rpm. The supernatant of some samples may be slightly turbid; however, this does not interfere with the test.

3. Insert the sample tubing into the supernatant, making sure that it is inserted into all tubes to the same level. The procedure may be semiautomated by aspirating the supernatant from an AutoAnalyzer Sampler II unit equipped with an adjustable cam (120/h) with a wash time of 8 s and an aspiration time of 22 s. In order to detect any possible drift it is recommended that standards be run frequently (e.g., after 10 samples).

CALCULATION

Prepare a standard curve and read the results either manually or with the aid of an AutoAnalyzer chart reader. Correct for displacement by the protein precipitate, where applicable (see Comments).

COMMENTS ON THE PROCEDURE AND SOURCES OF ERROR

1. Calcium values obtained with this procedure are, on the average, approximately 2.5 per cent higher (0.125 mmol $Ca^{2+}_{1/2}$/l for a specimen with 5.0 mmol $Ca^{2+}_{1/2}$/l) than those obtained with the reference method. This error is due to a 2 to 3 per cent contraction in volume caused by the removal of the dissolved protein (= displacement error) and can be corrected mathematically. The error is reduced by one-half if the 20-fold dilution of the sample is employed (see procedure step 1). Such an error, of course, does not occur with spinal fluid or urine which are low in protein content. Despite this limitation of the method, it is considered more practical for the routine laboratory than the alternative atomic absorption method discussed in the next paragraph. The precision is much better and burner maintenance much easier than with methods based on simple dilution of proteinaceous samples.

2. Some authors have claimed that precipitation of proteins before analysis is not necessary, and that serum dilutions (e.g., 1/50) containing 0.1 g La^{3+}/100 ml may be analyzed directly. It has been our experience that this approach will generally give results in good agreement with those obtained with the classic methods, but that accuracy and precision of these methods are less, with some samples differing by more than 5 per cent. It is believed that this is caused by protein which has precipitated in the burner, causing uneven flow or unstable flame characteristics. The most popular of these methods is that by Trudeau and Freier,[60] using a 50-fold dilution of the sample with an aqueous solution of La^{3+} (0.5 g La^{3+}/100 ml).

3. When aspirating the supernatant in step 3 of the procedure, the sample tubing should be held at approximately the same level for all determinations. Significant differences in the position of the sample tubing will produce different flow rates of supernatant into the burner and thus will affect the results.

4. The method presented may be applied to *urine*, provided that the urine has been acidified as outlined in connection with the manual methods that follow. No phosphorus interference is observed with levels up to 1.6 g/1000 ml. Presence of phosphate-phosphorus in a concentration of 3 g/1000 ml depresses calcium values by approximately 3 per cent.

5. If a calcium determination is to be performed on *stool*, the specimen must first be mixed. Approximately 1 g is transferred to a crucible, weighed, and ashed with a gentle flame. The residue is dissolved in a few drops of 6 molar HCl and quantitatively transferred to a 25 ml glass-stoppered graduated cylinder with several washes of distilled water. The flask is then filled to the mark and the solution analyzed. The average normal calcium excretion is about 16 mmol (0.64 g)/d.

NORMAL VALUES

In many publications, the range of normal values for calcium in *serum* is reported as 4.50 to 5.50 mmol $Ca^{2+}_{1/2}$/l (9.0 to 11.0 mg/100 ml). It is now widely agreed that these values are higher than those obtained by the better, modern methodology. However, there is still no close agreement as to the true normal range, whether or not there is a sex difference, or whether there is a decrease in values with age. Until completion of further studies tentative use of a range of 4.25 to 5.20 mmol $Ca^{2+}_{1/2}$/l (2.12 to 2.60 mmol/l) is recommended and the posture effect should be borne in mind. There is also considerable disagreement as to the excretion rate of calcium in *urine*, partially because of the significant effect of dietary intake of calcium. Values from 2.5 to 20 and from 25 to 75 mmol $Ca^{2+}_{1/2}$/d have been reported for Ca-free and low Ca diets, respectively.

REFERENCES

Tietz, N. W., Fiereck, E. A., and Green, A.: unpublished.

IONIZED CALCIUM

CLINICAL SIGNIFICANCE

The only form in which calcium is physiologically active is the ionized form, frequently referred to as free ionized calcium [$Ca^{2+}_{1/2}$ (f)]. Reasonably accurate methods for the determination of this fraction have been developed only recently; thus, relatively little information is available as to its clinical utility. It has been reasonably well documented, however, that the ionized calcium values reflect the functional status of the patient's calcium metabolism better, and that they show better correlation with the clinical condition, than do total calcium values.[25]

Hypercalcemia (increased total serum calcium), for example, may be seen not only in primary hyperparathyroidism, but also in sarcoidosis and in paraproteinemia due to an increase in levels of IgA or IgG (= increased protein binding of calcium) as well as in hypervitaminosis D (due to increased absorption). The ionized calcium, however, is elevated only in primary hyperparathyroidism[29] and in ectopic PTH-producing tumors. Such elevations are generally seen even if the total calcium is normal. Conversely, *hypocalcemia* (low total serum calcium) due to hypoproteinemia is generally associated with normal ionized calcium values, while primary hypoparathyroidism is associated with both low total calcium and low ionized calcium levels.

In view of the effect of pH on the ionized calcium fraction, any results obtained must be interpreted in light of the acid-base status of the patient. A significant decrease in the ionized calcium fraction, regardless of the total calcium level, may lead to an increase in neuromuscular irritability and thus to tetany.

DETERMINATION OF IONIZED CALCIUM BY ION-SELECTIVE ELECTRODES

SPECIMENS

Specimens must be collected anaerobically in a manner similar to that outlined for blood gases. Vacutainers have been found to be suitable if the stopper is left in place and if the specimen is immediately placed in ice water. The main reason for this precaution is the fact that pH changes greatly affect the protein binding of calcium. When the specimen reaches the laboratory, the sample is allowed to clot at room temperature. After centrifugation, the stopper is removed and an aliquot of the sample is immediately drawn into a 1 ml tuberculin syringe. If the assay is delayed, the syringe should be immediately capped and stored in the refrigerator. In this form, the ionized calcium fraction does not change for several hours; frozen, it will be suitable for analysis for several days.

Wybenga *et al.*,[65] as well as Henry,[17] have claimed that the ionized calcium values of fresh, anaerobically collected specimens are identical to results obtained from samples exposed to air if results are mathematically corrected for any pH changes.

Serum specimens are preferred by some over plasma, since heparin binds Ca^{2+} to a small degree. Specimens should be drawn without stasis and preferably in the fasting state.

Orion Research, Inc. (Cambridge, MA. 02139) suggests that whole blood specimens be used for analysis of ionized calcium with the recently released Ionized Calcium Analyzer, Model SS-20.

STANDARDIZATION

The procedure to be used for standardization of the method is still a strongly debated issue. Some authors claim that addition of trypsin and triethanolamine prevents coating of the electrode with proteins and increases the stability of the reading without introducing significant errors; others advise against addition of these agents since they are said to bind Ca^{2+}. Sodium is added to all standards, to correct for the response of the calcium selective electrode to sodium ions. Magnesium causes a decrease in the apparent free calcium and is therefore also added to standards. All standard solutions should be isotonic.

PROCEDURE

The determination is carried out by introducing respective standards into the flow-through electrode with the aid of a constant-speed pump. The electrode is then "conditioned" by repeatedly analyzing a serum pool until reproducible results are obtained. The sample is then treated in an identical manner. With most electrodes, the determination is carried out either at ambient temperature or at 25°C. A newly released thermostatted electrode (Orion) allows measurements at 37°C. It is felt that a recorder, attached to the voltmeter, increases the accuracy of the measurement and offers a convenience.

NORMAL VALUES

Normal values differ with the type of technique employed, the procedure for collection and storing the sample, the procedure used for standardization, and the temperature of the measurement. Yet, recently reported values for the technique employing different types of ion-selective electrodes show reasonably close agreement, namely 2.42 to 2.78,[17] 2.2 to 2.7,[27] 2.35 to 2.75,[24] and 2.24 to 2.61[14] mmol $Ca^{2+}_{1/2}$/l.

ALTERNATIVE METHODS FOR THE DETERMINATION OF IONIZED CALCIUM (Ca^{2+})

Several techniques have been utilized to determine the ionized calcium fraction in serum. All of these have significant limitations and are subject to a number of factors which affect the accuracy and reproducibility of the measurements. Although much progress in improving the methods has been made, they are still not suitable for use in the average clinical laboratory. The most suitable and convenient approach involves the use of ion-selective electrodes (see preceding section), of which the Orion flow-through electrode appears to be the most popular. There is still a considerable amount of disagreement, however, as to the details of the procedure and the technique for collection and storage of specimens. As a result, normal values differ and must be determined for each individual technique.

The ionized calcium concentration can also be determined by measuring the calcium levels before and after treatment with dextran gel,[51] by ultracentrifugation, or by ultrafiltration.[47] It has also been recommended that the level of total calcium in cerebrospinal fluid be used as an index of ionized calcium in serum, since this fluid may be considered an ultrafiltrate of plasma. Normal CSF calcium levels are 2.1 to 2.7 mmol $Ca^{2+}_{1/2}$/l. Results obtained on spinal fluid using acceptable calcium assay methods indeed correlate very closely with the ionized calcium level.

The approximate concentration of ionized calcium may also be calculated from a formula that is based on the nomogram by McLean and Hastings.[34]

$$\text{mg Ca}^{2+}/100\text{ ml} = \frac{(6 \times \text{total serum calcium in mg/100 ml}) - (\frac{1}{3} \times \text{g serum protein/100 ml})}{\text{g protein/100 ml} + 6}$$

To convert mg Ca^{2+}/100 ml to mmol $Ca^{2+}_{1/2}$/l, divide by 2.

It should be realized that this formula is based on the relationship between calcium and protein at a serum pH of 7.35 at 25°C. The formula also does not take into account the facts that there is a significant difference in the extent of binding of calcium by different proteins and that cephalin is able to bind calcium in a manner similar to protein. Consequently, any differences in the serum pH, in the proportions of serum proteins, or in the amount of phospholipids present may affect the results obtained by this formula and limit its usefulness.

DETERMINATION OF CALCIUM IN URINE BY CHLORANILATE PRECIPITATION

PRINCIPLE

Refer to the method for serum calcium by precipitation with chloranilic acid.

SPECIMEN

Values for urinary output of calcium are most meaningful if the patient has been on a low calcium, neutral ash diet for three days prior to the urine collection, and if a 24 h urine specimen is collected. Since urine may contain precipitated calcium salts, it is necessary to dissolve these by acidification to pH 1.0 prior to removal of an aliquot. Alternatively, the urine may be collected in a bottle containing 10 ml of 6 molar HCl.

PROCEDURE

Pipet 2.0 ml of urine into a 12 or 15 ml conical centrifuge tube and add 1 ml of sodium chloranilate as outlined under serum. If low calcium values are expected, use 4 ml of urine and 2 ml of chloranilate.

CALCULATION

Refer to the serum method. If 4.0 ml of urine was used, divide the values by 2.

NORMAL VALUES

The urinary excretion of calcium during a 24 h period varies greatly with the intake of calcium.

Type of Diet	Amount Excreted		
	mg/d	mmol $Ca^{2+}_{1/2}$/d	mmol Ca^{2+}/d
Free calcium	5–40	0.25–2.0	0.13–1.0
Low to average calcium	50–150	2.5–7.5	1.2–3.8
Average calcium (20 mmol/d)	100–300	5.0–15.0	2.5–7.5

Values are lower in persons above 70 years of age.

SEMIQUANTITATIVE DETERMINATION OF CALCIUM IN URINE (SULKOWITCH TEST)

PRINCIPLE

The Sulkowitch reagent contains oxalate and is buffered with acetate. When added to urine, it will produce a fine white precipitate of calcium oxalate without coprecipitation of other urine constituents. The amount of turbidity produced is the basis for the approximate quantitation.

REAGENTS

Sulkowitch reagent. Dissolve 2.5 g of oxalic acid and 2.5 g of ammonium oxalate in approximately 100 ml of water, add 5 ml of glacial acetic acid, and dilute to 150 ml with deionized water.

PROCEDURE

1. If urine is turbid, filter or centrifuge and use the clear supernatant for analysis.
2. Add 5 ml of protein-free urine to a test tube. Deproteinize the urine by acidifying the sample with 10 per cent acetic acid and by heating it to 100°C for several minutes. Restore the original volume by adding distilled water, and filter.
3. Add 5 ml of the Sulkowitch reagent to the urine or filtrate.
4. Mix and let stand for 2 to 3 min.

INTERPRETATION

No precipitate of calcium oxalate:	–
Faint turbidity visible against black background:	+
Turbidity visible without black background:	++
Opaque cloud:	+++
Flocculent precipitate:	++++

NORMAL VALUES

Healthy individuals on a normal diet show results equal to one plus (+) or two plus (++). Absence of a precipitate indicates that no calcium was present in urine and that the serum calcium levels are probably below 4.25 mmol $Ca^{2+}_{1/2}$/l. A four plus (++++) reaction indicates increased rate of excretion and suggests that the serum values may be above 6 mmol $Ca^{2+}_{1/2}$/l.

The test is only a screening test, and doubts have been raised whether the test is sufficiently accurate for clinical use. Its use is discouraged.

PHOSPHORUS IN BODY FLUIDS

DETERMINATION OF INORGANIC PHOSPHATE IN SERUM AND URINE

PRINCIPLE

A trichloroacetic acid filtrate of serum or urine is treated with molybdate reagent, which reacts with phosphate to form ammonium molybdophosphate (ammonium phosphomolybdate). This is thought to have the formula $(NH_4)_3[PO_4(MoO_3)_{12}]$. The addition of a suitable reducing agent such as aminonaphtholsulfonic acid produces a blue color of heteropolymolybdenum blue. A mild reducing agent is employed in order to avoid reduction of the excess of molybdate present. Other reducing agents, such as stannous chloride, ascorbic acid, Elon (*p*-methyl amino phenol) and N-phenyl-*p*-phenylenediamine (Semidine) have been utilized in this reaction, but aminonaphtholsulfonic acid is still widely used.

REAGENTS

1. Trichloroacetic acid, 5 g/100 ml. Place 50 g of trichloroacetic acid, AR, into a 1000 ml volumetric flask; dissolve in and fill to the mark with deionized water.
2. Sulfuric acid, 5 mol/l. Slowly add 300 ml of concentrated sulfuric acid, AR, to 750 ml of deionized water, mix well, and cool.
3. Molybdate reagent. Dissolve 25 g of ammonium molybdate, AR, in about 200 ml of deionized water. Into a 1 liter volumetric flask place 300 ml of 5 molar sulfuric acid, add the molybdate solution, dilute with washings to 1 liter with deionized water, and mix. Solution is stable indefinitely. Discard the reagent if blanks show a blue color.
4. Sodium bisulfite, 15 g/100 ml. Place 30 g of sodium bisulfite, AR, into a beaker, and dilute to 200 ml with deionized water from a graduated cylinder. Stir to dissolve; if the solution is turbid, allow to stand well stoppered for several days and then filter. Keep reagent well stoppered.
5. Sodium sulfite, 20 g/100 ml. Dissolve 20 g of sodium sulfite (anhydrous), AR, in deionized water and dilute to the 100 ml mark. Filter if necessary. Keep well stoppered.
6. Aminonaphtholsulfonic acid reagent. Place 195 ml of sodium bisulfite solution (15 g/100 ml) into a glass-stoppered cylinder or other suitable container. Add 0.5 g of 1,2,4-aminonaphtholsulfonic acid and 5 ml of sodium sulfite solution, 20 g/100 ml. Stopper and shake until the powder is dissolved. If solution is not complete, add, with continuous shaking, 1 ml of sodium sulfite at a time, until solution is complete. Avoid excess of sodium sulfite. Transfer the solution to a brown glass bottle and store in the cold. The solution is stable for about 1 month.
7. Stock standard, 0.4 mg phosphorus in 5 ml (2.61 mmol P/l). Place exactly 0.351 g of dry potassium dihydrogen phosphate, AR, into a 1 liter volumetric flask, dissolve in deionized water, add 10 ml of 5 molar sulfuric acid, and dilute to the mark with deionized water.
8. Working standard, 0.004 mg P/ml. Place 5.00 ml of the stock phosphate standard into a 100 ml volumetric flask and make up to the volume with trichloroacetic acid (5 g/100 ml).

PROCEDURE

1. Place 0.5 ml of serum into a 15 × 150 ml test tube or a 10 ml glass-stoppered cylinder.
2. Blow in 9.5 ml of trichloroacetic acid (5 g/100 ml), mix, and let stand for 5 min.
3. Centrifuge or filter through Whatman No. 42 filter paper.
4. Pipet 5 ml of clear filtrate into a test tube or glass-stoppered cylinder graduated

at 10 ml. Prepare a blank by using 5 ml of trichloroacetic acid (5 g/100 ml), and prepare a standard by using 5 ml of working standard (5 × 0.004 = 0.02 mg P).

5. Add 1 ml of molybdate reagent to all test tubes.
6. Add 0.4 ml of aminonaphtholsulfonic acid reagent; mix.
7. Dilute to the 10 ml mark with deionized water, mix, and allow to stand for 5 min.
8. Set blank at 100 per cent *T* or zero *A* and read standard and unknowns at 690 nm.

CALCULATIONS

Read results from a standard curve or calculate as follows:

$$\text{mg P/100 ml} = \frac{A_U}{A_S} \times 0.02 \times \frac{10}{5} \times \frac{100}{0.5}$$

$$= \frac{A_U}{A_S} \times 8$$

where

0.02 = mg P contained in 5 ml of working standard

10 = amount of filtrate prepared

5 = amount of filtrate used

100 = basis of expressing concentration (= 100 ml)

0.5 = amount of sample used

COMMENTS ON THE PROCEDURE

1. If *urine* is to be analyzed, follow steps 1 to 3 of the serum procedure. In step 4 use two test tubes, one with 0.5 and one with 2.5 ml of filtrate. Make up the missing volume with trichloroacetic acid (5 g/100 ml). Calculate as for serum and multiply the results by 10 or 2, respectively, to correct for the smaller amount of filtrate used. In case of a 24 h urine collection, express values in g P/d, using the following formula:

$$\text{g P/d} = \frac{\text{mg P/100 ml} \times \text{total 24 h urine volume (in ml)}}{100 \times 1000}$$

2. Phosphate in plasma (serum) at pH 7.4 is present as HPO_4^{2-} and $H_2PO_4^-$ in a ratio of 80/20. This ratio increases as the pH increases and decreases as the pH decreases. At a pH of 4.5, as may be found in urine, the ratio of HPO_4^{2-} to $H_2PO_4^-$ decreases to approximately 1/100. Thus, in any biological specimen the ratio of the two forms of phosphates will depend on the pH of the specimen. As a result, the amount of phosphate, expressed in mEq/l or mmol total ion charge/l, will change as the pH of the specimen changes. On the other hand, the total amount of phosphate-phosphorus is unaffected by changes in pH. It is for this reason that values are more commonly expressed in mg phosphate-phosphorus/100 ml. Reporting phosphates in mmol/l also resolves this problem.

To convert mg P/100 ml serum to mEq phosphate/l or total ion charge/l, the former is multiplied by an average factor of 0.58. This factor is derived from the following formula:

$$10\left(\frac{0.8}{1} \times \frac{2}{31}\right) + 10\left(\frac{0.2}{1} \times \frac{1}{31}\right) = 0.58$$

The figure 10 converts the volume of 100 ml to 1 liter; the figures 0.8/1 and 0.2/1 indicate the ratio of HPO_4^{2-} and $H_2PO_4^-$, respectively, at pH of 7.4; the figures 2 and 1 represent the valences of the two forms of phosphate; the number 31 represents the atomic mass of phosphorus.

3. Control of the pH during the color development is extremely important. At strongly acid pH, such as is provided by trichloroacetic acid and H_2SO_4, the reduction of the [P-Mo] complex is favored and reduction of the molybdate reagent itself is inhibited. At higher pH's, such as pH 2 or 3, reduction of the molybdate reagent takes place.

Labile organic phosphate esters, such as creatine phosphate or glucose-1-phosphate, are hydrolyzed at strongly acid pH. If inorganic phosphorus is to be measured in the presence of one of the esters, the pH must be adjusted to 4.5 to 5.0, and ascorbic acid must be used as a reducing agent. The latter is a mild reducing agent that will not reduce the molybdate reagent at this pH.

4. Lipid phosphorus and most of the phosphate from organic esters are not detected with this method.

NORMAL VALUES

Normal values for phosphorus are generally listed as follows:

Serum

	mg/100 ml	mmol ion charge/l*	mmol/l
First year of life	4.0–7.0	2.3–4.1	1.29–2.26
Children	4.5–6.5	2.5–3.6	1.45–2.09
Adults	3.0–4.5	1.7–2.5	0.96–1.45

The higher values seen in children are associated with increased levels of human growth hormone (HGH).

Phosphorus in serum is lower than normal during the menstrual period. Values are also lower after meals, making it important to perform phosphorus determinations on samples obtained in the fasting state.

Urine

The normal values for the excretion of urine phosphorus vary greatly with the dietary intake of phosphates, but average values most widely accepted are 0.4 to 1.3 g/d (12.9 to 42.0 mmol/d). There is a significant diurnal variation in the excretion of phosphorus in urine, with values in the afternoon being highest.

REFERENCE

Fiske, C. H. and Subbarow, Y.: J. Biol. Chem., *66*:375, 1925.

Magnesium

PHYSIOLOGY AND CLINICAL SIGNIFICANCE

The magnesium content of a 70 kg adult is about 20 g (820 mmol Mg^{2+}). More than 50 per cent of this is in bone, associated with calcium and phosphorus. Most of the remainder is in the soft tissues.

Among the intracellular cations, magnesium is second in quantity only to potassium. Like calcium, it is absorbed in the upper intestine, but unlike calcium its absorption is not dependent on vitamin D or any similar factor. About 65 to 70 per cent of the magnesium of blood serum is in the diffusible or free form; the remainder is bound to protein, mostly albumin.

Little is known about the factors regulating magnesium levels in plasma. It is believed,

* Based on a ratio of $HPO_4^{2-}/H_2PO_4^- = 4/1$

however, that PTH and aldosterone play a role. The latter is known to regulate the excretion of Mg^{2+} in a way similar to the excretion of K^+. A reciprocal relation between serum magnesium and serum calcium levels has been observed in some conditions and, in other conditions, between serum magnesium and serum phosphate; however, no details about the mechanism of these relationships are known.

Magnesium ions serve as activators for a number of important enzyme systems engaged in transfer and hydrolysis of phosphate groups,[61] such as hexokinase, creatine kinase, alkaline phosphatase (from red cells and bone), and prostatic acid phosphatase. Magnesium is required particularly in those enzyme reactions involving ATP, where it forms a Mg^{2+}–ATP^{4-} complex.

A "magnesium deficiency tetany" has been described. It is characterized by low serum magnesium and normal serum calcium levels and can be distinguished from "calcium deficiency tetany" in that it responds to magnesium but not to calcium administration. This is probably the most important application for magnesium determinations in serum. At the time of tetany, serum magnesium levels of 0.3 to 1.0 mmol $Mg^{2+}_{1/2}$/l* have been reported in the presence of normal serum calcium levels and normal pH. Treatment with magnesium sulfate resulted in all cases in a rise in the serum magnesium level and a concomitant disappearance of tetany and convulsions. As with Ca^{2+}, tetany reflects a deficiency in ionized Mg^{2+} and not in total Mg.

Decreased serum magnesium levels (*hypomagnesemia*) have been found in the malabsorption syndrome, acute pancreatitis, hypoparathyroidism, chronic alcoholism and delirium tremens, chronic glomerulonephritis, aldosteronism, digitalis intoxication, after protracted I.V. feeding, and after excessive loss of magnesium in the urine. This last situation has been observed in renal tubular reabsorption defects, in patients with congestive heart failure[22] treated with ammonium chloride and mercurial diuretics, and after treatment with chlorothiazides. Except during treatment with the mentioned drugs, the body has a great ability to preserve magnesium by tubular reabsorption. In states of magnesium deficiency, urine excretion is extremely low. Consequently, determination of the urinary excretion rate is most useful in confirming hypomagnesemia.

Serum levels of magnesium are, at times, a poor measure of cellular magnesium deficiency. Plasma levels do not decrease below 1.0 mmol $Mg^{2+}_{1/2}$/l until at least 25 per cent of the cellular magnesium has been lost. Symptoms of magnesium deficiency usually do not occur until serum magnesium levels decrease below 1.0 mmol $Mg^{2+}_{1/2}$/l.

Increased serum magnesium levels (*hypermagnesemia*) have been observed in dehydration, severe diabetic acidosis, and Addison's disease. Any condition interfering with glomerular filtration results in retention and thus elevation of serum magnesium levels. Some of the highest values have been observed in uremia. Hypermagnesemia leads to an increase in the atrioventricular conduction time of the electrocardiogram.

NORMAL VALUES

The normal concentration range of magnesium in *serum*, using the Titan Yellow method, is from 1.4 to 2.3 mmol $Mg^{2+}_{1/2}$/l. Values obtained with the atomic absorption method are slightly lower, from 1.3 to 2.1 mmol $Mg^{2+}_{1/2}$/l.

Serum levels in newborns are essentially the same as those in adults. Levels do not change appreciably throughout the day, but in females the highest serum values are obtained at the time of the menses.

Erythrocytes contain about 4.5 to 6.0 mmol $Mg^{2+}_{1/2}$/l.

* In analogy to the expression of the concentration of calcium (see p. 904), magnesium is best reported in terms of mmol of ion charges/l. Thus, 1.0 mmol $Mg^{2+}_{1/2}$/l = 1.0 mEq Mg^{2+}/l = 0.50 mmol Mg^{2+}/l.

The concentration ranges in *cerebrospinal fluid* for the Titan Yellow and atomic absorption methods are from 2.2 to 3.0 mmol $Mg^{2+}_{1/2}$/l and from 2.0 to 2.7 mmol $Mg^{2+}_{1/2}$/l, respectively. The urinary excretion is 6.0 to 10.0 mmol $Mg^{2+}_{1/2}$/d. Values for males are slightly higher than those for females. Earlier reports indicating a higher excretion rate are likely to be incorrect.

SPECIMENS

Serum magnesium determinations should be done on samples drawn without venous stasis. Since the magnesium concentration in red cells is substantially greater than that in serum, the specimen should be separated from the red cells as soon as possible and hemolyzed specimens should not be used for analysis. Magnesium levels are stable for several days if the serum is stored in the refrigerator, separated from the red cells.

If the test is performed on urine, the specimen should be acidified to pH 1.0 and mixed well prior to analysis. If the same 24 h collection is used for other tests, an aliquot of the well mixed urine should be acidified.

DETERMINATION OF SERUM MAGNESIUM BY THE TITAN YELLOW METHOD

PRINCIPLE

A trichloroacetic acid filtrate of serum is treated with the dye titan yellow (methyl benzothiazide-1,3-4,4′-diazo aminobenzol-2,2′-disulfonic acid) in alkaline solution. The red lake that forms is thought to be dye, adsorbed on the surface of colloidal particles of magnesium hydroxide, which are kept in solution with the aid of polyvinyl alcohol. This last reagent also increases the sensitivity of the method by a factor of approximately 2.

REAGENTS

1. Trichloroacetic acid, 5.0 g/100 ml. Dissolve 50.0 g of trichloroacetic acid, AR, in deionized water and make up to 1 liter. Store in a glass-stoppered borosilicate bottle.
2. Sodium hydroxide, 5.0 mol/l. Dissolve 200.0 g of NaOH in deionized water and dilute to 1 liter. The solution may also be prepared from a stock solution by dilution.
3. Polyvinyl alcohol, 0.1 g/100 ml. Suspend 1.0 g of polyvinyl alcohol (Elvanol, grade 70-05, E.I. Du Pont de Nemours & Co., Wilmington, DE. 19801) in 40 to 50 ml of 95 per cent ethanol and pour the mixture into 500 to 600 ml of swirling deionized water. Warm on a hot plate until the solution is clear. Allow to cool and dilute to 1 liter. Polyvinyl alcohol, dissolved directly in water, tends to form lumps; thus, the compound should be suspended in alcohol as outlined.
4. Titan yellow (stock solution). Dissolve 75 mg of titan yellow in and make up to 100 ml with polyvinyl alcohol, 0.1 g/100 ml; filter if not clear. The reagent is stable for 2 months if stored in a brown bottle at room temperature.
5. Titan yellow (working solution). Dilute 10 ml of the stock titan yellow solution to 100 ml with polyvinyl alcohol, 0.1 g/100 ml. The reagent is stable for 1 week if stored in a brown bottle.
6. Stock standard, 20 mmol $Mg^{2+}_{1/2}$/l. Dissolve 243.2 mg of bright magnesium metal turnings, AR, in a covered beaker with 50 ml of deionized water and 2 ml of concentrated hydrochloric acid. Avoid open flames since hydrogen is produced in this reaction. When the reaction subsides, add dropwise hydrochloric acid to dissolve the magnesium metal completely. Transfer the solution quantitatively to a 1 liter volumetric flask and make up to volume with deionized water. Accurately pipet 5, 10, 15, and 20 ml aliquots of this stock standard into 100 ml volumetric flasks and dilute to 100 ml with deionized water. These standards contain 1.0, 2.0, 3.0, and 4.0 mmol $Mg^{2+}_{1/2}$/l (0.5, 1.0, 1.5, and 2.0 mmol Mg^{2+}/l). The stock standard is stable if well stoppered.

PROCEDURE

1. Place 1 ml of the unknown serum into a 15 × 150 mm test tube and blow in 5 ml of trichloroacetic acid, 5 g/100 ml. Some prefer to use trichloroacetic acid, 7.5 g/100 ml.
2. Mix tubes gently but thoroughly, let stand for 5 min, and centrifuge for 5 min at 2000 rpm.
3. Transfer 3 ml of the clear supernatant to a Coleman or other suitable cuvet.
4. Prepare a standard set by pipetting 0.5 ml of each magnesium working standard into a separate cuvet, followed by 2.5 ml of trichloroacetic acid (5 g/100 ml). Prepare a reagent blank by substituting distilled water for the magnesium standard.
5. Add to all cuvets 2 ml of titan yellow working solution and 1 ml of 5.0 molar NaOH. Mix tubes thoroughly and read after 5 min, but not later than 30 min, at a wavelength of 540 nm with the reagent blank set to 100 per cent *T* or zero absorbance.

CALCULATION

Construct a standard curve or employ the following formula:

$$\frac{A_U}{A_S} \times C_S = \text{mmol } Mg^{2+}_{1/2}/l$$

where A = absorbance readings
C_S = concentration of magnesium standard most nearly corresponding to the value of the unknown

COMMENTS ON THE PROCEDURE

Since the procedure follows Beer's Law only over a short range, it is necessary to run a standard set with each determination; only in narrow band spectrophotometers will the procedure follow Beer's Law over the range observed in serum.

It is critical that the test be read at the absorption peak; thus, the maximum absorbance should be determined for the instrument used in this determination.

Results obtained with the procedure just outlined are invalid for sera from patients receiving calcium gluconate (low results) or mercurial diuretics. The latter interference may be removed by treatment of the sample with H_2S.

There have been unconfirmed claims by some investigators that the presence of Ca^{2+} intensifies the color. Thus, some authors have added Ca^{2+} to the standard or to both standard and unknown.

This method is simple and fast, but has the disadvantage that its accuracy is only within 10 per cent. The lack of greater accuracy is probably due to the "erratic and unsystematic variations presumably related to the colloidal nature of the material whose color is being measured."[16,58] The method should be used only if an atomic absorption spectrophotometer is not available.

REFERENCE

Basinski, D. H.: *In* Standard Methods of Clinical Chemistry. S. Meites, Ed. New York, Academic Press, Inc., 1965, vol. 5.

DETERMINATION OF SERUM MAGNESIUM BY FLUOROMETRIC AND COMPLEXOMETRIC TECHNIQUES

Magnesium ions and 8-hydroxy-5-quinoline sulfonic acid form a chelate compound that fluoresces if excited at a wavelength of 380 to 410 nm. The peak fluorescence occurs at 510 nm.

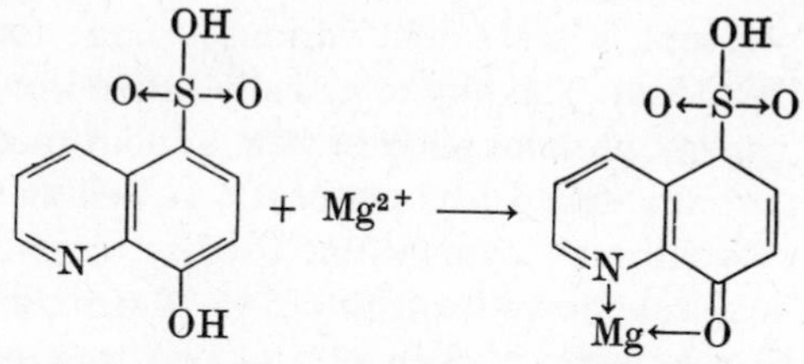

OH O←S→O N OH $+ Mg^{2+} \longrightarrow$ OH O←S→O N Mg←O

Other divalent ions will form similar complexes with this reagent. Because the binding capacity of the reagent for magnesium is greater than that for other ions and because the molar fluorescence of the magnesium compound is considerably greater, interferences by other ions in serum are negligible.

Although this method is extremely sensitive and simple, it is not widely used because of difficulties inherent in fluorometric methods. The enhancing or quenching effect of other compounds is the drawback that makes this method completely unsuitable for the determination of magnesium in urine.

Complexometric methods[1] employing Eriochrome, murexide, or other dyes are also rarely used because of the difficulties inherent in such procedures (see Determination of Serum Calcium by EDTA Titration).

DETERMINATION OF MAGNESIUM BY ATOMIC ABSORPTION SPECTROPHOTOMETRY

Atomic absorption spectrophotometry is the preferred technique for the determination of magnesium in biological fluids since it is fast, accurate, and reasonably simple. Some analytical difficulties observed are similar to those discussed under calcium analysis. Magnesium is released from proteins by treatment with acid (trichloroacetic acid; HCl). La^{3+} or Sr^{2+} is added to the reaction mixture to bind phosphates. The need for treating serum with acids and Sr^{2+} or La^{3+} has been questioned by some investigators but has been confirmed by others, especially for the analysis of urine specimens. Protein precipitation provides the additional advantage that precision is better and burner maintenance is easier.

In the method by Thiers, the sample is diluted with diluent (6.6 g La^{3+} and 40 g TCA/liter). After centrifugation, the sample is analyzed with an atomic absorption spectrophotometer using the 285.2 nm line of a Mg hollow cathode lamp. Results are calculated with the aid of a standard curve.

In the method by Hansen and Freier, samples are directly diluted with an aqueous solution of La^{3+} (0.2 g/100 ml) and analyzed without any further treatment, using the same Mg line used in the Thiers method.

Since methods are given in the manuals accompanying the atomic absorption spectrophotometers, no detailed method is presented here.

Magnesium has also been determined by flame emission spectrophotometry. This method is not recommended, however.

REFERENCES

Thiers, R. E.: *In* Standard Methods of Clinical Chemistry. S. Meites, Ed. New York, Academic Press, Inc., 1965, vol. 5.

Hansen, J. L., and Freier, E. F.: Am. J. Med. Technol., *33*:158, 1967.

Analytical Methods for Atomic Absorption Spectrophotometry. Manual, supplied by Perkin-Elmer, Norwalk, CN., 1971.

Serum Iron and Iron-Binding Capacity

The total amount of iron in an adult is approximately 4 to 5 g. About 70 to 75 per cent of this relatively small amount of iron has an active and vital physiological role, and the remaining 25 to 30 per cent is present in various storage forms that can readily be mobilized if needed (see Fig. 15–8). The physiologically active iron is mainly present in the form of oxygen-carrying chromoproteins such as hemoglobin (accounting for 65 per cent of Fe) and myoglobin (approximately 3 to 5 per cent), as well as in the form of a number of enzymes, such as the cytochromes, cytochrome oxidase, peroxidase, and catalase (less than 1 per cent). The main storage form for iron (15 to 20 per cent) is Fe_n^{3+}-ferritin, which is made up of ferric hydroxide-ferric phosphate attached to a protein called apoferritin (molecular mass 460,000). If the amount of apoferritin is not sufficient to bind the iron, the excess of iron is deposited in the form of small iron oxide granules, generally called hemosiderin (less than 0.1 per cent of Fe). Iron is stored mainly in the liver, spleen, and bone marrow.

The iron needs of the body are met by a dietary intake of approximately 5 to 20 mg/d. Since the body conserves iron extremely well, this rather minute intake of iron is sufficient to satisfy the normal adult requirement of approximately 12 mg/d. Less than 1 mg is lost per day through the skin, feces, and urine of adult males and nonmenstruating females. Loss through normal menstruation, however, may be as much as 80 mg per period. During pregnancy, approximately 400 mg iron are lost to the fetus and as a result of blood loss at the time of parturition.

Aside from a small amount of iron that is absorbed from the stomach, most of the absorption takes place in the duodenum and jejunum. Since only iron in its ferrous (Fe^{2+}) form can be absorbed, all dietary iron present in the Fe^{3+} form must first be reduced to the ferrous form. The rate of absorption is greatly dependent on pH. An acid pH, as provided by the gastric HCl, prevents precipitation of iron as phosphates, helps in solubilizing iron in food, and, finally, aids in the reduction of Fe^{3+} to Fe^{2+}. The iron absorbed from the intestinal tract is immediately oxidized in the mucosal cells to the ferric state and is temporarily stored there as Fe_n^{3+}-ferritin. When the storage capacity of the mucosal cell for Fe_n^{3+}-ferritin is exhausted, no further iron is absorbed. On demand, iron is released from the mucosal cell into the blood, where it circulates mainly as Fe_2^{3+}-transferrin, which in turn is in equilibrium with an extremely small amount of free Fe^{3+}. The mechanism for the Fe release is presently not understood. Transferrin (formerly called siderophilin) is the plasma iron transport protein, with a molecular mass of 90,000, that migrates electrophoretically with the β_1-globulin fraction and has the ability to form a complex with iron

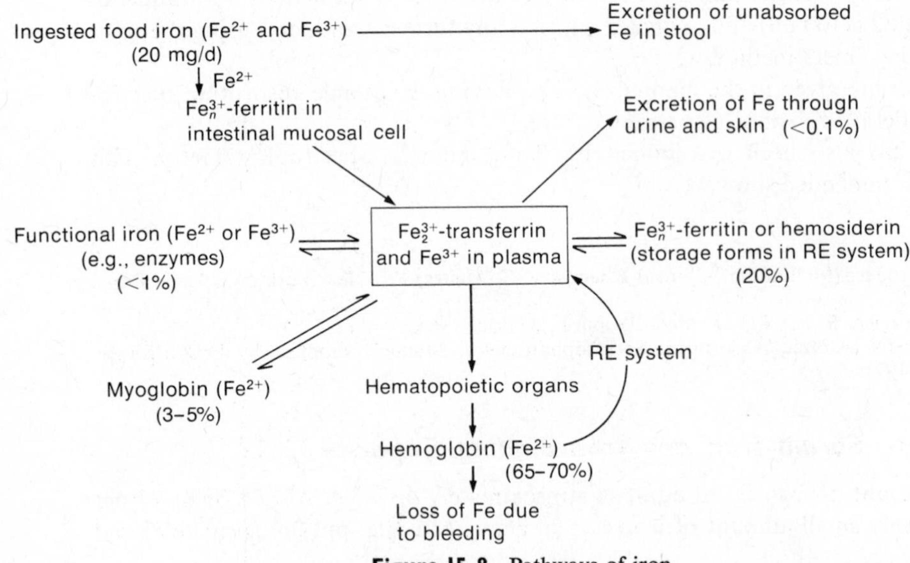

Figure 15–8 Pathways of iron.

through an ionic bond. Each molecule of transferrin combines with two molecules of Fe^{3+}. The iron is then carried in this form to the various body storage areas such as the liver and bone marrow, thus providing a dynamic equilibrium between the iron stores. Smaller amounts are carried to most other tissues, where iron is released from the protein. The total amount of circulating apo-transferrin (protein capable of binding iron) is generally only 25 to 30 per cent saturated with iron. The additional amount of iron (expressed in μg/100 ml) that can potentially be bound by apo-transferrin constitutes the latent or unsaturated iron-binding capacity (UIBC) of plasma. The total amount of iron circulating in the plasma plus the iron that could be bound constitutes the total iron-binding capacity (TIBC) of plasma.

At virtually each step of iron metabolism there is an interconversion of iron between the Fe^{2+} and Fe^{3+} forms. Examples are the oxidation of the Fe^{2+} absorbed from the intestinal tract into Fe_n^{3+}-ferritin, the reduction of Fe^{3+} for incorporation into hemoglobin, and the conversion of Fe^{2+} to Fe^{3+} in the degradation of hemoglobin. Ceruloplasmin, with its great facility to bind ferrous iron and its ferroxidase activity, is now thought to be significantly involved in the "ferrous-ferric cycle."

CLINICAL SIGNIFICANCE

Alterations in the level of serum iron and iron-binding capacity have been observed in a number of conditions (see Table 15–1). In most instances, it is not the value of either one, but the values of both the iron and the iron-binding capacity, that is of most clinical significance. In general, it can be stated that *increases in serum or plasma iron* may be seen: (1) in conditions characterized by increased red cell destruction (hemolytic anemia, decreased survival time of red cells); (2) in cases of decreased utilization of Fe (decreased formation of blood as in lead poisoning or pyridoxine deficiency); (3) in situations in which increased release of iron from body stores occurs (e.g., release of ferritin in necrotic hepatitis); (4) in states in which the process of iron storage is defective (as in pernicious anemia); and (5) in conditions in which there is an increased rate of absorption (e.g., hemochromatosis, hemosiderosis). *Decreases in serum or plasma iron* levels are generally due to a deficiency in the total amount of iron present in the body, which, in turn, may be caused by lack of sufficient intake or absorption of iron, increased loss of iron (chronic blood loss or nephrosis), or increased demand on the body stores (pregnancy). Diminished iron levels may also be caused by a decreased release of iron from body stores (reticuloendothelial

TABLE 15–1 SERUM IRON AND TIBC VALUES IN VARIOUS DISEASES

Disease	*Serum Iron*	*TIBC*
Iron deficiency anemia (dietary, malabsorption, chronic hemorrhage, late pregnancy)	↓↓	↑
Anemia of chronic infections	↓→	↓→
Anemia of neoplastic disease	↓→	↓→
Hemolytic anemia	↑	↓→
Pernicious anemia	↑→	↓
Hemochromatosis	↑↑	↓
Hemosiderosis	↑↑	→
Hepatitis	↑↑	↑
Chronic liver disease	↓→	↓→
Obstructive jaundice	→	→
Polycythemia	↓→	→↑
Nephrosis	↓	↓↓
Hodgkin's disease (terminal)	↑↑	

system), as seen in infections or with turpentine abscesses. *Increases in the total iron-binding capacity* of serum may be caused by increased production of the iron-binding protein apo-transferrin, as found in the various states of chronic iron deficiency, or may be caused by an increased release of ferritin, as in hepatocellular necrosis. *Decreases in the total iron-binding capacity* may be caused by a deficiency in ferritin, as found in cirrhosis and hemochromatosis, or as a result of an excessive loss of protein (apo-transferrin), as occurs in nephrosis.

DISCUSSION OF METHODS FOR THE DETERMINATION OF IRON IN SERUM

Most of the iron in serum is bound to protein; therefore, the first step in any iron procedure is the disruption of the iron-protein complex. An early approach was the wet digestion of the sample; however, the iron present in serum hemoglobin was released as well, resulting in too high serum iron values. More recent methods rely on the release of iron at strongly acid pH, which is provided by addition of either hydrochloric, sulfuric, or trichloroacetic acid. The last reagent acts simultaneously to release the iron and to precipitate the serum proteins, while the use of hydrochloric or sulfuric acids requires the additional use of a protein precipitating agent. Henry[16] has indicated, however, that cold trichloroacetic acid gives incomplete recovery of iron and, therefore, he recommends the use of hot trichloroacetic acid.

The next step in most procedures is the reaction of the iron in the protein-free filtrate with one of the following chromogens, listed in order of increasing sensitivity: thiocyanate, which with Fe^{3+} yields a strongly colored ferric thiocyanate, $Fe(SCN)_3$; α,α′-dipyridyl; 2,2′,2″-terpyridine (2,2′,2″-tripyridine); 1,10-phenanthroline (*o*-phenanthroline); 4,7-diphenyl-1,10-phenanthroline (bathophenanthroline) or its water soluble sulfonated form; TPTZ (2,4,6-tripyridyl-*s*-triazine); Ferrozine [3-(2-pyridyl)-5,6-*bis*(4-phenyl)sulfonic acid]; and Terosite [2,6-*bis*(4-phenyl-2,2-pyridyl)-4-phenyl pyridine]. The molar absorptivities (ε) for the iron complexes with the preceding reagents are 7000, 8600, 11000, 11000, 22400, 22600, 27900, and 30200, respectively.

These reagents, except thiocyanate, react only with Fe^{2+}, thus, requiring reduction of the Fe^{3+} with a suitable reducing agent. Hydrazine, dithionite, sulfite, and especially hydroxylamine, ascorbic acid, and thioglycolic acid have been used for this purpose (see also Henry[16]).

Ramsay[45] combined the protein precipitating agent (acetate buffer, pH 5.0), the reducing agent (hydroxylamine HCl), and the color reagent (2,2′,2″-tripyridine) and heated this mixture together with serum. Thus, protein precipitation, reduction of Fe^{3+}, and the color development are accomplished in one step. The procedure is extremely reliable and reproducible but, unfortunately, not very sensitive. A detailed modification of this method[31] was described in the first edition of this textbook.

A unique approach has been used by Schade.[50] In this method iron is released from its protein complex by adjusting the serum to approximately pH 6.0 with phosphate buffer of pH 5.3. Under these conditions the serum proteins remain in solution. The Fe^{3+} is then reduced by addition of ascorbic acid (stabilized with sodium metabisulfite) and sulfonated diphenylphenanthroline is added for color development:

$$Fe_2^{3+}\text{-transferrin} \xrightarrow{\text{pH } 6.0} Fe^{3+} + \text{apo-transferrin}$$

$$\downarrow + \text{ascorbic acid and sodium metabisulfite}$$

$$Fe^{2+} + \text{sulfonated diphenylphenanthroline} \rightarrow \text{color complex}$$

Correction for the interference by the nonspecific colors or opacity of serum, or both, is made by preparing individual serum blanks, which contain all the reagents except diphenylphenanthroline.

Webster,[62] as well as Askevold and Vellar,[2] used a similar approach but employed sodium dithionite as the reducing agent and added an anionic detergent, Teepol 710 or 610 (Norske Shell, A/S, or Shell Chemical Company, New York, N.Y.). The latter aids in the release of iron from its protein complex and also renders lipemic sera clear.

Iron has also been determined by atomic absorption spectrophotometry either directly after a 1/1 dilution of serum or after chelation of the iron with bathophenanthroline and extraction of the complex into methyl isobutyl ketone (4-methyl-2-pentanone, isopropyl acetone, MIBK). The use of atomic absorption spectrophotometry for measuring Fe in serum appears, at this time, to be impractical, especially because of the relatively low sensitivity. In the case of direct methods, matrix interference and inability to distinguish hemoglobin-Fe from transport-Fe are additional limitations. However, it is conceivable that further advances in technique and instrumentation design will increase the reliability of these methods. This may especially be true if the graphite furnace is employed to determine iron in the supernatant after precipitation of serum proteins with 10 per cent trichloroacetic acid.[4] However, the relatively small dilution (4-fold) causes a relatively large displacement error by precipitated proteins. Atomic absorption may successfully be used for urine iron determination if higher levels are measured, e.g., after administration of deferoxamine mesylate[3] for the evaluation of iron stores.

DISCUSSION OF METHODS FOR THE DETERMINATION OF SERUM IRON-BINDING CAPACITY

Only part of the apo-transferrin of serum is saturated with iron (see under serum iron). The amount of iron that can be bound, in addition to that already bound, is called the unsaturated (or latent) iron-binding capacity (UIBC or LIBC). Both values together represent the total iron-binding capacity (TIBC).

$$\text{Serum Fe} + \text{UIBC} = \text{TIBC}$$

The TIBC may be determined by one of two major approaches. (1) Excess iron equivalent to 500 μg Fe/100 ml in the form of ferric ammonium citrate is added to serum that was previously adjusted to a pH of 8.0 or above with a buffer solution. At this pH the iron transferrin complex is stable, and apo-transferrin will bind iron until all apo-transferrin has been saturated with iron (= Fe_2^{3+}-transferrin). The excess iron is then removed either by an ion exchange resin (which binds the citrate and, thus, the iron as well[18]) or by magnesium carbonate,[44] and the iron remaining in the supernatant (serum) is determined by one of the serum total iron methods. The value obtained represents the TIBC.

$$\text{UIBC} = \text{TIBC} - \text{Serum Fe}$$

(2) In the other possible approach,[50] excess iron is added as outlined under (1), but the iron in the supernatant is then determined at a pH above 7.5 at which only the excess (free) iron will react, since only this fraction can be reduced to Fe^{2+}. Iron bound by transferrin remains as Fe^{3+} and will give no color reaction. In this case:

$$\text{UIBC} = (\text{iron added}) - (\text{excess iron measured})$$

THE DETERMINATION OF IRON AND IRON-BINDING CAPACITY

PRINCIPLE

The serum sample is treated with buffer reagent which prevents precipitation of proteins, provides an acid medium for the dissociation of the Fe_2^{3+}-transferrin complex, and reduces Fe^{3+} to Fe^{2+}. Addition of color reagent results in formation of a deeply colored ferrozine-Fe^{2+} complex with an absorbance maximum at 562 nm. The presence of thiourea in the color reagent binds Cu and prevents formation of a ferrozine-Cu^{+} complex.

For the determination of the iron binding capacity, the serum is treated with excess iron. Part of the iron binds with the apo-transferrin and the remainder is removed by passing the treated sample through a column prepared with magnesium carbonate. The iron in the filtrate is determined in the same manner as the iron in serum and is a measure of the TIBC.

REAGENTS

Note: Use only deionized water, acid-washed glassware, and analytical grade reagents.

1. Ferrozine color reagent. Weigh out 400 mg of ferrozine (Hach Chemical Co., Ames, IA. 50010), and 2.5 g of thiourea and transfer to a 100 ml volumetric flask. Add deionized water to the mark and mix thoroughly. Filter out insoluble particles if necessary. Store in a dark brown bottle. Stable at least 2 months.

2. Sodium hydroxide solution, 12.5 mol/l. Weigh out 50 g of NaOH and transfer to a 100 ml volumetric flask containing about 70 ml of deionized water. Mix and allow to cool. Add water to the mark and mix. Store reagent in polyethylene bottle.

3. Stock acetate buffer, 1 mol/l. Add 60 ml of glacial acetic acid to a 1 liter volumetric flask containing about 500 ml of deionized water; while gently swirling the flask, add 28 ml of 12.5 molar NaOH. After cooling, add water to the mark and mix. Adjust the solution to pH 4.5 with either acetic acid or sodium hydroxide solution.

4. Iron reagent A. Transfer 50 ml of stock acetate buffer to a 500 ml volumetric flask containing about 300 ml of water. Add 30 g of sodium lauryl sulfate and mix until the powder is completely dissolved. Add water to the mark and mix. This reagent is stable for at least 4 weeks.

5. Iron reagent B. Transfer 50 ml of stock acetate buffer to a 500 ml volumetric flask. Add 30 g of ascorbic acid and 1.0 g of sodium metabisulfite. Dilute to the mark with deionized water. Store in a dark brown bottle in the refrigerator. This reagent is stable for 2 weeks.

6. Iron buffer reagent. This reagent is prepared fresh daily by mixing equal volumes of iron reagents A and B.

7. Filter columns. Use Quik-Sep columns with filter paper discs (#QS-P, Isolab Inc., Akron, OH. 44321), and add 0.15 to 0.20 g of $MgCO_3$ powder (Cat. #Mo-125, Sigma Chemical Co., St. Louis, MO.) and gently tap to pack the powder.

8. Stock iron standard, 1.0 g Fe/l. Weigh out 1.000 g of iron wire (Baker, AR, #36) and transfer to a 125 ml flask containing about 25 ml of 6 molar HCl. Heat the flask contents with gentle mixing on a hot plate under a hood. After the wire has completely dissolved, quantitatively transfer contents to a 1 liter volumetric flask. Dilute to the mark with deionized water and mix thoroughly. This reagent is stable indefinitely, if kept well-stoppered.

9. Working standard, 200 μg Fe/100 ml. Pipet 1.0 ml of stock standard (use a 1.0 ml transfer pipet) into a 500 ml volumetric flask and dilute to mark with deionized water. Mix. This standard is prepared fresh each day.

PROCEDURE

A. Serum Iron

1. Pipet 0.5 ml of serum, standard, control, and water, respectively, into appropriately labeled 19 × 150 mm Coleman cuvets and add 4.5 ml of iron buffer reagent to each. Run all determinations in duplicate.
2. Mix all tubes vigorously on a vortex-type mixer.
3. Incubate all tubes 15 min at 37°C in a water bath.
4. Read the absorbance (A_1) of each tube against the blank at 562 nm. *Note:* Since the total reaction volume is less than the minimum volume required for reading in a standard Coleman cuvet, a quarter-inch plug of Teflon is bonded to the bottom of the Coleman cuvet holder to elevate the cuvet.
5. Add 0.1 ml of ferrozine reagent to each tube (use a 1.0 ml serological pipet). Mix well and place in a 37°C water bath for 15 min.
6. Read the absorbance (A_2) of each tube against the blank at 562 nm.
7. Calculations:

$$\frac{A_2 - A_1 \text{ Unknown}}{A_2 - A_1 \text{ Standard}} \times 200 = \mu\text{g Fe/100 ml}$$

B. Determination of Total Iron Binding Capacity

1. Pipet 0.5 ml of serum and control into respective 12 × 75 mm acid washed tubes (use 0.5 ml Ostwald-Folin pipets). Add 1.0 ml of working iron standard to each tube and mix gently.
2. Let tubes stand at room temperature for 10 min.
3. Carefully transfer the contents of each tube into a filter column which is seated on a collection tube (12 × 75 mm tube). Avoid disrupting the packed magnesium carbonate powder during transfer.
4. Centrifuge the filter column–collection tube assembly for 10 min at approximately 500 × *g*.
5. Remove column-tubes from the centrifuge, and discard columns. If any particles of $MgCO_3$ have leaked into the filtrate, the tubes *must* be re-centrifuged to avoid erratic results.
6. Pipet 0.5 ml of filtrate from each collection tube into an appropriately labeled 19 × 150 mm calibrated Coleman cuvet.
7. Prepare "standard" and "blank" tubes by pipetting 0.5 ml of working standard and water, respectively, into labeled calibrated Coleman cuvets.
8. Add 4.5 ml of iron buffer reagent to all tubes and proceed as directed under Serum Iron Method, steps 2 to 6.
9. Calculations:

$$\frac{A_2 - A_1 \text{ Unknown}}{A_2 - A_1 \text{ Standard}} \times 600 = \text{TIBC (as } \mu\text{g Fe/100 ml)}$$

PROCEDURE NOTES

1. All glassware and pipets used for making reagents and performing the test must be acid washed in either diluted HCl (two-fold) or nitric acid (three-fold), and rinsed well with deionized water.
2. Specimens with barely visible hemolysis are suitable for this determination since

iron bound to hemoglobin is not detected with this procedure. Specimens showing definite hemolysis (pink to red appearance) are unsuitable.

3. No correction for bilirubin or turbidity is necessary with this method since calculation of Δabsorbance based on readings before and after addition of the color reagent corrects for any absorbance contributed by these factors.

4. Some authors have reported that $MgCO_3$ not only binds excess Fe^{2+} but also removes Fe^{3+} from transferrin. Results obtained with the method described above are similar to those obtained with methods employing resins instead of $MgCO_3$.

5. Since differences in the adsorption properties of $MgCO_3$ have been reported, it is advisable always to use products from the same manufacturer and to confirm the suitability of the reagent when a new lot is being used.

REFERENCES

White, J. M., and Flashka, H. A.: An automated procedure, with use of ferrozine for assay of serum iron and total iron-binding capacity. Clin. Chem., *19*:526–528, 1973.

Orynich, R. E., Tietz, N. W., and Fiereck, E. A.: Unpublished modification of the method by White and Flashka.

ALTERNATIVE METHOD FOR THE DETERMINATION OF IRON IN SERUM

[As Recommended by the International Committee for Standardization in Hematology (ICSH)]

PRINCIPLE

Serum proteins are precipitated with a reagent containing hydrochloric acid (to dissociate Fe^{3+}), thioglycolic acid (to reduce Fe^{3+} to Fe^{2+}), and trichloroacetic acid (to precipitate proteins). The ferrous iron is reacted with a solution of the disodium salt of bathophenanthroline sulfonate in acetate buffer, and the colored complex is measured spectrophotometrically at 535 nm.

REAGENTS

1. Protein precipitant. Dissolve 100 g trichloroacetic acid, AR, 30 ml thioglycolic acid, and 2 moles hydrochloric acid in deionized water and dilute to 1 liter. If stored in a dark brown bottle, the reagent is stable for at least 2 months.

2. Chromogen solution. Dissolve 2 moles of sodium acetate and 250 mg of the disodium salt of bathophenanthroline sulfonate in deionized water and dilute to 1 liter.

3. Iron standard, 200 μg/100 ml. (See preceding method.)

PROCEDURE

1. Place 2 ml of serum into an acid washed centrifuge tube and add 2 ml of protein precipitant solution. Mix thoroughly with a vortex mixer and allow the solution to stand for 5 min. Centrifuge until an optically clear supernatant is obtained.

2. To prepare a standard and reagent blank, add 2 ml of iron standard solution and 2 ml of iron-free water, respectively, into appropriately labeled test tubes. Add 2 ml of protein precipitant solution to each.

3. Centrifuge all "test" tubes and transfer 2 ml of the clear supernatant into a test tube marked "test."

4. Add 2 ml of chromogen solution to the filtrate, the standard, and the blank. Allow to stand for at least 5 min. Determine the absorbance of the standard and the unknown at 535 nm in a spectrophotometer set to zero absorbance against the blank.

CALCULATION

$$\text{Serum iron in } \mu g/100 \text{ ml} = \frac{A_U}{A_S} \times 200$$

COMMENTS

Rice and Fenner have evaluated this procedure and have suggested the following changes:

Optically clear filtrates are more consistently obtained if the HCl solution in the protein precipitant is reduced to 1.0 mol/l and if the mixture of serum and protein precipitant is heated for 15 to 20 min at 56°C.

The same authors also recommend substitution of the bathophenanthroline salt with the more sensitive and much more economical iron reagent ferrozine (see preceding iron method).

REFERENCES

Lewis, S. M.: International Committee for Standardization in Hematology: Proposed recommendations for measurement of serum iron in human blood. Am. J. Clin. Path., *56*:543–545, 1971.

Rice, E. W., and Fenner, H. E.: Study of the ICSH proposed reference method for serum iron assay: obtaining optically clear filtrates and substitution of ferrozine. Clin. Chim. Acta, *53*:391–393, 1974.

NORMAL VALUES

Average plasma and serum iron levels at birth approach 200 μg/100 ml, but fall off rapidly within hours to average values below 50 μg/100 ml and then increase to normal adult levels after the first 3 weeks of life. Adults males have a normal range of 60 to 150 μg/100 ml, and values for females range from 50 to 130 μg/100 ml. Serum iron levels decrease in elders to levels of 40 to 80 μg/100 ml.

Procedures for serum iron that do not require protein precipitation (see the section on serum iron determinations) give normal values that are approximately 10 to 20 μg/100 ml higher than those just listed. It is felt that these values are closer to the true serum values and that the lower values observed with the other methods are due to loss of iron with the protein precipitate.

Normal values for the TIBC in healthy adults are 270 to 380 μg/100 ml, with some authors reporting normal ranges of 300 to 360 μg/100 ml. TIBC values tend to decrease with age and are an average of 70 μg/100 ml lower in individuals above 70 years of age. Values for TIBC obtained with methods not requiring protein precipitation are 280 to 400 μg/100 ml.

Copper

PHYSIOLOGY AND CLINICAL SIGNIFICANCE

Copper belongs to the group of essential trace elements, with a daily requirement of 2.5 mg. The total copper present in an adult is about 100 mg, with a distribution of about 50 per cent in muscle tissue, 20 per cent in bones, 18 per cent in the liver, and the remainder in other body tissues, erythrocytes, and plasma. A deficiency in copper results in severe derangements in growth and metabolism. More specifically, there appears to occur a not clearly understood impairment of erythropoiesis resulting in a hypochromic, microcytic anemia; a decrease in erythrocyte survival time; an impaired mitochondrial function; and a decrease in the catalytic action of a number of copper-containing enzymes (e.g., tyrosinase, ceruloplasmin, and cytochrome oxidase).

Most of the copper that is ingested is lost through the stool; only a small portion is absorbed by the upper small intestine and reaches the blood. In plasma, most of it is first loosely bound to albumin, and then incorporated into a plasma α_2 globulin, ceruloplasmin, which is the copper transport protein (see p. 346 and p. 649). A variable but small amount (about 6 per cent) of copper is loosely bound to albumin (= direct reacting copper), and an even smaller quantity is dialyzable. The remainder of the body copper is mainly incorporated into various kinds of copper storage proteins, such as *erythrocuprein* in the erythrocytes, *cerebrocuprein* in brain, and *hepatocuprein* in the liver. All three copper storage proteins are colorless and have a molecular mass of 33,600 daltons, containing two atoms of Cu per molecule. All have similar amino acid compositions, identical immunochemical determinants, and similar sedimentation properties during ultracentrifugation. Copper that has once been absorbed is retained, and only very small amounts are excreted in urine.

The most significant clinical application of copper determinations is in the diagnosis of hepatolenticular degeneration (Wilson's disease). This disease is associated with a decrease in the synthesis of ceruloplasmin, which results in a low serum level of this enzyme ($< 20\ \mu g/100$ ml). The amount of free and albumin bound copper, however, is greater than normal, and this is attributed to a greater and uncontrolled rate of absorption of copper. Despite this fact, the total serum copper concentration is generally decreased (because of low ceruloplasmin values). The amount of copper deposited in tissues (e.g., liver, brain) is greatly increased and there is also an increased urinary excretion of copper, possibly due to the increase in free serum copper. Thus, the determination of serum ceruloplasmin, of total serum copper, and of the urinary copper are of great help in the diagnosis of this disease.

The low content of ceruloplasmin in Wilson's disease may possibly be explained by the synthesis of an abnormal molecule containing no sialic acid. Such defective molecules are rapidly picked up from serum by parenchymal cells of the liver. Wilson's disease is generally treated with penicillamine (β,β-dimethylcysteine), which chelates copper and, thus, promotes its excretion in the urine.

Low serum copper levels have also been observed in a number of hypoproteinemias as a result of malnutrition, malabsorption, and the nephrotic syndrome. In the last condition an appreciable amount of ceruloplasmin reaches the urine, but in normal urine there is essentially no ceruloplasmin.

Increased serum copper levels (hypercupremia) are found in a number of acute and chronic diseases, such as malignant diseases (including leukemia), hemochromatosis, biliary cirrhosis, thyrotoxicosis, and various infections. Serum copper levels are also high in patients who are taking contraceptives or estrogens. It is believed that the increase in these hormones during pregnancy is also responsible for the high copper and ceruloplasmin levels in this condition.

METHODS FOR THE DETERMINATION OF COPPER

DETERMINATION OF COPPER BY ATOMIC ABSORPTION SPECTROPHOTOMETRY

Atomic absorption spectrophotometry has been successfully applied to the determination of copper in various biological fluids. In one approach, the *serum* sample is diluted with an equal volume of deionized water and directly aspirated into the flame. Because of the high viscosity of the sample, its aspiration rate is slower than that of the standard, giving rise to false low results. It has therefore been recommended that the viscosity of the standards be changed by diluting them with an equal volume of a 6-fold dilution of ethylene glycol with deionized water[21] or by incorporating 10 ml of glycerol, AR, into each 100 ml of standard solution.[10] Ichida and Nobuoka[21] have also recommended the incorporation of L-histidine in a final concentration of 1.24 mmol/l, which compensates for the effect of other ions present in serum. The wavelength used is either 280 nm or the resonance line at 324.7 nm. The latter has no nearby interference and is in a region in which photomultipliers have high sensitivity and therefore operate at low noise levels.

In Henry's modification[17] of the method by Piper and Higgins,[42] serum is deproteinized with trichloroacetic acid, 25 g/100 ml, and the supernatant is aspirated into an atomic absorption spectrophotometer set at a wavelength of 324.7 nm. This concentration of TCA releases protein-bound copper, as does HCl, which is employed in several other methods. The standards employed contain copper in concentrations of 100 and 200 μg/100 ml, respectively, without any other additions.

Copper, liberated from proteins by acid treatment and protein precipitation, can also be complexed with ammonium pyrrolidine dithiocarbamate (APDC) and subsequently extracted into an organic solvent (e.g. *n*-butyl acetate).[5] This approach can also be used for the analysis of erythrocytes as well as urine specimens with or without prior wet ashing of the sample. This allows for concentration of specimens low in copper content, such as normal urine.

If undiluted, acidified *urine* (pH 2.0) is aspirated directly into the flame, the standard solution must contain a salt mixture comparable to the general "urine matrix" to compensate for salt effects. Such a solution can be prepared by placing 800 ml of deionized water into a 1 liter volumetric flask and dissolving in sequence: 0.67 ml concentrated H_2SO_4, 8.7 ml concentrated HCl, 0.312 g $CaCO_3$, 2.86 g KCl, 5.8 g NaCl, 0.418 g $MgCl_2{\cdot}6H_2O$, and 3.09 g $(NH_4)_2HPO_4$. The volume is then adjusted to 1 liter with deionized water. The copper working standard is then prepared by mixing equal volumes of stock standard with the salt mixture.[9]

DETERMINATION OF COPPER BY SPECTROPHOTOMETRIC METHODS

PRINCIPLE

Copper is released from serum proteins by treatment with dilute hydrochloric acid. The proteins are then precipitated by trichloroacetic acid and an aliquot of the filtrate is reacted with biscyclohexanoneoxalyldihydrazone (Cuprizone), which forms a stable blue-colored compound with cupric ions. Linearity holds for copper concentrations within the physiological range.

REAGENTS

1. Hydrochloric acid, 2 mol/l. Add 166 ml of HCl, AR, to deionized water, cool, and dilute to 1 liter.
2. Trichloroacetic acid, 20 g/100 ml. Dissolve 20.0 g of trichloroacetic acid (sulfate free, iron free, Eastman No. 259) in deionized water and dilute to 100 ml.
3. Buffer solution. Add 35.7 ml of saturated sodium pyrophosphate solution, 35.7 ml of saturated sodium citrate solution, and 80.3 ml of concentrated ammonium hydroxide into a 1000 ml volumetric flask and dilute to the mark with deionized water.
4. Biscyclohexanoneoxalyldihydrazone (Cuprizone, G. Frederick Smith Chemical Co., Columbus, OH.). Dissolve 0.5 g of the reagent in 100 ml of 50 per cent (v/v) ethanol.
5. Standard copper solution, 0.10 mg/ml. Dissolve 0.3928 g of copper sulfate pentahydrate, AR, in distilled water and dilute to the 1000 ml mark with deionized water.
6. Working standard, 200 μg/100 ml. Dilute 2 ml of the stock standard to 100 ml with deionized water.

PROCEDURE

1. Into 12 ml conical centrifuge tubes pipet 1.0 ml of water, 1.0 ml of working standard, and 1.0 ml of serum (or heparinized plasma), respectively.
2. Add 0.70 ml of 2.0 molar HCl to all tubes, mix, and let stand at room temperature for 10 min.
3. Add 1.0 ml of trichloroacetic acid (20 g/100 ml), mix with a thin stirring rod, and allow to stand for 10 min.

4. Cover the tubes with Parafilm, and centrifuge for 10 min at 2500 rpm.
5. Transfer 2.0 ml of the clear supernatant to another test tube.
6. Add 2.8 ml of buffer solution and 0.20 ml of Cuprizone reagent. Mix and allow to stand for 20 min.
7. Read absorbance at 620 nm.

CALCULATION

$$\frac{A_U}{A_S} \times 200 = \mu\text{g Cu/100 ml serum}$$

PROCEDURE NOTES

Syringes for the collection of blood as well as all glassware used in the test must be free of trace contamination. All reagents must be made up with chemicals of highest purity.

Some authors feel that protein precipitation by hot trichloroacetic acid extracts copper more quantitatively and thus gives better recovery results.

REFERENCE

Rice E. W.: Principles and Methods of Clinical Chemistry. Springfield IL. Charles C Thomas, Publisher, 1960, pp. 157-159.

ALTERNATE METHODS FOR THE DETERMINATION OF COPPER

In Henry's modification[16] of the method by Gubler et al.,[15] copper is dissociated from proteins by addition of hydrochloric acid, and proteins are then precipitated by hot trichloroacetic acid. The copper in the protein-free filtrate is reacted with diethyldithiocarbamate in the presence of citrate, which prevents interference from iron by forming a soluble iron complex.

In the determination of copper in *urine*, a sample is digested by wet ashing and the color produced after the addition of diethyldithiocarbamate is extracted into isoamyl alcohol.

Copper in serum can also be determined by treating the sample with a hydrochloric acid solution of oxalyldihydrazide. Addition of trichloroacetic acid precipitates the proteins, and the subsequent addition of ammonium hydroxide and acetaldehyde causes the formation of an intense lavender color as a result of the reaction of oxalyldihydrazide with Cu^{2+}.[46] The molar absorptivity of the resulting color complex is approximately 22,000 at 542 nm, in contrast to the molar absorptivity of 16,000 for the copper complex with Cuprizone and 8000 for the copper complex with diethyldithiocarbamate. Thus, the method offers the advantage of greater sensitivity, but it is slightly more complicated than the preceding method. The greatest sensitivity is obtained with 1,5-diphenylcarbohydrazide. Because of the high sensitivity of its copper complex (molar absorptivity 158,000), this reagent is suitable for use in a micro method.[35]

NORMAL VALUES

Birth to 6 months:	up to 70 μg/100 ml
6 years:	90 to 190 μg/100 ml
12 years:	80 to 160 μg/100 ml
Adult males:	70 to 140 μg/100 ml
Adult females:	80 to 155 μg/100 ml

Values in Blacks are about 8 to 12 per cent higher than in Caucasians. Values during pregnancy or in women on contraceptives may exceed 200 μg/100 ml, with increases in the total copper, the ceruloplasmin, and the direct reacting copper.

The copper concentration in *erythrocytes* is 90 to 150 μg/100 ml. *Urine* copper output in adults ranges from 15 to 30 μg/d.

Organic Anions

The organic fraction of the total anions of serum (exclusive of proteinate) is approximately 5 to 6 mmol ion charge/l (see Table 16-1). This fraction includes a great variety of individual anions, of which lactate, at a concentration of about 1 mmol/l (9 mg/100 ml), is normally the most abundant. All others (fatty acid, amino acid, β-hydroxybutyrate, as well as various other anions produced during metabolism) constitute collectively about 80 per cent of the total organic anion fraction, but the individual acids are normally present in relatively insignificant amounts. In various disease states, however, some of these acids may individually accumulate to rather significant concentrations. The levels of ketone bodies (acetoacetate and β-hydroxybutyrate) in diabetic acidosis, for example, may increase to more than 20 mmol/l (about 200 mg/100 ml). Seligson and associates[53] reported increases in the total organic anion fraction of as much as 26 mmol ion charges/l in severe renal disease. In acute methyl alcohol intoxication the quantity of formate (metabolite of methyl alcohol) may be as high as 15 mmol/l (68 mg/100 ml). In "lactic acidosis" one may see increases in the serum lactate concentration of up to 25 mmol/l (225 mg/100 ml).[53] In any of these conditions there may occur profound changes in the relative proportions of individual anions. This must be kept in mind when checking for the accuracy of electrolyte reports. (See also the discussion on the production of metabolic acidosis in Chapter 16 and refer to the calculations on page 970.)

KETONE BODIES

The metabolism of fatty acids results in the formation of a small amount of acetoacetate, which is subsequently metabolized in the peripheral tissues. In conditions in which there is carbohydrate deprivation (e.g., starvation) or decreased utilization of carbohydrates (e.g., diabetes mellitus), there is an increased production of acetoacetate, and the quantity present may exceed the capacity of the peripheral tissues to metabolize this compound (see Chapter 10, p. 485). Thus, the acetoacetate accumulates in the blood; a small part is converted to acetone by spontaneous decarboxylation, whereas the greater part is converted to β-hydroxybutyrate in accordance with the following reactions:

$$1.\quad \underset{\textbf{Acetoacetate}}{CH_3{-}CO{-}CH_2{-}COO^-} + H^+ \rightarrow \underset{\textbf{Acetone}}{CH_3{-}CO{-}CH_3} + CO_2$$

$$2.\quad \underset{\textbf{Acetoacetate}}{CH_3{-}CO{-}CH_2{-}COO^-} + NADH + H^+ \underset{\text{(in liver)}}{\overset{\beta\text{-hydroxybutyrate dehydrogenase}}{\rightleftharpoons}}$$

$$\underset{\boldsymbol{\beta}\textbf{-Hydroxybutyrate}}{CH_3{-}\overset{\overset{\large OH}{|}}{\underset{\underset{\large H}{|}}{C}}{-}CH_2{-}COO^-} + NAD^+$$

The relative proportions in which the three ketone bodies are present in blood may vary; average figures are 78 per cent β-hydroxybutyrate, 20 per cent acetoacetate, and 2 per cent acetone. Of the most commonly used methods for the detection and determination of ketone bodies in serum or urine, none reacts with all three ketone bodies. Gerhardt's ferric chloride test reacts with acetoacetate only, and the various tests employing nitroprusside are 15 to 20 times more sensitive for acetoacetate than for acetone and give no reaction at all with β-hydroxybutyrate. Thus, the tests to be described essentially detect or measure acetoacetate only. Tests for β-hydroxybutyrate are indirect; they require brief boiling of the urine to remove acetone and acetoacetate by evaporation (acetoacetate first breaks down spontaneously to acetone), which is followed by gentle oxidation of β-hydroxybutyrate to acetoacetate and acetone with peroxide, ferric ions, or dichromate. The acetoacetate thus formed may be detected with Gerhardt's tests or one of the procedures employing nitroprusside (see procedure).

CLINICAL SIGNIFICANCE

Excessive formation of ketone bodies results in increased blood levels (ketonemia) and increased excretion in the urine (ketonuria). This is observed in conditions associated with a decreased intake of carbohydrates, such as starvation, digestive disturbances, dietary imbalance, and frequent vomiting. Another, and possibly more frequent cause of increased production of ketone bodies is decreased utilization of carbohydrates, such as is found in diabetes mellitus. Both glycogen storage disease (Von Gierke's disease) and alkalosis (as a result of some obscure mechanism involving decreased carbohydrate utilization in the liver) may also result in excessive production of ketone bodies.

Semiquantitative determination of ketone bodies in blood is an extremely helpful guide (more so than the determination of these compounds in urine) in the treatment of ketonemia associated with diabetes. In fact, some authors (e.g., Lee and Duncan[26]) claim that knowledge of the degree of ketonemia is the most valuable guide to insulin therapy and offers more information than knowledge of the degree of ketonuria or of the blood glucose levels. Although this viewpoint is opposed by others, it still points out the great importance of detecting and measuring ketone bodies in serum.

DETERMINATION OF KETONE BODIES IN SERUM

Although a number of quantitative and semiquantitative methods for the estimation of ketone bodies have been devised, it is generally agreed that the use of the semiquantitative Acetest and Ketostix reagents (Ames Co., Div. Miles Laboratories, Elkhart, IN. 46514) offers information sufficient for clinical purposes.

SPECIMENS

The sera should be free of visible hemolysis, since discoloration of the tablet or reagent strip may occur if an excessive amount of hemoglobin is present. If there is any significant delay in performing the determination, the specimens should be kept well stoppered at refrigerator temperatures.

DETECTION OF KETONE BODIES BY ACETEST

PRINCIPLE

The Acetest tablets contain a mixture of glycine, sodium nitroprusside, disodium phosphate, and lactose. Acetoacetate or acetone in the presence of glycine will form a complex of lavender-purple color with nitroprusside. The disodium phosphate provides an optimum pH for the reaction and lactose enhances the color.

PROCEDURE

A detailed procedure for the detection of ketone bodies by Acetest is supplied by the manufacturer with each package of tablets, and the reader is referred to these instructions.

Acetest was designed mainly for the detection of ketone bodies in urine. If serum is used, the tablets should be crushed and a drop of serum should be added to the powder. Failure to do so will result in false low results. It has recently been reported that this procedure is more reliable than is the use of the Ketostix described later.[23]

A 1+ positive reaction (appearance of a purple-lavender color) indicates the presence of 5 to 10 mg of ketone bodies/100 ml. A color chart provided with the package may be used to estimate actual concentrations of the ketone bodies. A 4+ reaction (dark lavender color) corresponds to approximately 40 to 50 mg/100 ml. If desirable, dilutions of serum with saline can be prepared to measure levels of ketone bodies above 40 mg*. Since a 4+ reaction in an undiluted sample corresponds to approximately 40 mg/100 ml, a 4+ reaction in a 4-fold dilution corresponds to approximately 160 mg/100 ml. Similar calculations can be performed if other dilutions are used.

DETECTION OF KETONE BODIES BY KETOSTIX

Ketostix is a modification of the nitroprusside test in which a reagent strip is used instead of a tablet. The Ketostix test gives a positive reaction within 15 seconds with a specimen containing 5 to 10 mg of acetoacetate per 100 ml. Approximate serum acetoacetate values assigned to the color blocks representing increasingly more positive reactions are 10 mg/100 ml for "small," 30 mg/100 ml for "moderate," and 80 mg/100 ml for "large." Acetone reacts also, but to a considerably lesser extent.

DETERMINATION OF KETONE BODIES IN URINE

Acetest and Ketostix are also suitable for the detection of ketone bodies in urine. The sensitivity and specificity of the tests are the same as outlined for serum. The original test by Rothera has essentially been replaced by these two modifications.

Gerhardt's test is based on the reaction of ferric chloride with acetoacetate, resulting in the production of a wine red color. Other compounds[59] such as salicylates, phenol, and antipyrine give a similar color; thus, a positive reaction merely indicates the possible presence of acetoacetate. To confirm its presence, urine is heated to decompose acetoacetate to acetone and to drive off the latter. The test is then repeated. If it is now negative, it can be assumed that the original color was due to acetoacetate. This test also has been replaced by the Ketostix and Acetest procedures.

REFERENCES

Fraser, J., Fetter, M. C., Mast, R. L., and Free, A. H.: Clin. Chim. Acta, *11*:372, 1965.
Free, A. H., and Free, H. M.: Am. J. Clin. Path., *30*:7, 1958.

PROTEINATE

Plasma (serum) contains a number of electrolytes, including protein, that are normally not determined in connection with the customary "electrolyte pattern." The average amount of proteins (= 70 g/l) contributes approximately 16 mmol ion charge/l to the total anions and thus is the third largest fraction among the anions (see Chapter 16, Table 16–2). The protein concentration (in g/100 ml) times 2.25 gives the approximate number of

* Since the reaction is affected by proteins, any dilution with saline introduces a certain error.

mmols of ion charge/l (example: 7.0 g protein/100 ml × 2.25 = 15.75 mmol/l). Therefore, major changes in the protein content of plasma (serum) can cause significant proportional derangements in the relative anion composition.

The chemistry of proteins, the methods for their determination, and a discussion of their clinical significance are given in Chapter 7.

LACTATE

CLINICAL SIGNIFICANCE

Lactic acid, present in *blood* entirely as lactate ion ($pK_1 = 3.86$), is an intermediary product of carbohydrate metabolism and is derived mainly from muscle cells and erythrocytes. It is normally metabolized by the liver. The blood lactate concentration is, therefore, affected by the rate of production as well as the rate of metabolism. During exercise, lactate levels may increase significantly, from an average normal level of about 0.9 mmol/l to about 12 mmol/l. However, pyruvate levels increase under these conditions as well, and the normal ratio of lactate/pyruvate remains approximately 6 or 7/1. (Some investigators have reported higher normal ratios, e.g., 9/1 or even higher, owing to variations in technique and sample collection, especially when working under routine conditions).

Severe oxygen deprivation of tissues results in a blockage of aerobic oxidation of pyruvic acid in the tricarboxylic acid cycle and subsequent glycolytic reduction of pyruvate to lactate. This leads to a severe acidosis, called "lactic acidosis," which is associated with a significant increase in the lactate/pyruvate ratio in blood. Such extreme findings signal deterioration of the cellular oxidative process and are associated with marked hyperpnea, weakness, fatigue, stupor, and finally coma. Conditions of this type are frequently irreversible even when treatment for acidosis and hypoxia is instituted; examples are irreversible stage of shock, diabetic coma without ketosis, and a variety of illnesses in the terminal stage.

Hypoxia with hypoxemia is frequently seen in shock, cardiac decompensation, hematologic disorders, and pulmonary insufficiency. Such conditions are also associated with increases in blood lactate but are frequently reversible after treatment for hypoxia and the primary condition.

The liver can normally metabolize significantly more lactate than is produced. In the case of decreased perfusion of the liver, however, there may be significantly reduced removal or even generation of lactate by the liver. Thus, this organ may play an important role in the production of lactic acidosis.

Simultaneous determinations of lactate in *arterial and coronary sinus blood* after fast arterial pacing are sometimes requested to facilitate the diagnosis of doubtful angina pectoris. In such cases, high accuracy and precision are required, since differences in the lactate concentration in the two samples may be very small.

Lactate in *spinal fluid* normally parallels blood levels. In case of biochemical alterations in the central nervous system, however, CSF (Csf) lactate values change independently of blood values. Increased CSF levels are seen in cerebrovascular accidents (CVA), intracranial hemorrhage, meningitis, epilepsy, and other CNS disorders.

THE DETERMINATION OF LACTATE IN WHOLE BLOOD[28,30,32]

SPECIMENS

Collection of a satisfactory specimen for lactate analysis requires special procedures to prevent changes in lactate while and after the specimen is drawn. The patient should be fasting and at complete rest.

If a *venous specimen* is desired, the specimen is best drawn without the use of a tourniquet or immediately after the tourniquet has been applied. If there is any delay in obtaining the specimen, the tourniquet should be removed after the puncture has been performed, and the blood should be allowed to circulate for at least 2 min before the sample is withdrawn.

Both *venous* and *arterial blood* may conveniently be collected in heparinized syringes and immediately delivered into a premeasured amount of chilled protein precipitant, such as metaphosphoric acid or perchloric acid. Alternatively, the blood may be allowed to flow directly into the protein precipitant, preferably after discarding the first few ml of blood. The clear supernatant, after centrifugation, is stable at refrigerator temperature for up to 8 days.

Specimens collected as described are also suitable for the determination of pyruvate.

If blood is not preserved as indicated, lactate will increase rapidly in blood because of glycolysis. Increases may be as great as 20 per cent within 3 min and 70 per cent within 30 min at 25°C.

If *plasma* is required as specimen,[63] as for the DuPont ACA, the specimen is best collected in a container with 10 mg sodium fluoride (NaF) and 2 mg potassium oxalate ($K_2C_2O_4$)/ml of blood, followed by immediate chilling of the specimen and separation of the cells within 15 min.[49]

PRINCIPLE

Lactate in the presence of NAD^+ and LD is oxidized to pyruvate. The NADH formed in this reaction is measured spectrophotometrically at 340 nm, and serves as a measure of the lactate concentration.

$$\text{L-lactate} + NAD^+ \underset{\text{pH 9.0 to 9.6}}{\overset{\text{LD}}{\rightleftharpoons}} \text{pyruvate} + NADH + H^+$$

The equilibrium of the reaction normally lies far to the left. However, by using a pH of 9.0 to 9.6 and an excess of NAD^+, and by trapping the reaction product pyruvate with hydrazine, the equilibrium can be shifted to the right. Pyruvate can also be removed by reacting it with L-glutamate in the presence of alanine aminotransferase (ALT, GPT).

Use of Tris buffer results in faster completion of a side reaction between NAD^+ and hydrazine and, thus, prevents the "creeping" of blank values observed when glycine buffer is used.[28]

Metaphosphoric acid is preferred by some as the protein precipitating agent, since perchloric acid does not precipitate mucoproteins and it interferes with the enzymatic method for pyruvate (if this analyte is determined on the same filtrate as lactate). Also, the enzymatic LD reaction proceeds much more slowly in the presence of perchloric acid.

Because of its high specificity and simplicity, the enzymatic method is the method of choice for measuring lactate, although other methods may also be used (e.g., gas chromatography,[49] colorimetry[16,17]).

REAGENTS

1. Metaphosphoric acid (MPA), 5 g/100 ml. Dissolve 5 g of MPA in water and dilute to 100 ml. Prepare fresh daily. (See Comment.)

2. Metaphosphoric acid, 3 g/100 ml. Dissolve 3 g of MPA in water and dilute to 100 ml. Prepare fresh daily.

3. Tris–hydrazine buffer mixture, pH 9.6 (Tris, 79 mmol/l; hydrazine, 400 mmol/l). To about 700 ml of 1 molar NaOH, add 9.57 g of tris(hydroxymethyl)aminomethane, 52.05 g of hydrazine sulfate, and 1.85 g of ethylenediamine tetraacetate, disodium salt (to chelate metal ions and prevent inhibition of lactate dehydrogenase). Adjust the pH to 9.6 with 1 molar NaOH and dilute the solution to 1000 ml with distilled water. This buffer is stable for 8 days at 4°C.

4. NAD^+ solution, 27 mmol/l. Dissolve NAD^+ in distilled water to obtain a 20 mg/ml solution. This solution is stable for about 48 h at 4°C.

5. Lactate dehydrogenase solution. Dilute the stock solution (Sigma, grade III, crystalline beef heart LD, 10 mg protein/ml) with saline to a protein concentration of 3 mg/ml (0.3 ml stock diluted to 1.0 ml). Keep refrigerated and use within 48 h.

6. Lactate standard, 1 mmol/l (9 mg/100 ml). Dissolve 9.60 mg lithium L-lactate (Calbiochem, A Grade) in water in a 100 ml volumetric flask. Add 25 µl concentrated H_2SO_4 and bring to volume with water. Stable indefinitely at 4°C.

Lactate stock standards are also commercially available (Bio-Dynamics, Inc., Indianapolis, IN. 46250; Sigma Chemical Co., St. Louis, MO. 63178; and others).

COLLECTION OF SPECIMENS

To an appropriate number of tared test tubes (15 × 100 mm) add 6 ml of MPA (5 g/100 ml). Weigh and place the tubes into ice water. Run assays in duplicate. Collect about 5 ml of arterial or venous blood in a heparinized syringe without production of air bubbles and add 2 ml of the blood to each of two tubes within 15 s of withdrawal.

Mix the blood with the MPA (5 g/100 ml) solution by three inversions of the tubes. Once the proteins have been precipitated, there is no further need to chill the tubes. Allow the tubes to warm to room temperature and determine their weights. For complete protein precipitation, allow the tubes to stand for at least 15 min. Centrifuge at high centrifugal force (at least 1500 × *g* but preferably 3000 × *g*) for 15 to 30 min, depending on the centrifugal force used.[38] The supernatant *must* be clear.

Determine the dilution factor (D) as follows:

$$D = \frac{W_b - W_t}{W_b - W_m}$$

where W_t is the weight of the empty tube, W_m is the weight of the tube plus MPA solution, and W_b is the weight of the tube plus the MPA and blood.

If the method is followed as described, the dilution factor (D) is about 4.

PROCEDURE

1. To each of three cuvets labelled test, standard, and blank, respectively, add 2.0 ml of Tris–hydrazine buffer.

2. Add 0.1 ml of supernatant solution to the test cuvet, 0.1 ml of lactate standard to the standard cuvet, and 0.1 ml of MPA (3 g/100 ml) to the blank cuvet.

3. Mix the solutions and add 30 µl of LD solution and 0.2 ml of NAD^+ solution to each cuvet.

4. Mix the contents again; after approximately 15 min at room temperature, measure the absorbance at 340 nm against the blank. A separate reagent blank and working lactate standard should be determined with each run.

Versatol Acid-Base "Alkalosis" and Acid-Base "Normal" (General Diagnostic, Div. Warner-Lambert Co.) are suitable as quality control sera. The control sera are treated the same way as all unknowns.

Results obtained with the lactate method as described are linear up to concentrations of approximately 5.6 mmol/l (50 mg/100 ml).

CALCULATION

$$\text{Lactate (mmol/l)} = \frac{A_t}{A_s} \times 1 \times D$$

where t and s stand for the test and standard, respectively. and D is the dilution factor.

Alternatively, utilizing the extinction coefficient of NADH,

$$\text{Lactate (mmol/l)} = A_t \times \frac{2.33}{6.22} \times \frac{D}{0.1}$$

where

A_t = absorbance of the test

2.33 = total volume of the test

6.22 = millimolar extinction coefficient for NADH

0.1 = volume of supernatant

To convert mmol/l to mg/100 ml, multiply mmol/l by 9.

This calculation does not allow for the error due to the contraction of the liquid volume caused by the removal of dissolved solids, such as proteins, hemoglobin, and MPA (= displacement error). Some investigators have, therefore, applied average correction factors to compensate for this error. (See Alternative Procedures for Collection of Samples, and Comments.) Such corrections have the limitation that they do not allow for the differences in protein and hemoglobin concentration in various individuals.

ALTERNATIVE PROCEDURES FOR COLLECTION OF SAMPLES

Place 6.0 ml of MPA (5 g/100 ml) into graduated 12 ml centrifuge tubes and chill. Add 2 ml of arterial or venous blood collected in a heparinized syringe, and treat the mixture as described in the preceding section.

After centrifugation, carefully read from the tube scale the total volume and the volume of the precipitate, and calculate the dilution factor as follows:

$$D = \frac{V_t - V_p}{V_t - V_a}$$

where

D = dilution factor

V_t = total volume (ml)

V_p = volume of precipitate (ml)

V_a = volume of MPA added (ml)

Enter D into the formula given in the preceding section.

This method of correction for the displacement error caused by the volume of precipitate does not take into account any water trapped in the precipitate. Thus, the result is an underestimate of the actual lactate concentration.

COMMENTS

Mattenheimer[33] assumes that blood has a specific gravity of 1.06; accordingly, 2.0 ml of blood weighs 2.12 g. Furthermore, assuming an average water content of blood of 80 per cent, the volume (in ml) of the extract is (2.12 × 80/100) + 6.0 (= ml MPA) = 7.696. On this basis, the dilution factor is 7.696/2 or 3.848.

Metaphosphoric acid is generally composed of a variable mixture of HPO_3 and $NaPO_3$.[17] MPA in aqueous solution forms several polymers, $(HPO_3)_x$, and, catalyzed by hydrogen ions, it hydrolyzes to the ortho acid as follows: $HPO_3 + H_2O \rightarrow H_3PO_4$. Since orthophosphoric acid does not precipitate proteins, MPA solutions retain their protein precipitating activity for only about one week at refrigerator temperatures. By adjusting the pH to between 8 and 9 by addition of Na_2CO_3 or NaOH, a solution of sodium metaphosphate is made stable indefinitely at room temperature.

NORMAL VALUES

Fasting *venous blood* has a lactate concentration of 0.5 to 1.3 mmol/l (5 to 12 mg/100 ml). *Arterial blood* contains 0.36 to 0.75 mmol/l if the patient is in a state of complete rest. Patients under usual hospital conditions show a wider range, up to 0.9 to 1.7 mmol/l (8 to 15 mg/100 ml) for venous blood and up to 1.25 mmol/l for arterial blood. Sudden, severe exercise increases lactate levels dramatically. Even movement of leg muscles of patients at bed rest may result in significant increases above the normal. *Plasma* values are about 7 per cent higher than those seen in whole blood,[11] although differences are dependent on the procedure employed. CSF values are normally close to those found in blood but may change independently in CNS disorders. Normal 24 h urine output of lactate is 5.5 to 22 mmol/d.

PROCEDURE FOR THE DETERMINATION OF PYRUVATE IN BLOOD[30,32]

PRINCIPLE

The reaction involved in the determination of pyruvate is essentially the reverse of the reaction used in the lactate procedure.

$$\text{Pyruvate} + \text{NADH} + \text{H}^+ \xrightleftharpoons{\text{LD; pH 7.5}} \text{Lactate} + \text{NAD}^+$$

At about pH 7.5, the equilibrium constant strongly favors the reaction to the right. The method is very specific and α-oxoglutarate, oxalacetate, acetoacetate, and β-hydroxybutyrate do not interfere as is the case with some colorimetric methods.

SPECIMEN

Pyruvate in blood is extremely unstable and the same precautions detailed for lactate should be observed. A protein-free filtrate prepared with metaphosphoric acid is suitable for both lactate and pyruvate determinations.

The collection of blood and the preparation of a protein-free filtrate are discussed under lactate.

REAGENTS

1. Tris buffer, 0.75 mol/l. Dissolve 90.8 g tris(hydroxymethyl)aminomethane in distilled water and bring to a total volume of one liter.

2. NADH solution, 0.013 mol/l. Dissolve 10 mg of β-nicotinamide adenine dinucleotide, reduced form, disodium salt, in 1 ml of $NaHCO_3$ solution (1 g/100 ml). The solution should be refrigerated and used within 48 h.

3. Lactate dehydrogenase solution. Dilute the stock solution (Sigma Grade III, crystalline beef heart LD, 10 mg protein/ml) with 0.15 molar NaCl to a protein concentration of 3 mg/ml (e.g., 0.3 ml stock diluted to 1.0 ml). Keep refrigerated and use within 48 h.

4. Pyruvate standards.

(a) Stock pyruvate standard, 0.1 mol/l. Dissolve 1.101 g of sodium pyruvate (Sigma, Type II) in 0.1 molar HCl and dilute to a volume of 100 ml. Store in refrigerator.

(b) Working pyruvate standard, 0.05 mmol/l. Dilute the stock solution 2000-fold with MPA, 3 g/100 ml (e.g., 50 μl diluted to 100 ml). Prepare fresh daily.

PROCEDURE

1. To each of three appropriately labelled cuvets, add 1.0 ml of supernatant solution (test), pyruvate working standard, or MPA, 3 g/100 ml (blank) respectively, followed by 0.5 ml of Tris buffer, 0.75 mol/l, and 30 µl of NADH solution, 0.013 mol/l. Mix after each addition.
2. Measure absorbances at 340 nm against water.
3. Add to each cuvet 30 µl of LD solution (3 mg protein/ml), mix, and incubate at room temperature for 2 min.
4. Read absorbances again after 2 min and at additional 1 min intervals thereafter until stable readings are obtained.

CALCULATION

$$\text{Pyruvate (mmol/l)} = \frac{\Delta A_t - \Delta A_b}{\Delta A_s - \Delta A_b} \times 0.05 \times D$$

where t and s indicate test and working standard, respectively, and b indicates blank. D is the dilution factor.

Alternatively, utilizing the extinction coefficient of NADH,

$$\text{Pyruvate (mmol/l)} = (\Delta A_t - \Delta A_b) \times \frac{1.56}{6.22} \times \frac{D}{1.0}$$

where

ΔA_t = change in absorbance of test

1.56 = total volume in ml

6.22 = millimolar extinction coefficient of NADH

1.0 = volume of supernatant in ml

D = dilution factor

To convert mmol/1 to mg/100 ml, multiply mmol/l by 8.8.

NORMAL VALUES

Fasting *venous blood* withdrawn with the patient at rest:

0.034 to 0.102 mmol/l (0.3 to 0.9 mg/100 ml)

Normal 24 h *urine* output of pyruvate is approximately 1 mmol/d or less.
Pryce[43] gives values for CSF of 0.06 to 0.19 mmol/l.

NOTES

1. Pyruvate in blood is extremely unstable, a significant decrease being observable as early as 1 min after withdrawing the blood. Pyruvate in metaphosphoric acid filtrates of blood is stable for 6 days at room temperature and for 8 days at 4°C.
2. It is recommended that pyruvate standard solutions be freshly prepared. Pyruvic acid polymerizes in solution and may then behave differently in enzymatic reactions.[11,17]
3. If MPA with pH adjusted to between 8 and 9 is used, as explained under Comments in the lactate procedure, the pH of Tris buffer must be adjusted so that the pH of the reaction mixture is between 7.4 and 7.6.

REFERENCES

1. Alcock, N. W., and MacIntyre, I.: Methods for estimating magnesium in biological materials. *In*: Methods of Biochemical Analysis. D. Glick, Ed. New York, Interscience Publishers, Inc., 1966, vol. 14.
2. Askevold, R., and Vellar, O. D.: A simple method for the determination of iron in serum and other biological materials. Scand. J. Clin. Lab. Invest., *20*: 122–128, 1967.
3. Balcerzak, S. P., Westerman, M. P., Heinle, E. W., and Taylor, F. H.: Measurement of iron stores using deferoxamine. Ann. Int. Med., *68*: 518–525, 1968.
4. Berman, E.: Biochemical applications of flame emission and atomic absorption spectroscopy. Applied Spectroscopy, *29*: 1–9, 1975.
5. Blomfield, J., and MacMahon, R. A.: Micro determination of plasma and erythrocyte copper by atomic absorption spectrophotometry. J. Clin. Path., *22*: 136–143, 1969.
6. Burnett, R. W., and Noonan, D. C.: Calculations and correction factors used in determination of blood pH and blood gases. Clin. Chem., *20*: 1499–1506, 1974.
7. Cali, J. P., Mandel, J., Moore, L., and Young, D. S.: A referee method for the determination of calcium in serum. National Bureau of Standards Special Publication 260–36, 1972.
8. Council on Drugs: Evaluation of lithium carbonate for treatment of manic-depressive psychosis. J.A.M.A., *215*: 1486–1488, 1971.
9. Dawson, J. B., Ellis, D. J., and Newton-John, H.: Direct estimation of copper in serum and urine by atomic absorption spectroscopy. Clin. Chim. Acta, *21*: 33–42, 1968.
10. Determination of copper in serum. Clinical Methods for Atomic Absorption Spectroscopy. Manual, Perkin Elmer Corporation, Norwalk, CT., 1973.
11. Drewes, P. A.: Carbohydrate derivatives and metabolites. *In*: Clinical Chemistry—Principles and Technics. R. J. Henry, D. C. Cannon, and J. W. Winkelman, Eds. Hagerstown, Harper and Row, Publishers, 2nd ed., 1974, pp. 1327–1369.
12. Durst, R. A.: Temperature variation detection in a blood pH gas analyzer. Clin. Chem., *21*: 176–177, 1975.
13. Fingerhut, B.: A non-mercurimetric automated method for serum chloride. Clin. Chim. Acta, *41*: 247–253, 1972.
14. Fuchs, C., Paschen, K., Spieckermann, P. G., and v. Westberg, C.: Bestimmung des ionisierten Calciums im Serum mit einer ionenselektiven Durchflusselektrode: Methodik und Normalwerte. Klin. Wochenschr., *50*: 824–832, 1972.
15. Gubler, C. J., Lahey, M. E., Ashenbrucker, H., Cartwright, G. E., and Wintrobe, M. M.: Studies on copper metabolism. I. A method for the determination of copper in whole blood, red blood cells and plasma. J. Biol. Chem., *196*: 209–220, 1952.
16. Henry, R. J.: Clinical Chemistry—Principles and Technics. New York, Harper and Row, Publishers, 1964.
17. Henry, R. J., Cannon, D. C., and Winkelman, J. W., Eds.: Clinical Chemistry—Principles and Technics, 2nd ed. Hagerstown, Harper and Row, Publishers, 1974.
18. Henry, R. J., Sobel, C., and Chiamori, N.: On the determination of serum iron and iron binding capacity. Clin. Chim. Acta, *3*: 523–530, 1958.
19. Henry, R. J., and Szustkiewicz, C. P.: The preparation of protein-free filtrates. *In*: Clinical Chemistry—Principles and Technics. R. J. Henry, D. C. Cannon, and J. W. Winkelman, eds. Hagerstown, Harper and Row, Publishers, 2nd ed., 1974, pp. 389–404.
20. Husdan, H., Rapoport, A., Locke, S., and Oreopoulos, D.: Effect of venous occlusion of the arm on the concentration of calcium in serum and methods for its compensation. Clin. Chem., *20*: 529–532, 1974.
21. Ichida, T., and Nobuoka, M.: Determination of serum copper with atomic absorption spectrophotometry. Clin. Chim. Acta, *24*: 299–303, 1969.
22. Iseri, L. T., Freed, J., and Bures, A. R.: Magnesium deficiency and cardiac disorders. Am. J. Med., *58*: 837–846, 1975.
23. James, R. C., and Chase, G. R.: Evaluation of some commonly used semiquantitative methods for urinary glucose and ketone determinations. Diabetes, *23*: 474–479, 1974.
24. Ladenson, J. H., and Bowers, G. N., Jr.: Free calcium in serum. I. Determination with the ion-specific electrode, and factors affecting the results. Clin. Chem., *19*: 565–574, 1973.
25. Ladenson, J. H., and Bowers, G. N., Jr.: Free calcium in serum. II. Rigor of homeostatic control, correlations with total serum calcium, and review of data on patients with disturbed calcium metabolism. Clin. Chem., *19*: 575–582, 1973.
26. Lee, C. T., and Duncan, G. G.: Diabetic coma: the value of a simple test for acetone in the plasma—an aid to diagnosis and treatment. Metabolism, *5*: 144–149, 1956.
27. Li, T. K., and Piechocki, J. T.: Determination of serum ionic calcium with an ion-selective electrode: evaluation of methodology and normal values. Clin. Chem., *17*: 411–416, 1971.
28. Livesley, B., and Atkinson, L.: Accurate quantitative estimation of lactate in whole blood. Clin. Chem., *20*: 1478, 1974.
29. Low, J. C., Schaaf, M., Earll, J. M., Piechocki, J. T., and Li, T. K.: Ionic calcium determination in primary hyperparathyroidism. J. A. M. A., *223*: 152–155, 1973.
30. Lubran, M.: Measurement of lactic and pyruvic acid in biological fluids. *In*: Laboratory Diagnosis of Endocrine Diseases. F. W. Sunderman and F. W. Sunderman, Jr., Eds. St. Louis, Warren H. Green, Inc., 1971, pp. 401–408.

31. Mandel, E., and Niespodrzany, L.: Unpublished modification of method by W. Ramsay.
32. Marbach, E. P., and Weil, M. H.: Rapid enzymatic measurement of blood lactate and pyruvate. Clin. Chem., *13*: 314-325, 1967.
33. Mattenheimer, H.: Micromethods for the Clinical Biochemical Laboratory. Ann Arbor, Ann Arbor Science Publishers, Inc., 1970.
34. McLean, F. C., and Hastings, A. B.: The state of calcium in the fluids of the body. I. The conditions affecting the ionization of calcium. J. Biol. Chem., *108*: 285–322, 1935.
35. Mikac-Devic, D.: Micromethod for copper determination in serum and urine with 1,5-diphenylcarbohydrazide. Clin. Chim. Acta, *26*: 127–130, 1969.
36. Moorehead, W. R., and Biggs, H. G.: 2-Amino-2-methyl-1-propanol as the alkalizing agent in an improved continuous-flow cresolphthalein complexone procedure for calcium in serum. Clin. Chem., *20*: 1458–1460, 1974.
37. Natelson, S.: Techniques of Clinical Chemistry, 3rd ed. Springfield, IL., Charles C Thomas, Publisher, 1971.
38. Neville, J. F., Jr., and Gelder, R. L.: Modified enzymatic methods for the determination of L-(+)-lactic and pyruvic acids in blood. Amer. J. Clin. Path., *55*: 152–158, 1971.
39. Oliva, P. B.: Lactic acidosis. Am. J. Med., *48*: 209–225, 1970.
40. Peters, J. P., and Van Slyke, D. D.: Quantitative Clinical Chemistry, Methods. Baltimore, The Williams & Wilkins Co., 1932, vol. 2, pp. 245–283.
41. Pickup, J. F., Jackson, M. J., Price, E. M., and Brown, S. S.: Assessment of the reference method for determination of total calcium in serum. Clin. Chem., *20*: 1324–1330, 1974.
42. Piper, K. G., and Higgins, G.: Estimation of trace metals in biological materials by atomic absorption spectrophotometry. Proc. Assoc. Clin. Biochem., *4*: 190–197, 1967.
43. Pryce, J. D., Gant, P. W., and Saul, K. J.: Normal concentrations of lactate, glucose, and protein in cerebrospinal fluid. Clin. Chem., *16*: 562–565, 1970.
44. Ramsay, W. N. M.: *In*: Advances in Clinical Chemistry. H. Sobotka and C. P. Stewart, Eds. New York, Academic Press, Inc., 1958, vol. 1, pp. 2–39.
45. Ramsay, W. N. M.: The determination of iron in blood plasma or serum. Biochem. J., *53*: 227–231, 1953.
46. Rice, E. W.: Copper in serum. *In*: Standard Methods of Clinical Chemistry. D. Seligson, Ed. New York, Academic Press, Inc., 1963, vol. 4.
47. Robertson, W. G., and Peacock, M.: New techniques for the separation and measurement of the calcium fractions of normal human serum. Clin. Chim. Acta, *20*: 315–326, 1968.
48. Rosenthal, T. B.: The effect of temperature on the pH of blood and plasma in vitro. J. Biol. Chem., *173*: 25–30, 1948.
49. Savory J., and Kaplan, A.: A gas chromatographic method for the determination of lactic acid in blood. Clin. Chem., *12*: 559–569, 1966.
50. Schade, A. L., Oyama, J., Reinhart, R. W., and Miller J. R.: Bound iron and unsaturated iron-binding capacity of serum; rapid and reliable quantitative determination. Proc. Soc. Exp. Biol. Med., *87*: 443–448, 1954.
51. Schatz, B. C.: Thesis, Graduate School, University of Southern California, 1962.
52. Segal, M.: A rapid electrotitrimetric method for determining CO_2 combining power in plasma or serum. Am. J. Clin. Path., *25*: 1212–1216, 1955.
53. Seligson, D., Bluemle, L. W., Jr., Webster, G. D., Jr., and Senesky, D.: Organic acids in body fluids of the uremic patient. J. Clin. Invest., *38*: 1042–1043, 1959.
54. Severinghaus, J. W., Stupfel, M., and Bradley, A. F.: Accuracy of blood pH and pCO_2 determinations. J. Appl. Physiol., *9*: 189–196, 1956.
55. Siggaard-Andersen, O.: The Acid-Base Status of the Blood, 4th ed. Baltimore, William & Wilkins Company, 1974.
56. Skeggs, L. T., Jr: An automatic method for the determination of carbon dioxide in blood plasma. Am. J. Clin. Path., *33*: 181–185, 1960.
57. Skeggs, L. T., Jr., and Hochstrasser, H.: Multiple automatic sequential analysis. Clin. Chem., *10*: 918–936, 1964, and Technicon Method No. SE4–0005FD4, Technicon Instrument Corp., Tarrytown, NY, 1974.
58. Stewart, C. P., and Frazer, S. C.: Magnesium. *In:* Advances in Clinical Chemistry. H. Sobotka and C. P. Stewart, Eds. New York, Academic Press, Inc. 1963, vol. 6., pp. 29–65.
59. Thomas, G. H., and Howell, R. R.: Selected Screening Tests for Genetic Metabolic Diseases. Chicago, Yearbook Medical Publishers, Inc., 1973.
60. Trudeau, D. L., and Freier, E. F.: Determination of calcium in urine and serum by atomic absorption spectrophotometry (AAS). Clin. Chem., *13*: 101–114, 1967.
61. Ulmer, D. D.: Trace elements and clinical pathology. *In*: Progress in Clinical Pathology. New York, Grune & Stratton, Inc., 1966.
62. Webster, D.: The determination of serum iron after the intravenous injection of iron-dextran. J. Clin. Path., *13*: 246-248, 1960.
63. Westgard, J. O., Lahmeyer, B. L., and Birnbaum, M. L.: Use of the DuPont "Automatic Clinical Analyzer" in direct determination of lactic acid in plasma stabilized with sodium fluoride. Clin. Chem., *18*: 1334–1338, 1972.
64. Wills, M. R.: Value of plasma chloride concentration and acid-base status in the differential diagnosis of hyperparathyroidism from other causes of hypercalcaemia. J. Clin. Path., *24*: 219–227, 1971.

65. Wybenga, D. R., Cannon, D. C., and Ibbott, F. A.: Factor for correcting ionized calcium values obtained from serum collected and stored without special precautions. Clin. Chem., *18*: 715, 1972.
66. Zall, D. M., Fisher, D., and Garner, M. Q.: Photometric determination of chlorides in water. Anal. Chem., *28*: 1665–1668, 1956.

ADDITIONAL READING

1. Davenport, H. W.: The ABC of Acid-Base Chemistry, 6th ed. Chicago, University of Chicago Press, 1974.
2. Davidsohn, I., and Henry, J. B., Eds.: Todd-Sanford Clinical Diagnosis by Laboratory Methods, 15th ed. Philadelphia, W. B. Saunders Co., 1974.
3. Fleischer, W. R., and Gambino, S. R.: Blood pH, PO_2 and oxygen saturation. Commission on Continuing Education, Council on Clinical Chemistry, American Society of Clinical Pathologists, 1972, Chicago, IL.
4. Gambino, S. R.: pH and pCO_2. *In*: Standard Methods of Clinical Chemistry. S. Meites, Ed. New York, Academic Press, Inc., 1965, vol. 5, p. 169.
5. Gutman, A. B.: The biochemical significance of uric acid. The Harvey Lecture Series 60. New York, Academic Press, Inc., 1966.
6. Latner, A. L.: Cantarow & Trumper Clinical Biochemistry, 7th ed. Philadelphia, W. B. Saunders Co., 1975.
7. Peters, J. P., and Van Slyke, D. D.: Quantitative Clinical Chemistry, Interpretation. Baltimore, The Williams & Wilkins Co., 1932, vol. 1.
8. Peters, J. P., and Van Slyke, D. D.: Quantitative Clinical Chemistry, Methods. Baltimore, The Williams & Wilkins Co., 1932, vol. 2.
9. Ulmer, D. D.: Trace elements and clinical pathology. Prog. Clin. Path., *1*: 176–236, 1966.
10. Wacker, W. E. C.: Magnesium metabolism. J. Am. Diet. Assoc., *44*: 362–367, 1964

CHAPTER 16

ACID-BASE AND ELECTROLYTE BALANCE

by Norbert W. Tietz, Ph.D., and Ole Siggaard-Andersen, M. D., Ph.D.

The normal human diet is almost neutral and contains very little acid. However, metabolic processes in the body result in the production of relatively large amounts of carbonic, sulfuric, phosphoric, and other acids. Lactic acid and β-hydroxybutyric acid are intermediary products which are normally metabolized to carbon dioxide and water before excretion. In some abnormal conditions, however, such as diabetes mellitus, they may accumulate to a significant extent.

A person weighing 70 kg disposes daily of about 20 moles of carbon dioxide through the lungs and about 70 millimoles of titratable non-volatile acids, mainly sulfuric and phosphoric acids, through the kidneys. These products of metabolism are transported to the excretory organs (lungs and kidneys) via the extracellular fluid and the blood without producing any appreciable change in the plasma pH and with only minimal pH difference between the arterial and the venous blood plasma. This is accomplished by the combined functions of the *buffer systems* of the blood, the *respiratory system*, and the *renal mechanisms*.

Acid-base balance. In physiology, a component is *in balance* if the rates of input and output of this component are equal for a given time interval. A *positive balance* indicates that the mean rate of gain of the component in the body is positive (net gain), while a *negative balance* indicates that the mean rate of gain of the component in the body is negative (net loss). Imbalance results in either an increase or a decrease in the amount of the component in the body. A *balance account* is a detailed account of all input (intake plus production) and all output (excretion plus metabolic conversion) over a given time interval. A description of the *acid-base balance* involves a description of the CO_2 balance as well as the balance of noncarbonic acid and base. The *acid-base status* of the body fluids (i.e., the pH, the pCO_2, the concentration of titratable acid or base, and other acid-base variables) is the resultant of the input and output of acids and bases in the period preceding the sampling.

It is most difficult to evaluate accurately the acid-base status (and anion-cation composition) of the total body fluids on the basis of the analysis of plasma, since each body fluid compartment has a different composition (see p. 948) and movement of ions or other solutes between these compartments is not necessarily free. Also, in disease states, changes between compartments are not uniform. Analyses of acid-base parameters and ion composition of blood, plasma, or serum, therefore, does not necessarily give an accurate reflection of the situation in other compartments, particularly the intracellular compartment.[1] At best, results obtained from analysis of these fluids reflect the composition of the total extracellular fluid (plasma and interstitial fluid), which represents about 16 to 18 liters of water of the total body water of about 43 liters (extra- and intracellular). Of these 16 to 18 liters of water, plasma occupies about 2.6 to 3.5 liters. However, blood, plasma, and serum continue to be used for acid-base studies, since they are easily accessible.

Table 16–1 gives the acid-base status of normal arterial blood. The pH of the plasma may be considered to be a function of two independent variables: the pCO_2, which is regu-

TABLE 16–1 ACID-BASE STATUS OF NORMAL ARTERIAL BLOOD AT 37°C FOR ADULTS (95 PER CENT TOLERANCE LIMITS).

Quantity	*Male*	*Female*
pH of plasma	7.38–7.45	7.39–7.46
pCO_2 of plasma or blood (mm Hg)	35–45	32–43
excess concentration of base in the blood, ΔcBase(B) (unit: mmol/l)	−1.0–+3.1	−1.8–+2.8
concentration of bicarbonate in plasma, $cHCO_3^-$(P) (unit: mmol/l)	22.7–27.8	21.2–27.0
concentration of total CO_2 in plasma, $ctCO_2$(P) (unit: mmol/l)	23.8–29.0	22.3–28.3

lated by the respiratory mechanism, and the excess concentration of titratable base, which is regulated by the renal mechanism. For a pure bicarbonate solution, changes in the excess concentration of titratable base equal the changes in the bicarbonate concentration. Thus, the plasma bicarbonate concentration is often taken as an approximate measure of the base in plasma and extracellular fluid, although it is recognized that conditions in plasma differ greatly from those present in pure bicarbonate solutions.

The H^+-balance is closely associated with the balance of other electrolytes because H^+ cannot be introduced alone without a concomitant anion (e.g., Cl^-) or in exchange for a cation (e.g., Na^+).

Table 16–2 gives the concentrations of the cation and anion charges for normal plasma. (The term mmol of ion charge replaces the term mEq of ion, which is presently still widely used in the United States.)

Note that there is an exact equality of the concentrations of total anion charge and of total cation charge; i.e., the sum of negative and positive charges is zero. This law of electrical neutrality can be expressed mathematically as:

$$\sum_{i=1}^{n} z \times c\mathrm{I}_i^z = 0 \tag{1}$$

where z is the charge number of the ion I_i (positive for cations, negative for anions) and c is the substance concentration. Any increase in the concentration of one anion is accompanied either by a corresponding decrease of other anions, or by an increase of one or more cations, or both, so that total electrical neutrality is invariably maintained. Similarly, any decrease in the concentration of anions involves either a corresponding increase in other anions, or a decrease in cations, or both.

TABLE 16–2 CONCENTRATIONS OF CATION AND ANION CHARGES IN SERUM (EXPRESSED IN mmol/l).

Cation charges		*Anion charges*	
Na^+	142	Cl^-	103
K^+	4	HCO_3^-	27
Ca^{2+}	5	HPO_4^{2-}	2
Mg^{2+}	2	SO_4^{2-}	1
Others (trace elements)	1	Organic acids⁻	5
	154	Protein⁻	16
			154

The term electrical neutrality is not to be confused with acid-base neutrality (pH = 7.0, where the activity of H^+ equals the activity of OH^-), nor does it indicate that the pH is normal (7.4); it indicates only that there is an equality between the concentrations of total anion charge and total cation charge.

According to Brønsted's concept (1923), which is now generally employed, an acid is a substance that can donate protons (H^+) and a base is a substance that can accept protons (H^+).

$$\begin{aligned} \text{acid} &\rightleftharpoons H^+ + \text{base} \\ HCl &\rightleftharpoons H^+ + Cl^- \\ NH_4^+ &\rightleftharpoons H^+ + NH_3 \\ \text{glycine} &\rightleftharpoons H^+ + \text{glycinate}^- \\ \text{glycinium}^+ &\rightleftharpoons H^+ + \text{glycine (electrically neutral)} \end{aligned} \tag{2}$$

Thus, HCl, NH_4^+, and amine ions (glycinium$^+$) are acids, and NH_3, glycinate$^-$, and free (electrically neutral) amines are bases. Some (hydrogen-containing) anions, such as HCO_3^- and HPO_4^{2-}, as well as all amino acids (e.g., glycine) and proteins are both acids and bases (called ampholytes or amphoteric substances).

$$\begin{aligned} H_2PO_4^- &\rightleftharpoons H^+ + \underset{\text{(base)}}{HPO_4^{2-}} \\ \underset{\text{(acid)}}{HPO_4^{2-}} &\rightleftharpoons H^+ + PO_4^{3-} \end{aligned} \tag{3}$$

Concentrations of electrolytes in the body fluids are commonly expressed in units of millimoles per liter (mmol/l).* In the case of polyvalent ions it is important to distinguish between the (substance) concentration of the ion (cI^z) and the (substance) concentration of the ion charge ($cI^z_{1/z}$). Thus, the concentration (c) of the total calcium ions in normal plasma is $cCa^{2+}(P) = 2.5(0)$ mmol/l, while the concentration of the total calcium ion charge is $cCa^{2+}_{1/2} = 5.0$ mmol/l (equivalent to 5 mEq/l).

ELECTROLYTE COMPOSITION OF BODY FLUIDS

The body has two main fluid compartments, the *intracellular* and the *extracellular*, the latter being further divided into blood plasma (vascular compartment) and interstitial fluid. For simplicity, we will adhere to this division, although it should be realized that these compartments can be divided further into subcompartments (e.g., bones) with varying composition in regard to water, electrolytes, and other components. The masses and volumes of the different body compartments are summarized in Table 16–3.

* It is still the custom in the United States to express the electrolyte concentrations in terms of mEq/l (in case of calcium, also in mg/100 ml.) The use of these units is now discouraged.

TABLE 16–3 MASS (m) AND VOLUME (V) OF VARIOUS BODY COMPARTMENTS IN A 70 kg ADULT

	m (kg)	V (l)	mH_2O (kg)	mBone (kg)	mFat (kg)
Interstitial fluid (Isf) + cerebrospinal fluid (Csf) + bone	19	17	14	4.5	
Plasma (P)	3.4	3.2	3.0		
Erythrocyte fluid (E)	2.3	2.1	1.5		
Intracellular fluid (Icf)	45.5	44	24.5		12
Total	70	66	43		

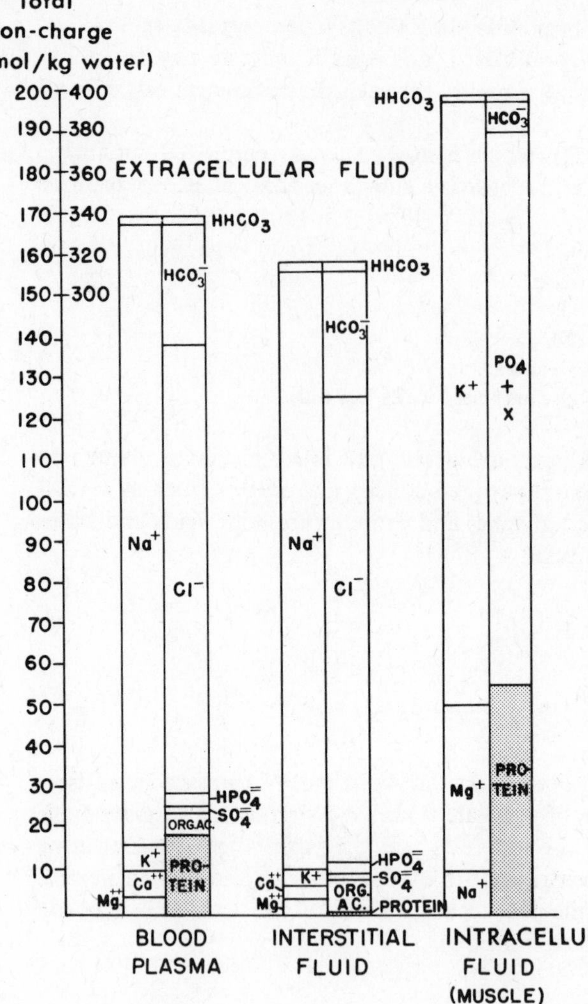

Figure 16–1 Electrolyte composition of blood plasma, interstitial fluid, and intracellular fluid. (After Gamble, J.: Chemical Anatomy, Physiology, and Pathology of Extracellular Fluid, Cambridge, Harvard University Press, 1950.)

Plasma, which is of main interest in our discussion, generally has a volume of 1300 to 1800 ml/m² of body surface and constitutes approximately 5 per cent of the body volume. Generally, body volume is derived from body mass by using a body density of 1.0 kg/l (the true value is about 1.06 kg/l). The electrolyte composition of plasma has been discussed earlier and is summarized in Table 16–2 and Figure 16–1.

Interstitial fluid, is essentially an ultrafiltrate of blood plasma. Its volume makes up approximately 15 per cent of the total body volume; however, when all extracellular spaces except plasma are included, the volume accounts for about 26 per cent of the total body volume (see Table 16–3). Plasma is separated from the interstitial fluid by the endothelial lining of the capillaries, which acts as a semipermeable membrane and allows passage of water and diffusible solutes, but not of compounds of large molecular mass such as proteins. However, this "impermeability" is not absolute, as is demonstrated by the varying (although low) concentration of protein in interstitial fluids.

The exact composition of ***intracellular fluid*** is extremely hard to measure because of the relative unavailability of cells free of contamination. Although erythrocytes are easily accessible, it would be incorrect to make any generalizations based on the composition of these highly specialized cells. Data on cell composition (Fig. 16–1), therefore, are considered only approximations. The volume of intracellular fluids contributes approximately 66 per cent to the total body volume (see Table 16–3).

Figure 16–1 illustrates the average composition of extracellular and intracellular fluids; note that the values are *molalities* of the ion charges. In clinical medicine, as in most analytical chemistry, where samples of solutions are measured volumetrically, it is customary to express concentrations of solutes as either mass concentration (i.e., mass of solute divided by volume of solution; unit: kg/l, g/l, or mg/l) or substance concentration (i.e., amount of substance of solute divided by volume of solution; unit: mol/l, mmol/l, or μmol/l). When comparing the concentrations of a solute in two aqueous solutions of widely different water content, however, it is usually more meaningful to use molalities (i.e., amount of substance of solute divided by the mass of the solvent), since it is the solvent only that provides the space in which solutes are free to move and be osmotically active. To convert concentration values into molalities (m) the following formula may be applied:

$$m\text{I(S)} = c\text{I(S)}/\rho\text{H}_2\text{O(S)} \tag{4}$$

This is immediately apparent when the definitions of m (molality), c (substance concentration), and ρ (mass concentration) are applied:

$$\text{Molality of ions in serum, } m\text{I(S)} = n\text{I(S)}/\boldsymbol{m}\text{H}_2\text{O(S)} \tag{5}$$

$$\text{Concentration of ions in serum, } c\text{I(S)} = n\text{I(S)}/V\text{(S)} \tag{6}$$

$$\text{Mass concentration of water in serum, } \rho\text{H}_2\text{O(S)} = \boldsymbol{m}\text{H}_2\text{O(S)}/V\text{(S)} \tag{7}$$

where n is amount of substance ("number of moles"), and $\boldsymbol{m}$ is mass.

The mass concentration of water in normal plasma is about 0.94 kg/l,* depending upon the protein and lipid content. Thus, a concentration of sodium in the plasma of 148 mmol/l would correspond to a molality of sodium in the plasma of 148 mmol $\times$ liter^{-1}/(0.94 kg $\times$ liter^{-1}) = 158 mmol/kg H_2O.

GIBBS-DONNAN EQUILIBRIUM

Two solutions separated by a semipermeable membrane will establish an equilibrium in such a way that all ions are equally distributed in both compartments, provided that the solution contains only solutes that can freely move through the membrane. At the state of equilibrium the total ion concentration and, therefore, the total concentration of osmotically active particles (osmolutes) are the same on both sides of the membrane.

If, on the other hand, the solution on one side of a membrane contains ions that cannot freely move through the membrane (e.g., proteins), distribution of the diffusible ions at the steady state is unequal.

Before steady state		*After steady state*		
Compartment I	Compartment II	Compartment I	Compartment II	
$c\text{Na}^+ = c_\text{I}$	$c\text{Na}^+ = c_\text{II}$	$c\text{Na}^+ = c_\text{I} - \Delta c$	$c\text{Na}^+ = c_\text{II} + \Delta c$	(8)
$c\text{Cl}^- = c_\text{I}$		$c\text{Cl}^- = c_\text{I} - \Delta c$	$c\text{Cl}^- = \Delta c$	
	$c\text{R}^- = c_\text{II}$		$c\text{R}^- = c_\text{II}$	

Let us assume that compartment I initially contains only Na^+ and Cl^- and that compartment II initially contains only Na^+ and R^-, where R^- is a nondiffusible ion such as protein. Let us further assume that the system is of fixed size and that the membrane is nonelastic. The initial concentrations are symbolized by c_I and c_II respectively. The change in the concentrations of Na^+ and Cl^- on the two sides of the membrane is symbolized

* The total weight (mass) of 1 liter of plasma is about 1.026 kg; thus, the total weight of solute in 1 liter of plasma is about 1.026 − 0.94 = 0.086 kg.

by Δc. In order to preserve electrical neutrality, Na^+ and Cl^- diffuse in pairs across the membrane until a steady state is established.

The unequal distribution of the diffusible ions (Na^+ and Cl^-) causes the development of a membrane potential, the magnitude of which is given by the Nernst equation (see Chap. 3, Section 3).

Applied to equation (8) this becomes

$$\Delta U(\mathrm{I} \mid \mathrm{II}) = U(\mathrm{II}) - U(\mathrm{I}) = -\frac{R \times T}{(+1) \times F} \times \ln \frac{a\mathrm{Na}^+(\mathrm{II})}{a\mathrm{Na}^+(\mathrm{I})}$$

$$\approx {}^* -\frac{R \times T}{(+1) \times F} \times \ln \frac{c\mathrm{Na}^+(\mathrm{II})}{c\mathrm{Na}^+(\mathrm{I})} \tag{9}$$

Similarly:

$$U(\mathrm{I} \mid \mathrm{II}) = \frac{R \times T}{(-1) \times F} \times \ln \frac{c\mathrm{Cl}^-(\mathrm{II})}{c\mathrm{Cl}^-(\mathrm{I})}$$

$$= \frac{R \times T}{(+1) \times F} \times \ln \frac{c\mathrm{Cl}^-(\mathrm{I})}{c\mathrm{Cl}^-(\mathrm{II})}$$

Since $U(\mathrm{I} \mid \mathrm{II})$ is identical for the Na^+ and Cl^-, therefore

$$\frac{R \times T}{(+1) \times F} \times \ln \frac{c\mathrm{Na}^+(\mathrm{II})}{c\mathrm{Na}^+(\mathrm{I})} = \frac{R \times T}{(+1) \times F} \times \ln \frac{c\mathrm{Cl}^-(\mathrm{I})}{c\mathrm{Cl}^-(\mathrm{II})}$$

This reduces to:

$$\frac{c\mathrm{Na}^+(\mathrm{II})}{c\mathrm{Na}^+(\mathrm{I})} = \frac{c\mathrm{Cl}^-(\mathrm{I})}{c\mathrm{Cl}^-(\mathrm{II})} \tag{10}$$

or:

$$c\mathrm{Na}^+(\mathrm{I}) \times c\mathrm{Cl}^-(\mathrm{I}) = c\mathrm{Na}^+(\mathrm{II}) \times c\mathrm{Cl}^-(\mathrm{II})$$

Consequently: *The product of the concentrations of the diffusible ions in one compartment is equal to the product of the concentrations of the diffusible ions in the other compartment* (Gibbs-Donnan law).

Application of the data from equation (8) to equation (10) allows calculation of Δc in terms of c_{I} and c_{II}:

$$\frac{c_{\mathrm{II}} + \Delta c}{c_{\mathrm{I}} - \Delta c} = \frac{c_{\mathrm{I}} - \Delta c}{\Delta c} \tag{11}$$

and rearrangement gives:

$$(c_{\mathrm{II}} + \Delta c)\, \Delta c = (c_{\mathrm{I}} - \Delta c)(c_{\mathrm{I}} - \Delta c)$$

$$(c_{\mathrm{II}} \times \Delta c) + \Delta c^2 = c_{\mathrm{I}}^2 - 2(c_{\mathrm{I}} \times \Delta c) + \Delta c^2$$

$$c_{\mathrm{II}} \times \Delta c = c_{\mathrm{I}}^2 - 2(c_{\mathrm{I}} \times \Delta c)$$

$$(c_{\mathrm{II}} \times \Delta c) + 2(c_{\mathrm{I}} \times \Delta c) = c_{\mathrm{I}}^2$$

$$\Delta c \times (c_{\mathrm{II}} + 2c_{\mathrm{I}}) = c_{\mathrm{I}}^2$$

$$\Delta c = \frac{c_{\mathrm{I}}^2}{c_{\mathrm{II}} + 2c_{\mathrm{I}}}$$

If, for example, we assume the same initial concentration for $c\mathrm{NaR(II)}$ and $c\mathrm{NaCl(I)}$ of

* The expression in terms of ion concentrations (or molalities) is accurate only if the activity coefficients of the ions are identical for compartments I and II.

3.0 mmol/l, then

$$\Delta c = cCl^-(II) = \frac{3^2}{3 + (2 \times 3)} = 9/9 = 1 \text{ mmol/l}$$

If we enter this figure into Equation 8, the following ion distribution results:

Before steady state		*After steady state*	
Compartment I	Compartment II	Compartment I	Compartment II
cNa^+ = 3 mmol/l	cNa^+ = 3 mmol/l	cNa^+ = 2 mmol/l	cNa^+ = 4 mmol/l
cCl^- = 3 mmol/l		cCl^- = 2 mmol/l	cCl^- = 1 mmol/l
	cR^- = 3 mmol/l		cR^- = 3 mmol/l

Note that after steady state has been established, the law of electrical neutrality is obeyed for both compartments.

The membrane potential according to the Na^+ distribution is

$$\Delta U(\text{I} \mid \text{II}) = -\frac{R \times T}{1 \times F} \times \ln\frac{4}{2}$$

or according to the Cl^- distribution:

$$\Delta U(\text{I} \mid \text{II}) = -\frac{R \times T}{-1 \times F} \times \ln\frac{1}{2}$$

which gives

$$\Delta U(\text{I} \mid \text{II}) = -18.5 \text{ mV}$$

when T = 310.15 K (=37°C); R (gas constant) = 8.314 Joules (J) × mol^{-1} × K^{-1}; and F (Faraday constant) = 96487 Coulombs (C) × mol^{-1}.

Note that the concentration of total particles in compartment II is greater than that of compartment I (compartment II: 8 mmol/l; compartment I: 4 mmol/l). By osmosis, water therefore tends to diffuse from compartment I into compartment II, resulting in an increased pressure in compartment II (since the volume in compartment II is fixed). This increase in pressure is due to the presence of a nondiffusible ion (colloid) and is called the *colloidal osmotic pressure*. This colloidal osmotic (oncotic) pressure due to proteins is especially important in biological systems because it aids in the proper distribution of water in the various fluid compartments. Decrease in the colloidal osmotic (oncotic) pressure of plasma (e.g., owing to loss of protein in renal disease) results in a loss of water from the vascular compartment to the interstitial fluid, causing the clinical condition known as edema.

The body cells which contain non-diffusible protein anions can withstand only a limited and temporary difference in osmotic pressure across the cell membrane. The osmotic pressure or the concentration of osmolutes is normally identical inside and outside the cells, since the cell membrane can correct concentration differences by excluding some small ions by active transport processes. If these processes cease, the cells gradually swell and eventually burst.

DISTRIBUTION OF IONS BY ACTIVE AND PASSIVE TRANSPORT

Examination of Figure 16–1 reveals that the electrolyte compositions of blood plasma and interstitial fluid (both extracellular fluids) are similar, but their compositions differ markedly from that of intracellular fluid. The major extracellular ions are Na^+, Cl^-, and HCO_3^-; in intracellular fluids the main ions are K^+, Mg^{2+}, organic phosphates, and protein. The mechanism for this unequal distribution of most electrolytes (or other constituents) between intracellular and extracellular fluids has so far not been adequately explained. According to the Nernst equation the ratio of the molalities of permeable ions should be directly predictable from the membrane potential, and changes in the latter should affect the ion distribution in a predictable manner. The sign and magnitude of the membrane potential can explain the uneven distribution of some ions (e.g., Cl^-, HCO_3^-, and H^+) between red cells and plasma. The membrane potential of the red cells (about −13 mV, inside negative) is due partly to the impermeability of the membrane to the negatively charged hemoglobin ions, and partly to its apparent impermeability to Na^+ and K^+. (Actually the red cell membrane is permeable to the cations, as can be shown by means of radioactive isotopes of Na^+ and K^+.) Nevertheless, the Na^+ distribution is greatly different from the distribution predicted by the membrane potential, according to which the concentration of sodium should be higher inside the red cells than outside. It is now established that this distribution is due to an active transport of Na^+ from inside to outside the cell against an electrochemical potential gradient. This process requires energy supplied by the metabolic processes in the red cell (e.g., glycolysis). An active sodium pump is now thought to be present in most cell membranes, frequently coupled with a transport of K^+ in the opposite direction. The actual chemical nature of the Na^+ pump is still unknown.

Other ions also appear to be transported actively, for instance H^+ in the muscle cells. The pH of the intracellular fluid of muscle cells has been measured to be about 6.9, while the pH of the interstitial fluid may be about 7.3. Therefore, on the basis of calculations using the Nernst equation, H^+ ions are thought to be actively pumped out of the muscle cells, possibly by the same mechanism which pumps Na^+ out of the cells.

THE HENDERSON-HASSELBALCH EQUATION

The Henderson-Hasselbalch equation aids in the understanding and explanation of pH control of body fluids; this will become clearer in the later discussions of the compensatory mechanisms of the body. The equation derived on page 893 can also be written as follows:

$$\text{pH} = \text{p}K' + \log \frac{c\text{HCO}_3^-}{c\text{dCO}_2} \qquad (12)$$

where $c\text{dCO}_2$ is the dissolved gas which by definition includes also the small concentration of undissociated carbonic acid. It is equal to $\alpha \times p\text{CO}_2$, where α is the solubility coefficient for CO_2.

The average normal ratio of the concentrations of bicarbonate and carbon dioxide in plasma, as explained previously, is 25 (mmol/l)/1.25 (mmol/l) = 20/1; the $\log_{10}$ of 20 is 1.3. Formula (12) applied to normal plasma can then be written:

$$\text{pH} = 6.1 + 1.3 = 7.4$$

It follows that any change in the concentration of either bicarbonate or dissolved CO_2 and, therefore, in the ratio of $c\text{HCO}_3^-/c\text{dCO}_2$ must be accompanied by a change in pH. Such changes in the ratio can occur through a change either in the numerator ($c\text{HCO}_3^-$) or in the denominator ($c\text{dCO}_2$), as will be discussed later.

Clinical conditions, as we shall see, are classified as primary disturbances in the HCO_3^- concentration or in the dCO_2 concentration. Various compensatory mechanisms that attempt

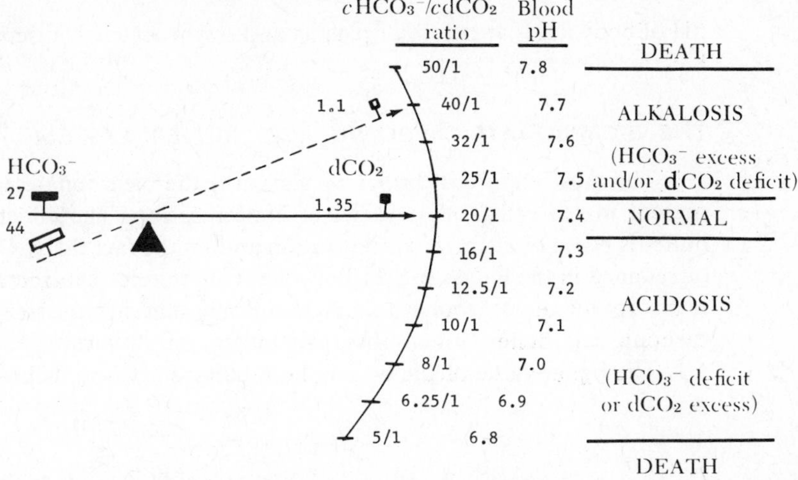

Figure 16–2 Scheme demonstrating the relation between pH and ratio of bicarbonate concentration to the concentration of dissolved CO_2. If the ratio in blood is 20/1 ($cHCO_3^- = 25$ mmol/l and $cdCO_2 = 1.25$ mmol/l) the resultant pH will be 7.4 as demonstrated by the solid beam. The dotted line shows a case of uncompensated alkalosis (bicarbonate excess) with a bicarbonate concentration of 44 mmol/l and a dCO_2 concentration of 1.1 mmol/l. The ratio therefore is 40/1 and the resultant pH 7.7. In a case of uncompensated acidosis, the pointer of the balance would point to a pH between 6.8 and 7.35, depending on the $cHCO_3^-/cdCO_2$ ratio. pH values below 6.8 or above 7.8 are incompatible with life. (After Weisberg, H. F.: Surg. Clin. N. Amer., *39*:93, 1959; Snively, W. D. and Wessner, M.: J. Ind. State Med. Assn., *47*:957, 1954.)

to correct these primary disturbances are geared toward reestablishment of the normal ratio of $cHCO_3^-/cdCO_2$ and hence normal pH. Here again, the compensatory mechanisms may result in changes in the bicarbonate and/or the dissolved CO_2 concentration.

Snively and Wessner, as well as Weisberg, illustrate the application of the Henderson-Hasselbalch equation by the lever-fulcrum (teeter-totter) diagram (Fig. 16–2).

BUFFER SYSTEMS AND THEIR ROLE IN REGULATING THE pH OF BODY FLUIDS

The action of buffers in the regulation of body pH can be explained by the bicarbonate buffer system as an example. (For definition of buffers, see Chapter 1, p. 35.) If we add a strong acid to a solution containing HCO_3^- and dCO_2, the H^+ will react with HCO_3^- to form more CO_2. The hydrogen ions are thereby bound, and the increase in the H^+ concentration will be minimal:

$$HCO_3^- + H^+ \rightarrow H_2CO_3 \rightarrow CO_2 + H_2O$$

On the other hand, if we add a base to the same buffer solution, the base will react with $HHCO_3$ directly (Equation 13) or OH^- will react with the H^+ of carbonic acid to form bicarbonate and water (Equations 14a and 14b). The pH change, therefore, will be small:

$$HPO_4^{2-} + H_2CO_3 \rightleftharpoons H_2PO_4^- + HCO_3^- \quad (13)$$

$$OH^- + H_2PO_4^- \rightleftharpoons HPO_4^{2-} + H_2O \quad (14a)$$

$$OH^- + H_2CO_3 \rightleftharpoons H_2O + HCO_3^- \quad (14b)$$

The buffer systems of most physiological interest in connection with regulation of the pH of body fluids are those of plasma and erythrocytes. A discussion of the most important buffers follows.

THE BICARBONATE/CARBONIC ACID BUFFER SYSTEM

The most important buffer of plasma is the bicarbonate/carbonic acid pair; it is also present in red cells, but at a lesser concentration. The effectiveness of the bicarbonate buffer is based on its high concentration and on the fact that CO_2 can readily be disposed of or retained in the lungs. In addition, the renal tubules can increase or decrease the rate of reabsorption of bicarbonate from the glomerular filtrate (see p. 964). The bicarbonate/carbonic acid buffer system obviously buffers only noncarbonic acid or base.

The buffer value of the bicarbonate buffer in plasma is by definition

$$\beta HCO_3^-(P) = \frac{\Delta cHCO_3^-(P)}{\Delta pH(P)} \tag{15}$$

i.e., the buffer value is defined as the amount of base required to cause a change in pH of one unit.

This coefficient can be derived by differentiating the Henderson-Hasselbalch equation. For a *closed system* where the concentration of total CO_2 ($ctCO_2$) is constant, the result is:

$$\beta HCO_3^-(P, \text{closed}) = \left(\frac{\partial cHCO_3^-(P)}{\partial pH(P)}\right)_{ctCO_2} = 2.303 \times ctCO_2(P) \times \frac{K' \times aH^+}{(K' + aH^+)^2} \tag{16}$$

For an *open system* where the pCO_2 is constant, the result is:

$$\beta HCO_3^-(P, \text{open}) = \left(\frac{\partial cHCO_3^-(P)}{\partial pH(P)}\right)_{pCO_2} = 2.303 \times cHCO_3^-(P) \tag{17}$$

Inserting $pK' = 6.10$, $pH = 7.40 \Leftrightarrow aH^+ = 40 \times 10^{-9}$, $ctCO_2 = 25.7$ mmol/l, and $cHCO_3^- =$ 24.5 mmol/l gives the buffer values:

$$\beta HCO_3^-(P, \text{closed}) = 2.7 \text{ mmol/l}$$

$$\beta HCO_3^-(P, \text{open}) = 56.6 \text{ mmol/l}$$

The above calculation illustrates the insignificance of the bicarbonate buffer in a closed system at pH 7.4 as against its great importance in an open system of constant pCO_2, which pertains to the situation in the living organism.

THE PLASMA PROTEIN BUFFER SYSTEM

The buffer value of the nonbicarbonate buffers of plasma is about 7.7 mmol/l at pH 7.40 for a plasma protein mass concentration of 72 g/l (7.2 g/100 ml). The value varies slightly in the physiological pH range with a maximum at pH $\approx$ 7.3. The proteins, especially albumin, account for the greatest portion (95 per cent) of the non-bicarbonate buffer value of the plasma. The most important buffer groups in the physiological pH range are the imidazole groups of the histidines (pK $\approx$ 7.3), of which 16 are present for each albumin molecule:

O H
C N
CH
CH_2
NH
H^+N
$\xrightleftharpoons[pK\,=\,7.3]{}$
O H
C N
CH
CH_2
NH
N
$+ H^+$

The significance of the nonbicarbonate buffers of the plasma can be illustrated by plotting the CO_2 equilibration curve of the plasma in a diagram of pH versus $cHCO_3^-$. The CO_2 equilibration curve is obtained by equilibrating the plasma with gas mixtures of varying pCO_2. The slope of the CO_2 equilibration curve in a pH-$cHCO_3^-$ diagram is equal to the buffer value of the nonbicarbonate buffers (with opposite sign). This is apparent when considering the chemical reactions during CO_2 equilibration:

$$CO_2 + H_2O \rightarrow H_2CO_3 \rightarrow H^+ + HCO_3^-,$$
$$HPr \leftarrow H^+ + Pr^-,$$

where the HPr/Pr$^-$ system represents all nonbicarbonate buffers. For each molecule of HCO_3^- which is generated, one molecule of nonbicarbonate buffer base disappears because the concentration of H^+ remains virtually constant compared to the changes in $cHCO_3^-$ and cPr^-. Mathematically this can be expressed as follows:

$$-\left(\frac{\partial cHCO_3^-(P)}{\partial pH(P)}\right)_{cB'(P)} \approx \left(\frac{\partial cPr^-(P)}{\partial pH(P)}\right)_{cB'(P)} = \beta Pr^-(P) = 7.7 \text{ mol/l} \quad (18)$$

which says that the negative value of the slope of the CO_2 equilibration curve in a pH-$cHCO_3^-$ diagram equals the buffer value of the nonbicarbonate buffers, which is approximately equal to the buffer value of the plasma proteins.

The equation for the CO_2 equilibration curve of plasma can therefore be written:

$$\Delta cHCO_3^-(P) = [-7.7 \text{ (mmol/l)} \times \Delta pH(P)] + \Delta cB'(P) \quad (19)$$

where $\Delta cHCO_3^-(P) = cHCO_3^-(P) - 24.5$ mmol/l, $\Delta pH(P) = pH(P) - 7.40$, and $\Delta cB'(P)$ is the concentration of titratable base minus the concentration of titratable acid when titrating the plasma with strong acid or base to pH(P) = 7.40 at pCO_2 = 40 mm Hg and 37°C.

THE PHOSPHATE BUFFER SYSTEM

At a plasma pH of 7.4, the ratio $cHPO_4^{2-}/cH_2PO_4^-$ is 80/20 ($pK' = 6.8$). The total concentration of this buffer in both erythrocytes and plasma is less than that of other major buffer systems. Inorganic phosphate accounts for only about 5 per cent of the nonbicarbonate buffer value of plasma. Organic phosphate, however, in the form of 2,3-diphosphoglycerate (present in erythrocytes in a concentration of about 4.5 mmol/l), accounts for about 16 per cent of the nonbicarbonate buffer value of erythrocyte fluid.

The phosphate buffer reacts with acids and with bases as follows:

$$HPO_4^{2-} + H^+ \rightarrow H_2PO_4^-$$
$$H_2PO_4^- + OH^- \rightarrow HPO_4^{2-} + H_2O$$

This system is important in the excretion of acids in the urine, as will be explained in the section on renal mechanism.

THE HEMOGLOBIN BUFFER SYSTEM

The buffer value of the nonbicarbonate buffers of erythrocyte fluid is about 63 mmol/l at pH 7.20 for an erythrocyte hemoglobin (Fe) concentration of 21 mmol/l (33.8 g/100 ml). Hemoglobin accounts for the major part (53 mmol/l), the remainder being due mainly to 2,3-diphosphoglycerate. The imidazole groups of hemoglobin are quantitatively the most important buffer groups.

The slope of the CO_2 equilibration curve of whole blood depends on the buffer value of nonbicarbonate buffers, i.e., mainly on the hemoglobin concentration of the blood. It is possible to derive an approximate equation for the CO_2 equilibration curve of whole blood:

$$\Delta c\mathrm{HCO_3^-(P)} = -\beta \times \Delta\mathrm{pH(P)} + \zeta^{-1} \times \Delta c\mathrm{B'(B)} \tag{20}$$

where

$$\Delta c\mathrm{HCO_3^-(P)} = c\mathrm{HCO_3^-(P)} - 24.5\ \mathrm{mmol/l}$$

$$\Delta\mathrm{pH(P)} = \mathrm{pH(P)} - 7.40$$

$\Delta c\mathrm{B'(B)} = c\mathrm{titr.\ base\ (B)} - c\mathrm{titr.\ acid\ (B)}$, titrating the blood with strong acid or strong base to pH(P) = 7.40 at pCO_2 = 40 mm Hg and 37°C

$$\beta = 2.3 \times c\mathrm{Hb(B)}^* + 7.7\ \mathrm{mmol/l}$$

$$\zeta = 1 - c\mathrm{Hb(B)}^* \times 0.023 \times \mathrm{l/mmol}$$

This equation, together with the Henderson-Hasselbalch equation, provides the simplest algorithm for calculation of the various acid-base variables by means of an electronic calculator. The derivation of Eq. (20) is complicated by the fact that blood is a two-phase system, and it is therefore not given here. When cHb(B) = 0 mmol/l, Eq. (19) and Eq. (20) are identical.

The buffer value of deoxyhemoglobin is slightly lower than that for oxyhemoglobin at pH $\approx$ 6.5 but higher at pH $\approx$ 7.8. This is due to a decrease in the pK-value of the so-called oxygen-linked acid-base groups when deoxyhemoglobin is oxygenated. This also causes a liberation of H^+ upon oxygenation of hemoglobin, a phenomenon called the *Haldane effect*. In a hemoglobin solution a close relationship exists between the Bohr effect (see Fig. 15–1) and the Haldane effect:

$$\left(\frac{\partial \log pO_2}{\partial \mathrm{pH}}\right)_{c\mathrm{HbO_2}} = \left(\frac{\partial c\mathrm{Hb\text{-}H^+}}{\partial c\mathrm{HbO_2}}\right)_{\mathrm{pH}} \tag{21}$$

The first coefficient is the Bohr coefficient, i.e., the change in log pO_2 with pH at constant $c\mathrm{HbO_2}$ (usually at a constant oxygen saturation of 0.5). The latter coefficient is the Haldane coefficient, i.e., the degree of change in the concentration of hemoglobin-bound hydrogen ion with changing concentration of hemoglobin-bound oxygen at constant pH. This is called the linkage equation. The value of the coefficients varies with pH, pCO_2, and the concentration of 2,3-diphosphoglycerate. For erythrocyte fluid with pH = 7.20, pCO_2 = 40 mm Hg, and a normal concentration of 2,3-diphosphoglycerate (4.5 mmol/l), the mean value of the Haldane coefficient when the oxygen saturation increases from zero to one is about −0.47. The value varies slightly with the oxygen saturation, with a maximum value at half-saturation. For whole blood, the heterogeneity of the blood must be taken into account and the "Haldane coefficient for whole blood," i.e., the rise in the concentration of titratable base of the whole blood (ΔcB′(B)), is only 0.3 mmol/l when oxyhemoglobin of a concentration of 1 mmol/l is converted to deoxyhemoglobin.

The oxygen-linked acid-base groups are the imino groups of the C-terminal histidines of the two β-chains (HC3(146)β), the amino groups of the N-terminal valines of the two α-chains (NA1(1)α), and possibly also the imino group of a histidine of the α-chains (H5(122)α). In deoxyhemoglobin these groups participate in the formation of salt bridges, and the pK values are thereby increased. In oxyhemoglobin the salt bridges are ruptured by changes in the tertiary and quaternary structures of the molecule, and the pK values thereby fall. The situation is considerably complicated by the fact that CO_2 and 2,3-diphosphoglycerate are bound to some of the oxygen-linked acid-base groups. CO_2 is bound as carbamino-CO_2 to the four terminal amino groups (valines) of the four peptide chains with a greater affinity to deoxyhemoglobin than to oxyhemoglobin. 2,3-DPG is bound in the central cavity of the deoxyhemoglobin molecule by salt bridges to the aminium groups of the two N-terminal valines (NA1(1)β) and the iminium groups of the two his-

* Expressed in mmol/l.

tidines (H21(147)α). In oxyhemoglobin the central cavity narrows and 2,3-DPG is expelled (see also Chap. 8).

RESPIRATION AND THE RESPIRATORY REGULATION OF THE ACID-BASE STATUS OF THE BLOOD

The respiratory mechanism is responsible for the adequate supply of oxygen to cells and, at the same time, for the removal of carbon dioxide produced during metabolic processes. Oxygen is transported from the lungs to the tissues, and carbon dioxide is brought from the place of production to the lungs with a minimum change in the pH of extracellular fluid. In fact, respiration and the respiratory mechanism are so closely involved in the maintenance of a constant pH of body fluids that the discussion of both processes in this chapter is justified.

The exchange of gases in the lungs takes place through the membranes of the alveoli. The average amount of oxygen that crosses the membrane and is absorbed by the blood per unit of time in an adult person is about 10 mmol/min ($\dot{V}O_2$* ≈ 240 ml/min), and the average amount of carbon dioxide eliminated per unit of time is about 8 mmol/min ($\dot{V}CO_2 \approx$ 190 ml/min). During exercise these figures increase significantly, possibly as much as tenfold. The gas exchange proceeds very rapidly, which is essential, since blood passes through the lungs in only 0.75 s in resting subjects and in 0.3 s during exercise. In some pathological conditions, the flow of blood through the lungs or the exchange of gas through the alveolar membranes is reduced, resulting in a decrease in both oxygenation and removal of carbon dioxide from the blood. This shall be discussed in more detail in connection with respiratory acidosis.

TRANSPORT OF OXYGEN

Although a small amount of oxygen (≈0.14 mmol/l or 0.314 ml/100 ml) may be transported in blood as physically dissolved oxygen (see Chap. 15, Section 2), most of the oxygen is carried in the form of oxyhemoglobin bound to the Fe^{2+} atom of the heme groups. Each mole of hemoglobin-Fe binds one mole of O_2 (Mol. mass = 32 daltons = 32 g/mol). One mole of hemoglobin tetramer (Mol. mass = 64 456) therefore binds four moles of O_2. The volume of oxygen (STP) which can be bound to hemoglobin divided by the mass of hemoglobin therefore is 4 × (22 400 ml/mol)/(64 456 g/mol) = 1.39 ml oxygen/g hemoglobin. A blood sample with a substance concentration of tetrameric hemoglobin of 2.3 mmol/l contains a mass concentration of hemoglobin of (2.3 mmol/l) × (64.456 g/mmol) = 148 g/l = 14.8 g/100 ml. When saturated with O_2, blood has a substance concentration of hemoglobin-bound O_2 of 9.2 (= 4 × 2.3) mmol/l. This corresponds to a "volume concentration" of O_2 (at STP) of (9.2 mmol/l) × (22.4 ml/mmol) = 206 ml/l.

The ability of hemoglobin to bind oxygen is influenced by a number of factors, such as pO_2, pCO_2, and pH. At the conditions of alveolar air (pO_2 = 100 mm Hg, pCO_2 = 40 mm Hg, and temperature = 37°C), there is an oxygen saturation of hemoglobin of approximately 95 to 98 per cent. With decrease in pO_2, there is initially only a slight decrease in the saturation of hemoglobin; e.g., at pO_2 values of 70 mm Hg, oxygen saturation is still more than 90 per cent. Below this level, however, decreases in the pO_2 will cause a significant drop in the degree of saturation, in accordance with the S-shaped curve shown in Figure 15–1. At pO_2 levels of 20 mm Hg, as it exists frequently in tissues, the oxygen saturation of hemoglobin drops below 30 per cent. The high degree of dissociation of oxygen from hemoglobin at low pO_2 values (e.g., 20 to 70 mm Hg) is an important factor in the supply of tissues with oxygen.

* $\dot{V}O_2$ refers to the *volume rate* of O_2 flow (vs. V = volume).

A decrease in pH in the blood, and, to a minor extent, an increase in pCO_2 (at constant pH) favors the dissociation of oxygen from oxyhemoglobin. This effect is especially pronounced at low pO_2 levels. Again, the factors that influence the dissociation of oxygen from hemoglobin are advantageous in regard to oxygen supply of tissues where the pCO_2 values are relatively high and the pH values are relatively low.

TRANSPORT OF CARBON DIOXIDE IN BLOOD

THE ISOHYDRIC AND CHLORIDE SHIFT

The carbon dioxide produced in metabolic processes is carried to the lungs in the blood (see Table 16–4). This is accomplished with a minimum of pH change between the arterial and the venous blood by a series of reactions generally referred to as the isohydric and chloride shift. Although most of these reactions are simultaneous, it is advantageous to discuss them step by step.

Owing to the continuous production of carbon dioxide within the tissue cells, there is a concentration gradient for carbon dioxide (or a pCO_2 gradient) from these cells to the plasma and the red cells, which causes a shift of physically dissolved carbon dioxide from the tissue cells into the plasma and the erythrocytes. A small portion of the carbon dioxide entering the plasma stays as dissolved carbon dioxide and another small portion reacts with water to form $HHCO_3$. The increased amount of H^+ is buffered by the plasma buffers, including the proteins (Fig. 16–3, reaction 1). Another small portion combines with the amino groups of proteins and forms carbamino compounds (Fig. 16–3, reaction 2):

$$R{-}NH_2 + CO_2 \rightleftharpoons R{-}NHCOO^- + H^+$$

The normal concentration of carbamate in the plasma is about 0.2 mmol/l, but the arterio-venous difference is negligible. Most of the carbon dioxide enters the erythrocytes and reacts with water to form $HHCO_3$. This reaction is catalyzed by the enzyme carbonic anhydrase and, therefore, proceeds at a relatively high speed (Fig. 16–3, reaction 3).

The $HHCO_3$ formed in reaction 3 contributes to the H^+-concentration. The pH change, however, is fully or partially compensated by the release of oxygen from oxyhemoglobin, which involves the conversion of the stronger acid ($HHbO_2$) into a weaker acid (HHb); the oxygen-linked acid-base groups will accept the H^+, as outlined in the discussion of the hemoglobin buffer. For each mole of O_2 given off, the hemoglobin binds about 0.5 mol of

TABLE 16-4 AVERAGE DISTRIBUTION OF CARBON DIOXIDE IN ONE LITER OF NORMAL BLOOD ASSUMING A HEMATOCRIT OF 40%

	Arterial		*Venous*		*Difference**	
Carbon Dioxide	mmol/l	% of total	mmol/l	% of total	mmol/l	% of total
In plasma (600 ml):						
as dissolved CO_2	0.72	3.32	0.81	3.47	0.09	5.38
as HCO_3^-†	15.27	70.56	16.26	69.75	0.99	59.28
In erythrocytes:						
as dissolved CO_2	0.36	1.66	0.40	1.71	0.04	2.39
as carbamino-CO_2	0.98	4.52	1.39	5.96	0.41	24.55
as HCO_3^-	4.31	19.91	4.45	19.09	0.14	8.38
Total	21.64		23.31		1.67	

* The difference between arterial and venous blood is considered to be that amount of carbon dioxide which is disposed of by the lungs.

† Plasma contains a small amount of carbamino-CO_2 (about 0.2 mmol/l), which is traditionally included in the HCO_3^- fraction.

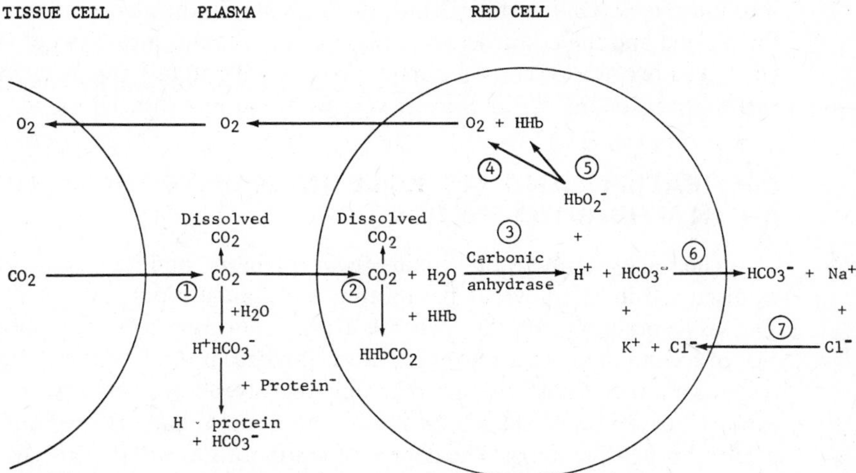

Figure 16–3 Scheme demonstrating the isohydric and chloride shift. The encircled numbers refer to the reactions described in the text.

H^+. Furthermore, the deoxyhemoglobin binds significantly more CO_2 in the form of carbamino-CO_2 than does oxyhemoglobin, and thus a significant fraction of CO_2 is transported in this form. The oxygen released from $HHbO_2$ moves from the red cells through the plasma into the tissue cells (see Fig. 16–3, reactions 3–5).

The remainder of the hydrogen ions formed in reaction 3 are buffered by the nonbicarbonate buffers of the erythrocyte fluid, while the concentration of HCO_3^- increases to the same extent that the concentration of Hb anions falls. The transformations described so far (Fig. 16–3, reactions 1–5) are referred to as the *isohydric shift* (= a shift in which the hydrogen ion concentration remains essentially unchanged).

The equilibrium between plasma and red cells has been disturbed by the reactions described so far. The concentration of HCO_3^- has increased relatively more in the red cells than in the plasma, the pH of plasma has fallen relatively more than the pH of the red cells, and, most important, the ion concentration of nondiffusible protein (Pr^-, Hb^-) in the erythrocytes has fallen. The membrane potential of the erythrocytes therefore falls numerically (becomes less negative) and the distribution of all the diffusible ions must change in accordance with the new membrane potential. The ion shifts that occur rapidly are a movement of HCO_3^- out of the erythrocytes and an electrically balancing movement of Cl^- into the erythrocytes. The Cl^- ratio [$mCl^-(E)/mCl^-(P)$ = molal concentration of Cl^- in erythrocytes/molal concentration of Cl^- in plasma] thereby increases in accordance with the change in the membrane potential. This shift of chloride ions is referred to as the *chloride shift* (see also Fig. 16–3, reactions 6 and 7). The replacement of polyvalent protein anions by monovalent anions causes an increase in the osmolality in the erythrocytes; therefore, a small amount of H_2O will pass from the plasma into the erythrocytes. The permeability of the erythrocyte membrane for Na^+ and K^+ is so low that the movements of these ions in and out of the erythrocytes during the circulation of the blood can be ignored; i.e., for a short time interval the erythrocyte membrane is practically impermeable to these ions. As a result of these ion and water fluxes, the concentration of chloride in the venous plasma is about 1 mmol/l lower than that in the arterial plasma, and the mean volume of an erythrocyte is about 1 femtoliter (fl) greater in the venous blood than in the arterial blood.

The HCO_3^-, the carbamino compounds, and the dissolved carbon dioxide are transported in the blood to the pulmonary capillaries and alveoli. The comparatively low pCO_2 in the alveoli will cause a shift of carbon dioxide from the erythrocytes and the plasma into the alveoli. On the other hand, the high pO_2 in the alveoli causes a shift of oxygen into the plasma and the erythrocytes. This exchange causes a reversal of reactions 1 to 7 in Fig. 16–3. The removal of carbon dioxide from the blood and the oxygenation of the blood are the major reactions which convert venous blood into arterial blood.

RESPIRATION AND ITS ROLE IN MAINTENANCE OF NORMAL pH IN THE BODY FLUIDS

The removal of carbon dioxide from the blood and the supply of tissues with oxygen, as discussed in the previous paragraph, are the main purposes of the respiratory process. The oxygen supply for this process arises from atmospheric air, which has an average pO_2 of 159 mm Hg. As atmospheric air is inhaled, it meets and mixes with air present in the respiratory tree; thus, the air reaching the alveoli is a mixture of atmospheric air and expired air. The alveolar air itself is in an approximate equilibrium with the blood that passes through the lungs. The degree of equilibration will depend on the factors discussed, namely, the concentration gradient of the gases, the speed with which the blood passes through the lungs, and the permeability of the alveolar membranes for blood gases. In normal individuals the equilibration is close to 100 per cent and is shown by a comparison of the pO_2 of alveolar air with arterial blood (mean 102 and 100 mm Hg, respectively) and the pCO_2 in alveolar air with arterial blood (mean, 36 and 40 mm Hg, respectively). As arterial blood passes through the tissues, oxygen diffuses from the blood into the cells and carbon dioxide is accepted by the blood, thus converting arterial blood into venous blood. Venous blood is again converted to arterial blood in its passage through the lungs, completing the cycle. Figure 16–4 presents this process in schematic form and gives the respective average values for pO_2 and pCO_2.

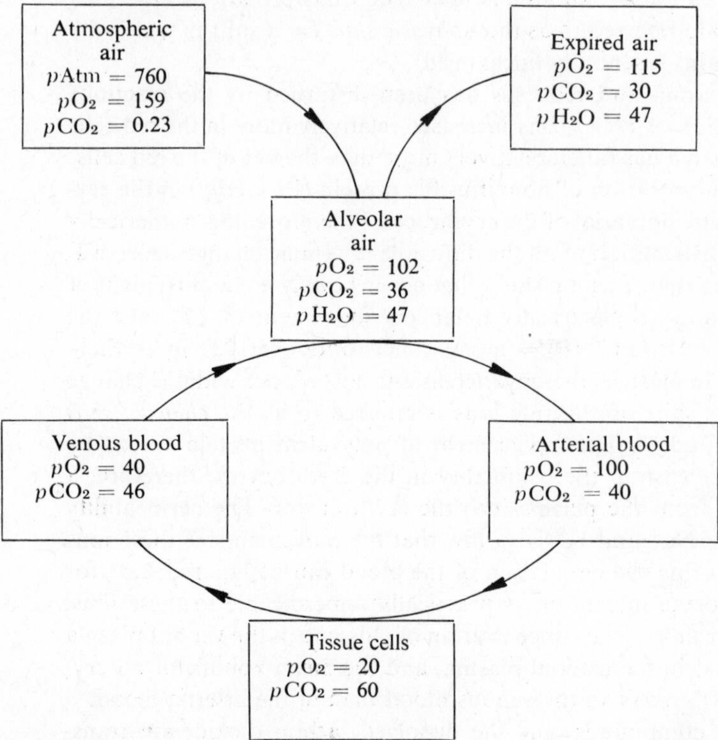

Figure 16–4 Average values for pO_2 and pCO_2 in air, blood, and tissue cells. Values are given in mm Hg.

The respiratory mechanism is regulated by the medullary respiratory center, which in turn is stimulated by the central chemoreceptors, located on the anterior surface of the medulla oblongata, and by the peripheral chemoreceptors, which include the carotid bodies and the aortic bodies. The peripheral chemoreceptors are stimulated by a fall in the arterial plasma pH, or more precisely, by a fall in the pH of the extracellular fluid surrounding the sensitive cells. The central chemoreceptors are stimulated by a fall in pH of the cerebrospinal fluid, which is in close contact with extracellular fluid bathing the sensitive cells in the brain. A fall in arterial pO_2 also stimulates the peripheral, but not the central chemoreceptors. Several other factors also influence the respiratory center. Apparently, reflexes from the working muscles are responsible for the initial hyperventilation during exercise. A rapid fall in skin temperature causes an inhibition of respiration, while increased temperatures accelerate the respiratory rate.

The regulation of the rate of pulmonary ventilation in response to changes in pCO_2 and pH is the basis of the pulmonary compensatory mechanism in acidosis and alkalosis. In metabolic acidosis (primary HCO_3^- deficit) the plasma pH is low, causing an increase in respiratory rate and depth (hyperventilation). This reduces the concentration of CO_2 in blood and helps to restore the ratio of $cHCO_3^-/cdCO_2$, although the total carbon dioxide concentration may be low.

Example: In a case of *metabolic acidosis* before compensation, the $cHCO_3^-/cdCO_2$ ratio may be 21.6/1.35 = 16/1, which represents a plasma pH of 7.3 (see Fig. 16–2). An increased rate and depth of respiration causes an elimination of carbon dioxide and thus a fall in $cdCO_2$ of plasma. The bicarbonate concentration falls slightly when the pCO_2 falls, but relatively less than does the pCO_2 (the change in plasma bicarbonate concentration will amount to about −1.5 mmol/l for a rise in plasma pH of 0.1). This changes the $cHCO_3^-/cdCO_2$ ratio, e.g., to 20.0/1.0 = 20/1, which results in a normal pH of 7.4. Such a condition is called a fully compensated metabolic acidosis. "Fully compensated" implies that the pH has returned to normal, and "acidosis" implies that there is still a primary $cHCO_3^-$ deficit. Those cases in which the $cHCO_3^-/cdCO_2$ ratio stays below the normal value of 18/1 to 22/1 after compensation are referred to as *partially compensated metabolic acidosis* (i.e., pH(P) below normal). Those cases in which the pCO_2 remains normal are referred to as *uncompensated metabolic acidosis.*

In *metabolic alkalosis* (primary $cHCO_3^-$ excess), the pH increase induces hypoventilation, which causes a retention of carbon dioxide and, therefore, an increase in $cdCO_2$ in plasma. This, again, aids in establishing the normal ratio of $cHCO_3^-/cdCO_2$, resulting in normal blood pH or in a pH approaching the normal; however, the total carbon dioxide may be above normal.

Example: A patient in metabolic alkalosis, before compensation, has a ratio of $cHCO_3^-/cdCO_2$ of 56/1.4 = 40/1, resulting in a plasma pH of 7.7. Decreased rate of respiration results in retention of carbon dioxide, an increase in plasma $cdCO_2$ and a slight increase in $cHCO_3^-$. The $cHCO_3^-/cdCO_2$ ratio changes to 57.5/1.80 = 32/1, resulting in a drop of pH to 7.6. Such a condition is called *partially compensated metabolic alkalosis.* The term "partially compensated" implies that the pH, despite some compensatory measures, is not in the normal range, and "metabolic alkalosis" implies that there is a primary excess of $cHCO_3^-$.

If the compensatory mechanisms had been successful in restoring a normal plasma pH, a *fully compensated metabolic alkalosis* would exist.

The rapid rate at which this compensation takes place makes the respiratory mechanism an extremely important link in the effort of the body to maintain a near normal plasma pH.

The respiratory compensation requires about 3 to 6 h to become maximal. Initially during an acute metabolic acidosis (e.g., that caused by lactic acid production during anaerobic muscular exercise), the pH falls in the arterial plasma but remains nearly normal in the cerebrospinal fluid owing to the rather slow equilibration of ions between blood and brain extracellular fluid or cerebrospinal fluid across the so-called blood-brain barrier. The peripheral chemoreceptors are immediately stimulated, causing a hyperventilation and a fall in pCO_2. Since CO_2 rapidly equilibrates between blood and brain extracellular fluid by nonionic diffusion, the pCO_2 immediately falls in brain extracellular fluid; the pH of the brain extracellular fluid therefore initially tends to rise rather than to fall. The central chemoreceptors are therefore initially inhibited rather than stimulated. The gradual fall in the plasma bicarbonate concentration in the course of 3 to 6 h leads to a fall in the bicarbonate concentration of the brain extracellular fluid. The pH of the brain extracellular fluid thereby returns towards normal or falls slightly below normal. When this has occurred, the stimulation of respiration becomes maximal.

The result of the maximal normal physiological compensation is that *the plasma (blood) pH returns about halfway toward normal* compared to the uncompensated situation. A fully compensated metabolic acidosis or alkalosis is a rare situation; it is the result of a combined metabolic and respiratory acid-base disturbance.

Besides the respiratory mechanism described here, there are, of course, the following renal mechanisms that aid in maintaining or restoring the normal $cHCO_3^-/cdCO_2$ ratio of 20/1 and hence a normal plasma pH.

THE RENAL COMPENSATORY MECHANISMS

The normal pH of plasma, as well as that of the glomerular filtrate, is about 7.4; the urinary pH of fasting individuals is about 6.0. This drop in pH is brought about by the kidney, which excretes the nonvolatile acids produced by metabolic processes. The various functions of the renal mechanism respond to the specific requirements: in case of acidosis, there is an increased excretion of acids and a conservation of base; in alkalosis, there is an increased excretion of base and conservation of acids. The pH of the urine changes correspondingly and may vary in random specimens from pH 4.5 to 8.2 (normal values from 4.8 to 7.8). This ability to excrete variable amounts of acid or base is of utmost importance and makes the kidney the final defense mechanism against any drastic changes in body pH and cation-anion composition.

The various acids produced during metabolic processes are buffered in the extracellular fluid, although at the expense of HCO_3^- (see bicarbonate buffer system). Therefore, the supply of HCO_3^- would finally be exhausted if the kidneys did not excrete the acids and restore the HCO_3^-.

EXCRETION OF ACIDS

None of the strong acids, such as sulfuric, hydrochloric, and many organic acids, can be excreted in their free form by the kidneys. Any H^+ ions derived from these acids are excreted only after they react with a buffer base (e.g., HPO_4^{2-} or HCO_3^-) or NH_3. Excretion of the acid anions is accompanied by the removal of an equal number of cations to provide electrical balance. Some weak acids, such as acetoacetic acid ($pK = 3.58$) and β-hydroxybutyric acid ($pK = 4.7$), are present in blood in almost entirely ionized form, but at an acid pH they are in part undissociated and thus may be partially excreted in the form of the free acids.

Hydrogen ions can be excreted into the tubular urine in exchange for sodium ions, which return into the tubular cell (see Fig. 16–5, reaction 3a).

The phosphate ions in plasma and in glomerular filtrate of pH 7.4 are present as HPO_4^{2-} and $H_2PO_4^-$ in a ratio of approximately 80/20. With increase in acidity, this ratio decreases

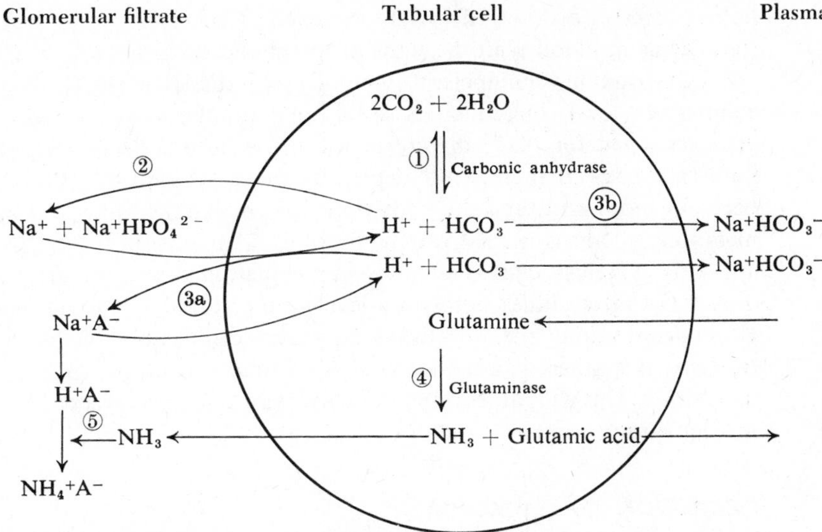

Figure 16–5 Scheme demonstrating acidification of urine by Na^+–H^+ exchange and formation of ammonia from amides. The encircled numbers refer to the reactions described in the text. (From Tietz, N. W.: The Chicago Medical School Quarterly, *22*:156, 1962.)

gradually to 1/99 at a pH of 4.8; at a urine pH of 4.5, essentially all phosphate is present in the form of $H_2PO_4^-$ in a ratio of 1/100. Each HPO_4^{2-} ion can accept one H^+ ion and can make one cation (mainly Na^+) available for reabsorption:

$$[2\ Na^+ + HPO_4^{2-}] + H^+ \leftrightharpoons [Na^+ + H_2PO_4^-] + Na^+$$

Again, the H^+ ions are probably derived from the tubular cells and are released in exchange for a cation, mainly Na^+. Since HPO_4^{2-} and $H_2PO_4^-$ are present in unmodified glomerular filtrate in a ratio of 80/20, each five molecules of phosphate are capable of accepting four H^+ ions, which are then excreted. At a pH of 4.5, nearly all phosphate (approximately 99 per cent) is present as $H_2PO_4^-$; thus, when this pH is attained, essentially no further H^+ can be accepted by this system.

The mechanism by which the Na^+ ions from the modified glomerular filtrate are exchanged for H^+ ions from the tubular cells (Na^+-H^+ exchange) is not exactly known, but the following *ion exchange theory* of Pitts[2] is currently most accepted:

1. Metabolic processes in the tubular cells produce carbon dioxide, which reacts with water in the presence of carbonic anhydrase to form $HHCO_3$ (Fig. 16–5, reaction 1).

2. The H^+ passes into the lumen of the tubule, where it forms either $H_2PO_4^-$ from HPO_4^{2-} (Fig. 16–5, reaction 2) or one of the previously named weak acids (Fig. 16–5, reaction 3a).

3. Na^+ ions equal in number to the H^+ ions, which passed into the tubular lumen, move into the tubular cell where they balance HCO_3^- formed in reaction 1, Figure 16–5. The $Na^+HCO_3^-$ then moves into the tubular blood, thus maintaining or restoring the HCO_3^- level. (It is now believed that there are a series of intermediary steps, but for the purpose of this discussion these will be omitted.)

The place where the Na^+-H^+ exchange takes place was formerly thought to be the distal portion of the renal tubule, but recently some evidence has been presented that it may take place in a larger portion of the tubule (collecting ducts). In renal tubular acidosis, the Na^+-H^+ exchange does not take place or is reduced; thus, the acids formed during metabolic processes are not effectively removed. This results in an increased accumulation of these acids in blood with the signs of metabolic acidosis.

Potassium ions compete in some way with the H^+ in the H^+-Na^+ exchange. If the intracellular K^+ level (including those of renal tubular cells) is high, more K^+ and less H^+ are exchanged for Na^+; therefore, the urine becomes less acid and the acidity of body fluids increases. If there is K^+ depletion, more H^+ ions are exchanged for Na^+, the urine becomes more acid, and the body fluids more alkaline. Since the body mechanism against metabolic alkalosis is relatively ineffective, K^+ depletion frequently results in a metabolic alkalosis. Alkalosis caused by potassium depletion, therefore, can be permanently corrected only if the intracellular potassium levels are restored. It should be emphasized that serum K^+ concentrations are not always an accurate indication of the intracellular K^+ concentrations, since there is a tendency of K^+ to move from the cells into the serum, especially in acidosis. This will, at least temporarily, maintain normal serum levels despite intracellular K^+ depletion.

EXCRETION OF AMMONIA

The renal tubular cells have the ability to form ammonia from amides (mainly glutamine) and some amino acids (Fig. 16–5, reaction 4). The process is greatly enhanced in acidosis and reduced in alkalosis. The ammonia (NH_3) produced diffuses into the tubular urine and combines there with H^+ to form NH_4^+ (Fig. 16–5, reaction 5). The H^+ ions thus bound in NH_4^+ do not contribute to the acidity of urine, making further exchange of Na^+ for H^+ possible. (As mentioned earlier, this latter process stops at a urine pH of about 4.5.) It follows that the formation of ammonia in the tubular cells permits increased excretion of H^+ ions and increased conservation of cations, mainly Na^+. Increased Na^+-H^+ exchange also increases HCO_3^- absorption as just explained (see Fig. 16–5).

REABSORPTION OF FILTERED BICARBONATE

The glomerular filtrate has the same concentration of HCO_3^- as does blood plasma; however, with increasing acidification of the urine, the urine HCO_3^- concentration decreases and may even approach zero, and the pCO_2 increases. Many theories have been postulated, but the most accepted involves the following reactions.

1. With decrease in urinary pH (owing to excretion of H^+ as just described), urinary HCO_3^- will be converted into $HHCO_3$, which in turn will be converted to water and carbon dioxide.

$$HCO_3^- + H^+ \rightleftharpoons HHCO_3$$

$$HHCO_3 \rightleftharpoons H_2O + CO_2$$

2. The increase in urinary pCO_2 causes carbon dioxide to diffuse across the tubule wall into the tubular cell. Therefore, the term "reabsorption of bicarbonate" is not quite correct since the bicarbonate is not reabsorbed in its original form but as carbon dioxide.

3. Along with the decrease of HCO_3^- in tubular urine, there is an increase in Cl^-. Both changes are accompanied by a simultaneous increase of HCO_3^- and decrease of Cl^- in blood.

The increased amount of bicarbonate reabsorbed helps restore the $cHCO_3^-/cdCO_2$ ratio, which is low in acidosis. In alkalosis, the reabsorption of bicarbonate is decreased (excretion in urine increased), which helps lower the $cHCO_3^-/cdCO_2$ ratio. Thus, it becomes apparent that the renal mechanism works with the other compensatory mechanisms either to retain or to restore the normal $cHCO_3^-/cdCO_2$ ratio and, thus, normal pH in body fluids.

CONDITIONS ASSOCIATED WITH ABNORMALITIES IN THE ACID-BASE STATUS AND ANION-CATION COMPOSITION OF BLOOD

Many pathological conditions are accompanied or caused by disturbances of the acid-base balance and electrolyte composition of blood. These changes are usually reflected in the acid-base status and anion-cation pattern of extracellular fluid. For this reason, the acid-base status (i.e., pH(P), pCO_2, $cHCO_3^-(P)$, and the excess concentration of base in the blood) is often determined in clinical medicine, and deviations from the normal are noted. It is important to realize that such results may not always be a true indication of the acid-base status of the intracellular fluid.

Abnormalities in the acid-base status of the blood are usually accompanied by characteristic changes in the electrolyte concentrations of the plasma. This is especially the case for the metabolic acid-base disturbances. Hydrogen ions cannot accumulate without a concomitant anion (e.g., Cl^-, SO_4^{2-}, lactate$^-$) or in exchange for a cation (e.g., Na^+ or K^+). For this reason, the electrolyte composition of blood plasma or serum is often determined in clinical medicine along with the acid-base status. Acid-base disturbances are usually classified in one of the following four groups: metabolic acidosis, metabolic alkalosis, respiratory acidosis, or respiratory alkalosis.

METABOLIC ACIDOSIS (PRIMARY BICARBONATE DEFICIT)

This condition is usually caused by one or more of the following processes:

1. Production of organic acids, which exceeds the rate of elimination (e.g., production of acetoacetic acid and β-hydroxybutyric acid in diabetic acidosis). Acidosis may be accompanied by the loss of cations that are excreted with the anions as explained earlier.

2. Reduced excretion of acids (e.g., renal failure, tubular acidosis).

3. Excessive loss of bicarbonate (base). This usually occurs if there is an excessive loss of duodenal fluid, as in diarrhea.

When any of these conditions exists, compensatory mechanisms act to restore the normal pH. If this restoration is complete (or nearly complete so that the pH remains between 7.35 and 7.45), the condition is called *fully compensated metabolic acidosis*. If, in spite of the compensatory mechanisms, the pH stays below 7.35, the condition is called *partially compensated metabolic acidosis* as explained by the example in the section on respiration. The ratio of $cHCO_3^-/cdCO_2$ is decreased because of the primary decrease in the bicarbonate.

COMPENSATORY MECHANISMS IN METABOLIC ACIDOSIS

Buffer systems. The buffer systems of the blood, mainly the bicarbonate/carbonic acid buffer, tend to minimize changes in pH (see page 953). The bicarbonate decreases and the ratio of $cHCO_3^-/cdCO_2$ will be less than 20/1. The respiratory mechanism and the renal mechanisms attempt to correct this ratio by increased excretion of $HHCO_3$ (as carbon dioxide) and by restoring the HCO_3^-, respectively.

Respiratory mechanism. The decrease in pH caused by any of the conditions just listed stimulates the respiratory mechanism and produces hyperventilation (Kussmaul respiration). The $HHCO_3$ diminishes and the ratio of $cHCO_3^-/cdCO_2$ approaches its normal value of 20/1 (e.g., HCO_3^-/dCO_2 before compensation $= 15/1.2$ or $12.5/1 \Rightarrow$ pH 7.2; after compensation $= 14.5/0.86$ or $16/1 \Rightarrow$ pH 7.3.)

Renal mechanism. The kidneys attempt to restore the original electrolyte composition and the pH by increased excretion of acid and preservation of base (increased rate of Na^+-H^+ exchange, increased ammonia formation, and increased reabsorption of bicarbonate). The total amount of H^+ excreted may be as much as 500 mmol/d.

LABORATORY FINDINGS IN METABOLIC ACIDOSIS

The plasma bicarbonate, carbonic acid, pCO_2, and also the total carbon dioxide concentration are decreased. The pH, in uncompensated cases, is decreased, the degree depending on the ratio of HCO_3^-/dCO_2. The remaining components of the electrolyte pattern vary, depending on the pathologic condition. In diabetic acidosis, for example, the fraction of organic acids is increased by the production of ketone bodies. The Na^+ and K^+ in serum are decreased because of the associated polyuria and their excretion as salts of acetoacetic acid and β-hydroxybutyric acid. (Serum levels of K^+ may be normal or even high despite a severe K^+ depletion. The actual level is determined by the amount of K^+ lost through the urine, by the amount of K^+ shifted from cells into extracellular fluid, and by the degree of dehydration.) In renal failure, organic acids as well as phosphate and sulfate ions are increased because of retention (Fig. 16–6).

Urinary acidity and urinary ammonia are increased, provided that the renal mechanism is functioning.

METABOLIC ALKALOSIS (PRIMARY BICARBONATE EXCESS)

This condition is most frequently caused by one of the following processes, which in turn cause a bicarbonate excess:

1. Administration of excess alkali, especially $NaHCO_3$.
2. Excessive loss of hydrochloric acid from the stomach as seen after prolonged vomiting, for example, in pyloric or high intestinal obstruction.
3. Potassium depletion as seen, for example, in Cushing's syndrome, after administration of ACTH or adrenocortical hormones, and in aldosteronism.

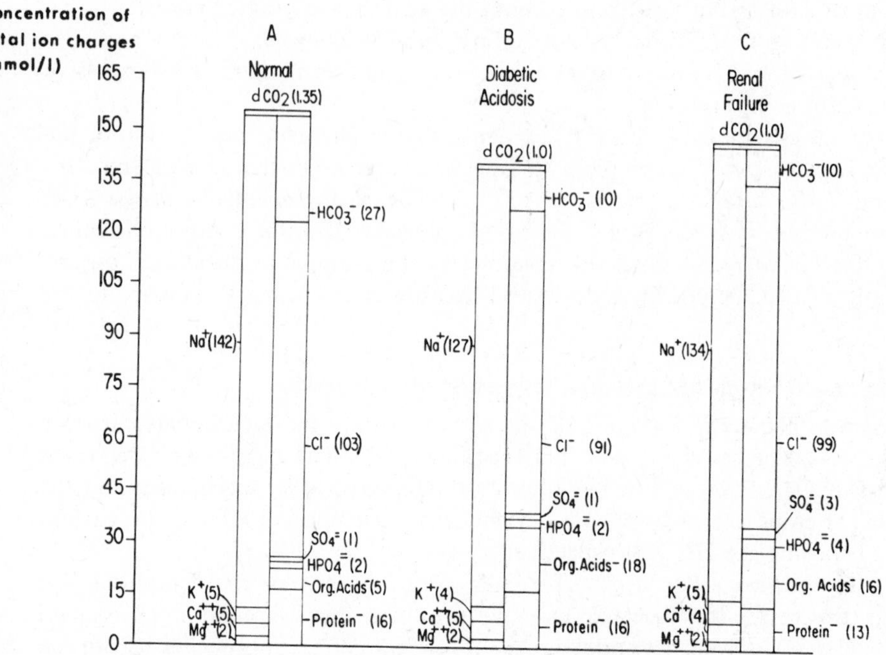

Figure 16–6 *A*, Gamblegram illustrating normal electrolyte composition of plasma. *B*, Example of anion-cation pattern as may be found in diabetic acidosis. Na^+ and Cl^- are decreased because of polyuria and excretion of Na^+ as salt of ketone acids. The organic acid fraction is increased because of excessive formation of ketone bodies. The ratio of HCO_3^-/dCO_2 is 10/1; thus, the plasma pH must be 7.10. *C*, Example of anion-cation pattern as may be seen in renal failure. Organic acids as well as phosphates and sulfates are retained owing to the decreased renal functional efficiency. The ratio of HCO_3^-/dCO_2 is 10/1, which results in a pH of 7.10 (From Tietz, N. W.: The Chicago Medical School Quarterly, *22*:156, 1962.)

4. Administration of certain diuretics, after prolonged usage.

If one of these conditions exists, the compensatory mechanisms of the body act to restore the normal plasma pH. If compensation is complete, we have a state of fully compensated metabolic alkalosis with a pH value within the normal range. With progression of the disturbance, the compensatory mechanisms are not effective enough and the pH will increase. In such a case, the ratio of HCO_3^-/dCO_2 is more than 20/1 because of a primary increase in bicarbonate, e.g., $48/1.5 = 32/1$, resulting in a pH of 7.6.

If the increase in pH is great enough, tetany may develop even in the presence of a normal serum total calcium concentration. The cause for this is thought to be increased binding of calcium ions by protein and other anions.

COMPENSATORY MECHANISMS IN METABOLIC ALKALOSIS

Buffer systems. As a result of loss of acid (e.g., HCl), excess base reacts with the carbonic acid of the HCO_3^- buffer system to form an increased amount of HCO_3^-, thereby minimizing pH change.

Respiratory mechanism. The increase in pH depresses the respiratory center, causing a retention of carbon dioxide, which in turn causes an increase in $HHCO_3$. Thus, the ratio of HCO_3^-/dCO_2, which was originally increased (see earlier section), approaches its normal value although the actual levels of both $cHCO_3^-$ and $cdCO_2$ are increased (e.g., the ratio of HCO_3^-/dCO_2 before compensation was 48/1.5 or $32/1 \Rightarrow$ pH 7.6; the ratio after partial compensation is 49.5/1.86 or $26.7/1 \Rightarrow$ pH 7.53 and the total CO_2 is $49.5 + 1.86 = 51.4$ mmol/l.

Renal mechanism. The kidneys respond to the state of alkalosis by decreased Na^+-H^+ exchange, decreased formation of ammonia, and decreased reabsorption of bicarbonate.

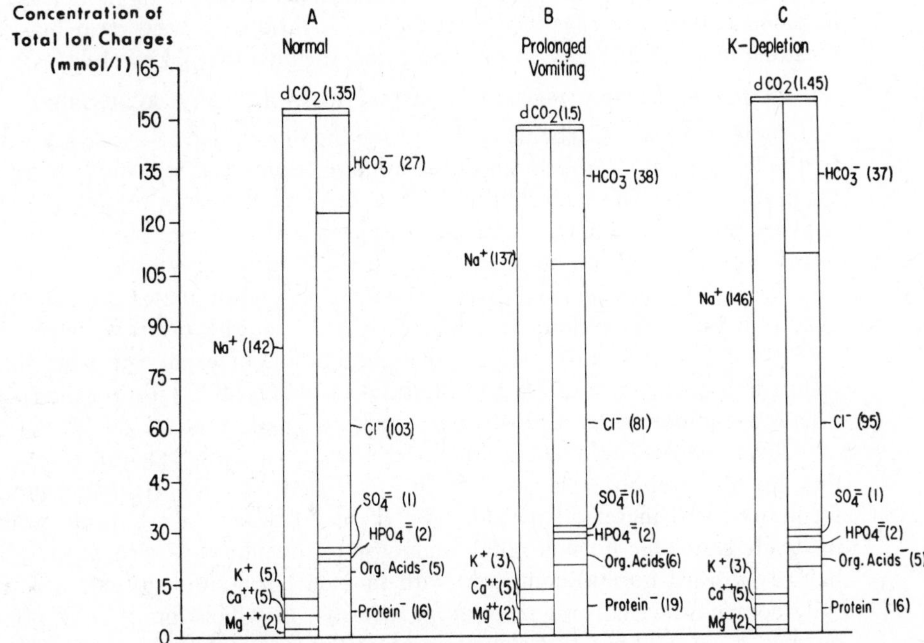

Figure 16-7 *A*, Electrolyte composition of normal plasma. *B*, Example of plasma electrolyte composition after prolonged vomiting, showing the decrease in K^+, Cl^- and the increase in protein⁻, organic acids⁻, HCO_3^-, and dCO_2. The ratio of the last two is $38/1.5 = 25.3$. The pH is therefore 7.50. *C*, Example of typical plasma electrolyte composition in a patient with intracellular K^+ depletion. There is a decrease in K^+ and Cl^- and an increase in HCO_3^- and dCO_2, resulting in a ratio of $37/1.45 = 25.4$ (pH = 7.50). (From Tietz, N. W.: The Chicago Medical School Quarterly, *22*:156, 1962.)

LABORATORY FINDINGS IN METABOLIC ALKALOSIS

Blood plasma values for $cHCO_3^-$, $cdCO_2$, and pCO_2 and, therefore, the total plasma CO_2 concentration increase with an increase in the ratio of HCO_3^-/dCO_2 (if uncompensated). The pH is increased, and the remaining ions of the electrolyte pattern vary depending on the condition. In cases of prolonged vomiting, Cl^- and possibly K^+ levels tend to be low because of the loss of these ions through the vomitus (Fig. 16–7). Protein values may be increased owing to dehydration and, if food intake is inadequate, formation of ketone bodies may increase the organic acid fraction. In cases of excessive administration of $NaHCO_3$, Na^+ levels are increased. In K^+ depletion, decreased concentrations of Cl^- are very common (Fig. 16–7). Serum K^+ concentrations are generally but not necessarily low. A total intracellular K^+ loss of about 100 to 200 mmol is required before serum K^+ concentrations decrease below normal.

Urinary pH values are usually increased because of the decreased excretion of acid and increased excretion of bicarbonate. Urinary ammonia values are decreased because of decreased formation of ammonia in the tubules. In K^+ depletion, the pH of the urine may be low in spite of a metabolic alkalosis; this is called paradoxical aciduria.

RESPIRATORY ACIDOSIS (PRIMARY dCO_2 EXCESS)

Any condition that results in a decreased elimination of carbon dioxide through the lungs results in a primary dCO_2 excess (respiratory acidosis). Inefficiency in carbon dioxide elimination may be mechanical, as in bronchopneumonia, pulmonary emphysema, and pulmonary fibrosis, or it may be caused by decreased circulation as in cardiac disease. Generally, however, cardiac disease results in a slight respiratory alkalosis because the hypoxemia stimulates hyperventilation. Rebreathing, or breathing of air high in CO_2 content, may also cause high pCO_2. Increase in pCO_2 results in an increase of dCO_2, which in turn causes a decrease in the HCO_3^-/dCO_2 ratio (e.g., the ratio may be $27/1.8 = 15$, resulting in a pH of 7.28). In respiratory conditions, the pH rarely goes below 7.20.

COMPENSATORY MECHANISMS IN RESPIRATORY ACIDOSIS

Buffer system. Carbonic acid entering the blood in excess is to a great extent buffered by the hemoglobin and protein buffer systems. Some Cl^- will move from the plasma into the erythrocytes. The buffering of CO_2 causes a slight rise in $cHCO_3^-(P)$. The relationship between the rise in $cHCO_3^-$ and the fall in plasma pH in acute respiratory acidosis has been found experimentally to be $\Delta cHCO_3^-(P)/\Delta pH(P) = -15$ mmol/l.

Respiratory mechanism. The increase in pCO_2 stimulates the respiratory center and results in increased pulmonary rate, provided that the primary defect is not decreased activity of the respiratory center. The elimination of carbon dioxide through the lungs results in a decrease of dCO_2 and the ratio of HCO_3^-/dCO_2 approaches the normal value. This is accompanied by a pH change toward normal.

Renal mechanism. The kidneys respond to respiratory acidosis in the same way that they do to metabolic acidosis, namely, increased Na^+-H^+ exchange, increased ammonia formation, and increased reabsorption of bicarbonate. In a chronic respiratory acidosis in steady state, the normal renal compensation results in return of the plasma pH about halfway toward normal compared with the acute (uncompensated) situation. A so-called fully compensated chronic respiratory acidosis, in which the pCO_2 is high but the pH is normal, is actually not a true compensation but is generally the result of a chronic respiratory acidosis with a superimposed metabolic alkalosis (e.g., because of prolonged administration of diuretics).

LABORATORY FINDINGS IN RESPIRATORY ACIDOSIS

Plasma levels of dCO_2, pCO_2, HCO_3^- and, therefore, the $ctCO_2$ are elevated. The ratio of HCO_3^-/dCO_2 is decreased (owing to an increase in dCO_2), resulting in a low pH.

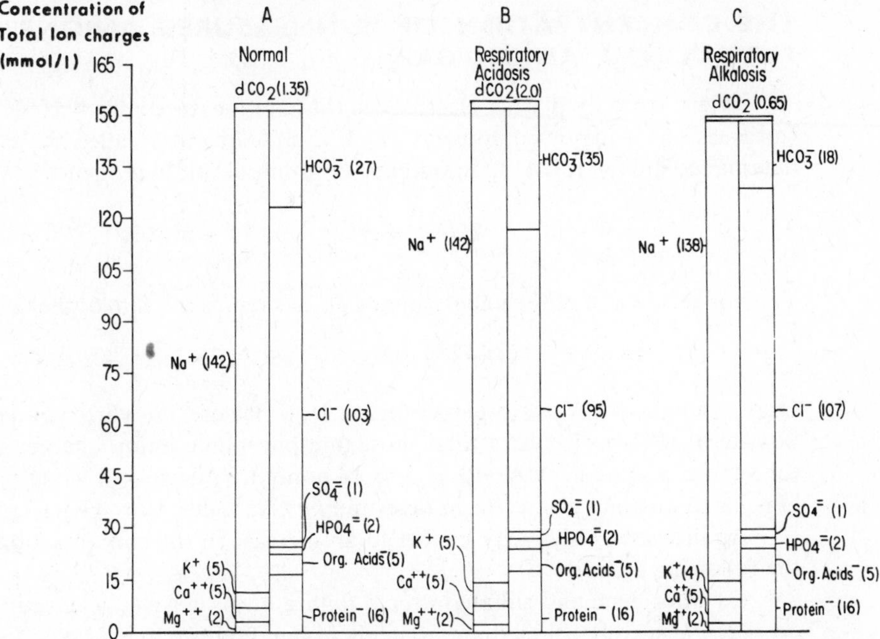

Figure 16–8 *A*, Electrolyte composition of normal plasma. *B*, Example of possible electrolyte pattern in a patient with respiratory acidosis. Note the increase in HCO_3^- and dCO_2. The increase in the latter fraction is more pronounced than that of HCO_3^-. The ratio therefore is decreased, and the chloride fraction shows a decrease. *C*, Theoretical electrolyte pattern of patient in respiratory alkalosis. There is a decrease in the HCO_3^- and especially dCO_2 fraction. Therefore, the ratio of HCO_3^-/dCO_2 and the pH are increased. The Na^+ is at the lower limit of normal. (From Tietz, N. W.: The Chicago Medical School Quarterly, *22*:156, 1962.)

The plasma chloride may be normal or slightly decreased, and urinary acidity and ammonia content are increased (see Fig. 16–8*B*).

RESPIRATORY ALKALOSIS (PRIMARY dCO_2 DEFICIT)

A primary deficit in dCO_2 occurs in all conditions that cause an increased rate or depth of respiration, or both (e.g., fever, high external temperatures, hysteria, hypoxia, and salicylate poisoning). The excessive elimination of carbon dioxide causes a decrease in $cdCO_2$ and, therefore, an increase in the $cHCO_3^-/cdCO_2$ ratio associated with an increase in pH (see Fig. 16–8*C*).

COMPENSATORY MECHANISMS IN RESPIRATORY ALKALOSIS

The compensatory mechanisms of the body to respiratory alkalosis are functions mainly of the kidneys, and these mechanisms correspond to those outlined in the discussion of metabolic alkalosis.

LABORATORY FINDINGS IN RESPIRATORY ALKALOSIS

In this condition, the $cdCO_2$, pCO_2, HCO_3^-, and, thus, the total CO_2 concentration are decreased. The ratio of HCO_3^-/dCO_2 is increased, causing an increase in pH, which, however, rarely exceeds 7.60. In prolonged severe alkalosis there may be an increase of ketone bodies due to decreased carbohydrate utilization, and phosphate levels may be significantly decreased. The concentration of lactate in the blood rises 2 to 4 mmol/l, probably owing to a decrease in hepatic blood flow.

THE CONCENTRATION OF "UNMEASURED ANIONS" IN PLASMA (THE ANION GAP)

Often, the only plasma electrolytes that are measured are Na^+, K^+, Cl^-, and HCO_3^- (measured as a major component of $ctCO_2$). A quantity called the concentration of undetermined anions (cUA^-), anion gap, or anion deficit may then be calculated:

$$cUA^- = cNa^+ - (cCl^- + cHCO_3^-) \tag{22}$$

This equals the sum of the undetermined anions minus the sum of the undetermined cations:

$$cUA^- = (cPr^- + cOrg^- + ctPO_4^- + 2\,cSO_4^{2-}) - (cK^+ + 2\,cCa^{2+} + 2\,cMg^{2+}) \tag{23}$$

Org^- symbolizes various organic anions (e.g., lactate, β-hydroxybutyrate, free fatty carboxylate); tPO_4^- symbolizes total inorganic phosphate anion charge. The reference range for venous plasma is $cUA^-(P) = 5$ to 14 mmol/l. A decreased value may be due to a low cPr^- or to an analytical error in determining cNa^+, cCl^-, or $cHCO_3^-$ (or $ctCO_2$).

An increased value may be due to an increase in the concentration of one or more of the following anions:

(1) Phosphate and sulfate in renal failure owing to insufficient renal excretion.

(2) Negatively charged amino acids in renal failure.

(3) Lactate in shock, muscular exercise, or the lactic acid syndrome.

(4) β-Hydroxybutyrate and acetoacetate in diabetic acidosis and during starvation.

(5) Formate during methanol poisoning, owing to oxidation of methanol to formic acid.

(6) Proteinate when the plasma protein concentration increases, e.g., in states of dehydration.

GRAPHICAL REPRESENTATION OF THE ACID-BASE STATUS OF THE BLOOD

Evaluation of acid-base data is often complicated by the fact that mixed respiratory and metabolic acid-base disturbances are present, and it is often difficult to memorize the various patterns of acid-base values characteristic of the different types of acid-base disturbances. For this reason the graphical diagram shown in Figure 16–9, illustrating the normal physiological relationship between the various acid-base variables, may be helpful in identifying the type of acid-base disturbance existing in a particular patient. The chart refers to the acid-base status determined in arterial blood or arterialized capillary blood at 37°C. Each point in the chart is characterized by four different coordinates representing the acid-base values in a given blood sample:

(1) *The* pH *and the hydrogen ion concentration of the plasma* are both indicated on the abscissa, with pH on a linear scale and the hydrogen ion concentration on a logarithmic scale. Hydrogen ion concentration is calculated as $cH^+ = \text{antilog}\,(9 - \text{pH})$ nmol/l; i.e., the activity coefficient of H^+ is tacitly assumed to be 1.0. The normal range (95 per cent tolerance limits) indicated on the abscissa for pH extends from 7.36 to 7.44. Any point in the left half of the chart indicates an increased plasma acidity, which is termed acidemia. Any point in the right half of the chart indicates a decreased plasma acidity, which is termed alkalemia.

(2) *The carbon dioxide partial pressure* (pCO_2) *of the blood* is indicated on the ordinate (on a logarithmic scale). The units are either mm Hg or kPa (kilopascal). The log scales on the abscissa and the ordinate use the same decade. The normal range indicated for pCO_2 is from 34 to 45 mm Hg or 4.5 to 6.0 kPa. Any point in the upper half of the chart indicates

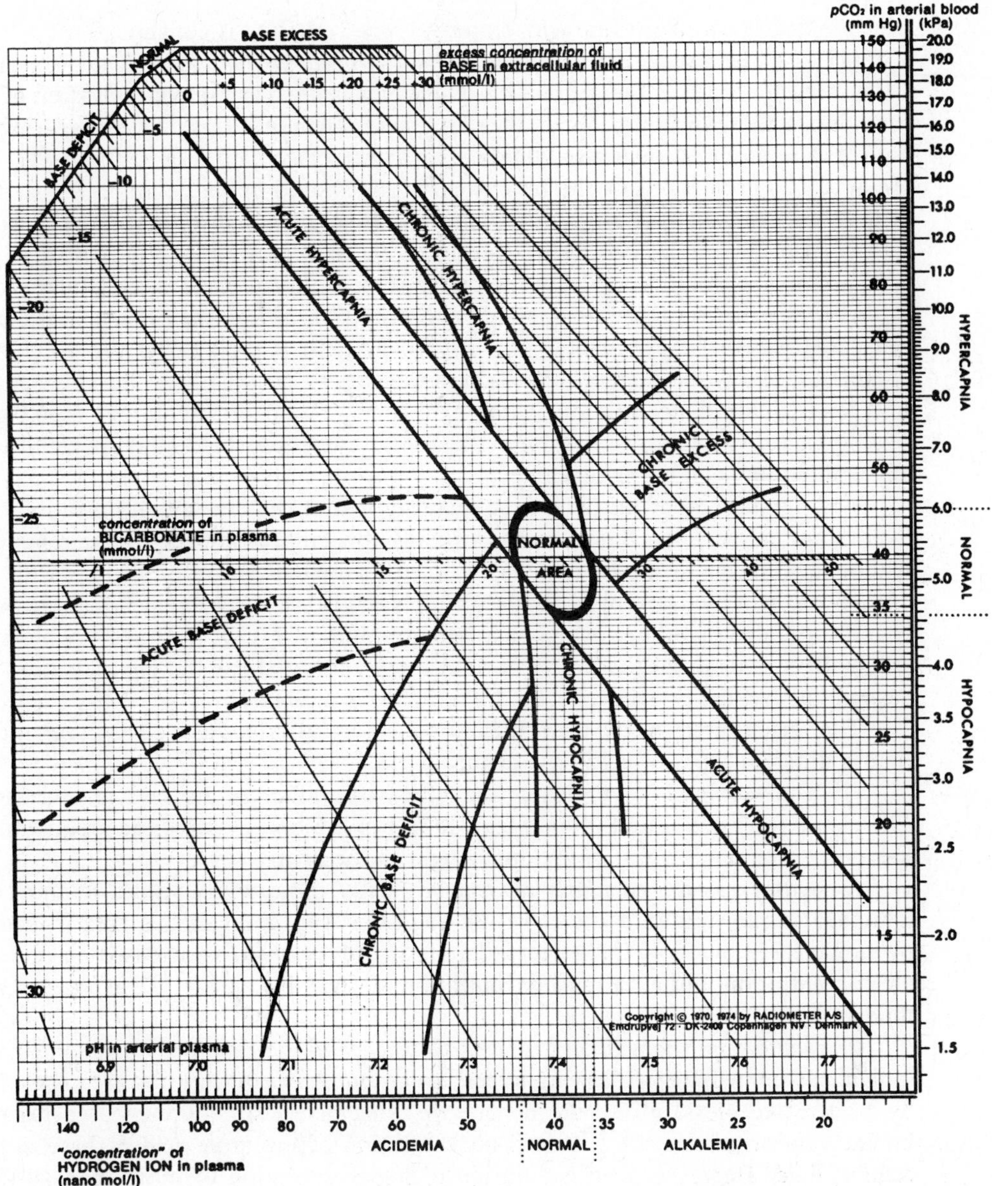

Figure 16–9 Acid-base chart for arterial blood with normal and pathophysiological reference areas. (From Siggaard-Andersen, O.: Scand. J. Clin. Lab. Invest., *27*:239–245, 1971, with permission from the Publisher.)

increased pCO_2, termed hypercapnia, which is generally caused by hypoventilation. Any point in the lower half of the chart indicates decreased pCO_2, or hypocapnia, which is generally caused by hyperventilation.

(3) The *bicarbonate concentration of the plasma* is indicated on the horizontal (logarithmic) scale in the middle of the chart. Projections to the bicarbonate scale should be made at an angle of $-45°$ to the scale. For this reason the divisions on the scale are slanting with slope -1 ($= -45°$). This is apparent when rearranging the Henderson-Hasselbalch equation as follows:

$$\log p\mathrm{CO_2} = -\mathrm{pH} + \mathrm{p}K' - \log \alpha + \log c\mathrm{HCO_3^-}$$

which indicates a straight line with slope -1 when $c\mathrm{HCO_3^-}$ is constant.

(4) *The base excess concentration of the extracellular fluid* is indicated on the scale in the upper left corner of the chart. This quantity is defined as the concentration of titratable base minus the concentration of titratable acid when titrating the average extracellular fluid (Ecf = blood + interstitial fluid) to an arterial plasma pH of 7.40 at $p\mathrm{CO_2} = 40$ mm Hg at 37 °C. Projections to the base excess scale should be made along the slanting lines of the chart. These "base excess lines" represent so-called *in vivo* CO_2 equilibration curves. The slope of these lines has been determined experimentally by CO_2-inhalation or hyperventilation. The slope depends on the buffer value of the average extracellular fluid, which is largely dependent on the hemoglobin concentration of the extracellular fluid, cHb(Ecf), calculated as cHb(Ecf) = cHb(B) × VB(Pt)/VEcf(Pt), where VB(Pt) and VEcf(Pt) are the volumes of blood and extracellular fluid in the patient. In the chart, the slope corresponds to a hemoglobin concentration in the extracellular fluid of 3.6 mmol/l (≈6.0 g/100 ml). The biological variations in the slope (due to polycythemia or anemia) are small and generally without any clinical significance.

The significance of this acid-base quantity is that the excess concentration of base in the average extracellular fluid remains virtually constant during acute changes in the $p\mathrm{CO_2}$. Although a $p\mathrm{CO_2}$ rise causes a rise in the bicarbonate concentration owing to the buffering of carbonic acid by the nonbicarbonate buffers, it also causes a slight fall in the base excess concentration of the blood owing to a redistribution of $\mathrm{HCO_3^-}$ between the blood and the less buffered interstitial fluid. The normal range indicated for the concentration of excess base in the average extracellular fluid is -3.0 to $+3.0$ mmol/l.

The various areas of the chart illustrate the normal values as well as the changes occurring in different types of acid-base disturbances.

(1) The *normal area* indicates the acid-base values of normal resting individuals. The values for women and infants tend to fall in the lower left of the area, while normal values for men tend to fall in the upper right of the area.

(2) *Acute respiratory acidosis* (*acute hypercapnia*) indicates the values obtained in normal individuals following an acute elevation of $p\mathrm{CO_2}$. Any point in this area is characterized by an increased blood $p\mathrm{CO_2}$, a decreased plasma pH, a slightly increased plasma bicarbonate concentration, but a normal concentration of titratable base in the average extracellular fluid.

(3) *Acute respiratory alkalosis* (*acute hypocapnia*) indicates values obtained in normal individuals immediately following hyperventilation. Any point in this area is characterized by a decreased blood $p\mathrm{CO_2}$, an increased plasma pH, a slightly decreased plasma bicarbonate concentration, but a normal concentration of titratable acid or base in the extracellular fluid. However, with a duration of hyperventilation of more than 10 to 15 min, the values tend to fall in the left side of the area or even outside and to the left of the area. This is due to a rapid formation of lactic acid in the liver and muscles, which causes a rise in the concentration of titratable acid in the extracellular fluid.

(4) *Chronic respiratory acidosis* (*chronic hypercapnia*) indicates the values obtained in patients with chronic respiratory insufficiency in a steady state with a normal renal function and a maximal renal compensation. The renal compensation is not maximal until several days after induction of hypercapnia. In the case of a concomitant potassium depletion, the values tend to fall in the right side or to the right of the indicated area. In this case the renal function cannot be said to be normal since the potassium depletion enhances the hydrogen ion excretion in the kidneys at a given plasma pH value. Any point in this area is characterized by increased blood $p\mathrm{CO_2}$, a compensatory elevated concentration of base in the extracellular fluid, a rise in the plasma bicarbonate concentration, and a slightly decreased plasma pH.

(5) *Chronic respiratory alkalosis* (*chronic hypocapnia*) indicates the values obtained in normal individuals acclimatized to high altitude. During the latter part of pregnancy the acid-base values also tend to fall in this area, characterized by a low blood pCO_2, a compensatory rise in the concentration of titratable acid in the extracellular fluid, a low plasma bicarbonate concentration, and a normal or slightly increased plasma pH.

(6) *Acute metabolic acidosis* (*acute base deficit*) indicates values obtained after acute production of non-carbonic acid in the organism, e.g., lactic acid in connection with severe anaerobic muscular exercise. The acidemia stimulates the peripheral chemoreceptors to hyperventilation, but in the most acute phase the acidity is not yet increased in the cerebrospinal fluid and the central chemoreceptors, so the respiratory compensation is only partial. "Acute" in this connection means a disturbance of less than 1 h duration. Any point in this area is characterized by an accumulation of titratable acid in the extracellular fluid (negative excess concentration of base), a slightly decreased pCO_2, a low plasma bicarbonate concentration, and a decreased plasma pH.

(7) *Chronic metabolic acidosis* (*chronic base deficit*) indicates values obtained in patients with a chronic base deficit but with a normal respiratory function, e.g., chronic renal insufficiency or diabetic acidosis. Maximal respiratory compensation develops when ionic equilibrium is reached between the blood plasma and the brain extracellular fluid. This requires 4 to 6 h. Any point in this area is characterized by a low plasma pH, a compensatory fall in blood pCO_2 (Kussmaul ventilation), a low plasma bicarbonate concentration, and an increased concentration of titratable acid in the extracellular fluid.

(8) *Metabolic alkalosis* (*chronic base excess*) indicates the values obtained in patients with a metabolic alkalosis but with a normal respiratory function. Any point in this area is characterized by a high plasma pH, a compensatory rise in blood pCO_2 (hypoventilation), a rise in plasma bicarbonate concentration, and a rise in the concentration of base in the extracellular fluid.

(9) *Mixed acid-base disturbances.* The areas of the chart outside the mentioned areas indicate intermediary situations or mixed acid-base disturbances.

Example: Plasma pH = 7.40, pCO_2 = 90 mm Hg. Projections to the bicarbonate scale (at an angle of $-45°$) gives 55 mmol/l. Projection to the base excess scale (along the slanting lines) gives an estimated concentration of titratable base in the extracellular fluid, ΔcBase(Ecf), of +27 mmol/l. This indicates a severe respiratory acidosis together with a metabolic alkalosis. The designation *fully compensated respiratory acidosis* is sometimes used because the plasma pH is within the normal range, but this designation can be misleading because "fully compensated" in a physiological sense would involve values within the area of chronic respiratory acidosis (chronic hypercapnia).

Example: Plasma pH = 7.10, pCO_2 = 29 mm Hg. Projections to the respective scales give $cHCO_3^-$(P) = 8.5 mmol/l and ΔcBase(Ecf) = -19 mmol/l. This indicates a metabolic acidosis with less than normal respiratory compensation, as in diabetic acidosis complicated by bronchopneumonia.

Example: Plasma pH = 7.18, pCO_2 = 70 mm Hg. Estimated values: $cHCO_3^-$(P) = 26 mmol/l, ΔcBase(Ecf) = -3 mmol/l. This indicates an acute respiratory acidosis; however, the values might also be due to a chronic respiratory acidosis complicated by a metabolic acidosis. The listed laboratory data alone allow only the conclusion that the acid-base values fall within the pattern of acute hypercapnia.

It must be emphasized that the laboratory report should not be considered an acid-base diagnosis in the clinical sense. The clinical acid-base diagnosis requires knowledge or a hypothesis as to the reasons for the shift to the present location in the diagram, i.e., knowledge of the path the acid-base values followed, and/or knowledge or an assumption concerning the acid-base *balance* (intake, production, and excretion of titratable acid or base).

REFERENCES

1. Baron, D. N.: Intracellular clinical chemistry. Clin. Chem. *18*:320, 1972.
2. Pitts, R. F.: Physiology of the Kidney and Body Fluids, 3rd ed. Chicago, Year Book Medical Publishers, 1974.

ADDITIONAL READING

Davenport, H. W.: The ABC of Acid-Base Chemistry, 5th ed. Chicago, The University of Chicago Press, 1969.

Siggaard-Andersen, O.: The Acid-Base Status of the Blood, 4th ed. Copenhagen, Munksgaard, 1974; Baltimore, Williams & Wilkins Company, 1974.

Weisberg, H. F.: Water, Electrolyte and Acid-Base Balance, 2nd ed. Baltimore, Williams & Wilkins Company, 1962.

CHAPTER 17

RENAL FUNCTION

by Willard R. Faulkner, Ph.D. and John W. King, M.D., Ph.D.

The human kidneys are paired, bean-shaped structures situated in the posterior part of the abdominal cavity. Imbedded in the perirenal fat and other supporting structures, the kidneys are firmly fixed to the posterior wall of the abdomen at an area between the upper border of the twelfth thoracic vertebra and the lower border of the third lumbar vertebra. Because the liver occupies the right upper quadrant of the abdominal cavity, the right kidney, which lies behind and slightly below it, is located somewhat lower than the left kidney. The right kidney is in immediate contact with the capsule of the liver and the descending loop of the duodenum. The left kidney lies behind and slightly below the spleen. In the adult human, each kidney weighs about 150 g and measures about 5 × 12 cm. The kidneys are slightly larger in the male than in the female. Their combined weight in proportion to the body weight is usually given as about 1:240. There is a mass of endocrine tissue on the superior pole of each kidney which is the adrenal or suprarenal gland. This is not part of the renal system (see Fig. 17–1*A*, *B*).

On the medial aspect of each kidney is a region called the hilus. If one thinks of the kidney as bean-shaped, the hilus of the kidney corresponds in location to the scar on the bean that marks the area of attachment to the pod. Blood vessels and nerves pass into and out of the kidney at this point. The upper end of the ureter forms a pocket in the hilar area into which the urine that is formed almost continuously by the kidney is collected. Aided by the peristaltic action of the ureter, the urine eventually reaches the bladder, which serves as a reservoir.

Malformations of the kidney occur. Sometimes only one kidney is present and functional, or the two kidneys are fused, usually at the superior pole, to form a horseshoe kidney. Double ureters or extra blood vessels are fairly common. Although many of these anomalies are compatible with life and health, this is not always the case, and many kidney problems, even those first manifesting themselves in later life, occur directly or indirectly as the result of congenital anatomic anomalies. One very serious developmental problem is the so-called polycystic kidney. In this condition, those parts of the kidney arising from the Wolffian duct and those parts arising from the metanephrogenic blastema fail to fuse normally. As a consequence, there is no proper connection between the part of the kidney that produces the urine and the part of the kidney that conducts the urine to the ureter. This condition is not compatible with health and these individuals usually die in renal failure. In other instances, when there is complete agenesis of the kidneys, there is death in utero.

Although the kidney has multiple functions, many of which are not yet well known, its principal role in the body metabolism is the *formation of urine*. This entails not only the excretion of waste products from the blood, but also the provision for the preservation of essential solutes and regulation of hydration and electrolyte balance. The vital importance of maintaining the body's internal environment has been recognized since the time of Claude Bernard. This nineteenth century French scientist was among the first to point out that the integrity of the body was dependent upon selective excretion of metabolites, which could not be allowed to accumulate within the body without causing harm to the

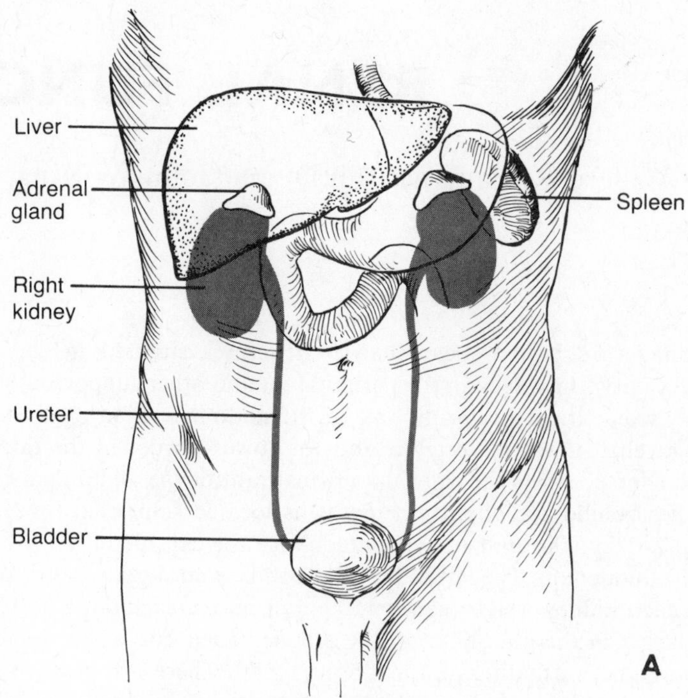

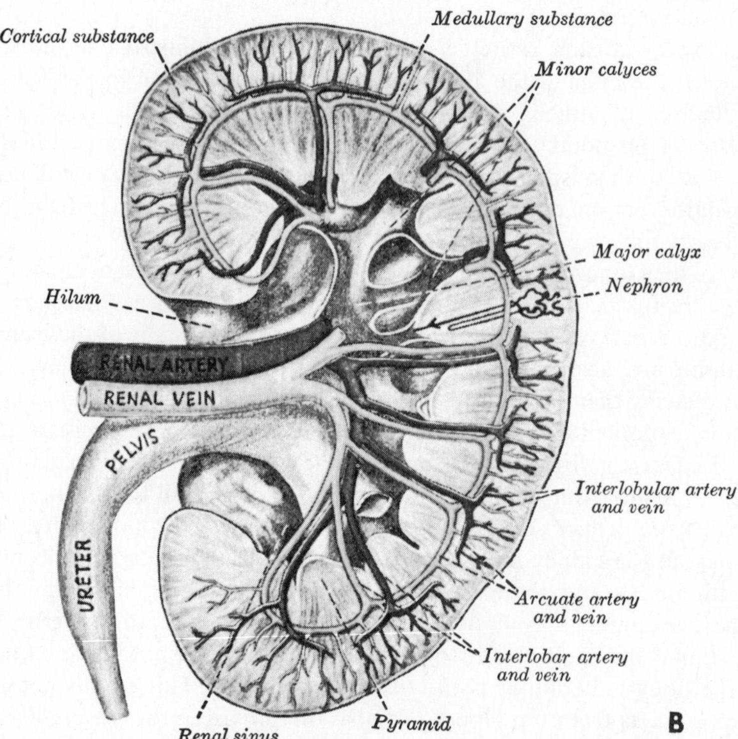

Figure 17–1 *A*, Schematic drawing of right and left kidneys with respect to liver, stomach, duodenal loop, spleen, adrenal gland, and bladder. *B*, Diagram of a vertical section through the kidney. Nephron and blood vessels greatly enlarged. (From Gray's Anatomy, 28th Edition, Edited by C. M. Goss, Lea & Febiger, 1966.)

individual. At the same time this excretion of metabolic products must be sufficiently selective so that substances that are utilized or required by the body are not lost. Homer Smith said, "The composition of the blood is determined not by what the mouth ingests but what the kidney keeps."

To accomplish its complex mission, the kidney must act first as a selective filter to remove water and filtrable solutes from the blood plasma. This process occurs in the glomeruli and the fluid formed is known as the glomerular filtrate. The volume of the glomerular filtrate is around 125 ml per minute and may exceed 180 liters in a single day. If the kidneys were to stop at this point and pass this filtrate to the bladder to be excreted, the individual could not replace the lost fluid and solutes and, therefore, life would be impossible. Among the components of the glomerular filtrate are many substances that are necessary for normal function. To prevent their loss, the kidney is designed to reabsorb water and useful solutes selectively. By this process the glomerular filtrate is converted to urine, which contains end products of metabolism that might be injurious if accumulated; it also contains any excess of essential solutes that might be present. Of course, sufficient water to keep these solutes in solution is also lost, but this volume is so small that the individual can usually replace this easily.

The filtration apparatus of the kidney is built around a functional unit called the *nephron*, which consists of the *glomerulus* and the *tubule*. There are about 1.2 million of these structures in each kidney. All of these structures are not working at any one time, but their very presence gives the kidney considerable reserve capacity in the event of stress, disease, or injury. Each nephron is supplied by a small blood vessel called the *afferent arteriole*. It carries blood from a branch of the renal artery into the nephron. These afferent arterioles carry blood to the nephrons at a rate of about 1200 ml per minute (total renal blood flow). The arteriole enters into an expanded portion of the renal tubule called *Bowman's capsule* (see Fig. 17–2). Within the capsule, the vessel breaks up into a plexus of capillaries, which ultimately recombine to form an *efferent arteriole*. This efferent arteriole then joins with other efferent arterioles to carry blood from the nephrons to the renal tubular area. The capillary plexus and its afferent and efferent arterioles are often referred to as the *glomerular tuft*. Filtration is accomplished through the thin walls of the capillaries that make up the plexus. The blood flows into the plexus from the relatively large afferent vessel and leaves the plexus through an efferent vessel that has a smaller lumen. This difference in size of the two vessels produces an increase in hydrostatic pressure within the capillaries that has been reported to be about 75 mm Hg. This pressure is almost twice as high as that found in capillaries in other parts of the body. In part because of the relatively high hydrostatic pressure, the filtrate is forced through the thin capillary epithelium and is caught in Bowman's capsule. As mentioned before, Bowman's capsule envelopes the glomerular tuft and is connected with the tubule where concentration and modification of the filtrate takes place. This is mainly an active process of selective secretion and reabsorption by the tubular epithelium of the kidney. By this process the kidney conserves solutes and metabolites that the body requires and, at the same time, disposes of molecules such as creatinine, urea, and metabolites that are not needed by the body and that must be eliminated.

Some of the substances that are present in the glomerular filtrate are known as *threshold* substances and appear in the urine only after they have reached certain minimum concentrations in the blood. Glucose is such a substance and does not appear in the urine until plasma glucose levels reach about 180 mg/100 ml. Other substances, such as creatinine, may be excreted without appreciable reabsorption. A foreign substance that behaves similarly is the polysaccharide inulin. Therefore, both creatinine and inulin are useful in measuring the glomerular filtration rate (see the next section).

About 20 per cent of the volume of the plasma that passes through the glomerular tuft passes into Bowman's capsule as glomerular filtrate. This filtrate has a specific gravity of 1.010 $\pm$ 0.002, a pH of approximately 7.4, and is isoosmotic with plasma. It contains

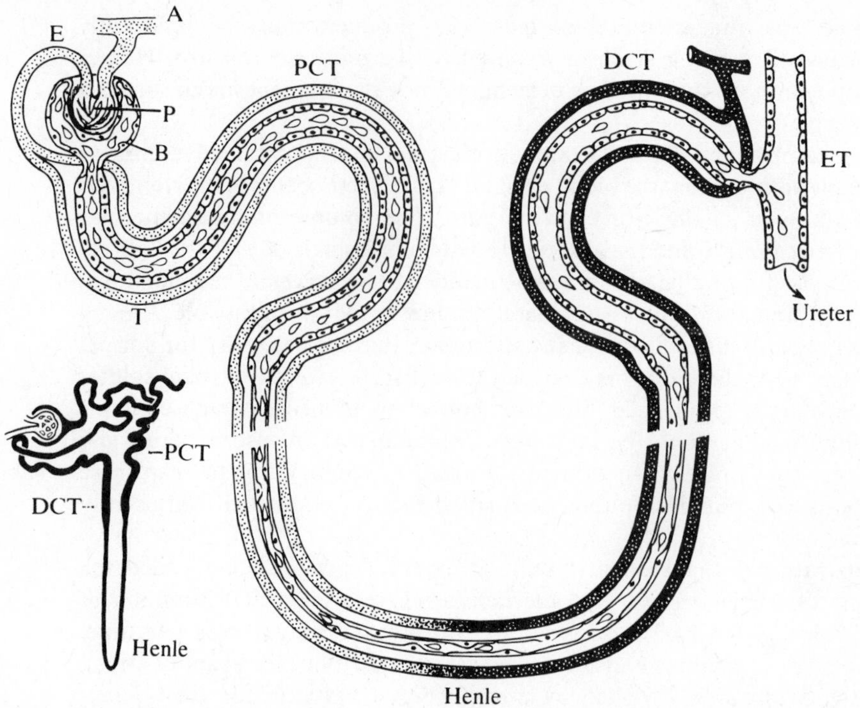

Figure 17–2 Schematic drawing of the glomerulus and the tubular system. *A*, Afferent arteriole; *E*, efferent arteriole; *P*, plexus of capillaries (glomerular tuft); *B*, Bowman's capsule; *T*, tubular blood supply; PCT, proximal convoluted tubule; Henle, loop of Henle; DCT, distal convoluted tubule; ET, excretory tubule or duct. The blood capillaries shown along the tubular system (T) gradually change to venous capillaries as they pass down the tubular system.

all the filtrate molecules present in plasma and at approximately the same relative concentrations. This filtrate passes into the tubular system, which is usually considered to be divided into four major sections (see Fig. 17–2).

The first section is a coiled structure that connects with Bowman's capsule and is known as the *proximal convoluted tubule*. This empties into a long narrow portion called the *loop of Henle*, which in turn changes into another convoluted tubule referred to as the *distal convoluted tubule*. The distal convoluted tubules join to form the *collecting tubules*, which terminate in the pelvis of the kidney, where the urine is emptied into the ureter.

Each section of the tubular system has a different histological structure and each appears to have a specific function. The proximal convoluted tubule is believed to be concerned mainly with the reabsorption of glucose and electrolytes. The passage of glucose through the tubular epithelium is a very complicated process and involves a series of enzymatic coupling and uncoupling reactions between glucose and the phosphate radical. Na^+ and Cl^- are also reabsorbed in the proximal convoluted tubule and, of course, large quantities of water are absorbed so that approximately 85 per cent of the volume of the glomerular filtrate is removed at this point of the tubular system. The modified filtrate passes into the loop of Henle, where more water and sodium are reabsorbed. The further modified filtrate then passes into the distal convoluted tubule, where again more water is reabsorbed. It is also at this point in the tubular system that exchange of Na^+, K^+, and H^+ occurs and the kidney forms ammonia from amides (see also Chap. 16). The urine has now assumed its final composition and passes, via the collecting tubules, to the pelvis of the kidney, to the ureter, and finally into the bladder. This remarkable process is subject to many types of disturbances, and it is sometimes possible to diagnose the type of lesion present from a knowledge of renal physiology and the results of some of the kidney function tests. Problems in proximal convoluted tubular activity may be measured by the phenolsulfonphthalein (PSP) excretion or the para-aminohippurate excretion tests, both of which will be discussed later in this chapter. Distal tubular activity affects the various concentration and dilution tests, and when damage is severe enough it may be reflected in alteration in serum electrolyte levels.

Normal values for urine show a much wider range than do normal values for constituents of plasma and other body fluids. The composition of urine is more markedly affected than corresponding values for plasma by factors such as fluid intake, diet, and environmental temperature. Thus, a patient may have a large decrease in daily urine volume merely because of an increase in the room temperature or because of a personal dislike for the fluids being served. A patient on a voluntary restriction of fluids may reduce his urinary output to a third of his normal output and at the same time raise his urinary specific gravity from his usual value to 1.025, 1.030, or higher. Similarly, the pH of the urine may be decreased by the administration of acidifying drugs or raised by the ingestion of foods that have an alkaline residue. Even after the urine has entered the bladder its pH may be altered because of bacterial action in the bladder, but normally, bladder urine is sterile. Significant urine pH changes *in vivo* are caused only by urea-splitting organisms such as *Proteus* and certain coliforms. Further changes may occur in the specimen bottle. For this reason, interpretation of pH values must be made with caution and changes in urinary pH cannot be ascribed to metabolic activity unless the urine is fresh and contains few or no bacteria.

Normal kidney function depends upon an adequate *blood flow* to the organ. Any situation that interferes with renal blood supply will result in diminished kidney function, will alter the amount and composition of the urine that is produced, and will result in an accumulation of metabolic products in the blood. Patients in cardiac failure frequently have diminished kidney function not due to any intrinsic lesion in the kidney, but rather due to an inadequate flow of blood to the kidney.

Intravascular changes that interfere with the flow of blood to the kidney may also produce increases in blood pressure. This condition is known as renal hypertension. Reduction of kidney function may also develop as a consequence of damage to the epithelium of the capillaries, which make up the glomerular tuft, as in glomerulonephritis. In this condition the endothelium is damaged to the extent that not only water and other small molecules are able to pass through the glomerular membrane, but also red blood cells and some plasma proteins that are not normally filtered. The relatively small molecules such as urea are subsequently (passively) partially reabsorbed by the tubules, but red cells and larger molecules (protein) cannot return to the blood and, therefore, are excreted with the urine. In chronic glomerulonephritis and other causes of the nephrotic syndrome, large amounts of protein are lost in urine.

Tubular as well as glomerular tuft damage may be the result of overloading the system with a toxic substance, e.g., following massive hemolytic reactions after transfusion with incompatible blood. In this situation, massive amounts of hemoglobin are released into the plasma and exert a toxic effect on the tubules. In addition, there may be an agglutination of red cells, resulting in anoxia and, consequently, damage to the nephron. Both factors result in an impairment of kidney function and possibly complete renal failure.

Administration of various poisons to the body may selectively interfere with certain functions of the tubule. Actual mechanical trauma to the kidney or the development of neoplastic growth, which either destroys tissue directly or produces pressure on the kidney tissue, may also disrupt its normal functioning.

Sometimes solutes in urine are present in such concentrations that precipitation of these substances occurs either within the pelvis of the kidney or in the ureter or bladder. These precipitates are referred to as *calculi* and have an appearance similar to natural stones (see Chap. 18). They may cause infections and produce injury by causing obstruction and subsequent back pressure, which reduces filtration rate and decreases urinary excretion. If the stone lodges in the ureter, reflex spasms occur which produce pain and obstruction. The pain is described as one of the most severe pains imaginable.

Occasionally patients are seen with inborn errors of metabolism involving the renal tubule. One such example is cystinuria. In this condition, the tubules are unable to reabsorb cystine and several other amino acids. As a consequence, these amino acids occur in urine in increased amounts (see Chap. 7).

Laboratory Tests Aiding in the Evaluation of Kidney Function

The kidney is a very complex organ; therefore, it is quite helpful to the physician to know the type and location of a possible pathological lesion. The diagnosis of functional renal disease, to a great extent, is made in the clinical laboratory. This is true because clinical signs and symptoms may be minimal or absent entirely and, even when present, will not always reflect the severity of the disease or the prognosis for the patient. It is indeed fortunate that the laboratory has a battery of kidney function tests available, which, when properly applied, can give valuable information about the status of the individual types of kidney functions and frequently about the location of the defect. It must be remembered, however, that the kidney has a considerable functional reserve and kidney function tests may be normal even in the presence of a relatively severe renal lesion.

Kidney function tests, in general, can be affected by *prerenal*, *renal*, *or postrenal* phenomena. Among the prerenal causes is dehydration, which may be found in pyloric and intestinal obstruction and in prolonged diarrhea. Conditions of shock and excessive loss of blood, such as are seen in severe intestinal bleeding or cardiac failure, are other prerenal causes for decreased kidney function. Decreased kidney function in these conditions may occur because of either decreased plasma volume or decreased blood flow.

Among the renal causes for decreased kidney function are diseases affecting the glomerular filtration rate, the tubular function, or any changes in the renal vascular system that decrease the blood flow.

A postrenal cause for decrease in kidney function is obstruction of the urine flow, which may be caused by enlargement of the prostate, stones in the urinary tract, or tumors of the bladder. In these cases the decreased function is due to reduction in effective filtration pressure of the glomeruli.

The comments on the causes for decreased renal function as well as the initial discussion on the physiology of the kidney make it most practical to separate kidney function tests into three major groups: (1) those measuring glomerular filtration, (2) those measuring renal blood flow, and (3) those measuring tubular function. Tests measuring the retention of nonprotein nitrogenous compounds in plasma are listed in the first category, although their plasma values may be affected by changes in any one of the three.

GENERAL CHARACTERISTICS OF URINE

Urine Volume

The volume of urine produced by a normal adult over a 24 h period is usually between 800 and 1800 ml. These values are subject to considerable variation because many factors determine the exact amount of urine any given individual will produce on any given day. In essence, the urinary output is the difference between total fluid intake and loss of fluid by other means (e.g., lung, perspiration, stool) or retention in the tissues. Thus, under some conditions the urinary output of normal individuals may be far outside the range just given. For example, voluntary restriction of fluid intake, as for the Fishberg concentration test, can reduce the daily urine volume to less than 500 ml. A moderate beer drinker can increase his daily output of urine well above the normal upper limit without much difficulty because not only is his urinary output affected by the volume of beer drunk, but also beer is believed to contain a diuretic factor, which further increases urine excretion. The prophylactic use of drugs for motion sickness may have the undesirable side effect of polyuria (increased urinary output). Nervousness, associated with stage fright, may be accompanied by polyuria that cannot be completely accounted for by increased intake of water.

Decrease of urinary output is known as *oliguria* and, in the absence of fluid deprivation, is usually an ominous sign. It may occur during the early development of edema or ascites or during shock. When occurring because of terminal uremia, it may be a prelude to *anuria* (no output of urine) and death.

Pathologic states associated with *polyuria* are diabetes mellitus, diabetes insipidus, and some types of central nervous system injuries. In some kidney diseases that ultimately result in oliguria, the initial effect will be polyuria as the kidney, unable to concentrate properly, seeks to compensate by increasing the urine volume.

Care must be taken to distinguish between polyuria and frequency. The person with a bladder infection may have frequency without significant increase in the total volume of urine produced; however, both conditions may be present at the same time, as in diabetes.

Appearance of Urine

Urine is ordinarily clear when voided, but may become turbid upon standing. Precipitation of solutes occurs as the urine cools and as the pH changes as a result of bacterial action. This may have little or no clinical significance; however, cloudiness in a *freshly* voided urine may have clinical significance. The most common reasons for this may be presence of blood, pus, or bacteria, or all three, in the urine. Blood rarely occurs to the extent that the urine is actually red except following surgery or trauma. (Some disturbances in porphyrin metabolism may produce red urine.) Usually *hematuria* (red cells in urine) is of the microscopic type and is said to produce a smoky urine, which some imaginative observers have described as looking like smoky Scotch whisky. Some foods may color the urine. Beets have been incriminated but the amount of beets necessary to produce red urine is greater than most people will consume, and this is true of most other foods. Riboflavin will impart to urine a yellow color. Many drugs, including several urinary tract antiseptics, change the color of the urine. The Diagnex Blue test, a technique for detecting the presence of hydrochloric acid in the stomach, may color the urine blue for several days. Other things being equal, concentrated urine of relatively high specific gravity will be darker than more dilute urine.

Odor of Urine

The odor of normal urine has been described in many ways, aromatic and nutty being two adjectives used; however, none has really described it properly. Let us simply state that fresh urine has a characteristic and not unpleasant odor. Abnormal odors can be produced by some foods (asparagus and garlic) and by various pathologic states. The odor of the urine of an uncontrolled diabetic is described as fruity because of the presence of acetone and acetoacetic acid. *Proteus* infections produce an ammoniacal urine, and less characteristic odors are produced by infections due to other bacteria.

Although modern laboratory workers, for obvious reasons, do not make a practice of tasting the specimens sent to the laboratory, many significant observations in the past were made by urine tasters. The one that is perhaps the most well known is the distinction between diabetes mellitus and diabetes insipidus. In the first condition, the urine tastes sweet; in the latter, it is virtually tasteless.

Urine pH

The average urine pH is approximately 6.0 (range: fasting, 5.5 to 6.5; random, 4.8 to 7.8). Night urine is generally more acid than day urine and this (besides the fact that night urine is more concentrated) is one of the reasons that specimens for examination for casts and other formed elements are best collected in the early morning. Metabolism of fats produces more acid residues than the metabolism of carbohydrates. Starvation, with consequent utilization of stored body fat, will also produce ketosis, and thus acid urine. In the treatment of urinary infection, acid producing drugs are sometimes used to reduce the pH of the urine. One of the more commonly used drugs of this type, and one which is also used as a mild diuretic, is ammonium chloride. Bacterial infections can alter the pH in either direction, depending upon the end products of bacterial metabolism. Ammonia producing organisms obviously produce alkaline urine, but certain other bacteria will cause the urine to become acid.

Formed Elements

Formed elements in urine are defined as objects that may be observed by direct microscopic examination of the urine specimen. This term usually refers to white and red blood cells, casts, crystals, bacteria, ova and parasites. A discussion of the clinical significance and the identification of these elements is beyond the scope of this text, and the reader is referred to special texts.

The problem of identifying crystals does not seem nearly as critical to the laboratory workers of today as it was to their counterparts of a century ago. It is still important to recognize a sulfa crystal, but we see sulfa crystals much less often than we did a few years ago. Tyrosine and leucine crystals do appear in urine, but generally we have much better ways of identifying amino acidurias than by the microscope and if we do this procedure today, it is usually an exercise in hindsight. Many crystals are merely a reflection of the patient's diet; it has been said that the presence of oxalic acid crystals in a patient's urine merely means that he has had rhubarb for breakfast. This is an oversimplification, but it is safe to say that many urine specimens contain crystals and, in general, their presence and their identification have relatively little clinical significance.

"Routine Urinalysis" Tests

Routine urinalysis procedures, as carried out in many hospitals, consist of determinations of urinary pH and specific gravity, tests for the detection of reducing sugars, protein, and ketone bodies, and, finally, a microscopic examination of the urinary sediment to detect the possible presence of red and white blood cells, casts, or any other formed elements. Some of these procedures are not tests of kidney function, since they detect abnormalities that are reflections of disease elsewhere in the body. An example of this is diabetes mellitus. In this disease, profound chemical abnormalities, such as presence of glucose and ketone bodies in the urine, may be noted as the kidney strives to correct abnormal physiological activities elsewhere in the body and to keep the internal environment within reasonable limits. The detailed laboratory procedures for urinary sugars, albumin, hemoglobin, bile, and ketone bodies are discussed in Chapters 6, 7, and 15.

Tests Measuring Glomerular Filtration Rate

CLEARANCE TESTS

The group of tests generally referred to as renal clearance tests are extremely useful in measuring the actual capacity of the kidney to eliminate certain substances present in plasma. The selection of the type of clearance test to be used is made by taking into consideration the aspect of kidney physiology which is to be evaluated. A substance may be excreted by glomerular filtration alone, or by filtration plus tubular excretion (depending on the substance, tubular excretion will more or less dominate), or it may be filtered but then subsequently reabsorbed by the tubules in whole or in part. If only a single facet of renal physiology (e.g., the glomerular filtration rate) is to be studied, a substance should be selected that is excreted (filtered) either completely or predominantly by the glomeruli without being either excreted or reabsorbed by the tubules. Inulin is such a substance; however, it is rarely employed because of the limited availability of procedures for the quantitation of inulin. Creatinine is a substance that behaves similarly to inulin. Since the procedure for determination of creatinine in plasma and urine is readily available in most clinical laboratories, the creatinine clearance test has become one of the most popular tests for measuring the glomerular filtration rate (Fig. 17–3). For the detailed discussion of the inulin and creatinine tests, the reader is referred to a later portion of this chapter.

Historically, the urea clearance test was the first clearance procedure commonly used. Urea, although cleared by the glomeruli, is subsequently partially reabsorbed by the tubules at an average rate of 40 per cent of that filtered. Furthermore, the rate of reabsorption, which is a process of passive diffusion, varies with the amount of water reabsorbed. Thus, the urea clearance test is not a measure of the glomerular filtration rate (although it has been used for this purpose), but it is an index of overall renal function. For the measurement of glomerular filtration, it has been widely replaced by the more accurate creatinine clearance test. The tests measuring renal blood flow are also based on the clearance concept, but they are discussed in a separate section.

Screening tests for measuring renal function involve the determination of blood (plasma, serum) constituents such as nonprotein nitrogen, urea nitrogen, and creatinine.

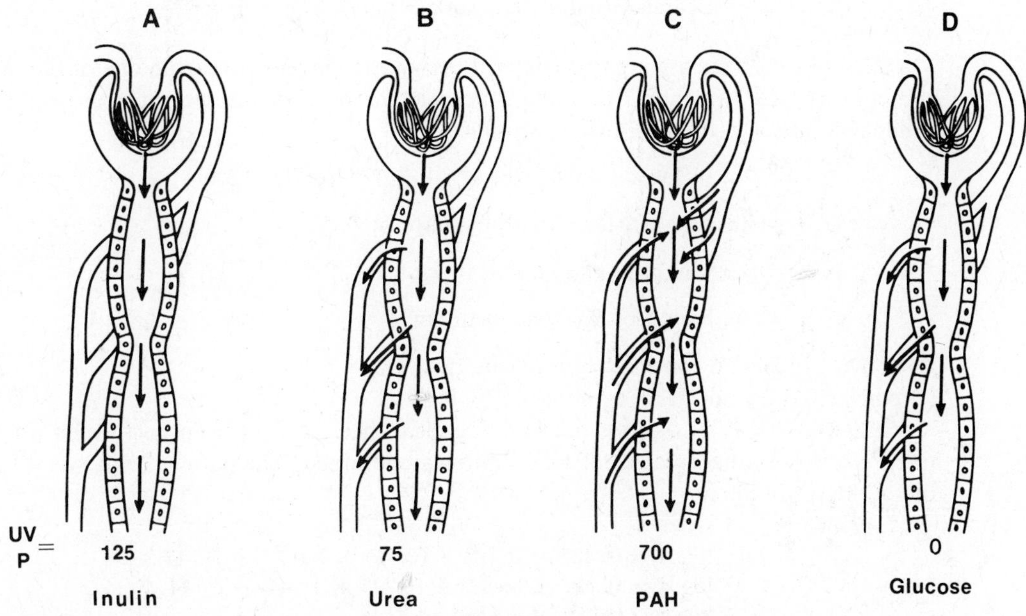

Figure 17–3 Schematic presentation of the excretion of various types of substances by the nephron. *A,* Inulin is excreted by glomerular filtration and passes through the tubular system without re-absorption. No inulin is removed by tubular secretion. Creatinine behaves very similarly to inulin. *B,* Urea is filtered through the glomerulus, but is subsequently partially reabsorbed by the tubular system. *C,* PAH (*p*-aminohippurate) is, to a limited extent, filtered by the glomerulus, but is mainly excreted by the tubules. Phenolsulfonphthalein behaves similarly. *D,* Glucose is filtered and subsequently reabsorbed by the tubules. (After Cantarow and Trumper.)

If the level of these substances increases, this implies impaired renal function. The sensitivity of these tests is limited, however, and kidney function may decrease to as much as 50 per cent of normal before these blood (plasma, serum) tests yield abnormal results.

The clearance tests, on the other hand, provide a much more sensitive measure of renal function. If the appropriate clearance test is used, its result is a measure of the functional capacity of a specific part of the nephron. The sensitivity and usefulness of these tests is due to the fact that we relate the quantity of a certain substance excreted in urine to the quantity of the same substance in plasma. The amount of substance cleared by the kidney is expressed as that volume of plasma which contains the quantity of the substance excreted in the urine in a period of one minute. This may be mathematically expressed as follows:

$$\text{clearance (ml/min)} = \frac{U}{P} \times V$$

where U = the concentration of the substance in urine

P = the concentration of the substance in plasma

V = the urine volume per minute, expressed in ml

The concentration of the substance in urine and plasma may be expressed in any convenient unit, but it is customary to use mg/100 ml. U and P must be expressed in the same unit. All other factors being equal, the clearance rate is roughly proportional to the size of the kidney and the body surface area of the individual. Therefore, the calculation for the clearance of any given substance should provide for correction for deviations from the average adult body surface. This is done by multiplying the clearance by the factor $1.73/A$, where 1.73 is the generally accepted average body surface in square meters and A is the body surface of the patient under investigation. The formula for calculating the renal clearance, therefore, expands as follows:

$$\text{clearance (ml/min/std. surface area)} = \frac{U}{P} \times V \times \frac{1.73}{A}$$

The body surface area may be determined more conveniently from one of the available nomograms (see Appendix) or it may be calculated from the weight and height of the patient by means of the following formula:

$$\log A = (0.425 \log W) + (0.725 \log H) - 2.144$$

where A = the body surface area in square meters

W = the weight of the patient in kg

H = the height of the patient in cm

To convert inches to centimeters multiply by 2.54.
To convert American lb to kg multiply by 0.45.

Example. Let us assume a creatinine clearance test has been performed on a child and a clearance (uncorrected) of 12 ml/min was obtained. The patient has a weight of 4 kg and a height of 35 cm.

Therefore:

$$\log A = (0.425 \times \log 4) + (0.725 \times \log 35) - 2.144$$
$$\log A = (0.425 \times 0.602) + (0.725 \times 1.544) - 2.144$$
$$\log A = 0.2559 + 1.1194 - 2.144$$
$$\log A = 1.3753 - 2.144$$
$$\log A = (-0.7687) \quad \text{or} \quad \begin{array}{r} 1.0000 \\ -0.7687 \\ \hline 0.2313 - 1 \end{array}$$
$$A = \text{antilog } (0.2313 - 1) = 0.1703 \text{ m}^2$$

The clearance of 12 ml/min corrected for the body surface of 0.1703 m², therefore, is:

$$12 \times \frac{1.73}{0.170} = 123 \text{ ml/min}$$

Correction for body surface is absolutely mandatory if the body surface of the patient differs greatly from that of the average person. The error otherwise introduced is substantial and in infants may be up to several hundred per cent, as shown by the previous example.

THE INULIN CLEARANCE TEST

The polysaccharide inulin, having a molecular mass of about 5100 daltons, is obtained from dahlias and artichokes. It has become the substance of choice for precise investigative work because it is filtered freely by the glomeruli but is neither secreted nor absorbed by the tubules.

Although the use of inulin for measuring the glomerular filtration rate cannot be regarded as a routine laboratory test (a fact that should be emphasized at the outset), a brief description of the procedure is given. Since inulin is not normally present in the plasma, it must be introduced at a suitable concentration in order to allow its clearance by the kidneys to be measured. This may be done by giving a sufficient quantity as a *priming* dose (25 ml of a 10 per cent solution of inulin) by intravenous injection to produce a satisfactory plasma level and then maintaining this level throughout the test period by a slow, continuous (*maintenance*) infusion of a less concentrated solution (500 ml of a 1.5 per cent solution).

An adequate fluid intake (1000 ml) is maintained during the hour before the test. It is not necessary for the patient to fast. A blood specimen is taken into a tube containing an anticoagulant and is used as the control. The patient empties his bladder and saves the sample of urine, which is also to be used as a control. The priming dose of inulin is then introduced slowly over a few minutes. This is followed by the maintenance solution, given at about 4 ml/min. After about one-half hour the bladder is emptied, and subsequently urine is collected in three accurately timed specimens of about 20 min each. Blood specimens are withdrawn at the beginning and at the end of the urine collection period. The mean value of the two plasma specimens is used in the calculation. Inulin is measured in all blood and urine specimens.

Inulin clearance is calculated for the three timed urine specimens, using the average value of the two plasma samples as follows:

$$\text{inulin clearance} = \frac{\text{mg inulin/100 ml urine}}{\text{mg inulin/100 ml plasma}} \times \text{ml urine excreted/min}$$

An average of the three clearance values is made; for a normal adult this is about 125 ml/min/std. surface area or about 70 ml/m² of body surface area. Goldring and Chasis reported a range of 110 to 152 ml/min for men and 102 to 132 ml/min for women.

CLINICAL SIGNIFICANCE

Inulin is almost completely cleared by glomerular filtration at an average rate of 125 ml/min. The amount of inulin, once filtered, is not reabsorbed by the tubules and, thus, is quantitatively excreted in the urine. The inulin clearance is the most accurate measure of glomerular filtration at the present time. Its clinical application is limited, however, by the fact that the test is cumbersome, expensive, time-consuming and uncomfortable to the patient. The test requires an intravenous infusion and constant attendance by the physician during the test period. For these reasons, the use of the inulin clearance test is restricted to research institutions or institutions specializing in the study of kidney diseases. Routine laboratories generally perform the creatinine clearance test, which is described next in this chapter.

REFERENCES

Dick, A., and Davies, C. E.: J. Clin. Path., *2*:67, 1949.
Goldring, W., and Chasis, H.: Hypertension and Hypertensive Disease. New York, The Commonwealth Fund, 1949.
Smith, H. W.: Kidney Structure and Function in Health and Disease. New York, Oxford University Press, 1951.
Gary, A., and Discombe, G.: Clinical Pathology. Philadelphia, F. A. Davis Co., 1966.

CREATININE CLEARANCE TEST (ENDOGENOUS)

CLINICAL SIGNIFICANCE

The creatinine clearance test is a relatively accurate and useful measure of the glomerular filtration rate, and it has largely replaced the less accurate urea clearance test.

When plasma creatinine levels increase considerably above the normal, creatinine is also secreted by the tubules; thus, the creatinine clearance value may be greater than the actual glomerular filtration rate. For this reason, the exogenous clearance test (in which creatinine is administered either orally or intravenously) is seldom employed.

The relationship between plasma creatinine and creatinine clearance values is shown in Table 17–1. In order to better demonstrate this relationship, the urine creatinine (in mg/100 ml) and the urine flow (in ml/min) are kept constant. It can be seen from this table that relatively minor changes in plasma creatinine are accompanied by changes in creatinine clearance which are more dramatic, especially in the early phase of kidney disease. For example, a change in plasma creatinine from 0.70 to 1.05 mg/100 ml (both in the normal range) is accompanied by a change in creatinine clearance values from 120 to 80 ml/min.

SPECIMEN

A precisely timed urine specimen and a serum sample are required for this test. The blood is generally collected in the middle or at the beginning *and* at the end of the urine collection period. Creatinine is subject to a slow equilibrium reaction with creatine, which is accelerated by hydrogen and hydroxyl ions. The best way to retard this process, as well as bacterial decomposition, is to store the specimen in a refrigerator until analyzed; this should be no longer than one working day after the collection.

PRINCIPLE

Creatinine occurring through metabolic production is eliminated from the plasma predominantly by glomerular filtration and, therefore, measurement of its rate of clearance affords a measure of this process.

TABLE 17–1. RELATIONSHIP BETWEEN PLASMA CREATININE AND CREATININE CLEARANCE VALUES*

Plasma creatinine (mg/100 ml)	28	16.8	8.4	4.2	2.1	1.40	1.05	0.84	0.70	0.60
Urine creatinine (mg/100 ml)	84	84	84	84	84	84	84	84	84	84
Urine flow (ml/min)	1	1	1	1	1	1	1	1	1	1
Creatinine clearance (ml/min)	3	5	10	20	40	60	80	100	120	140

A constant creatinine excretion of 1.21 g/d and a constant urine volume of 1440 ml/d are assumed.

* Modified from Bernstein, L. M.: Renal Function and Renal Failure. Baltimore, The Williams and Wilkins Company, 1965.

The general principle of clearance as described for urea (p. 988) is also applicable in this situation but with two modifications. (1) The clearance of creatinine is much less subject to the rate of urine flow than urea, and for this reason the distinction between standard and maximum clearance does not apply. (2) Creatinine clearance is stated only in terms of volume of plasma cleared per minute and not in per cent of normal. The test described here is an endogenous clearance test in which no creatinine is administered.

REAGENTS

These are the same as those used for the determination of plasma and urinary creatinine (see p. 996).

PROCEDURE

Preparation of the patient

(to be carried out by the nursing staff of the hospital ward)

1. Hydrate the patient by administering a minimum of 600 ml of water. Withhold tea, coffee, and drugs on the day of the test.
2. Have the patient void and discard the specimen.
3. Collect a 4, 12, or 24 h specimen, and record the exact times of starting and completing the collection. Keep patient well hydrated to insure a urine output of at least 2.0 ml/min.
4. Collect a specimen of clotted blood during the urine collection period. Creatinine values are relatively constant; hence, the blood specimen can be collected at any time during the urine collection.

Laboratory procedure

Determine the creatinine in an aliquot of the well-mixed urine specimen and in the plasma or serum according to the procedure described on page 997.

CALCULATION

$$\text{ml plasma cleared/min/std. surface area} = \frac{UV}{P} \times \frac{1.73}{A}$$

where U = concentration of creatinine in urine
P = concentration of creatinine in plasma
V = volume of urine in ml/min
A = body surface area in square meters

NORMAL VALUES

If the creatinine method described in this chapter is used, the normal creatinine clearance values for males are 105 $\pm$ 20 ml/min and for females 95 $\pm$ 20 ml/min. If a more specific method for the determination of creatinine is used (e.g., one employing Lloyd's reagent), the respective values for the clearance of creatinine by males and females are 117 $\pm$ 20 ml/min and 108 $\pm$ 20 ml/min. The reason for the difference in normal values is that values for urine creatinine obtained with both methods are relatively comparable, while serum values are significantly higher when measured with the nonspecific method.

SOURCES OF ERROR

1. Faulty timing or improper collection of the urine specimen is the most common source of error.
2. Vigorous exercise during the test may cause alteration in the clearance rate.
3. Proper hydration of the patient, insuring a urine flow of 2 ml/min or more, results

in a more precise measure of filtration rate and tends to eliminate retention of urine in the bladder as a source of negative error.

REFERENCES

Tobias, G. J., McLaughlin, R. F., and Hooper, J.: New England J. Med., *266*:317, 1962.
Edwards, K. D. G., and Whyte, H. M.: Aust. Ann. Med., *8*:218, 1959.

UREA CLEARANCE

CLINICAL SIGNIFICANCE

The urea clearance test has considerable historical significance, since it was the first of the clearance tests to be used widely. The test is a reasonably reliable measure of renal functional status if the rate of urinary excretion is 2 ml/min or more. It has been abandoned as a test for the measurement of glomerular function.

PRINCIPLE

The clearance of endogenous urea is most often measured. Urea clearances are affected by the rate of urine flow. When the excretion is 2 ml/min or more, the average normal clearance is 75 (64 to 99) ml/min and is known as the maximum clearance (C_m). When the urine flow is less than 2 ml/min, however, a larger portion of the urea in the glomerular filtrate passively diffuses into the tubules, thus lowering the clearance, which is then proportional to the square root of the excretory rate. The average normal is then 54 (41 to 68) ml/min and is termed the standard clearance (C_s). In an attempt to make the two clearances (C_m and C_s) comparable, urea clearance values are expressed as per cent of normal.

It is to be emphasized, however, that urea clearance values determined when urine flow is less than 2 ml/min are much less useful than those determined when the flow is 2 ml/min or more. Calculations on the basis of a flow between 1 and 2 ml/min have been included below for illustrative purposes but are not recommended.

PROCEDURE

Preparation of the patient

(to be carried out by the nursing staff of the hospital ward)

The collection of urine and blood samples is carried out in a way similar to that described under "creatinine clearance." In general, however, the urine collection period is only one or two hours.

Laboratory procedure

1. Measure and record the volume of the urine specimen.
2. Determine the urea nitrogen concentrations in the blood and in the urine specimen according to the procedure for urea nitrogen.

CALCULATIONS

1. Calculate the number of milliliters of urine excreted per minute by dividing the urine volume by the number of minutes over which the urine was collected.
2. Calculate urea clearance by substituting appropriate data in the following formula:

(a) When the urine volume is between 1.0 and 2.0 ml/min

$$\text{per cent } C_s = \frac{U\sqrt{V}}{B} \times 1.85 \times \frac{1.73}{A}$$

where C_s = standard clearance

1.85 = (100/54) converts the clearance (in ml/min) to per cent of normal clearance
54 = the average standard clearance in ml/min
B = blood concentration of urea in mg/100 ml

U, V, and the factor $1.73/A$ have been defined in the preceding general discussion.

(b) When the urine volume is 2.0 ml/min or more

$$\text{per cent } C_m = \frac{U}{B} \times V \times 1.33 \times \frac{1.73}{A}$$

where C_m = maximum clearance
1.33 = (100/75) converts the clearance (in ml/min) to per cent of normal
75 = the average maximum clearance in ml/min

NORMAL VALUES

	Mean	*Range*	*Range per cent of Normal*
Standard clearance	54 ml/min	41 to 68 ml/min	75 to 125
Maximum clearance	75 ml/min	64 to 99 ml/min	75 to 125

SOURCES OF ERRORS AND COMMENTS

1. The most common source of error in this test is the inaccurate timing of the urine collection periods. Also, incomplete emptying of the bladder will lead to falsely low results.

2. Values are most accurate, provided all other factors are equal, when the urine flow is 2.0 ml/min or more; accuracy tends to decrease with decreasing urine volumes. Some laboratories regard samples representing urine flows of less than 1 ml/min as unfit for analysis.

3. Historically, whole blood samples were used for the urea clearance test. In recent years, however, plasma (serum) has replaced blood as sample.

REFERENCES

Smith, H. W.: Kidney Structure and Function in Health and Disease. New York, Oxford University Press, 1951.
Austin, J. H., Stillman, E., and Van Slyke, D. D.: J. Biol. Chem., *46*:91, 1921.
Moller, E., McIntosh, J. F., and Van Slyke, D. D.: J. Clin. Invest., *6*:427, 1928.

NONPROTEIN NITROGEN IN SERUM (BLOOD)

The nonprotein nitrogen (NPN) fraction of serum (and blood) is composed of the nitrogen present in all nitrogenous compounds other than protein. Its major component is urea nitrogen, which constitutes approximately 45 per cent of the total. The other compounds included in the NPN fraction, listed in order of their quantitative distribution, are amino acids, uric acid, creatinine, creatine, and ammonia. Other nitrogenous compounds that are generally not identified are grouped together as "undetermined nitrogen."

The total NPN may be determined as a group, or the individual components may be measured in accordance with the procedures to be described.

THE DETERMINATION OF NONPROTEIN NITROGEN (NPN)

CLINICAL SIGNIFICANCE

Although the test for nonprotein nitrogen in whole blood and serum has for many years served as a test for kidney function, it has been replaced in recent years by the more useful and more convenient test for urea nitrogen. The routes of elimination of various nonprotein nitrogenous compounds of blood differ considerably (some are excreted by glomerular filtration only, some mainly by tubular excretion, and others are first excreted in the glomerular filtrate and are then partially absorbed by the tubules). Thus, the nonprotein nitrogen value is the result of many interacting factors. Increases in the NPN fraction are mainly a reflection of an increase in urea nitrogen, which normally makes up approximately 45 per cent of the total NPN. As the total NPN rises, the proportion of urea increases also.

The nonprotein nitrogen determination does not offer any information in addition to that provided by the urea nitrogen determination. An exception to this may be the simultaneous determination of NPN and urea nitrogen in patients with hepatic failure in the presence of renal disease. Under such conditions the ratio of nonprotein nitrogen to urea nitrogen may be substantially higher than that normally found. This change in the ratio is due to the decreased ability of the liver to synthesize urea and to deaminate amino acids.

Serum NPN levels above 35 mg/100 ml and blood levels above 50 mg/100 ml suggest renal insufficiency. Serum NPN levels may increase to more than 400 mg/100 ml, but these values are mostly seen in the terminal stages of renal failure.

SPECIMENS

Nonprotein nitrogen may be determined in whole blood, plasma, serum, or other biological fluids. Any anticoagulant that does not introduce nitrogen into the sample may be used. No preservative is required since the NPN is quite stable as long as gross bacterial contamination does not occur. Since a major portion of the NPN is distributed uniformly throughout the body water, hemolysis causes little difficulty. The nonprotein nitrogen compounds in filtrates of biological fluids vary, depending upon the protein precipitating agents used; however, the widely used tungstic acid protein-free filtrate employed in this procedure or trichloroacetic acid filtrates contain all NPN compounds present in the sample. Somogyi type filtrates should not be used, since some NPN compounds are adsorbed on the precipitate.

Since normal values differ depending upon the specimen (blood or serum), it is important to indicate in any report the type of specimen analyzed.

PRINCIPLE

The nitrogen, in whatever form, in a protein-free filtrate of the specimen is converted to the ammonium ion (NH_4^+) by Kjeldahl digestion with hot concentrated sulfuric acid in the presence of a catalyst such as copper sulfate. Catalysts such as mercury and selenium are used in other modifications. Also, a few procedures dispense with catalysts and introduce hydrogen peroxide to complete the digestion. The NH_4^+ formed is reacted with Nessler's reagent, either directly or after isolation by distillation or aeration (see p. 301). The color formed appears yellow when nitrogen is present in low to medium concentrations and orange-brown in high concentration. For most procedures Beer's law is valid up to concentrations of about 75 mg/100 ml. At very high levels (100 to 150 mg/100 ml) the colloidal material forms large aggregates and precipitates.

The following chemical reactions are involved:

1. $\text{N-compounds} + H_2SO_4 \xrightarrow[\text{heat}]{CuSO_4} NH_4^+ + HSO_4^-$

2. $(NH_4)HSO_4 + 2\,NaOH \longrightarrow Na_2SO_4 + NH_3 + 2\,H_2O$

The final reaction of ammonia with the double iodide is not definitely known, but it has been postulated to be:

$$2\,HgI_2 \cdot 2\,KI + 2\,NH_3 \rightarrow NH_2Hg_2I_3 + 4\,KI + NH_4I$$

Other formulas that have been proposed for the Nessler complex, based on x-ray diffraction studies, are:

$$Hg_2NI \cdot H_2O \quad \text{and} \quad HgOHNHHgI$$

UREA NITROGEN

Urea is synthesized in the liver from ammonia produced as a result of deamination of amino acids. This biosynthetic pathway is the chief means of excretion of surplus nitrogen by the body.

It is customary in most laboratories in the United States to express urea as urea nitrogen. This came about through the desire to compare the quantity of nitrogen in urea with that of other components included in the nonprotein (NPN) category. Since the NPN is seldom measured today, the expression of urea as urea nitrogen has little practical value or logic but continues to be used because of tradition.

The structure of urea is $NH_2\text{—}\overset{\overset{\displaystyle O}{\|}}{C}\text{—}NH_2$. Since its molecular mass is 60 daltons and it contains 2 nitrogen atoms, with a combined weight of 28, a urea nitrogen value can be converted to urea by multiplying by 60/28 or 2.14.

DETERMINATION OF UREA NITROGEN

CLINICAL SIGNIFICANCE

The determination of serum urea nitrogen is presently the most widely used screening test for the evaluation of kidney function. The test is frequently requested along with the serum creatinine test since simultaneous determination of these two compounds appears to aid in the differential diagnosis of prerenal, renal, and postrenal hyperuremia (see the discussion on creatinine). Serum urea determinations are considerably less sensitive than urea clearance (and creatinine clearance) tests, and levels may not be abnormal until the urea clearance has diminished to less than 50 per cent. As outlined in the preceding general discussion, increases in serum urea nitrogen may be due to prerenal causes (cardiac decompensation, water depletion due to decreased intake or excessive loss, increased protein catabolism). Among the renal causes are acute glomerulonephritis, in which only moderate increases are observed, chronic nephritis, polycystic kidney, nephrosclerosis, and tubular necrosis. Postrenal causes include all types of obstruction of the urinary tract (stone, enlarged prostate gland, tumors).

Urine urea nitrogen determinations are rarely done except as part of the urea clearance test.

SPECIMEN

Urea nitrogen may be determined directly in plasma, serum, urine, and most other biological fluids with the method presented here. Whole blood, however, must be deproteinized to eliminate the colorimetric interference of hemoglobin. Sodium fluoride in high concentrations inhibits the action of urease employed in the assay.

Since urea may be lost through bacterial action, the specimen should be analyzed within several hours of collection or should be preserved by refrigeration. Urine is particularly susceptible to loss as a result of bacterial decomposition of urea; therefore, in addition to refrigeration, crystals of thymol may be added to help to reduce the loss of urea.

PRINCIPLE

Urea is hydrolyzed to ammonium carbonate by urease, and the ammonia that is released from the carbonate by alkali reacts with phenol and sodium hypochlorite in an alkaline medium to form a blue indophenol. Sodium nitroprusside serves as a catalyst. The intensity of the blue color is proportional to the quantity of urea in the specimen. Ammonia already present in urine specimens is removed by adsorption on Lloyd's reagent (sodium aluminum silicate) or on Permutit.

The three reactions can be represented as follows:

1. $H_2N\text{—}\overset{\overset{\displaystyle O}{\|}}{C}\text{—}NH_2 + 2H_2O + H^+ \xrightarrow{\text{urease}} (NH_4)_2CO_3 + H^+ \longrightarrow 2NH_4^+ + HCO_3^-$

2. $NH_4^+ + OH^- \longrightarrow NH_3 + H_2O$

3. $NH_3 + NaOCl + 2\ C_6H_5OH \xrightarrow[Na_2Fe(CN)_5NO]{NaOH}$ $O{=}C_6H_4{=}N\text{—}C_6H_4\text{—}O^-$

Indophenol
(blue in dissociated form)

REAGENTS

1. Ammonia-free water. Allow distilled water to pass through a mixed cation-anion exchange resin bed and collect it in a glass-stoppered bottle.

2. Phenol-nitroprusside solution. Place 10 g of "pink-white" phenol and 0.050 g of sodium nitroprusside, $Na_2Fe(CN)_5NO \cdot 2\ H_2O$, AR, in a 1 liter volumetric flask containing about 500 ml of ammonia-free water. Dissolve the reagents, dilute to the mark, and mix thoroughly. Store this solution in a refrigerator at 5°C and discard after 2 months.

3. Alkaline-hypochlorite solution. Place 5 g of sodium hydroxide in about 500 ml of ammonia-free water in a 1 liter volumetric flask. Cool, add 0.42 g of sodium hypochlorite (commercial bleaches such as Clorox are satisfactory), dilute to the mark, and mix thoroughly. Store in an amber bottle in a refrigerator. Discard after 2 months.

4. Sodium ethylenediaminetetraacetate (EDTA), 1.0 g/100 ml, pH 6.5. Dissolve 10 g of the disodium salt of EDTA in about 800 ml of ammonia-free water. Adjust the pH to 6.5 with 1 molar sodium hydroxide and dilute to 1 liter. EDTA binds cations that might interfere with urease activity.

5. Urease stock solution (approximately 40 modified Sumner units/ml). Suspend 0.2 g of urease in 10 ml of water and add 10 ml of glycerol. Type V urease containing 3500 to 4100 units/g (obtainable from the Sigma Chemical Co., St. Louis, MO.) is suitable. Store in a refrigerator and discard after 4 months.

6. Urease working solution (0.4 units/ml). Dilute 1 ml of the urease stock to 100 ml with the EDTA solution. Store in a refrigerator and discard after 3 weeks.

7. Urea nitrogen stock standard (500 mg urea nitrogen/100 ml). Dissolve 1.0717 g of dry urea, AR, in about 50 ml of ammonia-free water in a 100 ml volumetric flask. Add 0.1 g of sodium azide, dilute to the mark, and mix. Sodium azide serves as a preservative that does not inhibit urease action. Store this standard in a refrigerator and discard after 6 months. Ammonium sulfate, $(NH_4)_2SO_4$, is sometimes used as a standard; it has the advantage of being stable, but the disadvantage of not controlling the enzymatic steps in the procedure.

8. Urea nitrogen working standard (50 mg/100 ml). Dilute 10 ml of the stock standard to about 90 ml with ammonia-free water in a 100 ml volumetric flask. Add 0.1 g of sodium azide, dilute to mark, and mix thoroughly. Store in a refrigerator and discard after 6 months.

9. Permutit, according to Folin, 40 to 60 mesh.

PROCEDURE

For plasma or serum

1. Label three tubes, blank, standard, and unknown, respectively.
2. Pipet 1.0 ml of urease working solution into each tube.

3. With a Kirk transfer or other micropipet add 10 μl of unknown serum and working standard to the appropriate tubes. Mix and incubate all tubes for 15 min at 37°C.

4. Add rapidly and successively, mixing after each addition, 5.0 ml of the phenol-nitroprusside solution and 5 ml of the alkaline hypochlorite.

5. Place the tubes in a water bath at 37°C for 20 minutes.

6. Measure the absorbance at 560 nm, using the blank as a reference.

For urine

1. Place about 0.5 g of Permutit in a 25 ml volumetric cylinder and wash twice with water. Drain completely.

2. Add 1.0 ml of urine and about 5 ml of water. Mix by swirling for 5 min. Add water to the mark, mix, and allow the Permutit to settle.

3. Label 15 × 120 mm tubes. Make the following additions and mix.

	Blank	*Standard*	*Unknown*
Urease working solution	1.0 ml	1.0 ml	1.0 ml
Diluted urine from step 2	0	0	10 μl
Urea working standard	0	10 μl	0

4. Incubate all tubes at 37°C for 15 min.

5. Add quickly, one after another, with mixing, 5 ml of the phenol-nitroprusside reagent and 5 ml of the alkaline hypochlorite reagent.

6. Incubate all tubes at 37°C for 20 min.

7. Measure the absorbances at 560 nm, using the blank as a reference.

CALCULATIONS

For plasma or serum

$$\frac{A \text{ unknown}}{A \text{ standard}} \times 50 = \text{mg urea nitrogen/100 ml}$$

For urine

1. $$\frac{A \text{ unknown}}{A \text{ standard}} \times 25 \times 50 = \text{mg urea nitrogen/100 ml}$$

2. $$\frac{\text{Urea nitrogen/100 ml}}{100} \times \frac{\text{24 h excretion (in ml)}}{1000} = \text{g urea nitrogen/d}$$

PROCEDURAL NOTES

The color produced is of such intensity that it is not practical to use the optimum wavelength of 628 nm. A wavelength of 560 nm allows a much wider range of concentrations to be measured.

NORMAL VALUES

Plasma or serum	7 to 18 mg urea nitrogen/100 ml (15 to 38 mg urea/100 ml or 2.5 to 6.3 mmol urea/l) Values may be higher in individuals on a high protein diet.
Urine	12 to 20 g urea nitrogen/d (25 to 43 g urea or 0.42 to 0.73 mol urea/d)

SOURCES OF ERROR

1. Ammonia in any of the reagents or in the atmosphere of the room in which the procedure is carried out will result in falsely high values.

2. Lipemic sera cause turbidity in the final colored solution. This may be corrected by extracting the final solution with several milliliters of ether.

DISCUSSION

Methods based on principles other than those just described are also widely used. Two of the most prevalent are nesslerization after urease hydrolysis and the direct reaction of urea with diacetyl.

In the former method, a urease suspension is added to the specimen (blood, plasma, serum). After enzymatic action is complete, the specimen is deproteinized with tungstic acid. The protein-free filtrate is treated with Nessler's reagent as described for nonprotein nitrogen.[3]

The diacetyl method is based on the reaction of urea with diacetyl to form a yellow compound.* Because diacetyl is unstable, it is replaced in most methods by the more stable diacetyl monoxime. The color is intensified by pentavalent arsenic, other polyvalent ions,[2] ferric ion, or Na thiocarbazide.

$$CH_3{-}\overset{\overset{O}{\|}}{C}{-}\overset{\overset{NOH}{\|}}{C}{-}CH_3 + H_2O \xrightarrow{H^+} CH_3{-}\overset{\overset{O}{\|}}{C}{-}\overset{\overset{O}{\|}}{C}{-}CH_3 + HONH_2$$

Diacetyl monoxime **Diacetyl** **Hydroxylamine**

$$\begin{array}{c} CH_3 \\ | \\ C{=}O \\ | \\ C{=}O \\ | \\ CH_3 \end{array} + \begin{array}{c} NH_2 \\ \quad\searrow \\ \quad C{=}O \\ \quad\nearrow \\ NH_2 \end{array} \xrightarrow{H^+} \begin{array}{c} CH_3 \\ | \\ C{=}N \\ | \\ C{=}N \\ | \\ CH_3 \end{array}\!\!\!\!\!\!\!\!\!\! >C{=}O + 2H_2O$$

Diacetyl **Urea** **Diazine derivative (yellow)**

Since diacetyl reacts directly with urea and not with ammonia, the latter compound does not need to be removed from urine specimens.

In the Technicon AutoAnalyzer procedure, diacetyl reacts with urea in the presence of thiosemicarbazide, which intensifies the color of the reaction product.[6]

REFERENCES

Kaplan, A.: Urea nitrogen and urinary ammonia. *In* Standard Methods of Clinical Chemistry. S. Meites, Ed. New York, Academic Press, Inc. 1965, vol. 5, pp. 245–256.

Chaney, A. L., and Marbach, E. P.: Clin. Chem., *8*:130, 1962.

CREATINE AND CREATININE

Creatine phosphate acts as a reservoir of high energy, readily convertible to ATP in the muscles and other tissues. Creatine is itself synthesized in the liver and pancreas from three amino acids, arginine, glycine, and methionine. After synthesis, creatine diffuses into the vascular system and is thus supplied to many kinds of cells, particularly those of muscle, where it becomes phosphorylated. Creatine and creatine phosphate total about 400 mg/100 g of fresh muscle. Both compounds are spontaneously converted into creatinine at the rate of about 2 per cent per day. Creatinine is a waste product derived from creatine and is excreted by the kidneys.

DETERMINATION OF CREATINE AND CREATININE

CLINICAL SIGNIFICANCE

Creatinine is removed from plasma by glomerular filtration and is then excreted in the urine without being reabsorbed by the tubules to any significant extent. This results

* Refer to Veniamin (Clin. Chem., *16*:3, 1970) for a discussion of this complex reaction.

in a relatively high clearance rate for creatinine, as compared, e.g., with urea (125 versus 70 ml/min). In addition, when plasma levels increase above the normal, the kidney can also excrete creatinine through the tubules. Consequently, serum or blood creatinine levels in renal disease generally do not increase until renal function is substantially impaired. In the presence of normal renal blood flow, any increase in creatinine values above 2 to 4 mg/100 ml is suggestive of moderate to severe kidney damage. This lack of sensitivity is in contrast to the creatinine clearance test, which is one of the most sensitive tests for measuring the glomerular filtration rate.

Simultaneous urea nitrogen and creatinine determinations (normal ratio between 15/1 and 24/1) appear to have some clinical significance. Elevations of serum urea nitrogen levels in renal disease (see the discussion on urea) are somewhat more pronounced than those of creatinine. In cases of retention of urea nitrogen due to prerenal causes (especially severe intestinal bleeding), the ratio between urea nitrogen and creatinine levels will be even higher, up to 40/1. Urea nitrogen levels of 35 or even 40 mg/100 ml in the presence of normal creatinine levels are not uncommon in these conditions. On the other hand, retention of nonprotein nitrogenous compounds due to obstruction of the urinary tract will cause almost simultaneous and proportional increases in both urea nitrogen and creatinine levels. These conditions will mechanically suppress the glomerular filtration rate and thus cause an increase in all compounds that are normally found in the glomerular filtrate. In severe tubular damage the ratio may be as low as 10/1.

Creatinine determinations have one advantage over urea determinations; they are not affected by a high protein diet as is the case for urea levels. Determination of *urine creatinine* levels is of little or no help in evaluating renal function unless it is done as a part of a creatinine clearance test. Since the excretion of creatinine in one given person is relatively constant, 24 h urine creatinine levels are used as a check on the completeness of a urine collection. Determinations of the excretion ratio of another compound under investigation to that of creatinine is considered advantageous in some cases. Such an example is the excretion of vanillylmandelic acid (3-methoxy-4-hydroxymandelic acid), which can be reported either in terms of the complete 24 h excretion or in terms of the amount excreted per milligram of creatinine.

Creatine in serum represents a small part of the nonprotein nitrogen fraction. No significant variations of this compound have as yet been associated with kidney diseases so that its determination has little clinical value in these conditions. Diseases associated with extensive muscle destruction may result in elevated levels of serum creatine as well as creatinuria. The test, in the diagnosis of muscle disease, has largely been replaced by creatine kinase measurements in serum.

SPECIMEN

Creatine and creatinine may be determined in any biological fluid, but plasma, serum, amniotic fluid, and urine are the specimens most commonly employed. Plasma and serum are preferred to whole blood since considerable amounts of noncreatinine chromogens are present in red cells. If kept for a few days, specimens for creatine and creatinine are best stored at refrigerator temperatures; if kept for longer periods, they should be frozen.

Aqueous solutions of creatine and creatinine very slowly approach a state of equilibrium with respect to each other, as indicated in the following:

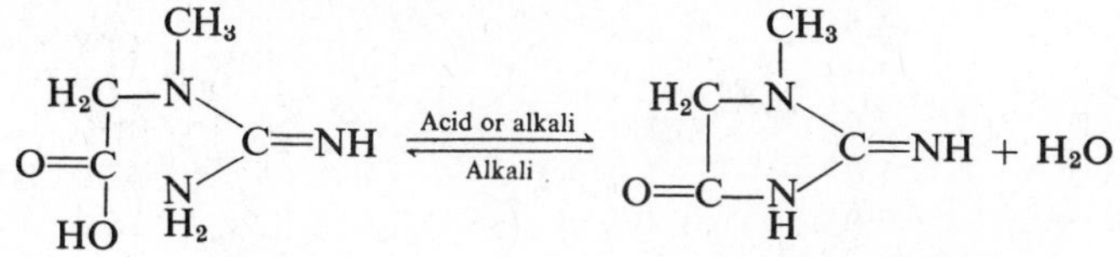

$$\underset{\textbf{Creatine}}{\left[\begin{array}{l} CH_3 \\ H_2C{-}N{-}C{=}NH \\ O{=}C{-}OH \quad NH_2 \end{array}\right]} \underset{\text{Alkali}}{\overset{\text{Acid or alkali}}{\rightleftharpoons}} \underset{\textbf{Creatinine (an anhydride of creatine)}}{\left[\begin{array}{l} CH_3 \\ H_2C{-}N{-}C{=}NH \\ O{=}C{-}NH \end{array}\right]} + H_2O$$

Although there are conflicting published reports about the speed with which this equilibrium is reached, it probably requires days or weeks at neutral pH. Creatinine, however, is formed rather quickly from creatine in either alkaline or acid solutions. Although this reversible reaction is catalyzed in both directions by hydroxyl ions, hydrogen ions promote the reaction only toward the right. Because of the lability of creatine and creatinine, it is advisable to carry out analysis for these two substances on fresh specimens. When this is not possible, adjustment of the pH to 7.0 or freezing, or both, may delay the change for indefinite periods.

Urine contains only small amounts of noncreatinine chromogens.

PRINCIPLE

Creatinine. This substance is determined in diluted urine or in a protein-free filtrate of plasma or serum, employing the Jaffe reaction which results in the production of a red tautomer of creatinine picrate after addition of an alkaline picrate solution.

Creatine. It is determined as the difference between the preformed creatinine and the total that results after the creatine present has been converted to creatinine by heating at an acid pH.

REAGENTS

1. Picric acid, 0.04 mol/liter. Dissolve about 9.3 g of picric acid, AR, in about 500 ml of water at 80°C. Cool to room temperature, dilute to 1 liter with water, and titrate with 0.1 molar NaOH, using phenolphthalein as the indicator. Dilute as necessary to make 0.04 molar.
2. Sodium hydroxide, 0.75 mol/liter. Dissolve 30 g of sodium hydroxide, AR, in water and when cool, dilute to 1 liter.
3. Creatinine stock standard, 1 mg/ml. Dissolve 0.100 g of creatinine, AR, in 100 ml of 0.1 molar HCl. Store in a refrigerator.
4. Creatinine working standard, 20 μg/ml. Dilute 2 ml of the stock solution to 100 ml with water in a volumetric flask. Add a few drops of chloroform as a preservative.
5. Sulfuric acid, 0.33 mol/liter. Add 18.8 ml of concentrated H_2SO_4, AR, to about 500 ml of water. When cool, dilute to 1 liter.
6. Sodium tungstate, 5.0 g/100 ml. Dissolve 50 g of $Na_2WO_4 \cdot 2\,H_2O$, AR, in water and dilute to 1 liter.

STANDARDIZATION

1. Place the following in cuvets, mixing after each addition:

	Creatinine working standard (20 μg/ml) (ml)	*Water* (ml)	*Picric acid* 0.04 mol/l (ml)	*Sodium hydroxide* (ml)	*Equivalent to: for plasma or serum* (mg/100 ml)	*Equivalent to: for urine* (mg/100 ml)
	0.25	3.75	1.0	1.0	1.0	100
	0.5	3.5	1.0	1.0	2.0	200
	1.0	3.0	1.0	1.0	4.0	400
	2.0	2.0	1.0	1.0	8.0	800
	3.0	1.0	1.0	1.0	12.0	1200
	4.0	0.0	1.0	1.0	16.0	1600
Blank	0	4.0	1.0	1.0	0	0

2. Allow tubes to stand for 15 min; then measure the absorbance of each at 500 nm, using the blank as a reference.

3. Construct an absorbance-concentration curve on rectangular coordinate paper, plotting absorbances as the ordinate.

PROCEDURE

Creatinine in plasma, serum, or urine

1. Add 1.0 ml of sodium tungstate (5.0 g/100 ml), 1.0 ml of 0.33 molar sulfuric acid and 1.0 ml of water to 1.0 ml of plasma or serum. Mix thoroughly and filter. If urine is analyzed, make a 1/400 dilution of urine with water.

2. Make the additions to labeled tubes as indicated below:

	Blank (ml)	*Standard* (ml)	*Unknown* (ml)
Water	4.0	3.5	2.0
Creatinine working standard	0	0.5	0
Diluted urine (1:400) or plasma (serum) filtrate	0	0	2.0
Picric acid, 0.04 mol/l	1.0	1.0	1.0
Sodium hydroxide, 0.75 mol/l	1.0	1.0	1.0

3. Allow to stand for 15 min and measure the absorbances at 500 nm.

4. Read the creatinine content from the preceding standard curve.

Note. The standard prepared in this run is equivalent to 2.0 mg creatinine/100 ml plasma (serum) or 200 mg of creatinine/100 ml of urine.

Creatine in plasma, serum or urine

1. Determine the preformed creatinine according to the procedure just described.

2. Measure the following into three 12 ml graduated centrifuge tubes, mixing after each addition:

	Blank (ml)	*Standard* (ml)	*Unknown* (ml)
Water	4.0	3.5	2.0
Creatinine standard	0	0.5	0
Protein-free filtrate of plasma or serum or diluted urine (1:400)	0	0	2.0
Picric acid, 0.04 mol/l	1.0	1.0	1.0

3. Heat for 1h in a 100°C bath.

4. Cool and make up the volume to 5.0 ml.

5. Add 1 ml of 0.75 molar sodium hydroxide, mix, and allow to stand for 15 min.

6. Measure the absorbances at 500 nm and read the values from the standard curve to obtain the total creatinine.

7. Subtract the preformed creatinine from the total and multiply the difference by 1.16 to obtain the concentration of creatine in mg/100 ml of plasma, serum or urine. The factor 1.16 is the ratio of the molecular mass of creatine to that of creatinine. Some laboratories use a factor such as 1.25 to allow for incomplete conversion of creatine to creatinine.

NORMAL VALUES

Plasma or serum

	Men	*Women*
Creatinine*	0.9–1.5 mg/100 ml	0.8–1.2 mg/100 ml
	0.08–0.13 mmol/l	0.07–0.11 mmol/l
Creatinine	0.6–1.2 mg/100 ml	0.5–1.0 mg/100 ml
Creatine	0.17–0.50 mg/100 ml	0.35–0.93 mg/100 ml

Urine

Creatinine	1.0–2.0 g/d	0.8–1.8 g/d
Creatine	0–40 mg/d	0–80 mg/d

DISCUSSION

The methods just presented, although not completely specific, are adequate for clinical purposes. A procedure for the isolation of creatinine from interfering substances has been presented by Owens and his associates.[7] After deproteinization of the specimen, creatinine is adsorbed from an acid medium onto Lloyd's reagent, an aluminum silicate, and subsequently desorbed in an alkaline solution. This type of method is highly specific but more time-consuming than that given here.

The color developed in the Jaffe reaction is relatively stable; but since the color caused by non-creatinine chromogens increases with time, spectrophotometric readings should be taken within 20 minutes. Although temperature and pH affect the color development, these effects are compensated by the inclusion of standards with each set of tests.

Slight hemolysis, although it does not affect the values for creatinine, can cause appreciable positive errors for creatine.

It has been proposed that creatinine be quantitated by measuring the rate of color development of the Jaffe reaction.[10] Although the mechanism of this reaction is not fully understood, it has been shown that color formation proceeds at a measurable rate which allows measurement of creatinine by a continuous monitoring technique. Noncreatinine substances in serum also react to produce color, but at a slower rate than does creatinine, thus allowing a possible way to avoid such interference. Because this method can be applied to a micro sample (100 μl) and can be completed quickly (1 minute), such methods are presently under investigation, especially in regard to their use in centrifugal analyzers.

It has also been proposed to measure creatinine by a coupled enzyme system. In the first approach, by F. Lim, creatinine is reacted with creatinase to form ammonia and *N*-methylhydantoin. The ammonia produced is then reacted with α-oxoglutarate and NADPH in the presence of glutamate dehydrogenase. The decrease in NADPH is followed fluorometrically.

In the second approach, G. A. Moss *et al.* use creatinine amidohydrolase to convert creatinine to creatine. The creatine formed is measured by an enzyme system involving CK, pyruvate kinase, and LDH at 340 nm (Clin. Chem., *20*:871, 1974, Abstracts 88 and 89).

The optimum pH conditions for the hydrolytic conversion of creatinine to creatine are carefully controlled in this procedure by the amount and concentration of picric acid used.

REFERENCES

Bonsnes, R. W., and Taussky, H. H.: J. Biol. Chem., *158*:581, 1945.
Brod, J., and Sirota, J. M.: J. Clin. Invest., *27*:645, 1948.

* Includes nonspecific chromogens.

Varley, H.: Practical Clinical Chemistry. New York, Interscience Publishers, Inc., 1967, pp. 197–200.
Henry, R. J.: Clinical Chemistry: *Principles and Technics*. New York, Hoeber Medical Div., Harper & Row, Publishers, 1964, p. 300.

URIC ACID

Uric acid is a waste product, derived from purines of the diet and those synthesized in the body. It has been shown that the healthy adult human body contains about 1.1 g of uric acid and that about one sixth of this is present in the blood, the remainder being in other tissues. Normally, about one half of the total uric acid is eliminated and replaced each day, partly by way of urinary excretion and partly through destruction in the intestinal tract by microorganisms. Uric acid is one of the components of the NPN fraction of plasma, which was discussed previously.

DETERMINATION OF URIC ACID

CLINICAL SIGNIFICANCE

Plasma uric acid is filtered by the glomeruli and is subsequently reabsorbed to about 90 per cent by the tubules. It is the end product of purine metabolism in man, in the anthropoid ape, and in the Dalmatian dog. Other mammals are able to metabolize the uric acid molecule to the more soluble end product, allantoin. Uric acid concentrations in serum are greatly affected by extrarenal as well as renal factors.

Determination of serum uric acid levels are most helpful in the diagnosis of gout, in which serum levels are frequently between 6.5 and 10 mg/100 ml. Occasional normal blood (serum) levels are found in this disease, but it is believed that repeated determinations will reveal hyperuricemia at some point of the disease. Serum uric acid levels are also increased whenever there is increased metabolism of nucleoproteins, such as in leukemia and polycythemia or after the intake of food rich in nucleoproteins, for example, liver, kidney, or sweetbread. Increased serum uric acid levels are also a constant finding in familial idiopathic hyperuricemia, of which there seem to be at least two types. In one type there is an overproduction of uric acid in the presence of normal excretion, and in the other there is a decreased rate of excretion in the presence of normal uric acid production.

Uric acid levels are elevated in conditions associated with decreased renal function. In severe renal impairment values up to 35 mg/100 ml have been observed, depending on the method employed. Although any decrease in renal function is generally accompanied by increases in serum uric acid levels, this test is rarely used in this connection because of the great effect of extrarenal factors on serum levels.

Urine uric acid levels are generally a reflection of the endogenous nucleic acid breakdown and of the amount of dietary purines (see normal values). Unless hyperuricemia is due to decreased excretion of uric acid, it is generally accompanied by increased levels of uric acid in urine.

SPECIMEN

Uric acid is stable in serum for several days at room temperature and for longer periods if refrigerated, but since it is susceptible to destruction by bacterial action, the addition of thymol may increase the stability. Any of the common anticoagulants, except potassium oxalate, can be used. Potassium phosphotungstates, which are insoluble, are formed when potassium salts are introduced into the analytical system, resulting in turbidity.

Uric acid has been reported to be stable in urine for several days at room temperature. If urine specimens are refrigerated, urates may precipitate, but may be brought into solution by adjustment of the pH to 7.5 to 8.0 and by warming the specimen to approximately 50°C.

PRINCIPLE

Uric acid is oxidized to allantoin and carbon dioxide by a phosphotungstic acid reagent in alkaline solution. Phosphotungstic acid is reduced in this reaction to tungsten blue, which is measured at 710 nm.

$$\text{Uric acid} + H_3PW_{12}O_{40} + O_2 \longrightarrow \text{Allantoin} + CO_2 + \text{tungsten blue}$$

Uric acid; Phosphotungstic acid; Allantoin

REAGENTS

1. Sodium tungstate, 10 g/100 ml. Dissolve 100 g of $Na_2WO_4{\cdot}2\ H_2O$, AR, in water and dilute to 1000 ml.
2. Sulfuric acid, 0.33 mol/l. Slowly add 18.8 ml of concentrated sulfuric acid, AR, to about 500 ml of water and dilute to 1000 ml.
3. Phosphotungstic acid reagent. Dissolve 40 g of sodium tungstate, AR, in 300 ml of water. Add 32 ml of 82 per cent orthophosphoric acid and several glass beads. Reflux gently for 2 h. Cool to room temperature and dilute to 1 liter. Mix thoroughly. Dissolve 32 g of $Li_2SO_4{\cdot}H_2O$, AR, in the reagent and mix thoroughly. Store in a refrigerator.
4. Sodium carbonate, 14 g/100 ml. Dissolve 70 g of anhydrous Na_2CO_3, AR, in water and dilute to 500 ml. Store in a polyethylene bottle.
5. Uric acid stock standard, 1 mg/ml. Measure 100 mg of uric acid, AR, and 60 mg of Li_2CO_3, AR, into a 100 ml volumetric flask. Add about 50 ml of water and warm to about 60°C to aid in the solution of the reagents. Cool to room temperature, dilute to the mark, and mix thoroughly. This reagent may be stable for several months in a refrigerator.
6. Uric acid working standards. Dilute 0.5, 1.0, and 1.5 ml of the 1 mg/ml stock standard to 100 ml with water. (Equivalent to 5, 10, and 15 mg of uric acid/100 ml plasma and 50, 100, and 150 mg of uric acid/100 ml urine.)

PROCEDURE

1. For plasma or serum, prepare a tungstic acid protein-free filtrate by mixing 1.0 ml of sample with 8.0 ml of water, 0.5 ml of 0.33 molar H_2SO_4 and 0.5 ml of sodium tungstate (10 g/100 ml). Mix and filter.
2. Measure the following into tubes in the order shown. Mix after each addition.

	Blank (ml)	*Standard* (ml)	*Unknown* (ml)
Protein-free filtrate (1:10) or urine (diluted 1:100 with water)	0	0	3.0
Water	3.0	0	0
Uric acid working standard	0	3.0	0
Sodium carbonate, 14 g/100 ml	1.0	1.0	1.0
Phosphotungstic acid	1.0	1.0	1.0

3. Allow tubes to stand 15 min.
4. Measure the absorbances at 690 to 710 nm, using the blank as a reference.

CALCULATIONS

For plasma or serum

$$\frac{A \text{ unknown}}{A \text{ standard}} \times 0.01^* \times 3 \times \frac{10}{3} \times \frac{100}{1}$$

or:

$$\frac{A \text{ unknown}}{A \text{ standard}} \times 10 = \text{mg uric acid/100 ml plasma or serum}$$

For urine

$$\frac{A \text{ unknown}}{A \text{ standard}} \times 0.01^* \times 3 \times \frac{100}{3} \times \frac{100}{1}$$

or:

$$\frac{A \text{ unknown}}{A \text{ standard}} \times 100^* = \text{mg uric acid/100 ml urine}$$

NORMAL VALUES

Plasma or serum

men:	3.5 to 7.2 mg/100 ml (0.21 to 0.43 mmol/l)
women:	2.6 to 6.0 mg/100 ml (0.16 to 0.36 mmol/l)

Urine

average diet:	250 to 750 mg/d
low purine diet:	up to 450 mg/d
high purine diet:	up to 1 g/d

SOURCES OF ERROR

This method has been reported to follow Beer's law up to an absorbance of 0.8 in a spectrophotometer; however, it is advisable to prepare several standards to check any deviation at higher concentrations.

DISCUSSION

Uric acid can be determined on the basis of another principle, that of differential spectrophotometry, in which uric acid is destroyed by the action of uricase. The decrease in absorbance after incubation with uricase is measured in the ultraviolet region (290 to 293 nm) and is proportional to the uric acid initially present. This method has great specificity because uricase acts only on uric acid. It is more time-consuming, however, and requires the use of an instrument capable of reading in the ultraviolet region; therefore, it is seldom used routinely.

$$\text{uric acid} \xrightarrow{\text{uricase}} \text{allantoin} + CO_2$$

Two automated adaptations of the manual uric acid procedures have been described, both using the AutoAnalyzer. One of these utilizes the Caraway carbonate-phosphotungstic acid method; the other is based on spectrophotometric measurements before and after destruction of uric acid in the sample by the action of uricase. These determinations can be made using the carbonate-phosphotungstic acid reaction (at 660 nm), or utilizing the uricase methods measuring $\Delta A_{290-293\text{ nm}}$.

In the Archibald modification[1] of the Kern and Stransky method, the specificity of the method is enhanced by pretreatment of the serum with sodium hydroxide, which causes an oxidative destruction of ascorbic acid and sulfhydryl compounds that otherwise would

* This factor applies if the 0.01 mg/ml standard is used.

lead to falsely high values. (In the Caraway method, sodium carbonate serves this purpose.) Phosphotungstic acid serves both as a protein precipitating agent and as a color agent. A glycerine-silicate reagent increases the sensitivity, and also provides alkalinity for the reduction of phosphotungstic acid to tungsten blue. Sodium polyanetholsulfonate (Liquoid, Hoffmann-La Roche Inc., Nutley, N.J.) is added to prevent turbidity.

REFERENCES

Henry, R. J., Sobel, C., and Kim, J.: Am. J. Clin. Path., *28*:152, 1957.
Henry, R. J.: Clinical Chemistry: Principles and Technics. New York, Hoeber Medical Div., Harper & Row, Publishers, 1964, pp. 278–280.
Caraway, W. T.: Uric acid. *In* Standard Methods of Clinical Chemistry. D. Seligson, Ed. New York, Academic Press, Inc., 1965, vol. 4, pp. 239–247.
Brown, R. A., and Freier, E. F.: Clin. Biochem., *1*:154, 1967.

AMMONIA IN BLOOD, SERUM, AND URINE

CLINICAL SIGNIFICANCE

Although ammonia is a part of the nonprotein nitrogen fraction of plasma and serum, it is present in both fluids only in small amounts and its determination is of very little or no value in the study of renal disease. Relatively significant increases in blood and serum ammonia are observed in impending and existing hepatic coma, which is discussed in Chapter 19, Liver Function.

The determination of ammonia in urine is used as an index of the ability of the kidney to produce ammonia. This process is discussed in more detail in Chapter 16. The determination of ammonia in urine has been replaced by more modern and accurate tests for kidney function.

AMINO ACIDS

CLINICAL SIGNIFICANCE

Although amino acids are a part of the nonprotein nitrogen fraction of blood and serum, their quantitative determination finds clinical application only in some selected congenital renal disorders. Amino acids in plasma are filtered by the glomerulus and appear in the glomerular filtrate in the same proportion as they do in plasma. A portion of the amino acids is subsequently reabsorbed by the proximal convoluted tubules. In some congenital disorders there is a defect in the reabsorption of amino acids, resulting in aminoaciduria. An example of such a condition is cystinuria, in which there is a failure to reabsorb dibasic amino acids (cystine, lysine, arginine, and ornithine). In Fanconi's syndrome there is a failure to reabsorb a wide variety of amino acids.

The reader is referred to Chapter 19, Liver Function, and especially to Chapter 7, Proteins and Amino Acids, in which more detailed information regarding the clinical significance of amino acid determinations can be found. Chapter 7 also contains detailed methodologies.

Tests Measuring Tubular Function

The renal tubules are engaged in a wide variety of activities and, consequently, there are several groups of tubular function tests that have been employed in the clinical laboratory.

The first group of these tests measures the excretory ability of the tubules. For this purpose substances are injected into the blood, which are cleared either exclusively or predominantly by the tubules. Such a compound is sodium *p*-aminohippurate (see *p*-Aminohippurate Clearance Test). The clearance test employing this substance is not widely used, however, because of technical difficulties in performance. The phenolsulfonphthalein test (to be described) is the test most widely used to evaluate the excretory capacity of the kidney. (The PSP test dose of 6 mg given in usual clinical tests does not, however, saturate the tubular capacity.) Like other renal function tests, both procedures are affected by the renal blood flow.

The second group of tubular function tests is concerned with the concentrating ability of the tubules. In cases of tubular damage, this is generally the first function to be decreased. Commonly used tests in this group are the measurement of the specific gravity, the measurement of the osmolality, and the Fishberg concentration tests (all to be described).

Other tubular function tests include the determination of ammonia in urine, which is a measure of the ability of the tubules to produce ammonia in states of acidosis. The test for "tubular reabsorption of phosphorus" is, strictly speaking, also a kidney function test; however, since this function of the tubule is greatly influenced by the parathyroid hormone, the test is mainly used to evaluate the status of the parathyroid gland.

p-AMINOHIPPURATE CLEARANCE TEST

Certain substances that are foreign to the human body, in addition to being filtered through the glomerulus, are also removed by the kidney via the tubules. Within certain limits of plasma concentration, up to 90 per cent of some of these substances (e.g., *p*-aminohippurate and Diodrast) are removed from the plasma in a single passage through the kidney. Since renal function is dependent on the renal blood flow, a clearance value of *p*-aminohippurate will provide a measure of the effective renal plasma flow in the absence of tubular functional impairment and vice versa. The clearance values of para-aminohippurate and Diodrast are 600 to 700 ml/min or 350 to 400 ml/min/m^2 of body surface area. Since about 10 per cent of the blood circulating through the kidney does not come into contact with functional cells, the renal plasma flow, as measured by this technique, will be about 650 ml/min or 390 ml/min/m^2 of body surface area. The actual renal blood flow is about 1200 ml/min and the plasma flow about 750 ml/min.

The *p*-aminohippurate clearance is performed in essentially the same way as the inulin clearance and is also to be regarded as a strictly investigative procedure, not a routine clinical one. Para-aminohippurate is given to the subject in a priming dose by intravenous injection, which is followed by a slow continuous administration of solution of low concentration to maintain a constant level. Blood and urine specimens are collected before the drug is given and are used for controls. Three timed urine specimens are collected, as well as blood samples at the beginning and at the end of each urine collection period. The mean of the two blood specimens is used in the calculation. An aliquot of each specimen is assayed for its *p*-aminohippurate concentration. The principle of this assay is a coupling of *p*-aminohippurate with diazotized N-(1-naphthyl)ethylenediamine dihydrochloride. The three clearance values are calculated and averaged. Values of plasma clearance are given in ml/min and also in ml/min/m^2 of body surface area.

REFERENCES

See the references under Glomerular Filtration and Inulin Clearance Test.

PHENOLSULFONPHTHALEIN (PSP) TEST

CLINICAL SIGNIFICANCE

Phenolsulfonphthalein is a dye that is removed from plasma to about 60 to 70 per cent during one passage through the kidney (renal clearance approximately 400 ml/min).

About 20 per cent is normally removed by the liver. Of that portion of the dye that is removed by the kidneys, about 6 per cent is excreted by glomerular filtration and about 94 per cent by tubular excretion. Thus, the test is mainly a measure of the secretory capacity of the tubules.

Traditionally, the test has been performed as a 2 h test. It has been well established, however, that the first specimen (15 min specimen) is of greatest clinical value since the challenge to the kidney is the greatest in the first interval and since the dye excretion during the first 15 min is affected to a lesser degree by extraneous factors such as the renal blood flow and the total urine flow. Therefore, many laboratories collect only one 15 min specimen.

SPECIMEN

Urine specimens, collected 15, 30, 60, and 120 min after the injection of dye, are used. If a split renal function (differential) test is ordered, urine specimens, collected from each kidney, are tested. Some laboratories use different collection times, for example only one 15 or 20 min sample.

PRINCIPLE

A standard dose of phenolsulfonphthalein (PSP) is injected intravenously and the quantity of PSP excreted is measured colorimetrically after alkalizing the specimen to convert the dye to the colored form.

REAGENTS

1. Sterile 1 ml vials of phenolsulfonphthalein (6 mg/ml) (Hynson, Westcott and Dunning, Baltimore, MD.).

2. Sodium hydroxide, 2.5 mol/l. Dissolve 100 g of reagent grade sodium hydroxide in water and, when cool, dilute to 1000 ml.

PROCEDURE

Patient preparation

(This is carried out by the nursing and clinical staffs on the hospital ward.)

1. Hydrate the patient by giving about 600 ml of water to drink.
2. After 10 min, inject intravenously 1 ml of sterile PSP solution (6 mg/ml).
3. Collect separate urine samples at 15, 30, 60, and 120 min after injection.

Laboratory procedure

1. Transfer each specimen into a 1000 ml graduated cylinder.

2. Add 10 ml of 2.5 molar NaOH to each cylinder. Filter, dilute to the 1 liter mark with water, and mix. If the color is pale, as it may be in the 60 and 120 min samples, dilute to only 250 or 500 ml and make corrections for this in the calculation.

3. Measure the per cent transmittance of each diluted specimen at 540 nm, using water as a reference.

4. Read the per cent of PSP excretion in each specimen from a calibration curve.

CALIBRATION CURVE

1. Pipet 1 ml of PSP solution (6 mg/ml) into a 1 liter volumetric flask containing 200 to 300 ml of water. Add 10 ml of 2.5 molar NaOH, dilute to the mark, and mix. This represents the 100 per cent standard.

2. Place in separate tubes the volumes of the 100 per cent standard and water indicated in the following table and mix thoroughly.

100% *Standard* (ml)	*Water* (ml)	*Standard* (%)
10.0	0	100
7	3	70
5	5	50
3	7	30
1	9	10

3. Measure the per cent transmittance of each standard at 540 nm, using water as a reference.

4. Plot the per cent transmittances of the standards against the corresponding concentrations on semilogarithmic paper. (Alternatively, plot absorbance values against concentration on rectangular coordinate paper.)

SOURCES OF ERROR

1. Incomplete urine collection resulting from retention in the bladder is the most frequent source of error in this test.

2. Injection of an incorrect amount of dye is another very common error. The PSP vial contains more than 1 ml; thus, not all the dye should be injected.

NORMAL VALUES

Specimen	*Per cent Excretion*
15 min	25–50
30 min	15–25
60 min	10–15
120 min	5
Total excretion	60–85

It should be noted that the results of the 15 and the 30 min specimens are the most significant, regardless of the outcome of the subsequent collections.

REFERENCES

Wells, B. B., and Halstead, J. A.: Clinical Pathology: Interpretation and Application. 4th ed. Philadelphia, W. B. Saunders Co., 1967, p. 288.

Roundtree, L. G., and Geraghty, J. T.: Arch. Int. Med., *9*:284, 1912.

SPECIFIC GRAVITY

Specific gravity is the ratio of the weight of a substance to the weight of an equal volume of water at a specified temperature. It is a direct but not proportional function of the number of particles in the urine. Since the work done by the kidney in eliminating substances is directly related to the number of particles present, specific gravity is one measure of the function of the kidney. Because each substance contributes differently to the specific gravity, this measure is not strictly a function of the number of particles as is the measurement of osmolality. Specific gravity is easily done, however, and furnishes information of considerable clinical value. It is still performed more frequently than the more informative osmolality measurement.

CLINICAL SIGNIFICANCE

The determination of the specific gravity of a urine sample is an important part of the routine urinalysis. Although there is a considerable fluctuation in the specific gravity values from day to day and also during the course of a day, the determination of the specific gravity of a randomly collected urine specimen has clinical value as a screening test to measure the concentrating ability of the renal tubules. In cases of renal tubular damage, this function is generally the first to be lost. Elevations in urine specific gravity in the absence of dehydration are most commonly seen in patients with uncontrolled diabetes with glycosuria; extremely high specific gravity values (above 1.050) may be seen in patients who have recently had urinary tract diagnostic studies that use iodine-containing x-ray contrast media.

Specific gravity measurements of urine are also a part of the various concentration tests (see Fishberg Concentration Test), which are more accurate measures of the concentrating ability of the kidney than a specific gravity measurement on a random urine specimen.

It has recently been felt that osmolality measurements should replace the specific gravity tests for both routine urinalysis and the Fishberg concentration test. Although such a development would be desirable, the specific gravity test probably will still be used widely because of its simplicity and speed.

NORMAL VALUES

Specific gravity values observed in normal individuals vary greatly with fluid intake and the state of hydration. Thus, the normal values for a 24 h specimen are usually considered to be from 1.015 to 1.025.

DETERMINATION OF SPECIFIC GRAVITY WITH THE URINOMETER

SPECIMEN

The specific gravity is most often performed as a part of a routine urinalysis on a random urine specimen. Less frequently, however, specific gravity is measured on timed specimens after water restriction, in which case more exact information is derived.

PRINCIPLE

The urinometer is a hydrometer designed for the measurement of urinary specific gravity. When placed in the specimen contained in a cylinder, it sinks to the level characteristic of the specific gravity of the specimen. The value may then be read directly from the calibrations on the stem. It is important that the urinometer float freely without sticking to the walls of the container.

PROCEDURE

1. Pour the specimen into the urinometer tube until it is about three-fourths full. Sufficient space should be allowed so that it will not overflow when the urinometer is floated in the sample.
2. Place the urinometer in the specimen with a slight twisting motion so that it will spin and have less tendency to stick to the sides of the tube.
3. Read the scale on the stem where it is intersected by the lowest line of the meniscus.
4. For greatest accuracy, measure the temperature and make the following correction: Add 0.001 to the specific gravity for each 3°C or 5.4°F that the temperature is above the urinometer calibration temperature. Subtract 0.001 for each 3°C or 5.4°F that it is below the calibration.

NOTES

1. The urinometer should be calibrated against distilled water and should read 1.000 at its calibration temperature. It should also be calibrated at a high value. This may be done by testing in a mixture of 75 ml of xylene and 28 ml of bromobenzene, which has a specific gravity of 1.030.

2. If the urine contains significantly large quantities of protein, a correction should be applied to compensate for this factor. Subtract 0.003 from the reading for each 1 g of protein/100 ml of urine.

3. Subtract 0.004 for each 1 g of glucose/100 ml of urine.

SOURCES OF ERROR

1. The urinometer must float freely, not adhering to the sides of the tube; otherwise, the reading may be in error. Also, there should be no bubbles clinging to the stem.

2. Failure to compensate for temperature or for gross proteinuria and glycosuria.

REFERENCES

Wells, B. B., and Halsted, J. A.: Clinical Pathology: Interpretation and Application. 4th ed. Philadelphia, W. B. Saunders Co., 1967, p. 273.

Monroe, L., and Hopper, J., Jr.: J. Lab. Clin. Med., *31*:934, 1946.

DETERMINATION OF SPECIFIC GRAVITY BY REFRACTOMETRY

SPECIMEN

See under Determination of Specific Gravity with the Urinometer.

PRINCIPLE

The refractive index and the specific gravity of a urine specimen are both related functions of the quantity and type of dissolved substance in the specimen. Each substance contributes differently to the refractive index and also to the specific gravity; however, because various urine specimens are likely to contain dissolved substances of similar types and proportions, the refractive index and the specific gravity may be correlated. Increased amounts of abnormal substances such as glucose and protein may partially invalidate the correlation and give specific values that are misleading.

The instrument most commonly used for this purpose is the TS (total solids) Meter (American Optical Corp., Scientific Instrument Div., Buffalo, N.Y.). This is a hand refractometer with two temperature compensated scales, which allows direct measurement of total solids of serum or the specific gravity of urine. Other refractometers have two scales, one calibrated in refractive index and the other in total serum protein concentration.

PROCEDURE

1. Place a small drop of sample on the lower glass plane surface of the TS Meter. Then bring the upper hinged surface down firmly on the drop so that the two glass planes are parallel.

2. Hold the meter toward a source of light so that the beam passes through the sample and the prisms.

3. Read the specific gravity from the proper scale at the sharp line of contrasting light and dark areas that falls across the scales.

NOTES

1. Accuracy is excellent as long as the protein content is low. Unfortunately there seems to be no published correction factor to compensate for the contribution of protein to the refractive index.

2. Wolf and his associates[9] describe a more accurate but more time-consuming method. The specific gravity is calculated from conversion tables provided with the instrument, through substitution of refractive index values of the untreated urine and of its supernatant after deproteinization with acetic acid.

REFERENCES

American Optical Co. Bull. N. SB 10400-1061.
Rubini, M. E., and Wolf, A. V.: J. Biol. Chem., *225*:869, 1957.

OSMOLALITY OF SERUM AND URINE

During the last decade there has been an increasing interest in the use of measurements of the osmotic concentration of body fluids such as serum and urine. Such measurements give clinical information regarding the physiological processes involved in the transport of solute and solvent across membranes which separate fluid compartments. Although osmotic concentration measurements may be clinically useful in the elucidation of certain metabolic, endocrinologic and renal disorders, such measurements should not replace but supplement electrolyte measurements.

As discussed in Chapter 3, page 154, osmotic concentration is best expressed in terms of osmoles or milliosmoles per kilogram of water, i.e., in terms of osmolality. The osmolar unit (mOsm/l fluid) is still used, but is less precise. There is still some inconsistency as to the unit in which the osmotic concentration is expressed. In fact, these terms are often used incorrectly and interchangeably, which is highly undesirable.

Assuming a water content of plasma of 93 per cent, the relation between mOsm/liter of fluid and mOsm/kg H_2O is as follows:

$$\text{mOsm/kg } H_2O = \text{mOsm/liter fluid} \times \frac{100}{93}$$

Thus, in this example, values expressed in mOsm/kg H_2O are approximately 7 per cent higher than those expressed in mOsm/liter plasma. In the new SI unit system, the concentration of osmotically active particles (osmolutes) is expressed in mol/l (see p. 154).

The principal solute components contributing to serum osmotic concentration are sodium, chloride, and bicarbonate ions, and to a lesser degree glucose and urea. Of the total osmotic concentration, ionic components make up over 95 per cent and the un-ionized solutes the small remainder. Serum sodium represents nearly half of the total osmolality because over 90 per cent of the total cation fraction is sodium. Thus, measurement of half of the ions should allow for a close estimate of the total osmolality.

CLINICAL SIGNIFICANCE

Measurement of the osmolality of serum or urine alone is of limited clinical usefulness. A more accurate and clinically more useful way of measuring the concentrating ability of the tubules is the simultaneous measurement of serum *and* urine osmolality. This allows for the calculation of the ratio of urine osmolality to serum osmolality and for the calculation of the osmotic clearance. Analogous to other clearance tests (see p. 982), here we measure the ratio of the concentration of the osmotically active particles in urine to that in serum. This ratio (or clearance) expresses the actual degree to which the kidney has concentrated the glomerular filtrate, which, in respect to osmolality, is very close to serum. The ratio is greatly affected by the volume of fluid intake. More meaningful data are therefore obtained if the fluid intake is restricted, as in the concentration test (see p. 1013), or if clearance is calculated taking into account the urine output.

URINE OSMOLALITY MEASUREMENTS

Measurements of the osmotic concentration of urine are considered more valid than specific gravity measurements in assessing the concentrating ability of the kidney, since the regulation of water excretion is, in part, determined by the osmolality of the fluid compartments of the body.* Consequently, measurement of the urine osmolality, especially as part of a concentration test, is preferred. Holmes performed a comparative study of specific gravity versus osmolality values in normal individuals and in unselected patients with renal disease. Although in the first group there appears to be a reasonably close relationship between specific gravity and osmolality, the relationship is maintained to a much lesser degree in patients with renal disease (see Figs. 17–4 and 17–5). This lack of correlation can be explained at least partially by the fact that the presence of heavy molecules, such as protein, glucose, or iodine-containing compounds, affects the specific gravity of urine substantially more than its osmolality.

The urine osmolality of normal individuals may vary widely depending on the state of hydration. After excessive intake of fluids, for example, the osmotic concentration may fall as low as 50 mOsm/kg, while in individuals with severely restricted fluid intake concentrations of up to 1400 mOsm/kg can be observed. In individuals on an average fluid intake, values of 300 to 900 mOsm/kg are most frequently seen.

If a *random urine specimen* of a patient has an osmolality of 600 mOsm/kg H_2O or

* The actual process of retention or excretion of water by the tubules is a complex process involving several mechanisms. One major mechanism is the secretion of anti-diuretic hormones (ADH) in response to high osmolality in the body fluid compartments. The action of ADH is a cyclic-AMP mediated process involving the tubular epithelial cells. J. T. Harrington and J. J. Cohen discussed this and related processes in a recent publication (Clinical Disorders of Urine Concentration and Dilution, Arch. Int. Med., *131*:810, 1973).

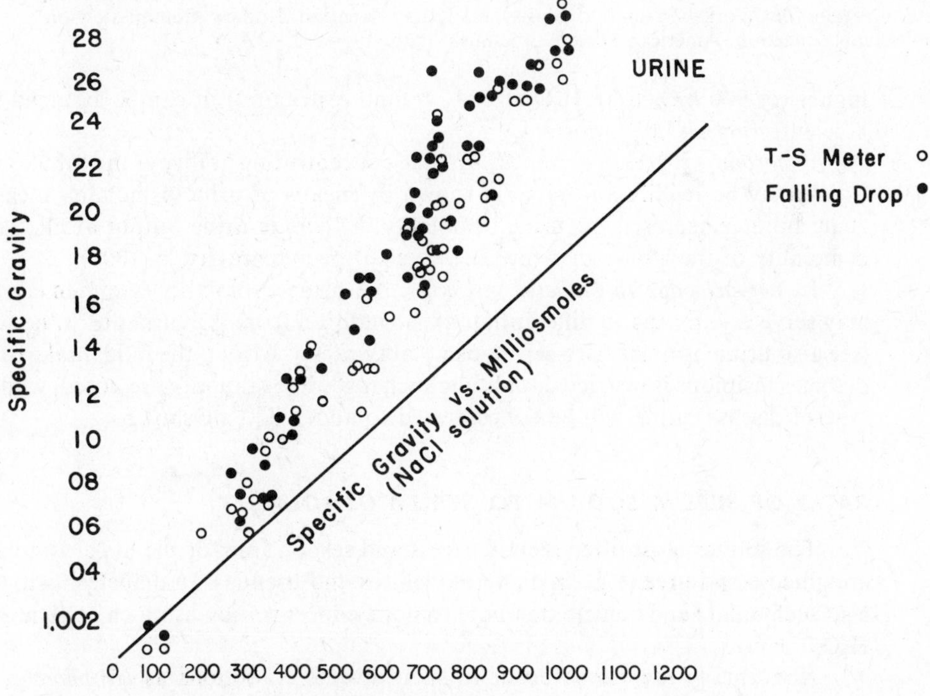

Figure 17–4 Comparison of the urinary specific gravity and urinary osmolality in a series of urines obtained from healthy medical students. The straight line represents comparative readings on various concentrations of sodium chloride solutions. (From Holmes, J. H.: Measurement of osmolality in serum, urine, and other biological fluids by the freezing point determination. *In:* Workshop on Urinalysis and Renal Function Studies. Commission on Continuing Education, American Society of Clinical Pathologists, 1962.)

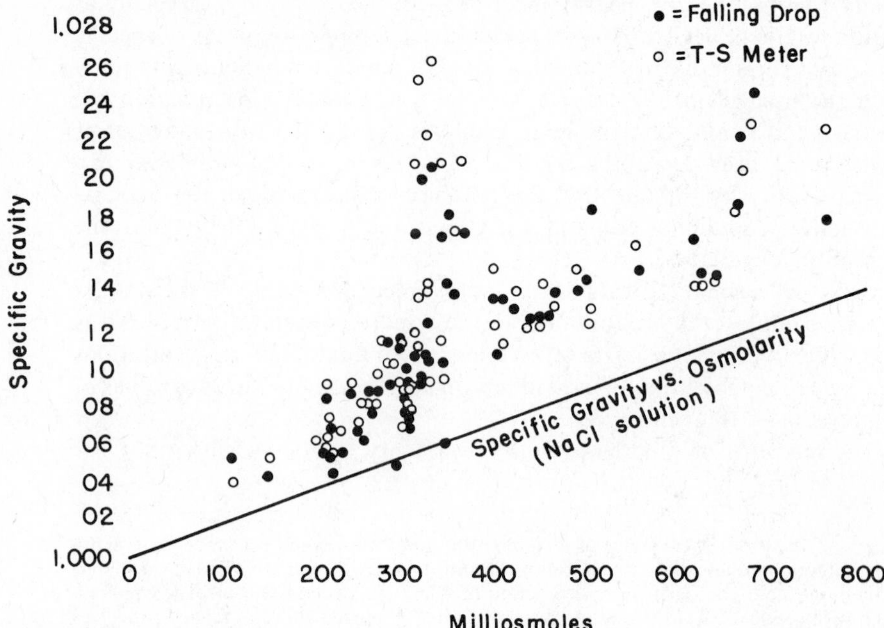

Figure 17–5 Comparison of the urinary specific gravity and urinary osmolality in a series of unselected urines obtained from patients on the renal service. The straight line represents comparative readings on various concentrations of sodium chloride solutions. (From Holmes, J. H.: Measurement of osmolality in serum, urine, and other biological fluids by the freezing point determination. *In:* Workshop on Urinalysis and Renal Function Studies. Commission on Continuing Education, American Society of Clinical Pathologists, 1962.)

higher (or 900 mOsm/kg H_2O after 12 h fluid restriction), it can be assumed that the renal concentrating ability is normal.

In *chronic progressive renal failure*, the concentrating ability of the tubules is diminished, which may be readily and reliably shown by means of urine osmolality measurements. In acute tubular necrosis, the urine osmolality, if there is urine output at all, approaches the osmolality of the glomerular filtrate (see also specific gravity, p. 1006).

In *polyuria due to diabetes insipidus*, the urine osmolality is extremely low, and this may serve as a means to differentiate this condition from polyuria due to neurogenic origin (see also urine osmolality to serum osmolality ratio). Also, if the fluid intake of patients with diabetes insipidus is restricted, only slight changes in osmolality are noted, while in the latter type of disease, urine will be concentrated to above 900 mOsm/kg.

RATIO OF SERUM SODIUM TO SERUM OSMOLALITY

The solutes most often seen in excess and responsible for the hyperosmolality of serum are glucose and urea (e.g., in diabetes mellitus and uremia). In diabetes, ketone bodies may also accumulate and contribute to the osmotic concentration as much as 10 mOsm/kg serum H_2O.

Abnormally *low ratios* occur in such diverse conditions as *lymphomas*, *cancer*, *liver failure*, *shock*, *acute infections* and *myocardial infarctions*, and indicate, in general, unfavorable prognosis. In special cases of *acute toxicity* due to overdose of drugs (e.g., salicylate poisoning) the finding of normal serum sodium and high serum osmolality may often be considered as an indication for dialysis.

In *dehydration*, where water loss exceeds salt loss, the osmolality and the serum sodium increase, but the ratio stays normal. Examples are *prolonged diarrhea* and *peritonitis* or *intestinal obstruction* in which fluid has been sequestered outside the vascular system.

RATIO OF URINE OSMOLALITY TO SERUM OSMOLALITY

In normal individuals on an average fluid intake, the ratio is most frequently between 1.0 and 3.0. After fluid restriction, as in the case of the concentration tests, the ratio is 3.0 or above and may rise in extremes to 4.7. In patients with *renal tubular deficiency*, the ratio will be below that observed in normal individuals. In polyuria of *diabetes insipidus*, the ratio will be between 0.2 and 0.7 even after fluid restriction. This allows for differentiation of this type of polyuria from that seen in other conditions such as diabetes mellitus. In polyuria of neurogenic origin, the ratio may be normal without fluid restriction and increases after fluid restriction.

DETERMINATION OF OSMOLALITY IN SERUM AND URINE

SPECIMEN

Blood should be collected by venipuncture with a minimum of stasis, and the serum should be separated by centrifugation soon after collection. In order to lessen the possible presence of particulate matter, a second centrifugation is recommended. If the serum is not to be analyzed soon after centrifuging, it should be refrigerated or frozen. Heparinized plasma is also satisfactory, but oxalated plasma is not.

Urine should be collected in clean, dry containers without preservatives and centrifuged at sufficiently high speed to remove all gross particulate material. If the analysis cannot be carried out soon after collection, the specimen should be refrigerated. Before analysis, refrigerated specimens should be warmed to aid the complete solution of any precipitated substances.

PROCEDURE

Several different osmometers are commercially available* and each one must be operated in a manner appropriate to the particular instrument, as described in its corresponding manual. However, the following procedure is general and may be applied to any instrument.

1. Centrifuge the specimen twice to eliminate any gross particulate matter.
2. Place the proper volume of specimen in a sample tube and position the tube in the instrument.
3. Supercool the specimen (about −7°C).
4. Initiate the freezing process and allow temperature equilibrium to take place.
5. Measure the freezing point and read the value in milliosmoles from the instrument.

STANDARDIZATION

Osmometers are generally standardized in accordance with the procedures supplied by the manufacturer, utilizing standard sodium chloride solutions.

Standard mOsm/kg H_2O	Expected freezing point depression (°C)	Concentration of NaCl (g/kg H_2O)
100	0.186	3.094
300	0.557	9.476
500	0.929	15.93
750	1.394	24.03
1000	1.858	32.12
1400	2.60	44.98

* Fiske Osmometer, Fiske Associates, Inc., Uxbridge, MA. Osmette 2007, Precision Systems Inc., 60 Union Ave., Sudbury, MA. Osmometer, Advanced Instruments, Inc., Needham Heights, MA.

TABLE 17–2 VALUES FOR OSMOLALITY AND OSMOLARITY OBSERVED UNDER VARIOUS CONDITIONS

Specimen	mOsm/liter Specimen	mOsm/kg H_2O
URINE		
Osmotic limits of renal dilution and concentration:	40–1370	50–1400
Normal random specimen (average fluid intake):		300–900
Normal range during maximum urine concentration (fluid restriction for 12 h):	855–1325	900–1400
SERUM		
Osmotic limits observed:	220–473	230–490
Normal range:	(275) 280–295	289–308
RATIO URINE OSMOLALITY / SERUM OSMOLALITY		
Normal random urine (average fluid intake):	1.0–3.0	1.0–3.0
Specimen after 12 h fluid restriction:	3.0–4.7	3.0–4.7
RATIO SERUM Na / SERUM OSMOLALITY	0.43–0.50	0.43–0.50

The sodium chloride should be dried at 110°C overnight and cooled to room temperature in a desiccator before weighing the amount designated in the table. Water of high quality (distilled-deionized) should be used.

The values given in the above table are calculated for 1 kg H_2O at 4°C. Since it is inconvenient in a routine laboratory to prepare such solution, it is customary to fill a 1 liter volumetric flask to the mark with water before adding the indicated amount of NaCl. In order to compensate for the decrease in density of water at room temperature (25 to 27°C), an additional 2.0 ml of water is added to the flask.

NORMAL VALUES

There is no close agreement regarding the values for osmolality in body fluids. The values given in Table 17–2 represent an average of those values found in the literature. One reason for the lack of agreement is that authors have not always made the distinction between *osmolality* and *osmolarity*; in fact, these terms have been used interchangeably by some.

PRECISION

It has been reported that a reproducibility of ±1 mOsm/kg H_2O with an equal accuracy can be obtained if the instrument is standardized correctly; however, under routine conditions an accuracy and precision of ±2 mOsm/kg H_2O is quite acceptable.

SOURCES OF ERROR

Faulty standardization, faulty use of the osmometer, and the presence of particulate matter in the specimen are the most common sources of inaccurate results.

REFERENCES

Osmometer Manual. Fiske Associates, Inc., Uxbridge, Mass.
Lobdell, D. H.: St. Vincent's Hosp. Med. Bull. (Bridgeport, Conn.), *8*:7, 1966.
Warhol, R. M., Eichenholz, A., and Mulhausen, R. O.: Arch. Int. Med., *116*:743, 1965.
Hendry, E. B.: Clin. Chem., *7*:156, 1961.
Lindemann, R. D., Van Buren, H. C., and Raisz, L. G.: New England J. Med., *262*:1306, 1960.

Wolf, A. V.: The Urinary Function of the Kidney. New York, Grune & Stratton, Inc., 1950.
Jacobson, M. H., Levy, S. E., Kaufman, R. M., Gallinek, W. E., and Donnelly, O. W.: Arch. Int. Med., *110*:83,1962.

CONCENTRATION AND DILUTION TESTS

PRINCIPLE

In the *concentration tests*, the ability of the kidneys to concentrate urine is tested by measuring the specific gravity or osmolality of urine voided at intervals in the morning, following an overnight period of fluid restriction. In the *dilution test*, the ability of the kidneys to excrete dilute urine is tested by administering a massive fluid load and testing the specific gravity and osmolality of urine specimens for a period of 3 to 4 h.

CLINICAL SIGNIFICANCE

Concentration tests for the assessment of renal tubular function are not used as frequently as they were several decades ago, since they have been gradually replaced by tests measuring plasma constituents. Of the many procedures available, the Fishberg test described in this section has perhaps had the widest application. Following the regimen described, the normal patient can concentrate his urine to a specific gravity of 1.025, often reaching 1.032. (The respective values for the osmotic concentration are >900 mOsm/kg H_2O or >855 mOsm/l plasma.) The ability to concentrate is lost to some extent with age, so that older patients usually show values in the lower portion of the normal range even if kidney function is apparently normal. *Tubular epithelial damage*, as may be seen after intake of *nephrotoxic drugs*, or in *severe alkalosis*, *shock syndrome*, or *impairment of tubular blood supply*, may cause impairment of the concentrating ability of the tubular cells.

In severe functional impairment, specific gravity values of 1.010 to 1.020 are observed; most often the value seen is near the lower figure, where it may persist for a certain time ("fixed specific gravity"). When healing occurs, e.g., after acute nephritis, the concentrating power of the kidneys is the last function to return to normal. This may reflect functional inadequacy of the newly regenerated tubular epithelial cells. Patients with *edema* who are receiving therapy and are losing their edema water may produce urine with low specific gravity. This has no relation with their renal status.

DILUTION TESTS. Production of a dilute urine, just as the formation of a concentrated urine, is a tubular function. Other physiologic processes, however, are also involved in the regulation of salt retention and limit the usefulness of the dilution tests as an index of tubular capacity. Among such non-tubular phenomena affecting the concentration of solids in urine are adrenal insufficiency, hepatic disease, and cardiac failure. The emotional state of the patient may also effect the result. Possible clinical contraindications to administering a large amount of fluid to a patient with decreased ability to excrete fluids must also be considered.

Although dilution tests are seldom used clinically, for the reasons mentioned, several procedures have been described. The subject drinks 1000 to 1200 ml of water in a 30 min period and all urine is collected at intervals over the next 3 to 4 h. Normally, the specific gravity of at least one specimen should be less than 1.003 (approximately 50 mOsm/kg H_2O) and at least half of the water ingested should have been excreted in the 3 to 4 h period.

THE FISHBERG CONCENTRATION TEST

PROCEDURE

Preparation of the patient

1. Allow the patient no more than 200 ml of total fluid intake during the evening meal on the day before the test and no fluid from 8 p.m. to 10 a.m. This period of water deprivation can be made more tolerable by reducing the salt intake during the meals.

2. Discard any urine voided by the patient during the night.

3. Collect separately all urine voided by the patient at 6 a.m., 8 a.m., and 10 a.m., or up to 10 a.m. if the patient cannot void at the specified times.

For the laboratory

1. Measure the specific gravity or the osmolality of each of the three urine specimens.

NORMAL VALUES

The specific gravity of one or more of the specimens should have a value of 1.025 or higher. The osmotic concentration should be >900 mOsm/kg H_2O or >855 mOsm/liter urine. Values below this figure indicate a decrease in the concentrating ability of the kidney.

SOURCES OF ERROR

If the patient ingests more fluid than allowed, the results will not be valid.

REFERENCE

Fishberg, A. M.: Hypertension and Nephritis. 5th ed. Philadelphia, Lea & Febiger, 1954.

REFERENCES

1. Archibald, R. M.: Colorimetric measurement of uric acid. Clin. Chem., *3*:102, 1957.
2. Friedman, H. S.: Modification of the determination of urea by the diacetyl monoxime method. Anal. Chem., *25*:662, 1953.
3. Gentzkow, C. J.: An accurate method for determination of blood urea nitrogen by direct nesslerization. J. Biol. Chem., *143*:531, 1942.
4. Henry, R. J.: Clinical Chemistry: Principles and Technics. New York, Hoeber Medical Division, Harper and Row, Publishers, 1964.
5. Johnson, R. B., and Hoch, H.: Osmolality of serum and urine. *In* Standard Methods of Clinical Chemistry. S. Meites, Ed. New York, Academic Press, Inc., 1965, vol. 5, p. 159.
6. Marsh, W. H., Fingerhut, B., and Miller, H.: Automated and manual direct methods for the determination of blood urea. Clin. Chem., *11*:624, 1965.
7. Owens, J. K., Iggo, B., Scandrette, F. J., and Stewart, C. P.: The determination of creatinine in plasma or serum and in urine; a critical examination. Biochem. J., *58*:426, 1954.
8. Vanselow, A. P.: Preparation of Nessler's reagent. Industr. Eng. Chem., Analyt. Ed., *12*:516, 1940.
9. Wolf, A. V., Fuller, J. B., Goldman, E. J., and Mahoney, T. D.: New refractometric methods for the determination of total protein in serum and urine. Clin. Chem., *8*:158, 1962.
10. Larsen, K.: Creatinine assay by a reaction-kinetic principle. Clin. Chim. Acta, *41*:209, 1972.

ADDITIONAL READINGS

Abbrecht, P. H.: An outline of renal structure and function. Chem. Eng. Progr., Symp. Ser., *64*:1, 1968.

Cantarow, A., and Trumper, M.: Renal function. *In* Clinical Biochemistry. 6th ed. A. Cantarow and M. Trumper, Eds. Philadelphia, W. B. Saunders Company, 1962, pp. 373–446.

Caraway, W. T.: Uric acid. *In* Standard Methods of Clinical Chemistry. D. Seligson, Ed. New York, Academic Press, Inc., 1963, vol. 4, pp. 239–247.

Corcoran, A. C., Hines, D. C., and Page, I. H.: Kidney Function in Health; Kidney Function in Disease. Indianapolis, Lilly Laboratory for Clinical Research.

Josephson, B., and Eck, J.: The assessment of the tubular function of the kidneys. *In* Advances in Clinical Chemistry. H. Sobotka and C. P. Stewart, Eds. New York, Academic Press, Inc., 1958, vol. 1.

Thurau, K., Valtin, H., and Schnermann, J.: Kidney. Ann. Rev. Physiol., *30*:441, 1968.

CHAPTER 18

ANALYSIS OF CALCULI

by Ermalinda A. Fiereck, M.S.

Calculi are deposited chemicals in compact form. These concretions are frequently found in the urinary tract and gallbladder and less frequently in the salivary gland, pancreas, prostate, and other locations. The cause of calculus formation is not known; however, the many physical factors which affect the metastable solubility or insolubility of crystalloids in complex biological fluids are involved in the phenomenon.

URINARY CALCULI

Urinary calculi have received the most attention because of the frequency of their occurrence and the variability of their composition. The occurrence of urinary "stones" dates back to antiquity. Evidence of renal calculi has been found in recent exploration of Egyptian tombs dated from 8000 B.C. Urinary calculus formation in children is frequent in the impoverished areas of the world; however, the incidence among European and American children has decreased in the past century. This decrease is thought to be related to an improvement in the dietary conditions among these populations. A wide geographic variation in the occurrence of renal calculi exists throughout the world. In the United States, the nationwide frequency is approximately 1 per 1000, and less than 1 per cent of these occur among children.[9] The majority of the patients requiring surgical removal of renal calculi in American hospitals are between the ages of 50 and 70.

Formation of Urinary Tract Calculi

Winer[16] lists seven factors that may contribute to the formation of calculi: (1) metabolic disturbances such as cystinuria and gout; (2) endocrinopathies such as hyperparathyroidism; (3) urinary obstruction; (4) infections; (5) mucosal metaplasia, which occurs in vitamin A deficiency; (6) extrinsic conditions such as dehydration, dietary excess, drug excess, or chemotherapy; and (7) isohydruria, which refers to the loss of the normal acid-alkaline tides ("fixation of pH"). The presence of two or more of these conditions is usually associated with stone formation, and, among the seven listed factors, isohydruria is found most often.

Many calculi contain a clearly defined nucleus around which the chemicals precipitate. Bacteria, blood clots, fibrin, or epithelial cells may serve as such nuclei. Calculi that have a definite center are thought to arise by localized precipitation of salts from a supersaturated urine. Some calculi have an indentation on one surface which suggests an initiation at a surface such as a renal papilla.[12]

A fibrous matrix, composed of mucoproteins and mucopolysaccharides, surrounds the center of the calculus at a series of spaced intervals, and constitutes approximately 3 per cent of its weight.[2] This matrix is thought to be an essential factor in the initial phase of calculus formation and a prime factor in determining the position and composition of the crystalline deposits.

The growth of a calculus is dependent upon the pH of the urine, the solubility of each substance, and the availability of the various solutes. The multiplicity of the crystalloids and colloids present and the interaction of these substances make it difficult, however, to predict the conditions under which crystallization of the individual chemicals occurs. Prolonged isohydruria, which may occur at almost any urinary pH within the physiological range, favors the formation of "pure"* calculi, whereas calculi of mixed composition are found in patients showing a temporary and wide fluctuation of urinary pH. Uric acid calculi are associated with a pH below 5.5, calcium oxalate calculi occur at the pH range of 5.5 to 6.0, and calcium phosphate and magnesium ammonium phosphate calculi occur at the alkaline range of 7.0 to 7.8.

Mixed calculi in which the center is composed of pure calcium oxalate, the intermediary layer of calcium oxalate and apatite,† and the peripheral layer of apatite and magnesium ammonium phosphate are rather frequently found. Periodic changes in the urinary environment, frequently associated with bacterial infections, cause the formation of these layered calculi.

Urea-splitting bacteria are thought to function in stone formation by enzymatic conversion of urea to ammonia and carbon dioxide. The resultant bicarbonate and hydroxyl ions raise the pH of the urine in the vicinity of the bacteria and thus favor the precipitation of magnesium ammonium phosphate. These bacteria or a small calculus may serve as the nucleus upon which the precipitation of magnesium ammonium phosphate occurs. Successful treatment of the bacterial infection, with the reversal of the urinary pH to an acidic pH, may inhibit further growth of the stone. After surgical removal of the calculus, no recurrence will result if the cause was only bacterial; however, if the cause was metabolic, continued stone formation may occur.

Excessive excretion of specific substances may also lead to the formation of urinary calculi. This excess may arise from a relative imbalance in the diet or as a consequence of a metabolic disorder. Cystinuria is a complex metabolic defect characterized by a decreased renal tubular reabsorption of arginine, cystine, ornithine, and lysine. Although greater quantities of lysine and arginine are excreted than cystine, cystine calculi result, owing to the low solubility of cystine in urine. Cystine is found in approximately 1 per cent of the calculi analyzed.

Uric acid calculi are frequently found and may occur in patients who show an excessive excretion of uric acid (urate). At a pH below 5.7, the predominant form is uric acid, which is nearly insoluble in urine and thus contributes to the growth of uric acid calculi. Excessive excretion of uric acid may occur after a high purine diet, as a result of extensive tissue breakdown, such as occurs in severe systemic or neoplastic diseases, or in patients with gout.

The average American diet contains considerably more calcium than is necessary for the maintenance of bones. This increased calcium intake is generally not harmful and the excess calcium is excreted in the urine or stool. In individuals who are predisposed to urinary calculi formation, hypercalciuria may favor the formation of calcium-containing stones. Hypercalciuria is found in hyperparathyroidism, and it is this condition that is associated most frequently with formation of calcium-containing calculi.

Inhibitors of Calculi Formation

Certain substances, such as pyrophosphates and polyphosphates, may inhibit calculi formation by forming chelate complexes. Individuals with the least amount of inhibitors would be most likely to develop calculi. No practical means, however, has been found to increase the concentration of the naturally occurring inhibitors.[7]

---

* The word "pure" should not be taken literally since each stone contains at least trace amounts of other constituents which are trapped during stone formation.

† Apatite is a mineral composed of complex calcium phosphate which sometimes contains carbonate.

Composition of Calculi

Approximately 30 to 40 per cent of the calculi analyzed are "pure" calculi; the remainder contain at least two principal constituents that may be of interest to the clinician for diagnostic and therapeutic purposes. The composition of a calculus is dependent to some degree on its location. Adult bladder stones frequently are composed of uric acid, whereas kidney stones generally contain calcium oxalate, apatite and triple phosphate.[9] Calcium is the chief cation found in calculi and is generally associated with oxalate, phosphate or carbonate. Apatite occurs frequently with calcium oxalate and uric acid in acid urine, and with magnesium ammonium phosphate in alkaline urine, which suggests that apatite may precipitate over the entire physiological urinary pH range. Xanthine, urostealith (fat and fatty acids) and fibrin calculi occur very rarely. Fibrin calculi are thought to arise from blood clots and may also serve as nuclei for other calculi.

Various other substances may be incorporated into the calculus during its formation. Trace amounts of many heavy metals have been found, as well as larger amounts of fluoride, amino acids, albumin, mucopolysaccharides, and mucoproteins.

The composition of calculi is also related to the geographic location of the patient. This and the other mentioned factors may explain the wide distribution of the constituents, noted by the various investigators, shown in Table 18–1.

TABLE 18–1 PERCENTAGE DISTRIBUTION OF CHEMICAL CONSTITUENTS OF URINARY CALCULI

(Components of mixed calculi are counted separately.)

Substance	*Prien and Frondel*[13]	*Herring*[5]	*Beeler et al.*[1]	*Kachmar*[6]
Calcium oxalate				
monohydrate	61.7	43.0	60.0	46.7% (includes $CaCO_3$)
dihydrate	46.0	61.0	37.0	
Apatite and tricalcium phosphate	20.8	61.8	46.0	88.8
Magnesium ammonium phosphate	17.0	15.7	15.0	18.7
Uric acid and urates	6.1	9.4	12.0	21.5
Cystine	3.8	0.89	5.0	0.93
Xanthine	0	0.04	1.0	0

Nonbiological materials such as sand, gravel, plaster, and other foreign material are occasionally submitted by malingerers or by drug addicts in an attempt to obtain attention or a prescription for narcotics. These stones are generally extremely hard and nearly impossible to crush.

OTHER CALCULI

Calculi may also occur in sites other than the urinary tract. The composition of these calculi, like that of urinary calculi, depends upon the concentrations and solubilities of the constituents that are locally available.

Prostatic calculi are composed of organic matter and inorganic constituents, such as carbonate, calcium, magnesium, ammonium, and phosphate. *Salivary calculi* are composed of "tartar," which contains magnesium, calcium, phosphate, carbonate, and some organic matter.

Pancreatic calculi are extremely rare and, if present, are generally associated with gallstones. Calcium, carbonate, phosphate, organic matter, and small amounts of magnesium and oxalate have been detected in these calculi.

Gallstones (biliary calculi) occur very frequently. They are composed chiefly of cholesterol, bile pigments, calcium, phosphate, and carbonate. Trace amounts of iron, copper, magnesium, manganese, and miscellaneous organic compounds are frequently found. Among the factors which are thought to contribute to the formation of biliary calculi are an increase in the cholesterol level in the bile, a change in pH, stagnation of bile, infections, and a change in the ratio of the bile acids to cholesterol.

DISCUSSION OF METHODS FOR STONE ANALYSIS

Various methods have been utilized in the analysis of calculi. Beeler *et al.*[1] compared the methods based on x-ray diffraction, infrared spectrophotometry, chemical analysis, optical analysis, electron microscopy, electron diffraction, and thermoluminescence. The last four techniques were reported as unsuitable for stone analyses; the three remaining techniques, namely, x-ray diffraction, infrared spectrophotometry, and chemical analyses, showed fairly good agreement. Infrared spectrophotometry and x-ray diffraction methods have the advantages that the results can be quantitated and a permanent record is obtained, but there are disadvantages. These instruments are not available in many clinical laboratories, the analyses are time-consuming, and much experience is needed in order to interpret the results correctly. Examples of infrared analysis of renal calculi containing a single constituent are shown in Figure 18–1. The spectra obtained from calculi containing numerous compounds, in contrast, are very complex and positive identification in such cases is extremely difficult, if not impossible. When infrared spectrophotometers and x-ray diffraction apparatus become readily available tools of clinical chemistry, these methods may supersede the chemical method for analysis of calculi. At the present time, however, the chemical method is still the method of choice since it is simple and accurate, requires no special equipment, and uses reagents that are readily available.

PROCEDURES FOR STONE ANALYSIS

Preliminary Examination of Calculi

The calculi submitted should be washed with deionized water and dried with moderate heat or ethanol-ether.* Included in the report should be a description stating the number of stones, the appearance, weight, and measurements. If the stone is large, it should be cut, using a bone saw, and a portion of each layer should be crushed and used for separate analysis. A small stone is crushed and chemically analyzed in its entirety. The external appearance of the calculi, the size, shape, and color, as well as the ashing of a portion of the calculus in an open flame, can be helpful in the identification of the constituents. For flame analysis, a small portion of the crushed calculus is heated in a platinum dish until it glows. If the original bulk is not appreciably changed by ashing and if there is little or no darkening, the calculus is mainly inorganic. If the sample chars and burns almost completely, the calculus is primarily organic.

Calcium oxalate calculi are hard calculi that are difficult to grind or crush. They vary from a small smooth type characterized as a hemp seed calculus to a large type possessing an extremely uneven surface. This rough surface may be due to the crystalline structure of calcium oxalate; a stone of this type is classed as a mulberry calculus. The hemp seed type is usually brown, whereas the mulberry type is usually light colored. After ashing a portion of a calcium oxalate stone, approximately two thirds of the bulk remains as a

* To wash small grain-sized stones conveniently and without loss, place them into a test tube, cover the tube with several layers of gauze secured with a rubber band, and allow water to flow over the stones.[14]

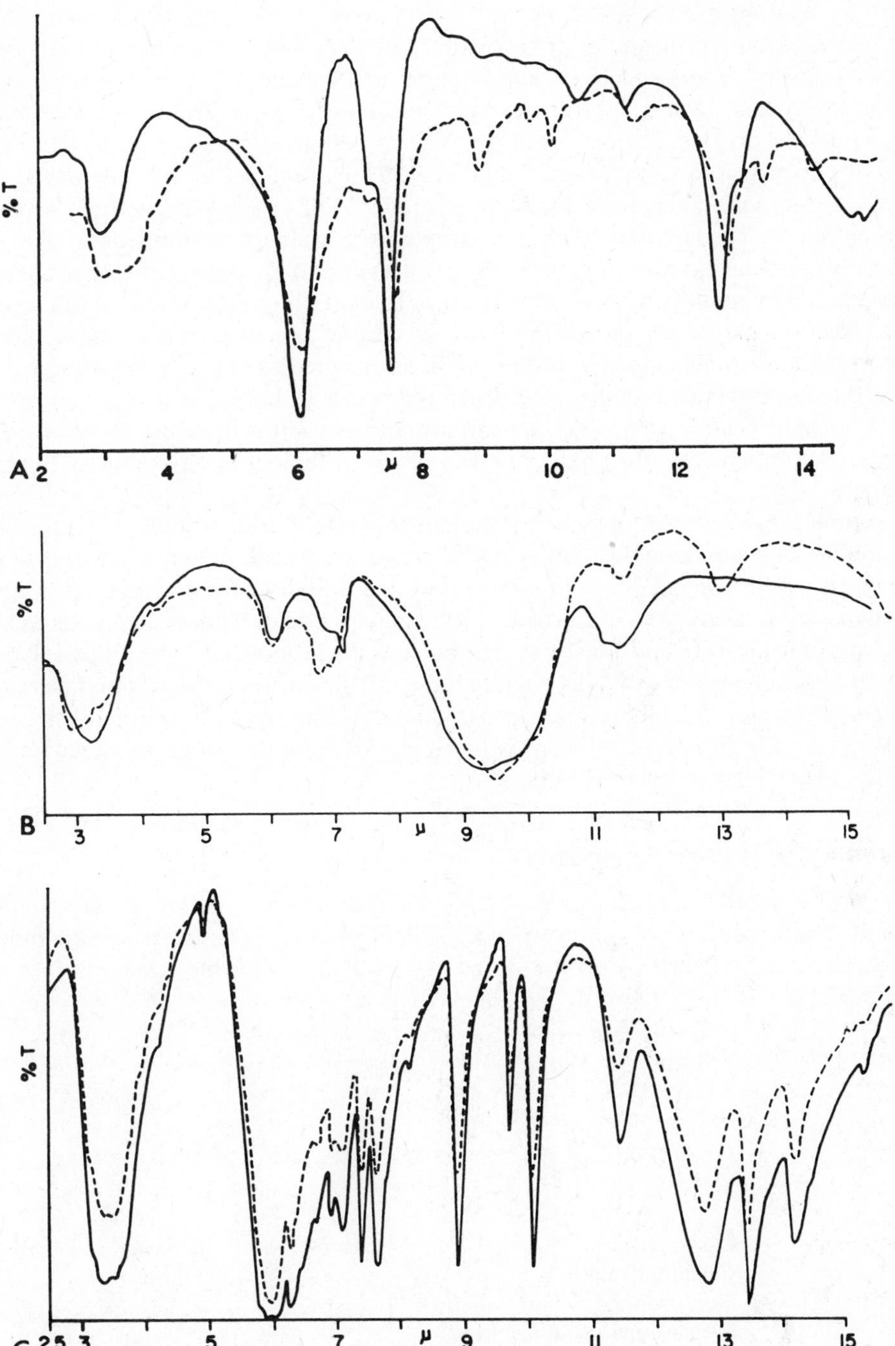

Figure 18–1 Comparison of infrared spectra of calculi (- - - -) with spectra of known compound (~). *A*, Calcium oxalate; *B*, magnesium ammonium phosphate; *C*, uric acid. (From Weissman et al.: Anal. Chem., *31*:1335, 1959.)

residue. (This is due to the conversion of calcium oxalate to calcium carbonate or oxide.)

Calcium carbonate calculi are small and round. They vary from white to gray in color and have a hard, smooth texture. Ashing of the sample leaves a white residue comparable in size to the original sample.

Uric acid (*urate*) *calculi* are often small, round, smooth, dull stones and are frequently found in groups. Occasionally a single large urate stone with an irregular crater-like appearance is seen. Uric acid (urate) calculi are usually colored yellow, brown, or reddish-brown, and the crushed powder is yellow. When ashed, they blacken rapidly with the formation of oily brown rings that burn completely as they progress up the side of the crucible.

Phosphate calculi may have a smooth or rough, chalk-like appearance and may be white gray, or yellow. The size and color of a sample after ashing remain unchanged.

Cystine calculi have a waxy, lustrous appearance and are pale yellow or white in color. They are usually round, and occasionally occur as multiple stones. When a cystine sample is ashed, the sample burns completely and leaves the crucible empty. A peculiar sharp penetrating odor, similar to that of burning hair, is noted as the sample burns. An odor of sulfur dioxide may also be noted as the sample begins to burn.

Xanthine calculi have a waxy appearance and are generally white to brownish-yellow in color. A xanthine calculus, like uric acid, blackens rapidly upon ashing and burns completely.

Fibrin calculi are small, black, lightweight stones that are insoluble in organic solvents, but soluble in potassium hydroxide in the presence of heat. When a portion of a fibrin calculus is ashed, the sample burns completely and an odor of burnt feathers is noted.

Urostealith calculi are composed of fat and fatty acid. These calculi are very brittle when dry but are soft and plastic when moist and exude an oil when squeezed. After a portion of the urostealith calculus is ignited, the sample burns completely and has an odor of shellac. Urostealith calculi are soluble in organic solvents. When a portion of this extract is evaporated on a glass slide, stained with Sudan III, and examined microscopically, urostealith is identified as red fat droplets.

QUALITATIVE CHEMICAL ANALYSIS

For a complete chemical analysis, place small portions of the powdered stone into three 10 × 75 mm test tubes and two porcelain evaporating dishes or crucibles, and proceed to check for the individual constituents according to the following scheme:*

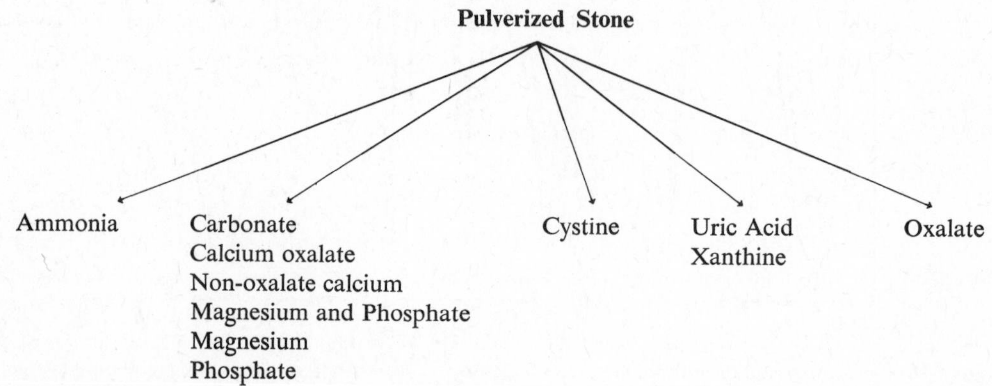

* Use only small portions of the stone; the use of large portions may lead to false positive results because of the precipitation of some constituents at their saturation point.

If the calculus submitted is very small, the stone can be analyzed according to the procedure given under Analysis of Small Calculi, included later in the chapter.

AMMONIA

To a small amount of powdered calculus in a 10 × 75 mm test tube, add a few drops of H_2O. Carefully add 3 or 4 drops of 1.8 molar KOH (10 g/100 ml) into the bottom of the tube. Place a piece of water-moistened pH paper over the mouth of the tube. (When adding the KOH to the tube, be careful not to wet the top of the tube; contamination of the pH paper with KOH will give a false positive result.) If the moistened pH paper indicates an increase in alkalinity, *ammonia is present*. The volatile ammonia released by the KOH combines with the H_2O to form NH_4OH, which reacts with the indicator.

ALTERNATIVE PROCEDURE FOR AMMONIA

Heat a small amount of the powdered stone with 2 ml of 0.6 molar HCl (5 ml conc. acid/100 ml). Cool and neutralize the solution with 2.5 molar NaOH (10 g/100 ml). Add 0.5 ml of Nessler's reagent (see serum nonprotein nitrogen procedure). If an orange-brown precipitate forms, *ammonia is present*.

CARBONATE

To a small amount of the powdered calculus in a 10 × 75 mm test tube, add approximately 3 ml of 0.6 molar HCl. If effervescence occurs, *carbonate is present*.

The addition of a strong acid (e.g., HCl) to a carbonate results in the release of CO_2, noted as small bubbles.

$$CO_3^{2-} + 2\,H^+ \rightarrow H_2CO_3 \rightarrow H_2O + CO_2\uparrow$$

CALCIUM OXALATE

Heat the HCl solution from the carbonate procedure in order to solubilize the constituents. Cool and filter through Whatman No. 1 filter paper (5.5 cm in diameter) into another tube. Add 0.5 ml of saturated sodium acetate and adjust the pH to approximately 5 with 1.7 molar acetic acid (10 ml/100 ml). If a white precipitate forms, *calcium oxalate is present*.

NON-OXALATE CALCIUM

If a precipitate formed in the calcium oxalate procedure, filter using Whatman No. 1 filter paper; if no precipitate was noted, continue without filtering. Add 0.5 ml of 0.3 molar potassium oxalate (5.5 g/100 ml) and adjust the pH to approximately 5.0 using 1.7 molar acetic acid. If a white precipitate forms, *non-oxalate calcium is present*.

$$Ca^{2+} + C_2O_4^{2-} \xrightarrow{\text{pH } 5.0} CaC_2O_4$$

At a pH of 5, oxalate combines with calcium to form an insoluble white precipitate of calcium oxalate. It is important that the pH be approximately 5 because phosphates, if present, will coprecipitate at alkaline pH, while the precipitation of CaC_2O_4 is incomplete at acid pH.*

MAGNESIUM *AND* PHOSPHATE

If a precipitate formed in the non-oxalate calcium procedure, filter through Whatman No. 1 paper; if no precipitate was noted, continue without filtering. Add NH_4OH until a pH of 8.0 is obtained. If a white precipitate forms, *magnesium and phosphate are present*. At pH > 8.0, the addition of ammonium ions to a solution containing magnesium and phosphate ions results in the precipitation of NH_4MgPO_4.

$$Mg^{2+} + PO_4^{3-} \xrightarrow[\text{pH}>8.0]{NH_4^+} NH_4MgPO_4$$

---

* If only a small amount of calcium is present, a slowly forming cloudiness will be noted. Allow the test to stand 10 min before filtering to insure complete precipitation.

If no precipitate forms (as is the case when either magnesium *or* phosphate is absent), divide the solution into two portions and proceed as follows:

MAGNESIUM

To one portion, add 0.5 ml of 0.3 molar Na_2HPO_4 (5.5 g/100 ml). If a white precipitate forms, *magnesium is present*. The addition of phosphate and ammonium ions to a solution containing magnesium ions, at pH > 8.0, results in the precipitation of NH_4MgPO_4.

PHOSPHATE

To the second portion, add 0.5 ml of 0.4 molar $MgSO_4$ (5 g/100 ml). If a white precipitate forms, *phosphate is present*. The addition of magnesium and ammonium ions to a solution containing phosphate ions, at pH > 8.0, results in the precipitation of NH_4MgPO_4.

ALTERNATIVE PROCEDURES FOR MAGNESIUM AND PHOSPHATE

Heat a small amount of the powdered stone with 4 ml of 0.6 molar HCl. Cool and divide the solution into two parts. Analyze one portion for magnesium and the other for phosphate.

Neutralize one portion with NH_4OH and add 1 ml of 0.02 molar alcoholic *p*-nitrobenzeneazoresorcinol (0.5 g/100 ml ethanol). If a blue color forms, *magnesium is present.*

Another test for magnesium is to add titan yellow and NaOH (see serum magnesium procedure) to a 0.6 molar HCl solution of the calculus. If an orange-red color or precipitate forms, *magnesium is present.*

Neutralize the second portion with NaOH and add 0.5 ml of molybdate solution and 0.2 ml of aminonaphtholsulfonic acid (see serum phosphorus procedures). If a blue color forms, *phosphate is present.*

CYSTINE

To a small amount of pulverized stone in an evaporating dish, add 1 or 2 drops of 2.5 molar NaOH, and heat. Add several drops of 0.3 molar lead acetate (10 g/100 ml) and heat again. If a black precipitate forms, *cystine is probably present.* Cystine decomposes in the presence of heat and NaOH to yield sodium sulfide, which combines with lead to form a black precipitate of lead sulfide. Sulfides from other sources such as organic matter give a positive lead sulfide test; therefore, the presence of cystine should be confirmed by the nitroprusside test or by a microscopic examination for crystals.

ALTERNATIVE PROCEDURE FOR CYSTINE (NITROPRUSSIDE TEST)

Boil a small amount of the powdered stone with 2 ml of H_2O. Add 2 ml of 1.0 molar NaCN (5 g/100 ml). After 5 min, add 3 drops of a freshly prepared solution of sodium nitroprusside (approximately 25 mg/ml H_2O). If a red-wine color forms, *cystine is present.*

NaCN converts cystine to cysteine, which then reacts with sodium nitroprusside to form a red color.

$$\begin{array}{l} \quad\quad NH_2 \\ \quad\quad | \\ S{-}CH_2C{-}COOH \\ | \quad\quad\;\, | \\ | \quad\quad\;\, H \\ | \\ S{-}CH_2CH{-}COOH \\ \quad\quad | \\ \quad\quad NH_2 \end{array} \xrightarrow{NaCN} HS{-}CH_2{-}\underset{\displaystyle H}{\overset{\displaystyle NH_2}{\underset{|}{\overset{|}{C}}}}{-}COOH \xrightarrow{Na_2[Fe(NO)(CN)_5]\cdot 2H_2O}$$

$$(COOH{-}\underset{\displaystyle H}{\overset{\displaystyle NH_2}{\underset{|}{\overset{|}{C}}}}{-}CH_2{-}S)_2[Fe(NO)(CN)_5]$$

Red Color

URIC ACID AND XANTHINE (MUREXIDE TEST)

To a small amount of pulverized stone in an evaporating dish, add 1 or 2 drops of concentrated HNO_3 and evaporate the solution slowly just to dryness (a pink-orange color forms in the presence of uric acid). Cool and add several drops of concentrated NH_4OH. If a purple color develops, *uric acid is present.* The reaction of uric acid with HNO_3 results in an oxidation of the uric acid to dialuric acid and alloxan, which condense to form alloxantin. The addition of NH_4OH to alloxantin results in the formation of ammonium purpurate, murexide.

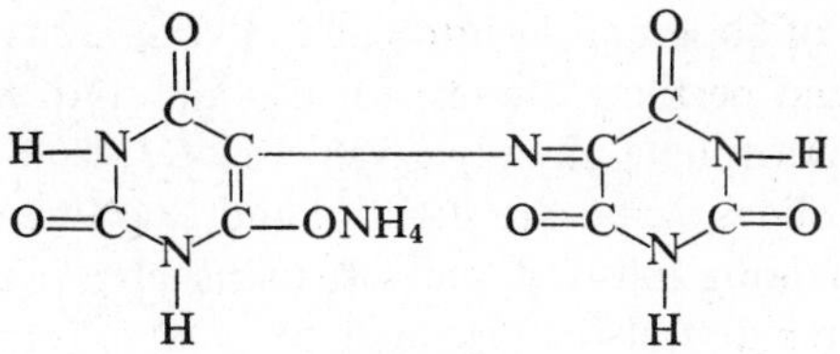

```
          O                    O
          ‖                    ‖
          C                    C
     H—N     C———————————N=C      N—H
        |    ‖             |      |
      O=C    C—ONH4      O=C      C=O
          N                    N
          |                    |
          H                    H
```

Murexide

If a yellow color develops, which turns red after the addition of several drops of 2.5 molar NaOH and heating, *xanthine is present.*

TEST FOR DIFFERENTIATING BETWEEN XANTHINE AND URIC ACID[10]

Although this test is not totally specific for xanthine, it can be used to distinguish xanthine from uric acid.

Heat a small amount of the powdered calculus in 2 ml of freshly prepared Ehrlich's diazo reagent (see serum bilirubin procedure). Add 2 drops of 2.5 molar NaOH. Xanthine couples with the reagent to form a red-wine color; uric acid gives a faint yellow color or no color.

ALTERNATIVE PROCEDURE FOR URIC ACID

Heat a small amount of the powdered stone in 2 ml of 0.6 molar HCl. Cool and neutralize with 2.5 molar NaOH. Add 1 ml of phosphotungstic acid (see serum uric acid procedure). If a blue color forms, *uric acid is present.*

ALTERNATIVE PROCEDURES FOR OXALATE

1. Heat a small amount of pulverized stone briefly in a porcelain dish. Cool and place the residue in a 12 × 75 mm test tube. Add approximately 2 ml of 0.6 molar HCl. If effervescence occurs, and there was *no* effervescence in the carbonate procedure, *oxalate is present.** Moderate heat converts oxalate to carbonate, which upon acidification yields CO_2. Prolonged intense heat will convert the oxalate to oxide, and will invalidate this test.

2. Add 2 ml of 0.6 molar HCl to a small portion of the powdered stone. Add a pinch of MnO_2. Do not shake the tube. MnO_2 oxidizes the oxalate to CO_2, which can be seen as tiny bubbles rising from the sediment. It may be necessary to heat the tube slightly to obtain the reaction if only trace quantities of oxalate are present.

ANALYSIS OF SMALL CALCULI

Small stones, many less than 2 mm in diameter, are often submitted for analysis. If the sample is not adequate for a complete stone analysis, the tests for composition should be performed either according to the frequency of their occurrence as noted in Table 18–1, or on

* If effervescence was obtained in the carbonate procedure, effervescence would also occur here; therefore, oxalate cannot be determined by this method if carbonate is present.

the basis of clinical judgment (e.g., in a patient with a history of triple phosphate stones, check for $MgNH_4PO_4$; in patients with hyperparathyroidism, check for calcium; in patients with hyperuricemia, check for uric acid). When small amounts of sample are used, the volume of the reagents should be reduced accordingly, as false negative results may be obtained if the dilution is too great.

An alternative approach for the analysis of small stones is to solubilize the crushed calculus in approximately 0.5 ml of 0.6 molar HCl. If there is effervescence, *carbonate is present*. Heat the HCl solution to solubilize the constituents. Centrifuge. Use a microdropper to place 2 drops of supernatant into each of four wells of a spot plate. Use drop portions of the reagents and perform the test for calcium and the alternative tests for ammonia, magnesium, and phosphate already given. Place 1 or 2 drops of the HCl calculus solution on a microscope slide, evaporate slightly, and examine microscopically for cystine crystals. Transfer the remaining solution, and any undissolved calculus, into an evaporating dish and perform the murexide test for uric acid.

Several kit procedures for stone analysis are commercially available. These procedures may be useful for the analysis of small stones since, in some kit methods, drop amounts of the reagents are used with very small portions of the calculus.

MICROSCOPIC EXAMINATION OF CRYSTALS

The microscopic examination of crystals has been used to a limited extent for the identification of some constituents of calculi. This technique is perhaps most useful as a confirmatory test for cystine. Triple phosphate, calcium carbonate, and calcium oxalate crystals may sometimes be recognized; however, the various calcium salts are difficult to identify and may appear as amorphous material.

The method is as follows: a small portion of the powdered stone is dissolved in 6 molar HCl on a microscope slide. A portion of the liquid is evaporated and the residue is examined immediately under the low power of a microscope. The crystals are identified on the basis of a comparison to known crystalline preparations. When this technique is used as a confirmatory test for cystine, a small portion of the powdered calculus is dissolved in 0.75 molar NH_4OH (10 ml conc. NH_4OH/100 ml) on a microscope slide. Cystine crystals are seen as hexagonal plates (six-sided flat crystals).

BILIARY CALCULI

The following tests for cholesterol and bilirubin are performed only on gallstones.

CHOLESTEROL

Extract a portion of the stone with approximately 3 ml of an absolute alcohol–ether mixture (1/1). Add 0.5 ml of the sulfuric acid–acetic anhydride reagent (see serum cholesteryl–ester procedure). If a green color develops, *cholesterol is present*. Cholesterol reacts with sulfuric acid and acetic anhydride to form a green colored compound.

BILIRUBIN

Extract a portion of the stone with several ml of methanol. Add 0.5 ml of diazotized sulfanilic acid (see serum bilirubin procedure). If a violet color develops, *bilirubin is present*. Bilirubin couples with diazotized sulfanilic acid to form azobilirubin.

REFERENCES

1. Beeler, M., Veith, D., Morriss, R., and Biskind, G.: Analysis of urinary calculus. Am. J. Clin. Path., *41*:553, 1964.

2. Boyce, W., and King, J. S.: Crystal-matrix interrelations in calculi. J. Urol., *81*:351, 1959.
3. Cantarow, A., and Trumper, M.: Clinical Biochemistry. 6th ed. Philadelphia, W. B. Saunders Company, 1962.
4. Henry, R. J.: Clinical Chemistry. Principles and Technics. New York, Harper and Row, Publishers, 1964.
5. Herring, L.: Observation on the analysis of ten thousand urinary calculi. J. Urol., *88*:545, 1962.
6. Kachmar, J.: Personal communication.
7. King, J. S., Jr.: Currents in renal stone research. Clin. Chem., *17*:971, 1971.
8. Kolmer, J., Spaulding, E., and Robinson, H.: Approved Laboratory Technic. 5th ed. New York, Appleton-Century-Crofts, Inc., 1951.
9. Lonsdale, K.: Human stones. Science, *159*:3820, 1968.
10. McIntosh, J. F., and Salter, R. W.: The qualitative examination of urinary calculi. J. Clin. Invest., *21*:751, 1942.
11. Oser, B. L.: Hawk's Physiological Chemistry. 14th ed. New York, McGraw-Hill Book Co., Inc., 1965.
12. Prien, E. L.: Studies in urolithiasis: II Relationship between pathogenesis, structure and composition of calculi. J. Urol., *61*:821, 1949.
13. Prien, E. L., and Frondel, C.: Studies in urolithiasis: Composition of urinary calculi. J. Urol., *57*:949, 1947.
14. Reiner, M., Cheung, H. L., and Thomas, J. L.: Calculi. *In* Standard Methods of Clinical Chemistry. R. P. MacDonald, Ed. New York, Academic Press, 1970, pp. 193–214.
15. Weissman, M., Klein, B., and Berkowitz, J.: Clinical applications of infrared spectroscopy, analysis of renal tract calculi. Anal. Chem., *31*:1334, 1959.
16. Winer, J.: Practical value of analysis of urinary calculi. J.A.M.A., *169*:1715, 1959.
17. Zinsser, H.: Urinary calculi. J.A.M.A., *174*:116, 1960.

CHAPTER 19

LIVER FUNCTION

by Joseph I. Routh, Ph.D.

The liver is a large organ (1200 to 1600 g) shaped like a wedge with its base to the right. It has two major lobes, the larger right one being about six times the size of the left in the normal adult. The right lobe lies behind the rib cage with its upper border at about the level of the fifth rib. The smaller left lobe is tongue-shaped and its apex reaches to the dome of the left diaphragm (Fig. 19–1). The circulatory system of the liver is characterized by a dual blood supply. Approximately 80 per cent of this blood supply enters from the portal vein and 20 per cent from the hepatic artery. This rich flow of blood, amounting to about 1500 ml per minute, brings oxygen and nutrient material to the liver. There is also an extensive system of lymph vessels that carries the interstitial fluid or liver lymph from that organ into the general circulation. In addition, a network of bile ducts transports substances secreted or excreted by the liver cells for storage in the gallbladder and subsequent use in the digestive process. The circulatory system will be examined in greater detail in connection with the function of liver cells and the biliary system.

The liver is composed of a multitude of functional units called *lobules*. A branch of the hepatic vein forms the central core of the lobule, which is surrounded by *parenchymal cells* arranged in plates or sheets one cell thick (Fig. 19–2). These plates or sheets form the walls of spaces or *lacunae*, which are divided into two systems of irregular tunnels containing the portal tracts and the central hepatic vein. In most instances the two systems of tunnels run in planes perpendicular to each other. The portal tracts contain branches of the portal vein, the hepatic artery, bile ducts, and lymph vessels. The portal vein and the hepatic

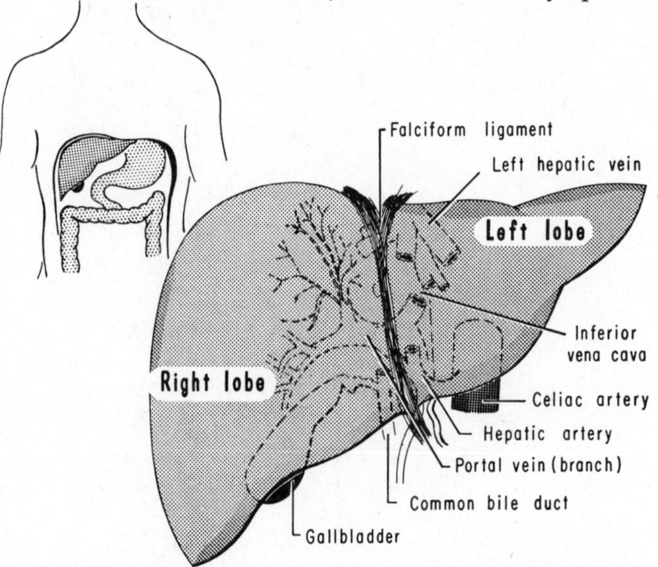

Figure 19–1 Anterior view of the liver, showing major blood and bile vessels.

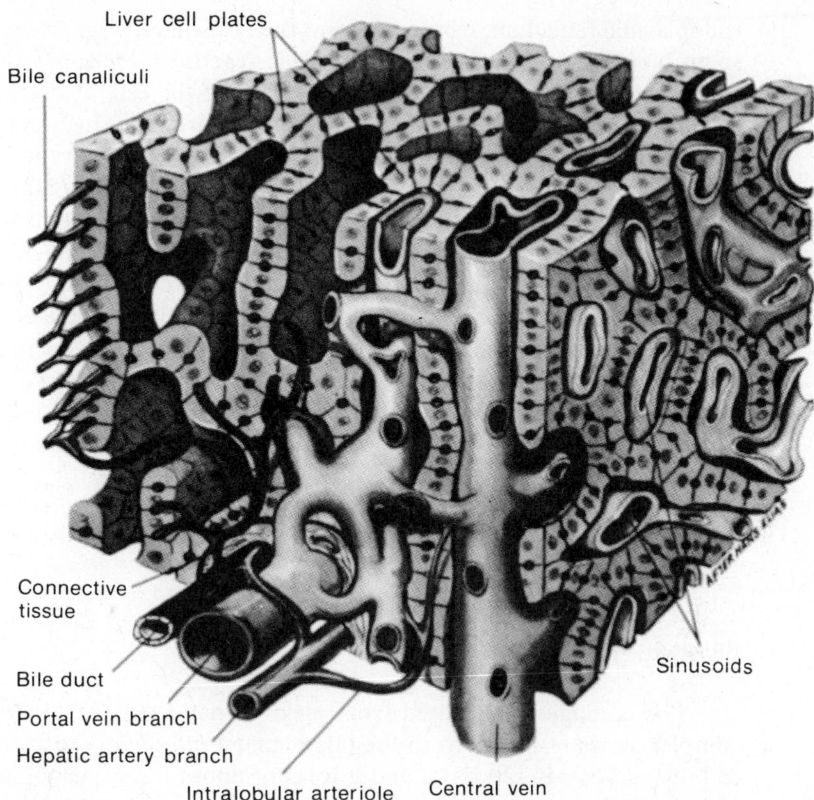

Figure 19–2 Illustration of a section of a lobule of the liver, showing the plates of parenchymal cells, the lacunae, sinusoids, space of Disse, and bile canaliculi, and their relation to the central vein and portal tract.

artery supply the blood that flows through special capillaries called *sinusoids*, which empty into the central hepatic vein at right angles. The walls of the sinusoids are lined with reticuloendothelial cells, called *Kupffer cells*, which are phagocytic. Between the cell plates and the sinusoids are spaces, the *space of Disse*, which contain interstitial fluid that functions in the transfer of nutrients and waste products between the liver cells and the blood. Small bile *canaliculi* or bile capillaries are formed by grooves in the contact surfaces of liver cells that fit together to form a cylindrical tube. The canaliculi form polygonal networks that surround each liver cell and carry the bile to small ductules; the ductules combine to make up the larger bile ducts.

The gross appearance of a cut section of the liver would be a pattern of red dots representing the central veins of the lobules, separated by greyish dots that represent the portal tracts, all against the brown background of liver tissue. The central veins are from 0.3 to 1.0 mm apart and are a measure of the diameter of the lobule. The liver cells are approximately polyhedral in shape, with a diameter of about 30 μm. Each gram of liver tissue contains about 200 million cells, of which 170 million are parenchymal and 30 million are sinusoidal epithelial cells (Kupffer cells).

In recent years, attention has been focused on the specific function of the subcellular components of the *parenchymal cells* of the liver. As is well known, the liver is the major site of intermediary metabolism and synthesis of many important compounds, the site of conjugation and detoxication of natural and potentially toxic foreign substances, and the site of storage of glycogen. In addition, synthesis and destruction of many of the factors involved in the blood clotting mechanism take place in the liver. These functions are carried out by subcellular components such as the nucleus, mitochondria, Golgi apparatus, endoplasmic reticulum, ribosomes, and the lysosomes (Fig. 19–3). The *mitochondria* are responsible for a large number of metabolic reactions, especially those that provide energy, and are thus often called the "powerhouses" of the cell. They contain many enzymes and coenzymes that are essential in the pathways of carbohydrate, lipid, and protein metabolism. A very important function of the mitochondria is their ability to transform energy released from food into the energy-rich bonds of ATP by the process of oxidative phosphorylation. The *Golgi apparatus* is believed to function in the storage, transportation, and secretion of molecules such as proteins and bilirubin formed by the microsomal components. The *microsomal fraction* (obtained by centrifugation) is composed of free and bound ribosomes and pieces of rough and smooth endoplasmic reticulum. These components are responsible for the activation of amino acids prior to protein synthesis, the process of protein synthesis itself, cholesterol synthesis, the conjugation of bilirubin, and the detoxication of drugs. *Lysosomes* contain several hydrolytic enzymes that digest cellular material or debris brought into the cell by phagocytosis. Lysosomes are also involved in the metabolism of iron, bile pigments, and copper. The *Kupffer cells* (Fig. 19–3) contain a nucleus, mitochondria, and lysosomes, and sometimes phagocytized material. They function as part of the reticuloendothelial system of the body and are especially involved in bilirubin metabolism.

Bilirubin Metabolism

The hemoglobin released from aged or damaged red blood cells is converted by a complex series of reactions to the bile pigment *bilirubin*. The average life of the red blood cell in the body is 120 days, and a total of about 6 g of hemoglobin is released each day from the disintegration of overaged cells. The cells of the reticuloendothelial system, especially in the spleen, liver, and bone marrow, first phagocytize the red cells and then convert the released hemoglobin into bilirubin. Hemoglobin consists of a protein, *globin*, attached to *heme* groups, which are porphyrin compounds containing iron. Each heme molecule is composed of four pyrrole rings joined by methene bridges and containing a central iron atom (see Chap. 8). One of the methene bridges of heme in hemoglobin appears to be split first by the reticuloendothelial cells to produce a biliverdin-iron-protein complex

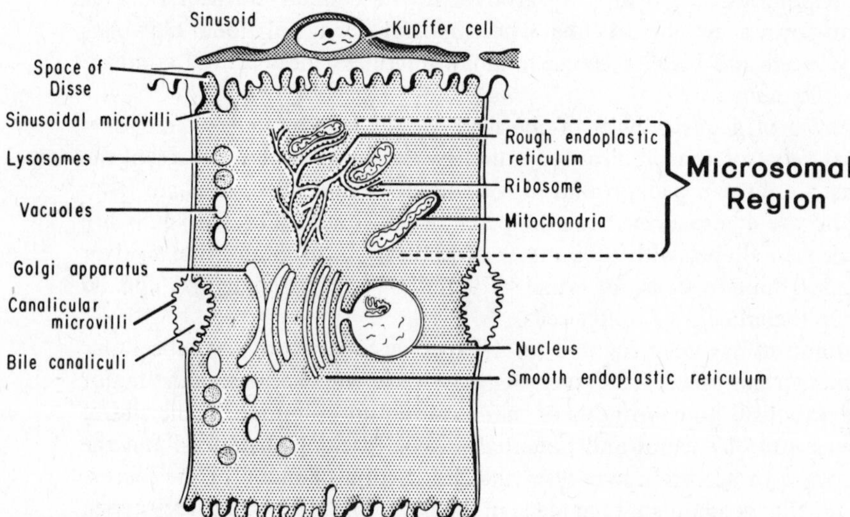

Figure 19–3 Structure and subcellular components of a normal liver cell.

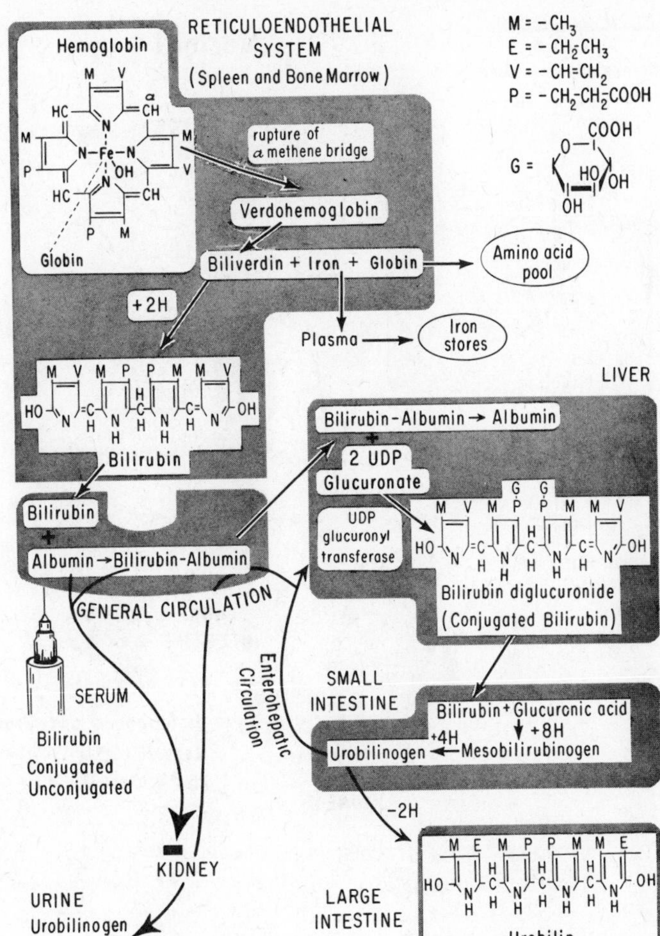

Chart 19–1 The normal metabolism of bilirubin.

known as *verdohemoglobin* or choleglobin. The iron is then removed, the globin is released, and the biliverdin is reduced to form bilirubin (see Chart 19–1). The bilirubin is then transported from the extrahepatic sources as a *bilirubin-albumin complex* into the hepatic sinusoids. From the sinusoids, the bilirubin complex passes through the sinusoidal microvilli into the liver cell. The protein is separated from the complex, and bilirubin is converted into *bilirubin diglucuronide* by a reaction with uridine diphosphate glucuronate catalyzed by the enzyme system *UDP-glucuronyl transferase.* Unconjugated bilirubin and bilirubin glucuronides are loosely bound to albumin and transported in this form in the blood (see Chart 19–1 and Figs. 19–4 and 19–5).

It was concluded from the original experiments involving electrophoresis and chromatography of fresh bile that both mono- and diglucuronides of bilirubin were present in normal bile. Subsequent studies have shown that "Pigment I," the monoglucuronide, is a complex of unconjugated bilirubin and bilirubin diglucuronide, and that the monoglucuronide is not formed in the liver in the reaction catalyzed by UDP-glucuronyl transferase.

Conjugation to the glucuronide is a prerequisite for excretion of bilirubin into the bile and may be a limiting factor in transportation because conjugated bilirubin, when injected into the blood, is excreted in the bile much more rapidly than the unconjugated pigment. There are some clinical conditions in which excretion is abnormal despite a normal conjugating system, which suggests that an active transport system may be required for excretion. From experimental work with fluorescent dyes similar to bilirubin, it has been

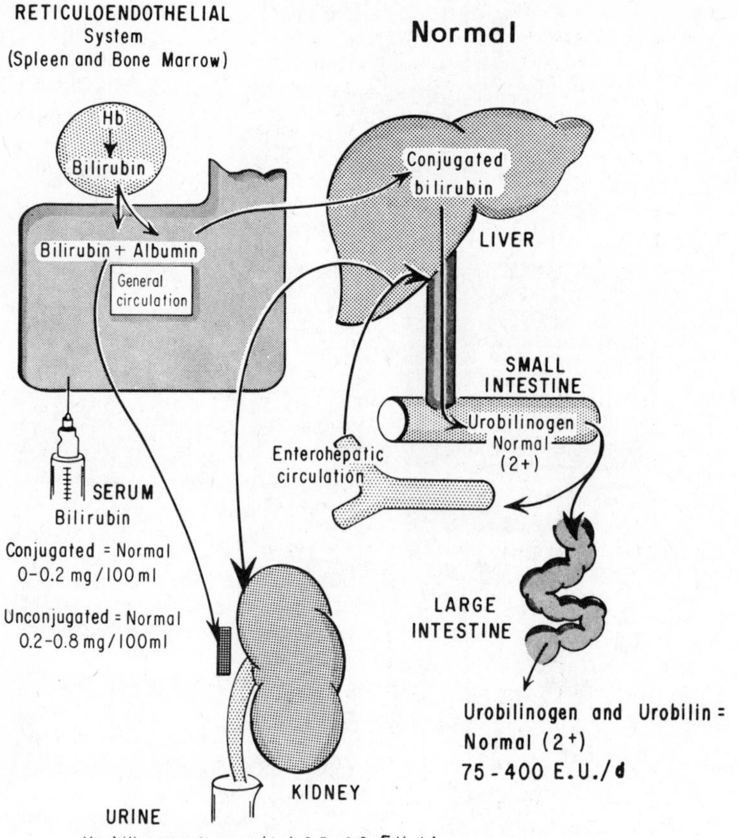

Figure 19-4 Metabolism and excretion of bilirubin in a normal individual.

postulated that after passing into the liver cell, bilirubin is first concentrated at the sinusoidal membrane surface and is then moved across the cell and concentrated at the canalicular membrane surface before excretion into the bile. The unidirectional nature of this process is vital; disturbances of these mechanisms produce leaks from the bile canaliculi back into

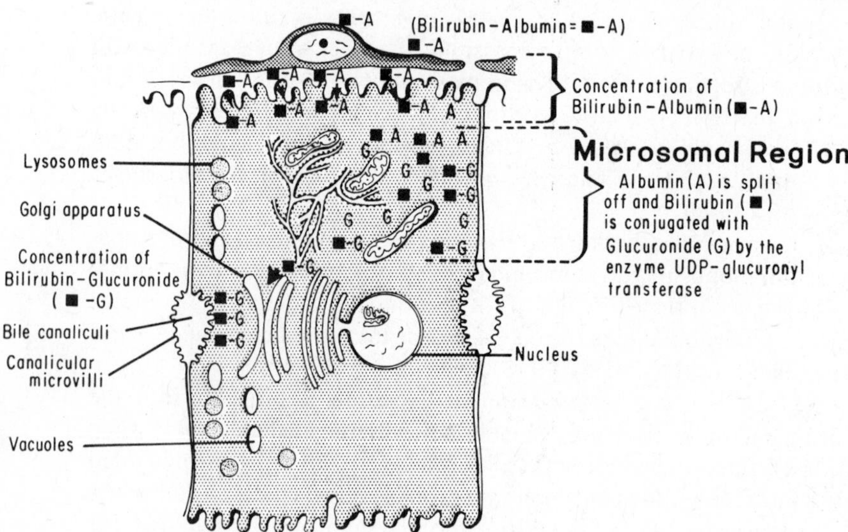

Figure 19-5 Normal transport of bilirubin from the sinusoids to the bile canaliculi.

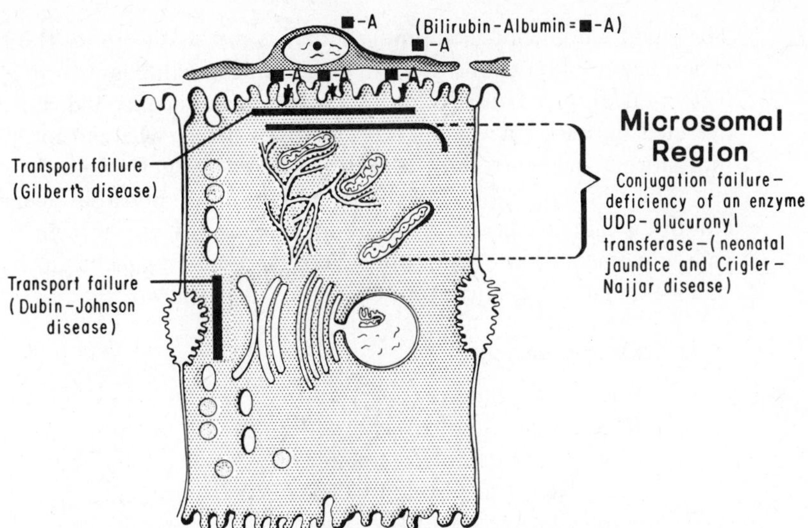

Figure 19–6 Impairment of transport and conjugation of bilirubin in disease.

the sinusoids, resulting in jaundice. Pinocytotic vacuoles, lysosomes, and the Golgi apparatus have all been implicated in the transportation of bilirubin from the sinusoids to the bile canaliculi. A representation of this process is illustrated in Figure 19–5 (see also Fig. 19–6).

The glucuronide, along with some free bilirubin, is excreted into the bile, passes into the small intestine, and is exposed to the reducing action of the enzymes of anaerobic bacteria. The reduction products, mesobilirubinogen, stercobilinogen, and urobilinogen, known collectively as *urobilinogen*, are first formed in the colon; then up to 50 per cent of the urobilinogen is absorbed into the portal circulation and returned to the liver. The normal liver removes all but a small amount of the urobilinogen from the blood, probably oxidizes some of it to bilirubin, and excretes both into the bile for a return trip to the colon. A portion of the urobilinogen remaining in the colon is oxidized by microorganisms to form urobilin (stercobilin), an orange-brown colored pigment which is excreted in the stool along with the unchanged urobilinogen. The scheme of bilirubin metabolism from the breakdown of hemoglobin to the formation of urobilin is shown in Chart 19–1.

JAUNDICE

Before consideration of the three major types of jaundice, the normal fate of bilirubin and urobilinogen should be illustrated. Figure 19–4 is a representation of the events that occur in the metabolism of bilirubin and the disposal of its compounds by the normal individual. The reticuloendothelial cells in the spleen and bone marrow convert the hemoglobin of red blood cells into bilirubin, which is loosely bound to albumin and carried to the liver, where the Kupffer cells may also produce more bilirubin from hemoglobin. After passage of the bilirubin-protein complex into the parenchymal liver cells, the protein is split from the complex and bilirubin is converted into the diglucuronide by enzyme systems in the microsomal region and is then passed into the bile canaliculi. The bile is transported to the common bile duct and then to the gallbladder, where it is concentrated and emptied into the small intestine. The bilirubin glucuronide is thus carried to the large intestine and, under the influence of the reducing enzymes of the bacteria in the colon, is changed into urobilinogen. Most of this urobilinogen is oxidized to D-, I-, and L-urobilin and excreted in the stool. The other part is carried back to the liver by the portal vein and may be either converted by the liver cells to bilirubin glucuronide or excreted into the bile canaliculi unchanged. The glucuronides and any urobilinogen that escapes transformation are carried back to the intestine by the bile and thus complete the enterohepatic circulation.

The small amount of urobilinogen that is not taken up by the liver cells is carried in the general circulation to the kidney. Urobilinogen will pass through the kidney, and normally 0.5 to 4.0 Ehrlich Units (0.5 to 3.5 mg) per day is excreted. Unconjugated bilirubin is not excreted by the kidney and is absent from urine, whereas conjugated bilirubin, if present in abnormal concentration, is excreted by the kidney tubules.

The abnormal metabolism or retention of bilirubin usually results in *jaundice*, a condition that is characterized by an increase of bilirubin in the blood and a brownish-yellow pigmentation of the skin, sclera, and mucous membranes. One of several suggested schemes for the classification of jaundice is as follows:

Types of jaundice

A. Prehepatic jaundice
   (Hemolytic jaundice)
   - a. Acute hemolytic anemia
   - b. Neonatal physiologic jaundice
   - c. Chronic hemolytic anemia

B. Hepatic jaundice
   1. Conjugation failure
      - a. Neonatal physiologic jaundice
      - b. Crigler-Najjar disease*
   2. Bilirubin transport disturbances
      - a. Preconjugation transport failure (Gilbert's disease)*
      - b. Postconjugation transport failure (Dubin-Johnson disease)
   3. Diffuse hepatocellular damage or necrosis
      - a. Viral hepatitis
      - b. Toxic hepatitis
      - c. Cirrhosis
   4. Intrahepatic obstruction (e.g., edema)

C. Posthepatic jaundice: obstruction of the common bile duct
   1. By stones
   2. By neoplasms
   3. By spasms or stricture

The *prehepatic type of jaundice* is commonly caused by hemolytic anemia. The increased destruction of red cells brings a larger load of bilirubin-protein complex to the liver than the organ can handle (Fig. 19–7). When this additional bilirubin is converted to the glucuronide and excreted into the intestinal tract, an increased amount of urobilinogen is formed in the colon. A portion of this is excreted in the stool; the other portion is reabsorbed and returned to the liver. The liver cannot pick up the large quantities of urobilinogen; thus, increased amounts escape into the general circulation and are excreted into the urine by the kidneys. The increased breakdown of hemoglobin also causes an increase in the amount of unconjugated bilirubin in the blood without a corresponding increase in conjugated bilirubin.

The determination of urobilinogen in urine and stool are useful clinical tests in the diagnosis of *hemolytic anemia*. In this disorder, the 24 h urine specimen may contain from 5 to as much as 350 Ehrlich Units (E.U.) of urobilinogen, and the amount excreted in the feces may range from 350 to 1800 E.U./d. These values are considerably higher than the normal 0.5 to 4 E.U. for urine and 75 to 400 E.U./d for feces.

---

* Some authors have classified conditions in which there is an inability to convert bilirubin to glucuronides as "prehepatic"; thus, the Crigler-Najjar disease and Gilbert's disease are, in some classifications, listed in this group.

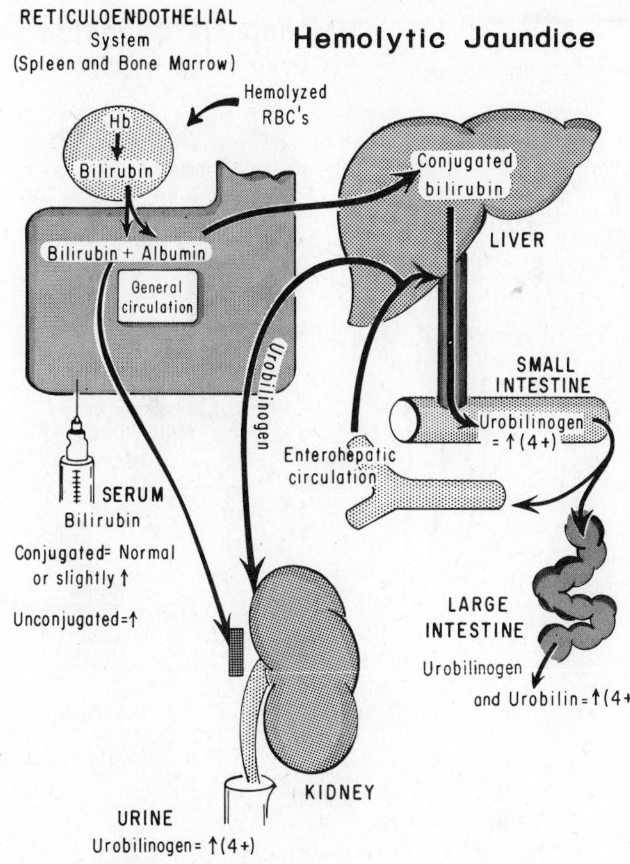

Figure 19–7 Metabolism and excretion of bilirubin in hemolytic jaundice.

There are many cases of jaundice that involve the liver cell directly. There may be specific defects, as will be mentioned, or there may be diffuse damage such as in viral hepatitis, drug toxicity, and intrahepatic obstruction. The posthepatic type of jaundice usually involves the blockage of the common bile duct by a stone, stricture, or neoplasm.

Impairment of the liver cell mechanisms can result in several types of hepatic jaundice. As shown in Figure 19–6, the *transportation of bilirubin* or its protein complex from the sinusoidal membrane to the microsomal region may be impaired in the familial type of nonhemolytic jaundice known as *Gilbert's disease*. A deficiency of the enzyme system involving UDP-glucuronyl transferase is the cause of the hyperbilirubinemia of the *Crigler-Najjar disease*. This enzyme deficiency is also responsible for the lack of conjugation of bilirubin in the *physiologic jaundice* of the newborn, especially premature infants. This last mentioned type of jaundice is ordinarily short-lived; the deficient enzymes appear soon after the first few days of life, and there is also a concomitant decrease in hemoglobin destruction. Levels of unconjugated bilirubin may reach 10 mg/100 ml of serum at the height of the physiologic jaundice. A disturbance of the transportation system that carries the bilirubin glucuronides from the microsomes to the bile canaliculi is involved in the *Dubin-Johnson* type of hyperbilirubinemia, whereas the jaundice caused by the drug chlorpromazine results from intrahepatic cholestasis.

Viral hepatitis, *toxic hepatitis*, and *cirrhosis* involve overall damage and necrosis of the liver cells and produce jaundice as illustrated in Figure 19–8. The injured cells lose their ability to remove bilirubin from the serum and to conjugate bilirubin with glucuronate; thus, an increase in unconjugated bilirubin occurs in the serum. The parenchymal cells

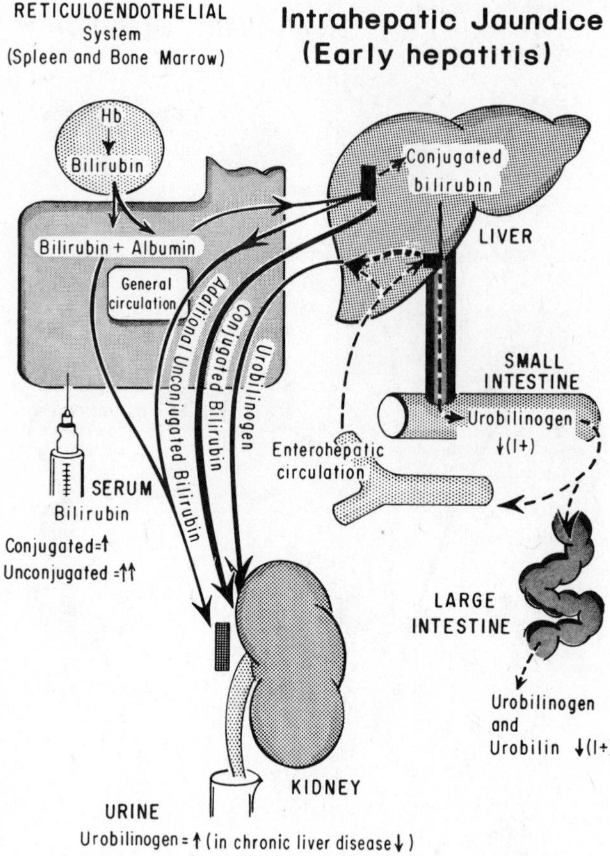

Figure 19–8 Metabolism and excretion of bilirubin in intrahepatic jaundice.

that are damaged provide a path for the leakage of bilirubin glucuronide back into the sinusoids, which causes an increase in conjugated bilirubin as well. Furthermore, the damaged cells are not capable of removing all of the urobilinogen from the portal blood in the enterohepatic circulation and more of this compound reaches the general circulation. The kidney excretes the excess urobilinogen and the bilirubin glucuronides, and appreciable quantities of these compounds are found in the urine. Urine urobilinogen determinations are useful in the detection of early viral hepatitis, whereas in later stages of the disease the formation and passage of bilirubin glucuronide into the bile is decreased to such an extent that urobilinogen formation, reabsorption, and urinary excretion also decrease.

Obstruction of the common bile duct, if complete, produces a condition illustrated in Figure 19–9. A gallstone, spasm, or neoplasm that blocks the passage of bile into the intestine causes regurgitation of bile into the sinusoids. The bilirubin glucuronides cannot reach the intestine and, therefore, no urobilinogen is produced for recirculation to the liver or excretion in the stool. Urobilin, the pigment of the stool, is lacking and the feces vary from a light brown to a chalky white color. The urine does not contain any urobilinogen, but does contain appreciable amounts of bilirubin glucuronides. Since there are conditions in which the posthepatic blockage is intermittent or incomplete, the stools may contain decreased amounts of urobilinogen, and small quantities of this compound may appear in the urine. The blood contains increased amounts of bilirubin and bilirubin glucuronides caused by regurgitation and impairment of liver cell function. In a severe case of *obstructive jaundice*, the serum may contain 12 mg/100 ml of conjugated bilirubin and a total bilirubin level of 18 mg/100 ml or higher.

Methods for the Evaluation of Liver Function

ICTERUS INDEX

A patient with jaundice is sometimes referred to as icteric. The icterus index was developed by Meulengracht in 1919 as a measure of the degree of icterus in a plasma or serum specimen in cases of jaundice. In this test, serum or plasma is diluted with sodium citrate solution and the yellow color is read spectrophotometrically at a wavelength of 460 nm. The results are expressed in units obtained by comparison with a standard potassium dichromate solution.

Substances other than bilirubin in the serum (e.g., carotene, xanthophyll, hemoglobin) may contribute to the icterus index, thus limiting its clinical usefulness; in fact, this test is now considered obsolete.

REFERENCE

Henry, R. J., Golub, O. J., Berkman, S., and Segalove, M.: Am. J. Clin. Path., *23*:841, 1953.

SERUM BILIRUBIN

In 1883, Ehrlich described a reaction by which bilirubin is coupled with diazotized sulfanilic acid to form a blue pigment in strongly acid or alkaline solutions. Van den Bergh applied this color reaction to the quantitative determination of serum bilirubin and reported the effect of alcohol on the rate of the coupling reaction. Malloy and Evelyn in

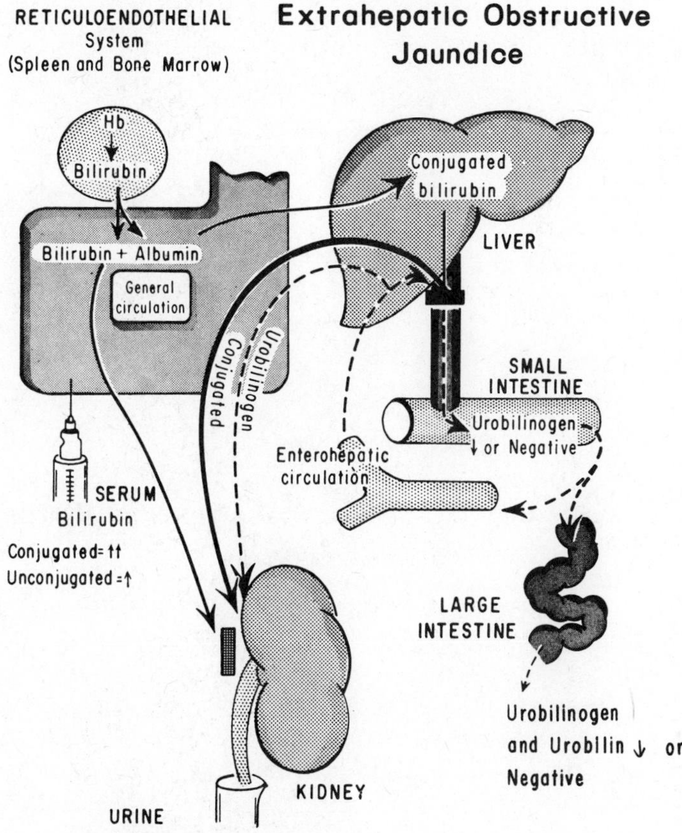

Figure 19–9 Metabolism and excretion of bilirubin in extrahepatic obstructive jaundice.

1937 introduced the use of 50 per cent methanol in the serum determination, a concentration that avoided precipitation of the proteins. These authors developed the first useful quantitative method. For many years, results of bilirubin determinations in serum were reported as values for direct bilirubin (the fraction that produced a color in the van den Bergh method in aqueous solution) and indirect bilirubin (the fraction that produced a color only after alcohol was added). Since 1956, it has been known that the "direct" reaction is given by the diglucuronide of bilirubin, so-called *conjugated bilirubin*, which is water soluble. The "indirect" reaction, on the other hand, is given by *unconjugated bilirubin*, which is water insoluble but dissolves in alcohol to couple with the diazotized sulfanilic acid.

The reactions involved in the coupling of bilirubin with Ehrlich's reagent are shown in Chart 19–2.

The literature is filled with discussions and modifications of the van den Bergh method, especially that proposed by Malloy and Evelyn. Compounds other than methanol have been used to "accelerate" the color reactions of bilirubin, and considerable attention has been paid to the time period required to assure complete reaction of the glucuronide in aqueous solution. Also, it has been established that hemoglobin interferes with the diazo reaction and depresses the values for total bilirubin in hemolyzed serum specimens commonly obtained from newborn infants. Therefore, it is desirable to have available a method for total bilirubin that is not affected by hemolysis and that can be carried out on the small specimens of serum obtained from newborns.

It should be noted that azobilirubin has indicator properties and in strongly acid or strongly alkaline solution its color is blue. This blue color is more intense than the familiar purple color produced at moderately acid pH and is less affected by nonbilirubin pigments. Stoner and Weisberg[18] developed a micromethod that employs a strongly acid pH, while in the method of Jendrassik and Grof the final blue color is developed at a strongly alkaline pH.

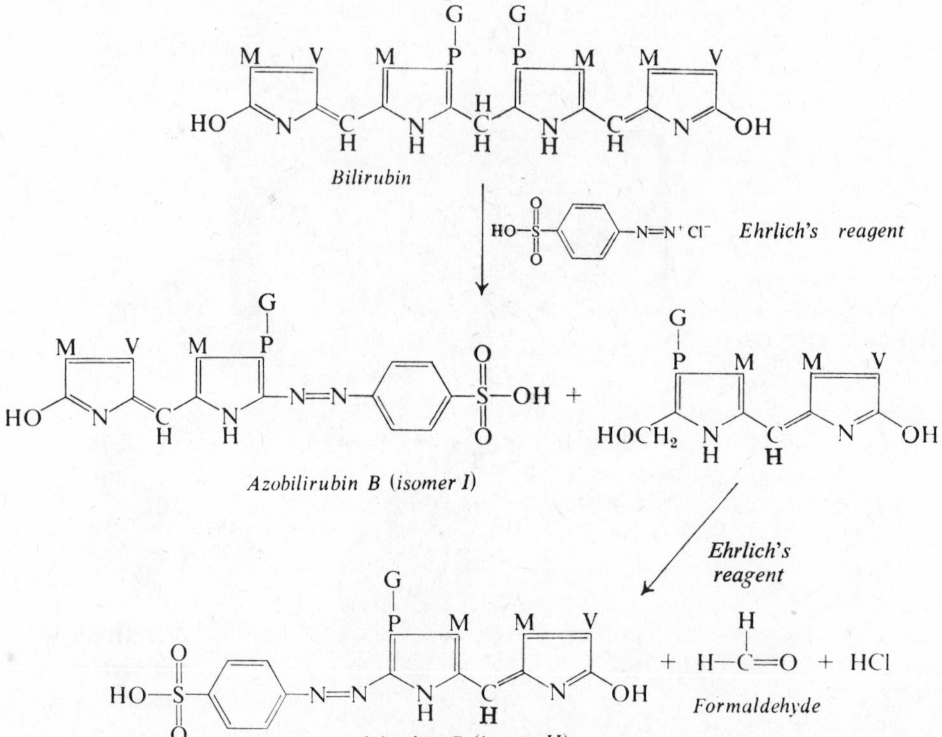

Bilirubin

Ehrlich's reagent

Azobilirubin B (isomer I)

Ehrlich's reagent

$+ \ H{-}\overset{H}{\underset{|}{C}}{=}O \ + \ HCl$

Formaldehyde

Azobilirubin B (isomer II)

Chart 19–2 The formation of isomers I and II of azobilirubin B through the reaction of bilirubin glucuronide with Ehrlich's reagent. Unconjugated bilirubin reacts in the same way, resulting in isomers I and II of azobilirubin A.

Results of a recent bilirubin proficiency testing program conducted by the Center for Disease Control indicated that when bilirubin was added to cord blood only 50% recovery was obtained by the Malloy and Evelyn method, compared to over 90% by the Jendrassik and Grof method. Both methods will be described in this chapter since they are commonly used in clinical laboratories. The Jendrassik and Grof method, however, should be considered the method of choice.

CLINICAL SIGNIFICANCE

As discussed earlier in the chapter, in *obstructive jaundice* there is an increase in the total bilirubin, but this increase is primarily in the bilirubin glucuronide. In *hemolytic jaundice* and in *neonatal jaundice*, there is an increase primarily in the unconjugated ("indirect") bilirubin fraction. In the newborn, jaundice may be caused by *Rh*, *ABO*, or other *blood group incompatibilities*, by *hepatic immaturity*, or by *hereditary defects* in bilirubin conjugation (see Types of Jaundice). Both conjugated and unconjugated bilirubin are increased in the serum in *hepatitis*. The relative proportion of the conjugated fraction increases with progression of the disease until eventually the liver loses its ability to carry out the conjugation reaction.

DETERMINATION OF CONJUGATED AND TOTAL BILIRUBIN IN SERUM

SPECIMEN

Serum is preferably obtained from a blood specimen collected in the postabsorptive state to avoid lipemia. Hemolysis must also be avoided, since it causes falsely low results in the diazo method. Before analysis the serum should be kept in the dark, and the determination should be carried out as soon as possible and not later than 2 to 3 h after clotting of the blood. Direct sunlight may cause up to 50 per cent decrease in bilirubin within 1 h, especially when serum is kept in capillary tubes. Serum specimens may be kept in the dark in a refrigerator for up to a week and in the freezer for 3 months without appreciable change in the bilirubin content.

The Jendrassik and Grof Method

SPECIMEN

Either serum or plasma may be used, and the precautions suggested above should be followed.

PRINCIPLE

For the determination of *total bilirubin* in serum or plasma, the specimen is added to a solution of sodium acetate and caffeine-sodium benzoate. The sodium acetate buffers the pH of the diazotization reaction, while the caffeine-sodium benzoate accelerates the coupling of bilirubin with diazotized sulfanilic acid. The diazotization reaction is terminated by the addition of ascorbic acid, which destroys the excess diazo reagent. A strongly alkaline tartrate solution is then added to convert the purple azobilirubin to blue azobilirubin, and the intensity of the color is read at 600 nm.

For the determination of *bilirubin glucuronide* (conjugated or direct bilirubin) in serum or plasma, the specimen is first added to a dilute acid solution, followed by the diazo reagent. The absence of accelerator allows only for the reaction of conjugated bilirubin with diazotized sulfanilic acid. The reaction is terminated by ascorbic acid, and the azobilirubin is made alkaline by the addition of the tartrate solution. In some other methods, after the reaction is terminated by ascorbic acid, caffeine reagent is added to shift the absorption peak of the azobilirubin from 585 nm to the 600 nm wavelength at which the color is measured.

The azobilirubin color produced by bilirubin glucuronides is developed for 1 min, whereas that from total bilirubin is essentially completely developed in 10 min. After the addition of ascorbic acid and the tartrate solution, the color is stable for 30 min. By reading the blue azobilirubin color at 600 nm, the effect of the nonbilirubin yellow, red, and brown pigments in the specimen is minimized.

REAGENTS

1. Hydrochloric acid, 0.05 mol/l.
2. Caffeine reagent. Add 50 g of caffeine, 75 g of sodium benzoate and 125 g of sodium acetate, $CH_3COONa{\cdot}3H_2O$, to 400 to 500 ml of warm (50 to 60°C) distilled water. When cool, add 1 g Na_2 EDTA and dilute to 1 liter. The reagent is stable for at least 6 months at room temperature.
3. Reagent A. Add 5.0 g of sulfanilic acid to 15 ml of concentrated hydrochloric acid in a 1 liter volumetric flask and dilute to the mark with distilled water. The solution is stable indefinitely.
4. Reagent B. Dissolve 500 mg of sodium nitrite, $NaNO_2$, in distilled water and dilute to 100 ml. This reagent is stable up to 2 weeks in the refrigerator.
5. Diazo reagent. Mix 10.0 ml of Reagent A with 0.25 ml of Reagent B. This reagent should be prepared within 30 min of the time of use.
6. Ascorbic acid solution, 4 g/100 ml. Dissolve 200 mg of ascorbic acid in 5 ml of water. Prepare fresh daily and keep in refrigerator.
7. Tartrate solution. Add 100 g of sodium hydroxide and 350 g of sodium tartrate, $Na_2C_4H_4O_6{\cdot}2H_2O$, to 400 to 500 ml of distilled water. When cool, dilute to 1 liter. The reagent is stable for at least 6 months at room temperature.

PROCEDURE

Conjugated bilirubin

1. Add 1.0 ml of 0.05 molar HCl to a cuvet (12 × 75 mm round or 1.00 cm rectangular cuvet) labeled C (for conjugated) and 1.0 ml of caffeine reagent to a cuvet labeled B (for blank).
2. Add 100 μl of serum or plasma to each cuvet, followed by 150 μl of water.
3. Add 250 μl of sulfanilic acid solution to cuvet B and mix.
4. Add 250 μl of fresh diazo reagent to cuvet C and mix.
5. Exactly 1 min* later, add 50 μl of ascorbic acid solution to cuvets B and C; then immediately add 0.5 ml of tartrate solution to each cuvet and mix.
6. Read the absorbance of cuvet C at 600 nm against cuvet B at zero absorbance.

Total bilirubin

1. Add 1.0 ml of caffeine reagent to a cuvet labeled T (for total).
2. Add 100 μl of serum or plasma to the cuvet, followed by 150 μl of water. (If the total bilirubin is expected to be in the normal range, up to 250 μl of serum or plasma may be used for more accurate results. If the latter procedure is followed, adjust the dilution with water to give a total volume of 250 μl.)
3. Add 250 μl of fresh diazo reagent to the cuvet and mix.
4. Exactly 10 min after the addition of diazo reagent in step 3, add 0.5 ml of tartrate solution to the cuvet and mix.
5. Read the absorbance of cuvet T at 600 nm against cuvet B (from step 5 in the conjugated bilirubin method) set at zero absorbance.

---

* Some investigators prefer a 5 or 10 min waiting period for color development in the direct method. Doumas, for example, found a 1 min waiting period satisfactory for normal or slightly elevated conjugated bilirubin values, but suggested 5, 10, or even 20 min periods for abnormally elevated concentrations.

CALIBRATION CURVE

It has been recommended by a committee composed of representatives from the American Academy of Pediatrics, the College of American Pathologists, the American Association of Clinical Chemists, and the National Institutes of Health that bilirubin acceptable for the preparation of calibration curves should give $E^{25°C}_{453nm} = 60{,}700 \pm 1600$ in chloroform. This means that a 1 molar solution of bilirubin should yield an absorbance (A) reading of 60,700 or, for practical spectrophotometry, a 1/60,700 molar solution (9.63 mg/l) should have an A value of 1.0. In addition, the committee recommended the preparation of *bilirubin standards in serum*. (Others have recommended that the azobilirubin formed with this standard must also meet specified absorbance values.) Bilirubin standards in serum require an acceptable serum diluent, which is prepared by pooling approximately 150 ml of fresh (less than 4 hours old), nonhemolyzed, nonicteric, nonlipemic serum. The acceptability is checked by diluting 1 ml of the pooled serum to 25 ml with 0.85 g/100 ml sodium chloride solution. The absorbance of this diluted serum read against a 0.85 g/100 ml sodium chloride blank set at zero A must be less than 0.100 at 414 nm and less than 0.040 at 460 nm to be acceptable as a bilirubin diluent. The bilirubin standard to be used for the preparation of the calibration curve should be obtained as follows:

1. Weigh out exactly 20.0 mg of an acceptable bilirubin, transfer to a 100 ml volumetric flask, and dissolve by adding 2 ml of 0.1 molar Na_2CO_3 and 1.5 ml of 0.1 molar NaOH. If standards more concentrated than 20 mg/100 ml are prepared, add 0.5 ml of 0.1 molar NaOH for each additional 5 mg of bilirubin. The solution should be clear and red in color. Dilute this solution to 100 ml with the acceptable pooled serum and mix well. Since the bilirubin is light sensitive, the flask should immediately be wrapped with aluminum foil and protected from light as much as possible. The stability of the bilirubin standard also depends on the storage temperature. Deterioration in a regular freezer compartment is about 1.5% per week; at −30°C, about 1.5% per month; and at −70°C, about 1% in six months (B. T. Doumas, personal communication).

2. Standard solutions for the preparation of the calibration curve are obtained as follows:

Total bilirubin (mg/100 ml)	*Pooled serum* (ml)	20 mg/100 ml *standard* (ml)
0	4	0
2	9	1
5	3	1
10	2	2
15	1	3
20	0	4

3. Determine the absorbance of each of the preceding solutions using the procedure for the analysis of serum. Subtract the absorbance value of the respective pooled serum blank from the absorbance of each of the five standard solutions. The corrected absorbance values may then be used to plot absorbance versus concentration to prepare the calibration curve for total bilirubin.

4. A consistent use of the SI System would require reporting analyses in mol/l. However, as an interim solution, the unit of mg/l has been adapted by some journals in the USA to replace mg/100 ml.

Note: The details of the methods for both total and conjugated bilirubin are presently under intensive study. In the preparation of the bilirubin standard solution, the use of dimethyl sulfoxide (DMSO) and Na_2CO_3 instead of Na_2CO_3 and NaOH to dissolve the bilirubin has been suggested. Furthermore, serum albumin (HSA), fraction V, has been found to have certain advantages as a diluent compared to the pooled serum suggested in the text: (1) it has a very low absorbance and stays clear during storage; and (2) no suppression of the azobilirubin color has been observed with HSA. The procedure involves dissolving the 20 mg of bilirubin in 1 ml of DMSO, followed by the addition of 2 ml of 0.1 molar Na_2CO_3 and immediate addition of about 80 ml of 4 per cent HSA V. The Na_2CO_3 is neutralized by the addition of 2 ml of 0.1 molar HCl, and the solution is diluted to 100 ml with the HSA solution. The pH of the HSA solution should be adjusted to between 7.3 and 7.4 for maximum stability of the bilirubin standard (Doumas, B. T., Perry, B. W., Sasse, E. A., and Straumfjord, J. V., Jr.: Clin. Chem., *19*:984, 1973; and Doumas, B. T.: Personal communication).

NORMAL VALUES

Normal values for serum bilirubin content of newborns, infants, and adults are shown in the following tabulation:

Age	*Premature* (mg/100 ml)	*Full term* (mg/100 ml)
Up to 24 h	1–8	2–6
Up to 48 h	6–12	6–10
Days 3 to 5	10–14	4–8
Infants after 1 month and adults:		
Conjugated:	0–0.2	
Unconjugated:	0.2–0.8	
Total:	0.2–1.0	

REFERENCES

Jendrassik, L., and Grof, P.: Biochem. Z., *297*:81, 1938.
Gambino, S. R.: *In* Standard Methods of Clinical Chemistry. S. Meites, Ed. New York, Academic Press, 1965, vol. 5, pp. 55–64.

The Malloy and Evelyn Method

PRINCIPLE

Bilirubin in the serum is coupled with diazotized sulfanilic acid to form *azobilirubin*. The intensity of the purple color that is formed is proportional to the bilirubin concentration in the serum. Bilirubin glucuronide, the conjugated or direct bilirubin, reacts with the diazo reagent in aqueous solution to form a colored diazo compound within 1 min. The subsequent addition of alcohol accelerates the reaction of unconjugated bilirubin in the serum, and a value for total bilirubin is obtained after letting the specimen stand for 30 min. The total bilirubin value represents the sum of the bilirubin glucuronide (direct) and the unconjugated (indirect) bilirubin.

REAGENTS

1. Methanol, absolute, A.R.
2. Reagent A. Dissolve 5 g of sulfanilic acid in 60 ml of concentrated hydrochloric acid in a 1 liter volumetric flask, and dilute to the mark with water. The solution is stable indefinitely.
3. Sodium nitrite solution, 20 g/100 ml (stock). Dissolve 20 g of $NaNO_2$ in water and dilute to a volume of 100 ml. The solution is stable up to 4 weeks when stored in the refrigerator.
4. Reagent B. Sodium nitrite solution, 2 g/100 ml. Prepare fresh daily by diluting the 20 g/100 ml $NaNO_2$ solution tenfold.
5. Diazo reagent. Add 0.3 parts of Reagent B to 10 parts of Reagent A and mix. This reagent should be prepared within 30 min of the time of use.

6. Diazo blank. Dilute 60 ml of concentrated hydrochloric acid to a volume of 1 liter with water. The reagent is stable indefinitely at room temperature.

PROCEDURE

1. Into a test tube labeled B (blank), pipet 0.5 ml of clear, unhemolyzed serum and 9.5 ml of distilled water and mix by gentle inversion.
2. Transfer 5.0 ml of the diluted serum to a second tube labeled U (unknown).
3. Add 1.0 ml of diazo blank solution to tube B and 1.0 ml of diazo reagent to tube U and mix immediately.
4. Exactly 1 min after the addition of the diazo reagent, read the absorbance of tube U at 540 nm against tube B, which was set at zero absorbance or at 100 per cent *T*. When several specimens are run at the same time, the addition of the diazo reagent (step 3) should be spaced to enable readings to be carried out exactly 1 min after mixing.
5. Blow 6.0 ml of methanol into each tube, mix by gentle inversion, let stand for 30 min, and read the absorbance of tube U at 540 nm against tube B.

CALCULATIONS

1. Values for total bilirubin may be obtained directly from the calibration curve for total bilirubin.
2. Values for conjugated bilirubin may be obtained by dividing by 2 the values that are read from the calibration curve of absorbance versus concentration of total bilirubin. (The solution is not diluted with methanol and is thus twice as concentrated.)
3. The values for unconjugated (indirect) bilirubin are obtained by subtracting the conjugated (direct) bilirubin from the total bilirubin.
4. In laboratories where microcuvets are used, the method may be scaled down by using one tenth of the volumes given in the procedure.

REFERENCES

Malloy, H. T., and Evelyn, K. A.: J. Biol. Chem., *119*:481, 1937 (modified).
Recommendation on a Uniform Bilirubin Standard: *In* Standard Methods of Clinical Chemistry. S. Meites, Ed. New York, Academic Press, 1965, vol. 5, p. 75.

DIRECT SPECTROPHOTOMETRIC METHOD FOR TOTAL BILIRUBIN IN SERUM

SPECIMENS

Serum obtained from capillary blood of infants is drawn from the heel or fingertip. Use of such specimens poses two major problems: the necessity of a micro- or ultra-micromethod because of the small amount of blood that is drawn, and the fact that there is frequently hemolysis of the serum sample.

PRINCIPLE

The absorbance of bilirubin in the serum at 455 nm is proportional to its concentration. The serum of newborn infants does not contain lipochromes, such as carotene, and other pigments that increase the absorbance at 455 nm; however, these pigments may be present in serum from adults. The absorbance of hemoglobin at 455 nm is corrected by subtracting the absorbance at 575 nm.

REAGENT

1. Phosphate buffer, pH 7.4. Weigh out 7.65 g of $Na_2HPO_4 \cdot 7H_2O$ and 1.74 g of anhydrous KH_2PO_4. Dissolve in water and dilute to a volume of 1000 ml in a volumetric flask. Check the pH with a pH meter.

PROCEDURE

1. With a micropipet, add 20 μl of serum to a microcuvet (1.0 cm light path) containing 1 ml of phosphate buffer. Rinse the pipet several times with the buffer.

2. Add 1 ml of phosphate buffer to a microcuvet and, using this as a blank, set at zero absorbance; read the absorbances of the diluted serum at 455 and 575 nm. To obtain the accurate values of absorbances required in this method, a spectrophotometer capable of transmitting bandwidths of 10 nm or less should be used.

CALCULATIONS

The absorbance values of the serum at 455 nm or 575 nm are due to both bilirubin and hemoglobin. Two equations are required to express the effect of both pigments.

$$A_{455} = (K_{b455} \times C_b) + (K_{h455} \times C_h) \tag{1}$$

$$A_{575} = (K_{b575} \times C_b) + (K_{h575} \times C_h) \tag{2}$$

In equations (1) and (2), C_b is concentration of bilirubin in the sample used, A_{455} and A_{575} are the absorbances at 455 and 575 nm, and K_{b455}, K_{b575}, K_{h455}, and K_{h575} are the absorption constants of bilirubin and hemoglobin solutions at 455 and 575 nm, respectively. These two equations may be solved for C_b by multiplying (1) by K_{h575} and (2) by K_{h455}

$$A_{455} \times K_{h575} = (K_{b455} \times K_{h575} \times C_b) + (K_{h455} \times K_{h575} \times C_h) \tag{3}$$

$$A_{575} \times K_{h455} = (K_{b575} \times K_{h455} \times C_b) + (K_{h575} \times K_{h455} \times C_h) \tag{4}$$

Subtracting (4) from (3), the expression $K_{h455} \times K_{h575} \times C_h$ cancels and C_b may be expressed as

$$C_b = \frac{(A_{455} \times K_{h575}) - (A_{575} \times K_{h455})}{(K_{b455} \times K_{h575}) - (K_{b575} \times K_{h455})} \tag{5}$$

The absorption constants for bilirubin and hemoglobin are determined with the same spectrophotometer that is used in the procedure with diluted serum specimens. Pure bilirubin dissolved in Na_2CO_3 solution or hemoglobin prepared from lysed erythrocytes (previously washed with saline) is added to a 5 per cent albumin solution or a hemoglobin-free, low bilirubin ($<$0.3 mg/100 ml) pooled serum to prepare two series of standard solutions (bilirubin 2 to 20 mg/100 ml, and hemoglobin 5 to 100 mg/100 ml). These solutions are treated as serum specimens (see under Procedure) and the A value for each solution is determined. The absorption constants at 455 and 575 nm are obtained by dividing A by the milligram/100 ml for each solution. The K values are averaged to obtain the four values required in equation (5). An example using reasonable K values will illustrate the simplification of equation (5).

$$K_{b455} = 0.780$$

$$K_{b575} = 0.0115$$

$$K_{h455} = 0.0103$$

$$K_{h575} = 0.0098$$

$$C_b = \frac{0.0098A_{455} - 0.0103A_{575}}{(0.780 \times 0.0098) - (0.0115 \times 0.0103)} = \frac{0.0098A_{455} - 0.0103A_{575}}{0.00753}$$

$$= 1.30A_{455} - 1.37A_{575}$$

$$C_b\text{, mg /100 ml} = \text{dilution} \times (1.30A_{455} - 1.37A_{575})$$

REFERENCES

Meites, S., and Hogg, C. K.: Clin. Chem., *6*:421, 1960 (modified).

White, D., Haidar, G. A., and Reinhold, J. G.: Clin. Chem., *4*:211, 1958.

DIRECT-READING BILIRUBINOMETERS

The principle of the direct spectrophotometric method for measuring total bilirubin in serum has been applied in the design of direct-reading bilirubinometers. These instruments are in essence differential spectrophotometers, either with two light paths passing at right angles to each other through the same cuvet or with a single light path that is split into two beams after passing through the cuvet. The resulting light beams are passed through two separate narrow-bandpass filters to separate photodetection cells. One filter has a wavelength of 454 (or 461) nm, which is the absorption peak wavelength for bilirubin in serum. The other filter wavelength is 540 (or 561) nm, chosen as the point at which oxyhemoglobin has the same absorbance as at 454 (or 461) nm. The instrument is set to "zero" using an oxyhemoglobin solution.

When bilirubin is present in the sample, the absorbance at 454 (or 461) nm increases and unbalances the photodetection system. The bilirubin concentration in the sample is read directly on a meter, or the null point needle is returned to zero and the concentration is read on a scale calibrated in mg/100 ml. If oxyhemoglobin is present in the sample, no interference will occur because the absorbances of oxyhemoglobin at 540 (or 561) nm and at 454 (or 461) nm are equal and will therefore have no effect on the reading.

The instruments can be standardized with bilirubin solutions of known concentrations or with secondary standards, such as a methyl orange solution at pH 7.4 or a multilayer colored glass standard.

These instruments are available commercially from Advanced Instruments, Inc. (Newton Highlands, MA.), and American Optical Corp. (Buffalo, N.Y.).

URINE BILIRUBIN

It has been recognized for many years that the presence of conjugated bilirubin in the urine suggests hepatocellular disease or obstructive jaundice. Generally, one is concerned only with its presence, not with the exact amount; for this reason, many qualitative tests have been devised. These tests have been based on the observation of the color of the urine, the characteristic colors formed on oxidation of bilirubin, the addition of methylene blue until the yellow-brown color of the urine becomes blue, and diazotization of the bilirubin. The sensitivity and specificity of a method are of prime importance in the choice of a method. In general, methods based on diazotization are most satisfactory.

CLINICAL APPLICATION

Any method that consistently gives a negative test for bilirubin with normal urines, and yet is sensitive enough to detect slightly increased quantities of bilirubin, is valuable in clinical diagnosis. In any form of hepatitis that involves impairment or destruction of liver cells, in transportation defects such as the Dubin-Johnson syndrome, and in obstructive jaundice, conjugated bilirubin is excreted in the urine.

METHODS FOR THE DETECTION OF BILIRUBIN IN URINE

Although many attempts have been made to modify the various qualitative methods, as yet there has not been an acceptable quantitative method developed for the routine laboratory.

Fouchet's Test

A test that is often used in the laboratory is based on the green color produced by the reaction of bilirubin with Fouchet's reagent (1.0 g ferric chloride in a solution of 25.0 g trichloroacetic acid/100 ml). A common modification of Fouchet's test is carried out as follows: A strip of thick filter paper, impregnated with a saturated solution of barium chloride and dried, is inserted for about half its length into the urine specimen. (Barium chloride forms barium phosphate, which adsorbs and concentrates the bilirubin.) The strip is removed from the urine and a few drops of Fouchet's reagent are added at the boundary between the wet and dry portions of the strip. A green color is produced if bilirubin is present in the urine; the intensity of the color varies with the amount of bilirubin. The color response is usually graded from 0 to 4+.

Ictotest

A qualitative test employing diazotization that is both sensitive and specific has been developed by Free and Free. Semiquantitative results may also be obtained by this method.

SPECIMEN

Bilirubin in urine is unstable and will decrease in concentration in light and at room temperature. Therefore, a fresh specimen of urine is most satisfactory for this test. Specimens may be kept in the dark in a refrigerator for 1 day.

PRINCIPLE

Diazotization of bilirubin in the urine specimen under acid conditions produces a blue to purple color. The diazo agent is *p*-nitrobenzenediazonium *p*-toluenesulfonate, which produces a color within 30 s. The speed of color development and the intensity of the color are related to the amount of bilirubin in the urine.

REAGENTS

1. A test mat composed of asbestos and cellulose fibers is used to adsorb and concentrate the urine bilirubin and other bile pigments on its surface.
2. A powdered mixture composed of 0.6 parts of *p*-nitrobenzendiazonium *p*-toluenesulfonate, 10 parts of $NaHCO_3$, 100 parts of sulfosalicylic acid, and 20 parts of boric acid is used to carry out the diazotization reaction.

The proper test mats and tablets containing the diazo mixture are commercially available under the trade name Ictotest (Ames Co., Div. Miles Laboratories, Elkhart, IN.).

PROCEDURE

1. Place 5 drops of fresh urine in the center of one square of the test mat.
2. Place an Ames Ictotest reagent tablet in the center of the moistened area.
3. Add 2 drops of water onto the tablet, making sure that the water flows from the tablet onto the mat.
4. A positive test is indicated by a blue or purple color around the tablet within 30 s. A pink to red color or any color that develops after 30 s should be ignored.
5. Semiquantitation of the test can be carried out by performing the test on serial dilutions of the urine specimen. The highest dilution that still gives a positive test will have a bilirubin concentration of 0.1 mg/100 ml. A simple calculation (multiplication by the dilution factor) will yield the approximate bilirubin concentration of the original urine.

SOURCES OF ERROR

The specificity of many of the qualitative tests for bilirubin in urine is questionable. The Ictotest just outlined has a high degree of specificity; it seldom (see the following) yields false positive tests and is sensitive enough to pick up weak positive tests sometimes missed by other methods. The sensitivity of the test ranges from 0.05 to 0.1 mg of bilirubin per 100 ml of urine. High levels of indican, urobilin, or salicylate will produce a red color in the test; Pyridium and Serenium, drugs used in the treatment of urinary tract infections, color the urine red and also yield an atypical color in the test.

REFERENCES

Free, A. H., and Free, H. M.: Gastroenterology *24*:414, 1953.
Watson, C. V., and Hawkinson, V.: J. Lab. Clin. Med., *31*:914, 1946.

UROBILINOGEN IN URINE AND FECES

As already mentioned under Bilirubin Metabolism, urobilinogen is the name given to the end products of bilirubin metabolism: mesobilirubinogen, stercobilinogen, and urobilinogen. These compounds, known collectively as urobilinogen, are colorless reduction products of bilirubin, and are oxidized by intestinal microorganisms to the brown pigments D-, I- and L-urobilin (stercobilin).

In the normal individual, part of the urobilinogen that is formed by the reduction of bilirubin in the colon is excreted in the feces, and the remainder is reabsorbed into the portal blood and returned to the liver. A small amount escapes re-excretion into the intestine by the liver and is excreted in the urine.

CLINICAL SIGNIFICANCE

In various forms of *hepatitis* involving impairment or destruction of liver cells, the liver is unable to remove an appreciable fraction of the urobilinogen from the portal blood; thus, increasing amounts enter the general circulation and are excreted by the kidneys. The determination of urine urobilinogen is, in fact, a very sensitive and useful test for detecting the early stages of hepatitis. In *obstructive jaundice*, bilirubin excretion into the intestine stops or is greatly decreased. This decreases the formation of urobilinogen; consequently, its excretion in the urine is decreased. Fecal excretion of urobilinogen in the presence of a normally functioning liver is dependent on the rate of breakdown of hemoglobin. *Hemolytic anemia* increases the amount of urobilinogen excreted, but anemias not related to destruction of red blood cells cause a decrease. Liver disease, in general, reduces the flow of bilirubin glucuronides to the intestine and thus decreases fecal excretion of urobilinogen. Complete *obstruction* of the bile duct reduces the urobilinogen of the stool to very low values. The clay-colored or chalky white stool in obstructive jaundice reflects the exclusion of bile pigments from the intestine.

DETERMINATION OF URINE UROBILINOGEN (SEMIQUANTITATIVE)

SPECIMEN

A fresh urine specimen is collected over a 2 h period from 2 to 4 p.m. (or 1 to 3 p.m.). The patient is asked to empty his bladder at 2 p.m. and this specimen is discarded. He is then given a glass of water. All urine specimens voided during the next 2 h are collected and pooled. The specimen should be kept cool and protected from light. To prevent oxidation of urobilinogen to urobilin, there must be no delay in running the analysis once the sample is collected. If a 24 h sample is collected, it should be collected in a dark bottle that contains 5 g of sodium bicarbonate to minimize oxidation of urobilinogen and 100 ml of toluene to minimize bacterial growth and to form a protective layer against oxygen from the air.

PRINCIPLE

The majority of the quantitative methods for urobilinogen are based on the reaction of this substance with *p*-dimethylaminobenzaldehyde to form a red color. Although some doubt exists concerning the exact chemical structure of the red compound produced, the reaction may be represented as shown in Chart 19–3. This reaction was first described by Ehrlich in 1901, and methods involving the Ehrlich reagent have been modified over the years to improve their specificity. Major improvements were made in 1925 by Terwen, who used alkaline ferrous hydroxide to reduce urobilin to urobilinogen and added sodium acetate to eliminate interference from indole and skatole. In 1936, Watson introduced the use of petroleum ether rather than diethyl ether for the extraction of urobilinogen to assist in the removal of other interfering substances. Studies by Henry *et al.* in 1964 indicated that even the so-called quantitative method that involves extraction with petroleum ether is not capable of complete recovery of added urobilinogen from urine. These investigators suggest that the more rapid semiquantitative method yields values of sufficient clinical significance.

The principle of the method chosen, therefore, involves the spectrophotometric determination of urobilinogen that has reacted with *p*-dimethylaminobenzaldehyde. Ascorbic acid is added as a reducing agent to maintain urobilinogen in a reduced state and

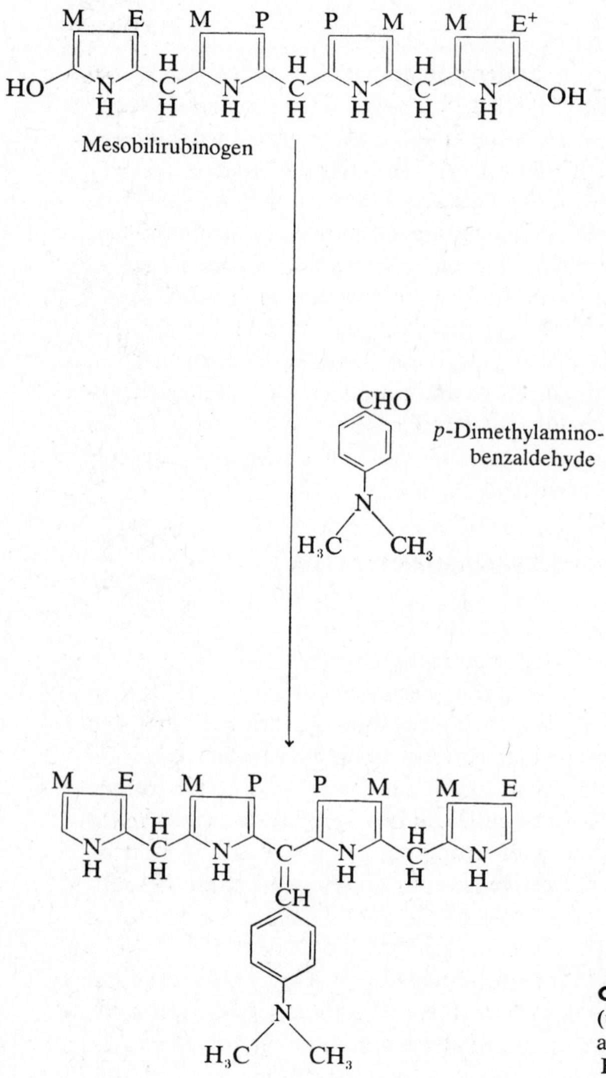

Mesobilirubinogen

p-Dimethylamino-
benzaldehyde

Chart 19–3 The reaction of mesobilirubinogen (urobilinogen and stercobilinogen) with *p*-dimethylaminobenzaldehyde. (M= $—CH_3$; E= $—CH_2CH_3$; P = $—CH_2CH_2COOH$.)

prevent the re-formation of urobilin. Sodium acetate is used to reduce the acidity after the reaction of urobilinogen with Ehrlich's reagent. The sodium acetate also inhibits color formation from indole and skatole, and intensifies the color with Ehrlich's reagent. The results of the method are expressed in Ehrlich units (E.U.), where 1 Ehrlich unit is equivalent to the color produced by 1 mg of urobilinogen.

REAGENTS

1. Ehrlich's reagent, modified. Dissolve 0.7 g of *p*-dimethylaminobenzaldehyde in 150 ml of concentrated HCl, A.R.; add to 100 ml of distilled water and mix. This reagent is stable.

2. Saturated sodium acetate solution. Saturate 1 liter of distilled water with either A.R. grade anhydrous or triple-hydrated sodium acetate, maintaining extra crystals in the solution to insure saturation. Two to three pounds of sodium acetate will be required, depending on the form that is used. This reagent is stable at room temperature.

3. Ascorbic acid, powder, A.R.

4. Standard PSP dye solution. Dissolve 20.0 mg of phenolsulfonphthalein (phenol red) in 100 ml of 0.05 per cent NaOH. Use the acid form of the dye, which may be obtained from Eastman Kodak Company (Rochester, N.Y.) and other suppliers.

5. Working standard. Dilute the stock solution (step 4) 1:100 with 0.05 per cent NaOH. This solution contains 0.2 mg/100 ml and has a color equivalent to that of a solution of urobilinogen-aldehyde containing 0.346 mg of urobilinogen per 100 ml of final colored solution in the method. PSP salts or solutions for intravenous injection should not be used to prepare the standard. The working standard should have an absorbance of 0.384 at 562 nm in a cuvet with a 1.0 cm light path on a high resolution spectrophotometer. Other methods use a mixture of Pontacyl Violet and Pontacyl Carmine 2B as the standard dye solution. This mixture has an absorption spectrum curve different from that of urobilinogen-aldehyde, and when it is employed as a standard one must be aware that the analytical results vary with the spectral resolution of the spectrophotometer used.

PROCEDURE

1. Measure the volume of the 2 h urine sample.

2. Test the urine for bilirubin. If more than a trace is present, mix 2.0 ml of 10 g/100 ml $BaCl_2$ solution with 8.0 ml of urine, and filter. The final result must be multiplied by 1.25 to correct for this step.

3. Dissolve 100 mg of ascorbic acid in 10 ml of urine (centrifuge if cloudy) and place 1.5 ml aliquots in each of two tubes labeled B for blank and U for unknown.

4. To the B tube add 4.5 ml of a freshly prepared mixture of 1 volume of Ehrlich's reagent and 2 volumes of saturated sodium acetate; mix.

5. To the U tube add 1.5 ml of Ehrlich's reagent, mix well, and immediately add 3.0 ml of saturated sodium acetate; mix.

6. Within 5 min, measure the absorbance of the tubes marked B and U at 562 nm against water set at zero. Measure the absorbance of the PSP working standard against water at the same wavelength.

CALCULATION

$$\text{Ehrlich units/100 ml urine} = \frac{A_U - A_B}{A_S} \times 0.346 \times \frac{6.0}{1.5} = \frac{A_U - A_B}{A_S} \times 1.38$$

and

$$\text{Ehrlich units/2 h} = \frac{A_U - A_B}{A_S} \times 0.0138 \times \text{urine volume in ml}$$

Note. Multiply answer by 1.25 if bilirubin was removed by $BaCl_2$ (step 2).

NORMAL VALUES

The normal range for this method is 0.1 to 1.0 E.U./2 h or 0.5 to 4.0 E.U./d.

SOURCES OF ERROR

1. The specimen should be fresh and the procedure should be carried out without delay in the absence of sunlight, bright fluorescent, or other light.

2. Pure solutions of urobilinogen develop color immediately when mixed with Ehrlich's reagent. In the urine, Ehrlich's reagent produces color with nonurobilinogen substances on standing. For this reason it is important to stop these slower color reactions by the addition of saturated sodium acetate immediately after the urine is mixed with Ehrlich's reagent.

3. Since the urobilinogen-aldehyde color slowly decreases in intensity, the spectrophotometric readings should be made within 5 min after the production of the color.

4. Compounds other than urobilinogen that may be present in the urine, such as porphobilinogen, sulfonamides, procaine, and 5-hydroxyindoleacetic acid, may also react with Ehrlich's reagent. Bilirubin, when present, will form a green color and must, therefore, be removed before analyzing for urobilinogen.

REFERENCES

Henry, R. J., Fernandez, A. A., and Berkman, S.: Clin. Chem., *10*:440, 1964.
Watson, C. J., and Hawkinson, V.: Am. J. Clin. Path., *17*:108, 1947.

DETERMINATION OF FECAL UROBILINOGEN (SEMIQUANTITATIVE)

SPECIMEN

Any single fresh specimen may be used and the analysis should be carried out in the absence of direct sunlight or bright artificial light.

PRINCIPLE

This method involves the same principles described earlier for the urine procedure. It is carried out on an aqueous extract of fresh feces, and any urobilin present is reduced to urobilinogen by treatment with alkaline ferrous hydroxide before Ehrlich's reagent is added.

REAGENTS

1. Ferrous sulfate solution, 20 g/100 ml. Prepare just before use.
2. NaOH solution, 2.5 mol/l (100 g/l).
3. Ehrlich's reagent (for reagents 3, 4, and 6, see under urine urobilinogen).
4. Saturated sodium acetate solution (see p. 1047).
5. Ascorbic acid powder.
6. Stock and working standard of PSP dye solution (see p. 1047).

PROCEDURE

1. Transfer 10 g of a blended or homogenized sample of fresh feces to a large mortar. Add water to a 250 ml graduated cylinder to the 190 ml mark. Add 20 ml of water from the cylinder to the mortar and grind the feces to a paste. Add 80 ml more of water and grind again. To a 500 ml Erlenmeyer flask, add 100 ml of the ferrous sulfate solution; then add the supernatant suspension of feces from the mortar. Add another 50 ml of water to the residue in the mortar, grind, and transfer the supernatant material to the flask. Repeat this process with the remaining water in the cylinder. Slowly add 100 ml of 2.5 molar NaOH to the flask with swirling. Stopper the flask, shake, and allow to stand in the dark at room temperature for 1 to 3 h.

2. Mix the contents of the flask and filter a portion of the contents. Dilute 5.0 ml of the filtrate to 50 ml with water. This solution should be nearly colorless. If the filtrate is highly colored, measure 50 ml into a flask, add 25 ml of the ferrous sulfate solution and 25 ml of 2.5 molar NaOH, and mix well. Stopper the flask and allow to stand in the dark for 1 to 3 h. Mix the contents and filter as before; dilute10 ml of the filtrate to 50 ml with water to correct for the additional dilution before proceeding to step 3.

3. Dissolve 100 mg of ascorbic acid in 10 ml of the diluted filtrate and place 1.5 ml aliquots into each of two tubes labeled B for blank and U for unknown.

4. Carry the tubes through steps 4, 5, and 6 as described in the urine procedure.

CALCULATION

$$\text{Ehrlich units/100 g wet feces} = \frac{A_U - A_B}{A_S} \times 0.346 \times \frac{6.0}{1.5} \times \frac{50}{5} \times \frac{400}{10} = \frac{A_U - A_B}{A_S} \times 552$$

NORMAL VALUES

A range of 75 to 275 Ehrlich units/100 g of fresh feces, or 75 to 400 E.U./24 h specimen is considered normal.

REFERENCES

See references under Determination of Urine Urobilinogen, page 1048.

CARBOHYDRATE METABOLISM AND LIVER FUNCTION

The liver is essential for normal carbohydrate metabolism. Monosaccharides such as fructose and galactose are converted to glucose; the glucose is stored as glycogen, or may enter various metabolic pathways to be converted into amino acids or fatty acids or be broken down to carbon dioxide and water with the release of energy. The liver serves as a source of readily available glucose, either from its store of glycogen or by the process of gluconeogenesis. It plays an important role in the metabolism of lactic and pyruvic acids and produces ketone bodies in the process of oxidation of fatty acids (see Chap. 6, Carbohydrates). The liver is involved in so many essential metabolic reactions that one would think it would be a simple matter to devise tests to study abnormalities in these reactions. Although many investigators have attempted to provide such sensitive tests, in general, they have failed, since extensive impairment of liver cells or destruction of liver tissue is required before there is significant interference with metabolic function. For this reason, the clinical laboratory seldom carries out function tests based on impairment of the metabolic activities of the liver.

It is reasonable to assume that the glucose tolerance test (see Chap. 6) would be grossly abnormal in the presence of liver disease, but attempts at correlation of this test with various forms of liver disease have been disappointing. At times even normal curves are seen in serious liver disease, and it must be concluded that this test is of limited value in liver function assessment. The most common abnormality in acute liver disease is the occurrence of a rapid sharp peak of about 200 mg/100 ml in the blood glucose level within the first hour, followed by a return to normal or hypoglycemic levels within 3 to 5 h. Of all the tests of metabolic function that have been devised, the galactose tolerance test is probably the most valuable, but it is rarely used.

THE GALACTOSE TOLERANCE TEST

The galactose that is carried to the liver cells from the intestinal tract is normally converted to glucose, which is then further converted into glycogen. If galactose is given orally (40 g in 200 ml water) or injected into the blood (1 ml of a 50 per cent solution per kg of body weight), the speed of removal of this sugar is related to the integrity of the liver and the normal functioning of its cells.

Although the test is of some value in assessing liver function, its major limitation is its insensitivity. Abnormal tolerance is not evident until rather severe impairment or destruction of liver cells occurs. A progressive decrease of tolerance, however, does parallel a developing liver cell necrosis and may assist in distinguishing obstructive jaundice from hepatitis. Nevertheless, the galactose tolerance test has been discarded by many clinicians.

In the oral test the blood galactose level should reach a peak of 40 to 60 mg/100 ml in 30 to 60 min. The total galactose excreted in the urine in the 5 h period should not exceed 3 g. In the intravenous test the blood level should not exceed 42 mg/100 ml after 60 min.

THE DETERMINATION OF GALACTOSE IN BLOOD

PRINCIPLE

Methods for the determination of galactose in the blood or urine ordinarily involve removal of glucose by fermentation with yeast or by treatment with glucose oxidase. The concentration of the remaining reducing sugar is determined and is expressed as galactose. Several modifications of such methods have been proposed and they all suffer from a lack of specificity. With the recent commercial availability of *galactose oxidase*, a more specific method can be devised. The enzymatic reaction is similar to that used for the determination of glucose with glucose oxidase.

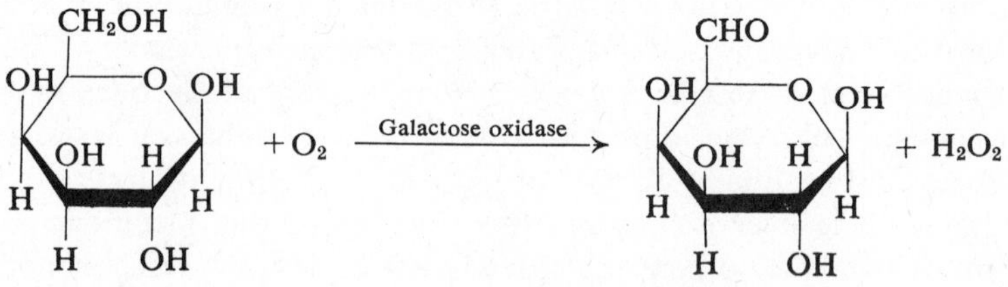

$$\text{D-galactose} + O_2 \xrightarrow{\text{Galactose oxidase}} \text{D-galacto-hexoaldose} + H_2O_2$$

D-*galactose* → D-*galacto-hexoaldose*

$$H_2O_2 + \text{Reduced chromogen} \xrightarrow{\text{Peroxidase}} \text{Oxidized chromogen}$$

REFERENCES

Sempere, J. M., Gancedo, D., and Asensio, C.: Anal. Biochem., *12*:509, 1965.

Worthington Enzyme Manual, Galactostat Brochure. Freehold, N. J., Worthington Biochemical Corp., 1967.

PROTEIN METABOLISM AND LIVER FUNCTION

Normally functioning liver cells are essential for normal protein metabolism. Deamination and transamination of amino acids, urea formation, and synthesis of prothrombin and of many of the plasma proteins are dependent on normal liver function. In protein metabolism, as in carbohydrate metabolism, an extensive impairment or destruction of liver cells is required before abnormal function can be clearly demonstrated. In general, liver function tests that evaluate overall protein metabolism are less sensitive than those that depend, for example, on abnormal concentrations of specific plasma proteins such as albumin and gamma (γ) globulin.

PLASMA PROTEINS

The plasma proteins such as albumin, fibrinogen, and the majority of the globulins, with the exception of γ-globulin, are synthesized by the liver. The decrease in albumin and increase in γ-globulin and lipoproteins in the β-globulin fraction are characteristic of *chronic liver disease.* In *obstructive jaundice*, the albumin is only slightly decreased and the γ-globulin slightly increased, but a definite increase in the lipoprotein fractions, α_2- and β-globulins, is observed. In general, the plasma protein changes in *parenchymal liver disease* consist of a decrease in albumin and an overall increase of the globulin fractions. Total protein determinations alone are not particularly helpful, since normal values may be obtained by the combination of low albumin and high globulins. The albumin/globulin (A/G) ratio is often reversed in liver disease, but this is not specific and indicates a need for further fractionation of the plasma proteins. Changes in the plasma proteins occur in chronic hepatitis and cirrhosis and are readily demonstrated by electrophoretic analysis (see Chap. 7, Proteins and Amino Acids). It should be emphasized that changes in the plasma proteins as measured by the methods discussed in this section are the result of prolonged or extensive liver cell impairment and, unlike some enzyme tests, are of little assistance in the detection of early liver disease.

MUCOPROTEINS

The mucoproteins of the serum are found in the α_1- and β-globulin fractions and range in concentration from 75 to 135 mg/100 ml in normal individuals. In *viral hepatitis, toxic hepatitis*, and *cirrhosis*, the level decreases, whereas a progressive increase above normal levels occurs in *extrahepatic obstructive jaundice.* The determination of mucoproteins in the serum has been suggested as a valuable diagnostic aid in the differentiation of jaundice in hepatitis from that in extrahepatic obstruction (see the discussion on jaundice and Chapter 7, Proteins and Amino Acids).

AMINO ACIDS

Amino acids undergo deamination in the liver. Thus, an increase in the serum level could result from extensive impairment or destruction of liver cells. In cases of *acute hepatic necrosis* caused by chemical agents such as phosphorus or carbon tetrachloride, and in the terminal stages of liver disease, the level may increase three- to fivefold. In early hepatitis or mild cirrhosis, the changes are too slight to be of diagnostic assistance (see Chap. 7).

AMMONIA

The level of circulating ammonia in the blood is extremely low in normal individuals and ranges from 10 to 70 μg N/100 ml. This low concentration is surprising when one considers the continuous processes of oxidative deamination and transamination of dietary and tissue amino acids. Since any appreciable level of ammonia in the blood would adversely affect acid-base balance and brain function, a major mechanism for its removal is essential. Although a small amount of ammonia is used for the synthesis of *glutamine*, the synthesis of *urea* is mainly responsible for the removal of ammonia from the blood. In the first step of urea synthesis, ammonia combines with carbon dioxide and water under the influence of carbamyl phosphate synthetase to form carbamyl phosphate.

$$NH_3 + CO_2 + H_2O + 2\,ATP \xrightarrow[\text{synthetase}]{\text{Carbamyl phosphate}} \underset{\textit{Carbamyl phosphate}}{H_2N{-}CO{-}OPO_3H_2} + 2\,ADP + PO_4$$

This important reaction requires the expenditure of two molecules of ATP and is irreversible.

The complete urea cycle involves (1) the formation of *citrulline* from *ornithine* and carbamyl phosphate, (2) the addition of another amino group from aspartate in the formation of *argininosuccinate*, (3) oxidation to *arginine* and *fumarate*, and (4) hydrolysis of arginine by arginase, yielding *urea* and the starting material, ornithine.

Glutamine is also synthesized by the brain, and this synthesis increases when the blood ammonia is elevated. In terminal liver disease, the ammonia not removed by the liver is used for glutamine synthesis by the brain, resulting in an elevated blood concentration of glutamine. Since glutamate is used in glutamine synthesis, this would cause a drain on the glutamic acid and on the citric acid cycle intermediates of the brain, decreasing oxidative metabolism. Respiration decreases and coma may result. The relationship between glutamine synthesis, glutamic acid, and the citric cycle intermediates in the brain may be represented as follows:

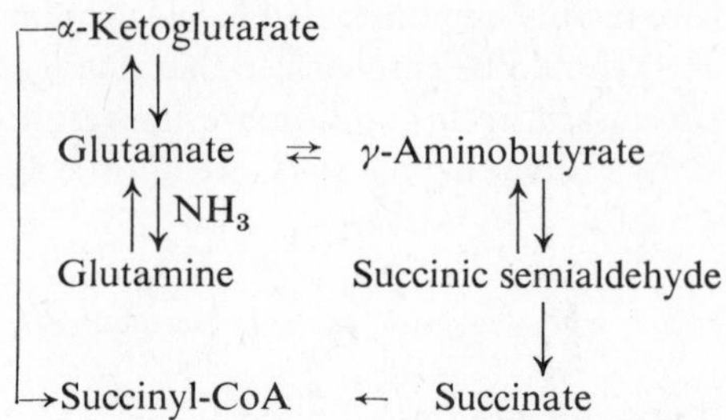

Hepatic coma and the terminal stages of cirrhosis are often marked by an increase in blood ammonia. In fact, the determination is sometimes requested to assist in establishing a diagnosis of impending or existing hepatic coma. Hepatic coma, however, has been observed in the presence of normal ammonia levels, while some patients with high ammonia levels do not exhibit such a coma. The test is therefore of questionable value; in this context, however, it has recently been found to be useful in the diagnosis of Reye's syndrome.

DETERMINATION OF AMMONIA IN BLOOD, PLASMA, AND SERUM

The ammonia content of freshly drawn normal blood rises rapidly on standing (owing to enzymatic deamination of labile amides like glutamine) to two or three times its original value in the course of several hours at room temperature; it rises less rapidly at refrigerator temperature, and remains constant for several days if kept in a deep freeze at −20°C. No change in ammonia content is noted if the blood is placed immediately in an ice bath and analyzed within 20 min. Some authors suggest immediate preparation of trichloroacetic acid or tungstic acid filtrates, since such filtrates show no changes in ammonia content on standing.

The problem of ammonia formation affects the procedure itself. Conway and Cooke[1] published a diffusion method in which the ammonia is reabsorbed in hydrochloric acid containing a suitable acid-base indicator. The amount of ammonia is determined by titration. Other authors recommend that the amount of ammonia absorbed by the acid be determined by nesslerization or by the Berthelot reaction (see serum urea nitrogen method). Claims have been made that under the conditions of this test (pH of potassium carbonate-serum mixture = 12.5 to 13.0) ammonia formation continues. Faulkner and Britton,[2] therefore, used a solution of potassium bicarbonate and potassium carbonate that gives a final pH of 9.9, which apparently reduces ammonia formation.

Forman[3] recommends mixing plasma with a strongly acidic cation exchange resin (sulfonated polystyrene cation exchanger, sodium form, 60 to 80 mesh), which captures the ammonium ion. Addition of sodium phenoxide in the presence of hypochlorite and nitroprusside will simultaneously elute the ammonium ion and colorimetrically react with it to produce a stable blue color. This procedure gives values lower than those obtained with other procedures; it is assumed that there is a loss of ammonia on the column. Kurahasi, Ishihara, and Uehara[9] passed diluted blood through a small column of Dowex 50-X12 (50–100 mesh); following elution with a 4 molar sodium chloride solution, they measured the blue color after indophenol formation. In a study involving elution through a sequence of three columns, the amount of ammonia recovered from the second and third columns was less than 0.3% of that recovered from the first column, which approximated the original value for plasma ammonia.

A recent and possibly promising approach to the determination of plasma ammonia has been the use of an enzymatic method based on the reaction of ammonia and α-ketoglutarate in the presence of glutamate dehydrogenase:

$$\alpha\text{-Ketoglutarate} + NH_4^+ + NADH \xrightarrow[\text{dehydrogenase}]{\text{glutamate}} \text{Glutamate} + H_2O + NAD^+$$

Quantitation is provided by kinetic measurement of the decrease in absorbance at 340 nm, as NADH is converted to NAD^+.[7]

Another promising approach is the use of an ammonia selective electrode. (See also section on Electrochemistry.) Such an electrode has recently been made available by Orion Research, Inc., Cambridge, MA.

Neither the resin method nor the enzymatic method eliminates the problems related to collection of specimens. The enzymatic method is also affected by the liberation of ammonia during the test procedure. In view of the many unresolved problems related to the determination of ammonia, and because of the questionable value of this test, no procedure is given here in detail. The interested reader is referred to the references given.

NORMAL VALUES

(Values are given as ammonia nitrogen.)

Enzymatic method: 40 to 80 μg/100 ml
Resin method: 15 to 45 μg/100 ml
Conway diffusion methods: 40 to 110 μg/100 ml
Levels in impending or actual hepatic coma: up to 400 μg/100 ml

THE BLOOD CLOTTING PROCESS

The clotting of blood is a very complex process that involves several plasma protein components, all of which are synthesized by the liver. Investigators are still involved in establishing the exact mechanism of blood coagulation and many precursors, activators, and inhibitors have been implicated in the series of chemical reactions. The difficulties encountered in the isolation of pure, active complex protein clotting factors have led to considerable confusion. The essential steps involved in the clotting process, including many of the factors proposed by current investigators, are outlined in the following scheme. To conserve space, the clotting factors in the scheme are represented by abbreviations or by the names of investigators associated with their isolation. Each factor will be described in the paragraphs that follow.

Step IA. Blood is shed; contact with a rough surface releases platelet factors and activates the Hageman factor.

B. PTC activated by the Hageman factor + platelet factors + AHG + PTA → thromboplastin intermediate.

C. Thromboplastin intermediate + Ac-globulin + Stuart-Prower factor + Ca^{2+} → thromboplastin.

Step IIA. Prothrombin + Ac-globulin + SPCA + Stuart-Prower factor + Ca^{2+} + thromboplastin → prothrombin-cephalin-calcium complex.

B. Prothrombin-cephalin-calcium complex → thrombin.
Step IIIA. Fibrinogen + thrombin → fibrin.
B. Fibrin polymerization → fibrin clot.

Several of the factors responsible for the production of active thromboplastin were discovered in the course of studies of patients with hemophilia and other diseases involving coagulation defects. One such plasma factor, *antihemophilic globulin* (AHG or Factor VIII) reacts with tissue or platelet thromboplastin to initiate the clotting process. This component is lacking in the plasma of patients with classical hereditary hemophilia. Other factors present in plasma are *plasma thromboplastin component* or PTC (Factor IX or Christmas factor), *plasma thromboplastin antecedent* or PTA (Factor XI), a labile factor called *accelerator globulin* or Ac-globulin (Factor V), the *Hageman factor* (Factor XII), and the Stuart-Prower factor (Factor X). The *platelet thromboplastin factor* is released from platelets on contact with a rough surface.

Prothrombin is a glycoprotein in the plasma which has an electrophoretic mobility similar to that of α_2-globulin. In the clotting process, it is converted to the enzyme *thrombin.* The series of reactions involved in the conversion requires thromboplastin and Ca^{2+}, the Stuart-Prower factor, and a labile as well as a stable factor. The labile factor, accelerator globulin (Factor V), is consumed during the clotting process and is not found in the serum. The stable factor, *serum prothrombin converting factor* or SPCA (Factor VII) is not consumed during the process and is stable during long periods of storage. Deficiencies in any of the components required in Step II result in a prolonged prothrombin time. Prothrombin itself is synthesized by the liver and requires adequate amounts of vitamin K for its synthesis.

The condition of *hypoprothrombinemia* is not uncommon, but it leads to severe bleeding only when liver damage is so extensive that the prothrombin synthesis is significantly decreased. In obstructive jaundice, the impairment of absorption of vitamin K from the intestines, owing to lack of bile salts, may result in hypoprothrombinemia. The liver requires vitamin K also for the production of Ac-globulin and SPCA. A deficiency in either of these factors, in the Stuart-Prower factor, or in vitamin K results in a prolonged prothrombin time. In acute liver cell damage, SPCA (Factor VII) is decreased first, followed by prothrombin and then the Stuart-Prower factor (Factor X).

Fibrinogen is a plasma globulin that is also synthesized by the liver. It is changed into fibrin by the action of thrombin. This reaction involves splitting off two small polypeptides to produce a smaller fibrin monomer, which undergoes three-dimensional polymerization to form the fibrin clot. On standing, the gel-like clot contracts into a harder mass and extrudes serum. This process is called *syneresis* or clot retraction. It can be observed in the laboratory in tubes containing clotted blood.

Although the liver is the sole source of fibrinogen, liver disease itself seldom leads to such low levels as to cause bleeding, which occurs at fibrinogen levels below 100 mg/100 ml. *Hypofibrinogenemia* may be observed, however, in severe liver injury, in some cases of carcinoma of the prostate, and in such complications of pregnancy as premature separation of the placenta and retention of a dead fetus. Hypofibrinogenemia is generally caused not by deficient production but by increased consumption owing to clot formation and breakdown by the action of fibrinolytic enzymes. A method for the determination of fibrinogen is described in Chapter 7.

In addition to the scheme outlined for the clotting process, there is also a mechanism concerned with clot lysis. This system includes precursors such as prefibrinolysin (plasminogen), and activators present in plasma, such as urokinase, that lead to the formation of *fibrinolysin* (*plasmin*), which in turn causes lysis of the fibrin clot. This process is of obvious importance for the dissolution of intravascular clots. Normally, the two systems of blood clotting and clot lysis are under a well-balanced control.

FLOCCULATION AND TURBIDITY TESTS

Both qualitative and quantitative alterations in the plasma protein fractions in disease have been used as the basis for diagnostic tests for many years. Although many tests were devised and some were applied to liver disease, at present the best known are the *cephalin-cholesterol flocculation*, *thymol turbidity*, and *zinc sulfate turbidity* tests. In general, the response to these tests depends on the state of balance between the stabilizing and precipitating factors in the serum. The precipitating factors include γ-globulins and lipoproteins such as the β-globulins. The stabilizing factors are albumin and α_1-globulin. In normal serum, the distribution of the protein components is such that the stabilizing factors prevent turbidity or flocculation when any of the three tests are carried out. In the design of each test, therefore, the concentration of the precipitating reagent is adjusted to produce minimal or negative response with normal serum, while exhibiting a reaction to changes in the precipitating and stabilizing factors. In varying degrees, the three tests are useful in distinguishing the several types of hepatitis from extrahepatic biliary obstruction, but are gradually being replaced by more specific tests such as protein electrophoresis and enzyme activity measurements.

CLINICAL SIGNIFICANCE

The *cephalin-cholesterol flocculation test*[4,13] responds readily to serum specimens containing increased γ-globulin and decreased albumin. This condition is common in a high percentage of cases of viral hepatitis, cirrhosis, and hepatic necrosis, and occurs less frequently in posthepatic obstruction. The test is very sensitive to *qualitative* changes in serum albumin, which may explain its rapid response to acute hepatitis. More specifically, a positive test is obtained under any of the following conditions: decrease in albumin, increase in γ-globulin, or production of an abnormal albumin with less stabilizing power. The last condition is frequently evident in patients with subclinical hepatitis and patients recovering from hepatitis. Serum specimens from patients with various types of hepatitis react with a cephalin-cholesterol suspension to produce a flocculant precipitate (graded from 0 to 4+) which is probably an α- and *β-globulin-cholesterol complex.*

The *thymol turbidity and flocculation test*[15,16] is affected by a decrease in albumin and an increase in γ-globulin as are the other tests, but it is mainly affected by an increase in lipids and β-globulins. The test does not respond as rapidly to viral hepatitis as the cephalin-cholesterol flocculation, but the increase in thymol turbidity persists at times longer than an abnormal cephalin-cholesterol test. Hyperglobulinemic states and any condition that results in high serum lipoprotein levels, such as nephrosis, may result in an increase in thymol turbidity. In the test, serum specimens from patients with hepatitis will produce definite turbidity when mixed with a thymol solution in barbiturate buffer of pH 7.55. The turbidity is caused by the precipitation of a *globulin-thymol-phospholipid complex*. The normal adult range is 0 to 5 Shank-Hoagland units. The test should not be performed on lipemic sera, since such specimens will give false high results. In the screening of blood donors for possible hepatitis, postprandial hyperlipemia interferes with the interpretation of the test. In these cases, a thymol flocculation test, which involves allowing the serum-thymol mixture to stand overnight, may be carried out, since this test is not affected by lipemia.

The zinc sulfate turbidity test[8,11] is more affected by changes in γ-globulin alone than are the other two tests. The high values for γ-globulin characteristic of chronic hepatitis and cirrhosis are readily detected by the zinc sulfate turbidity test. One of the outstanding characteristics of this test is the low value obtained in extrahepatic obstructive jaundice; thus, the test is most useful in distinguishing between hepatitis and obstructive jaundice. In the test, serum from patients with high γ-globulins will produce varying degrees of turbidity when mixed with a dilute solution of zinc sulfate in a barbiturate buffer of pH 7.5. The normal adult range is 2 to 12 turbidity units. This covers a mixture of races, since the normal range for Caucasians, 2 to 9, is lower than that for blacks, 5 to 12.

LIPID METABOLISM AND LIVER FUNCTION

The liver plays a major role in lipid metabolism. It is involved in the complex transportation of lipid material between the blood and the bile. The liver is an important site of synthesis of fatty acids, bile acids, ketone bodies, cholesterol and cholesterol esters, phosphatides, and lipoproteins. Fatty acids are continually produced in the liver from acetyl-CoA, resulting from glucose metabolism. The acetyl-CoA is converted to malonyl-CoA to initiate this synthesis. The fatty acids are used in the synthesis of triglycerides, cholesterol esters, and phosphatides. Oxidation of fatty acids to produce energy and the removal of cholesterol and phosphatides from plasma are carried out by the liver (see Chap. 10, Lipids).

CHOLESTEROL AND CHOLESTERYL ESTERS

The liver is the key organ in the synthesis and excretion of cholesterol; therefore, diseases of the liver or biliary tract affect the plasma concentration of the free and ester forms. The *esterification of cholesterol* with polyunsaturated fatty acids by the liver cells depends upon the presence of a transferase enzyme and involves a different mechanism than does the synthesis of cholesterol. Impairment of the synthesis of transferase, as in parenchymal liver disease, is reflected in a decrease in the absolute and relative amounts (percentages) of cholesterol esters.

In *obstructive jaundice* there is a disproportionate increase in serum levels of free cholesterol, since free cholesterol and bile acids are normally excreted in the bile. Any type of obstruction, either intra- or extrahepatic, will cause an increase in total cholesterol levels in the serum, but the ratio of the esterified to the free form is less than that in normal serum. Acute obstruction may be characterized by total cholesterol levels of 300 to 400 mg/100 ml, whereas chronic biliary obstruction may result in levels around 800 mg/100 ml of serum. *Intrahepatic obstruction* tends to increase free cholesterol levels, and an overall increase in the total level may be seen in *early hepatitis*. In chronic conditions, such as *cirrhosis*, that involve considerable destruction of liver cells, the cholesterol level eventually falls below normal levels since decreased synthesis is taking place. Total cholesterol levels may be below 100 mg/100 ml, and the ester level may be less than 20 per cent of the total. In other conditions, such as *hypothyroidism* and *nephrosis*, an increase in total serum cholesterol is observed, but the percentages of free and ester forms are normal in these diseases.

For details of the methods for cholesterol and cholesterol esters and their relation to other lipids in the body, see Chapter 10.

BILE ACIDS AND SALTS

Bile acids such as *cholic acid* are catabolic products of cholesterol formed in the liver. The acids are conjugated with either glycine or the sulfur-containing compound taurine (sulfate analog of cysteine) to form bile salts that are called glycocholates or taurocholates. The bile salts are excreted in the bile and recirculate back to the liver through an enterohepatic pathway. About 0.8 g of the salts is excreted per day in the feces. The bile salts normally function in the emulsification of dietary fat, the activation of the lipases, and the absorption of lipids through the intestinal mucosa.

CLINICAL SIGNIFICANCE

The normal level of bile acids in the serum is 0.3 to 3.0 mg/100 ml. Increased levels are commonly observed in hepatitis and obstructive jaundice with no clear-cut differentiation between intra- and extrahepatic obstruction. The increased levels in the serum are caused by a decrease in the rate of conjugation with glycine and taurine in liver cell damage, and by an impairment of the enterohepatic circulation in obstruction. In the normal individual only small amounts of bile salts are excreted in the urine. As the level in the serum increases in liver disease, the urinary output increases.

Methods

Unfortunately, complex methods are required for the determination of bile acids in serum. These methods involve extraction with organic solvents, partition chromatography, gas chromatography-mass spectroscopy, and spectrophotometry or ultraviolet light absorption or fluorescence measurements in concentrated sulfuric acid. Investigators employing a recently developed complicated fluorescence method concluded that even when a sensitive quantitative method was available, the results added nothing to the diagnostic assistance gained from other tests for liver function.

Modifications of the Pettenkofer test may be used to detect the increased urinary excretion of bile acids in liver disease. The Pettenkofer reaction results in the development of a red-purple color when cholic acids react with fructose or furfural after the addition of sulfuric acid. The Mylius modification for the detection of bile salts in urine is carried out as follows: To 5 ml of urine in a test tube, add 3 drops of 0.1 per cent aqueous furfural solution, then add 2 to 3 ml of concentrated H_2SO_4 to form a lower layer. Cool under running water and shake carefully. The presence of appreciable amounts of bile acids will produce a red color.

CONJUGATION, DETOXICATION, AND EXCRETION

In addition to the role played by the liver in bilirubin, carbohydrate, protein, and lipid metabolism, another essential function involves conjugation and detoxication. The liver, by means of conjugation, is able to convert many toxic substances into nontoxic compounds, change active drugs into inactive conjugates, and alter the solubility of metabolites by esterification or conjugation to assist in normal excretion. Phenols, menthol, camphor, salicylates, indole, hormones, bromsulfophthalein (BSP dye), and bilirubin are common examples of compounds that are detoxified or conjugated in the liver prior to excretion. Glycine, glucuronates, and sulfates are the most common *conjugating agents*. Benzoic, nicotinic, and salicylic acids may be conjugated with glycine to form hippuric, nicotinuric, and salicyluric acids. Other conjugation mechanisms involve the formation of glucuronides of drugs such as salicylates and phenacetin, phenols, benzoic acid, and sterols. One of the essential steps in the metabolism of bilirubin is the formation of bilirubin diglucuronides by the liver cell before excretion in the bile. So many compounds are excreted in combination with glucuronic acid that the increase in glucuronides after the administration of a test substance has been proposed as a test of liver function.

Although the conjugation processes of the liver fulfill many useful functions for the body, the presence of conjugated forms often complicates the determination of drugs and metabolites in the urine. For example, to measure the total excretion of salicylates or of steroids such as 17-hydroxycorticosteroids or 17-ketosteroids, the conjugated forms must be hydrolyzed before analysis.

THE HIPPURIC ACID TEST

Of all the tests that have been proposed to evaluate the ability of the liver to detoxify or conjugate, the hippuric acid test remains the most practical. Benzoic acid in the form of sodium benzoate is conjugated with glycine to form *hippuric acid* for excretion by the kidney. Since the test requires the presence of both the conjugating agent and the enzyme system involved in the conjugation, it should provide a measure of liver function unlike tests described earlier. Nevertheless, it is little used today in liver function testing.

In the oral test, after emptying the bladder, the patient is given 6.0 g Na benzoate dissolved in about 200 ml of water. The urine is collected for a 4 h period. In the intravenous test, a sterile solution containing 1.77 g of sodium benzoate in 20 ml of water is injected by a physician. The patient empties his bladder and drinks a glass of water just prior to the injection. One hour after the injection the patient empties his bladder completely, collecting the urine. The normal excretion of hippuric acid expressed as benzoic acid in the oral test is 3.0 to 3.5 g/4 h, while the normal excretion in the intravenous test is 0.6 to 0.9 g/1 h.

REFERENCES

Quick, A. J.: Am. J. Med. Sci., *185*:630, 1933.
Weichselbaum, T. E., and Probstein, J. G.: J. Lab. Clin. Med., *24*:636, 1939 (modified).

THE BROMSULFOPHTHALEIN TEST

Phenolphthalein and several of its derivatives are excreted by the kidney and the liver. The brominated derivative of phenolsulfophthalein, called *bromsulfophthalein* or *BSP*, was found to be excreted almost entirely by the liver. The excretion of this dye has been used as a general liver function test since 1925. The rate of removal of BSP and its excretion into the bile depends on several factors: the blood level of the dye, the hepatic blood flow, the condition of the liver cells, and the patency of the bile ducts.

CLINICAL SIGNIFICANCE

In the absence of jaundice, the bromsulfophthalein test provides a simple, sensitive test of liver function that is capable of detecting early lesions of the liver cells. In the presence of jaundice from both intra- and posthepatic disorders including fatty liver, there is increased retention of the dye. In cirrhosis, in the absence of jaundice, an increased retention of 25 to 45 per cent of BSP is frequently observed. Significant retention of BSP is observed in some *space occupying lesions* of the liver, even in the absence of jaundice. In the presence of hepatic jaundice due to liver cell disease or due to obstruction, the test is of little value since it is abnormal in both cases.

SPECIMEN

The serum specimen should be free from hemolysis or lipemia. To avoid lipemia, the specimen is usually collected in the morning while the patient is in the fasting state.

PRINCIPLE

A serum specimen drawn after the injection of the dye is diluted with an alkaline buffer, and the absorbance of the purple color is measured at 580 nm. An acid reagent is then added to convert the dye to the colorless form, and the absorbance is read again. The difference in absorbance is due to the BSP dye present in the serum. The strong anion, *p*-toluenesulfonate, is added to the alkaline buffer to release the BSP dye from the serum albumin.

REAGENTS

1. Alkaline buffer, pH 10.6 to 10.7. Dissolve 6.46 g of Na_2HPO_4, 1.77 g of $Na_3PO_4 \cdot 12\ H_2O$, and 3.2 g of sodium *p*-toluenesulfonate in water and dilute to volume in a 500 ml volumetric flask. Check the pH and adjust if necessary with 1 molar NaOH or 1 molar HCl.
2. Acid reagent, 2 molar NaH_2PO_4. Dissolve 27.6 g of $NaH_2PO_4 \cdot H_2O$ in water and dilute to volume in a 100 ml volumetric flask.
3. BSP dye standard, 5 mg/100 ml; equivalent to 50 per cent retention. The intravenous BSP solution, 50 mg/ml, is diluted 1/1000 (for example, 0.5 ml of the dye diluted to 500 ml with water). The diluted standard is stable for 1 week.

PROCEDURE

1. To avoid lipemia, the test is usually run in the morning on a patient in the fasting state. The dye, 5 mg/kg of body weight, is injected by a physician under proper clinical conditions and a blood specimen is obtained 45 minutes after the injection of the dye.
2. To 1.0 ml of the serum add 7.0 ml of alkaline buffer and mix.
3. Read the absorbance (A_1) at 580 nm against H_2O.
4. Add 0.2 ml of the acid reagent, mix, and read absorbance (A_2) at 580 nm.
5. To 1.0 ml of the BSP dye standard add 7.0 ml of alkaline buffer, mix, and read the absorbance (A_3) at 580 nm.

CALCULATIONS

$$\text{per cent retention of BSP} = \frac{A_1 - A_2}{A_3} \times 50$$

NORMAL VALUES

With the dosage of 5 mg/kg, normal adults will show less than 6 per cent retention of BSP dye at 45 min. Very obese or markedly underweight people have different normal values, and the following corrections for body weight have been suggested: from 110 to 149 lb, add 1 per cent retention; from 170 to 189 lb, subtract 1 per cent retention; and from 190 to 279 lb, subtract 3 per cent retention.

PROCEDURE NOTES

A standard curve may be constructed by diluting the 50 mg/ml BSP solution 1/500 to prepare a stock 10 mg/100 ml solution representing 100 per cent retention. Dilute the stock solution to prepare 5 mg, 2.5 mg, 1 mg, and 0.5 mg/100 ml standards respectively, representing 50, 25, 10, and 5 per cent retention. To 1.0 ml of each of the five standard solutions, add 7.0 ml of alkaline buffer, mix, and read the absorbance at 580 nm against a water blank. Plot absorbance against per cent retention. Since the curve follows Beer's law, the 50 per cent retention standard may be used as in the preceding procedure.

REFERENCES

Seligson, D., Marino, J., and Dodson, D.: Clin. Chem., *3*:638, 1957 (modified).
Zieve, L., and Hill, E.: Gastroenterology, *28*:766, 1955.

ENZYMES IN SERUM AS AN AID IN THE DIAGNOSIS OF LIVER DISEASE

The search for useful liver function tests has naturally included a study of the activity of a number of enzymes in the serum. Many of these enzymes are involved in intermediary metabolism and are present in the liver cell in high concentration. When the cells are injured or disrupted, as occurs in acute liver disease, these enzymes are released into the serum and their increased levels are often of diagnostic significance. The relation of specific enzymes to liver function will be discussed in this section; however, detailed information on these and other enzymes, as well as methods for their determination, is given in Chapter 12.

The list of enzymes related to liver function is rather long and includes the transaminases, alkaline phosphatase, isocitrate dehydrogenase (ICD), α-hydroxybutyrate dehydrogenase (HBD), and lactate dehydrogenase (LDH). Other enzyme activity measurements of interest in liver disease include ornithine carbamyl transferase (OCT), 5′-nucleotidase, γ-glutamyl transpeptidase (γ-GT), sorbitol dehydrogenase (SDH), aldolase (ALD), and guanase.

LIVER FUNCTION TESTS IN THE DIFFERENTIAL DIAGNOSIS OF LIVER DISEASE

Table 19–1 lists various commonly used liver function tests and compares the changes observed in acute hepatitis, obstructive jaundice, and chronic hepatitis.

In *acute hepatitis* and liver cell injury, there is an increase in conjugated and unconjugated bilirubin (in the early stages, the unconjugated bilirubin is the predominant form), flocculation tests are positive, and alkaline phosphatase levels in serum increase one- to three-fold. Significant increases in the activities of the transaminases in serum (>8 × normal) are observed, L-alanine aminotransferase (GPT) being predominant over aspartate

TABLE 19–1 SUMMARY OF FUNCTION TESTS

Normal values for these function tests vary with the method employed; therefore, only relative increases or decreases from normal are given in the table. The normal values for each test may be found in the text and in the table of normal values in the Appendix.

Test	*Acute Hepatitis*	*Obstructive Jaundice*	*Chronic Hepatitis*
Enzyme Measurements in Serum			
Aldolase	Marked increase	Normal	Normal
Alkaline phosphatase	Slight increase	Marked increase	Slight increase
Cholinesterase	Decreased	Normal	Decreased
Aspartate aminotransferase (GOT)	Marked increase	Slight increase	Slight increase
L-Alanine aminotransferase (GPT)	Marked increase	Slight increase	Slight increase
γ-Glutamyl transpeptidase	Slight increase	Marked increase	Slight increase
Guanase	Marked increase	Normal	Slight increase
α-Hydroxybutyrate dehydrogenase	Normal	Normal	Normal
Isocitrate dehydrogenase	Marked increase	Slight increase	Increased
Lactate dehydrogenase	Moderate increase	Normal or slight increase	Normal
5′-Nucleotidase	Slight increase	Marked increase	Slight increase
Ornithine carbamyl transferase	Marked increase	Slight increase	Increased
Sorbitol dehydrogenase	Marked increase	Near zero	Near zero
Flocculation Tests			
Cephalin-cholesterol	Increased	Normal	Increased
Thymol turbidity	Increased	Normal (often)	Increased
Zinc sulfate turbidity	Increased	Normal	Increased
Miscellaneous Tests			
(tests measuring conjugation, detoxication, excretion, carbohydrate metabolism, and/or synthesis)			
Bilirubin, serum conjugated	Increased	Increased	Variable
unconjugated	Increased	Increased	Variable
Bilirubin, urine	Increased	Increased	Increased
Bromsulfophthalein	Increased	Increased	Increased
retention	(Test not carried out in presence of jaundice)		
Cholesterol, serum total	Normal	Increased	Normal or decreased
esters %	Decreased	Normal or decreased	Decreased
Galactose tolerance	Decreased	Normal	Decreased
Hippuric acid	Decreased	Normal	Decreased
Icterus index	Increased	Increased	Variable
Protein, serum total	Normal	Normal	Normal or decreased
albumin	Decreased	Normal	Decreased
globulin	Increased	Normal	Increased or normal
Prothrombin time	Increased	Increased	Increased
Urobilinogen, stool	Slight increase	Decreased	Slight decrease
Urobilinogen, urine	Increase (in early stage)	Decreased	Decreased

aminotransferase (GOT). Carbamyl transferase, sorbitol dehydrogenase, aldolase, and guanase are also markedly elevated in acute hepatitis compared to the normal or mildly increased serum levels found in obstructive jaundice.

LDH is found in many types of tissues and is a rather insensitive and unspecific index of liver disease. Electrophoretic separation of the five LDH isoenzymes, however, will show a characteristic increase in LDH_4 and LDH_5 in acute hepatic disease. Increased presence of the LDH_5 may also be demonstrated by heat inactivation of this heat labile isoenzyme (see Chap. 12). Determinations of α-hydroxybutyrate dehydrogenase (HBD) activity in serum reflects closely the activities of the LDH_1 and LDH_2 (heart muscle) isoenzymes and is, therefore, normal in liver necrosis. The normal ratio of HBD/LDH decreases markedly in acute hepatitis owing to an increase in LDH without a simultaneous increase in HBD. Thus, the measurement of the activity ratio differentiates between the LDH increase in hepatitis and that in myocardial infarction without the necessity of LDH isoenzyme determinations.

BSP retention is increased to varying degrees depending on the extent of cell necrosis, the degree of intrahepatic obstruction, and the conjugating ability by the liver. The total cholesterol level is essentially normal but the esterified fraction is decreased. The total protein level is normal with a decrease in the albumin (decreased synthesis) and an increase in the globulin, especially the gamma globulin (immune response). The urobilinogen is increased in the early stages, but decreases as intrahepatic obstruction increases and general liver function decreases. Total serum iron levels increase owing to release of iron and transferrin from the necrotic liver cell.

Patients with *obstructive jaundice* exhibit, early in the disease, a predominant absolute increase in the conjugated bilirubin; but later in the disease, possibly owing to secondary liver cell damage, significant increases in both conjugated and unconjugated bilirubin are seen. The flocculation tests are generally normal, unless there is significant secondary liver damage. Serum levels of alkaline phosphatase activity increase more than 3 × normal (compared to slight increases in liver cell damage), and significant elevation in the activities of 5′-nucleotidase and γ-glutamyl transpeptidase are seen. Although the activity of the last two mentioned enzymes parallels that of alkaline phosphatase, there is an additional advantage in that no increase in the activity of these enzymes is seen in bone disease. Thus, these tests are valuable in differentiating the increased levels of alkaline phosphatase observed in liver disease and liver malignancies from those found in skeletal conditions. Activities of the transaminases, LDH, HBD, SDH, ALD, and guanase in serum are only mildly increased compared with the marked elevations in acute hepatic injury (necrosis).

BSP retention is increased, although this test should not be done if jaundice is present. The total cholesterol level is increased and the percentage of esterified cholesterol is normal or decreased. Urobilinogen in stool and urine is low or absent, depending on the degree of obstruction. Total protein, albumin, globulin, and total serum iron values are normal.

In *cirrhosis*, which is characterized by the presence of large areas of liver necrosis and fibrous tissue formation, there is less active liver tissue and blood flow is diminished. This results in a general decrease in liver function associated with a very large increase in BSP retention and a severe decrease in albumin, accompanied by a characteristic increase in the β- and γ-globulins. Abnormal flocculation tests are caused by these serum protein changes. The total bilirubin may be normal or moderately increased, with the conjugated bilirubin essentially normal. The total serum cholesterol is markedly decreased and a low percentage of cholesterol esters is observed. (In biliary cirrhosis, on the other hand, total cholesterol values are very high.) Increases in enzymes in serum are marginal; aspartate transferase is generally higher than alanine transferase. The alkaline phosphatase is essentially normal; the total serum iron and the iron binding capacity are low, owing to a decrease in the synthesis of transferrin.

The diagnosis of *hemolytic jaundice* is facilitated by the results of liver function tests. Increased levels of unconjugated (and total) bilirubin, in the presence of essentially normal conjugated bilirubin levels in serum and increased urobilinogen levels in urine and stool, are generally found in this condition. Activities of enzymes which are present in red cells (transaminases, LDH) may be slightly elevated owing to the release of erythrocyte enzymes into the serum. Other enzyme tests, as well as the flocculation tests, are usually normal in this condition. Total serum iron is increased.

REFERENCES

1. Conway, E. J., and Cooke, R.: Biochem. J., *33*:457,1939.
2. Faulkner, W. R., and Britton, R. C.: Cleveland Clinic Quart., *27*:202, 1960.
3. Forman, D. T.: Clin. Chem., *10*:497, 1964.
4. Hanger, F. J.: Trans. Assoc. Am. Physicians, *53*:148, 1938; J. Clin. Invest., *18*:261, 1939 (modified).
5. Henry, R. J.: Clinical Chemistry: Principles and Technics. New York, Hoeber Medical Div., Harper & Row, Publishers, 1964.
6. Hougie, C.: Fundamentals of Blood Coagulation in Clinical Medicine. New York, McGraw-Hill Book Co., Inc., 1963.
7. Ishihara, A., Kurahasi, K., and Uehara, H.: Clin. Chim. Acta, *41*:255, 1972.
8. Kunkel, H. G.: Proc. Soc. Exp. Biol. Med., *66*:217, 1947 (modified).
9. Kurahasi, K., Ishihara, A., and Uehara, H.: Clin. Chim. Acta, *42*:141, 1972.
10. Leevy, C. M.: Evaluation of Liver Function. Indianapolis, Eli Lilly Research Laboratories, 1965.
11. Maclagen, N. F.: Brit. J. Exp. Path., *25*:334, 1944.
12. Meites, S., and Hogg, C. K.: Clin. Chem., *5*:470, 1959.
13. Neefe, J. R., and Reinhold, J. G.: Science, *100*:83, 1944.
14. Popper, H., and Schaffner, F.: Liver: Structure and Function. New York, McGraw-Hill Book Co., Inc., 1957.
15. Reinhold, J. G.: Advances in Clinical Chemistry. New York, Academic Press, 1960, vol. 3, p. 84.
16. Reinhold, J. G., and Yonan, V. L.: Am. J. Clin. Path., *26*:669, 1956 (modified).
17. Sherlock, S.: Diseases of the Liver and Biliary System. Springfield, Ill., Charles C Thomas, Publisher, 1965.
18. Stoner, R. A., and Weisberg, H. F.: Clin. Chem., *3*:22, 1957.

CHAPTER 20

GASTRIC, PANCREATIC, AND INTESTINAL FUNCTION

by Norbert W. Tietz, Ph.D.

Although it has long been assumed that there is a functional relationship between the stomach, the intestinal tract and the pancreas, little evidence had been available to support this. In recent years, however, the availability of highly sensitive laboratory assays for hormones regulating gastrointestinal functions[3] has made it possible to demonstrate clearly that such an interplay indeed exists.[10] It is therefore appropriate to discuss gastric, pancreatic, and intestinal physiology in one chapter.

The **human stomach** consists of three major zones, the cardiac zone, the body, and the pyloric zone (Fig. 20–1). The upper *cardiac zone* contains mucus-secreting *surface epithelial cells*. The body of the stomach contains cells or cell groups of four different types: (1) the surface epithelial cells, which secrete mucus; (2) the *parietal cells*, which are the main, and possibly only, source of hydrochloric acid; (3) the *chief* or *peptic cells*, which secrete a considerable amount of pepsinogen; and, finally, (4) the *neck chief cells* or mucus cells, which secrete mucus and pepsinogen. The third portion of the stomach, the *pyloric zone*, may be subdivided into the *antrum*, the *pyloric canal*, and the *sphincter*. Its cells secrete mucus,

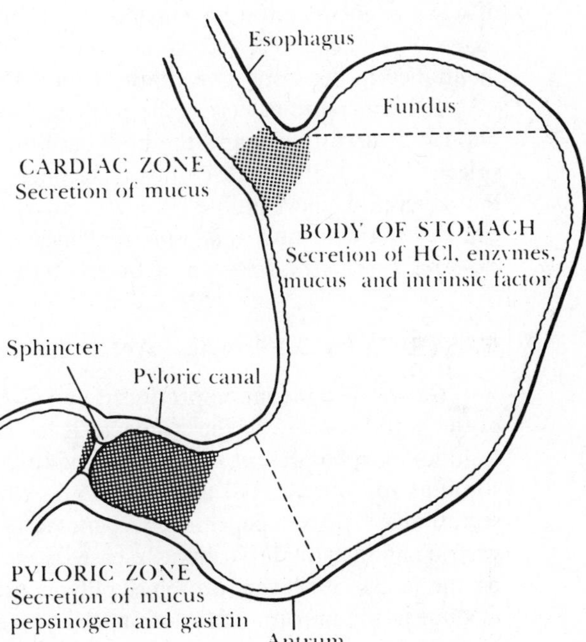

Figure 20–1 Schematic drawing of the stomach, with major zones.

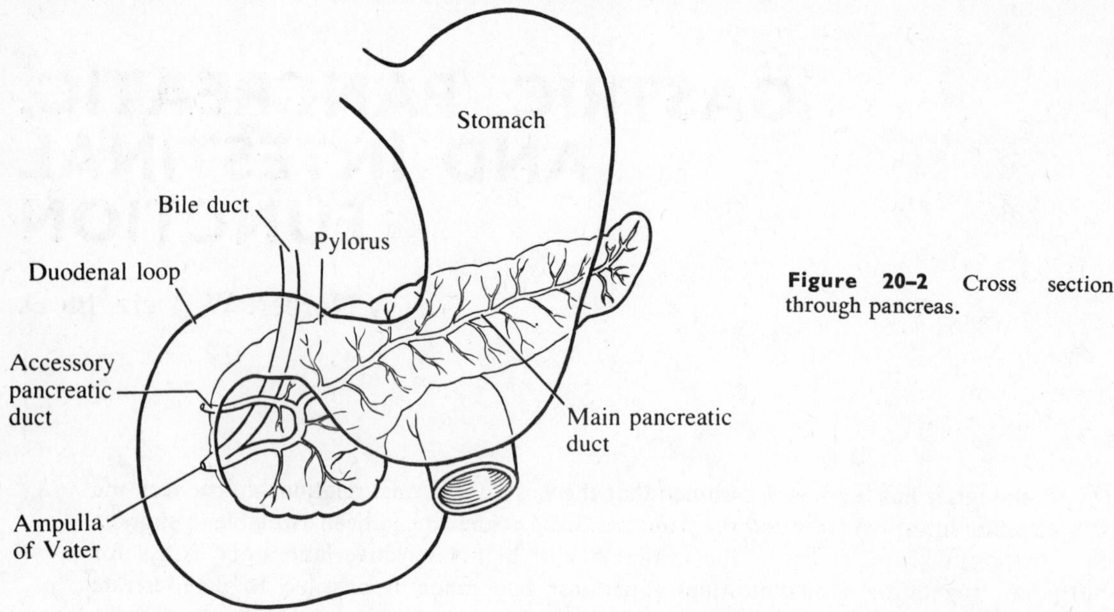

Figure 20–2 Cross section through pancreas.

some pepsinogen, and gastrin but no HCl. The pyloric zone also plays a major role in the process of emptying food into the **duodenum** by virtue of its strong musculature, which moves chyme through the pylorus. In the second portion of the duodenal loop (Fig. 20–2) is the ampulla of Vater with the sphincter of Oddi, through which the pancreatic juice and the bile enter the duodenum and mix with the food.

The food bolus is moved from the duodenum into the jejunum and ileum (= small intestines) and finally into the cecum, colon, and rectum (= large intestines). The intestinal tract is lined with mucosal cells which fulfill a variety of functions. They are the source of hormones (such as secretin, cholecystokinin-pancreozymin, and enterogastrone) and of a multitude of enzymes (e.g., disaccharidases), and they are involved in the absorption of water, minerals, and various other food materials. Most of these topics are beyond the scope of this chapter, but some will be briefly discussed later, while others have been mentioned in the respective chapters on carbohydrates, lipids, and electrolytes.

The **pancreas** lies across the posterior wall of the abdomen. The head is located in the duodenal curve (loop) and the body and tail are directed toward the left, extending to the spleen (Fig. 20–2). Because of its position deep in the retroperitoneal space in the abdomen, the pancreas is inaccessible for direct physical examination. The functions of the pancreas can be separated into *endocrine functions* (mainly secretion of glucagon and insulin) and *exocrine functions* (secretion of bicarbonate and digestive enzymes into the duodenum).

GASTRIC, INTESTINAL, AND PANCREATIC HORMONES[2,10]

Gastrin is a hormone produced and stored mainly by special endocrine cells (G-cells) of the antral mucosa of the stomach. It has also been demonstrated by immunofluorescent techniques to be present in the intestinal mucosa and in the pancreas.[44] Gastrin is a strong stimulus for gastric HCl and pancreatic enzyme secretion, but it also weakly stimulates secretion of gastric pepsinogen, pancreatic HCO_3^-, and hepatic bile. It also increases gastric and intestinal motility. Release of gastrin by the antral granular mucosa is mediated by the vagus; however, proteins, polypeptides, amino acids (glycine and β-alanine), and alcohol in the antrum of the stomach, as well as gastric and antral distention, also cause release of gastrin.

Gastrin is a polypeptide comprising a chain of 17 amino acids. Two forms, gastrin I and II, have been identified; they differ only by the presence of a sulfate ($—OSO_3H$) group on the tyrosine residue located in position 12 in the gastrin II molecule (see Fig. 20–3). There is evidence of the coexistence of a "little gastrin" with a molecular mass of 2100 and a "big gastrin" with a molecular mass of 7000 daltons. These can be separated by starch gel or paper electrophoresis or by Sephadex G-50. Both forms have identical immunoreactivity.

It has been established that the C-terminal tetrapeptide residue in the gastrin molecule is responsible for all its physiologic actions. Although derivatives of gastrin containing the same terminal tetrapeptide residue are physiologically active, the potency for each derivative differs. Synthetic *pentagastrin* (see Fig. 20–3 and discussion on gastric stimuli) is an example of such a derivative.

Maximum secretion of gastrin from the fundus was found to occur at an antral pH range of 5 to 7. Acidification of the antrum decreases gastrin secretion. This appears to be a safeguard against over-acidification.

Gastrin has been assayed in the past by biological *assays*, which are tedious, insensitive, and imprecise. Newer radioimmunoassays, however, are highly sensitive, reasonably reproducible, and can be performed in a hospital laboratory, as will be explained later.

Cholecystokinin-pancreozymin (CCK-PZ) is a polypeptide composed of 33 amino acid residues. The carboxy (C)-terminal pentapeptide sequence is identical to that of gastrin (see Fig. 20–3). It is believed that the distribution of CCK-PZ in the intestinal mucosa corresponds closely to that of the hormone secretin. For a long time it was thought that CCK and PZ were separate hormones, the former causing the contraction of the gall bladder and the latter stimulating the secretion of pancreatic enzymes. It is now well established that one single polypeptide hormone is responsible for both actions.

CCK-PZ is mainly produced by the mucosa of the duodenum and upper small bowel. Its secretion into the blood is stimulated primarily by peptones, but also by amino acids (methionine, valine, phenylalanine), fatty acids, and gastric hydrochloric acid entering the duodenum.

CCK-PZ also increases the motility of the duodenum and the small intestine. Since it possesses the same terminal amino acid tetrapeptides as gastrin, it also stimulates to a slight degree gastric HCl and pepsinogen secretion, antral motility, and pancreatic HCO_3^- secretion. Since CCK-PZ competes with gastrin for the receptor sites on the HCl secreting cells, secretion of this less potent hormone results in a decreased output of HCl and may contribute to the termination of gastric secretion after a meal. On the other hand, gastrin and CCK-PZ are additive in their stimulation of the pancreas and both increase the effect of secretin on pancreatic function.

Secretin is a hormone secreted by the S-cells located in large numbers in the duodenum and upper jejunum. It contains 27 amino acid residues, of which the positions of 14 residues are identical with those found in glucagon (see Fig. 20–3). The hormone is released primarily on contact of the S-cells with gastric HCl, but also by fatty acids and amino acids. As pancreatic juice flows into the duodenum, it neutralizes gastric acid and thereby removes one stimulus for its own secretion.

Secretin strongly stimulates the pancreas and the liver to secrete HCO_3^- and it also causes some pepsinogen secretion by the chief cells of the stomach. In addition, it slightly increases enzyme secretion from the pancreas and may provide a weak stimulus for gall bladder contraction. Secretin inhibits gastrin-stimulated acid secretion and gastric motility.

There is evidence of the presence of yet another hormone, **enterogastrone,** whose release is stimulated by fats and fatty acids as they are absorbed by the small intestine and the jejunum. This hormone has not yet been isolated, nor has its structure been elucidated. It is said to inhibit gastric motility and gastric secretion.

Insulin and glucagon are also pancreatic hormones; they have been discussed in the chapters on Carbohydrates and Endocrine Function.

	Approximate Molecular Mass	Structure
Gastrin	*I* *II* 2096 2176	1 2 3 4 5 6 7 8 9 10 11 12 13 14 15 16 17 Glu-Gly-Pro-Trp-Leu-Glu-Glu-Glu-Glu-Glu-Ala-Tyr-Gly-Trp-Met-Asp-Phe-NH_2 | (Tyr) SO_3H
Pentagastrin	768	*N-t*-butyloxycarbonyl-β-Ala-Trp-Met-Asp-Phe-NH_2
Cholecystokinin-pancreozymin (CCK-PZ)	3884	Lys-(Ala-Gly-Pro-Ser)-Arg-Val- (Ileu-Met-Ser)-Lys-Asp-(Asp-Glu-His-Leu_2-Pro-Ser_2)-Arg-Ileu-(Asp-Ser)-Arg-Asp-Tyr-Met-Gly-Trp-Met-Asp-Phe-NH_2 | (Tyr) SO_3H
Secretin:	3000	1 2 3 4 5 6 7 8 9 10 11 12 13 14 15 16 17 18 19 20 21 22 23 24 25 26 27 *His*-*Ser*-Asp-*Gly*-*Thr*-*Phe*-*Thr*-*Ser*-Glu-Leu-*Ser*-Arg-Leu-Arg-*Asp*-*Ser*-Ala-*Arg*-Leu-*Gln*-Arg-Leu-Leu-*Gln*-Gly-*Leu*-Val-NH
Glucagon:	3550	1 2 3 4 5 6 7 8 9 10 11 12 13 14 15 16 17 18 19 20 21 22 23 24 25 26 27 28 29 *His*-*Ser*-Gln-*Gly*-*Thr*-*Phe*-*Thr*-*Ser*-Asp-Tyr-*Ser*-Lys-Tyr-Leu-*Asp*-*Ser*-Arg-*Arg*-Ala-*Gln*-Asp-Phe-Val-*Gln*-Tyr-*Leu*-Met-Asn-Thr

Figure 20-3 Structure of some gastrointestinal hormones.

ENZYMES OF THE GASTROINTESTINAL TRACT

ENZYMES IN GASTRIC CONTENT

The *chief* and *peptic cells* and the *neck chief cells* or mucus cells secrete *pepsinogen*, which is activated to pepsin at a strongly acid pH provided by gastric hydrochloric acid. The active enzyme pepsin effects partial digestion of proteinaceous foods during their passage through the stomach. Some of the pepsinogen enters the blood directly and is subsequently excreted in the urine as *uropepsinogen*.

A small amount of *lipase* is also secreted by the stomach, but this contributes only slightly to the digestion of fats. Gastric content also includes some *amylase*, which is derived from swallowed saliva. Other enzymes are also present in gastric content, but their measurement does not seem to have any diagnostic significance. The activity of the multitude of enzymes present in the gastric cells has so far not been measured routinely for diagnostic purposes.

ENZYMES DERIVED FROM THE PANCREAS

The normal pancreas secretes a number of enzymes that pass almost entirely into the duodenum. Only a fraction of these enzymes reaches the blood directly, where they can be demonstrated. Enzymes of most clinical interest are *amylase*, *lipase*, and a group of *proteolytic enzymes*. Pancreatic enzymes are synthesized by the acinar cells and stored there in the form of zymogen granules (see Fig. 20–4). The relative proportions of lipase, amylase, and proteolytic enzymes in pancreatic juice change if the diet is predominantly of one type for long periods of time (e.g., proteolytic enzymes are secreted in larger amounts if there is a high protein intake.)

Other enzymes secreted in the pancreatic juice (in the form of precursors) include chymotrypsin A and B, carboxypeptidases A and B, phospholipase A, ribonucleases, elastase, and collagenase.

Upon stimulation by CCK-PZ and to a lesser extent by gastrin and secretin, these enzymes are released by the acinar cells in a fluid containing HCO_3^- and are secreted into the lumen of the acinus. From there they pass through the ductules into the main pancreatic duct, which empties into the duodenum through the ampulla of Vater. Frequently, the bile duct joins the pancreatic duct just proximal to or at the ampulla of Vater (see Fig. 20–2). The pancreas may have a second duct (duct of Santorini) which empties separately or jointly with the main duct into the duodenum. In the duodenum, the pancreatic juice mixes with

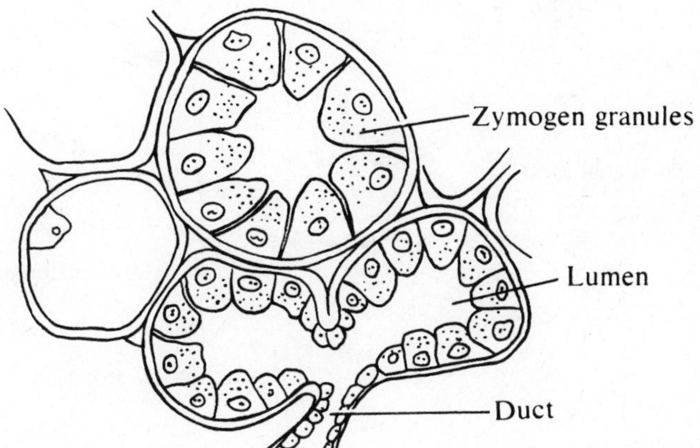

Figure 20–4 Section through acinar cells.

the food material coming from the stomach. The combination of enzymes secreted by the pancreas, at the proper pH provided by the bicarbonate, can digest virtually any food material.

In some disorders, such as pancreatitis or obstruction of the pancreatic duct due to stones or carcinoma, the flow of enzymes and bicarbonate into the duodenum is impeded (Fig. 20–5). This results in a decreased secretion of pancreatic juice into the duodenum, as can be demonstrated by direct examination of the duodenal content (see Secretin and CCK-PZ Test). At the same time, however, an increased amount of pancreatic enzymes reaches the blood stream under such conditions (back flooding). The mechanism by which the enzyme enters the circulation is not exactly known, but it is thought to be due to changes in the pressure in the pancreatic duct and ductules, changes in the permeability of acinar cells, or disruption of the acinar limiting membrane.

The events which lead to pancreatitis and necrosis of pancreatic tissues are presently under intensive debate.[41] Implicated in this process are lysosomal hydrolases and kallikrein, both heavily concentrated in pancreatic tissues. Kallikrein, as is true for all kinins, can by itself induce the cardinal features of inflammation. Activated proteases are also assumed to penetrate pancreatic tissue, resulting in edema and necrosis. Necrosed acini may release elastase and collagenase, resulting in damage to the walls of capillaries in the gland. Attention has recently also been directed to phospholipase A, which acts on phospholipids to release lysolecithin, a strong cytotoxic agent. Bile, refluxed into the pancreas by obstructive mechanisms, can act as a substrate for lysolecithin formation.

Enzymes released by the necrotic process spread not only through but also beyond the confines of the pancreatic gland into surrounding structures. Lipase, for example, causes fat necrosis both in the pancreas and in a wide area of contiguous structures. In this process, fatty acids are formed, which withdraw Ca^{2+} from the extracellular fluid by formation of calcium soaps (see pp. 903 and 1094).

The clinically most important enzymes in serum, in regard to evaluation of pancreatic function, are amylase and lipase. Measurement of proteolytic enzymes in serum has been less successful owing to the presence of inhibitors in plasma and serum.

Amylase and lipase activity in serum. The activities of both amylase and lipase in serum are significantly elevated in cases of *acute pancreatitis* and *obstruction of the pancreatic duct.* Values of **amylase** activity frequently exceed 600 Somogyi units (normal: 40 to 180 Somogyi units/100 ml or 65 to 290 U/1). Values above 1000 Somogyi units suggest pancreatitis due to obstruction by a gall stone. Amylase activity in serum tends to increase rapidly after an attack and may be demonstrated as early as six to eight hours after its onset. Levels generally

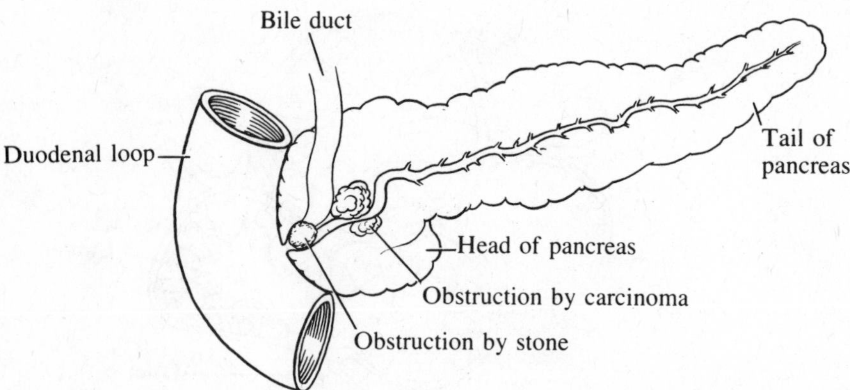

Figure 20–5 Cross section through pancreas with obstruction of the pancreatic duct by stone and carcinoma respectively.

stay elevated for one to three days and then return rapidly to normal levels, reflecting the efficient renal clearance of the enzyme. Thus, blood samples for amylase activity measurements must be collected as soon as possible if they are to be of diagnostic value. Between 70 and 90 per cent of the increase in amylase activity in acute pancreatitis is due to an increase in isoenzymes derived from the pancreas. Persistently high levels of amylase activity in serum indicate a *pseudocyst*.

Serum amylase activity is frequently elevated sooner than lipase activity, but the latter may stay elevated for a longer period. Such findings, however, are not consistent and it appears that amylase and lipase activity measurements in serum complement rather than exclude each other.[42]

In *chronic pancreatitis*, values for serum amylase activity are generally lower than those seen in acute pancreatitis and are limited to periods of exacerbation. No elevation may be found if significant amounts of acinar tissue have been destroyed.

In *intra-abdominal diseases* such as perforated peptic, gastric, or duodenal ulcers, intestinal obstruction, and acute peritonitis, and after surgical procedures in the abdomen, elevated amylase values are frequently observed. Although values in these disorders rarely exceed 500 units, values of more than 1000 units have been observed in a few cases. Amylase elevations have also been observed in acute diseases of the salivary glands such as *mumps*, in *renal disease* with diminished clearance, and after administration of *drugs* such as morphine, codeine, and meperidine (Demerol). These drugs cause a transient spasm of the duodenal musculature and Oddi's sphincter and thus obstruct the flow of bile and pancreatic juice.

Mumps and *bacterial parotitis*, which block the secretion of salivary amylase, are associated with mild elevation of serum amylase (200 to 600 Somogyi units). Hyperamylasemia due to presence of *macroamylase* is unrelated to pancreatic disease and is discussed in more detail in Chapter 12.

Low serum amylase values have been found in abcesses of the liver, acute hepatocellular damage, cirrhosis, cancer of the liver and bile duct, and cholecystitis. Jaundice alone and chronic diseases of the gall bladder do not influence amylase levels.

Measurements of **lipase activity in serum,** in general, have the same clinical significance as amylase activity measurements. At times, lipase activity may increase to a greater degree and may be found to be elevated for much longer periods (as long as 14 days in some cases).[45]

Lipase activity in acute pancreatitis or obstruction of the pancreatic duct may reach levels of more than 10 times the upper normal with two-point assays[46] or 30 to 50 times the upper normal with recently developed kinetic assays.[47] Lipase activity in serum is normal in acute diseases of the salivary gland and is less affected by intra-abdominal diseases than is amylase activity.

The literature referring to the clinical significance of lipase activity measurements in serum is very confusing and unsatisfactory, because most studies were based on experiments employing methods that are either unsatisfactory for lipase activity measurements or nonspecific in that they measure esterase rather than lipase activity.

Techniques for the measurement of amylase and lipase activity in serum are described in detail in Chapter 12.

Amylase activity in urine.[7,23,29] Amylase is cleared from the plasma by the kidneys at a rate of 2.9 ± 0.2 ml/min and is present in the urine of all healthy individuals. In acute pancreatitis the quantity of amylase cleared may increase 3- to 5-fold, resulting in a significantly increased amylase activity in urine. For this reason, it is felt by some investigators that urine amylase measurement is a more sensitive test in the diagnosis of pancreatitis than the determination of amylase activity in serum. It has been claimed that amylase activity in urine increases sooner and stays elevated for longer periods than does serum amylase activity. Other investigators have found results in some cases to be inconsistent and have reported false positive results. Amylase activity in the urine of normal adults is 1000 to 5000 (6000) Somogyi units/d. The average daytime output of amylase is 45 to 275 Somogyi units/h. Values for females are slightly lower.[7]

Levitt *et al.*[23] have suggested that the diagnostic value of urinary amylase measurements may be enhanced if amylase excretion is related to creatinine excretion. The ratio of amylase clearance (C_{Am}) to creatinine clearance (C_{Cr}) is markedly increased in acute pancreatitis, but it is decreased in renal insufficiency and extremely low in patients with macroamylasemia.

There have been reports that urine also contains *pancreatic lipase*. With newer techniques, however, no lipase activity can be demonstrated in urine. It is not established whether this is due to the absence of lipase or due to the presence of inhibitors. No practical diagnostic test is presently available.

Leucine aminopeptidase (LAP) in serum. LAP is found in almost all human tissues and also in serum, urine, and bile. It was once believed that activity measurements of LAP in serum could be helpful in the diagnosis of carcinoma of the pancreas. It has subsequently been shown that such elevations are not consistently present. Also, elevations have been reported in a number of other unrelated conditions such as obstructive jaundice, infectious hepatitis, metastatic cancer, acute pancreatitis, and pregnancy. Thus, it appears that the clinical usefulness of this test in the diagnosis of pancreatic carcinoma is rather limited.

Proteolytic activity. The proteolytic activity present in pancreatic juice was originally attributed to one enzyme, trypsin. It is now well established that pancreatic juice not only contains trypsinogen, which is activated by intestinal enterokinase or by autocatalysis to trypsin, but it also contains other proteases and peptidases. The exopeptidases (carboxypeptidases A and B) hydrolyze the terminal peptide bond at the carboxyl (—COOH) end of the peptide chain, and endopeptidases (trypsin, chymotrypsin, and elastase) hydrolyze the interior peptide bonds. Trypsin attacks the peptide chains specifically at a point where basic amino acids (arginine and lysine) remain at the —COOH end of the peptide chain after hydrolysis by the peptidases (see Fig. 20–6).

Trypsin in serum. Although trypsin activity has been demonstrated in serum, there is presently no simple test for trypsin in serum available. This is mainly due to the presence in serum of several trypsin inhibitors.[35]

Trypsin in duodenal juice. Trypsinogen, the precursor of active trypsin, is secreted by the pancreas and subsequently activated by the action of intestinal enterokinase. Subsequent activation proceeds autocatalytically (newly formed trypsin activates trypsinogen). Trypsin is an endopeptidase which preferentially hydrolyzes the peptide bonds involving the carboxyl group of arginine and lysine. For most clinical applications a screening procedure, such as that described in Chapter 12, is adequate. This test is not specific, however, and measures other proteases as well. For a specific, quantitative determination of tryptic activity in duodenal content, one of the synthetic substrates, such as *p*-toluenesulfonyl-L-arginine methyl ester (TAME), is the preferred substrate.[35] The reaction can be followed by measuring the absorbance at 247 nm, which increases as the ester bond is hydrolyzed. The reaction can

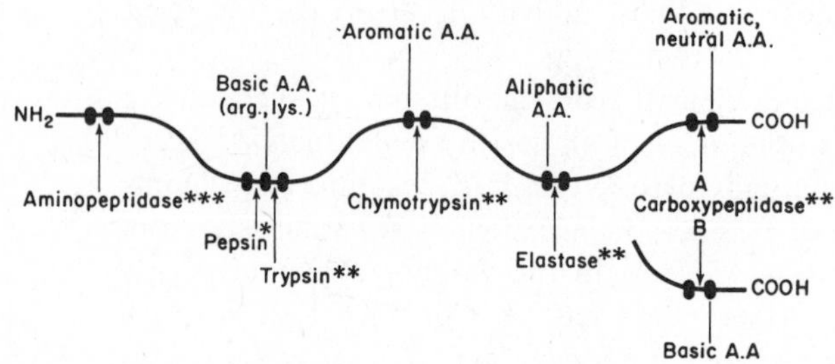

Figure 20–6 Sites of action of gastric (*), pancreatic (**), and small bowel (***) proteolytic enzymes on hypothetical protein molecules. (From Newcomer, A. D.: Digestion and absorption of proteins. Mayo Clin. Proc., *48*:614–621, 1973.)

also be monitored with a pH-stat by measuring the newly formed H^+.[35] Benzoylarginine-*p*-nitroanilide can also be used as a substrate. It is preferred by some, since it is chromogenic and the reaction may be followed by measuring the absorbance at 405 nm. Some commercial kit procedures are based on this principle. Proteins such as hemoglobin or casein should not be used as substrates for trypsin activity measurements, since chymotrypsin also hydrolyzes these substrates.

Trypsin and chymotrypsin in stool. The trypsin and chymotrypsin present in duodenal content mix with food material and are excreted in the stool (except for the fraction which is being digested or inactivated in the intestinal tract). Activity measurements of these enzymes in stool have therefore been used in the diagnosis of obstruction of the pancreatic duct by stone or tumor and in the diagnosis of cystic fibrosis, in which a large portion of the acinar tissues has been replaced by fibrous tissue. Results should be interpreted carefully, because false normal results may be obtained in the presence of proteolytic bacteria in stool and false negative results have been reported in some individuals with normal pancreatic secretory capacity.

Most frequently, a screening test, described in Chapter 12, is employed. Some investigators recommend that the activity of chymotrypsin rather than trypsin be measured, since chymotrypsin appears to be more stable during passage through the intestine.[1]

ENZYMES IN THE INTESTINAL MUCOSA

Although the intestinal mucosa contains a wide variety of enzymes, most of these are not measured for routine diagnostic purposes. The importance of disaccharidases for the absorption of disaccharides will be discussed in a later section of this chapter.

THE DIGESTIVE PROCESS AND THE INTERACTION OF GASTROINTESTINAL HORMONES[2,10,11]

The availability of radioimmunoassays for measuring most of the gastrointestinal hormones has greatly advanced our understanding of the digestive process. It is now well documented that there is a very delicate and complex interplay between the various gastrointestinal hormones which can briefly be summarized as follows:

The intake of food stimulates the afferent nerves in the brain as a result of stimuli from taste buds and smell sensors. This initiates the *cephalic phase* of the digestive process, which is mediated by the vagus. As a result, there is direct stimulation of the gastric chief and oxyntic (parietal) cells to secrete pepsinogen and HCl. Acetylcholine appears to be the mediator in this process, possibly sensitizing the parietal cells to the action of histamine. In addition, vagal fibers to the pyloric glandular mucosa cause the release of gastrin, which also stimulates pepsinogen and HCl secretion. The actions of acetylcholine and gastrin on the secretory cell are additive and result in a greater stimulation than vagal or gastrin stimulation alone.

As soon as food enters the stomach, the resulting distention initiates the *gastric phase* of secretion, which is mediated by local and vagal reflexes. Gastrin secretion is also promoted by distention of the stomach, especially the pyloric gland area. The greatest stimulation of gastrin, however, is produced by partially digested meat products, amino acids, and alcohol. The secretion of gastrin is optimal between pH 5 and 7, which may be obtained in the pyloric zone as a result of neutralization of gastric HCl by the ingested food. This additional secretion of gastrin fully stimulates the gastric secretory process. Moreover, gastrin stimulates antral motility and initiates secretion of a fluid from the pancreas that is rich in enzymes. It also stimulates the release of secretin by the duodenal mucosa and the release of insulin by the pancreatic islet cells.

As the weakly acidic digestive products enter the duodenum, CCK-PZ is released mainly by digestion products of protein and fat. This results in secretion of enzymes and in contraction of the gall bladder. In addition, insulin and glucagon are released by pancreatic islet cells. The acidic digestion products entering the duodenum also cause the release of secretin by the intestinal mucosa, which in turn stimulates secretion of bicarbonate-containing fluid by the pancreas and the liver (*intestinal phase of digestion*). At the same time, however, CCK-PZ and secretin inhibit gastrin-stimulated acid secretion and gastric motility and thus possibly initiate the end of the gastric phase of secretion.

GASTRIC FUNCTION TESTS

The functions of the stomach include accepting, mixing, storing, and discharging food into the duodenum as well as secreting enzymes, intrinsic factor, and especially hydrochloric acid. This has been discussed in more detail in the preceding sections. The various motor and secretory activities of the stomach can be evaluated by a number of clinical laboratory tests, among which are the analysis of gastric residue (the content of the stomach after fasting) and the determinations of the secretion rate in the basal state and after stimulation with appropriate stimuli. These tests, however, are more helpful in evaluating specific functions of the stomach, than in providing specific diagnostic data.[6,8] Detection of compounds not normally seen in gastric contents (e.g., blood or lactic acid) is also of clinical significance.

GASTRIC RESIDUE AND ITS MAJOR CONSTITUENTS

Gastric residue is defined as the content of the stomach after a fast of approximately 12 h. The specimen is obtained by aspiration through a gastric tube. The *normal total volume* is between 20 and 100 ml, usually being below 50 ml; volumes above 100 ml may be considered abnormal. Among the causes for *increased volume* are: delayed emptying, as in pyloric obstructions; increase in gastric secretion, as in duodenal ulcer or Zollinger-Ellison syndrome; and admixture of regurgitated material containing bile from the duodenum. In this last case, the gastric residue generally contains an excessive amount of bile, which can be confirmed by chemical tests for bilirubin (e.g., Ictotest test). The *consistency* of gastric residue is rather fluid but may be viscous in the presence of excessive amounts of mucus. The normal *odor* of gastric juice is sharply sour; a foul smelling gastric juice generally indicates putrefaction or fermentation. The residue is usually colorless, but in the presence of regurgitated bile its color may be slightly yellow or green. This is the case in approximately 25 per cent of normal individuals and in the majority of patients after partial gastrectomy or gastroenterostomy. A red or brown color in the gastric residue is usually due to blood, which can readily be confirmed by chemical tests, such as the Hematest (see Chap. 8).

Free Hydrochloric Acid. Hydrochloric acid is secreted by the parietal cells at a constant concentration of approximately 155 mmol/l. The pH of parietal secretion is approximately 0.9. As the secretion mixes with other gastric constituents such as mucus, saliva, regurgitated material, and ingested food, the final concentration of hydrogen ions decreases to approximately 0 to 40 mmol/l and the pH increases to 1.5 to 3.5. It is believed that the variations in the concentration of hydrochloric acid and in the pH of gastric residue are due to changes in the proportions of hydrochloric acid relative to other stomach contents. The secretion of hydrochloric acid by the mucosa is continuous, but the volume fluctuates considerably depending upon the degree and type of stimulation. As previously discussed, the flow is regulated by a neural mechanism mediated through the vagus (*psychic or cephalic phase*) and is stimulated by gastrin.

The *concentration* of free acid in gastric residue of normal individuals varies from approximately 0 to 40 mmol/l (without stimulation). Approximately 4 per cent of young normal individuals may have no free hydrochloric acid in the fasting stomach. This percentage increases with increase in age and is about 25 per cent in individuals at age 60. Absence of free hydrochloric acid in gastric residue is considered abnormal only if the condition also persists after maximal stimulation with one of the stimuli listed later. It follows that patients without free hydrochloric acid should always be subjected to gastric stimulation before a diagnosis of *achlorhydria* is made. *False achlorhydria* is a term used for those cases in which free hydrochloric acid is secreted, but in which the hydrogen ions are subsequently partially or fully neutralized by either saliva, food, or regurgitated materials. A determination of chloride might help avoid misinterpretation in these cases (normal: 45 to 155 mmol/l).

The determination of the concentration of free hydrochloric acid is generally done by titration. Measurement of the pH of the gastric residue will also give an indication of the amount of hydrogen ions present and is the preferred method according to some investigators (see Method).[28]

The hydrochloric acid secreted into the lumen of the stomach derives ultimately from the blood. Although the mechanism for this phenomenal process is still not understood, it is presently believed that there is an active transport of these ions across the membrane of the parietal cells, and that the energy for this process is supplied by ATP which is generated by oxidative metabolism and glycolysis. The hydrogen ion concentration is raised from 4×10^{-5} mmol/l plasma to 155 mmol/l in parietal secretions. The chloride concentration is increased from 103 mmol to about 170 mmol/l. The rapid secretion of gastric hydrochloric acid is accompanied by a decrease in blood chlorides and an increase in pH and HCO_3^- of the blood, the so-called *alkaline tide*.

Total Acidity. This component includes hydrogen ions occurring as (1) free HCl, (2) mucoprotein, (3) acid salts, and (4) organic acids such as lactic and butyric acid. Lactate and butyrate are normally not found in gastric juice. The total acid concentration is usually between 10 and 50 mmol/l, but may occasionally be slightly higher.

Combined Acidity. The difference between free and total acidity is due to acids other than free hydrochloric acid (see Total Acidity) and amounts to about 10 to 20 mmol/l. The old term "combined acidity" stems from the fact that some of the acid is bound to or reacted with other material, especially proteins (acids or bases with a pK of less than 1×10^{-4}).

Moore[28] has recommended that the term "combined acidity" be replaced by the term "nonionized acids." There is considerable doubt whether the determination of the "combined acidity" has any clinical significance, and many investigators suggest that this test be discontinued.

Organic Acids. Lactate and butyrate are formed in gastric content by bacterial action when food is retained in the stomach for long periods (over 6 h) at neutral or slightly alkaline pH. Such a condition is usually associated with carcinoma of the stomach or pyloric stenosis. If free hydrochloric acid is present, these organic acids are absent.

Enzymes. Gastric residue contains a number of enzymes such as pepsin, lipase, and salivary amylase. The most important of these is pepsin, which has been discussed elsewhere.

Mucus. Mucus is produced by the surface epithelial cells and the neck chief cells of the stomach. It is present in only small amounts in normal gastric contents; however, increased amounts are sometimes found in gastric carcinoma, in gastritis, and in cases of mechanical irritation due to passage of a stomach tube. Mucus has a pH of approximately 7.4 to 8.2 and is chemically composed of mucopolysaccharides and protein moieties. There is generally no clinical interest in analyzing gastric content for mucus.

Blood. In some pathologic conditions, such as carcinoma of the stomach, peptic ulcer, gastritis, or bleeding gums, blood may be present in gastric residue. Its appearance will vary with the pH of the gastric content. At the strongly acid pH in the stomach, acid hematin is formed; it has a brownish appearance, resembling coffee grounds. Fresh red blood may be due to accidental trauma from the gastric tube; however, in some instances, underlying lesions, such as ulcer or carcinoma, may be responsible for its presence.

Food. Normal gastric residue does not contain any appreciable amount of food. Excessive accumulation of food particles indicates decreased motor activity of the stomach or pyloric obstruction.

Miscellaneous Materials. Gastric residue may contain mucus cells, chief cells, parietal cells, and regurgitated materials such as pancreatic juice, bile, and duodenal secretions. It also contains the intrinsic factor, which combines in some way with vitamin B_{12} and makes absorption of this vitamin possible. Lack of this factor causes vitamin B_{12} deficiency with resultant arrest in the development of red cells. It is now believed that the intrinsic factor is a mucoprotein with a molecular mass of about 50,000 and a terminal structure similar to that of specific blood group carbohydrates. It is secreted by the cells of the fundus and behaves electrophoretically as a β-globulin. The determination of the intrinsic factor is now feasible, but it is rarely done except indirectly by the Schilling test, which will be described in a later portion of this chapter.

STIMULI OF GASTRIC SECRETION

Test Meals, Caffeine, and Alcohol. During the past decades of gastric testing, stimulation of gastric secretion was performed by administration of meals such as toast with water or tea (Ewald meal). Caffeine sodium benzoate (500 mg in 200 ml of water) admitted into the stomach through a gastric tube or in the form of tablets (see Diagnex Test) as well as ethanol (3.5 ml ethanol in 50 ml water) has also been used to stimulate gastric secretion. These stimuli have the advantage that they give essentially no undesirable side reactions; however, the degree of stimulation is relatively weak. Thus, other stimuli capable of eliciting maximum response, such as Histalog or pentagastrin, are now generally preferred.

Histalog. Histalog (3-beta-aminoethylpyrazol dihydrochloride, Eli Lilly & Co., Indianapolis, IN.) is presently the most widely used stimulus for gastric HCl secretion. It stimulates the parietal cells to secrete hydrochloric acid, without provoking the chief cells to secrete pepsinogen. In achlorhydria and achylia there is no secretion of hydrochloric acid after Histalog stimulation. Histalog has replaced histamine as a stimulus since it shows fewer side effects in patients. (Cases of acute erosive gastritis and gastric bleeding have been reported, in addition to anaphylactic shock). Nevertheless, all patients should be questioned about any history of allergic reactions, and the drug should not be administered if the systolic blood pressure is below 110 mm Hg. The drug should always be administered by a physician. Simultaneous administration of an antihistamine 30 min before the Histalog or histamine injection is recommended by some and will not affect hydrochloric acid secretion. The most commonly used dose for Histalog is 1.5 mg/kg of body weight. Maximum stimulation is obtained with 1.7 to 2.0 mg/kg;[22] however, such high levels may not be advisable for routine use because of the higher incidence of side reactions. The peak of hydrochloric acid secretion occurs 90 to 120 min after Histalog stimulation.

Gastrin and Pentagastrin.[44] *Gastrin I and II* are the natural and most powerful stimuli for gastric HCl secretion (see p. 1064). Gastrin has been prepared in pure form and is administered by a single subcutaneous dose of 2 μg of gastrin/kg of body weight or 1 μg/kg intramuscularly. Gastrin may also be administered by continuous intravenous infusion (67 μg/h) or by a single intraveneous injection (50 μg). The extent of stimulation is nearly 20 per cent greater than that produced by the maximal dose of histamine; on a molar basis, it is 500 times as potent as histamine. Although this agent is not yet being used routinely, it is conceivable that it will become the stimulus of choice in the future.

The synthetic product *pentagastrin* contains the same terminal sequence of four amino acids that is responsible for the physiological action of gastrin. This compound is also a potent stimulus for gastric HCl secretion, although a higher dose is required for maximal stimulation (6 μg subcutaneously/kg body weight). The response to the stimulus is highly reproducible, and side reactions are extremely rare and usually limited to epigastric discomfort. The drug has not been released by the FDA for regular use in the USA, and thus may presently be used only for investigative purposes.

Insulin. Hypoglycemia, as a result of insulin administration, causes stimulation of the vagal medullary nuclei, resulting in increased gastrin release. The amount of insulin administered should result in a drop in blood glucose level to below 45 mg/100 ml; this is usually obtained with a dose of 20 units of insulin (see Hollander Test).

PROCEDURES FOR GASTRIC ANALYSIS

ANALYSIS OF GASTRIC RESIDUE

PREPARATION OF THE PATIENT

The patient should be in the fasting state (no food or liquids for 12 h). The patient is not permitted to smoke the morning of or during the test and should avoid any form of exercise.

COLLECTION OF GASTRIC CONTENT

A lubricated gastric Levin tube is inserted orally; nasal intubation is used if the patient has a hyperactive gag reflex. Proper positioning of the tube should be confirmed by X-ray or fluoroscopy. When the patient has become calm and adjusted to the presence of the tube (after about 10 to 15 min), aspirate all gastric residue while the patient is in an upright position and leaning slightly to the left. Place the specimen into an appropriate container. Measure the pH or place a drop of Toepfer's reagent into the gastric residue. A pH below 3.0 or a red color after the addition of Toepfer's reagent indicates the presence of free acid. A pH above 3.0 or a yellow color of Toepfer's reagent indicates absence of free acid. For some purposes (e.g., exclusion of pernicious anemia), the test may be discontinued at this point, as presence of HCl excludes this diagnosis. If no HCl is found, a gastric stimulus is administered (see Gastric Stimuli).

Currently, many clinicians and scientists feel strongly that analysis of gastric residue gives inadequate information and that this test should be replaced by tests measuring the *secretion rate* of HCl. Directions for determinations of the basal secretion rate and maximum or peak acid output after stimulation are given in a later section.

DETERMINATION OF FREE HYDROCHLORIC ACID IN GASTRIC RESIDUE

PRINCIPLE

A known amount of gastric residue is titrated with 0.10 molar NaOH to a pH of 3.5 using a pH meter, or until Toepfer's reagent (0.5 g diethylaminoazobenzene/100 ml ethanol) added to gastric juice takes on a salmon color.

Titration to a pH of 3 to 3.5 detects essentially all free hydrochloric acid, as can be shown by titration curves established by Michaelis[27] and Lubran.[25] (Since HCl is the only strongly ionized acid in gastric content, this test is essentially a test for free HCl.) Titration beyond a pH of 3.5, as recommended by some, will overestimate the HCl concentration to varying degrees, depending on the composition of the gastric residue. On the other hand, titration to pH 3.5 may underestimate the amount of free H^+ secreted if some of these H^+ ions are bound to or have reacted with other constituents of gastric content. Thus, there is presently no fully satisfactory procedure available which measures accurately the true total amount of free acid secreted by the gastric mucosa.

Moore[28] maintains that titration to a preselected pH is an erroneous concept since the activity coefficients for hydrogen ions in hydrochloric acid and in gastric juice are different. The author recommends pH measurements and calculation of the H^+ concentration using an apparent activity coefficient. It is doubtful whether it is possible to accurately convert pH to H^+ concentration in a material as complex as mucus-containing gastric content, and whether such an approach can yield results of greater accuracy than that of sample titration. It is also questionable whether such an additional effort is warranted in view of the wide overlap of results in various clinical conditions.

PROCEDURE

1. Determine the pH of the gastric specimen with a pH meter. If the pH is above 3.5, no free acid is present. Such a specimen need not be titrated, and the patient should be administered a gastric stimulus.
2. Pipet a convenient volume of gastric juice (e.g., 5.0 ml, but not more than 10.0 ml) into a clean titration jar. If the gastric juice contains food particles or mucus, centrifuge the sample or filter it through gauze.
3. Titrate the sample with 0.10 molar NaOH to a pH of 3.5, using a pH meter. If a pH meter is not available, add 2 drops of Toepfer's reagent and titrate to a salmon color.

CALCULATION

$$\text{Free HCl in mmol/l} = \frac{\text{ml of 0.10 molar NaOH} \times 0.1 \times 1000}{\text{ml of gastric content used}}$$

where 0.1 = mmol NaOH/ml titrant,

or, if 5 ml of gastric content was used:

$$\text{mmol/l} = \text{ml of 0.10 molar NaOH} \times 20$$

REMARKS AND SOURCES OF ERROR

Normal gastric content is colorless and has a sour odor; its total volume should not exceed 100 ml. If these criteria are not met, the appropriate observation should be noted on the report. The appearance of the specimen may suggest performance of additional tests such as the following: a test for blood (e.g., Hematest) in case of brownish gastric content; a test for bile (e.g., Ictotest) if the gastric content appears yellowish or green; a test for lactic acid if a foul smell is observed.

DETERMINATION OF pH OF GASTRIC JUICE

Intragastric pH measurements with electrodes, mainly glass electrodes, represent the most recent approach to pH measurement of gastric juice. Technical difficulties, such as problems in proper placement of the electrode, and pH fluctuations due to swallowed air, limit this technique mainly to research work, and make it impractical for routine work.

Determinations of pH on aspirated gastric juice are best done with a pH meter. Use of pH papers may introduce significant errors.

There is no agreement about whether the gastric juice should be centrifuged before performance of the test. Many investigators recommend this technique, but others feel that some H^+ may be adsorbed to the sediment and thus may not be detected. Clear gastric content with little mucus may be analyzed without centrifugation.

DETERMINATION OF TOTAL TITRATABLE ACIDITY IN GASTRIC RESIDUE

Many investigators have recommended the determination of total acidity in gastric residue (or other gastric specimens such as basal secretions or secretions after stimulation) by titration to a pH of 7.0 using a pH meter or Phenol Red as indicator. Others have titrated gastric juice using phenolphthalein as indicator (pH 8.4).

It must be realized that titration to either pH 7.0 or 8.4 is *not* a measure of free hydrochloric acid and that values thus obtained are less consistent owing to the variable admixture of other constituents of gastric content (see p. 1073).

Calculation of total acidity is done in the same way as the calculation of free hydrochloric acid shown previously.

The *normal total acidity* of gastric residue (without stimulation) is 10 to 60 mmol/l.

DETERMINATION OF BASAL ACID OUTPUT (BAO) AND MAXIMUM ACID OUTPUT (MAO) OF GASTRIC ACID[4]

DETERMINATION OF THE BASAL SECRETION RATE OF GASTRIC HCl

A basal condition, in the context of gastric analysis, is one in which the patient is at complete rest and not exposed to any visual, auditory, or olfactory stimuli. Such a condition is maintained during sleep. It is for this reason that many clinicians prefer a 12 h nocturnal collection for the determination of the basal acid output. Such an approach has the disadvantage, however, that the patient is exposed to significant discomfort since he has to retain the tube overnight, and in addition he must sit upright and slightly turned to the left to avoid a loss of gastric content into the duodenum. In order to assure that all these conditions are met, the patient needs close supervision throughout the entire night.

A reasonably satisfactory alternative is the collection of a specimen for a shorter period of time (e.g., 30 min) after the patient has had a satisfactory night's sleep in a quiet separate room. This is the procedure most frequently followed.

In either case, the tube has to be placed and the position confirmed as outlined under "Collection of Gastric Residue." The entire residue is then discarded and the specimen collected as outlined in the following section.

12 h Nocturnal Secretion Rate. The patient should be given a clear, liquid meal for dinner on the day of the test. The collection of the 12 h gastric specimen should be started at approximately 8 p.m. The patient should not eat any food or drink liquid after dinner and throughout the entire 12 h collection period. Medication influencing gastric secretion, such as antacids, anticholinergic drugs, reserpine, alcohol, adrenergic drugs, or adrenocorticosteroids, should be withheld 12 h prior to the test period.

After aspirating and discarding the gastric residue, the gastric tube is connected to an aspirator (low suction, GUMCO Pump) fitted with a 1000 ml specimen reservoir. After a 12 h collection, the gastric tube is disconnected from the pump and carefully removed from the patient. The specimen is sent to the laboratory immediately for analysis.

PROCEDURE

The free acid in the specimen is determined by the procedure outlined on p. 1076.

CALCULATION

$$\text{free acid in mmol/l} = \frac{\text{ml of 0.10 molar NaOH} \times 0.1 \times 1000}{\text{ml of sample titrated}}$$

$$\text{free acid in mmol/h} = \frac{\text{mmol free acid/l} \times \text{total volume of specimen (ml)}}{1000} \times \frac{60}{\text{collection period of specimen (min)}}$$

The basal acid output (BAO), regardless of the collection period, is generally reported as acid output/h.

Thirty minute basal secretion rate test. Prepare the patient as outlined in the preceding section. Then collect gastric fluid for a 30 min period and place specimen into a properly labeled container fitted with a cover. Request the pátient to expectorate all saliva during the collection period. Send the specimen to the laboratory and analyze for free HCl as outlined under "Determination of Free HCl." Calculate the BAO/h by the formula given under "12 h Nocturnal Secretion Rate."

MAXIMUM ACID OUTPUT (MAO) AND PEAK ACID OUTPUT (PAO)

After collection of a basal specimen (12 h nocturnal specimen or 30 min basal specimen) the patient is given a stimulus for gastric secretion, most frequently Histalog (1.5 mg/kg). If the patient has a history of asthma or paroxysmal hypertension, or a systolic blood pressure below 110 mm Hg, Histalog should not be given. It is also recommended that an ampule of adrenaline chloride be readily available in case of an unsuspected reaction to Histalog. The stimulus should always be administered by a physician, and the patient should be observed during the entire test period.

Begin immediate collection of four 15 min specimens for a total of 60 min. Place each 15 min specimen into a separate container, cover, and send immediately to the laboratory for analysis. Some investigators prefer collection of six 15 min specimens, since PAO and MAO are dependent on maximal stimulation, which may in some individuals start after 30 min and continue beyond the 1 h period.

Determine the amount of HCl in each specimen by the method described on p. 1076.

CALCULATION

Maximum Acid Output. Calculate the free acid output of each 15 min post-stimulation specimen in mmol/h, using the formula given under "12 h Nocturnal Acid Output." Average the acid output values for all four post-stimulation specimens. This gives the MAO/h.

Peak Acid Output. Calculate the acid output for each 15 min post-stimulation specimen as described for the MAO/h. Select the two specimens with the highest acid output. Add these two values and divide by two. This gives the peak acid output (PAO/h).

Some investigators prefer the PAO/h instead of the MAO/h value. Blackman[4] has shown that the data for PAO and MAO have a correlation coefficient of 0.998.

BAO/MAO Ratio. $$\frac{\text{BAO}}{\text{MAO}} \times 100 = \text{per cent BAO/MAO}$$

INTERPRETATION AND NORMAL VALUES[4,5,11,36]

Gastric residue (after 12 h fast)

Volume:	20 to 100 ml (generally below 50 ml)
pH:	1.5 to 3.5
Free acid (without stimulation):	0 to 40 mmol/l
Free acid (after stimulation with Histalog):	10 to 130 mmol/l
12 h Gastric Secretion	
Volume:	150 to 1000 ml
pH:	1.5 to 3.5
Gastric Acid Secretion Rate	
1. *Basal Acid Output* (*BAO*)	
Normal; gastric ulcer	0 to 5 mmol/h
Possible duodenal ulcer	5 to 15 mmol/h
Zollinger-Ellison (Z-E) syndrome	>20 mmol/h (at times more than 60 mmol/h)

2. *Maximum Acid Output (MAO)*

Normal; gastric ulcer	1 to 20 mmol/h
High normal; duodenal ulcer; possible Z-E syndrome	20 to 60 mmol/h
Z-E syndrome	>60 mmol/h (generally not more than twice basal acid output)

3. *BAO/MAO*

Normal; gastric ulcer or gastric carcinoma	<20 per cent
Gastric or duodenal ulcer	20 to 40 per cent
Possible duodenal ulcer or Z-E syndrome	40 to 60 per cent
Z-E syndrome	>60 per cent

NOTE: Data related to the acid output are not diagnostic by themselves. A high acid output is compatible with duodenal and prepyloric ulcer or Z-E syndrome. Achlorhydria is compatible with pernicious anemia. All other intermediate values are of little diagnostic significance, owing to wide overlap of results.

CONFIRMATION OF COMPLETE VAGOTOMY[5,11,19] (HOLLANDER TEST)

A hypoglycemia of 45 mg/100 ml or less produced by insulin administration (0.15 to 0.2 units/kg body weight) strongly stimulates gastric HCl and pepsinogen secretion. It is believed that the hypoglycemia is sensed by the cells in the hypothalamus[10] with subsequent stimulation of the vagal medullary nuclei. Such a response to hypoglycemia is not observed in *completely* vagotomized subjects. The insulin (Hollander) test is therefore performed to confirm the completeness of the nerve division (vagotomy).

The optimal time for performance of the Hollander test is between 3 and 6 months after the surgical procedure. The test should not be performed in the first 2 weeks immediately after surgery.

PROCEDURE

Patient preparation and placement of the gastric tube are the same as outlined under "Collection of Gastric Residue." A glucose solution for I.V. injection should be kept at hand in case of severe hypoglycemic reactions. The test is contraindicated in any patient with a condition that already predisposes him to hypoglycemia.

1. Aspirate the gastric residue completely and discard the specimen.
2. Collect a 30 min basal gastric secretion.
3. Draw blood samples for glucose determination before and 30, 60, and 90 min after insulin injection.
4. Administer insulin intravenously (0.15 to 0.2 unit/kg body weight) immediately after the collection of the basal gastric sample (this must be done by a physician). Observe the patient for any possible side reactions.
5. Collect eight 15 min post-insulin gastric specimens (total 120 min). Place specimens in an appropriate container and send them together with the basal secretions to the laboratory for analysis.

INTERPRETATION

If the response to insulin within the first 45 min is negative, vagotomy is complete. Vagotomy is considered incomplete if (1) there is an increase in acid concentration greater than 20 mmol/l or more than 10 mmol/l if basal secretion was not acid, or (2) if the total acid output is more than 2 mmol/h. Test results are valid only if serum glucose values decreased below 45 mg/100 ml.

DETECTION OF LACTIC ACID IN GASTRIC CONTENTS

Lactic acid in gastric contents (and also butyric and acetic acid) is most likely the product of bacterial action and fermentation as a result of food stagnation or absence of free hydrochloric acid, or both. The testing for lactic acid is of relatively little clinical value, but it is occasionally requested because the presence of lactic acid, together with gastric retention and hypochlorhydria, is a rather common finding in carcinoma of the stomach.

PRINCIPLE

A portion of gastric content is extracted with ether. An aliquot of this ether extract is treated with ferric chloride, which gives a slight yellow-greenish color with low concentrations of lactic acid and an intense yellow-green color with high concentrations.

METHOD

1. Introduce 5 ml of strained or centrifuged stomach contents into a separatory funnel.
2. Add 20 ml of ether (AR) and extract the lactic acid by vigorous shaking for 1 min. Permit the ether phase to separate.
3. Transfer 5 ml of the upper ether layer to another separatory funnel, and add 20 ml of distilled water and 2 drops of a ferric chloride solution (10 g/100 ml). Mix gently.

INTERPRETATION

A slight yellow-greenish color is observed if lactic acid is present in concentrations of more than 50 mg/100 ml of gastric content. Concentrations of more than 100 mg/100 ml give an intense yellow-greenish color. Lack of color development indicates absence of lactic acid.

TUBELESS GASTRIC ANALYSIS (DIAGNEX BLUE TEST)[37,38]

The passage of a stomach tube is an unpleasant experience for the patient and, in some conditions, it is even contraindicated; therefore, much effort has been expended to develop a tubeless form of gastric analysis. Although the techniques now available have their limitations (see Comments on the Method and Sources of Error), they are useful as screening procedures. The most popular of the various tests devised is the Diagnex Blue Test.

PRINCIPLE

After an overnight fast, the patient is given a stimulant, generally caffeine or Histalog, which is followed 1 h later by administration of an azure A resin. The resin has been prepared by exchanging the hydrogen ions of a carboxylic cation exchange resin with azure A (3-amino-7-dimethylaminophenazathionium chloride). When this resin indicator compound comes in contact with free hydrochloric acid, the azure A indicator material on the resin is exchanged with H^+. This release begins at a pH below 3.0 and has its maximum at a pH of 1.5. The dye that is thus released is subsequently absorbed by the intestinal tract and excreted through the urine. The dye may be excreted in its blue form or as a colorless compound. It is not known whether the colorless compound is a reduced leucoform, or a conjugated form, or a combination of both, but it can be converted into the blue form by boiling with acid. The amount of dye excreted is determined either semiquantitatively with the aid of a comparator block or quantitatively by spectrophotometry.

PREPARATION OF THE PATIENT

The patient should be kept fasting overnight and during the test and should refrain from smoking on the morning of the test since nicotine stimulates adrenalin secretion, which in turn increases gastric secretion. Intake of water in limited quantity is permissible.

On the morning of the test day, the patient is given 500 mg of caffeine sodium benzoate with one glass of water. A Diagnex Blue test kit containing caffeine sodium benzoate and the resin is available commercially. Since caffeine is a weak stimulant, some investigators have used Histalog instead. After 1 h the patient is asked to void, and this urine specimen is discarded. The patient should now swallow the azure A resin suspended in water. From this time on, all urine voided for a 2 h period is collected and the entire urine specimen is sent to the laboratory for examination.

REAGENTS

1. HCl, 6 mol/l. Place 50 ml of concentrated HCl into a 100 ml volumetric flask containing approximately 30 ml of distilled water. Bring to mark with distilled water.
2. Diagnex Blue test kit (E. R. Squibb & Sons, New York, N.Y.).

PROCEDURE

1. Place the entire 2 h urine sample into a graduated cylinder and dilute to 300 ml with distilled water. If the total urine volume is greater than 300 ml, use the urine undiluted as is.
2. Place approximately 10 ml of the diluted urine specimen into each of three test tubes. Label two tubes "control" and the other tube "test."
3. Add 30 mg ascorbic acid (1 capsule supplied with kit) to both control tubes. The ascorbic acid will reduce the blue color due to azure A, but not most other pigments present in the urine.
4. Place the two controls into the slots of the Squibb Diagnex Comparator block that are marked 0.3 and 0.6 mg. The 0.3 and 0.6 mg standard slots are prepared with a blue screen of a color intensity that is comparable to that of urine containing 0.3 or 0.6 mg azure A dye, respectively. The tube marked "test" is placed into the middle slot labeled "test."
5. The Diagnex Comparator block with the three test tubes is held against a suitable light source, and the color intensity of the test is compared with those of the controls in the respective standard slots. If the color intensity of the test urine is equal to or exceeds that of the 0.6 mg standard, the test is completed at this point and "presence of free gastric hydrochloric acid" may be presumed. The two control urines in the 0.3 and 0.6 mg slot help prevent misinterpretations due to presence of unspecific chromogens.
6. If the color intensity of the test is less than that of the 0.6 mg standard, the test tube with urine (step 2) should be acidified with 1 drop of Squibb acid–copper sulfate solution or 2 drops of 6 molar HCl and then placed into a 100°C heating bath for 10 min.
7. Remove all tubes from the heating bath and allow to cool for 2 h. This time is needed to bring out the full color of the azure A dye. Determine the color intensity of the test as outlined in step 5. A color intensity of the test which exceeds that of the 0.6 mg standard suggests that "free hydrochloric acid is present." If the color of the test falls between the 0.6 and 0.3 mg standards, there is "presumptive evidence for hypochlorhydria." A color intensity of the test less than that of the 0.3 mg standard is "presumptive evidence for achlorhydria."

COMMENTS ON THE METHOD AND SOURCES OF ERROR

Although the tubeless gastric analysis eliminates the use of the stomach tube, it creates the need for a rather lengthy procedure for the patient and it has many potential sources of error (in proper timing, in complete urine collection, and so on).

The test gives only an indication of the presence or absence of hydrochloric acid. It provides no information as to the concentration of hydrochloric acid present, nor does it indicate the total volume of gastric residue.

If hydrochloric acid was indeed secreted by the gastric mucosa, but was subsequently neutralized by either regurgitated material, food particles, or saliva, a *false negative* result will be obtained. (In a regular gastric analysis, one has a sample on hand for observation of the presence of food particles or bile.)

The outcome of the tubeless gastric analysis depends not only on the presence of hydrochloric acid but also on such factors as rate of emptying of the stomach, rate of intestinal absorption of the free azure A dye, and rate of excretion of the dye through the kidneys. Abnormalities in any of these steps could affect the amount of dye appearing in the urine and thus the test result; therefore, the test is unreliable in patients with subtotal gastrectomy, pyloric obstruction, malabsorption, marked congestive heart failure, severe liver and kidney disorders, and urinary retention.

The replacement of the azure A dye from the resin is accomplished not only by hydrogen ions but also by a number of other cations such as sodium, potassium, barium, and iron. Segal[37] states, however, that this accounts for the excretion of less than 0.35 mg of azure A in the first 2 h.

The control generally compensates for pigments that might be present in urine; however, false positive results may be obtained if dyes other than azure A are present, which decolorize after addition of ascorbic acid. In this case, the nonspecific colors contribute to the absorbance of the unknown without proper compensation for this color by the control. It is therefore best that patients to be tested receive no colored medication or colored foods (e.g., beets) for at least 24 h before the test.

SUMMARY

The tubeless gastric analysis has provided useful information in selected cases but should be considered as a screening test only. The various sources of error just listed are responsible for approximately 5 to 10 per cent false negative and 2 to 5 per cent false positive results.

REFERENCES

Segal, A. L., Miller, L. I., and Plumb, E. J.: Gastroenterology, *28*:402, 1955.
Diagnex Blue test kit package insert, E. R. Squibb & Sons, New York, N.Y.

DETERMINATION OF GASTRIN IN SERUM[14,43]

SPECIMEN

Following a complete overnight fast, collect a blood sample in a tube containing *no* additives. (Heparin will interfere with the test procedure). Centrifuge the blood and freeze the separated serum promptly until analysis. Plasma is unsuitable for analysis.

Serum specimens must be frozen immediately to prevent destruction of gastrin by proteolytic enzymes in serum.

PRINCIPLE

An aliquot of serum (or standard) is incubated with gastrin antiserum and ^{125}I-labeled gastrin in polyethylene vials. The gastrin in the sample will compete with the labeled gastrin for the sites on the antiserum.

After incubation, the bound ^{125}I-labeled gastrin is separated from the unbound gastrin either by precipitating the antigen-antibody complex with polyethylene glycol followed by centrifugation, or by solid phase adsorption of the free antigen to charcoal or anionic exchange resins. The radioactivity in either the precipitate (bound ^{125}I) or the supernatant (free ^{125}I) is counted and compared with the radioactivity observed in tests in which standards were used instead of serum.

Depending on the type of procedure employed, various interferences have been reported. Heparin interferes especially when resin separation is used, while separation with polyethylene glycol or charcoal requires a uniform protein concentration. NaCl in high concentration has also been reported to interfere. The immunological activity of gastrin appears to be linked to the integrity of an at least partially rigid three-dimensional structure of gastrin. This structure is present in "little gastrin" and "big gastrin."

McGuigan and Trudeau[26] have developed a double antibody radioimmunoassay technique for measuring gastrin in human serum. In this procedure, a sample of ^{125}I SHG: 2-17 (serum human gastrin I residues 2 through 17 covalently conjugated to bovine serum albumin) is incubated in the presence of rabbit anti-SHG antibodies and either human gastrin I (standard) or human serum.

After 24 h at 4°C, goat antiserum to rabbit gamma globulin is added and the mixture is incubated for an additional 24 h at 4°C. Radioactivity in the immune precipitate after centrifugation is inversely proportional to the amount of gastrin in the sample. (The procedure is based on the competition of gastrin in the sample with the ^{125}I SHG:2-17 for antigastrin binding sites).

NORMAL VALUES

Normal values for gastrin in serum range up to 300 pg/ml non-fasting. Values in elderly individuals are higher and reach 500 $\pm$ 300 pg/ml in patients 80 years of age.

Serum values between 300 and 500 pg/ml in young adults constitute a "gray zone." Such values may be seen not only in some normal young and older individuals but also in patients with duodenal ulcers,[49] in patients having disorders associated with achlorhydria or hypochlorhydria (e.g., atrophic gastritis), and in patients with decreased renal function. Values in pernicious anemia are 300 to 3000 pg/ml, and significant elevations of gastrin are found in patients with Zollinger-Ellison (Z-E) syndrome[43,44] with values ranging from 3500 to 62,000 pg/ml. Patients with pernicious anemia may have similar elevations; however, in this condition, gastrin levels can readily be reduced by infusion of HCl into the stomach, since gastrin levels are regulated by the negative feedback mechanism. In Z-E syndrome, gastrin secretion is essentially independent of gastric pH.

If gastrin levels are in the range between 200 and 500 pg/ml, an intravenous *calcium infusion test* may be performed. In normal individuals, serum gastrin levels increase only slightly, while in patients with Zollinger-Ellison syndrome, calcium infusion provokes a marked rise in serum gastrin levels.[48]

TEST FOR PRESENCE OF INTRINSIC FACTOR (SCHILLING TEST)

Absorption of vitamin B_{12} requires the presence of the intrinsic factor (see Chapter 11). Since this factor is missing in pernicious anemia, the Schilling test, which provides a measure of vitamin B_{12} absorption, is used to detect this disorder.

The rate of absorption of orally administered ^{57}Co- or ^{58}Co-labeled vitamin B_{12} can be measured by determining the radioactivity in feces, in urine, or in serum, or by externally scanning the liver. The most frequently applied procedure is the measurement of radioactivity in a 24 h urine sample, which is collected after oral administration of 0.5 or 1 μCi of radioactive cobalt-labeled vitamin B_{12} following an overnight fast. (Initially, a ^{60}Co-labeled vitamin B_{12} was used for this test, but ^{57}Co or ^{58}Co is now preferred, since they have shorter half-lives and the counting efficiency is greater.) The patient also receives a 1 mg dose of unlabeled vitamin B_{12} intramuscularly to saturate the binding capacity of plasma and liver for vitamin B_{12} and to assure adequate excretion of the vitamin. In normal individuals, more than 7.5 per cent of the dose administered will be excreted in the urine.

In individuals with pernicious anemia, significantly less than 7.5 per cent (about 0 to 2.5 per cent) of the administered labeled vitamin B_{12} dose will be excreted. Oral administration of the intrinsic factor simultaneously with the vitamin B_{12}, however, will significantly increase the rate of absorption by the intestinal tract and therefore the excretion in the urine. In patients with malabsorption, oral administration of intrinsic factor will essentially not affect the absorption rate.

The Schilling test is mostly used in studying patients with achlorhydria in whom latent pernicious anemia is suspected but hematologic manifestations have not yet developed. The test is also used to confirm the diagnosis of pernicious anemia.

REFERENCE

Silver, S.: Radioactive Nuclides in Medicine and Biology. 3rd ed. Philadelphia, Lea & Febiger, 1968, p. 341.

CONDITIONS ASSOCIATED WITH ABNORMAL GASTRIC FUNCTION

Carcinoma of the stomach

Achlorhydria is found in about 50 per cent of cases; a large portion of the remainder of this group have hypochlorhydria, and only a small percentage have hyperacidity. In some cases, there is blood in the stomach. The lactic acid test is positive in a high percentage of cases. Also, the lactobacillus of Boas-Oppler can frequently be demonstrated.

Gastric ulcers

In some cases, gastric acidity is high, but in the majority of cases it is normal. Hypoacidity, which is sometimes found, is thought to be due to chronic gastritis. Blood is constantly or intermittently found in many cases, and the same is true for occult blood in the stool. The diagnosis is made chiefly on the basis of the clinical and x-ray findings.

Peptic ulcers are associated with an increased secretion rate of gastric HCl. These types of ulcers are rarely found in the stomach but if they occur, they are generally found in the antral region. Peptic ulcers are most often found in the duodenum. (Peptic ulcers, by definition, are produced by gastric juice high in pepsin content and HCl concentration.)

No patient with common peptic ulcer disease has a fasting serum level of gastrin greater than 400 pg/ml.

The variable concentration of gastric secretion in patients with gastric ulcers may be due to the fact that the etiology of the ulcer may be quite variable. The integrity of the gastric mucosa depends on the balance between "attack" and "defense."[9] In the case of peptic ulcers, the high pepsin concentration in the presence of a high concentration of HCl provides a strong "attack" which may overwhelm the defense (namely, the gastric mucosal area). The nature of the area is not fully elucidated, but it is believed that the apical plasma membrane of the surface epithelial cells plays a major role. It is also believed that this membrane, like all cell membranes, is composed of lipids and thus may be destroyed by substances which disturb the arrangement of lipid layers. Substances which are known to have the ability to penetrate the barrier are detergents such as bile salts and lysolecithin (which may be regurgitated from the duodenum into the stomach), ethanol, aliphatic nonionized acids, eugenol, salicylic acid, and acetylsalicylic acid. If the integrity of the mucosal barrier has been affected, acid may diffuse from the lumen of the stomach back into the mucosa, and sodium ions can diffuse from the mucosa into the lumen.[9] Such back diffusion can, at least in part, account for low gastric acidity.

Duodenal ulcers

Hyperchlorhydria is observed in more than 70 per cent of cases; however, a small percentage of cases with hypochlorhydria have been reported. Histamine-fast achlorhydria excludes the diagnosis of duodenal ulcer. The volume of gastric secretion is usually increased and may be twice the normal volume. The basal hourly secretion may be more than 200 ml, and the 12 h nocturnal secretion may increase to as much as 5 to 10 liters.

Atrophy of the stomach (atrophic gastritis)

Achlorhydria and low volume are rather constant findings in this disease state. Gastrin levels are frequently moderately elevated.

Pernicious anemia

True achylia (no HCl or pepsinogen secretion) and a decreased volume of gastric secretions are constant findings. Demonstration of free acid in a patient with megaloblastic anemia provides strong evidence that the diagnosis is *not* pernicious anemia.

Serum gastrin levels may be significantly elevated. Infusion of HCl into the stomach, however, will result in a prompt and significant decrease in gastrin.

Pyloric obstruction

The volume of gastric residue is usually large, and an excessive amount of food can be demonstrated in gastric content. Although hydrochloric acid may have been secreted, it frequently will have been neutralized by the food. Lactate and butyrate, as well as yeast cells, may be demonstrated when pyloric obstruction is accompanied by a high pH of gastric content (as in gastric carcinoma).

Zollinger-Ellison syndrome

The Zollinger-Ellison syndrome is caused by a functioning pancreatic cell adenoma or carcinoma which secretes gastrin continuously or intermittently. The tumor is a non-beta cell (probably a delta-islet cell) non-insulin producing tumor of the pancreas. Secretion of gastrin causes continuous and prolonged stimulation of the gastric mucosa. Thus, the BAO is more than 20 mmol/h and at times more than 60 mmol/h. The MAO is at least 20 to 60 but most often more than 60 mmol/h with a BAO/MAO of over 40 and generally over 60 per cent. Serum gastrin levels are significantly elevated, with values of 3500 to 62,000 pg/ml.

Emotional states

Gastric secretions may be inhibited under various circumstances, such as emotional disturbances and depressions, presumably via inhibitory fibers in the vagus and splanchnic nerves. Conditions normally referred to as "stress" may stimulate the hypothalamus and result in increased gastric secretions.

TESTS MEASURING THE EXOCRINE FUNCTION OF THE PANCREAS AND INTESTINAL ABSORPTION

The predominant exocrine function of the pancreas is the production and secretion of pancreatic juice that is rich in enzymes and bicarbonate. Normal pancreatic juice is colorless and odorless; it has a pH of 8 to 8.3 and a specific gravity of 1.007 to 1.042. The total 24 h secretion volume is approximately 800 to 3000 ml.

A number of laboratory tests are available to measure these functions either directly (secretin and CCK-PZ tests, trypsin in stool) or indirectly (fat, vitamin A, and carotene absorption tests). All of these tests must be viewed mainly as measuring the pancreatic function and not as tests designed to diagnose a specific pancreatic disorder.[7] When interpreting results, allowance must always be made for other disorders that may give rise to similar manifestations (e.g., steatorrhea due to intestinal malabsorption). There is also great overlap between results observed in normal individuals and those found in patients with pancreatic disorders. This is partly the result of the large functional reserve of the pancreas. It has been estimated, for example, that pancreatic insufficiency cannot be clearly demonstrated until at least 50 per cent of the acinar cells have been destroyed.

SECRETION OF BICARBONATE AND PANCREATIC ENZYMES IN RESPONSE TO SECRETIN AND CCK-PZ

SECRETIN TEST[5,7,50,51,52]

A direct measure of pancreatic function is the determination of the rate of bicarbonate output and the secretion rate of pancreatic juice in response to stimulation with secretin. Secretin is administered either at a dose of 1 clinical (C) unit/kg of body weight intramuscularly or 2 C-units/kg per hour by continuous intravenous infusion in 0.15 molar sodium chloride. Although intramuscular injections are resorted to more frequently, this technique may produce powerful contractions of the duodenum and is therefore to be discouraged. It is also felt that the results obtained with the steady rate of secretion produced by intravenous infusion are easier to interpret. Various preparations of secretin differ greatly in their potency, but secretin obtained from the GIH Laboratory (Karolinska Institute, Stockholm, Sweden) has been found to be most reliable. A synthetic preparation can be obtained in the United States from Squibb. The Boots preparation (England), which is marketed in this country by WarrenTeed Pharmaceuticals, Inc. (Columbus, OH), has considerably less potency (required dose: up to 8 Boots units/kg body weight).

INTERPRETATION

Subjects with normal pancreatic function secrete more than 15 mmol bicarbonate/30 min. The concentration of bicarbonate is normally more than 100 mmol/l but may be lower in normal individuals if the total volume secreted is high. In general, the measurement of HCO_3^- output is clinically more discriminatory than the measurement of the HCO_3^- concentration.

The total volume for an 80 min period should be more than 80 ml or more than 2.0 ml/kg body weight.

Abnormally low concentrations of constituents of pancreatic juice after secretin stimulation have been reported in cases of chronic pancreatitis, pancreatic cysts, cystic fibrosis, calcification, carcinoma of the pancreas, and edema of the pancreas. Findings in these conditions are very similar and are of relatively little value in the differential diagnosis. The greatest value of the secretin test is probably in excluding pancreatic dysfunction, as seen in fibrocystic disease, and in differentiating steatorrhea of pancreatic origin from that caused by sprue or celiac disease.

In the presence of obstruction (carcinoma, stone) or if an obstructive action is produced by drugs (morphine sulfate, methacholine chloride, or Urecholine), the outflow of pancreatic juice into the duodenum is prevented and significant increases in serum amylase and lipase will be observed (see p. 1068). Patients with substantially decreased amounts of functioning acinar tissue (i.e., patients with cystic fibrosis or chronic pancreatitis) show significantly smaller increases in serum enzyme levels. A normal average response to combined secretin-morphine administration is a serum amylase elevation of 390 Somogyi units. Use of this test in pancreatitis may be contraindicated because of the possible risk involved.

PREPARATION OF THE PATIENT AND COLLECTION OF THE SPECIMENS

The patient should be in the fasting state. A double-lumen gastroduodenal tube is introduced under fluoroscopic guidance in such a way that the shorter end of the tube lies in the stomach and the longer end lies in the duodenum distal to the ampulla of Vater. Proper positioning of the tube must be confirmed by fluoroscopy. Constant suction is then applied and aspiration is continued until the duodenal content becomes clear and not contaminated with gastric juice. At this time, secretin is administered and three to eight consecutive 10 min specimens are collected for 30 to 80 min. The pH of each specimen is measured. A sudden increase in pH of the gastric content indicates contamination by duodenal fluid; a sudden decrease in pH of the duodenal content indicates contamination by gastric juice.

At the end of the total collection period, all duodenal specimens are pooled, the total volume is measured, and the specimen is tested for blood and assayed for bicarbonate content either by a manometric technique (e.g., Natelson Microgasometer; see Chap. 15) or by a titrimetric technique.[21] This is followed by calculation of the bicarbonate secretion rate (mmol HCO_3^-/collection period). Detection of occult or overt blood (see Chap.8) in any one of the specimens may be helpful in pinpointing gastrointestinal bleedings.

THE SECRETIN-CCK-PZ TEST[5,34,50,51,52]

This test is designed to measure the capacity of the pancreas to secrete enzymes in response to CCK-PZ administration. Since this hormone does not stimulate volume output, secretin must be administered in combination with CCK-PZ. Hormones are administered by continuous intravenous infusion of 0.25 C-unit secretin/kg/h plus 16 *C*rick-*H*arper-*R*aper (CHR-) units CCK-PZ (Boots)/kg/h in 0.15 molar sodium chloride. CCK-PZ from the Karolinska Institute is measured in Ivy (I-) units. For the measurement of enzymes, it is important that the indicated lower dose of secretin be given. The high dose used in the secretin test may inhibit pancreatic enzyme secretion in response to CCK-PZ. On the other hand, this smaller dose may not provide the maximum stimulus for bicarbonate secretion. Thus, if borderline results for bicarbonate output are obtained, a second test using the high dose of secretin may be required.[51]

The combined secretin-CCK-PZ test is rarely used in the United States. CCK-PZ has not been released by the FDA for routine use. Also, information gained from this test generally does not add significantly to the information gained from the measurement of volume and bicarbonate output after secretin administration alone.

The preparation of the patient and collection of the specimen, as well as the procedure for administering the hormones, are the same as those outlined under "Secretin Test." For greater reliability it is recommended that at least two enzymes be measured. Determination of *trypsin* activity is clinically most informative, but *amylase* is most often determined in adults. In infants, amylase production is extremely low, and therefore contamination by salivary amylase is a serious source of error.

Trypsin is readily inactivated at room temperature. It is therefore recommended that duodenal fluid for enzyme analysis be stored in ice water. *Lipase* activity measurements can readily be done, but uniform normal values have not been established. Most authors have used their own methods and their own sets of normals. Techniques for the determination of trypsin, amylase, and lipase are discussed in Chapter 12.

INDIRECT TESTS OF PANCREATIC FUNCTION

This group of tests is based on the fact that proper pancreatic function is essential for normal intestinal absorption of certain substances. Since the pancreas is the major source of amylase, lipase, and proteolytic enzymes, significant decrease in pancreatic function causes decreased absorption of starch, fats, and protein and a resultant increased excretion of these food materials in the stool. Demonstrated normal absorption will generally rule out severe pancreatic deficiency. Careful interpretation of the results is necessary, however, since the pancreas has a large functional reserve and absorption may be normal even in the presence of pancreatic disorders. Only if a significant amount of pancreatic tissue is destroyed, or if the flow of pancreatic juice into the duodenum is greatly interfered with, will these tests yield abnormal results. The greatest value of these tests seems to be in the exclusion of pancreatic pathology, especially cystic fibrosis.

FAT ABSORPTION TEST

Normal absorption of fat from the intestinal tract can take place only in the presence of an adequate amount of pancreatic lipase and bile. Consequently, a number of fat absorption tests have been devised in which the patient is given a high-fat meal. Serum lipid levels are measured before and at certain periods after administration of the meal (e.g., 3 and 6 h). Serum lipids may be determined either by quantitative analysis (see Chap. 10) or by measurement of the turbidity of the serum. These tests have largely been replaced by those listed below. It has been stated that fat absorption may not be abnormal until lipase secretion has decreased to 10 per cent of normal. Also, fat absorption tests give abnormal results in intestinal malabsorption (such as in celiac disease and tropical sprue) and do not discriminate between these causes of malabsorption.

MEASUREMENT OF ABSORPTION OF ^{131}I-LABELED TRIOLEIN AND ^{125}I-LABELED OLEIC ACID

A clinically more useful modification of the fat absorption test is based on oral administration of ^{131}I-labeled triolein and ^{125}I-labeled oleic acid and subsequent determination of both forms of radioactivity *in blood serum.* While absorption of triolein is dependent on pancreatic lipase, oleic acid absorption is independent of the presence of this enzyme. Consequently, normal absorption of labeled triolein indicates that pancreatic function is probably normal or not greatly impaired. Normal or near normal absorption of labeled oleic acid in the presence of abnormal triolein uptake suggests impaired pancreatic function (see Fig. 20–7). Failure of absorption of both triolein and oleic acid indicates a malabsorption syndrome of nonpancreatic origin.

Use of different isotopes for triolein and oleic acid allows measurement of both isotopes in the same sample. Use of the same isotope would require that the tests for triolein and oleic acid uptake be performed on different days. It has also been suggested that the ratio of absorbed triolein/absorbed oleic acid be calculated and used as a means to differentiate between malabsorption of pancreatic origin and that of nonpancreatic origin. A ratio of approximately 1/1 indicates normal absorption, whereas a low ratio indicates malabsorption of pancreatic origin. In malabsorption of nonpancreatic origin, the ratio may be normal or slightly decreased, but the absolute amounts of both triolein and oleic acid absorbed are significantly decreased.

The determination of total radioactivity *in stool* after administration of labeled triolein and oleic acid is considered more accurate than plasma determinations alone. The remaining radioactivity in a 72 h stool specimen should be less than 5 per cent of the administered dose. In pancreatic insufficiency, ^{131}I-triolein values in stool may be as high as 20 to 30 per cent, but ^{125}I-oleic acid values are normal or only slightly increased. Hydrolysis of triolein by bacteria in the colon limits the usefulness of the test.

PROCEDURE

The evening before the test, the patient receives iodine (e.g., 10 drops Lugol's solution) to saturate the thyroid gland. On the test day, ^{131}I-labeled triolein and ^{125}I-labeled oleic acid (50 μCi each) are given with milk and the radioactivity of blood plasma is measured after 4 and 6 h. Appearance of more than 1.7 per cent of the administered dose/liter of plasma is considered normal for both oleic and triolein. In the presence of decreased absorption of triolein, oleic acid levels may also be slightly decreased because of increased intestinal motility or other causes. In such a case, oleic acid values approaching the normal range are considered normal (see Fig. 20-7) as long as the oleic acid absorption is substantially greater than that of triolein. For more detail on the procedure, consult Silver.[40]

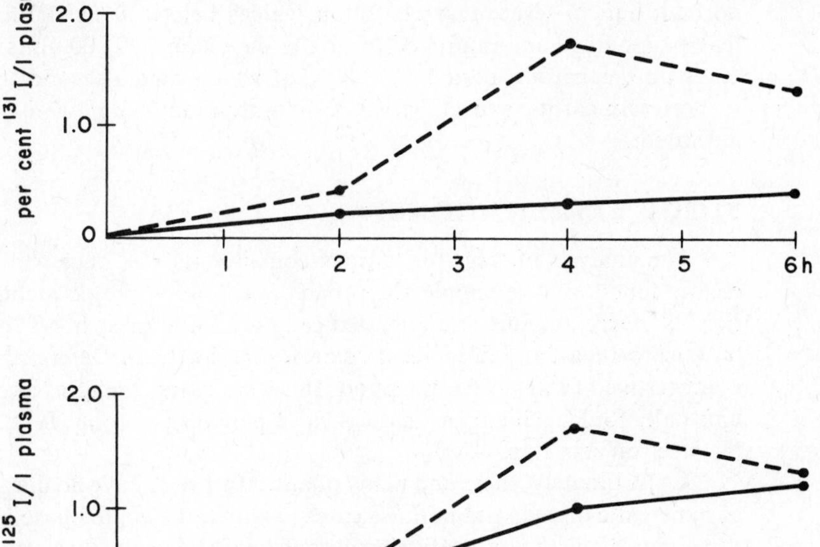

Figure 20–7 Examples of absorption of ^{131}I-labeled triolein (top) and ^{125}I-labeled oleic acid (bottom) in a patient with pancreatic insufficiency; (- - - -) indicates lower limit of normal range.

SERUM CAROTENE, VITAMIN A, AND VITAMIN A TOLERANCE TEST

Serum vitamin A levels in children with cystic fibrosis of the pancreas and in patients with pancreatic insufficiency are usually below the normal range of 30 to 65 μg/100 ml (1.0 to 2.3 μmol/l). In these patients, oral administration of 5000 units of vitamin A in oil/kg body weight causes only a slight increase in the serum vitamin A level. Persons with normal absorption show a peak increase from 200 to 600 μg of vitamin A/100 ml (7 to 21 μmol/l) of serum in the 3 or 6 h specimen, or both.

Blood specimens are drawn before as well as 3 and 6 h after administration of the vitamin. The patient may drink water or consume light meals during the test period. The procedure for determining vitamin A has been described in Chapter 11.

A more standardized vitamin A absorption test was proposed by Kahan.[15] In this test, a patient in the fasting state is given 7500 IU vitamin A palmitate/kg of body weight (maximally 350,000 IU), and vitamin A determinations are carried out before and 4, 5, and 6 h after administration of the vitamin. *Normal values* for vitamin A in serum show a log normal distribution with a mean value of 28 μg/100 ml and a range of 20 to 110 μg/100 ml (0.8 to 4.0 μmol/l, mean 1.0 μmol/l) up to three years of age; and a mean of 25 μg/100 ml (range 15 to 95 μg/100 ml or 0.6 to 3.3 μmol/l) in the age group of 4 to 15 years. Maximum increments of the serum vitamin A level (over the fasting level) after ingestion of the vitamin are about 180 to 2100 μg/100 ml (6.5 to 75 μmol/l) in the age group of 0 to 3 years and approximately 320 to 2200 μg/100 ml (11.5 to 80 μmol/l) for the age group up to 15 years.

The test is based on the fact that vitamin A, a fat-soluble vitamin, can be absorbed to any significant extent only if fats are first hydrolyzed (by pancreatic lipase). The test is subject to the same limitations as other absorption tests (see previous discussion).

Measurement of the fasting serum carotene level has a similar clinical application, although it is not as reliable. Values of 50 to 200 μg/100 ml (1.0 to 4.0 μmol/l) are considered normal, but, in severe malabsorption, values below 30 μg/100 ml are observed. This test may be modified and improved by giving the patient 15,000 units of carotene three times a day with meals for a period of 3 days, at which time a second serum sample is obtained. If the serum carotene level increased by more than 35 μg/100 ml, normal fat absorption is indicated.[32]

STOOL EXAMINATIONS

The analysis of feces for various constituents was once widely used to measure pancreatic function. The simplest approach is the microscopic identification and demonstration of a large amount of undigested cell nuclei and meat fibers (creatorrhea), of increased fat (steatorrhea), and of increased starch (amylorrhea). Demonstration of these conditions is suggestive of impaired absorption. In severe cases, the feces are usually pale, bulky, and unusually foul smelling. In steatorrhea of pancreatic origin, fat droplets usually appear in the feces on standing.

Unfortunately, these and many quantitative tests have no diagnostic value in mild cases of pancreatic disease and in those cases in which the acute phase has subsided. In addition, some patients with pancreatic insufficiency may have almost normal digestion owing to the activity of microorganisms and intestinal enzymes. On the other hand, patients with celiac disease or sprue may have stools similar to those of patients with pancreatic insufficiency because increased intestinal motility decreases the time that food remains in the intestinal tract, during which digestive enzymes can act. The quantitative assay of fat and nitrogen in the stool is clinically more helpful, but it also has its limitations. (Normal: <2.0 g nitrogen/d.)

FAT IN STOOL

Ingested fat is normally split by pancreatic lipase into fatty acids, glycerol, and monoesters of glycerol, and the products of hydrolysis are absorbed by the intestinal tract. Therefore, the stool content of neutral fat, free fatty acids, and soaps is relatively low. The total content of fat in feces should not exceed 6 g/d, provided that the patient has been on a diet containing 60 to 150 g fat/d. In pancreatic insufficiency, the fat content of stool will increase rather substantially and may reach 20 g/d or even more. Children up to 6 years excrete up to 2 g fat/d. Free fatty acids and soaps normally constitute more than 40 per cent of the total fat fraction. Stool contains approximately 25 per cent dry solids (dry matter), and up to 25 per cent of this dry matter may be total lipids.

A procedure for the quantitative determination of fat in feces is described in detail in Chapter 10.

TESTS MEASURING DISACCHARIDASE DEFICIENCY[20,24,31]

The brush border region of the epithelial cells of the small intestine contains a series of enzymes that hydrolyze disaccharides such as lactose, sucrose, and maltose. Hydrolysis of these disaccharides by the respective enzymes is a prerequisite for their absorption as monosaccharides.

Disaccharidase deficiences are characterized either by a lack of a single enzyme (e.g., lactase) or by a generalized deficiency secondary to mucosal disease such as celiac sprue.

Enzyme deficiency causes an intolerance in the patient for the respective disaccharide, which frequently results in cramps, bloating, and diarrhea that may be accompanied by metabolic acidosis. Fatalities from this condition during the first year of life have been reported. Lactose intolerance has been observed most frequently. This appears to be quite important because of the relatively high intake of milk (and therefore lactose) by infants.

Lactase deficiency states can be classified into several groups.[31] The congenital form is autosomal recessive and is extremely rare. It becomes manifest shortly after birth. Symptoms are generally developed in the first week of life. The racial-ethnic type of lactase deficiency is most common, and occurs in a large percentage of blacks, Orientals, Jews, Arabs, Eskimos, and Indians during childhood. This deficiency often develops with decreased intake of lactose (milk). Caucasians of Northern and Western Europe, the United States, and Australia generally maintain lactase activity throughout adulthood regardless of diet. Secondary lactase deficiency may be acquired as a result of diseases of the small bowel mucosa such as tropical sprue, celiac sprue, protein malnutrition, and infections.

Sucrase-isomaltase deficiency is an autosomal recessive defect which has been reported primarily in children and so far only in a few adults. There is intolerance to sucrose and a less prominent intolerance to starch, which on hydrolysis yields some isomaltose.

The most direct diagnostic test for the listed enzyme deficiencies is a biopsy and examination of the brush border region of the small intestine; however, this is difficult and is done only rarely. A more practical approach is the administration of one disaccharide at a time, followed by determinations of serum glucose levels. If the respective disaccharidase is present, the disaccharides will be hydrolyzed and absorbed as monosaccharides, and the glucose component can be determined in blood by employing one of the standard procedures for glucose. In cases of disaccharidase deficiency, serum glucose levels will increase only insignificantly. Changes of less than 20 mg/100 ml should be considered as abnormal. The dose of disaccharides given is generally 50 g/m^2 body surface, or 1 g/kg body weight.

Appearance of the monosaccharide in serum obviously depends not only upon the presence of the enzyme but also on the ability of the intestines to absorb monosaccharides. In order to eliminate misinterpretations due to malabsorption, a glucose tolerance test may be performed on a separate day. Appearance of this monosaccharide in serum in normal amounts would exclude this condition. Alternatively, the respective enzyme (usually 500 mg) may be given simultaneously with the disaccharide. If the glucose appears in serum to a greater extent than after administration of the disaccharide alone, a deficiency of the respective enzyme may be suspected. Some investigators feel that glucose values obtained on capillary blood are clinically more informative than those obtained on venous blood.

D-XYLOSE ABSORPTION TESTS* [33]

Xylose is a pentose not normally present in significant amounts in blood. When given orally, it is passively absorbed in the proximal small (duodeno-jejunal) intestine, passes unchanged through the liver, and is excreted by the kidneys. The xylose absorption test is considered a reliable index of the functional integrity of the jejunum (in the absence of massive bacterial overgrowth). Low absorption of xylose is observed in intestinal malabsorption, but not in malabsorption due to pancreatic insufficiency. In the latter condition, the absorption of xylose will be essentially normal provided that there is no significant increase in intestinal motility. Therefore, the test is of some help in distinguishing between these two types of malabsorption.

PRINCIPLE

A 25 g dose of xylose (or 5 g if so specified by the physician) is given to the patient, and all of the urine voided over the next 5 h is collected. A blood specimen is taken approximately 2 h after the xylose is given (in children after 1 h). The xylose concentrations in the 5 h urine specimen and in the blood specimen are determined by treating the diluted urine and a

* Xylose has not yet been released by the Food and Drug Administration for oral administration; thus, the test may be used only for investigative purposes.

protein-free filtrate of the blood with *p*-bromoaniline in an acid medium. Xylose, when heated with acids, will form furfural, which in turn reacts with *p*-bromoaniline. A pink color is produced in this reaction. The concentrations of xylose in blood and urine are calculated on the basis of a standard xylose solution.

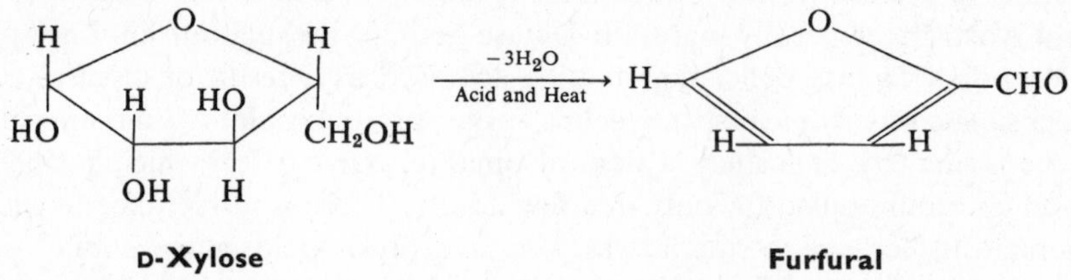

H O H
H HO
HO CH$_2$OH
OH H

$\xrightarrow[\text{Acid and Heat}]{-3H_2O}$

O
H— —CHO
H— —H

D-Xylose Furfural

SAMPLE COLLECTION

Adults

1. Keep patient fasting overnight and during test period.
2. In the morning (e.g., at 7 a.m.) have the patient empty his bladder completely and discard this urine.*
3. Give 25 g of D(+)-xylose (or 5 g if so specified by the physician) dissolved in 250 ml of water, followed immediately by an additional 250 ml water.
4. Collect and pool all urine specimens voided during the next 5 h period, including the 5 h specimen.
5. After approximately 2 h, collect a blood sample, preferably in a tube containing sodium fluoride.

Children

1. Keep patient fasting overnight and during test period.
2. In the morning (e.g., at 7 a.m.) have the patient empty his bladder completely.
3. Give 0.5 g of D(+)-xylose/lb of body weight, up to 25 g, cutting the amount of water according to the weight of the patient.
4. Collect and pool all urine specimens during the next 5 h period.
5. Collect a blood sample approximately 60 min after xylose administration. The time of blood collection is different from that in adults because the peak of absorption in children is reached sooner.

NORMAL VALUES

Adults, after a 25 g dose of xylose, should excrete more than 4.0 g of xylose/5 h urine specimen. In patients over 65 years of age, the minimum excretion is 3.5 g/5 h. Blood values are normally more than 25 mg xylose/100 ml blood. Normal children excrete 16 to 33 per cent of the ingested xylose and their blood xylose level is more than 30 mg/100 ml.

REAGENTS

1. Zinc sulfate ($ZnSO_4 \cdot 7H_2O$), 5 g/100 ml water.
2. Barium hydroxide, 0.15 mol/l. Dissolve 25 g of $Ba(OH)_2 \cdot 8H_2O$ in water and dilute to 500 ml with water. Boil for a few minutes, stopper, cool, and filter.
3. *p*-Bromoaniline reagent. Prepare a saturated solution of thiourea in glacial acetic acid by shaking about 4 g of thiourea/100 ml of glacial acetic acid, and decant the supernatant. Dissolve 2 g of *p*-bromoaniline in 100 ml of the decanted thiourea solution.

* Some investigators recommend checking a fasting urine and blood sample for nonspecific chromogens that react with *p*-bromoaniline; such interferences have been reported in some cases. Correction may be made by subtracting the base value from that found in the test specimen.

4. Stock standard, 200 mg of xylose/100 ml. Place 200 mg D(+)-xylose into a 100 ml volumetric flask. Dissolve in, and fill up to volume with, benzoic acid (0.3 g/100 ml).

5. Working standard (0.1 mg/ml). Dilute stock standard 20-fold with saturated benzoic acid.

6. Working standard (0.2 mg/ml). Dilute stock standard 10-fold with saturated benzoic acid.

PROCEDURE

1. Urine. If 5 to 15 g of xylose is given, dilute the urine to 500 ml; if 15 g or more of xylose is given, dilute the urine to 1000 ml (freeze a part of the first diluted urine for possible repeat). Further dilute a convenient portion of the urine 40-fold.

2. Blood. Deproteinize the blood by using: 1 vol blood, 7 vol H_2O, 1 vol $ZnSO_4$ solution (5 g/100 ml), and 1 vol $Ba(OH)_2$ (0.15 mol/l). Mix after each addition and centrifuge.

3. Add the following amounts (in ml) of blood filtrate, urine, standard solutions, and *p*-bromoaniline to a set of test tubes as indicated:

	Blood		*Urine*		*Standard (0.1 mg/ml)*		*Standard (0.2 mg/ml)*	
	Test	*Blank*	*Test*	*Blank*	*Test*	*Blank*	*Test*	*Blank*
Supernatant	0.5	0.5	—	—	—	—	—	—
Diluted urine	—	—	0.5	0.5	—	—	—	—
0.1 mg/ml Xylose	—	—	—	—	0.5	0.5	—	—
0.2 mg/ml Xylose	—	—	—	—	—	—	0.5	0.5
p-Bromoaniline	2.5	2.5	2.5	2.5	2.5	2.5	2.5	2.5

Place all tests into a 70°C water bath for 10 min, and all blanks in the dark at room temperature.

Cool the incubated tubes and place all test tubes in the dark. After 70 min, read the test against the blank, using a spectrophotometer or colorimeter at a wavelength of 520 nm. Alternatively, all tests and blanks may be read against a water blank.

CALCULATION

Blood

$$\text{mg xylose/100 ml} = \frac{A(\text{test})}{A(\text{standard, 0.1 mg/ml})} \times 10 \times 10$$

If all test and blank tubes are read against water:

$$\text{mg xylose/100 ml} = \frac{A(\text{test}) - A(\text{blank})}{A(\text{standard}) - A(\text{blank})} \times 10 \times 10$$

Urine

If urine was diluted to 1000 ml:

$$\text{g xylose excreted} = \frac{A(\text{test})}{A(\text{standard, 0.2 mg/ml})} \times \frac{20 \times 40}{100} \times \frac{1000}{1000}$$

If urine was diluted to 500 ml:

$$\text{g xylose excreted} = \frac{A(\text{test})}{A(\text{standard, 0.2 g/100 ml})} \times \frac{20 \times 40}{100} \times \frac{500}{1000}$$

If all test and blank tubes were read against water, the absorbance of the blank must be subtracted from the absorbance of the test as indicated under blood.

COMMENTS ON METHODS AND SOURCES OF ERROR

The accuracy of the procedure depends not only on the rate of absorption but also on the rate of excretion of xylose by the kidneys. Thus, patients with renal insufficiency will excrete a decreased amount of xylose. In order to eliminate misinterpretations because of renal retention, a blood determination of xylose is carried out along with the determination of xylose in urine. Normal blood xylose level in the presence of decreased urine xylose levels would suggest renal retention, myxedema, or incomplete urine collection, and thus invalidate the test.

The test findings are abnormal in 80 per cent of patients with malabsorption syndrome.

Aspirin decreases the excretion of xylose by the kidneys, and indomethacin depresses intestinal absorption.

Some abdominal discomfort or slight diarrhea may be observed in some patients. These symptoms can be minimized by use of the 5 g xylose dose. Normal urinary excretion of xylose after a 5 g dose is more than 1.2 g/5 h.

Kendall *et al.*[17] recommend the 5 g dose because it is easier to take, and much less likely to cause abdominal distension and diarrhea. The author also suggests reducing the collection period to 2 h to minimize any possible metabolism of xylose by microorganisms.

REFERENCE

Reiner, M., and Cheung, H. L.: Xylose. *In* Standard Methods of Clinical Chemistry. S. Meites, Ed. New York, Academic Press, 1965, vol. 5, p. 257.

MISCELLANEOUS FINDINGS IN PANCREATIC DISEASES

A significant number of patients with acute and chronic pancreatitis, as well as those with carcinoma of the body and tail of the pancreas, show decreased *glucose tolerance*. Fasting hyperglycemia is observed in acute pancreatitis, but only in 15 to 30 per cent of cases of chronic pancreatitis. Glycosuria has been reported in 5 to 25 per cent of cases of acute pancreatitis.

Serum *calcium* levels are decreased to the range of 8.0 to 8.5 mg/100 ml in the majority of cases of acute pancreatitis and are occasionally as low as 7.0 mg/100 ml. These changes have been observed from 24 to 72 h after the attack, and are attributed to a sudden withdrawal of calcium from extracellular fluids in the formation of calcium soaps. (See also p. 903.) Some authors believe that the extent of the decrease of calcium levels parallels somewhat the severity of the condition.

ELECTROLYTE COMPOSITION OF SWEAT AND SALIVA IN CYSTIC FIBROSIS[16]

THE SWEAT TEST

Cystic fibrosis is a generalized disorder of the exocrine glands with changes in the lungs, upper respiratory tract, liver, gall bladder, pancreas, and other mucus-producing glands throughout the body. Pathological changes appear to be secondary to accumulation of excess mucous secretions rich in glycoproteins, which precipitate and then obstruct all organ passages. Sweat glands show no morphologic changes, but the chemical composition of their secretion is abnormal.[13] In 98 per cent of patients, the secretion of chloride in sweat is increased by two to five times the normal level (see Table 20–1). In fact, the determination of the electrolyte composition of sweat is considered the most reliable single test in the diagnosis of cystic fibrosis.

TABLE 20–1 SODIUM AND CHLORIDE VALUES OF SWEAT IN NORMAL AND ABNORMAL INDIVIDUALS IN MMOL/L

	Normal Value (Homozygotes)	*Normal Value (Heterozygotes)*	*Values in Cystic Fibrosis (Homozygotes)*
Na^+	10–40	40–70	80–190
Cl^-	5–35	35–60	60–160

Electrolyte elevations in sweat are seen even in the absence of gastrointestinal or respiratory symptoms or when pancreatic insufficiency cannot be demonstrated by other tests. In most affected infants, the test becomes positive between 3 and 5 weeks of age.

Similar elevations of sodium, potassium, and chloride have been found in the *saliva* of patients with fibrocystic diseases of the pancreas,[16] but the results of these determinations are inconsistent and by no means as reliable as those obtained from the sweat test.

Normal values for saliva without stimulation (resting), expressed in mmol/l, are: sodium, 14.8 (6.5 to 21.7); potassium, 22.1 (19 to 23); chloride, 12.5 (5 to 20). After stimulation of saliva production by chewing paraffin for 1 h, the respective values for sodium, potassium, and chloride are: Na, 44.6 (43 to 46); K, 18.3 (18 to 19); and Cl, 44 mmol/l.

COLLECTION OF SWEAT BY IONTOPHORESIS[12]

1. With the use of forceps, place two prewashed 2 inch square gauzes (or two Whatman No. 42 filter papers, diameter 5.5 cm) into a weighing bottle and determine the combined weight accurately. The weighing bottle should be handled with tissue or gauze to avoid direct contact with the fingers.

2. Place two other 2 inch square gauzes (not the preweighed gauze) on the anterior (inside) surface of the forearm and moisten them well with a freshly prepared solution of 0.2 g pilocarpine nitrate/100 ml. (Pilocarpine is a drug that, when introduced into the skin, induces sweating.) Place two gauzes saturated with saline on the posterior (outside) surface of the arm. The positive electrode is placed over the gauze with pilocarpine, and the negative electrode is placed on the saline-soaked gauze. Insure good contact and secure the electrodes with rubber bands or similar means. If the arm is too small to secure the electrodes, as in small children, use the thigh of the patient.

3. Apply a current of 2.5 mA for 5 min. The current will tend to increase during this time interval and should be maintained at approximately 2.5 mA. After 5 min, take off the electrodes and clean the area well with distilled water; dry the area.

If the patient complains about discomfort, discontinue the test. A tickling sensation at the site of the electrode is a common finding and should be disregarded. After the test, the skin may be somewhat reddish, but this will disappear within a few hours.

4. With the aid of forceps, place the preweighed gauze or filter paper over the skin area that was exposed to pilocarpine. Place an approximately 4 inch square plastic sheet over this area and seal with surgical tape. Allow the sweat to accumulate on the gauze or filter paper. This usually takes approximately 20 to 30 min, but the time of sweat collection may be extended as long as necessary. In general, the appearance of droplets on the plastic sheet indicates that enough sweat has accumulated.

5. Remove the gauze or filter paper with forceps, place it immediately into the weighing bottle, and stopper. Send the weighing bottle to the laboratory.

RECOVERY OF SWEAT

Accurately determine the weight of the bottle or Petri dish containing the gauze or filter paper and calculate the amount of sweat by difference. One gram of sweat is assumed to be 1 ml of sweat. (Generally 0.2 to 0.5 ml of sweat is obtained.)

Many methods have been suggested to recover the sweat from the gauzes or filter paper. The method of choice will depend on the particular laboratory setup. The gauzes, for example, may be placed in a funnel and washed off with deionized water. The weighing container should also be rinsed with water and the washings should be combined and brought up to a predetermined volume (e.g., 50 ml). To avoid such an elaborate procedure, it has been recommended that a golf tee be placed point down into a 15 ml conical centrifuge tube and that the gauzes be put on the top of the tee. The tube, tightly sealed with a rubber cap, is then centrifuged at moderate speed. The sweat will accumulate in the tip of the tube in its original concentration. If this procedure is used, weighing of the gauzes is not necessary.

ALTERNATE COLLECTION PROCEDURE

An alternative procedure for inducing sweating is local heating of the skin. For this procedure, a selected area of the forearm is washed with distilled water and thoroughly dried. An aluminum cylinder is heated to 48°C in a water bath, dried, and covered with Parafilm. The Parafilm-covered side of the cylinder is held for a 5 min period firmly against the cleansed skin area of the patient. At the end of the 5 min, the heated cylinder is removed and the sweat (collected between Parafilm and skin) is analyzed directly with an ion specific electrode; or the sweat is collected in a capillary tube for chemical analysis.

DETERMINATION OF SWEAT CHLORIDE

A detailed example of sweat recovery for a chloride determination by the Schales and Schales method is now outlined.

1. Add deionized water to the weighing container with the gauze so that the total volume of sweat plus water will be 5 ml (e.g., if 0.3 ml of sweat was collected, add 4.7 ml of water). Stir well to elute the chloride from the paper or gauze.
2. Place 2.5 ml of the extract into a suitable titration vessel and determine the chloride content as outlined in Chapter 15.

CALCULATION

$$\text{chloride in mmol/l} = \frac{\text{reading of unknown} \times 0.2 \times 100 \times 5.0}{\text{reading of standard} \times \text{total amount sweat in ml} \times 2.5}$$

The formula reduces to:

$$\text{chloride in mmol/l} = \frac{\text{reading of unknown} \times 40}{\text{reading of standard} \times \text{total amount of sweat in ml}}$$

Note: The figure 0.2 in the preceding formula indicates the amount of sample for which the chloride procedure has been set up; 100 is the factor for the concentration of the standard; 5.0 is the final volume of the dilution in the weighing container; 2.5 is the amount of extract used in the titration.

MEASUREMENT OF SWEAT COMPOSITION BY CONDUCTIVITY MEASUREMENTS OR ION SPECIFIC ELECTRODES

The electrolyte concentration of sweat can also be assessed by determining its *electrical conductivity*.[39] (Equipment available from Heat Technology Laboratory, Inc., 4308 Governors Drive, Huntsville, AL.) Results obtained with this technique are a measure of total electrolyte content, rather than chloride or sodium concentration.

Other investigators[18] have recommended the use of ion specific electrodes for Na^+ and Cl^- measurement, respectively. Such electrodes measure the *ion activity*, rather than the ion concentration, but a good correlation between these two measured quantities has been observed. The principle of ion specific electrodes has been discussed in the section "Electrochemistry."

REMARKS AND SOURCES OF ERROR

False high results can be obtained if the skin of the child is not cleaned properly or if contaminated sponges are used. Thus, gauzes have to be cleaned and dried carefully before use or a blank determination has to be done to assure that the gauze or filter paper is low in chloride content.

False normal results can be obtained in patients with pure salt depletion, which is common in the affected group of patients during hot weather periods. Additional electrolyte studies on serum help to avoid misinterpretation of results in such instances. Elevated levels of electrolytes in sweat have also been reported in meconium ileus, in adrenal insufficiency, and in some cases of renal disease.

It has been found that determination of the chloride concentration is the most reliable index for diagnosing cystic fibrosis; therefore, many laboratories use only this test.

AGAR PLATE TEST

The palm of the hand of the patient is brought in contact with an agar plate or a commerically available paper prepared with silver salts. An increased amount of chloride in the sweat results in a white precipitate of silver chloride which can be graded semiquantitatively. Sweating is induced by placing the hand into a plastic bag. Such tests should be used only for screening purposes, since results are not as reliable as those obtained from direct measurement of sweat electrolyte contents.

LABORATORY TESTS USEFUL IN THE DIAGNOSIS OF PANCREATIC DISORDERS

Acute Pancreatitis. Amylase and lipase activity in serum is generally significantly increased. Amylase activity in urine may be increased even if serum amylase is normal. Serum calcium levels are frequently decreased. The secretin-CCK-PZ test shows relatively normal exocrine secretory capacity during the early phase of the disease. The sweat test is normal.

Chronic Pancreatitis. Activity of serum amylase and lipase is most often normal or only slightly elevated. Amylase values in urine may be increased. The secretin-CCK-PZ test shows a low output of total volume and a significant decrease in the bicarbonate output. Enzyme secretion is somewhat better maintained than bicarbonate secretion. The sweat test is normal.

Carcinoma of the Pancreas. In carcinoma of the *head* of the pancreas with obstruction, the secretin-CCK-PZ test shows only negligible response. Serum lipase and amylase values may be elevated if a sufficient amount of acinar tissue remains. Leucine aminopeptidase (LAP) activity in serum is frequently increased, and serum bilirubin determinations show the values generally observed in obstructive jaundice. Fat absorption tests and ^{131}I-triolein uptake are abnormal (low). Cytological examinations of pancreatic juice (if secreted) may show cancer cells.

In cancer of the *body or tail* of the pancreas, amylase and lipase activity in serum is most often normal, but LAP activity is frequently increased. The CCK-PZ test shows a decrease in total volume of pancreatic juice, but the composition is relatively normal with enzyme concentration somewhat more impaired than bicarbonate output. Absorption tests (except ^{125}I-oleic acid) are abnormal. The sweat test is normal.

Cystic Fibrosis. A sweat test shows an increased concentration of Na^+ and Cl^- in sweat. The secretin-CCK-PZ test shows little response. If there is residual pancreatic function, the enzyme secretion is better maintained than the bicarbonate output. Absorption tests (except ^{125}I-oleic acid) are abnormal. Stool fat excretion is increased, and stool trypsin is absent.

Malabsorption of Non-pancreatic Origin. Stool fat excretion is increased. Absorption tests, including ^{125}I-oleic acid uptake, are abnormal. Xylose absorption is decreased. Microscopic examination of stool shows digested meat fibers and starch. The CCK-PZ test and the sweat test are normal.

REFERENCES

1. Ammann, R. W., Tagwercher, E., Kashiwagi, H., and Rosenmund, H.: Diagnostic value of fecal chymotrypsin and trypsin assessment for detection of pancreatic disease. Am. J. Digest. Dis., *13*:123–146, 1968.
2. Anderson, S.: Secretion of gastric intestinal hormones. Ann. Rev. Physiol., *35*:431–452, 1973.
3. Berson, S. A., and Yalow, R. S.: Radioimmunoassay in gastroenterology. Gastroenterology, *62*:1061–1084, 1972.
4. Blackman, A. H., Lambert, D. L., Thayer, W. R., and Martin, H. F.: Computed normal values for peak acid output based on age, sex, and body weight. Am. J. Digest. Dis., *15*:783–789, 1970.
5. Bouchier, I. A. D.: Gastric and pancreatic function tests. Ann. Clin. Biochem., *7*:122–125, 1970.
6. Brooks, F. P.: Clinical usefulness of gastric acid secretory tests. Postgrad. Med., *51*:189–193, 1972.
7. Brooks, F. P.: Tests of pancreatic function. *In:* Practice of Medicine, Vol. 2, Hagerstown, Md., Harper & Row, 1973.
8. Brooks, F. P., and O'Neill, F.: Gastric analysis for hydrochloric acid. *In:* Practice of Medicine, Vol. 2, Hagerstown, Md., Harper & Row, 1973.
9. Davenport, H. W.: The gastric mucosal barrier. Digestion, *5*:162–165, 1972.
10. Davenport, H. W.: Control of secretion. *In:* Physiology of the Digestive Tract. 3rd ed. H. W. Davenport, ed. Chicago, Yearbook Medical Publishers, Inc., 1971.
11. Farrar, G. E., and Bower, R. J.: Gastric juice and secretion: physiology and variations in disease. Ann. Rev. Physiol., *29*:141–168, 1967.
12. Gibson, L. E., and Cooke, R. E.: A test for concentration of electrolytes in sweat in cystic fibrosis of the pancreas utilizing pilocarpine by iontophoresis. Pediatrics, *23*:545–549, 1959.
13. Jones, J. D., Steige, H., and Logan, G. B.: Variations of sweat sodium values in children and adults with cystic fibrosis and other diseases. Mayo Clin. Proc., *45*:768–773, 1970.
14. Kaess, H.: The radioimmunoassay of gastrin. Acta Hepato-Gastroenterol., *20*:1–5, 1973.
15. Kahan, J.: The vitamin A absorption test. I. Studies on children and adults without disorders in the alimentary tract. Scand. J. Gastroenterol., *4*:313–324, 1969.
16. Kaiser, E., Kunstadter, R. H., and Mendelsohn, R. S.: Electrolyte concentrations in sweat and saliva. A.M.A. J. Dis. Child., *92*:369–373, 1956.
17. Kendall, M. J., Nutter, S., and Hawkins, C. F.: Bacteria and the xylose test. Lancet, *1*:1017–1018, 1972.
18. Kopito, L., and Shwachman, H.: Studies in cystic fibrosis: Determination of sweat electrolytes in situ with direct reading electrodes. Pediatrics, *43*:794–798, 1969.
19. Korman, M. G., Soveny, C., and Hansky, J.: Radioimmunoassay of gastrin. The response of serum gastrin to insulin hypoglycaemia. Scand. J. Gastroenterol., *6*:71–75, 1971.
20. Kretcher, N.: Lactose and lactase. Scientific American, *227*:70–78, 1972.
21. Lagerlöf, H. O.: Pancreatic function and pancreatic disease: studied by means of secretin. Acta Med. Scand. Supplement, *128*:1–289, 1942.
22. Laudano, O. M., and Roncoroni, E. C.: Determination of the dose of histalog that provokes maximal gastric secretory response. Gastroenterology, *49*:372–374, 1965.
23. Levitt, M. D., Rapoport, M., and Cooperband, S. R.: The renal clearance of amylase in renal insufficiency, acute pancreatitis, and macroamylasemia. Ann. Int. Med., *71*:919–925, 1969.
24. Lifshitz, F., and Holman, G. H.: Disaccharidase deficiencies with steatorrhea. J. Pediatr., *64*:34–44 1964.
25. Lubran, M.: Measurement of gastric acidity. Lancet, *2*:1070–1071, 1966.
26. McGuigan, J. E., and Trudeau, W. L.: Studies with antibodies to gastrin. Radioimmunoassay in human serum and physiological studies. Gastroenterology, *58*:139–150, 1970.
27. Michaelis, L.: Harvey Lectures 1926–1927. p. 59–89. Academic Press, Inc., New York, 1928.
28. Moore, E. W.: The terminology and measurement of gastric acidity. Ann. N.Y. Acad. Sci., *140*:866–874, 1967.
29. Mulhausen, R., Brown, D. C., and Onstad, G.: Renal clearance of amylase during pancreatitis. Metabolism, *18*:669–674, 1969.
30. Newcomer, A. D.: Digestion and absorption of proteins. Mayo Clin. Proc., *48*:624–629, 1973.
31. Newcomer, A. D.: Disaccharidase deficiencies. Mayo Clin. Proc., *48*:648–652, 1973.
32. Onstand, G. R., and Zieve, L.: Carotene absorption. A screening test for steatorrhea. J.A.M.A., *221*:677–679, 1972.

33. Reiner, M., and Cheung, H. L.: Xylose. Standard Methods of Clin. Chem., *5*:257–268, 1965.
34. Rick, W.: Der Secretin-pankreozymin-test in der Diagnostik der Pankreasinsuffizienz. Internist, *11*:110–117, 1970.
35. Rick, W.: Chemical methods in the diagnosis of pancreatic disease. *In:* Clinics in Gastroenterology, Vol. 1, The Exocrine Pancreas. H. T. Howat, Ed. London, Philadelphia, Toronto, W. B. Saunders Co., 1972.
36. Segal, H. L.: Gastric analysis. J.A.M.A., *196*:655, 1966.
37. Segal, H. L.: Tubeless gastric analysis as a tool to measure gastric secretory activity. Ann. N.Y. Acad. Sci., *140*:896–903, 1967.
38. Segal, H. L., Miller, L. L., and Plumb, E. J.: Tubeless gastric analysis with an azure A ion-exchange compound. Gastroenterology, *28*:402–408, 1955.
39. Shwachman, H., Dunham, B. S., and Phillips, W. R.: Electrical conductivity of sweat: A simple diagnostic test in children. Pediatrics, *32*:85–88, 1963.
40. Silver, S.: Radioactive Nuclides in Medicine and Biology. 3rd ed. Philadelphia, Lea & Febiger, 1968.
41. Sodeman, W. A., Jr., and Sodeman, W. A., Eds. Pathologic Physiology: Mechanisms of Disease. 5th ed. London, Philadelphia, Toronto, W. B. Saunders Co., 1974.
42. Song, H., Tietz, N. W., and Tan, C.: Usefulness of serum lipase, esterase, and amylase estimation in the diagnosis of pancreatitis—a comparison. Clin. Chem., *16*:264–268, 1970.
43. Stadil, F., and Rehfeld, J. F.: Determination of gastrin in serum. An evaluation of the reliability of a radio-immunoassay. Scand. J. Gastroenterol., *8*:101–112, 1973.
44. Thompson, J. C.: Gastrin and gastric secretion. Ann. Rev. Med., *20*:291–314, 1969.
45. Tietz, N. W., Borden, T., and Stepleton, J. D.: An improved method for the determination of lipase in serum. Am. J. Clin. Path., *31*:148–154, 1959.
46. Tietz, N. W., and Fiereck, E.: A specific method for serum lipase determination. Clin. Chim. Acta, *13*:352–358, 1966, and Serum Lipase. *In:* Standard Methods of Clinical Chemistry. G. R. Cooper, Ed. New York, Academic Press, 1972, Vol. 7, p. 19–31.
47. Tietz, N. W., and Repique, E.: Proposed standard method for measuring lipase activity in serum by a continuous sampling technique. Clin. Chem., *19*:1268–1275, 1973.
48. Trudeau, W. L., and McGuigan, J. E.: Effects of calcium on serum gastrin levels in the Zollinger-Ellison syndrome. New Eng. J. Med., *281*:862–866, 1969.
49. Trudeau, W. L., and McGuigan, J. E.: Relations between serum gastrin levels and rates of gastric hydrochloric acid secretion. New Eng. J. Med., *284*:408–412, 1971.
50. Wormsley, K. G.: Further studies of the response to secretin and pancreozymin in man. Scand. J. Gastroenterol., *6*:343–350, 1971.
51. Wormsley, K. G.: Tests of pancreatic function. Br. J. Clin. Pract., *24*:271–275, 1970.
52. Wormsley, K. G.: Pancreatic function tests. *In:* Clinics in Gastroenterology, Vol. 1, The Exocrine Pancreas. H. T. Howat, Ed. London, Philadelphia, Toronto, W. B. Saunders Co., 1972.

ADDITIONAL READING

1. Carey, L. C., Ed.: The Pancreas. St. Louis, C. V. Mosby Co., 1973.
2. Gambill, E. E.: Pancreatitis. St. Louis, C. V. Mosby Co., 1973.
3. Henry, R. J., Cannon, D. C., and Winkelman, J. W.: Clinical Chemistry, Principles and Techniques. 2nd ed. Hagerstown, Harper and Row, 1974.
4. Howat, H. T., Ed.: The Exocrine Pancreas. Clinics in Gastroenterology, Vol. 1. London, Philadelphia, Toronto, W. B. Saunders Co., 1972.
5. Paulson, M., Ed.: Gastroenterologic Medicine. Philadelphia, Lea & Febiger, 1969.

CHAPTER 21

ANALYSIS OF DRUGS AND TOXIC SUBSTANCES

by Robert V. Blanke, Ph.D.

The incidence of poisoning, both accidental and suicidal, continues to increase in this country. In 1968, accidental poisoning was responsible for 4109 deaths in the United States. In the same year, 5684 suicides by poison occurred.[45] Valid statistics concerning the number of non-fatal poisonings are not available, but the National Clearing House for Poison Control Centers received reports on 136,051 incidents of poison ingestion during 1972.[25] If one accepts the frequently cited estimate that five to eight unsuccessful suicide attempts occur for each successful one, then 28,000 to 45,000 attempts occur annually.[4,61]

The high incidence of poisonings of all types has led most hospitals and clinical laboratories to recognize the need for laboratory services that give rapid and reliable information about the type and quantity of poison ingested by the patient.

The purpose of this chapter is to describe a number of laboratory procedures that can be carried out using equipment common to most clinical laboratories. These procedures require careful attention to detail, but not more than would be necessary in any other clinical test conducted by a well trained technologist. Some of the tests described are qualitative in nature; others enable an accurate, quantitative estimation of serum and urine levels to be made. Some of these tests are of the "screening" type; that is, a negative result indicates that the toxic substance is not present in significant amounts, and a positive result indicates that toxic substances may be present in significant quantities. Other procedures are very specific and yield positive results with only one or a limited group of substances.

The modern toxicological specialist uses common, as well as other more sophisticated tools of science in isolating, identifying, and quantitating toxic substances in biological material. Activation analysis, x-ray diffraction, crystallography, optical rotatory dispersion, and other specialized or complex techniques may be required for specific problems. In addition, the toxicological specialist should be in a position to interpret the laboratory results, give advice about treatment of the patient, or give an opinion as to the effect, if any, of the toxic substance on the condition of the patient.

Although toxicological specialists, with their unique methodology and expertise, are not generally part of the clinical laboratory staff, every clinical laboratory director should be informed about the special problems relating to toxicological tests. He should also know the closest toxicological specialist to whom he can turn for advice when problems arise that are beyond the capabilities of his staff and facilities.

In medicolegal cases, that is, cases in which the possibility of criminal poisoning exists or cases in which it is probable that legal action may be necessary in order to determine liability, the advice and services of an experienced toxicologist should be sought. In these cases, special precautions are necessary in collecting, transporting, and storing specimens to insure that loss, tampering, or contamination cannot occur. Also, the specimens and other evidence must be in the custody of a responsible individual at all times in order to establish a proper chain of custody and avoid legal objections during the presentation of evidence in court.

Ideally the analyst will collect the specimen and keep it in his custody until the analysis can be completed. If this is impractical, the individual collecting the specimen should place it in a proper container labeled with the name of the patient, the time and date it was collected, and the nature of the specimen. Many ingenious sealing devices have been described to make the container tamperproof. One of the most practical is a simple gummed-paper strip placed across the cap or stopper of the container so that the cap cannot be removed without tearing the strip. The paper strip should carry the signature or initials of the person who collected the specimen. The analyst can then note whether the seal was intact at the time he received the specimen, and he can also preserve the seal for later identification by the person who prepared it.

Other aspects of specimen handling of concern in legal cases are: the use of clean containers, the use of proper preservatives or anticoagulants, the procurement of receipts when specimens are transferred from one individual to another, the mailing of specimens by registered mail, preparation of a written description of the container and its contents as received in the laboratory (photography is useful in this situation), and the preservation of unused specimens by freezing for duplicate analysis by an independent laboratory. These and other problems of a similar nature are considered in Stewart and Stolman's book.[54] It is best to obtain the advice of a forensic toxicologist on these matters before collecting and sending him specimens.

The costly and complex laboratory procedures, together with the legal problems and inconveniences of court testimony, have had the result that most clinical laboratory directors are reluctant even to contemplate the performance of toxicological tests. Such considerations need not, and indeed should not, prevent clinical laboratories from carrying out those tests that are within the limitations of their personnel and facilities. The useful information gained can be rapidly communicated to the clinician, which results in more intelligent and faster treatment and possibly the saving of lives.

Clinical toxicology laboratories generally have two main functions. The first and most obvious function is the identification and quantitation of toxic substances in specimens from patients brought to the emergency room after possible acute toxic episodes. In general, the clinician needs the laboratory results promptly in order to answer questions such as "What toxic substance, if any, is present?", "Is the quantity of any toxic substance present consistent with the signs and symptoms displayed by the patient?", "Will the quantity of toxic substance present influence the choice of the treatment procedure to be utilized?", and "Do the laboratory findings confirm the history, diagnosis, or clinical impression?"

The second general function of the clinical toxicology laboratory is to monitor certain chemical substances in patients. This may be done for the purpose of maintaining therapeutically effective drug levels (as in epileptic patients), or it may be done to determine the effectiveness of a treatment procedure in reducing levels of toxic agents. A relatively recent area of interest may also include the determination of levels of metallic ions in cases of industrial exposure or when alterations in metal metabolism may be associated with a clinical condition.

Both of these general functions utilize common laboratory facilities and personnel, so that it is reasonable that the clinical toxicology laboratory be engaged in both activities. In general, in order to fulfill its role in dealing with emergency situations and to be of maximum use to the clinician, the toxicology laboratory must provide 24 hour service and utilize reliable, rapid methods resulting in a short turnaround time. The second general function, monitoring chemical substances in patients, can usually be achieved in a normal workday time period using proven, sometimes automated, procedures on a relatively continuous basis.

Since there are literally thousands of toxic substances which may be encountered in clinical toxicology, it is absolutely essential that these substances be examined realistically at the outset with a view toward selecting those which are commonly encountered in the community served by the laboratory. Some substances are common to almost all treatment facilities, while others may be unique to a particular locality. In rural and urban communities, for example, the relative importances of problems relating to pesticides and narcotics may differ. An important industry in one locality may present specific toxicological problems which would not usually be encountered elsewhere.

It is commonly acknowledged that good communication between the clinician and the laboratory facility is essential for the production of meaningful laboratory results in all areas of activity. Unfortunately, the gap between acknowledging this truism and implementing it is difficult to bridge. Lines of communication between the clinical toxicology laboratory and the clinician can be laid down initially by establishing the priority with which new toxicological procedures should be set up. It is essential that the clinical toxicology laboratory understand the problems of the clinician and that the clinician understand the problems of the laboratory. For example, it may be important in a specific case to determine whether a patient has ingested an overdose of lysergic acid diethylamide (LSD). It is now possible to assay LSD levels by a radioimmunoassay procedure. But is the frequency of this determination high enough to justify the expenditure of funds for equipment and personnel to make this assay procedure available on a round-the-clock basis? Conversely, the laboratory may have an elegant method for measuring fluoride levels in serum, but is it of practical use to the clinician when the test requires 24 hours before results are obtained?

It is better for a clinical toxicology laboratory to conduct a relatively small number of tests by reliable methods in reasonable time periods than to attempt to handle all types of situations which might arise. Once a basic nucleus of tests is agreed upon, the laboratory can adapt and evaluate published procedures to give the clinician results which are meaningful both in reliability and in turnaround time.

The "unknown" toxic agent is always a difficult problem to handle. Again, good communication can help to resolve these problems, at least partially. Most clinical toxicology laboratories have the ability to conduct screening tests of a variety of types. None of these are all-inclusive, and most of them are chiefly of value when they are negative. Some screening tests are of the spot-test type, which can be conducted on urine or gastric contents after a minimum of handling. Others can be used for screening for metals, alcohol or other volatile substances, or narcotic drugs. The simple request for "toxicological screening" or "test for poisons" presents a dilemma to the clinical toxicology laboratory. Should the screening tests be directed toward detecting drugs of abuse, or heavy metals? Should pesticides be considered, or is the patient a known alcoholic? Valuable time can be saved and costs minimized by a brief history or clinical summary of the patient's condition. If nothing else, this would indicate what substances may not be present.

Thoughtful and well-planned coordination of the laboratory's efforts with those of the clinician can be provided by establishing and maintaining good communication between both parties. This is particularly true in the application of quick screening tests. These tests are appropriate in certain circumstances, provided their limitations are recognized *and understood by the clinician.*[5]

In the procedures to be described, an effort has been made to keep them as simple and economical as possible without sacrificing accuracy and specificity. With each procedure, additional references are given to more specific or more rapid methods that require special equipment or training.

The analytical toxicologist, realizing that some overlapping of categories always occurs, tends to classify poisons according to the following groups: (1) gases, (2) volatile substances (i.e., substances, usually liquids, that are separable by steam distillation), (3) corrosives, (4) metals, (5) nonmetals (elemental or combined forms), and (6) nonvolatile organic substances.

No attempt has been made here to present a systematic approach to the "general unknown" type of problem, i.e., one in which the type of poison present, if any, is not known. Neither has any attempt been made to include procedures for the detection of all toxic substances. Space does not permit such an all-inclusive approach. Only the most common toxic substances in each group are considered.

For those interested in other toxicological tests useful in the clinical laboratory, the excellent small book by Curry[12] or the manual edited by Sunshine[57] are recommended. Gleason's book with its supplements[30] serves as a useful source of information relating to diagnosis and treatment.

GASES

CARBON MONOXIDE

Carbon monoxide, the product of incomplete combustion of organic substances, is the most common of the gaseous poisons. It is present in the free state in manufactured gas (coal gas), but not in natural gas. Both types of gas, when used as fuel for stoves, furnaces, and other appliances, release carbon monoxide as one of the combustion products. This is true, of course, of coal, oil, and other types of fuel as well. Malfunctioning or poorly ventilated heating appliances, therefore, are frequently causes of carbon monoxide poisoning. Since this gas is also a component of exhaust fumes from internal combustion engines, accidental poisonings can occur when gasoline powered tools or outboard motors, as well as automobile engines, are used under conditions of poor ventilation or improper operation.

Carbon monoxide combines reversibly with hemoglobin in a manner almost identical to oxygen. Thus, carbon monoxide occupies the sites on the hemoglobin molecule which normally bind with oxygen. This results in a decrease in the amount of oxygen carried by hemoglobin. In addition, the presence of carbon monoxide on the hemoglobin molecule produces a shift in the oxyhemoglobin dissociation curve. Thus, not only does the amount of oxygen bound by hemoglobin decrease, but the oxygen which is carried is not released as effectively to be utilized for cellular respiration. The primary toxic effect of carbon monoxide therefore is cellular hypoxia or anoxia. The binding force of carbon monoxide to hemoglobin is about 210 times as strong as that of oxygen. As a result, carbon monoxide is not displaced from hemoglobin (except at high oxygen tension), and accidental poisonings can occur even at low levels of carbon monoxide in the atmosphere with prolonged exposure. (See Table 21-1.) In these instances the carbon monoxide level in blood builds up slowly until toxic levels are reached.

TABLE 21-1 CARBON MONOXIDE TOXICITY*

% (v/v) *in Air*	*Response*
0.01	Allowable for an exposure of several hours.
0.04–0.05	Can be inhaled for 1 hour without appreciable effect.
0.06–0.07	Causing a just noticeable effect after 1 hour's exposure.
0.1–0.12	Causing unpleasant but not dangerous symptoms after 1 hour's exposure.
0.15–0.20	Dangerous for exposure of 1 hour.
0.4 and above	Fatal in exposure of less than 1 hour.

* From Deichmann, W. B., and Gerarde, H. W.: Symptomatology and Therapy of Toxicological Emergencies. New York, Academic Press Inc., 1964. © Copyright Academic Press Inc.

The detection and estimation of carbon monoxide in biological specimens can be approached in two general ways: (1) release of the gas from the hemoglobin complex with subsequent direct or indirect measurement of the gas, or (2) estimation of carboxyhemoglobin by its typical color or absorption bands. The first approach can be carried out by gasometric techniques,[63] gas chromatography,[31] microdiffusion,[22] or infrared spectrophotometry.[55] The second approach utilizes spectrophotometric[59] or spectrographic analysis or simple color comparison. An example of each approach will be described.

Regardless of the analytical method used, the specimen to be analyzed must contain hemoglobin. Relatively little carbon monoxide dissolves in the aqueous or lipid fractions of tissue, compared to that bound to hemoglobin.

This rather obvious statement is made since some clinicians request serum, spinal fluid, or even urine carbon monoxide levels. The most satisfactory specimen is whole blood. Clotted blood is less desirable and must be homogenized with a minimum exposure to air before analysis. This can be done using a Ten Broeck hand homogenizer. The clot is gently disintegrated, avoiding the introduction of air into the specimen or "whipping" the specimen. In those fatalities in which the victim is so badly burned or mutilated that blood is not available, tissue rich in hemoglobin, such as bone marrow or spleen, can be used. In these cases, estimation of carbon monoxide, rather than carboxyhemoglobin, is done.

DETERMINATION OF CARBON MONOXIDE BY MICRODIFFUSION

The principle of diffusion can be used to separate a number of toxic substances from biological material. It is particularly applicable for separating gases and volatile substances. The specimen, containing a substance to be separated, and a "trapping" solution are placed in separate containers inside a third sealed container. The specimen and the trapping solution are in contact with the same atmosphere. The substance to be separated, because of its vapor pressure, leaves the specimen and enters the atmosphere, from which it is absorbed by the trapping solution. Thus by gaseous diffusion the substance to be separated is transferred from the specimen to the trapping solution until an equilibrium is reached. If the trapping solution contains a reagent that converts the separated substance into a different compound, equilibrium does not occur and a quantitative transfer results. The entire operation can be carried out with a small specimen by using a Conway unit. This unit consists of two round, concentric chambers molded into a porcelain or glass dish that can be sealed by a glass plate. The trapping solution is placed in the center well and the specimen in the outer compartment.

The time required for completion of the diffusion process is variable, depending on the vapor pressure of the substance to be separated, the volume of the specimen solution, the nature of the trapping solution, and the temperature at which the process is carried out. In general, the diffusion time is shortened by higher temperatures, by small volumes of specimen solutions, and by substances of high vapor pressure. It should be emphasized that procedures worked out for a specific diffusion assembly may require different time periods for completion in an assembly of different dimensions. Also, if a diffusion is carried out at an elevated temperature, precautions must be taken to insure against loss of expanding gases, which tend to lift the glass plate. Even at room temperature this can occur if cold solutions are used. It is advisable to place a 10 ounce weight on the lid at room temperature, and at elevated temperatures the lid should be clamped in place. For a mathematical consideration of these variables, see Conway.[9]

PRINCIPLE

The microdiffusion technique can be utilized to detect carbon monoxide gas that has been released from hemoglobin by the action of dilute sulfuric acid or by lactic acid-ferricyanide solution. Carbon monoxide will reduce Pd^{2+} to metallic palladium, which appears as a silvery-black film on the surface of the reagent:

$$PdCl_2 + CO + H_2O \rightarrow Pd + CO_2 + 2HCl$$

The amount of carbon monoxide can be estimated indirectly by determining the amount of palladium reduced or the amount of hydrochloric acid produced in the reaction. The former is a colorimetric method, and the latter is titrimetric. We have found that microdiffusion is a very useful screening test, since the amount of metallic palladium produced is proportional to the amount of carbon monoxide in the specimen. Although this procedure can be carried out quantitatively, it is generally more convenient to use the spectrophotometric method, described later, for this purpose.

In the original method[22] on which this procedure is based, the authors recommend the use of 10 per cent sulfuric acid as an agent to liberate carbon monoxide from hemoglobin. We have found that some substances, particularly formic acid, may be converted to carbon monoxide by this treatment. In order to avoid such false positive results, we recommend the use of lactic acid-ferricyanide solution as the liberating agent.

REAGENTS

1. Palladium chloride solution. Dissolve 0.22 g of $PdCl_2$ in 250 ml of 0.01 molar HCl. The solution is stable if protected from the carbon monoxide in the atmosphere.
2. Hemolyzing solution. Mix equal parts of $K_3Fe(CN)_6$ solution (3.2 g/100 ml) and 0.1 molar lactic acid solution.
3. Sealing compound. Either stopcock grease or petrolatum can be used.

PROCEDURE

Place 1.0 ml of blood into the outer compartment of a Conway diffusion dish and 2.0 ml of $PdCl_2$ solution into the center compartment. Add 2.0 ml of hemolyzing solution into the outer compartment opposite the blood, and cover the dish with the ground glass plate treated with sealing compound. Carefully mix the blood and hemolyzing solution by swirling the Conway dish gently. Allow diffusion to proceed for 1 hour at room temperature. In the presence of carbon monoxide, a mirror of metallic palladium will be noted on the surface of the palladium chloride solution in the inner compartment.

The only interference in this test is from the presence of sulfides in putrified blood. In the absence of such interference, a small spot of reduced palladium is consistent with the amount of carbon monoxide due to heavy smoking or subtoxic exposures. A bright mirror covering the entire compartment is typical of a lethal level of carbon monoxide in the blood.

For those interested in carrying out this procedure quantitatively, the original procedure should be consulted.[22] Much more dilute solutions are used in the titrimetric method, together with dilute, standardized acidic and basic solutions.

REFERENCE

Feldstein, M., and Klendshoj, N.: J. Forensic Sci., *2*:39, 1957.

CARBON MONOXIDE BY SPECTROPHOTOMETRIC DETERMINATION

Hemoglobin and its derivatives have characteristic absorption bands in the visible region that can be utilized to detect carboxyhemoglobin and to measure the quantity present. Oxygenated hemoglobin and carboxyhemoglobin have similar double bands in alkaline solution. The absorption maxima for oxygenated hemoglobin are 576 to 578 and 540 to 542 nm; for carboxyhemoglobin they are 568 to 572 and 538 to 540 nm. Deoxygenated hemoglobin has a single broad band at 555 nm (see Fig. 21–1).

If a weakly alkaline dilution of blood is treated with sodium hydrosulfite, oxygenated hemoglobin (and any methemoglobin present) is converted to deoxygenated hemoglobin. Carboxyhemoglobin is unaffected by such treatment.

$$\left.\begin{array}{c}HbO_2\\ \text{or}\\ MetHb\end{array}\right. + Na_2S_2O_4 \rightarrow Hb$$

$$HbCO + Na_2S_2O_4 \rightarrow \text{No reaction}$$

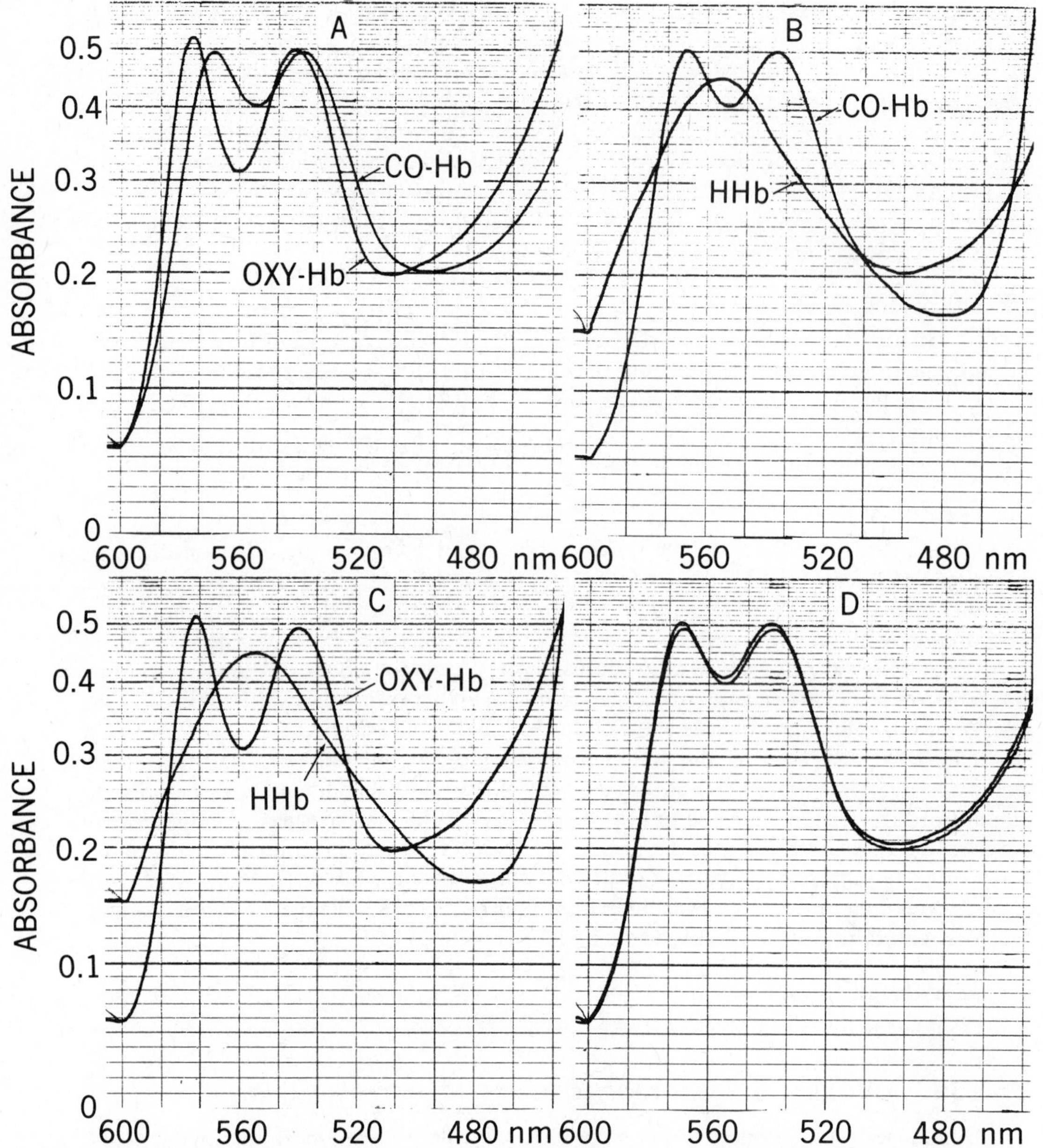

Figure 21–1 *A*, Spectral curve of 100 per cent carboxyhemoglobin and 100 per cent oxyhemoglobin.

B, Spectral curve of 100 per cent carboxyhemoglobin and 100 per cent oxyhemoglobin after treatment with sodium dithionite.

C, Spectral curve of 100 per cent oxyhemoglobin before and after treatment with sodium dithionite.

D, Spectral curve of 100 per cent carboxyhemoglobin before and after treatment with sodium dithionite. (The Beckman DB Recording Spectrophotometer was used to make these recordings. From Tietz, N. W., and Fiereck, E.A.: Ann. Clin. Lab. Sci., *3*:36, 1973.)

This is the basis of several methods for the determination of per cent saturation of hemoglobin by carbon monoxide. The method to be described[59] works satisfactorily with fresh, whole blood, but is not satisfactory with postmortem blood or specimens containing denatured hemoglobin.

PRINCIPLE

A dilute hemolysate of blood is treated with sodium hydrosulfite, which reduces methemoglobin and oxyhemoglobin but does not affect carboxyhemoglobin. The absorbance of this solution is measured at 541 and 555 nm, the absorbance ratio A_{541}/A_{555} is calculated, and the per cent carboxyhemoglobin is determined from the standard curve.

REAGENTS

1. NH_4OH, 0.12 mol/l. Dilute 15.9 ml of concentrated NH_4OH to 1.0 liter with deionized water. This solution is stable.
2. Sodium hydrosulfite (sodium dithionite), AR. Preweigh 10 mg portions of sodium dithionite into individual small test tubes. Stopper the test tubes or cover with Parafilm.
3. Carbon monoxide. Lecture bottle (Matheson Gas Products, Div. of Will Ross, Inc., East Rutherford, N.J. 07073).
4. Oxygen, CP.

SPECIAL APPARATUS

A narrow band pass ($<$2 nm) spectrophotometer with 1.00 cm cuvettes is required, although the use of a recording spectrophotometer with the same specifications is desirable. The procedure listed below can be performed on a Beckman DB Recording Spectrophotometer or other similar instrument.

It is imperative that the spectrophotometer be checked regularly for wavelength and spectrophotometric accuracy with appropriate calibrating filters (e.g., NBS Reference Material 930) and with liquid photometric standards (e.g., NBS Reference Material 931).

PROCEDURE

1. Add 100.0 μl of whole heparinized blood to 25 ml of 0.12 molar NH_4OH. Mix the solution and allow it to stand for 2 min.
2. Transfer 3.0 ml of NH_4OH (blank) and 3.0 ml of the hemolysate (test) respectively into 1.0 cm cuvettes. (Analyze the sample in triplicate.)
3. Add 10 mg of sodium dithionite to each of the cuvettes. Cover the cuvettes with Parafilm and invert gently 10 times.
4. Exactly 5 min after the addition of dithionite to the sample, read the absorbance at 541 and 555 nm against the NH_4OH blank. (If a number of samples are analyzed, space the addition of the reducing agent so that each can be read after exactly 5 min.)
5. Calculate the ratio of the absorbance at 541 nm to that at 555 nm, and determine the per cent carboxyhemoglobin from the calibration chart.

Note: For confirmation and for the purpose of record, the sample without and with dithionite (steps 1 and 3, respectively) may be scanned between 450 and 600 nm. (See Fig. 21–1.)

PREPARATION OF THE STANDARD CURVE

Caution—Use a fume hood when working with carbon monoxide gas.

1. Collect 20 ml of heparinized blood from a healthy person who does not smoke.
2. Transfer a 4.0 ml portion of the fresh, heparinized blood sample into each of two 125 ml separatory funnels. Treat one sample with pure oxygen and the other with pure carbon monoxide for 15 min while the funnels are gently rotated. After the addition of the

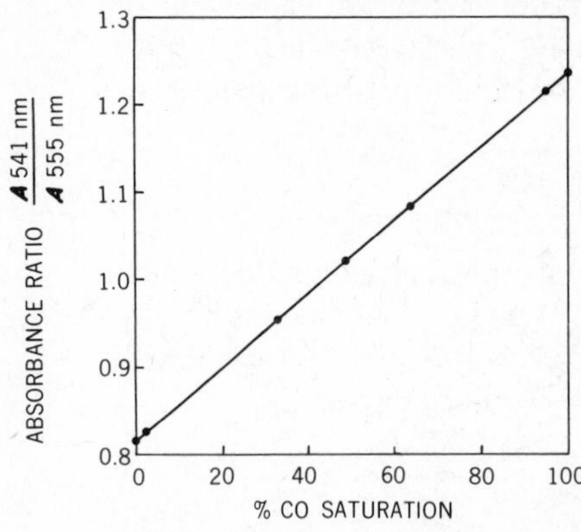

Figure 21–2 Example of a standard curve for the conversion of the 541/555 nm absorbance ratio to per cent carboxyhemoglobin saturation.

gases, close the separatory funnels and rotate them gently for an additional 15 min. Analyze the fully saturated samples immediately, in triplicate, according to the procedure given above. Use these results for the establishment of the 0 and 100 per cent carboxyhemoglobin calibration points. These samples may not be used to establish the intermediate calibration points.

Plot the ratio of the absorbance at 541 nm to that at 555 nm for the 0 per cent and for the 100 per cent carboxyhemoglobin samples and draw a line between the two points. (See Fig. 21–2.)

3. Fill the funnel containing the 100 per cent carboxyhemoglobin sample with nitrogen gas and rotate it for 5 min. Treatment with nitrogen removes the physically dissolved CO from the sample, but a small amount of CO will also dissociate from hemoglobin. Determine the exact carboxyhemoglobin content of this sample by the method described, using the standard curve just prepared. Prepare intermediate standards by mixing appropriate proportions of the nitrogen-treated sample with the oxygen-treated sample.

4. Analyze each of the diluted blood samples from step 3 in triplicate, according to the procedure given above.

5. Plot the calculated concentrations against the absorbance ratios obtained. These points should fall on the line drawn for the fully saturated samples, since the curve is linear over the entire range. (See Fig. 21–2.)

INTERPRETATION

Normal carbon monoxide levels depend on the degree of exposure to this gas, without signs and symptoms of poisoning being produced. For example, in smokers the following levels may occur:

Smokers (one to two packs per day): up to 4 to 5 per cent saturation of hemoglobin with carbon monoxide.

Heavy smokers (more than two packs per day): up to 8 to 9 per cent saturation of hemoglobin with carbon monoxide.

Nonsmokers: 0.5 to 1.5 per cent saturation of hemoglobin with carbon monoxide.

In our experience, patients can survive brief periods of 70 to 75 per cent saturation. Prolonged periods at these high levels can, of course, be fatal. Interpretation of lethal levels must be related to other factors in each case, i.e., time of exposure, normal hemoglobin level of the patient, age and general health of the patient, degree of activity, and so on. Nevertheless, Table 21–2 is a useful guide for interpretation of results in the average patient.

In patients treated with oxygen, carbon monoxide is fairly rapidly released from hemoglobin. Frequently, a patient is treated with oxygen while being transported to the hospital. By the time a blood sample is drawn and analyzed, the carbon monoxide level may be

TABLE 21-2 CARBON MONOXIDE TOXICITY*

Carboxyhemoglobin (%)	*Effect*
10	Shortness of breath on vigorous muscular exertion.
20	Shortness of breath on moderate exertion; slight headache.
30	Decided headache; irritation; ready fatigue; disturbance of judgement.
40–50	Headache, confusion, collapse, and fainting on exertion.
60–70	Unconsciousness; respiratory failure and death if exposure is long continued.
80	Rapidly fatal.
Over 80	Immediately fatal.

* From Deichmann, W. B., and Gerarde, H. W.: Symptomatology and Therapy of Toxicological Emergencies. New York, Academic Press Inc., 1964. © Copyright Academic Press Inc.

close to normal. These patients should be kept quiet with good oxygenation to insure that all tissue-bound carbon monoxide (e.g., with myoglobin and heme-containing enzymes) is dissipated before the patient is discharged.

REFERENCE

Tietz, N. W., and Fiereck, E. A.: Ann. Clin. Lab. Science, *3*:36–42, 1973.

VOLATILE SUBSTANCES

This group of toxic compounds consists mainly of liquids that have boiling points of 100°C or lower. For this reason, they can be separated from biological specimens by steam distillation. Members of this group include almost all types of chemical compounds, and many are solvents commonly used in industry or in household products.

Steam distillation, although a useful procedure, necessitates assembly of glass apparatus and, particularly for some very volatile substances, great care in conducting the separation. The principle of microdiffusion can also be used to separate many members of this group (e.g., alcohols, aldehydes, ketones, chlorinated aliphatic hydrocarbons, and aromatic hydrocarbons[21]) with the advantages of small sample size, minimal equipment, and simplicity of operation. The use of the Conway dish with its various modifications greatly simplifies this type of separation.

Although only a few substances are considered in detail here, many other toxic, volatile substances can be determined using the Conway dish. For further details see Conway[9] or Feldstein.[21]

Volatile substances are ideally suited for analysis by gas chromatography. Numerous methods have been described which are both convenient and accurate. One involves mixing the specimen with potassium carbonate in a tube closed with a Vacutainer stopper with needle septum. After equilibration, a syringe is used to remove a portion of the "air" above the specimen and to inject it into the gas chromatograph. This method, known as head space analysis, is useful for rapid screening for the presence of many volatiles.[12] Quantitative analyses can also be done by the head space technique, but pressure and temperature variables must be carefully controlled. Direct injection of blood or urine can also be done.[15] On-column injection of specimens diluted with an internal standard is rapid, accurate, and precise.

Using the following procedure adapted from that of Curry,[15] the three most common alcohols (ethanol, methanol, and 2-propanol) together with acetone, a metabolite of

2-propanol, can be determined simultaneously. This is particularly appropriate, since "alcohol" ingestion may frequently be missed if only ethanol is measured.

DETERMINATION OF ALCOHOLS BY GAS CHROMATOGRAPHY

PRINCIPLE

Poisoning by alcohols almost always results in blood or serum levels which are quite high compared to those of most toxic agents. This fact, coupled with the low boiling points of alcohols, permits gas chromatographic analysis at low temperatures using specimens which can be injected directly or after simple dilution. Normal constituents of biological specimens do not interfere. If a flame ionization detector is used, the large excess of water in the specimen does not interfere significantly since it will elicit only a minimal response by the detector.

If large numbers of specimens are to be analyzed, an automatic diluter such as the one cited below is useful; occasional specimens can be diluted manually. For greatest accuracy, the use of a recorder with an integrator is recommended. If ± 10 per cent accuracy is tolerable, peak height can be used for quantitation. 1-Propanol is used in this procedure as an internal standard.

APPARATUS

1. Diluter (Hobbs No. 015-Spec), designed to sample 0.40 ml and then deliver the sample with 3.60 ml of diluent (10-fold dilution). A wash cycle is included to insure that there is no carry-over from one sample to the next. This diluter is not necessary, but is very useful if a large number of samples are to be analyzed.

2. Gas chromatograph, equipped with a flame ionization detector, a 1 mV recorder with attached Disc Integrator, and a glass column, 5 feet long and $\frac{1}{4}$ inch O.D. packed with 10 per cent PEG 400 on 100-120 Anakrom SD.

Operating conditions

Gas Flow Rates: carrier—70 ml/min; hydrogen—70 ml/min; air—600 ml/min.

Temperature: column 85°C; injection port 110°C; detector 125°C.

Place a small plug of glass wool in the end of the column nearest the injection port to trap solids. The plug is easily removed and should be replaced periodically with a clean plug.

REAGENTS

All chemicals used should be anhydrous, analytical reagent grade.

1. Diluent stock solution, 23.4 g/l. Dilute 30.0 ml of 1-propanol to 1 liter with water at 20°C.

2. Diluent working solution, 23.4 mg/100 ml. Dilute 10.0 ml of the diluent stock solution to 1.00 liter with water.

3. Alcohol stock solutions. Dilute 3.00 ml each of anhydrous methanol, ethanol, and 2-propanol respectively to 100.0 ml with water. At 20°C, this will result in concentrations as follows:

(a) Methanol, 2.37 g/100 ml
(b) Ethanol, 2.37 g/100 ml
(c) 2-Propanol, 2.36 g/100 ml

4. Alcohol standard solutions. Dilute 1.00, 2.00, 4.00, 8.00, and 16.00 ml of each stock solution to 100.0 ml with water. This will result in standard solutions as follows:

(a) Methanol, 23.7, 47.4, 94.8, 190, and 379 mg/100 ml
(b) Ethanol, 23.7, 47.4, 94.8, 190, and 379 mg/100 ml

(c) 2-Propanol, 23.6, 47.2, 94.4, 189, and 378 mg/100 ml

PROCEDURE

1. Dilute the specimen 10-fold with diluent working solution. The Hobbs diluter may be used, or 1.0 ml of specimen may be added to a 10 ml volumetric flask and diluted to volume with the diluent working solution.
2. Inject duplicate 1 μl aliquots of the prepared samples into the gas chromatograph, making sure that the stated operating conditions are met.
3. Inject 1 μl aliquots of each alcohol standard solution, suitably diluted with the diluent working solution.

CALCULATION

Compare the peak-area or peak-height ratio with respect to 1-propanol obtained from the specimen with that obtained from the corresponding alcohol standard solution of comparable concentration. The ratios obtained from the reference solutions are linear functions of their respective concentrations from 0 to 400 mg/100 ml.

ACCURACY AND PRECISION

This method can easily detect concentrations of 10.0 mg alcohol/100 ml blood. Lower concentrations can be detected by injecting larger volumes of diluted specimen into the gas chromatograph. The standard deviation of the method is $\pm$3.0 mg/100 ml when using peak-area ratios.

INTERPRETATION

Under the stated conditions, 1-propanol has a retention time of about 6 minutes. The retention times for other volatile substances relative to 1-propanol are as follows: acetone, 0.418; methanol, 0.585; 2-propanol, 0.609; ethanol, 0.654.

REFERENCE

Curry, A. S., Walker, G. W., and Simpson G. S.: Analyst, *91*:742, 1966.

ETHANOL

Ethanol (ethyl alcohol) is the most common toxic substance encountered in medicolegal cases. Not only is it lethal in its own right, but it is commonly a contributory factor in accidents of all types. In the case of a patient brought to the hospital in coma, the effect of alcohol, if any, must be ruled out in a differential diagnosis of the cause of coma.

There are probably more published methods for the determination of ethanol in blood than for any other toxic substance. In general they can be divided into those methods that are simple but nonspecific, and those that are specific but complex. The ideal method, as far as specificity and rapidity are concerned, is based on the use of gas chromatography.[15] Even this procedure is not absolutely specific by itself, although it is extremely unlikely that an interfering substance would be encountered in biological samples. A specific but time-consuming method is an enzymatic one utilizing alcohol dehydrogenase.[40] Besides the time factor, an ultraviolet spectrophotometer is necessary for this test; however, this method lends itself well to automation.

Two methods that are easily adaptable for use in most clinical laboratories use the principle of diffusion for separation of ethanol from the specimen. The Conway microdiffusion method has already been referred to, but macrodiffusion can also be useful.[46]

DETERMINATION OF ETHANOL BY DIFFUSION

PRINCIPLE

The ethanol released from the sample is absorbed in an acid dichromate solution. Ethanol is oxidized by this solution to acetic acid. During this process, the yellow dichromate is reduced to the green chromic ion, and this color change constitutes a rough qualitative indication of the progress of the reaction:

$$3C_2H_5OH + 2K_2Cr_2^{6+}O_7 + 8H_2SO_4 \rightarrow 3CH_3COOH + 2K_2SO_4 + 2Cr_2^{3+}(SO_4)_3 + 11H_2O$$

The amount of unreacted dichromate is measured by reacting it with potassium iodide to form iodine, which is estimated by titration with thiosulfate:

$$K_2Cr_2^{6+}O_7 + 6KI + 7H_2SO_4 \rightarrow 4K_2SO_4 + Cr_2^{3+}(SO_4)_3 + 3I_2 + 7H_2O$$
$$I_2 + 2Na_2S_2O_3 \rightarrow 2NaI + Na_2S_4O_6$$

INTERPRETATION

Any volatile reducing substance will interfere with this method. The possibility of interference by methanol, isopropanol, and aldehydes must be ruled out by separate tests for these substances.

ETHANOL DETERMINATION WITH ALCOHOL DEHYDROGENASE (ADH)

PRINCIPLE

Ethanol is oxidized in the presence of ADH to acetaldehyde. In the course of this reaction, NAD, a coenzyme, is reduced:

$$C_2H_5OH + NAD^+ \xrightleftharpoons{ADH} CH_3CHO + NADH + H^+$$

The increase in NADH can be measured by the increase in absorbance at its absorption maximum of 340 nm. The equilibrium for this reaction lies strongly to the left. At neutral pH and at normal NAD concentrations, less than 1 per cent of the ethanol present is oxidized to acetaldehyde. However, the reaction can be driven almost completely to the right by maintaining a high pH and removing the acetaldehyde as it is formed by reacting it with semicarbazide.

Both yeast and mammalian liver ADH can be used for this reaction. At a pH of 8, the Michaelis constant is 30 times greater for yeast than for liver ADH. The turnover number, however, is greater for yeast than for liver ADH. These differences are only of slight importance in analytical work when a large excess of the enzyme is used.

It was once thought that this was a specific method, but it is now evident that other alcohols can interfere at high concentrations. Despite this drawback, it is still used in many clinical laboratories because of its convenience and the relatively rare occurrence of other alcohol intoxications.

Many modifications of this method have been reported. Several commercial kits are available. The procedure to be described[40] uses the Stat-Pak kit marketed by Calbiochem (see reagents list).

SPECIMEN

Blood, serum, urine, and saliva are all appropriate specimens for this test. The site of venipuncture, finger puncture, or ear puncture should be cleansed and disinfected with aqueous Zephiran (benzalkonium chloride), aqueous Merthiolate (thimerosal), or other suitable disinfectant. Never use alcohol or other volatile disinfectants. If reusable containers, syringes, and needles are used, they must not be cleaned or stored with alcohol or other volatile solvents. Fluoride is the best preservative but citrate, oxalate, and heparin can be used as anticoagulants. All specimens must be stored under refrigeration until they are analyzed.

REAGENTS

1. Ethyl alcohol Stat-Pak kit with reagents and standard (Calbiochem, 10933 N. Torrey Pines Rd., La Jolla, CA. 92037).

2. Sodium chloride solution, 0.15 mol/l. Dissolve 9.0 g NaCl, AR, in distilled water and dilute to 1.0 liter.

3. Perchloric acid, 0.33 mol/l. Dilute 29 ml of 70 per cent perchloric acid, AR, to 1.0 liter with distilled water.

Caution: Perchloric acid oxidizes organic materials. See warning on p. 54.

PROCEDURE

1. Preparation of the activated substrate: Add 13.5 ml distilled water to vial B (NAD^+) and swirl gently to dissolve. Open vial A (alcohol dehydrogenase with trapping agent) and add entire contents of vial B to it. Cap and dissolve by gentle inversion. DO NOT SHAKE.

2. Sample preparation:

a. Dilute 100 μl of urine, saliva, or unhemolyzed serum or plasma with 4.9 ml of 0.15 molar NaCl.

b. Alternatively, add 100 μl of whole blood or hemolyzed serum to 4.9 ml of 0.33 molar perchloric acid. Mix well. Centrifuge at 3000 rpm for 5 min. The supernatant must be assayed within 24 h. Keep stoppered.

c. Treat 100 μl of standard and a separate 100 μl of distilled water in a manner identical to the sample, step 2a or 2b.

3. To separate test tubes labelled "standard," "sample," and "blank," add 2.6 ml of activated substrate.

4. To the appropriate tubes, add 100 μl of diluted standard, sample, or water blank from step 2. Mix and incubate at 30°C for 10 min.

5. With the ultraviolet spectrophotometer set to read zero absorbance at 340 nm with the blank, read the absorbances of the sample and of the standard, using 1.00 cm cuvettes.

6. Do not discard solutions until after the calculations have been performed. If the standard is measured to be less than 200 mg ethanol/100 ml, incubate the solutions an additional 10 min and repeat step 5.

7. If the concentration in the sample is calculated to be greater than 200 mg ethanol/100 ml, repeat the analysis after diluting the specimen appropriately.

CALCULATION

$$A_{340} \times 1000 \times \text{dilution factor} = \text{mg ethanol/100 ml}$$

Normally the dilution factor is 1. If, for example, the initial result is approximately 400 mg/100 ml for blood, 1.00 ml of supernatant from step 2b is diluted to 2.00 ml with 0.33 molar perchloric acid and carried through the procedure. The dilution factor then becomes 2.

COMMENTS

This is an extremely sensitive test. Alcohol from outside sources must be carefully avoided as a contaminant.

Temperature and reaction time are not critical since the reaction proceeds to completion. However, if the temperature is lower than 30°C the reaction may not be complete after 10 min. This is one reason for checking the standard with each group of tests.

For interpretation of results (see below) whole blood is the specimen of choice. Serum or plasma will have much higher ethanol levels than blood. The saliva ethanol concentration is 1.1 to 1.2 times greater than that of blood, while the urine concentration is quite variable.

Although ethanol is the preferred substrate for this enzyme, the rate of dehydrogenation of 2-propanol is 6 per cent, that for 1-propanol is 36 per cent, and that for 1-butanol is 17.5 per cent of that for ethanol. The rate of dehydrogenation of methanol is negligible.

INTERPRETATION

With blood ethanol levels of 50 to 100 mg/100 ml, various signs of intoxication may be observed: flushing, loquaciousness, slowing of reflexes, impairment of visual acuity, and so on; however, there is much individual variation in this regard. Above 100 mg/100 ml all individuals are under the influence of alcohol and depression of the central nervous system is more apparent. Because of impairment of good judgment and visual acuity, as well as slowing of reflexes, driving a motor vehicle or operating machinery is hazardous when an individual is under the influence of alcohol.

With higher blood alcohol levels, central nervous system impairment is more pronounced and true coma may appear at levels of 300 mg/100 ml. Death may occur with levels above 400 mg/100 ml.

In many areas, state laws require that blood alcohol levels, when measured for legal purposes, be stated in terms of weight per cent. In these cases, weighing the sample is mandatory. For clinical work pipetting the sample is more convenient. Interpretation of blood levels is the same in each case.

REFERENCE

Lundguist, F.: *In* Methods of Biochemical Analysis. D. Glick, ed. Interscience Publishers Inc., New York, 1959, vol. 7, p. 217.

METHANOL

Methanol (methyl or wood alcohol) is a widely used solvent in paints, varnishes, and paint removers. It is used alone as an antifreeze fluid and with ethanol and soap as a solid canned fuel. Poisonings are usually due to accidental ingestion by children or by alcoholics. In some areas, methanol may be a contaminant in "moonshine."

DETERMINATION OF METHANOL

The most convenient and reliable method for methanol determination is gas chromatography. Such a method has already been described (see p. 1110). Methanol can also be measured by a variety of other methods, most of which involve measuring the color intensity after oxidation of methanol to formaldehyde, followed by the development of a color by reacting formaldehyde with chromotropic acid (CTA):

$$CH_3OH + MnO_4^- + 2H^+ \rightarrow CH_2O + MnO_2 + 2H_2O$$

CH$_2$O + [naphthalene ring with OH, OH and HO$_3$S, SO$_3$H] → CH$_2$O – CTA (Unknown structure)

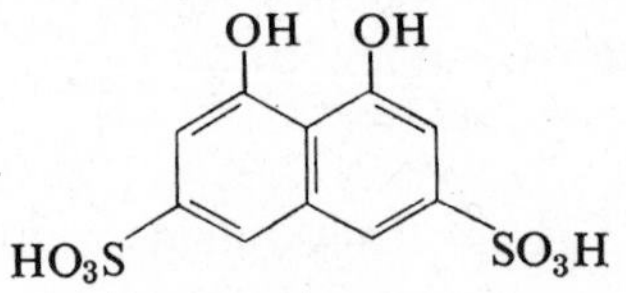

Violet color

These methods work well since chromotropic acid is specific for formaldehyde and, hence, for methanol after oxidation. The microdiffusion method referred to earlier[21] is useful for the determination of methanol, and it also utilizes CTA for color development.

The CTA colorimetric procedure for methanol has two major drawbacks. First, methanol is not quantitatively oxidized to formaldehyde. It is readily apparent that after formation of formaldehyde by the oxidation reaction just noted, the formaldehyde itself can be oxidized to formic acid and further to carbon dioxide as follows:

$$CH_2O + MnO_4^- \rightarrow CO_2 + MnO_2 + H_2O$$

This means that before a quantitative procedure can be devised, conditions must be chosen such that constant proportions of methanol are oxidized. Thus, the method is empirical and the set conditions must be established and adhered to rigidly before quantitative results can be achieved.

Second, the presence of reducing substances other than methanol will affect the system so that the procedure can no longer be applied quantitatively. The most common interference in cases of methanol poisoning is ethanol. It is not generally appreciated that the presence of ethanol invalidates a methanol procedure based on oxidation followed by CTA color development, if the calibration curve has been set up using pure methanol standards.

The procedure of Hindberg and Wieth[34] obviates both drawbacks. First, the procedure must be carried out identically for standards and unknowns. Second, an excess of ethanol is added to *both* standards and unknowns. This results in a constant "interference" of a magnitude much greater than would ever be encountered in practice.

A third, but minor, drawback of any CTA procedure for determining methanol is the use of concentrated sulfuric acid for development of the final color. The dehydrating effect of concentrated sulfuric acid can produce formaldehyde from appropriate organic compounds and there will be false high results. We have encountered this interference occasionally in patients with severe acidosis. Apparently, some substances may appear in a trichloroacetic acid filtrate, such as glycolic acid, which reacts as follows:

$$H_2C(OH)COOH \xrightarrow[\text{Heat}]{H_2SO_4} HCHO + CO + H_2O$$

This type of interference can be detected by running a blank for comparison. A portion of filtrate is carried through the procedure except that the oxidation step is omitted. Any color developed in the unoxidized specimen is due to formaldehyde contaminating the original specimen, or to glycolic acid. The following is a convenient, qualitative screening test for methanol.

REAGENTS

All chemicals used should be analytical reagent grade.

1. Trichloroacetic acid, 20 g/100 ml. Dissolve 20 g of trichloroacetic acid in water and dilute to 100 ml.
2. Potassium permanganate, 3 g/100 ml. Add 3.0 g of potassium permanganate to a solution of 15 ml of 85 per cent phosphoric acid and dilute to 100 ml with water.
3. Sodium bisulfite.
4. Chromotropic acid (1,8-dihydroxynaphthalene-3,6-disulfonic acid).
5. Sulfuric acid, concentrated.

PROCEDURE

1. To 2.0 ml of blood, serum, or urine, add 4.0 ml of the trichloroacetic acid solution.

Do not use heparin or EDTA as an anticoagulant for blood specimens. Shake the mixture thoroughly, and then centrifuge.

2. Into each of two tubes labelled "sample" and "sample blank," pipet 1.0 ml of supernatant. Add 2 drops of the potassium permanganate solution to the *sample tube only*. Wait exactly 2 min.

3. Add approximately 10 mg of sodium bisulfite to both tubes to decolorize (reduce) the excess permanganate. Mix thoroughly. If some permanganate color remains, add more sodium bisulfite, but avoid a large excess.

4. Add approximately 2 mg of chromotropic acid to both tubes and mix the solution by swirling.

5. Carefully underlay this solution with 3.0 ml of concentrated sulfuric acid by inclining the test tube and flowing the acid down the side of the inclined tube. A purple ring at the interface may be considered a positive test for methanol.

6. Shake the tube to diffuse the purple color. The color is fully developed after about 20 minutes.

COMMENTS

This method is very sensitive and will detect about 10 mg methanol/100 ml sample.

An aqueous control should always be processed in parallel with each batch of samples for comparison with the specimen. Formalin, heparin, methenamine, and EDTA also give positive tests. If the "sample blank" is also positive, one of the interferences discussed above may be present.

INTERPRETATION

Methanol poisoning is considerably more dangerous than that due to ethanol. Methyl alcohol is metabolized in man to formaldehyde and formic acid. The accumulation of formic, and other acids, severely reduces the alkali reserve, resulting in a metabolic acidosis (see Chapters 15 and 16). In addition, necrosis of the pancreas and serum amylase elevations have been demonstrated. Therefore, in addition to blood methanol levels, plasma carbon dioxide content, serum amylase determinations, and electrolyte studies are useful laboratory tests for determining the severity of the poisoning and following the progress of treatment.

Metabolites of methyl alcohol can damage the optic nerve, resulting in either temporary or permanent blindness. The mechanism of this effect is not well understood, nor is it a constant finding; nevertheless, prompt treatment of these cases may not only be lifesaving but may also preserve the eyesight.

As little as 2 teaspoonsful (10 ml) of methanol are considered toxic; fatal results have been reported with dosages between 2 and 8 ounces. A blood level greater than 80 mg/100 ml is dangerous to life.

Treatment is twofold. First, the acidosis is treated, generally with sodium bicarbonate, both intravenously and orally. Second, it has also been proposed that in severe cases, ethyl alcohol be administered to saturate the alcohol dehydrogenase enzyme system. Since ethyl alcohol is the preferred substrate for this enzyme, this prevents the conversion of methanol to its toxic metabolites.

REFERENCE

Kaye, S.: Handbook of Emergency Toxicology. 3rd ed. Springfield, IL., Charles C Thomas, 1970, p. 313.

CYANIDE

Cyanide inhibits cellular respiration because of its combination with important respiratory enzymes. This mechanism of action is the same whether cyanide is inhaled as the gas, hydrocyanic acid, or ingested as the potassium or sodium salt, or other combined form.

Since death follows very quickly if sufficient cyanide is absorbed, the patient rarely survives long enough for treatment. Despite this fact, it is desirable to have a test for this poison available in order to confirm a suspected cyanide death.

DETERMINATION OF CYANIDE

PRINCIPLE

In the method to be described,[22] cyanide is separated from the specimen by microdiffusion, trapped in dilute alkali, and converted to cyanogen chloride:

$$CN^- + \text{Chloramine-T} \rightarrow ClCN$$

The ClCN is then reacted with pyridine to form N-cyanopyridinium chloride (König reaction):

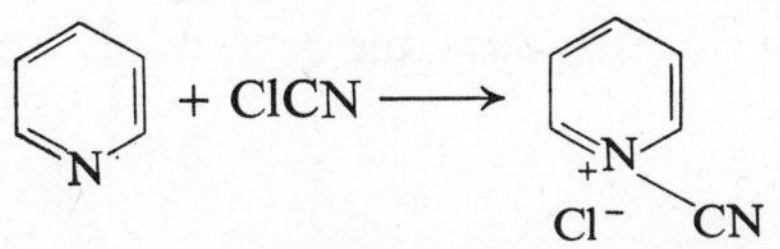

Pyridine + ClCN ⟶ N-Cyanopyridinium chloride

Pyridine *Cyanogen chloride* *N-Cyanopyridinium chloride*

This is followed by a modified Aldrich reaction in which the N-cyanopyridinium chloride is cleaved to form an anil of glutaconic aldehyde. The aldehyde is coupled with barbituric acid to form a colored, highly resonant system:

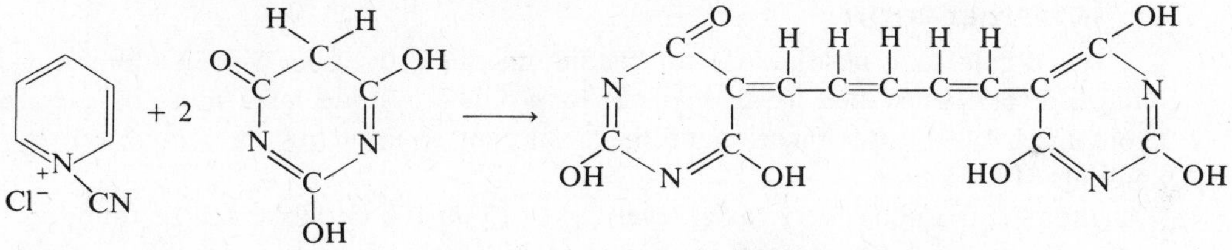

N-Cyanopyridinium chloride + 2 Barbituric acid ⟶ Postulated colored product

N-Cyanopyridinium chloride *Barbituric acid* *Postulated colored product*

The preceding reaction cannot be carried out on blood or tissues that contain formalin, since formaldehyde reacts with cyanide to form cyanohydrin, which is readily hydrolyzed to glycolic acid and ammonia:

$$HCN + HCHO \longrightarrow H_2C(OH)CN \xrightarrow{H_2O} H_2C(OH)COOH + NH_3$$

REAGENTS

1. Sulfuric acid, 1.9 mol/l.
2. Sodium hydroxide, 0.10 mol/l.
3. Sodium phosphate, monobasic, 1.0 mol/l.
4. Chloramine-T, 0.25 g/100 ml aqueous solution. This is stable when kept in the refrigerator.
5. Pyridine-barbituric acid reagent. Add 15 ml of pyridine to 3.0 g of barbituric acid in a 50 ml volumetric flask; mix. Add 3.0 ml of concentrated HCl; mix. Dilute to volume with distilled water. Mix thoroughly, since the ingredients dissolve slowly. Let stand for 30 min and filter if necessary. Prepare fresh as needed.

PROCEDURE

1. Place 4.0 ml of blood or urine into the outer compartment of a Conway diffusion dish. If tissue is used, homogenize a portion in an equal weight of saline (0.9 g NaCl/100 ml) and use a measured quantity of homogenate. Place 2.0 ml of 0.10 molar NaOH in the center compartment and prepare the cover with silicone grease for a tight seal. Add 6 drops of 1.9 molar H_2SO_4 to the outer compartment, seal the top quickly, and swirl gently to mix. Allow diffusion to proceed for 3 to 4 h at room temperature.
2. After diffusion is complete, transfer 1.0 ml of absorbing solution (center well) to a test tube.
3. Prepare a blank consisting of 1.0 ml of 0.10 molar NaOH in a second test tube.
4. To each tube add 2.0 ml of NaH_2PO_4 solution and 1.0 ml of Chloramine-T solution. Mix and let stand 2 to 3 min.
5. To each tube add 3.0 ml of pyridine-barbituric acid solution. Mix and allow to stand 10 min.
6. Observe the color, if any, or read the absorbance at 580 nm in a spectrophotometer and compare with standards run through the same procedure.

CALCULATIONS

Prepare a series of standards ranging from 5 to 100 μg cyanide/100 ml of 0.10 molar NaOH and carry these standards through the entire procedure, including the diffusion step. Construct a standard curve by plotting absorbance readings against concentration. The calibration curve will be a straight line up to values of 200 μg/100 ml.

INTERPRETATION

A red color is a positive test for cyanide in this procedure. Since Chloramine-T can oxidize certain substances (e.g., glycine) to produce cyanide, care must be exercised to avoid mechanical contamination of the absorbing solution in the center well by trace amounts of specimen.

Although cyanide is very toxic, levels up to 15 μg/100 ml blood can be found in adults without symptoms. In cases of death due to ingestion of an overdose of a cyanide salt, levels of 1.0 mg/100 ml or higher may be found. A lethal level is about 0.1 mg/100 ml.

REFERENCE

Feldstein, M., and Klendshoj, N.: J. Forensic Sci., *2*:39, 1957.

KEROSINE

Kerosine and other petroleum hydrocarbons are frequently ingested accidentally by children. Any hydrocarbon aspirated into the lungs or absorbed material excreted into the lungs can produce a dangerous chemical pneumonitis. Frequently, the odor of the breath is an indication that this material has been ingested.

Kerosine is used as a fuel in some areas, and this material is a common solvent for household products.

The aliphatic hydrocarbons are chemically inert. This makes the problem of testing for them exceedingly difficult. Generally, one can rely on the characteristic odor of this substance in gastric contents as a positive test. If a sufficient quantity of the hydrocarbon is present and can be physically separated, it can be identified by its physical constants. No simple laboratory tests for detecting the presence of kerosine are available.

CORROSIVES

This group includes those strong mineral acids or fixed alkalies that produce chemical burns on contact. There are no good tests that can be carried out on blood, serum, or urine by which the type of acid or alkali can be detected and the ingested quantity estimated.

The only specimen that can be examined profitably is gastric contents. Frequently, this specimen is not available unless the patient has vomited, since gastric lavage is contraindicated in this type of poisoning. If gastric contents are available, the pH should be measured. Ions such as Na^+, K^+, Cl^-, SO_4^{2-}, and PO_4^{3-} can be demonstrated by methods used in the laboratory for routine analysis. Obviously, most of the common ions would be present normally in gastric contents. To be of significance in this type of case, a large excess must be present. Since many compounds in the group of corrosives can cause major disturbances in acid-base balance, it is advisable to perform electrolyte studies on blood. Usually the clinician has evidence from lesions in the mouth and esophagus that a corrosive substance has been ingested.

METALS

All metals are toxic if a sufficient quantity is absorbed. Generally they are not encountered in their toxic form in the elemental or free state, but rather in the form of salts. The degree of toxicity of a given metal is dependent on the solubility of the salt; the greater the solubility, the more likely it is that it will be absorbed and the greater will be its toxicity. For example, barium chloride is soluble and extremely toxic, but barium sulfate is insoluble enough to be used as a radiopaque medium for the gastrointestinal tract.

In general, metals can be detected after burning away the organic material in the specimen and measuring the metallic ions in the inorganic residue by some standard procedure. The combustion process may be either a wet digestion with strong, oxidizing acids, or a dry ashing procedure in a furnace.

Some metals and their salts are volatile at high temperatures (e.g., mercury) and special precautions must be taken to avoid loss during the ashing step. Some metal ions, such as sodium, potassium, calcium, magnesium, and iron, are commonly analyzed in clinical laboratories and thus will not be considered here. The so-called trace metals are present in biological material in only minute amounts, even after ingestion of toxic amounts. For this reason most metal determinations require analytical techniques used in trace analysis. These are difficult and require considerable experience if they are to be carried out with validity. Instrumental analysis, particularly atomic absorption spectrophotometry, makes it possible to conduct trace metal analyses more easily.

ARSENIC AND RELATED METALS

Arsenic, despite its reputation, is not a common poison. It is still a favorite homicidal poison, but homicidal poisonings are rare. Since arsenic is an ingredient in some herbicides and insecticides, accidental poisonings, both acute and chronic, may still be encountered on occasion.

Clinically, the symptoms of both acute and chronic arsenic poisonings can easily be confused with a variety of other conditions. It is not uncommon, therefore, for a clinician to request that the presence of arsenic be ruled out as an aid in the differential diagnosis. The specimen of choice in this case is urine, even if 2 to 3 weeks have elapsed after ingestion of the poison. In long-term chronic cases, analysis of hair and nails may be informative, but this is subject to difficulties in interpretation (see under Interpretation).

In the older literature, arsenic is frequently described as a "protoplasmic poison." This term is as good as any for describing the mode of action of the metal. Arsenic combines readily with proteins because of its great affinity for sulfhydryl groups. This results in the precipitation of proteins, producing gastrointestinal irritation and irreversible inhibition of important enzyme systems, which are important toxic effects of arsenic. The great affinity of arsenic for tissue proteins is also responsible for the rapid removal of arsenic from the blood. Blood, therefore, is not a good specimen except in cases in which a large overdose of this substance has been ingested.

DETECTION OF ARSENIC

The test to be described is commonly referred to as the *Reinsch test*. It depends on the fact that metallic copper in the presence of acid will reduce arsenic to the elemental form. The arsenic is deposited on the copper as a visible, dark film:

$$3Cu^0 + 2As^{3+} \xrightarrow{HCl} 3Cu^{2+} + 2As^0$$

The oxidized forms of antimony, bismuth, mercury, and selenium can also be reduced by metallic copper under these conditions. Thus, the same test constitutes an exclusion test for these metals as well. The test was applied by Gettler[29] in a systematic way to biological material, and its modification by Rieders[49] will be described.

REAGENTS

1. Hydrochloric acid, concentrated, AR.
2. Copper spiral. Wind bright, clean copper wire around a 3 mm glass rod about eight to ten times to make a tight spiral. A 1.0 cm square copper foil may also be used.

PROCEDURE

To 100 ml of urine in a shallow dish, add 10 ml of concentrated HCl. Add a copper spiral and gently boil the solution until the volume is reduced to about 20 ml. Remove the copper, rinse gently with distilled water, examine, and note any color change.

INTERPRETATION

If the copper is still bright, arsenic (25 μg/l or more), mercury (50 μg/100 ml or more) and selenium (50 μg/100 ml or more) have been ruled out. In the presence of arsenic or selenium, the surface of the copper will be gray to black; in the presence of mercury, the film will be light gray to silvery and will become shiny on rubbing. Some sulfur compounds, antimony, bismuth, or tellurium, also give gray to black deposits.

In the case of a positive test, the nature of the deposit on the copper must be verified by further tests.[29] Arsenic can be quantitated after wet digestion of another specimen, by an excellent colorimetric method.[47] Recently, the technique of atomic absorption spectroscopy has been used in a practical determination of arsenic in biological material. The principle of atomic absorption spectroscopy is discussed in Chapter 3.

Normal arsenic levels in urine are less than 50 μg/l. In cases of chronic poisoning, arsenic levels in urine will rise to 100 μg/l; in acute poisoning, 1.0 mg/l or more may be present.

Since arsenic is readily bound by sulfhydryl groups in protein, considerable arsenic is bound by keratin and subsequently deposited in hair and nails. This phenomenon has led to the analysis of hair and nails in an effort to determine whether a previous exposure to arsenic has occurred. Interpretation of these analyses is difficult because of the problem of differentiating between surface contamination of the hair and endogenous arsenic. If such an examination is required, a minimum of 1.0 g of clean hair (a large handful), clipped close to the scalp, should be submitted to a toxicological specialist.

LEAD

Lead is still one of the most serious of the metallic poisons. In adults, inorganic and organic lead compounds may be encountered in industrial exposures. An increasing awareness of this danger has promoted the use of prophylactic measures. Education of workers about the hazards of lead intoxication has also been of help in minimizing industrial poisonings.

Unfortunately, children are particularly sensitive to lead poisoning and the exposure of children to lead-containing paint and plaster, particularly in lower class housing, has continued despite regulations, labeling laws, and attempts to educate the public. Severe poisoning in a child can cause lead encephalopathy, the mortality rate for which is high. Those children who survive frequently show evidence of permanent central nervous system damage.

The diagnosis of lead poisoning is difficult, and the demonstration of an elevated lead level in blood or urine constitutes the most positive indication of absorption of a lead compound. Being a ubiquitous element, lead is normally present in trace amounts in biological material. Analytical procedures must be extremely sensitive and conducted with great care in order to achieve valid results. These requirements generally make lead analyses the function of a special laboratory, particularly one which has experience with trace metal analyses and their special problems. This can be illustrated by some facts relating to lead analysis. An average normal lead level in blood is 30 μg/100 ml and an amount of 100 μg/100 ml represents a toxic level. Thus, 5 ml of the normal blood specimen contains 1.5 μg, and the abnormal sample contains 5 μg. It is obvious that any method used must not only be extremely sensitive, but it must also have an excellent accuracy and precision in order to discriminate between the ends of the 3.5 μg range separating the normal and toxic lead levels in blood. In addition, all of the glassware and reagents used in the analysis contain traces of lead. Thus, after careful selection of reagents and cleaning of glassware, the analyst must still exercise meticulous technique in order to keep blank values of lead low.

DETECTION OF LEAD OR LEAD POISONING

The actual analysis may follow one of many techniques: colorimetric analysis with diphenylthiocarbazone,[27] polarography,[3] or atomic absorption spectrophotometry,[37] to name some of the most reliable ones.

The clinical laboratory performs two very important functions that aid in the diagnosis of lead poisoning, even if the lead analysis is done by others. First, the specimens to be analyzed must be collected in a valid way, that is, free of contamination. Second, other diagnostic tests can be done, for screening purposes or for confirmation. These tests are based on the effects of lead on erythropoiesis. Lead interferes in the biosynthesis of hemoglobin, which results in anemia. Although the precise mechanism of this interference is still under investigation, one result is a buildup of precursors of hemoglobin. Two precursors that accumulate in lead poisoning are δ-aminolevulinic acid and coproporphyrin III, and urinary excretion of these substances increases markedly. Methods for the detection of these substances and related enzymes are discussed in Chapter 9.

Elevated urinary porphyrin levels can occur in conditions other than lead poisoning; however, after several hundred comparisons in the author's laboratory, it has been noted that in all cases of proven lead poisoning a corresponding elevation of urinary copropor-

TABLE 21–3 COMPARISON OF URINARY LEAD AND COPROPORPHYRIN LEVELS IN 140 CASES OF ESTABLISHED LEAD POISONING

	Number of Cases (total = 140)			
Urinary Lead Range μg/l	*Test for Coproporphyrin Positive*	*Test for Coproporphyrin Doubtful*	*Test for Coproporphyrin Negative*	
			A*	B†
0–80 (Normal)	5	3	68	—
80–100	4	0	3	2
100–200	19	3	5	1
200–300	10	0	0	2
300–400	3	0	0	1
400–500	2	0	1	0
500–600	2	0	0	0
600–700	2	0	0	0
700–800	2	0	0	0
800–900	2	0	0	0

* A—Single specimen analyzed for lead.
† B—First specimen, Pb abnormal; other specimens, Pb normal.

phyrin levels occurred. Table 21–3 shows the correlation of urinary lead and elevated coproporphyrin levels in 140 cases of lead poisoning.

SPECIMEN COLLECTION

A 24 h urine specimen is the specimen of choice. The patient should void directly into a lead-free container (a borosilicate glass or polyethylene container from which surface lead has been removed by washing, then rinsing with hot 1 molar nitric acid, and rinsing twice with metal-free water). A preservative should not be added because it might contaminate the specimen. After noting the total volume, the entire specimen, or a minimum of 100 ml, is submitted to the toxicological laboratory for analysis. Catheterized specimens should not be used unless it is unavoidable. In this case, the catheter should be cleansed (as just noted) to remove surface lead before sterilization. In some cases we have found that an indwelling catheter through which urine has been flowing freely for 24 to 48 h is usually free from surface lead. The possibility of contamination should always be borne in mind when catheterized specimens are submitted for analysis. In an emergency, it may be necessary to analyze a random urine specimen rather than a 24 h urine specimen. In such a case, the specimen must be collected with the same care as just outlined. Interpretation of the result is subject to the same difficulty as discussed next in connection with blood specimens.

Blood specimens can be analyzed as readily as urine, but lead levels may fluctuate widely in different blood specimens from the same patient. We have had the experience of seeing normal lead levels in occasional blood specimens from lead-poisoned patients in well-documented cases. For this reason, properly collected 24 h urine specimens are preferable. In very young or acutely ill patients, blood may be the only practical specimen available. In these cases it may be necessary to run several specimens before lead poisoning can be ruled out.

If the test is to be performed on blood, an anticoagulant or preservative should not be used unless the exact lead content of these agents is known so that proper correction can be made. The needle, syringe, test tube, and stopper should be of lead-free material, cleaned as previously described. Special tubes for blood-lead collection are commercially available. Since most of the lead is in the erythrocytes, a serum lead level is of little value.

As with any analysis of trace substances, the sensitivity of the analysis and the expected level of substance controls the amount of specimen to be collected. For example, if a lead method is used that is sensitive to 1 μg of lead and in which the known reagent blank is also 1 μg, and the expected blood level is within normal limits, or about 30 μg/100 ml, then a minimum of 10.0 ml of blood must be collected. This quantity of specimen would contain 3 μg of lead, a level that can be differentiated from a blank with some degree of validity.

NORMAL VALUES

Normal lead levels range up to 80 μg/l of urine or 80 μg/100 ml of blood, with an average of 30 μg/100 ml. Levels higher than normal indicate elevated absorption of lead compounds; levels greater than 100 μg/l of urine, or 100 μg/100 ml of blood, are usually associated with signs and symptoms of lead poisoning. Normal blood lead levels in children are 15 to 20 μg/100 ml. In this age group, levels of 40 μg/100 ml may represent an abnormal exposure to lead compounds. Some clinicians prefer urine lead levels to be reported on a per diem basis. In the author's opinion, it is preferable to report these levels in μg/l together with the total volume of the 24 h specimen. This allows the clinician to correlate the 24 h excretion of lead with other factors that may be related to an excessively high or low urinary output.

COLORIMETRIC DETERMINATION OF LEAD

PRINCIPLE

Lead (as Pb^{2+}) forms a red complex with diphenylthiocarbazone (dithizone) that is soluble in a number of organic solvents. Interference by other metal ions, such as zinc, cobalt, nickel, cadmium, silver, copper, mercury, stannous tin, bismuth, and thallous thallium is eliminated by use of complexing agents and performance of extractions at controlled pH levels.

Specimens are either digested with nitric acid or ashed in a muffle furnace at 500°C. The residue, dissolved in HCl, is transferred to a separatory funnel. Citrate is added to complex the iron, the pH is made alkaline with NH_4OH, and the solution is extracted with a CCl_4 solution of dithizone. Pb, Zn, Cu, Ni, Co, Cd, Hg, and Bi are extracted quantitatively and Ag partially. The dithizone solution is then shaken with dilute HCl, of pH 2. This step removes Pb, Zn, and Cd from the dithizone solution, leaving the other metals in the CCl_4-dithizone layer. The dilute HCl solution is now treated with citrate, it is made alkaline with NH_4OH, cyanide is added to complex zinc and cadmium, and the solution is extracted with dithizone in toluene. Unreacted dithizone is extracted into the alkaline aqueous phase, leaving the red lead dithizonate in the toluene. After separating and filtering the toluene layer, the color intensity is read in a spectrophotometer at 520 nm. The absorbance is compared with those of standards carried through the same procedure. It is essential to run blank determinations and correct the final result for any trace quantities of lead present in the reagents and glassware.

REFERENCE

Gant, V. A.: Industr. Med., 7:608, 1938.

DETERMINATION OF LEAD BY ATOMIC ABSORPTION SPECTROSCOPY

PRINCIPLE

Determination of the lead concentration in a biological specimen is the most definitive diagnostic test for lead poisoning. The problem is of particular concern in children, where the need exists for methods requiring small volumes of blood. Classical colorimetric methods in which dithizone is used require very large sample volumes, and even conventional atomic absorption spectrophotometry requires samples of a milliliter or more.

Conventional atomic absorption spectrophotometry necessitates processing a blood specimen by any of a variety of methods prior to actual measurement. These include ashing the specimen,[48] precipitating protein with trichloroacetic acid followed by direct measurement of lead in the supernatant,[10] or measuring absorption after concentration of lead by solvent extraction.[70] All of these suffer from poor sensitivity, necessitating rather large blood samples, and are prone to reagent contamination.

The recent development of the carbon rod atomizer promises to circumvent many of these problems. Application of this method to 0.5 μl samples of packed red cells[41] or 1 μl of diluted whole blood[37] illustrate the remarkable sensitivity of the technique.

Delves conceived a microsampling atomic absorption method in which a nickel cup and an absorption tube mounted in the flame are used, which offers great potential in screening programs and in routine clinical analyses because it accepts 10 μl volumes and is very rapid. Procedures based on this technique require that the analysis be performed by the method of additions to compensate for the matrix effects of blood, and, in the Delves technique, to compensate for poor response to aqueous standards. This increases the time and effort required for the analysis and increases pipetting errors.

Olsen and Jatlow modified Delves' original procedure to gain improved precision and the ability to use aqueous standards. Specifically, a dilute albumin solution is used to precoat the bottoms of the nickel cups before aqueous standards are added, which then provide the same response as lead in blood. Preliminary studies suggest that the technique may also be used for analysis of urine.

EQUIPMENT

The Delves sampling cup system is mounted on an atomic absorption spectrophotometer (Perkin-Elmer, Model 305) equipped with a 3-slot Boling burner, a lead hollow-cathode lamp operated at 9 mA, and a strip-chart recorder (Perkin-Elmer, Model 165) operated on the 5 mV range (2 × scale expansion). Use the resonance line for lead at 283.3 nm and a slit setting of 4 (0.7 nm). The silica absorption tube and "Inconel" cup holder supplied by the manufacturer are used. The loop height is adjusted to 2 to 3 mm below the absorption tube. For acceptable precision, it is imperative to secure the Delves cup accessory firmly to the burner by tightening the thumb screws with pliers, and to secure the burner firmly to the base of the spectrophotometer by a wooden wedge and a spring-loaded wire strap going up over the top of the mixing-chamber part of the burner. This mechanical stabilization of the burner interferes with any positioning adjustments of the burner, so these adjustments must be made before the stabilization.

An air-acetylene flame (air, 24 liters/min, with acetylene set slightly fuel-rich to give faint yellow streaks at the base of the flame) is used. The hot plate is a "Thermolyne" Model HP-A1915B, operated at 140°C, and equipped with a 20° to 180°C surface thermometer (Model 311C, Pacific Transducer Corp., Los Angeles, CA. 90064).

Eppendorf microliter pipets, 10 μl and 20 μl, are used.

All glassware is acid washed overnight in dilute HNO_3 (3 parts acid to 7 parts water).

REAGENTS

Note: All water is doubly distilled and de-ionized. All stock standard solutions are stored in acid-washed polyethylene bottles. All reagents are reagent grade unless otherwise specified.

1. Stock standard, 400 μg of lead/ml. Dissolve 320 mg of dried $Pb(NO_3)_2$ in 500 ml of dilute HNO_3 (0.5 ml/100 ml).

2. Working stock standard. Dilute stock standard with water to 4 μg of lead/ml. Prepare working standards (20, 40, 80, 120, and 160 μg/100 ml) in 50 millimolar NaCl from the working stock solution on the day of analysis.

3. Albumin. Dilute normal human serum albumin (USP, salt-poor, 250 g/l) or bovine albumin (300 g/l) with an equal volume of water and store in a 3 ml plastic disposable syringe equipped with a stainless-steel needle.

COLLECTION OF SAMPLES

Collect venous blood samples in lead-free heparinized tubes and refrigerate.

PROCEDURE

1. With a 5 μl syringe, add one small drop (about 2 μl) of the diluted albumin to each cup that will contain standards. (This volume is not critical, but should be just sufficient to coat the bottom of the cup.) Dry the albumin by warming the cups on the hot plate for about 30 s. Remove the cups from the hot plate; allow them to cool a few seconds.
2. Pipet 10 μl of blank (50 millimolar saline) and 10 μl of each aqueous standard into the centers of the albumin-coated cups. Ten microliters of the blood sample is similarly added in triplicate to untreated cups. Place the cups on the hot plate for about 20 s.
3. Remove the cups from the hot plate, cool a few seconds, and add 20 μl of 30 per cent hydrogen peroxide to each cup. Dry the cups on the hot plate (about 1.5 to 2 min).
4. Place each cup in the Delves sampling apparatus, and ignite in the flame, recording the smoke and lead peaks.

Results are read from the standard curve, and the mean value of the three samples is used.

COMMENTS

In any method such as this, when extremely small samples are used and the sensitivity of the method is high, considerable attention must be given to details. The cups must be exactly centered, preferably 2 to 3 mm below the orifice of the quartz tube. Decreasing this distance results in poorer resolution of the smoke peak from that of the lead, and poor precision; increasing the distance decreases sensitivity. The precision of the micro-scale pipetting is critical to the total precision of the procedure.

The cups used for any one run must be of uniform age because sensitivity gradually diminishes. After about 60 determinations per cup, sensitivity and precision are so low that the cups should be replaced.

Lead standards deteriorate rapidly in polystyrene or borosilicate glass containers exposed to light. This is probably due to adsorption of lead ions on the glass or plastic surface. Aqueous solutions containing 0.5 or 1.0 μg Pb/l are stable for several days in polystyrene or polyethylene containers wrapped in several layers of carbon paper. Upon exposure to light, a solution containing 0.5 μg Pb/l lost 20 per cent of the lead within 3 h, 50 per cent within 6 h, and 90 per cent within 24 h. More concentrated solutions saturate the container surface and, after losing about 0.5 μg Pb/l, remain constant for several days. Borosilicate adsorbs more lead than polystyrene. To prevent errors, solutions containing less than 0.2 μg Pb/l should be prepared in a darkened room and analyzed immediately.

No appreciable lead loss occurs in urine stored under conditions similar to those described for aqueous solutions. Urine stored without a preservative in clear polystyrene containers, exposed to light at room temperature, showed no significant lead change for a period of 10 days. These observations of Kopito and Schwachman emphasize the problems associated with trace metal analysis.[35]

REFERENCE

Olsen, E. D., and Jatlow, P. I.: Clin. Chem., *18*:1312–1317, 1972.

THALLIUM

This metal is rarely encountered. Thallium salts are used as rodenticides, usually by professional exterminators. Cases have been reported of children eating fruit slices or other foods used as bait and treated with thallium salts, which had been carelessly scattered by exterminators intending to kill rats. Formerly, thallium salts were used, both internally and externally, as a depilatory. Fortunately, because of the high toxicity of these substances, this use has been virtually abandoned.

In many respects, the toxic effects of thallium are similar to those of lead. One characteristic feature of poisoning by this metal is loss of hair or occasionally loss of nails and the skin of the feet. In some intoxications in children, loss of hair was the only sign. A lethal dose can be as little as 0.2 g of a soluble thallium salt. Since this metal is not present in biological material, except in extreme trace quantities, any amount demonstrated in the urine is significant.

The following procedure is simple and useful for the detection of this substance.[49]

DETECTION OF THALLIUM

PRINCIPLE

Thallium is converted to its oxidized form by the action of bromine water. After destruction of excess bromine, the thallic ion is complexed with methyl violet to form a blue to violet compound of unknown structure that is soluble in benzene.

REAGENTS

1. Hydrochloric acid, concentrated, AR.
2. Bromine water. Into a glass-stoppered bottle containing 50 ml of water, add 2.0 ml of liquid bromine (*Caution:* Do this in a fume hood! Avoid contact with skin!) Stopper the bottle and shake thoroughly. Some undissolved liquid bromine should remain in the bottom of the bottle. Store in the fume hood.
3. Sulfosalicylic acid, 20 g/100 ml in water.
4. Methyl violet, AR, 20 g/100 ml in water.
5. Benzene, AR.

PROCEDURE

1. Into a glass-stoppered tube, place 1.0 ml of urine and 3 drops of concentrated HCl; mix.
2. Add 5 drops of bromine water; mix thoroughly.
3. Add 5 drops of sulfosalicylic acid solution to decolorize the bromine.
4. Add 1 drop of methyl violet solution and mix.
5. Add 1.0 ml of benzene, stopper the tube and shake thoroughly. After separation of the layers, decant or aspirate off the benzene and observe its color, if any.

INTERPRETATION

A colorless benzene layer rules out the presence of 0.8 μg or more of thallium. A positive test imparts a blue to violet color to the benzene. No interference to the test is seen with levels up to 1.0 mg of borate, oxalate, chlorate, nitrate, phosphate, sulfate, chloride, bromide, perchlorate, or EDTA. The color formation is inhibited by 1.0 mg quantities of nitrite, sulfite, sulfide, thiosulfate, and thiocyanate; 0.2 mg or more of iodide gives a false positive test. Of all the metals tested, the only one that gives a false positive is 0.01 mg of Hg^{2+}. Alkyl aryl sulfonate detergents give false positives, but these also give a color if the bromine and sulfosalicylic acid are omitted. This test can be made quantitative by measuring the absorption at 610 nm.

Other procedures that can be used for the quantitative determination of thallium in blood, tissues, and other specimens involve emission spectrography or various colorimetric methods following digestion of the specimen. Atomic absorption spectroscopy (see under Lead) can also be applied for the determination of this metal.

REFERENCE

Rieders, F.: Ann. N.Y. Acad. Sci., *111*:591, 1964.

NONMETALS

The toxic nonmetals are usually encountered as compounds with other elements, or as sodium and potassium salts. They are infrequently found in the free elemental form. Perhaps a more descriptive heading for this group would be toxic anions.

BORON

Boric acid and borate salts are commonly found in the home laundry or medicine cabinet. Accidental poisonings occur chiefly in children who ingest these preparations, or in infants treated with talcum powders containing borates, which may be absorbed through abraded or irritated skin.

Quantitative analysis of biological material for boron is a difficult problem. Not the least of the difficulties is that boron-free glassware must be used. The following simple test does not require special equipment and is convenient for screening purposes.[49]

DETECTION OF BORON

PRINCIPLE

Turmeric or curcuma is a plant native to the East Indies and China. It is used as a condiment (curry powder) in the tropical East. From the root of the plant is obtained an orange-yellow coloring matter, which is curcumin or turmeric yellow. By an unknown reaction, this substance combines with borates to form a characteristic color. This has been used as a qualitative test for boron for many years, and papers treated with turmeric are readily available.

REAGENTS

1. Hydrochloric acid, concentrated, AR.
2. Turmeric paper (commercially available), AR.
3. Ammonium hydroxide, concentrated, AR.

PROCEDURE

1. Mix 5 drops of urine with 1 drop of concentrated HCl.
2. Place 1 drop of the acidified urine on turmeric paper. Observe the color.
3. Let the paper dry and again observe the color.
4. Hold the paper over concentrated NH_4OH and observe the color.

INTERPRETATION

A positive test is indicated by a brownish-red color of the acidified urine on the wet or dry paper. A green-black or blue color results after exposure to ammonia fumes.

If the wet acidified spot does not change color, the urine contains less than 10 mg borate/l.

If, after drying, the spot still does not change color, less than 5 mg borate/l is present. If, after exposure to ammonia, no color is produced, less than 3 mg borate/l is present.

Normally, less than 2 mg borate/l is present in urine. After a patient has been exposed to borate, the urine level increases sharply, but even at levels of 10 mg/l no particular signs and symptoms of borate poisoning are seen. This test, although not sensitive to normal urinary borate levels, can detect elevated levels before toxic effects are displayed.

REFERENCE

Rieders, F.: Ann. N.Y. Acad. Sci., *111*:591, 1964.

BROMIDES

Bromides are used in both organic and inorganic forms in medicine, chiefly for the purpose of sedation. These drugs are sometimes abused or may be taken in overdosage accidentally. The nonprescription status of drugs containing bromide makes them easily available to the patient prone to drug abuse.

DETERMINATION OF BROMIDE IN SERUM

PRINCIPLE

The procedure to be described[33] measures free Br^- only; thus, the bromine in most of the organic compounds is not detected. When organic bromides are ingested, however, they are metabolized eventually to inorganic bromide (see Interpretation).

The bromide anion readily displaces chloride from gold trichloride, forming gold tribromide:

$$AuCl_3 + 3Br^- \rightarrow AuBr_3 + 3Cl^-$$

The formation of gold tribromide may also be accompanied by the formation of $AuBrCl_2$ and $AuBr_2Cl$. The resulting brown color is very stable in acid solution and can be read quantitatively at 440 nm.

REAGENTS

1. Trichloroacetic acid, 10 g/100 ml aqueous solution.
2. Gold (auric) chloride solution. Wash the contents of a 1.0 g ampule of gold chloride into a 200 ml volumetric flask and dilute to the mark with water. The solution is stable.
3. Trichloroacetic acid (10 g/100 ml)–sodium chloride (0.06 g/100 ml) mixture. Place 0.6 g of NaCl in a 1 liter volumetric flask and add 500 ml of water. Add 100 g of trichloroacetic acid and dilute to volume with water.
4. Standards.

a. Stock, 10 mg/ml. Weigh exactly 1.000 g of NaBr, AR, dissolve in water, and dilute to 100 ml.

b. Dilute standard, 0.5 mg/ml. Pipet 10.0 ml of stock standard into a 200 ml volumetric flask and dilute to volume with the trichloroacetic acid–NaCl mixture.

PROCEDURE

1. Prepare a 1:10 trichloroacetic acid filtrate of serum.
2. Pipet 5.0 ml of clear filtrate (sample) into one tube and 5.0 ml of 10 g/100 ml trichloroacetic acid solution (blank) into a second tube.
3. Prepare standards as follows:

a. Pipet 0.5 ml of dilute standard into a labeled tube and add 4.5 ml of 10 g/100 ml trichloroacetic acid–NaCl mixture. Mix well (corresponds to 50 mg NaBr/100 ml).

b. Pipet 2.0 ml of dilute standard into a labeled tube and add 3 ml of 10 g/100 ml trichloroacetic acid–NaCl mixture. Mix well (corresponds to 200 mg NaBr/100 ml).

4. Add 0.5 ml of 0.5 g/100 ml $AuCl_3$ solution to all tubes. Mix well.

5. Read at 440 nm.

CALCULATION

$$\frac{\text{A unknown}}{\text{A standard*}} \times \text{concentration of standard*} = \text{mg NaBr/100 ml}$$

INTERPRETATION

Although normal bromide levels in serum are 0.8 to 1.5 mg/100 ml, this method may occasionally give results up to 5 mg/100 ml even with normal serum. It has been suggested that this is due to a slight turbidity that may at times develop. Therapeutic levels may be in the order of 100 mg/100 ml, and toxic levels are usually greater than 150 mg/100 ml. With a single overdose of an organic bromide compound, serum levels of inorganic bromide do not rise above normal levels. After prolonged therapy with these drugs, serum levels of inorganic bromide may increase to more than 100 mg/100 ml. At these levels, mental disturbances may be elicited.

REFERENCE

Hepler, O. E.: Manual of Clinical Laboratory Methods. 4th ed. Springfield, IL., Charles C Thomas, Publisher, 1963, p. 325.

FLUORIDE

This element is accessible to the public in the form of the sodium salt. Sodium fluoride is a common ingredient in roach and ant poisons and, as such, it is frequently kept around the house and even in the kitchen. For this reason, accidental poisonings have occurred, especially since the white crystalline material can be mistaken for ordinary salt or baking powder. In recent years a blue dye is usually added to these preparations to avoid this type of accident.

The fatal dose of sodium fluoride is 5 to 10 g. Once the compound reaches the stomach, the acidity of the gastric contents converts the salt to free hydrofluoric acid, which produces a dark red corrosion of the mucous membrane. For this reason, inorganic fluorides could also be classified with the corrosives.

A number of organic fluoride compounds are extremely toxic, and one of these, sodium fluoroacetate (sometimes called "1080"), has been used as a rat poison. The toxicity of this substance is due to its competition with acetate in the tricarboxylic acid cycle with the eventual formation of fluorocitric acid. It is estimated that a lethal oral dose in man is about 50 mg.

Despite the marked toxicity of these substances, poisonings of this type have not been common in the past. This was fortunate for the analyst because the difficulty and length of time needed for fluoride analysis led him to avoid it, if at all possible. Now, with the aid of microdiffusion in plastic dishes, this analysis has been greatly simplified. Plastic containers are ideal for collecting specimens as well as for conducting the analysis, since silica in glassware reacts with fluoride to form a volatile product, resulting in loss of fluoride.

Rieders[49] has described a screening test using modified polypropylene Conway cells. The method takes about 1 hour for completion; however, the test to be described,[50] although somewhat lengthy, yields somewhat better quantitative results.

* Use the standard whose absorbance is closest to that of the unknown.

DETERMINATION OF FLUORIDE IN BIOLOGICAL SAMPLES

PRINCIPLE

Fluoride is separated from the specimen by diffusing it into solid NaOH, at an elevated temperature for 20 h. The separated fluoride is then estimated by developing a color with a cerium or lanthanum complex with alizarin complexone:

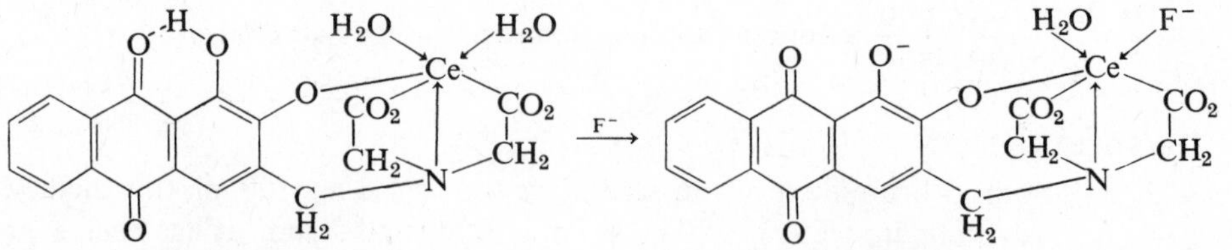

(Magenta) **(Blue)**

(Cerium and lanthanum function similarly in these complexes)

REAGENTS

1. Perchloric acid, 70 g/100 g.
2. Silver perchlorate, AR.
3. Sodium hydroxide, 0.50 molar. Dissolve 2.00 g of NaOH in 50 ml of water and dilute to 100 ml with 95 per cent ethanol.
4. Dye solution. Dissolve 10.0 g of Amadac-F (Burdick & Jackson Laboratories, Inc., Muskegon, MI.) in 60 per cent isopropyl alcohol and make up to 100 ml with the same solvent. Amadac-F contains all the necessary components of the dye reagent.
5. Fluoride standard. Dissolve 0.2210 g of sodium fluoride, AR, in water and make up to 100 ml; 1.0 ml of this solution contains 1.0 mg of fluoride. A 1:100 dilution of this solution will result in a useful working standard in which 1.0 ml contains 0.010 mg of fluoride.

PROCEDURE

1. Place 0.10 ml of 0.50 molar NaOH in the center of the inside top of a plastic Petri dish (Millipore Filter Corp., Bedford, MA., Cat. No. PD 10 047 00). Distribute the solution evenly and evaporate to dryness with a fan.
2. Transfer 1.0 ml of blood, urine, or gastric contents to the bottom of the plastic Petri dish diffusion unit. Add about 0.2 g of silver perchlorate to the specimen.
3. Add 2.0 ml of perchloric acid to the specimen and cover immediately with the prepared receiver top. Mix by gentle swirling.
4. Place the unit in an oven, previously heated to 50°C, and allow it to remain at this temperature for 20 h.
5. Carefully remove the unit from the oven and allow to cool. Remove the receiver top and add 1.0 ml of distilled water to the NaOH residue in the receiver top.
6. Gently stir with the tip of a dropping pipet, aspirate the solution, and transfer to a 5 ml volumetric flask. Repeat last part of step 5 and first part of step 6 twice more.
7. Add 1.0 ml of Amadac-F reagent and dilute the solution to 5.00 ml.
8. After 1 h, read the color at 620 nm and compare with a water blank and standard carried through the entire procedure.

INTERPRETATION

Normal blood fluoride levels range up to 0.050 mg/100 ml. In fatal cases, blood levels may be as high as 0.2 to 0.3 mg/100 ml. Excretion of fluoride in the urine is about 1.0 mg/d. Even subtoxic doses of fluoride can result in sharp increases in urine levels, and in severe poisoning the urine concentration can be several milligrams/100 ml.

REFERENCE

Rowley, R. J., and Farrah, G. H.: Am. Industr. Hyg. Assn. J., *23*:314, 1962.

NONVOLATILE ORGANIC SUBSTANCES

This, the largest group of substances, includes most drugs and alkaloids. The problems associated with analysis of this group are quite complex. Extraction methods must usually be employed to separate the drug from the specimen. Extractions are frequently not quantitative or may result in troublesome emulsions or may be pH dependent. Some drugs are rapidly metabolized, excreted, or bound to protein, and this makes their detection difficult. Many drugs are chemically similar to naturally occurring substances and must be differentiated from them by purification steps or highly specific chemical tests.

Those drugs that are acids or bases are usually water soluble when they are in the form of salts. By reconverting the drug back to the free acid or base, it is made less water soluble but more soluble in solvents such as chloroform or ether. This property of organic acids and bases is used in separation and purification steps. By adjusting the pH of the aqueous phase and extracting with less polar immiscible solvents, separation is usually successful.

Those drugs that are neutral or amphoteric are more difficult to extract and purify. Chromatography, electrophoresis, sublimation, and countercurrent liquid-liquid extraction are some of the techniques that have been used.[54]

The various extracts can be rapidly screened for ultraviolet absorbing compounds by scanning with a recording spectrophotometer. Organic acids can be extracted from the immiscible solvent by a small quantity of 0.5 molar NaOH. Organic bases are then extracted by a small quantity of 0.5 molar HCl. The neutral compounds are recovered by evaporating the solvent to dryness and taking up the residue in alcohol. Both aqueous solutions and the alcohol solution are scanned, individually, from 220 to 350 nm. In general, aromatic compounds or compounds with conjugated double bonds can be detected in this manner if their extinction coefficient is great enough. Spectral curves thus obtained can be compared with similar curves obtained with known compounds.[54]

The advent of thin-layer chromatography has been a great help in detecting organic drugs in extracts of biological material. Although the extraction technique may not be quantitative, the speed and sensitivity of thin-layer chromatography enables the extract to be easily screened for the presence or absence of certain drugs or groups of drugs.

A number of books and articles discuss the technique more comprehensively, and these should be consulted by anyone planning to use this extremely valuable tool more fully.[7,53,56] The technique of thin-layer chromatography in general is also discussed in Chapter 3, Analytical Procedures, and Chapter 10, Lipids.

Gas liquid chromatography (GLC) has also proved to be a useful tool in toxicological analysis. At elevated temperatures many members of the "nonvolatile" organic group are volatile enough and sufficiently stable to be separated and detected by this method. The availability of detectors sensitive to picogram quantities of organic compounds makes this technique particularly appropriate to analytical toxicology.

Detailed discussions of the theory and many applications of GLC can be found in Chapter 3 and elsewhere.[44] It is pertinent however, to emphasize that with most GLC methods, extractions of biological specimens are still necessary. Frequently methods are reported which give striking results with pure standards or spiked samples but which fail to live up to this promise when applied to biological material. This may be due to many factors: protein binding, interfering substances, rapid metabolism, and others.

Another point for emphasis is the obvious but frequently ignored fact that a peak on a GLC chart does not constitute an unequivocal identification of a substance. As in thin layer chromatography, positive results strongly suggest the presence of a specific substance, but interferences can occur. Plasticizers, like pesticides, are present in almost all tissues and body fluids in detectable amounts and may be a source of interference in GLC methods. Normal constituents, such as steroids, lipids, and fatty acids, can interfere. The purity of reagents, solvents, and water is also a critical issue. The utility of some plastic ware is marred by the nuisance of imparting GLC-sensitive components to reagents.

Despite these problems, the technique is of practical importance. Since many drugs are detected under similar conditions, it is not unusual that unsuspected drugs are uncovered. Simultaneous determination of several drugs and their metabolites is also possible.

Application of GLC to analytical toxicology has been reviewed extensively.[7,56] Owing to interferences and lack of specificity in both TLC and GLC, it is well for any toxicology laboratory to maintain back-up facilities in order to confirm questionable cases by an independent method.

Both of these techniques will be illustrated by practical problems for which each is appropriately designed. For TLC, a urine drug screening procedure will be described. For GLC, a method by which anticonvulsant drugs and their metabolites can be detected is appropriate in view of the interest in this problem shown by neurologists.

URINE SCREENING FOR DRUGS

PRINCIPLE

There is a great interest in discovering and controlling drug abuse and addiction. The ease with which urine can be collected, and the fact that most drugs or their metabolites are excreted in part by the kidney, has led to the development of screening procedures utilizing this specimen.

A variety of techniques have evolved. Drugs are separated from the specimen by solvent extraction or by adsorption on resins, or they are assayed directly by immunoassay techniques. After separation, drugs are detected by TLC, GLC, fluorimetry, spectrophotometry, or chemical methods. Depending on workload, methods may be automated, semiautomated, or performed manually. Many descriptions of various combinations of these methods are available, and Finkle[24] has reviewed these recently. The following TLC procedure is commonly used and is quite appropriate for clinical laboratories.

REAGENTS

The solvents and chemicals are reagent grade and are used without further purification.

1. Extracting solvent. Chloroform, 960 ml; isopropanol, 40 ml.
2. Buffer solution. To 100 ml of saturated ammonium chloride solution, add concentrated ammonium hydroxide solution to make the pH 9.5.
3. Developing solvent. Ethyl acetate, 170 ml; methyl alcohol, 20 ml; and concentrated ammonium hydroxide, 10 ml.
4. Spray reagents.
   a. Ninhydrin, 0.1 g/100 ml acetone.
   b. Diphenylcarbazone, 0.01 g/100 ml, in equal parts of acetone and water.
   c. Mercuric sulfate solution, 0.25 g/100 ml in 1 molar sulfuric acid. Dissolve 0.50 g mercuric oxide in 20 ml concentrated sulfuric acid; add the acid solution slowly to water and dilute with water to 200 ml.
   d. Iodoplatinate solution. Add 1 g of platinic chloride in 10 ml of water to 60 g of potassium iodide in 200 ml of water. Dilute this mixture to 250 ml with water. Store in the refrigerator. Prepare fresh solution every 2 weeks.
   e. Dragendorff solution. Add 1.3 g of bismuth subnitrate in 60 ml of water, as well as 15 ml of glacial acetic acid, to 12 g of potassium iodide in 30 ml of water. Dilute the mixture with 100 ml of water and 25 ml of glacial acetic acid.
5. Drug standards (available from a variety of commercial sources).

PROCEDURE

1. Add 10 ml of a urine sample and 1 ml of buffer to 50 ml of the extracting solvent in a 125 ml glass-stoppered bottle (or separatory funnel). Prepare a control urine sample containing 10 μg each of morphine, codeine, and quinine per 10 ml of urine, and 50 μg each of amphetamine, phenobarbital, pentobarbital, glutethimide, and chlorpromazine per 10 ml of urine, and analyze it in the same way as the urine sample.

2. Agitate the bottles manually or in a shaking machine (3-inch stroke, 180 cycles per min) for 5 min.

3. Aspirate and discard the aqueous layer.

4. Filter the organic layer through filter paper into a 100 ml beaker.

5. Boil off the ammonia; then add one drop of 0.1 molar HCl. Evaporate the solvent to dryness on a steam bath.

6. Wash the sides and the bottom of the beaker with a fine stream of methyl alcohol. Evaporate the alcohol to approximately 25 μl while the beaker is tilted at an angle of 45 degrees.

7. Transfer the entire extract in 5 μl aliquots to a 250 μm silica gel thin-layer plate with a disposable capillary pipet. The diameter of the spot at the point of application should not exceed 1.0 cm. Normally, seven or eight spots can be conveniently applied to each plate. Dry the spots with warm air (hair dryer).

8. Add 200 ml of developing solvent to the developing tank.

9. Insert the plates into the developing tank and allow the solvent to travel 10 cm past the point of application.

10. Remove the plates and air dry to eliminate the more volatile solvents. Then heat in an oven at 75°C for 10 min.

11. Spray the plate, while still hot, with the ninhydrin solution and place under ultraviolet light for 2 min.

12. Remove the plate from the ultraviolet light and mark the pink spot due to amphetamine.

13. Spray the plate first with diphenylcarbazone solution and then with the mercuric sulfate reagent. Apply the mercuric sulfate rather heavily until the barbiturates in the positive control specimen are clearly visible. After air drying, the barbiturate and glutethimide spots appear blue to pink in color.

14. Reheat the plate in the 75°C oven for 2 min. After removal from the oven, violet to orange-red colored spots due to phenothiazine drugs are noted.

15. View the plate under ultraviolet light for fluorescence due to quinine.

16. Spray the cooled plates with the iodoplatinate reagent, air dry, and spray with the Dragendorff solution. After 3 to 5 min the position and color of the spots obtained are noted and compared with those of the positive control sample. Iodoplatinate followed by Dragendorff solution results in characteristic colors with basic drugs such as morphine, codeine, quinine, and phenothiazine derivatives. Table 21–4 shows R_f values and color reactions to the sprays described for a number of commonly encountered drugs.

COMMENTS

The sensitivity of this procedure ranges from 0.5 to 1.0 μg/ml of urine, depending on the drug. Since no attempt is made to break down complexes, such as morphine glucuronide, the method detects only free drug rather than total drug present. In addition, metabolites of some drugs may be present but not identified by this procedure. Cocaine and methadone are metabolized readily and may be missed. Thus, false negative results may be found.

False positive results, although infrequent, may also be a problem. Some drugs have similar R_f values, and metabolites may also lead to confusion. It is difficult to reproduce R_f values accurately, thus including standards on each TLC plate is always necessary.

TABLE 21–4 THIN-LAYER CHROMATOGRAPHIC DATA FOR VARIOUS DRUGS EXTRACTABLE FROM URINE*

Compound	R_f	*Ninhydrin*	*Diphenyl-carbazone,* $HgSO_4$	*Oven*	*Ultraviolet Light*	*Iodoplatinate, Dragendorff*
ANALGESICS						
Acetaminophen	NR					
Cocaine	0.96					OR
Phenacetin	0.98					Br
Propoxyphene (Darvon)	0.90					O
Salicylic acid	NR					
ANTIHISTAMINES						
Chlorpheniramine (Chlor-Trimeton)	0.88					RV
Diphenhydramine (Benadryl)	0.90					V
Dimenhydrinate (Dramamine)	0.90					OR
Methapyrilene (Histadyl)	0.94					Pu
Promethazine (Phenergan)	0.95					BV
HYPNOTICS AND SEDATIVES						
Amobarbital	0.75		P			
Butabarbital	0.73		P			
Carbromal	NR					
Diphenylhydantoin (Dilantin)	0.75		P			
Glutethimide (Doriden)	0.99		P			
Pentabarbital	0.75		Pu			
Phenobarbital	0.46		P			
Secobarbital	0.75		P			
OPIUM ALKALOIDS						
Codeine	0.54					RV
Morphine	0.32					DB
SYNTHETIC NARCOTICS						
Meperidine (Demerol)	0.90					RV
Methadone	0.99					OR
STIMULANTS						
Amphetamine	0.78	P	P	P		Br
Caffeine	0.80					B
Methamphetamine	NR					
Phenylpropanolamine	0.60	V	V			
TRANQUILIZERS						
Amitriptyline (Elavil)	0.98					Br
Chlordiazepoxide (Librium)	0.88				B	RV
Chlorpromazine (Thorazine)	0.96			P		BV
Diazepam (Valium)	0.98					V
Imipramine (Tofranil)	0.95				B	Pu
Meprobamate (Miltown)	0.75		PF			Br
Perphenazine (Trilafon)	0.84			P		BV
Thioridazine (Mellaril)	0.97			P		OBr
Trifluoperazine (Stelazine)	0.99			P		Pu
MISCELLANEOUS						
Atropine	0.43					Pu
Nicotine	0.90					B
Quinidine	0.65				LB	RV
Quinine	0.65				LB	RV

* NR, no reaction; B, blue; Br, brown; D, dark; L, light; P, pink; Pu, purple; R, red; O, orange; V, violet; Y, yellow; F, fades.

For these reasons, interpreting the results of urine drug screening is hazardous. Reports based on screening tests such as this should not state flatly that "urine is positive (or negative) for drug X." It is better to use a reporting format, such as "urine screen is consistent with the presence of drug X. Drugs Y and Z were not detected." If decisions affecting the life (e.g., suspected overdose), liberty (e.g., parole violation), or pursuit of happiness (e.g., job applicant screening) of an individual depend on laboratory results, all positive findings must be confirmed by an independent, more specific test.

REFERENCE

Davidow, B., Petri, N. L. and Quame, B.: Tech. Bulletin Reg. Med. Tech., *38*:714–719, 1968.

DETERMINATION OF ACID AND NEUTRAL DRUGS BY GLC

PRINCIPLE

Although acid and neutral drugs undergo protein binding to some extent, direct extraction of biological specimens with an organic solvent gives reasonably good recoveries. This ease of extraction, and the sensitivity and resolution of GLC, make a good combination for rapid detection of these drugs in emergencies. Also, when used as anticonvulsants, many of these drugs are used in therapeutic combinations. Thus, it is useful to be able to determine them simultaneously.

The following procedure is a combination of several which have been published. The

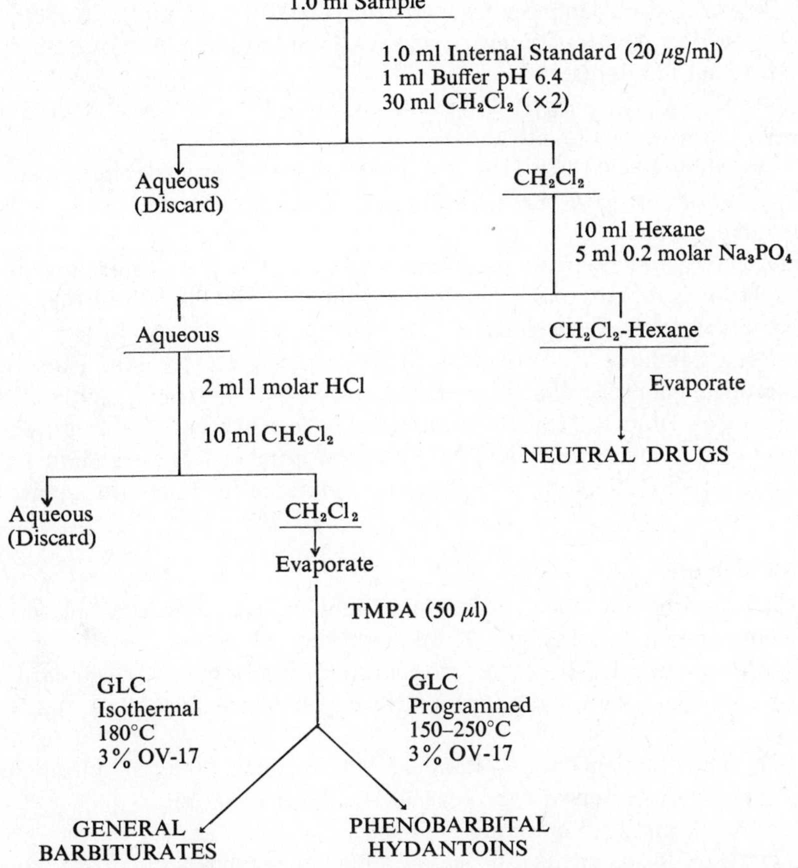

Figure 21–3 Extraction scheme for barbiturates, anticonvulsant drugs, and neutral drugs.

extraction is done in a uniform way, and the choice of which extract to examine, as well as the GLC conditions to use, depend on the drug or group of drugs to be determined (Fig. 21–3).

EQUIPMENT

1. Gas chromatograph with flame ionization detector equipped for either isothermal or linear temperature programmed operation.

2. Water bath or heating block arranged for low temperature (approximately 60°C) heating of 15 ml conical centrifuge tubes under a current of nitrogen, for rapid evaporation of solvent.

REAGENTS

1. Extracting solvent. Methylene chloride, pesticide grade (ethylene dichloride, spectral grade, or chloroform, spectral grade, may be substituted).

2. Hexane, pesticide grade.

3. Methanol, anhydrous, AR.

4. Phosphate buffer, pH 6.4. Mix 13.3 ml Na_2HPO_4 (0.2 molar) and 36.8 ml KH_2PO_4 (0.2 molar). Check with pH meter.

5. Sodium phosphate, tribasic, 0.2 mol/l. Dissolve 7.6 g $Na_3PO_4 \cdot 12\ H_2O$ in water and make up to 100 ml.

6. Hydrochloric acid, 1 mol/l. Dilute 82.6 ml concentrated reagent grade HCl to 1 liter with water.

7. Trimethylphenylammonium hydroxide (TMPA), 0.1 mol/l, in methanol (Eastman No. A10943). Refrigerate. Dilute 1.0 ml with 2.0 ml of methanol for each day's use.

8. Internal standards.

(a) 5-(*p*-Methylphenyl)-5-phenylhydantoin (Aldrich No. 16,145–4). Used for hydantoin-phenobarbital combination.

(b) Methohexital (Eli Lilly & Co.). Used for general barbiturate analysis.

(c) α,α-Dimethyl-β-methylsuccinimide (Aldrich No. 16,350–3). Used for neutral, anticonvulsant drugs.

Each internal standard is prepared in methanol at a concentration of 20 mg/100 ml (stock solution); 10.0 ml of stock solution diluted to 100 ml with methanol gives working solutions of 20 μg/ml. Refrigerate.

9. Drug standards. The various drugs of interest are obtained from the appropriate pharmaceutical manufacturer. (See Physicians' Desk Reference, 29th Edition, Medical Economics Co., Oradell, N.J., 1975. Purchase of certain scheduled drugs requires a Bureau of Narcotics and Dangerous Drugs registration number.) The standard for each drug is made up to contain 20 μg/ml in methanol as described for internal standards. Refrigerate.

PROCEDURE

1. In an extraction tube or separatory funnel marked "patient," place 1.0 ml of specimen (blood, serum, or urine) and 30 ml methylene chloride.

2. In a second extraction tube or separatory funnel marked "standard," place 1.0 ml of drug-free serum, 1.0 ml of the appropriate drug standard (20 μg/ml), and 30 ml methylene chloride.

3. To both "patient" and "standard" tubes, add 1.0 ml of phosphate buffer, pH 6.4, and 1.0 ml of the appropriate internal standard (20 μg/ml).

4. Shake or mix on Vortex mixer for 60 s.

5. Draw off lower, organic phases and filter through Whatman #3 paper into second tubes or separatory funnels marked "patient" and "standard."

6. Repeat the extraction of sample and standard with an additional 30 ml methylene chloride. Repeat step 5. Discard aqueous phase.

7. To each of the combined, filtered extracts add 10 ml of hexane and 5.0 ml of 0.2 molar Na_3PO_4. Shake or mix on Vortex mixer for 60 s.

8. Draw off lower organic phases. Discard, or save for neutral drug analysis.

9. To the aqueous phases from step 7, add 2 ml of 1 molar HCl and 10 ml of methylene chloride. Shake or mix on Vortex mixer for 60 s.

10. Filter the lower, organic phases through Whatman #3 paper into marked 15 ml conical centrifuge tubes. Evaporate at 60°C under a stream of nitrogen to about 50 μl.

11. With an Eppendorf pipet, add 50 μl of diluted TMPA to each tube, washing down the constricted tip of the tubes in the process.

12. Promptly inject 2 μl of the resulting solution into the gas chromatograph, using the isothermal mode for general barbiturates and the programmed mode for diphenylhydantoin-phenobarbital combinations.

Isothermal mode

Glass column, 5 ft, $\frac{1}{8}$ inch ID, 3 per cent OV-17 on Gas Chrom Q.

Temperature: Injection port, 250°C; column, 180°C; detector, 250°C.

Carrier gas: N_2, 40 ml/min. FID:H_2, 40 ml/min; air, 400 ml/min.

Programmed mode

Glass column, 5 ft, $\frac{1}{8}$ inch ID, 3 per cent OV-17 on Gas Chrom Q.

Temperature: Injection port, 250°C; detector, 290°C; column 150–250°C programmed at 10°/min. Begin program immediately after injection.

Carrier gas: N_2, 40 ml/min. FID:H_2, 40 ml/min; air, 400 ml/min.

CALCULATIONS

1. Draw a base line for each peak. Measure the peak height or peak area of the drug and of the internal standard.

2. Calculate the ratio of the peak height or area of the drug to that of the internal standard for various concentrations of drug standards, and construct a standard curve.

3. Calculate drug levels of an unknown sample from the standard curve when samples are treated exactly like the standards.

4. Alternatively, concentrations can be calculated as follows:

$$\frac{Dp}{Ip} \times \frac{Is}{Ds} \times C_I = C_p$$

where Dp and Ds are the peak heights or areas of the drug in the patient's serum or standard, respectively

Ip and Is are the peak heights or areas of the internal standard in patient's serum or standard, respectively

C_I is concentration of internal standard used (20 μg/ml)

C_p is concentration of drug in patient's serum (μg/ml)

COMMENTS

A new standard curve should be prepared each time a new column is used or conditions such as temperature or flow rates are changed.

A standard should be run daily. If retention times or drug/internal standard ratios change, all parameters should be rechecked. This may indicate that a new column should be used.

Adding the internal standard before extraction and running parallel standards compensates for losses and improves accuracy and reproducibility.

The TMPA methylating reagent is simple and convenient to use. It is strongly alkaline, however, and prolonged contact of this reagent with barbiturate extracts may decrease recoveries owing to degradation of the barbiturate. The GLC analysis should be done within 5 to 10 min after adding TMPA reagent.

Phenobarbital and mephobarbital form identical dimethyl derivatives and cannot be separated by this technique. This is also true of barbital and metharbital. In this event, a separate GLC analysis can be done without the addition of TMPA reagent. Barbiturate peaks without methylation are broader and show longer retention times. Measuring peak area results in greater accuracy and reproducibility. If peaks are sharp and symmetrical, peak height measurements are satisfactory.

If hydantoin-phenobarbital or other drug combinations are measured frequently, it is more convenient to make up the appropriate standards combined in one solution.

REFERENCES

General Barbiturates—Fiereck, E. A., and Tietz, N. W.: Clin. Chem., *17*:1024–1027, 1971.
Phenobarbital and Diphenylhydantoin—Kupferberg, H. J.: Clin. Chim. Acta, *29*:283–288, 1970.
Other Anticonvulsant Drugs—*Antiepileptic Drugs*, D. M. Woodbury, J. K. Penry, and R. P. Schmidt, eds. Raven Press, New York, 1972.

BARBITURATES

In cases associated with drug overdose, barbiturates are the leading offenders. These drugs are extremely useful in modern medicine and are commonly prescribed as treatment for a variety of conditions. Because of their availability, they are frequently the cause of accidental poisoning. Since they are hypnotics (sleep-producing drugs), they are commonly

TABLE 21-5 COMMON BARBITURATES AND THEIR TOXIC BLOOD LEVELS

Generic Name	R_1	R_2	*Blood Level When Consciousness Regained* (mg/100 ml)
Barbital (Veronal)	$-CH_2-CH_3$	$-CH_2-CH_3$	8
Phenobarbital (Luminal)	$-CH_2-CH_3$	$-C_6H_5$	5
Butabarbital (Butisol)	$-CH_2-CH_3$	$-C(CH_3)(H)-CH_2-CH_3$	3
Amobarbital (Amytal)	$-CH_2-CH_3$	$-CH_2-CH_2-C(CH_3)(CH_3)-H$	3
Pentobarbital (Nembutal)	$-CH_2-CH_3$	$-C(CH_3)(H)-CH_2-CH_2-CH_3$	1
Secobarbital (Seconal)	$-CH_2-CH{=}CH_2$	$-C(CH_3)(H)-CH_2-CH_2-CH_3$	1

used for suicide or in suicide attempts. There are many individual drugs in this group, but all are chemically similar, and produce sleep and, upon overdosage, coma and death. They are chemically characterized by a pyrimidine ring with two substitutions on carbon atom 5 (see Table 21–5).

Some barbituric acid derivatives may have a methyl group substituted on one of the N atoms of the ring (e.g., mephobarbital and hexobarbital); others have an S atom instead of O at carbon atom 2 (e.g., thiopental and thiamylal). The sulfur derivatives are used as anesthetic agents and are not likely to be encountered in cases of overdosage.

Since these drugs can be readily extracted from blood, serum, or urine, and since they all have good ultraviolet absorption properties, some reliable methods for determining barbiturates in biological materials require the availability of an ultraviolet spectrophotometer, preferably one that records automatically. The qualitative test for barbiturates that uses a color reaction produced by cobaltous acetate and an organic base under anhydrous conditions (the Koppanyi reaction) is unreliable. False positive results and, when the test is improperly carried out, false negative results may be observed.

Another colorimetric method has been reported in which a complex is formed between mercury(II) and barbituric acid derivatives.[2] The complex is soluble in organic solvents, and the barbiturate can be estimated by using diphenylthiocarbazone to determine the amount of mercury present in the organic solvent.[68] A similar method, which is rapid and does not require a colorimeter, has been described by Curry.[11,14] This method is useful for blood levels greater than 2.0 mg/100 ml.

BARBITURATE DETERMINATION BY ULTRAVIOLET SPECTROPHOTOMETRY

PRINCIPLE

Ultraviolet spectrophotometric methods for the determination of 5,5-disubstituted barbiturates in biological material are based on the fact that these compounds exist in three forms in solution: a nonionized form in acid solution, with almost no absorption in the range 230 to 270 nm; the first ionized form at pH 9.8 to 10.5 with an absorption maximum at 240 nm; and the second ionized form at pH 13 to 14 with an absorption maximum at 252 to 255 nm and a minimum at 234 to 237 nm. The three forms of the drug can be represented as follows:

Acid form $\underset{H^+}{\overset{OH^-}{\rightleftharpoons}}$ *First ionized form* $\underset{H^+}{\overset{OH^-}{\rightleftharpoons}}$ *Second ionized form*

The 1,5,5-trisubstituted barbiturates exist only in two forms in solution, since they lack one enolizable hydrogen:

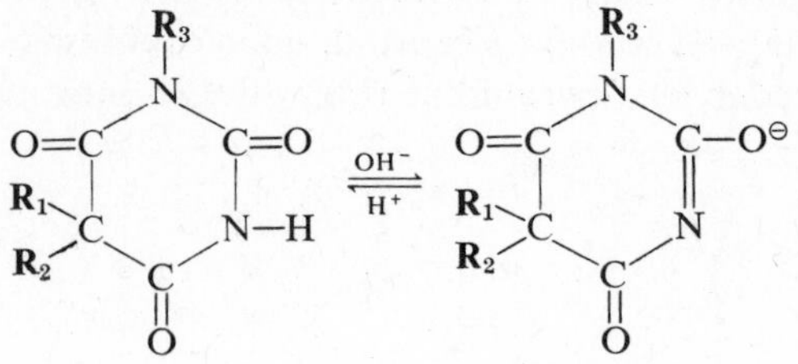

Acid form $\underset{H^+}{\overset{OH^-}{\rightleftharpoons}}$ *Ionized form*

The nonionized form in acid solution has almost no absorption in the range of 230 to 270 nm; the first and only ionized form at pH 9.8 to 14 has an absorption maximum at 245 nm.

In the acid form, barbiturates are relatively water insoluble, but they are soluble in organic solvents. In both the first and second ionized forms these drugs are very water soluble, but insoluble in organic solvents. Thus, they can be extracted from blood or serum at physiological pH values, or from acidified urine, by organic solvents. By washing the organic solvent with an aqueous phosphate buffer of pH 7.4, some interfering impurities, e.g., salicylates, can be removed, although some loss of barbiturates occurs. Shaking the organic solvent with dilute alkaline solution converts the free acid form of barbiturate into its salt, resulting in the transfer of the barbiturate into the aqueous phase. This aqueous extract is used for scanning in the ultraviolet spectrophotometer.

Proper interpretation of a blood or serum barbiturate level cannot be done unless the type of barbiturate present is known. For example, a 1 mg/100 ml level of barbital is not too serious, but the same level of secobarbital is close to a lethal level (see Table 21–5). Also, 2 mg/100 ml of phenobarbital in blood is not generally considered a dangerous level, but the same level of secobarbital can be lethal.

Since the prognosis in a given case of overdosage is influenced by the type of barbiturate involved, it is important to identify the drug or to determine the type of barbiturate present. This is done either by treatment of the barbiturate with hot alkaline solution, by thin-layer chromatography, or by GLC (see pp. 1132, 1135, and 1141).

About 29 different barbituric acid derivatives are, or have been, used clinically. Of these, only about six or eight are commonly prescribed and available to the general public.

Pharmacologically, the barbiturates can be classified according to their duration of action. Four groups are commonly described: long, intermediate, short, and ultrashort acting.[30]

Long acting: barbital, phenobarbital, mephobarbital, diallylbarbituric acid.
Intermediate acting: amobarbital, aprobarbital, butabarbital, hexethal.
Short acting: cyclobarbital, pentobarbital, secobarbital.
Ultrashort acting: hexobarbital, thiamylal, thiopental.

The ultrashort acting barbiturates are used exclusively as anesthetic agents and consequently are rarely encountered in cases of accidental or intentional overdosage. Thus, the balance of this discussion is restricted to the classification of the first three groups: long acting, intermediate acting, and short acting barbiturates.

Classification of an unknown barbiturate into one of these major groups may be accomplished by treatment of the alkaline extract (step 9 of spectrophotometric procedure) with heat. Fast acting barbiturates are more stable to this treatment (alkaline hydrolysis at high temperatures) than the intermediate or slow acting barbiturates. Under this treatment barbiturate derivatives are hydrolyzed to malonic acid derivatives and urea. These compounds show little absorption in the ultraviolet region and, therefore, the decrease in ultraviolet absorption can be taken as a direct measure of the degree of hydrolysis of barbiturates (for more detail see under Procedure).

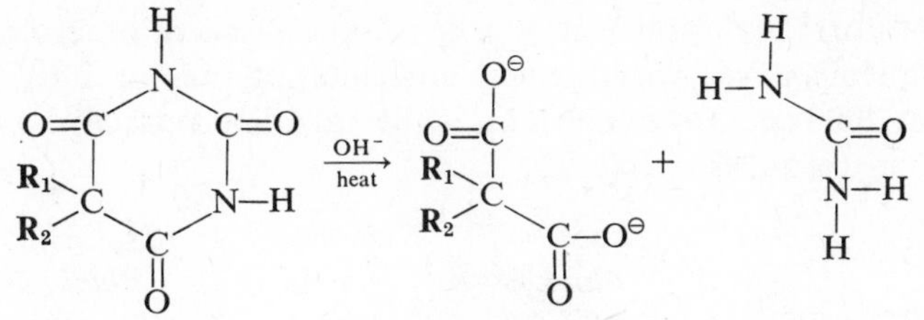

Barbiturate derivative $\xrightarrow[\text{heat}]{OH^-}$ *Malonic acid derivative* + *Urea*

REAGENTS

1. Chloroform. Prepare a daily supply by washing sufficient U.S.P. chloroform with 1/10 vol of 1 molar NaOH followed by two washings with 1/10 vol of distilled water. (Washing the $CHCl_3$ may be omitted if a blank determination shows the absence of interfering materials in the $CHCl_3$.)

2. Boric acid, 0.6 mol/l—potassium chloride solution. Dissolve 37.1 g of H_3BO_3 and 44.7 g of KCl in distilled water and make up to the mark in a 1 liter volumetric flask. The solution is stable.

3. Sodium hydroxide, 0.45 mol/l. Dissolve 18.0 g of NaOH in distilled water and make up to the mark in a 1 liter volumetric flask. This solution need not be standardized, but when equal volumes of 0.45 molar NaOH and 0.6 molar H_3BO_3–KCl are mixed, a pH of about 9.9 should result. Check with a pH meter. It may be necessary to add acid or alkali to the 0.45 molar NaOH to get the correct pH on mixing with 0.6 molar H_3BO_3–KCl. The solution is stable.

4. Potassium phosphate, monobasic, 0.5 mol/l. Dissolve 17.0 g of KH_2PO_4 in distilled water and dilute to the mark in a 250 ml volumetric flask. The solution is stable.

5. Sodium phosphate, dibasic, 0.5 mol/l. Dissolve 179 g of $Na_2HPO_4 \cdot 12\,H_2O$ (or equivalent weight of anhydrous or other hydrated forms) in distilled water and dilute to the mark in a 1 liter volumetric flask. The solution is stable.

6. Phosphate buffer, pH 7.4. Mix 19.2 ml of 0.5 molar KH_2PO_4 (reagent 4) with 80.8 ml of 0.5 molar Na_2HPO_4 (reagent 5). Check the pH with a pH meter and adjust, if necessary, with reagent 4 or 5, respectively.

PROCEDURE

For the extractions use separatory funnels with Teflon stopcocks. Stopcock grease should not be used since some of the components of stopcock grease interfere with ultraviolet measurements.

1. Measure 10.0 ml of blood, serum, urine, or gastric contents into a 125 ml separatory funnel. With urine or gastric contents, the pH should be checked with indicator paper and adjusted to a pH of 7 or less by the addition of dilute HCl. Volumes other than 10.0 ml can be used with appropriate adjustment of the calculation. If the type of barbiturate need not be identified, 5.0 ml of specimen are sufficient.

2. Add 30 ml of chloroform (reagent 1) and extract the specimen by shaking for 1 min.

3. Draw off the chloroform and filter through Whatman No. 3 filter paper into a second 125 ml separatory funnel. Repeat the extraction two more times, using 30 ml of chloroform each time. If an emulsion occurs at this point, add a few milliliters of chloroform in excess and gently invert the separatory funnel several times. The layers will separate easily with most specimens. If not, centrifuge at 2000 rpm for 5 min.

4. Wash the combined filtered chloroform extracts twice with 5 ml of phosphate buffer. After the second phosphate wash, filter the chloroform through fresh Whatman No. 3 filter paper into a third 125 ml separatory funnel.

5. Extract the chloroform by shaking for 3 min with 10.0 ml of 0.45 molar NaOH. Draw off the chloroform and discard it or save it for analysis of other drugs.

6. Run the aqueous phase, together with any emulsion present, into a 15 ml centrifuge tube. Centrifuge at 2000 rpm for 5 min. This is the "alkaline extract."

7. Prepare four test tubes as follows:

	Borate Blank	*Borate Sample*	NaOH *Blank*	NaOH *Sample*
0.6 molar H_3BO_3–KCl	2.0 ml	2.0 ml	—	—
0.45 molar NaOH	2.0 ml	—	4.0 ml	2.0 ml
Alkaline extract	—	2.0 ml	—	2.0 ml

8. Scan these solutions in the ultraviolet spectrophotometer from 220 to 300 nm as follows (if only a manual instrument is available, make readings at 5 nm intervals): Read the borate sample against the borate blank as reference.* Read the NaOH sample against the NaOH blank as reference. If typical barbiturate curves are obtained (see Interpretation) and the type of barbiturate is known, proceed to Calculations. If the type of barbiturate is unknown, proceed as follows.

9. Pipet 5.0 ml of alkaline extract from step 6 into a tube calibrated at 5.0 ml and place into a boiling water bath for 15 min (time exactly!).

10. After 15 min, immediately transfer the tube to an ice-water bath. After cooling to room temperature, adjust the volume to 5.0 ml with distilled water and mix thoroughly. This is the hydrolyzed alkaline extract.

11. Prepare two test tubes as follows:

	NaOH-Hydrolyzed Sample	*Borate-Hydrolyzed Sample*
0.6 molar H_3BO_3–KCl	—	2.0 ml
0.45 molar NaOH	2.0 ml	—
Hydrolyzed alkaline extract	2.0 ml	2.0 ml

12. Scan these solutions in the ultraviolet spectrophotometer in the same manner as in step 8, using the same blank solutions as reference solutions.

CALCULATIONS

Table 21–6 shows the calibration data for 5,5-disubstituted barbiturates commonly encountered. In this table, K_N and K_B are extinction coefficients of the particular barbiturate at a concentration of 1 mg/l in 0.45 molar NaOH and in borate solution of pH 9.9, re-

* After completion of the reading of the borate sample, adjust the same recorder paper in such a way that the scan of the NaOH sample will exactly overlay the scan of the borate sample.

TABLE 21–6 CALIBRATION DATA FOR 5,5-DISUBSTITUTED BARBITURATES

Barbiturate	*Duration of Action*	K_N	K_B	*F*	*R* %
Phenobarbital	Slow	0.0315	0.0086	43.7	31.8
Barbital	Slow	0.0329	0.0062	37.5	42.5
Butabarbital	Intermediate	0.0314	0.0056	38.8	49.6
Amobarbital	Intermediate	0.0292	0.0052	41.7	55.8
Pentobarbital	Fast	0.0268	0.0060	48.1	98.2
Secobarbital	Fast	0.0288	0.0078	47.5	97.6

spectively, at 260 nm. F is a calibration constant and is equal to $1/(K_N - K_B)$. R is the percentage of barbiturate remaining after alkaline hydrolysis for 15 min.

In the calculations to follow, these symbols are used:

A_1 = absorbance of NaOH sample at 260 nm.
A_2 = absorbance of borate sample at 260 nm.
A_3 = absorbance of NaOH hydrolyzed sample at 260 nm.
A_4 = absorbance of borate hydrolyzed sample at 260 nm.
C = concentration of barbiturate in sample.

$$(A_1 - A_2) \times F = C \text{ in } \mu\text{g/ml of solution in cuvet} \tag{a}$$

or

$$(A_1 - A_2) \times F \times 0.2 = C \text{ in mg/100 ml of specimen} \tag{b}$$

If specimen volumes other than 10.0 ml are used, and the chloroform is extracted with volumes of 0.45 molar NaOH other than 10.0 ml, then equation (c) is used:

$$(A_1 - A_2) \times F \times 0.2 \times \frac{\text{ml 0.45 molar NaOH}}{\text{ml sample}} = C \text{ in mg/100 ml specimen} \tag{c}$$

If the barbiturate is known to be one of the six shown in Table 21–6, the appropriate F value is used. If the specific barbiturate is unknown, but it is found by hydrolysis to be slow acting, use an F value of 40.6; if intermediate acting, use $F = 40.3$; and if fast acting use $F = 47.8$. If the type of barbiturate is unknown, and insufficient specimen is available for hydrolysis, an approximate level can be estimated by using the mean F value of 42.9.

In order to determine the type of barbiturate present, the following calculation is made:

$$R = \frac{A_3 - A_4}{A_1 - A_2} \times 100$$

If: $R = 30$ to 45, a slow acting barbiturate is present.
$R = 45$ to 56, an intermediate acting barbiturate is present.
$R = 90$ to 98, a fast acting barbiturate is present.

INTERPRETATION

The criteria for the identification of barbiturates are as follows (see Fig. 21–4):
In pH 9.9 solution—a maximum absorbance at 238 to 240 nm.
In pH 14 solution—a maximum absorbance at 252 to 255 nm.
In pH 14 solution—a minimum absorbance at 234 to 237 nm.
Isosbestic points*—227 to 230 nm and 247 to 250 nm.

Dilute solutions of barbiturates may not give all of these characteristic points, and the absorption peaks of salicylates and sulfonamides may obscure some of them; however, the phosphate wash should remove most of these interferences.

The value of R is valid only if a single barbiturate is present. For example, if a mixture of slow and fast acting barbiturates is present, R may indicate an intermediate acting barbiturate.

Blood barbiturate levels must be interpreted cautiously for the following reasons: (1) there is variation in the response of different individuals to a given dose of any drug; (2) if the blood level is rising, a given concentration of barbiturate has a more profound

* Isosbestic points are points on a spectral curve at which two substances have equal absorbance at the same wavelength.

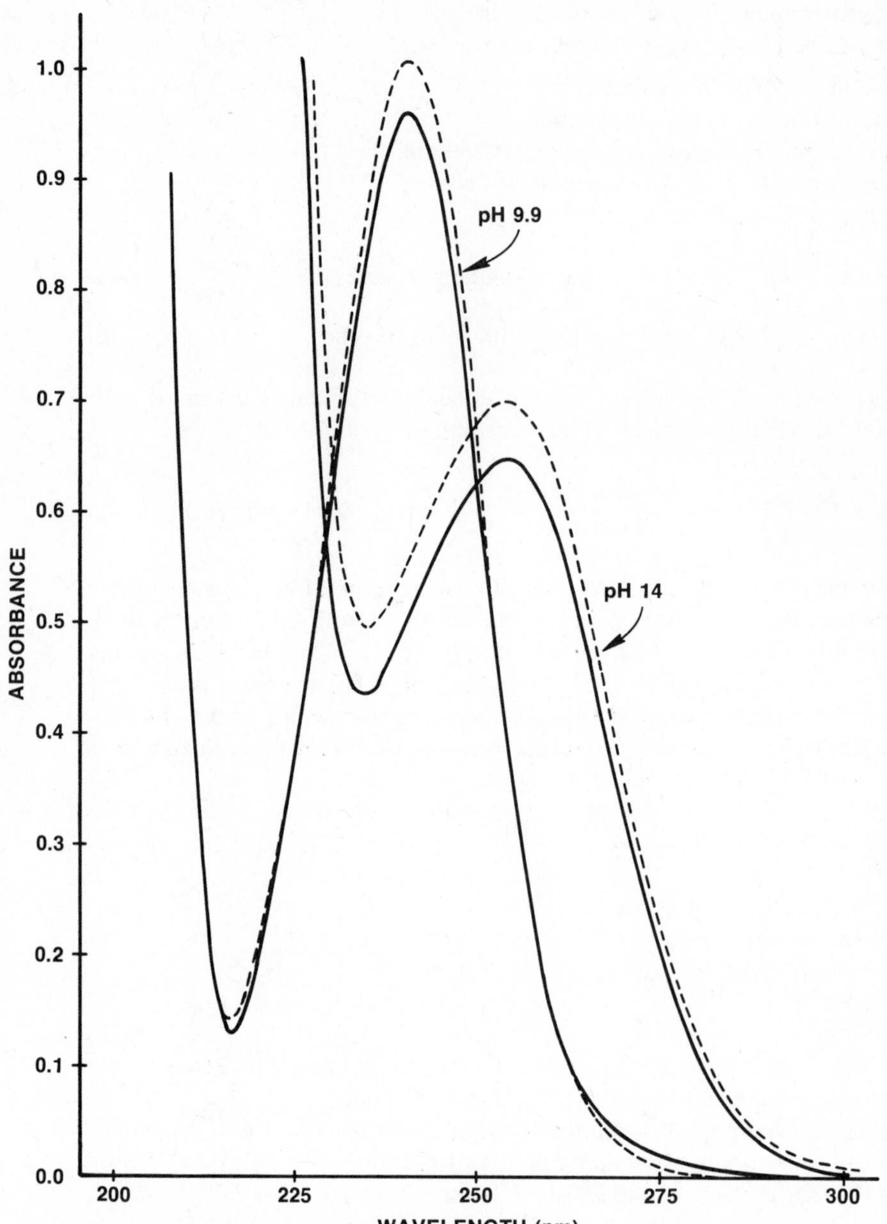

Figure 21–4 Typical ultraviolet absorption curve of secobarbital, a fast acting barbiturate. Dashed curve, alkaline extract; solid curve, hydrolyzed alkaline extract.

effect than if the blood level is falling; (3) if other depressant drugs, particularly alcohol, are present in addition to the barbiturates, the effects observed in the patient will be more severe; (4) some individuals, who are tolerant to or addicted to barbiturates, may have high blood levels without serious effects.

Table 21–5 shows approximate blood barbiturate levels at which coma disappears (falling blood levels). At levels higher than those indicated in Table 21–5, the prognosis is poor since fatalities have occurred at these levels. Except in epileptics treated with phenobarbital, normal therapeutic blood levels rarely rise higher than 0.1 to 0.2 mg/100 ml.

REFERENCE

Broughton, P. M. G.: Biochem. J., *63*:207, 1956.

ALTERNATIVE METHODS FOR THE IDENTIFICATION OF BARBITURATES

Numerous ultraviolet spectrophotometric methods have been published.[7,57] Barbiturates may be identified by infrared spectrophotometry,[62] gas chromatography,[23] paper chromatography,[12] and thin-layer chromatography.[7,56]

GLUTETHIMIDE (DORIDEN)

Glutethimide, a nonbarbiturate hypnotic, has been prescribed frequently in recent years. As a result, its use in attempted and successful suicides, as well as its abuse, have been reported with increasing frequency.

Chemically this drug is an imide and therefore is a very weak acid. Its acidic property is so weak, however, that it is usually extracted in the neutral fraction.

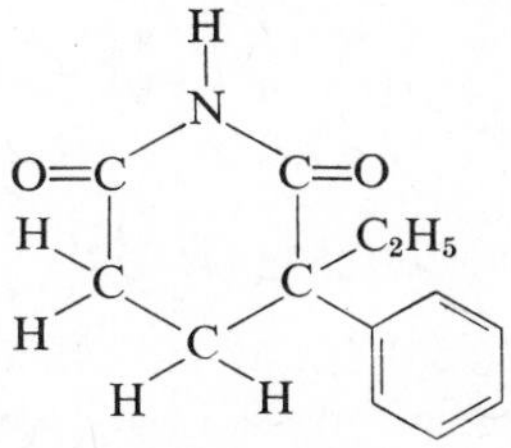

Glutethimide
2-ethyl-2-phenylglutarimide

Glutethimide has an absorption maximum in alkaline solution at 235 nm. Unfortunately, it is very unstable in alkaline solution and is easily hydrolyzed to the substituted monoamide of glutaric acid, thereby losing its ultraviolet absorbing property:

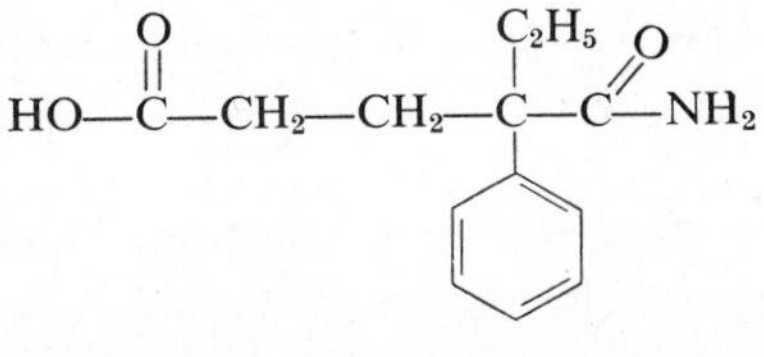

$$HO{-}\overset{O}{\overset{\|}{C}}{-}CH_2{-}CH_2{-}\overset{C_2H_5}{\overset{|}{C}}(C_6H_5){-}\overset{O}{\overset{\|}{C}}{-}NH_2$$

Goldbaum[32] utilized this property of glutethimide and, under rigidly controlled conditions, measured the rate of hydrolysis by plotting ultraviolet absorption at 235 nm against time. By extrapolating back to zero time, it was possible to calculate the original concentration of the drug. This method, with modifications, was the only good procedure available for many years. It is somewhat time-consuming, however, and requires some experience in order to obtain reproducible results. Korzun *et al.*[36] reported a simple chromatographic procedure to identify glutethimide in blood. This procedure is rapid, does not require any elaborate equipment, is sensitive, and uses 2.0 ml of blood. It is capable of producing semiquantitative results by visually comparing the spot size of the unknown drug in the blood with the spot size of a known graded series of glutethimide standards.

INTERPRETATION

Therapeutic blood levels of glutethimide range up to about 0.5 mg/100 ml. Levels higher than this usually indicate that an overdose has been ingested. Deep unconsciousness or coma may be produced at levels of 3.0 mg/100 ml, although higher levels than this have been reported. Blood levels at death have been reported between 3 and 20 mg/100 ml.

Frequently, patients may have several central nervous system depressant drugs prescribed. The possibility of overdosage of drug combinations thus presents itself, and interpretation of analytical results is very difficult in these situations. Combinations of glutethimide and alcohol or glutethimide and barbiturates are most common, but almost every type of combination has been encountered. As a general rule, if the depth of coma cannot be reasonably explained by the analytical results, a drug combination should be suspected. If drugs available to the patient are not known, alcohol should be ruled out followed by barbiturates, chlordiazepoxide, meprobamate, and phenothiazine derivatives.

PHENOTHIAZINE DERIVATIVES

Presently, there are a number of tranquilizers and antihistamines in clinical use that have in common the phenothiazine ring structure:

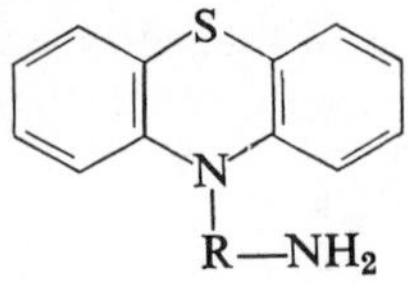

S

N

R—NH_2

DETECTION OF PHENOTHIAZINE DERIVATIVES

These substances can be detected easily in urine by the use of "FPN (ferric-perchloric-nitric) reagent." Although not specific for this ring structure, the test is sensitive, and can detect the drug at therapeutic levels. It is a useful screening test since false negative results do not occur. Elevated urinary levels of bile metabolites, particularly urobilinogen, yield color reactions similar to the drugs. Phenylpyruvic acid and *p*-aminosalicylic acid, as well as natural conjugated or synthetic estrogens, may give false positive results. These false positive reactions are generally of very weak intensity.

REAGENTS

1. Ferric chloride, 5 g/100 ml. Not stable, use immediately to prepare FPN reagent.
2. Perchloric acid, 2 mol/l.
3. Nitric acid, 5 mol/l.
4. FPN reagent. Mix 5.0 ml of the ferric chloride, 45 ml of the perchloric acid, and 50 ml of the nitric acid solutions. The reagent is stable.

PROCEDURE

Add 1.0 ml of FPN reagent to 1.0 ml of urine. The color is observed immediately. The stability of the color varies from 10 s at low dosages up to several minutes for high dosages.

INTERPRETATION

With a drug intake of 5 to 20 mg/day, a light pink-orange is observed. A drug intake of 25 to 70 mg/day produces more intense shades of pink. Doses of 75 to 120 mg/day produce gradually more intense shades of violet, and with doses above 125 mg/day deep purple colors are noted.

The value of this test is that a negative result effectively eliminates this group of drugs as a factor in diagnosing the condition of the patient. If the test is positive, several possibilities are suggested. First, the original author's interpretation (see preceding paragraph) is valid.[26] Second, if considerable time has elapsed between possible drug ingestion and collection of the urine specimen (e.g., 24 to 48 h), an overdosage is indicated since a therapeutic dose would have been excreted in this time period. Finally, the presence of an interfering substance is possible. To resolve these possibilities, one must use ultraviolet spectrophotometry[7] or thin-layer chromatography.[8]

REFERENCE

Forrest, I. S., and Forrest, F. M.: Clin. Chem., *6*:11, 1960.

SALICYLATES

Aspirin is responsible for more cases of accidental poisonings in children than any other substance. This extremely useful analgesic is so widely used and readily available (and carelessly handled) that children frequently ingest a toxic quantity by eating the flavored tablets like candy or mimicking adults. Toxic doses of salicylates initially produce a stimulation of the central nervous system. This may be reflected by hyperventilation, flushing, and fever. Unfortunately, an unrecognized case of salicylate poisoning may be thought to be a case of infection and further aspirin given in a vain attempt to control the fever. Central nervous system stimulation is followed by depression.

A complex disturbance of acid-base balance results from severe hyperventilation. Initially a respiratory alkalosis occurs, but this may be followed, especially in infants, by a metabolic acidosis. The net effect may be a decrease in the blood pH.

DETERMINATION OF SALICYLATES IN BIOLOGICAL FLUIDS

The procedure to be described for determining salicylate in urine, serum, or other specimen[60] is based on the formation of a violet colored complex between ferric iron and phenols.

$$C_6H_4(OH)(COOH) + Fe^{3+} = \text{Violet color complex}$$

This test is not specific for salicylates, but false negative results do not occur. The color developing solution contains acid and mercuric ions to precipitate protein.

REAGENTS

1. Color reagent. Dissolve 40 g of mercuric chloride, AR, in 850 ml of water by heating. Cool the solution and add 120 ml of 1 molar HCl and 40 g of ferric nitrate ($Fe(NO_3)_3 \cdot 9\,H_2O$). When all the ferric nitrate has dissolved, dilute the solution to 1 liter with water. It is stable indefinitely.
2. Salicylate standard, stock. Dissolve 580.0 mg of sodium salicylate (500 mg salicylic acid) in water and dilute to 250 ml. Add a few drops of chloroform as a preservative. This solution contains 2.0 mg salicylic acid/ml. Store in refrigerator; it is stable for about 6 months.
3. Working standard. Dilute 25.0 ml of stock salicylate solution to 100.0 ml with water. Add a few drops of chloroform as a preservative. This solution contains 0.5 mg of salicylic acid/ml. Store in refrigerator; the solution is stable for about 6 months.

PROCEDURE FOR SALICYLATE DETERMINATION IN CEREBROSPINAL FLUID, SERUM, OR WHOLE BLOOD

1. Pipet 1.0 ml of specimen into a centrifuge tube and add 5.0 ml of color reagent while shaking the tube. Continue shaking until the precipitate is finely dispersed.
2. Centrifuge at 2000 rpm for 2 minutes or filter through Whatman No. 42 filter paper.
3. Transfer the clear supernatant or filtrate to a 1.0 cm cuvet and read at 540 nm

against a blank consisting of 1.0 ml water mixed with 5.0 ml of color reagent. The color is stable for one hour.

4. Prepare a series of standards as follows:

	A	*B*	*C*	*D*	*E*
ml Working standard	0.2	0.4	0.6	0.8	1.0
ml Water	0.8	0.6	0.4	0.2	0.0
mg Salicylate/100 ml	10.0	20.0	30.0	40.0	50.0

Run these standards as described above (steps 1 to 3), reading the absorbances at 540 nm against the reagent blank, and plot a calibration curve. Beer's law is followed over this range.

5. If the unknown absorbance is greater than 0.7, repeat the analysis using a smaller portion of specimen diluted to 1.0 ml with water.

PROCEDURE FOR SALICYLATE DETERMINATION IN URINE

1. Follow the procedure as outlined under serum. If the urine contains more than 50 mg/100 ml salicylic acid, make an appropriate dilution of urine with distilled water and repeat the test.

2. After reading the absorbance of the unknown, obtain a sample blank reading by setting the instrument with water and reading the absorbance of a solution prepared by mixing 1.0 ml of urine or diluted urine with 5.0 ml of color reagent and 0.1 ml of syrupy phosphoric acid (sp. gr. 1.75), using the same cuvets as before. Urine solutions may not require centrifuging. If it is necessary to clarify the solutions, centrifuge both unknown and urine sample blank.

CALCULATION

Urine salicylic acid (mg/100 ml) = (mg/100 ml in diluted unknown − mg/100 ml in diluted sample blank) × dilution factor.

INTERPRETATION

Blank values for serum, cerebrospinal fluid, and plasma are less than 1.1 mg/100 ml as salicylic acid. For blood, the blank is less than 2.0 mg/100 ml and for urine it is less than 4.5 mg/100 ml. Recoveries of added salicylate are quantitative and the following substances in the indicated concentrations do not interfere: phosphate (100 mg/100 ml), bilirubin (20 mg/100 ml), phenol (25 mg/100 ml), heparin (10,000 U), glucose (1000 mg/100 ml), and urea (1000 mg/100 ml). Acetoacetic acid forms a pink color with ferric iron, and at a level of 50 mg/100 ml gives a value of 1 mg/100 ml as salicylate. The procedure can easily be adapted to micro- or ultramicroscale, a useful feature in pediatric cases.

Therapeutic levels of salicylic acid rarely rise above 20 mg/100 ml in blood or serum. Above 30 mg/100 ml, toxic symptoms such as headache, tinnitus, flushing, and hyperventilation may be seen. Serum electrolytes should be followed and any imbalance corrected. Lethal salicylate levels are usually greater than 60 mg/100 ml.

REFERENCE

Trinder, P.: Biochem. J., *57*:301, 1954.

DRUGS DETERMINED BY DERIVATIVE FORMATION

Some drugs require special techniques for their detection. It is beyond the scope of this chapter to list special techniques in detail, but an interesting approach to this problem has been the work of Wallace and co-workers.[65] This approach has been applied to drugs that can easily be extracted or separated from biological specimens, but which have weak ultraviolet absorption maxima and can not be readily detected. Wallace and his group found ways in which these drugs can be converted to derivatives that have strong, well defined absorption maxima. The drugs can therefore be determined indirectly by ultraviolet spectrophotometry of their derivatives.

In general, the procedures can be outlined as follows:

Separation of drug from specimen (extraction or steam distillation)	→	Derivative formation (usually an oxidation reaction)	→	
Separation of derivative (extraction or steam distillation)	→	Concentration if necessary (vacuum evaporation)	→	Spectrophotometric scan

Drugs for which this approach has been successfully used are: diphenylhydantoin, phenylbutazone, amphetamine, propoxyphene, ephedrine, phenaglycodol, ethchlorvynol, and imipramine. Two examples of this general technique will be given here. One, the determination of imipramine or its analogs,[64] involves oxidation of the extracted drug to an intensely colored derivative which can be measured by a colorimetric method. The other, a method for propoxyphene,[42,66] hydrolyzes the dιug to a derivative which, after exposure to ultraviolet light, is converted to a product with high ultraviolet absoιbance.

DETERMINATION OF IMIPRAMINE

PRINCIPLE

The antidepressant drug imipramine is strongly protein bound and never reaches very high levels in blood and serum. For these reasons, it has been a problem drug from the analytical point of view. Its primary metabolite, desipramine, has been found to be active and is now marketed as a separate drug. A third analog is trimipramine. The tricyclic structure of these drugs was found to respond to oxidizing agents by the formation of an intense blue color.[64] This property has been utilized by several investigators for qualitative and quantitative detection of these drugs. Although there is some speculation as to the mechanism of this reaction,[16] the precise nature of the oxidation product is not known.

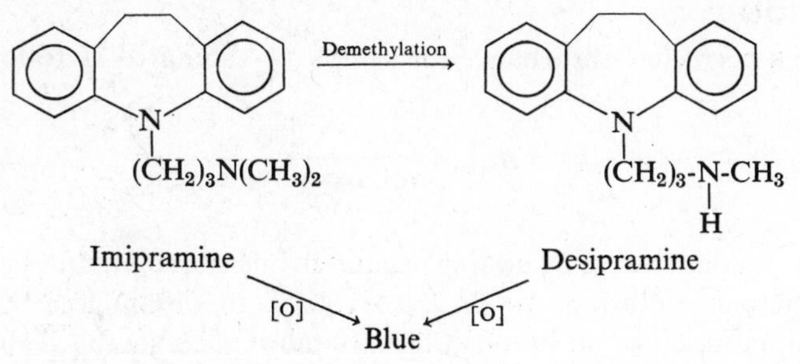

Demethylation

N
$(CH_2)_3N(CH_3)_2$

Imipramine

N
$(CH_2)_3$-N-CH_3
H

Desipramine

[O] → Blue ← [O]

REAGENTS

1. Hydrochloric acid, concentrated, AR.

2. Potassium hydroxide, 60 g/100 ml. Cautiously dissolve 60.0 g of potassium hydroxide, AR, in about equal weight of water. Cool the solution to room temperature. Dilute to 100 ml with water.

3. *n*-Hexane, spectral grade.

4. Sulfuric acid, 3 mol/l. Cautiously add 168 ml of concentrated sulfuric acid, AR, to 600 ml of water. Cool the solution to room temperature, then dilute it to 1 liter with water.

5. Sulfuric acid, 0.5 mol/l. Cautiously add 28.0 ml of concentrated sulfuric acid, AR, to 100 ml of water. Cool the solution to room temperature, then dilute it to 1 liter with water.

6. Cerium (IV) sulfate, 200 mg/100 ml. Dissolve 200 mg of cerium sulfate in 0.5 molar sulfuric acid. Dilute the solution to 100 ml with additional 0.5 molar sulfuric acid.

7. Color-developing reagent. Cautiously mix 30.0 ml of the cerium sulfate solution and 70 ml of concentrated sulfuric acid, AR.

8. Imipramine stock solution, 100 mg/100 ml. Dissolve 113 mg of imipramine hydrochloride (Geigy Pharmaceuticals) in water and dilute to 100 ml.

9. Imipramine reference solutions. Dilute 1.0, 5.0, and 10.0 ml of the imipramine stock solution to 100 ml with water to obtain solutions containing 10, 50, and 100 μg imipramine per ml, respectively.

PROCEDURE

1. In a 50 ml test tube, mix 10 ml of blood with 2 ml of water and 8 ml of concentrated hydrochloric acid. Heat the mixture in a boiling water bath for not longer than 5 min.

2. Remove the tube from the water bath and cool it in an ice bath.

3. When cool, make the mixture distinctly alkaline by adding 12 ml of potassium hydroxide solution; then transfer it to a 500 ml separatory funnel (check alkalinity with pH paper).

4. Add 200 ml of *n*-hexane and shake the mixture for 3 min. Allow the layers to separate (centrifuge if necessary).

5. Filter the organic phase through Whatman No. 1 filter paper into a graduated cylinder. Note the volume of recovered hexane.

6. Transfer the hexane to a 250 ml separatory funnel and wash it with 10 ml of water. Discard the water wash.

7. Add 5 ml of the 3 molar sulfuric acid and extract the washed *n*-hexane for 3 min.

8. Separate the acidic layer and centrifuge it.

9. Pipet 1.5 ml of the acidic extract into a test tube and add 1.5 ml of the color-developing reagent. Read within 2 h.

10. Measure the absorbance of the resulting blue color at 620 nm, using a mixture containing equal volumes of 3 molar sulfuric acid and color-developing reagent as the reference solution.

CALCULATION

1. Calculate a corrected absorbance for sample or standards as follows:

$$A_{\text{corr.}} = A_{620} \times \frac{200}{\text{vol. recovered hexane}}$$

2. Prepare a standard curve by adding accurately measured volumes of the imipramine hydrochloride reference solutions to drug-free blood to obtain final concentrations of 1 μg to 10 μg imipramine per ml blood. Carry 10 ml of each standard through the above procedures.

3. Determine the imipramine concentration in the specimen by comparing its corrected absorbance with the standard curve.

COMMENTS

The blue color produced by imipramine in this procedure is stable for 2 h, provided the color-developing reagent is stabilized with concentrated sulfuric acid. Recovery of the drug added to specimens is about 90 per cent for blood, 75 per cent for urine, and 85 per cent for homogenized tissue. As mentioned earlier, this method does not distinguish between imipramine, desipramine and trimipramine. At high concentrations thioridazine will interfere, but a large number of other similar drugs have been shown not to interfere.

INTERPRETATION

Therapeutic levels of imipramine range from 0.05 to 0.6 $\mu g/ml$ of blood. Toxic ingestions may result in blood levels of 1 to 5 $\mu g/ml$. Liver levels in fatalities are 5 to 10 times greater than blood levels. Since renal failure often occurs in overdose cases with this drug, urine levels may be surprisingly low.

REFERENCES

Wallace, J. E., and Biggs, J. D.: J. Forensic Sciences, *14*:528, 1969.
Cimbura, G.: *In* Manual of Analytical Toxicology. I. Sunshine, ed. Cleveland, Ohio, Chemical Rubber Company, 1971.

DETERMINATION OF PROPOXYPHENE

PRINCIPLE

Propoxyphene exists in two optically active forms. The dextrorotatory form is an analgesic marketed as Darvon, while the levorotatory form is available as Novrad, an antitussive. The isomers cannot be differentiated by usual chemical means; therefore, levels are reported as propoxyphene. Quantitative methods for assaying blood and tissue levels of this drug have been plagued by many difficulties. Unlike most organic bases, the hydrochloride salt of propoxyphene is soluble in certain organic solvents. Normal metabolism of this drug converts it to norpropoxyphene. Although this metabolite would not be expected to differ greatly from the parent drug, strong alkalinization of norpropoxyphene causes it to rearrange, forming norpropoxyphene amide. Once the amide is extracted from an alkaline medium, it cannot be re-extracted from the organic solvent by acid solutions.[43] These factors, plus protein binding of the drug, have contributed to the difficulties associated with assays of propoxyphene in biological specimens.

$CH_3CH_2COOC(C_6H_5)(CH_2C_6H_5)—CH(CH_3)—CH_2N(CH_3)_2$ —Demethylation (Metabolism)→ $CH_3CH_2COOC(C_6H_5)(CH_2C_6H_5)—CH(CH_3)—CH_2NH—CH_3$

Propoxyphene — Norpropoxyphene

↓ OH^-, pH 11

$HO—C(C_6H_5)(CH_2C_6H_5)—CH(CH_3)CH_2N(CH_3)—C(=O)—C_2H_5$

Norpropoxyphene amide

The procedure to be described circumvents many of these problems by subjecting the specimen initially to strong acid hydrolysis. This not only breaks some drug-protein binding sites but converts propoxyphene and its metabolites to forms which are easily extracted and have increased ultraviolet absorbing properties.

$$CH_3CH_2COOC(C_6H_5)(CH_2C_6H_5){-}CH(CH_3){-}CH_2N(CH_3)_2 \xrightarrow[\text{Heat}]{\text{Acid}} (C_6H_5)(C_6H_5CH_2)C{=}C(CH_3){-}CH_2N(CH_3)_2 \rightleftharpoons C_6H_5CH{=}C(C_6H_5){-}CH(CH_3)CH_2N(CH_3)_2$$

Propoxyphene

The products of the above reaction are subsequently treated with ultraviolet irradiation, which further increases their ultraviolet absorbing properties.

REAGENTS

1. Hydrochloric acid, concentrated, AR.

2. Hydrochloric acid, 0.2 mol/l. Dilute 16.5 ml of concentrated HCl to 1.0 liter with water.

3. Sodium hydroxide, 12.5 mol/l. Cautiously add 500 ml of water to 500 g of NaOH pellets, AR, in a 2.0 liter beaker. Stir with a glass rod, cool, and dilute to 1.0 liter with water. Store in a polyethylene bottle.

4. Diethyl ether, AR.

5. Propoxyphene standard. (Note: 2-naphthalene sulfonic acid in propoxyphene napsylate may interfere in this method and should not be used as a standard.)

(a) Stock solution, 500 μg/ml. Dissolve 55.4 mg of propoxyphene HCl (Eli Lilly and Co.) in water and make up to 100.0 ml with water.

(b) Working standard, 5 μg/ml. Dilute 1.00 ml of stock solution to 100.0 ml with water.

SPECIMEN REQUIREMENTS

Owing to low blood levels of propoxyphene, 10 ml of blood are required. The choice of anticoagulant is not critical, but if clotted blood is submitted, the clot and serum must be homogenized.

PROCEDURE

1. Into a 125 ml Erlenmeyer flask labeled "patient," pipet 10.0 ml of the patient's blood. Into a separate 125 ml Erlenmeyer flask labeled "standard," pipet 10.0 ml of propoxyphene working standard.

2. Add 15 ml of concentrated HCl to both flasks and heat them in a boiling water bath for 20 min with occasional stirring.

3. Cool the flasks in an ice-water bath and add 15 ml of 12.5 molar NaOH to each flask. Check the pH with indicator paper to ensure that alkaline conditions prevail. Cool thoroughly.

4. Transfer the contents of each flask to similarly labeled 250 ml separatory funnels.

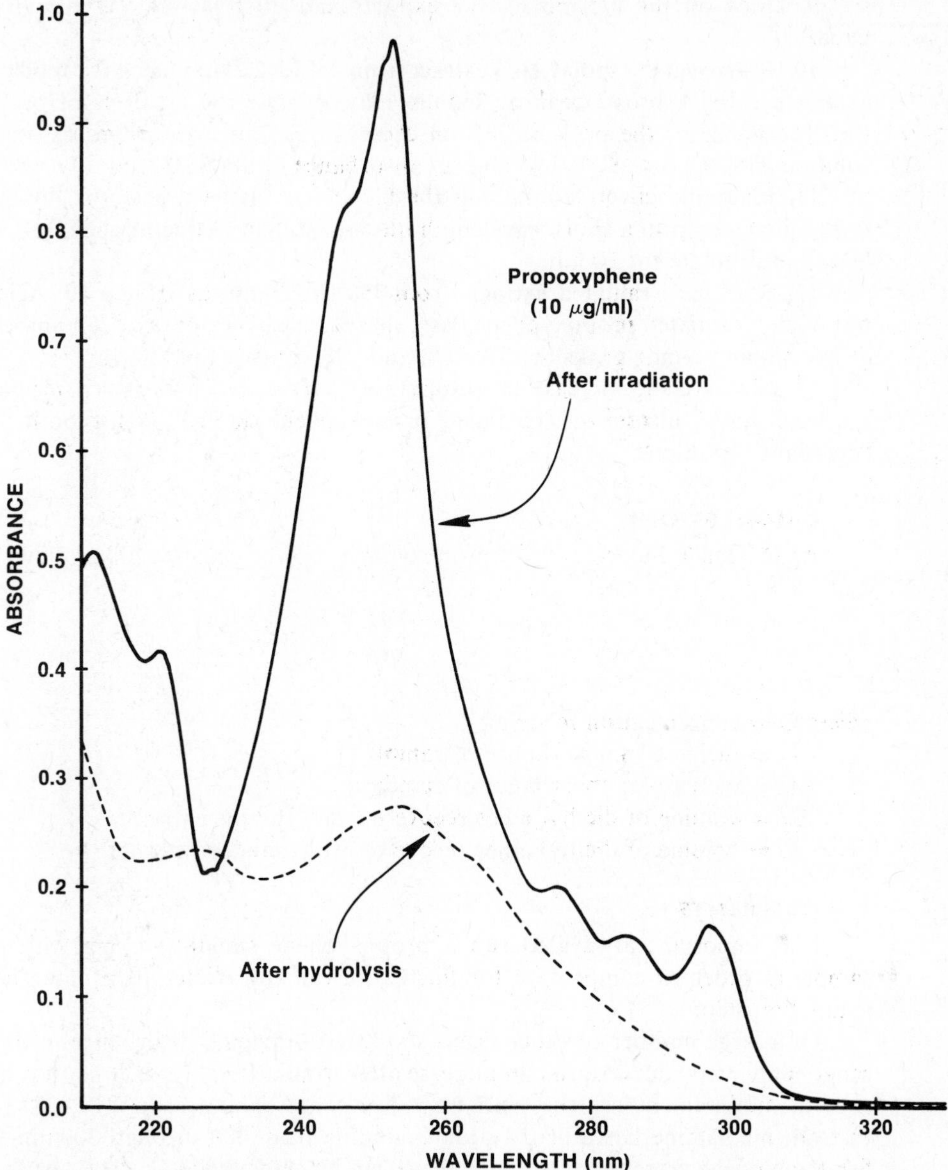

Figure 21–5 Ultraviolet absorbance characteristics of the hydrolysis product (dashed curve) and ultraviolet irradiated product (solid curve) of equivalent amounts of propoxyphene.

Rinse each flask with a few ml of water from a wash bottle, adding the washings to the separatory funnels.

5. Add 150 ml of diethyl ether to each separatory funnel and shake vigorously for 1 min. After separation of the phases, draw off and discard the lower aqueous phase in each funnel.

6. Add 10 ml water to the ether phase in each funnel. Shake vigorously for 1 min. Draw off and discard the aqueous washes.

7. Filter the ether from each funnel through Whatman No. 1 filter paper into graduated cylinders labeled "patient" and "standard." Note and record the volume of ether recovered.

8. Transfer the ether extracts to clean, labeled 250 ml separatory funnels. Add 5.0 ml of 0.2 molar HCl to each funnel and shake the funnels vigorously for 1 min.

9. Draw off the 0.2 molar HCl extracts and filter through Whatman No. 41 filter paper.

10. Scan each 0.2 molar HCl extract from 350 to 220 nm using 0.2 molar HCl solution as a reference. A broad peak at 255 nm may indicate the presence of propoxyphene. If the absorbance of the peak at 255 nm exceeds 0.3 (rare with blood extracts), dilute the solution with 0.2 molar HCl to give an absorbance of about 0.3.

11. Place the cuvets containing the 0.2 molar HCl extracts (or diluted extracts) a few millimeters from a short wavelength ultraviolet light (Mineralight UVSL-25 or equivalent) and irradiate for 30 min.

12. Scan the irradiated extracts from 350 to 220 nm as in step 10. Acid hydrolyzed, ultraviolet irradiated propoxyphene has a sharp absorbance peak at 253 nm with a shoulder at 247 nm and minor peaks at 277, 287, and 298 nm (see Fig. 21–5).

13. Measure the increase in absorbance (ΔA) of the hydrolyzed propoxyphene and the hydrolyzed, ultraviolet irradiated propoxyphene at 253 nm for both "patient" and "standard" solutions.

CALCULATIONS

Approximate blood levels of propoxyphene can be calculated as follows:

$$C = \frac{\Delta A_p \times E_s}{\Delta A_s \times E_p} \times 5$$

where C = concentration in μg/ml
ΔA_p = increase in absorbance of sample
ΔA_s = increase in absorbance of standard
E_p = volume of diethyl ether recovered with sample extract
E_s = volume of diethyl ether recovered with standard extract

COMMENTS

It is important always to run a propoxyphene standard along with the unknown sample in order to compensate for fluctuations of ultraviolet light intensity during the irradiation step.

Of a large number of other drugs showing ultraviolet absorbance in this wavelength range, only methadone shows an increase after irradiation. Methadone, however, does not show a minimum before the peak at 253 nm, nor shoulders at 277, 287, and 298 nm. Instead, methadone has a broad peak extending from 280 through 300 nm. Further confirmation of the presence of propoxyphene can be made by TLC, GLC, or spectrofluorimetry. The original article should be consulted for these details.[42]

INTERPRETATION

Therapeutic administration of propoxyphene results in blood levels of 30 μg/100 ml or less. Fatalities resulting from propoxyphene overdose show blood levels greater than 200 μg/100 ml.

REFERENCE

McBay, A. J., Turk, R. F., Corbett, B. W., and Hudson, P.: J. Forensic Sciences, *19*:81–89, 1974.

IMMUNOASSAY TECHNIQUES

The application of immunoassay procedures to the detection of drugs in biological specimens has opened an exciting new field in toxicological analysis. Most wet chemical or instrumental methods for drug detection require prior separation of the drug from the specimen by extraction. This is the weakest link in the chain of steps in the analytical procedure. Even when spiked specimens show essentially one hundred per cent recovery of the added drug, this does not duplicate a specimen from an overdosed patient in which the drug is distributed between erythrocytes and plasma, between lipid and polar fractions, or between free and bound forms of the drug. Moreover, extraction procedures frequently separate the drug together with interfering substances.

The remarkable sensitivity and specificity of immunoassay procedures enables the investigator to add serum or urine directly to an appropriate medium containing the necessary components for the reaction, and then read the final end point. Only 20 to 100 μl of specimen are used and, since few manipulations are involved, the methods are easily adaptable to automation for screening large numbers of specimens.

Substances that can be determined by immunoassay include steroids, vitamins, hormones, and cardiac glycosides. Of these, insulin and cardiac glycoside overdoses are likely to be the only toxicological problems encountered; for practical purposes, they can be detected in serum by immunoassay. Among the more commonly encountered drugs, methods are now available for morphine[51] (and opiates in general), barbiturates,[52] amphetamines, methadone, and lysergic acid diethylamide (LSD).[58]

Many approaches to practical utilization of immunoassay procedures have been made. All are useful in toxicology, and the choice of which approach to use depends on such factors as availability of equipment, workload or frequency of testing, and cost.

Radioimmunoassay (RIA) employs the principle of isotopic dilution as the end point measurement. The disadvantages of this approach are the initial cost of equipment, the necessity of handling radioactive materials, and the increased cost of trained personnel. These may be offset by the advantage that there are a great many other clinically useful applications of RIA, enabling the same personnel and equipment to perform toxicological and clinical procedures. The general principles of RIA have been discussed in Chapter 7.

Spin immunoassay (SIA) is similar to RIA, except that spin-labeled (free-radical-labeled) antibody is used.[39] Thus, an electron spin resonance spectrometer can be used to measure the end point. Equipment cost is still high, but since the use of radioisotopes is avoided, other problems are minimized. Results are obtained quickly.

In SIA, the drug is labeled with a free radical such as nitroxide. The antibody–spin-labeled-drug complex produces very broad spectral peaks in an electron spin resonance spectrometer, owing to the immobilization of the free radical at the antibody sites. Once the labeled drug has been displaced by free drug in the test sample, an immediate sharp peak appears, reflecting the number of spin-labeled molecules now tumbling freely in solution. Thus, SIA gives an immediate result with fewer manipulations than RIA.[39]

Hemagglutination-inhibition assay (HIA) is still another approach.[1] No expensive equipment or special training are necessary for this method, so costs are low. The necessity for a long incubation step, which lengthens turn-around time for a single analysis, and the availability of fewer clinical applications are possible disadvantages.

In HIA, a small quantity of test sample is mixed first with antibody and subsequently with antigen (drug) coated erythrocytes. In the presence of free drug in the test sample, the added antibody binds preferentially and completely with the free drug. Thus, no antibody is available to react with the coated erythrocytes, leaving them to settle as a pellet in the bottom of the microtiter cup. If there is no drug in the sample, the antibody can bind with the coated erythrocytes (= agglutinate), causing them to adhere to each other and to the sides of the cup in a diffuse pattern. Test samples are compared with positive and negative controls. The method requires a 90 min incubation period at room temperature but is simple, sensitive, and semiquantitative.[1]

It should be emphasized that, despite the excellent sensitivity of the methods described, they are not absolutely specific. Thus, a method designed to detect morphine will respond to codeine and other opiates. When absolute identification is required, additional confirming tests must be done. Nevertheless, these methods are valuable as screening procedures and hold promise of development into more specific tests for a broad spectrum of drugs.

REFERENCES

RIA
Spector, S.: Ann. Rev. Pharmacology, *13*:359–370, 1973.
SIA
Lente, R., Ullman, E. G., and Goldstein, A.: J. Amer. Med. Assoc., *221*:1231–1234, 1972.
HIA
Adler, F. L., and Liu, C-T.: J. Immunology, *106*:1684–1685, 1971.

DETERMINATION OF DIGOXIN BY RIA

Until the advent of RIA, toxicity of cardiac glycosides could be judged only by the clinical state of the patient. Serum levels could not be measured. Even in cases of death following ingestion of a large overdose of these drugs, demonstration of toxic levels in tissues was a formidable undertaking. Thus, the use of RIA has opened new approaches to analytical toxicology.

The procedure described uses the RIA kit for digoxin supplied by Schwarz/Mann (Division of Becton, Dickinson, and Co., Orangeburg, N.Y. 10962) and contains digoxin[^{3}H], digoxin antiserum, digoxin standard solution, and UNOGEL sufficient for 240 tubes. The volume of radioactive digoxin is sufficient to include it as an internal standard in all serum assays. Corrections can then be made for quenching owing to components of serum samples.

PRINCIPLE

The assay involves the competition between ^{3}H-labeled digoxin and circulating digoxin for binding by the antibody to digoxin. The digoxin-antibody complex which forms at room temperature during an incubation time of 30 min has a level of radioactivity which is inversely related to the concentration of digoxin in the blood serum or standard.

The separation of antibody-bound digoxin from free digoxin is achieved by adsorption of the free digoxin, both labeled and unlabeled, on dextran-coated charcoal. After centrifugation, the bound radioactivity is determined upon decantation of the supernatant into Schwarz/Mann UNOGEL. The supernatant fluid is totally soluble in UNOGEL.

A standard curve is prepared when each group of serum samples is analyzed. The curve is obtained by incubating known levels of digoxin with a constant amount of antibody and a specific amount of tritiated digoxin. The percentage of digoxin[^{3}H] bound is plotted against nanograms digoxin/ml serum on semi-logarithmic paper, with ng/ml as the logarithmic function. A linear curve is obtained in the range from 1 to 10 ng. Although levels of digoxin are plotted per ml serum, the technique described requires only 200 μl serum per assay.

This protocol is a modification of the procedure for the radioimmunoassay of digoxin described previously by Smith, Butler, and Haber.

REAGENTS

1. Blood serum or plasma. Serum or plasma is used in the preparation of the standard curve. It should be free of radioactivity and of digoxin. Store in a refrigerator.

2. Phosphate buffered saline. Dissolve 1.392 g of potassium phosphate, K_2HPO_4, AR; 0.276 g of sodium phosphate, $NaH_2PO_4 \cdot H_2O$, AR; and 8.77 g of sodium chloride, NaCl, AR, in about 900 ml of distilled water. If necessary, add 0.003 molar phosphoric acid or 0.01 molar KOH to bring the solution to pH 7.4. Add water to make 1000 ml, giving 0.01 molar phosphate in 0.15 molar NaCl, pH 7.4. Refrigerate.

3. Dextran-coated charcoal (DCC) stock suspension. Dissolve 62.5 mg of dextran, radioimmunoassay grade, in 25 ml of phosphate buffered saline (reagent 2). Use a magnetic stirrer. Add 2.5 g of Norit A (neutral) charcoal. Mix well. Refrigerate. Remix thoroughly in the cold before use.

4. DCC working suspension. Resuspend reagent 3 as indicated above. Pipet 2 ml of reagent 3 into 18 ml of phosphate buffered saline (reagent 2). The working charcoal suspension, a 10–fold dilution of the stock suspension, is mixed as it is used by magnetic stirring.

5. UNOGEL, Schwartz/Mann.

OPERATING PROCEDURE

The assays are carried out in polystyrene test tubes. Polystyrene tubes have shown no affinity for the antibody under the conditions of the assay and also can be safely taken through the centrifugation step. In the following protocol, the tests are described in duplicate.

PREPARATION OF STANDARD CURVE

1. Into each of 16 numbered tubes, pipet 200 μl of serum or plasma (reagent 1). The tubes are kept at room temperature.
2. Add 800 μl of phosphate buffered saline (reagent 2) to each tube.
3. Add 10 μl of 95 per cent ethanol to tubes 1 to 4.
4. Add digoxin standard solution as follows:

Tube No.	*Digoxin Standard*	*Digoxin as* ng/ml *Serum*
5, 6	2 μl	0.4
7, 8	5 μl	1.0
9, 10	10 μl	2.0
11, 12	15 μl	3.0
13, 14	25 μl	5.0
15, 16	50 μl	10.0

5. Add 10 μl of digoxin[^{3}H] to tubes 1 to 16. Mix well.
6. Add 10 μl of digoxin[^{3}H] to 200 μl of serum and 1.3 ml of phosphate buffered saline (reagent 2) in each of two scintillation vials numbered 17 and 18.
7. Add 10 μl of digoxin antiserum to tubes 3 to 16.
8. Each tube is mixed well after each of the above additions. Shake the rack of tubes to mix all reagents thoroughly.
9. Incubate at room temperature for 30 min from the time of the last addition in step 7.
10. Add 0.5 ml of DCC working suspension (reagent 4) to tubes 1 to 16. The reagent is "squirted" into each tube to obtain a uniform suspension of charcoal in the reaction mixture.
11. Keep the tubes at room temperature for 10 min from the time of the last addition in step 10.
12. Centrifuge the tubes at about 5000 rpm in the cold for 20 min.
13. Decant the clear supernatant into 10 ml of UNOGEL using a numbered sequence of scintillation vials. Maximal transfer is obtained by striking the rims together. Cap each vial and shake well. Discard the charcoal residues.
14. The scintillations in the vials are counted in sequences for 1 to 10 min. Include in the sequence vials 17 and 18, to which 10 ml of UNOGEL has been added. The counts in these vials should be 1500 to 2500 cpm; these vials give the Total Count per assay. (See Calculations, step 2c.)
15. A standard curve is drawn on semi-logarithmic paper, showing percentage digoxin[^{3}H] bound against ng/ml serum. Calculations are described in a following section.

PROCEDURE

1. To each of two tubes, add 200 μl of the patient's serum or plasma. The tubes are kept at room temperature.
2. Add 800 μl of phosphate buffered saline (Reagent 2) to each tube.
3. Add 10 μl of digoxin[^3H] to each tube.
4. Add 10 μl of digoxin antiserum to each tube. Mix well.
5. Incubate at room temperature for 30 min.
6. Add 0.5 ml of DCC working suspension (reagent 4) to each tube. The reagent is "squirted" into each tube to obtain a uniform suspension of charcoal in the reaction mixture.
7. Keep the tubes at room temperature for 10 min from the time of the last addition in step 6.
8. Centrifuge the tubes at about 5000 rpm in the cold for 20 min.
9. Decant the clear supernatant into 10 ml of UNOGEL using a numbered sequence of scintillation vials. Maximal transfer is obtained by striking the rims together. Cap each vial and shake well. Discard the charcoal residues.
10. Count each vial for the same time period taken under "Preparation of Standard Curve," step 14.
11. If the liquid scintillation counter does not provide a system for automatic external standardization, corrections for quenching owing to components of the sera are made by adding 10 μl of digoxin[^3H], as an internal standard, to each sample for assay.
12. Shake each vial and recount in sequence for the same time period as before.

CALCULATIONS

1. The "Blank" or "Background" is the average count found in vials 1 and 2.
2. Under "Preparation of Standard Curve":

   a. The counts found in vials 3 to 16 are corrected by subtracting the "Blank" counts.

   b. The average of the counts found in vials 17 and 18, the Total Count per assay, is corrected by subtracting the "Blank" counts.

TABLE 21–7 TYPICAL COUNTING DATA IN DETERMINATION OF DIGOXIN BY RIA

Vial No.	*CPM*	*Corrected CPM*	*Ave. CPM*	*Percentage Bound*	*Digoxin ng/ml*
1	192			—	—
			184		
2	176			—	—
3	1245	1061		50.5	Trace
4	1245	1061		50.5	Trace
5	1134	950		45.3	0.4
6	1139	955		45.5	0.4
7	953	769		36.6	1.0
8	971	787		37.5	1.0
9	749	565		26.9	2.0
10	743	559		26.6	2.0
11	622	438		20.9	3.0
12	633	449		21.4	3.0
13	506	322		15.3	5.0
14	498	314		15.0	5.0
15	379	193		9.3	10.0
16	389	183		9.8	10.0
17	2293	2109		—	—
			2099		
18	2272	2089		—	—

c. $\text{Percentage Bound} = \dfrac{\text{Standard Count (Step 2a)}}{\text{Total Count (Step 2b)}} \times 100.$

d. The Percentage Bound for tubes 3 and 4 indicates the binding of digoxin[^{3}H] in the absence of unlabeled digoxin.

e. Plot Percentage Bound against ng/ml serum on semi-logarithmic paper, with ng/ml as the logarithmic function. Typical counting data and the calculated Percentage Bound are tabulated in Table 21–7.

Caution: The plotted data included in this protocol must not be employed by the user in lieu of a standard curve obtained by the user in his own laboratory with a Digoxin Kit.

3. Under "Procedure":

a. The counts found in step 10 are corrected for the "Blank" counts, as above.

b. The counts found in step 12, in which the internal standard was added, are corrected by subtracting the counts found in step 10 after subtracting the "Blank" counts. This gives the Total Sample Counts corrected for quenching.

c. $\text{Percentage Bound} = \dfrac{\text{Serum Sample Count (step 3a)}}{\text{Total Sample Count (step 3b)}} \times 100.$

d. Determine ng/ml serum from the standard curve.

COMMENTS

Under "Preparation of Standard Curve," the counts found in the vials for the Total Counts may vary in the range from 1500 to 2500 cpm. This count will depend upon the total efficiency of the scintillation counter in use. In turn, this efficiency will actually determine the counting time. A longer counting time is required for an instrument with a low efficiency. The counting time required for an accumulated count of 1500 to 2500 in the vials prepared for the Total Count will indicate the time required for all assays with the user's scintillation counter. For most instruments, this time will not exceed 10 min.

Since efficiency varies with the temperature of the scintillation counter, it is important that a uniform temperature be maintained during the counting sequence. Chilling of vials before counting is required for refrigerated units.

In the case of counters with systems for automatic external standardization, the user should follow the instructions contained in the operating manual supplied by the manufacturer of his counting system. With this type of instrument, an alternate method for plotting the data can be used. The reciprocal of the corrected bound cpm $\times 10^3$ can be plotted against concentration of unlabeled digoxin on square grid paper. A rectilinear relationship is obtained which lends itself to computer usage.

INTERPRETATION OF RESULTS

Smith, Butler, and Haber have defined digoxin toxicity in patients in terms of disturbances of impulse formation or conduction, followed by the disappearance of the rhythm disturbance when digoxin is withheld. The difference between the mean values for patients with nontoxic and toxic levels was found to be statistically significant. The mean value for serum digoxin concentrations was 1.1 ± 0.3 ng/ml (range, 0.8 to 1.6) for patients on nontoxic oral doses of 0.25 mg/d. A mean value of 1.4 ± 0.4 ng/ml (range 0.9 to 2.4) was found for patients on nontoxic doses of 0.50 mg/d. The mean value for patients with toxic levels was 3.3 ± 1.5 ng/ml (range, 2.1 to 8.7).

Knowledge of serum digoxin levels is particularly useful when an accurate history of dosage is not available, when the extent of absorption of orally administered doses is uncertain, and in cases of hemodynamically unstable patients with fluctuating renal function.

REFERENCE

Smith, T. W., Butler, Jr., V. P., and Haber, E.: New Eng. J. Med., *281*:1212, 1969.

MISCELLANEOUS SUBSTANCES

There are always those substances that cannot be categorized. Many toxic substances (e.g., snake and insect venoms) are complex mixtures that have not been completely characterized chemically. Others (e.g., cholinesterase inhibitors) act by inhibiting enzyme systems, and their effects can be more reliably estimated by determining enzyme activity. Many pesticides, some of the so-called "nerve gases," and some drugs fall into this category.

On the other hand, there are many substances that could be put into one or the other of the categories because of their chemical type, but for some other reason cannot be tested for in the usual way and must be sought by special, specific tests. Organic substances that are extremely inert or extremely insoluble are examples of this type.

Finally, the miscellaneous group includes the phenomenon of drug sensitivity or allergy. In this situation, an individual may succumb to a normal therapeutic dose of a drug, or less. In this case it is necessary to demonstrate that true sensitivity to a chemical agent did exist. For this the techniques and tools of serology are brought into play.

Fortunately, cases involving poisonings in this group are extremely rare. Considerable study and research are needed in this area since the difficulties encountered have discouraged many workers. The problems associated with members of the miscellaneous category are too numerous to be discussed here. The interested student is referred to more comprehensive texts.[15,54]

SUMMARY

It is somewhat frustrating to write a chapter consisting chiefly of methods used in toxicological analysis. The field is changing so rapidly, technological improvements are so frequent, and new drugs with their clinical problems are so variable that it is rare to find a toxicology laboratory that is not constantly seeking better methods. Although attempts were made to include a variety of analytical approaches in this chapter, it was necessary to omit many excellent methods because of space restrictions; the author truly regrets the omissions.

Since toxicology impinges on so many fields, advances in methodology, new knowledge relating to mechanisms of action, diagnosis and treatment of toxic episodes, and interpretation of laboratory results are found in publications directed to many areas. Analytical and clinical chemistry, pathology, pediatrics, cancer research, general medicine, criminalistics, veterinary medicine, entomology, and many other disciplines have contributions to make. Fortunately, several publications have now appeared which attempt to draw together information from diverse sources in a form useable in the clinical toxicology laboratory.[7,15,57] It is hoped that these will continue to appear at practical intervals.

Finally, even the best equipped and staffed laboratory encounters problems which are new and novel. These, together with the "general unknown" problem, constantly present a challenge to the toxicologist. Screening tests and spot tests[18] still play an important role in these cases when coupled with sound training and the ingenuity of the toxicologist.

REFERENCES

1. Adler, F. L., and Liu, C-T.: J. Immunology, *106*:1684, 1971.
2. Björling, C. O., Berggren, A., and Nygord, B.: Acta Chem. Scand., *16*:1481, 1962.
3. Blanke, R. V.: J. Forensic Sci., *1*:79, 1956.
4. Blanke, R. V.: J. Forensic Sci., *19*:284–291, 1974.

5. Blanke, R. V.: Med. Coll. Va. Quart., *9*:301–303, 1973.
6. Broughton, P. M. G.: Biochem. J., *63*:207, 1956.
7. Clarke, E. G. C. (Ed.): Isolation and Identification of Drugs. London, The Pharmaceutical Press, 1969.
8. Cochin, J., and Daly, J. W.: J. Pharmacol. Exp. Ther., *139*:160, 1963.
9. Conway, E. J.: Microdiffusion Analysis and Volumetric Error. 5th ed. London, Crosby Lockwood & Sons Ltd., 1962.
10. Cramer, S., and Selander, K.: Brit. J. Ind. Med., *25*:209, 1968.
11. Curry, A. S.: Brit. Med. J., *2*:1040, 1963.
12. Curry, A. S.: Poison Detection in Human Organs. 2nd ed. Springfield, Ill., Charles C Thomas, 1969.
13. Curry, A. S.: *In* Toxicology—Mechanisms and Analytical Methods. C. P. Stewart and A. Stolman, Eds. New York, Academic Press, Inc., 1961, vol. 2, p. 185.
14. Curry, A. S.: Brit. Med. J., *1*:354, 1964.
15. Curry, A. S.: Advances in Forensic and Clinical Toxicology. Cleveland, Ohio, Chemical Rubber Co., 1972.
16. Dal Cortivo, L. A., Giaquinta, P. and Umberger, C. J.: J. Forensic Sci., *8*:526, 1963.
17. Davidow, B., Petri, N. L., and Quame, B.: Tech. Bull. Reg. Med. Tech., *38*:714, 1968.
18. Decker, W. J., and Treuting, J. J.: Clin. Tox., *4*:89, 1971.
19. Deichmann, W. B., and Gerarde, H. W.: Symptomatology and Therapy of Toxicological Emergencies. New York, Academic Press, Inc., 1964.
20. Farrelly, R. O., and Pybus, J.: Clin. Chem., *15*:566, 1969.
21. Feldstein, M.: *In* Toxicology—Mechanisms and Analytical Methods. C. P. Stewart and A. Stolman, Eds. New York, Academic Press, Inc., 1960, vol. 1, chap. 16.
22. Feldstein, M., and Klendshoj, N.: J. Forensic Sci., *2*:39, 1957.
23. Fiereck, E. A., and Tietz, N. W.: Clin. Chem., *17*:1024, 1971.
24. Finkle, B. S.: Anal. Chem., *44*:18A, 1972.
25. Food and Drug Administration: Poison Control Statistics, 1972.
26. Forrest, I. S., and Forrest, F. M.: Clin. Chem., *6*:11, 1960.
27. Gant, V. A: Industr. Med., *7*:608, 1938; *7*:679, 1938.
28. Gerarde, H. W., and Skiba, P.: Clin. Chem., *6*:327, 1960.
29. Gettler, A. O., and Kaye, S.: J. Lab. Clin. Med., *35*:146, 1950.
30. Gleason, M. N., Gosselin, R. E., Hodge, H. C., and Smith, R. P.: Clinical Toxicology of Commercial Products. 3rd ed. Baltimore, The Williams & Wilkins Co., 1969.
31. Goldbaum, L. R., Schloegel, E. L., and Dominguez, A. M.: *In* Progress in Chemical Toxicology. A. Stolman, Ed. New York, Academic Press, Inc., 1963, vol. 1, chap. 1.
32. Goldbaum, L. R., Williams, M. D., and Koppanyi, T.: Anal. Chem. *32*:81, 1960.
33. Hepler, O. E.: Manual of Clinical Laboratory Methods. 4th ed. Springfield, Ill., Charles C Thomas, 1963, p. 325.
34. Hindberg, J., and Wieth, J. O.: J. Lab. Clin. Med., *61*:355, 1963.
35. Kopito, L., and Schwachman, H.: J. Lab. Clin. Med., *70*:326, 1967.
36. Korzun, B. P., Brody, S. M., Keegan, P. G., Luders, R. C., and Rehm, C. R.: J. Lab. Clin. Med., *68*:333, 1966.
37. Kubasik, N. P., Volosin, M. T., and Murray, M. H.: Clin. Chem., *18*:410, 1972.
38. Kupferberg, H. J.: Clin. Chim. Acta, *29*:283, 1970.
39. Leute, R., Ullman, E. F., and Goldstein, A.: J. Amer. Med. Assoc. *221*:1231, 1972.
40. Lundquist, F.: *In* Methods of Biochemical Analysis. D. Glick, Ed. New York, Interscience Publishers, Inc., 1959, vol. 7, p. 217.
41. Matousek, J. P., and Stevens, B. J.: Clin. Chem. *17*:363, 1971.
42. McBay, A. S., Turk, R. F., Corbett, B. W., and Hudson, P.: J. Forensic Sci., *19*:81–89, 1974.
43. McMahon, R. E., Ridolfo, A. S., Culp, H. W., Wolen, R. L., and Marshall, F. J.: Tox. Applied Pharmacol., *19*:427, 1971.
44. McNair, H. M., and Bonelli, E. J.: Basic Gas Chromatography. 5th ed. Varian Aerograph, 1969.
45. National Center for Health Statistics: Vital Statistics of the United States—1968. Volume II—Mortality. U.S. Dept. H.E.W., P.H.S., 1972.
46. Nickolls, L. C.: Analyst, *85*:840, 1960.
47. Official Methods of Analysis of the Association of Official Agricultural Chemists. 10th ed. 1965, p. 357. American Association of Official Agricultural Chemists (Publ.).
48. Pierce, J. O., and Cholak, J.: Arch. Environ. Health, *13*:208, 1966.
49. Rieders, F.: Ann. N.Y. Acad. Sci., *111*:591, 1964.
50. Rowley, R. J., and Farrah, G. H.: Am. Industr. Hyg. Assn. J., *23*:314, 1962.
51. Spector, S., and Parker, C. W.: Science, *168*:1347, 1970.
52. Spector, S., and Flynn, E. J.: Science, *174*:1036, 1971.
53. Stahl, E.: Thin Layer Chromatography. Berlin, Springer-Verlag, 1965.
54. Stewart, C. P., and Stolman, A. (Eds.): Toxicology—Mechanisms and Analytical Methods. New York, Academic Press, Inc., 1960, vol. 1; 1961, vol. 2.
55. Stewart, R. D., and Erley, D. S.: J. Forensic Sci., *8*:31, 1963.
56. Sunshine, I. (Ed.): Handbook of Analytical Toxicology. Cleveland, Ohio, Chemical Rubber Co., 1969.
57. Sunshine, I. (Ed.): Manual of Analytical Toxicology. Cleveland, Ohio, Chemical Rubber Co., 1971.

58. Taunton-Rigby, A., Sher, S. E., and Kelley, P. R.: Science, *181*:165, 1973.
59. Tietz, N. W., and Fiereck, E. A.: Ann. Clin. Lab. Sci., *3*:36, 1973.
60. Trinder, P.: Biochem. J., *57*:301, 1954.
61. Tuckman, J., Youngman, W. F., and Bleiburg, B. M.: Pub. Health Reports, *77*:605, 1962.
62. Umberger, C. J., and Adams, G.: Anal. Chem. *24*:1309, 1952.
63. Van Slyke, D. D., and Salvasen, H. A.: J. Biol. Chem., *40*:103, 1919.
64. Wallace, J. E., and Biggs, J. D.: J. Forensic Sci., *14*:528, 1969.
65. Wallace, J. E., and Ladd, S. L.: Ind. Med., *39*:23, 1970.
66. Wallace, J. E., Ladd, S. L., and Blum, K.: J. Forensic Sci., *17*:164, 1972.
67. Woodbury, D. M., Penry, J. K., and Schmidt, R. P. (Eds.): Antiepileptic Drugs. New York, Raven Press, 1972.
68. Zaar, B., and Gronwall, A.: Scand. J. Lab. Invest., *13*:225, 1961.
69. de Zeeuw, R. A.: Anal. Chem., *40*:915, 1968.
70. Zinterhofer, L. J. M., Jatlow, P. I., and Fappiano, A.: J. Lab. Clin. Med., *78*:664, 1971.

CHAPTER 22

AMNIOTIC FLUID

by Ermalinda A. Fiereck, M.S.

Amniotic fluid was at one time considered a stagnant pool which merely functions to protect and cushion the fetus. In the past 20 years, extensive studies have shown that the amniotic fluid is a very complex and dynamic fluid. Amniotic fluid functions to equalize the pressure and thus protects the fetus from external trauma, and it assures fetal mobility. It is active in the disposal of fetal secretions and excretions, it may provide nutrients by fetal swallowing, and in addition it probably has innumerable immunological and biochemical functions. The mechanism of action and the control of these various functions are presently not clearly understood.

Origin. The origin of amniotic fluid is a matter of conjecture. Ostergard[18] reports that it may arise as a transudate of maternal serum across the placental and fetal membranes, or as a transudate across the umbilical cord. A portion of the amniotic fluid also arises from fetal urine, from tracheal and bronchial secretions, and possibly from the fetal skin. Each of these sources may contribute to the formation of amniotic fluid to varying degrees at specific times during gestation.

According to Lind *et al.*,[14] the composition of amniotic fluid early in pregnancy resembles more closely fetal plasma than maternal plasma; this suggests that the amniotic fluid found early in gestation is probably an extension of fetal extracellular fluid. At the early stage of gestation, the outer limit of the extracellular fluid is probably the amnion rather than the fetal skin. Later in gestation, as the skin becomes more stratified, the movement of fluid by this route is restricted. As gestation proceeds, there is an increased participation by the fetal renal and respiratory systems in amniotic fluid formation. Fetal urine is formed as early as 14 weeks; however, its contribution to amniotic fluid is probably not significant until approximately the 20th week of gestation. The contributions of the secretions from the tracheobronchial tract, salivary glands, and buccal mucosa to the content of amniotic fluid have not been fully clarified. The increase in concentration of certain amniotic fluid lipids late in pregnancy, when at the same time the developing lung tissue shows a similar lipid change, suggests that respiratory tissue secretions also contribute to the amniotic fluid composition.[3]

Volume. The volume of amniotic fluid increases as pregnancy progresses, from approximately 25 ml of amniotic fluid at 10 weeks gestation to approximately 800 to 1000 ml near term.[18]

Increased volumes are seen in fetuses with severe hydrops fetalis, congenital obstruction of the esophagus or upper gastrointestinal tract, anencephaly, and central nervous system defects. *Low* amniotic fluid volumes occur in conditions such as fetal bilateral renal agenesis and in placental insufficiency.

Disposal. The fate of amniotic fluid, like its formation, is not clearly understood.[1,18] Amniotic fluid is most likely exchanged across the placental and umbilical membranes; fetal swallowing is thought to play a major role in the removal of fluid. This is substantiated by the excessive accumulation of amniotic fluid when there is a mechanical or neurological disturbance of this function. It has been estimated that, near term, approximately 500 to

1500 ml of fluid/h are swallowed. The total water exchange between the fetus and mother at term is approximately 3500 ml/h.

Composition. The composition of amniotic fluid varies considerably with the duration of gestation.[2,3,4,14,18] In the early stages of pregnancy, the composition of the electrolytes is similar to that of plasma. The amniotic fluid becomes progressively hypotonic, owing mainly to a decrease in *sodium* and *chloride*. This is accompanied by a marked increase in *urea*, *creatinine*, and *uric acid* to levels approximately 2 to 3 times those found in maternal serum. These changes result from admixture of fetal urine, which dilutes amniotic fluid and adds nonprotein nitrogenous constituents.

Potassium levels remain relatively stable throughout gestation (approximately 4.4 mmol/l), and no significant change in the values for calcium, phosphorus, or glucose occurs between the second and third trimesters. Amniotic fluid glucose values near term have been reported as 10 to 61 mg/100 ml for normal nondiabetic mothers and from 23 to 139 mg/100 ml for diabetic mothers.

Variable values for pH, pCO_2, and pO_2 have been reported but have not been linked consistently to any abnormal state of pregnancy.

Both total *protein* and *albumin* decrease with gestational age (1/10 of normal serum level during second trimester and 1/20 near term). *Globulins* may increase slightly throughout gestation, and *fibrinogen* is absent. *Amino acid* concentrations are similar to or slightly lower than those of maternal serum. Abnormalities in amino acid metabolism are associated with a number of genetic disorders (see Table 22–1).

Some changes in enzyme activities have been observed during gestation. *Aspartate amino transferase* activity (GOT), for example, increases approximately twofold from the second to the third trimester. *Lactate dehydrogenase* activity is normally lower in amniotic fluid than in maternal serum; however, highly elevated levels (2 to 20 times normal) have been associated with fetal death. *Alkaline phosphatase* activity increases with the length of gestation and may be due to the increase in maternal alkaline phosphatase noted with advancing pregnancy. This increase is in the 65°C heat stable isoenzyme, which is probably of placental origin[11].

Studies of enzyme *activities* in amniotic fluid have not shown any clear association between activity and the fetal pathological state. *Absence of a specific enzyme activity* may, however, be clinically extremely useful in the diagnosis of particular genetic metabolic defects (see Table 22–1).

Oxyhemoglobin is frequently found in amniotic fluid and is thought to result primarily from the hemolysis of erythrocytes which enter the specimen during collection. *Methemalbumin* is a common constituent in amniotic fluid when there is impending intrauterine death. The *non-heme iron* concentration near term is 8 to 30 μg/100 ml. A rapid increase in iron indicates a poor prognosis[13]. A small amount of *bilirubin* is present in amniotic fluid normally and decreases in concentration as gestation continues, perhaps owing to the dilution of amniotic fluid by fetal urine, or to the increased ability of the placenta to remove bilirubin. In hemolytic disease owing to Rh incompatibility, the bilirubin is present in an increased concentration. (See Amniotic Fluid Analysis in Erythroblastosis, p. 1168). Amniotic fluid bilirubin is also increased when there is an increase in maternal serum bilirubin, such as occurs in mothers with sickle-cell crisis.[7]

A number of *hormones* have been detected in amniotic fluid, and their quantitation may be of potential value in the diagnosis of certain congenital disorders. *Serotonin* itself has not been detected in amniotic fluid, but the serotonin metabolite 5-*hydroxyindoleacetic acid* has been found in increased concentration in patients with toxemia of pregnancy. *Estriol* determinations in amniotic fluid can be used to evaluate the placento-fetal integrity; however, maternal urine is currently more often used for this evaluation (see Chap. 13). *Prostaglandins* have been measured in amniotic fluid; the E series is found earlier in gestation

TABLE 22–1 AMNIOTIC FLUID IN THE DIAGNOSIS OF HEREDITARY DISORDERS (AFTER MILUNSKY *et al.*[15])

Disorder	*Major Clinical Manifestation*	*Accumulated Product*	*Deficient Enzyme*
LIPID METABOLISM:			
Tay-Sachs disease (G_{M2} gangliosidosis)	Degenerative neurological disorder.	G_{M2} ganglioside and its asialo derivative	Hexosaminidase A
Metachromatic leukodystrophy	Degenerative neurological disorder. Mental retardation.	Sulfatide	Arylsulfatase A
Niemann-Pick disease	Hepatosplenomegaly. Variable skeletal and neurologic involvement.	Sphingomyelin	Sphingomyelinase
CARBOHYDRATE METABOLISM:			
Hurler's syndrome	Gargoyle-like facies. Early psychomotor retardation. Early clouding of cornea. Dwarfism.	Dermatan sulfate, heparin sulfate	Specific β-galactosidase
Pompe's disease (Glycogen storage disease, Type II)	Failure to thrive. Cardiomegaly. Hepatomegaly.	Glycogen	α-1,4-Glucosidase
Galactosemia	Failure to thrive. Mental retardation. Cirrhosis.	Galactose	Galactose 1, phosphouridyl transferase
AMINO ACID METABOLISM:			
Maple syrup urine disease	Ketoacidosis. Neurologic abnormality. Mental retardation.	Valine, leucine, isoleucine, alloisoleucine	Branched-chain keto acid decarboxylase
Methyl malonic aciduria	Acidosis. Failure to thrive.	Methylmalonic acid, glycine, hemocystine, cystathionine	Methylmalonyl CoA isomerase or vitamin B_{12} coenzyme
MISCELLANEOUS:			
Adrenogenital syndrome	Adrenal insufficiency.	17-Ketosteroids, pregnanetriol	Failure of C-21 or C-11 hydroxylation
I-cell disease	Gargoyle-like facies. Dwarfism. Psychomotor retardation.	Acid mucopolysaccharide, glycolipids	β-glucuronidase, excess acid phosphatase
Lesch-Nyhan syndrome*	Self-mutilation. Mental retardation.	Uric acid	Hypoxanthine-guanine-phosphoribosyltransferase
Lysosomal acid phosphatase deficiency	Failure to thrive. Progressive neuromuscular involvement.	Unknown	Lysosomal acid phosphatase

* All disorders are autosomal recessive with the exception of Lesch-Nyhan syndrome, which is sex-linked.

than the F series. With the onset of labor, however, the F group is present in greater proportion. Prostaglandins of the F series are thought to function in the initiation of labor, although their exact action is not known.

The *lipid* concentration of centrifuged amniotic fluid was studied by Biezenski *et al.*,[3] and the observed values were tabulated in three groups according to weeks of gestation, namely 27 to 33 weeks, 34 to 40 weeks, and at labor. The mono-, di-, and triglycerides, free fatty acids, cholesterol, cholesterol esters, and hydrocarbons remained approximately unchanged. The only noticeable change was an approximate twofold increase in the *phospholipids* between 27 to 33 weeks of gestation and term. The similarity of the lecithin-fatty acids of amniotic fluid to those from fetal lung effluents at various stages of gestation suggests that amniotic fluid lipids originate from developing respiratory tract tissue. The *ratio of amniotic fluid lecithin to sphingomyelin* (L/S ratio) was used by Gluck *et al.*[10] for the evaluation of fetal lung maturity. Prior to 34 weeks, the amniotic fluid lecithin and sphingomyelin concentrations are approximately equal. After this time there is a marked increase in lecithin (Fig. 22–1) and the lecithin/sphingomyelin ratio increases to greater than 5[5a] at term. Amniotic fluid *lecithin* measurements, instead of L/S ratio measurements, have also been used to estimate fetal lung maturity. Since a change in the ratio may occur without corresponding increase in lecithin itself, the estimation of lecithin (or lecithin phosphorus) is thought to be clinically more useful.[17]

Adequate amounts of lecithin are needed for *normal surface activity** of the lungs. The lungs of infants with respiratory distress syndrome have a markedly lower concentration of lecithin. Delivery of a low weight (premature) fetus, whose amniotic fluid lecithin-phosphorus concentration is less than 0.050 mg/100 ml (or L/S ratio of approximately 1.0[10]), is contraindicated since the infant will probably develop respiratory distress or hyaline membrane disease. A lecithin-phosphorus value of 0.100 mg/100 ml (or L/S ratio of >2[10] or >5[5a]) indicates adequate fetal lung maturity and suggests that respiratory distress after delivery is not likely.

Besides the normal changes in lipids that occur in the later stages of gestation, several other changes have been observed in pathological conditions of pregnancy. A high *total lipid* concentration, with the increase chiefly in the free fatty acids, has been reported to be associated with intrauterine death. Toxemia of pregnancy is associated with an increase in lipids which is primarily in the triglyceride fraction.

* Surface activity is that phenomenon whereby the surface tension of the air-alveolar lining interface is lowered with expiration. In the absence of normal surface activity, the alveolar radius becomes smaller with expiration, the wall tension rises, and the alveoli collapse.

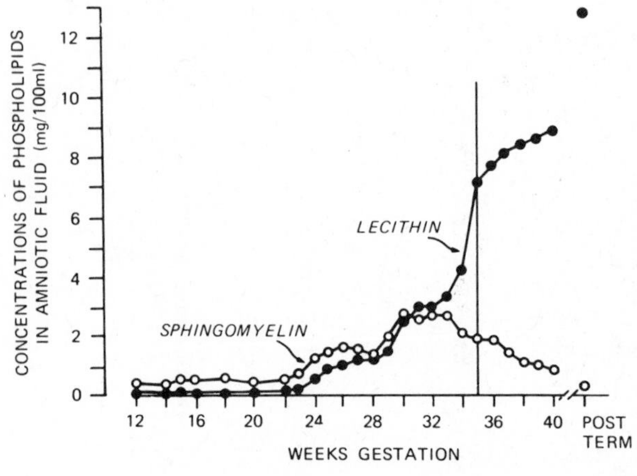

Figure 22–1 Changes in mean concentrations of lecithin and sphingomyelin in amniotic fluid during gestation in normal pregnancy. (From Gluck, L., and Kulovich, M.: Lecithin/sphingomyelin ratios in amniotic fluid in normal and abnormal pregnancy. Am. J. Obstet. Gynecol., *115*:539, 1973.)

AMNIOTIC FLUID ANALYSES

The determination of specific substances in amniotic fluid aids in the *in utero* diagnosis and treatment of many fetal pathological conditions. Results obtained must be interpreted cautiously, however, since the *normal range* for amniotic fluid constituents is wide and constituents vary with the gestational age. Useful physiological information is often more accurately evaluated by serially sampling the same subject and relating these results to the normal *trend.*

Normal values for each constituent at each period of gestation must still be established. Useful normal ranges are available only for bilirubin as it relates to hemolytic disease and for the tests used to evaluate fetal maturity.

Since a collection of amniotic fluid itself entails certain risks, amniotic fluid analyses are presently limited to specimens from patients with suspected fetal pathology. Maternal complications following amniocentesis include sepsis, hemorrhage, Rh sensitization, and abortion. Fetal complications include sepsis, fetal injury, abortion, and perhaps the induction of malformations. Additional unsuspected long term effects may also be initiated by amniocentesis.

Specimens for amniotic fluid analysis can be placed into three distinct groups: those collected prior to 20 weeks of gestation, primarily used for genetic counseling; those collected after 28 weeks, used to evaluate hemolysis in Rh incompatible pregnancies; and those collected near term, used to evaluate fetal maturity in high-risk pregnancies. The last group of disorders, including diabetes, hypertension, renal disease, Rh disease, and toxemia, are those in which early delivery may be indicated.

AMNIOTIC FLUID ANALYSIS IN GENETIC COUNSELING

In the past, genetic counseling was based solely on statistical probabilities in accordance with the genetic history of the parents and their offspring. Genetic counseling is usually sought after the birth of one child with a genetic disease. Several fairly recent advances have increased the accuracy of genetic counseling: (1) the identification of fetal sex by examination of fetal amniotic fluid cells, (2) the demonstration of fetal karyotypes from cultured amniotic fluid cells, and (3) the demonstration that cells retain their specific enzyme patterns in tissue culture.[6,15,16,18]

Significant *chromosome abnormalities* occur with an overall frequency of approximately 1 in 200 pregnancies; in mothers over 40 years of age, the incidence is much higher and approximates 1 in 40. Gross chromosomal abnormalities, e.g., *Down's syndrome*, can be diagnosed *in utero* by an examination of the chromosomal pattern of cultured (and in some cases, uncultured) amniotic fluid cells.

Identification of *fetal sex* using cultured amniotic fluid cells aids in the prediction of possible sex-linked inherited diseases. If the mother is a known carrier of a sex-linked disease, there is a 50 per cent chance that each male fetus has the particular sex-linked hereditary disease. Differentiation between the normal and the affected male fetus is possible in some cases in which a known biochemical excess or deficiency is associated with the disease (e.g., uric acid in Lesch-Nyhan syndrome).

Specific biochemical testing of the cultured amniotic fluid cells (and to a limited extent the amniotic fluid itself) can be used to diagnose some *sex-linked diseases* and a number of autosomal genetic diseases. Approximately 45 metabolic diseases with recognized phenotypes in cultured fibroblasts are known and may possibly be identified by studies on the cultured amniotic fluid cells. At least 12 genetic diseases have been diagnosed *in utero* by noting the specific enzyme deficiency or the excess metabolite associated with the disease, in an extract from the cultured amniotic fluid cells (Table 22–1).

Since the overall incidence of most hereditary diseases is very low, these studies are requested only for high-risk fetuses (i.e., when the metabolic disease is suspected from the family history). The biochemical study is restricted to the particular constituent associated with the disease in question; e.g., if *Tay-Sachs disease* is suspected, the cultured cell extract (or amniotic fluid) is assayed for hexosaminidase A and B. The disease is identified by a low activity or absence of hexosaminidase A and a normal or increased activity of hexosaminidase B.

Amniotic fluid *samples used in genetic counseling* are collected during the second trimester. The preferred time for the collection of a sample for cell culture is at about 16 to 17 weeks of gestation. Chromosomal studies are usually possible after 2 weeks of culture growth, but successful biochemical determinations require a greater number of cells and at least 3 to 6 weeks of culture time.

Various problems have been associated with cell culture assays, among which are (1) the possibility of mistakenly culturing maternal rather than fetal cells, (2) the difference in enzyme activities of the two types of fetal cells which are present in amniotic fluid (fibroblasts and epithelial cells), (3) the considerable time (sometimes 6 weeks) required to grow sufficient cells for reliable assay, (4) the variability of enzyme activities during different stages of the cell cycle (this could lead to an inaccurate diagnosis if the culture is merely at a low enzyme phase but the results are reported as showing an enzyme deficiency), (5) the methods used may lack the sensitivity and specificity needed for accurate measurement, and (6) the normal values may not be well established for either cultured skin fibroblasts or amniotic fluid cells, leading to errors in the interpretation of the results.

Development of more sensitive microbiochemical assays may permit accurate quantitation of specific substances in the uncultured cells or, in some cases, in the amniotic fluid itself. Currently, these assays are usually performed in specialized laboratories since they require preliminary tissue culture by methods often not available in routine clinical laboratories.

Although the reliability of genetic counseling has been improved markedly by the use of amniotic fluid cell assays, more basic information is still needed if an accurate prenatal diagnosis is to be possible. Normal values for both cultured and uncultured amniotic fluid cells must be clearly defined for each constituent associated with the defect. Pathological changes associated with each defect must be verified in specimens from patients known to have the disease, and possible artifacts caused by culture effects must be ruled out. Improvements in the various technical aspects may ultimately lead to accurate diagnosis *in utero*; however, the important ethical questions arising from these amniotic fluid assays will undoubtedly remain unresolved.

AMNIOTIC FLUID ANALYSIS IN ERYTHROBLASTOSIS

Neonatal or intrauterine deaths occur in approximately 20 per cent of pregnancies of Rh-negative, sensitized women. Many fatalities could possibly have been avoided by early induced delivery or by intrauterine blood transfusions if there had been an accurate means of assessing the status of these fetuses *in utero*. Until recently, the determination of the Rh antibody titer, the obstetrical history, and radiographic analysis were the chief tools for evaluating the condition of the fetus *in utero* and for determining which patients could benefit most by preterm delivery. These indirect methods often gave too little or wrong information. Many severely erythroblastotic infants who may have benefited by preterm delivery were not so managed, owing to low antibody titers and a previous normal obstetrical history in the mother. Conversely, in the presence of high antibody titers a normal or slightly affected infant was occasionally found, who would in all probability have progressed better had the pregnancy been allowed to continue. A more decisive means of evaluating the *in utero* condition of fetuses with hemolytic disease is by the examination of the amniotic fluid periodically throughout gestation.

ABSORBANCE CURVE OF AMNIOTIC FLUID

In 1953, Bevis observed that the amniotic fluid from some Rh-incompatible pregnancies had a yellow-green coloration. This coloration was reported to be due to bilirubin or other intermediate products of heme metabolism. When the amniotic fluid absorbance curves of normal and erythroblastotic fetuses were compared, a distortion was noted in the absorption curve of the amniotic fluid obtained from the erythroblastotic fetuses. The degree of distortion at 450 nm was used to predict the severity of the hemolytic condition.

The absorbance curve of amniotic fluid from normal infants, at term, shows an approximately linear increase in absorbance* with decrease in wavelength between 550 and 365 nm (see Fig. 22–2*A*). This uniform increase in absorbance is probably due to the scattering of light by proteins. Any increment above this uniform increase can be determined by noting the extent to which the observed absorbance lies above a straight line, drawn between the observed absorbance minima at approximately 550 nm and 365 nm (see Fig. 22–2*B*). Liley[13] used the differences between the observed absorbance at 450 nm and that indicated by this straight line at 450 nm to assess the extent of hemolytic disease. This difference is called "the absorbance peak at 450 nm."

* The term absorbance will be used throughout this text, although in the medical literature related to amniotic fluid analysis, the older term, optical density (O.D.), is often used.

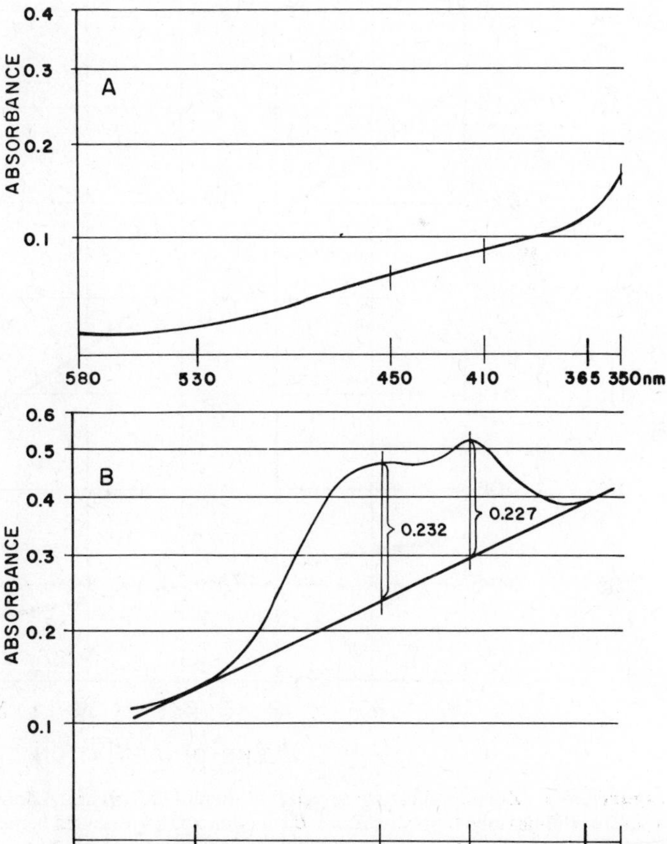

Figure 22–2 *A*, Normal amniotic fluid. Note near linearity of curve. *B*, Amniotic fluid showing bilirubin peak at 450 nm and oxyhemoglobin peak at approximately 410 nm. Note baseline drawn between linear parts of curve, from 550 to 365 nm.

RELATIONSHIP OF ABSORBANCE PEAK AT 450 nm TO WEEKS OF GESTATION

Amniotic fluid from normal fetuses, or those with mild hemolytic disease, shows a moderate absorbance peak at 450 nm early in pregnancy, which decreases with approaching maturity. This decrease is perhaps due to the continually increasing ability of the placenta to remove bilibrubin,[21] or to the increased contribution of fetal urine to amniotic fluid. At 28 weeks of gestation, the 450 nm absorbance peak of amniotic fluid from normal fetuses or fetuses with mild hemolytic disease may be as high as 0.06, and at 40 weeks it may range from zero to 0.02. In severe hemolytic disease, the absorbance peak at 450 nm is greatly increased and does not always decrease with approaching maturity; in fact, there is frequently an increase in absorbance. Generally the initial amniocentesis is performed between the 28th and 32nd weeks of gestation and, if indicated, amniocentesis is repeated every 2 weeks. For correct evaluation, the absorbance at 450 nm must be related to the weeks of gestation. Liley[13] plotted the 450 nm absorbance obtained on amniotic fluid specimens against the weeks of gestation. He noted a grouping of the results into zones indicative of the severity of the hemolytic disease. A modified graph of this type is shown in Figure 22-3.

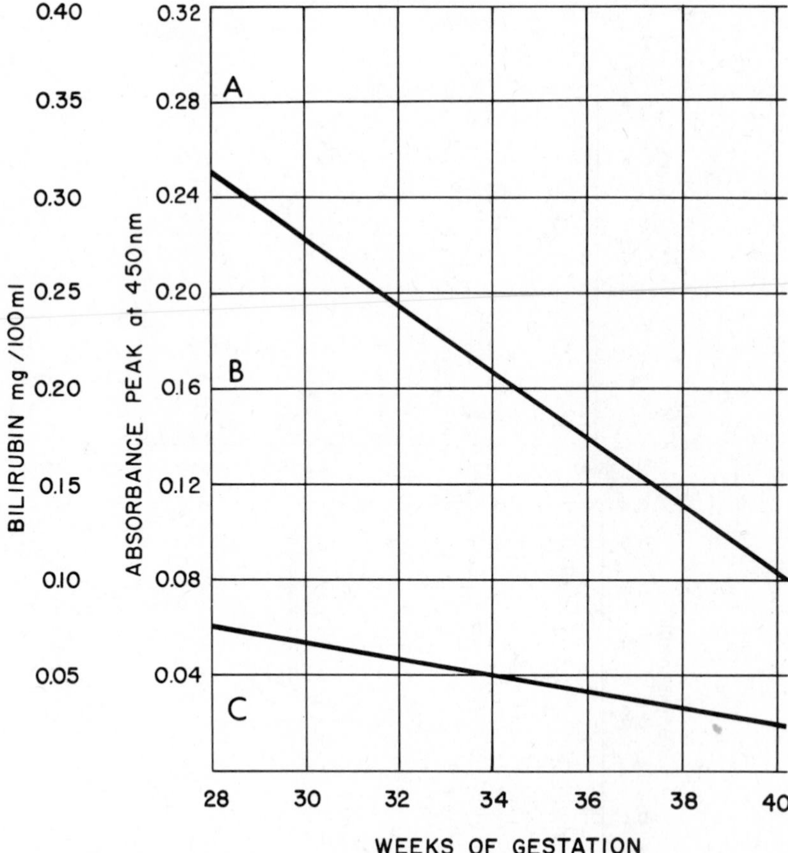

Figure 22-3 Zones indicating severity of hemolytic process. Zone *A* suggests a severe fetal hemolytic process. Zone *B* suggests a mild to severe hemolytic process. Zone *C* suggests the absence of hemolytic disease *or* the presence of a mild hemolytic process. (After Liley, A.: Am. J. Obstet. Gynecol., *82*:139, 1961.)

BILIRUBIN CONCENTRATION AND ITS RELATIONSHIP TO THE ABSORBANCE PEAK AT 450 nm

Attempts have been made to determine the bilirubin concentration of amniotic fluid chemically and to relate this to the severity of the hemolytic disease. Considerable disagreement in values exists in the literature, some perhaps due to difficulties in measuring the low bilirubin concentrations in amniotic fluid. Further disagreement was encountered when the bilirubin concentration was not related to the weeks of gestation.

Bilirubin concentrations reported by Watson *et al.*[22] at 32 to 33 weeks ranged upward from 0.01 mg/100 ml and were, in general, directly related to the severity of the hemolytic condition. These workers reported that bilirubin glucuronide is not present in either normal amniotic fluid or in amniotic fluid obtained from erythroblastotic fetuses; however, the possible presence of other bilirubin conjugates was not excluded.

A modified Evelyn and Malloy method[22] and the Jendrassik and Grof method[8] have been used for amniotic fluid bilirubin quantitation. No detailed chemical bilirubin method will be presented here since the analysis of amniotic fluid by its absorbance curve is generally preferred to that of chemical bilirubin measurement. In cases in which a suitable spectrophotometer for the analysis of amniotic fluid by its absorbance at 450 nm is not available, the Jendrassik and Grof procedure for bilirubin as presented by Gambino and Freda[8] could be successfully used.

The addition of bilirubin in known concentrations* to amniotic fluids with no absorbance peak at 450 nm, followed by the determination of the absorbance curve, has permitted us to relate absorbance at 450 nm to bilirubin concentration. A direct linear relationship was obtained in which each 0.10 mg of bilirubin/100 ml corresponded to an absorbance of 0.08 at 450 nm. (In other words, absorbance peak $\times$ 1.25 = mg bilirubin/100 ml amniotic fluid.)

A similar correlation was reported by Gambino and Freda,[8] using the chemical method of Jendrassik and Grof. The level of bilirubin in amniotic fluid can, therefore, be assessed by determining the absorbance peak at 450 nm and expressing the results as mg bilirubin/100 ml amniotic fluid. The analysis of amniotic fluid by its absorbance curve avoids the difficulties of chemically measuring low concentrations of bilirubin and demonstrates not only the presence of bilirubin in amniotic fluid but also the possible presence of other interfering substances.

The *normal* amniotic fluid bilirubin values, based upon the relationship of weeks of gestation to the absorbance peak at 450 nm as reported by Liley,[13] and the relationship of absorbance at 450 nm to bilirubin concentration, are shown in Figure 22–3. The bilirubin concentration of amniotic fluid from normal fetuses or those with mild hemolytic disease, at 28 weeks, may be as high as 0.075 mg/100 ml and at 40 weeks ranges from zero to 0.025 mg/100 ml.

Increased amniotic fluid bilirubin concentrations have been reported in several conditions besides erythroblastosis. An increase was noted in the amniotic fluid of a fetus with duodenal atresia,[12] possibly owing to the regurgitation of bile into the amniotic fluid. Maternal infectious hepatitis[20] and maternal sickle cell crisis[7] cause an elevation of bilirubin in the maternal serum which is reflected in the amniotic fluid.

METHOD FOR SPECTRAL SCAN OF AMNIOTIC FLUID

COLLECTION OF SPECIMEN AND SAMPLE PREPARATION

Approximately 10 ml of amniotic fluid is withdrawn by transabdominal amniocentesis. Since bilirubin is unstable in light, the specimen is immediately placed in a sterile tube and

* The appropriate amounts of bilirubin are added in the form of an aqueous solution of bilirubin in Na_2CO_3 and NaOH as described in the procedurc for serum bilirubin (Chap. 19).

continually protected from light. (This can be accomplished conveniently by using foil-wrapped sterile tubes.) Liley[13] reported that the 450 nm peak had a half-life of 10 hours in laboratory daylight and 12 to 18 minutes in winter sunlight. Storage of a sterile specimen in the dark preserves the specimen for 9 months in a refrigerator and 30 days at room temperature.[13]

Amniotic fluid specimens are generally turbid because of the presence of proteins and particulate matter. Immediate centrifugation, using a centrifuge such as the International Model (size 2, International Equipment Co., Needham, MA.) at approximately 3000 rpm for 10 minutes, partially clears the sample by removing the larger particulate matter, such as erythrocytes and epithelial cells. Further clarification may be necessary if the sample is extremely turbid; this can be accomplished by centrifugation at 12,000 rpm, using a high speed centrifuge (e.g., Lourdes Model LRA, Lourdes Instrument Corp., Old Bethpage, N.Y.), at 0°C for 30 minutes. Centrifugation does not eliminate the turbidity completely, but it usually clears the specimen adequately for spectral analysis.

Undiluted specimens are generally used; however, it may be necessary to dilute heavily pigmented specimens before analysis. If a dilution is necessary, 0.15 molar NaCl (0.9 g/100 ml) is used as the diluent.

Amniotic fluid specimens should be analyzed immediately. If there is a delay in analysis, the specimen should be centrifuged, protected from light, and stored in a refrigerator until analyzed.

MEASUREMENT OF THE 450 nm ABSORBANCE PEAK

The instrument used for this analysis must have a narrow band-pass (preferably <4 nm) and, in addition, the wavelength scale must correspond with the actual band of energy passed through the photometer, since inaccuracies in the spectral output may result in inaccurate values for the "bilirubin and oxyhemoglobin peaks." The instrument should be checked with the appropriate calibrating filters prior to the analyses.

PROCEDURE

1a. If a recording spectrophotometer with a 1.0 cm light path is employed, adjust the instrument and standardize as recommended by the manufacturer. Scan the amniotic fluid sample between 580 and 350 nm, using 0.15 molar NaCl in the reference cuvet.

1b. If a double beam spectrophotometer is used (without scanning attachment and recorder), take readings manually at 5 nm intervals between 580 and 350 nm, using 0.15 molar NaCl in the reference cuvet.

1c. If a single beam spectrophotometer is used, set the instrument to zero absorbance with 0.15 molar NaCl at each change in wavelength before measuring the absorbance of the sample. Take readings every 5 nm between 580 and 350 nm.

2. Remove the plotted absorbance curve from the recorder. If the readings were taken manually, prepare the curve by plotting the absorbance against the specific wavelength.

3. Draw a line connecting the lower portions of the curve at approximately 550 and 365 nm. The absorbance peak at 450 nm is calculated as the difference in absorbance between the top of the peak and the baseline (see Fig. 22–2*B*).

4. A peak at approximately 410 nm indicates the presence of oxyhemoglobin, which causes an error in the 450 nm peak. This error can be corrected by subtracting 5 per cent of the absorbance peak at 410 nm from the bilirubin absorbance peak at 450 nm.[13]

Example (based on values from Fig. 22–2*B*):

$$A_{450} \text{ corrected for oxyhemoglobin} = A_{450} - (0.05 \times A_{410})$$

where $A_{450} = 0.232$ (absorbance at 450 nm measured from baseline to peak)

A_{410} = 0.227 (absorbance at approximately 410 nm measured from baseline to peak)
0.05 = correction factor (5 per cent) for the oxyhemoglobin error in the 450 nm absorbance peak.

Substituting these values into the previous formula:

absorbance peak at 450 nm corrected for oxyhemoglobin

$$= 0.232 - (0.05 \times 0.227)$$

$$= 0.221$$

5. The corrected absorbance at 450 nm can be converted to mg bilirubin/100 ml amniotic fluid by multiplying the corrected 450 nm absorbance by 1.25. In the example given, the 0.221 absorbance at 450 nm is equivalent to 0.28 mg bilirubin/100 ml.

6. If a dilution was necessary, the results must be multiplied by the dilution factor.

7. The absorbance or bilirubin concentration may be recorded in a graph such as that shown in Figure 22–3. This relates the absorbance or bilirubin concentrations, or both, to the weeks of gestation and suggests the severity of the hemolytic disease.

FACTORS AFFECTING THE ACCURACY OF THE SPECTRAL SCAN

Contamination of the sample. Amniotic fluid may be contaminated with bilirubin from maternal or fetal blood as a result of trauma during aspiration. If the contamination is due to fetal blood from an erythroblastotic fetus, a substantial error may be introduced in the measurement of amniotic fluid bilirubin. For example, the inclusion of 0.05 ml of fetal serum (approximately 0.1 ml blood), with a bilirubin concentration of 10 mg/100 ml, in 5 ml of amniotic fluid, would raise the amniotic fluid results by 0.1 mg bilirubin/100 ml amniotic fluid (or an absorbance of 0.08 at 450 nm). This contamination would result in a very serious error.

Contamination of the amniotic fluid with whole blood is readily recognized because of the red color of hemoglobin, and an observant technician is warned that such a sample is unsuitable for analysis. If a specimen containing blood is centrifuged and only the clear portion submitted (e.g., a referral specimen), an error due to added bilirubin may not be suspected and incorrect results may be reported. A small amount of whole blood (noted as a button of red cells after centrifugation) does not necessarily invalidate the test; however, the hemolysis of even a few cells does cause contamination from oxyhemoglobin.

Oxyhemoglobin is frequently present in amniotic fluids. It may occasionally be a true constituent of the fluid, but generally results from the hemolysis of contaminating erythrocytes, which enter at the time of amniocentesis. The presence of oxyhemoglobin is noted by a peak in the region of 408 to 415 nm as previously discussed (see Fig. 22–2*B*).[13,22] The close proximity of the oxyhemoglobin peak to that for bilirubin (450 nm) introduces a positive error into the 450 nm reading. By experimental contamination Liley[12] has shown this error to be approximately 5 per cent; that is, 5 per cent of the observed oxyhemoglobin absorbance at approximately 410 nm will appear as added absorbance at 450 nm. If oxyhemoglobin is present to a significant extent, noted by a deviation of absorbance at 410 nm from the baseline, the 450 nm absorbance peak, from which bilirubin is calculated, must be corrected (subtract 5 per cent of the 410 nm absorbance from the reading at 450 nm). The method for applying this correction is illustrated in detail earlier in this chapter. This correction applies only to oxyhemoglobin and does not correct for the bilirubin derived from the serum portion of blood (see previous section).

Repeated amniocentesis at short intervals (more than once a week) should be avoided since trauma may cause blood to enter the amniotic sac. The bilirubin formed from the hemoglobin would temporarily raise the bilirubin level in the amniotic fluid. Two or three

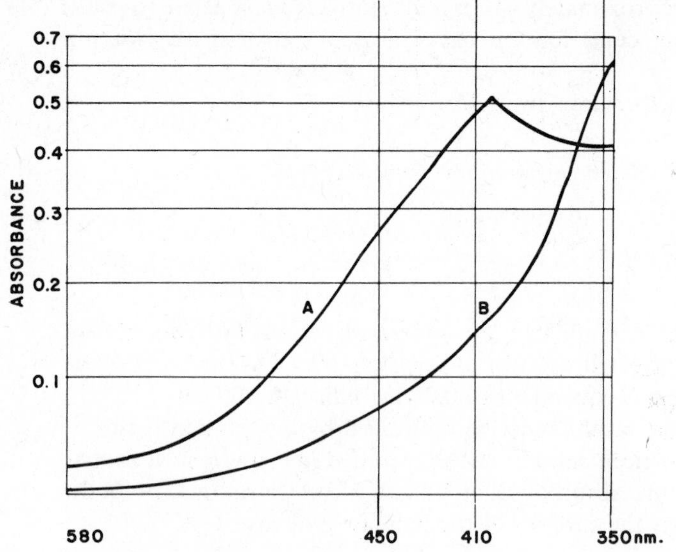

Figure 22–4 *A*, Absorbance curve of amniotic fluid containing unidentified pigment, perhaps meconium.[21] *B*, Absorbance curve of urine from a normal pregnant woman.

weeks[12] may be required for the elevated bilirubin value of the amniotic fluid, following such traumatic contamination, to return to its precontamination level.

Incorrect fluid. The specimen submitted is assumed to be amniotic fluid; however, on occasion maternal urine, fluid from amniotic cysts, fetal ascitic fluid, or meconium has been submitted.[12,21] Any atypically shaped absorbance curve should be viewed with this in mind. A scan of a urine specimen from a normal pregnant woman is shown in Figure 22–4*B*.

A specimen of meconium (fetal intestinal contents) has a green-black color and is generally very turbid. If meconium is present as a contaminant of amniotic fluid, the specimen may resemble the amniotic fluid of a fetus with severe hemolytic disease. Liley[13] states that a specimen containing meconium has a high 450 nm absorbance and a greater absorbance peak at 408 to 415 nm, with no inflection in the curve between these points. Figure 22–4*A* shows an absorbance curve of an amniotic fluid containing an unknown substance, perhaps meconium;[21] analysis of meconium, obtained by autopsy, showed an identical spectral curve.

Multiple amniotic sacs. The sample of amniotic fluid aspirated will assess only the condition of the fetus or fetuses included in the amniotic sac; therefore, if more than one amniotic sac is present, a sample from each should be submitted for analysis. Results obtained on samples from different amniotic sacs could lead to misinterpretation of the results.[12]

INTERPRETATION OF THE RESULTS

This relationship of the bilirubin concentrations (or absorbance at 450 nm) at specific weeks of gestation to the severity of the hemolytic disease is shown in Figure 22–3. Values falling in the *A* zone of the graph suggest a severe fetal hemolytic process, those falling in the *B* zone suggest a mild to severe fetal hemolytic disease, and values falling in the *C* zone suggest the absence of hemolytic disease or the presence of a mild hemolytic reaction.[13]

AMNIOTIC FLUID ANALYSES FOR THE EVALUATION OF FETAL MATURITY

The evaluation of fetal maturity is needed in certain high-risk patients, e.g., those with premature rupture of membranes, diabetes, preeclampsia, Rh disease, or other conditions in which early delivery may be indicated. If the fetus is to survive in the extrauterine environment, a certain stage of fetal maturity is necessary. Tests used to evaluate fetal maturity include the quantitative determination of creatinine, lecithin, and the lecithin-sphingomyelin ratio. A cytological evaluation of the number and types of cells (basal, precornified, anucleated squamous cells) present in the fluid provides additional information.[9,14,18]

CREATININE

As gestation progresses to term, the sodium and chloride ion concentrations and the osmolality of the amniotic fluid decrease, whereas the levels of urea, uric acid, and creatinine steadily increase. Creatinine is the only constituent among these that has been used to any extent successfully for the measurement of fetal maturity.[2,9,18,19] Statistically there is good correlation between an amniotic fluid creatinine of more than 2.0 mg/100 ml (in the presence of a normal maternal creatinine) and a period of gestation greater than 37 weeks, and also between an amniotic fluid creatinine of less than 1.5 mg/100 ml and a period of gestation of less than 37 weeks.[19] The latter group showed a death rate of approximately 40 per cent, with death resulting predominantly from respiratory failure.

The level of amniotic fluid creatinine, although useful, is not a totally reliable indicator of fetal maturity. A number of near-term pregnancies are associated with amniotic fluid creatinine concentrations normally seen in immature fetuses. On the other hand, amniotic fluid obtained from diabetic mothers and from mothers with preeclampsia generally shows a rise in creatinine values earlier during gestation, thus implying a degree of maturity not actually reached. The analytical procedure used for the measurement of amniotic fluid creatinine is the same as that for serum creatinine (see p. 994).

LIPIDS

Since the majority of postnatal deaths following premature delivery result from respiratory complications, the determination of amniotic fluid lecithin or the L/S ratio, as previously discussed, may be more useful for the determination of fetal lung maturity (see p. 1166). For a specific procedure for the measurement of the lipid components of amniotic fluid, see specific references[5,10,17] or Chapter 10, Lipids. The normal amniotic fluid lipid values indicating adequate fetal lung development are a lecithin-phosphorus concentration greater than 0.100 mg/100 ml[17] and an L/S ratio greater than 2 [10] or greater than 5[5a].

REFERENCES

1. Abramovich, D. R.: Fetal factors influencing the volume and composition of liquor amnii. J. Obstet. & Gynaecol. Br. Commonw., *77*:865, 1970.
2. Abdul-Karim, R. W., and Beydoun, S. N.: Amniotic fluid: the value of prenatal analysis. Postgrad. Med., *52*:147, 1972; *52*:236, 1972.
3. Biezenski, J. J., Pomerance, W., and Goodman, J.: Studies on the origin of amniotic fluid lipids, 1. normal composition. Am. J. Obstet. Gynecol., *102*:853, 1968.
4. Bosnes, R. W.: Composition of amniotic fluid. Clin. Obstet. Gynecol., *9*:440, 1966.
5. Clements, J. A.: Assessment of the risk of respiratory distress syndrome by a rapid test for surfactant in amniotic fluid. N. Engl. J. Med., *286*:1077, 1972.

5a. Coch, E. H., Kessler, G., and Meyer, J. S.: Rapid thin-layer chromatographic method for assessing the lecithin/sphingomyelin ratio in amniotic fluid. Clin. Chem. *20*:1368, 1974.

6. Davidson, R., and Rattazzi, M.: Prenatal diagnosis of genetic disorders. Clin. Chem., *18*:179, 1972.
7. Fort, A. T., Morrison, J. C., Ragland, J. B., Morgan, B. S., and Fish, S. A.: Correlation of maternal serum and amniotic fluid bilirubin in gravid patients with sickle cell anemia who were actively hemolyzing. Am. J. Obstet. Gynecol., *112*:227, 1972.
8. Gambino, S. R., and Freda, V. J.: The measurement of amniotic fluid bilirubin by the method of Jendrassik and Grof. Am. J. Clin. Pathol., *46*:198, 1966.
9. Gauthier, C., Desjardins, P., and McLean, F.: Fetal maturity: Amniotic fluid analysis correlated with neonatal assessment. Am. J. Obstet. Gynecol., *112*:344, 1972.
10. Gluck, L., and Kulovich, M.: Lecithin/sphingomyelin ratios in amniotic fluid in normal and abnormal pregnancy. Am. J. Obstet. Gynecol., *115*:539, 1973.
11. Hunter, R. J.: Serum heat stable alkaline phosphatase: an index of placental function. Obstet. Gynecol. Survey, *25*:544, 1970.
12. Liley, A.: Errors in the assessment of hemolytic disease from amniotic fluid. Am. J. Obstet. Gynecol., *86*:485, 1963.

13. Liley, A.: Liquor amnii analysis in the management of the pregnancy complicated by rhesus sensitisation. Am. J. Obstet. Gynecol., *82*:139, 1961.
14. Lind, T., Billewicz, W. Z., and Cheyne, G. A.: Composition of amniotic fluid and maternal blood through pregnancy. J. Obstet. Gynaecol. Br. Commonw., *78*:505, 1971.
15. Milunsky, A., Littlefield, J., Kanfer, J., Kolodny, E., Shih, V., and Atkins, L.: Prenatal genetic diagnosis. N. Engl. J. Med., *283*:1370, 1441, and 1498, 1970.
16. Nadler, H. L., and Gerbie, A. B.: Role of amniocentesis in the intrauterine detection of genetic disorders. N. Engl. J. Med., *282*:596, 1970.
17. Nelson, G. H.: Relationship between amniotic fluid lecithin concentration and respiratory distress syndrome. Am. J. Obstet. Gynecol., *112*:82, 1972.
18. Ostergard, D. R.: The physiology and clinical importance of amniotic fluid. A review. Obstet. Gynecol. Survey, *25*:297, 1970.
19. Pitkin, R. M. and Zwirek, S. J.: Amniotic fluid creatinine. Am. J. Obstet. Gynecol., 98:1135, 1967.
20. Stewart, A., and Taylor, W.: Amniotic fluid analysis as an aid to antepartum diagnosis of hemolytic disease. J. Obstet. Gynaecol. Br. Commonw., *71*:604, 1964.
21. Tietz, N. W.: Spectrophotometry—application. *In*: Workshop Manual on Instrumentation. Sponsored by Commission on Continuing Education, A.S.C.P./A.S.M.T., March 25–26, 1966.
22. Watson, D., Mackay, E., and Trevella, W.: Amniotic fluid analysis and fetal erythroblastosis. Clin. Chim. Acta, *12*:500, 1965.

ADDITIONAL READING

Natelson, S., Scommegna, A., and Epstein, M. B.: Amniotic Fluid. Physiology, Biochemistry, and Clinical Chemistry. New York, John Wiley & Sons, 1974.

Appendix

Compiled by Ermalinda A. Fiereck, M.S.

METRIC UNITS

Prefix Name	*Prefix Symbol*	*Equivalent*	*Units of Length*	*Units of Mass*	*Units of Capacity*
kilo-	k	10^3	kilometer (km)	kilogram (kg)	kiloliter (kl)
		1	meter (m)	gram (g)	liter (l)
deci-	d	10^{-1}	decimeter (dm)	decigram (dg)	deciliter (dl)
centi-	c	10^{-2}	centimeter (cm)	centigram (cg)	centiliter (cl)
milli-	m	10^{-3}	millimeter (mm)	milligram (mg)	milliliter (ml)
micro-	μ	10^{-6}	micrometer (μm)	microgram (μg)	microliter (μl)
nano-	n	10^{-9}	nanometer (nm)	nanogram(ng)	nanoliter (nl)
pico-	p	10^{-12}	picometer (pm)	picogram (pg)	picoliter (pl)

CONVERSION CHARTS

I. UNITS OF LENGTH

Kilometer km	*Meter* m	*Decimeter* dm	*Centimeter* cm	*Millimeter* mm	*Micrometer* μm	*Nanometer* nm	*Angstrom* Å	*Picometer* pm	*Inch* in.
1	10^3	10^4	10^5	10^6	10^9	10^{12}	10^{13}	10^{15}	39.37×10^3
10^{-3}	1	10	10^2	10^3	10^6	10^9	10^{10}	10^{12}	39.37
10^{-4}	10^{-1}	1	10	10^2	10^5	10^8	10^9	10^{11}	39.37×10^{-1}
10^{-5}	10^{-2}	10^{-1}	1	10	10^4	10^7	10^8	10^{10}	39.37×10^{-2}
10^{-6}	10^{-3}	10^{-2}	10^{-1}	1	10^3	10^6	10^7	10^9	39.37×10^{-3}
10^{-9}	10^{-6}	10^{-5}	10^{-4}	10^{-3}	1	10^3	10^4	10^6	39.37×10^{-6}
10^{-12}	10^{-9}	10^{-8}	10^{-7}	10^{-6}	10^{-3}	1	10	10^3	39.37×10^{-9}
10^{-13}	10^{-10}	10^{-9}	10^{-8}	10^{-7}	10^{-4}	10^{-1}	1	10^2	39.37×10^{-10}
10^{-15}	10^{-12}	10^{-11}	10^{-10}	10^{-9}	10^{-6}	10^{-3}	10^{-2}	1	39.37×10^{-12}
2.54×10^{-5}	2.54×10^{-2}	2.54×10^{-1}	2.54	2.54×10	2.54×10^4	2.54×10^7	2.54×10^8	2.54×10^{10}	1

2. UNITS OF MASS

Kilogram kg	*Gram* g	*Decigram* dg	*Centigram* cg	*Milligram* mg	*Microgram* μg	*Nanogram* ng	*Picogram* pg	*Ounce (Av.)* oz.	*Pound (Av.)* lb.
1	10^3	10^4	10^5	10^6	10^9	10^{12}	10^{15}	35.27	2.2
10^{-3}	1	10	10^2	10^3	10^6	10^9	10^{12}	35.27×10^{-3}	2.2×10^{-3}
10^{-4}	10^{-1}	1	10	10^2	10^5	10^8	10^{11}	35.27×10^{-4}	2.2×10^{-4}
10^{-5}	10^{-2}	10^{-1}	1	10	10^4	10^7	10^{10}	35.27×10^{-5}	2.2×10^{-5}
10^{-6}	10^{-3}	10^{-2}	10^{-1}	1	10^3	10^6	10^9	35.27×10^{-6}	2.2×10^{-6}
10^{-9}	10^{-6}	10^{-5}	10^{-4}	10^{-3}	1	10^3	10^6	35.27×10^{-9}	2.2×10^{-9}
10^{-12}	10^{-9}	10^{-8}	10^{-7}	10^{-6}	10^{-3}	1	10^3	35.27×10^{-12}	2.2×10^{-12}
10^{-15}	10^{-12}	10^{-11}	10^{-10}	10^{-9}	10^{-6}	10^{-3}	1	35.27×10^{-15}	2.2×10^{-15}
28.35×10^{-3}	28.35	28.35×10	28.35×10^2	28.35×10^3	28.35×10^6	28.35×10^9	28.35×10^{12}	1	0.0625
0.454	454	454×10	454×10^2	454×10^3	454×10^6	454×10^9	454×10^{12}	16	1

3. UNITS OF CAPACITY

Kiloliter kl	*Liter* l	*Deciliter* dl	*Centiliter* cl	*Milliliter* ml	*Microliter* μl	*Nanoliter* nl	*Picoliter* pl	*Ounce* oz.	*Quart* qt.
1	10^3	10^4	10^5	10^6	10^9	10^{12}	10^{15}	33.81×10^3	1.06×10^3
10^{-3}	1	10	10^2	10^3	10^6	10^9	10^{12}	33.81	1.06
10^{-4}	10^{-1}	1	10	10^2	10^5	10^8	10^{11}	33.81×10^{-1}	1.06×10^{-1}
10^{-5}	10^{-2}	10^{-1}	1	10	10^4	10^7	10^{10}	33.81×10^{-2}	1.06×10^{-2}
10^{-6}	10^{-3}	10^{-2}	10^{-1}	1	10^3	10^6	10^9	33.81×10^{-3}	1.06×10^{-3}
10^{-9}	10^{-6}	10^{-5}	10^{-4}	10^{-3}	1	10^3	10^6	33.81×10^{-6}	1.06×10^{-6}
10^{-12}	10^{-9}	10^{-8}	10^{-7}	10^{-6}	10^{-3}	1	10^3	33.81×10^{-9}	1.06×10^{-9}
10^{-15}	10^{-12}	10^{-11}	10^{-10}	10^{-9}	10^{-6}	10^{-3}	1	33.81×10^{-12}	1.06×10^{-12}
29.57×10^{-6}	29.57×10^{-3}	29.57×10^{-2}	29.57×10^{-1}	29.57	29.57×10^3	29.57×10^6	29.57×10^9	1	3.125×10^{-2}
0.946×10^{-3}	0.946	0.946×10	0.946×10^2	0.946×10^3	0.946×10^6	0.946×10^9	0.946×10^{12}	32	1

GREEK ALPHABET (UPRIGHT AND SLOPING TYPES)

alpha	Α	α	*A*	*α*	nu	Ν	ν	*N*	*ν*
beta	Β	β	*B*	*β*	xi	Ξ	ξ	*Ξ*	*ξ*
gamma	Γ	γ	*Γ*	*γ*	omicron	Ο	ο	*O*	*ο*
delta	Δ	δ	*Δ*	*δ*	pi	Π	π	*Π*	*π*
epsilon	Ε	ε, ϵ	*E*	*ε, ϵ*	rho	Ρ	ρ	*P*	*ρ*
zeta	Ζ	ζ	*Z*	*ζ*	sigma	Σ	σ	*Σ*	*σ*
eta	Η	η	*H*	*η*	tau	Τ	τ	*T*	*τ*
theta	Θ	ϑ, θ	*Θ*	*ϑ, θ*	upsilon	Υ	υ	*Υ*	*υ*
iota	Ι	ι	*I*	*ι*	phi	Φ	φ, ϕ	*Φ*	*φ, ϕ*
kappa	Κ	ϰ, κ	*K*	*ϰ, κ*	chi	Χ	χ	*X*	*χ*
lambda	Λ	λ	*Λ*	*λ*	psi	Ψ	ψ	*Ψ*	*ψ*
mu	Μ	μ	*M*	*μ*	omega	Ω	ω	*Ω*	*ω*

TEMPERATURE CONVERSIONS

(Celsius-Fahrenheit)

$$°C = 5/9 \times (°F - 32)$$
$$°F = (9/5 \times °C) + 32$$

Temp. °C	0	1	2	3	4	5	6	7	8	9
−10	14.0	12.2	10.4	8.6	6.8	5.0	3.2	1.4	−0.4	−2.2
−0	32.0	30.2	28.4	26.6	24.8	23.0	21.2	19.4	17.6	15.8
0	32.0	33.8	35.6	37.4	39.2	41.0	42.8	44.6	46.4	48.2
10	50.0	51.8	53.6	55.4	57.2	59.0	60.8	62.6	64.4	66.2
20	68.0	69.8	71.6	73.4	75.2	77.0	78.8	80.6	82.4	84.2
30	86.0	87.8	89.6	91.4	93.2	95.0	96.8	98.6	100.4	102.2
40	104.0	105.8	107.6	109.4	111.2	113.0	114.8	116.6	118.4	120.2
50	122.0	123.8	125.6	127.4	129.2	131.0	132.8	134.6	136.4	138.2
60	140.0	141.8	143.6	145.4	147.2	149.0	150.8	152.6	154.4	156.2
70	158.0	159.8	161.6	163.4	165.2	167.0	168.8	170.6	172.4	174.2
80	176.0	177.8	179.6	181.4	183.2	185.0	186.8	188.6	190.4	192.2
90	194.0	195.8	197.6	199.4	201.2	203.0	204.8	206.6	208.4	210.2
100	212.0	213.8	215.6	217.4	219.2	221.0	222.8	224.6	226.4	228.2

ATOMIC MASSES *

Name	Atomic Number	Symbol	International Atomic Mass	Oxidative States	Name	Atomic Number	Symbol	International Atomic Mass	Oxidative States
Aluminum	13	Al	26.98	+3	Manganese	25	Mn	54.94	+2, +3, +4, +7
Antimony, stibium	51	Sb	121.8	+3, +5, −3	Mercury	80	Hg	200.6	+1, +2
Argon	18	Ar	39.95	0	Molybdenum	42	Mo	95.94	+6
Arsenic	33	As	74.92	+3, +5, −3	Neon	10	Ne	20.18	0
Barium	56	Ba	137.3	+2	Nickel	28	Ni	58.71	+2, +3
Beryllium	4	Be	9.012	+2	Nitrogen	7	N	14.01	+1, +2, +3, +4, +5, −1, −2, −3
Bismuth	83	Bi	209.0	+3, +5					
Boron	5	B	10.81	+3	Oxygen	8	O	16.00	−2
Bromine	35	Br	79.91	+1, +5, −1	Palladium	46	Pd	106.4	+2, +4
Cadmium	48	Cd	112.4	+2	Phosphorus	15	P	30.97	+3, +5, −3
Calcium	20	Ca	40.08	+2	Platinum	78	Pt	195.1	+2, +4
Carbon	6	C	12.01	+2, +4, −4	Potassium, kalium	19	K	39.10	+1
Cerium	58	Ce	140.1	+3, +4	Selenium	34	Se	78.96	+4, +6, −2
Cesium	55	Cs	132.9	+1	Silicon	14	Si	28.09	+2, +4, −4
Chlorine	17	Cl	35.45	+1, +5, +7, −1	Silver, argentum	47	Ag	107.9	+1
Chromium	24	Cr	52.00	+2, +3, +6	Sodium, natrium	11	Na	22.99	+1
Cobalt	27	Co	58.93	+2, +3	Strontium	38	Sr	87.62	+2
Copper	29	Cu	63.54	+1, +2	Sulfur	16	S	32.06	+4, +6, −2
Fluorine	9	F	19.00	−1	Tellurium	52	Te	127.6	+4, +6, −2
Gold, aurum	79	Au	197.0	+1, +3	Thallium	81	Tl	204.4	+1, +3
Helium	2	He	4.003	0	Thorium	90	Th	232.0	+4
Hydrogen	1	H	1.008	+1, −1	Tin, stantium	50	Sn	181.7	+2, +4
Iodine	53	I	126.9	+1, +5, +7, −1	Titanium	22	Ti	47.90	+2, +3, +4
Iron, ferrum	26	Fe	55.85	+2, +3	Tungsten, wolfram	74	W	183.8	+6
Lanthanum	57	La	138.9	+3	Uranium	92	U	238.0	+3, +4, +5, +6
Lead, plumbum	82	Pb	207.2	+2, +4	Vanadium	23	V	50.94	+2, +3, +4, +5
Lithium	3	Li	6.939	+1	Xenon	54	Xe	131.3	0
Magnesium	12	Mg	24.31	+2	Zinc	30	Zn	65.37	+2

* Values as of 1963, based on carbon-12 and rounded off to four significant figures.

BOILING POINTS OF COMMONLY USED SOLVENTS

Name	*Additional Names*	*Molecular Mass*	*Boiling Point** (°C)
Acetic acid	Ethanoic acid	60.05	118.5^{760}
Acetoacetic acid	3-Oxobutanoic acid	102.09	<100 *d*†
Acetone	2-Propanone	58.08	56.2
Aniline	Aminobenzene	93.13	184.3^{760}
Benzene		78.11	80.1
n-Butanol	1-Butanol	74.12	117.5^{760}
Carbon disulfide		76.14	45^{760}
Carbon tetrachloride	Tetrachloromethane	153.82	76.8^{760}
Chloroform	Trichloromethane	119.38	61.2^{760}
Ethanol	Ethyl alcohol	46.07	78.5
Ethyl acetate	Acetic acid ethyl ester	88.11	77.1^{760}
Ethyl ether	Diethyl ether	74.12	34.6
Ethylene dichloride	1,2-Dichloroethane	98.96	84^{760}
Heptane		100.21	98.4
Isoamyl acetate	Acetic acid 3-methylbutyl ester	130.2	142
Isoamyl alcohol	3-Methyl-1-butanol	88.15	131^{760}
Isobutyl alcohol	2-Methyl-1-propanol	74.12	108.4
Isopropyl alcohol	2-Propanol	60.09	82.4
Methanol	Carbinol; Methyl alcohol	32.04	65.0^{760}
Methyl isobutyl ketone	4-Methyl-2-propanone	100.16	116.9
Methylene chloride	Dichloromethane	84.93	40
Nitrobenzene		123.11	210.8^{760}
Petroleum ether		Varies with fraction	Appr. 40 to 120 Varies with fraction
Pyridine		79.10	115.5
Toluene	Methylbenzene	92.13	110.6
p-Xylene	1,4-Dimethylbenzene	106.16	138
m-Xylene	1,3-Dimethylbenzene	106.16	139
o-Xylene	1,2-Dimethylbenzene	106.16	144

* Superscript indicates the barometric pressure at which the boiling point was measured. If no figure is given, the barometric pressure was measured at approximately 1 atmosphere.

† *d* = decomposes.

PRIMARY AND SECONDARY STANDARDS

PRIMARY STANDARDS

	Formula	*Molecular Mass (also, g quantity needed for 1 liter of solution containing 1 mole)*	*Equivalent Mass (also, g quantity needed for 1 liter of solution containing 1 equivalent)*
Sodium carbonate	Na_2CO_3	105.989	52.994
Sodium oxalate	$Na_2C_2O_4$	134.000	67.000
Sodium chloride	$NaCl$	58.443	58.443
Potassium iodate	KIO_3	214.005	35.67
Potassium dichromate	$K_2Cr_2O_7$	294.192	49.04
Potassium hydrogen phthalate	$KHC_8H_4O_4$	204.229	204.229
Succinic acid	$HOOC—(CH_2)_2—COOH$	118.090	59.045
Sodium tetraborate, decahydrate	$Na_2B_4O_7 \cdot 10\ H_2O$	381.373	190.686
Tris(hydroxymethyl)-aminomethane (Tromethamine, THAM, or Tris)	$NH_2C(CH_2OH)_3$	121.14	121.14

SECONDARY STANDARDS

	Formula	*Molecular Mass (also, g quantity needed for 1 liter of solution containing 1 mole)*	*Equivalent Mass (also, g quantity needed for 1 liter of solution containing 1 equivalent)*
Oxalic acid	$H_2C_2O_4$	90.035	45.018
Oxalic acid, dihydrate	$H_2C_2O_4 \cdot 2\ H_2O$	126.067	63.038
Nitric acid	HNO_3	63.013	*
Hydrochloric acid	HCl	36.461	*
Sulfuric acid	H_2SO_4	98.077	*

* Calculate from specific gravity and assay of acid and establish exact concentration by titration against primary standard.

INDICATORS

Common Name	Chemical Name and Formula	pH *Range*	Color Change	Commonly Used Concentration
Bromcresol green	3,3′,5,5′-Tetrabromo-*m*-cresolsulfonphthalein 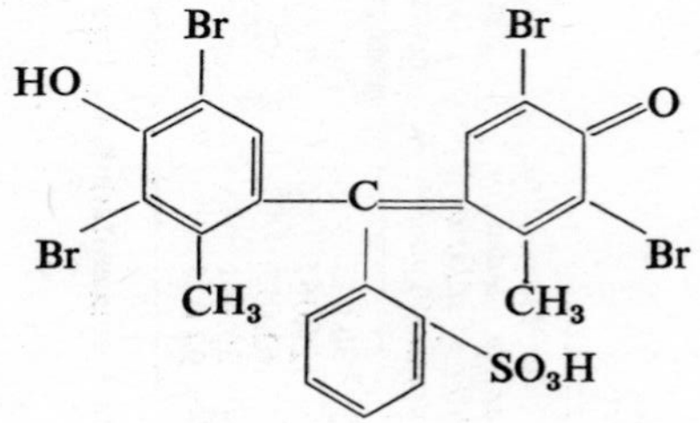	3.8–5.4	yellow to green	0.04 g/100 ml of 0.6 mmolar NaOH
Bromphenol blue	3,3′,5,5′-Tetrabromophenolsulfonphthalein 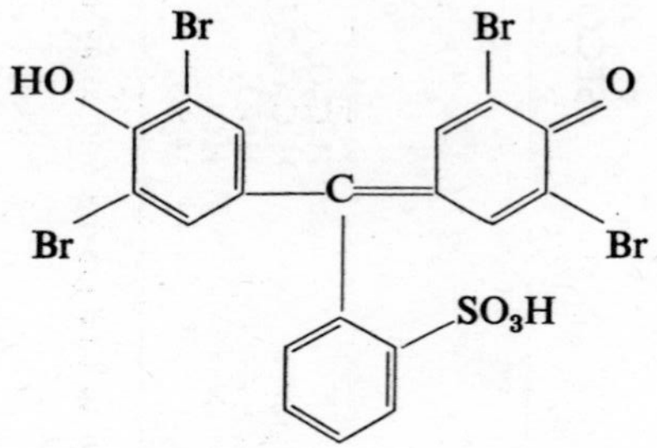	3.0–4.6	yellow to blue	0.04 g/100 ml of 0.6 mmolar NaOH

Bromthymol blue	3,3′-Dibromothymolsulfonphthalein 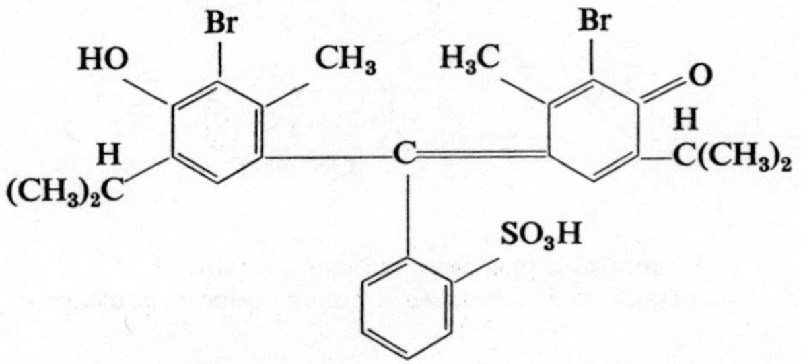	6.0–7.6	yellow to blue	0.04 g/100 ml of 0.6 mmolar NaOH
Congo red	Sodium diphenyldiazo-bis-α-naphthylaminesulfonate 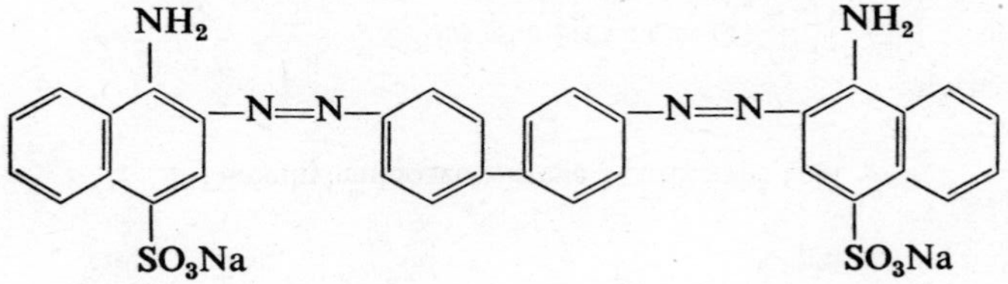	3.0–5.0	blue–violet to red	0.1 g/100 ml of H_2O
Cresol red	*o*-Cresolsulfonphthalein; α-hydroxy-α,α-bis(4-hydroxy-*m*-tolyl)-*o*-toluenesulfonic acid γ-sultone 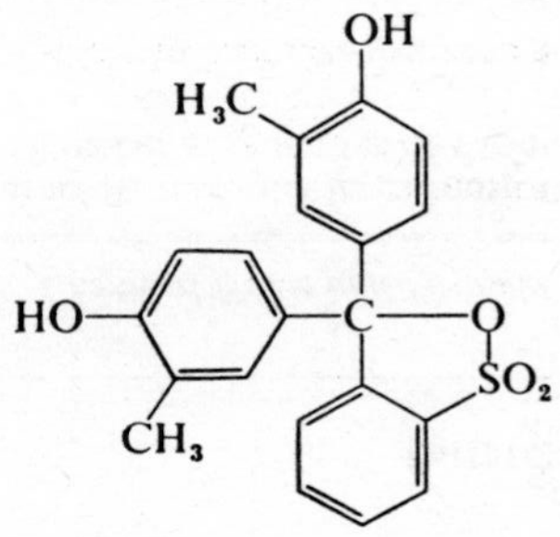	7.2–8.8	yellow to red (orange to amber at pH 2–3)	0.04 g/100 ml of 1.1 mmolar NaOH

Table continued on the following page

INDICATORS (Continued)

Common Name	Chemical Name and Formula	pH Range	Color Change	Commonly Used Concentration
Litmus	Lacmus; tournesol; turnsole; lacca musica; lacca coerulea. (Blue coloring matter of various lichens.)	4.5–8.3	red to blue	
Methyl orange	Sodium 4′-dimethylaminoazobenzene 4-sulfonate 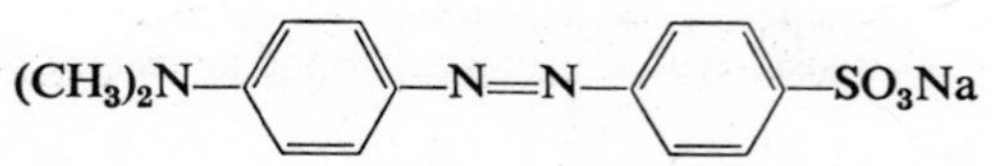 $(CH_3)_2N$—N=N—SO_3Na	3.0–4.4	red to yellow	0.1 g/100 ml of H_2O
Methyl red	4′-Dimethylaminoazobenzene 2-carboxylic acid 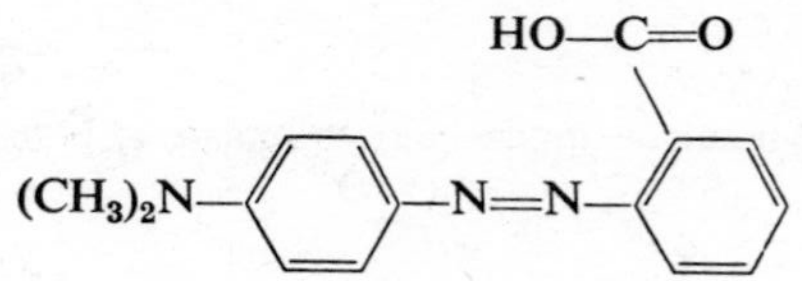 HO—C=O $(CH_3)_2N$—N=N—	4.2–6.3	red to yellow	0.05 g/100 ml of 50 per cent ethanol
Phenol red	Phenolsulfonphthalein; α-hydroxy-α,α-bis (*p*-hydroxyphenyl)-*o*-toluenesulfonic acid γ-sultone 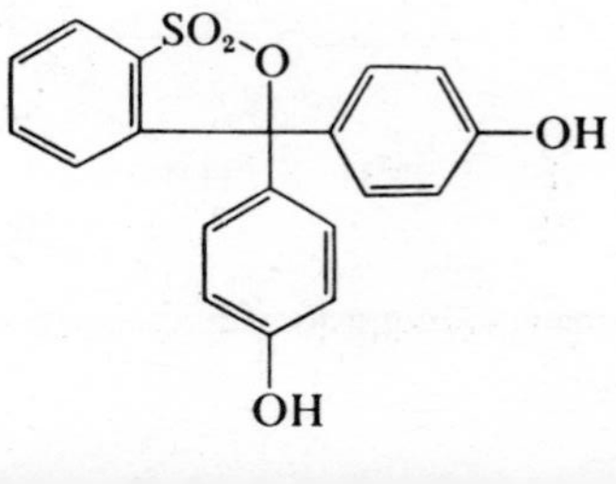 SO_2-O OH OH	6.8–8.4	yellow to red	0.04 g/100 ml of 1.1 mmolar NaOH

Phenolphthalein	3,3-Bis(*p*-hydroxyphenyl)phthalide 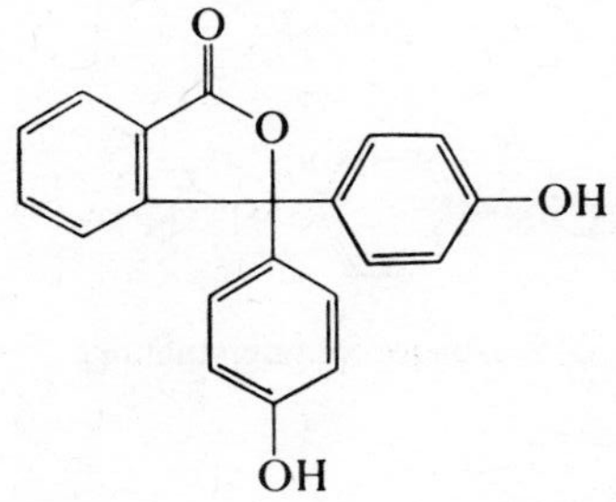	8.3–10.0	colorless to red	0.1 g/100 ml of 95 per cent ethanol
Thymol blue	Thymolsulfonephthalein; α-hydroxy-α,α-bis(5-hydroxy-carvacryl)-*o*-toluenesulfonic acid γ-sultone 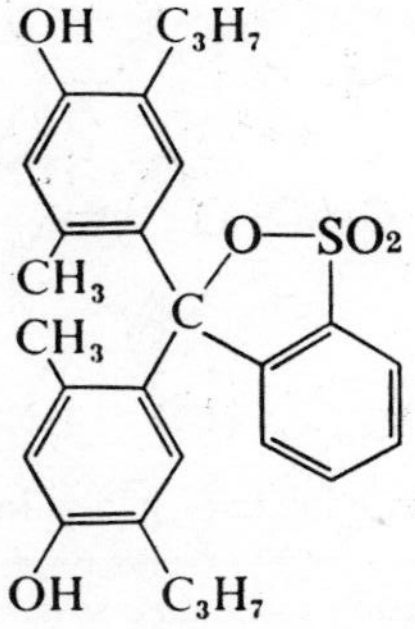	Acid Range: 1.2–2.8	red to yellow	If sodium salt is used: 0.04 g/100 ml of 95 per cent ethanol
		Alkaline Range: 8.0–9.6	yellow to blue	If acid form is used: 0.04 g/100 ml of 1.0 mmolar NaOH

Table continued on the following page

INDICATORS (Continued)

Common Name	*Chemical Name and Formula*	pH *Range*	*Color Change*	*Commonly Used Concentration*
Thymolphthalein	5′,5″-Diisopropyl-2′,2″-dimethylphenolphthalein	9.3–10.5	colorless to blue	0.1 g/100 ml of 95 per cent ethanol
Toepfer's reagent	Dimethylaminoazobenzene $C_6H_5-N{=}N-C_6H_4-N(CH_3)_2$	2.9–4.0	red to yellow	0.5 g/100 ml of 95 per cent ethanol

IONIZATION CONSTANTS (K) FOR COMMON ACIDS AND BASES IN WATER*

	K	pK_a
Acetic acid	1.75×10^{-5}	4.76
Acetoacetic acid	2.62×10^{-4}	3.58 (18°C)
Ammonia	5.6×10^{-10}	9.25
Boric acid†	6.4×10^{-10}	9.19
Carbonic acid	4.47×10^{-7}	6.35
	4.68×10^{-11}	10.34
Citric acid	7.4×10^{-4}	3.13
	1.7×10^{-5}	4.77
	4.0×10^{-7}	6.40
Diethylbarbituric acid (Veronal)	3.7×10^{-8}	7.43
Ethylenediamine	1.4×10^{-7}	6.85
	1.12×10^{-10}	9.93
Ethylenediamine tetraacetate	1.00×10^{-2}	2.00
	2.16×10^{-3}	2.67
	6.92×10^{-7}	6.16
	5.50×10^{-11}	10.26
Formic acid	1.76×10^{-4}	3.75
Glycine	4.5×10^{-3}	2.35
	1.7×10^{-10}	9.77
Glycylglycine	7.24×10^{-4}	3.14
	5.62×10^{-9}	8.25
Hydroxylamine	9.1×10^{-9}	8.04
Imidazole	1.01×10^{-7}	6.95
Isocitric acid	5.13×10^{-4}	3.29
	1.99×10^{-5}	4.70
	3.98×10^{-7}	6.40
p-Nitrophenol	7×10^{-8}	7.15
Oxalacetic acid	2.75×10^{-3}	2.56
	4.27×10^{-5}	4.37
Oxalic acid	6.5×10^{-2}	1.19
	6.1×10^{-5}	4.21
Phosphoric acid	7.5×10^{-3}	2.12
	6.2×10^{-8}	7.21
	4.8×10^{-13}	12.32
Phosphorous acid	5×10^{-2}	1.30
	2.6×10^{-7}	6.59
Pyruvic acid	3.23×10^{-3}	2.49
Succinic acid	6.2×10^{-5}	4.21
	2.3×10^{-6}	5.64
Sulfuric acid	$\gg 1$	—
	1.2×10^{-2}	1.92
Tartaric acid	1.1×10^{-3}	2.96
	6.9×10^{-5}	4.16
Triethanolamine	1.26×10^{-8}	7.90
Tris(hydroxymethyl)-amino methane	8.32×10^{-9}	8.08

* Temperature at or near room temperature (25°C) unless otherwise indicated.

† Boric acid acts as a monotropic acid in aqueous solution.

DATA ON COMPRESSED GASES*

	Relative Specific Gravity (Air = 1)	*State of Gas in Cylinder*	*Approximate Pressure in Cylinder (psi)*	*Purity of Best Grade in Cylinder (Molfr.)*	*Flammability and Toxicity*	*Flammability Limit in Air by Volume (Per Cent)*	*Special Notes*	*Leak Detection*
Acetylene	0.9073	Dissolved in acetone	250 (21°C)	>0.996	Highly flammable. Nontoxic but asphyxiant and anesthetic	2.5–81	1. Use only at pressures less than 30 psi 2. Never use with unalloyed Cu, Ag, and Hg 3. Store upright—acetone is included in cylinder	Soap solution
Air	1.000	Nonliquefied	2200–2400 (21°C)	Mixture	Nonflammable. Nontoxic		Dry air is inert to metals and plastics at ambient temperatures	Soap solution
Ammonia	0.5870	Liquefied	114 (21°C)	>0.9995	Flammable. Toxic in concentrations greater than 100 μl/l for 8 h	15–28	1. Cu, Sn, Zn, and their alloys are attacked by moist ammonia 2. Never use mercury; ammonia can combine with Hg to form explosive compounds	1. Open bottle of HCl 2. Wet red litmus paper 3. Wet phenol-phthalein paper
Argon	1.38	Nonliquefied	2200 (21°C)	>0.999995	Inert. Nontoxic but acts as a simple asphyxiant		Inert	Soap solution
Butane	2.076 (15°C, 1 atm.)	Liquefied	16.3 (21°C)	>0.9999	Extremely flammable. Nontoxic but acts as a simple asphyxiant and has anesthetic effect	1.9–8.5	Noncorrosive	Soap solution
Carbon dioxide	1.5289	Liquefied	830 (21°C)	>0.99995	Nonflammable. Toxic in concentrations greater than 5000 μl/l for 8 h		Dry gas is relatively inert	1. Soap solution 2. Aqueous ammonia

Carbon monoxide	0.9678 (21°C, 1 atm.)	Nonliquefied	1500 (21°C)	>0.998	Extremely flammable. Extremely toxic; maximum allowable concentration for 8 h exposure is 100 μl/l	12.5–74	Corrosion by pure CO at low pressure can be considered negligible. At high pressures it will react with Fe, Ni, and other metals forming carbonyls	Soap solution
Helium	0.137	Nonliquefied	2200 (21°C)	>0.999995	Inert. Nontoxic but acts as a simple asphyxiant		Inert	Soap solution
Hydrogen	0.06952	Nonliquefied	2000 (21°C)	>0.999995	Extremely flammable. Nontoxic but can act as asphyxiant	4.0–75	Noncorrosive	Soap solution
Hydrogen chloride	1.268 (gas, 0°C)	Gas over liquid	613 (21°C)	>0.990	Nonflammable. Highly toxic; 0.13–0.2 per cent lethal in a few minutes. Maximum accepted concentration for 8 h is 5 μl/l		1. Hydrogen chloride is essentially inert and does not attack metals; however, in the presence of moisture, hydrogen chloride will corrode most metals except silver, platinum, and tantalum 2. *Always* shut off hydrogen chloride from the use end, backward to the cylinder	1. Large leaks evident by dense white fumes on contact with the atmosphere 2. Open bottle of conc. NH_4OH 3. Wet blue litmus paper
Hydrogen sulfide	1.1895 (gas, 15°C)	Liquefied	250 (21°C)	>0.995	Highly flammable. Highly toxic. Maximum allowable concentration for 8 h is 20 μl/l	4.3–45		1. Soap solution 2. Cadmium chloride solution (turns yellow upon contact with H_2S) 3. Moist lead acetate paper (turns black upon contact with H_2S)
Methane	0.5549 (15°C, 1 atm.)	Nonliquefied	2265	>0.99993	Extremely flammable. Nontoxic but acts as a simple asphyxiant	5.3–14	Noncorrosive	Soap solution

* Compiled from *Matheson Gas Data Book*, 4th Ed., The Matheson Company Inc., Herst Litho Inc., New York, 1966.

Table continued on the following page

DATA ON COMPRESSED GASES* (Continued)

	Relative Specific Gravity (Air = 1)	*State of Gas in Cylinder*	*Approximate Pressure in Cylinder (psi)*	*Purity of Best Grade in Cylinder (Molfr.)*	*Flammability and Toxicity*	*Flammability Limit in Air by Volume (Per Cent)*	*Special Notes*	*Leak Detection*
Neon	0.6964 (gas, 21°C, 1 atm.)	Nonliquefied	1800	>0.99995	Inert. Nontoxic but acts as an asphyxiant		Inert	Soap solution
Nitrogen	0.9670	Nonliquefied	2200	>0.999999	Inert. Nontoxic but acts as an asphyxiant		Inert	Soap solution
Nitrous oxide	1.530 (gas, 15°C)	Liquefied	745 (21°C)	>0.980	Nonflammable. Nontoxic but acts as asphyxiant and anesthetic		Noncorrosive	Soap solution
Oxygen	1.1053	Nonliquefied	2200 (21°C)	>0.9999	Nonflammable but may cause explosive oxidation of organic substances such as oil, grease, etc. Nontoxic		Noncorrosive. Extreme caution should be taken to avoid contact with oil, grease, or other readily combustible substances as explosion may occur	Soap solution
Propane	1.5503 (15°C, 1 atm.)	Liquefied	110 (21°C)	>0.9999	Extremely flammable. Nontoxic but acts as an asphyxiant and in high concentration has an anesthetic action	2.2–9.5	Noncorrosive	Soap solution

* Compiled from *Matheson Gas Data Book*, 4th Ed., The Matheson Company Inc., Herst Litho Inc., New York, 1966.

DESIRABLE WEIGHTS (MASSES*) FOR MEN AND WOMEN†

ACCORDING TO HEIGHT AND FRAME. AGES 25 AND OVER

Height (In Shoes) ft. + in. ...... cm	*Weight (Mass) (In Indoor Clothing)*					
	Small Frame lb.	kg	*Medium Frame* lb.	kg	*Large Frame* lb.	kg
	Men					
5′ 2″......157.5	112–120	50.9–54.5	118–129	53.6–58.6	126–141	57.3–64.1
3″......160.0	115–123	52.3–55.9	121–133	55.0–60.4	129–144	58.6–66.5
4″......162.6	118–126	53.6–57.2	124–136	56.4–61.8	132–148	60.0–67.3
5″......165.1	121–129	55.0–58.6	127–139	57.7–63.2	135–152	61.4–69.1
6″......167.6	124–133	56.4–60.4	130–143	59.0–65.0	138–156	62.7–70.9
7″......170.2	128–137	58.2–62.3	134–147	60.9–66.8	142–161	64.5–73.2
8″......172.7	132–141	60.0–64.1	138–152	62.7–69.1	147–166	66.8–75.4
9″......175.3	136–145	61.8–65.9	142–156	64.5–70.9	151–170	68.6–77.2
10″......177.8	140–150	63.6–68.2	146–160	66.4–72.7	155–174	70.4–79.1
11″......180.3	144–154	65.4–70.0	150–165	68.2–75.0	159–179	72.3–81.4
6′ 0″......182.9	148–158	67.3–71.8	154–170	70.0–77.3	164–184	74.5–83.6
1″......185.4	152–162	69.1–73.6	158–175	71.8–79.5	168–189	76.4–85.9
2″......188.0	156–167	70.9–75.9	162–180	73.6–81.8	173–194	78.6–88.2
3″......190.5	160–171	72.7–77.7	167–185	75.9–84.1	178–199	80.9–90.4
4″......193.0	164–175	74.5–79.5	172–190	78.2–86.4	182–204	82.7–92.7
	Women					
4′ 10″......147.3	92–98	41.8–44.5	96–107	43.6–48.6	104–119	47.3–54.1
11″......149.9	94–101	42.7–45.9	98–110	44.5–50.0	106–122	48.2–55.4
5′ 0″......152.4	96–104	43.6–47.3	101–113	45.9–51.4	109–125	49.5–56.8
1″......154.9	99–107	45.0–48.6	104–116	47.3–52.7	112–128	50.9–58.2
2″......157.5	102–110	46.4–50.0	107–119	48.6–54.1	115–131	52.3–59.5
3″......160.0	105–113	47.7–51.4	110–122	50.0–55.4	118–134	53.6–60.9
4″......162.6	108–116	49.1–52.7	113–126	51.4–57.3	121–138	55.0–62.7
5″......165.1	111–119	50.4–54.1	116–130	52.7–59.1	125–142	56.8–64.5
6″......167.6	114–123	51.8–55.9	120–135	54.5–61.4	129–146	58.6–66.4
7″......170.2	118–127	53.6–57.7	124–139	56.4–63.2	133–150	60.4–68.2
8″......172.7	122–131	55.4–59.5	128–143	58.2–65.0	137–154	62.3–70.0
9″......175.3	126–135	57.3–61.4	132–147	60.0–66.8	141–158	64.1–71.8
10″......177.8	130–140	59.1–63.6	136–151	61.8–68.6	145–163	65.9–74.1
11″......180.3	134–144	60.9–65.4	140–155	63.6–70.4	149–168	67.7–76.4
6′ 0″......182.9	138–148	62.7–67.3	114–159	65.4–72.3	153–173	69.5–78.6

* According to the SI system, mass is the preferred term; however, weight is more commonly used in the United States.

† Prepared by the Metropolitan Life Insurance Company. Derived primarily from data of the *Build and Blood Pressure Study, 1959*, Society of Actuaries. Reproduced with permission (modified).

BROMSULPHALEIN DOSAGE SCHEDULE FOR BSP LIVER FUNCTION TEST

Vial concentration: 50 mg/ml
Dose: 5 mg/kg (5 mg/2.2 lb.)
Calculation:

$$\text{ml needed for injection} = \frac{1\ \text{ml}}{50\ \text{mg}} \times \frac{5\ \text{mg}}{2.2\ \text{lb.}} \times \text{mass (weight*) of patient in lb.}$$

or 0.0454 × mass (weight*) of patient in lb.
or 0.1 × mass (weight*) of patient in kg

Mass (Weight)* lb.	kg	*ml of Dye Solution*
60	27.3	2.7
70	31.8	3.2
80	36.4	3.6
90	40.9	4.1
100	45.4	4.5
110	50.0	5.0
120	54.5	5.5
130	59.1	5.9
140	63.6	6.4
150	68.2	6.8
160	72.7	7.3
170	77.3	7.7
180	81.8	8.2
190	86.4	8.6
200	90.9	9.1
210	95.4	9.5
220	100.0	10.0

* See first footnote on p. 1193.

NOMOGRAM FOR THE DETERMINATION OF BODY SURFACE AREA OF CHILDREN*

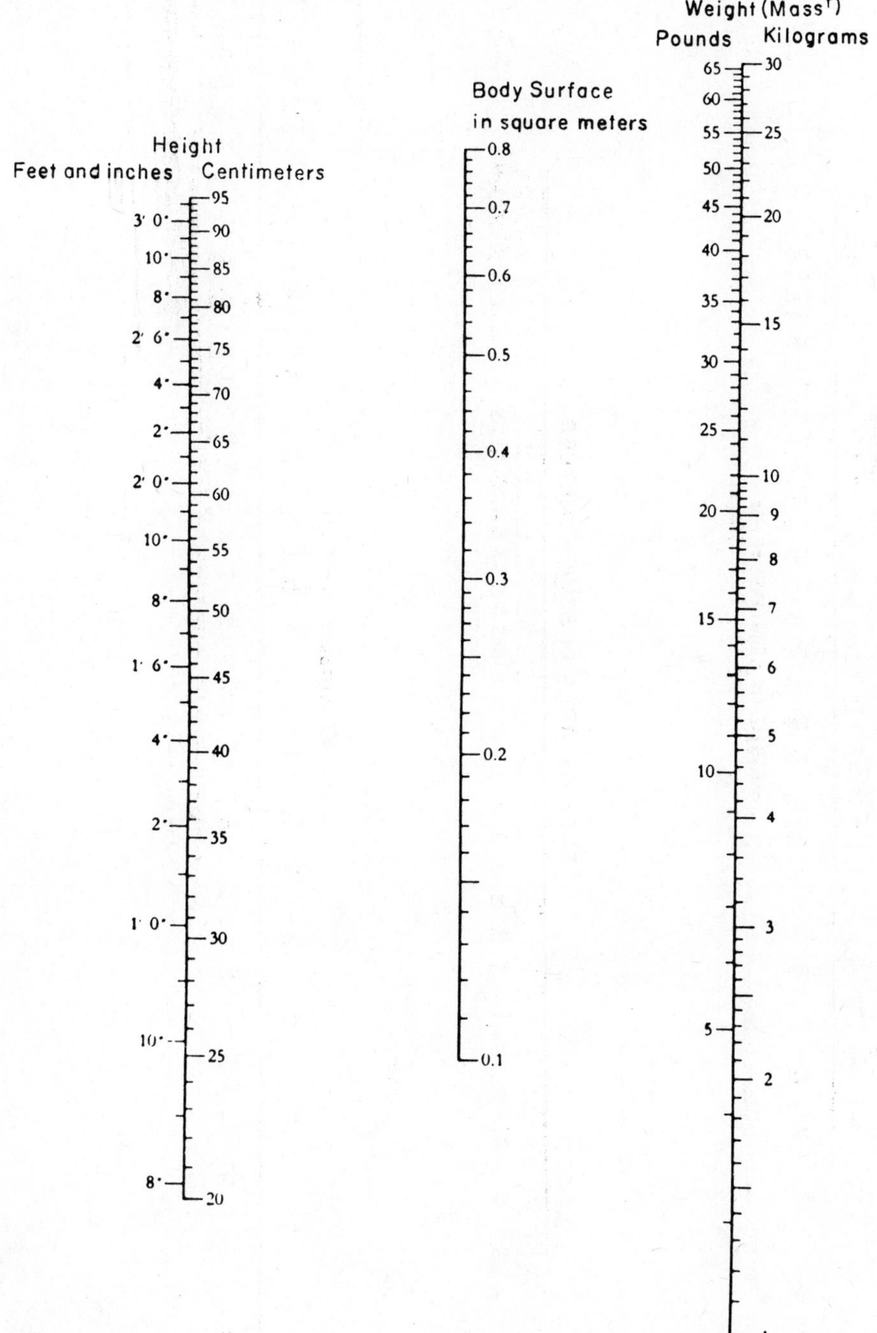

* From DuBois, E. F.: *Basal Metabolism in Health and Disease*, Philadelphia, Lea & Febiger, 1936.

† See first footnote on p. 1193.

NOMOGRAM FOR THE DETERMINATION OF BODY SURFACE OF CHILDREN AND ADULTS*

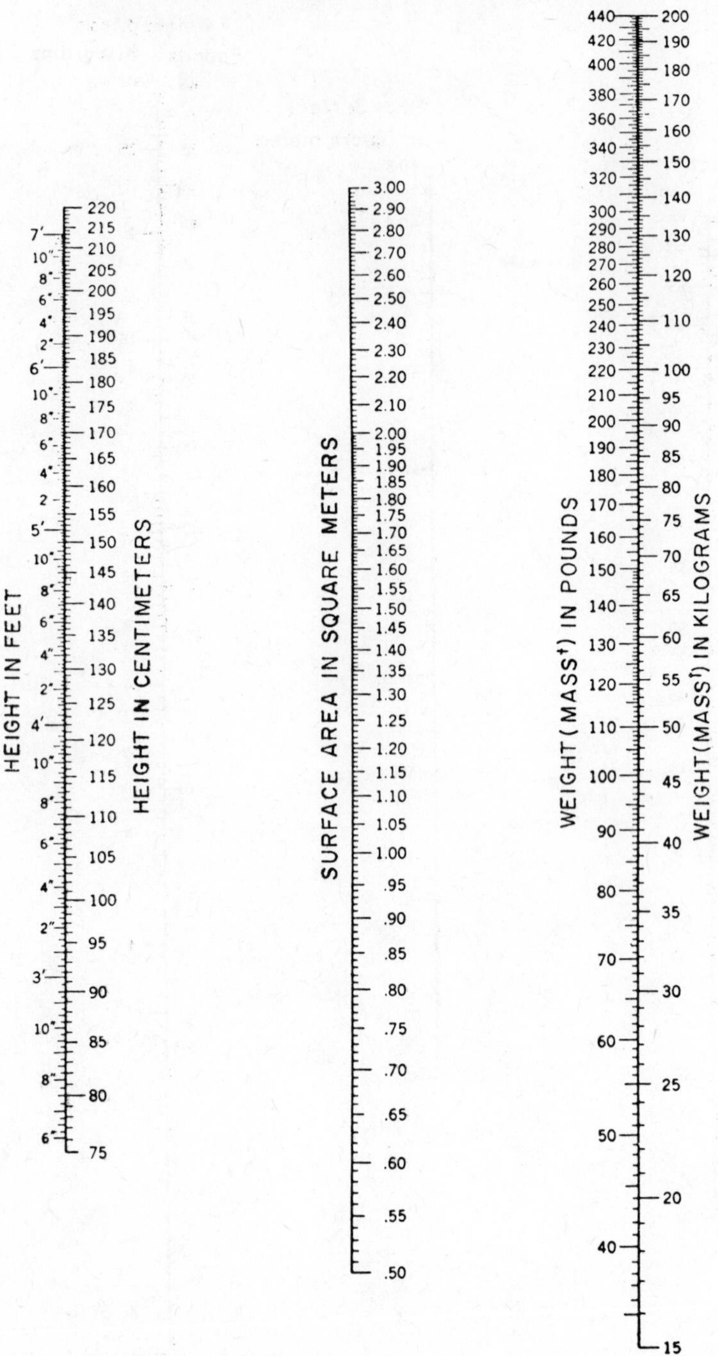

* From Boothby, W. M., and Sandiford, R. B.: Boston M. & S.J. *185*:337, 1921.

† See first footnote on p. 1193.

THE IEC RELATIVE CENTRIFUGAL FORCE NOMOGRAPH

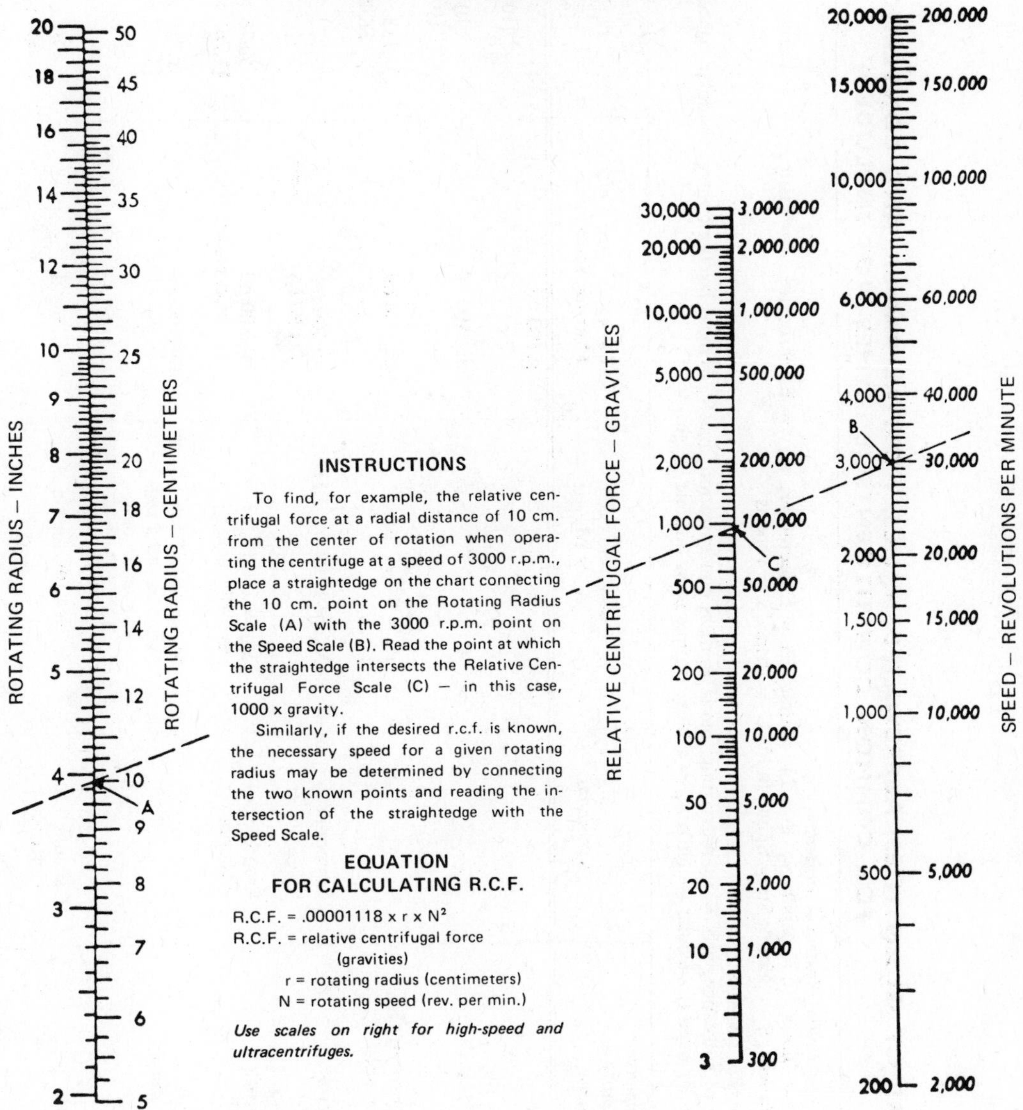

INSTRUCTIONS

To find, for example, the relative centrifugal force at a radial distance of 10 cm. from the center of rotation when operating the centrifuge at a speed of 3000 r.p.m., place a straightedge on the chart connecting the 10 cm. point on the Rotating Radius Scale (A) with the 3000 r.p.m. point on the Speed Scale (B). Read the point at which the straightedge intersects the Relative Centrifugal Force Scale (C) — in this case, 1000 x gravity.

Similarly, if the desired r.c.f. is known, the necessary speed for a given rotating radius may be determined by connecting the two known points and reading the intersection of the straightedge with the Speed Scale.

EQUATION FOR CALCULATING R.C.F.

R.C.F. = $.00001118 \times r \times N^2$

R.C.F. = relative centrifugal force (gravities)

r = rotating radius (centimeters)

N = rotating speed (rev. per min.)

Use scales on right for high-speed and ultracentrifuges.

LIST OF QUANTITIES, ABBREVIATED NAMES, NORMAL RANGE, AND CONVERSION FACTORS FOR COMMON CONSTITUENTS (ENZYMES NOT INCLUDED)

Abbreviations for Description of System Analyzed:
a = arterial; B = Blood; c = capillary; d = 24 hours; Ex = Expectorate; Ery = Erythrocyte; f = fasting; F = Feces; Hb = Hemoglobin; n = night
P = Plasma; Pt = Patient; S = Serum; Sp = Spinal fluid; U = Urine; v = venous

Abbreviations for Kinds of Quantity:
ams. = amount of substance; arb. = arbitrary unit; masc. = mass concentration; masfr. = mass fraction; molc. = molar concentration; molfr. = mole fraction
M_r = relative molecular mass (molecular weight); A_r = relative atomic mass (atomic weight)

System-component, Kind of Quantity, Normal Range	*Synonyms*	*Comments & A_r or M_r*	*Currently Used Units (Not Recommended)*	*Conversion Factors* → [←]	*Recommended Unit*
S-Albumin, masc., 35–47 g/l			g/100 ml	10 [10^{-1}]	g/l
Sp-Albumin, masc., 0.10–0.30 g/l			mg/100 ml	10^{-2} [10^{2}]	g/l
fS-Amino acid nitrogen (N), molc., 2.1–4.2 mmol/l		A_r = 14.01 for N	mg/100 ml	7.138×10^{-1} [1.401]	mmol/l
P-Ammonium, molc., 23.5–47.0 μmol/l	Ammonia	M_r = 17.03 for NH_3	μg/100 ml (as NH_3)	5.872×10^{-1} [1.703]	μmol/l
fP-Ascorbate, molc., 28.4–85.2 μmol/l	Ascorbic acid Vitamin C	M_r = 176.12 for ascorbic acid	mg/100 ml (as ascorbic acid)	5.678×10^{1} [1.761×10^{-2}]	μmol/l
fS-Bilirubin (total), molc., 3.4–17.1 μmol/l		M_r = 584.65 for unconjugated bilirubin as undissociated acid	mg/100 ml	1.710×10^{1} [5.847×10^{-2}]	μmol/l
dU-Calcium (Ca), mole rate, 1.25–3.74 mmol/d	Calcium excretion	A_r = 40.08 for Ca	mg/d	2.495×10^{-2} [4.008×10^{1}]	mmol/d

fS-Calcium (Ca, total), molc., 2.12–2.62 mmol/l	Calcium	A_r = 40.08 for Ca	mg/100 ml ↑ 2.004	2.495×10^{-1} [4.008]	mmol/l
			mEq/l	5.000×10^{-1} [2.000]	mmol/l
Alveolar gas-Carbon dioxide, partial pressure	Alveolar carbon dioxide tension	See Next Entry			
Gas-Carbon dioxide, partial pressure (aB equil.)	Carbon dioxide pressure pCO_2		mmHg	1.333×10^{-1} [7.502]	kPa
P-Carbonate (total + CO_2), molc., 23–30 mmol/l	Alkali reserve Bicarbonate Total carbon dioxide		Volume percent (as carbon dioxide at 0°C, 1013 mbar, dry, having a molar volume of 22.26 liters/mol)	4.492×10^{-1} [2.226]	mmol/l
dU-Chloride, mole rate, 110–250 mmol/d	Chloride excretion	A_r = 35.46 for Cl	mg/d	2.820×10^{-2} [3.546×10^{1}]	mmol/d
			mEq/d	1 [1]	mmol/d
S-Chloride, molc., 98–108 mmol/l		A_r = 35.46 for Cl	mg/100 ml	2.820×10^{-1} [3.546]	mmol/l
U-Chloride, molc. (See dU-Chloride above)		A_r = 35.46 for Cl	mg/100 ml ↑ 3.546	2.820×10^{-1} [3.546]	mmol/l
			mEq/l	1 [1]	mmol/l
fS-Cholesterol (total), molc., 3.36–6.46 mmol/l	Total cholesterol	M_r = 386.64 for cholesterol	mg/100 ml	2.586×10^{-2} [3.866×10^{1}]	mmol/l
S-Copper (Cu, total), molc., 11.0–22.0 μmol/l		A_r = 63.54 for Cu	μg/100 ml	1.574×10^{-1} [6.354]	μmol/l

Table continued on the following page.

System-component, Kind of Quantity, Normal Range	*Synonyms*	*Comments & A_r or M_r*	*Currently Used Units (Not Recommended)*	*Conversion Factors* → [←]	*Recommended Unit*
dU-Coproporphyrins (I & III), mole rate, 91.6–274.8 mmol/d	Coproporphyrin excretion	M_r = 654.73 for coproporphyrins	μg/d	1.527 [6.547×10^{-1}]	nmol/d
dU-Creatininium, mole rate, 8.8–17.7 mmol/d	Creatinine excretion	M_r = 113.12 for creatinine	g/d as creatinine	8.840 [1.131×10^{-1}]	mmol/d
fS-Creatininium, molc., 0.053–0.106 mmol/l	Creatinine	M_r = 113.12 for creatinine	mg/100 ml as creatinine	8.840×10^{-2} [1.131×10^{1}]	mmol/l
Kidneys-Creatininium, clear. (in: plasma; out: urine, vol. rate >33 μl/s), 1.62–2.28 ml/s			ml/min	1.667×10^{-2} [60]	ml/s
P-Fibrinogen, masc., 2.0–4.0 g/l			mg/100 ml	10^{-2} [10^{2}]	g/l
B(Hb)-Fetal hemoglobin (Fe), molfr. of Hb(Fe) total, <0.01 mol/mol	Alkali-stable hemoglobin Hemoglobin F	Dimensionless quantity. Symbol mol/mol may be omitted.	per cent	10^{-2} [10^{2}]	(mol/mol)
dU-Glucose, mole rate, 2.78–8.34 mmol/d	Glucose excretion	M_r = 180.16 for glucose	g/d	5.551 [1.802×10^{1}]	mmol/d
U-Glucose, arb. molc. (Clinistix), 0 arb. unit	Glucose qualitative sugar	Use 0 arb. unit instead of −; use 0 or 1 arb. unit for ±.			

U-Glucose, molc., <1.67 mmol/l		M_r = 180.16 for glucose	mg/100 ml	5.551 × 10^{-2} [1.802 × 10^{1}]	mmol/l
			per cent (presumed that this incorrect unit = g/100 ml)	5.551 × 10^{1} [1.802 × 10^{-2}]	mmol/l
fP-Glucose, molc., 3.89–5.83 mmol/l		M_r = 180.16 for glucose	mg/100 ml	5.551 × 10^{-2} [1.802 × 10^{1}]	mmol/l
S-Haptoglobin (Hb(Fe)), molc. (Hb sat.), 43.4–86.8 μmol/l		M_r = 16115, an average for 1/4 of hemoglobin molecule	mg/100 ml of hemoglobin bound	6.205 × 10^{-1} [1.612]	μmol/l
B-Hemoglobin (Fe), molc., 8.07–11.17 mmol/l	Hemoglobin per cent	M_r = 16115, an average for 1/4 of hemoglobin molecule	g/100 ml	6.205 × 10^{-1} [1.612]	mmol/l
			per cent (a value of 100 per cent is assumed to correspond to 148 g/l)	9.183 × 10^{-2} [1.089 × 10^{1}]	mmol/l
(B) Ery-Hemoglobin (Fe), molc. (mean), appr. 21 mmol/l	MCHC Mean cell hemoglobin concentration Saturation index		g/100 ml	6.205 × 10^{-1} [1.612]	mmol/l
dU-17-Hydroxycorticosteroids (as cortisol), mole rate, 8.3–27.7 μmol/d	17-ketogenic steroids	M_r = 361.5 for cortisol	mg/d	2.766 [3.615 × 10^{-1}]	μmol/d
fS-Iodine (I, protein bound), molc., 0.32–0.63 μmol/l	PBI, Protein bound iodine	A_r = 126.9 for I	μg/100 ml	7.880 × 10^{-2} [1.269 × 10^{1}]	μmol/l
fS-Iron (Fe, transferrin bound), molc., 10.7–26.9 μmol/l	Iron	A_r = 55.85 for Fe	μg/100 ml	1.791 × 10^{-1} [5.585]	μmol/l

Table continued on the following page.

System-component, Kind of Quantity, Normal Range	*Synonyms*	*Comments & A_r or M_r*	*Currently Used Units (Not Recommended)*	*Conversion Factors* → [←]	*Recommended Unit*
dU-17-Ketosteroids (as androsterone), mole rate, 27.5–68.8 μmol/d		$M_r = 344.3$ for androsterone	mg/d	3.443 [2.904×10^{-1}]	μmol/d
B-Lead (Pb), molc., <2.43 μmol/l		$A_r = 207.2$ for Pb	μg/100 ml	4.826×10^{-2} [2.072×10^{1}]	μmol/l
fS-Lipids (total), masc., 4.0–7.0 g/l	Fat		mg/100 ml	10^{-2} [10^{2}]	g/l
dry F-Lipids (total), masfr., <250 g/kg	Fat		per cent	10^{1} [10^{-1}]	g/kg
S-Magnesium (Mg), molc., 0.7–1.3 mmol/l	Magnesium	$A_r = 24.32$ for Mg	mg/100 ml ↑ 1.216	4.112×10^{-1} [2.432]	mmol/l
			mEq/l	5.000×10^{-1} [2.000]	mmol/l
S-Nitrogen (N), non-protein, molc., 14.3–25.0 mmol/l	NPN Non-protein nitrogen	$A_r = 14.01$ for N	mg/100 ml	7.138×10^{-1} [1.401]	mmol/l
(aB) Hb-Oxyhemoglobin (Fe), molfr. (of Hb(Fe) Total), 0.90–0.96 (mol/mol)	Oxygen saturation	Dimensionless quantity. Symbol mol/mol may be omitted.	per cent	10^{-2} [10^{2}]	(mol/mol)
S-Phosphate (P, inorganic), molc., 0.97–1.45 mmol/l	Phosphorus	$A_r = 30.97$ for P Conversion to equivalent concentrations of phosphate requires measurement of pH	mg/100 ml as P	3.229×10^{-1} [3.097]	mmol/l

dU-Phosphate (P, total), mole rate, 29–42 mmol/d	Phosphorus excretion	A_r = 30.97 for P	g/d	3.229×10^1 [3.097×10^{-2}]	mmol/d
dU-Porphobilinogen, mole rate, 0–8.8 μmol/d	Porphobilinogen excretion	M_r = 226.24 for porphobilinogen	μg/d	4.420×10^{-3} [2.262×10^2]	μmol/d
S-Potassium, molc., 3.5–5.3 mmol/l	Potassium	A_r = 39.10 for K	mg/100 ml	2.558×10^{-1} [3.910]	mmol/l
			↑ 3.910 mEq/l	1 [1]	mmol/l
dU-Potassium, mole rate, 30–90 mmol/d	Potassium excretion	A_r = 39.10 for K	mg/d	2.558×10^{-1} [3.910]	mmol/d
			mEq/d	1	mmol/d
U-Protein, arb. masc. (Albustix), 0 arb. units	Protein qualitative	Use 0 arb. unit instead of −; use 0 or 1 arb. unit for ±.			
fS-Protein, masc., 60–82 g/l	Protein		g/100 ml	10^1	g/l
			per cent (presumed that this incorrect unit = g/100 ml)	10^1 [10^{-1}]	g/l
dU-Protein, mass rate, 50–100 mg/d	Protein excretion		mg/d	1	mg/d
B-Serum, rel. density $\left(\frac{S, 20°C}{H_2O, 20°C}\right)$, appr. 1.026	Specific gravity				

Table continued on the following page.

System-component, Kind of Quantity, Normal Range	Synonyms	Comments & A_r or M_r	Currently Used Units (Not Recommended)	Conversion Factors → [←]	Recommended Unit
S-Sodium, molc., 135–148 mmol/l	Sodium	$A_r = 22.99$ for Na	mg/100 ml	4.350×10^{-1} [2.299]	mmol/l
			↑ 2.299 mEq/l	1	mmol/l
dU-Sodium, mole rate, 40–220 mmol/d	Sodium excretion	$A_r = 22.99$ for Na	mg/d	4.350×10^{-1} [2.299]	mmol/d
			mEq/d	1	mmol/d
fS-Transferrin (Fe), molc. (Fe sat.), 48.4–68.1 μmol/l	TIBC Total iron binding capacity	$A_r = 55.85$ for Fe Probably, S-Transferrin (Fe), molc. (Fe sat.) × 0.5 = S-Transferrin, molc., as one transferrin molecule seems to bind two iron atoms	μg/100 ml	1.791×10^{-1} [5.585]	μmol/l
fS-Urate, molc., 0.15–0.42 mmol/l	Uric acid	$M_r = 168.11$ for uric acid	mg/100 ml	5.948×10^{-2} [1.681×10^{1}]	mmol/l
dU-Urate (total), mole rate, 1.49–4.46 mmol/d	Uric acid excretion	$M_r = 168.11$ for uric acid	mg/d	5.948×10^{-3} [1.681×10^{2}]	mmol/d
fS-Urea, molc., 2.50–6.41 mmol/l	Blood urea	$M_r = 60.06$ for urea	mg/100 ml	1.665×10^{-1} [6.006]	mmol/l
dU-Urea, mole rate, 430–716 mmol/d	Urea excretion	$M_r = 60.06$ for urea	g/d	1.665×10^{1} [6.006×10^{-2}]	mmol/d
fS-Urea nitrogen (N), molc., 2.50–6.41 mmol/l	Serum urea nitrogen	$M_r = 28.02$ for N in urea	mg/100 ml	3.569×10^{-1} [2.802]	mmol/l

dU-Urea nitrogen (N), mole rate, 430–716 mmol/d	Urea nitrogen excretion	$M_r = 28.02$ for N in urea	g/d	3.569×10^1 [2.802×10^{-2}]	mmol/d
Kidneys-Urea, clear. (in: plasma; out: urine, vol rate >33 μl/s), 1.07–1.65 ml/s	Urea clearance Maximum urea clearance	See above	ml/min	1.667×10^{-2} [6.0×10^1]	ml/s
Kidneys-Urea nitrogen (N), clear. (in: plasma; out: urine, vol rate >33 μl/s), 1.07–1.65 ml/s	Urea nitrogen clearance	See above	ml/min	1.667×10^{-2} [6.0×10^1]	ml/s
dU-Urine, rel. density $\left(\frac{\text{dU, 20°C}}{\text{H}_2\text{O, 20°C}}\right)$, appr. average 1.016	Specific gravity				
fU-Urobilinogen, arb. molc. (Ehrlich 1884), 0–1 arb. unit	Ehrlich reaction	Use 0 arb. unit instead of −; use 0 or 1 arb. unit for ±.			

Compiled from Dybkaer, R., and Jorgensen, K., *Quantities and Units in Clinical Chemistry*, Williams and Wilkins Company, Baltimore, 1967 (modified). Normal ranges were taken from the Normal Value Table immediately following this table.

TABLE OF NORMAL VALUES*

Test	Specimen	Value	Special Instructions and Interfering Substances
Acetoacetic acid, qualitative	Serum or urine	Negative	
Acetone, qualitative	Serum or urine	Negative	Analyze without delay.
quantitative	Serum or plasma	0.3–2.0 mg/100 ml	
quantitative (acetone and acetoacetic acid)	Serum or plasma	0.5–3.0 mg/100 ml	
Adrenocorticotrophic hormone (ACTH)	Plasma	8 A.M.: 36–168 pg/ml (lower in afternoon)	
Alanine aminotransferase, ALT (Glutamate pyruvate transaminase, GPT)	Serum	Males: 6–21 U/l at 30°C Females: 4–17 U/l at 30°C (conditions in many methods not optimal; 7–33 U/l at 30°C proposed by one optimized method using P-5-P)	Do not use specimen with visible hemolysis.
	Spinal fluid	None	
Albumin/globulin ratio	Serum	1.1–1.8 by salt fractionation (27%) or electrophoresis	
	Spinal fluid	1.6–2.2	Presence of blood invalidates the test.
Albumin, quantitative	Serum	3.8–5.0 g/100 ml for ambulatory patients (approx. 0.5 g lower for recumbent individuals); varies with technique used	
	Spinal fluid	10–30 mg/100 ml	Presence of blood invalidates the test.
	Urine	Average: 10 mg/d	
Alcohol	Serum or plasma	see Ethanol and Methanol	
Aldolase	Serum	1.0–7.5 U/l at 30°C (1.5–12.0 U/l at 37°C); lower in patients at bed rest (0.7–3.0 U/l at 30°C)	Keep at room temperature.
Aldosterone	Plasma	Recumbent (2 h), 0.3–2.5 ng/100 ml; upright (4 h), 4.0–19.6 ng/100 ml	
	Urine, 24 h	2–26 μg/d	Refrigerate during collection. Acidify to pH 4–5 with hydrochloric, acetic, or boric acid.
Alkapton bodies	Urine, random	Negative	
α-Amino acid nitrogen	Plasma	2.5–4.5 mg/100 ml or 3.5–7.0 mg/100 ml depending on specificity of method used	
	Urine, 24 h	Adults: 50–200 mg/d (by naphthoquinone method)	Preserve with thymol or HCl. Keep refrigerated.
p-Amino hippurate clearance test	Urine and plasma	600–700 ml/min or 350–400 ml/min/m² body surface	
δ-Aminolevulinate dehydratase	Red cells	139–211 units/ml erythrocytes	

* The normal values listed in this table, and normal values in general, are not to be taken as absolute but only as guidelines. The normal values vary with the particular procedure employed, and may indeed vary from laboratory to laboratory. Thus, ideally, normal values should be established for each constituent under the particular method and conditions of the laboratory (see Chapter 2). Values for various blood constituents, in general, are those obtained in the fasting state. The recommended SI units for the common constituents are given in the preceding table, along with conversion factors to and from the commonly used units. Since the SI units are not used extensively at the present time, the common units are generally used in this normal value table. The values listed in this table were compiled chiefly from those given in this text, and are supplemented by values obtained from the sources[2,3,4] listed at the end of this table. Other references[1,5,6] for normal values are also noted. Enzyme values are given in conventional units (units) or in international units (U). The names given indicate the authors of the methods.

TABLE OF NORMAL VALUES (*continued*)

Test	*Specimen*	*Value*	*Special Instructions and Interfering Substances*
δ-Aminolevulinic acid	Serum	15–23 μg/100 ml; lower in children	
	Urine, 24 h	1.5–7.5 mg/d	Collect 24 h specimen in bottle containing 10 ml glacial acetic acid. Refrigerate.
Ammonia nitrogen	Serum or plasma	Enzymatic method: 40–80 μg/100 ml Resin method: 15–45 μg/100 ml Conway diffusion method: 40–110μg/100 ml Newborns: 90–150 μg/100 ml (higher in premature or jaundiced infants) Older children: 40–80 μg/100 ml (Resin method)	For plasma, collect with sodium heparinate. Analyze immediately, as concentration increases rapidly on standing.
	Urine, 24 h	500–1200 mg/d	Avoid contamination with ammonia and bacteria.
Amniotic fluid analysis	Amniotic fluid		Protect from light. Centrifuge and analyze immediately. If analysis is delayed, refrigerate the specimen after centrifugation.
absorbance at 450 nm		28 weeks: 0–0.048 *A* 40 weeks: 0–0.02 *A*	
bilirubin		28 weeks: 0–0.075 mg/100 ml 40 weeks: 0–0.025 mg/100 ml	
L/S ratio		Varies with method 2.0–5.0 indicates probable fetal lung maturity	
Amylase	Serum	60–180 Somogyi units/100 ml or 95–290 U/l (saccharogenic method) 45–200 dye units/100 ml at 37°C	Inhibited by oxalate or citrate.
	Urine, 24 h or timed	1500 (1800)*–6000 (7500)* Somogyi units/d or 70–275 Somgyi units/h (saccharogenic method) 40–330 dye units/h at 37°C	Refrigerate during collection.
α_1-Antitrypsin	Serum	150–275 mg/100 ml	
Arsenic	Whole blood	<3 μg/100 ml	Collect in acid-cleaned container.
	Urine, 24 h	<50 μg/l	
Ascorbic acid	Plasma	0.6–2.0 mg/100 ml	Unstable; analyze immediately. Stable in acid.
Aspartate aminotransferase, AST (Glutamate oxalacetate transaminase, GOT)	Serum	Without P-5-P Males: 7–21 U/l at 30°C Females: 6–18 U/l at 30°C With P-5-P 12–29 U/l at 30°C (females slightly lower)	Do not use specimen with visible hemolysis.
	Spinal fluid	3–10 U/l at 30°C	
Barbiturate	Whole blood	Negative Therapeutic concentration: 0.1–0.2 mg/100 ml (higher for epileptics treated with phenobarbital)	Collect with sodium heparinate.

* Value in parentheses represents another reported value for the designated lower or upper limit.

TABLE OF NORMAL VALUES (*continued*)

Test	*Specimen*	*Value*	*Special Instructions and Interfering Substances*
Barbiturate, *cont'd*		Approx. coma concentration: Short acting: 1 mg/100 ml Intermediate acting: 3 mg/100 ml Long acting: 5–8 mg/100 ml	
	Urine, random	Negative	
Base excess	Whole blood	−3 to +3 mmol/l	
Bence Jones protein	Urine, first morning	Negative	Analyze fresh specimen.
Beryllium	Urine, 24 h	<0.05 μg/d	Collect in acid-washed container.
Bicarbonate	Plasma, arterial	21–28 mmol/l	Collect anaerobically.
	venous	22–29 mmol/l	
Bile, unaltered, qualitative	Feces, fresh random	Negative	
Bile acids	Serum	0.3–3.0 mg/100 ml	Analyze immediately or freeze.
Bile salts	Feces, 24 h	Approx. 0.8 g/d	
Bilirubin	Amniotic fluid	28 weeks: 0–0.075 mg/100 ml (or 0–0.048 *A* units at 450 nm) 40 weeks: 0–0.025 mg/100 ml (or 0–0.02 *A* units at 450 nm) Varies with method	Protect specimen from light.
	Serum	Total: 0.2–1.0 mg/100 ml	Avoid exposure of specimen to direct light.
		Conjugated: 0–0.2 mg/100 ml	
		Unconjugated: 0.2–0.8 mg/100 ml	
		Infants <1 week of age, total: 1–12 mg/100 ml	See Chapter 19 for details.
Bilirubin, qualitative	Urine, fresh random	Negative	Protect specimen from light.
Blood, occult	Urine, random	Negative	
	Feces, random	Negative	Place patient on meat-free diet for 2 days.
Blood volume	Whole blood	Adults: 60–70 ml/kg	Varies with hematocrit and sex.
	Plasma	Adults: 33–45 ml/kg	
Borate	Urine, 24 h	<2 mg/l	
Bromide	Serum	Values given as NaBr Adult: 0.8–1.5 mg/100 ml Children <1.0 mg/100 ml Therapeutic concentration: approx. 100 mg/100 ml Toxic concentration: >200 mg/100 ml	
BSP (Bromsulfophthalein)	Serum	<10% after 30 minutes <5% after 45 minutes	Dose: 5 mg/kg.
Calcium	Feces, 24 h	Average: 0.64 g/d (varies with diet)	
	Serum	Ionized: 4.2–5.4 mg/100 ml (2.1–2.7 mmol $Ca^{2+}_{1/2}$/l) Total: 8.5–10.4 mg/100 ml (4.25–5.2 mmol $Ca^{2+}_{1/2}$/l or 2.12–2.60 mmol/l)	
	Spinal fluid	4.2–5.4 mg/100 ml (2.1–2.7 mmol $Ca^{2+}_{1/2}$/l)	
	Urine, 24 h	Average: 50–150 mg/d (25–75 mmol $Ca^{2+}_{1/2}$/d or 12.5–37 mmol/d) (varies with intake)	Acidify during collection.

TABLE OF NORMAL VALUES (*continued*)

Test	*Specimen*	*Value*	*Special Instructions and Interfering Substances*
Calcium, *cont'd*			
qualitative (Sulkowitch test)	Urine, random	1+ to 2+	
Carbon dioxide (total CO_2)	Whole blood, arterial	Children: (1–3 years) 12–20 mmol/l (increases gradually to adult level) Adults: 19–25 mmol/l	Collect anaerobically.
	Whole blood, venous	22–27 mmol/l	
	Plasma or serum, capillary	22–29 mmol/l	
	Plasma or serum, venous	23–30 mmol/l	
Carbon dioxide pressure (pCO_2)	Whole blood, arterial	Males: 35–45 mm Hg Females: 32–43 mm Hg	Collect with heparinized syringe, and seal. Handle anaerobically.
	Whole blood, venous	38–50 mm Hg	
Carbon monoxide hemoglobin (carboxyhemoglobin)	Whole blood	Nonsmokers: 0.5–1.5% saturation of Hb Average smokers: 4–5% saturation Heavy smokers: 8–9 (15)*% saturation	Avoid exposure to light and air.
Carbonic acid (CO_2 gas + H_2CO_3)	Plasma, arterial venous	0.97–1.38 mmol/l 1.16–1.53 mmol/l	
β-Carotene	Serum	Adult: 60–200 μg/100 ml Newborn: Approx. 70 μg/100 ml	
Catecholamines, fractionation	Plasma	Epinephrine: 18–26 ng/100ml Norepinephrine: 47–69 ng/100 ml	
total	Urine, random	0–14 μg/100 ml	Refrigerate and acidify during collection.
	Urine, 24 h	<100 μg/d Borderline: 150–230 μg/d (varies with muscular activity)	
Cephalin cholesterol	Serum	Negative to 1+	
Ceruloplasmin	Serum	25–43 mg/100 ml	
Chloride	Gastric residue	45–155 mmol/l	
	Saliva	Without stimulation: 5–20 mmol/l With stimulation: <44 mmol/l	
	Serum or plasma	98–106 mmol/l	
	Spinal fluid	118–132 mmol/l	
	Sweat	Normal homozygotes: 5–35 mmol/l Heterozygotes: 35–60 mmol/l	
	Urine, 24 h	110–250 mmol/d (varies with intake)	
Cholesterol, total	Serum	Adult: 140–250 mg/100 ml (Varies with diet, age, and seasons. See Chapter 10.) Lower in children Birth–1 month: 45–100 mg/100 ml 1 month–20 years: 100–230 mg/100 ml	
in low density lipids		66–178 mg/100 ml	
Cholesterol esters	Serum	65–75% of total	
Cholesterol/phosphatide ratio	Serum	0.55–1.05	

* Value in parentheses represents another reported value for the designated lower or upper limit.

TABLE OF NORMAL VALUES (*continued*)

Test	*Specimen*	*Value*	*Special Instructions and Interfering Substances*
Cholinesterase, pseudo	Serum or plasma	5–12 U/ml at 37°C (PTC as substrate). Values vary with substrate used; see Chapter 12.	Refrigerate. Collect plasma with sodium heparinate.
	Spinal fluid	13–21 mU/ml	
Concentration test (Fishberg)	Urine, after fluid restriction	Specific gravity: >1.025 Osmolality: >900 mOsm/kg Osmolarity: >850 mOsm/l	
Congo red test	Plasma	After 1 h: 65–75% retention	See Chapter 7.
Copper	Serum	Males: 70–140 μg/100 ml Females: 80–155 μg/100 ml Higher in children, and in women on contraceptive pills and during pregnancy	Avoid hemolysis. Use acid-cleaned collection tubes.
	Urine, 24 h	15–30 μg/d	Use acid-cleaned collection vessel.
Coproporphyrins, quantitative	Feces, 24 h	<30 μg/g dry weight	Refrigerate (or freeze) sample if analysis is delayed.
qualitative	Urine, fresh random	Negative or trace	Avoid exposure to direct light. Collect in a dark bottle.
quantitative	Urine, fresh random	3–15 μg/100 ml	Avoid exposure to direct light.
	Urine, 24 h	Adults: <200 μg/d Children: <2 μg/kg	Collect in a dark bottle with 5 g Na_2CO_3 to maintain pH 6.5–9.0. Refrigerate during collection.
Corticosteroids, 17-OH	Urine, 24 h	Males: 3–10 mg/d Females: 2–8 mg/d Lower in children	Refrigerate during collection. Use 1 g boric acid/l of urine. For values after stimulation or suppression tests and drug interference, see Chapter 13.
Cortisol	Plasma	9–10 A.M.: 6–26 μg/100 ml 4 P.M.: 2–18μg/100 ml	For values after stimulation or suppression tests and drug interference, see Chapter 13.
Cortisol, free	Urine, 24 h	0–108 μg/d	
Creatine	Serum or plasma	Males: 0.2–0.5 mg/100 ml Females: 0.4–0.9 mg/100 ml	
	Urine, 24 h	Males: 0–40 mg/d Females: 0–80 (100) mg/d Higher in pregnancy (>12% of creatinine) Children: >30% of creatinine	Refrigerate during collection.
Creatine kinase, CK (Creatine phosphokinase, CPK)	Serum	Males: 12–65 U/l at 30°C Females: 10–50 U/l at 30°C Values lower in patients at bed rest Values higher in blacks	Some isoenzymes are unstable; therefore, analyze as soon as possible.
Creatinine	Amniotic fluid	>2.0 mg/100 ml generally indicates fetal maturity when maternal serum creatinine normal	

TABLE OF NORMAL VALUES (*continued*)

Test	*Specimen*	*Value*	*Special Instructions and Interfering Substances*
Creatinine, *cont'd*	Serum or plasma	Using nonspecific method: Males: 0.9–1.5 mg/ 100 ml Females: 0.8–1.2 mg/100 ml Using specific method: Males: 0.6–1.2 mg/100 ml Females: 0.5–1.0 mg/ 100 ml	Refrigerate or freeze.
	Urine, 24 h	Males: 1.0–2.0 g/d Females: 0.8–1.8 g/d	Refrigerate.
Creatinine clearance (endogenous)	Urine and serum	Using nonspecific method: Males: 85–125 ml/min Females: 75–115 ml/min Using specific method: Males: 97–137 ml/min Females: 88–128 ml/min	Refrigerate or freeze. Correct for body surface; see Chapter 17.
Cryoglobulins	Serum	Negative	Keep specimen at 37°C.
Cyanide	Blood	Negative Lethal concentration: >0.1 mg/100 ml	
Cystine, qualitative	Urine	Negative or trace	
Cystine and cysteine	Urine, 24 h	10–100 mg/d	
Diagnex (tubeless gastric analysis)	Urine, 2 h	Presumptive evidence for presence of free HCl	Run test in postabsorptive state.
Digoxin	Serum	Negative Therapeutic concentration: 0.8–1.6 ng/ml Toxic concentration: >2.0 ng/ml	
Disaccharide tolerance	Serum	Change of >20 mg/100 ml in glucose concentration	Dose: 50 g/m² or 1 g/kg.
Duodenal drainage	Duodenal juice	For secretin test, see Chapter 20	
Electrophoresis, hemoglobin	Whole blood	See Chapter 8	
Electrophoresis	Serum	% / g/100 ml alb. 53–65 / 3.5–4.7 α_1 2.5–5.0 / 0.2–0.3 α_2 7.0–13.0 / 0.4–0.9 β 8.0–14.0 / 0.5–1.1 γ 12.0–22.0 / 0.7–1.5	Hemolysis interferes.
	Spinal fluid	Average % of Total Protein pre-alb. 2.9–5.3 alb. 56.8–68.0 α_1 4.1–6.4 α_2 6.2–10.2 β 10.8–14.8 γ 6.1–8.3	Concentrate specimen prior to analysis, unless protein content is significantly increased.
	Urine, 24 h	Average % of Total Protein alb. 37.9 α_1 27.3 α_2 19.5 β 8.8 γ 3.3	Concentrate specimen prior to analysis (see above).
Epinephrine	Plasma	Values not well established 18–26 ng/100 ml	
	Urine, 24 h	0–20 μg/d	
Epinephrine tolerance test	Plasma	Fasting: normal glucose 40–60 min: increase of 35–45 mg/100 ml 2 h: return to approx. fasting level	Dose: 1 ml of 1/1000 solution of epinephrine HCl.

TABLE OF NORMAL VALUES (*continued*)

Test	*Specimen*	*Value*	*Special Instructions and Interfering Substances*
Estriol	Plasma	During pregnancy at term: 9.3–56 μg/100 ml	
	Urine, 24 h	During pregnancy 30 wks: 6.3–17.5 mg/d 35 wks: 9.7–27.0 mg/d 40 wks: 14.5–41.2 mg/d	
Estrogens, total	Urine, 24 h	Males: 5–18 μg/d Children: <1 μg/d Females: Postmenopausal: 1.4–19.6 μg/d Pregnant: up to 45,000 μg/d Nonpregnant: preovulatory phase: 4–25 μg/d ovulation peak: 28–99 μg/d luteal peak: 22–105 μg/d	Keep refrigerated. Preserve with boric acid.
Estrogens, fractionation	Urine, 24 h	Nonpregnant females: (ovulation peak) Estradiol: 4–14 μg/d Estriol: 13–54 μg/d Estrone: 11–31 μg/d	
Ethanol	Whole blood or serum	Negative	Avoid exposure to air. Keep container tightly stoppered and refrigerate.
Fat	Feces, 72 h	Older children and adults: <6 g/d Children up to 6 years: <2 g/d Total fat: 15–25% of dry weight	Diet: 60–150 g of fat/d.
Fat, neutral (triglycerides)	Serum	Males: 40–160 mg/100 ml Females: 35–135 mg/100 ml See Chapter 10	Use fasting specimen. Remove serum from clot soon after drawing. Hemolysis interferes.
Fatty acids, "free" (NEFA)	Serum	0.30–0.90 mmol/l, children and obese individuals slightly higher, with upper limit of 1.10 mmol/l	Patient should be fasting. Remove serum from clot soon after drawing. Analyze promptly.
α_1-Fetoprotein	Serum	Adults: Negative	
Fibrinogen	Plasma	0.20–0.40 g/100 ml	Use oxalate, citrate, or EDTA as anticoagulant.
FIGLU test (*N*-Formiminoglutamic acid)	Urine, 24 h	<3 mg/d (after 15 g L-histidine: approx. 4 mg/8 h)	
Fluoride	Whole blood	<0.05 mg/100 ml	
	Urine, 24 h	<1 mg/d	
Folic acid	Serum	3.0–25.0 ng/ml	
Follicle stimulating hormone (FSH), measured as total gonadotrophins	Urine, 24 h	See Gonadotrophin	
radioimmunoassay	Plasma	Males: 4–25 mIU/ml Females: Premenopausal: 4–30 mIU/ml Midcycle peak: 2 × baseline Postmenopausal: 40–250 mIU/ml	
Fructose	Serum	<7.5 mg/100 ml	
	Urine, 24 h	Approx. 60 mg/d	
Galactose	Blood	Children: <20 mg/100 ml	

TABLE OF NORMAL VALUES (*continued*)

Test	*Specimen*	*Value*	*Special Instructions and Interfering Substances*
Galactose tolerance	Plasma	**I.V.:** 60 min: <42 mg/100 ml	Dose: 0.5 g/kg. Collect blood with potassium oxalate and sodium fluoride or analyze without delay.
	Plasma and urine	**Oral:** 30–60 min: 40–60 mg/100 ml plasma 5 h urine: <3 g/5 h	Dose: 40 g in 200 ml H_2O. Collect blood with potassium oxalate and sodium fluoride or analyze without delay.
Gamma globulin	Serum	0.7–1.5 g/100 ml See also Electrophoresis	
Gamma glutamyltransferase, GGT	Serum	Males: 6–28 U/l at 25°C 8–37 U/l at 30°C 9–54 U/l at 37°C Females: 4–18 U/l at 25°C 6–24 U/l at 30°C 8–35 U/l at 37°C	
Gastric analysis	Gastric residue	Without stimulation	
acids, free		0–40 mmol/l	
acids, combined		10–20 mmol/l	
acids, total		10–50 mmol/l	
pH		1.5–3.5	
volume		20–100 ml	
Gastric analysis, secretion rate	Gastric secretion	Basal acid output (BAO): 0–5 mmol/h Maximum acid output (MAO): 1–20 mmol/h BAO/MAO: <20%	Histalog dose: 1.5 mg/kg.
	Gastric secretion	12 h Volume: 150–1000 ml pH: 1.5–3.5	
Gastric analysis, tubeless (Diagnex)	Urine, 2 h	Presumptive evidence for presence of free HCl	Run test in postabsorptive state.
Gastrin	Serum	<300 pg/ml Elders: 200–800 pg/ml	
Globulins, total	Serum	2.3–3.5 g/100 ml	
qualitative (Pandy)	Spinal fluid	Negative or trace	Presence of blood causes false high values.
quantitative	Spinal fluid	See Chapter 7	
	Urine, 24 h	Approx. 20 mg/d	
Globulin-β_{1c}	Serum	70–160 mg/100 ml	
Glomerular selectivity test (IgG/albumin clearance)	Urine	<0.16 indicates high selectivity	
Glucose	Serum or plasma	70–105 mg/100 ml ("True" glucose method) 75–110 mg/100 ml (AutoAnalyzer ferricyanide method) 80–120 mg/100 ml (Folin-Wu method)	Collect with potassium oxalate and sodium fluoride, or analyze without delay. Serum or plasma glucose has greater stability than whole blood glucose.
	Whole blood	65–95 mg/100 ml ("True" glucose method) Capillary blood higher; see Chapter 6	
	Spinal fluid	40–70 mg/100 ml ("True" glucose method) 45–75 mg/100 ml (AutoAnalyzer ferricyanide method) Approx. 60–70% of plasma glucose	

TABLE OF NORMAL VALUES (*continued*)

Test	*Specimen*	*Value*	*Special Instructions and Interfering Substances*
Glucose, *cont'd*			
qualitative	Urine, random	<30 mg/100 ml (Clinitest negative)	
quantitative	Urine, 24 h	0.5–1.5 g/d (Total reducing substances) <0.5 g/d as glucose (average 0.13 g/d)	
2 h postprandial	Plasma	<120 mg/100 ml (Diabetics: >140 mg/100 ml)	Collect with potassium oxalate and sodium fluoride, or analyze without delay.
Glucose tolerance	Plasma	**I.V.:** Fasting: normal glucose level 5 min: maximum of 250 mg/100 ml 60 min: significant decrease 120 min: <140 mg/100 ml 180 min: fasting level	I.V. dose: 0.5 g/kg. Collect specimens with potassium oxalate and sodium fluoride or analyze without delay.
		Oral: Fasting: normal glucose level 30–60 min: peak <60% above fasting level but should not exceed 170 mg/100 ml 2 h: <120 mg/100 ml (Diabetics: >140 mg/100 ml)	Oral adult dose: 100 g. Children (below 12 years): 2 g/kg. Collect specimens with potassium oxalate and sodium fluoride or analyze without delay.
Glucose-6-phosphate dehydrogenase (G-6-P-D)	Red cells	120–280 U/10^{12} cells or 3.4–8.0 U/g Hb See Chapter 12	Analyze promptly. If specimen must be stored, preserve the specimen with citrate.
	Serum	0.3–3.0 U/l at 30°C	
Glutethimide	Serum or whole blood	Negative Therapeutic concentration: approx. 0.5 mg/100 ml Coma concentration: >3.0 mg/100 ml	
Gonadotrophin, pituitary (FSH & LH)	Urine, 24 h	Adults: approx. 10–50 Mouse Uterine (M.U.) units/d Children, prepuberty: <6 M.U. units/d Postmenopausal: 50–190 M.U. units/d	Preserve with 2 ml of 1% thymol in glacial acetic acid or adjust pH to 5–6.5 with glacial acetic acid.
	Plasma	See FSH	
Growth hormone	Plasma	Males: 0–8 ng/ml Females: 0–30 ng/ml	
Haptoglobin	Serum or plasma	40–180 mg/100 ml (as hemoglobin binding capacity)	
Hematocrit	Whole blood	Males: 42–50 ml/100 ml Females: 37–47 ml/100 ml Higher at birth	
Hemoglobin	Whole blood	Males: 14–17 g/100 ml Females: 12–15 g/100 ml Newborn: 16–20 g/100 ml Increased values obtained at high altitudes	Collect with EDTA.
electrophoresis	Whole blood	See Chapter 9	
free, qualitative	Serum or plasma	Negative (by spectroscopy)	Hemolysis invalidates this test.
	Urine, random	Negative (by spectroscopy)	
free, quantitative	Plasma	<2.5 mg/100 ml <1 mg/100 ml when special blood collection technique is used	Hemolysis invalidates this test. Serum is unsuitable as a specimen.

TABLE OF NORMAL VALUES (*continued*)

Test	*Specimen*	*Value*	*Special Instructions and Interfering Substances*
Hemoglobin A_2	Whole blood	1.6–3.2% of Hb	Collect with EDTA.
Hemoglobin F by acid elution (Slide test)	Whole blood	Complete elution of hemoglobin	
Hemoglobin F (alkali-resistant)	Whole blood	1 year and up: <2% of total Hb	
Hemoglobin solubility test	Whole blood	Genotype A/A: 90–105% of total hemoglobin	
Hollander test (confirmation of complete vagotomy)	Gastric contents	No increase in HCl within 45 min after insulin indicates complete vagotomy or HCl: <20 mmol/l above fasting if free acid present in fasting specimen; <10 mmol/l above fasting if free acid absent in fasting specimen	Insulin dose: 0.15–0.2 units/kg. Test valid only if serum glucose falls below 45 mg/100 ml.
Homogentisic acid	Urine, random	Negative	
Homovanillic acid (HVA)	Urine, 24 h	<15 mg/d 1–40 μg/mg creatinine	Use boric acid as preservative.
Human chorionic gonadotrophin (HCG)	Serum or urine	Pregnancy: 1st trimester peak: up to 163,000 mIU/ml	
Human placental lactogen (HPL)	Serum	At term: 5.5–6.5 μg/ml	
α-Hydroxybutyric dehydrogenase	Serum	61–155 U/ml at 30°C (Henry's modification of Rosalki and Wilkinson Method)	
17-Hydroxycorticosteroids (as cortisol)	Plasma	9–10 A.M.: 6–26 μg/100 ml 4 P.M.: 2–18 μg/100 ml	For values after stimulation or suppression tests and drug interference, see Chapter 13.
17-Hydroxycorticosteroids	Urine, 24 h	Males: 3–10 mg/d Females: 2–8 mg/d Lower in children	Refrigerate during collection. Preserve with 1 g boric acid/l of urine. For values after stimulation or suppression tests and drug interference, see Chapter 13.
5-Hydroxyindole acetic acid	Urine, random	Negative	Some drugs and fruits interfere.
	Urine, 24 h	2–8 mg/d	
Hydroxyproline	Urine, 24 h	Total: 10–50 mg/d Free: 0.2–1.0 mg/d	Low collagen diet required.
Imipramine	Whole blood	Negative Therapeutic concentration: 0.05–0.6 μg/ml Toxic concentration: 1–5 μg/ml	
Immunoglobulins	Serum	IgG: 0.8–1.4 g/100 ml IgA: 0.09–0.45 g/100 ml IgM: 0.06–0.25 g/100 ml Lower in children; see Chapter 7	
	Spinal fluid	IgG: 0.8–6.4 mg/100 ml IgA: Trace IgM: Nil	
Indican	Urine, 24 h	10–20 mg/d	
Insulin	Plasma	Fasting: 11–240 μIU/ml (bioassay) 4–24 μIU/ml (radioimmunoassay)	

TABLE OF NORMAL VALUES (*continued*)

Test	*Specimen*	*Value*	*Special Instructions and Interfering Substances*
Insulin tolerance test	Plasma	Fasting: normal plasma glucose 30 min: decrease to 50% of fasting level 90–120 min: approaches fasting level	Diet containing 300 g carbohydrate 2–3 days prior to test. Dose: 0.1 unit insulin/kg. Collect specimens with potassium oxalate and sodium fluoride or analyze immediately. Hypoglycemic reaction may occur.
Inulin clearance	Urine and whole blood	Males: 110–152 ml/min Females: 102–132 ml/min	Dose: 25 ml of a 10% inulin solution (priming dose) and 500 ml of 1.5% inulin solution (maintenance infusion).
Iodine, radioactive uptake (RAI)		After 24 h: 7–33%	
Iodine, protein bound (PBI)	Serum	(3.5)* 4.0–8.0 μg/100 ml	Test not reliable if iodine-containing drugs or x-ray contrast media were given prior to test; see Chapter 14.
Iron, total	Serum	Using protein precipitation method: Males: 60–150 μg/100 ml Females: 50–130 μg/100 ml Elders: 40–80 μg/100 ml Birth: approx. 200 μg/100 ml (within 1 h falls to 50 μg/100 ml) Children: adult level after 3 weeks Without protein precipitation: Values 10–20 μg/100 ml higher than those above	Avoid hemolysis. For diurnal variation, see Chapter 15.
Iron-binding capacity (See also Transferrin)	Serum	Using protein precipitation methods: Adults: 270–380 μg/100 ml >70 years: approx. 70 μg/100 ml lower Children: 180–575 μg/100 ml Without protein precipitation: Adults: 280–400 μg/100 ml	Avoid hemolysis.
Isocitrate dehydrogenase	Serum	1.2–7.0 U/l at 30°C	Separate serum from clot soon after drawing to avoid erratic blank readings.
17-Ketogenic steroids, total	Urine, 24 h	Males: 5–23 mg/d Females: 3–15 mg/d Children: 2.3–3.8 mg/d	Refrigerate during collection. Preserve with 1 g boric acid/l of urine. For values after stimulation or suppression tests and drug interference, see Chapter 13.
Ketone bodies, qualitative	Serum	Negative	
	Urine, random	Negative	
quantitative	Serum	0.5–3.0 mg/100 ml	

* Value in parentheses represents another reported value for the designated lower or upper limit.

TABLE OF NORMAL VALUES (*continued*)

Test	*Specimen*	*Value*	*Special Instructions and Interfering Substances*
11-Ketopregnanetriol	Urine, 24 h	0–0.3 mg/d	
17-Ketosteroids (Zimmermann reaction)	Urine, 24 h	Males: 8–20 mg/d Young males: 9–22 mg/d Females: (5)* 6–15 mg/d Decline after 60 years Children: up to 1 yr: <1 mg/d 1–4 years: <2 mg/d 5–8 years: <3 mg/d 8–12 years: 3–10 mg/d 13–16 years: 5–12 mg/d	Refrigerate during collection. Preserve with 1 g boric acid/l of urine. For values after stimulation or suppression tests and drug interference, see Chapter 13.
17-Ketosteroids (Gas chromatographic fractionation)	Urine, 24 h	Total: Males: 5–12 mg/d Females: 3–10 mg/d Androsterone M: 2.0–5.0 mg/d F: 0.5–3.0 mg/d Etiocholanolone M: 1.4–5.0 mg/d F: 0.8–4.0 mg/d Dehydroepiandrosterone M: 0.2–2.0 mg/d F: 0.2–1.8 mg/d 11-Ketoandrosterone M: 0.2–1.0 mg/d F: 0.2–0.8 mg/d 11-Ketoetiocholanolone M: 0.2–1.0 mg/d F: 0.2–0.8 mg/d 11-Hydroxyandrosterone M: 0.1–0.8 mg/d F: 0.0–0.5 mg/d 11-Hydroxyetiocholanolone M: 0.2–0.6 mg/d F: 0.1–1.1 mg/d	Refrigerate during collection. For values after stimulation and suppression tests, see Chapter 13.
17-Ketosteroids			
alpha/beta ratio	Urine, 24 h	Above 5	
beta/alpha ratio	Urine, 24 h	Below 0.2	
Lactate	Whole blood, arterial	0.36–0.75 mmol/l (3–7 mg/100 ml)	Collect blood without tourniquet. Use iodoacetate as a preservative or analyze immediately. Patient must be fasting and at complete bed rest.
	Whole blood, venous	0.5–1.3 mmol/l (4–12 mg/100 ml)	
Lactate dehydrogenase, LD	Serum	P → L: 200–425 Wroblewski units/ml at 30°C or 85–190 U/l at 30°C L → P: 35–88 U/l at 30°C	Hemolysis causes elevated results. Store at room temperature.
	Spinal fluid	P → L: 7–30 U/l	
LD/HBD	Serum	1.2–1.6 (when LD analyzed by P → L method)	
Lactate dehydrogenase isoenzymes	Serum		Value for each isoenzyme depends on method and support media used. See Chapter 12.
Lactic acid	Gastric contents	Negative	

* Value in parentheses represents another reported value for the designated lower or upper limit.

TABLE OF NORMAL VALUES (*continued*)

Test	*Specimen*	*Value*	*Special Instructions and Interfering Substances*
Lactose	Plasma Urine, random Urine, 24 h	<5 mg/100 ml Children: <1.5 mg/100 ml Adults: 12–40 mg/d	
Lactose tolerance	Plasma	Results similar to glucose tolerance curve of patient	Oral adult dose: 100 g lactose
Lead	Urine, 24 h	<80 μg/l	Use lead-free container for collection.
	Whole blood	<80 μg/100 ml Children: <40 μg/100 ml 50–80 μg/100 ml borderline	Use glassware and anticoagulant (heparin) that are lead-free.
Lecithin/sphingomyelin (L/S) ratio	Amniotic fluid	2.0–5.0 indicates probable fetal lung maturity	
Lecithin phosphorus	Amniotic fluid	>0.10 mg/100 ml indicates probable adequate fetal lung maturity	
Leucine amino peptidase, LAP (aminopeptidase)	Serum	Males: 84–200 GR units/ml at 37°C Females: 76–184 GR units/ml at 37°C (Goldbarg and Rutenberg)	
Lipase	Serum	0.10–1.0 units/ml at 37°C 28–280 U/l at 37°C (Tietz and Fiereck) Up to 200 U/l with triolein as substrate (30°C) (Tietz and Repique)	
Lipids	Feces, 72 h	Older children and adults: <6 g/d Children up to 6 years: <2 g/d Total: 15–25% of dry weight	Diet: 60–150 g of fat/d.
	Serum or plasma	Total: 400–700 mg/100 ml	Draw blood after a 12 h fast.
Lipid phosphorus	Serum	7–11 mg/100 ml	See also phospholipids and phosphatides. Hemolysis interferes. See Chapter 10.
Lipoprotein electrophoresis (paper or agarose)	Serum	Distinct beta band; negligible chylomicron and pre-beta bands. See Chapter 10	Collect blood with EDTA after 12 h fast.
Lithium	Serum	Negative Therapeutic concentration: 0.5–1.5 mmol/l Toxic concentration: >2.0 mmol/l and rarely <1.5 mmol/l	
Long-acting thyroid stimulator, LATS	Serum	Below detectable limits	
Luteinizing hormone, LH	Plasma	Males: <11 mIU/ml Females: Premenopausal: <25 mIU/ml Midcycle: >3 × base level Postmenopausal: >25 mIU/ml	
Macroglobulins ($\alpha_2 + \gamma$)	Serum	0.07–0.43 g/100 ml	

TABLE OF NORMAL VALUES (*continued*)

Test	*Specimen*	*Value*	*Special Instructions and Interfering Substances*
Magnesium	Serum	1.3–2.1 mmol $Mg^{2+}_{1/2}$/l (essentially the same in newborns)	Avoid hemolysis.
	Erythrocytes	4.5–6.0 mmol $Mg^{2+}_{1/2}$/l	
	Spinal fluid	2.2–3.0 mmol $Mg^{2+}_{1/2}$/l	Presence of blood invalidates the test.
	Urine, 24 h	6–8.5 mmol $Mg^{2+}_{1/2}$/d	
Melanogens, qualitative	Urine, random	Negative	
Mercury	Urine, 24 h	<20 μg/d	Collect in acid-cleaned container.
Metanephrines, total	Urine, 24 h	0.3–0.9 mg/d	
Methanol	Serum	Negative	
Methemoglobin	Whole blood	0–0.25 g/100 ml or $<1.5\%$ of total hemoglobin	Analyze immediately. Collect with heparin, EDTA, or ACD solution.
3-Methoxy-4-hydroxy-mandelic acid (Vanilmandelic acid, VMA)	Urine, 24 h	Adults: 1.8–7.1 mg/d (1.5–7.0 μg/mg creatinine) Infants: <83 μg/kg/d	Refrigerate and acidify specimen during collection Excessive ingestion of coffee, desserts, or fruit may cause an increase in results by some methods but not with Pisano's method.
Mucoproteins	Serum	75–135 mg/100 ml (80–200 mg/100 ml by less specific technique)	
	Urine, 24 h	Approx. 70 mg/d	
Myoglobin	Urine, 24 h	<1.6 mg/l	
qualitative	Urine, random	Negative	
Nitrogen, total	Feces, 24 h	<2 g/d	
Nonprotein nitrogen, NPN	Serum or plasma	20–35 mg/100 ml	
Norepinephrine	Plasma	47–69 ng/100 ml	
5′-Nucleotidase	Serum	2–17 U/ml at 37°C (Campbell) Lower in children	
Occult blood, qualitative	Urine, random	Negative	
	Feces, random	Negative, if patient on meat-free diet for two days	
Oleic acid-I^{131} absorption test	Plasma	1.7% of administered dose/l after 4–6 h	Dose: 50 μCi in milk.
	Feces, 72 h	Less than 5% of administered dose in 72 h specimen	
Osmolality	Serum	289–308 mOsm/kg	
	Urine, random	>600 mOsm/kg >900 mOsm/kg after 12 h fluid restriction	
Osmolality, urine/serum ratio	Serum and urine	1.0–3.0 with average fluid intake >3.0 after 12 h fluid restriction	
Oxygen capacity	Whole blood, arterial	16–24 vol. % or 1.34 ml/g hemoglobin	
Oxygen content	Whole blood, arterial	15–23 vol. %	
Oxygen pressure, pO_2	Whole blood, arterial	12–40 years: 83–108 mm Hg 40–80 years: 72–104 mm Hg	Collect in heparinized syringe Store on ice. Handle anaerobically.

TABLE OF NORMAL VALUES (*continued*)

Test	*Specimen*	*Value*	*Special Instructions and Interfering Substances*
Oxygen saturation	Whole blood, arterial	95–99%	Collect in heparinized syringe. Analyze promptly. Lipemia and bilirubin interfere.
	Whole blood, venous	60–85%	
pCO_2 (Pressure of carbon dioxide)	Whole blood, arterial	Males: 35–45 mm Hg Females: 32–43 mm Hg	Collect in heparinized syringe and seal. Handle anaerobically.
	Whole blood, venous	38–50 mm Hg (varies considerably depending on skin temp., duration of stasis, and muscular pumping)	
pO_2 (Pressure of oxygen)	Whole blood, arterial	12–40 years: 83–108 mm Hg 40–80 years: 72–104 mm Hg	Collect in heparinized syringe. Store on ice. Handle anaerobically.
pO_2 at half saturation ($pO_2(0.5)$ or P_{50})	Whole blood, arterial	Adults: 25–29 mm Hg Newborns: 18–24 mm Hg	
Pandy test	Spinal fluid	Negative	Presence of blood causes false high results.
Pentoses, total	Urine, 24 h	Adult: 2–5 mg/kg/d on fruit-free diet	
pH	Gastric contents	1.5–3.5	
	Serum or plasma, venous	7.35–7.45	Handle anaerobically. See Chapter 15 for blood collection.
	Whole blood, venous	7.33–7.43	
	Whole blood, capillary	7.35–7.45	
	Urine, fasting	5.5–6.5	
	Urine, random	4.8–7.8	
Phenolsulfonphthalein	Urine	% Excretion 15 min: 25–50 30 min: 15–25 60 min: 10–15 120 min: 5 60–85 (Total)	I.V. dose: 6 mg.
Phenylalanine	Serum	Adults: 0.8–1.8 mg/100 ml Full-term, normal weight newborns: 1.2–3.4 mg/100 ml Premature or low weight newborns: 2.0–7.5 mg/100 ml (drops to the normal full-term newborn range within 7–20 days)	
Phenylalanine tolerance	Serum	Non-carrier of phenylketonuria: Fasting: normal phenylalanine concentration 1–2 h: approx. 9 mg/100 ml 4 h: <5 mg/100 ml Carrier: Fasting: normal 1–2 h: approx. 19 mg/100 ml 4 h: delayed fall	Dose: 100 mg phenylalanine/kg.
Phenylpyruvic acid, qualitative	Urine, fresh random	Negative by $FeCl_3$ test	

TABLE OF NORMAL VALUES (*continued*)

Test	*Specimen*	*Value*	*Special Instructions and Interfering Substances*
Phosphatase, acid	Serum	Total: Males: 0.15–0.65 BLB units at 37°C (2.5–11.7 U/l) Females: 0.02–0.55 BLB units at 37°C (0.3–9.2 U/l) Prostatic fraction (tartrate labile): Males: 0.01–0.03 BLB units at 37°C (0.2–5.0 U/l) Females: 0–0.05 BLB units at 37°C (0–0.8 U/l) (BLB = Bessey, Lowry, and Brock using PNPP) Prostatic: 0.11–0.60 U/l at 37°C with TMP (Roy, Brower, and Hayden)	Avoid hemolysis. Perform test without delay. If there is a delay, add disodium hydrogen citrate (10 mg/ml serum). Fluoride and oxalate inhibit the reaction. See Chapter 12.
Phosphatase, alkaline	Serum	Adults: 3.5–13 units/100 ml at 37°C with phenylphosphate, or 25–92 U/l (Kind and King) 25–90 U/l at 30°C with PNPP and MAP buffer (Bowers and McComb) 70–220 U/l at 30°C with PNPP and DEA buffer (Tietz, Weinstock, and Wills) Children: 10–30 units/100 ml at 37°C with phenylphosphate (Kind and King) 20 (40)*–150 U/l at 30°C with PNPP and MAP buffer (Bowers and McComb)	
Phosphatides	Plasma	Birth: 29–93 mg/100 ml (as lecithin) Adults up to 65 years: Males: 175–275 mg/100 ml Females, nonpregnant: 158–232 mg/100 ml Females, early pregnant: 205–291 mg/100 ml Males (and females over 65 years): 196–366 mg/100 ml	Avoid hemolysis.
Phospholipid-P	Serum	7–11 mg/100 ml	Avoid hemolysis.
Phosphorus, inorganic	Serum	Adults: 3.0–4.5 mg/100 ml (0.96–1.45 mmol/l) Children: 4.5–6.5 mg/100 ml (1.45–2.09 mmol/l) Birth–1 year: 4.0–7.0 mg/100 ml (1.29–2.26 mmol/l)	Avoid hemolysis. Separate serum from cells promptly.

* Value in parentheses represents another reported value for the designated lower or upper limit.

TABLE OF NORMAL VALUES (*continued*)

Test	*Specimen*	*Value*	*Special Instructions and Interfering Substances*
Phosphorus, inorganic, *cont'd*	Urine, 24 h	0.4–1.3 g/d (12.9–42.0 mmol/d) (varies greatly with intake)	
Porphobilinogen, qualitative	Urine, random	Negative	Analyze promptly.
Porphobilinogen, quantitative	Urine, 24 h	<1.0 mg/d	Avoid exposure to direct light. Collect in a dark bottle with 5 g Na_2CO_3 to maintain pH between 6.5–9.0. Refrigerate during collection.
Porphyrins	Blood	Up to 50–60 μg/100 ml	Collect specimen with heparin.
Potassium	Serum	Adults: 3.5–5.3 mmol/l Newborn: 4.0–5.9 mmol/l All K^+ values 0.2–0.5 mmol/l lower in plasma	Avoid hemolysis.
	Saliva	Without stimulation: 19–23 mmol/l With stimulation: 18–19 mmol/l	
	Sweat	5–17 mmol/l	
	Urine, 24 h	Average diet: 25–120 mmol/d (varies with diet)	
Pregnanediol	Urine, 24 h	Males: 0–1.0 mg/d Females: Proliferative phase: 0.1–1.3 mg/d Luteal phase: 1.2–9.5 mg/d Pregnant: 27–47 mg/d at peak (28th–32nd weeks) and remains steady until delivery	Refrigerate during collection.
Pregnanetriol	Urine, 24 h	0.5–2.0 mg/d Children: <0.5 mg/d	Refrigerate during collection.
Δ^5-Pregnanetriol	Urine, 24 h	Females: 0.2–0.9 mg/d	
Progesterone	Plasma	Females: Follicular phase: 40–60 ng/100 ml Luteal phase: 1000–2000 ng/100 ml During pregnancy increases steadily from 9th–32nd week	
Protein	Serum	6.0–8.0 g/100 ml in recumbent patients. Approx. 0.5 g higher when patients ambulatory	Based on the protein nitrogen factor of 6.25.
	Spinal fluid	15–45 mg/100 ml	
	Ventricular fluid	10–15 mg/100 ml	
	Lumbar fluid	20–45 mg/100 ml	
	Urine, 24 h	50–100 mg/d	
	Urine, random	<10 mg/100 ml	
	Urine, first morning	<20 mg/100 ml	

TABLE OF NORMAL VALUES *(continued)*

Test	*Specimen*	*Value*	*Special Instructions and Interfering Substances*
Protein fractionation	Serum	Albumin: 3.8–5.0 g/100 ml Globulin: 2.3–3.5 g/100 ml (Albumin and globulin values are for ambulatory patients. Approx. 0.5 g lower values for recumbent patients.) Electrophoresis: See Table 7–7	Hemolysis interferes.
Protein-bound iodine, PBI	Serum	(3.5)* 4.0–8.0 μg/100 ml	Test not reliable if iodine-containing drugs or x-ray contrast media were given prior to test.
Propoxyphene	Whole blood	Negative Therapeutic concentration: <30 μg/100 ml	
Reducing substances, total	Urine, 24 h	<150 mg/100 ml (as glucose)	
Renal blood flow	Whole blood	Approx. 1200 ml/min	
	Plasma	Approx. 650 ml/min (or 390 ml/min/m² body surface)	
Riboflavin activity coefficient	Red cell hemolysate	<1.2	
Salicylates	Serum	Negative (<2.0 mg/100 ml considered negative) Therapeutic concentration: <20 mg/100 ml Toxic concentration: >30 mg/100 ml	
	Urine, random	Negative (<4.5 mg/100 ml considered negative)	
Secretin test	Duodenal contents	See Chapter 20	
Secretin-CCK-PZ test	Duodenal contents	See Chapter 20	
Schilling test (Intrinsic factor test)	Urine, 24 h	>7.5% of dose	Dose: 0.5–1 μCi ^{58}Co-Vitamin B_{12}.
Sia test	Serum	Negative	
Sodium	Saliva	Without stimulation: 6.5–21.7 mmol/l After stimulation: 43–46 mmol/l	
	Serum	135–148 mmol/l	
	Spinal fluid	138–150 mmol/l	
	Sweat	Normal homozygotes: 10–40 mmol/l Heterozygotes: 40–70 mmol/l	
	Urine, 24 h	40–220 mmol/d (varies with diet)	
Solids, total	Urine, 24 h	45–70 g/d	Refrigerate during collection.
Sorbitol dehydrogenase	Serum	<1.0 U/l at 25°C <1.3 U/l at 30°C	Very unstable. Anticoagulants other than heparin interfere.
Specific gravity	Urine, 24 h	1.015–1.025	
	Urine, random	1.002–1.030 (varies greatly with fluid intake and state of hydration)	
	Urine, first morning	>1.025 after fluid restriction	

* Value in parentheses represents another reported value for the designated lower or upper limit.

TABLE OF NORMAL VALUES (*continued*)

Test	*Specimen*	*Value*	*Special Instructions and Interfering Substances*
Stool analysis	Feces, 24 h		
Dry matter (See also individual constituent requested)		Up to 25% of total weight	
Sulfhemoglobin	Whole blood	<1% of total hemoglobin	Collect with heparin, EDTA, or ACD solution.
Sulfonamides	Serum or whole blood	Negative Therapeutic concentration: Approx. 10 mg/100 ml	False positive caused by other aromatic amines.
T_3, free	Serum	160–270 ng/100 ml	
T_3 resin uptake (RT_3U)	Serum	25–35% resin uptake	
Testosterone	Plasma	Males: 500–860 ng/100 ml Females: 26–54 ng/100 ml	Collect with heparin. Analyze immediately or freeze.
Thymol flocculation	Serum	Negative	
Thymol turbidity	Serum	0–5 units (Shank and Hoagland)	Lipemia interferes.
Thyroid stimulating hormone, TSH	Serum	1.5–8.0 μUnits/ml	
Thyroid uptake of ^{131}I		After 24 h: 7–33%	
Thyroid uptake of $^{99m}TcO_4^-$		After 24 h: 0.4–3.0%	
Thyroxine, free	Serum	3.0–5.1 ng/100 ml, as thyroxine	
Thyroxine (T_4) by column	Serum	2.8–6.4 μg/100 ml, as iodine (4.3–9.8 μg/100 ml, as thyroxine)	Organic iodine–containing drugs and contrast media interfere.
Thyroxine (Murphy-Pattee method)	Serum	2.9–7.0 μg/100 ml, as iodine (4.4–10.7 μg/100 ml, as thyroxine)	No interference by organic iodine-containing compounds except excessive T_3, diiodothyronine, phenytoin (diphenylhydantoin), and choloxin (*d*-isomer of T_4).
Thyroxine binding globulin	Serum	10–26 ng/100 ml as thyroxine	
Titratable acidity	Urine, 24 h	10–60 mol ion charge/d	Refrigerate and preserve with toluene.
Tolbutamide (Orinase) tolerance test	Plasma	Fasting: Glucose normal; 30 min: glucose decrease of approx. 50% 2–3 h: >75% of fasting glucose concentration	I.V. dose: 1 g sodium tolbutamide in 20 ml H_2O Collect specimens with potassium oxalate and sodium fluoride. Severe hypoglycemic reaction may occur; therefore, glucose for I.V. administration should be readily available.
Transaminase, glutamate oxalacetate, GOT (Aspartate aminotransferase, AST)	Serum	Without P-5-P Males: 7–21 U/l at 30°C Females: 6–18 U/l at 30°C With P-5-P 12–29 U/l at 30°C (females slightly lower)	Do not use specimen with visible hemolysis.
	Spinal fluid	3–10 U/l at 30°C	
Transaminase, glutamate pyruvate, GPT (alanine aminotransferase, ALT)	Serum	Males: 6–21 U/l at 30°C Females: 4–17 U/l at 30°C (conditions in many methods not optimal; 7–33 U/l at 30°C proposed by one optimized method using P-5-P)	Do not use specimen with visible hemolysis.

TABLE OF NORMAL VALUES (*continued*)

Test	*Specimen*	*Value*	*Special Instructions and Interfering Substances*
Transaminase, glutamate pyruvate, GPT (alanine aminotransferase, ALT), *cont'd*	Spinal fluid	None	
Transferrin	Serum	200–400 mg/100 ml See also Iron-binding capacity	
Triglycerides	Serum	Males: 40–160 mg/100 ml Females: 35–135 mg/100 ml (see Chapter 10)	Use fasting specimen. Remove serum from clot soon after drawing. Hemolysis interferes.
	Urine	<20 μg/d	
Triolein-^{131}I, absorption test	Plasma	1.7% of administered dose/l after 4–6 h	Dose: 50 μCi in milk.
	Feces, 72 h	<5% of administered dose in 72 h specimen	
Trypsin	Feces, fresh random	Infants: Positive in >80-fold dilution. (Semiquantitative method using x-ray film.) Older children: Positive in 20–40 fold dilution	Patient must be off enzyme preparations 3 days prior to test. Test unreliable in adults.
	Duodenal fluid	Positive in 64–256 fold dilution	
Tyrosine	Serum	Adult: 0.8–1.3 mg/100 ml Full-term, normal weight newborns: 1.6–3.7 mg/100 ml (drops to normal adult level in 4–8 weeks) Premature and low weight full-term newborns: 7.0–24.0 mg/100 ml (drops to normal full-term newborn range in 7–20 days)	
	Urine	8–20 mg/d	
Urea clearance	Urine and blood	Standard clearance: 41–68 ml/min Maximum clearance: 64–99 ml/min or 75–125% of normal clearance	Correct for body surface.
Urea nitrogen	Serum or plasma	7–18 mg/100 ml (15–38.5 mg urea/100 ml or 2.5–6.3 mmol/l) 1–2 yrs: 5–15 mg/100 ml (11–32 mg urea/100 ml or 1.8–5.3 mmol/l)	Values higher in individuals on high protein diet.
	Urine, 24 h	12–20 g/d (25–43 g urea/d or 0.42–0.73 mol/d)	Refrigerate during collection.
Uric acid	Serum or plasma	Males: 3.5–7.2 mg/100 ml Females: 2.6–6.0 mg/100 ml Children: 2.0–5.5 mg/100 ml	Separate serum from cells.
	Urine, 24 h	Average diet: 250–750 mg/d Low purine diet: <450 mg/d High purine diet: <1.0 g/d	Refrigerate during collection.

TABLE OF NORMAL VALUES (*continued*)

Test	*Specimen*	*Value*	*Special Instructions and Interfering Substances*
Urobilinogen, quantitative	Feces, 24 h Feces, random Urine, 2 h Urine, 24 h	100–400 Ehrlich units/d 75–275 Ehrlich units/100 g 0.1–1.0 Ehrlich units/2 h 0.5–3.5 mg/d or 0.5–4.0 Ehrlich units/d	Refrigerate during collection. Protect from light. Refrigerate during collection. Preserve with 5 g $NaHCO_3$. Protect from light. Bilirubin, sulfonamide, procaine, and 5-HIAA interfere.
Uromucoid	Urine, 24 h	Approx. 70 mg/d	
Uropepsin	Urine, 24 h or timed specimen	15–45 U/h at 37°C (West) 1500–5000 U/d at 25°C (Anson)	Stable 2–3 days at pH 5.0–6.5.
Uroporphyrins			
qualitative	Urine, random	Negative	Analyze promptly.
quantitative	Urine, 24 h	<40 μg/d	Avoid exposure to direct light. Collect in a dark bottle with 5 g Na_2CO_3 to maintain pH 6.5–9.0
Vanilmandelic acid, VMA (3-Methoxy-4-hydroxy-mandelic acid)	Urine, 24 h	Adults: 1.8–7.1 mg/d (1.5–7.0 μg/mg creatinine) Infants: <83 μg/kg/d (2–12 μg/mg creatinine)	Refrigerate and acidify during collection. Excessive ingestion of coffee, desserts, or fruit may cause an increase in results by some methods but not with Pisano's method.
Vitamin A	Serum	30–65 μg/100 ml (lower at birth)	
Vitamin A tolerance	Serum	Fasting: 30–65 μg/100 ml 3 or 6 h: Increase to 200–600 μg/100 ml	Dose: 5000 U.S.P. Units Vit. A (in oil)/kg.
Vitamin B_{12}	Serum	200–1600 pg/ml	
Vitamin C (Ascorbic acid)	Plasma	0.6–2.0 mg/100 ml	Analyze immediately, unstable. Stable in acid solution. Collect with oxalate.
Volume	Amniotic fluid	10 weeks: approx. 25 ml 40 weeks: approx. 900 ml	
	Whole blood	60–70 ml/kg	
	Plasma	33–45 ml/kg	
	Gastric residue	After 12 h fast: 20–100 ml (generally <50 ml)	
	Urine, 24 h	Males: 800–1800 ml/d Females: 600–1600 ml/d (varies with intake and other factors)	
Xanthochromia	Spinal fluid	Absent	
Xylose	Urine, 24 h	Average: 49 mg/d	
Xylose absorption test	Whole blood and urine	Adults: Blood: >25 mg/100 ml after 2 h Urine: >4 g/5 h (after 65 yr: >3.5 g/5 h) Children: Blood: >30 mg/100 ml after 1 h Urine: 16–33% of ingested xylose	Dose: Adult, 25 g; children, 0.23 g/kg (0.5 g/lb). Analyze urine immediately after collection or preserve with 100 mg NaF/ml. Collect blood with potassium oxalate and sodium fluoride.
Zinc turbidity	Serum	Caucasians: 2–9 units Blacks: 5–12 units	Lipemia interferes.

REFERENCES FOR NORMAL VALUE TABLE

1. Bergmeyer, H. U.: Methods of Enzymatic Analysis. 2nd English Ed. New York, Academic Press, 1974.
2. Handbook of Specialized Diagnostic Laboratory Tests. 10th Ed. Van Nuys, California, Bioscience Laboratories, 1973.
3. Henry, R. J., Cannon, D. C., and Winkelman, J. W.: Clinical Chemistry. Principles and Technics. 2nd Ed. New York, Hoeber Medical Division, Harper & Row Publishers, 1974.
4. O'Brien, D., Ibbott, F. A., and Rodgerson, D. O.: Laboratory Manual of Pediatric Micro-Biochemical Techniques. 4th Ed. New York, Hoeber Medical Division, Harper & Row Publishers, 1968.
5. Page, L. B., and Culver, P. J.: A Syllabus of Laboratory Examinations in Clinical Diagnosis. Revised Ed. Cambridge, Massachusetts, Harvard University Press, 1960.
6. Robinson, H. W.: Appendix, normal blood values. *In* Textbook of Pediatrics, 8th Ed. W. E. Nelson, Ed. Philadelphia, W. B. Saunders Co., 1964, pp. 1583–1587.

INDEX

Page numbers in *italics* indicate figures; page numbers followed by (t) indicate tables; those followed by (n) indicate footnotes.

TABLE OF FOUR

	0	1	2	3	4	5	6	7	8	9
1.0	.0000	.0043	.0086	.0128	.0170	.0212	.0253	.0294	.0334	.0374
1.1	.0414	.0453	.0492	.0531	.0569	.0607	.0645	.0682	.0719	.0755
1.2	.0792	.0828	.0864	.0899	.0934	.0969	.1004	.1038	.1072	.1106
1.3	.1139	.1173	.1206	.1239	.1271	.1303	.1335	.1367	.1399	.1430
1.4	.1461	.1492	.1523	.1553	.1584	.1614	.1644	.1673	.1703	.1732
1.5	.1761	.1790	.1818	.1847	.1875	.1903	.1931	.1959	.1987	.2014
1.6	.2041	.2068	.2095	.2122	.2148	.2175	.2201	.2227	.2253	.2279
1.7	.2304	.2330	.2355	.2380	.2405	.2430	.2455	.2480	.2504	.2529
1.8	.2553	.2577	.2601	.2625	.2648	.2672	.2695	.2718	.2742	.2765
1.9	.2788	.2810	.2833	.2856	.2878	.2900	.2923	.2945	.2967	.2989
2.0	.3010	.3032	.3054	.3075	.3096	.3118	.3139	.3160	.3181	.3201
2.1	.3222	.3243	.3263	.3284	.3304	.3324	.3345	.3365	.3385	.3404
2.2	.3424	.3444	.3464	.3483	.3502	.3522	.3541	.3560	.3579	.3598
2.3	.3617	.3636	.3655	.3674	.3692	.3711	.3729	.3747	.3766	.3784
2.4	.3802	.3820	.3838	.3856	.3874	.3892	.3909	.3927	.3945	.3962
2.5	.3979	.3997	.4014	.4031	.4048	.4065	.4082	.4099	.4116	.4133
2.6	.4150	.4166	.4183	.4200	.4216	.4232	.4249	.4265	.4281	.4298
2.7	.4314	.4330	.4346	.4362	.4378	.4393	.4409	.4425	.4440	.4456
2.8	.4472	.4487	.4502	.4518	.4533	.4548	.4564	.4579	.4594	.4609
2.9	.4624	.4639	.4654	.4669	.4683	.4698	.4713	.4728	.4742	.4757
3.0	.4771	.4786	.4800	.4814	.4829	.4843	.4857	.4871	.4886	.4900
3.1	.4914	.4928	.4942	.4955	.4969	.4983	.4997	.5011	.5024	.5038
3.2	.5051	.5065	.5079	.5092	.5105	.5119	.5132	.5145	.5159	.5172
3.3	.5185	.5198	.5211	.5224	.5237	.5250	.5263	.5276	.5289	.5302
3.4	.5315	.5328	.5340	.5353	.5366	.5378	.5391	.5403	.5416	.5428
3.5	.5441	.5453	.5465	.5478	.5490	.5502	.5514	.5527	.5539	.5551
3.6	.5563	.5575	.5587	.5599	.5611	.5623	.5635	.5647	.5658	.5670
3.7	.5682	.5694	.5705	.5717	.5729	.5740	.5752	.5763	.5775	.5786
3.8	.5798	.5809	.5821	.5832	.5843	.5855	.5866	.5877	.5888	.5899
3.9	.5911	.5922	.5933	.5944	.5955	.5966	.5977	.5988	.5999	.6010
4.0	.6021	.6031	.6042	.6053	.6064	.6075	.6085	.6096	.6107	.6117
4.1	.6128	.6138	.6149	.6160	.6170	.6180	.6191	.6201	.6212	.6222
4.2	.6232	.6243	.6253	.6263	.6274	.6284	.6294	.6304	.6314	.6325
4.3	.6335	.6345	.6355	.6365	.6375	.6385	.6395	.6405	.6415	.6425
4.4	.6435	.6444	.6454	.6464	.6474	.6484	.6493	.6503	.6513	.6522
4.5	.6532	.6542	.6551	.6561	.6571	.6580	.6590	.6599	.6609	.6618
4.6	.6628	.6637	.6646	.6656	.6665	.6675	.6684	.6693	.6702	.6712
4.7	.6721	.6730	.6739	.6749	.6758	.6767	.6776	.6785	.6794	.6803
4.8	.6812	.6821	.6830	.6839	.6848	.6857	.6866	.6875	.6884	.6893
4.9	.6902	.6911	.6920	.6928	.6937	.6946	.6955	.6964	.6972	.6981
5.0	.6990	.6998	.7007	.7016	.7024	.7033	.7042	.7050	.7059	.7067
5.1	.7076	.7084	.7093	.7101	.7110	.7118	.7126	.7135	.7143	.7152
5.2	.7160	.7168	.7177	.7185	.7193	.7202	.7210	.7218	.7226	.7235
5.3	.7243	.7251	.7259	.7267	.7275	.7284	.7292	.7300	.7308	.7316
5.4	.7324	.7332	.7340	.7348	.7356	.7364	.7372	.7380	.7388	.7396
5.5	.7404	.7412	.7419	.7427	.7435	.7443	.7451	.7459	.7466	.7474
5.6	.7482	.7490	.7497	.7505	.7513	.7520	.7528	.7536	.7543	.7551
5.7	.7559	.7566	.7574	.7582	.7589	.7597	.7604	.7612	.7619	.7627
5.8	.7634	.7642	.7649	.7657	.7664	.7672	.7679	.7686	.7694	.7701
5.9	.7709	.7716	.7723	.7731	.7738	.7745	.7752	.7760	.7767	.7774

PLACE LOGARITHMS

	0	1	2	3	4	5	6	7	8	9
6.0	.7782	.7789	.7796	.7803	.7810	.7818	.7825	.7832	.7839	.7846
6.1	.7853	.7860	.7868	.7875	.7882	.7889	.7896	.7903	.7910	.7917
6.2	.7924	.7931	.7938	.7945	.7952	.7959	.7966	.7973	.7980	.7987
6.3	.7993	.8000	.8007	.8014	.8021	.8028	.8035	.8041	.8048	.8055
6.4	.8062	.8069	.8075	.8082	.8089	.8096	.8102	.8109	.8116	.8122
6.5	.8129	.8136	.8142	.8149	.8156	.8162	.8169	.8176	.8182	.8189
6.6	.8195	.8202	.8209	.8215	.8222	.8228	.8235	.8241	.8248	.8254
6.7	.8261	.8267	.8274	.8280	.8287	.8293	.8299	.8306	.8312	.8319
6.8	.8325	.8331	.8338	.8344	.8351	.8357	.8363	.8370	.8376	.8382
6.9	.8388	.8395	.8401	.8407	.8414	.8420	.8426	.8432	.8439	.8445
7.0	.8451	.8457	.8463	.8470	.8476	.8482	.8488	.8494	.8500	.8506
7.1	.8513	.8519	.8525	.8531	.8537	.8543	.8549	.8555	.8561	.8567
7.2	.8573	.8579	.8585	.8591	.8597	.8603	.8609	.8615	.8621	.8627
7.3	.8633	.8639	.8645	.8651	.8657	.8663	.8669	.8675	.8681	.8686
7.4	.8692	.8698	.8704	.8710	.8716	.8722	.8727	.8733	.8739	.8745
7.5	.8751	.8756	.8762	.8768	.8774	.8779	.8785	.8791	.8797	.8802
7.6	.8808	.8814	.8820	.8825	.8831	.8837	.8842	.8848	.8854	.8859
7.7	.8865	.8871	.8876	.8882	.8887	.8893	.8899	.8904	.8910	.8915
7.8	.8921	.8927	.8932	.8938	.8943	.8949	.8954	.8960	.8965	.8971
7.9	.8976	.8982	.8987	.8993	.8998	.9004	.9009	.9015	.9020	.9026
8.0	.9031	.9036	.9042	.9047	.9053	.9058	.9063	.9069	.9074	.9079
8.1	.9085	.9090	.9096	.9101	.9106	.9112	.9117	.9122	.9128	.9133
8.2	.9138	.9143	.9149	.9154	.9159	.9165	.9170	.9175	.9180	.9186
8.3	.9191	.9196	.9201	.9206	.9212	.9217	.9222	.9227	.9232	.9238
8.4	.9243	.9248	.9253	.9258	.9263	.9269	.9274	.9279	.9284	.9289
8.5	.9294	.9299	.9304	.9309	.9315	.9320	.9325	.9330	.9335	.9340
8.6	.9345	.9350	.9355	.9360	.9365	.9370	.9375	.9380	.9385	.9390
8.7	.9395	.9400	.9405	.9410	.9415	.9420	.9425	.9430	.9435	.9440
8.8	.9445	.9450	.9455	.9460	.9465	.9469	.9474	.9479	.9484	.9489
8.9	.9494	.9499	.9504	.9509	.9513	.9518	.9523	.9528	.9533	.9538
9.0	.9542	.9547	.9552	.9557	.9562	.9566	.9571	.9576	.9581	.9586
9.1	.9590	.9595	.9600	.9605	.9609	.9614	.9619	.9624	.9628	.9633
9.2	.9638	.9643	.9647	.9652	.9657	.9661	.9666	.9671	.9675	.9680
9.3	.9685	.9689	.9694	.9699	.9703	.9708	.9713	.9717	.9722	.9727
9.4	.9731	.9736	.9741	.9745	.9750	.9754	.9759	.9763	.9768	.9773
9.5	.9777	.9782	.9786	.9791	.9795	.9800	.9805	.9809	.9814	.9818
9.6	.9823	.9827	.9832	.9836	.9841	.9845	.9850	.9854	.9859	.9863
9.7	.9868	.9872	.9877	.9881	.9886	.9890	.9894	.9899	.9903	.9908
9.8	.9912	.9917	.9921	.9926	.9930	.9934	.9939	.9943	.9948	.9952
9.9	.9956	.9961	.9965	.9969	.9974	.9978	.9983	.9987	.9991	.9996

W9-CCQ-099
Teeth
T-rex had **massive jaws** and 6-inch teeth, and it never ran out of teeth. As soon as it lost one tooth, another would replace it.
Skin
Tyrannosaurus' skin was **bumpy**, like alligator skin. It had **good eyesight** for spotting its prey.
Short Arms
T-rex had very short arms—just **over 3 feet** long. The arms didn't even reach up to its mouth. It had **two fingers** on each hand.

Mosasaur

(say MOES-ah-SAWR)

After mosasaurs died out, sharks soon took over the sea.

DiscoveryFact™

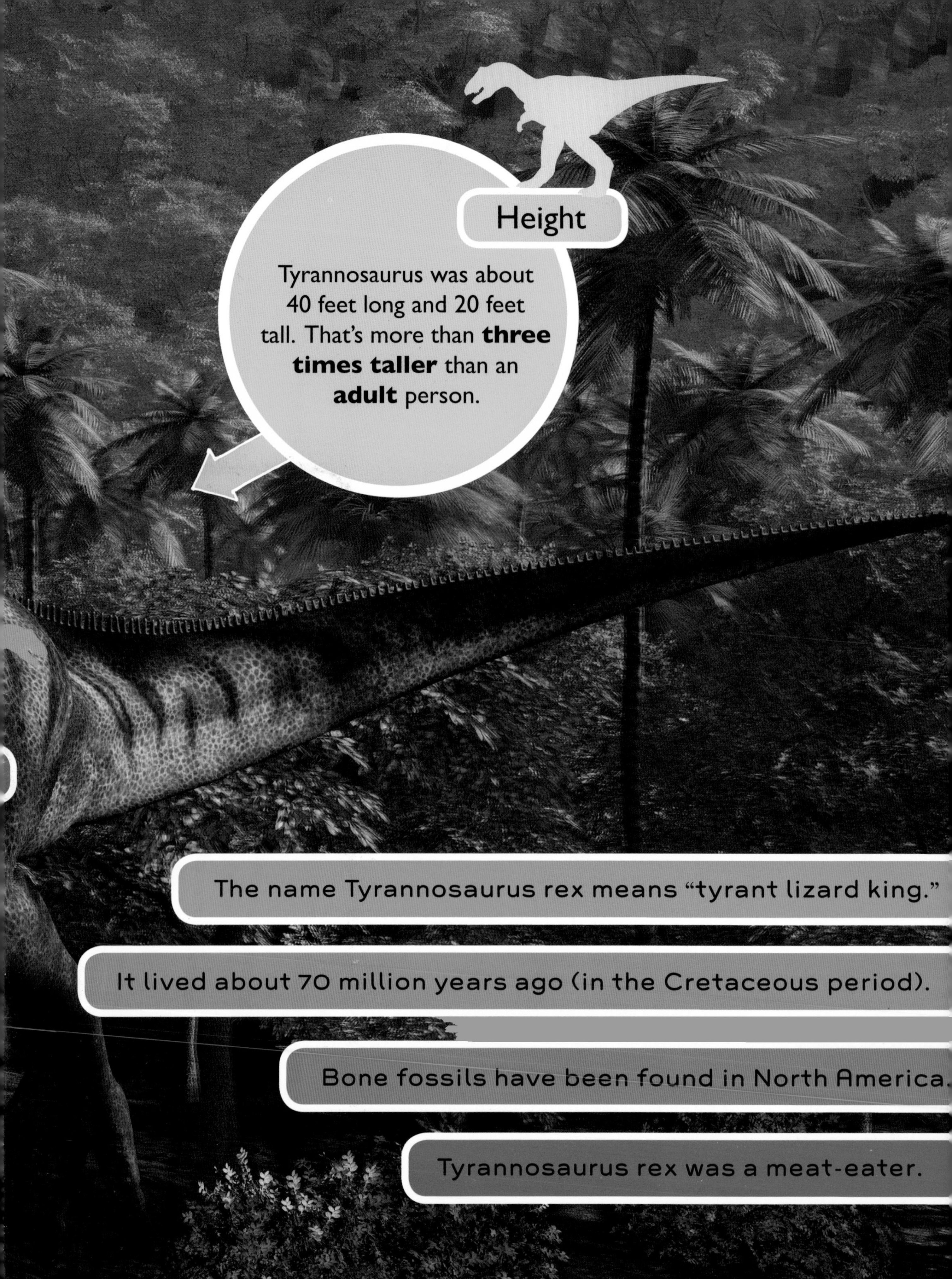
Height
Tyrannosaurus was about 40 feet long and 20 feet tall. That's more than **three times taller** than an **adult** person.
The name Tyrannosaurus rex means "tyrant lizard king."
It lived about 70 million years ago (in the Cretaceous period).
Bone fossils have been found in North America.
Tyrannosaurus rex was a meat-eater.

(say brak-ee-oh-SORE-us)

Nose

Its nose was on the **top** of its **head**, and it had **big nostrils**, so it probably had a **good** sense of **smell**.

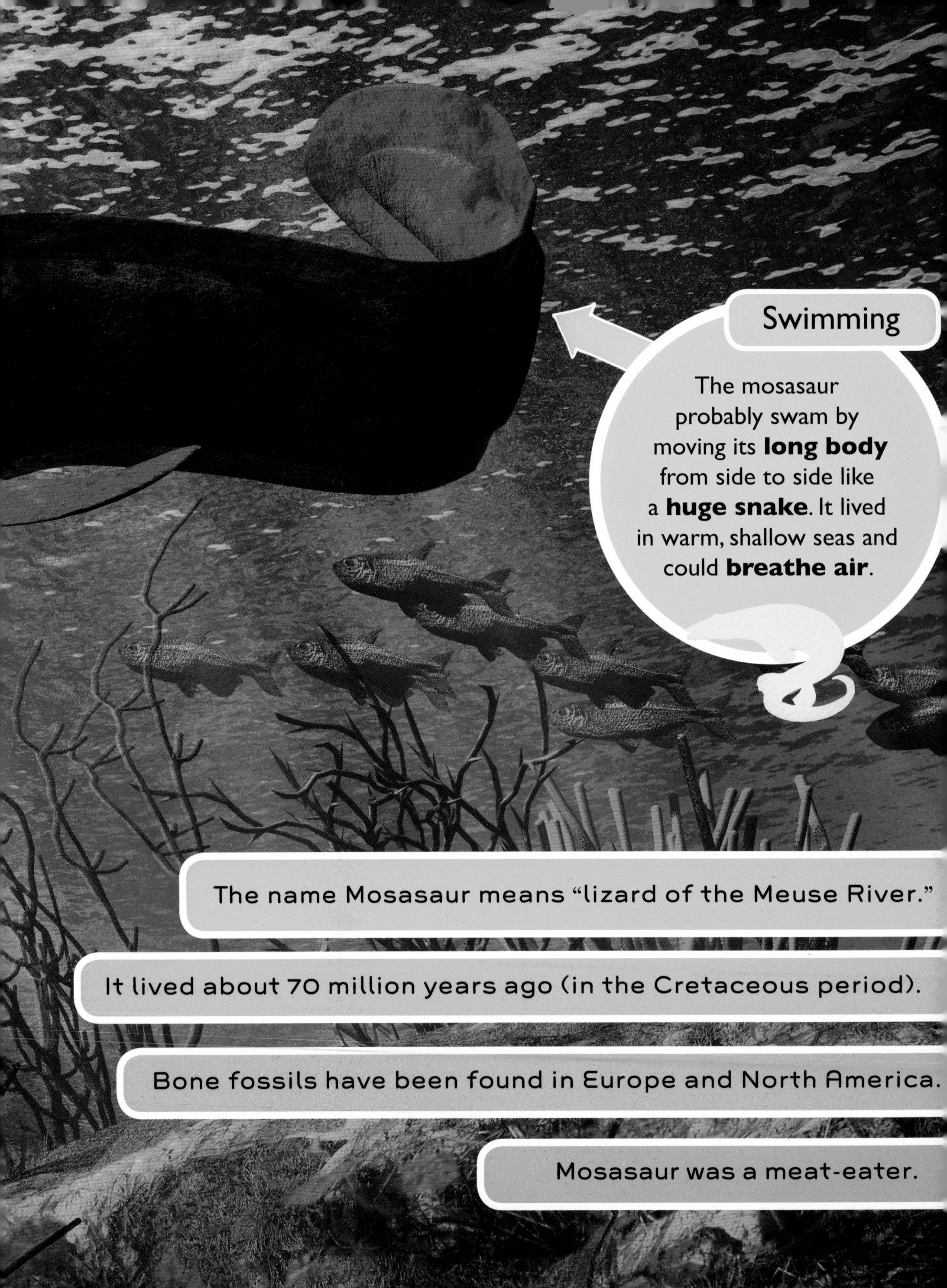

Swimming

The mosasaur probably swam by moving its **long body** from side to side like a **huge snake**. It lived in warm, shallow seas and could **breathe air**.

The name Mosasaur means "lizard of the Meuse River."

It lived about 70 million years ago (in the Cretaceous period).

Bone fossils have been found in Europe and North America.

Mosasaur was a meat-eater.

The huge Brachiosaurus was one of the **largest dinosaurs** found. It was about 85 feet long and as tall as a four-story **building**. It had a very **long neck** and **tail**. This meant that it could reach out and **munch** on lots of **different plants** and use its tail for balance.

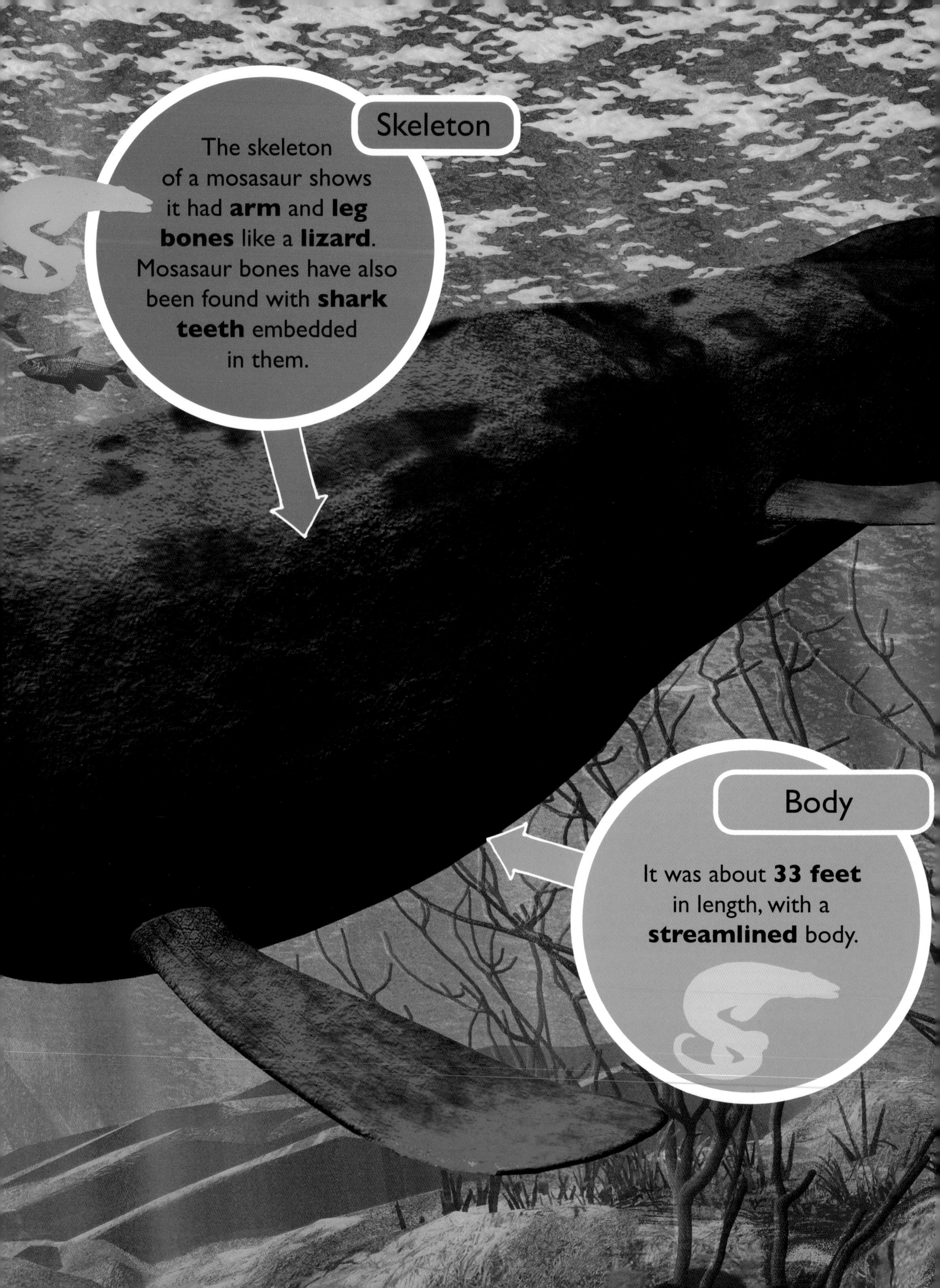
Skeleton
The skeleton of a mosasaur shows it had **arm** and **leg bones** like a **lizard**. Mosasaur bones have also been found with **shark teeth** embedded in them.
Body
It was about **33 feet** in length, with a **streamlined** body.

Herd
Brachiosaurus lived in a **group**, or **herd**, of animals. They spent most of their time **looking for food** and eating.

Velociraptor was about **6** feet long and **3** feet tall.
Height

Velociraptor

(say vel-AH-si-RAP-tor)

Running

It ran **quickly** on its thin back legs. It held its **big foot claws** off the ground when it wasn't attacking prey, so they didn't get **worn down** and blunt.

Legs
Brachiosaurus' **front legs** were longer than its back legs. This would have meant it stood like a **giraffe** and could have reared up on its legs to strip leaves from trees.
This big animal needed lots of food. It may have eaten 440 pounds of plants every day.
DiscoveryFact™
The name Brachiosaurus means "arm lizard."
It lived about 150 million years ago (in the Jurassic period).
Bone fossils have been found in North America and Africa.
Brachiosaurus was a plant-eater.

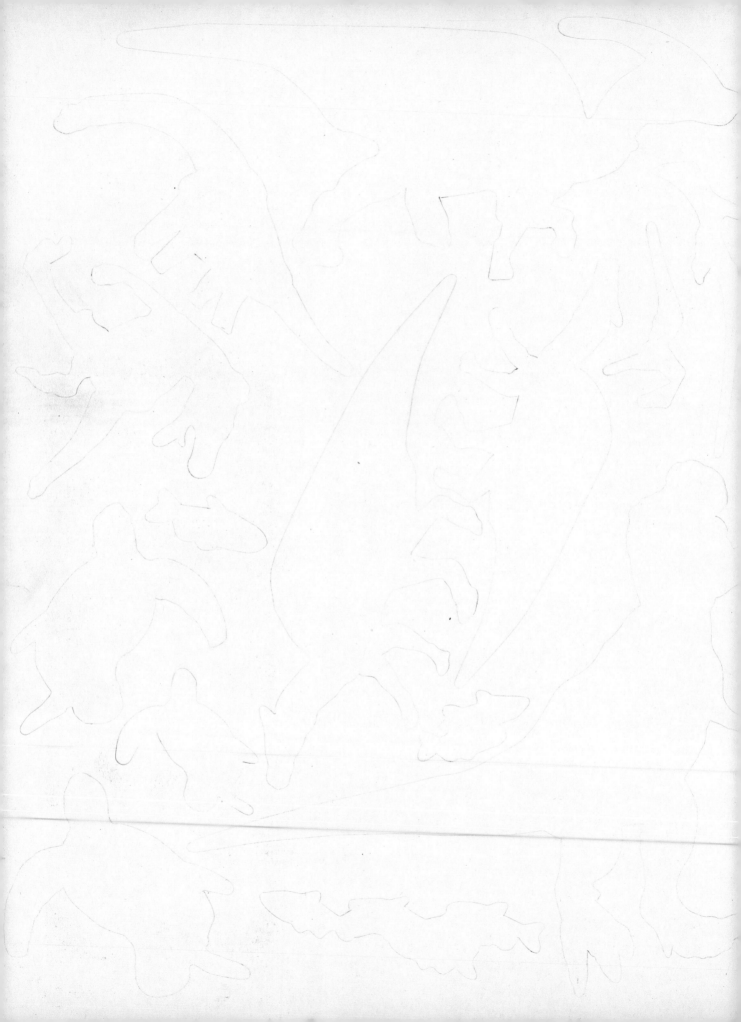

Pteranodon

(say ter-AN-oh-DON)

Beak

It had a long, **pointed** beak, similar to that of a modern bird, which was probably **good for fishing**. It also had lots of small teeth.

The pteranodon had very long wings that measured more than four adults lying head to toe.

DiscoveryFact™

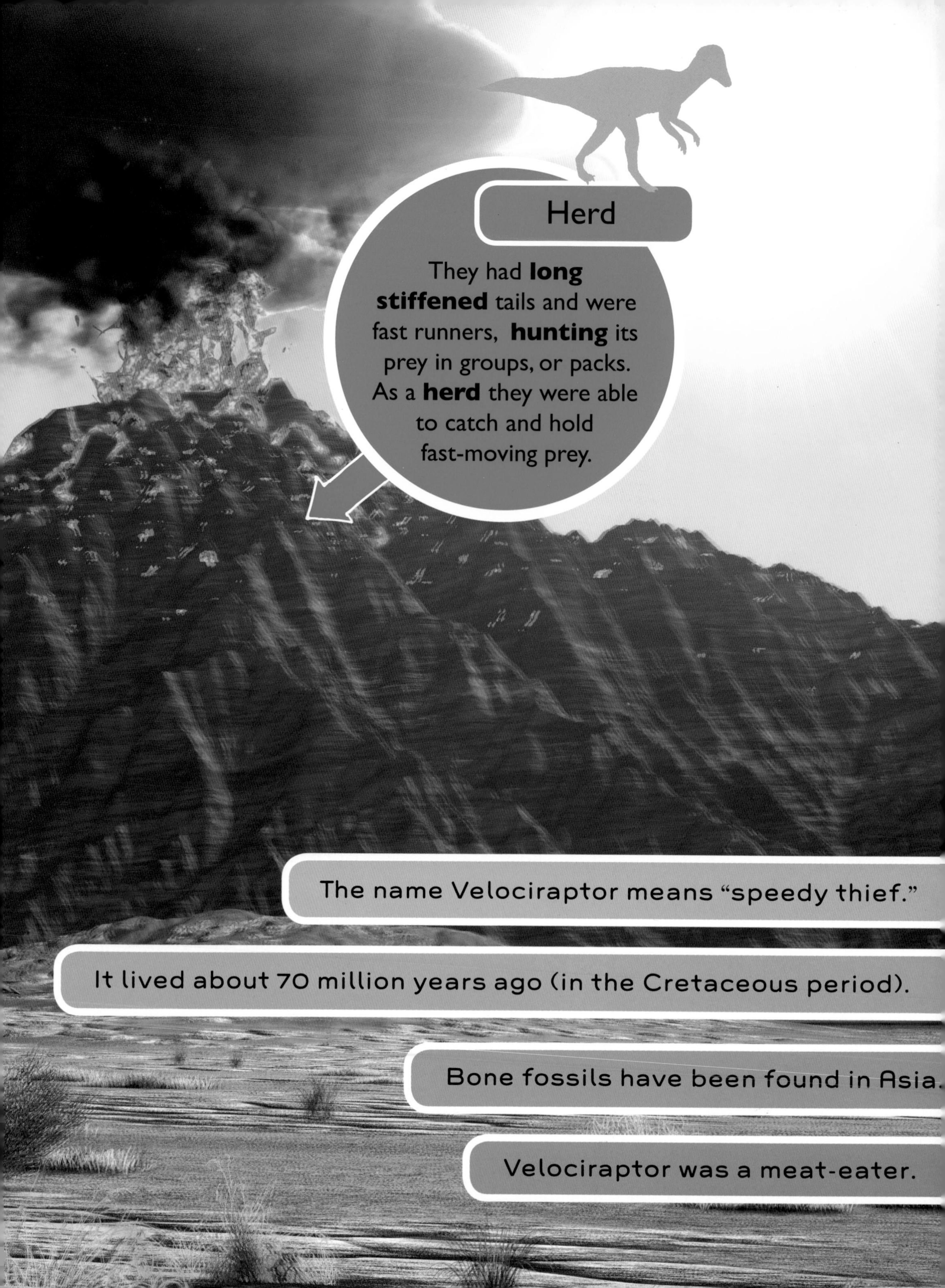
Herd
They had **long stiffened** tails and were fast runners, **hunting** its prey in groups, or packs. As a **herd** they were able to catch and hold fast-moving prey.
The name Velociraptor means “speedy thief.”
It lived about 70 million years ago (in the Cretaceous period).
Bone fossils have been found in Asia.
Velociraptor was a meat-eater.

The velociraptor was a very **fierce hunter** and one of the Dromaeosaurs. This family of dinosaurs had **large brains** in relation to their body size and were **intelligent** animals. They had **big sharp claws** on their feet and strong hands for grabbing hold of prey.

The velociraptor was made famous when it appeared in the film *Jurassic Park* although it appeared bigger than it would have been in real life.
DiscoveryFact™

Pteranodon was not really a dinosaur but a prehistoric **flying reptile** and part of the pterosaur family. Although it could fly, its **wings** were **not feathery.** Some pterosaurs were as small as a blackbird of today. The biggest was the size of a small plane.

Crest
Pteranodon had a crest on its head made up of **skull bones**. The **size** and **shape** of the crest **varied** depending on the species. It had long hair that grew along its neck.

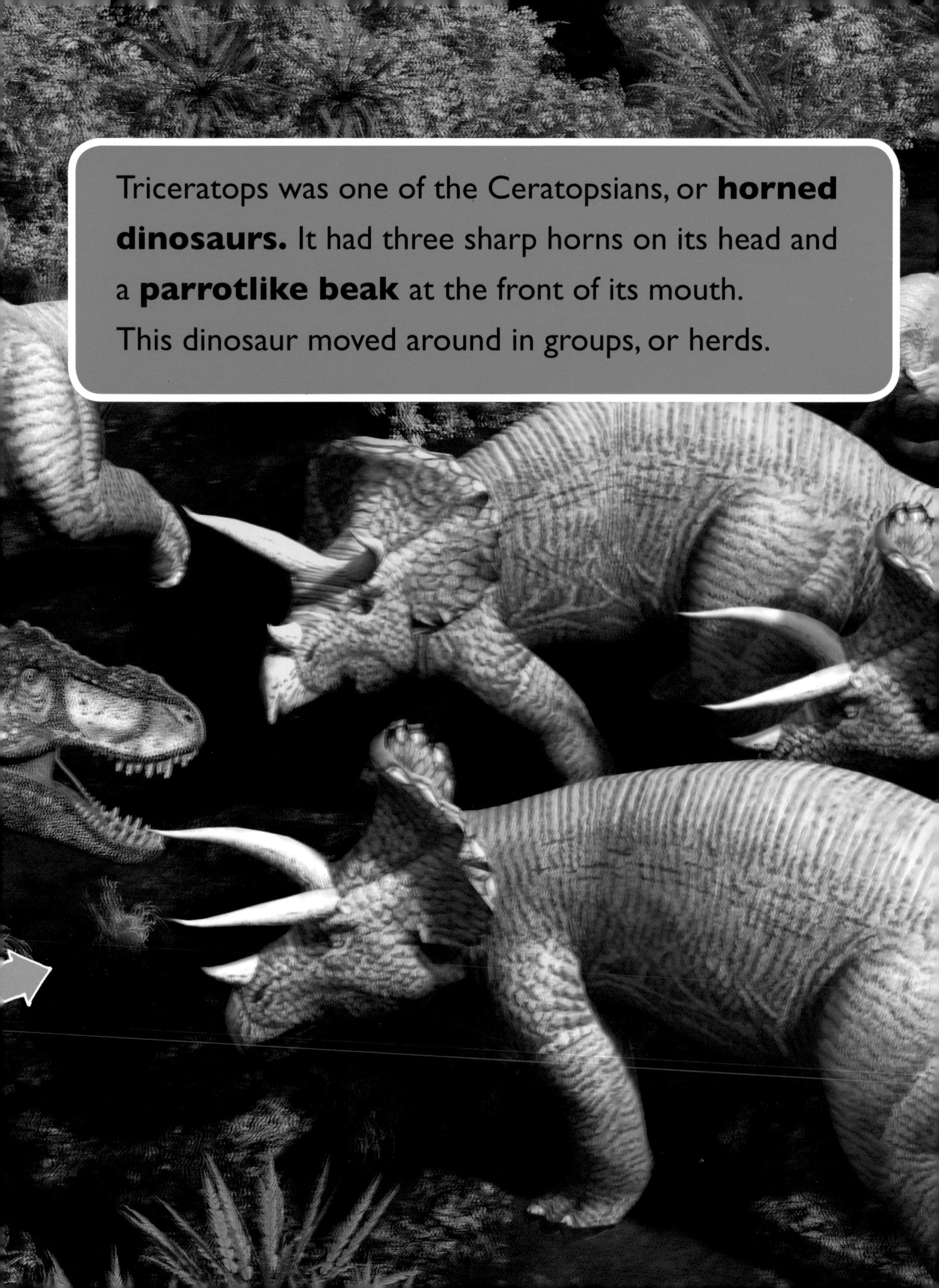

Triceratops was one of the Ceratopsians, or **horned dinosaurs.** It had three sharp horns on its head and a **parrotlike beak** at the front of its mouth. This dinosaur moved around in groups, or herds.

Triceratops

(say try-SERRA-tops)

Triceratops probably fought over mates or who would lead the herd. They crashed their heads together and locked horns in combat.

DiscoveryFact™

Horns

It may have used the horns on its head to **knock** down **small trees** so that it could eat the leaves. But horns would also have been useful in **defending itself** from predators, such as T-rex.

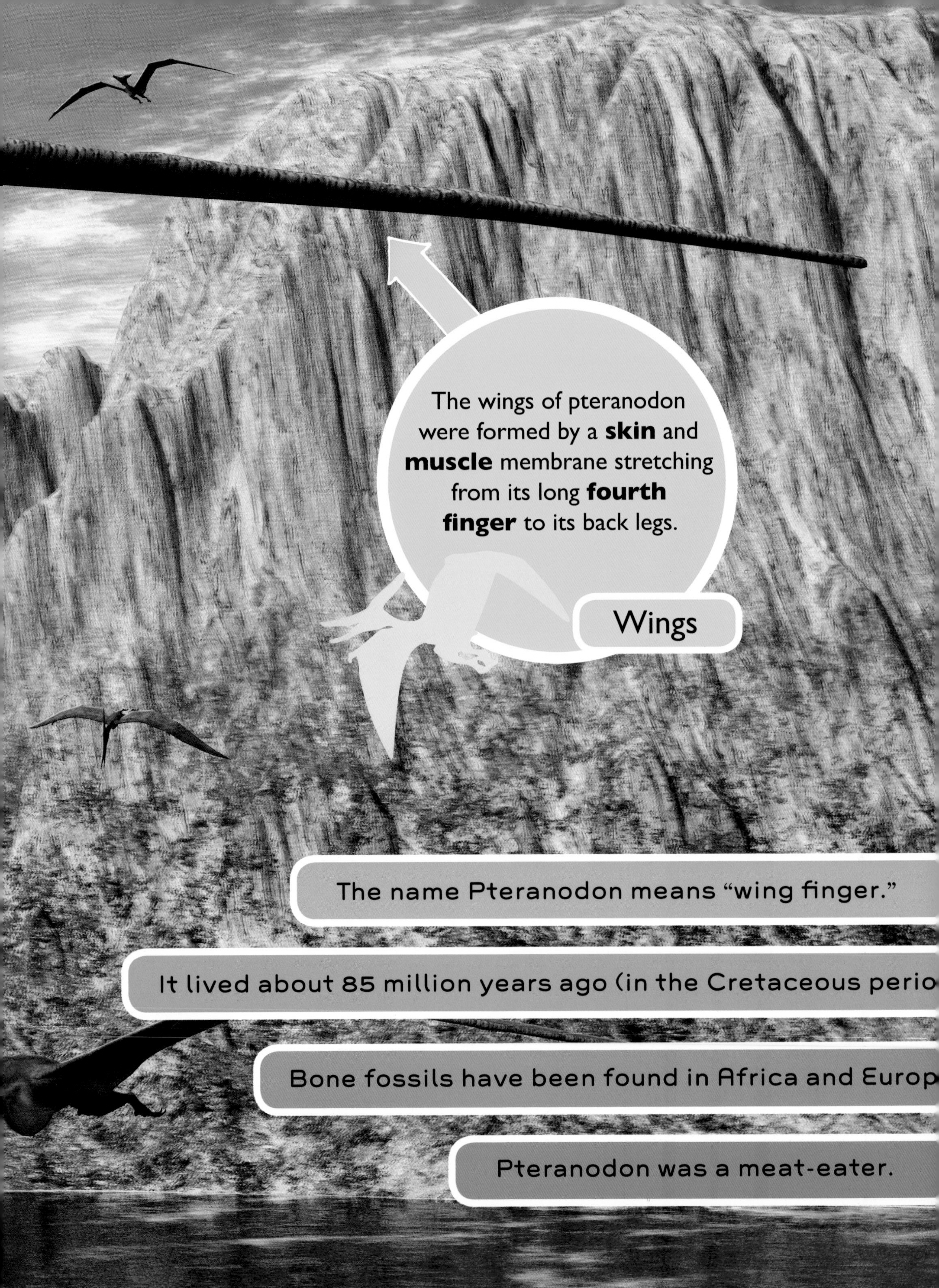
The wings of pteranodon were formed by a **skin** and **muscle** membrane stretching from its long **fourth finger** to its back legs.
Wings
The name Pteranodon means "wing finger."
It lived about 85 million years ago (in the Cretaceous perio
Bone fossils have been found in Africa and Europ
Pteranodon was a meat-eater.

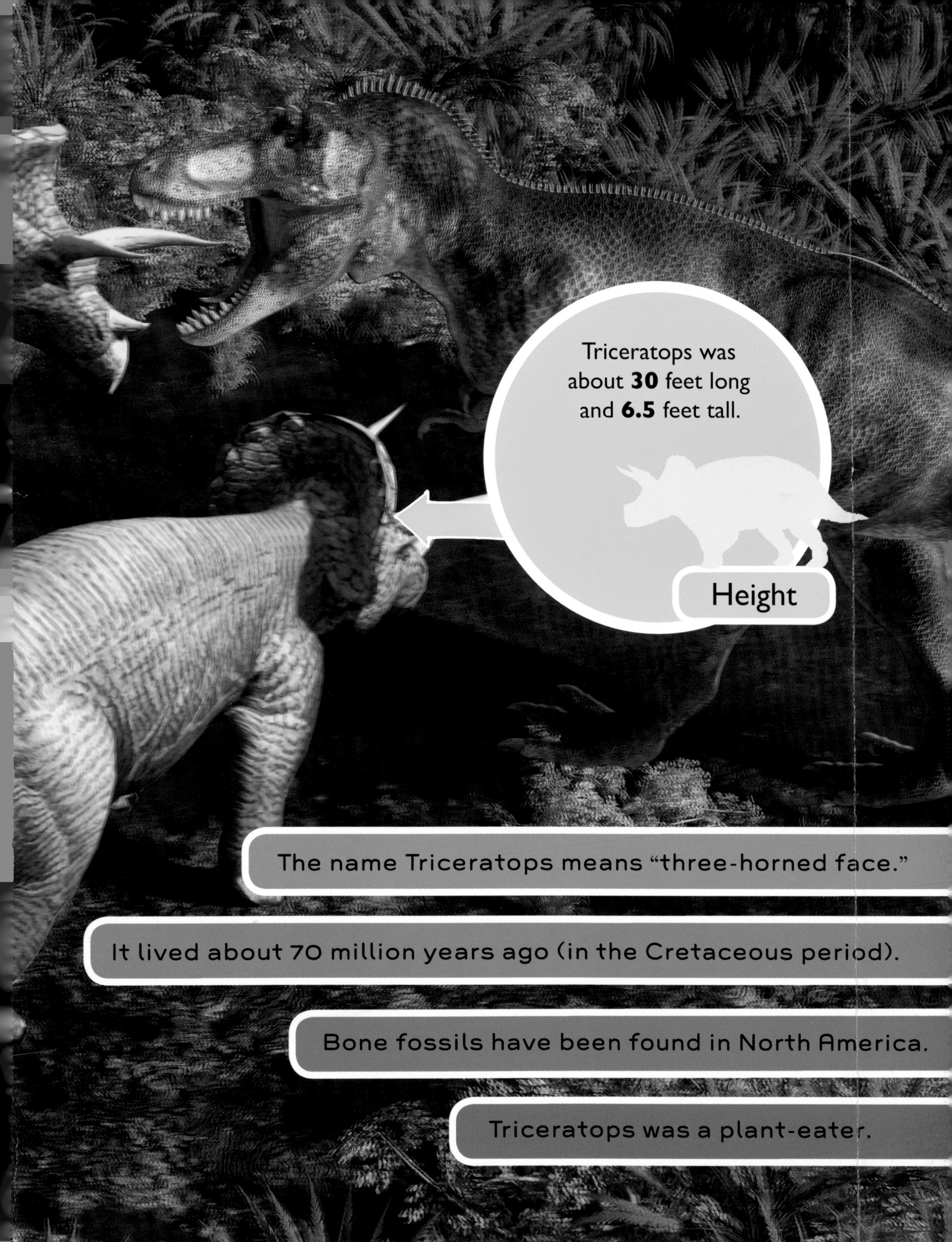
Triceratops was
about **30** feet long
and **6.5** feet tall.
Height
The name Triceratops means "three-horned face."
It lived about 70 million years ago (in the Cretaceous period).
Bone fossils have been found in North America.
Triceratops was a plant-eater.

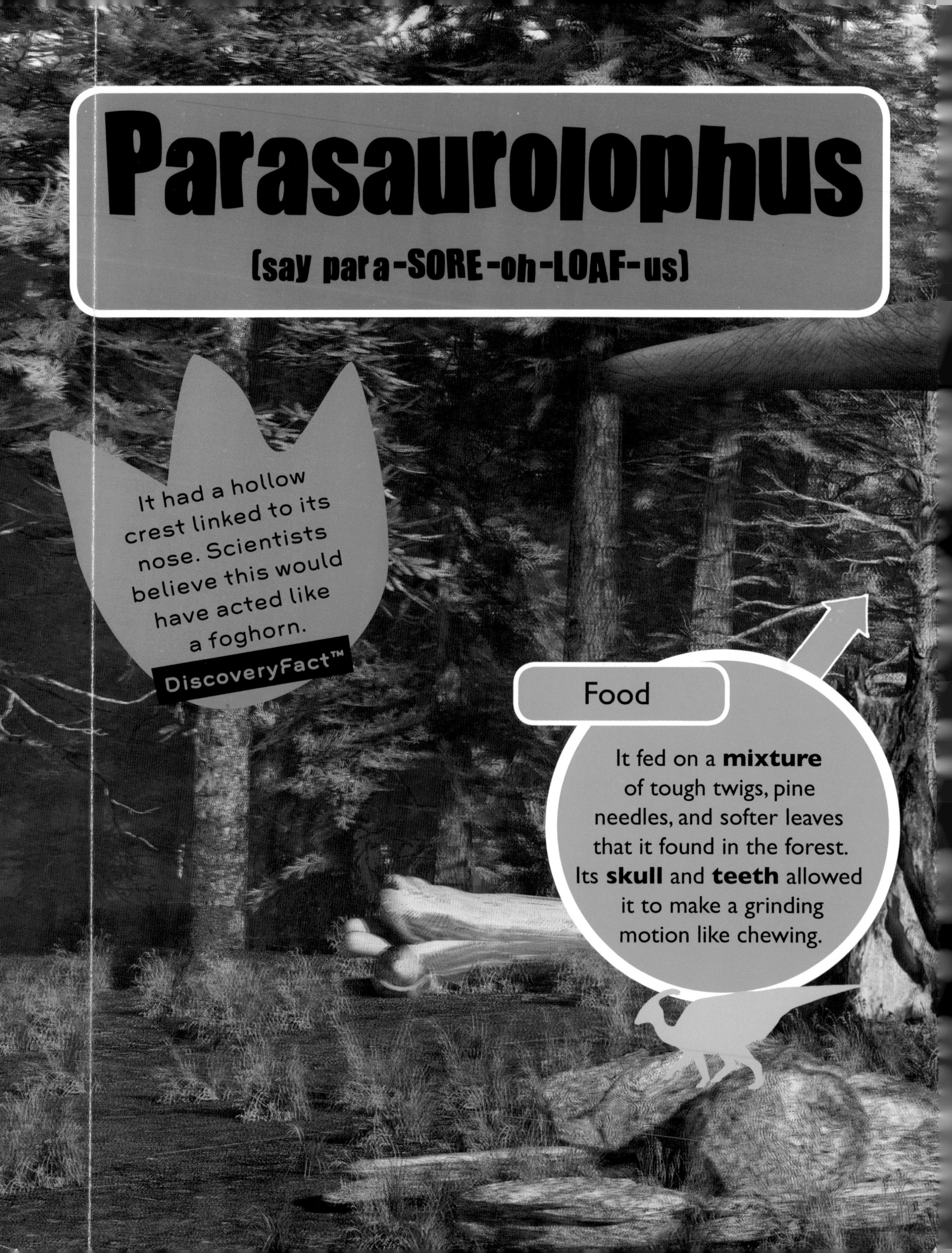
Parasaurolophus
(say para-SORE-oh-LOAF-us)
It had a hollow crest linked to its nose. Scientists believe this would have acted like a foghorn.
DiscoveryFact™
Food
It fed on a **mixture** of tough twigs, pine needles, and softer leaves that it found in the forest. Its **skull** and **teeth** allowed it to make a grinding motion like chewing.

Body
Triceratops had a **chunky body**, short tail, and **thick legs**. This would have made it **good** at **charging** and **fighting** off fierce hunters.

Parasaurolophus was a type of hadrosaur. This family of dinosaurs are also called "**duckbills**" because they have a beak like a duck's. They spent most of their time on all fours, but they could **rear up** on their **strong back legs** to run and to feed on tall trees.